2015

VOLUME **2**

CODE AND
COMMENTARY

The complete IRC with
commentary after each
section

INTERNATIONAL
CODE COUNCIL®

2015 International Residential Code® Commentary—Volume II

First Printing: August 2015
Second Printing: March 2016

ISBN: 978-1-60983-285-8 (soft-cover edition)

COPYRIGHT © 2015
by
INTERNATIONAL CODE COUNCIL, INC.

T019522

PREFACE

The principal purpose of the Commentary is to provide a basic volume of knowledge and facts relating to building construction as it pertains to the regulations set forth in the 2015 *International Residential Code*®. The person who is serious about effectively designing, constructing and regulating buildings and structures will find the Commentary to be a reliable data source and reference to almost all components of the built environment.

As a follow-up to the *International Residential Code*, we offer a companion document, the *International Residential Code*® *Commentary—Volume II*. Volume II covers Chapters 12 through Appendix Q of the 2015 *International Residential Code*. The basic appeal of the Commentary is thus: it provides, in a small package and at reasonable cost, thorough coverage of many issues likely to be dealt with when using the *International Residential Code* — and then supplements that coverage with historical and technical background. Reference lists, information sources and bibliographies are also included.

Throughout all of this, strenuous effort has been made to keep the vast quantity of material accessible and its method of presentation useful. With a comprehensive yet concise summary of each section, the Commentary provides a convenient reference for regulations applicable to the construction of buildings and structures. In the chapters that follow, discussions focus on the full meaning and implications of the code text. Guidelines suggest the most effective method of application, and the consequences of not adhering to the code text. Illustrations are provided to aid understanding; they do not necessarily illustrate the only methods of achieving code compliance.

The format of the Commentary includes the full text of each section, table and figure in the code, followed immediately by the commentary applicable to that text. At the time of printing, the Commentary reflects the most up-to-date text of the 2015 *International Residential Code*. Each section's narrative includes a statement of its objective and intent, and usually includes a discussion about why the requirement commands the conditions set forth. Code text and commentary text are easily distinguished from each other. All code text is shown as it appears in the *International Residential Code*, and all commentary is indented below the code text and begins with the symbol ❖. All code figures and tables are reproduced as they appear in the IRC. Commentary figures and tables are identified in the text by the word "Commentary" (as in "see Commentary Figure 704.3"), and each has a full border.

Readers should note that the Commentary is to be used in conjunction with the *International Residential Code* and not as a substitute for the code. **The Commentary is advisory only;** the code official alone possesses the authority and responsibility for interpreting the code.

Comments and recommendations are encouraged, for through your input, we can improve future editions. Please direct your comments to the Codes and Standards Development Department at the Central Regional Office.

TABLE OF CONTENTS

Part V—Mechanical

CHAPTER 12   MECHANICAL ADMINISTRATION .... 12-1 – 12-2

CHAPTER 13   GENERAL MECHANICAL SYSTEM REQUIREMENTS .... 13-1 – 13-18

CHAPTER 14   HEATING AND COOLING EQUIPMENT AND APPLIANCES .... 14-1 – 14-16

CHAPTER 15   EXHAUST SYSTEMS .... 15-1 – 15-14

CHAPTER 16   DUCT SYSTEMS .... 16-1 – 16-20

CHAPTER 17   COMBUSTION AIR .... 17-1 – 17-2

CHAPTER 18   CHIMNEYS AND VENTS .... 18-1 – 18-14

CHAPTER 19   SPECIAL APPLIANCES, EQUIPMENT AND SYSTEMS .... 19-1 – 19-4

CHAPTER 20   BOILERS AND WATER HEATERS .... 20-1 – 20-8

CHAPTER 21   HYDRONIC PIPING .... 21-1 – 21-10

CHAPTER 22   SPECIAL PIPING AND STORAGE SYSTEMS .... 22-1 – 22-4

CHAPTER 23   SOLAR THERMAL ENERGY SYSTEMS .... 23-1 – 23-8

Part VI—Fuel Gas

CHAPTER 24   FUEL GAS .... 24-1 – 24-264

Part VII—Plumbing

CHAPTER 25   PLUMBING ADMINISTRATION .... 25-1 – 25-6

CHAPTER 26   GENERAL PLUMBING REQUIREMENTS .... 26-1 – 26-22

CHAPTER 27   PLUMBING FIXTURES .... 27-1 – 27-24

CHAPTER 28   WATER HEATERS .... 28-1 – 28-12

CHAPTER 29   WATER SUPPLY AND DISTRIBUTION .... 29-1 – 29-78

CHAPTER 30   SANITARY DRAINAGE .... 30-1 – 30-38

CHAPTER 31   VENTS .... 31-1 – 31-26

CHAPTER 32   TRAPS . . . . 32-1 – 32-4

CHAPTER 33   STORM DRAINAGE . . . . 33-1 – 33-4

Part VIII—Electrical

CHAPTER 34   GENERAL REQUIREMENTS . . . . 34-1 – 34-20

CHAPTER 35   ELECTRICAL DEFINITIONS . . . . 35-1 – 35-18

CHAPTER 36   SERVICES . . . . 36-1 – 36-32

CHAPTER 37   BRANCH CIRCUIT AND FEEDER REQUIREMENTS . . . . 37-1 – 37-22

CHAPTER 38   WIRING METHODS . . . . 38-1 – 38-10

CHAPTER 39   POWER AND LIGHTING DISTRIBUTION . . . . 39-1 – 39-48

CHAPTER 40   DEVICES AND LUMINAIRES . . . . 40-1 – 40-14

CHAPTER 41   APPLIANCE INSTALLATION . . . . 41-1 – 41-6

CHAPTER 42   SWIMMING POOLS . . . . 42-1 – 42-20

CHAPTER 43   CLASS 2 REMOTE-CONTROL, SIGNALING AND
POWER-LIMITED CIRCUITS . . . . 43-1 – 43-4

Part IX—Referenced Standards

CHAPTER 44   REFERENCED STANDARDS . . . . 44-1 – 44-28

APPENDIX A   SIZING AND CAPACITIES OF GAS PIPING . . . . A1 – A12

APPENDIX B   SIZING OF VENTING SYSTEMS SERVING
APPLIANCES EQUIPPED WITH DRAFT HOODS,
CATEGORY I APPLIANCES, AND APPLIANCES
LISTED FOR USE WITH TYPE B VENTS . . . . B-1 – B-10

APPENDIX C   EXIT TERMINALS OF MECHANICAL DRAFT
AND DIRECT-VENT VENTING SYSTEMS . . . . C-1 – C-2

APPENDIX D   RECOMMENDED PROCEDURE FOR SAFETY
INSPECTION OF AN EXISTING APPLIANCE INSTALLATION . . . . D-1 – D-6

APPENDIX E   MANUFACTURED HOUSING USED AS DWELLINGS . . . . E-1 – E-14

APPENDIX F   RADON CONTROL METHODS . . . . F-1 – F-10

APPENDIX G   PIPING STANDARDS FOR VARIOUS APPLICATIONS . . . . G-1 – G-6

APPENDIX H   PATIO COVERS . . . . H-1 – H-6

APPENDIX I   PRIVATE SEWAGE DISPOSAL. . . . . I-1 – I-2

APPENDIX J   EXISTING BUILDINGS AND STRUCTURES . . . . J-1 – J-10

APPENDIX K   SOUND TRANSMISSION . . . . K-1 – K-2

APPENDIX L   PERMIT FEES . . . . L-1 – L-2

APPENDIX M   HOME DAY CARE—R-3 OCCUPANCY . . . . M-1 – M-2

APPENDIX N   VENTING METHODS . . . . N-1 – N-6

APPENDIX O   AUTOMATIC VEHICULAR GATES. . . . . O-1 – O-2

APPENDIX P   SIZING OF WATER PIPING SYSTEM. . . . . P-1 – P-20

APPENDIX Q   RESERVED . . . . Q1 – Q-2

APPENDIX R   LIGHT STRAW-CLAY CONSTRUCTION. . . . . R1 – R10

APPENDIX S   STRAWBALE CONSTRUCTION. . . . . S1 – S28

APPENDIX T   RECOMMENDED PROCEDURE FOR WORST-CASE TESTING
OF ATMOSPHERIC VENTING SYSTEMS UNDER N1102.4 OR
N1105 CONDITIONS $\leq 5ACH_{50}$ . . . . T1 – T4

APPENDIX U   SOLAR-READY PROVISIONS—DETACHED ONE- AND
TWO-FAMILY DWELLINGS, MULTIPLE SINGLE-FAMILY
DWELLINGS (TOWNHOUSES). . . . . U1 – U2

INDEX . . . . INDEX-1 – INDEX-14

Chapter 12:
Mechanical Administration

User note: Code change proposals to this chapter will be considered by the IRC – Plumbing and Mechanical Code Development Committee during the 2015 (Group A) Code Development Cycle.

General Comments

Chapter 12 provides regulations for the administration of the mechanical provisions of the code. Though this may be the smallest chapter in the code, it is very important in that it defines the application of the mechanical provisions to both existing and new construction. It also relates this chapter to the administrative provisions in Chapter 1.

Section M1201 addresses this set of mechanical regulations' relationship with Chapter 1 and the validity of the standards that are referenced. Section M1202 provides the applicability to existing mechanical systems. While the code mainly deals with new systems, this section indicates that some existing situations may fall under the requirements in the chapter.

Purpose

A set of mechanical regulations is intended to be adopted as a legally enforceable document that can safeguard health, safety, property and public welfare. Such regulations cannot be effective without adequate provisions for their administration and enforcement. The official charged with the administration and enforcement of mechanical regulations has a great responsibility, and with this responsibility goes authority. No matter how detailed the mechanical regulations may be, the building official must, to some extent, exercise judgement in determining code compliance. She or he has the responsibility to establish that the homes in which the citizens of the community reside are designed and constructed to be reasonably free from hazards associated with the presence and use of mechanical equipment, appliances and systems.

SECTION M1201
GENERAL

M1201.1 Scope. The provisions of Chapters 12 through 24 shall regulate the design, installation, maintenance, *alteration* and inspection of mechanical systems that are permanently installed and used to control environmental conditions within buildings. These chapters shall also regulate those mechanical systems, system components, *equipment* and *appliances* specifically addressed in this code.

❖ This section lists the chapters in the code that regulate mechanical systems. It indicates that the design, installation, and maintenance of mechanical equipment used to control the environmental conditions within the building are regulated by these chapters. It also states that other mechanical systems specifically addressed within these chapters are so regulated. Other provisions in the code reference the *International Mechanical Code®* (IMC®) and the *International Fuel Gas Code®* (IFGC®). This regulates virtually all mechanical systems and equipment within a dwelling in some form or another.

M1201.2 Application. In addition to the general administration requirements of Chapter 1, the administrative provisions

of this chapter shall also apply to the mechanical requirements of Chapters 13 through 24.

❖ This section makes reference to the administrative requirements of Chapter 1 to include those administrative provisions and make them applicable to the mechanical chapters.

SECTION M1202
EXISTING MECHANICAL SYSTEMS

M1202.1 Additions, alterations or repairs. *Additions, alterations,* renovations or repairs to a mechanical system shall conform to the requirements for a new mechanical system without requiring the existing mechanical system to comply with all of the requirements of this code. *Additions, alterations* or repairs shall not cause an existing mechanical system to become unsafe, hazardous or overloaded. Minor *additions, alterations* or repairs to existing mechanical systems shall meet the provisions for new construction, unless such work is done in the same manner and arrangement as was in the existing system, is not hazardous, and is *approved.*

❖ Major alterations or additions to existing mechanical systems must comply with the provisions of the code.

However, the code does not require existing systems to be upgraded to comply with new requirements unless the alteration or addition renders the existing system unsafe, overloaded or hazardous.

Minor additions, alterations or repairs to existing mechanical systems may be made following the same manner and arrangement as was found for the existing systems, as long as the work does not render the existing and new system unsafe, hazardous or overloaded. The building official has to make a judgement call on the extent of the addition, alteration or repair and determine whether the work is "minor," thereby allowing the use of the provisions of the old code, or that it is "major," thereby requiring the new work to be in compliance with the new code. Only when the new work makes the existing system overloaded, unsafe or hazardous does the building official need to require bringing the existing system up to the new provisions of the code.

M1202.2 Existing installations. Except as otherwise provided for in this code, a provision in this code shall not require the removal, *alteration* or abandonment of, nor prevent the continued use and maintenance of, an existing mechanical system lawfully in existence at the time of the adoption of this code.

❖ An existing mechanical system is generally "grandfathered" with code adoption, provided that the system meets a minimum level of safety. Frequently the criteria for determining this level of safety are the regulations (or code) under which the existing building was originally constructed. If there are no previous code criteria that apply, the building official is to apply those provisions of the code that are reasonably applicable to existing buildings. Provisions dealing with hazard abatement in existing buildings and provisions dealing with maintenance, as contained in the property maintenance and fire prevention codes, dictate a specific level of safety.

M1202.3 Maintenance. Mechanical systems, both existing and new, and parts thereof shall be maintained in proper operating condition in accordance with the original design and in a safe and sanitary condition. Devices or safeguards that are required by this code shall be maintained in compliance with the code edition under which installed. The owner or the owner's designated agent shall be responsible for maintenance of the mechanical systems. To determine compliance with this provision, the *building official* shall have the authority to require a mechanical system to be reinspected.

❖ This section allows the continued use of an existing mechanical system if it was originally lawfully permitted and installed. Periodic adoption of new codes or new provisions that may affect an existing system does not require upgrading the system to comply with the new provisions. However, this is predicated on the system being maintained by the owner in proper operating condition, in accordance with the manufacturer's requirements. The building official has the authority, as provided for in this section, to require the inspection of a mechanical system to determine whether work was

performed to maintain the equipment in a safe operating condition. If the system becomes hazardous or unsafe, the building official has the authority to require repair, modification or even replacement to render the installation safe.

Bibliography

The following resource materials were used in the preparation of the commentary for this chapter of the code:

IFGC–15, *International Fuel Gas Code*. Washington, DC: International Code Council, 2014.

IMC–15, *International Mechanical Code*. Washington, DC: International Code Council, 2014.

Legal Aspects of Code Administration. Washington, DC: International Code Council, 2002.

Chapter 13:
General Mechanical System Requirements

User note: Code change proposals to this chapter will be considered by the IRC – Plumbing and Mechanical Code Development Committee during the 2015 (Group A) Code Development Cycle.

General Comments

This chapter contains the provisions that apply to various types of mechanical appliances. The approval of appliances and their proper installation is the main theme. Section M1301 states the scope of the chapter and addresses its relationship with the *International Mechanical Code®* (IMC®) and the *International Fuel Gas Code®* (IFGC®). Section M1302 indicates that all mechanical appliances must be listed and labeled by an approved agency. Section M1303 addresses the information that is needed on the labels. Section M1304 discusses the proper design of appliances considering the appliances' type of fuel and the geographical location of the installed appliance. Section M1305 addresses access to installed appliances for servicing and potential replacement. Section M1306 contains the allowance for reduced clearances between appliances and combustible construction. Section M1307 contains the criteria for the safe installation of appliances. Section M1308 is a cross reference to the proper sections in the building portion of the code for the drilling and notching of structural members of the building.

Purpose

This chapter contains requirements for the safe and proper installation of mechanical equipment and appliances.

SECTION M1301
GENERAL

M1301.1 Scope. The provisions of this chapter shall govern the installation of mechanical systems not specifically covered in other chapters applicable to mechanical systems. Installations of mechanical *appliances, equipment* and systems not addressed by this code shall comply with the applicable provisions of the *International Mechanical Code* and the *International Fuel Gas Code*.

❖ This section provides general requirements for mechanical systems not specifically covered in other chapters of the code. In addition, it refers to the IMC and the IFGC for regulations governing equipment not addressed by the code.

M1301.1.1 Flood-resistant installation. In flood hazard areas as established by Table R301.2(1), mechanical *appliances, equipment* and systems shall be located or installed in accordance with Section R322.1.6.

❖ The local jurisdiction must fill in Table R301.2(1) upon adoption of the code, including the flood hazards information. Mechanical appliances, equipment, and systems that are located in flood hazard areas must be installed above the design flood elevation or must be designed and installed to prevent the entrance of water into the components and to resist the forces of the flood waters on the components (see commentary, Section R322.1.6).

M1301.2 Identification. Each length of pipe and tubing and each pipe fitting utilized in a mechanical system shall bear the identification of the manufacturer.

❖ The manufacturer is given the option of determining the type of marking for the material. If there is no applicable standard or the applicable standard does not require that the material be identified, identification of the manufacturer is still required by the code. Where the code indicates compliance with an approved standard, the manufacturer must comply with the requirements for marking in accordance with the applicable standard.

M1301.3 Installation of materials. Materials shall be installed in strict accordance with the standards under which the materials are accepted and approved. In the absence of such installation procedures, the manufacturer's instructions shall be followed. Where the requirements of referenced standards or manufacturer's instructions do not conform to minimum provisions of this code, the provisions of this code shall apply.

❖ Mechanical components and materials are to be installed in accordance with the installation requirements of the applicable standard listed in the code. Where a standard is not provided, the manufacturer's instructions must be followed. For example, because there are very few standards available that regulate the installation of valves, the manufacturer's instructions must be used to install these components. The code trumps where a referenced standard or manufacturer's instructions are less stringent than the code. It is rare, but the code may contain requirements that are more restrictive than the installation instructions or product listing.

M1301.4 Plastic pipe, fittings and components. Plastic pipe, fittings and components shall be third-party certified as conforming to NSF 14.

❖ Plastic piping, fittings and plastic pipe-related components, including solvent cements, primers, tapes, lubricants and seals used in mechanical systems, must be tested and certified as conforming to NSF 14. This includes hydronic piping and fittings and plastic piping system components including but not limited to pipes, fittings, valves, joining materials, gaskets and appurtenances. This section does not apply to components that only include plastic parts such as brass valves with a plastic stem.

M1301.5 Third-party testing and certification. Piping, tubing and fittings shall comply with the applicable referenced standards, specifications and performance criteria of this code and shall be identified in accordance with Section M1301.2. Piping, tubing and fittings shall either be tested by an approved third-party testing agency or certified by an approved third-party certification agency.

❖ This section requires that all piping, tubing and fittings comply with the referenced standards. However, the provisions contained in Section 104.11 regarding the evaluation and approval of alternative materials, methods and equipment are still applicable (see commentary, Section 104.11). Additionally, the code has been revised to include requirements for third-party certification and testing of such products. "Third-party certified" indicates that the minimum level of quality required by the applicable standard is maintained and the product is often referred to as "listed." "Third-party tested" indicates a product that has been tested by an approved testing laboratory and found to be in compliance with the standard. Although the code does not specifically state the identification or marking requirements, except for the manufacturer's identification, the applicable referenced standard states the minimum information required. The identification or marking requirements typically include the name of the manufacturer, product name or serial number, installation specifications, applicable tests and standards, testing agency and labeling agency.

SECTION M1302
APPROVAL

M1302.1 Listed and labeled. *Appliances* regulated by this code shall be *listed* and *labeled* for the application in which

they are installed and used, unless otherwise *approved* in accordance with Section R104.11.

❖ Mechanical appliances must be listed and labeled by an approved agency to show that they comply with applicable national standards. The code requires listing and labeling for appliances such as boilers, furnaces, space heaters, cooking appliances and clothes dryers. The code also requires listing for system components. The label is the primary, if not the only, assurance to the installer, the inspector and the end user that a similar appliance has been tested and evaluated by an approved agency and performed safely and efficiently when installed and operated in accordance with its listing.

The label is part of the information that the code official is to consider in the approval of appliances. The only exception to the labeling requirement occurs when the code official approves a specific appliance in accordance with the authority granted in Section R104.11.

The requirement that appliances are to be used only in accordance with their listing is intended to prevent the use of products that have a listing for some application but are being used in a different application for which they have not been tested. An example would be a fan that is listed for use only as a bathroom exhaust fan but is installed for use as a kitchen exhaust hood fan or as a clothes dryer booster fan. Another potential misapplication could be an appliance that has been tested and listed for indoor installation only, but is installed outdoors. Such misapplications have the potential to create hazardous situations.

The code official should exercise extreme caution when considering the approval of unlisted appliances.

Approval of unlisted appliances must be based on some form of documentation that demonstrates compliance with the applicable standards or equivalence with an appliance that is listed and labeled to the applicable standards. Where no product standards exist, documentation must be provided to demonstrate that the appliance is appropriate for the intended use and will provide the same level of performance as would be expected from a similar appliance that is listed and labeled. Sometimes appliances are listed in the field on a case-by-case basis using requirements or outlines of investigation derived from relative appliance standards. One fundamental principle of the code is the reliance on the listing and labeling process to ensure appliance performance. Approvals granted in accordance with Section R104.11 must be justified with supporting documentation. To the code official, installer and end-user very little is known about the performance of an appliance that is not tested and built to an appliance standard.

SECTION M1303
LABELING OF APPLIANCES

M1303.1 Label information. A permanent factory-applied nameplate(s) shall be affixed to *appliances* on which shall appear, in legible lettering, the manufacturer's name or trademark, the model number, a serial number and the seal or *mark* of the testing agency. A *label* also shall include the following:

1. Electrical *appliances.* Electrical rating in volts, amperes and motor phase; identification of individual electrical components in volts, amperes or watts and motor phase; and in Btu/h (W) output and required clearances.

2. Absorption units. Hourly rating in Btu/h (W), minimum hourly rating for units having step or automatic modulating controls, type of fuel, type of refrigerant, cooling capacity in Btu/h (W) and required clearances.

3. Fuel-burning units. Hourly rating in Btu/h (W), type of fuel *approved* for use with the *appliance* and required clearances.

4. Electric comfort-heating appliances. The electric rating in volts, amperes and phase; Btu/h (W) output rating; individual marking for each electrical component in amperes or watts, volts and phase; and required clearances from combustibles.

5. Maintenance instructions. Required regular maintenance actions and title or publication number for the operation and maintenance manual for that particular model and type of product.

❖ This section requires that appliances have a label that is a permanent nameplate. In general, labels other than metal tags or plates usually consist of material that is similar in appearance to a decal, and the label, its adhesive and the printed information are all durable and water resistant. Because of the important information on a label, the label must be permanent, not susceptible to damage and legible for the life of the appliance. The standards appliances are tested to usually specify the required label material, the method of attachment and the required label information. The code requires that the label be affixed permanently and prominently on the appliance or equipment and specifies the information that must appear on the label. The manufacturer may be required by the relevant standard or may voluntarily provide additional information on the label (see Commentary Figure M1303.1).

AMERICAN STANDARD INC.
THE TRANE COMPANY
TRENTON, N.J. 08619   MADE IN U.S.A.

FORCED AIR FURNACE   CATEGORY I

ANSI Z21.47 - 1990 CENTRAL FURN

FOR INDOOR INSTALLATION IN A BUILDING
CONSTRUCTED ON SITE.   **NRTL**

MODEL NO. TUD1C0C948A1	SERIAL NO. G36520130	EQUIPPED FOR NAT. GAS
INPUT 100,000 BTU/HR.	LIMIT SETTING 200 °F	MFRD. 09/92
TEMP. RISE °F FROM 35 TO 65	MAX. EXT. STATIC PRESS .50 INCHES WATER	MAX. DESIGN AIR TEMP. 165 °F
VOLTS/PHASE/HERTZ 115/1/60	TOTAL AMPS 9.8	SERVICE CODE 1

MANIFOLD PRESSURE
(IN INCHES OF WATER)
NAT. 3.5 **LP** 10.5
SUPPLY PRESSURE
(IN INCHES OF WATER)
MAX. NAT. 10.5, **LP** 13.0
MIN. NAT. 4.5**L P** 11.0 **FOR PURPOSE OF INPUT ADJUSTMENT.**

FLAME ROLLOUT SWITCH - REPLACE
IF BLOWN WITH CATALOG NO.
WG09X0533 (333 F CUTOFF TEMP.)
ONE TIME THERMAL FUSE.

MINIMUM CLEARANCE COMBUSTIBLE MATERIALS:

FOR	CLOSET	INSTALLATION AS FOLLOWS:		
SIDES	0	IN. W/SINGLE WALL VENT		
FLUE	6	IN. W/SINGLE WALL VENT	1 IN. W/TYPE B-1 VENT	
FRONT	6	IN.	BACK 0 IN.	TOP 1 IN.

UPFLOW UNITS. FOR INSTALLATION COMBUSTIBLE FLOORING.   21D340159 P01

Figure M1303.1
TYPICAL LABEL FOR A CATEGORY I GAS-FIRED FURNACE
(Courtesy of The Trane Company and American Standard Company)

SECTION M1304
TYPE OF FUEL

M1304.1 Fuel types. Fuel-fired *appliances* shall be designed for use with the type of fuel to which they will be connected and the altitude at which they are installed. *Appliances* that comprise parts of the building mechanical system shall not be converted for the use of a different fuel, except where *approved* and converted in accordance with the manufacturer's instructions. The fuel input rate shall not be increased or decreased beyond the limit rating for the altitude at which the *appliance* is installed.

❖ An element of information used for the approval of appliances is the label, which ensures that the appliance has been tested in accordance with a valid standard and performed acceptably when installed and operated in accordance with the appliance listing. Manufacturers usually design mechanical appliances to operate on a specific type of fuel. Thus, the fuel used in the appliance test must be the type of fuel specified by the manufacturer on the label. When an appliance is converted to a different type of fuel, the original label that appears on the appliance is no longer valid. Because the original approval of the appliance was based in part on the label, the appliance is no longer approved for use.

Field conversions will more likely allow for the safe operation of the appliance if, as required, the conversion is approved by the code official and done in accordance with the manufacturer's installation instructions. Fuel conversions that are not performed correctly can cause serious malfunctions and hazardous operation. Before a fuel conversion is performed, the manufacturer must be contacted for installation instructions outlining the procedures to follow for proper operation of the appliance. In most cases, conversion kits from the manufacturer are available along with the installation instructions. Once a conversion has been completed, a supplemental label must be installed to update the information contained on the original label, thereby alerting any service personnel of the modifications that have been made.

All fuel-fired appliances are designed to operate with a maximum and minimum British thermal units per hour (Btu/h) input capacity. This capacity is field adjusted to suit the elevation because of the change in air density at different elevations. Alteration of Btu/h input beyond the allowable limits can result in hazardous over-firing or under-firing. Either condition can cause operation problems that include overheating, vent failure, corrosion, poor draft and poor combustion.

SECTION M1305
APPLIANCE ACCESS

M1305.1 Appliance access for inspection service, repair and replacement. *Appliances* shall be accessible for inspection, service, repair and replacement without removing per-

manent construction, other *appliances*, or any other piping or ducts not connected to the *appliance* being inspected, serviced, repaired or replaced. A level working space not less than 30 inches deep and 30 inches wide (762 mm by 762 mm) shall be provided in front of the control side to service an *appliance*.

❖ Because mechanical equipment and appliances require routine maintenance, repair and possible replacement, access is required. Additionally, manufacturer's installation instructions usually contain access recommendations or requirements. As a result, the provisions stated herein supplement the manufacturer's installation instructions.

The provisions of this section specify that access must be provided to components that require observation, inspection, adjustment, servicing, repair or replacement. Access is also necessary for operating procedures such as startup or shutdown. The level working space in front of the control side of the appliance must be 30 inches (762 mm) wide and 30 inches (762 mm) deep to provide adequate space for the technician or inspector to safely perform the work.

The code states that "accessible" means "access that might require the removal of an access panel or similar removable obstruction." An appliance or piece of equipment is not accessible if any portion of the structure's permanent finish materials, such as drywall, plaster, paneling, built-in furniture or cabinets or any other similar permanently affixed building component, must be removed before access is achieved. In addition, removal of all or part of another appliance or the piping or duct or serving other appliances must not be necessary to perform the service, replacement or inspection of an appliance. Such an installation could result in unnecessarily high costs to the homeowner and improper or unsafe reassembly of other appliance and system components. This could also result in service personnel having to perform disassembly and reassembly of appliances and system components that are not within the personnel's area of expertise or licensed work.

The intent is to provide access to all components such as controls, gauges, burners, filters, blowers and motors that require observation, inspection, adjustment, servicing, repair or replacement.

M1305.1.1 Furnaces and air handlers. Furnaces and air handlers within compartments or alcoves shall have a minimum working space clearance of 3 inches (76 mm) along the sides, back and top with a total width of the enclosing space being not less than 12 inches (305 mm) wider than the furnace or air handler. Furnaces having a firebox open to the atmosphere shall have not less than a 6-inch (152 mm) working space along the front combustion chamber side. Combustion air openings at the rear or side of the compartment shall comply with the requirements of Chapter 17.

Exception: This section shall not apply to replacement *appliances* installed in existing compartments and alcoves where the working space clearances are in accordance with the *equipment* or *appliance* manufacturer's installation instructions.

❖ Furnaces and air handlers installed in compartments or alcoves must have clearances from the enclosure so that they can be removed, maintained or repaired as necessary. The minimum clearances specified in the code apply even though the manufacturer's instructions might permit a lesser clearance. Clearances provide access, ventilation, cooling of the appliance and equipment and protection for surrounding combustibles. The front (firebox) clearance helps protect combustibles against flame rollout and allows free movement of combustion air. The exception exempts replacement appliances and equipment installed in existing compartments or alcoves if the installation complies with the manufacturer's instructions. The manufacturer's installation instructions usually contain minimum workspace requirements and minimum clearances to surrounding construction that must be maintained.

M1305.1.2 Appliances in rooms. *Appliances* installed in a compartment, alcove, *basement* or similar space shall be accessed by an opening or door and an unobstructed passageway measuring not less than 24 inches (610 mm) wide and large enough to allow removal of the largest *appliance* in the space, provided there is a level service space of not less than 30 inches (762 mm) deep and the height of the *appliance*, but not less than 30 inches (762 mm), at the front or service side of the *appliance* with the door open.

❖ This section specifies an access opening and passageway to afford service personnel reasonable access to appliances and to allow for the passage of system components. Quite often appliances such as furnaces, boilers and water heaters are installed in spaces with little or no forethought about future access for maintenance or replacement.

M1305.1.3 Appliances in attics. *Attics* containing *appliances* shall be provided with an opening and a clear and unobstructed passageway large enough to allow removal of the largest *appliance*, but not less than 30 inches (762 mm) high and 22 inches (559 mm) wide and not more than 20 feet (6096 mm) long measured along the centerline of the passageway from the opening to the *appliance*. The passageway shall have continuous solid flooring in accordance with Chapter 5 not less than 24 inches (610 mm) wide. A level service space not less than 30 inches (762 mm) deep and 30 inches (762 mm) wide shall be present along all sides of the *appliance* where access is required. The clear access opening dimensions shall be not less than of 20 inches by 30 inches (508 mm by 762 mm), and large enough to allow removal of the largest appliance.

Exceptions:

1. The passageway and level service space are not required where the *appliance* can be serviced and removed through the required opening.

2. Where the passageway is unobstructed and not less than 6 feet (1829 mm) high and 22 inches (559 mm)

wide for its entire length, the passageway shall be not more than 50 feet (15 250 mm) long.

❖ There is not always sufficient room for mechanical equipment and appliances to be installed in spaces such as basements, alcoves, utility rooms and furnace rooms. In an effort to save floor space or simplify an installation, designers often locate appliances and mechanical equipment on roofs, in attics or in similar remote locations. Access to appliances and equipment could be difficult because of roof slope, stone roof ballast or the lack of a walking surface, such as might occur in an attic or similar space with exposed ceiling joists. The intent of this section is to require a suitable access opening, passageway and workspace that will allow reasonably easy access without endangering the service person (see Commentary Figure M1305.1.3). The longer the attic passageway, the more the service person will be exposed to extreme temperatures and the risk of injury. The attic access opening (typically a scuttle) must be large enough to allow the largest appliance in the attic to pass through such opening. For example, if an attic furnace is the largest appliance, the furnace itself should be able to be removed from the attic without having to disassemble the furnace. Of course, it is understood that ducts, plenums, cooling coil cabinets and other attachments might have to be disconnected from the furnace before removal is possible.

The first exception allows the passageway and level service space to be eliminated if the technician can reach the appliance through the access opening without having to step into the attic. The second exception allows the length of the passageway to be extended to 50 feet (15 250 mm) if there is at least 6 feet (1829 mm) of clear headroom for the entire length of the passageway. This is allowed because there is less danger of lengthy exposure to extreme temperatures if the service personnel can walk erect and unimpeded to the equipment rather than crawling.

Note that some appliances might not be listed for attic installation or might otherwise be unsuitable for such conditions.

M1305.1.3.1 Electrical requirements. A luminaire controlled by a switch located at the required passageway opening and a receptacle outlet shall be installed at or near the *appliance* location in accordance with Chapter 39. Exposed lamps shall be protected from damage by location or lamp guards.

❖ An appliance located in an attic is generally not easy to access. A lighting outlet and receptacle outlet encourage and facilitate appliance maintenance. The receptacle will accommodate power tools, drop lights and diagnostic instruments. Also, these provisions negate the need for extension cords, which can be hazardous to service personnel. The lighting outlet is to allow the attic space to be safely navigated and is not intended to provide the necessary lighting for servicing and repair of the appliances. Where exposed lamps (naked light bulbs) are installed as the required lighting outlets, they must be located out of harm's way or must be provided with a suitable lamp guard. If service personnel hit and break the lamp with their bodies, tools, parts or other objects, the result could be a shock and/or fire hazard, with the additional hazard of sudden darkness in a dangerous location.

M1305.1.4 Appliances under floors. Underfloor spaces containing *appliances* shall be provided with an unobstructed passageway large enough to remove the largest *appliance*, but not less than 30 inches (762 mm) high and 22 inches (559 mm) wide, nor more than 20 feet (6096 mm) long measured along the centerline of the passageway from the opening to the *appliance*. A level service space not less than 30 inches (762 mm) deep and 30 inches (762 mm) wide shall be present at the front or service side of the *appliance*. If the depth of the passageway or the service space exceeds 12 inches (305 mm) below the adjoining grade, the walls of the passageway shall be lined with concrete or masonry extending 4 inches (102 mm) above the adjoining grade in accordance with Chapter 4. The rough-framed access opening dimensions shall be not less than 22 inches by 30 inches (559 mm by 762 mm), and large enough to remove the largest *appliance*.

Exceptions:

1. The passageway is not required where the level service space is present when the access is open, and the *appliance* can be serviced and removed through the required opening.

2. Where the passageway is unobstructed and not less than 6 feet high (1929 mm) and 22 inches (559 mm) wide for its entire length, the passageway shall not be limited in length.

❖ This section, which applies to crawl spaces, has concepts similar to those of Section M1305.1.3. The more difficult the access to appliances and equipment is, the less likely that the appliance or equipment will be inspected and serviced on a regular basis. Attic and crawl space installations suffer from the "out-of-sight, out-of-mind" syndrome.

The first exception has the same intent as the first exception of Section M1305.1.3. The second exception allows unlimited length of the passageway if there is at least 6 feet (1829 mm) of clear headroom for the entire length of the passageway (see commentary, Section M1305.1.3).

M1305.1.4.1 Ground clearance. *Equipment* and *appliances* supported from the ground shall be level and firmly supported on a concrete slab or other *approved* material extending not less than 3 inches (76 mm) above the adjoining ground. Such support shall be in accordance with the manufacturer's installation instructions. *Appliances* suspended from the floor shall have a clearance of not less than 6 inches (152 mm) from the ground.

❖ This section's requirement provides a buffer from the corrosive effects of an appliance's contact with the

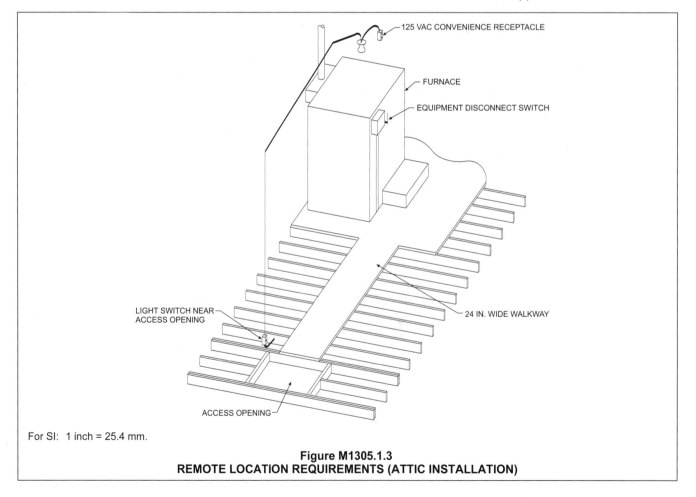

125 VAC CONVENIENCE RECEPTACLE

FURNACE

EQUIPMENT DISCONNECT SWITCH

LIGHT SWITCH NEAR ACCESS OPENING

24 IN. WIDE WALKWAY

ACCESS OPENING

For SI: 1 inch = 25.4 mm.

Figure M1305.1.3
REMOTE LOCATION REQUIREMENTS (ATTIC INSTALLATION)

ground. If supported on the ground, the appliance is to rest on a material that will be a barrier between the appliance and the ground. Concrete is the material that is prescribed, but other approved materials could be used if they provide the same level of protection as the concrete. If the appliance is suspended from the floor assembly above the ground, a minimum separation of 6 inches (152 mm) is called out.

The slab or other support surface must be at least 3 inches (76 mm) above grade so that the appliance being supported will be well above grade and protected from prolonged exposure to moisture and soil. The 3-inch (76 mm) minimum also provides some protection against settling of the slab or support base.

M1305.1.4.2 Excavations. Excavations for *appliance* installations shall extend to a depth of 6 inches (152 mm) below the *appliance* and 12 inches (305 mm) on all sides, except that the control side shall have a clearance of 30 inches (762 mm).

❖ This section allows for appliances to be located below the level of the ground surface in underfloor areas. Again, the concern is to provide a separation from the ground itself. Minimum clearances are provided in this section. The 30-inch-deep (762 mm) workspace is required for adequate access for servicing the appliance.

M1305.1.4.3 Electrical requirements. A luminaire controlled by a switch located at the required passageway opening and a receptacle outlet shall be installed at or near the *appliance* location in accordance with Chapter 39. Exposed lamps shall be protected from damage by location or lamp guards.

❖ This is the same requirement as found in Section M1305.1.3.1. Lighting and a power supply are required for the service of mechanical appliances. Note that Section E3902 requires ground fault circuit interrupter (GFCI) protection for crawl space receptacle outlets. Where exposed lamps (naked light bulbs) are installed as the required lighting outlets, they must be located out of harm's way or must be provided with a suitable

lamp guard. If service personnel hit and break the lamp with their bodies, tools, parts or other objects, the result could be a shock and/or fire hazard, with the additional hazard of sudden darkness in a dangerous location.

SECTION M1306
CLEARANCES FROM COMBUSTIBLE CONSTRUCTION

M1306.1 Appliance clearance. *Appliances* shall be installed with the clearances from unprotected combustible materials as indicated on the *appliance label* and in the manufacturer's installation instructions.

❖ Requirements for clearances to combustibles are emphasized because of the potential fire hazard posed when those clearances are not observed. Maintaining an appropriate distance from the outer surface of an appliance or piece of equipment to combustible materials reduces the possibility of ignition of combustible materials. This section requires appliances to be installed with clearances from unprotected combustibles as indicated on the label for the listed appliance and the manufacturer's installation instructions. The minimum clearances to combustibles are specified in the manufacturer's installation instructions for a labeled appliance. Because an approved agency tests appliances in accordance with these instructions, the clearances required are necessary for the correct installation and operation of the appliance. Note, however, that this section does not include provisions or guidelines for the installation of unlisted appliances. Figure M1306.1 is a graphic image of the reduced clearances allowed by the code. It works in conjunction with Table M1306.2.

FIGURE M1306.1. See below.

❖ Figure M1306.1 is a graphic image of the reduced clearances allowed by the code. It works in conjunction with Table M1306.2.

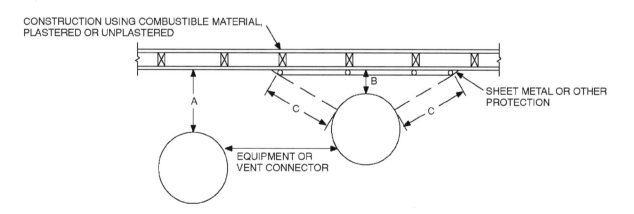

Note: "A" equals the required clearance with no protection. "B" equals the reduced clearance permitted in accordance with Table M1306.2. The protection applied to the construction using combustible material shall extend far enough in each direction to make "C" equal to "A."

FIGURE M1306.1
REDUCED CLEARANCE DIAGRAM

GENERAL MECHANICAL SYSTEM REQUIREMENTS

M1306.2 Clearance reduction. The reduction of required clearances to combustible assemblies or combustible materials shall be based on Section M1306.2.1 or Section M1306.2.2.

❖ See the commentary to Sections M1306.2.1 and M1306.2.2.

TABLE M1306.2. See below.

❖ The column headings of Table M1306.2 list required clearances without protection. The numbers to the right of each method indicate the permissible reduced clearance measured from the heat-producing appliances to the face of the combustible surface.

The rationale behind the methods of protection listed in Table M1306.2 is based on the ability of the protection to reduce radiant heat transmission from the appliance and equipment to the combustible material so that the temperature rise of the combustible material will remain below the maximum allowed.

TABLE M1306.2
REDUCTION OF CLEARANCES WITH SPECIFIED FORMS OF PROTECTION[a, c, d, e, f, g, h, i, j, k, l]

TYPE OF PROTECTION APPLIED TO AND COVERING ALL SURFACES OF COMBUSTIBLE MATERIAL WITHIN THE DISTANCE SPECIFIED AS THE REQUIRED CLEARANCE WITH NO PROTECTION (See Figures M1306.1 and M1306.2)	WHERE THE REQUIRED CLEARANCE WITHOUT PROTECTION FROM APPLIANCE, VENT CONNECTOR, OR SINGLE WALL METAL PIPE IS:									
	36 inches		18 inches		12 inches		9 inches		6 inches	
	Allowable clearances with specified protection (Inches)[b] Use column 1 for clearances above an appliance or horizontal connector. Use column 2 for clearances from an appliance, vertical connector and single-wall metal pipe.									
	Above column 1	Sides and rear column 2	Above column 1	Sides and rear column 2	Above column 1	Sides and rear column 2	Above column 1	Sides and rear column 2	Above column 1	Sides and rear column 2
3½-inch-thick masonry wall without ventilated air space	—	24	—	12	—	9	—	6	—	5
½-inch insulation board over 1-inch glass fiber or mineral wool batts	24	18	12	9	9	6	6	5	4	3
Galvanized sheet steel having a minimum thickness of 0.0236-inch (No. 24 gage) over 1-inch glass fiber or mineral wool batts reinforced with wire or rear face with a ventilated air space	18	12	9	6	6	4	5	3	3	3
3½-inch-thick masonry wall with ventilated air space	—	12	—	6	—	6	—	6	—	6
Galvanized sheet steel having a minimum thickness of 0.0236-inch (No. 24 gage) with a ventilated air space 1-inch off the combustible assembly	18	12	9	6	6	4	5	3	3	2
½-inch-thick insulation board with ventilated air space	18	12	9	6	6	4	5	3	3	3
Galvanized sheet steel having a minimum thickness of 0.0236-inch (No. 24 gage) with ventilated air space over 24 gage sheet steel with a ventilated space	18	12	9	6	6	4	5	3	3	3
1-inch glass fiber or mineral wool batts sandwiched between two sheets of galvanized sheet steel having a minimum thickness of 0.0236-inch (No. 24 gage) with a ventilated air space	18	12	9	6	6	4	5	3	3	3

For SI: 1 inch = 25.4 mm, 1 pound per cubic foot = 16.019 kg/m³, °C = [(°F)-32/1.8], 1 Btu/(h × ft² × °F/in.) = 0.001442299 (W/cm² × °C/cm).

a. Reduction of clearances from combustible materials shall not interfere with combustion air, draft hood clearance and relief, and accessibility of servicing.

b. Clearances shall be measured from the surface of the heat producing appliance or equipment to the outer surface of the combustible material or combustible assembly.

c. Spacers and ties shall be of noncombustible material. Spacers and ties shall not be used directly opposite appliance or connector.

d. Where all clearance reduction systems use a ventilated air space, adequate provision for air circulation shall be provided as described. (See Figures M1306.1 and M1306.2.)

e. There shall be not less than 1 inch between clearance reduction systems and combustible walls and ceilings for reduction systems using ventilated air space.

f. If a wall protector is mounted on a single flat wall away from corners, adequate air circulation shall be permitted to be provided by leaving only the bottom and top edges or only the side and top edges open with not less than a 1-inch air gap.

g. Mineral wool and glass fiber batts (blanket or board) shall have a minimum density of 8 pounds per cubic foot and a minimum melting point of 1,500°F.

h. Insulation material used as part of a clearance reduction system shall have a thermal conductivity of 1.0 Btu inch per square foot per hour °F or less. Insulation board shall be formed of noncombustible material.

i. There shall be not less than 1 inch between the appliance and the protector. The clearance between the appliance and the combustible surface shall not be reduced below that allowed in this table.

j. All clearances and thicknesses are minimum; larger clearances and thicknesses are acceptable.

k. Listed single-wall connectors shall be permitted to be installed in accordance with the terms of their listing and the manufacturer's instructions.

l. For limitations on clearance reduction for solid-fuel-burning appliances see Section M1306.2.3.

Although the materials referred to in Table M1306.2 are common construction materials, confusion often arises over what satisfies the requirement for "insulation board" (Item 6 in the table), sometimes referred to as inorganic insulating board, noncombustible mineral board or noncombustible insulating board. These products are not made of carbon-based compounds.

Carbon-based compounds are those found in cellulose (wood), plastics and other materials manufactured from raw materials that once existed in living organisms. Cement board materials must have a specified maximum "C" (conductance) value in addition to being noncombustible.

Note h specifies a maximum thermal conductivity of Btu/(h · ft · °F). Conductivity is the amount of heat in Btus that will flow each hour through a 1-foot-square (0.0929 m²) slab of material, 1-inch (25 mm) thick with a 1°F temperature difference between both sides and is usually identified by the symbol k. Tables of k values usually do not include the area term in the dimensions for conductivity, and it must be understood that the value must be multiplied by the area to obtain the total Btu value.

Thermal conductance (overall) is the time rate of heat flow through a body not taking thickness into account and is usually identified by the symbol C Btu/(h · ft² · °F).

Thermal resistance (overall) is the reciprocal of overall thermal conductance and is usually identified by the symbol R (hour · ft² · °F/Btu).

This translates into a minimum required insulation R-value of 1.0 (square foot · hour · °F)/Btu per inch of insulating material. The methods in Table M1306.2 control heat transmission by reflecting heat radiation, retarding thermal conductance and providing convective cooling. Where sheet metal materials or metal plates are specified, the effectiveness of the protection can be enhanced by the reflective surface of the metal. Painting or otherwise covering the surface would reduce the metal's ability to reflect radiant heat and, depending on the color, could increase heat absorption. The airspace between the protected surface and the clearance-reduction assembly allows convection air currents to cool the protection assembly by carrying away heat that has been conducted through the assembly. Where a clearance-reduction assembly must be spaced 1 inch (25 mm) off the wall, the top, bottom and sides of the assembly must remain open as required by Notes d and f to permit unrestricted airflow (convection currents). If the openings were not provided, the air-cooling effect would not take place, and the protection assembly would not be as effective in limiting the temperature rise on the protected surfaces. Ideally, the protection assembly should be open on all sides to provide maximum ventilation.

Spacers must be noncombustible. Spacers should not be placed directly behind the heat source because the location would increase the amount of heat conduction through the spacer, thus creating a "hot spot." Figure M1306.2 specifically shows a noncombustible spacer arrangement.

The performance of a protective assembly when applied to a horizontal surface, such as a ceiling, will differ substantially from the same assembly placed in a vertical plane. Obviously, temperatures at a ceiling surface will be higher because of natural convection and because the air circulation between the method of protection and the protected ceiling surface will be substantially reduced or nonexistent. It is for these reasons that Table M1306.2 is divided into two application groups.

The manufacturer's instructions or label for many appliances will state an absolute minimum clearance, regardless of any clearance reduction method used. Those clearance requirements take precedence over Table M1306.2. For example, a typical wood-burning room heater will require in all cases an airspace clearance of at least 12 inches (305 mm), with no further reduction allowed.

The methods in Table M1306.2 are intended to be permanent installations properly supported to prevent displacement or deformation. Movement could adversely affect the performance of the protection method, thus posing a potential fire hazard.

The assemblies in Table M1306.2 are the product of experience and testing. To achieve predictable and dependable performance, the components of the various assemblies cannot be mixed, matched, combined, substituted or otherwise rearranged to comprise new assemblies, and materials cannot be substituted for those prescribed in the table. Any alterations or substitutions could have an effect on the assembly, and its performance would have to be tested and approved.

As stated in Note b to the table, the reduced clearance is measured from the heat source to the combustible material, disregarding any intervening protection assembly.

Note l serves to remind the table user that Section M1306.2.1 puts severe limitations on reducing clearances to solid-fuel appliances.

FIGURE M1306.2. See page 13-10.

❖ See the commentary to Section M1306.2.

M1306.2.1 Labeled assemblies. The allowable clearance shall be based on an approved reduced clearance protective assembly that is listed and labeled in accordance with UL 1618.

❖ Listed and labeled clearance reduction assemblies are available that can be used to reduce clearances based on the testing and listing of the assemblies. Such assemblies may or may not require some field assembly. The manufacturer's instructions must be closely followed for assembly and installation in order to safely achieve the desired clearance reduction.

M1306.2.2 Reduction table. Reduction of clearances shall be in accordance with the *appliance* manufacturer's instructions and Table M1306.2. Forms of protection with ventilated air space shall conform to the following requirements:

1. Not less than 1-inch (25 mm) air space shall be provided between the protection and combustible wall surface.

2. Air circulation shall be provided by having edges of the wall protection open not less than 1 inch (25 mm).

3. If the wall protection is mounted on a single flat wall away from corners, air circulation shall be provided by having the bottom and top edges, or the side and top edges not less than 1 inch (25 mm).

4. Wall protection covering two walls in a corner shall be open at the bottom and top edges not less than 1 inch (25 mm).

❖ Heat-producing appliances and mechanical equipment must be installed with the required minimum clearances to combustible materials indicated by their listing label. It is not uncommon to encounter practical or structural difficulties in maintaining clearances. Therefore, clearance-reduction methods have been developed to allow, in some cases, reduction of the minimum prescribed clearance distance while achieving equivalent protection. An important understanding is that all prescribed clearances to combustibles are airspace clearances measured from the heat source to the face of the nearest combustible surface, even if that combustible surface is not visible. An example of that would be a wood stud wall located behind a metal panel. If installation of an appliance with the required clearances from combustibles is not possible, this section allows reduction of the clearances in accordance with Table M1306.2 if the appliance manufacturer's instructions allow the reductions to the extent desired. The table includes several forms of protection,

depending on the extent of reduction. This section requires that a 1-inch (25 mm) air space be maintained between the protection and the combustible wall surface to allow unimpeded circulation of convection air needed to keep the temperature rise within acceptable limits. Most of the methods in Table M1306.2 depend on convective cooling as an essential part of the system (see Commentary Figure G2409.2).

When using the assemblies described in Table M1306.2, the clearance is measured as described in Note b. The required clearances are intended to be clear airspace, and therefore the space is not to be filled with insulation or any other material other than an assembly intended to allow clearance reduction. This is especially important where clearances are required from appliances and equipment that rely on the airspace for convection cooling to maintain their proper operating temperature.

The provisions contained in this section are based on the principles of heat transfer. Mechanical equipment or appliances producing heat can become hot, and many appliances have hot exterior surfaces by design. The heat energy is then radiated to objects surrounding the appliances or equipment. When mechanical equipment and appliances are tested, the minimum clearances are established so that radiant and, to a lesser extent, convective heat transfer do not represent an ignition hazard to adjacent surfaces and objects. This distance is called the "required clearance" to combustible materials. Appliance and equip-

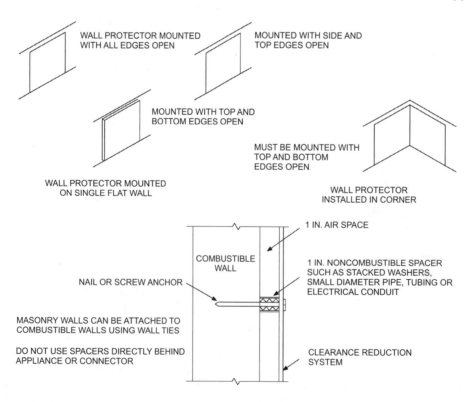

For SI: 1 inch = 25.4 mm.

FIGURE M1306.2
WALL PROTECTOR CLEARANCE REDUCTION SYSTEM

ment labels must specify minimum clearances in all directions.

This section permits the use of materials and systems as radiation shields, decreasing the amount of heat energy transferred to surrounding objects and reducing the required clearances between mechanical appliances and equipment to combustibles.

Plaster and gypsum by themselves are classified as noncombustible materials. Under continued exposure to heat, however, these materials will gradually decompose as water molecules are driven out of the material. Plaster on wood lath, plasterboard, sheetrock and drywall are all considered to be combustible materials when the code provisions are applied.

Additionally, gypsum wallboard has a paper face that has a flame spread index that is measurable in the ASTM E84 test. This alone identifies the need to classify gypsum wallboard as a combustible material for the purpose of requiring a separation from heat-producing equipment and appliances.

M1306.2.3 Solid-fuel appliances. Table M1306.2 shall not be used to reduce the clearance required for solid-fuel *appliances* listed for installation with minimum clearances of 12 inches (305 mm) or less. For *appliances listed* for installation with minimum clearances greater than 12 inches (305 mm), Table M1306.2 shall not be used to reduce the clearance to less than 12 inches (305 mm).

❖ Because solid-fuel-burning appliances can produce high-intensity radiant heat and have wide variations in heat output, this section restricts the use of Table M1306.2 for types of solid-fuel-burning appliances that are listed for clearances of 12 inches (305 mm) or less. With small clearances, the protection method could be inadequate because of the radiation intensity, reradiation from the protection assembly and inability of the protection method to dissipate the heat energy at the rate received from the appliance. If the listed clearance is greater than 12 inches (305 mm), Table M1306.2 is applicable provided that the clearance is not reduced to less than 12 inches (305 mm). The bottom line is that clearance for solid-fuel appliances can never be less than 12 inches (305 mm).

SECTION M1307
APPLIANCE INSTALLATION

M1307.1 General. Installation of *appliances* shall conform to the conditions of their *listing* and *label* and the manufacturer's instructions. The manufacturer's operating and installation instructions shall remain attached to the *appliance*.

❖ Manufacturer's installation instructions are thoroughly evaluated by the listing agency verifying that a safe installation is prescribed. The listing agency can require that the manufacturer alter, delete or add information to the instructions as necessary to achieve compliance with applicable standards and code requirements. Manufacturer's installation instructions are an enforceable extension of the code and must be in the hands of the code official when an inspection

takes place. Without access to the instructions, the code official would be unable to complete an inspection.

When an appliance is tested to obtain a listing and label, the approved agency installs the appliance in accordance with the manufacturer's installation instructions. The appliance is tested under these conditions; thus, the installation instructions become an integral part of the labeling.

The listing and labeling process ensures that the appliance and its installation instructions are in compliance with applicable standards. Therefore, an installation in accordance with the manufacturer's instructions is required, except where the code requirements are more stringent. An inspector must carefully and completely read and comprehend the manufacturer's instructions to properly perform an installation inspection.

In some cases, the code will specifically address an installation requirement that is also addressed in the manufacturer's installation instructions. The code requirements may be the same or may exceed the requirements in the manufacturer's installation instructions or the manufacturer's installation instructions could contain requirements that exceed those in the code. In all cases, the more restrictive requirements would apply.

Even if an installation appears to be in compliance with the manufacturer's instructions, the installation cannot be complete or approved until all associated components, connections and systems that serve the appliance or equipment are also in compliance with the applicable provisions of the code. For example, an oil-fired boiler installation must not be approved if the boiler is connected to a deteriorated, undersized or otherwise unsafe chimney or vent. Likewise, the same installation must not be approved if the existing oil piping is in poor condition or if the electrical supply circuit is inadequate or unsafe.

In the case of replacement installations, the intent of this section is to require new work associated with the installation to comply with the code without necessarily requiring full compliance for the existing, unchanged portions of the related ductwork, piping, electrical, venting and similar mechanical systems. For example, if a furnace is replaced in an existing building, the new work and connections involved with the replacement would be treated as new construction and the existing unaltered system components would be considered as existing mechanical systems. Existing mechanical systems are accepted on the basis that they are free from hazard although not necessarily compliant with current codes. The code is not retroactive except where specifically stated that it applies to existing systems.

Manufacturer's installation instructions are often updated and changed for various reasons, such as changes in the appliance or equipment design, revisions to the product standard and as a result of field experience related to existing installations. The code official should stay abreast of any changes by reviewing the manufacturer's instructions for every installation.

Equipment and appliances must be installed in accordance with the manufacturer's installation instructions. The manufacturer's label and installation instructions must be consulted in determining whether or not an appliance or piece of equipment can be installed and operated in a particular hazardous location. The manufacturer's installation instructions must be available on the job site when the equipment is being inspected.

M1307.2 Anchorage of appliances. *Appliances* designed to be fixed in position shall be fastened or anchored in an *approved* manner. In Seismic Design Categories D_0, D_1 and D_2, and in townhouses in Seismic Design Category C, water heaters and thermal storage units shall be anchored or strapped to resist horizontal displacement caused by earthquake motion in accordance with one of the following:

1. Anchorage and strapping shall be designed to resist a horizontal force equal to one-third of the operating weight of the water heater storage tank, acting in any horizontal direction. Strapping shall be at points within the upper one-third and lover one-third of the *appliance's* vertical dimensions. At the lower point, the strapping shall maintain a minimum distance of 4 inches (102 mm) above the controls.

2. The anchorage strapping shall be in accordance with the appliance manufacturer's recommendations.

❖ This section requires the anchorage of fixed appliances to prevent movement or the possible tipping over of the appliance. In areas of high seismic risk,

water heaters must be secured directly to the building. Water heaters can weigh several hundred pounds, and if not properly secured, the water, gas and electrical connections may be damaged or severed, creating a hazardous situation. For example, a 40 gallon water heater will have a total weight that includes the weight of the empty heater plus the weight of the water (40 x 8.333 pounds). Commentary Figure M1307.2 shows a method of anchorage of a residential water heater that is known to have been used. Note that the required strapping must be substantial to resist lateral movement. Water heater manufacturers can provide details for such support and possibly support kits. The use of plumber's perforated hanger strap, also known as "holy iron," is commonly encountered and is probably quite inadequate.

M1307.3 Elevation of ignition source. *Appliances* having an *ignition source* shall be elevated such that the source of ignition is not less than 18 inches (457 mm) above the floor in garages. For the purpose of this section, rooms or spaces that are not part of the *living space* of a *dwelling unit* and that communicate with a private garage through openings shall be considered to be part of the garage.

Exception: Elevation of the ignition source is not required for appliances that are listed as flammable-vapor-ignition resistant.

❖ To reduce the hazard of fire or explosion from gasoline and other chemical vapors that are likely to be stored in a garage, appliances having an ignition source must be elevated to keep the ignition source a safe distance above garage floors (see Commentary Figure M1307.3). It is not the intent of this section for the

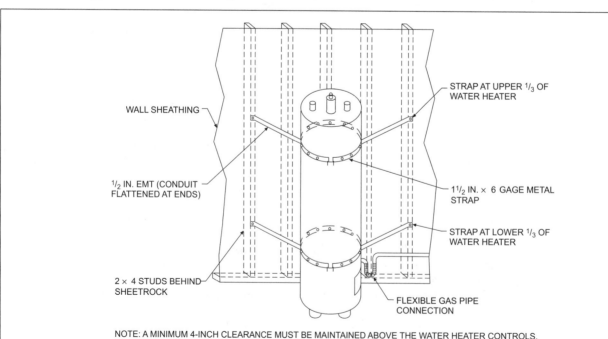

NOTE: A MINIMUM 4-INCH CLEARANCE MUST BE MAINTAINED ABOVE THE WATER HEATER CONTROLS.

For SI: 1 inch = 25.4 mm.

Figure M1307.2
ANCHORAGE OF WATER HEATER

installer to measure the elevation from the surface of a dedicated appliance stand or other structure built to support the appliance where that stand or structure does not afford room for the storage of flammable liquids. Some flammable and combustible liquids typically associated with hazardous locations give off vapors that are denser than air and tend to collect near the floor. The 18-inch (457 mm) height requirement is intended to reduce the possibility of flammable vapor ignition by keeping the ignition sources elevated above the anticipated level of accumulated vapors. The 18-inch (457 mm) value is a minimum requirement and must be increased when required by the manufacturer's installation instructions.

This section will effectively prohibit the installation of most furnaces, boilers, space heaters, clothes dryers and water heaters directly on the floor of residential garages.

The accumulation of flammable vapors more than 18 inches (457 mm) deep is unlikely in most ventilated locations; therefore, maintaining all possible sources of ignition at least 18 inches (457 mm) above the floor will substantially reduce the risk of explosion and fire [see Commentary Figures G2408.2(1) and G2408.2(2)].

In the context of this section, a source of ignition could be a pilot flame, burner, burner igniter or electrical component capable of producing a spark. The term "ignition source" is defined and can be interpreted as

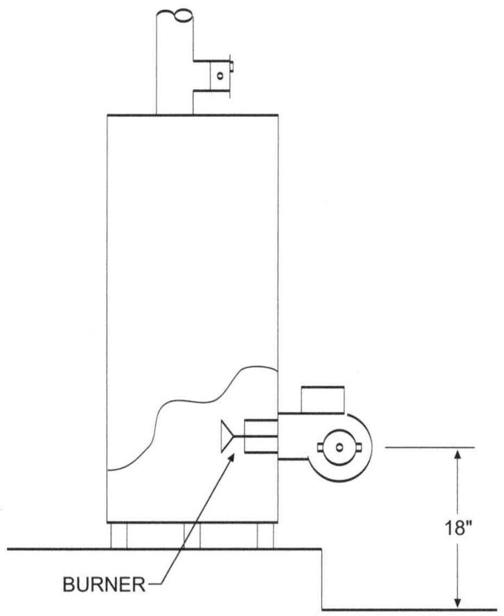

For SI: 1 inch = 25.4 mm.

**Figure M1307.3
OIL-FIRED WATER HEATER INSTALLATION
IN A GARAGE**

meaning an intentional source of ignition in a fuel-fired appliance or an unintentional source of ignition for any flammable vapors that may be present in the space (see the definition of "Ignition source").

An appliance installed in a closet or room that is accessible only from the garage must be considered as part of the garage for application of this section. Even though the room may be separated from the garage by walls and a door, there are no practical means of making the door vapor tight nor is there any assurance that the door will remain closed during normal use. An appliance room that is accessed only from the outdoors or only from the living space would not be considered as part of the garage. Rooms such as utility rooms and laundry rooms that communicate with both the garage and the living space and that are part of the living space, are not considered to be part of the garage (see definition of "Living space").

Appliances in the garage must be protected from impact by automobiles and elevation of the appliance may not be sufficient to guard against damage from impact unless the code official determines that the appliance platform is of substantial construction and the necessary height to be capable of protecting the appliance from impact.

The exception recognizes that new technology exists that allows appliances, such as water heaters, to be tested and listed as being flammable-vapor-ignition resistant and suitable for installation without the 18-inch (457 mm) elevation requirement. Note that while new gas-fired water heaters have flammable vapor ignition-resistant technology that allows them to be installed on the floor without the 18-inch (457 mm) elevation (see commentary, Section G2408.2), to the author's knowledge, oil-fired and electric water heaters are not listed as flammable-vapor-ignition resistant.

M1307.3.1 Protection from impact. *Appliances* shall not be installed in a location subject to vehicle damage except where protected by *approved* barriers.

❖ Mechanical appliances installed in garages, carports and any other location where motorized vehicles are present must be protected from vehicular impact.

Although the code does not specify methods of protection, the most apparent method would be to locate the appliance where it could not be struck by a vehicle. A practical method of protection would be to place a formidable and permanent barrier between the motor vehicles and the appliance. This barrier could include such items as an effectively located vehicle wheel stop that is anchored in place, an elevated platform higher than the vehicle's bumpers or one or more concrete-filled steel pipes. Final approval of the method of protection is left to the local code official. See Commentary Figure M1307.3.1 for some commonly used methods of protection.

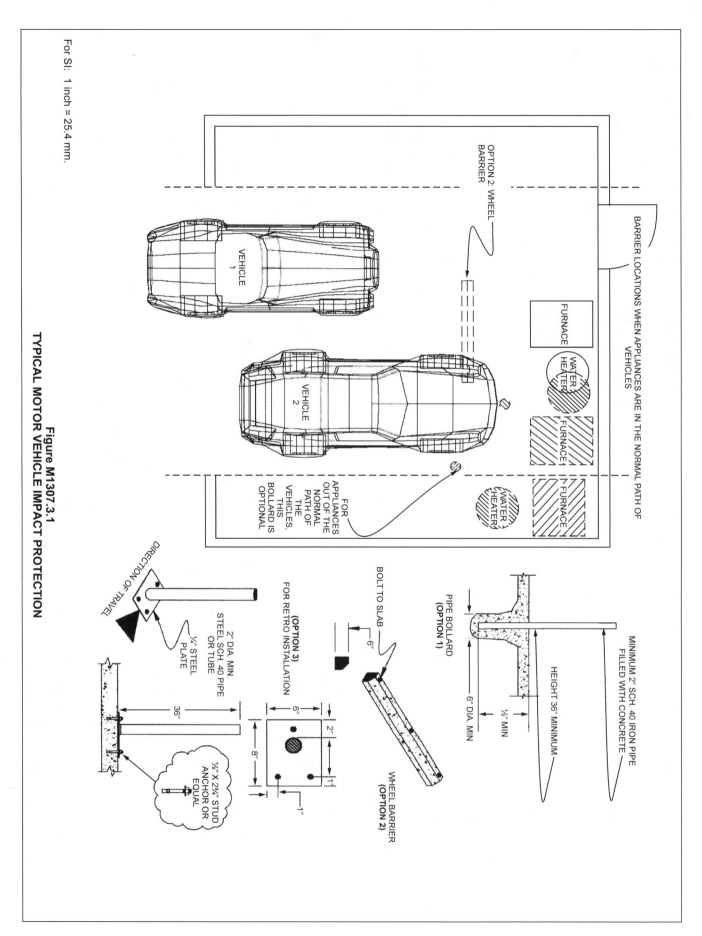

Figure M1307.3.1
TYPICAL MOTOR VEHICLE IMPACT PROTECTION

For SI: 1 inch = 25.4 mm.

M1307.4 Hydrogen generating and refueling operations. *Ventilation* shall be required in accordance with Section M1307.4.1, M1307.4.2 or M1307.4.3 in private garages that contain hydrogen-generating *appliances* or refueling systems. For the purpose of this section, rooms or spaces that are not part of the *living space* of a *dwelling unit* and that communicate directly with a private garage through openings shall be considered to be part of the private garage.

❖ The use of hydrogen to fuel vehicles and generate electricity in fuel cell appliances to replace petroleum-based fuels is a rapidly developing technology. Key factors in the increased use of hydrogen are the reduced atmospheric emissions associated with hydrogen and the nation's shift to renewable sources of energy.

  Typically, the code official will encounter two classes of equipment—those that generate hydrogen for use by other equipment, such as vehicles, and those that use hydrogen as their energy input, such as fuel cell appliances. This section intends to minimize the potential for explosions by limiting the source of hydrogen gas and by requiring sufficient ventilation to dissipate any leakage.

M1307.4.1 Natural ventilation. Indoor locations intended for hydrogen-generating or refueling operations shall be limited to a maximum floor area of 850 square feet (79 m^2) and shall communicate with the outdoors in accordance with Sections M1307.4.1.1 and M1307.4.1.2. The maximum rated output capacity of hydrogen-generating *appliances* shall not exceed 4 standard cubic feet per minute (1.9 L/s) of hydrogen for each 250 square feet (23 m^2) of floor area in such spaces. The minimum cross-sectional dimension of air openings shall be 3 inches (76 mm). Where ducts are used, they shall be of the same cross-sectional area as the free area of the openings to which they connect. In those locations, *equipment* and *appliances* having an *ignition source* shall be located so that the source of ignition is not within 12 inches (305 mm) of the ceiling.

❖ The 850-square-foot (79 m^2) maximum floor space used for hydrogen generation or refueling and the maximum rated output capacity of such appliances are intended to limit the amount of hydrogen gas that can accumulate in the space, thus minimizing the potential for explosions.

  The location of an ignition source parallels the intent of Section M1307.3, but with respect to the ceiling instead of the floor. This section does not apply to spaces that house only vehicles and that do not contain hydrogen-generating or refueling operations. Because it is more buoyant, hydrogen will dissipate more quickly than natural gas and much more quickly than either propane or gasoline, both of which have vapors that are heavier than air and will linger at an accident site. However, hydrogen and natural gas can both accumulate in unventilated pockets at the top of indoor structures and could represent a risk in such situations.

  Similarly, gasoline fumes can accumulate at the floor level in unventilated spaces, posing a different risk. Thus, ignition sources must be avoided at the top of any unventilated spaces for hydrogen gas. Also, hydrogen is odorless, colorless and burns with a flame that is not generally visible to the human eye. This means that it is unlikely that people will be able to detect unsafe conditions without appropriate instrumentation [similar to carbon monoxide (CO) accumulation in a structure].

M1307.4.1.1 Two openings. Two permanent openings shall be constructed within the garage. The upper opening shall be located entirely within 12 inches (305 mm) of the ceiling of the garage. The lower opening shall be located entirely within 12 inches (305 mm) of the floor of the garage. Both openings shall be constructed in the same exterior wall. The openings shall communicate directly with the outdoors and shall have a minimum free area of $^1/_2$ square foot per 1,000 cubic feet (1.7 m^2/1000 m^3) of garage volume.

❖ The location requirement will prevent the openings from being more than 12 inches (305 mm) tall because the required openings must be entirely within the 12 inches (305 mm) of wall space measured down from the ceiling and up from the floor. The openings must be in the same wall to help create a gravity flow of gases driven by the natural buoyancy of hydrogen gas. The bottom opening is an air inlet and the top opening is an air outlet.

M1307.4.1.2 Louvers and grilles. In calculating free area required by Section M1307.4.1, the required size of openings shall be based on the net free area of each opening. If the free area through a design of louver or grille is known, it shall be used in calculating the size opening required to provide the free area specified. If the design and free area are not known, it shall be assumed that wood louvers will have a 25-percent free area and metal louvers and grilles will have a 75-percent free area. Louvers and grilles shall be fixed in the open position.

❖ This section recognizes that louvers and grilles are usually installed over air inlets and outlets to prevent rain, snow and animals from entering the building. When louvers or grilles are used, the solid portion of the louver or grille must be considered when determining the unobstructed (net clear) area of the opening.

  Air openings are sized based on a free, unobstructed area for the passage of air. Louvers or grilles placed over these openings reduce the area of the openings because of the area occupied by the solid portions of the grille or louver. The reduction in area must be considered because only the unobstructed area can be credited toward the required opening size.

  The reduction in opening area caused by the presence of grilles or louvers will always require openings to be larger than determined from the sizing ratios of this chapter and larger than any duct of the minimum required size that might connect to these openings.

M1307.4.2 Mechanical ventilation. Indoor locations intended for hydrogen-generating or refueling operations shall be ventilated in accordance with Section 502.16 of the *International Mechanical Code*. In these locations, *equip-*

ment and *appliances* having an *ignition source* shall be located so that the source of ignition is below the mechanical *ventilation* outlet(s).

❖ Section 502.16 of the IMC provides criteria for the design and operation of a mechanical ventilation system in repair garages for natural-gas-fueled and hydrogen-fueled vehicles. This section makes Section 502.16 of the IMC applicable to indoor spaces containing hydrogen-generating and/or refueling operations. Although the IMC does not specify that the outlet must be within 12 inches (305 mm) of the ceiling as is required for natural ventilation, it does require the opening to be located at the high point of the space. This section requires that any ignition source be located below the mechanical exhaust outlet opening.

M1307.4.3 Specially engineered installations. As an alternative to the provisions of Sections M1307.4.1 and M1307.4.2, the necessary supply of air for *ventilation* and dilution of flammable gases shall be provided by an *approved* engineered system.

❖ The code is not intended to inhibit innovative ideas or technological advances. A comprehensive regulatory document such as a fuel gas code cannot envision and then address all future innovations in the industry. As a result, a code must be applicable to and provide a basis for the approval of an increasing number of newly developed, innovative materials, systems and methods for which no code text or referenced standards yet exist. The fact that a material, product or method of construction is not addressed in the code is not an indication that prohibition of the material, product or method is intended. The code official is expected to apply sound technical judgment in accepting materials, systems or methods that, while not anticipated by the drafters of the current code text, can be demonstrated to offer equivalent performance. By virtue of its text, the code regulates new and innovative construction practices while addressing the relative safety of building occupants. The code official is responsible for determining whether a requested alternative provides a level of protection of the public health, safety and welfare as required by the code.

M1307.5 Electrical appliances. Electrical *appliances* shall be installed in accordance with Chapters 14, 15, 19, 20 and 34 through 43 of this code.

❖ Electrical appliances must be installed in accordance with the electrical provisions of Chapters 34 through 43 and the installation requirements for specific appliances in Chapters 14, 15, 19 and 20.

M1307.6 Plumbing connections. Potable water and drainage system connections to *equipment* and *appliances* regulated by this code shall be in accordance with Chapters 29 and 30.

❖ Many appliances addressed in the mechanical part of the code have potable water connections and/or drain connections. Chapters 29 and 30 contain provisions for such connections to protect the potable water supply from contamination and to protect the occupants

from health hazards associated with improper connections to the sanitary drainage system.

SECTION M1308
MECHANICAL SYSTEMS INSTALLATION

M1308.1 Drilling and notching. Wood-framed structural members shall be drilled, notched or altered in accordance with the provisions of Sections R502.8, R602.6, R602.6.1 and R802.7. Holes in load-bearing members of cold-formed steel light-frame construction shall be permitted only in accordance with Sections R505.2.6, R603.2.6 and R804.2.6. In accordance with the provisions of Sections R505.3.5, R603.3.4 and R804.3.3, cutting and notching of flanges and lips of load-bearing members of cold-formed steel light frame construction shall not be permitted. Structural insulated panels (SIPs) shall be drilled and notched or altered in accordance with the provisions of Section R610.7.

❖ During mechanical system installation, it is usually necessary to penetrate structural members. The building chapters of the code regulate how these penetrations should be made so as not to weaken the structural members. The provisions of this section refer to those requirements, which are especially strict regarding the cutting and notching of steel members and prohibit the cutting and notching of flanges and lips of cold-formed, steel-framed, load-bearing members. See the commentary to the sections listed in the code text and also to Section P2603.2.

M1308.2 Protection against physical damage. Where piping will be concealed within light-frame construction assemblies, the piping shall be protected against penetration by fasteners in accordance with Sections M1308.2.1 through M1308.2.3.

   Exception: Cast iron piping and galvanized steel piping shall not be required to be protected.

❖ This section is intended to minimize the possibility of damage to refrigerant piping and other mechanical system piping from nails, screws or other fasteners. Because nails and screws sometimes miss the stud, rafter joist or top or sole plate, the shield must protect the pipe through the full width of the member and must extend not less than 2 inches (51 mm) above the sole plates and 2 inches (51 mm) below the top plates. Commentary Figure M1308.2 shows typical shield plates. Cast-iron and galvanized steel pipe have wall thicknesses greater than the required thickness of the shield plate, which makes them inherently resistant to nail and screw penetrations. The same requirement exists in Section P2603.2.1 for plumbing piping and Section G2415.7 for gas piping (see commentary, Sections P2603.2.1 and G2415.7). Piping such as refrigerant and gas piping should not be run in close proximity to the underside of roof deck sheathing because there have been cases where roofing nails have penetrated such piping, especially during reroofing. See Section M1411.7.

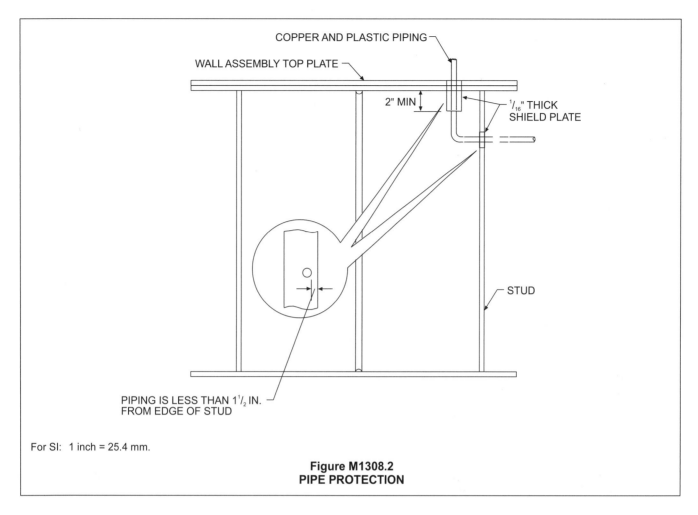

For SI: 1 inch = 25.4 mm.

Figure M1308.2
PIPE PROTECTION

M1308.2.1 Piping through bored holes or notches. Where *piping* is installed through holes or notches in framing members and is located less than $1^1/_2$ inches (38 mm) from the framing member face to which wall, ceiling or floor membranes will be attached, the pipe shall be protected by shield plates that cover the width of the pipe and the framing member and that extend 2 inches (51 mm) to each side of the framing member. Where the framing member that the piping passes through is a bottom plate, bottom track, top plate or top track, the shield plates shall cover the framing member and extend 2 inches (51 mm) above the bottom framing member and 2 inches (51 mm) below the top framing member.

❖ See the commentary to Section G2415.7.1.

M1308.2.2 Piping in other locations. Where piping is located within a framing member and is less than $1^1/_2$ inches (38 mm) from the framing member face to which wall, ceiling or floor membranes will be attached, the piping shall be protected by shield plates that cover the width and length of the piping. Where piping is located outside of a framing member and is located less than $1^1/_2$ inches (38 mm) from the nearest edge of the face of the framing member to which the membrane will be attached, the piping shall be protected by shield plates that cover the width and length of the piping.

❖ See the commentary to Section G2415.7.2.

M1308.2.3 Shield plates. Shield plates shall be of steel material having a thickness of not less than 0.0575 inch (1.463 mm) (No. 16 gage).

❖ See the commentary to Section G2415.7.3.

Bibliography

The following resource materials were used in the preparation of the commentary for this chapter of the code:

IFGC–15, *International Fuel Gas Code*. Washington, DC: International Code Council, 2014.

NFPA 70–14, *National Electrical Code*. Quincy, MA: National Fire Protection Association, 2013.

Chapter 14:
Heating and Cooling
Equipment and Appliances

User note: Code change proposals to this chapter will be considered by the IRC – Plumbing and Mechanical Code Development Committee during the 2015 (Group A) Code Development Cycle.

General Comments

This chapter is concerned with the installation and hazards associated with heating and cooling appliances and equipment. These types of appliances and equipment have risks either because of the potential for fire or the accidental release of refrigerants. Both situations are undesirable in dwellings that are covered by the code.

The chapter is a collection of requirements for various heating and cooling appliances, dedicated to single topics by section. The common theme is that all of these types of appliances use energy in one form or another, and the improper installation of the appliances would present a hazard to the occupants of the dwellings.

Section M1401 consists of the general requirements that apply to all such appliances and equipment. Section M1402 contains the provisions for central furnaces. Section M1403 addresses heat pump equipment. Section M1404 refers the reader to Section M1411 for the criteria for refrigeration cooling equipment. Section M1405 contains the provisions for baseboard convectors. Sec-

tion M1406 addresses radiant heating systems. Section M1407 consists of the criteria for duct heaters. Section M1408 contains the provisions for vented floor furnaces. Section M1409 contains provisions for vented wall furnaces. Section M1410 addresses vented room heaters. Section M1411 consists of the provisions for refrigeration cooling equipment. Section M1412 contains the provisions for absorption cooling equipment. Section M1413 addresses evaporative cooling equipment, and Section M1414 contains the provisions for fireplace stoves.

Purpose

This chapter provides, in a simple format, the requirements for heating and cooling appliances and equipment commonly found in dwellings covered by the code. These regulations pertain to the safety of the occupants by establishing minimum requirements for the design, construction, installation and operation of these systems.

SECTION M1401
GENERAL

M1401.1 Installation. Heating and cooling *equipment* and *appliances* shall be installed in accordance with the manufacturer's instructions and the requirements of this code.

❖ As discussed in the commentary to Section M1307.1, the manufacturer's installation instructions are an enforceable extension of the code. The manufacturer's instructions must be available at the job site when the equipment and appliances are being inspected.

M1401.2 Access. Heating and cooling *equipment* and appliances shall be located with respect to building construction and other *equipment* and appliances to permit maintenance, servicing and replacement. Clearances shall be maintained to permit cleaning of heating and cooling surfaces; replacement of filters, blowers, motors, controls and vent connections; lubrication of moving parts; and adjustments.

    Exception: Access shall not be required for ducts, piping, or other components approved for concealment.

❖ Because mechanical equipment and appliances require routine maintenance, repairs and possible

replacement, access is required. Additionally, access recommendations or requirements are usually stated in the manufacturer's installation instructions. As a result, the provisions stated here are intended to supplement the manufacturer's installation instructions. The intent is that clearances be sufficient to permit cleaning of the heating and cooling heat exchange surfaces and permit replacement and maintenance of filters, motors, controls, etc. Section M1305.1 contains similar requirements.

The exception recognizes that providing access to some components is not necessary or practical. This is especially true for ducts and piping which are often concealed in walls, floors and ceilings. Components that do not require observation, inspection, adjustment, servicing, repair or replacement are not required to be provided with access when approved by the code official.

M1401.3 Equipment and appliance sizing. Heating and cooling *equipment* and *appliances* shall be sized in accordance with ACCA Manual S or other approved sizing methodologies based on building loads calculated in accordance with ACCA Manual J or other *approved* heating and cooling calculation methodologies.

Exception: Heating and cooling equipment and appliance sizing shall not be limited to the capacities determined in accordance with Manual S where either of the following conditions applies:

1. The specified equipment or appliance utilizes multi-stage technology or variable refrigerant flow technology and the loads calculated in accordance with the approved heating and cooling calculation methodology are within the range of the manufacturer's published capacities for that equipment or appliance.

2. The specified equipment or appliance manufacturer's published capacities cannot satisfy both the total and sensible heat gains calculated in accordance with the approved heating and cooling calculation methodology and the next larger standard size unit is specified.

❖ The Air Conditioning Contractors of America's Manual J, *Load Calculations for Residential Winter and Summer Air Conditioning*, contains a simplified method of calculating heating and cooling loads. It includes a room-by-room calculation method that allows the designer to determine the required capacity of the heating and cooling equipment. In addition, it provides for an estimate of the airflow requirements for each of the areas in the house. The estimate can be used in sizing the duct system for the types of heating and cooling units that use air as the medium for heat transfer. Other approved methods may be used with the code official's approval.

ACCA Manual S or an approved equivalent is used to accurately choose the size of HVAC equipment after the building's heat gains and losses are determined in accordance with Manual J. Manual S takes into account the varying performance of cooling equipment based on the climate zone it will be installed in. The manual also allows calculation of heating equipment output British thermal units per hour (Btu/h) based on the manufacturer's given input ratings. Fuel-fired heating equipment is typically rated based on Btu/h input, but the input rating cannot be used alone to size the appliance. The actual output rating must be known to match the output to the heat loss of the building. Actual appliance output depends on the efficiency of the appliance and the altitude at which the appliance is installed. For example, an 80 percent efficient appliance with an input rating of 100,000 Btu/h will have an output of 80,000 Btu/h at sea level and that output will decrease as the appliance is elevated above sea level. Appliances must be derated based on the altitude at which they are installed in accordance with the manufacturer's instructions. At higher elevations, the air is less dense and this will affect the fuel and oxygen mixture entering the burners. Some appliances will require adjustments to the burner manifold gas pressure, replacement of orifices and possibly other modifications to the burners, controls and/or venting components. A commonly used derating prescription calls for

the input rating to be reduced 4 percent for each 1,000 feet (304 800 mm) that the appliance is elevated above sea level with no reduction required for an elevation of up to 2,000 feet (609 600 mm) above sea level.

The exception, condition 1 addresses variable capacity equipment and would allow the equipment capacity to exceed the calculated equipment size. The calculated equipment size must fall within the range of the specified (chosen) variable capacity equipment. Condition 2 would allow the next larger standard size of the specified (chosen) equipment where the calculated size of the equipment is inadequate for both the sensible and the sensible and latent heat gains. In other words, if XYZ brand of equipment is specified and the size calculated in accordance with this section won't handle the loads, the next larger size of that specific equipment can be used.

M1401.4 Exterior installations. *Equipment* and *appliances* installed outdoors shall be *listed* and *labeled* for outdoor installation. Supports and foundations shall prevent excessive vibration, settlement or movement of the *equipment*. Supports and foundations shall be in accordance with Section M1305.1.4.1.

❖ Appliances installed outdoors must be specifically listed and labeled for outdoor installation. The concern is for weather and ambient temperatures. The manufacturer's instructions and listing must be consulted to determine whether a particular appliance is designed for or can be made suitable for outdoor installation. For example, furnaces cannot be installed outdoors in cold climates regardless of any weatherproof enclosure, unless the heat exchanger, burner assemblies and venting system are designed for exposure to temperatures below normal indoor temperatures. Cold ambient temperatures can cause harmful condensation to occur on heat exchanger surfaces of fuel-fired appliances.

Additionally, there may be local ordinances that govern the outdoor installation of appliances. Before installing an appliance outdoors, consult local zoning regulations, ordinances and subdivision covenants. Many of these regulations strictly limit the location of outdoor mechanical equipment and appliances.

Supports and foundations must be adequate for the loads imposed such that there is no movement that could put stress and strains on connections and attachments to the equipment and appliances. Supports and foundations must also comply with the requirements in Section M1305.1.4.1 (see commentary, Section M1305.1.4.1).

M1401.5 Flood hazard. In flood hazard areas as established by Table R301.2(1), heating and cooling *equipment* and *appliances* shall be located or installed in accordance with Section R322.1.6.

❖ Mechanical appliances, equipment and systems for dwellings that are located in flood hazard areas must be installed at or above the elevation required by the code or must be designed and installed to prevent the entry of water into the components and to resist the forces of the flood waters on the components (see commentary, Section R322.1.6).

SECTION M1402
CENTRAL FURNACES

M1402.1 General. Oil-fired central furnaces shall conform to ANSI/UL 727. Electric furnaces shall conform to UL 1995.

❖ The term "central furnace" refers to a furnace in a central location of the home that heats the air and then distributes it through ducts to the rooms of the home. The term originates from the days when homes were heated by local sources within a room such as a fireplace or a stove. This section requires oil-fired central furnaces to meet the requirements in UL 727, and it requires electric furnaces to meet the requirements of UL 1995. Furnaces can come in many configurations, including upflow, counterflow (downflow), horizontal flow and indoor and outdoor units (see Chapter 24 for gas-fired furnaces).

M1402.2 Clearances. Clearances shall be provided in accordance with the *listing* and the manufacturer's installation instructions.

❖ The clearances indicated on the equipment label and in the manufacturer's installation instructions must always be followed, as indicated throughout these mechanical provisions.

M1402.3 Combustion air. *Combustion air* shall be supplied in accordance with Chapter 17. *Combustion air* openings shall be unobstructed for a distance of not less than 6 inches (152 mm) in front of the openings.

❖ This section refers to Chapter 17 for combustion air requirements and specifically requires a minimum 6-inch (152 mm) unobstructed clearance in front of the firebox to ensure adequate distribution of combustion air to the appliance. Fuel-gas-fired furnaces must meet the combustion air requirements in Section G2407.

SECTION M1403
HEAT PUMP EQUIPMENT

M1403.1 Heat pumps. Electric heat pumps shall be listed and labeled in accordance with UL 1995 or UL/CSA/ANCE 60335-2-40.

❖ Heat pumps are central heating units that use the rejection heat from the condenser of a refrigeration circuit to heat the conditioned space. They are often backed up by auxiliary electric resistance heating elements which supplement the heat pump during defrost cycles and during times of very low outdoor ambient temperatures. The source of heat energy is the outdoor air for air-to-air units and bodies of water or wells for water-to-air units. Heat pumps utilize reversing valves that cause the evaporator to become a condenser and vice versa, thus allowing the unit to both heat and cool conditioned spaces.

SECTION M1404
REFRIGERATION COOLING EQUIPMENT

M1404.1 Compliance. Refrigeration cooling *equipment* shall comply with Section M1411.

❖ This section is a cross reference to the actual provisions found in Section M1411.

SECTION M1405
BASEBOARD CONVECTORS

M1405.1 General. Electric baseboard convectors shall be installed in accordance with the manufacturer's instructions and Chapters 34 through 43 of this code. Electric baseboard heaters shall be listed and labeled in accordance with UL 1042.

❖ Electric baseboard convectors must be listed and labeled in accordance with UL 1042 and be installed in accordance with manufacturers' installation instructions and with the electrical requirements of the code. The latter are based on the electrical provisions of the *National Electrical Code*, published by the National Fire Protection Association. See Commentary Figure M1405.1 for an illustration of a typical electric baseboard convector.

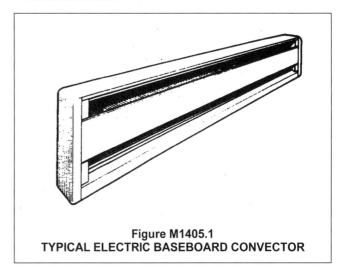

Figure M1405.1
TYPICAL ELECTRIC BASEBOARD CONVECTOR

SECTION M1406
RADIANT HEATING SYSTEMS

M1406.1 General. Electric radiant heating systems shall be installed in accordance with the manufacturer's instructions and Chapters 34 through 43 of this code and shall be listed for the application.

❖ Radiant heating systems contain electrical elements that create heat when an electric current passes through them. Radiant heating systems consist of heating cables and radiant heating panels that have been evaluated for contact with specific surfaces. If the system is not installed properly, electricity and the heat emitted could create hazards for the occupants of the dwelling. This section refers to the manufacturer's instructions and the electrical provisions in Chapters

34 through 43 for safe installations. The manufacturer's installation instructions will have details on the different type of surfaces and the temperature limits for each surface. These requirements are in addition to the requirements found in Sections M1406.2 through M1406.5.

M1406.2 Clearances. Clearances for radiant heating panels or elements to any wiring, outlet boxes and junction boxes used for installing electrical devices or mounting luminaires shall comply with Chapters 34 through 43 of this code.

❖ This section references certain minimum clearances located in the electrical chapters of the code so that the heating elements are not damaged or the electrical circuit continuity is not interrupted, thus rendering portions or all of the radiant heating systems inoperable. This section is also intended to protect the electrical systems, equipment, and components from damage from the heating elements.

M1406.3 Installation of radiant panels. Radiant panels installed on wood framing shall conform to the following requirements:

1. Heating panels shall be installed parallel to framing members and secured to the surface of framing members or mounted between framing members.

2. Mechanical fasteners shall penetrate only the unheated portions provided for this purpose. Panels shall not be fastened at any point closer than $^1/_4$ inch (6.4 mm) to an element. Other methods of attachment of the panels shall be in accordance with the panel manufacturer's instructions.

3. Unless *listed* and *labeled* for field cutting, heating panels shall be installed as complete units.

❖ Radiant heat panels mounted on wood framing may act as an ignition source if improperly installed. Because these systems generate heat, they must be mounted to wood framing by fastening through the unheated portions of each panel. Metallic fasteners, such as nails or screws, are excellent conductors of heat as well as electricity; therefore, they must not be driven through any portion of a panel closer than $^1/_4$ inch (6.4 mm) to a heating element. The manufacturer's installation instructions may contain additional methods that can be used to attach the panels.

M1406.4 Installation in concrete or masonry. Radiant heating systems installed in concrete or masonry shall conform to the following requirements:

1. Radiant heating systems shall be identified as being suitable for the installation, and shall be secured in place as specified in the manufacturer's installation instructions.

2. Radiant heating panels or radiant heating panel sets shall not be installed where they bridge expansion joints unless protected from expansion and contraction.

❖ Fire hazards created by electric radiant heating systems installed in masonry construction are minimal, but these installations require special mounting hardware and must be installed in accordance with the manufacturer's instructions. Radiant panels must be protected from expansion and contraction when mounted over masonry expansion joints to avoid structural damage.

M1406.5 Finish surfaces. Finish materials installed over radiant heating panels or systems shall be installed in accordance with the manufacturer's instructions. Surfaces shall be secured so that nails or other fastenings do not pierce the radiant heating elements.

❖ Finish materials must be carefully installed over electric radiant heating systems so that nails or screws do not penetrate the heating elements. Fasteners that are accidentally driven through these elements could cause short circuits, ground faults, destroy the elements or cause the surfaces to overheat and start a fire. The finish surfaces must be installed in accordance with the instructions supplied by the manufacturer of the radiant heating system.

SECTION M1407
DUCT HEATERS

M1407.1 General. Electric duct heaters shall be installed in accordance with the manufacturer's instructions and Chapters 34 through 43 of this code. Electric duct heaters shall comply with UL 1996.

❖ Electric duct heaters are designed to be installed "inline" in an air duct system and use electric resistance heating elements to heat the air passing through them. The code requires that their installation comply with the manufacturer's instructions and the electrical provisions in the code.

M1407.2 Installation. Electric duct heaters shall be installed so that they will not create a fire hazard. Class 1 ducts, duct coverings and linings shall be interrupted at each heater to provide the clearances specified in the manufacturer's installation instructions. Such interruptions are not required for duct heaters *listed* and *labeled* for zero clearance to combustible materials. Insulation installed in the immediate area of each heater shall be classified for the maximum temperature produced on the duct surface.

❖ These heaters could be installed in ducts that have combustible components, coverings or insulating materials. This section requires that these heaters be listed for zero clearance to combustible material or that the duct materials be kept at the specified distance from the heaters.

M1407.3 Installation with heat pumps and air conditioners. Duct heaters located within 4 feet (1219 mm) of a heat pump or air conditioner shall be *listed* and *labeled* for such installations. The heat pump or air conditioner shall additionally be *listed* and *labeled* for such duct heater installations.

❖ Heat pumps are typically unable to provide all of the heat required in a building at design temperature and also may have to shut down periodically for a defrost cycle. Thus, they are usually supplemented with electrical resistance heating to meet the occasional periods of peak load requirements and carry the load during defrost.

M1407.4 Access. Duct heaters shall be accessible for servicing, and clearance shall be maintained to permit adjustment, servicing and replacement of controls and heating elements.

❖ Section M1305 has additional criteria for access to the installations of appliances in specific locations.

M1407.5 Fan interlock. The fan circuit shall be provided with an interlock to prevent heater operation when the fan is not operating.

❖ A fan interlock is required to prevent the duct heater from operating when there is no air flow. Without air flow, the duct heater would overheat, possibly creating a fire hazard.

SECTION M1408
VENTED FLOOR FURNACES

M1408.1 General. Oil-fired vented floor furnaces shall comply with UL 729 and shall be installed in accordance with their *listing*, the manufacturer's instructions and the requirements of this code.

❖ Oil-fired floor furnaces are vented appliances that are installed in an opening in the floor. Such units supply heat to the room by gravity convection and direct radiation and typically serve as the sole source of space heating. Such furnaces are common in cottages, small homes, seasonally occupied structures and rural homes. Because the floor grille can become very hot, extreme care must be exercised to prevent occupants, especially children, from contacting the grille by walking or falling on it. Also, care must be taken to avoid a fire hazard caused by placement of materials or furnishings on or near the furnace floor grille.

M1408.2 Clearances. Vented floor furnaces shall be installed in accordance with their listing and the manufacturer's instructions.

❖ Vented floor furnaces must be installed in accordance with the manufacturer's instructions, which are an enforceable extension of the code.

M1408.3 Location. Location of floor furnaces shall conform to the following requirements:

1. Floor registers of floor furnaces shall be installed not less than 6 inches (152 mm) from a wall.

2. Wall registers of floor furnaces shall be installed not less than 6 inches (152 mm) from the adjoining wall at inside corners.

3. The furnace register shall be located not less than 12 inches (305 mm) from doors in any position, draperies or similar combustible objects.

4. The furnace register shall be located not less than 5 feet (1524 mm) below any projecting combustible materials.

5. The floor furnace burner assembly shall not project into an occupied under-floor area.

6. The floor furnace shall not be installed in concrete floor construction built on grade.

7. The floor furnace shall not be installed where a door can swing within 12 inches (305 mm) of the grille opening.

❖ Because of the heat generated in the combustion chamber that is emitted through the floor register, minimum clearances are necessary from the face of the air registers to surrounding walls, overhead projections and other combustible items, such as doors and draperies. See Commentary Figure M1408.3 for required clearances.

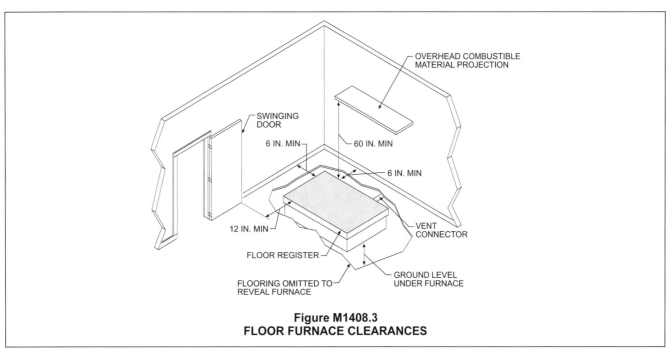

Figure M1408.3
FLOOR FURNACE CLEARANCES

M1408.4 Access. An opening in the foundation not less than 18 inches by 24 inches (457 mm by 610 mm), or a trap door not less than 22 inches by 30 inches (559 mm by 762 mm) shall be provided for access to a floor furnace. The opening and passageway shall be large enough to allow replacement of any part of the *equipment*.

❖ Either a floor opening (trap door) or a foundation opening must be provided to allow for maintenance or replacement of the furnace. Without such access, appliances are not as likely to receive the maintenance necessary for efficient and safe operation.

M1408.5 Installation. Floor furnace installations shall conform to the following requirements:

1. Thermostats controlling floor furnaces shall be located in the room in which the register of the floor furnace is located.

2. Floor furnaces shall be supported independently of the furnace floor register.

3. Floor furnaces shall be installed not closer than 6 inches (152 mm) to the ground. The minimum clearance shall be 2 inches (51 mm), where the lower 6 inches (152 mm) of the furnace is sealed to prevent water entry.

4. Where excavation is required for a floor furnace installation, the excavation shall extend 30 inches (762 mm) beyond the control side of the floor furnace and 12 inches (305 mm) beyond the remaining sides. Excavations shall slope outward from the perimeter of the base of the excavation to the surrounding *grade* at an angle not exceeding 45 degrees (0.79 rad) from horizontal.

5. Floor furnaces shall not be supported from the ground.

❖ Floor furnaces must have supports independent of the floor register and the ground below. A 6-inch (152 mm) clearance must also be maintained from the ground to protect against moisture, rust and corrosion of the furnace casing. This may be reduced to 2 inches (51 mm) if the unit is sealed against flooding. When the ground must be excavated to provide space for a floor furnace, a 30-inch (762 mm) clearance must be provided for access to the control side and a 12-inch (305 mm) clearance is required on all other sides.

SECTION M1409
VENTED WALL FURNACES

M1409.1 General. Oil-fired vented wall furnaces shall comply with UL 730 and shall be installed in accordance with their *listing*, the manufacturer's instructions and the requirements of this code.

❖ Vented, oil-fired wall furnaces are a type of room heater typically designed to be installed within a 2 × 4 stud cavity in frame construction. Wall furnaces are typically used in cottages, room additions and homes in mild climates. Some units are designed to serve a single room, and others are designed as through-the-wall units to serve adjacent rooms. Wall furnaces are ductless; however, some units are listed for use with a

surface-mounted supply outlet extension. Wall furnaces can be either gravity or forced-air type.

M1409.2 Location. The location of vented wall furnaces shall conform to the following requirements:

1. Vented wall furnaces shall be located where they will not cause a fire hazard to walls, floors, combustible furnishings or doors. Vented wall furnaces installed between bathrooms and adjoining rooms shall not circulate air from bathrooms to other parts of the building.

2. Vented wall furnaces shall not be located where a door can swing within 12 inches (305 mm) of the furnace air inlet or outlet measured at right angles to the opening. Doorstops or door closers shall not be installed to obtain this clearance.

❖ Wall furnaces, like all room heaters, can present a fire hazard if improperly located. The heat discharged or directly radiated from such units can ignite nearby wall or floor surfaces, furniture, trim items, window treatments and doors. A through-the-wall unit serving both a bathroom and an adjacent room must not recirculate air be-tween the two spaces. A combustible door that swings close to a wall furnace could be a fire hazard if at some position the door would be within 12 inches (305 mm) of an air inlet or outlet. A door could also interfere with air-flow through the furnace, thereby causing appliance overheating. Because door closures and doorstops are easily defeated, they must not be depended on to secure the required clearance. If the door swing cannot comply with this section, the door would have to be removed or rehung to change its swing direction.

M1409.3 Installation. Vented wall furnace installations shall conform to the following requirements:

1. Required wall thicknesses shall be in accordance with the manufacturer's installation instructions.

2. Ducts shall not be attached to a wall furnace. Casing extensions or boots shall be installed only where listed as part of a *listed* and *labeled appliance*.

3. A manual shut off valve shall be installed ahead of all controls.

❖ Most wall furnaces are not designed to force air through ducts, especially the gravity types that do not use fans. Attachment of ducts would add resistance to the flow of air through the furnace, thereby causing abnormally high temperature rises across the heat exchanger and abnormally high temperatures on furnace surfaces. Some wall furnaces are listed and designed for use with supply duct extensions designed for wall mounting that are intended to improve heat distribution. Such duct extensions are factory-built and supplied only by the furnace manufacturer.

The fuel-line connection to every appliance must be equipped with an individual shutoff valve to permit maintenance, repair or replacement of the appliance or its components.

M1409.4 Access. Vented wall furnaces shall be provided with access for cleaning of heating surfaces; removal of burners; replacement of sections, motors, controls, filters and other working parts; and for adjustments and lubrication of parts requiring such attention. Panels, grilles and access doors that must be removed for normal servicing operations shall not be attached to the building construction.

❖ Proper access for servicing and replacement of vented wall furnaces is very important, as it is with the other mechanical appliances regulated by the code (see commentary, Section M1408.4).

SECTION M1410
VENTED ROOM HEATERS

M1410.1 General. Vented room heaters shall be tested in accordance with ASTM E1509 for pellet-fuel burning, UL 896 for oil-fired or UL 1482 for solid fuel-fired and installed in accordance with their *listing*, the manufacturer's installation instructions and the requirements of this code.

❖ This section addresses vented space/room heaters, including direct-vent and vent-connected or chimney-connected appliances. As required by the appliance standard, such appliances are equipped with safety controls that will prevent fuel flow to the burners in the event of ignition system failure. UL 896 is the governing standard for oil-fired room heaters, UL 1432 is the standard for solid-fuel heaters and ASTM E1509 is the standard for pellet-fuel-burning heaters. Solid-fuel-burning room heaters in accordance with UL 1482 address solid fuel-burning, free-standing fire chamber assemblies that heat space by direct radiation, circulated heated air or a combination of both. Solid-fuel room heaters are chimney connected, and are designed for operation with the fire chamber (firebox) closed (see commentary, Section M1414.1). Mechanical appliances and equipment must be installed in accordance with the manufacturer's instructions. This is important because the manufacturer's instructions are used by the testing agency when the test installation is constructed. Therefore, a unit tested in compliance with the specified standard and installed in accordance with the manufacturer's instructions is expected to perform the same as the representative sample unit that was tested.

M1410.2 Floor mounting. Room heaters shall be installed on noncombustible floors or *approved* assemblies constructed of noncombustible materials that extend not less than 18 inches (457 mm) beyond the *appliance* on all sides.

Exceptions:

1. *Listed* room heaters shall be installed on noncombustible floors, assemblies constructed of noncombustible materials or floor protectors *listed* and *labeled* in accordance with UL 1618. The materials and dimensions shall be in accordance with the *appliance* manufacturer's instructions.

2. Room heaters *listed* for installation on combustible floors without floor protection shall be installed in

accordance with the *appliance* manufacturer's instructions.

❖ Room heaters must be installed on noncombustible floors or approved floor assemblies constructed of noncombustible materials that extend 18 inches (457 mm) beyond the appliance on all sides or installed on assemblies or listed protectors in accordance with the appliance listing and instructions, or they must be listed for installation on combustible floors. They must also be installed following the manufacturer's instructions.

SECTION M1411
HEATING AND COOLING EQUIPMENT

M1411.1 Approved refrigerants. Refrigerants used in direct refrigerating systems shall conform to the applicable provisions of ANSI/ASHRAE 34.

❖ The classification of refrigerants is based on ASHRAE 34, which numbers and classifies refrigerants in accordance with their potential hazards. Refrigerants are identified by their number preceded by the letter R, for example, R-134a. Manufacturer trademark names such as "Freon" are not used in the code.

M1411.2 Refrigeration coils in warm-air furnaces. Where a cooling coil is located in the supply plenum of a warm-air furnace, the furnace blower shall be rated at not less than 0.5-inch water column (124 Pa) static pressure unless the furnace is *listed* and *labeled* for use with a cooling coil. Cooling coils shall not be located upstream from heat exchangers unless *listed* and *labeled* for such use. Conversion of existing furnaces for use with cooling coils shall be permitted provided the furnace will operate within the temperature rise specified for the furnace.

❖ If a cooling coil is not a part of a listed furnace or air conditioning system and is added to a warm-air furnace, the following requirements must be met:

1. The furnace blower must be rated at 0.5-inch (12.7 mm) of water-column minimum static pressure. The furnace blower must be capable of moving the larger flow rates of air required for cooling.

2. Unless the furnace heat exchanger is corrosion resistant (usually stainless steel), the cooling coil unit must be installed downstream of the heat exchanger.

Moisture in the air could condense on the surface of a heat exchanger that is cooled by being downstream of a cooling coil and such condensation can corrode metal parts.

M1411.3 Condensate disposal. Condensate from cooling coils and evaporators shall be conveyed from the drain pan outlet to an *approved* place of disposal. Such piping shall maintain a minimum horizontal slope in the direction of discharge of not less than $^1/_8$ unit vertical in 12 units horizontal

(1-percent slope). Condensate shall not discharge into a street, alley or other areas where it would cause a nuisance.

❖ Appliances and equipment containing evaporators or cooling coils, including refrigeration, dehumidification and comfort cooling equipment, can produce condensate from the water vapor in the atmosphere. A drainage system is necessary to dispose of the condensate and prevent damage to the structure. Condensate disposal is a local issue dependent on specific local conditions such as soil conditions, contour of the area, sewer loading and water treatment plant capacity. Because of this, the approval of the disposal place is the responsibility of the local jurisdiction. Some of the locations that could be accepted are storm or sanitary sewers, rooftops, French drains, drainage ditches, collection ponds or simply into the yard, as is typical for most homes.

Some water treatment facilities operate near maximum capacity and do not allow condensate disposal into the sanitary sewer because condensate does not require treatment. If connected to the sanitary sewer, the connection must be indirect in compliance with the plumbing provisions of the code to prevent sewage and gases from entering the equipment or system. French drains, or seepage pits, are effective depending on the permeability of the soil, the rate of discharge and the size of the pit.

The piping conveying the condensate to its disposal point must be properly sloped to ensure drainage just the same as any drainage system. Sagging pipe could trap water and air that would impede the flow of the condensate.

M1411.3.1 Auxiliary and secondary drain systems. In addition to the requirements of Section M1411.3, a secondary drain or auxiliary drain pan shall be required for each cooling or evaporator coil where damage to any building components will occur as a result of overflow from the *equipment* drain pan or stoppage in the condensate drain piping. Such piping shall maintain a minimum horizontal slope in the direction of discharge of not less than $^1/_8$ unit vertical in 12 units horizontal (1-percent slope). Drain piping shall be not less than $^3/_4$-inch (19 mm) nominal pipe size. One of the following methods shall be used:

1. An auxiliary drain pan with a separate drain shall be installed under the coils on which condensation will occur. The auxiliary pan drain shall discharge to a conspicuous point of disposal to alert occupants in the event of a stoppage of the primary drain. The pan shall have a minimum depth of 1.5 inches (38 mm), shall be not less than 3 inches (76 mm) larger than the unit or the coil dimensions in width and length and shall be constructed of corrosion-resistant material. Galvanized sheet steel pans shall have a minimum thickness of not less than 0.0236-inch (0.6010 mm) (No. 24 Gage). Nonmetallic pans shall have a minimum thickness of not less than 0.0625 inch (1.6 mm).

2. A separate overflow drain line shall be connected to the drain pan installed with the *equipment*. This overflow drain shall discharge to a conspicuous point of disposal to alert occupants in the event of a stoppage of the primary drain. The overflow drain line shall connect to the drain pan at a higher level than the primary drain connection.

3. An auxiliary drain pan without a separate drain line shall be installed under the coils on which condensation will occur. This pan shall be equipped with a water level detection device conforming to UL 508 that will shut off the *equipment* served prior to overflow of the pan. The pan shall be equipped with a fitting to allow for drainage. The auxiliary drain pan shall be constructed in accordance with Item 1 of this section.

4. A water level detection device conforming to UL 508 shall be installed that will shut off the *equipment* served in the event that the primary drain is blocked. The device shall be installed in the primary drain line, the overflow drain line or the *equipment*-supplied drain pan, located at a point higher than the primary drain line connection and below the overflow rim of such pan.

❖ An auxiliary (redundant) drain pan or a secondary drain is required for equipment locations where condensate overflow would cause damage to any building components. The auxiliary drain pan is installed to catch condensate spilling from the primary condensate removal system in the equipment. This backup pan protects the building from structural and finish damage. A secondary (overflow) drain is a second drain connection to the primary pan and is at a higher elevation than the primary drain. These "back-up provisions" help protect the building from structural and finish damage.

Condensate drains are notorious for clogging because of debris (lint, dust) from air-handling systems and the natural affinity to produce slime growths in drain pans and pipes. It is relatively common for condensate overflows to cause damage to buildings. The specified minimum horizontal slope is intended to maintain flow to prevent the blockage of the condensate pipe. However, properly sloping the drain will not ensure that blockage will not occur, therefore, this section provides four options for preventing such damage where the equipment is located in spaces such as attics, above suspended ceilings and furred spaces and locations on upper stories. One of the four methods must be used.

The first method uses an auxiliary drain pan below the coils on which condensate will occur with an independent drain line that discharges to a location that is easily observable to notify the building occupants that a problem with the primary pan exists. The code prescribes the depth of the pan and specific material thicknesses to ensure that the pan will be corrosion resistant and will have sufficient holding capacity and capture ability.

The second method uses an independent overflow drain line connected to the primary drain pan at a point higher than the primary drain line. Most evaporator coil pans are factory provided with an overflow drain tapping that can be used for this purpose. As in the first method, the point of discharge must be readily observable.

The third method uses a water-level detection device, usually a float switch or electronic sensor that must conform to the requirements of UL 508, located in the auxiliary drain pan [see Commentary Figure M1411.3.1(1)]. These detection devices will shutdown the equipment before the pan can overflow. There is no requirement for a separate drain line for the auxiliary pan in this method; however, there is a requirement for a drain fitting with a plug or cap that can be used to manually drain water from the pan to prevent water from standing in the pan for long periods of time.

The fourth method uses a water-level detection device installed in the primary drain line, in the overflow drain tapping on the primary pan or in the primary drain pan supplied by the manufacturer [see Commentary Figures M1411.3.1(2) and (3)]. This detection device shuts down the equipment before the pan can overflow. It also serves as a very noticeable warning to the homeowner that there is a problem with the HVAC system. Note that Item 4 is an acceptable method of protection, despite the fact that the main paragraph speaks only of secondary drains and auxiliary drain pans and Item 4 is neither.

M1411.3.1.1 Water-level monitoring devices. On downflow units and other coils that do not have secondary drain or provisions to install a secondary or auxiliary drain pan, a water-level monitoring device shall be installed inside the primary drain pan. This device shall shut off the equipment

served in the event that the primary drain becomes restricted. Devices shall not be installed in the drain line.

❖ The intent of this section is to provide adequate overflow protection on all coils that do not have a secondary drain and have no provisions for providing a secondary or auxiliary drain pan. Typical down-flow appliances and equipment used in dwelling units will have cooling

Photo courtesy of SMD Research, Inc.

Figure M1411.3.1(2)
CONDENSATE OVERFLOW SHUTOFF SWITCH
INSTALLED IN THE OUTLET OF THE AUXILIARY PAN

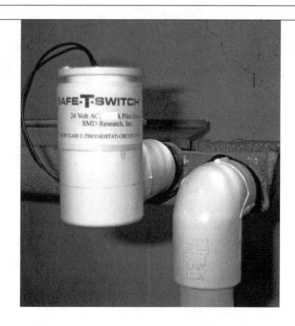

Photo courtesy of SMD Research, Inc.

Figure M1411.3.1(3)
CONDENSATE OVERFLOW SHUTOFF SWITCH
INSTALLED IN THE OVERFLOW DRAIN LINE

Photo courtesy of SMD Research, Inc.

Figure M1411.3.1(1)
CONDENSATE OVERFLOW
SHUTOFF SWITCH INSTALLED IN A PAN

coil primary drain pans that are equipped with secondary over-flow drain outlets; therefore, this section would rarely have to be applied in residential occupancies. The addition of a water level detection device inside the drain pan [see Commentary Figure M1411.3.1(1)] will prevent overflow caused by stoppages that occur anywhere in the drain, including at the site of the drain opening inside the primary drain pan provided by the manufacturer. When algae, airborne debris and sometimes foreign objects create a stoppage at the drain opening of the primary pan, overflow will not be prevented by a device that is installed externally in the drain line. Devices installed in the drain lines can only monitor stoppages that occur downstream of such devices and are in no way affected by the rising water level from a blockage inside the drain pan provided with the equipment.

When the water overflows from the pan, it typically runs into the duct, causing mold and mildew problems. It will eventually leak out through the joints or seams and cause damage to the building structure.

M1411.3.2 Drain pipe materials and sizes. Components of the condensate disposal system shall be ABS, cast iron, copper, cross-linked polyethylene, CPVC, galvanized steel, PE-RT, polyethylene, polypropylene or PVC pipe or tubing. Components shall be selected for the pressure and temperature rating of the installation. Joints and connections shall be made in accordance with the applicable provisions of Chapter 30. Condensate waste and drain line size shall be not less than $^3/_4$-inch (19 mm) nominal diameter from the drain pan connection to the place of condensate disposal. Where the drain pipes from more than one unit are manifolded together for condensate drainage, the pipe or tubing shall be sized in accordance with an *approved* method.

❖ Condensate drains must be constructed of one of the materials listed in this section, which are corrosion resistant. Such drains must be straight runs of pipe without sags or dips that would trap liquid. If coiled stock tubing is used, it would be difficult to maintain a uniform slope, and the low points (dips) in the tubing would trap water, causing the drain to be air bound, thereby blocking flow. When drains are merged, the piping must be properly sized for the aggregate flow. Note that the minimum size drain is required to be $^3/_4$-inch (19 mm) nominal diameter (I.D.) meaning that the drain line would be referred to or labeled as $^3/_4$-inch size, even though the actual diameter will be slightly more or less than $^3/_4$ inch, depending on the material. The $^3/_4$ inch refers to the trade size of the material, not an actual dimension.

M1411.3.3 Drain line maintenance. Condensate drain lines shall be configured to permit the clearing of blockages and performance of maintenance without requiring the drain line to be cut.

❖ Drains that convey condensate water from cooling coils and evaporators are known to develop blockages as a result of debris and biological growth in the system. These drains are commonly cleared of blockages by a compressed gas such as air or nitrogen being forced through the drain. It is inherently hazardous to pressurize plastic piping such as PVC and CPVC with a compressed gas because of the potential for violent rupture and propelled shards of plastic. The drains are seldom large enough to accommodate mechanical drain cleaning (rodding) equipment. The code permits any arrangement that provides access to the drain interior without the drain being severed or cut. This includes capped or plugged tees and cross fittings, unions, removable mechanical couplings and specialty devices made specifically for the attachment of compressed-gas hoses. The intent is to prevent the spillage of condensate that would cause damage to the structure (see Commentary Figure M1411.3.3).

M1411.3.4 Appliances, equipment and insulation in pans. Where *appliances, equipment* or insulation are subject to

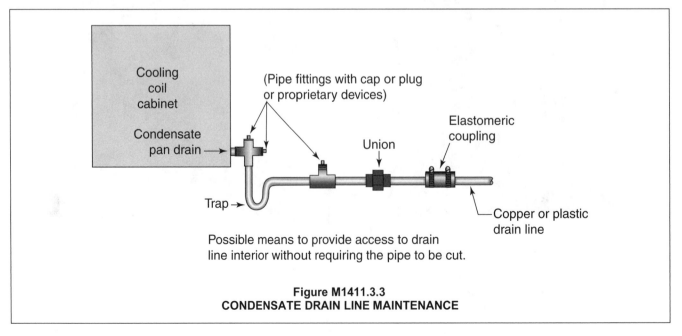

Possible means to provide access to drain line interior without requiring the pipe to be cut.

Figure M1411.3.3
CONDENSATE DRAIN LINE MAINTENANCE

water damage when auxiliary drain pans fill, those portions of the *appliances, equipment* and insulation shall be installed above the flood level rim of the pan. Supports located inside of the pan to support the *appliance* or *equipment* shall be water resistant and *approved*.

❖ If an auxiliary pan receives water and the appliance/ equipment is sitting directly on the bottom of the pan, water damage can occur to the appliance/equipment. Electrical components, metal items subject to corrosion and insulation within the appliance are examples of items that are subject to damage when submerged in water. This must be prevented by elevating the appliance/equipment above the rim of the pan such that the pan would overflow before any water reached the appliance/equipment. The means of elevating the appliance/equipment must be unaffected by repeated wettings [see Commentary Figures M1411.3.4(1) and M1411.3.4(2)].

M1411.4 Condensate pumps. Condensate pumps located in uninhabitable spaces, such as attics and crawl spaces, shall be connected to the appliance or equipment served such that when the pump fails, the appliance or equipment will be prevented from operating. Pumps shall be installed in accordance with the manufacturer's instructions.

❖ Condensate pumps are often located in attics and crawl spaces and above ceilings where they are not readily observable. If they fail, the condensate overflow can cause structural damage to the building, especially where the overflow will not be noticed immediately. The majority of such pumps are equipped with simple float controls that can be wired in series with the appliance/ equipment control circuit. When the pump system fails, the float will rise in the reservoir and open a switch before the condensate starts to overflow the reservoir. These float controls are commonly not connected, and in other cases the pump might not be equipped with an overflow switch. This new code section requires the installation of condensate pumps that have this overflow shutoff capability and requires that the appliance/ equipment served be connected to take advantage of that feature (see Commentary Figure M1411.4).

M1411.5 Auxiliary drain pan. Category IV condensing *appliances* shall have an auxiliary drain pan where damage to any building component will occur as a result of stoppage in the condensate drainage system. These pans shall be installed in accordance with the applicable provisions of Section M1411.3.

Exception: Fuel-fired *appliances* that automatically shut down operation in the event of a stoppage in the condensate drainage system.

❖ Evaporators and cooling coils are not the only appliances/equipment that produce condensate. Condensing-type fuel-fired appliances such as Category IV furnaces, boilers and condensing water heaters produce water condensate from the combustion of fuel and are considered to be high-efficiency appliances.

Figure M1411.3.4(1)
UNACCEPTABLE APPLIANCE AND PAN INSTALLATION

Manufacturers of condensing appliances may recommend that a pan be installed under their appliances; however, this section requires it where leakage would result in damage. The exception allows installation of appliances without an auxiliary drain pan if the appli-

ances will automatically shut down in the event of a condensate drain stoppage. In accordance with the appliance standards, if a drain blockage occurs, some appliances have the option of continuing to operate safely or automatically shutting down.

Figure M1411.3.4(2)
ACCEPTABLE APPLIANCE AND PAN INSTALLATION

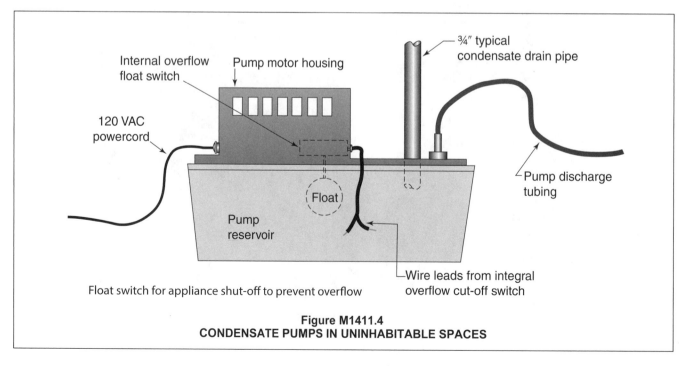

Figure M1411.4
CONDENSATE PUMPS IN UNINHABITABLE SPACES

M1411.6 Insulation of refrigerant piping. Piping and fittings for refrigerant vapor (suction) lines shall be insulated with insulation having a thermal resistivity of not less than R-4 and having external surface permeance not exceeding 0.05 perm [2.87 ng/(s · m² · Pa)] when tested in accordance with ASTM E 96.

❖ To improve the efficiency of the refrigeration cycle, to allow for proper compressor cooling and to control the formation of condensation on the piping, the refrigerant vapor (suction) lines must be insulated. Premade refrigerant line sets are commonly used and the suction line is factory insulated. Great care must be taken during installation to prevent the insulation from being torn, ripped or compressed. Damaged insulation will allow condensation to form on the tubing, so the insulation must be repaired or replaced.

M1411.7 Location and protection of refrigerant piping. Refrigerant piping installed within $1^1/_2$ inches (38 mm) of the underside of roof decks shall be protected from damage caused by nails and other fasteners.

❖ Refrigerant line sets for cooling systems and heat pumps are often routed up exterior walls and into attic spaces, which results in the lines being close to the roof deck at some points. When the roof covering is applied initially or during a reroofing installation, roofing nails and staples have been known to pierce the refrigerant lines. This results in loss of the system's refrigerant charge and results in environmental harm. The system will have to be repaired, evacuated (vacuum pumped) and recharged at significant expense. Protection of such lines is difficult to achieve; therefore, the best approach is to keep the lines away from the roof deck for a distance greater than $1^1/_2$ inches.

M1411.8 Locking access port caps. Refrigerant circuit access ports located outdoors shall be fitted with locking-type tamper-resistant caps or shall be otherwise secured to prevent unauthorized access.

❖ This provision intends to address a relatively new type and method of substance abuse where people intentionally inhale refrigerant gases for the intoxicating effect. In some cases, this inhalant abuse has resulted in the death of the individual. The typical condensing unit or heat pump unit is located outdoors and is equipped with access ports on the vapor and liquid refrigerant lines. These access ports are necessary for several purposes, such as to allow the connection of diagnostic gauges and to allow refrigerant to be added to or taken from the unit during servicing of the unit. Some of these access ports require back-seated valves to be opened with a wrench to allow refrigerant to escape and many of these access ports are equipped with simple "Schraeder" valves that are similar to the valve cores used on car and truck tires. All access ports are provided with threaded caps to keep out debris and moisture and to also guard against valve leakage. Individuals intent on inhaling refrigerants have found that they can withdraw refrigerant from these units by using simple hand tools or their bare hands to remove the

caps and open the valves.

The purpose of this section is to prevent this dangerous form of substance abuse by making it difficult if not impossible to remove these access port caps. Some of the available locking-type caps are designed such that wrenches, pliers and fingers cannot be used to remove the caps without a special tool/key.

All newly installed equipment, including replacement equipment, must be fitted with locking-type caps as part of such installations or other means must be provided to prevent access to the service ports. Such other means might include locked cages, locked access panels and locating the service access ports where they would be inaccessible to unauthorized persons [see Commentary Figures M1411.8(1) through M1411.8(4)].

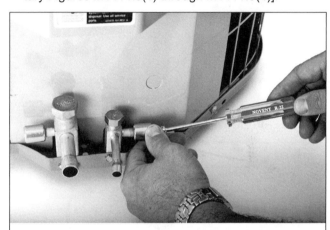

Photo courtesy of Novent L.L.C. and Airtec Products Corporation Inc.

Figure M1411.8(1)
INSTALLATION OF LOCKING CAPS FOR
REFRIGERANT ACCESS PORT

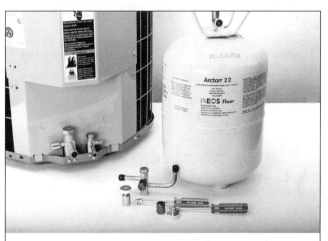

Photo courtesy of Novent L.L.C and Airtec Products Corporation Inc.

Figure M1411.8(2)
REFRIGERANT LOCKING CAPS AND TOOLS

Photo courtesy of Novent L.L.C and Airtec Products Corporation, Inc.

Figure M1411.8(3)
REFRIGERANT LABELS FOR LOCKING CAPS

Photo courtesy of Novent L.L.C and Airtec Products Corporation Inc.

Figure M1411.8(4)
SPECIAL REMOVAL TOOLS FOR
REFRIGERANT LOCKING CAPS

SECTION M1412
ABSORPTION COOLING EQUIPMENT

M1412.1 Approval of equipment. Absorption systems shall be installed in accordance with the manufacturer's instructions. Absorption equipment shall comply with UL 1995 or UL/CSA/ANCE 60335-2-40.

❖ Absorption refrigeration systems use two fluids (an absorbent and a refrigerant) and a heat source to remove heat through evaporation at a lower pressure and then reject the heat through condensation at a higher pressure. Typical absorption systems use ammonia as the refrigerant and water as the absorbent or lithium bromide as the absorbent and water as the refrigerant.

This section states that absorption systems must be installed in accordance with manufacturers' installation instructions, which are an enforceable extension of the code. Absorption equipment must be in compliance with UL 1995.

M1412.2 Condensate disposal. Condensate from the cooling coil shall be disposed of as provided in Section M1411.3.

❖ This section and Section M1411.3 provide for the proper and safe disposal of condensate.

M1412.3 Insulation of piping. Refrigerant piping, brine piping and fittings within a building shall be insulated to prevent condensation from forming on piping.

❖ Condensation will deteriorate piping and cause serious damage to other parts of buildings, so the piping must be insulated. Note that insulation alone cannot prevent condensation on piping, because a vapor barrier is necessary to prevent moisture from reaching the pipe surface. Some insulation materials also function as vapor barriers.

M1412.4 Pressure-relief protection. Absorption systems shall be protected by a pressure-relief device. Discharge from the pressure-relief device shall be located where it will not create a hazard to persons or property.

❖ In addition to requiring that the installation of this type of equipment be in compliance with the manufacturer's instructions, this section specifically requires the installation of pressure-relief devices that are to be discharged to a safe location.

SECTION M1413
EVAPORATIVE COOLING EQUIPMENT

M1413.1 General. Evaporative cooling equipment and appliances shall comply with UL 1995 or UL/CSA/ANCE 60335-2-40 and shall be installed:

1. In accordance with the manufacturer's instructions.

2. On level platforms in accordance with Section M1305.1.4.1.

3. So that openings in exterior walls are flashed in accordance with Section R703.4.

4. So as to protect the potable water supply in accordance with Section P2902.

5. So that air intake opening locations are in accordance with Section R303.5.1.

❖ This section requires installation of exterior equipment on a level platform at least 3 inches (76 mm) above the surrounding grade to protect the equipment from corrosion. This section states that evaporative cooling equipment must be installed in accordance with the manufacturer's instructions, which are an enforceable extension of the code.

SECTION M1414
FIREPLACE STOVES

M1414.1 General. Fireplace stoves shall be *listed*, *labeled* and installed in accordance with the terms of the listing. Fireplace stoves shall be tested in accordance with UL 737.

❖ A fireplace stove is a free-standing, chimney-connected, solid-fuel-burning heater designed to be operated with the fire chamber doors in either the open or closed position. The difference between a fireplace stove and a solid-fuel-type room heater is that a room heater is chimney connected and designed for operation with the fire chamber (firebox) closed. A fireplace stove must be tested in accordance with UL 737 whereas a solid-fuel-type room heater is tested in accordance with UL 1482 (see Section M1410). Note that some appliances are listed as both a stove and a room heater and, therefore, can be operated with the fire chamber in either the open or closed position. Fireplace stoves must be installed in accordance with the manufacturer's instructions. Such instructions are an integral part of the product listing since these instructions are used by the testing agency when the test installation is constructed. Therefore, a unit tested in compliance with UL 737 and installed in accordance with the manufacturer's instructions is expected to perform the same as the representative sample unit that was tested.

M1414.2 Hearth extensions. Hearth extensions for fireplace stoves shall be installed in accordance with the *listing* of the fireplace stove. The supporting structure for a hearth extension for a fireplace stove shall be at the same level as the supporting structure for the fireplace unit. The hearth extension shall be readily distinguishable from the surrounding floor area.

❖ Fireplace stoves must be installed in strict accordance with the manufacturer's instructions. These instructions will include specifications for the materials, dimensions and installation of a hearth extension. The hearth extension must be of noncombustible materials and must be supported at the same level as the fireplace stove. The floor supporting the hearth extension can be of combustible construction. Because a fireplace stove is a free-standing heater, in most installations the fireplace stove will be installed on the hearth extension. The dimensions of the hearth extension will be dictated by the manufacturer's installation instructions and will include distances that the hearth extension must extend beyond the sides, back and front of the fireplace stove.

A fireplace stove can be operated with the firebox doors in the open position, similar to a traditional masonry fireplace. The hearth extension will protect the floor structure on all sides from radiant heat, and it will protect the floor structure in front of the stove from sparks, embers, coals and burning solid fuel that might escape from the firebox. The nature of solid fuel can cause burning material to be expelled from the fire box, and large pieces of the fuel can fall from the grate or roll out of the firebox. In addition to the required hearth extensions, screens should be used as the initial and primary protection against sparks and embers from the firebox.

SECTION M1415
MASONRY HEATERS

M1415.1 General. Masonry heaters shall be constructed in accordance with Section R1002.

❖ Masonry heaters are a type of solid-fuel-burning appliance and are covered in Section R1002, which requires installation in accordance with ASTM E1602 or the terms of the listing. Such heaters are assembled on the job site from locally obtained raw materials or from factory components shipped to the job site. Masonry heaters are designed to be much more efficient than conventional fireplaces by incorporating immense thermal mass (masonry) and multiple pass flueways that provide greater heat exchange surface area and longer combustion gas retention time. Masonry heaters provide heat to the conditioned space by direct radiation of the heat energy stored in the immense thermal mass. In addition to having high thermal efficiencies, masonry heaters have the ability to provide space heating for many hours after the fire has burned out. Some factory-built kits are listed. Clearances to combustibles and convection cooling airflow are extremely important because these heaters store heat at high temperatures for long periods of time.

Bibliography

The following resource materials were used in the preparation of the commentary for this chapter of the code:

NFPA 70–14, *National Electrical Code*. Quincy, MA: National Fire Protection Association, 2013.

Chapter 15:
Exhaust Systems

User note: Code change proposals to this chapter will be considered by the IRC – Plumbing and Mechanical Code Development Committee during the 2015 (Group A) Code Development Cycle.

General Comments

As the title states, this chapter is a compilation of all exhaust system-related code requirements applicable to those uses covered by the code. The code regulates the materials used for constructing and installing such duct systems. Air brought into the building for ventilation, combustion or makeup is protected from contamination by the provisions found in this chapter.

Section M1501 requires that all mechanically exhausted air must discharge to the outdoors. Section M1502 pertains to the exhaust ducts from clothes dryers. Section M1503 has the criteria for exhaust ducts for range hoods. Section M1504 requires that microwave ovens be listed and labeled and properly installed. Section M1505 contains the regulations for exhausting domestic open-top broilers, and Section M1506 indicates that if a particular exhaust duct construction is not found in this chapter, the reader must use the criteria for ducts in Chapter 16. Section M1507 contains the minimum ventilation rates for bathroom and kitchen exhaust systems and some restrictions on recirculation of exhaust air.

The code official must be aware of the effect that exhaust systems can have on space conditioning and ventilation systems. For example, improperly designed exhaust systems can create negative pressure conditions, which can adversely affect the operation of fuel-burning appliances and the operation of other exhaust systems.

Purpose

Chapter 15 presents provisions associated with exhaust systems so hazards associated with exhaust systems and air contaminants can be avoided.

SECTION M1501
GENERAL

M1501.1 Outdoor discharge. The air removed by every mechanical exhaust system shall be discharged to the outdoors in accordance with Section M1506.3. Air shall not be exhausted into an attic, soffit, ridge vent or crawl space.

> **Exception:** Whole-house *ventilation*-type *attic* fans that discharge into the *attic* space of *dwelling units* having private *attics* shall be permitted.

❖ The primary intent of this section is to avoid exhausting contaminants into areas that may be occupied by people or into concealed spaces such as attics and crawl spaces where moisture can damage the building components. To prevent the introduction of contaminants into the ventilation air of a building, exhaust openings must not direct exhaust so that it could be readily drawn in by a ventilating system. To achieve this, the requirement for termination points for exhaust ducts must be in accordance with the location and placement of intake and exhaust openings in accordance with Section M1506.3.

Attics and crawl spaces are not considered to be outdoors. Exhaust ducts cannot terminate in these spaces. Exhaust ducts must connect directly to terminals that pass through the building envelope to the outside atmosphere. Pointing, aiming or similarly directing an exhaust duct at an opening in the envelope of the building such as an attic louver, ridge vent, gable vent, roof vent, eave vent or soffit vent, in no way ensures that all or any of the exhaust will reach the outdoors. In fact, it is possible that the majority, if not all, of the exhaust vapors and gases will discharge to the attic space rather than to the outdoors (see Commentary Figure M1501.1).

In the case of a duct that terminates in a ventilated soffit, the exhaust can rise into the attic as opposed to falling through the perforated soffit. The flow of air through an opening for attic ventilation is dependent on attic temperature, wind direction, wind speed and opening configuration and location, making attic air movement unpredictable and often in a direction not intended. Additionally, grilles and louvers offer resistance and interfere with the exhaust flow directed at them. This may cause deflection of exhaust back into the attic.

The exception addresses a mechanical alternative to natural ventilation of a dwelling unit. A whole-house ventilation-type attic fan is less source-specific than a dryer, bathroom or kitchen exhaust system. Whole-house ventilation systems are installed to provide a fresh air flow for the occupants either continuously or at timed intervals. The exception is specific to whole-house "ventilation-type" attic fans that create an airflow through open windows. Source-specific fans (bathroom exhaust or kitchen exhaust) would not qualify for the exception because their primary function is to exhaust moisture, odors, vapors or products of combustion. The contaminants associated with these functions might be harmful to the occupants, detrimental to the structural integrity of the building or even impact the fire safety performance of the building, and must discharge directly outdoors. The exception applies only to whole-house fans dedicated solely to ventilation and comfort cooling. Within the context of the code, a "private attic" is one in which air discharged from one dwelling unit cannot enter another dwelling unit's attic space. The exception is not related to the whole-house ventilation systems covered in Sections R303.4 and M1507.

SECTION M1502
CLOTHES DRYER EXHAUST

M1502.1 General. Clothes dryers shall be exhausted in accordance with the manufacturer's instructions.

❖ Clothes dryers must be exhausted in compliance with the dryer manufacturer's installation instructions. The manufacturer's installation instructions apply and where the code provisions are more strict, the code applies. Note that some of the provisions in the code are redundant with the appliance installation instructions. Dryers are designed and built to meet industry safety standards. The manufacturer's installation instructions are evaluated by the agency responsible for testing the appliance and, therefore, those instructions will prescribe an installation that is consistent with the appliance installation that was tested. This requirement is consistent with Section M1307.1, which addresses the installation of all appliances. The correct terminology is "dryer exhaust duct," as the term "vent" is used in a different context elsewhere in the code.

M1502.2 Independent exhaust systems. Dryer exhaust systems shall be independent of all other systems and shall convey the moisture to the outdoors.

Exception: This section shall not apply to *listed* and *labeled* condensing (ductless) clothes dryers.

❖ Clothes dryer exhaust systems are a potential fire hazard because of the combustible lint and debris that can accumulate in the duct and the heat source of the appliance. This combination of combustible lint and elevated temperature has been the cause of fires. The dryer exhaust system must be independent of other systems to prevent the hazard associated with one exhaust entering into or affecting other systems or areas in the building.

Clothes dryer exhaust systems must convey the moisture and any products of combustion directly to the exterior of the building (outdoors); therefore, any devices that allow all or part of the exhaust to discharge to the indoors are in violation of the code. The requirement to convey the exhaust to the outdoors applies to both fuel-fired and nonfuel-fired (i.e., electric) appliances. Clothes dryer exhaust systems cannot terminate in or discharge to any enclosed space, such as an attic or crawl space, regardless of whether or not the space is ventilated through openings to the outdoors. The high levels of moisture in the exhaust air

Figure M1501.1
PROHIBITED EXHAUST TERMINATION

can cause condensation to form on exposed surfaces or in insulation materials. Water vapor condensation can cause structural damage, deterioration of building materials and contribute to the growth of mold and fungus. Clothes dryer exhausts that discharge to enclosed spaces will also cause an accumulation of combustible lint and debris, creating a significant fire hazard. An improperly installed clothes dryer exhaust system not only reduces dryer efficiency and increases running time, but also can cause a significant increase in exhaust temperature, causing the dryer to cycle on its high limit control, which is an unsafe operating condition.

The exception specifically excludes condensing clothes dryers from the requirement to exhaust to the outdoors. Condensing dryers are electrically heated and recirculate the same air through the clothes drum. An air-to-air heat exchanger condenses the water vapor in the clothes drum air by using room air as the condensing medium. The condensate is collected in a reservoir that is either manually emptied or automatically emptied by an integral condensate pump. These dryers do not discharge heated air to the outdoors and have the effect of slightly elevating the indoor ambient temperature.

These "ductless" dryers must be listed for this application and installed to comply with the listing. They are commonly installed in applications where a conventional dryer exhaust duct cannot be used because of distance or aesthetics problems.

M1502.3 Duct termination. Exhaust ducts shall terminate on the outside of the building. Exhaust duct terminations shall be in accordance with the dryer manufacturer's installation instructions. If the manufacturer's instructions do not specify a termination location, the exhaust duct shall terminate not less than 3 feet (914 mm) in any direction from openings into buildings. Exhaust duct terminations shall be equipped with a backdraft damper. Screens shall not be installed at the duct termination.

❖ The clothes dryer exhaust discharge must terminate outdoors because of the high levels of moisture, combustible lint and, for gas-fired units, combustion products in the exhaust. If discharged indoors, such exhaust could present a health and fire hazard, could cause structural damage and deterioration of building material and could contribute to the growth of mold and fungus. The exhaust also contains highly combustible clothes fibers and, in the case of gas-fired units, products of combustion.

The requirement to locate the exhaust outlet at least 3 feet (914 mm) from any openings into the building is intended to minimize the possibility of drawing the dryer exhaust into the house through windows, doors or fresh air intake openings. Note that the dryer installation instructions may specify a greater distance. The 3-foot (914 mm) dimension is consistent with similar requirements in the *International Mechanical Code®* (IMC®) and has been successfully used in the housing industry for years.

Backdraft dampers must be installed in dryer

exhaust ducts to avoid outdoor air infiltration during periods when the dryer is not operating and to prevent the entry of animals. These dampers should be designed and installed to provide an adequate seal when in a closed position to minimize air leakage (infiltration). Backdraft dampers are typically of the gravity type, which are opened by the energy of the exhaust discharge. Some dryer manufacturers prohibit the use of magnetically held backdraft dampers because of the extra resistance that the exhaust flow must overcome.

Exhaust terminal opening size is also governed by the dryer manufacturer's instructions. Full opening terminals present less resistance to flow and might be mandated by the dryer manufacturer. A "full opening" is considered to be an opening having no dimension less than the diameter of the exhaust duct. Screens are prohibited on the duct termination because lint from the dryer will clog the screen creating flow resistance and eventually a complete blockage. This will result in a fire hazard. Consider that the filter screen integral with the appliance becomes restricted with lint in each cycle of operation. The code does not recognize any type or size of screen as being acceptable. Some dryer exhaust terminals are manufactured with a "basket" type of animal screen and the code makes no exception for such screens. Clothes fibers have the ability to coalesce and clog any screen.

M1502.4 Dryer exhaust ducts. Dryer exhaust ducts shall conform to the requirements of Sections M1502.4.1 through M1502.4.7.

❖ See the commentary to subsequent sections.

M1502.4.1 Material and size. Exhaust ducts shall have a smooth interior finish and be constructed of metal having a minimum thickness of 0.0157 inches (0.3950 mm) (No. 28 gage). The duct shall be 4 inches (102 mm) nominal in diameter.

❖ This section requires that the diameter of the exhaust duct be 4 inches (102 mm), no larger and no smaller. A 4-inch (102 mm) duct is the basis for the design of the appliance exhaust system. Ducts that are too small would restrict air movement and would cause the dryer to consume more energy because of the longer drying time. They could also cause appliance overheating. Ducts that are too large will cause a decrease in the flow velocity, causing lint and debris to drop out of the exhaust air stream. For example, increasing the duct size from 4 inches to 5 inches (102 mm to 127 mm) will result in a reduction of duct velocity of approximately 37 percent for the typical dryer. The exhaust duct material must be smooth-wall metal with the specified minimum wall thickness. Plastic ducts, corrugated ducts and flexible ducts are all prohibited with the exception of transition ducts in accordance with Section M1502.4.3.

M1502.4.2 Duct installation. Exhaust ducts shall be supported at intervals not to exceed 12 feet (3658 mm) and shall be secured in place. The insert end of the duct shall extend into the adjoining duct or fitting in the direction of airflow. Exhaust duct joints shall be sealed in accordance with Section

M1601.4.1 and shall be mechanically fastened. Ducts shall not be joined with screws or similar fasteners that protrude more than $^1/_8$ inch (3.2 mm) into the inside of the duct.

❖ Clothes dryer ducts must be well supported to prevent deformation and joint separation. The male end of the duct and fittings must point in the direction of airflow to lessen flow resistance, joint leakage and collection of lint. This is common practice for all ducts, regardless of application. Previous editions of the code did not permit duct fasteners to protrude into the inside of the ducts. This created a problem since all ducts have been required by tradition, codes or installation standards to be mechanically fastened at all joints and joint sealing is not equivalent to or a substitute for mechanical fastening (see Section M1601.4.1 applicable to HVAC ducts). This also created a problem when duct cleaning firms attempted to clean dryer exhaust ducts and they would come apart within concealed wall and ceiling spaces. The code permits mechanical fasteners to penetrate up to $^1/_8$ inch (3.2 mm) into the inside of the dryer exhaust duct. A maximum penetration of $^1/_8$ inch (3.2 mm) is expected to create only an insignificant amount of lint buildup. The $^1/_8$-inch (3.2 mm) limitation was intended to allow the use of short pop-rivets and $^1/_8$-inch-long (3.2 mm) screws. Dryer exhaust ducts must be exceptionally well supported and sealed so that the joints will not separate. This is especially important where the ducts are to be concealed within the building construction and where they will be mechanically cleaned by duct cleaning apparatus.

M1502.4.3 Transition duct. Transition ducts used to connect the dryer to the exhaust *duct system* shall be a single length that is *listed* and *labeled* in accordance with UL 2158A. Transition ducts shall be not greater than 8 feet (2438 mm) in length. Transition ducts shall not be concealed within construction.

❖ Within the context of this section, a transition duct is a flexible connector used as a transition between the dryer outlet and the connection point to the exhaust duct system. Transition duct connectors must be listed and labeled as transition ducts for clothes dryer application. Transition ducts must be listed to comply with UL 2158A. Such transition ducts are not considered as flexible air ducts or flexible air connectors.

Transition ducts are flexible ducts constructed of a metalized (foil) laminated fabric supported on a spiral wire frame. They are more fire resistant than the typical plastic spiral duct. Transition duct connectors are necessary for domestic dryers because of appliance movement, vibration and outlet location. In many cases, connecting a domestic clothes dryer directly to rigid duct would be difficult.

Transition duct connectors are limited to 8 feet (2438 mm) in length and must be installed in accordance with their listing and the manufacturer's instructions. These duct connectors must not be concealed by any portion of the structure's permanent finish materials, such as drywall, plaster, paneling, built-in furniture or cabinets or any other similar permanently affixed building component; they must remain entirely within the room in which the appliance is installed. Transition duct connectors cannot be joined to extend beyond the 8-foot (2438 mm) maximum length limit. Transition ducts are to be cut to length as needed to avoid excess duct and unnecessary bends. Note that in the application of Section M1502.4.5.1, the length of the transition duct is not counted in the overall duct system length, despite the fact that the transition duct does contribute considerable airflow resistance to the exhaust duct system.

M1502.4.4 Dryer exhaust duct power ventilators. Domestic dryer exhaust duct power ventilators shall conform to UL 705 for use in dryer exhaust duct systems. The dryer exhaust duct power ventilator shall be installed in accordance with the manufacturer's instructions.

❖ The previous code editions did not recognize dryer exhaust duct power ventilators (DEDPVs) as an option for clothes dryer installations. DEDPVs are loosely referred to as "dryer booster fans" in the marketplace. Prior to the 2015 code, the designer's choices for exhaust duct length were: 1. To limit the duct length to 35 feet; 2. To follow the length limits in the clothes dryer manufacturer's instructions; or, if neither of those choices work, 3. To relocate the dryer. A fourth option was to get the code official to approve the installation of a DEDPV under the alternative approval provision in Section 105. Exhaust ducts that exceed the developed length allowed by the code are a potential fire hazard, create maintenance problems, increase drying times and cause the dryer to be inefficient and waste energy. Dryer exhaust systems are commonly installed improperly with excessive lengths, too many elbows and the wrong duct materials. Because of the high incidence of reported dryer fires, the code strictly regulates the installation. DEDPVs are listed to a revised version of UL 705 that now contains tests and construction requirements that are specific to these devices. DEDPVs have been around for years but until recently were not listed to a national consensus standard that was specific to these devices. The UL 705 standard contains requirements for the construction, testing and installation of DEDPVs and requires them to be equipped with features such as interlocks, limit controls, monitoring controls and enunciator devices to make certain that the dryers or dryer operators are aware of the operating status of the DEDPVs (see Commentary Figure M1502.4.4).

M1502.4.5 Duct length. The maximum allowable exhaust duct length shall be determined by one of the methods specified in Sections M1502.4.5.1 through M1502.4.5.3.

❖ The duct length, the number and angle of fittings and the smoothness of the duct interior all contribute to the total friction loss of the airflow. When the friction loss is excessive, a reduction of the air velocity occurs, allowing lint and debris to accumulate in the duct and thus creating a fire hazard. Impeded airflow will also increase drying time, decrease appliance efficiency and raise system temperatures. The manufacturer's instructions and the code limitations concerning duct length, the number of elbows installed and the type of termination fittings must be adhered to in order to help prevent a fire hazard and allow proper appliance operation.

Note that this section introduces two distinct choices in its two subsections.

M1502.4.5.1 Specified length. The maximum length of the exhaust duct shall be 35 feet (10 668 mm) from the connection to the transition duct from the dryer to the outlet terminal. Where fittings are used, the maximum length of the exhaust duct shall be reduced in accordance with Table M1502.4.5.1. The maximum length of the exhaust duct does not include the transition duct.

❖ The maximum exhaust duct length of 35 feet (7620 mm) is the requirement for domestic clothes dryers. The 35-foot (7620 mm) limit is based on the worst case scenario where the dryer is rated for a maximum duct

length of 35 feet (7620 mm). The intent was to ensure that the least capable dryer available would be compatible with a 35-foot (7620 mm) duct system. If the exhaust duct system is designed for worst case, it should work properly in all cases. Note that there may be some dryers on the market that specify a duct length of less than 35 feet (10 668 mm), and the manufacturer's instructions would prevail in such case.

Because the maximum length is based on equivalent length, all fittings must be accounted for in accordance with Table M1502.4.5.1. The length is measured from the point where the transition duct (if used) connects to the rigid exhaust duct system to the terminal outdoors (see Table M1502.4.5.1). The transition duct is excluded from the calculation of equivalent length of the duct system.

Figure M1502.4.4
DRYER EXHAUST DUCT POWER VENTILATOR

TABLE M1502.4.5.1
DRYER EXHAUST DUCT FITTING EQUIVALENT LENGTH

DRYER EXHAUST DUCT FITTING TYPE	EQUIVALENT LENGTH
4 inch radius mitered 45 degree elbow	2 feet 6 inches
4 inch radius mitered 90 degree elbow	5 feet
6 inch radius smooth 45 degree elbow	1 foot
6 inch radius smooth 90 degree elbow	1 foot 9 inches
8 inch radius smooth 45 degree elbow	1 foot
8 inch radius smooth 90 degree elbow	1 foot 7 inches
10 inch radius smooth 45 degree elbow	9 inches
10 inch radius smooth 90 degree elbow	1 foot 6 inches

For SI: 1 inch = 25.4 mm, 1 foot = 304.8 mm, 1 degree = 0.0175 rad.

Table M1502.4.5.1 assigns an equivalent length to the various types of elbow fittings that are available. The first two rows of the table cover the traditional elbows that the code and dryer manufacturer's installation instructions have always addressed. The other rows in the table cover newer, high-performance elbows that have longer turning radii and smoother walls (i.e., no mitered joints); thus, the equivalent length is substantially reduced. When an installer discovers that the length of an exhaust duct will exceed the limits of the code, he or she may find that using higher performance elbows will keep the length within the allowable limits.

M1502.4.5.2 Manufacturer's instructions. The size and maximum length of the exhaust duct shall be determined by the dryer manufacturer's installation instructions. The code official shall be provided with a copy of the installation instructions for the make and model of the dryer at the concealment inspection. In the absence of fitting equivalent length calculations from the clothes dryer manufacturer, Table M1502.4.5.1 shall be used.

❖ This section allows the 35-foot (7620 mm) limit of the previous section to be exceeded where longer exhaust duct lengths are allowed by the appliance manufacturer's instructions. Today's appliances often, but not always, permit longer distances, therefore this section allows the installer to take advantage of those longer distances where specified by the manufacturer. The make and model of the dryer must be provided to the code official, along with the respective installation instructions, to permit the code official to inspect the duct installation based on the manufacturer's instructions. Because the installation is specific to and dependent on a certain appliance make and model, the code official must follow through to make sure that the appliance did get installed. Remember that the prime objective is to prevent a dangerous mismatch between a dryer and its exhaust system.

Where not otherwise specified by the dryer manufacturer, the flow resistance contributed by fittings installed in the duct system must be in accordance with Table M1502.4.5.1. Each type of fitting and its angle is equated to a certain length of straight duct and this "equivalent length" is added to the actual straight duct length in the system to obtain the total length. The table recognizes newer fitting designs that have smooth bore interiors and much larger turning radii than traditionally used fittings. Such designs greatly reduce friction loss as can be seen by their much smaller equivalent lengths. If an installer is unable to stay within the allowable length on a particular job, he or she may find that substituting better fittings will correct the problem (see Section M1502.4.6).

M1502.4.5.3 Dryer exhaust duct power ventilator. The maximum length of the exhaust duct shall be determined in accordance with the manufacturer's instructions for the dryer exhaust duct power ventilator.

❖ See the commentary to Section M1502.4.4.

M1502.4.6 Length identification. Where the exhaust duct equivalent length exceeds 35 feet (10 668 mm), the equivalent length of the exhaust duct shall be identified on a permanent label or tag. The label or tag shall be located within 6 feet (1829 mm) of the exhaust duct connection.

❖ A point to consider is the situation that will exist when the original dryer is replaced or the original occupant moves out and the new occupant installs a different dryer. In either case, the replacement dryer might not be compatible with the existing duct length. If the duct system equivalent length is not identified in a conspicuous manner, the new resident will be unaware of the potentially hazardous situation that has been created by the dryer and exhaust duct mismatch. It is assumed that the installer will be aware of the equivalent length of exhaust ducts that are entirely exposed and observable (see Commentary Figure M1502.4.6). If the exhaust duct equivalent length is 35 feet or less, there is no point in identifying it because it is assumed to be compatible with all dryers. See Section M1502.4.5.1.

M1502.4.7 Exhaust duct required. Where space for a clothes dryer is provided, an exhaust *duct system* shall be installed. Where the clothes dryer is not installed at the time of occupancy the exhaust duct shall be capped or plugged in the space in which it originates and identified and marked "future use."

Exception: Where a *listed* condensing clothes dryer is installed prior to occupancy of the structure.

❖ Where a space is provided for the future installation of a clothes dryer, an exhaust duct system must be installed so that such system can be inspected and approved (see Commentary Figure M1502.4.7). Otherwise, it is likely that the dryer and exhaust duct system will be installed later without the code official's knowledge and it will not be inspected for proper installation. Note that Section M1502.4.5.2 does not appear to be an option when an exhaust duct system is "roughed-in" for future use because the make, model and exhaust capability of the dryer will be unknown. The code official could approve identifying the equivalent length in accordance with Section M1502.4.6 under the alternative approval provisions of Section R104.11.

The requirement to cap the exhaust duct where a dryer is not installed at the time of occupancy is intended to keep the duct from becoming a conduit for air to travel between the inside of the structure and the outdoors. While a backdraft damper is required at the outdoor termination, capping the duct further reduces the chance of air movement. The capping will also prevent the entry of debris and foreign objects. The cap is required inside at the future dryer's location so that it will be readily apparent when the future dryer is installed. Requiring that a listed condensing dryer be installed prior to occupancy will prevent someone from taking advantage of this exception to avoid installing an exhaust duct system by indicating that a future condensing dryer will be installed (see the commentary for the exception to Section M1502.2).

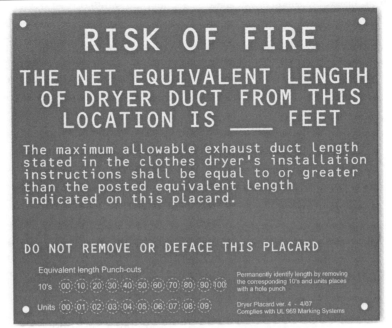

Figure M1502.4.6
EXAMPLE OF A DRYER EXHAUST LENGTH LABEL

Figure M1502.4.7
SPACE FOR CLOTHES DRYER INSTALLATION

M1502.5 Protection required. Protective shield plates shall be placed where nails or screws from finish or other work are likely to penetrate the clothes dryer exhaust duct. Shield plates shall be placed on the finished face of framing members where there is less than $1^1/_4$ inches (32 mm) between the duct and the finished face of the framing member. Protective shield plates shall be constructed of steel, shall have a minimum thickness of 0.062-inch (1.6 mm) and shall extend not less than 2 inches (51 mm) above sole plates and below top plates.

❖ Protection from nail and screw penetration has been required for wiring, vent systems, plumbing piping, fuel piping, hydronic piping and refrigerant piping for some time in the code, and now it is also required for dryer exhaust ducts (see Commentary Figure M1502.5). Depending on the depth of penetration, a nail or screw could create a serious obstruction problem and potential fire hazard.

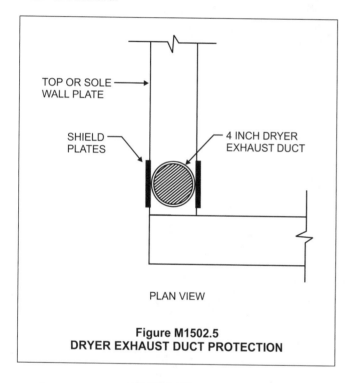

Figure M1502.5
DRYER EXHAUST DUCT PROTECTION

SECTION M1503
RANGE HOODS

M1503.1 General. Range hoods shall discharge to the outdoors through a duct. The duct serving the hood shall have a smooth interior surface, shall be air tight, shall be equipped with a back-draft damper and shall be independent of all other exhaust systems. Ducts serving range hoods shall not terminate in an attic or crawl space or areas inside the building.

Exception: Where installed in accordance with the manufacturer's instructions, and where mechanical or natural *ventilation* is otherwise provided, *listed* and *labeled* ductless range hoods shall not be required to discharge to the outdoors.

❖ A domestic kitchen exhaust system is one that serves appliances typically found in residential occupancies

such as within dwelling units. When compared to commercial cooking operations, residential cooking operations are far less frequent, of shorter duration, have lower heat output and produce fewer grease-laden vapors. However, airborne contaminant control may be even more important in residential cooking operations because of the lower or nonexistent ventilation rates typical of dwelling units. If natural or mechanical ventilation is otherwise provided for a space, this section does not require that the range hoods discharge to the outdoors. This section does not require range hoods to be installed, but regulates their installation. Section M1505.1 does require hoods for open-top broiler units and requires them to exhaust to the outdoors.

By requiring the ducts to have smooth inner walls, this section prohibits the use of flexible and semirigid corrugated ducts for range hood exhaust. As with all exhaust systems, the exhaust must discharge to the outdoors, not to an attic or crawl space or ridge, soffit, gable or roof vent (see commentary, Section M1501.1).

The ducts must be sealed air tight to prevent leakage of air and grease into wall and ceiling cavities. Ducts should be sealed with a material that is suitable for long-term exposure to elevated temperatures. Backdraft dampers prevent the infiltration of outdoor air when the exhaust system is not operating. The hood manufacturer's instructions may require that the ducts be installed with a minimum clearance to combustibles (see commentary, Section M1307.1).

The exception to this section allows the use of listed ductless (recirculating) range hoods as an alternative to the ducted type. However, exhaust to the outdoors would be required if natural or mechanical ventilation was not provided for the kitchen area. Note that many residential-type range hoods are designed to be exhausted to the outdoors and are also designed to be used as recirculating hoods, depending on how they are installed. Ductless residential hoods can remove some grease particulate by means of filters, but otherwise discharge the cooking effluent and combustion products back into the living space.

M1503.2 Duct material. Ducts serving range hoods shall be constructed of galvanized steel, stainless steel or copper.

Exception: Ducts for domestic kitchen cooking *appliances* equipped with down-draft exhaust systems shall be permitted to be constructed of schedule 40 PVC pipe and fittings provided that the installation complies with all of the following:

1. The duct is installed under a concrete slab poured on grade.

2. The underfloor trench in which the duct is installed is completely backfilled with sand or gravel.

3. The PVC duct extends not more than 1 inch (25 mm) above the indoor concrete floor surface.

4. The PVC duct extends not more than 1 inch (25 mm) above grade *outside of the building*.

5. The PVC ducts are solvent cemented.

❖ Ducts used for dwelling unit kitchen exhaust must be constructed of noncombustible materials to reduce the potential fire hazard associated with the collection of grease or the absorption of grease into a more porous duct material. This section requires domestic kitchen exhaust ducts to be constructed of galvanized or stainless steel or copper. Section M1503.1 effectively requires rigid ducts.

The exception allows the use of plastic ducts to exhaust the downdraft-type domestic kitchen cooking appliances if all five of the specified installation criteria are met. Plastic ducts are not allowed in above-ground applications because the plastic cannot contain fire within the confines of the duct. This exception allows plastic for the duct material because the duct must be below a slab on grade, and the trench for the duct must be completely backfilled with sand or gravel to: 1. Limit the amount of combustion air available to sustain a duct fire; and 2. Prevent a fire within the duct from spreading to the rest of the building. Additionally, the combustion gases and particulate from the burning plastic will be exhausted to the outdoors, not introduced into the occupied space. The plastic material also has the advantage of corrosion resistance. Metal ducts below a slab are susceptible to deterioration.

M1503.3 Kitchen exhaust rates. Where domestic kitchen cooking *appliances* are equipped with ducted range hoods or down-draft exhaust systems, the fans shall be sized in accordance with Section M1507.4.

❖ This section does not require installation of a hood that exhausts to the outdoors for domestic kitchen cooking appliances. It only provides guidance for sizing the exhaust fans for hoods that do exhaust to the outdoors. Mechanical ventilation would be required only where natural ventilation is not provided (see Section R303).

M1503.4 Makeup air required. Exhaust hood systems capable of exhausting in excess of 400 cubic feet per minute (0.19 m³/s) shall be mechanically or naturally provided with makeup air at a rate approximately equal to the exhaust air rate. Such makeup air systems shall be equipped with not less than one damper. Each damper shall be a gravity damper or an electrically operated damper that automatically opens when the exhaust system operates. Dampers shall be accessible for inspection, service, repair and replacement without removing permanent construction or any other ducts not connected to the damper being inspected, serviced, repaired or replaced.

❖ It is becoming more common for residential kitchens to resemble commercial kitchens, both aesthetically and functionally. As a result, much larger-capacity range hoods are being installed which aggravates the already existing problem of a lack of adequate makeup air for all exhaust systems in the dwelling. Homes typically suffer multiple ills because of indoor negative pressures caused by an imbalance of exhaust air to makeup air. The typical home relies solely on infiltra-

tion for makeup air and homes are being built much more airtight than in the past. Compounding this issue is the trend for much larger exhaust fans, and the pressure imbalance problem has a very real safety concern. The primary concern is depressurization that can cause chimneys and vents to fail. This section requires a makeup air supply system at the prescribed threshold and further requires it to be tied to the exhaust system controls such that both systems operate simultaneously. The intent is to not allow the homeowner the option of operating the exhaust fan without also operating the makeup air system.

This section states that the makeup air system can be a passive or active system (i.e., natural or mechanical). The makeup air is supposed to be provided "at a rate approximately equal to the exhaust air rate," which implies an active makeup air system that utilizes a fan; however, the text specifically allows a natural means of providing makeup air. The typical natural means is an opening to the outdoors with a motorized damper that operates in parallel with the exhaust fan. Whether active or passive (mechanical or natural), the makeup supply must have a damper to close the opening when not in use. This section allows a gravity-type damper that opens by the force of air entering the building caused by the negative pressure created by the exhaust fan. In other words, gravity dampers depend on a negative pressure in the building with respect to the outdoors. This negative pressure must overcome the resistance of the inlet opening/duct and also the resistance created by the weight of the damper and its support bearings. The primary concern of this section is that negative pressures in a home invite unwanted infiltration air and can cause natural draft fuel-fired appliances to backdraft and spill combustion products into the home. If the dwelling is allowed to depressurize because of exhaust fans that lack sufficient makeup air, the drafting forces in the vents for natural draft fuel-fired appliances will be overwhelmed and the vents will flow backwards, bringing in makeup air from the outdoors instead of conveying combustion products to the outdoors. Opening a window or relying on infiltration is not an acceptable means of providing makeup air.

M1503.4.1 Location. Kitchen exhaust makeup air shall be discharged into the same room in which the exhaust system is located or into rooms or *duct systems* that communicate through one or more permanent openings with the room in which such exhaust system is located. Such permanent openings shall have a net cross-sectional area not less than the required area of the makeup air supply openings.

❖ This section simply clarifies how makeup air is delivered to the kitchen area. Some readers have interpreted the code to require the makeup air to be delivered directly to the kitchen; however, this has never been required by the code. The makeup air can be delivered to any space that freely communicates with the kitchen as well as to the kitchen or any combination of such locations.

SECTION M1504
INSTALLATION OF MICROWAVE OVENS

M1504.1 Installation of a microwave oven over a cooking appliance. The installation of a *listed* and *labeled* cooking *appliance* or microwave oven over a *listed* and *labeled* cooking *appliance* shall conform to the terms of the upper *appliance's listing* and *label* and the manufacturer's installation instructions. The microwave oven shall conform to UL 923.

❖ A cooking appliance or microwave oven installed above other cooking appliances must be listed for use above the lower appliance. This section addresses the fire hazard created by having hot surfaces, open flames and, possibly, a cooking fire located under a cooking appliance or microwave oven. Installation clearances must comply with the manufacturer's instructions for the upper appliance. UL 923 is the standard applicable to microwave ovens. Combination units consisting of an oven with an integral exhaust hood are commonly installed above ranges and cooktops.

SECTION M1505
OVERHEAD EXHAUST HOODS

M1505.1 General. Domestic open-top broiler units shall have a metal exhaust hood, having a minimum thickness of 0.0157-inch (0.3950 mm) (No. 28 gage) with $^{1}/_{4}$ inch (6.4 mm) clearance between the hood and the underside of combustible material or cabinets. A clearance of not less than 24 inches (610 mm) shall be maintained between the cooking surface and the combustible material or cabinet. The hood shall be not less than the width of the broiler unit, extend over the entire unit, discharge to the outdoors and be equipped with a backdraft damper or other means to control infiltration/exfiltration when not in operation. Broiler units incorporating an integral exhaust system, and *listed* and *labeled* for use without an exhaust hood, need not have an exhaust hood.

❖ The surface material of the hood must not come in direct contact with combustible material, therefore, a $^{1}/_{4}$-inch (6.4 mm) clearance between the hood and the underside of combustible material and cabinets is required. Otherwise, a grease fire or hot gases from the broiler unit may heat the hood to temperatures that could cause ignition of combustible material mounted above the hood.

A clearance of at least 24 inches (610 mm) must be maintained between the cooking surface and overhead cabinets constructed of combustible materials. These clearances help ensure that combustible material will not be ignited in the event of a fire on the broiler. Note that this section mandates that the hood exhaust to the outdoors and there is no option for recirculating-type hoods.

SECTION M1506
EXHAUST DUCTS AND EXHAUST OPENINGS

M1506.1 Duct construction. Where exhaust duct construction is not specified in this chapter, construction shall comply with Chapter 16.

❖ This section refers the reader to Chapter 16 for any exhaust duct construction not specified in Chapter 15. For example, clothes dryer exhaust duct construction is specified in this chapter, but toilet/bathroom exhaust duct construction is addressed only in Chapter 16.

M1506.2 Duct length. The length of exhaust and supply ducts used with ventilating equipment shall not exceed the lengths determined in accordance with Table M1506.2.

Exception: Duct length shall not be limited where the duct system complies with the manufacturer's design criteria or where the flow rate of the installed ventilating equipment is verified by the installer or approved third party using a flow hood, flow grid or other airflow measuring device.

❖ Commonly, exhaust fans, especially bathroom exhaust fans, are installed and ducts are run from the fans to the outlet terminals with little regard for the pressure losses in the duct run. A fan rated at 100 cfm, for example, will not move 100 cfm of air if the exhaust duct creates a static pressure drop greater than that for which the fan was rated. Flexible duct offers greater flow resistance than smooth-wall duct, making the problem worse in many cases. The use of duct fittings greatly adds to the flow resistance. Section M1507.4 sets the required exhaust rates and fans are chosen to satisfy that section, however, poorly designed exhaust ducts can prevent the systems from complying with the intent of Section M1507.4 because the fans cannot deliver the required airflow through poorly designed ducts. The exception allows the ducts to be designed by the fan manufacturer, where such information is provided. Often the manufacturer of the fan is silent on the allowable length of duct and number of fittings. Compliance can also be achieved by field testing the installed system and verifying the actual airflow rates.

The table addresses smooth-wall and flexible ducts and sets the maximum duct length based on the diameter of the duct. Looking at the table, one can quickly see that many commonly installed fan and duct combinations are substandard and will not perform as intended. For example, the common 50 cfm fan can never utilize 3-inch flex duct and can use 3-inch smooth-wall duct only in 5-foot runs with no elbows. The size of the fan outlet cannot dictate the size of the duct as, quite often, the duct will have to be larger than the fan outlet. Note c in Table M1506.2 explains how duct fittings are addressed in the table. For example, a 4-inch flexible duct with an 80 cfm fan could not have any elbows, and a 4-inch smooth-wall duct with an 80 cfm fan could have only two elbows and 1 foot of duct, or one elbow and 16 feet of duct.

Note that Table M1506.2 states that the fans were rated with an assumed static pressure loss of 0.25 inch water column in the duct. Some fans might be rated with an assumed static pressure of 0.10 inch w.c. and if they are, the table would allow ducts that are too long for such fans, resulting in fans not meeting their design airflow rate. When using the table, the static pressure at which the fan was tested should be taken into consideration.

M1506.3 Exhaust openings. Air exhaust openings shall terminate not less than 3 feet (914 mm) from property lines; 3 feet (914 mm) from operable and nonoperable openings into the building and 10 feet (3048 mm) from mechanical air intakes except where the opening is located 3 feet (914 mm) above the air intake. Openings shall comply with Sections R303.5.2 and R303.6.

❖ These provisions for termination points are for air that is exhausted from a dwelling unit. Such air will include kitchen exhaust, bathroom exhaust and clothes dryer exhaust and these exhaust terminations would be required to be 3 feet (914 mm) from operable openings into a building (i.e., windows and doors). The exhaust terminations that penetrate through the exterior walls or roof will need to be located 3 feet (914 mm) from windows that are providing the natural ventilation to the dwelling unit. Considering that the exhaust from a dwelling unit is not considered to be hazardous or noxious and is of low volume, the 3-foot (914 mm) separation from windows is deemed to be reasonable. Mechanical air intakes must be located not less than 10 feet (3048 mm) from exhaust from a bathroom, kitchen and domestic clothes dryer. In the case where the required 10-foot (3048 mm) separation cannot be met, the intake must be located at least 3 feet (914 mm) below the exhaust termination. It is assumed that the exhaust that is present will be buoyant in air because of its temperature or specific gravity and such exhaust will rise above and away from the intake opening.

SECTION M1507
MECHANICAL VENTILATION

M1507.1 General. Where local exhaust or whole-house mechanical ventilation is provided, the equipment shall be designed in accordance with this section.

❖ Mechanical ventilation can consist of either local exhaust or whole-house mechanical ventilation. Local exhaust is defined as an exhaust system that uses one or more fans to exhaust air from a specific room or rooms within a dwelling. Examples of local exhaust include bathroom and kitchen exhaust. Section R303 requires either natural or mechanical ventilation of toilet rooms and bathrooms. This section brings mechanical ventilation into the exhaust chapter for convenience. It should be noted that the choice of using either natural or mechanical ventilation is one the designer must make except as required by Section R303.4.

Whole-house mechanical ventilation is defined as an exhaust system, supply system or combination thereof, that is designed to mechanically exchange indoor air for outdoor air. The system can either operate continuously or it can be programmed to operate intermittently.

M1507.2 Recirculation of air. Exhaust air from bathrooms and toilet rooms shall not be recirculated within a residence or to another *dwelling unit* and shall be exhausted directly to the outdoors. Exhaust air from bathrooms and toilet rooms shall not discharge into an *attic*, crawl space or other areas inside the building.

❖ This section prohibits the recirculation of exhaust air from toilet rooms and bathrooms within a dwelling or to another dwelling. In some cases, installers connect the outlets of bathroom exhaust fans together to a com-

TABLE M1506.2
DUCT LENGTH

DUCT TYPE	FLEX DUCT								SMOOTH-WALL DUCT							
Fan airflow rating (CFM @ 0.25 inch wc[a])	50	80	100	125	150	200	250	300	50	80	100	125	150	200	250	300
Diameter[b] (inches)	Maximum length[c, d, e] (feet)															
3	X	X	X	X	X	X	X	X	5	X	X	X	X	X	X	X
4	56	4	X	X	X	X	X	X	114	31	10	X	X	X	X	X
5	NL	81	42	16	2	X	X	X	NL	152	91	51	28	4	X	X
6	NL	NL	158	91	55	18	1	X	NL	NL	NL	168	112	53	25	9
7	NL	NL	NL	NL	161	78	40	19	NL	NL	NL	NL	NL	148	88	54
8 and above	NL	NL	NL	NL	NL	189	111	69	NL	NL	NL	NL	NL	NL	198	133

For SI: 1 foot = 304.8 mm.

a. Fan airflow rating shall be in accordance with ANSI/AMCA 210-ANSI/ASHRAE 51.

b. For noncircular ducts, calculate the diameter as four times the cross-sectional area divided by the perimeter.

c. This table assumes that elbows are not used. Fifteen feet of allowable duct length shall be deducted for each elbow installed in the duct run.

d. NL = no limit on duct length of this size.

e. X = not allowed. Any length of duct of this size with assumed turns and fittings will exceed the rated pressure drop.

mon duct to avoid multiple roof, wall or soffit penetrations. This will result in some exhaust air flowing backwards through the fan that is not running because the backdraft dampers in such fans will allow some leakage. This problem is much worse where the common duct serves fans that are located in different dwelling units, such as for a duplex. The code does not expressly prohibit the discharge side of exhaust fans from being connected to a common duct, but the fan manufacturer might prohibit such installations. Recirculating (ductless) exhaust fans are not allowed if mechanical ventilation of a space is required, but such fans are allowed in naturally ventilated spaces.

Recirculation to other dwelling units is prohibited because odors, disease-causing organisms, tobacco smoke, moisture and other contaminants could be transported from one dwelling unit to another.

The requirement for directing air exhausted from bathrooms directly to the outdoors is the same fundamental requirement as expressed in Section M1501 (see commentary, Section M1501.1).

M1507.3 Whole-house mechanical ventilation system. Whole-house mechanical ventilation systems shall be designed in accordance with Sections M1507.3.1 through M1507.3.3.

❖ All habitable rooms are required by Section 303.1 to be provided with natural ventilation through windows, doors or other approved openings, with an option for mechanical ventilation in accordance with this section. For years natural ventilation has been used which relied on the occupants to open windows and doors to provide ventilation to dwellings. While natural ventilation is still allowed, concerns over indoor air quality have raised questions over the effectiveness of naturally ventilating a dwelling. There are provisions in the code that allow mechanical ventilation to be provided instead of natural ventilation, and, in some cases, require mechanical ventilation to be provided. Section R303.4 will require a dwelling to be provided with whole-house mechanical ventilation when natural ventilation alone is not sufficient for dwellings that are tightly sealed such that their infiltration rate is 5 or fewer air changes per hour at an indoor/outdoor pressure differential of 50 Pa (ACH_{50}) (see commentary, Section R303.4).

In the case where the air infiltration is 5 ACH or greater, these provisions are not mandatory, but they can be used by the designer as another tool or option to provide ventilation for a home in lieu of using natural ventilation.

M1507.3.1 System design. The whole-house ventilation system shall consist of one or more supply or exhaust fans, or a combination of such, and associated ducts and controls. Local exhaust or supply fans are permitted to serve as such a system. Outdoor air ducts connected to the return side of an air handler shall be considered as providing supply ventilation.

❖ The proponent of the change that is responsible for the whole-house mechanical ventilation requirements indicated that these provisions came from the 2004 edition of ASHRAE 62.2. The intent was to reduce the standard

down to the nuts and bolts of mechanical ventilation and provide requirements that are simple and straight forward. Therefore, if a mechanical ventilation system is installed in a dwelling, the designer or builder will have prescriptive requirements on which the system can be based and the code official has provisions on which he or she can evaluate the system for compliance.

The system is made up of supply or exhaust fans or a combination of supply and exhaust fans along with associated ducts and controls for the system. Local bathroom exhaust fans and kitchen fans can be used as part of the system and outdoor air can be brought in through passive openings, dedicated supply fans or outdoor air ducts connected to the return side of the air-handling unit. Note that if the dwelling's air-handling unit or furnace is used to supply outdoor air, the unit must run continuously or have controls to operate intermittently in accordance with Section M1507.3.3. If the outdoor air duct is connected to the air-handler return air duct system and exhaust fans are used as the whole-house mechanical ventilation system (WHMVS), it would not matter if the airhandler is running because the outdoor air will be able to move through the duct system because of the negative pressure created by the exhaust fans. This scenario has the advantage of mixing the outdoor air with indoor air and distributing it in a manner that will not cause discomfort for the occupants.

There are generally three types of whole-house exhaust systems consisting of exhaust ventilation, supply ventilation and balanced ventilation. Exhaust ventilation systems are better suited for cold climates since they work by depressurizing the dwelling. Positively pressurizing a dwelling in a cold climate can force moist air into wall cavities causing condensation to occur in the wall assemblies. In warmer climates with high humidity, reducing the inside pressure will cause air to migrate from the outdoors to inside the dwelling units' wall cavities, which can cause moisture damage if it condenses. Exhaust fans are used to remove air from the dwelling, usually in bathrooms and the kitchen where the most contaminates are located, and then makeup air replaces the air that is exhausted. While infiltration might account for some of the makeup air, outdoor air connections to the return side of the air-handling unit can be used to supply the outdoor makeup air. Since the dwelling is under negative pressure, extreme care must be taken to make sure that fuel-fired appliances and appliance vents are not affected (see Sections M1503.4 and G2407.4).

Supply ventilation systems are better suited for warmer climates since they work by pressurizing the dwelling. In colder climates, when the dwelling is pressurized, warm air can migrate from inside the dwelling into the building cavities which could cause condensation and structural damage to the dwelling. Supply fans are used to supply air to the dwelling with outdoor air into areas of the dwelling which are usually occupied (i.e., family rooms and bedrooms). For this reason, the air should be located so that it mixes with the conditioned air so that it does not create drafts and make the occupants uncomfortable. Relief vents and kitchen and bath fan ducts are used to relieve air from the dwelling.

Balanced ventilation systems consist of supply and exhaust fans which provide approximately equal amounts of outdoor air and exhaust air. Exhaust is usually taken from bathrooms and kitchen areas and the outdoor air is usually supplied to the occupied areas of the dwelling, like family rooms and the bedrooms. A balanced ventilation system using supply and exhaust fans in combination is well suited for an energy recovery ventilation system, thus allowing the energy penalty associated with ventilation to be minimized.

M1507.3.2 System controls. The whole-house mechanical ventilation system shall be provided with controls that enable manual override.

❖ The whole-house mechanical ventilation system must run continuously or have controls to operate intermittently in accordance with Section M1507.3.3. A manual override switch/control must be provided to allow the occupants an option to override the system. Note that an override function could mean manually turning off the system or manually turning on the system. Many systems are also designed with a "vacation" setting which allows the occupants to shut down the system when they will be away. See Section M1507.3.3 for more discussion on system control schemes.

M1507.3.3 Mechanical ventilation rate. The whole-house mechanical ventilation system shall provide outdoor air at a continuous rate of not less than that determined in accordance with Table M1507.3.3(1).

Exception: The whole-house mechanical ventilation system is permitted to operate intermittently where the system has controls that enable operation for not less than 25-percent of each 4-hour segment and the ventilation rate prescribed in Table M1507.3.3(1) is multiplied by the factor determined in accordance with Table M1507.3.3(2).

❖ The amount of continuous outdoor air required for a WHMVS is indicated in Table M1507.3.3(1). The rates in this table are based on ASHRAE 62.2 and are tied to the area of the conditioned space and the number of bedrooms. Basements, utility areas, laundry rooms, and similar rooms would be considered as part of the dwelling unit's floor area if these spaces are conditioned. For example, a 2,500-square-foot dwelling with four bedrooms would require 75 cfm of continuous outdoor airflow based on Table M1507.3.3(1).

Intermittent operation is permitted in accordance with the exception. The exception permits intermittent operation based on the run time percentage in each 4-hour time segment. When there is anything less than 100-percent continuous operation, the amount of outdoor air required in Table M1507.3.3(1) is multiplied by the factors in Table M1507.3.3(2). The factors are based on the percentage of time that the system runs in every 4-hour segment of a day. While intermittent operation is permitted, the tradeoff is that more outdoor air is required. For example, in the 2,500-square-foot dwelling with four bedrooms, 75 cfm of continuous outdoor air is required. If intermittent operation was used and 50 percent operation during a 4-hour period was chosen, then the amount of airflow would have to be multiplied by a factor of 2. That would mean that in each 4-hour segment of a day, the ventilation system must run for 2 hours (50 percent of 4 hours) and the amount of airflow provided would be 150 cfm (75 cfm × 2). The minimum run time permitted is 25 percent or 1 hour of run time, which would require the amount of outdoor air to be multiplied by 4. In the above example, for 1 hour of operation in a 4-hour period, 300 cfm of outdoor air is required. The only option to having a

TABLE M1507.3.3(1)
CONTINUOUS WHOLE-HOUSE MECHANICAL VENTILATION SYSTEM AIRFLOW RATE REQUIREMENTS

DWELLING UNIT FLOOR AREA (square feet)	NUMBER OF BEDROOMS				
	0 – 1	2 – 3	4 – 5	6 – 7	> 7
	Airflow in CFM				
< 1,500	30	45	60	75	90
1,501 – 3,000	45	60	75	90	105
3,001 – 4,500	60	75	90	105	120
4,501 – 6,000	75	90	105	120	135
6,001 – 7,500	90	105	120	135	150
> 7,500	105	120	135	150	165

For SI: 1 square foot = 0.0929 m², 1 cubic foot per minute = 0.0004719 m³/s.

TABLE M1507.3.3(2)
INTERMITTENT WHOLE-HOUSE MECHANICAL VENTILATION RATE FACTORS[a, b]

RUN-TIME PERCENTAGE IN EACH 4-HOUR SEGMENT	25%	33%	50%	66%	75%	100%
Factor[a]	4	3	2	1.5	1.3	1.0

a. For ventilation system run time values between those given, the factors are permitted to be determined by interpolation.

b. Extrapolation beyond the table is prohibited.

timed control system is to run the fans continuously.

Where fans are designed to run continuously, such fans should not be controlled by the typical wall switch to which all occupants are accustomed because the occupants will inadvertently turn off the fans when they leave the room, as they have done most of their lives. For example, bathroom/toilet room exhaust fans can be used as a component of the WHMVS in a dwelling. The occupants of such dwelling would turn off the fans when they left the bathroom if they were provided with the typical wall switch that they are accustomed to using, thus disabling the WHMVS. Although not explicitly stated in the code, the implied intent is to require the WHMVS to have controls that the occupants will recognize as being part of a WHMVS. For example, a wall switch can be located and labeled such that it is obviously not intended to be turned off unless an occupant purposely intends to shut off the WHMVS. Alternatively, the fans could be controlled solely by the branch circuit breaker that supplies them, with the circuit identified as bath fans for the WHMVS.

M1507.4 Local exhaust rates. *Local exhaust* systems shall be designed to have the capacity to exhaust the minimum air flow rate determined in accordance with Table M1507.4.

❖ The mechanical ventilation rates for kitchens, toilet rooms and bathrooms are specified in Table M1507.4. See Section G2407.4 regarding required makeup air where appliance operation can be affected by negative building pressures.

TABLE M1507.4
MINIMUM REQUIRED LOCAL EXHAUST RATES FOR ONE- AND TWO-FAMILY DWELLINGS

AREA TO BE EXHAUSTED	EXHAUST RATES
Kitchens	100 cfm intermittent or 25 cfm continuous
Bathrooms-Toilet Rooms	Mechanical exhaust capacity of 50 cfm intermittent or 20 cfm continuous

For SI: 1 cubic foot per minute = 0.0004719 m³/s.

❖ The exhaust rates specified in this table are identical to those in Table 403.3.1.1 of the IMC. The continuous rates are lower than the intermittent rates, reflecting the fact that an exhaust fan that runs continuously will provide some degree of overventilation during periods in which contaminants are not being produced. An intermittent exhaust fan will run for short periods of time only, requiring a higher exhaust rate to effectively control odors, cooking effluent and moisture.

Bibliography

The following resource materials were used in the preparation of the commentary for this chapter of the code:

IMC–15, *International Mechanical Code.* Washington, DC: International Code Council, 2014.

NFPA 70–14, *National Electrical Code.* Quincy, MA: National Fire Protection Association, 2013.

Chapter 16:
Duct Systems

User note: Code change proposals to this chapter will be considered by the IRC – Plumbing and Mechanical Code Development Committee during the 2015 (Group A) Code Development Cycle.

General Comments

This chapter addresses duct systems by instituting requirements for the protection of the occupants of the building and the building itself. This chapter regulates the materials and methods of construction. Duct construction affects the performance of the entire air distribution system.

Section M1601 contains the material requirements that are generally applied to all duct construction and installation. Section M1602 contains regulations for return air.

Purpose

Chapter 16 contains requirements for the installation of supply, return and exhaust air systems. The chapter contains no information on the design of these systems from the standpoint of air movement, but it is concerned with the structural integrity of the systems and the overall impact of the systems on the fire-safety performance of the building. Design considerations such as duct sizing, maximum efficiency, cost effectiveness, occupant comfort and convenience are the responsibility of the designer.

SECTION M1601
DUCT CONSTRUCTION

M1601.1 Duct design. *Duct systems* serving heating, cooling and *ventilation equipment* shall be installed in accordance with the provisions of this section and ACCA Manual D, the appliance manufacturer's installation instructions or other *approved* methods.

❖ Material and construction requirements for duct systems can be divided into two major categories: those for above-ground ducts and those for underground ducts. The demands placed on each of these types of duct systems are different; therefore, they are treated separately in the code and in the following commentary.

Adequately sized supply air and return air ductwork is essential for efficient and proper circulation of conditioned air. Ducts must be large enough to allow sufficient airflow through appliances to satisfy refrigerant coil airflow demands and to avoid excessive temperature rise. As a practical matter, restriction of the supply duct system is an infrequently encountered problem. However, inadequately sized return systems are often found in the field, particularly on heat pump installations. The code does not provide specific requirements for sizing duct systems, but relies on ACCA *Manual D*, the appliance manufacturer or other sizing criteria approved by the code official.

M1601.1.1 Above-ground duct systems. Above-ground *duct systems* shall conform to the following:

1. *Equipment* connected to *duct systems* shall be designed to limit discharge air temperature to not greater than 250°F (121°C).

2. Factory-made ducts shall be listed and labeled in accordance with UL 181 and installed in accordance with the manufacturer's instructions.

3. Fibrous glass duct construction shall conform to the SMACNA *Fibrous Glass Duct Construction Standards* or NAIMA *Fibrous Glass Duct Construction Standards*.

4. Field-fabricated and shop-fabricated metal and flexible duct constructions shall conform to the SMACNA HVAC *Duct Construction Standards—Metal and Flexible* except as allowed by Table M1601.1.1. Galvanized steel shall conform to ASTM A653.

5. The use of gypsum products to construct return air ducts or plenums is permitted, provided that the air temperature does not exceed 125°F (52°C) and exposed surfaces are not subject to condensation.

6. *Duct systems* shall be constructed of materials having a flame spread index of not greater than 200.

7. Stud wall cavities and the spaces between solid floor joists to be used as air plenums shall comply with the following conditions:

    7.1. These cavities or spaces shall not be used as a plenum for supply air.

    7.2. These cavities or spaces shall not be part of a required fire-resistance-rated assembly.

    7.3. Stud wall cavities shall not convey air from more than one floor level.

    7.4. Stud wall cavities and joist-space plenums shall be isolated from adjacent concealed spaces by tight-fitting fireblocking in accordance with Section R602.8.

    7.5. Stud wall cavities in the outside walls of building envelope assemblies shall not be utilized as air plenums.

❖ Metal ducts are usually constructed of galvanized sheet steel. Duct size is based on required airflow, system pressure, flow velocity and pressure losses caused by friction. Duct material thickness is determined by duct size, static pressure of the system, distance between supports and whether the duct is reinforced. Metal ducts must be constructed with the minimum thicknesses specified in Table M1601.1.1, which bases the minimum required duct thickness on the geometry of the duct, the material used and the major dimension of the duct (the diameter for round ducts and the widest side for rectangular ducts).

Metallic ducts must be constructed to comply with the requirements contained in SMACNA *HVAC Duct Construction Standards - Metal and Flexible*. This standard contains information on duct reinforcement, joints, fittings, hangers and supports and other pertinent design information needed to achieve a stable, efficient and durable installation of ductwork. While this standard contains the minimum required thickness of duct materials, ducts that are installed within a single dwelling are allowed to be constructed with the minimum thicknesses specified in Table M1601.1.1.

Factory-made ducts must be tested and classified in accordance with the provisions of UL 181. Only Class 0 and Class 1 are recognized by UL 181. Class 0 indicates flame spread and smoke-developed indices of zero; Class 1 indicates a flame spread index not greater than 25 and a smoke-developed index not greater than 50, when tested to ASTM E84. The exception to the 25 flame spread index for nonmetallic ducts is Item 6 of this section, which is intended to allow the installation of plastic ducts above ground where such ducts have a flame spread index of 200 or less. However, it should be noted that there is an ICC committee formal interpretation that opines that plastic ducts are allowed to be installed above ground only if they meet the Class I flame spread index requirement of 25 and smoke development index of 50. Plastic rigid ducts are typically made of PVC that cannot meet a smoke development index of 50.

UL 181 requires that a nonmetallic duct be tested to determine its fire-performance characteristics, corrosion resistance, mold-growth resistance, humidity resistance, leakage resistance, temperature resistance, erosion resistance and structural performance. Air ducts that conform to the requirements of UL 181 are identified by the manufacturer's or vendor's name, rated velocity, negative and positive pressure classification and duct material class.

Fibrous ducts are constructed of a composite material of rigid (high-density) fiberglass board and a factory-applied facing, typically reinforced aluminum. The surface of the fibrous duct that is exposed to the airflow is sealed with a fiber-bonding adhesive that prevents erosion of the fiberglass material. The factory-applied exterior duct-board facing contributes to the strength and rigidity of the composite material, acts as a heat reflector, serves as a vapor barrier and is an integral component of the joining method used to construct fibrous ducts. The material is available in board form for shop or field fabrication into rectangular sections or into 10-sided duct form, which approximates a circular cross section. Factory-built round fibrous glass ducts are also available.

Fibrous ducts take advantage of the inherent insulating qualities of the glass fiber material. The air friction factors for fibrous ducts are higher than those for sheet metal because of the relatively rough surface finish of the former.

Construction of fibrous glass ducts must conform to the requirements of the SMACNA Fibrous Glass Duct Construction Standards or NAIMA Fibrous Glass Duct Construction Standards, which provides details for the design and fabrication of air distribution systems using fibrous glass ducts. The SMACNA and NAIMA standards referenced in this section and the previous section are enforceable extensions of the code.

The maximum discharge temperature permitted by industry standards for warm-air heating systems is 250°F (121°C). This section prohibits nonmetallic ducts from being used in applications in which the air temperature would exceed this maximum because the material has not been tested to withstand higher temperatures, and high temperatures will cause accelerated aging of the duct material.

Gypsum board is a composite material commonly used for the construction of air plenums and shafts. Gypsum board can reduce construction costs because it is a common component of building construction assemblies. By serving a dual purpose, gypsum board eliminates the need for independent duct construction. The use of gypsum board to form ducts and plenums is specifically regulated to prevent deterioration of the gypsum board material. Air temperatures that exceed 125°F (52°C) will, over time, dry both the paper facing and the gypsum of the gypsum board, leading to deterioration of the panel.

Gypsum board can also deteriorate when exposed to moisture, which will happen if the surface temperature of the gypsum board is lower than the airstream dew-point temperature, causing water to condense on the surface of the gypsum board. For these reasons, gypsum board cannot be used for air distribution systems using evaporative cooling equipment. It is further restricted to return-air system applications only, a maximum airstream temperature of 125°F (52° C) and an airstream dew-point temperature continuously below the temperature of the gypsum board surface. Evaporative cooling equipment, such as "swamp coolers," uses water as a refrigerant. The resulting addition of moisture to the airstream could cause deterioration of the gypsum board.

Stud spaces and joist spaces have commonly been used as plenums in residential construction (see Section N1103.3.5). These spaces are limited to return air plenums because the negative pressures within the return air plenum with respect to surrounding spaces will decrease the likelihood of spreading smoke to other spaces via the plenum. Also, the temperature and moisture content of heated, cooled and conditioned supply air could cause a fire hazard or deterioration of the construction materials exposed in the spaces.

Note that Section N1103.3.5 prohibits the use of all framing cavities as ducts or plenums.

The space must not be a part of a fire-resistance-rated assembly because the ASTM E119 test does not consider the impact of air movement within the assembly on the fire-resistance rating. This restriction is a concession to the convenience and cost-saving potential of this method of moving air. The use of stud spaces inherently means the interconnection of different floor levels by the concealed space. Because of the hazard of such an interconnection, the use of this type of plenum is limited to return air from one floor level only for each independent stud cavity. All cavities not used for air movement must be isolated from the plenum by fireblocking constructed and installed in accordance with the code. The use of stud wall cavities as a plenum in outside walls is prohibited due to the difficulty in sealing the space to prevent outdoor air from infiltrating the cavity, which would increase the heating and cooling loads on the system. Note that Section N1103.3.5 prohibits the use of all framing cavities as ducts or plenums.

Commentary Figure M1601.1.1(1) shows an example of an acceptable stud and joist space installation. The bottom plate of the wall is cut away for the plenum to function, and fireblocking is installed in the joist and stud space to limit communication of the plenum with other spaces. The stud cavity shown is being used to return air from one floor level only, conducting it to the space below, where the air-handling equipment is

located. Commentary Figure M1601.1.1(2) shows an example of an unacceptable stud and joist space installation.

Whether viewed as a shaft or as a duct, a stud-cavity plenum penetrates floor assemblies and is, therefore, subject to the floor penetration protection requirements of Section R302.4.1.

Because air is returned from more than one floor level through the same stud cavity, this section prohibits this type of installation because it creates a direct connection from one floor level to another by means of a concealed cavity. Such direct connections would act as a chase or chimney, allowing fire and smoke to spread quickly upward through the building. Thus, stud-cavity return-air plenums are subject to the same restrictions and requirements as floor openings and penetrations regulated by Section R302.4. Stud-space and joist-space plenums must be viewed as an exception to the fireblocking provisions of Section R302.11 because one or more fireblocks (wall plates) must be removed or relocated to construct the plenums. The code does not mention the type of materials allowed for "panning" the bottom of open joists to create joist-space plenums. Traditionally, sheet metal has been used; however, composite materials are also used. The building official must determine what materials are acceptable for joist panning.

TABLE M1601.1.1.

❖ The minimum thicknesses indicated are the low end of the tolerance range for the nominal thicknesses of

TABLE M1601.1.1
DUCT CONSTRUCTION MINIMUM SHEET METAL THICKNESS FOR SINGLE DWELLING UNITS[a]

ROUND DUCT DIAMETER (inches)	STATIC PRESSURE			
	1/2 inch water gage		1 inch water gage	
	Thickness (inches)		Thickness (inches)	
	Galvanized	Aluminum	Galvanized	Aluminum
≤ 12	0.013	0.018	0.013	0.018
12 to 14	0.013	0.018	0.016	0.023
15 to 17	0.016	0.023	0.019	0.027
18	0.016	0.023	0.024	0.034
19 to 20	0.019	0.027	0.024	0.034

RECTANGULAR DUCT DIMENSION (inches)	STATIC PRESSURE			
	1/2 inch water gage		1 inch water gage	
	Thickness (inches)		Thickness (inches)	
	Galvanized	Aluminum	Galvanized	Aluminum
≤ 8	0.013	0.018	0.013	0.018
9 to 10	0.013	0.018	0.016	0.023
11 to 12	0.016	0.023	0.019	0.027
13 to 16	0.019	0.027	0.019	0.027
17 to 18	0.019	0.027	0.024	0.034
19 to 20	0.024	0.034	0.024	0.034

For SI: 1 inch = 25.4 mm, 1 inch water gage = 249 Pa.

a. Ductwork that exceeds 20 inches by dimension or exceeds a pressure of 1 inch water gage (250 Pa) shall be constructed in accordance with SMACNA *HVAC Duct Construction Standards—Metal and Flexible*.

commonly available sheet materials according to SMACNA *HVAC Duct Construction Standards*. The minimum thickness allows the code official to verify that the proper nominal thickness of material is being used. The dimension indicated for rectangular sizes is the largest dimension. The indicated thicknesses for aluminum sheet are for aluminum alloy 3003-H 14. Note a was added to the table to indicate that for ducts having static pressures greater than 1-inch water gauge, and for ducts with a dimension greater than 20 inches (508 mm), the ducts must be designed in accordance with the material thickness and reinforcing requirements of the SMACNA *HVAC Duct Construction Standard*.

M1601.1.2 Underground duct systems. Underground *duct systems* shall be constructed of *approved* concrete, clay, metal or plastic. The maximum duct temperature for plastic ducts shall not be greater than 150°F (66°C). Metal ducts shall be protected from corrosion in an *approved* manner or shall be completely encased in concrete not less than 2 inches (51 mm) thick. Nonmetallic ducts shall be installed in accordance with the manufacturer's instructions. Plastic pipe and fitting materials shall conform to cell classification 12454-B of ASTM D1248 or ASTM D1784 and external loading properties of ASTM D2412. Ducts shall slope to an accessible point for drainage. Where encased in concrete, ducts shall be sealed and secured prior to any concrete being poured. Metallic ducts having an *approved* protective coating and nonmetallic ducts shall be installed in accordance with the manufacturer's instructions.

❖ Ducts installed underground must be able to resist the forces imposed on them by the materials that encase them, the forces created by floodwaters in and around

them and corrosion. The ASHRAE *Handbook of HVAC Systems and Equipment* recommends that all underground ducts and fittings be round to provide optimum structural performance. Unlike round ducts, square or rectangular ducts offer little resistance to deformation or collapse caused by the structural loads associated with burial.

Metal ducts must have either a protective coating to resist corrosion or be completely encased in concrete a minimum of 2 inches (52 mm) thick all around. Concrete-encased ducts may eventually corrode; however, the air passageway will be maintained because of the remaining concrete enclosure. All nonmetallic ducts and metallic ducts with factory-applied protective coatings must be approved, and all such duct installations must be in accordance with the manufacturer's installation instructions. Application of any field-applied protective coating must be approved. Great care must be taken to protect underground ducts from damage prior to placing the concrete or installing the permanent structure above them. Plastic duct and fitting systems designed for underground applications are available and allow corrosion-resistant and waterproof installations.

Plastic ducts have the advantage of being corrosion resistant. The required external loading properties of ASTM D2412 ensure that the pipe has the ability to resist deformation from the loads associated with direct burial. Plastic ducts rapidly lose strength as their temperature approaches their maximum service temperature. At temperatures above 150°F (65.5°C), PVC pipe is substantially weakened and deformation/collapse is possible.

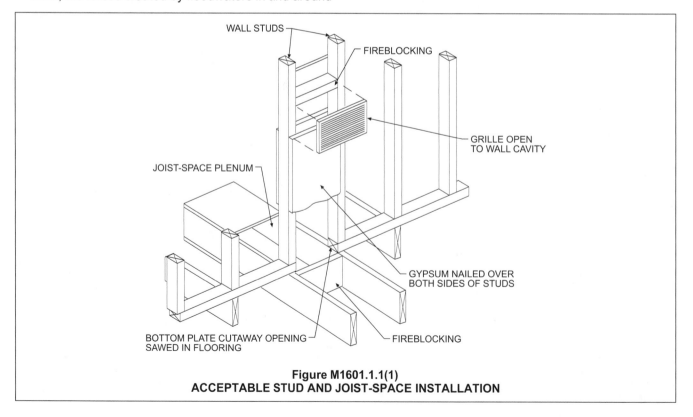

Figure M1601.1.1(1)
ACCEPTABLE STUD AND JOIST-SPACE INSTALLATION

Underground ducts must be sloped to drain to an accessible point in the event that water enters the duct through duct openings or from the subsoil. Water can cause corrosion, deterioration of duct materials and duct blockages; therefore, sloping the duct to drain to a collection point will allow removal of water. The code does not require that ducts be watertight. This section requires that the duct be sealed before the concrete is poured to prevent concrete from entering the duct.

M1601.2 Vibration isolators. Vibration isolators installed between mechanical *equipment* and metal ducts shall be fabricated from *approved* materials and shall not exceed 10 inches (254 mm) in length.

❖ Isolators must be built from materials that will withstand the temperatures and pressures of typical conditioned air passing through the duct. The 10-inch (254 mm) length limitation prevents installation of long lengths of nonrigid duct material, which could adversely affect the integrity of the duct system.

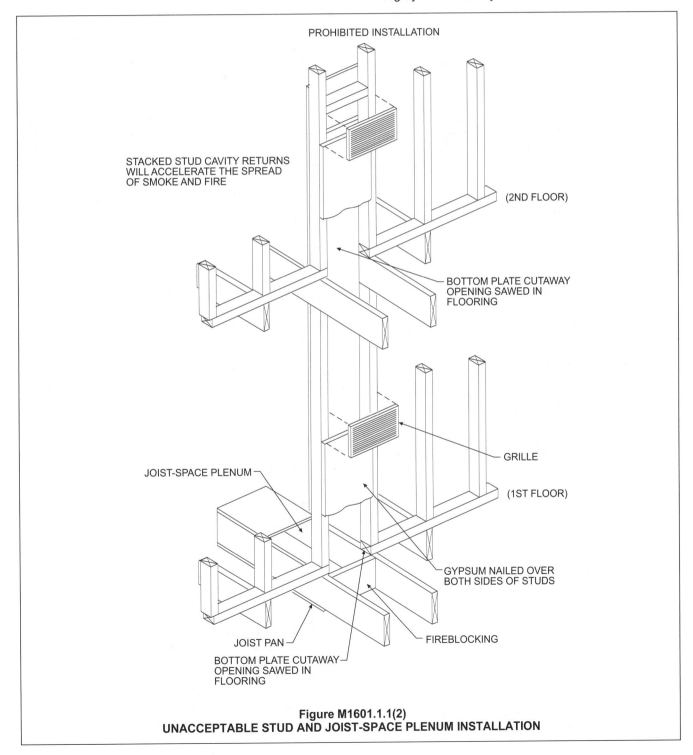

Figure M1601.1.1(2)
UNACCEPTABLE STUD AND JOIST-SPACE PLENUM INSTALLATION

M1601.3 Duct insulation materials. Duct insulation materials shall conform to the following requirements:

1. Duct coverings and linings, including adhesives where used, shall have a flame spread index not higher than 25, and a smoke-developed index not over 50 when tested in accordance with ASTM E84 or UL 723, using the specimen preparation and mounting procedures of ASTM E2231.

   Exception: Spray application of polyurethane foam to the exterior of ducts in *attics* and crawl spaces shall be permitted subject to all of the following:

   1. The flame spread index is not greater than 25 and the smoke-developed index is not greater than 450 at the specified installed thickness.

   2. The foam plastic is protected in accordance with the ignition barrier requirements of Sections R316.5.3 and R316.5.4.

   3. The foam plastic complies with the requirements of Section R316.

2. Duct coverings and linings shall not flame, glow, smolder or smoke when tested in accordance with ASTM C411 at the temperature to which they are exposed in service. The test temperature shall not fall below 250°F (121°C). Coverings and linings shall be listed and labeled.

3. External reflective duct insulation shall be legibly printed or identified at intervals not greater than 36 inches (914 mm) with the name of the manufacturer, the product R-value at the specified installed thickness and the flame spread and smoke-developed indices. The installed thickness of the external duct insulation shall include the enclosed air space(s). The product R-value for external reflective duct insulation shall be determined in accordance with ASTM C1668.

4. External duct insulation and factory-insulated flexible ducts shall be legibly printed or identified at intervals not longer than 36 inches (914 mm) with the name of the manufacturer, the thermal resistance R-value at the specified installed thickness and the flame spread and smoke-developed indexes of the composite materials. Spray polyurethane foam manufacturers shall provide the same product information and properties, at the nominal installed thickness, to the customer in writing at the time of foam application. Nonreflective duct insulation product R-values shall be based on insulation only, excluding air films, vapor retarders or other duct components, and shall be based on tested C-values at 75°F (24°C) mean temperature at the installed thickness, in accordance with recognized industry procedures. The installed thickness of duct insulation used to determine its R-value shall be determined as follows:

   4.1. For duct board, duct liner and factory-made rigid ducts not normally subjected to compression, the nominal insulation thickness shall be used.

   4.2. For ductwrap, the installed thickness shall be assumed to be 75 percent (25-percent compression) of nominal thickness.

   4.3. For factory-made flexible air ducts, The installed thickness shall be determined by dividing the difference between the actual outside diameter and nominal inside diameter by two.

   4.4. For spray polyurethane foam, the aged R-value per inch measured in accordance with recognized industry standards shall be provided to the customer in writing at the time of foam application. In addition, the total R-value for the nominal application thickness shall be provided.

❖ Because duct systems connect most rooms in a building, they can provide a path for fire and smoke to travel throughout that building. Duct coverings and linings are exposed to the surrounding environment and to the airstream in the duct. To reduce the possible spread of fire and smoke, duct coverings and linings must be tested to ASTM E84 or UL 723. These materials are limited to a maximum flame spread index of 25 and a smoke-developed index of 50, which correspond to Class 1 material. In addition, duct coverings and linings must be rated for the air temperatures expected; this avoids degradation of these materials.

An exception to the requirement for meeting the flame spread and smoke-developed indices was added to allow spray polyurethane insulation to be applied on the exterior of ducts in attics and crawl spaces as long as all of the requirements of the exception are met. Although the exception requires this foam plastic to be protected by an ignition barrier, Sections 316.5.3 and 316.5.4 allow for deletion of the ignition barrier if the foam plastic material is specifically approved in accordance with Section 316.6.

To verify that duct coverings and linings will not present a fire hazard, these materials must be tested in the form in which they will be installed at their rated temperatures but to not less than 250°F (121°C) in accordance with the procedures of ASTM C411. This minimum temperature for testing represents the maximum temperature that industry standards will permit in the airstream of a warm-air heating appliance. It is important that the duct coverings and linings are tested in their composite form rather than having tests conducted on each component that comprises the product (i.e., the insulation, the facing and the adhesive). Each component could pass the ASTM E84 or UL 723 tests individually, but could fail when combined into the final assembly to be installed in the field. Requiring the product to be listed and labeled ensures that the product installed is the same one that was tested.

ASTM E84 and UL 723 indicate the test methods required for duct coverings and linings, including a requirement for testing of systems representative of the actual field installation. ASTM E2231 was added to the code because it contains the specimen preparation and mounting procedures necessary to ensure that the specimen tested in the laboratory is as close as possible to the actual field installation. This will result in a safer field installation where the actual performance of the material can be more accurately predicted.

To assist inspectors, the code requires that duct insulation have a label with the manufacturer's name, thermal resistance (*R*), and the flame/smoke indexes. However, because labeling of the exterior of spray polyurethane insulation is not feasible, the code requires that the same information be given, in writing, to the customer at the time of application.

A third-party agency must provide quality control inspections at the manufacturer's facility in accordance with the requirements for labeling. Testing performed by an independent agency must determine the insulating *R*-value, the flame spread index and the smoke-developed index. The 36-inch (914 mm) label intervals are intended to increase the likelihood that every cut piece of insulation and flexible duct will have a label.

Reflective duct insulation is a defined term in Chapter 2. It is typically constructed of aluminum foil or metalized film membranes having closed air cells sandwiched between the membranes. The air cells provide *R*-value and the foil or metalized film acts as a reflector of radiant energy. *Reflective duct insulation* is often installed with spacers that create an airspace (not to be confused with air films) between the duct surface and the inside surface of the insulation, thereby increasing the overall *R*-value of the assembly. Items 4.1 through 4.4 of Item 4 do not apply to *reflective duct insulation*.

The thermal performance of duct insulation is dependent on its "installed" condition including the compressed condition, as is the case for duct wrap insulations other than *reflective duct insulation*. This section is intended to provide manufacturers, installers and inspectors with specific guidance for meeting the intent of the code. For example, installed fiberglass duct wrap is assumed to have a thickness of 75 percent of the nominal uninstalled thickness. The *R*-value reduction for the product will account for the decreased thermal resistance caused by compression of the product.

M1601.4 Installation. Duct installation shall comply with Sections M1601.4.1 through M1601.4.10.

❖ See Sections M1601.4.1 through M1601.4.10.

M1601.4.1 Joints, seams and connections. Longitudinal and transverse joints, seams and connections in metallic and nonmetallic ducts shall be constructed as specified in SMACNA HVAC *Duct Construction Standards—Metal and Flexible* and NAIMA *Fibrous Glass Duct Construction Standards*. Joints, longitudinal and transverse seams, and connections in ductwork shall be securely fastened and sealed with welds, gaskets, mastics (adhesives), mastic-plus-embedded-fabric systems, liquid sealants or tapes. Tapes and mastics used to seal fibrous glass ductwork shall be *listed* and *labeled* in accordance with UL 181A and shall be marked "181A-P" for pressure-sensitive tape, "181 A-M" for mastic or "181 A-H" for heat-sensitive tape.

Tapes and mastics used to seal metallic and flexible air ducts and flexible air connectors shall comply with UL 181B and shall be marked "181 B-FX" for pressure-sensitive tape or "181 BM" for mastic. Duct connections to flanges of air distribution system equipment shall be sealed and mechanically fastened. Mechanical fasteners for use with flexible nonmetallic air ducts shall comply with UL 181B and shall be marked 181B-C. Crimp joints for round metallic ducts shall have a contact lap of not less than 1 inch (25 mm) and shall be mechanically fastened by means of not less than three sheet-metal screws or rivets equally spaced around the joint.

Closure systems used to seal all ductwork shall be installed in accordance with the manufacturers' instructions.

Exceptions:

1. Spray polyurethane foam shall be permitted to be applied without additional joint seals.

2. Where a duct connection is made that is partially inaccessible, three screws or rivets shall be equally spaced on the exposed portion of the joint so as to prevent a hinge effect.

3. For ducts having a static pressure classification of less than 2 inches of water column (500 Pa), additional closure systems shall not be required for continuously welded joints and seams and locking-type joints and seams of other than the snap-lock and button-lock types.

❖ Duct sealing is commonly overlooked or poorly performed. The U.S. Environmental Protection Agency (EPA) estimates that 20 percent of the energy efficiency of heating and cooling systems can be lost due to duct air leaks located outside of the conditioned space. With the increased focus on lowering energy use in all types of buildings, substantial gains in energy efficiency can be obtained by making sure that ducts are "substantially" airtight. While approved duct materials have low permeability, the joints in these materials as well as the joints in the connections of these materials to fittings/equipment must be carefully sealed to achieve a "substantially" air-tight condition.

Field experience has shown that poor workmanship and failure to follow manufacturer's instructions has resulted in a high incidence of joint failure in attics and crawl spaces. It is not uncommon to find that air is supplied to or air is returned from an attic because of a separated duct joint. This is obviously a tremendous waste of energy.

In general, joints must be sealed using tapes, mastics, liquid sealants, gasketing or other approved closure systems. A "closure system" consists of the materials and an installation method used to make the joint substantially airtight. This section specifically addresses sealing for three types of duct material which can be accomplished as follows:

- Rigid fibrous glass ductwork must be sealed in accordance with UL 181A. Where pressure-sensitive tape is used, the exterior of the tape must be factory marked with the designation "181 A-P." Where mastic is used, the container label must indicate "181 A-M." Where heat-sensitive tape is used, the exterior of the tape must be factory marked with the designation "181 A-H."

- Metallic ducts, flexible air ducts and flexible air connectors must be sealed in accordance with

UL 181B. Where pressure-sensitive tape is used, the exterior of the tape must be factory marked with the designation "181B F-X." Where mastic is used, the container label must indicate "181 B-M."

In addition to sealing, joints must also be mechanically fastened. In other words, sealing by itself is not a substitute for the mechanical fastening of a duct joint. Two examples are taping a flexible air duct to a fitting and "gluing" a round metal duct joint with caulking. Both practices are not acceptable fastening methods. Mechanical fasteners must be used. This section specifically addresses the mechanical fastening of three types of duct material as follows:

- Rigid fibrous glass ductwork must be mechanically attached to flanges of air distribution equipment or sheet metal fittings. For example, where a rigid fibrous glass duct connects to a sheet metal flange of an air handler, sheet metal screws (with flat washers) can be used to attach the duct board to the metal flange. Where fittings for other ducts must attach to the duct, a "spin in," "twist lock" or "tabbed" collar provides for the required mechanical fastening of the joint.

- Flexible air ducts and flexible air connectors must be fastened to collar stubs or duct connector sleeves. Worm gear band clamps and plastic cable ties are most commonly used for fastening. Where cable ties are used, they must comply with UL 181B and be marked "181-C." The collar stubs should have a convex bead formed on the circumference of the fitting to prevent the flexible duct from being pulled off of the connector. Commentary Figure M1601.4.1 shows installation guidelines for flexible air ducts and flexible air connectors.

- Round metal duct with crimp joints must have an overlap of at least 1 inch (25.4 mm). At least three sheet metal screws or rivets must be installed, equally spaced around the duct, for mechanical fastening of the joint. A minimum of 3 fasteners creates a rigid joint that will not allow a hinge movement.

Exception 1 allows spray polyurethane foam insulation to serve as a duct joint sealing method. Spray foam insulation has strong adhesive properties and low porosity such that its application provides an equivalent air-tight condition as compared to other closure systems. Exception 2 allows the spacing of the screws or rivets on round metal ducts to be equally spaced on the portion of duct joint that can be accessed, if the entire perimeter of the joint cannot be accessed. The installation of duct work in close quarters such as in joint spaces and chases, often limits access to the complete circumference of the joint. The installation of three screws or rivets in the portion of joint that can be accessed is a reasonable accommo-

dation for such conditions. For air systems with a static pressure of less than 2 inches water column (0.50 kPa), Exception 3 allows for longitudinal (including spiral-type) welded joints and locking-type joints and seams to not be sealed. A welded or locking-type (e.g., Pittsburgh lock) joint is sufficiently air tight at low pressures such that joint sealing would be of limited value. Snap-lock and button-lock type joints do not fall under Exception 3 and must be sealed because such joints will leak. Note that some duct manufacturers provide snap-lock joints that contain a factory-applied sealant in the locking groove that serves to seal the joint when assembled. S-slip joints and lap joints must always be sealed.

M1601.4.2 Duct lap. Crimp joints for round and oval metal ducts shall be lapped not less than 1 inch (25 mm) and the male end of the duct shall extend into the adjoining duct in the direction of airflow.

❖ The crimped end of a round duct is the male end and it points in the direction of flow. The minimum required lap creates joint stability and allows for proper mechanical fastening. Crimped joints can be exceptionally leaky depending on the length of lap. Section M1601.4.1 requires such joints to be sealed.

M1601.4.3 Plastic duct joints. Joints between plastic ducts and plastic fittings shall be made in accordance with the manufacturer's installation instructions.

❖ Plastic ducts are typically installed underground. If plastic ducts are installed, this section requires the installer to follow the duct manufacturer's instructions for making joints in the duct. Some manufacturers require joints to be solvent cemented while others require the joints to be caulked. Section M1601.4.1 does not address plastic ducts.

M1601.4.4 Support. Factory-made ducts listed in accordance with UL 181 shall be supported in accordance with the manufacturer's installation instructions. Field- and shop-fabricated fibrous glass ducts shall be supported in accordance with the SMACNA *Fibrous Glass Duct Construction Standards* or the NAIMA *Fibrous Glass Duct Construction Standards*. Field- and shop-fabricated metal and flexible ducts shall be supported in accordance with the SMACNA *HVAC Duct Construction Standards—Metal and Flexible*.

❖ These material requirements and the 10-foot (3048 mm) maximum spacing requirement prescribe structural support for metal ductwork to limit deflection and maintain alignment. Nonmetallic ducts must be supported in accordance with the manufacturer's installation instructions because these ducts are produced in various configurations. See Commentary Figure M1601.4.1 for information on nonmetallic duct support.

M1601.4.5 Fireblocking. Duct installations shall be fireblocked in accordance with Section R602.8.

❖ Section R602.8 refers to Section R302.11 and requires that all openings around ducts at ceiling and floor levels must be fireblocked. Fireblocking retards the spread of fire to other areas in the dwelling.

4

Installation Guidelines

4.1 Code Reference

The "authority having jurisdiction" should be referenced to determine what law, ordinance or code shall apply in the use of flexible duct.

Ducts conforming to NFPA 90A or 90B shall meet the following requirements:

    a. Shall be tested in accordance with Sections 5-21 of Underwriters Laboratories Standard for Factory-Made Air Ducts and Air Connectors, UL 181.

    b. Shall be installed in accordance with the conditions of their listing.

    c. Shall be installed within the limitations of the applicable NFPA 90A or 90B Standard.

4.2 General

The routing of flexible duct, the number of bends, the number or degrees in each bend and the amount of sag allowed between support joints will have serious effects on system performance due to the increased resistance each introduces. Use the minimum length of flexible duct to make connections. It is not recommended that excess length of ducts be installed to allow for possible future relocations of air terminal devices.

Avoid installations where exposure to direct sunlight can occur, e.g. turbine vents, sky lights, canopy windows, etc. Prolonged exposure to sunlight will cause degradation of the vapor barrier. Direct exposure to UV light from a source lamp installed within the HVAC system will cause degradation of some inner core/liner materials.

Terminal devices shall be supported independently of the flexible duct.

Repair torn or damaged vapor barrier/jacket with duct tape listed and labeled to Standard UL 181B. If internal core is penetrated, replace flexible duct or treat as a connection.

4.3 Installation and Usage

Install duct fully extended, do not install in the compressed state or use excess lengths. This will noticeably increase friction losses.

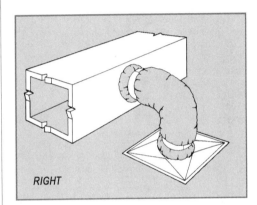

RIGHT

Figure 6

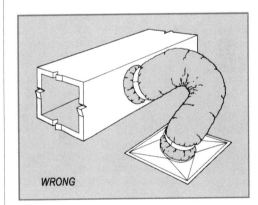

WRONG

Figure 7

Figure M1601.4.1
INSTALLATION INSTRUCTIONS FOR FLEXIBLE AIR DUCTS AND AIR CONNECTORS
(Courtesy of Air Diffusion Council)

(continued)

12

ADC Flexible Duct Performance & Installation Standards, 4th Edition

Installation Guidelines . . . continued

Avoid bending ducts across sharp corners or incidental contact with metal fixtures, pipes or conduits. Radius at center line shall not be less than one duct diameter.

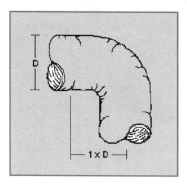

Figure 8

Do not install near hot equipment (e.g. furnaces, boilers, steam pipes, etc.) that is above the recommended flexible duct use temperature.

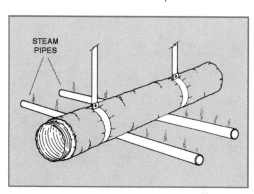

Figure 9

4.4 Connecting, Joining and Splicing Flexible Ducts

All connections, joints and splices shall be made in accordance with the manufacturer's installation instructions.

For flexible ducts with plain ends, standardized installation instructions conforming to this standard are shown in Sections 4.5 "Nonmetallic With Plain Ends" (uses tape and clamp to seal/secure the core to the fitting), 4.6 "Alternate Nonmetallic With Plain Ends" (uses mastic and clamp to seal/secure the core to the fitting), and 4.7 "Metallic With Plain Ends (optional use of tape or mastic and metal screws to seal/secure the core to the fitting).

Due to the wide variety of ducts and duct assemblies with special end treatments, e.g. factory installed fittings, taped ends, crimped metal ends, etc., no standardized installation instructions are shown. Reference manufacturer's installation instructions.

All tapes, mastics, and nonmetallic clamps used for field installation of flexible ducts shall be listed and labeled to Standard UL 181B - Closure Systems for Use With Flexible Air Ducts and Air Connectors.

Sheet metal fittings to which flexible ducts with plain ends are attached shall be beaded and have a minimum of 2 inches [50 mm] collar length. Beads are optional for fittings when attaching *metallic* flexible ducts.

Sheet metal sleeves used for joining two sections of flexible duct with plain ends shall be a minimum of 4 inches [100 mm] in length and beaded on each end. Beads are optional for sleeves when joining *metallic* flexible ducts.

Flexible ducts secured with nonmetallic clamps shall be limited to 6 inches w.g. [1500 Pa] positive pressure.

13

ADC Flexible Duct Performance & Installation Standards, 4th Edition

Figure M1601.4.1—continued
INSTALLATION INSTRUCTIONS FOR FLEXIBLE AIR DUCTS AND AIR CONNECTORS
(Courtesy of Air Diffusion Council)

(continued)

Installation Guidelines . . . continued

4.5 Installation Instructions for Air Ducts and Air Connectors - Nonmetallic with Plain Ends

Connections

1. After desired length is determined, cut completely around and through duct with knife or scissors. Cut wire with wire cutters. Fold back jacket and insulation.

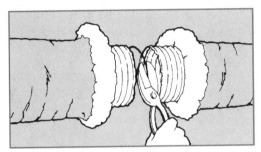

2. Slide at least 1" [25 mm] of core over fitting and past the bead. Seal core to collar with at least 2 wraps of duct tape. Secure connection with clamp placed over the core and tape and past the bead.

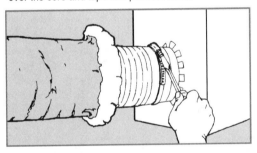

3. Pull jacket and insulation back over core. Tape jacket with at least 2 wraps of duct tape. A clamp may be used in place of or in combination with the duct tape.

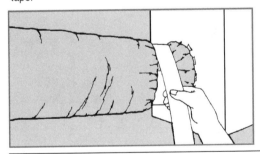

Splices

1. Fold back jacket and insulation from core. Butt two cores together on a 4" [100 mm] length metal sleeve.

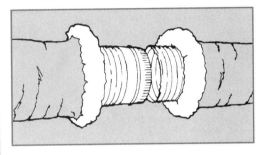

2. Tape cores together with at least 2 wraps of duct tape. Secure connection with 2 clamps placed over the taped core ends and past the beads.

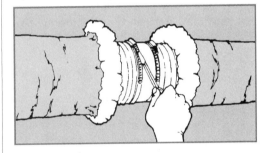

3. Pull jacket and insulation back over cores. Tape jackets together with at least 2 wraps of duct tape.

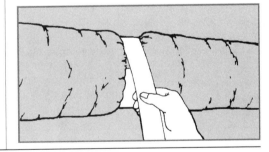

NOTES:
1. For uninsulated air ducts and air connectors, disregard references to insulation and jacket.
2. Use beaded sheet metal fittings and sleeves.
3. Use tapes listed and labeled in accordance with Standard UL 181B and marked "181B-FX".
4. Nonmetallic clamps shall be listed and labeled in accordance with Standard UL 181B and marked "181B-C". Use of nonmetallic clamps shall be limited to 6 in. w.g. [1500 Pa] positive pressure.

Figure M1601.4.1—continued
INSTALLATION INSTRUCTIONS FOR FLEXIBLE AIR DUCTS AND AIR CONNECTORS
(Courtesy of Air Diffusion Council)

(continued)

Installation Guidelines . . . continued

4.6 Alternate Installation Instructions for Air Ducts and Air Connectors - Nonmetallic with Plain Ends

Connections and Splices

Step 1
After desired length is determined, cut completely around and through duct with knife or scissors. Cut wire with wire cutters. Pull back jacket and insulation from core.

Step 2
Apply mastic approximately 2" [50 mm] wide uniformly around the collar of the metal fitting or over the ends of a 4" [100 mm] metal sleeve. Reference data on mastic container for application rate, application thickness, cure times and handling information.

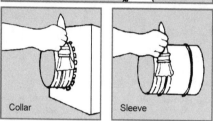

Collar   Sleeve

Step 3
Slide at least 2" [50 mm] of core over the fitting or sleeve ends and past the bead.

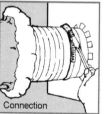 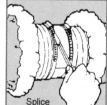

Connection   Splice

Step 4
Secure core to collar with a clamp applied past the bead. Secure cores to sleeve ends with 2 clamps applied past the beads.

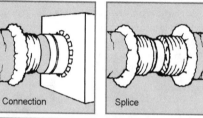

Connection   Splice

Step 5
Pull jacket and insulation back over core ends. Tape jacket(s) with at least 2 wraps of duct tape. A clamp may be used in place of or in combination with the duct tape.

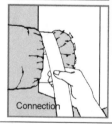 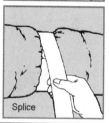

Connection   Splice

15

ADC Flexible Duct Performance & Installation Standards, 4th Edition

NOTES:
1. For uninsulated air ducts and air connectors, disregard references to insulation and jacket.
2. Use beaded sheet metal fittings and sleeves.
3. Use mastics listed and labeled in accordance with Standard UL 181B and marked "181B-M" on container.
4. Use tapes listed and labeled in accordance with Standard UL 181B and marked "181B-FX".
5. Nonmetallic clamps shall be listed and labeled in accordance with standard UL 181B and marked "181B-C". Use of nonmetallic clamps shall be limited to 6 in. w.g. [1500 Pa] positive pressure.

Figure M1601.4.1—continued
INSTALLATION INSTRUCTIONS FOR FLEXIBLE AIR DUCTS AND AIR CONNECTORS
(Courtesy of Air Diffusion Council)

(continued)

Installation Guidelines . . . continued

4.7 Installation Instruction for Air Ducts and Air Connectors - Metallic with Plain Ends

Connections and Splices

1. After cutting duct to desired length, fold back jacket and insulation exposing core. Trim core ends squarely using suitable metal shears. Determine optional sealing method (Steps 2 or 5) before proceeding.

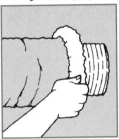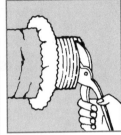

4. Secure to collar/sleeve using #8 sheet metal screws spaced equally around circumference. Use 3 screws for diameters under 12" [300 mm] and 5 screws for diameters 12" [300 mm] and over.

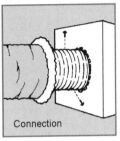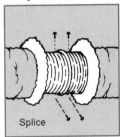

Connection | Splice

2. When mastics are required and for pressures 4" w.g. [1000 Pa] and over, seal joint with mastic applied uniformly to the outside surface of collar/sleeve. (Disregard this step when not using mastics and proceed to Step 3).

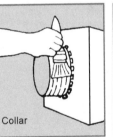

Collar | Sleeve

5. For pressures under 4" w.g. [1000 Pa] seal joint using 2 wraps of duct tape applied over screw heads and spirally lapping tape to collar/sleeve. (Disregard this step when using mastics per Step 2).

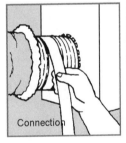

Connection | Splice

3. Slide at least 1" [25 mm] of core over metal collar for attaching duct to take off or over ends of a 4" [100 mm] metal sleeve for splicing 2 lengths of duct.

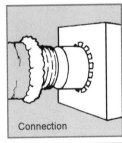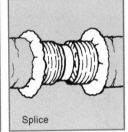

Connection | Splice

6. Pull jacket and insulation back over core. Tape jacket with 2 wraps of duct tape. A clamp may be used in place of or in combination with the duct tape.

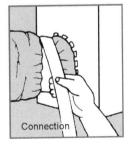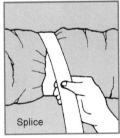

Connection | Splice

NOTES:
1. For uninsulated air ducts and air connectors, disregard references to insulation and jacket.
2. Use mastics listed and labeled to Standard UL 181B and marked "181B-M" on container.
3. Use tapes listed and labeled to Standard UL 181B and marked "181B-FX".
4. Nonmetallic clamps shall be listed and labeled in accordance with Standard UL 181B and marked "181B-C".

Figure M1601.4.1—continued
INSTALLATION INSTRUCTIONS FOR FLEXIBLE AIR DUCTS AND AIR CONNECTORS
(Courtesy of Air Diffusion Council)

(continued)

Installation Guidelines . . . continued

4.8 Supporting Flexible Duct

Flexible duct shall be supported at manufacturer's recommended intervals, but at no greater distance than 5' [1.5 m]. Maximum permissible sag is ½" per foot [42 mm per meter] of spacing between supports.

A connection to rigid duct or equipment shall be considered a support joint. Long horizontal duct runs with sharp bends shall have additional supports before and after the bend approximately one duct diameter from the center line of the bend.

Hanger or saddle material in contact with the flexible duct shall be of sufficient width to prevent any restriction of the internal diameter of the duct when the weight of the supported section rests on the hanger or saddle material. In no case will the material contacting the flexible duct be less than 1½" [38 mm] wide.

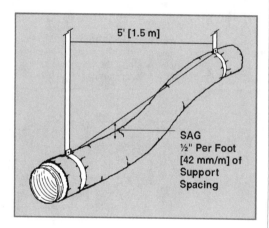

Figure 10

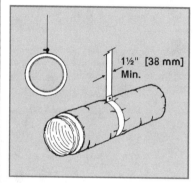

Figure 11

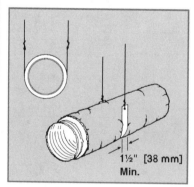

Figure 12

Figure M1601.4.1—continued
INSTALLATION INSTRUCTIONS FOR FLEXIBLE AIR DUCTS AND AIR CONNECTORS
(Courtesy of Air Diffusion Council)

(continued)

Installation Guidelines . . . continued

Factory installed suspension systems integral to the flexible duct are an acceptable alternative hanging method when manufacturer's recommended procedures are followed.

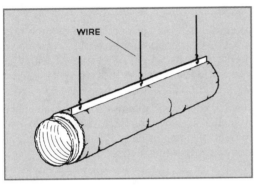

Figure 13

Flexible ducts may rest on ceiling joists or truss supports. Maximum spacing between supports shall not exceed the maximum spacing per manufacturer's installation instruction.

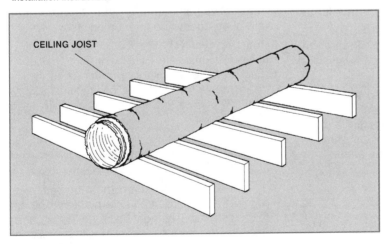

CEILING JOIST

Figure 14

Note:
Factory-made air ducts may not be used for vertical risers in air duct systems serving more than two stories.

Support the duct between a metal connection and bend by allowing the duct to extend straight for a few Inches before making the bend. This will avoid possible damage of the flexible duct by the edge of the metal collar.

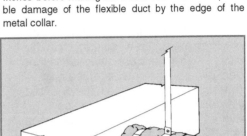

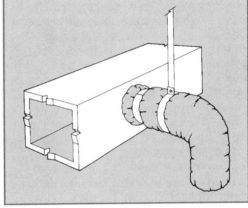

Figure 15

19

Vertically installed duct shall be stabilized by support straps at a max. of 6' [1.8 m] on center.

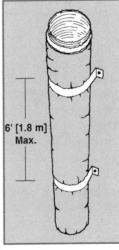

6' [1.8 m] Max.

Figure 16

ADC Flexible Duct Performance & Installation Standards, 4th Edition

Figure M1601.4.1—continued
INSTALLATION INSTRUCTIONS FOR FLEXIBLE AIR DUCTS AND AIR CONNECTORS
(Courtesy of Air Diffusion Council)

M1601.4.6 Duct insulation. Duct insulation shall be installed in accordance with the following requirements:

1. A vapor retarder having a maximum permeance of 0.05 perm [2.87 ng/(s · m^2 · Pa)] in accordance with ASTM E96, or aluminum foil with a minimum thickness of 2 mils (0.05 mm), shall be installed on the exterior of insulation on cooling supply ducts that pass through unconditioned spaces conducive to condensation except where the insulation is spray polyurethane foam with a maximum water vapor permeance of 3 perm per inch [1722 ng/(s · m^2 · Pa)] at the installed thickness.

2. Exterior *duct systems* shall be protected against the elements.

3. Duct coverings shall not penetrate a fireblocked wall or floor.

❖ Where a duct is installed outdoors or in an unconditioned area (such as an attic or crawl space), it can be exposed to humidity and temperature differentials that can create condensation on the outside of the duct. A vapor retarder must be installed on the exterior of the duct to protect the duct and/or insulation from damage caused by moisture. Duct insulation and coverings must not penetrate fireblocked assemblies. If the exterior insulation and vapor retarder burn in a fire, a pathway will open for the fire to spread through the penetration in the structure for the duct.

The exterior of cooling supply ducts that are insulated with spray polyurethane foam which has a maximum water vapor permeance of 3 perm per inch does not require a vapor retarder or aluminum foil. Spray polyurethane foam has been successfully used without a vapor barrier in other applications, such as exterior walls. The natural low permeance of the installed material makes a vapor barrier over the spray foam insulation redundant.

M1601.4.7 Factory-made air ducts. Factory-made air ducts shall not be installed in or on the ground, in tile or metal pipe, or within masonry or concrete.

❖ Factory-made air ducts must be listed and labeled for underground installations. Ducts not listed for in-ground installations might not have sufficient strength to withstand the loads applied by these types of installations and might not be suitable for high-moisture areas.

M1601.4.8 Duct separation. Ducts shall be installed with not less than 4 inches (102 mm) separation from earth except where they meet the requirements of Section M1601.1.2.

❖ A physical separation from earth is the best method of protecting metallic and nonmetallic ducts from the effects of corrosion and moisture.

M1601.4.9 Ducts located in garages. Ducts in garages shall comply with the requirements of Section R302.5.2.

❖ Ducts in garages and ducts penetrating separation walls or ceilings between garages and living spaces must be designed to prevent fire and smoke from easily entering the living spaces (see commentary, Sections R302.5.2 and M1601.6). Section R302.5.2 is intended to protect the penetration of the fire separa-

tion between a dwelling and an attached garage. The code assumes that the 26-gage steel ductwork and the furnace/air-handler casing will offer a significant impediment to the spread of fire from the garage to the dwelling interior. Placing furnaces, airhandlers and ductwork in unconditioned spaces is becoming less common because of energy saving efforts. Concern has been expressed by some that having a furnace/air-handler with nonmetallic access doors to the blower or filter compartments creates a weak spot in the steel duct barrier. The code official can approve materials other than steel in accordance with Sections R302.5.2 and R104.11.

M1601.4.10 Flood hazard areas. In flood hazard areas as established by Table R301.2(1), *duct systems* shall be located or installed in accordance with Section R322.1.6.

❖ In buildings and structures located in flood hazard areas, ducts must be installed above the design flood elevation or must be capable of preventing water from entering the ducts and capable of withstanding the forces of buoyancy and moving water (see commentary, Section R322.1.6).

M1601.5 Under-floor plenums. Under-floor plenums shall be prohibited in new structures. Modification or repairs to under-floor plenums in existing structures shall conform to the requirements of this section.

❖ New structures must not be designed to use crawl spaces as plenums. This practice is quite rare and such arrangements can waste energy and negatively affect indoor air quality. Because some existing structures might still utilize under-floor plenums, this section allows those arrangements to continue to exist even if modifications or repairs are performed to those systems. See Commentary Figure M1601.5 for an example of an under-floor plenum system.

M1601.5.1 General. The space shall be cleaned of loose combustible materials and scrap, and shall be tightly enclosed. The ground surface of the space shall be covered with a moisture barrier having a minimum thickness of 4 mils (0.1 mm). Plumbing waste cleanouts shall not be located within the space.

Exception: Plumbing waste cleanouts shall be permitted to be located in unvented crawl spaces that receive *conditioned air* in accordance with Section R408.3.

❖ Loose combustible scrap must be removed from an under-floor space used as a plenum. The space must be substantially air tight because the furnace will place the plenum under a slight positive pressure to force conditioned air out of the registers into the rooms above. If there are leaks in the plenum walls, the conditioned air will escape to the outdoors and result in energy loss and the creation of negative pressure in the living space.

Efforts to rod drains through cleanouts located in under-floor plenums could create spills and splashes of sewage in the plenum and cleanout openings might be carelessly left open allowing sewer gases to enter the plenum. Because this would create an insanitary air condition for the building occupants, plumbing cleanouts are prohibited in these spaces. An issue that often comes up is whether fuel gas piping is prohibited in under-floor plenums. Chapter 24 contains the requirements for fuel-gas piping and does not prohibit gas piping in such spaces.

M1601.5.2 Materials. The under-floor space, including the sidewall insulation, shall be formed by materials having flame spread index values not greater than 200 when tested in accordance with ASTM E84 or UL 723.

❖ The enclosing materials of an under-floor plenum, including the sidewall insulation, are limited to materials having a flame spread index less than or equal to 200 because the plenum space could introduce smoke and fire to the living space.

M1601.5.3 Furnace connections. A duct shall extend from the furnace supply outlet to not less than 6 inches (152 mm) below the combustible framing. This duct shall comply with the provisions of Section M1601.1. A noncombustible receptacle shall be installed below any floor opening into the plenum in accordance with the following requirements:

1. The receptacle shall be securely suspended from the floor members and shall be not more than 18 inches (457 mm) below the floor opening.

2. The area of the receptacle shall extend 3 inches (76 mm) beyond the opening on all sides.

3. The perimeter of the receptacle shall have a vertical lip not less than 1 inch (25 mm) in height at the open sides.

❖ A noncombustible receptacle, (i.e., a pan) must be installed below each of the supply openings in the floor. This receptacle (pan) must be suspended from the floor members and must be located not greater than 18 inches (457 mm) below the floor opening so that items and debris dropped into floor openings can be easily retrieved by reaching down through the opening. The receptacle (pan) must extend 3 inches (76 mm) beyond the opening on all sides and have a vertical lip at least 1-inch (25 mm) high to keep items from sliding or rolling off the edges of the pan. The requirement for the pan material to be noncombustible anticipates the potential for burning material being dropped into the opening.

M1601.5.4 Access. Access to an under-floor plenum shall be provided through an opening in the floor with minimum dimensions of 18 inches by 24 inches (457 mm by 610 mm).

❖ An 18-inch by 24-inch (457 mm by 610 mm) or larger opening in the floor is required to provide access to the under-floor plenum to permit maintenance and repairs in the crawl space.

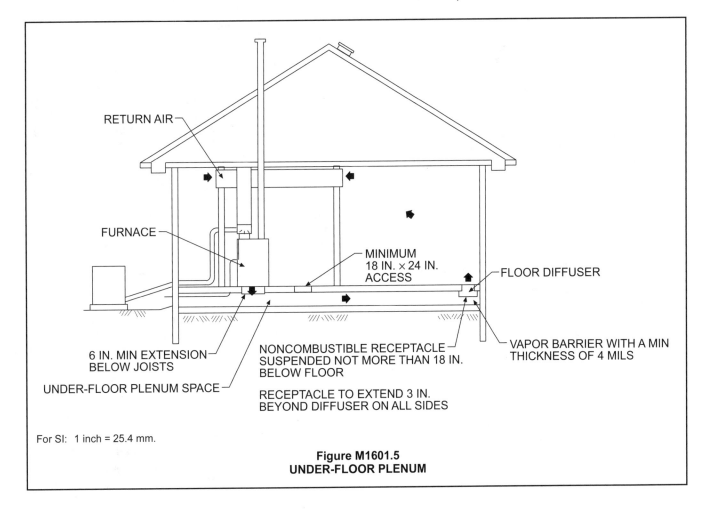

For SI: 1 inch = 25.4 mm.

Figure M1601.5
UNDER-FLOOR PLENUM

M1601.5.5 Furnace controls. The furnace shall be equipped with an automatic control that will start the air-circulating fan when the air in the furnace bonnet reaches a temperature not higher than 150°F (66°C). The furnace shall additionally be equipped with an *approved* automatic control that limits the outlet air temperature to 200°F (93°C).

❖ Because exposed wood joists and flooring often form under-floor plenums, the outlet air temperature from the furnace must be limited to 200°F (93°C).

Furnaces must also be equipped with an automatic control that starts the air circulation fan before the air temperature in the furnace at the heat exchanger exceeds 150°F (66°C). This requirement serves to limit plenum air temperature. Note that replacement furnaces will not likely be available that can meet these control requirements and altering the furnaces would violate the listing of the appliance.

M1601.6 Independent garage HVAC systems. Furnaces and air-handling systems that supply air to living spaces shall not supply air to or return air from a garage.

❖ This section prohibits a furnace or air-handler from serving both living spaces and garages. If a garage is conditioned, it must be done so by an independent system. For example, if a furnace supplies air to a garage and also to a dwelling interior, the air to the garage will be lost because return air cannot be taken from a garage. The result will be negative pressure in the dwelling with respect to the garage, and the condition will cause contaminants in the garage to migrate into the dwelling (see M1602.2, Item 4).

Garages could contain contaminants that would negatively affect air quality in the living spaces. Pressure differentials and duct leakage could cause contaminants to migrate into the living space. Requiring that the living space system be completely separate ensures that contaminants will not be introduced into the living spaces.

SECTION M1602
RETURN AIR

M1602.1 Outdoor air openings. Outdoor intake openings shall be located in accordance with Section R303.5.1. Opening protection shall be in accordance with Section R303.6.

❖ See Sections R303.5.1 and R303.6.

M1602.2 Return air openings. Return air openings for heating, ventilation and air conditioning systems shall comply with all of the following:

1. Openings shall not be located less than 10 feet (3048 mm) measured in any direction from an open combustion chamber or draft hood of another appliance located in the same room or space.

2. The amount of return air taken from any room or space shall be not greater than the flow rate of supply air delivered to such room or space.

3. Return and transfer openings shall be sized in accordance with the appliance or equipment manufacturers'

installation instructions, Manual D or the design of the registered design professional.

4. Return air shall not be taken from a closet, bathroom, toilet room, kitchen, garage, mechanical room, boiler room, furnace room or unconditioned attic.

Exceptions:

1. Taking return air from a kitchen is not prohibited where such return air openings serve the kitchen only, and are located not less than 10 feet (3048 mm) from the cooking appliances.

2. Dedicated forced-air systems serving only the garage shall not be prohibited from obtaining return air from the garage.

5. Taking return air from an unconditioned crawl space shall not be accomplished through a direct connection to the return side of a forced-air furnace. Transfer openings in the crawl space enclosure shall not be prohibited.

6. Return air from one dwelling unit shall not be discharged into another dwelling unit.

❖ Return air is defined and is typically partially or completely recirculated, thus it is important to control the locations from which return air is taken to prevent contamination. Item 1 expresses the concern for drawing combustion products into the return air intakes. Item 4 addresses locations where contaminants would be expected to be present. Item 2 is intended to prevent the creation of negative pressures that would interfere with the function of vents, chimneys and fuel-fired appliances. Such interference can cause appliance malfunction, unsafe operation and spillage of combustion products into the space served by the return air system. Item 3 defers the sizing of return air openings and transfer openings to the design professional, the appliance manufacturer or the ACCA *Manual D* for residential duct systems. The aggregate area of all ducts or openings that convey return air back to the furnace must be adequate to allow the required airflow through the furnace. A furnace that is "starved" for return air will produce an abnormal temperature rise across the heat exchanger, which is a fire hazard and likely detrimental to the furnace. Item 6 prohibits a return system from serving more than one dwelling unit. Any arrangement in which dwelling units share all or part of an air distribution system would allow a communication of atmospheres in the units. This communication would spread odors, smoke, contaminants and disease-causing organisms from one dwelling unit to another and therefore must be avoided. Item 5 addresses the practice of indirectly conditioning a crawl space to the extent that it is considered to be within the thermal envelope of the building. Transfer openings are allowed for this purpose, but locating a ducted return air opening in the crawl space is prohibited because of the potential for contaminants such as moisture, dust, radon, mold, etc., to be present in such space. Item 4 addresses potential sources of air contaminants, and the creation of pressure differentials that can draw combustion products and contaminants from garages, chimneys, vents and appliances. Return air openings might interfere with the ability of exhaust systems and hoods to capture the target contaminants or effluent. Taking return air from an unconditioned attic would also cause significant energy waste.

Exception 1 recognizes that return air could be necessary for conditioning a kitchen and contamination of the return air is avoided by the required minimum separation from cooking appliances. Exception 2 recognizes that return air from a garage cannot be avoided where the return system is part of a space conditioning system that serves the garage.

Chapter 17:
Combustion Air

User note: Code change proposals to this chapter will be considered by the IRC – Plumbing and Mechanical Code Development Committee during the 2015 (Group A) Code Development Cycle.

General Comments

Complete combustion of fuel is essential for the proper operation of appliances, for control of harmful emissions and for achieving maximum fuel efficiency. Combustion air supplies, among other things, the oxygen necessary for the complete and efficient burning of fuel. If insufficient quantities of oxygen are supplied, the combustion process will be incomplete, creating dangerous byproducts and wasting energy in the form of unburned fuel (hydrocarbons). The byproducts of incomplete combustion are poisonous, corrosive and combustible and can cause serious appliance or equipment malfunctions that pose fire or explosion hazards.

Combustion air also serves other purposes in addition to supplying oxygen. It ventilates and cools appliances and the rooms or spaces that enclose them, and it plays an important role in producing and controlling draft in vents and chimneys. An insufficient combustion air supply could cause an appliance to overheat and discharge combustion byproducts into the building.

A combustion air supply is also necessary to prevent oxygen depletion, which threatens the safety of the building occupants. Both building-occupant respiration and the combustion process of fuel-burning appliances consume the available oxygen in a room or space. The depletion of available oxygen promotes incomplete fuel combustion.

Adequate combustion air supply is extremely important, but it is one aspect of mechanical installation that is often overlooked, ignored or compromised. Depending on the appliance type, location and building construction, supplying combustion air can be easy, or it can involve complex designs and extraordinary methods. In any case, the importance of a proper combustion air supply cannot be overemphasized.

This chapter does not actually contain any combustion air requirements. It was recognized in previous editions of the code that the requirements in this chapter were based on combustion air requirements that were developed for fuel-gas burning appliances. Liquid and solid fuel-burning appliances do not have the same requirements for combustion air. Therefore, all of the requirements in this chapter were deleted and replaced with instructions on where to find fuel-specific combustion air requirements for solid, liquid and fuel gas-fired appliances. Chapter 24 covers gas appliances.

Purpose

This chapter directs the reader to appropriate combustion air requirements provided in other publications or chapters of the code.

SECTION M1701
GENERAL

M1701.1 Scope. Solid fuel-burning *appliances* shall be provided with *combustion air* in accordance with the *appliance* manufacturer's installation instructions. Oil-fired *appliances* shall be provided with *combustion air* in accordance with NFPA 31. The methods of providing *combustion air* in this chapter do not apply to fireplaces, fireplace stoves and direct-vent *appliances*. The requirements for combustion and dilution air for gas-fired *appliances* shall be in accordance with Chapter 24.

❖ Combustion air is necessary for complete fuel combustion. In some situations, air is necessary for ventilation of the appliance enclosure. A lack of combustion air results in incomplete fuel combustion, which causes soot production, increased carbon monoxide production, serious appliance malfunction and the risk of fire or explosion. Incomplete fuel combustion and improper draft and venting compound each other and

greatly increase the risk of the release of carbon monoxide, thereby endangering the occupants of the space or building. Lack of the air necessary for ventilation of the appliance enclosure can result in excessive temperatures in that enclosure, thus introducing the risks of overheating the appliance and fire.

Potential sources of combustion air are (1) the indoor atmosphere in which the appliance is located, (2) outdoor openings and ducts, (3) a combination of the two and (4) any other approved arrangement for the dependable introduction of outdoor air into the building or directly to the fuel-burning appliance.

Combustion air requirements for a fuel-burning appliance are dependent on the type of fuel burned by the appliance. As such, this section directs the reader to other publications or Chapter 24 of the code for the requirements specific to the type of fuel that the appliance burns.

As stated in the general comments at the beginning of this chapter, this combustion air chapter in previous code editions did not specifically address the combustion air requirements for solid and liquid fuel burning appliances as this chapter was originally developed long ago (1992 CABO, *One- and Two-family Dwelling Code*) based on the combustion air requirements for gas-fired appliances. Up until the 2000 edition (first edition) of the code, this chapter was applicable to liquid, solid and gas-fired appliances. In the 2000 edition of the code, a new fuel gas chapter was introduced which included combustion air requirements for gas-fired appliances. With the introduction of the new fuel gas chapter, it was intended that the combustion air chapter be applicable only to liquid and solid fired appliances. Even though the first section of Chapter 17 from the 2000 edition forward did attempt to redefine the scope of the chapter to be only for liquid and solid fuel-fired appliances, many users of the code continued to apply this chapter to gas-fired appliance installations. As subsequent revisions to Chapter 24 occurred (and those same revisions were not carried over into Chapter 17), the requirements of the two chapters began to diverge creating confusion for code users who attempted to apply both chapters to gas-fired appliances. Replacing all of the combustion air requirements in Chapter 17 with instructions on where to find combustion air requirements for the specific type of fuel-burning appliance eliminated the confusion about gas-fired appliance combustion air requirements and provided the location for appropriate combustion air requirements for solid and liquid fuel-fired appliances.

Because of the unique configurations and variety of solid fuel-burning appliances (e.g., wood, pellet, biomass), no one set of combustion air requirements could adequately cover all solid fuel-fired appliance installations. Therefore, the best information must be obtained from the manufacturer's installation instructions for the specific appliance.

NFPA 31, *Installation of Oil-burning Equipment*, has long been recognized for many aspects of oil-fired appliance installations that were not covered by the code. Because NFPA 31 has combustion air requirements, the standard is the appropriate source for this information. The reader is advised to consult NFPA 31 directly.

M1701.2 Opening location. In flood hazard areas as established in Table R301.2(1), *combustion air* openings shall be located at or above the elevation required in Section R322.2.1 or R322.3.2.

❖ Appropriately locating ventilation and combustion air openings in buildings that are subject to flooding will substantially reduce the amount of damage to appliances as well as to the building structure in the event of flooding.

Chapter 18:
Chimneys and Vents

User note: Code change proposals to this chapter will be considered by the IRC – Plumbing and Mechanical Code Development Committee during the 2015 (Group A) Code Development Cycle.

General Comments

Chapter 18 regulates the design, construction, installation, maintenance, repair and approval of chimneys, vents and their connections to fuel-burning appliances. A properly designed chimney or vent system is needed to conduct the flue gases produced by a fuel-burning appliance to the outdoors. In the case of natural draft appliances, the chimney or vent system serving the appliance is expected to produce a draft at the appliance connection. This chapter does not cover venting requirements for fuel gas-fired appliances (see Chapter 24).

Draft is produced by the temperature difference between combustion gases (flue gases) and the ambient atmosphere. It is measured as pressure in inches of water column (kPa) and is a negative pressure-meaning that it is less than the atmospheric pressure at the point of measurement. Because hotter gases are less dense, they are buoyant and will rise in the chimney, causing negative pressures to develop that are directly proportional to the height of the chimney or vent and directly proportional to the temperature difference between the flue gases and the ambient air. Draft produced by a chimney or vent is the motivating force that conveys combustion products from natural draft appliances to the outside atmosphere.

The type of vent or chimney used in a given installation is generally based on the type of fuel burned by the appliance, the temperature of the flue gases produced by the combustion process and the appliance category. Because the code classifies appliances into three temperature categories (low, medium and high heat), the fuel type and characteristics of the flue gas must be analyzed for the design of a chimney or vent system. Higher efficiency appliances continue to produce lower flue gas temperatures, and the diversity of mechanical venting systems continues to expand.

Because of new technology and higher efficiency appliances/equipment, venting means are becoming increasingly diversified and complex. Many of today's appliances are designed to be vented by more than one type of venting system.

Purpose

This chapter contains provisions that would minimize the hazards associated with the products of combustion from fuel-burning appliances.

SECTION M1801
GENERAL

M1801.1 Venting required. Fuel-burning *appliances* shall be vented to the outdoors in accordance with their *listing* and *label* and manufacturer's installation instructions except *appliances* listed and *labeled* for unvented use. Venting systems shall consist of *approved* chimneys or vents, or venting assemblies that are integral parts of *labeled appliances*. Gas-fired *appliances* shall be vented in accordance with Chapter 24.

❖ Every fuel-burning appliance must be vented by a chimney or vent, unless the appliance is listed and labeled for unvented use. This chapter does not apply to unvented fuel gas-fired appliances or vented fuel gas-fired appliances. Quite often there is confusion resulting from users of the code trying to apply this chapter to gas-fired appliances. This chapter does not apply to gas-fired appliances.

M1801.2 Draft requirements. A venting system shall satisfy the draft requirements of the *appliance* in accordance with the manufacturer's installation instructions, and shall be con-structed and installed to develop a positive flow to convey combustion products to the outside atmosphere.

❖ Every fuel-burning appliance must be vented by a chimney or vent unless the appliance is direct vented, integrally vented or mechanically vented. Although chimneys and vents share a basic purpose, they are different creatures with different code requirements. Some code sections pertain to chimneys only, some to vents only and some to both. The code is frequently misapplied because people refer to vents as chimneys and vice versa.

The combustion byproducts produced by the burning of fuels consist primarily of carbon dioxide; nitrogen; water vapor and small amounts of carbon monoxide, nitric oxide, nitrogen dioxide and other compounds that vary with the type and purity of the fuel. Some of the compounds, especially carbon monoxide (CO), are poisonous, and some are corrosive. The amount of carbon monoxide in combustion gases varies with the type of fuel, the type of burner and the condition and adjustment of the burner.

Carbon monoxide poisoning as a result of malfunctioning fuel-burning appliances or malfunctioning chimneys or vents, or the lack thereof, has caused many fatalities. The products of combustion must be conducted to the outdoors so that operation of the appliance does not pose a threat to the building occupants. Fuel-burning appliances have widely varying operating characteristics; the method of venting the combustion byproducts must therefore be designed for the particular type of appliance served. Flue gases produced by appliances differ primarily in temperature and chemical composition. Methods of venting differ in the temperature and pressure ranges over which they operate and whether or not they can accommodate condensation.

A mismatch between an appliance and a chimney or vent can result in a hazardous operating condition. For example, an appliance that produces high-temperature flue gases can cause vent material to degrade or melt. It can also cause excessively high temperatures on the outer surface of the vent. On the other hand, an appliance that produces low-temperature flue gases can cause vent material to corrode and deteriorate because of condensation.

M1801.3 Existing chimneys and vents. Where an *appliance* is permanently disconnected from an existing chimney or vent, or where an *appliance* is connected to an existing chimney or vent during the process of a new installation, the chimney or vent shall comply with Sections M1801.3.1 through M1801.3.4.

❖ Existing chimneys and vents must be reevaluated for continued suitability whenever the "conditions of use change." A "condition of use change" is the permanent removal of one appliance or the installation of an appliance (either an additional appliance or the replacement for an existing appliance). Such changes could result in failure of the venting system to produce adequate draft for the appliances and could also cause the venting system to produce harmful condensation. The chimney or vent reevaluation must be in accordance with the following four code sections. First, the size of the existing chimney or vent must be verified (see Section M1801.3.1). Second, if the existing chimney or vent is of the proper size, the exterior of the chimney or vent must be inspected to verify that there is proper clearance to combustibles (see Section M1801.3.4). Third, the internal flue passageway must be inspected and cleaned (see Section M1801.3.2). Finally, masonry chimneys that are to be reused must have a cleanout installed if one does not already exist (see Section M1801.3.3).

M1801.3.1 Size. The chimney or vent shall be resized as necessary to control flue gas condensation in the interior of the chimney or vent and to provide the *appliance*, or *appliances* served, with the required draft. For the venting of oil-fired *appliances* to masonry chimneys, the resizing shall be done in accordance with NFPA 31.

❖ The combined input from multiple appliances, especially older lower efficiency appliances, can maintain chimney or vent temperatures that are high enough to

provide the necessary draft and avoid harmful condensation. Changing an existing configuration by disconnecting and eliminating an appliance or by substituting a higher efficiency appliance can cause a decrease in flue gas temperature, resulting in poor draft and/or condensation. The elimination of one or more draft hood equipped appliances will reduce the amount of dilution air in a venting system, increasing the likelihood of condensation.

The requirements of the code for the design of a new chimney or vent must be used to determine the required chimney/vent size to serve all of the appliances that will be connected. If the required size is different (either smaller or larger) than the existing system, the dimensions of the chimney or vent will likely need to be changed. If the required size is smaller, a liner system complying with UL 1777 could be installed in an existing chimney or vent (see the exception for Section M1801.3.4). If the required size is larger, the existing chimney or vent must be replaced with one of the proper size; or abandoned, if a new chimney or vent system is to be installed in another location.

Where a masonry chimney vents an oil-fired appliance, the requirements of NFPA 31 must be followed when resizing a masonry chimney.

A common scenario involves removing a chimney-vented furnace or boiler and leaving a water heater as the only appliance vented to the chimney. In such cases, the chimney is grossly oversized for the water heater which typically results in poor draft and the production of continuous condensation within the chimney. This scenario has received much attention and has caused those in the industry to use the phrase "orphaned water heater" to describe the arrangement.

M1801.3.2 Flue passageways. The flue gas passageway shall be free of obstructions and combustible deposits and shall be cleaned if previously used for venting a solid or liquid fuel-burning *appliance* or fireplace. The flue liner, chimney inner wall or vent inner wall shall be continuous and free of cracks, gaps, perforations, or other damage or deterioration that would allow the escape of combustion products, including gases, moisture and creosote.

❖ The chimney or vent must be inspected, cleaned (if needed), reinspected (if cleaned), and repaired or replaced (if needed). Where a chimney or vent previously conveyed flue gases from solid or liquid fuel, the inside surfaces of the flue gas passageway must be cleaned regardless of what an initial inspection reveals. Cleaning after solid or liquid fuel flue gas exposure is the only way to ensure proper conditions for a thorough examination of the flue gas passageway. The cleaning also eliminates fuel deposits that might cause a chimney or vent fire, reduces frictional resistance to flow and prevents any residual deposits from falling into lower parts of the venting system.

Use of a chimney or vent without proper inspection (and cleaning, if required) is a code violation. If the inspection shows that the flue gas passageway of the vent or chimney is incapable of preventing leakage of combustion gases and liquids, the chimney or vent must be either repaired or relined (see commentary, Section M1801.3.1). Where an oil-fired appliance is vented through the chimney, flue repairs must comply with the requirements of NFPA 31.

M1801.3.3 Cleanout. Masonry chimneys shall be provided with a cleanout opening complying with Section R1003.17.

❖ If an existing masonry chimney is to be reused but is found to not have a cleanout, a cleanout must be installed.

M1801.3.4 Clearances. Chimneys and vents shall have air-space clearance to combustibles in accordance with this code and the chimney or vent manufacturer's installation instructions.

> **Exception:** Masonry chimneys equipped with a chimney lining system tested and *listed* for installation in chimneys in contact with combustibles in accordance with UL 1777, and installed in accordance with the manufacturer's instructions, shall not be required to have a clearance between combustible materials and exterior surfaces of the masonry chimney. Noncombustible firestopping shall be provided in accordance with this code.

❖ An existing chimney or vent that is to be reused must be inspected to verify that proper external clearances to combustibles exist for the entire length of the chimney or vent, as required by the code for a new chimney or vent installation. If proper clearances do not exist, either the building structure (within allowable structural limitations) must be altered to provide the required clearances or the chimney/vent system realigned to provide the required clearances. Where proper clearances cannot be provided by either method, the chimney/vent system must be replaced with a system installed with the proper clearances. Note that where a chimney/vent system passes through floors or ceilings of a building, the code requires the installation of noncombustible fireblocking between the exterior of the chimney/vent system and the structure (see Chapter 10).

Clearances for vents and factory-built chimneys are typically indicated on the exterior of the vent or chimney. If the required clearances are not indicated on the vent or chimney, the manufacturer must be contacted and the original installation instructions consulted.

Existing masonry chimneys commonly do not have the proper external clearances to combustibles required by the code, especially if the structure is fairly old. The exception allows for listed chimney lining systems to be installed in masonry chimneys. Listed chimney liners are designed to provide insulation value to compensate for the lack of clearance of the masonry chimney to combustibles. Such chimney liner systems must comply with the requirements of UL 1777.

M1801.4 Space around lining. The space surrounding a flue lining system or other vent installed within a masonry chimney shall not be used to vent any other *appliance*. This shall not prevent the installation of a separate flue lining in accor-

dance with the manufacturer's installation instructions and this code.

❖ When a chimney lining system is installed within a chimney, an annular space will be created between the liner and the chimney. Such a space usually has an irregular shape and is not conducive to the proper flow of flue gases. Also, vent gases could degrade the lining system.

M1801.5 Mechanical draft systems. A mechanical draft system shall be used only with *appliances listed* and *labeled* for such use. Provisions shall be made to prevent the flow of fuel to the *equipment* when the draft system is not operating. Forced draft systems and portions of induced draft systems under positive pressure during operation shall be designed and installed to prevent leakage of flue gases into a building.

❖ Commonly used chimney and vent systems are not designed for positive pressure (above atmospheric pressure). Such chimneys and vents are intended to produce a draft and thus operate with negative (below atmospheric) internal pressures. Specialized vents and chimneys are available that are designed for positive pressure applications. The misapplication of chimney and venting materials can result in leakage of combustion gases and can cause chimney or vent deterioration due to condensation corrosion.

M1801.6 Direct-vent appliances. Direct-vent *appliances* shall be installed in accordance with the manufacturer's instructions.

❖ Direct-vent appliances must be listed and labeled and installed in accordance with the manufacturer's instructions. These appliances are designed to draw combustion air through a duct that is connected directly to the outdoors and expel combustion products though another duct to the outdoors. Some direct vent systems operate by natural draft while others are blower assisted. The vent and intake pipes are often terminated outdoors in a single vent/air intake assembly constructed in a concentric arrangement.

M1801.7 Support. Venting systems shall be adequately supported for the weight of the material used.

❖ Vent manufacturers supply support parts, and manufacturers' instructions provide detailed requirements for support of vent systems. Improper support can cause strain on vent components, appliance connectors and fittings resulting in vent, appliance or connector damage and loss of required clearance to combustibles and proper pitch. Frequently, venting installations suffer from lack of or improperly installed supports, brackets and hangers (see Commentary Figure M1801.7).

M1801.8 Duct penetrations. Chimneys, vents and vent connectors shall not extend into or through supply and return air ducts or plenums.

❖ In residential construction, stud and joist spaces are sometimes used instead of ducts for return air and combustion air. Areas under a floor are sometimes

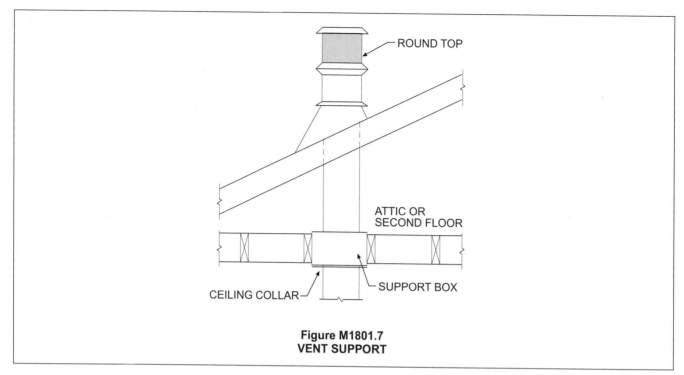

Figure M1801.7
VENT SUPPORT

used as plenums. A combustion products vent, vent connector or exhaust duct is not permitted to extend into or through an air plenum or space used as an air duct. The combustion vent could leak flue gases into the circulating air, creating a health hazard. The restriction does not apply to a space that is being used to convey combustion air.

M1801.9 Fireblocking. Vent and chimney installations shall be fireblocked in accordance with Section R602.8.

❖ Fireblocking with noncombustible materials must be installed in wood-frame construction at all openings around vents, pipes, ducts, chimneys and fireplaces at ceiling and floor levels. See the commentary to Section R602.8 (R302.11).

M1801.10 Unused openings. Unused openings in any venting system shall be closed or capped.

❖ Unused openings in chimneys or vents can allow vent gases to escape into the building, cause loss of draft or allow the entry of birds and rodents. The method used to close an unused opening must not create a protrusion that can restrict flow through the vent or chimney passageway.

M1801.11 Multiple-appliance venting systems. Two or more *listed* and *labeled appliances* connected to a common natural draft venting system shall comply with the following requirements:

1. *Appliances* that are connected to common venting systems shall be located on the same floor of the *dwelling*.

    Exception: Engineered systems as provided for in Section G2427.

2. Inlets to common venting systems shall be offset such that no portion of an inlet is opposite another inlet.

3. Connectors serving *appliances* operating under a natural draft shall not be connected to any portion of a mechanical draft system operating under positive pressure.

❖ If there is more than one opening into a single chimney or vent passageway, the openings must be located at different heights or otherwise be arranged so that the flow of combustion products from one connector cannot enter another connector instead of going up the flueway. This provision intends to prevent the flow or spilling of flue gases back into the building through an adjacent appliance flue connection.

Each solid-fuel-burning appliance or fireplace must have a dedicated independent chimney or vent. Solid-fuel appliances cannot share a common chimney with any other appliance or fireplace. Solid-fuel-burning appliances produce creosote deposits on the interior walls of chimney liners. The creosote is highly combustible. In the event of a chimney fire, other connections to the chimney could allow fire to break out of the chimney and into the building. Also, chimney passageways can become partially blocked by creosote formation, thus forcing vent gases from other appliances to discharge back into the building.

M1801.12 Multiple solid fuel prohibited. A solid fuel-burning *appliance* or fireplace shall not connect to a chimney passageway venting another *appliance*.

❖ Each solid-fuel-burning appliance or fireplace must have a dedicated independent chimney or vent. Solid-fuel appliances cannot share a common chimney with any other appliance or fire-place. Solid-fuel-burning appliances produce creosote deposits on the interior walls of chimney liners. The creosote is highly combustible. In the event of a chimney fire, other connections to the chimney could allow fire to break out of the chimney and into the building. Also, chimney passageways can become partially blocked by creosote formation, thus forcing vent gases from other appliances to discharge back into the building.

SECTION M1802
VENT COMPONENTS

M1802.1 Draft hoods. Draft hoods shall be located in the same room or space as the *combustion air* openings for the *appliances*.

❖ Draft hoods are installed in gas-venting systems to regulate the draft intensity and to prevent downdrafts from entering gas appliances. Draft hoods also allow dilution air to be drawn into the vents, inhibiting the formation of moisture condensed from the flue gases. Air drawn into the venting system must be replaced to prevent a negative pressure that could produce spillage of vent gases into the dwelling. Draft hoods must therefore be located in the same room or space as combustion air openings. This provision will ensure that the venting system will be in the same pressure zone as the combustion chamber.

M1802.2 Vent dampers. Vent dampers shall comply with Sections M1802.2.1 and M1802.2.2.

❖ Dampers control the rate of combustion by regulating the amount of draft. When the appliance is not operating, they allow flue gas passageways to be closed to reduce the loss of conditioned air through the chimney. This section addresses the use of both manually operated and automatic dampers.

M1802.2.1 Manually operated. Manually operated dampers shall not be installed except in connectors or chimneys serving solid fuel-burning *appliances*.

❖ Manual dampers can be installed only in solid-fuel-burning appliances. Such dampers control the rate of combustion by regulating the amount of draft and also reduce the loss of conditioned air through the vent or chimney when the appliance is not operating by allowing the vent or chimney to be closed. The damper is usually opened manually during the process of starting a fire. If the damper is not opened, smoke will spill back into the room, alerting the user. For gas- or liquid-fuel-fired appliances, however, the user may not be aware of a partially or completely closed damper, and a hazardous condition could develop. The code therefore prohibits the use of manual dampers for all but solid-fuel-burning appliances. Note that this section does not prohibit the use of fixed baffles that are components of approved appliance venting systems.

M1802.2.2 Automatically operated. Automatically operated dampers shall conform to UL 17 and be installed in accordance with the terms of their *listing* and *label*. The installa-

tion shall prevent firing of the burner when the damper is not opened to a safe position.

❖ Automatic dampers can be used with gas- or liquid-fuel-fired natural draft appliances. Because damper failure can result in a hazardous condition, automatic dampers must be listed and labeled, and installed in strict accordance with the manufacturer's installation instructions and the terms of the listing. An improperly installed or malfunctioning damper can cause the appliance to malfunction and discharge the combustion products back into the building. Automatic dampers are designed to prove that they are in the open position before allowing the appliance burners to fire.

M1802.3 Draft regulators. Draft regulators shall be provided for oil-fired *appliances* that must be connected to a chimney. Draft regulators provided for solid fuel-burning *appliances* to reduce draft intensity shall be installed and set in accordance with the manufacturer's installation instructions.

❖ Oil-fired appliances must be fitted with a draft regulator unless the appliance is listed and labeled for use without the draft regulator. Draft regulators may be installed on solid-fuel-burning appliances to reduce the draft intensity. When required, draft regulators must be installed in the same room or enclosure as the appliance so that no difference in pressure between the air at the regulator and the combustion air supply will exist. Draft regulators are of the barometric type and are somewhat comparable to the draft hood used with gas appliances. See Commentary Figure M1802.3 for illustrations of draft hoods and draft regulators.

M1802.3.1 Location. Where required, draft regulators shall be installed in the same room or enclosure as the *appliance* so that a difference in pressure will not exist between the air at the regulator and the *combustion air* supply.

❖ See the commentary to Section M1802.1.

SECTION M1803
CHIMNEY AND VENT CONNECTORS

M1803.1 General. Connectors shall be used to connect fuel-burning *appliances* to a vertical chimney or vent except where the chimney or vent is attached directly to the *appliance*.

❖ Unless the chimney or vent is connected directly to the appliance, a connector is necessary. A connector is piping or a factory-built assembly that is used to connect a fuel-burning appliance to a vent or chimney, including any necessary fittings to make the connection or a change in direction.

M1803.2 Connectors for oil and solid fuel appliances. Connectors for oil and solid fuel-burning *appliances* shall be constructed of factory-built chimney material, Type L vent material or single-wall metal pipe having resistance to corrosion and heat and thickness not less than that of galvanized steel as specified in Table M1803.2.

❖ Connectors for oil- and solid-fuel-burning appliances must be constructed of the materials listed in this section to ensure that they are capable of tolerating the high temperatures and corrosive nature of the vent gases being conveyed.

TABLE M1803.2
THICKNESS FOR SINGLE-WALL METAL PIPE CONNECTORS

DIAMETER OF CONNECTOR (inches)	GALVANIZED SHEET METAL GAGE NUMBER	MINIMUM THICKNESS (inch)
Less than 6	26	0.019
6 to 10	24	0.024
Over 10 through 16	22	0.029

For SI: 1 inch = 25.4 mm.

M1803.3 Installation. Vent and chimney connectors shall be installed in accordance with the manufacturer's instructions and within the space where the *appliance* is located. *Appliances* shall be located as close as practical to the vent or chimney. Connectors shall be as short and straight as possible and installed with a slope of not less than $^1/_4$ inch (6 mm) rise per foot of run. Connectors shall be securely supported and joints shall be fastened with sheet metal screws or rivets. Devices that obstruct the flow of flue gases shall not be installed in a connector unless *listed* and *labeled* or *approved* for such installations.

❖ A connector must be supported for the design and weight of the material used. Proper support is necessary to maintain the clearances required by Section M1803.3.4, to maintain the required pitch, and to prevent physical damage and separation of joints. The joints between connectors and appliance draft hood outlets and flue collars must be secured with screws, rivets or other approved means. The joints between pipe sections also must be secured with screws, rivets or other approved means. Some connectors are available with a proprietary-type fastening method. The special proprietary joints do not require screws, rivets or other fasteners if they provide equivalent resistance to disengagement or displacement. This section addresses connectors, not chimneys and vents, and most chimney and vent manufacturers do not require or might even prohibit the use of screws or rivets for securing joints for chimney and vent pipe sections and fittings.

The ideal chimney or vent configuration is a vertical system, even though it is not always practical. This section requires all portions of a chimney or vent connector to rise vertically a minimum $^1/_4$ inch per each foot (0.02 mm/m) of its horizontal length. The connector slope is intended to induce the flow of flue gases by taking advantage of the natural buoyancy of the hot gases. Connector slope can promote the priming of a cold venting system and can partially compensate for short connector vertical rise.

M1803.3.1 Floor, ceiling and wall penetrations. A chimney connector or vent connector shall not pass through any floor or ceiling. A chimney connector or vent connector shall not pass through a wall or partition unless the connector is *listed* and *labeled* for wall pass-through, or is routed through a device *listed* and *labeled* for wall pass-through and is installed in accordance with the conditions of its *listing* and *label*. Connectors for oil-fired *appliances listed* and *labeled* for Type L vents, passing through walls or partitions shall be in accordance with the following:

1. Type L vent material for oil *appliances* shall be installed with not less than *listed* and *labeled* clearances to combustible material.

2. Single-wall metal pipe shall be *guarded* by a ventilated metal thimble not less than 4 inches (102 mm) larger in diameter than the vent connector. A minimum 6 inches (152 mm) of clearance shall be maintained between the thimble and combustibles.

❖ Walls and partitions in dwellings are often constructed of combustible materials and must be protected from exposure to high temperatures from venting compo-

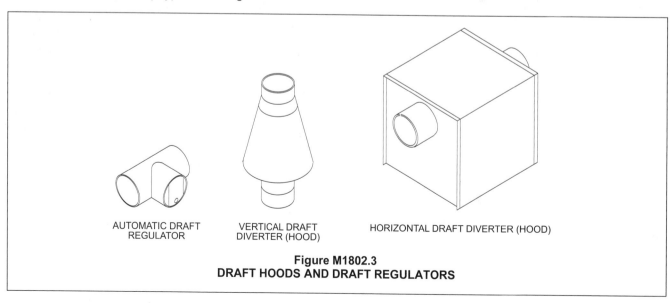

AUTOMATIC DRAFT REGULATOR

VERTICAL DRAFT DIVERTER (HOOD)

HORIZONTAL DRAFT DIVERTER (HOOD)

Figure M1802.3
DRAFT HOODS AND DRAFT REGULATORS

nents. Type B and Type L vents must be installed with clearances to combustibles in accordance with their listing and label. Because single-wall metal pipe does not have the insulating qualities of Type B and Type L vents, it must be installed within a ventilated metal thimble when passing through combustible walls and partitions. The thimble must be at least 4 inches (102 mm) larger in diameter than the connector. Note that this section prohibits all penetrations of ceilings and floors.

M1803.3.2 Length. The horizontal run of an uninsulated connector to a natural draft chimney shall not exceed 75 percent of the height of the vertical portion of the chimney above the connector. The horizontal run of a *listed* connector to a natural draft chimney shall not exceed 100 percent of the height of the vertical portion of the chimney above the connector.

❖ The appliance and its chimney or vent should be located in close proximity to keep the connector length to a minimum. A connector must be limited in length because of flow resistance of the connector and heat loss through the connector wall. A listed connector is assumed to be an air insulated double-wall type, therefore a longer length is permitted because it will help maintain sufficiently high vent gas temperatures to produce draft.

M1803.3.3 Size. A connector shall not be smaller than the flue collar of the *appliance*.

  Exception: Where installed in accordance with the *appliance* manufacturer's instructions.

❖ Unless otherwise specified in the manufacturer's instructions, the size of the connector must be at least equal to the diameter of the appliance flue collar. A lesser size could restrict the flow of flue gases and result in spillage of flue gases into the dwelling.

M1803.3.4 Clearance. Connectors shall be installed with clearance to combustibles as set forth in Table M1803.3.4. Reduced clearances to combustible materials shall be in accordance with Table M1306.2 and Figure M1306.1.

❖ Connectors, vents and chimneys must be installed with minimum clearances to combustibles. These clearances can be reduced if the combustible material is protected by one of the methods of protection given in Table M1306.2. The clearance requirements in Table M1803.3.4 apply to unlabeled single-wall connectors serving a listed appliance. Connectors that are listed and labeled must be installed with the clearance to combustibles required by the connector manufacturer's installation instructions. If the appliance or connector manufacturer's installation instructions specify larger clearances than those prescribed by this section, the manufacturer's installation instructions govern.

TABLE M1803.3.4
CHIMNEY AND VENT CONNECTOR CLEARANCES TO COMBUSTIBLE MATERIALS[a]

TYPE OF CONNECTOR	MINIMUM CLEARANCE (inches)
Single-wall metal pipe connectors: Oil and solid-fuel appliances Oil appliances listed for use with Type L vents	 18 9
Type L vent piping connectors: Oil and solid-fuel appliances Oil appliances listed for use with Type L vents	 9 3[b]

For SI: 1 inch = 25.4 mm.

a. These minimum clearances apply to unlisted single-wall chimney and vent connectors. Reduction of required clearances is permitted as in Table M1306.2.

b. Where listed Type L vent piping is used, the clearance shall be in accordance with the vent listing.

M1803.3.5 Access. The entire length of a connector shall be accessible for inspection, cleaning and replacement.

❖ Because connectors are exposed to high temperatures and possible condensation, the entire length of the connector must be accessible to allow observation of any degradation of the connector or lack of clearances and to allow replacement.

M1803.4 Connection to fireplace flue. Connection of *appliances* to chimney flues serving fireplaces shall comply with Sections M1803.4.1 through M1803.4.4.

❖ This section regulates installations where fireplace chimneys are used to vent appliances, such as room heaters and fireplace-insert heaters.

M1803.4.1 Closure and accessibility. A noncombustible seal shall be provided below the point of connection to prevent entry of room air into the flue. Means shall be provided for access to the flue for inspection and cleaning.

❖ Without an air-tight connection between the chimney flue and the appliance chimney connector, chimney draft will draw room air into the flue, thus weakening the draft drawn through the appliance. Because the chimney flue is no longer open to the fireplace firebox, the chimney connection must be designed to provide access to the chimney for inspection and cleaning.

M1803.4.2 Connection to factory-built fireplace flue. A different *appliance* shall not be connected to a flue serving a factory-built fireplace unless the *appliance* is specifically *listed* for such an installation. The connection shall be made in conformance with the *appliance* manufacturer's instructions.

❖ Factory-built fireplaces may or may not be tested for uses other than as a traditional fire-place. Connecting an appliance or fireplace-insert heater could result in abnormally high fire-place or chimney temperatures, leading to component deterioration and hazardous operation. The fireplace manufacturer will provide specific details stating what appliance connections are allowed. Usually, the manufacturer's instructions will prohibit any modifications and any use of the fireplace other than that for which it was designed.

M1803.4.3 Connection to masonry fireplace flue. A connector shall extend from the *appliance* to the flue serving a masonry fireplace to convey the flue gases directly into the flue. The connector shall be accessible or removable for inspection and cleaning of both the connector and the flue. *Listed* direct-connection devices shall be installed in accordance with their *listing*.

❖ The appliance chimney connector must extend up through the fireplace damper and smoke chamber and terminate in the chimney flue. This helps produce an adequate draft and reduces the possibility of combustible creosote deposits forming within the fireplace throat and smoke chamber.

M1803.4.4 Size of flue. The size of the fireplace flue shall be in accordance with Section M1805.3.1.

❖ The size of the connector must be equal to the diameter of the appliance to prevent a restriction in the flow of the flue gases. See the commentary to Section M1805.3.1.

SECTION M1804
VENTS

M1804.1 Type of vent required. *Appliances* shall be provided with a *listed* and *labeled* venting system as set forth in Table M1804.1.

❖ Vents and chimneys are distinct systems. The provisions of this section apply only to vents. The vents addressed in this section are natural draft venting systems that produce draft using the same principles that produce draft in chimneys.

Vents regulated by this section are factory fabricated and must be listed and labeled. The labeling requirement applies to all components of the system, including the sections of pipe, fittings, terminal caps, supports and spacers.

A vent system can be used only with oil-burning and pellet-burning appliances that are listed and labeled for use with a vent. Solid-fuel-burning appliances must not be connected to a vent system, because such appliances produce flue gas temperatures that are much higher than the temperatures for which vents are designed. The burning of solid fuels also produces creosote. Creosote formation leaves a combustible deposit on the surface of the flue passageway that, if ignited, can produce a fire with temperatures higher than 2,000°F (1093°C). Such temperatures far exceed the maximum safe temperature of a vent system.

Type L vents and pellet vents must be tested in accordance with UL 641 and are designed to vent oil-burning and pellet-fuel-burning appliances. Oil-burning appliances produce a higher flue gas temperature than gas-fired appliances. Type L vents are typically double-wall, air-insulated vent piping systems constructed of galvanized and stainless steel. They are designed for natural draft applications only and must not be used to convey combustion gases under a positive pressure. For example, a Type L vent must not be used with Category III or IV gas-fired appliances and must not be used on the discharge side of an exhauster or power-vented appliance. Positive pressures will cause a Type L vent to leak combustion gases into the building, and the leakage can cause condensation to form within the vent system pipe, fittings and joints, resulting in corrosion damage to the vent and eventual vent failure.

TABLE M1804.1
VENT SELECTION CHART

VENT TYPES	APPLIANCE TYPES
Type L oil vents	Oil-burning appliances listed and labeled for venting with Type L vents
Pellet vents	Pellet fuel-burning appliances listed and labeled for use with pellet vents

M1804.2 Termination. Vent termination shall comply with Sections M1804.2.1 through M1804.2.6.

❖ The standards for vents and appliances require that venting instructions be supplied by the manufacturer of the appliance and the manufacturer of the vent system. Such instructions are part of the labeling requirements. Any deviation from them is a violation of the code.

The clearance to combustibles for a vent system is determined by the testing agency and stated on the component labels and in the manufacturer's instructions. Not all vents have the same required clearances to combustibles. The clearances are determined by vent performance during testing in accordance with the applicable standard. Different vent materials and designs impact the vent's ability to control the amount of heat transmitted to surrounding combustibles. Installation of the vent system in accordance with the clearances listed on the vent's label and in the manufacturer's installation instructions is critical.

The termination of a natural draft vent must comply with the requirements of the manufacturer's installation instructions and Sections M1804.2.1 through M1804.2.5. A vent used in conjunction with a mechanical exhauster must meet the termination requirements established in Section M1804.2.6.

The vent system must be physically protected to prevent damage to the vent and to prevent combustibles from coming into contact with or being placed too close to the vents. This protection is usually provided by enclosing the vent in chases, shafts or cavities during building construction. Physical protection is not required in the room or space where the vent originates (at the appliance connection) and is not required in such locations as attics that are not occupied or used for storage. For example, assume that a vent is installed in an existing building and that it extends from the basement, through a first-floor closet, through the attic and through the roof. The portion of the vent passing through the closet must be protected from damage.

The means of physical damage protection must also be designed to maintain separation between the vent and any combustible storage. To prevent the passage of fire and smoke through the annular space around a vent penetration through a floor or ceiling, the vent must be fireblocked with a noncombustible material in accordance with the code. Vent manufacturers provide factory-built components and installation instructions for fireblocking penetrations.

M1804.2.1 Through the roof. Vents passing through a roof shall extend through flashing and terminate in accordance with the manufacturer's installation requirements.

❖ Manufacturers' installation instructions contain the requirements for the termination of vents through the roof (see Commentary Figure M1804.2.1).

M1804.2.2 Decorative shrouds. Decorative shrouds shall not be installed at the termination of vents except where the shrouds are *listed* and *labeled* for use with the specific venting system and are installed in accordance with the manufacturer's instructions.

❖ Decorative shrouds have become a popular architectural feature that, in some cases, have proven to be a fire hazard. They are designed to be aesthetically pleasing and to hide chimney and vent terminations. These shrouds have caused fires resulting from overheated combustible construction and the accumulation of debris and animal nesting. Shrouds can also interfere with the function of a chimney or vent system. Shrouds are allowed only where they are listed and labeled for use with a specific factory-built chimney system. Unlisted, field-constructed shrouds are prohibited.

M1804.2.3 Natural draft appliances. Vents for natural draft *appliances* shall terminate not less than 5 feet (1524 mm) above the highest connected *appliance* outlet, and natural draft gas vents serving wall furnaces shall terminate at an elevation not less than 12 feet (3658 mm) above the bottom of the furnace.

❖ To develop and maintain adequate venting action (draft), the termination of gravity vents must be at least 5 feet (1524 mm) above the vent collar of the appliance and at least 12 feet (3658 mm) above the bottom of a wall furnace.

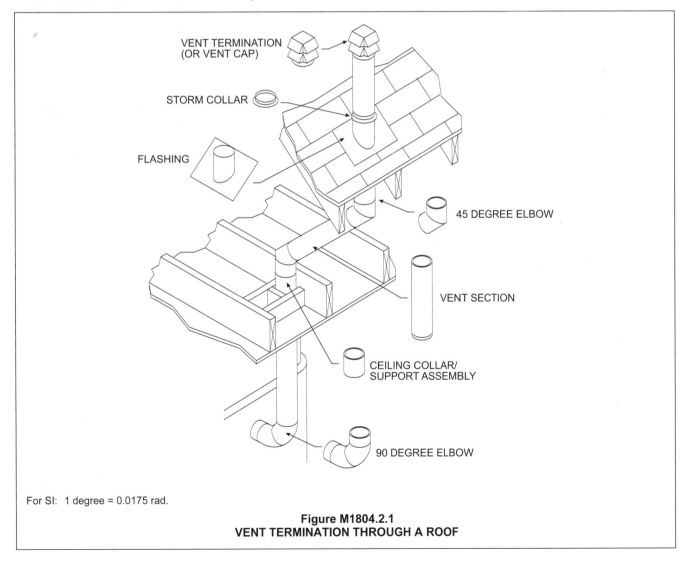

For SI: 1 degree = 0.0175 rad.

Figure M1804.2.1
VENT TERMINATION THROUGH A ROOF

M1804.2.4 Type L vent. Type L venting systems shall conform to UL 641 and shall terminate with a *listed* and *labeled* cap in accordance with the vent manufacturer's installation instructions not less than 2 feet (610 mm) above the roof and not less than 2 feet (610 mm) above any portion of the building within 10 feet (3048 mm).

❖ The termination heights are consistent with the requirements of NFPA 31.

Consideration must be given to vent location in the design/installation phase of a building so that the vent termination will not require long extensions of vent pipe exposed to the outdoors. Vent piping exposed to the outdoors encourages condensation in cold weather, could require guy wires or braces depending on height, and is considered aesthetically unattractive. Another good reason to plan for vent location during the design phase of a building is to allow straight vertical runs of vent and avoid difficult offsets. Good planning can also prevent vents from passing through the roof on the street side of the building. Quite often, difficult vent offsets in attics are needed for vents to penetrate the roof on the back side of a building.

M1804.2.5 Direct vent terminations. Vent terminals for direct-vent *appliances* shall be installed in accordance with the manufacturer's instructions.

❖ Direct-vent appliances include appliances that have a closed combustion chamber and a conduit for bringing all combustion air directly from the outdoors. Direct-vent appliances are often referred to as "sealed combustion" or "separated combustion" appliances.

The code does not provide prescriptive termination requirements for these appliances and relies on the manufacturer's installation instructions.

M1804.2.6 Mechanical draft systems. Mechanical draft systems shall comply with UL 378 and shall be installed in accordance with their *listing*, the manufacturer's instructions and, except for direct-vent *appliances*, the following requirements:

1. The vent terminal shall be located not less than 3 feet (914 mm) above a forced air inlet located within 10 feet (3048 mm).

2. The vent terminal shall be located not less than 4 feet (1219 mm) below, 4 feet (1219 mm) horizontally from, or 1 foot (305 mm) above any door, window or gravity air inlet into a *dwelling*.

3. The vent termination point shall be located not closer than 3 feet (914 mm) to an interior corner formed by two walls perpendicular to each other.

4. The bottom of the vent terminal shall be located not less than 12 inches (305 mm) above finished ground level.

5. The vent termination shall not be mounted directly above or within 3 feet (914 mm) horizontally of an oil tank vent or gas meter.

6. Power exhauster terminations shall be located not less than 10 feet (3048 mm) from *lot lines* and adjacent buildings.

7. The discharge shall be directed away from the building.

❖ Mechanical draft systems must be listed to UL 378 which contains a comprehensive set of construction and performance requirements that are used to evaluate and list draft equipment. The appliance installations addressed in this section use auxiliary or integral fans and blowers to force the flow of combustion products to the outdoors. This section applies to nonintegral power exhausters, integrally power-exhausted appliances, and venting systems equipped with draft inducers. This section does not address direct-vent appliances.

The seven requirements in this section provide guidance in locating the vent terminal to prevent drawing the vented products of combustion back into the building and creating a hazardous condition.

M1804.3 Installation. Type L and pellet vents shall be installed in accordance with the terms of their *listing* and *label* and the manufacturer's instructions.

❖ The standards for vents and appliances require that venting instructions be supplied by the manufacturer of the appliance and the manufacturer of the vent system. Such instructions are part of the labeling requirements. Any deviation from them is a violation of the code.

The clearance to combustibles for a vent system is determined by the testing agency and stated on the component labels and in the manufacturer's instructions. Not all vents have the same required clearances to combustibles. The clearances are determined by vent performance during the testing in accordance with the applicable standard. Different vent materials and designs impact the vent's ability to control the amount of heat transmitted to surrounding combustibles. Installation of the vent system in accordance with the clearances listed on the vent's label and in the manufacturer's installation instructions is critical.

Physical protection of the vent system is required to prevent damage to the vent and to prevent combustibles from coming into contact with or being placed too close to the vents. This protection is typically provided by enclosing the vent in chases, shafts or cavities in the building construction. Physical protection is not required in the room or space where the vent originates (at the appliance connection) and would not be required in locations such as attics that are not occupied or used for storage. For example, assume that a vent is installed in an existing building and that it extends from the basement, through a first-floor closet, through the attic and through the roof. The portion of the vent passing through the closet must be protected from damage. The means of physical damage protection must also be designed to maintain separation between the vent and any combustible storage. To prevent the passage of fire and smoke through annular space around a vent penetration through a floor or ceiling, the vent must be fireblocked with a noncombustible material in accordance with the code. Vent manufacturers provide installation instructions and factory-built components for fireblocking penetrations.

M1804.3.1 Size of single-appliance venting systems. An individual vent for a single *appliance* shall have a cross-sectional area equal to or greater than the area of the connector to the *appliance*, but not less than 7 square inches (4515 mm^2) except where the vent is an integral part of a *listed* and *labeled appliance*.

❖ This section sets the minimum vent area allowed on gravity vent appliances at 7 square inches (4515 mm^2) [which calculates to a minimum 3-inch (76 mm) diameter vent]. Obviously, if the vent is an integral part of the listed appliance, lesser sizes may be accepted. Section M1803.3.3 requires that the size of the vent or connector be not less than the appliance flue collar. Therefore, when the appliance flue collar is larger than 3 inches (76 mm) in diameter, the connector and vent must be increased in size accordingly.

M1804.4 Door swing. Appliance and equipment vent terminals shall be located such that doors cannot swing within 12 inches (305 mm) horizontally of the vent terminals. Door stops or closers shall not be installed to obtain this clearance.

❖ Vent terminals for sidewall-vented appliances (such as direct-vent gas fireplaces and fireplace heaters, direct-vent room heaters, direct-vent water heaters, furnaces and boilers and also nondirect-vent appliances) are sometimes located where a side-swinging door can impact the vent terminal or swing close to the terminal. The results can be damage to the vent terminal, a fire hazard and interference with the appliance venting and combustion air intake. Also, if the door blocks or deflects the vent discharge such that the combustion products are pulled back into the combustion air intake, the result would be excessive CO production and serious appliance malfunction and sooting.

A damaged vent terminal could cause appliance malfunction and carbon monoxide poisoning. It could also cause damage to nearby combustible surfaces and create a fire hazard. Door stops and closer devices cannot be depended on because they are easily defeated or removed [see Commentary Figures G1804.4(1) and G1804.4(2)].

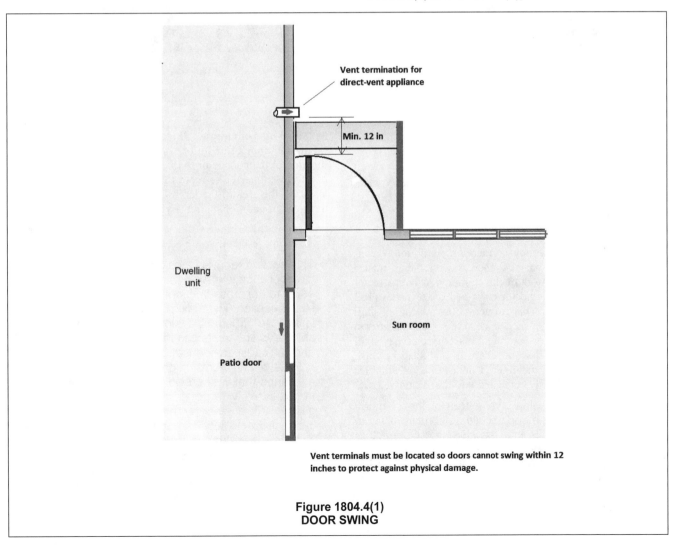

Vent terminals must be located so doors cannot swing within 12 inches to protect against physical damage.

Figure 1804.4(1)
DOOR SWING

Figure G1804.4(2)
DOOR SWING

SECTION M1805
MASONRY AND FACTORY-BUILT CHIMNEYS

M1805.1 General. Masonry and factory-built chimneys shall be built and installed in accordance with Sections R1003 and R1005, respectively. Flue lining for masonry chimneys shall comply with Section R1003.11.

❖ See Sections R1003, R1003.11 and R1005.

M1805.2 Masonry chimney connection. A chimney connector shall enter a masonry chimney not less than 6 inches (152 mm) above the bottom of the chimney. Where it is not possible to locate the connector entry at least 6 inches (152 mm) above the bottom of the chimney flue, a cleanout shall be provided by installing a capped tee in the connector next to the chimney. A connector entering a masonry chimney shall extend through, but not beyond, the wall and shall be flush with the inner face of the liner. Connectors, or thimbles where used, shall be firmly cemented into the masonry.

❖ The 6-inch space below the bottom of the chimney connector serves as a trap to collect debris from within the chimney that falls into the chimney. Connectors must not extend beyond the inner face of the chimney liner to prevent complete or partial obstruction of the chimney. Connectors must be sealed and attached to the chimney to prevent leakage and dislocation of the connector.

M1805.3 Size of chimney flues. The effective area of a natural draft chimney flue for one *appliance* shall be not less than the area of the connector to the *appliance*. The area of chimney flues connected to more than one *appliance* shall be not less than the area of the largest connector plus 50 percent of the areas of additional chimney connectors.

Exception: Chimney flues serving oil-fired *appliances* sized in accordance with NFPA 31.

❖ This section requires a chimney or vent serving a single appliance to have a minimum cross-sectional area equal to the cross-sectional area of the appliance outlet connection. The maximum size of a chimney or vent is also an important design consideration. Oversized chimneys and vents can fail to produce sufficient draft and can cause condensation of the flue gases. Existing chimneys and vents might become oversized if an appliance that was previously connected is removed. In excessively large chimneys and vents, the flue gases are cooled by expansion and because of increased heat loss through the larger flue passage surface area. The cooling effect reduces the temperature differences necessary to produce draft and can cause the flue gases to reach the dew point temperature. The resulting condensation is highly corrosive and can cause deterioration of the chimney, vent, connected appliances and the building.

For connection to more than one appliance, this section requires that the area of the flue be increased by 50 percent of the area of the additional connector. The reason is based on a diversity factor arising from the fact that all appliances are not likely to be firing simultaneously at their maximum rate. Any design conflicting with the sizing criteria of this section must be considered as an engineered system. Note that this sizing rule does not apply to gas-fired appliances, however, an extremely limited version of the rule is found in Sections G2427.5.4 and G2427.6.8.1.

M1805.3.1 Size of chimney flue for solid-fuel appliance. Except where otherwise specified in the manufacturer's installation instructions, the cross-sectional area of a flue connected to a solid-fuel-burning *appliance* shall be not less than the area of the flue collar or connector, and not larger than three times the area of the flue collar.

❖ So that flue gases exhaust properly, the cross-sectional area of the flue connected to a solid-fuel-burning appliance must not be less than the area of the flue collar or connector. For proper draft, and to allow the chimney to be heated sufficiently to produce and maintain this draft, this section puts a maximum size limitation on the chimney, not to exceed three times the area of the flue collar (see commentary, Section M1805.3). As with all chimneys and all vents, they can fail from being too small and from being too big. In either case, the chimney or vent can fail to convey the combustion products to the outdoors and can fail to prevent condensation damage within the chimney or vent.

Chapter 19:
Special Appliances,
Equipment and Systems

User note: Code change proposals to this chapter will be considered by the IRC – Plumbing and Mechanical Code Development Committee during the 2015 (Group A) Code Development Cycle.

General Comments

This chapter regulates the installation of fuel-burning appliances that are not covered in other chapters. Section M1901 covers the installation of and clearances to ranges and ovens. Section M1902 addresses the proper installation and controls necessary for safe operation of sauna heaters. Section M1903 addresses the testing and installation of stationary fuel cell power plants. Section M1904 directs the reader to the sections of the code and other *International Codes*® (I-Codes®) that have installation requirements for hydrogen systems.

The chapters of the code are dedicated to single sub-jects, however, because the two subjects in this chapter do not contain the volume of text necessary to warrant individual chapters, they have been combined into a single chapter. This chapter is a collection of requirements for various appliances. The only commonality is that the subjects use energy to perform some task or function.

Purpose

This chapter provides regulations for the proper installation of ranges, ovens, fuel cells and sauna heaters. The intent is to provide a reasonable level of protection for the occupants of the dwelling.

SECTION M1901
RANGES AND OVENS

M1901.1 Clearances. Freestanding or built-in ranges shall have a vertical clearance above the cooking top of not less than 30 inches (762 mm) to unprotected combustible material. Reduced clearances are permitted in accordance with the *listing* and *labeling* of the range hoods or *appliances*. The installation of a listed and labeled cooking appliance or microwave oven over a listed and labeled cooking appliance shall be in accordance with Section M1504.1. The clearances for a domestic open-top broiler unit shall be in accordance with Section M1505.1.

❖ Freestanding and built-in ranges must be located for required clearances to unprotected combustible materials. Cooking range tops can produce intense heat. The vertical clearance called out will help prevent the ignition of any unprotected combustible materials that are commonly found in residential kitchens. A cooking appliance or microwave oven and an open top broiler unit have additional requirements found in Sections M1504.1 and M1505.1, respectively (see commentary, Sections M1503, M1507, M1504.1, M1505.1 and G2447.5).

Ducted or ductless range hoods are often installed over ranges. Note, however, that the code does not require such hoods. When installed, range hoods must conform to the manufacturer's installation instructions. Reduced clearances to these hoods are allowed if the hoods are listed and labeled for such.

M1901.2 Cooking appliances. Cooking *appliances* shall be *listed* and *labeled* for household use and shall be installed in accordance with the manufacturer's instructions. The installation shall not interfere with *combustion air* or access for operation and servicing. Electric cooking appliances shall comply with UL 1026 or UL 858. Solid-fuel-fired fireplace stoves shall comply with UL 737.

❖ This section requires the listing and labeling of all cooking appliances and installation in accordance with the manufacturer's installation instructions. Cooking appliances listed only for commercial use must not be installed in residential kitchens. Sections G2447.2 and G2447.3 prohibit gas-fired commercial cooking appliances from being installed in dwelling units, except where the appliances are listed as both commercial and domestic (household) types. Commercial cooking appliances do not have the same safety features as residential appliances, such as push-to turn safety knobs and insulated doors. Commercial appliances allow much higher surface temperatures on the enclosure components, have higher energy output, higher effluent output and higher clearances to combustibles, all of which can be unsuitable or unsafe for domestic environments. There are cooking appliances on the market that have the size, features and appearance of commercial appliances, but they are listed for domestic/residential installation or are listed for both commercial and domestic use. The referenced standards listed contain a comprehensive set of construction and performance requirements that are used to evaluate and list cooking equipment specifically for use in dwelling units.

SECTION M1902
SAUNA HEATERS

M1902.1 Locations and protection. Sauna heaters shall be protected from accidental contact by persons with a guard of material having a low thermal conductivity, such as wood. The guard shall not have a substantial effect on the transfer of heat from the heater to the room.

❖ Sauna heaters produce steam by evaporating water that passes through or over a heating element. Guards must be installed around the source of steam because occupants of the space of the sauna are usually unclothed. These guards must be made of a material that has low thermal conductivity, such as wood, to protect the occupants from burns.

M1902.2 Installation. Sauna heaters shall be installed in accordance with the manufacturer's instructions. Sauna heaters shall comply with UL 875.

❖ Sauna heaters must be listed, labeled and installed in accordance with the manufacturer's installation instructions. The listing and label must indicate compliance with UL 875, which contains construction and performance requirements that are used to evaluate sauna heaters.

M1902.3 Combustion air. *Combustion air* and venting for a nondirect vent-type heater shall be provided in accordance with Chapters 17 and 18, respectively.

❖ Combustion air for fuel-burning appliances must not be taken from within the sauna room itself. Sauna rooms are usually very small, confined spaces that do not contain an adequate volume of combustion air. For other than direct vent appliances, combustion air must be supplied in accordance with Chapter 17.

M1902.4 Controls. Sauna heaters shall be equipped with a thermostat that will limit room temperature to not greater than 194°F (90°C). Where the thermostat is not an integral part of the heater, the heat-sensing element shall be located within 6 inches (152 mm) of the ceiling.

❖ The room must be kept at a temperature not to exceed 194°F (90°C). This temperature has been determined to be the maximum safe level, beyond which occupants can be injured. So that the thermostat will measure the highest temperature of the room and thus maintain an acceptable and safe temperature range throughout the space, the thermostat for the appliance must either be integral with the heater or be installed within the top 6 inches (76 mm) of the room.

SECTION M1903
STATIONARY FUEL CELL POWER PLANTS

M1903.1 General. Stationary fuel cell power plants having a power output not exceeding 1,000 kW, shall comply with ANSI/CSA America FC 1 and shall be installed in accordance with the manufacturer's instructions and NFPA 853.

❖ The private generation of electricity for commercial and residential use is an emerging technology that is expected to gain widespread acceptance as future

technological improvements result in lower costs. Fuel cells have been included in the code to provide guidance to the building community for those future installations.

The code requires that fuel cell power plants with an output of 1,000 kW or less be listed in accordance with ANSI/CSA America FC 1, and requires that the units be installed in accordance with the manufacturer's instructions and NFPA 853. NFPA 853 provides guidance on location of the power plants, fuel system supplies, ventilation and exhaust of the space containing the unit and fire protection requirements. Although fuel cell technology has been available since the 1800s, its first practical use was by NASA to generate electricity and water for the Gemini and Apollo spacecraft and the space shuttles. Because the installation of fuel cells in privately owned buildings is relatively new, the following discussion of the fuel cell process is included to provide a basic understanding of the technology.

A fuel cell is an electrochemical device that converts the chemical energy of a hydrocarbon fuel (hydrogen, natural gas, coal-bed methane, methanol, gasoline, etc.) and an oxidant (air or oxygen) into useable electricity. The fuel cell consists of a fuel electrode (anode) and an oxidant electrode (cathode) separated by an ion-conducting membrane (electrolyte). When a hydrocarbon fuel is introduced into the fuel cell, the catalyst surface of the membrane converts the hydrogen gas molecules into hydrogen ions and electrons. The hydrogen ions pass through the membrane to react with oxygen and electrons to form water as a reaction byproduct. The electrons, which cannot pass through the membrane, must travel around it through an external circuit, thereby creating direct current (dc) electricity [see Commentary Figure M1903.1(1)].

When fueled by pure hydrogen, the only byproducts of the fuel cell process are heat and water. Low levels of CO_2 are produced when other fuels are used. Unlike traditional fossil-fuel electric generating stations that burn the fuels, the electrochemical fuel cell process produces no particulate matter, nitrogen oxides or sulfur oxides.

When the waste heat is captured for use in heating water or a cogeneration process, fuel cell efficiency can reach 80 percent.

A typical fuel cell consists of three sections: a fuel processor, a fuel stack (power generator) and a power conditioner [see Commentary Figure M1903.1(2)].

The fuel processor section has two components: (1) the fuel reformer, which processes a hydrocarbon fuel, such as methane, into a hydrogen-rich formate gas and (2) a carbon monoxide cleanup system to reduce CO concentrations to acceptable levels.

The fuel cell stack consists of many individual fuel cells that operate as previously described to generate electricity and usable heat. The power output of the unit increases based on the number and arrangement of individual cells stacked in series and/or in parallel in the unit. There are several types of fuel cell technologies, including phosphoric acid (PA), molten carbonate

(MC), solid oxide (SO) and proton exchange membrane (PEM), which are named based on the type of electrolyte used in the membrane section of the cell.

The power conditioner (inverter) section converts the direct current (dc) produced in the fuel cell stack into alternating current (ac). Batteries are used to smooth out power surge demands, such as HVAC start-ups, and to supply additional power when demand exceeds the peak out-put of the cell stack.

Although currently available, fuel cell units are not cost competitive with established electricity suppliers, but the initial and operating costs are expected to decrease as more manufacturers enter the market. Where there are constraints on transmission expansion, in remote locations or in areas where reduced emission of pollutants is critical, fuel cells could be seriously considered as an energy source. Fuel cells can also be attractive where the quality and dependability of commercial power is a concern.

SECTION M1904
GASEOUS HYDROGEN SYSTEMS

M1904.1 Installation. Gaseous hydrogen systems shall be installed in accordance with the applicable requirements of Sections M1307.4 and M1903.1 and the *International Fuel Gas Code*, the *International Fire Code* and the *International Building Code*.

❖ Systems that use or produce hydrogen gas can create potentially explosive situations where installation and ventilation instructions are not adhered to properly. Chapter 7 of the *International Fuel Gas Code®* (IFGC®) is the primary source of regulations related to gaseous hydrogen systems, but the *International Building Code®* (IBC®), the *International Fire Code®* (IFC®) and Sections M1307.4 and M1903.1 of this code also contain code requirements. Chapters 4 and 5 of the IBC state requirements for hydrogen cut-off rooms and Chapter 23 of the IFC provide requirements for hydrogen fueling stations.

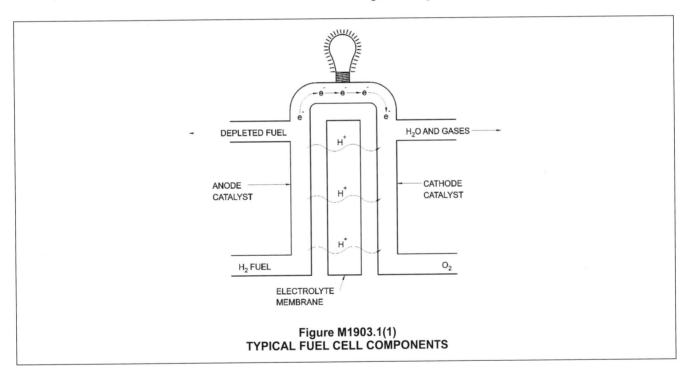

Figure M1903.1(1)
TYPICAL FUEL CELL COMPONENTS

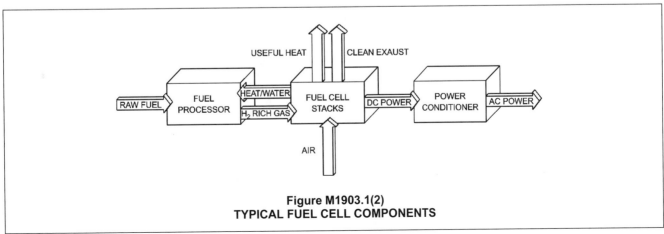

Figure M1903.1(2)
TYPICAL FUEL CELL COMPONENTS

Bibliography

The following resource materials were used in the preparation of the commentary for this chapter of the code:

IBC–15, *International Building Code*. Washington, DC: International Code Council, 2014.

IFC–15, *International Fire Code*. Washington, DC: International Code Council, 2014.

IFGC–15, *International Fuel Gas Code*. Washington, DC: International Code Council, 2014.

Chapter 20:
Boilers and Water Heaters

User note: Code change proposals to this chapter will be considered by the IRC – Plumbing and Mechanical Code Development Committee during the 2015 (Group A) Code Development Cycle.

General Comments

This chapter applies to all boilers and water heaters. A water heater is any appliance that heats potable water and supplies it to the plumbing hot water distribution system. However, some water heaters are also used for space heating. These are listed for combination potable water heating and space heating applications. A boiler either heats water or generates steam for space heating and is generally a closed system.

Purpose

This chapter regulates the installation of boilers and water heaters. Its purpose is to protect the occupants of the dwelling from the potential hazards associated with such appliances.

SECTION M2001
BOILERS

M2001.1 Installation. In addition to the requirements of this code, the installation of boilers shall conform to the manufacturer's instructions. The manufacturer's rating data, the nameplate and operating instructions of a permanent type shall be attached to the boiler. Boilers shall have their controls set, adjusted and tested by the installer. A complete control diagram together with complete boiler operating instructions shall be furnished by the installer. Solid and liquid fuel-burning boilers shall be provided with *combustion air* as required by Chapter 17.

❖ This section governs the installation and commissioning of boilers and their control systems. The mechanical equipment requirements for approval, labeling, installation, maintenance, repair and alteration are regulated by Chapters 12 and 13.

A durable copy of the complete operating instructions must be attached to the boiler upon completion of installation by the contractor. Boiler systems can be complex and generally require coordinated operation among several pieces of equipment. The proper operating procedures, set points, etc., must be specified in the operating instructions. The control system design documentation in the operating instructions typically includes diagrams, system schematics and control sequence descriptions.

M2001.1.1 Standards. Packaged oil-fired boilers shall be listed and labeled in accordance with UL 726. Packaged electric boilers shall be listed and labeled in accordance with UL 834. Solid fuel-fired boilers shall be listed and labeled in accordance with UL 2523. Boilers shall be designed, constructed and certified in accordance with the ASME *Boiler and Pressure Vessel Code*, Section I or IV. Controls and safety devices for boilers with fuel input ratings of 12,500,000 Btu/hr (3 663 388 watts) or less shall meet the requirements of ASME CSD-1. Gas-fired boilers shall conform to the requirements listed in Chapter 24.

❖ The history of boiler accidents shows that they are potentially dangerous if not properly designed, constructed and installed. Along with the code, several industry standards are referenced for the design and construction of boilers, and equipment manufacturers produce boilers in accordance with these referenced standards. The design and construction of boilers are regulated by these references. ASME CSD-1 specifies what safety devices and controls are required for boilers, such as pressure and temperature limit controls and low-water cutoffs. Design and construction requirements for specific types of boilers are found in the applicable sections of these referenced standards.

M2001.2 Clearance. Boilers shall be installed in accordance with their *listing* and *label*.

❖ Because boilers, related appliances and mechanical equipment require inspection, observation, routine maintenance, repairs and possible replacement, working clearance around such equipment and appliances is required. The manufacturer's installation instructions state the access requirements and clearances to combustibles. See Commentary Figure M2001.2 for an illustration of a typical installation.

M2001.3 Valves. Every boiler or modular boiler shall have a shutoff valve in the supply and return piping. For multiple boiler or multiple modular boiler installations, each boiler or modular boiler shall have individual shutoff valves in the supply and return piping.

Exception: Shutoff valves are not required in a system having a single low-pressure steam boiler.

❖ Shutoff valves are required to facilitate maintenance and repairs by isolating a boiler from the system that it serves. Without isolation valves, the entire system of piping and components would have to be taken out of service, relieved of pressure and drained before a boiler or water-side/steam-side component could be removed for servicing, repair or replacement. Where systems have multiple boilers, isolation valves permit one or more boilers to be removed from service without affecting the remaining units, thus preserving partial system operation. A modular boiler unit is a field-assembled boiler composed of multiple interdependent modules. Modular units are designed to permit close matching of boiler capacity to building loads and to allow for future expansion of capacity by the addition of modules to a boiler "package." Each independent boiler or modular boiler, whether single or in groups, must have individual shutoff valves (see Commentary Figure M2001.3).

The exception states that shutoff valves are not required for systems with a single low-pressure steam boiler. Shutoff valves would not be of value for a single low-pressure steam boiler installation because of a negligible amount of heat transfer medium to relieve from the system. Also, maintaining a partial capacity system operation is impossible with only a single unit.

M2001.4 Flood-resistant installation. In flood hazard areas established in Table R301.2(1), boilers, water heaters and their control systems shall be located or installed in accordance with Section R322.1.6.

❖ The local jurisdiction must fill in Table R301.2(1) upon adoption of the code, including the flood hazards information. Boilers, water heaters and associated control systems that are located in flood hazard areas must be installed above the design flood elevation or must be designed and installed to prevent the entrance of water into the components and to resist the forces of the flood waters on the components (see commentary, Section R322.1.6).

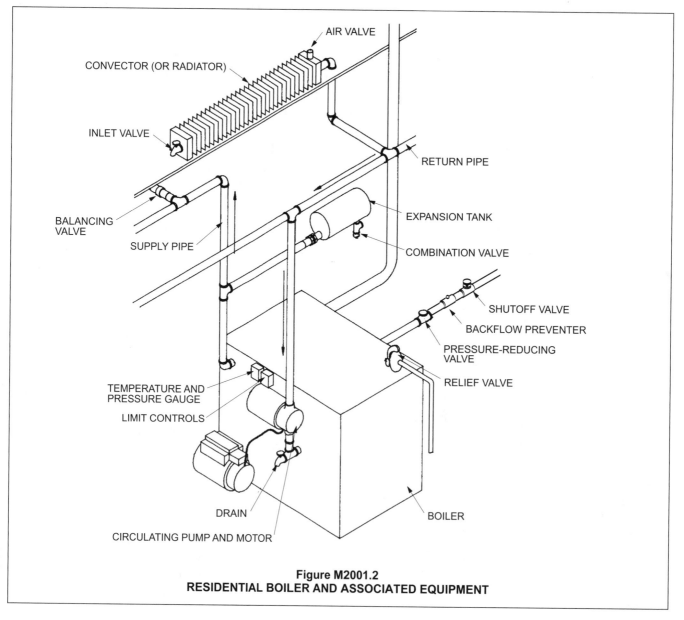

Figure M2001.2
RESIDENTIAL BOILER AND ASSOCIATED EQUIPMENT

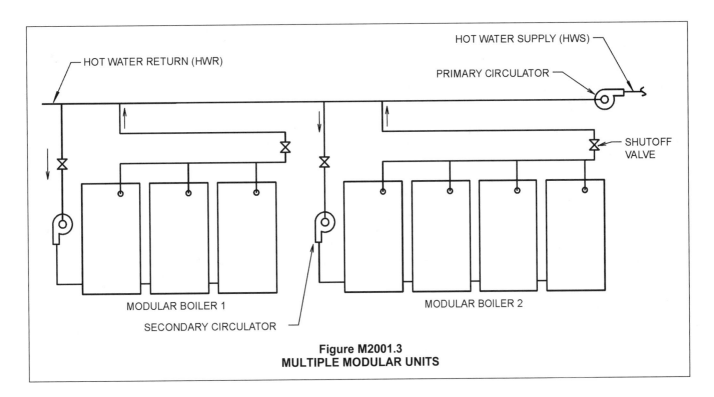

Figure M2001.3
MULTIPLE MODULAR UNITS

SECTION M2002
OPERATING AND SAFETY CONTROLS

M2002.1 Safety controls. Electrical and mechanical operating and safety controls for boilers shall be *listed* and *labeled*.

❖ This section requires the listing and labeling of electrical and mechanical operating and safety controls to help ensure their safety and dependability.

M2002.2 Hot water boiler gauges. Every hot water boiler shall have a pressure gauge and a temperature gauge, or combination pressure and temperature gauge. The gauges shall indicate the temperature and pressure within the normal range of the system's operation.

❖ All boilers operate within certain pressure and temperature limitations. Problems and hazards can develop if these limitations are exceeded. Pressure and temperature gauges are necessary to allow boiler owners and service technicians to check and monitor system operating conditions. Gauges indicate hazardous conditions and causes of problems or malfunctions. They also help prevent future hazards or problems from developing by signaling the need for system repairs and service. To allow reasonably accurate readings, gauges must have a range and scale of values that are proportioned to the boiler's normal operating range. For example, a pressure gauge with a range of 0 to 250 psi (0 to 1724 kPa) would barely deflect the needle when installed on a 10-psi (69 kPa) system, and pressure readings are inaccurate and not readily discerned at the extreme ends of a gauge's range. The gauges must have the resolution and increments to allow an accurate reading to be taken. Hot water boiler gauges are often equipped with combina-

tion pressure and temperature gauges (see Commentary Figure M2002.2).

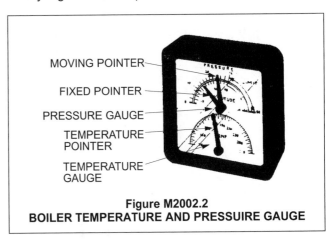

Figure M2002.2
BOILER TEMPERATURE AND PRESSUIRE GAUGE

M2002.3 Steam boiler gauges. Every steam boiler shall have a water-gauge glass and a pressure gauge. The pressure gauge shall indicate the pressure within the normal range of the system's operation. The gauge glass shall be installed so that the midpoint is at the normal water level.

❖ All steam boilers operate within certain pressure and water level limits that must be observed. However, unlike with hot water boilers, water temperature is not a concern for steam systems. Whenever steam is present, the water temperature in a steam boiler will be a function of steam pressure and is easily determined when the steam pressure is known. A steam pressure gauge indicates both normal and abnormal operating conditions and the presence of a hazardous condition and is necessary to observe and monitor boiler operation. The water level in a boiler fluctuates within a narrow margin between too low and too high. Because proper boiler operation is dependent on the water level being maintained within the high and low parameters, a water-gauge glass is necessary to visually indicate the actual water level in the boiler.

A water-gauge glass, commonly referred to as a "sight glass," indicates the following: boiler flooding, low water conditions, condensate return operation, unstable water levels resulting from foaming and surging, boiler water condition and the amount of automatic or manual makeup water feed (see Commentary Figure M2002.3). If either or both of the gauge-glass cocks are closed, the water-level reading will be erroneous, and the boiler could be empty even though the gauge glass reads "normal."

Ideally, a steam boiler pressure gauge should indicate the normal system operating pressure at approximately the first quarter to midpoint of the scale. For example, a 0 to 100 psi (0 to 690 kPa) pressure gauge would be a very poor indicator for a low-pressure [15 psi (103 kPa) or less] steam heating boiler (see commentary, Section M2002.2). Low-pressure steam gauges are often designed to read both pressure and vacuum because some steam systems will develop a partial vacuum as steam condenses in the distribution system.

A water-gauge glass is always installed on a boiler with the bottom of the glass connected below the normal water level and the top of the glass connected above the normal water level. When connected in this manner, the boiler water level will be accurately represented in the column of water in the gauge glass. It has been an industry tradition to equate the midpoint of a gauge glass with the "normal" water level in a boiler. The coincidence of the gauge-glass mid-point and normal water level also allows the greatest amount of indicator range above and below normal, thus allowing clear indication of abnormal water levels.

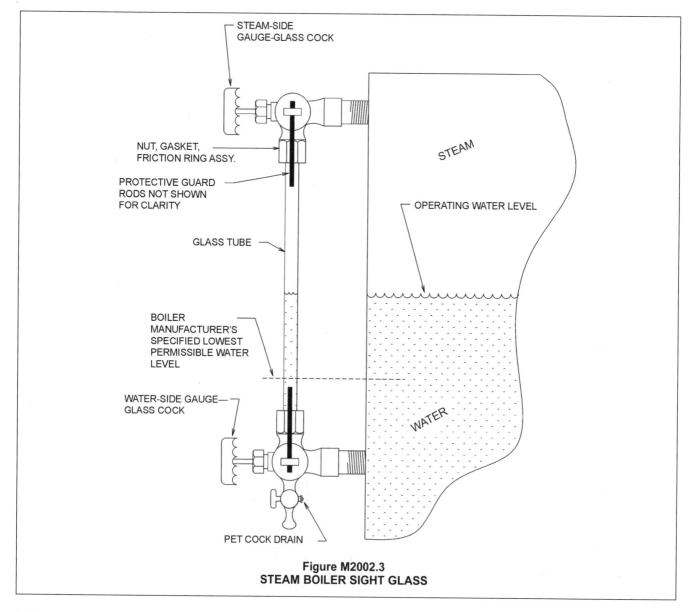

Figure M2002.3
STEAM BOILER SIGHT GLASS

M2002.4 Pressure-relief valve. Boilers shall be equipped with pressure-relief valves with minimum rated capacities for the *equipment* served. Pressure-relief valves shall be set at the maximum rating of the boiler. Discharge shall be piped to drains by gravity to within 18 inches (457 mm) of the floor or to an open receptor.

❖ Boilers must be protected from overpressure by one or more relief valves. A relief valve is designed to open in direct proportion to the boiler pressure force acting on its closure disk. The higher the pressure, the greater the force, and the wider the valve opens. Boilers installed without a relief valve can produce devastating explosions and have been responsible for loss of life and damage to property. Relief valves also play an important role in protecting aging equipment that may have deteriorated or weakened because of prolonged use. Relief valves must be installed directly into the openings provided for that purpose by the boiler manufacturer and must be discharged to a safe location. The discharge is a threat to the occupants and the building because of the high temperature produced when the valves discharge. Because relief valve discharge is a symptom of a boiler malfunction, the discharge must be observable to provide a warning to initiate remedial action. Relief valve discharge piping must never be located where it is subject to freezing or installed in a such a way that it functions as a "trap" by allowing discharge water to remain in the pipe.

M2002.5 Boiler low-water cutoff. Steam and hot water boilers shall be protected with a low-water cutoff control.

Exception: A low-water cutoff is not required for coil-type and water-tube type boilers that require forced circulation of water through the boiler and that are protected with a flow sensing control.

❖ If a steam or hot water boiler is operated (fired) without water or with water below the minimum level, overheating and severe boiler damage can occur. In addition to damage or destruction of the boiler, a severe fire or explosion hazard can result from the overheating or failure of the combustion chamber enclosure. Burner malfunction can also occur as a result of leakage and destruction of the boiler heat exchange surfaces. The heat exchange surfaces of fuel-fired boilers absorb radiated heat from the flames and, by conduction, absorb heat from the combustion gases produced by the combustion of fuel. This heat is conducted to the water at a rate that prevents the temperature of the heat exchanger material from exceeding the maximum for which it was designed. As the water level drops in a boiler, a greater area of heat exchange surface becomes dry and thus subject to overheating. All boiler manufacturers specify and mark on the unit the minimum water level to be maintained in the boiler, below which damage would be possible. All boilers are subject to a loss of water. A leak anywhere in the piping system or in the boiler itself can result in a dangerous condition caused by water loss. This is especially true for steam boilers because they are dependent on the timely return of condensate water and the proper oper-

ation and sequencing of controls for the return and makeup water feed systems.

A low-water cutoff is an essential control device designed to prevent boiler operation when the water level is too low. Because steam boilers (as opposed to hot water boilers, which are completely filled) maintain a water level, a low-water condition is much more likely to occur in a steam system. However, the hazards resulting from low water levels are the same for both boilers. Also, a flash steam explosion can occur if makeup water is introduced into an over-heated steam or hot water boiler. For these reasons, hot water boilers are also required to be protected by a low-water cutoff control. This section requires low-water cutoffs without regard to heat energy input. Some low-water cutoff controls are intended to be field installed as part of the boiler "trim" (appurtenances), and others are factory installed directly into tapped openings provided on the boiler.

Traditional low-water cutoff controls use float-actuated mechanisms. Electronic probe-type controls, which are commonly used for hot water boiler protection, detect the water level with an electrode that relies on the slight conductivity of the boiler water to complete a relay circuit (see Commentary Figure M2002.5). In the case of coil-type (water-tube) boilers requiring forced circulation to prevent coil or tube overheating, a flow-sensing device, which detects flow and verifies that the boiler and system are full of water, is required. Low-water cutoff controls do not sense flow and, therefore, cannot protect a forced circulation coil-type/water-tube boiler from overheating caused by loss of circulation.

Photo courtesy of the Hydro Level Company

Figure M2002.5
BOILER LOW-WATER CUTOFF CONTROL DEVICE

Electric boilers are generally protected with factory-installed water-level sensing devices to prevent heating element failure, which could occur if the elements were not submerged.

The sole purpose of a low-water cutoff control is to stop (cut off) the heat input to the boiler whenever the water level is dangerously low. These devices automatically interrupt the power supply to the burner controls or heating elements to cause boiler shutdown. For forced-circulation boilers, such as hot water, water-tube-type boilers, the flow-sensing control must shut off the energy input when circulation has stopped or is less than the required flow rate as established by the manufacturer of the boiler.

M2002.6 Operation. Low-water cutoff controls and flow sensing controls required by Section M2002.5 shall automatically stop the combustion operation of the appliance when the water level drops below the lowest safe water level as established by the manufacturer or when the water circulation flow is less than that required for safe operation of the appliance, respectively.

❖ See the commentary to Section M2002.5.

SECTION M2003
EXPANSION TANKS

M2003.1 General. Hot water boilers shall be provided with expansion tanks. Nonpressurized expansion tanks shall be securely fastened to the structure or boiler and supported to carry twice the weight of the tank filled with water. Provisions shall be made for draining nonpressurized tanks without emptying the system.

❖ Hot water heating systems are closed systems that are completely filled with water. When the water is heated, it expands and, because it is in a closed system, quickly develops hydrostatic pressure. Expansion tanks are used to absorb additional system water volume caused by expansion, thus avoiding wide variation in system pressure and possible discharge of pressure relief valves. Expansion tanks are either open tank reservoirs (nonpressurized) or sealed vessels (pressurized).

M2003.1.1 Pressurized expansion tanks. Pressurized expansion tanks shall be consistent with the volume and capacity of the system. Tanks shall be capable of withstanding a hydrostatic test pressure of two and one-half times the allowable working pressure of the system.

❖ Because they are subjected to the same pressures as the boiler system, closed-type expansion tanks must have a pressure rating equal to or exceeding that of the boiler system operating pressure. The code requires that pressurized tanks withstand a hydrostatic test pressure of $2^1/_2$ times the allowable working pressure of the system. Closed-type expansion tanks contain a cushion of air that compresses as water expands into the tank. The pressure in a hot water system will increase several pounds per square inch as the temperature increases. However, the compression of the air in a properly sized expansion tank will prevent the pressure from exceeding the maximum system operating pressure. Open-type expansion tanks are elevated reservoirs holding the expanded hot water at atmospheric pressure.

M2003.2 Minimum capacity. The minimum capacity of expansion tanks shall be determined from Table M2003.2.

❖ The capacity of nonpressurized and pressurized expansion tanks is very important to the proper operation of the boiler system. Refer to the table showing minimum capacities of expansion tanks as they relate to the hot water system volume. Note that the system volume in gallons includes the volume of the water in the boiler, piping, radiators and convectors. The volume of expansion tanks is not included in the system volume.

TABLE M2003.2
EXPANSION TANK MINIMUM CAPACITY[a]
FOR FORCED HOT-WATER SYSTEMS

SYSTEM VOLUME[b] (gallons)	PRESSURIZED DIAPHRAGM TYPE	NONPRESSURIZED TYPE
10	1.0	1.5
20	1.5	3.0
30	2.5	4.5
40	3.0	6.0
50	4.0	7.5
60	5.0	9.0
70	6.0	10.5
80	6.5	12.0
90	7.5	13.5
100	8.0	15.0

For SI: 1 gallon = 3.785 L, 1 pound per square inch gauge = 6.895 kPa, °C = [(°F)-32]/1.8.

a. Based on average water temperature of 195°F (91°C), fill pressure of 12 psig and a maximum operating pressure of 30 psig.

b. System volume includes volume of water in boiler, convectors and piping, not including the expansion tank.

❖ This table gives the minimum capacities of two types of expansion tanks. One is the pressurized diaphragm type, which is described in Section M2003.1.1. The other is the nonpressurized type. The capacities of the tanks vary depending on the volume of the system under consideration.

SECTION M2004
WATER HEATERS USED FOR SPACE HEATING

M2004.1 General. Water heaters used to supply both potable hot water and hot water for space heating shall be installed in accordance with this chapter, Chapter 24, Chapter 28 and the manufacturer's instructions.

❖ Water heaters serving the dual purpose of supplying potable hot water and being a heat source for a hot water space-heating system must be listed and labeled for that dual application. This section does not address water heaters used solely for space-heating applications, but rather addresses water heaters that serve a primary purpose of domestic water heating and a secondary purpose of space heating. Water heater labels indicate whether or not the appliance is suitable for space heating. Water heaters are listed and labeled as water heaters which are defined in Chapter 2. A water heater used solely for space heating applications would not meet the definition of a water heater and would likely violate the conditions of its listing.

Chapter 28 contains additional requirements for the proper installation of such water heaters.

SECTION M2005
WATER HEATERS

M2005.1 General. Water heaters shall be installed in accordance with Chapter 28, the manufacturer's instructions and the requirements of this code. Water heaters installed in an attic shall comply with the requirements of Section M1305.1.3. Gas-fired water heaters shall comply with the requirements in Chapter 24. Domestic electric water heaters shall comply with UL 174. Oiled-fired water heaters shall comply with UL 732. Thermal solar water heaters shall comply with Chapter 23 and UL 174. Solid fuel-fired water heaters shall comply with UL 2523.

❖ This section recognizes that water heaters and hot water storage tanks must be considered as both mechanical appliances (equipment) and plumbing appliances; therefore, they must comply with both the mechanical and plumbing sections of this code. Manufacturer's installation instructions are thoroughly evaluated by the listing and labeling agency to establish that a safe installation is prescribed. The listing and labeling agency can require the manufacturer to alter, delete or add information in the installation instructions as necessary to achieve compliance with the applicable standards and code requirements.

When a water heater is tested to obtain a listing and label, the approved agency installs the water heater in accordance with the manufacturer's installation instructions. The water heater is then tested under these conditions; thus, the installation instructions become an integral part of the listing and labeling process.

The manufacturer's installation instructions must be available to the code official because they are an enforceable extension of the code and are necessary for determining that the water heater has been properly installed. Simply put, the listing and labeling process indicates that the water heater and its installation instructions are in compliance with applicable standards. Therefore, an installation in accordance with the manufacturer's installation instructions is required, except where the code requirements are more stringent. It is necessary for an inspector to carefully and completely read and comprehend the manufacturer's instructions in order to properly perform an installation inspection.

Even if an installation appears to be in compliance with the manufacturer's instructions, the installation cannot be complete or approved until all associated components, connections and systems that serve the water heater are also in compliance with the applicable code provisions.

A water heater installation is complex in that it has a fuel or power supply; a chimney or vent connection, if fuel fired; a combustion air supply, if fuel fired; connections to the plumbing potable water distribution system; and controls and devices to prevent a multitude of potential hazards from conditions such as excessively high temperatures, pressures and ignition failure. In addition to the requirements of Section M1305.1.3, the structural provisions of the code must be observed for attic installations because of the large point load contributed by the water heater. The typical 40-gallon water heater will weigh over 400 pounds when filled with water.

M2005.2 Prohibited locations. Fuel-fired water heaters shall not be installed in a room used as a storage closet. Water heaters located in a bedroom or bathroom shall be installed in a sealed enclosure so that *combustion air* will not be taken from the living space. Installation of direct-vent water heaters within an enclosure is not required.

❖ The intent of this section is to prevent fuel-fired appliances from being installed in rooms and spaces where the combustion process could pose a threat to the occupants. Potential threats include depleted oxygen levels; elevated levels of carbon dioxide, nitrogen oxides and carbon monoxide, and other combustion gases; ignition of combustibles; and elevated levels of flammable vapors. See Section G2406.2 for the same concerns regarding gas-fired water heaters.

In small rooms such as bedrooms and bathrooms, the doors are usually closed when the room is occupied, and combustion gases could build up to life-threatening levels. Occupants would not be aware of impending danger. The last sentence of this section relates only to the second sentence.

M2005.2.1 Water heater access. Access to water heaters that are located in an *attic* or underfloor crawl space is permitted to be through a closet located in a sleeping room or bathroom where *ventilation* of those spaces is in accordance with this code.

❖ Mechanical equipment and appliances require regular maintenance. Proper access to enable repairs or replacement of equipment is a must. See Sections M2005.2 and G2406.2.

All water heaters have a far shorter working life than the building in which they are located; therefore, they must be replaced one or more times during the lifetime of the building. This future need has often been overlooked, and water heaters have been installed in locations that would necessitate the dismantling or destruction of a permanent portion of the structure to install a new water heater.

M2005.3 Electric water heaters. Electric water heaters shall also be installed in accordance with the applicable provisions of Chapters 34 through 43.

❖ This section refers the user back to the electrical chapters in the code for electric water heaters.

M2005.4 Supplemental water-heating devices. Potable water heating devices that use refrigerant-to-water heat exchangers shall be *approved* and installed in accordance with the manufacturer's instructions.

❖ Waste heat from refrigeration systems can be used to heat potable water and can provide substantial energy savings overtime. This section allows the use of such systems if the equipment is approved and installed in accordance with the manufacturer's installation instructions. Heat pump water heaters are becoming more common and provide a high coefficient of performance and also provide the beneficial side effect of free space cooling in the location of the heat pump appliance.

SECTION M2006
POOL HEATERS

M2006.1 General. Pool and spa heaters shall be installed in accordance with the manufacturer's installation instructions. Oil-fired pool heaters shall comply with UL 726. Electric pool and spa heaters shall comply with UL 1261.

❖ This section addresses specialized types of water heaters used with swimming pools, recreational or therapeutic spas and hot tubs. Pool and spa heaters are similar in design to hot water boilers. Usually, these heaters are of the water-tube type, are open nonpressurized systems and are designed for either indoor or outdoor installation.

M2006.2 Clearances. The clearances shall not interfere with *combustion air*, draft hood or flue terminal relief, or accessibility for servicing.

❖ The manufacturer's installation instructions can contain recommendations that are either more or less restrictive than the code for combustion air requirements, draft hood or terminal relief and accessibility for servicing the appliance. In either case, the more restrictive requirement governs.

M2006.3 Temperature-limiting devices. Pool heaters shall have temperature-relief valves.

❖ Temperature relief devices must be installed on pool heaters to protect against the development of high temperatures that could cause scalding injuries. Pool heaters are defined by ANSI Z2.56 as "An appliance designed for heating non-potable water stored at atmospheric pressure, such as water in swimming pools, spas, hot tubs and similar applications." This section is not specific as to the type of pool heater, thus it would appear to apply to electric, fuel-fired and heat pump type pool heaters. The code is silent on the set-

point of the relief device, therefore such temperature needs to be approved by the code official and also be consistent with the heater manufacturer's instructions.

M2006.4 Bypass valves. Where an integral bypass system is not provided as a part of the pool heater, a bypass line and valve shall be installed between the inlet and outlet piping for use in adjusting the flow of water through the heater.

❖ Unless a similar system is integrated within the heater, a bypass line and valve must be installed to allow control of the flow rate of water through the heater, which in turn allows control of the temperature rise of the water circulated through the unit.

Chapter 21:
Hydronic Piping

User note: Code change proposals to this chapter will be considered by the IRC – Plumbing and Mechanical Code Development Committee during the 2015 (Group A) Code Development Cycle.

General Comments

Hydronic piping includes piping, fittings and valves used in building space conditioning systems. Applications include hot water, chilled water, steam, steam condensate, brines and water/antifreeze mixtures. This chapter regulates installation, alteration and repair of all hydronic piping systems.

Section M2101 contains the installation requirements that apply to all hydronic systems. Section M2102 discusses baseboard convectors. Section M2103 addresses floor heating systems. Section M2104 covers low temperature piping applications. Section M2105 contains the provisions for ground source heat pump system loop piping.

Purpose

The requirements herein are intended to affect the reliability, serviceability, energy efficiency and safety of hydronic systems.

SECTION M2101
HYDRONIC PIPING SYSTEMS INSTALLATION

M2101.1 General. Hydronic piping shall conform to Table M2101.1. *Approved* piping, valves, fittings and connections shall be installed in accordance with the manufacturer's instructions. Pipe and fittings shall be rated for use at the operating temperature and pressure of the hydronic system. Used pipe, fittings, valves or other materials shall be free of foreign materials.

❖ Table M2101.1 lists the piping materials that are permitted for use in hydronic systems. This section does not prohibit the use of other materials that provide equivalent performance as approved by the building official. Materials that have previously been used for other than hydronic applications can contain residual contaminants that are harmful to the hydronic system. Before any used pipe, fitting, valve or other material is used, it must be examined to determine that its performance will be equivalent to that required for new materials for the intended application.

M2101.2 System drain down. Hydronic piping systems shall be installed to permit draining of the system. Where the system drains to the plumbing drainage system, the installation shall conform to the requirements of Chapters 25 through 32 of this code.

> **Exception:** The buried portions of systems embedded underground or under floors.

❖ To facilitate system repairs and maintenance, hydronic piping systems must be sloped and arranged to allow the transfer-medium fluids or condensate to be drained from the system. Each trapped section of the system piping must be fitted with drain cocks, unions or some other means of opening the system for draining. Drainage discharge to the plumbing system must be by indirect connection in accordance with the code.

Where portions of hydronic systems are buried below grade or installed below floors, such as a slab on grade, for which there is no access to drain piping by gravity, an exception allows those portions to not have a gravity drain point. System fluid can be removed by other means such as vacuuming or blowing out with compressed air.

M2101.3 Protection of potable water. The potable water system shall be protected from backflow in accordance with the provisions listed in Section P2902.

❖ Hydronic systems require means of supplying initial fill water and makeup water to replace any water lost because of evaporation, leakage or intentional draining. Where direct connections are made to the potable water supply, the connections must be isolated from the potable water source in compliance with the plumbing provisions of the code. This provision protects the potable water system from contamination where a direct connection is made to a hydronic system. Because hydronic systems are usually pressurized and can contain nonpotable water and fluids, antifreeze and conditioning or cleaning chemicals, the potable water system must be protected so that backflow does not occur.

M2101.4 Pipe penetrations. Openings through concrete or masonry building elements shall be sleeved.

❖ Concrete and masonry penetrations must be sleeved to protect the pipe from abrasion.

M2101.5 Contact with building material. A hydronic piping system shall not be in direct contact with any building material that causes the piping material to degrade or corrode.

❖ Piping must be protected to avoid external degradation resulting from corrosion, abrasion and chemical reactions. Allowing the pipe to come into contact with certain materials can cause corrosion or degradation. For example, some plastic pipe materials can interact with petroleum-based materials, and concrete can be corrosive to some metals. If hydronic piping is to be installed in an environment or soils that are corrosive, the piping must be protected. Coating, wrappings and sleeves are forms of protection for piping installed in such an environment.

TABLE M2101.1
HYDRONIC PIPING MATERIALS

MATERIAL	USE CODE[a]	STANDARD[b]	JOINTS	NOTES
Acrylonitrile butadiene styrene (ABS) plastic pipe	1, 5	ASTM D1527; ASTM F2806; ASTM F2969	Solvent cement joints	
Brass pipe	1	ASTM B43	Brazed, welded, threaded, mechanical and flanged fittings	
Brass tubing	1	ASTM B135	Brazed, soldered and mechanical fittings	
Chlorinated poly (vinyl chloride) (CPVC) pipe and tubing	1, 2, 3	ASTM D2846	Solvent cement joints, compression joints and threaded adapters	
Copper pipe	1	ASTM B42, B302	Brazed, soldered and mechanical fittings threaded, welded and flanged	
Copper tubing (type K, L or M)	1, 2	ASTM B75, B88, B251, B306	Brazed, soldered and flared mechanical fittings	Joints embedded in concrete
Cross-linked polyethylene (PEX)	1, 2, 3	ASTM F876, F877	(See PEX fittings)	Install in accordance with manufacturer's instructions
Cross-linked polyethylene/ aluminum/cross-linked polyethylene-(PEX-AL-PEX) pressure pipe	1, 2	ASTM F1281 or CAN/ CSA B137.10	Mechanical, crimp/insert	Install in accordance with manufacturer's instructions
PEX fittings		ASTM F877 ASTM F1807 ASTM F1960 ASTM F2098 ASTM F2159 ASTM F2735	Copper-crimp/insert fittings, cold expansion fittings, stainless steel clamp, insert fittings	Install in accordance with manufacturer's instructions
Polybutylene (PB) pipe and tubing	1, 2, 3	ASTM D3309	Heat-fusion, crimp/insert and compression	Joints in concrete shall be heat-fused
Polyethylene/aluminum/polyethylene (PE-AL-PE) pressure pipe	1, 2, 3	ASTM F1282 CSA B 137.9	Mechanical, crimp/insert	
Polypropylene (PP)	1, 2, 3	ISO 15874 ASTM F2389	Heat-fusion joints, mechanical fittings, threaded adapters, compression joints	
Raised temperature polyethylene (PE-RT)	1, 2, 3	ASTM F2623 ASTM F2769	Copper crimp/insert fitting stainless steel clamp, insert fittings	
Raised temperature polyethylene (PE-RT) fittings	1, 2, 3	ASTM F1807 ASTM F2159 ASTM F2735 ASTM F2769 ASTM F2098	Copper crimp/insert fitting stainless steel clamp, insert fittings	
Steel pipe	1, 2	ASTM A53, A106	Brazed, welded, threaded, flanged and mechanical fittings	Joints in concrete shall be welded. Galvanized pipe shall not be welded or brazed.
Steel tubing	1	ASTM A254	Mechanical fittings, welded	

For SI: °C = [(°F)-32]/1.8.
a. Use code:
   1. Above ground.
   2. Embedded in radiant systems.
   3. Temperatures below 180°F only.
   4. Low temperature (below 130°F) applications only.
   5. Temperatures below 160°F only.
b. Standards as listed in Chapter 44.

M2101.6 Drilling and notching. Wood-framed structural members shall be drilled, notched or altered in accordance with the provisions of Sections R502.8, R602.6, R602.6.1 and R802.7. Holes in load bearing members of cold-formed steel light-frame construction shall be permitted only in accordance with Sections R505.2.6, R603.2.6 and R804.2.6. In accordance with the provisions of Sections R505.3.5, R603.3.4 and R804.3.3, cutting and notching of flanges and lips of load-bearing members of cold-formed steel light-frame construction shall not be permitted. Structural insulated panels (SIPs) shall be drilled and notched or altered in accordance with the provisions of Section R610.7.

❖ The code recognizes that cutting or notching framing members may sometimes be necessary. Experience and testing has shown that proper drilling and notching techniques must be followed when mechanical equipment, ducts, piping and related electrical wiring and cables are installed through framing members. Notches that are too deep or holes that are too large or too close together will significantly weaken these members. The code refers to the building chapters (Chapters 5: Floors, 6: Wall Construction, and 8: Roof/Ceiling Construction) that elaborate on such requirements and limitations. Reference to these sections is imperative for these installations so that the structure will not be weakened beyond its design parameters. This section is limited in scope to those framing members cited in the code and does not cover engineered framing members, such as trusses, laminated veneer beams and premanufactured I-joists.

M2101.7 Prohibited tee applications. Fluid in the supply side of a hydronic system shall not enter a tee fitting through the branch opening.

❖ Commentary Figure M2101.7 depicts a prohibited "bullhead" tee arrangement. Installation of such a connection creates excessive pressure drop, poor system performance and possible cavitation (air bubbles) that cause erosion of metal pipes. Supply side flow must not enter a tee from the branch opening. This section does not apply to tees in return piping arrangements because the flow entering a tee branch opening merges with the unidirectional flow in the run of the tee.

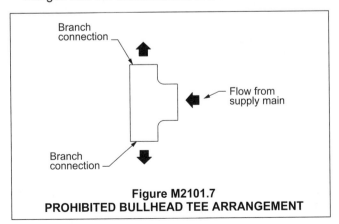

Figure M2101.7
PROHIBITED BULLHEAD TEE ARRANGEMENT

M2101.8 Expansion, contraction and settlement. Piping shall be installed so that piping, connections and *equipment*

shall not be subjected to excessive strains or stresses. Provisions shall be made to compensate for expansion, contraction, shrinkage and structural settlement.

❖ Changes in temperatures cause dimensional changes in all materials to varying degrees. The greatest amount of expansion and contraction in piping occurs along its length. For systems operating at high temperatures, such as steam and hot water, the amount of expansion is substantial, and significant movements can occur even in short runs of piping. Larger movements can be expected in longer lengths of piping. Accordingly, larger forces can develop in restrained piping. Piping systems must be capable of accommodating the forces resulting from thermal expansion and contraction. Inadequate provisions to accommodate these forces can result in failure of pipe and supports, joint damage and leakage.

M2101.9 Piping support. Hangers and supports shall be of material of sufficient strength to support the piping, and shall be fabricated from materials compatible with the piping material. Piping shall be supported at intervals not exceeding the spacing specified in Table M2101.9.

❖ As with all piping systems, the support of the piping is as important as any other part of the overall design. Proper supports are necessary to maintain piping alignment and slope, to support the weight of the piping and fluid within, to control movement and to resist imposed loads. Table M2101.9 provides hanger spacing intervals required for different types of piping material.

TABLE M2101.9
HANGER SPACING INTERVALS

PIPING MATERIAL	MAXIMUM HORIZONTAL SPACING (feet)	MAXIMUM VERTICAL SPACING (feet)
ABS	4	10[a]
CPVC ≤ 1-inch pipe or tubing	3	5[a]
CPVC ≥ 1$^1/_4$ inches	4	10[a]
Copper or copper alloy pipe	12	10
Copper or copper alloy tubing	6	10
PB pipe or tubing	2.67	4
PE pipe or tubing	2.67	4
PE-RT ≤ 1 inch	2.67	10[a]
PE-RT ≥ 1$^1/_4$ inches	4	10[a]
PEX tubing	2.67	4
PP < 1-inch pipe or tubing	2.67	4
PP > 1$^1/_4$ inches	4	10[a]
PVC	4	10[a]
Steel pipe	12	15
Steel tubing	8	10

For SI: 1 inch = 25.4 mm, 1 foot = 304.8 mm.

a. For sizes 2 inches and smaller, a guide shall be installed midway between required vertical supports. Such guides shall prevent pipe movement in a direction perpendicular to the axis of the pipe.

❖ See the commentary to Section M2101.9. Note a requires a guide to be installed to prevent lateral movement of vertical plastic piping. The axis of the pipe is along its length and movement perpendicular to that axis is lateral movement in a horizontal plane. Where the piping is plastic and the distance between vertical supports is allowed to be up to 10 feet, it is expected that the pipe will need the guide to maintain alignment and help prevent stress, abrasion and noise.

M2101.10 Tests. Hydronic piping systems shall be tested hydrostatically at a pressure of one and one-half times the maximum system design pressure, but not less than 100 pounds per square inch (689 kPa). The duration of each test shall be not less than 15 minutes and not more than 20 minutes.

❖ Hydronic piping systems must be tested to determine that the system is leak free and capable of withstanding system operating pressures. The piping under test must be isolated from any component that cannot safely tolerate the test pressure. The prescribed test pressure of 100 psi (690 kPa) will typically exceed the maximum operating pressure of boilers and related components as well as the setting of the system's pressure relief devices. The code official will normally observe required tests. Maintaining the test pressure for at least 15 minutes allows time for any leaks to be detected.

SECTION M2102
BASEBOARD CONVECTORS

M2102.1 General. Baseboard convectors shall be installed in accordance with the manufacturer's instructions. Convectors shall be supported independently of the hydronic piping.

❖ Convectors must be installed in accordance with the manufacturer's installation instructions. This section requires that convectors be supported independently of the hydronic piping to prevent strain on the piping system and to allow for independent expansion and contraction.

SECTION M2103
FLOOR HEATING SYSTEMS

M2103.1 Piping materials. Piping for embedment in concrete or gypsum materials shall be standard-weight steel pipe, copper and copper alloy pipe and tubing, cross-linked polyethylene/aluminum/cross-linked polyethylene (PEX-AL-PEX) pressure pipe, chlorinated polyvinyl chloride (CPVC), polybutylene, cross-linked polyethylene (PEX) tubing, polyethylene of raised temperature (PE-RT) or polypropylene (PP) with a minimum rating of 100 psi at 180°F (690 kPa at 82°C).

❖ Piping embedded in concrete or covered by permanent construction materials or assemblies is not accessible. Therefore, this section is more restrictive than Section M2101.1, in that it allows only specific materials for embedded hydronic piping.

M2103.2 Thermal barrier required. Radiant floor heating systems shall have a thermal barrier in accordance with Sections M2103.2.1 through M2103.2.4.

Exception: Insulation shall not be required in engineered systems where it can be demonstrated that the insulation

will decrease the efficiency or have a negative effect on the installation.

❖ In order for the maximum amount of heat to be transferred to the space intended to be heated, the hydronic piping system must be insulated to limit the amount of heat transferred to spaces or materials that are not intended to be heated. Sections M2103.2.1 and M2103.2.2 describe two applications where this is a concern and provides the insulation requirements.

M2103.2.1 Slab-on-grade installation. Radiant piping used in slab-on-grade applications shall have insulating materials having a minimum *R*-value of 5 installed beneath the piping.

❖ Typically, rigid foam board insulation is used for insulating slabs on grade that will be heated by hydronic systems. For most manufacturers of this product, an *R*-value of 5 equates to 1-inch (25 mm) material thickness. Other types of insulation products that are suitable for under-slab installation are also available and can be used if marked in accordance with Section M2103.2.4.

M2103.2.2 Suspended floor installation. In suspended floor applications, insulation shall be installed in the joist bay cavity serving the heating space above and shall consist of materials having a minimum *R*-value of 11.

❖ Typically, fiberglass insulation is used for insulating hydronically heated floors that are above an unheated space, such as a crawl space or basement. Insulation is placed between the joists (or floor trusses) and should be held in place.

M2103.2.3 Thermal break required. A thermal break consisting of asphalt expansion joint materials or similar insulating materials shall be provided at a point where a heated slab meets a foundation wall or other conductive slab.

❖ The exterior edges of heated concrete slabs must be protected against heat loss. Placement of a thermal barrier at the edge of the slab where it would contact the foundation wall breaks the heat conduction path to the exterior.

M2103.2.4 Thermal barrier material marking. Insulating materials used in thermal barriers shall be installed so that the manufacturer's *R*-value mark *is readily observable upon inspection.*

❖ Insulation materials must be marked so that the inspector can verify that the correct *R*-value is being used for the application. "Readily observable" means that the inspector will not be required to remove insulation in order to see the marking. The *R*-value marking requirement might be a problem where unfaced fiberglass insulation is used or where the facing of the fiberglass insulation is required to be located such that the marking is not observable without removal of the insulation.

M2103.3 Piping joints. Copper and copper alloy systems shall be soldered in accordance with ASTM B828. Fluxes for soldering shall be in accordance with ASTM B813. Brazing fluxes shall be in accordance with AWS A5.31. Piping joints that are embedded shall be installed in accordance with the following requirements:

1. Steel pipe joints shall be welded.
2. Copper tubing shall be joined by brazing complying with Section P3003.6.1.
3. Polybutylene pipe and tubing joints shall be installed with socket-type heat-fused polybutylene fittings.
4. CPVC tubing shall be joined using solvent cement joints.
5. Polypropylene pipe and tubing joints shall be installed with socket-type heat-fused polypropylene fittings.
6. Cross-linked polyethylene (PEX) tubing shall be joined using cold expansion, insert or compression fittings.
7. Raised temperature polyethylene (PE-RT) tubing shall be joined using insert or compression fittings.

❖ Joints for these types of installations must be of the type that is least susceptible to failures. For that reason, steel pipe joints must be welded, copper pipe joints brazed, polypropylene heat-fused, etc.

M2103.4 Testing. Piping or tubing to be embedded shall be tested by applying a hydrostatic pressure of not less than 100 psi (690 kPa). The pressure shall be maintained for 30 minutes, during which, the joints shall be visually inspected for leaks.

❖ Hydronic piping systems must be tested to determine that the system is leak free and capable of withstanding the pressures to which it will be subjected. It is customary for the code official to observe required tests. This testing is especially important where the piping is to be encased and inaccessible. See Section M2101.10.

SECTION M2104
LOW TEMPERATURE PIPING

M2104.1 Piping materials. Low temperature piping for embedment in concrete or gypsum materials shall be as indicated in Table M2101.1.

❖ Table M2101.1 lists piping materials that can be used in low temperature piping, along with their applicable standards. This section does not prohibit the use of other approved materials that provide equivalent performance.

M2104.2 Piping joints. Piping joints, other than those in Section M2103.3, that are embedded shall comply with the following requirements:

1. Cross-linked polyethylene (PEX) tubing shall be installed in accordance with the manufacturer's instructions.
2. Polyethylene tubing shall be installed with heat fusion joints.
3. Polypropylene (PP) tubing shall be installed in accordance with the manufacturer's instructions.
4. Raised temperature polyethylene (PE-RT) shall be installed in accordance with the manufacturer's instructions.

❖ All joining methods for PEX tubing are considered to be mechanical joints that must be installed in accor-

dance with the manufacturer's installation instructions. Polyethylene pipe and fittings are joined by heat-fusion joining methods, which provide dependability for underground piping. Embedded polypropylene tubing must be installed in accordance with the manufacturer's instructions for underground installations. Heat-fusion joining methods are considered to be highly dependable.

M2104.3 Raised temperature polyethylene (PE-RT) plastic tubing. Joints between raised temperature polyethylene tubing and fittings shall conform to Sections M2104.3.1, M2104.3.2 and M2104.3.3. Mechanical joints shall be installed in accordance with the manufacturer's instructions.

❖ The fitting manufacturer's installation instructions must be followed when making connections of raised temperature polyethylene (PE-RT) tubing.

M2104.3.1 Compression-type fittings. Where compression-type fittings include inserts and ferrules or O-rings, the fittings shall be installed without omitting such inserts and ferrules or O-rings.

❖ All parts of a fitting assembly must be utilized as intended by the manufacturer.

M2104.3.2 PE-RT-to-metal connections. Solder joints in a metal pipe shall not occur within 18 inches (457 mm) of a transition from such metal pipe to PE-RT pipe.

❖ The required heat for soldering fittings to copper pipe could damage the PE-RT pipe if the soldering operation was performed too close to a PE-RT-to-copper transition. This section applies to solder joints that occur in the same section of pipe as a PE-RT-to-copper transition joint.

M2104.3.3 PE-RT insert fittings. PE-RT insert fittings shall be installed in accordance with the manufacturer's instructions.

❖ Insert fittings are mechanical joints that involve a metal insert and a crimp ring or that involve a compression nut assembly. The piping manufacturer will specify what fittings are to be used with their product.

M2104.4 Polyethylene/Aluminum/Polyethylene (PE-AL-PE) pressure pipe. Joints between polyethylene/aluminum/polyethylene pressure pipe and fittings shall conform to Sections M2104.4.1 and M2104.4.2. Mechanical joints shall be installed in accordance with the manufacturer's instructions.

❖ The fitting manufacturer's installation instructions must be followed when making connections of polyethylene/aluminum/polyethylene (PE-AL-PE) tubing.

M2104.4.1 Compression-type fittings. Where compression-type fittings include inserts and ferrules or O-rings, the fittings shall be installed without omitting such inserts and ferrules or O-rings.

❖ All parts of a fitting assembly must be utilized as intended by the manufacturer. It is hard to imagine that an installer would purposely omit parts of a fitting assembly, but, this code text is testament to the fact that this must have been an issue for some types of fittings.

M2104.4.2 PE-AL-PE to metal connections. Solder joints in a metal pipe shall not occur within 18 inches (457 mm) of a transition from such metal pipe to PE-AL-PE pipe.

❖ The required heat for soldering fittings to copper pipe could damage the PE-AL-PE pipe if the soldering operation was performed too close to a PE-AL-PE-to-copper transition. This section applies to solder joints that occur in the same section of pipe as a PE-AL-PE-to-copper transition joint.

SECTION M2105
GROUND-SOURCE HEAT-PUMP
SYSTEM LOOP PIPING

M2105.1 Plastic ground-source heat-pump loop piping. Plastic piping and tubing material used in water-based ground-source heat-pump ground-loop systems shall conform to the standards specified in this section.

❖ Section M2105 applies to ground-source heat-pump loop systems, specifically, the piping for the loops. These systems are mistakenly referred to as Geo-thermal systems, but Geo-thermal systems involve underground sources of heat associated with hot springs, lava domes and other areas in the earth's crust where inner earth heat energy rises close to the earth's surface. Ground-source heat-pump systems simply use the soil, water wells and surface bodies of water as a heat source and as a heat sink. See Sections M2105.4 and M2105.5.

M2105.2 Used materials. Reused pipe, fittings, valves, and other materials shall not be used in ground-source heat-pump loop systems.

❖ For the greatest reliability, the piping system must be constructed of new (unused) materials.

M2105.3 Material rating. Pipe and tubing shall be rated for the operating temperature and pressure of the ground-source heat-pump loop system. Fittings shall be suitable for the pressure applications and recommended by the manufacturer for installation with the pipe and tubing material installed. Where used underground, materials shall be suitable for burial.

❖ All materials must be rated for the pressures and temperatures to which they will be exposed. Fittings have

to be a type that is specifically recommended by the fitting manufacturer for use with the piping being installed. Components that are to be buried must be intended for burial.

M2105.4 Piping and tubing materials standards. Ground-source heat-pump ground-loop pipe and tubing shall conform to the standards listed in Table M2105.4.

❖ See Table M2105.4.

M2105.5 Fittings. Ground-source heat-pump pipe fittings shall be approved for installation with the piping materials to be installed, shall conform to the standards listed in Table M2105.5 and, where installed underground, shall be suitable for burial.

❖ See Table M2105.5.

M2105.6 Joints and connections. Joints and connections shall be of an approved type. Joints and connections shall be tight for the pressure of the ground-source loop system. Joints used underground shall be approved for such applications.

❖ This section requires code official approval of the types of joints and connections. Such approval would be based on the compliance with Table M2105.5, fitting manufacturer's instructions and recommendations, product literature and consideration for the conditions to which the joints and connections will be exposed.

M2105.6.1 Joints between different piping materials. Joints between different piping materials shall be made with approved transition fittings.

❖ See Section M2105.6.

M2105.7 Preparation of pipe ends. Pipe shall be cut square, reamed, and shall be free of burrs and obstructions. CPVC, PE and PVC pipe shall be chamfered. Pipe ends shall have full-bore openings and shall not be undercut.

❖ This section requires proper workmanship for preparing pipe to be joined. These requirements are very often disregarded. Cutting pipe ends squarely, de-burring and chamfering are necessary to achieve a reliable joint for many reasons related to structural strength and leak-tightness. With some types of joints, it is possible to inspect for compliance, but with others, inspection is not possible without disassembling the

TABLE M2105.4
GROUND-SOURCE LOOP PIPE

MATERIAL	STANDARD
Chlorinated polyvinyl chloride (CPVC)	ASTM D2846; ASTM F437; ASTM F438; ASTM F439; ASTM F441; ASTM F442; CSA B137.6
Cross-linked polyethylene (PEX)	ASTM F876; ASTM F877 CSA B137.5
Polyethylene/aluminum/polyethylene (PE-AL-PE) pressure pipe	ASTM F1282; CSA B137.9; AWWA C 903
High-density polyethylene (HDPE)	ASTM D2737; ASTM D3035; ASTM F714; AWWA C901; CSA B137.1; CSA C448; NSF 358-1
Polypropylene (PP-R)	ASTM F2389; CSA B137.11
Polyvinyl chloride (PVC)	ASTM D1785; ASTM D2241; CSA 137.3
Raised temperature polyethylene (PE-RT)	ASTM F2623; ASTM F2769

joint or connection. The inspector may determine compliance by requiring a sample joint to be opened, by observing discarded parts of joints that have been removed or by interviewing the installers about their procedures. There are requirements for installation that are very difficult to inspect and this sort of difficulty has led to code requirements such as purple primer for PVC pipe joints. The purple color is used because the use of a primer cannot be visually verified without the distinguishing color.

M2105.8 Joint preparation and installation. Where required by Sections M2105.9 through M2105.11, the preparation and installation of mechanical and thermoplastic-welded joints shall comply with Sections M2105.8.1 and M2015.8.2.

❖ Some of the material specific subsections could refer back to Sections M2105.8.1 and M1205.8.2 relative to mechanical and thermoplastic welded joints where necessary. Thermoplastic welding is similar to metal welding and involves a heat source and welding rod (spline). The plastic pipe and fittings and the welding rod (spline) are heated typically by hot air and the materials fuse together.

M2105.8.1 Mechanical joints. Mechanical joints shall be installed in accordance with the manufacturer's instructions.

❖ Mechanical joints can be of many different types. See the definition of "Mechanical joint."

M2105.8.2 Thermoplastic-welded joints. Joint surfaces for thermoplastic-welded joints shall be cleaned by an approved procedure. Joints shall be welded in accordance with the manufacturer's instructions.

❖ See section M2105.8.

M2105.9 CPVC plastic pipe. Joints between CPVC plastic pipe or fittings shall be solvent-cemented in accordance with Section P2906.9.1.2. Threaded joints between fittings and CPVC plastic pipe shall be in accordance with Section M2105.9.1.

❖ See Sections M2105.9.1 and P2906.9.1.2.

M2105.9.1 Threaded joints. Threads shall conform to ASME B1.20.1. The pipe shall be Schedule 80 or heavier plastic pipe and shall be threaded with dies specifically designed for plastic pipe. Thread lubricant, pipe-joint compound or tape shall be applied on the male threads only and shall be approved for application on the piping material.

❖ Schedule 40 pipe cannot be threaded because the pipe wall is too thin to provide the necessary strength. It is very important to make sure that pipe thread compounds (pipe dope) are compatible with the pipe material. Some compounds can degrade plastic piping.

M2105.10 Cross-linked polyethylene (PEX) plastic tubing. Joints between cross-linked polyethylene plastic tubing and fittings shall comply with Sections M2105.10.1 and M2105.10.2. Mechanical joints shall comply with Section M2105.8.1.

❖ See Sections M2105.8.1, M2105.10.1 and M2105.10.2.

M2105.10.1 Compression-type fittings. Where compression-type fittings include inserts and ferrules or O-rings, the fittings shall be installed without omitting the inserts and ferrules or O-rings.

❖ See Section M2104.3.1.

M2105.10.2 Plastic-to-metal connections. Solder joints in a metal pipe shall not occur within 18 inches (457 mm) of a transition from such metal pipe to plastic pipe or tubing.

❖ Solder joints involve temperatures of 400°F and higher. Copper pipe is an excellent conductor of heat and the heat of soldering could be conducted to nearby plastic piping, possibly damaging the plastic pipe and fittings. Where plastic pipe connects to metal pipe, soldering must occur at a safe distance from the plastic components.

M2105.11 Polyethylene plastic pipe and tubing. Joints between polyethylene plastic pipe and tubing or fittings for ground-source heat-pump loop systems shall be heat-fusion joints complying with Section M2105.11.1, electrofusion

TABLE M2105.5
GROUND-SOURCE LOOP PIPE FITTINGS

PIPE MATERIAL	STANDARD
Chlorinated polyvinyl chloride (CPVC)	ASTM D2846; ASTM F437; ASTM F438; ASTM F439; ASTM F1970; CSA B137.6
Cross-linked polyethylene (PEX)	ASTM F877; ASTM F1807; ASTM F1960; ASTM F2080; ASTM F2159; ASTM F2434; CSA B137.5
Polyethylene/aluminum/polyethylene (PE-AL-PE)	ASTM F2434; ASTM F1282; CSA B137.9
High-density polyethylene (HDPE)	ASTM D2683; ASTM D3261; ASTM F1055; CSA B137.1; CSA C448; NSF 358-1
Polypropylene (PP-R)	ASTM F2389; CSA B137.11; NSF 358-2
Polyvinyl chloride (PVC)	ASTM D2464; ASTM D2466; ASTM D2467; ASTM F1970, CSA B137.2; CSA B137.3
Raised temperature polyethylene (PE-RT)	ASTM D3261; ASTM F1807; ASTM F2159; F2769; B137.1

joints complying with Section M2105.11.2, or stab-type insertion joints complying with Section M2105.11.3.

❖ See Sections M2105.11.1, M2105.11.2 and M2105.11.3.

M2105.11.1 Heat-fusion joints. Joints shall be of the socket-fusion, saddle-fusion or butt-fusion type, and joined in accordance with ASTM D2657. Joint surfaces shall be clean and free of moisture. Joint surfaces shall be heated to melt temperatures and joined. The joint shall be undisturbed until cool. Fittings shall be manufactured in accordance with ASTM D2683 or ASTM D3261.

❖ Heat fusion joints are very reliable if created using the proper procedures and using the proper tools.

M2105.11.2 Electrofusion joints. Joints shall be of the electrofusion type. Joint surfaces shall be clean and free of moisture, and scoured to expose virgin resin. Joint surfaces shall be heated to melt temperatures for the period of time specified by the manufacturer. The joint shall be undisturbed until cool. Fittings shall be manufactured in accordance with ASTM F1055.

❖ Electrofusion joining utilizes special fittings that have integral heating elements. The fittings are connected to a power supply that provides the necessary current for the correct amount of time to heat the mating parts to the fusion temperature.

M2105.11.3 Stab-type insert fittings. Joint surfaces shall be clean and free of moisture. Pipe ends shall be chamfered and inserted into the fittings to full depth. Fittings shall be manufactured in accordance with ASTM F1924.

❖ Stab-type fittings are referred to as such because of the action that inserts (roughly analogous to stabbing) the tubing into the fitting. Such fittings utilize O-rings and gripping devices that retain the tubing in the fitting. Chamfering is necessary to remove the sharp angular outside edge that could damage the O-ring seals when the tubing is inserted.

M2105.12 Polypropylene (PP) plastic. Joints between PP plastic pipe and fittings shall comply with Sections M2105.12.1 and M2105.12.2.

❖ See Sections M2105.12.1 and M2105.12.2.

M2105.12.1 Heat-fusion joints. Heat-fusion joints for polypropylene (PP) pipe and tubing joints shall be installed with socket-type heat-fused polypropylene fittings, electrofusion polypropylene fittings or by butt fusion. Joint surfaces shall be clean and free from moisture. The joint shall be undisturbed until cool. Joints shall be made in accordance with ASTM F2389.

❖ Heat fusion and mechanical joints are used with polypropylene (PP) because it cannot be solvent welded. Specialized tools must be used to heat the pipe/tubing and the fittings to the point where the mating parts fuse together.

M2105.12.2 Mechanical and compression sleeve joints. Mechanical and compression sleeve joints shall be installed in accordance with the manufacturer's instructions.

❖ Mechanical joints come in many varieties and they all rely on sealing elements and mechanical compression. Strict compliance with the manufacturer's instructions is imperative to create dependable joints.

M2105.13 Raised temperature polyethylene (PE-RT) plastic tubing. Joints between raised temperature polyethylene tubing and fittings shall comply with Sections M2105.13.1 and M2105.13.2. Mechanical joints shall comply with Section M2105.8.1.

❖ See Sections M2105.8.1, M2105.13.1 and M2105.13.2.

M2105.13.1 Compression-type fittings. Where compression-type fittings include inserts and ferrules or O-rings, the fittings shall be installed without omitting the inserts and ferrules or O-rings.

❖ See Section M2104.3.1.

M2105.13.2 PE-RT-to-metal connections. Solder joints in a metal pipe shall not occur within 18 inches (457 mm) of a transition from such metal pipe to PE-RT pipe or tubing.

❖ See Section M2105.10.2.

M2105.14 PVC plastic pipe. Joints between PVC plastic pipe or fittings shall be solvent-cemented in accordance with Section P2906.9.1.4. Threaded joints between fittings and PVC plastic pipe shall be in accordance with Section M2105.9.1.

❖ See Section M2105.9.1. Solvent welding is performed as dictated in Chapter 29 for water service piping.

M2105.15 Shutoff valves. Shutoff valves shall be installed in ground-source loop piping systems in the locations indicated in Sections M2105.15.1 through M2105.15.6.

❖ Shutoff valves are required to allow the system to be serviced and repaired and to allow components and buildings to be isolated from the system.

M2105.15.1 Heat exchangers. Shutoff valves shall be installed on the supply and return side of a heat exchanger.

Exception: Shutoff valves shall not be required where heat exchangers are integral with a boiler or are a component of a manufacturer's boiler and heat exchanger packaged unit and are capable of being isolated from the hydronic system by the supply and return valves required by Section M2001.3.

❖ Heat exchangers include water-to-air, refrigerant-to-water and water-to-water coils. Heat exchangers are sometimes integral to storage tanks and heat pump units.

M2105.15.2 Central systems. Shutoff valves shall be installed on the building supply and return of a central utility system.

❖ Multiple buildings could be served by a single heat pump system and each building must be capable of being isolated.

M2105.15.3 Pressure vessels. Shutoff valves shall be installed on the connection to any pressure vessel.

❖ Pressure vessels include expansion tanks and storage tanks.

M2105.15.4 Pressure-reducing valves. Shutoff valves shall be installed on both sides of a pressure-reducing valve.

❖ Pressure-reducing valves need to be isolated to allow replacement and repair.

M2105.15.5 Equipment and appliances. Shutoff valves shall be installed on connections to mechanical equipment and appliances. This requirement does not apply to components of ground-source loop systems such as pumps, air separators, metering devices, and similar equipment.

❖ This section exempts water pumps; however, replacing or servicing a water pump without pump isolation valves can be difficult. Heat pump equipment isolation valves facilitate servicing and replacement, and can also allow specific equipment to be taken in and out of service.

M2105.15.6 Expansion tanks. Shutoff valves shall be installed at connections to nondiaphragm-type expansion tanks.

❖ Diaphragm-type expansion tanks do not require maintenance or servicing, so isolation valves are required only for tanks that do not have a diaphragm membrane. Not having a shutoff valve makes it impossible to inadvertently isolate the piping system from the expansion tank.

M2105.16 Reduced pressure. A pressure relief valve shall be installed on the low-pressure side of a hydronic piping system that has been reduced in pressure. The relief valve shall be set at the maximum pressure of the system design. The valve shall be installed in accordance with Section M2002.

❖ A relief valve must be installed downstream of a pressure reducing valve (PRV) to protect the lower pressure system from pressure excursions or PRV failure.

M2105.17 Installation. Piping, valves, fittings, and connections shall be installed in accordance with the manufacturer's instructions.

❖ See Sections M1301.3, M1307.1 and M1401.1.

M2105.18 Protection of potable water. Where ground-source heat-pump ground-loop systems have a connection to a potable water supply, the potable water system shall be protected from backflow in accordance with Section P2902.

❖ The potable water system must be protected against backflow because the heat pump system contains nonpotable water and can contain antifreeze mixtures and other chemicals.

M2105.19 Pipe penetrations. Openings for pipe penetrations in walls, floors and ceilings shall be larger than the penetrating pipe. Openings through concrete or masonry building elements shall be sleeved. The annular space surrounding pipe penetrations shall be protected in accordance with Section P2606.1.

❖ See Section P2606.1. Larger openings and sleeves protect the piping against abrasion damage caused by expansion, contraction and vibration.

M2105.20 Clearance from combustibles. A pipe in a ground-source heat pump piping system having an exterior surface temperature exceeding 250°F (121°C) shall have a clearance of not less than 1 inch (25 mm) from combustible materials.

❖ Such temperature would be unlikely in a heat pump system, even where a supplemental source of heat such as a boiler might be involved.

M2105.21 Contact with building material. A ground-source heat-pump ground-loop piping system shall not be in direct contact with building materials that cause the piping or fitting material to degrade or corrode, or that interfere with the operation of the system.

❖ Piping must be separated or otherwise protected from contact with building materials that can cause detrimental chemical reactions with the piping.

M2105.22 Strains and stresses. Piping shall be installed so as to prevent detrimental strains and stresses in the pipe. Provisions shall be made to protect piping from damage resulting from expansion, contraction and structural settlement. Piping shall be installed so as to avoid structural stresses or strains within building components.

❖ Piping must be supported such that stresses are not created in the piping. Offsets, loops and expansion fitting must be provided as necessary to compensate for the expansion and contraction of the piping. Plastic piping, in general, has a high coefficient of expansion and the forces generated are substantial.

M2105.22.1 Flood hazard. Piping located in a flood hazard area shall be capable of resisting hydrostatic and hydrodynamic loads and stresses, including the effects of buoyancy, during the occurrence of flooding to the *design flood elevation*.

❖ Buried ground-source loop piping that is filled with liquid is likely to be immune from damage caused by area flooding, however, such immunity must be demonstrated.

M2105.23 Pipe support. Pipe shall be supported in accordance with Section M2101.9.

❖ See Section M2101.9.

M2105.24 Velocities. Ground-source heat-pump ground-loop systems shall be designed so that the flow velocities do not exceed the maximum flow velocity recommended by the pipe and fittings manufacturer. Flow velocities shall be controlled to reduce the possibility of water hammer.

❖ The flow velocity in piping of all materials (plastic and metal) must be kept within the recommendations of the material manufacturer to protect the piping from erosion and cavitation damage. Water hammer results from the sudden deceleration of flow velocity and the forces generated can be destructive to piping and system components.

M2105.25 Labeling and marking. Ground-source heat-pump ground-loop system piping shall be marked with tape, metal tags or other methods where it enters a building. The marking shall state the following words: "GROUND-SOURCE HEAT-PUMP LOOP SYSTEM." The marking shall indicate if antifreeze is used in the system and shall indicate the chemicals by name and concentration.

❖ Proper labeling will help prevent mistaken identity of piping systems when any system in the building is to be worked on. The nature of the heat transfer medium must be identified to help prevent the introduction of the wrong chemicals and incompatible chemicals, and to maintain the required freeze protection. The label also serves as a warning about potentially toxic chemicals.

M2105.26 Chemical compatibility. Antifreeze and other materials used in the system shall be chemically compatible with the pipe, tubing, fittings and mechanical systems.

❖ Plastic pipe manufacturers publish chemical compatibility information for their products. Chemicals must not be introduced into a piping system without verifying that the chemicals and the piping system materials are compatible. Damage to underground piping would be quite expensive to repair and could result in hazardous operation of equipment in the building.

M2105.27 Makeup water. The transfer fluid shall be compatible with the makeup water supplied to the system.

❖ Makeup water is water added to the system to replace any fluid that is lost through leakage and intentional draining for repairs and maintenance. Makeup water is limited to the amount absolutely necessary to maintain a completely filled system because makeup water contains dissolved minerals and gasses that are undesirable in the system.

M2105.28 Testing. Before connection header trenches are backfilled, the assembled loop system shall be pressure tested with water at 100 psi (689 kPa) for 15 minutes without observed leaks. Flow and pressure loss testing shall be performed and the actual flow rates and pressure drops shall be compared to the calculated design values. If actual flow rate or pressure drop values differ from calculated design values by more than 10 percent, the cause shall be identified and corrective action taken.

❖ Groundwater heat pump systems are either closed loop or open loop. Closed-loop systems circulate water or an antifreeze solution through plastic pipes buried in the ground. Some loop systems, although not as common, use surface water bodies or underground water as the source of heat energy. The heat-transfer fluid absorbs heat from the ground (earth) and transfers it to the building in the heating season, then reverses to absorb heat from the building and transfer it to the ground in the cooling season.

Ground source loop systems use significant lengths of piping installed in multiple circuits (loops) to provide the required area of heat-exchange surface. Also, multiple loops have the advantage over a single loop system because a single piping failure will not cause failure of the entire heat pump system.

The significant difference between testing ground source loop systems and other hydronic systems is that ground source loops are allowed to be buried prior to testing. The connection headers (manifolds) that connect the loops (circuits) of piping must remain exposed until completion of testing. Because loops (circuits) are typically installed without joints, all of the potential sources of leaks will likely be located at the point of connection between the ends of the loops and the header. For this reason, the loops can be covered during testing while the headers remain exposed. Also, it is difficult to dig trenches for large loop fields without backfilling each trench before digging the adjacent trenches.

Depending on the lot, it may be necessary to backfill "as you go" to allow room for the excavator machine.

Requiring that each loop be inspected prior to covering could require time-consuming multiple inspections at the same property if the excavator had to backfill each trench before digging the next.

M2105.29 Embedded piping. Ground-source heat-pump ground-loop piping to be embedded in concrete shall be pressure tested prior to pouring concrete. During pouring, the pipe shall be maintained at the proposed operating pressure.

❖ Embedding piping in concrete is generally avoided because the loop piping will be inaccessible and because the piping is easily damaged during the concrete placement process. If the piping is damaged during the concrete pour, the leak will be apparent if the system is kept under pressure.

Chapter 22:
Special Piping and Storage Systems

User note: Code change proposals to this chapter will be considered by the IRC – Plumbing and Mechanical Code Development Committee during the 2015 (Group A) Code Development Cycle.

General Comments

This chapter regulates the design and installation of fuel oil storage and piping systems. The regulations include reference to construction standards for above-ground and underground storage tanks, material standards for piping systems (both above-ground and underground) and extensive requirements for the proper assembly of system piping and components.

The importance of adequate regulations for the storage, handling and use of combustible liquid fuel cannot be overstated. A fuel-oil piping system includes piping, valves and fittings. If installed, pumps, reservoirs, regulators, strainers, filters, relief valves, oil preheaters, controls and gauges are also included in the fuel-oil piping system. The piping materials and the method of joining sections of pipe must be approved, and all other components included in the piping system must be labeled and installed in accordance with the manufacturer's instructions. Over the past 10 years, public scrutiny of flammable and combustible liquid storage installations has increased with the public's awareness of the consequences of release of these liquids into the environment. Environmental studies have shown that improper installation of these storage systems is a major contributing factor in system failure. Improper use of such systems also accounts for many fire losses. Although the hazards are well known, accidents involving combustible liquids remain one of the most common fire scenarios in the United States.

Fuel oil is considered a Class II combustible liquid, having a flash point of 100°F (38°C) or higher. Though the classification boundaries are somewhat arbitrary, flammable and combustible liquids are distinguished by their flash points. The flash point is that temperature at which the liquid produces sufficient vapor to form an ignitable vapor-air mixture above its surface. Because Class I flammable liquids all have a flash point below 100°F (38°C), it is prudent to assume that such liquids may be capable of igniting when unconfined under normal environmental conditions. On the other hand, combustible liquids, including fuel oil, are materials with flash points above 100°F (38°C) and must usually be heated above their flash points or, in the case of extremely high flash-point liquids, above their boiling points before they will ignite.

Combustible liquids possess significant characteristics other than their flash points. These include ignition temperature, autoignition temperature, flammable (explosive) range, viscosity, vapor density, vapor pressure, boiling point, evaporation rate, specific gravity and water solubility. Once the liquid is ignited, these variables have little influence over the material's heat release rate. Factors such as evaporation rate, viscosity and water solubility may profoundly affect how these fires are extinguished.

Generally, combustible liquids have low specific gravities, high vapor densities and narrow flammable ranges. These characteristics mean that liquids usually float on water, vapors most often hug the ground and ignitable vapor-air mixtures are generally confined to a range between 6 and 15 percent in air. Thus, smothering is difficult and ignition sources near the ground are more likely to pose a hazard.

Purpose

The requirements in this chapter are intended to prevent fires, leaks and spills involving fuel oil storage and piping systems, whether inside or outside structures and above or below ground.

SECTION M2201
OIL TANKS

M2201.1 Materials. Supply tanks shall be *listed* and *labeled* and shall conform to UL 58 for underground tanks and UL 80 for indoor tanks.

❖ Shop-fabricated tank(s) must be listed and labeled by an approved agency. The code specifies the national standards applicable to the manufacture of these tanks.

M2201.2 Above-ground tanks. The maximum amount of fuel oil stored above ground or inside of a building shall be 660 gallons (2498 L). The supply tank shall be supported on rigid noncombustible supports to prevent settling or shifting.

Exception: The storage of fuel oil, used for space or water heating, above ground or inside buildings in quantities exceeding 660 gallons (2498 L) shall comply with NFPA 31.

❖ The quantity of fuel oil is limited to lessen the hazard associated with leakage and storage system failures. Tanks must have good support bases to support large point loads and to prevent the tanks from moving and stressing piping connected to the tanks.

M2201.2.1 Tanks within buildings. Supply tanks for use inside of buildings shall be of such size and shape to permit installation and removal from *dwellings* as whole units. Supply tanks larger than 10 gallons (38 L) shall be placed not less than 5 feet (1524 mm) from any fire or flame either within or external to any fuel-burning *appliance*.

❖ This section limits the size of a tank, in addition to the capacity limitation of Section M2201.2, by taking into consideration the maintenance aspect of replacing a tank. If a tank must be cut apart to be removed from a residence, it could create a hazardous condition.

In addition, this section limits the location of a tank with respect to an open flame to prevent ignition of spilled or leaking fuel.

M2201.2.2 Outside above-ground tanks. Tanks installed outdoors above ground shall be not less than 5 feet (1524 mm) from an adjoining property line. Such tanks shall be suitably protected from the weather and from physical damage.

❖ This section takes into account the location of tanks installed outside of a building and above ground. Protection of a tank from physical damage and deterioration caused by local weather conditions is the main concern. The location of the tank must be evaluated with respect to possible impact damage (vehicles, equipment, etc.) and protected in an approved manner. Such protection usually consists of high-quality exterior grade paint. In areas subject to extreme weather conditions, additional protection may be required.

M2201.3 Underground tanks. Excavations for underground tanks shall not undermine the foundations of existing structures. The clearance from the tank to the nearest wall of a *basement*, pit or property line shall be not less than 1 foot (305 mm). Tanks shall be set on and surrounded with noncorrosive inert materials such as clean earth, sand or gravel well tamped in place. Tanks shall be covered with not less than 1 foot (305 mm) of earth. Corrosion protection shall be provided in accordance with Section M2203.7.

❖ These requirements exist so that a tank will not shift or move because of settlement of the surrounding earth. Shifting stresses a tank and piping and can cause leakage of oil.

M2201.4 Multiple tanks. Cross connection of two supply tanks shall be permitted in accordance with Section M2203.6.

❖ A maximum of two interconnected tanks are allowed in place of one, as long as the total capacity does not exceed 660 gallons (2498 L). See the commentary to Section M2203.6.

M2201.5 Oil gauges. Inside tanks shall be provided with a device to indicate when the oil in the tank has reached a predetermined safe level. Glass gauges or a gauge subject to breakage that could result in the escape of oil from the tank shall not be used. Liquid-level indicating gauges shall comply with UL 180.

❖ This section applies to indoor tanks and prohibits the use of any gauging device, including sight glasses, that

would release tank contents in the event of the gauging device breaking. This helps prevent a hazardous material spill and possible fuel oil vapor ignition. Liquid-level gauges must be in compliance with UL 180, which contains construction and performance requirements for gauges used in fuel-oil storage systems. A predetermined safe level is related to the filling of tanks.

M2201.6 Flood-resistant installation. In flood hazard areas as established by Table R301.2(1), tanks shall be installed in accordance with Section R322.2.4 or R322.3.7.

❖ The local jurisdiction must fill in Table R301.2(1) upon adoption of the code, including the "Flood Hazards" information. Tanks that are located in flood hazard areas must be installed above the design flood elevation or must be designed and installed to resist the lateral and buoyant forces of the flood waters. See the commentary to Sections 322.1.6, R322.2.4 and R322.3.7.

M2201.7 Tanks abandoned or removed. Exterior above-grade fill piping shall be removed when tanks are abandoned or removed. Tank abandonment and removal shall be in accordance with the *International Fire Code*.

❖ The *International Fire Code®* (IFC®) contains the requirements for abandoning or removing fuel oil tanks. The IFC does not, however, require the removal of the above-grade fill piping associated with the tanks. Exterior fill and vent piping has to be removed because of the potential danger that hundreds of gallons of fuel oil could be accidentally delivered to the wrong address. There have been instances of accidental filling where tanks have been removed but the fill pipe has remained. The oil delivery service sees only the exterior connection and may not be aware that the tank has been removed or disconnected. Piping must be removed, not just capped off; accidental filling has occurred when piping systems have been "capped off." When fuel oil contamination occurs, the cost of repairs and cleanup is extremely high and in some cases contamination has led to condemnation of such structures.

SECTION M2202
OIL PIPING, FITTING AND CONNECTIONS

M2202.1 Materials. Piping shall consist of steel pipe, copper and copper alloy pipe and tubing or steel tubing conforming to ASTM A539. Aluminum tubing shall not be used between the fuel-oil tank and the burner units.

❖ Aluminum is prohibited because it corrodes more readily when it is exposed to moisture or is adjacent to some types of construction material.

M2202.2 Joints and fittings. Piping shall be connected with standard fittings compatible with the piping material. Cast iron fittings shall not be used for oil piping. Unions requiring gaskets or packings, right or left couplings, and sweat fittings employing solder having a melting point less than 1,000°F (538°C) shall not be used for oil piping. Threaded joints and connections shall be made tight with a lubricant or pipe thread compound.

❖ Joining and connecting methods and materials must be compatible with the pipe and the fuel oil used. Lubricants and tape are designed to lubricate the threads for proper mating and to fill in small imperfections within these threads. Lubricants must be compatible with the pipe and fuel oil used or they will deteriorate and allow fuel oil leaks. Gaskets and packing materials in unions can lose their resiliency and deteriorate over time, which can also lead to fuel oil leaks. Solder with a low melting point can quickly fail in a fire. Cast-iron fittings are very brittle and can be damaged by impact.

M2202.3 Flexible connectors. Flexible metallic hoses shall be *listed* and *labeled* in accordance with UL 536 and shall be installed in accordance with their *listing* and *labeling* and the manufacturer's installation instructions. Connectors made from combustible materials shall not be used inside of buildings or above ground outside of buildings.

❖ Flexible metal hoses are used to reduce the transfer of vibrations and to compensate for movements in the system. These connectors must be listed and labeled in accordance with UL 536 for their intended use so that they are compatible with the environment and the type of fuel used. Combustible connectors are prohibited for installations inside the building and are also prohibited anywhere above ground because of concern for fire damage and longevity of the material.

SECTION M2203
INSTALLATION

M2203.1 General. Piping shall be installed in a manner to avoid placing stresses on the piping, and to accommodate expansion and contraction of the piping system.

❖ Changes in temperatures can result in substantial longitudinal movement of piping, especially in long runs. Provisions must be made to allow expansion or contraction without creating stresses on the system. Otherwise, leakage or failure of pipe, pipe supports and pipe joints can occur.

M2203.2 Supply piping. Supply piping used in the installation of oil burners and *appliances* shall be not smaller than $^3/_8$-inch (9 mm) pipe or $^3/_8$-inch (9 mm) outside diameter tubing. Copper tubing and fittings shall be a minimum of Type L.

❖ This section requires that the fuel oil supply piping be of a size capable of delivering the maximum fuel demand of the appliance it supplies at the recommended working pressure of the appliance burner. For purposes of flow volume and pressure control, pumps may supply more fuel oil than an appliance consumes, in which case the excess oil must be routed back to the tank through a return line. Because the amount returning to the tank is less than what was supplied, the return line pipe diameter may be smaller than that of the supply pipe.

M2203.3 Fill piping. Fill piping shall terminate outside of buildings at a point not less than 2 feet (610 mm) from any building opening at the same or lower level. Fill openings shall be equipped with a tight metal cover.

❖ Requiring fill pipe terminations to be outside buildings and at least 2 feet (610 mm) from building openings that are at the same level or lower than the fill pipe reduces the possibility that liquid or vapor spillage during delivery will enter the building and come into contact with a source of ignition. Likewise, fill pipes must be arranged to reduce the possibility of liquid spilling onto the ground or pavement upon completion of the filling operation, which could result in vapors traveling to an ignition source. One method of managing such spills is to install a spill containment device at the end of the fill pipe. These devices are usually equipped with a water-tight cover and are essentially reservoirs designed to catch any spilled fuel oil and retain it for proper disposal. Some models of these devices have a manual drain valve that allows the captured liquid to drain through the fill pipe and into the tank. Other models have a liquid level alarm to warn responsible parties on the premises that the reservoir needs to be emptied.

M2203.4 Vent piping. Vent piping shall be not smaller than $1^1/_4$-inch (32 mm) pipe. Vent piping shall be laid to drain toward the tank without sags or traps in which the liquid can collect. Vent pipes shall not be cross connected with fill pipes, lines from burners or overflow lines from auxiliary tanks. The lower end of a vent pipe shall enter the tank through the top and shall extend into the tank not more than 1 inch (25 mm).

❖ Because vent piping can have oil in it under certain circumstances, this section requires that it be laid to drain back to the tank without sags or traps. If oil reaches the vent, it can drain back to the tank without leaving pockets of oil that could block the vent. The vent must enter the tank through the top and not extend more than 1 inch (25.4 mm) into the tank, again to protect against the vent being blocked by fuel oil.

M2203.5 Vent termination. Vent piping shall terminate outside of buildings at a point not less than 2 feet (610 mm), measured vertically or horizontally, from any building opening. Outer ends of vent piping shall terminate in a weatherproof cap or fitting having an unobstructed area at least equal to the cross-sectional area of the vent pipe, and shall be located sufficiently above the ground to avoid being obstructed by snow and ice.

❖ The 2-foot (610 mm) minimum clearance reduces the possibility of vapors entering the building. Protection of the open end of the vent pipe with an approved vent cap or fitting is necessary to prevent rainwater from entering the tank through the vent.

The vent pipe must terminate above ground at a level that will reduce the possibility of the opening being obstructed by snow accumulation.

M2203.6 Cross connection of tanks. Cross connection of two supply tanks, not exceeding 660 gallons (2498 L) aggregate capacity, with gravity flow from one tank to another, shall be acceptable providing that the two tanks are on the same horizontal plane.

❖ Cross connection of two supply tanks, with gravity flow from one tank to another, is acceptable if the two tanks are installed on the same horizontal plane. The capacity of both tanks combined must not exceed 660 gallons (2498 L). Two smaller supply tanks connected together, instead of a single larger tank, allow more flexibility of location, such as with indoor storage tanks.

M2203.7 Corrosion protection. Underground tanks and buried piping shall be protected by corrosion-resistant coatings or special alloys or fiberglass-reinforced plastic.

❖ To minimize the possibility of a hazardous fuel oil spill, underground tanks and buried piping must be protected by corrosion-resistant coating, use of special alloys or fiberglass-reinforced plastic. Materials installed underground and left unprotected are subject to moisture, soil chemistry and stresses that can cause rapid deterioration.

SECTION M2204
OIL PUMPS AND VALVES

M2204.1 Pumps. Oil pumps shall be positive displacement types that automatically shut off the oil supply when stopped. Automatic pumps shall be *listed* and *labeled* in accordance with UL 343 and shall be installed in accordance with their *listing*.

❖ Pumps must be of the positive-displacement type to prevent gravity or siphon delivery of oil when the pump is not operating. Otherwise, oil could drain through the pump and burner and into the combustion chamber, creating a hazard.

M2204.2 Shutoff valves. A *readily accessible* manual shutoff valve shall be installed between the oil supply tank and the burner. Where the shutoff valve is installed in the discharge line of an oil pump, a pressure-relief valve shall be incorporated to bypass or return surplus oil. Valves shall comply with UL 842.

❖ Valves can be used to stop the flow of oil if a leak occurs and can also be used during maintenance or repair of the burners. When the shutoff valve is installed downstream of a pump, the pump becomes capable of overpressurizing the piping system if it pumps against a closed valve. Therefore, if the pump has no built-in bypass, a pressure relief valve must be installed. The fuel oil discharged by the relief valve must be routed back to the tank through a return line to prevent a spill.

M2204.3 Maximum pressure. Pressure at the oil supply inlet to an *appliance* shall be not greater than 3 pounds per square inch (20.7 kPa).

❖ Higher pressures could cause malfunctioning of the appliance and/or leaks if the internal parts of the appliance are not designed to withstand higher pressures.

M2204.4 Relief valves. Fuel-oil lines incorporating heaters shall be provided with relief valves that will discharge to a return line when excess pressure exists.

❖ Fuel-oil systems are sometimes designed to incorporate a heater that decreases the viscosity of the oil and improves flow within the system. Because oil will expand when heated, it can reach pressures beyond those of the design operating pressure. Relief valves are necessary to prevent overpressures.

Chapter 23:
Solar Thermal Energy Systems

User note: Code change proposals to this chapter will be considered by the IRC – Plumbing and Mechanical Code Development Committee during the 2015 (Group A) Code Development Cycle.

General Comments

Chapter 14 contains requirements for the construction, alteration and repair of all systems and components of solar energy systems used for space heating or cooling and domestic hot water heating or processing. The provisions of this chapter are limited to those necessary to achieve safe installations that are relatively hazard free.

A solar energy system can be designed to handle 100 percent of the energy load of a building, although this is rarely accomplished. Because solar energy is a low-intensity energy source and dependent on the weather, it is usually necessary to supplement a solar energy system with traditional energy sources.

Solar heating and cooling systems are classified as either passive or active systems. A passive solar energy system uses solar collectors, a gas or liquid heat-transfer medium and distribution piping or ductwork. This system does not use circulators or fans but relies on natural (gravity) flow. The flow of energy (heat) is by natural convection, conduction, and radiation. Commentary Figures 23(1) and 23(2) show typical passive solar heating applications.

A complete active solar energy system includes solar collectors, a gas or liquid heat transfer medium, piping, ducts, circulators, fans and controls to move the medium to the load or to an energy storage system. Commentary Figure 23(3) shows a typical active solar system used for space and domestic water heating, supplemented by an auxiliary furnace and water heater.

Many design variations exist, such as hybrid systems with components of both active and passive systems. Solar systems can act in harmony with the architectural features of the building and can be very complex, or they can be as simple as glazing [see Commentary Figure 23(4)]. Solar designs can be used to work with, and improve the performance of, conventional space heating systems. For example, a simple passive collection system can greatly improve the performance of an air-to-air or water-to-air heat pump system.

The energy collected by a solar system can be used as it is collected or transferred to some form of thermal storage system. Typical thermal storage systems consist of large quantities of a dense mass, such as stone, masonry or water. Storage mass can also consist of materials in which a phase change occurs, such as chemical compounds that store the latent heat required to liquefy the compounds that are solid at room temperature.

If the energy collected by a solar system is not used or stored for later use, and unless the collection system is designed to withstand the higher temperatures and pressures resulting from the static (no-flow) mode of operation, the collection system must either stop collecting or dump the energy to the building exterior.

The structural loading effect that the collectors and supports have on the building's roof system also must be considered. Collectors can add considerable weight, affect snow accumulation and increase wind loads and uplift forces. As with any roof-mounted equipment or appliances, the installation also must comply with the applicable provisions of the building portions of this code.

Purpose

This chapter establishes provisions for the safe installation, operation and repair of solar energy systems. Although these systems use components similar to those of conventional mechanical equipment, many of these provisions are unique to solar energy systems.

SECTION M2301
THERMAL SOLAR ENERGY SYSTEMS

M2301.1 General. This section provides for the design, construction, installation, *alteration* and repair of *equipment* and systems using thermal solar energy to provide space heating or cooling, hot water heating and swimming pool heating.

❖ The provisions within the this section cover the design, installation, construction, alteration and repair of equipment for thermal solar systems that heat water for domestic hot water use, space heating and, on rare occasions, space cooling. The *International Residential Code*® (IRC®) also covers the installation of collector systems that provide hot water for swimming pools. There are five major components in active solar heating systems:

- Collector(s) to capture solar energy. Usually, systems used for space heating will have many more collectors than systems designed for domestic water heating only.

- Circulation system to move a fluid between the collectors to a storage tank. The circulation system will either be a direct (open loop) system or an indirect (closed loop) system. A direct system pumps domestic water directly from the storage tank (usually a hot water heater) through the collectors and back to the storage tank. An indirect system circulates an antifreeze fluid, oil, refrigerant or other fluid with a high boiling temperature and low freezing temperature through the collectors and back through a heat exchanger. The heat exchanger keeps the collection fluid separate from the domestic water.

- Storage tank(s) to collect heated water. This may be a standard water heater or a separate unfired hot water storage tank located next to the water heater.
- Backup heating system. For domestic water heating this will usually be a hot water heater. Backup space-heating systems may consist of electric baseboard heating or a gas or electric heater.
- Control system to regulate overall system operation. The control system will include valves and shutoff devices to protect from excessively high temperatures.

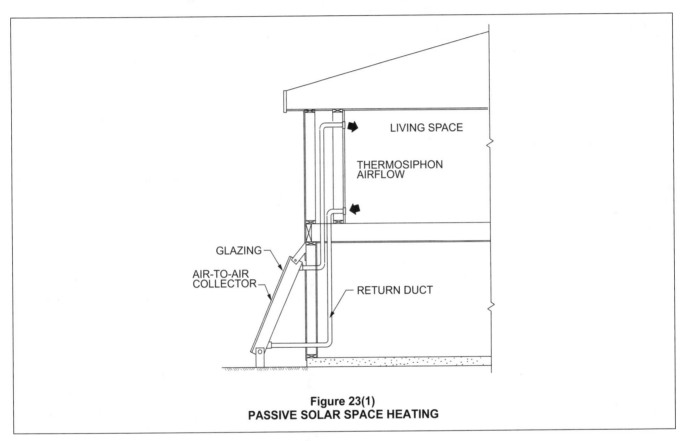

Figure 23(1)
PASSIVE SOLAR SPACE HEATING

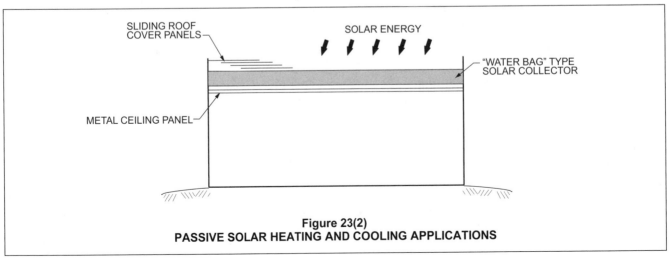

Figure 23(2)
PASSIVE SOLAR HEATING AND COOLING APPLICATIONS

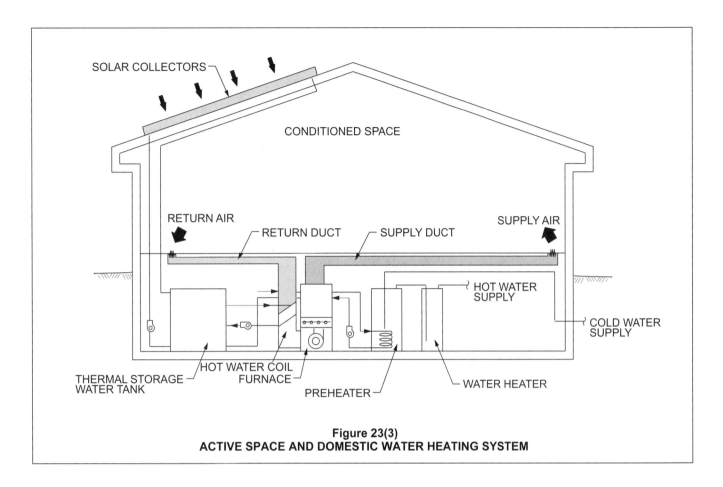

Figure 23(3)
ACTIVE SPACE AND DOMESTIC WATER HEATING SYSTEM

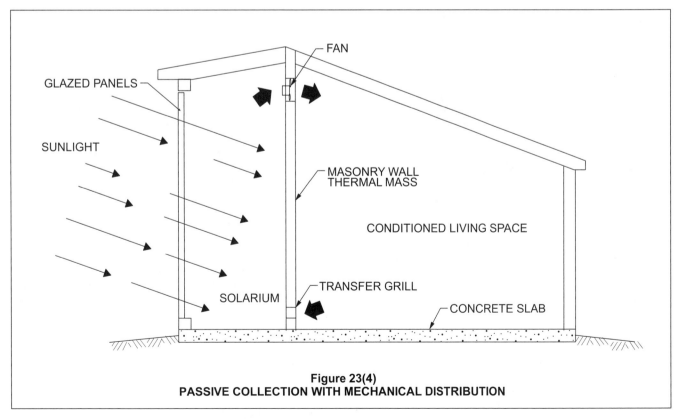

Figure 23(4)
PASSIVE COLLECTION WITH MECHANICAL DISTRIBUTION

M2301.2 Design and installation. The design and installation of thermal solar energy systems shall comply with Sections M2301.2.1 through M2301.2.13.

❖ The requirements of this chapter address the potential hazards to life and property associated with solar installations. The systems must be installed in accordance with the manufacturer's installation instructions. Components of the solar system must also conform to other applicable sections of the code.

M2301.2.1 Access. Solar energy collectors, controls, dampers, fans, blowers and pumps shall be accessible for inspection, maintenance, repair and replacement.

❖ This section requires that solar collectors, controls, dampers, fans, blowers and pumps associated with the solar system be accessible for inspection, maintenance, repair and replacement. The collection system pumps and controls are often located in the garage or pool equipment building, allowing easy access for the inspector, and are mounted in close proximity to each other. While not a code requirement, it is beneficial to have operation and maintenance manuals available to the owner once the system is installed.

M2301.2.2 Collectors and panels. Solar collectors and panels shall comply with Sections M2301.2.2.1 and M2301.2.2.2.

❖ See Sections M2301.2.2.1 and M2301.2.2.2.

M2301.2.2.1 Roof-mounted collectors. The roof shall be constructed to support the loads imposed by roof-mounted solar collectors. Roof-mounted solar collectors that serve as a roof covering shall conform to the requirements for roof coverings in Chapter 9 of this code. Where mounted on or above the roof coverings, the collectors and supporting structure shall be constructed of noncombustible materials or fire-retardant-treated wood equivalent to that required for the roof construction.

❖ Roof mounted solar collectors can be designed to be installed integral with the roof, thereby serving as the roof covering. They can also be mounted on or above the roof covering. Systems that are installed as a roof covering must be watertight and properly flashed, in addition to meeting the other requirements applicable to the installation in Chapter 9 of this code. Systems installed on or above the roof covering must be securely fastened to the roof/rafter system. Often, solar collector systems are installed with manufactured metal mounting brackets that can be attached to the roofing system or wood blocking, or with sleepers using lag bolts or other secure mounting systems. The manufacturer's mounting specifications must be followed so that the panels are properly mounted. This is especially important in areas with high wind loads. Also, the weight of the collector system must be checked (when operational and filled with fluid) to confirm that the roof structure is designed to meet the increased load.

Most manufactured panel systems used for domestic hot water are constructed using a metal housing to hold the collection system. This housing must meet the requirements for noncombustible materials. Panel systems constructed on site by the owner/builder that use a housing made of wood may or may not meet the requirements for noncombustible materials.

M2301.2.2.2 Collector sensors. Collector sensor installation, sensor location and the protection of exposed sensor wires from ultraviolet light shall be in accordance with SRCC 300.

❖ The referenced standard contains installation instructions for sensor devices used in collectors and collector arrays. Sunlight will degrade many materials, especially certain plastics.

M2301.2.3 Pressure and temperature relief valves and system components. System components containing fluids shall be protected with temperature and pressure relief valves or pressure relief valves. Relief devices shall be installed in sections of the system so that a section cannot be valved off or isolated from a relief device. Direct systems and the potable water portion of indirect systems shall be equipped with a relief valve in accordance with Section P2804. For indirect systems, pressure relief valves in solar loops shall comply with SRCC 300. System components shall have a working pressure rating of not less than the setting of the pressure relief device.

❖ Pressure and temperature relief valves are required to prevent collector, piping and equipment failures caused by excessive temperatures and pressures. Such failures could involve violent explosions, similar to those associated with boilers and water heaters. The collection surface, or absorber plate, can easily heat up to 200°F (93°C) on a freezing winter day and up to 400°F (204°C) on a summer day when the water is stagnant and not circulated through the collector. These relief valves are often installed as part of the collector system. Multiple relief valves could be necessary where there are separate piping system circuits.

M2301.2.4 Vacuum relief. System components that might be subjected to pressure drops below atmospheric pressure during operation or shutdown shall be protected by a vacuum-relief valve.

❖ System components that are subjected to pressure drops below atmospheric pressure during operation or shutdown can be damaged by the force of atmospheric pressure acting on the external side. Partial vacuums can also interfere with system drain-down. Vacuum relief valves can be used in conjunction with freeze protection valves that drain the collector panels when the temperature reaches close to 32°F (0°C) or a drain-down system that automatically drains the collector fluid when the system is shut off.

M2301.2.5 Piping insulation. Piping shall be insulated in accordance with the requirements of Chapter 11. Exterior insulation shall be protected from ultraviolet degradation. The entire solar loop shall be insulated. Where split-style insulation is used, the seam shall be sealed. Fittings shall be fully insulated.

Exceptions:

1. Those portions of the piping that are used to help prevent the system from overheating shall not be required to be insulated.

2. Those portions of piping that are exposed to solar radiation, made of the same material as the solar collector absorber plate and are covered in the same manner as the solar collector absorber, or that are used to collect additional solar energy, shall not be required to be insulated.

3. Piping in thermal solar systems using unglazed solar collectors to heat a swimming pool shall not be required to be insulated.

❖ Piping insulation will improve the efficiency of the system. Sunlight will degrade many types of pipe insulation, such as closed cell plastic insulation. Split-type insulation refers to insulation tubes that are split down their length to allow them to be fitted over installed piping. The split seam must be permanently sealed. The exceptions recognize portions of the piping that are meant to absorb solar energy or dissipate heat energy.

M2301.2.6 Protection from freezing. System components shall be protected from damage resulting from freezing of heat-transfer liquids at the winter design temperature provided in Table R301.2(1). Freeze protection shall be provided by heating, insulation, thermal mass and heat transfer fluids with freeze points lower than the winter design temperature, heat tape or other *approved* methods, or combinations thereof.

Exception: Where the winter design temperature is greater than 32°F (0°C).

❖ In climates where the design winter temperature is 32°F (0°C) or less, the system must be protected from freezing. The greatest damage that can occur to the collector system is for water to freeze in the collection system, causing the collectors, components and piping to rupture. To meet the intent of the code, an indirect system can be protected with antifreeze heat transfer fluids or the system can be a type that drains the outdoor system components when the temperature drops to a specific temperature.

M2301.2.7 Storage tank sensors. Storage tank sensors shall comply with SRCC 300.

❖ Storage tanks store the energy collected from the sun for eventual use external to the storage tank. The storage might be domestic potable water or it might be nonpotable heat transfer fluid. Tank sensors communicate the storage temperatures to the system's pumps and controllers.

M2301.2.8 Expansion tanks. Expansion tanks in solar energy systems shall be installed in accordance with Section M2003 in solar collector loops that contain pressurized heat transfer fluid. Where expansion tanks are used, the system shall be designed in accordance with SRCC 300 to provide an expansion tank that is sized to withstand the maximum operating pressure of the system.

Exception: Expansion tanks shall not be required in *drain-back systems*.

❖ This section requires the installation of pressure expansion tanks in closed loop systems. The text is stating two different requirements; one being that an expansion tank must be installed and the other being that the installation must comply with Section M2003. Closed loop systems cannot accommodate the increased volume of expanded liquid that results from a rise in temperature. Without an expansion tank, the expanding fluid would rupture a system component or a pressure relief device would have to routinely open and discharge heat transfer fluid. Manufactured systems that are preengineered based on the number of panels in the system, storage tank size and circulation pump size often come with a specified expansion tank engineered for the system. The expansion tank is usually installed next to the pumps, storage tank and controls. Expansion tanks are installed in closed loop hydronic heating systems for the same reasons.

M2301.2.9 Roof and wall penetrations. Roof and wall penetrations shall be flashed and sealed in accordance with Chapter 9 of this code to prevent entry of water, rodents and insects.

❖ Roof and wall penetrations must be sealed during and after installation of the solar systems. At a minimum, there will be roof penetrations for the supply and return piping going to and from the collector panels. Systems mounted above the roof covering will require penetrations into the rafter system or blocking to attach the collectors to the roof, and care must be taken to seal the holes with a waterproof sealant. Chapter 9 states flashing and waterproofing requirements for several roof types. Wall penetrations must also be sealed to prevent the entry of water, rodents and insects.

M2301.2.10 Description and warning labels. Solar thermal systems shall comply with description label and warning label requirements of Section M2301.2.11.2 and SRCC 300.

❖ Labeling of system components provides valuable guidance to homeowners, occupants and service personnel and also provides necessary warnings relative to high temperatures and pressures and required actions to prevent hazards.

M2301.2.11 Solar loop. Solar loops shall be in accordance with Sections M2301.2.11.1 and M2301.2.11.2.

❖ See Sections M2301.2.11.1 and M2301.2.11.2.

M2301.2.11.1 Solar loop isolation. Valves shall be installed to allow the solar collectors to be isolated from the remainder of the system.

❖ Isolation of the collectors could be necessary for servicing, repair or replacement. A loop is a circuit of piping and components that circulates a fluid in a closed circuit (loop). In an indirect system, the collector loop circulates between the collectors and a heat exchanger or storage tank. This section requires isolation of the collector loop, not necessarily isolation of the collectors from their collector loop. If the collectors themselves were isolated from loop piping, they might be isolated from the required expansion tanks and pressure and temperature relief devices, thereby creating a serious hazard.

M2301.2.11.2 Drain and fill valve labels and caps. Drain and fill valves shall be labeled with a description and warning that identifies the fluid in the solar loop and a warning that the fluid might be discharged at high temperature and pressure. Drain caps shall be installed at drain and fill valves.

❖ Access points into a collector loop, such as fill valve ports and drain valve ports, must be labeled to warn of the potential hazards relative to chemical content of the loop and the pressures and temperatures in the loop. Personnel and occupants could be injured by opening valves if they are unaware of the dangers and fail to take the necessary precautions. All access valves, such as hose bibb drain valves, must be capped to prevent accidental discharge. Discharge of system fluid would require two actions; removing a cap and opening one or more valves.

M2301.2.12 Maximum temperature limitation. Systems shall be equipped with means to limit the maximum water temperature of the system fluid entering or exchanging heat with any pressurized vessel inside the *dwelling* to 180°F (82°C). This protection is in addition to the required temperature- and pressure-relief valves required by Section M2301.2.3.

❖ To prevent potential problems with clearances to combustibles, to prevent burn injuries resulting from human contact with piping or pressure vessels and to prevent overheating of the loads served by the solar system, the system fluid temperature is limited to 180°F (82°C) where the fluid enters or conveys heat energy to any pressurized vessel inside of a dwelling. The method of limiting the temperature is to be determined by the designer or system installer. Possible methods include a system to bypass the solar collectors and/or a rejection heat exchanger outside of the dwelling.

M2301.2.13 Thermal storage unit seismic bracing. In Seismic Design Categories D_0, D_1 and D_2 and in townhouses in Seismic Design Category C, thermal storage units shall be anchored in accordance with Section M1307.2.

❖ See the commentary to Section M1307.2.

M2301.3 Labeling. *Labeling* shall comply with Sections M2301.3.1 and M2301.3.2.

❖ The designer, installer, owner and building official must know the quality, design and application limitations of the collectors and thermal storage units. This section requires that they be labeled. The label is a form of quality assurance and lists specifications that apply to the equipment, as well as any necessary information for its installation and application.

M2301.3.1 Collectors and panels. Solar thermal collectors and panels shall be listed and labeled in accordance with SRCC 100 or SRCC 600. Collectors and panels shall be *listed* and *labeled* to show the manufacturer's name, model number, serial number, collector weight, collector maximum allowable temperatures and pressures, and the type of heat transfer fluids that are compatible with the collector or panel. The

label shall clarify that these specifications apply only to the collector or panel.

❖ Solar collectors must be listed and labeled to the specified standards and must be labeled to indicate the manufacturer's name, model number, serial number, collector weight, collector maximum allowable temperatures and pressures and the type of heat transfer fluids that are compatible with the collector. The label must state that the specifications on the label do not apply to components external to the collectors, otherwise, dangerous assumptions could be implied about the other portions of the solar system. The Solar Rating and Certification Corporation (SRCC) is one entity that provides a third party certification that will meet the requirements of this provision. The *Directory of SRCC Certified Solar Collector System Ratings, Directory of SRCC Certified Water Heating System Ratings*, and *Summary of SRCC Certified Solar Collector and Water Heating System Ratings* are published by the SRCC annually and list all the information required for this provision except for the serial number, which can be found on the collector.

M2301.3.2 Thermal storage units. Pressurized thermal storage units shall be *listed* and *labeled* to show the manufacturer's name, model number, serial number, storage unit maximum and minimum allowable operating temperatures and pressures, and the type of heat transfer fluids that are compatible with the storage unit. The *label* shall clarify that these specifications apply only to the thermal storage unit.

❖ Pressurized thermal storage units must bear a listing and label that discloses the manufacturer's name, model, serial number and important operating characteristics such as operating temperatures and pressures, and the type of heat transfer fluid. The information is important for installation, application, servicing, repairs and inspection. The label must state that the specifications on the label do not apply to components external to the thermal storage unit, otherwise dangerous assumptions could be implied about the other portions of the solar system.

M2301.4 Heat transfer gasses or liquids and heat exchangers. *Essentially toxic transfer fluids*, ethylene glycol, flammable gases and flammable liquids shall not be used as heat transfer fluids. Heat transfer gasses and liquids shall be rated to withstand the system's maximum design temperature under operating conditions without degradation. Heat exchangers used in solar thermal systems shall comply with Section P2902.5.2 and SRCC 300.

Heat transfer fluids shall be in accordance with SRCC 300. The flash point of the heat transfer fluids utilized in solar thermal systems shall be not less than 50°F (28°C) above the design maximum nonoperating or no-flow temperature attained by the fluid in the collector.

❖ Flammable gases, flammable liquids, toxic chemicals and ethylene glycol must not be used as heat transfer fluids because they each pose significant risks. Ethylene glycol is automotive antifreeze, is very toxic and is an unacceptable risk as a heat transfer fluid. Chemicals introduced into a solar system must be compatible with the piping and system components and must also be capable of tolerating, without degradation, the temperatures to which they will be exposed.

M2301.5 Backflow protection. Connections from the potable water supply to solar systems shall comply with Section P2902.5.5.

❖ The potable water supply connected to a solar system must be protected from backflow because this is a direct connection between a potable and a nonpotable piping system, which poses significant health risks. See Section P2902.5.5.

M2301.6 Filtering. Air provided to occupied spaces that passes through thermal mass storage systems by mechanical means shall be filtered for particulates at the outlet of the thermal mass storage system.

❖ Some thermal storage systems utilize mass, such as stone aggregate, to store heat energy and force air through the stone bed to condition the air. Before such conditioned air is delivered to an occupied space, it must filtered to remove any particulate matter picked up from the mass, whatever the mass may be.

M2301.7 Solar thermal systems for heating potable water. Where a solar thermal system heats potable water to supply a potable hot water distribution system, the solar thermal system shall be in accordance with Sections M2301.7.1, M2301.7.2 and P2902.5.5.

❖ See Sections M2301.7.1, M2301.7.2 and P2902.5.5.

M2301.7.1 Indirect systems. Heat exchangers that are components of indirect solar thermal heating systems shall comply with Section P2902.5.2.

❖ See Section P2902.5.2. Indirect systems consist of a collector loop that conveys the energy from the collectors to a heat exchanger that transfers that energy to the potable water. It is considered as indirect because there is a heat exchange coil or plate type exchanger that isolates the potable water system from the solar collector system. A direct system would pass the potable water through the collectors without a heat exchanger. See Section M2301.7.2.

M2301.7.2 Direct systems. Where potable water is directly heated by a solar thermal system, the pipe, fittings, valves and other components that are in contact with the potable water in the solar heating system shall comply with the requirements of Chapter 29.

❖ Because a direct solar system heats the potable water without an isolating heat exchanger, the potable water is in contact with the collectors, the collector loop piping, the pumps, expansion tanks, etc. Anything in contact with the potable water must not contaminate the water, therefore, the components of a direct system have to meet the same requirements as potable water piping and potable water system components. See Chapter 29.

Chapter 24:
Fuel Gas

User note: Code change proposals to this chapter will be considered by the IRC – Plumbing and Mechanical Code Development Committee during the 2015 (Group A) Code Development Cycle.

The text of this chapter is extracted from the 2015 edition of the *International Fuel Gas Code* and has been modified where necessary to conform to the scope of application of the *International Residential Code for One- and Two-Family Dwellings*. The section numbers appearing in parentheses after each section number are the section numbers of the corresponding text in the *International Fuel Gas Code.*

General Comments

This chapter covers all installations of gas piping, gas appliance installation and gas appliance venting systems. It is extracted from the *International Fuel Gas Code®* (IFGC®) and is identical in intent. This chapter contains its own gas-specific coverage of combustion air, clearance reduction methods, chimneys and vents, and appliance installation. The dual section numbering system allows this text to be cross-referenced with the IFGC. Chapters 12, 13, 14 and 20 also contain requirements applicable to gas appliance installations.

The IFGC itself is segregated by section number into two categories: code and standard. In that document, code sections are identified as IFGC; standards sections are identified as IFGS. The IFGS is a copyrighted work of the American Gas Association.

The commentary text of this chapter is produced and copyrighted by the International Code Council® (ICC®).

Purpose

Chapter 24 intends to protect occupants and their property from fire, explosion and health hazards that could result from the improper installation of gas piping systems, gas appliances and appliance venting systems.

SECTION G2401 (101)
GENERAL

G2401.1 (101.2) Application. This chapter covers those fuel gas *piping systems*, fuel-gas *appliances* and related accessories, *venting systems* and *combustion air* configurations most commonly encountered in the construction of one- and two-family dwellings and structures regulated by this *code*.

Coverage of *piping systems* shall extend from the *point of delivery* to the outlet of the *appliance* shutoff *valves* (see definition of *"Point of delivery"*). *Piping systems* requirements shall include design, materials, components, fabrication, assembly, installation, testing, inspection, operation and maintenance. Requirements for gas *appliances* and related accessories shall include installation, combustion and ventilation air and venting and connections to *piping systems*.

The omission from this chapter of any material or method of installation provided for in the *International Fuel Gas Code* shall not be construed as prohibiting the use of such material or method of installation. Fuel-gas *piping systems*, fuel-gas *appliances* and related accessories, *venting systems* and *combustion air* configurations not specifically covered in these chapters shall comply with the applicable provisions of the *International Fuel Gas Code*.

Gaseous hydrogen systems shall be regulated by Chapter 7 of the *International Fuel Gas Code.*

This chapter shall not apply to the following:

1. Liquified natural gas (LNG) installations.

2. Temporary LP-*gas piping* for buildings under construction or renovation that is not to become part of the permanent *piping system*.

3. Except as provided in Section G2412.1.1, gas *piping*, meters, gas pressure regulators, and other appurtenances used by the serving gas supplier in the distribution of gas, other than undiluted LP-gas.

4. Portable LP-gas *appliances* and *equipment* of all types that is not connected to a fixed fuel *piping system*.

5. Portable fuel cell *appliances* that are neither connected to a fixed *piping system* nor interconnected to a power grid.

6. Installation of hydrogen gas, LP-gas and compressed natural gas (CNG) systems on vehicles.

❖ This section describes the types of fuel gas systems to which the code is intended to apply and specifically lists those systems to which the code does not apply.

The applicability of the code spans from the initial

design of fuel gas systems, through the installation and construction phases, and into the maintenance of operating systems. Chapter 24 of the *International Residential Code®* (IRC®) covers fuel gas systems and is a duplication of the applicable IFGC text. The provisions of Chapter 24 of the IRC are identical to those in the IFGC. Chapter 24 contains only the IFGC text that is applicable to one- and two-family dwellings and townhouses. If something is encountered on plans or in the field that is not covered by Chapter 24, the provisions of the IFGC take over. Note that each Chapter 24 section number is followed by the corresponding IFGC section number. This makes it easy to cross-reference Chapter 24 to its source, the IFGC.

SECTION G2402 (201)
GENERAL

G2402.1 (201.1) Scope. Unless otherwise expressly stated, the following words and terms shall, for the purposes of this chapter, have the meanings indicated in this chapter.

❖ In the application of the code, the terms used have the meanings given in Section G2403.

G2402.2 (201.2) Interchangeability. Words used in the present tense include the future; words in the masculine gender include the feminine and neuter; the singular number includes the plural and the plural, the singular.

❖ Although the definitions contained in Section G2403 are to be taken literally, gender and tense are considered to be interchangeable; thus, any grammatical inconsistencies within the code text will not hinder the understanding or enforcement of the requirements.

G2402.3 (201.3) Terms defined in other codes. Where terms are not defined in this code and are defined in the *International Building Code, International Fire Code, International Mechanical Code, International Fuel Gas Code* or *International Plumbing Code*, such terms shall have meanings ascribed to them as in those *codes*.

❖ When a word or term appears in the code that is not defined in this chapter, other references may be used to find its definition. These include the *International Building Code®* (IBC®), the *International Fire Code®* (IFC®), the *International Mechanical Code®* (IMC®) and the *International Plumbing Code®* (IPC®). These codes contain additional definitions (some parallel and duplicative) that may be used in the enforcement of this code or in the enforcement of the other codes by reference.

SECTION G2403 (202)
GENERAL DEFINITIONS

❖ Definitions of terms can help in the understanding and application of the code requirements. The purpose for including here those definitions that are most closely associated with the subject matter of this chapter is to provide more convenient access to them without having to refer back to Chapter 2. For convenience, these terms are also listed in Chapter 2 with a cross reference to this section. The use and application of all defined terms, including those defined in this section, are set forth in Section 201.

ACCESS (TO). That which enables a device, *appliance* or *equipment* to be reached by ready *access* or by a means that first requires the removal or movement of a panel, door or similar obstruction (see also "Ready *access*").

AIR CONDITIONER, GAS-FIRED. A gas-burning, automatically operated *appliance* for supplying cooled and/or dehumidified air or chilled liquid.

❖ Gas-fired air conditioners are not specifically referred to in the code, but would include systems fueled with natural or propane gas including absorption type units and reciprocating units powered by internal combustion engines.

AIR CONDITIONING. The treatment of air so as to control simultaneously the temperature, humidity, cleanness and distribution of the air to meet the requirements of a conditioned space.

AIR, EXHAUST. Air being removed from any space or piece of *equipment* or *appliance* and conveyed directly to the atmosphere by means of openings or ducts.

❖ Exhaust air may be from a space, an appliance or a piece of equipment. Exhaust air systems are terminated outside of the building, in some cases after the exhaust air has been treated to remove any harmful emissions. Exhaust air is not recirculated.

AIR-HANDLING UNIT. A blower or fan used for the purpose of distributing supply air to a room, space or area.

❖ In addition to blowers, air-handling units may contain heat exchangers, filters and means to control air volume.

AIR, MAKEUP. Any combination of outdoor and transfer air intended to replace exhaust air and exfiltration.

❖ Makeup air is not to be confused with combustion air. Makeup air replaces the air being exhausted through such systems as bathroom and toilet exhausts, kitchen exhaust hoods, and clothes dryer exhaust systems. Refer to Section G2439.5 for specific requirements for makeup air. Exhaust systems cannot function at design capacity without adequate volumes of makeup air to replace the air being exhausted.

ALTERATION. A change in a system that involves an extension, addition or change to the arrangement, type or purpose of the original installation.

❖ An alteration is any modification or change made to an existing installation. For example, increasing the size of piping for a portion of the system to accommodate different appliances would be an alteration.

ANODELESS RISER. A transition assembly in which plastic *piping* is installed and terminated above ground outside of a building.

❖ As the name implies, these riser assemblies protect the steel riser from corrosion by means other than

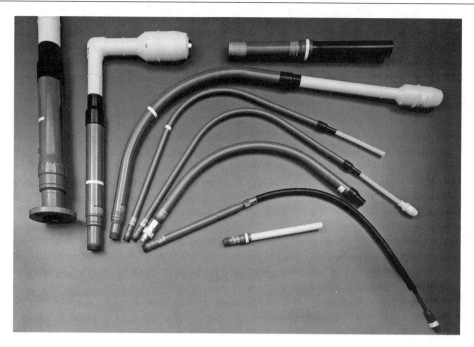

Photo courtesy of Perfection Corporation

Figure G2403(1)
ANODELESS RISERS

cathodic protection that involves a sacrificial anode. Some anodeless risers allow the termination of plastic piping above ground by encasing the piping in a steel conduit equipped with a plastic-to-steel transition fitting [see Commentary Figure 2403(1)].

APPLIANCE. Any apparatus or device that utilizes a fuel or raw material to produce light, heat, power, refrigeration or air conditioning.

❖ An appliance is a manufactured component or assembly of components that converts one source of energy into a different form of energy to serve a specific purpose. The term "appliance" generally refers to residential- and commercial-type utilization equipment that is manufactured in standardized sizes or types. The term "appliance" is generally not associated with industrial-type equipment. For the application of the code provisions, the terms "appliance" and "equipment" are not interchangeable.

Examples of appliances regulated by the code include furnaces; boilers; water heaters; room heaters; decorative gas log sets; cooking equipment; clothes dryers; pool, spa and hot tub heaters; unit heaters; ovens and similar gas-fired appliances (see the definition of "Equipment"). Note that the definition intends to state that a fuel can be used as a raw material to produce energy rather than the apparent statement that an appliance utilizes raw materials that are not fuels. For example, a fuel-cell system could use natural gas as a raw material for the production of hydrogen for use in the fuel cell. The definition was revised to recognize that in a few instances in the code, the term "appliance" is used to describe appliances that are fueled by other than gaseous fuels, whereas generally, the term describes gas-fired appliances.

APPLIANCE, AUTOMATICALLY CONTROLLED. Appliances equipped with an automatic *burner* ignition and safety shut-off device and other automatic devices, which accomplish complete turn-on and shut-off of the gas to the *main burner* or *burners*, and graduate the gas supply to the *burner* or *burners*, but do not affect complete shut-off of the gas.

❖ With respect to the code, an automatically controlled appliance is an appliance that is cycled through its operation by controls, such as thermostats, pressure switches and timers, without manual intervention. A residential water heater is automatically controlled: a four-burner cooktop is not.

APPLIANCE, FAN-ASSISTED COMBUSTION. An *appliance* equipped with an integral mechanical means to either draw or force products of combustion through the combustion chamber or heat exchanger.

❖ Fan-assisted appliances are a specific type of Category I appliance, typically furnaces and boilers. They are not to be confused with Category II, III or IV appliances; power-vented appliances; or appliances served by exhausters. As the name implies, these appliances use a fan (blower) to "assist" the combustion process by helping the flue gases to overcome the internal flow resistance of the appliance heat exchanger. Some fan-assisted appliances are field-convertible to Category III appliances.

APPLIANCE, UNVENTED. An *appliance* designed or installed in such a manner that the products of combustion are not conveyed by a vent or *chimney* directly to the outside atmosphere.

❖ Direct-fired makeup air heaters, gas-fired infrared radiant unit heaters, gas-fired cooking equipment and small gas-fired room heaters are examples of appliances that are sometimes unvented. With the exception of direct-fired equipment, the appliances listed herein can also be of the vented type, depending on the design and type of appliance and the particular application. Unvented appliances must be designed, installed and operated to avoid both the depletion of the oxygen supply and the accumulation of toxic combustion by products (see Sections G2425.8, G2445 and G2451).

APPLIANCE, VENTED. An *appliance* designed and installed in such a manner that all of the products of combustion are conveyed directly from the *appliance* to the outside atmosphere through an *approved chimney* or vent system.

❖ Most fuel-fired appliances are designed to vent the products of combustion to the outdoors through one or more specific types of vent or chimney (see Section G2425.2).

APPROVED. Acceptable to the *code official*.

❖ As related to the process of acceptance of fuel gas related installations, including materials, equipment and construction systems, this definition identifies where ultimate authority rests. Whenever this term is used, it means that only the enforcing authority can accept a specific installation or component as complying with the code. Research reports prepared and published by the ICC may be used by code officials to aid in their review and approval of the material or method described in the report. Publishing a report does not indicate automatic "approval" for the material or method described in the report. When the code states that an item or method "shall be approved," it does not mean that the code official is obligated to allow or accept it. Rather, it means that the code official must determine whether the item or method is acceptable; that is, the code official must make the decision to allow or disallow; accept or reject.

APPROVED AGENCY. An established and recognized agency that is regularly engaged in conducting tests or furnishing inspection services, where such agency has been *approved* by the *code official*.

ATMOSPHERIC PRESSURE. The pressure of the weight of air and water vapor on the surface of the earth, approximately 14.7 pounds per square inch (psia) (101 kPa absolute) at sea level.

❖ This is the "standard" atmospheric pressure at sea level. The actual atmospheric pressure varies with climate conditions. Barometric pressure is a measure of the actual atmospheric pressure, usually given as inches of mercury. One standard atmosphere (14.7 psi) is equal to 29.9 inches of mercury (101.4 kPa) at 32°F (0°C). Absolute pressure includes atmospheric pressure. Gauge pressure is the pressure above atmo-

spheric pressure. If not otherwise defined, stated pressure is the pressure above atmospheric pressure (gauge pressure). A vacuum (partial or complete) is a pressure below atmospheric pressure.

AUTOMATIC IGNITION. Ignition of gas at the *burner(s)* when the gas controlling device is turned on, including reignition if the flames on the *burner(s)* have been extinguished by means other than by the closing of the gas controlling device.

❖ As opposed to manual ignition, automatic ignition is accomplished by pilot burners, spark ignitors, hot surface ignitors and similar means.

BAROMETRIC DRAFT REGULATOR. A balanced *damper* device attached to a *chimney,* vent *connector,* breeching or flue gas manifold to protect combustion *appliances* by controlling *chimney* draft. A double-acting *barometric draft regulator* is one whose balancing *damper* is free to move in either direction to protect combustion *appliances* from both excessive *draft* and backdraft.

❖ These units automatically open or close depending on the difference between the internal vent pressure and the atmospheric pressure. See the commentary to the definition of "Atmospheric pressure." Excessive negative pressure in the vent will cause excessive draft; therefore, the draft regulator will open as a result of the higher atmospheric pressure on the exterior of the damper, allowing air to enter the vent and thereby increase the internal vent pressure. With excessive positive internal vent pressure, a double-acting regulator will open to relieve the pressure, lowering the vent internal pressure [see Commentary Figure G2403(2)].

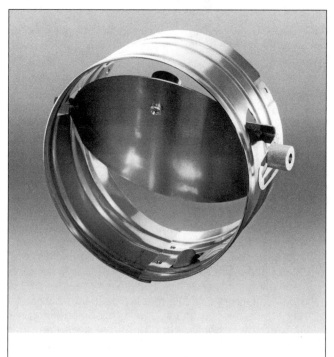

Photo courtesy of Tjernlund Products, Inc.

**FIGURE G2403(2)
BAROMETRIC DRAFT REGULATOR**

BOILER, LOW-PRESSURE. A self-contained *appliance* for supplying steam or hot water.

❖ Low-pressure boilers operate at pressures less than or equal to 15 pounds per square inch (psi) (103 kPa) for steam and 160 psi (1103 kPa) for water. High-pressure boilers operate at pressures exceeding those pressures.

Boilers are usually manufactured of steel, cast iron or copper and are used to transfer heat from the combustion of a fuel or from an electric-resistance element to water to make steam or pressurized hot water for heating or other process or power purposes.

Boilers are usually installed in closed systems where the heat transfer medium is recirculated and retained within the system.

Boilers must be labeled and installed in accordance with the manufacturer's installation instructions and the applicable sections of the code. Boilers are rated in accordance with standards published by the American Society of Mechanical Engineers (ASME), the Hydronics Institute, the Steel Boiler Institute (SBI), the Canadian Standards Association (CSA) and the American Boiler Manufacturers Association (ABMA). Boilers can be classified by working temperature, working pressure, type of fuel used (or electric boilers), materials of construction and whether or not the heat transfer medium changes phase from a liquid to a vapor [see Commentary Figure G2403(3)].

Hot water heating boiler. A boiler in which no steam is generated, from which hot water is circulated for heating purposes and then returned to the boiler, and that operates at water pressures not exceeding 160 pounds per square inch gauge (psig) (1100 kPa gauge) and at water tempera-

WEIL-McLAIN
MODEL LGB SERIES 2 BOILER

TO DETERMINE BOILER SIZE, COUNT THE NUMBER OF SECTIONS OR MEASURE THE JACKET LENGTH. CHECK BOX NEXT TO BOILER SIZE INSTALLED.

MODEL NUMBER	NUMBER OF SECTIONS	JACKET LENGTH INCHES	MIN. RELIEF VALVE CAP. LBS/HR OR MBH	INPUT BTU/HR	MINIMUM INPUT BTU/HR	CSA GROSS OUTPUT BTU/HR	Steam Sq. Ft.	Steam MBH	Water MBH
☐ LGB-4	4	21	325	400,000	---	324,000	1013	243	282
☐ LGB-5	5	26	422	520,000	---	421,200	1317	316	366
☐ LGB-6	6	31	527	650,000	325,000	526,500	1646	395	458
☐ LGB-7	7	36	632	780,000	390,000	631,800	1975	474	549
☐ LGB-8	8	41	738	910,000	455,000	737,100	2304	553	641
☐ LGB-9	9	46	843	1,040,000	520,000	842,400	2633	632	733
☐ LGB-10	10	51	948	1,170,000	585,000	947,700	2965	711	824
☐ LGB-11	11	56	1053	1,300,000	650,000	1,053,000	3292	790	916
☐ LGB-12	12	61	1159	1,430,000	715,000	1,158,300	3621	869	1007
☐ LGB-13	13	66	1264	1,560,000	780,000	1,263,600	3954	949	1099
☐ LGB-14	14	71	1369	1,690,000	845,000	1,368,900	4313	1035	1190
☐ LGB-15	15	76	1475	1,820,000	910,000	1,474,200	4679	1123	1282
☐ LGB-16	16	81	1580	1,950,000	975,000	1,579,500	5046	1211	1373
☐ LGB-17	17	86	1685	2,080,000	1,040,000	1,684,800	5408	1298	1465
☐ LGB-18	18	91	1791	2,210,000	1,105,000	1,790,100	5775	1386	1557
☐ LGB-19	19	96	1896	2,340,000	1,170,000	1,895,400	6125	1470	1648
☐ LGB-20	20	101	2001	2,470,000	1,235,000	2,000,700	6471	1553	1740
☐ LGB-21	21	106	2106	2,600,000	1,300,000	2,106,000	6813	1635	1831
☐ LGB-22	22	111	2212	2,730,000	1,365,000	2,211,300	7155	1717	1923
☐ LGB-23	23	116	2317	2,860,000	1,430,000	2,316,600	7496	1799	2014

CERTIFIED BY
WEIL-McLAIN
523 S New Street
Eden, North Carolina
27288-3623

MAWP, WATER 50 PSI
MAWP, STEAM 15 PSI
MAX. WATER TEMP. 250°F

Electrical Input Less Than
12 Amperes, 120 Volts, & 60 Hertz.
Canadian Registration No. H7268.51234679T
MEA Number: 333-85-E
Design Certified Under ANSI Z21.13a 2005• CSA 4.9a 2005
Low Pressure Boiler

IMPORTANT
Installation not complete unless gas information label attached here.
Labels are attached to gas train in gas control carton.

- For installation on noncombustible flooring.
- Provide service clearances and minimum 24" between jacket and any combustible wall(s) and ceiling. Install in space large in comparison to size of boiler.
- Vent Category I.

550-223-880 (0606)

Figure courtesy of Weil-McLain

Figure G2403(3)
LABEL FOR LOW-PRESSURE BOILER

tures not exceeding 250°F (121°C) at or near the boiler outlet.

❖ Hot water heating boilers are normally part of a closed system in which the heated water is circulated through fan coils or radiators of various types, including convectors, finned tube units, radiant floor tubing systems and baseboard units.

Hot water supply boiler. A boiler, completely filled with water, which furnishes hot water to be used externally to itself, and that operates at water pressures not exceeding 160 psig (1100 kPa gauge) and at water temperatures not exceeding 250°F (121°C) at or near the boiler outlet.

❖ Hot water supply boilers are normally part of open systems in which the heated water is supplied and used externally to the boiler. Large domestic (potable) water heating systems often use hot water supply boilers. Such boilers will not likely be found in dwellings.

Steam heating boiler. A boiler in which steam is generated and that operates at a steam pressure not exceeding 15 psig (100 kPa gauge).

❖ Steam heating boilers are normally part of a closed system in which low-pressure steam can be circulated through a variety of terminal units, including natural convection units, forced-convection units and radiant panel systems.

BONDING JUMPER. A conductor installed to electrically connect metallic gas *piping* to the grounding electrode system.

❖ This term is used in subsections of G2411.1.1 relative to the bonding of gas piping systems. It is a number 6 AWG or larger copper conductor or equivalent that is used to electrically bond gas piping to the grounding electrodes that are a part of the electrical service for the structure.

BRAZING. A metal-joining process wherein coalescence is produced by the use of a nonferrous filler metal having a melting point above 1,000°F (538°C), but lower than that of the base metal being joined. The filler material is distributed between the closely fitted surfaces of the joint by capillary action.

❖ Brazing is the act of producing a brazed joint and is often referred to as silver soldering. Silver soldering is more accurately described as silver brazing and uses high-silver-bearing alloys primarily composed of silver, copper and zinc. Silver soldering (brazing) typically requires temperatures in excess of 1,000°F (538°C), and these solders are classified as "hard" solders.

Confusion has always existed concerning the distinction between "silver solder" and "silver-bearing solder." Silver solders are unique and can be further subdivided into soft and hard categories that are determined by the percentages of silver and the other component elements of the particular alloy. The distinction is that silver-bearing solders [melting point below 600°F (316°C)] are used in soft-soldered joints, and silver solders [melting point above 1,000°F (538°C)] are used in silver-brazed joints.

Brazed joints are considered to have superior strength and stress resistance and, because of the high melting point, are less likely to fail when exposed to fire.

Brazing with a filler metal conforming to AWS A5.8 produces a strong joint that will perform under extreme service conditions. The surfaces to be brazed must be cleaned to be free from oxides and impurities. Flux should be applied as soon as possible after the surfaces have been cleaned. Flux helps to remove residual traces of oxides, to promote wetting and to protect the surfaces from oxidation during heating. Care should be taken to prevent flux from entering the piping system during the brazing operation because flux that remains may corrode the pipe or contaminate the system.

Air and any other residual products should be removed from the pipe being brazed by purging the piping with a nonflammable gas such as carbon dioxide or nitrogen. Purging the system has several benefits, such as preventing oxidation from occurring on the inside of the pipe and preventing the creation of toxic gases that can result from the chemical breakdown of other products. Additionally, purging will eliminate the possibility of an explosion from any other flammable gas in the pipe that could ignite when mixed with air.

BTU. Abbreviation for British thermal unit, which is the quantity of heat required to raise the temperature of 1 pound (454 g) of water 1°F (0.56°C) (1 *Btu* = 1055 J).

BURNER. A device for the final conveyance of the gas, or a mixture of gas and air, to the combustion zone.

❖ There are several types of burners that vary in the method of providing combustion air and the pressures of fuel/air mixtures. Burner types include atmospheric ribbon, ported, porous ceramic, slotted and cone types and atmospheric or powered inshot and upshot gun burners.

Induced-draft. A *burner* that depends on *draft* induced by a fan that is an integral part of the *appliance* and is located downstream from the *burner*.

❖ See the commentary to the definition of "Draft/mechanical or induced draft."

Power. A *burner* in which gas, air or both are supplied at pressures exceeding, for gas, the line pressure, and for air, atmospheric pressure, with this added pressure being applied at the *burner*.

❖ Power burners are typically used on very large non-residential-type appliances and use a blower to introduce combustion air under pressure and mix it with the fuel.

CHIMNEY. A primarily vertical structure containing one or more flues, for the purpose of carrying gaseous products of *combustion* and air from an *appliance* to the outside atmosphere.

❖ Chimneys differ from vents in their materials of construction and the type of appliance they are designed to serve. Chimneys are capable of venting much higher temperature flue gases than vents [see Commentary Figure G2403(4)].

Factory-built chimney. A *listed* and *labeled* chimney composed of factory-made components, assembled in the field in accordance with manufacturer's instructions and the conditions of the listing.

❖ A factory-built chimney is a manufactured listed and labeled chimney that has been tested by an approved agency to determine its performance characteristics. Factory-built chimneys are manufactured in two basic designs: a double-wall mineral fiber insulated design and a triple-wall air-cooled design. Some chimneys use both convection cooling and mineral-fiber insulation. Both designs use stainless steel inner liners to resist the corrosive effects of combustion products [see Commentary Figure G2403(5)].

Masonry chimney. A field-constructed chimney composed of solid masonry units, bricks, stones or concrete.

❖ Masonry chimneys can have one or more flues within them and are field constructed of brick, stone, concrete or fire-clay materials. Masonry chimneys can stand alone or be part of a masonry fireplace.

CLEARANCE. The minimum distance through air measured between the heat-producing surface of the mechanical *appliance*, device or *equipment* and the surface of the *combustible material* or *assembly*.

❖ Clearances between sources of heat and combustibles are always airspace clearances.

**Figure G2403(4)
FACTORY-BUILT CHIMNEYS**

Photo courtesy of Selkirk L.L.C.

CLOTHES DRYER. An *appliance* used to dry wet laundry by means of heated air.

❖ Type 1 clothes dryers are used in dwelling units and in residential settings.

  Type 1. Factory-built package, multiple production. Primarily used in the family living environment. Usually the smallest unit physically and in function output.

❖ Type 1 clothes dryers are used in dwelling units and in residential settings.

CODE. These regulations, subsequent amendments thereto, or any emergency rule or regulation that the administrative authority having jurisdiction has lawfully adopted.

❖ The adopted regulations are generally referred to as "the code" and include not only this code, but also any adopted modifications to the code and all other related rules and regulations promulgated and enacted by the jurisdiction.

CODE OFFICIAL. The officer or other designated authority charged with the administration and enforcement of this code, or a duly authorized representative.

❖ The statutory power to enforce the code is normally vested in a building department (or the like) of a state, county or municipality whose designated enforcement officer is termed the "code official."

COMBUSTIBLE ASSEMBLY. Wall, floor, ceiling or other assembly constructed of one or more component materials that are not defined as noncombustible.

❖ A combustible assembly is made up of one or more combustible components. For example, a wall assem-

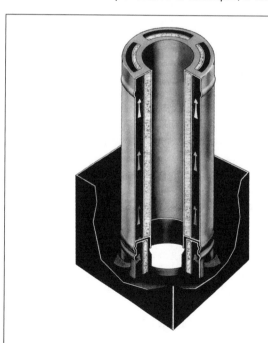

Figure courtesy of Simpson Dura-Vent Co., Inc.

**Figure G2403(5)
AIR-COOLED FACTORY-BUILT CHIMNEY SECTION**

bly with wood studs is combustible regardless of the nature of the sheathing attached to the studs. Likewise, a metal stud wall with gypsum board sheathing is a combustible assembly because the gypsum board is not noncombustible. A noncombustible assembly has no combustible elements (see the definition of "Noncombustible materials").

COMBUSTIBLE MATERIAL. Any material not defined as noncombustible.

❖ See the definition of "Noncombustible materials."

COMBUSTION. In the context of this code, refers to the rapid oxidation of fuel accompanied by the production of heat or heat and light.

❖ The primary components of combustion are fuel, oxygen and heat. The code regulates many aspects of combustion technology, including the process of combustion and providing sufficient air; the use of energy produced from combustion; the safe venting of the products of combustion; combustion efficiency; the containment and control of combustion; and the fuel supplies for combustion equipment and appliances.

COMBUSTION AIR. Air necessary for complete combustion of a fuel, including theoretical air and excess air.

❖ The process of combustion requires a specific amount of oxygen to initiate and sustain the combustion reaction. Combustion air includes primary air, secondary air, draft hood dilution air and excess air. Combustion air includes the amount of atmospheric air required for complete combustion of a fuel and is related to the molecular composition of the fuel being burned, the design of the fuel-burning equipment and the percentage of oxygen in the combustion air. Too little combustion air will result in incomplete combustion of a fuel and the possible formation of carbon deposits (soot), carbon monoxide, toxic alcohols, ketones, aldehydes, nitrous oxides and other by-products. The required amount of combustion air is usually stated in terms of cubic feet per minute (m^3/s) or pounds per hour (kg/h).

COMBUSTION CHAMBER. The portion of an *appliance* within which combustion occurs.

❖ Combustion chambers are either open to the atmosphere or are isolated (sealed) from the atmosphere in which the appliance is installed. Combustion chambers are often referred to as "fireboxes."

COMBUSTION PRODUCTS. Constituents resulting from the combustion of a fuel with the oxygen of the air, including the inert gases, but excluding excess air.

❖ Such products include water, carbon dioxide, carbon monoxide, nitrous oxides and various trace compounds.

CONCEALED LOCATION. A location that cannot be accessed without damaging permanent parts of the building structure or finish surface. Spaces above, below or behind

readily removable panels or doors shall not be considered as concealed.

❖ The space above a "drop-in" tile suspended ceiling system, for example, would not be considered as a concealed location.

CONCEALED PIPING. *Piping* that is located in a *concealed location* (see "Concealed location").

❖ This refers to gas piping in locations meeting the definition of "Concealed location."

CONDENSATE. The liquid that condenses from a gas (including flue gas) caused by a reduction in temperature or increase in pressure.

❖ Condensate forms when the temperature of a vapor is lowered to its dew point temperature. Air conditioning systems produce condensate when an airstream contacts cooling coils. The moisture in the air condenses on the cold surface of the coils and the air is "dehumidified." High-efficiency (84 percent and up) fuel burning appliances produce condensate from the combustion gases. Condensate also forms within improperly designed chimneys and vents when the products of combustion (which contain water vapor) contact the colder inner walls of the vent or chimney. If the temperature of the products of combustion is lowered to the dew point temperature of the water vapor, condensate will form on the inside walls of the vent or chimney. Condensed steam in hydronic systems is also referred to as "condensate."

CONNECTOR, APPLIANCE (Fuel). Rigid metallic *pipe* and fittings, semirigid metallic *tubing* and fittings or a *listed* and *labeled* device that connects an *appliance* to the *gas piping system*.

❖ An appliance fuel connector is the piping, tubing or factory-built device that is installed downstream of the equipment shutoff to connect the appliance to the gas piping system. See the definition of "Piping system." Section G2422.1 lists the types of appliance connectors which include both factory-built assemblies and field-constructed assemblies of pipe, tube and fittings. Note that rigid metallic piping, such as Schedule 40 black steel, used to connect an appliance to the gas piping system is considered to be an appliance connector because it is placed between the appliance and the piping system, downstream of the appliance shutoff valve. As defined, the gas piping ends at the appliance/equipment shutoff valve, therefore, everything downstream of that point is defined, sized and installed as a connector except as required by Sections G2420.5.2, G2420.5.3 and the exception to Section G2422.1.2.1.

CONNECTOR, CHIMNEY OR VENT. The *pipe* that connects an *appliance* to a chimney or vent.

❖ In most cases, appliances are not located directly in line with the vertically rising chimney or vent; therefore, a vent or chimney connector is necessary to connect the appliance flue outlet to the vent or chimney. Vent and chimney connectors can be single- or double-wall pipes and are usually made from steel, galvanized steel, stainless steel or aluminum sheet metal, depending on the application. Vent connectors are also made of corrugated aluminum. In many installations, the vent connectors must be constructed of the same material as the vent, as is typically done with Type B vent systems. There are many requirements in the code (Sections G2427 and G2428) regulating the location, length, size and type of connector.

CONTROL. A manual or automatic device designed to regulate the gas, air, water or electrical supply to, or operation of, a mechanical system.

❖ A control is a device designed to respond to changes in temperature, pressure, liquid or gas flow rates, current, voltage, resistance, humidity or liquid levels.

CONVERSION BURNER. A unit consisting of a *burner* and its *controls* for installation in an *appliance* originally utilizing another fuel.

❖ Typical conversion burners are designed to convert an appliance, such as a boiler, from solid-fuel fired to oil fired or from oil fired to gas fired.

CUBIC FOOT. The amount of gas that occupies 1 cubic foot (0.02832 m³) when at a temperature of 60°F (16°C), saturated with water vapor and under a pressure equivalent to that of 30 inches of mercury (101 kPa).

❖ This definition applies specifically to a measurement of fuel gas.

DAMPER. A manually or automatically controlled device to regulate *draft* or the rate of flow of air or combustion gases.

❖ Dampers act as restrictors or valves in a gaseous flow stream (see Section G2427).

DECORATIVE APPLIANCE, VENTED. A *vented appliance* wherein the primary function lies in the aesthetic effect of the flames.

❖ Such appliances include so-called "gas fireplaces" that are designed to simulate a wood-burning fireplace. The title of the standard for vented decorative gas-fired appliances (ANSI Z21.50) has been changed to "Vented Gas Fireplaces."

DECORATIVE APPLIANCES FOR INSTALLATION IN VENTED FIREPLACES. A *vented appliance* designed for installation within the fire chamber of a vented *fireplace*, wherein the primary function lies in the aesthetic effect of the flames.

❖ These appliances include gas log sets for installation in solid-fuel-burning fireplaces.

DEMAND. The maximum amount of gas input required per unit of time, usually expressed in cubic feet per hour, or *Btu*/h (1 *Btu*/h = 0.2931 W).

❖ Demand refers to the load that appliances place on the fuel-gas distribution system.

DESIGN FLOOD ELEVATION. The elevation of the "design flood," including wave height, relative to the datum specified on the community's legally designated flood hazard map. In areas designated as Zone AO, the *design flood elevation* shall be the elevation of the highest existing grade of the *building's* perimeter plus the depth number, in feet, specified on the flood hazard map. In areas designated as Zone AO where a depth number is not specified on the map, the depth number shall be taken as being equal to 2 feet (610 mm).

❖ This term is used in Section G2404.7. See the commentary to Sections R322.1.4 and R322.1.4.1.

DILUTION AIR. Air that is introduced into a *draft hood* and is mixed with the *flue gases*.

❖ Dilution air is associated only with draft-hood equipped Category I appliances. Fan-assisted Category I appliances do not involve dilution air because all of the air brought in by the fan passes directly through the combustion chamber of the appliance. Dilution air lowers the dew point of vent gases, reducing the possibility that condensation will occur inside the vent or chimney.

DIRECT-VENT APPLIANCES. *Appliances* that are constructed and installed so that all air for combustion is derived directly from the outside atmosphere and all *flue gases* are discharged directly to the outside atmosphere.

❖ These appliances have independent exhaust and intake pipes or have concentric pipes that vent combustion gases through the inner pipe and convey combustion air in the annular space between the inner and outer pipe walls [see Commentary Figures G2403(6) and G2403(7)].

DRAFT. The pressure difference existing between the *appliance* or any component part and the atmosphere, that causes a continuous flow of air and products of combustion through the gas passages of the *appliance* to the atmosphere.

❖ Draft is the negative static pressure measured relative to atmospheric pressure that is developed in chimneys and vents and in the flue-ways of fuel-burning appliances. Draft can be produced by hot flue-gas buoyancy ("stack effect"), mechanically by fans and exhausters or by a combination of both natural and mechanical means.

   Mechanical or induced draft. The pressure difference created by the action of a fan, blower or ejector that is located between the *appliance* and the chimney or vent termination.

❖ Induced draft systems use a fan or blower to boost or "induce" draft in a venting system that produces insufficient natural draft. Draft induction implies that the combustion gases are "pulled" through a passageway or conduit. Draft inducers produce negative pressures on the inlet (upstream) side of the fan or blower. They are separate field-installed units located between an appliance and its venting system. Draft inducers are used with natural draft venting systems to overcome the resistance of vent or chimney connectors and to compensate for the inability of the chimney or vent to produce sufficient and reliable draft.

Draft induction is also a design principle used in many fan-assisted appliances. Draft-inducer fans or blowers that are integral with fuel-fired appliances are necessary to overcome the internal resistance of the heat exchanger flue passageways. To attain higher thermal transfer efficiencies, some heat exchanger designs use flue passageways that retain the combustion gases longer over a greater surface area, thus extracting more heat. Such heat exchanger designs cannot rely on natural (gravity) venting to overcome the higher resistance to flow and, therefore, must rely on mechanical means to pull the combustion gases through the heat exchanger.

Induced draft systems are not to be confused with systems using power exhausters or other self-venting equipment.

Mechanical draft systems produce positive pressures on the discharge side of the fan or blower and can be said to "push" combustion gases through a passageway or conduit.

Natural draft. The pressure difference created by a vent or chimney because of its height, and the temperature difference between the *flue gases* and the atmosphere.

❖ Natural draft systems do not use mechanical devices such as fans or blowers, but instead rely on the principle of buoyancy to carry the products of combustion to the atmosphere. Because of the difference in temperature and the resultant difference in density between the hot products of combustion and the ambient atmosphere, the gases within the chimney or flue will rise, creating a buoyant "draft." The phenomenon of natural draft is sometimes referred to as "stack effect" and is measured in inches of water column (kPa). The amount of draft is affected by the height of the chimney or vent and also by the ability of the chimney or vent to maintain the temperature differential between the combustion gases and the ambient air.

DRAFT HOOD. A nonadjustable device built into an *appliance*, or made as part of the *vent connector* from an *appliance*, that is designed to (1) provide for ready escape of the *flue gases* from the *appliance* in the event of no *draft,* backdraft, or stoppage beyond the *draft hood*, (2) prevent a backdraft from entering the *appliance*, and (3) neutralize the effect of stack action of the chimney or gas vent upon operation of the *appliance*.

❖ Draft hoods are integral to or supplied with natural draft atmospheric-burner gas-fired appliances other than fan-assisted appliances. When classified, appliances equipped with draft hoods are classified by the manufacturer as Category I appliances. Because of minimum efficiency standards and the popularity of mechanical and other special proprietary venting systems, draft-hood-equipped appliances are becoming increasingly rare in the marketplace [see Commentary Figure G2403(8)].

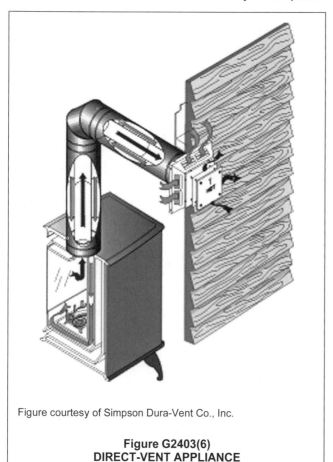

Figure courtesy of Simpson Dura-Vent Co., Inc.

Figure G2403(6)
DIRECT-VENT APPLIANCE

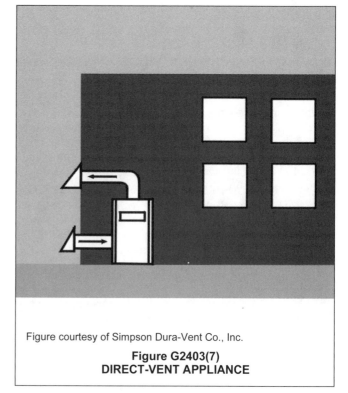

Figure courtesy of Simpson Dura-Vent Co., Inc.

Figure G2403(7)
DIRECT-VENT APPLIANCE

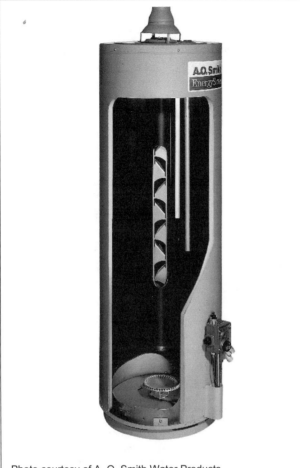

Photo courtesy of A. O. Smith Water Products

Figure G2403(8)
DRAFT-HOOD-EQUIPPED APPLIANCE

DRAFT REGULATOR. A device that functions to maintain a desired *draft* in the *appliance* by automatically reducing the *draft* to the desired value.

❖ Excessive draft reduces the combustion and thermal efficiency of an appliance. Draft regulators automatically adjust the draft by admitting air into the vent, thereby reducing the draft [see Commentary Figure G2403(2)].

DRIP. The container placed at a low point in a system of *piping* to collect *condensate* and from which the *condensate* is removable.

❖ These reservoirs, also referred to as "drip legs," are made up of pipe and fittings and are intended to collect liquid in piping systems where condensables are possible. A "drip leg" is distinct from a "sediment trap" even though they may be constructed identically.

DUCT FURNACE. A warm-air *furnace* normally installed in an air-distribution duct to supply warm air for heating. This definition shall apply only to a warm-air heating *appliance*

that depends for air circulation on a blower not furnished as part of the *furnace*.

❖ Duct furnaces are vented appliances and consist of burners, heat exchangers and related controls. Duct furnaces depend on an external air handler or blower.

DWELLING UNIT. A single unit providing complete, independent living facilities for one or more persons, including permanent provisions for living, sleeping, eating, cooking and sanitation.

❖ Contrast this definition with the definition of "Sleeping unit." All of the listed provisions must be present for the occupancy to be considered as a dwelling unit.

EQUIPMENT. Apparatus and devices other than *appliances*.

❖ The term "equipment" does not refer to or describe anything that is defined as an "Appliance," despite the fact that some appliances are commonly referred to in the field as "pieces of equipment." The term "equipment" includes control devices, pressure regulators, valves, appliance appurtenances, gas connectors, power exhausters, fans, air handlers and a multitude of other devices that do not fit the description of "appliance."

EXCESS FLOW VALVE (EFV). A valve designed to activate when the fuel gas passing through it exceeds a prescribed flow rate.

❖ Some utility companies install excess flow valves (EFVs) in gas service lines to provide protection against high-pressure leaks resulting from damage to the piping caused by excavation, hole boring or hand digging. The device is a type of shutoff valve that abruptly closes when the gas flow through it reaches an abnormal flow rate corresponding to its design setting. The excess flow pushes a spring-loaded element against a seat in the EFV body. The upstream gas pressure holds the valve in the closed position. After the gas line is repaired and closed off on the load end, the EFV will reset itself by allowing a small flow of gas to pass through an orifice in the valve into the downstream piping, thereby equalizing the pressure on both sides of the valve and allowing the valve to reopen by the force of the spring. Service line EFVs do not fall within the scope of the code because the service line is upstream of the point of delivery. Appliance-type EFVs are addressed by the code, are designed to protect individual appliances and are installed in the supply piping to the appliance. Some types are installed on the supply end of a gas appliance connector. The basic operating principle is the same for all types, although some do not allow bypass gas flow and are, therefore, not self-resetting. The moving component of EFVs, depending on the type, is held open by gravity, a spring or a magnet. The gravity type must be installed in a vertical position.

EFVs can be compared to electrical circuit breakers in that they stop gas flow when a predetermined flow rate is exceeded, just as circuit breakers interrupt electrical current flow when the amperage setting is exceeded. Imagine a scenario where an appliance gas connector is severed or a gas line breaks loose at a fitting and gas would be flowing into the space at a high rate. An EFV device installed upstream of the point of leakage would recognize the abnormal flow and either shut off the gas flow completely or reduce the flow to a small bypass flow (see Section G2421.4 and Figure G2421.4).

EXTERIOR MASONRY CHIMNEYS. Masonry chimneys exposed to the outdoors on one or more sides below the roof line.

❖ The part of a chimney that extends above the roof will obviously be exposed to the outdoors, but the sizing methods in Section G2428 are based on the chimney being indoors below the roof. If any one wall or multiple walls of a chimney are exposed to the outdoors, the chimney will lose heat more rapidly, will likely produce condensation internally and may fail to produce the required draft.

Exterior masonry chimneys are not addressed in Chapter 24 because such chimneys are rarely used to vent gas appliances today and the code attempts to limit its coverage to those installations that are common and relevant. These chimneys are addressed in the IFGC and if one is encountered on a job site, the provisions of the IFGC would apply (see Section G2401.1).

FIREPLACE. A fire chamber and hearth constructed of *noncombustible material* for use with solid fuels and provided with a chimney.

❖ Fireplaces burn solid fuels (wood, coal, etc.) and are not referred to as appliances in the code. The IBC and the code cover construction of masonry fireplaces.

Factory-built fireplace. A *fireplace* composed of *listed* factory-built components assembled in accordance with the terms of listing to form the completed *fireplace*.

❖ Factory-built fireplaces are solid-fuel-burning units having a fire chamber that is intended to be either open to the room or, if equipped with doors, operated with the doors either open or closed. Fireplaces are not referred to as appliances. The term "fireplace" describes a complete assembly that includes the hearth, the fire chamber and a chimney. A factory-built fireplace is composed of factory-built components representative of the prototypes tested and is installed in accordance with the manufacturer's installation instructions to form the completed fireplace.

Masonry fireplace. A hearth and fire chamber of solid masonry units such as bricks, stones, *listed* masonry units or reinforced concrete, provided with a suitable chimney.

❖ Masonry fireplaces must be constructed in accordance with the requirements found in Chapter 10. These specific requirements are based on tradition and field experience and describe the conventional fireplace that has proven to be reliable where properly constructed, used and maintained.

FLAME SAFEGUARD. A device that will automatically shut off the fuel supply to a *main burner* or group of *burners* when the means of ignition of such *burners* becomes inoperative, and when flame failure occurs on the *burner* or group of *burners*.

❖ These devices are primary safety controls and are provided on all automatically operated fuel-fired appliances.

FLASHBACK ARRESTOR CHECK VALVE. A device that will prevent the backflow of one gas into the supply system of another gas and prevent the passage of flame into the gas supply system.

❖ See the commentary to Section G2421.5.

FLOOD HAZARD AREA. The greater of the following two areas:

1. The area within a floodplain subject to a 1 percent or greater chance of flooding in any given year.

2. This area designated as a *flood hazard area* on a community's flood hazard map, or otherwise legally designated.

❖ See the commentary to Sections R322.1.4 and R322.1.4.1.

FLOOR FURNACE. A completely self-contained *furnace* suspended from the floor of the space being heated, taking air for combustion from outside such space and with means for observing flames and lighting the *appliance* from such space.

❖ These units supply heat through a floor grille placed directly over the unit's heat exchanger.

FLUE, APPLIANCE. The passage(s) within an *appliance* through which *combustion products* pass from the *combustion chamber* of the *appliance* to the *draft hood* inlet opening on an *appliance* equipped with a *draft hood* or to the outlet of the *appliance* on an *appliance* not equipped with a *draft hood*.

❖ Appliance flues are the passages through an appliance from the combustion chamber, through the heat exchanger and through the vent outlet.

FLUE COLLAR. That portion of an *appliance* designed for the attachment of a *draft hood*, *vent connector* or venting system.

❖ The flue collar size will be a determining factor in vent connector sizing.

FLUE GASES. Products of combustion plus excess air in *appliance flues* or heat exchangers.

❖ The exact composition of flue gases will depend on the fuel being burned. The primary components of flue gases are nitrogen, carbon dioxide, water vapor, particulates and a myriad of compounds and trace elements that vary with the nature of the fuel and purity of the combustion air. Carbon monoxide is also a component of flue gas.

FLUE LINER (LINING). A system or material used to form the inside surface of a flue in a *chimney* or vent, for the purpose of protecting the surrounding structure from the effects of *combustion products* and for conveying *combustion products* without leakage to the atmosphere.

❖ Flue liners must be resistant to heat and the corrosive action of the products of combustion. They provide insulation value to retard the transfer of heat to the chimney structure and limit exposure of the chimney structure to the harmful effects of combustion products. Flue liners are generally made of fire-clay tile, refractory brick, poured-in-place refractory materials, stainless steel alloys and aluminum. Flue liners used with gas-fired appliances are typically made of stainless steel or aluminum [see Commentary Figure G2403(9)].

FUEL GAS. A natural gas, manufactured gas, *liquefied petroleum gas* or mixtures of these gases.

❖ The nature of fuel gases makes proper design, installation and selection of materials and devices necessary to minimize the possibility of fire or explosion. Bringing fuel gases into a building is in itself a risk. The provisions of the code are intended to reduce that risk to a level comparable to that associated with other energy sources such as electricity.

The two most commonly used fuel gases are natural gas and liquefied petroleum gas (LP-gas or LPG). These fuel gases have the following characteristics or properties.

Natural gas: The principal constituent of natural gas is methane (CH_4). It can also contain small quantities of nitrogen, carbon dioxide, hydrogen sulfide, water vapor, other hydrocarbons (such as ethane and propane) and various trace elements. Natural gas is colorless, tasteless and odorless; however, an odorant (methyl mercaptan) is added to the gas so that it can be readily detected. Natural gas is lighter than air (specific gravity of 0.60 typical) and has the tendency to rise when escaping to the atmosphere.

Natural gas has a rather narrow flammability range (approximately 3 to 15 percent volume in air) above and below which the gas-to-air mixture ratio will be too rich or too lean to support combustion. The heating value of natural gas is approximately 1,050 Btu per cubic foot (39 MJ/m^3).

LP-gas: Liquefied petroleum gases include commercial propane and commercial butane. LP-gas vapors are heavier than air (specific gravity of 1.52 typical) and tend to accumulate in low areas and near the floor; however, a gas-air mixture that is within the flammability range will have a specific gravity very close to that of air. The ranges of flammability for LP-gases are narrower than those of natural gas (approximately 2 to 10 percent volume in air). Like natural gas, LP-gases are odorized to make them detectable. The heating value of propane is approximately 2,500 Btu per cubic foot (93 MJ/m^3) of gas. The heating value of butane is approximately 3,300 Btu per cubic foot (123 MJ/m^3) of gas.

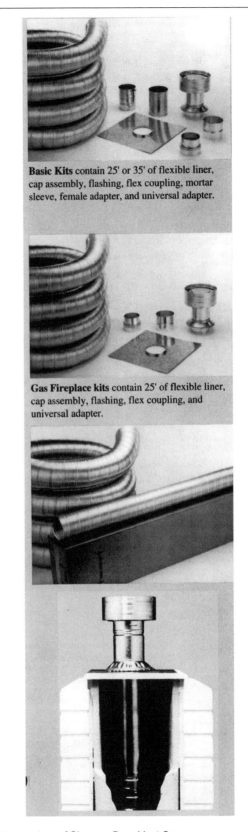

Basic Kits contain 25' or 35' of flexible liner, cap assembly, flashing, flex coupling, mortar sleeve, female adapter, and universal adapter.

Gas Fireplace kits contain 25' of flexible liner, cap assembly, flashing, flex coupling, and universal adapter.

Photos courtesy of Simpson Dura-Vent Company
For SI: 1 foot = 304.8 mm.

Figure G2403(9)
CHIMNEY FLUE LINER SYSTEM FOR GAS APPLIANCE

FURNACE. A completely self-contained heating unit that is designed to supply heated air to spaces remote from or adjacent to the *appliance* location.

❖ The single most distinguishing characteristic of furnaces is that they use air as the heat transfer medium. Furnaces can be fueled by gas, oil, solid fuel or electricity and use fans, blowers or gravity (convection) to circulate air to and from the unit. In the context of the code, the primary usage of the term "furnace" refers to heating appliances that combine a combustion chamber with related burner and control components, one or more heat exchangers, a flue gas conveying system and air-handling fans/blowers.

FURNACE, CENTRAL. A self-contained *appliance* for heating air by transfer of heat of *combustion* through metal to the air, and designed to supply heated air through ducts to spaces remote from or adjacent to the *appliance* location.

❖ The term "central furnace" has been adopted to identify a furnace that supplies conditioned air to remote locations from a central location. There are many types of furnaces. Most of these variations are a result of architectural design related to available space, desired location of the unit and size of the area to be conditioned.

FURNACE PLENUM. An air compartment or chamber to which one or more ducts are connected and which forms part of an air distribution system.

❖ The term "plenum" is no longer defined or used in this chapter. A furnace plenum is a box made of sheet metal or other duct material. Such a box is attached directly to a furnace to facilitate connection of supply and return ducts to the furnace air inlet and outlet. This term does not refer to cavities within a building, such as stud and joist spaces and spaces above dropped ceilings.

GAS CONVENIENCE OUTLET. A permanently mounted, manually operated device that provides the means for connecting an *appliance* to, and disconnecting an *appliance* from, the supply *piping*. The device includes an integral, manually operated valve with a nondisplaceable valve member and is designed so that disconnection of an *appliance* only occurs when the manually operated valve is in the closed position.

❖ These devices are listed and labeled and designed to mate with specialized appliance connectors. A gas outlet is the point at which the gas distribution system connects to a gas appliance. The outlets are intended for use with portable appliances such as space heaters and outdoor cooking appliances [see Commentary Figures G2403(10) and G2403(11)].

GAS PIPING. An installation of pipe, valves or fittings installed on a premises or in a building and utilized to convey fuel gas.

❖ Gas piping includes all the components, fittings and piping needed to deliver the fuel gas from the point of delivery to the appliance or equipment connection. The point of delivery may be a regulator or meter that is typically installed by the gas utility. The point of delivery may be located at the user's property line, immediately outside the structure or in some instances in the structure.

Photo courtesy of Gastite Division/Titflex Corporation

Figure G2403(10)
GAS CONVENIENCE OUTLET

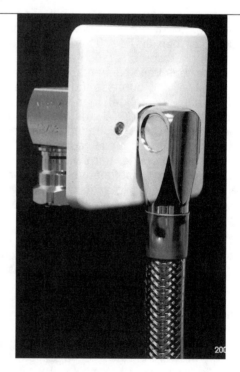

Photo courtesy of Maxitrol Company

Figure G2403(11)
GAS CONVENIENCE OUTLET

HAZARDOUS LOCATION. Any location considered to be a fire hazard for flammable vapors, dust, combustible fibers or other highly combustible substances. The location is not necessarily categorized in the *International Building Code* as a high-hazard use group classification.

❖ The environment in which mechanical equipment and appliances operate plays a significant role in the safe performance of the equipment installation. Locations that may contain ignitable or explosive atmospheres are classified as hazardous locations for the installation of mechanical equipment and appliances. For example, repair garages can be classified as hazardous locations because they can contain gasoline vapors from vehicles stored within them as well as other volatile chemicals. Public and private garages are not considered hazardous locations even though the presence of motor vehicles and the storage of fuel, paint, varnish, thinner, lawn- and home-maintenance products and other chemicals are a concern, as evidenced by Sections G2408.2 and G2408.3.

HOUSE PIPING. See *"Piping system."*

❖ House piping is the distribution piping downstream of the point of delivery. House piping is an antiquated term.

IGNITION PILOT. A *pilot* that operates during the lighting cycle and discontinues during *main burner* operation.

❖ Commonly referred to as "interrupted pilots," such pilots are ignited by a spark or hot surface device. Generally an electronic circuit controls the ignition system. When main burner ignition has been proven, the ignition pilot is shut off, and the main burner flame is supervised for the entire burn cycle.

IGNITION SOURCE. A flame spark or hot surface capable of igniting flammable vapors or fumes. Such sources include *appliance burners, burner* ignitors and electrical switching devices.

❖ This definition is important in the application of Section G2408.2 regarding the elevation of ignition sources. By means of this definition, Section G2408.2 applies to unintentional ignition sources, such as electrical switching devices, as well as intentional ignition sources, such as pilot lights, spark ignitors and hot surface ignitors.

In the context of Section G2408.2, an "ignition source" is something capable of igniting flammable vapors that are present in the atmosphere in the locations listed in Section G2408.2.

INFRARED RADIANT HEATER. A heater which directs a substantial amount of its energy output in the form of infrared radiant energy into the area to be heated. Such heaters are of either the vented or unvented type.

❖ These heaters include low-intensity tubular types and high-intensity ceramic burner element types. They heat objects (such as people) by direct radiation and do not heat the ambient air.

JOINT, FLARED. A metal-to-metal compression joint in which a conical spread is made on the end of a tube that is compressed by a flare nut against a mating flare.

❖ Because the pipe end is expanded in a flared joint, only annealed and bending-tempered soft-drawn copper tubing may be flared. Commonly used flaring tools use a screw yoke and block assembly or an expander tool that is driven into the tube with a hammer. The flared tubing end is compressed between a fitting seat and a threaded nut to form a metal-to-metal seal.

JOINT, MECHANICAL. A general form of gas-tight joints obtained by the joining of metal parts through a positive-holding mechanical construction, such as press joint, flanged joint, threaded joint, flared joint or compression joint.

❖ Mechanical joints can take many forms, but most share a common characteristic, which is applying a radial pressure to the pipes or fittings they join. Mechanical joining means can be proprietary, meaning that they are specific to one manufacturer of the fittings and devices and such manufacturer must provide adequate instructions for assembling the joint. Press-joint fittings must be assembled with a special tool that is unique to the fitting. Press joints are irreversible because the fitting and pipe are permanently deformed by the pressing operation. Some press-joint systems are manufactured for use with copper piping and some are manufactured for use with steel piping.

JOINT, PLASTIC ADHESIVE. A joint made in thermoset plastic *piping* by the use of an adhesive substance which forms a continuous bond between the mating surfaces without dissolving either one of them.

❖ Unlike solvent welded joints, adhesive (glue) joints are surface bonded. Often, people refer to solvent welded joints, such as used for PVC, CPVC and ABS piping, as "glued" joints, when actually they are chemically welded, not glued.

LABELED. Equipment, materials or products to which have been affixed a label, seal, symbol or other identifying mark of a nationally recognized testing laboratory, inspection agency or other organization concerned with product evaluation that maintains periodic inspection of the production of the above-labeled items and whose labeling indicates either that the *equipment*, material or product meets identified standards or has been tested and found suitable for a specified purpose.

❖ When a product is labeled, the label indicates that the material has been tested for conformance to an applicable standard and that the component is subject to third-party inspection to verify that the minimum level of quality required by the standard is maintained. Labeling provides a readily available source of information that is useful for field inspection of installed products. The label identifies the product or material and provides other information that can be further investigated if there is question concerning the suitability of the product or material for the specific installation.

The labeling agency performing the third-party inspection must be approved by the code official. The basis for this approval may include, but is not necessarily limited to, the capacity and capability of the agency to perform the specific testing and inspection.

The referenced standard often states the minimum identifying information that must be on a label. The data contained on a label typically includes, but is not necessarily limited to, name of the manufacturer; product name or serial number; installation specifications; applicable tests and standards; and the approved testing and labeling agency.

LEAK CHECK. An operation performed on a gas *piping system* to verify that the system does not leak.

❖ A leak check is not the same as the pressure test required by Sections G2417.1 through G2417.5.2 (see Section G2417.6).

LIQUEFIED PETROLEUM GAS or LPG (LP-GAS). *Liquefied petroleum gas* composed predominately of propane, propylene, butanes or butylenes, or mixtures thereof that is gaseous under normal atmospheric conditions, but is capable of being liquefied under moderate pressure at normal temperatures.

❖ LP-gases are usually obtained as a by product of oil refinery operations or by stripping natural gas and are commercially available as butane, propane or a mixture of the two. Propane has a boiling point of -40°F (-40°C) at atmospheric pressure and a heating value of approximately 2,500 Btu per cubic foot (93 MJ/m²). Butane has a boiling point of 32°F (0°C) at atmospheric pressure and a heating value of approximately 3,300 Btu per cubic foot (123 MJ/m³). LP-gas is odorized so that gas leakage can be detected.

LISTED. Equipment, materials, products or services included in a list published by an organization acceptable to the code official and concerned with evaluation of products or services that maintains periodic inspection of production of *listed equipment* or materials or periodic evaluation of services and whose listing states either that the *equipment*, material, product or service meets identified standards or has been tested and found suitable for a specified purpose.

❖ When a product is listed, it indicates that it has been tested for conformance to an applicable standard and is subject to a third-party inspection quality assurance (QA) program. The QA verifies that the minimum level of quality required by the appropriate standard is maintained. The agency performing the third-party inspection must be approved by the code official, and the basis for this approval may include, but is not limited to, the capacity and capability of the agency to perform the specified testing and inspection.

LIVING SPACE. Space within a *dwelling unit* utilized for living, sleeping, eating, cooking, bathing, washing and sanitation purposes.

❖ The code uses this term to define locations that are not to be considered as part of a private garage for the purpose of requiring the elevation of ignition sources (see Section G2408.2).

LOG LIGHTER. A manually operated solid-fuel ignition *appliance* for installation in a vented solid-fuel-burning *fireplace*.

❖ Log lighters are simple, manually operated burners used to start wood fires. Log lighters are not considered as decorative but rather as functional appliances. The heat produced by the log lighter flame raises the temperature of the wood fuel to its ignition temperature. A log lighter is designed to be turned off manually after the wood fire is capable of sustaining combustion.

MAIN BURNER. A device or group of devices essentially forming an integral unit for the final conveyance of gas or a mixture of gas and air to the combustion zone, and on which combustion takes place to accomplish the function for which the *appliance* is designed.

❖ With the exception of small amounts of fuel for the pilot or ignition burners, the main burner consumes the entire fuel input into an appliance. Common types of atmospheric burners are the drilled port, slotted port, ribbon and single port.

METER. The instrument installed to measure the volume of gas delivered through it.

❖ The meter is the actual point of commerce and is usually associated with the point of delivery. This code does not require or regulate meters. The gas-supplying utility generally installs meters for natural gas. Occasionally in locations where there are several tenants (large office buildings, mobile home parks and apartment complexes, for example), the gas utility installs one meter for the property owner and the owner installs individual tenant meters. Liquefied petroleum may be metered to individual tenants from a storage tank in the same manner [see Commentary Figure G2403(12)].

MODULATING. Modulating or throttling is the action of a *control* from its maximum to minimum position in either predetermined steps or increments of movement as caused by its actuating medium.

❖ In the context of this code, modulating is associated with burner control valves. One type of automatic control is a modulating control that can infinitely vary the controlled variable over the range being controlled. Another common type of control is a two-position action control, which simply turns the variable being controlled, such as fuel gas, on and off. Another type of control is the floating action control. It regulates the controlled device by adjusting the controlled variable between the open or closed position using a sensing element, which must react faster than the actuating drive time or the system would be essentially the same as a two-position control.

Photo courtesy of Washington Gas

Figure G2403(12)
TYPICAL RESIDENTIAL GAS METER SETTING

NONCOMBUSTIBLE MATERIALS. Materials that, when tested in accordance with ASTM E136, have at least three of four specimens tested meeting all of the following criteria:

1. The recorded temperature of the surface and interior thermocouples shall not at any time during the test rise more than 54°F (30°C) above the furnace temperature at the beginning of the test.

2. There shall not be flaming from the specimen after the first 30 seconds.

3. If the weight loss of the specimen during testing exceeds 50 percent, the recorded temperature of the surface and interior thermocouples shall not at any time during the test rise above the furnace air temperature at the beginning of the test, and there shall not be flaming of the specimen.

❖ If the material is not determined to be noncombustible, it is, by default, a combustible material. ASTM E136 is a test standard used to determine the combustion characteristics of building materials. This test method in the standard is intended to indicate whether a material will assist in combustion or add appreciable heat to an ambient fire. The three listed criteria are taken from ASTM E136. Items 1 and 2 are applicable if the weight loss of the specimen during the test is 50 percent or less and Item 3 applies if such weight loss is greater

than 50 percent. The standard intends for three out of four samples to comply with either Items 1 and 2 or Item 3.

OFFSET (VENT). A combination of *approved* bends that make two changes in direction bringing one section of the vent out of line, but into a line parallel with the other section.

❖ An offset results in the lateral displacement of a vertical vent. Vent offsets are measured in degrees of angle from vertical and are limited in venting systems because they restrict the flow of flue gas. An offset angle is measured between a line that is an extension of the original vertical vent and the new line of direction of the vent [see Commentary Figure G2403(13)]. Note that an offset is more than just a single change in direction and always involves two or more fittings and two changes of direction.

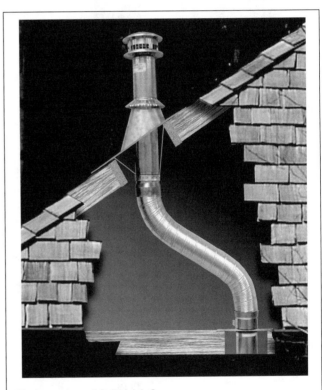

Photo courtesy of Selkirk L.L.C.

Figure G2403(13)
VENT OFFSET

OUTLET. The point at which a gas-fired *appliance* connects to the gas *piping system*.

❖ A gas outlet is analogous to an electrical receptacle outlet. See Section G2415.13 for requirements covering the location of gas piping outlets.

OXYGEN DEPLETION SAFETY SHUTOFF SYSTEM (ODS). A system designed to act to shut off the gas supply to the main and *pilot burners* if the oxygen in the surrounding atmosphere is reduced below a predetermined level.

❖ An oxygen depletion safety shutoff system (ODS) is intended to protect the occupants from carbon monoxide build-up. The ODS measures the oxygen level in the air and shuts off the appliance if the level falls below a preset level, typically 18 percent. Monitoring oxygen levels will inversely monitor carbon monoxide levels because of the natural relationship between a decreasing oxygen level and an increasing carbon monoxide level [see Commentary Figure G2403(14)].

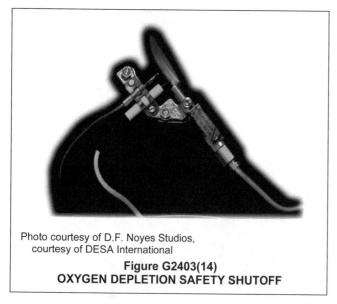

Photo courtesy of D.F. Noyes Studios, courtesy of DESA International

Figure G2403(14)
OXYGEN DEPLETION SAFETY SHUTOFF

PILOT. A small flame that is utilized to ignite the gas at the *main burner* or *burners.*

❖ Pilot burners (pilot lights) can be continuously burning or can be intermittent or interrupted depending upon the type of appliance. Continuously burning (standing) pilots are becoming more rare and are typically found only in certain hearth appliances, water heaters and commercial ranges. Intermittent pilots burn only while the main burner(s) are firing and interrupted pilots burn only until the main burner(s) are ignited and proven.

PIPING. Where used in this code, *"piping"* refers to either *pipe* or *tubing,* or both.

    Pipe. A rigid conduit of iron, steel, copper, brass or plastic.

    Tubing. Semirigid conduit of copper, aluminum, plastic or steel.

❖ Piping includes tubing and pipe used to convey fuel gases (see definition of "Piping system").

PIPING SYSTEM. All fuel *piping,* valves and fittings from the outlet of the *point of delivery* to the outlets of the *appliance* shutoff valves.

❖ A piping system includes tubing, pipe, fittings, valves and line pressure regulators used to convey fuel gas from the point of delivery to the appliance. This definition was revised to state that the piping system ends at the outlets of the appliance shutoff valves. This revision allows the last 6 feet (1829 mm) or less of gas conduit between the shutoff valve and the appliance inlet connection to be sized as a connector (see Section 409.5 and the definition for "Point of delivery").

PLASTIC, THERMOPLASTIC. A plastic that is capable of being repeatedly softened by increase of temperature and hardened by decrease of temperature.

❖ Polyvinyl chloride (PVC) is a type of thermoplastic.

POINT OF DELIVERY. For natural gas systems, the *point of delivery* is the outlet of the service meter assembly or the outlet of the service regulator or service shutoff valve where a meter is not provided. Where a valve is provided at the outlet of the service meter assembly, such valve shall be considered to be downstream of the *point of delivery.* For undiluted liquefied petroleum gas systems, the point of delivery shall be considered to be the outlet of the service pressure regulator, exclusive of line gas regulators, in the system.

❖ It is necessary to understand where the point of delivery begins with respect to fuel gas systems because that is the location where the enforcement of Chapter 24 begins. The point of delivery is a point of demarcation that determines what is or is not covered by this chapter. For natural gas systems, the utility-owned service piping, service shutoff valve, service pressure regulator and meter are all upstream of the point of delivery and thus are not covered by the code. Any valve located on the outlet of the service meter assembly is considered to be part of the piping system and under the scope of the code.

    For LP-gas applications, the point of delivery is the outlet of the service pressure regulator (see definition). NFPA 58 regulates the piping and components upstream of the outlet of the service pressure regulator, including the regulator itself. In typical two-stage LP-gas systems, the service pressure regulator is the second-stage regulator. Two-stage systems use a first-stage regulator located at the tank to reduce tank pressure down to 10 psig, and a second-stage regulator to reduce the 10 psig pressure down to 11 inches water column (w.c.). Some two-stage systems use separate first- and second-stage regulators and some use integral two-stage regulators, located at the tank, in which both the first- and second-stage regulators are contained within a single unit. Two-psig systems are becoming more popular in LP-gas applications and these systems use three stages of regulation, which include a first-stage regulator to reduce tank pressure to 10 psig, a 2-psi regulator (the service pressure regulator) to reduce the 10 psig pressure to 2 psig and line gas pressure regulators (see definition) used to reduce the 2-psig pressure to 11 inches w.c. for delivery to the appliances. In these systems, the point of delivery is the outlet of the 2-psi regulator.

PRESSURE DROP. The loss in pressure due to friction or obstruction in pipes, valves, fittings, *regulators* and *burners.*

❖ The pressure in a fuel gas system is reduced as the gas flows in the system. This is caused by a loss of energy resulting from friction and turbulence. Fuel gas systems are designed so that the pressure drop does not result in the system pressure falling below the minimum pressure required for proper equipment operation.

PRESSURE TEST. An operation performed to verify the gas-tight integrity of *gas piping* following its installation or modification.

❖ Installed piping systems are tested on the job site to verify that the system is free of leaks.

PURGE. To free a gas conduit of air or gas, or a mixture of gas and air.

❖ See Section G2417.7.

READY ACCESS (TO). That which enables a device, *appliance* or *equipment* to be directly reached, without requiring the removal or movement of any panel, door or similar obstruction. (See "Access.")

❖ Ready access can be described as the capability of being quickly reached or approached for the purpose of operation, inspection, observation or emergency action. Ready access does not require the removal or movement of any access door, panel or similar obstruction, nor does it require surmounting physical obstructions or obstacles, including differences in elevation. Note that this definition does not refer to the egress/ingress door through which a person enters a room or building.

REGULATOR. A device for controlling and maintaining a uniform gas supply pressure, either pounds-to-inches water column (MP regulator) or inches-to-inches water column (*appliance regulator*).

❖ In the context of the code, these devices are fuel-gas pressure regulating devices that reduce a higher pressure to a lower pressure.

REGULATOR, GAS APPLIANCE. A *pressure regulator* for controlling pressure to the manifold of the gas *appliance*.

❖ These regulators are integral to the appliance and serve to stabilize the appliance input pressure and supply fuel gas to the appliance at the desired pressure. Most appliances come equipped with appliance regulators because the appliances are designed to operate at a pressure lower than the delivery pressure, even when low pressure gas [0.5 psig or lower (3.4 kPa)] is supplied. Commonly, appliance regulators reduce a pressure of 5 to 8 inches w.c. down to a manifold pressure of 3.5 inches w.c.

REGULATOR, LINE GAS PRESSURE. A device placed in a gas line between the *service pressure regulator* and the *appliance* for controlling, maintaining or reducing the pressure in that portion of the *piping system* downstream of the device.

❖ Line gas pressure regulators are used to reduce the gas pressure supplied by the service (point of delivery) regulator. They are typically located at the connection to each gas appliance and are commonly used as "pounds to inches" regulators in systems having elevated [14 inches w.c. (3.5 kPa) to 5 psi typical (34.5 kPa)] supply pressures.

REGULATOR, MEDIUM-PRESSURE (MP Regulator). A line *pressure regulator* that reduces gas pressure from the range of greater than 0.5 psig (3.4 kPa) and less than or equal to 5 psig (34.5 kPa) to a lower pressure.

❖ With the increased use of high-pressure gas distribution systems, systems are designed with one or more regulators in addition to the service regulators. Medium-pressure regulators (MP regulators) are most commonly associated with 2 psi manifold distribution systems used with copper tubing or corrugated stainless steel tubing (CSST) systems. Appliances are also equipped with factory-installed regulators that control burner input (manifold) pressure. MP regulators are used to reduce the service pressure to a pressure suitable for delivery to the gas appliances where the appliance regulator may further reduce the pressure (see Section G2421.2). MP regulators are used where the distribution line pressure is in the range of 5 psig (34.5 kPa) down to 0.5 psig (3.4 kPa) and such pressure needs to be reduced to be compatible with the input pressure needs of the load.

REGULATOR, PRESSURE. A device placed in a gas line for reducing, controlling and maintaining the pressure in that portion of the *piping system* downstream of the device.

❖ See the definition and commentary to "Regulator."

REGULATOR, SERVICE PRESSURE. For natural gas systems, a device installed by the serving gas supplier to reduce and limit the service line pressure to delivery pressure. For undiluted liquefied petroleum gas systems, the regulator located upstream from all line gas pressure regulators, where installed, and downstream from any first stage or a high pressure regulator in the system.

❖ The utility company (gas supplier) installs this regulator to reduce the gas pressure from the utility distribution mains to that required for delivery to the meter and customer. For the typical house service, the pressure is reduced from 60 psig to 2 psig or less. The devices are usually located near the meter and outdoors [see Commentary Figure G2403(15)]. For LP-gas systems, the service regulator is the regulator between the first-stage regulator and the line-pressure regulators or between the first-stage regulator and the appliances served where there are no line-pressure regulators. The service regulator is upstream of the point of delivery for natural and LP-gases (see the definition of "Point of delivery").

RELIEF OPENING. The opening provided in a *draft hood* to permit the ready escape to the atmosphere of the flue products from the *draft hood* in the event of no *draft*, backdraft or stoppage beyond the *draft hood*, and to permit air into the *draft hood* in the event of a strong chimney updraft.

❖ Dilution air, a component of combustion air, enters the venting system through the draft hood opening. Combustion gases spilling from (exiting) a draft hood relief opening create an abnormal and potentially hazardous condition except for brief periods of spillage that might occur at the start of an appliance firing cycle. Typical periods of spillage at start-up last only a few seconds.

Photo courtesy of Washington Gas

Figure G2403(15)
SERVICE REGULATOR

RELIEF VALVE (DEVICE). A safety valve designed to forestall the development of a dangerous condition by relieving either pressure, temperature or vacuum in the hot water supply system.

❖ A relief valve is a safety device designed to prevent the development of a potentially damaging or dangerous condition in a closed system. This is usually associated with a closed system that has a heat source as part of the system. Requirements for the installation of relief valves in water distribution systems and water heaters are given in Section 504 of the IPC.

RELIEF VALVE, PRESSURE. An *automatic valve* that opens and closes a relief vent, depending on whether the pressure is above or below a predetermined value.

❖ Pressure relief valves are intended to prevent harm to life and property. These valves are safety devices used to relieve abnormal pressures and to prevent the over-pressurization of the vessel or system to which the valves are connected. Pressure relief valves are designed to operate (discharge) only when an abnormal pressure exists in a system. They are not to be used as regulating valves, nor are they intended to control or regulate flow or pressure. Pressure relief valves are set at the factory to begin opening at a predetermined pressure and are rated according to the maximum energy discharge per unit of time, usually in units of British thermal units per hour (Btu/h) (W).

Pressure relief valves must be properly sized and rated for the boiler, pressure vessel or system served; otherwise, a system malfunction could cause the pressure within the system to continue to rise, thereby creating a hazardous condition. Pressure relief valves do not open fully when they reach the factory-preset pressure; rather, the valves open an amount proportional to the forces produced by the pressure in the system. The valves open fully at a certain percentage above the

preset pressure, then close when the pressure drops below the preset pressure. Pressure relief valves are sometimes combined in the same body with temperature relief valves, creating combination temperature and pressure relief valves.

RELIEF VALVE, TEMPERATURE.

Manual reset type. A valve that automatically opens a *relief* vent at a predetermined temperature and that must be manually returned to the closed position.

Reseating or self-closing type. An *automatic valve* that opens and closes a relief vent, depending on whether the temperature is above or below a predetermined value.

❖ Temperature relief valves are designed to open in response to excessive temperatures and to discharge heated water to limit the temperature of the water in the vessel, tank or system. Water heater control failure or improper adjustments could allow excessive temperature rises, which can heat water above the temperature at which it would vaporize at atmospheric pressure. Water heated to above 212°F (100°C) is referred to as superheated water and will flash to steam when the pressure is reduced below its vapor pressure. Very hot water and the presence of steam are hazardous; however, more importantly, higher temperatures produce higher pressures and superheated water increases the potential for and the magnitude of an explosion.

RELIEF VALVE, VACUUM. A valve that automatically opens and closes a vent for relieving a vacuum within the hot water supply system, depending on whether the vacuum is above or below a predetermined value.

❖ The vacuum relief valve is intended to prevent the possible reduction in pressure within the system (water heater tanks being of main concern) to below atmospheric pressure (that is, a partial vacuum). Many tanks are not designed to resist external pressures exceeding internal pressures; therefore, a vacuum relief valve is necessary to prevent atmospheric pressure from possibly collapsing or otherwise damaging the tank. A partial vacuum could also cause water to be siphoned from the tank, thereby creating an overheating hazard in a water heater.

The vacuum relief valve operates by automatically opening the system to the atmosphere when a partial vacuum is created, thereby permitting air to enter and maintaining the internal pressure at atmospheric pressure.

RISER, GAS. A vertical *pipe* supplying fuel gas.

❖ A vertical gas pipe that distributes gas from a building main to upper floor levels is commonly referred to as a "riser."

ROOM HEATER, UNVENTED. See *"Unvented room heater."*

❖ See the commentary to "Unvented room heater" and Commentary Figure G2403(16).

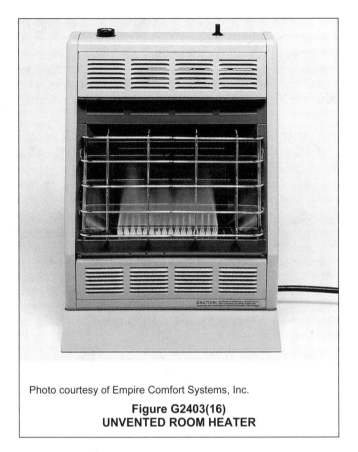

Photo courtesy of Empire Comfort Systems, Inc.

Figure G2403(16)
UNVENTED ROOM HEATER

ROOM HEATER, VENTED. A free-standing heating unit used for direct heating of the space in and adjacent to that in which the unit is located. (See also *"Vented room heater."*)

❖ These heaters are typically small space-heating appliances designed to heat a single room.

SAFETY SHUTOFF DEVICE. See *"Flame safeguard."*

❖ See the commentary to "Flame safeguard."

SHAFT. An enclosed space extending through one or more stories of a building, connecting vertical openings in successive floors, or floors and the roof.

❖ A shaft is an enclosed, vertical passage that passes through stories and floor levels within a building. Shafts are sometimes referred to as "chases" and are usually constructed of fire-resistance-rated assemblies.

SPECIFIC GRAVITY. As applied to gas, *specific gravity* is the ratio of the weight of a given volume to that of the same volume of air, both measured under the same condition.

❖ Analogous to density, specific gravity values are used to describe how a gaseous substance will behave in, for example, air or piping systems. "Specific gravity" indicates the density of a gas with respect to air, with air being assigned a value of 1. For example, natural gas has a specific gravity of less than 1; therefore, it is less dense and lighter than air. LP-gas has a specific gravity of 1.5; therefore, it is one and one-half times the density of air.

THERMOSTAT.

Electric switch type. A device that senses changes in temperature and controls electrically, by means of separate components, the flow of gas to the *burner(s)* to maintain selected temperatures.

Integral gas valve type. An automatic device, actuated by temperature changes, designed to control the gas supply to the *burner(s)* in order to maintain temperatures between predetermined limits, and in which the thermal actuating element is an integral part of the device.

1. Graduating thermostat. A thermostat in which the motion of the *valve* is approximately in direct proportion to the effective motion of the thermal element induced by temperature change.

2. Snap-acting thermostat. A thermostat in which the thermostatic valve travels instantly from the closed to the open position, and vice versa.

❖ Thermostats most often consist of one or more thermally actuated on/off switches, but also include types that transmit varying voltage and signals. A thermostat is also referred to as a controller. The set point or desired unit of temperature is adjusted or set at the thermostat. The thermostat transmits a signal to the controlled device or appliance.

THIRD-PARTY CERTIFICATION AGENCY. An approved agency operating a product or material certification system that incorporates initial product testing, assessment and surveillance of a manufacturer's quality control system.

❖ These agencies evaluate products, verify compliance with specific standards or criteria and conduct periodic inspections of the manufacturer's production facilities to verify ongoing compliance with such standards or criteria (see Section 401.10).

THIRD-PARTY CERTIFIED. Certification obtained by the manufacturer indicating that the function and performance characteristics of a product or material have been determined by testing and ongoing surveillance by an approved third-party certification agency. Assertion of certification is in the form of identification in accordance with the requirements of the third-party certification agency.

❖ A third-party certification agency certifies that a product meets specific standards or criteria and identifies the product with a certification mark. This mark is an indication to the installer and inspector that the product has been evaluated and tested and the manufacturing process has been monitored so as to verify compliance with the relative product standards or criteria. Additionally, the agency typically provides a list of the products that it certifies, which provides information relative to the certification (see Section 401.10).

THIRD-PARTY TESTED. Procedure by which an approved testing laboratory provides documentation that a product, material or system conforms to specified requirements.

❖ A third-party testing laboratory or agency tests a product to verify that it meets specific standards or criteria and provides documentation to that fact. Having a product tested is not the same as having it listed, as listing entails additional requirements such as ongoing monitoring of the manufacturing process. Third-party tested products can be marked to indicate the testing agency.

TRANSITION FITTINGS, PLASTIC TO STEEL. An adapter for joining plastic *pipe* to steel *pipe*. The purpose of this fitting is to provide a permanent, pressure-tight connection between two materials that cannot be joined directly one to another.

❖ Joining of piping of dissimilar materials such as plastic and steel requires specialized transition fittings. For example, such fittings are used with underground plastic fuel-gas piping and steel meter setting risers [see Commentary Figure G2403(17)].

Photo courtesy of Perfection Corporation

Figure G2403(17)
TRANSITION FITTINGS

UNIT HEATER.

High-static pressure type. A self-contained, automatically controlled, vented *appliance* having integral means for circulation of air against 0.2 inch w.c. (50 Pa) or greater static pressure. Such *appliance* is equipped with provisions for attaching an outlet air duct and, where the *appliance* is for indoor installation remote from the space to be heated, is also equipped with provisions for attaching an inlet air duct.

Low-static pressure type. A self-contained, automatically controlled, vented *appliance*, intended for installation in the space to be heated without the use of ducts, having integral means for circulation of air. Such units are allowed to be equipped with louvers or face extensions made in accordance with the manufacturer's specifications.

❖ Unit heaters are similar to warm-air furnaces, except that ducts are not usually associated with unit heaters.

These heaters are typically suspended from ceilings or roof structures.

UNVENTED ROOM HEATER. An unvented heating *appliance* designed for stationary installation and utilized to provide comfort heating. Such *appliances* provide radiant heat or convection heat by gravity or fan circulation directly from the heater and do not utilize ducts.

❖ As opposed to vented heaters, unvented heaters discharge all products of combustion into the space being heated [see Commentary Figure G2403(16)].

VALVE. A device used in *piping* to control the gas supply to any section of a system of *piping* or to an *appliance*.

❖ The code requires valves to perform various functions in a fuel-gas system as described in the definitions below. These valves are designed for the specific use intended and many are factory installed on equipment and are part of the listed equipment. However, there are valves that are not listed or manufactured in accordance with the standards listed in the code and if allowed must be evaluated in accordance with Section R104.11.

Appliance shutoff. A *valve* located in the *piping system*, used to isolate individual *appliances* for purposes such as service or replacement.

❖ The definition makes it clear that the primary purpose of an appliance shutoff valve is not related to the emergency shutoff of the appliance.

Automatic. An automatic or semiautomatic device consisting essentially of a *valve* and an operator that control the gas supply to the *burner(s)* during operation of an *appliance*. The operator shall be actuated by application of gas pressure on a flexible diaphragm, by electrical means, by mechanical means or by other *approved* means.

Automatic gas shutoff. A *valve* used in conjunction with an automatic gas shutoff device to shut off the gas supply to a water-heating system. It shall be constructed integrally with the gas shutoff device or shall be a separate assembly.

Individual main burner. A *valve* that controls the gas supply to an individual *main burner*.

Main burner control. A *valve* that controls the gas supply to the *main burner* manifold.

Manual main gas-control. A manually operated *valve* in the gas line for the purpose of completely turning on or shutting off the gas supply to the *appliance*, except to *pilot* or pilots that are provided with independent shutoff.

Manual reset. An automatic shutoff valve installed in the gas supply *piping* and set to shut off when unsafe conditions occur. The device remains closed until manually reopened.

Service shutoff. A valve, installed by the serving gas supplier between the service meter or source of supply and the customer *piping system*, to shut off the entire *piping system*.

VENT. A *pipe* or other conduit composed of factory-made components, containing a passageway for conveying *combustion products* and air to the atmosphere, *listed* and *labeled* for use with a specific type or class of *appliance*.

❖ In code terminology, vents are distinguished from chimneys and usually are constructed of factory-made listed and labeled components intended to function as a system. Type B and BW vents are constructed of galvanized steel and aluminum sheet metal and are double wall and air insulated. Such vents are designed to vent gas-fired appliances and equipment that are equipped with draft hoods or are specifically listed (labeled) for use with Type B or BW vents. Type L vents are typically constructed of sheet steel and stainless steel. They are double wall and air insulated and are designed to vent gas- and oil-fired appliances and equipment. Some appliances are designed for use with corrosion-resistant special vents, such as those made of plastic pipe and special alloys of stainless steel [see Commentary Figure G2403(18)].

Photo courtesy of Selkirk, L.L.C.

Figure G2403(18)
TYPE B VENT

Special gas vent. A vent *listed* and *labeled* for use with *listed* Category II, III and IV gas *appliances*.

❖ These vents include high-temperature plastic pipe, stainless steel pipe and low-temperature (PVC and CPVC) plastic pipes.

Type B vent. A vent *listed* and *labeled* for use with *appliances* with *draft hoods* and other Category I *appliances* that are *listed* for use with Type B vents.

❖ See the commentary to "Vent."

Type BW vent. A vent *listed* and *labeled* for use with wall *furnaces.*

❖ See the commentary to "Vent."

Type L vent. A vent *listed* and *labeled* for use with *appliances* that are *listed* for use with Type L or Type B vents.

❖ See the commentary to "Vent."

VENT CONNECTOR. See "Connector."

❖ See the commentary to "Connector."

VENT PIPING.

Breather. *Piping* run from a pressure-regulating device to the outdoors, designed to provide a reference to *atmospheric pressure*. If the device incorporates an integral pressure *relief* mechanism, a breather vent can also serve as a *relief* vent.

❖ A breather vent is distinct from a relief vent and allows the cavity on the air side of the regulator diaphragm to "breathe" by letting air in and out as the diaphragm moves within the regulator housing (see commentary, Section G2421.3). Although not addressed in the definition, breather vents are also part of diaphragm-operated gas controls such as high- and low-pressure gas cutoffs.

Relief. *Piping* run from a pressure-regulating or pressure-limiting device to the outdoors, designed to provide for the safe venting of gas in the event of excessive pressure in the *gas piping system.*

❖ Relief-vent piping discharges fuel gas from a pressure relief valve or pressure regulator with an integral relief valve. Unlike breather vents that would discharge fuel gas only in the event of diaphragm failure, relief vents will always discharge fuel gas, if they discharge at all. Relief devices release fuel gas under emergency conditions to prevent overpressurizing the piping system being monitored and protected.

VENTED APPLIANCE CATEGORIES. *Appliances* that are categorized for the purpose of vent selection are classified into the following four categories:

❖ Gas appliances are categorized for the purpose of matching the appliance to the required type of venting. Not all gas appliances are assigned a category.

Category I. An *appliance* that operates with a nonpositive vent static pressure and with a vent gas temperature that avoids excessive *condensate* production in the vent.

❖ These appliances include draft-hood-equipped and fan-assisted types.

Category II. An *appliance* that operates with a nonpositive *vent* static pressure and with a vent gas temperature that is capable of causing excessive *condensate* production in the vent.

❖ The condition of nonpositive pressure (draft dependent) combined with low flue-gas temperatures makes it difficult to design such an appliance. At the time of this publication, no such appliances are known to exist in the marketplace.

Category III. An *appliance* that operates with a positive vent static pressure and with a vent gas temperature that avoids excessive *condensate* production in the vent.

❖ These appliances include mid-efficiency (approximately 78 to 83 percent) appliances that are typically side-wall-vented with stainless steel special vents.

Category IV. An *appliance* that operates with a positive vent static pressure and with a vent gas temperature that is capable of causing excessive *condensate* production in the vent.

❖ These appliances include high-efficiency (84 percent and higher) condensing-type furnaces, boilers and water heaters.

VENTED ROOM HEATER. A vented self-contained, free-standing, nonrecessed *appliance* for furnishing warm air to the space in which it is installed, directly from the heater without duct connections.

❖ These appliances are typically small space heaters that discharge all products of combustion to the outdoors.

VENTED WALL FURNACE. A self-contained vented *appliance* complete with grilles or equivalent, designed for incorporation in or permanent attachment to the structure of a building, mobile home or travel trailer, and furnishing heated air circulated by gravity or by a fan directly into the space to be heated through openings in the casing. This definition shall exclude *floor furnaces, unit heaters* and *central furnaces* as herein defined.

❖ Wall furnaces are designed to occupy very little area of the room and provide space heating for one or more rooms in small occupancies such as dwelling units, cottages and the like.

VENTING SYSTEM. A continuous open passageway from the *flue collar* or *draft hood* of an *appliance* to the outdoor atmosphere for the purpose of removing flue or vent gases. A venting system is usually composed of a vent or a chimney and *vent connector*, if used, assembled to form the open passageway.

❖ Venting systems fall into one of two categories: natural draft or mechanical draft systems. Natural draft chimneys and vents do not rely on any mechanical means to convey combustion products to the outdoors. Draft is produced by the temperature difference between the vent gases (combustion gases) and the ambient atmosphere. Hot gases are less dense and more buoyant; therefore, they are displaced by cooler (more dense) ambient gases, causing the hotter gases to rise and produce a draft.

Mechanical draft systems use fans or other mechanical means to cause the removal of flue or vent gases and also fall into one of two categories, based on positive or nonpositive static vent pressure. The positive vent pressure systems are referred to as forced-draft or power venting systems, and the nonpositive vent pressure systems are referred to as induced-draft venting systems.

WALL HEATER, UNVENTED TYPE. A room heater of the type designed for insertion in or attachment to a wall or partition. Such heater does not incorporate concealed venting arrangements in its construction and discharges all products of *combustion* through the front into the room being heated.

❖ See the definition and commentary to "Unvented room heater."

WATER HEATER. Any heating *appliance* or *equipment* that heats potable water and supplies such water to the potable hot water distribution system.

❖ A water heater is a closed pressure vessel or heat exchanger that has a heat source and that supplies potable (drinkable) water to the building's hot water distribution system. Large commercial water heaters are sometimes referred to as "hot water supply boilers." Water heaters can be of the storage type with an integral storage vessel, circulating type for use with an external storage vessel, tankless instantaneous type without storage capacity and point-of-use type with or without storage capacity. This chapter and the plumbing chapters regulate water heaters because these appliances have elements and installation requirements related to both mechanical and plumbing systems (see Chapter 28).

SECTION G2404 (301)
GENERAL

G2404.1 (301.1) Scope. This section shall govern the approval and installation of all *equipment* and *appliances* that comprise parts of the installations regulated by this *code* in accordance with Section G2401.

❖ This section states that this chapter governs the approval and installation of all gas-fired equipment and appliances that are regulated by the code. Section G2401.1 establishes the scope of application of the code (see commentary, Section G2401.1).

G2404.2 (301.1.1) Other fuels. The requirements for *combustion* and *dilution air* for gas-fired *appliances* shall be governed by Section G2407. The requirements for *combustion* and *dilution air* for *appliances* operating with fuels other than fuel gas shall be regulated by Chapter 17.

❖ Chapters 17 and 24 each have combustion air provisions that are specific to the fuels addressed in the respective chapters.

G2404.3 (301.3) Listed and labeled. *Appliances* regulated by this *code* shall be *listed* and *labeled* for the application in which they are used unless otherwise *approved* in accordance with Section R104.11. The approval of unlisted *appliances* in accordance with Section R104.11 shall be based upon *approved* engineering evaluation.

❖ Gas-fired appliances must be listed and labeled by an approved agency to show that they comply with the applicable national standards. The code requires listing and labeling for appliances such as boilers, furnaces, space heaters, direct-fired heaters, cooking appliances, clothes dryers, rooftop heating, ventilating and air-conditioning (HVAC) units, etc. The code also requires listing for system components as specifically stated in the text addressing those components. The label is the primary, if not the only, assurance to the installer, the inspector and the end user that a representative sample of an appliance model has been tested and evaluated by an approved agency and has been determined to perform safely and efficiently when installed and operated in accordance with its listing.

Appliances must be listed and labeled for the application in which they are used, otherwise the installation would be a misapplication of the appliance. For example, if an appliance is listed for indoor use only and is installed outdoors, this installation is a misapplication of the appliance and serious malfunctions and/or conditions could result. An appliance might be marketed and installed for a particular purpose for which it was not tested and listed and this is what this section intends to prohibit. Verifying that an appliance has a testing agency label is only part of the code official's responsibility. He or she must also verify that the listing from the testing agency includes the application at hand. The bottom line is, the use of an appliance must match the use for which the appliance was tested.

The presence of a label is part of the information that the code official is to consider in the approval of appliances. The only exception to the labeling requirement occurs when the code official approves a specific appliance in accordance with the authority granted in Section R104.11.

Approval of unlabeled appliances must be based on documentation that demonstrates compliance with applicable standards or, where no product standards exist, that the appliance is appropriate for the intended use and will provide the same level of performance as would be provided by listed and labeled appliances. A fundamental principle of the code is the reliance on the listing and labeling process to ensure appliance performance; approvals granted in accordance with Section 105 must be well justified with supporting documentation. To the code official, the installer and the end-user, very little is known about the performance of an appliance that is not tested and built to an appliance standard.

G2404.4 (301.8) Vibration isolation. Where means for isolation of vibration of an *appliance* is installed, an *approved* means for support and restraint of that *appliance* shall be provided.

❖ Where vibration isolation connections are used in ducts and piping and where equipment is mounted with vibration dampers, support is required for the ducts, piping and equipment to maintain positioning and alignment and to prevent stress and strain on the vibration connectors and dampers.

G2404.5 (301.9) Repair. Defective material or parts shall be replaced or repaired in such a manner so as to preserve the original approval or listing.

❖ Repair work must not alter the nature of appliances and equipment in a way that would invalidate the listing

or conditions of approval. For example, replacement of safety control devices with different devices could alter the design and operation of an appliance from that intended by the manufacturer and the listing agency.

G2404.6 (301.10) Wind resistance. *Appliances* and supports that are exposed to wind shall be designed and installed to resist the wind pressures determined in accordance with this *code*.

❖ Installations of equipment and appliances that are subject to wind forces must be designed to resist those forces. The wind pressures must be based on the wind provisions in this code. The wind pressure requirements are based on the exposure of the building and wind speeds for that region.

G2404.7 (301.11) Flood hazard. For structures located in flood hazard areas, the appliance, equipment and system installations regulated by this code shall be located at or above the elevation required by Section R322 for utilities and attendant equipment.

Exception: The appliance, equipment and system installations regulated by this code are permitted to be located below the elevation required by Section R322 for utilities and attendant equipment provided that they are designed and installed to prevent water from entering or accumulating within the components and to resist hydrostatic and hydrodynamic loads and stresses, including the effects of buoyancy, during the occurrence of flooding to such elevation.

❖ Fuel gas appliances, equipment and installations in flood hazard areas must be installed at or above the elevation specified in the code or must be designed and installed to prevent the entry of water into components and to resist flood loads on the components (see commentary, Section R322.1.6). Exposure to water can cause deterioration of system components and serious appliance and equipment malfunctions. For example, most appliance manufacturers require the replacement of appliances that have been submerged in floodwaters. Components such as pressure regulators, gas controls, motors and electronic circuitry can be ruined by exposure to water. The exception allows appliances and system components to be located below the design flood elevation if they are designed for submersion in floodwaters. Appliances and system components designed for submersion might be difficult or impossible to obtain and it might be extremely difficult to install effective barriers to keep out floodwaters.

G2404.8 (301.12) Seismic resistance. When earthquake loads are applicable in accordance with this code, the supports shall be designed and installed for the seismic forces in accordance with this code.

❖ This code requires building systems to be designed for specified seismic forces. This section references the detailed equipment and component seismic support requirements contained in this code to bring these requirements to the attention of the design professional and the permit applicant.

Equipment piping must be braced for earthquake loads as stated in this code. The failure of the supports for these components has been shown to be a threat to health and safety in geographical areas where moderate- to high-magnitude earthquakes occur. The code specifies the geographical locations where earthquake design is required for certain piping and equipment and the size of the components that must be braced.

G2404.9 (301.14) Rodentproofing. Buildings or structures and the walls enclosing habitable or occupiable rooms and spaces in which persons live, sleep or work, or in which feed, food or foodstuffs are stored, prepared, processed, served or sold, shall be constructed to protect against the entry of rodents.

❖ This section states the requirements to prevent rodent infestation of a building. Efforts must be made to protect the annular spaces around openings in exterior walls. The annular spaces can be sealed by any effective method, but the primary methods used are to fill the annular space with a sealant material or to place a metal collar around the penetrating pipe, duct, etc. The collar material must be durable for the weather exposure and strong enough to prevent the rodents from chewing through and entering the building.

Effort must also be made to cover ventilation and combustion air openings with wire cloth to prevent the entry of rodents through the opening.

G2404.10 (307.5) Auxiliary drain pan. Category IV condensing *appliances* shall be provided with an auxiliary drain pan where damage to any building component will occur as a result of stoppage in the *condensate* drainage system. Such pan shall be installed in accordance with the applicable provisions of Section M1411.

Exception: An auxiliary drain pan shall not be required for *appliances* that automatically shut down operation in the event of a stoppage in the *condensate* drainage system.

❖ This section treats condensing (Category IV) appliances the same as the mechanical chapters treat appliances and equipment having evaporators and cooling coils. Category IV appliances produce condensate and, like cooling equipment and evaporators, can cause structural damage to a building if the condensate drainage system fails. This section parallels Section M1411.3. Section M1411.3.1 is applicable; however, only Items 1 and 3 apply because this section requires an auxiliary pan in all cases. In many cases, this provision will require nothing more than what is already being provided for the cooling coil or evaporator that is part of the gas appliance installation (e.g., the typical gas-fired furnace with cooling coil/ evaporator). For example, an up-flow Category IV furnace with an "A" coil on its outlet is installed in an attic and it is sitting in an auxiliary drain pan. This pan can also serve as protection for the furnace, therefore, no additional installation is required because the furnace happened to be a condensing type (see Commentary Figure G2404.10).

The exception recognizes that some condensing appliances are designed to shut down automatically upon failure of the condensate drainage system. For example, the design/test standards for some appliances allow a choice of appliance reaction to a failed condensate drain. The appliance may either continue to function safely with a failed condensate drain or the appliance must shut down upon failure of the condensate drain.

G2404.11 (307.6) Condensate pumps. Condensate pumps located in uninhabitable spaces, such as attics and crawl spaces, shall be connected to the *appliance* or *equipment* served such that when the pump fails, the *appliance* or *equipment* will be prevented from operating. Pumps shall be installed in accordance with the manufacturer's instructions.

❖ Condensate pumps are often located in attics and crawl spaces and above ceilings, where they are not readily observable. If they fail, the condensate overflow can cause structural damage to the building, especially where the overflow will not be immediately noticed. The majority of such pumps are equipped with simple float controls that can be wired in series with the appliance/equipment control circuit. When the pump system fails, the float will rise in the reservoir and open a switch before the condensate starts to overflow the reservoir. These float controls are commonly not connected, and in other cases the pump might not be equipped with an overflow switch. This new code section requires the installation of condensate pumps that have this overflow shutoff capability and requires that the appliance/equipment served be connected to take advantage of that feature (see Commentary Figure G2404.11).

SECTION G2405 (302)
STRUCTURAL SAFETY

G2405.1 (302.1) Structural safety. The building shall not be weakened by the installation of any gas *piping*. In the process of installing or repairing any gas *piping*, the finished floors, walls, ceilings, tile work or any other part of the building or premises which is required to be changed or replaced shall be left in a safe structural condition in accordance with the requirements of this code.

❖ The installation of fuel-gas systems must not adversely affect the structural integrity of the building components. This code dictates the structural safety requirements that must be applied to any structural portion of the building that is penetrated, altered or removed during the installation, replacement or repair of fuel-gas systems.

G2405.2 (302.4) Alterations to trusses. Truss members and components shall not be cut, drilled, notched, spliced or otherwise altered in any way without the written concurrence and approval of a registered design professional. *Alterations* resulting in the addition of loads to any member, such as HVAC equipment and water heaters, shall not be permitted without verification that the truss is capable of supporting such additional loading.

❖ A truss is an engineered system of components that function as a structural unit. Trusses are susceptible to weakening by almost any type of alteration, such as boring, notching and cutting. Unlike dimensional solid lumber members, such as studs, rafters and joists, trusses cannot be altered in a code-prescribed way. The only exception is obtaining written approval of a specific alteration from a registered design professional.

The placement of a water heater in an attic, for example, can result in abnormal loads on roof trusses that must be accounted for structurally. A simple 40-gallon water heater weighs well over 400 pounds and is a point load. Subjecting trusses to loads that they were not designed to carry is considered to be an alter-

ation and is prohibited without a structural analysis to verify that the truss can bear the additional weight.

G2405.3 (302.3.1) Engineered wood products. Cuts, notches and holes bored in trusses, structural composite lumber, structural glued-laminated members and I-joists are prohibited except where permitted by the manufacturer's recommendations or where the effects of such *alterations* are specifically considered in the design of the member by a registered design professional.

❖ This section applies to engineered lumber products which are manufactured structural members. These products often have installation instructions that specifically state if, where and how the member can be altered, drilled, notched, etc. What appears to be a harmless cut, notch or boring might actually be structurally damaging to the member (see Section G2405.2).

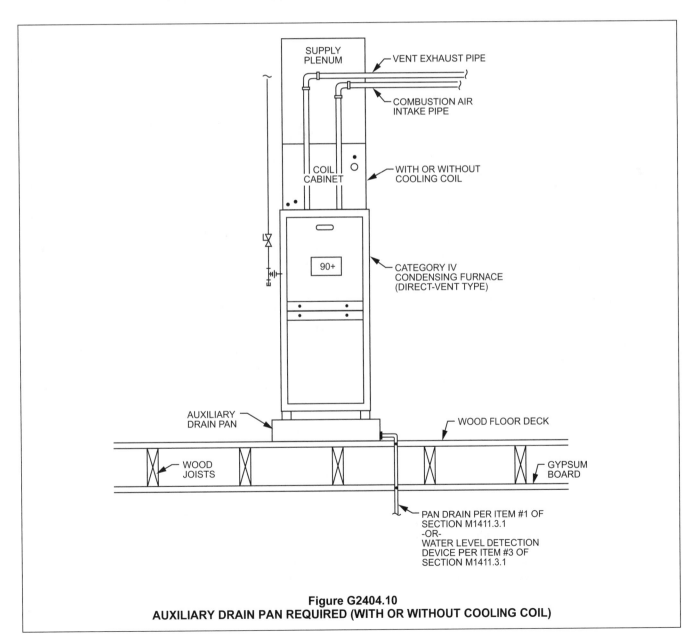

Figure G2404.10
AUXILIARY DRAIN PAN REQUIRED (WITH OR WITHOUT COOLING COIL)

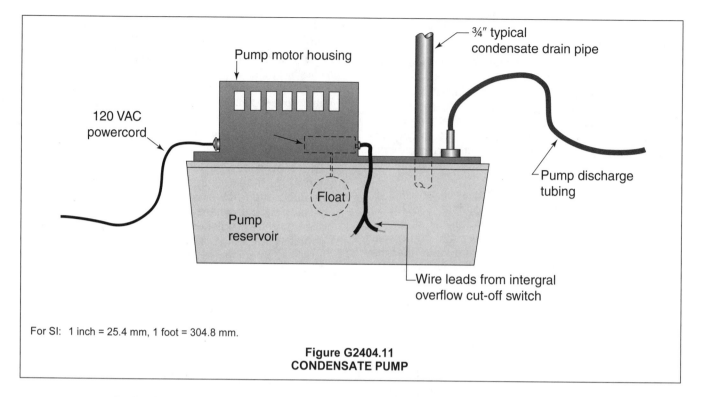

120 VAC
powercord

Pump motor housing

¾" typical
condensate drain pipe

Float

Pump
reservoir

Pump discharge
tubing

Wire leads from intergral
overflow cut-off switch

For SI: 1 inch = 25.4 mm, 1 foot = 304.8 mm.

**Figure G2404.11
CONDENSATE PUMP**

SECTION G2406 (303)
APPLIANCE LOCATION

G2406.1 (303.1) General. *Appliances* shall be located as required by this section, specific requirements elsewhere in this code and the conditions of the *equipment* and *appliance* listing.

❖ Section G2406 is a consolidation of the code's generally applicable location requirements and limitations. The listing for an appliance or equipment will often contain location requirements parallel with or in addition to these sections.

G2406.2 (303.3) Prohibited locations. *Appliances* shall not be located in sleeping rooms, bathrooms, toilet rooms, storage closets or surgical rooms, or in a space that opens only into such rooms or spaces, except where the installation complies with one of the following:

1. The *appliance* is a direct-vent *appliance* installed in accordance with the conditions of the listing and the manufacturer's instructions.

2. *Vented room heaters*, wall *furnaces*, vented decorative *appliances*, vented gas *fireplaces*, vented gas *fireplace* heaters and decorative *appliances* for installation in vented solid fuel-burning *fireplaces* are installed in rooms that meet the required volume criteria of Section G2407.5.

3. A single wall-mounted *unvented room heater* is installed in a bathroom and such *unvented room heater* is equipped as specified in Section G2445.6 and has an input rating not greater than 6,000 *Btu*/h (1.76 kW). The bathroom shall meet the required volume criteria of Section G2407.5.

4. A single wall-mounted *unvented room heater* is installed in a bedroom and such *unvented room heater* is equipped as specified in Section G2445.6 and has an input rating not greater than 10,000 *Btu*/h (2.93 kW). The bedroom shall meet the required volume criteria of Section G2407.5.

5. The *appliance* is installed in a room or space that opens only into a bedroom or bathroom, and such room or space is used for no other purpose and is provided with a solid weather-stripped door equipped with an *approved* self-closing device. All *combustion air* shall be taken directly from the outdoors in accordance with Section G2407.6.

❖ The intent of this section is to prevent fuel-fired appliances from being installed in rooms and spaces where the combustion process could pose a threat to the occupants. Potential threats include depleted oxygen levels; elevated levels of carbon dioxide, nitrous oxides, carbon monoxide, and other combustion gases; ignition of combustibles and elevated levels of flammable gases.

In small rooms such as bedrooms and bathrooms, the doors are typically closed when the room is occupied, which could allow combustion gases to build up to life-threatening levels. In bedrooms, sleeping occupants would not be alert to or aware of impending danger.

If an appliance obtains combustion air from a room or space, it communicates with the atmosphere in that room or space whether or not it is installed in that room or space. An appliance might be in a room, closet or alcove and obtain combustion air from an adjacent room, so Section G2406.2 is worded to address the location of an appliance in the rooms listed and in spaces that open only into such rooms. In other words, an appliance in a closet accessed from a bedroom is no different from an appliance located within the bedroom. It is not the intent of this section to prevent combustion air from being taken from a bedroom, bathroom, etc., as evidenced in Items 2, 3 and 4. For example, the volume of a bedroom could be added to the volume of other rooms for the purpose of providing indoor combustion air for an appliance not installed in a location prohibited by this section if openings are installed to conjoin the space volumes in accordance with Section G2407.5.3. If an appliance obtains combustion air from a room, the appliance combustion chamber would be open to the room and the appliance must be considered to be in that room.

Item 1 recognizes that direct-vent appliances have sealed combustion chambers and obtain all combustion air directly from the outdoors. The appliance combustion chambers do not communicate with the room atmosphere.

Item 2 requires that the room be able to supply the necessary combustion air by infiltration as specified in Section G2407.5. Note that Item 2 includes both vented decorative appliances and vented gas fireplaces which are both addressed by ANSI Z21.50 and are essentially the same appliance, known by different names. Vented gas fireplace heaters (ANSI Z21.88) were also added to Item 2.

Items 3 and 4 allow the installation of a single wall-mounted unvented room heater in bathrooms and bedrooms if the heaters are equipped with oxygen depletion safety shutoff systems, are limited in Btu input rating and the space is capable of supplying indoor combustion air in accordance with Section G2407.5. These exceptions specify "wall-mounted" heaters, which are fully enclosed and less susceptible to tampering and other conditions that might affect the combustion process. Items 3 and 4 would not apply to room heaters that stand on the floor or fasten to a fireplace hearth or ventless firebox hearth [see Commentary Figure G2406.2(1)].

Item 5 would allow installation of fuel-fired appliances within a separate dedicated space that is accessed from the rooms and spaces listed in this section. A separated space containing the appliance must be open to the outdoors in accordance with Section G2407.6, and the access door to the space must be solid and weather-stripped to prevent communication between atmospheres in the separated spaces. The door must also be self-closing and not rely on occupants to keep it closed. The enclosure must not be used for storage or any other purpose. The intent is to isolate the appliance(s) from the rooms listed in this section and to obtain all combustion air directly from the outdoors [see Commentary Figure G2406.2(2)]. This item can be used to avoid relocating an appliance when an existing appliance installed in a prohibited location needs to be replaced [see Commentary Figure G2406.2(2)].

G2406.3 (303.6) Outdoor locations. *Appliances* installed in outdoor locations shall be either listed for outdoor installation or provided with protection from outdoor environmental factors that influence the operability, durability and safety of the *appliance*.

❖ Appliances installed outdoors must be specifically listed for outdoor installation or be protected from outdoor conditions that will affect the "operability, durability and safety of the equipment."

The concern is for weather and ambient temperatures. The manufacturer's instructions and listing must be consulted to determine whether a particular appliance is designed for or can be made suitable for outdoor installation. For example, furnaces cannot be installed outdoors in cold climates regardless of any weather-proof enclosure, unless the heat exchanger, burner assemblies and venting system are designed for exposure to temperatures below normal indoor temperatures. Cold ambient temperatures can cause harmful condensation to occur on heat exchanger surfaces of fuel-fired appliances.

Additionally, there may be local ordinances that govern the outdoor installation of appliances. Before installing an appliance outdoors, consult local zoning regulations, ordinances and subdivision covenants. Many of these regulations strictly limit the location of outdoor mechanical equipment and appliances. Also, for roof installations, the roof structure must be able to support all imposed static and dynamic structural loads. Substantiating data on the structural adequacy of the entire installation must be submitted and approved.

Photo by D.F. Noyes Studios, courtesy of DESA International

Figure G2406.2(1)
UNVENTED WALL-MOUNTED ROOM HEATER

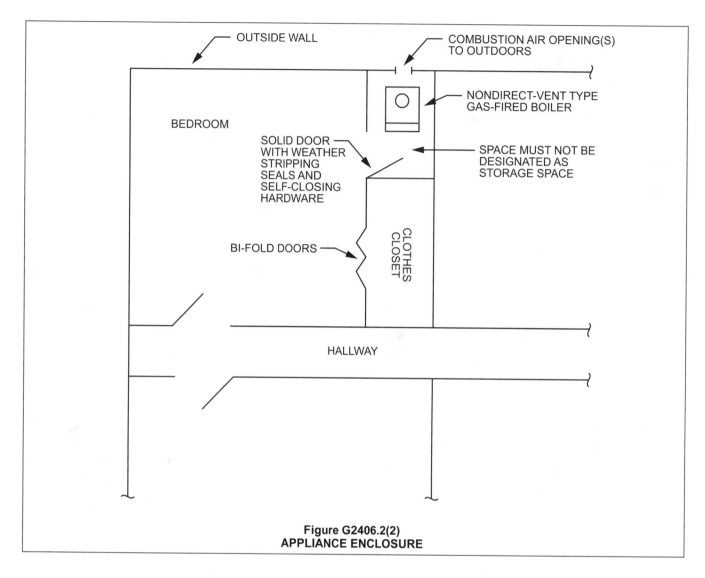

Figure G2406.2(2)
APPLIANCE ENCLOSURE

SECTION G2407 (304)
COMBUSTION, VENTILATION AND DILUTION AIR

G2407.1 (304.1) General. Air for *combustion*, ventilation and dilution of *flue gases* for *appliances* installed in buildings shall be provided by application of one of the methods prescribed in Sections G2407.5 through G2407.9. Where the requirements of Section G2407.5 are not met, outdoor air shall be introduced in accordance with one of the methods prescribed in Sections G2407.6 through G2407.9. *Direct-vent appliances*, gas *appliances* of other than *natural draft* design, vented gas *appliances* not designated as Category I and *appliances* equipped with power burners, shall be provided with *combustion*, ventilation and *dilution air* in accordance with the *appliance* manufacturer's instructions.

Exception: *Type 1 clothes dryers* that are provided with *makeup air* in accordance with Section G2439.5.

❖ The provisions of Section G2407 describe requirements for the combustion air necessary for the complete combustion of fuel gas, dilution of flue gases, and ventilation of gas-fired appliances and the space in which they are installed. An inadequate combustion air supply to gas-fired appliances can compromise safety by causing in-complete combustion, resulting in appliance malfunction and production of excess carbon monoxide.

Complete combustion of fuel gas is essential for the proper operation of gas-fired appliances. If insufficient quantities of oxygen are supplied, the combustion process will be incomplete, creating hazardous by-products [see Commentary Figures G2407.1(1) and G2407.1(2)].

Although not implied in the term, combustion air also serves other purposes in addition to supplying oxygen. Combustion air ventilates and cools appliances and the rooms or spaces that enclose them. Combustion air also plays an important role in producing and controlling draft in vents and chimneys.

Despite the fact that an adequate combustion air supply is extremely important, it is one aspect of gas-fired equipment installations that is often overlooked, ignored or compromised. Depending on the appliance type, location and building construction, supplying combustion air can either be easy or can involve complex designs and extraordinary methods. In any case, the importance of a proper combustion air supply cannot be overemphasized.

The methods of supplying combustion air range from simple (inherently more dependable) methods to more complex methods. This section offers five methods for providing combustion air [see Commentary Figure G2407.1(3)]:

1. All indoor air.

2. All outdoor air.

3. Combination indoor and outdoor air.

4. Mechanical combustion air supply.

5. Engineered design.

The last sentence of this section clarifies that the section is intended to apply only to natural-draft atmospheric-burner-design appliances, Category I appliances, nondirect-vent appliances and appliances not equipped with power burners (see commentary, Section G2407.8). For example, the application of Section G2407.6 with large power-burner-equipped (nonnatural draft, nonatmospheric-burner designs) boilers could result in excessively large openings in an outside wall, causing environmental problems within the boiler room.

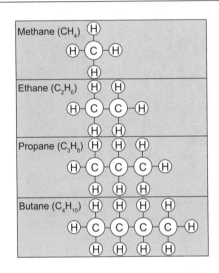
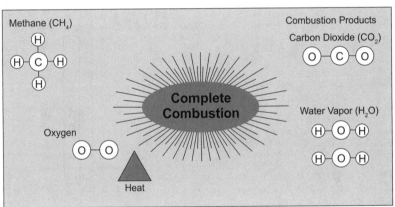

Figure courtesy of Reznor/Thomas & Betts Corporation

Figure G2407.1(1)
THE CHEMISTRY OF IDEAL COMBUSTION

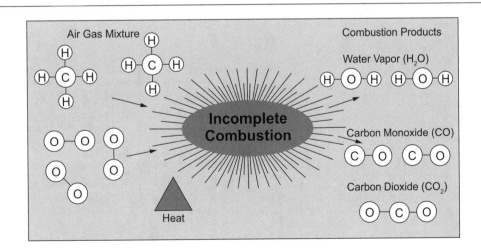

Figure courtesy of Reznor/Thomas & Betts Corporation

Figure G2407.1(2)
THE CHEMISTRY OF INCOMPLETE COMBUSTION

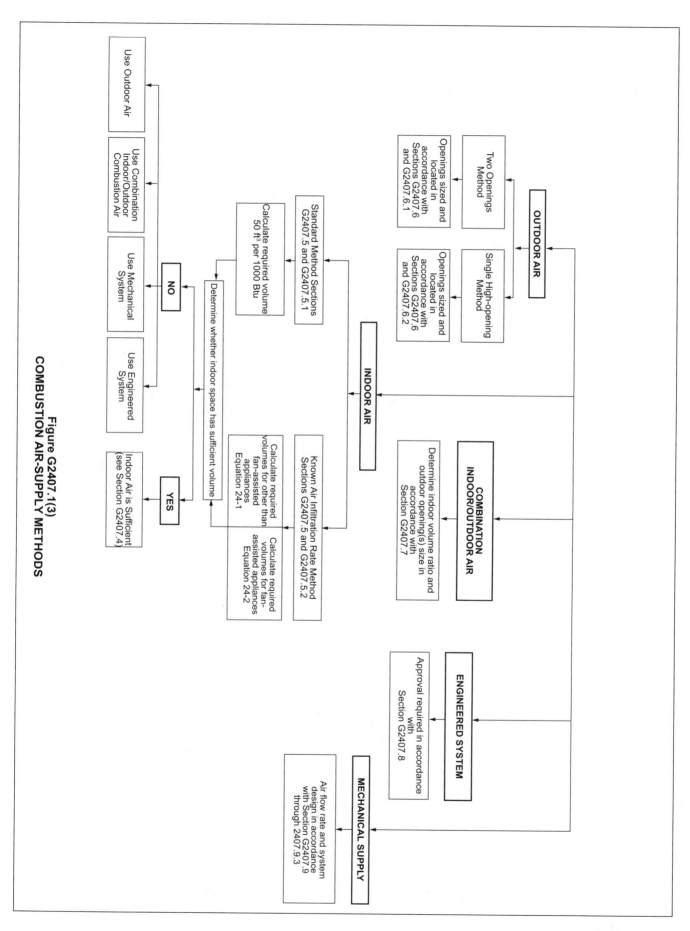

Figure G2407.1(3)
COMBUSTION AIR-SUPPLY METHODS

Section G2407.1 states that the methods of providing combustion air as prescribed in Sections G2407.5 through G2407.9 are applicable only to certain types of appliances, namely natural-draft appliances with draft hoods and fan-assisted appliances. The other types of appliances, including direct-vent appliances (direct-vent appliances can be natural-draft type but have no draft hoods), power-vented appliances and Categories II, III and IV appliances, must be provided with combustion air as dictated in the appliance manufacturer's instructions. Another type of appliance has been added to the types that are subject only to the requirements in the manufacturer's instructions—appliances equipped with power burners. Power burners pull in combustion air under fan power and mix it with the fuel gas before conveying it to the burner.

Natural-draft appliances take in combustion air by means of the weak force produced by the natural draft in the venting system only. Clearly there is a significant difference between the mechanical force of power burners with combustion air fans and natural-draft appliances that must rely on the venting system draft to cause combustion air to enter the appliance. The methods in Sections G2407.5 through G2407.9 were based on the physics of natural-draft, fan-assisted and Category I appliances and therefore failed to account for the nature of powered systems like power burners. This fact has perplexed designers for many years as they tried to design combustion air openings for large input boilers and similar appliances that are commonly equipped with power burners. When they used the sizing rules in the code for combustion air openings in the exterior wall of a boiler room, for example, the opening size would be so large that it would be impossible to prevent the freezing of pipes and equipment in colder climates. The openings to the outdoors were prohibitively large for such large input power-burner appliances and out of proportion with the need when considering that power burners mechanically draw in combustion air. When natural draft is depended on, larger openings are necessary to reduce the pressure drop across the openings, but power burners can tolerate much larger pressure drops, thus allowing smaller openings. The manufacturer's instructions for large input power-burner appliances specify the size of combustion air openings by their own methods. Typically those methods result in much smaller openings than prescribed in the code; however, some manufacturers simply refer back to the code instead of offering their own method. A research study demonstrated that the code-prescribed opening sizing methods were overkill for power-burner appliances [see Commentary Figure G2407.1(6)]. Direct-vent appliances supply their own combustion air through an outdoor air intake pipe or duct [see Commentary Figures G2407.1(4) and G2407.1(5)].

The exception recognizes that clothes dryers receive the required combustion air from the makeup air that compensates for the exhaust air flow from the appliance.

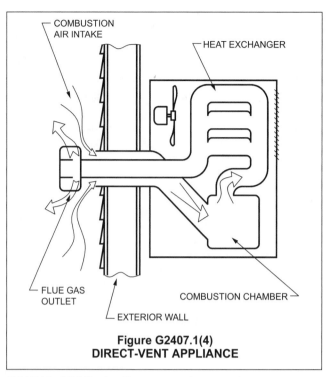

Figure G2407.1(4)
DIRECT-VENT APPLIANCE

G2407.2 (304.2) Appliance location. *Appliances* shall be located so as not to interfere with proper circulation of *combustion*, ventilation and *dilution air*.

❖ The provisions of Section G2407 rely on the natural (gravity) movement of air; therefore, installations must allow unimpeded flow of combustion air from the source to the appliances being supplied. Other mechanical systems, appliances or equipment in the same room or building can adversely affect the combustion air supply. For example, when placed too close to the combustion air openings, gas-fired appliances or equipment will restrict the free circulation of air.

G2407.3 (304.3) Draft hood/regulator location. Where used, a *draft hood* or a *barometric draft regulator* shall be installed in the same room or enclosure as the *appliance* served to prevent any difference in pressure between the hood or regulator and the *combustion air* supply.

❖ It is important that draft hoods and barometric draft regulators be located in the same pressure zone as the appliance combustion chamber.

Draft hoods and barometric regulators both serve to stabilize draft through an appliance and in some cases act to relieve backpressure/downdraft in the venting system. It is the draft in a Category I appliance that moves combustion air into the appliance and conveys the combustion gases to the venting system. Pressure differences between the atmospheres in which the draft hood/regulator and the combustion chamber are located could interfere with the function of the draft hood/regulator, resulting in poor combustion, excess draft, insufficient draft, combustion gas spillage and/or hazardous appliance operation. A draft hood/regulator is installed under the assumption that the combustion chamber it serves sees the same atmospheric pressure.

G2407.4 (304.4) Makeup air provisions. Where exhaust fans, *clothes dryers* and kitchen ventilation systems interfere with the operation of *appliances, makeup air* shall be provided.

❖ The introduction of makeup air is critical to the proper operation of all fuel-burning appliances located in areas subject to the effects of exhaust systems. Too little makeup air will cause excessive negative pressures to develop, not only reducing the exhaust airflow but also interfering with appliance venting. Too little makeup air can cause loss of draft in appliance vents and chimneys or cause combustion byproducts to discharge into the building. It is the draft produced in an appliance that causes combustion air to enter the combustion chamber of natural-draft atmospheric-burner appliances. It is the draft in the venting system that causes the combustion gases and dilution air to move through the venting system toward the vent terminal. A lack of combustion air can result in incomplete fuel combustion, appliance malfunction and flue gas spillage.

A significant reduction in building pressure could be created by fireplaces, exhaust fans, ventilation systems, clothes dryers and similar appliances and equipment. Gas-fired appliances are often in competition

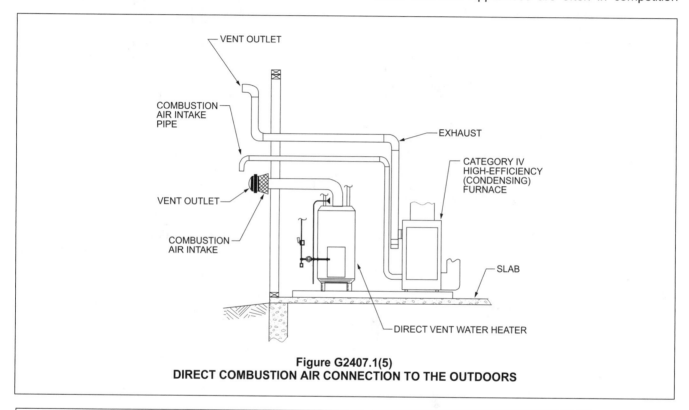

Figure G2407.1(5)
DIRECT COMBUSTION AIR CONNECTION TO THE OUTDOORS

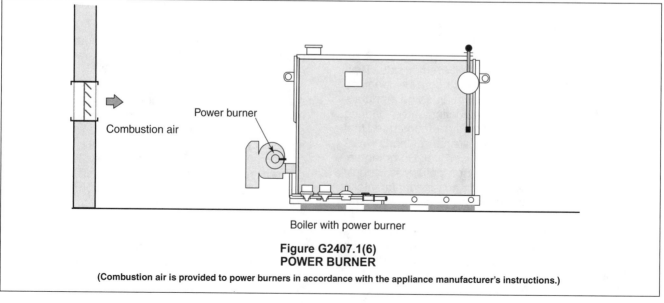

Figure G2407.1(6)
POWER BURNER

(Combustion air is provided to power burners in accordance with the appliance manufacturer's instructions.)

with other equipment or systems for the available combustion air and makeup air that flows through (infiltrates) the building envelope or is otherwise introduced into a building, room or space. The competition between powered exhaust equipment and natural-draft gas-fired appliance venting systems is an unfair contest at best. Eventually, the powered equipment will starve the natural-draft appliances unless provisions are made to compensate for the effect of the powered exhaust equipment/appliances. Supplying makeup air will prevent negative pressures within a space, thus negating the effect of exhaust fans and clothes dryers. In many cases, makeup air will be required because this is the only way to prevent exhaust systems from negatively affecting the appliance venting systems that draw combustion air from the same space from which the exhaust systems remove air.

Natural draft appliances also compete among themselves for combustion air. The appliance venting system that produces the strongest draft, such as a solid-fuel appliance or fireplace, can cause combustion air shortages for the appliance venting systems that produce a weaker draft.

Exhaust fans and similar equipment and appliances can produce significant negative building pressures that can interfere with the operation of vents and chimneys. This interference can cause reverse flow as outdoor air enters the building through the vents and chimneys be cause of the pressure difference. Any such interference with vents or chimneys could cause products of combustion to be discharged into the building and, therefore, must be avoided. The provisions of Section G2407 could be woefully inadequate or nullified where compensating features are not installed to mitigate the effect of appliances or equipment removing air from any room or space.

When any system depends on infiltration as the source of combustion air, it will probably be necessary to supply makeup air to offset the deficiency in many buildings. It is obvious that natural draft appliance installations cannot compete with exhaust fans and clothes dryers for the available infiltration air. Without makeup air, the gas-fired appliance venting systems and the mechanical exhaust systems will indeed compete for the same infiltration air, which could be insufficient to satisfy either need (see Commentary Figure G2407.4). Note that exhaust fans, clothes dryers and similar equipment that remove air from a space will have no negative impact on unvented appliances because unvented appliances have no venting system. Direct-vent appliances are also immune in most cases.

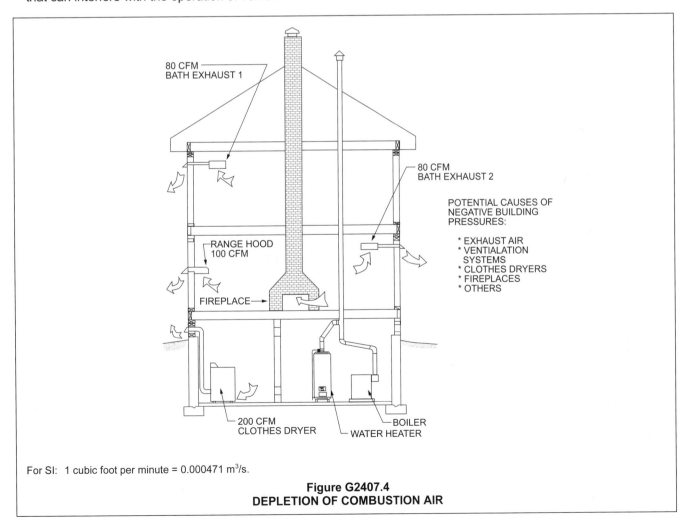

For SI: 1 cubic foot per minute = 0.000471 m³/s.

Figure G2407.4
DEPLETION OF COMBUSTION AIR

G2407.5 (304.5) Indoor combustion air. The required volume of indoor air shall be determined in accordance with Section G2407.5.1 or G2407.5.2, except that where the air infiltration rate is known to be less than 0.40 air changes per hour (ACH), Section G2407.5.2 shall be used. The total required volume shall be the sum of the required volume calculated for all *appliances* located within the space. Rooms communicating directly with the space in which the *appliances* are installed through openings not furnished with doors, and through *combustion air* openings sized and located in accordance with Section G2407.5.3, are considered to be part of the required volume.

❖ The terms "confined space," "unconfined space" and "unusually tight construction" are no longer used in the code. The required room volume is determined by two methods: one method based on the actual air infiltration rate of the building and the other based on the familiar fixed ratio of 50 ft³ per 1000 Btu/h (4.8 m³/kW). Air taken directly from inside the building is an acceptable source of combustion air if a sufficient volume of air is available for the appliances served and the building construction allows sufficient infiltration. The provisions of this section rely on building envelope infiltration as the only source of combustion air.

The method given in Section G2407.5.1 is based on a conservative assumed air infiltration rate of at least 0.40 air change per hour (ACH); therefore, if the ACH is known to be less than 0.40, this method cannot be used. The air infiltration rate for most buildings is not known, and it is possible for the ACH rate of a building to be actually less than 0.40; however, research suggests that buildings having less than 0.40 ACH are rare. It is believed that 80 to 95 percent of U.S. homes have infiltration rates of 0.35 ACH or greater; however, those numbers are changing because of more stringent energy codes and sustainable "green" building practices. A rate of 0.40 ACH or less is the quantitative expression of "tight construction." Testing also indicates that the 50 ft³/1 000 Btu/h (4.8 m³/kW) convention is somewhat liberal; thus, a built-in safety factor exists. If the designer suspects that the building in question is extraordinarily tight (less than 0.40 ACH), calculations and/or testing should be done to verify the ACH rate. A conservative approach could be to use the method in Section G2407.5.2 with a conservative ACH rate such as 0.25, 0.30 or 0.35. Commentary Figure G2407.5.2 shows calculated required volumes for appliances up to 300,000 Btu/h (87.9 kW) input rating based on the standard method and known ACH method for multiple ACH rates.

The standard method can be calculated for each appliance and then all the volumes are added or can be calculated by adding all of the appliance inputs first and then calculating the total volume. The known ACH-rate method must be performed in at least two distinct calculations where both fan-assisted and draft-hood-equipped appliances are present.

To increase the available volume, the code allows rooms and spaces to be coupled together by doorways without doors and by openings installed in accordance with Section G2407.5.3.

G2407.5.1 (304.5.1) Standard method. The minimum required volume shall be 50 cubic feet per 1,000 *Btu*/h (4.8 m³/kW) of the appliance input rating.

❖ This is the optional default method intended for use when the ACH rate is unknown. If the actual ACH rate is known to be greater than 0.40, the method of Section G2407.5.2 will yield smaller required volumes.

If the ACH rate is known to be less than 0.40, this method must not be used because it is based on an assumed ACH rate of at least 0.40. The rate of 50 ft³/1,000 Btu/h (4.8 m³/kW) is a carryover from the previous editions of the code and was used for buildings that were not of unusually tight construction.

G2407.5.2 (304.5.2) Known air-infiltration-rate method. Where the air infiltration rate of a structure is known, the minimum required volume shall be determined as follows:

For *appliances* other than fan-assisted, calculate volume using Equation 24-1.

$$Required\ Volume_{other} \geq \frac{21\,\text{ft}^3}{ACH}\left(\frac{I_{other}}{1{,}000\ \text{Btu/h}}\right)$$

(Equation 24-1)

For fan-assisted *appliances*, calculate volume using Equation 24-2.

$$Required\ Volume_{fan} \geq \frac{15\,\text{ft}^3}{ACH}\left(\frac{I_{fan}}{1{,}000\ \text{Btu/hr}}\right)$$

(Equation 24-2)

where:

I_{other} = All *appliances* other than fan assisted (input in *Btu*/h).

I_{fan} = Fan-assisted *appliance* (input in *Btu*/h).

ACH = Air change per hour (percent of volume of space exchanged per hour, expressed as a decimal).

For purposes of this calculation, an infiltration rate greater than 0.60 *ACH* shall not be used in Equations 24-1 and 24-2.

❖ This method considers the actual or calculated ACH rate and requires space volumes to be commensurate with that rate. This method can also be used when the ACH rate is unknown by simply picking a conservative ACH rate (0.40 ACH or less) representing the lowest anticipated ACH rate for the given building. Equation 24-1 is for draft-hood-equipped appliances and reflects recent research and test results that show that such appliances need less air flow than previously assumed. The standard ratio of 50 ft³/1,000 Btu/h (4.8 m³/kW) that has been in codes for many years was based on an assumed combustion air flow need of 25 ft³/1,000 Btu/h (2.4 m³/kW) with an assumed ACH rate of 0.50. The 25 ft³/1,000 Btu/h air flow consisted of the air required for stoichiometric combustion [10 ft³/ft³ of natural gas] plus excess air to assure complete combustion, draft hood dilution air and a safety factor allowance. The volume of 25 cubic feet (0.7 m3) divided by an ACH rate of 0.50 yields the familiar 50 ft3/1,000 Btu/h ratio. Equation 24-1 is based on a total airflow need of 21 cubic feet (0.6 m3) instead of 25(0.7 m3) because research supports a less conservative revised volume.

Equation 24-2 is for fan-assisted appliances only and accounts for the fact that fan-assisted appliances have no draft hood and, therefore, do not need dilution air. Fan-assisted appliances need only air for stoichiometric combustion [(10 ft³/ft³) of natural gas] and excess air [5 ft³/ft³ of natural gas]. Dividing by the ACH rate adjusts the required volume to correspond to the available infiltration rate.

For example, if the ACH rate is 0.35, 42.8 cubic feet (1.2 m³) would be required for each 1,000 Btu/h (ft³ natural gas), and if the ACH rate is 0.50, 30 cubic feet (0.8 m³) would be required for each 1,000 Btu/h. The greater the ACH rate, the lower the room volume needed.

Research and testing results indicate that the absolute minimum volume for draft-hood-equipped appliances must be 35 ft³/1,000 Btu/h (3.4 m³/kW) and for fan-assisted appliances it must be 25 ft³/1,000 Btu/h (2.4 m³/kW). This is why this section limits the infiltration rate to a maximum of 0.60 ACH. Limiting the ACH rate to 0.60 results in a lower limit safety factor for the required volume, thereby preventing appliances from being installed in exceedingly small spaces. Commentary Figure G2407.5.2 is based on the methodology of Sections G2407.5.1 and G2407.5.2 and is provided for convenience.

G2407.5.3 (304.5.3) Indoor opening size and location. Openings used to connect indoor spaces shall be sized and located in accordance with Sections G2407.5.3.1 and G2407.5.3.2 (see Figure G2407.5.3).

❖ Sections G2407.5.3.1 and G2407.5.3.2 prescribe the size and location of openings used to conjoin (couple) spaces for the purpose of increasing the available volume. The last sentence of Section G2407.5 speaks of this. These openings must be permanently open, except where interlocked motorized dampers are used in accordance with Section G2407.10.

G2407.5.3.1 (304.5.3.1) Combining spaces on the same story. Each opening shall have a minimum free area of 1 square inch per 1,000 *Btu*/h (2,200 mm²/kW) of the total input rating of all *appliances* in the space, but not less than 100 square inches (0.06 m²). One opening shall commence within 12 inches (305 mm) of the top and one opening shall commence within 12 inches (305 mm) of the bottom of the enclosure. The minimum dimension of air openings shall be not less than 3 inches (76 mm).

❖ This section is applicable to adjacent spaces on the same floor level and provides for the familiar high and low openings. The opening configuration creates a thermosiphon air flow with the bottom opening acting as an inlet and the top opening acting as an outlet. This provision originated as a means to conjoin a small appliance enclosure (such as a furnace or boiler room) with other spaces, thereby effectively increasing the volume of the enclosures.

G2407.5.3.2 (304.5.3.2) Combining spaces in different stories. The volumes of spaces in different stories shall be considered as communicating spaces where such spaces are connected by one or more openings in doors or floors having a total minimum free area of 2 square inches per 1,000 *Btu*/h (4402 mm²/kW) of total input rating of all *appliances*.

❖ This section is a new concept that allows spaces on different floor levels to be conjoined, thereby effectively increasing the available volume for supplying combustion air. Installing a high and low opening configuration in accordance with the previous section would serve no purpose when vertically connecting adjacent stories; thus, a single opening is permitted. The opening or openings can be in a floor or in a door opening to an unenclosed stairway that connects the two stories. This method of conjoining spaces is particularly useful for dwelling units where the volume of a basement in which the appliances are located can be conjoined with the open spaces on upper stories by means of a louvered door at the basement stairs. The opening size requirement is simply the addition of the areas of the two openings required by Section G2407.5.3.1. A combustion air opening in a floor would be considered as a transfer opening in a horizontal assembly.

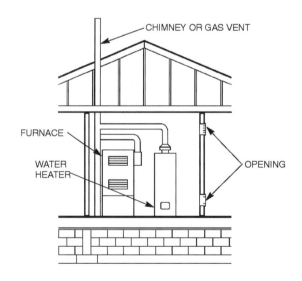

FIGURE G2407.5.3 (304.5.3)
ALL AIR FROM INSIDE THE BUILDING
(see Section G2407.5.3)

Standard Method Required Volume All Appliances		Known Air Infiltration Rate Method, Minimum Space Volume for Appliance Other Than Fan-Assisted, For Specified Infiltration Rates (ACH*)				Known Air Infiltration Rate Method, Minimum Space Volume for Fan-Assisted Appliance, For Specified Infiltration Rates (ACH*)			
Appliance Input Btu/h	Space Volume (ft³)	Appliance Input Btu/h	Space Volume (ft³) 0.25 ACH	Space Volume (ft³) 0.30 ACH	Space Volume (ft³) 0.35 ACH	Appliance Input Btu/h	Space Volume (ft³) 0.25 ACH	Space Volume (ft³) 0.30 ACH	Space Volume (ft³) 0.35 ACH
5,000	250	5,000	420	350	300	5,000	300	250	214
10,000	500	10,000	840	700	600	10,000	600	500	429
15,000	750	15,000	1,260	1,050	900	15,000	900	750	643
20,000	1,000	20,000	1,680	1,400	1,200	20,000	1,200	1,000	857
25,000	1,250	25,000	2,100	1,750	1,500	25,000	1,500	1,250	1,071
30,000	1,500	30,000	2,520	2,100	1,800	30,000	1,800	1,500	1,286
35,000	1,750	35,000	2,940	2,450	2,100	35,000	2,100	1,750	1,500
40,000	2,000	40,000	3,360	2,800	2,400	40,000	2,400	2,000	1,714
45,000	2,250	45,000	3,780	3,150	2,700	45,000	2,700	2,250	1,929
50,000	2,500	50,000	4,200	3,500	3,000	50,000	3,000	2,500	2,143
55,000	2,750	55,000	4,620	3,850	3,300	55,000	3,300	2,750	2,357
60,000	3,000	60,000	5,040	4,200	3,600	60,000	3,600	3,000	2,571
65,000	3,250	65,000	5,460	4,550	3,900	65,000	3,900	3,250	2,786
70,000	3,500	70,000	5,800	4,900	4,200	70,000	4,200	3,500	3,000
75,000	3,750	75,000	6,300	5,250	4,500	75,000	4,500	3,750	3,214
80,000	4,000	80,000	6,720	5,600	4,800	80,000	4,800	4,000	3,429
85,000	4,250	85,000	7,140	5,950	5,100	85,000	5,100	4,250	3,643
90,000	4,500	90,000	7,500	6,300	5,400	90,000	5,400	4,500	3,857
95,000	4,750	95,000	7,900	6,650	5,700	95,000	5,700	4,750	4,071
100,000	5,000	100,000	8,400	7,000	6,000	100,000	6,000	5,000	4,286
105,000	5,250	105,000	8,820	7,350	6,300	105,000	6,300	5,250	4,500
110,000	5,500	110,000	9,240	7,700	6,600	110,000	6,600	5,500	4,714
115,000	5,750	115,000	9,660	8,050	6,900	115,000	6,900	5,750	4,929
120,000	6,000	120,000	10,080	8,400	7,200	120,000	7,200	6,000	5,143
125,000	6,250	125,000	10,500	8,750	7,500	125,000	7,500	6,250	5,357
130,000	6,500	130,000	10,920	9,100	7,800	130,000	7,800	6,500	5,571
135,000	6,750	135,000	11,340	9,450	8,100	135,000	8,100	6,750	5,786
140,000	7,000	140,000	11,760	9,800	8,400	140,000	8,400	7,000	6,000
145,000	7,250	145,000	12,180	10,150	8,700	145,000	8,700	7,250	6,214
150,000	7,500	150,000	12,600	10,500	9,000	150,000	9,000	7,500	6,429
160,000	8,000	160,000	13,440	11,200	9,600	160,000	9,600	8,000	6,857
170,000	8,500	170,000	14,280	11,900	10,200	170,000	10,200	8,500	7,286
180,000	9,000	180,000	15,120	12,600	10,800	180,000	10,800	9,000	7,714
190,000	9,500	190,000	15,960	13,300	11,400	190,000	11,400	9,500	8,143
200,000	10,000	200,000	16,800	14,000	12,000	200,000	12,000	10,000	8,571
210,000	10,500	210,000	17,640	14,700	12,600	210,000	12,600	10,500	9,000
220,000	11,000	220,000	18,480	15,400	13,200	220,000	13,200	11,000	9,429
230,000	11,500	230,000	19,320	16,100	13,800	230,000	13,800	11,500	9,857
240,000	12,000	240,000	20,160	16,800	14,400	240,000	14,400	12,000	10,286
250,000	12,500	250,000	21,000	17,500	15,000	250,000	15,000	12,500	10,714
260,000	13,000	260,000	21,840	18,200	15,600	260,000	15,600	13,000	11,143
270,000	13,500	270,000	22,680	18,900	16,200	270,000	16,200	13,500	11,571
280,000	14,000	280,000	23,520	19,600	16,800	280,000	16,800	14,000	12,000
290,000	14,500	290,000	24,360	20,300	17,400	290,000	17,400	14,500	12,429
300,000	15,000	300,000	25,200	21,000	18,000	300,000	18,000	15,000	12,857

*ACH=Air Change per Hour

For SI: 1 British thermal unit per hour = 0.293 W, 1 cubic foot = 0.028 m³.

Table courtesy of American Gas Association.

Figure G2407.5.2
CALCULATED VOLUMES

G2407.6 (304.6) Outdoor combustion air. Outdoor *combustion* air shall be provided through opening(s) to the outdoors in accordance with Section G2407.6.1 or G2407.6.2. The minimum dimension of air openings shall be not less than 3 inches (76 mm).

❖ This section describes two methods for supplying combustion air from the outdoors: the traditional method of two direct openings or ducts to the outdoors and a newer method using one opening or duct to the outdoors.

Openings to spaces that are naturally ventilated with outdoor air, such as attic or crawl spaces, are considered as an acceptable alternative to a direct connection to the outdoors. Attic and crawl spaces can be acceptable sources of combustion air only where such spaces have adequate natural ventilation openings directly to the outdoors. Attic and crawl spaces ventilated by mechanical means are not an acceptable source of combustion air.

Combustion air ducts and openings that penetrate components of wall, floor, ceiling and roof assemblies must be installed as required by this code.

When designing combustion air installations, the effect that openings to the outdoors can have on appliances, plumbing systems and building occupants must be considered. For example, depending on the location, openings to the outdoors can:

• Cause drafts that can blow out pilot lights or otherwise interfere with appliance ignition and operation.

• Cause freezing of plumbing piping or other water-containing components.

• Cause objectionable cold drafts that encourage the occupants to block or cover the openings.

In all cases, combustion air openings should be located to reduce the likelihood they will be accidentally or intentionally blocked or covered. A ceiling transfer opening that connects a furnace room with an attic is an example of a combustion air opening that is likely to be intentionally blocked by an occupant because of the drafts that can occur in cold climates.

Building occupants typically do not understand the need for or the importance of combustion air openings.

No side dimension of a square or rectangular opening and no diameter of a round opening can be less than 3 inches (76 mm). The smallest allowed square opening area would be 9 square inches (5806 mm^2) and the smallest allowed round opening area would be 7 square inches (4516 mm^2) (see commentary, Section G2407.10).

G2407.6.1 (304.6.1) Two-permanent-openings method. Two permanent openings, one commencing within 12 inches (305 mm) of the top and one commencing within 12 inches (305 mm) of the bottom of the enclosure, shall be provided. The openings shall communicate directly or by ducts with the outdoors or spaces that freely communicate with the outdoors.

Where directly communicating with the outdoors, or where communicating with the outdoors through vertical ducts, each opening shall have a minimum free area of 1 square inch per 4,000 *Btu*/h (550 mm^2/kW) of total input rating of all *appliances* in the enclosure [see Figures G2407.6.1(1) and G2407.6.1(2)].

Where communicating with the outdoors through horizontal ducts, each opening shall have a minimum free area of not less than 1 square inch per 2,000 *Btu*/h (1,100 mm^2/kW) of total input rating of all *appliances* in the enclosure [see Figure G2407.6.1(3)].

❖ Two openings located as prescribed in this section are intended to induce a convective air current in the room or space by admitting cooler, denser air in the lower opening and allowing the escape of warmer, less dense air through the upper opening. The farther apart the openings, the greater the temperature differential and the greater the convective force behind the current. A component of combustion air is cooling (ventilation) air for the appliance enclosure. The two-opening method was created to ventilate the appliance enclosure in addition to supplying combustion air. This ventilation cools the appliances and would help remove any combustion gases that spilled from the appliances.

G2407.6.2 (304.6.2) One-permanent-opening method. One permanent opening, commencing within 12 inches (305 mm) of the top of the enclosure, shall be provided. The *appliance* shall have *clearances* of at least 1 inch (25 mm) from the sides and back and 6 inches (152 mm) from the front of the *appliance*. The opening shall directly communicate with the outdoors or through a vertical or horizontal duct to the outdoors, or spaces that freely communicate with the outdoors (see Figure G2407.6.2) and shall have a minimum free area of 1 square inch per 3,000 *Btu*/h (734 mm^2/kW) of the total input rating of all *appliances* located in the enclosure and not less than the sum of the areas of all *vent connectors* in the space.

❖ Research has shown that for modern appliances, a single opening to the outdoors will perform as well as the traditional two-opening method. The one-opening method described in this section depends on a reduced pressure being created in the enclosure by the draft created by the venting system. This reduced pressure causes combustion air to enter the enclosure through the single opening. The opening must be properly sized considering both sizing criteria: the square-inch-area-per-Btu/h ratio and the area minimum based on the sum of the areas of all vent connectors in the enclosure. This method allows for fewer openings, fewer ducts and fewer objections by the owners/occupants.

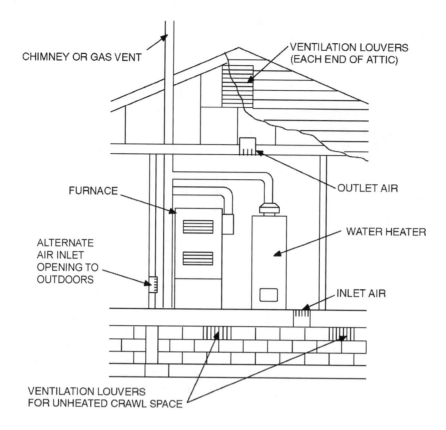

FIGURE G2407.6.1(1) [304.6.1(1)]
ALL AIR FROM OUTDOORS—INLET AIR FROM VENTILATED CRAWL SPACE AND OUTLET AIR TO VENTILATED ATTIC
(see Section G2407.6.1)

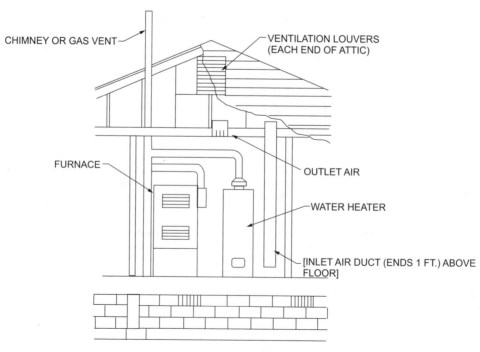

For SI: 1 foot = 304.8 mm.

FIGURE G2407.6.1(2) [304.6.1(2)]
ALL AIR FROM OUTDOORS THROUGH VENTILATED ATTIC
(see Section G2407.6.1)

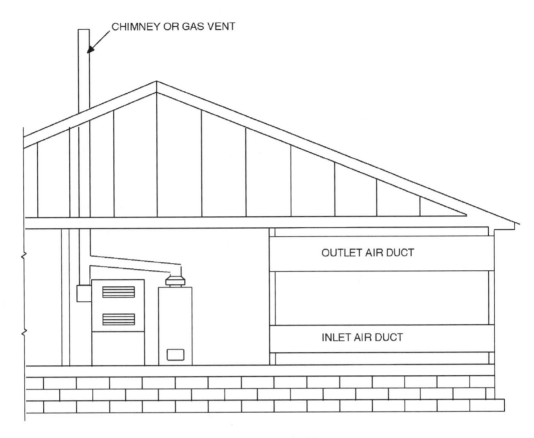

FIGURE G2407.6.1(3) [304.6.1(3)]
ALL AIR FROM OUTDOORS (see Section G2407.6.1)

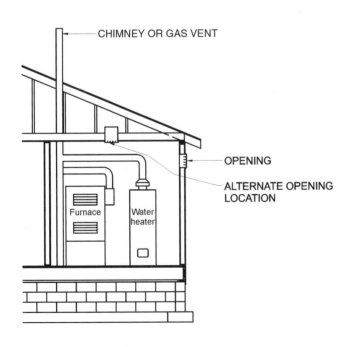

FIGURE G2407.6.2 (304.6.2)
SINGLE COMBUSTION AIR OPENING,
ALL AIR FROM OUTDOORS

G2407.7 (304.7) Combination indoor and outdoor combustion air. The use of a combination of indoor and outdoor *combustion air* shall be in accordance with Sections G2407.7.1 through G2407.7.3.

❖ This method of supplying combustion air is a combined application of Section G2407.5 and Section G2407.6. This method allows credit for the amount of infiltration that exists and makes up for the shortage with supplemental outdoor air. In other words, in addition to obtaining combustion air directly from the outdoors, this method relies on building infiltration for a portion of the total combustion air. In spaces where the volume is insufficient to satisfy the method of Section G2407.5 or where smaller outdoor air openings than required by Section G2407.6 are desired, this method allows infiltration and outdoor openings to supplement each other. If the appliances are enclosed in a small room such as a closet, openings to adjacent spaces as prescribed by Section G2407.5.3 must be provided to couple the appliance room volume with any other space volume counted on to provide combustion air. Frequently, sufficient volume cannot be obtained in the appliance enclosure or by opening the appliance enclosure to adjacent spaces. This method of combining indoor and outdoor air is an alternative solution. Simply stated, this section uses ratios of what is required to what is actually supplied so that when combined, the indoor air component and the outdoor air component add up to the whole required.

G2407.7.1 (304.7.1) Indoor openings. Where used, openings connecting the interior spaces shall comply with Section G2407.5.3.

❖ See the commentary to Section G2407.5.3.

G2407.7.2 (304.7.2) Outdoor opening location. Outdoor opening(s) shall be located in accordance with Section G2407.6.

❖ See the commentary to Section G2407.6.

G2407.7.3 (304.7.3) Outdoor opening(s) size. The outdoor opening(s) size shall be calculated in accordance with the following:

1. The ratio of interior spaces shall be the available volume of all communicating spaces divided by the required volume.

2. The outdoor size reduction factor shall be one minus the ratio of interior spaces.

3. The minimum size of outdoor opening(s) shall be the full size of outdoor opening(s) calculated in accordance with Section G2407.6, multiplied by the reduction factor. The minimum dimension of air openings shall be not less than 3 inches (76 mm).

❖ Although the principle has not changed, this method has been simplified compared to the same provision in previous editions of the code. The intent is still the same; that is, the fraction of the required indoor volume plus the fraction of outdoor openings must be equal to or greater than 1. The indoor volume method and the out-door air method can both be stand-alone methods; therefore, half of one and half of the other will work, as will three fourths of one and a quarter of the other, etc.

This section is expressed in the following equation:

$$\left[1 - \left(\frac{\text{available indoor volume}}{\text{volume required by Section G2407.5}}\right)\right] \times \begin{array}{c}\text{full size opening}\\\text{required by}\\\text{Section G2407.6}\end{array} = \begin{array}{c}\text{reduced size}\\\text{outdoor air}\\\text{openings}\end{array}$$

Example:

Given:

7,500 ft^3 of indoor volume is required if all indoor air is to be used.

4,950 ft^3 of indoor volume is available.

One opening to the outdoors is desired and must be 50 in.^{2} if all outdoor air is to be used.

$$\left[1 - \left(\frac{4,950 \text{ ft}^3}{7,500 \text{ ft}^3}\right)\right] \times 50 \text{ in.}^2 = \text{reduced outdoor opening size}$$

$0.34 \times 50 \text{ in}^2 = 17 \text{ in}^2$ outdoor opening size

Either of the provisions of Section G2407.6 when combined with the provision of this section can be used to satisfy the equation.

Example 1:

Room A is within a dwelling unit and contains a fuel-fired furnace and a fuel-fired water heater with input ratings of 150,000 and 50,000 Btu/h, respectively.

The size of Room A is 20 feet by 30 feet, and adjacent Room B is also 20 feet by 30 feet. Both rooms have 8-foot-high ceilings. Rooms A and B communicate through two openings located in accordance with Section G2407.5.3, each having dimensions of 10 inches by 20 inches. Two 3-inch-diameter (round) direct openings to the outdoors are installed in Room A and are located in accordance with Section G2407.6.1.

Question: Do the openings to the outdoors, when combined with the volumes of Rooms A and B, meet the combustion air demand of the appliances?

For SI: 1 inch = 25.4 mm, 1 foot = 305 mm, 1 Btu/h = 0.2931 W, 1 cubic foot = 0.0283 m^3, 1 square inch = 645 mm^2.

Given:

The air infiltration rate of the building is unknown.

Volume of Room A = 4,800 ft^3

Volume of Room B = 4,800 ft^3

Area of 3-inch-diameter direct openings = 7.07 in^2

Area of each opening between Rooms A and B = 200 in^2

Total input rating of appliances = 200,000 Btu/h

Step 1: Determine whether Rooms A and B meet the volume requirements of Section G2407.5.

Available volume of Rooms A and B = 4,800 ft^3 + 4,800 ft^3 = 9,600 ft^3

The required volume is:

$$\frac{200,000 \text{ Btu/h}}{1,000 \text{ Btu/h}} \times 50 \text{ ft}^3 = 10,000 \text{ ft}^3$$

Rooms A and B combined do not meet the volume requirements of Section G2407.5.

Step 2: Determine whether the openings between Rooms A and B meet the requirements of Section G2407.5.3.

The minimum required area is:

$$\frac{200,000 \text{ Btu/h}}{1,000 \text{ Btu/h}} \times 1 \text{ in}^2 = 200 \text{ in}^2$$

Actual area of each opening = 200 in^2

The area of each opening meets the size requirement of Section G2407.5.3.

Step 3: Determine the required area of each direct outdoor opening in accordance with Section G2407.6.1.

Required area of each opening:

$$\frac{200,000 \text{ Btu/h}}{4,000 \text{ Btu/h}} \times 1 \text{ in}^2 = 50 \text{ in}^2$$

Step 4: Determine whether the 3-inch-diameter direct openings combined with the volumes of Rooms A and B comply with:

$$\left[1 - \left(\frac{9,600 \text{ ft}^3}{10,000 \text{ ft}^3}\right)\right] \times 50 = (1 - 0.96) \times 50 = 0.04 \times 50 = 2 \text{ in.}^2$$

The actual area of each direct opening (7.07 in²) exceeds the required area of 2 in².

Therefore, the combination of the 3-inch-diameter direct openings and the volumes of Rooms A and B satisfies the combustion air demand of the appliances.

Example 2:

A combination of indoor and outdoor combustion air is to be supplied for two natural draft boilers in a room as shown in Commentary Figure G2407.7.3 (the air infiltration rate is not known).

1. In accordance with Section G2407.5.1, the required volume of the room containing the boilers would be:

$$\frac{90,000 \text{ Btu/h} + 90,000 \text{ Btu/h}}{1,000} \times 50 \text{ ft}^2$$

$$= \frac{180,000}{1,000} \times 50 \text{ ft}^3 = 180 \times 50 \text{ ft}^3 = 9,000 \text{ ft}^3$$

2. The actual room volume is 20ft × 50ft × 8ft = 8,000 ft³.

3. In accordance with Section G2407.7.3, Item 1, determine the ratio of available (actual) volume to the required volume.

$$\frac{\text{Actual Volume}}{\text{Required Volume}} \quad \frac{8,000 \text{ ft}^3}{9,000 \text{ ft}^3} = 0.89$$

4. In accordance with Section G2407.7.3, Item 2, determine outdoor opening size reduction factor.

1 - 0.89 = 0.11

5. In accordance with Section G2407.7.3, Item 3, determine the size of the outdoor opening required as if all combustion air is to be supplied through the outdoor opening. Because a single high opening is built into an exterior wall, Section G2407.6.2 applies and would require an area of 1 in²/3000 Btu/h of total appliance input rating (Assuming 6-in. vent connectors, the total area of the boiler vent connectors is 56.6 in²).

$$\frac{90,000 \text{ Btu/h} + 90,000 \text{ Btu/h}}{3,000} = 60 \text{ in}^2$$

6. The required size of the outdoor air opening is 60 in² × 0.11 = 6.6 in². Because the minimum dimensions must be not less than 3 inches, the minimum area of a square opening must be 9 in². This combination satisfies the combustion air demand of the appliances; however, note that Section G2407.6.2 requires an area not less than the sum of the vent connector areas, which is 56.6 in² in this example.

G2407.8 (304.8) Engineered installations. Engineered *combustion air* installations shall provide an adequate supply of *combustion*, ventilation and *dilution air* and shall be *approved*.

❖ Potential sources of combustion air are the indoor atmosphere in which the appliance is located, outdoor openings and ducts that communicate with the outdoor atmosphere, a combination of these or any other

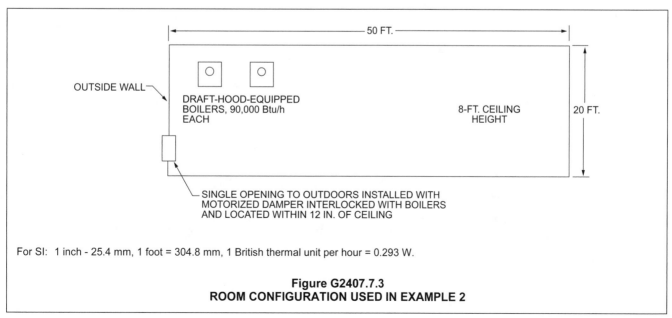

For SI: 1 inch - 25.4 mm, 1 foot = 304.8 mm, 1 British thermal unit per hour = 0.293 W.

Figure G2407.7.3
ROOM CONFIGURATION USED IN EXAMPLE 2

approved arrangement for the dependable introduction of outdoor air into the building or directly to the fuel-burning appliance.

The provisions of Section G2407 are intended for application only to appliances and equipment using natural draft atmospheric burners. Gas-fired appliances, especially those that are not of the atmospheric-burner type, are commonly supplied with combustion air by engineered systems. For example, special engineered systems are usually designed for appliances equipped with power burners. Power burners force combustion air into the burner under pressure and are, therefore, not dependent on gravity airflow. In some applications, engineered combustion air supplies have advantages and can be preferable to large gravity (natural) openings to the outdoors. Engineered designs must be evaluated for code compliance, approved and installed in accordance with the appliance manufacturer's instructions [see Commentary Figures G2407.8(1) through G2407.8(3)].

G2407.9 (304.9) Mechanical combustion air supply. Where all *combustion air* is provided by a mechanical air supply system, the *combustion air* shall be supplied from the outdoors at a rate not less than 0.35 cubic feet per minute per 1,000 *Btu*/h (0.034 m³/min per kW) of total input rating of all *appliances* located within the space.

❖ This method supplies combustion air by means of a fan/blower that runs when any of the served appliances are in a firing cycle. The fan/blower must be sized to supply the required airflow based on the simultaneous operation of all appliances that are served by the fans/blowers. The total Btu/h input rating of all fuel-burning appliances located in the room or space must be used because the potential exists for all of the appliances to be operating at the same time. This method is used where gravity openings to the outdoors are impractical or undesirable. A small fan-powered intake opening can substitute for comparatively large gravity openings, especially where freezing temperatures are a problem or there is no room for the required size gravity openings. Commentary Figures G2407.8(1), (2) and (3) illustrate factory-engineered mechanical combustion air supply units (see Commentary Figure G2407.9).

The prescribed air flow rate of 0.35 cfm per 1,000 Btu/h is equivalent to 1 cfm per 2,867 Btu/h and is based on a combustion air need of 21 cubic feet of volume for each cubic foot of natural gas (1,000 Btu/h). The rate is derived as follows:

$$\frac{2857 \text{ Btu/h}}{1000 \text{ Btu/ft}^3} = 2.857 \text{ ft}^3 \text{ of gas per hour}$$

Thus, 1 cfm is required for each 2,857 Btu/h or 0.35 cfm per 1,000 Btu/h.

G2407.9.1 (304.9.1) Makeup air. Where exhaust fans are installed, *makeup air* shall be provided to replace the exhausted air.

❖ This section expresses the exact same concern as Section G2407.4 (see commentary, Section G2407.4). If the air that is supplied by the combustion air fan becomes makeup air for an exhaust fan, the combustion air will not be available to the appliance(s) for which the air was intended.

Figure courtesy of Tjernlund Products, Inc.

Figure G2407.8(1)
MECHANICAL COMBUSTION AIR/MAKEUP AIR-SUPPLY SYSTEM

Photo courtesy of Tjernlund Products, Inc.

Figure G2407.8(2)
MECHANICAL COMBUSTION AIR-SUPPLY SYSTEM

Photo courtesy of Tjernlund Products, Inc.

Figure G2407.8(3)
MECHANICAL COMBUSTION AIR/MAKEUP AIR-SUPPLY SYSTEM

G2407.9.2 (304.9.2) Appliance interlock. Each of the *appliances* served shall be interlocked with the mechanical air supply system to prevent *main burner* operation when the mechanical air supply system is not in operation.

❖ This section requires an interlock circuit as opposed to simple parallel operation wiring. In other words, the combustion air fan/blower must start when initiated by the means that controls the appliance operation cycles (e.g., thermostat), and only after the combustion air fan/blower is proven to be operating will the appliance be allowed to start a firing cycle. If the combustion air fan/blower fails during a firing cycle of an appliance, the appliance firing cycle must be terminated (see Commentary Figure G2407.9.2).

The following steps must occur in this order:

1. Call for heat.

2. Combustion air fan/blower starts.

3. Combustion air fan/blower is proven to be operating.

4. Proving controls allow appliance to fire.

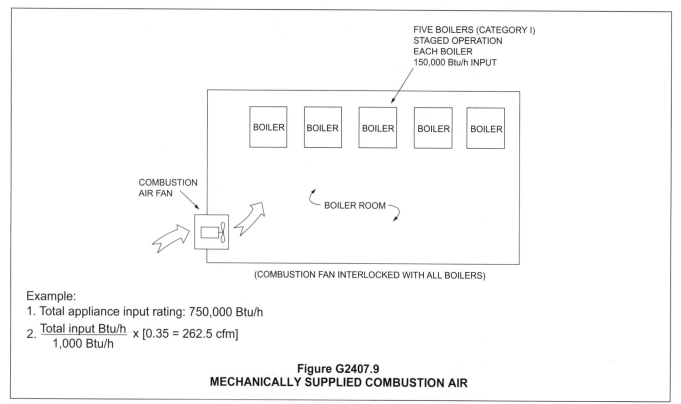

Example:
1. Total appliance input rating: 750,000 Btu/h

2. $\dfrac{\text{Total input Btu/h}}{1,000 \text{ Btu/h}}$ x [0.35 = 262.5 cfm]

Figure G2407.9
MECHANICALLY SUPPLIED COMBUSTION AIR

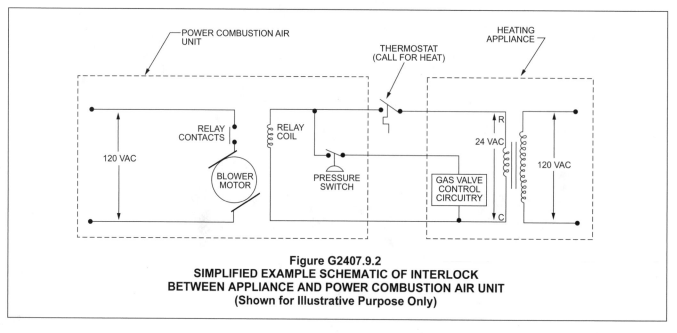

Figure G2407.9.2
SIMPLIFIED EXAMPLE SCHEMATIC OF INTERLOCK
BETWEEN APPLIANCE AND POWER COMBUSTION AIR UNIT
(Shown for Illustrative Purpose Only)

G2407.9.3 (304.9.3) Combined combustion air and ventilation air system. Where *combustion air* is provided by the building's mechanical ventilation system, the system shall provide the specified *combustion air* rate in addition to the required ventilation air.

❖ If combustion air is mechanically supplied through an HVAC system, the system must deliver the required combustion air rate above and beyond any outdoor air rate supplied for ventilation of the occupied spaces. The HVAC system must also be interlocked with the appliance(s) as required by Section G2407.9.2. Neither continuous operation nor parallel operation of the HVAC system satisfies the requirement for a true interlock as required by Section G2407.9.2.

G2407.10 (304.10) Louvers and grilles. The required size of openings for *combustion*, ventilation and *dilution air* shall be based on the net free area of each opening. Where the free area through a design of louver, grille or screen is known, it shall be used in calculating the size opening required to provide the free area specified. Where the design and free area of louvers and grilles are not known, it shall be assumed that wood louvers will have 25-percent free area and metal louvers and grilles will have 75-percent free area. Screens shall have a mesh size not smaller than $^1/_4$ inch (6.4 mm). Nonmotorized louvers and grilles shall be fixed in the open position. Motorized louvers shall be interlocked with the *appliance* so that they are proven to be in the full open position prior to *main burner* ignition and during *main burner* operation. Means shall be provided to prevent the *main burner* from igniting if the louvers fail to open during *burner* start-up and to shut down the *main burner* if the louvers close during operation.

❖ This section recognizes that louvers and grilles are usually installed over air inlets and outlets to prevent rain, snow and animals from entering the building. When louvers or grilles are used, the solid portion of the louver or grille must be considered when determining the unobstructed (net clear) area of the opening.

Combustion air openings are sized based on there being a free, unobstructed area for the passage of air into the space where the fuel-burning appliances are located. Louvers or grilles placed over these openings reduce the area of the openings because of the area occupied by the solid portions of the grille or louver. The reduction in area must be considered because only the unobstructed area can be credited toward the required opening size.

The reduction in opening area caused by the presence of grilles or louvers will always require openings to be larger than determined from the sizing ratios of this chapter and larger than any duct of the minimum required size that might connect to these openings. Once the required size of a combustion air duct or opening is determined, any increase in opening size that must be made as a result of the effect of louvers or grilles must be determined. If a metal louver is used and based on the default opening area of 75 percent, the designer must calculate what size opening times

75 percent will yield the previously determined combustion air opening or duct area.

For example, assume that a combustion air duct needs to have a cross-sectional area of 200 square inches and this duct connects to an exterior wall opening with a metal louver protecting it.

x = The required opening area adjusted to account for the louver.

$$0.75x = 200 \text{ in}^2$$
$$x = \frac{200}{75}$$
$$x = 266.7 \text{ in}^2$$

This can be simplified to multiplying by the reciprocal of the percentage as follows:

For metal louvers and grilles, the required combustion air duct or opening size multiplied by 1.33 equals the adjusted opening size.

For wooden louvers and grilles, the required combustion air duct or opening size multiplied by 4 equals the adjusted opening size.

The actual free opening of a louver or grille should be obtained from the manufacturer rather than relying on default assumptions.

This section does not apply to grilles, louvers or screens that are an integral component of a labeled appliance that is installed in accordance with the manufacturer's instructions. Where the manufacturer states the free, unobstructed area of a grille or louver, it is acceptable to consider that area as actual without requiring further calculation.

The code also requires the free area of screens to be considered in the opening sizing calculations. If it is unknown, it will have to be calculated because there is no stated default area for screens as there is for louvers and grilles. The code prohibits screens having mesh sizes smaller than $^1/_4$ inch (6 mm). Screens with a mesh size of $^1/_4$ inch (6 mm) and larger are not insect screens and would be suitable only for keeping out large debris, rodents and other small animals. Smaller size mesh is likely to become obstructed by lint, plant fibers, cottonwood seeds, etc. At best, placing screens in combustion air openings is risky, and grilles and louvers are preferred to keep out the unwanted.

Metal louvers are constructed so that the solid portion of the louver occupies approximately 25 percent of the opening area, leaving 75 percent of this area unobstructed. Wood louvers occupy approximately 75 percent of the opening area, leaving 25 percent of the area unobstructed. When the required amount of unobstructed area is determined, the effect that a louver has on that area must be taken into account to make certain the minimum amount of unobstructed opening is achieved. The material type, thickness and spacing all have a significant effect on the actual net free area of any louver or grille.

Note that the 75-percent and 25-percent default numbers might be too liberal for some louvers or grilles and every effort should be made to consult the louver or grille manufacturer's specifications.

To prevent freezing of pipes, to save energy and maintain comfortable temperatures in appliance rooms, outdoor openings often have motorized dampers that close during the appliance(s) off cycle. The damper motors and/or damper blades must be supervised so that an interlock can be established between the dampers and the appliances, thereby preventing appliance operation when the dampers are closed.

Combustion air openings must not be closable unless they open and close automatically using a mechanism that is interlocked with the appliances (see Commentary Figure G2407.10). Damper supervision is easily accomplished by damper motor end switches or blade position travel switches that monitor the position of the drive motor shaft, the damper/motor linkage hardware or the damper blades. These switches provide feedback to the appliance control circuitry to allow firing only when the dampers are proven to be fully open.

G2407.11 (304.11) Combustion air ducts. *Combustion air* ducts shall comply with all of the following:

1. Ducts shall be constructed of galvanized steel complying with Chapter 16 or of a material having equivalent corrosion resistance, strength and rigidity.

Exception: Within dwellings units, unobstructed stud and joist spaces shall not be prohibited from conveying *combustion air*, provided that not more than one required fireblock is removed.

2. Ducts shall terminate in an unobstructed space allowing free movement of *combustion air* to the *appliances*.

3. Ducts shall serve a single enclosure.

4. Ducts shall not serve both upper and lower *combustion air* openings where both such openings are used. The separation between ducts serving upper and lower *combustion air* openings shall be maintained to the source of *combustion air*.

5. Ducts shall not be screened where terminating in an attic space.

6. Horizontal upper *combustion air* ducts shall not slope downward toward the source of *combustion air*.

7. The remaining space surrounding a *chimney* liner, gas vent, special gas vent or plastic *piping* installed within a masonry, metal or factory-built *chimney* shall not be used to supply *combustion air*.

Exception: Direct-vent gas-fired *appliances* designed for installation in a solid fuel-burning *fireplace* where installed in accordance with the manufacturer's instructions.

8. *Combustion air* intake openings located on the exterior of a building shall have the lowest side of such open-

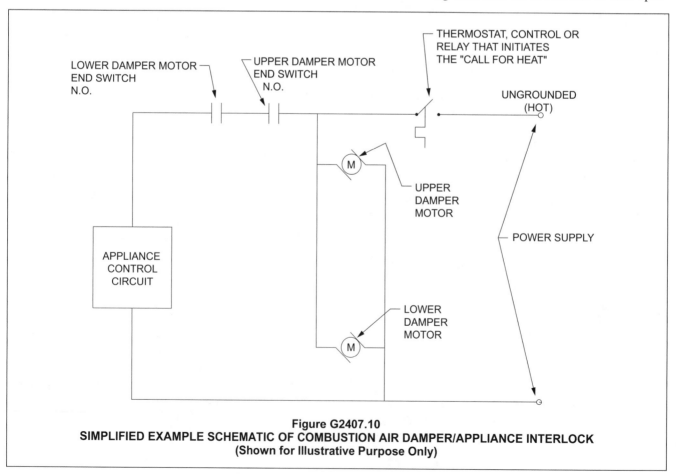

Figure G2407.10
SIMPLIFIED EXAMPLE SCHEMATIC OF COMBUSTION AIR DAMPER/APPLIANCE INTERLOCK
(Shown for Illustrative Purpose Only)

ings located not less than 12 inches (305 mm) vertically from the adjoining finished ground level.

❖ This section addresses the construction and installation of combustion air ducts.

Item 1 requires that combustion air ducts be constructed of galvanized sheet steel or an approved equivalent material. The intent is to cause ducts to be constructed of a material that is resistant to corrosion and physical damage, which would allow the ducts to remain in place undamaged for the life of the installation. It is doubtful that the authors of Item 1 ever anticipated the use of flexible ducts and the higher friction losses associated with such rough-wall ducts. It is also reasonable to assume that all of the sizing criteria in Section G2407 are based on smooth-wall ducts. The strength and rigidity of typical flexible ducts would not be equivalent to rigid steel ducts. The exception expresses a concern for fire safety where stud cavities are used as a duct.

Item 2 is intended to prevent blockages and minimize resistance (friction loss) to airflow.

Item 3 prohibits a duct from serving more than one appliance enclosure because the duct and opening sizing criteria are based on a single duct serving a single opening in an appliance location. Multiple openings supplied by a single duct could result in unpredictable airflow and might not create the convective movements of air intended by the combustion air methods.

Item 4 recognizes that where both upper and lower combustion air openings are required, air circulation cannot occur if a single duct serves both combustion air openings. The lower opening functions as an air inlet and the upper opening as an air outlet, thus independent ducts are required (see Commentary Figure G2407.11).

Item 5 is intended to prevent obstruction or blockage by insulation materials.

Item 6 is intended to prevent trapping or sloping of the upper combustion air opening in a way that would impede the convective movement of air. The horizontal duct for the upper openings must slope upward as it

travels away from the appliance(s) and toward the outdoors.

When a vent or chimney liner is installed within a chimney, an annular space will exist between the vent or liner and the interior chimney walls. According to Item 7, this space cannot be used as a conduit for conveying combustion air because of poor flow characteristics, the tendency for reverse flow to occur as the combustion air is warmed by the liner or vent and the adverse cooling effect on the vent or liner. The exception recognizes that there could be appliances that are designed and listed for the type of installation that this item intends to prohibit.

Item 8 intends to prevent combustion air openings from being obstructed by snow, leaves and vegetation. No part of a combustion air opening is allowed to be less than 12 inches (305 mm) above the finished ground level under such opening.

G2407.12 (304.12) Protection from fumes and gases. Where corrosive or flammable process fumes or gases, other than products of *combustion*, are present, means for the disposal of such fumes or gases shall be provided. Such fumes or gases include carbon monoxide, hydrogen sulfide, ammonia, chlorine and halogenated hydrocarbons.

In barbershops, beauty shops and other facilities where chemicals that generate corrosive or flammable products, such as aerosol sprays, are routinely used, nondirect vent-type *appliances* shall be located in a mechanical room separated or partitioned off from other areas with provisions for *combustion air* and *dilution air* from the outdoors. *Direct-vent appliances* shall be installed in accordance with the *appliance* manufacturer's instructions.

❖ In many occupancies, the routine use of chemicals contaminates the indoor combustion air. These contaminants can combine with water vapor in the combustion gases to produce acids and other corrosive compounds that can destroy appliance components, vents, chimneys and connectors. For example, it is common to see vents and vent connectors for draft-hood-equipped water heaters in beauty shops that have deteriorated to the point of structural failure. Other examples are swimming pool heaters exposed to chlorine, boilers exposed to laundry and dry cleaning chemicals and boilers exposed to refrigerant leakage. Some contaminants are more toxic to life after they have passed through an appliance combustion chamber. The highly poisonous chemical "phosgene" (carbonyl chloride) can be produced when common refrigerants contaminate combustion air.

Obviously, choosing direct-vent appliances will eliminate this problem in most cases and will not require isolation enclosures for the appliances. It is possible that some atmospheres are unsuitable for any appliance, direct-vent or otherwise. The code provides an alternative to direct-vent appliances that requires the appliances to be isolated from the contaminated atmosphere by reasonably air-tight enclosures and requires that all combustion air be supplied for the appliance(s) from the outdoors.

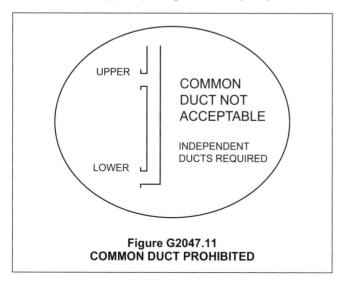

UPPER

LOWER

COMMON DUCT NOT ACCEPTABLE

INDEPENDENT DUCTS REQUIRED

Figure G2047.11
COMMON DUCT PROHIBITED

SECTION G2408 (305)
INSTALLATION

G2408.1 (305.1) General. *Equipment* and *appliances* shall be installed as required by the terms of their approval, in accordance with the conditions of listing, the manufacturer's instructions and this code. Manufacturer's installation instructions shall be available on the job site at the time of inspection. Where a code provision is less restrictive than the conditions of the listing of the *equipment* or *appliance* or the manufacturer's installation instructions, the conditions of the listing and the manufacturer's installation instructions shall apply.

Unlisted *appliances approved* in accordance with Section G2404.3 shall be limited to uses recommended by the manufacturer and shall be installed in accordance with the manufacturer's instructions, the provisions of this code and the requirements determined by the *code official*.

❖ Manufacturers' installation instructions are evaluated by the listing agency verifying compliance with the applicable standard. The listing agency can require that the manufacturer alter, delete or add information to the instructions as necessary to achieve compliance with applicable standards and code requirements. Manufacturers' installation instructions are an enforceable extension of the code and must be in the hands of the code official when an inspection takes place.

When an appliance is tested to obtain a listing and label, the approved agency installs the appliance in accordance with the manufacturer's installation instructions. The appliance is tested under these conditions; thus, the installation instructions become an integral part of the labeling process.

The listing and labeling process assures that the appliance and its installation instructions are in compliance with applicable standards. Therefore, an installation in accordance with the manufacturer's instructions is required, except where the code requirements are more stringent. An inspector must carefully and completely read and comprehend the manufacturer's instructions to properly perform an installation inspection.

In some cases, the code will specifically address an installation requirement that is also addressed in the manufacturer's installation instructions. The code requirements may be the same or may exceed the requirements in the manufacturer's installation instructions, or the manufacturer's installation instructions could contain requirements that exceed those in the code. In all such cases, the more restrictive requirements would apply.

In addition to the installation requirements for the appliance or equipment itself, this section also regulates the connections to appliances and equipment by requiring compliance with all other applicable sections of this code and other *International Codes*®. These connections include but are not limited to fuel and electrical supplies, control wiring, hydronic piping, chimneys, vents and ductwork. Some overlap or coincidence of connection requirements may occur in the code and

the manufacturer's installation instructions. In the unlikely event that the code was less restrictive (more lenient) than the manufacturer's instructions regarding an installation issue, the manufacturer's instructions must prevail. Where differences or conflicts occur, the requirements that provide the greatest level of safety always apply.

Even if an installation appears to be in compliance with the manufacturer's instructions, the installation cannot be complete or approved until all associated components, connections, and systems that serve the appliance or equipment are also in compliance with the applicable provisions of the code. For example, a gas-fired boiler installation must not be approved if the boiler is connected to a deteriorated, undersized, improper or otherwise unsafe chimney or vent. Likewise, the same installation must not be approved if the existing gas piping is in poor condition or if the electrical supply circuit is inadequate or unsafe.

In the case of replacement installations, the intent of this section is to require all new work associated with the installation to comply with the code without necessarily requiring full compliance for the existing, unchanged portions of the related ductwork, piping, electrical, venting and similar mechanical systems. In the case of an appliance replacement, such as a furnace, it is the expectation of both the manufacturer's instructions and the code that the new furnace be connected to a code-complying venting system, fuel supply system and electrical circuit, all of which are part of the furnace installation. If the ductwork system was such that the furnace temperature rise across the heat exchanger exceeded the maximum allowable temperature rise, the ductwork system might have to be altered as part of the furnace installation. Existing mechanical systems are accepted on the basis that they are free from hazard although not necessarily compliant with current codes. The code is not retroactive except where specifically stated that it applies to existing systems.

Manufacturer's installation instructions are often updated and changed for various reasons, such as changes in the appliance or equipment design, revisions to the product standards and as a result of field experience related to existing installations. The code official should stay abreast of any changes by reviewing the manufacturer's instructions for every installation.

Equipment and appliances must be installed in accordance with the manufacturer's installation instructions. The manufacturer's label and installation instructions must be consulted in determining whether or not an appliance or piece of equipment can be installed and operated in a particular hazardous location. Without the manufacturer's installation instructions, an inspector cannot fully perform his or her job, which is why the manufacturer's installation instructions must be available on the job site when the equipment is being inspected.

G2408.2 (305.3) Elevation of ignition source. *Equipment* and *appliances* having an *ignition source* shall be elevated such that the source of ignition is not less than 18 inches (457 mm) above the floor in *hazardous locations* and public garages, private garages, repair garages, motor fuel-dispensing facilities and parking garages. For the purpose of this section, rooms or spaces that are not part of the *living space* of a *dwelling unit* and that communicate directly with a private garage through openings shall be considered to be part of the private garage.

Exception: Elevation of the *ignition source* is not required for *appliances* that are *listed* as flammable-vapor-ignition resistant.

❖ To reduce the possibility of ignition of flammable vapors in hazardous areas, the potential sources of ignition must be elevated above the surface supporting the equipment or appliance. It is not the intent of this section for the installer to measure the elevation from the surface of a dedicated appliance stand or other structure built to support the appliance where that stand or structure does not afford room for the storage of flammable liquids. Some flammable and combustible liquids typically associated with hazardous locations give off vapors that are denser than air and tend to collect near the floor. The 18-inch (457 mm) height requirement is intended to reduce the possibility of flammable vapor ignition by keeping the ignition sources elevated above the anticipated level of accumulated vapors. The 18-inch (457 mm) value is a minimum requirement and must be increased when required by the manufacturer's installation instructions.

This section will effectively prohibit the installation of most furnaces, boilers, space heaters, clothes dryers and water heaters directly on the floor of residential garages.

The accumulation of flammable vapors more than 18 inches (457 mm) deep is unlikely in most ventilated locations; therefore, maintaining all possible sources of ignition at least 18 inches (457 mm) above the floor will substantially reduce the risk of explosion and fire [see Commentary Figures G2408.2(1) and G2408.2(2)].

In the context of this section, a source of ignition could be a pilot flame, burner, burner igniter or electrical component capable of producing a spark. The term "ignition source" is defined and can be interpreted as meaning an intentional source of ignition, such as for a burner, or an unintentional source of ignition for any flammable vapors that may be present (see definition of "Ignition source").

An appliance installed in a closet or room that is accessible only from the garage must be considered as part of the garage for application of this section. Even though the room may be separated from the garage by walls and a door, there is no practical means of making the door vapor tight, nor is there any assurance that the door will remain closed during normal use. An appliance room that is accessed only from the outdoors or only from the living space would not be considered as part of the garage (see Section G2408.2.1). Rooms

---

GAS-FIRED
WATER HEATER

GAS CONTROL

PILOT BURNER

COMBUSTION
CHAMBER

MAIN
BURNER

18" MIN

RAISED PLATFORM

GARAGE FLOOR

For SI: 1 inch = 25.4 mm.

Figure G2408.2(1)
GAS-FIRED WATER HEATER
INSTALLATION IN A GARAGE

Figure G2408.2(2)
SOURCES OF IGNITION ELEVATED IN PRIVATE GARAGE

such as utility rooms or laundry rooms that communicate with both the garage and the living space and that are considered as part of the living space are not part of the garage (see definition of "Living space"). If a room opens to the garage and that room is not living space, it is part of the garage for the purpose of this section [see Commentary Figure G2408.2(3)].

The exception recognizes that new technology exists that allows appliances, such as water heaters, to be tested and listed as being flammable-vapor-ignition resistant and suitable for installation without the 18-inch (457 mm) elevation requirement. This new technology was developed for conventional draft-hood-equipped water heaters, and now the standard for these appliances, ANSI Z21.10.1, mandates that by specified effective dates, all water heaters must comply with the flammable-vapor-ignition criteria in the standard. The technology involves variously configured flame arrestor screens having precisely engineered openings (slots). All combustion air for the appliance is passed through the flame arrestor screen. To pass through the small slots in the screen, the combustion air velocity is increased (Venturi principle) such that it exceeds the velocity at which flames travel through a

flammable vapor. If the combustion air contains flammable vapors, the vapors will enter the appliance and be ignited within the combustion chamber. The flames will be contained (trapped) within the combustion chamber because the arrestor screen will not allow flames to exit the combustion chamber. The result is that vapor ignition is limited to the appliance interior and is not allowed to spread to the atmosphere surrounding the appliance. During such an event, a thermal sensing device shuts down and locks out the appliance. Some appliance manufacturers might require that the appliance be replaced after a flammable-vapor-ignition event. Others require that a sensor be replaced or reset. The appliance standard includes testing requirements to determine whether the arrestor screen is subject to blockage by debris, including lint, plant fibers, dust particles, etc. If the arrestor screen openings become restricted by debris, the appliance will be deprived of combustion air and incomplete combustion and serious appliance malfunction can result. If an appliance shuts down as a result of the arrestor screen being blocked by debris, thus restricting the combustion air supply, typically a thermal device can be reset or replaced to restore operation of the appliance.

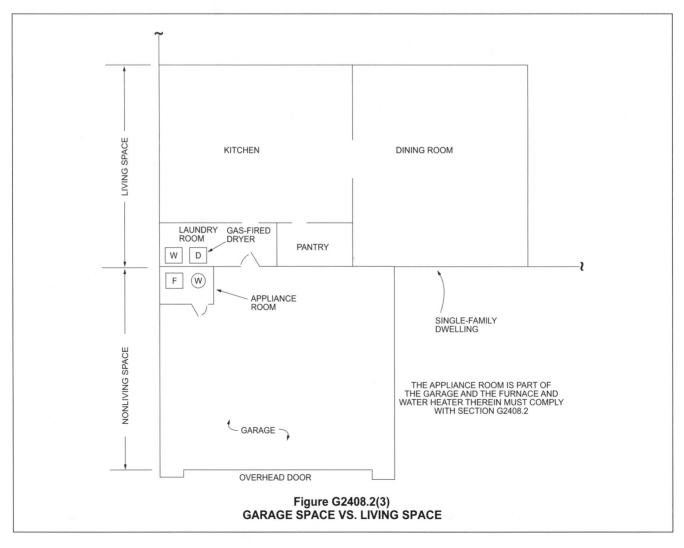

Figure G2408.2(3)
GARAGE SPACE VS. LIVING SPACE

If the installation instructions or the label of a water heater equipped with flammable-vapor-ignition-resistant technology state that the appliance must be elevated so that the source of ignition is at least 18 inches (457 mm) above the floor, it must be installed as directed [see Commentary Figures G2408.2(1) through G2408.2(6)].

G2408.2.1 (305.3.1) Installation in residential garages. In residential garages where *appliances* are installed in a separate, enclosed space having access only from outside of the garage, such *appliances* shall be permitted to be installed at floor level, provided that the required *combustion air* is taken from the exterior of the garage.

❖ See Commentary Figure G2408.2.1 and Section G2408.2.

G2408.3 (305.5) Private garages. *Appliances* located in private garages shall be installed with a minimum *clearance* of 6 feet (1829 mm) above the floor.

Exception: The requirements of this section shall not apply where the *appliances* are protected from motor vehicle impact and installed in accordance with Section G2408.2.

❖ Appliances located in a private garage or carport must be protected from vehicle impact (also see Section M1307.3.1). This section is applicable to appliances located in an area where motor vehicles can be operated and includes appliances under which a vehicle can pass and those located anywhere in a vehicle's path where impact is possible. The 6-foot (1829 mm) minimum height requirement is intended to provide adequate clearance above the typical automobile.

Figure courtesy of American Water Heater Company

Figure G2408.2(4)
EXAMPLE OF FLAMMABLE-VAPOR-IGNITION-RESISTANT WATER HEATER TECHNOLOGY

Photo courtesy of Bradford White Corporation
(Bradford White Defender Safety System™)

Figure G2408.2(6)
EXAMPLE OF FLAMMABLE-VAPOR-IGNITION-RESISTANT WATER HEATER TECHNOLOGY

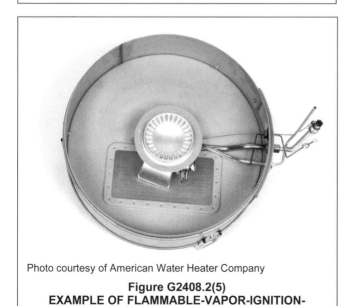

Photo courtesy of American Water Heater Company

Figure G2408.2(5)
EXAMPLE OF FLAMMABLE-VAPOR-IGNITION-RESISTANT WATER HEATER TECHNOLOGY

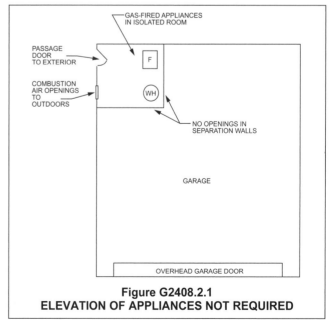

Figure G2408.2.1
ELEVATION OF APPLIANCES NOT REQUIRED

With the popularity of conversion vans and recreational vehicles, which can be much higher than other automobiles, the 6-foot (1829 mm) minimum installation height above the floor may not provide adequate clearance; additional height might be necessary. The garage door height can be used as a guide in determining the maximum vehicle height [see Commentary Figures G2408.3(1) through G2408.3(5)].

G2408.4 (305.7) Clearances from grade. *Equipment* and *appliances* installed at grade level shall be supported on a level concrete slab or other *approved* material extending not less than 3 inches (76 mm) above adjoining grade or shall be suspended not less than 6 inches (152 mm) above adjoining grade. Such supports shall be installed in accordance with the manufacturer's instructions.

❖ This section is consistent with Section M1305.1.4.1 and addresses outdoor and crawl space installations where an appliance is resting on the earth and not supported by a building. The intent is to maintain level support and protect the appliance from corrosion and deterioration. The slab should be concrete or a material having comparable strength and longevity. Outdoor appliance installations must be designed to tolerate movement where frost heave will cause a slab to rise and fall.

G2408.5 (305.8) Clearances to combustible construction. Heat-producing *equipment* and *appliances* shall be installed to maintain the required clearances to combustible construction as specified in the listing and manufacturer's instructions. Such *clearances* shall be reduced only in accordance with Section G2409. *Clearances* to combustibles shall include such considerations as door swing, drawer pull, overhead projections or shelving and window swing. Devices, such as door stops or limits and closers, shall not be used to provide the required *clearances*.

❖ Section G2409 allows clearances to be reduced where protective assemblies are installed to decrease the transfer of heat from the source to the combustible material/assembly. Clearances must be measured from shelving, overhead projections and the possible positions of doors, windows and drawers. Doorstops and closers are easily defeated and subject to failure.

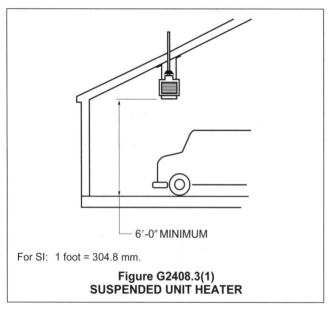

For SI: 1 foot = 304.8 mm.

Figure G2408.3(1)
SUSPENDED UNIT HEATER

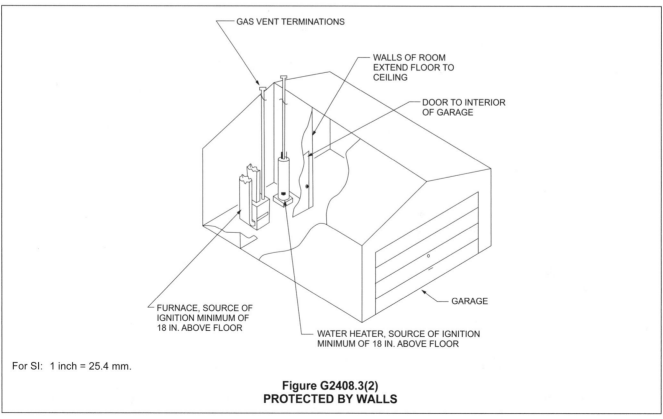

For SI: 1 inch = 25.4 mm.

Figure G2408.3(2)
PROTECTED BY WALLS

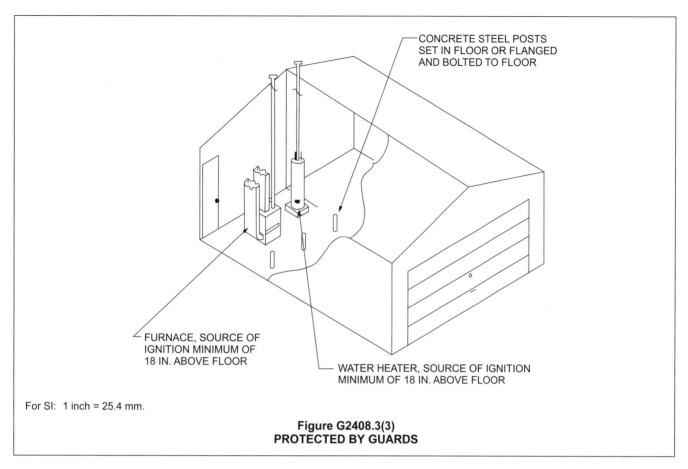

CONCRETE STEEL POSTS
SET IN FLOOR OR FLANGED
AND BOLTED TO FLOOR

FURNACE, SOURCE OF
IGNITION MINIMUM OF
18 IN. ABOVE FLOOR

WATER HEATER, SOURCE OF IGNITION
MINIMUM OF 18 IN. ABOVE FLOOR

For SI: 1 inch = 25.4 mm.

Figure G2408.3(3)
PROTECTED BY GUARDS

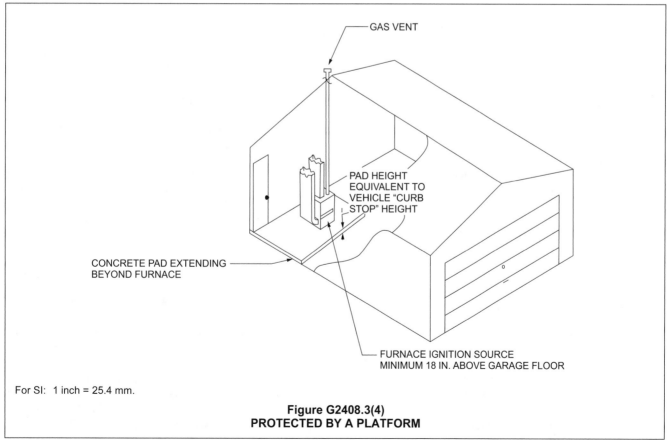

GAS VENT

PAD HEIGHT
EQUIVALENT TO
VEHICLE "CURB
STOP" HEIGHT

CONCRETE PAD EXTENDING
BEYOND FURNACE

FURNACE IGNITION SOURCE
MINIMUM 18 IN. ABOVE GARAGE FLOOR

For SI: 1 inch = 25.4 mm.

Figure G2408.3(4)
PROTECTED BY A PLATFORM

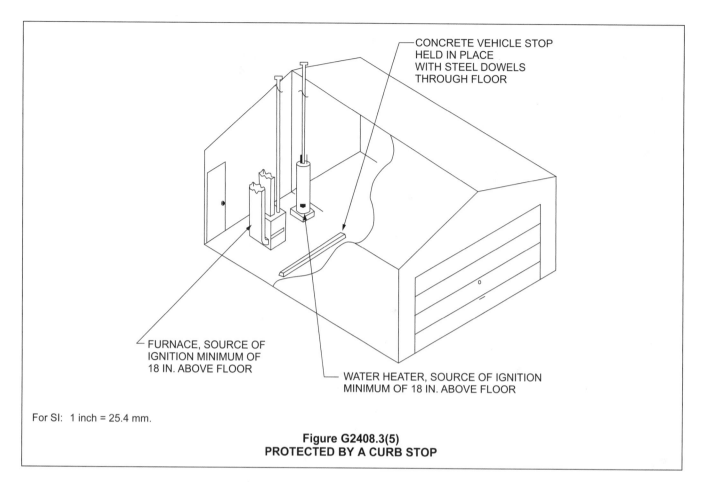

CONCRETE VEHICLE STOP
HELD IN PLACE
WITH STEEL DOWELS
THROUGH FLOOR

FURNACE, SOURCE OF
IGNITION MINIMUM OF
18 IN. ABOVE FLOOR

WATER HEATER, SOURCE OF IGNITION
MINIMUM OF 18 IN. ABOVE FLOOR

For SI: 1 inch = 25.4 mm.

Figure G2408.3(5)
PROTECTED BY A CURB STOP

G2408.6 (305.12) Avoid strain on gas piping. *Appliances* shall be supported and connected to the *piping* so as not to exert undue strain on the connections.

❖ Connections under strain can transmit forces to gas controls that can distort valve/control bodies and put strain on burner manifolds and other components. Gas piping must not exert forces on the appliance connection and vice versa. To accomplish this intent, the appliance and the gas piping must be supported independently. "Undue" strain is not specific, but is to be taken to mean any strain that is reasonably capable of threatening the integrity of the appliance connection over any period of time. That being very hard to determine, it would be wise to make the point of connection a neutral point where no forces are transmitted to or from the appliance. For example, where steel gas piping connects to an appliance's gas control, it can be determined if the piping is exerting strain on the gas control by simply observing if the piping moves when the union is opened. If the piping moves in some direction, it is exerting a force on the appliance gas control in that same direction.

SECTION G2409 (308)
CLEARANCE REDUCTION

G2409.1 (308.1) Scope. This section shall govern the reduction in required clearances to *combustible materials*, includ-

ing gypsum board, and *combustible assemblies* for chimneys, vents, appliances, devices and equipment. Clearance requirements for air-conditioning equipment and central heating boilers and furnaces shall comply with Sections G2409.3 and G2409.4.

❖ Heat-producing appliances and mechanical equipment must be installed with the required minimum clearances to combustible materials indicated by their listing label. It is not uncommon to encounter practical or structural difficulties in maintaining clearances. Therefore, clearance reduction methods have been developed to allow, in some cases, reduction of the minimum prescribed clearance distance while achieving equivalent protection. An important understanding is that all prescribed clearances to combustibles are air-space clearances measured from the heat source to the face of the nearest combustible surface (see Commentary Figure G2409.1). When using the assemblies described in Table G2409.2, the clearance is measured as described in Note b. In the case of listed equipment, the required clearances are intended to be clear airspace, and therefore the space is not to be filled with insulation or any other material other than an assembly intended to allow clearance reduction. This is especially important where clearances are required from appliances and equipment that rely on the airspace for convection cooling to maintain their proper operating temperature.

The provisions contained in this section are based on the principles of heat transfer. Mechanical equipment or appliances producing heat can become hot, and many appliances have hot exterior surfaces by design. The heat energy is then radiated to objects surrounding the appliances or equipment. When mechanical equipment and appliances are tested, the minimum clearances are established so that radiant and, to a lesser extent, convective heat transfer do not represent an ignition hazard to adjacent surfaces and objects. This distance is called the "required clearance" to combustible materials. Appliance and equipment labels must specify minimum clearances in all directions.

This section permits the use of materials and systems as radiation shields, decreasing the amount of heat energy transferred to surrounding objects and reducing the required clearances between gas-fired appliances and equipment to combustibles.

Plaster and gypsum by themselves are classified as noncombustible materials. Under continued exposure to heat, however, these materials will gradually decompose as water molecules are driven out of the material. Plaster on wood lath, plasterboard, sheetrock and drywall are all considered to be combustible materials when the code provisions are applied.

Additionally, gypsum wallboard is faced with paper that has a flame spread index that is measurable in the ASTM E84 test. This alone identifies the need to classify gypsum wallboard as a combustible material for the purpose of requiring a separation from heat-producing equipment and appliances.

The code specifically prohibits the reduction of clearances in certain applications. For examples, see Sections G2409.2 through G2409.4.

G2409.2 (308.2) Reduction table. The allowable *clearance* reduction shall be based on one of the methods specified in Table G2409.2 or shall utilize a reduced *clearance* protective assembly *listed* and *labeled* in accordance with UL 1618. Where required *clearances* are not listed in Table G2409.2, the reduced clearances shall be determined by linear interpolation between the distances listed in the table. Reduced *clearances* shall not be derived by extrapolation below the range of the table. The reduction of the required *clearances* to combustibles for *listed* and *labeled appliances* and *equipment* shall be in accordance with the requirements of this section, except that such *clearances* shall not be reduced where reduction is specifically prohibited by the terms of the *appliance* or *equipment listing* [see Figures G2409.2(1) through 2409.2(3)].

❖ Another option for reducing the required clearances to combustible materials is to use one of the on-site field-constructed methods specified in Table G2409.2.

FIGURES G2409.2(1) and G2409.2(2). See page 24-58.

❖ See the commentary to Section G2409.2

TABLE G2409.2. See page 24-60.

❖ The column headings of Table G2409.2 list required clearances without protection. The numbers to the right of each method indicate the permissible reduced

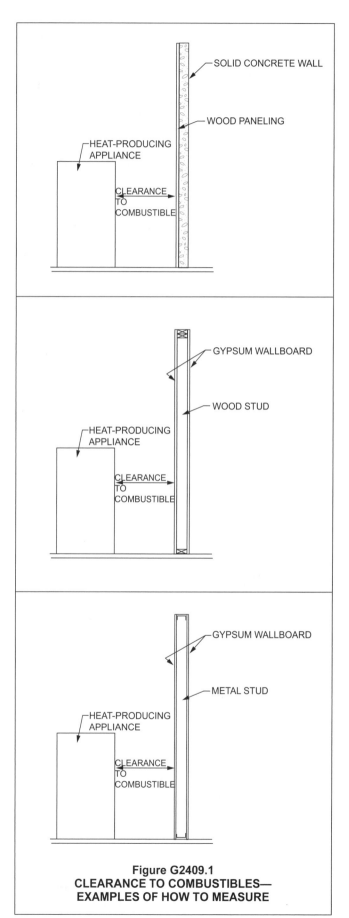

Figure G2409.1
CLEARANCE TO COMBUSTIBLES—
EXAMPLES OF HOW TO MEASURE

clearance measured from the heat-producing appliances/equipment to the face of the combustible material.

The rationale behind the methods of protection listed in Table G2409.2 is based on the ability of the protection to reduce radiant heat transmission from the appliance and equipment to the combustible material so that the temperature rise of the combustible material will remain below the maximum allowed.

Although the materials referred to in Table G2409.2 are common construction materials, confusion often arises over what satisfies the requirement for "insula-

tion board" (Item 6 in the table), sometimes referred to an inorganic insulating board, noncombustible mineral board or noncombustible insulating board. These products are not made of carbon-based compounds.

Carbon-based compounds are those found in cellulose (wood), plastics and other materials manufactured from raw materials that once existed in living organisms. Cement board materials must have a specified maximum "C" (conductance) value in addition to being noncombustible.

Note h specifies a maximum thermal conductivity of $Btu/in^2 \cdot h \cdot °F$. Conductivity is the amount of heat in

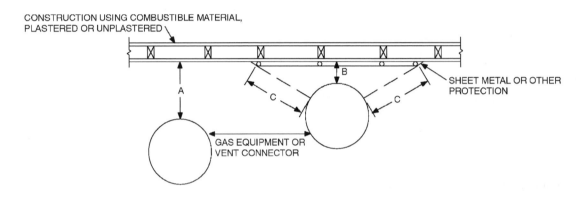

NOTES:
"A" equals the *clearance* without protection.
"B" equals the reduced *clearance* permitted in accordance with Table G2409.2. The protection applied to the construction using *combustible material* shall extend far enough in each direction to make "C" equal to "A."

FIGURE G2409.2(1) [308.2(1)]
EXTENT OF PROTECTION NECESSARY TO REDUCE CLEARANCES FROM GAS EQUIPMENT OR VENT CONNECTORS

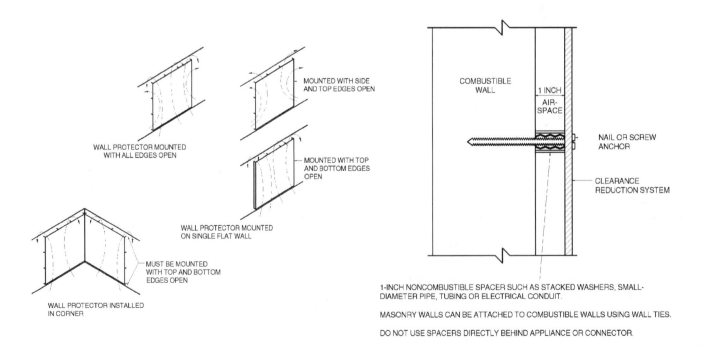

For SI: 1 inch = 25.4 mm.

FIGURE G2409.2(2) [308.2(2)]
WALL PROTECTOR CLEARANCE REDUCTION SYSTEM

Btus that will flow each hour through a 1-foot-square (0.0929 m^2) slab of material, 1-inch (25.4 mm) thick with a 1°F temperature difference between both sides and is usually identified by the symbol "k." Tables of k values usually do not include the area term in the dimensions for conductivity, and it must be understood that the value must be multiplied by the area to obtain the total Btu value.

Thermal conductance (overall) is the time rate of heat flow through a body not taking thickness into account and is usually identified by the symbol "C" (Btu/h · ft^2 · °F).

Thermal resistance (overall) is the reciprocal of overall thermal conductance and is usually identified by the symbol "R" (hr · ft^2 · °F/Btu).

This translates into a minimum required insulation R-value of 1.0 (ft^2 · hr · °F)/Btu per inch of insulating material. The methods in Table G2409.2 control heat transmission by reflecting heat radiation, retarding thermal conductance and providing convective cooling. Where sheet metal materials or metal plates are specified, the effectiveness of the protection can be enhanced by the reflective surface of the metal. Painting or otherwise covering the surface would reduce the metal's ability to reflect radiant heat and, depending on the color, could increase heat absorption. The airspace between the protected surface and the clearance-reduction assembly allows convection air currents to cool the protection assembly by carrying away heat that has been conducted through the assembly. Where a clearance-reduction assembly must be spaced 1 inch off the wall, the top, bottom and sides of the assembly must remain open as required by Notes d and f to permit unrestricted airflow (convection currents). If the openings were not provided, the air-cooling effect would not take place, and the protection assembly would not be as effective in limiting the temperature rise on the protected surfaces. Ideally, the protection assembly should be open on all sides to provide maximum ventilation. Figure G2409.2(2) and Commentary Figure G2409.2 show assemblies incorporating airspace.

Spacers must be noncombustible. Spacers should not be placed directly behind the heat source because the location would increase the amount of heat conduction through the spacer, thus creating a "hot spot." Figure G2409.2(2) specifically shows a noncombustible spacer arrangement.

The performance of a protective assembly when applied to a horizontal surface, such as a ceiling, will

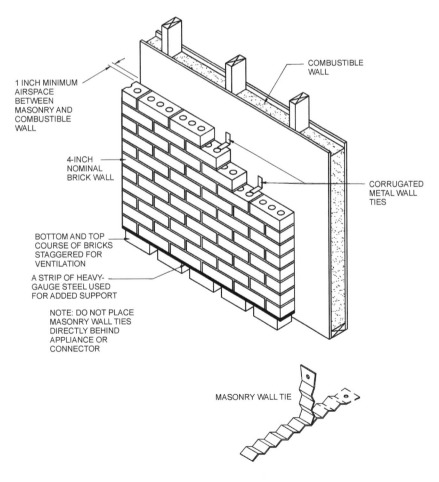

For SI: 1 inch = 25.4 mm.

FIGURE G2409.2(3) [308.2(3)]
MASONRY CLEARANCE REDUCTION SYSTEM

differ substantially from the same assembly placed in a vertical plane. Obviously, temperatures at a ceiling surface will be higher because of natural convection and because the air circulation between the method of protection and the protected ceiling surface will be substantially reduced or nonexistent. It is for these reasons that Table G2409.2 is divided into two application groups.

The manufacturer's instructions or label for many appliances will state an absolute minimum clearance, regardless of any clearance reduction method used.

Those clearance requirements take precedence over Table G2409.2.

The methods in Table G2409.2 are intended to be permanent installations properly supported to prevent displacement or deformation. Movement could adversely affect the performance of the protection method, thus posing a potential fire hazard.

The assemblies in Table G2409.2 are the product of experience and testing. To achieve predictable and dependable performance, the components of the various assemblies cannot be mixed, matched, combined

TABLE G2409.2 (308.2)[a through k]
REDUCTION OF CLEARANCES WITH SPECIFIED FORMS OF PROTECTION

TYPE OF PROTECTION APPLIED TO AND COVERING ALL SURFACES OF COMBUSTIBLE MATERIAL WITHIN THE DISTANCE SPECIFIED AS THE REQUIRED CLEARANCE WITH NO PROTECTION [see Figures G2409.2(1), G2409.2(2), and G2409.2(3)]	WHERE THE REQUIRED CLEARANCE WITH NO PROTECTION FROM APPLIANCE, VENT CONNECTOR, OR SINGLE-WALL METAL PIPE IS: (inches)									
	36		18		12		9		6	
	Allowable clearances with specified protection (inches)									
	Use Column 1 for clearances above appliance or horizontal connector. Use Column 2 for clearances from appliance, vertical connector and single-wall metal pipe.									
	Above Col. 1	Sides and rear Col. 2	Above Col. 1	Sides and rear Col. 2	Above Col. 1	Sides and rear Col. 2	Above Col. 1	Sides and rear Col. 2	Above Col. 1	Sides and rear Col. 2
1. 3¹/₂-inch-thick masonry wall without ventilated airspace	—	24	—	12	—	9	—	6	—	5
2. ¹/₂-inch insulation board over 1-inch glass fiber or mineral wool batts	24	18	12	9	9	6	6	5	4	3
3. 0.024-inch (nominal 24 gage) sheet metal over 1-inch glass fiber or mineral wool batts reinforced with wire on rear face with ventilated airspace	18	12	9	6	6	4	5	3	3	3
4. 3¹/₂-inch-thick masonry wall with ventilated airspace	—	12	—	6	—	6	—	6	—	6
5. 0.024-inch (nominal 24 gage) sheet metal with ventilated airspace	18	12	9	6	6	4	5	3	3	2
6. ¹/₂-inch-thick insulation board with ventilated airspace	18	12	9	6	6	4	5	3	3	3
7. 0.024-inch (nominal 24 gage) sheet metal with ventilated airspace over 0.024-inch (nominal 24 gage) sheet metal with ventilated airspace	18	12	9	6	6	4	5	3	3	3
8. 1-inch glass fiber or mineral wool batts sandwiched between two sheets 0.024-inch (nominal 24 gage) sheet metal with ventilated airspace	18	12	9	6	6	4	5	3	3	3

For SI: 1 inch = 25.4 mm, °C = [(°F - 32)/1.8], 1 pound per cubic foot = 16.02 kg/m³, 1 Btu per inch per square foot per hour per °F = 0.144 W/m² · K.

a. Reduction of *clearances* from *combustible materials* shall not interfere with combustion air, draft hood *clearance* and relief, and accessibility of servicing.

b. All *clearances* shall be measured from the outer surface of the *combustible material* to the nearest point on the surface of the *appliance*, disregarding any intervening protection applied to the *combustible material*.

c. Spacers and ties shall be of *noncombustible material*. A spacer or tie shall not be used directly opposite an *appliance* or *connector*.

d. For all clearance reduction systems using a ventilated airspace, adequate provision for air circulation shall be provided as described [see Figures G2409.2(2) and G2409.2(3)].

e. There shall be at least 1 inch between *clearance* reduction systems and combustible walls and ceilings for reduction systems using ventilated airspace.

f. Where a wall protector is mounted on a single flat wall away from corners, it shall have an air gap of not less than 1 inch. To provide air circulation, the bottom and top edges, or only the side and top edges, or all edges shall be left open.

g. Mineral wool batts (blanket or board) shall have a density of not less than 8 pounds per cubic foot and a melting point of not less than 1500°F.

h. Insulation material used as part of a *clearance* reduction system shall have a thermal conductivity of 1.0 Btu per inch per square foot per hour per °F or less.

i. There shall be not less than 1 inch between the *appliance* and the protector. The *clearance* between the *appliance* and the combustible surface shall not be reduced below that allowed in this table.

j. All *clearances* and thicknesses are minimum; larger *clearances* and thicknesses are acceptable.

k. *Listed* single-wall connectors shall be installed in accordance with the manufacturer's instructions.

or otherwise rearranged to comprise new assemblies, and materials cannot be substituted for those prescribed in the table. Any alterations or substitutions could have an effect on the assembly, and its performance must be tested and approved.

The reduced clearance is measured from the heat source to the combustible material, disregarding any intervening protection assembly.

Factory-built clearance reduction assemblies are available that are listed to UL1618. Such assemblies may or may not require field assembly of components.

G2409.3 (308.3) Clearances for indoor air-conditioning appliances. *Clearance* requirements for indoor air-conditioning *appliances* shall comply with Sections G2409.3.1 through G2409.3.4.

❖ Requirements for clearances to combustibles are emphasized because of the potential fire hazard posed when those clearances are not observed. Maintaining an appropriate distance from the outer surfaces of an appliance or piece of equipment to combustible materials reduces the possibility of ignition of combustible materials.

The minimum clearances to combustibles are specified in the manufacturer's installation instructions for a labeled appliance. Because an approved agency tests appliances in accordance with these instructions, the clearances required are necessary for correct installation and operation of the appliance.

Reduction of the required clearances to combustibles is allowed only when the combustibles are protected by one of the methods outlined in Section

G2409 and clearance reduction is not prohibited by Section G2409 or the appliance and equipment listing (see commentary, Section G2409). The manufacturer's specified minimum clearances and the clearances specified in Section G2409 are all airspace clearances, and such spaces cannot be filled with insulation or any other material, even if the material is noncombustible. In some cases, the manufacturer's installation instructions will specify absolute minimum clearances that must not be reduced by any clearance reduction method.

The most common wall covering material-gypsum wallboard-is a combustible finish material for the purpose of the code. As a result, gypsum wallboard as well as all other combustible wall finishes must be separated from an appliance or equipment in accordance with the prescribed clearance to combustibles. Clearances to combustibles also apply to furnishings, window treatments and moveable items that can be placed within the required clearance range of appliances and equipment.

G2409.3.1 (308.3.1) Appliances clearances. Air-conditioning *appliances* shall be installed with clearances in accordance with the manufacturer's instructions.

❖ The concept of "rooms large or not large in comparison with the appliance" is no longer part of the code. The code has been simplified by no longer tying rules for clearances to whether or not an appliance is installed in a room having a specific volume. The dated concept of room volume in comparison to appliance volume was meant to determine when the room was consid-

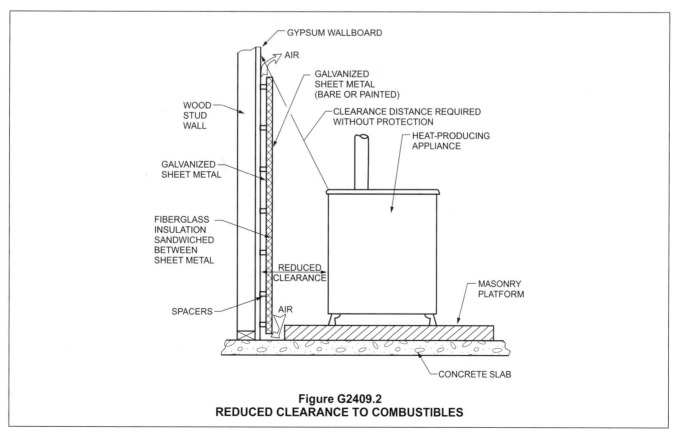

Figure G2409.2
REDUCED CLEARANCE TO COMBUSTIBLES

ered to be a closet or alcove. If the room was determined to be a closet or alcove, the code would then prohibit the use of clearance reduction methods and require the appliances to be listed for installation in closets and alcoves because the small room volume would not allow adequate dissipation of heat. This concept has outlived its intent because modern appliances are more efficient, have less jacket heat losses and are listed to current appliance standards. Now, the code simply defers to the appliance manufacturer's instructions and the appliance listing to establish the clearance requirements. In some cases, the installation instructions for an appliance might specifically prohibit the application of a clearance reduction method (see Section G2409.2).

G2409.3.2 (308.3.2) Clearance reduction. Air-conditioning appliances shall be permitted to be installed with reduced clearances to *combustible material*, provided that the *combustible material* or *appliance* is protected as described in Table G2409.2 and such reduction is allowed by the manufacturer's instructions.

❖ The clearance reduction methods in the code are always allowed unless the manufacturer's instructions specifically prohibit the reduction of clearances or specify different clearance reduction methods (see commentary, Section G2409.3.1). Note that some appliances specify clearances that are absolute minimums that cannot be reduced by any means.

G2409.3.3 (308.3.3) Plenum clearances. Where the *furnace plenum* is adjacent to plaster on metal lath or *noncombustible material* attached to *combustible material*, the *clearance* shall be measured to the surface of the plaster or other noncombustible finish where the *clearance* specified is 2 inches (51 mm) or less.

❖ Where the air-conditioning equipment plenum is adjacent to surfaces covered by noncombustible material as described in this section and the required clearance from combustibles is 2 inches (51 mm) or less, the required clearance must be measured to the surface of the adjacent material. Measuring to the combustible material would reduce the clearance between the equipment plenum and adjacent surface. Reducing this clearance would affect the cooling of the equipment and could cause overheating.

G2409.3.4 (308.3.4) Clearance from supply ducts. Supply air ducts connecting to listed central heating furnaces shall have the same minimum clearance to combustibles as required for the furnace supply plenum for a distance of not less than 3 feet (914 mm) from the supply plenum. Clearance is not required beyond the 3-foot (914 mm) distance.

❖ Portions of ducts that are within 3 feet (914 mm) of the supply plenum to which they connect will reach nearly the same temperature as the plenum.

G2409.4 (308.4) Central heating boilers and furnaces. *Clearance* requirements for central-heating boilers and *fur-*

naces shall comply with Sections G2409.4.1 through G2409.4.5. The *clearance* to these *appliances* shall not interfere with *combustion air*; *draft hood clearance* and relief; and accessibility for servicing.

❖ See the commentary to Section G2409.3.

G2409.4.1 (308.4.1) Appliances clearances. Central-heating furnaces and low-pressure boilers shall be installed with clearances in accordance with the manufacturer's instructions.

❖ See the commentary to Section G2409.3.1.

G2409.4.2 (308.4.2) Clearance reduction. Central-heating furnaces and low-pressure boilers shall be permitted to be installed with reduced clearances to *combustible material* provided that the *combustible material* or *appliance* is protected as described in Table G2409.2 and such reduction is allowed by the manufacturer's instructions.

❖ See the commentary to Section G2409.3.2.

G2409.4.3 (308.4.4) Plenum clearances. Where the *furnace plenum* is adjacent to plaster on metal lath or *noncombustible* material attached to *combustible material*, the *clearance* shall be measured to the surface of the plaster or other noncombustible finish where the *clearance* specified is 2 inches (51 mm) or less.

❖ See the commentary to Section G2409.3.3.

G2409.4.4 (308.4.5) Clearance from supply ducts. Supply air ducts connecting to listed central heating furnaces shall have the same minimum clearance to combustibles as required for the furnace supply plenum for a distance of not less than 3 feet (914 mm) from the supply plenum. Clearance is not required beyond the 3-foot (914 mm) distance.

❖ See the commentary to Section G2409.3.4.

G2409.4.5 (308.4.3) Clearance for servicing appliances. Front *clearance* shall be sufficient for servicing the *burner* and the *furnace* or boiler.

❖ Because mechanical equipment and appliances require routine maintenance, repairs and possible replacement, access is required. Additionally, access recommendations or requirements are usually stated in the manufacturer's installation instructions.

The intent of this section is to secure access to components that require observation, inspection, adjustment, servicing, repair or replacement such as controls, gauges, burners, filters, blowers, and motors. Access is also necessary to conduct operating procedures such as startup or shut down.

The code defines access as being able to be reached, but first may require the removal of a panel, door or similar obstruction. An appliance or piece of equipment is not accessible if any portion of the structure's permanent finish materials, such as drywall, plaster, paneling, built-in furniture, cabinets or any other similar permanently affixed building component must be removed.

SECTION G2410 (309)
ELECTRICAL

G2410.1 (309.1) Grounding. *Gas piping* shall not be used as a grounding electrode.

❖ Section E3508.6 prohibits metal underground gas piping from being used as a grounding electrode. Metal gas piping must be bonded to the electrical service grounding electrode system, but it cannot serve as a grounding electrode. This is somewhat contradictory and this section has been the center of much confusion relative to the "equal-potential" bonding of gas piping. Underground gas piping cannot be used as an earth electrode because it is not recognized as such in the *National Electric Code®* (NEC®), also known as NFPA 70, and because of the perceived risk of potential current flow on piping that conveys flammable gas. There is also the obvious fact that underground piping is typically plastic or metal, with a dielectric fitting isolating the metal piping from above-ground piping. However, bonding underground metal gas piping to the grounding electrode system will cause the piping to act as an additional electrode whether or not this is intended or desired. This is unavoidable, unless dielectric fittings are used to electrically isolate the underground portion of the metal piping from the above-ground portion. Any fault current will take all possible paths to earth with the intensity dictated by the impedance of the path. The key to remember is that this section does not negate the requirement to bond all metal above-ground gas piping, as required by the NEC and Section G2411. See Sections G2411.1 and G2415.10 for clarification of gas piping bonding means.

G2410.2 (309.2) Connections. Electrical connections between *appliances* and the building wiring, including the grounding of the *appliances,* shall conform to Chapters 34 through 43.

❖ Field-installed power wiring and control wiring for appliances and equipment must be installed in accordance with Chapters 34 through 43.

The power wiring includes all the wiring, disconnects, overcurrent protection devices, starters and related hardware used to supply electrical power to the appliance or equipment. The control wiring includes all the wiring, devices and related hardware that connect the main unit to all external controls and accessories, such as temperature and pressure sensors, thermostats, exhausters, equipment contractors, interlock controls and remote damper motors. The internal factory wiring of appliances and equipment is not covered by this section unless it is specifically addressed in Chapters 34 through 43; however, the wiring is covered by the testing and review performed by an approved agency as part of the listing and labeling process.

The mechanical or electrical code official responsible for the inspection of appliances and equipment must be familiar with the applicable sections of Chapters 34 through 43.

SECTION G2411 (310)
ELECTRICAL BONDING

G2411.1 (310.1) Pipe and tubing other than CSST. Each above-ground portion of a *gas piping system* other than corrugated stainless steel tubing (CSST) that is likely to become energized shall be electrically continuous and bonded to an effective ground-fault current path. *Gas piping* other than CSST shall be considered to be bonded where it is connected to *appliances* that are connected to the *equipment* grounding conductor of the circuit supplying that *appliance.*

❖ This section is consistent with Section 250.104(B) of the 2014 edition of the NEC (NFPA 70). Although difficult to define, gas piping that is likely to become energized is typically considered to be any gas piping that connects to one or more appliances that are electrically powered. The metal components of the appliances are required to be grounded; therefore, the metal gas piping that connects to the appliances should be as well because the piping has the same likelihood of being energized as the appliance components. The term "bonding" simply means to electrically tie metal parts together by means of conductors to create a low impedance path for electricity to flow between them. Metallic objects that are bonded together will have no voltage difference between them and they will all have the same potential with respect to ground.

In the majority of cases, the gas piping system will be connected to at least one, if not several, appliances that use electrical power. The gas connections involve metal-to-metal joints that ensure electrical continuity with the appliance. Those appliances will be connected to an equipment grounding conductor that will serve to bond the gas piping to the grounding electrode system for the building. Where the appliance branch circuits include grounding conductors properly connected to the appliances and one or more such appliances are connected to the gas piping system, the gas-piping system is automatically bonded as required by this section (see Commentary Figure G2411.1). This method of bonding is not allowed as the sole means of bonding for CSST gas piping systems. Note that the manufacturer's installation instructions for some gas tubing systems will require additional bonding beyond what is required by this section. This is true for CSST systems and is also mandated by Section G2411.1.1 (see Section G2411.1.1).

The purpose of the bonding required by this section is to make sure that there is no electrical potential (voltage difference) between gas piping and appliances, between gas piping and any other piping and between any piping and the earth so as to eliminate electrical shock hazards. If all of the metal piping, systems and appliances are bonded together, there cannot be a voltage difference between them; therefore, human contact across any such piping systems and appliances will not result in electrical shock. This is referred to as "equal-potential" bonding, the primary purpose of which is to prevent shock. Section G2411.1.1 addresses bonding that has an additional purpose.

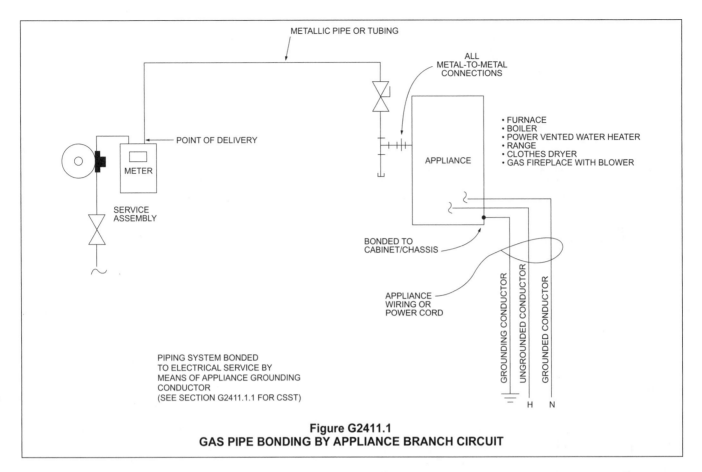

Figure G2411.1
GAS PIPE BONDING BY APPLIANCE BRANCH CIRCUIT

G2411.1.1 (310.1.1) CSST. Corrugated stainless steel tubing (CSST) gas *piping* systems and piping systems containing one or more segments of CSST shall be bonded to the electrical service grounding electrode system or, where provided, the lightning protection electrode system.

❖ This section addresses gas pipe bonding specific to CSST and does so for the purpose of lessening the possibility of damage to the CSST caused by the electrical energy created by indirect lightning strikes.

It is assumed that nothing is capable of fully protecting anything subjected to a direct strike by lightning because of the extremely high energy levels involved and unpredictability of the paths of flow. It is more common for building components to be energized by nearby lightning strikes to the earth, trees, poles, towers, etc. This intense electrical energy can travel through the earth or other objects and energize gas piping systems, among other metal systems and components. This can happen by simple conduction to the piping or by induction through forms of magnetic field coupling. The electrical energy can travel along a pipe or tube and unpredictably jump off and onto another metal pipe, wire, duct or other metal object. This "jumping off" point will experience an arc that has enough energy to burn through a pipe, tube or connector wall. The resulting exit wound caused by the arcing is a perforation that will result in gas leakage and possibly a fire or explosion. This damage can happen to any piping material, including steel pipe, CSST and copper

tube, and also gas connectors. Relatively recent experience has shown that CSST is susceptible to this type of lightning damage and this section intends to mitigate the possibility of such damage.

Recall that the CSST piping will already be bonded to some degree by the appliance grounding conductor(s), as explained in the commentary to Section G2411.1, assuming that it is connected to appliances that are electrically powered. This bonding, however, is not considered to be adequate to protect CSST, therefore, Section G2411.1.1 mandates a more effective bonding means. Directly bonding the CSST piping to the electrical service grounding electrode system or the lightning protection electrode system has been shown, by laboratory experiments, to substantially reduce the risk of lightning damage. The minimum size bonding jumper was chosen based on the maximum size grounding electrode conductor that is required to connect to "made" electrodes, such as ground rods and plates as addressed in the NEC (NFPA 70) (see Section E3608.3). A number 6 AWG copper bonding jumper provides an electrical path with considerably less impedance than the typical equipment grounding conductor found in an appliance branch circuit. The gas service and electrical service are often in close proximity, thus allowing for a short bonding jumper between the point of delivery and the electrical service. The shorter the bonding jumper, the better it will perform. In cases where the electrical service and gas service are not in close proximity, the bonding jumper should be routed to take the shortest path, noting the length limit of Section G2411.1.1.3.

The requirement of this section is also a requirement in the manufacturer's installation instructions for CSST.

As stated in the manufacturer's instructions, the bonding clamp must connect only to rigid steel pipe or a brass CSST fitting and it must never connect directly to the tubing itself. Clamping directly to the corrugated tubing would result in a poor electrical connection and could promote failure from current flow at that location. Bonding clamps must be listed for the application, for example, clamps used outdoors must be a type identified for outdoor use. The typical installation will have Schedule 40 steel pipe extending from the point of delivery (typically the meter outlet) and passing through the outside wall into the building for some distance. The bonding clamp will be attached to this steel

pipe either indoors or outdoors or attached to any other section of steel pipe or any downstream CSST fitting. Note that this section applies to gas-piping systems that have any CSST in them. For example, if a piping system is constructed mostly of steel pipe, but contains one or more sections of CSST, such a system is treated as if it was constructed entirely of CSST and this code section applies [see Commentary Figures G2411.1.1(1) and G2411.1.1(2)].

Some CSST manufacturers have designed and are marketing a type of CSST that has an electrically conductive outer jacket or jacket assembly intended to conduct stray currents and dissipate them in such a manner that arcing is prevented. The code allows these conductive jacketed types, but the bonding provisions of this section still apply unless the code official approves otherwise in accordance with Section R104.11. See Sections G2411.1.1.1 through G2411.1.1.5.

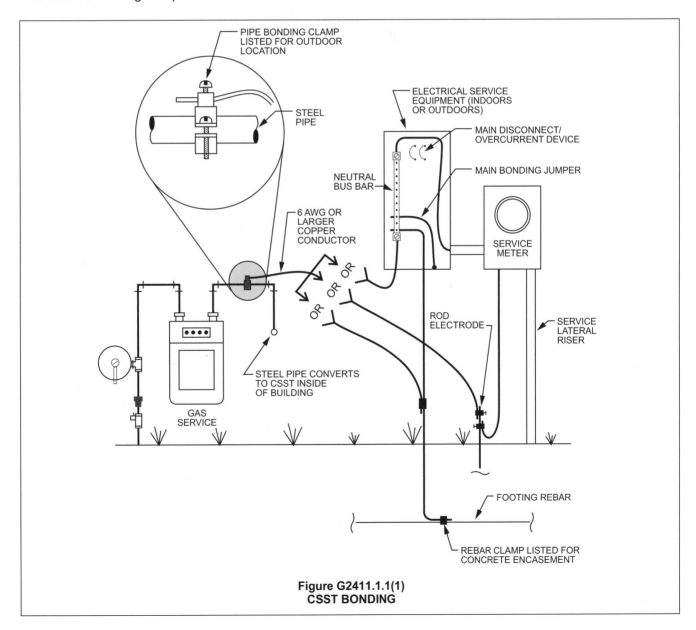

Figure G2411.1.1(1)
CSST BONDING

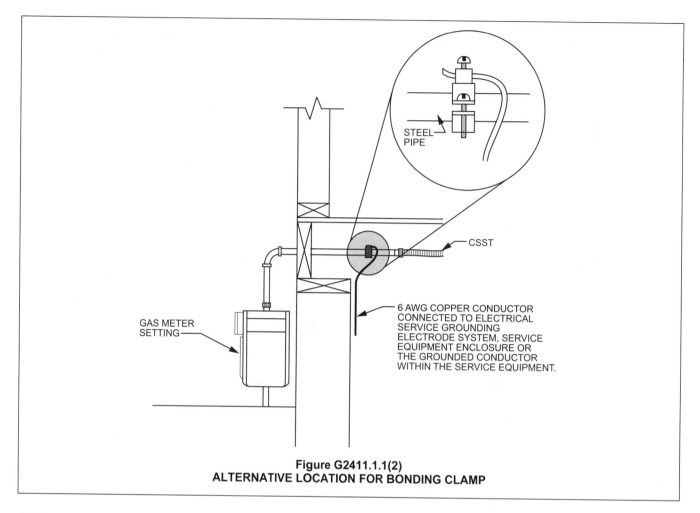

Figure G2411.1.1(2)
ALTERNATIVE LOCATION FOR BONDING CLAMP

G2411.1.1.1 (310.1.1.1) Point of connection. The boding jumper shall connect to a metallic pipe, pipe fitting or CSST fitting.

❖ The bonding jumper must connect to metal pipe or a CSST fitting, but such connection point can be anywhere in the distribution system. The bonding connection does not have to be between the point of delivery and the first downstream CSST fitting [see Commentary Figure G2411.1.1.3(1)].

G2411.1.1.2 (310.1.1.2) Size and material of jumper. The bonding jumper shall be not smaller than 6 AWG copper wire of equivalent.

❖ See the commentary to Section G2411.1.1.

G2411.1.1.3 (310.1.1.3) Bonding jumper length. The length of the bonding jumper between the connection to a gas piping system and the connection to a grounding electrode system shall not exceed 75 feet (22 860 mm). Any additional grounding electrodes used shall be bonded to the electrical service grounding electrode system or, where provided, the lightning protection grounding electrode system.

❖ It is a well-known fact that the longer a bonding jumper is the less effective it is because of the increasing impedance to electrical flow on the wire. Therefore, the shorter the better for jumper effectiveness. Extensive

testing was performed by the CSST industry to determine how well the bonding protects the CSST from indirect lightning strikes and lightning-induced currents. The testing concluded that the bonding was effective in preventing perforations in the CSST under the conditions of the predicted lightning events. The testing also determined that the bonding jumper was functionally adequate up to approximately 100 feet in preventing arcing, thus suggesting the need for a length limit. A length limit of 75 feet was chosen to provide a safety factor and also because it was believed that 75 feet would accommodate the majority of building designs and utility service entrances [see Commentary Figure G2411.1.1.3(1)].

Bonding the CCST to an independent grounding electrode is prohibited; however, the code does not prevent a designer or installer from installing a supplemental grounding electrode ("additional," as stated in the code text) for perceived additional protection. Where such supplemental electrodes are installed, the code requires that they be bonded back to the electrical service grounding electrode system, as this is consistent with NFPA 70 requirements for a common grounding electrode system. The author believes that the code is not intended to allow the length limit to be circumvented by the installation of supplemental electrodes. Where supplemental electrodes are installed by choice, the code implies that the bonding jumpers that connect to the electrical service grounding electrode system are still limited to 75 feet in combined length. An opposing interpretation is that the length of the bonding jumper between the CSST and the supplemental grounding electrode is limited to 75 feet, and the length of the jumper that connects the supplemental grounding electrode back to the electrical service grounding electrode system is limited only by the NEC, NFPA 70. The new code text implies that the more conservative interpretation is intended.

The points of connection to the electrical service grounding electrode system, the methods of connections and the protection of the bonding conductors must be in accordance with NFPA 70 (NEC). The devices, such as clamps, that are used to connect the bonding jumper on both ends must be listed for the application and environment in which they are installed. For example, clamps used outdoors must be listed for exposure to the elements. Some commonly used bonding clamps are suitable only for indoor use, and some are suitable for both indoor and outdoor use [see Commentary Figure G2411.1.1.3(2)].

G2411.1.1.4 (310.1.1.4) Bonding connections. Bonding connections shall be in accordance with NFPA 70.

❖ See the commentary to Section G2411.1.1.3.

G2411.1.1.5 (310.1.1.5) Connection devices. Devices used for making the bonding connections shall be *listed* for the application in accordance with UL 467.

❖ See the commentary to Section G2411.1.1.3.

SECTION G2412 (401)
GENERAL

G2412.1 (401.1) Scope. This section shall govern the design, installation, modification and maintenance of *piping systems*. The applicability of this *code* to *piping systems* extends from the *point of delivery* to the connections with the *appliances* and includes the design, materials, components, fabrication, assembly, installation, testing, inspection, operation and maintenance of such *piping systems*.

❖ This section regulates aspects of fuel-gas distribution systems, including design, installation, testing, repair and maintenance. The applicability of this chapter is limited to the consumer side of the public utility company's gas distribution system.

The code governs all piping and system components from the end point of the gas purveyor's (utility company) service line to the supplied appliances and equipment. Typically, the gas utility company service line terminates at the service pressure regulator and meter setting. In other words, the code does not apply to piping and components that are owned by the gas utility company. The construction of utility-owned gas piping is governed by the federal Department of Transportation (DOT) regulations wherever the piping is located. However, in cases where the utility-owned piping runs through, in or on a building, it must do so in a way that does not jeopardize the structural integrity or fire safety of the building. Therefore, the relationship of the utility-owned piping to the building (prohibited locations, penetrations, etc.) is governed by the building portion of this code (see Section G2412.1.1). No

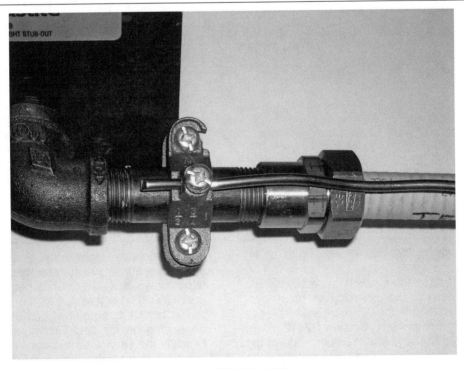

Figure G2411.1.1.3(2)
BONDING CLAMP

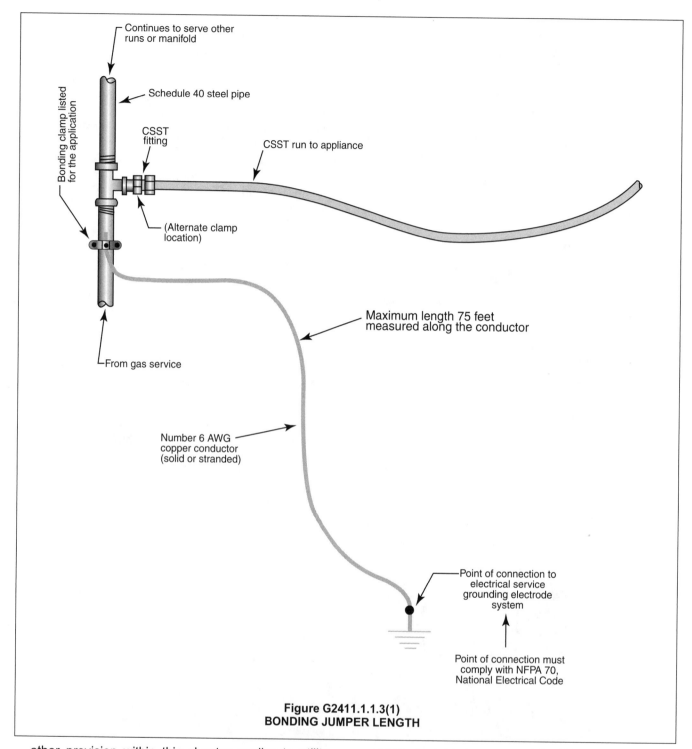

Figure G2411.1.1.3(1)
BONDING JUMPER LENGTH

other provision within this chapter applies to utility-owned gas piping.

Although LP-gas storage systems generally remain under the ownership of the gas supplier, these systems are governed by the code through the reference to NFPA 58 (see commentary, Section G2412.2). An LP-gas piping system begins at the outlet side of the first-stage pressure regulator. The piping between the first- and second-stage (final) regulator is governed by NFPA 58, and the piping downstream of the second-stage regulator is governed by this code. The first-stage pressure regulator reduces the storage container pressure to 10 pounds per square inch gauge (psig) (69 kPa) or less.

The termination point of gas utility services varies depending on the particular company's policies, but typically the service terminates at the property line or outdoors, immediately adjacent to the building or structure served by the service [see Commentary Figures G2412.1(1) and G2412.1(2)].

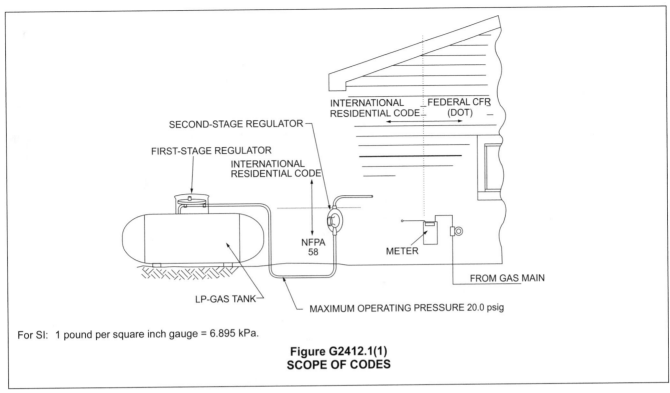

SECOND-STAGE REGULATOR

FIRST-STAGE REGULATOR

INTERNATIONAL
RESIDENTIAL CODE

INTERNATIONAL
RESIDENTIAL CODE

FEDERAL CFR
(DOT)

NFPA
58

METER

FROM GAS MAIN

LP-GAS TANK

MAXIMUM OPERATING PRESSURE 20.0 psig

For SI: 1 pound per square inch gauge = 6.895 kPa.

**Figure G2412.1(1)
SCOPE OF CODES**

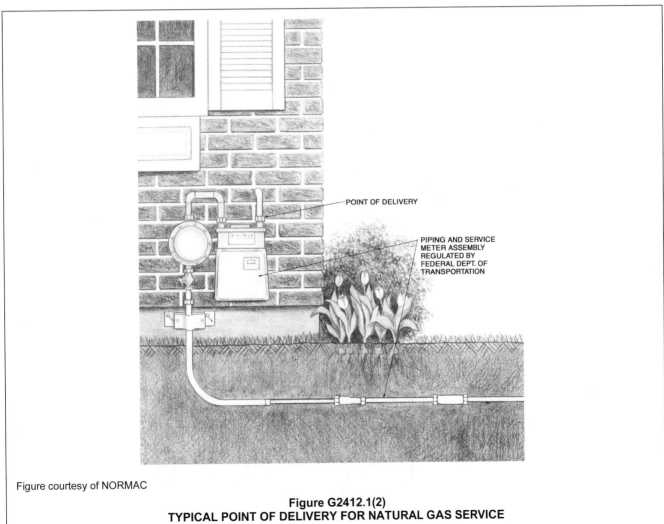

POINT OF DELIVERY

PIPING AND SERVICE
METER ASSEMBLY
REGULATED BY
FEDERAL DEPT. OF
TRANSPORTATION

Figure courtesy of NORMAC

**Figure G2412.1(2)
TYPICAL POINT OF DELIVERY FOR NATURAL GAS SERVICE**

G2412.1.1 (401.1.1) Utility piping systems located within buildings. Utility service *piping* located within buildings shall be installed in accordance with the structural safety and fire protection provisions of this code.

❖ Piping and components upstream of the point of delivery are designed, installed, tested, owned and maintained by the gas utility company and are regulated by federal DOT regulations, not this code. This code does, however, regulate two aspects of the installation related to protection of the structure from a structural and fire safety standpoint. By reference, the building portion of this code would regulate the piping penetrations through structural components and through fire-resistance-rated assemblies. The building portion of this code would also regulate piping support to handle structural loads and seismic forces.

G2412.2 (401.2) Liquefied petroleum gas storage. The storage system for *liquefied petroleum gas* shall be designed and installed in accordance with the *International Fire Code* and NFPA 58.

❖ LP-gas tanks, containers, cylinders, storage vessels and related components must comply with NFPA 58. The storage system includes storage vessels of any type and the vessel appurtenances such as shutoff valves, pressure gauges, liquid level indicators, pressure regulators, pressure relief valves and filling connections. The first- and second-stage pressure regulators and interconnecting piping are also addressed by the standard and could, therefore, be considered part of the LP-gas storage system. If the piping between the first-stage regulator and the point of delivery [second-stage or 2-psi (13.7 kPa) regulator] is not placed under the control of NFPA 58, it would be unregulated because Section G2412.1 states that the code coverage begins at the point of delivery (see definition of "Point of delivery"). Second-stage 2-psi regulators reduce the outlet pressure of the first-stage regulator to 2-psi (13.7 kPa) and a line pressure regulator downstream of the 2-psi regulator reduces the pressure from 2-psi to a maximum of 14 inches w.c. (3.5 kPa). Second-stage regulators reduce the outlet pressure of the first-stage regulator to a maximum of 14 inches w.c. NFPA 58 addresses all aspects of LP-gas storage systems such as container design and construction, container appurtenances, safety devices, container location and installation. A referenced standard is enforceable only in the context in which it is referenced. Any portions of NFPA 58 that extend beyond the scope of the storage system are not applicable because of the context of the reference in this section.

G2412.3 (401.3) Modifications to existing systems. In modifying or adding to existing *piping systems*, sizes shall be maintained in accordance with this chapter.

❖ The gas volume demand on any portion of a distribution system cannot be increased beyond the capacity of the piping serving that portion. Modifications and additions to existing piping systems can have a detrimental effect on the existing piping system's ability to serve the connected loads. Commonly, new gas-fired equipment or appliances will be added to existing distribution piping without any regard for the effect of the extra load. To prevent dangerously low gas pressures, existing piping must be able to adequately supply additional loads, or the piping size must be increased. For example, a substantial load increase would occur where a storage-type (tank) water heater is replaced by an instantaneous tankless water heater.

G2412.4 (401.4) Additional appliances. Where an additional *appliance* is to be served, the existing *piping* shall be checked to determine if it has adequate capacity for all *appliances* served. If inadequate, the existing system shall be enlarged as required or separate *piping* of adequate capacity shall be provided.

❖ This section parallels Section G2412.8 and makes it clear that when any appliances are added to an existing gas piping system, the system must be carefully evaluated to verify that adding an appliance will not create a dangerous low-gas-pressure condition within the gas piping system. The evaluation must be conducted on the completed system as it will be configured after the appliance is added, and if the results indicate that the existing piping system cannot adequately serve the gas demand of the new appliance, the system must be revised accordingly. Undersized (overloaded) piping will cause pressure losses that can result in serious appliance malfunction.

G2412.5 (401.5) Identification. For other than steel *pipe*, exposed *piping* shall be identified by a yellow label marked "Gas" in black letters. The marking shall be spaced at intervals not exceeding 5 feet (1524 mm). The marking shall not be required on *pipe* located in the same room as the *appliance* served.

❖ The intent of this section is to prevent gas piping from being mistaken for other piping. A case of mistaken identity could lead to a dangerous condition if gas piping were to be cut or opened unintentionally. It is not uncommon for fuel-gas piping systems and various other gas or liquid piping systems in a building to be constructed of the same materials and thus have the same appearance. The method of identifying fuel gas piping must be permanent, legible and conspicuous. Steel pipe (typically what is referred to as "black iron") is exempt from the identification requirements because it is the traditional gas pipe material and is generally recognized as such. CSST is manufactured in compliance with the marking requirements of this section, and some copper tubing is manufactured with a similar yellow plastic jacket [see Commentary Figures G2414.5 and G2414.5.3(1)]. Steel pipe does not have to be identified, regardless of its color and regardless of any applied paint. Note that steel pipe that has been painted another color is still steel pipe.

G2412.6 (401.6) Interconnections. Where two or more *meters* are installed on the same premises but supply separate consumers, the *piping systems* shall not be interconnected on the outlet side of the *meters*.

❖ If gas distribution systems with independent meters were connected downstream of the meters, it would be impossible to determine the actual consumption of each of the interconnected systems. As the result of building renovations, remodeling and additions, or as the result of intentionally valved cross-connections between separately metered systems, it is possible for a consumer to be paying for fuel gas consumed by another consumer. Accurate consumption metering can occur only if each metered distribution system is independent of all other metered distribution systems on the premises. Also, a system thought to be shut off could be backfed from an interconnected system, thereby maintaining pressure in a system that was intended to be shut off.

G2412.7 (401.7) Piping meter identification. *Piping* from multiple *meter* installations shall be marked with an *approved* permanent identification by the installer so that the *piping system* supplied by each *meter* is readily identifiable.

❖ This provision allows service and emergency personnel to readily locate the shutoff valve that serves each piping system. The identification also helps to prevent interconnections between systems and between spaces when additions and/or alterations are made in the piping systems.

G2412.8 (401.8) Minimum sizes. All *pipe* utilized for the installation, extension and *alteration* of any *piping system* shall be sized to supply the full number of outlets for the intended purpose and shall be sized in accordance with Section G2413.

❖ Undersized gas piping systems are not capable of delivering the required volume of fuel at the required pressure. Inadequate gas pressure can cause hazardous operation of appliances. Appliances depend on proper gas pressure to maintain the design Btu (J) input rate, to maintain the required minimum manifold pressure and to produce the required gas velocity in the burners. If the gas pressure to an appliance is too low, the result can be incomplete combustion, burner malfunction and flashback, soot production and appliance malfunction and damage. Gas piping must be sized to maintain the required minimum gas pressures at the appliance inlet.

When designing a gas piping distribution system, all connected appliances and equipment are assumed to operate simultaneously, except as provided for in Section G2413.2. In other words, the system is designed for maximum demand. The code contains provisions for sizing gas piping systems using an exact equation or design tables. The tabular method is most often used because it is simpler and less likely to produce errors. Factors that affect the sizing of gas piping systems include the specific gravity of the gas, the length of the pipe and the number of fittings installed, the

maximum gas demand, the allowable pressure loss through the system and any diversity factor that is applicable.

The design of gas piping installations should also take into account the possibility of future increases in gas demand (load). The farther the gas meter or point of delivery is from the building, the more extensive will be the disruption of the property in the event that a larger yard line between the meter and the building becomes necessary in the future as a result of increased gas demand within the building, a change in fuel gas type or a change in service pressure.

G2412.9 (401.9) Identification. Each length of pipe and tubing and each pipe fitting, utilized in a fuel gas system, shall bear the identification of the manufacturer.

❖ All pipe, tubing and fittings for the pipe and tubing must be able to be traced back to their source in case there is an issue with the material or its listing. Fittings of all sizes are easily marked with the symbol or logo of the manufacturer, but this requirement is more difficult for close and shoulder pipe nipples because there is little or no surface area suitable for such marking. Pipe nipples constructed in the field must be cut and threaded using pipe that meets the code. The pipe is marked, but the field-cut nipples might not be.

G2412.10 (401.10) Third-party testing and certification. Piping, tubing and fittings shall comply with the applicable referenced standards, specifications and performance criteria of this code and shall be identified in accordance with Section G2412.9. Piping, tubing and fittings shall either be tested by an approved third-party testing agency or certified by an approved *third-party certification agency*.

❖ These materials have always been required to comply with the standards prescribed by the code and this section requires verification by a third-party entity. The materials, or in some cases their packaging, must bear proof that a third-party entity has verified that the materials comply. This requirement will make it easier for builders, designers and code officials to determine that materials comply with the appropriate product standards.

SECTION G2413 (402)
PIPE SIZING

G2413.1 (402.1) General considerations. *Piping systems* shall be of such size and so installed as to provide a supply of gas sufficient to meet the maximum *demand* and supply gas to each *appliance* inlet at not less than the minimum supply pressure required by the *appliance*.

❖ This section is a summation of the provisions of Sections G2412.8 and G2413.5, and is intended to give guidance to the designer (see commentary, Sections G2412.8 and G2413.5). The pressure loss (drop) in a gas piping system must not exceed the allowable loss described in Section G2413.5. Pressure loss must be controlled to make sure that the supply pressure at each appliance inlet is not less than the minimum required by the appliance. Inadequate supply pressure can result in hazardous appliance operation or malfunction.

G2413.2 (402.2) Maximum gas demand. The volumetric flow rate of gas to be provided shall be the sum of the maximum input of the *appliance*s served.

The total connected hourly load shall be used as the basis for pipe sizing, assuming that all appliances could be operating at full capacity simultaneously. Where a diversity of load can be established, pipe sizing shall be permitted to be based on such loads.

The volumetric flow rate of gas to be provided shall be adjusted for altitude where the installation is above 2,000 feet (610 m) in elevation.

❖ To determine the demand volume required by an appliance in cubic feet (m³) of gas per hour, the input rating of the appliance in British thermal units per hour (Btu/h) (W) as specified by the appliance manufacturer must be known. This information will be on the appliance nameplate. Note that, depending on the elevation above sea level at which the appliance is installed, the manufacturer of the appliance might also require that the input be derated to compensate for the effect of altitude. If a derating adjustment of input is required, this adjusted input rate would be used in the calculation. If the average heating value of the gas is known [typically about 1,000 Btu/ft³ (37.3 MJ/m³) for natural gas], the volume of gas required per hour can be calculated. The heating value can be obtained for the gas supplier. The input rating of all appliances must be shown on the appliance label. When the input rating of the appliances is unknown at the time of piping system design, the gas demand will have to be estimated based on input from appliance manufacturers, gas utilities or another source. However, when the actual appliance demands are determined upon appliance selection, any approximate calculations must be checked to verify that the piping system is, in fact, adequate to supply the known demand for all appliances. The designer must be aware of the fact that a system designed in accordance with an estimated load might be undersized relative to the actual connected load.

In all cases, the piping must have the capacity to supply the actual connected load of appliances. As also stated in this section and Section G2412.8, the load on the piping system must be based on the simultaneous operation of all appliances at full output. The only exception to this would be the demonstration of actual load diversity. For example, the designer could have access to data that demonstrates that the total load for cooking ranges or water heaters in a multiple-family dwelling (apartment building) is less than the sum of all of the appliance demands at full output. The code official would have to be convinced that any such diversity factor had been established. Any known diversity factors would likely come from the serving gas supplier.

The following is an example of how gas demand is calculated for a representative appliance: an appliance installed at sea level has a nameplate (label) input rating of 120,000 Btu/h (35.2 kW). Using natural gas with an average heating value of 1,000 Btu/ft³ (37.3 MJ/m³), the volume of gas required for the appliance is: 120,000/1,000 = 120 cubic feet per hour (3.4 m³/h).

G2413.3 (402.3) Sizing. *Gas piping* shall be sized in accordance with one of the following:

1. *Pipe* sizing tables or sizing equations in accordance with Section G2413.4.

2. The sizing tables included in a *listed piping* system's manufacturer's installation instructions.

3. Other *approved* engineering methods.

❖ Using the sizing tables is generally simpler and less time consuming than using the equations found in Section G2413.4. The gas system pressure, allowable pressure drop, maximum gas demand and the specific gravity of the fuel gas must be known in order to use tables. The allowable pressure drop is the designer's choice because the intent of the code is to provide the appliance or equipment with the required rate of gas flow at the required minimum pressure. The tables are based on a given pressure drop and cannot be used if the designer wishes to use any other pressure drop. The sizing tables are based on using smooth-bore pipe or tube and can be used with the piping materials listed in Section G2414. Pipe materials with higher resistance to flow cannot be sized using these tables. CSST is sized using CSST-specific tables. The pipe sizes in the pipe sizing tables are based on the traditional steel pipe size designation as used in the material standards. More information may be obtained from a variety of pipe design publications, including the *Piping Handbook* by Sabin Crocker and the ASHRAE *Handbook of HVAC Systems and Equipment*. Each of the tables presents different piping material and design parameters and represents the most common system design applications. The tables assume that the system being sized is constructed of a single type pipe or tubing material throughout; thus, a single table cannot be used for systems constructed of multiple materials. For example, it is common to find hybrid systems composed of both steel pipe and CSST. Such systems would have to be sized from two different tables based on the longest run of the entire system.

The code does not limit sizing methodology to the tables or equations. Manufacturers of gas piping/tubing may provide sizing charts, tables or slide calculators that, in effect, become part of the installation instructions. This is the case for CSST, which is a listed piping system.

Section G2413.4 describes a sizing method that uses an exact thermodynamic flow equation to determine the required pipe sizes. Although this is an exact method, the degree of accuracy it affords may not always prove valuable, because pipe sizes are standardized and the pipe diameter calculated may not correspond to an available pipe size. For example, a calculated pipe diameter of 0.80 inch (20.3 mm) will require the selection of the next larger standard pipe size [1-inch (25.4 mm) pipe].

There are two equations in Section G2413.4 to be used, depending on the gas pressure within the piping system.

As stated in Appendix A, Sections A.2.2 and A.2.3, an equivalent length of pipe should be added to the length of any piping run having four or more fittings (see Appendix A, Table A.2.2). Some designers simply add 50 percent of the actual length of piping as an all-inclusive fitting allowance (i.e., actual pipe length x 1.5). This is a conservative estimate and is used because the necessary number and type of fittings is usually not known when the system is being designed.

Steel pipe sizes $^1/_4$ inch and $^3/_8$ inch have been deleted from the tables because the use of such sizes is very rare. The pressure of "0.50 or less" that appeared in some table titles has been replaced by "less than 2 psi" because such tables can be used with any pressure less than 2 psi (13.8 kPa), including 0.50 psi (3.5 kPa) and less.

G2413.4 (402.4) Sizing tables and equations. Where Tables G2413.4(1) through G2413.4(21) are used to size *piping* or *tubing*, the *pipe* length shall be determined in accordance with Section G2413.4.1, G2413.4.2 or G2413.4.3.

Where Equations 24-3 and 24-4 are used to size *piping* or *tubing*, the *pipe* or *tubing* shall have smooth inside walls and the pipe length shall be determined in accordance with Section G2413.4.1, G2413.4.2 or G2413.4.3.

1. Low-pressure gas equation [Less than $1^1/_2$ pounds per square inch (psi) (10.3 kPa)]:

$$D = \frac{Q^{0.381}}{19.17\left(\dfrac{\Delta H}{C_r \times L}\right)^{0.206}} \quad \textbf{(Equation 24-3)}$$

2. High-pressure gas equation [1.5 psi (10.3 kPa) and above]:

$$D = \frac{Q^{0.381}}{18.93\left[\dfrac{(P_1^2 - P_2^2) \times Y}{C_r \times L}\right]^{0.206}} \quad \textbf{(Equation 24-4)}$$

where:

D = Inside diameter of *pipe*, inches (mm).

Q = Input rate *appliance(s)*, cubic feet per hour at 60°F (16°C) and 30-inch mercury column.

P_1 = Upstream pressure, psia (P_1 + 14.7).

P_2 = Downstream pressure, psia (P_2 + 14.7).

L = Equivalent length of *pipe*, feet.

ΔH = *Pressure drop*, inch water column (27.7 inch water column = 1 psi).

❖ The length of pipe or tube to be used with the tables and equations is determined by the sizing method being used. Section G2413.4.1 involves only one length whereas Sections G2413.4.2 and G2413.4.3 involve multiple lengths.

TABLE G2413.4 (402.4)
C_r AND Y VALUES FOR NATURAL GAS AND UNDILUTED PROPANE AT STANDARD CONDITIONS

GAS	EQUATION FACTORS	
	C_r	Y
Natural gas	0.6094	0.9992
Undiluted propane	1.2462	0.9910

For SI: 1 cubic foot = 0.028 m³, 1 foot = 305 mm,
1-inch water column = 0.249 kPa,
1 pound per square inch = 6.895 kPa,
1 British thermal unit per hour = 0.293 W.

❖ **Example 1:**

Determine the required pipe sizes for the gas distribution system in Commentary Figure G2413.4 using the equation method outlined in this section. The system is designed to operate at less than 0.5 pound per square inch (psig) (3.5 kPa), using steel pipe. Assume the following:

Fuel gas	Natural
Specific gravity	0.60
Heating value	1,000 Btu per cubic foot
Maximum allowable pressure drop	$^1/_2$ inch of water column

Negligible number of fittings

For SI: 1 inch = 25.4mm, 1 foot = 304.8mm, 1 Btu/h = 0.2931 W, 1 Btu/ft³ = 0.0373 MJ/m³, 1 inch water column = 0.2488 kPa, 1 cubic foot = 0.02832 m³, 1 pound per square inch gauge = 6.895 kPa.

To use equations in this section, the gas volume of each appliance must be known.

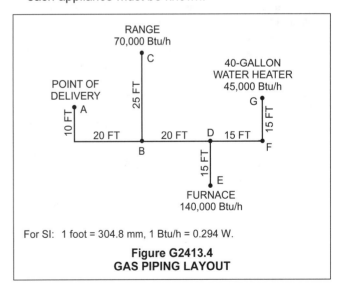

For SI: 1 foot = 304.8 mm, 1 Btu/h = 0.294 W.

Figure G2413.4
GAS PIPING LAYOUT

Gas volume of an appliance = input rating of an appliance (Btu/h)/Heating value of the fuel gas (Btu/ft³).

Using the equation, we can solve for the minimum pipe diameter (D) as follows:

$D = Q^{0.381}/19.17 \ [(\Delta H)/(C_r \times L]^{0.206}$

APPLIANCE	INPUT RATING (Btu/h)	GAS VOLUME (ft³/h)
Range	70,000	70
Furnace	140,000	140
Water heater	45,000	45

Starting at the most remote appliance from the point of delivery (Point G - water heater), determine the minimum pipe size for pipe segment D.

$$D = (45)^{0.381}/19.17 \ [(^1/_2)/(0.609 \times 80)]^{0.206} = 0.574$$

Repeat the calculation for each pipe segment. The gas piping system is sized as follows:

PIPE SEGMENT	FLOW RATE (Q)	CALCULATED PIPE DIAMETER (D) (in.)	NOMINAL PIPE SIZE (in.)
A-B	255	1.11	$1^1/_4$
B-C	70	0.679	$^3/_4$
B-D	185	0.983	1
D-E	140	0.884	1
D-G	45	0.574	$^1/_2$

Most gas pipe sizing will be determined by using the sizing tables. In the event that circumstances warrant it, however, pipe sizing may be done by calculation using the provisions of Appendix A or other approved methods. If in a particular design, the piping length, desired pressure loss, system supply pressure, specific gravity or pipe size does not fall within the parameters of a table, an approved alternate sizing procedure must be used.

Instead of using the methods specified in this section, piping sizes may be determined by the use of accurate gas flow, computer or pressure drop charts. Such sizing tools must be acceptable to the code official. Note that calculators, charts and computer programs can be designed to size gas piping for any pressure drop because the intent of the code is to make sure that the supply pressure at the appliance is greater than or equal to the minimum pressure required for proper operation of the appliance (see Section G2413.5).

TABLES G2413.4(1) through G2413.4(21). See pages 24-76 through 24-96.

❖ See the commentary to Section G2413.4.

G2413.4.1 (402.4.1) Longest length method. The *pipe* size of each section of *gas piping* shall be determined using the longest length of *piping* from the *point of delivery* to the most remote *outlet* and the load of the section.

❖ This section provides a step-by-step approach to proper application of the tables using the traditional longest-length method. When using the tables, the maximum pipe length from the point of delivery to the farthest outlet must be determined, including any allowance for the equivalent length of fittings. The designer will use the row in the table that equals the determined length or the next higher row if the determined length is between table values. The system will be designed using only values taken from this row. This is known as the longest-run method. Basing the sizing on the most demanding circuit (longest run) compensates for pressure losses throughout the entire system. Even though the appliances are not typical for dwellings, Commentary Figure G2413.4.1 and the example serve to illustrate the longest length method.

Longest length method example: Basing the sizing on the most demanding circuit (longest run using Table G2413.4(1), size the piping system in Commentary Figure G2413.4.1. The longest run is from the meter (point of delivery) to Roof Top Unit (RTU) 5 and is 210 feet (64 008 mm), therefore, all pipe sizes will be chosen from the 250 foot (76 200 mm) row of the table.

PIPE SECTION	LOAD (MBH)	SIZE (in.)
A	690	2
B	690	2
C	540	2
D	450	2
E	250	$1^1/_2$
F	250	$1^1/_2$
G	150	$1^1/_4$
H	90	1
I	200	$1^1/_4$
J	100	1
K	150	$1^1/_4$

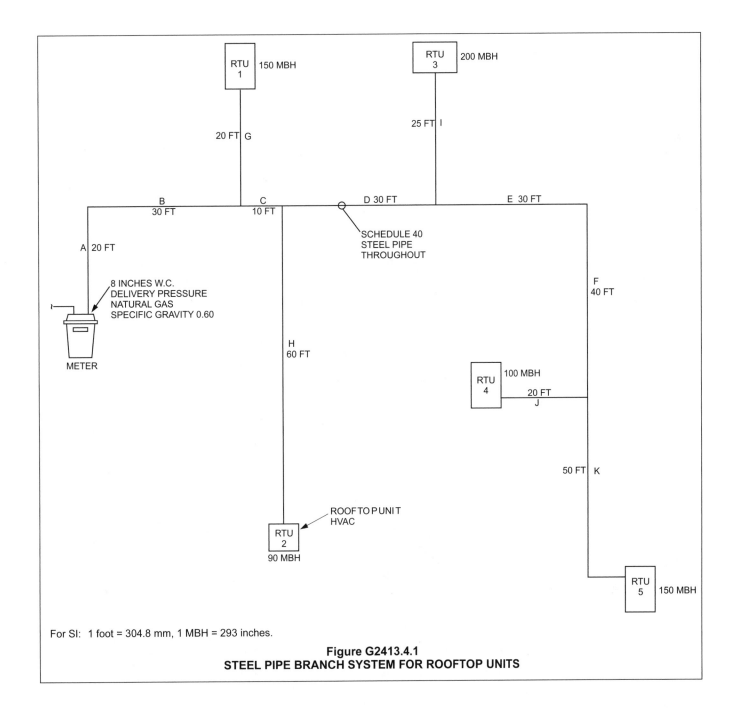

For SI: 1 foot = 304.8 mm, 1 MBH = 293 inches.

Figure G2413.4.1
STEEL PIPE BRANCH SYSTEM FOR ROOFTOP UNITS

TABLE G2413.4(1) [402.4(2)]
SCHEDULE 40 METALLIC PIPE

Gas	Natural
Inlet Pressure	Less than 2 psi
Pressure Drop	0.5 in. w.c.
Specific Gravity	0.60

	PIPE SIZE (inches)													
Nominal	1/2	3/4	1	1 1/4	1 1/2	2	2 1/2	3	4	5	6	8	10	12
Actual ID	0.622	0.824	1.049	1.380	1.610	2.067	2.469	3.068	4.026	5.047	6.065	7.981	10.020	11.938
Length (ft)	Capacity in Cubic Feet of Gas per Hour													
10	172	360	678	1,390	2,090	4,020	6,400	11,300	23,100	41,800	67,600	139,000	252,000	399,000
20	118	247	466	957	1,430	2,760	4,400	7,780	15,900	28,700	46,500	95,500	173,000	275,000
30	95	199	374	768	1,150	2,220	3,530	6,250	12,700	23,000	37,300	76,700	139,000	220,000
40	81	170	320	657	985	1,900	3,020	5,350	10,900	19,700	31,900	65,600	119,000	189,000
50	72	151	284	583	873	1,680	2,680	4,740	9,660	17,500	28,300	58,200	106,000	167,000
60	65	137	257	528	791	1,520	2,430	4,290	8,760	15,800	25,600	52,700	95,700	152,000
70	60	126	237	486	728	1,400	2,230	3,950	8,050	14,600	23,600	48,500	88,100	139,000
80	56	117	220	452	677	1,300	2,080	3,670	7,490	13,600	22,000	45,100	81,900	130,000
90	52	110	207	424	635	1,220	1,950	3,450	7,030	12,700	20,600	42,300	76,900	122,000
100	50	104	195	400	600	1,160	1,840	3,260	6,640	12,000	19,500	40,000	72,600	115,000
125	44	92	173	355	532	1,020	1,630	2,890	5,890	10,600	17,200	35,400	64,300	102,000
150	40	83	157	322	482	928	1,480	2,610	5,330	9,650	15,600	32,100	58,300	92,300
175	37	77	144	296	443	854	1,360	2,410	4,910	8,880	14,400	29,500	53,600	84,900
200	34	71	134	275	412	794	1,270	2,240	4,560	8,260	13,400	27,500	49,900	79,000
250	30	63	119	244	366	704	1,120	1,980	4,050	7,320	11,900	24,300	44,200	70,000
300	27	57	108	221	331	638	1,020	1,800	3,670	6,630	10,700	22,100	40,100	63,400
350	25	53	99	203	305	587	935	1,650	3,370	6,100	9,880	20,300	36,900	58,400
400	23	49	92	189	283	546	870	1,540	3,140	5,680	9,190	18,900	34,300	54,300
450	22	46	86	177	266	512	816	1,440	2,940	5,330	8,620	17,700	32,200	50,900
500	21	43	82	168	251	484	771	1,360	2,780	5,030	8,150	16,700	30,400	48,100
550	20	41	78	159	239	459	732	1,290	2,640	4,780	7,740	15,900	28,900	45,700
600	19	39	74	152	228	438	699	1,240	2,520	4,560	7,380	15,200	27,500	43,600
650	18	38	71	145	218	420	669	1,180	2,410	4,360	7,070	14,500	26,400	41,800
700	17	36	68	140	209	403	643	1,140	2,320	4,190	6,790	14,000	25,300	40,100
750	17	35	66	135	202	389	619	1,090	2,230	4,040	6,540	13,400	24,400	38,600
800	16	34	63	130	195	375	598	1,060	2,160	3,900	6,320	13,000	23,600	37,300
850	16	33	61	126	189	363	579	1,020	2,090	3,780	6,110	12,600	22,800	36,100
900	15	32	59	122	183	352	561	992	2,020	3,660	5,930	12,200	22,100	35,000
950	15	31	58	118	178	342	545	963	1,960	3,550	5,760	11,800	21,500	34,000
1,000	14	30	56	115	173	333	530	937	1,910	3,460	5,600	11,500	20,900	33,100
1,100	14	28	53	109	164	316	503	890	1,810	3,280	5,320	10,900	19,800	31,400
1,200	13	27	51	104	156	301	480	849	1,730	3,130	5,070	10,400	18,900	30,000
1,300	12	26	49	100	150	289	460	813	1,660	3,000	4,860	9,980	18,100	28,700
1,400	12	25	47	96	144	277	442	781	1,590	2,880	4,670	9,590	17,400	27,600
1,500	11	24	45	93	139	267	426	752	1,530	2,780	4,500	9,240	16,800	26,600
1,600	11	23	44	89	134	258	411	727	1,480	2,680	4,340	8,920	16,200	25,600
1,700	11	22	42	86	130	250	398	703	1,430	2,590	4,200	8,630	15,700	24,800
1,800	10	22	41	84	126	242	386	682	1,390	2,520	4,070	8,370	15,200	24,100
1,900	10	21	40	81	122	235	375	662	1,350	2,440	3,960	8,130	14,800	23,400
2,000	NA	20	39	79	119	229	364	644	1,310	2,380	3,850	7,910	14,400	22,700

For SI: 1 inch = 25.4 mm, 1 foot = 304.8 mm, 1 pound per square inch = 6.895 kPa, 1-inch water column = 0.2488 kPa,
1 British thermal unit per hour = 0.2931 W, 1 cubic foot per hour = 0.0283 m^3/h, 1 degree = 0.01745 rad.

Notes:
1. NA means a flow of less than 10 cfh.
2. All table entries have been rounded to three significant digits.

**TABLE G2413.4(2) [402.4(5)]
SCHEDULE 40 METALLIC PIPE**

Gas	Natural
Inlet Pressure	2.0 psi
Pressure Drop	1.0 psi
Specific Gravity	0.60

PIPE SIZE (inches)									
Nominal	$^1/_2$	$^3/_4$	1	$1^1/_4$	$1^1/_2$	2	$2^1/_2$	3	4
Actual ID	0.622	0.824	1.049	1.380	1.610	2.067	2.469	3.068	4.026
Length (ft)	Capacity in Cubic Feet of Gas per Hour								
10	1,510	3,040	5,560	11,400	17,100	32,900	52,500	92,800	189,000
20	1,070	2,150	3,930	8,070	12,100	23,300	37,100	65,600	134,000
30	869	1,760	3,210	6,590	9,880	19,000	30,300	53,600	109,000
40	753	1,520	2,780	5,710	8,550	16,500	26,300	46,400	94,700
50	673	1,360	2,490	5,110	7,650	14,700	23,500	41,500	84,700
60	615	1,240	2,270	4,660	6,980	13,500	21,400	37,900	77,300
70	569	1,150	2,100	4,320	6,470	12,500	19,900	35,100	71,600
80	532	1,080	1,970	4,040	6,050	11,700	18,600	32,800	67,000
90	502	1,010	1,850	3,810	5,700	11,000	17,500	30,900	63,100
100	462	934	1,710	3,510	5,260	10,100	16,100	28,500	58,200
125	414	836	1,530	3,140	4,700	9,060	14,400	25,500	52,100
150	372	751	1,370	2,820	4,220	8,130	13,000	22,900	46,700
175	344	695	1,270	2,601	3,910	7,530	12,000	21,200	43,300
200	318	642	1,170	2,410	3,610	6,960	11,100	19,600	40,000
250	279	583	1,040	2,140	3,210	6,180	9,850	17,400	35,500
300	253	528	945	1,940	2,910	5,600	8,920	15,800	32,200
350	232	486	869	1,790	2,670	5,150	8,210	14,500	29,600
400	216	452	809	1,660	2,490	4,790	7,640	13,500	27,500
450	203	424	759	1,560	2,330	4,500	7,170	12,700	25,800
500	192	401	717	1,470	2,210	4,250	6,770	12,000	24,400
550	182	381	681	1,400	2,090	4,030	6,430	11,400	23,200
600	174	363	650	1,330	2,000	3,850	6,130	10,800	22,100
650	166	348	622	1,280	1,910	3,680	5,870	10,400	21,200
700	160	334	598	1,230	1,840	3,540	5,640	9,970	20,300
750	154	322	576	1,180	1,770	3,410	5,440	9,610	19,600
800	149	311	556	1,140	1,710	3,290	5,250	9,280	18,900
850	144	301	538	1,100	1,650	3,190	5,080	8,980	18,300
900	139	292	522	1,070	1,600	3,090	4,930	8,710	17,800
950	135	283	507	1,040	1,560	3,000	4,780	8,460	17,200
1,000	132	275	493	1,010	1,520	2,920	4,650	8,220	16,800
1,100	125	262	468	960	1,440	2,770	4,420	7,810	15,900
1,200	119	250	446	917	1,370	2,640	4,220	7,450	15,200
1,300	114	239	427	878	1,320	2,530	4,040	7,140	14,600
1,400	110	230	411	843	1,260	2,430	3,880	6,860	14,000
1,500	106	221	396	812	1,220	2,340	3,740	6,600	13,500
1,600	102	214	382	784	1,180	2,260	3,610	6,380	13,000
1,700	99	207	370	759	1,140	2,190	3,490	6,170	12,600
1,800	96	200	358	736	1,100	2,120	3,390	5,980	12,200
1,900	93	195	348	715	1,070	2,060	3,290	5,810	11,900
2,000	91	189	339	695	1,040	2,010	3,200	5,650	11,500

For SI: 1 inch = 25.4 mm, 1 foot = 304.8 mm, 1 pound per square inch = 6.895 kPa, 1-inch water column = 0.2488 kPa,
    1 British thermal unit per hour = 0.2931 W, 1 cubic foot per hour = 0.0283 m³/h, 1 degree = 0.01745 rad.

Note: All table entries have been rounded to three significant digits.

TABLE G2413.4(3) [402.4(9)]
SEMIRIGID COPPER TUBING

	Gas	Natural
	Inlet Pressure	Less than 2 psi
	Pressure Drop	0.5 in. w.c.
	Specific Gravity	0.60

		TUBE SIZE (inches)								
Nominal	K & L	$^1/_4$	$^3/_8$	$^1/_2$	$^5/_8$	$^3/_4$	1	$1^1/_4$	$1^1/_2$	2
	ACR	$^3/_8$	$^1/_2$	$^5/_8$	$^3/_4$	$^7/_8$	$1^1/_8$	$1^3/_8$	—	—
Outside		0.375	0.500	0.625	0.750	0.875	1.125	1.375	1.625	2.125
Inside		0.305	0.402	0.527	0.652	0.745	0.995	1.245	1.481	1.959
Length (ft)		Capacity in Cubic Feet of Gas per Hour								
10		27	55	111	195	276	590	1,060	1,680	3,490
20		18	38	77	134	190	406	730	1,150	2,400
30		15	30	61	107	152	326	586	925	1,930
40		13	26	53	92	131	279	502	791	1,650
50		11	23	47	82	116	247	445	701	1,460
60		10	21	42	74	105	224	403	635	1,320
70		NA	19	39	68	96	206	371	585	1,220
80		NA	18	36	63	90	192	345	544	1,130
90		NA	17	34	59	84	180	324	510	1,060
100		NA	16	32	56	79	170	306	482	1,000
125		NA	14	28	50	70	151	271	427	890
150		NA	13	26	45	64	136	245	387	806
175		NA	12	24	41	59	125	226	356	742
200		NA	11	22	39	55	117	210	331	690
250		NA	NA	20	34	48	103	186	294	612
300		NA	NA	18	31	44	94	169	266	554
350		NA	NA	16	28	40	86	155	245	510
400		NA	NA	15	26	38	80	144	228	474
450		NA	NA	14	25	35	75	135	214	445
500		NA	NA	13	23	33	71	128	202	420
550		NA	NA	13	22	32	68	122	192	399
600		NA	NA	12	21	30	64	116	183	381
650		NA	NA	12	20	29	62	111	175	365
700		NA	NA	11	20	28	59	107	168	350
750		NA	NA	11	19	27	57	103	162	338
800		NA	NA	10	18	26	55	99	156	326
850		NA	NA	10	18	25	53	96	151	315
900		NA	NA	NA	17	24	52	93	147	306
950		NA	NA	NA	17	24	50	90	143	297
1,000		NA	NA	NA	16	23	49	88	139	289
1,100		NA	NA	NA	15	22	46	84	132	274
1,200		NA	NA	NA	15	21	44	80	126	262
1,300		NA	NA	NA	14	20	42	76	120	251
1,400		NA	NA	NA	13	19	41	73	116	241
1,500		NA	NA	NA	13	18	39	71	111	232
1,600		NA	NA	NA	13	18	38	68	108	224
1,700		NA	NA	NA	12	17	37	66	104	217
1,800		NA	NA	NA	12	17	36	64	101	210
1,900		NA	NA	NA	11	16	35	62	98	204
2,000		NA	NA	NA	11	16	34	60	95	199

For SI: 1 inch = 25.4 mm, 1 foot = 304.8 mm, 1 pound per square inch = 6.895 kPa, 1-inch water column = 0.2488 kPa,
     1 British thermal unit per hour = 0.2931 W, 1 cubic foot per hour = 0.0283 m³/h, 1 degree = 0.01745 rad.

Notes:

1. Table capacities are based on Type K copper tubing inside diameter (shown), which has the smallest inside diameter of the copper tubing products.

2. NA means a flow of less than 10 cfh.

3. All table entries have been rounded to three significant digits.

2015 INTERNATIONAL RESIDENTIAL CODE® COMMENTARY

TABLE G2413.4(4) [402.4(12)]
SEMIRIGID COPPER TUBING

	Gas	Natural
	Inlet Pressure	2.0 psi
	Pressure Drop	1.0 psi
	Specific Gravity	0.60

Nominal	K & L	¹/₄	³/₈	¹/₂	⁵/₈	³/₄	1	1¹/₄	1¹/₂	2
	ACR	³/₈	¹/₂	⁵/₈	³/₄	⁷/₈	1¹/₈	1³/₈	—	—
Outside		0.375	0.500	0.625	0.750	0.875	1.125	1.375	1.625	2.125
Inside		0.305	0.402	0.527	0.652	0.745	0.995	1.245	1.481	1.959
Length (ft)		Capacity in Cubic Feet of Gas per Hour								
10		245	506	1,030	1,800	2,550	5,450	9,820	15,500	32,200
20		169	348	708	1,240	1,760	3,750	6,750	10,600	22,200
30		135	279	568	993	1,410	3,010	5,420	8,550	17,800
40		116	239	486	850	1,210	2,580	4,640	7,310	15,200
50		103	212	431	754	1,070	2,280	4,110	6,480	13,500
60		93	192	391	683	969	2,070	3,730	5,870	12,200
70		86	177	359	628	891	1,900	3,430	5,400	11,300
80		80	164	334	584	829	1,770	3,190	5,030	10,500
90		75	154	314	548	778	1,660	2,990	4,720	9,820
100		71	146	296	518	735	1,570	2,830	4,450	9,280
125		63	129	263	459	651	1,390	2,500	3,950	8,220
150		57	117	238	416	590	1,260	2,270	3,580	7,450
175		52	108	219	383	543	1,160	2,090	3,290	6,850
200		49	100	204	356	505	1,080	1,940	3,060	6,380
250		43	89	181	315	448	956	1,720	2,710	5,650
300		39	80	164	286	406	866	1,560	2,460	5,120
350		36	74	150	263	373	797	1,430	2,260	4,710
400		33	69	140	245	347	741	1,330	2,100	4,380
450		31	65	131	230	326	696	1,250	1,970	4,110
500		30	61	124	217	308	657	1,180	1,870	3,880
550		28	58	118	206	292	624	1,120	1,770	3,690
600		27	55	112	196	279	595	1,070	1,690	3,520
650		26	53	108	188	267	570	1,030	1,620	3,370
700		25	51	103	181	256	548	986	1,550	3,240
750		24	49	100	174	247	528	950	1,500	3,120
800		23	47	96	168	239	510	917	1,450	3,010
850		22	46	93	163	231	493	888	1,400	2,920
900		22	44	90	158	224	478	861	1,360	2,830
950		21	43	88	153	217	464	836	1,320	2,740
1,000		20	42	85	149	211	452	813	1,280	2,670
1,100		19	40	81	142	201	429	772	1,220	2,540
1,200		18	38	77	135	192	409	737	1,160	2,420
1,300		18	36	74	129	183	392	705	1,110	2,320
1,400		17	35	71	124	176	376	678	1,070	2,230
1,500		16	34	68	120	170	363	653	1,030	2,140
1,600		16	33	66	116	164	350	630	994	2,070
1,700		15	31	64	112	159	339	610	962	2,000
1,800		15	30	62	108	154	329	592	933	1,940
1,900		14	30	60	105	149	319	575	906	1,890
2,000		14	29	59	102	145	310	559	881	1,830

For SI: 1 inch = 25.4 mm, 1 foot = 304.8 mm, 1 pound per square inch = 6.895 kPa, 1-inch water column = 0.2488 kPa,
  1 British thermal unit per hour = 0.2931 W, 1 cubic foot per hour = 0.0283 m³/h, 1 degree = 0.01745 rad.

Notes:
1. Table capacities are based on Type K copper tubing inside diameter (shown), which has the smallest inside diameter of the copper tubing products.
2. All table entries have been rounded to three significant digits.

TABLE G2413.4(5) [402.4(15)]
CORRUGATED STAINLESS STEEL TUBING (CSST)

Gas	Natural
Inlet Pressure	Less than 2 psi
Pressure Drop	0.5 in. w.c.
Specific Gravity	0.60

	TUBE SIZE (EHD)													
Flow Designation	13	15	18	19	23	25	30	31	37	39	46	48	60	62
Length (ft)	Capacity in Cubic Feet of Gas per Hour													
5	46	63	115	134	225	270	471	546	895	1,037	1,790	2,070	3,660	4,140
10	32	44	82	95	161	192	330	383	639	746	1,260	1,470	2,600	2,930
15	25	35	66	77	132	157	267	310	524	615	1,030	1,200	2,140	2,400
20	22	31	58	67	116	137	231	269	456	536	888	1,050	1,850	2,080
25	19	27	52	60	104	122	206	240	409	482	793	936	1,660	1,860
30	18	25	47	55	96	112	188	218	374	442	723	856	1,520	1,700
40	15	21	41	47	83	97	162	188	325	386	625	742	1,320	1,470
50	13	19	37	42	75	87	144	168	292	347	559	665	1,180	1,320
60	12	17	34	38	68	80	131	153	267	318	509	608	1,080	1,200
70	11	16	31	36	63	74	121	141	248	295	471	563	1,000	1,110
80	10	15	29	33	60	69	113	132	232	277	440	527	940	1,040
90	10	14	28	32	57	65	107	125	219	262	415	498	887	983
100	9	13	26	30	54	62	101	118	208	249	393	472	843	933
150	7	10	20	23	42	48	78	91	171	205	320	387	691	762
200	6	9	18	21	38	44	71	82	148	179	277	336	600	661
250	5	8	16	19	34	39	63	74	133	161	247	301	538	591
300	5	7	15	17	32	36	57	67	95	148	226	275	492	540

For SI: 1 inch = 25.4 mm, 1 foot = 304.8 mm, 1 pound per square inch = 6.895 kPa, 1-inch water column = 0.2488 kPa, 1 British thermal unit per hour = 0.2931 W, 1 cubic foot per hour = 0.0283 m³/h, 1 degree = 0.01745 rad.

Notes:

1. Table includes losses for four 90-degree bends and two end fittings. Tubing runs with larger numbers of bends or fittings shall be increased by an equivalent length of tubing to the following equation: $L = 1.3n$, where L is additional length (feet) of tubing and n is the number of additional fittings or bends.
2. EHD—Equivalent Hydraulic Diameter, which is a measure of the relative hydraulic efficiency between different tubing sizes. The greater the value of EHD, the greater the gas capacity of the tubing.
3. All table entries have been rounded to three significant digits.

TABLE G2413.4(6) [402.4(18)]
CORRUGATED STAINLESS STEEL TUBING (CSST)

Gas	Natural
Inlet Pressure	2.0 psi
Pressure Drop	1.0 psi
Specific Gravity	0.60

TUBE SIZE (EHD)														
Flow Designation	13	15	18	19	23	25	30	31	37	39	46	48	60	62
Length (ft)	Capacity in Cubic Feet of Gas Per Hour													
10	270	353	587	700	1,100	1,370	2,590	2,990	4,510	5,037	9,600	10,700	18,600	21,600
25	166	220	374	444	709	876	1,620	1,870	2,890	3,258	6,040	6,780	11,900	13,700
30	151	200	342	405	650	801	1,480	1,700	2,640	2,987	5,510	6,200	10,900	12,500
40	129	172	297	351	567	696	1,270	1,470	2,300	2,605	4,760	5,380	9,440	10,900
50	115	154	266	314	510	624	1,140	1,310	2,060	2,343	4,260	4,820	8,470	9,720
75	93	124	218	257	420	512	922	1,070	1,690	1,932	3,470	3,950	6,940	7,940
80	89	120	211	249	407	496	892	1,030	1,640	1,874	3,360	3,820	6,730	7,690
100	79	107	189	222	366	445	795	920	1,470	1,685	3,000	3,420	6,030	6,880
150	64	87	155	182	302	364	646	748	1,210	1,389	2,440	2,800	4,940	5,620
200	55	75	135	157	263	317	557	645	1,050	1,212	2,110	2,430	4,290	4,870
250	49	67	121	141	236	284	497	576	941	1,090	1,890	2,180	3,850	4,360
300	44	61	110	129	217	260	453	525	862	999	1,720	1,990	3,520	3,980
400	38	52	96	111	189	225	390	453	749	871	1,490	1,730	3,060	3,450
500	34	46	86	100	170	202	348	404	552	783	1,330	1,550	2,740	3,090

For SI: 1 inch = 25.4 mm, 1 foot = 304.8 mm, 1 pound per square inch = 6.895 kPa, 1-inch water column = 0.2488 kPa, 1 British thermal unit per hour = 0.2931 W, 1 cubic foot per hour = 0.0283 m³/h, 1 degree = 0.01745 rad.

Notes:

1. Table does not include effect of pressure drop across the line regulator. Where regulator loss exceeds $^3/_4$ psi, DO NOT USE THIS TABLE. Consult with the regulator manufacturer for pressure drops and capacity factors. Pressure drops across a regulator can vary with flow rate.
2. CAUTION: Capacities shown in the table might exceed maximum capacity for a selected regulator. Consult with the regulator or tubing manufacturer for guidance.
3. Table includes losses for four 90-degree bends and two end fittings. Tubing runs with larger numbers of bends or fittings shall be increased by an equivalent length of tubing to the following equation: $L = 1.3n$ where L is additional length (feet) of tubing and n is the number of additional fittings or bends.
4. EHD—Equivalent Hydraulic Diameter, which is a measure of the relative hydraulic efficiency between different tubing sizes. The greater the value of EHD, the greater the gas capacity of the tubing.
5. All table entries have been rounded to three significant digits.

TABLE G2413.4(7) [402.4(21)]
POLYETHYLENE PLASTIC PIPE

Gas	Natural
Inlet Pressure	Less than 2 psi
Pressure Drop	0.5 in. w.c.
Specific Gravity	0.60

PIPE SIZE (inches)						
Nominal OD	$^1/_2$	$^3/_4$	1	$1^1/_4$	$1^1/_2$	2
Designation	SDR 9	SDR 11	SDR 11	SDR 10	SDR 11	SDR 11
Actual ID	0.660	0.860	1.077	1.328	1.554	1.943
Length (ft)	Capacity in Cubic Feet of Gas per Hour					
10	201	403	726	1,260	1,900	3,410
20	138	277	499	865	1,310	2,350
30	111	222	401	695	1,050	1,880
40	95	190	343	594	898	1,610
50	84	169	304	527	796	1,430
60	76	153	276	477	721	1,300
70	70	140	254	439	663	1,190
80	65	131	236	409	617	1,110
90	61	123	221	383	579	1,040
100	58	116	209	362	547	983
125	51	103	185	321	485	871
150	46	93	168	291	439	789
175	43	86	154	268	404	726
200	40	80	144	249	376	675
250	35	71	127	221	333	598
300	32	64	115	200	302	542
350	29	59	106	184	278	499
400	27	55	99	171	258	464
450	26	51	93	160	242	435
500	24	48	88	152	229	411

For SI: 1 inch = 25.4 mm, 1 foot = 304.8 mm, 1 pound per square inch = 6.895 kPa, 1-inch water column = 0.2488 kPa, 1 British thermal unit per hour = 0.2931 W, 1 cubic foot per hour = 0.0283 m³/h, 1 degree = 0.01745 rad.

Note: All table entries have been rounded to three significant digits.

TABLE G2413.4(8) [402.4(22)]
POLYETHYLENE PLASTIC PIPE

Gas	Natural
Inlet Pressure	2.0 psi
Pressure Drop	1.0 psi
Specific Gravity	0.60

PIPE SIZE (inches)						
Nominal OD	$^1/_2$	$^3/_4$	1	$1^1/_4$	$1^1/_2$	2
Designation	SDR 9	SDR 11	SDR 11	SDR 10	SDR 11	SDR 11
Actual ID	0.660	0.860	1.077	1.328	1.554	1.943
Length (ft)	Capacity in Cubic Feet of Gas per Hour					
10	1,860	3,720	6,710	11,600	17,600	31,600
20	1,280	2,560	4,610	7,990	12,100	21,700
30	1,030	2,050	3,710	6,420	9,690	17,400
40	878	1,760	3,170	5,490	8,300	14,900
50	778	1,560	2,810	4,870	7,350	13,200
60	705	1,410	2,550	4,410	6,660	12,000
70	649	1,300	2,340	4,060	6,130	11,000
80	603	1,210	2,180	3,780	5,700	10,200
90	566	1,130	2,050	3,540	5,350	9,610
100	535	1,070	1,930	3,350	5,050	9,080
125	474	949	1,710	2,970	4,480	8,050
150	429	860	1,550	2,690	4,060	7,290
175	395	791	1,430	2,470	3,730	6,710
200	368	736	1,330	2,300	3,470	6,240
250	326	652	1,180	2,040	3,080	5,530
300	295	591	1,070	1,850	2,790	5,010
350	272	544	981	1,700	2,570	4,610
400	253	506	913	1,580	2,390	4,290
450	237	475	856	1,480	2,240	4,020
500	224	448	809	1,400	2,120	3,800
550	213	426	768	1,330	2,010	3,610
600	203	406	733	1,270	1,920	3,440
650	194	389	702	1,220	1,840	3,300
700	187	374	674	1,170	1,760	3,170
750	180	360	649	1,130	1,700	3,050
800	174	348	627	1,090	1,640	2,950
850	168	336	607	1,050	1,590	2,850
900	163	326	588	1,020	1,540	2,770
950	158	317	572	990	1,500	2,690
1,000	154	308	556	963	1,450	2,610
1,100	146	293	528	915	1,380	2,480
1,200	139	279	504	873	1,320	2,370
1,300	134	267	482	836	1,260	2,270
1,400	128	257	463	803	1,210	2,180
1,500	124	247	446	773	1,170	2,100
1,600	119	239	431	747	1,130	2,030
1,700	115	231	417	723	1,090	1,960
1,800	112	224	404	701	1,060	1,900
1,900	109	218	393	680	1,030	1,850
2,000	106	212	382	662	1,000	1,800

For SI: 1 inch = 25.4 mm, 1 foot = 304.8 mm, 1 pound per square inch = 6.895 kPa, 1-inch water column = 0.2488 kPa,
1 British thermal unit per hour = 0.2931 W, 1 cubic foot per hour = 0.0283 m³/h, 1 degree = 0.01745 rad.

Note: All table entries have been rounded to three significant digits.

TABLE G2413.4(9) [402.4(25)]
SCHEDULE 40 METALLIC PIPE

Gas	Undiluted Propane
Inlet Pressure	10.0 psi
Pressure Drop	1.0 psi
Specific Gravity	1.50

INTENDED USE	Pipe sizing between first stage (high-pressure regulator) and second stage (low-pressure regulator).								
	PIPE SIZE (inches)								
Nominal	1/2	3/4	1	1 1/4	1 1/2	2	2 1/2	3	4
Actual ID	0.622	0.824	1.049	1.380	1.610	2.067	2.469	3.068	4.026
Length (ft)	Capacity in Thousands of Btu per Hour								
10	3,320	6,950	13,100	26,900	40,300	77,600	124,000	219,000	446,000
20	2,280	4,780	9,000	18,500	27,700	53,300	85,000	150,000	306,000
30	1,830	3,840	7,220	14,800	22,200	42,800	68,200	121,000	246,000
40	1,570	3,280	6,180	12,700	19,000	36,600	58,400	103,000	211,000
50	1,390	2,910	5,480	11,300	16,900	32,500	51,700	91,500	187,000
60	1,260	2,640	4,970	10,200	15,300	29,400	46,900	82,900	169,000
70	1,160	2,430	4,570	9,380	14,100	27,100	43,100	76,300	156,000
80	1,080	2,260	4,250	8,730	13,100	25,200	40,100	70,900	145,000
90	1,010	2,120	3,990	8,190	12,300	23,600	37,700	66,600	136,000
100	956	2,000	3,770	7,730	11,600	22,300	35,600	62,900	128,000
125	848	1,770	3,340	6,850	10,300	19,800	31,500	55,700	114,000
150	768	1,610	3,020	6,210	9,300	17,900	28,600	50,500	103,000
175	706	1,480	2,780	5,710	8,560	16,500	26,300	46,500	94,700
200	657	1,370	2,590	5,320	7,960	15,300	24,400	43,200	88,100
250	582	1,220	2,290	4,710	7,060	13,600	21,700	38,300	78,100
300	528	1,100	2,080	4,270	6,400	12,300	19,600	34,700	70,800
350	486	1,020	1,910	3,930	5,880	11,300	18,100	31,900	65,100
400	452	945	1,780	3,650	5,470	10,500	16,800	29,700	60,600
450	424	886	1,670	3,430	5,140	9,890	15,800	27,900	56,800
500	400	837	1,580	3,240	4,850	9,340	14,900	26,300	53,700
550	380	795	1,500	3,070	4,610	8,870	14,100	25,000	51,000
600	363	759	1,430	2,930	4,400	8,460	13,500	23,900	48,600
650	347	726	1,370	2,810	4,210	8,110	12,900	22,800	46,600
700	334	698	1,310	2,700	4,040	7,790	12,400	21,900	44,800
750	321	672	1,270	2,600	3,900	7,500	12,000	21,100	43,100
800	310	649	1,220	2,510	3,760	7,240	11,500	20,400	41,600
850	300	628	1,180	2,430	3,640	7,010	11,200	19,800	40,300
900	291	609	1,150	2,360	3,530	6,800	10,800	19,200	39,100
950	283	592	1,110	2,290	3,430	6,600	10,500	18,600	37,900
1,000	275	575	1,080	2,230	3,330	6,420	10,200	18,100	36,900
1,100	261	546	1,030	2,110	3,170	6,100	9,720	17,200	35,000
1,200	249	521	982	2,020	3,020	5,820	9,270	16,400	33,400
1,300	239	499	940	1,930	2,890	5,570	8,880	15,700	32,000
1,400	229	480	903	1,850	2,780	5,350	8,530	15,100	30,800
1,500	221	462	870	1,790	2,680	5,160	8,220	14,500	29,600
1,600	213	446	840	1,730	2,590	4,980	7,940	14,000	28,600
1,700	206	432	813	1,670	2,500	4,820	7,680	13,600	27,700
1,800	200	419	789	1,620	2,430	4,670	7,450	13,200	26,900
1,900	194	407	766	1,570	2,360	4,540	7,230	12,800	26,100
2,000	189	395	745	1,530	2,290	4,410	7,030	12,400	25,400

For SI: 1 inch = 25.4 mm, 1 foot = 304.8 mm, 1 pound per square inch = 6.895 kPa, 1-inch water column = 0.2488 kPa,
   1 British thermal unit per hour = 0.2931 W, 1 cubic foot per hour = 0.0283 m³/h, 1 degree = 0.01745 rad.
Note: All table entries have been rounded to three significant digits.

TABLE G2413.4(10) [402.4(26)]
SCHEDULE 40 METALLIC PIPE

Gas	Undiluted Propane
Inlet Pressure	10.0 psi
Pressure Drop	3.0 psi
Specific Gravity	1.50

INTENDED USE	Pipe sizing between first stage (high-pressure regulator) and second stage (low-pressure regulator).								
	PIPE SIZE (inches)								
Nominal	$^1/_2$	$^3/_4$	1	$1^1/_4$	$1^1/_2$	2	$2^1/_2$	3	4
Actual ID	0.622	0.824	1.049	1.380	1.610	2.067	2.469	3.068	4.026
Length (ft)	**Capacity in Thousands of Btu per Hour**								
10	5,890	12,300	23,200	47,600	71,300	137,000	219,000	387,000	789,000
20	4,050	8,460	15,900	32,700	49,000	94,400	150,000	266,000	543,000
30	3,250	6,790	12,800	26,300	39,400	75,800	121,000	214,000	436,000
40	2,780	5,810	11,000	22,500	33,700	64,900	103,000	183,000	373,000
50	2,460	5,150	9,710	19,900	29,900	57,500	91,600	162,000	330,000
60	2,230	4,670	8,790	18,100	27,100	52,100	83,000	147,000	299,000
70	2,050	4,300	8,090	16,600	24,900	47,900	76,400	135,000	275,000
80	1,910	4,000	7,530	15,500	23,200	44,600	71,100	126,000	256,000
90	1,790	3,750	7,060	14,500	21,700	41,800	66,700	118,000	240,000
100	1,690	3,540	6,670	13,700	20,500	39,500	63,000	111,000	227,000
125	1,500	3,140	5,910	12,100	18,200	35,000	55,800	98,700	201,000
150	1,360	2,840	5,360	11,000	16,500	31,700	50,600	89,400	182,000
175	1,250	2,620	4,930	10,100	15,200	29,200	46,500	82,300	167,800
200	1,160	2,430	4,580	9,410	14,100	27,200	43,300	76,500	156,100
250	1,030	2,160	4,060	8,340	12,500	24,100	38,400	67,800	138,400
300	935	1,950	3,680	7,560	11,300	21,800	34,800	61,500	125,400
350	860	1,800	3,390	6,950	10,400	20,100	32,000	56,500	115,300
400	800	1,670	3,150	6,470	9,690	18,700	29,800	52,600	107,300
450	751	1,570	2,960	6,070	9,090	17,500	27,900	49,400	100,700
500	709	1,480	2,790	5,730	8,590	16,500	26,400	46,600	95,100
550	673	1,410	2,650	5,450	8,160	15,700	25,000	44,300	90,300
600	642	1,340	2,530	5,200	7,780	15,000	23,900	42,200	86,200
650	615	1,290	2,420	4,980	7,450	14,400	22,900	40,500	82,500
700	591	1,240	2,330	4,780	7,160	13,800	22,000	38,900	79,300
750	569	1,190	2,240	4,600	6,900	13,300	21,200	37,400	76,400
800	550	1,150	2,170	4,450	6,660	12,800	20,500	36,200	73,700
850	532	1,110	2,100	4,300	6,450	12,400	19,800	35,000	71,400
900	516	1,080	2,030	4,170	6,250	12,000	19,200	33,900	69,200
950	501	1,050	1,970	4,050	6,070	11,700	18,600	32,900	67,200
1,000	487	1,020	1,920	3,940	5,900	11,400	18,100	32,000	65,400
1,100	463	968	1,820	3,740	5,610	10,800	17,200	30,400	62,100
1,200	442	923	1,740	3,570	5,350	10,300	16,400	29,000	59,200
1,300	423	884	1,670	3,420	5,120	9,870	15,700	27,800	56,700
1,400	406	849	1,600	3,280	4,920	9,480	15,100	26,700	54,500
1,500	391	818	1,540	3,160	4,740	9,130	14,600	25,700	52,500
1,600	378	790	1,490	3,060	4,580	8,820	14,100	24,800	50,700
1,700	366	765	1,440	2,960	4,430	8,530	13,600	24,000	49,000
1,800	355	741	1,400	2,870	4,300	8,270	13,200	23,300	47,600
1,900	344	720	1,360	2,780	4,170	8,040	12,800	22,600	46,200
2,000	335	700	1,320	2,710	4,060	7,820	12,500	22,000	44,900

For SI: 1 inch = 25.4 mm, 1 foot = 304.8 mm, 1 pound per square inch = 6.895 kPa, 1-inch water column = 0.2488 kPa,
1 British thermal unit per hour = 0.2931 W, 1 cubic foot per hour = 0.0283 m³/h, 1 degree = 0.01745 rad.
Note: All table entries have been rounded to three significant digits.

TABLE G2413.4(11) [402.4(27)]
SCHEDULE 40 METALLIC PIPE

Gas	Undiluted Propane
Inlet Pressure	2.0 psi
Pressure Drop	1.0 psi
Specific Gravity	1.50

INTENDED USE	Pipe sizing between 2 psig service and line pressure regulator.								
	PIPE SIZE (inches)								
Nominal	¹/₂	³/₄	1	1¹/₄	1¹/₂	2	2¹/₂	3	4
Actual ID	0.622	0.824	1.049	1.380	1.610	2.067	2.469	3.068	4.026
Length (ft)	Capacity in Thousands of Btu per Hour								
10	2,680	5,590	10,500	21,600	32,400	62,400	99,500	176,000	359,000
20	1,840	3,850	7,240	14,900	22,300	42,900	68,400	121,000	247,000
30	1,480	3,090	5,820	11,900	17,900	34,500	54,900	97,100	198,000
40	1,260	2,640	4,980	10,200	15,300	29,500	47,000	83,100	170,000
50	1,120	2,340	4,410	9,060	13,600	26,100	41,700	73,700	150,000
60	1,010	2,120	4,000	8,210	12,300	23,700	37,700	66,700	136,000
70	934	1,950	3,680	7,550	11,300	21,800	34,700	61,400	125,000
80	869	1,820	3,420	7,020	10,500	20,300	32,300	57,100	116,000
90	815	1,700	3,210	6,590	9,880	19,000	30,300	53,600	109,000
100	770	1,610	3,030	6,230	9,330	18,000	28,600	50,600	103,000
125	682	1,430	2,690	5,520	8,270	15,900	25,400	44,900	91,500
150	618	1,290	2,440	5,000	7,490	14,400	23,000	40,700	82,900
175	569	1,190	2,240	4,600	6,890	13,300	21,200	37,400	76,300
200	529	1,110	2,080	4,280	6,410	12,300	19,700	34,800	71,000
250	469	981	1,850	3,790	5,680	10,900	17,400	30,800	62,900
300	425	889	1,670	3,440	5,150	9,920	15,800	27,900	57,000
350	391	817	1,540	3,160	4,740	9,120	14,500	25,700	52,400
400	364	760	1,430	2,940	4,410	8,490	13,500	23,900	48,800
450	341	714	1,340	2,760	4,130	7,960	12,700	22,400	45,800
500	322	674	1,270	2,610	3,910	7,520	12,000	21,200	43,200
550	306	640	1,210	2,480	3,710	7,140	11,400	20,100	41,100
600	292	611	1,150	2,360	3,540	6,820	10,900	19,200	39,200
650	280	585	1,100	2,260	3,390	6,530	10,400	18,400	37,500
700	269	562	1,060	2,170	3,260	6,270	9,990	17,700	36,000
750	259	541	1,020	2,090	3,140	6,040	9,630	17,000	34,700
800	250	523	985	2,020	3,030	5,830	9,300	16,400	33,500
850	242	506	953	1,960	2,930	5,640	9,000	15,900	32,400
900	235	490	924	1,900	2,840	5,470	8,720	15,400	31,500
950	228	476	897	1,840	2,760	5,310	8,470	15,000	30,500
1,000	222	463	873	1,790	2,680	5,170	8,240	14,600	29,700
1,100	210	440	829	1,700	2,550	4,910	7,830	13,800	28,200
1,200	201	420	791	1,620	2,430	4,680	7,470	13,200	26,900
1,300	192	402	757	1,550	2,330	4,490	7,150	12,600	25,800
1,400	185	386	727	1,490	2,240	4,310	6,870	12,100	24,800
1,500	178	372	701	1,440	2,160	4,150	6,620	11,700	23,900
1,600	172	359	677	1,390	2,080	4,010	6,390	11,300	23,000
1,700	166	348	655	1,340	2,010	3,880	6,180	10,900	22,300
1,800	161	337	635	1,300	1,950	3,760	6,000	10,600	21,600
1,900	157	327	617	1,270	1,900	3,650	5,820	10,300	21,000
2,000	152	318	600	1,230	1,840	3,550	5,660	10,000	20,400

For SI: 1 inch = 25.4 mm, 1 foot = 304.8 mm, 1 pound per square inch = 6.895 kPa, 1-inch water column = 0.2488 kPa,
    1 British thermal unit per hour = 0.2931 W, 1 cubic foot per hour = 0.0283 m³/h, 1 degree = 0.01745 rad.

Note: All table entries have been rounded to three significant digits.

2015 INTERNATIONAL RESIDENTIAL CODE® COMMENTARY

TABLE G2413.4(12) [402.4(28)]
SCHEDULE 40 METALLIC PIPE

Gas	Undiluted Propane
Inlet Pressure	11.0 in. w.c.
Pressure Drop	0.5 in. w.c.
Specific Gravity	1.50

INTENDED USE	Pipe sizing between single- or second-stage (low pressure) regulator and appliance.								
PIPE SIZE (inches)									
Nominal	$^1/_2$	$^3/_4$	1	$1^1/_4$	$1^1/_2$	2	$2^1/_2$	3	4
Actual ID	0.622	0.824	1.049	1.380	1.610	2.067	2.469	3.068	4.026
Length (ft)	Capacity in Thousands of Btu per Hour								
10	291	608	1,150	2,350	3,520	6,790	10,800	19,100	39,000
20	200	418	787	1,620	2,420	4,660	7,430	13,100	26,800
30	160	336	632	1,300	1,940	3,750	5,970	10,600	21,500
40	137	287	541	1,110	1,660	3,210	5,110	9,030	18,400
50	122	255	480	985	1,480	2,840	4,530	8,000	16,300
60	110	231	434	892	1,340	2,570	4,100	7,250	14,800
80	101	212	400	821	1,230	2,370	3,770	6,670	13,600
100	94	197	372	763	1,140	2,200	3,510	6,210	12,700
125	89	185	349	716	1,070	2,070	3,290	5,820	11,900
150	84	175	330	677	1,010	1,950	3,110	5,500	11,200
175	74	155	292	600	899	1,730	2,760	4,880	9,950
200	67	140	265	543	814	1,570	2,500	4,420	9,010
250	62	129	243	500	749	1,440	2,300	4,060	8,290
300	58	120	227	465	697	1,340	2,140	3,780	7,710
350	51	107	201	412	618	1,190	1,900	3,350	6,840
400	46	97	182	373	560	1,080	1,720	3,040	6,190
450	42	89	167	344	515	991	1,580	2,790	5,700
500	40	83	156	320	479	922	1,470	2,600	5,300
550	37	78	146	300	449	865	1,380	2,440	4,970
600	35	73	138	283	424	817	1,300	2,300	4,700
650	33	70	131	269	403	776	1,240	2,190	4,460
700	32	66	125	257	385	741	1,180	2,090	4,260
750	30	64	120	246	368	709	1,130	2,000	4,080
800	29	61	115	236	354	681	1,090	1,920	3,920
850	28	59	111	227	341	656	1,050	1,850	3,770
900	27	57	107	220	329	634	1,010	1,790	3,640
950	26	55	104	213	319	613	978	1,730	3,530
1,000	25	53	100	206	309	595	948	1,680	3,420
1,100	25	52	97	200	300	578	921	1,630	3,320
1,200	24	50	95	195	292	562	895	1,580	3,230
1,300	23	48	90	185	277	534	850	1,500	3,070
1,400	22	46	86	176	264	509	811	1,430	2,930
1,500	21	44	82	169	253	487	777	1,370	2,800
1,200	24	50	95	195	292	562	895	1,580	3,230
1,300	23	48	90	185	277	534	850	1,500	3,070
1,400	22	46	86	176	264	509	811	1,430	2,930
1,500	21	44	82	169	253	487	777	1,370	2,800
1,600	20	42	79	162	243	468	746	1,320	2,690
1,700	19	40	76	156	234	451	719	1,270	2,590
1,800	19	39	74	151	226	436	694	1,230	2,500
1,900	18	38	71	146	219	422	672	1,190	2,420
2,000	18	37	69	142	212	409	652	1,150	2,350

For SI: 1 inch = 25.4 mm, 1 foot = 304.8 mm, 1 pound per square inch = 6.895 kPa, 1-inch water column = 0.2488 kPa,
   1 British thermal unit per hour = 0.2931 W, 1 cubic foot per hour = 0.0283 m³/h, 1 degree = 0.01745 rad.

Note: All table entries have been rounded to three significant digits.

TABLE G2413.4(13) [402.4(29)]
SEMIRIGID COPPER TUBING

Gas	Undiluted Propane
Inlet Pressure	10.0 psi
Pressure Drop	1.0 psi
Specific Gravity	1.50

INTENDED USE		Sizing between first stage (high-pressure regulator) and second stage (low-pressure regulator).								
		TUBE SIZE (inches)								
Nominal	K & L	$^1/_4$	$^3/_8$	$^1/_2$	$^5/_8$	$^3/_4$	1	$1^1/_4$	$1^1/_2$	2
	ACR	$^3/_8$	$^1/_2$	$^5/_8$	$^3/_4$	$^7/_8$	$1^1/_8$	$1^3/_8$	—	—
Outside		0.375	0.500	0.625	0.750	0.875	1.125	1.375	1.625	2.125
Inside		0.305	0.402	0.527	0.652	0.745	0.995	1.245	1.481	1.959
Length (ft)		Capacity in Thousands of Btu per Hour								
10		513	1,060	2,150	3,760	5,330	11,400	20,500	32,300	67,400
20		352	727	1,480	2,580	3,670	7,830	14,100	22,200	46,300
30		283	584	1,190	2,080	2,940	6,290	11,300	17,900	37,200
40		242	500	1,020	1,780	2,520	5,380	9,690	15,300	31,800
50		215	443	901	1,570	2,230	4,770	8,590	13,500	28,200
60		194	401	816	1,430	2,020	4,320	7,780	12,300	25,600
70		179	369	751	1,310	1,860	3,980	7,160	11,300	23,500
80		166	343	699	1,220	1,730	3,700	6,660	10,500	21,900
90		156	322	655	1,150	1,630	3,470	6,250	9,850	20,500
100		147	304	619	1,080	1,540	3,280	5,900	9,310	19,400
125		131	270	549	959	1,360	2,910	5,230	8,250	17,200
150		118	244	497	869	1,230	2,630	4,740	7,470	15,600
175		109	225	457	799	1,130	2,420	4,360	6,880	14,300
200		101	209	426	744	1,060	2,250	4,060	6,400	13,300
250		90	185	377	659	935	2,000	3,600	5,670	11,800
300		81	168	342	597	847	1,810	3,260	5,140	10,700
350		75	155	314	549	779	1,660	3,000	4,730	9,840
400		70	144	292	511	725	1,550	2,790	4,400	9,160
450		65	135	274	480	680	1,450	2,620	4,130	8,590
500		62	127	259	453	643	1,370	2,470	3,900	8,120
550		59	121	246	430	610	1,300	2,350	3,700	7,710
600		56	115	235	410	582	1,240	2,240	3,530	7,350
650		54	111	225	393	558	1,190	2,140	3,380	7,040
700		51	106	216	378	536	1,140	2,060	3,250	6,770
750		50	102	208	364	516	1,100	1,980	3,130	6,520
800		48	99	201	351	498	1,060	1,920	3,020	6,290
850		46	96	195	340	482	1,030	1,850	2,920	6,090
900		45	93	189	330	468	1,000	1,800	2,840	5,910
950		44	90	183	320	454	970	1,750	2,750	5,730
1,000		42	88	178	311	442	944	1,700	2,680	5,580
1,100		40	83	169	296	420	896	1,610	2,540	5,300
1,200		38	79	161	282	400	855	1,540	2,430	5,050
1,300		37	76	155	270	383	819	1,470	2,320	4,840
1,400		35	73	148	260	368	787	1,420	2,230	4,650
1,500		34	70	143	250	355	758	1,360	2,150	4,480
1,600		33	68	138	241	343	732	1,320	2,080	4,330
1,700		32	66	134	234	331	708	1,270	2,010	4,190
1,800		31	64	130	227	321	687	1,240	1,950	4,060
1,900		30	62	126	220	312	667	1,200	1,890	3,940
2,000		29	60	122	214	304	648	1,170	1,840	3,830

For SI: 1 inch = 25.4 mm, 1 foot = 304.8 mm, 1 pound per square inch = 6.895 kPa, 1-inch water column = 0.2488 kPa,
    1 British thermal unit per hour = 0.2931 W, 1 cubic foot per hour = 0.0283 m³/h, 1 degree = 0.01745 rad.

Notes:

1. Table capacities are based on Type K copper tubing inside diameter (shown), which has the smallest inside diameter of the copper tubing products.

2. All table entries have been rounded to three significant digits.

2015 INTERNATIONAL RESIDENTIAL CODE® COMMENTARY

TABLE G2413.4(14) [402.4(30)]
SEMIRIGID COPPER TUBING

Gas	Undiluted Propane	
Inlet Pressure	11.0 in. w.c.	
Pressure Drop	0.5 in. w.c.	
Specific Gravity	1.50	

INTENDED USE		Sizing between single- or second-stage (low-pressure regulator) and appliance.								
		TUBE SIZE (inches)								
Nominal	K & L	$^1/_4$	$^3/_8$	$^1/_2$	$^5/_8$	$^3/_4$	1	$1^1/_4$	$1^1/_2$	2
	ACR	$^3/_8$	$^1/_2$	$^5/_8$	$^3/_4$	$^7/_8$	$1^1/_8$	$1^3/_8$	—	—
Outside		0.375	0.500	0.625	0.750	0.875	1.125	1.375	1.625	2.125
Inside		0.305	0.402	0.527	0.652	0.745	0.995	1.245	1.481	1.959
Length (ft)		Capacity in Thousands of Btu per Hour								
10		45	93	188	329	467	997	1,800	2,830	5,890
20		31	64	129	226	321	685	1,230	1,950	4,050
30		25	51	104	182	258	550	991	1,560	3,250
40		21	44	89	155	220	471	848	1,340	2,780
50		19	39	79	138	195	417	752	1,180	2,470
60		17	35	71	125	177	378	681	1,070	2,240
70		16	32	66	115	163	348	626	988	2,060
80		15	30	61	107	152	324	583	919	1,910
90		14	28	57	100	142	304	547	862	1,800
100		13	27	54	95	134	287	517	814	1,700
125		11	24	48	84	119	254	458	722	1,500
150		10	21	44	76	108	230	415	654	1,360
175		NA	20	40	70	99	212	382	602	1,250
200		NA	18	37	65	92	197	355	560	1,170
250		NA	16	33	58	82	175	315	496	1,030
300		NA	15	30	52	74	158	285	449	936
350		NA	14	28	48	68	146	262	414	861
400		NA	13	26	45	63	136	244	385	801
450		NA	12	24	42	60	127	229	361	752
500		NA	11	23	40	56	120	216	341	710
550		NA	11	22	38	53	114	205	324	674
600		NA	10	21	36	51	109	196	309	643
650		NA	NA	20	34	49	104	188	296	616
700		NA	NA	19	33	47	100	180	284	592
750		NA	NA	18	32	45	96	174	274	570
800		NA	NA	18	31	44	93	168	264	551
850		NA	NA	17	30	42	90	162	256	533
900		NA	NA	17	29	41	87	157	248	517
950		NA	NA	16	28	40	85	153	241	502
1,000		NA	NA	16	27	39	83	149	234	488
1,100		NA	NA	15	26	37	78	141	223	464
1,200		NA	NA	14	25	35	75	135	212	442
1,300		NA	NA	14	24	34	72	129	203	423
1,400		NA	NA	13	23	32	69	124	195	407
1,500		NA	NA	13	22	31	66	119	188	392
1,600		NA	NA	12	21	30	64	115	182	378
1,700		NA	NA	12	20	29	62	112	176	366
1,800		NA	NA	11	20	28	60	108	170	355
1,900		NA	NA	11	19	27	58	105	166	345
2,000		NA	NA	11	19	27	57	102	161	335

For SI: 1 inch = 25.4 mm, 1 foot = 304.8 mm, 1 pound per square inch = 6.895 kPa, 1-inch water column = 0.2488 kPa,
     1 British thermal unit per hour = 0.2931 W, 1 cubic foot per hour = 0.0283 m³/h, 1 degree = 0.01745 rad.

Notes:

1. Table capacities are based on Type K copper tubing inside diameter (shown), which has the smallest inside diameter of the copper tubing products.

2. NA means a flow of less than 10,000 Btu/hr.

3. All table entries have been rounded to three significant digits.

TABLE G2413.4(15) [402.4(31)]
SEMIRIGID COPPER TUBING

Gas	Undiluted Propane
Inlet Pressure	2.0 psi
Pressure Drop	1.0 psi
Specific Gravity	1.50

INTENDED USE		Tube sizing between 2 psig service and line pressure regulator.								
		TUBE SIZE (inches)								
Nominal	K & L	$^1/_4$	$^3/_8$	$^1/_2$	$^5/_8$	$^3/_4$	1	$1^1/_4$	$1^1/_2$	2
	ACR	$^3/_8$	$^1/_2$	$^5/_8$	$^3/_4$	$^7/_8$	$1^1/_8$	$1^3/_8$	—	—
Outside		0.375	0.500	0.625	0.750	0.875	1.125	1.375	1.625	2.125
Inside		0.305	0.402	0.527	0.652	0.745	0.995	1.245	1.481	1.959
Length (ft)		Capacity in Thousands of Btu per Hour								
10		413	852	1,730	3,030	4,300	9,170	16,500	26,000	54,200
20		284	585	1,190	2,080	2,950	6,310	11,400	17,900	37,300
30		228	470	956	1,670	2,370	5,060	9,120	14,400	29,900
40		195	402	818	1,430	2,030	4,330	7,800	12,300	25,600
50		173	356	725	1,270	1,800	3,840	6,920	10,900	22,700
60		157	323	657	1,150	1,630	3,480	6,270	9,880	20,600
70		144	297	605	1,060	1,500	3,200	5,760	9,090	18,900
80		134	276	562	983	1,390	2,980	5,360	8,450	17,600
90		126	259	528	922	1,310	2,790	5,030	7,930	16,500
100		119	245	498	871	1,240	2,640	4,750	7,490	15,600
125		105	217	442	772	1,100	2,340	4,210	6,640	13,800
150		95	197	400	700	992	2,120	3,820	6,020	12,500
175		88	181	368	644	913	1,950	3,510	5,540	11,500
200		82	168	343	599	849	1,810	3,270	5,150	10,700
250		72	149	304	531	753	1,610	2,900	4,560	9,510
300		66	135	275	481	682	1,460	2,620	4,140	8,610
350		60	124	253	442	628	1,340	2,410	3,800	7,920
400		56	116	235	411	584	1,250	2,250	3,540	7,370
450		53	109	221	386	548	1,170	2,110	3,320	6,920
500		50	103	209	365	517	1,110	1,990	3,140	6,530
550		47	97	198	346	491	1,050	1,890	2,980	6,210
600		45	93	189	330	469	1,000	1,800	2,840	5,920
650		43	89	181	316	449	959	1,730	2,720	5,670
700		41	86	174	304	431	921	1,660	2,620	5,450
750		40	82	168	293	415	888	1,600	2,520	5,250
800		39	80	162	283	401	857	1,540	2,430	5,070
850		37	77	157	274	388	829	1,490	2,350	4,900
900		36	75	152	265	376	804	1,450	2,280	4,750
950		35	72	147	258	366	781	1,410	2,220	4,620
1,000		34	71	143	251	356	760	1,370	2,160	4,490
1,100		32	67	136	238	338	721	1,300	2,050	4,270
1,200		31	64	130	227	322	688	1,240	1,950	4,070
1,300		30	61	124	217	309	659	1,190	1,870	3,900
1,400		28	59	120	209	296	633	1,140	1,800	3,740
1,500		27	57	115	201	286	610	1,100	1,730	3,610
1,600		26	55	111	194	276	589	1,060	1,670	3,480
1,700		26	53	108	188	267	570	1,030	1,620	3,370
1,800		25	51	104	182	259	553	1,000	1,570	3,270
1,900		24	50	101	177	251	537	966	1,520	3,170
2,000		23	48	99	172	244	522	940	1,480	3,090

For SI: 1 inch = 25.4 mm, 1 foot = 304.8 mm, 1 pound per square inch = 6.895 kPa, 1-inch water column = 0.2488 kPa,
    1 British thermal unit per hour = 0.2931 W, 1 cubic foot per hour = 0.0283 m³/h, 1 degree = 0.01745 rad.

Notes:

1. Table capacities are based on Type K copper tubing inside diameter (shown), which has the smallest inside diameter of the copper tubing products.

2. All table entries have been rounded to three significant digits.

TABLE G2413.4(16) [402.4(32)]
CORRUGATED STAINLESS STEEL TUBING (CSST)

Gas	Undiluted Propane
Inlet Pressure	11.0 in. w.c.
Pressure Drop	0.5 in. w.c.
Specific Gravity	1.50

INTENDED USE: SIZING BETWEEN SINGLE OR SECOND STAGE (Low Pressure) REGULATOR AND THE APPLIANCE SHUTOFF VALVE.													
TUBE SIZE (EHD)													
Flow Designation	13	15	18	19	23	25	30	31	37	39	46	48	60	62
Length (ft)	Capacity in Thousands of Btu per Hour												
5	72	99	181	211	355	426	744	863	1,420	1,638	2,830	3,270	5,780	6,550
10	50	69	129	150	254	303	521	605	971	1,179	1,990	2,320	4,110	4,640
15	39	55	104	121	208	248	422	490	775	972	1,620	1,900	3,370	3,790
20	34	49	91	106	183	216	365	425	661	847	1,400	1,650	2,930	3,290
25	30	42	82	94	164	192	325	379	583	762	1,250	1,480	2,630	2,940
30	28	39	74	87	151	177	297	344	528	698	1,140	1,350	2,400	2,680
40	23	33	64	74	131	153	256	297	449	610	988	1,170	2,090	2,330
50	20	30	58	66	118	137	227	265	397	548	884	1,050	1,870	2,080
60	19	26	53	60	107	126	207	241	359	502	805	961	1,710	1,900
70	17	25	49	57	99	117	191	222	330	466	745	890	1,590	1,760
80	15	23	45	52	94	109	178	208	307	438	696	833	1,490	1,650
90	15	22	44	50	90	102	169	197	286	414	656	787	1,400	1,550
100	14	20	41	47	85	98	159	186	270	393	621	746	1,330	1,480
150	11	15	31	36	66	75	123	143	217	324	506	611	1,090	1,210
200	9	14	28	33	60	69	112	129	183	283	438	531	948	1,050
250	8	12	25	30	53	61	99	117	163	254	390	476	850	934
300	8	11	23	26	50	57	90	107	147	234	357	434	777	854

For SI: 1 inch = 25.4 mm, 1 foot = 304.8 mm, 1 pound per square inch = 6.895 kPa, 1-inch water column = 0.2488 kPa,
  1 British thermal unit per hour = 0.2931 W, 1 cubic foot per hour = 0.0283 m³/h, 1 degree = 0.01745 rad.

Notes:

1. Table includes losses for four 90-degree bends and two end fittings. Tubing runs with larger numbers of bends or fittings shall be increased by an equivalent length of tubing to the following equation: $L = 1.3n$ where L is additional length (feet) of tubing and n is the number of additional fittings or bends.

2. EHD—Equivalent Hydraulic Diameter, which is a measure of the relative hydraulic efficiency between different tubing sizes. The greater the value of EHD, the greater the gas capacity of the tubing.

3. All table entries have been rounded to three significant digits.

TABLE G2413.4(17) [402.4(33)]
CORRUGATED STAINLESS STEEL TUBING (CSST)

Gas	Undiluted Propane
Inlet Pressure	2.0 psi
Pressure Drop	1.0 psi
Specific Gravity	1.50

INTENDED USE: SIZING BETWEEN 2 PSI SERVICE AND THE LINE PRESSURE REGULATOR.													
TUBE SIZE (EHD)													
Flow Designation	13	15	18	19	23	25	30	31	37	39	46	48	60	62
Length (ft)	Capacity in Thousands of Btu per Hour													
10	426	558	927	1,110	1,740	2,170	4,100	4,720	7,130	7,958	15,200	16,800	29,400	34,200
25	262	347	591	701	1,120	1,380	2,560	2,950	4,560	5,147	9,550	10,700	18,800	21,700
30	238	316	540	640	1,030	1,270	2,330	2,690	4,180	4,719	8,710	9,790	17,200	19,800
40	203	271	469	554	896	1,100	2,010	2,320	3,630	4,116	7,530	8,500	14,900	17,200
50	181	243	420	496	806	986	1,790	2,070	3,260	3,702	6,730	7,610	13,400	15,400
75	147	196	344	406	663	809	1,460	1,690	2,680	3,053	5,480	6,230	11,000	12,600
80	140	189	333	393	643	768	1,410	1,630	2,590	2,961	5,300	6,040	10,600	12,200
100	124	169	298	350	578	703	1,260	1,450	2,330	2,662	4,740	5,410	9,530	10,900
150	101	137	245	287	477	575	1,020	1,180	1,910	2,195	3,860	4,430	7,810	8,890
200	86	118	213	248	415	501	880	1,020	1,660	1,915	3,340	3,840	6,780	7,710
250	77	105	191	222	373	448	785	910	1,490	1,722	2,980	3,440	6,080	6,900
300	69	96	173	203	343	411	716	829	1,360	1,578	2,720	3,150	5,560	6,300
400	60	82	151	175	298	355	616	716	1,160	1,376	2,350	2,730	4,830	5,460
500	53	72	135	158	268	319	550	638	1,030	1,237	2,100	2,450	4,330	4,880

For SI: 1 inch = 25.4 mm, 1 foot = 304.8 mm, 1 pound per square inch = 6.895 kPa, 1-inch water column = 0.2488 kPa, 1 British thermal unit per hour = 0.2931 W, 1 cubic foot per hour = 0.0283 m³/h, 1 degree = 0.01745 rad.

Notes:

1. Table does not include effect of pressure drop across the line regulator. Where regulator loss exceeds $^1/_2$ psi (based on 13 in. w.c. outlet pressure), DO NOT USE THIS TABLE. Consult with the regulator manufacturer for pressure drops and capacity factors. Pressure drops across a regulator can vary with flow rate.

2. CAUTION: Capacities shown in the table might exceed maximum capacity for a selected regulator. Consult with the regulator or tubing manufacturer for guidance.

3. Table includes losses for four 90-degree bends and two end fittings. Tubing runs with larger numbers of bends or fittings shall be increased by an equivalent length of tubing to the following equation: $L = 1.3n$ where L is additional length (feet) of tubing and n is the number of additional fittings or bends.

4. EHD—Equivalent Hydraulic Diameter, which is a measure of the relative hydraulic efficiency between different tubing sizes. The greater the value of EHD, the greater the gas capacity of the tubing.

5. All table entries have been rounded to three significant digits.

TABLE G2413.4(18) [402.4(34)]
CORRUGATED STAINLESS STEEL TUBING (CSST)

Gas	Undiluted Propane
Inlet Pressure	5.0 psi
Pressure Drop	3.5 psi
Specific Gravity	1.50

	TUBE SIZE (EHD)													
Flow Designation	13	15	18	19	23	25	30	31	37	39	46	48	60	62
Length (ft)	Capacity in Thousands of Btu per Hour													
10	826	1,070	1,710	2,060	3,150	4,000	7,830	8,950	13,100	14,441	28,600	31,200	54,400	63,800
25	509	664	1,090	1,310	2,040	2,550	4,860	5,600	8,400	9,339	18,000	19,900	34,700	40,400
30	461	603	999	1,190	1,870	2,340	4,430	5,100	7,680	8,564	16,400	18,200	31,700	36,900
40	396	520	867	1,030	1,630	2,030	3,820	4,400	6,680	7,469	14,200	15,800	27,600	32,000
50	352	463	777	926	1,460	1,820	3,410	3,930	5,990	6,717	12,700	14,100	24,700	28,600
75	284	376	637	757	1,210	1,490	2,770	3,190	4,920	5,539	10,300	11,600	20,300	23,400
80	275	363	618	731	1,170	1,450	2,680	3,090	4,770	5,372	9,990	11,200	19,600	22,700
100	243	324	553	656	1,050	1,300	2,390	2,760	4,280	4,830	8,930	10,000	17,600	20,300
150	196	262	453	535	866	1,060	1,940	2,240	3,510	3,983	7,270	8,210	14,400	16,600
200	169	226	393	464	755	923	1,680	1,930	3,050	3,474	6,290	7,130	12,500	14,400
250	150	202	352	415	679	828	1,490	1,730	2,740	3,124	5,620	6,390	11,200	12,900
300	136	183	322	379	622	757	1,360	1,570	2,510	2,865	5,120	5,840	10,300	11,700
400	117	158	279	328	542	657	1,170	1,360	2,180	2,498	4,430	5,070	8,920	10,200
500	104	140	251	294	488	589	1,050	1,210	1,950	2,247	3,960	4,540	8,000	9,110

For SI: 1 inch = 25.4 mm, 1 foot = 304.8 mm, 1 pound per square inch = 6.895 kPa, 1-inch water column = 0.2488 kPa,
   1 British thermal unit per hour = 0.2931 W, 1 cubic foot per hour = 0.0283 m^3/h, 1 degree = 0.01745 rad.

Notes:

1. Table does not include effect of pressure drop across line regulator. Where regulator loss exceeds1 psi, DO NOT USE THIS TABLE. Consult with the regulator manufacturer for pressure drops and capacity factors. Pressure drop across regulator can vary with the flow rate.

2. CAUTION: Capacities shown in the table might exceed maximum capacity of selected regulator. Consult with the tubing manufacturer for guidance.

3. Table includes losses for four 90-degree bends and two end fittings. Tubing runs with larger numbers of bends or fittings shall be increased by an equivalent length of tubing to the following equation: $L = 1.3n$ where L is additional length (feet) of tubing and n is the number of additional fittings or bends.

4. EHD—Equivalent Hydraulic Diameter, which is a measure of the relative hydraulic efficiency between different tubing sizes. The greater the value of EHD, the greater the gas capacity of the tubing.

5. All table entries have been rounded to three significant digits.

TABLE G2413.4(19) [402.4(35)]
POLYETHYLENE PLASTIC PIPE

Gas	Undiluted Propane
Inlet Pressure	11.0 in. w.c.
Pressure Drop	0.5 in. w.c.
Specific Gravity	1.50

INTENDED USE	PE pipe sizing between integral 2-stage regulator at tank or second stage (low-pressure regulator) and building.					
	PIPE SIZE (inches)					
Nominal OD	$^1/_2$	$^3/_4$	1	$1^1/_4$	$1^1/_2$	2
Designation	SDR 9	SDR 11	SDR 11	SDR 10	SDR 11	SDR 11
Actual ID	0.660	0.860	1.077	1.328	1.554	1.943
Length (ft)	**Capacity in Thousands of Btu per Hour**					
10	340	680	1,230	2,130	3,210	5,770
20	233	468	844	1,460	2,210	3,970
30	187	375	677	1,170	1,770	3,180
40	160	321	580	1,000	1,520	2,730
50	142	285	514	890	1,340	2,420
60	129	258	466	807	1,220	2,190
70	119	237	428	742	1,120	2,010
80	110	221	398	690	1,040	1,870
90	103	207	374	648	978	1,760
100	98	196	353	612	924	1,660
125	87	173	313	542	819	1,470
150	78	157	284	491	742	1,330
175	72	145	261	452	683	1,230
200	67	135	243	420	635	1,140
250	60	119	215	373	563	1,010
300	54	108	195	338	510	916
350	50	99	179	311	469	843
400	46	92	167	289	436	784
450	43	87	157	271	409	736
500	41	82	148	256	387	695

For SI: 1 inch = 25.4 mm, 1 foot = 304.8 mm, 1 pound per square inch = 6.895 kPa, 1-inch water column = 0.2488 kPa, 1 British thermal unit per hour = 0.2931 W, 1 cubic foot per hour = 0.0283 m³/h, 1 degree = 0.01745 rad.

Note: All table entries have been rounded to three significant digits.

TABLE G2413.4(20) [402.4(36)]
POLYETHYLENE PLASTIC PIPE

Gas	Undiluted Propane
Inlet Pressure	2.0 psi
Pressure Drop	1.0 psi
Specific Gravity	1.50

INTENDED USE	PE pipe sizing between 2 psig service regulator and line pressure regulator.					
	PIPE SIZE (inches)					
Nominal OD	$^1/_2$	$^3/_4$	1	$1^1/_4$	$1^1/_2$	2
Designation	SDR 9	SDR 11	SDR 11	SDR 10	SDR 11	SDR 11
Actual ID	0.660	0.860	1.077	1.328	1.554	1.943
Length (ft)	**Capacity in Thousands of Btu per Hour**					
10	3,130	6,260	11,300	19,600	29,500	53,100
20	2,150	4,300	7,760	13,400	20,300	36,500
30	1,730	3,450	6,230	10,800	16,300	29,300
40	1,480	2,960	5,330	9,240	14,000	25,100
50	1,310	2,620	4,730	8,190	12,400	22,200
60	1,190	2,370	4,280	7,420	11,200	20,100
70	1,090	2,180	3,940	6,830	10,300	18,500
80	1,010	2,030	3,670	6,350	9,590	17,200
90	952	1,910	3,440	5,960	9,000	16,200
100	899	1,800	3,250	5,630	8,500	15,300
125	797	1,600	2,880	4,990	7,530	13,500
150	722	1,450	2,610	4,520	6,830	12,300
175	664	1,330	2,400	4,160	6,280	11,300
200	618	1,240	2,230	3,870	5,840	10,500
250	548	1,100	1,980	3,430	5,180	9,300
300	496	994	1,790	3,110	4,690	8,430
350	457	914	1,650	2,860	4,320	7,760
400	425	851	1,530	2,660	4,020	7,220
450	399	798	1,440	2,500	3,770	6,770
500	377	754	1,360	2,360	3,560	6,390
550	358	716	1,290	2,240	3,380	6,070
600	341	683	1,230	2,140	3,220	5,790
650	327	654	1,180	2,040	3,090	5,550
700	314	628	1,130	1,960	2,970	5,330
750	302	605	1,090	1,890	2,860	5,140
800	292	585	1,050	1,830	2,760	4,960
850	283	566	1,020	1,770	2,670	4,800
900	274	549	990	1,710	2,590	4,650
950	266	533	961	1,670	2,520	4,520
1,000	259	518	935	1,620	2,450	4,400
1,100	246	492	888	1,540	2,320	4,170
1,200	234	470	847	1,470	2,220	3,980
1,300	225	450	811	1,410	2,120	3,810
1,400	216	432	779	1,350	2,040	3,660
1,500	208	416	751	1,300	1,960	3,530
1,600	201	402	725	1,260	1,900	3,410
1,700	194	389	702	1,220	1,840	3,300
1,800	188	377	680	1,180	1,780	3,200
1,900	183	366	661	1,140	1,730	3,110
2,000	178	356	643	1,110	1,680	3,020

For SI: 1 inch = 25.4 mm, 1 foot = 304.8 mm, 1 pound per square inch = 6.895 kPa, 1-inch water column = 0.2488 kPa,
1 British thermal unit per hour = 0.2931 W, 1 cubic foot per hour = 0.0283 m³/h, 1 degree = 0.01745 rad.

Note: All table entries have been rounded to three significant digits.

TABLE G2413.4(21) [402.4(37)]
POLYETHYLENE PLASTIC TUBING

Gas	Undiluted Propane
Inlet Pressure	11.0 in. w.c.
Pressure Drop	0.5 in. w.c.
Specific Gravity	1.50

INTENDED USE: PE PIPE SIZING BETWEEN INTEGRAL 2-STAGE REGULATOR AT TANK OR SECOND STAGE (low-pressure regulator) AND BUILDING.		
Plastic Tubing Size (CTS) (inch)		
Nominal OD	$\frac{1}{2}$	1
Designation	SDR 7	SDR 11
Actual ID	0.445	0.927
Length (ft)	Capacity in Cubic Feet of Gas per Hour	
10	121	828
20	83	569
30	67	457
40	57	391
50	51	347
60	46	314
70	42	289
80	39	269
90	37	252
100	35	238
125	31	211
150	28	191
175	26	176
200	24	164
225	22	154
250	21	145
275	20	138
300	19	132
350	18	121
400	16	113
450	15	106
500	15	100

For SI: 1 inch = 25.4 mm, 1 foot = 304.8 mm, 1 pound per square inch = 6.895 kPa, 1-inch water column = 0.2488 kPa,
    1 British thermal unit per hour = 0.2931 W, 1 cubic foot per hour = 0.0283 m³/h, 1 degree = 0.01745 rad.
Note: All table entries have been rounded to three significant digits.

G2413.4.2 (402.4.2) Branch length method. *Pipe* shall be sized as follows:

1. *Pipe* size of each section of the longest *pipe* run from the *point of delivery* to the most remote *outlet* shall be determined using the longest run of *piping* and the load of the section.

2. The *pipe* size of each section of branch *piping* not previously sized shall be determined using the length of *piping* from the *point of delivery* to the most remote *outlet* in each branch and the load of the section.

❖ This sizing method is a variation of the longest-length method, and the results can be less conservative. Because less headroom is built into this method, it is especially important to account for the equivalent length of fittings installed in the system [see Commentary Figures G2413.4.2(1) and (2) and examples].This method involves multiple piping lengths within a system for application of the tables or equations, whereas the longest-length method involves only one piping length per system.

G2413.4.3 (402.4.3) Hybrid pressure. The *pipe* size for each section of higher pressure *gas piping* shall be determined using the longest length of *piping* from the *point of delivery* to the most remote line *pressure regulator*. The *pipe* size from the line *pressure regulator* to each *outlet* shall be determined using the length of *piping* from the *regulator* to the most remote outlet served by the *regulator*.

❖ Hybrid pressure piping systems convey gas at different pressures in different parts of the system. One or more pressure regulators are used to reduce the pressure in portions of the system. In some cases, a single gas piping system must deliver widely varying pressure to loads throughout a building. The most common hybrid pressure systems allow small economically sized piping to carry large loads over long distances by conveying gas at high pressures. Near the load being supplied, pressure regulators reduce the higher pressure to suit the load. For example, rooftop units on

buildings are typically designed for a natural gas pressure of 14 inches w.c. (3.5 kPa) or less. Running hundreds of feet of pipe across expansive roofs would require large expensive pipe at such low pressure. At 5-psi (34.4 kPa) pressure, small, less expensive pipe can be run over long distances, and regulators at each rooftop unit reduce the 5-psi (34.4 kPa) pressure to less than $^1/_2$ psi (3.4 kPa), which is typically 8 inches w.c. (2 kPa) [see Commentary Figures G2413.4.3(1) and (2) and examples.

Branch length method [Commentary Figure G2413.4.2(1)] Example 1:

1. In accordance with Section G2413.4.2 Item 1, determine the size of Sections A, B, C, D and E (constructed of Schedule 40 steel pipe) based on the load of each section and the longest run length of 85 feet (25 908 mm). Because the system pressure is less than 0.5 psi [(14 in. w.c.) (3.5 kPa), Table G2413.4(1) is chosen. Because the longest run length is between rows in the table, the 90-foot (27 432 mm) row must be chosen.

PIPE SECTION	LOAD (MBH)	SIZE (in.)
A	230	$1^1/_4$
B	170	1
C	135	1
D	95	$^3/_4$
E	20	$^1/_2$

2. In accordance with Section G2413.4.2 Item 2, determine the size of Branch Sections F, G, H, I and J (constructed of CSST) based on the load of each section and the length of piping and tubing from the point of delivery to the outlet on that section. Table G2413.4(5) is chosen because the branch sections are constructed of CSST

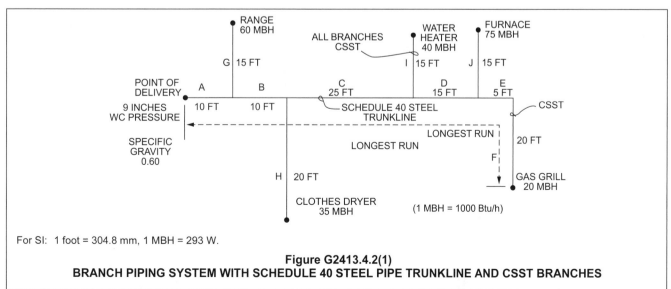

For SI: 1 foot = 304.8 mm, 1 MBH = 293 W.

Figure G2413.4.2(1)
BRANCH PIPING SYSTEM WITH SCHEDULE 40 STEEL PIPE TRUNKLINE AND CSST BRANCHES

and the pressure is less than 0.5 psi [(14 in. w.c) (3.5 kPa)]. Where a length falls between entries in the table, use the next longer length row.

CSST BRANCH	LOAD (MBH)	LENGTH PIPING AND TUBING (ft)	SIZE EHD
F	20	85	18
G	60	25	19
H	35	40	18
I	40	60	23
J	75	75	30

Branch length method [Commentary Figure G2413.4.2(2)] Example 2:

Determine the minimum required size of each piping section using branch length method.

Given:

- Piping arranged in parallel.
- Main section and individual appliance runs constructed of CSST.
- Supply pressure is 6 inches wc downstream of meter.
- Natural gas with specific gravity of 0.60.
- Chosen pressure drop 0.5 inch w.c. Minimum appliance supply pressure requirements are 5 inches wc.

- Heating value of gas is 1,000 Btu per cubic foot.

Sizing Section A:

- Size Section A based on appliance total load and length from meter to furthest appliance.
- Total load = 80 + 36 + 52 + 28 + 25 = 221 cfh.
- Section length = 45 + 20 + 5 = 70 feet.
- Use Table G2413.4(5).

Sizing Section B:

- Size Section B based on appliance total load and length from meter to appliance.
- Total load = 80 cfh.
- Section length (furnace) = 45 + 15 = 60 feet.
- Use Table G2413.4(5).

Sizing Section C:

- Size Section C based on appliance total load and length from meter to appliance.
- Total load = 36 cfh.
- Section length (water heater) = 45 + 10 = 55.
- Use Table G2413.4(5).

Sizing Section D:

- Size Section D based on appliance total load and length from meter to appliance.
- Total load = 52 cfh.
- Section Length (range) = 45 + 20 = 65 feet.
- Use Table G2413.4(5).

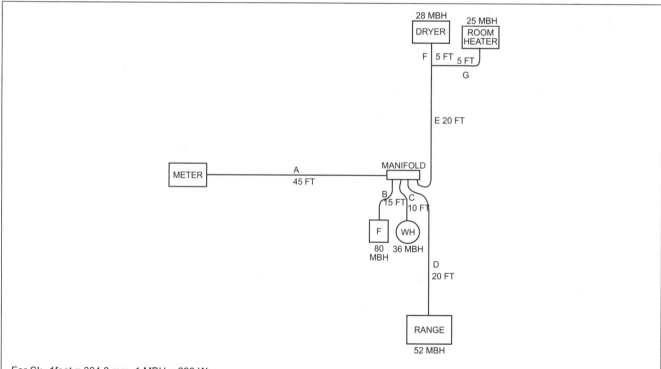

For SI: 1foot = 304.8 mm, 1 MBH = 293 W.

Figure G2413.4.2(2)
SINGLE PRESSURE CSST MANIFOLD SYSTEM

Sizing Section E:

- Size Section E base on appliance total load and length from meter to furthest appliance.
- Total load = 25 + 28 = 53 cfh.
- Section length (dryer/room heater) = 45 + 20 + 5 = 70 feet.
- Use Table G2413.4(5).

Sizing Section F:

- Size Section F based on appliance total load and length from meter to furthest appliance.
- Total load = 28 cfh.
- Section length (dryer) = 45 + 20 + 5 = 70 feet.
- Use Table G2413.4(5).

Sizing Section G:

- Size Section G based on appliance total load and length from meter to furthest appliance.
- Total load (room heater) = 25 cfh.
- Section length = 45 + 20 + 5 = 70 feet.
- Use Table G2413.4(5).

Branch length method [Commentary Figure G2413.4.3(1)] Example 1:

Note: This example illustrates hybrid system design, but is not typical for dwellings.

1. Using Table 402.4(7) of the IFGC, determine the minimum required size of piping Sections A through D. The longest run of piping from the point of delivery to the most remote regulator is 175 feet.

   - Section A serves a load of 625 MBH and in the 175 foot row of the table, $^1/_2$-inch pipe is shown to have a capacity of 728 ft^3/hr or 728,000 Btu/h for gas with a heating value of 1,000 Btu/ft^3.
   - Because Sections B, C and D all serve lesser loads than Section A, they too can be $^1/_2$ inch in size.
   - Because Table 402.4(7) of the IFGC is based upon a pressure drop of 3.5 psi, the available pressure at the inlets of the "pounds-to-inches" regulators under full load condition will be at least 1.5 psig.

2. Determine size of piping Sections E, F and G, using Table 402.4(2) of the IFGC and the longest length of piping from the pounds-to-inches regulator to the most remote outlet. In this case, rooftop unit 1 is the most remote at 65 feet from the regulator.

3. Determine size of piping sections H, I and J in the same manner described in Item 2.

NOTE: MBH = 1,000 Btu/h

Hybrid pressure [Commentary Figure G2413.4.3(2)] Example 2:

Given:

- All piping is CSST.
- Piping branches run in parallel.
- Individual CSST appliance branch runs supplied by manifold.
- Supply pressure is 2 psi downstream of meter.
- Chosen pressure drop is 1 psi for Section A.
- Natural gas with specific gravity of 0.60.
- Heating value of gas is 1,000 Btu/ft^3.
- Supply pressure at manifold is 8 inches w.c.
- Chosen pressure drop is 3 inches w.c. for Sections B through G.
- Appliances need 5 inches w.c. minimum supply pressure.
- Pressure drop across regulator is less than $^1/_2$ psi at full load of this system.

Sizing Section A:

- Size Section A based on total load and length from meter to regulator.
- Total load = 80 + 36 + 52 + 28 + 25 = 221 cfh.
- Section length = 45 feet.
- Use Table G2413.4(6).

Sizing Section B:

- Size Section B based on appliance load and length from regulator to appliance.

SECTION	LOAD (cfh)	SECTION LENGTH (feet)	SECTION LENGTH (feet) USED WITH TABLE	REQUIRED SIZE (CSST)
A	221	45	70	
B	80	15	60	
C	36	10	55	
D	52	20	65	
E	53	20	70	
F	28	5	70	
G	25	5	70	

SUMMARY TABLE FOR BRANCH LENGTH METHOD EXAMPLE 2

- Appliance load = 80 cfh.
- Section length (furnace) = 15 feet.
- Use Table 402.4(16) of the IFGC.

Sizing Section C:

- Size Section C based on appliance load and length from regulator to appliance.
- Total load = 36 cfh.
- Section length (water heater) = 10 feet.
- oUse Table 402.4(16) of the IFGC.

Sizing Section D:

- Size Section D based on appliance load and length from regulator to appliance.
- Total load = 52 cfh.
- Section length (range) = 20 feet
- Use Table 402.4(16) of the IFGC

Sizing Section E:

- Size Section E based on appliance total load and length from regulator to furthest appliance.
- Total load = 25 + 28 = 53 cfh.

SECTION	LOAD (cfh)	ACTUAL SECTION LENGTH (feet) (including equivalent length of fittings and bends)	SECTION LENGTH (FT) USED WITH TABLE	REQUIRED SIZE (CSST)
A	221	45	45[a]	
B	80	15	15[b]	
C	36	10	10[b]	
D	52	20	20[b]	
E	53	20	25[b]	
F	28	5	25[b]	
G	25	5	25[b]	

SUMMARY TABLE FOR HYBRID PRESSURE EXAMPLE 2

a. Distance from meter to regulator is the run length for the main section.

b. Distance from regulator to each appliance is the run length for appliance branch run sections.

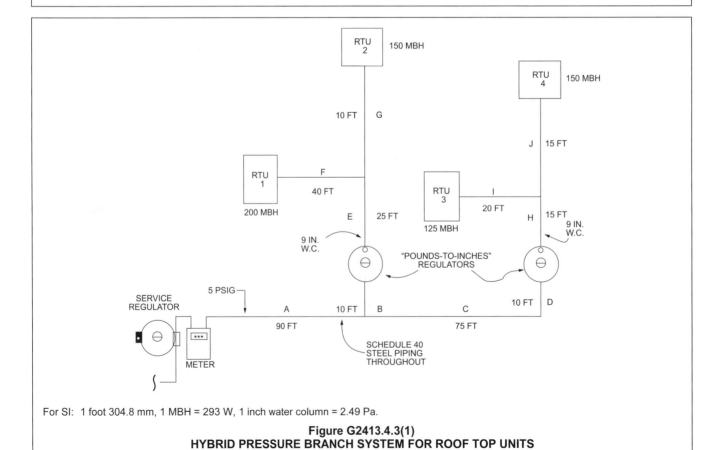

For SI: 1 foot 304.8 mm, 1 MBH = 293 W, 1 inch water column = 2.49 Pa.

Figure G2413.4.3(1)
HYBRID PRESSURE BRANCH SYSTEM FOR ROOF TOP UNITS

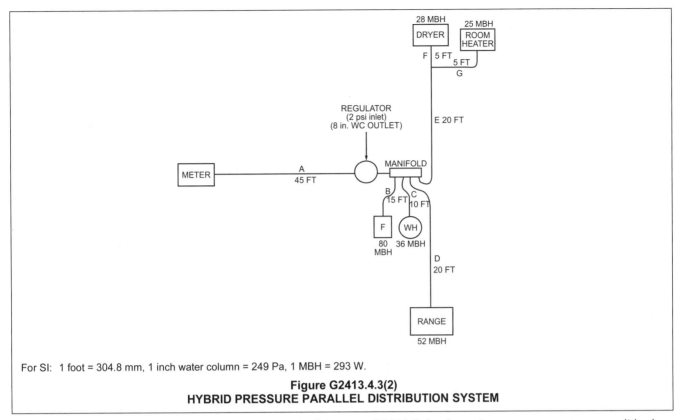

For SI: 1 foot = 304.8 mm, 1 inch water column = 249 Pa, 1 MBH = 293 W.

Figure G2413.4.3(2)
HYBRID PRESSURE PARALLEL DISTRIBUTION SYSTEM

- Section length (dryer/room heater) = 20 + 5 =25 feet.
- Use Table 402.4(16) of the IFGC.

Sizing Section F:
- Size Section F based on appliance total load and length from regulator to furthest appliance.
- Total load (dryer) = 28 cfh (0.8 m³/h).
- Section length = 20 + 5 = 25 feet (7620 mm).
- Use Table 402.4(16) of the IFGC.

Sizing Section G:
- Size Section G based on appliance total load and length from regulator to furthest appliance.
- Total load (room heater) = 25 cfh.
- Section length = 20 + 5 = 25 feet.
- Use Table 402.4(16) of the IFGC.

G2413.5 (402.5) Allowable pressure drop. The design pressure loss in any *piping system* under maximum probable flow conditions, from the *point of delivery* to the inlet connection of the *appliance*, shall be such that the supply pressure at the *appliance* is greater than or equal to the minimum pressure required by the *appliance*.

❖ Between the point of delivery and the load, gas pressure losses will occur because of pipe friction. The design of the system must control these losses to ensure that the pressure at the load (appliance) connection point will be equal to or greater than the minimum required pressure as dictated by the appliance manufacturer. As stated in the commentary to Section

G2412.8, inadequate gas pressures can result in dangerous appliance operation. This section speaks of "maximum probable flow conditions," which must be taken to mean the total connected load of all appliances operating at full capacity except where a diversity-of-load factor is applied.

The designer can choose any pressure loss for a piping system if the goal of this section is met. The code does not specify any maximum pressure drop other than what is required by this section. For example, consider an appliance that has a minimum required inlet pressure of 5 inches w.c. (1.25 kPa), and the gas pressure at the point of delivery is 8 inches w.c. (2 kPa). The maximum allowable pressure drop through the piping system would be 3 inches w.c. (0.75 kPa) because the pressure at the appliance connection must be greater than or equal to 5 inches w.c. Some sizing tables such as Tables G2413.4(1) and G2413.4(3) are based on very small pressure drops [0.5 inch w.c. (125 Pa)] and, thus are quite conservative sizing methods that account for the friction losses through a limited number of fittings used in a piping system. The tables are optional sizing methods placed in the code for convenience. The code allows piping systems to be designed for the exact pressure losses allowed by this section. Some tables are based on much larger pressure losses and can be used where the designer wants to keep pipe sizes to their minimum. The code intends to allow the designer to have the option of choosing the system pressure losses, provided that such losses meet the requirement of this section.

G2413.6 (402.6) Maximum design operating pressure. The maximum design operating pressure for *piping systems* located inside buildings shall not exceed 5 pounds per square inch gauge (psig) (34 kPa gauge) except where one or more of the following conditions are met:

1. The *piping system* is welded.

2. The *piping* is located in a ventilated chase or otherwise enclosed for protection against accidental gas accumulation.

3. The *piping* is a temporary installation for buildings under construction.

❖ This section is specific to indoor piping systems. The pressure is limited to 5 psig (34 kPa) to minimize the consequences of a leak. Also, the higher the pressure, the more difficult it is to contain any gas and, thus leakage is more likely. Item 1 recognizes the superior strength and integrity of steel piping with welded joints. Item 2 allows the option of higher pressures where the piping is enclosed to prevent leakage from entering the building. Item 3 allows higher pressures for temporary piping for construction site applications, such as temporary space heating, because of low occupant density, the presence of trained personnel and the short duration of use. Item 3 recognizes the unique nature of temporary piping on construction sites where small piping is desirable, thus making higher pressures necessary.

This section applies to all fuel gases including LP-gas (see Section G2413.6.1).

G2413.6.1 (402.6.1) Liquefied petroleum gas systems. LP-gas systems designed to operate below -5°F (-21°C) or with butane or a propane-butane mix shall be designed to either accommodate liquid LP-gas or prevent LP-gas vapor from condensing into a liquid.

❖ In low-temperature environments, LP-gases might condense into the liquid state within the piping system. If the liquid fuel enters system components or appliances, dangerous malfunctions could result. The temperature and pressure of a gas both affect the point at which a gas will change states to a liquid phase. LP-gas can be made to condense to liquid by applying pressure, by cooling the gas and by a combination of both. At atmospheric pressure, 14.7 psia at sea level, propane will condense at -44°F (-42°C) and butane will condense at 15°F (-9°C) meaning that LP liquid would remain as a liquid in an open container at temperatures of minus 44 degrees F and lower. As the liquid warms to temperatures above the condensing temperature, the liquid will begin to boil producing gas vapor and the resultant vapor pressure will have to be held in a closed container to prevent vaporization of all of the liquid. This is what happens in an LP-gas storage tank. The tank holds both liquid and gaseous LP under pressure that will vary with the ambient temperature. Within a piping system, the pressure could be higher than one atmosphere and LP-gas would, therefore, condense at a higher temperature. For example, at 20 psig (138 kPa), propane will condense at -5°F (-21°C). The -5°F threshold requirement of this section was chosen because 20 psi system piping was commonly installed outdoors.

SECTION G2414 (403)
PIPING MATERIALS

G2414.1 (403.1) General. Materials used for *piping systems* shall comply with the requirements of this chapter or shall be *approved*.

❖ This section dictates what materials and components can be used to construct gas distribution systems and also specifies the allowable applications of those materials.

G2414.2 (403.2) Used materials. *Pipe*, fittings, *valves* or other materials shall not be used again unless they are free of foreign materials and have been ascertained to be adequate for the service intended.

❖ Ideally, gas pipe installations should be constructed of new materials. However, this section recognizes that there are occasions when used piping materials may be perfectly acceptable for an installation if they meet a number of very strict criteria that intend to reduce the potential for poor performance. To judge equivalency of reused piping materials, compliance with approval criteria for new materials should be used.

G2414.3 (403.3) Other materials. Material not covered by the standards specifications listed herein shall be investigated and tested to determine that it is safe and suitable for the proposed service, and, in addition, shall be recommended for that service by the manufacturer and shall be *approved* by the *code official*.

❖ This section echoes the intent of Section R104.11.

G2414.4 (403.4) Metallic pipe. Metallic *pipe* shall comply with Sections G2414.4.1 and G2414.4.2.

❖ This section addresses the traditional metal pipes used in piping installations.

G2414.4.1 (403.4.1) Cast iron. Cast-iron *pipe* shall not be used.

❖ Because cast iron pipe is brittle compared to the malleable metals, it is more likely to fail under stress and therefore is not suitable for conveying fuel gases. Steel pipe will deform under extreme stress, but cast iron pipe will break.

G2414.4.2 (403.4.2) Steel. Steel and wrought-iron *pipe* shall be at least of standard weight (Schedule 40) and shall comply with one of the following standards:

1. ASME B36.10, 10M.

2. ASTM A53/A 53M.

3. ASTM A106.

❖ Steel pipe must be Schedule 40 or heavier, must comply with one of the listed standards and can be black iron or galvanized. Contrary to popular belief, natural gas does not adversely react with the zinc coating on galvanized pipe.

G2414.5 (403.5) Metallic tubing. Seamless copper, aluminum alloy and steel *tubing* shall not be used with gases corrosive to such materials.

❖ Aluminum and steel tubing are used in the manufacturing of appliances but are rarely used for gas distribution. Copper tubing is widely used in gas distribution systems and is used as an alternative to CSST in parallel distribution manifold systems. As with copper pipe, copper tubing cannot be used with gases that will cause it to corrode. Commentary Figure G2414.5 illustrates a type of sheathed copper tubing that is designed for gas distribution. The corrosiveness of the gas can be determined by contacting the gas supplier. Depending on the chemical composition of the gas, copper can form scale deposits on the interior pipe wall, and such scale can find its way into gas controls, regulators and valves causing malfunctions.

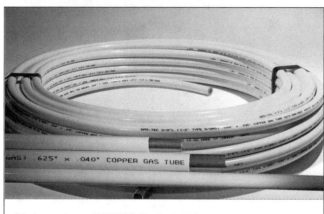

Photo courtesy of KAMCO Products Ltd.

Figure G2414.5
PLASTIC-SHEATHED COPPER TUBING

G2414.5.1 (403.5.1) Steel tubing. Steel *tubing* shall comply with ASTM A254.

❖ Steel tubing is used only in industrial applications and in the construction of appliances.

G2414.5.2 (403.5.2) Copper and copper alloy tubing. Copper *tubing* shall comply with Standard Type K or L of ASTM B88 or ASTM B280.

Copper and copper alloy *tubing* shall not be used if the gas contains more than an average of 0.3 grains of hydrogen sulfide per 100 standard cubic feet of gas (0.7 milligrams per 100 liters).

❖ ASTM B88 applies to copper water tube, and Type K and L refers to wall thickness. Type M tubing is the thinnest wall and is not allowed for gas service.

G2414.5.3 (403.5.4) Corrugated stainless steel tubing. Corrugated stainless steel *tubing* shall be *listed* in accordance with ANSI LC 1/CSA 6.26.

❖ ANSI LC-1 contains material and performance criteria and installation requirements for corrugated stainless steel tubing (CSST) gas distribution systems. CSST systems are used as an alternative to more traditional

gas piping systems for the distribution of natural gas within buildings.

Corrugated stainless steel tubing is a semirigid stainless steel tubing with a plastic jacket and is available in coil lengths varying from 100 to 250 feet (30 480 to 76 200 mm) and in the following standard sizes: $^3/_8$, $^1/_2$, $^3/_4$, 1, $1^1/_4$, $1^1/_2$ and 2 inch (larger sizes may be available) [see Commentary Figure G2414.5.3(1)]. Note that CSST sizes are designated by equivalent hydraulic diameters (EHD) not standard tube sizes in inches. Any standard size markings in inches provided by the manufacturer are approximations only. Special proprietary fittings are required for connection of the tubing [see Commentary Figures G2414.5.3(2) through G2414.5.3(4)]. CSST can be used in low-pressure systems [up to 0.5 psig (14 inches w.c.) (3.45 kPa)] and in higher pressure systems operating at pressures up to 5 psig (34.5 kPa). Commentary Figure G2414.5.3(5) shows a typical manifold installation using CSST and shows a multiport manifold with a pressure regulator and shutoff valve installed. To prevent damage or puncture by screws and nails, specialized steel shield plates and protective flexible conduit are used to protect CSST systems that pass through or near structural members. Such protection must employ the parts specified by the

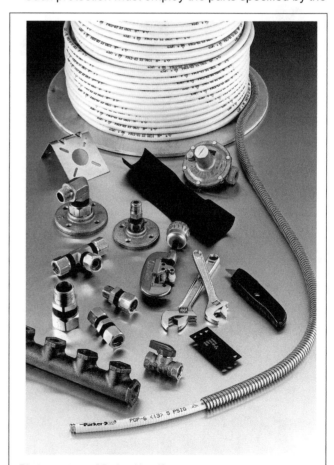

Photo courtesy of Parker Hannifin

Figure G2414.5.3(1)
CSST AND SYSTEM COMPONENTS

CSST manufacturer as opposed to standard off-the-shelf protection devices [see Commentary Figures G2414.5.3(1) and G2414.5.3(6)].

CSST systems are usually designed as parallel distribution (manifold) systems, whereas traditional gas distribution systems are typically designed as branch (series) systems [see Commentary Figure G2414.5.3(7)].

Medium-pressure CSST systems, typically 2 psig (13.8 kPa) maximum, are designed to allow the use of small diameter tubing for long runs. Significant distribu-

tion pressure losses can be compensated for by providing sufficiently high supply pressures. Medium-pressure distribution systems require the installation of pressure regulators, commonly called pounds-to-inches regulators, to reduce the supply pressure to the pressure required for delivery to the gas-fired appliances [see commentary, Section G2421 and Commentary Figure G2414.5.3(8)]. Note that systems with pressures over 2 psi (13.8 kPa) that supply appliances designed for a maximum inlet pressure of 14 inches

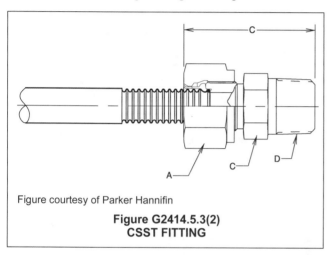

Figure courtesy of Parker Hannifin

Figure G2414.5.3(2)
CSST FITTING

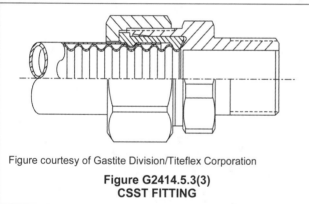

Figure courtesy of Gastite Division/Titeflex Corporation

Figure G2414.5.3(3)
CSST FITTING

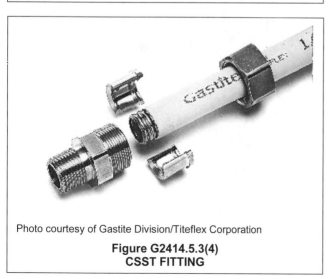

Photo courtesy of Gastite Division/Titeflex Corporation

Figure G2414.5.3(4)
CSST FITTING

Photo courtesy of Gastite Division/Titeflex Corporation

Figure G2414.5.3(5)
MULTIPORT MANIFOLD INSTALLATION USING CSST

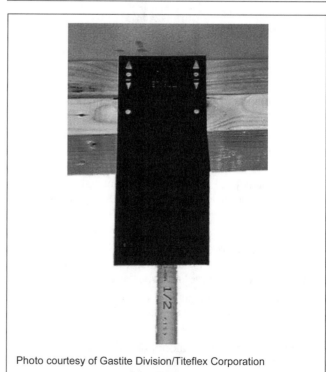

Photo courtesy of Gastite Division/Titeflex Corporation

Figure G2414.5.3(6)
TOP PLATE TUBING PROTECTION SHIELD

w.c. (3.5 kPa) must have overpressure protection devices (OPD) in addition to the pounds-to-inches line pressure regulators. Such OPDs are factory installed on pressure regulators shipped for use in systems having pressures greater than 2 psi (13.8 kPa).

Section G2415.5 prohibits tubing fittings in concealed locations but does not apply to CSST fittings. In accordance with the manufacturer's installation instructions and the material listing, CSST fittings can be concealed in the building construction. The tubing, fittings, shields and outlet terminals are all part of a system and must be installed together as a system. The manufacturer's installation instructions are to be considered as a complete system installation manual, which also includes the proprietary sizing criteria. Except for those tables that are specific to CSST, the sizing tables in Section G2413 do not apply to and cannot be used for CSST systems.

Installation of CSST is not allowed underground (directly buried) unless it is of a type specially designed and listed for underground applications [see Commentary Figure G2414.5.3(9)]. Some manufacturers' instal-

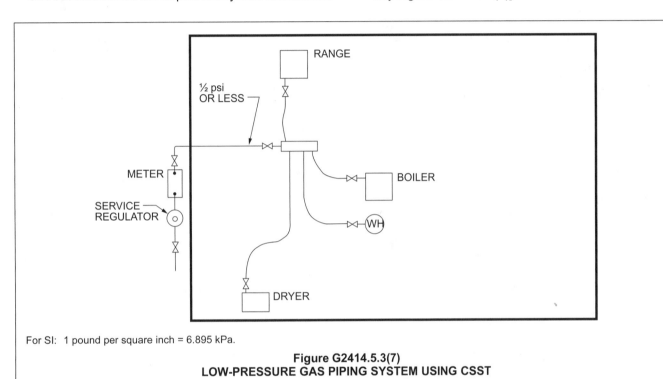

For SI: 1 pound per square inch = 6.895 kPa.

Figure G2414.5.3(7)
LOW-PRESSURE GAS PIPING SYSTEM USING CSST

For SI: 1 pound per square inch = 6.896 kPa, 1 inch water column = 249 Pa.

Figure G2414.5.3(8)
MEDIUM-PRESSURE GAS PIPING SYSTEM USING CSST

lation instructions allow CSST to be "indirectly" buried by installing the tubing within a nonmetallic, water-tight conduit, such as PVC, PE or ABS plastic pipe. Other manufacturers may offer CSST with an integral casing and sealing fittings specifically designed for underground installation. See Sections G2415.6 and G2415.12 through G2415.14 regarding piping buried in slabs and beneath buildings.

CSST is allowed by the manufacturers' installation instructions to connect directly to fixed-in-place, nonportable, nonmovable appliances such as furnaces, boilers, gas fireplaces and water heaters. CSST is not intended for direct connection to moveable appliances such as clothes dryers or ranges nor is it intended to connect to appliances suspended on chains or rods such as radiant heaters and unit heaters. Also, connecting directly to log lighters installed in solid-fuel burning fireplaces is not allowed. Direct connection is allowed for gas fireplace appliances in accordance with the CSST and fireplace manufacturers' instructions. At range and clothes dryer locations, the CSST must terminate at a CSST termination fitting assembly, and a gas appliance connector must be used to connect the appliance to the termination fitting assembly. CSST must not be exposed to impact, vibration or repeated movement. "Drops" to fixed-in-place nonmovable appliances must be well supported and routed to avoid being disturbed. In other words "flying runs" of CSST are prohibited. CSST, like copper tubing, is not flexible, but rather is semirigid, allowing it to be formed into shape as it is installed. CSST, like any metallic tubing, must not be repeatedly bent in the same location because this can result in kinking, work hardening and structural failure.

Generally, the sale of CSST materials is restricted and only individuals who have been specially trained and certified are allowed to purchase and install CSST systems.

Traditionally, the utility company (gas supplier) service regulator delivers natural gas to the building at a pressure of approximately 6 to 10 inches w.c. (1.49 to 2.49 kPa). Where medium-pressure [2 psig (13.8 kPa)] systems are used, the service regulator is set to deliver gas at a pressure of 2 psig (13.8 kPa). A medium-pressure gas distribution system must be properly designed to deliver gas to the connected appliances at pressures not exceeding the maximum inlet (supply) pressure for which the appliance is designed. A hazardous condition could develop if gas is supplied to appliances at pressures greater than the design inlet pressure of the appliances. See Section 416 of the IFGC.

ANSI LC-1 /CSA 6.26 addresses CSST systems and contains installation requirements. Installers and inspectors must have copies of the standard and the CSST manufacturer's installation instructions to properly install and inspect CSST systems. Manufacturers' instructions contain detailed requirements for protection hardware designed to prevent damage to installed CSST systems. As with most tubing systems, the most common installation errors involve abuse of the CSST tubing, lack of support and improperly installed puncture protection devices or the omission of protection devices. CSST manufacturers publish design and installation guides that are very helpful and should be on the book shelves of inspectors, installers and designers [see Commentary Figures G2414.5.3(10) and G2414.5.3(11)]. Some CSST manufacturers offer preassembled manifolds in a rough-in cabinet to sim-

Photo courtesy of Omega Flex

Figure G2414.5.3(9)
CSST MANUFACTURED FOR UNDERGROUND INSTALLATION

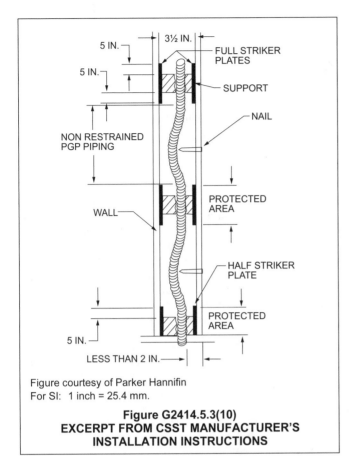

Figure courtesy of Parker Hannifin
For SI: 1 inch = 25.4 mm.

Figure G2414.5.3(10)
EXCERPT FROM CSST MANUFACTURER'S
INSTALLATION INSTRUCTIONS

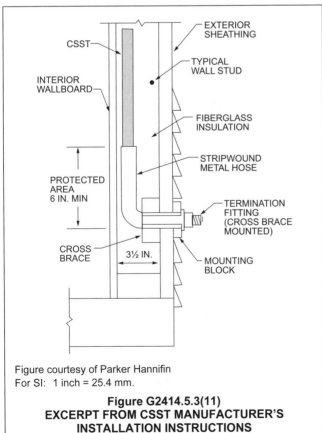

Figure courtesy of Parker Hannifin
For SI: 1 inch = 25.4 mm.

Figure G2414.5.3(11)
EXCERPT FROM CSST MANUFACTURER'S
INSTALLATION INSTRUCTIONS

plify installation [see Commentary Figure G2414.5.3(12)]. CSST transitions to rigid pipe appliance connections and listed appliance connectors must be made with fittings that prevent stress and rotational forces from being applied to the CSST. Commentary Figures G2414.5.3(13) through (15) illustrate typical fitting assemblies used to make the transition from CSST to appliance connections or connectors. It is a code violation to construct appliance outlets with a standard pipe elbow and rigid pipe nipple stubbed out of a wall or up through a floor because any rotational forces applied by tools to the outlet nipple will cause the CSST to rotate behind the wall or below the floor. The same is true for tee-handle valves concealed in a wall. The proper terminal fittings and brackets must be used to prevent forces from being transferred to the CSST.

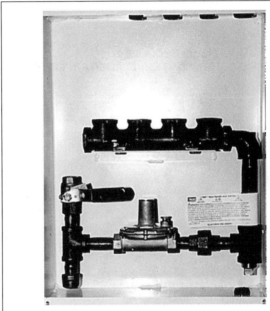

Photo courtesy of Parker Hannifin

Figure G2414.5.3(12)
MANIFOLD FOR 2 PSI PARALLEL
DISTRIBUTION SYSTEM

Photo courtesy of Parker Hannifin

Figure G2414.5.3(13)
"STUB-OUT" TERMINATION TRANSITION FITTING

Photo courtesy of Gastite Division/Titeflex Corporation

Figure G2414.5.3(14)
"VALVE BOX" TERMINATION TRANSITION FITTING

Photo courtesy of Gastite Division/Titeflex Corporation

Figure G2414.5.3(15)
TERMINATION TRANSITION FITTING

G2414.6 (403.6) Plastic pipe, tubing and fittings. Polyethylene plastic pipe, tubing and fittings used to supply fuel gas shall conform to ASTM D2513. Such pipe shall be marked "Gas" and "ASTM D2513."

Plastic pipe, tubing and fittings, other than polyethylene, shall be identified and conform to the 2008 edition of ASTM D2513. Such pipe shall be marked "Gas" and "ASTM D2513."

Polyvinyl chloride (PVC) and chlorinated polyvinyl chloride (CPVC) plastic pipe, tubing and fittings shall not be used to supply fuel gas.

❖ The installation of plastic pipe and tubing is limited to areas that are both outside the building (outdoors) and underground because of the potential hazard associated with the use of a material that has lower resistance to physical damage and heat compared to metallic pipe. Plastic pipe and tubing must never be installed in or under any building. Plastic pipe and tubing are widely used for underground gas distribution systems because of their ease of installation and inherent resistance to corrosion (see Section G2415.17.1). The scope of ASTM D2513 was recently changed to limit its application to polyethylene (PE) pipe; therefore, an older edition is referenced for other types of plastic. This section specifically prohibits the use of PVC and CPVC pipe, tubing and fittings for conveying fuel gas.

The code now references the 2013 edition of ASTM D2513, which has been revised to address polyethylene (PE) plastic pipe, tubing and fittings only, whereas the 2008 edition addressed all plastic materials. The code had to maintain a reference to the 2008 edition of the standard in order to address plastics other than PE such as polyamide (nylon). It was determined that polyamide pipe is currently used to supply fuel gas; however, PVC and CPVC are not. Further, it was decided that because of the brittle nature of PVC and CPVC, especially at low temperatures, these materials are not suitable for conveying fuel gas. Rather than being silent, the code now prohibits what the marketplace has failed to embrace as a viable material for fuel gas (see Commentary Figure G2414.6).

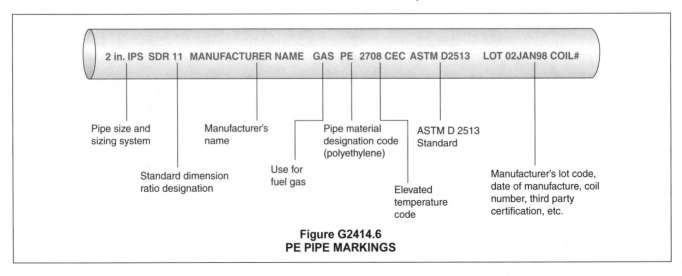

2 in. IPS SDR 11   MANUFACTURER NAME   GAS   PE   2708 CEC   ASTM D2513   LOT 02JAN98 COIL#

Pipe size and sizing system

Manufacturer's name

Standard dimension ratio designation

Use for fuel gas

Pipe material designation code (polyethylene)

Elevated temperature code

ASTM D 2513 Standard

Manufacturer's lot code, date of manufacture, coil number, third party certification, etc.

Figure G2414.6
PE PIPE MARKINGS

G2414.6.1 (403.6.1) Anodeless risers. Plastic pipe, tubing and anodeless risers shall comply with the following:

1. Factory-assembled anodeless risers shall be recommended by the manufacturer for the gas used and shall be leak tested by the manufacturer in accordance with written procedures.

2. Service head adapters and field-assembled anodeless risers incorporating service head adapters shall be recommended by the manufacturer for the gas used, and shall be designed and certified to meet the requirements of Category I of ASTM D2513, and U.S. Department of Transportation, Code of Federal Regulations, Title

49, Part 192.281(e). The manufacturer shall provide the user with qualified installation instructions as prescribed by the U.S. Department of Transportation, Code of Federal Regulations, Title 49, Part 192.283(b).

❖ This section is an exception to the absolute prohibition of plastic gas piping above ground. Plastic pipe is allowed to rise from underground within a steel riser conduit constructed as an anodeless riser assembly. Such risers are coated to resist corrosion and do not use a sacrificial anode for cathodic corrosion protection, hence the name "anodeless riser." The steel riser protects the PE plastic pipe from physical damage and terminates the plastic pipe with a threaded steel pipe transition fitting. A sweeping 90-degree (1.57 rad) bend is placed on the underground end of rigid risers to prevent them from rotating in the ground when threaded connections are made at the terminal end. The plastic pipe could be damaged by rotational and bending stresses transmitted by the riser; therefore, a horizontal portion of steel piping is installed at the bottom of the riser to limit such movement. These risers can eliminate the need for an underground plastic-to-metal transition fitting and simplify installation. They are used extensively by utility companies for service laterals and can be used for other installations such as supply runs to out-buildings, pool heaters, gas grills and yard lights.

Service head adapters are also a type of terminal transition fitting that allow plastic pipe to be pulled through a retired steel gas service lateral and terminated at the point where the concentric piping passes through the building foundation wall. The adapter makes the transition from plastic pipe to steel pipe and seals the end of the retired steel service lateral [see Commentary Figures G2403.(1), G2414.6.1(1) G2414.6.1(2) and G2414.6.1(3)].

Photo courtesy of Perfection Corporation

Figure G2414.6.1(1)
SERVICE HEAD ADAPTERS

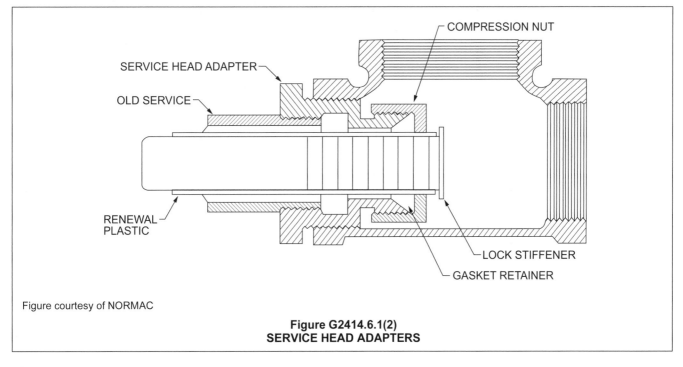

Figure courtesy of NORMAC

Figure G2414.6.1(2)
SERVICE HEAD ADAPTERS

Photo courtesy of NORMAC

Figure G2414.6.1(3)
FLEXIBLE ANODELESS RISER

G2414.6.2 (403.6.2) LP-gas systems. The use of plastic pipe, tubing and fittings in undiluted liquefied petroleum gas *piping* systems shall be in accordance with NFPA 58.

❖ NFPA 58 limits plastic piping for LP-gas service to only polyethylene pipe and tubing. The PE pipe and tubing must be marked as being in compliance with ASTM D2513 and is allowed for LP-gas service only if so recommended by the material manufacturer.

G2414.6.3 (403.6.3) Regulator vent piping. Plastic pipe and fittings used to connect *regulator* vents to remote vent terminations shall be of PVC conforming to ANSI/UL 651. PVC vent *piping* shall not be installed indoors.

❖ The referenced standard is for Schedule 40 and 80 PVC electrical conduit. This material was chosen because it is required to be sunlight-resistant for use above ground and the typical application for regulator vent piping is above ground, outdoors where it would be subject to the degrading effects of ultraviolet radiation in sunlight. Consistent with other provisions in this code, plastic pipe that does or could convey fuel gas is prohibited indoors. Regulator vent outlets are piped to maintain separation between vent outlets and sources of ignition and air intake openings in building envelopes.

G2414.7 (403.7) Workmanship and defects. *Pipe, tubing* and fittings shall be clear and free from cutting burrs and defects in structure or threading, and shall be thoroughly brushed, and chip and scale blown.

Defects in *pipe* or *tubing* or fittings shall not be repaired. Defective *pipe, tubing* or fittings shall be replaced. (See Section G2417.1.2.)

❖ This section stresses that proper installation procedures are necessary to secure a safe installation. In the context of this section, good workmanship is an enforceable requirement. Burrs result from the cutting of pipe or tube and cause flow restriction. Pipe scale, dirt, construction site debris, metal chips from cutting and threading operations and excess thread compound are potentially harmful contaminates that might cause damage to valves, regulators and appliances. Damaged and/or improperly cut threads can cause leakage or joint failure. Defective materials must be replaced because attempts to repair them are always makeshift at best, and any cost savings would not be worth the risk.

G2414.8 (403.8) Protective coating. Where in contact with material or atmosphere exerting a corrosive action, metallic *piping* and fittings coated with a corrosion-resistant material shall be used. External or internal coatings or linings used on *piping* or components shall not be considered as adding strength.

❖ Protective coatings in the form of wrappings, tapes, enamels, epoxies, sleeves, casings and factory-applied coverings are a common method of isolating metallic piping materials from the atmosphere or from the earth in which they are buried. Application of these materials under strictly controlled factory conditions improves the quality control process and reduces the likelihood of coating defects. Corrosion can be caused by weather exposure, burial in the soil or construction materials that are in contact with the pipe. Cinders (also known as dross) are the waste residue from steel production and carry the waste products skimmed from the molten metal during the smelting process. Cinders are also the waste residue of coal-fired appliances and equipment. These materials are extremely corrosive to metallic piping and may not be used in any form (backfill or cinder blocks, for example) where in direct contact with the pipe. Protection is usually provided by a factory-applied coating or by field wrapping the pipe with a protective covering, such as a coal-tar-based or plastic wrapping. Galvanized piping in contact with soil would require additional coating or protection. The zinc coating does not usually provide long-term protection in underground applications. Where possible, a piping material is chosen that is not subject to the type of corrosion of the application.

Corrosion can also be caused by galvanic action that takes place where dissimilar metals are joined in a current-carrying medium, such as soil or water. For example, if steel and copper pipe are joined in a medium that conducts electrical current, the steel pipe will corrode at an accelerated rate because of the electro-chemical process between the dissimilar metals. In this case, the soil acts as an electrolyte, the steel will be a sacrificial anode, and the copper will be a cathode. Because a cell is created, current will flow through the metal junction and the soil, resulting in the gradual deterioration of the steel. To protect against galvanic corrosion, dielectric fittings and couplings are used to join the piping, thus breaking the circuit of the cell.

This section is commonly interpreted as requiring priming and painting of black steel pipe that is exposed to the weather as the minimum form of protection.

G2414.9 (403.9) Metallic pipe threads. Metallic *pipe* and fitting threads shall be taper *pipe* threads and shall comply with ASME B 1.20.1.

❖ Threads must be taper cut in accordance with the stated standard. The standard regulates tapered and straight threads, but this section specifies tapered threads. Tapered pipe threads, when made up, form a metal-to-metal (interference fit) seal. Pipe-joint compound or PTFE (Teflon) tape is to be applied to male threads only to decrease the possibility of tape fragments or compound entering the piping system. Such debris in piping systems can block orifices, restrict flow or interfere with the operation of controls. The primary purpose of pipe-thread compounds is to act as a lubricant to allow proper tightening and to achieve a metal-to-metal seal. They also fill in small imperfections on the threaded surfaces. This section prohibits the use of straight tapped fittings such as couplings because the threaded joint will not provide the same seal as a joint made with taper-tapped fittings. Steel taper-threaded couplings are easily identified because they look exactly like the run of a tee fitting (see Commentary Figure G2414.9). The common designation "NPT" is misinterpreted as "nominal pipe threads" and actually stands for "national standard, pipe, tapered."

G2414.9.1 (403.9.1) Damaged threads. *Pipe* with threads that are stripped, chipped, corroded or otherwise damaged shall not be used. Where a weld opens during the operation of cutting or threading, that portion of the *pipe* shall not be used.

❖ Damaged or improperly cut threads and defective welded seam pipe must be eliminated from any gas piping system because of the potential for leaks.

G2414.9.2 (403.9.2) Number of threads. Field threading of metallic *pipe* shall be in accordance with Table G2414.9.2.

❖ The table sets forth the required number of threads per unit length of pipe based on the pipe size. These specifications are intended to be both maximums and minimums because both undercutting and overcutting can result in faulty joints. Proper threading of pipe requires skill, practice, and the proper well-maintained tools.

TABLE G2414.9.2 (403.9.2)
SPECIFICATIONS FOR THREADING METALLIC PIPE

IRON PIPE SIZE (inches)	APPROXIMATE LENGTH OF THREADED PORTION (inches)	APPROXIMATE NO. OF THREADS TO BE CUT
$^1/_2$	$^3/_4$	10
$^3/_4$	$^3/_4$	10
1	$^7/_8$	10
$1^1/_4$	1	11
$1^1/_2$	1	11

For SI: 1 inch = 25.4 mm.

G2414.9.3 (403.9.3) Thread joint compounds. Thread joint compounds shall be resistant to the action of liquefied petroleum gas or to any other chemical constituents of the gases to be conducted through the *piping*.

❖ Joint compounds in both paste and tape forms are commonly misapplied and used for the wrong application. Some compounds react chemically with the gas

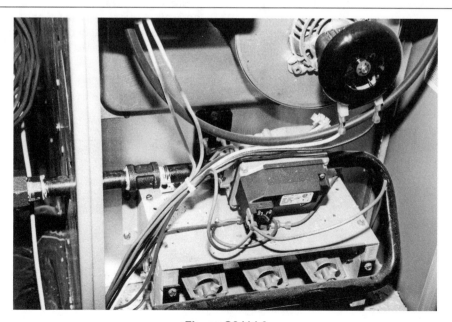

Figure G2414.9
TAPERED THREAD COUPLING IN CONNECTION TO FURNACE

being conveyed, which could result in leakage. The label on the compound container will specify the applications for which the compound is suitable. Thread compounds act as a lubricant for the threads during assembly and also act as a sealant for the life of the joint. Care must be taken to keep all compounds out of the piping system interior because the contamination can cause damage to components and appliances. Pipe-joint compound or tape is limited to application on the male threads only to decrease the possibility of tape fragments or compound entering the piping system. Such debris in piping systems can block orifices, restrict flow or interfere with the operation of safety controls. It is important to leave the first one or two threads on the end of the male threads bare to help prevent the compound from entering the piping system.

Once made from lead compounds and linseed oil, joint compound formulations now commonly contain PTFE (Teflon). The primary purpose of pipe-thread compounds is to act as a lubricant to allow proper tightening and to achieve a metal-to-metal seal. They also fill in small imperfections on the threaded surfaces. Pipe-thread compounds and tapes must be compatible with both the piping material and the contents of the piping.

LP-gases can be a solvent for some pipe joint compound formulations. Therefore, a type must be chosen that will not react with the gas.

G2414.10 (403.10) Metallic piping joints and fittings. The type of *piping* joint used shall be suitable for the pressure-temperature conditions and shall be selected giving consideration to joint tightness and mechanical strength under the service conditions. The joint shall be able to sustain the maximum end force caused by the internal pressure and any additional forces due to temperature expansion or contraction, vibration, fatigue, or to the weight of the *pipe* and its contents.

❖ This section is in performance language and requires the designer and/or installer to choose the type of join-

ing means that is appropriate for the application. Besides being leak-tight, the joints must be able to withstand the conditions of service. Many joining methods are addressed in the code, but not all of them are suitable for all applications. For example, PE pipe for LP-gas use is limited to a maximum pressure of 30 psig (206.8 kPa), and most tubing materials would be unsuitable for uses involving equipment vibration that is transferred to the tubing. Also, some fittings might not be suitable for exposure to extreme changes in temperature.

G2414.10.1 (403.10.1) Pipe joints. *Pipe* joints shall be threaded, flanged, brazed or welded. Where nonferrous *pipe* is brazed, the *brazing* materials shall have a melting point in excess of 1,000°F (538°C). *Brazing* alloys shall not contain more than 0.05-percent phosphorus.

❖ Joints in piping materials can be made by four methods: threaded (screwed), brazed, welded and flanged. Flanged joints must also involve threading or welding as the means of attaching the flanges to the pipe. The limit on phosphorus content is related to corrosion that can occur from the reaction of phosphorus with a trace contaminant (sulphur) in the gas (see commentary, Section G2414.10.2).

G2414.10.2 (403.10.2) Tubing joints. *Tubing* joints shall be made with *approved gas tubing* fittings or be brazed with a material having a melting point in excess of 1,000°F (538°C) or made with press-connect fittings complying with ANSI LC-4. *Brazing alloys* shall not contain more than 0.05-percent phosphorus.

❖ Tubing joints can be made by three methods: brazing, tubing fittings and press-connect fittings. Tubing fittings include flare, compression and CSST. Press-connect fittings require special tools that make an irreversible joint (see Commentary Figure G2414.10.2).

Brazing is the act of producing a brazed joint. Brazing is often referred to as silver soldering. Silver soldering is more accurately described as silver brazing and employs high-silver-bearing alloys primarily composed

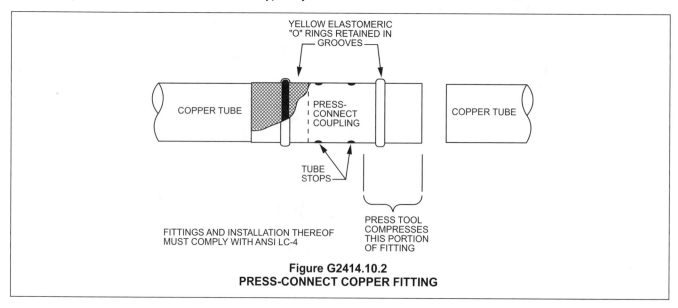

Figure G2414.10.2
PRESS-CONNECT COPPER FITTING

of silver, copper and zinc. Silver soldering (brazing) typically requires temperatures in excess of 1,000°F (538°C), and such solders are classified as "hard" solders (see commentary to "Brazed joint").

Confusion has always been present with respect to the distinction between "silver solder" and "silver-bearing" solder. Silver solders are unique and can be further subdivided into soft and hard categories, which are determined by the percentages of silver and the other component elements of the particular alloy. The distinction is that silver-bearing solders [melting point below 600° F (316°C)] are used in soft-soldered joints, and silver solders [melting point higher than 1,000°F (538°C)] are used in silver-brazed joints.

Brazed joints are considered to have superior strength and stress resistance and, because of the high melting point, are less likely to fail when exposed to fire.

Brazing with a filler metal conforming to AWS A5.8 produces a strong joint that will perform under extreme service conditions. The surfaces to be brazed must be cleaned free of oxides and impurities. Flux should be applied as soon as possible after the surfaces have been cleaned. Flux helps to remove residual traces of oxides, to promote wetting and to protect the surfaces from oxidation during heating. Care should be taken to prevent flux from entering the piping system during the brazing operation because flux that remains may corrode the pipe or contaminate the system.

Air should be removed from the pipe being brazed by purging the piping with a nonflammable gas such as carbon dioxide or nitrogen. Purging the system has several benefits, such as preventing oxidation from occurring on the inside of the pipe. Mechanical joints are usually proprietary joints that are developed and marketed by individual manufacturers. Many types of mechanical joints use a sleeve or ferrule that is compressed around the circumference of the pipe or tube. Mechanical joints must be specifically designed for and compatible with the type of pipe or tube to be joined.

G2414.10.3 (403.10.3) Flared joints. *Flared joints* shall be used only in systems constructed from nonferrous *pipe* and *tubing* where experience or tests have demonstrated that the joint is suitable for the conditions and where provisions are made in the design to prevent separation of the joints.

❖ Flared joints are typically used with copper and aluminum tubing and are prohibited for steel tubing. Flared joints require the use of specialized tools. Because the pipe end is expanded in a flared joint, only annealed and bending tempered (drawn) copper tubing may be flared. Commonly used flaring tools use a screw yoke and block assembly or an expander tool that is driven into the tube with a hammer. The flared tubing end is compressed between a fitting seat and a threaded nut to form a metal-to-metal seal.

G2414.10.4 (403.10.4) Metallic fittings. Metallic fittings, shall comply with the following:

1. Fittings used with steel or wrought-iron *pipe* shall be steel, copper alloy, malleable iron or cast iron.

2. Fittings used with copper or copper alloy *pipe* shall be copper or copper alloy.

3. Cast-iron bushings shall be prohibited.

4. Special fittings. Fittings such as couplings, proprietary-type joints, saddle tees, gland-type compression fittings, and flared, flareless and compression-type *tubing* fittings shall be: used within the fitting manufacturer's pressure-temperature recommendations; used within the service conditions anticipated with respect to vibration, fatigue, thermal expansion and contraction; and shall be *approved*.

5. Where pipe fittings are drilled and tapped in the field, the operation shall be in accordance with all of the following:

   5.1. The operation shall be performed on systems having operating pressures of 5 psi (34.5 kPa) or less.

   5.2. The operation shall be performed by the gas supplier or the gas supplier's designated representative.

   5.3. The drilling and tapping operation shall be performed in accordance with written procedures prepared by the gas supplier.

   5.4. The fittings shall be located outdoors.

   5.5. The tapped fitting assembly shall be inspected and proven to be free of leakage.

❖ Item 1 allows use of cast iron fittings with steel, but cast iron pipe is prohibited. Malleable iron (steel) fittings are used almost exclusively.

Item 2 requires fittings for nonferrous pipes to be of a material consistent with the pipe material.

Item 3 places limitations on cast iron fittings because of the brittle nature of cast iron. Bushings have been known to split from overtightening.

Item 4 speaks of special fittings of the mechanical type.

Item 5 addresses a practice that is allowed only for gas utility (gas supplier) companies.

Occasionally, an installer will drill a hole in a Schedule 40 or heavier pipe fitting and tap threads in the hole for the purpose of making a test connection or branch connection. The pipe fittings were never designed to be drilled and tapped, and the thickness of the fitting wall does not allow for even one complete thread in Schedule 40 fittings. The resulting pipe connection or plugged opening would not have sufficient threads to make a joint that is both strong and free of leakage. Lacking the required number of complete threads, such joints would rely solely on the thread-sealant paste or tape to effect a seal. Normal threaded joints rely on an interference fit of several threads and metal-to-metal sealing, while depending on pipe-joint pastes and tapes only as thread lubricants and for sealing tiny imperfections in the threads. Field drilling and tapping of outdoor fittings on the outlet side of the meter is performed by some gas suppliers (utilities) for a unique purpose. If the gas supplier needs to replace a customer's gas meter, and it is not convenient to shut off the gas supply to the customer in order to replace the meter, the gas supplier may use a special apparatus to tap into the customer's piping near the meter for the purpose of temporarily supplying gas to the system at the same time the meter is isolated, removed and replaced. The gas is temporarily supplied from tanks of compressed gas on the utility's service trucks. After replacing the meter, the tapped opening is plugged. The five conditions in the new text limit this practice to gas-supplier personnel and authorized contractors employed by the gas supplier. The practice is limited to outdoor fittings, and the entire operation must be performed with specialized tools and equipment in accordance with the prescribed procedures. In cases where a leak-free plugged assembly cannot be achieved, the gas supplier will cover the affected area of the fitting with specialized devices or leak-repair methods used in the gas industry.

The drilling and tapping of pipe fittings is prohibited in all cases except where performed in accordance with all five conditions that are intended to mitigate the risk [see Commentary Figures G2414.10.4(1) and (2)].

G2414.11 (403.11) Plastic piping, joints and fittings. Plastic *pipe*, *tubing* and fittings shall be joined in accordance with the manufacturers' instructions. Such joints shall comply with the following:

1. The joints shall be designed and installed so that the longitudinal pull-out resistance of the joints will be at least equal to the tensile strength of the plastic *piping* material.

2. Heat-fusion joints shall be made in accordance with qualified procedures that have been established and proven by test to produce gas-tight joints at least as strong as the *pipe* or *tubing* being joined. Joints shall be made with the joining method recommended by the *pipe* manufacturer. Heat fusion fittings shall be marked "ASTM D2513."

3. Where compression-type *mechanical joints* are used, the gasket material in the fitting shall be compatible with the plastic *piping* and with the gas distributed by the system. An internal tubular rigid stiffener shall be used in conjunction with the fitting. The stiffener shall be flush with the end of the *pipe* or *tubing* and shall extend at least to the outside end of the compression fitting when installed. The stiffener shall be free of rough or sharp edges and shall not be a force fit in the plastic. Split tubular stiffeners shall not be used.

4. Plastic *piping* joints and fittings for use in *liquefied petroleum gas piping systems* shall be in accordance with NFPA 58.

❖ This section contains general requirements for the proper joining of plastic gas pipe materials. Although plastic pipe materials offer many advantages such as ease of installation, corrosion resistance, lighter weight and lower cost, these materials require different handling and installation methods than steel or other metal pipe or tubing. In all cases, the pipe material manufacturer's installation instructions must be strictly adhered to in order to reduce the likelihood of creating improper

**Figure 2414.10.4(1)
DRILLED AND
TAPPED FITTING**

**Figure 2414.10.4(2)
OUTDOOR METER SETTING**

joints, which could lead to gas leakage or piping or joint failure. Dissimilar types of plastic materials may not be joined by heat fusion methods. Because of the resulting reduction in wall thickness, plastic pipe and tubing must not be field-threaded as a joining method. All approved joints must be designed to account for the forces acting upon the piping system, such as expansion, contraction and external loads imposed by burial in underground installations. The longitudinal pull-out resistance of the joint should be approximately the same as the tensile strength of the materials being joined.

Heat-fusion joints for plastic pipe are analogous to the welding of steel pipe. Heat fusion must be performed in accordance with the pipe manufacturer's

instructions. The process involves heating the pipe and fittings with a special iron. When the parts to be joined reach their melting points, they are assembled and allowed to fuse. Electrical fusion fittings are available and employ an integral heating element that is connected to a power supply unit that allows the fitting to be heated at the required rate and duration.

Mechanical compression joints employ an elastomeric seal that must be compatible not only with the piping material but also with the gas that is conveyed in the system. The manufacturers' instructions normally require an internal stiffener insert used in conjunction with the fitting to provide additional support. The stiffener insert prevents the pipe from deforming under the compression force exerted by the seal and compression nut assembly.

Category 1 fittings in accordance with ASTM D2513 are designed to provide a gas-tight seal and resist axial (pull-out) forces equal to the yield strength of the piping material [see Commentary Figures G2414.11(1) through G2414.11(4)].

Photo courtesy of Perfection Corporation

Figure G2414.11(1)
MECHANICAL JOINTS

SECTION G2415 (404)
PIPING SYSTEM INSTALLATION

G2415.1 (404.1) Installation of materials. Materials used shall be installed in strict accordance with the standards under which the materials are accepted and approved. In the absence of such installation procedures, the manufacturer's instructions shall be followed. Where the requirements of referenced standards or manufacturer's instructions do not conform to minimum provisions of this code, the provisions of this code shall apply.

❖ Piping materials and related system components must be installed in accordance with the standards to which

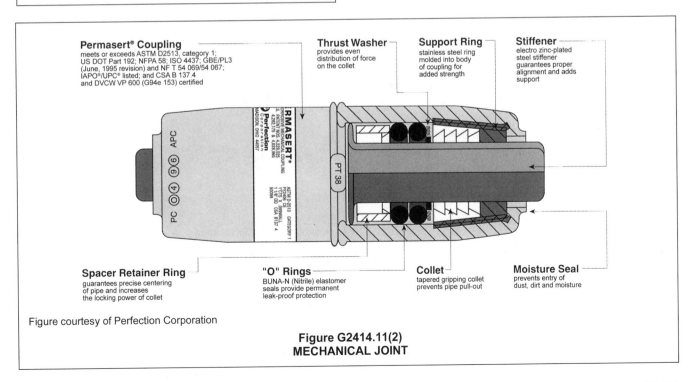

Figure courtesy of Perfection Corporation

Figure G2414.11(2)
MECHANICAL JOINT

they are listed. Improper installation can result in substandard performance, product failures and hazardous conditions. In some cases, the product standard will contain installation requirements, such as found in ANSI LC-1 for CSST. In other cases, the product standards might not contain any installation requirements; therefore, the manufacturer of the product supplies the only installation requirements. The manufacturer could also have installation requirements that are in addition to or in excess of those in the standard or vice versa. If the installation requirements in the product standards or manufacturer's instructions are somehow less stringent than a provision of the code, the code must prevail.

G2415.2 (404.2) CSST. CSST piping systems shall be installed in accordance with the terms of their approval, the conditions of listing, the manufacturer's instructions and this code.

❖ The CSST product standard dictates what the manufacturer's instructions must include at a minimum. This section was added because Section G2414.5.3 only required that CSST be listed and failed to clarify that the CSST system must also comply with the standard, the manufacturer's instructions, the code and any conditions established in the approval of such systems.

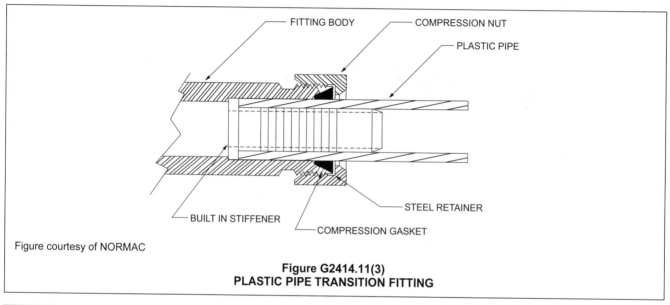

Figure courtesy of NORMAC

Figure G2414.11(3)
PLASTIC PIPE TRANSITION FITTING

NORMAC LOCK Stiffener and Ring

Tightening the nut compresses the gasket against the plastic, gripping its exterior while forming a gas-tight seal.

As the end of the plastic approaches the gasket, the lock ring engages the flange of the stiffener and the brass beads of the gasket, transmitting the internal friction of the grooved stiffener through the gasket to the fitting. The more tensile pull force exerted, the more gasket pressure transferred, thus forming a force link.

Figure courtesy of NORMAC

Figure G2414.11(4)
PLASTIC PIPE TRANSITION FITTING

G2415.3 (404.3) Prohibited locations. *Piping* shall not be installed in or through a ducted supply, return or exhaust, or a clothes chute, *chimney* or gas vent, dumbwaiter or elevator shaft. *Piping* installed downstream of the *point of delivery* shall not extend through any townhouse unit other than the unit served by such *piping*.

❖ An air supply duct or areas that create a shaft that provides a path for the gas to travel are considered to be locations where the problem of a gas leak could be compounded by spreading the gas or any resultant fire throughout the building. Also, in some locations, the gas piping could be subject to deterioration because of temperature or corrosive conditions such as in air ducts or chimneys.

Locating fuel gas piping in certain areas or atmospheres may cause corrosion of the pipe, which in turn may cause a leak in the piping system. Within a supply air duct, the conditioned air may cause moisture to condense on or within the pipe, thereby causing corrosion of the pipe. Fuel-gas piping located within a chimney or a vent will be subjected to high temperatures and the corrosive effects of flue gases.

Additionally, piping located in a dumbwaiter or elevator shaft may be subject to an additional hazard resulting in mechanical damage to the pipe. Damage from a dumbwaiter or elevator impacting the fuel-gas piping may not only cause a leak that would allow the gas to escape and travel up the shaft but may also put any occupants within an elevator in immediate danger. The elevator codes also intend to prohibit nonessential piping within elevator shafts.

Gas piping is not prohibited in concealed spaces used to convey environmental air; for example, above ceiling and below floor plenums, stud cavities and joist spaces. Gas piping is often installed in return air plenums above a suspended ceiling with no history of problems. The piping system is assumed not to leak in a ceiling air plenum or any other location, for that matter. In fact, when viewed logically, it would be more hazardous for a leak to occur in a location where the air is static and an explosive gas-air mixture could be created. Return, supply and exhaust air ducts are not considered to be a plenum. These ducts are supply, return and exhaust air conduits constructed as components of a ductwork system. A furnace plenum is part of a ductwork system and would be treated as an air duct (see the definition of "Furnace plenum").

This section now prohibits what was once a common practice, that is, the practice of placing a bank of gas meters on one end of a townhouse complex and running the load side gas line from each meter through the attic or crawl spaces of the townhouse units to reach the units served. For example, the gas supply line for the unit farthest away from the meter bank will pass through all of the other units to reach the unit it serves. This is a concern because of what might be done to the piping within the units through which it passes and because of the potential for one of the units to be destroyed by fire, thereby damaging the gas piping that passes through the fire walls/separation walls between units. Townhouses are considered to be separate buildings in the code and are constructed to prevent fire from spreading beyond the unit of origin. This is similar to the provision in the code that prevents electrical service entrance conductors from passing through dwelling units other than the unit served (see Commentary Figure G2415.3).

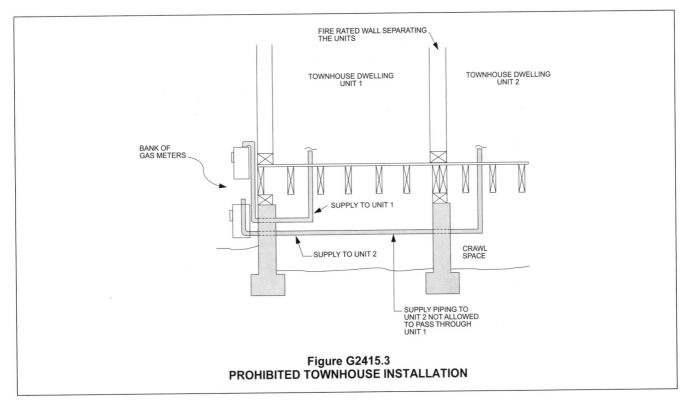

Figure G2415.3
PROHIBITED TOWNHOUSE INSTALLATION

G2415.4 (404.4) Piping in solid partitions and walls. *Concealed piping* shall not be located in solid partitions and solid walls, unless installed in a chase or casing.

❖ As with the alternative installation requirements in Section G2415.8, this section allows installation of gas piping within solid walls or partitions only if the piping is installed within a chase or casing to protect the pipe from stress and from corrosive effects of wall materials such as concrete.

G2415.5 (404.5) Fittings in concealed locations. Fittings installed in concealed locations shall be limited to the following types:

1. Threaded elbows, tees and couplings.

2. Brazed fittings.

3. Welded fittings.

4. Fittings *listed* to ANSI LC-1/CSA 6.26 or ANSI LC-4.

❖ A concealed location is a location that requires the removal of permanent construction in order to gain access (see the definition of "Concealed location"). The space above a dropped ceiling having readily removable lay-in panels or other locations that have removable access panels are not considered concealed locations for the purposes of this section. Concealed locations include wall, floor and ceiling cavities bounded by permanent finish materials such as gypsum board, masonry or paneling. This section lists what is allowed to be concealed; therefore, by omission, all other types of fittings are not allowed in concealed locations. Note that threaded plugs and caps are not listed, but, the author is not aware of any reason to prohibit them. The National Fuel Gas Code Z223.1, 2015 edition, has been revised to specifically allow threaded plugs and caps in concealed locations. Unions and mechanical joint tubing fittings are not permitted in concealed locations because they are more likely to loosen and leak than other joining means and fittings. Tubing fittings include flare, compression and similar proprietary-type fittings, all of which are mechanical joints.

Joints for CSST systems are mechanical tubing joints; however, these joints use specialized proprietary fittings that are listed for concealment as part of the CSST system. Such fittings undergo stringent testing to determine their ability to remain leak-tight.

Right and left couplings are somewhat archaic and were used as a union is now used. The coupling is made with right-hand threads on one end and left-hand threads on the other, and the corresponding male pipe threads can be used to connect pipe and fittings within a section of existing piping. The code does not specifically prohibit right and left couplings, as it refers only to "couplings" generically; however, if a right and left coupling was concealed, people would not be aware of such a fitting in a piping system and would not realize that they were loosening a joint as they applied the traditional tightening torque to an exposed section of piping. Piping should always be backed up with wrenches to prevent rotational forces from extending beyond the work piece. When the code refers to couplings, those

fittings are tapered-thread fittings. See Section G2414.9. Bushings have been known to split after assembly and are not trusted in concealed locations.

Brazing is a commonly used method of joining copper tubing and is as reliable as welded joints for steel pipe.

This section does not prohibit NPT threaded (screwed) elbow, tee and coupling fittings from being concealed; however, it does prohibit threaded unions and bushings from being concealed.

G2415.6 (404.6) Underground penetrations prohibited. Gas *piping* shall not penetrate building foundation walls at any point below grade. Gas *piping* shall enter and exit a building at a point above grade and the annular space between the *pipe* and the wall shall be sealed.

❖ The preferred and required method for bringing outdoor gas piping into or out of a building is to penetrate the exterior wall at a point above grade. In other words, gas piping that comes into or leaves a building should not provide a path for underground gas leakage to enter the interior of a building. This section does not apply to utility-company-owned piping upstream of the point of delivery. See Section G2401.1.

The penetration of the pipe and any protective sleeve must be adequately sealed to prevent the possibility of water or insect entry into the building. A gas pipe is not permitted to enter or exit a building below grade because any gas leakage in the underground piping will be channeled into the building through the surrounding soil and the wall penetration. Gas leakage can travel in the annular space that exists between the outside wall of the pipe and the soil that surrounds it. With an above-ground wall penetration, any outdoor piping leakage will vent to the outdoors with very little possibility of it finding its way into the building. Underground penetrations of building foundations are very difficult to seal against gas migration and the best way to eliminate the possibility of gas entering the building is to simply require the piping to enter and exit above grade. There have been cases of explosions that were believed to be caused by gas entering basements and crawl spaces through underground wall penetrations (see Commentary Figure G2415.6).

G2415.7 (404.7) Protection against physical damage. Where *piping* will be concealed within light-frame construction assemblies, the *piping* shall be protected against penetration by fasteners in accordance with Sections G2415.7.1 through G2415.7.3.

Exception: Black steel *piping* and galvanized steel *piping* shall not be required to be protected.

❖ This section is intended to minimize the possibility that nails or screws will be driven into the gas pipe or tube. Because nails and screws sometimes miss the stud, rafter, joist or sole or top plates, the shield plates must extend parallel to the pipe or tube not less than 4 inches (102 mm) beyond the member on each side or not less than 4 inches (102 mm) above or below sole or top wall plates, respectively. Commentary Figures

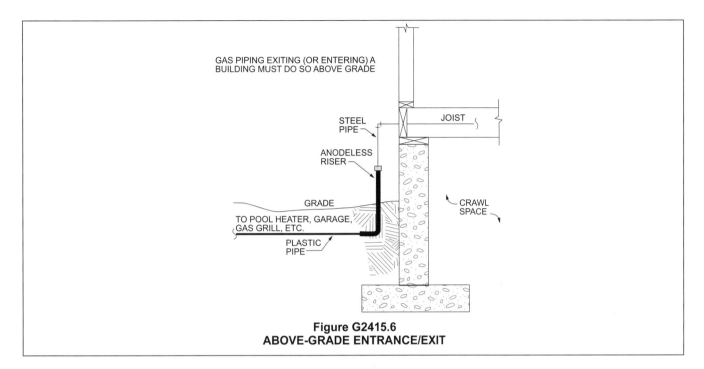

Figure G2415.6
ABOVE-GRADE ENTRANCE/EXIT

G2415.7 and G2414.5.3(6) show typical shield plates. Black and galvanized steel pipe (Schedule 40) each have wall thicknesses greater than the required thickness for the shield plates, which makes these piping materials inherently resistant to nail and screw penetrations. This section does not apply to CSST tubing, because CSST systems have their own protection requirements dictated by the manufacturer's instructions and ANSI LC-1/CSA 6.26. The protection requirements for CSST systems are, overall, more stringent than those of this section. Before studs, rafters, joists or other structural members are drilled or notched, the building portion of this code should be consulted.

The section on protection of piping has been completely rewritten to address more than just bored holes and notches in structural members. It now addresses piping parallel to framing members and piping within framing members. The new text requires that the protection extend well beyond the edge of members that are bored or notched.

Piping and tubing other than Schedule 40 steel pipe must be protected from penetration by nails and screws where the pipe or tubing is less than $1^1/_2$ inches from the face of the member where membranes will be attached. This protection is necessary whether the pipe or tube is perpendicular or parallel to the framing member. If a pipe or tube is run inside of the "C" channel of a $3^1/_2$-inch metal stud, it will almost certainly be penetrated by screws unless the pipe or tube is $^1/_2$ inch or less in diameter and located dead center in the stud channel. Where pipes and tubing are attached to and run parallel with the side of a framing member, penetration by a nail or screw is likely if the installer misses the member with the nail or screw and the pipe or tube is less than $1^1/_2$ inches from either face of the stud.

Extending the protection shield 4 inches beyond the edges of the framing member is intended to protect against fasteners that miss the member or that exit the member on an angle. To avoid having protection plates run parallel with a member, the installer could simply place the pipe or tube on "standoffs" such that the pipe/tube is not less than $1^1/_2$ inches from the nearest edge of the face of the member. As always, a designer's careful planning of the routing of gas piping or tubing can avoid the need for installation of protection plates by simply relying on a distance of at least $1^1/_2$ inches between the pipe/tubing and the fastener face of the member. Pipes and tubes are sometimes penetrated by screws or nails, and a leak does not develop immediately. It may take years for the fastener to corrode enough for a leak to eventually develop.

This section pertains to piping and tubing that will be concealed within wood or steel light-frame construction assemblies, which is the same scope as the original text.

G2415.7.1 (404.7.1) Piping through bored holes or notches. Where *piping* is installed through holes or notches in framing members and the *piping* is located less than $1^1/_2$ inches (38 mm) from the framing member face to which wall, ceiling or floor membranes will be attached, the pipe shall be protected by shield plates that cover the width of the pipe and the framing member and that extend not less than 4 inches (51 mm) to each side of the framing member. Where the framing member that the *piping* passes through is a bottom plate, bottom track, top plate or top track, the shield plates shall cover the framing member and extend not less than 4 inches (51 mm) above the bottom framing member and not less than 4 inches (51 mm) below the top framing member.

❖ See the commentary to Section G2415.7.

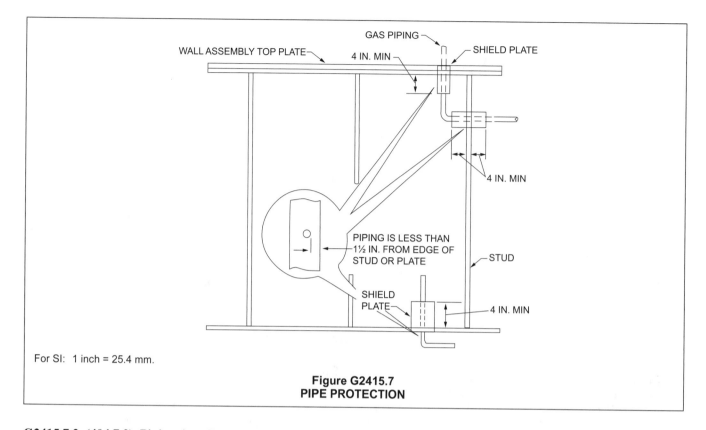

For SI: 1 inch = 25.4 mm.

Figure G2415.7
PIPE PROTECTION

G2415.7.2 (404.7.2) Piping installed in other locations. Where the *piping* is located within a framing member and is less than 1¹/₂ inches (38 mm) from the framing member face to which wall, ceiling or floor membranes will be attached, the *piping* shall be protected by shield plates that cover the width and length of the *piping*. Where the *piping* is located outside of a framing member and is located less than 1¹/₂ inches (38 mm) from the nearest edge of the face of the framing member to which the membrane will be attached, the *piping* shall be protected by shield plates that cover the width and length of the *piping*.

❖ This section addresses piping that is run along the side of framing members and also piping that is run inside the channel of metal framing members. Where the piping is within 1¹/₂ inches of the face of the framing member where fasteners will penetrate the member, there is the danger that fasteners will miss the target or be driven at an angle such that the piping is penetrated. Piping is sometimes run within the "C" channel of metal studs, which practically begs for the pipe/tube to be penetrated by fasteners that attach wall sheathing to the framing members (see commentary, Section G2415.7).

2415.7.3 (404.7.3) Shield plates. Shield plates shall be of steel material having a thickness of not less than 0.0575 inch (1.463 mm) (No. 16 gage).

❖ See the commentary to Section G2415.7.

G2415.8 (404.8) Piping in solid floors. *Piping* in solid floors shall be laid in channels in the floor and covered in a manner that will allow access to the *piping* with a minimum amount of damage to the building. Where such *piping* is subject to exposure to excessive moisture or corrosive substances, the *piping* shall be protected in an *approved* manner. As an alternative to installation in channels, the *piping* shall be installed in a conduit of Schedule 40 steel, wrought iron, PVC or ABS pipe in accordance with Section G2415.6.1 or G2415.6.2.

❖ Piping must not be installed in any solid concrete or masonry floor construction. The potential for pipe damage from slab settlement, cracking or the corrosive action of the floor material makes it imperative that one of the installation methods in this section be used. This section does not intend to allow any direct encasement of gas piping in solid concrete or masonry floor systems.

Gas piping installed within a solid floor system must be safeguarded by installation in a sealed casing or in a floor channel with a removable cover for pipe access (see Commentary Figure G2415.8). Either of these methods should provide reasonable protection of the pipe from the effects of settling, cracking and being in contact with corrosive materials.

Because casings constructed of metal could corrode where installed within a concrete slab on grade, consideration should be given to corrosion protection for the steel casing, or an alternate material should be chosen.

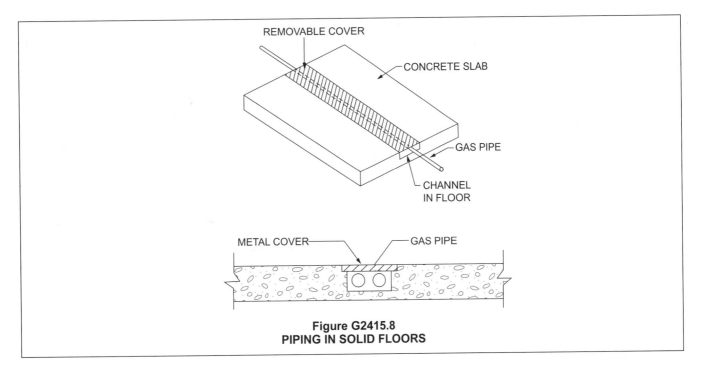

Figure G2415.8
PIPING IN SOLID FLOORS

G2415.8.1 (404.8.1) Conduit with one end terminating outdoors. The conduit shall extend into an occupiable portion of the building and, at the point where the conduit terminates in the building, the space between the conduit and the *gas piping* shall be sealed to prevent the possible entrance of any gas leakage. The conduit shall extend not less than 2 inches (51 mm) beyond the point where the *pipe* emerges from the floor. If the end sealing is capable of withstanding the full pressure of the gas *pipe,* the conduit shall be designed for the same pressure as the *pipe.* Such conduit shall extend not less than 4 inches (102 mm) outside the building, shall be vented above grade to the outdoors and shall be installed to prevent the entrance of water and insects.

❖ This method could be used in any application, but is more logically applied where piping enters the building from the outdoors or exits the building to the outdoors. The intent is to direct any gas leakage to the outdoors by means of the secondary containment conduit and the outdoor vent terminal.

  Note that the application of this section is extremely limited because of Section G2415.6, which prohibits gas piping from entering or exiting a building below grade. Clearly, Section G2415.8.1 was originally intended to address installations involving gas piping that entered or exited a building; however, this section now has only one application and that would be for gas piping that runs from point to point all within the building. The requirements of this section are more stringent than those of Section G2415.8.2; therefore, it is not likely that an installer would choose the more difficult method prescribed in this section. Because of Section G2415.6, the methods prescribed in both Sections G2415.8.1 and G2415.8.2 are applicable only to gas piping that does not enter or exit the building. Note that this discussion also applies to Sections G2415.14.1 and G2415.14.2.

G2415.8.2 (404.8.2) Conduit with both ends terminating indoors. Where the conduit originates and terminates within the same building, the conduit shall originate and terminate in an accessible portion of the building and shall not be sealed. The conduit shall extend not less than 2 inches (51 mm) beyond the point where the pipe emerges from the floor.

❖ This method allows piping to run between any two points within the building without a vent run to the outdoors and without sealing the ends of the conduit. The minimum rise of 2 inches (51 mm) above the floor is intended to prevent liquids and debris from falling into the conduit. The ends of the conduit are not to be sealed so as to allow any leakage to be noticed by the occupants, the same as leakage would be noticed from piping located anywhere else in the structure. If the conduit ends were sealed, there would be no evidence of a piping defect, no matter how severe the defect, possibly leading to a major leak in the event of conduit or end-seal failure (see Commentary Figure G2415.8.2).

G2415.9 (404.9) Above-ground piping outdoors. *Piping* installed outdoors shall be elevated not less than 3¹/₂ inches (152 mm) above ground and where installed across roof surfaces, shall be elevated not less than 3¹/₂ inches (152 mm) above the roof surface. *Piping* installed above ground, outdoors, and installed across the surface of roofs shall be securely supported and located where it will be protected from physical damage. Where passing through an outside wall, the *piping* shall also be protected against corrosion by coating or wrapping with an inert material. Where *piping* is encased in a protective pipe sleeve, the annular space between the *piping* and the sleeve shall be sealed.

❖ Gas piping in any location must be properly supported and protected from physical damage. Protection from damage is especially important where piping is run outdoors near grade or across roof surfaces. See Sections G2418 and G2424. Piping passing through an outside wall must be protected where the material of the wall could corrode or abrade the piping (concrete, masonry or stucco, for example). To help protect piping from corrosion resulting from exposure to moisture, the piping must always be located at least $3^1/_2$ inches (89 mm) above the earth and roof surfaces. The distance of $3^1/_2$ inches (89 mm) was chosen because it corresponds to the size of nominal 4-inch by 4-inch (102 mm by 102 mm) pressure-treated lumber which is commonly used to support piping run across roof surfaces.

Where piping is supported on 4-inch by 4-inch (102 mm by 102 mm) wood blocks or similar supports, the piping should be attached to the blocking with suitable clamps or straps. It is common to see wood blocking or piping that has been displaced because the piping was not properly fastened to the blocking. Section G2424.1 dictates the minimum required spacing for blocking.

G2415.10 (404.10) Isolation. Metallic *piping* and metallic *tubing* that conveys *fuel gas* from an LP-gas storage container shall be provided with an *approved* dielectric fitting to electrically isolate the underground portion of the pipe or tube from the above ground portion that enters a building. Such dielectric fitting shall be installed above ground outdoors.

❖ Such isolation is common practice for natural gas utility service risers. The required isolation fitting is intended to isolate the piping/tubing in the building from piping extending underground. This section does not require isolation between the LP-gas storage tank and the piping/tubing running from it to below ground. In other words, in the typical installation, piping/tubing enters the soil near the tank and rises out of the soil near the building served and the isolation is required where the piping/tubing rises to enter the building. This section is related to Sections G2410 and G2411. There has been concern raised about bonding metal gas piping/tubing that extends underground and this section is intended to alleviate such concern (see Commentary Figure G2415.10).

G2415.11 (404.11) Protection against corrosion. Metallic pipe or *tubing* exposed to corrosive action, such as soil condition or moisture, shall be protected in an *approved* manner. Zinc coatings (galvanizing) shall not be deemed adequate protection for *gas piping* underground. Where dissimilar metals are joined underground, an insulating coupling or fitting shall be used. *Piping* shall not be laid in contact with cinders.

❖ See the commentary to Section G2414.8.

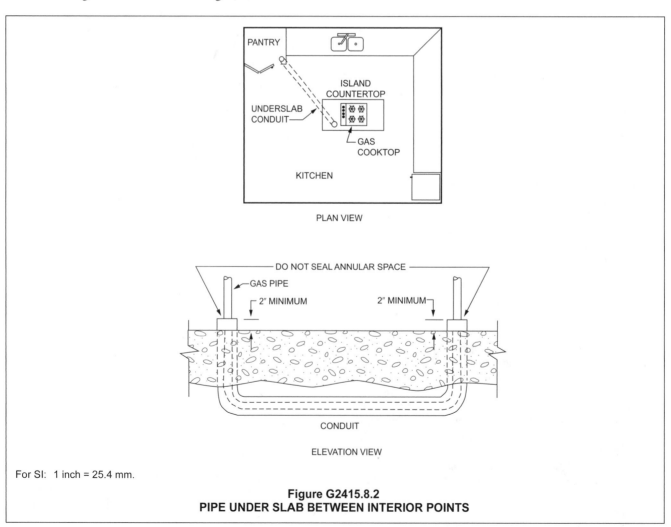

For SI: 1 inch = 25.4 mm.

Figure G2415.8.2
PIPE UNDER SLAB BETWEEN INTERIOR POINTS

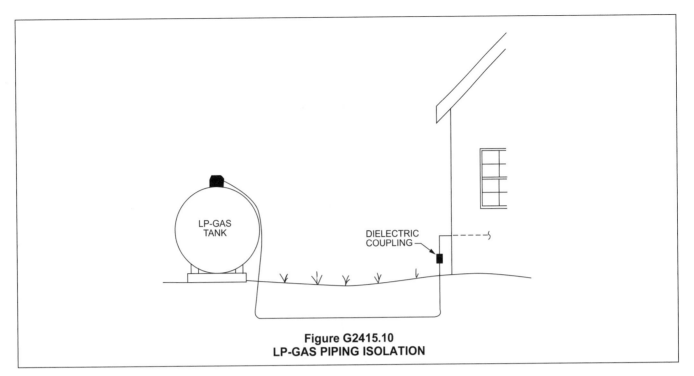

Figure G2415.10
LP-GAS PIPING ISOLATION

G2415.11.1 (404.11.1) Prohibited use. Uncoated threaded or socket-welded joints shall not be used in *piping* in contact with soil or where internal or external crevice corrosion is known to occur.

❖ Pits, crevices and other surface imperfections can create concentrated corrosion cells where the corrosion effect is amplified. Pipe or fitting failure could occur more quickly than if the corrosion was uniformly distributed over the entire surface of the pipe or fitting (see commentary, Section G2415.9.2).

G2415.11.2 (404.11.2) Protective coatings and wrapping. Pipe protective coatings and wrappings shall be *approved* for the application and shall be factory applied.

Exception: Where installed in accordance with the manufacturer's instructions, field application of coatings and wrappings shall be permitted for pipe nipples, fittings and locations where the factory coating or wrapping has been damaged or necessarily removed at joints.

❖ Protective coatings in the form of wrappings, tapes, enamels, epoxies, sleeves, casings and factory-applied coverings are a common method of isolating metallic piping materials from the atmosphere or from the earth in which they are buried. Corrosion can be caused by soil or construction materials that are in contact with the pipe. Protection is usually provided by a factory-applied coating or by field wrapping the pipe with a protective covering such as a coal-tar-based or plastic wrapping. Where possible, a piping material is chosen that is not subject to the type of corrosion typical of the application. Application of these materials under strictly controlled factory conditions improves

the quality control process and reduces the likelihood of coating defects. In the event that coatings or wrappings must be applied in the field, the exception states that they must be installed in accordance with the manufacturer's installation instructions. Coatings and wrappings should be applied only by persons trained and experienced in such work in order to improve reliability of the coating/wrapping process. Field application is permitted only in cases where the factory-applied material may have been damaged during transit or installation of the pipe, where the applied coating was removed for pipe welding or threading of the pipe or for short sections (nipples) of pipe used in the installation. Corrosion can be concentrated on the pipe or fitting where there are flaws in the coating or wrapping; therefore, improperly applied coatings and wrappings can cause pipe or fitting failures to occur faster than failures would occur in unprotected metals. It is evident why plastic piping is used extensively in underground installations.

G2415.12 (404.12) Minimum burial depth. Underground *piping systems* shall be installed a minimum depth of 12 inches (305 mm) below grade, except as provided for in Section G2415.12.1.

❖ The depth of 12 inches (305 mm) is considered to be sufficient to avoid possible harm to the pipe from the use of hand tools such as spades and shovels. However, if the piping is located in an area subject to surface loads such as vehicular traffic, the 12-inch (305 mm) depth may not be sufficient to protect the piping from those loads (see Commentary Figure G2415.12).

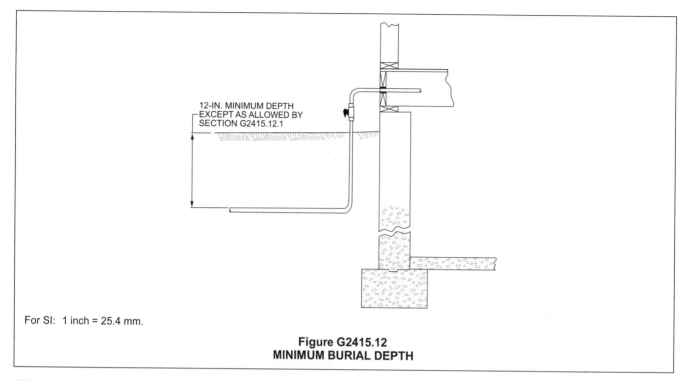

For SI: 1 inch = 25.4 mm.

Figure G2415.12
MINIMUM BURIAL DEPTH

G2415.12.1 (404.12.1) Individual outside appliances. Individual lines to outdoor lights, grills or other *appliances* shall be installed not less than 8 inches (203 mm) below finished grade, provided that such installation is *approved* and is installed in locations not susceptible to physical damage.

❖ Gas piping may be installed within 8 inches (203 mm) of the ground surface where it serves individual outdoor appliances and is not likely to be subjected to damage such as might occur from vehicular traffic, gardening, future excavation, etc. Each individual installation must be reviewed and approved by the code official. The intent is to allow shallow installations for small distribution lines that serve outdoor gas lights, cooking appliances, pool heaters and similar loads only where the piping is unlikely to be disturbed.

G2415.13 (404.13) Trenches. The trench shall be graded so that the pipe has a firm, substantially continuous bearing on the bottom of the trench.

❖ Where trenches have nonuniform depth or peaks and valleys in the bottom, the piping could lack continuous support and could be subjected to stresses from the backfill and surface loads.

G2415.14 (404.14) Piping underground beneath buildings. *Piping* installed underground beneath buildings is prohibited except where the *piping* is encased in a conduit of wrought iron, plastic pipe, steel pipe or other *approved* conduit material designed to withstand the superimposed loads. The conduit shall be protected from corrosion in accordance with Section G2415.11 and shall be installed in accordance with Section G2415.14.1 or G2415.14.2.

❖ This section prohibits the installation of gas piping beneath buildings to reduce the potential for an inaccessible pipe failure caused by settling of the structure.

The prohibition also reduces the potential for corrosion-caused failure of piping embedded in soil or fill. See Section G2415.8 for gas pipe installations in floor slabs. Where underground installation under a building is unavoidable, the piping must be encased in another pipe (conduit) to act as a secondary containment system or as a protective shield from the fill under the floor (see Commentary Figure G2415.8.2).

G2415.14.1 (404.14.1) Conduit with one end terminating outdoors. The conduit shall extend into an occupiable portion of the building and, at the point where the conduit terminates in the building, the space between the conduit and the *gas piping* shall be sealed to prevent the possible entrance of any gas leakage. The conduit shall extend not less than 2 inches (51 mm) beyond the point where the *pipe* emerges from the floor. Where the end sealing is capable of withstanding the full pressure of the gas pipe, the conduit shall be designed for the same pressure as the pipe. Such conduit shall extend not less than 4 inches (102 mm) outside the building, shall be vented above grade to the outdoors and shall be installed so as to prevent the entrance of water and insects.

❖ See the commentary to Section G2415.8.1.

G2415.14.2 (404.14.2) Conduit with both ends terminating indoors. Where the conduit originates and terminates within the same building, the conduit shall originate and terminate in an accessible portion of the building and shall not be sealed. The conduit shall extend not less than 2 inches (51 mm) beyond the point where the pipe emerges from the floor.

❖ See the commentary to Section G2415.8.2.

G2415.15 (404.15) Outlet closures. Gas *outlets* that do not connect to *appliances* shall be capped gas tight.

Exception: *Listed* and *labeled* flush-mounted-type quick-disconnect devices and *listed* and *labeled* gas *convenience* *outlets* shall be installed in accordance with the manufacturer's instructions.

❖ Unused fuel gas outlets must be capped or plugged gas tight, regardless of whether a shutoff valve is installed at the outlet. A closed valve alone is not dependable and poses an unnecessary risk of leakage, accidental opening and vandalism. The exception recognizes that listed gas outlet devices are built with an inherent safety feature that will automatically shut off the gas flow if the mating connector is disengaged or that will not allow the appliance connector to be disengaged until the integral shutoff valve is manually closed.

G2415.16 (404.16) Location of outlets. The unthreaded portion of *piping outlets* shall extend not less than 1 inch (25 mm) through finished ceilings and walls and where extending through floors, outdoor patios and slabs, shall not be less than 2 inches (51 mm) above them. The *outlet* fitting or *piping* shall be securely supported. *Outlets* shall not be placed behind doors. *Outlets* shall be located in the room or space where the *appliance* is installed.

Exception: *Listed* and *labeled* flush-mounted-type quick-disconnect devices and *listed* and *labeled gas convenience outlets* shall be installed in accordance with the manufacturer's instructions.

❖ This section regulates the location, installation and termination of gas piping system outlets to reduce the likelihood of physical damage and to provide clearances for the use of tools. When making connections to piping outlets, "back-up" wrenches are used to prevent piping from rotating, loosening or being damaged. Sufficient pipe length is necessary to allow the application of tools. Gas outlet devices that have been tested and labeled for installation methods other than those addressed in this section must be installed in accordance with the terms of their testing and listing as contained in the manufacturer's installation instructions. Gas outlets for CSST systems must be installed with the termination fitting designed specifically for that purpose and provided by the CSST manufacturer.

G2415.17 (404.17) Plastic pipe. The installation of plastic *pipe* shall comply with Sections G2415.17.1 through G2415.17.3.

❖ This section places restrictions on the location and operating pressures for plastic pipe in addition to stating installation requirements specific to plastic pipe.

G2415.17.1 (404.17.1) Limitations. Plastic pipe shall be installed outdoors underground only. Plastic pipe shall not be used within or under any building or slab or be operated at pressures greater than 100 psig (689 kPa) for natural gas or 30 psig (207 kPa) for LP-gas.

Exceptions:

1. Plastic pipe shall be permitted to terminate above ground outside of buildings where installed in premanufactured *anodeless risers* or service head adapter risers that are installed in accordance with the manufacturer's instructions.

2. Plastic pipe shall be permitted to terminate with a wall head adapter within buildings where the plastic pipe is inserted in a *piping* material for *fuel gas* use in buildings.

3. Plastic pipe shall be permitted under outdoor patio, walkway and driveway slabs provided that the burial depth complies with Section G2415.10.

❖ Because of the potential hazard associated with the use of a material that has lower resistance to physical damage and heat as compared to metallic pipe, installation of plastic pipe and tubing is limited to areas that are both outside of the building and underground. Plastic pipe and tubing are widely used for underground gas distribution systems because of their ease of installation and inherent resistance to corrosion. Polyethylene (PE) and polyamide (PA) SDR 11 pipe are the only allowable plastic pipes for use with LP-gas, and the code user is directed to the referenced standard, NFPA 58, which requires that these piping materials comply with ASTM D2513. The exception makes it clear that if an equivalent level of physical protection is installed, plastic pipe may terminate above ground outside of the building. It is common practice for gas utility companies to install premanufactured riser assemblies at their meter settings. Such risers typically consist of a steel pipe with a corrosion-resistant coating in which a length of PE pipe is preinstalled for coupling to the service lateral. A 90-degree (1.57 rad) sweeping bend is built into the underground end of the riser assembly. Plastic piping is not allowed in or under a concrete slab except as permitted by Exception 3. It is not the intent of this section to prevent an outdoor slab from being placed over customer-owned (private) underground gas lines, such as in the case of walkways, driveways and open patios. The prohibition on piping within a slab is applicable to all slabs without exception. Exceptions 1 and 2 relate to Section G2414.6.1; see the commentary to that section (see Commentary Figure G2415.17.1).

G2415.17.2 (404.17.2) Connections. Connections outdoors and underground between metallic and plastic *piping* shall be made only with transition fittings conforming to ASTM D2513 Category I or ASTM F1973.

❖ Because installation of plastic piping is allowed only where both outside and underground, the same requirement holds for joining plastic and metallic piping. Mechanical compression joints employ an elastomeric seal that must be compatible not only with the piping material but also with the gas that is conveyed in the system. The manufacturer's instructions normally require use of an internal stiffener insert in conjunction with the fitting for additional support. The stiffener insert prevents the pipe from deforming under the compression force exerted by the seal and compression nut assembly.

Category 1 fittings are designed to make a gas-tight seal and resist axial (pull-out) forces equal to the yield strength of the piping material [see Commentary Figure G2414.11(3)].

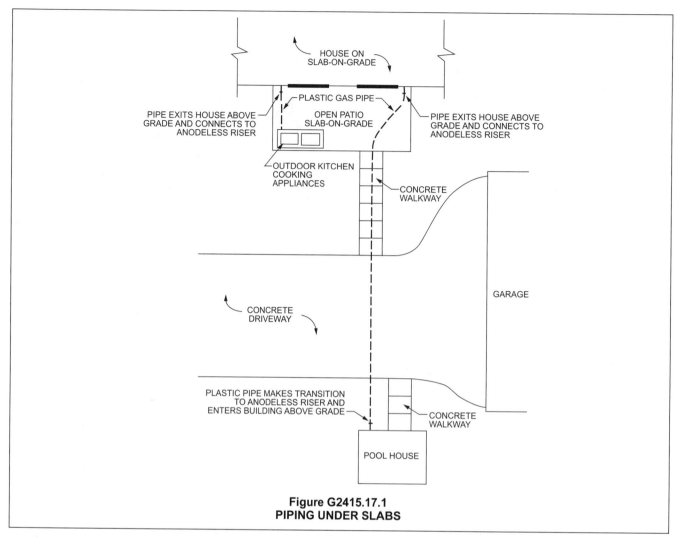

Figure G2415.17.1
PIPING UNDER SLABS

G2415.17.3 (404.17.3) Tracer. A yellow insulated copper tracer wire or other *approved* conductor shall be installed adjacent to underground nonmetallic *piping*. *Access* shall be provided to the tracer wire or the tracer wire shall terminate above ground at each end of the nonmetallic *piping*. The tracer wire size shall not be less than 18 AWG and the insulation type shall be suitable for direct burial.

❖ To avoid piping damage and a hazardous condition, the location of underground piping must be known prior to excavation work in the vicinity of such piping.

In the past, gas utility companies, installers and contractors relied on metal detectors to locate gas piping underground. With the advent of nonmetallic (plastic) piping, this is no longer possible. A tracer wire provides a means for locating nonmetallic gas piping without having to excavate, thereby avoiding the possibility of damaging the pipe. An electrical current can be passed through the wire to allow a metal detector to locate the piping. Insulated wire is required because copper wire or other approved conductors may be susceptible to corrosion in some soils.

Yellow is the standardized color used to identify gas piping. Specialized tracer systems made of proprietary materials are available. Tapes that contain a conduct-

ing strip and also serve as a visual warning are also available.

G2415.18 (404.18) Pipe cleaning. The use of a flammable or combustible gas to clean or remove debris from a *piping* system shall be prohibited.

❖ Such cleaning operations were once used on industrial sites using high-pressure fuel gas. Regardless of the location, fuel gas must never be used as the means of flushing debris from gas piping because the discharge of gas from the piping poses a serious threat of explosion.

G2415.19 (404.19) Prohibited devices. A device shall not be placed inside the *piping* or fittings that will reduce the cross-sectional area or otherwise obstruct the free flow of gas.

Exceptions:

1. Approved gas filters.

2. An approved fitting or device where the gas piping system has been sized to accommodate the pressure drop of the fitting or device.

❖ Devices such as filters and flow-measuring instruments could create considerable pressure drop and are prohibited except where the piping system is designed to tolerate such losses. An unaccounted for restriction in the piping could make an otherwise properly designed system fail to supply the required pressure. This section is not intended to apply to shutoff valves and EFV devices because it specifically addresses devices installed in the interior of piping as opposed to devices installed "in" piping systems. Exception 2 addresses devices such as EFVs that have internal components that can be viewed as being placed inside of pipe or fittings. An EFV will act as a restriction in the piping and the resultant pressure loss from this restriction must be considered to determine if such loss is tolerable and that the load will be supplied with the required minimum inlet pressure.

G2415.20 (404.20) Testing of piping. Before any system of *piping* is put in service or concealed, it shall be tested to ensure that it is gas tight. Testing, inspection and purging of *piping systems* shall comply with Section G2417.

❖ A pressure test is required after every installation, alteration, addition or repair to the fuel-gas piping system. The location of a leak may be difficult to determine, especially if it is concealed in the building construction. If a leak is found, the leaking component must be repaired or replaced before the system is concealed or put into operation. Section G2417 specifies testing pressures based on the type of system, the design working pressure or other parameters. The piping system must sustain the test pressure for the test duration period without exhibiting any sign of leakage.

SECTION G2416 (405)
PIPING BENDS AND CHANGES IN DIRECTION

G2416.1 (405.1) General. Changes in direction of pipe shall be permitted to be made by the use of fittings, factory bends or field bends.

❖ This section permits making changes in direction of gas piping through the use of pipe fittings compatible with the piping material or, if recommended by the pipe manufacturer, by bending the pipe. If bending is the chosen method, it must be accomplished in accordance with Sections G2416.2 and G2416.3. In practice, steel gas piping is rarely bent to accomplish changes in direction because pipe failure at a seam is a possibility for steel pipe that is bent.

G2416.2 (405.2) Metallic pipe. Metallic pipe bends shall comply with the following:

1. Bends shall be made only with bending tools and procedures intended for that purpose.

2. All bends shall be smooth and free from buckling, cracks or other evidence of mechanical damage.

3. The longitudinal weld of the pipe shall be near the neutral axis of the bend.

4. Pipe shall not be bent through an arc of more than 90 degrees (1.6 rad).

5. The inside radius of a bend shall be not less than six times the outside diameter of the pipe.

❖ Pipe is bent to accomplish changes in direction without the use of fittings. Pipe must be bent with the appropriate bending tools and materials. Some materials are not tempered or intended to be bent; therefore, consideration must be given to choosing the proper materials. Bending tools are designed to make bends without damaging the pipe. Pipe intended to be bent has a minimum bend radius which must be observed to avoid damage to the pipe. For example, bending welded seam pipe with the seam located outside of the neutral axis of the bend may result in a split seam because of the stresses induced from the bend. In other words, the welded seam must not be oriented in a bend that would place it in tension or compression. If pipe is to be bent, it should be stated in the pipe specifications that the pipe is suitable for bending. Rigid gas piping is not commonly bent because of the perceived risk of pipe stress failures at the bend and because bending can be more labor intensive than using fittings.

G2416.3 (405.3) Plastic pipe. Plastic pipe bends shall comply with the following:

1. The pipe shall not be damaged and the internal diameter of the pipe shall not be effectively reduced.

2. Joints shall not be located in pipe bends.

3. The radius of the inner curve of such bends shall not be less than 25 times the inside diameter of the pipe.

4. Where the *piping* manufacturer specifies the use of special bending tools or procedures, such tools or procedures shall be used.

❖ Because of the material characteristics of plastic pipe, the code defers to the manufacturer's installation and bending instructions to achieve the performance level contemplated by this section. For example, coiled plastic pipe and tubing have limitations regarding bending beyond or against the natural curvature of the coil. A joint in a bend would be subject to bending stresses and pipe/fitting misalignments that could cause joint structural failure and/or leakage. Thus, a 1-inch (25.4 mm) ID plastic pipe would have a minimum bending radius of 25 inches (635 mm) measured to the inner curve of the bend.

SECTION G2417 (406)
INSPECTION, TESTING AND PURGING

G2417.1 (406.1) General. Prior to acceptance and initial operation, all *piping* installations shall be visually inspected and pressure tested to determine that the materials, design, fabrication and installation practices comply with the requirements of this code.

❖ Before any gas piping system is put to use, it must be inspected, tested and approved. See Sections G2417.1.1 through G2417.6.4. Pressure testing is distinct from leakage checking as covered in Section G2417.6.

G2417.1.1 (406.1.1) Inspections. Inspection shall consist of visual examination, during or after manufacture, fabrication, assembly or *pressure tests.*

❖ The inspection of piping installations is intended to be a visual observation of the system and the testing procedure. The designer may require that welded joints in, for example, very high pressure applications be examined by a method that is capable of discovering internal defects that are not detectable by visual observation.

G2417.1.2 (406.1.2) Repairs and additions. In the event repairs or additions are made after the *pressure test*, the affected *piping* shall be tested.

Minor repairs and additions are not required to be *pressure tested* provided that the work is inspected and connections are tested with a noncorrosive leak-detecting fluid or other *approved* leak-detecting methods.

❖ Some time after the initial test of the piping, a repair could be made or a branch could be added or extended. The repaired, added or extended portion of the piping system needs to be tested without requiring the unchanged portions to be tested again. Minor work on an existing system is allowed without pressure testing the minor work if the work is visually inspected and tested by a leak-detecting method. Leak-detecting methods include bubble test fluids and electronic sensors. Some bubble test fluids could be corrosive to some piping; therefore, a noncorrosive fluid designed for this purpose should always be used. Methods other than bubble test fluids must be acceptable to the code official.

G2417.1.3 (406.1.3) New branches. Where new branches are installed to new *appliances*, only the newly installed branches shall be required to be *pressure tested.* Connections between the new *piping* and the existing *piping* shall be tested with a noncorrosive leak-detecting fluid or other *approved* leak-detecting methods.

❖ This section applies to all new branches regardless of where in the piping system they connect. Often, because it is convenient or because the existing piping

is fully loaded, a new piping branch will be run from the point of delivery to serve a new appliance installation. The new branch is typically taken from a tee fitting installed immediately downstream of the meter. The entire new branch piping must be tested, including the tee fitting in the existing piping from which the new branch is supplied. The point of connection to the existing piping can be tested by a means other than pressure testing, although all other piping in the new branch must be isolated from the existing piping and pressure tested as required for new work. Section G2417.1.4 requires that the new branch piping be disconnected from the existing piping during pressure testing, except where an assembly consisting of two closed valves in series with an intermediate open valved port is installed to isolate the new piping from the existing piping (see Commentary Figure G2417.1.4).

G2417.1.4 (406.1.4) Section testing. A *piping system* shall be permitted to be tested as a complete unit or in sections. Under no circumstances shall a *valve* in a line be used as a bulkhead between gas in one section of the *piping system* and test medium in an adjacent section, except where a double block and bleed valve system is installed. A valve shall not be subjected to the test pressure unless it can be determined that the valve, including the valve closing mechanism, is designed to safely withstand the test pressure.

❖ Depending on the progression of a job, it may be desirable to test portions of a system as they are completed. It is also possible that portions of a system will be put in service before the entire system is completed. To prevent a test medium from leaking into piping containing fuel gas or vice versa, portions of piping under test must be isolated from portions that are in service. A single valve cannot be depended on to establish this isolation because all valves have some "leak-through" allowance, and also the valve could be accidentally opened. To eliminate this risk, two valves in series must be used, and a valved open port (tell-tale) must be installed between the valves. This arrangement is called a double-block and bleed-valve system. The

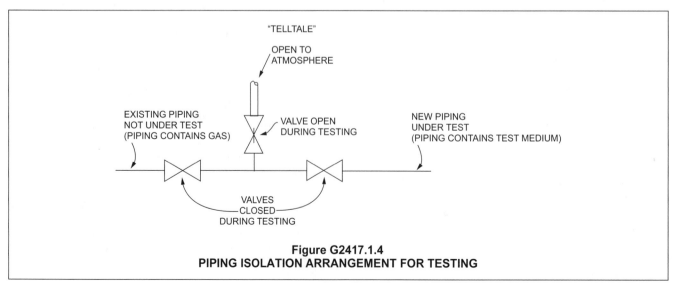

Figure G2417.1.4
PIPING ISOLATION ARRANGEMENT FOR TESTING

open port will release any gases that have passed through either isolation valve and also will serve to indicate any leakage or accidental opening of an isolation valve (see Commentary Figure G2417.1.4).

G2417.1.5 (406.1.5) Regulators and valve assemblies. *Regulator* and valve assemblies fabricated independently of the *piping system* in which they are to be installed shall be permitted to be tested with inert gas or air at the time of fabrication.

❖ Assemblies of valves, regulators and/or OPDs might be factory or shop assembled and may be tested independently of the piping system in which they will be installed.

G2417.1.6 (406.1.6) Pipe clearing. Prior to testing, the interior of the pipe shall be cleared of all foreign material.

❖ Piping will contain debris from pipe storage, handling and assembly, including soil, construction site trash, pipe-joint compound, welding debris, threading and reaming scrap and even tools. Clearing the piping of foreign materials will prevent such materials from entering system controls, devices and appliances and will also reduce the chances of injury to personnel from ejected debris and objects (see Section G2415.18).

G2417.2 (406.2) Test medium. The test medium shall be air, nitrogen, carbon dioxide or an inert gas. Oxygen shall not be used.

❖ The gas that is forced into the piping system for pressure testing must be chemically unreactive to prevent undesirable reactions with the piping system components and the fuel to be conveyed in the piping. Oxygen is not inert, and any residual amount could form an explosive mixture with the fuel gas. The typical test gas is air, which is also not inert but is readily available and free. Nitrogen and carbon dioxide are considered to be inert for the purpose of testing piping systems and are inexpensive compared to most truly inert gases.

G2417.3 (406.3) Test preparation. *Pipe* joints, including welds, shall be left exposed for examination during the test.

Exception: Covered or *concealed pipe* end joints that have been previously tested in accordance with this *code*.

❖ This section is consistent with Section R109.1.2 and requires piping joints to remain exposed during testing to allow locations of defects. This section speaks only of joints, but Section R109.1.2 requires that the piping and joints be left exposed until they are inspected and approved.

G2417.3.1 (406.3.1) Expansion joints. Expansion joints shall be provided with temporary restraints, if required, for the additional thrust load under test.

❖ If the piping system uses expansion joints to accommodate piping movement, those joints might need to be restrained to resist the additional thrust loading caused by the movement of the pressurized test gas. Generally, testing does not produce noticeable dynamic thrust forces because of the slow rate at which the test gas is put in and taken out of the piping system. The thrust force produced by the static pres-

sure of the test gas is of concern. Expansion joints are rarely used because expansion loops and offsets perform the same function.

G2417.3.2 (406.3.2) Appliance and equipment isolation. *Appliances* and *equipment* that are not to be included in the test shall be either disconnected from the *piping* or isolated by blanks, blind flanges or caps.

❖ Some equipment can be damaged by test pressure and is therefore disconnected from the piping system or is otherwise isolated by blank-offs placed between the flanges.

G2417.3.3 (406.3.3) Appliance and equipment disconnection. Where the *piping system* is connected to *appliances* or *equipment* designed for operating pressures of less than the test pressure, such *appliances* or *equipment* shall be isolated from the *piping system* by disconnecting them and capping the *outlet(s)*.

❖ This requirement is often overlooked in the field, which could lead to appliance damage and serious consequences. Appliances are typically designed for a maximum inlet gas pressure of $^{1}/_{2}$ psig (3.5 kPa) or less, above which damage to controls and regulators could occur. The test pressure will be at least 3 psig (20.7 kPa) and oftentimes much higher in accordance with general practice or local traditions or requirements. Installers routinely pressurize piping systems to pressures between 50 and 100 psig (345 and 690 kPa), often at the insistence of the local authority. It is obvious that there is a good potential to expose appliances to extreme pressures during a test, which is why the code requires the precaution of disconnecting the appliances from the supply piping. Disconnection is normally accomplished by opening a union or disconnecting the appliance connector. The appliance shutoff valve must not be used as a means of appliance isolation because the valve can leak through and/or can be accidentally opened or left open (see Section G2417.3.4 and Commentary Figure G2417.3.3).

G2417.3.4 (406.3.4) Valve isolation. Where the *piping system* is connected to *appliances* or *equipment* designed for operating pressures equal to or greater than the test pressure, such *appliances* or *equipment* shall be isolated from the *piping system* by closing the individual *appliance* or *equipment* shutoff valve(s).

❖ This section will likely never apply to dwellings because the test pressure (see Section G2417.4.1) will exceed the operating pressure of the appliances. If the appliances and other system components are not subject to damage by the test pressure, the equipment shutoff valve can serve to isolate the equipment during the test. Unlike Section G2417.3.3, this section does not require disconnection of the appliance/equipment. This section would probably be applicable only in an industrial occupancy. It is believed that no residential appliance can be safely subjected to even the minimum test pressure of 3 psi (207 kPa); therefore, it is always prudent to disconnect all appliances prior to applying test pressure (see Section G2417.3.3).

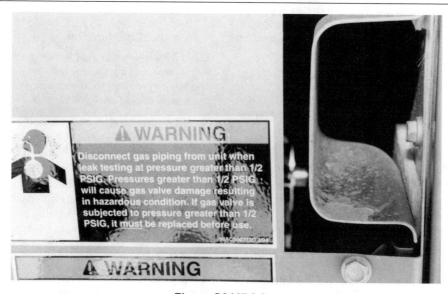

Figure G2417.3.3
MANUFACTURER'S WARNING REGARDING PRESSURE TESTING OF GAS PIPING

G2417.3.5 (406.3.5) Testing precautions. Testing of *piping* systems shall be performed in a manner that protects the safety of employees and the public during the test.

❖ Compressed gases store energy in direct proportion to the pressure. This stored energy can be destructive and dangerous to personnel if released quickly because of joint, component or pipe failure or an intentional action. Debris will be found in all piping systems after installation. This debris can be harmful to appliances and system components and can act as projectiles during testing and purging operations. Piping is typically purged before testing to flush out foreign materials.

G2417.4 (406.4) Test pressure measurement. Test pressure shall be measured with a manometer or with a pressure-measuring device designed and calibrated to read, record, or indicate a pressure loss caused by leakage during the *pressure test* period. The source of pressure shall be isolated before the *pressure tests* are made. Mechanical gauges used to measure test pressures shall have a range such that the highest end of the scale is not greater than five times the test pressure.

❖ Test pressures must be measured with a manometer or other pressure-measuring device that is designed and calibrated to indicate a pressure loss caused by leakage during the pressure test period. Pressure-measuring devices must be graduated so that small variations in test pressure will be readily detectable. Small pressure variations might go undetected in instruments having broader ranges. Gauges should have scale increments that are small enough to detect small changes in pressure. Mechanical gauges generally achieve their best accuracy near mid-range of their scale. All test instruments should have been calibrated recently and their accuracy verified before starting the test. Although no piping system can be absolutely leak free, the intent of the code is to make the system free

of leaks that are measurable with the testing procedures and instruments prescribed in the code and attainable in the field. Spring-type mechanical pressure gauges must have a high-end scale reading of no greater than 5 times the test pressure. If the test pressure was 3 psig (20.7 kPa), for example, the gauge would have to have a maximum range of 0 to 15 psig (0 to 104 kPa). Such gauges are a wise choice because the test reading will be higher in the range of the gauge where the gauge accuracy is better. Also, the gauge can have much smaller increments than the typical 100 psig (690 kPa) gauge, thus allowing greater resolution and discernability in detecting a leak [see Commentary Figure G2417.4(1)]. All spring-type mechanical gauges are susceptible to inaccuracies caused by conditions of use and mishandling and

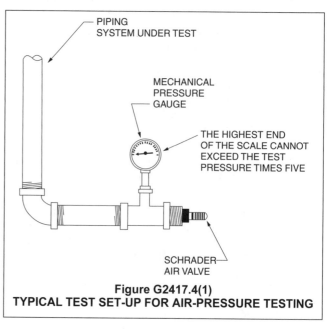

Figure G2417.4(1)
TYPICAL TEST SET-UP FOR AIR-PRESSURE TESTING

should be calibrated regularly. Manometers are very accurate and do not need to be calibrated. In all cases, the intent of this section is to require an instrument that is capable of indicating the very small pressure changes that occur as a result of a detectable leak. Test instrument sensitivity is especially important considering the short duration of the test.

The source of pressure in the piping system under test must be disconnected from the system to prevent the source from invalidating the test. For example, the compressed air tank or inert gas cylinder could be adding gas to the piping system at the same rate as a leak is letting gas escape; the test would then be meaningless.

Commentary Figures G2417.4(2), (3) and (4) show examples of test instruments that can be considerably more accurate than mechanical spring gauges. Electronic instruments can provide high resolution readings, making very small leaks as well as regular leaks detectable within short time periods. Commentary Figures G2417.4(3) and (4) show instruments that are designed to give accurate pressure readings and give positive indication of a leak. The liquid manometer-type instrument, shown in Commentary Figure G2417.4(3), functions as a manometer that readily indicates test pressure and leakage. A separate feature of the instrument allows small leaks to be recognized quickly. While the system is under test pressure, an integral valve can be closed that will isolate the liquid reservoir from the piping system, thereby trapping the test pressure in the manometer's liquid reservoir. A small diameter vertical dip tube, having its open end submerged in the liquid, rises above the liquid level in the reservoir. The top end of the dip tube is connected to the piping system under test. If there is no leakage, the pressures at the upper end and lower (submerged) end of the dip tube are equal and liquid will not rise. If leakage occurs, the pressure in the piping system will fall below the pressure trapped in the reservoir at the

start of the test procedure and liquid will be forced up into the small diameter dip tube, indicating the leak. The use of a small diameter dip tube allows a small change in liquid level in the reservoir to be seen as a comparatively large movement of liquid in the dip tube.

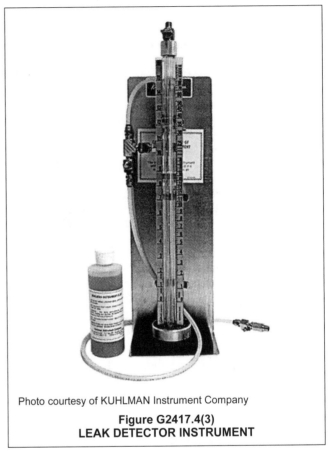

Photo courtesy of KUHLMAN Instrument Company

Figure G2417.4(3)
LEAK DETECTOR INSTRUMENT

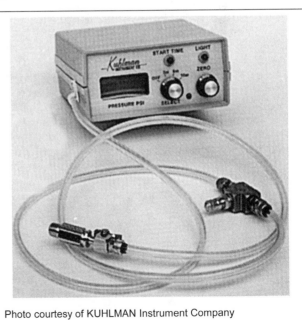

Photo courtesy of KUHLMAN Instrument Company

Figure G2417.4(4)
ELECTRONIC LEAK DETECTOR INSTRUMENT

Photo courtesy of Dwyer Instruments, Inc.

Figure G2417.4(2)
ELECTRONIC PRESSURE GAUGE

G2417.4.1 (406.4.1) Test pressure. The test pressure to be used shall be not less than $1^{1}/_{2}$ times the proposed maximum working pressure, but not less than 3 psig (20 kPa gauge), irrespective of design pressure. Where the test pressure exceeds 125 psig (862 kPa gauge), the test pressure shall not exceed a value that produces a hoop stress in the *piping* greater than 50 percent of the specified minimum yield strength of the pipe.

❖ The minimum test pressure will never be less than 3 psig (20 kPa). The majority of residential and small commercial occupancies are served by piping systems with a pressure of less than $^{1}/_{2}$ psig [(14 inches water column) (3.5 kPa)], and for such systems the test pressure must be 3 psig (20 kPa) or higher. The test pressure must also be not less than 1.5 times the pressure at which the piping system is designed to operate. If the piping system is designed to operate at 5 psig (34.5 kPa), the test pressure must be not less than 1.5 × 5 = 7.5 psig (51.7 kPa). In very high pressure applications, the test pressure might approach the structural limits of the piping material, and, therefore, the test pressure must be limited. Hoop stress is the result of internal pressure that tends to expand the pipe walls outward along the circumference of the pipe.

G2417.4.2 (406.4.2) Test duration. The test duration shall be not less than 10 minutes.

❖ As the piping system becomes larger, the test duration must be longer so that any leak can be detected. For example, a leakage rate of 1 cubic foot per hour (472 cm³/min) would produce a more easily detectable pressure drop in a small 10 cubic foot (0.3 m³) volume system than it would in a large 500 cubic foot (14.2 m³) volume system. It would require a longer test period to detect the leak in the large volume system because the leakage of 1 cubic foot represents a much smaller fraction of the total gas volume in the system. If a system has a volume of 501 cubic feet, the minimum test duration would be 1 hour as required by the IFGC. To put this in perspective, 500 cubic feet of piping system internal volume would equate to 83,333 feet (25 400 m) of 1-inch (25 mm) Schedule 40 steel pipe or 9,747 feet (2971 m) of 3-inch (76 mm) Schedule 40 steel pipe.

Because 10 cubic feet (0.3 m³) equates to 1,667 feet (508.1 m) of 1-inch (25 mm) pipe, it is obvious that single-family dwellings fall in the 10-cubic-foot or less volume category. A 10-minute test period is not much time to detect a pressure drop, which makes it even more important to use a pressure instrument that is capable of indicating very small drops in pressure (see commentary, Section G2417.4).

G2417.5 (406.5) Detection of leaks and defects. The *piping system* shall withstand the test pressure specified without showing any evidence of leakage or other defects. Any reduction of test pressures as indicated by pressure gauges shall be deemed to indicate the presence of a leak unless such reduction can be readily attributed to some other cause.

❖ Any pressure drop, no matter how small, is considered a failure of the test, except where the drop can be shown to result from a change in temperature or other cause. This would be practically impossible to demonstrate in the field, especially for short test durations. A true leak should show up as a continuous drop in test pressure. If the rate of pressure drop slows as the test progresses, it may be caused by cooling of the test medium, which causes it to contract. For example, warm air coming from a compressor will contract when introduced into a cold piping system, thereby causing a test instrument to indicate a drop in pressure. To help eliminate any guess work, the temperature of the test medium should be allowed to stabilize within the piping system before the test begins.

G2417.5.1 (406.5.1) Detection methods. The leakage shall be located by means of an *approved* gas detector, a noncorrosive leak detection fluid or other *approved* leak detection methods.

❖ Once the pressure measuring instrument indicates leakage, the leak or leaks must be located for repair or replacement. Electronic sensors and bubble fluids are used to find leaks that are not readily found by human senses. Many people who have worked with and/or inspected gas piping installations can recall instances of personal injury, fires and close calls that were caused by searching for leaks with an open flame. Open flames must never be used for leak detection because of the explosion danger and because open flames will not detect small leaks.

G2417.5.2 (406.5.2) Corrections. Where leakage or other defects are located, the affected portion of the *piping system* shall be repaired or replaced and retested.

❖ Where a leak is detected by a method named in Section G2417.5.1, the defect must be corrected. Corrections include tightening of fittings or threaded joints and replacing defective pipe, tubing or fittings.

G2417.6 (406.6) Piping system and equipment leakage check. Leakage checking of systems and *equipment* shall be in accordance with Sections G2417.6.1 through G2417.6.4.

❖ Leakage checking is different from pressure testing. Sections G2417.1 through G2417.5 address the pressure testing of newly installed or altered piping systems for the purpose of locating defects in the installation or materials. This section addresses leakage checks that are intended to discover open outlets, defective appliance connections and defects that have developed since the initial installation (see Section G2417.6.2).

G2417.6.1 (406.6.1) Test gases. Leak checks using fuel gas shall be permitted in *piping systems* that have been pressure tested in accordance with Section G2417.

❖ Piping systems are pressure tested with air or an inert gas, and leakage checking is done with the fuel gas at whatever pressure the system is designed to operate (see Section G2417.6.2).

G2417.6.2 (406.6.2) Before turning gas on. During the process of turning gas on into a system of new *gas piping*, the entire system shall be inspected to determine that there are no open fittings or ends and that all *valves* at unused outlets are closed and plugged or capped.

❖ This section requires that a precautionary measure be taken to prevent fuel gas from escaping from the piping system when the gas is turned on in the system for the first time. Because the piping system has never been used, it is not to be fully trusted as being gas tight. Because anything could have been done to the system since it was pressure tested, an additional visual check is justified before the gas is turned on. The condition of appliances may also be unknown, therefore, it may be wise to make sure that all appliance shutoff valves are closed until the appliances are ready to be put in service. Often, parties other than those who installed and tested the piping system are involved with connecting to the outlets in the system. This makes it likely that what was gas tight at the initial pressure testing is no longer gas tight as a result of incomplete or improper connections at the system outlets. Consistent with Section G2415.15, all unused outlets must be capped or plugged.

G2417.6.3 (406.6.3) Leak check. Immediately after the gas is turned on into a new system or into a system that has been initially restored after an interruption of service, the *piping system* shall be checked for leakage. Where leakage is indicated, the gas supply shall be shut off until the necessary repairs have been made.

❖ This section is a logical progression from the previous section. After the visual check for the status of outlets and equipment shutoff valves, the gas is introduced into the system, and the system is checked for leakage. This checking is typically accomplished by examining all the outlets, observing a gas meter for indication of flow, monitoring the system pressure for any drop or any combination of these. Because all outlet valves are closed and unused outlets are capped, any gas flow or pressure drop indicates leakage, and the gas must be turned off. This section would also allow leakage checking with the appliance shutoff valves open and the appliances shut down to prevent their operation. Testing in this manner would verify that the appliance gas controls were not faulty. A leakage check is also required after gas service is restored.

G2417.6.4 (406.6.4) Placing appliances and equipment in operation. *Appliances* and *equipment* shall not be placed in operation until after the *piping* system has been checked for leakage in accordance with Section G2417.6.3, the *piping system* has been purged in accordance with Section G2417.7 and the connections to the *appliances* have been checked for leakage.

❖ Gas appliances must not be put into service until the gas-piping system that supplies them has been checked for leakage and purged of air and the appliance connections have been checked for leakage (see Section G2417.7.3).

G2417.7 (406.7) Purging. The purging of piping shall be in accordance with Sections G2417.7.1 through 2417.7.3.

❖ Purging is intended to prevent a flammable gas/air mixture from being created in the piping. A flammable mixture could be a fire/explosion hazard in the piping, room or space, or in an appliance combustion chamber. There have been documented accidents involving explosions caused by failure to take the necessary precautions during piping purging operations. It has been purported that some accidents were caused by a condition referred to as "odor fade." It is a fact that steel pipe can absorb the chemical additive in fuel gas that gives the gas (natural or LP) its distinctive odor. This is evidenced by the fact that steel gas piping will smell like gas for a long time after it is no longer exposed to gas. Odorant fade appears to happen when gas is introduced into a new piping system that has never before contained gas. Once a piping system has been conditioned (seasoned) by exposure to gas, its ability to absorb the odorant diminishes with time. If personnel purge piping into an enclosed area and depend on their sense of smell to detect when gas has reached the purge discharge outlet, they could be fooled if the gas has lost its odor and a serious accident could result from such a dangerous practice. Also, it is known that not all humans have the same ability to detect the odorants in fuel gases. The sense of smell should never be relied upon to detect the presence of fuel gas. Using one's sense of smell to detect fuel gas is subjective and unreliable at best; therefore, this section prescribes methods that do not depend on the sense of smell. Combustible gas indicators and detectors are commonly used to detect gas during purging operations (see commentary, Sections G2417.7.1 through G2417.7.3).

G2417.7.1 (406.7.1) Piping systems required to be purged outdoors. The purging of piping systems shall be in accordance with the provisions of Sections G2417.7.1.1 through G2417.7.1.4 where the *piping system* meets either of the following:

1. The design operating gas pressure is greater than 2 psig (13.79 kPa).

2. The piping being purged contains one or more sections of pipe or tubing meeting the size and length criteria of Table G2417.7.1.1.

❖ This section requires that piping systems meeting either or both of the stated criteria comply with the listed subsections. Note that Section G2417.7.1.3 requires that the piping be purged only to the outdoors. The two criteria (and those of Section G2417.7.2) were chosen to distinguish the piping systems typically found in industrial, large commercial and large multiple-family residential buildings from those piping systems typically found in small commercial and small residential buildings. The gas-piping systems typically installed in industrial, large commercial and large multiple-family occupancies have larger internal pipe volumes and/or higher flow rates that justify special procedures to avoid creating explosive gas-air mixtures within buildings and within the piping. Combustible gas-air mixtures can be the source of explosions in the building or within the piping itself. Because there is no correlation between the occupancy groups established by the IBC and the size and nature of gas-piping systems installed in buildings, occupancy groups could not be used to distinguish between the larger, more complex piping systems and the smaller, less complex systems. In other words, the operating pressure and size of piping systems are much more dependent upon the type of appliances to be served than upon the occupancy classification of the building. Thus, the criteria of this section and Section G2417.7.2 represent the best attempt to establish a threshold for requiring piping systems to be purged to the outdoors. Note that the gas venting and gas displacement requirements of Sections G2417.7.1.1 and G2417.7.1.2 also apply to piping systems meeting the criteria or criterion of this section.

Criterion 2 is met if any portion of the piping system is larger than 2-inch nominal size and also longer than 50 feet for $2^1/_2$-inch pipe, 30 feet for 3-inch pipe, 15 feet for 4-inch pipe, 10 feet for 6-inch pipe and zero feet for 8-inch and larger pipe. In other words, if any portion of the piping system meets the criteria of Table G2417.7.1.1, the piping system meets Criterion 2 of this section. For example, if a piping system is constructed of $1^1/_2$- and 2-inch piping, but contains one section of 3-inch pipe that is over 30 feet in length, that system meets Criteria 2; however, if that same section of 3-inch pipe is 30 feet or less in length, the system does not comply with Criteria 2 (i.e., 30 feet is not > 30 feet).

If the operating pressure of the system is greater than 2 psig, Sections G2417.7.1.1 through G2417.7.1.4 apply regardless of the sizes of pipe in the system.

G2417.7.1.1 (406.7.1.1) Removal from service. Where existing gas piping is opened, the section that is opened shall be isolated from the gas supply and the line pressure vented in accordance with Section G2417.7.1.3. Where gas *piping* meeting the criteria of Table G2417.7.1.1 is removed from service, the residual fuel gas in the *piping* shall be displaced with an inert gas.

❖ Where a piping system to which this section applies is to be opened for any reason, it must be isolated from the gas supply and then the pressure in the piping system must be relieved by venting the gas to the outdoors in accordance with Section G2417.7.1.3. If the piping being opened or retired meets the criteria of Table G2417.7.1.1, the unpressurized gas in the piping must be displaced with an inert gas (i.e., a gas that will not support combustion such as nitrogen or argon). The displacement with inert gas will prevent an explosive mixture from forming inside of the pipe by eliminating any residual gas that could mix with the ambient

air. The larger and longer the pipe, the greater the amount of residual gas in the pipe and the greater the potential for the formation of an explosive mixture.

TABLE G2417.7.1.1 (406.7.1.1)
SIZE AND LENGTH OF PIPING

NOMINAL PIPE SIZE (inches)[a]	LENGTH OF PIPING (feet)
$\geq 2^1/_2 < 3$	> 50
$\geq 3 < 4$	> 30
$\geq 4 < 6$	> 15
$\geq 6 < 8$	> 10
≥ 8	Any length

For SI: 1 inch = 25.4 mm, 1 foot = 304.8 mm.

a. CSST EHD size of 62 is equivalent to nominal 2-inch pipe or tubing size.

G2417.7.1.2 (406.7.1.2) Placing in operation. Where gas *piping* containing air and meeting the criteria of Table G2417.7.1.1 is placed in operation, the air in the *piping* shall first be displaced with an inert gas. The inert gas shall then be displaced with fuel gas in accordance with Section G2417.7.1.3.

❖ Where new piping or piping that has been left open for some time is to be put into service, the air that exists in such piping must be removed to prevent an explosive gas-air mixture from forming inside of the piping. This is accomplished by first displacing the air with a gas that does not support combustion (inert gas) and then introducing fuel gas to displace the inert gas. This procedure prevents fuel gas from ever coming into contact with air inside of the piping. As the size and length of the piping increases, so does the volume of air in it and the potential for forming explosive mixtures within the piping. To help determine when sufficient inert gas has been introduced during the process of displacing air with an inert gas, the oxygen levels in the piping system could be monitored.

G2417.7.1.3 (406.7.1.3) Outdoor discharge of purged gases. The open end of a *piping* system being pressure vented or purged shall discharge directly to an outdoor location. Purging operations shall comply will all of the following requirements:

1. The point of discharge shall be controlled with a shutoff valve.

2. The point of discharge shall be located not less than 10 feet (3048 mm) from sources of ignition, not less than 10 feet (3048 mm) from building openings and not less than 25 feet (7620 mm) from mechanical air intake openings.

3. During discharge, the open point of discharge shall be continuously attended and monitored with a combustible gas indicator that complies with Section G2417.7.1.4.

4. Purging operations introducing fuel gas shall be stopped when 90 percent fuel gas by volume is detected within the pipe.

5. Persons not involved in the purging operations shall be evacuated from all areas within 10 feet (3048 mm) of the point of discharge.

❖ This section applies to piping systems meeting the criteria of Section G2417.7.1 and mandates that the purged gases discharge directly to the outdoors. Discharging purged gases to the building interior is prohibited by this section. Purging to the outdoors can be accomplished by either running temporary piping or hoses to the outdoors from the piping being purged or by installing permanent piping to the outdoors in locations where purging will be repeated regularly. It could also be accomplished by using existing piping that extends to the outdoors for some other purpose, but that is suitable for and able to be reconfigured for the purpose of gas-pipe purging. Other configurations are possible. The concept has been proposed by gas industry pundits and, some day, the code might contain provisions for purging air from piping by means of evacuation of the air with vacuum pumps and then breaking the vacuum with fuel gas such that the piping will contain nearly 100-percent fuel gas by volume. Purging by vacuum evacuation is only conceptual at this point in time.

Item 1 requires a means to control the purge gas discharge in the event of an emergency. Item 2 intends to prevent purged gases from reaching an ignition source and from entering a building. Item 3 requires that the point of discharge be constantly attended by one or more individuals who are monitoring the discharge gases with an instrument meeting the requirements of Section G2417.7.1.4. Item 4 requires purging operations that are introducing fuel gas into the piping to be stopped immediately after determining that the concentration of fuel gas within the purge gas discharge pipe is 90 percent of the purge-gas volume. This item intends to limit the volume of purged gases that discharge to the outdoor atmosphere. When the content of the piping is 90-percent fuel gas by volume, an explosive mixture is not possible within the piping interior and the piping content upstream of the discharge point is expected to be 90-percent fuel gas or higher. Item 5 intends to protect personnel and passersby from the potential hazard of purge gas discharge and also intends to prevent an individual from bringing an ignition source near the point of discharge.

G2417.7.1.4 (406.7.1.4) Combustible gas indicator. Combustible gas indicators shall be listed and shall be calibrated in accordance with the manufacturer's instructions. Combustible gas indicators shall numerically display a volume scale from zero percent to 100 percent in 1 percent or smaller increments.

❖ Note that a combustible gas indicator (CGI) is a different instrument than the instrument addressed in Section G2417.7.2.2. A CGI is required to be used by Item 3 of Section G2417.7.1.3. This instrument must be capable of indicating the percentage of fuel gas in the purge-gas discharge and do so by a number scale of zero to 100 percent. The display resolution requirement means that the instrument must be able to display percentages across the entire scale with not more than 1 percent between successive numbers. The increments on the 0 to 100 percent scale must be 1 percent or smaller. In other words, the percentage display must be able to read out "1%, 2%, 3%, 4%, 5%," etc., all the way up to 100%. These instruments need to be calibrated routinely because the nature of the sensors used in these instruments is such that they drift over time and with use.

G2417.7.2 (406.7.2) Piping systems allowed to be purged indoors or outdoors. The purging of *piping systems* shall be in accordance with the provisions of Section G2417.7.2.1 where the *piping system* meets both of the following:

1. The design operating gas pressure is 2 psig (13.79 kPa) or less.

2. The *piping* being purged is constructed entirely from pipe or tubing not meeting the size and length criteria of Table G2417.7.1.1.

❖ This section addresses purging procedures that allow the purged gases to discharge to the outdoors or to the indoors. Items 1 and 2 represent the division between the larger, more complex piping systems and the smaller, less complex piping systems (see commentary, Section G2417.7.1). In this section, the piping system must meet both criteria, meaning that the operating pressure of the system must be 2 psig or less and the piping must not meet the criteria of Table G2417.7.1.1. Not meeting the criteria of Table G2417.7.1.1 means that the piping system is either constructed entirely of 2-inch or smaller nominal size pipe, or the piping is 50 feet or less in length for 2$^1/_2$-inch pipe, 30 feet or less in length for 3-inch pipe, 15 feet or less in length for 4-inch pipe, 10 feet or less in length for 6-inch pipe and zero feet in length for 8-inch and longer pipe. For example, if a piping system is constructed of 2-inch and smaller pipe but contains one section of 4-inch pipe 15 feet long, that system does not meet the criteria of Table G2417.7.1.1 and thus meets Criteria 2 of this section (i.e., 15 feet is shorter than > 15 feet). If both criteria of this section are met, the piping system is allowed to be purged in accordance with Section G2417.7.2.1; however, if both criteria are not met, purging must comply with Sections G2417.7.1.1 through G2417.7.1.4.

Other examples of their criteria are as follows: A piping system that operates at 2 psig and is constructed of all 2-inch pipe meets both Criteria 1 and 2. A piping system that operates at 2 psig and is constructed of 2-inch and smaller pipe with one 55-foot section of 2$^1/_2$-inch pipe meets Criteria 1, but does not meet Criteria 2 because the 55-foot section of 2$^1/_2$-inch falls within the size and length criteria of the table.

G2417.7.2.1 (406.7.2.1) Purging procedure. The p*iping system* shall be purged in accordance with one or more of the following:

1. The *piping* shall be purged with fuel gas and shall discharge to the outdoors.

2. The *piping* shall be purged with fuel gas and shall discharge to the indoors or outdoors through an *appliance* burner not located in a combustion chamber. Such burner shall be provided with a continuous source of ignition.

3. The *piping* shall be purged with fuel gas and shall discharge to the indoors or outdoors through a burner that has a continuous source of ignition and that is designed for such purpose.

4. The *piping* shall be purged with fuel gas that is discharged to the indoors or outdoors, and the point of discharge shall be monitored with a listed combustible gas detector in accordance with Section G2417.7.2.2. Purging shall be stopped when fuel gas is detected.

5. The *piping* shall be purged by the gas supplier in accordance with written procedures.

❖ The purging provisions of this section are applicable only where the piping system meets both criteria of Section G2417.7.2. These provisions are intended for the piping systems typically found in light commercial and small residential occupancies (see commentary, Section G2417.7.1). Item 1 allows the simple procedure of purging air in a pipe by displacing it with fuel gas and discharging the purged gases to the outdoors. Item 2 allows a procedure used by some installers and gas suppliers in which air in the piping is displaced with fuel gas and the purged gases are discharged through an open appliance burner, typically on a range or cooktop. Any such open burner must have a continuous ignition source that is either integral with the burner or that is temporarily provided by the personnel performing the purging operation. If the purged gases were discharged into an enclosed combustion chamber such as in a furnace, boiler or water heater, an explosive gas/air mixture could form in the presence of an ignition source, resulting in a hazard. Item 3 is basically the same procedure as specified in Item 2, except that the burner is not an appliance burner. Some installers and gas suppliers utilize special purge burners designed for this procedure. Item 4 allows the purged gases to discharge to the indoors or to the outdoors, provided that the point of discharge is continuously monitored by the personnel performing the purging operation using a combustible gas detector (CGD) that complies with Section G2417.7.2.2. The purging operation must be immediately stopped when the CGD indicates the presence of fuel gas at the point of discharge. Item 5 allows personnel employed by the gas supplier to perform purging operations in accordance with detailed procedures that are in writing and originating from the gas supplier. This item assumes that the written procedures are based on practices that have been proven by experience to be safe.

G2417.7.2.2 (406.7.2.2) Combustible gas detector. Combustible gas detectors shall be listed and shall be calibrated or tested in accordance with the manufacturer's instructions.

Combustible gas detectors shall be capable of indicating the presence of fuel gas.

❖ A CGD is different than the CGI required by Section G2417.7.1.4; however, both of these instruments need to be calibrated or tested routinely because the nature of the sensors used in these instruments is such that they drift over time and with use.

G2417.7.3 (406.7.3) Purging appliances and equipment. After the *piping system* has been placed in operation, *appliances* and *equipment* shall be purged before being placed into operation.

❖ This section simply requires that after the gas piping has been properly purged and charged with fuel gas, the appliances served be purged of air before startup.

SECTION G2418 (407)
PIPING SUPPORT

G2418.1 (407.1) General. *Piping* shall be provided with support in accordance with Section G2418.2.

❖ Often, piping support is inadequate, poorly installed or overlooked entirely, despite the fact that support is critical in preventing stresses and strains in piping, fittings, joints, connectors and appliance gas trains (see Section G2418.2).

G2418.2 (407.2) Design and installation. *Piping* shall be supported with metal pipe hooks, metal pipe straps, metal bands, metal brackets, metal hangers or building structural components suitable for the size of *piping*, of adequate strength and quality, and located at intervals so as to prevent or damp out excessive vibration. *Piping* shall be anchored to prevent undue strains on connected *appliances* and shall not be supported by other *piping*. Pipe hangers and supports shall conform to the requirements of MSS SP-58 and shall be spaced in accordance with Section G2424. Supports, hangers and anchors shall be installed so as not to interfere with the free expansion and contraction of the *piping* between anchors. All parts of the supporting *equipment* shall be designed and installed so that they will not be disengaged by movement of the supported *piping*.

❖ As with all piping systems, the support of the fuel gas system is as important as any other part of the overall design. Proper supports are necessary to maintain piping alignment and slope, to support the weight of the pipe and its contents, to control movement and to resist dynamic loads, such as thrust. Inadequate support can cause piping to fail under its own weight, resulting in fire, explosion or property damage. Building design must take into consideration the structural loads created by the support of piping systems.

Hangers or supports must not react with or be detrimental to the pipe they support. Hangers or supports for metallic pipe must be of a material that is compatible with the pipe to prevent any corrosive action. For example, copper, copper-clad or specially coated hangers are required if the piping system is constructed of copper tubing. Hangers and supports must be constructed of noncombustible materials to prevent premature failure in a fire. Also, nonmetallic materials can elongate, deform and "creep" under load. This section prohibits combustible hooks, straps, bands, brackets and hangers. Plastic hooks, straps, hangers, etc., such as used with plumbing piping, are prohibited for use with gas piping. Building structural components are listed as a means of support because piping often rests on wall plates, beams, joists and similar structural members where such members are capable of carrying the additional load of the piping. Gas piping must not depend on other piping for support; thus, it must be supported from the structure independently. Gas controls for appliances can be stressed and damaged by unsupported piping drops that bear on or produce bending moments in the controls. A simple check for proper support at an appliance connected with rigid piping is to open the union and observe any movement of the piping. If the piping drops or moves in any direction, it is evident that stress is being applied to the appliance controls. Eliminating such stress is one of the advantages of using flexible gas connectors (see Section G2424).

SECTION G2419 (408)
DRIPS AND SLOPED PIPING

G2419.1 (408.1) Slopes. *Piping* for other than dry gas conditions shall be sloped not less than $^1/_4$ inch in 15 feet (6.4 mm in 4572 mm) to prevent traps.

❖ The required minimum slope (0.14-percent slope) is intended for gas known to contain enough water vapor to cause condensate to form inside the piping. Such conditions are rarely, if ever, encountered today.

G2419.2 (408.2) Drips. Where wet gas exists, a *drip* shall be provided at any point in the line of pipe where *condensate* could collect. A *drip* shall also be provided at the outlet of the *meter* and shall be installed so as to constitute a trap wherein an accumulation of *condensate* will shut off the flow of gas before the *condensate* will run back into the *meter*.

❖ Drips, often referred to as "drip legs," are distinct from sediment traps. Modern gas supplies and distribution systems are typically dry; thus, drips would be required only when recommended by the gas supplier.

G2419.3 (408.3) Location of drips. *Drips* shall be provided with *ready access* to permit cleaning or emptying. A *drip* shall not be located where the *condensate* is subject to freezing.

❖ A drip is not the same as a sediment trap; they are required for different reasons (see Section G2419.4).

G2419.4 (408.4) Sediment trap. Where a sediment trap is not incorporated as part of the appliance, a sediment trap shall be installed downstream of the appliance shutoff valve as close to the inlet of the appliance as practical. The sediment trap shall be either a tee fitting having a capped nipple of any length installed vertically in the bottommost opening of the tee as illustrated in Figure G2419.4 or other device approved as an effective sediment trap. Illuminating appliances, ranges,

clothes dryers, decorative vented appliances for installation in vented fireplaces, gas fireplaces and outdoor grills need not be so equipped.

❖ In addition to the code requirement, most appliance manufacturers require the installation of a sediment trap (dirt leg) to protect the appliance from debris in the gas. Note that a drip leg is not the same as a sediment trap (see Section G2419.2). Sediment traps are necessary to protect appliance gas controls from the dirt, soil, pipe chips, pipe-joint tapes and compounds and construction site debris that enter the piping during storage, handling, installation and repairs. Hazardous appliance operation could result from foreign matter entering gas controls and burners. Despite the fact that utilities supply clean gas, foreign matter can enter the piping prior to and during installation both on the utility side of the system and on the customer side.

Sediment traps are designed to cause the gas flow to change direction 90 degrees (1.57 rad) at the sediment collection point, thus causing the solid or liquid contaminants to drop out of the gas flow [see Commentary Figure G2419.4(1)]. The nipple and cap should not be placed in the branch opening of a tee fitting because this would not create a change in direction of flow and would allow debris to simply pass/jump over the capped nipple collection point. Commentary Figure G2419.4(2) illustrates a relatively ineffective sediment trap; however, such configurations are not expressly prohibited by the wording of this section except that the text does state "as illustrated in Figure G2419.4." The text speaks of the bottommost opening

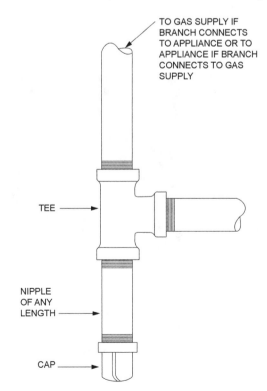

FIGURE G2419.4 (408.4)
METHOD OF INSTALLING A TEE FITTING SEDIMENT TRAP

of the tee without prohibiting such opening from being the branch opening. Note that code Figure G2419.4 illustrates the intent to have the nipple and cap in the run of the tee, thereby suggesting that it is not the intent to have it in the branch of the tee. The code does not specify a minimum length for the capped nipple; therefore, it could be from a close nipple on up. Three to 6 inches (76 to 152 mm) is the customary length. The capped nipple must be in a vertical plane to allow

the sediments to fall in by gravity. The sediment trap must be as close to the appliance inlet as practical to be able to capture sediment from all of the piping upstream of the appliance connection. The sediment trap must be downstream of the appliance shutoff valve to allow the trap to be serviced after closing the upstream shutoff valve. Manufactured sediment traps are available that have the configuration of a straight section of pipe and are equipped with cleanout openings. Although it would be wise to install sediment traps at all appliance connections, they are not mandated by code for gas lights, ranges, clothes dryers and outdoor grills. These appliances are also susceptible to harm from debris in gas, especially ranges and clothes dryers, and the appliance manufacturer may require sediment traps where the code does not. The code's logic is that these exempt appliances are manually operated rather than automatically operated; therefore, the user would be in attendance and aware of a problem.

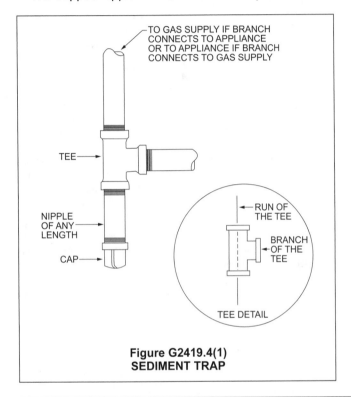

Figure G2419.4(1)
SEDIMENT TRAP

SECTION G2420 (409)
SHUTOFF VALVES

G2420.1 (409.1) General. *Piping systems* shall be provided with shutoff valves in accordance with this section.

❖ This section addresses shutoff valves for meters, buildings, tenant spaces, appliances and fireplaces.

G2420.1.1 (409.1.1) Valve approval. Shutoff valves shall be of an *approved* type; shall be constructed of materials compatible with the *piping*; and shall comply with the standard that is applicable for the pressure and application, in accordance with Table G2420.1.1.

❖ The code official must approve the type of shutoff valve used. For example, he or she might not approve a

Figure G2419.4(2)
RELATIVELY INEFFECTIVE SEDIMENT TRAP CONFIGURATION

valve that requires a wrench to operate. Shutoff valves must be tested to the specified standard and so identified. ANSI Z21.15 is applicable to appliance shutoff valves and appliance connector shutoff valves. Ball valves are used almost exclusively today because of their low pressure losses, ease of operation and reliability [see Commentary Figures G2420.1.1(1) and G2420.1.1(2)].

TABLE G2420.1.1. See below.

❖ Table G2420.1.1 lists two valve application categories: appliance shutoff application and all other applications. The application of a valve must be consistent with the listing of the valve. For example, a valve used ahead of an MP regulator (see Section G2420.4) in a 2-psi system must comply with ASME B16.44 and be marked "2G." It could also comply with ASME B16.33 and without being marked "2G."

G2420.1.2 (409.1.2) Prohibited locations. Shutoff valves shall be prohibited in *concealed locations* and *furnace plenums*.

❖ This section does not prohibit shutoff valves in spaces above ceilings and below floors used for conveying environmental air (see definition of "Concealed location"). A valve in a concealed location would be of no value and a potential leak source. The concern for

valves in furnace plenums is temperature extremes and lack of access.

G2420.1.3 (409.1.3) Access to shutoff valves. Shutoff *valves* shall be located in places so as to provide access for operation and shall be installed so as to be protected from damage.

❖ As expressed in the previous section, a shutoff valve has no utility if it is installed where it cannot be accessed. Location is especially important for valves installed for emergency use, such as a ball valve installed indoors upstream of all appliances to allow the building owner to shut off all gas in the event of an emergency. Outdoor valves are particularly subject to damage.

G2420.2 (409.2) Meter valve. Every *meter* shall be equipped with a shutoff valve located on the supply side of the *meter*.

❖ This valve is typically installed and owned by the gas supplier because it is upstream of the point of delivery for utility-owned meters. These valves require a wrench to operate and are capable of being locked off.

G2420.3 (409.3.2) Individual buildings. In a common system serving more than one building, shutoff valves shall be installed outdoors at each building.

❖ This section requires the shutoff valve for each building to be located outside of the building, thus allowing quick access in an emergency.

TABLE G2420.1.1 (409.1.1)
MANUAL GAS VALVE STANDARDS

VALVE STANDARDS	APPLIANCE SHUTOFF VALVE APPLICATION UP TO $^1/_2$ psig PRESSURE	OTHER VALVE APPLICATIONS			
		UP TO $^1/_2$ psig PRESSURE	UP TO 2 psig PRESSURE	UP TO 5 psig PRESSURE	UP TO 125 psig PRESSURE
ANSI Z21.15	X	—	—	—	—
ASME B 16.44	X	X	X[a]	X[b]	—
ASME B 16.33	X	X	X	X	X

For SI: 1 pound per square inch gauge = 6.895 kPa.
a. If labeled 2G.
b. If labeled 5G.

Photo courtesy of Dormont Manufacturing Company
Figure G2420.1.1(1)
BALL-TYPE GAS SHUTOFF VALVE

Photo courtesy of Dormont Manufacturing Company
Figure G2420.1.1(2)
BALL-TYPE GAS SHUTOFF VALVE
DESIGNED TO MATE WITH A CONNECTOR

G2420.4 (409.4) MP regulator valves. A listed shutoff valve shall be installed immediately ahead of each MP *regulator*.

❖ See the definition of "Regulator, Medium-pressure." This regulator (also known as "pounds-to-inches" regulator) is placed between the service regulator and the appliance regulator and is typically associated with manifold parallel distribution systems using CSST or copper tubing. A shutoff valve ahead of the MP regulator will allow the regulator to be conveniently serviced, adjusted or replaced. The valve also serves to isolate the higher pressure system from the lower pressure system. For purposes of sizing, the piping downstream of an MP regulator is considered as a separate system, with the MP regulator being the source of supply for that system.

G2420.5 (409.5) Appliance shutoff valve. Each *appliance* shall be provided with a shutoff valve in accordance with Section G2420.5.1, G2420.5.2 or G2420.5.3.

❖ Each appliance, regardless of type, size or location, must have its own dedicated shutoff valve to isolate the appliance from its gas supply. It is not the intent of this section to allow a single shutoff valve to control more than one appliance. Appliance shutoff valves are not considered to be emergency valves, but rather, are considered to be service valves that allow the appliance to be serviced, repaired, replaced or shut down (see definition of "Valve, Appliance shutoff"). A shutoff valve is a valve other than the valve that is integral with the appliance gas control. The intent is that the separate shutoff valve be located ahead of all appliance control valves. The location requirements of this section are intended to make the valve conspicuous and easy to operate. Note that the three subsections to this section are distinct choices and the shutoff valves have to meet one of these choices.

G2420.5.1 (409.5.1) Located within same room. The shutoff valve shall be located in the same room as the *appliance*. The shutoff valve shall be within 6 feet (1829 mm) of the *appliance*, and shall be installed upstream of the union, connector or quick disconnect device it serves. Such shutoff *valves* shall be provided with *access*. *Appliance shutoff valves* located in the firebox of a *fireplace* shall be installed in accordance with the *appliance* manufacturer's instructions.

❖ This section contains the traditional appliance shutoff valve requirements such as within 6 feet (1829 mm) of and in the same room as the appliance served. This section does not require the appliance shutoff valve to be provided with "ready access." Because such a valve is not an emergency valve, there is no reason to require ready access, and simply providing access is sufficient. Common practice has been to install the appliance shutoff valve behind ranges and clothes dryers and in the control compartment of decorative gas fireplace appliances. This section allows this practice because the valves are accessible. To avoid the use of an exposed valve or concealed "key-handle" valve, it is accepted practice to locate the shutoff valve for gas fireplace appliances in the control compartment in the bottom of the unit, behind the control access panel (see Commentary Figure G2420.5.1).

The code does not indicate how the 6-foot (1829 mm) limit is to be measured, but traditionally, it is measured along the piping such that no more than 6 feet (1829 mm) of piping extends between the outlet of the shutoff valve and the inlet of the appliance.

This section works with the definition of "Piping system" and Section G2413.3 to allow the piping down-

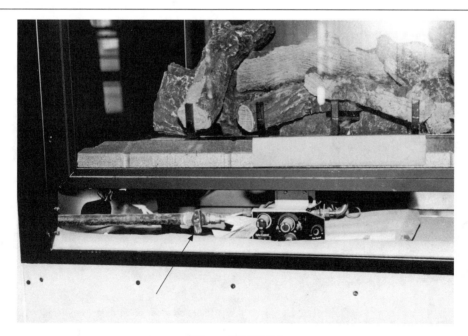

Figure G2420.5.1
SHUTOFF VALVE IN CONTROL COMPARTMENT OF GAS FIREPLACE

stream of the shutoff valve to be considered and sized as a connector. The code does not state that the required shutoff valve has to be located external to an appliance cabinet, enclosure or housing; therefore, the valve could be located behind an appliance access door/panel.

The last sentence of this section addresses appliances, such as decorative gas log sets, that are installed in solid fuel-burning fireplaces. The manual shut off may be in the firebox of the fireplace if it is allowed by the appliance manufacturer's instructions and the valve is located as directed by those instructions. There is typically an area adjacent to the appliance controls or inlet connection that is shielded from radiant heat, and this "cool zone" is where the manual shutoff valve must be located. A shutoff valve must never be located in the firebox if solid fuel is burned.

G2420.5.2 (409.5.2) Vented decorative appliances and room heaters. Shutoff valves for vented decorative *appliances*, room heaters and decorative *appliances* for installation in vented fireplaces shall be permitted to be installed in an area remote from the *appliances* where such valves are provided with *ready access*. Such *valves* shall be permanently identified and shall not serve another *appliance*. The *piping* from the shutoff valve to within 6 feet (1829 mm) of the *appliance* shall be designed, sized and installed in accordance with Sections G2412 through G2419.

❖ This section allows locating shutoff valves for the stated appliances remote from the appliance if the valves are permanently tagged as to their function, serve only the single appliance and have ready access. The shutoff valve can be located anywhere, indoors or outdoors, as long as it is provided with ready access and is permanently identified. Ready access is required because the shutoff valve will be more difficult to locate, being remote from the appliance served. This section is an alternative to using concealed "key-handle-type" valves located in the wall or floor adjacent to the appliance. The code intends for such valves to be permanently identified by means of a durable tag suitable for the environment, such as brass medallions on chains. The reference to room heaters in this section is not specific as to the type, thus, it would include both vented and unvented room heaters.

The shutoff valve for a log lighter is also the appliance operating valve; therefore, that valve cannot be located remotely.

Because the remote shutoff valve allowed by this section could be any distance [more than 6 feet (1829 mm)] from the appliance, the piping from the shutoff valve outlet to within 6 feet of the appliance inlet must be treated and sized as gas piping, not as a connector. Recall that the definition of "Piping system" says that the piping system ends at the appliance shutoff valves, therefore, the piping between the shutoff valve and the appliance becomes a connector, no matter how long it is. This section accounts for the potential pressure loss through lengthy connectors by requiring them to be sized as required for the piping system, except for the last 6 feet (1829 mm) before the appliance [see the

exception to Section G2422.1.2.1 and Commentary Figure G2422.1.2.1(2)].

G2420.5.3 (409.5.3) Located at manifold. Where the *appliance* shutoff valve is installed at a manifold, such shutoff valve shall be located within 50 feet (15 240 mm) of the *appliance* served and shall be *readily accessible* and permanently identified. The *piping* from the manifold to within 6 feet (1829 mm) of the *appliance* shall be designed, sized and installed in accordance with Sections G2412 through G2419.

❖ Similar to Section G2420.5.2, this section allows remote appliance shutoff valves, but, unlike Section G2420.5.2, this section specifies the remote location and imposes a distance restriction. Manifolds are part of parallel distribution systems commonly installed with CSST and copper tubing gas distribution systems. A manifold can be a manufactured fitting with multiple branch openings or it can be an assembly of close connected tee fittings. The individual "run-outs" of tubing to each appliance all connect to this manifold and the appliance shutoff valves can be installed at the manifold in a group configuration. Such valves must have ready access and must be labeled as to their function, the same as required in the previous section.

The code does not specify how the 50-foot (15 240 mm) distance limitation is to be measured. Taken literally, it would be measured in a straight line through walls, floors and ceilings. Clearly, such measurement could be difficult to perform in the field and should be approached liberally, keeping in mind the intent to simply have the valve within a reasonable distance.

Because the remote shutoff valve allowed by this section could be any distance [up to 50 feet (12 240 mm)] from the appliance, the piping from the shutoff valve outlet to within 6 feet of the appliance inlet must be treated and sized as gas piping, not as a connector. Recall that the definition of "Piping system" says that the piping system ends at the appliance shutoff valves, therefore, the piping between the shutoff valve and the appliance becomes a connector, no matter how long it is. This section accounts for the potential pressure loss through lengthy connectors by requiring them to be sized as required for the piping system, except for the last 6 feet before the appliance [see the exception to Section G2422.1.2.1 and Commentary Figure G2422.1.2.1(2)].

SECTION G2421 (410)
FLOW CONTROLS

G2421.1 (410.1) Pressure regulators. A line *pressure regulator* shall be installed where the *appliance* is designed to operate at a lower pressure than the supply pressure. *Line gas pressure regulators* shall be *listed* as complying with ANSI Z21.80. *Access* shall be provided to *pressure regulators*. *Pressure regulators* shall be protected from physical damage. *Regulators* installed on the exterior of the building shall be *approved* for outdoor installation.

❖ All gas supply systems employ regulators at some point in the distribution system, and the type and location of these devices depends on the design pressures of the system and utilization equipment. A line pressure regulator reduces the pressure from the service delivery pressure to the pressure of the distribution system. Additional pressure regulators could be required at other points in the system or at the appliance to reduce the pressure even further [see Commentary Figure G2421.1(1)]. With the increased use of medium-pressure gas distribution systems, systems are designed with one or more regulators in addition to the service regulators. Appliances and gas utilization equipment are also equipped with factory-installed regulators that control burner input (manifold) pressure.

Pressure regulators are extremely important devices that are relied on to prevent overpressurization of appliances and equipment. Typical gas appliance controls are not designed to tolerate a pressure of more than 14 inches w.c. (3.5 kPa). The standards that regulate the design and function of appliance gas controls require that the controls be tested for resistance to permanent damage resulting from exposure to excessive gas supply pressure. Each function of the gas controls must be tested for exposure to $2^{1}/_{2}$ psi (17 kPa) pressure for a period of 1 minute, after which, the control must function normally. Excess pressures could lead to severe appliance malfunction. Depending on the application, gas pressure regulators may contain relief mechanisms that are designed to vent gas from the downstream side in the event that the downstream (load) side pressure exceeds a predeter-mined maximum. This feature is a safeguard intended to prevent the overpressurization of the piping and utilization equipment supplied by the regulator. Regulators with relief devices must be installed outdoors, or the relief vent must be piped to the outdoors to prevent gas from being discharged into the building.

Access to pressure regulators is required for the maintenance, adjustment and servicing of the devices. The definition of "Access" appears in Section G2403. Protecting regulators from physical damage essentially means that regulators must be installed in locations that are not prone to impact from equipment or vehicles, or they must be shielded from impact. Because of the range of environmental conditions encountered outside the building, regulators must be evaluated and approved for installation outdoors. Regulators exposed to the weather must be protected from moisture and precipitation. Commentary Figure G2421.1(2) shows a service regulator with discharge piping connected to the integral relief valve port. This regulator establishes the pressure at the point of delivery. Commentary Figure G2421.1(3) illustrates the basic operating principle of gas pressure regulators. Outlet pressure acts upon the flexible diaphragm and adjustment spring, causing the valve to close on a rise in outlet pressure and open on a fall in outlet pressure. The vent allows the diaphragm to move freely by admitting or emitting air above the diaphragm. Atmospheric pressure and the adjustment spring force act together to establish the outlet pressure setting of the regulator [see Commentary Figure G2421.1(4)].

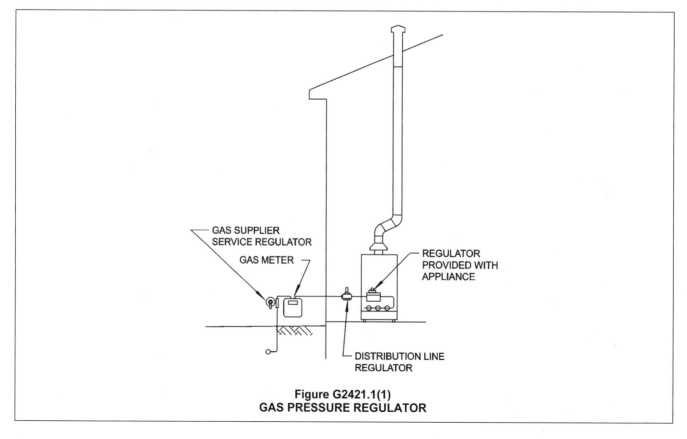

Figure G2421.1(1)
GAS PRESSURE REGULATOR

G2421.2 (410.2) MP regulators. MP *pressure regulators* shall comply with the following:

1. The MP *regulator* shall be *approved* and shall be suitable for the inlet and outlet gas pressures for the application.

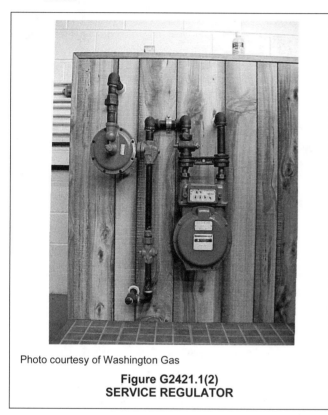

Photo courtesy of Washington Gas

Figure G2421.1(2)
SERVICE REGULATOR

2. The MP *regulator* shall maintain a reduced outlet pressure under lock-up (no-flow) conditions.

3. The capacity of the MP *regulator*, determined by published ratings of its manufacturer, shall be adequate to supply the *appliances* served.

4. The MP *pressure regulator* shall be provided with *access*. Where located indoors, the *regulator* shall be vented to the outdoors or shall be equipped with a leak-limiting device, in either case complying with Section G2421.3.

5. A tee fitting with one opening capped or plugged shall be installed between the MP *regulator* and its upstream shutoff valve. Such tee fitting shall be positioned to allow connection of a pressure-measuring instrument and to serve as a sediment trap.

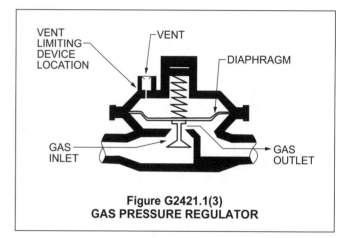

Figure G2421.1(3)
GAS PRESSURE REGULATOR

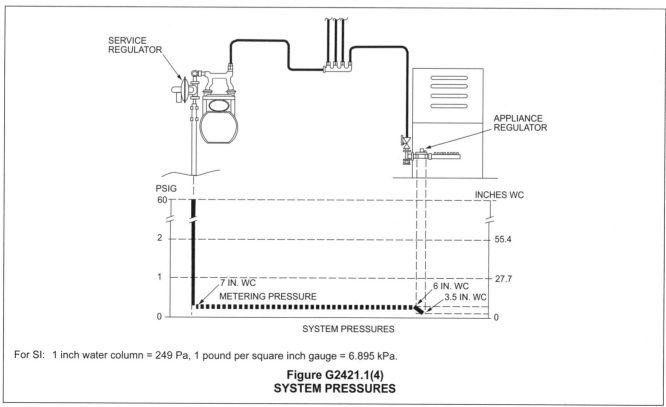

For SI: 1 inch water column = 249 Pa, 1 pound per square inch gauge = 6.895 kPa.

Figure G2421.1(4)
SYSTEM PRESSURES

6. A tee fitting with one opening capped or plugged shall be installed not less than 10 pipe diameters downstream of the MP *regulator* outlet. Such tee fitting shall be positioned to allow connection of a pressure-measuring instrument.

7. Where connected to rigid *piping*, a union shall be installed within 1 foot (304 mm) of either side of the MP *regulator*.

❖ This section states the minimum requirements for medium-pressure (MP) gas regulators installed in 2-psi (13.8 kPa) gas piping systems or portions of systems. These requirements intend to keep the regulators accessible for service and repair, protect them from physical damage, require that they are suitable for the use intended, are properly sized and installed, and, if installed indoors, are properly vented to the outdoors (see commentary, Section G2421.3). Item 2 requires a "lockup"-type regulator capable of holding the approximate outlet pressure setpoint indefinitely under no-flow conditions. Section G2420.4 requires a gas shutoff valve immediately upstream of the MP regulator to allow isolation of the regulator during testing, repair or replacement of the regulator. Items 5 and 6 require a tee fitting upstream and downstream of the MP regulator to facilitate the connection of pressure-measuring instruments for calibrating the regulator. These fittings should have a $^{1}/_{8}$-inch (3.2 mm) threaded tapping for connection of common gauges and manometers. Item 5 also requires a sediment trap on the inlet side of the regulator. Because a lot of faith is placed in the integrity of this regulator, it deserves the protection from debris afforded by a simple tee, nipple and cap arrangement (see Section G2419.4).

Where a line pressure regulator serves appliances rated for $^{1}/_{2}$ psi (3.5 kPa) or less inlet pressure and the regulator's inlet pressure is over 2 psi (13.8 kPa), a downstream over-pressure protection device must be installed in accordance with ANSI Z21.80, the standard for line pressure regulators [see Commentary Figures G2421.2(1) and G2421.2(2)]. The OPD is designed to limit the downstream (outlet) pressure to 2 psi (13.8 kPa).

MP regulators are line pressure regulators that serve to reduce pressures that are above 0.5 psi and

Photo courtesy of MAXITROL Company

Figure G2421.2(2)
PRESSURE REGULATOR/OVERPRESSURE PROTECTION DEVICE ASSEMBLIES

For SI: 1 inch water column = 249 Pa, 1 pound per square inch gauge = 6.895 kPa.

Figure G2421.2(1)
SYSTEM PRESSURES

less than or equal to 5 psi to some lower pressure. They are typically installed in 2-psi and 5-psi gas-distribution systems that serve appliances having a maximum input pressure of 0.5 psi (14 inches water column). If such regulators are installed with steel piping on the inlet and outlet side, the regulators cannot be removed or isolated without disassembly of the piping system for some distance or cutting the piping. To facilitate removal or isolation of the regulator, a union fitting must be placed near the inlet or outlet side of the regulator. This is simply a common-sense provision [see Commentary Figure G2421.2(3)].

G2421.3 (410.3) Venting of regulators. *Pressure regulators* that require a vent shall be vented directly to the outdoors. The vent shall be designed to prevent the entry of insects, water and foreign objects.

> **Exception:** A vent to the outdoors is not required for *regulators* equipped with and *labeled* for utilization with an *approved* vent-limiting device installed in accordance with the manufacturer's instructions.

❖ All regulators use synthetic or natural rubber diaphragms, one side of which is vented to the atmo-sphere. The devices must vent one side of the diaphragm to the atmosphere to allow unimpeded movement of the diaphragm within its enclosure. Thus, these devices, in effect, see atmospheric pressure as a reference point for controlling the fuel-gas pressure. Regulators cannot operate without the vent. If the diaphragm were to rupture or perforate, gas would be vented to the exterior of the device. For this reason, regulators installed indoors must be vented to the outdoors or must comply with the exception to this section. See the definition of "Vent piping, breather." The vent termination outdoors must be protected from blockage and the entry of foreign objects, insects and water, which could cause regulator failure and a severe hazard [see Commentary Figures G2421.3(1) and G2421.3(2)]. If a regulator vent is blocked by ice, water, insect nests, debris or other foreign material, the air above the diaphragm will be trapped and the regulator port valve could be held in the open position thus disabling the regulator and causing overpressurization of the piping system.

The exception permits installation of regulators indoors without a vent to the outdoors where the regu-

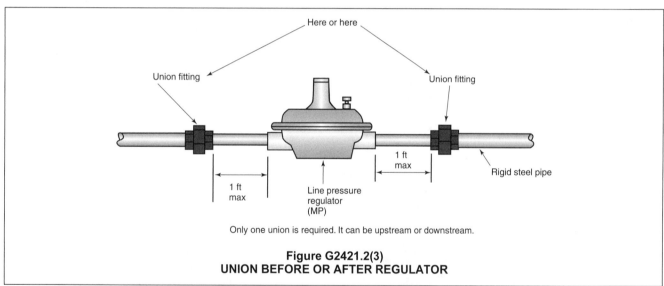

Only one union is required. It can be upstream or downstream.

Figure G2421.2(3)
UNION BEFORE OR AFTER REGULATOR

Photo courtesy of Washington Gas

Figure G2421.3(1)
REGULATOR WITH VENT PROTECTION

Photo courtesy of MAXITROL Company

Figure G2421.3(2)
PRESSURE REGULATOR WITH VENT-LIMITING DEVICE

lator is equipped with and labeled for use with a factory-installed vent-limiting device. Vent-limiting devices, as the name implies, are designed to allow the vent to "breathe" under normal operating conditions. In the event of diaphragm failure, the vent-limiting devices will limit gas escapement to a rate near 1 cubic foot per hour (0.0283 m³/h) for natural gas at 2-psi (13.8 kPa) pressure. A vent-limiting device is a ball check valve with a fixed orifice and is designed to allow air inhalation and escapement while restricting gas leakage to a small rate. Even small leakage rates can be hazardous if inadequate ventilation exists to prevent buildup of explosive concentrations. Although a standard is not referenced in this section, ANSI Z21.18 is appropriate for evaluating such vent-limiting devices. The standard limits the volume of gas that a vent-limiting device is al-lowed to discharge into a room or space. Vent-limiting devices should be installed in locations that are ventilated sufficiently to dissipate any gas discharged from the device.

G2421.3.1 (410.3.1) Vent piping. Vent *piping* for relief vents and breather vents shall be constructed of materials allowed for *gas piping* in accordance with Section G2414. Vent *piping* shall be not smaller than the vent connection on the pressure regulating device. Vent *piping* serving relief vents and combination relief and breather vents shall be run independently to the outdoors and shall serve only a single device vent. Vent *piping* serving only breather vents is permitted to be connected in a manifold arrangement where sized in accordance with an *approved* design that minimizes backpressure in the event of diaphragm rupture. *Regulator* vent *piping* shall not exceed the length specified in the *regulator* manufacturer's installation instructions.

❖ See the definitions of relief vents and breather vents. Relief vents must not be combined with any other relief vents or breather vents because of the potential to restrict relief capacity and/or cause regulator malfunction. Only breather vents are allowed to be manifolded together and only where in compliance with this section.

In all cases, the minimum size of vent piping is the size of the vent connection on the device served. Although not stated in this section, there is always a limit on how far any size vent piping can be run. Vent piping may have to be increased in size to compensate for long vent runs. The longer the vent piping, the greater the flow resistance and flow resistance must be limited so that the air flow necessary for unimpeded movement of the regulator diaphragm is not restricted. The regulator manufacturer's installation instructions may include vent piping length limitations and vent piping lengths must be in accordance with such instructions. Manifold piping for breather vents must be engineered based on the predicted rise in internal vent pressure in the event of a diaphragm failure. If regulator vents are blocked or the airflow is restricted in any way, including by undersized or excessively long piping runs, the result will be impaired regulator operation, causing

retarded regulator response time, underpressure during appliance start-ups and overpressurization of the downstream system, all of which can be hazardous.

This section also requires regulators to be vented individually to the outdoors to prevent a potentially hazardous situation that may arise if more than one regulator is served by a single vent. The hazard could occur if the diaphragm of one regulator failed and the gas leakage into the vent caused an increase of pressure within the vent, which would then act on the other regulator diaphragms, causing all of the other regulators using the common vent to operate at a higher discharge pressure. This section permits use of engineered vent manifold systems to vent more than one regulator if the manifold is designed and sized to allow unimpeded air movement and to minimize back pressure in the event of a diaphragm failure.

G2421.4 (410.4) Excess flow valves. Where automatic *excess flow valves* are installed, they shall be listed for the application and shall be sized and installed in accordance with the manufacturer's instructions.

❖ The code does not mandate the installation of EFVs; rather, it provides requirements for installations where such installations are chosen. EFVs must be listed for the application in which they are to be used. The manufacturer's installation instructions provide sizing and installation requirements that must be adhered to. Some devices have to be installed in certain positions to work properly. Sizing is very important because the EFV will cause a gas pressure drop and, if this drop is not accounted for in the system, the appliance might be supplied with inadequate pressure, causing a hazardous condition (see commentary, Section G2413.5). Proper sizing for EFVs is also important to ensure that they will function as intended. An undersized EFV might trip unnecessarily, resulting in the appliance being inoperative, and an oversized EFV might not trip at all. See Commentary Figure G2421.4 and the commentary to the definition of "Excess flow valve (EFV)."

G2421.5 (410.5) Flashback arrestor check valve. Where fuel gas is used with oxygen in any hot work operation, a listed protective device that serves as a combination flashback arrestor and backflow check valve shall be installed at an *approved* location on both the fuel gas and oxygen supply lines. Where the pressure of the piped fuel gas supply is insufficient to ensure such safe operation, *approved* equipment shall be installed between the gas meter and the appliance that increases pressure to the level required for such safe operation.

❖ This section relates to commercial and industrial installations that use fuel gas and oxygen for welding and cutting operations. The devices required by this section are designed to prevent the flow of one gas into the piping for another gas caused by differences in line pressures. The result of any such backflow would be potentially hazardous because an explosive mixture could be created in either piping system and such mixture could be delivered to a burner head.

3. Listed and labeled *appliance connectors* in compliance with ANSI Z21.24 and installed in accordance with the manufacturer's instructions and located entirely in the same room as the *appliance*.

4. *Listed* and *labeled* quick-disconnect devices used in conjunction with *listed* and *labeled appliance connectors*.

5. *Listed* and *labeled* convenience outlets used in conjunction with *listed* and *labeled appliance connectors*.

6. *Listed* and *labeled* outdoor *appliance connectors* in compliance with ANSI Z21.75/CSA 6.27 and installed in accordance with the manufacturer's instructions.

7. *Listed* outdoor gas hose connectors in compliance with ANSI Z21.54 used to connect portable outdoor *appliances*. The gas hose connection shall be made only in the outdoor area where the *appliance* is used, and shall be to the gas *piping* supply at an *appliance* shutoff valve, a *listed* quick-disconnect device or *listed* gas convenience outlet.

❖ Connectors are used to connect the appliance or equipment to the gas distribution system outlet. The pipe sizing methods in this chapter size distribution piping up to but not beyond the appliance shutoff valve outlet and this section sizes the connection between the shutoff valve outlet and the appliance [see definition of "Piping system" and Commentary Figure G2422.1(1)]. The choice of a connection type must take into consideration such factors as appliance movement, vibration, ambient conditions and susceptibility to physical damage. Between the shutoff valve and the appliance, a union fitting or similar arrangement must allow a means of disconnecting the piping. Flared or ground-joint connections that are part of an approved, labeled semirigid ("flexible") connector can serve as a union [see Commentary Figures G2422.1(2) and (3)].

Item 2 addresses CSST connected directly to appliances. Although CSST is allowed to directly connect to specific appliances (fixed/nonmoveable), CSST is not to be used as a substitute for a connector where one is

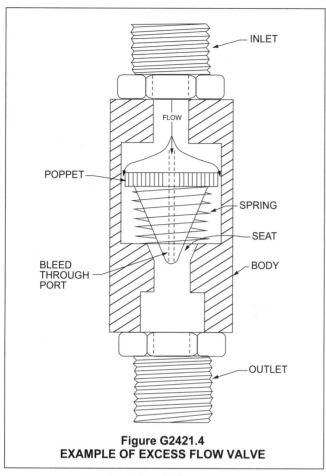

Figure G2421.4
EXAMPLE OF EXCESS FLOW VALVE

SECTION G2422 (411)
APPLIANCE CONNECTIONS

G2422.1 (411.1) Connecting appliances. *Appliances* shall be connected to the *piping system* by one of the following:

1. Rigid metallic pipe and fittings.

2. Corrugated stainless steel *tubing* (CSST) where installed in accordance with the manufacturer's instructions.

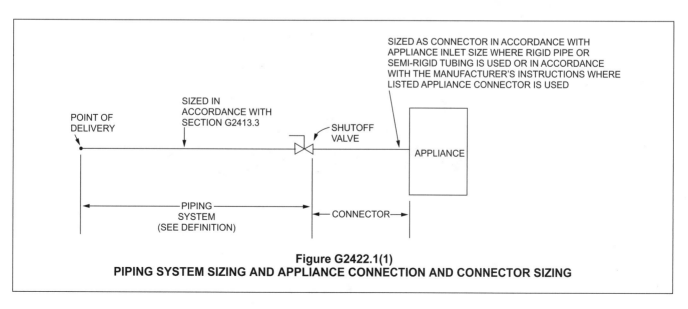

Figure G2422.1(1)
PIPING SYSTEM SIZING AND APPLIANCE CONNECTION AND CONNECTOR SIZING

required. CSST manufacturers' instructions specify where CSST can directly connect to appliances and where it cannot [see Commentary Figures G2422.1(4) and (5)].

Item 3 refers to the commonly used, so called "flexible" connectors typically made of corrugated stainless steel and previously made of brass. Appliance connectors are listed and labeled with the label typically being wrapped around the connector tubing.

Most appliance connectors manufactured today have attached plastic labels that state the installation and sizing instructions. These connectors are not designed for repeated movement that causes bending of the metal and should not be reused after the initial installation. Reuse is typically prohibited by the manufacturers' instructions. Repeated bending and/or vibration can cause metal fatigue, stress cracking and gas leak-age [see Commentary Figures G2422.1(6) and (7)].

The piping system is sized up to the outlet of the appliance shutoff valve. The connector occurs between the shutoff valve outlet and the appliance and is sized by the connector manufacturer or is based on the appliance input rating, gas inlet size and length of the connector [see Commentary Figure G2422.1(1)]. For example, a boiler is supplied by branch piping required to be 1 inch (25 mm) based on the sizing methods in Section G2413. The branch piping is 1 inch (25 mm) up to the boiler shutoff valve, but the boiler inlet opening is only $^1/_2$ inch (13 mm). The piping (connector) downstream of the shutoff valve is sized as a connector

because it is 6 feet (1829 mm) or less in length (see commentary, Section G2420.5.1).

Item 7 addresses connections for portable outdoor appliances [see Commentary Figure G2422.1(8)].

Because portable appliances can be moved, methods 1 through 7 of this section are not suited for connecting portable appliances to the gas distribution piping system. Most appliance connectors are not designed to be used with any appliance that can be readily moved. All appliance connectors must be used within the parameters of their listings. For outdoor portable appliances, new method 7 is the only apparent option. Outdoor gas-hose connectors have to be resistant to mechanical damage, possible heat exposure and the harmful effects of exposure to the weather. Connectors listed to ANSI Z21.54 are evaluated and tested for the particularly harsh environment of outdoor use. The gas-hose connector must be located entirely outdoors and must be connected to the gas piping system at a point outdoors. The point of connection to the gas distribution system piping must be through a listed device that allows the hose to be readily manually disconnected or through an appliance shutoff valve. Quick-disconnect devices have safety features such as thermal shutoffs that will close the valve when exposed to high temperatures and interlocking systems that will not allow the hose to be removed until the manual gas valve is closed.

The intent of new Item 7 is to address portable outdoor appliance connections and to mandate that such connectors be listed to a specific safety standard.

Photo courtesy of Dormont Manufacturing Company

Figure G2422.1(2)
QUICK-DISCONNECT DEVICE FOR USE WITH COMMERICAL APPLIANCE CONNECTORS

Photo courtesy of Dormont Manufacturing Company

Figure G2422.1(3)
QUICK-DISCONNECT DEVICE FOR USE WITH COMMERCIAL APPLIANCE CONNECTORS

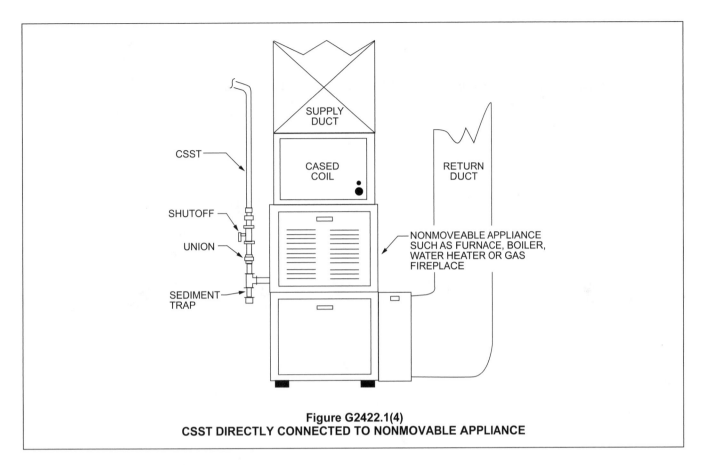

Figure G2422.1(4)
CSST DIRECTLY CONNECTED TO NONMOVABLE APPLIANCE

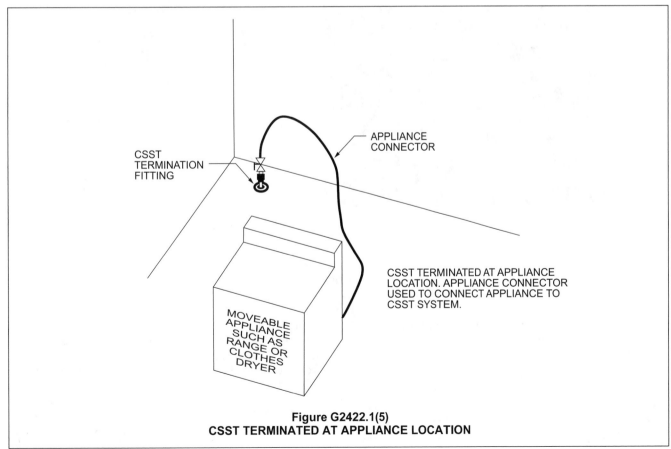

Figure G2422.1(5)
CSST TERMINATED AT APPLIANCE LOCATION

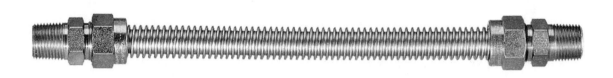

Photo courtesy of Dormont Manufacturing Company

Figure G2422.1(6)
SEMIRIGID (FLEXIBLE) APPLIANCE CONNECTOR

Photo courtesy of Dormont Manufacturing Company

Figure G2422.1(7)
OUTDOOR MOBILE HOME CONNECTOR

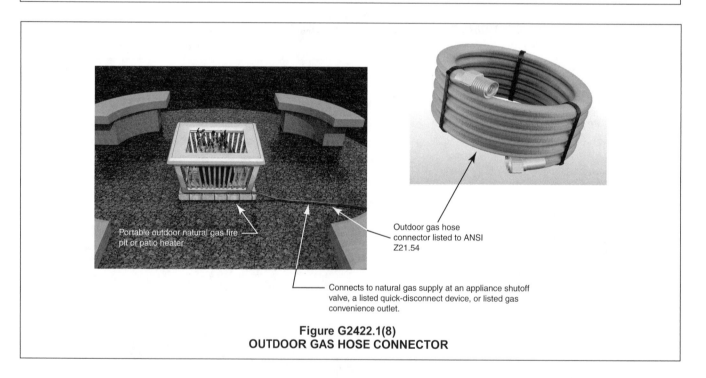

Portable outdoor natural gas fire pit or patio heater

Outdoor gas hose connector listed to ANSI Z21.54

Connects to natural gas supply at an appliance shutoff valve, a listed quick-disconnect device, or listed gas convenience outlet.

Figure G2422.1(8)
OUTDOOR GAS HOSE CONNECTOR

G2422.1.1 (411.1.2) Protection from damage. Connectors and *tubing* shall be installed so as to be protected against physical damage.

❖ Appliance connectors, although sound, are not constructed or tested to withstand the same rigors of service as the gas piping system without physical damage. Therefore, they must be installed to minimize the potential for damage. For example, "flexible" and semirigid tubing connectors should not be used where subject to excessive vibration or impact by occupants, vehicles, animals, doors, stored materials, etc. Impact, repeated movement and vibration can cause connector failure.

It is not the intent of this section to require that appliance connectors be hidden from view behind or under an appliance to consider them as protected from damage. For typical range and clothes dryer installations, the connectors are physically protected by the appliances themselves. However, for water heaters, furnaces, boilers, unit heaters, infrared heaters, etc., the connector will be located in the open and thus will need protection from damage by proper location, placement, elevation or guards. Protection from damage is commonly afforded by locating the connector out of harm's way, such as between a wall and an appliance or 8 feet (2438 mm) or more above the floor.

G2422.1.2 (411.1.3) Connector installation. *Appliance* fuel connectors shall be installed in accordance with the manufacturer's instructions and Sections G2422.1.2.1 through G2422.1.2.4.

❖ Flexible (semirigid) appliance fuel connectors are primarily intended for use with cooking ranges, clothes dryers and similar appliances where the gas connection is located behind the appliance and some degree of flexibility is necessary to facilitate the hook-up; however, flexible connectors may be used with any appliance where they are not subject to damage (see commentary, Section G2422.1.1). Rigid connections are not practical for movable appliances. Flexible connectors are typically constructed of stainless steel and are labeled with tags or metal rings placed over the tubing. The manufacturer's installation instructions dictate the installation requirements that are intended to protect the connector from damage and prevent leakage. Most flexible connectors are intended to be installed only once and are not designed for repeated bending and forming. For other than commercial flexible connections that are designed for repeated move-

ment, a new connector should be used each time an appliance is reinstalled. Repeated flexing or bending can cause metal fatigue and connector failure. Flexible connectors must be protected from physical damage and are not allowed to pass through any walls, floors, ceilings or appliance housings except as allowed by Section G2422.1.2.3.

G2422.1.2.1 (411.1.3.1) Maximum length. Connectors shall have an overall length not to exceed 6 feet (1829 mm). Measurement shall be made along the centerline of the connector. Only one connector shall be used for each *appliance*.

> **Exception:** Rigid metallic *piping* used to connect an *appliance* to the *piping system* shall be permitted to have a total length greater than 6 feet (1829 mm) provided that the connecting pipe is sized as part of the *piping system* in accordance with Section G2413 and the location of the *appliance* shutoff valve complies with Section G2420.5.

❖ Flexible connectors must bear the label of an approved agency and are limited to a maximum of 6 feet (1829 mm) in length. Six-foot (1829 mm) lengths have traditionally been used for range and dryer installations to allow the connector to be looped or otherwise arranged to allow limited movement of the appliance without stressing the connector. The safety record for such connectors has been good, therefore, the use of 6-foot-length connectors is not limited to specific appliances. Flexible connectors are evaluated in accordance with ANSI Z21.24 or ANSI Z21.69. Multiple connectors must not be connected together to circumvent the maximum length requirement [see Commentary Figures G2422.1.2.1(1) and (2)].

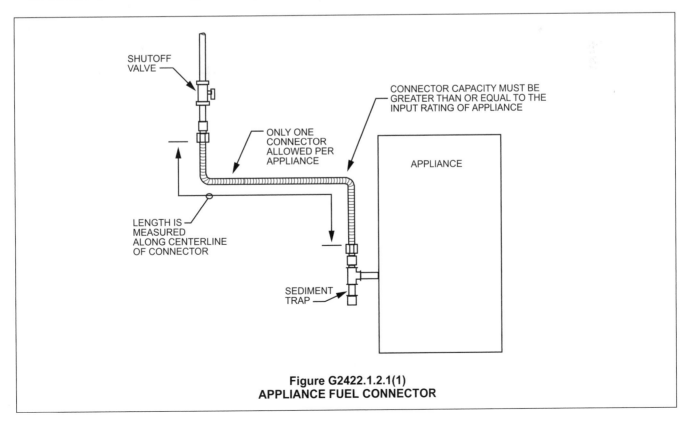

Figure G2422.1.2.1(1)
APPLIANCE FUEL CONNECTOR

The exception allows rigid metallic piping (steel) to exceed 6 feet (914 mm) in length for connection to any appliance if the shutoff valve is located in accordance with Section G2420.5 and the connector is sized as gas piping. If the shutoff valve must be within 6 feet (1829 mm) of the appliance, the connector length is also limited to 6 feet (1829 mm). If the shutoff valve is remote in accordance with Section G2420.5.2 or G2420.5.3, the connector length is unlimited if sized as gas piping (see Sections G2420.5.1 through G2420.5.3).

G2422.1.2.2 (411.1.3.2) Minimum size. Connectors shall have the capacity for the total *demand* of the connected *appliance.*

❖ Connectors must be chosen on the basis that they will provide the total gas demand of the appliance served at the required minimum pressure, as indicated in the manufacturer's instructions. Listed connectors provide capacity information on the tag attached to the connector.

G2422.1.2.3 (411.1.3.3) Prohibited locations and penetrations. Connectors shall not be concealed within, or extended through, walls, floors, partitions, ceilings or *appliance* housings.

Exceptions:

1. Connectors constructed of materials allowed for *piping systems* in accordance with Section G2414 shall be permitted to pass through walls, floors, partitions and ceilings where installed in accordance with Section G2420.5.2 or G2420.5.3.

2. Rigid steel pipe connectors shall be permitted to extend through openings in *appliance* housings.

3. *Fireplace* inserts that are factory equipped with grommets, sleeves or other means of protection in accordance with the listing of the *appliance.*

4. Semirigid *tubing* and *listed* connectors shall be permitted to extend through an opening in an *appliance* housing, cabinet or casing where the tubing or connector is protected against damage.

❖ This section requires connectors to be located entirely in the same room as the appliance served to protect them from damage and because the connector begins at the appliance shutoff valve, which must also be in the same room as the appliance served in accordance with Section G2420.5.1. Exception 1 applies to piping system materials, not semirigid (flexible) appliance connectors, and recognizes that Sections G2420.5.2 and G2420.5.3 both permit remote shutoff valves, meaning that in many cases, the piping between the shutoff valve and the appliance will necessarily have to pass through walls, floors, etc., in order to reach the appliance served. For example, a gas fireplace on the first floor has its shutoff valve in the basement and the piping (connector) must pass through the floor to connect to the fireplace. With respect to passing through appliance housings, this section is intended to address listed so-called "flexible" appliance connectors and metal tubing used as a connector. As addressed in Exception 2, rigid steel piping has always been allowed to pass through appliance housings and is often required by appliance installation instructions. CSST installed upstream of the appliance shutoff valve is not considered to be a connector; rather, it is part of the piping system. CSST installed downstream of an appliance shutoff valve, like all other piping materials, is treated as a connector (see Commentary Figure G2422.1.2.3).

Exception 2 recognizes that rigid steel pipe is not likely to be damaged by passing through an appliance housing/cabinet opening designed for that purpose.

Recall that all piping downstream from a shutoff valve is connector piping by the definition of "Piping system."

Exception 3 allows any connector of any material to pass through fireplace insert appliance housings where

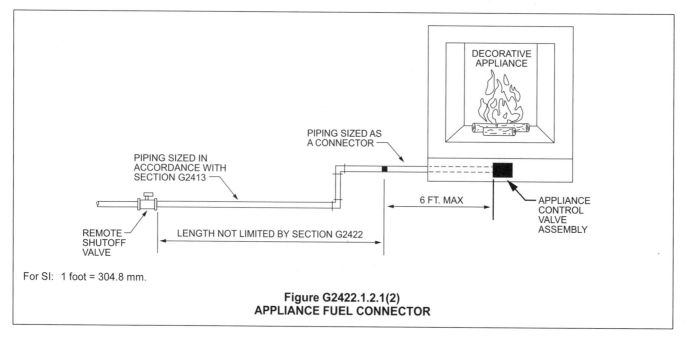

**Figure G2422.1.2.1(2)
APPLIANCE FUEL CONNECTOR**

For SI: 1 foot = 304.8 mm.

means integral with the appliance provide adequate protection from abrasion, cutting and vibration.

Exception 4 allows listed appliance connectors and tubing used as a connector to pass through any appliance housing/cabinet/casing provided that the connector is adequately protected from abrasion, cutting and vibration.

G2422.1.2.4 (411.1.3.4) Shutoff valve. A shutoff valve not less than the nominal size of the connector shall be installed ahead of the connector in accordance with Section G2420.5.

❖ Connectors are installed downstream of the shutoff valve and that valve must be located in the same room as the appliance except as allowed by Sections G2420.5.2 and G2420.5.3. All of the connectors addressed in Section G2422.1 must be sized for the connected load.

G2422.1.3 (411.1.5) Connection of gas engine-powered air conditioners. Internal combustion engines shall not be rigidly connected to the gas supply *piping*.

❖ Internal combustion engines produce significant vibrations and will move on their mounting during operation. Such movement could weaken and damage gas connections and/or gas controls. Some type of "flexible" connector designed for this purpose must be used to allow for movement.

G2422.1.4 (411.1.6) Unions. A union fitting shall be provided for *appliances* connected by rigid metallic pipe. Such unions shall be accessible and located within 6 feet (1829 mm) of the *appliance*.

❖ Although unions have always been installed in rigid pipe connections to appliances by necessity, previous code editions did not specifically require them. Regardless of where the shutoff valve is located, the union must be within 6 feet (1829 mm) of the appliance. Note that this section applies only to rigid piping because tubing systems and appliance connectors are joined with fittings that can serve as a union. A union is a special type of threaded coupling fitting consisting of three mating components and it is used "inline" to join two sections of piping that could not otherwise be threaded together. Unions also allow an appliance to be easily disconnected for repairs and during gas piping system testing (see Section G2415.5).

G2422.1.5 (411.1.4) Movable appliances. Where *appliances* are equipped with casters or are otherwise subject to periodic movement or relocation for purposes such as routine cleaning and maintenance, such *appliances* shall be connected to the supply system *piping* by means of an *appliance connector listed* as complying with ANSI Z21.69 or by means of Item 1 of Section G2422.1. Such flexible connectors shall be

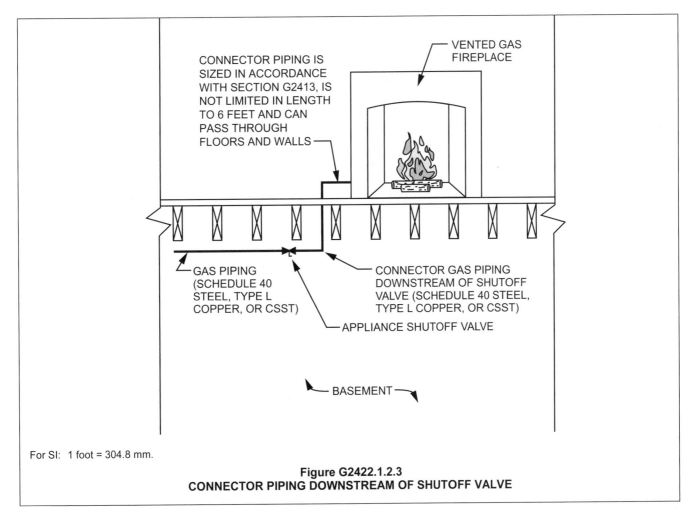

For SI: 1 foot = 304.8 mm.

Figure G2422.1.2.3
CONNECTOR PIPING DOWNSTREAM OF SHUTOFF VALVE

installed and protected against physical damage in accordance with the manufacturer's instructions.

❖ Listed residential appliances on casters are not likely to be found in a dwelling. Appliances, such as cooking appliances, commonly use heavy-duty, commercial-type flexible gas connectors. These connectors are designed to tolerate repeated movement to allow appliances on wheels or casters to be moved for cleaning operations or relocation.

Such connectors are commonly equipped with quick-disconnect couplings for ease of disconnection and reconnection and restraint cables to limit the amount of appliance movement, thereby preventing stress and damage to the connector.

Flexible connectors designed for specific applications are also used for the connection of equipment and appliances that produce significant vibrations. For example, gas-burning combustion engines in generator sets and HVAC equipment are connected with specially designed vibration dampening flexible connectors. Some pulse-combustion heating appliances employ flexible connectors to control the transmission of noise to gas distribution piping.

G2422.2 (411.3) Suspended low-intensity infrared tube heaters. Suspended low-intensity infrared tube heaters shall be connected to the building *piping system* with a connector *listed* for the application complying with ANSI Z21.24/CGA 6.10. The connector shall be installed as specified by the tube heater manufacturer's instructions.

❖ Such heaters are used in garages, shops and recreation/hobby spaces where the desire is to partially condition the space by heating the occupants rather than the ambient air in the space. Although uncommon, radiant tube heaters could be found within dwelling accessory structures and garages. This type of appliance will move each time it cycles on and off because of linear expansion and contraction of the radiant tube heating element. This movement has been known to cause stress failures of the appliance connectors that serve these appliances. The appliance manufacturer's installation instructions will provide very specific details on how the appliance connector is to be installed, configured and oriented. Commentary Figures G2422.2(1) and G2422.2(2) are examples of one manufacturer's installation instructions. The instructions will vary from manufacturer to manufacturer, but the theme and intent are the same, that is, to orient and configure the connector such that it will flex with the movement of the appliance without causing stress in the connector.

SECTION G2423 (413) COMPRESSED NATURAL GAS MOTOR VEHICLE FUEL-DISPENSING FACILITIES

G2423.1 (413.1) General. Motor fuel-dispensing facilities for CNG fuel shall be in accordance with Section 413 of the *International Fuel Gas Code*.

❖ Section 2308 of the IFC regulates CNG service stations.

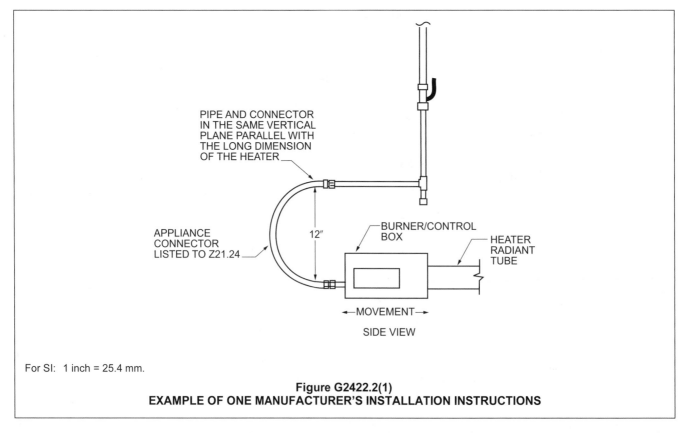

For SI: 1 inch = 25.4 mm.

Figure G2422.2(1)
EXAMPLE OF ONE MANUFACTURER'S INSTALLATION INSTRUCTIONS

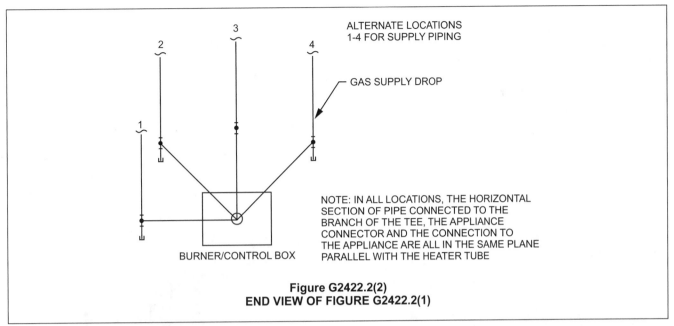

Figure G2422.2(2)
END VIEW OF FIGURE G2422.2(1)

ALTERNATE LOCATIONS
1-4 FOR SUPPLY PIPING

GAS SUPPLY DROP

NOTE: IN ALL LOCATIONS, THE HORIZONTAL
SECTION OF PIPE CONNECTED TO THE
BRANCH OF THE TEE, THE APPLIANCE
CONNECTOR AND THE CONNECTION TO
THE APPLIANCE ARE ALL IN THE SAME PLANE
PARALLEL WITH THE HEATER TUBE

BURNER/CONTROL BOX

SECTION G2424 (415)
PIPING SUPPORT INTERVALS

G2424.1 (415.1) Interval of support. *Piping* shall be supported at intervals not exceeding the spacing specified in Table G2424.1. Spacing of supports for CSST shall be in accordance with the CSST manufacturer's instructions.

❖ See the commentary to Table G2424.1.

TABLE G2424.1 (415.1)
SUPPORT OF PIPING

STEEL PIPE, NOMINAL SIZE OF PIPE (inches)	SPACING OF SUPPORTS (feet)	NOMINAL SIZE OF TUBING SMOOTH-WALL (inch O.D.)	SPACING OF SUPPORTS (feet)
$1/2$	6	$1/2$	4
$3/4$ or 1	8	$5/8$ or $3/4$	6
$1 1/4$ or larger (horizontal)	10	$7/8$ or 1 (horizontal)	8
$1 1/4$ or larger (vertical)	Every floor level	1 or larger (vertical)	Every floor level

For SI: 1 inch = 25.4 mm, 1 foot = 304.8 mm.

❖ This table is two tables side-by-side. The first two columns are for steel pipe, and the second two columns are for tubing materials (see Section G2418). Support for CSST is dictated by the manufacturers' installation instructions.

SECTION G2425 (501)
GENERAL

G2425.1 (501.1) Scope. This section shall govern the installation, maintenance, repair and approval of factory-built *chimneys, chimney* liners, vents and connectors and the utilization of masonry chimneys serving gas-fired *appliances*.

❖ This section contains requirements for the installation, maintenance, repair and approval of residential, commercial and industrial chimney and venting systems that convey the products of combustion from a gas-fired appliance to the outside atmosphere. Venting systems for fuel-fired appliances other than gas-fired appliances, such as oil and solid-fuel appliances, are covered in Chapter 18. The construction of masonry chimneys is regulated by Chapter 10 (see Section G2425.3).

G2425.2 (501.2) General. Every *appliance* shall discharge the products of combustion to the outdoors, except for *appliances* exempted by Section G2425.8.

❖ Appliances other than those listed in Section G2425.8 must be vented to convey the potentially harmful combustion byproducts to the outdoor atmosphere (see Section G2425.8).

G2425.3 (501.3) Masonry chimneys. *Masonry chimneys* shall be constructed in accordance with Section G2427.5 and Chapter 10.

❖ A masonry chimney is a field-constructed assembly that can consist of solid masonry units, reinforced concrete, rubble stone, fire-clay liners and mortars. A masonry chimney is permitted to serve low-, medium- and high-heat appliances. Chapter 10 outlines the general code requirements regarding the construction details for masonry chimneys, including those serving masonry fireplaces.

G2425.4 (501.4) Minimum size of chimney or vent. *Chimneys* and vents shall be sized in accordance with Sections G2427 and G2428.

❖ Sections G2427 and G2428 contain the requirements for sizing Category I appliance venting systems. Sizing of venting systems for other categories is dictated by the appliance manufacturer.

G2425.5 (501.5) Abandoned inlet openings. Abandoned inlet openings in *chimneys* and vents shall be closed by an *approved* method.

❖ Abandoned inlet openings result from appliances being disconnected or reconnected at different elevations. Unused openings in chimneys and vents can allow combustion gases to enter the building; can cause loss of draft at other appliances connected to the chimney or vent; can allow conditioned air to escape, resulting in energy losses; and can allow the entry of birds and rodents. For example, consider an appliance that was connected to a chimney and has been replaced with an appliance that vents through an exterior wall. The opening in the chimney left by the disconnection of the appliance must be properly closed to prevent a hazardous condition. Combustion gases from appliances connected to chimneys and vents can escape through openings in the chimney and vent. Those openings can affect the draft of the remaining appliances, causing them to spill combustion gases into the building.

The method used to close an unused opening must not create a protrusion in the flue passageway, which could restrict vent gas flow. It is not uncommon to find multiple openings in chimneys that were used at one time but have since been abandoned and covered or closed in some makeshift fashion, or left open. Inlet openings that are not closed could allow the escape of vent gases and could affect the draft produced by the chimney or vent. Excess dilution air will be admitted through unused openings, which could weaken draft by cooling the vent gases and bypassing the flow through appliance connectors. It is rare to find unused openings in vents because vent openings are easily closed or adapted to different appliance connections; however, the hazard of abandoned openings is the same as for chimneys. Vent and factory-built chimney openings are easily closed by eliminating tees and wyes or by using metal caps. Closure of masonry chimney openings may require the replacement of masonry, especially where the chimney has been structurally weakened by the openings.

G2425.6 (501.6) Positive pressure. Where an *appliance* equipped with a mechanical forced *draft* system creates a positive pressure in the venting system, the venting system shall be designed for positive pressure applications.

❖ Commonly used chimney and vent systems are not designed for positive pressure (above atmospheric). Chimneys and vents are intended to produce a draft and thus operate with negative internal pressures. Specialized vents and chimneys are available that are designed for positive pressure applications (see Commentary Figure G2425.6). The misapplication of materials can result in leakage of combustion gases and can cause chimney or vent deterioration from condensate corrosion. Masonry chimneys, Type B vents, Type L vents and most factory-built chimneys are designed for negative pressure applications, meaning that the venting system is producing draft. Negative-

pressure venting systems will leak if subjected to positive internal pressures. Leakage can allow combustion gases to enter the building, and condensation can cause rapid deterioration of the chimney or vent.

G2425.7 (501.7) Connection to fireplace. Connection of *appliances* to *chimney* flues serving *fireplaces* shall be in accordance with Sections G2425.7.1 through G2425.7.3.

❖ This section regulates installations where fireplace chimneys are used as the venting means for appliances such as room heaters and fireplace-insert-type heaters.

Figure courtesy of Selkirk L.L.C.

Figure G2425.6
VENT DESIGNED FOR GATEGORY III AND IV (POSITIVE PRESSURE) APPLICATIONS

G2425.7.1 (501.7.1) Closure and access. A noncombustible seal shall be provided below the point of connection to prevent entry of room air into the flue. Means shall be provided for *access* to the flue for inspection and cleaning.

❖ Without an airtight connection between the chimney flue and the appliance chimney connector, chimney draft will draw room air into the flue, thus weakening the draft drawn through the appliance. The chimney connection must be designed for access for inspection and cleaning of the chimney because the chimney flue is no longer open to the fireplace firebox.

G2425.7.2 (501.7.2) Connection to factory-built fireplace flue. An *appliance* shall not be connected to a flue serving a *factory-built fireplace* unless the *appliance* is specifically *listed* for such installation. The connection shall be made in accordance with the *appliance* manufacturer's installation instructions.

❖ Factory-built fireplaces may or may not be tested for any use other than as a traditional fireplace. Installing an appliance or fireplace-insert-type heater could result in abnormally high fireplace or chimney temperatures, component deterioration and hazardous operation. The fireplace manufacturer will provide specific instructions covering what is or is not allowed regarding appliance installations. Typically, the manufacturer's instructions will prohibit any modifications and any use of the fireplace other than for what it was designed.

G2425.7.3 (501.7.3) Connection to masonry fireplace flue. A connector shall extend from the *appliance* to the flue serving a *masonry fireplace* such that the *flue gases* are exhausted directly into the flue. The connector shall be accessible or removable for inspection and cleaning of both the connector and the flue. *Listed* direct connection devices shall be installed in accordance with their listing.

❖ The appliance chimney connector must extend up through the fireplace damper and smoke chamber and terminate in the chimney flue. This helps produce an adequate draft.

G2425.8 (501.8) Appliances not required to be vented. The following *appliances* shall not be required to be vented:

1. Ranges.

2. Built-in domestic cooking units *listed* and marked for optional venting.

3. Hot plates and laundry stoves.

4. *Type 1 clothes dryers* (*Type 1 clothes dryers* shall be exhausted in accordance with the requirements of Section G2439).

5. Refrigerators.

6. Counter *appliances*.

7. Room heaters *listed* for unvented use.

Where the *appliances* listed in Items 5 through 7 above are installed so that the aggregate input rating exceeds 20 Btu per hour per cubic foot (207 W/m³) of volume of the room or space in which such *appliances* are installed, one or more

shall be provided with venting *systems* or other *approved* means for conveying the *vent gases* to the outdoor atmosphere so that the aggregate input rating of the remaining *unvented appliances* does not exceed 20 Btu per hour per cubic foot (207 W/m³). Where the room or space in which the *appliance* is installed is directly connected to another room or space by a doorway, archway or other opening of comparable size that cannot be closed, the volume of such adjacent room or space shall be permitted to be included in the calculations.

❖ The specified appliances do not have to be vented unless required by the last paragraph of this section, which addresses Items 5 through 7 and not Items 1 through 4. Section G2404.3 requires that all appliances be listed.

Item 1 recognizes that residential-type cooking appliances are listed for unvented operations.

Item 4 clarifies that clothes dryers are not vented, but instead, are exhausted. The products of combustion for gas-fired clothes dryers are discharged to the outdoors along with the moisture exhaust; in effect, the dryer is "vented," but not in the sense that other appliances are vented. Clothes dryer exhaust ducts are not referred to as vents because they are exhaust ducts, not a type of vent. Clothes dryer exhaust ducts are erroneously referred to as "vents." Clothes dryers, both gas and electric, must exhaust to the outdoors.

Twenty Btu per hour per cubic foot of room volume (207 w/m³) is the same ratio as 50 cubic feet per 1000 Btu per hour (4.8 m³/w) on which Section G2407.5.1 is based. In other words, this section requires that unvented appliances be located in a space that meets the volume requirement for indoor combustion air specified in Section G2407.5.1. If the total unvented appliance input rating exceeds the ratio of 20 Btu/h per cubic foot of space volume (207 w/m³), one or more appliances would have to be vented or removed to keep the total at or below the limit. Maintaining a minimum space volume will help control the level of combustion products in the space by dilution with infiltration air. The appliances listed in Items 5 through 7 are allowed to be unvented, only where the total Btu/h input is at or below the limit set for the space volume.

G2425.9 (501.9) Chimney entrance. Connectors shall connect to a *masonry chimney* flue at a point not less than 12 inches (305 mm) above the lowest portion of the interior of the *chimney* flue.

❖ This requirement is intended to prevent blockage of the connector opening by debris that has collected at the bottom of the flue passage. The 12-inch deep (305 mm) space below the connector functions as a trap (dirt leg) for debris such as dead animals, leaves, flaking tile, mortar and soot. This section is not intended to apply to factory-built chimneys because those chimneys are protected from the entry of debris by the required cap.

G2425.10 (501.10) Connections to exhauster. *Appliance* connections to a *chimney* or vent equipped with a power exhauster shall be made on the inlet side of the exhauster. Joints on the positive pressure side of the exhauster shall be sealed to prevent flue-gas leakage as specified by the manufacturer's installation instructions for the exhauster.

❖ When a mechanical draft device or power exhauster is installed, the resulting "forced draft" creates a positive pressure inside the chimney or vent on the discharge or outlet side of the fan. A hazardous condition will result if an appliance is connected on the outlet or discharge side of a power exhauster. The appliance connector provides an alternate path for the flue gases to travel back into the building. Also, the appliance connected on the discharge side (outlet) of the exhauster will not be supplied with draft, and its combustion products will spill into the building interior.

A hazardous condition can also occur when a vent connector serving an appliance vented by natural draft is connected to the vent of an appliance equipped with integral power-venting means. This section does not prohibit Category I fan-assisted appliances from being common-vented with draft-hood (natural draft) appliances as addressed in Section G2428.

Any vent, chimney or connector piping into which an exhauster discharges must be rated and approved for positive pressure and properly installed to prevent leakage of combustion gases. For example, Type B vents cannot be used on the outlet side of an exhauster or a power-vented appliance, regardless of any appliance manufacturer's recommendations. Type B vent material cannot be sealed to prevent leakage when exposed to positive pressure. Positive pressure can force combustion products into the annular space between the pipe walls where condensation could occur, causing deterioration of the piping.

G2425.11 (501.11) Masonry chimneys. *Masonry chimneys* utilized to vent *appliances* shall be located, constructed and sized as specified in the manufacturer's installation instructions for the *appliances* being vented and Section G2427.

❖ Manufacturers' instructions for today's gas-fired appliances specify very limited conditions under which the appliance is allowed to vent to a masonry chimney. The conditions include chimney size, condition, location and construction. Because many existing masonry chimneys are oversized, unlined, in poor structural condition or partially exposed to the outdoors, they might not be suitable to serve gas-fired appliances. Masonry chimneys serving to vent fan-assisted gas-fired appliances are becoming rare. In many cases, a masonry chimney must be relined with a retrofit metal liner system, or a vent such as Type B vent will have to be installed within the flueway of the chimney. Masonry chimneys have a large thermal mass that has to be heated to above the dew point of the combustion gases to avoid condensation. The flue gas tempera-

tures discharged from many appliances will not be high enough to sufficiently warm the chimney mass, especially where the chimney is oversized and/or exposed to the exterior.

G2425.12 (501.12) Residential and low-heat appliances flue lining systems. *Flue lining* systems for use with residential-type and low-heat *appliances* shall be limited to the following:

1. Clay *flue lining* complying with the requirements of ASTM C315 or equivalent. Clay *flue lining* shall be installed in accordance with Chapter 10.

2. *Listed chimney* lining systems complying with UL 1777.

3. Other *approved* materials that will resist, without cracking, softening or corrosion, *flue gases* and *condensate* at temperatures up to 1,800°F (982°C).

❖ Flue lining systems for masonry chimneys include clay tile, poured-in-place refractory materials and stainless steel pipe.

G2425.13 (501.13) Category I appliance flue lining systems. *Flue lining* systems for use with Category I *appliances* shall be limited to the following:

1. *Flue lining* systems complying with Section G2425.12.

2. *Chimney* lining systems *listed* and *labeled* for use with gas *appliances* with *draft hoods* and other Category I gas *appliances listed* and *labeled* for use with Type B vents.

❖ Chimney lining systems for chimneys serving gas-fired appliances include stainless steel and aluminum piping (uninsulated or insulated) designed for both new construction and for relining existing chimneys.

Type B vent is also used as a chimney liner, although it is not considered to be a lining system, and in such cases the chimney is actually a masonry chase containing a vent. Relining of existing chimneys is quite common now because of the incompatibility of mid-efficiency appliances and masonry chimneys. Relining systems allow reuse of existing chimneys, thereby saving expense and retaining desirable architectural features of buildings [see Commentary Figures G2425.13(1) through G2425.13(4)].

G2425.14 (501.14) Category II, III and IV appliance venting systems. The design, sizing and installation of vents for Category II, III and IV *appliances* shall be in accordance with the *appliance* manufacturer's instructions.

❖ Venting systems for these appliances are special vent systems specified by the appliance manufacturer and are not addressed further in the code (see Commentary Figure G2425.14).

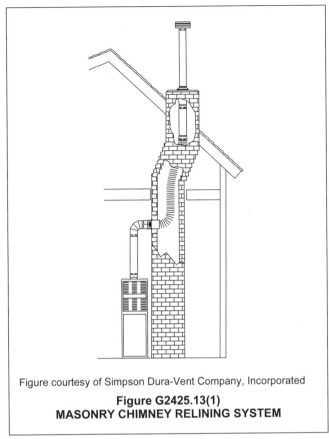

Figure courtesy of Simpson Dura-Vent Company, Incorporated

Figure G2425.13(1)
MASONRY CHIMNEY RELINING SYSTEM

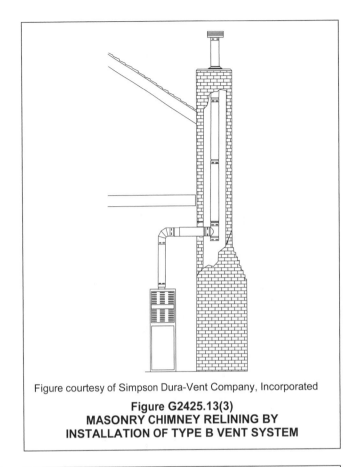

Figure courtesy of Simpson Dura-Vent Company, Incorporated

Figure G2425.13(3)
MASONRY CHIMNEY RELINING BY
INSTALLATION OF TYPE B VENT SYSTEM

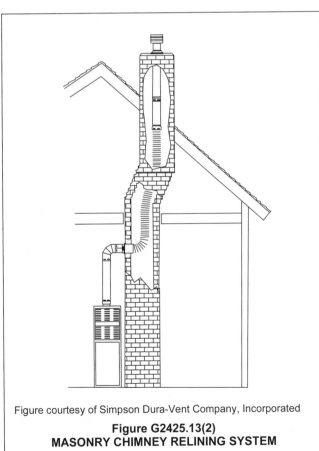

Figure courtesy of Simpson Dura-Vent Company, Incorporated

Figure G2425.13(2)
MASONRY CHIMNEY RELINING SYSTEM

Photo courtesy of Simpson Dura-Vent Company, Incorporated

Figure G2425.13(4)
MASONRY CHIMNEY RELINING SYSTEM

Photo courtesy of Selkirk L.L.C.

Figure G2425.14
SPECIAL VENT SYSTEM FOR
CATEGORY II, III AND IV APPLIANCES

G2425.15 (501.15) Existing chimneys and vents. Where an *appliance* is permanently disconnected from an existing *chimney* or vent, or where an *appliance* is connected to an existing *chimney* or vent during the process of a new installation, the *chimney* or vent shall comply with Sections G2425.15.1 through G2425.15.4.

❖ Existing chimneys and vents must be reevaluated for continued suitability whenever the conditions of use change. Size, which is covered in Section G2425.15.1, is of primary importance. Other considerations are the presence of liner obstructions, combustible deposits in the liner, the structural condition of the liner, the inclusion of a cleanout and clearances to combustibles, which are addressed in Sections G2425.15.2 through G2425.15.4. It is the intent of this section that whenever a new appliance is connected or an existing appliance is disconnected from a chimney or vent, the chimney or vent is subject to the requirements in Sections G2425.15.2 through G2425.15.4, which include requirements for inspection, cleaning, possible repair, the installation of a cleanout if one is not already there and the establishment of clearances to combustibles in accordance with the requirements for new chimneys or vent installations.

Chimneys, vents and the appliances served are all designed to function together as a system. Any change to an existing chimney or vent system will have an impact on the performance of that system. Something as simple as disconnecting an appliance from a chimney can upset the system balance and cause the vent-

ing system to fail to produce a draft for the remaining appliances, and could cause the venting system to produce harmful condensation (see Commentary Figure G2425.15).

G2425.15.1 (501.15.1) Size. The *chimney* or vent shall be resized as necessary to control flue gas condensation in the interior of the *chimney* or vent and to provide the *appliance* or *appliances* served with the required *draft*. For Category I *appliances*, the resizing shall be in accordance with Section G2426.

❖ The combined input from multiple appliances, especially older lower-efficiency appliances, can maintain sufficiently high chimney or vent temperatures to provide the necessary draft and to avoid condensation. Changing an existing configuration by disconnecting and eliminating an appliance or by substituting a higher-efficiency appliance can cause a decrease in flue gas temperature, resulting in condensation or poor draft. Also, the elimination of one or more draft-hood-equipped appliances will reduce the amount of dilution air in a venting system, thus increasing the likelihood of condensation. Often, it is necessary to resize a chimney or vent by replacing it with a smaller-size chimney or vent or by installing a liner system (see Commentary Figure G2425.15).

A common scenario involves removing chimney-vented furnaces or boilers and leaving a water heater as the only appliance vented to the chimney. In such cases, the chimney would typically be grossly oversized for the water heater, could fail to produce adequate draft and could be subject to continuous condensation. This scenario has received much attention and has created the phrase "orphaned water heaters."

G2425.15.2 (501.15.2) Flue passageways. The flue gas passageway shall be free of obstructions and combustible deposits and shall be cleaned if previously used for venting a solid or liquid fuel-burning *appliance* or *fireplace*. The *flue liner*, *chimney* inner wall or vent inner wall shall be continuous and shall be free of cracks, gaps, perforations, or other damage or deterioration that would allow the escape of *combustion products*, including gases, moisture and creosote.

❖ A chimney or flue must be cleaned to examine the flue liner surface, to eliminate fuel for a possible chimney fire, to reduce frictional resistance to flow and to prevent any residual deposits from falling into lower parts of the venting system. Installation or removal of an appliance without an inspection of the existing flue is prohibited. If the flue served a solid- or liquid-fuel-burning appliance or fireplace, cleaning is required in order to make the required inspection and to remove any combustible deposits. In that case, the addition or removal of an appliance without both cleaning and inspection of the existing flue would be a code violation. If inspection shows that the flueway of the vent or chimney is incapable of preventing the leakage of combustion gases, the chimney or vent would have to be repaired or relined.

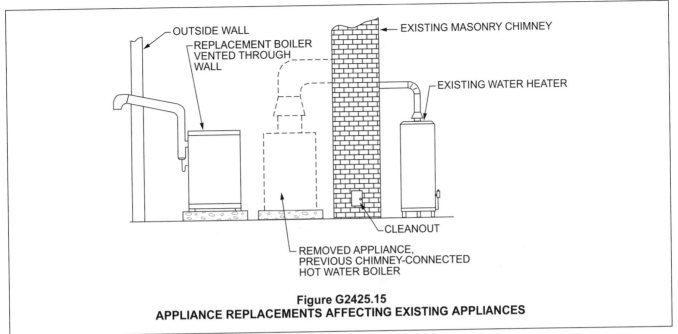

Figure G2425.15
APPLIANCE REPLACEMENTS AFFECTING EXISTING APPLIANCES

G2425.15.3 (501.15.3) Cleanout. *Masonry chimney* flues shall be provided with a cleanout opening having a minimum height of 6 inches (152 mm). The upper edge of the opening shall be located not less than 6 inches (152 mm) below the lowest *chimney* inlet opening. The cleanout shall be provided with a tight-fitting, noncombustible cover.

❖ This section requires installation of a cleanout in a chimney to facilitate cleaning and inspection. A fireplace inherently provides access to its chimney through the firebox, throat and smoke chamber. The cleanout cover and opening frame must be of an approved material such as cast iron, precast cement or other noncombustible material and must be arranged to remain tightly closed. A loose fitting or unsecured cleanout can allow air to flow into the chimney, affecting draft at the appliance or fireplace, and can result in an energy loss in the building. Under certain conditions, combustion products could escape into the building through an ill-fitting cleanout door or cover. The requirement for placing the cleanout at least 6 inches (152 mm) below the lowest connection to the chimney is intended to minimize the possibility of combustion products exiting the chimney through the cleanout. It is the intent of this section that a cleanout opening be installed, if one is not already provided, for existing masonry chimney flues whenever another appliance is installed or an existing appliance is removed.

G2425.15.4 (501.15.4) Clearances. *Chimneys* and vents shall have airspace *clearance* to combustibles in accordance with Chapter 10 and the *chimney* or vent manufacturer's installation instructions.

Exception: *Masonry chimneys* without the required airspace *clearances* shall be permitted to be used if lined or relined with a *chimney* lining system *listed* for use in *chimneys* with reduced *clearances* in accordance with UL 1777. The *chimney clearance* shall be not less than that

permitted by the terms of the *chimney* liner listing and the manufacturer's instructions.

❖ This section requires that the existing chimney or vent be inspected for clearances to combustibles for the entire length of the chimney or vent. Clearances to combustibles for masonry chimneys are found in Chapter 10. The required clearances for a listed factory-built chimney or vent would be in accordance with the existing chimney or vent manufacturer's installation instructions. The clearances to combustibles for vents and factory-built chimneys are typically indicated on the vent or chimney. If they are not, the manufacturer must be contacted and the original installation instructions consulted.

Existing masonry chimneys commonly do not have the clearances required by the code, especially if the structure is fairly old. Listed chimney lining systems, which provide insulation value to compensate for the lack of clearance, are available for use in such chimneys. Lining systems listed to UL 1777 include poured-in-place insulating refractory cement and insulated metal liners. The lining system listing will specify what the required clearance must be from the relined chimney to combustibles.

G2425.15.4.1 (501.15.4.1) Fireblocking. Noncombustible fireblocking shall be provided in accordance with Chapter 10.

❖ See Section R1003.19.

SECTION G2426 (502)
VENTS

G2426.1 (502.1) General. Vents, except as provided in Section G2427.7, shall be *listed* and *labeled*. Type B and BW vents shall be tested in accordance with UL 441. Type L vents shall be tested in accordance with UL 641. Vents for Category II and III *appliances* shall be tested in accordance with UL 1738. Plastic vents for Category IV *appliances* shall not be required to be *listed* and *labeled* where such vents are as specified by the *appliance* manufacturer and are installed in accordance with the *appliance* manufacturer's instructions.

❖ Vents regulated by this section are factory fabricated and must be listed and labeled. The labeling requirement applies to all components of the system such as the sections of pipe, fittings, terminal caps, supports and spacers.

The provision for unlisted plastic vents is necessary to allow for the venting systems commonly specified in the installation instructions for Category IV condensing appliances. Unlisted plastic pipe (commonly PVC or CPVC plumbing pipe) vents are distinct from listed high-temperature plastic special gas vents that were designed primarily for Category III appliances. Unlisted plastic pipe can be used only when specified by the appliance manufacturer and must be installed as directed in the appliance installation instructions. Unlisted plastic pipe is used only for high-efficiency Category IV gas-fired appliances. The pipe functions as an exhaust pipe for combustion gases and as a combustion air intake pipe.

Gas appliances (furnaces and boilers) are categorized so that the proper venting system can be selected. Four categories have been developed based on the pressure produced in the vent passageway and whether or not the flue gas temperatures approach the dew point, making condensation likely (see the definition for "Vented gas appliance categories").

Category I appliances operate under a negative or neutral vent pressure, are noncondensing appliances and may generate vent gases with temperatures up to 550°F (288°C). A Category I appliance can use a draft hood or an integral vent blower or draft inducer (fan assisted) [see Commentary Figure G2426.1(1)].

Category II appliances also operate under negative or neutral vent pressure, but the vent gas temperature and percent of CO_2 yield a flue loss of less than 17 percent, meaning that condensation of flue gases can occur. Vent systems serving Category II appliances fall under the special vent provisions of Sections G2425.14, G2427.4 and G2427.6.8.3.

Category III appliances operate with a positive vent pressure and can produce condensation, although they are not considered to be "condensing appliances." The positive pressure in the vent system necessitates airtight flue passageways to avoid leakage. Vent systems serving Category III appliances fall under the special vent provisions of Sections G2425.14, G2427.4, and G2427.6.8.3. They are typically constructed of high-temperature plastic or stainless steel pipe. The type of stainless steel specified today is typically AL29-4C, a special alloy known for its corrosion resistance.

Category IV appliances operate with a positive vent pressure and condense flue gases. Condensing (high-efficiency) appliances require airtight vent passageways and a method of collecting and draining condensation. The vent system serving this type of appliance is usually constructed of plastic, such as PVC or CPVC. Vent systems serving Category IV appliances also fall under the special vent provisions of Sections G2425.14, G2427.4 and G2427.6.8.3 [see Commentary Figure G2426.1(2)].

Figure G2426.1(1)
CATEGORY I FAN-ASSISTED FURNACE

Figure courtesy of Carrier Corporation

Figure G2426.1(2)
CATEGORY IV APPLIANCE (FURNACE)

A vent system can be used only in conjunction with gas-burning appliances that are listed and labeled for use with a vent. Appliances equipped with draft hoods and Category I (including fan-assisted) appliances both depend on the vent to produce a draft to convey the combustion products to the outdoors. Solid-fuel-burning appliances must not be connected to a vent system because they produce flue-gas temperatures that are much higher than the temperatures for which vents are designed. The burning of solid fuels also produces creosote. Creosote formation leaves a combustible deposit on the surface of the flue passageway that, if ignited, can produce a fire with temperatures exceeding 2,000°F (1093°C), far exceeding the safe temperature limitations of a vent system.

Type B vents are designed for venting noncondensing gas appliances equipped with a draft hood and fan-assisted appliances that operate with a nonpositive vent pressure and that are listed and labeled for use with a Type B vent. Type B vents are capable of withstanding continuous firing at flue temperatures not exceeding 400°F (222°C) above room temperature. This section requires that a Type B vent be tested in accordance with UL 441. Type B vents must not be used with incinerators, appliances readily converted to the use of solid or liquid fuels or combination gas/oil burning appliances. Type B vents can serve multiple gas-fired appliance installations [see Commentary Figure G2426.1(3)].

Type BW vents must also be tested in accordance with UL 441 and can serve only labeled gas-fired wall furnaces. A Type BW vent is an oval-shaped vent designed and tested for installation in a stud wall cavity. This type of vent is capable of withstanding continuous firing at flue temperatures not exceeding 400°F (222°C) above room temperature. Type BW vents must not be used to vent incinerators, appliances readily converted to the use of solid or liquid fuels or combination gas/oil-burning appliances.

Type L vents must be tested in accordance with UL 641 and are designed to vent oil-burning appliances, which produce a higher flue gas temperature than gas-fired appliances. Type L vents can also be used with appliances listed and labeled for Type B vent systems, meaning that Type L vents can substitute for Type B vents, but Type B vents cannot serve appliances that require Type L vents.

Type B and BW vents are double-walled, air-insulated vent piping systems constructed of galvanized steel outer walls and aluminum inner walls [see Commentary Figure G2426.1(4)]. Type L vents are typically double-wall, air-insulated vent piping systems constructed of galvanized and stainless steel.

Type B, BW and L vents are designed for natural draft applications only and must not be used to convey combustion gases under positive pressure. For example, a Type B vent must not be used with Category III or IV gas-fired appliances and must not be used on the

Photo courtesy of Simpson Dura-Vent Company, Incorporated

**Figure G2426.1(3)
TYPE B GAS VENT SERVING
TWO CATEGORY I APPLIANCES**

Figure courtesy of Simpson Dura-Vent Company, Incorporated

**Figure G2426.1(4)
TYPE B VENT SECTIONS WITH
AIR SPACE BETWEEN PIPE WALLS**

discharge side of an exhauster or power-vented appliance. Positive pressures will cause a Type B vent to leak combustion gases into the building, and the leakage can cause condensation to form within the vent system pipe, fittings and joints, resulting in corrosion damage to the vent and eventual vent failure. Fan-assisted Category I appliances are not considered power vented and do not produce positive vent pressures when vented vertically to a properly sized draft-producing vent such as a Type B vent system (see commentary, Section G2428). UL 1738 covers factory-built vents for Category II, III and IV appliances. These vents can be constructed of metallic or plastic materials including aluminum alloys, steel (coated, galvanized, stainless or aluminized), PVC and CPVC.

G2426.2 (502.2) Connectors required. Connectors shall be used to connect *appliances* to the vertical *chimney* or vent, except where the *chimney* or vent is attached directly to the *appliance*. Vent *connector* size, material, construction and installation shall be in accordance with Section G2427.

❖ Connectors designed, constructed and installed in accordance with Section G2427 are required except where a chimney or vent serves a single appliance and connects directly to that appliance. In such cases, a connector does not exist. The use of single-wall connectors for fan-assisted appliances is extremely limited by the sizing tables because of the high heat input necessary (minimum capacity column) to offset the heat loss of single-wall pipe.

G2426.3 (502.3) Vent application. The application of vents shall be in accordance with Table G2427.4.

❖ See Section G2427.4.

G2426.4 (502.4) Insulation shield. Where vents pass through insulated assemblies, an insulation shield constructed of steel having a minimum thickness of 0.0187 inch (0.4712 mm) (No. 26 gage) shall be installed to provide *clearance* between the vent and the insulation material. The *clearance* shall not be less than the *clearance* to combustibles specified by the vent manufacturer's installation instructions. Where vents pass through attic space, the shield shall terminate not less than 2 inches (51 mm) above the insulation materials and shall be secured in place to prevent displacement. Insulation shields provided as part of a *listed* vent system shall be installed in accordance with the manufacturer's instructions.

❖ Loose insulation in attic floor assemblies or roof assemblies can fall against vents, creating a fire hazard, and can also cause abnormally high vent temperatures. Even though clearances to combustible construction such as wood framing are maintained, the continued heating of combustible insulation materials that have fallen against a vent could be a source of ignition. This section applies regardless of the combustibility of the insulation. Shields constructed in the field must be sufficient to serve their purpose and must be securely attached to building construction because shields that are merely resting in place could be inad-

vertently dislodged or removed during maintenance activities.

G2426.5 (502.5) Installation. Vent systems shall be sized, installed and terminated in accordance with the vent and *appliance* manufacturer's installation instructions and Section G2427.

❖ The standards with which vents and appliances comply require that venting instructions be supplied by the manufacturer of the appliance and the manufacturer of the vent system. These instructions are part of the labeling requirements. Any deviation from them is in violation of the code.

The instructions for Type B vents and for gas appliances approved for use with Type B vents may be specific to a certain appliance or manufacturer's B-vent material. The clearance to combustibles for a vent system is determined by the testing agency and is stated on the component labels and in the manufacturer's instructions. Not all vents have the same required clearances to combustibles. The clearances are determined by the vent's performance during testing in accordance with the applicable standard. Different vent materials and designs impact the vent's ability to control the amount of heat transmitted to surrounding combustibles. It is critical that the vent system be installed in accordance with the clearances listed on the vent's label and in the manufacturer's installation instructions.

The termination of a natural draft vent must comply with the requirements of the manufacturer's installation instructions and Sections G2427.6.3, G2427.6.6 and G2427.6.7. A vent used in conjunction with a mechanical exhauster must meet the termination requirements established in Sections G2427.3.3 and G2427.8.

Physical protection of the vent system is required to prevent damage to the vent and to prevent combustibles from coming into contact with or being placed too close to the vents. Such protection is typically provided by enclosing the vent in chases, shafts or cavities in the building construction. Physical protection is not required in the room or space where the vent originates (at the appliance connection) and would not be required in such locations as attics that are not occupied or used for storage. For example, assume that a vent is installed in an existing building and that it extends from the basement, through a first-floor closet, through the attic and through the roof. The portion of the vent passing through the closet must be protected from damage. The means of physical damage protection should also be designed to maintain separation between the vent and any combustible storage. To prevent the passage of fire and smoke through the annular space around a vent penetration through a floor or ceiling, the vent must be fireblocked with a non-combustible material in accordance with this code.

Vent manufacturers provide installation instructions and factory-built components for fireblocking penetrations.

G2426.6 (502.6) Support of vents. All portions of vents shall be adequately supported for the design and weight of the materials employed.

❖ Vent manufacturers supply support parts, and manufacturers' instructions contain detailed requirements for support of vent systems. Improper support can cause strain on vent components, appliance connectors and fittings, resulting in vent, appliance or connector damage, as well as loss of required clearance to combustibles and proper slope. Frequently, venting installations suffer from lack of or improperly installed supports, brackets and hangers.

G2426.7 (502.7) Protection against physical damage. In *concealed locations*, where a vent is installed through holes or notches in studs, joists, rafters or similar members less than $1^1/_2$ inches (38 mm) from the nearest edge of the member, the vent shall be protected by shield plates. Protective steel shield plates having a minimum thickness of 0.0575-inch (1.463 mm) (16 gage) shall cover the area of the vent where the member is notched or bored and shall extend a minimum of 4 inches (102 mm) above sole plates, below top plates and to each side of a stud, joist or rafter.

❖ This section requires protection from nail and screw penetration for vents in the same manner that the code protects gas piping and tubing (see Commentary Figure G2426.7). Plastic and metal vent systems typically pass through or are enclosed by construction assemblies, thereby making them highly susceptible to penetration by fasteners. Such penetrations can damage the vent and the damage would not be visible in con-

cealed locations. Possible results from vent penetration include cracking/splitting of pipe, leakage of condensate and/or combustion gases, corrosion failure of metals, structural decay, interference with required clearances and restraint of expansion and contraction movement.

G2426.7.1 (502.7.1) Door swing. Appliance and equipment vent terminals shall be located such that doors cannot swing within 12 inches (305 mm) horizontally of the vent terminal. Door stops or closures shall not be installed to obtain this clearance.

❖ Vent terminals for sidewall-vented appliances (such as direct-vent gas fireplaces and fireplace heaters, direct-vent room heaters, direct-vent water heaters, furnaces and boilers and also nondirect-vent appliances) are sometimes located where a side-swinging door can impact the vent terminal or swing close to the terminal. The results can be damage to the vent terminal, a fire hazard and interference with the appliance venting and combustion air intake. Also, if the door blocks or deflects the vent discharge such that the combustion products are pulled back into the combustion air intake, the result would be excessive CO production and serious appliance malfunction and sooting.

A damaged vent terminal could cause appliance malfunction and carbon monoxide poisoning. It could also cause damage to nearby combustible surfaces and create a fire hazard. Door stops and closer devices cannot be depended on because they are easily defeated or removed [see Commentary Figures G2426.7.1(1) and (2)].

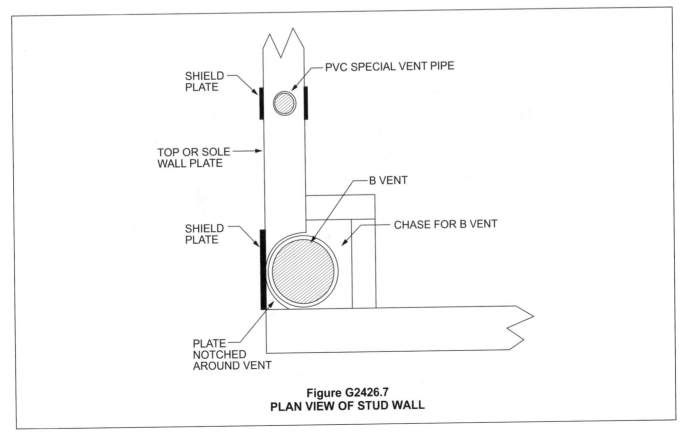

Figure G2426.7
PLAN VIEW OF STUD WALL

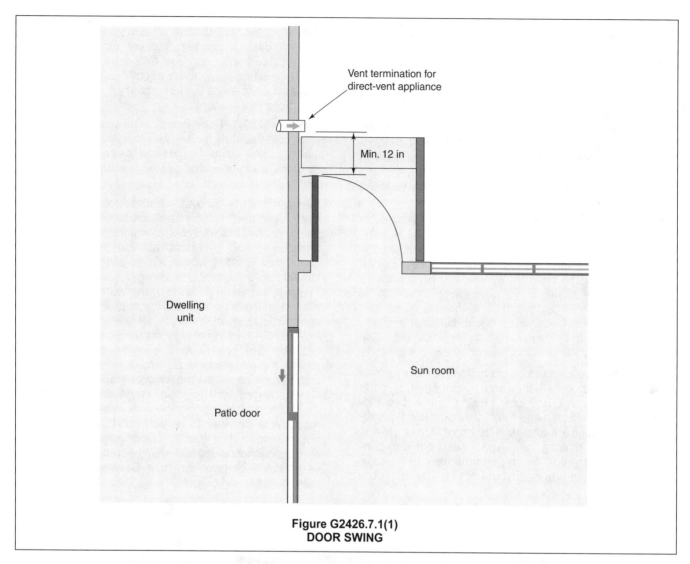

**Figure G2426.7.1(1)
DOOR SWING**

**Figure G2426.7.1(2)
DOOR SWING**

SECTION G2427 (503)
VENTING OF APPLIANCES

G2427.1 (503.1) General. The venting of appliances shall be in accordance with Sections G2427.2 through G2427.16.

❖ Section G2427 contains general sizing, design and installation requirements for venting systems. Section G2428 covers vent system sizing for Category I appliances.

G2427.2 (503.2) Venting systems required. Except as permitted in Sections G2427.2.1, G2427.2.2 and G2425.8, all *appliances* shall be connected to *venting systems*.

❖ All appliances must be vented by a venting system except those allowed to be unvented by Section G2425.8.

 The appliances addressed in Sections G2427.2.1 and G2427.2.2 must have a venting system.

G2427.2.1 (503.2.3) Direct-vent appliances. *Listed direct-vent appliances* shall be installed in accordance with the manufacturer's instructions and Section G2427.8, Item 3.

❖ Direct-vent appliances are vented by a natural or mechanical draft system that is designed as part of the appliance and that usually requires some assembling on the job site. Except for the vent terminal requirements of Section G2427.8, Item 3, the code depends on the appliance manufacturer's installation instructions for all aspects of venting direct-vent appliances. Direct-vent appliances include some Category IV condensing furnaces and boilers and power- and gravity-vented direct-vent water heaters, unit heaters, space heaters and packaged terminal units.

Direct-vent appliances have the distinct advantage of providing their own combustion air supply in addition to their own dedicated venting means [see definition of "Direct-vent appliance" and Commentary Figures G2403(6) and G2427.2.1(1) through G2427.2.1(6)].

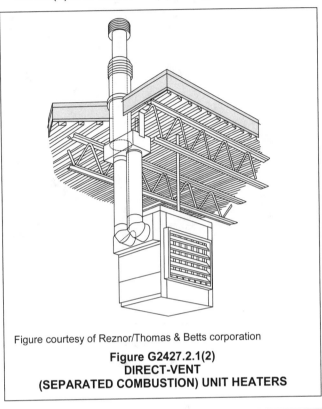

Figure courtesy of Reznor/Thomas & Betts corporation

Figure G2427.2.1(2)
DIRECT-VENT
(SEPARATED COMBUSTION) UNIT HEATERS

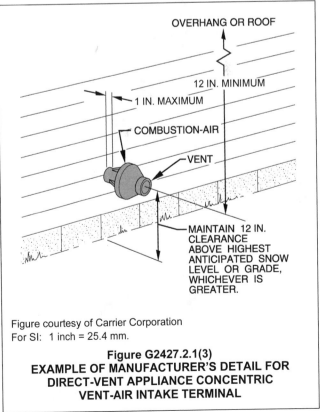

Figure courtesy of Carrier Corporation
For SI: 1 inch = 25.4 mm.

Figure G2427.2.1(3)
EXAMPLE OF MANUFACTURER'S DETAIL FOR
DIRECT-VENT APPLIANCE CONCENTRIC
VENT-AIR INTAKE TERMINAL

Figure courtesy of Carrier Corporation

Figure G2427.2.1(1)
DIRECT-VENT APPLIANCE

Photo courtesy of Lochinvar Corporation

Figure G2427.2.1(4)
DIRECT-VENT APPLIANCES

Photo courtesy of A.O. Smith Water Products

Figure G2427.2.1(5)
GRAVITY DIRECT-VENT WATER HEATER

Photo courtesy of Simpson Dura-Vent Company, Incorporated

Figure G2427.2.1(6)
DIRECT-VENT SYSTEM COMPONENTS

G2427.2.2 (503.2.4) Appliances with integral vents. *Appliances* incorporating integral venting means shall be installed in accordance with the manufacturer's instructions and Section G2427.8, Items 1 and 2.

❖ As it does with direct-vent appliances, the code trusts the manufacturer's installation instructions for the installation of appliances having an integral venting means. Integral-vent appliances such as rooftop HVAC units, outdoor makeup air heaters, and outdoor heaters for swimming pools have a built-in natural or mechanical venting means. The natural-draft integral venting means is simply a short vertical vent with a factory supplied vent cap assembly. The mechanical integral-vent appliances use supervised blowers to discharge the combustion gases directly to the atmosphere. The typical means of blower supervision is one or more pressure switches or a centrifugal switch mounted to the blower motor shaft. Integral-vent appliances must not be confused with unvented appliances and appliances that need not be vented in accordance with Section G2425.8. Integral-vent appliances must be vented to the outside atmosphere and thus are always designed for outdoor use only [see Section G2427.8, Items 1 and 2, and Commentary Figures G2427.2.2(1) through G2427.2.2(3)].

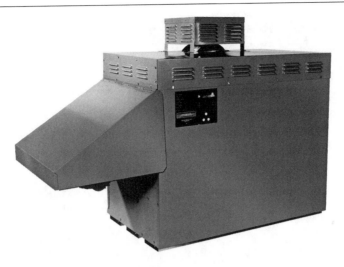

Figure courtesy of A.O. Smith Water Products

Figure G2427.2.2(1)
HOT WATER BOILER WITH INTEGRAL VENT FOR OUTDOOR INSTALLATION

Figure courtesy of Reznor/Thomas & Betts Corporation

Figure G2427.2.2(2)
OUTDOOR DUCT FURNACE WITH INTEGRAL POWER VENTING

G2427.3 (503.3) Design and construction. Venting systems shall be designed and constructed so as to convey all flue and *vent gases* to the outdoors.

❖ This section is stated in performance language and requires venting systems to convey the products of combustion to the outdoors. The reference to a "positive flow" does not imply that the venting system is under a positive pressure, but rather implies that the vent produces a dependable flow at its design operating pressure.

G2427.3.1 (503.3.1) Appliance draft requirements. A venting system shall satisfy the *draft* requirements of the *appliance* in accordance with the manufacturer's instructions.

❖ This section is stated in performance language and requires venting systems to produce the draft neces-

sary for the operation of the appliance being vented. An appliance manufacturer will specify the required type of venting means and, to assure the required amount of draft, may specify the size and height of a vent or chimney or may specify a mechanical draft system.

G2427.3.2 (503.3.2) Design and construction. *Appliances* required to be vented shall be connected to a venting system designed and installed in accordance with the provisions of Sections G2427.4 through G2427.16.

❖ As defined in Section G2403, a venting system includes natural-induced- and forced-draft systems, which are addressed in Sections G2427.3.3 through G2427.16.

Photo courtesy of Carrier Corporation

Figure G2427.2.2(3)
ROOFTOP HVAC UNIT WITH INTEGRAL POWER VENT

G2427.3.3 (503.3.3) Mechanical draft systems. Mechanical *draft* systems shall comply with the following:

1. Mechanical *draft* systems shall be *listed* and shall be installed in accordance with the manufacturer's instructions for both the *appliance* and the mechanical *draft* system.

2. *Appliances* requiring venting shall be permitted to be vented by means of mechanical *draft* systems of either forced or induced *draft* design.

3. Forced *draft* systems and all portions of induced *draft* systems under positive pressure during operation shall be designed and installed so as to prevent leakage of flue or *vent gases* into a building.

4. *Vent connectors* serving *appliances* vented by natural *draft* shall not be connected into any portion of mechanical *draft* systems operating under positive pressure.

5. Where a mechanical *draft* system is employed, provisions shall be made to prevent the flow of gas to the *main burners* when the *draft* system is not performing so as to satisfy the operating requirements of the *appliance* for safe performance.

6. The exit terminals of mechanical *draft* systems shall be not less than 7 feet (2134 mm) above finished ground level where located adjacent to public walkways and shall be located as specified in Section G2427.8, Items 1 and 2.

❖ The appliances and equipment installations addressed in this section use auxiliary or integral fans and blowers to force the flow of combustion products to the outdoors.

This section applies to externally installed power exhausters, integrally power-exhausted appliances and venting systems equipped with draft inducers. This section does not address fan-assisted Category I appliances and does not address direct-vent appliances. Category III and IV appliances are subject to the requirements of this section if they are not listed as direct-vent appliances. Some Category I appliances are field convertible to Category III. Item 4 also applies to Category I fan-assisted appliances because they are vented by natural draft.

Power exhausters are field-installed pieces of equipment that are independent of, but used in conjunction with, other appliances [see Commentary Figures G2427.3.3(1) through G2427.3.3(7)].

Power exhausters are typically used where other means of venting are impractical, impossible or uneconomical. Power exhausters are typically designed for use with gas-fired and oil-fired natural draft appliances. Some are also designed for use with fan-assisted appliances and appliances equipped with integral forced-draft (power-vent) fans. Power exhausters produce negative pressures at their inlet connection and positive pressures at their outlet (discharge connection). The most common installation locates the exhauster at the point of termination of the vent or chimney. In such installations, the vent or chimney between the appliance and the exhauster operates under negative pressure [see Commentary Figures G2427.3.3(1) through G2427.3.3(7)].

Induced draft systems employ separate field-installed units designed to boost draft in a natural (gravity) draft chimney or vent [see Commentary Figures G2427.3.3(8) and G2427.3.3(9)].

Power-vented (self-venting) appliances are equipped with factory-installed integral blowers that force the combustion products through the special venting systems that are addressed in Sections G2425.14 and G2427.4. Power-vented appliances include mid- to high-efficiency furnaces, boilers, water heaters and heating units that are designed for through-the-wall or through-the-roof venting [see

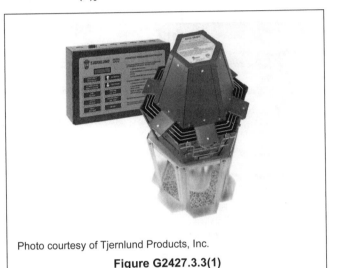

Photo courtesy of Tjernlund Products, Inc.

Figure G2427.3.3(1)
VERTICAL DRAFT INDUCER

Photo courtesy of Tjernlund Products, Inc.

Figure G2427.3.3(2)
HORIZONTAL POWER EXHAUSTER

Photo courtesy of Exhausto, Inc.

Figure G2427.3.3(3)
HORIZONTAL POWER EXHAUSTERS

Commentary Figure G2427.3.3(10)]. Appliances for outdoor installation are typically power vented and discharge directly to the atmosphere. They are addressed in Section G2427.2.2.

Although not specifically referred to as an appliance, a power exhauster (mechanical draft system) is considered to be a mechanical appliance appurtenance and therefore must bear the label of an approved agency in accordance with Item 1 (see definition of "Appliance"). A power exhauster is an essen-

tial component of the appliance installation it serves and must be installed in accordance with the manufacturer's installation instructions for both the power exhauster and the appliance served. Mechanical draft devices that are an integral part of an appliance are covered by the appliance listing.

Vent or chimney systems installed downstream (discharge side) of a power exhauster must be designed and approved for positive-pressure applica-

Photo courtesy of Field Controls Company

Figure G2427.3.3(4)
POWER EXHAUSTER OUTDOOR TERMINAL-TYPE UNIT

Photo courtesy of Tjernlund Products, Inc.

Figure G2427.3.3(6)
TYPICAL POWER EXHAUSTER

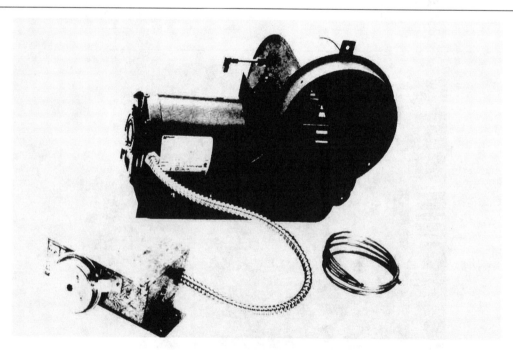

Photo courtesy of Tjernlund Products, Inc.

Figure G2427.3.3(5)
EXHAUSTER AND CONTROL PACKAGE

tions. For example, Type B vent cannot be used on the discharge side of an exhauster because such pipe is not designed for positive pressure applications. The application and installation of exhausters (power venters) must comply with the manufacturers' installation instructions for the exhausters and the appliance(s) served by the exhausters (see commentary, Section G2425.6).

There are three distinct variations in the use of blowers at the combustion chamber inlet working in conjunction with the fuel burner.

1. Blowers that supply turbulent combustion air to aid fuel-air mixing in a combustion chamber that is under negative pressure. To obtain proper negative overfire draft (which optimizes combustion) also requires steady negative (below atmospheric) pressure at the flue outlet. This will be produced by a natural draft chimney and may be controlled by a barometric draft regulator.

2. Blowers that supply sufficient combustion air and pressure to produce flow through the combustion chamber, but the combustion process does not need additional vent or chimney draft. This permits use of gravity or neutral draft venting products such as Type B vent for gas conversion burners. A draft regulator may be used for such equipment to prevent excess draft from affecting combustion efficiency.

Figure courtesy of Tjernlund Products, Inc.

**Figure G2427.3.3(7)
TYPICAL EXHAUSTER APPLICATIONS**

Photo courtesy of Tjernlund Products, Inc.

**Figure G2427.3.3(8)
DRAFT INDUCER**

Photo courtesy of Field Controls Company

**Figure G2427.3.3(9)
DRAFT INDUCER**

3. Blowers with enough power to overcome internal flue passage pressure losses (i.e., in fire-tube boilers) that also produce positive pressure at the outlet. This outlet pressure must be

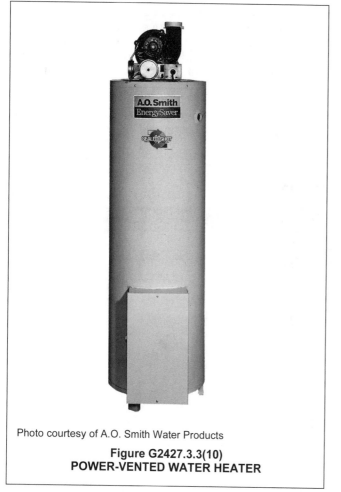

Photo courtesy of A.O. Smith Water Products

Figure G2427.3.3(10)
POWER-VENTED WATER HEATER

added to gravity draft as the motive force for flow in a chimney. If positive outlet pressure exists, the use of a sealed pressure-tight chimney is required.

These three types of equipment usually have integral blower/burner systems, and all could be considered as forced combustion systems. Only those described in the third paragraph truly produce forced chimney draft. When power-venting equipment is installed, it becomes an essential part of the appliance it serves. The appliance relies on the power-venting equipment (exhauster) to provide sufficient draft for proper appliance operation and for venting of the combustion products. Power exhausters are electrically interlocked with the appliance or appliances they serve to ensure that the appliances will not operate if there is insufficient draft to vent the products of combustion. If an exhauster fails, the appliances served by the exhauster would discharge the products of combustion into the building unless shut off by controls that monitor draft and draft-hood spillage. An improper or lacking interlock can cause the appliance(s) to malfunction and spill harmful flue gases into the building space. The electrical interlock (interconnection) is typically accomplished with controls provided by the exhauster manufacturer [see Commentary Figure G2427.3.3(11)].

The usual sequence of operation is as follows: The call for heat from the appliance operating control starts the exhauster, and pressure controls start the appliance only after adequate draft has been proven to exist. In some cases, temperature sensors are used in addition to pressure controls to sense draft-hood spillage or unusual vent system temperatures and shut off the appliance.

Item 6 addresses the concern for pedestrian safety relative to exposure to combustion gases (see Section G2427.8).

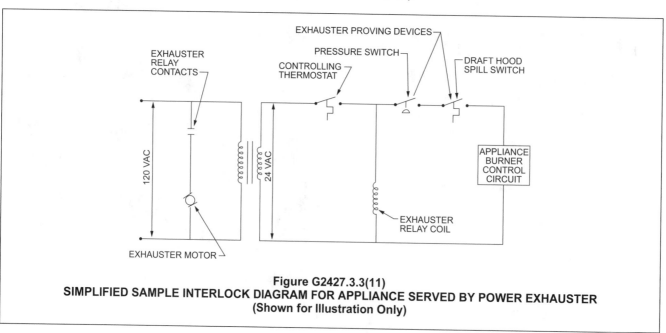

Figure G2427.3.3(11)
SIMPLIFIED SAMPLE INTERLOCK DIAGRAM FOR APPLIANCE SERVED BY POWER EXHAUSTER
(Shown for Illustration Only)

G2427.3.4 (503.3.5) Air ducts and furnace plenums. *Venting systems* shall not extend into or pass through any fabricated air duct or *furnace plenum*.

❖ If a vent or chimney or a connector passes through or extends into a duct or furnace plenum, it is possible to subject the venting system to negative or positive pressures and/or temperature extremes that could cause the venting system to deteriorate, fail to produce the required draft, produce condensation and/or leak combustion gases. A furnace plenum is a component of the ductwork system constructed as a junction box for multiple duct connections. The term "furnace plenum" is defined in Section G2403 and the term "fabricated air duct" refers to supply and return air ducts and does not refer to nonducted building cavities and plenums, such as spaces above ceilings used to convey air. Section G2427.3.5 addresses nonducted interstitial spaces used as plenums.

G2427.3.5 (503.3.6) Above-ceiling air-handling spaces. Where a venting system passes through an above-ceiling air-handling space or other nonducted portion of an air-handling system, the venting system shall conform to one of the following requirements:

1. The venting system shall be a *listed* special gas vent; other venting system serving a Category III or Category IV *appliance*; or other positive pressure vent, with joints sealed in accordance with the *appliance* or vent manufacturer's instructions.

2. The venting system shall be installed such that fittings and joints between sections are not installed in the above-ceiling space.

3. The venting system shall be installed in a conduit or enclosure with sealed joints separating the interior of the conduit or enclosure from the ceiling space.

❖ Vents are allowed to pass through interstitial spaces of a building used to convey environmental air such as above-ceiling return air-handling spaces (plenums), if the penetration/pass-through conforms to one of the three described installations (see Commentary Figure G2427.3.5). All three installations are intended to prevent pressure differentials between the air-handling space and the vent interior from causing leakage of combustion products from the vent.

G2427.4 (503.4) Type of venting system to be used. The type of venting system to be used shall be in accordance with Table G2427.4.

❖ The table lists various types of vents and the corresponding types of appliances that can be served by the vents. The vent system must be tested and specifically approved for use with the approved appliance. If the vent system is not a tested and labeled component of the appliance, the material must be approved for use with the appliance and installed in accordance with the manufacturer's installation instructions.

Gas-burning appliances (listed as Category II, III or IV) require special vent systems that are specific to the type of appliance. Special vent systems are typically associated with mid- to high-efficiency appliances and include vent materials such as polyvinyl chloride (PVC), chlorinated polyvinyl chloride (CPVC) and special alloys of stainless steel.

A primary consideration for the design of a special vent system that serves a high-efficiency appliance is

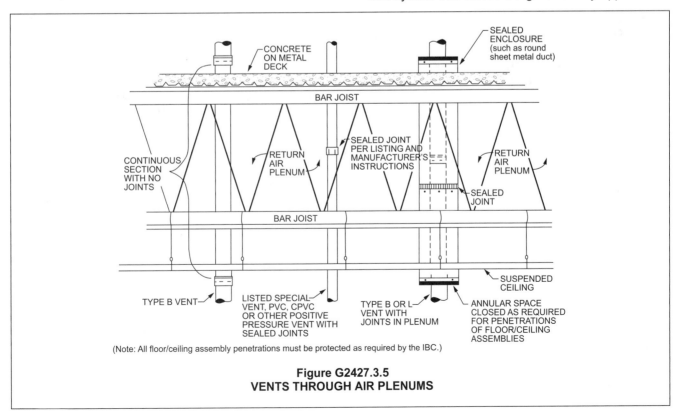

Figure G2427.3.5
VENTS THROUGH AIR PLENUMS

the selection of a material that is capable of withstanding the corrosive effects of condensate. Flue gas condensate is slightly acidic and corrosive and can deteriorate many vent materials. Sulfur in the fuel and halogens carried in the combustion air can combine with the combustion products to create acids. To prevent the corrosion of a vent and possible escape of flue gas, metallic flue passageways can be protected with appropriate coatings; however, any imperfections, pinholes or discontinuities in the coating can create areas for condensate to collect and accelerate the deterioration of the chimney or vent. The solution to this problem has been to use a material such as PVC, PP, CPVC or high-temperature plastic that is resistant to acidic condensation and compatible with the temperature of the appliance flue gases. Plastic pipe is a suitable material because it is impervious to corrosive condensation and can easily be designed to drain off accumulated condensate. The flue-gas temperatures of most condensing-type appliances are not high enough to affect such materials adversely.

Special vent systems must be designed and installed in compliance with the manufacturers' instructions, which will specify installation requirements that are specific to that type of vent and appliance. For example, a maximum developed length, a maximum number of directional fittings, fitting turn radius and specific support requirements will be specified for special vents using plastic pipe. Special vent systems also have special vent termination requirements, and typically incorporate condensate collection and drainage fittings and connections.

Gas- and oil-burning appliances can be used with a vent system if the appliance is so labeled and approved.

The type and size of the vent must be as dictated by the manufacturer's installation instructions for the appliance. The design and installation instructions provided by the vent manufacturer must also be consulted when designing a vent system for any particular application. Commentary Figure G2427.4 shows some of the variables that affect vent system sizing. In addition to Section G2428, the vent manufacturers' design and application handbooks and installation instructions contain information and tables for the sizing and application of the vent systems.

TABLE G2427.4. See below.

❖ Table G2427.4 lists the currently available types of appliances and the types of venting systems to be used with such appliances. In the first row, the three types of appliances can be matched up with any of the five types of venting systems, if allowed by the appliance manufacturer's installation instructions. The fourth row up from the bottom addresses unlisted appliances that are not allowed by the code except under the alternative approval provisions of Section R104.11.

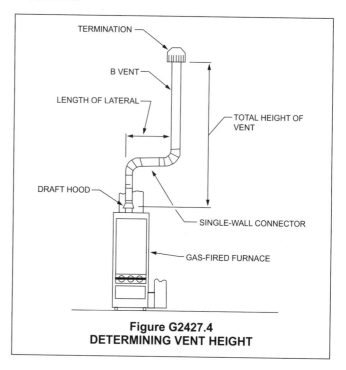

Figure G2427.4
DETERMINING VENT HEIGHT

TABLE G2427.4 (503.4)
TYPE OF VENTING SYSTEM TO BE USED

APPLIANCES	TYPE OF VENTING SYSTEM
Listed Category I *appliances* *Listed appliances* equipped with draft hood *Appliances* listed for use with Type B gas vent	Type B gas *vent* (Section G2427.6) *Chimney* (Section G2427.5) Single-wall metal pipe (Section G2427.7) *Listed* chimney lining system for gas venting (Section G2427.5.2) Special gas *vent* listed for these appliances (Section G2427.4.2)
Listed vented wall furnaces	Type B-W gas vent (Sections G2427.6, G2436)
Category II *appliances*	As specified or furnished by manufacturers of *listed appliances* (Sections G2427.4.1, G2427.4.2)
Category III *appliances*	As specified or furnished by manufacturers of *listed appliances* (Sections G2427.4.1, G2427.4.2)
Category IV *appliances*	As specified or furnished by manufacturers of *listed appliances* (Sections G2427.4.1, G2427.4.2)
Unlisted *appliances*	Chimney (Section G2427.5)
Decorative *appliances* in vented fireplaces	Chimney
Direct-vent *appliances*	See Section G2427.2.1
Appliances with integral vent	See Section G2427.2.2

2015 INTERNATIONAL RESIDENTIAL CODE® COMMENTARY

G2427.4.1 (503.4.1) Plastic piping. Where plastic piping is used to vent an appliance, the appliance shall be listed for use with such venting materials and the appliance manufacturer's installation instructions shall identify the specific plastic piping material.

❖ Commonly used plastic venting materials include PVC, PP and CPVC (see Commentary Figure G2427.4.1). This section no longer requires code official approval of vents because the code official is not in a position to determine what venting materials are suitable and safe for venting various appliances. Such decisions should be left to design engineers, manufacturers and the testing and listing agencies that list the appliances. The previous code text did not actually require that the appliance be listed for use with specific venting system materials, although this was implied by the text and it is required in the appliance standards.

The appliance manufacturer's installation instructions must clearly specify what plastic materials are required or allowed for venting an appliance. The installation instructions will be consistent with how the appliance was tested by the listing agency. The product standards for gas appliances contain various testing procedures for plastic venting systems. The appliance manufacturer determines the type of plastic vent that is suitable for venting its product, and the testing and listing agency tests the appliance with that venting system for compliance with the product standards. There must not be any uncertainty about what type of venting system is required for any appliance, so that venting system failures can be avoided. The definition of "Vent" does not include plastic pipe such as PVC, ABS and CPVC because such pipes are not currently listed as factory-built venting systems. Section 503.4.1 refers to the plastic pipe as a material used for venting; it is not referred to as a vent conforming to the definition. PVC, ABS and CPVC pipe manufacturers do not recommend that their pipe be used for appliance venting because such products are not currently listed for such applications. There are polypropylene venting systems on the market that are listed to UL 1738 as appliance venting systems, and they would fall under the definition of "Vent" (see Commentary Figure G2427.4.1).

G2427.4.1.1 (503.4.1.1) (IFGS) Plastic vent joints. Plastic *pipe* and fittings used to vent *appliances* shall be installed in accordance with the *appliance* manufacturer's instructions. Where a primer is required, it shall be of a contrasting color.

❖ Experience has shown that plastic pipe and fittings used for venting appliances are not always joined together properly and joint failures occur resulting in leakage of combustion products. Piping materials such as PVC and CPVC usually require the use of a primer in the solvent welding process. The color of the primer, typically purple, must contrast with the color of the pipe, fittings and solvent cement so as to be visible for inspection. The use of a clear primer will not be evident on a completed solvent weld joint.

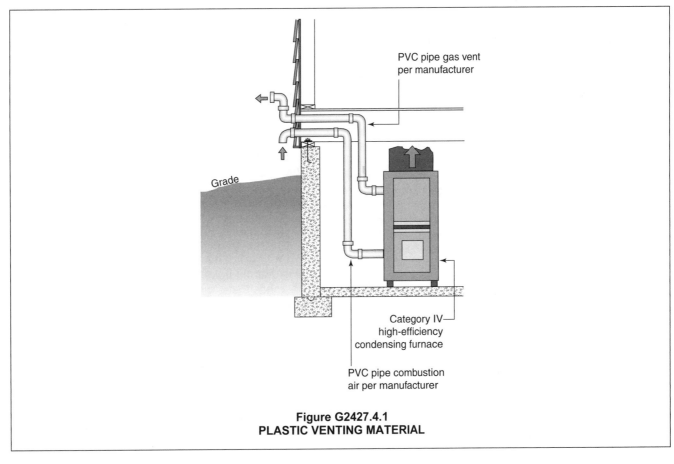

Figure G2427.4.1
PLASTIC VENTING MATERIAL

G2427.4.2 (503.4.2) Special gas vent. Special gas *vent* shall be *listed* and installed in accordance with the special gas *vent* manufacturer's instructions.

❖ Special gas vents include high-temperature plastic pipe, stainless steel alloy pipe and positive-pressure pipe that are specified by the appliance manufacturer for use with a specific appliance. This section does not necessarily require that all plastic pipe be listed as a venting system component. For example, the installation instructions for Category IV appliances often specify PVC, CPVC and ABS plastic plumbing pipe for venting the appliances, and such pipe is not listed as a component of a venting system (see Section G2427.4.1).

G2427.5 (503.5) Masonry, metal and factory-built chimneys. Masonry, metal and factory-built *chimneys* shall comply with Sections G2427.5.1 through G2427.5.9.

❖ See Sections G2427.5.1 through G2427.5.9.

G2427.5.1 (503.5.1) Factory-built chimneys. Factory-built *chimneys* shall be installed in accordance with the manufacturer's instructions. Factory-built *chimneys* used to vent *appliances* that operate at a positive vent pressure shall be *listed* for such application.

❖ Prefabricated chimney systems must bear the label of an approved agency. A label is required on all components of the chimney system such as the pipe sections, shields, fireblocks, fittings, termination caps and supports. The label states information such as the type of appliance the chimney was tested for use with, a reference to the manufacturer's installation instructions and the minimum required clearances to combustibles [see Commentary Figures G2403(4) and G2403(5)].

Most factory-built chimneys are either of the double-walled fiber-insulated design or the triple-walled air-cooled design. Factory-built chimneys are constructed of stainless steel inner liners with stainless steel outer walls. Factory-built chimneys, like vent systems, are composed of components that must be installed as a complete system. Components from different manufacturers are not designed to be mixed and installed together.

The manufacturer's instructions contain sizing criteria and requirements for every aspect of a factory-built chimney installation. The requirements include component assembly, clearances to combustibles, support, terminations, connections, protection from damage and fireblocking.

Any vent or chimney that conveys vent gases under positive pressure must be factory designed for that application. There is no allowable method of taping or sealing a chimney or vent that will convert a nonpositive-pressure chimney or vent into a positive-pressure chimney.

G2427.5.2 (503.5.3) Masonry chimneys. Masonry *chimneys* shall be built and installed in accordance with NFPA 211 and shall be lined with *approved* clay *flue lining*, a *listed chimney* lining system or other *approved* material that will resist corrosion, erosion, softening or cracking from vent gases at temperatures up to 1,800°F (982°C).

Exception: Masonry *chimney* flues serving *listed* gas *appliances* with *draft hoods*, Category I *appliances* and other gas *appliances listed* for use with Type B vents shall be permitted to be lined with a *chimney* lining system specifically *listed* for use only with such *appliances*. The liner shall be installed in accordance with the liner manufacturer's instructions. A permanent identifying label shall be attached at the point where the connection is to be made to the liner. The label shall read: "This *chimney* liner is for *appliances* that burn gas only. Do not connect to solid or liquid fuel-burning appliances or incinerators."

❖ A chimney liner might be listed for use with solid-, liquid- and gas-fuel-fired appliances. The label required by this section is intended to warn the unknowing installer who sees a masonry chimney and thinks that it can serve a solid- or liquid-fuel-fired appliance. The label would be applicable to chimney liner systems that are designed for gas-fired appliances only (see Section G2425.3).

G2427.5.3 (503.5.4) Chimney termination. *Chimneys* for residential-type or low-heat *appliances* shall extend not less than 3 feet (914 mm) above the highest point where they pass through a roof of a building and not less than 2 feet (610 mm) higher than any portion of a building within a horizontal distance of 10 feet (3048 mm). *Chimneys* for medium-heat *appliances* shall extend not less than 10 feet (3048 mm) higher than any portion of any building within 25 feet (7620 mm). *Chimneys* shall extend not less than 5 feet (1524 mm) above the highest connected *appliance draft hood* outlet or *flue collar*. Decorative shrouds shall not be installed at the termination of factory-built *chimneys* except where such shrouds are *listed* and *labeled* for use with the specific factory-built *chimney* system and are installed in accordance with the manufacturer's instructions.

❖ Low-heat and residential-type chimneys must extend at least 3 feet (914 mm) above the roof, measured from the highest point of the roof penetration. They must also be at least 2 feet (610 mm) higher than any portion of the roof within a 10-foot (3048 mm) horizontal distance [see Commentary Figures G2427.5.3(1) and G2427.5.3(2)].

The 2-foot (610 mm) termination requirement is intended to prevent wind and pressure zones from reducing the amount of draft produced by the chimney. Wind and wind-induced eddy currents can react with building structural surfaces, thereby creating air pressure zones that can diminish chimney draft and cause reverse flow (backdraft) in the chimney. Loss of draft will cause the appliance or fireplace served by the chimney to discharge combustion products into the building interior. Locating the chimney outlet well into the undisturbed wind stream and away from the cavity and wake (eddy) zones around the building can counteract the adverse effects and also prevent the reentry of vent gases into the building through openings and fresh air intakes. Terminating a chimney in the eddy current area recirculates the combustion products and allows them to enter the building via infiltration, wall openings and air intakes. A chimney terminal properly located above the eddy current area allows the wind to carry the combustion products away from the building. Factory-built chimneys have termination caps that are part of the labeled chimney system and that must be installed. Like vent systems, the chimney cap is designed to keep out precipitation and animals and minimize the adverse effects of wind.

Decorative shrouds are designed to be aesthetically pleasing and to hide chimney terminations from view.

These shrouds have, in some cases, caused fires resulting from overheated combustible construction and the accumulation of debris and animal nesting. Shrouds can also interfere with the functioning of a vent or chimney. The listing of the shroud must state the specific make and model of chimney with which the shroud is intended to be used. Some manufacturers are manufacturing shrouds that are listed for use with a specific make and model of chimney. Most often, shrouds are field-constructed of sheet metal, masonry or stucco on frame.

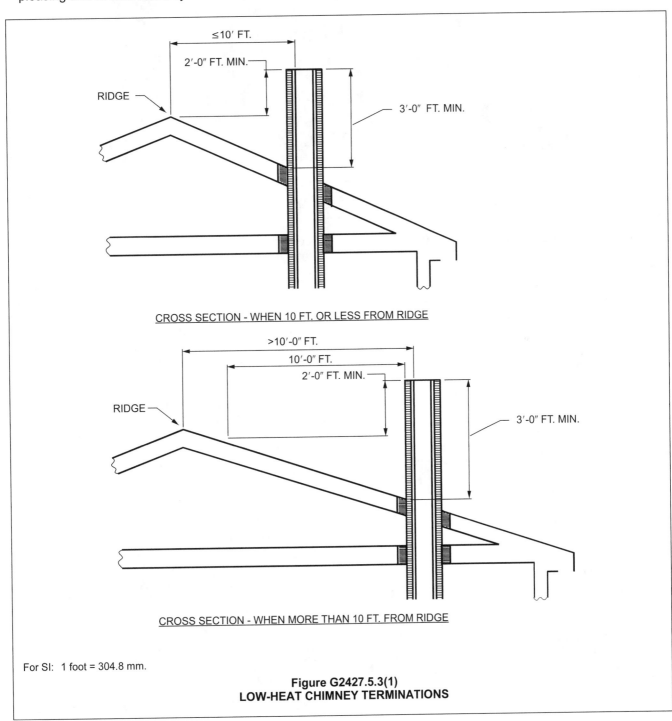

For SI: 1 foot = 304.8 mm.

Figure G2427.5.3(1)
LOW-HEAT CHIMNEY TERMINATIONS

Figure G2427.5.3(2)
DECORATIVE SHROUDS

G2427.5.4 (503.5.5) Size of chimneys. The effective area of a *chimney* venting system serving *listed appliances* with *draft hoods*, Category I *appliances*, and other *appliances listed* for use with Type B vents shall be determined in accordance with one of the following methods:

1. The provisions of Section G2428.

2. For sizing an individual *chimney* venting system for a single *appliance* with a *draft hood*, the effective areas of the *vent connector* and *chimney* flue shall be not less than the area of the *appliance flue collar* or *draft hood* outlet, nor greater than seven times the *draft hood* outlet area.

3. For sizing a *chimney* venting system connected to two *appliances* with *draft hoods*, the effective area of the *chimney* flue shall be not less than the area of the larger *draft hood* outlet plus 50 percent of the area of the smaller *draft hood* outlet, nor greater than seven times the smallest *draft hood* outlet area.

4. *Chimney venting systems* using mechanical *draft* shall be sized in accordance with *approved* engineering methods.

5. Other *approved* engineering methods.

❖ Chimneys can be sized in accordance with Section G2428, by Item 2, 3 or 4 of this section or by an engineering method acceptable to the code official. Some appliances are not categorized; however, appliances with draft hoods and appliances listed for use with Type B vents would fit in Category I if they were categorized. The manufacturer's instructions for some Category I appliances, such as fan-assisted furnaces and

boilers, may not allow connecting the appliance to a masonry chimney except under specific limited conditions.

Item 2 applies only to a single appliance that is factory equipped with a draft hood. The "seven times rule" prevents the chimney from being too large, which could result in poor draft and condensation problems.

Like vents, chimneys that are too large in area can be as problematic as chimneys that are too small in area.

Item 3 is limited to only one venting arrangement. It can only be used for a chimney that serves two appliances, both of which must be factory-equipped with draft hoods. The method is based on the areas of the chimneys and draft hood outlets, not on the diameters.

Item 4 recognizes that some manufacturers of mechanical draft equipment (exhausters) have their own sizing methodology specific to their product. Recall that Item 1 of Section G2427.3.3 requires listing of such equipment.

G2427.5.5 (503.5.6) Inspection of chimneys. Before replacing an existing *appliance* or connecting a vent *connector* to a *chimney*, the *chimney* passageway shall be examined to ascertain that it is clear and free of obstructions and it shall be cleaned if previously used for venting solid or liquid fuel-burning appliances or *fireplaces*.

❖ If a chimney is going to serve a new appliance installation or a replacement appliance installation, the chimney must be inspected to determine whether it is still serviceable and free of deposits. Chimneys can become obstructed by debris such as leaves, animal carcasses, loose mortar and pieces of masonry and liner. Combustible deposits can accumulate on chimney walls used to vent liquid- and solid-fuel-fired appliances, and this section mandates that the chimney be cleaned if it served such appliances in the past. Masonry chimneys are especially susceptible to internal deterioration and should be inspected regularly, in addition to when a new or replacement appliance is connected.

G2427.5.5.1 (503.5.6.1) Chimney lining. *Chimneys* shall be lined in accordance with NFPA 211.

Exception: Where an existing chimney complies with Sections G2427.5.5 through G2427.5.5.3 and its sizing is in accordance with Section G2427.5.4, its continued use shall be allowed where the *appliance* vented by such *chimney* is replaced by an *appliance* of similar type, input rating and efficiency.

❖ The liner forms the flue passageway and is the actual conductor of all products of combustion. The chimney liner must be able to withstand exposure to high temperatures and corrosive chemicals. The chimney lining protects the masonry construction of the chimney walls and allows the chimney to be constructed gas tight. This section regulates liners and relining systems. Liners are often used to reline an existing chimney to salvage a masonry chimney or allow connection of higher-efficiency appliances.

Flue lining systems for masonry chimneys include clay tile, poured-in-place refractory materials and stainless steel pipe. Chimney lining systems for chimneys serving gas-fired appliances include stainless steel and aluminum piping (uninsulated or insulated) designed for both new construction and for relining existing chimneys [see Commentary Figures G2427.5.5.1(1) and G2427.5.5.1(2)]. Type B vent is also used as a chimney liner, although it is not considered a lining system, and in such cases the chimney is actually a masonry chase containing a vent.

Relining of existing chimneys is quite common now because of the incompatibility of mid-efficiency appliances and masonry chimneys. Relining systems allows reuse of existing chimneys, thereby saving expense and retaining desirable architectural features of buildings.

The exception is intended to allow an unlined chimney to serve a new appliance that is installed to replace a previously served appliance if the new appliance does not create any different operating conditions in the chimney; that is, the volume, water vapor content and dilution air content of the new appliance flue gases are the same as those of the old appliance.

The exception assumes that the existing chimney and the existing appliances operated as intended by this code. If there is evidence that the previous chimney and appliance system was not working as required by this code, the exception is not intended to be appli-

Photo courtesy of Selkirk L.L.C.

Figure G2427.5.5.1(1)
CHIMNEY LINING SYSTEM FOR GAS APPLIANCES

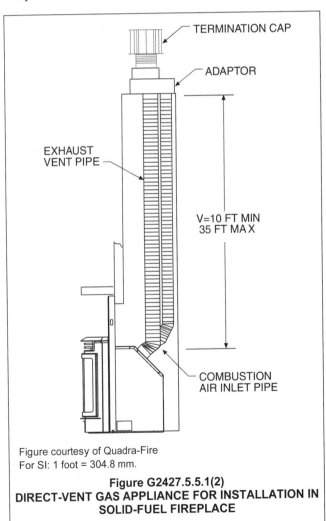

Figure courtesy of Quadra-Fire
For SI: 1 foot = 304.8 mm.

Figure G2427.5.5.1(2)
DIRECT-VENT GAS APPLIANCE FOR INSTALLATION IN SOLID-FUEL FIREPLACE

cable. Because this exception would allow an installation that is in conflict with the typical appliance manufacturer's installation instructions, it is in effect negated because the manufacturer's instructions will prevail in this case (see Sections G2408.1 and G2427.5.5.3). Before the exception could ever be applied, the chimney would have to be inspected and determined to be clear, clean, free of obstructions, safe for the intended use and properly sized for the appliances served. Also, the replacement appliance(s) would have to be the same type with nearly the same input rating and efficiency to make sure that the same amount of heat and dilution air enters the chimney. This will rarely, if ever, be encountered in the field, therefore, for multiple reasons, the exception will likely never be applicable.

G2427.5.5.2 (503.5.6.2) Cleanouts. Cleanouts shall be examined to determine if they will remain tightly closed when not in use.

❖ A cleanout door or panel must be capable of securely sealing the opening to prevent air from entering the chimney, which would reduce draft for the appliances served. Under certain conditions, a faulty cleanout could also cause vent gases to leak from the chimney.

G2427.5.5.3 (503.5.6.3) Unsafe chimneys. Where inspection reveals that an existing *chimney* is not safe for the intended application, it shall be repaired, rebuilt, lined, relined or replaced with a vent or *chimney* to conform to NFPA 211 and it shall be suitable for the *appliances* to be vented.

❖ If a chimney is inspected and found to be unsafe for any reason, it must be made safe or be replaced. See Section G2425.3. Some chimneys and appliances are not compatible as dictated by the appliance and/or factory-built chimney manufacturer. Masonry chimneys usually become unsafe as a result of deteriorated mortar joints and liners. Other causes include structural instability, cracks, faulty cleanouts, deposit formation and moisture damage.

G2427.5.6 (503.5.7) Chimneys serving appliances burning other fuels. *Chimneys* serving *appliances* burning other fuels shall comply with Sections G2427.5.6.1 through G2427.5.6.4.

❖ See the commentary to Sections G2427.5.6.1 through G2427.5.6.4.

G2427.5.6.1 (503.5.7.1) Solid fuel-burning appliances. An *appliance* shall not be connected to a *chimney* flue serving a separate *appliance* designed to burn solid fuel.

❖ Each solid fuel-burning appliance or fireplace must be connected to a dedicated independent chimney, or a dedicated independent flue in multiple-flue chimney constructions. Solid-fuel-burning appliances and fireplaces cannot share a common chimney or flue-way with any other appliance or fireplace.

Solid-fuel-burning appliances produce creosote deposits on the interior walls of chimney liners. The creosote formation is highly combustible and creates a potential fire hazard. Because of the potential for a chimney fire, other connections to the chimney can allow fire to break out of the chimney into the building.

Also, chimney passageways can become restricted by the creosote formations. If other appliances were vented to such chimneys, combustion products could be discharged into the building and the other appliances might seriously interfere with the draft for the solid-fuel-burning appliance.

This section is also intended to prevent the possibility of creosote leaking out of the chimney through appliance connectors.

Combination (dual fuel) gas- or oil- and solid-fuel-burning appliances are designed to be connected to a single chimney passageway. Such dual-fuel appliances must be listed, labeled and installed in accordance with the manufacturers' instructions.

G2427.5.6.2 (503.5.7.2) Liquid fuel-burning appliances. Where one *chimney* flue serves gas *appliances* and liquid fuel-burning appliances, the appliances shall be connected through separate openings or shall be connected through a single opening where joined by a suitable fitting located as close as practical to the *chimney*. Where two or more openings are provided into one *chimney* flue, they shall be at different levels. Where the appliances are automatically controlled, they shall be equipped with *safety shutoff devices*.

❖ Gas-fired and oil-fired appliances are allowed to share a chimney or flue. The code does not include a sizing method for such arrangements; therefore, the system would have to be engineered or otherwise approved by the code official. Placing the openings at different levels in the chimney flue will minimize interference between the flow of the chimney connectors. The last sentence is consistent with the design of all automatically controlled appliances listed to today's appliance standards as required by this code.

G2427.5.6.3 (503.5.7.3) Combination gas- and solid fuel-burning appliances. A combination gas- and solid fuel-burning *appliance* shall be permitted to be connected to a single chimney flue where equipped with a manual reset device to shut off gas to the *main burner* in the event of sustained backdraft or flue gas spillage. The *chimney* flue shall be sized to properly vent the *appliance*.

❖ A dual fuel appliance, gas and solid fuel, can be served by a single chimney or flue if the appliance is equipped with a safety control that monitors chimney spillage. A solid-fuel fire cannot be turned on and off like other fuel fires; thus, it is possible that the gas burner could be operated while the solid-fuel fire is still burning, and the chimney could be overloaded, causing spillage of combustion gases. The chimney size would have to be engineered or would have to comply with the appliance manufacturer's instructions.

G2427.5.6.4 (503.5.7.4) Combination gas- and oil fuel-burning appliances. A *listed* combination gas- and oil fuel-burning *appliance* shall be permitted to be connected to a single *chimney* flue. The *chimney* flue shall be sized to properly vent the *appliance*.

❖ See the commentary to Section G2427.5.6.3.

G2427.5.7 (503.5.8) Support of chimneys. All portions of *chimneys* shall be supported for the design and weight of the materials employed. Factory-built *chimneys* shall be supported and spaced in accordance with the manufacturer's installation instructions.

❖ Chimneys, including factory-built, are very heavy compared to vents and need substantial support to carry their weight and prevent displacement. Factory-built chimneys require special support fittings at offsets to prevent the elbows from being damaged by bearing the weight of the chimney sections above the offset. Chimneys, like piping systems, often suffer from lack of adequate support (see Section G2426.6).

G2427.5.8 (503.5.9) Cleanouts. Where a *chimney* that formerly carried flue products from liquid or solid fuel-burning appliances is used with an *appliance* using *fuel gas*, an accessible cleanout shall be provided. The cleanout shall have a tight-fitting cover and be installed so its upper edge is at least 6 inches (152 mm) below the lower edge of the lowest *chimney* inlet opening.

❖ If an existing chimney has no cleanout and was used for oil or solid-fuel appliances, a cleanout must be added. The cleanout allows access for a person to inspect and clean the chimney and monitor deposits on the interior walls. The cleanout must be located so that vent gases will not be in contact with the cleanout door/cover.

G2427.5.9 (503.5.10) Space surrounding lining or vent. The remaining space surrounding a *chimney* liner, gas vent, special gas *vent* or plastic *piping* installed within a masonry *chimney* flue shall not be used to vent another *appliance*. The insertion of another liner or vent within the *chimney* as provided in this *code* and the liner or vent manufacturer's instructions shall not be prohibited.

The remaining space surrounding a *chimney* liner, gas vent, special gas vent or plastic *piping* installed within a masonry, metal or factory-built *chimney* shall not be used to supply *combustion air*. Such space shall not be prohibited from supplying *combustion air* to *direct-vent appliances* designed for installation in a solid fuel-burning *fireplace* and installed in accordance with the manufacturer's instructions.

❖ If a vent or a chimney liner is installed within a chimney flue, there will be space between the vent or liner and the chimney walls. This space must not be used to convey vent gases because of its irregular size and geometry and because the vent gases could damage the vent or liner installed in the chimney. A metallic vent or liner could be corroded by surrounding vent gases, and plastic pipes could be overheated. Liner systems and vent systems are not listed for use within an atmosphere of vent gases. Installing multiple liners or vents within a chimney is not the same as using the annular space between a vent or liner and the inside chimney walls.

The space between a liner, vent or pipe and the interior chimney walls cannot be used as a duct for conveying combustion air. That space would have poor air-flow characteristics, and moving cold combustion air over a liner or vent can cause condensation in the venting system. An exception is made for direct-vent appliances that are listed for installation in solid-fuel-burning fireplaces.

G2427.6 (503.6) Gas vents. Gas vents shall comply with Sections G2427.6.1 through G2427.6.11. (See Section G2403, Definitions.)

❖ See the commentary to Sections G2427.6.1 through G2427.6.11.

G2427.6.1 (503.6.1) Installation, general. Gas vents shall be installed in accordance with the manufacturer's instructions.

❖ Gas vents, such as Type B vents, come with installation instructions available at the place of purchase or enclosed within the boxed items. The manufacturers also will provide, upon request, very detailed and informative design manuals that discuss sizing and venting fundamentals. These design manuals are invaluable for someone who installs, inspects or designs Type B vent systems (see Sections G2426.1 and G2426.5).

G2427.6.2 (503.6.2) Type B-W vent capacity. A Type B-W gas vent shall have a listed capacity not less than that of the *listed vented wall furnace* to which it is connected.

❖ A Type B-W vent is an oval shaped version of a Type B vent designed to fit within a nominal 2-inch by 4-inch framed wall and serve a wall furnace (see Section G2436).

G2427.6.3 (503.6.4) Gas vent terminations. A gas vent shall terminate in accordance with one of the following:

1. Gas vents that are 12 inches (305 mm) or less in size and located not less than 8 feet (2438 mm) from a vertical wall or similar obstruction shall terminate above the roof in accordance with Figure G2427.6.3.

2. Gas vents that are over 12 inches (305 mm) in size or are located less than 8 feet (2438 mm) from a vertical wall or similar obstruction shall terminate not less than 2 feet (610 mm) above the highest point where they pass through the roof and not less than 2 feet (610 mm) above any portion of a building within 10 feet (3048 mm) horizontally.

3. As provided for direct-vent systems in Section G2427.2.1.

4. As provided for *appliances* with integral vents in Section G2427.2.2.

5. As provided for mechanical *draft* systems in Section G2427.3.3.

❖ This section duplicates typical vent manufacturers' instructions and emphasizes that a vent is a system of components that are all necessary for proper functioning. The cap must be as specified by the vent manufacturer and is a listed component. A vent cap not only keeps out moisture, debris and animals, it also serves to prevent wind interference that could negatively affect the vent's ability to produce the required draft. Vent caps are tested for flow resistance and performance in wind.

This section is consistent with typical vent manufacturers' instructions and applies only to vents equipped with listed caps as required by Section G2427.6.5. The relationship between vent terminal height and roof pitch is based on reducing the effects of wind and maintaining a minimum separation between the vent terminal and the roof surface.

See Section G2428.2.9 for a discussion of vent terminations that extend above the roof higher than required by the code.

Figure G2427.6.3 requires greater vent height above the roof as the roof approaches being a vertical surface. The greater the roof pitch, the greater the effect of wind striking the roof surface and the closer the cap becomes (horizontally) to the roof surface (see Figure G2427.6.3).

It is a common misapplication for code users to apply chimney termination height requirements to vents, thereby causing vents to extend above roofs much higher than required in many cases. For example, a 6-inch (152 mm) diameter Type B vent must extend above a simple gabled roof only 1 foot (305 mm) for roofs having up to 6/12 pitch, if the terminal is not within 8 feet (2438 mm) of a vertical surface such as an upper-story exterior wall. Consideration should be given to vent location in a building in the design/installation phase so that the vent termination will not be difficult or require long extensions of vent pipe exposed to the outdoors. Vent piping exposed to the outdoors encourages condensation in cold weather, could require guy wires or braces depending on height and is considered aesthetically unattractive. Another good reason to plan for vent location during the design phase of a building is to allow straight vertical runs of vent and avoid offsets. Good planning can also prevent vents from passing through the roof on the street side of the building. Quite often, vent offsets in attics are installed for the sole purpose of penetrating the roof on the back side of a building.

Vent termination heights are affected by the pitch of the roof and the proximity of walls and other vertical or near vertical structural components. Where vent terminal locations within 8 feet (2438 mm) of a gable end, upper story or other vertical surface cannot be avoided, vent terminations must be regulated in a manner similar to chimneys [i.e., 2-feet (610 mm) minimum height and not less than 2 feet (610 mm) above any part of building within 10 feet (3048 mm)] (see Commentary Figure G2427.6.3). Commentary Figure G2427.6.3 does not apply to vent sizes larger than 12 inches (305 mm). This is consistent with manufacturers' instructions and relates to the testing of gas vents to the referenced standard.

The intent of Section G2428.2.9 is for vents exposed to the outdoors below the roof penetration to be sized by a method other than the tables in Section G2428. This means that if a vent is within 8 feet (2438 mm) of a vertical wall or is larger in size than 12 inches (305 mm), it must extend 2 feet (3048 mm) above any part of the building within 10 feet horizontally, and this could

expose a substantial run of vent to the outdoors. In such cases, the tables of Section G2428 would not apply, and the vent system would have to be engineered.

FIGURE G2427.6.3. See page 24-185.

❖ See the commentary to Section G2427.6.3

G2427.6.3.1 (503.6.4.1) Decorative shrouds. Decorative shrouds shall not be installed at the termination of gas vents except where such shrouds are *listed* for use with the specific gas venting system and are installed in accordance with manufacturer's instructions.

❖ Decorative shrouds have become a popular architectural feature. They are designed to be aesthetically pleasing and to hide chimney and vent terminations. These shrouds have, in some cases, caused fires resulting from overheated combustible construction and the accumulation of debris and animal nesting. Shrouds can also interfere with the functioning of a chimney or vent system. Shrouds are allowed only where they are listed and labeled for use with the specific factory-built chimney system or gas venting system (see Commentary Figure G2427.6.3.1).

G2427.6.4 (503.6.5) Minimum height. A Type B or L gas vent shall terminate at least 5 feet (1524 mm) in vertical

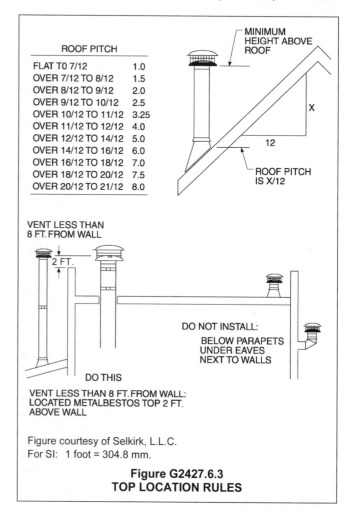

Figure courtesy of Selkirk, L.L.C.
For SI: 1 foot = 304.8 mm.

Figure G2427.6.3
TOP LOCATION RULES

height above the highest connected *appliance draft hood* or *flue collar*. A Type B-W gas vent shall terminate not less than 12 feet (3658 mm) in vertical height above the bottom of the wall *furnace*.

❖ The amount of draft produced by a vent is directly related to the height of the vent. A minimum height must be established to produce the minimum draft necessary for the appliance served. This is made evident by looking at the tables in Section G2428 that indicate increasing vent capacity as the vent height increases. This is a result of the increase in draft and vent flow velocity. The tables in Section G2428 and some vent manufacturers' instructions indicate a minimum height of 6 feet (1829 mm) because this is the lowest entry for height in the sizing tables.

This section is often violated for appliances installed near the roof such as suspended unit heaters and appliances installed under shed roofs.

G2427.6.5 (503.6.6) Roof terminations. Gas vents shall extend through the roof flashing, roof jack or roof thimble and terminate with a *listed* cap or *listed* roof assembly.

❖ This section duplicates the typical vent manufacturer's instructions and emphasizes that a vent is a system of components that are all necessary for proper functioning. The cap must be as specified by the vent manufacturer and is a listed component. A vent cap not only keeps out moisture, debris and animals, it also serves to prevent wind interference that could negatively affect the vent's ability to produce the required draft. Vent caps are tested for flow resistance and perfor-

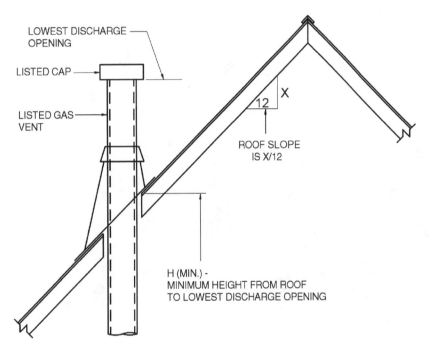

ROOF SLOPE	H (minimum) ft
Flat to $^6/_{12}$	1.0
Over $^6/_{12}$ to $^7/_{12}$	1.25
Over $^7/_{12}$ to $^8/_{12}$	1.5
Over $^8/_{12}$ to $^9/_{12}$	2.0
Over $^9/_{12}$ to $^{10}/_{12}$	2.5
Over $^{10}/_{12}$ to $^{11}/_{12}$	3.25
Over $^{11}/_{12}$ to $^{12}/_{12}$	4.0
Over $^{12}/_{12}$ to $^{14}/_{12}$	5.0
Over $^{14}/_{12}$ to $^{16}/_{12}$	6.0
Over $^{16}/_{12}$ to $^{18}/_{12}$	7.0
Over $^{18}/_{12}$ to $^{20}/_{12}$	7.5
Over $^{20}/_{12}$ to $^{21}/_{12}$	8.0

For SI: 1 foot = 304.8 mm.

FIGURE G2427.6.3 (503.6.4)
TERMINATION LOCATIONS FOR GAS VENTS WITH LISTED CAPS 12 INCHES OR LESS
IN SIZE AT LEAST 8 FEET FROM A VERTICAL WALL

mance in wind. The emphasis here is that the vent system must extend through the roof all the way to the cap using components that are listed as part of the listed venting system.

G2427.6.6 (503.6.7) Forced air inlets. Gas vents shall terminate not less than 3 feet (914 mm) above any forced air inlet located within 10 feet (3048 mm).

❖ If the vent termination is 10 feet (3048 mm) or more from the forced-air (mechanical intake opening), this section does not apply. The intent is to prevent combustion gases from being drawn into the building through powered air intakes of any type (see Commentary Figure G2427.6.6).

G2427.6.7 (503.6.8) Exterior wall penetrations. A gas *vent* extending through an exterior wall shall not terminate adjacent to the wall or below eaves or parapets, except as provided in Sections G2427.2.1 and G2427.3.3.

❖ The termination locations prohibited by this section would result in poor draft or back draft because of the effects of wind. Also, termination in such locations could result in a fire hazard, damage to the structure and entry of vent gases into the building. If a gas vent (typically Type B) penetrates an exterior wall, the vent would have to extend vertically above the roof in accordance with Section G2427.6.3; however, this exterior extension of vent presents a problem in itself. Gas vents, such as Type B vents, are not designed for out-

Figure G2427.6.3.1
DECORATIVE SHROUDS

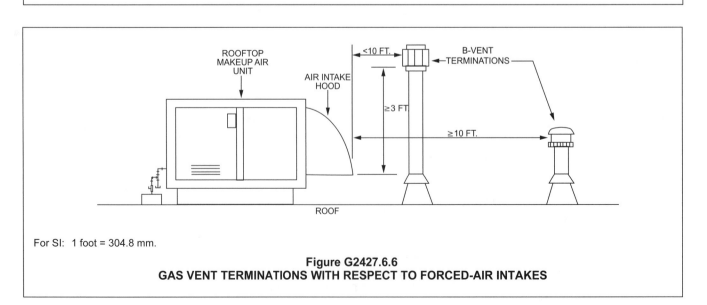

For SI: 1 foot = 304.8 mm.

Figure G2427.6.6
GAS VENT TERMINATIONS WITH RESPECT TO FORCED-AIR INTAKES

door exposure, except for the short section of vent that passes through the roof. Sections G2428.2.9 and G2428.3.16 reinforce this by stating that the vent sizing tables in Section G2428 do not apply to vents exposed to the outdoors below the roof line. Vents exposed to the outdoors will suffer from poor draft and condensation production in most climates because of the heat loss through the exposed vent walls. The bottom line is that vents sized in accordance with Section G2428 cannot extend up the exterior side of a building as implied by this section and Section G2427.6.3. This type of installation could occur only if the vent system was engineered to account for the outdoor exposure. Note that this is also consistent with typical vent manufacturers' design manuals and installation instructions. Even though it is common to see gas vents run up the exterior wall of a building, such installations are not allowed unless they are engineered and allowed by the vent manufacturer, neither of which is likely. Typically, gas vents run outdoors up the side of a building will show signs of severe corrosion, rusting and degradation.

G2427.6.8 (503.6.9) Size of gas vents. *Venting systems* shall be sized and constructed in accordance with Section G2428 or other *approved* engineering methods and the gas vent and *appliance* manufacturer's installation instructions.

❖ Gas vents are to be sized by either Section G2428 or an engineering method acceptable to the code official and, in all cases, in accordance with the vent manufacturer's instructions (see Sections G2427.6.8.1 through G2427.6.8.3).

G2427.6.8.1 (503.6.9.1) Category I appliances. The sizing of *natural draft venting systems* serving one or more *listed appliances* equipped with a *draft hood* or *appliances listed* for use with Type B gas vent, installed in a single story of a building, shall be in accordance with one of the following methods:

1. The provisions of Section G2428.

2. For sizing an individual gas vent for a single, draft-hood-equipped *appliance*, the effective area of the vent *connector* and the gas vent shall be not less than the area of the *appliance draft hood* outlet, nor greater than seven times the *draft hood* outlet area.

3. For sizing a gas vent connected to two *appliances* with *draft hoods*, the effective area of the vent shall be not less than the area of the larger *draft hood* outlet plus 50 percent of the area of the smaller *draft hood* outlet, nor greater than seven times the smaller *draft hood* outlet area.

4. *Approved* engineering practices.

❖ This section reiterates the intent of the preceding section, is specific to Category I appliances, and applies only to appliances installed within the same story. Multiple-story applications are addressed in the IFGC. Category I appliances are vented by natural draft and, depending upon the design and efficiency, are either equipped with a draft hood or a fan-assisted combus-

tion system. Because of the higher thermal efficiencies and lack of dilution air, vent designs for fan-assisted appliances are different from those for draft hood-equipped appliances.

Item 3 allows a very limited application of the old "50-percent rule." Item 2 applies only to a single draft-hood-equipped appliance. Item 3 applies only to a common vent system serving two draft-hood equipped appliances. Neither Item 2 nor 3 can be applied to the venting of fan-assisted appliances, because the old alternate method (50-percent rule) was not created with fan-assisted appliances in mind and will not work with such appliances. Besides being limited to draft hood appliances, Items 2 and 3 should be further limited to simple venting system configurations having short laterals; few, if any, changes in direction (fittings); short connector lengths with maximum vertical rise and tall vent heights. Vent manufacturers emphatically warn against the use of the alterative sizing rule in Item 3 because this method does not account for many factors now known to be important in vent sizing.

G2427.6.8.2 (503.6.9.2) Vent offsets. Type B and L vents sized in accordance with Item 2 or 3 of Section G2427.6.8.1 shall extend in a generally vertical direction with offsets not exceeding 45 degrees (0.79 rad), except that a vent system having not more than one 60-degree (1.04 rad) *offset* shall be permitted. Any angle greater than 45 degrees (0.79 rad) from the vertical is considered horizontal. The total horizontal distance of a vent plus the horizontal vent *connector* serving *draft hood*-equipped *appliances* shall be not greater than 75 percent of the vertical height of the vent.

❖ This section addresses vent offsets differently than Section G2428 and is in no way related to Section G2428. This section would be applied only if Section G2428 is not applied; therefore, this section is rarely applied, because Section G2428 is necessary for sizing the majority of today's Category I venting systems. The only time that a Type B vent system would not be designed in accordance with Section G2428 is if it were an engineered system or if it were designed in accordance with Section G2427.6.8.1, Item 2 or 3, applicable only to draft-hood-equipped appliances.

This section regulates vent offsets, but without the tables and design requirements of Section G2428, there are only very limited design requirements in Section G2427 to accompany the offset provisions. In other words, this section would be only one of many parts necessary to design a vent system. This section is not intended for application in conjunction with the requirements of Section G2428.

This section is not applicable to fan-assisted appliances and applies only to Type B and L vents sized in accordance with Item 2 or 3 of Section G2427.6.8.1.

G2427.6.8.3 (503.6.9.3) Category II, III and IV appliances. The sizing of gas vents for Category II, III and IV appliances shall be in accordance with the appliance manufacturer's instructions. The sizing of plastic pipe that is specified by the appliance manufacturer as a venting material for

Category II, III and IV appliances, shall be in accordance with the manufacturer's instructions.

❖ Venting systems for Category II, III and IV appliances are special vent systems specified by the appliance manufacturer and are not addressed further in the code (see Section G2427.4).

Because plastic pipes such PVC, ABS and CPVC plumbing pipes are not listed and labeled as appliance vents (see the definition of "Vent"), the code was silent on how to size such pipes. The sizing is covered in the appliance manufacturer's instructions, and the code requires compliance with such instructions. However, for consistency, this section has been modified to address both listed vents and unlisted materials used as vents. For example, PVC pipe that vents a Category IV furnace is not listed as a vent; rather, it is a material that is used as a vent, and the appliance must be listed for use with the PVC pipe.

G2427.6.8.4 (503.6.9.4) Mechanical draft. *Chimney venting systems* using mechanical *draft* shall be sized in accordance with *approved* engineering methods.

❖ The manufacturers of exhausters (mechanical draft systems) provide sizing criteria for the vents and chimneys served by such equipment. The code's sizing criteria are based on natural draft chimneys and vents; therefore, engineered sizing methods must be used for other than natural draft (see Section G2427.3.3).

G2427.6.9 (503.6.11) Support of gas vents. Gas vents shall be supported and spaced in accordance with the manufacturer's installation instructions.

❖ Type B vents must be supported to prevent the weight of the system from bearing on appliances and vent fittings. Fittings such as adjustable elbows are not designed to carry the weight of upper or lower sections of vent pipe, nor are appliance draft hoods and flue collars. Support bases, brackets and spacers are available from the vent manufacturer and are used to support the weight of the vent and maintain alignment and clearance to combustibles. On most job sites, the vents will be supported by various field-constructed brackets and straps made of sheet metal scraps, duct slips and drives and plumber's perforated straps. Gas vents are often poorly supported by methods that allow the vent fittings to rotate, that cause excessive loads on fittings and that allow the vent to sag or lose its required clearance. The installations should be carefully examined to determine that such field-constructed supports are adequate. Sheet metal screws must not penetrate the inner liner of any Type B vent component. Type B vents are designed for installation without the use of any additional fasteners that penetrate the vent walls (see Section G2427.6.11).

G2427.6.10 (503.6.12) Marking. In those localities where solid and liquid fuels are used extensively, gas vents shall be permanently identified by a label attached to the wall or ceiling at a point where the *vent connector* enters the gas vent. The determination of where such localities exist shall be made by the *code official*. The label shall read:

"This gas vent is for *appliances* that burn gas. Do not connect to solid or liquid fuel-burning appliances or incinerators."

❖ Type B vents installed in localities where oil- and solid-fuel-fired appliances are common must be labeled to warn the installers and occupants that Type B vents cannot be used to vent any appliances other than gas fired. The label is not placed on the vent; rather, it is to be placed on a wall or ceiling where it would be conspicuous to someone attempting to connect to the vent. The code official determines whether their locality is the intended target for this provision. It is assumed that ignorance of gas vents will be more likely in regions where oil and solid fuel have been commonly used for fuel.

G2427.6.11 (503.6.13) Fastener penetrations. Screws, rivets and other fasteners shall not penetrate the inner wall of double-wall gas vents, except at the transition from an *appliance draft hood* outlet, a *flue collar* or a single-wall metal connector to a double-wall vent.

❖ Gas vents are not designed to be joined with screws or other fasteners, yet installers insist on using screws or rivets to augment the proprietary joint methods provided by the vent manufacturer. Manufacturers might not prohibit this practice, but they do not encourage it and definitely do warn against penetrating the inner wall. Screws are also commonly used to attach support (stand-off) brackets and hangers to vent pipe. It is typical to find duct slips and S-drives used as supports for vents with screws used to attach to the vent.

When fasteners penetrate the inner wall, condensation in the vent will corrode the fastener and leak into the air space between the vent walls causing vent deterioration. Vent manufacturers offer adapter fittings to transition between a vent and an appliance outlet or flue collar; however, installers usually ignore these fittings and connect the vent pipe or vent elbow directly to the appliance. This section permits the inner wall to be penetrated only at the point of transition between single-wall pipe and the vent, and between an appliance outlet collar and the vent (see Commentary Figure G2427.6.11).

G2427.7 (503.7) Single-wall metal pipe. Single-wall metal *pipe* vents shall comply with Sections G2427.7.1 through G2427.7.13.

❖ Although discouraged or prohibited by designers, code officials and appliance and vent manufacturers alike, the code still recognizes the use of unlisted single-wall metal pipe as a vent. All other types of vents are listed systems as required by Section G2426.1. As evidenced by Sections G2427.7.1 through G2427.7.13, single-wall metal pipe is restricted to very limited applications, and extraordinary installation precautions are necessary. Single-wall metal pipe was once common, but it will not be found in residential construction today except in extremely rare circumstances in existing buildings. Appliance manufacturers' instructions will typically prohibit the use of a single-wall metal vent (see Section G2427.7.8).

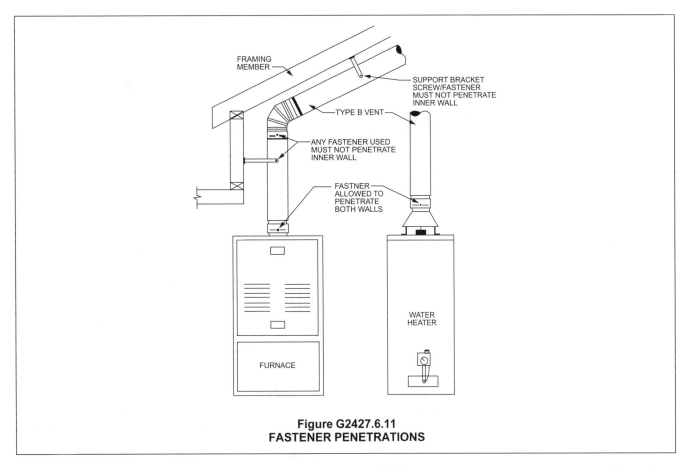

Figure G2427.6.11
FASTENER PENETRATIONS

G2427.7.1 (503.7.1) Construction. Single-wall metal pipe shall be constructed of galvanized sheet steel not less than 0.0304 inch (0.7 mm) thick, or other *approved*, noncombustible, corrosion-resistant material.

❖ The galvanized sheet steel thickness specified is equivalent to 22 gage, which is heavier than the pipe commonly used for connectors for most residential solid-fuel-fired appliances.

G2427.7.2 (503.7.2) Cold climate. Uninsulated single-wall metal pipe shall not be used outdoors for venting *appliances* in regions where the 99-percent winter design temperature is below 32°F (0°C).

❖ This section limits the use of single-wall vents to regions such as the southwestern and southeastern parts of the United States, unless the pipe is an insulated type to compensate for the high heat loss of single-wall pipe. The code contains no guidance for the type of insulation required or what *R*-value is required. At the minimum, insulation applied directly to the pipe would have to be a noncombustible material, the *R*-value would have to exceed that afforded by a double-wall vent, and the insulation would have to be permanently attached and protected from the weather. Insulating single-wall pipe could be considerably more difficult and expensive than using a venting system designed for outdoor use. Bear in mind that air-insulated double-wall vent pipe (Type B) is not designed for use outdoors in cold climates. When applying Section

G2428, single-wall vents are also prohibited from being exposed to the outdoors below the point of roof penetration.

G2427.7.3 (503.7.3) Termination. Single-wall metal pipe shall terminate at least 5 feet (1524 mm) in vertical height above the highest connected *appliance draft hood* outlet or *flue collar*. Single-wall metal pipe shall extend at least 2 feet (610 mm) above the highest point where it passes through a roof of a building and at least 2 feet (610 mm) higher than any portion of a building within a horizontal distance of 10 feet (3048 mm). An *approved* cap or roof assembly shall be attached to the terminus of a single-wall metal pipe.

❖ Single-wall metal vents are not allowed to terminate as allowed by Section G2427.6.3 and must comply with the termination requirements for chimneys, except that the minimum height above the roof penetration is 2 feet (610 mm), whereas it is 3 feet (914 mm) for chimneys [see Commentary Figure G2427.5.3(1)]. The minimum height and vent cap requirements are consistent with Sections G2427.6.3 and G2427.6.4.

G2427.7.4 (503.7.4) Limitations of use. Single-wall metal pipe shall be used only for runs directly from the space in which the *appliance* is located through the roof or exterior wall to the outdoor atmosphere.

❖ Single-wall metal pipe is further restricted by this section to only two applications. It can run from the appliance directly to the outdoors through an exterior wall or it can run from the appliance directly to the outdoors through a roof. Single-wall metal pipe cannot pass through any floors, interior walls or partitions and cannot pass through attics or concealed spaces. This means that the appliance served must be in a space with an exterior wall or in the same story as a cathedral or shed-type roof.

G2427.7.5 (503.7.5) Roof penetrations. A pipe passing through a roof shall extend without interruption through the roof flashing, roof jack or roof thimble. Where a single-wall metal pipe passes through a roof constructed of combustible material, a noncombustible, nonventilating thimble shall be used at the point of passage. The thimble shall extend not less than 18 inches (457 mm) above and 6 inches (152 mm) below the roof with the annular space open at the bottom and closed only at the top. The thimble shall be sized in accordance with Section G2427.7.7.

❖ A thimble is a sheet metal assembly designed to provide clearance between a vent and combustible materials, analogous to a thimble that protects one's fingers from a sewing needle. It is essentially a spacer that provides an annular space around the vent. Annular space is usually ventilated by punched holes in the faces of the thimble assembly. Some thimbles are constructed with insulation in addition to an annular space.

A roof jack is an assembly that passes through the roof and contains a thimble and weather protection all in one. Roof jacks are flashing/thimble combinations and typically mount to a roof curb.

G2427.7.6 (503.7.6) Installation. Single-wall metal pipe shall not originate in any unoccupied attic or concealed space and shall not pass through any attic, inside wall, concealed space, or floor. The installation of a single-wall metal pipe through an exterior combustible wall shall comply with Section G2427.7.7.

❖ A single-wall metal vent must be exposed for its entire length, except for the section of pipe that is within a thimble or roof jack assembly. Single-wall vents can penetrate only exterior walls and roofs. A roof/ceiling assembly cannot be penetrated if there is an attic or a concealed space between the ceiling and the roof (see Section G2427.7.4).

G2427.7.7 (503.7.7) Single-wall penetrations of combustible walls. Single-wall metal pipe shall not pass through a combustible exterior wall unless guarded at the point of passage by a ventilated metal thimble not smaller than the following:

1. For *listed appliances* with *draft hoods* and *appliances listed* for use with Type B gas vents, the thimble shall be not less than 4 inches (102 mm) larger in diameter than the metal pipe. Where there is a run of not less than 6 feet (1829 mm) of metal pipe in the open between the *draft hood* outlet and the thimble, the thimble shall be permitted to be not less than 2 inches (51 mm) larger in diameter than the metal pipe.

2. For unlisted *appliances* having *draft hoods*, the thimble shall be not less than 6 inches (152 mm) larger in diameter than the metal pipe.

3. For residential and low-heat *appliances*, the thimble shall be not less than 12 inches (305 mm) larger in diameter than the metal pipe.

Exception: In lieu of thimble protection, all *combustible material* in the wall shall be removed a sufficient distance from the metal pipe to provide the specified *clearance* from such metal pipe to *combustible material*. Any material used to close up such opening shall be noncombustible.

❖ The only single-wall vent component that is allowed to penetrate an exterior wall is a single-wall vent. Single-wall vent connectors cannot penetrate an exterior wall because if they did, they would be outdoors and exposed to the ambient outdoor temperatures. See Sections G2427.7.2 and G2427.10.2.2. This section is located in the single-wall vent section because this is the only place that it is applicable.

Items 1 through 3 dictate the size of thimbles used to protect the combustible wall. For example, in accordance with the first sentence of Item 1, a 4-inch (102 mm) metal pipe would require an 8-inch (203 mm) diameter thimble. Ventilated thimbles have holes in the faces of the thimble to allow air to cool the thimble by convection. Item 2 would rarely apply because unlisted appliances are not allowed by this code unless specifically approved by the code official under Section R104.11. Item 3 would rarely apply because it is addressing appliances other than those addressed in Items 1 and 2. The exception requires that the combustible wall be made noncombustible at the point of penetration. The noncombustible portion of the wall must extend from the metal pipe in all directions a distance equal to the required airspace clearance for the metal pipe.

G2427.7.8 (503.7.8) Clearances. Minimum *clearances* from single-wall metal pipe to *combustible material* shall be in accordance with Table G2427.10.5. The *clearance* from single-wall metal pipe to *combustible material* shall be permitted to be reduced where the *combustible material* is protected as specified for *vent connectors* in Table G2409.2.

❖ Consideration should be given to the fact that single-wall vents will have high surface temperatures compared to insulated vents such as Type B vents. High surface temperatures could be an ignition or burn injury hazard.

G2427.7.9 (503.7.9) Size of single-wall metal pipe. A venting system constructed of single-wall metal pipe shall be sized in accordance with one of the following methods and the *appliance* manufacturer's instructions:

1. For a draft-hood-equipped *appliance*, in accordance with Section G2428.

2. For a venting system for a single *appliance* with a *draft hood*, the areas of the connector and the pipe each shall be not less than the area of the *appliance flue collar* or *draft hood* outlet, whichever is smaller. The vent area

shall not be greater than seven times the *draft hood* outlet area.

3. Other *approved* engineering methods.

❖ As can be seen from Items 1, 2 and 3, and Tables 504.2(5) and 504.3(5) of the IFGC, single-wall metal pipe vents are limited to use with draft-hood-equipped appliances, except where the system is engineered. Note that the appliance manufacturer's instructions will generally require connection to a listed vent system.

G2427.7.10 (503.7.10) Pipe geometry. Any shaped single-wall metal pipe shall be permitted to be used, provided that its equivalent effective area is equal to the effective area of the round pipe for which it is substituted, and provided that the minimum internal dimension of the pipe is not less than 2 inches (51 mm).

❖ Single-wall metal pipe does not have to be round; it could be oval, square, or rectangular. Round pipe has the best flow characteristics as evidenced by the fact that a round chimney liner will have a greater capacity than a rectangular liner of equal area.

G2427.7.11 (503.7.11) Termination capacity. The vent cap or a roof assembly shall have a venting capacity of not less than that of the pipe to which it is attached.

❖ Section G2427.7.3 requires an approved vent cap or roof termination assembly, and this section requires that the terminus be sized and designed to not reduce the capacity of the vent. An unlisted, untested cap or roof assembly will have an unknown capacity unless it is engineered.

G2427.7.12 (503.7.12) Support of single-wall metal pipe. All portions of single-wall metal pipe shall be supported for the design and weight of the material employed.

❖ Single-wall vents, like all vents and chimneys, must be supported to prevent structural failure and joint separation and to maintain the required clearances.

G2427.7.13 (503.7.13) Marking. Single-wall metal pipe shall comply with the marking provisions of Section G2427.6.10.

❖ See the commentary to Section G2427.6.10.

G2427.8 (503.8) Venting system termination location. The location of venting system terminations shall comply with the following (see Appendix C):

1. A mechanical *draft* venting system shall terminate not less than 3 feet (914 mm) above any forced-air inlet located within 10 feet (3048 mm).

    Exceptions:

    1. This provision shall not apply to the *combustion air* intake of a direct-vent *appliance*.

    2. This provision shall not apply to the separation of the integral outdoor air inlet and flue gas discharge of *listed* outdoor *appliances*.

2. A mechanical *draft* venting system, excluding *direct*-vent *appliances*, shall terminate not less than 4 feet (1219 mm) below, 4 feet (1219 mm) horizontally from,

or 1 foot (305 mm) above any door, operable window or gravity air inlet into any building. The bottom of the vent terminal shall be located not less than 12 inches (305 mm) above finished ground level.

3. The vent terminal of a *direct*-vent *appliance* with an input of 10,000 *Btu* per hour (3 kW) or less shall be located not less than 6 inches (152 mm) from any air opening into a building. Such an *appliance* with an input over 10,000 *Btu* per hour (3 kW) but not over 50,000 *Btu* per hour (14.7 kW) shall be installed with a 9-inch (230 mm) vent termination *clearance*, and an *appliance* with an input over 50,000 Btu per hour (14.7 kW) shall have not less than a 12-inch (305 mm) vent termination *clearance*. The bottom of the vent terminal and the air intake shall be located not less than 12 inches (305 mm) above grade finished ground level.

4. Through-the-wall vents for Category II and IV *appliances* and noncategorized condensing *appliances* shall not terminate over public walkways or over an area where *condensate* or vapor could create a nuisance or hazard or could be detrimental to the operation of *regulators, relief valves* or other *equipment*. Where local experience indicates that *condensate* is a problem with Category I and III *appliances*, this provision shall also apply. Drains for *condensate* shall be installed in accordance with the appliance and vent manufacturer's installation instructions.

5. Vent systems for Category IV appliances that terminate through an outside wall of a building and discharge flue gases perpendicular to the adjacent wall shall be located not less than 10 feet (3048 mm) horizontally from an operable opening in an adjacent building. This requirement shall not apply to vent terminals that are 2 feet (607 mm) or more above or 25 feet (7620 mm) or more below operable openings.

❖ This section addresses the terminations of mechanical draft systems and direct-vent appliances. To prevent vent gases from entering the building, Item 1 requires forced air (mechanical) intakes to be at least 10 feet (3048 mm) away from a mechanical draft termination, or the termination must be at least 3 feet (914 mm) above the air intake. The natural buoyancy of vent gases is the justification for allowing a closer distance where the terminal is above the air intake. A combustion air intake for a direct-vent appliance is not what Item 1 intends to address (see Exception 1). Direct-vent appliance exhaust and intake opening locations are dictated by the appliance listing and manufacturer's instructions. Note the distinction between direct-vent and other than direct-vent appliances in the manufacturer's instructions. Item 1 also does not apply to appliances such as rooftop units that have built-in outdoor air intakes and combustion gas exhaust outlets. The code assumes that the design and listing of the outdoor appliance accounts for the location of those openings (see Exception 2).

Item 2 addresses gravity air inlets in the exterior envelope of a building, including doors, operable (openable) windows and intake louvers and grilles. The appliances addressed by this item use auxiliary or integral fans and blowers to force the flow of combustion products to the outdoors. This item applies to externally installed power exhausters, integrally power-exhausted appliances and venting systems equipped with draft inducers. This item does not address fan-assisted Category I appliances and does not address direct-vent appliances. Category III and IV appliances are subject to the requirements of this item if those appliances are not listed as direct-vent appliances. Some Category I appliances are field convertible to Category III.

Power exhausters are field-installed pieces of equipment that are independent of, but used in conjunction with, other appliances [see Commentary Figure G2427.8(1)].

This item is consistent with appliance installation instructions that will contain additional details for the location of the venting system terminals, including but not limited to clearance to grade, inside corners, outside comers, decks, porches, balconies, soffits, roof

eaves, utility meters, combustion air intakes, other appliance vents, service regulators, sidewalks and driveways.

There are significant differences between the requirements of Items 2 and 3 as they relate to nondirect-vent and direct-vent appliances, respectively. A Category IV, 90-percent efficient furnace or boiler would be subject to Item 2 if installed without a combustion air intake pipe to the outdoors but would be subject to Item 3, not Item 2, if installed with a combustion air intake pipe to the outdoors. This is because such a furnace can be listed as both a direct-vent and nondirect-vent appliance, and as such can be installed either way. Less restrictive clearances, improved operating efficiency, longer appliance life, immunity from the effects of building exhaust systems and ease of providing combustion air are some of the advantages of direct-vent appliances. For example, according to Item 2, a side-wall vent terminal must be 4 feet (1219 mm) or more to either side of an openable window, or must be 4 feet (1219 mm) or more below the bottom of

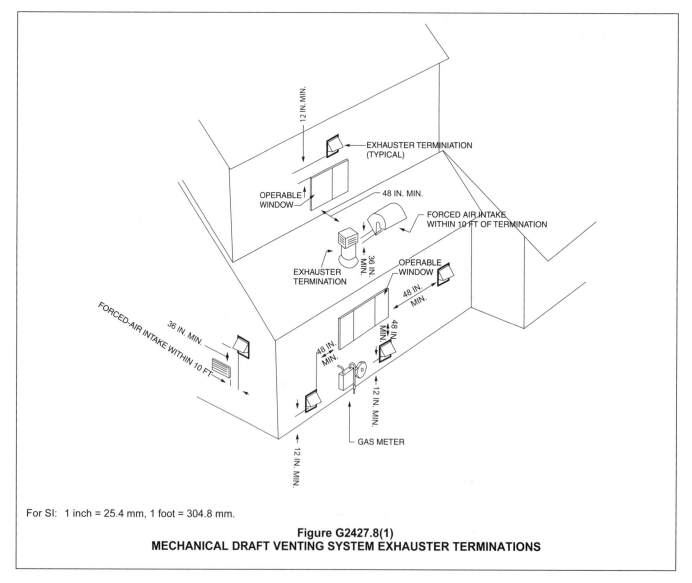

For SI: 1 inch = 25.4 mm, 1 foot = 304.8 mm.

Figure G2427.8(1)
MECHANICAL DRAFT VENTING SYSTEM EXHAUSTER TERMINATIONS

the window or must be 1 foot (305 mm) or more above the top of the window.

Item 3 is consistent with appliance manufacturers' instructions. Research has shown that the distances associated with the input ranges are necessary to allow the combustion gases to dissipate in the atmosphere, thus avoiding entry into the building through openings. This item is applicable only to direct-vent appliances. For example, a Category IV, 90-percent efficient furnace installed without an outdoor air intake

pipe for combustion air would be subject to Item 2, not 3 [see Commentary Figures G2427.8(2), G2427.8(3) and G2427.8(4)].

Item 4 addresses the possible effect of condensation on pedestrians, service equipment, etc. Condensate is corrosive and could create a hazard when it freezes. Condensate ice could be a slip hazard and could obstruct regulator and relief valve vents. Category I and III appliances do not produce as much condensate as appliances in the other two categories, but they can also cause condensate problems in cold cli-

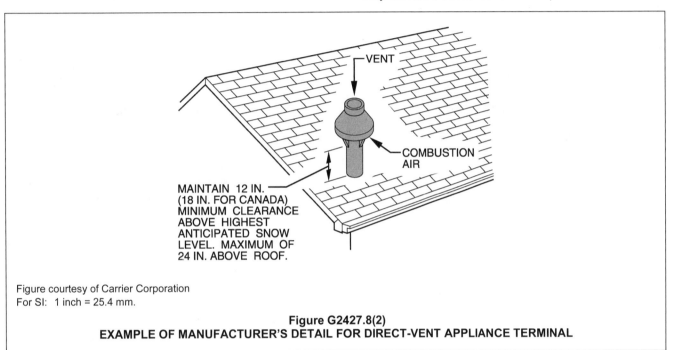

Figure courtesy of Carrier Corporation
For SI: 1 inch = 25.4 mm.

Figure G2427.8(2)
EXAMPLE OF MANUFACTURER'S DETAIL FOR DIRECT-VENT APPLIANCE TERMINAL

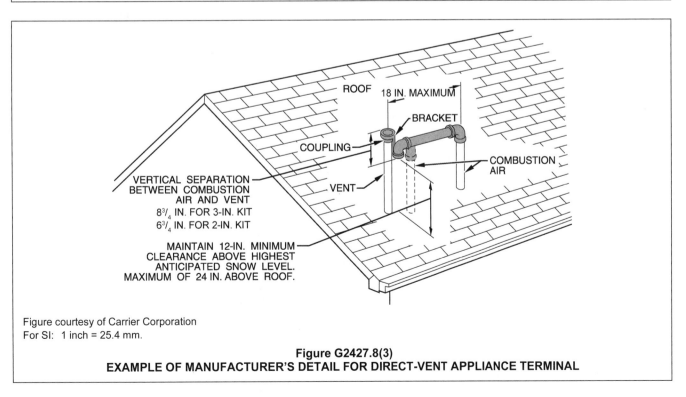

Figure courtesy of Carrier Corporation
For SI: 1 inch = 25.4 mm.

Figure G2427.8(3)
EXAMPLE OF MANUFACTURER'S DETAIL FOR DIRECT-VENT APPLIANCE TERMINAL

mates. It is common to see large masses of ice build up on or near the vent terminals of Category I and III vents. Such ice masses can fall or slide from roofs and injure someone or cause damage to a building, vehicle or equipment.

Item 5 is specific to sidewall terminations of vents for Category IV appliances such as furnaces and boilers [see Commentary Figure G2427.8(5)].

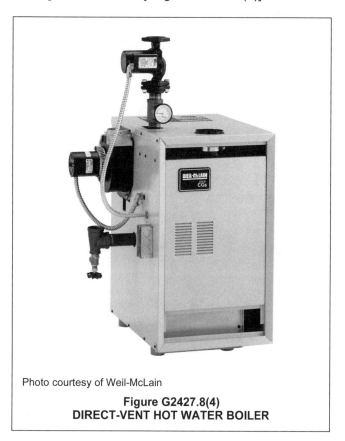

Photo courtesy of Weil-McLain

Figure G2427.8(4)
DIRECT-VENT HOT WATER BOILER

A common issue arises where buildings, especially homes, are located very close to each other, and sidewall-vented appliances are installed with the vent terminals directed toward the neighboring home. The concern is that combustion gases will enter the adjacent building through openings in the exterior walls that face the appliance vent terminal. This section applies only to Category IV (condensing) appliances that are sidewall-vented with stainless steel or plastic vents.

Computer simulations were conducted as part of a research project, and the results indicated that in many scenarios the combustion products would impinge on the neighboring building. Many factors impact the simulated scenarios including wind speed and direction, the height of the adjacent buildings and the type of vent terminal (i.e., straight pipe, tee fitting, deflector cap, directional fitting, etc.). The worst-case scenario that this code section addresses is a straight open-ended pipe used as the appliance vent terminal that is perpendicular to the wall it passes through. This scenario is the most common and the most likely to project combustion gases far enough to be a potential danger to the neighbors. The research project suggested that vent terminals that utilize a tee-fitting outlet or a deflector cap or that are directed at some angle downward are much less likely to interfere with the neighbors because with these terminals the combustion gases disperse and lack the velocity to impinge on the adjacent building.

G2427.9 (503.9) Condensation drainage. Provisions shall be made to collect and dispose of *condensate* from *venting systems* serving Category II and IV *appliances* and noncategorized condensing *appliances* in accordance with Section G2427.8, Item 4. Where local experience indicates that condensation is a problem, provisions shall be made to drain off and dispose of *condensate* from *venting systems* serving Cat-

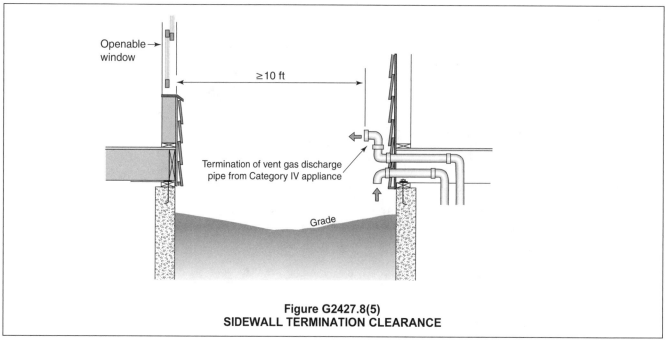

Figure G2427.8(5)
SIDEWALL TERMINATION CLEARANCE

egory I and III *appliances* in accordance with Section G2427.8, Item 4.

❖ See Section G2427.8, Item 4.

G2427.10 (503.10) Vent connectors for Category I appliances. Vent *connectors* for Category I *appliances* shall comply with Sections G2427.10.1 through G2427.10.13.

❖ See Sections G2427.10.1 through G2417.10.5.

G2427.10.1 (503.10.1) Where required. A vent *connector* shall be used to connect an *appliance* to a gas vent, *chimney* or single-wall metal pipe, except where the gas vent, *chimney* or single-wall metal pipe is directly connected to the *appliance*.

❖ Unless the chimney or vent is connected directly to the appliance, a connector, defined in Section G2403 as pipe used to connect an approved fuel-burning appliance to a chimney or vent, is necessary. This includes the fittings necessary to make a connection or change in direction. This is usually accomplished with a single-wall metal pipe, but it is also common practice to use listed and labeled chimney and vent pipe or listed factory-built single- or double-wall bendable connectors.

Many factors affect the design and configuration of a connector. The most important of these is the appliance location with respect to the chimney or vent system. This impacts the connector size, length and rise. Another important factor is the number of appliances being vented. The appliance manufacturer's installation instructions may prohibit the use of single-wall connectors. For example, a single-wall vent connector may not be appropriate for fan-assisted appliances because of the possibility of condensation, corrosion and leakage. They are also not permitted in attic or crawl space installations that are subject to cold temperatures. The cold temperatures increase the possibility of condensation, which leads to accelerated material failure. If the vent or chimney extends all the way to a single appliance vent outlet (collar), no connector is needed.

G2427.10.2 (503.10.2) Materials. *Vent connectors* shall be constructed in accordance with Sections G2427.10.2.1 through G2427.10.2.4.

❖ See Sections G2427.10.2.1 through G2427.10.2.4.

G2427.10.2.1 (503.10.2.1) General. A *vent connector* shall be made of noncombustible corrosion-resistant material capable of withstanding the vent gas temperature produced by the *appliance* and of sufficient thickness to withstand physical damage.

❖ See Sections G2427.10.2.2 through G2427.10.2.4.

G2427.10.2.2 (503.10.2.2) Vent connectors located in unconditioned areas. Where the *vent connector* used for an *appliance* having a *draft hood* or a Category I *appliance* is located in or passes through attics, crawl spaces or other unconditioned spaces, that portion of the *vent connector* shall

be *listed* Type B, Type L or listed vent material having equivalent insulation properties.

Exception: Single-wall metal pipe located within the exterior walls of the building in areas having a local 99-percent winter design temperature of 5°F (-15°C) or higher shall be permitted to be used in unconditioned spaces other than attics and crawl spaces.

❖ Compared to double-wall pipe, single-wall pipe has a much higher heat loss, and when installed in an unconditioned space, condensation can occur, deteriorating the pipe. A typical scenario is an attic, garage or crawl space appliance installation where a connector would be exposed to cold temperatures, practically ensuring the formation of condensation in the connector. Except as allowed in the exception, this section effectively prohibits the use of single-wall connectors in garages, attics and crawl spaces when the spaces are unheated.

The term "unconditioned" is not defined and the debate over the intent created the text in the exception, which is climate related. Attics and crawl spaces are usually unconditioned spaces; however, the uncertainty has been with residential basements and garages, which are common locations for gas-fired appliances. In many climates, the garage winter temperatures will be cold enough to cause the combustion gases in a connector to reach their dew point, meaning that condensate will form and linger in the connector, causing its rapid destruction. Most residential garages and some basements are considered to be unconditioned insofar as they are not directly or indirectly heated as is the living space.

The typical residential garage is not heated (conditioned) and, in most climates, will certainly have low ambient temperatures. The exception allows single-wall metal pipe in unconditioned spaces (other than attics and crawl spaces) if the climate is as described. The winter design temperatures can be found in the ASHRAE *Handbook of Fundamentals*. To put the climate criterion in perspective, Philadelphia, Pennsylvania; Louisville, Kentucky; Fall River, Massachusetts; and Providence, Rhode Island all have 99-percent winter design temperatures of 5°F (-15°C) or higher. The exception is applicable only to vent connectors that are inside a building (i.e., located within the exterior walls) (see Commentary Figure G2427.10.2.2).

G2427.10.2.3 (503.10.2.3) Residential-type appliance connectors. Where *vent connectors* for residential-type *appliances* are not installed in attics or other unconditioned spaces, connectors for *listed appliances* having *draft hoods*, *appliances* having *draft hoods* and equipped with *listed conversion burners* and Category I *appliances* shall be one of the following:

1. Type B or L vent material.

2. Galvanized sheet steel not less than 0.018 inch (0.46 mm) thick.

3. Aluminum (1100 or 3003 alloy or equivalent) sheet not less than 0.027 inch (0.69 mm) thick.

4. Stainless steel sheet not less than 0.012 inch (0.31 mm) thick.

5. Smooth interior wall metal pipe having resistance to heat and corrosion equal to or greater than that of Item 2, 3 or 4.

6. A *listed* vent *connector*.

Vent connectors shall not be covered with insulation.

Exception: *Listed* insulated *vent connectors* shall be installed in accordance with the manufacturer's instructions.

❖ This section applies to draft-hood-equipped appliances only and lists the allowable materials, including sheet metals, listed connectors and listed vent system pipe. The minimum thickness for single-wall galvanized steel pipe is equivalent to 26 gage. Installers have field-installed insulation on vent connectors in an attempt to control heat loss and the resultant condensation; however, this is prohibited because of the combustibility of most insulation materials and the possibility of excessive connector temperatures. Listed double-wall (air insulated) Type B connectors are available as an option to using Type B vent pipe (see Commentary Figure G2427.10.2.2).

Photo courtesy of Selkirk L.L.C.

Figure G2427.10.2.2
LISTED BENDABLE DOUBLE-WALL CONNECTOR

G2427.10.2.4 (503.10.2.4) Low-heat appliance. A *vent connector* for a nonresidential, low-heat *appliance* shall be a factory-built *chimney* section or steel *pipe* having resistance to heat and corrosion equivalent to that for the appropriate galvanized pipe as specified in Table G2427.10.2.4. Factory-built *chimney* sections shall be joined together in accordance with the *chimney* manufacturer's instructions.

❖ Residential appliances are also low-heat by definition, and the provisions of Section G2427.10.2.3 apply as well. The table is intended for nonresidential-type appliances, which is apparent when compared to the materials allowed by Section G2427.10.2.3.

TABLE G2427.10.2.4 (503.10.2.4)
MINIMUM THICKNESS FOR GALVANIZED STEEL VENT CONNECTORS FOR LOW-HEAT APPLIANCES

DIAMETER OF CONNECTOR (inches)	MINIMUM THICKNESS (inch)
Less than 6	0.019
6 to less than 10	0.023
10 to 12 inclusive	0.029
14 to 16 inclusive	0.034
Over 16	0.056

For SI: 1 inch = 25.4 mm.

❖ See the commentary to Section G2427.10.2.4.

G2427.10.3 (503.10.3) Size of vent connector. *Vent connectors* shall be sized in accordance with Sections G2427.10.3.1 through G2427.3.5.

❖ See the commentary to Sections G2427.10.3.1 through G2427.10.3.5.

G2427.10.3.1 (503.10.3.1) Single draft hood and fan-assisted. A *vent connector* for an *appliance* with a single *draft hood* or for a Category I fan-assisted *combustion* system *appliance* shall be sized and installed in accordance with Section G2428 or other *approved* engineering methods.

❖ The sizing tables in Section G2428 will determine connector size with the fundamental requirement that a connector not be smaller than the appliance vent connection, except as allowed by Sections G2428.2.2 and G2428.3.17. The tables will often require that a connector be larger than the appliance vent outlet (see Section G2428).

G2427.10.3.2 (503.10.3.2) Multiple draft hood. For a single *appliance* having more than one *draft hood* outlet or *flue collar*, the manifold shall be constructed according to the instructions of the *appliance* manufacturer. Where there are no instructions, the manifold shall be designed and constructed in accordance with *approved* engineering practices. As an alternate method, the effective area of the manifold shall equal the combined area of the *flue collars* or *draft hood* outlets and the *vent connectors* shall have a minimum 1-foot (305 mm) rise.

❖ Some appliances are designed with more than one venting (flue) outlet. For example, some large furnaces were built with dual heat exchangers and burner sections, and thus have two draft hoods or flue collar outlets. The manufacturer's instructions must be followed; however, in the event that the manufacturer's instructions are not available, this section specifies required sizing criteria. Section G2408.1 requires the manufacturer's instructions to be on the job site.

G2427.10.3.3 (503.10.3.3) Multiple appliances. Where two or more *appliances* are connected to a common *vent* or *chimney*, each *vent connector* shall be sized in accordance with Section G2428 or other *approved* engineering methods.

As an alternative method applicable only when all of the *appliances* are *draft hood* equipped, each *vent connector* shall have an effective area not less than the area of the *draft hood* outlet of the *appliance* to which it is connected.

❖ The sizing tables in Section G2428 will determine connector size with the fundamental requirement that a connector not be smaller than the appliance vent connection, except as allowed by Sections G2428.2.2 and G2428.3.17. The tables will often require a connector that is larger than the appliance vent outlet (see Section G2428). The alternative method paragraph of this section is incompatible with Section G2428 and is intended only for use with the alternative sizing method of Section G2427.6.8.1.

G2427.10.3.4 (503.10.3.4) Common connector/manifold. Where two or more *appliances* are vented through a common *vent connector* or vent manifold, the common *vent connector* or vent manifold shall be located at the highest level consistent with available headroom and the required *clearance* to *combustible materials* and shall be sized in accordance with Section G2428 or other *approved* engineering methods.

As an alternate method applicable only where there are two *draft hood*-equipped *appliances*, the effective area of the common *vent connector* or vent manifold and all junction fittings shall be not less than the area of the larger *vent connector* plus 50 percent of the area of the smaller *flue collar* outlet.

❖ Vent manufacturers' instructions contain detailed design criteria for manifolds. Connectors serving gas-fired appliances are allowed to join to form a common connector manifold rather than connecting independently to the vertical common vent [see Commentary Figures G2427.10.3.4(1) and G2427.10.3.4(2)]. Various manufacturers' instructions require a 10-percent to 20-percent reduction in the capacity of a common vent where a manifold is used (see commentary, Section G2428.3.4). A manifold is defined as a vent or connector that is a lateral (horizontal) extension of the

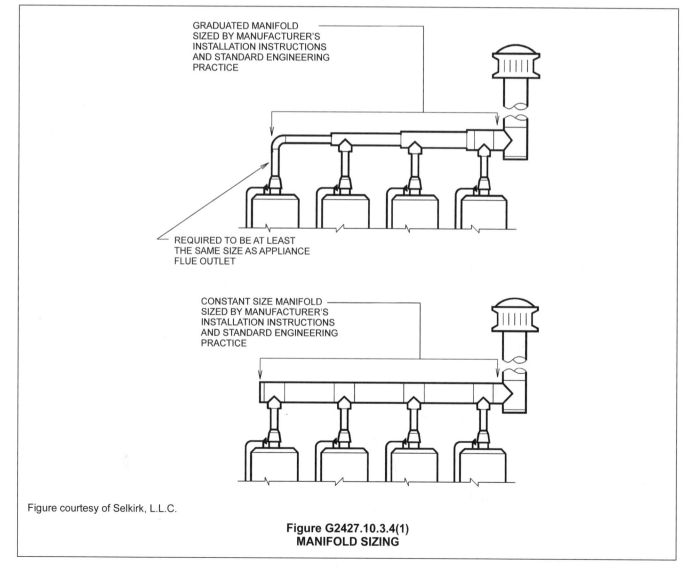

GRADUATED MANIFOLD SIZED BY MANUFACTURER'S INSTALLATION INSTRUCTIONS AND STANDARD ENGINEERING PRACTICE

REQUIRED TO BE AT LEAST THE SAME SIZE AS APPLIANCE FLUE OUTLET

CONSTANT SIZE MANIFOLD SIZED BY MANUFACTURER'S INSTALLATION INSTRUCTIONS AND STANDARD ENGINEERING PRACTICE

Figure courtesy of Selkirk, L.L.C.

Figure G2427.10.3.4(1)
MANIFOLD SIZING

lower end of a common vent. Because a manifold serves multiple appliances and thus conveys flue gases from two or more appliances, manifolds must be sized by the common vent tables. This section requires that manifolds be installed as high as possible, respective of ceiling height and clearance-to combustibles requirements. This will result in the most appliance connector rise, which is always beneficial for venting performance. Connector rise takes advantage of the energy of hot gases discharging directly from the appliance and develops flow velocity and draft in addition to that produced by the common vent [see Commentary Figures G2427.10.3.4(1) and G2427.10.3.4(2)]. The alternative method in the second paragraph is intended for use only with the alternative sizing method of Section G2427.6.8.1 and is not compatible with Section G2428.

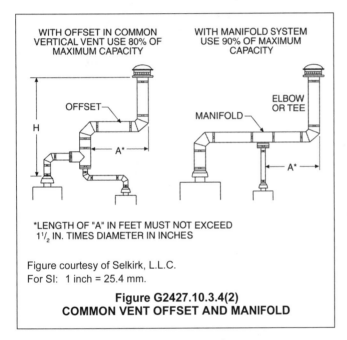

WITH OFFSET IN COMMON VERTICAL VENT USE 80% OF MAXIMUM CAPACITY

WITH MANIFOLD SYSTEM USE 90% OF MAXIMUM CAPACITY

OFFSET

H

A*

MANIFOLD

ELBOW OR TEE

A*

*LENGTH OF "A" IN FEET MUST NOT EXCEED 1¹/₂ IN. TIMES DIAMETER IN INCHES

Figure courtesy of Selkirk, L.L.C.
For SI: 1 inch = 25.4 mm.

**Figure G2427.10.3.4(2)
COMMON VENT OFFSET AND MANIFOLD**

G2427.10.3.5 (503.10.3.5) Size increase. Where the size of a *vent connector* is increased to overcome installation limitations and obtain connector capacity equal to the *appliance* input, the size increase shall be made at the *appliance draft hood* outlet.

❖ If a connector must be larger than the appliance vent outlet, the size increase must occur at the appliance connection by use of an increaser fitting. For example, it is common for a water heater with a 3-inch (76 mm) draft-hood outlet to be connected to a 4-inch (102 mm) connector by a 3-inch by 4-inch (76 mm by 102 mm) increaser fitting, as dictated by the vent tables in Section G2428.

G2427.10.4 (503.10.4) Two or more appliances connected to a single vent or chimney. Where two or more *vent connectors* enter a common gas vent, *chimney* flue, or single-wall metal pipe, the smaller connector shall enter at the highest level consistent with the available headroom or *clearance* to *combustible material*. *Vent connectors* serving Category I

appliances shall not be connected to any portion of a *mechanical draft* system operating under positive static pressure, such as those serving Category III or IV *appliances*.

❖ A common vent is somewhat oversized when only a single appliance is operating; therefore, every effort is made to achieve proper vent/chimney operation during all possible operating circumstances. Placing the smallest appliance connection at the highest elevation in the common vent/chimney will allow the greatest connector rise for the smaller appliance connector and will take advantage of any draft priming effect caused by the lower connector. For example, it has been common practice to connect a domestic water heater connector above the furnace or boiler connector in a chimney or vent system. Category I appliances require a vent that produces a draft and therefore cannot be connected to any venting system that produces a positive pressure. For example, a Category I appliance and a Category III appliance cannot share the same vent unless they both connect to the negative pressure (inlet) side of a mechanical draft system in accordance with the appliance and draft system installation instructions (see Section G2425.10).

G2427.10.4.1 (503.10.4.1) Two or more openings. Where two or more openings are provided into one *chimney* flue or vent, the openings shall be at different levels, or the connectors shall be attached to the vertical portion of the *chimney* or vent at an angle of 45 degrees (0.79 rad) or less relative to the vertical.

❖ The intent is to prevent turbulence and flow interference that could impede flow from a chimney or vent connector into the chimney or common vent. Openings at the same elevation, such as those directly opposite of each other, could negatively affect the flow from either or both of the connectors (see Commentary Figure G2427.10.4.1).

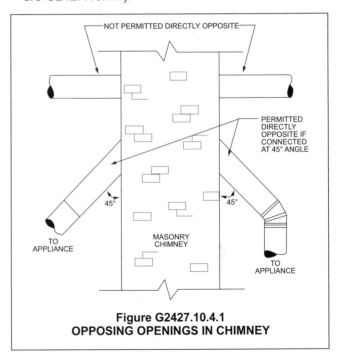

NOT PERMITTED DIRECTLY OPPOSITE

PERMITTED DIRECTLY OPPOSITE IF CONNECTED AT 45° ANGLE

45°

45°

TO APPLIANCE

MASONRY CHIMNEY

TO APPLIANCE

**Figure G2427.10.4.1
OPPOSING OPENINGS IN CHIMNEY**

G2427.10.5 (503.10.5) Clearance. Minimum *clearances* from *vent connectors* to *combustible material* shall be in accordance with Table G2427.10.5.

> **Exception:** The *clearance* between a *vent connector* and *combustible material* shall be permitted to be reduced where the *combustible material* is protected as specified for *vent connectors* in Table G2409.2.

❖ Flue gas passageways must have minimum clearances to ignitable materials. The single-wall metal pipe clearance requirements in Table G2427.10.5 apply to unlisted single-wall connectors. Connectors that are listed and labeled for this use must be installed with a clearance to combustibles as required by the connector manufacturer's installation instructions. If the appliance or connector manufacturer's installation instructions specify larger clearances than those prescribed by this section, the manufacturer's installation instructions govern. Table G2427.10.5 dictates the required air-space clearances between vent connectors and combustible materials and assemblies. The provisions of Section G2409 would allow reduction of connector clearances where the prescribed clearance cannot be provided or is impractical. Lack of vent connector clearance is a common code violation. Connectors usually lack the required clearances to wood joists, plastic plumbing piping, building and piping insulation materials and gypsum board. A major disadvantage of single-wall connectors is the large clearance required to combustibles. Listed connectors usually require far less clearance, in some cases only 1 inch (25.4 mm).

G2427.10.6 (503.10.6) Joints. Joints between sections of connector *piping* and connections to *flue collars* and *draft hood* outlets shall be fastened by one of the following methods:

1. Sheet metal screws.

2. *Vent connectors* of *listed* vent material assembled and connected to *flue collars* or *draft hood* outlets in accordance with the manufacturers' instructions.

3. Other *approved* means.

❖ Single-wall connectors (unlisted) have been traditionally fastened with sheet metal screws and rivets. All connector joints must be fastened, including the joint at the appliance draft hood or flue collar. A displaced connector could result in a life-threatening condition; therefore, connectors must be fastened and supported well. Appliance connectors are often located where they can be impacted or otherwise disturbed by building occupants, making proper fastening even more important. Item 2 speaks of listed connectors such as Type B vent material and factory-built corrugated (bendable) connectors, which use adapter fittings or integral collars that are mechanically fastened to the appliance with screws or rivets (see Commentary Figure G2427.10.2.2).

G2427.10.7 (503.10.7) Slope. A *vent connector* shall be installed without dips or sags and shall slope upward toward the vent or *chimney* at least $^1/_4$ inch per foot (21 mm/m).

> **Exception:** *Vent connectors* attached to a mechanical *draft* system installed in accordance with the *appliance* and *draft* system manufacturers' instructions.

❖ The ideal chimney or vent configuration is a totally vertical system, even though it is not always practical. This section requires all portions of a chimney or vent connector to rise vertically a minimum of a $^1/_4$ inch per each foot (21 mm/m) of its horizontal length. The connector slope is intended to induce the flow of flue gases using the natural buoyancy of the hot gases. Connector slope can promote the priming of a cold venting system and can partially compensate for short connector vertical rise. Low points, dips and sags could also trap condensate and accelerate corrosion of the connector. The exception allows connectors without slope where connected to mechanical draft systems because the connector is on the negative pressure (inlet) side of the exhauster and slope would provide no benefit.

TABLE G2427.10.5 (503.10.5)[a]
CLEARANCES FOR CONNECTORS

APPLIANCE	MINIMUM DISTANCE FROM COMBUSTIBLE MATERIAL			
	Listed Type B gas vent material	Listed Type L vent material	Single-wall metal pipe	Factory-built chimney sections
Listed appliances with draft hoods and appliances listed for use with Type B gas vents	As listed	As listed	6 inches	As listed
Residential boilers and furnaces with listed gas conversion burner and with draft hood	6 inches	6 inches	9 inches	As listed
Residential appliances listed for use with Type L vents	Not permitted	As listed	9 inches	As listed
Listed gas-fired toilets	Not permitted	As listed	As listed	As listed
Unlisted residential appliances with draft hood	Not permitted	6 inches	9 inches	As listed
Residential and low-heat appliances other than above	Not permitted	9 inches	18 inches	As listed
Medium-heat appliances	Not permitted	Not permitted	36 inches	As listed

For SI: 1 inch = 25.4 mm.

a. These clearances shall apply unless the manufacturer's installation instructions for a listed appliance or connector specify different clearances, in which case the listed clearances shall apply.

G2427.10.8 (503.10.8) Length of vent connector. The maximum horizontal length of a single-wall connector shall be 75 percent of the height of the *chimney* or vent except for engineered systems. The maximum horizontal length of a Type B double-wall connector shall be 100 percent of the height of the *chimney* or vent except for engineered systems.

❖ This section is applied in conjunction with Section G2428.3.2. The appliance and chimney or vent must be located to keep the connector length as short as practicable. This section establishes the maximum allowable length for uninsulated chimney and vent connectors, insulated chimney and vent connectors and individual connectors for a chimney or vent system serving multiple appliances. These requirements are based on the heat loss of the connector and the ability of the vent or chimney system to produce a draft.

An insulated connector (double wall) reduces the amount of heat transfer through the connector pipe; thus, the flue gas is maintained at a higher temperature (see commentary, Section G2428.3.2). The amount of draft is directly related to the chimney or vent height and the difference in temperature between the flue gases and the ambient air. An uninsulated connector is a run of single-wall pipe. Commentary Figure G2427.10.8 shows that the total chimney or vent height is measured from the top of the highest appliance flue outlet connection to the termination point of the chimney or vent. The length limitations of this section are based on the total vertical rise of the chimney or vent and on the developed length of horizontal connectors within the chimney or vent system. The venting tables will often prohibit the use of single-wall connectors with fan-assisted appliances, and most appliance and vent manufacturers recommend against the use of

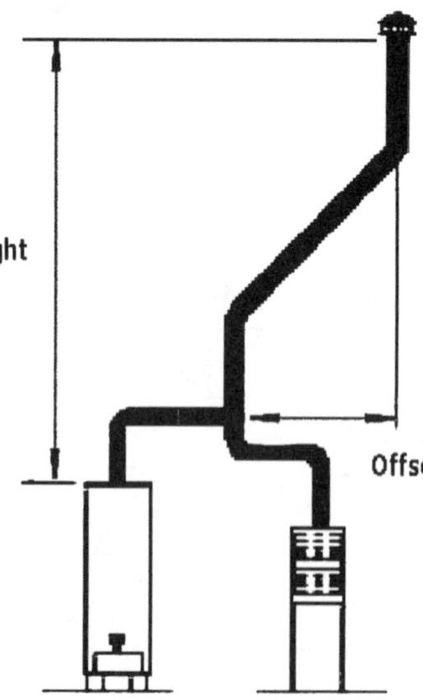

Figure courtesy of Simpson Dura-Vent Company, Incorporated

Figure G2427.10.8
VENT HEIGHT

single-wall (uninsulated) connectors.

Connectors are limited in length because of the flow resistance of the connector pipe and because heat loss through the connector is directly related to its length. For a venting system to work, the draft produced by the vertical vent or chimney must be able to overcome the resistance to flow created by the connector. The longer the connector, the longer it takes to prime a cold venting system and develop draft. As stated in Section G2428.3.2, connectors should always be as short as the installation conditions will permit. The sizing methodology and tables in Section G2428 for connectors in single-appliance installations might allow connector lengths greater than the general limits given in this section or might require a shorter connector length than would be allowed by these limits, depending on the combination of appliance type and input, connector size and type and vent height.

Because the sizing tables in Section G2428.2 are based on specific computer modeling of the installation configuration, the limitations or allowances in the single-appliance tables supersede what would be required in this section. For multiple-appliance systems, the sizing methods and tables do not contain the same limitations on connector length. This section establishes an absolute length limitation for connectors in multiple-appliance systems (100 percent of the height of the chimney or vent), which still provides liberal installation flexibility without permitting excessively long connectors. This section functions as an absolute cap for the connector length provisions of Sections G2428.3.2 and G2428.3.3.

The appliance and vent location should be well planned to allow the shortest horizontal run of connector piping (see Sections G2428.3.2 and G2428.3.3).

The intent of the reference to engineered systems is to allow connector length to be determined by design as part of an engineered venting system.

G2427.10.9 (503.10.9) Support. A *vent connector* shall be supported for the design and weight of the material employed to maintain *clearances* and prevent physical damage and separation of joints.

❖ A connector must be supported for the design and weight of the material used. Proper support is necessary to maintain the clearances required by Section G2427.10.5, to maintain the required slope and to prevent physical damage and separation of joints. The joints between connectors and appliance draft-hood outlets and flue collars must be secured with screws, rivets or other approved means. The joints between pipe sections also must be secured with screws, rivets or other approved means. Some connectors are available with a proprietary-type fastening method, which does not require screws, rivets or other fasteners if it provides equivalent resistance to disengagement or displacement. This section addresses connectors, not chimneys and vents, and most chimney and vent manufacturers do not require or might even prohibit the use of screws or rivets for securing joints for chimney and vent pipe sections and fittings.

G2427.10.10 (503.10.10) Chimney connection. Where entering a flue in a masonry or metal *chimney*, the *vent connector* shall be installed above the extreme bottom to avoid stoppage. Where a thimble or slip joint is used to facilitate removal of the connector, the connector shall be firmly attached to or inserted into the thimble or slip joint to prevent the connector from falling out. Means shall be employed to prevent the connector from entering so far as to restrict the space between its end and the opposite wall of the *chimney* flue (see Section G2425.9).

❖ All of the precautions taken to provide adequate chimney design may be ineffective if the connection between the appliance and the chimney is not properly accomplished. Improper connections can lead to appliance or connector failure and the leakage of vent gases into the building. This section contains requirements for the connection between a chimney and the appliance connector. The connection to a vent is properly accomplished by vent fittings and likewise for factory-built chimneys. Chimney connectors are required to pass through a masonry chimney wall to the inner face of the liner, but not beyond. A connector that extends into a chimney passageway can restrict the flow of vent gases and provide a ledge on which debris can accumulate. The joint between the connector and the chimney must be fastened in a manner that will prevent separation. If the connector enters a masonry chimney, it must be cemented in place with an approved material such as refractory mortar or other heat-resistant cement.

This section also permits the use of thimbles at a masonry chimney opening to provide for easy removal of the connector to facilitate cleaning. When a thimble is installed, it is to be permanently cemented in place with an approved high-temperature cement. A connector must be attached to a thimble in an approved manner to prevent displacement. A collar on the connector or a similar arrangement must be provided to limit the penetration of the connector into the chimney. Otherwise, it is quite possible that a connector will be shoved blindly into a chimney, reaching to the back wall of the chimney, and thus be obstructed.

G2427.10.11 (503.10.11) Inspection. The entire length of a *vent connector* shall be provided with *ready access* for inspection, cleaning and replacement.

❖ A connector cannot be concealed by any form of building construction, whether or not such construction is removable. Connectors must be inspected periodically for corrosion damage, structural integrity, joint separation and clearances to combustibles (see Section G2427.10.13).

G2427.10.12 (503.10.12) Fireplaces. A *vent connector* shall not be connected to a *chimney* flue serving a *fireplace* unless the *fireplace* flue opening is permanently sealed.

❖ A fireplace chimney cannot serve any appliance unless the fireplace is retired by permanently sealing off the chimney from the fireplace. If this is not done, the fireplace can affect draft for the connected appli-

ance, and/or vent gases could escape into the building through the fireplace opening.

G2427.10.13 (503.10.13) Passage through ceilings, floors or walls. Single-wall metal pipe connectors shall not pass through any wall, floor or ceiling except as permitted by Section G2427.7.4.

❖ This section no longer prohibits double-wall, Type B connectors from passing through walls, floors and ceilings. The prohibition on pass-through is limited to single-wall metal pipe connectors, recognizing that Section G2427.7.4 allows single-wall metal pipe vents to extend from the appliance to the outdoors by passing through a roof or exterior wall. A single-wall connector has more than double the heat loss of a listed metal vent or chimney system. This produces high temperatures on the outside surface and cooling of the flue gases on the inside of the connector. The cooling of flue gases produces condensation, which can cause connector deterioration, while the excessive surface temperatures pose a potential fire hazard. For these reasons, a single-wall connector can be located only within the room or space in which the appliance is located. Single-wall connectors are prohibited from passing through any ceiling, floor or wall. If single-wall connectors were allowed to pass through walls, floors or ceilings, they might not be readily observable, and a potential fire hazard or connector deterioration could go undetected. Also, pass-through increase the likelihood that the connector will be excessively long, exposed to low ambient temperatures or subject to contact with combustibles. Connectors that pass through a wall, floor or ceiling would be out of sight and therefore out of mind, meaning that they would not get the attention needed to avoid hazardous conditions (see Commentary Figure G2427.10.13).

G2427.11 (503.11) Vent connectors for Category II, III and IV appliances. *Vent connectors* for Category II, III and IV *appliances* shall be as specified for the *venting systems* in accordance with Section G2427.4.

❖ See Sections G2427.4, G2427.4.1 and G2427.4.2.

G2427.12 (503.12) Draft hoods and draft controls. The installation of *draft hoods* and draft controls shall comply with Sections G2427.12.1 through G2427.12.7.

❖ See Sections G2427.12.1 through G2427.12.7.

G2427.12.1 (503.12.1) Appliances requiring draft hoods. *Vented appliances* shall be installed with *draft hoods*.

Exception: Dual oven-type combination ranges; *direct-vent appliances*; fan-assisted *combustion* system *appliances*; *appliances* requiring *chimney draft* for operation; single firebox boilers equipped with *conversion burners* with inputs greater than 400,000 *Btu* per hour (117 kW); *appliances* equipped with blast, power or pressure *burners* that are not *listed* for use with *draft hoods*; and *appliances* designed for forced venting.

❖ Draft hoods are integral to or are supplied with natural draft atmospheric-burner gas-fired appliances (see Commentary Figure G2427.12.1). With the development of higher appliance efficiencies, fan-assisted combustion, direct-vent appliances and Category III and IV appliances, fewer draft-hood appliances will be used. Draft hoods are still common on conventional tank-type and tankless water heaters and 80-percent-efficient boilers. A draft-hood-equipped appliance will allow a continuous flow of air into the venting system, which represents an energy loss except where an automatic vent damper is installed to close off the vent in the off cycle. Without an automatic vent damper, conditioned air is taken from the space in which the appliance is located and residual heat is taken away from the appliance itself, both of which contribute to energy loss. Not all appliances are categorized, but if they were, draft-hood-equipped appliances would fall under Category I. Draft hoods have multiple functions. They serve as a relief opening in the event of vent or chimney backdraft or blockage, they stabilize/regulate the amount of draft that occurs in the appliance combustion chamber and they allow the introduction of dilution air into the chimney or vent.

G2427.12.2 (503.12.2) Installation. A *draft hood* supplied with or forming a part of a *listed vented appliance* shall be installed without *alteration*, exactly as furnished and specified by the *appliance* manufacturer.

❖ Draft hoods are often shipped with appliances as a part that must be field installed. Installers have been known to shorten or otherwise modify the draft hoods to compensate for installation constraints and have also installed them in an orientation other than vertical.

G2427.12.2.1 (503.12.2.1) Draft hood required. If a *draft hood* is not supplied by the *appliance* manufacturer where one is required, a *draft hood* shall be installed, shall be of a *listed* or *approved* type and, in the absence of other instructions, shall be of the same size as the *appliance flue* collar.

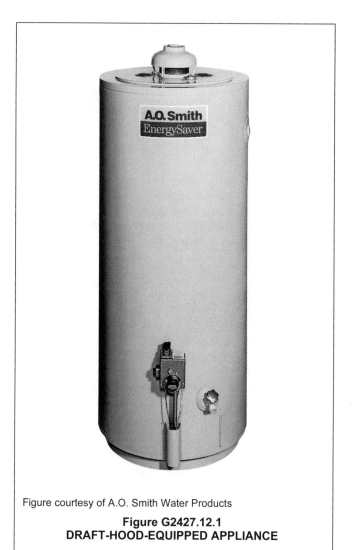

Figure courtesy of A.O. Smith Water Products

**Figure G2427.12.1
DRAFT-HOOD-EQUIPPED APPLIANCE**

COMMON B VENT SYSTEM WITH B VENT CONNECTORS

COMMON B VENT SYSTEM WITH B VENT CONNECTORS

IN OLDER EDITIONS OF THE CODE, TYPE B CONNECTORS NOT ALLOWED TO PASS THROUGH CEILINGS

IN LATER EDITIONS OF THE CODE TYPE B CONNECTORS ALLOWED TO PASS THROUGH CEILINGS

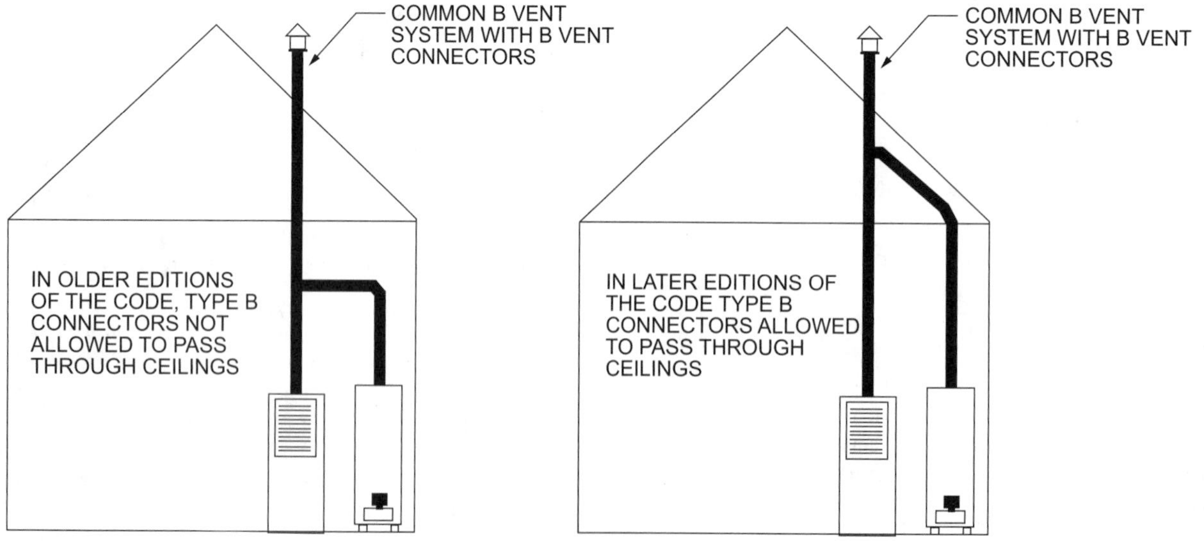

**Figure G2427.10.13
TYPE B VENT CONNECTOR PENETRATIONS**

Where a *draft hood* is required with a *conversion burner*, it shall be of a *listed* or *approved* type.

❖ In the unlikely event that an appliance manufacturer does not supply the required draft hood, a listed or approved draft hood must be obtained and installed.

G2427.12.2.2 (503.12.2.2) Special design draft hood. Where it is determined that a *draft hood* of special design is needed or preferable for a particular installation, the installation shall be in accordance with the recommendations of the *appliance* manufacturer and shall be *approved*.

❖ There could be unusual circumstances, such as where a horizontal type draft hood is needed, because of low ceilings and/or insufficient connector rise.

G2427.12.3 (503.12.3) Draft control devices. Where a *draft control* device is part of the *appliance* or is supplied by the *appliance* manufacturer, it shall be installed in accordance with the manufacturer's instructions. In the absence of manufacturer's instructions, the device shall be attached to the *flue collar* of the *appliance* or as near to the *appliance* as practical.

❖ Draft hoods also serve to control draft; however, this section would be redundant with Section G2427.12.2 if it were not addressing draft controls such as barometric dampers. Two types of draft controls are installed in the appliance vent outlet or vent connector: the passive draft hood and the active barometric damper. Barometric dampers consist of a damper plate mounted on an axle that rotates in low friction bearing points such as needle or knife-edge bearings. The device is mounted in a tee fitting arrangement and connects with the appliance vent connector. It admits air in varying amounts into the vent connector to control the amount of draft through the appliance combustion chamber. These devices respond to varying chimney or vent draft caused by changes in indoor and outdoor temperatures, changes in atmospheric (barometric) pressure and changes in wind speed and direction. Barometric dampers are tuned by adjustable balancing weights attached to the damper plate. The damper manufacturer will provide instructions on how and where to install the device [see Commentary Figure G2403(2)].

G2427.12.4 (503.12.4) Additional devices. *Appliances* requiring a controlled *chimney draft* shall be permitted to be equipped with a *listed* double-acting barometric-*draft regulator* installed and adjusted in accordance with the manufacturer's instructions.

❖ Like Section G2427.12.3, this section addresses barometric dampers, specifically, the double-acting type. These dampers open inward to admit air into the vent system like all other barometric dampers, but can also open outward in the event of backdraft to allow vent gases to be spilled into the room or space in which the appliance is located, just as a draft hood would do. The manufacturer's installation instructions may require or recommend that a sensor (spill switch) be installed that will sense prolonged vent-gas spillage from the

damper and shut down the appliance. In such cases, a manufacturer's recommendation should be interpreted as an enforceable requirement.

Some Category I fan-assisted furnaces can be fitted with a factory-supplied conversion part that is, in effect, a retrofit draft hood. This will lower the efficiency of the appliance by converting it to a draft-hood equipped appliance that will now introduce dilution air (conditioned room air) into the vent. This is done solely to allow fan-assisted furnaces to connect to existing masonry chimneys that would otherwise not be allowed to serve such appliances. These conversion parts use a sensor (spill switch) in the relief/air inlet opening. The manufacturer of the appliance and the draft hood conversion kit will provide specific installation instructions that must be followed for the conversion.

Vent gas spillage must not occur from an incinerator venting system; therefore, a draft regulating device must be of the single-acting type.

G2427.12.5 (503.12.5) Location. *Draft hoods* and *barometric draft regulators* shall be installed in the same room or enclosure as the *appliance* in such a manner as to prevent any difference in pressure between the hood or *regulator* and the *combustion air* supply.

❖ Draft-regulation devices are designed to maintain a constant draft through the appliance combustion chamber and flue passages for the purpose of maintaining proper combustion conditions and achieving the high-est possible efficiency. If the draft control sees a different ambient pressure than the appliance, it could allow too much or too little draft through the appliance. The positive, negative or neutral pressures within a building must act equally on both the appliance and the draft control to allow the draft control to be adjusted and perform its intended function.

G2427.12.6 (503.12.6) Positioning. *Draft hoods* and *draft regulators* shall be installed in the position for which they were designed with reference to the horizontal and vertical planes and shall be located so that the *relief opening* is not obstructed by any part of the *appliance* or adjacent construction. The *appliance* and its *draft hood* shall be located so that the *relief opening* is accessible for checking *vent* operation.

❖ Draft hoods and barometric dampers are both affected by their orientation. There are two basic types of draft hoods: vertical draft hoods and horizontal draft hoods. These types of draft hoods are not interchangeable. Barometric dampers can be adjusted to connect to either vertical or horizontal vent connectors; however, they are always positioned so that the closed damper plate is in a vertical plane. The manufacturer's instructions must always be followed. The relief opening on draft controls is also the dilution air inlet, and it must be unobstructed to allow air to enter under normal conditions, allow vent gases to spill in the event of backdraft and allow testing of the vent system for proper draft. For example, for many years, appliance venting/draft has been checked by observing the movement of smoke introduced near the relief/inlet opening.

G2427.12.7 (503.12.7) Clearance. A *draft hood* shall be located so its *relief opening* is not less than 6 inches (152 mm) from any surface except that of the *appliance* it serves and the venting system to which the *draft hood* is connected. Where a greater or lesser *clearance* is indicated on the *appliance* label, the *clearance* shall be not less than that specified on the label. Such *clearances* shall not be reduced.

❖ The specified clearance is necessary to assure the unimpeded flow of air into the draft hood and flow of vent gas spillage from the hood. The clearance also protects materials from the hot gases that can spill from the draft hood and allows access for testing and observation.

G2427.13 (503.13) Manually operated dampers. A manually operated *damper* shall not be placed in the vent *connector* for any *appliance*. Fixed baffles shall not be classified as manually operated *dampers*.

❖ Manual dampers are associated with solid-fuel appliances but not with gas-fired appliances. A manual damper requires manual operation by a human, which cannot be relied on because humans forget. If a damper is left closed or partially closed during appliance operation, a severe hazard could result from vent-gas spillage and/or appliance malfunction (see Section G2427.14).

G2427.14 (503.14) Automatically operated vent dampers. An automatically operated vent damper shall be of a *listed* type.

❖ Automatic vent dampers are intended for use with gas-fired natural-draft appliances. An automatic vent damper must be installed in strict compliance with the manufacturer's installation instructions. Because automatic vent-damper failure can result in a hazardous condition, automatic dampers must be listed and labeled. An automatic vent damper is installed on the draft hood outlet of an individual gas-fired appliance (see Commentary Figure G2427.14). These dampers must not serve more than one appliance. The manufacturer's installation instructions require that the damper be installed by a qualified installer in accordance with the terms of the listing and the manufacturer's instructions.

Because the purpose of a vent damper is to close or restrict the flue passageway of an appliance, it is imperative that the device be properly installed to minimize the possibility of failure. A malfunctioning or improperly installed vent-damper device can cause the

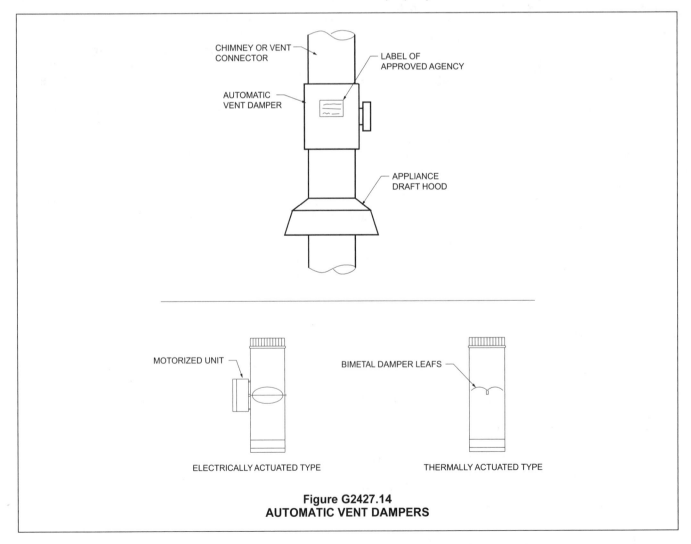

Figure G2427.14
AUTOMATIC VENT DAMPERS

appliance to malfunction and could cause the discharge of products of combustion directly into the building interior. Automatic vent dampers are energy-saving devices designed to close off or restrict an appliance flue passageway when the appliance is not operating and is in its "off" cycle. These devices save energy by trapping residual heat in a heat exchanger after the burners shut off and by preventing the escape of conditioned room air up the chimney or vent. Thus, appliance efficiency can be boosted, and building air infiltration can be reduced. Such devices can be field-installed additions to existing equipment and can be a factory-supplied component of an appliance (see Commentary Figure G2428.2.1).

A common application for vent dampers today is for mid-efficiency (80 to 83 percent) hot water boilers, which are draft-hood equipped. There are three types of automatic vent dampers: electrical-, mechanical- and thermal-actuating. The electric-motor-operated type is the commonly encountered type today (see Commentary Figures G2427.14 and G2428.2.1). Electrical- and mechanical-actuating automatic dampers must open automatically prior to main burner ignition and must remain open until completion of the burner cycle. They must also be interlocked electrically with the appliance control circuitry to prevent operation of the appliance in the event of damper failure. Thermal-actuating dampers open in response to flue gas temperature after burner ignition and close after burner shutdown. Thermal-actuated dampers use bimetal components that convert heat energy into mechanical energy. Some thermal-actuated dampers require the installation of a draft-hood spillage sensor that would shut off the appliance if the damper failed to open. The most common type of vent damper uses an electric motor to rotate a damper blade. The device uses switches that prove the damper position and allow the appliance to start only after the damper is opened and verified. The sequence of operation is the same as that described in the commentary to Section G2427.3.3.

G2427.15 (503.15) Obstructions. Devices that retard the flow of *vent gases* shall not be installed in a *vent connector*, *chimney*, or vent. The following shall not be considered as obstructions:

1. *Draft regulators* and safety *controls* specifically listed for installation in *venting systems* and installed in accordance with the manufacturer's instructions.

2. *Approved draft regulators* and safety *controls* that are designed and installed in accordance with *approved* engineering methods.

3. Listed heat reclaimers and automatically operated vent dampers installed in accordance with the manufacturer's instructions.

4. *Approved* economizers, heat reclaimers and recuperators installed in *venting systems* of *appliances* not required to be equipped with *draft hoods*, provided that the *appliance* manufacturer's instructions cover the installation of such a device in the venting system and performance in accordance with Sections G2427.3 and G2427.3.1 is obtained.

5. Vent dampers serving *listed appliances* installed in accordance with Sections G2428.2.1 and G2428.3.1 or other *approved* engineering methods.

❖ Manual dampers, flow restricting orifices and similar devices are not accounted for in the vent design methods of this code. The tables in Section G2428 assume that no such devices are installed in the vent. Item 3 addresses the automatic vent dampers covered in Section G2427.14. These dampers open fully before the appliance is allowed to fire and offer negligible flow resistance because the damper plate is parallel to the vent gas flow. Item 4 addresses devices that extract heat from the vent gases to save energy that would otherwise be lost to the outside atmosphere. Appliances without draft hoods, excluding Category I fan-assisted, Category II, III and IV appliances, usually have higher vent gas temperatures than those with draft hoods; therefore, waste heat can be reclaimed with less chance of diminishing draft (see Section G2428.2.1).

G2427.16 (503.16) (IFGS) Outside wall penetrations. Where vents, including those for *direct-vent appliances*, penetrate outside walls of buildings, the annular spaces around such penetrations shall be permanently sealed using *approved* materials to prevent entry of *combustion products* into the building.

❖ Vents that terminate through outside walls are subject to being covered (buried) in snow in many climates and experience has shown that combustion products can enter the building through the annular space between the vent and the wall assembly. In some cases, snow cover can trap combustion gases and cause them to reenter a direct-vent appliance combustion chamber via the combustion air intake. This will generate excessive carbon monoxide levels as a result of oxygen depletion. The snow cover can trap and channel these toxic gases into the building through any unsealed annular space in the wall penetration. The required sealing must be permanent and able to tolerate movement of the vent caused by expansion and contraction of the vent. Plastic pipe used for vents will expand and contract much more than metal pipe, and the annular space seal will be broken if allowances are not made for this movement.

Although not explicitly stated, this section should also be applied to penetrations by combustion air intake pipes that are associated with direct-vent appliances because such intakes are in close proximity to the vent discharge terminal and are affected by the same snow cover.

This section applies to all vents, not just those in regions subject to snowfall. Wind, fences, visual screens, vegetation and other causes could also create the conditions where combustion gases could enter the building.

SECTION G2428 (504)
SIZING OF CATEGORY I
APPLIANCE VENTING SYSTEMS

G2428.1 (504.1) Definitions. The following definitions apply to the tables in this section.

❖ The definitions in this section are necessary for application of the vent sizing tables and are exclusively related to Section G2428. As stated in the main section title, this section applies to Category I appliances. See the definition of "Vented appliance categories." However, it also applies to draft-hood-equipped appliances, which may or may not be categorized at all, even though they could be Category I because they meet the definition of Category I appliances. For example, a draft-hood water heater need not be categorized by its governing standard, but if it were categorized it would be a Category I appliance. A gas-fired appliance with a draft hood depends on gravity flow for supply of combustion air, for flow of combustion products through the heat exchanger and for proper gas venting. The draft hood is designed to maintain proper combustion in the event of draft fluctuations. Its relief opening serves as a flue product exit in the event of a blocked vent or downdraft. The relief opening, however, also allows dilution air to enter the vent during normal appliance/vent operation. When this dilution air is obtained from within the heated space, there is a loss of seasonal efficiency, which can be reduced by installation of a vent damper designed to reduce the flow of air through the vent when the burner is not firing.

To improve annual fuel utilization efficiency (AFUE), many appliance manufacturers have designed fan-assisted combustion using mechanical means (blowers or fans) to obtain either induced or forced flow of combustion air and combustion products. Greater efficiency results from three major effects. First, heat exchange improves because of higher internal flow velocity, which enhances heat transfer by creating turbulent flow and the "scrubbing" of heat exchanger surfaces. Second, induced draft through the appliance combustion chambers and heat exchangers allows the heat exchangers to be constructed with longer passes, which results in greater surface area and thus longer flue gas retention time. This allows extraction of more heat from the combustion gases. Third, there is no longer a draft hood to cause heated air loss up the vent, both when the appliance is operating and when it is not [see Commentary Figures G2428.1(1) through G2428.1(4)].

A fan-assisted combustion appliance uses pressure-sensing and heat-sensing controls to monitor proper venting. The controls must prove that the appliance flue outlet is at neutral or negative pressure and that adequate flow of flue gases exists. Without a draft hood and dilution air to dilute the combustion products, gases entering the vent will have a higher water vapor content than those from a draft-hood appliance. [see Commentary Figure G2428.1(3)]. There will likely be a longer period of condensation in the vent (wet time), particularly in the upper and/or colder portions of the venting system. These differences between fan-assisted and draft-hood appliances are reflected in the tables by the lack of minimum capacities for draft hood appliances. The minimums for fan-assisted appliances

Figure courtesy of Simpson Dura-Vent Company, Incorporated
For SI: 1 cubic foot = 0.0283 m³.

Figure G2428.1(1)
COMBUSTION PROCESS FOR NATURAL GAS

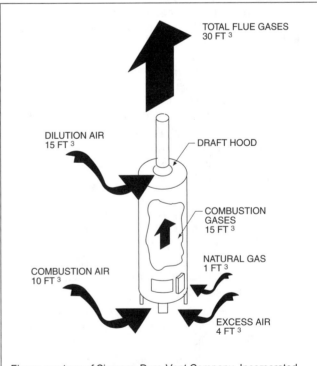

Figure courtesy of Simpson Dura-Vent Company, Incorporated
For SI: 1 cubic foot = 0.0283 m³.

Figure G2428.1(2)
FLUE GAS COMPOSITION

are based on the heat input needed to control the duration of wet time or condensation during appliance startup and operation.

The tables do not apply to:

1. Wall furnaces (recessed heaters), which require Type BW vents.

2. Decorative gas appliances, which generally require a specific vent size and are best individually vented.

3. Category II, III or IV gas appliances. If permitted by the appliance manufacturer, some Category III appliances can also be vented in accordance with Category I conditions. Specifically, if the vent for a Category III appliance is sized and configured (connector rise, total height, etc.) for the input, the vent will operate under nonpositive pressure. This allows the use of a Type B gas vent. For a Category III appliance to operate as a Category I, appliance manufacturers' instructions must be followed, and the vent size may need to be larger than the appliance flue collar.

4. Gas-fired appliances listed for use only with chimneys or dual-fuel appliances such as oil/gas, wood/gas, or coal/gas. Dual-fuel appliances require chimneys sized in accordance with the appliance manufacturer's instructions and in accordance with the chimney manufacturer's instructions.

Capacities in the tables assume cold starts such as with an intermittent ignition system (no standing pilot). The type of ignition system does not affect maximum capacities, but it does affect minimums. A standing pilot on an appliance will keep vent gas temperatures slightly above outdoor ambient temperature, whereas a tank of hot water plus water-heater pilot operation will maintain flue temperature ahead of the draft hood at approaching the same temperature as the stored water. This flow of heat into a vent aids in priming as well as reducing wet time. In some cases, the fan-assisted appliance manufacturer and/or the sizing tables in the code will require connection of a draft-hood-type appliance to the venting system because the relatively inefficient draft-hood appliance contributes heat to the venting system necessary to make it function.

The tables also assume that there are no adverse or building depressurization effects. A strong wind or mechanical ventilation/exhaust might cause a downdraft with draft-hood appliances that will prevent the vent from priming properly.

The input capacity values in the tables are computed for typical natural gas. They can be used for LP-gases, such as propane and butane, and mixtures of these with air. With LP-gases, the maximum capacity remains the same, but because these have less hydrogen and produce less water vapor, the possibility of condensation is somewhat lower. To simplify matters, assume that minimum capacities are the same regardless of fuel type. Oversized or excessively long vents can cause condensation and should be avoided. The tables define limits that minimize wet time in the indoor portions as well as in the upper exposed end of the vent.

The maximum vent lateral lengths for FAN appliances are calculated on the basis of several assumed conditions, including flue gas temperature and com-

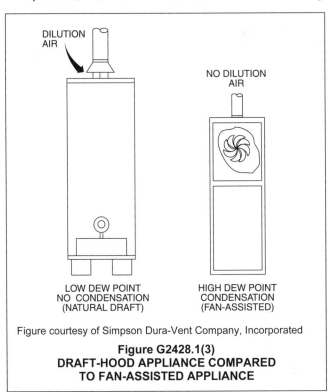

Figure courtesy of Simpson Dura-Vent Company, Incorporated

Figure G2428.1(3)
DRAFT-HOOD APPLIANCE COMPARED
TO FAN-ASSISTED APPLIANCE

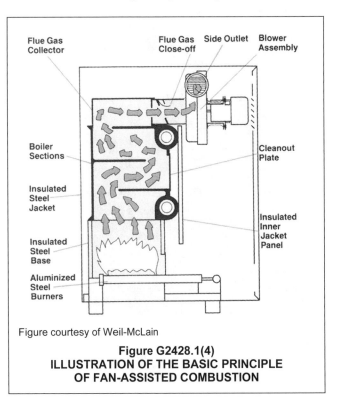

Figure courtesy of Weil-McLain

Figure G2428.1(4)
ILLUSTRATION OF THE BASIC PRINCIPLE
OF FAN-ASSISTED COMBUSTION

position, and an outdoor ambient temperature of 42°F (6°C), chosen as a representative value. Colder assumed temperatures or greater outdoor exposure of the vent would lead to shorter maximum allowable lengths or to greater possibilities of condensation. To minimize condensation, it is essential to operate closer to maximum than minimum capacity and also to use the smallest allowable vent size.

APPLIANCE CATEGORIZED VENT DIAMETER/ AREA. The minimum vent area/diameter permissible for Category I *appliances* to maintain a nonpositive vent static pressure when tested in accordance with nationally recognized standards.

❖ The minimum Category I appliance vent area/diameter necessary to maintain a nonpositive vent static pressure when tested in accordance with the applicable standard. The appliance categorized vent diameter/ area is determined by the appliance manufacturer and is typically the vent outlet collar installed on the appliance.

FAN-ASSISTED COMBUSTION SYSTEM. An *appliance* equipped with an integral mechanical means to either draw or force products of *combustion* through the *combustion chamber* or heat exchanger.

❖ An appliance equipped with an integral mechanical means to either draw or force products of combustion through the combustion chamber or heat exchanger [see Commentary Figures G2428.1(4) and (5)].

FAN Min. The minimum input rating of a Category I fan-assisted *appliance* attached to a vent or connector.

❖ The minimum appliance input rating of a Category I appliance with a fan-assisted combustion system that is capable of being attached to a vent.

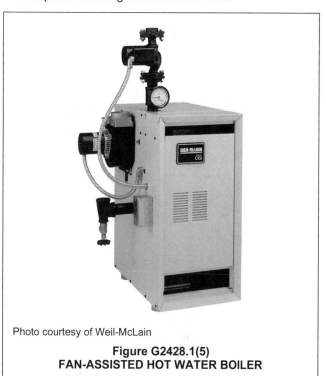

Photo courtesy of Weil-McLain

Figure G2428.1(5)
FAN-ASSISTED HOT WATER BOILER

FAN Max. The maximum input rating of a Category I fan-assisted *appliance* attached to a vent or connector.

❖ The maximum appliance input rating of a Category I appliance with a fan-assisted combustion system that is capable of being attached to a vent.

NAT Max. The maximum input rating of a Category I draft-hood-equipped *appliance* attached to a vent or connector.

❖ The maximum input rating of a Category I appliance equipped with a draft hood that is capable of being attached to a vent. There are no minimum appliance input ratings for draft-hood-equipped appliances.

FAN + FAN. The maximum combined *appliance* input rating of two or more Category I fan-assisted *appliances* attached to the common vent.

FAN + NAT. The maximum combined *appliance* input rating of one or more Category I fan-assisted *appliances* and one or more Category I draft-hood-equipped *appliances* attached to the common vent.

NA. Vent configuration is not permitted due to potential for *condensate* formation or pressurization of the venting system, or not applicable due to physical or geometric restraints.

❖ Vent configuration is prohibited.

NAT + NAT. The maximum combined *appliance* input rating of two or more Category I draft-hood-equipped *appliances* attached to the common vent.

G2428.2 (504.2) Application of single appliance vent Tables G2428.2(1) and G2428.2(2). The application of Tables G2428.2(1) and G2428.2(2) shall be subject to the requirements of Sections G2428.2.1 through G2428.2.17.

❖ Note that all of the table titles have been reformatted to make them easier to use and to lessen the possibility for error in choosing the appropriate table. The tables do not apply to Type B-W vents; vents for decorative gas appliances; vents for Category II, III or IV appliances; vents for dual-fuel appliances; vents for appliances not listed for use with Type B vent and appliances listed only for connection to chimneys. Category I appliances can be a fan-assisted design or can be a draft-hood-equipped design, and each design has different vent system design considerations. A noteworthy difference between fan-assisted and draft-hood equipped designs is that the vents for fan-assisted appliances have both a maximum and a minimum vent capacity. In the past, lower-efficiency draft-hood-type appliances were dominant, and vent sizing was based only on the maximum venting capacity of the vent. In other words, the only concern was to make sure that the vent was big enough. This is no longer the case with the advent of mid-efficiency fan-assisted-type appliances. Today, it is possible for a vent to be either too small or too large. An undersized vent will have insufficient capacity and can allow flue gas spillage or cause positive pressure to occur within the vent. An oversized vent can fail to produce sufficient draft and can be subject to the continuous formation of water vapor condensation on the interior vent surfaces. A typical individual vent is shown in Commentary Figure G2428.2(1) for a FAN furnace with an input rating of 150,000 Btu/h (44 kW) and a 5-inch (127 mm) outlet connection.

Procedure for a Type B vent, Table G2428.2(1): Go down the height column to 20 feet (6096 mm) and across on the 10-foot (3048 mm) lateral line.

For this furnace, the MAX capacity under the 5-inch (127 mm) size heading is 229,000 Btu/h (67.1 kW) and the MIN capacity is 50,000 Btu/h (14.7 kW). A 5-inch (127 mm) size Type B vent will be correct. See Commentary Figure G2428.2(1). In applying the tables, extreme care must be taken to observe the title of the table and to apply the conditions and instructions set forth in Sections G2428.2.1 through G2428.2.17.

To determine the proper size for an individual vent, apply the table as follows:

1. Determine the total vent height and length of lateral, based on the appliance and vent location and the height to the top of the vent, as indicated in Commentary Figure G2428.2(1). If gas appliances, such as a furnace, boiler or water heater, have not been chosen or installed, estimate the height beginning at 6 feet (1829 mm) above the floor. For attic or horizontal furnaces, floor furnaces, room heaters and small boilers, the height location of the draft hood outlet or vent collar should be known.

2. Read down the height column to a height equal to or less than the estimated total height.

3. Select the horizontal row for the appropriate lateral (L) length.

4. Read across to the first column under the type of appliance (FAN or NAT) that shows an appliance in-put rating equal to or greater than the name plate sea level input rating of the appliance to be vented.

Commentary Figure G2428.2(2) illustrates the measurement of height and laterals for draft-hood appliances.

Some vent manufacturers include special requirements regarding certain types of appliances in their installation/design instructions.

G2428.2.1 (504.2.1) Vent obstructions. These venting tables shall not be used where obstructions, as described in Section G2427.15, are installed in the venting system. The installation of vents serving *listed appliances* with vent dampers shall be in accordance with the *appliance* manufacturer's instructions or in accordance with the following:

1. The maximum capacity of the vent system shall be determined using the "NAT Max" column.

2. The minimum capacity shall be determined as if the *appliance* were a fan-assisted *appliance*, using the "FAN Min" column to determine the minimum capacity of the vent system. Where the corresponding "FAN Min" is "NA," the vent configuration shall not be permitted and an alternative venting configuration shall be utilized.

❖ An engineered-vent sizing method or appliance and vent/chimney manufacturer's instructions must be used to size vents where devices (obstructions) are installed in a venting system. The tables do not account for the flow resistance caused by such devices, nor do they account for the reduction of flue gas temperatures resulting from heat reclaim exchangers. Section G2427.15 lists several items that are not to be viewed

Figure courtesy of Selkirk, L.L.C.
For SI: 1 foot = 304.8 mm.

Figure G2428.2(1)
EXAMPLE OF INDIVIDUAL VENT DIMENSIONS

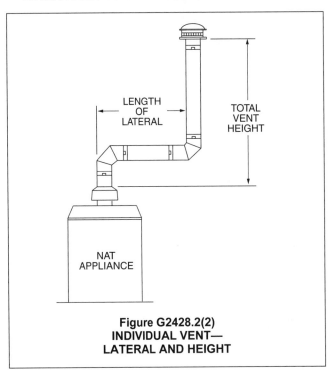

Figure G2428.2(2)
INDIVIDUAL VENT—
LATERAL AND HEIGHT

as obstructions, including automatic vent dampers. This section recognizes the use of the vent sizing tables with appliances equipped with automatic vent-damper devices. By referring to the appliance manufacturer's installation instructions, this section implies that the appliances are factory equipped with automatic vent dampers. Item 1 recognizes that during the firing cycle, an appliance with an automatic vent damper affects a venting system no differently than a draft-hood appliance without one. Item 2 addresses the effect that appliances equipped with automatic vent dampers have on venting systems during the appliance "off" cycle. The appliance acts as if it were an idle fan-assisted appliance because a closed vent damper will allow little or no air to enter the vent system, just as an idle fan-assisted appliance allows no air to enter the vent system. The lack of dilution air entering the venting system is a primary reason that the minimum input columns were created in the vent sizing tables. Boilers and water heaters with draft hoods commonly use automatic vent dampers to boost the efficiency of the appliance by trapping residual heat of the firing cycle, reducing standby losses and reducing the amount of conditioned air that escapes up the vent through the draft hood (see Commentary Figure G2428.2.1).

Photo courtesy of Weil-McLain

Figure G2428.2.1
CATEGORY I NATURAL DRAFT HOT WATER BOILER WITH DRAFT HOOD AND AUTOMATIC VENT DAMPER

G2428.2.2 (504.2.2) Minimum size. Where the vent size determined from the tables is smaller than the *appliance draft hood outlet* or *flue collar*, the smaller size shall be permitted to be used provided all of the following requirements are met:

1. The total vent height (H) is at least 10 feet (3048 mm).

2. Vents for *appliance draft hood* outlets or *flue collars* 12 inches (305 mm) in diameter or smaller are not reduced more than one table size.

3. Vents for *appliance draft hood* outlets or *flue collars* larger than 12 inches (305 mm) in diameter are not reduced more than two table sizes.

4. The maximum capacity listed in the tables for a fan-assisted *appliance* is reduced by 10 percent (0.90 × maximum table capacity).

5. The *draft hood* outlet is greater than 4 inches (102 mm) in diameter. Do not connect a 3-inch-diameter (76 mm) vent to a 4-inch-diameter (102 mm) *draft hood* outlet. This provision shall not apply to fan-assisted *appliances*.

❖ If a vent complies with all of Items 1 through 5 and is sized in accordance with the appropriate table, it is allowed to be smaller than the appliance connection, primarily because of the increase in capacity of the vent resulting from its height and because vent capacity is based on the Btu/h input rating of the appliance and not the size of the appliance flue collar. This code section suggests that flue collars are sized for worst case vent designs and are thus somewhat oversized for vents with higher venting capacities. Item 4 is necessary to prevent positive pressure from developing in the vent serving fan-assisted appliances. A 6-inch (152 mm) draft hood connected to a 5-inch (127 mm) Type B vent, for example, might look like an installation error, but this section could allow it in some cases, and operating a vent nearer its maximum capacity is always desirable (see Section G2428.2.10).

G2428.2.3 (504.2.3) Vent offsets. Single-*appliance* venting configurations with zero (0) lateral lengths in Tables G2428.2(1) and G2428.2(2) shall not have elbows in the *venting system*. Single-*appliance* venting configurations with lateral lengths include two 90-degree (1.57 rad) elbows. For each additional elbow up to and including 45 degrees (0.79 rad), the maximum capacity listed in the venting tables shall be reduced by 5 percent. For each additional elbow greater than 45 degrees (0.79 rad) up to and including 90 degrees (1.57 rad), the maximum capacity listed in the venting tables shall be reduced by 10 percent. Where multiple *offsets* occur in a vent, the total lateral length of all *offsets* combined shall not exceed that specified in Tables G2428.2(1) and G2428.2(2).

❖ A vent system with a zero lateral length is an ideal system because it is entirely vertical and thus has the least flow resistance and greatest draft. To create a lateral run of vent or vent connector, one or two elbows (adjustable fittings) are necessary [see Commentary Figure G2428.2(2)]. The tables account for two 90-degree (1.57 rad) changes in direction, which will offer considerable resistance to the flow of flue gases. Depending on the appliance flue collar or draft hood outlet orientation, a single elbow can create a lateral, but the tables assume that a vertical outlet orientation and two elbows are used. Two 45-degree (0.79 rad) fittings joined together, for example, are the equivalent of a single 90-degree (1.57 rad) fitting but would offer slightly less resistance to flow because of the larger turning radius. The capacity reduction penalty for each change in direction in excess of the two "free" 90-degree (1.57 rad) fittings accounts for the increased flow resistance caused by the additional fittings (elbows). For example, a vent using two 90-degree (1.57 rad) fittings and two 45-degree (0.79 rad) fittings would have a 10 percent capacity reduction penalty. A vent using two 90-degree (1.57 rad) fittings and two 30-degree (0.52 rad) fittings would also have a 10 percent capacity penalty. All changes of direction (fittings), regardless of the angle, that are in addition to the two "free" 90-degree fittings are accounted for by a capacity reduction penalty. A fitting adjusted to 45-degrees (0.79 rad) or less will cause a 5 percent penalty and a fitting adjusted to greater than 45-degrees (0.79 rad) and up to 90 degrees (1.57 rad) will cause a 10 percent penalty.

Whenever possible, an individual vent should be located directly over the appliance outlet. If the appliance flue outlet is horizontal, one 90-degree (1.57 rad) elbow should be used with the vent directly over it. A straight vertical vent is more easily supported and has less flow resistance.

There is no need to offset the vertical vent to include a tee and bracket. The use of a tee for cleanout or inspection purposes is quite unnecessary for three reasons. First, using a B-vent cap keeps debris and birds out. Second, clean-burning gas does not produce any deposits needing removal. Third, Type B vent joints are easily opened for inspection of the inside of the piping.

Should an offset be needed, the use of two elbows will provide somewhat greater capacity than an elbow and a tee. A tee has greater flow resistance than an elbow. The two 90-degree turns that are accounted for by the tables for laterals can occur anywhere within the vent system. This means that if the turns are used where the appliance connects to the system, the rest of the vent must be entirely vertical, or else the penalties of this section must be applied. The two "free" 90-degree (1.57 rad) turns can be composed of any number and angle of fittings that total not more than 180 degrees (3.1 rad). Simply put, there are free turns up to a maximum of 180 degrees (3.1 rad), and after that each additional turn will cause a capacity penalty to apply (see definition of "Offset").

Vent offsets should always be avoided where possible, especially offsets in the upper portion of the vent and in cold spaces such as attics and truss spaces. Offsets can be avoided with simple planning of the vent system and appliance locations. The attic offset is prevalent because the vent and appliance location is often not planned, and the building owners and designers do not want the vent to terminate on the street side of a roof for aesthetic reasons. This type of offset causes the most problems for several reasons. The typical attic will have winter temperatures close to the outdoor temperature; therefore, offsets in attics will cause more vent pipe to be exposed to the cold attic (see commentary, Section G2428.2.9). This can increase the production of condensate in the vent, and the fittings used will be points at which the condensate can leak from the inner vent lining. Most condensate leakage from vents can be traced to a fitting used in an offset. The code (see Sections G2428.2.9 and G2428.3.16) assumes that a vent will not be exposed to outdoor temperatures except for the necessary portion that extends above the roof. The temperature in a typical attic today is nearly equivalent to the outdoors.

Studies have shown that offsets in the upper portion of vent systems can produce slight positive pressures because of the flow resistance, and this encourages water vapor to escape from the vent and condense between the vent walls. Offsets in attics usually involve long runs of pipe that increase the surface area of vent pipe exposed to cold temperature. It is quite common to find that attic offsets exceed the horizontal length allowed by Section G2428.3.5.

Each offset introduces a horizontal component to a vent and this horizontal component is limited by the "lateral" column of the single appliance tables. The horizontal component of all offsets in a vent must be totaled and the total cannot exceed the lateral length allowance in the single appliance tables.

G2428.2.4 (504.2.4) Zero lateral. Zero (0) lateral (L) shall apply only to a straight vertical vent attached to a top outlet *draft hood* or *flue collar*.

❖ The zero lateral rows in the tables refer only to an entirely vertical vent run without offsets, lateral runs and directional fittings. This would be the ideal vent. Unfortunately, it is seldom seen because of lack of planning and coordination of vent and appliance locations. Architectural features of buildings usually dictate that vents take less than desirable paths to get from the appliances to the outdoors.

G2428.2.5 (504.2.5) High-altitude installations. Sea-level input ratings shall be used when determining maximum capacity for high-altitude installation. Actual input, derated for altitude, shall be used for determining minimum capacity for high-altitude installation.

❖ Appliances installed at elevations above sea level must have their Btu/h (W) input reduced (derated) because the atmosphere is less dense and oxygen levels decrease as altitude increases. The actual volume of flue gases entering a vent (i.e., primary and secondary combustion air, excess air and combustion byproducts) does not significantly change for different elevations; therefore, the maximum vent capacity is based on the sea-level input rating. The minimum capacity of a vent depends on heat input to the vent to control condensation and maintain the required draft. Therefore, the actual heat input (as derated) is used to determine the minimum capacity of a vent.

TABLE G2428.2(1) [504.2(1)]
TYPE B DOUBLE-WALL GAS VENT

		Number of Appliances	Single
		Appliance Type	Category I
		Appliance Vent Connection	Connected directly to vent

VENT DIAMETER—(D) inches

APPLIANCE INPUT RATING IN THOUSANDS OF BTU/H

HEIGHT (H) (feet)	LATERAL (L) (feet)	3 FAN Min	3 FAN Max	3 NAT Max	4 FAN Min	4 FAN Max	4 NAT Max	5 FAN Min	5 FAN Max	5 NAT Max	6 FAN Min	6 FAN Max	6 NAT Max	7 FAN Min	7 FAN Max	7 NAT Max	8 FAN Min	8 FAN Max	8 NAT Max	9 FAN Min	9 FAN Max	9 NAT Max
6	0	0	78	46	0	152	86	0	251	141	0	375	205	0	524	285	0	698	370	0	897	470
	2	13	51	36	18	97	67	27	157	105	32	232	157	44	321	217	53	425	285	63	543	370
	4	21	49	34	30	94	64	39	153	103	50	227	153	66	316	211	79	419	279	93	536	362
	6	25	46	32	36	91	61	47	149	100	59	223	149	78	310	205	93	413	273	110	530	354
8	0	0	84	50	0	165	94	0	276	155	0	415	235	0	583	320	0	780	415	0	1,006	537
	2	12	57	40	16	109	75	25	178	120	28	263	180	42	365	247	50	483	322	60	619	418
	5	23	53	38	32	103	71	42	171	115	53	255	173	70	356	237	83	473	313	99	607	407
	8	28	49	35	39	98	66	51	164	109	64	247	165	84	347	227	99	463	303	117	596	396
10	0	0	88	53	0	175	100	0	295	166	0	447	255	0	631	345	0	847	450	0	1,096	585
	2	12	61	42	17	118	81	23	194	129	26	289	195	40	402	273	48	533	355	57	684	457
	5	23	57	40	32	113	77	41	187	124	52	280	188	68	392	263	81	522	346	95	671	446
	10	30	51	36	41	104	70	54	176	115	67	267	175	88	376	245	104	504	330	122	651	427
15	0	0	94	58	0	191	112	0	327	187	0	502	285	0	716	390	0	970	525	0	1,263	682
	2	11	69	48	15	136	93	20	226	150	22	339	225	38	475	316	45	633	414	53	815	544
	5	22	65	45	30	130	87	39	219	142	49	330	217	64	463	300	76	620	403	90	800	529
	10	29	59	41	40	121	82	51	206	135	64	315	208	84	445	288	99	600	386	116	777	507
	15	35	53	37	48	112	76	61	195	128	76	301	198	98	429	275	115	580	373	134	755	491
20	0	0	97	61	0	202	119	0	349	202	0	540	307	0	776	430	0	1,057	575	0	1,384	752
	2	10	75	51	14	149	100	18	250	166	20	377	249	33	531	346	41	711	470	50	917	612
	5	21	71	48	29	143	96	38	242	160	47	367	241	62	519	337	73	697	460	86	902	599
	10	28	64	44	38	133	89	50	229	150	62	351	228	81	499	321	95	675	443	112	877	576
	15	34	58	40	46	124	84	59	217	142	73	337	217	94	481	308	111	654	427	129	853	557
	20	48	52	35	55	116	78	69	206	134	84	322	206	107	464	295	125	634	410	145	830	537

(continued)

TABLE G2428.2(1) [504.2(1)]—continued
TYPE B DOUBLE-WALL GAS VENT

Number of Appliances	Single
Appliance Type	Category I
Appliance Vent Connection	Connected directly to vent

VENT DIAMETER—(D) inches

APPLIANCE INPUT RATING IN THOUSANDS OF BTU/H

HEIGHT (H) (feet)	LATERAL (L) (feet)	3 FAN Min	3 FAN Max	3 NAT Max	4 FAN Min	4 FAN Max	4 NAT Max	5 FAN Min	5 FAN Max	5 NAT Max	6 FAN Min	6 FAN Max	6 NAT Max	7 FAN Min	7 FAN Max	7 NAT Max	8 FAN Min	8 FAN Max	8 NAT Max	9 FAN Min	9 FAN Max	9 NAT Max
30	0	0	100	64	0	213	128	0	374	220	0	587	336	0	853	475	0	1,173	650	0	1,548	855
	2	9	81	56	13	166	112	14	283	185	18	432	280	27	613	394	33	826	535	42	1,072	700
	5	21	77	54	28	160	108	36	275	176	45	421	273	58	600	385	69	811	524	82	1,055	688
	10	27	70	50	37	150	102	48	262	171	59	405	261	77	580	371	91	788	507	107	1,028	668
	15	33	64	NA	44	141	96	57	249	163	70	389	249	90	560	357	105	765	490	124	1,002	648
	20	56	58	NA	53	132	90	66	237	154	80	374	237	102	542	343	119	743	473	139	977	628
	30	NA	NA	NA	73	113	NA	88	214	NA	104	346	219	131	507	321	149	702	444	171	929	594
50	0	0	101	67	0	216	134	0	397	232	0	633	363	0	932	518	0	1,297	708	0	1,730	952
	2	8	86	61	11	183	122	14	320	206	15	497	314	22	715	445	26	975	615	33	1,276	813
	5	20	82	NA	27	177	119	35	312	200	43	487	308	55	702	438	65	960	605	77	1,259	798
	10	26	76	NA	35	168	114	45	299	190	56	471	298	73	681	426	86	935	589	101	1,230	773
	15	59	70	NA	42	158	NA	54	287	180	66	455	288	85	662	413	100	911	572	117	1,203	747
	20	NA	NA	NA	50	149	NA	63	275	169	76	440	278	97	642	401	113	888	556	131	1,176	722
	30	NA	NA	NA	69	131	NA	84	250	NA	99	410	259	123	605	376	141	844	522	161	1,125	670

For SI: 1 inch = 25.4 mm, 1 foot = 304.8 mm, 1 British thermal unit per hour = 0.2931 W.

TABLE G2428.2(2) [504.2(2)]
TYPE B DOUBLE-WALL GAS VENT

	Number of Appliances	Single
	Appliance Type	Category I
	Appliance Vent Connection	Single-wall metal connector

VENT DIAMETER—(D) inches

APPLIANCE INPUT RATING IN THOUSANDS OF BTU/H

HEIGHT (H) (feet)	LATERAL (L) (feet)	3 FAN Min	3 FAN Max	3 NAT Max	4 FAN Min	4 FAN Max	4 NAT Max	5 FAN Min	5 FAN Max	5 NAT Max	6 FAN Min	6 FAN Max	6 NAT Max	7 FAN Min	7 FAN Max	7 NAT Max	8 FAN Min	8 FAN Max	8 NAT Max	9 FAN Min	9 FAN Max	9 NAT Max	10 FAN Min	10 FAN Max	10 NAT Max	12 FAN Min	12 FAN Max	12 NAT Max
6	0	38	77	45	59	151	85	85	249	140	126	373	204	165	522	284	211	695	369	267	894	469	371	1,118	569	537	1,639	849
	2	39	51	36	60	96	66	85	156	104	123	231	156	159	320	213	201	423	284	251	541	368	347	673	453	498	979	648
	4	NA	NA	33	74	92	63	102	152	102	146	225	152	187	313	208	237	416	277	295	533	360	409	664	443	584	971	638
	6	NA	NA	31	83	89	60	114	147	99	163	220	148	207	307	203	263	409	271	327	526	352	449	656	433	638	962	627
8	0	37	83	50	58	164	93	83	273	154	123	412	234	161	580	319	206	777	414	258	1,002	536	360	1,257	658	521	1,852	967
	2	39	56	39	59	108	75	83	176	119	121	261	179	155	363	246	197	482	321	246	617	417	339	768	513	486	1,120	743
	5	NA	NA	37	77	102	69	107	168	114	151	252	171	193	352	235	245	470	311	305	604	404	418	754	500	598	1,104	730
	8	NA	NA	33	90	95	64	122	161	107	175	243	163	223	342	225	280	458	300	344	591	392	470	740	486	665	1,089	715
10	0	37	87	53	57	174	99	82	293	165	120	444	254	158	628	344	202	844	449	253	1,093	584	351	1,373	718	507	2,031	1,057
	2	39	61	41	59	117	80	82	193	128	119	287	194	153	400	272	193	531	354	242	681	456	332	849	559	475	1,242	848
	5	52	56	39	76	111	76	105	185	122	148	277	186	190	388	261	241	518	344	299	667	443	409	834	544	584	1,224	825
	10	NA	NA	34	97	100	68	132	171	112	188	261	171	237	369	241	296	497	325	363	643	423	492	808	520	688	1,194	788
15	0	36	93	57	56	190	111	80	325	186	116	499	283	153	713	388	195	966	523	244	1,259	681	336	1,591	838	488	2,374	1,237
	2	38	69	47	57	136	93	80	225	149	115	337	224	148	473	314	187	631	413	232	812	543	319	1,015	673	457	1,491	983
	5	51	63	44	75	128	86	102	216	140	144	326	217	182	459	298	231	616	400	287	795	526	392	997	657	562	1,469	963
	10	NA	NA	NA	95	116	79	128	201	131	182	308	203	228	438	284	284	592	381	349	768	501	470	966	628	664	1,433	928
	15	NA	NA	NA	NA	NA	72	158	186	124	220	290	192	272	418	269	334	568	367	404	742	484	540	937	601	750	1,399	894
20	0	35	96	60	54	200	118	78	346	201	114	537	306	149	772	428	190	1,053	573	238	1,379	750	326	1,751	927	473	2,631	1,346
	2	37	74	50	56	148	99	78	248	165	113	375	248	144	528	344	182	708	468	227	914	611	309	1,146	754	443	1,689	1,098
	5	50	68	47	73	140	94	100	239	158	141	363	239	178	514	334	224	692	457	279	896	596	381	1,126	734	547	1,665	1,074
	10	NA	NA	NA	NA	NA	86	125	223	146	177	344	224	222	491	316	277	666	437	339	866	570	457	1,092	702	646	1,626	1,037
	15	NA	NA	NA	NA	NA	80	155	208	136	216	325	210	264	469	301	325	640	419	393	838	549	526	1,060	677	730	1,587	1,005
	20	NA	NA	NA	NA	NA	NA	186	192	126	254	306	196	309	448	285	374	616	400	448	810	526	592	1,028	651	808	1,550	973

(continued)

TABLE G2428.2(2) [504.2(2)]—continued
TYPE B DOUBLE-WALL GAS VENT

Number of Appliances	Single
Appliance Type	Category I
Appliance Vent Connection	Single-wall metal connector

VENT DIAMETER—(D) inches

APPLIANCE INPUT RATING IN THOUSANDS OF BTU/H

HEIGHT (H) (feet)	LATERAL (L) (feet)	3 FAN Min	3 FAN Max	3 NAT Max	4 FAN Min	4 FAN Max	4 NAT Max	5 FAN Min	5 FAN Max	5 NAT Max	6 FAN Min	6 FAN Max	6 NAT Max	7 FAN Min	7 FAN Max	7 NAT Max	8 FAN Min	8 FAN Max	8 NAT Max	9 FAN Min	9 FAN Max	9 NAT Max	10 FAN Min	10 FAN Max	10 NAT Max	12 FAN Min	12 FAN Max	12 NAT Max
30	0	34	99	63	53	211	127	76	372	219	110	584	334	144	849	472	184	1,168	647	229	1,542	852	312	1,971	1,056	454	2,996	1,545
30	2	37	80	56	55	164	111	76	281	183	109	429	279	139	610	392	175	823	533	219	1,069	698	296	1,346	863	424	1,999	1,308
30	5	49	74	52	72	157	106	98	271	173	136	417	271	171	595	382	215	806	521	269	1,049	684	366	1,324	846	524	1,971	1,283
30	10	NA	NA	NA	91	144	98	122	255	168	171	397	257	213	570	367	265	777	501	327	1,017	662	440	1,287	821	620	1,927	1,234
30	15	NA	NA	NA	115	131	NA	151	239	157	208	377	242	255	547	349	312	750	481	379	985	638	507	1,251	794	702	1,884	1,205
30	20	NA	NA	NA	NA	NA	NA	181	223	NA	246	357	228	298	524	333	360	723	461	433	955	615	570	1,216	768	780	1,841	1,166
30	30	NA	NA	NA	NA	NA	NA	NA	NA	NA	NA	NA	NA	389	477	NA	461	670	426	541	895	574	704	1,147	720	937	1,759	1,101
50	0	33	99	66	51	213	133	73	394	230	105	629	361	138	928	515	176	1,292	704	220	1,724	948	295	2,223	1,189	428	3,432	1,818
50	2	36	84	61	53	181	121	73	318	205	104	495	312	133	712	443	168	971	613	209	1,273	811	280	1,615	1,007	401	2,426	1,509
50	5	48	80	NA	70	174	117	94	308	198	131	482	305	164	696	435	204	953	602	257	1,252	795	347	1,591	991	496	2,396	1,490
50	10	NA	NA	NA	89	160	NA	118	292	186	162	461	292	203	671	420	253	923	583	313	1,217	765	418	1,551	963	589	2,347	1,455
50	15	NA	NA	NA	112	148	NA	145	275	174	199	441	280	244	646	405	299	894	562	363	1,183	736	481	1,512	934	668	2,299	1,421
50	20	NA	NA	NA	NA	NA	NA	176	257	NA	236	420	267	285	622	389	345	866	543	415	1,150	708	544	1,473	906	741	2,251	1,387
50	30	NA	NA	NA	NA	NA	NA	NA	NA	NA	315	376	NA	373	573	NA	442	809	502	521	1,086	649	674	1,399	848	892	2,159	1,318

For SI: 1 inch = 25.4 mm, 1 foot = 304.8 mm, 1 British thermal unit per hour = 0.2931 W.

TABLE G2428.3(1) [504.3(1)]
TYPE B DOUBLE-WALL VENT

Number of Appliances	Two or more
Appliances Type	Category I
Appliances Vent Connection	Type B double-wall connector

VENT CONNECTOR CAPACITY

VENT HEIGHT (H) (feet)	CONNECTOR RISE (R) (feet)	3 FAN Min	3 FAN Max	3 NAT Max	4 FAN Min	4 FAN Max	4 NAT Max	5 FAN Min	5 FAN Max	5 NAT Max	6 FAN Min	6 FAN Max	6 NAT Max	7 FAN Min	7 FAN Max	7 NAT Max	8 FAN Min	8 FAN Max	8 NAT Max	9 FAN Min	9 FAN Max	9 NAT Max	10 FAN Min	10 FAN Max	10 NAT Max
6	1	22	37	26	35	66	46	46	106	72	58	164	104	77	225	142	92	296	185	109	376	237	128	466	289
	2	23	41	31	37	75	55	48	121	86	60	183	124	79	253	168	95	333	220	112	424	282	131	526	345
	3	24	44	35	38	81	62	49	132	96	62	199	139	82	275	189	97	363	248	114	463	317	134	575	386
8	1	22	40	27	35	72	48	49	114	76	64	176	109	84	243	148	100	320	194	118	408	248	138	507	303
	2	23	44	32	36	80	57	51	128	90	66	195	129	86	269	175	103	356	230	121	454	294	141	564	358
	3	24	47	36	37	87	64	53	139	101	67	210	145	88	290	198	105	384	258	123	492	330	143	612	402
10	1	22	43	28	34	78	50	49	123	78	65	189	113	89	257	154	106	341	200	125	436	257	146	542	314
	2	23	47	33	36	86	59	51	136	93	67	206	134	91	282	182	109	374	238	128	479	305	149	596	372
	3	24	50	37	37	92	67	52	146	104	69	220	150	94	303	205	111	402	268	131	515	342	152	642	417
15	1	21	50	30	33	89	53	47	142	83	64	220	120	88	298	163	110	389	214	134	493	273	162	609	333
	2	22	53	35	35	96	63	49	153	99	66	235	142	91	320	193	112	419	253	137	532	323	165	658	394
	3	24	55	40	36	102	71	51	163	111	68	248	160	93	339	218	115	445	286	140	565	365	167	700	444
20	1	21	54	31	33	99	56	46	157	87	62	246	125	86	334	171	107	436	224	131	552	285	158	681	347
	2	22	57	37	34	105	66	48	167	104	64	259	149	89	354	202	110	463	265	134	587	339	161	725	414
	3	23	60	42	35	110	74	50	176	116	66	271	168	91	371	228	113	486	300	137	618	383	164	764	466
30	1	20	62	33	31	113	59	45	181	93	60	288	134	83	391	182	103	512	238	125	649	305	151	802	372
	2	21	64	39	33	118	70	47	190	110	62	299	158	85	408	215	105	535	282	129	679	360	155	840	439
	3	22	66	44	34	123	79	48	198	124	64	309	178	88	423	242	108	555	317	132	706	405	158	874	494

COMMON VENT CAPACITY

VENT HEIGHT (H) (feet)	4 FAN+FAN	4 FAN+NAT	4 NAT+NAT	5 FAN+FAN	5 FAN+NAT	5 NAT+NAT	6 FAN+FAN	6 FAN+NAT	6 NAT+NAT	7 FAN+FAN	7 FAN+NAT	7 NAT+NAT	8 FAN+FAN	8 FAN+NAT	8 NAT+NAT	9 FAN+FAN	9 FAN+NAT	9 NAT+NAT	10 FAN+FAN	10 FAN+NAT	10 NAT+NAT
6	92	81	65	140	116	103	204	161	147	309	248	200	404	314	260	547	434	335	672	520	410
8	101	90	73	155	129	114	224	178	163	339	275	223	444	348	290	602	480	378	740	577	465
10	110	97	79	169	141	124	243	194	178	367	299	242	477	377	315	649	522	405	800	627	495
15	125	112	91	195	164	144	283	228	206	427	352	280	556	444	365	753	612	465	924	733	565
20	136	123	102	215	183	160	314	255	229	475	394	310	621	499	405	842	688	523	1,035	826	640
30	152	138	118	244	210	185	361	297	266	547	459	360	720	585	470	979	808	605	1,209	975	740
50	167	153	134	279	244	214	421	353	310	641	547	423	854	706	550	1,164	977	705	1,451	1,188	860

For SI: 1 inch = 25.4 mm, 1 foot = 304.8 mm, 1 British thermal unit per hour = 0.2931 W.

TABLE G2428.3(2) [504.3(2)]
TYPE B DOUBLE-WALL VENT

Number of Appliances	Two or more
Appliances Type	Category I
Appliances Vent Connection	Single-wall metal connector

VENT CONNECTOR CAPACITY

VENT HEIGHT (H) (feet)	CONNECTOR RISE (R) (feet)	3 FAN Min	3 FAN Max	3 NAT Max	4 FAN Min	4 FAN Max	4 NAT Max	5 FAN Min	5 FAN Max	5 NAT Max	6 FAN Min	6 FAN Max	6 NAT Max	7 FAN Min	7 FAN Max	7 NAT Max	8 FAN Min	8 FAN Max	8 NAT Max	9 FAN Min	9 FAN Max	9 NAT Max	10 FAN Min	10 FAN Max	10 NAT Max
6	1	NA	NA	26	NA	NA	46	NA	NA	71	NA	NA	102	207	223	140	262	293	183	325	373	234	447	463	286
	2	NA	NA	31	NA	NA	55	NA	NA	85	168	182	123	215	251	167	271	331	219	334	422	281	458	524	344
	3	NA	NA	34	NA	NA	62	121	131	95	175	198	138	222	273	188	279	361	247	344	462	316	468	574	385
8	1	NA	NA	27	NA	NA	48	NA	NA	75	NA	NA	106	226	240	145	285	316	191	352	403	244	481	502	299
	2	NA	NA	32	NA	NA	57	125	126	89	184	193	127	234	266	173	293	353	228	360	450	292	492	560	355
	3	NA	NA	35	NA	NA	64	130	138	100	191	208	144	241	287	197	302	381	256	370	489	328	501	609	400
10	1	NA	NA	28	NA	NA	50	119	121	77	182	186	110	240	253	150	302	335	196	372	429	252	506	534	308
	2	NA	NA	33	84	85	59	124	134	91	189	203	132	248	278	183	311	369	235	381	473	302	517	589	368
	3	NA	NA	36	89	91	67	129	144	102	197	217	148	257	299	203	320	398	265	391	511	339	528	637	413
15	1	NA	NA	29	79	87	52	116	138	81	177	214	116	238	291	158	312	380	208	397	482	266	556	596	324
	2	NA	NA	34	83	94	62	121	150	97	185	230	138	246	314	189	321	411	248	407	522	317	568	646	387
	3	NA	NA	39	87	100	70	127	160	109	193	243	157	255	333	215	331	438	281	418	557	360	579	690	437
20	1	49	56	30	78	97	54	115	152	84	175	238	120	233	325	165	306	425	217	390	538	276	546	664	336
	2	52	59	36	82	103	64	120	163	101	182	252	144	243	346	197	317	453	259	400	574	331	558	709	403
	3	55	62	40	87	107	72	125	172	113	190	264	164	252	363	223	326	476	294	412	607	375	570	750	457
30	1	47	60	31	77	110	57	112	175	89	169	278	129	226	380	175	296	497	230	378	630	294	528	779	358
	2	51	62	37	81	115	67	117	185	106	177	290	152	236	397	208	307	521	274	389	662	349	541	819	425
	3	54	64	42	85	119	76	122	193	120	185	300	172	244	412	235	316	542	309	400	690	394	555	855	482

COMMON VENT CAPACITY

VENT HEIGHT (H) (feet)	4 FAN+FAN	4 FAN+NAT	4 NAT+NAT	5 FAN+FAN	5 FAN+NAT	5 NAT+NAT	6 FAN+FAN	6 FAN+NAT	6 NAT+NAT	7 FAN+FAN	7 FAN+NAT	7 NAT+NAT	8 FAN+FAN	8 FAN+NAT	8 NAT+NAT	9 FAN+FAN	9 FAN+NAT	9 NAT+NAT	10 FAN+FAN	10 FAN+NAT	10 NAT+NAT
6	NA	78	64	NA	113	99	200	158	144	304	244	196	398	310	257	541	429	332	665	515	407
8	NA	87	71	NA	126	111	218	173	159	331	269	218	436	342	285	592	473	373	730	569	460
10	NA	94	76	163	137	120	237	189	174	357	292	236	467	369	309	638	512	398	787	617	487
15	121	108	88	189	159	140	275	221	200	416	343	274	544	434	357	738	599	456	905	718	553
20	131	118	98	208	177	156	305	247	223	463	383	302	606	487	395	824	673	512	1,013	808	626
30	145	132	113	236	202	180	350	286	257	533	446	349	703	570	459	958	790	593	1,183	952	723
50	159	145	128	268	233	208	406	337	296	622	529	410	833	686	535	1,139	954	689	1,418	1,157	838

For SI: 1 inch = 25.4 mm, 1 foot = 304.8 mm, 1 British thermal unit per hour = 0.2931 W.

TABLE G2428.3(3) [504.3(3)]
MASONRY CHIMNEY

Number of Appliances	Two or more
Appliances Type	Category I
Appliances Vent Connection	Type B double-wall connector

VENT CONNECTOR CAPACITY

VENT HEIGHT (H) (feet)	CONNECTOR RISE (R) (feet)	3 FAN Min	3 FAN Max	3 NAT Max	4 FAN Min	4 FAN Max	4 NAT Max	5 FAN Min	5 FAN Max	5 NAT Max	6 FAN Min	6 FAN Max	6 NAT Max	7 FAN Min	7 FAN Max	7 NAT Max	8 FAN Min	8 FAN Max	8 NAT Max	9 FAN Min	9 FAN Max	9 NAT Max	10 FAN Min	10 FAN Max	10 NAT Max
6	1	24	33	21	39	62	40	52	106	67	65	194	101	87	274	141	104	370	201	124	479	253	145	599	319
6	2	26	43	28	41	79	52	53	133	85	67	230	124	89	324	173	107	436	232	127	562	300	148	694	378
6	3	27	49	34	42	92	61	55	155	97	69	262	143	91	369	203	109	491	270	129	633	349	151	795	439
8	1	24	39	22	39	72	41	55	117	69	71	213	105	94	304	148	113	414	210	134	539	267	156	682	335
8	2	26	47	29	40	87	53	57	140	86	73	246	127	97	350	179	116	473	240	137	615	311	160	776	394
8	3	27	52	34	42	97	62	59	159	98	75	269	145	99	383	206	119	517	276	139	672	358	163	848	452
10	1	24	42	22	38	80	42	55	130	71	74	232	108	101	324	153	120	444	216	142	582	277	165	739	348
10	2	26	50	29	40	93	54	57	153	87	76	261	129	103	366	184	123	498	247	145	652	321	168	825	407
10	3	27	55	35	41	105	63	58	170	100	78	284	148	106	397	209	126	540	281	147	705	366	171	893	463
15	1	24	48	23	38	93	44	54	154	74	72	277	114	100	384	164	125	511	229	153	658	297	184	824	375
15	2	25	55	31	39	105	55	56	174	89	74	299	134	103	419	192	128	558	260	156	718	339	187	900	432
15	3	26	59	35	41	115	64	57	189	102	76	319	153	105	448	215	131	597	292	159	760	382	190	960	486
20	1	24	52	24	37	102	46	53	172	77	71	313	119	98	437	173	123	584	239	150	752	312	180	943	397
20	2	25	58	31	39	114	56	55	190	91	73	335	138	101	467	199	126	625	270	153	805	354	184	1,011	452
20	3	26	63	35	40	123	65	57	204	104	75	353	157	104	493	222	129	661	301	156	851	396	187	1,067	505

COMMON VENT CAPACITY

VENT HEIGHT (H) (feet)	12 FAN+FAN	12 FAN+NAT	12 NAT+NAT	19 FAN+FAN	19 FAN+NAT	19 NAT+NAT	28 FAN+FAN	28 FAN+NAT	28 NAT+NAT	38 FAN+FAN	38 FAN+NAT	38 NAT+NAT	50 FAN+FAN	50 FAN+NAT	50 NAT+NAT	63 FAN+FAN	63 FAN+NAT	63 NAT+NAT	78 FAN+FAN	78 FAN+NAT	78 NAT+NAT	113 FAN+FAN	113 FAN+NAT	113 NAT+NAT
6	NA	74	25	NA	119	46	NA	178	71	NA	257	103	NA	351	143	NA	458	188	NA	582	246	1,041	853	NA
8	NA	80	28	NA	130	53	NA	193	82	NA	279	119	NA	384	163	NA	501	218	724	636	278	1,144	937	408
10	NA	84	31	NA	138	56	NA	207	90	NA	299	131	NA	409	177	606	538	236	776	686	302	1,226	1,010	454
15	NA	NA	36	NA	152	67	NA	233	106	NA	334	152	523	467	212	682	611	283	874	781	365	1,374	1,156	546
20	NA	NA	41	NA	NA	75	NA	250	122	NA	368	172	565	508	243	742	668	325	955	858	419	1,513	1,286	648
30	NA	NA	NA	NA	NA	NA	NA	270	137	NA	404	198	615	564	278	816	747	381	1,062	969	496	1,702	1,473	749
50	NA	NA	NA	NA	NA	NA	NA	NA	NA	NA	NA	NA	NA	620	328	879	831	461	1,165	1,089	606	1,905	1,692	922

For SI: 1 inch = 25.4 mm, 1 square inch = 645.16 mm², 1 foot = 304.8 mm, 1 British thermal unit per hour = 0.2931 W.

TABLE G2428.3(4) [504.3(4)]
MASONRY CHIMNEY

Number of Appliances	Two or more
Appliances Type	Category I
Appliances Vent Connection	Single-wall connector

VENT CONNECTOR CAPACITY

VENT HEIGHT (H) (feet)	CONNECTOR RISE (R) (feet)	3 FAN Min	3 FAN Max	3 NAT Max	4 FAN Min	4 FAN Max	4 NAT Max	5 FAN Min	5 FAN Max	5 NAT Max	6 FAN Min	6 FAN Max	6 NAT Max	7 FAN Min	7 FAN Max	7 NAT Max	8 FAN Min	8 FAN Max	8 NAT Max	9 FAN Min	9 FAN Max	9 NAT Max	10 FAN Min	10 FAN Max	10 NAT Max
6	1	NA	NA	21	NA	NA	39	NA	NA	66	179	191	100	231	271	140	292	366	200	362	474	252	499	594	316
	2	NA	NA	28	NA	NA	52	NA	NA	84	186	227	123	239	321	172	301	432	231	373	557	299	509	696	376
	3	NA	NA	34	NA	NA	61	134	153	97	193	258	142	247	365	202	309	491	269	381	634	348	519	793	437
8	1	NA	NA	21	NA	NA	40	NA	NA	68	195	208	103	250	298	146	313	407	207	387	530	263	529	672	331
	2	NA	NA	28	NA	NA	52	137	139	85	202	240	125	258	343	177	323	465	238	397	607	309	540	766	391
	3	NA	NA	34	NA	NA	62	143	156	98	210	264	145	266	376	205	332	509	274	407	663	356	551	838	450
10	1	NA	NA	22	NA	NA	41	130	151	70	202	225	106	267	316	151	333	434	213	410	571	273	558	727	343
	2	NA	NA	29	NA	NA	53	136	150	86	210	255	128	276	358	181	343	489	244	420	640	317	569	813	403
	3	NA	NA	34	97	102	62	143	166	99	217	277	147	284	389	207	352	530	279	430	694	363	580	880	459
15	1	NA	NA	23	NA	NA	43	129	151	73	199	271	112	268	376	161	349	502	225	445	646	291	623	808	366
	2	NA	NA	30	92	103	54	135	170	88	207	295	132	277	411	189	359	548	256	456	706	334	634	884	424
	3	NA	NA	34	96	112	63	141	185	101	215	315	151	286	439	213	368	586	289	466	755	378	646	945	479
20	1	NA	NA	23	87	99	45	128	167	76	197	303	117	265	425	169	345	569	235	439	734	306	614	921	347
	2	NA	NA	30	91	111	55	134	185	90	205	325	136	274	455	195	355	610	266	450	787	348	627	986	443
	3	NA	NA	35	96	119	64	140	199	103	213	343	154	282	481	219	365	644	298	461	831	391	639	1,042	496

COMMON VENT CAPACITY

MINIMUM INTERNAL AREA OF MASONRY CHIMNEY FLUE (square inches)

COMBINED APPLIANCE INPUT RATING IN THOUSANDS OF BTU/H

VENT HEIGHT (H) (feet)	12 FAN+FAN	12 FAN+NAT	12 NAT+NAT	19 FAN+FAN	19 FAN+NAT	19 NAT+NAT	28 FAN+FAN	28 FAN+NAT	28 NAT+NAT	38 FAN+FAN	38 FAN+NAT	38 NAT+NAT	50 FAN+FAN	50 FAN+NAT	50 NAT+NAT	63 FAN+FAN	63 FAN+NAT	63 NAT+NAT	78 FAN+FAN	78 FAN+NAT	78 NAT+NAT	113 FAN+FAN	113 FAN+NAT	113 NAT+NAT
6	NA	NA	25	NA	118	45	NA	176	71	NA	255	102	NA	348	142	NA	455	187	NA	579	245	NA	846	NA
8	NA	NA	28	NA	128	52	NA	190	81	NA	276	118	NA	380	162	NA	497	217	NA	633	277	1,136	928	405
10	NA	NA	31	NA	136	56	NA	205	89	NA	295	129	NA	405	175	NA	532	234	171	680	300	1,216	1,000	450
15	NA	NA	36	NA	NA	66	NA	230	105	NA	335	150	NA	400	210	677	602	280	866	772	360	1,359	1,139	540
20	NA	NA	NA	NA	NA	74	NA	247	120	NA	362	170	NA	503	240	765	661	321	947	849	415	1,495	1,264	640
30	NA	NA	NA	NA	NA	NA	NA	NA	135	NA	398	195	NA	558	275	808	739	377	1,052	957	490	1,682	1,447	740
50	NA	NA	NA	NA	NA	NA	NA	NA	NA	NA	NA	NA	NA	612	325	NA	821	456	1,152	1,076	600	1,879	1,672	910

For SI: 1 inch = 25.4 mm, 1 square inch = 645.16 mm², 1 foot = 304.8 mm, 1 British thermal unit per hour = 0.2931 W.

G2428.2.6 (504.2.6) Multiple input rate appliances. For *appliances* with more than one input rate, the minimum vent capacity (FAN Min) determined from the tables shall be less than the lowest *appliance* input rating, and the maximum vent capacity (FAN Max/NAT Max) determined from the tables shall be greater than the highest *appliance* rating input.

❖ To improve efficiency, some appliances are equipped with modulating input or high-low two-stage input. Such appliances typically use two-stage thermostats or timers to change over from low-fire to high-fire operation, thus more closely matching the appliance input to the heating load. The venting system must be capable of operating properly throughout the entire firing range of the appliance. The venting system must be capable of handling the maximum input of the appliance and also must be compatible with the appliance when the appliance is operating at its lowest input. Such appliances will usually operate at the "low-fire" input most of their total operating time. To satisfy the minimum input requirements for fan-assisted appliance vents, the vent design must be based on the lowest Btu/h rating of the appliance because this is what the vent senses as the connected load. The lowest appliance input rate is used for minimum vent capacity, and the highest input rate must be used for maximum vent capacity. This requirement can be easily overlooked if the appliance label is not carefully studied because the appliance will usually be described by its maximum input, and any minimum input will have to obtained from the label.

G2428.2.7 (504.2.7) Liner system sizing and connections. *Listed* corrugated metallic *chimney* liner systems in masonry *chimneys* shall be sized by using Table G2428.2(1) or G2428.2(2) for Type B vents with the maximum capacity reduced by 20 percent (0.80 × maximum capacity) and the minimum capacity as shown in Table G2428.2(1) or G2428.2(2). Corrugated metallic liner systems installed with bends or offsets shall have their maximum capacity further reduced in accordance with Section G2428.2.3. The 20-percent reduction for corrugated metallic *chimney* liner systems includes an allowance for one long-radius 90-degree (1.57 rad) turn at the bottom of the liner.

Connections between *chimney* liners and listed double-wall connectors shall be made with listed adapters designed for such purpose.

❖ A liner system installed within a masonry chimney is sized and designed as if the chimney/liner system combination were a Type B vent. Corrugated metal liners are constructed of semirigid piping having a higher flow resistance than smooth-wall rigid pipe. A capacity reduction penalty compensates for the poorer flow characteristics of corrugated pipe. Bends in semirigid liners count as elbows and are regulated by Section G2428.2.3. Note that the sweeping 90-degree (1.57 rad) bend in the liner at the point where it exits the bottom of the chimney and connects to the appliance(s) is already accounted for in the 20-percent capacity penalty and no further capacity adjustment is required for such turn/bend. Listed adapter fittings must be used to connect double-wall connectors to the liner to avoid "make shift" field connections that can result in poor fit and leakage.

G2428.2.8 (504.2.8) Vent area and diameter. Where the vertical vent has a larger diameter than the *vent connector*, the vertical vent diameter shall be used to determine the minimum vent capacity, and the connector diameter shall be used to determine the maximum vent capacity. The flow area of the vertical vent shall not exceed seven times the flow area of the listed *appliance* categorized vent area, *flue collar* area, or *draft hood* outlet area unless designed in accordance with *approved* engineering methods.

❖ The maximum capacity of a venting system must be based on the element of the system having the least capacity (i.e., the weakest link in the chain). Likewise, the minimum Btu/h (W) input required for proper vent system operation must be based on the largest element of the system because the largest element will naturally have the highest minimum input requirement. A connector that is smaller than the vent will set the maximum capacity limit for the system because it has the greater flow resistance and smaller cross-sectional area. The minimum Btu/h input is based on the larger vent size because the vent will require more heat to produce adequate draft and avoid condensation. The ratio of appliance vent-connection area to vertical vent area is intended to prevent oversizing of the vent, which can result in poor draft and condensation problems. Because the tables will allow vents to be larger than needed, this section serves as a limit to prevent gross oversizing. For example, in Table G2428.2(1) it can be seen that a fan-assisted appliance having an input rating near the maximum end of the minimum/maximum range for any 4-inch (102 mm) venting system configuration can be served by not only a 4-inch (102 mm) vent but also a 5-inch (127 mm) and sometimes a 6-inch (152 mm) or 7-inch (178 mm) vent. Consistent with vent manufacturers' instructions, a venting system should always be designed to use the smallest vent allowed by the tables. Vents that are larger than they need to be usually will not perform as well as accurately sized vents and are more likely to suffer from poor draft and condensate problems.

G2428.2.9 (504.2.9) Chimney and vent locations. Tables G2428.2(1) and G2428.2(2) shall be used only for chimneys and vents not exposed to the outdoors below the roof line. A Type B vent or listed chimney lining system passing through an unused masonry chimney flue shall not be considered to be exposed to the outdoors. Where vents extend outdoors above the roof more than 5 feet (1524 mm) higher than required by Figure G2427.6.3 and where vents terminate in accordance with Section G2427.6.3, Item 2, the outdoor portion of the vent shall be enclosed as required by this section for vents not considered to be exposed to the outdoors or such venting system shall be engineered. A Type B vent shall not be considered to be exposed to the outdoors where it passes through an unventilated enclosure or chase insulated to a value of not less than R8.

❖ Chimneys that have one or more walls exposed to the outdoors and vents exposed to the outdoors can develop condensation and weak draft problems in cold weather. Commentary Figure G2428.2.9(1) illustrates why masonry chimneys, including interior chimneys, are more prone to draft and condensation problems. The tables assume that the only portion of the chimney or vent that is exposed to the outdoors is the portion that extends above the roof penetration. Even though an attic is technically not outdoors, the temperature in an attic can be very close to the outdoor temperature. In some climates, the actual attic temperature can be far lower than the assumed temperature used in the computer modeling on which the vent sizing tables are based. Therefore, the portion of a venting system that passes through an attic should be kept to an absolute minimum. If at all possible, the installation should be planned to avoid the need for vent offsets in an attic. Section G2428.2.7 states that a chimney lined with a listed liner or Type B vent is to be sized as if it were a Type B vent. The walls of a chimney will provide some insulating effect for a properly installed Type B vent or lining system installed within the chimney. A chimney lined with a Type B vent is not considered to be a chimney, but is treated as a Type B vent installed within a masonry chase (see Section G2428.2.3).

This section intends that the listed tables be applicable only to vents and chimneys that are not exposed to the outdoors below the roof penetration. If the vent or chimney is exposed to the outdoors below the roof penetration, the listed sizing tables do not apply and the venting system would have to be engineered. The intent is that vents and chimneys be enclosed within the building envelope to lessen exposure to low ambient temperatures and wind. It is these conditions that were considered in the development of the vent sizing tables. The sizing tables are based on an assumed outdoor temperature of 42°F (6°C) and the assumption that all portions of a vent or chimney below the roof penetration are exposed to an ambient temperature of 60°F (16°C). This makes it obvious that long runs of vent piping through attics and long extensions above roofs should be avoided in cold climates (see Section G2428.3.5).

This section prohibits the use of the vent sizing tables for the installation of a Type B vent run up the out-side of an exterior wall of a building. This is consistent with vent manufacturers' instructions. Outdoor vents obviously suffer from poorer draft and increased condensation formation because they are exposed to the elements, low temperatures being the main factor. "Stack effect" in the building interior can cause back drafts (reverse flow) to occur in any vent or chimney, particularly vents exposed to the outdoors. For the same reasons, vents should extend above the roof no more than required by the code because the more vent pipe that extends above the roof, the more vent pipe will be exposed to the cold and wind. In some cases, designers and installers will intentionally raise a vent above a roof more than Section G2427.6.3 requires just to increase the vent capacity. In doing so, additional vent pipe is exposed to the outdoors, thus inviting condensation and possibly diminishing draft. Section G2427.6.3 could also require a vent to extend well above a lower roof penetration, thus causing the sizing tables of this section to be inapplicable to such a vent. A Type B vent that is enclosed in an airtight (unventilated) chase insulated to a value of at least R8 is protected from the weather and is therefore considered as an indoor vent. R8 is the same insulation value required by the *International Energy Conservation Code®* (IECC®) for ducts located outside of the building thermal envelope. To be effective, the chase enclosure would have to be built air tight and with an internal volume no larger than necessary to accommodate the vent with the required clearance to combustibles.

Where vents extend outdoors above the roof more than 5 feet higher than required by Figure G2427.6.3 and also where vents extend outdoors above the roof as required by Section G2427.6.3, Item 2, the outdoor portion of such vents must be enclosed in an insulated and sealed chase/enclosure. This requirement is necessary because such vents are not within the parameters upon which the vent sizing tables were calculated; thus, these vents would be expected to not perform as intended [see Commentary Figures G2428.2.9(1) and (2)]. If the owner, designer, builder or installer wants to avoid the expense and any displeasing aesthetics of these outdoor enclosures, the location of vent terminations will have to be thoughtfully planned, or already installed vents may have to be relocated. The allowance for 5 feet higher than required by Figure G2427.6.3 is a necessary tolerance to account for the fact that sections of vent pipe are available only in certain lengths. Therefore, in many cases, vent pipes extending above the roof some distance higher than required by the code will be unavoidable.

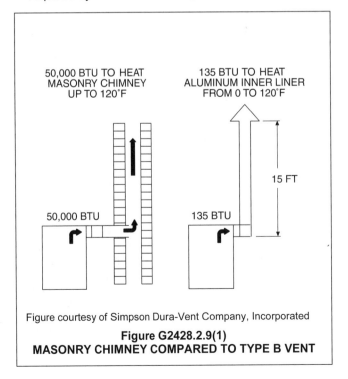

Figure courtesy of Simpson Dura-Vent Company, Incorporated

Figure G2428.2.9(1)
MASONRY CHIMNEY COMPARED TO TYPE B VENT

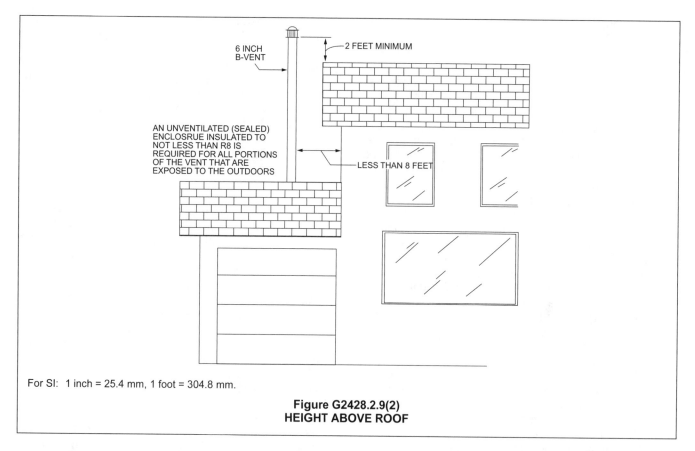

6 INCH
B-VENT

2 FEET MINIMUM

AN UNVENTILATED (SEALED)
ENCLOSRUE INSULATED TO
NOT LESS THAN R8 IS
REQUIRED FOR ALL PORTIONS
OF THE VENT THAT ARE
EXPOSED TO THE OUTDOORS

LESS THAN 8 FEET

For SI: 1 inch = 25.4 mm, 1 foot = 304.8 mm.

Figure G2428.2.9(2)
HEIGHT ABOVE ROOF

G2428.2.10 (504.2.10) Corrugated vent connector size. Corrugated *vent connectors* shall be not smaller than the listed *appliance* categorized *vent* diameter, *flue collar* diameter, or *draft hood* outlet diameter.

❖ Corrugated vent connectors are becoming more common because they can save installation time and avoid the use of fittings. This section is consistent with Section G2428.3.17 and applies to both draft-hood equipped and fan-assisted appliances. Corrugated materials have higher resistance to flow than smooth wall materials; therefore, to ensure the required capacity, full size connectors must be used.

G2428.2.11 (504.2.11) Vent connector size limitation. *Vent connectors* shall not be increased in size more than two sizes greater than the listed *appliance* categorized vent diameter, *flue collar* diameter or *draft hood* outlet diameter.

❖ The sizing tables of this section will often require that a connector be larger than the appliance vent connection. The appliance manufacturer's instructions may also require that the size of a connector be increased. The increase in size compensates for flow resistance in the connector and prevents positive pressurization. A grossly oversized connector will allow the vent gases to expand and cool too much, thus negatively affecting draft and contributing to the formation of condensation. Connector size is sometimes increased for the sole purpose of gaining more allowable length (Section G2428.3.2), and this section will limit that practice.

G2428.2.12 (504.2.12) Component commingling. In a single run of vent or *vent connector*, different diameters and types of vent and connector components shall be permitted to be used, provided that all such sizes and types are permitted by the tables.

❖ In the unlikely event that an installer wants to mix, for example, Type B vent with single-wall pipe in a single connector (for example, Type B vent from a draft hood, switching to single wall and connecting to Type B vent again), the mixed system must comply with the code based on each type of material as if the run were constructed entirely of that material. There may be some unique situations that would justify commingling of different types and/or sizes of components, but clearances, heat loss, ambient temperatures, joining methods and flow characteristics must be considered.

G2428.2.13 (504.2.13) Draft hood conversion accessories. *Draft hood* conversion accessories for use with *masonry chimneys* venting listed Category I fan-assisted *appliances* shall be listed and installed in accordance with the manufacturer's instructions for such listed accessories.

❖ Some appliance manufacturers offer an accessory kit that allows the appliance to be converted from a fan-assisted Category I appliance to something that performs closely to a draft-hood-equipped Category I appliance. This conversion will result in some loss in appliance efficiency (AFUE) and its sole purpose is to allow the appliance to vent to a masonry chimney, whereas without the conversion, the appliance would be prohibited from venting to a masonry chimney. The net result of the conversion is to put dilution air into the chimney, thus helping to warm it and reduce the possibility of condensation. Such conversion kits come with their own installation instructions and are specific to a particular make and model of appliance.

G2428.2.14 (504.2.14) Table interpolation. Interpolation shall be permitted in calculating capacities for vent dimensions that fall between the table entries.

❖ Because Tables G2428.2(1) and G2428.2(2) do not contain values for every possible height of a vent or length of a lateral, interpolation could be necessary. The maximum and minimum capacity of vents having heights not listed in the tables can be determined using the following equations.

Interpolated Maximum Vent Capacity = $(NB \times C) + D$
Interpolated Minimum Vent Capacity = $E - (AB \times F)$

where:

A = The difference between the design height entry and the next lower height entry in the applicable table.

B = The difference between the closest consecutive height entries in the applicable table.

C = The difference between the maximum capacity column consecutive entries in the applicable table.

D = The lower maximum capacity entry in the applicable table.

E = The higher minimum capacity entry in the applicable table.

F = The difference between the minimum capacity column consecutive entries in the applicable table.

For SI: 1 inch = 25.4 mm, 1 foot = 304.8 mm, 1 Btu/h = 0.2931 W.

Example 1:

Determine the minimum and maximum vent capacity for a single fan-assisted appliance given the following:

Vertical design height of vent	17 feet
Diameter of vent	6 inches
Vent type	Type B
Length of lateral	10 feet

Using Table G2428.2(1):

Interpolated Maximum Vent Capacity = [(17 ft - 15 ft)/(20 ft - 15 ft) × (351,000 Btu/h - 315,000 Btu/h)] + 315,000 Btu/h = (2/5 × 36,000 Btu/h) + 315,000 Btu/h)] 329,400 Btu/h

Interpolated Minimum Vent Capacity = 64,000 Btu/h - [(17 ft - 15 ft)/(20 ft - 15 ft) × (64,000 Btu/h - 62,000 Btu/h)] = 64,000 Btu/h - (2/5 × 2,000 Btu/h) = 63,200 Btu/h

Example 2:

Determine the maximum vent capacity for a single fan assisted furnace given the following:

Input rating of furnace	210,000 Btu/h
Vertical design height of vent	12½ feet
Diameter of vent	6 inches
Vent type	Type B
Length of lateral	2 feet

Using Table G2428.2(1):

Step 1: Find the maximum vent capacity (MAX) at the first height entry greater than 12½ feet, i.e., 15 feet.

MAX_{15} = 226,000 Btu/h

Step 2: Find the maximum vent capacity (MAX) at the next height entry lower than 12½ feet, i.e., 10 feet.

MAX_{10} = 194,000 Btu/h

Step 3: Determine the difference between the two maximum vent capacities.

$MAX_{15} - MAX_{10}$ = 226,000 Btu/h - 194,000 Btu/h = 32,000 Btu/h

Step 4: Determine the maximum vent capacity for a 12½-foot high vent using:

$MAX\ 12\frac{1}{2}$ = [(NB) × ($MAX_{15}\ MAX_{10}$)] + MAX_{10}

where:

A = The difference between the next higher table vent height and the design vent height.

B = The difference between the consecutive closest table vent heights.

$MAX\ 12\frac{1}{2}$ = [(15 ft - 12½ ft)/(15 ft -10 ft) × 32,000 Btu/h] + 194,000 Btu/h = [(2½ ft)/(5 ft) × 32,000 Btu/h] + 194,000 Btu/h = 16,000 Btu/h + 194,000/h = 210,000 Btu/h

When a vent height falls between height entries in a table, the code user can choose to interpolate or use the closest table height entry that is lower than the actual height of the vent. These are the only allowable options.

This example also illustrates how the tables may be used to reduce vent size if there is adequate height and capacity to do so (see commentary, Section G2428.2.2).

It is important to consider interpolation before assuming that an "in between" height vent installation is not in compliance with the code. For example, a vent with a 25-foot (7620 mm) height could fail to comply with a table when it is viewed as a 20-foot (6096 mm) vent, but that same vent could quite possibly comply with the table when viewed as what it actually is, a 25-foot (7620 mm) high vent.

This interpolation process can also be used to estimate in-between capacities for intermediate lengths of laterals, as well as in-between minimum capacities.

There is no way to estimate in-between capacity between a zero lateral (straight vertical vent) and a 2-foot (610 mm) lateral (which has two elbows). If the vent has just one 90-degree (1.57 rad) turn, the 2-foot (610 mm) lateral capacity applies.

G2428.2.15 (504.2.15) Extrapolation prohibited. Extrapolation beyond the table entries shall not be permitted.

❖ Projecting capacities below the lowest or above the highest boundary entries of the tables is not allowed because the tables are based on the known operational limits of venting systems. The tables cannot predict the operating characteristics of venting systems having heights shorter or taller than the table limits; therefore, any such system must be an engineered system.

G2428.2.16 (504.2.16) Engineering calculations. For *vent* heights less than 6 feet (1829 mm) and greater than shown in the tables, engineering methods shall be used to calculate *vent* capacities.

❖ This section parallels the previous section by stating that vents having heights not within the tables must be designed by an engineered method. In other words, extrapolation above or below the table height entries is not allowed (see Section G2428.2.14).

G2428.2.17 (504.2.17) Height entries. Where the actual height of a vent falls between entries in the height column of the applicable table in Tables G2428.2(1) and G2428.2(2), either interpolation shall be used or the lower appliance input rating shown in the table entries shall be used for FAN Max and NAT Max column values and the higher appliance input rating shall be used for the FAN MIN column values.

❖ This section makes it very clear what is to be done where the vent being designed is of a height that does not correspond to a height row in the applicable table. For example, it is given that a Type B vent has a height of 17 feet and the venting system is subject to Table G2428.2(2). The height of 17 feet falls between the 15-foot row and the 20-foot row in the table, which means that neither the 15- nor 20-foot row provides accurate values for this vent. Often times, the code user will mistakenly choose the 20-foot row because it is close to 17, but this is incorrect because a 17-foot-high vent will have less capacity than a 20-foot-high vent, as can be seen by comparing the FAN MAX and NAT MAX entries for the 15- and 20-foot rows, and because a 17-foot-high vent requires a greater minimum input than a 20-foot-high vent, as can be seen by comparing the FAN MIN entries. Clearly, the 20-foot row is inappropriate. Because the 20-foot row is not a choice, either interpolation must be used to mathematically construct entries for a 17-foot row or the 15-foot row must be used. When the 15-foot row is used, the FAN MAX and NAT MAX values will be lower than the corresponding values in the 20-foot row and the FAN MIN value will be higher than the corresponding value in the 20-foot row. This approach is a safe alternative to interpolation because the 17-foot vent will perform better than a 15-foot vent. In other words, the 17-foot vent has a higher maximum appliance input capacity and a lower minimum appliance input requirement than a 15-foot vent. Using the 15-foot row corresponds to the instructions of this section to use the lower appliance input rating given in each of the MAX columns and to use the higher appliance input rating for the FAN MIN column.

G2428.3 (504.3) Application of multiple appliance vent Tables G2428.3(1) through G2428.3(4). The application of Tables G2428.3(1) through G2428.3(4) shall be subject to the requirements of Sections G2428.3.1 through G2428.3.23.

❖ The application of Tables G2428.3(1) through G2428.3(4) shall be subject to the requirements of Sections G2428.3.1 through G2428.3.23. This part of Section G2428 regulates vents and chimneys that serve more than one appliance. These venting systems can be described as combined vents and include multiple connectors. A "combined vent" is a vent for two or more appliances at one level served by a common vent. "Least total height" is the vertical distance from the highest appliance outlet (draft hood or flue collar) to the lowest discharge opening of the vent top. This height dimension is illustrated in Commentary Figure G2428.3(1) for a typical (FAN + NAT) system. The same height measurement applies to FAN + FAN and NAT + NAT systems. Least total height is used for vent sizing of all connected appliances on one level. "Connector rise" for any appliance is the vertical distance from its outlet connection to the level at which it joins the common vent, as shown in Commentary Figure G2428.3(1).

A "connector" for purposes of designing a combined vent is that part of the vent piping between the appliance outlet and its junction with or interconnection to the rest of the system. A chimney is referred to as a venting means, although a chimney and a vent are distinct. An improperly sized common vent or chimney could fail to vent the combustion products of the appliances served. Many factors influence the sizing of a vent or chimney, including total height, offsets, lateral lengths, the type of appliance, the appliance energy input and the appliance connector configurations. The "common vent" is that portion of the system serving two or more connected appliances. If connectors are joined before reaching the vertical vent, the run between the last entering connector and the vertical portion is also treated as part of the common vent. In Commentary Figure G2428.3(1), the vertical common vent is cross-hatched beginning at the interconnection tee.

For each connector, the correct size must be found from the applicable tables based on its appliance input, rise and least total vent height. For draft-hood (NAT) appliances, the outlet size may be too small if there is not enough rise; therefore, connector design involves choosing the correct size and verifying that the use of a connector the same size as the outlet is within input rating limits.

For the common vent, the capacity table shows maximum combined ratings only. The size of the common vent is thus based on least total height and the combination of attached appliances: FAN + FAN, FAN + NAT

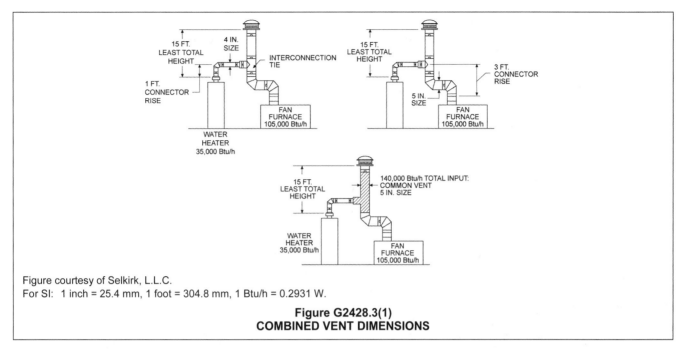

Figure courtesy of Selkirk, L.L.C.
For SI: 1 inch = 25.4 mm, 1 foot = 304.8 mm, 1 Btu/h = 0.2931 W.

**Figure G2428.3(1)
COMBINED VENT DIMENSIONS**

or NAT + NAT. To find each connector size, use Table G2428.3(1) for Type B vent connectors or Table G2428.3(2) for single-wall metal connectors and proceed as follows:

1. Determine "least total height" for the system.

2. Determine connector rise for each appliance.

3. Enter the applicable vent connector table at the least total height. Continue across on the line for appliance connector rise to a MAX input rating equal to or greater than that of the appliance. For a FAN appliance, this input rating should also be greater than shown in the MIN column. Read the connector size at the top of the column. If Table G2428.3(2) for single wall shows NA, use a double-wall connector in accordance with Table G2428.3(1). In some cases, the table will dictate a connector size that is larger than the appliance vent outlet (flue collar).

To find the common vent size, proceed as follows:

1. Add all appliance Btu input ratings to get the total Btu input.

2. If one or both connectors are single-wall metal, use Table G2428.3(2) for the common vent.

3. Enter the common vent table at the same least total height used for connectors.

4. Continue across and stop at the first applicable column of combined appliance input rating equal to or greater than the total. If Table G2428.3(2) for the common vent shows NA, the entire system must be double-wall Type B vent.

5. Read the size of the common vent at the top of the applicable column (NAT + NAT, FAN + NAT or FAN + FAN).

a. Regardless of table results for size, the common vent must be at least as large as the largest connector (see Section G2428.3.8).

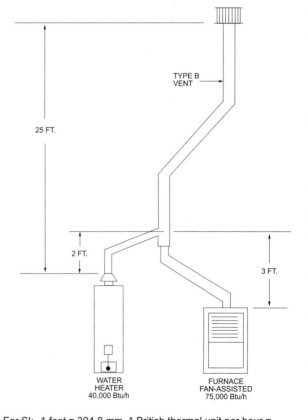

For SI: 1 foot = 304.8 mm, 1 British thermal unit per hour = 0.2931 W.

**Figure G2428.3(2)
FAN AND NAT VENT SIZING EXAMPLE**

b. In accordance with vent manufacturer's instructions, for NAT + NAT appliance combinations, if both connectors are the same size, the common vent must be at least one size larger.

Example 1:

Commentary Figure G2428.3(1) shows a two-appliance FAN + NAT system, combining a draft-hood water heater (NAT appliance) with a fan-assisted combustion Category I (FAN) furnace. The system is designed in the following steps using Table G2428.3(1) for double-wall connectors:

1. For the water heater, enter the vent connector table at a least total height of 15 feet (4572 mm) and a connector rise of 1 foot (305 mm). Read across to the MAX Btu/h rating for a NAT appliance vent higher than 35,000 Btu/h (10.3 kW). The table shows that a 4-inch (102 mm) connector is needed. This size must be used beginning at the draft hood, regardless of draft hood size, which might be 3 inches (76 mm) [see Commentary Figure G2428.3(1)].

2. For the 105,000 Btu/h (30.8 kW) furnace, enter the vent connector table at the same least total height (15 feet) at a connector rise of 3 feet (914 mm). Read across to 163,000 under the FAN MAX column. The MIN is 51,000. Therefore, 5 inches (127 mm) is the correct connector size [see Commentary Figure G2428.3(1)].

3. For the common vent, the sum of the two ratings is 140,000 Btu/h (41.0 kW). Enter the common vent table at 15 foot (4572 mm) least total height.

For a FAN + NAT combination, the maximum input of 5-inch (127 mm) vent is 164,000 Btu/h (48.1 kW), so 5 inches (127 mm) is the proper size, as shown in Commentary Figure G2428.3(1). Both draft-hood and fan-assisted appliances may be common vented in any combination, as indicated by the headings in the common vent table. Appliance types include the following:

- Central heating furnaces.
- Central heating boilers (hot water and steam).
- Water heaters.
- Unit heaters.
- Duct furnaces.
- Room heaters*.
- Floor furnaces*.

*If these have draft hoods, a design input increase of 40 percent is recommended by some vent manufacturers in order to use the tables (see commentary, Section G2428.2).

The common vent tables do not apply to the following:

1. Gas cooking appliances, which should be vented into an exhaust hood or vented individu-

ally in accordance with the appliance manufacturer's instructions.

2. Forced draft and commercial or industrial hot water or steam boilers without draft hoods. For this equipment, see the boiler and vent/chimney manufacturers' instructions.

3. Categories II, III and IV gas-burning equipment, for which the equipment manufacturer's venting instructions must be used.

4. Gas-fired incinerators.

The manufacturers of chimney and vent systems provide installation and design information as part of their installation instructions. The common chimney or vent must be designed to properly vent the products of combustion when any one, any combination or all of the connected appliances operate. [see Commentary Figure G2428.3(2)]. Solid-fuel-burning appliances must not connect to common chimneys (see Section G2427.5.6.1).

Combined vent systems for two or more gas appliances of either type (FAN or NAT) must be designed to prevent draft-hood spillage for natural draft (NAT) appliances and to avoid positive pressure for fan-assisted (FAN) appliances. The connector and common vent tables have been computed by examining the most critical situations for any operating combination. A common vent must function properly when all, one or any combination of the connected appliances are firing. It is particularly demanding of a vent system to operate properly when only a single appliance having the lowest input rating is firing, thus creating the greatest potential for poor draft and condensation. Oversizing of venting systems should be avoided.

The connector tables are based on the most critical condition for a particular appliance when operating by itself, whereas the common vent tables show sizes that assure adequate capacity and draft, whether one or all appliances are operating simultaneously.

All parts of a combined vent must be checked for capacity. For connectors, the size must be determined from the tables, particularly for low-height vents or where headroom restricts available connector rise.

The connector in a combined-vent system is defined as the piping from a draft hood or flue collar to the junction of the common vent or to a junction in a vent manifold. Proper connector design is vital to obtaining adequate capacity. The connector must produce its share of the total draft for its NAT or FAN appliance and must deliver enough heat to the common vent so that it can contribute the balance of draft needed.

For application of the code's installation provisions, the Type B gas vent connector is a "gas vent." It is essential, however, for system design purposes to use the word "connector" so that its rise and configuration may be explained and tabulated. The connector tables for combined vents show MIN and MAX capacities only for FAN appliances because no minimums are needed for NAT appliances with Type B gas-vent connectors. Important factors in connector design include the following:

- Connector material.
- Connector length.
- Connector rise.
- Number.
- Appliance location as it affects the piping arrangement.
- Number of attached appliances or different connector sizes.
- Connection to an offset or manifold rather than directly to the vertical common vent.

Example 2:

This example is a common scenario [see Commentary Figure G2428.3(2)].

Given:

- Single-family house with Type B vent serving two appliances;
- Category I fan-assisted furnace, 80-percent efficient, 75,000 Btu/h input (4-inch vent collar);
- Furnace connector single-wall pipe with 3-foot rise;
- Draft-hood-equipped water heater, 40,000 Btu/h (3-inch draft hood outlet);
- Water heater connector single-wall with 2-foot rise;
- Vent height: 25 feet;
- A 45-degree offset occurs in the common vent in the attic.

Step 1: Choose the applicable table. Table G2428.3(2) addresses two or more Category I appliances with single-wall connectors.

Step 2: Using the top portion of the table, size the appliance connectors. Because the table has no row for the 25-foot height, the 20-foot row must be used, or interpolation must be used to calculate the capacities at 25 feet. To simplify the procedure, first proceed using the 20-foot row until interpolation is necessary.

a. Looking in the "NAT" columns, see that a 4-inch connector is required for the water heater because the water heater input of 40,000 Btu/h exceeds the maximum capacity (36,000) of a 3-inch connector with a 2-foot rise.

b. Looking in the "FAN" columns, see that the furnace input of 75,000 Btu/h is less than the minimum capacity of 4-inch and larger connectors with a 3-foot rise. This means that single-wall pipe cannot be used for the furnace connector. The furnace connector must be double-wall (Type B or equivalent) and must be sized using Table G2428.3(1). In Table G2428.3(1), see that a 4-inch double-wall connector is suitable

for the furnace because the 75,000 Btu/h input fits within the range of 35,000 to 110,000 Btu/h for a 3-foot rise.

The furnace connector has been determined to be 4-inch double-wall pipe, and the water heater connector has been determined to be 4-inch single-wall pipe.

Step 3: Using the bottom portion of Table G2428.3(2), the common vent is sized. (Because both single- and double-wall connectors are used within the same system, Section G2428.3.18 requires use of the single-wall table.) The water heater and furnace inputs total 115,000 Btu/h. Under the 4-inch vent heading, FAN + NAT column, see that 4-inch Type B vent has a capacity of 118,000 Btu/h, which exceeds the appliance total of 115,000. However, the system contains an offset, which will reduce this capacity by 20 percent in accordance with Section G2428.3.5 (0.80 x 118,000) = 94,400 Btu/h.

The actual capacity of a 4-inch common vent is 94,400 Btu/h, which is less than the 115,000 appliance total; therefore, a 5-inch common vent is required.

Verify 5-inch capacity: $(0.80 \times 177) = 141,600$ Btu/h.

From the 30-foot height rows, it is apparent that interpolation for the actual height of 25 feet would not have changed the outcome of this system design.

Appendix B contains additional sizing examples.

G2428.3.1 (504.3.1) Vent obstructions. These venting tables shall not be used where obstructions, as described in Section G2427.15, are installed in the venting system. The installation of vents serving listed *appliances* with vent dampers shall be in accordance with the *appliance* manufacturer's instructions or in accordance with the following:

1. The maximum capacity of the *vent connector* shall be determined using the NAT Max column.

2. The maximum capacity of the vertical vent or *chimney* shall be determined using the FAN+NAT column when the second *appliance* is a fan-assisted *appliance*, or the NAT+NAT column when the second *appliance* is equipped with a *draft hood*.

3. The minimum capacity shall be determined as if the *appliance* were a fan-assisted *appliance*.

  3.1. The minimum capacity of the *vent connector* shall be determined using the FAN Min column.

  3.2. The FAN+FAN column shall be used when the second *appliance* is a fan-assisted *appliance*, and the FAN+NAT column shall be used when the second *appliance* is equipped with a *draft hood*, to determine whether the vertical vent or *chimney* configuration is not permitted (NA). Where the vent configuration is NA, the vent configuration shall not be permitted and an alternative venting configuration shall be utilized.

❖ See the commentary to Section G2428.2.1. This section is a bit more complex than Section G2428.2.1

because the tables for multiple appliances are two-part tables. The difference lies in Items 2 and 3.2 dealing with the common vent portion of the sizing tables. Item 2 determines the maximum capacity of the common vent by treating the vent-damper-equipped appliance as a draft-hood appliance coupled with either another draft-hood appliance or a fan-assisted appliance. Item 3.2 is a test wherein the vent-damper-equipped appliance is treated as a fan-assisted appliance. If the table says NA, the vent system must be redesigned, regardless of the fact that it may have been allowed by the table under Item 2.

G2428.3.2 (504.3.2) Connector length limit. The *vent connector* shall be routed to the vent utilizing the shortest possible route. Except as provided in Section G2428.3.3, the maximum *vent connector* horizontal length shall be $1^1/_2$ feet for each inch (18 mm per mm) of connector diameter as shown in Table G2428.3.2.

❖ Connectors are necessary for chimney and vent systems that serve multiple appliances. To reduce friction loss and flue-gas heat loss and to conserve available draft, connectors must be kept to a minimum length. Thoughtful placement of appliances and chimneys or vents can reduce connector lengths and achieve optimum venting performance. Unlike Tables G2428.2(1) and G2428.2(2), Tables G2428.3(1) through G2428.3(4) do not provide a range of permissible lateral lengths. Therefore, Section G2427.10.8 places an overall maximum limit on connector length. Section G2427.10.8 provides a necessary limiting function because Tables G2428.3(1) through G2428.3(4) would appear to allow (unintentionally) a connector to be considerably longer than the height of the vent.

Increasing the length of the connector increases the required minimum capacity of a vent and decreases its maximum capacity. Some vent manufacturers appear to measure only the horizontal (lateral) run of a connector in their design manuals. The more conservative approach would be to consider the friction loss and heat loss for the entire run of a connector throughout its developed length (see commentary, Section G2427.10.8). The vent sizing tables were developed with the assumption that vent connectors are no longer than 18 inches per inch of connector diameter. This section establishes the basic rule, which is modified by Sections G2428.3.3 and G2427.10.8. Simply put, connector length is limited by this section to 18 inches per inch of diameter without capacity penalties or is limited in length by Section G2427.10.8 if the capacity penalties of Section G2428.3.3 are applied and the minimum capacity is met for fan-assisted appliances. Section G2428.3.3 allows longer connector lengths, but this section still requires connectors to take the shortest route to the vent.

TABLE G2428.3.2 (504.3.2)
MAXIMUM VENT CONNECTOR LENGTH

CONNECTOR DIAMETER (inches)	CONNECTOR MAXIMUM HORIZONTAL LENGTH (feet)
3	$4^1/_2$
4	6
5	$7^1/_2$
6	9
7	$10^1/_2$
8	12
9	$13^1/_2$

For SI: 1 inch = 25.4 mm, 1 foot = 304.8 mm.

❖ See the commentary to Section G2428.3.2.

G2428.3.3 (504.3.3) Connectors with longer lengths. Connectors with longer horizontal lengths than those listed in Section G2428.3.2 are permitted under the following conditions:

1. The maximum capacity (FAN Max or NAT Max) of the *vent connector* shall be reduced 10 percent for each additional multiple of the length allowed by Section G2428.3.2. For example, the maximum length listed in Table G2428.3.2 for a 4-inch (102 mm) connector is 6 feet (1829 mm). With a connector length greater than 6 feet (1829 mm) but not exceeding 12 feet (3658 mm), the maximum capacity must be reduced by 10 percent (0.90 × maximum vent *connector* capacity). With a connector length greater than 12 feet (3658 mm), but not exceeding 18 feet (5486 mm), the maximum capacity must be reduced by 20 percent (0.80 × maximum vent capacity).

2. For a connector serving a fan-assisted *appliance*, the minimum capacity (FAN Min) of the connector shall be determined by referring to the corresponding single-*appliance* table. For Type B double-wall connectors, Table G2428.2(1) shall be used. For single-wall connectors, Table G2428.2(2) shall be used. The height (*H*) and lateral (*L*) shall be measured according to the procedures for a single-*appliance* vent, as if the other *appliances* were not present.

❖ Section G2427.10.8 still applies and serves to limit the application of this section and Section G2428.3.2. For fan-assisted appliances, both Items 1 and 2 apply. Item 1 allows the maximum length permitted by Section G2428.3.2 to be increased (doubled, tripled, etc.) if the vent connector capacity is reduced by 10 percent when the maximum connector length is doubled, by 20 percent when tripled, by 30 percent when quadrupled, etc. Regardless of the length permitted, connectors should always be as short as practicable.

Item 2 applies to fan-assisted appliances and allows connector lengths to be as long as the lateral entries in Table G2428.2(1) or G2428.2(2) if the limits of Section

G2427.10.8 are not exceeded and the appliance complies with the minimum input requirement of the respective table.

When horizontal lengths in excess of those stated in Section G2428.3.2 are necessary, the minimum capacity of the system is determined by referring to the corresponding single-appliance table. In this case, for each appliance the entire vent connector and common vent from appliance to vent termination is treated as a single-appliance vent (of the same size as the common vent), as if the others were not present. Any appliance failing to meet the Min input may be prone to creating excessive condensation and/or insufficient draft within the vent system if the appliance is operated by itself. In that case, options may include relocation of appliances, selective sequential or simultaneous operation of appliances or separate vent installations. This section is intended for use only where a longer connector cannot be avoided as required by Section G2427.10.8 and the connector takes the shortest possible route to the vent. The longer the connector, the greater the flow resistance and the greater the heat loss, both of which negatively affect vent system operation. The capacity penalties account for the increased flow resistance and Item 2 accounts for the increased heat loss. Item 2 addresses the worst case condition when a single appliance is operating and treats a connector serving a fan-assisted appliance as if the appliance was being independently vented, hence, the reference to the single appliance tables. Recall that the single appliance tables limit lateral length, which can be, in effect, a connector length limitation.

In many cases, the sizing tables will prevent someone from taking advantage of the allowances of this section for connectors. For example, a single-wall connector serving a fan-assisted appliance may not be able to meet the minimum input requirement (FAN Min) if the capacity penalty of this section forces increasing the connector size. Note that Section G2427.10.8 will not allow a connector length to exceed the vent or chimney height in any case.

G2428.3.4 (504.3.4) Vent connector manifold. Where the *vent connectors* are combined prior to entering the vertical portion of the common vent to form a common vent manifold, the size of the common vent manifold and the common vent shall be determined by applying a 10-percent reduction (0.90 × maximum common vent capacity) to the common vent capacity part of the common vent tables. The length of the common *vent connector* manifold (L_m) shall not exceed $1^1/_2$ feet for each inch (18 mm per mm) of common *vent connector* manifold diameter (D).

❖ This section inflicts a 10-percent capacity penalty on the common vent as a result of the lateral (offset) effect of a manifold. A connector manifold is a horizontal extension of the lower end of a common vent [see Commentary Figures G2427.10.3.4(1) and (2)]. The capacity reduction accounts for the additional turn in the common vent. The manifold portion of a vent is sized based on the common vent portion of the tables. The length of a connector manifold is measured from

the vertical common vent to the most upstream interconnection fitting.

G2428.3.5 (504.3.5) Common vertical vent offset. Where the common vertical vent is *offset*, the maximum capacity of the common vent shall be reduced in accordance with Section G2428.3.6. The horizontal length of the common vent *offset* (L_o) shall not exceed $1^1/_2$ feet for each inch (18 mm per mm) of common vent diameter (D). Where multiple *offsets* occur in a common vent, the total horizontal length of all *offsets* combined shall not exceed $1^1/_2$ feet for each inch (18 mm/mm per) of the common vent diameter (D).

❖ Offsets, as opposed to a single change in direction, require two fittings (elbows) [see Commentary Figures G2403(13) and G2427.10.3.4(2)]. An offset offers resistance to flow as does any change of direction in a vent chimney or connector, and the capacity reduction penalty compensates for this flow resistance. Common vent locations should be well planned in a building to avoid the need for an offset. Offsets also introduce the need for additional supports to prevent stress on fittings (elbows). The term Offset (Vent) is defined in Section G2403 [see commentary, Sections G2428.2.3 and G2428.2.9].

Offsets should be avoided because of the flow resistance, capacity reduction, additional support requirements and possibility of condensation leakage. The length limitation of this section is absolute with no allowance for greater lengths. This provision is commonly violated in attic space offsets with some violations resulting in offsets that are as long or longer than the vent height. The length of an offset is measured in a horizontal plane, not along the developed length of the vent pipe. For example, a 45-degree offset with 7 feet (2134 mm) of pipe between the fittings will have an offset length of 5 feet (1524 mm) for application of this section. (For a 45-degree (0.79 rad) offset, actual pipe developed length is approximately 1.4 times the horizontal distance between the parallel vertical sections of pipe) (see Commentary Figure G2428.3.5).

The total length of all horizontal offsets combined cannot exceed the maximum limit of $1^1/_2$ feet (457 mm) per inch of vent diameter. In other words, horizontal offsets are accumulative.

G2428.3.6 (504.3.6) Elbows in vents. For each elbow up to and including 45 degrees (0.79 rad) in the common vent, the maximum common vent capacity listed in the venting tables shall be reduced by 5 percent. For each elbow greater than 45 degrees (0.79 rad) up to and including 90 degrees (1.57 rad), the maximum common vent capacity listed in the venting tables shall be reduced by 10 percent.

❖ This section applies to the common vent portion of a vent system (see commentary, Section G2428.2.3). For example, a common vent may have two 30-degree (0.52 rad) changes of direction at its base and might also have an offset constructed with 45-degree (0.79 rad) fittings in an attic space, thus having a total capacity reduction of 20 percent. If that same vent system had two 45-degree (0.79 rad) changes of direction at the base and an offset constructed with 60-degree

(1.05 rad) fittings in the attic, the total capacity penalty would be 30 percent. In Commentary Figure G2428.3.6(1), an offset is created in the common vent by the use of two fittings (elbows) adjusted to 45 degree (0.79 rad) angles. The common vent capacity would be reduced by 10 percent total (5 percent for each fitting/elbow). If the fittings were adjusted to angles greater than 45 degrees (0.79 rad), the total capacity reduction would be 10 percent per fitting for a total of 20 percent (see Commentary Figures G2428.3.6(2) and G2428.3.6(3)].

G2428.3.7 (504.3.7) Elbows in connectors. The *vent connector* capacities listed in the common vent sizing tables include allowance for two 90-degree (1.57 rad) elbows. For each additional elbow up to and including 45 degrees (0.79 rad), the maximum *vent connector* capacity listed in the venting tables shall be reduced by 5 percent. For each elbow greater than 45 degrees (0.79 rad) up to and including 90 degrees (1.57 rad), the maximum *vent connector* capacity listed in the venting tables shall be reduced by 10 percent.

❖ In Commentary Figure G2428.3.6(1), the connector for the fan-assisted appliance is offset with two 60-degree (1.05 rad) fittings (elbows). Because the tables include an allowance for two "free" 90-degree (1.57 rad) fittings, no capacity reduction would be required for this connector. The draft-hood-equipped appliance has a connector with a total of four 90-degree (1.57 rad) fittings, including the tee in the common vent. This would require a total capacity reduction of 20 percent. Note that Section G2427.10.8 intends to prohibit connectors with excessive length and fittings/elbows. Excess fittings and length are often the result of poor planning, improper appliance location and obstructions, such as ducts, structural members and other appliances. Capacity reduction penalties can force an increase in connector size, and in the case of fan-assisted appliances, increasing the connector size might raise the minimum required input rating above the appliance rating, thus forcing redesign of the system.

G2428.3.8 (504.3.8) Common vent minimum size. The cross-sectional area of the common vent shall be equal to or greater than the cross-sectional area of the largest connector.

❖ This requirement is consistent with vent manufacturers' instructions and is necessary to prevent overloading of the common vent. It is apparent that a vent's maximum capacity could be exceeded where multiple connectors are involved, especially if one of the connectors is larger than the common vent itself. Vent manufacturers may also require that the common vent be increased one size where more than one appliance connector and the common vent are the same size.

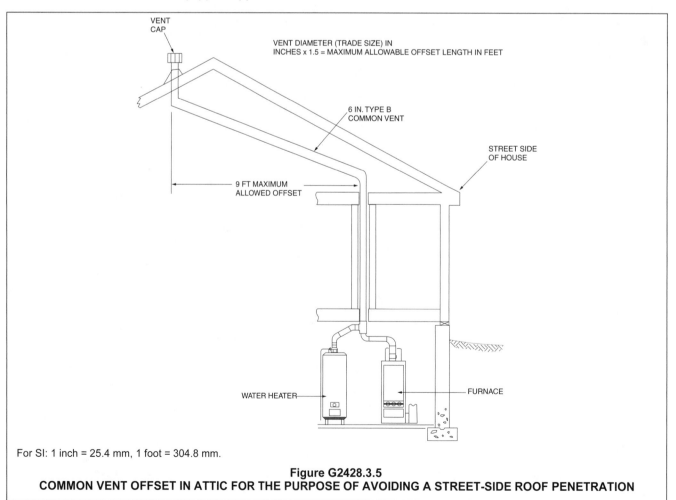

For SI: 1 inch = 25.4 mm, 1 foot = 304.8 mm.

Figure G2428.3.5
COMMON VENT OFFSET IN ATTIC FOR THE PURPOSE OF AVOIDING A STREET-SIDE ROOF PENETRATION

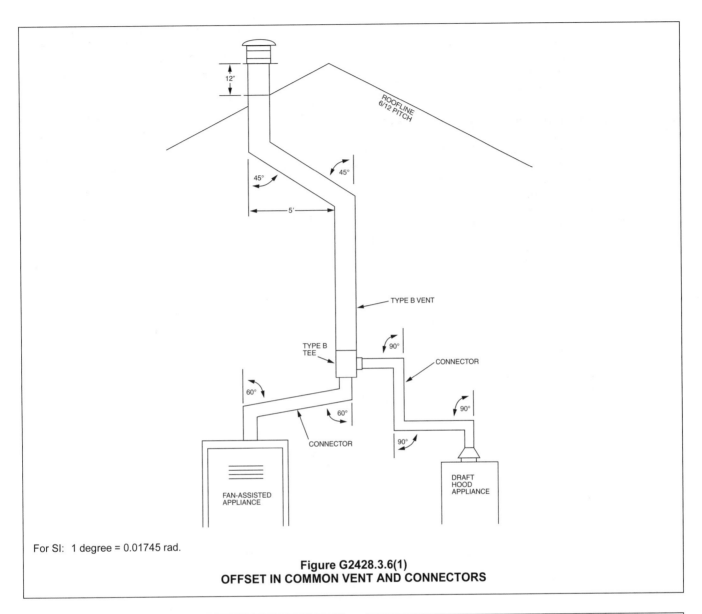

For SI: 1 degree = 0.01745 rad.

Figure G2428.3.6(1)
OFFSET IN COMMON VENT AND CONNECTORS

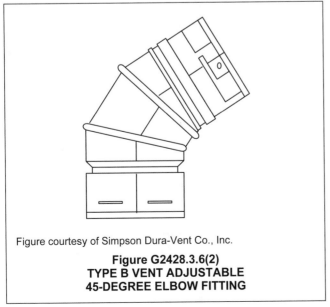

Figure courtesy of Simpson Dura-Vent Co., Inc.

Figure G2428.3.6(2)
TYPE B VENT ADJUSTABLE
45-DEGREE ELBOW FITTING

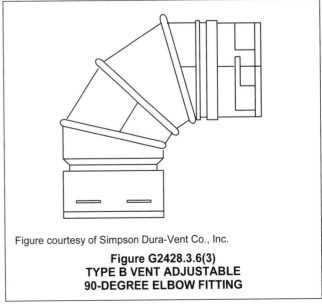

Figure courtesy of Simpson Dura-Vent Co., Inc.

Figure G2428.3.6(3)
TYPE B VENT ADJUSTABLE
90-DEGREE ELBOW FITTING

G2428.3.9 (504.3.9) Common vent fittings. At the point where tee or wye fittings connect to a common vent, the opening size of the fitting shall be equal to the size of the common vent. Such fittings shall not be prohibited from having reduced-size openings at the point of connection of *appliance vent connectors.*

❖ This section specifically allows use of tee and wye fittings with reduced-size openings to join appliance vent connectors to the common vent. The fittings must be the same size as the common vent at the point where they join, but are allowed to have smaller branch openings and smaller openings on the opposite end of the run of the fitting. A common question regarding interconnection fittings is whether or not the code intends to allow single-wall interconnection fittings such as tees and wyes. These fittings have commonly been used with single-wall connectors to join two appliances to a common Type B vent 5 × 4 × 3-inch wyes, 6 × 6 × 4-inch tees, etc. Vent manufacturers' opinion on this issue is that because the interconnection fitting is part of the common vent, it should be constructed as required for the common vent, meaning Type B vent tees and wyes. Certainly, double-wall vent connectors are not allowed to discharge into a single-wall fitting. The code was previously silent with regard to single-wall connectors discharging into single-wall fittings. However, Section G2428.3.9.1 has settled this issue. Double-wall tees and wyes are preferable because they have far less heat loss and less required clearance to combustibles [see Commentary Figures G2428.3.9(1) and (2) and Section G2428.3.9.1].

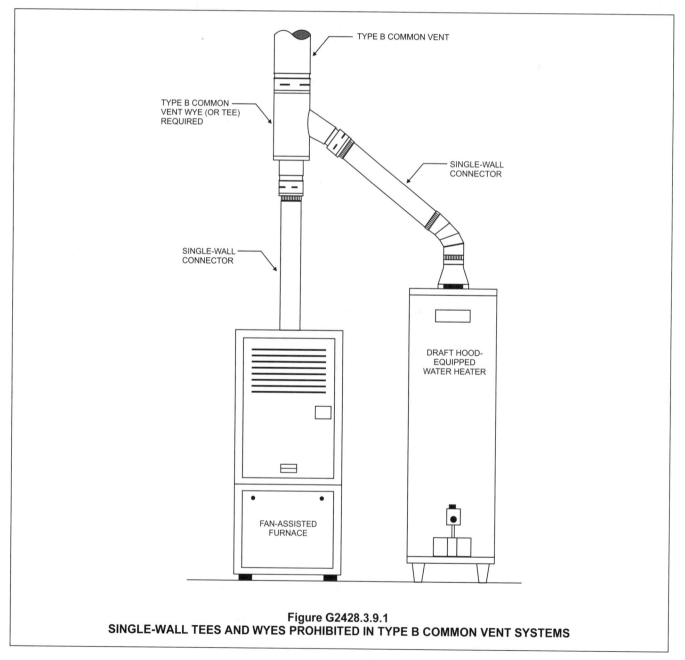

Figure G2428.3.9.1
SINGLE-WALL TEES AND WYES PROHIBITED IN TYPE B COMMON VENT SYSTEMS

G2428.3.9.1 (504.3.9.1) Tee and wye fittings. Tee and wye fittings connected to a common gas vent shall be considered as part of the common gas vent and shall be constructed of materials consistent with that of the common gas vent.

❖ Because the tee and wye fittings are part of the common vent, they must be of the same construction as the common vent. This is consistent with the vent manufacturer's recommendations [see Commentary Figures G2428.3.9.1, G2428.3.9(1) and G2428.3.9(2)]. Tee and wye fittings used with Type B common vents are not allowed to be single wall fittings.

G2428.3.10 (504.3.10) High-altitude installations. Sea-level input ratings shall be used when determining maximum capacity for high-altitude installation. Actual input, derated for altitude, shall be used for determining minimum capacity for high-altitude installation.

❖ See the commentary to Section G2428.2.5.

G2428.3.11 (504.3.11) Connector rise measurement. Connector rise (R) for each *appliance connector* shall be measured from the *draft hood* outlet or *flue collar* to the centerline where the vent gas streams come together.

❖ It is desirable to achieve the greatest connector rise that conditions will permit [see Commentary Figures G2428.3.11 and G2428.3(2)].

G2428.3.12 (504.3.12) Vent height measurement. For multiple *appliances* all located on one floor, available total height (H) shall be measured from the highest *draft hood* outlet or *flue collar* up to the level of the outlet of the common vent.

❖ The vent design is based on the "worst case" (most conservative) height so that a vent is not given credit for being taller than it actually is [see commentary Figure G2428.3(2)].

G2428.3.13 (504.3.17) Vertical vent maximum size. Where two or more *appliances* are connected to a vertical vent or *chimney*, the flow area of the largest section of vertical vent or *chimney* shall not exceed seven times the smallest listed *appliance* categorized vent areas, *flue collar* area, or *draft hood* outlet area unless designed in accordance with *approved* engineering methods.

❖ Vent manufacturers' instructions address ratios of areas of appliance connectors connecting to a com-

Photo courtesy of Simpson Dura-Vent Company, Incorporated

Figure G2428.3.9(2)
TYPE B VENT REDUCING WYE

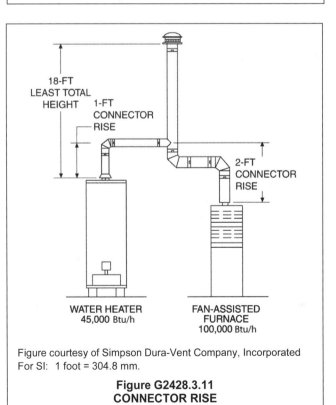

Figure courtesy of Simpson Dura-Vent Company, Incorporated
For SI: 1 foot = 304.8 mm.

Figure G2428.3.11
CONNECTOR RISE

Photo courtesy of Simpson Dura-Vent Company, Incorporated

Figure G2428.3.9(1)
TYPE B VENT REDUCING TEE

mon vent. This section relates the area ratio limitation to the smallest appliance connection and the area of the vertical common vent. The concern is that the smaller or smallest appliance may not be able to provide enough heat to the common vent to develop sufficient draft and prevent a condensation problem. It can be seen that the common vent tables would allow gross oversizing if it were not for the limit of this section. For example, if the table permits a 5-inch common vent, it would also permit a 6-, 7-, 8-, 9- and 10-inch common vent. This section serves to limit the size of the common vent, whereas the table does not. The larger the common vent, the more difficult it becomes to maintain draft and avoid condensation, especially when the smallest appliance is firing by itself.

G2428.3.14 (504.3.18) Multiple input rate appliances. For *appliances* with more than one input rate, the minimum *vent connector* capacity (FAN Min) determined from the tables shall be less than the lowest *appliance* input rating, and the maximum *vent connector* capacity (FAN Max or NAT Max) determined from the tables shall be greater than the highest *appliance* input rating.

❖ See the commentary to Section G2428.2.6.

G2428.3.15 (504.3.19) Liner system sizing and connections. Listed, corrugated metallic *chimney* liner systems in masonry *chimneys* shall be sized by using Table G2428.3(1) or G2428.3(2) for Type B vents, with the maximum capacity reduced by 20 percent (0.80 × maximum capacity) and the minimum capacity as shown in Table G2428.3(1) or G2428.3(2). Corrugated metallic liner systems installed with bends or offsets shall have their maximum capacity further reduced in accordance with Sections G2428.3.5 and G2428.3.6. The 20-percent reduction for corrugated metallic *chimney* liner systems includes an allowance for one long-radius 90-degree (1.57 rad) turn at the bottom of the liner. Where double-wall connectors are required, tee and wye fittings used to connect to the common vent *chimney* liner shall be listed double-wall fittings. Connections between *chimney* liners and listed double-wall fittings shall be made with listed adapter fittings designed for such purpose.

❖ See the commentary to Section G2428.2.7.

G2428.3.16 (504.3.20) Chimney and vent location. Tables G2428.3(1), G2428.3(2), G2428.3(3) and G2428.3(4) shall be used only for chimneys and vents not exposed to the outdoors below the roof line. A Type B vent or *listed* chimney lining system passing through an unused masonry chimney flue shall not be considered to be exposed to the outdoors. Where vents extend outdoors above the roof more than 5 feet (1524 mm) higher than required by Figure G2427.6.3 and where vents terminate in accordance with Section G2427.6.3, Item 2, the outdoor portion of the vent shall be enclosed as required by this section for vents not considered to be exposed to the outdoors or such venting system shall be engineered. A Type B vent shall not be considered to be exposed to the outdoors where it passes through an unventilated enclosure or chase insulated to a value of not less than R8.

❖ See the commentary to Section G2428.2.9.

G2428.3.17 (504.3.21) Connector maximum and minimum size. *Vent connectors* shall not be increased in size more than two sizes greater than the listed *appliance* categorized vent diameter, *flue collar* diameter or *draft hood* outlet diameter. *Vent connectors* for draft-hood-equipped *appliances* shall not be smaller than the *draft hood* outlet diameter. Where a *vent connector* size(s) determined from the tables for a fan-assisted *appliance(s)* is smaller than the *flue collar* diameter, the use of the smaller size(s) shall be permitted provided that the installation complies with all of the following conditions:

1. *Vent connectors* for fan-assisted *appliance flue collars* 12 inches (305 mm) in diameter or smaller are not reduced by more than one table size [e.g., 12 inches to 10 inches (305 mm to 254 mm) is a one-size reduction] and those larger than 12 inches (305 mm) in diameter are not reduced more than two table sizes [e.g., 24 inches to 20 inches (610 mm to 508 mm) is a two-size reduction].

2. The fan-assisted *appliance(s)* is common vented with a draft-hood-equipped *appliance(s)*.

3. The vent *connector* has a smooth interior wall.

❖ Using a connector that is larger than the appliance outlet connection may sometimes be necessary to comply with the requirements of the sizing tables for multiple-appliance systems. However, having a connector significantly larger than the appliance outlet connection could cause reduced flow velocity, excessive loss of heat and could result in the allowance of excessively long connectors.

In accordance with Section G2428.2.2, a venting system serving a single appliance can, under certain conditions, be smaller than the appliance outlet connection. However, this is prohibited by this section for draft-hood appliances in common venting systems serving more than one appliance (see commentary, Section G2428.2.11). The reason for this prohibition is that a common vent or chimney, being large enough for multiple appliances, may not be able to produce adequate draft or avoid condensation where a single appliance is operating with a reduced-size vent connector.

Recent research involving computerized modeling has shown that connectors for fan-assisted appliances in multiple-appliance systems can be reduced under conditions similar to those for single-appliance installations. Therefore, this section makes an exception for systems with fan-assisted appliances commonly vented with appliances equipped with a draft hood (see commentary, Section G2428.2.2). The allowance for a connector to be smaller than an appliance flue collar is applicable only to fan-assisted appliances. A reduction in size is not allowed for corrugated metal connectors.

G2428.3.18 (504.3.22) Component commingling. All combinations of pipe sizes, single-wall and double-wall metal pipe shall be allowed within any connector run(s) or within the common vent, provided that all of the appropriate tables permit all of the desired sizes and types of pipe, as if they were used for the entire length of the subject connector or vent. Where single-wall and Type B double-wall metal pipes are used for *vent connectors* within the same venting system, the common vent must be sized using Table G2428.3(2) or G2428.3(4), as appropriate.

❖ Because this section parallels Section G2428.2.12, see the commentary to that section. This section addresses the installation that uses both single-wall connectors and double-wall connectors in the same venting system. The penalty for mixing is that the common vent must be sized from the more restrictive single-wall connector tables. Double-wall pipe is not designed to be interchangeable with single-wall pipe, and all such transitions would have to made with fittings designed for the transition.

G2428.3.19 (504.3.23) Draft hood conversion accessories. *Draft hood* conversion accessories for use with *masonry chimneys* venting listed Category I fan-assisted *appliances* shall be listed and installed in accordance with the manufacturer's instructions for such listed accessories.

❖ See the commentary to Section G2428.2.13.

G2428.3.20 (504.3.24) Multiple sizes permitted. Where a table permits more than one diameter of pipe to be used for a connector or vent, all the permitted sizes shall be permitted to be used.

❖ This section broadly states that any size pipe permitted by the sizing tables may be used. However, do not overlook Sections G2428.3.8, G2428.3.13 and G2428.3.17, which contain provisions that can override this section. Vent manufacturers stress that the smallest size pipe allowed by the code should be used rather than any larger sizes that may be permitted. Operating a vent or chimney at nearer to its maximum capacity will produce stronger draft and result in better protection against the formation of condensation.

G2428.3.21 (504.3.25) Table interpolation. Interpolation shall be permitted in calculating capacities for vent dimensions that fall between table entries.

❖ See Section G2428.2.14.

G2428.3.22 (504.3.26) Extrapolation prohibited. Extrapolation beyond the table entries shall not be permitted.

❖ See Section G2428.2.15.

G2428.3.23 (504.3.27) Engineering calculations. For vent heights less than 6 feet (1829 mm) and greater than shown in the tables, engineering methods shall be used to calculate vent capacities.

❖ See Section G2428.2.16.

G2428.3.24 (504.3.28) Height entries. Where the actual height of a vent falls between entries in the height column of the applicable table in Tables G2428.3(1) through G2428.3(4), either interpolation shall be used or the lower appliance input rating shown in the table shall be used for FAN Max and NAT Max column values and the higher appliance input rating shall be used for the FAN Min column values.

❖ See the commentary to Section G2428.2.17.

SECTION G2429 (505)
DIRECT-VENT, INTEGRAL VENT, MECHANICAL VENT AND VENTILATION/EXHAUST HOOD VENTING

G2429.1 (505.1) General. The installation of direct-vent and integral vent *appliances* shall be in accordance with Section G2427. Mechanical *venting systems* shall be designed and installed in accordance with Section G2427.

❖ See Sections G2427.2.1, G2427.2.2, G2427.3.3 and G2427.8.

SECTION G2430 (506)
FACTORY-BUILT CHIMNEYS

G2430.1 (506.1) Listing. Factory-built *chimneys* for building heating *appliances* producing *flue gases* having a temperature not greater than 1,000ºF (538ºC), measured at the entrance to the *chimney*, shall be listed and *labeled* in accordance with UL 103 and shall be installed and terminated in accordance with the manufacturer's instructions.

❖ Factory-built chimneys can be used with gas-fired appliances if allowed by the chimney manufacturer. The sizing tables in Section G2428 do not apply to factory-built chimneys; therefore, sizing must be engineered or as specified by the chimney and appliance manufacturers' instructions. Factory-built chimneys will be too large for many residential gas appliances [see Commentary Figures G2430.1(1) and G2430.1(2)].

Photo courtesy of Selkirk L.L.C.

Figure G2430.1(1)
FACTORY-BUILT CHIMNEY

Photo courtesy of Simpson Dura-Vent Co., Inc.

Figure G2430.1(2)
FACTORY BUILT AIR-COOLED CHIMNEY SYSTEM

G2430.2 (506.2) Support. Where factory-built *chimneys* are supported by structural members, such as joists and rafters, such members shall be designed to support the additional load.

❖ Factory-built chimneys are supported by the building structure and can impose considerable weight on structural components. Joists, rafters and other structural members must be designed to support the additional loading. Structural evaluation may be necessary, especially where framing for a chimney requires cutting and heading of joists and rafters.

SECTION G2431 (601)
GENERAL

G2431.1 (601.1) Scope. Sections G2432 through G2454 shall govern the approval, design, installation, construction, maintenance, *alteration* and repair of the *appliances* and *equipment* specifically identified herein.

❖ This chapter regulates all aspects of appliances and equipment to the extent found in each of Sections G2432 through G2454.

SECTION G2432 (602)
DECORATIVE APPLIANCES FOR
INSTALLATION IN FIREPLACES

G2432.1 (602.1) General. Decorative *appliances* for installation in *approved* solid fuel-burning *fireplaces* shall be tested in accordance with ANSI Z21.60 and shall be installed in accordance with the manufacturer's instructions. Manually lighted natural gas decorative *appliances* shall be tested in accordance with ANSI Z21.84.

❖ These appliances include gas log sets that are designed to simulate wood fires (see Commentary Figure G2432.1). Sections G2404 through G2411 address the requirements for testing, labeling and installing mechanical equipment and appliances. The gas-burning appliance must be tested to the standard or standards appropriate for the equipment. The testing agency is responsible for determining the standard to be used to test the equipment. In the case of decorative gas log sets, ANSI Z21.60 or Z21.84 is the applicable test standard.

Labeling is the code official's assurance that the subject product is a representative duplication of the product that the testing agency tested in the laboratory. The label indicates that an independent agency has conducted inspections at the plant to verify that all units conform to the specifications that the quality control manual sets forth for fabricating the gas appliances. Information that must be contained on the label is described in Section M1303.1 and includes the manufacturer's identification, the third-party inspection agency's identification, the model number, the serial number, the input ratings and the type of fuel the appliance is designed to burn.

The code requires that the appliance be installed in accordance with the manufacturer's installation instructions. This requirement is also linked to the laboratory testing of the appliance because the laboratory used the same installation instructions to install the prototype appliance being tested. When the appliance has been tested and evaluated for code compliance and judged to meet the performance and construction requirements of the applicable standard, the installation instructions become an integral part of the labeling requirements and must be strictly adhered to.

Photography by D.F. Noyes Studio, Courtesy of DESA International

Figure G2432.1
VENTED GAS LOG SET

The intent of this section is to regulate gas-burning appliances that are accessory to, and designed for installation in, vented solid-fuel-burning fireplaces. Gas-fired decorative log sets and log lighters are examples of accessory appliances that are designed for installation in solid-fuel-burning fireplaces. Gas log sets provide some radiant heat; however, their primary function is to create an aesthetically pleasing simulation of a wood log fire. This section addresses vented appliances and does not address "unvented gas log sets" (room heaters).

Decorative gas-burning appliances, such as some gas log sets, are designed to simulate wood fires by intentionally causing incomplete combustion as necessary to yield yellow or yellow-tipped flames. This incomplete combustion results in an increase in the amount of carbon monoxide produced. The highly toxic carbon monoxide is odorless and colorless, and the accompanying products of combustion produced by a gas-burning appliance do not have a strong odor or an odor that is readily recognized by an untrained person.

If a decorative gas appliance were operated with the fireplace damper closed, the carbon monoxide levels could be dangerously high before the building occupants became aware of the hazard. To prevent harm and the possible asphyxiation of the occupants, it is imperative that the fireplace damper be open whenever the appliance is burning. The manufacturer's installation instructions will specify the minimum free area of damper opening required to vent the appliance combustion products. The damper area is proportional to the appliance's input rating. The fireplace damper plate must be removed or permanently fixed in a position that provides the opening area required by the appliance manufacturer's installation instructions.

G2432.2 (602.2) Flame safeguard device. Decorative *appliances* for installation in *approved* solid fuel-burning *fireplaces*, with the exception of those tested in accordance with ANSI Z21.84, shall utilize a direct ignition device, an ignitor or a *pilot* flame to ignite the fuel at the *main burner*, and shall be equipped with a *flame safeguard* device. The *flame safeguard* device shall automatically shut off the fuel supply to a *main burner* or group of *burners* when the means of ignition of such *burners* becomes inoperative.

❖ To eliminate the hazards associated with manually igniting a decorative gas appliance, the code requires that the main burner be ignited by a supervised means of direct ignition, such as a standing pilot. The exception to this requirement is for manually lighted appliances listed to ANSI Z21.84. The use of automatic ignition controls prevents the operator of the appliance from being exposed to the potential hazard of trying to ignite the main burner with a match or other manual lighting device. The flame safeguard device consists of a control valve assembly with an integral means of pilot or igniter supervision and is designed to prevent or shut off the flow of gas to the appliance main burner in the event that the source of ignition is extinguished or otherwise fails.

The typical flame safeguard device used with gas log set appliances is a combination manual control valve, pressure regulator, pilot feed and magnetic pilot safety mechanism with a thermocouple generator. If the pilot flame is extinguished, the drop in thermocouple output voltage will cause the control valve to "lock out" in the closed position, thereby preventing the flow of gas to the main burner and the pilot burner. It is not the intent of this section to require flame safeguard devices for manually operated log lighter appliances used to kindle wood fires.

A flame safeguard, also known as a safety shutoff device, functions to automatically shut off the fuel supply to the main burner or burners of an appliance when the source of ignition becomes inoperative. Gas-burning equipment may operate using standing pilot ignition, which is a small flame kept constantly burning for the sole purpose of igniting the main burner when the thermostat calls for heat or a manual gas valve is opened by the operator. If the pilot is not lit, for whatever reason, the flame safeguard will automatically activate to prevent the flow of fuel to the main burner or burners of the appliance. Appliances that use automatic sparking devices or other means for ignition are subject to the same requirements for providing flame safeguards, and the method of operation of the flame safeguard will differ depending on the type of ignition system in use.

Automatically operated gas-burning equipment and appliances are equipped with some means of monitoring the means of ignition or the main burner flame because fuel introduced into a combustion chamber and not ignited and burned could cause a dangerous condition.

The most commonly used type of flame safeguard device consists of a thermocouple and an electromechanical device that function together to supervise a standing pilot flame. In the event of pilot failure, the thermocouple and flame safeguard device will function to prevent the flow of fuel to the main burners, and if the safeguard control is of the 100-percent shutoff type, it will also shut off the flow of fuel to the pilot burner.

Modern appliances seldom have standing pilots and instead use electric-spark-ignited intermittent pilots or direct ignition devices such as hot surface igniters and glow coils. The methods of detecting the presence of a pilot flame include thermocouples, millivolt (power pile) generators, liquid-filled capillary tubes, bimetal mechanisms and flame rectification circuits. Where only the main burner is monitored to verify ignition and continued combustion, the methods of detection include flame rectification, infrared detectors, ultraviolet detectors and bimetal stack sensors. The majority of today's appliances rely on electronic flame rectification or radiant energy detectors (infrared or ultraviolet) as the means of proving ignition of a pilot burner or the main burner(s). The flame rectification method of sensing the presence of flames is very common today and involves two electrodes, those being a flame rod and the burner itself. An AC voltage is applied between the flame rod and the burner. If a flame is present and impinges on both of these electrodes, a few micro-amps of current will flow through the ionized gases in the flame, but almost entirely in one direction because of the relative size of the surface areas of the two elec-trodes. This unidirectional flow rectifies the ac current to dc current, which is then recognized by electronic circuitry, thereby proving the presence of a flame. Manually lighted natural gas decorative appliances tested and listed to ANSI Z21.84 do not have to be equipped with a flame safeguard device.

Flame safeguard devices are commonly integral with a combination gas control. These controls typically incorporate multiple solenoid valves, a manual valve, an appliance gas-pressure regulator, a pressure operated valve and flame safeguard devices all in one device body. It is also common for flame safeguard devices to be incorporated into an electronic integrated circuit control module that is separate from the combination gas control. For example, spark ignition and hot surface ignition systems incorporate the means of ignition and ignition supervision in an electronic control module that may also contain components and circuitry for control of combustion fans, furnace blowers, combustion chamber purging and monitoring of multiple system pressure and temperature sensors located throughout an appliance. Commentary Figure G2432.2(1) illustrates a type of combination gas control designed for use with an outboard intermittent ignition control module. Commentary Figure G2432.2(2) illustrates the internal workings that are typical of combination gas controls. These controls have redundant valve mechanisms as an additional safety feature.

Photo courtesy of Emerson Climate Technologies, White Rodgers

Figure G2432.2(1)
COMBINATION GAS CONTROL

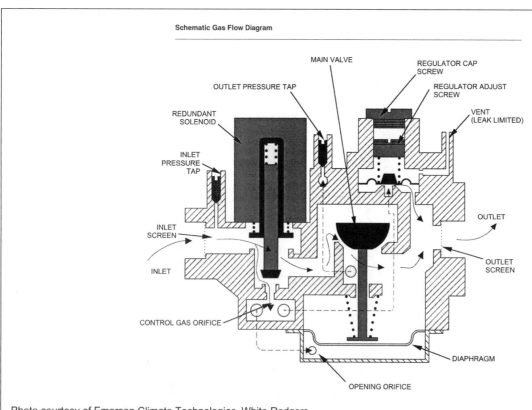

Photo courtesy of Emerson Climate Technologies, White Rodgers

Figure G2432.2(2)
COMBINATION GAS CONTROL FOR APPLIANCES
(Shown for Illustration Only)

G2432.3 (602.3) Prohibited installations. Decorative *appliances* for installation in *fireplaces* shall not be installed where prohibited by Section G2406.2.

❖ Section G2406.2 lists, with exceptions, various locations where gas appliances cannot be installed, such as bedrooms and bathrooms. Those prohibitions also apply to decorative appliances.

SECTION G2433 (603)
LOG LIGHTERS

G2433.1 (603.1) General. Log lighters shall be tested in accordance with CSA 8 and shall be installed in accordance with the manufacturer's instructions.

❖ Log lighters are simple manually operated burners used to start wood fires. Log lighters are functional rather than decorative appliances. The heat produced by the log lighter flames raises the temperature of the wood fuel to its ignition temperature. Log lighters are designed to be turned off manually after the wood fire is capable of sustaining combustion. The code, rather than providing prescriptive installation requirements, requires that log lighters be installed in accordance with the manufacturer's installation instructions. Log lighters must comply with the standard CSA 8. Note that log lighters are often packaged with a T-Handle shutoff valve and the package will indicate that the device is listed, referring to the shutoff valve. The code requires that the log lighter itself be listed. The lighter must be designed for the type of gas, i.e., natural or propane.

SECTION G2434 (604)
VENTED GAS FIREPLACES
(DECORATIVE APPLIANCES)

G2434.1 (604.1) General. Vented gas *fireplaces* shall be tested in accordance with ANSI Z21.50, shall be installed in accordance with the manufacturer's instructions and shall be designed and equipped as specified in Section G2432.2.

❖ Unlike the appliances addressed in Section G2432, these appliances are self-contained and do not rely on a fireplace to contain or vent them. Such appliances are referred to as gas fireplaces because they are designed to simulate a solid-fuel-burning fireplace. The standard, ANSI Z21.50, that regulates these appliances has recently been retitled as "Vented Gas Fireplaces." Therefore, these appliances will be referred to as gas fireplaces even though they do not fall under the definition of "Fireplace" [see Commentary Figures G2434.1(1), G2434.1(2) and G2434.1(3)]. These appliances are designed for various methods of venting, including direct-venting through the wall or roof and conventional venting with Type B vent or factory-supplied vent material. Section G2406.2 controls the type of appliance that is allowed in bathrooms, toilet rooms and bedrooms. Direct-vent appliances have the advantage of a closed combustion chamber that does not communicate with the room in which they are installed, as well as the advantage of an outdoor combustion air supply.

Photography by D.F. Noyes Studio, courtesy of DESA International

Figure G2434.1(1)
VENTED DECORATIVE APPLIANCE

Photography by D.F. Noyes Studio, courtesy of DESA International

Figure G2434.1(2)
DIRECT-VENT DECORATIVE APPLIANCE

Figure G2434.1(3)
DIRECT-VENT DECORATIVE APPLIANCE

G2434.2 (604.2) Access. Panels, grilles and access doors that are required to be removed for normal servicing operations shall not be attached to the building.

❖ Access to the appliance for maintenance and repair may be through panels, grilles or doors, but such panels must be removable as intended by the manufacturer of the appliance and not a permanent part of the structure (see definition of "Access" in Section G2403).

SECTION G2435 (605)
VENTED GAS FIREPLACE HEATERS

G2435.1 (605.1) General. Vented gas *fireplace* heaters shall be installed in accordance with the manufacturer's instructions, shall be tested in accordance with ANSI Z21.88 and shall be designed and equipped as specified in Section G2432.2.

❖ These appliances are similar to those addressed in Section G2434, the main difference being that vented gas fireplace heaters are designed with more emphasis on space heating while maintaining the decorative features. These heaters must comply with minimum thermal efficiency requirements (see Commentary Figure G2435.1).

SECTION G2436 (608)
VENTED WALL FURNACES

G2436.1 (608.1) General. *Vented wall furnaces* shall be tested in accordance with ANSI Z21.86/CSA 2.32 and shall be installed in accordance with the manufacturer's instructions.

❖ Wall furnaces are a type of room heater usually designed to be installed within a 2-inch by 4-inch (51 mm by 102 mm) stud cavity in frame construction. They are typically used in cottages, room additions and homes in mild climates. Some units are designed to serve a single room, and others are designed as through-the-wall units to serve adjacent rooms. Wall furnaces are ductless; however, some units are listed for use with a surface-mounted supply outlet extension. Wall furnaces can be either gravity or forced air type.

G2436.2 (608.2) Venting. *Vented wall furnaces* shall be vented in accordance with Section G2427.

❖ Because wall furnaces are designed to fit within a wall cavity, a special type of oval-shaped vent is required to fit within the same wall. Type BW vent is specially designed for wall furnace applications (see commentary, Section G2427).

G2436.3 (608.3) Location. *Vented wall furnaces* shall be located so as not to cause a fire hazard to walls, floors, combustible furnishings or doors. *Vented wall furnaces* installed

Photo courtesy of Quadra-Fire
Figure G2435.1
VENTED GAS FIREPLACE HEATER

between bathrooms and adjoining rooms shall not circulate air from bathrooms to other parts of the building.

❖ Wall furnaces, like all room heaters, can present a fire hazard if improperly located. The heat discharged or directly radiated from these units can ignite nearby wall or floor surfaces, furniture, trim items, window treatments and doors (see commentary, Section G2436.4). A through-the-wall unit serving both a bathroom and an adjacent room must not be capable of recirculating air between those spaces.

G2436.4 (608.4) Door swing. *Vented wall furnaces* shall be located so that a door cannot swing within 12 inches (305 mm) of an air inlet or air outlet of such *furnace* measured at right angles to the opening. Doorstops or door closers shall not be installed to obtain this *clearance*.

❖ A combustible door that swings close to a wall furnace could be a fire hazard if at some position the door would be within 12 inches (305 mm) of an air inlet or outlet. Because door closers and door stops are easily defeated, they must not be depended upon to secure the required clearance. If the door swing cannot comply with this section, the door would have to be removed or rehung to change its swing direction.

G2436.5 (608.5) Ducts prohibited. Ducts shall not be attached to wall *furnaces*. Casing extension boots shall not be installed unless listed as part of the *appliance*.

❖ Most wall furnaces are not designed to force air through ducts, especially those gravity types that do not use fans. Attachment of ducts would add resistance to the flow of air through the furnace, thereby causing an abnormally high temperature rise across the heat exchanger and abnormally high temperatures on furnace surfaces. Some wall furnaces are listed and designed for use with supply duct extensions intended for wall mounting and intended to improve heat distribution. Such duct extensions are factory-built and supplied only by the furnace manufacturer.

G2436.6 (608.6) Access. *Vented wall furnaces* shall be provided with *access* for cleaning of heating surfaces, removal of *burners*, replacement of sections, motors, *controls*, filters and other working parts, and for adjustments and lubrication of parts requiring such attention. Panels, grilles and access doors that are required to be removed for normal servicing operations shall not be attached to the building construction.

❖ All access panels and doors must be removable as intended by the manufacturer of the furnace.

SECTION G2437 (609)
FLOOR FURNACES

G2437.1 (609.1) General. *Floor furnaces* shall be tested in accordance with ANSI Z21.86/CSA 2.32 and shall be installed in accordance with the manufacturer's instructions.

❖ Floor furnaces are vented appliances that are installed in an opening in the floor. These units heat the room by gravity convection and direct radiation and usually serve as the sole source of space heating. Such fur-

naces are common in cottages, small homes, seasonally occupied structures and rural homes. Because the floor grille can become hot, extreme care must be exercised to prevent occupants, especially children, from contacting the grille by walking or falling on it. Also, care must be taken to avoid a fire hazard caused by placement of materials or furnishings on or near the furnace floor grille (see Commentary Figure G2437.1).

Photo courtesy of Empire Comfort Systems Inc.

Figure G2437.1
FLOOR FURNACE

G2437.2 (609.2) Placement. The following provisions apply to *floor furnaces*:

1. Floors. *Floor furnaces* shall not be installed in the floor of any doorway, stairway landing, aisle or passageway of any enclosure, public or private, or in an exitway from any such room or space.

2. Walls and corners. The register of a *floor furnace* with a horizontal warm air outlet shall not be placed closer than 6 inches (152 mm) to the nearest wall. A distance of at least 18 inches (457 mm) from two adjoining sides of the *floor furnace* register to walls shall be provided to eliminate the necessity of occupants walking over the warm-air discharge. The remaining sides shall be permitted to be placed not closer than 6 inches (152 mm) to a wall. Wall-register models shall not be placed closer than 6 inches (152 mm) to a corner.

3. Draperies. The *furnace* shall be placed so that a door, drapery, or similar object cannot be nearer than 12 inches (305 mm) to any portion of the register of the *furnace*.

4. Floor construction. *Floor furnaces* shall not be installed in concrete floor construction built on grade.

5. *Thermostat.* The controlling *thermostat* for a *floor furnace* shall be located within the same room or space as the *floor furnace* or shall be located in an adjacent room

or space that is permanently open to the room or space containing the *floor furnace.*

❖ 1. Floor furnaces must not be located where they would interfere with or impede egress. In an emergency egress situation, a floor furnace could be a tripping hazard and could collapse under the live load of many occupants.

2. The floor furnace register must be at least 6 inches (152 mm) from walls to avoid creating a fire hazard by raising the wall temperature to a combustible level. The same reasoning applies to the 6-inch (152 mm) distance from wall registers to a corner. The 18-inch (457 mm) clearance from adjoining sides of a register to walls provides space for the occupants to walk around the grille, which can reach temperatures high enough to cause burns.

3. As stated in Section G2436.3, the furnace location must not create a fire hazard by being too close to wall surfaces, trim items, furnishings and window treatments.

4. A floor furnace installed in a slab on grade would have to be in a pit, would be subject to flooding and corrosion, and would be inaccessible for service and inspection.

5. If the controlling thermostat does not sense the air temperature in the room in which the furnace is installed, dangerous overheating could result. A thermostat isolated from the source of heat it controls would not respond to the condition in the space served by the furnace.

G2437.3 (609.3) Bracing. The floor around the *furnace* shall be braced and headed with a support framework designed in accordance with Chapter 5.

❖ The framing around the floor opening must be capable of supporting the floor system, the anticipated floor loads and the weight of the furnace. The structural requirements of the IBC must be complied with.

G2437.4 (609.4) Clearance. The lowest portion of the *floor furnace* shall have not less than a 6-inch (152 mm) *clearance* from the grade level; except where the lower 6-inch (152 mm) portion of the *floor furnace* is sealed by the manufacturer to prevent entrance of water, the minimum *clearance* shall be reduced to not less than 2 inches (51 mm). Where such *clearances* cannot be provided, the ground below and to the sides shall be excavated to form a pit under the *furnace* so that the required *clearance* is provided beneath the lowest portion of the *furnace.* A 12-inch (305 mm) minimum clearance shall be provided on all sides except the *control* side, which shall have an 18-inch (457 mm) minimum *clearance.*

❖ This section specifies clearances between the ground and the furnace, which apply to crawl space installations. The clearances allow access for service and inspection and help prevent corrosion of the furnace assembly.

G2437.5 (609.5) First floor installation. Where the basement story level below the floor in which a *floor furnace* is installed is utilized as habitable space, such *floor furnaces*

shall be enclosed as specified in Section G2437.6 and shall project into a nonhabitable space.

❖ Where a floor furnace is installed above a habitable basement space, the furnace must project into an uninhabitable space and be separated from that space by noncombustible construction (see commentary, Section G2437.6).

G2437.6 (609.6) Upper floor installations. *Floor furnaces* installed in upper stories of buildings shall project below into nonhabitable space and shall be separated from the nonhabitable space by an enclosure constructed of *noncombustible materials.* The *floor furnace* shall be provided with *access, clearance* to all sides and bottom of not less than 6 inches (152 mm) and *combustion air* in accordance with Section G2407.

❖ Where the floor furnace is installed in an upper floor of a building, the furnace must project into a nonhabitable space, similar to the installation in Section G2437.5, and must be separated from that space by noncombustible construction. Access and clearance must be provided for maintenance and servicing. An adequate source of combustion air must also supply the furnace.

SECTION G2438 (613)
CLOTHES DRYERS

G2438.1 (613.1) General. *Clothes dryers* shall be tested in accordance with ANSI Z21.5.1 and shall be installed in accordance with the manufacturer's instructions.

❖ This section addresses clothes dryer appliances, Types 1 and 2 (see definition of "Clothes dryer"). Dryers are tested to the applicable safety standard for the appliance, and the manufacturer's installation instructions convey the information needed to duplicate the installation configuration that was tested and found to meet the requirements of the safety standard. The manufacturer's installation instructions are evaluated by the agency responsible for testing, listing and labeling the appliance, and will therefore prescribe an installation that is consistent with the appliance installation that was tested. Clothes dryers must be exhausted in compliance with Section G2439 and the manufacturer's installation instructions (see Commentary Figure G2438.1).

SECTION G2439 (614)
CLOTHES DRYER EXHAUST

G2439.1 (614.1) Installation. *Clothes dryers* shall be exhausted in accordance with the manufacturer's instructions. Dryer exhaust systems shall be independent of all other systems and shall convey the moisture and any products of *combustion* to the outside of the building.

❖ Clothes dryer exhaust systems must convey the moisture and any products of combustion directly to the exterior of the building (outdoors). The code does not use the term "dryer vent." Gas-fired clothes dryers are not "vented" as are other fuel-fired appliances, but rather are exhausted.

CLOTHES DRYER MAYTAG NEWTON, IOWA U.S.A.

MODEL NO. MDG5500AWW   SERIAL NO. 10000001XX

OUTSIDE EXHAUST SEE INSTALLATION INSTRUCTIONS. ALSO SUITABLE FOR MFD. HOME (MOBILE HOME) INSTALLATION. CERTIFIED FOR USE WITH NATURAL OR PROPANE GAS. TESTED FOR: NAT. MFG. LP GASES. MINIMUM GAS SUPPLY PRESSURE FOR THE PURPOSE OF INPUT ADJUSTMENT - 4.5 in. wc (1.12 KPa)(1.8 mbar). MAXIMUM PERMISSIBLE GAS SUPPLY PRESSURE 10.5 in. wc (2.62 KPa)(4.2 mbar). ANSI Z21.5.1a - 1999 - CSA 7.1a - M99 - CLOTHES DRYER VOL.1 (1999 SÉCHEUSES VOL.1.)

120V 60HZ 6A   MANIFOLD PRESSURE 3.5in wc (0.87 Kpa)(1.4 mbar)   EQUIPPED FOR NAT. GAS INPUT 22,000 BTU/HR (6.4 kVv)

Figure courtesy of Maytag Corporation

Figure G2438.1
TYPE I CLOTHES DRYER LABEL

Clothes dryer exhaust ducts must be installed to comply with the dryer manufacturer's installation instructions and the requirements of this section. Dryers are designed and built to meet industry safety standards.

The clothes dryer manufacturer's installation instructions control the type of exhaust duct material allowed and the method of installation. For example, typical dryer installation instructions will require metallic duct materials and will impose more stringent length limitations for flexible metallic ducts than for rigid ducts because of the poorer flow characteristics of flexible duct materials.

Because clothes dryer exhaust contains high concentrations of combustible lint, debris and water vapor, dryer exhaust systems must be independent of all other systems. This requirement prevents the fire hazards associated with such an exhaust system from extending into or affecting other systems or other areas in the building. Additionally, this section intends to prevent products of combustion from entering the building through other systems.

Dryer exhaust ducts must be independent of other dryer exhaust ducts unless connected to an engineered exhaust system specifically designed to serve multiple dryers. Type I domestic dryers are not designed for connection to a common exhaust duct serving multiple dryers. For example, connecting multiple Type I dryers to a common duct riser would pressurize the riser and cause exhaust to back up into any dryer that was not operating. Also, the low-flow velocities and duct temperature losses could result in water vapor condensation and the buildup of lint and debris at low points.

Clothes dryer exhaust systems cannot terminate in or discharge to any enclosed space such as an attic or crawl space, regardless of whether the space is ventilated through openings to the outdoors. The high levels of moisture in the exhaust air can cause condensation to form on exposed surfaces or in insulation materials. Water vapor condensation can cause structural damage and deterioration of building materials and contribute to the growth of mold and fungus. Clothes dryer exhausts that discharge to enclosed spaces will also cause an accumulation of combustible lint and debris, creating a significant fire hazard. An improperly installed clothes dryer exhaust system not only reduces dryer efficiency and increases running time but can also cause a significant increase in exhaust temperature, in turn causing the dryer to cycle on its

high limit control, which is an unsafe operating condition.

G2439.2 (614.2) Duct penetrations. Ducts that exhaust *clothes dryers* shall not penetrate or be located within any fireblocking, draftstopping or any wall, floor/ceiling or other assembly required by this *code* to be fire-resistance rated, unless such duct is constructed of galvanized steel or aluminum of the thickness specified in the mechanical provisions of this *code* and the fire-resistance rating is maintained in accordance with this *code*. Fire dampers shall not be installed in *clothes dryer* exhaust duct systems.

❖ Rigid clothes dryer exhaust ducts are permitted to penetrate assemblies that are not fire-resistance-rated and building elements not used as fireblocking or draftstopping. In all other cases, ducts must be constructed of galvanized steel or aluminum of the thickness specified in the mechanical provisions of this code, and the penetration must be protected to maintain the fire-resistance rating and integrity of the assembly or element being penetrated. Because of the strength and rigidity differences between steel and aluminum, aluminum ducts generally must be of a heavier (thicker) gage than steel ducts for a given application.

The metal thickness requirements of Chapter 16 necessitate rigid pipe and rule out the use of flexible duct where the duct must penetrate fireblocking, draftstopping or a fire-resistance-rated assembly. Where penetrating fireblocking or draftstopping, the exhaust duct must be constructed of galvanized steel or aluminum, and the annular space around the duct must be fireblocked in accordance with the IBC (see Section G2439.7.1).

Where penetrating a fire-resistance-rated assembly, the penetration must be protected. The requirements of this section, combined with Section G2439.7.4, make a compelling case for always placing clothes dryers against outside walls to avoid long duct runs and penetrations of other than exterior walls.

G2439.3 (614.4) Exhaust installation. Exhaust ducts for *clothes dryers* shall terminate on the outside of the building and shall be equipped with a backdraft *damper*. Screens shall not be installed at the duct termination. Ducts shall not be connected or installed with sheet metal screws or other fasteners that will obstruct the flow. *Clothes dryer* exhaust ducts shall not be connected to a *vent connector*, vent or *chimney*. *Clothes dryer* exhaust ducts shall not extend into or through ducts or plenums.

❖ Exhaust ducts must connect directly to terminals that pass through the building envelope to the outdoor atmosphere. Attics and crawl spaces are not considered to be outdoors, and exhaust ducts cannot terminate in those spaces (see commentary, Section G2439.1). Backdraft dampers must be installed in dryer exhaust ducts to avoid outdoor air infiltration during periods when the dryer is not operating and to prevent the entry of animals. These dampers should be designed and installed to provide an adequate seal when in a closed position to minimize air leakage (infiltration). A backdraft damper is usually of the gravity type that is opened by the energy of the exhaust discharge. Some dryer manufacturers prohibit the use of magnetic backdraft dampers because of the extra resistance that the exhaust flow must overcome.

Exhaust terminal opening size is also governed by the dryer manufacturer's instructions. Full-opening terminals present less resistance to flow and might be mandated by the dryer manufacturer. A "full opening" is considered to be an opening having no dimension less than the diameter of the exhaust duct. Dryer exhaust flow must not be restricted by screens or fastening devices such as sheet metal screws. See the allowance in Section G2439.7.2 for fasteners. Any type of screen, including so-called bird screens, could become completely blocked with fibers in a very short time. Consider that the filter screen integral with the appliance becomes restricted with lint in each cycle of operation. These restrictions and projections will promote the accumulation of combustible lint and debris in the exhaust duct, thereby creating a potential fire hazard and causing flow resistance. Duct tape should not be relied upon as the sole means of joining dryer exhaust ducts because adhesives can deteriorate with age and when exposed to high temperatures, causing joints to separate. However, sealing joints is still desirable to limit leakage of lint, fibers, moisture vapor and combustion products into the occupied space. Ducts can also separate during duct-cleaning operations using brushes.

G2439.4 (614.5) Dryer exhaust duct power ventilators. Domestic dryer exhaust duct power ventilators shall be listed and labeled to UL 705 for use in dryer exhaust duct systems. The dryer exhaust duct power ventilator shall be installed in accordance with the manufacturer's instructions.

❖ The previous code editions did not recognize dryer exhaust duct power ventilators (DEDPVs) as an option for clothes dryer installations. DEDPVs are loosely referred to as "dryer booster fans" in the marketplace. Prior to the 2015 code, the designer's choices for exhaust duct length were to: 1. Limit the duct length to 35 feet; 2. Follow the length limits in the clothes dryer manufacturer's instructions; or 3. Relocate the dryer. A fourth option was to get the code official to approve the installation of a DEDPV under the alternative approval provision in Section 105.

Exhaust ducts that exceed the developed length allowed by the code are a potential fire hazard, create maintenance problems, increase drying times and

cause the dryer to be inefficient and waste energy. Dryer exhaust systems are commonly installed improperly with excessive lengths, too many elbows and the wrong duct materials. Because of the high incidence of reported dryer fires, the code strictly regulates the installation. DEDPVs are listed to a revised version of UL 705 that now contains tests and construction requirements that are specific to these devices. DEDPVs have been around for years, but until recently, were not listed to a national consensus standard that was specific to these devices. The UL 705 standard contains requirements for the construction, testing and installation of DEDPVs and requires them to be equipped with features such as interlocks, limit controls, monitoring controls and enunciator devices to make certain that the dryers or dryer operators are aware of the operating status of the DEDPVs [see Commentary Figures G2439.4(1) and (2).]

G2439.5 (614.6) Makeup air. Installations exhausting more than 200 cfm (0.09 m³/s) shall be provided with *makeup air*. Where a closet is designed for the installation of a *clothes dryer*, an opening having an area of not less than 100 square inches (0.0645 m²) for *makeup air* shall be provided in the closet enclosure, or *makeup air* shall be provided by other *approved* means.

❖ Makeup air must be supplied to compensate for the air exhausted by the dryer exhaust system. A typical domestic clothes dryer will exhaust less than 200 cubic feet per minute (0.09 m³/s). For closet installations, the closet door or the closet enclosure must have an opening with a minimum area of 2 square inches (0.0645 m²). Where louvers or grilles are used, the solid portion of the louver or grille should be evaluated in accordance with Section G2407.10. Makeup air is necessary to prevent the room or space housing the dryer(s) from developing a negative pressure with respect to adjacent spaces or the outdoors, which could result in the improper and dangerous operation of the dryer and other fuel-burning appliances. The required amount of makeup air should be approximately equal to the amount exhausted and is normally supplied by infiltration of air from outdoors or through openings to the outdoors.

The makeup air not only supplies the air that is to be exhausted from the dryer, but also supplies combustion air for gas-fired appliances.

G2439.6 (614.7) Protection required. Protective shield plates shall be placed where nails or screws from finish or other work are likely to penetrate the *clothes dryer* exhaust duct. Shield plates shall be placed on the finished face of all framing members where there is less than 1¼ inches (32 mm) between the duct and the finished face of the framing member. Protective shield plates shall be constructed of steel, shall have a minimum thickness of 0.062 inch (1.6 mm) and shall extend a minimum of 2 inches (51 mm) above sole plates and below top plates.

❖ See the commentary to Section M1502.5 (see Commentary Figure G2439.6).

Figure G2439.4(1)
DRYER EXHAUST DUCT POWER VENTILATORS

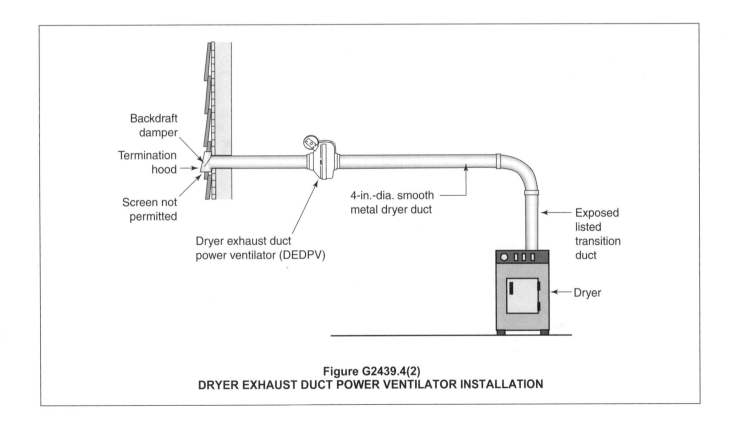

Backdraft
damper

Termination
hood

Screen not
permitted

Dryer exhaust duct
power ventilator (DEDPV)

4-in.-dia. smooth
metal dryer duct

Exposed
listed
transition
duct

Dryer

Figure G2439.4(2)
DRYER EXHAUST DUCT POWER VENTILATOR INSTALLATION

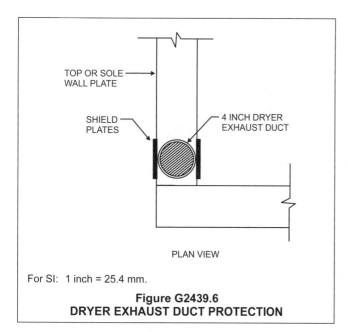

TOP OR SOLE
WALL PLATE

SHIELD
PLATES

4 INCH DRYER
EXHAUST DUCT

PLAN VIEW

For SI: 1 inch = 25.4 mm.

**Figure G2439.6
DRYER EXHAUST DUCT PROTECTION**

G2439.7 (614.8) Domestic clothes dryer exhaust ducts.
Exhaust ducts for domestic *clothes dryers* shall conform to
the requirements of Sections G2439.7.1 through G2439.7.6.

❖ Section G2408.1 states that gas-fired equipment and
appliances must be installed in accordance with the
manufacturer's installation instructions for the listed
and labeled equipment. An installation complying with
the manufacturer's installation instructions is required,
except where the code requirements are more stringent. The code specifically addresses dryer exhaust
ducts, an installation requirement that is also
addressed in the dryer manufacturer's installation
instructions. The code requirements for the items
addressed in these sections, such as maximum total
developed length of dryer exhaust ducts, equivalent
lengths of directional fittings and minimum nominal
size to name a few, may parallel or exceed the applicable requirements in the manufacturer's installation
instructions. The manufacturer's installation instructions could also contain requirements that exceed
those in the code.

In all cases, the more restrictive requirements would
apply. Therefore, the clothes dryer exhaust duct size
must be determined by the clothes dryer manufacturer
but must not be less than 4 inches (102 mm) in diameter.

Dryer exhaust ducts cannot be larger than 4 inches
(102 mm) in diameter unless specifically allowed by
the manufacturer. Note that enlarging a duct will cause
a reduction in flow velocity, and dryer exhaust ducts
must maintain sufficient flow velocity to transport lint
and fibers through the duct to the discharge terminal.

The duct length, the number and degree of directional fittings, the smoothness of the duct interior wall
and the type of exhaust outlet terminal all contribute to

the overall friction loss of a clothes dryer exhaust system. When the friction loss is high enough to restrict
the required exhaust flow, the duct system must be
redesigned to allow the required flow rate. An exhaust
duct may have to be shortened or rerouted to compensate for the flow resistance.

Excessive friction losses will also result in reduced
flow velocities, which means that the exhaust ducts
would be much more likely to collect debris. An
improperly designed exhaust system will result in poor
dryer performance and poor energy efficiency and can
cause the appliance to cycle on its limit control, which
can be hazardous.

Joints in the dryer exhaust system must be reasonably air tight, must have a smooth interior finish and
must run in the direction of airflow. For example, the
male end of each section of duct must point away from
the dryer. This permits the dryer exhaust system to
function as intended so that joints and connections do
not serve as collection points for lint and debris.

Proper support is required and necessary for maintaining alignment of the dryer exhaust duct system and
to prevent excessive stress on ducts and duct joints. A
sagging duct will increase the internal resistance to
airflow, reduce the efficiency of the system and cause
the accumulation of lint and debris.

Section G2439.7.3 specifically addresses transition
duct connectors. Within the context of this section, a
transition duct is a flexible connector used as a transition between the dryer outlet and the connection point
to the exhaust duct system. Transition duct connectors must be listed and labeled as transition ducts for
clothes dryer application. Transition ducts are currently listed to comply with UL 2158A and are not considered to be flexible air ducts or flexible air
connectors subject to the material requirements of
Chapter 16.

Transition ducts are flexible ducts constructed of a
metalized (foil) fabric supported on a spiral wire frame.
They are more fire resistant than the typical plastic spiral duct. Transition duct connectors are limited to 8
feet (1829 mm) in and length and must be installed in
compliance with their listing and the manufacturer's
instructions. These duct connectors must not be concealed by any portion of the structure's permanent finish materials such as drywall, plaster, paneling, built-in
furniture or cabinets or any other similar permanently
affixed building component; they must remain entirely
within the room in which the appliance is installed.
Transition duct connectors cannot be joined together
to extend beyond the 8-foot (1829 mm) maximum
length limit. Transition ducts are to be cut to length as
needed to avoid excess duct and unnecessary bends.

Transition duct connectors are necessary for
domestic dryers because of appliance movement,
vibration and outlet location. In many cases, connecting a domestic clothes dryer directly to rigid duct would
be difficult.

G2439.7.1 (614.8.1) Material and size. Exhaust ducts shall have a smooth interior finish and shall be constructed of metal a minimum 0.016-inch (0.4 mm) thick. The exhaust duct size shall be 4 inches (102 mm) nominal in diameter.

❖ This section requires that the diameter of the exhaust duct be 4 inches (102 mm), no larger and no smaller. Round duct is intended. A 4-inch (102 mm) duct is the basis for the design of the appliance exhaust system. Ducts that are too small would restrict air movement and would cause the dryer to consume more energy because of the longer drying time. They could also cause appliance overheating. Ducts that are too large will cause a decrease in the flow velocity, causing lint and debris to drop out of the exhaust air stream. For example, increasing the duct size from 4 inches to 5 inches (102 mm to 127 mm) will result in a reduction of duct velocity of approximately 37 percent for the typical dryer. The exhaust duct material must be smooth-wall metal with the specified minimum wall thickness. Plastic ducts, corrugated ducts and flexible ducts are all prohibited with the exception of transition ducts in accordance with Section M1502.4.3.

G2439.7.2 (614.8.2) Duct installation. Exhaust ducts shall be supported at 4-foot (1219 mm) intervals and secured in place. The insert end of the duct shall extend into the adjoining duct or fitting in the direction of airflow. Ducts shall not be joined with screws or similar fasteners that protrude more than $^1/_8$ inch (3.2 mm) into the inside of the duct.

❖ Section G2439.3 states that fasteners used to join fittings and sections of dryer exhaust ducts together must not obstruct the airflow. This could be interpreted as a prohibition of screws and rivets, or it could mean that such fasteners must not penetrate too far into the duct. The revision to Section G2439.7.2 now makes it clear how Section G2439.3 is to be interpreted. A fastener protrusion of $^1/_8$ inch or less will collect some lint, but it is believed to be insignificant. Actually, smooth duct walls also collect lint. The trade-off for allowing tiny amounts of lint to collect is improved duct construction. If dryer exhaust ducts are not allowed to be mechanically fastened, the only method to prevent separation of joints is duct tape. Duct tape should never be depended on as the sole means of securing duct sys-

tems because it is a sealing means, not a fastening means. Now ducts can be properly and securely fastened and then sealed with tapes or mastics. Section M1502 requires dryer exhaust ducts to be mechanically fastened and allows the same $^1/_8$-inch maximum penetration. See the commentary to Section M1502.4.2 and Commentary Figure G2439.7.2.

G2439.7.3 (614.8.3) Transition ducts. Transition ducts used to connect the dryer to the exhaust duct system shall be a single length that is *listed* and *labeled* in accordance with UL 2158A. Transition ducts shall be not more than 8 feet (2438 mm) in length and shall not be concealed within construction.

❖ See the commentary to Section M1502.4.3.

G2439.7.4 (614.8.4) Duct length. The maximum allowable exhaust duct length shall be determined by one of the methods specified in Sections G2439.7.4.1 through G2439.7.4.3.

❖ See the commentary to Section M1502.4.5.

G2439.7.4.1 (614.8.4.1) Specified length. The maximum length of the exhaust duct shall be 35 feet (10 668 mm) from the connection to the transition duct from the dryer to the outlet terminal. Where fittings are used, the maximum length of the exhaust duct shall be reduced in accordance with Table G2439.7.4.1.

❖ The maximum length of 35 feet (7620 mm) and the associated reductions for changes in direction can be very restrictive and make it necessary, in most cases, to locate the clothes dryer close to an exterior wall. Locating the dryer in the middle of a house or apartment creates a design problem for the architect and the contractor. Exceeding the length requirement of this section will contribute to excessive pressure loss, poor dryer performance and potential fire hazards (see commentary, Section G2439.7.4).

The length reduction for changes in direction applies to all changes in direction in the exhaust duct, including the first 90-degree (1.57 rad) turn required to run the duct inside a wall cavity. Any change in direction made with the 8-foot (1829 mm) transition duct connector does not require a reduction in length. The length penalties for bends account for the flow resistance created by the change in direction. See Table G2439.7.4.1, which allows lesser length penalties for bends (elbows)

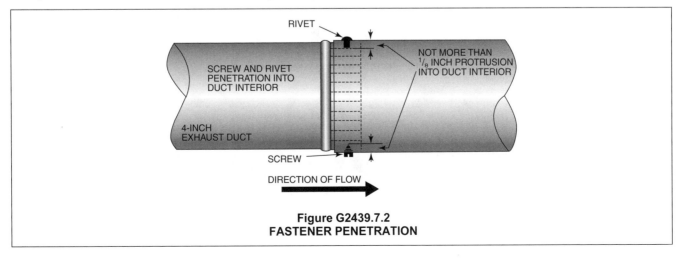

Figure G2439.7.2
FASTENER PENETRATION

used in a system if such bends are long turn radius (long sweep) bends. Section G2439.7.4.2 allows the 35-foot (7620 mm) duct length to be exceeded if the manufacturer's installation instructions specify a longer duct allowance. The manufacturer's instructions must be made available to the code official to allow verification of the allowable length and to allow inspection of the rough-in installation for compliance with those instructions. The code official will also have to verify that the specific clothes dryer is installed. The dryer chosen for the application of the exception to this section must be installed before final approval can be given.

Note that Section G2439.7.4.2 could create a problem when the occupants move and take the dryer with them. The next occupant must either buy a dryer with an equivalent duct length allowance or install a duct that is compatible with the dryer to be used. If the new occupant is not aware of the duct requirements, he or she might install a standard dryer designed for only 35 feet (7620 mm) of duct length. This could result in a build-up of lint in the duct and a potential fire hazard. The code official is usually not aware of the occupancy change and has no method of ensuring that the proper dryer and duct are installed. See Section G2439.7.5.

See Sections G2439.4 and G2439.7.4.3 regarding "booster fans" (DEDPV) that are marketed for the purpose of extending exhaust duct lengths. Clothes dryer exhaust duct length problems could be eliminated if designers would give more thought to the location of laundry facilities.

TABLE G2439.7.4.1 (TABLE 614.8.4.1)
DRYER EXHAUST DUCT FITTING EQUIVALENT LENGTH

DRYER EXHAUST DUCT FITTING TYPE	EQUIVALENT LENGTH
4 inch radius mitered 45-degree elbow	2 feet, 6 inches
4 inch radius mitered 90-degree elbow	5 feet
6 inch radius smooth 45-degree elbow	1 foot
6 inch radius smooth 90-degree elbow	1 foot, 9 inches
8 inch radius smooth 45-degree elbow	1 foot
8 inch radius smooth 90-degree elbow	1 foot, 7 inches
10 inch radius smooth 45-degree elbow	9 inches
10 inch radius smooth 90-degree elbow	1 foot, 6 inches

For SI: 1 inch = 25.4 mm, 1 foot = 304.8 mm, 1 degree = 0.0175 rad.

G2439.7.4.2 (614.8.4.2) Manufacturer's instructions. The maximum length of the exhaust duct shall be determined by the dryer manufacturer's installation instructions. The *code official* shall be provided with a copy of the installation instructions for the make and model of the dryer. Where the exhaust duct is to be concealed, the installation instructions shall be provided to the *code official* prior to the concealment inspection. In the absence of fitting equivalent length calculations from the clothes dryer manufacturer, Table G2439.5.5.1 shall be utilized.

❖ See the commentary to Section M1502.4.5.2.

G2439.7.4.3 (614.8.4.3) Dryer exhaust duct power ventilator length. The maximum length of the exhaust duct shall be determined by the dryer exhaust duct power ventilator manufacturer's installation instructions.

❖ See the commentary to Section G2439.3.

G2439.7.5 (614.8.5) Length identification. Where the exhaust duct equivalent length exceeds 35 feet (10 668 mm), the equivalent length of the exhaust duct shall be identified on a permanent label or tag. The label or tag shall be located within 6 feet (1829 mm) of the exhaust duct connection.

❖ See the commentary to Section M1502.4.6 (see Commentary Figure G2439.7.5).

Figure G2439.7.5
LENGTH LABEL

G2439.7.6 (614.8.6) Exhaust duct required. Where space for a *clothes dryer* is provided, an exhaust duct system shall be installed.

Where the *clothes dryer* is not installed at the time of occupancy, the exhaust duct shall be capped at location of the future dryer.

Exception: Where a *listed* condensing *clothes dryer* is installed prior to occupancy of the structure.

❖ See the commentary to Section M1502.4.7.

SECTION G2440 (615)
SAUNA HEATERS

G2440.1 (615.1) General. Sauna heaters shall be installed in accordance with the manufacturer's instructions.

❖ Sauna heaters are used in steam baths and similar rooms to generate heat and steam. They must be installed in compliance with the manufacturer's instructions to ensure they are installed as designed and tested. The code relies on these installation instructions rather than stating prescriptive requirements that might contradict the manufacturer.

G2440.2 (615.2) Location and protection. Sauna heaters shall be located so as to minimize the possibility of accidental contact by a person in the room.

❖ Sauna heaters produce steam by passing water over a heated surface. The choice of locations must consider the fact that a heat-producing appliance will be exposed within a small, limited-visibility room with unclothed occupants.

G2440.2.1 (615.2.1) Guards. Sauna heaters shall be protected from accidental contact by an *approved* guard or barrier of material having a low coefficient of thermal conductivity. The guard shall not substantially affect the transfer of heat from the heater to the room.

❖ Guards must be installed to protect the occupants from being burned. The guards must be constructed of a material that is a poor conductor of heat (e.g., wood) so that the guard itself will not present a burn hazard. The design of the guard must protect the occupants but impede the flow of heat into the room from the heater.

G2440.3 (615.3) Access. Panels, grilles and access doors that are required to be removed for normal servicing operations, shall not be attached to the building.

❖ Access panels, covers and doors must not be made unusable by trim, woodwork or room enclosures that impede or interfere with access.

G2440.4 (615.4) Combustion and dilution air intakes. Sauna heaters of other than the direct-vent type shall be installed with the *draft hood* and *combustion air* intake located outside the sauna room. Where the *combustion air* inlet and the *draft hood* are in a dressing room adjacent to the sauna room, there shall be provisions to prevent physically blocking the *combustion air* inlet and the *draft hood* inlet, and to prevent physical contact with the *draft hood* and vent assembly, or warning notices shall be posted to avoid such contact. Any warning notice shall be easily readable, shall contrast with its background and the wording shall be in letters not less than $1/4$ inch (6.4 mm) high.

❖ Combustion air and dilution air must not be taken from the sauna room because of the excessive water vapor in the air and the fact that the sauna room would probably be incapable of providing the required volume of combustion air. If a draft hood or combustion air inlet is obstructed, the appliance could malfunction, which could threaten the occupants. For example, a combustion-air-starved heater would produce high levels of carbon monoxide that could enter the sauna room in the event of venting failure. In general, the code expresses concern for the use of fuel-fired appliances in small closed rooms, especially where the occupants are sleeping or would have impaired senses or a diminished ability to recognize danger.

G2440.5 (615.5) Combustion and ventilation air. *Combustion air* shall not be taken from inside the sauna room. *Combustion* and ventilation air for a sauna heater not of the direct-vent type shall be provided to the area in which the *combustion air* inlet and *draft hood* are located in accordance with Section G2407.

❖ This section would require the heater to be either a direct-vent type or a separated-combustion type in which the combustion chamber does not communicate with the sauna room (see commentary, Section G2440.4).

G2440.6 (615.6) Heat and time controls. Sauna heaters shall be equipped with a *thermostat* which will limit room temperature to 194°F (90°C). If the *thermostat* is not an integral part of the sauna heater, the heat-sensing element shall be located within 6 inches (152 mm) of the ceiling. If the heat-sensing element is a capillary tube and bulb, the assembly shall be attached to the wall or other support, and shall be protected against physical damage.

❖ A thermostat is required to limit the temperature in the sauna for fire safety. The control must sense the warmest air near the ceiling and must be protected from physical damage.

G2440.6.1 (615.6.1) Timers. A timer, if provided to *control main burner* operation, shall have a maximum operating time of 1 hour. The *control* for the timer shall be located outside the sauna room.

❖ Timers are not required by the code but provide an extra level of protection when installed. To protect both the occupants and the building, timers must limit the heater operating time to 1 hour. Resetting the timer would require the occupant to exit the sauna, thus lessening the chances of overexposure.

G2440.7 (615.7) Sauna room. A ventilation opening into the sauna room shall be provided. The opening shall be not less than 4 inches by 8 inches (102 mm by 203 mm) located near the top of the door into the sauna room.

❖ A ventilation opening is required to allow the escape of steam and heat and to supply ventilation for the occupants. The opening is required near the ceiling to take advantage of the natural tendency of heat and steam to rise, facilitating the exhaust.

SECTION G2441 (617)
POOL AND SPA HEATERS

G2441.1 (617.1) General. Pool and spa heaters shall be tested in accordance with ANSI Z21.56/CSA 4.7 and shall be installed in accordance with the manufacturer's instructions.

❖ Pool and spa heaters are specialized water heaters very similar in design to hot water supply boilers and are used with swimming pools, recreational or therapeutic spas and hot tubs. These heaters are usually of the water-tube type and are designed for either indoor or outdoor installation.

SECTION G2442 (618)
FORCED-AIR WARM-AIR FURNACES

G2442.1 (618.1) General. Forced-air warm-air *furnaces* shall be tested in accordance with ANSI Z21.47 or UL 795 and shall be installed in accordance with the manufacturer's instructions.

❖ Forced-air warm-air furnaces are considered to be central heating units and consist of burners or heating elements, heat exchangers, blowers and associated controls. Forced-air furnaces are made in many different configurations, including upflow, counterglow (down flow), horizontal flow, multiposition flow and indoor and outdoor units. Commentary Figure G2426.1(2) is an example of a multiposition flow (i.e., up, down and horizontal flow) forced-air warm-air furnace.

G2442.2 (618.2) Forced-air furnaces. The minimum unobstructed total area of the outside and return air ducts or openings to a forced-air warm-air *furnace* shall be not less than 2 square inches for each 1,000 *Btu*/h (4402 mm²/W) output rating capacity of the *furnace* and not less than that specified in the *furnace* manufacturer's installation instructions. The minimum unobstructed total area of supply ducts from a forced-air warm-air *furnace* shall be not less than 2 square inches for each 1,000 *Btu*/h (4402 mm²/W) output rating capacity of the *furnace* and not less than that specified in the *furnace* manufacturer's installation instructions.

> **Exception:** The total area of the supply air ducts and outside and return air ducts shall not be required to be larger than the minimum size required by the *furnace* manufacturer's installation instructions.

❖ The aggregate area of all ducts or openings that convey supply air from the furnace or return air back to the furnace must be adequate to allow the required airflow through the furnace. A furnace that is "starved" for return air or is restricted by an inadequate supply air duct size will produce an abnormal temperature rise across the heat exchanger, which is both a fire hazard and detrimental to the furnace. The furnace output rating is not usually indicated on the label and would be determined as approximately the input rating in Btu/h (W) times the efficiency rating of the furnace. For example, 100,000 Btu/h (29 310 W) input times 0.80 (80-percent efficiency) is 80,000 Btu/h (23 448 W) output. Return or supply air openings required by this section must not be less than that specified by the furnace manufacturer's installation instructions.

The exception intends to clarify that if the furnace installation instructions specify a lesser return or supply area than this section, that lesser area is permitted. The code-specified minimum area applies where the furnace manufacturer does not specify a minimum area.

G2442.3 (618.3) Dampers. Volume dampers shall not be placed in the air inlet to a *furnace* in a manner that will reduce the required air to the *furnace*.

❖ Dampers are usually avoided in return air ducts and openings because of the risk of starving the furnace for return air. If dampers are installed, the total unrestricted return air duct or opening area must be as required by the code with all of the dampers in the fully closed position.

G2442.4 (618.4) Prohibited sources. Outdoor or return air for forced-air heating and cooling systems shall not be taken from the following locations:

1. Closer than 10 feet (3048 mm) from an *appliance* vent outlet, a vent opening from a plumbing drainage system or the discharge outlet of an exhaust fan, unless the outlet is 3 feet (914 mm) above the outside air inlet.

2. Where there is the presence of objectionable odors, fumes or flammable vapors; or where located less than 10 feet (3048 mm) above the surface of any abutting public way or driveway; or where located at grade level by a sidewalk, street, alley or driveway.

3. A hazardous or insanitary location or a refrigeration machinery room as defined in the *International Mechanical Code.*

4. A room or space, the volume of which is less than 25 percent of the entire volume served by such system. Where connected by a permanent opening having an area sized in accordance with Section G2442.2, adjoining rooms or spaces shall be considered as a single room or space for the purpose of determining the volume of such rooms or spaces.

   > **Exception:** The minimum volume requirement shall not apply where the amount of return air taken from a room or space is less than or equal to the amount of supply air delivered to such room or space.

5. A room or space containing an *appliance* where such a room or space serves as the sole source of return air.

   > **Exception:** This shall not apply where:
   >
   > 1. The *appliance* is a direct-vent *appliance* or an *appliance* not requiring a vent in accordance with Section G2425.8.
   >
   > 2. The room or space complies with the following requirements:
   >
   >   2.1. The return air shall be taken from a room or space having a volume exceeding 1 cubic foot for each 10 Btu/h (9.6L/W) of combined input rating of all fuel-burning appliances therein.

2.2. The volume of supply air discharged back into the same space shall be approximately equal to the volume of return air taken from the space.

2.3. Return-air inlets shall not be located within 10 feet (3048 mm) of a draft hood in the same room or space or the combustion chamber of any atmospheric burner *appliance* in the same room or space.

3. Rooms or spaces containing solid fuel-burning appliances, provided that return-air inlets are located not less than 10 feet (3048 mm) from the firebox of such appliances.

6. A closet, bathroom, toilet room, kitchen, garage, boiler room, furnace room or unconditioned attic.

Exceptions:

1. Where return air intakes are located not less than 10 feet (3048 mm) from cooking appliances and serve only the kitchen area, taking return air from a kitchen area shall not be prohibited.

2. Dedicated forced air systems serving only a garage shall not be prohibited from obtaining return air from the garage.

7. A crawl space by means of direct connection to the return side of a forced-air system. Transfer openings in the crawl space enclosure shall not be prohibited.

❖ This section prohibits outdoor air and return air from being taken from locations that are potential sources of contamination, odor, flammable vapors or toxic substances and also from locations that would negatively affect the operation of the furnace itself or other fuel-burning appliances.

Items 1, 2 and 3 are contaminant related.

Part of the intent of Items 4 and 5 is to prevent the system from being starved for return air by the placement of the main or only return-air intake in an area not meeting the volume requirements. Also there is concern for the possibility of combustion gases leaking from fuel-fired appliances because of negative pressure in the space produced by an air handler drawing return air from such space. It is not the intent of this section to prohibit the common practice of installing return-air intakes in bedrooms and similarly sized rooms that typically have a volume that is far less than 25 percent of the total volume of the space served by the furnace. The return-air system must be able to convey the required air flow to the furnace regardless of the position of any doors to any rooms in the building served by the furnace. Item 6 is contaminate related and space pressurization related. Exception 1 recognizes that taking return air from a kitchen is of no con-

cern if the air-handling system serves only the kitchen and the intake does not interfere with cooking exhaust systems or entrain cooking effluent from cooking appliances. Exception 2 recognizes that it is illogical to prohibit return air from being taken from a garage where the air handler is serving only the garage.

Item 7 is contaminate related and recognizes that some crawl spaces are designed to be within the thermal envelope of a building.

It is the intent of this section to avoid arrangements that cause an air pressure imbalance, which can cause fuel-fired appliances to spill combustion products into the occupied space. Pressure imbalances can be avoided by making sure that the amount of supply air discharge to a room or space is approximately equal to the amount of return-air taken from the room or space and by locating return air inlets where they will not cause depressurization of an enclosed space.

G2442.5 (618.5) Screen. Required outdoor air inlets shall be covered with a screen having $^1/_4$-inch (6.4 mm) openings.

❖ The inlet openings for outdoor air must be covered with screen or mesh material with a mesh opening size as specified. The screen openings must be small enough to keep out insects, rodents, birds, etc., while being large enough to prevent blockage of air flow by lint, debris and plant fibers.

G2442.6 (618.6) Return-air limitation. Return air from one *dwelling unit* shall not be discharged into another *dwelling unit*.

❖ This section prohibits a forced-air heating/cooling system from serving more than one dwelling unit. Any arrangement in which dwelling units share all or part of an air distribution system would allow a communication of atmospheres between the units. This type of communication would spread odors, smoke, allergens, contaminants and disease-causing organisms from one dwelling unit to another and, therefore, must be avoided.

G2442.7 (618.7) Furnace plenums and air ducts. Where a *furnace* is installed so that supply ducts carry air circulated by the *furnace* to areas outside of the space containing the *furnace*, the return air shall be handled by a duct(s) sealed to the *furnace* casing and terminating outside of the space containing the *furnace*.

❖ This section is somewhat redundant with Section G2442.4 and has the same intent. That is, to prevent a furnace room, closet, alcove, etc., from developing a negative pressure with respect to the outside of the enclosure (see commentary, Section G2442.4). It is a typical design to have a furnace or air handler in a closet with a water heater and the closet door is louvered or a grille is placed in the closet wall. If none of the appliances in the closet are fuel fired, this design is harmless, but a real hazard is created if any are fuel fired.

SECTION G2443 (619)
CONVERSION BURNERS

G2443.1 (619.1) Conversion burners. The installation of *conversion burners* shall conform to ANSI Z21.8.

❖ The referenced standard is an installation standard that, in addition to the manufacturers' instructions, would govern the installation of conversion burners. Conversion burners are an assembly of components including burners, gas controls, blowers, safety devices and supporting means. These units are designed to convert an existing appliance from another fuel to gas, commonly from fuel oil or coal. Conversion of an existing appliance to a different fuel can involve much more than installation of a conversion burner. It can also include the addition of safety controls and limits, combustion air and secondary air supplies, fuel gas piping installation, chimney or vent alterations and other modifications to the existing appliance and its control system.

SECTION G2444 (620)
UNIT HEATERS

G2444.1 (620.1) General. *Unit heaters* shall be tested in accordance with ANSI Z83.8 and shall be installed in accordance with the manufacturer's instructions.

❖ Unit heaters are ductless warm-air space heaters that are self-contained and usually suspended from a ceiling or roof structure. Garages, workshops, warehouses, factories, gymnasiums, mercantile spaces and similar large, open spaces are the most common locations for unit heaters [see Commentary Figures G2444.1(1) through G2444.1(5)].

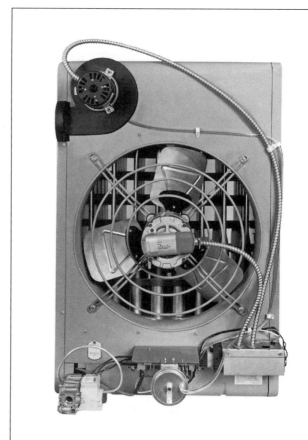

Photo courtesy of Modine Manufacturing Company

Figure G2444.1(2)
POWER-VENTED UNIT HEATER

Photo courtesy of Modine Manufacturing Company

Figure G2444.1(1)
DIRECT-VENT UNIT HEATER

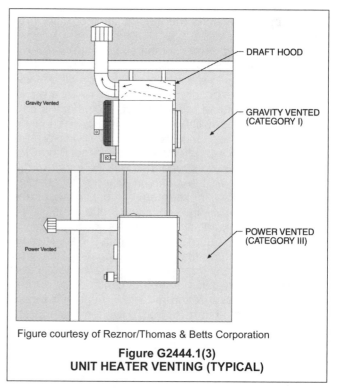

Figure courtesy of Reznor/Thomas & Betts Corporation

Figure G2444.1(3)
UNIT HEATER VENTING (TYPICAL)

SERIES 100
MODEL F

PHOTO I.D. 1997

Photo courtesy of Reznor/Thomas & Betts Corporation

**Figure G2444.1(4)
DRAFT-HOOD-EQUIPPED UNIT HEATER**

G2444.2 (620.2) Support. Suspended-type *unit heaters* shall be supported by elements that are designed and constructed to accommodate the weight and dynamic loads. Hangers and brackets shall be of noncombustible material.

❖ As with all suspended fuel-fired appliances, a support failure can result in a fire, explosion or injury to building occupants. The supports themselves must be properly designed. Equally important are the structural members to which the supports are attached, such as rafters, beams, joists and purlins. Brackets, pipes, rods, angle iron, structural members and fasteners must be designed for the dead and dynamic loads of the suspended appliance.

G2444.3 (620.3) Ductwork. Ducts shall not be connected to a unit heater unless the heater is *listed* for such installation.

❖ Unit heaters are usually not designed to move air through ductwork. Unless specifically listed for the application, the fans or blowers on unit heaters are designed only for moving air across the heat exchanger without the added friction of ductwork. The addition of ductwork could create a hazard by restricting airflow through the heater.

G2444.4 (620.4) Clearance. Suspended-type *unit heaters* shall be installed with *clearances* to *combustible materials* of not less than 18 inches (457 mm) at the sides, 12 inches (305 mm) at the bottom and 6 inches (152 mm) above the top where the unit heater has an internal *draft hood* or 1 inch (25 mm) above the top of the sloping side of the vertical *draft hood*.

Floor-mounted-type *unit heaters* shall be installed with *clearances* to *combustible materials* at the back and one side only of not less than 6 inches (152 mm). Where the *flue gases*

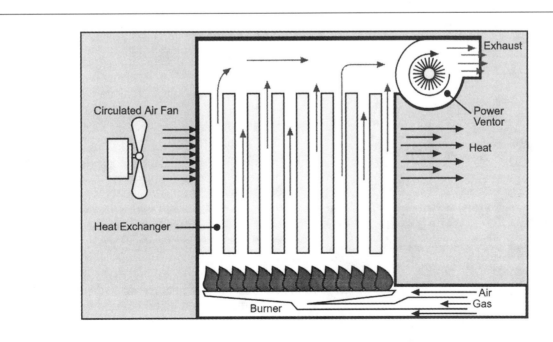

Figure courtesy of Reznor/Thomas & Betts Corporation

**Figure G2444.1(5)
FUNDAMENTAL COMPONENTS OF UNIT HEATER**

are vented horizontally, the 6-inch (152 mm) *clearance* shall be measured from the *draft hood* or *vent* instead of the rear wall of the unit heater. Floor-mounted-type *unit heaters* shall not be installed on combustible floors unless *listed* for such installation.

Clearances for servicing all *unit heaters* shall be in accordance with the manufacturer's installation instructions.

> **Exception:** *Unit heaters listed* for reduced *clearance* shall be permitted to be installed with such *clearances* in accordance with their listing and the manufacturer's instructions.

❖ The required clearances to combustible materials for suspended and floor-mounted unit heaters are specified in this section and must be adhered to unless the unit is listed for less clearance. The clearance reduction methods detailed in Section G2409 may be applied where applicable. See the commentary to Section G2409 for a complete discussion concerning clearance requirements and clearance reduction methods.

SECTION G2445 (621)
UNVENTED ROOM HEATERS

G2445.1 (621.1) General. *Unvented room heaters* shall be tested in accordance with ANSI Z21.11.2 and shall be installed in accordance with the conditions of the listing and the manufacturer's instructions.

❖ Unvented room heaters are limited-size gas-fired space heaters that discharge the combustion byproducts into the space being heated. Like all appliances regulated by this code, unvented gas-fired room heaters must be listed and labeled, and their installation must comply with the manufacturer's installation instructions [see Commentary Figures G2445.1(1) and G2445.1(2)].

Photo courtesy of Empire Comfort Systems Inc.

Figure G2445.1(1)
UNVENTED ROOM HEATER

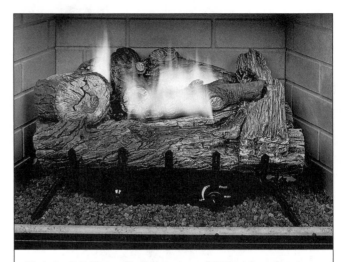

Photo by D.F. Noyes Studio, courtesy of DESA International

Figure G2445.1(2)
UNVENTED ROOM HEATER

G2445.2 (621.2) Prohibited use. One or more *unvented room heaters* shall not be used as the sole source of comfort heating in a *dwelling unit*.

❖ Unvented room heaters are designed to supplement a central heating system to allow zone heating of particular rooms and spaces. Unvented room heaters are not intended for continuous use as would occur if they were the only source of heat in a building. One or more unvented room heaters used as the sole source of heat would not provide adequate heat distribution in most building arrangements and, depending on the heating load, could require continuous operation of the appliance. Unvented gas-log heaters are listed as room heaters and their installation in factory-built fireplaces is addressed in UL 127 [see Commentary Figure G2445.1(2)]. Fireplaces built to the current edition of UL 127 will have a label and installation instructions that will either allow or disallow the installation of an unvented gas log in the fireplace firebox (see Section G2445.2).

G2445.3 (621.3) Input rating. *Unvented room heaters* shall not have an input rating in excess of 40,000 *Btu*/h (11.7 kW).

❖ The input rating limitation is consistent with the industry standard for such appliances and allows unvented room heaters to be categorized in the supplemental room heater classification.

G2445.4 (621.4) Prohibited locations. The location of *unvented room heaters* shall comply with Section G2406.2.

❖ In accordance with this section and Section G2406.2, unvented gas-fired heaters are prohibited in assembly, educational and institutional occupancies and in sleeping rooms, bathrooms and toilet rooms in all occupancies (see commentary, Section G2406.2 for exceptions). These heaters are considered to be an unacceptable risk in such occupancies because of occupant density, occupant age, occupant physical condition and awareness and small room volumes.

G2445.5 (621.5) Room or space volume. The aggregate input rating of all *unvented appliances* installed in a room or space shall not exceed 20 *Btu*/h per *cubic foot* (207 W/m^3) of volume of such room or space. Where the room or space in which the *appliances* are installed is directly connected to another room or space by a doorway, archway or other opening of comparable size that cannot be closed, the volume of such adjacent room or space shall be permitted to be included in the calculations.

❖ The Btu/h (W) input to room volume ratio limits the accumulation of combustion byproducts in the building interior. Combustion byproducts include carbon monoxide, nitrogen oxides and water vapor. The required room volume would allow dilution of the combustion byproducts by infiltration (see commentary, Section G2425.8).

G2445.6 (621.6) Oxygen-depletion safety system. *Unvented room heaters* shall be equipped with an oxygen-depletion-sensitive safety shutoff system. The system shall shut off the gas supply to the main and *pilot burners* when the oxygen in the surrounding atmosphere is depleted to the percent concentration specified by the manufacturer, but not lower than 18 percent. The system shall not incorporate field adjustment means capable of changing the set point at which the system acts to shut off the gas supply to the room heater.

❖ Because unvented heaters are not vented to the outdoors, the appliance standard requires them to incorporate this extra safety feature. This safety system consists of a special pilot burner device that is incorporated with the appliance's flame safeguard device (see commentary, Section G2432.2). The oxygen-depletion sensor is basically a pilot burner that is extremely sensitive to the oxygen content in the combustion air. See definition of oxygen depletion safety shutoff system. If the oxygen content (approximately 20 percent normal) drops to a predetermined level, the pilot flame will become "lazy," shifting from a stable horizontal flame to a less stable, more vertical flame, which is incapable of sufficiently heating the thermocouple or thermopile generator, resulting in main gas control valve shutdown and lockout. The predetermined oxygen level that activates burner shutdown is above the level at which incomplete combustion would start to occur.

The monitoring of oxygen levels in this manner is an indicator of approaching insufficient combustion air and is designed to prevent the formation of excess carbon monoxide. If the oxygen level reaches a low enough percentage of air, the level of oxygen will have an inverse relationship with the level of carbon monoxide. This means that as burners are increasingly starved for oxygen, the amount of carbon monoxide produced by the flames is increased. This relationship is valid for appliances in good working order. An improperly maintained, maladjusted or defective burner could produce abnormal (elevated) amounts of carbon monoxide regardless of the oxygen level in the combustion air, and in such cases, the oxygen-depletion sensor device will react only to the oxygen level because it cannot directly detect burner malfunction or abnormal carbon monoxide production.

G2445.7 (621.7) Unvented decorative room heaters. An unvented decorative room heater shall not be installed in a *factory-built fireplace* unless the *fireplace* system has been specifically tested, *listed* and *labeled* for such use in accordance with UL 127.

❖ Because the fireplace chimney damper can be closed while an unvented decorative (log) heater (room heater) is operating, the firebox might reach surface temperatures higher than would occur with the damper open. The lack of chimney draft with the resultant dilution air introduction into the firebox would allow higher temperatures to occur in the fireplace components. To prevent surface temperatures from exceeding that allowed by UL 127, that standard now requires factory-built fireplace manufacturers to either test their units for use with unvented log heaters and meet the standard criteria or provide instructions and labels that prohibit such use.

If a factory-built fireplace has not been tested for use with an unvented appliance, the installation instructions and labels must state that the use is prohibited.

Fireplaces manufactured before the UL standard added coverage for unvented decorative (log) heaters could not have been tested for that use, and the manufacturers did not have the opportunity to include instructions and labels covering unvented log heaters for use with their units. The installation instructions for fireplace units that were built prior to the unvented heater coverage being added to the UL 127 standard stated that the fireplace was permitted for use only with solid fuel and decorative gas logs listed to ANSI Z21.60. Therefore, regardless of when the fireplace unit was manufactured, unvented decorative heaters can be installed only in factory-built fireplaces that specifically state that they are tested for use with unvented decorative heaters. The fireplace manufacturers will currently indicate whether or not their units are intended for use with unvented decorative heaters. If listed for the application, an unvented decorative (log) heater can be used as a vented decorative appliance by opening the fireplace chimney damper.

G2445.7.1 (621.7.1) Ventless firebox enclosures. Ventless firebox enclosures used with unvented decorative room heaters shall be *listed* as complying with ANSI Z21.91.

❖ Ventless firebox enclosures are designed and tested to accommodate the appliances addressed in Section G2445.7. Such enclosures have no chimney or vent connection (see Commentary Figure G2445.7.1).

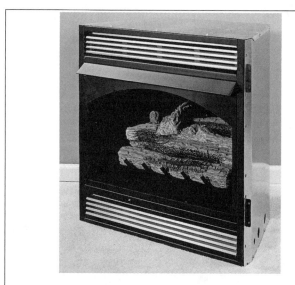

Photography by D.F. Noyes Studio, courtesy of DESA International

Figure G2445.7.1
FIREBOX FOR UNVENTED ROOM HEATER

SECTION G2446 (622)
VENTED ROOM HEATERS

G2446.1 (622.1) General. *Vented room heaters* shall be tested in accordance with ANSI Z21.86/CSA 2.32, shall be designed and equipped as specified in Section G2432.2 and shall be installed in accordance with the manufacturer's instructions.

❖ This section addresses vented gas-fired space/room heaters, including direct-vent and vent- or chimney-connected appliances, which are limited-size room heaters designed to connect to a vent or chimney. As required by the appliance standard, these appliances are equipped with safety controls that will prevent gas flow to the burners in the event of ignition system failure (see Commentary Figure G2446.1).

Photo courtesy of Maytag Corporation

Figure G2446.1
VENTED ROOM HEATER

SECTION G2447 (623)
COOKING APPLIANCES

G2447.1 (623.1) Cooking appliances. Cooking *appliances* that are designed for permanent installation, including ranges, ovens, stoves, broilers, grills, fryers, griddles, hot plates and barbecues, shall be tested in accordance with ANSI Z21.1 or ANSI Z21.58 and shall be installed in accordance with the manufacturer's instructions.

❖ This section addresses cooking appliances in all occupancies that are designed for permanent installation, including, but not limited to, ranges, ovens, stoves, broilers, grills, fryers, griddles and barbecues. These appliances must be installed in accordance with the listing and the manufacturer's installation instructions.

The code intends to regulate the design, construction and installation of cooking appliances that are designed for permanent installation-heated countertop appliances. Appliances that are not readily moveable to another location because of a gas-fuel supply connection would be considered permanently installed, even if they were on casters. Line equipment under a Type I hood, for example, is usually on casters and connected with quick-disconnect-type fuel supply lines to allow movement for routine cleaning. This kind of equipment would be considered permanently installed.

G2447.2 (623.2) Prohibited location. Cooking appliances designed, tested, *listed* and *labeled* for use in commercial occupancies shall not be installed within dwelling units or within any area where domestic cooking operations occur.

Exception: Appliances that are also listed as domestic cooking appliances.

❖ Commercial cooking appliances are tested and labeled to different standards than those listed for domestic use. Commercial cooking appliances generally are not insulated to the same level, have higher surface operating temperatures and require a much greater clearance to combustible material. The safety measures inherent to household cooking appliances, such as child-safe push-to-turn knobs and insulated oven doors, are not usually found in commercial cooking appliances. Commercial cooking appliances also have a greater ventilation air requirement for safe operation than household-type cooking appliances. For these reasons, the installation of commercial-type cooking appliances in dwellings is prohibited. Note that many cooking appliances are dual-listed as both commercial and domestic (household) cooking appliances. The code does not prohibit such dual-listed appliances in dwelling units because such appliances have the safety features that are necessary for the domestic environment (see Commentary Figure G2447.2).

G2447.3 (623.3) Domestic appliances. Cooking *appliances* installed within *dwelling units* and within areas where domestic cooking operations occur shall be *listed* and *labeled* as household-type *appliances* for domestic use.

❖ Cooking appliances used in dwelling units or in areas where domestic cooking operations occur require a greater degree of user protection and must be listed and labeled as household-type appliances for domestic use (see commentary, Section G2447.2). To satisfy residential consumer demand for commercial appliances in the home, some manufacturers are producing listed household-type appliances that have the appearance of commercial application cooking appliances [see Commentary Figures G2447.3(1) and G2447.3(2)].

G2447.4 (623.4) Range installation. Ranges installed on combustible floors shall be set on their own bases or legs and shall be installed with *clearances* of not less than that shown on the label.

❖ This section requires the installation of a domestic range using the legs or base provided by the manufacturer where the range is resting on a combustible surface. These supports were part of the test set-up when the range was tested and must be included in the installation to ensure compliance with the listing. Clearances behind and to the sides of the range must also comply with the listing.

Figure G2447.2
COOKING RANGE IN DWELLING

G2447.5 (623.7) Vertical clearance above cooking top. Household cooking *appliances* shall have a vertical *clearance* above the cooking top of not less than 30 inches (760 mm) to *combustible material* and metal cabinets. A minimum *clearance* of 24 inches (610 mm) is permitted where one of the following is installed:

1. The underside of the *combustible material* or metal cabinet above the cooking top is protected with not less than

Photo courtesy of Maytag Corporation

Figure G2447.3(1)
DOMESTIC (HOUSEHOLD) RANGE

Figure courtesy of Maytag Corporation

Figure G2447.3(2)
HOUSEHOLD-TYPE COOKING APPLIANCE LABEL

$^1/_4$-inch (6 mm) insulating millboard covered with sheet metal not less than 0.0122 inch (0.3 mm) thick.

2. A metal ventilating hood constructed of sheet metal not less than 0.0122 inch (0.3 mm) thick is installed above the cooking top with a *clearance* of not less than $^1/_4$ inch (6 mm) between the hood and the underside of the *combustible material* or metal cabinet. The hood shall have a width not less than the width of the *appliance* and shall be centered over the *appliance*.

3. A *listed* cooking *appliance* or microwave oven is installed over a *listed* cooking *appliance* and in compliance with the terms of the manufacturer's installation instructions for the upper *appliance*.

❖ This section addresses the fire hazard created by having hot surfaces, open flames and, possibly, a cooking fire located under combustible cabinets, soffits, etc., and also under metal cabinets with combustible contents (see Commentary Figure G2447.5).

SECTION G2448 (624)
WATER HEATERS

G2448.1 (624.1) General. Water heaters shall be tested in accordance with ANSI Z21.10.1 and ANSI Z21.10.3 and shall be installed in accordance with the manufacturer's instructions.

❖ Water heaters are recognized by the codes as both a plumbing and a mechanical appliance. Chapter 5 of the IPC and Chapter 10 of the IMC contain more detailed information concerning the testing and installation of water heaters than is given in this code. This section identifies those aspects of water heaters specifically related to fuel gas. The IMC addresses those aspects related to other fuel or power sources.

The standards listed are specific to gas-fired water heaters. ANSI Z21.10.1 is for water heaters with an input rating less than or equal to 75,000 Btu/h (22 kw). ANSI Z21.10.3 is for water heaters with an input rating greater than 75,000 Btu/h (22 kW) and for circulating and instantaneous water heaters.

G2448.1.1 (624.1.1) Installation requirements. The requirements for *water heaters* relative to sizing, *relief valves*, drain pans and scald protection shall be in accordance with this code.

❖ This section refers the user to the IPC for generic requirements related to water heaters not related to the fuel source. These include connections to the potable water system, safety devices such as relief valves, drain pans for protection of the structure and sizing of the water heaters.

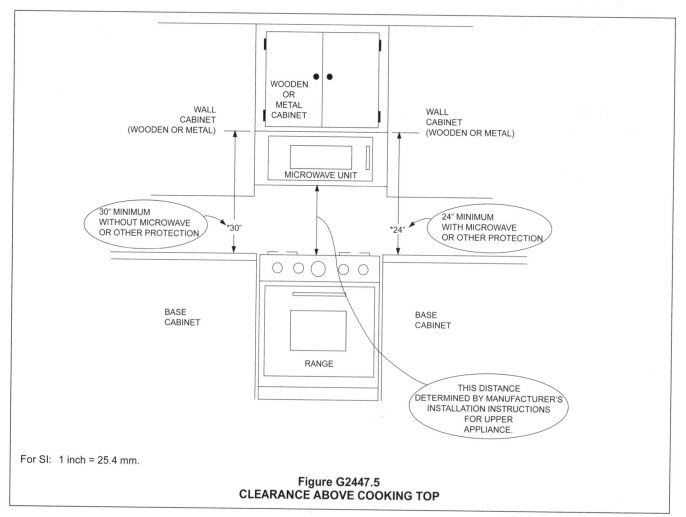

For SI: 1 inch = 25.4 mm.

Figure G2447.5
CLEARANCE ABOVE COOKING TOP

G2448.2 (624.2) Water heaters utilized for space heating. *Water heaters* utilized both to supply potable hot water and provide hot water for space-heating applications shall be *listed* and *labeled* for such applications by the manufacturer and shall be installed in accordance with the manufacturer's instructions and this code.

❖ Water heaters serving the dual purpose of supplying potable hot water and serving as a heat source for a space-heating system must be listed and labeled for that dual application. This section does not address water heaters used solely for space-heating applications, but rather addresses water heaters that serve a secondary purpose of space heating. The label will indicate whether the water heater is suitable for space heating [see Commentary Figures G2448.2(1) and (2)].

The plumbing chapters contain additional requirements for the proper installation of these appliances.

SECTION G2449 (627)
AIR-CONDITIONING APPLIANCES

G2449.1 (627.1) General. Gas-fired air-conditioning *appliances* shall be tested in accordance with ANSI Z21.40.1 or ANSI Z21.40.2 and shall be installed in accordance with the manufacturer's instructions.

❖ Gas-fired air conditioning systems include absorption types and internal-combustion-engine-driven machines. The referenced American National Standards Institute (ANSI) standards contain requirements for gas-fired air conditioning equipment. ANSI Z21.40.1 regulates gas-fired absorption systems; ANSI Z21.40.2 regulates work-activated gas-fired air conditioning and heat pump systems. The code relies on the manufacturer's installation instructions and the referenced standards for installation requirements.

Figure G2448.2(1)
DIRECT-VENT WATER HEATER DESIGNED FOR BOTH POTABLE WATER HEATING AND SPACE HEATING APPLICATIONS

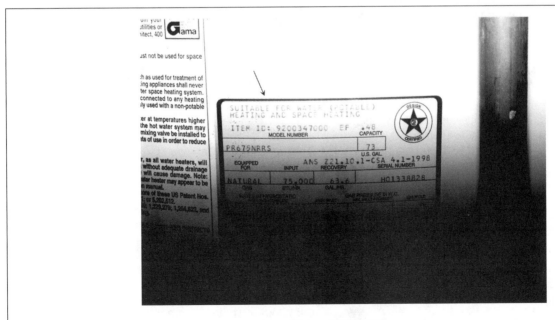

Figure G2448.2(2)
WATER HEATER LABEL INDICATING SUITABILITY FOR BOTH POTABLE WATER HEATING AND SPACE HEATING

G2449.2 (627.2) Independent piping. *Gas piping* serving heating *appliances* shall be permitted to also serve cooling *appliances* where such heating and cooling *appliances* cannot be operated simultaneously (see Section G2413).

❖ Where gas supply piping serves both heating and cooling equipment, this section permits the piping to be sized based on the larger gas demand of the two systems rather than sizing based on the total demand of the two systems. This provision is not applicable unless the heating and cooling systems are incapable of operating simultaneously.

G2449.3 (627.3) Connection of gas engine-powered air conditioners. To protect against the effects of normal vibration in service, gas engines shall not be rigidly connected to the gas supply *piping*.

❖ Because of the inherent vibration of a gas engine, a connector must be installed between the engine and the gas supply pipe so that the vibration does not place undue stress on the connector or have an adverse effect on the shutoff valve or the gas piping. Vibration transmitted to the connector and piping could eventually cause joint failure and/or connector failure. The connector must be designed and approved for this application.

G2449.4 (627.6) Installation. Air conditioning *appliances* shall be installed in accordance with the manufacturer's instructions. Unless the *appliance* is *listed* for installation on a combustible surface such as a floor or roof, or unless the surface is protected in an *approved* manner, the *appliance* shall be installed on a surface of noncombustible construction with *noncombustible material* and surface finish and with no *combustible material* against the underside thereof.

❖ The intent of this section is to prohibit the installation of gas-fired air-conditioning equipment on combustible surfaces unless specifically listed for such an installation or the surface is protected in a manner approved by the code official.

SECTION G2450 (628)
ILLUMINATING APPLIANCES

G2450.1 (628.1) General. Illuminating *appliances* shall be tested in accordance with ANSI Z21.42 and shall be installed in accordance with the manufacturer's instructions.

❖ Illuminating appliances must be listed and labeled as required for other appliances regulated by this code. These appliances include gas lamps designed for outdoor use.

G2450.2 (628.2) Mounting on buildings. Illuminating *appliances* designed for wall or ceiling mounting shall be securely attached to substantial structures in such a manner that they are not dependent on the *gas piping* for support.

❖ Gas piping is not designed to act as a support for any appliance. All appliances must be independently sup-

ported to prevent stresses and strains on fuel supply connections.

G2450.3 (628.3) Mounting on posts. Illuminating *appliances* designed for post mounting shall be securely and rigidly attached to a post. Posts shall be rigidly mounted. The strength and rigidity of posts greater than 3 feet (914 mm) in height shall be at least equivalent to that of a 2$\frac{1}{2}$-inch-diameter (64 mm) post constructed of 0.064-inch-thick (1.6 mm) steel or a 1-inch (25 mm) Schedule 40 steel *pipe*. Posts 3 feet (914 mm) or less in height shall not be smaller than a $\frac{3}{4}$-inch (19.1 mm) Schedule 40 steel *pipe*. Drain openings shall be provided near the base of posts where there is a possibility of water collecting inside them.

❖ Regardless of the intended method of mounting and support of an appliance, the installation must be secure and must not transmit any loading to the fuel supply connection. A requirement is added to provide a drain at the base of the post to prevent collection of water that can deteriorate the post material.

G2450.4 (628.4) Appliance pressure regulators. Where an *appliance pressure regulator* is not supplied with an illuminating *appliance* and the service line is not equipped with a *service pressure regulator*, an *appliance pressure regulator* shall be installed in the line to the illuminating *appliance*. For multiple installations, one *regulator* of adequate capacity shall be permitted to serve more than one illuminating *appliance*.

❖ Outdoor lighting appliances generally require low-pressure gas flow. Some of these lighting appliances are supplied by the manufacturer with a built-in pressure regulator set for the proper operating pressure. If, however, the regulator is not provided, this section requires the installation of one in the gas line serving the appliance. This regulator must be appropriate for the pressure required by the manufacturer. One pressure regulator with the required capacity may be installed to serve more than one lighting appliance if the piping pressure losses permit.

SECTION G2451 (630)
INFRARED RADIANT HEATERS

G2451.1 (630.1) General. Infrared radiant heaters shall be tested in accordance with ANSI Z83.19 or Z83.20 and shall be installed in accordance with the manufacturer's instructions.

❖ This section addresses radiant heaters including ceramic element and steel tube-type designs. Infrared heaters are produced in both vented and unvented types and function by creating a very hot surface area from which heat energy is directly radiated. Infrared radiant heaters are usually suspended from ceilings or roofs. These heaters are typically used for "spot" heating in spaces that are otherwise unconditioned or that are not conditioned to human comfort levels. Radiant heaters have the advantage of being able to heat objects and personnel without having to heat the surrounding air [see Commentary Figures G2451.1(1) through G2451.1(3)].

The code relies on the manufacturers' installation instructions for installation requirements. Maintaining the clearance to combustibles for radiant heaters is of paramount importance. Direct radiation is a very effective method of transferring heat energy, and improperly located combustible materials can be readily ignited [see Commentary Figure G2451.1(4)].

G2451.2 (630.2) Support. *Infrared radiant heaters* shall be fixed in a position independent of gas and electric supply lines. Hangers and brackets shall be of *noncombustible material.*

❖ Supports must prevent radiant heaters from falling. A fall would mean losing the required clearance to combustibles; putting tension or pressure on electrical, fuel and vent connections; and dislocating that redirects the radiant output to where a fire hazard would result. Often, such heaters are improperly hung and restrained, allowing heater movement (swinging) to stress flexible (semirigid) gas connectors. Gas connectors must be designed and approved for applications in which the appliance moves because of expansion and contraction or lack of restraints. To accommodate expansion/contraction movement, some appliance manufacturers require flexible connectors to be coiled or shaped in a manner that allows limited movement without stressing the metal of the connector, similar to expansion loops and offsets in rigid piping systems.

SECTION G2452 (631)
BOILERS

G2452.1 (631.1) Standards. Boilers shall be *listed* in accordance with the requirements of ANSI Z21.13 or UL 795. If applicable, the boiler shall be designed and constructed in accordance with the requirements of ASME CSD-1 and as applicable, the ASME *Boiler and Pressure Vessel Code*, Sections I, II, IV, V and IX and NFPA 85.

❖ The scope of this section includes boilers in all occupancies including power plants; factories; industrial plants; schools; and institutional occupancies such as hospitals, commercial laundries, hotels and residential structures. Boilers are defined in Section G2403 of this code (see Commentary Figure G2452.1).

Boilers are potentially dangerous if not properly designed, constructed and operated, more so than many other appliances because of the potential explosion hazard associated with pressure vessels. In addition to this code, several industry standards are referenced. Manufactured gas-fired boilers must be listed to either ANSI Z21.13 or UL 795. Commentary Figure G2403(3) shows a typical label with the required information for a listed boiler. Only the design and construction requirements of boilers are regulated by the referenced standards. The requirements for specific types of boilers can be found in the respective sections of the referenced standards.

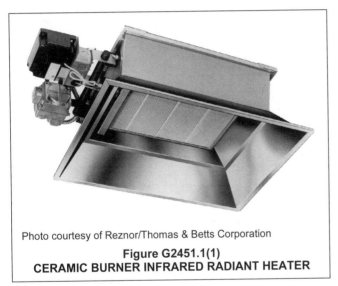

Photo courtesy of Reznor/Thomas & Betts Corporation

Figure G2451.1(1)
CERAMIC BURNER INFRARED RADIANT HEATER

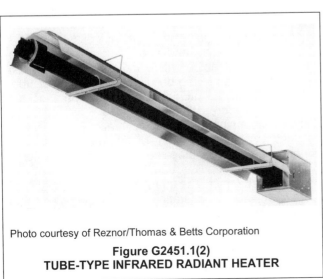

Photo courtesy of Reznor/Thomas & Betts Corporation

Figure G2451.1(2)
TUBE-TYPE INFRARED RADIANT HEATER

Photo courtesy of Reznor/Thomas & Betts Corporation

Figure G2451.1(3)
TUBE-TYPE INFRARED RADIANT HEATER

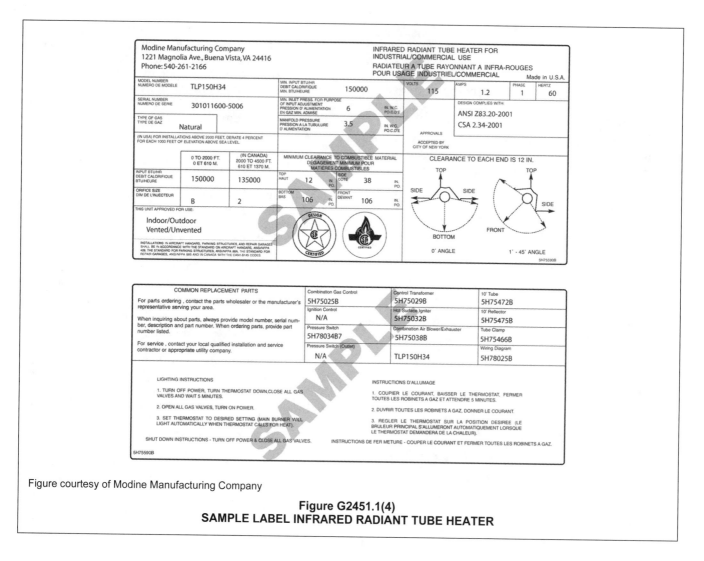

Figure courtesy of Modine Manufacturing Company

Figure G2451.1(4)
SAMPLE LABEL INFRARED RADIANT TUBE HEATER

G2452.2 (631.2) Installation. In addition to the requirements of this code, the installation of boilers shall be in accordance with the manufacturer's instructions. Operating instructions of a permanent type shall be attached to the boiler. Boilers shall have all *controls* set, adjusted and tested by the installer. A complete *control* diagram together with complete boiler operating instructions shall be furnished by the installer. The manufacturer's rating data and the nameplate shall be attached to the boiler.

❖ This section governs the installation and commissioning of boilers and their control systems. The mechanical equipment requirements for approval, labeling, installation, maintenance, repair and alteration are regulated by the IMC and the manufacturer's instructions. The IMC contains installation requirements for boilers including provisions for shutoff valves, pressure relief valves, safety valves, electrical control wiring, blowoff valves, expansion tanks and low water cutoff controls [see Commentary Figures G2452.1 and G2452.2(1) through G2452.2(3)].

Complete operating instructions must be permanently affixed to the boiler upon completion of the installation. Boiler systems can be complex and generally require coordination of several pieces of equipment. The proper operating procedures, set points, etc., must be specified and included in these instructions. This is usually done in the control system design documentation, including such things as diagrams, system schematics and control sequence descriptions. Typically, an operating and maintenance (O&M) manual is given to the building owner/operator upon completion of the project. Along with operating and control procedures, the O&M manual will clearly identify routine maintenance and calibration information. It is also typical for operation sequences and calibration information to be displayed under glass in a conspicuous place in the boiler room for use by operating and service personnel. The intent is to provide the owner/operator with all operating and control information necessary to properly operate and maintain the boiler and its associated controls and equipment.

Photo courtesy of Weil-McLain

**Figure G2452.1
STEAM HEATING BOILER WITH GAUGE GLASS,
LOW WATER CUTOFF CONTROL, PRESSURE
GAUGE, PRESSURE LIMIT CONTROL
AND AUTOMATIC VENT DAMPER**

Photo courtesy of McDonnell & Miller

**Figure G2452.2(1)
ELECTRONIC PROBE-TYPE LOW-WATER
CUTOFF CONTROL WITH GAUGE GLASS
AND TRI-COCKS FOR STEAM BOILER APPLICATION**

Photo courtesy of McDonnell & Miller

**Figure G2452.2(2)
ELECTRONIC PROBE-TYPE LOW-WATER CUTOFF
CONTROL FOR HOT WATER BOILER APPLICATION**

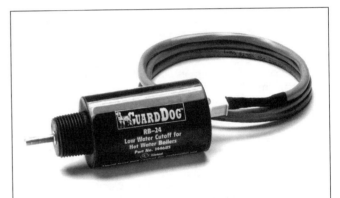

Photo courtesy of McDonnell & Miller

**Figure G2452.2(3)
ELECTRONIC PROBE-TYPE LOW-WATER CUTOFF
CONTROL FOR HOT WATER BOILER APPLICATION**

G2452.3 (631.3) Clearance to combustible material. *Clearances* to *combustible materials* shall be in accordance with Section G2409.4.

❖ Boilers, like all gas-fired equipment and appliances, operate at high temperatures that require adequate separation from combustibles. This section refers the user to Section G2409 of this code for requirements for clearances to combustibles and allowable clearance reduction methods.

**SECTION G2453 (634)
CHIMNEY DAMPER OPENING AREA**

G2453.1 (634.1) Free opening area of chimney dampers. Where an unlisted decorative *appliance* for installation in a vented *fireplace* is installed, the *fireplace damper* shall have a permanent free opening equal to or greater than specified in Table G2453.1.

❖ This section intends to ensure that an adequate fireplace damper opening is installed for all unlisted decorative appliances. The minimum damper opening size is part of the listing for listed appliances, but the installer and the code official might not have access to this information for an unlisted appliance. Section G2404.3 requires all appliances to be listed.

The free opening size is a function of the height of the chimney and the appliance input rating. As stated in note "a," the opening areas in the table correspond to the cross-sectional area of round chimney flues having nominal sizes from 3 to 8 inches (76 to 203 mm), and the 64 square inches (41 290 mm^2) represents the cross-sectional area of a standard 8- by 8-inch chimney tile.

SECTION G2454 (636)
OUTDOOR DECORATIVE APPLIANCES

G2454.1 (636.1) General. Permanently fixed-in-place outdoor decorative appliances shall be tested in accordance with ANSI Z21.97 and shall be installed in accordance with the manufacturer's instructions.

❖ There are many different types of gas-fired decorative appliances that are for outdoor use only, such as simulated camp fire units and table-top decorative display burner units. Such appliances are popular for outdoor living areas, patios, outdoor kitchens, etc. (see Commentary Figure G2454.1).

Bibliography

The following resource materials were used in the preparation of the commentary for this chapter of the code:

IBC–15, *International Building Code*. Washington, DC: International Code Council, 2014.

IFGC–15, *International Fuel Gas Code*. Washington, DC: International Code Council, 2014.

IMC–15, *International Mechanical Code*. Washington DC: International Code Council, 2014.

IPC–15, *International Plumbing Code*. Washington, DC: International Code Council, 2014.

Photo courtesy of Hearth and Home Technologies. Inc.

Figure G2454.1
OUTDOOR (CAMPFIRE) DECORATIVE APPLIANCE

TABLE G2453.1 (634.1)
FREE OPENING AREA OF CHIMNEY DAMPER FOR VENTING FLUE GASES
FROM UNLISTED DECORATIVE APPLIANCES FOR INSTALLATION IN VENTED FIREPLACES

CHIMNEY HEIGHT (feet)	MINIMUM PERMANENT FREE OPENING (square inches)[a]						
	8	13	20	29	39	51	64
	Appliance input rating (Btu per hour)						
6	7,800	14,000	23,200	34,000	46,400	62,400	80,000
8	8,400	15,200	25,200	37,000	50,400	68,000	86,000
10	9,000	16,800	27,600	40,400	55,800	74,400	96,400
15	9,800	18,200	30,200	44,600	62,400	84,000	108,800
20	10,600	20,200	32,600	50,400	68,400	94,000	122,200
30	11,200	21,600	36,600	55,200	76,800	105,800	138,600

For SI: 1 inch = 25.4 mm, 1 foot = 304.8 mm, 1 square inch = 645.16 mm^2, 1,000 Btu per hour = 0.293 kW.

a. The first six minimum permanent free openings (8 to 51 square inches) correspond approximately to the cross-sectional areas of chimneys having diameters of 3 through 8 inches, respectively. The 64-square-inch opening corresponds to the cross-sectional area of standard 8-inch by 8-inch chimney tile.

Chapter 25:
Plumbing Administration

User note: Code change proposals to this chapter will be considered by the IRC – Plumbing and Mechanical Code Development Committee during the 2015 (Group A) Code Development Cycle.

General Comments

Chapter 25 contains administrative requirements that apply to the installation of plumbing systems and to the ensuing plumbing related chapters of the code. Chapter 25 refers the user to the applicable provisions of Chapter 1 and addresses the code requirements' applicability to additions, alterations and repairs. This chapter also outlines the building official's authority and discretion regarding existing construction that is used in conjunction with new construction.

Chapter 25 lists the types and phases of various inspections as well as detailing the permit holder's responsibilities for scheduling and performing tests.

Purpose

The requirements of Chapter 25 do not supersede the administrative provisions of Chapter 1. Rather, the administrative guidelines of Chapter 25 pertain to plumbing installations that are best referenced and located within the plumbing chapters. This chapter addresses how to apply the plumbing provisions of this code to specific types or phases of construction. This chapter also outlines the responsibilities of the applicant, the installer and the inspector with regard to testing plumbing installations.

SECTION P2501
GENERAL

P2501.1 Scope. The provisions of this chapter shall establish the general administrative requirements applicable to plumbing systems and inspection requirements of this code.

❖ This section states that the provisions of Chapter 25 set forth the general administration requirements for plumbing systems.

P2501.2 Application. In addition to the general administration requirements of Chapter 1, the administrative provisions of this chapter shall also apply to the plumbing requirements of Chapters 25 through 32.

❖ Section P2501.2 refers to the applicability of the administrative requirements of Chapter 1 and of plumbing Chapters 25 through 33.

SECTION P2502
EXISTING PLUMBING SYSTEMS

P2502.1 Existing building sewers and building drains. Where the entire sanitary drainage system of an existing building is replaced, existing *building drains* under concrete slabs and existing *building sewers* that will serve the new system shall be internally examined to verify that the piping is sloping in the correct direction, is not broken, is not obstructed and is sized for the drainage load of the new plumbing drainage system to be installed.

❖ A home could be so significantly damaged from fire or wind that the home must be completely removed in order for another home to be constructed. Or, the

repair of an older home might be so cost prohibitive that it is more cost effective to tear it down and build a new home on the same lot. The routing of the existing building sewer might be complicated by existing driveways, trees or accessory buildings. In some jurisdictions, the homeowner is responsible for their building sewer as it passes into a right-of-way, such as crossing a street en route to the public sewer main on the other side. In other situations, the homeowner's building sewer might cross another owner's property.

Another situation might be that a concrete slab-on-grade building is completely replaced using the existing slab as the foundation for the new building. The existing building drain system could still be intact and in good condition such that the slab would not have to be cut up to replace the building drain.

Although the code does not specifically state that new-built homes must be served by a new building sewer (and a new building drain), some code officials might interpret the code as requiring an all new building drain and building sewer. In the scenarios given in the previous two paragraphs, replacement of building drains and building sewers might be very costly. Furthermore, the replacement might be completely unnecessary. Why tear out good, serviceable building drains and building sewers just for sake of replacing with new material? The only way to know if existing building drains and existing building sewers are serviceable is to internally examine the piping for problems. The only way to properly examine this piping is by an internal video camera survey.

Video pipe surveys can identify reverse slopes, sags, breaks, tree root intrusions, illegal taps, diameter changes and the overall condition of the piping so that an evaluation can be made as to whether to repair or replace the piping.

P2502.2 Additions, alterations or repairs. Additions, *alterations*, renovations or repairs to any plumbing system shall conform to that required for a new plumbing system without requiring the existing plumbing system to comply with the requirements of this code. Additions, *alterations* or repairs shall not cause an existing system to become unsafe, insanitary or overloaded.

Minor additions, *alterations*, renovations and repairs to existing plumbing systems shall be permitted in the same manner and arrangement as in the existing system, provided that such repairs or replacement are not hazardous and are *approved*.

❖ Simply stated, new work must comply with current code requirements. Any alteration or addition to an existing system involves some new work, and therefore is subject to the requirements of the code. Additions or alterations to an existing system can place additional loads or different demands on the system, which could necessitate changing all or part of the existing system. For example, the addition of plumbing fixtures to an existing system may necessitate an increase in drain piping size and water distribution piping size. Additions and alterations must not cause an existing system to be any less in compliance with the code than it was before the changes.

Repair of an existing nonconforming plumbing system is permitted without having to completely replace the nonconforming portion. This typically occurs when repairing a fixture or piping. Although some types of fixtures or piping arrangements are no longer permitted, existing fixtures or piping can be repaired and remain in service if a health hazard or insanitary condition is not maintained or created. This section distinguishes between alterations (subject to applicable provisions of the code) and ordinary repairs (maintenance activities not requiring a permit). The intent of this section is to allow the continued use of existing plumbing systems and equipment that might not be designed and constructed as required for new installations.

Existing plumbing systems and equipment will normally require repair and component replacement to remain operational. This section permits repair and component replacements without requiring the redesign, alteration or replacement of the entire system. In other words, the plumbing system is allowed to stay as it was if it is not hazardous. It is important to note that the word "minor" in this section is intended to modify "additions," "alterations," "renovations" and "repairs." It is not the intent of this section to waive code requirements for the replacement of all or major portions of systems under the guise of repair. Any work other than minor repairs or replacement of minor portions of a system must be considered as new work subject to all applicable provisions of the code. Repairs and minor component replacements are permitted in a manner that is consistent with the existing system if those repairs or replacements are approved by the code official; are not hazardous; do not cause the system or equipment to be any less in compliance with the code than before; and are, to the extent practicable, in compliance with the provisions of the code applicable to new work.

SECTION P2503
INSPECTION AND TESTS

P2503.1 Inspection required. New plumbing work and parts of existing systems affected by new work or *alterations* shall be inspected by the *building official* to ensure compliance with the requirements of this code.

❖ This section establishes the building official's authority to conduct inspections of plumbing work to determine code compliance. The code requires inspection and approval of new work and existing systems that could be affected by new work. This provision allows the building official to inspect existing systems as part of the overall approval process.

P2503.2 Concealment. A plumbing or drainage system, or part thereof, shall not be covered, concealed or put into use until it has been tested, inspected and *approved* by the *building official*.

❖ It is the responsibility of the contractor, the builder, the owner or other authorized party to arrange for the required inspections and to coordinate them to prevent work from being concealed or put into use before the work is inspected, tested and approved by the authority having jurisdiction.

P2503.3 Responsibility of permittee. Test equipment, materials and labor shall be furnished by the permittee.

❖ The permit holder is responsible for supplying all of the labor, equipment and apparatus necessary to conduct such tests. The code official only observes the tests being performed.

P2503.4 Building sewer testing. The *building sewer* shall be tested by insertion of a test plug at the point of connection with the public sewer, filling the *building sewer* with water and pressurizing the sewer to not less than 10-foot (3048 mm) head of water. The test pressure shall not decrease during a period of not less than 15 minutes. The *building sewer* shall be watertight at all points.

A forced sewer test shall consist of pressurizing the piping to a pressure of not less than 5 psi (34.5 kPa) greater than the pump rating and maintaining such pressure for not less than 15 minutes. The forced sewer shall be water tight at all points.

❖ This section requires that the gravity building sewer be plugged at the point of connection to the public sewer and tested with 10 feet (3048 mm) of water for 15 minutes without leaking. Even though such a test applies a relatively small amount of pressure [approximately 5 pounds per square inch (psi) (34 kPa)], it is sufficient for unpressurized piping applications.

The second paragraph of this section is for forced sewers (i.e., pressurized building sewers). The piping must be pressurized to not less than 5 psi greater than the rating of the pump that creates the pressure to move the sewage to the public sewer. The pump curve (or data from the manufacturer) must be consulted to determine the greatest pressure that the pump can generate.

P2503.5 Drain, waste and vent systems testing. Rough-in and finished plumbing installations of drain, waste and vent systems shall be tested in accordance with Sections P2503.5.1 and P2503.5.2.

❖ Section P2503.5 indicates that drainage and venting systems are to be tested once during the rough-in phase and again after the plumbing fixtures have been set and traps installed.

P2503.5.1 Rough plumbing. DWV systems shall be tested on completion of the rough piping installation by water or, for piping systems other than plastic, by air, without evidence of leakage. Either test shall be applied to the drainage system in its entirety or in sections after rough-in piping has been installed, as follows:

1. Water test. Each section shall be filled with water to a point not less than 5 feet (1524 mm) above the highest fitting connection in that section, or to the highest point in the completed system. Water shall be held in the section under test for a period of 15 minutes. The system shall prove leak free by visual inspection.

2. Air test. The portion under test shall be maintained at a gauge pressure of 5 pounds per square inch (psi) (34 kPa) or 10 inches of mercury column (34 kPa). This pressure shall be held without introduction of additional air for a period of 15 minutes.

❖ This section describes the procedures for performing a water or air test of the drainage, waste and vent (DWV) system. Note that air cannot be used for testing plastic piping systems. The water test is performed during the rough-in phase to determine whether the plumbing system complies with the code and operates as intended. This test allows identification and repair of any defects before the system is concealed within the building. Except for the top 5 feet (1524 mm) of the system, each section must be tested with not less than a 5-foot (1524 mm) head of water for a period of 15 minutes. The system must be leak free at all points.

As an alternative, this section allows an air-pressure test of the DWV system. Note that air cannot be used for testing plastic piping systems. The system must hold 5 psi (34 kPa) pressure throughout for a period of at least 15 minutes without introducing additional air. Testing DWV systems with air is inherently more dangerous to personnel than testing with water because of the energy release that can occur if the system fails or ruptures. Increasing the test pressure beyond the material's safety limits poses a hazard.

P2503.5.2 Finished plumbing. After the plumbing fixtures have been set and their traps filled with water, their connec-

tions shall be tested and proved gas tight or water tight as follows:

1. Water tightness. Each fixture shall be filled and then drained. Traps and fixture connections shall be proven water tight by visual inspection.

2. Gas tightness. Where required by the local administrative authority, a final test for gas tightness of the DWV system shall be made by the smoke or peppermint test as follows:

   2.1. Smoke test. Introduce a pungent, thick smoke into the system. When the smoke appears at vent terminals, such terminals shall be sealed and a pressure equivalent to a 1-inch water column (249 Pa) shall be applied and maintained for a test period of not less than 15 minutes.

   2.2. Peppermint test. Introduce 2 ounces (59 mL) of oil of peppermint into the system. Add 10 quarts (9464 mL) of hot water and seal the vent terminals. The odor of peppermint shall not be detected at any trap or other point in the system.

❖ This section requires a visual inspection of the finished plumbing system. When there is any doubt that the finished system is not water tight, this section allows the code official to require the entire system to be subjected to a smoke or peppermint test. The smoke test involves filling all traps with water and then introducing a thick pungent smoke produced by a smoke machine [see Commentary Figure P2503.5.2(1)]. When smoke appears at stack openings, they must be capped and a pressure equivalent to 1-inch water column (w.c.) (249 Pa) must be maintained for a period of 15 minutes. The pressure developed for the smoke test must not exceed 1 inch (249 Pa) of w.c. because any higher pressure could affect the water seal in the traps, thereby rendering the test invalid. Detection of any smoke indicates a leak that must be repaired [see Commentary Figure P2503.5.2(2)].

Figure P2503.5.2(1)
SMOKE GENERATING EQUIPMENT
(Photo courtesy of Hurco Technologies, Inc.)

Figure P2503.5.2(2)
SMOKE INDICATING PRESENCE OF LEAK
(Photo courtesy of Hurco Technologies, Inc.)

P2503.6 Shower liner test. Where shower floors and receptors are made water tight by the application of materials required by Section P2709.2, the completed liner installation shall be tested. The pipe from the shower drain shall be plugged water tight for the test. The floor and receptor area shall be filled with potable water to a depth of not less than 2 inches (51 mm) measured at the threshold. Where a threshold of not less than 2 inches (51 mm) in height does not exist, a temporary threshold shall be constructed to retain the test water in the lined floor or receptor area to a level not less than 2 inches (51 mm) in depth measured at the threshold. The water shall be retained for a test period of not less than 15 minutes and there shall not be evidence of leakage.

❖ The resulting water damage, increased difficulty in making repairs and the additional expense involved associated with discovering a leak after a building has been completed requires that field-fabricated shower liners be tested at the rough-in stage.

This section clearly spells out how to perform the test. The evidence of leakage could be either the lowering of the water level from the full threshold level or water drips/seepage outside of the receptor area [see Commentary Figures P2503.6(1) and P2503.6(2)].

P2503.7 Water-supply system testing. Upon completion of the water-supply system or a section of it, the system or portion completed shall be tested and proved tight under a water pressure of not less than the working pressure of the system or, for piping systems other than plastic, by an air test of not less than 50 psi (345 kPa). This pressure shall be held for not

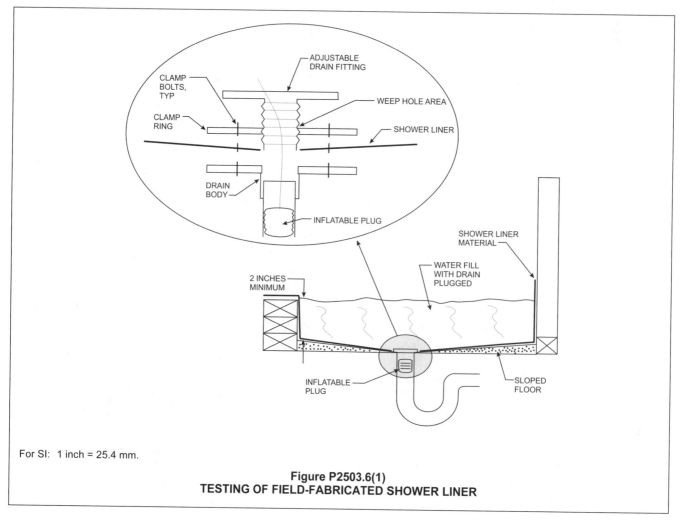

For SI: 1 inch = 25.4 mm.

Figure P2503.6(1)
TESTING OF FIELD-FABRICATED SHOWER LINER

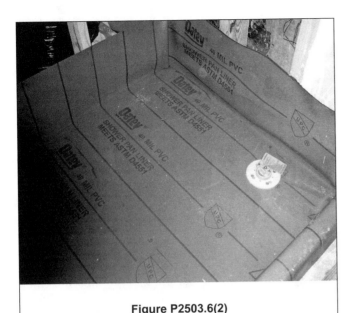

Figure P2503.6(2)
INSTALLED SHOWER LINER READY FOR TESTING

less than 15 minutes. The water used for tests shall be obtained from a potable water source.

❖ The water supply system consists of piping from the water meter at the curb, the stop valve at the curb or the well system to the ends of the water distribution piping in the building.

The term "working pressure" is not defined in the code. It is understood to be the maximum pressure in a water supply system under normal operating conditions. The working pressure can be different depending on where the pressure is measured in a system. For example, the working pressure in a water service line from the curb stop valve to the entry point into the building might be 120 psi (827 kPa). As this pressure is greater than 80 psi (551 kPa), in accordance with Section P2903.3.1, a pressure-reducing valve must be installed to limit the pressure in the building's water distribution system to no greater than 80 psi (551 kPa). Therefore, the working pressure in the water distribution system portion of the water supply system is 80 psi (551 kPa).

The phrase "the system shall be proved tight," although a somewhat archaic expression, means that by visual inspection, no evidence of leakage from the piping system is observed. Evidence of leakage is typically determined by attaching a pressure gauge to the system, pressurizing the system to the test pressure and, without further addition of test water or air to the system, verifying after 15 minutes that the pressure gauge indication has not changed from the reading taken at the beginning of the test (note the test gauge requirements of Section P2503.9). Where minor repairs or modifications are made to a water-supply system and the system is tested with water, the pressure gauge method is not necessary as evidence of leakage could be simply determined by observing

each piping connection for the presence of leaking water.

If air is used to test a water supply system of other than plastic material, the test pressure need only be 50 psi (344 kPa). Where testing with compressed air, it is advisable to allow the system under test pressure to thermally stabilize before the start of the observation period. Warm air from a compressor introduced into cold piping will cool and result in a decrease in the test pressure, falsely indicating a leak. Gaskets and O-rings in shower mixing valves are intended to seal against water pressure and typically do not initially seat or seal well when first pressurized with air. In most cases, leaks from these locations will eventually cease after a "seating in" period of time. Adjustments for changes in pressure resulting from ambient temperature fluctuations or the seating of gaskets must be made prior to the start of the 15-minute test period.

Air testing of water supply systems of plastic material is prohibited by the code and many plastic piping manufacturers due to the risk of personal injury caused by flying shards of plastic should the piping rupture. Compressed air stores potential energy not unlike a compressed coil spring stores energy. If the piping ruptures, the stored potential energy becomes kinetic energy that can propel plastic pieces with great force and velocity.

Sections P2503.5.1 and P2503.5.2 specify two general construction stages that require inspection: (1) rough-in, including underground and above ground and (2) final (completion of plumbing work). Depending on the size and required sequence of a project, multiple inspections may be necessary for each stage. The intent of this section is to ensure that all under-ground and above-ground water supply system piping is tested prior to concealment so that any leaks can be readily located and repaired. The final inspection and "test" (see Section P2503.5.2) provides for visual inspection to find leaks in exposed water supply system connections, such as water heater piping, stop valves, supply tube connections and fixture/faucet assemblies.

P2503.8 Inspection and testing of backflow prevention devices. Inspection and testing of backflow prevention devices shall comply with Sections P2503.8.1 and P2503.8.2.

❖ A device must be tested in accordance with Section P2503.8.2 to determine whether it is operating properly. A visual inspection will determine whether the unit is installed according to its listing and the manufacturer's instructions (see commentary, Section P2902).

P2503.8.1 Inspections. Inspections shall be made of backflow prevention assemblies to determine whether they are operable.

❖ This section requires inspection of all backflow assemblies to determine whether they are installed correctly and are operating in accordance with the manufacturer's instructions. The commentary for Section P2902 provides further discussion of installation.

P2503.8.2 Testing. Reduced pressure principle, double check, double check detector and pressure vacuum breaker backflow preventer assemblies shall be tested at the time of installation, immediately after repairs or relocation and every year thereafter.

❖ Four types of assemblies require testing: reduced-pressure principle backflow prevention assemblies; double-check-valve backflow prevention assemblies; double-detector check-valve backflow prevention assemblies and pressure vacuum breaker assemblies (see commentary, Section P2902). Other backflow prevention devices, such as atmospheric vacuum breakers and air gaps, are used within a water distribution system to protect against cross contamination. For the purpose of testing their adequacy under certain working conditions, the four assemblies referenced in this section are equipped with test cocks (or ports) that are used in conjunction with specific testing equipment and procedures. These tests are required at the time of installation, repair and relocation, and at least annually thereafter.

P2503.9 Test gauges. Gauges used for testing shall be as follows:

1. Tests requiring a pressure of 10 psi or less shall utilize a testing gauge having increments of 0.10 psi (0.69 kPa) or less.

2. Tests requiring a pressure higher than 10 psi (0.69 kPa) but less than or equal to 100 psi (690 kPa) shall use a testing gauge having increments of 1 psi (6.9 kPa) or less.

3. Tests requiring a pressure higher than 100 psi (690 kPa) shall use a testing gauge having increments of 2 psi (14 kPa) or less.

❖ It is common to see test gauges with 1-, 2-, 5- or even 10-psi (0.1, 0.2, 0.5 or 1.0 kg/cm²) increments being used in a DWV inspection that requires only a 5-psi (34.5 kPa) test. These gauges are difficult to read and interpret and lack the resolution necessary to detect all but the largest of leaks.

In each case, the user needs to ensure that the proper pressure gauge is selected with respect to indicating range and design (see Commentary Figure P2503.9). Most standard dial-type mechanical pressure gauges use a bourdon tube sensing element generally made of a copper alloy (brass) or stainless steel. The construction is simple, and operation does not require any additional power source. The C-shaped or spiral wound bourdon tube flexes when pressure is applied, producing a rotational movement that in turn causes the pointer to indicate the measured pressure. Installation of shut-off devices between the measuring point and the pressure gauge is recommended. This will allow an exchange of the pressure gauge and checks on the gauge's zero setting while the system remains under pressure.

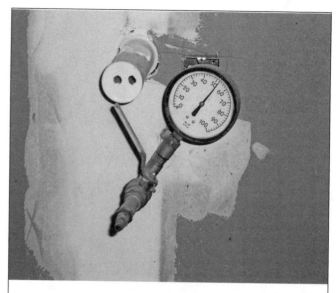

**Figure P2503.9
IMPROPER GAUGE FOR TEST
(DIAL INCREMENTS TOO LARGE)**

Chapter 26:
General Plumbing Requirements

User note: Code change proposals to this chapter will be considered by the IRC – Plumbing and Mechanical Code Development Committee during the 2015 (Group A) Code Development Cycle.

General Comments

This chapter provides for the proper installation of piping systems used in conjunction with plumbing fixtures, drains and appliances. It mandates the removal of liquid wastes and sewage within a building and specifies approved sources of potable water and proper collection of wastes.

Chapter 26 outlines approved methods for maintaining structural integrity when installing the plumbing system as well as for protecting the piping system itself. This chapter also provides guidelines for installing and inspecting underground piping and for preventing the undermining of foundation footings.

Chapter 26 includes a table as well as specifications for providing proper horizontal and vertical support to piping systems using diverse materials. This chapter also addresses the water tightness of a structure's exterior membrane penetrations. Chapter 26 includes provisions for adhering to standards that regulate materials used in piping systems for durability and health effects.

Purpose

Chapter 26 contains requirements for plumbing systems that do not correlate to the provisions set forth in the remaining plumbing related chapters. In other words, if specific provisions do not demand that a requirement be located in another chapter, the requirement is located in this chapter.

SECTION P2601
GENERAL

P2601.1 Scope. The provisions of this chapter shall govern the installation of plumbing not specifically covered in other chapters applicable to plumbing systems. The installation of plumbing, *appliances, equipment* and systems not addressed by this code shall comply with the applicable provisions of the *International Plumbing Code.*

❖ Chapter 26 covers the items that are not specifically addressed in the other chapters of Part VII-Plumbing. In other words, this chapter contains subject matter that is generally applicable to all plumbing systems covered by the code. Rather than repeat such general requirements in each chapter, they are collected together in this chapter for easy reference and brevity of the code.

As stated in the introduction section of the preface of the code, the *International Residential Code®* (IRC®) is a comprehensive, stand-alone code that applies only to one- and two-family dwellings and townhouses. The content of the code was developed and has been maintained in a manner that provides the homebuilder a single code book that covers all aspects of typical home construction. If the subject matter is covered in the IRC, then it is intended that the IRC will provide all of the needed coverage for that subject. In other words, it is not intended that the IRC be applied with additional requirements plucked from the other codes. Therefore, users of the IRC expect that all of the applicable requirements are in the IRC and do not expect to have to determine what additional unknown requirements might apply from some other code. If this was not the case, the IRC user would never be sure that he

or she had all of the applicable requirements in front of him or her. Other codes apply only where the IRC is silent on a subject; however, the fact that the IRC does not contain some particular requirement from another code does not mean that the IRC is silent on a subject. For example, the IRC covers the design of drainage system piping, as does the *International Plumbing Code®* (IPC®). If the IPC contains an additional requirement that is not also in the IRC, such additional requirement is not to be applied to an installation regulated by the IRC. In this case, the IRC contains all of the coverage for drainage piping that was intended. In other words, it is not intended that the IRC provisions be "mixed and matched" with provisions of other codes and it is not intended that code users "shop" in or "cherry pick" from other codes when applying the IRC. It is likely that where the IRC does not contain a particular requirement that is found in another code, such requirement was intentionally left out of the IRC.

A comparison between the IPC and Part VII-Plumbing of this code will reveal many instances where the IPC is different, perhaps more stringent, than this code. These differences are not oversights, but are the result of the cost/benefit and technical analysis decision-making process that occurs in the development of the IRC. Thus, the IPC may have more regulations, or more stringent regulations, relative to a particular aspect of plumbing systems, but the intent is for the IRC to be used as a stand-alone code that is separate and distinct from the regulations of other codes, such as the IPC.

The majority of one- and two-family dwellings and townhouses are, in general, simplistic in design so the regulations contained in the IRC are able to completely cover all aspects of the design and installation of such structures. In rare cases, a homebuilder may come across a situation or feature where the IRC is silent on a subject. For example, imagine that a homeowner wishes that a urinal be installed. Because the IRC is silent on urinals, the homebuilder and code official would look to the IPC where urinals are covered in order to find regulations that would be useful for application in a home environment. The intent of the last line of this code section is to provide a link to the IPC where it is necessary to address plumbing, appliances, equipment and systems that the IRC does not address. It is not the intent of this code section to provide the link to the IPC so that additional or more stringent regulations of the IPC be applied to applications that are already covered by the IRC.

P2601.2 Connections to drainage system. Plumbing fixtures, drains, appurtenances and *appliances* used to receive or discharge liquid wastes or sewage shall be directly connected to the sanitary drainage system of the building or premises, in accordance with the requirements of this code. This section shall not be construed to prevent indirect waste connections where required by the code.

Exception: Bathtubs, showers, lavatories, clothes washers and laundry trays shall not be required to discharge to the sanitary drainage system where such fixtures discharge to systems complying with Sections P2910 and P2911.

❖ Plumbing fixtures, drains, appurtenances and appliances that receive or discharge liquid waste or sewage must be directly connected to the sanitary drainage system unless specifically required to be indirectly connected. An example of an indirect waste connection would be a standpipe receiving the discharge of a clothes washer (see commentary, Section P2718).

The exception to this section allows for the indicated fixtures (producing gray water) to be connected to a nonpotable water reuse system or a subsurface irrigation system instead of being connected to the sanitary drainage system.

P2601.3 Flood hazard areas. In flood hazard areas as established by Table R301.2(1), plumbing fixtures, drains, and *appliances* shall be located or installed in accordance with Section R322.1.6.

❖ See the commentary to Section R322.1.6.

SECTION P2602
INDIVIDUAL WATER SUPPLY AND
SEWAGE DISPOSAL

P2602.1 General. The water-distribution and drainage system of any building or premises where plumbing fixtures are installed shall be connected to a public water supply or sewer system, respectively, if available. Where either a public water-supply or sewer system, or both, are not available, or connection to them is not feasible, an individual water supply or individual (private) sewage-disposal system, or both, shall be provided.

❖ For compliance with Section P2601, this section requires that both the water distribution system and the drainage system be connected to a public water supply and a public sewer system, respectively. The code allows for connection to an individual water supply and a private sewage disposal when a public connection is not feasible. Usually, the local jurisdictional authorities (i.e., building department, water purveyor, sewer utility district, etc.) determine feasibility. The most common individual water supply is a private well.

P2602.2 Flood-resistant installation. In flood hazard areas as established by Table R301.2(1):

1. Water supply systems shall be designed and constructed to prevent infiltration of floodwaters.

2. Pipes for sewage disposal systems shall be designed and constructed to prevent infiltration of floodwaters into the systems and discharges from the systems into floodwaters.

❖ This section requires that certain plumbing systems be designed to minimize or eliminate infiltration of floodwaters into them. These requirements are intended to prevent damage to the system, guard against contamination of the water supply through floodwater entering the system and prevent sewage from escaping and further contaminating floodwaters.

Potable water service pipes, both in the lower levels of the building and those buried below ground, are to be water tight to prevent floodwater from entering them. Although a water pipe under normal conditions does not leak, it could develop a leak when it is subjected to the additional stress from floodwater. This means that the entire piping system subject to floodwater contact must be designed accordingly by, for example, having hangers that are capable of resisting not only the normal downward load of the pipe and its contents, but also lateral or upward forces on the pipe caused by floodwaters.

Sanitary drainage and vent piping must be designed so that the additional stress of the floodwater does not cause it to fail and thus contaminate the floodwater.

SECTION P2603
STRUCTURAL AND PIPING PROTECTION

P2603.1 General. In the process of installing or repairing any part of a plumbing and drainage installation, the finished floors, walls, ceilings, tile work or any other part of the building or premises that must be changed or replaced shall be left in a safe structural condition in accordance with the requirements of the building portion of this code.

❖ Where new plumbing components are installed within existing construction, structural components that are modified must comply with the minimum structural requirements of the code's building chapters

P2603.2 Drilling and notching. Wood-framed structural members shall not be drilled, notched or altered in any manner except as provided in Sections R502.8, R602.6, R802.7 and R802.7.1. Holes in load-bearing members of cold-formed steel light-frame construction shall be made only in accordance with Sections R505.2.6, R603.2.6 and R804.2.6. In accordance with the provisions in Sections R505.3.5,

R603.3.3 and R804.3.4, cutting and notching of flanges and lips of load-bearing members of cold-formed steel light-frame construction shall be prohibited. Structural insulated panels (SIPs) shall be drilled and notched or altered in accordance with the provisions of Section R613.7.

❖ The code recognizes that it is occasionally necessary to modify framing members by boring holes, notching or cutting in the course of installing plumbing piping or plumbing fixture supports. While the original intent of this section was to address modifications to solid sawn wood members (commonly called "dimensional lumber"), modern building construction includes a mix of "engineered" wood members and assemblies, cold-formed steel light-frame members, structural steel members and log members. To maintain a building's structural integrity, the load-carrying capacity of these members must not be reduced. If a structural member is modified beyond what is specifically permitted by the code, the member must either be replaced with an unmodified member or a structural analysis of that member must be performed by a registered design professional (and approved by the code official) to verify that the modified member will have the required structural capacity. As either solution could be costly, the plumbing installer should carefully consider the specified limitations for modification of all framing members before such alterations are performed.

Bearing Walls versus Nonbearing Walls

The terms "bearing" and "nonbearing" are used in the prescriptive allowances for the alteration of studs. In order to apply the allowances in the least restrictive manner, the plumbing installer must understand how the roof and floor loads of a building are supported by the wall framing. Exterior walls are always considered to be (load) bearing walls. Where floor joists, trusses or I-joists are supported by an interior wall, the wall is a (load) bearing wall. Roof trusses might also require support by interior walls and as such, those walls would be (load) bearing. However, not all walls located under the bottom chord of a roof truss are intended to support roof trusses. Where a wall only supports itself and no other framing member, it is non(load) bearing. If the plumbing installer has doubt about whether a wall is bearing or nonbearing, he or she should either consult with the party responsible for the framing design or assume that the wall is (load) bearing.

Sawn Wood Members

The code offers prescriptive allowances for cutting, notching and boring of sawn wood studs, joists and rafters. These allowances apply to wood-constructed buildings that have been designed in accordance with the conventional light-frame wood construction design method.

Figure R502.8 illustrates the limitations for sawn lumber (wood) joists.

Figures R602.6(1) and R602.6(2) illustrate the limitations for wood studs in bearing and nonbearing walls, respectively.

Section R802.7.1 indicates the limitations for sawn lumber (wood) ceiling joists, rafters and beams. Because these limitations are the same as for wood joists, Figure 502.8 can be used for illustration of the limitations.

In Commentary Figure P2603.2(1), the bearing wall studs have $2^1/_8$-inch (54 mm) holes drilled in 2 by 4 studs for passage of the $1^1/_2$-inch (38 mm) drain and vent pipes. The holes exceed 40 percent of the $3^1/_2$-inch (89 mm) actual depth of the stud and therefore, according to the *International Building Code*® (IBC®), the modified studs should have been evaluated by a design professional for structural adequacy. However, in this situation, the code official allowed the use of metal reinforcing brackets (often called "stud shoes") to reinforce the studs at the oversized hole locations. An exception in Section R602.6 allows for approved metal "stud shoes" to be used for reinforcing of a stud that has been modified beyond the prescribed limitations. The stud shoes must be installed in accordance with the manufacturer's recommendations. The proper fasteners, in accordance with the manufacturer's instructions, must be used to attach the bracket to the stud. The code official must approve the type of bracket being used.

Commentary Figures P2603.2(2) through P2603.2(4) also show notching on the face of the studs. The allowable notch depth for a bearing stud is 25 percent of the stud depth or $7/_8$ inch (22 mm). The notches in these studs are deeper than $7/_8$ inch (22 mm) and thus are in violation of the notching rules.

Section R802.7.2 prohibits any modifications to engineered wood products, such as trusses, structural composite lumber, structural glue-laminated members or I-joists, unless specifically allowed by the manufacturer or where approved by a registered design professional.

Sawn Wood Wall Plate Reinforcement

The top plate of a (load) bearing wood stud wall is a structural member that has limitations for cutting, notching or boring. See the commentary to Section R602.6.1 for those limitations and required reinforcement. Commentary Figures P2603.2(5) and P2603.2(6) illustrate two commonly occurring violations.

Steel Joists

Figure R505.2.5.1 illustrates the hole limitations for cold-formed steel floor joists. Section R505.3.5 prohibits the cutting or notching of flanges or lips of steel joists. Section R505.1.3 prohibits any alteration of steel floor trusses.

Figure P2603.2(1)
NOTCHING AND BORING IN A WOOD STUD WALL

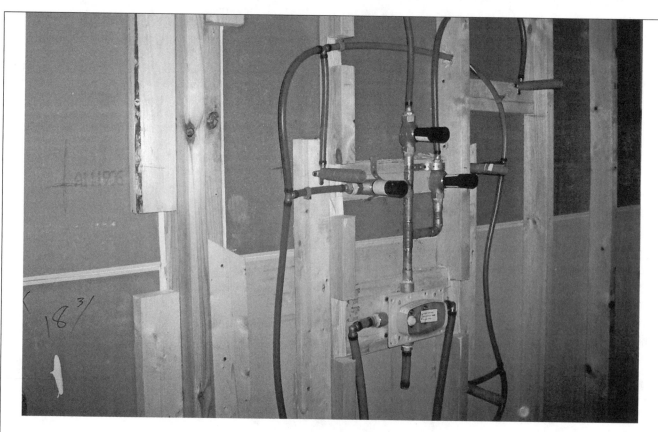

Figure P2603.2(2)
VIOLATION: NOTCHES TOO DEEP IN STUDS

Figure P2603.2(3)
VIOLATION: EXTERIOR STUD NOTCHED TOO DEEP

Figure P2603.2(4)
VIOLATION: EXTERIOR STUD COMPLETELY SEVERED

Figure P2603.2(5)
VIOLATION: WALL TOP PLATE NOT REINFORCED

Figure P2603.2(6)
VIOLATION: WALL TOP PLATE NOT REINFORCED

Steel Roofing Members

Limitations for holes in webs of steel roofing members are indicated in Section 804.2.5.1 and illustrated by Figure 804.2.5.1. Section 804.3.4 prohibits notching or cutting of flanges and lips of steel roofing members.

Engineered Wood Members

Section R502.8.2 prohibits any modifications to engineered wood products, such as trusses, structural composite lumber, structural glue-laminated members or I-joists, unless specifically allowed by the manufacturer or where approved by a registered design professional. Engineered wood members are manufactured in many ways. The most common products used are laminated veneer lumber (LVL) beams [see Commentary Figure P2603.2(7)], trusses of dimensional lumber, and I-joists of oriented strand board (OSB) webs with dimensional lumber top and bottom flanges (chords). LVL members must not be cut or notched. Because most LVL manufacturers either prohibit the installation of holes or severely limit the size, location and number of holes in LVL members, as a general rule, holes should never be bored in LVL members. Truss members must not be altered in any manner.

The top and bottom chords of I-joist members must never be cut or notched. Although most I-joist manufacturers allow holes to be cut in the web of the member, the maximum size and location of holes must be in accordance with the specific manufacturer's product instructions. The cutting, notching and boring limitations indicated for sawn lumber joists (see Figure R502.8) do not apply to I-joists. The installer and inspector must have access to the product instructions in order to determine if hole sizes and locations are acceptable. A review of several I-joist manufacturers' installation instructions provides some generalizations that can help the plumbing installer plan piping routes as well as alert the inspector that a violation might exist. The generalizations are:

1. The closer the hole is located to the bearing locations (supports) of the I-joist, the smaller the allowable size of the hole becomes. Typically, holes near the bearing ends are limited to $1^1/_2$ inches (38 mm) in diameter with a spacing (hole edge to hole edge) of not less than 12 inches (305 mm).

2. As the hole location is placed farther from the bearing locations of the I-joist, the allowable size of the hole becomes larger. Large holes should be spaced apart (hole edge to hole edge) not less than the largest diameter or longest dimension of the largest hole.

3. Holes should be cut round, square or rectangular. The edge of any hole should not be closer than $^1/_4$ inch (6.4 mm) to a top or bottom chord of the I-joist.

Always refer to the I-joist manufacturer's installation instructions for specific hole limitations. Commentary Figures P2603.2(8) through P2603.2(11) show prohibited boring, cutting and notching of trusses and I-joists.

Figure P2603.2(7)
LAMINATED VENEER LUMBER BEAM

Figure P2603.2(8)
VIOLATION: TOP CHORD OF TRUSS NOTCHED

Figure P2603.2(9)
VIOLATION: HOLE IN BOTTOM CHORD OF ROOF TRUSS

Figure P2603.2(10)
VIOLATION: TOP CHORD OF I-JOIST COMPLETELY SEVERED

Figure P2603.2(11)
VIOLATION: I-JOIST WEB HOLES IRREGULAR SHAPE, TOO LARGE, TOO CLOSE TOGETHER

Cold-formed Steel Studs

Figure R603.2.5.1 illustrates the hole limitations for steel studs, both load bearing and nonload bearing. Section R603.3.4 prohibits notching or cutting of flanges and lips of steel studs. Section R603.2.5.3 covers hole patching requirements if a hole in a steel stud is greater than the hole size limitation, but less than or equal to 70 percent of the stud width.

Structural Insulated Panels (SIPs)

As the name implies, the entire panel is a structural member. These panels are primarily used for exterior walls and roofs, but can also be used for interior bearing walls. Section R613.7 allows one rectangular or circular hole with the largest dimension no greater than 12 inches (305 mm) to be cut in each panel section. For example, a 2-inch (51 mm) vent pipe that is required to pass through a sloped SIP roof might require a 4-inch-diameter (102 mm) hole. Otherwise, SIPs must not be cut or notched in any manner. Therefore, plumbing water supply, drainage or vent piping cannot be installed within a structural insulated panel.

Log Members

The code references ICC 400 for the construction of log structures. Section 302.2.4 of ICC 400 covers the limitations for notching and boring of holes in logs used in structural applications.

P2603.2.1 Protection against physical damage. In concealed locations, where piping, other than cast-iron or galvanized steel, is installed through holes or notches in studs, joists, rafters or similar members less than $1\frac{1}{4}$ inches (31.8 mm) from the nearest edge of the member, the pipe shall be protected by steel shield plates. Such shield plates shall have a thickness of not less than 0.0575 inch (1.463 mm) (No. 16 Gage). Such plates shall cover the area of the pipe where the member is notched or bored, and shall extend not less than 2 inches (51 mm) above sole plates and below top plates.

❖ This section is intended to minimize the possibility of drainage and water pipe damage caused by nails, screws or other fasteners used to affix wall panels. A shield plate must be installed to protect the pipe through the full width of the member and must extend not less than 2 inches (51 mm) above or below sole or top wall plates. Commentary Figure P2603.2.1(1) shows locations for shield plates. Commentary Figure

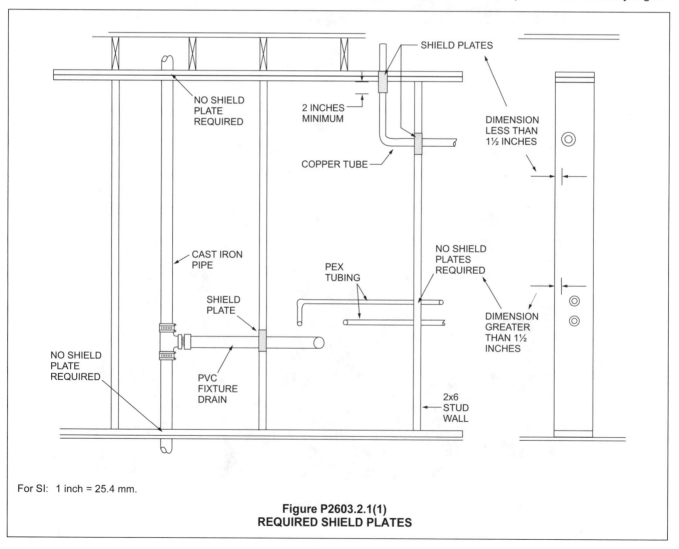

For SI: 1 inch = 25.4 mm.

Figure P2603.2.1(1)
REQUIRED SHIELD PLATES

P2603.2.1(2) shows shield plates improperly installed because the plate(s) do not extend 2 inches (51 mm) below the bottom member of the double top plate. Cast-iron and galvanized steel pipe each have wall thicknesses greater than the required thicknesses for the shield plate, which makes them inherently resistant to nail and screw penetrations.

The shield plate minimum thickness specified represents the "low end" of the thickness tolerance range for No. 16 gage galvanized sheet metal in accordance with Sheet Metal and Air Conditioning Contractors' National Association (SMACNA) standards. The "low end" thickness dimension is stated so that if a field measurement of shield plate thickness is necessary, the absolute minimum thickness will allow all possible thicknesses of No. 16 gage shield plates to be acceptable. No. 16 gage has been proven to be the minimum thickness to prevent penetration by commonly used fasteners without causing noticeable "bulge" appearance problems in the covering materials.

P2603.3 Protection against corrosion. Metallic piping, except for cast iron, ductile iron and galvanized steel, shall not be placed in direct contact with steel framing members, concrete or masonry. Metallic piping shall not be placed in direct contact with corrosive soil. Where sheathing is used to prevent direct contact, the sheathing material thickness shall be not less than 0.008 inch (8 mil) (0.203 mm) and shall be made of plastic. Where sheathing protects piping that penetrates concrete or masonry walls or floors, the sheathing shall be installed in a manner that allows movement of the piping within the sheathing.

❖ Metallic piping must be protected from external corrosion in locations where the piping is in direct contact with materials known to be corrosive.

Steel framing members can be source of corrosion to metallic piping, especially to copper piping.

Concrete, in a wet slurry state, is caustic (pH level of 12 to 13) due to the lime-stone-based portland cement in the concrete reacting with water. Because concrete (and masonry units made from concrete) could contain acid-producing cinders and cured concrete is still somewhat caustic, metallic piping must be protected from direct contact with concrete or concrete masonry units.

The exceptions for cast iron, ductile iron and galvanized steel are provided because these pipe materials are heavy walled products. External corrosion as the result of being in contact with steel framing members, concrete and masonry has minimal impact on the piping.

Corrosive soils such as found in swamps, peat bogs or tidal marshes can corrode metallic piping. Fill materials containing furnace slag, cinders, ash or spent industrial byproducts can also be corrosive to metallic piping. The designer or installer should consult with soils engineers, piping manufacturers as well as local code officials concerning the need for corrosion protection of metallic piping in contact with soil or fill materials of potentially corrosive nature.

Direct exposure of metallic piping to corrosive materials can be prevented by sheathing the piping. Sheathing involves inserting the piping into a flexible plastic sleeve of material. The sheathing must allow for pipe movement. In other words, the sheathing should not be adhered to the pipe so that the pipe can expand and contract within the protective material. This section requires the thickness of the sheathing or wrapping to be at least 0.008 inch (8 mil) (0.203 mm).

Figure P2603.2.1(2)
PIPE PROTECTION VIOLATION

P2603.4 Pipes through foundation walls. A pipe that passes through a foundation wall shall be provided with a relieving arch, or a pipe sleeve shall be built into the foundation wall. The sleeve shall be two pipe sizes greater than the pipe passing through the wall.

❖ Piping installed through a foundation wall must be structurally protected from any transferred loading from the foundation wall. This protection must be provided by either a relieving arch or a pipe sleeve.

Where a pipe sleeve is used, it must be sized to be two pipe sizes larger than the pipe that passes through the pipe sleeve. For example, a 4-inch (102 mm) drainage pipe would require a 6-inch (152 mm) pipe sleeve. The annular space between the pipe sleeve and the pipe that passes through the sleeve will allow for any differential movement between the pipe and the wall. The pipe sleeve or relieving arch protects the pipe so that it will not be subjected to undue stresses that could cause it to rupture and leak [see Commentary Figures P2603.4(1) and P2603.4(2)].

In rare circumstances, it is sometimes necessary for a pipe to pass though a footing. For example, the slope of a building drain might not be able to be changed, resulting in the drainage pipe needing to pass through the footing. Although the code does not specifically address pipes passing through footings, the need for protecting pipes passing though footings is no different than for pipes passing through foundation walls. In these situations, protection of the pipe is not the only concern as the strength of the footing could be compromised by the installation of a pipe sleeve or relieving arch. Any footing to be altered, either before or after placement of the footing, should be reviewed by a design professional to determine what footing design changes might be necessary to maintain the required footing strength.

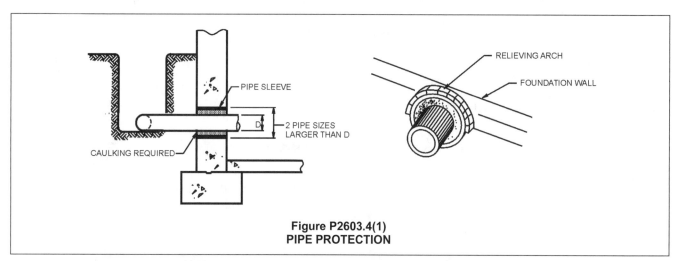

Figure P2603.4(1)
PIPE PROTECTION

Figure P2603.4(2)
BUILDING DRAIN PENETRATION THROUGH FOUNDATION WALL

Prior editions of the code required a relieving arch or a pipe sleeve for a pipe that passes under a footing. This was determined to be unnecessary since the footing already acts as a relieving arch to "span over" the area where the pipe passes under the footing. This is not to say that routing any size or number of pipes under a footing is without consequences. Footings are designed to be uniformly supported by undisturbed or highly compacted soil. A trench for a pipe passing under the footing disrupts the uniform support for the footing. If the trench can be backfilled and compacted to the same level of compaction as the surrounding undisturbed or highly compacted soil, then the footing will be uniformly supported. But if the trench cannot be compacted sufficiently, such as where the trench is dug after the footing is in place, then the footing will not be uniformly supported by the soil. The footing will experience more stress than it was designed for. In actual practice, in a low-rise building where one or two "small" diameter pipes [less than 8 inches (203 mm)] pass under a footing in the same area, the footing will most likely have sufficient reserve capacity to span the trench. Larger buildings, typically requiring larger pipes, will have footings that are already designed for spanning over the trenches of these larger pipes. There are no "rules of thumb" to know when a footing design should be analyzed for adequacy. Installers and code officials need to be aware of the possible consequences to the footings and, where necessary, consult with a design professional concerning the footing design.

P2603.5 Freezing. In localities having a winter design temperature of 32°F (0°C) or lower as shown in Table R301.2(1) of this code, a water, soil or waste pipe shall not be installed outside of a building, in exterior walls, in *attics* or crawl spaces, or in any other place subjected to freezing temperature unless adequate provision is made to protect it from freezing by insulation or heat or both. Water service pipe shall be installed not less than 12 inches (305 mm) deep and not less than 6 inches (152 mm) below the frost line.

❖ Water, soil and waste pipes must be protected from freezing. Although vent piping is not required to be protected from freezing, it may be prudent to insulate vent pipes in extreme freezing climates. Where a water or drain pipe is installed in an exterior wall or ceiling of a space intended for occupancy (in other words, a heated space), some degree of freeze protection can be achieved by making sure that the thermal insulation for the wall (or ceiling) is installed between the exterior wall surface and the piping [see Commentary Figures P2603.5(1) and P2603.5(2)]. Whether or not this arrangement will prevent freezing temperatures at the piping location depends on the climatic conditions, the thickness of insulation (between the exterior wall surface and the piping) and the room temperature.

Note that there must always be a heat source along with an appropriate insulation thickness in order to protect pipes from freezing conditions. Insulation by itself (without a heat source) cannot protect a pipe from freezing; insulation only slows the rate of heat loss. The closer a pipe is to the heat source and the less insulation there is between the pipe and the heat source the warmer the pipe will be. Commentary Figure P2603.5(3) shows piping in an exterior wall of a building. The water distribution piping has been covered with foam tube insulation. On the left of the photo, note the close proximity of the horizontal water pipe to the exterior sheathing. While the insulation on the pipe between the pipe and the sheathing does help protect the pipe from outdoor temperatures, the insulation on the pipe between the pipe and the room will insulate the pipe from the heat coming from the room. In all likelihood, the wall insulation will be installed over the insulated pipe, resulting in further insulation of the pipe from the room heat. On the right side of this photo, note the multitude of water distribution pipes behind the fixture location (probably a laundry tub). Again, while the pipe insulation between the pipes and the sheathing is beneficial for freeze protection of the pipes, the insulation on the room side of the pipes is detrimental because it insulates the pipes from the room heat. The congestion in this area will undoubtedly cause difficulty for installing wall insulation, especially between the piping and the sheathing. In the center of this photo,

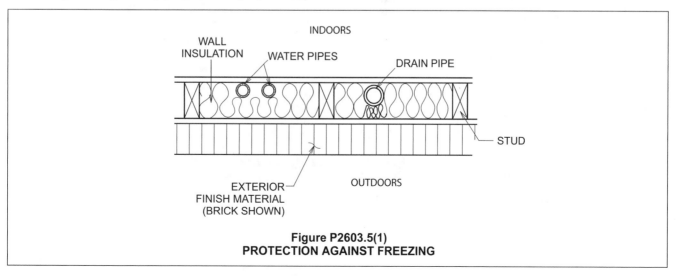

Figure P2603.5(1)
PROTECTION AGAINST FREEZING

note that the trap for the clothes washer standpipe is in close proximity to the sheathing, so much so that it will be difficult to install any insulation between the trap and the sheathing. The arrangement and insulation of pipes shown in this figure would most likely only be suitable for a building located where freezing climatic conditions are unlikely or are of limited duration.

Where piping is not directly adjacent to heated spaces in a building, electric-resistance heat tapes or cables can be used to supply heat to the piping. Generally, heat tapes should not be used on piping in concealed spaces since the tapes can burn out and require replacement. Since the code is silent with regard to the use of heat tape in a concealed area, however,

Figure P2603.5(2)
WATER DISTRIBUTION PIPING IN EXTERIOR WALL

Figure P2603.5(3)
PIPING IN EXTERIOR WALL

the only recourse the code official has to prohibit such installations is to determine if the manufacturer's instructions limit the product use to unconcealed areas. There are heavy-duty electric heat-tracing products that are suitable for concealed- and earth-burial applications. Note that plastic piping requires self-limiting-type heat tape to prevent overheating of the pipe.

The code does not provide any guidelines for insulating methods to ensure that piping will not be exposed to freezing temperatures. The code official is responsible for determining whether water and drain piping is adequately protected for the climatic conditions expected in the geographic area.

Freezeproof Hose Faucets

Hose bibs and wall hydrants located on the exterior wall must be protected where installed in areas subject to freezing temperatures. This can be accomplished by installing devices, such as freezeproof hose bibs, that locate the valve seat within the heated space and allow residual water within the hydrant to drain after the valve is closed. A freezeproof hose bibb cannot be installed with the valve seat located in an unheated garage or storage room. The valve seat must extend through to the heated side of the exterior wall. Valve casings on these devices are available in various lengths to accommodate a variety of wall thicknesses [see Commentary Figure P2603.5(4)].

Underground Water Service Piping

Water service piping that is installed exterior to a building and underground must be buried at least 12 inches (305 mm) below grade or 6 inches (152 mm) below the established frost penetration depth for the geographic area [see Commentary Figure P2603.5(5)], whichever is the greater depth. For example, if the frost depth for the region is 24 inches

(610 mm) below grade, the water service pipe must be installed so that the top surface of the pipe is at least 30 inches (762 mm) below grade. If the frost depth is 5 inches (127 mm) below grade, the water service pipe must be installed such that the top surface of the pipe is at least 12 inches (305 mm) below grade. Where frost penetration depth controls the burial depth, the 6-inch (152 mm) buffer protects the pipe from forces caused by the freezing and expansion of moisture in the above soil (i.e., "frost heave") [see Commentary Figure P2603.5(6)]. The minimum burial depth of 12 inches (305 mm) protects the pipe from the most common accidental damage: shallow hand digging for landscape plantings.

Plumbing in Seasonal-use Buildings and Outdoor-use Areas

The potable water plumbing systems for poolside bathhouses, summer-use-only cabins and other unheated areas not used during freezing temperature conditions are typically designed in some way to be "winterized" so that water in the piping does not freeze and break the piping. Two common ways of accomplishing winterization of plumbing systems are: (1) gravity draining the potable water system to an outlet at the system's lowest point or (2) pumping an antifreeze solution into the potable water system to displace the water. Although the code is silent about winterization, there are several code sections to consider: (1) stop and waste valves below grade (Section P2903.9.5) and (2) introduction of nonpotable liquids into a potable water system (Section P2902.1).

Consider a poolside bathhouse built on a slab-on-grade foundation. Gravity draining of the water from the potable water system would require a drain outlet located below grade so that water in the system can drain down to an elevation below the frost line. However, Section P2903.9.5 prohibits stop and waste

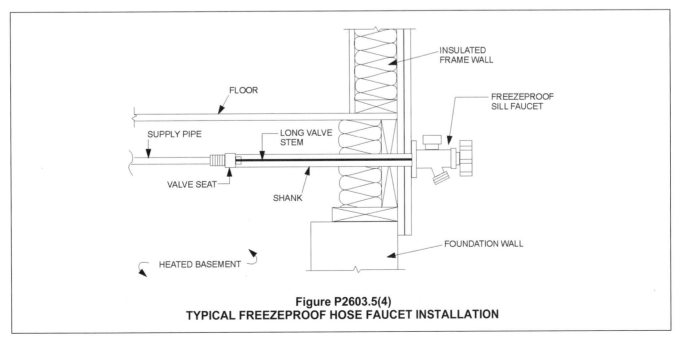

Figure P2603.5(4)
TYPICAL FREEZEPROOF HOSE FAUCET INSTALLATION

valves below grade because there is a high probability for the drain outlet valve to be left open during the off season such that ground water could enter and contaminate the potable water system. Therefore, gravity draining of the bathhouse potable water system through the below-grade outlet is not an acceptable method for winterizing. One possible solution for evacuating the water from the potable water system is to design the system so that the water could be vacuumed out from one or more specifically designed above-grade openings that are connected directly to the low points in the system. Other parts of the potable water system would require careful piping installation to ensure that the water in the system gravity drains to the low points of the system.

A potable water system could also be winterized by pumping "antifreeze" into the system to displace the water throughout all of the piping. In general, plastic piping manufacturers recommend using only a glycerin-based antifreeze solution since it is generally regarded as nontoxic and is chemically compatible with all types of plastic pipe materials listed in the code. The use of glycerin-based antifreeze for nonplastic piping should be equally acceptable. Waste glycerin antifreeze could be captured for reuse or, if disposed of, should be done so in an environmentally safe manner.

P2603.5.1 Sewer depth. *Building sewers* that connect to private sewage disposal systems shall be a not less than **[NUMBER]** inches (mm) below finished grade at the point of septic tank connection. *Building sewers* shall be not less than **[NUMBER]** inches (mm) below grade.

❖ A building sewer is not subject to the same constraints as a water pipe because a sewer conveys heat to the surrounding soil as the drainage flows through it. A

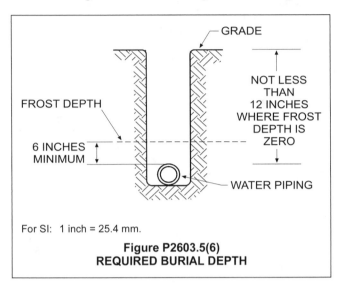

For SI: 1 inch = 25.4 mm.

Figure P2603.5(6)
REQUIRED BURIAL DEPTH

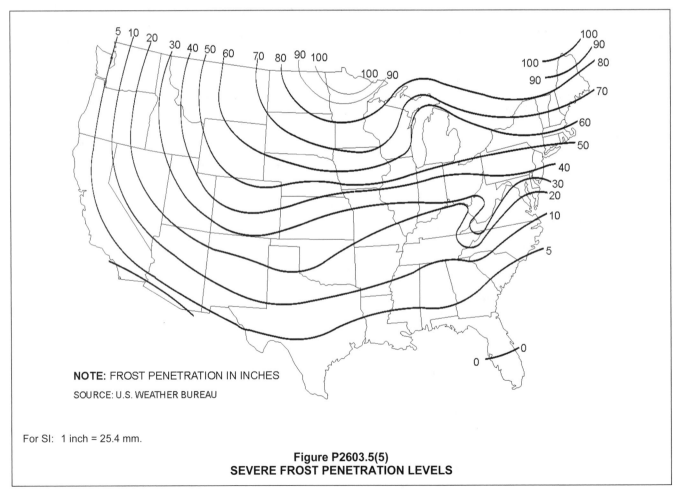

NOTE: FROST PENETRATION IN INCHES
SOURCE: U.S. WEATHER BUREAU

For SI: 1 inch = 25.4 mm.

Figure P2603.5(5)
SEVERE FROST PENETRATION LEVELS

building sewer also has an intermittent flow. As a result, the frost depth is higher in the immediate area of the building sewer; therefore, the burial depth of a building sewer does not have to be below the recorded frost depth. A typical burial depth for a building sewer ranges from 6 to 18 inches (152 to 457 mm). The installation depth of the sewer pipe connecting to a septic tank could be different to facilitate the proper operation of the private sewage disposal system. The connection at the septic tank is typically 18 inches (457 mm) below grade; however, the code official must determine the depths necessary for the locality.

SECTION P2604
TRENCHING AND BACKFILLING

P2604.1 Trenching and bedding. Where trenches are excavated such that the bottom of the trench forms the bed for the pipe, solid and continuous load-bearing support shall be provided between joints. Where over-excavated, the trench shall be backfilled to the proper grade with compacted earth, sand, fine gravel or similar granular material. Piping shall not be supported on rocks or blocks at any point. Rocky or unstable soil shall be over-excavated by two or more pipe diameters and brought to the proper grade with suitable compacted granular material.

❖ A trench must be sufficiently wide to allow proper pipe alignment. Piping must be supported on solid, continuous bedding for its entire length. Improper backfill material will subject the pipe to uneven earth loading that can lead to damage to the pipe or joints. Where a trench contains unstable material, the code requires overexcavation by at least two pipe diameters below the pipe and placement of suitable, compacted granular material below the pipe to provide proper and even support. Manufacturer's installation instructions often include material-specific information on burial techniques.

P2604.2 Water service and building sewer in same trench. Where the water service piping and *building sewer* piping is installed in same trench, the installation shall be in accordance with Section P2906.4.1.

❖ See Section P2906.4.1.

P2604.3 Backfilling. Backfill shall be free from discarded construction material and debris. Backfill shall be free from rocks, broken concrete and frozen chunks until the pipe is covered by not less than 12 inches (305 mm) of tamped earth. Backfill shall be placed evenly on both sides of the pipe and tamped to retain proper alignment. Loose earth shall be carefully placed in the trench in 6-inch (152 mm) layers and tamped in place.

❖ This section requires that backfill material be devoid of construction materials, trash, debris, rocks, broken concrete, frozen chunks and other rubble. The provisions of this section are intended to protect buried piping from all of the rubbish, scraps, rubble, etc., that all too commonly find their way into piping excavations. Piping excavations are not to be used as a waste disposal "landfill." Twelve inches (305 mm) of clean tamped earth, sand or fine gravel must be used to cover the pipe. Loose earth must be placed carefully in the trench in 6-inch (152 mm) layers and tamped in place to achieve stability and proper compaction (see Commentary Figure P2604.3).

P2604.4 Protection of footings. Trenching installed parallel to footings and walls shall not extend into the bearing plane of a footing or wall. The upper boundary of the bearing plane is a line that extends downward, at an angle of 45 degrees from horizontal, from the outside bottom edge of the footing or wall.

❖ This section provides for the installation of pipes parallel to a footing. This is intended to protect the piping from building loads and to prevent the disturbance of the soil in the bearing plane of the footing (see Commentary Figure P2604.4).

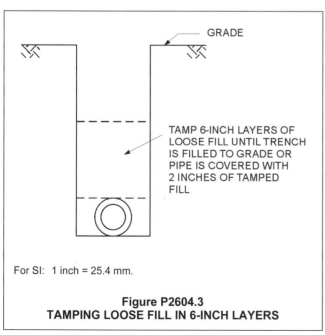

For SI: 1 inch = 25.4 mm.

Figure P2604.3
TAMPING LOOSE FILL IN 6-INCH LAYERS

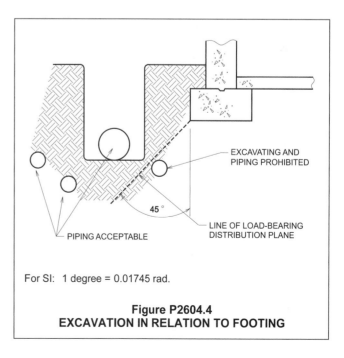

For SI: 1 degree = 0.01745 rad.

Figure P2604.4
EXCAVATION IN RELATION TO FOOTING

SECTION P2605
SUPPORT

P2605.1 General. Piping shall be supported in accordance with the following:

1. Piping shall be supported to ensure alignment and prevent sagging, and allow movement associated with the expansion and contraction of the piping system.

2. Piping in the ground shall be laid on a firm bed for its entire length, except where support is otherwise provided.

3. Hangers and anchors shall be of sufficient strength to maintain their proportional share of the weight of pipe and contents and of sufficient width to prevent distortion to the pipe. Hangers and strapping shall be of *approved* material that will not promote galvanic action. Rigid support sway bracing shall be provided at changes in direction greater than 45 degrees (0.79 rad) for pipe sizes 4 inches (102 mm) and larger.

4. Piping shall be supported at distances not to exceed those indicated in Table P2605.1.

❖ The intent of supporting the pipe is to maintain proper slope and alignment and to prevent sagging, while allowing movement of the pipe resulting from expansion and contraction [see Commentary Figure P2605.1(2)]. Hangers and supports for piping must be capable of supporting the load imposed by the piping system and must not be detrimental to the pipes they support. Using hanger or strapping material that is not compatible with the piping material can result in corrosion caused by galvanic action. Galvanic action occurs when dissimilar metals are in contact in the presence of an electrolyte. Section P2605.1 also contains lateral support requirements. Pipes are subject to the momentum resulting from the flow of waste. Hangers alone may not be sufficient to resist these forces, and, thus, rigid support sway bracing is required for pipe sizes 4 inches (102 mm) or larger at changes of direction greater than 45 degrees (0.79 rad) [see Commentary Figures P2605.1(1) and P2605.1(3)].

TABLE P2605.1. See below.

❖ The code refers to Table P2605.1 for piping support requirements. The maximum amount of sag occurs at a point equidistant from the supports and when the pipe is full of water. The recommended spacing of supports in Table P2605.1 is designed to limit sag and pipe stress and to maintain slope and alignment. As the pipe's diameter increases, so does the pipe's strength and resistance to sag. The spacing in the table is calculated for a small-diameter pipe [see Commentary Figures P2605.1(2) and P2605.1(3)]. Consult the pipe manufacturer's installation instructions for more specific spacing instructions.

TABLE P2605.1
PIPING SUPPORT

PIPING MATERIAL	MAXIMUM HORIZONTAL SPACING (feet)	MAXIMUM VERTICAL SPACING (feet)
ABS pipe	4	10[b]
Aluminum tubing	10	15
Cast-iron pipe	5[a]	15
Copper or copper alloy pipe	12	10
Copper or copper alloy tubing (1¼ inches in diameter and smaller)	6	10
Copper or copper alloy tubing (1½ inches in diameter and larger)	10	10
Cross-linked polyethylene (PEX) pipe, 1 inch and smaller	2.67 (32 inches)	10[b]
Cross-linked polyethylene (PEX) pipe, 1¼ inch and larger	4	10[b]
Cross-linked polyethylene/aluminum/cross-linked polyethylene (PEX-AL-PEX) pipe	2.67 (32 inches)	4[b]
CPVC pipe or tubing (1 inch in diameter and smaller)	3	10[b]
CPVC pipe or tubing (1¼ inches in diameter and larger)	4	10[b]
Lead pipe	Continuous	4
PB pipe or tubing	2.67 (32 inches)	4
Polyethylene of raised temperature (PE-RT) pipe, 1 inch and smaller	2.67 (32 inches)	10[b]
Polyethylene of raised temperature (PE-RT) pipe, 1¼ inch and larger	4	10[b]
Polypropylene (PP) pipe or tubing (1 inch and smaller)	2.67 (32 inches)	10[b]
Polypropylene (PP) pipe or tubing (1¼ inches and larger)	4	10[b]
PVC pipe	4	10[b]
Stainless steel drainage systems	10	10[b]
Steel pipe	12	15

For SI: 1 inch = 25.4 mm, 1 foot = 304.8 mm.

a. The maximum horizontal spacing of cast-iron pipe hangers shall be increased to 10 feet where 10-foot lengths of pipe are installed.

b. For sizes 2 inches and smaller, a guide shall be installed midway between required vertical supports. Such guides shall prevent pipe movement in a direction perpendicular to the axis of the pipe.

The spacing of supports indicated in Table P2605.1 determines the minimum number of supports required. Extra supports should be used as necessary because of fittings, connections, change in direction or connection to a vertical pipe.

SECTION P2606
PENETRATIONS

P2606.1 Sealing of annular spaces. The annular space between the outside of a pipe and the inside of a pipe sleeve or between the outside of a pipe and an opening in a building envelope wall, floor, or ceiling assembly penetrated by a pipe shall be sealed with caulking material or foam sealant or closed with a gasketing system. The caulking material, foam sealant or gasketing system shall be designed for the conditions at the penetration location and shall be compatible with the pipe, sleeve and building materials in contact with the sealing materials. Annular spaces created by pipes penetrating fire-resistance-rated assemblies or membranes of such assemblies shall be sealed or closed in accordance with the building portion of this code.

❖ The annular space created between a pipe sleeve and a pipe or an opening in the building envelope and a pipe must be sealed to prevent airflow though the space and to block the space from insects. The material used must be somewhat elastic to prevent structural loading of the pipe should there be some movement between the pipe and the opening. Caulking, expanding foam sealant or a gasketing system can be used to seal the ends of the annular space. It is not required to fill the entire annular space with material. The materials used to seal the annular space must be compatible with the pipe as well as the pipe sleeve material or building envelope materials.

Annular spaces created by penetrations in fire-resistance-rated assemblies must only be sealed in accordance with materials and methods specific to the firestop assembly chosen for the through penetration (see Section R302).

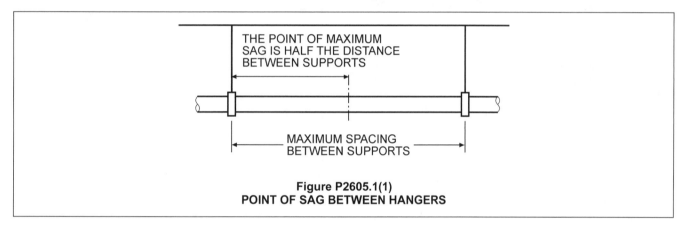

Figure P2605.1(1)
POINT OF SAG BETWEEN HANGERS

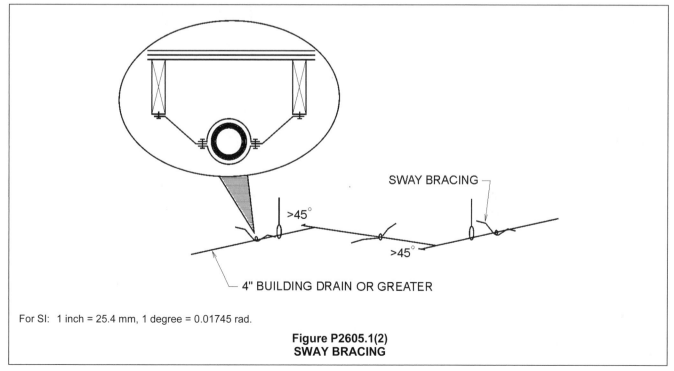

For SI: 1 inch = 25.4 mm, 1 degree = 0.01745 rad.

Figure P2605.1(2)
SWAY BRACING

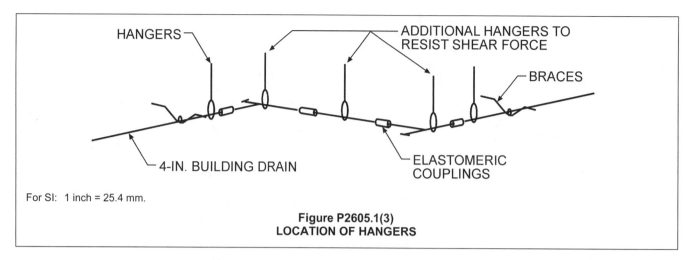

For SI: 1 inch = 25.4 mm.

Figure P2605.1(3)
LOCATION OF HANGERS

SECTION P2607
WATERPROOFING OF OPENINGS

P2607.1 Pipes penetrating roofs. Where a pipe penetrates a roof, a flashing of lead, copper, galvanized steel or an *approved* elastomeric material shall be installed in manner that prevents water entry into the building. Counterflashing into the opening of pipe serving as a vent terminal shall not reduce the required internal cross-sectional area of the vent pipe to less than the internal cross-sectional area of one pipe size smaller.

❖ Roof penetrations must be watertight where penetrated by piping. Use of lead, copper, plastic or galvanized steel flashing is a common method of waterproofing roof penetrations.

   Where using flashing materials, such as lead that wraps over the top of the pipe, significant reduction of the required cross-sectional area of the vent terminal should be avoided (see Commentary Figure P2607.1).

P2607.2 Pipes penetrating exterior walls. Where a pipe penetrates an exterior wall, a waterproof seal shall be made on the exterior of the wall by one of the following methods:

   1. A waterproof sealant applied at the joint between the wall and the pipe.

   2. A flashing of an *approved* elastomeric material.

❖ Exterior walls can be subject to driving rain and wind. Pipe penetrations need to be water tight. Sealing with a waterproof sealant or installation of an approved wall flashing can provide the necessary sealing.

SECTION P2608
WORKMANSHIP

P2608.1 General. Valves, pipes and fittings shall be installed in correct relationship to the direction of the flow. Burred ends shall be reamed to the full bore of the pipe.

❖ Drainage piping and fittings must be installed in the correct direction to assist the flow of waste. Table P3005.1 regulates the installation of fittings according to the change of direction. Pipe ends that are burred because of cutting procedures must be reamed to the full bore of the pipe to eliminate turbulence in water piping and rough edges that can cause blockage in drainage piping.

   Burred ends on the pipe's outer edges must be removed to create the desired joint. A chamfering tool can be used to remove the burred ends (see Commentary Figure P2608.1).

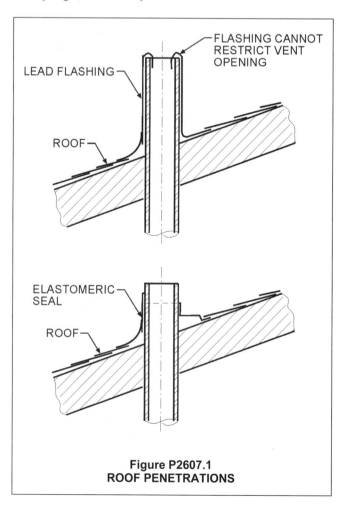

Figure P2607.1
ROOF PENETRATIONS

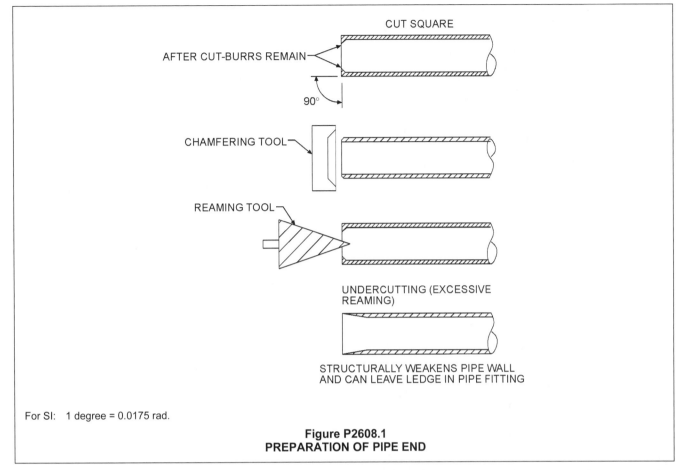

For SI: 1 degree = 0.0175 rad.

Figure P2608.1
PREPARATION OF PIPE END

SECTION P2609
MATERIALS EVALUATION AND LISTING

P2609.1 Identification. Each length of pipe and each pipe fitting, trap, fixture, material and device utilized in a plumbing system shall bear the identification of the manufacturer and any markings required by the applicable referenced standards. Nipples created from the cutting and threading of *approved* pipe shall not be required to be identified.

Exception: Where the manufacturer identification cannot be marked on pipe fittings and pipe nipples because of the small size of such fittings, the identification shall be printed on the item packaging or on documentation provided with the item.

❖ The manufacturer is given the option of determining the type of marking for the material. If there is no applicable standard or the applicable standard does not require identification of a material, identification of the manufacturer is still required by the code. Where the code indicates compliance with an approved standard, the manufacturer must comply with the requirements for marking in accordance with that standard.

The exception recognizes that nipples fabricated in the field from stock pipe lengths will not always have the required marking because the stock length markings occur at such wide intervals. Also, for small fittings and short factory-made pipe nipples, it is sometimes impractical for the manufacturer to apply the standard markings on each piece. In those situations, packaging or carton indications of the standards will suffice to comply with this section.

P2609.2 Installation of materials. Materials used shall be installed in strict accordance with the standards under which the materials are accepted and *approved*. In the absence of such installation procedures, the manufacturer's instructions shall be followed. Where the requirements of referenced standards or manufacturer's instructions do not conform to the minimum provisions of this code, the provisions of this code shall apply.

❖ Plumbing components and materials are to be installed in accordance with the installation requirements of the applicable standard listed in the code.

Where a standard is not applicable, the manufacturer's instructions and the code provide the only requirements. The code prevails where there are conflicts or inconsistencies between the code and the referenced standards and the manufacturer's instructions.

P2609.3 Plastic pipe, fittings and components. Plastic pipe, fittings and components shall be third-party certified as conforming to NSF 14.

❖ All plastic piping, fittings and plastic-pipe-related components, including solvent cements, primers, tapes, lubricants and seals used in plumbing systems, must be tested and certified as conforming to NSF 14. This includes all water service, water distribution and drainage piping and fittings as well as plastic piping system components, including but not limited to pipes, fittings, valves, joining materials, gaskets and appurtenances. This section does not apply to components that include plastic parts such as brass valves with a plastic stem or to fixture fittings such as fixture stop valves. Plastic piping systems, fittings and related components intended for use in the potable water supply system must comply with NSF 61 in addition to NSF 14.

P2609.4 Third-party certification. Plumbing products and materials required by the code to be in compliance with a referenced standard shall be *listed* by a third-party certification agency as complying with the referenced standards. Products and materials shall be identified in accordance with Section P2609.1.

❖ To simplify inspections and approvals, the code requires that all plumbing products and materials that are to be in compliance with a referenced standard be third-party certified. The code official only has to confirm that the product or material has the mark of the third-party certifying agency. Commentary Figure P2609.4 shows an example of typical markings on PVC plastic pipe.

P2609.5 Water supply systems. Water service pipes, water distribution pipes and the necessary connecting pipes, fittings, control valves, faucets and appurtenances used to dispense water intended for human ingestion shall be evaluated and listed as conforming to the requirements of NSF 61.

❖ Piping material that comes in contact with the potable water source must comply with the requirements of NSF 61. The purpose is to control potential adverse health effects produced by indirect additives, products, and materials that come in contact with potable water. Plastic piping systems, fittings and related components intended for use with the potable water supply system must comply with NSF 14 in addition to NSF 61.

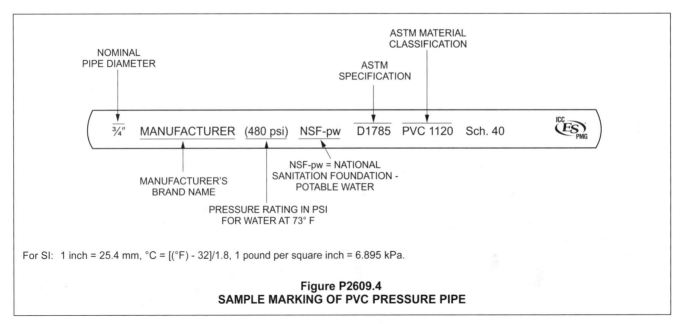

For SI: 1 inch = 25.4 mm, °C = [(°F) - 32]/1.8, 1 pound per square inch = 6.895 kPa.

Figure P2609.4
SAMPLE MARKING OF PVC PRESSURE PIPE

Chapter 27:
Plumbing Fixtures

User note: Code change proposals to this chapter will be considered by the IRC – Plumbing and Mechanical Code Development Committee during the 2015 (Group A) Code Development Cycle.

General Comments

Chapter 27 addresses each type of fixture individually, including the accessory parts or installation criteria specific to a given fixture.

Chapter 27 outlines the fixtures, faucets and fixture fittings and the standards for their manufacture. In particular, it covers the materials used in the manufacture of tailpieces and traps, their fittings and the strainer requirements for the fixture outlet. This chapter also regulates fixture tailpieces and specifies the minimum tailpiece size required for each fixture. Chapter 27 lists access requirements for inspection and repair of slip-joint connections.

This chapter also details installation criteria for material, methods and location of fixtures and their parts.

The majority of the chapter is devoted to detailing specific provisions relating to waste receptors, kitchen fixtures and appliances, bathroom and toilet room fixtures, laundry facilities, valves and fixture fittings and macerating toilet systems.

Purpose

Fixtures must be of the proper type, approved for the purpose and installed properly to promote usability and a safe, sanitary condition. Chapter 27 contains the standards to which fixtures must be manufactured and installed.

SECTION P2701
FIXTURES, FAUCETS AND FIXTURE FITTINGS

P2701.1 Quality of fixtures. Plumbing fixtures, faucets and fixture fittings shall have smooth impervious surfaces, shall be free from defects, shall not have concealed fouling surfaces, and shall conform to the standards indicated in Table P2701.1 and elsewhere in this code.

❖ This section requires plumbing fixtures, faucets and fixture fittings to be in compliance with the appropriate referenced standards found in Table P2701.1 and other locations in the code. The products cannot have pervious surfaces, defects or concealed fouling surfaces that might lead to insanitary conditions that the user cannot resolve.

SECTION P2702
FIXTURE ACCESSORIES

P2702.1 Plumbing fixtures. Plumbing fixtures, other than water closets, shall be provided with *approved* strainers.

Exception: Hub drains receiving only clear water waste and standpipes shall not require strainers.

❖ This section requires the drain outlet(s) of plumbing fixtures, except water closets, to be equipped with approved strainers. This requirement prevents items such as toothbrushes, rings, cosmetic devices and large food waste particles from entering the drainage system and creating clogs. The exception recognizes that hub drains receiving clear water waste and standpipes do not require strainers because these are not the type of fixtures where items will be inadvertently dropped into them.

TABLE P2701.1. See page 27-2.

P2702.2 Waste fittings. Waste fittings shall conform to ASME A112.18.2/CSA B125.2, ASTM F409 or shall be made from pipe and pipe fittings complying with any of the standards indicated in Tables P3002.1(1) and P3002.3.

❖ Waste fittings are the tail pieces, traps, waste and overflow assemblies and continuous waste assemblies used to connect the outlet of a fixture to the fixture drain pipe. The American Society of Mechanical Engineers (ASME) and ASTM International (ASTM) standards regulate the material quality, wall thicknesses and dimensional uniformity of tubular size waste fittings, among other criteria. The ASME standard covers metal and plastic fittings; the ASTM standard covers polypropylene plastic fittings.

The installer is not confined to using tubular waste fittings to connect a fixture outlet to a fixture drain since this section also allows any pipe or fittings that comply with above-ground drainage and vent pipe and fittings to be used [see Table P3002.1(1)].

P2702.3 Plastic tubular fittings. Plastic tubular fittings shall conform to ASTM F409 as indicated in Table P2701.1.

❖ Literally, this section requires plastic tubular waste fittings to be made of only polypropylene plastic. However, Section P2702.2 allows tubular waste fittings to conform to ASME A112.18.2/CSA B125.2, which allows other plastic materials such as PVC and ABS.

TABLE P2701.1
PLUMBING FIXTURES, FAUCETS AND FIXTURE FITTINGS

MATERIAL	STANDARD
Air gap fittings for use with plumbing fixtures, appliances and appurtenances	ASME A 112.1.3
Bathtub/whirlpool pressure-sealed doors	ASME A 112.19.15
Diverters for faucets with hose spray, anti-syphon type, residential application	ASTM A112.18.1/CSA B125.1
Enameled cast-iron plumbing fixtures	ASME A 112.19.1M/CSA B45.2
Floor drains	ASME A 112.6.3
Floor-affixed supports for off-the-floor plumbing fixtures for public use	ASME A 112.6.1M
Framing-affixed supports for off-the-floor water closets with concealed tanks	ASME A 112.6.2
Hose connection vacuum breaker	ASSE 1052
Hot water dispensers, household storage type, electrical	ASSE 1023
Household disposers	ASSE 1008
Hydraulic performance for water closets and urinals	ASME A 112.19.2/CSA B45.1
Individual automatic compensating valves for individual fixture fittings	ASME A 112.18.1/CSA B125.1
Individual shower control valves anti-scald	ASSE 1016/ASME A 112.1016/CSA B125.16
Macerating toilet systems and related components	ASME A 112.3.4/CSA B45.9
Nonvitreous ceramic plumbing fixtures	ASME A 112.19.2/CSA B45.1
Plastic bathtub units	CSA B45.5/IAPMO Z124, ASME A112.19.2/CSA B45.1
Plastic lavatories	CSA B45.5/IAPMO Z124
Plastic shower receptors and shower stall	CSA B45.5/IAPMO Z124
Plastic sinks	CSA B45.5/IAPMO Z124
Plastic water closet bowls and tanks	CSA B45.5/IAPMO Z124
Plumbing fixture fittings	ASME A 112.18.1/CSA B125.1
Plumbing fixture waste fittings	ASME A 112.18.2/CSA B125.2, ASTM F409
Porcelain-enameled formed steel plumbing fixtures	ASME A 112.19.1/CSA B45.2
Pressurized flushing devices for plumbing fixtures	ASSE 1037, CSA B125.3
Specification for copper sheet and strip for building construction	ASTM B370
Stainless steel plumbing fixtures	ASME A 112.19.3/CSA B45.4
Suction fittings for use in whirlpool bathtub appliances	ASME A 112.19.7 /CSA B45.10
Temperature-actuated, flow reduction valves to individual fixture fittings	ASSE 1062
Thermoplastic accessible and replaceable plastic tube and tubular fittings	ASTM F409
Trench drains	ASME A 112.6.3
Trim for water closet bowls, tanks and urinals	ASME A 112.19.5/CSA B45.15
Vacuum breaker wall hydrant-frost-resistant, automatic-draining type	ASSE 1019
Vitreous china plumbing fixtures	ASME A 112.19.2/CSA B45.1
Wall-mounted and pedestal-mounted, adjustable and pivoting lavatory and sink carrier systems	ASME A 112.19.12
Water closet flush tank fill valves	ASSE 1002, CSA B125.3
Whirlpool bathtub appliances	ASME A 112.19.7 /CSA B45.10

P2702.4 Carriers for wall-hung water closets. Carriers for wall-hung water closets shall conform to ASME A112.6.1 or ASME A112.6.2.

❖ Contrary to what the name implies, a wall-hung water closet is not actually supported by the wall that the water closet appears to be bolted to. The weight of the water closet and of the person sitting on the water closet is transferred through bolts or studs that are screwed into a robust metal carrier assembly. The carrier assembly is bolted to the floor behind the wall. The referenced standards, ASME A112.6.1 or ASME A112.6.2, regulate the design and test requirements for the strength of the hanger (see Commentary Figure P2702.4).

SECTION P2703
TAIL PIECES

P2703.1 Minimum size. Fixture tail pieces shall be not less than $1^1/_2$ inches (38 mm) in diameter for sinks, dishwashers, laundry tubs, bathtubs and similar fixtures, and not less than $1^1/_4$ inches (32 mm) in diameter for bidets, lavatories and similar fixtures.

❖ This section regulates the tail piece size for specific types of fixtures. The tail piece of a fixture is the drain tube immediately downstream of the fixture waste outlet opening and upstream of the fixture trap. Lavatories and bidets must have a tail piece diameter that is not less than $1^1/_4$ inches (32 mm). Sinks, dishwashers, laundry tubs and bathtubs must have a tail piece diameter that is not less than $1^1/_2$ inches (38 mm). These requirements apply to similar types of fixtures.

SECTION P2704
ACCESS TO CONNECTIONS

P2704.1 General. Slip joints shall be made with an *approved* elastomeric gasket and shall be installed only on the trap outlet, trap inlet and within the trap seal. Fixtures with concealed slip-joint connections shall be provided with an access panel or utility space not less than 12 inches (305 mm) in its smallest dimension or other *approved* arrangement so as to provide access to the slip connections for inspection and repair.

❖ Slip joints are found on tubular waste fittings (including tubular waste traps) that are used to connect fixtures with $1^1/_4$-inch and $1^1/_2$-inch (32 mm and 38 mm) tubular waste outlet tail pieces to the fixture drain (see Commentary Figure P2704.1). Because slip-joint connections can become loose or start leaking over time, they must be located where they can be accessed without the need for removing permanent portions of the building, such as drywall or cabinetry. Where access panels (or cabinet doors) are provided for such access, the opening must be not less than 12 inches (305 mm) in the least dimension so that service personnel have adequate space to work on the joint.

One of the most common violations of this section is where a bathtub is located above a finished ceiling and tubular waste fittings with slip joints have been used for the waste outlet connections. This section would require an access panel in the ceiling below. However, the need for an access panel could be eliminated if the waste overflow assembly and the slip-jointed trap were replaced with a polyvynil chloride (PVC) overflow assembly and trap having solvent welded joints. The code does not require access panels in finished walls where joints are accessible from an under-floor space area (crawl space or unfinished basement ceiling).

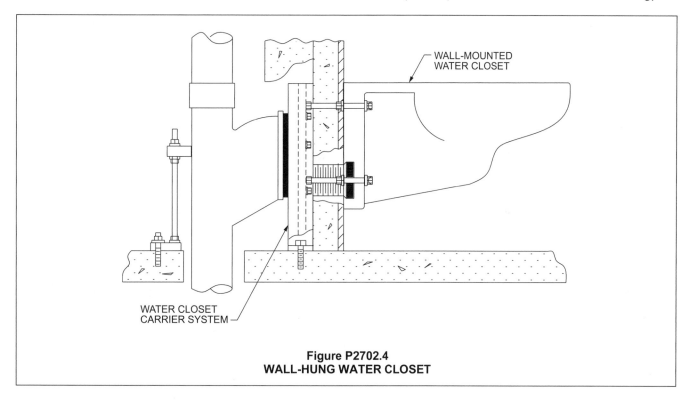

Figure P2702.4
WALL-HUNG WATER CLOSET

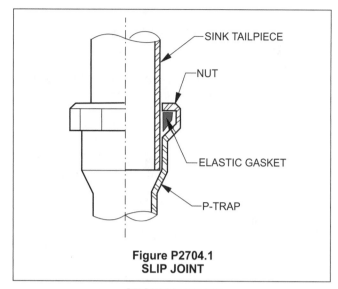

Figure P2704.1
SLIP JOINT

Labels: SINK TAILPIECE, NUT, ELASTIC GASKET, P-TRAP

SECTION P2705
INSTALLATION

P2705.1 General. The installation of fixtures shall conform to the following:

1. Floor-outlet or floor-mounted fixtures shall be secured to the drainage connection and to the floor, where so designed, by screws, bolts, washers, nuts and similar fasteners of copper, copper alloy or other corrosion-resistant material.

2. Wall-hung fixtures shall be rigidly supported so that strain is not transmitted to the plumbing system.

3. Where fixtures come in contact with walls and floors, the contact area shall be water tight.

4. Plumbing fixtures shall be usable.

5. Water closets, lavatories and bidets. A water closet, lavatory or bidet shall not be set closer than 15 inches (381 mm) from its center to any side wall, partition or vanity or closer than 30 inches (762 mm) center-to-center between adjacent fixtures. There shall be a clearance of not less than 21 inches (533 mm) in front of a water closet, lavatory or bidet to any wall, fixture or door.

6. The location of piping, fixtures or equipment shall not interfere with the operation of windows or doors.

7. In flood hazard areas as established by Table R301.2(1), plumbing fixtures shall be located or installed in accordance with Section R322.1.6.

8. Integral fixture-fitting mounting surfaces on manufactured plumbing fixtures or plumbing fixtures constructed on site, shall meet the design requirements of ASME A112.19.2/CSA B45.1 or ASME A112.19.3/CSA B45.4.

❖ Most water closets utilize a floor-flange drainage connection where the water closet bolts to the flange and the flange is attached to the floor [see Commentary Figure P2705.1(1)]. Other designs of water closets are fastened directly to the floor. The fasteners must be corrosion resistant because they could be repeatedly exposed to moisture. The required screws, bolts, washers, nuts and similar fasteners are usually made of copper, copper alloy (previously referred to as brass) or plastic.

Wall-hung water closets must be rigidly supported using specially designed carriers that are securely fastened to the floor structure behind the wall (see commentary, Section P2702.4). The piping to the water closet must not provide any support for the water closet. Wall-hung lavatories must be adequately supported and fastened to the wall so the lavatory is not applying loads to the piping systems that connect to the lavatory.

Where fixtures contact walls or floors, those joints must be caulked so that water does not enter the concealed space between the fixture and the wall or floor [see Commentary Figure P2705.1(2)].

Plumbing fixtures require space around them for use and cleaning purposes. This section item requires that centerline of water closets, lavatories and bidets be not less than 15 inches (381 mm) from an adjacent sidewall, partition or vanity and not be less than 30 inches (762 mm) from the centerline to any adjacent fixture. A clearance of not less than 21 inches (533 mm) is required in front of a water closet, bidet and lavatory to any wall, fixture or door [see Commentary Figure P2705.1(3)]. The 21-inch (533 mm) clearance does not apply to the swing arc of a door.

Plumbing fixture locations must not interfere with the operation of any window or door. For example, if opening a window or door causes it to swing inward, the swing of the window or door must not be obstructed by a plumbing fixture. For the purposes of interpreting this section, the maximum window opening is the mechanical limit of the window assembly or 90 degrees (1.57 rad) from the fully closed position, whichever is less. The maximum door opening is considered to be 90 degrees (1.57 rad) from the fully closed position [see Commentary Figure P2705.1(4)]. Also see Section R307, which regulates fixture location.

Plumbing fixtures must be located at or above the flood elevation required in Section R322.1.6 unless specially designed to prevent floodwaters from entering during conditions of the design flood. Plumbing fixtures should not be necessary below elevated buildings, given the use of enclosed areas below elevated buildings is limited to parking of vehicles, building access and storage (see commentary, Section R322). See Section R307, which also regulates fixture location.

Fixture fittings are faucets or waste outlet assemblies that are intended to be mounted integrally to plumbing fixtures, such as lavatories or sinks. The design of the fixture fittings is in accordance with fixture-fitting standards that are coordinated with plumbing-fixture standards. For example, a 4-inch (102 mm) centerset lavatory faucet will fit a lavatory that is drilled in accordance with one of the plumbing-fixture standards. Where plumbing fixtures are field constructed, the mounting-surface dimensions required must be in accordance with the same standards.

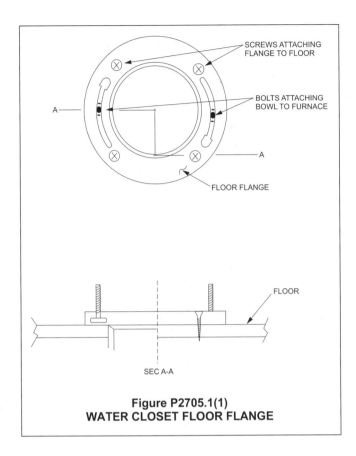

**Figure P2705.1(1)
WATER CLOSET FLOOR FLANGE**

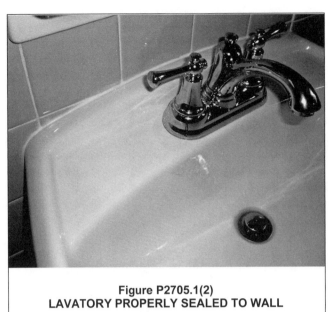

**Figure P2705.1(2)
LAVATORY PROPERLY SEALED TO WALL**

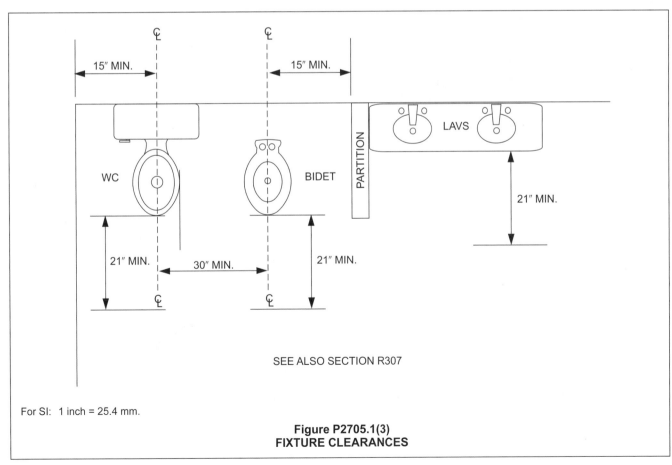

SEE ALSO SECTION R307

For SI: 1 inch = 25.4 mm.

**Figure P2705.1(3)
FIXTURE CLEARANCES**

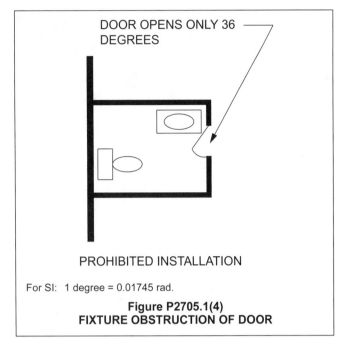

DOOR OPENS ONLY 36
DEGREES

PROHIBITED INSTALLATION

For SI: 1 degree = 0.01745 rad.

Figure P2705.1(4)
FIXTURE OBSTRUCTION OF DOOR

SECTION P2706
WASTE RECEPTORS

P2706.1 General. For other than hub drains that receive only clear-water waste and standpipes, a removable strainer or basket shall cover the waste outlet of waste receptors. Waste receptors shall not be installed in concealed spaces. Waste receptors shall not be installed in plenums, attics, crawl spaces or interstitial spaces above ceilings and below floors. Waste receptors shall be readily accessible.

❖ Waste receptors include floor sinks, hub drains, laundry tubs, standpipes and funnel-equipped floor drains. Waste receptors are typically used to receive discharge from fixtures or appliances that are required to discharge through an air break or air gap, such as heating, ventilating and air-conditioning (HVAC) condensate drains, water heater T&P valve discharge pipes, water heater drip pan drains, clothes washer drains and dishwasher drains. Waste receptors must be equipped with a strainer unless the receptor is an open hub receptor or standpipe. Note that open hub receptors can only be installed in locations where the surrounding floor is water impervious.

Waste receptors cannot be concealed in an interstitial space such as inside of a wall, above a ceiling and below a floor, or a plenum. They also cannot be installed in areas not occasionally visited by the building occupant. These spaces would be attics and crawl spaces (see Commentary Figure P2706.1).

Waste receptors must be readily accessible, meaning that access does not require removal or opening of a panel or panel door.

P2706.1.1 Hub drains. Hub drains shall be in the form of a hub or a pipe that extends not less than 1 inch (25 mm) above a water-impervious floor.

❖ Hub drain drains are commonly used for a disposal point for condensate from HVAC systems and water heater T&P valve discharges.

P2706.1.2 Standpipes. Standpipes shall extend not less than 18 inches (457 mm) and not greater than 42 inches (1067 mm) above the trap weir.

❖ A standpipe is typically used for capturing the waste flow from a pumped discharge plumbing appliance such as a dishwasher or a clothes washer. An 18-inch (457 mm) minimum height standpipe provides a small volume for accumulation of waste flow above the entrance to the trap. If the appliance discharges at a greater rate than the trap is discharging to the fixture drain, the accumulated volume will not spill out the top of the standpipe. The accumulation of waste volume above the entry of the trap also creates a small amount of head pressure on the trap inlet to make the trap pass the waste flow at a higher rate. The limitation on the maximum height of a standpipe is so the falling waste water does not gain too much velocity, which might cause the trap to self-siphon.

Standpipes and the traps for standpipes can be concealed within construction, but the inlet to the standpipe must be accessible to allow drain cleaning.

P2706.1.2.1 Laundry tray connection to standpipe. Where a laundry tray waste line connects into a standpipe for an automatic clothes washer drain, the standpipe shall extend not less than 30 inches (762 mm) above the standpipe trap weir and shall extend above the flood level rim of the laundry tray. The outlet of the laundry tray shall not be greater than 30 inches (762 mm) horizontally from the standpipe trap.

❖ A laundry tray is more commonly known as a laundry tub or a laundry sink and typically has a capacity of at least 20 gallons (76 L). A common laundry room arrangement places the laundry tray adjacent to the clothes washer and thus adjacent to the standpipe for the clothes washer discharge.

The discharge from a laundry tray (laundry tub) is allowed to connect to the clothes washer standpipe if the top of the standpipe is at least 30 inches (762 mm) above the trap weir of the clothes washing standpipe trap and the laundry tray outlet is not more than 30 inches (762 mm) horizontally from the standpipe trap. To prevent overflow of the standpipe, the top of the standpipe must be higher than the flood level rim of the sink. The 30-inch (762 mm) horizontal distance limitation is intended to maintain a sanitary condition by minimizing the length of piping on the inlet side (house side) of the trap (see Commentary Figure P2706.1.2.1).

P2706.2 Prohibited waste receptors. Plumbing fixtures that are used for washing or bathing shall not be used to receive the discharge of indirect waste piping.

Exceptions:

1. A kitchen sink trap is acceptable for use as a receptor for a dishwasher.

2. A laundry tray is acceptable for use as a receptor for a clothes washing machine.

❖ For sanitary reasons, this section prohibits discharge of indirect waste piping into a sink used for washing or bathing purposes. However, this provision does not apply to the indirect connection of a dishwasher to a kitchen sink because a dishwasher essentially performs the same function as a kitchen sink. By the same logic, this section also permits a clothes washing machine to discharge into a laundry tub (see Commentary Figure 2706.1).

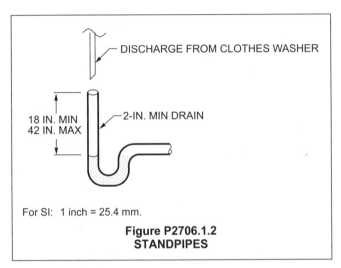

For SI: 1 inch = 25.4 mm.

Figure P2706.1.2
STANDPIPES

SECTION P2707
DIRECTIONAL FITTINGS

P2707.1 Directional fitting required. *Approved* directional-type branch fittings shall be installed in fixture tailpieces receiving the discharge from food-waste disposer units or dishwashers.

❖ Waste disposal units and dishwashers connected to a tailpiece of a kitchen sink must use a directional-type branch fitting to guide the flow, thus helping to prevent blockages and the backup of discharge into the sink. One option is to use a tee equipped with a baffle, a wye branch or a similar type of directional fitting that would guide the discharge in the direction of flow.

SECTION P2708
SHOWERS

P2708.1 General. Shower compartments shall have not less than 900 square inches (0.6 m^2) of interior cross-sectional area. Shower compartments shall be not less than 30 inches (762 mm) in minimum dimension measured from the finished interior dimension of the shower compartment, exclusive of fixture valves, shower heads, soap dishes, and safety grab bars or rails. The minimum required area and dimension shall be measured from the finished interior dimension at a height equal to the top of the threshold and at a point tangent to its centerline and shall be continued to a height of not less than 70 inches (1778 mm) above the shower drain outlet. Hinged

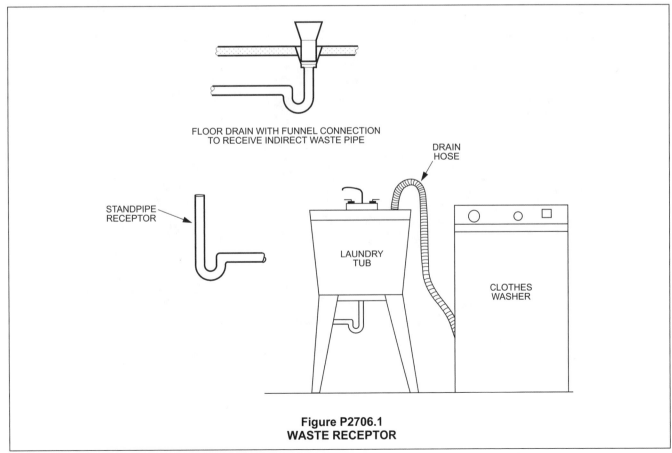

Figure P2706.1
WASTE RECEPTOR

shower doors shall open outward. The wall area above built-in tubs having installed shower heads and in shower compartments shall be constructed in accordance with Section R702.4. Such walls shall form a water-tight joint with each other and with either the tub, receptor or shower floor.

Exceptions:

1. Fold-down seats shall be permitted in the shower, provided the required 900-square-inch (0.6 m²) dimension is maintained when the seat is in the folded-up position.

2. Shower compartments having not less than 25 inches (635 mm) in minimum dimension measured from the finished interior dimension of the compartment provided that the shower compartment has a cross-sectional area of not less than 1,300 square inches (0.838 m²).

❖ In plan view, from the elevation of the top of the threshold to an elevation 70 inches (1778 mm) above the top of the drain, the interior dimension from the back wall to the centerline of the threshold must not be less than 30 inches (762 mm) and the area of the compartment must not be less than 900 square inches (0.6 m²). When measuring the interior dimension of the compartment, minor encroachments into the area, such as handrails, shower valves, shower heads and soap

dishes, are to be ignored (see Commentary Figure P2708.1).

Because of the limited area within the typical size of most shower compartments, shower doors must open outward. This allows for the user to easily step out of the shower without having to navigate on a wet shower floor around a shower door.

Section R307.2 requires that the walls of shower compartments be nonabsorbent and extend to a point that is not less than 72 inches (1829 mm) above the finished floor outside the shower. To protect concealed areas from water damage, this section (Section P2708.1) requires all intersections of the nonabsorbent walls to each other and to adjacent finished surfaces be water tight.

Exception 1 allows a wall-mounted folding seat to be installed in a shower as long as the minimum required area or minimum required interior dimension is not violated with the seat in the folded-up position.

Exception 2 allows for a shower to have an interior dimension of less than 30 inches (762 mm), but not less than 25 inches (635 mm) as long as the area of the shower is not less than 1,300 square inches (0.838 m²). This exception is necessary for replacing a standard size bathtub [typically 30 inches by 60 inches (762 mm by 1524 mm)] with a shower compartment.

For SI: 1 inch = 25.4 mm.

Figure P2706.1.2.1
LAUNDRY TUB CONNECTION TO STANDPIPE

For SI: 1 inch = 25.4 mm.

PLUMBING FIXTURES

shower doors shall open outward. The wall area above built-in tubs having installed shower heads and in shower compartments shall be constructed in accordance with Section R702.4. Such walls shall form a water-tight joint with each other and with either the tub, receptor or shower floor.

Exceptions:

1. Fold-down seats shall be permitted in the shower, provided the required 900-square-inch (0.6 m²) dimension is maintained when the seat is in the folded-up position.

2. Shower compartments having not less than 25 inches (635 mm) in minimum dimension measured from the finished interior dimension of the compartment provided that the shower compartment has a cross-sectional area of not less than 1,300 square inches (0.838 m²).

❖ In plan view, from the elevation of the top of the threshold to an elevation 70 inches (1778 mm) above the top of the drain, the interior dimension from the back wall to the centerline of the threshold must not be less than 30 inches (762 mm) and the area of the compartment must not be less than 900 square inches (0.6 m²). When measuring the interior dimension of the compartment, minor encroachments into the area, such as handrails, shower valves, shower heads and soap

dishes, are to be ignored (see Commentary Figure P2708.1).

Because of the limited area within the typical size of most shower compartments, shower doors must open outward. This allows for the user to easily step out of the shower without having to navigate on a wet shower floor around a shower door.

Section R307.2 requires that the walls of shower compartments be nonabsorbent and extend to a point that is not less than 72 inches (1829 mm) above the finished floor outside the shower. To protect concealed areas from water damage, this section (Section P2708.1) requires all intersections of the nonabsorbent walls to each other and to adjacent finished surfaces be water tight.

Exception 1 allows a wall-mounted folding seat to be installed in a shower as long as the minimum required area or minimum required interior dimension is not violated with the seat in the folded-up position.

Exception 2 allows for a shower to have an interior dimension of less than 30 inches (762 mm), but not less than 25 inches (635 mm) as long as the area of the shower is not less than 1,300 square inches (0.838 m²). This exception is necessary for replacing a standard size bathtub [typically 30 inches by 60 inches (762 mm by 1524 mm)] with a shower compartment.

For SI: 1 inch = 25.4 mm.

Figure P2706.1.2.1
LAUNDRY TUB CONNECTION TO STANDPIPE

27-8

2015 INTERNATIONAL RESIDENTIAL CODE® COMMENTARY

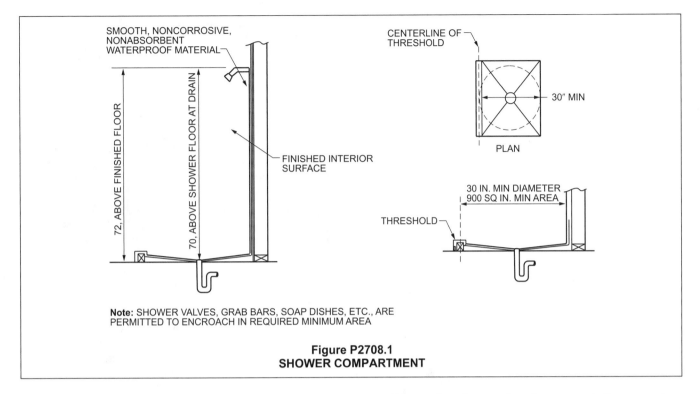

Figure P2708.1
SHOWER COMPARTMENT

P2708.1.1 Access. The shower compartment access and egress opening shall have a clear and unobstructed finished width of not less than 22 inches (559 mm).

❖ Twenty-two inches (559 mm) corresponds to the approximate width of the shoulders of the average adult.

P2708.2 Shower drain. Shower drains shall have an outlet size of not less than $1^1/_2$ inches [38 mm] in diameter.

❖ The minimum size of a shower drain outlet is $1^1/_2$ inches.

P2708.3 Water supply riser. Water supply risers from the shower valve to the shower head outlet, whether exposed or concealed, shall be attached to the structure using support devices designed for use with the specific piping material or fittings anchored with screws.

❖ The shower head riser must be attached to a building structural element to protect the pipe and shower valve from damage caused by shower head movement [see Commentary Figures P2708.3(1) and P2708.3(2)]. Users commonly put force on the shower head arm when they adjust the spray pattern, add hose attachments and hang bathing product organizers. A standard practice for installing the riser is to place a drop-ear elbow at the top of the riser. A drop-ear elbow has two wing connections, permitting the elbow to be screwed into a structural member. Where drop-ear elbows are not used, one or more pipe straps are installed at the top of the riser. A support system specifically designed for riser support can also be used. This section intends to prohibit shoddy practice such as supporting the riser with nails driven beside the pipe and bent over the pipe.

Where the shower riser is exposed, the manufacturer typically provides a clamp or other means of support matching the finish of the exposed piping. The clamp or other support is firmly secured to a structural member and attached to the riser.

P2708.4 Shower control valves. Individual shower and tub/shower combination valves shall be equipped with control valves of the pressure-balance, thermostatic-mixing or combination pressure-balance/thermostatic-mixing valve types with a high limit stop in accordance with ASSE 1016/ASME A112.1016/CSA B125.16. The high limit stop shall be set to limit the water temperature to not greater than 120°F (49°C). In-line thermostatic valves shall not be used for compliance with this section.

❖ Commentary Figure P2708.4 shows exposure times at various water temperatures that cause third-degree burns to human skin. Industry safety experts have chosen 120°F (49°C) to be the maximum safe water temperature for showering and bathing purposes. In a showering application where the shower head is fixed, a user subjected to a surge of high-temperature water usually reacts by abruptly moving away from the hot spray. This abrupt movement often causes falls, resulting in the stunned user not being able to escape the scalding water stream. Children and some elderly persons often have delayed or no physical reaction to the scalding water such that they continue to stand in the stream. Therefore, showers and combination tub/showers must have a shower control valve that is capable of protecting an individual from being exposed to water temperatures in excess of 120°F (49°C). The control valve must be installed at the point of use. In other words, the person in the shower must have access to the control handle(s) of the valve.

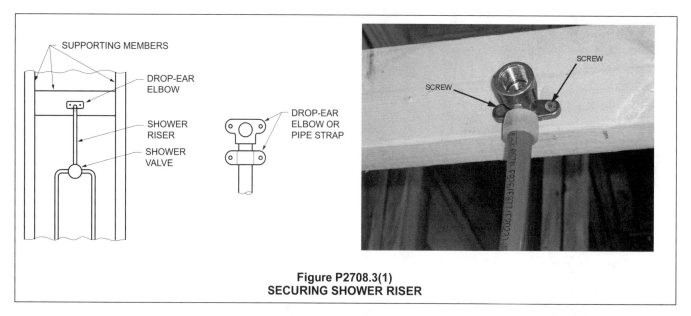

Figure P2708.3(1)
SECURING SHOWER RISER

Figure P2708.3(2)
SPECIAL BRACKET FOR SHOWER RISER
(Photo courtesy of Holdrite-Hubbard Enterprises)

The control valve must be designed to protect against two types of events: (1) extreme temperature fluctuations from the user's set temperature caused by changes in hot or cold water distribution line pressures and (2) extreme temperature conditions caused by the user either purposely or accidentally adjusting the control valve to deliver the hottest water available from the hot water distribution system. Where water inlet pressures or temperatures fluctuate during shower use, control valves complying with ASSE 1016/ASME A112.1016/CSA B125.16 must automatically and rapidly adjust to maintain the water discharge temperature to ±3.6°F (±2°C) of the user-selected temperature.

The standards also require that shower control valves have a maximum temperature-limit device that is not adjustable by the user at the point of use. The high-limit stop is typically an adjustable set screw or cam that is manually set to limit the travel of the control valve handle. The high-limit stop must be field adjusted at the time of installation to limit the delivered water temperature to a maximum of 120°F (49°C). Most control valve manufacturers' instructions require that a thermometer be used to verify the maximum discharge temperature from the valve.

The three types of shower control valves available are pressure balancing, thermostatic and combination thermostatic/pressure balancing. A pressure-balancing valve senses changes in pressure of the hot and cold water supplies (up to 50-percent pressure change). If the pressure on one side changes, the valve reacts so that the flow from each side of the valve is adjusted to maintain the user's temperature selection. The thermostatic valve senses the discharge water temperature and adjusts flows of supply water to the valve to maintain the user's set temperature. A thermostatic valve provides some limited protection against hot and cold supply pressure fluctuations (up to 20-percent pressure change). A combination thermostatic/pressure-balancing valve adjusts to changes in supply pressures (up to 50-percent pressure change) and discharge temperature.

Pressure-balancing control valves are used in a majority of shower applications. One slight disadvantage of the pressure-balanced-type control valve is that any change in the temperature setting of the water heating system will affect the maximum discharge temperature available from the control valve. If the system temperature is lowered, some users might

	TIME AND TEMPERATURE RELATIONSHIP TO SERIOUS BURNS		
		Adults (skin thickness of 2.5 mm)	Children (skin thickness of 0.56 mm)
WATER TEMPERATURE		Time required for a third-degree burn to occur	
155°F	(68°C)	1 second	0.5 second
148°F	(64°C)	2 seconds	1 second
140°F	(60°C)	5 seconds	1 second
133°F	(56°C)	15 seconds	4 seconds
127°F	(52°C)	1 minute	10 seconds
124°F	(51°C)	3 minutes	1.5 minutes
120°F	(48°C)	5 minutes	2.5 minutes
100°F	(37°C)	Safe temperature for bathing	Safe temperature for bathing

For SI: 1 inch = 25.4 mm.

Figure P2708.4
TEMPERATURE BURN CHART

complain that they are not receiving hot enough water. If the system temperature is raised, the maximum water temperature available from the valve might be in excess of 120°F (49°C). Seasonal fluctuations in the cold water supply temperature can also affect the maximum discharge temperature. Thus, pressure-balancing-type control valves might require periodic readjustment of the high-limit setting.

The maximum discharge temperature from a thermostatic-type control valve is not affected by changes in the temperatures of the hot and cold water supplies to the valve except for when the hot water supply temperature falls below the high-limit setting. Although this type of valve might significantly reduce the need for readjustments to the high-limit setting, due to its inherent limited protection against supply-side pressure changes, a thermostatic valve might not be suitable for applications where large pressure fluctuations (greater than 20-percent change in pressure) are expected.

A combination pressure balanced/thermostatic control valve offers the greatest degree of protection for the shower user. As the name implies, this type of valve offers the best features of both the pressure-balanced-type valve and the thermostatic-type valve designs.

This section does not apply to control valves that supply water only for filling tubs (see Section P2713.3). Inline-type thermostatic and pressure-balancing valves do not satisfy the requirements of this section because they are not the final control valve that the user adjusts at the point of use and they do not have a high-limit setting feature.

P2708.5 Hand showers. Hand-held showers shall conform to ASME A112.18.1/CSA B125.1. Hand-held showers shall provide backflow protection in accordance with ASME A112.18.1/CSA B125.1 or shall be protected against backflow by a device complying with ASME A112.18.3.

❖ Because it is possible for a hand-held shower to be submerged in the bathtub or shower compartment base, a cross connection to waste water could be created. The hand-held shower must have adequate protection against backflow. The backflow protection may be integral to the hand-held shower or may be provided by an external backflow device that meets ASME A112.18.3. These types of showers are commonly installed in accessible shower enclosures. Their use, however, has increased in popularity in showers within a dwelling unit. The referenced standards specify the requirements for backflow protection of the hand-held shower.

A hand-held shower, used in a hand-held mode, is not considered to be the type of shower that requires a means to protect the user from scalding. The reason for this comes from the justifications that eventually led to the code requirement for antiscald shower valves (see commentary, Section P2708.3). Therefore, if the hand-held shower is mounted for use in a hands-free mode with the user in a standing position, a hand-held shower is no different than any other fixed-type shower head and must be supplied through an antiscald shower valve in accordance with Section P2708.3.

However, where a hand-held shower is used in a hand-held mode, most users, when subjected to a high-temperature surge of water spray, will either drop or simply redirect the hand-held shower. If the user is standing, the potential for a fall injury is minimal. If the user is sitting, fall injuries cannot occur. Therefore, where a hand-held shower is used only in a hand-held mode, the water discharged from a hand-held-only shower is not required to be supplied through an antiscald shower valve.

SECTION P2709
SHOWER RECEPTORS

P2709.1 Construction. Where a shower receptor has a finished curb threshold, it shall be not less than 1 inch (25 mm) below the sides and back of the receptor. The curb shall be not less than 2 inches (51 mm) and not more than 9 inches (229 mm) deep when measured from the top of the curb to the top of the drain. The finished floor shall slope uniformly toward the drain not less than $^1/_4$ unit vertical in 12 units horizontal (2-percent slope) nor more than $^1/_2$ unit vertical per 12 units horizontal (4-percent slope) and floor drains shall be flanged to provide a water-tight joint in the floor.

❖ This section regulates three aspects of shower receptors: (1) threshold height, (2) finished floor slope and (3) connection of the floor drain to the receptor. For factory-fabricated shower receptors, the requirements of the section are automatically covered by the applicable standards, ANSI Z124.2 and CSA B45.5. This section provides the necessary information for field-fabricated shower receptors.

The first part of the first sentence of this section allows for a shower receptor to not have a curb. This is necessary to enable "zero height threshold" shower receptors to be installed for wheelchair access or for ease of access for disabled users. If the shower receptor is to be designed with a threshold, this section requires that the top of the threshold be not less than 1 inch (25 mm) below the sides and back of the receptor. This is required so that if the drain should clog, water will overflow the threshold before overflowing the back or sides of the receptor. With respect to the top of the drain, the top of the curb cannot be less than 2 inches (51 mm) above the drain. This allows for some amount of water storage in the receptor so that if the drain is sluggish or a washcloth is accidently dropped over the drain, the user will notice the accumulation of water on the floor before water begins flowing over the threshold. There is also a maximum height of the top of the threshold above the drain of 9 inches (229 mm) so that the user does not have to step too far down into the receptor.

The required minimum slope of the finished floor is to keep shower water moving towards the drain. The maximum slope of the finished floor is to keep the shower floor from being too steep so that the user could slip and fall too easily.

Obviously, a floor drain in the shower receptor must be connected to the receptor in a water-tight manner. This requires a specially flanged drain design that enables the installer to make a proper water-tight seal. There are two types of special floor drains for showers: (1) flanged type and (2) clamp-ring type. Flanged-type drains are used on factory-fabricated receptors such as one-piece fiberglass or molded plastic receptors. Clamp-ring-type drains are used for field-fabricated receptors ("built up") where a liner or a coating that acts like a liner is used to make the shower floor waterproof.

P2709.2 Lining required. The adjoining walls and floor framing enclosing on-site built-up shower receptors shall be lined with one of the following materials:

1. Sheet lead.

2. Sheet copper.

3. Plastic liner material that complies with ASTM D4068 or ASTM D4551.

4. Hot mopping in accordance with Section P2709.2.3.

5. Sheet-applied load-bearing, bonded waterproof membranes that comply with ANSI A118.10.

The lining material shall extend not less than 2 inches (51 mm) beyond or around the rough jambs and not less than 2 inches (51 mm) above finished thresholds. Sheet-applied load bearing, bonded waterproof membranes shall be applied in accordance with the manufacturer's instructions.

❖ There are a variety of materials that can be used for a waterproof lining of shower receptors that are field built. Extending the material at least 2 inches (51 mm) beyond rough jambs and 2 inches (51 mm) above the finished threshold creates a water-tight pan beneath the finished floor of the shower receptor.

For the 2012 edition, the amount of liner material required above the finished threshold was reduced from 3 inches (76 mm) to 2 inches (51 mm) to be in alignment with *International Plumbing Code®* (IPC®) requirements.

P2709.2.1 PVC sheets. Plasticized polyvinyl chloride (PVC) sheet shall meet the requirements of ASTM D4551. Sheets shall be joined by solvent welding in accordance with the manufacturer's instructions.

❖ PVC shower liner membranes are an alternative to lead, copper and chlorinated polyethylene (CPE) liner materials.

P2709.2.2 Chlorinated polyethylene (CPE) sheets. Non-plasticized chlorinated polyethylene sheet shall meet the requirements of ASTM D4068. The liner shall be joined in accordance with the manufacturer's instructions.

❖ CPE shower liner membranes are an alternative to lead, copper and PVC liner materials.

P2709.2.3 Hot-mopping. Shower receptors lined by hot-mopping shall be built-up with not less than three layers of standard grade Type 15 asphalt-impregnated roofing felt. The bottom layer shall be fitted to the formed subbase and each succeeding layer thoroughly hot-mopped to that below. Corners shall be carefully fitted and shall be made strong and water tight by folding or lapping, and each corner shall be reinforced with suitable webbing hot-mopped in place. Folds, laps and reinforcing webbing shall extend not less than 4 inches (102 mm) in all directions from the corner and webbing shall be of *approved* type and mesh, producing a tensile strength of not less than 50 pounds per inch (893 kg/m) in either direction.

❖ Hot-mopping is permitted if three separate layers of Type 15 asphalt-impregnated roofing felt are thoroughly hot-mopped in succession. Corners must be carefully fitted, with folds extending at least 4 inches (102 mm) and webbing that has a minimum tensile strength of 50 pounds per inch (893 kg/m) in either direction.

P2709.2.4 Liquid-type, trowel-applied, load-bearing, bonded waterproof materials. Liquid-type, trowel-applied, load-bearing, bonded waterproof materials shall meet the requirements of ANSI A118.10 and shall be applied in accordance with the manufacturer's instructions.

❖ Liquid-type, trowel-applied, load-bearing, bonded waterproof material can be applied to various substrates in accordance with the manufacturer's instructions. No reinforcing materials are necessary as the cured product forms a flexible water-tight barrier that prevents water from penetrating to the substrate materials.

P2709.3 Installation. Lining materials shall be sloped one-fourth unit vertical in 12 units horizontal (2-percent slope) to weep holes in the subdrain by means of a smooth, solidly formed subbase, shall be properly recessed and fastened to *approved* backing so as not to occupy the space required for the wall covering, and shall not be nailed or perforated at any point less than 1 inch (25.4 mm) above the finished threshold.

❖ This section requires that the lining material (just below the finished floor) maintain a gradient to the drain of at least one fourth unit vertical in 12 units horizontal (2-percent slope) to the weep holes in the subdrain (see Commentary Figure P2709.1). No punctures or perforations in the lining material are permitted at any point less than 1 inch (25.4 mm) above finished threshold.

P2709.3.1 Materials. Lead and copper linings shall be insulated from conducting substances other than the connecting drain by 15-pound (6.80 kg) asphalt felt or its equivalent. Sheet lead liners shall weigh not less than 4 pounds per square foot (19.5 kg/m²). Sheet copper liners shall weigh not less than 12 ounces per square foot (3.7 kg/m²). Joints in lead and copper pans or liners shall be burned or silver brazed, respectively. Joints in plastic liner materials shall be joined in accordance with the manufacturer's instructions.

❖ The specified weights determine the thickness of lead and copper sheet materials.

P2709.4 Receptor drains. An *approved* flanged drain shall be installed with shower subpans or linings. The flange shall be placed flush with the subbase and be equipped with a clamping ring or other device to make a water-tight connection between the lining and the drain. The flange shall have weep holes into the drain.

❖ As illustrated in Commentary Figure P2709.1, a flanged drain must be installed flush with the sub-base and equipped with a clamping ring to create a water-tight connection between the lining and the drain. The purpose of this is to drain any water that seeps down to the liner from the shower floor.

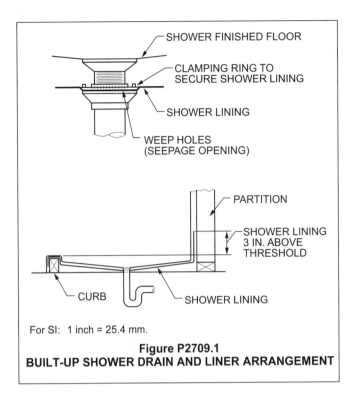

For SI: 1 inch = 25.4 mm.

Figure P2709.1
BUILT-UP SHOWER DRAIN AND LINER ARRANGEMENT

SECTION P2710
SHOWER WALLS

P2710.1 Bathtub and shower spaces. Walls in shower compartments and walls above bathtubs that have a wall-mounted showerhead shall be finished in accordance with Section R307.2.

❖ Section P2710.1 directs the reader to Section R307.2, which addresses wall construction. Also see Section R702.4.

SECTION P2711
LAVATORIES

P2711.1 Approval. Lavatories shall conform to ASME A112.19.1/CSA B45.2, ASME A112.19.2/CSA B45.1, ASME A112.19.3/CSA B45.4 or CSA B45.5/IAPMO Z124.

❖ Lavatories are available in enameled cast iron, vitreous china, stainless steel, porcelain-enameled-formed steel, plastic and nonvitreous ceramic. Lavatories also come in a variety of shapes and sizes, such as wall-mounted, hanger-mounted, pedestal, rimmed, above-counter basins and undermounted types. Referenced standards previously required lavatories to have an overflow; however, that is not currently the case. The provision of an overflow and its location is an option of the manufacturer. The reason for eliminating the overflow requirement was the lack of use, which resulted in the growth of bacteria and microorganisms.

P2711.2 Cultured marble lavatories. Cultured marble vanity tops with an integral lavatory shall conform to CSA B45.5/IAPMO Z124.

❖ Vanity tops with an integral lavatory that is all plastic must conform to the standard. Lavatory units can be constructed of a variety of synthetic materials including ABS, PVC, gel-coated fiberglass-reinforced plastic, acrylic, cultured marble cast-filled fiberglass, polyester and cultured marble acrylic. Cultured marble vanity tops are impregnated with fire-retardant chemicals, reducing the fuel contribution of the unit during a fire.

P2711.3 Lavatory waste outlets. Lavatories shall have waste outlets not less than $1^1/_4$ inch (32 mm) in diameter. A strainer, pop-up stopper, crossbar or other device shall be provided to restrict the clear opening of the waste outlet.

❖ The first sentence of this section requires that lavatories have a waste outlet assembly that is not less than $1^1/_4$ inches (32 mm). The second sentence requires that lavatory drain openings have a restriction device that prevents foreign items such as jewelry, toiletry items and cosmetics from inadvertently entering the drainage system.

P2711.4 Movable lavatory systems. Movable lavatory systems shall comply with ASME A112.19.12.

❖ Movable and pivoting lavatories allow the user to adjust the location of a lavatory upward and downward, from side-to-side and front-to-back to maximize the ease of use of the fixture. Some products also pivot.

SECTION P2712
WATER CLOSETS

P2712.1 Approval. Water closets shall conform to the water consumption requirements of Section P2903.2 and shall conform to ASME A112.19.2/CSA B45.1, ASME A112.19.3/CSA B45.4 or CSA B45.5/IAPMO Z124. Water closets shall conform to the hydraulic performance requirements of ASME A112.19.2/CSA B45.1. Water closet tanks shall conform to ASME A112.19.2/CSA B45.1, ASME A112.19.3/CSA B45.4 or CSA B45.5/IAPMO Z124. Water closets that have an invisible seal and unventilated space or walls that are not thoroughly washed at each discharge shall be prohibited. Water closets that allow backflow of the contents of the bowl into the flush tank shall be prohibited. Water closets equipped with a dual flushing device shall comply with ASME A112.19.14.

❖ A water closet is referred to by many other names, including "toilet," "stool" and "commode." The term "water closet" is the technical expression that derives from the first attempt to bring plumbing indoors. Early water closets consisted of a room that was the size of a closet, containing a cistern and a chair. Today, the term "water closet" applies to the name of the fixture itself. The most common word used in the United States to identify a water closet is "toilet;" however, the word "toilet" more accurately describes the room in which a water closet is installed.

The water closet is the fixture primarily responsible for the modernization of plumbing. The fixture effectively disposes of human waste in a sanitary manner. A water closet that does not thoroughly scour the bowl surface when flushed will harbor bacteria and germs that pose health risks to occupants. This would also be the case if the contents of the bowl were to enter the flush tank.

In accordance with Table P2903.2, water closets must be of the water-conservation type. The fixture standard limits the water consumption to an average of 1.6 gallons (6.1 L) of water per flushing cycle. No cycle during the testing can exceed 2.0 gallons (7.6 L) per flush. To determine conformance, a representative model of the water closet must be tested for its flushing efficiency. The flushing test uses polypropylene balls and polyethylene granule discs. There is also a test for the ability to remove an ink stain on the inside surface of the water closet bowl and a dye test measuring the exchange rate of water in the bowl. Another test measures the capacity of the drain line attached to the water closet to carry away the solids from the bowl after being flushed.

Dual flush water closets are becoming very popular and, in some locales, mandatory, so that less water is used to flush the bowl contents. These water closets must comply with the standard for dual flush water closets.

P2712.2 Flushing devices required. Water closets shall be provided with a flush tank, flushometer tank or flushometer valve designed and installed to supply water in sufficient quantity and flow to flush the contents of the fixture, to cleanse the fixture and refill the fixture trap in accordance with ASME A112.19.2/CSA B45.1.

❖ Generally, there are three physical configurations of water closets: flushometer valve equipped (bowl only), close coupled and one piece. A flushometer valve water closet is a bowl only (no tank) that requires the installation of a separate flushometer valve to the bowl. A close-coupled water closet has a bowl with a separate gravity-type flush tank or flushometer tank that is mounted to the bowl. A one-piece water closet is constructed with a gravity-type flush tank or flushometer tank integral with the bowl construction. The most common type installed in residential buildings is the close-coupled, gravity-type tank water closet. Regardless of the type of flushing mechanism used, the resulting effect must be the evacuation of the contents of the bowl, cleansing of the bowl's wetted surfaces and restoration of water in the trapway to maintain a 2-inch (51 mm) liquid seal. These performance parameters are ensured for each model of water closet by the testing protocols of ASME A112.19.2/CSA B45.1.

P2712.3 Water supply for flushing devices. An adequate quantity of water shall be provided to flush and clean the fixture served. The water supply to flushing devices equipped for manual flushing shall be controlled by a float valve or other automatic device designed to refill the tank after each discharge and to completely shut off the water flow to the tank when the tank is filled to operational capacity. Provision shall be made to automatically supply water to the fixture so as to refill the trap after each flushing.

❖ In a more general way, the first sentence of this section states the same thing as required by Section P2712.2. The remainder of this section describes the required operation of a gravity-type flush tank. A gravity-type flush tank has an automatic fill valve (sometimes called a "float valve") that allows water to fill the tank to a prescribed height. A valve at the bottom of the tank (called the "flush valve") is triggered to open by the user activating a lever or button on the outside of the tank. The water drains from the flush tank into the internal waterways of the water closet bowl to start a flushing action within the bowl. Once the water in the flush tank depletes to a certain level, the flush valve closes and the automatic fill valve begins filling the tank with water to be ready for the next flush cycle. Depending on the design of the water closet, replenishment of the water in the trap of the bowl could be accomplished by a bowl internal waterway design or a small tube from the fill valve (called the "refill tube"), directing water into the tank overflow tube, which drains into the bowl.

The design of gravity-flush tanks varies considerably even within the same manufacturer's product line. Flush tanks are engineered specifically for each water closet bowl design in order to produce the required flushing action using not more than a specified amount of water required by the fixture standard (ASME A112.19.2/CSA B45.1). In order to ensure that the water closet functions as originally intended by the manufacturer, identically designed replacement parts should be used when making repairs.

As Section P2712.2 indicates, two other major devices for supplying flushing water to a water closet bowl are the flushometer tank and the flushometer valve. From the exterior, a flushometer tank looks like a gravity-flush tank; however, the flushing mechanism is quite different. This design uses a pressure vessel to accumulate a specific volume of water with a volume of compressed air above the water. When an external lever or button is actuated by the user, a flush valve opens, allowing the pressurized water to exit the pressure vessel into the internal waterways of the bowl to initiate flushing action. The flushing water exits with significant velocity so as to create a rapid and highly effective flushing action in the bowl.

The flushometer valve flushing device is a special type of water control valve that is mounted above the water closet bowl. A 1$^1/_2$-inch (38 mm) vertical tube connects the outlet of the flushometer valve to a connector (called the "spud") on the top or the top rear of the water closet bowl. When the user actuates a lever

on the flushometer valve, a diaphragm mechanism in the valve moves to open an internal water valve to discharge water into the outlet tube of the valve. The valve stays open for a predetermined amount of time to discharge a specific volume of water and then automatically closes. The discharged water enters the internal waterways of the water closet bowl to initiate flushing action in the bowl.

P2712.4 Flush valves in flush tanks. Flush valve seats in tanks for flushing water closets shall be not less than 1 inch (25 mm) above the flood-level rim of the bowl connected thereto, except an *approved* water closet and flush tank combination designed so that when the tank is flushed and the fixture is clogged or partially clogged, the flush valve will close tightly so that water will not spill continuously over the rim of the bowl or backflow from the bowl to the tank.

❖ This section pertains only to gravity-type flush tanks. The design of the flush valve in the bottom of the tank must have the seat of the valve located not less than 1 inch (25 mm) above the flood-level rim of the water closet bowl. This is to protect the flush tank from contamination caused by waste from the water closet bowl from backing up into the tank should the outlet of the water closet bowl become clogged.

An exception is provided for those designs of gravity-type flush tanks that are integral to the water closet bowl where the flush valve seat cannot be above the flood-level rim of the water closet bowl. In these designs, a special mechanical arrangement must be used that keeps the flush valve tightly closed should the water closet bowl become partially or fully clogged. This arrangement satisfies the requirement of preventing backflow of the bowl contents into the tank and prevents water from continuously spilling into the bowl when the bowl is clogged.

P2712.5 Overflows in flush tanks. Flush tanks shall be provided with overflows discharging to the water closet connected thereto and such overflow shall be of sufficient size to prevent flooding the tank at the maximum rate at which the tanks are supplied with water according to the manufacturer's design conditions.

❖ This section pertains only to gravity-type flush tanks. The design of the flush tank must have an overflow pipe that drains into the water closet bowl to prevent the water in the tank from rising above a certain level should the automatic fill valve fail to shut off the water flow. The overflow pipe must be designed to accommodate draining of the maximum water discharge rate from a fully opened fill valve.

Automatic fill valves for gravity-type flushing water closets must be of the antisiphon type to protect the water distribution system from contamination that might be in the water of the flush tank (see Table P2701.1). Antisiphon fill valves must have a "critical level" marking that, when the fill valve is installed in the tank, must be above the top of the overflow pipe. Proper location of the critical level of the fill valve ensures that the backflow protection of the fill valve is not compromised. Fill valve manufacturer's instructions will indicate the location of the critical level of the valve and how far that location needs to be above the top of the overflow pipe. Typically, this dimension is 1 inch (25 mm).

In certain designs of water closet bowls with integral flush tanks, the top of the overflow pipe is below the water closet flood-level rim. To ensure that the fill valve backflow protection is not compromised should the tank water level rise above the top of the overflow pipe, overflow openings are provided in the back wall of the tank to serve as a secondary overflow.

P2712.6 Access. Parts in a flush tank shall be accessible for repair and replacement.

❖ The components in both gravity-type and flushometer tank-type flush tanks for water closets will periodically require repair and replacement. Permanent construction must not block access to the flush tank so that service personnel can work on the components without needing to disassemble the flush tank from the water closet or remove the water closet from its mounted location.

P2712.7 Water closet seats. Water closets shall be equipped with seats of smooth, nonabsorbent material and shall be properly sized for the water closet bowl type.

❖ A water closet seat must be designed to be easily cleaned. The seat must also be designed to suit the bowl. For example, an elongated bowl requires an elongated seat.

P2712.8 Flush tank lining. Sheet copper used for flush tank linings shall have a weight of not less than 10 ounces per square foot (3 kg/m^2).

❖ Long ago, sheet copper was commonly used to line wooden flush tanks so that they would hold water without leaking.

P2712.9 Electro-hydraulic water closets. Electro-hydraulic water closets shall conform to ASME A112.19.2/CSA B45.1.

❖ An electrohydraulic water closet is a siphonic or wash-down water closet with a nonmechanical trap seal. The water closet incorporates electrical motor(s), pump(s) and controllers to facilitate the flushing action.

SECTION P2713
BATHTUBS

P2713.1 Bathtub waste outlets and overflows. Bathtubs shall be equipped with a waste outlet and an overflow outlet. The outlets shall be connected to waste tubing or piping not less than 1$^1/_2$ inches (38 mm) in diameter. The waste outlet shall be equipped with a water-tight stopper.

❖ An overflow opening must be provided in a bathtub. Waste overflows are sometimes used to access the drain piping for the purpose of rodding to clear blockages in the trap and fixture drain piping. Waste overflow drain assemblies are typically designed only for the removal of displaced water when a bather enters a full tub. They are not designed to prevent overfilling of

a bathtub should the tub filler faucet be left open and unattended.

A device for blocking the flow of tub water from the outlet (tub drain) must be provided in order to allow the user to fill the tub with water for bathing. The device could be as simple as a "pull out"-type rubber plug that tightly fits in the opening of the waste assembly or other approved methods such as trip lever waste units, cable-actuated closure devices or metal push/pull or pop-up devices.

P2713.2 Bathtub enclosures. Doors within a bathtub enclosure shall conform to ASME A112.19.15.

❖ Bathtub/whirlpool bathtubs with pressure-sealed doors allow the bather to enter the fixture at approximately the same level as the floor. The door can be opened, allowing the bather easier access to the bathtub. Once inside the fixture, the door is closed and the water is turned on. Sensor switches, which note the presence of water entering the vessel, activate an air pump that pressurizes the door seal, thereby keeping the water within the bathing vessel. Upon completion of bathing, the bather can open the door only when the bath water is drained from the vessel.

P2713.3 Bathtub and whirlpool bathtub valves. Hot water supplied to bathtubs and whirlpool bathtubs shall be limited to a temperature of not greater than 120°F (49°C) by a water-temperature limiting device that conforms to ASSE 1070 or CSA B125.3, except where such protection is otherwise provided by a combination tub/shower valve in accordance with Section P2708.4.

❖ This section addresses faucets used only for the filling of whirlpool tubs or bathtubs. The water discharged from these "tub-filler" faucets must be limited to a maximum temperature of 120°F (49°C) to prevent skin burns (see commentary, Section P2708.3). It is not uncommon for a user to fill a bathtub in advance, with water hotter than he or she could stand, in an attempt to compensate for the cooling effect caused by the tub walls or a planned delay in entering the tub. These users typically avoid skin burns by testing the temperature of the water in the tub and adding cold water to temper the hot water before getting into the tub; however, there have been cases in which the user was sitting on the side of the tub, reached to adjust the control for cold water, accidentally slipped into the tub and received third-degree burns. In other cases, children placed in a tub being filled have been scalded because the water became hotter some time after an adult tested the flowing water temperature at the beginning of tub filling. While it would seem that hot water supply temperatures might be simply controlled by setting the water heater thermostat to 120°F (49°C), experience has indicated that this is an unreliable means for limiting the temperature of hot water in any type of building because thermostats are inaccurate and are too easy for laypersons to adjust. The adjustment to a higher temperature setting is usually in response to users' complaints that there is not enough hot water during periods of high demand. Adjusting a water heater to a higher temperature effectively creates more hot water volume. This is because less, higher temperature, hot water flow is required for any particular mixed hot/cold water temperature and, thus, there is effectively more hot water available to the users. Because the code does not regulate the temperature to which water heater thermostats must be set and the thermostats are not tamperproof or have the necessary repeatable accuracy, water temperature control devices exterior to the water heater unit must be used to positively limit the discharge water temperature to 120°F (49°C).

The temperature-limiting device required by this section is a control valve that might require periodic adjustment, maintenance or replacement. Obviously, these valves should be located so that they can be accessed. Access panels in finished walls, floors or ceilings might be required in order to provide access.

This code section allows for a combination tub/shower valve (complying with ASSE 1016/ASME A112.1016/CSA B125.16), with or without a fixed shower head, to be used for filling a tub, including a large whirlpool tub. In actual practice, however, the installation of a combination tub/shower valve as a tub filler for a large tub (such as a garden or whirlpool tub) is rarely done, for several reasons: (1) faucet manufacturers generally do not provide or offer "extra long" wall-mount tub spouts to enable the discharge of the spout to be located over the tub well so as to not splash onto the tub deck and (2) the placement of the water control on the wall makes it difficult for a user sitting in a large tub to add more water if needed. This "allowance" for a combination tub/shower valve for a tub application might lead the reader to assume that any device meeting ASSE 1016/ASME A112.1016/CSA B125.16 is also acceptable for meeting the requirements of this section, but this is not the case. Combination tub/shower valves are the only ASSE 1016/ASME A112.1016/CSA B125.16 device that can be used to meet the requirements of this section.

This is a reason why ASSE 1016 valves (other than combination tub/shower valves) cannot be used for the ASSE 1070 or CSA B125.3 device required by this section. ASSE 1016 or CSA B125.3 devices, other than combination tub/shower valves, have a limited useful "maximum flow" rating. In other words, for flows in excess of 2.5 gallons per minute (gpm) [9.5 liters per minute (lpm)], the pressure drop through these devices severely limits the user's ability to fill a large tub in short order. ASSE 1070 and CSA B125.3 devices have a much higher usable flow rate, typical of what is needed to rapidly fill a large bathtub or whirlpool tub. The "allowance" to use a combination tub/shower valve complying with Section 424.3 for a tub-filler faucet is a modest compromise directed at the most commonly installed, minimum size [30 inches (762 mm) x 60 inches (1524 mm) x 16 inches (406 mm) deep], standard tub found in most residential bathrooms. The filling time of this standard tub size

using a combination tub/shower valve through a standard tub spout is deemed "reasonable" based upon years of user experience with filling tubs through anti-scald-type combination tub/shower valves. However, where tub sizes are larger, higher faucet flow rates are required to fill those tubs within an acceptable length of time. Requiring the use of an ASSE 1070 device with the higher flow-through rating ensures that larger tubs can be filled quickly so that the tub water temperature has less time to "cool off" since these devices are limiting the discharge temperature to 120°F (49°C).

Where a hand-held shower is connected to a tub-filler faucet, the hand shower is not required to be provided with water supplied through an ASSE 1016/ASME A112.1016/CSA B125.16 device as is required for a standup-type shower application. The hand-held shower also does not "reclassify" a tub-filler faucet as a "combination tub/shower valve" such that it would be required to comply with ASSE 1016/ASME A112.1016/CSA B125.16.

SECTION P2714
SINKS

P2714.1 Sink waste outlets. Sinks shall be provided with waste outlets not less than $1^1/_2$ inches (38 mm) in diameter. A strainer, crossbar or other device shall be provided to restrict the clear opening of the waste outlet.

❖ This section requires sinks such as dishwashing sinks, food prep sinks and handwashing sinks (but not lavatories) be provided with waste outlet assemblies that are not less than $1^1/_2$ inches (38 mm) in diameter. The drain opening in the sink can be larger than $1^1/_2$ inches (38 mm) so that a $3^1/_2$-inch (89 mm) basket strainer assembly or a food waste disposer can be installed. Waste outlet assemblies must be provided with strainers or crossbars to prevent large items from entering the drainage system.

P2714.2 Movable sink systems. Movable sink systems shall comply with ASME A112.19.12.

❖ Movable sinks allow the user to adjust the elevation of the sink for easier use.

SECTION P2715
LAUNDRY TUBS

P2715.1 Laundry tub waste outlet. Each compartment of a laundry tub shall be provided with a waste outlet not less than $1^1/_2$ inches (38 mm) in diameter. A strainer or crossbar shall restrict the clear opening of the waste outlet.

❖ This section requires laundry tubs be provided with waste outlet assemblies that are not less than $1^1/_2$ inches (38 mm) in diameter. Waste outlet assemblies must be provided with strainers or crossbars to prevent large items from entering the drainage system.

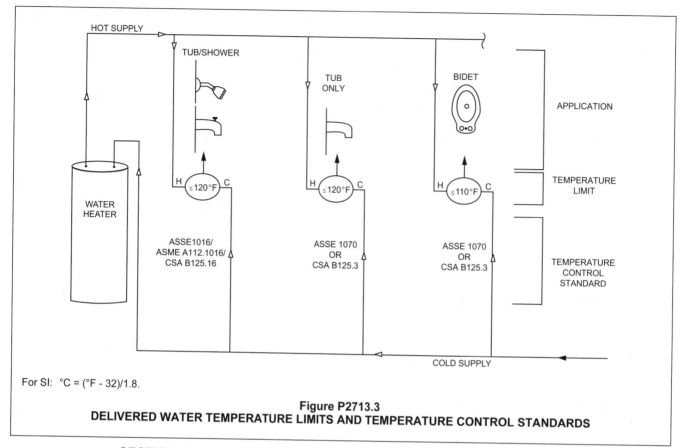

For SI: °C = (°F - 32)/1.8.

Figure P2713.3
DELIVERED WATER TEMPERATURE LIMITS AND TEMPERATURE CONTROL STANDARDS

SECTION P2716
FOOD-WASTE DISPOSER

P2716.1 Food-waste disposer waste outlets. Food-waste disposers shall be connected to a drain of not less than $1\frac{1}{2}$ inches (38 mm) in diameter.

❖ The outlet, or discharge, of the food-waste disposer must be at least $1\frac{1}{2}$ inches (38 mm) in diameter, which is consistent with the size of the waste outlet assembly tubing from a kitchen sink (see Section P2714.1).

P2716.2 Water supply required. A sink equipped with a food-waste disposer shall be provided with a faucet.

❖ A supply of water is necessary to promote proper food waste disposer operation and to carry food waste into and through the drain.

SECTION P2717
DISHWASHING MACHINES

P2717.1 Protection of water supply. The water supply to a dishwasher shall be protected against backflow by an *air gap* complying with ASME A112.1.3 or A112.1.2 that is installed integrally within the machine or a backflow preventer in accordance with Section P2902.

❖ This section requires protection of the water distribution system against backflow from a dishwasher water

supply connection. An air gap or air gap device inside of the machine (i.e., integral with the machine) is what the vast majority of manufacturers provide. Where an air gap is not provided integral to the machine, a backflow preventer external to the machine must be installed in the water supply line to the dishwasher.

P2717.2 Sink and dishwasher. The combined discharge from a dishwasher and a one- or two-compartment sink, with or without a food-waste disposer, shall be served by a trap of not less than $1\frac{1}{2}$ inches (38 mm) in outside diameter. The dishwasher discharge pipe or tubing shall rise to the underside of the counter and be fastened or otherwise held in that position before connecting to the head of the food-waste disposer or to a wye fitting in the sink tailpiece.

❖ In the majority of residential kitchen arrangements, a dishwasher is often located directly adjacent to the kitchen sink. Where the kitchen sink is not equipped with a food grinder, the discharge tube from the dishwasher must connect to the sink waste tailpiece through an angled tube (see the right panel of Commentary Figure P2717.2). This section also requires that the discharge tube be fastened to the underside of the countertop in order to reduce the possibility of waste water from the sink backing up into the dishwasher should a drain clog occur.

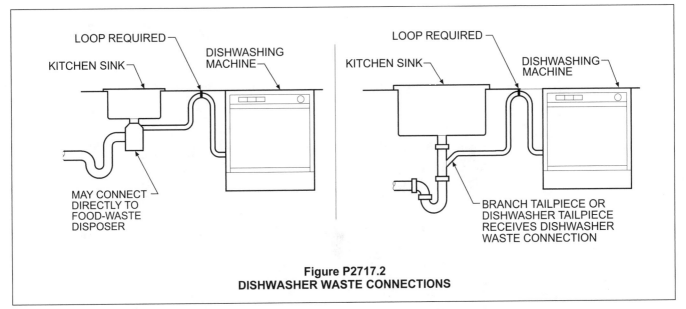

Figure P2717.2
DISHWASHER WASTE CONNECTIONS

SECTION P2718
CLOTHES WASHING MACHINE

P2718.1 Waste connection. The discharge from a clothes washing machine shall be through an *air break*.

❖ A clothes-washing machine must not be directly connected to the sanitary drainage system in order to prevent contamination of the washer with sewage should the drainage system become clogged. Because the discharge from a clothes washer is waste water, the indirect connection is only required to be an air break. The air-break connection also prevents siphoning of the liquids within the clothes washer where the discharge is to a standpipe waste receptor [see the commentary to Section P2706.2 for further information on standpipes and Section P2706.3, Exception 2, for use of a laundry tray (tub) as a waste receptor].

SECTION P2719
FLOOR DRAINS

P2719.1 Floor drains. Floor drains shall have waste outlets not less than 2 inches (51 mm) in diameter and a removable strainer. Floor drains shall be constructed so that the drain can be cleaned. Access shall be provided to the drain inlet. Floor drains shall not be located under or have their access restricted by permanently installed appliances.

❖ Although the code does not specifically require floor drains, designers often include floor drains in single-family dwellings to provide for emergency drainage or to meet specific drainage needs (e.g., relief valve drains for water heaters, condensate drains, etc.). Floor drains must conform to ASME A112.6.3 as listed in Table P2701.1. Unless used as a receptor, floor drains are viewed as emergency drains and Table P3004.1 does not assign them a fixture unit value. Section P2719.1 clearly states that the outlet (hence,

the trap and drain also) must be not less than 2 inches (51 mm) in diameter.

It is important to not have floor drains concealed under furnaces, water heaters or other permanently installed appliances so that the drain strainer can be inspected and cleaned periodically and the floor drain and connected piping can be rodded to clear clogs.

SECTION P2720
WHIRLPOOL BATHTUBS

P2720.1 Access to pump. Access shall be provided to circulation pumps in accordance with the fixture or pump manufacturer's installation instructions. Where the manufacturer's instructions do not specify the location and minimum size of field-fabricated access openings, an opening of not less than 12-inches by 12-inches (305 mm by 305 mm) shall be installed for access to the circulation pump. Where pumps are located more than 2 feet (610 mm) from the access opening, an opening of not less than 18 inches by 18 inches (457 mm by 457 mm) shall be installed. A door or panel shall be permitted to close the opening. The access opening shall be unobstructed and be of the size necessary to permit the removal and replacement of the circulation pump.

❖ Whirlpool bathtubs are often ordered, located or installed without consideration for the location of the circulation pump and the requirement for access for servicing and replacement. The fixture or pump manufacturer's instructions usually specify the access requirements for the serviceable components of the whirlpool tub. But if the instructions are silent, this section requires the minimum size of openings for access to the pumps. In no case can such openings be obstructed by piping, framing, etc. [see Commentary Figures P2720.1(1) and (2)].

Figure P2720.1(1)
NONCOMPLIANT WHIRLPOOL BATHTUB PUMP ACCESS

Figure P2720.1(2)
NONCOMPLIANT WHIRLPOOL BATHTUB PUMP ACCESS

P2720.2 Piping drainage. The circulation pump shall be accessibly located above the crown weir of the trap. The pump drain line shall be properly graded to ensure minimum water retention in the volute after fixture use. The circulation piping shall be installed to be self-draining.

❖ The pump and suction lines for the whirlpool bathtub must be designed to drain completely after each use. Waste water left standing in the pump body or in the suction and discharge piping will promote bacterial growth. Therefore, the pump must be located above the crown weir of the trap, and the piping must be properly graded to ensure minimum water retention.

P2720.3 Leak testing. Leak testing and pump operation shall be performed in accordance with the manufacturer's instructions.

❖ A whirlpool bathtub is the combination of a plumbing fixture and a plumbing appliance. As whirlpool tubs are installed at the plumbing rough-in stage and then the sides are covered with finish construction, leak and operational testing of the pump, jets and recirculation piping must be performed prior to concealment to verify that the unit is water tight and operational. Repair of a leak or replacement of a pump is far easier to accomplish at the plumbing rough-in stage. It also prevents water damage to finished construction should a leak not be discovered until the occupant starts using the tub.

P2720.4 Manufacturer's instructions. The product shall be installed in accordance with the manufacturer's instructions.

❖ The installation requirements of whirlpool tubs vary from manufacturer to manufacturer. One of the most important aspects of whirlpool tub installations is making sure that the supporting structure is capable of the significant loads caused by the weight of the tub, its pump/piping, the water and the occupant(s). Although a whirlpool tub might appear to be "deck mounted," that is, supported at the outer edge of the deck flange, in most cases, the tub must be solidly supported under the basin of the tub. Some manufacturers require that the bottom of the tub be set in a bed of mortar in order to provide proper support and dampen the vibrations from operation. Manufacturer's instructions must be followed to ensure that the tub is not damaged by improper installation.

SECTION P2721
BIDET INSTALLATIONS

P2721.1 Water supply. The bidet shall be equipped with either an air-gap-type or vacuum-breaker-type fixture supply fitting.

❖ The outlet end of the water supply nozzle of a bidet is typically located below the flood-level rim of the fixture. To protect the water distribution system from backflow of contaminants, the water supply to the nozzle must pass through either an air gap or a vacuum breaker. Many bidet faucet manufacturers offer designs that

have integral vacuum breaker protection built into the faucet.

P2721.2 Bidet water temperature. The discharge water temperature from a bidet fitting shall be limited to not greater than 110°F (43°C) by a water-temperature-limiting device conforming to ASSE 1070 or CSA B125.3.

❖ Users of bidets would be at significant risk for scalding of sensitive areas of the body if not for this section's requirement for a water temperature-limiting device. Temperature-limiting devices meeting ASSE 1070 or CSA B125.3 provide for temperature-limiting control for flow rates as small as 0.5 gpm (1.8 lpm). Note that because these devices are valves and might require periodic adjustment and repair, access must be provided to the valves.

SECTION P2722
FIXTURE FITTING

P2722.1 General. Fixture supply valves and faucets shall comply with ASME A112.18.1/CSA B125.1 as indicated in Table P2701.1. Faucets and fixture fittings that supply drinking water for human ingestion shall conform to the requirements of NSF 61, Section 9. Flexible water connectors shall conform to the requirements of Section P2905.7.

❖ The design and manufacture of faucets and fixture supply valves must be controlled by a standard so that interchangeability exists between the various models produced for the market. Furthermore, items that convey water for human consumption must be tested in compliance with NSF 61 to make sure that the internal wetted surfaces of these items do not impart any significant level of toxic substances. Flexible water connectors are not covered by the ASME/CSA standard but are covered by ASME A112.18.6, which requires compliance with Section 9 of NSF 61.

P2722.2 Hot water. Fixture fittings supplied with both hot and cold water shall be installed and adjusted so that the left-hand side of the water temperature control represents the flow of hot water when facing the outlet.

Exception: Shower and tub/shower mixing valves conforming to ASSE 1016/ASME A112.1016/CSA B125.16, where the water temperature control corresponds to the markings on the device.

❖ For safety reasons, the code requires that the handle location for hot water corresponds to the left side of the fixture when facing the fixture fitting. It is traditional and is, therefore, human nature to equate the left side of the faucet or control with hot water. If the fixture was incorrectly connected to result in hot water being controlled by the right handle location, a user might not realize that hot water was being discharged before inserting his or her hands into the flow stream. This could result in scalding, especially for children or the elderly as they might not have the reaction speed to avoid a skin burn.

The exception is provided to allow single-handle shower valves complying with ASSE 1016 or ASME

A112.18.1/CSA B125.1 to have identifiable "hot" designations on the control valve that do not correspond with the left side of the valve. These devices operate by rotating a single handle or knob away from the "off" position to initiate water flow. Moving the handle further in the same direction increases the temperature of flowing water. Because the rotating operation is limited to one direction, the user cannot mistakenly activate water flow at the maximum hot water setting without first rotating the handle or knob through the cold water setting. Both the ASSE 1016 and ASME A112.18.1/CSA B125.1 standards require control valves to have markings that indicate clearly the direction or means of adjusting the temperature.

P2722.3 Hose-connected outlets. Faucets and fixture fittings with hose-connected outlets shall conform to ASME A112.18.3 or ASME A112.18.1/CSA B125.1.

❖ A hose connected to a faucet could have the end of the hose submerged in waste water or other nonpotable fluids. To prevent a backsiphonage of contaminant into the water distribution system, the faucet must comply with either ASME A112.18.3 or ASME A112.18.1/CSA B125.1, which ensure that the faucet hose assembly has proper backflow protection. A kitchen faucet with a side spray is an example of a faucet with a hose connected.

P2722.4 Individual pressure-balancing in-line valves for individual fixture fittings. Individual pressure-balancing in-line valves for individual fixture fittings shall comply with ASSE 1066. Such valves shall be installed in an accessible location and shall not be used as a substitute for the balanced pressure, thermostatic or combination shower valves required in Section P2708.3.

❖ For faucets (fixture fittings) where the code does not require control of the temperature of the water discharge, there still may be a need (but not a code requirement) to keep the user's set, faucet outlet temperature from fluctuating due to pressure changes. For example, the flushing of a water closet might cause significant pressure changes in the water distribution system such that the water temperature discharged from a lavatory faucet might change considerably. This could be annoying and perhaps dangerous. An inline pressure-balancing valve that complies with ASSE 1066 has no adjustments and no upper temperature limit. The valve simply adjusts the flow rate through one side of the valve in order to compensate for a pressure differential between the sides (hot and cold) of up to 50 percent.

ASSE 1066 valves do not meet the requirements for balanced-pressure, thermostatic or combination balanced-pressure/thermostatic mixing valves for showers (see Section P2708.3.). An ASSE 1066 valve is only intended for connection to a single-fixture fitting.

P2722.5 Water closet personal hygiene devices. Personal hygiene devices integral to water closets or water closet seats shall conform to ASME A112.4.2.

❖ These devices can be integral to some water closet designs or part of a water closet seat. Typical functions of personal hygiene devices are bidet spray and blow drying. The standard ensures that products have proper potable water backflow prevention, strength and dimensional conformity to fit a variety of water closets. Some models have wireless programmable wall controls for ease of use.

SECTION P2723
MACERATING TOILET SYSTEMS

P2723.1 General. Macerating toilet systems shall be installed in accordance with manufacturer's instructions.

❖ A macerating toilet system collects the waste from a single water closet, lavatory and bathtub located in the same room. The system then grinds the waste and discharges it through a small discharge line to the drainage system. The concept of this system is similar to that of a building drain installation using a sewage ejector pump [see Commentary Figures P2723.1(1) and P2723.1(2)].

P2723.2 Drain. The size of the drain from the macerating toilet system shall be not less than $^3/_4$ inch (19 mm) in diameter.

❖ Because a macerating toilet "liquefies" the solids in the waste water, a $^3/_4$-inch (19.1 mm) discharge line is adequate for handling the pumped discharge.

SECTION P2724
SPECIALTY TEMPERATURE CONTROL DEVICES
AND VALVES

P2724.1 Temperature-actuated mixing valves. Temperature-actuated mixing valves, which are installed to reduce water temperatures to defined limits, shall comply with ASSE 1017. Such valves shall be installed at the hot water source.

❖ There could be hot water system applications where it is necessary to reduce the temperature of the hot water generated to "reasonable" temperature levels. For example, a water heater could also be supplying water for a space heating system that requires the water temperature to be greater than 140°F (60°C). Section P2802.2 requires that a thermostatic mixing valve (also known as a temperature-actuated mixing valve) be installed to limit the water temperature to the plumbing system. Section P2724.1 requires the valve to be located at the source (water heater) and not at some distant point near plumbing fixtures.

Figure P2723.1(1)
MACERATING TOILET SYSTEM
(Photo courtesy of SFA SANIFLO)

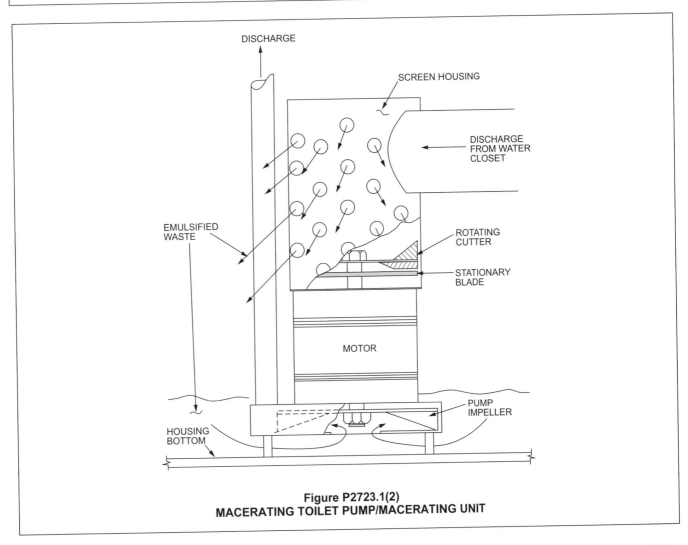

Figure P2723.1(2)
MACERATING TOILET PUMP/MACERATING UNIT

P2724.2 Temperature-actuated, flow-reduction devices for individual fixtures. Temperature-actuated, flow-reduction devices, where installed for individual fixture fittings, shall conform to ASSE 1062. Such valves shall not be used as a substitute for the balanced pressure, thermostatic or combination shower valves required for showers in Section P2708.3.

❖ Temperature-actuated, flow-reduction valves for individual fixture fittings must comply with the provisions in ASSE 1062. These valves reduce flow to 0.25 gpm (0.946 L/m) or less automatically in response to outlet temperatures higher than a preset temperature not to exceed 120°F (49°C). This limits a person's exposure to high-temperature water discharge from an individual fixture fitting, such as a bath and utility faucet, lavatory or sink faucet. These devices may not be used as a substitute for shower or tub/shower valves required in Section P2708.3.

SECTION P2725
NONLIQUID SATURATED TREATMENT SYSTEMS

P2725.1 General. Materials, design, construction and performance of nonliquid saturated treatment systems shall comply with NSF 41.

❖ The standard covers composting toilet systems.

Bibliography

The following resource materials were used in the preparation of the commentary for this chapter of the code:

ASSE 1016/ASME A112.1016/CSA B125.16–2011, *Performance Requirements for Automatic Compensating Valves for Individual Showers and Tub/Shower Combinations*. Westlake, OH: American Society of Sanitary Engineering, 2011.

Chapter 28:
Water Heaters

User note: Code change proposals to this chapter will be considered by the IRC – Plumbing and Mechanical Code Development Committee during the 2015 (Group A) Code Development Cycle.

General Comments

A water heater is any appliance that heats potable water and supplies it to the potable hot water distribution system of the building. Some water heaters might also provide hot water for space heating purposes in addition to supplying hot water to the water distribution system. This chapter does not apply to appliances, equipment and devices that provide hot water for the sole purpose of space heating (hydronic heating systems) because, by definition in this code, a water heater supplies water to the potable water distribution system. Hydronic heating systems are regulated by Chapters 20 and 21 of the code.

Historically, the main focus of this chapter has been to regulate the installation and safety features for automatic storage tank water heaters and unfired hot water storage tanks. Other water heating methods and equipment, such as solar, electric heat pump, ground source thermal transfer, boiler water recirculation, steam heat transfer, solid fuel-fired technology and the necessary controls/safety devices for those methods, are not specifically addressed in this chapter.

Questions often arise as to how the regulations in this chapter are intended to apply to the installation of alternative water-heating methods. Where the code does not specifically address those questions, manufacturers' instructions must serve as the basis for the installation of these products.

Wherever hot water is stored in a closed vessel (tank), there is a potential for explosion. There are documented cases where improperly installed (or poorly maintained) water heaters have exploded with such force that they have been propelled through floor and roof assemblies and over 100 feet (30 480 mm) in the air. Thus, the intent of this chapter is to regulate the materials, design and installation of water heaters and related safety devices in order to protect property and life.

Purpose

Chapter 28 regulates the design, approval and installation of water heaters and related safety devices. The intent is to minimize the hazards associated with the installation and operation of water heaters.

SECTION P2801
GENERAL

P2801.1 Required. Hot water shall be supplied to plumbing fixtures and plumbing appliances intended for bathing, washing or culinary purposes.

❖ A source of hot water must be supplied to plumbing fixtures and appliances that are intended to be used for bathing, washing, food preparation or for cleanup of pots, pans, dishes and flatware. This requirement corresponds to Section R306.4, which requires that both hot and cold water be supplied to plumbing fixtures. Hot water helps to melt oils, allowing soaps and detergents to reach the soils and dirt. The need for hot water is also a human comfort issue in most situations.

For many decades, the typical source of hot water for a dwelling was either a self-contained storage tank water heater or a hot water storage tank where the water was heated by an outside heat source such as a boiler. Tanks for storing hot water must be made of noncorrosive metal (e.g., stainless steel) or be made with a noncorrosive liner (usually glass).

The code does not provide criteria for determining the "minimum required" capacity of a storage tank water heater for a particular application. The sizing of storage-type water-heating equipment for domestic hot water applications is difficult because user behavior patterns are not highly predictable. There have been many studies of hot water demand for a variety of residential, institutional and commercial applications. At best, these studies have concluded that without significant oversizing (i.e., more storage volume) of the storage water heater, users will experience a shortage of hot water. Customer satisfaction surveys in one study revealed that in general, single-family home occupants are satisfied with their hot water availability even if they run out of hot water up to 12 times per year. Most residential users understand that there is a limit to hot water availability and if they run out of hot water frequently, they learn to adjust their living patterns to accommodate the existing storage tank capacity.

In recent years, tankless water heaters have become increasingly popular because they occupy less floor space than a storage-type tank water heater, use zero energy in the standby mode and, if properly maintained, can last considerably longer than a storage tank water heater. But perhaps their most advertised feature is that a tankless water heater can provide an "unlimited" flow of hot water. This is opposed to a storage tank water heater that will eventually be depleted of its stored hot water, given enough time of hot water flowing from a hot water outlet. A tankless water heater instantly and continuously heats water as long as hot water is demanded at a fixture.

But if tankless water heaters can provide an "unlimited" flow of hot water, could installing a tankless water heater instead of a storage tank water heater be a solution to not running out of hot water? The answer depends on the flow rate demanded of the tankless water heater.

Tankless water heaters are rated to provide for a specific temperature increase (rise) of the incoming cold water at a specific flow rate. For example, one rating might be 50°F (10°C) temperature rise at 2.6 gallons per minute (gpm) (9.8 L/m). If the cold water temperature flowing into the tankless heater is 60°F (16°C) and the flow through the heater is 2.6 gpm (9.8 L/m), the outlet water temperature will be 50°F (10°C) + 60°F (16°C) = 110°F (43°C). Therefore, if the hot water demand at the fixture is 2.6 gpm (9.8 L/m), hot water at a temperature of 110°F (43°C) can be expected to be discharged at the hot water outlet (neglecting heat losses from the hot water piping sys-

tem). This flow of hot water can continue indefinitely so, indeed, the user "never runs out of hot water."

One important characteristic of tankless water heaters is that if the flow through the water heater becomes greater, the temperature rise becomes less. Conversely, as the flow becomes less, the temperature rise becomes greater. Commentary Figure P2801.1(1) shows typical capacity graphs for a Model A and a Model B of electric tankless water heaters. Although some manufacturers might choose to show the same information in table form as shown in Commentary Table P2801.1, the concept is the same as that shown in a capacity graph. Commentary Figure P2801.1(2) illustrates the operating point of 50°F (10°C) temperature rise at 2.6 gpm (9.8 L/m) for a Model A unit. If the flow through the unit is 3 gpm (11 L/m), the temperature rise would only be approximately 40°F (4°C). If the flow through the unit is only 1.5 gpm (5.7 L/m), the temperature rise would be 100°F (38°C) (assuming that high outlet water temperature controls do not exist). What does this information mean in terms of the operation of plumbing fixtures in a building?

FLOW	TEMPERATURE RISE, °F		
	1.0 gpm	2.0 gpm	4.0 gpm
Model A (9.5 kW)	65	32	16
Model B (19 kW)	130	65	32

For SI: 1 gallon per minute = 3.785 L/m, °C = [(°F) - 32]/1.8.

Figure P2801.1
TYPICAL CAPACITY TABLE FOR TANKLESS WATER HEATERS

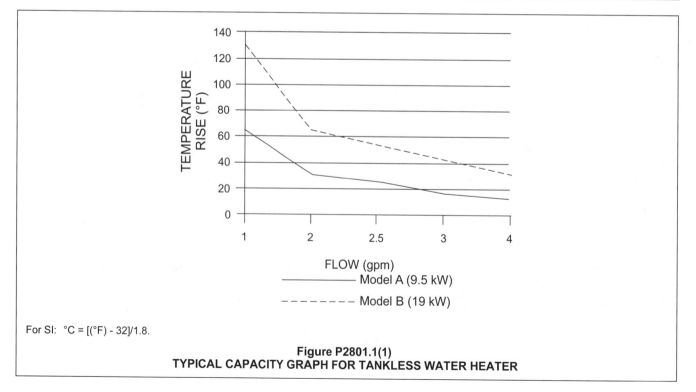

For SI: °C = [(°F) - 32]/1.8.

Figure P2801.1(1)
TYPICAL CAPACITY GRAPH FOR TANKLESS WATER HEATER

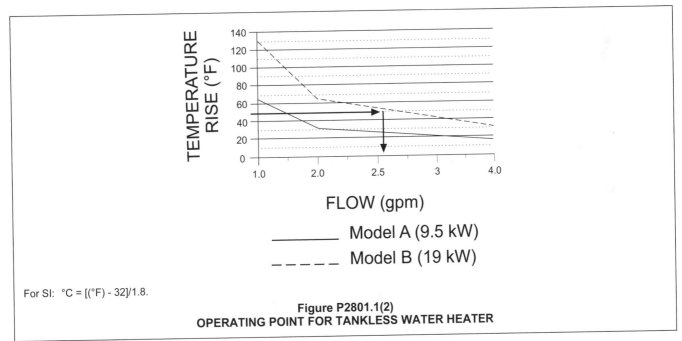

For SI: °C = [(°F) - 32]/1.8.

Figure P2801.1(2)
OPERATING POINT FOR TANKLESS WATER HEATER

Consider a shower with one shower head that discharges 2.5 gpm (9.8 L/m). Then consider a shower user desiring a shower water temperature of 110°F (43°C). If a Model A tankless water heater unit (as described previously) supplies hot water to the plumbing system, the shower user could turn the shower mixing valve almost all the way to the full hot position and enjoy a 110°F (43°C) shower, indefinitely. Now consider that another user desires to wash his or her hands at a lavatory [flowing at 1.5 gpm (5.7 L/m)] connected to the same hot water system as the shower. The lavatory user turns on only the hot water side of the faucet to get hot water flowing to the faucet before turning on the cold water to temper the hot water to a usable temperature. What is the result of this action? Now the flow rate through the tankless water heater unit is 2.5 + 1.5 = 4 gpm (15 L/m). According to the graph shown in Commentary Figure P2801.1(2), at 4 gpm (15 L/m), the Model A unit has a temperature rise of only 30°F (-1°C)—a drop of 20°F (-7°C) from the 50°F (10°C) temperature rise at 2.5 gpm (9.8 L/m). The result is that both the shower user and the lavatory user will only receive 110°F (38°C) - 20°F (-7°C) = 90°F (32°C) hot water. The shower user will most likely notice the change in temperature and will wonder what happened to his or her "hot water that never runs out."

The previous example is an "extreme" case to illustrate the effect of multiple simultaneous users of hot water generated by a tankless water heater. In reality, the effects of multiple simultaneous uses of hot water are a little more complex because of temperature-controlled shower valves, the variety of possible settings of the shower valve, incoming cold water temperature and the characteristics of the tankless water heater. However, the basic concept for tankless water heaters that more hot water demand causes less temperature rise is substantially different than a storage tank water heater that can provide the same water temperature to all fixtures operating simultaneously, but only for a limited time.

Because showers typically require the largest flow rate of hot water in a residential building, some tankless water heater manufacturers have made selection of a "properly" sized tankless water heater simpler by providing unit sizing/selection charts based upon the number of showers (and sometimes "plus" other fixtures) that the designer believes will be operating simultaneously. While this selection method is simple, the designer must be aware of the fixture flow rates and incoming water temperature conditions that were assumed in creation of the sizing/selection chart. For example, a flow rate of 2.0 gpm (7.6 L/m) for a shower and 0.5 gpm (1.9 L/m) for a lavatory and an incoming cold water temperature of 55°F (13°C) might be used to develop the manufacturer's chart. If the actual operating conditions are different, then the tankless water heater may not deliver the flow rate of hot water at the desired temperature as anticipated. Where greater flow rates, different cold water temperatures or multiple shower heads per shower are involved, the designer must use the capacity graphs or tables to determine the appropriate size of tankless water heaters.

There are many different models of electric and fuel-gas-powered tankless water heaters, each having a specific capacity graph or table. If the largest tankless water heater cannot accommodate the total demand, additional tankless units can be added to meet the total demand. For electric units, each can be installed to serve a specific fixture or group of fixtures that can be accommodated by that unit. Where additional fuel-gas units are required to meet the total demand, each unit could be connected to serve only specific fixture groups within the building. However, most fuel-gas-powered unit controls are sophisticated enough so that multiple units can be connected to a common header and the controls interconnected so as to cause additional units to "come online" when hot water demand increases beyond the capacity of the first unit.

Depending on how systems were originally designed, the users of tankless water heater systems might be required to adjust their expectations concerning when and to what extent simultaneous hot water demands can be expected before temperature (and flow) of the hot water is affected. This adjustment of expectations is no different than for most users of storage tank-type water heater systems. Users implicitly understand that there is a limit to how much hot water can be used before the temperature of the hot water becomes unacceptable for the purposes intended and how much time it takes for the system to recover sufficiently in order to provide an acceptable supply and temperature level of hot water.

P2801.2 Drain valves. Drain valves for emptying shall be installed at the bottom of each tank-type water heater and *hot water* storage tank. The drain valve inlet shall be not less than $^3/_4$-inch (19.1 mm) nominal iron pipe size and the outlet shall be provided with a male hose thread.

❖ Eventually, hot water storage tanks and tank-type water heaters will require replacement because of tank failure (leaks in the walls of the tank). Also, some maintenance procedures require draining of the tank to be able to perform the maintenance. The drain valve is needed for both purposes. The requirement for the valve to be on the bottom of the tank is not to be construed as meaning to be on the underside of the tank. For a vertical tank, a valve on the side of the tank, near the bottom of the side, is typical for vertical storage tanks. The intent of the valve is to drain most of the water from the tank so the tank is not so heavy that it is unnecessarily difficult to move.

P2801.3 Installation. Water heaters shall be installed in accordance with this chapter and Chapters 20 and 24.

❖ Chapters 20, "Boilers and Water Heaters," and 24, "Fuel Gas," provide installation requirements for water heaters, in addition to this chapter.

P2801.4 Location. Water heaters and storage tanks shall be installed in accordance with Section M1305 and shall be located and connected to provide access for observation, maintenance, servicing and replacement.

❖ This section requires installation compliance with Section M1305, as well as access to all components of a water heater that require observation, inspection, adjustment, servicing, repair and replacement. Access is also required to conduct startup and shutdown operations.

The manufacturer's installation instructions typically provide clearance requirements for accessing controls and performing component removal, as well as the minimum acceptable clearances between the unit and

building surfaces, such as walls and ceilings. As provided by Sections M2005.1 and G2448.1, the manufacturer's instructions are part of the code.

P2801.5 Prohibited locations. Water heaters shall be located in accordance with Chapter 20.

❖ This section refers to Chapter 20 for water heater location requirements with respect to attics, storage closets, bedrooms, bathrooms and crawl spaces (see Sections M2005 and G2406).

P2801.6 Required pan. Where a storage tank-type water heater or a hot water storage tank is installed in a location where water leakage from the tank will cause damage, the tank shall be installed in a pan constructed of one of the following:

1. Galvanized steel or aluminum of not less than 0.0236 inch (0.6010 mm) in thickness.

2. Plastic not less than 0.036 inch (0.9 mm) in thickness.

3. Other approved materials.

A plastic pan shall not be installed beneath a gas-fired water heater.

❖ Where leaks from a storage-type water heater or hot water storage tank would cause damage to the building structure, the tank must be installed in a pan (also known as a "safe" pan or "drip" pan) (see Commentary Figure P2801.6). The pans must be made of galvanized steel, aluminum or plastic in the thicknesses indicated in the code text. The list also allows for other materials to be used but those materials would have to be approved by the code official. Note that plastic pans cannot be installed under gas-fired water heaters because of the high temperatures that might occur under those types of water heaters.

The required thickness of galvanized sheet metal is the minimum thickness tolerance for 24-gauge galvanized sheet metal so that a field measurement of thickness would allow for all thicknesses within the tolerance range of 24 gauge to be acceptable.

Note that tankless water heaters are not required to have a pan.

P2801.6.1 Pan size and drain. The pan shall be not less than $1^1/_2$ inches (38 mm) deep and shall be of sufficient size and shape to receive dripping or condensate from the tank or water heater. The pan shall be drained by an indirect waste pipe of not less than $^3/_4$ inch (19 mm) diameter. Piping for safety pan drains shall be of those materials indicated in Table P2906.5. Where a pan drain was not previously installed, a pan drain shall not be required for a replacement water heater installation.

❖ The pan required by this section is intended to catch only dripping water from tank or flue condensation, a small leak in the tank, a small leak at a connection to the tank, or a leaking temperature and pressure relief valve. The typical pan design having a flat bottom, a depth of $1^1/_2$ inches (38 mm) and a side outlet drain of $^3/_4$ inch (19 mm) is not capable of accommodating large flows, such as would occur if the water heater tank ruptured or the temperature and pressure relief valve discharged at its full capacity. In other words, a water heater pan meeting the minimum requirements of this section is not a waste receptor as defined by the code. Although a storage tank could rupture and a temperature and pressure relief valve could "trip," causing large flows of water, such occurrences happen so infrequently that requiring a true waste receptor pan design for every installation requiring a pan is not justified.

The pan drain pipe must be made of one of the materials listed for piping material suitable for water distribution system use. The concern is that leaking hot water could occur for extended periods of time until discovered. If the pan drain pipe was not made of a material suitable for hot water, the small diameter drain piping might sag between the supports to cause sluggish draining and clogging, which ultimately could cause the drain pan to overflow. Although many commercially available pans have a PVC adapter fitting for connecting the pan drain piping, PVC pipe material is not included in Table P2905.5 because it is not rated for hot water temperatures.

Where Section P2801.6 requires a pan under a water heater, the replacement of an existing water heater in those locations requires a pan, even though the code might not have required a pan for the previous water heater installation. Adding a pan to a replacement water heater job is not usually a significant issue, but providing a drain pipe and a suitable discharge point for the drain pipe can be an obstacle. The water heater might be in a basement where the building drain is above the elevation of the bottom of the water heater or the water heater might be in the middle of a slab-on-grade building, without a nearby floor drain and not adjacent to an outside wall. The last sentence of this code section allows the pan to be installed without a drain. Although it might seem illogical to require a pan that does not have a drain, the intent is that when a tank leak occurs, it is contained until the point at which the leaking water overflows the pan. Perhaps an occupant might notice water in the pan that is normally dry, or if water does overflow the pan and someone is investigating the cause for dampness in adjacent areas, water overflowing the pan is a sure sign that the tank is leaking. Otherwise, a leaking tank could leak for months without any visible evidence until after the leaking water has started to rot wood or create mold conditions.

The presence of a water heater drain pan (whether it is required or not) also allows the T & P valve drain pipe of the water heater to terminate at the drain pan in accordance with Item 5 of Section P2804.6.1.

If a pan drain is not required for the pan, then the implication is that the pan does not have a drain opening. As many factory-made pans already have a hole punched for a drain adapter, this means that the punched hole needs to be blocked off. One way to do this would be to install the male adapter provided with the pan in the opening and then install a plug into the adapter; or cap a short piece of pipe installed in the adapter.

The adapter (and plug, or cap materials) might not need to be of Table P2905.5 materials, as the hot water pipe material requirement for the pan drain is mainly concerned about sagging of drain piping on its way to the pipe termination. The "plug assembly" on the pan would not have this problem. The point is, with or without a prepunched hole in the pan, the pan should have the full $1^1/_2$-inch-deep capacity for water.

Another advantage of a drain pan (whether it is required or not) without a pan drain is that it allows for the installation of water leak detection alarms. Such alarms are not required by the code, however, many homeowners after having experienced a leaking water heater, would have liked to have known when leakage began so that they could have rescued personal possessions and shut off the water supply to the water heater.

P2801.6.2 Pan drain termination. The pan drain shall extend full-size and terminate over a suitably located indirect waste receptor or shall extend to the exterior of the building and terminate not less than 6 inches (152 mm) and not more than 24 inches (610 mm) above the adjacent ground surface.

❖ Although a pan drain is not required to follow the same rules as for gravity flow sanitary drainage, it should be obvious that any reductions in pan drain pipe size would restrict flow and be a potential clogging point, especially where pans might have collected years of dust and possible rust flakes from corroding steel tanks. Since the drainage from the pan is basically clear water, the drain line is not required to have a trap and the drain can terminate outdoors, at a waste receptor or at a floor drain. The discharge at a waste receptor or floor drain does not have to be through an air gap as might be assumed by the term "over." Where the pan drain terminates outdoors, a termination point of at least 6 inches (152 mm) above the ground protects the open end from being blocked by careless landscaping work and vegetation growth, which might allow for insects to build nests in the opening. The 6 inches (152 mm) minimum provides for clearance above grade to prevent blockage by ground cover. The 24 inches (610 mm) maximum provides a level of safety to keep hot water flow away from passers-by face and hands.

P2801.7 Water heaters installed in garages. Water heaters having an *ignition source* shall be elevated such that the source of ignition is not less than 18 inches (457 mm) above the garage floor.

Exception: Elevation of the *ignition source* is not required for appliances that are *listed* as flammable vapor ignition-resistant.

❖ An ignition source could be many things including an open flame, such as a pilot light, an electrical switch, open resistance heating coils or an electrical igniter unit (such as hot surface or spark ignition types). See the definition of "Ignition source" in Chapter 2. Residential garages have a high potential for volatile liquids, such as gasoline and paint thinners, to be spilled

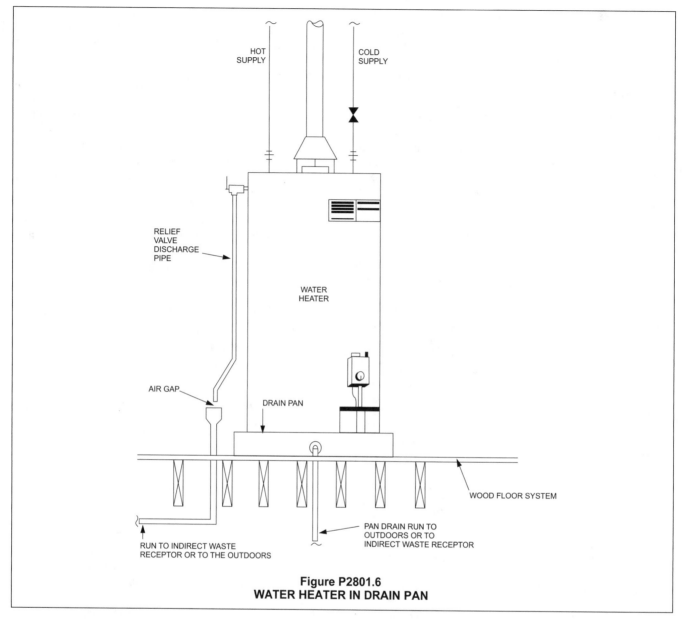

Figure P2801.6
WATER HEATER IN DRAIN PAN

or leak from their containers. Because the vapors from these liquids are heavier than air, they concentrate just above floor level, posing an explosion hazard. Testing has indicated that locating ignition sources at an elevation of 18 inches (457 mm) above the floor in these areas provides a degree of safety due to the time delay for the vapors from a spilled combustible liquid to build up to ignitable concentrations at the elevation of the ignition source. However, given large enough spills and enough time, an ignition source at any elevation in an enclosed area could cause an explosion.

Many electric water heater thermostats have enclosed contacts, but they are not sealed gas tight. Therefore, if an electric water heater having a thermostat is located less than 18 inches (457 mm) from the bottom of the unit, both this section and Section M1307.3 require that the unit be elevated so the ignition source (thermostat) is at least 18 inches (457 mm)

above the garage floor. Electric water heaters having all switching controls located above 18 inches (457 mm) from the bottom of the water heater are not required to be elevated.

Section G2408.2 for gas-fired appliances reiterates the requirement for elevation above the garage floor, but has an exception to allow gas-fired appliances having flammable ignition vapor resistant (FVIR) design to be installed without elevating the unit. Note that in this section and Section M1307.3, rooms or spaces that are not part of the living space of a dwelling unit and that communicate directly with the garage are considered to be part of the private garage. For example, a storage room on the same level as the garage having a door opening into the garage is in "direct communication" (in terms of the exchange of air) with the garage.

A frequently asked question is how this section, as well as Sections M1307.3.1 and G2408.2, should be applied to a replacement water heater that originally was not required by the adopted code at that time to be elevated. The code makes no distinction between an installation in new construction and a replacement installation in existing construction. Where installers are confronted with obstacles that appear to prevent strict compliance with the code, they should consult with the code official before performing the replacement work.

P2801.8 Water heater seismic bracing. In Seismic Design Categories D_0, D_1 and D_2 and townhouses in Seismic Design Category C, water heaters shall be anchored or strapped in the upper one-third and in the lower one-third of the appliance to resist a horizontal force equal to one-third of the operating weight of the water heater, acting in any horizontal direction, or in accordance with the appliance manufacturer's recommendations.

❖ The seismic design category for a building is determined by Section R301. The required strapping is intended to prevent movement of the appliance when subjected to lateral forces generated during an earthquake. The weight of the heater plus the weight of the water in it equals the operating weight (see Commentary Figure P2801.8).

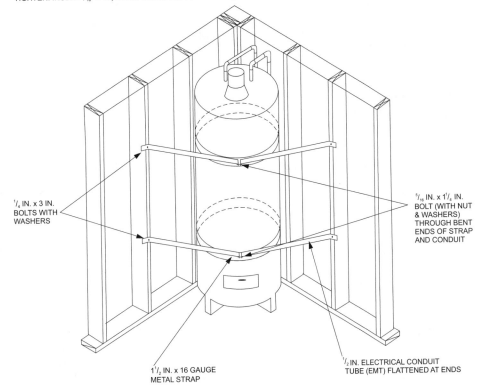

NOTE: THIS IS AN EXAMPLE OF AN EARTHQUAKE RESTRAINT DESIGN FOR WATER HEATERS UNDER 100 GALLONS (EXAMPLE TAKEN FROM A SPECIFICATION ISSUED BY A WEST COAST CITY IN U.S.A.)

THE ILLUSTRATION BELOW SHOWS THE STRAPPING REQUIREMENTS FOR WATER HEATERS UP TO 40 GALLONS WITHIN 12 INCHES OF A STUD WALLS. FOR WATER HEATERS OVER 40 GALLONS BUT LESS THAN 100 GALLONS, USE $^3/_4$ INCH EMT CONDUIT AND $1^1/_2$ INCH METAL STRAP. WATER HEATERS OVER 100 GALLONS AND/OR WATER HEATERS MORE THAN 12 INCHES FROM A WALL REQUIRE A DESIGNED SYSTEM.

NOTE: APPROVED MANUFACTURED WATER HEATER SUPPORT KITS SOLD AT BUILDING AND PLUMBING SUPPLY STORES MAY BE USED AS AN ALTERNATE TO THESE REQUIREMENTS.

INSTALLATION CHECKLIST
• LOCATE WALL STUDS ON BOTH SIDES OF THE WATER HEATER. PRE-DRILL HOLES FOR ANCHOR BOLTS IN STUD CENTER.
• MEASURE DISTANCE AROUND WATER HEATER. CUT TWO $1^1/_2$" x 16 GAUGE STRAPS TO ENCOMPASS WATER HEATER.
• DRILL HOLES IN THE ENDS OF THE STRAPS AND BEND THEM 45 DEGREES.
• CUT FOUR PIECES OF $^1/_2$" EMT CONDUIT TO THE PROPER LENGTH AS SHOWN IN DIAGRAM AND FLATTEN ENDS.
• DRILL HOLES IN FLATTENED ENDS OF CONDUIT. BEND FLATTENED ENDS 45 DEGREES ON ONE SIDE.
• INSTALL STRAPPING. STRAPPING SHALL BE LOCATED AT POINTS WITHIN THE UPPER AND LOWER ($^1/_2$) ONE THIRD OF THE VERTICAL DIMENSION OF THE WATER HEATER. AT THE LOWER POINT, A MINIMUM DISTANCE OF 4 INCHES SHALL BE MAINTAINED ABOVE THE CONTROLS. WRAP THE STRAPS AROUND THE HEATER AND INSERT A $^5/_{16}$"X $1^1/_4$" BOLT WITH WASHER INTO THE BENT ENDS.
• INSTALL EMT CONDUIT. INSERT $^1/_4$" x 3" LAG BOLTS THROUGH HOLE IN ENDS OF CONDUIT INTO THE WALL STUD AND TIGHTEN. INSERT $^5/_{16}$"X $1^1/_4$" BOLT, WHICH IS IN STRAPPING, THROUGH OTHER END OF CONDUIT AND TIGHTEN.

$^1/_4$ IN. x 3 IN. BOLTS WITH WASHERS

$^5/_{16}$ IN. x $1^1/_4$ IN. BOLT (WITH NUT & WASHERS) THROUGH BENT ENDS OF STRAP AND CONDUIT

$1^1/_2$ IN. x 16 GAUGE METAL STRAP

$^1/_2$ IN. ELECTRICAL CONDUIT TUBE (EMT) FLATTENED AT ENDS

For SI: 1 gallon = 3.785 L, 1 inch = 25.4 mm, 1 degree = 0.0175 rad.

Figure P2801.8
EXAMPLE OF SEISMIC RESTRAINT FOR SMALL WATER HEATERS
(Shown for illustration only)

SECTION P2802
SOLAR WATER HEATING SYSTEMS

P2802.1 Water temperature control. Where heated water is discharged from a solar thermal system to a *hot water* distribution system, a thermostatic mixing valve complying with ASSE 1017 shall be installed to temper the water to a temperature of not greater than 140°F (60°C). Solar thermal systems supplying *hot water* for both space heating and domestic uses shall comply with Section P2803.2. A temperature-indicating device shall be installed to indicate the temperature of the water discharged from the outlet of the mixing valve. The thermostatic mixing valve required by this section shall not be a substitute for water temperature limiting devices required by Chapter 27 for specific fixtures.

❖ Solar thermal water-heating systems can generate very high water temperatures. Water of a temperature greater than 140°F must not be released into the potable water distribution system because of the scalding implications that such high temperature water could cause. Therefore, a thermostatic mixing valve in accordance with ASSE 1017 must be installed to limit the temperature to not more than 140°F.

P2802.2 Isolation valves. Isolation valves in accordance with P2903.9.2 shall be provided on the cold water feed to the water heater. Isolation valves and associated piping shall be provided to bypass solar storage tanks where the system contains multiple storage tanks.

❖ Solar water-heating systems need repair and servicing from time to time, so a shutoff valve is required on the water supply to the system. Where multiple storage tanks are installed for a solar water-heating system, each tank must have a bypass piping arrangement to bypass the tank for maintenance and replacement.

SECTION P2803
WATER HEATERS USED FOR SPACE HEATING

P2803.1 Protection of potable water. Piping and components connected to a water heater for space heating applications shall be suitable for use with potable water in accordance with Chapter 29. Water heaters that will be used to supply potable water shall not be connected to a heating system or components previously used with nonpotable-water heating *appliances*. Chemicals for boiler treatment shall not be introduced into the water heater.

❖ There are combined systems that serve the dual purpose of providing both space heating and domestic hot water. In this case, precautions must be taken to protect the potable water from contamination. Thus, any piping that is used as part of the space heating system must meet the material and installation requirements of Chapter 29.

P2803.2 Temperature control. Where a combination water heater-space heating system requires water for space heating at temperatures exceeding 140°F (60°C), a master thermostatic mixing valve complying with ASSE 1017 shall be installed to temper the water to a temperature of not greater than 140°F (60°C) for domestic uses.

❖ Where a water heater has a dual purpose of supplying hot water and serving as a heat source for a hot water space heating system, the maximum outlet water temperature for the potable hot water distribution system is limited to 140°F (60°C). A master thermostatic mixing valve conforming to ASSE 1017 must be installed to limit the water temperature. These valves are used extensively in applications for domestic service to mix hot and cold water to reduce high-service water temperature to the building distribution system (see Commentary Figure P2803.2). These devices are not intended for final temperature control at fixtures and appliances (see commentary, Section P2708.3).

A water heater used as part of a space heating system must be protected from any conditions that can cause contamination of the potable water supply system. A typical installation might be an under-floor radiant heating system or baseboard fin-tube radiation system. Because the water heater is part of the potable water system, materials used in the heating system must be approved for use in a potable water system, and all connections must be protected against contamination. In the summer months when the heating system is inactive, a method to prevent stagnation of the water can be employed. A small orifice in the isolation valves to permit a small amount of water to circulate through the system may be provided for this purpose. Chemicals of any type must not be added to the heating system because this would directly contaminate the potable water supply. The potable water supply must be protected in accordance with Section P2902.

SECTION P2804
RELIEF VALVES

P2804.1 Relief valves required. Appliances and equipment used for heating water or storing hot water shall be protected by one of the following:

1. A separate pressure-relief valve and a separate temperature-relief valve.

2. A combination pressure-and-temperature relief valve.

❖ A combination pressure- and temperature-relief valve or separate temperature-relief and pressure-relief valves are needed to protect the water heater against damage and potential violent explosion. Pressure-relief valves are designed to relieve excessive pressures that can develop in a closed system (the water heater and piping). Temperature-relief valves are designed to open in response to excessive temperatures and to discharge heated water, thereby limiting the water temperature in the water heater. In the event of water heater control malfunction, it is possible for the water pressure to exceed the maximum pressure for which the water heater and piping system were

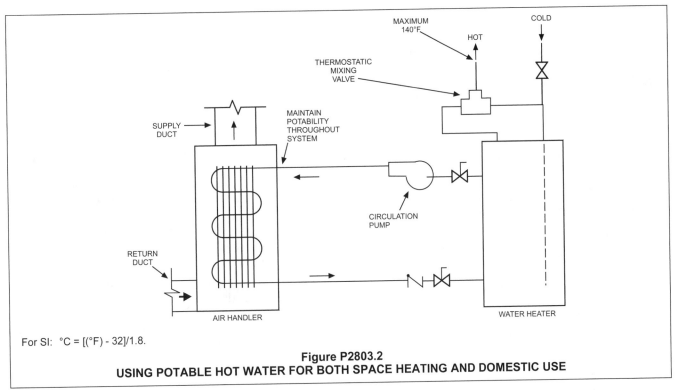

For SI: °C = [(°F) - 32]/1.8.

Figure P2803.2
USING POTABLE HOT WATER FOR BOTH SPACE HEATING AND DOMESTIC USE

designed. Also, water heater control failures could allow excessive temperature rises that could superheat the water (i.e., heated above the temperature at which it would vaporize at atmospheric pressure).

P2804.2 Rating. Relief valves shall have a minimum rated capacity for the equipment served and shall conform to ANSI Z21.22.

❖ To perform as intended, the relief valve capacity must be equal to or greater than the energy input capacity of the appliance it protects. Relief valves must be manufactured and tested in accordance with ANSI Z21.22. Conformance to the standard indicates to the installer, the inspector and the occupant that a similar relief valve has been tested and evaluated to demonstrate that this valve will perform as required when properly installed.

P2804.3 Pressure-relief valves. Pressure-relief valves shall have a relief rating adequate to meet the pressure conditions for the appliances or equipment protected. In tanks, they shall be installed directly into a tank tapping or in a water line close to the tank. They shall be set to open at not less than 25 psi (172 kPa) above the system pressure and not greater than 150 psi (1034 kPa). The relief-valve setting shall not exceed the rated working pressure of the tank.

❖ The code requires an independent pressure-relief valve for a water heater or hot water storage tank to be set to open at least 25 pounds per square inch (psi) (172 kPa) above the system pressure, but not over 150 psi (1034 kPa). Regardless, the valve setting must not exceed the tank's rated working pressure. Because the valve reacts to pressure, not temperature, its location in the system is not as critical as for temperature-relief valves. However, the location must

be in a tank opening or immediately adjacent to the tank in a water pipe connected to the tank. All tank-type water heaters manufactured today provide a tapping in the tank for installation of relief valves. To prevent the relief valve from being rendered useless, valves cannot be located between a relief valve and the vessel it protects (see Section P2804.6).

P2804.4 Temperature-relief valves. Temperature-relief valves shall have a relief rating compatible with the temperature conditions of the appliances or equipment protected. The valves shall be installed such that the temperature-sensing element monitors the water within the top 6 inches (152 mm) of the tank. The valve shall be set to open at a temperature of not greater than 210°F (99°C).

❖ The code requires that temperature-relief valves be set to open at a maximum temperature of 210°F (99°C). The highest temperature water rises to the top of the tank in tank-type water heaters. The thermal sensing element of the temperature relief valve must be installed within 6 inches (152 mm) of the top of the tank. This measure allows the valve to sense the hot-test water temperature accurately and to provide an early response to afford maximum protection. The water heater manufacturer's installation instructions usually specify the location, which is the tank tapping provided and so identified on the heater (see Commentary Figure P2804.4).

P2804.5 Combination pressure-and-temperature relief valves. Combination pressure and temperature-relief valves shall comply with the requirements for separate pressure- and temperature-relief valves.

❖ All of the provisions of Sections P2804.3 and P2804.4 that apply to separate relief valves apply to combination relief valves. Because combination valves operate by sensing changes in temperature and pressure, the location requirements of Section P2804.4 apply.

P2804.6 Installation of relief valves. A check or shutoff valve shall not be installed in any of the following locations:

1. Between a relief valve and the termination point of the relief valve discharge pipe.

2. Between a relief valve and a tank.

3. Between a relief valve and heating appliances or equipment.

❖ A check valve or shutoff valve can restrict or completely cut off the physical connection between the water heater and the relief valve, thus rendering the relief valve ineffective for its designed purpose and thereby causing the installation to be an extreme hazard.

P2804.6.1 Requirements for discharge pipe. The discharge piping serving a pressure-relief valve, temperature-relief valve or combination valve shall:

1. Not be directly connected to the drainage system.

2. Discharge through an air gap located in the same room as the water heater.

3. Not be smaller than the diameter of the outlet of the valve served and shall discharge full size to the air gap.

4. Serve a single relief device and shall not connect to piping serving any other relief device or equipment.

5. Discharge to the floor, to the pan serving the water heater or storage tank, to a waste receptor or to the outdoors.

6. Discharge in a manner that does not cause personal injury or structural damage.

7. Discharge to a termination point that is readily observable by the building occupants.

8. Not be trapped.

9. Be installed to flow by gravity.

10. Terminate not more than 6 inches (152 mm) and not less than two times the discharge pipe diameter above the floor or waste receptor flood level rim.

11. Not have a threaded connection at the end of the piping.

12. Not have valves or tee fittings.

13. Be constructed of those materials indicated in Section P2906.5 or materials tested, rated and *approved* for such use in accordance with ASME A112.4.1.

14. Be one nominal size larger than the size of the relief-valve outlet, where the relief-valve discharge piping is constructed of PEX or PE-RT tubing. The outlet end of such tubing shall be fastened in place.

❖ The discharge pipe from a water heater temperature and pressure relief valve (T&P valve) is an extension of the potable water distribution system. Because Section P2902.1 prohibits cross connections between the potable water supply and any source of contamination (such as a drainage system), a T&P valve discharge pipe that discharges to a drain system must connect indirectly to that drainage system.

Regardless of where the T&P valve discharge pipe terminates, an air gap is required to protect the potable water supply system. Section P2902.3.1 requires a minimum air gap dimension of twice the diameter of the effective opening of the discharge pipe. There are three reasons for the air gap to be in the same room as the water heater: (1) it prevents a direct connection to concealed discharge piping that might be bent, flattened, plugged, reverse-sloped or inadvertently capped off; (2) it provides a location for observing discharge when testing the relief valve and (3) it provides a readily accessible location to observe valve leakage indicating a defective T&P valve, a water distribution

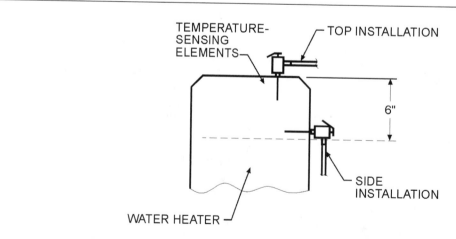

For SI: 1 inch = 25.4 mm.

Figure P2804.4
TEMPERATURE AND PRESSURE RELIEF VALVE INSTALLATION

system overpressure problem, or a water heater operation problem.

The size of the discharge pipe must be not less than the size of the T&P valve outlet to ensure that the valve can discharge at its full capacity. The pipe size must not be reduced as this would create a restriction that might prevent full-capacity discharge in an emergency condition.

The discharge pipe cannot be combined with any other discharge pipe or connect with any other piping before terminating at the required air gap. Connection of other piping could introduce flow that would interfere with the relief valve discharge flow. Also, full-capacity discharge could damage other connected equipment or cause the discharge to exit at other points where persons could be injured by the escaping hot water.

Water discharged from the T&P valve must be directed to one of four locations: 1. The floor below the water heater; 2. The water heater or storage tank drip pan, if present; 3. A waste receptor, such as a floor drain; or 4. The outdoors. The choice of discharge location must consider the potential for personal injury and structural damage that water discharge might cause. For example, a floor discharge might be suitable in a concrete-floored and curbed garage, but where the garage walls are of wood and rest directly on the floor, this discharge point might be unsuitable. Another suitable floor discharge example might be the tiled and sloped floor of a laundry/utility room that has a floor drain. Discharge to laundry trays/tubs and sinks would not be a suitable location as it violates the intent of Item 6 of this section, which is to protect the person using the fixture from hot water and steam that could come from the discharge pipe.

As discharge from a T&P valve is an indicator of a problem, the discharge point should be readily observable by the occupants so they can take action to correct the problem.

T&P relief valve pipes must not have traps and must be so installed to cause water to completely flow out of the pipe by gravity. If a trap was in the line or the line was not sloped to completely drain, hot water drying out in the trapped location could cause deposits to build up and eventually block the flow. Where the pipe is exposed to freezing temperatures, an ice block could form in the pipe and block flow.

Where the T&P discharge pipe discharges to a floor or waste receptor, such as a floor drain, the opening of the pipe must not be any higher than 6 inches (152 mm) above the floor or receptor. This precaution is necessary to keep any hot water discharge from splashing and possibly causing injury to someone nearby. However, the installer must be careful to allow for an air gap above the floor or flood level rim of the waste receptor. For arrangements where the T&P valve discharge pipe is routed to the water heater drip pan (see Item 5), the intent is that the air gap at the end of the pipe be above the flood level rim of the pan, even though the pan is not considered to be a waste receptor.

Threads on the end of a T&P discharge pipe are an invitation for someone to install a threaded cap to stop a nuisance relief valve drip. What would be perceived as a fix would actually be creating a very serious hazard. Because the installation of valves or tees in the relief valve would only invite someone to close the valve (thus blocking the flow) or connect another drain line to the discharge pipe (creating another hazard), their installation is prohibited.

Relief valve discharge piping must be one of the piping materials listed in Table P2906.5. Although the pipe materials in this table have a pressure rating of at least 100 psi (690 kPa) at 180°F (82°C), they have the capability to survive the limited time exposures (e.g., 15 minutes) at a temperature of 210°F (99°C) to carry a full output relief valve discharge. A discharge tube assembly complying with ASME A112.4.1 can also be installed.

The use of PEX or PE-RT tubing for T & P discharge piping is permitted as those are piping types indicated in Table P2906.5. However, the typical fittings for these tubing types are insert fittings. The insert fittings have a smaller inside diameter than the tubing and as such, would create additional resistance to flow. This would not be a good situation for a "full trip" event of the T&P valve. Therefore, Item 14 requires that the tubing size be one nominal size larger than the relief valve opening so that the insert fittings for the one size larger tubing are not a significant restrictor of flow. PEX and PE-RT tubing is flexible and, after uncoiling, has the inclination to want to return to its former coiled shape. Therefore, the free end of PEX and PE-RT tubing T&P discharge pipe must be fastened in the intended position.

A frequently asked question is how this section should be applied to a relief valve discharge pipe termination that, originally, terminated just above the floor without a drain or waste receptor to capture the flow, regardless of the potential for damage. The water heater might be in a basement where the building drain is above the elevation of the water heater T&P valve or the water heater might be in the middle of a slab-on-grade building, without a nearby floor drain and not adjacent to an outside wall. Where water heater replacement installers are confronted with obstacles that appear to prevent strict compliance with this section, they should consult with the code official before performing the replacement work.

P2804.7 Vacuum-relief valve. Bottom fed tank-type water heaters and bottom fed tanks connected to water heaters shall have a vacuum-relief valve installed that complies with ANSI Z21.22.

❖ A vacuum-relief valve must be installed in bottom fed water heaters and bottom fed tanks connected to water heaters. The vacuum relief valve is intended to prevent a possible pressure reduction within the tank to below atmospheric pressure (i.e., a partial vacuum). Many tanks are not designed to resist external pressures exceeding internal pressures; therefore, a vacuum-relief valve is necessary to prevent atmospheric pressure from possibly collapsing the tank or damaging an internal flue.

The vacuum-relief valve operates by automatically allowing air into the supply piping to the tank when a partial vacuum is created, (see Commentary Figure P2804.7). Bottom fed tanks are also susceptible to the creation of a vacuum by tank drainage resulting from negative supply pipe pressure or leakage.

Vacuum-relief valves must be constructed and tested in accordance with ANSI Z21.22.

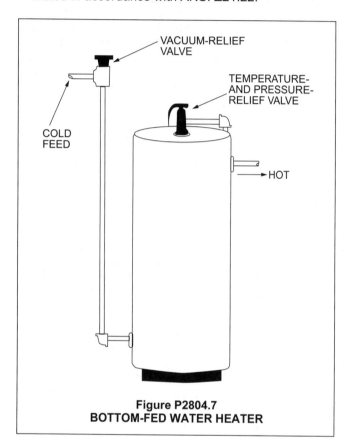

Figure P2804.7
BOTTOM-FED WATER HEATER

Chapter 29:
Water Supply and Distribution

User note: Code change proposals to this chapter will be considered by the IRC – Plumbing and Mechanical Code Development Committee during the 2015 (Group A) Code Development Cycle.

General Comments

Chapter 29 regulates the design and installation of water supply systems. The code requires potable water to be supplied to dwelling units, and contains specific guidelines for maintaining the potability of the water supply. These provisions for protecting the water system from contamination are without question the most important part of this chapter and perhaps all of the plumbing chapters.

The chapter further outlines the methods that can be used in sizing water supply systems, along with specific precautions that must be adhered to in order to maintain the integrity of the material and its contents. Information specific to the types of materials that are approved for use within a potable water supply system are included, along with provisions relating to the types of connecting components or methods specific to each material.

The type and installation of various drinking water treatment units, such as water softeners and water filters, are also included.

Purpose

This chapter contains the minimum requirements necessary to design and install a water distribution system. The provisions include materials, installation guidelines and methods for maintaining the potability of the water system.

SECTION P2901
GENERAL

P2901.1 Potable water required. Potable water shall be supplied to plumbing fixtures and plumbing *appliances* except where treated rainwater, treated gray water or municipal reclaimed water is supplied to water closets, urinals and trap primers. The requirements of this section shall not be construed to require signage for water closets and urinals.

❖ A dwelling unit must be supplied with potable water to provide for the sanitation, hygiene, cooking and drinking needs of the occupants. An increasing number of homes use nonpotable water for irrigation and for flushing of water closets.

P2901.2 Identification of nonpotable water systems. Where *nonpotable* water systems are installed, the piping conveying the nonpotable water shall be identified either by color marking, metal tags or tape in accordance with Sections P2901.2.1 through P2901.2.2.3.

❖ Where nonpotable water supply systems are installed in a building, the piping of the nonpotable system must be identified to prevent an inadvertent cross connection between different systems, especially the potable water system, during repairs or renovations. For example, a building could be provided with a recycled gray-water supply system that supplies nonpotable water for flushing urinals and water closets. Without identification of the recycled gray-water piping, the piping system might be assumed to convey potable water because it is commonplace for urinals and water closets to be supplied with potable water. A connection to an unmarked nonpotable piping system for the purpose of obtaining potable water could lead to significant health risks for the building's occupants.

P2901.2.1 Signage required. Nonpotable water outlets such as hose connections, open-ended pipes and faucets shall be identified with signage that reads as follows: "Nonpotable water is utilized for [application name]. CAUTION: NONPOTABLE WATER. DO NOT DRINK." The words shall be legibly and indelibly printed on a tag or sign constructed of corrosion-resistant water-proof material or shall be indelibly printed on the fixture. The letters of the words shall be not less than 0.5 inches (12.7 mm) in height and in colors in contrast to the background on which they are applied. In addition to the required wordage, the pictograph shown in Figure P2901.2.1 shall appear on the required signage.

❖ Some nonpotable water systems have outlets such as hose connections, open-ended pipes and faucets that could be mistaken for potable water outlets. Therefore, at these locations, the code requires signage to warn the building's occupants that water obtained from these outlets is not safe for drinking. The pictograph in Figure P2901.2.1 conveys the intent in a visual manner.

FIGURE P2901.2.1
PICTOGRAPH—DO NOT DRINK

P2901.2.2 Distribution pipe labeling and marking. Nonpotable distribution piping shall be purple in color and shall be embossed or integrally stamped or marked with the words: "CAUTION: NONPOTABLE WATER—DO NOT DRINK" or the piping shall be installed with a purple identification tape or wrap. Pipe identification shall include the contents of the piping system and an arrow indicating the direction of flow. Hazardous piping systems shall contain information addressing the nature of the hazard. Pipe identification shall be repeated at intervals not exceeding 25 feet (7620 mm) and at each point where the piping passes through a wall, floor or roof. Lettering shall be readily observable within the room or space where the piping is located.

❖ Distribution piping conveying nonpotable water must be purple in color and have the caution lettering as indicated. Many types of water distribution pipe materials are offered with the external purple color and caution lettering. For those materials that are not offered with purple color or lettering, or if the installer wishes to use standard pipe (not purple), the code allows for pipe identification using tapes or wraps. The tapes and wraps would need to be of purple color (primarily) and have the cautionary lettering. The intent of this section is not to require that elbows, tees, and valves in the piping be purple colored or have the cautionary lettering. Identification of the pipe is all that is required as the pipe is where someone would be most likely to attempt to "tap into."

Piping identification must indicate the contents and the direction of flow, and any hazardous nature of the flow such as temperature, pressure, flammability and toxicity. The repetition of the identification locations is necessary so that those performing maintenance and modifications can easily find the pipe identification to determine the contents of the piping. Obviously, the pipe identification must be in a position such that a person within the room or space can easily see the identification markings.

P2901.2.2.1 Color. The color of the pipe identification shall be discernable and consistent throughout the building. The color purple shall be used to identify reclaimed, rain and gray water distribution systems.

❖ Where color is used for pipe identification, it must be clearly recognizable from any other pipe identification colors and be the same color for the same type of piping in the entire building. Where identifying reclaimed, rain and gray water distribution systems with color, the color purple must be used.

P2901.2.2.2 Lettering size. The size of the background color field and lettering shall comply with Table P2901.2.2.2.

❖ See Table P2901.2.2.2 for the required size of pipe identification.

P2901.2.2.3 Identification Tape. Where used, identification tape shall be not less than 3 inches (76 mm) wide and have white or black lettering on a purple field stating "CAUTION: NONPOTABLE WATER—DO NOT DRINK." Identification tape shall be installed on top of nonpotable rainwater distribution pipes and fastened not greater than every 10 feet (3048 mm) to each pipe length, and run continuously the entire length of the pipe.

❖ This section provides requirements for identification tapes for nonpotable water distribution piping. Also, this section requires that identification tape used for rainwater distribution piping be located on top of the piping and fastened at 10-foot intervals (and that the tape be continuously applied along the entire length of the pipe).

It is believed that this text comes from one of the many rainwater reuse "best practices" manuals that have been created over the last decade and that the text was selected primarily for the requirements for the identification tape. The part about use of the tape for identifying rainwater distribution piping, while important for buried rainwater distribution piping, does not seem to be logical to apply to rainwater piping inside of buildings (unless that piping is buried under the building).

The code does not specify how to use/apply identification tapes, other than the buried application for rainwater piping. Where tapes are used for identifying buried piping for any type of nonpotable water distribution piping, the tape requirements for buried rainwater distribution piping appear to be reasonable, although this is not a code requirement for other than rainwater piping. Where nonpotable water distribution piping is in a building (and not buried below the building), nonself-adhering tapes could be (spiral) wrapped around nonpotable water distribution piping because simply fastening at 10-foot intervals might not result in a "permanent" marking. For example, if the tape were fastened at 10-foot intervals on a horizontal pipe, the tape could sag and eventually be torn off during other work. Spiral wrapping of tape is not a code requirement but seems to be a logical method for maintaining identification.

Identification tapes could also be self-adhering, however, application of such tapes should consider

TABLE P2901.2.2.2
SIZE OF PIPE IDENTIFICATION

PIPE DIAMETER (inches)	LENGTH OF BACKGROUND COLOR FIELD (inches)	SIZE OF LETTERS (inches)
$^3/_4$ to $1^1/_4$	8	0.5
$1^1/_2$ to 2	8	0.75
$2^1/_2$ to 6	12	1.25
8 to 10	2	2.5
over 10	32	3.5

For SI: 1 inch = 25.4 mm.

whether the adhesive will have a negative impact on the pipe material, whether the adhesive will hold up in the expected environmental conditions (heat, condensation, etc.) and whether the adhesive can be removed without damage to the piping, should a repair or alteration be necessary.

SECTION P2902
PROTECTION OF POTABLE WATER SUPPLY

P2902.1 General. A potable water supply system shall be designed and installed as to prevent contamination from nonpotable liquids, solids or gases being introduced into the potable water supply. Connections shall not be made to a potable water supply in a manner that could contaminate the water supply or provide a cross-connection between the supply and a source of contamination except where *approved* backflow prevention assemblies, backflow prevention devices or other means or methods are installed to protect the potable water supply. Cross-connections between an individual water supply and a potable public water supply shall be prohibited.

❖ History is filled with localized and widespread occurrences of sicknesses, diseases and deaths caused by contamination of potable water supplies. It is therefore imperative that potable water supply systems be designed and maintained in a safe-for-drinking condition at all times. Although the requirements of this section appear straightforward, piping arrangements, fixture designs and appliance operations can be complex. Adding to this complexity is the potential for upset conditions to occur with regard to pressures within piping systems. As a result, designers and installers of potable water systems must be aware of hazardous conditions that might occur because of connections made to potable water systems or changing pressure conditions within piping systems that would not be normally expected.

A potable water "cross connection" is an arrangement where there is a potential for contamination to flow into the potable water supply. Because potable water supply systems normally flow in one direction, (i.e., from the source to the point of delivery), a flow back into a potable water system is called a "backflow." To prevent backflows from occurring, specific piping arrangements or specially designed devices must be installed. These are called "backflow prevention methods" and "backflow prevention devices," respectively. A discussion on backflow methods and devices is presented in the commentary to Section P2902.3.

The last sentence of this section prohibits the interconnection of a public water supply system and a private water system, such as a well. Even though a private well might be a good source of potable water, there is no guarantee that a private well will continue to produce potable water as wells are subject to contamination from numerous sources beyond the control of the well owner. Therefore, private water systems must not be connected to public water systems regardless of whether a backflow prevention method or device is installed at the connection.

To protect the potable water supply, identify potential threats and determine the appropriate backflow prevention method; an understanding of how backflow can occur is necessary. Backflow depends on atmospheric pressure, water pressures and variations in the water supply at any moment. The following discusses pressure and backflow:

ATMOSPHERIC PRESSURE: The pressure exerted by the weight of the atmosphere. Atmospheric pressure at sea level is 14.7 pounds per square inch (psi) (101 kPa) and varies with altitude.

GAUGE PRESSURE: The pressure read on a gauge. The reference point for the gauge is atmospheric pressure equal to zero. In other words, gauge pressure is the amount of pressure above or below atmospheric pressure.

ABSOLUTE PRESSURE: The sum of atmospheric pressure and gauge pressure.

WATER PRESSURE: A function of the weight and height of a column of water. A cubic foot of water weights 62.4 pounds (28.3 kg); the pressure exerted at the base of the cube is 62.4 pounds per square foot (psf) or 0.433 psi (2.987 kPa). Therefore, the height of a column of water determines the pressure that is exerted at its base. The diameter of the pipe or vessel has no bearing on this pressure. For example, a 100-foot high (30 480 mm) column of water that is 20 feet (6096 mm) in diameter has the same pressure at its base as a 100-foot high (30 480 mm) column of water that is 1 inch (25 mm) in diameter.

Columns of water with a free surface (exposed to atmosphere) will have atmospheric pressure in addition to pressure that is caused by the weight of water exerted at its base.

BACKFLOW: Flow of liquids in potable water distribution piping that is in reverse of its intended direction. Two types of pressure conditions cause backflow: backsiphonage and backpressure.

BACKSIPHONAGE: Backflow of water that is caused by system pressure falling below atmospheric pressure. Atmosphere supplies the force that reverses flow. Any gauge reading of pressure below atmospheric pressure will be negative.

To illustrate the effect atmospheric pressure has on water, assume that a 50-foot-long (15 240 mm) tube open on both ends is inserted vertically into a container of water. Atmospheric pressure acts equally on the surface of the water within the tube and outside the tube.

If the top of the tube is sealed and a vacuum pump is used to evacuate all the air from the tube, a vacuum with a pressure of 0 pounds per square inch gauge (psig) (0 Pa) is created within the tube. Atmospheric pressure will force water up into the tube and the water in the tube will rise to 33.9 feet (10 333 mm) above the surface of the water in the container. This height is equal to the height that can be supported by the atmospheric pressure. Standard (sea level) atmospheric pressure is 14.7 psi (101 kPa) and will support a column of water 33.9 feet (10 333 mm) high [33.9 feet × 0.433 psi = 14.7 psi (101 kPa)].

Backsiphonage occurs in the water distribution system when supply pressure falls below atmospheric pressure. The following are practical examples of conditions that can cause back-siphonage.

A simple siphon occurs when pressure is reduced as a result of a difference in water levels at two separate points within a continuous fluid system [see Commentary Figure P2902.1(1)].

Aspiration or venturi principle is achieved when reduced pressure is created within a fluid system as a result of fluid motion. Intentional or unintentional con-

striction of flow through an opening or pipe increases velocity and decreases pressure. If this point of reduction is linked to a source of pollution, backsiphonage of the pollutant can occur [see Commentary Figure P2902.1(2)].

Supply pressure can be reduced at the pump intake or suction side. Insufficient water pressure or undersized piping supplying the pump can cause a negative pressure and allow contaminated water to be drawn into the system [see Commentary Figure P2902.1(3)].

BACKPRESSURE: Pressure created in a nonpotable system in excess of the water supply mains, causing backflow. Backpressure may be created by mechanical means, such as a pump; by static head pressure including an elevated tank; or by thermal expansion from a heat source, such as a water heater.

P2902.2 Plumbing fixtures. The supply lines and fittings for every plumbing fixture shall be installed so as to prevent backflow. Plumbing fixture fittings shall provide backflow protection in accordance with ASME A112.18.1/CSA B125.1.

❖ Faucets and fixture fittings are required to comply with ASME A112.18.1/CSA B125.1, which addresses backflow protection. These components must be installed in a manner that does not allow backflow conditions to occur.

**Figure P2902.1(1)
EXAMPLE OF SIMPLE SIPHON**

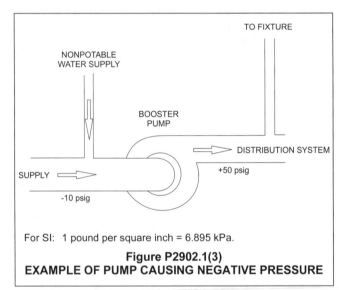

For SI: 1 pound per square inch = 6.895 kPa.

**Figure P2902.1(3)
EXAMPLE OF PUMP CAUSING NEGATIVE PRESSURE**

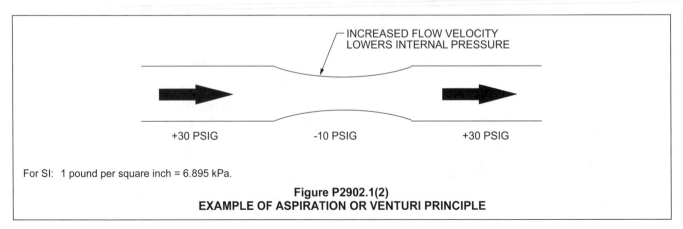

For SI: 1 pound per square inch = 6.895 kPa.

**Figure P2902.1(2)
EXAMPLE OF ASPIRATION OR VENTURI PRINCIPLE**

P2902.3 Backflow protection. A means of protection against backflow shall be provided in accordance with Sections P2902.3.1 through P2902.3.6. Backflow prevention applications shall conform to Table P2902.3, except as specifically stated in Sections P2902.4 through P2902.5.5.

❖ This section refers the reader to six subsections, each detailing a type of approved backflow preventer that can be used, depending on the type of installation and commensurate with the degree of hazard.

TABLE P2902.3. See below.

❖ Where the code does not specifically address backflow prevention methods or devices for a specific application within Sections P2902.4 through P2902.5.5, this table must be used to select a suitable method or device for the application. To use the table, first identify whether the potential hazard substance that might enter the potable water system is a low or high hazard to the potable water supply. Note a directs the reader to the Chapter 2 definitions of "Pollution" and "Contamination" for determining whether the hazard is low or high, respectively. Once the hazard is selected, this should eliminate some entries in the table as being viable for the application. Next, determine if the application will be subject to backpressure conditions. If so, this will eliminate additional table entries as being viable. (If not, then no table entries will be eliminated because all methods and devices are suitable for backsiphonage conditions). Of the remaining entries to choose from, the reader must then select a method or device using the description based upon the specific application for which the description fits. Of the final list of viable entries, the best selection is typically the one that is most economical to purchase and install.

TABLE P2902.3
APPLICATION FOR BACKFLOW PREVENTERS

DEVICE	DEGREE OF HAZARD[a]	APPLICATION[b]	APPLICABLE STANDARDS
Backflow Prevention Assemblies			
Double check backflow prevention assembly and double check fire protection backflow prevention assembly	Low hazard	Backpressure or backsiphonage Sizes $^3/_8$" – 16"	ASSE 1015, AWWA C510, CSA B64.5, CSA B64.5.1
Double check detector fire protection backflow prevention assemblies	Low hazard	Backpressure or backsiphonage Sizes 2" – 16"	ASSE 1048
Pressure vacuum breaker assembly	High or low hazard	Backsiphonage only Sizes $^1/_2$" – 2"	ASSE 1020, CSA B64.1.2
Reduced pressure principle backflow prevention assembly and reduced pressure principle fire protection backflow prevention assembly	High or low hazard	Backpressure or backsiphonage Sizes $^3/_8$" – 16"	ASSE 1013, AWWA C511, CSA B64.4, CSA B64.4.1
Reduced pressure detector fire protection backflow prevention assemblies	High or low hazard	Backsiphonage or backpressure (Fire sprinkler systems)	ASSE 1047
Spill-resistant vacuum breaker	High or low hazard	Backsiphonage only Sizes $^1/_4$" – 2"	ASSE 1056, CSA B64.1.3
Backflow Preventer Plumbing Devices			
Antisiphon-type fill valves for gravity water closet flush tanks	High hazard	Backsiphonage only	ASSE 1002, CSA B125.3
Backflow preventer with intermediate atmospheric vents	Low hazard	Backpressure or backsiphonage Sizes $^1/_4$" – $^3/_8$"	ASSE 1012, CSA B64.3
Dual-check-valve-type backflow preventers	Low hazard	Backpressure or backsiphonage Sizes $^1/_4$" – 1"	ASSE 1024, CSA B64.6
Hose-connection backflow preventer	High or low hazard	Low head backpressure, rated working pressure backpressure or backsiphonage Sizes $^1/_2$" – 1"	ASSE 1052, CSA B64.2.1.1
Hose-connection vacuum breaker	High or low hazard	Low head backpressure or backsiphonage Sizes $^1/_2$", $^3/_4$", 1"	ASSE 1011, CSA B64.2, B64.2.1
Laboratory faucet backflow preventer	High or low hazard	Low head backpressure and backsiphonage	ASSE 1035, CSA B64.7
Pipe-applied atmospheric-type vacuum breaker	High or low hazard	Backsiphonage only Sizes $^1/_4$" – 4"	ASSE 1001, CSA B64.1.1
Vacuum breaker wall hydrants, frost-resistant, automatic-draining type	High or low hazard	Low head backpressure or backsiphonage Sizes $^3/_4$" – 1"	ASSE 1019, CSA B64.2.2
Other Means Or Methods			
Air gap	High or low hazard	Backsiphonage only	ASME A112.1.2
Air gap fittings for use with plumbing fixtures, appliances and appurtenances	High or low hazard	Backsiphonage or backpressure	ASME A112.1.3

For SI: 1 inch = 25.4 mm.

a. Low hazard—See Pollution (Section R202). High hazard—See Contamination (Section R202).
b. See Backpressure (Section R202). See Backpressure, Low Head (Section R202). See Backsiphonage (Section R202).

P2902.3.1 Air gaps. *Air gaps* shall comply with ASME A112.1.2 and *air gap* fittings shall comply with ASME A112.1.3. An *air gap* shall be measured vertically from the lowest end of a water outlet to the flood level rim of the fixture or receptor into which the water outlets discharges to the floor. The required *air gap* shall be not less than twice the diameter of the effective opening of the outlet and not less than the values specified in Table P2902.3.1.

❖ The table referred to in this section has been a mainstay in the plumbing industry since 1940. The values specified have not changed in more than 70 years. The minimum air gap distance must be twice the diameter of the effective opening [see Commentary Figure P2902.3.1(1)].

This section refers the reader to Table P2902.3.1 for distance requirements from side walls (see commentary, Table P2902.3.1).

Further provisions specify minimum air gap requirements. An air gap is required at the discharge point of a relief valve, and within dishwashers and clothes washers (see commentary, Sections P2717.1, P2718.1 and P2803.6.1).

TABLE P2902.3.1. See below.

❖ The middle column of the table provides minimum air gap dimensions for outlets located away from walls or obstructions. The third column provides more stringent air gap dimensions for outlets located near walls or obstructions [see Commentary Figures P2902.3.1(2) and P2902.3.1(3)].

ASME A112.1.2 is the standard for air gaps that require specific dimensions; these are listed in Table P2902.3.1. An air gap is not a device and has no moving parts. It is, therefore, the most effective and dependable method of preventing backflow and should be used where feasible. However, when this method of protection is used, there must be an assurance or at least a level of awareness by the user that the air gap cannot be bypassed or defeated by the addition of a hose or extension from the end of the potable water outlet to the source of contamination. The potable water opening or outlet must terminate at an elevation above the level of the source of contamination. The potable water could be contaminated only if the entire area or room is flooded to a depth that would submerge the potable water opening or outlet. An air gap installation must always be constructed so that the air gap remains permanently fixed.

ASME A112.1.3 is the standard for air gap fittings for use with plumbing fixtures, appliances and appurtenances. This standard was developed for applications where an air gap fitting is manufactured for appliance applications.

TABLE P2902.3.1
MINIMUM AIR GAPS

FIXTURE	MINIMUM AIR GAP	
	Away from a wall[a] (inches)	Close to a wall (inches)
Effective openings greater than 1 inch	Two times the diameter of the effective opening	Three times the diameter of the effective opening
Lavatories and other fixtures with effective opening not greater than $^1/_2$ inch in diameter	1	1.5
Over-rim bath fillers and other fixtures with effective openings not greater than 1 inch in diameter	2	3
Sink, laundry trays, gooseneck back faucets and other fixtures with effective openings not greater than $^3/_4$ inch in diameter	1.5	2.5

For SI: 1 inch = 25.4 mm.

a. Applicable where walls or obstructions are spaced from the nearest inside edge of the spout opening a distance greater than three times the diameter of the effective opening for a single wall, or a distance greater than four times the diameter of the effective opening for two intersecting walls.

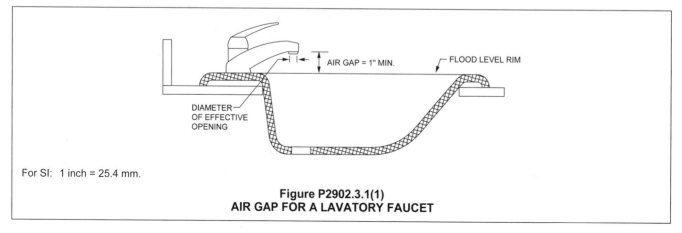

For SI: 1 inch = 25.4 mm.

Figure P2902.3.1(1)
AIR GAP FOR A LAVATORY FAUCET

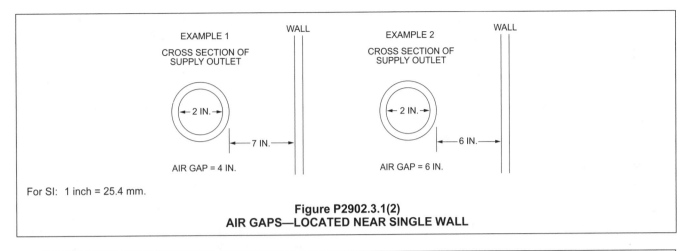

For SI: 1 inch = 25.4 mm.

Figure P2902.3.1(2)
AIR GAPS—LOCATED NEAR SINGLE WALL

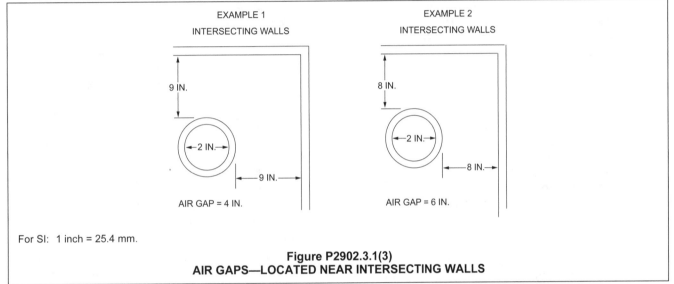

For SI: 1 inch = 25.4 mm.

Figure P2902.3.1(3)
AIR GAPS—LOCATED NEAR INTERSECTING WALLS

P2902.3.2 Atmospheric-type vacuum breakers. Atmospheric-type vacuum breakers shall conform to ASSE 1001 or CSA B64.1.1. Hose-connection vacuum breakers shall conform to ASSE 1011, ASSE 1019, ASSE 1035, ASSE 1052, CSA B64.2, CSA B64.2.1, CSA B64.2.1.1, CSA B64.2.2 or CSA B64.7. Both types of vacuum breakers shall be installed with the outlet continuously open to the atmosphere. The critical level of the atmospheric vacuum breaker shall be set at not less than 6 inches (152 mm) above the highest elevation of downstream piping and the flood level rim of the fixture or device.

❖ An atmospheric vacuum breaker is designed to prevent siphonic action from occurring down-stream of the device. When a negative pressure (full or partial vacuum) occurs in the water supply, the vacuum breaker opens to the atmosphere, allowing air to enter the piping system, thus breaking the vacuum. This prevents contamination of the potable water system by stopping the back-siphonage backflow. Under backsiphonage conditions the disc float (poppet) drops over the inlet seat opening and air enters the atmospheric port opening allowing the remaining water in the piping downstream from the vacuum breaker to drain.

These devices are not effective in preventing backflow resulting from backpressure. The outlet of atmospheric vacuum breakers must remain open to the atmosphere by terminating with a pipe, spout or similar unobstructed opening. Valves must not be installed downstream of this device because this would subject the device to supply pressure, which could cause the disc float (poppet) to stick in the closed position, thereby rendering it inoperative.

An atmospheric-type vacuum breaker:

- Provides protection against low-hazard or high-hazard contamination.

- Provides protection against backsiphonage only.

- Cannot be installed where it is under continuous pressure from the water supply (12-hour or less intervals).

- Has a critical level of installation of 6 inches (152 mm) above the highest point of use downstream.

- Must be accessible for inspection and replacement.

Hose-connection backflow preventers are designed to be installed on the discharge side of a hose-threaded outlet on a potable water system. This two-check device protects against backflow caused by backsiphonage and low-head backpressure, under the high-hazard conditions present at a hose-threaded outlet. This device must be used only on systems where sources of backpressure are not introduced and where low-head backpressure does not exceed that generated by an elevated hose equal to or less than 10 feet (3048 mm) in height. For example, using a garden hose while standing on a ladder could create a backpressure of 10 feet (3048 mm) of head.

P2902.3.3 Backflow preventer with intermediate atmospheric vent. Backflow preventers with intermediate atmospheric vents shall conform to ASSE 1012 or CSA B64.3. These devices shall be permitted to be installed where subject to continuous pressure conditions. These devices shall be prohibited as a means of protection where any hazardous chemical additives are introduced downstream of the device. The relief opening shall discharge by *air gap* and shall be prevented from being submerged.

❖ A backflow preventer with intermediate atmospheric vent is a nontestable device consisting of two spring-loaded check valves in the closed position and a vent to atmosphere between the two check valves. The design and principle of operation of these backflow preventers are similar to those of a reduced pressure principle backflow preventer. Supply pressure keeps the atmospheric vent closed, but zero supply pressure or backsiphonage will open the inner chamber to the atmosphere.

With a flow through the valve, the primary check opens away from the diaphragm seal. The atmospheric vent remains closed by the deflection of the diaphragm seal. The secondary check opens away from the downstream seal, permitting the flow of water through the valve.

A backflow preventer with intermediate atmospheric vent:

• Provides protection against low-hazard contamination.

• Provides protection against backsiphonage and backpressure backflow.

• Must have its relief vent opening discharge through an air gap.

• Can be used where it is under continuous pressure from the water supply.

• Cannot be installed below grade where it may be subject to submersion.

• Must be accessible for inspection and replacement.

Provisions must be made at or near the location of the installation to prevent drainage from the relief vent opening causing damage to the structure.

P2902.3.4 Pressure vacuum breaker assemblies. Pressure vacuum breaker assemblies shall conform to ASSE 1020 or CSA B64.1.2. Spill-resistant vacuum breaker assemblies shall comply with ASSE 1056. These assemblies are designed for installation under continuous pressure conditions where the critical level is installed at the required height. The critical level of a pressure vacuum breaker and a spill-resistant vacuum breaker assembly shall be set at not less than 12 inches (304 mm) above the highest elevation of downstream piping and the flood level rim of the fixture or device. Pressure vacuum breaker assemblies shall not be installed in locations where spillage could cause damage to the structure.

❖ A pressure vacuum breaker assembly contains one or two independently operating spring-loaded check valves and an independently operating spring-loaded air inlet valve located on the discharge side of the check valve. This device is testable because it has a shutoff valve and test cock at each end (see Commentary Figure P2902.3.4). Although a pressure vacuum breaker assembly operates by the same principle as an atmospheric vacuum breaker, a pressure vacuum breaker assembly is spring activated. The water pressure acting against the spring closes off the air opening. Because of the spring activation, a valve can be located downstream of a pressure vacuum breaker assembly.

A pressure-type vacuum breaker:

• Provides protection against low-hazard and high-hazard contamination.

• Provides protection against backsiphonage only.

• Can be used where it is under continuous pressure from the water supply (valving permitted downstream).

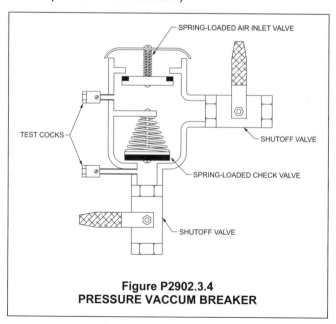

**Figure P2902.3.4
PRESSURE VACCUM BREAKER**

- Has a critical level of installation of 6 inches (152 mm) above flood-level rim.
- Must be accessible for testing and inspection.

Provisions must be made at or near the location of the installation to prevent leakage from the air vent opening from causing damage to the structure when the device is installed in a building or structure.

ASSE 1056 was developed to specifically address indoor applications offering the same vacuum breaker capabilities as ASSE 1020; however, water discharges each time the valve is pressurized. As with ASSE 1020 valves, backflow protection against back-siphonage is achieved by a check valve backed up with an air inlet vent that opens in response to a loss of supply pressure. Such valves are not for use in any system where backpressure is applied to the device. Where the system is pressurized, the vent closes to prevent flow through the upstream check valve and to eliminate vent spillage.

P2902.3.5 Reduced pressure principle backflow prevention assemblies. Reduced pressure principle backflow prevention assemblies and reduced pressure principle fire protection backflow prevention assemblies shall conform to ASSE 1013, AWWA C511, CSA B64.4 or CSA B64.4.1. Reduced pressure detector fire protection backflow prevention assemblies shall conform to ASSE 1047. These devices shall be permitted to be installed where subject to continuous pressure conditions. The relief opening shall discharge by *air gap* and shall be prevented from being submerged.

❖ A reduced pressure principle backflow prevention assembly is considered to be the most reliable mechanical method to prevent backflow. These devices consist of dual, independently acting, spring-loaded check valves that are separated by a chamber or "zone" equipped with a relief valve (see Commentary Figure P2902.3.5). The pressure downstream of the device and in the central chamber between the check valves is maintained at a minimum of 2 psi (13.8 kPa) less than the potable water supply pressure at the device inlet; hence the name "reduced pressure principle." The relief valve located in the central chamber is held closed by the pressure differential between the inlet supply pressure and the central chamber pressure. If the inlet supply pressure decreases, the device will discharge water from the central chamber, thus maintaining a lower pressure in the chamber. In the event of backpressure or negative supply pressure, the relief valve will open to the atmosphere and drain any backflow that has leaked through the check valve. The relief vent will also allow air to enter to prevent any siphonage.

During a static or no-flow condition, both check valves and the relief vent remain closed. When there is normal flow, the two check valves open and the relief vent port remains closed. The supply pressure drops across the valve because of the force required to open the spring-loaded check valves. If backsiphonage or backpressure occurs, both check valves close and the relief vent opens to the atmosphere, discharging the water in the intermediate zone. If both check valves are fouled and held in the open position, and there is either a backsiphonage or backpressure, or both, the relief vent port opens, discharging the possibly contaminated water to the atmosphere and allowing air to enter to break the siphon. This gives a visual indication of a malfunction of one or both of the check valves or relief valve.

ASSE 1013 is the standard for two types of backflow prevention assemblies, identified as "reduced pressure principle backflow preventers" (RP) and "reduced

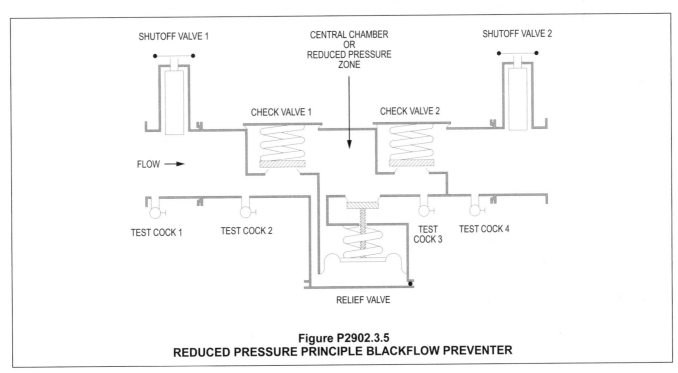

Figure P2902.3.5
REDUCED PRESSURE PRINCIPLE BLACKFLOW PREVENTER

pressure principle fire protection backflow preventers" (RPF). An RP device is designed for a working pressure of 150 psi (1024 kPa) and an RPF device is designed for a working pressure of 175 psi (1207 kPa). The RP and RPF are identical in their backflow protection. The RPF has specific performance requirements relating to its use on fire protection systems.

ASSE 1047 is the standard for devices known as "reduced pressure detector fire protection backflow prevention assemblies" (RPDF). This device is primarily used for the protection of the public water supply from contaminants found in fire sprinkler systems, and it is used to detect low rates of flow up to 2 gallons per minute (gpm) (0.126 L/s) within the sprinkler system caused by leakage or unauthorized use. This standard also has provisions for allowing inclusion of alarm signaling devices in the assembly.

An RP:

- Provides protection against low- or high-hazard contamination.
- Provides protection against backsiphonage and backpressure backflow.
- Can be used where it is under continuous pressure from the water supply.
- Must have its relief vent opening discharge through an air gap.
- Must be accessible for testing and inspection.
- Cannot be installed below grade where it may be subject to submersion.

Provisions must be made at or near the location of the installation to prevent drainage from the relief vent opening causing damage to the structure when the device is installed in a building or structure.

P2902.3.6 Double check backflow prevention assemblies. Double check backflow prevention assemblies shall conform to ASSE 1015, CSA B64.5, CSA B64.5.1 or AWWA C510. Double check detector fire protection backflow prevention assemblies shall conform to ASSE 1048. These assemblies shall be capable of operating under continuous pressure conditions.

❖ These devices are designed for low-hazard applications subject to backpressure and backsiphonage applications. The devices consist of two independent spring-loaded check valves in series. Test cocks are provided to permit testing of the devices. Note that these devices must not be confused with dual check-valve devices or two single check valves placed in series.

Double-detector check-valve assemblies are always permitted as an option where double check-valve assemblies are required because they offer identical protection. Double-detector devices provide the user the ability to monitor water flow through the device to determine whether there is a system leak or water is being intentionally withdrawn from the system downstream of the device.

P2902.3.7 Dual check backflow preventer. Dual check backflow preventers shall conform with ASSE 1024 or CSA B64.6.

❖ Some water purveyors/utilities require that the building water supply system be isolated from the public water main by means of dual check-valve backflow preventers. The code does not require this. Typically, this type of backflow preventer is installed along with the water meter "set," by the water purveyor/utility. This device is not field testable.

P2902.4 Protection of potable water outlets. Potable water openings and outlets shall be protected by an *air gap*, a reduced pressure principle backflow prevention assembly, an atmospheric vent, an atmospheric-type vacuum breaker, a pressure-type vacuum breaker assembly or a hose connection backflow preventer.

❖ The potable water system comes in contact with the atmosphere at openings or outlets. These include faucets, hose bibbs and plumbing appliances. The openings can be protected by an air gap, RP principle backflow preventer with atmospheric vent, atmospheric-type vacuum breaker, pressure-type vacuum breaker or hose connection backflow preventer.

P2902.4.1 Fill valves. Flush tanks shall be equipped with an antisiphon fill valve conforming to ASSE 1002 or CSA B125.3. The critical level of the fill valve shall be located not less than 1 inch (25 mm) above the top of the flush tank overflow pipe.

❖ The fill valve is the refilling component that controls the flow of water into the flush tank. Because the fill valve connects the potable water to the flush tank, it must have adequate protection against backflow. The referenced standards require an antisiphon device in the fill valve. An antisiphon device is a type of vacuum breaker that must be located a minimum of 1 inch (25 mm) above the overflow pipe to prevent its submersion. The 1-inch (25 mm) minimum is equivalent to the vacuum breaker critical level.

P2902.4.2 Deck-mounted and integral vacuum breakers. *Approved* deck-mounted or equipment-mounted vacuum breakers and faucets with integral atmospheric vacuum breakers or spill–resistant vacuum breaker assemblies shall be installed in accordance with the manufacturer's instructions and the requirements for labeling. The critical level of the breakers and assemblies shall be located at not less than 1 inch (25 mm) above the *flood level rim*.

❖ These devices are installed on the decks of plumbing fixtures, such as custom bathtubs, whirlpools, bidets or sinks. They need to be installed only 1 inch (25 mm), instead of 6 inches (152 mm), above the flood-level rim unless required otherwise by the manufacturer. However, these devices must be tested for that reduced elevation so that they function properly and prevent backsiphonage. Faucets with integral atmospheric vacuum breakers or spill-resistant vacuum breaker assemblies must also be installed 1 inch (25 mm) above the flood-level rim of the fixture served. The manufacturer's instructions must be followed so that the devices will function as intended.

P2902.4.3 Hose connection. Sillcocks, hose bibbs, wall hydrants and other openings with a hose connection shall be protected by an atmospheric-type or pressure-type vacuum breaker, a pressure vacuum breaker assembly or a permanently attached hose connection vacuum breaker.

Exceptions:

1. This section shall not apply to water heater and boiler drain valves that are provided with hose connection threads and that are intended only for tank or vessel draining.

2. This section shall not apply to water supply valves intended for connection of clothes washing machines where backflow prevention is otherwise provided or is integral with the machine.

❖ Sillcocks, hose bibbs, wall hydrants and other openings with hose connections must be protected by either an atmospheric, or pressure-type vacuum breaker or a permanently attached hose connection vacuum breaker (see Commentary Figure P2902.4.3). There is danger of backsiphonage with any outlet that has a hose connected to it if the hose can be submerged. Therefore, a device that will stop the siphonage action is required. The type of device will depend on whether a shutoff valve is installed upstream or downstream of the device. Where a shutoff valve is installed upstream of the device, and the device is not subjected to prolonged flow, an atmospheric-type vacuum breaker or a hose connection vacuum breaker can be installed. Where a shutoff valve is installed downstream, or the device is subjected to prolonged flow, a pressure-type vacuum breaker is required.

Backflow prevention devices are not required for fixtures that already include integral backflow prevention

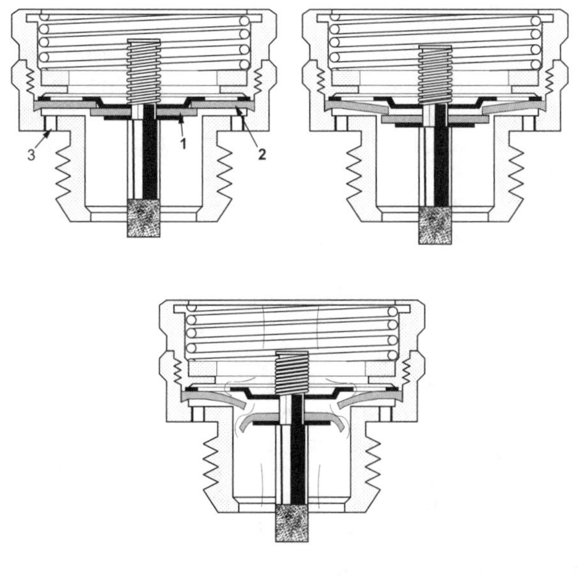

Figure P2902.4.3
HOSE CONNECTION VACUUM BREAKER

devices, such as dishwashers, clothes washers and ice machines.

P2902.5 Protection of potable water connections. Connections to the potable water shall conform to Sections P2902.5.1 through P2902.5.5.

❖ This section regulates direct connections between potable and nonpotable systems that result in a cross connection. There are instances where specific provisions override the previous general provisions, as indicated in Sections P2902.5.1 through P2902.5.5.

P2902.5.1 Connections to boilers. Where chemicals will not be introduced into a boiler, the potable water supply to the boiler shall be protected from the boiler by a backflow preventer with an intermediate atmospheric vent complying with ASSE 1012 or CSA B64.3. Where chemicals will be introduced into a boiler, the potable water supply to the boiler shall be protected from the boiler by an *air gap* or a reduced pressure principle backflow prevention assembly complying with ASSE 1013, CSA B64.4 or AWWA C511.

❖ Because boilers are pressurized vessels, the potential for backflow caused by backpressure is high. If the water supply pressure drops below the boiler pressure, backflow can occur. When the boiler contains untreated water (without additives), a backflow preventer with an intermediate atmospheric vent can be installed. However, where chemicals are added to the boiler, the potable water supply connection must be protected by a reduced pressure zone (RPZ) device or by an air gap.

If potable water is used for domestic use in conjunction with space heating, such systems must be protected from conditions that can cause contamination. Because the water heater or boiler is part of the potable water system, all materials used in the heating system must be approved for use in a potable water system. Note the temperature-protection requirements of Section P2803.2.

P2902.5.2 Heat exchangers. Heat exchangers using an essentially toxic transfer fluid shall be separated from the potable water by double-wall construction. An *air gap* open to the atmosphere shall be provided between the two walls. Single-wall construction heat exchangers shall be used only where an *essentially nontoxic transfer fluid* is utilized.

❖ Heat exchangers are used to transfer energy to or from potable water without mixing with the medium carrying the energy to or from the heat exchanger. Because a wall separates the medium from the potable water, care must be exercised so that mixing between the two will not occur, especially when the medium is toxic. If the medium is toxic, double-wall heat exchangers with an air gap between the walls vented to atmosphere are required so that any leak from either side will leak to the outside, alerting the occupant that a problem exists. Single-wall heat exchangers are only allowed where the heat transfer fluid is not toxic.

P2902.5.3 Lawn irrigation systems. The potable water supply to lawn irrigation systems shall be protected against backflow by an atmospheric vacuum breaker, a pressure vacuum breaker assembly or a reduced pressure principle backflow prevention assembly. Valves shall not be installed downstream from an atmospheric vacuum breaker. Where chemicals are introduced into the system, the potable water supply shall be protected against backflow by a reduced pressure principle backflow prevention assembly.

❖ Lawn irrigation system outlets can be installed at or below grade and are not regulated by the code; however, the potable water supply point of connection to the system must be protected against backflow by an atmospheric vacuum breaker, a pressure vacuum breaker assembly or a reduced pressure principle backflow prevention assembly. The most common method of protection uses pressure vacuum breakers. These devices are permitted for all irrigation systems that use valved sprinklers or zone controls that could render atmospheric vacuum breakers inoperative. Where chemicals can be introduced into the irrigation system, the potable water system must be protected by a reduced pressure principle backflow prevention assembly. RPZ devices are typically installed in cases where it is not known at the time of installation whether or not chemicals will be introduced into the irrigation system.

P2902.5.4 Connections to automatic fire sprinkler systems. The potable water supply to automatic fire sprinkler systems shall be protected against backflow by a double check backflow prevention assembly, a double check fire protection backflow prevention assembly, a reduced pressure principle backflow prevention assembly or a reduced pressure principle fire protection backflow prevention assembly.

> **Exception:** Where systems are installed as a portion of the water distribution system in accordance with the requirements of this code and are not provided with a fire department connection, backflow protection for the water supply system shall not be required.

❖ Automatic fire sprinkler systems installed in dwellings are either installed as a portion of the water distribution system in accordance with the requirements of the code, in which case there are usually no stagnant portions, or they are completely isolated from the potable water system, in which case there could be areas where water can stagnate and possibly develop pollutants and residue from the piping system. If the entire piping system, including sprinkler piping, is constructed of materials that are approved for potable water, a backflow preventer is not required. However, if the water distribution system and fire sprinkler system share only the main supply (water service) in common, a double check-valve assembly is required to isolate the sprinkler system from the potable water supply main.

P2902.5.4.1 Additives or nonpotable source. Where systems contain chemical additives or antifreeze, or where systems are connected to a nonpotable secondary water supply, the potable water supply shall be protected against backflow by a reduced pressure principle backflow prevention assembly or a reduced pressure principle fire protection backflow prevention assembly. Where chemical additives or antifreeze

is added to only a portion of an automatic fire sprinkler or standpipe-system, the reduced pressure principle fire protection backflow preventer shall be permitted to be located so as to isolate that portion of the system.

❖ Some fire sprinkler systems incorporate the use of chemicals or additives that can be used to protect the piping system from corrosion or protect it from freezing. In such cases, a reduced pressure principle backflow prevention assembly is required. In cases where sections of the system are chemically treated, only those treated sections need to be isolated.

P2902.5.5 Solar thermal systems. Where a solar thermal system heats potable water to supply a potable *hot water* distribution or any other type of heating system, the solar thermal system shall be in accordance with Section P2902.5.5.1, P2902.5.5.2 or P2902.5.5.3 as applicable.

❖ There are two basic types of solar thermal water-heating systems: direct and indirect. The following sections cover the plumbing requirements for these systems.

P2902.5.5.1 Indirect systems. Water supplies of any type shall not be connected to the solar heating loop of an indirect solar thermal *hot water* heating system. This requirement shall not prohibit the presence of inlets or outlets on the solar heating loop for the purposes of servicing the fluid in the solar heating loop.

❖ An indirect solar thermal water-heating system heats a heat transfer fluid and then passes that heated fluid through a heat exchanger that is in contact with the water to be heated for distribution in the building's potable hot water distribution system. The heat exchanger construction must comply with Section P2902.5.2.

P2902.5.5.2 Direct systems for potable water distribution systems. Where a solar thermal system directly heats potable water for a potable water distribution system, the pipe, fittings, valves and other components that are in contact with the potable water in the system shall comply with the requirements of Chapter 29.

❖ A direct solar thermal water-heating system heats the potable water that is to be distributed to the building's potable hot water distribution system. All components in a direct solar thermal water-heating system must comply with requirements for potable water distribution system piping and components.

P2902.5.5.3 Direct systems for other than potable water distribution systems. Where a solar thermal system directly heats water for a system other than a potable water distribution system, a potable water supply connected to such system shall be protected by a backflow preventer with an intermediate atmospheric vent complying with ASSE 1012. Where a solar thermal system directly heats chemically treated water for a system other than a potable water distribution system, a potable water supply connected to such system shall be protected by a reduced pressure principle backflow prevention assembly complying with ASSE 1013.

❖ Solar thermal water-heating systems can be used for purposes other than heating potable water for distribution of hot water to the building's potable hot water distribution system. One purpose might be for space heating. To protect the potable water system, a backflow preventer must be installed between the solar thermal water-heating system and the potable water supply.

P2902.6 Location of backflow preventers. Access shall be provided to backflow preventers as specified by the manufacturer's installation instructions.

❖ Because backflow preventers require periodic inspection, testing and repair, they must be installed and oriented in locations that permit these activities to be performed. The manufacturer's installation instructions must be followed to provide the proper access.

P2902.6.1 Outdoor enclosures for backflow prevention devices. Outdoor enclosures for backflow prevention devices shall comply with ASSE 1060.

❖ A typical residential backflow preventer that is located outdoors is associated with a lawn irrigation system. As these units are commonly located near a curb-located water meter or in landscaped areas outside of the residence, residential owners desire that these units be camouflaged by a variety of methods. Because some makeshift structures might compromise the performance of the unit, the code requires enclosures to meet the requirements of ASSE 1060. Enclosures meeting this standard are required to resist at least 100 psf (4.79 kPa) vertical loading and allow for water that is discharged from the backflow preventer to readily drain to the exterior.

P2902.6.2 Protection of backflow preventers. Backflow preventers shall not be located in areas subject to freezing except where they can be removed by means of unions, or are protected by heat, insulation or both.

❖ A backflow preventer located outdoors in freezing climates might be damaged by freezing conditions. If the backflow preventer is used only when seasonal temperatures are above freezing, then one solution is to remove the unit and store it in an above freezing environment. Unions in the piping to the unit allow for ease of removal. If the backflow preventer is to remain in place during freezing weather, it must be protected from freezing. Where outdoor temperatures dip to just below freezing for only a few hours each day, wrapping the unit with insulating material might successfully protect the unit. However, any insulation covering must not block the discharge port. A better solution would be to use an insulated enclosure complying with ASSE 1060. Where temperatures are below freezing for extended periods of time, the enclosure must have a heat source to keep the unit from freezing.

P2902.6.3 Relief port piping. The termination of the piping from the relief port or air gap fitting of the backflow preventer shall discharge to an *approved* indirect waste receptor or to the outdoors where it will not cause damage or create a nuisance.

❖ Wherever a backflow preventer is located, provisions must be made to dispose of the water that could dis-

charge from the unit. Backflow preventers that are located indoors must have the relief discharge port located over a suitable waste receptor or an area that drains to the out-doors. Where piping from the relief port is required to route discharges to a suitable disposal location, the piping must be connected to an air gap fitting that is an integral part of the back-flow preventer, or the termination of the relief port piping at the discharge point must provide for an air gap.

SECTION P2903
WATER SUPPLY SYSTEM

P2903.1 Water supply system design criteria. The water service and water distribution systems shall be designed and pipe sizes shall be selected such that under conditions of peak demand, the capacities at the point of outlet discharge shall not be less than shown in Table P2903.1.

❖ Each plumbing fixture requires a given flow rate and minimum water supply pressure to function properly. The minimum values are shown in Table P2903.1. For fixtures not listed in the table, manufacturer's specifications for the specific fixture must be used. Minimum flow rate and pressure requirements must be satisfied during periods of peak demand. Peak demand is the maximum flow rate of water at any given time during the day. Usually, peak demand for residential occupancies is in the morning before 8 a.m. and at night between 9 p.m. and 10 p.m. Because it is unrealistic to consider the maximum flow rate based on all fixtures in a building operating at the same time, the code uses water supply demand factors for determining the maximum probable flow at peak conditions. These factors are provided in Table P2903.6 and are called water supply fixture units (w.s.f.u.). An w.s.f.u. is a dimensionless factor that is the result of combining probability of use with the water demand for each type of plumbing fixture. The w.s.f.u values only have meaning when used with a table that converts the value into gallons per minute flow [see commentary, Section P2903.6 and Table P2903.6(1)].

TABLE P2903.1. See Page 29-14.

❖ This table specifies the design parameters required for a water distribution system. These minimum values must be satisfied in the design of the system for proper functioning of each plumbing fixture. The design professional must use the table when calculating pipe sizes as required in Section P2903.7 (see commentary, Section P2903.7).

P2903.2 Maximum flow and water consumption. The maximum water consumption flow rates and quantities for plumbing fixtures and fixture fittings shall be in accordance with Table P2903.2.

❖ This section conserves water resources by limiting the consumption of water by plumbing fixtures. Water conservation is important not only to conserve our limited water supplies, but also to conserve the energy required to pump, transport and treat water. The relationship between water use and energy use is evidenced by the fact that the federal government water conservation requirements are part of the National Energy Policy Act. Table P2903.2 contains the maximum allowable flow rates for various types of fixtures.

TABLE P2903.2. See below.

❖ The fixtures and faucets listed in this table are those that function in a cyclical manner or that are allowed to discharge water continuously for the duration of fixture use. Water is conserved by limiting the flow rate for manually controlled fixtures and by limiting the volume-per-cycle usage for cyclically operating fixtures.

The maximum flow rates for nonmetering showers, lavatories and sink faucets are dependent on the supply pressure; therefore, a reference pressure is stated for each item in the table. Where the actual supply pressure is less than the pressure in the table, the actual flow rate of the faucet or shower head will be less than the maximum flow rate in the table.

P2903.3 Minimum pressure. Where the water pressure supplied by the public water main or an individual water supply system is insufficient to provide for the minimum pressures and quantities for the plumbing fixtures in the building, the pressure shall be increased by means of an elevated water tank, a hydropnuematic pressure booster system or a water pressure booster pump.

❖ Table P2903.1 indicates the minimum flow pressures that are required for fixtures, appliances and outlets. Note that Section P2903.1 requires these pressures be available at "peak demand" of the system. See the commentary to Section P2903.1 concerning peak demand. Where the pressures at the fixtures cannot be provided by the public or private water supply system, additional equipment to increase the water pressure needs to be installed.

P2903.3.1 Maximum pressure. The static water pressure shall be not greater than 80 psi (551 kPa). Where the main pressure exceeds 80 psi (551 kPa), an *approved* pressure-reducing valve conforming to ASSE 1003 or CSA B356 shall be installed on the domestic water branch main or riser at the connection to the water service pipe.

❖ The piping system, components and fixtures in a water distribution system are designed to withstand a given pressure. The code establishes 80 psi (55 kPa) as the maximum working pressure of the system. Where the pressure from the public water system or the individual private water supply system exceeds 80 psi (55 kPa), a pressure-reducing valve (PRV) conforming to ASSE 1003 must be installed in the water service line before

TABLE P2903.1
REQUIRED CAPACITIES AT POINT OF OUTLET DISCHARGE

FIXTURE SUPPLY OUTLET SERVING	FLOW RATE (gpm)	FLOW PRESSURE (psi)
Bathtub, balanced-pressure, thermostatic or combination balanced-pressure/thermostatic mixing valve	4	20
Bidet, thermostatic mixing valve	2	20
Dishwasher	2.75	8
Laundry tray	4	8
Lavatory	0.8	8
Shower, balanced-pressure, thermostatic or combination balanced-pressure/thermostatic mixing valve	2.5[a]	20
Sillcock, hose bibb	5	8
Sink	1.75	8
Water closet, flushometer tank	1.6	20
Water closet, tank, close coupled	3	20
Water closet, tank, one-piece	6	20

For SI: 1 pound per square inch = 6.895 kPa, 1 gallon per minute = 3.785 L/m.

a. Where the shower mixing valve manufacturer indicates a lower flow rating for the mixing valve, the lower value shall be applied.

TABLE P2903.2
MAXIMUM FLOW RATES AND CONSUMPTION FOR PLUMBING FIXTURES AND FIXTURE FITTINGS[b]

PLUMBING FIXTURE OR FIXTURE FITTING	MAXIMUM FLOW RATE OR QUANTITY
Lavatory faucet	2.2 gpm at 60 psi
Shower head[a]	2.5 gpm at 80 psi
Sink faucet	2.2 gpm at 60 psi
Water closet	1.6 gallons per flushing cycle

For SI: 1 gallon per minute = 3.785 L/m,
1 pound per square inch = 6.895 kPa.

a. A handheld shower spray shall be considered a shower head.

b. Consumption tolerances shall be determined from referenced standards.

it connects to the water distribution system of the building.

P2903.4 Thermal expansion control. A means for controlling increased pressure caused by thermal expansion shall be installed where required in accordance with Sections P2903.4.1 and P2903.4.2.

❖ As the water in a water heater is heated to higher temperatures, it expands in volume. Usually, the water expands into the water service and the public water main. This volume relief prevents an increase in water pressure in the building distribution system. If the water cannot expand into the water service, high pressures can develop in the water distribution system, which can damage the piping, fixtures, appliances and water heaters, and cause the water heater relief valve to discharge. The code, therefore, mandates that an approved device for thermal expansion control be installed.

P2903.4.1 Pressure-reducing valve. For water service system sizes up to and including 2 inches (51 mm), a device for controlling pressure shall be installed where, because of thermal expansion, the pressure on the downstream side of a pressure-reducing valve exceeds the pressure-reducing valve setting.

❖ A pressure reducing valve (PRV) is installed where the pressure in the water main and water service is greater than the 80 psi (552 kPa) limit indicated in Section P2903.3.1. A PRV can act like a check valve and prevent water from expanding through it back out into the water service. If the pressure on the outlet (downstream) side of the PRV exceeds the setpoint of the PRV, either the PRV has failed or thermal expansion has caused the elevated pressure to occur. Either way, the purpose of the PRV has been defeated. Some PRVs have an integral bypass valve that will allow reverse flow through the PRV when excessive back-pressure occurs on the outlet side of the valve; however, this feature is useless when the inlet pressure (street pressure) is higher than the outlet pressure. For example, a bypass valve cannot allow reverse flow if the outlet pressure is 85 psi (586 kPa) and the inlet

pressure is 120 psi (827 kPa). The street pressure being higher than allowed for the building is typically the reason why the PRV was installed and, therefore, an internal bypass feature will not serve as a means of thermal expansion control. A thermal expansion tank is commonly used to control thermal expansion. Such tanks must be sized based on the volume of the water heater, the temperature rise across the heater and the maximum allowable system pressure. Recall that Section P2903.3.1 limits the maximum pressure to 80 psi (552 kPa) at all times under all conditions (see commentary, Section P2903.4.2 and Commentary Figure P2903.4).

P2903.4.2 Backflow prevention device or check valve. Where a backflow prevention device, check valve or other device is installed on a water supply system using storage water heating equipment such that thermal expansion causes an increase in pressure, a device for controlling pressure shall be installed.

❖ A backflow preventer, check valve or PRV installed in a water service or anywhere else up-stream of the water heater creates a closed system, and thermal expansion can cause destructive and hazardous pressures to develop in the piping system. This section requires expansion control for closed piping systems served by storage-type (tank) water heaters. The typical solution to this problem is an expansion (compression) tank properly sized for the application. Thermal expansion of water is not addressed for tankless (nonstorage) water heaters because such heaters have a negligible water volume and are operational only when water is flowing through them to an outlet. Storage-type (tank) water heaters can heat water when no water usage is occurring and the piping system is, therefore, a closed system (see commentary, Section P2902.4.1). Note that expansion tanks are commonly, but incorrectly, sized only to prevent the system pressure from approaching the setpoint of the relief valve. For example, for a system with a 150-psi (1034 kPa) relief valve, the expansion tank will often be sized to limit system pressure to approximately 90 percent of the relief valve setting, or 135 psi (931 kPa). In such a system, pressure excursions of up to 135 psi (931 kPa) violate Section P2903.3.1 because the system pressure should never exceed 80 psi (552 kPa).

P2903.5 Water hammer. The flow velocity of the water distribution system shall be controlled to reduce the possibility of water hammer. Water-hammer arrestors shall be installed in accordance with the manufacturer's instructions. Water-hammer arrestors shall conform to ASSE 1010.

❖ "Water hammer" is a term used to describe the effect of hydraulic shock within a piping system. Hydraulic shock happens when fluid flowing through a pipe is subjected to a rapid decrease in flow velocity. Flowing water has kinetic energy. When the flow is abruptly stopped, the kinetic energy must dissipate. The dissipation of energy occurs in the form of a pressure surge that produces a hydraulic shockwave traveling at extremely high velocity. Depending on the speed at

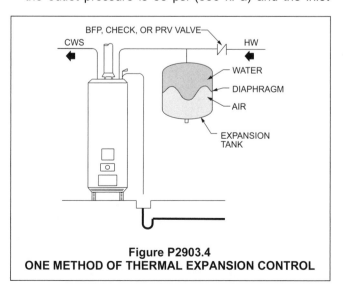

Figure P2903.4
ONE METHOD OF THERMAL EXPANSION CONTROL

which the flow is stopped, the pressure surge can be significant. A rule of thumb for determining the intensity of this pressure is 60 times the change in velocity. In other words, if a 10-foot-per-second (3 m/s) (fps) flow is stopped, the pressure surge could be as great as 600 psi (4137 kPa). The pressure surge is sometimes accompanied by a banging noise within the piping system, hence the expression "water hammer." Any time flowing water is abruptly stopped, water hammer occurs, whether heard or not. Because water hammer creates a pressure spike within the piping system, the potential for piping, fixture or appliance damage exists.

The primary methods for keeping water hammer from causing damage or noise are designing the piping for low-flow velocities the and the use of slow-closing faucets and valves. Another method for controlling water hammer is the installation of a water hammer arrestor in the piping near faucet and valve locations that are known to have a high probability for causing water hammer.

The code does not mandate the use of water hammer arrestors for quick-closing valves. There has been great debate over what constitutes a quick-closing valve and because other methods can be used to reduce water hammer.

A common, but rarely effective, practice is the installation of air chambers in the piping system in order to control water hammer. Unfortunately, air chambers quickly become waterlogged and after a short time period, the air chambers become ineffective in controlling water hammer.

Mechanical water hammer arrestors provide for permanent and reliable control of water hammer. Typically, a mechanical water hammer arrestor has a pressurized gas chamber separated by either a moving piston or diaphragm. When a piping system pressure surge occurs, the piston or diaphragm moves to compress the captive gas and, thus, provides a damping effect that absorbs the pressure surge. Another type of mechanical water hammer uses an elastic metal bellows to absorb the pressure surges (see Commentary Figure P2903.5). Where mechanical water hammer arrestors are used, they must be in conformance with ASSE 1010. Typical locations where water hammer might cause problems are the supply piping to dishwashers, icemakers, washing machines and self-closing faucets or water controls (see commentary for the definition of "Quick-closing valves" in Chapter 2).

P2903.6 Determining water supply fixture units. Supply loads in the building water distribution system shall be determined by total load on the pipe being sized, in terms of water supply fixture units (w.s.f.u.), as shown in Table P2903.6, and gallon per minute (gpm) flow rates [see Table P2903.6(1)]. For fixtures not listed, choose a w.s.f.u. value of a fixture with similar flow characteristics.

❖ As stated in the commentary for Section P2903.1, the w.s.f.u. for determining flow has evolved through the years and has been accepted as a standard method for determining water demands within a building. Table P2903.6 shows w.s.f.u. assigned for various fixtures. However, in order to apply hydraulic formulas for sizing of piping, these dimensionless values must be converted into flow rates. Table P2903.6(1) provides the information for this conversion. Where a building fixture is not listed in Table P2903.6, the w.s.f.u. value of a similar fixture must be chosen and allocated.

TABLE P2903.6. See page 29-17.

❖ This table shows w.s.f.u. values for various fixtures and groups of fixtures that are designed to work in conjunction with the prescriptive sizing methods such as those found in Appendix P. Values are given for hot water and cold water use, along with a combined hot and cold use, the combined use value being somewhat less than the total of the hot and cold w.s.f.u. values. To estimate the supply demand of the building main and branches of

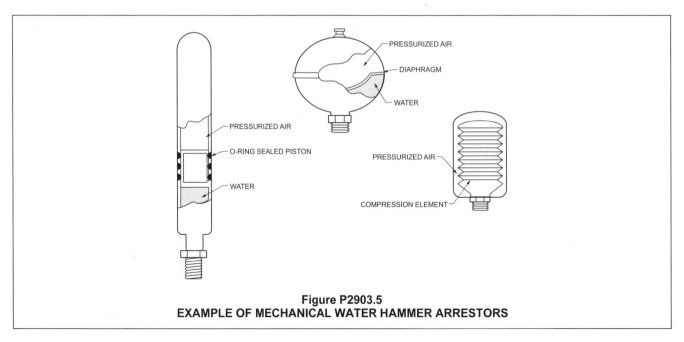

Figure P2903.5
EXAMPLE OF MECHANICAL WATER HAMMER ARRESTORS

the system, total the corresponding demands in the right column of the table.

Section P2903.7 does not contain provisions for the minimum size of a fixture supply pipe (see the definition of "Fixture supply" in Chapter 2). To size the fixture supply pipe when connecting faucets, sinks, lavatories, tubs, dishwashers and hose bibbs, the referenced standard under fixture fittings in Section P2722.1 is used.

TABLE P2903.6(1). See below.

❖ This table is the basis for conversion from w.s.f.u. to gallons per minute (gpm) flow in order to determine water service and water distribution pipe sizing when using w.s.f.u. to estimate water supply demand.

P2903.7 Size of water-service mains, branch mains and risers. The size of the water service pipe shall be not less than

TABLE P2903.6
WATER-SUPPLY FIXTURE-UNIT VALUES FOR VARIOUS PLUMBING FIXTURES AND FIXTURE GROUPS

TYPE OF FIXTURES OR GROUP OF FIXTURES	WATER-SUPPLY FIXTURE-UNIT VALUE (w.s.f.u.)		
	Hot	Cold	Combined
Bathtub (with/without overhead shower head)	1.0	1.0	1.4
Clothes washer	1.0	1.0	1.4
Dishwasher	1.4	—	1.4
Full-bath group with bathtub (with/without shower head) or shower stall	1.5	2.7	3.6
Half-bath group (water closet and lavatory)	0.5	2.5	2.6
Hose bibb (sillcock)[a]	—	2.5	2.5
Kitchen group (dishwasher and sink with or without food-waste disposer)	1.9	1.0	2.5
Kitchen sink	1.0	1.0	1.4
Laundry group (clothes washer standpipe and laundry tub)	1.8	1.8	2.5
Laundry tub	1.0	1.0	1.4
Lavatory	0.5	0.5	0.7
Shower stall	1.0	1.0	1.4
Water closet (tank type)	—	2.2	2.2

For SI: 1 gallon per minute = 3.785 L/m.

a. The fixture unit value 2.5 assumes a flow demand of 2.5 gpm, such as for an individual lawn sprinkler device. If a hose bibb or sill cock will be required to furnish a greater flow, the equivalent fixture-unit value may be obtained from this table or Table P2903.6(1).

TABLE P2903.6(1)
CONVERSIONS FROM WATER SUPPLY FIXTURE UNIT TO GALLON PER MINUTE FLOW RATES

SUPPLY SYSTEMS PREDOMINANTLY FOR FLUSH TANKS			SUPPLY SYSTEMS PREDOMINANTLY FOR FLUSHOMETER VALVES		
Load	Demand		Load	Demand	
(Water supply fixture units)	(Gallons per minute)	(Cubic feet per minute)	(Water supply fixture units)	(Gallons per minute)	(Cubic feet per minute)
1	3.0	0.04104	—	—	—
2	5.0	0.0684	—	—	—
3	6.5	0.86892	—	—	—
4	8.0	1.06944	—	—	—
5	9.4	1.256592	5	15.0	2.0052
6	10.7	1.430376	6	17.4	2.326032
7	11.8	1.577424	7	19.8	2.646364
8	12.8	1.711104	8	22.2	2.967696
9	13.7	1.831416	9	24.6	3.288528
10	14.6	1.951728	10	27.0	3.60936
11	15.4	2.058672	11	27.8	3.716304
12	16.0	2.13888	12	28.6	3.823248
13	16.5	2.20572	13	29.4	3.930192
14	17.0	2.27256	14	30.2	4.037136
15	17.5	2.3394	15	31.0	4.14408
16	18.0	2.90624	16	31.8	4.241024
17	18.4	2.459712	17	32.6	4.357968
18	18.8	2.513184	18	33.4	4.464912
19	19.2	2.566656	19	34.2	4.571856
20	19.6	2.620128	20	35.0	4.6788

(continued)

TABLE P2903.6(1)—continued
CONVERSIONS FROM WATER SUPPLY FIXTURE UNIT TO GALLON PER MINUTE FLOW RATES

SUPPLY SYSTEMS PREDOMINANTLY FOR FLUSH TANKS			SUPPLY SYSTEMS PREDOMINANTLY FOR FLUSHOMETER VALVES		
Load	Demand		Load	Demand	
(Water supply fixture units)	(Gallons per minute)	(Cubic feet per minute)	(Water supply fixture units)	(Gallons per minute)	(Cubic feet per minute)
25	21.5	2.87412	25	38.0	5.07984
30	23.3	3.114744	30	42.0	5.61356
35	24.9	3.328632	35	44.0	5.88192
40	26.3	3.515784	40	46.0	6.14928
45	27.7	3.702936	45	48.0	6.41664
50	29.1	3.890088	50	50.0	6.684

For SI: 1 gallon per minute = 3.785 L/m, 1 cubic foot per minute = 0.4719 L/s.

$^3/_4$ inch (19 mm) diameter. The size of water service mains, branch mains and risers shall be determined from the water supply demand [gpm (L/m)], available water pressure [psi (kPa)] and friction loss caused by the water meter and *developed length* of pipe [feet (m)], including *equivalent length* of fittings. The size of each water distribution system shall be determined according to design methods conforming to acceptable engineering practice, such as those methods in Appendix P and shall be *approved* by the code official.

❖ There are numerous methods that can be used to size piping for a building and the code leaves it up to the user of the code and the code official to agree on what method is approved. Regardless of the method, the code requires that water demand, pipe friction loss, water meter friction loss, equivalent length of fittings and developed length of pipe be factors in determining pipe size.

P2903.8 Gridded and parallel water distribution systems. Hot water and cold water manifolds installed with parallel-connected individual distribution lines and cold water manifolds installed with gridded distribution lines to each fixture or fixture fitting shall be designed in accordance with Sections P2903.8.1 through P2903.8.5. Gridded systems for hot water distribution systems shall be prohibited.

❖ A gridded water supply system involves distribution pipes that are fed from each end. Such systems allow greater design flexibility and smaller diameter piping. A manifold with parallel water distribution lines is a system that provides a centralized distribution point for all fixture supply piping. Each fixture is supplied with an individual supply line that extends from a common manifold, thus eliminating concealed joints and simplifying the method of installation. This type of system allows for smaller pipe sizes than are used for a conventional trunk and branch system if the system is installed in accordance with Sections P2903.8.1 through P2903.8.7. Commentary Figure P2903.8 shows typical details of a manifold system.

The prohibition for gridded distribution systems for hot water distribution is because these types of systems delay the arrival of hot water to the outlets, wasting water and energy.

P2903.8.1 Sizing of manifolds. Manifolds shall be sized in accordance with Table P2903.8.1. Total gallons per minute is the demand for all outlets.

❖ The size of parallel water distribution system manifolds is regulated by Table P2903.8.1. The total w.s.f.u. must be converted into gallons per minute (gpm) demand [see Table P2903.6(1)].

TABLE P2903.8.1
MANIFOLD SIZING

PLASTIC		METALLIC	
Nominal Size ID (inches)	Maximum[a] gpm	Nominal Size ID (inches)	Maximum[a] gpm
$^3/_4$	17	$^3/_4$	11
1	29	1	20
$1^1/_4$	46	$1^1/_4$	31
$1^1/_2$	66	$1^1/_2$	44

For SI: 1 inch = 25.4 mm, 1 gallon per minute = 3.785 L/m,
1 foot per second = 0.3048 m/s.

Note: See Table P2903.6(1) for w.s.f.u. and Table 2903.6(1) for gallon-per-minute (gpm) flow rates.

a. Based on velocity limitation: plastic-12 fps; metal-8 fps.

❖ For example, a typical two-story single-family residence with three full baths, an extra shower, kitchen sink, dishwasher, laundry tray, clothes washer and two hose bibbs would have a total w.s.f.u. count of 22.2 and a 20.5 gallon (81 L) per minute demand based on flush tank water closets. In accordance with Table P2903.8.1, a 1 inch (25 mm) plastic manifold size could serve this residence.

P2903.8.2 Minimum size. Where the *developed length* of the distribution line is 60 feet (18 288 mm) or less, and the available pressure at the meter is not less than 40 pounds per square inch (276 kPa), the size of individual distribution lines shall be not less than $^3/_8$ inch (10 mm) diameter. Certain fixtures such as one-piece water closets and whirlpool bathtubs shall require a larger size where specified by the manufacturer. If a water heater is fed from the end of a cold water manifold, the manifold shall be one size larger than the water heater feed.

❖ The minimum size of individual distribution lines is determined by the required gallons per minute (gpm) flow rate of each fixture and the developed length. This section allows a minimum individual distribution line size of $^3/_8$ inch (10 mm) if the minimum pressure and maximum length criteria are met. However, one-piece water closets and whirlpool bathtubs might require a larger size based on the manufacturer's instructions.

P2903.8.3 Support and protection. Plastic piping bundles shall be secured in accordance with the manufacturer's instructions and supported in accordance with Section P2605. Bundles that have a change in direction equal to or greater than 45 degrees (0.79 rad) shall be protected from chafing at the point of contact with framing members by sleeving or wrapping.

❖ This section requires that plastic pipe bundles be secured in accordance with the manufacturer's installation instructions and the requirements of Section P2605. In addition, changes in direction can create movement of the branches because of expansion and contraction, and thrust forces resulting from water flow (see commentary, Section P2605.1 and Table P2605.1).

P2903.8.4 Valving. Fixture valves, when installed, shall be located either at the fixture or at the manifold. Valves installed at the manifold shall be labeled indicating the fixture served.

❖ Fixture valves can be installed either at the fixture or at the manifold. However, where valves are installed at the manifold, they must be labeled to identify the fixture served.

P2903.8.5 Hose bibb bleed. A *readily accessible* air bleed shall be installed in hose bibb supplies at the manifold or at the hose bibb exit point.

❖ This section requires installation of a readily accessible air bleed to allow the purging of any trapped air that might be in the system.

P2903.9 Valves. Valves shall be installed in accordance with Sections P2903.9.1 through P2903.9.5.

❖ This section addresses requirements for fixture shutoff valves. Where such valves are located at the manifold, they must be labeled to identify the fixture served.

P2903.9.1 Service valve. Each *dwelling unit* shall be provided with an accessible main shutoff valve near the entrance of the water service. The valve shall be of a full-open type having nominal restriction to flow, with provision for drainage such as a bleed orifice or installation of a separate drain valve. Additionally, the water service shall be valved at the curb or lot line in accordance with local requirements.

❖ Each dwelling unit must have a full-open valve at the curb or property line as required by the local water supplier. In addition, this section requires an accessible full-open valve for the dwelling unit at or near the entrance of the water service. In multiple-family dwellings, the full-open valve must be located at the point where the water distribution pipe enters each dwelling unit so that in the event of a leak within the unit, the unit can be isolated [see Commentary Figures P2903.9.1(1), P2903.9.1(2) and P2903.9.1(3)]. The two valves allow isolation of the meter for the water supplier and shutoff of the entire building for service.

Full-open valves, such as gate valves and ball valves, reduce the amount of friction loss, as compared to other valves, such as globe valves. Therefore, this section requires full-open-type valves to minimize friction loss and to coincide with the prescriptive sizing methods of water distribution systems as outlined in this chapter. The bleed orifice or termination of the drain from a drain valve must be above grade to prevent contamination of the potable water system.

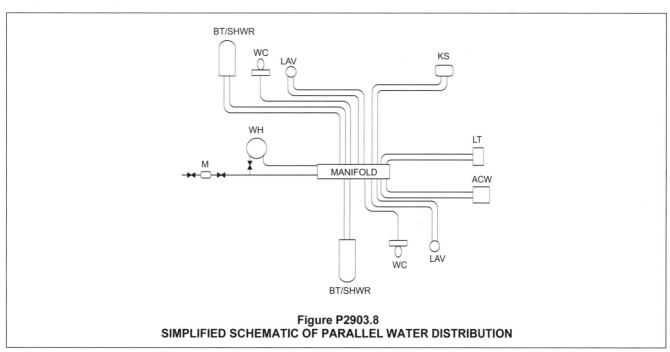

Figure P2903.8
SIMPLIFIED SCHEMATIC OF PARALLEL WATER DISTRIBUTION

P2903.9.2 Water heater valve. A *readily accessible* full-open valve shall be installed in the cold-water supply pipe to each water heater at or near the water heater.

❖ To allow for maintenance, replacement or emergency shutoff of a water heater, a readily accessible full-open valve is required on the cold water supply to the water heater. The valve must be located at or near the water heater so that it is readily recognized as serving the water heater.

P2903.9.3 Fixture valves and access. Shutoff valves shall be required on each fixture supply pipe to each plumbing appliance and to each plumbing fixture other than bathtubs and showers. Valves serving individual plumbing fixtures, *plumbing appliances,* risers and branches shall be *accessible*.

❖ Within the dwelling, a shutoff valve is required on the fixture supply pipe to each plumbing fixture except for bathtubs and showers. Bathtubs and showers typically do not have shutoff valves because it could be difficult

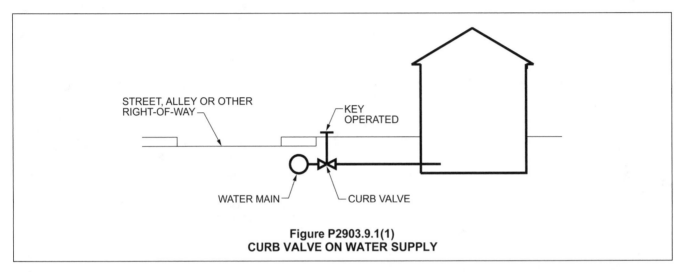

Figure P2903.9.1(1)
CURB VALVE ON WATER SUPPLY

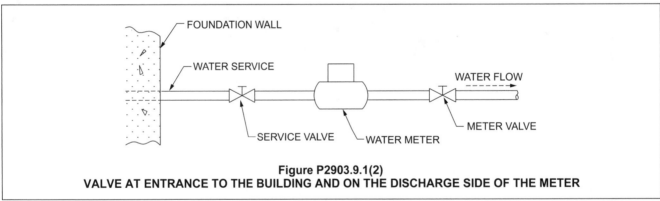

Figure P2903.9.1(2)
VALVE AT ENTRANCE TO THE BUILDING AND ON THE DISCHARGE SIDE OF THE METER

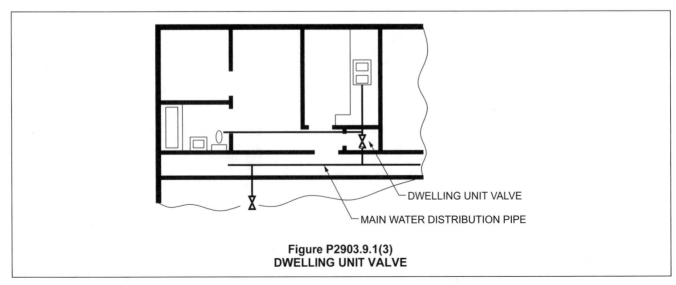

Figure P2903.9.1(3)
DWELLING UNIT VALVE

to provide access to valves for locations such as upper-story bathrooms. Some tub and shower controls are available with integral shutoff valve (stops) that can be accessed by removing a trim piece or escutcheon.

P2903.9.4 Valve requirements. Valves shall be compatible with the type of piping material installed in the system. Valves shall conform to one of the standards listed in Table P2903.9.4 or shall be *approved*. Valves intended to supply drinking water shall meet the requirements of NSF 61.

❖ Valves must comply with one of the standards indicated in Table P2903.4 so that a level of quality for valves is ensured. Where a valve does not meet one of the standards, the building official is required to review the information about the valve and determine if the valve meets the overall intent of the code (see Section R104.11).

Any valve through which water flows for consumption by living creatures must meet NSF 61. This standard evaluates the valves for their affect on the potability of the water.

P2903.9.5 Valves and outlets prohibited below grade. Potable water outlets and combination stop-and-waste valves shall not be installed underground or below grade. Freeze-proof yard hydrants that drain the riser into the ground are considered to be stop-and-waste valves.

Exception: Installation of freezeproof yard hydrants that drain the riser into the ground shall be permitted if the potable water supply to such hydrants is protected upstream of the hydrants in accordance with Section P2902 and the hydrants are permanently identified as non-potable outlets by *approved* signage that reads as follows: "Caution, Nonpotable Water. Do Not Drink."

❖ A potable water outlet or a valve with a waste drain outlet located below grade has the potential for becoming submerged by either ground water or rainwater. If the valve or the valve waste port does not completely shut off or develops a leak during a loss in system pressure, backsiphonage could occur and contaminate the pota-

ble water supply line connected to the valve. Also, contaminated water can enter piping downstream of the valve through the open waste opening when the valve is closed. A common example of a violation is a hose bibb valve located in a pit below grade or in a recessed box in a concrete slab.

Any fixture or device that incorporates a stop-and-waste feature is prohibited if the waste opening is underground or in any location where waste water or water-borne contaminants could enter the device, or water supply from the ground or other source by reversal of flow.

Many water utilities routinely use stop-and-waste valves on curb stops and fire hydrants. An-other common location for stop-and-waste valves is in yard hydrants equipped with a below-grade self-draining weep hole [see Commentary Figures P2903.9.5(1) and P2903.9.5(2)].

Sanitary yard hydrants are available that protect against ground-water contamination by eliminating the waste opening below grade. These devices use a sealed reservoir to collect water draining from the hydrant head and riser, and empty the reservoir upon opening of the valve the next time the hydrant is used [see Commentary Figure P2903.9.5(3)].

The exception allows for installations of frostproof yard hydrants (having a combination stop-and-waste valve below grade) for applications where nonpotable water is acceptable for the use intended. Examples of nonpotable water use applications are hydrants used for farm area wash down, public park area cleaning, and garbage area cleanup. Yard hydrants having stop-and-waste valves below grade must have the potable water supply supplying the hydrant protected with backflow protection as specified in Section P2902. The hydrant must be marked permanently with a sign having the code specified words. The sign must meet the approval of the code official [see Commentary Figure P2903.9.5(4)].

TABLE P2903.9.4
VALVES

MATERIAL	STANDARD
Chlorinated polyvinyl chloride (CPVC) plastic	ASME A112.4.14, ASME A112.18.1/CSA B125.1, ASTM F1970, CSA B125.3
Copper or copper alloy	ASME A112.4.14, ASME A112.18.1/CSA B125.1, ASME B16.34, CSA B125.3, MSS SP-67, MSS SP-80, MSS SP-110
Gray and ductile iron	ASTM A126, AWWA C500, AWWA C504, AWWA C507, MSS SP-42, MSS SP-67, MSS SP-70, MSS SP-71, MSS SP-72, MSS SP-78
Cross-linked polyethylene (PEX) plastic	ASME A112.4.14, ASME A112.18.1/CSA B125.1, CSA B125.3, NSF 359
Polypropylene (PP) plastic	ASME A112.4.14, ASTM F2389
Polyvinyl chloride (PVC) plastic	ASME A112.4.14, ASTM F1970

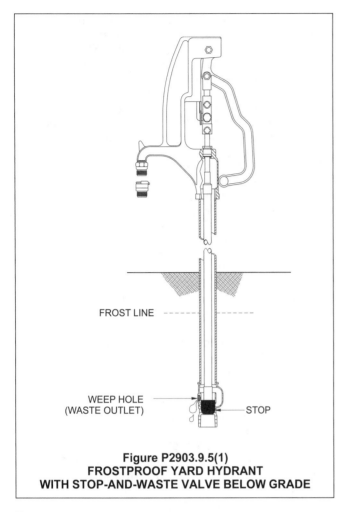

Figure P2903.9.5(1)
FROSTPROOF YARD HYDRANT
WITH STOP-AND-WASTE VALVE BELOW GRADE

Figure P2903.9.5(3)
SANITARY FROSTPROOF YARD HYDRANT
(Figure courtesy of Murdock-Super Secure)

P

Figure P2903.9.5(2)
VIOLATION: FROSTPROOF YARD HYDRANT CONNECTED TO WATER SERVICE LINE

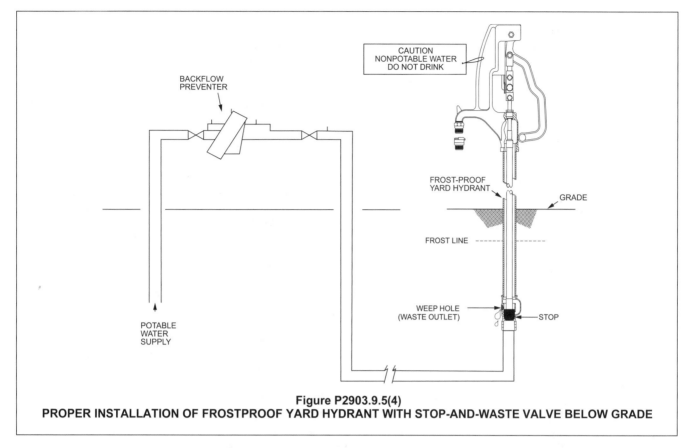

Figure P2903.9.5(4)
PROPER INSTALLATION OF FROSTPROOF YARD HYDRANT WITH STOP-AND-WASTE VALVE BELOW GRADE

2903.10 Hose bibb. Hose bibbs subject to freezing, including the "frostproof" type, shall be equipped with an accessible stop-and-waste-type valve inside the building so that they can be controlled and drained during cold periods.

Exception: Frostproof hose bibbs installed such that the stem extends through the building insulation into an open heated or semi*conditioned space* need not be separately valved (see Figure P2903.10).

❖ This section requires that hose bibbs that are subject to freezing be supplied through an accessible stop-and-waste type valve. This feature allows drainage of water from the piping and hose bibb. An exception to this requirement exists for frostproof hose bibbs that have stems extending beyond the building insulation to an area that will not subject the water to freezing conditions. To prevent cross contamination of the potable water system, the bleed orifice or termination of the drain must be above grade. The exception applies only to frostproof hose bibbs where the point of shutoff is not subject to freezing (see commentary, Figure P2903.10). If the bibb is subject to freezing, the exception does not apply.

FIGURE P2903.10. See Page 29-24.

❖ This figure shows a detail of a proper installation of a frostproof hose bibb through an insulated exterior wall separating a conditioned space from the exterior of the building. The point of shutoff (valve seat) is within the conditioned (heated) space. The hose bibb is installed so that the shank portion drains automatically when

the valve seat is in the closed position. This type of installation would not require a separate shutoff valve as specified in Section P2903.10.

P2903.11 Drain water heat recovery units. Drain water heat recovery units shall be in accordance with Section N1103.5.4.

❖ As the recovery of waste energy is becoming very popular, recovery of heat from warm waste water has become beneficial in the overall goal to reduce energy consumption of a building. This code section does not require drain water heat recovery units. This section is simply a pointer to Chapter 11, the energy conservation chapter, to alert the plumbing community that standards for these units are available and where drain water heater recovery (DWHR) units are installed, they must comply with those standards.

SECTION P2904
DWELLING UNIT FIRE SPRINKLER SYSTEMS

P2904.1 General. The design and installation of residential fire sprinkler systems shall be in accordance with NFPA 13D or Section P2904, which shall be considered equivalent to NFPA 13D. Partial residential sprinkler systems shall be permitted to be installed only in buildings not required to be equipped with a residential sprinkler system. Section P2904 shall apply to stand-alone and multipurpose wet-pipe sprinkler systems that do not include the use of antifreeze. A multipurpose fire sprinkler system shall provide domestic water to both fire sprinklers and plumbing fixtures. A stand-alone sprinkler system shall be separate and independent from the water distribution system. A backflow preventer shall not be required to separate a stand-alone sprinkler system from the water distribution system.

❖ This section does not require fire sprinkler systems to be installed in buildings covered by the code, rather, it provides a specification for the design and installation of a sprinkler system (or portion of a sprinkler system) where such systems are installed. Other sections of the code dictate which buildings are required to have sprinkler systems.

The terms "sprinkler system" and "sprinklers" do not have the same meaning. A sprinkler system is the complete piping system that provides water to the sprinklers. A sprinkler is the individual device that sprays the water. The term "sprinklers" means multiple sprinkler devices, not a sprinkler system.

NFPA 13D is a widely accepted standard for the design and installation of residential fire sprinkler systems. Although not as complicated or stringent as its parent, NFPA 13 and 13D requires hydraulic calculations that essentially limit the design and installation of such systems to qualified sprinkler system contractors with experience with such calculations. Section P2904 was developed to provide a design and installation method for a fire sprinkler system that will provide an equivalent level of protection as provided by an NFPA 13D system, but without the need for sophisticated calculations and other installation complexities. Thus, the residential plumbing contractor, and even the homeowner doing his or her own plumbing work, can design and install a sprinkler system in conjunction with the installation of plumbing systems.

This section can also be used for the design and installation of a stand-alone sprinkler system, that is, a system that does not supply water for purposes other than fire protection. Because the piping used for systems designed in accordance with this section is approved for potable water distribution, a backflow preventer is not required at the connection of the stand-alone system to the potable water supply.

Because the required flow rate for residential sprinkler systems designed in accordance with this section is much greater than the flow rate required for most residential plumbing systems, the pipe sizes for the sprinkler system are more than adequate for supplying water to plumbing fixtures. Thus, a system designed in accordance with Section 2904 is suitable as a multipurpose system design without adding any additional flow for the plumbing fixture operation. Although it is possible that one or even two sprinklers could activate at the same time as one or more plumbing fixtures would be in operation, it is assumed that the operation of plumbing fixtures generally imposes so little demand on the sprinkler system piping that flow to any sprinkler is not significantly comprised.

Systems designed in accordance with this section must not be designed to have portions of the system freeze protected by the use of antifreeze solutions. The reason for this is that the owner of buildings covered by the code might not reliably check the condition of antifreeze solution to avoid freeze damage to the piping.

P2904.1.1 Required sprinkler locations. Sprinklers shall be installed to protect all areas of a *dwelling unit*.

Exceptions:

1. Attics, crawl spaces and normally unoccupied concealed spaces that do not contain fuel-fired appliances do not require sprinklers. In *attics*, crawl spaces and normally unoccupied concealed spaces

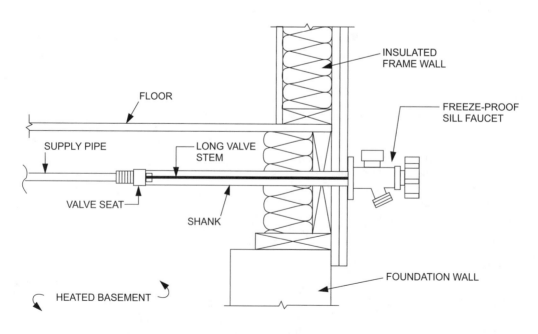

FIGURE P2093.10
TYPICAL FREEZE-PROOF HOSE BIBB INSTALLATION NOT REQUIRING SEPARATE VALVE

that contain fuel-fired equipment, a sprinkler shall be installed above the equipment; however, sprinklers shall not be required in the remainder of the space.

2. Clothes closets, linen closets and pantries not exceeding 24 square feet (2.2 m²) in area, with the smallest dimension not greater than 3 feet (915 mm) and having wall and ceiling surfaces of gypsum board.

3. Bathrooms not more than 55 square feet (5.1 m²) in area.

4. Garages; carports; exterior porches; unheated entry areas, such as mud rooms, that are adjacent to an exterior door; and similar areas.

❖ The section does not mandate sprinkler systems for all dwelling units, but requires sprinklers for all areas of dwelling units that are required to have a sprinkler system installed. Although it might seem odd that this section has a requirement to install sprinklers in all areas of dwelling unit, followed by exceptions that allow areas not to have sprinklers, the language arrangement ensures that areas not specifically mentioned in the exception will have sprinklers installed.

Studies have shown that most fires in residential buildings have a high probability of occurring in the living areas and at locations of fuel-fired equipment. The installation of sprinklers beyond those rooms and locations offers only a minimal increase in safety since fires rarely begin in those areas. Note that garages and carports are not required to have sprinklers even though, statistically, there is a moderate probability of fire occurrence in those areas. There are several reasons for this exception: 1. Dwelling units are required to be fire protected from attached garages by the application of gypsum board to the garage wall (and ceiling, in most cases); and 2. Unheated garages are difficult to protect with a sprinkler system that does not allow the use of antifreeze or preaction-type sprinklers.

P2904.2 Sprinklers. Sprinklers shall be new listed residential sprinklers and shall be installed in accordance with the sprinkler manufacturer's instructions.

❖ Sprinklers are required to pass tests by an independent agency to prove that they: 1. Open within a specified time when exposed to the required temperature range; 2. Discharge the required water flow rate at a rated pressure; and 3. Produce the spray coverage pattern as indicated by the manufacturer. Only sprinklers listed for residential use can be used in a residential sprinkler system because residential sprinklers have a faster response time than sprinklers used in other occupancies.

Once a sprinkler has been activated or has been removed from service, it cannot be refurbished or reused in another system. In other words, used or recycled sprinklers must not be installed in a sprinkler system.

The manufacturer's installation instructions must be followed so: 1. The appropriate type of sprinkler is installed for each application; 2. The proper thread sealing methods are used; 3. The approved wrenches for tightening are used; and 4. The correct escutcheon plates or decorative covers are properly installed.

There are two types of residential sprinklers: pendant type and horizontal sidewall type. Pendant-type sprinklers release a jet of water downward towards a horizontal deflector that distributes the water in droplets into the covered area [see Commentary Figure P2904.2(1)]. A horizontal sidewall sprinkler directs a water jet horizontally towards both a horizontal and vertical deflector assembly to distribute the water in droplets into the covered area [see Commentary Figure P2904.2(2)].

Figure P2904.2(1)
PENDANT SPRINKLER

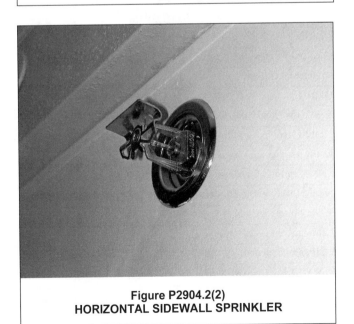

Figure P2904.2(2)
HORIZONTAL SIDEWALL SPRINKLER

P2904.2.1 Temperature rating and separation from heat sources. Except as provided for in Section P2904.2.2, sprinklers shall have a temperature rating of not less than 135°F (57°C) and not more than 170°F (77°C). Sprinklers shall be separated from heat sources as required by the sprinkler manufacturer's installation instructions.

❖ When a fire starts, the heat from the combustion process raises the temperature of that area until the sensor in the sprinkler is heated to within the range of 135°F to 170°F (57°C to 77°C). The sprinkler opens and sprays the area with water. If there are nearby heat sources, such as a fireplace or skylight, the heat from those sources could possibly activate a sprinkler even though no fire exists. Therefore, the manufacturer's installation instructions must be followed for locating "ordinary" temperature range sprinklers near heat-generating sources (see commentary, Section P2904.2.2).

P2904.2.2 Intermediate temperature sprinklers. Sprinklers shall have an intermediate temperature rating not less than 175°F (79°C) and not more than 225°F (107°C) where installed in the following locations:

1. Directly under skylights, where the sprinkler is exposed to direct sunlight.

2. In *attics*.

3. In concealed spaces located directly beneath a roof.

4. Within the distance to a heat source as specified in Table P2904.2.2.

❖ Certain areas in a building are known to have elevated temperatures that could prematurely activate ordinary temperature sprinklers. Sprinklers that are directly under skylights, in attics or in concealed spaces located directly beneath a roof must be intermediate temperature sprinklers. Where there are heat sources, as listed in Table P2904.2.2, and sprinklers are located within the dimension range of those heat sources as indicated in Table P2904.2.2, those sprinklers must be intermediate temperature sprinklers.

TABLE P2904.2.2. See below.

❖ As stated in the commentary to Section P2904.2.1, ordinary temperature range sprinklers must not be located too near to heat sources. This section identifies three specific locations where intermediate temperature sprinklers, instead of ordinary temperature sprinklers, must be installed. The upper end of the distance range to heat sources in Table P2904.2.2 indicates the smallest distance that standard sprinklers can be to specific heat sources. Where the sprinkler location is less than the upper end of the distance range, an intermediate temperature sprinkler must be installed at that location. Note that intermediate temperature sprinklers cannot be located any closer to the heat source than the lower end of the distance range in Table P2904.2.2, unless the specific sprinkler listing allows a closer distance (see Note a). For example, a sprinkler cannot be located less than 9 inches (229 mm) from a chimney connector, regardless of its rating, unless listed otherwise. Commentary Figure P2904.2.2 illustrates allowable distances for the two temperature ranges of sprinklers to a ceiling-mounted warm air register.

TABLE P2904.2.2
LOCATIONS WHERE INTERMEDIATE TEMPERATURE SPRINKLERS ARE REQUIRED

HEAT SOURCE	RANGE OF DISTANCE FROM HEAT SOURCE WITHIN WHICH INTERMEDIATE TEMPERATURE SPRINKLERS ARE REQUIRED[a, b] (inches)
Fireplace, side of open or recessed fireplace	12 to 36
Fireplace, front of recessed fireplace	36 to 60
Coal and wood burning stove	12 to 42
Kitchen range top	9 to 18
Oven	9 to 18
Vent connector or chimney connector	9 to 18
Heating duct, not insulated	9 to 18
Hot water pipe, not insulated	6 to 12
Side of ceiling or wall warm air register	12 to 24
Front of wall mounted warm air register	18 to 36
Water heater, furnace or boiler	3 to 6
Luminaire up to 250 watts	3 to 6
Luminaire 250 watts up to 499 watts	6 to 12

For SI: 1 inch = 25.4 mm.

a. Sprinklers shall not be located at distances less than the minimum table distance unless the sprinkler listing allows a lesser distance.

b. Distances shall be measured in a straight line from the nearest edge of the heat source to the nearest edge of the sprinkler.

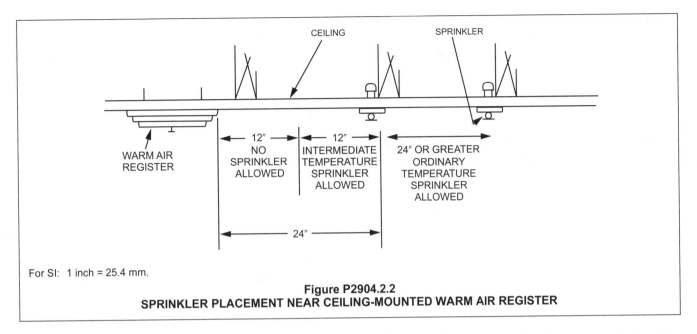

For SI: 1 inch = 25.4 mm.

Figure P2904.2.2
SPRINKLER PLACEMENT NEAR CEILING-MOUNTED WARM AIR REGISTER

P2904.2.3 Freezing areas. Piping shall be protected from freezing as required by Section P2603.6. Where sprinklers are required in areas that are subject to freezing, dry-sidewall or dry-pendent sprinklers extending from a nonfreezing area into a freezing area shall be installed.

❖ Sprinkler piping containing water must be protected from freezing in the same manner as the water distribution piping is protected so as to prevent freeze damage to the piping and sprinklers, and to prevent frozen piping from disabling the system. Where a sprinkler is required to be located in an area subject to freezing conditions, dry-type sprinklers can be installed. Dry-type sprinklers have an extended body to place the water stopping device of the sprinkler in a location where the water piping is not subject to freezing. Although Section P2904.2 requires that sprinklers be a listed residential type, dry-pendant or dry-horizontal-sidewall sprinklers having a residential listing may not be commonly available from all sprinkler manufacturers.

P2904.2.4 Sprinkler coverage. Sprinkler coverage requirements and sprinkler obstruction requirements shall be in accordance with Sections P2904.2.4.1 and P2904.2.4.2.

❖ The following two sections prescribe limitations for sprinkler coverage area and requirements relative to obstructions to a sprinkler water spray pattern.

P2904.2.4.1 Coverage area limit. The area of coverage of a single sprinkler shall not exceed 400 square feet (37 m²) and shall be based on the sprinkler listing and the sprinkler manufacturer's installation instructions.

❖ Every type of sprinkler is rated to cover a maximum area size based upon a specific flow and pressure; however, not more than 400 square feet (37 m²) of floor area can be protected by a single sprinkler. The area

limit is dictated by the manufacturer, and the 400-square-foot (37 m²) limit is simply an overall cap on area. If the sprinkler piping is designed to deliver the manufacturer's flow rate at the manufacturer's stated pressure, the sprinkler will have a very high probability of extinguishing a fire in that covered area. Some sprinklers may have a smaller area of coverage.

P2904.2.4.2 Obstructions to coverage. Sprinkler discharge shall not be blocked by obstructions unless additional sprinklers are installed to protect the obstructed area. Additional sprinklers shall not be required where the sprinkler separation from obstructions complies with either the minimum distance indicated in Figure P2904.2.4.2 or the minimum distances specified in the sprinkler manufacturer's instructions where the manufacturer's instructions permit a lesser distance.

❖ A sprinkler's water spray pattern could be blocked by various building elements, such as lintels, beams, soffits, protruding walls, chases, cabinets, surface-mounted ceiling luminaires and ceiling fans thereby causing gaps in spray coverage. The manufacturer's installation instructions provide the necessary sprinkler location requirements to ensure that all areas requiring sprinklers are covered.

P2904.2.4.2.1 Additional requirements for pendent sprinklers. Pendent sprinklers within 3 feet (915 mm) of the center of a ceiling fan, surface-mounted ceiling luminaire or similar object shall be considered to be obstructed, and additional sprinklers shall be installed.

❖ Ceiling-mounted fixtures could block the spray pattern of a ceiling-mounted sprinkler if located within 3 feet (915 mm) of the center of a sprinkler. If the location of the sprinkler cannot be moved outside of the 3-foot (915 mm) proximity, extra sprinklers must be installed to cover the blocked area or the fixtures must be relocated.

P2904.2.4.2.2 Additional requirements for sidewall sprinklers. Sidewall sprinklers within 5 feet (1524 mm) of the center of a ceiling fan, surface-mounted ceiling luminaire or similar object shall be considered to be obstructed, and additional sprinklers shall be installed.

❖ Ceiling-mounted fixtures could also block the spray pattern of sidewall-mounted sprinklers if located within 5 feet (1524 mm) of the center of a sprinkler. If the location of the sprinkler cannot be moved outside of the 5-foot (1524 mm) proximity, extra sprinklers must be installed to cover the blocked area or the fixtures must be relocated.

P2904.2.5 Sprinkler installation on systems assembled with solvent cement. The solvent cementing of threaded adapter fittings shall be completed and threaded adapters for sprinklers shall be verified as being clear of excess cement prior to the installation of sprinklers on systems assembled with solvent cement.

❖ Because all sprinklers are designed to thread into a fitting that is connected to the sprinkler system piping, adapter fittings are required where the sprinkler piping is of a material that is solvent cemented to fittings and adapters. Excess solvent cement must be removed from the inside of the adapter fittings so that the sprinkler orifice does not become plugged upon sprinkler activation.

P2904.2.6 Sprinkler modifications prohibited. Painting, caulking or modifying of sprinklers shall be prohibited. Sprinklers that have been painted, caulked, modified or damaged shall be replaced with new sprinklers.

❖ Painting or decorating of sprinklers after factory shipment could affect their operation and the temperature at which they activate. Sprinklers are available from the manufacturer in a variety of colors to coordinate with interior finish colors. Sprinklers must not be field painted or modified in any manner. Field-painted, modified or damaged sprinklers must be replaced with new sprinklers.

P2904.3 Sprinkler piping system. Sprinkler piping shall be supported in accordance with requirements for cold water distribution piping. Sprinkler piping shall comply with the requirements for cold water distribution piping. For multipurpose piping systems, the sprinkler piping shall connect to and be a part of the cold water distribution piping system.

Exception: For plastic piping, it shall be permissible to follow the manufacturer's installation instructions.

❖ The support of sprinkler piping must be in accordance with the requirements of Table P2605.1. The support spacing requirements in this section might be more stringent than what is stated in design and installation manuals from sprinkler piping manufacturers. Because of the reaction force of an activated sprinkler, special consideration must be given to supporting the piping near sprinklers that are supported only by the piping. For example, when a ceiling sprinkler activates, the downward force of the water flow creates an oppo-

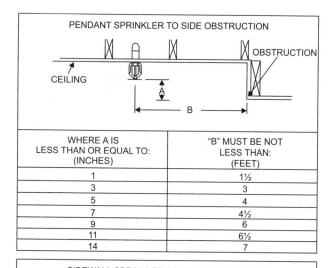

WHERE A IS LESS THAN OR EQUAL TO: (INCHES)	"B" MUST BE NOT LESS THAN: (FEET)
1	1½
3	3
5	4
7	4½
9	6
11	6½
14	7

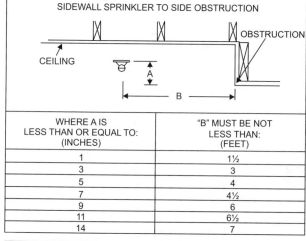

WHERE A IS LESS THAN OR EQUAL TO: (INCHES)	"B" MUST BE NOT LESS THAN: (FEET)
1	1½
3	3
5	4
7	4½
9	6
11	6½
14	7

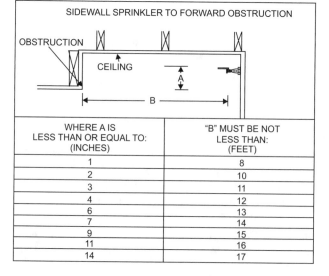

WHERE A IS LESS THAN OR EQUAL TO: (INCHES)	"B" MUST BE NOT LESS THAN: (FEET)
1	8
2	10
3	11
4	12
6	13
7	14
9	15
11	16
14	17

For SI: 1 inch = 25.4 mm, 1 foot = 304.8 mm.

FIGURE P2904.2.4.2
MINIMUM ALLOWABLE DISTANCE BETWEEN
SPRINKLER AND OBSTRUCTION

site reaction in the sprinkler. If the sprinkler piping at the sprinkler is not properly fastened to the structure, the sprinkler, upon activation, could be forced above the ceiling membrane, causing the sprinkler to be ineffective. Sprinkler manufacturer's installation instructions, also sprinkler piping manufacturer's installation instructions, will typically indicate the required support arrangements for piping that supports sprinklers. The exception attempts to clarify that piping manufacturer and sprinkler manufacturer instructions can also be used for piping support instructions.

P2904.3.1 Nonmetallic pipe and tubing. Nonmetallic pipe and tubing, such as CPVC, PEX, and PE-RT shall be listed for use in residential fire sprinkler systems.

❖ Where nonmetallic piping is used for a multipurpose sprinkler system, the piping from the service valve to every sprinkler must be "listed" for use as fire sprinkler piping. The term "listed" is defined in Chapter 2 of this code as an approved agency that has evaluated and tested the product and has put the product on a list of products that are intended for use in residential fire sprinkler systems. Although not required by this section, most products that are listed will be labeled with an identifying mark of the agency that lists the product. The majority of sprinkler products are listed and labeled by either Underwriters Laboratories (UL) or Factory Mutual (FM). Other agencies may meet the approval of the authority having jurisdiction.

Because sprinklers, sprinkler piping and fittings are in contact with potable water that flows to the plumbing fixtures, these components are also required to be certified to NSF 61.

For cross-linked polyethylene (PEX) material, listed tubing must comply with ASTM F877 or CSA B137.10M. PEX fittings must comply with the fittings standards as recommended by the PEX pipe manufacturer.

For chlorinated polyvinyl chloride (CPVC) material, listed pipe must comply with ASTM F442, and listed fittings must comply with ASTM F437 (threaded fittings in Schedule 80 wall thickness), ASTM F438 (socket fittings in Schedule 40 wall thickness) or ASTM F439 (socket fittings in Schedule 80 wall thickness). Although there are listed resins that are black in color, the majority of CPVC tubing, pipe and fittings for sprinkler system service is orange in color.

CPVC pipe and fittings that are listed for residential sprinkler service are iron pipe size (ips), not CPVC copper tube size (CTS). CTS CPVC, which is normally used for residential plumbing, is not listed for residential sprinkler use. Where CPVC is used for a multipurpose residential sprinkler system, listed CPVC pipe must be used for all piping supplying water to sprinklers. Piping branches from the multipurpose sprinkler system piping for supplying water to plumbing fixtures can be CTS CPVC. Special adapter fittings are available to convert from CPVC pipe size to CPVC CTS to allow for a solvent-cemented connection. However, the code does not address a conflict between the requirements for solvent cement for CPVC plumbing piping (or tubing) and the solvent cement required by manufacturer's listed CPVC residential sprinkler piping. Section P2905.9.1.2 requires that solvent cement for CPVC materials be yellow or orange in color and comply with ASTM F493. The listing for CPVC sprinkler piping requires a solvent cement that is red in color and in compliance with ASTM F493. The code official will have to determine if solvent cementing of these different types of CPVC is acceptable with the solvent cement the installer chooses for this connection. Otherwise, the connection should be made using threaded adapters.

P2904.3.1.1 Nonmetallic pipe protection. Nonmetallic pipe and tubing systems shall be protected from exposure to the living space by a layer of not less than $^3/_8$-inch-thick (9.5 mm) gypsum wallboard, $^1/_2$-inch-thick (13 mm) plywood, or other material having a 15-minute fire rating.

Exceptions:

1. Pipe protection shall not be required in areas that do not require protection with sprinklers as specified in Section P2904.1.1.

2. Pipe protection shall not be required where exposed piping is permitted by the pipe listing.

❖ Nonmetallic pipe must be protected from the potential heat from a fire that could occur in a protected area. Nonmetallic piping is allowed to be exposed in areas not protected by sprinklers. For example, sprinkler piping could be exposed in a small [i.e., less than 24-square-feet (2.2 m²)] closet, but not exposed in the adjacent room requiring sprinklers. The assumption is that because the area is not required to be protected, it will be unlikely that a fire will occur in that area and expose the pipe to heat.

Where pipe protection is required, the protection must consist of gypsum board not less than $^3/_8$-inch (9.5 mm) thick, $^1/_2$-inch (13 mm) plywood or a 15-minute fire-rated material.

The listing for nonmetallic pipe might allow exposed piping in protected areas given certain conditions, such as depth of joists above the sprinkler location, sprinkler spacing and type of area. Although it might appear that the use of nonmetallic piping in a protected area could result in a piping failure before a fire in that area is extinguished, consider that the pipe is full of water, which helps to dissipate heat that would otherwise soften the pipe if it held no water.

P2904.3.2 Shutoff valves prohibited. With the exception of shutoff valves for the entire water distribution system, valves shall not be installed in any location where the valve would isolate piping serving one or more sprinklers.

❖ Having only one shutoff valve in a multipurpose fire sprinkler system ensures that whenever the valve is in the "off" position, no water is available at the plumbing fixture, thereby, indicating that water is also not available for the sprinkler system.

P2904.3.3 Single dwelling limit. Piping beyond the service valve located at the beginning of the water distribution system shall not serve more than one *dwelling*.

❖ Once the water service pipe enters into a two-family dwelling, the water distribution piping for each unit must be completely separate so that only the sprinkler system for a single unit will be shut off when the water to that same unit's plumbing fixtures is shut off. This prevents one valve from controlling more than one dwelling unit.

P2904.3.4 Drain. A means to drain the sprinkler system shall be provided on the system side of the water distribution shut-off valve.

❖ A residential sprinkler system might require maintenance, therefore, a means for draining down the system must be provided. In most cases, the sprinkler system can be drained sufficiently by opening plumbing fixture faucets and hose bibs.

P2904.4 Determining system design flow. The flow for sizing the sprinkler piping system shall be based on the flow rating of each sprinkler in accordance with Section P2904.4.1 and the calculation in accordance with Section P2904.4.2.

❖ A sprinkler will discharge a specified water flow rate at a specified pressure. As long as the piping system is designed to maintain the pressure at the sprinkler during flow, the system will operate as intended.

P2904.4.1 Determining required flow rate for each sprinkler. The minimum required flow for each sprinkler shall be determined using the sprinkler manufacturer's published data for the specific sprinkler model based on all of the following:

1. The area of coverage.

2. The ceiling configuration.

3. The temperature rating.

4. Any additional conditions specified by the sprinkler manufacturer.

❖ The selection of which sprinkler model to be used for protecting an area begins with: (1) evaluating the floor dimensions of the space to be protected, (2) evaluating the ceiling condition (e.g., the presence of ceiling slope greater than 9$\frac{1}{2}$ degrees or beams) and (3) deciding whether use of a pendent-type or sidewall-type sprinkler results in better piping economy or protection from freezing. Generally, the sprinkler model having the lowest K-value that will provide coverage for the area, given the ceiling condition, is the optimum selection. The K-factor is a dimensionless flow coefficient that is used by sprinkler manufacturers to easily determine flow rates at different pressures. The following sprinkler manufacturer data information and examples illustrate the selection and design process:

Sprinkler model A, Pendent, K = 3.1

COVERAGE	MINUMUM WATER SUPPLY REQUIREMENTS
12 feet by 12 feet	9 gpm @ 8.4 psi
14 feet by 14 feet	10 gpm @ 10.4 psi

Sprinkler model B, Pendent, K = 4.3

COVERAGE	MINUMUM WATER SUPPLY REQUIREMENTS
12 feet by 12 feet	12 gpm @ 7.8 psi
14 feet by 14 feet	13 gpm @ 9.1 psi
16 feet by 16 feet	13 gpm @ 9.1 psi
18 feet by 18 feet	17 gpm @ 15.6 psi
20 feet by 20 feet	21 gpm @ 23.9 psi

Sprinkler model C, Horizontal Sidewall, Deflector 4-6 inches from ceiling, K = 4.0

COVERAGE	MINUMUM WATER SUPPLY REQUIREMENTS
12 feet by 12 feet	11 gpm @ 7.6 psi
14 feet by 14 feet	12 gpm @ 9 psi
16 feet by 16 feet	15 gpm @ 14.1 psi

Sprinkler model D, Horizontal Sidewall, Deflector 4-6 inches from ceiling, K = 4.3

COVERAGE	MINUMUM WATER SUPPLY REQUIREMENTS
12 feet by 12 feet	11 gpm @ 7.6 psi
14 feet by 14 feet	12 gpm @ 9 psi
16 feet by 16 feet	16 gpm @ 16 psi
16 feet by 18 feet	16 gpm @ 16 psi
16 feet by 20 feet	23 gpm @ 33.1 psi
18 feet by 18 feet	19 gpm @ 22.6 psi
18 feet by 20 feet	24 gpm @ 36 psi

Where the coverage area is rectangular, the first number is the width of the room (i.e., the dimension that is parallel to the wall that the sprinkler is located on) and the second number is the length of the room (i.e., the dimension perpendicular to the wall that the sprinkler is located on).

Example 1

Given: Room floor dimensions of 16 feet by 16 feet.

Ceiling slope is zero degrees (i.e., a flat ceiling).

A horizontal sidewall sprinkler is chosen for best piping economy.

Sprinkler C is chosen because the required flow is 15 gpm versus the required flow of 16 gpm for Model D.

Example 2

Given: Room floor dimensions of 14 feet by 20 feet.

Ceiling slope is zero degrees.

A pendent sprinkler is chosen for best piping economy.

Sprinkler B is chosen because Sprinkler A does not have a rating for coverage of 14-foot by 20-foot. The flow rate and pressure (21 gpm @ 23.9 psi) for the 20-

foot by 20-foot area for Sprinkler B must be used. One alternative to using one Sprinkler B to cover the area would be to use two Sprinkler As, at the minimum spacing of 8 feet apart, at the flow rate and pressure for a coverage of a 14-foot by 14-foot area (requiring 10 gpm @ 10.4 psi). Note that by using two of Sprinkler A, the flow required and pressure required for the room (10 + 10 = 20 gpm versus 21 gpm and 10.4 psi versus 23.9 psi) is reduced.

Another alternative would be to use a sidewall sprinkler. One Sprinkler D covering 16-foot by 20-foot (23 gpm @ 33.1 psi) would be suitable. While all selections will provide adequate protection to the area, if available supply pressure is low, the two Sprinkler A option would be better than the one Sprinkler B. If there is plenty of supply pressure, one Sprinkler B would be preferable over one Sprinkler D due to the lesser flow and pressure requirements.

P2904.4.2 System design flow rate. The design flow rate for the system shall be based on the following:

1. The design flow rate for a room having only one sprinkler shall be the flow rate required for that sprinkler, as determined by Section P2904.4.1.

2. The design flow rate for a room having two or more sprinklers a shall be determined by identifying the sprinkler in that room with the highest required flow rate, based on Section P2904.4.1, and multiplying that flow rate by 2.

3. Where the sprinkler manufacturer specifies different criteria for ceiling configurations that are not smooth, flat and horizontal, the required flow rate for that room shall comply with the sprinkler manufacturer's instructions.

4. The design flow rate for the sprinkler system shall be the flow required by the room with the largest flow rate, based on Items 1, 2 and 3.

5. For the purpose of this section, it shall be permissible to reduce the design flow rate for a room by subdividing the space into two or more rooms, where each room is evaluated separately with respect to the required design flow rate. Each room shall be bounded by walls and a ceiling. Openings in walls shall have a lintel not less than 8 inches (203 mm) in depth and each lintel shall form a solid barrier between the ceiling and the top of the opening.

❖ Because of the "light" fire hazard nature of residential interiors and the required fast sprinkler response time for residential sprinklers, it is assumed that if there is one sprinkler in a room, a fire in that room will be extinguished before the heat of the fire can travel to other rooms and activate other sprinklers. If a room has more than one sprinkler, it is assumed that only two sprinklers in that room would be activated before a fire is extinguished. Thus, if there is only one sprinkler per room in a building, the sprinkler piping system design only needs to be based upon one sprinkler discharging. If there is more than one sprinkler per room, the sprinkler piping design must be based on twice the flow of the room sprinkler having the largest required flow.

Where ceilings have slopes greater than 9.5 degrees or have beams, the chosen model of sprinkler might require a higher flow rate and pressure for those conditions. This section requires the higher flow rate and pressure to be used for the design of the sprinkler piping system. Use of the higher flow rate does not permit the designer to consider only one sprinkler in a room having multiple sprinklers. The higher flow rate must be doubled for sprinkler piping design.

A large room with multiple sprinklers can be considered as separate rooms if lintels divide the room into two or more areas. The lintel must extend from the ceiling down at least 8 inches (203 mm) to be considered as a separation for the areas. The lintel must be solid so that heat from a fire in one room cannot flow to another to activate sprinklers not over the fire area.

P2904.5 Water supply. The water supply shall provide not less than the required design flow rate for sprinklers in accordance with Section P2904.4.2 at a pressure not less than that used to comply with Section P2904.6.

❖ The source of the water for the fire sprinkler system must have the capability to provide the flow and pressure to the sprinkler piping to result in the design flow rate and pressure at the design sprinkler location.

P2904.5.1 Water supply from individual sources. Where a *dwelling unit* water supply is from a tank system, a private well system or a combination of these, the available water supply shall be based on the minimum pressure control setting for the pump.

❖ For sprinkler systems supplied from individual water supplies, such as wells or tanks, the sprinkler piping design must be based upon the pump "cut-on" (low) pressure setting and not the pump "cut-off" (high) pressure setting. This accounts for the worst-case pressure condition.

P2904.5.2 Required capacity. The water supply shall have the capacity to provide the required design flow rate for sprinklers for a period of time as follows:

1. Seven minutes for *dwelling units* one *story* in height and less than 2,000 square feet (186 m^2) in area.

2. Ten minutes for *dwelling units* two or more stories in height or equal to or greater than 2,000 square feet (186 m^2) in area.

Where a well system, a water supply tank system or a combination thereof is used, any combination of well capacity and tank storage shall be permitted to meet the capacity requirement.

❖ The other purpose of a sprinkler system is to provide the occupants ample time to realize that they must exit the building and then actually do so. Occupants can exit a smaller dwelling [one story and less than 2,000 square feet (186 m^2)] much quicker than they can a larger [2,000 square feet (186 m^2) or greater] or multi-story building (two or more stories). The requirements of this section are identical to that required by NFPA 13D.

P2904.6 Pipe sizing. The piping to sprinklers shall be sized for the flow required by Section P2904.4.2. The flow required to supply the plumbing fixtures shall not be required to be added to the sprinkler design flow.

❖ For multipurpose piping systems, because the sprinkler demand on the piping is much greater than the plumbing demand and because the plumbing fixtures' water demand on a system during a fire will likely be minimal or zero, the design of the piping system only needs to consider the sprinkler design flow rate.

P2904.6.1 Method of sizing pipe. Piping supplying sprinklers shall be sized using the prescriptive method in Section P2904.6.2 or by hydraulic calculation in accordance with NFPA 13D. The minimum pipe size from the water supply source to any sprinkler shall be $^3/_4$ inch (19 mm) nominal. Threaded adapter fittings at the point where sprinklers are attached to the piping shall be not less than $^1/_2$ inch (13 mm) nominal.

❖ Once the sprinkler design flow and pressure is known, the piping system can be designed in accordance with Section P2904.6.2 or by the methods as required by NFPA 13D. Where Section P2904.6.2 is used, the sprinkler design flow value is used in Tables P2904.6.2(4) through P2904.6.2(9) to determine the allowable length of the water distribution pipe to the most remote sprinkler creating the design sprinkler flow.

Most residential sprinklers have $^1/_2$-inch (12.7 mm) male pipe threads. Because the piping to each sprinkler must not be less than $^3/_4$ inch (19.1 mm), an adapter fitting must be used to couple the piping to the sprinkler. For copper piping or CPVC piping, the adapter must have a socket on one end that is sized for the sprinkler piping and $^1/_2$-inch (12.7 mm) female pipe threads on the other end. The female threaded end of CPVC adapter fittings must have either metal reinforcement around the outside of the body of the fitting where the female threads are located or have a metal insert in the fitting to prevent the fitting from cracking when the sprinkler is tightened into the fitting. Fittings for PEX tubing are made of metal construction and have barbed ends that insert into the sprinkler tubing. The tubing connection to the barbed end of the fitting is with a clamped ring. Depending on the piping material, there are various configurations of adapter fittings that can be used, such as a straight, 90-degree (1.57 rad) elbow, back-to-back tee and back-to-back cross.

P2904.6.2 Prescriptive pipe sizing method. Pipe shall be sized by determining the available pressure to offset friction loss in piping and identifying a piping material, diameter and length using the equation in Section P2904.6.2.1 and the procedure in Section P2904.6.2.2.

❖ The rationale of this section's prescriptive design is simply to subtract all known pressure losses (except for piping pressure losses), and the design sprinkler flow pressure, from the available supply pressure. The remainder is the pressure available for overcoming pipe friction losses.

P2904.6.2.1 Available pressure equation. The pressure available to offset friction loss in the interior piping system (P_t) shall be determined in accordance with the Equation 29-1.

$$P_t = P_{sup} - PL_{svc} - PL_m - PL_d - PL_e - P_{sp} \quad \textbf{(Equation 29-1)}$$
where:

P_t = Pressure used in applying Tables P2904.6.2(4) through P2904.6.2(9).

P_{sup} = Pressure available from the water supply source.

PL_{svc} = Pressure loss in the water-service pipe.

PL_m = Pressure loss in the water meter.

PL_d = Pressure loss from devices other than the water meter.

PL_e = Pressure loss associated with changes in elevation.

P_{sp} = Maximum pressure required by a sprinkler.

❖ This formula indicates the factors that must be considered in the calculation of the pressure that is remaining and available for overcoming pipe friction losses.

TABLE P2904.6.2(1). See page 29-33.

❖ This table is applicable for all types of piping listed in Table P2905.4.

P2904.6.2.2 Calculation procedure. Determination of the required size for water distribution piping shall be in accordance with the following procedure:

Step 1—Determine P_{sup}

Obtain the static supply pressure that will be available from the water main from the water purveyor, or for an individual source, the available supply pressure shall be in accordance with Section P2904.5.1.

❖ If the water supply is from the public water main, the supplier of the water can provide the static pressure at the tap location. As this pressure will vary over a 24-hour period, as well as seasonally, worse case pressure should be used. For individual water sources (wells), the pressure will be the pump "cut-on" pressure setting.

Step 2—Determine PL_{svc}

Use Table P2904.6.2(1) to determine the pressure loss in the water service pipe based on the selected size of the water service.

❖ The table indicates pressure loss caused by friction, and includes friction factors for fittings and shutoff valves that are typically used in service lines. If the design sprinkler flow falls between table values, select the next higher value in the table. Note that if the service line supplies water for two dwellings, 5 gpm (0.3 L/s) must be added to the design sprinkler flow before using this table.

For individual water sources (wells), there may not be a water service line that requires determination of pressure loss because the water distribution system usually begins at the point where the pressure switch is located (i.e., the pressure tank).

TABLE P2904.6.2(1)
WATER SERVICE PRESSURE LOSS (PL_{svc})[a, b]

FLOW RATE[c] (gpm)	3/4-INCH WATER SERVICE PRESSURE LOSS (psi)				1-INCH WATER SERVICE PRESSURE LOSS (psi)				1 1/4-INCH WATER SERVICE PRESSURE LOSS (psi)			
	Length of water service pipe (feet)				Length of water service pipe (feet)				Length of water service pipe (feet)			
	40 or less	41 to 75	76 to 100	101 to 150	40 or less	41 to 75	76 to 100	101 to 150	40 or less	41 to 75	76 to 100	101 to 150
8	5.1	8.7	11.8	17.4	1.5	2.5	3.4	5.1	0.6	1.0	1.3	1.9
10	7.7	13.1	17.8	26.3	2.3	3.8	5.2	7.7	0.8	1.4	2.0	2.9
12	10.8	18.4	24.9	NP	3.2	5.4	7.3	10.7	1.2	2.0	2.7	4.0
14	14.4	24.5	NP	NP	4.2	7.1	9.6	14.3	1.6	2.7	3.6	5.4
16	18.4	NP	NP	NP	5.4	9.1	12.4	18.3	2.0	3.4	4.7	6.9
18	22.9	NP	NP	NP	6.7	11.4	15.4	22.7	2.5	4.3	5.8	8.6
20	27.8	NP	NP	NP	8.1	13.8	18.7	27.6	3.1	5.2	7.0	10.4
22	NP	NP	NP	NP	9.7	16.5	22.3	NP	3.7	6.2	8.4	12.4
24	NP	NP	NP	NP	11.4	19.3	26.2	NP	4.3	7.3	9.9	14.6
26	NP	NP	NP	NP	13.2	22.4	NP	NP	5.0	8.5	11.4	16.9
28	NP	NP	NP	NP	15.1	25.7	NP	NP	5.7	9.7	13.1	19.4
30	NP	NP	NP	NP	17.2	NP	NP	NP	6.5	11.0	14.9	22.0
32	NP	NP	NP	NP	19.4	NP	NP	NP	7.3	12.4	16.8	24.8
34	NP	NP	NP	NP	21.7	NP	NP	NP	8.2	13.9	18.8	NP
36	NP	NP	NP	NP	24.1	NP	NP	NP	9.1	15.4	20.9	NP

For SI: 1 inch = 25.4 mm, 1 foot = 304.8 mm, 1 gallon per minute = 0.063 L/s, 1 pound per square inch = 6.895 kPa.

NP = Not permitted. Pressure loss exceeds reasonable limits.

a. Values are applicable for underground piping materials listed in Table P2905.4 and are based on an SDR of 11 and a Hazen Williams C Factor of 150.

b. Values include the following length allowances for fittings: 25% length increase for actual lengths up to 100 feet and 15% length increase for actual lengths over 100 feet.

c. Flow rate from Section P2904.4.2. Add 5 gpm to the flow rate required by Section P2904.4.2 where the water-service pipe supplies more than one dwelling.

Step 3—Determine PL_m

Use Table P2904.6.2(2) to determine the pressure loss from the water meter, based on the selected water meter size.

❖ The table indicates pressure loss from friction through the meter. If the sprinkler demand flow rate is between table values, select the next higher value in the table. Note that if the service line gpm supplies water for two dwellings, 5 gpm (0.34 L/s) must be added to the sprinkler design flow before using this table.

Step 4—Determine PL_d

Determine the pressure loss from devices other than the water meter installed in the piping system supplying sprinklers, such as pressure-reducing valves, backflow preventers, water softeners or water filters. Device pressure losses shall be based on the device manufacturer's specifications. The flow rate used to determine pressure loss shall be the rate from Section P2904.4.2, except that 5 gpm (0.3 L/s) shall be added where the device is installed in a water-service pipe that supplies more than one *dwelling*. As an alternative to deducting pressure loss for a device, an automatic bypass valve shall be installed to divert flow around the device when a sprinkler activates.

❖ Devices other than water meters are often installed in water service lines. All devices will create a pressure loss based upon the water flow rate through the

device. Device manufacturers can provide pressure loss values for the sprinkler design flow. Where the pressure loss for a device is not available or the pressure loss is known and too high, this section allows for an automatic bypass valve to be installed so that flow can be diverted around the device so that the device pressure loss does not have to be included in the calculations. Pressure loss due to the automatic bypass valve must be included in the calculation. Note that if the service line supplies water for two dwellings, 5 gpm (0.34 L/s) must be added to the sprinkler design flow before using this table.

Step 5—Determine PL_e

Use Table P2904.6.2(3) to determine the pressure loss associated with changes in elevation. The elevation used in applying the table shall be the difference between the elevation where the water source pressure was measured and the elevation of the highest sprinkler.

❖ Note that the difference in elevation is to the highest sprinkler in the building, but not necessarily will that sprinkler be the one creating the sprinkler design flow. The table uses the conversion of 1 foot (305 mm) of water head is equal to 0.433 psi (2.99 kPa). Where the difference in elevation falls between values, use the next higher value in the table or interpolate between the values.

TABLE P2904.6.2(2)
MINIMUM WATER METER PRESSURE LOSS (PL_m)[a]

FLOW RATE (gallons per minute, gpm)[b]	5/8-INCH METER PRESSURE LOSS (pounds per square inch, psi)	3/4-INCH METER PRESSURE LESS (pounds per square inch, psi)	1-INCH METER PRESSURE LOSS (pounds per square inch, psi)
8	2	1	1
10	3	1	1
12	4	1	1
14	5	2	1
16	7	3	1
18	9	4	1
20	11	4	2
22	NP	5	2
24	NP	5	2
26	NP	6	2
28	NP	6	2
30	NP	7	2
32	NP	7	3
34	NP	8	3
36	NP	8	3

For SI: 1 inch = 25.4 mm, 1 pound per square inch = 6.895 kPa, 1 gallon per minute = 0.063 L/s.

NP—Not permitted unless the actual water meter pressure loss is known.

a. Table P2904.6.2(2) establishes conservative values for water meter pressure loss or installations where the water meter loss is unknown. Where the actual water meter pressure loss is known, P_m shall be the actual loss.

b. Flow rate from Section P2904.4.2. Add 5 gpm to the flow rate required by Section P2904.4.2 where the water-service pipe supplies more than one dwelling.

TABLE P2904.6.2(3)
ELEVATION LOSS (PL_e)

ELEVATION (feet)	PRESSURE LOSS (psi)
5	2.2
10	4.4
15	6.5
20	8.7
25	10.9
30	13
35	15.2
40	17.4

For SI: 1 foot = 304.8 mm, 1 pound per square inch = 6.895 kPa.

Step 6—Determine P_{sp}

Determine the maximum pressure required by any individual sprinkler based on the flow rate from Section P2904.4.1. The required pressure is provided in the sprinkler manufacturer's published data for the specific sprinkler model based on the selected flow rate.

❖ For the sprinkler creating the design flow, obtain the required sprinkler pressure from the manufacturer's published literature.

Step 7—Calculate P_t

Using Equation 29-1, calculate the pressure available to offset friction loss in water-distribution piping between the service valve and the sprinklers.

❖ The value, P_t, will be necessary for using Tables P2904.6.2(4) through P2904.6.2(9).

Step 8—Determine the maximum allowable pipe length

Use Tables P2904.6.2(4) through P2904.6.2(9) to select a material and size for water distribution piping. The piping material and size shall be acceptable if the *developed length* of pipe between the service valve and the most remote sprinkler does not exceed the maximum allowable length specified by the applicable table. Interpolation of P_t between the tabular values shall be permitted.

The maximum allowable length of piping in Tables P2904.6.2(4) through P2904.6.2(9) incorporates an adjustment for pipe fittings. Additional consideration of friction losses associated with pipe fittings shall not be required.

❖ From the building drawings, determine the developed length of the water distribution piping from the water service valve (at the discharge end of the water service line) to the most remote sprinkler. This actual length is compared to an allowable length that is derived from Tables P2904.6.2(4) through P2904.6.2(9).

Sample Problem 1

Refer to the single floor plan shown in Commentary Figure P2904.6.2.2(1). The length of the water service line from the public water main to the entry point into the building is 100 feet. The public water main's minimum expected pressure at the tap location is 105 psi.

The building is located on a hill such that the building floor elevation is 30 feet above the public water main. A $^5/_8$-inch water meter will be provided by the water utility company. The water service pipe will enter the building at 4 feet below the floor level of the building. The available sprinklers are the generic Models A, B, C and D indicated in the commentary to Section P2902.5.1. The heating, ventilating and air-conditioning (HVAC) unit is electric (not fuel fired) and is located in the attic. All ceilings are flat. The water dis-

TABLE P2904.6.2(4)
ALLOWABLE PIPE LENGTH FOR $^3/_4$-INCH TYPE M COPPER WATER TUBING

SPRINKLER FLOW RATE[a] (gpm)	WATER DISTRIBUTION SIZE (inch)	AVAILABLE PRESSURE—P_t (psi)									
		15	20	25	30	35	40	45	50	55	60
		Allowable length of pipe from service valve to farthest sprinkler (feet)									
8	$^3/_4$	217	289	361	434	506	578	650	723	795	867
9	$^3/_4$	174	232	291	349	407	465	523	581	639	697
10	$^3/_4$	143	191	239	287	335	383	430	478	526	574
11	$^3/_4$	120	160	200	241	281	321	361	401	441	481
12	$^3/_4$	102	137	171	205	239	273	307	341	375	410
13	$^3/_4$	88	118	147	177	206	235	265	294	324	353
14	$^3/_4$	77	103	128	154	180	205	231	257	282	308
15	$^3/_4$	68	90	113	136	158	181	203	226	248	271
16	$^3/_4$	60	80	100	120	140	160	180	200	220	241
17	$^3/_4$	54	72	90	108	125	143	161	179	197	215
18	$^3/_4$	48	64	81	97	113	129	145	161	177	193
19	$^3/_4$	44	58	73	88	102	117	131	146	160	175
20	$^3/_4$	40	53	66	80	93	106	119	133	146	159
21	$^3/_4$	36	48	61	73	85	97	109	121	133	145
22	$^3/_4$	33	44	56	67	78	89	100	111	122	133
23	$^3/_4$	31	41	51	61	72	82	92	102	113	123
24	$^3/_4$	28	38	47	57	66	76	85	95	104	114
25	$^3/_4$	26	35	44	53	61	70	79	88	97	105
26	$^3/_4$	24	33	41	49	57	65	73	82	90	98
27	$^3/_4$	23	30	38	46	53	61	69	76	84	91
28	$^3/_4$	21	28	36	43	50	57	64	71	78	85
29	$^3/_4$	20	27	33	40	47	53	60	67	73	80
30	$^3/_4$	19	25	31	38	44	50	56	63	69	75
31	$^3/_4$	18	24	29	35	41	47	53	59	65	71
32	$^3/_4$	17	22	28	33	39	44	50	56	61	67
33	$^3/_4$	16	21	26	32	37	42	47	53	58	63
34	$^3/_4$	NP	20	25	30	35	40	45	50	55	60
35	$^3/_4$	NP	19	24	28	33	38	42	47	52	57
36	$^3/_4$	NP	18	22	27	31	36	40	45	49	54
37	$^3/_4$	NP	17	21	26	30	34	38	43	47	51
38	$^3/_4$	NP	16	20	24	28	32	36	40	45	49
39	$^3/_4$	NP	15	19	23	27	31	35	39	42	46
40	$^3/_4$	NP	NP	18	22	26	29	33	37	40	44

For SI: 1 inch = 25.4 mm, 1 foot = 304.8 mm, 1 pound per square inch = 6.895 kPa, 1 gallon per minute = 0.963 L/s.

NP—Not permitted.

a. Flow rate from Section P2904.4.2.

tribution piping material will be CPVC. The building is located in a freezing climate such that sprinkler piping must not be installed above the ceiling (in the attic), or in outside walls. Design a multipurpose sprinkler system for this building.

Problem Approach

Locate the sprinkler positions for every area of the building exempting those areas listed in Section P2904.1.1. Determine the length of piping to the most remote sprinkler location. Determine which room has the greatest sprinkler flow. Where a room has two sprinklers, the sprinkler demand flow is double the flow of the sprinkler having the greater flow. Calculate the pressure available for pipe friction loss, P_t, to be used in the pipe sizing tables. From the pipe sizing tables, identify the size of the water service pipe to the building and the size of the water distribution piping to all sprinklers.

TABLE P2904.6.2(5)
ALLOWABLE PIPE LENGTH FOR 1-INCH TYPE M COPPER WATER TUBING

SPRINKLER FLOW RATE[a] (gpm)	WATER DISTRIBUTION SIZE (inch)	AVAILABLE PRESSURE—P_t (psi)									
		15	20	25	30	35	40	45	50	55	60
		Allowable length of pipe from service valve to farthest sprinkler (feet)									
8	1	806	1075	1343	1612	1881	2149	2418	2687	2955	3224
9	1	648	864	1080	1296	1512	1728	1945	2161	2377	2593
10	1	533	711	889	1067	1245	1422	1600	1778	1956	2134
11	1	447	586	745	894	1043	1192	1341	1491	1640	1789
12	1	381	508	634	761	888	1015	1142	1269	1396	1523
13	1	328	438	547	657	766	875	985	1094	1204	1313
14	1	286	382	477	572	668	763	859	954	1049	1145
15	1	252	336	420	504	588	672	756	840	924	1008
16	1	224	298	373	447	522	596	671	745	820	894
17	1	200	266	333	400	466	533	600	666	733	799
18	1	180	240	300	360	420	479	539	599	659	719
19	1	163	217	271	325	380	434	488	542	597	651
20	1	148	197	247	296	345	395	444	493	543	592
21	1	135	180	225	270	315	360	406	451	496	541
22	1	124	165	207	248	289	331	372	413	455	496
23	1	114	152	190	228	267	305	343	381	419	457
24	1	106	141	176	211	246	282	317	352	387	422
25	1	98	131	163	196	228	261	294	326	359	392
26	1	91	121	152	182	212	243	273	304	334	364
27	1	85	113	142	170	198	226	255	283	311	340
28	1	79	106	132	159	185	212	238	265	291	318
29	1	74	99	124	149	174	198	223	248	273	298
30	1	70	93	116	140	163	186	210	233	256	280
31	1	66	88	110	132	153	175	197	219	241	263
32	1	62	83	103	124	145	165	186	207	227	248
33	1	59	78	98	117	137	156	176	195	215	234
34	1	55	74	92	111	129	148	166	185	203	222
35	1	53	70	88	105	123	140	158	175	193	210
36	1	50	66	83	100	116	133	150	166	183	199
37	1	47	63	79	95	111	126	142	158	174	190
38	1	45	60	75	90	105	120	135	150	165	181
39	1	43	57	72	86	100	115	129	143	158	172
40	1	41	55	68	82	96	109	123	137	150	164

For SI: 1 inch = 25.4 mm, 1 foot = 304.8 mm, 1 pound per square inch = 6.895 kPa, 1 gallon per minute = 0.963 L/s.
a. Flow rate from Section P2904.4.2.

The solution is broken down into Parts A, B, and C:

Solution—Part A

Because every closet is less than 24 square feet with the smallest dimension less than 3 feet, and the bathroom is less than 55 square feet, these areas do not require sprinklers. Because the HVAC unit is all electric, a sprinkler is not required at the HVAC unit. As sprinkler piping cannot be installed above the ceiling, horizontal sidewall sprinklers will be selected for all applications.

Note that Bedroom 1 has a ceiling fan, Bedroom 2 has an alcove entry area, the living room has a fireplace, and the kitchen has a surface-mounted luminaire (light fixture) and range. The presence of these features could affect placement of sprinklers or could require intermediate temperature sprinklers to be used in those areas.

Review the living room area to locate a sprinkler. Considering that sprinkler piping cannot be in exterior walls because of the potential for freezing, the two possible locations for a single sprinkler to cover the

TABLE P2904.6.2(6)
ALLOWABLE PIPE LENGTH FOR $^3/_4$-INCH CPVC PIPE

SPRINKLER FLOW RATE[a] (gpm)	WATER DISTRIBUTION SIZE (inch)	AVAILABLE PRESSURE—P_t (psi)									
		15	20	25	30	35	40	45	50	55	60
		Allowable length of pipe from service valve to farthest sprinkler (feet)									
8	$^3/_4$	348	465	581	697	813	929	1045	1161	1278	1394
9	$^3/_4$	280	374	467	560	654	747	841	934	1027	1121
10	$^3/_4$	231	307	384	461	538	615	692	769	845	922
11	$^3/_4$	193	258	322	387	451	515	580	644	709	773
12	$^3/_4$	165	219	274	329	384	439	494	549	603	658
13	$^3/_4$	142	189	237	284	331	378	426	473	520	568
14	$^3/_4$	124	165	206	247	289	330	371	412	454	495
15	$^3/_4$	109	145	182	218	254	290	327	363	399	436
16	$^3/_4$	97	129	161	193	226	258	290	322	354	387
17	$^3/_4$	86	115	144	173	202	230	259	288	317	346
18	$^3/_4$	78	104	130	155	181	207	233	259	285	311
19	$^3/_4$	70	94	117	141	164	188	211	234	258	281
20	$^3/_4$	64	85	107	128	149	171	192	213	235	256
21	$^3/_4$	58	78	97	117	136	156	175	195	214	234
22	$^3/_4$	54	71	89	107	125	143	161	179	197	214
23	$^3/_4$	49	66	82	99	115	132	148	165	181	198
24	$^3/_4$	46	61	76	91	107	122	137	152	167	183
25	$^3/_4$	42	56	71	85	99	113	127	141	155	169
26	$^3/_4$	39	52	66	79	92	105	118	131	144	157
27	$^3/_4$	37	49	61	73	86	98	110	122	135	147
28	$^3/_4$	34	46	57	69	80	92	103	114	126	137
29	$^3/_4$	32	43	54	64	75	86	96	107	118	129
30	$^3/_4$	30	40	50	60	70	81	91	101	111	121
31	$^3/_4$	28	38	47	57	66	76	85	95	104	114
32	$^3/_4$	27	36	45	54	63	71	80	89	98	107
33	$^3/_4$	25	34	42	51	59	68	76	84	93	101
34	$^3/_4$	24	32	40	48	56	64	72	80	88	96
35	$^3/_4$	23	30	38	45	53	61	68	76	83	91
36	$^3/_4$	22	29	36	43	50	57	65	72	79	86
37	$^3/_4$	20	27	34	41	48	55	61	68	75	82
38	$^3/_4$	20	26	33	39	46	52	59	65	72	78
39	$^3/_4$	19	25	31	37	43	50	56	62	68	74
40	$^3/_4$	18	24	30	35	41	47	53	59	65	71

For SI: 1 inch = 25.4 mm, 1 foot = 304.8 mm, 1 pound per square inch = 6.895 kPa, 1 gallon per minute = 0.963 L/s.
a. Flow rate from Section P2904.4.2.

area are Points A and B [see Commentary Figure P2904.6.2.2(2)]. Because the one dimension of the room is 17 feet, one Model C sprinkler will not cover the area. One Model D sprinkler located at Point A will cover the room if it is provided with 16 gpm at a pressure of 16 psi. If the Model D sprinkler is located at Point B, it would have to be supplied with 19 gpm at 22.6 psi in order to cover the room. Note that both Points A and B are located farther than 60 inches from the front of the fireplace, thus an immediate tempera-

ture sprinkler is not required for either location. Because Point B is adjacent to the possible Point C location for a kitchen sprinkler, Point B might be the better choice, in order to save piping costs.

Review the kitchen area to locate a sprinkler. The two possible locations for a single sprinkler to cover the area are Points C and D. However, because of the proximity of the surface-mounted luminaire, with the sprinkler located at Point D, Point C is the only acceptable location. A Model C sprinkler is chosen, requiring

TABLE P2904.6.2(7)
ALLOWABLE PIPE LENGTH FOR 1-INCH CPVC PIPE

SPRINKLER FLOW RATE[a] (gpm)	WATER DISTRIBUTION SIZE (inch)	AVAILABLE PRESSURE—P_t (psi)									
		15	20	25	30	35	40	45	50	55	60
		Allowable length of pipe from service valve to farthest sprinkler (feet)									
8	1	1049	1398	1748	2098	2447	2797	3146	3496	3845	4195
9	1	843	1125	1406	1687	1968	2249	2530	2811	3093	3374
10	1	694	925	1157	1388	1619	1851	2082	2314	2545	2776
11	1	582	776	970	1164	1358	1552	1746	1940	2133	2327
12	1	495	660	826	991	1156	1321	1486	1651	1816	1981
13	1	427	570	712	854	997	1139	1281	1424	1566	1709
14	1	372	497	621	745	869	993	1117	1241	1366	1490
15	1	328	437	546	656	765	874	983	1093	1202	1311
16	1	291	388	485	582	679	776	873	970	1067	1164
17	1	260	347	433	520	607	693	780	867	954	1040
18	1	234	312	390	468	546	624	702	780	858	936
19	1	212	282	353	423	494	565	635	706	776	847
20	1	193	257	321	385	449	513	578	642	706	770
21	1	176	235	293	352	410	469	528	586	645	704
22	1	161	215	269	323	377	430	484	538	592	646
23	1	149	198	248	297	347	396	446	496	545	595
24	1	137	183	229	275	321	366	412	458	504	550
25	1	127	170	212	255	297	340	382	425	467	510
26	1	118	158	197	237	276	316	355	395	434	474
27	1	111	147	184	221	258	295	332	368	405	442
28	1	103	138	172	207	241	275	310	344	379	413
29	1	97	129	161	194	226	258	290	323	355	387
30	1	91	121	152	182	212	242	273	303	333	364
31	1	86	114	143	171	200	228	257	285	314	342
32	1	81	108	134	161	188	215	242	269	296	323
33	1	76	102	127	152	178	203	229	254	280	305
34	1	72	96	120	144	168	192	216	240	265	289
35	1	68	91	114	137	160	182	205	228	251	273
36	1	65	87	108	130	151	173	195	216	238	260
37	1	62	82	103	123	144	165	185	206	226	247
38	1	59	78	98	117	137	157	176	196	215	235
39	1	56	75	93	112	131	149	168	187	205	224
40	1	53	71	89	107	125	142	160	178	196	214

For SI: 1 inch = 25.4 mm, 1 foot = 304.8 mm, 1 pound per square inch = 6.895 kPa, 1 gallon per minute = 0.963 L/s.
a. Flow rate from Section P2904.4.2.

15 gpm at 14.6 psi to cover a 16-foot by 16-foot area. Because Point C was chosen, it is more economical to locate the living room sprinkler at Point B.

For Bedrooms 2 and 3, it is more economical to locate the sprinkler for each room adjacent to one another on the wall common to both rooms (Points E and F). A sprinkler location at Point E provides coverage for the room entry alcove in Bedroom 2. Therefore, a Model C sprinkler is chosen for coverage of a 16-foot by 16-foot area requiring 15 gpm at 14.1 psi.

For Bedroom 3, either a Model C or D sprinkler can be chosen for a coverage area of 14 feet by 14 feet, which is more than adequate to cover the 12-foot, 6-inch by 7-foot area. The required flow for both models of sprinklers is 12 gpm at 9 psi.

Bedroom 1 has only one logical location for a sprinkler and that is at Point G. As the ceiling fan is located in the center of the room, it is more than 5 feet from the sprinkler and, therefore, the ceiling fan is not considered to be an obstruction to the sprinkler spray pat-

TABLE P2904.6.2(8)
ALLOWABLE PIPE LENGTH FOR $^3/_4$-INCH PEX AND PE-RT TUBING

SPRINKLER FLOW RATE[a] (gpm)	WATER DISTRIBUTION SIZE (inch)	AVAILABLE PRESSURE—P_t(psi)									
		15	20	25	30	35	40	45	50	55	60
		Allowable length of pipe from service valve to farthest sprinkler (feet)									
8	$^3/_4$	93	123	154	185	216	247	278	309	339	370
9	$^3/_4$	74	99	124	149	174	199	223	248	273	298
10	$^3/_4$	61	82	102	123	143	163	184	204	225	245
11	$^3/_4$	51	68	86	103	120	137	154	171	188	205
12	$^3/_4$	44	58	73	87	102	117	131	146	160	175
13	$^3/_4$	38	50	63	75	88	101	113	126	138	151
14	$^3/_4$	33	44	55	66	77	88	99	110	121	132
15	$^3/_4$	29	39	48	58	68	77	87	96	106	116
16	$^3/_4$	26	34	43	51	60	68	77	86	94	103
17	$^3/_4$	23	31	38	46	54	61	69	77	84	92
18	$^3/_4$	21	28	34	41	48	55	62	69	76	83
19	$^3/_4$	19	25	31	37	44	50	56	62	69	75
20	$^3/_4$	17	23	28	34	40	45	51	57	62	68
21	$^3/_4$	16	21	26	31	36	41	47	52	57	62
22	$^3/_4$	NP	19	24	28	33	38	43	47	52	57
23	$^3/_4$	NP	17	22	26	31	35	39	44	48	52
24	$^3/_4$	NP	16	20	24	28	32	36	40	44	49
25	$^3/_4$	NP	NP	19	22	26	30	34	37	41	45
26	$^3/_4$	NP	NP	17	21	24	28	31	35	38	42
27	$^3/_4$	NP	NP	16	20	23	26	29	33	36	39
28	$^3/_4$	NP	NP	15	18	21	24	27	30	33	36
29	$^3/_4$	NP	NP	NP	17	20	23	26	28	31	34
30	$^3/_4$	NP	NP	NP	16	19	21	24	27	29	32
31	$^3/_4$	NP	NP	NP	15	18	20	23	25	28	30
32	$^3/_4$	NP	NP	NP	NP	17	19	21	24	26	28
33	$^3/_4$	NP	NP	NP	NP	16	18	20	22	25	27
34	$^3/_4$	NP	NP	NP	NP	NP	17	19	21	23	25
35	$^3/_4$	NP	NP	NP	NP	NP	16	18	20	22	24
36	$^3/_4$	NP	NP	NP	NP	NP	15	17	19	21	23
37	$^3/_4$	NP	NP	NP	NP	NP	NP	16	18	20	22
38	$^3/_4$	NP	NP	NP	NP	NP	NP	16	17	19	21
39	$^3/_4$	NP	NP	NP	NP	NP	NP	NP	16	18	20
40	$^3/_4$	NP	NP	NP	NP	NP	NP	NP	16	17	19

For SI: 1 inch = 25.4 mm, 1 foot = 304.8 mm, 1 pound per square inch = 6.895 kPa, 1 gallon per minute = 0.963 L/s.

NP— Not permitted.

a. Flow rate from Section P2904.4.2.

tern. A Model C sprinkler is chosen to cover the 12-foot by 12-foot area. The sprinkler will require 11 gpm at 7.6 psi.

The sprinkler at Point B has the greatest demand flow at 19 gpm at 22.6 psi. The sprinkler at Point B also happens to be the most remote sprinkler location in the building. In this example, it is just a coincidence that the sprinkler having the greatest demand is also the sprinkler that is most remote. If the example was a two-story building and the living room still had the

sprinkler with the greatest demand, the most remote sprinkler would be at a sprinkler location on the second floor.

Solution—Part B

The supply pressure, P_{sup}, is 105 psi. To calculate pressure loss of the water service line, PL_{svc}, find the sprinkler demand flow of 19 gpm, in the far left column of Table P2904.6.2(1) and select the row that is equal to or greater than 19 gpm. The row for 20 gpm is chosen.

TABLE P2904.6.2(9)
ALLOWABLE PIPE LENGTH FOR 1-INCH PEX AND PE-RT TUBING

SPRINKLER FLOW RATE[a] (gpm)	WATER DISTRIBUTION SIZE (inch)	AVAILABLE PRESSURE—P_t (psi)									
		15	20	25	30	35	40	45	50	55	60
		Allowable length of pipe from service valve to farthest sprinkler (feet)									
8	1	314	418	523	628	732	837	941	1046	1151	1255
9	1	252	336	421	505	589	673	757	841	925	1009
10	1	208	277	346	415	485	554	623	692	761	831
11	1	174	232	290	348	406	464	522	580	638	696
12	1	148	198	247	296	346	395	445	494	543	593
13	1	128	170	213	256	298	341	383	426	469	511
14	1	111	149	186	223	260	297	334	371	409	446
15	1	98	131	163	196	229	262	294	327	360	392
16	1	87	116	145	174	203	232	261	290	319	348
17	1	78	104	130	156	182	208	233	259	285	311
18	1	70	93	117	140	163	187	210	233	257	280
19	1	63	84	106	127	148	169	190	211	232	253
20	1	58	77	96	115	134	154	173	192	211	230
21	1	53	70	88	105	123	140	158	175	193	211
22	1	48	64	80	97	113	129	145	161	177	193
23	1	44	59	74	89	104	119	133	148	163	178
24	1	41	55	69	82	96	110	123	137	151	164
25	1	38	51	64	76	89	102	114	127	140	152
26	1	35	47	59	71	83	95	106	118	130	142
27	1	33	44	55	66	77	88	99	110	121	132
28	1	31	41	52	62	72	82	93	103	113	124
29	1	29	39	48	58	68	77	87	97	106	116
30	1	27	36	45	54	63	73	82	91	100	109
31	1	26	34	43	51	60	68	77	85	94	102
32	1	24	32	40	48	56	64	72	80	89	97
33	1	23	30	38	46	53	61	68	76	84	91
34	1	22	29	36	43	50	58	65	72	79	86
35	1	20	27	34	41	48	55	61	68	75	82
36	1	19	26	32	39	45	52	58	65	71	78
37	1	18	25	31	37	43	49	55	62	68	74
38	1	18	23	29	35	41	47	53	59	64	70
39	1	17	22	28	33	39	45	50	56	61	67
40	1	16	21	27	32	37	43	48	53	59	64

For SI: 1 inch = 25.4 mm, 1 foot = 304.8 mm, 1 pound per square inch = 6.895 kPa, 1 gallon per minute = 0.963 L/s.

a. Flow rate from Section P2904.4.2.

The length of the water service is given as 100 feet. Follow across the 20-gpm row and identify the first 100-foot column, which has a value. The pressure loss of 18.7 psi is identified. Note that the water service line must be at least a 1-inch (25.4 mm) pipe or tube size.

To calculate the water meter pressure loss, PLm, find the sprinkler demand flow of 19 gpm, in the far left column of Table P2904.6.2(2) and select the row that is equal to or greater than 19 gpm. The row for 20 gpm is chosen. Follow across the 20 gpm row and identify the first column that has a number. A $^5/_8$-inch meter has an 11-psi pressure loss.

The next unknown to calculate is the pressure loss for all devices, such as PRVs, filters, water softeners, etc., that are in the water supply or in the main water distribution line supplying water to the sprinklers. Because the public main has a pressure of 105 psi, a PRV might be required. However, due to the location of the building on a hill, the static pressure at the building entry point must first be calculated. Note that the elevation difference is given between the public water main and the floor elevation and not to the sprinkler elevation. Therefore, assuming that the ceiling height is 8 feet, the sprinkler elevation is approximately 7.5 feet above the floor elevation since sidewall sprinklers are located at 4 to 6 inches from the ceiling. Enter the left hand column of Table P2904.6.2(3) and select a row that is equal to or greater than $37^1/_2$ feet. The 40-psi row is chosen. Move across that row to the right-hand column to identify a pressure loss of 17.4 psi due to the elevation difference. Subtracting 17.4 psi from 105 psi equals 87.6 psi. Because the code requires (see Section 2903.3.1) that the supply pressure of the building plumbing system not exceed 80-psi static pressure, a PRV is required to be installed in order to limit the pressure to 80 psi. As there are so many different sizes and models of PRVs, the code does not provide values for pressure losses for PRVs. The values must be obtained from the manufacturer of the valve to be used on the project. In this example, it is determined (from manufacturer's data sheets) that a $^3/_4$-inch (19.1 mm) PRV is required and at the design sprinkler flow of 19 gpm, the pressure loss is 22 psi.

The design sprinkler pressure, P_{sp}, is stated as 22.6 psi. The pressure available for pipe friction, P_t is calculated as follows:

$$P_t = P_{sup} - PL_{svc} - PL_m - PL_d - PL_e - P_{sp}$$
$$P_t = 105 - 18.7 - 11 - 22 - 17.4 - 22.6$$
$$P_t = 13.3 \text{ psi}$$

A total of 13.3 psi is available for pipe friction losses from the point of water service entry into the building to the most remote sprinkler.

Looking at the allowable pipe length tables [see Tables P2904.6.2(4) through (9)], it is realized that the lowest available pressure in every table is 15 psi. In this example, with the sizes chosen to this point, there

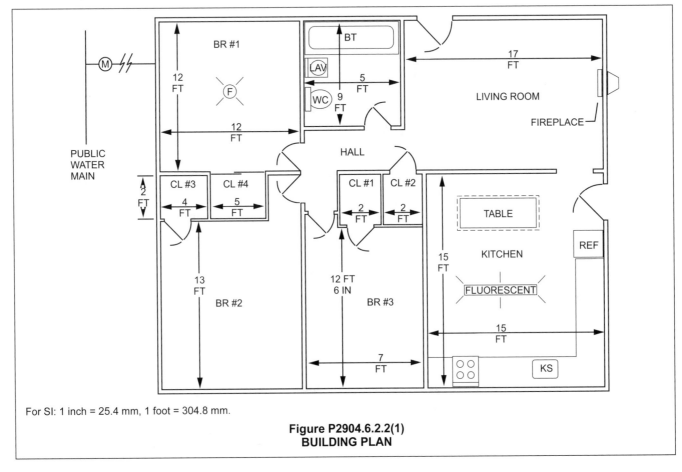

For SI: 1 inch = 25.4 mm, 1 foot = 304.8 mm.

Figure P2904.6.2.2(1)
BUILDING PLAN

is not enough pressure in order to use the tables. Therefore, some recalculation is necessary in order to increase the pressure remaining to accommodate pipe friction losses.

In review of the previous calculations, the water service pressure loss at 18.7 psi and the PRV pressure loss at 22 psi are big losses that could be reduced by choosing larger sizes. Therefore, a $1^1/_4$-inch water service line is chosen, which has a pressure loss of only 7 psi and a 1-inch PRV is chosen, which has a pressure loss (according to the manufacturer) of only 16 psi. Substituting the revised values into the equation:

$P_t = 105 - 7 - 11 - 16 - 17.4 - 22.6$

$P_t = 31$ psi

Solution—Part C

From the building drawings, determine the length of piping from the water service line entry point to most remote sprinkler [see Commentary Figure P2904.6.2.2(3)]. Starting from the below-grade entry of the water service line into the structure, the line runs up 4 feet to the bottom of the joists, over 26 feet, over 10 feet and up 9 feet for a total developed length of 49 feet. Because CPVC piping will be used, Table P2904.6.2(6) will be the starting point for determining what size pipe is required. Enter in the left-hand column of the table and select a row that is equal to or

greater than 19 gpm (the required flow rate or the highest demand sprinkler). The 19 gpm row is chosen. Enter the topmost row having the available pressure values (15, 20, 30, etc.) and identify the column that is equal to or less than 31 psi. The 30 psi column is chosen. At the intersection of the chosen row and column, identify the value. In this example, the value of 141 feet is found. This value represents the maximum length of $^3/_4$-inch (19 mm) CPVC piping allowed from the water service entrance point into the building to the most remote sprinkler. Because 141 feet (42 977 mm) is greater than the actual 49 feet of pipe, the design is acceptable using $^3/_4$-inch (19 mm) piping for the piping to every sprinkler.

P2904.7 Instructions and signs. An owner's manual for the fire sprinkler system shall be provided to the owner. A sign or valve tag shall be installed at the main shutoff valve to the water distribution system stating the following: "Warning, the water system for this home supplies fire sprinklers that require certain flows and pressures to fight a fire. Devices that restrict the flow or decrease the pressure or automatically shut off the water to the fire sprinkler system, such as water softeners, filtration systems and automatic shutoff valves, shall not be added to this system without a review of the fire sprinkler system by a fire protection specialist. Do not remove this sign."

❖ A fire sprinkler system is not unlike other types of mechanical systems in the home that require maintenance. An owner's manual might include details as to

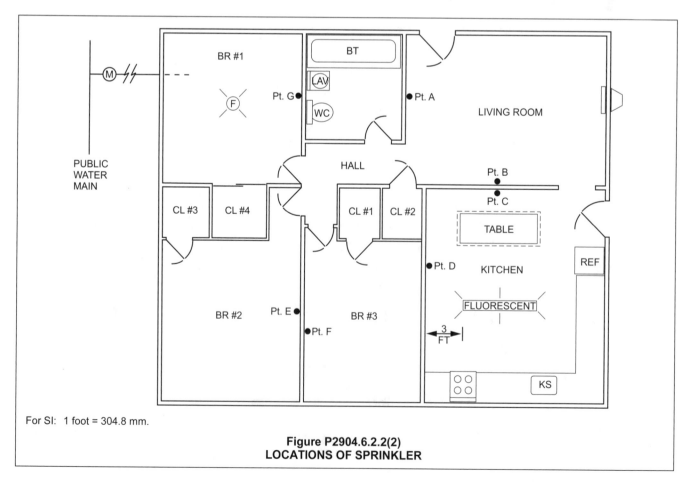

For SI: 1 foot = 304.8 mm.

Figure P2904.6.2.2(2)
LOCATIONS OF SPRINKLER

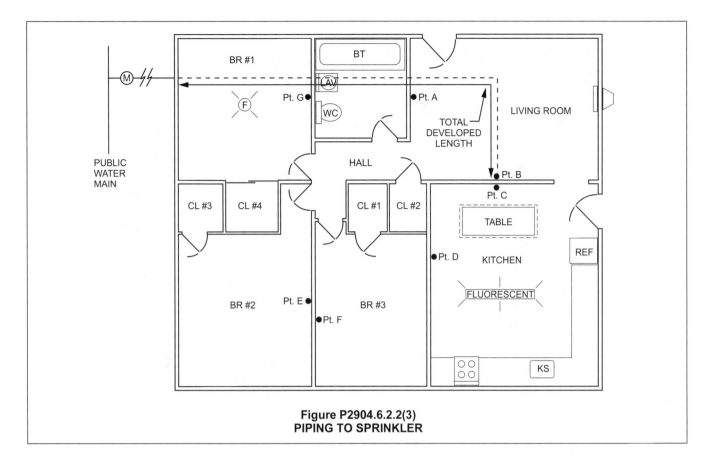

Figure P2904.6.2.2(3)
PIPING TO SPRINKLER

the model of sprinklers used for each location, specification sheets for the sprinklers used, calculation sheets, sprinkler replacement procedures and warnings. The required warning tag is necessary to alert service personnel that modifications to the system might require that the system design be reviewed by someone who is familiar with the design of fire sprinkler systems.

P2904.8 Inspections. The water distribution system shall be inspected in accordance with Sections P2904.8.1 and P2904.8.2.

❖ In order for sprinkler systems to be effective when needed, the installation must be inspected prior to concealment and at completion.

P2904.8.1 Preconcealment inspection. The following items shall be verified prior to the concealment of any sprinkler system piping:

1. Sprinklers are installed in all areas as required by Section P2904.1.1.

2. Where sprinkler water spray patterns are obstructed by construction features, luminaires or ceiling fans, additional sprinklers are installed as required by Section P2904.2.4.2.

3. Sprinklers are the correct temperature rating and are installed at or beyond the required separation distances from heat sources as required by Sections P2904.2.1 and P2904.2.2.

4. The pipe size equals or exceeds the size used in applying Tables P2904.6.2(4) through P2904.6.2(9) or, if the piping system was hydraulically calculated in accordance with Section P2904.6.1, the size used in the hydraulic calculation.

5. The pipe length does not exceed the length permitted by Tables P2904.6.2(4) through P2904.6.2(9) or, if the piping system was hydraulically calculated in accordance with Section P2904.6.1, pipe lengths and fittings do not exceed those used in the hydraulic calculation.

6. Nonmetallic piping that conveys water to sprinklers is listed for use with fire sprinklers.

7. Piping is supported in accordance with the pipe manufacturer's and sprinkler manufacturer's installation instructions.

8. The piping system is tested in accordance with Section P2503.7.

❖ This section provides a checklist of items to verify before the sprinkler system piping is concealed [see Commentary Figures P2904.8.1(1) and P2904.8.1(2)].

Figure P2904.8.1(1)
SPRINKLER SYSTEM ROUGH-IN

Figure P2904.8.1(2)
SPRINKLER SHOWN WITHOUT TEMPORARY PROTECTIVE COVER INSTALLED

P2904.8.2 Final inspection. The following items shall be verified upon completion of the system:

1. Sprinkler are not painted, damaged or otherwise hindered from operation.

2. Where a pump is required to provide water to the system, the pump starts automatically upon system water demand.

3. Pressure-reducing valves, water softeners, water filters or other impairments to water flow that were not part of the original design have not been installed.

4. The sign or valve tag required by Section P2904.7 is installed and the owner's manual for the system is present.

❖ This section provides a checklist of items to verify at final inspection.

SECTION P2905
HEATED WATER DISTRIBUTION SYSTEMS

P2905.1 Heated water circulation systems and heat trace systems. Circulation systems and heat trace systems that are installed to bring heated water in close proximity to one or more fixtures shall meet the requirements of Section N1103.5.1.

❖ This section is meant to alert the plumbing community that there are requirements (and prohibitions) in Section N1103.5.1 for hot water circulation systems.

P2905.2 Demand recirculation systems. Demand recirculation water systems shall be in accordance with Section N1103.5.2.

❖ This section is meant to alert the plumbing community that there are requirements in Section N1103.5.2 for demand circulation water systems. See the Chapter 11 definition of "Demand recirculation water systems."

SECTION P2906
MATERIALS, JOINTS AND CONNECTIONS

P2906.1 Soil and groundwater. The installation of water service pipe, water distribution pipe, fittings, valves, appurtenances and gaskets shall be prohibited in soil and groundwater that is contaminated with solvents, fuels, organic compounds or other detrimental materials that cause permeation, corrosion, degradation or structural failure of the water service or water distribution piping material.

❖ This section requires that soil be investigated for contaminants, based on the material selected. When a pipe is buried, the surrounding soil conditions may cause a component of the system to corrode or degrade at an accelerated rate. The soil must be evaluated, or historical knowledge of the soil applied, to determine whether additional requirements are necessary to protect the piping material. Most soils are corrosive to galvanized steel pipe; therefore, additional protection would be required. The pipe is either coated or wrapped with coal tar, or an elastomeric or epoxy coating is applied to the exterior of the pipe. Some soils affect copper tubing, and soil with cinders will attack unprotected copper pipe or tubing. The pipe or tube must therefore be coated with a protective layer. Thermoplastic piping can degrade quickly in soil that contains heavy concentrations of hydrocarbons.

P2906.1.1 Investigation required. Where detrimental conditions are suspected by or brought to the attention of the *building official*, a chemical analysis of the soil and groundwater conditions shall be required to ascertain the acceptability of the water service material for the specific installation.

❖ If historical knowledge of the soil or a visual inspection reveals the possibility of contaminants being present, a chemical analysis is required. Based on the results of such analysis, approved materials must be installed.

P2906.1.2 Detrimental condition. Where a detrimental condition exists, *approved* alternate materials or alternate routing shall be required.

❖ If a detrimental soil condition exists, either an alternative material must be used or the piping must bypass the detrimental soil.

P2906.2 Lead content. The lead content in pipe and fittings used in the water supply system shall be not greater than 8 percent.

❖ An amendment of the Safe Drinking Water Act (SDWA) occurred in 2011 to further reduce the amount of lead in plumbing products. The amendment became law in January of 2014. The amendment changes the meaning of "lead free" from 8 percent or less to 0.25 percent or less. However, the law also changes the determination of lead content to a "weighted average of wetted surfaces method." Therefore, the SDWA no longer requires laboratory testing of water exposed to products to determine the lead content. However, this section remains in the code.

P2906.2.1 Lead content of drinking water pipe and fittings. Pipe, pipe fittings, joints, valves, faucets and fixture fittings utilized to supply water for drinking or cooking purposes shall comply with NSF 372 and shall have a weighted average lead content of 0.25 percent lead or less.

❖ In recent years, additional health study data has been gathered to indicate that limiting the lead content to not greater than 8 percent might not be restrictive enough to avoid human health problems. Initially, two states (California and Vermont) enacted legislation to require the lead content of pipe, pipe fittings, plumbing fittings and fixture fittings (faucets) to be not greater than 0.25 percent (by wetted surface weighted average method). The United States Environmental Protection Agency followed up with a similar recommendation for all states and the "Low Lead" amendment to the Safe Drinking Water Act became law in January, 2014. See commentary to Section 605.2.

Neither NSF 372 nor the federal law requires low-lead compliant products to be marked or identified in any particular manner. Identification markings of "low-lead" products are not standardized among manufacturers or third-party certification agencies. Some manufacturers might make both low-lead compliant products and products that are intended to be used in systems such as hydronic heating systems and nonpotable water systems that do not require low-lead compliant products. Contractors will need to be aware of the difference to avoid using the wrong product, such as a noncompliant valve, in an application where a low-lead compliant valve is required. Code officials will need to be knowledgeable about various product markings and vigilant for improper application errors when making inspections. Other manufacturers that make products such as kitchen faucets, lavatory faucets and drinking fountains that are intended for sale in areas covered by federal law will have only the low-lead compliant versions available for sale in those areas. However, with the increasing global access to plumbing products, contractors and code officials will need to be careful about the products that are being installed to ensure that federal law is not violated. This code's requirement for third-party certification to NSF 372 will make such verifications easier. Evaluation of products for compliance with the low-lead federal law (and NSF 372) is by calculation method only.

P2906.3 Polyethylene plastic piping installation. Polyethylene pipe shall be cut square using a cutter designed for plastic pipe. Except where joined by heat fusion, pipe ends shall be chamfered to remove sharp edges. Pipe that has been kinked shall not be installed. For bends, the installed radius of pipe curvature shall be greater than 30 pipe diameters or the coil radius where bending with the coil. Coiled pipe shall not be bent beyond straight. Bends within 10 pipe diameters of any fitting or valve shall be prohibited. Joints between polyethylene plastic pipe and fittings shall comply with Section P2906.3.1 or P2906.3.2.

❖ Any properly made pipe joint requires the pipe ends be cut square for proper alignment, proper insertion depth and adequate surface area for joining. Because polyethylene plastic (PE) pipe and tubing can be kinked and stress weakened, a minimum pipe bending radius is specified so the structural capacity of the material is not exceeded. The minimum bending radius must not be less than the radius of the pipe coil as it came from the manufacturer or less than 30 pipe diameters, whichever is larger (see Commentary Figure P2906.3).

The pipe can be straightened as it is taken from the coil, but it cannot be bent in a direction opposite the bending direction of the coil (see Commentary Figure P2906.3). To avoid stress on connected fittings or valves, the location of pipe bends is restricted to specific distances from such components.

P2906.3.1 Heat-fusion joints. Joint surfaces shall be clean and free from moisture. Joint surfaces shall be heated to melting temperature and joined. The joint shall be undisturbed until cool. Joints shall be made in accordance with ASTM D2657.

❖ Polyethylene (PE) pipe cannot be solvent welded; therefore, it is joined by thermal welding (heat fusion) or by mechanical joints.

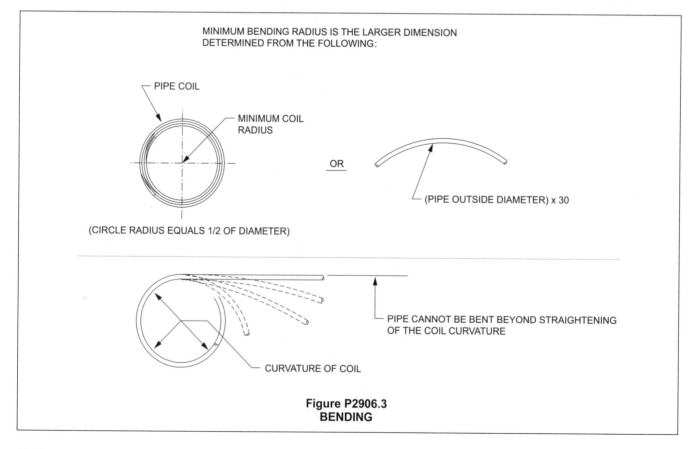

Figure P2906.3
BENDING

P2906.3.2 Mechanical joints. Mechanical joints shall be installed in accordance with the manufacturer's instructions.

❖ Mechanical joints include insert fittings and compression fittings.

P2906.4 Water service pipe. Water service pipe shall conform to NSF 61 and shall conform to one of the standards indicated in Table P2906.4. Water service pipe or tubing, installed underground and outside of the structure, shall have a working pressure rating of not less than 160 pounds per square inch at 73°F (1103 kPa at 23°C). Where the water pressure exceeds 160 pounds per square inch (1103 kPa), piping material shall have a rated working pressure equal to or greater than the highest available pressure. Water service piping materials not third-party certified for water distribution shall terminate at or before the full open valve located at the entrance to the structure. Ductile iron water service piping shall be cement mortar lined in accordance with AWWA C104/A21.4.

❖ Water service pipe must conform to the requirements of NSF 61 and to the requirements for one of the materials listed in Table P2906.4. See the definition of "Water service pipe" in Chapter 2. Materials that will be underground and outside of the building must have a working pressure rating of at least 160 psi (1103 kPa) and not less than the highest available pressure in the supply main to which they connect [see Commentary Figure P2906.4(1)].

TABLE P2906.4. See below.

❖ The referenced table specifies allowable piping materials for water service. It also identifies the standards with which these materials must comply.

ACRYLONITRILE BUTADIENE STYRENE (ABS) PLASTIC PIPE: ABS pipe is rarely manufactured for water service. However, it is approved for use outside the structure. The pipe is black and must be rated for distribution of potable water. ASTM D1527 is the referenced standard for Schedules 40 and 80 ABS pipe. The schedule number indicates the wall thickness; the higher the schedule number, the thicker the pipe wall. ASTM D2282 is the referenced standard for SDR-PR pipe. SDR is an abbreviation for "standard (pipe) dimension ratio," which is the ratio of pipe diameter to wall thickness. The PR designation indicates that the pipe is "pressure rated."

COPPER AND COPPER ALLOY PIPE: Copper and copper alloy pipe is approved for both water service and water distribution. The piping material is composed of approximately 85 percent copper and 15 percent zinc. The pipe is available in both standard and extra strong weight. Extra strong weight pipe has the same outside diameter as standard weight pipe, but with a thicker pipe wall.

Standard weight pipe is often referred to as "Schedule 40," and extra strong pipe is often referred to as "Schedule 80." Brass pipe dimensions are very similar to steel pipe dimensions.

Section P2904.5 requires copper and copper alloy pipe to comply with NSF 61, which restricts lead content to very small concentrations.

COPPER OR COPPER-ALLOY TUBING (TYPE K, WK, L, WL, M OR WM): This material is approved for both water service and water distribution. The tubing

TABLE P2906.4
WATER SERVICE PIPE

MATERIAL	STANDARD
Acrylonitrile butadiene styrene (ABS) plastic pipe	ASTM D1527; ASTM D2282
Chlorinated polyvinyl chloride (CPVC) plastic pipe	ASTM D2846; ASTM F441; ASTM F442; CSA B137.6
Chlorinated polyvinyl chloride/aluminum/chlorinated polyvinyl chloride (CPVC/AL/CPVC) plastic pipe	ASTM F2855
Copper or copper-alloy pipe	ASTM B42; ASTM B43; ASTM B302
Copper or copper-alloy tubing (Type K, WK, L, WL, M or WM)	ASTM B75; ASTM B88; ASTM B251; ASTM B447
Cross-linked polyethylene/aluminum/cross-linked polyethylene (PEX-AL-PEX) pipe	ASTM F1281; ASTM F2262; CSA B137.10
Cross-linked polyethylene/aluminum/high-density polyethylene (PEX-AL-HDPE) pipe	ASTM F1986
Cross-linked polyethylene (PEX) plastic tubing	ASTM F876; ASTM F877; CSA B137.5
Ductile iron water pipe	AWWA C115/A21.15; AWWA C151/A21.51
Galvanized steel pipe	ASTM A53
Polyethylene/aluminum/polyethylene (PE-AL-PE) pipe	ASTM F1282; CSA B137.9
Polyethylene (PE) plastic pipe	ASTM D2104; ASTM D2239; AWWA C901; CSA B137.1
Polyethylene (PE) plastic tubing	ASTM D2737; AWWA C901; CSA B137.1
Polyethylene of raised temperature (PE-RT) plastic tubing	ASTM F2769
Polypropylene (PP) plastic tubing	ASTM F2389; CSA B137.11
Polyvinyl chloride (PVC) plastic pipe	ASTM D1785; ASTM D2241; ASTM D2672; CSA B137.3
Stainless steel (Type 304/304L) pipe	ASTM A312; ASTM A778
Stainless steel (Type 316/316L) pipe	ASTM A312; ASTM A778

is manufactured in two different tempers, which are identified as drawn (called hard copper with the designation "H") and annealed (called soft copper with the designation "O"). Hard (drawn) copper tubing comes in straight lengths. Soft (annealed) copper tubing comes in both straight lengths and coils.

Annealed copper and tempered drawn copper may be formed by bending. Bending tempered copper is identified on the pipe with the designation "BT." Only bending tempered copper tubing may be joined by flared connections.

Seamless copper water tubing is available in Type K, L or M. The tubing type indicates the wall thickness. Type K has the greatest wall thickness, followed by Types L and M. The outside diameter of the tubing is the same for all three types; only the inside diameter varies with wall thickness. Welded copper tubing has the designation "WK," "WL" or "WM." The "W" indicates that the tubing is welded-seam. The pipe dimensions correspond to Type K, L or M seamless tubing.

The types of copper tubing are continuously identified with a color marking for ease of identification. These color markings are as follows:

- Type K (WK)-Green
- Type L (WL)-Blue
- Type M (WM)-Red

CHLORINATED POLYVINYL CHLORIDE (CPVC) PLASTIC PIPE: CPVC is approved for both water service and water distribution. CPVC raw material used to produce pipe is a polyvinyl chloride (PVC) that has been chlorinated to improve the material characteristics. The resulting pipe is more resistant to temperature extremes than PVC pipe.

CPVC plastic pipe may be either a white or a milky white (cream) color. The pipe listed is continuously marked with the manufacturer's name, the ASTM standard, and CPVC 4120 or CPVC 41, followed by two additional numbers. The designation CPVC 4120 identifies the quality of the material used to produce the pipe. The pipe is made with Grade 23447 plastic material. The grade number is a code for the quality of the pipe. The first digit indicates that the base resin is of CPVC. The remaining four digits indicate the impact strength, tensile strength, modulus of elasticity and deflection temperature.

Grade 23445 was previously designated Type IV, Grade 1, which was shortened to CPVC 41. The last two digits in CPVC 4120 indicate the hydrostatic design stress in hundreds of pounds per square inch. CPVC 4120 has a hydrostatic design stress of 2,000 psi (13 790 kPa).

ASTM D2846 is the referenced standard for SDR 11 CPVC plastic pipe and CPVC socket fittings. ASTM F441 is the referenced standard for Schedules 40 and 80 plastic pipe. The dimensions of the pipe are the same as Schedule 40 and 80 steel pipe. The referenced standard for SDR 13.5, 17, 21, 26 and 32.5 CPVC pipe is ASTM F442. The pipe is designated "PR," indicating that it is pressure rated.

CHLORINATED POLYVINYL CHLORIDE/ALUMINUM/CHLORINATED POLYVINYL CHLORIDE (CPVC/AL/CPVC) TUBING: The CPVC material used to produce CPVC/AL/CPVC tubing has similar characteristics as CPVC tubing. This pipe is a three-layer assembly that has aluminum as the middle layer. The outside color is off-white. The three layers are bonded together with melt adhesive. The main advantage of

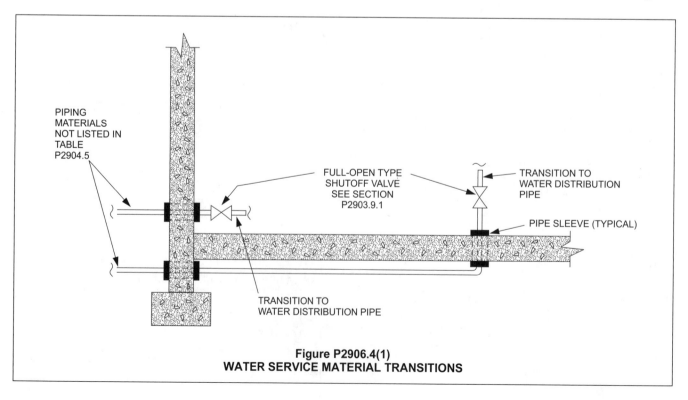

Figure P2906.4(1)
WATER SERVICE MATERIAL TRANSITIONS

this type of pipe is that it is "form stable," which means that the piping can be field bent to make permanent bends.

Bends can be made as small as 6 times the outside diameter of the tubing. The tubing can be formed to make a 90-degree bend (requiring a 135-degree "over bend" to make the bend). Bending tools should be used for making such "tight" bends to avoid collapse of the pipe.

The tubing must be cut with a roll-type cutting tool and not with a scissor-type cutter. The cut ends of the tube must have a bushing (marked F2855) cemented into the end of the tubing to prevent delamination at the tube end.

The thermal expansion is approximately 2.2×10^5 inches/(inch · °F), which is less than for CPVC tubing.

The outside diameter (OD) dimensions of the tubing product are copper tube size (CTS) $1/_2$, $3/_4$ and 1 inch. The inside diameters (IDs) of the tubing are 0.448, 0.691 and 0.932 inch, respectively.

Section 303.3 requires identification of all plastic pipe and fittings with the mark of an approved agency establishing conformance to NSF 14 (see commentary, Section P2609.3). Where used for drinking water, the product must be certified to NSF 61.

The product is marked at intervals of not more than 5 feet with the following information:

- The manufacturer's name.
- CPVC/AL/CPVC TUBING.
- 0.079 minimum wall.
- CAUTION: USER SHALL FIRST SOLVENT CEMENT F2855 BUSHINGS INTO CUT ENDS OF TUBING PRIOR TO JOINING.
- The standard designation (ASTM F2855).
- Certification for conformance to NSF 61.
- Nominal size.
- Type of plastic material (CPVC 23447).
- Pressure rating (100 psi at 180 psi).

DUCTILE-IRON WATER PIPE: Piping material is identified as either cast iron or ductile iron. Both materials are within the classification of cast iron. Ductile iron is a cast iron that is higher in ductility, hence the name. It is not as brittle as cast-iron pipe. Cement-mortar lining is necessary to provide a protective barrier between the potable water supply and the ductile-iron pipe to prevent impurities and contaminants from leaching into the water supply.

GALVANIZED STEEL PIPE: There is a wide variety of pipe classified as galvanized steel pipe. The grade of steel (thickness) can differ greatly. These thicknesses are referred to as Schedules 40, 80 and 160. As the schedule number increases, the wall thickness of the pipe increases. The outside diameter, however, remains the same for each pipe size.

Galvanized steel pipe is sometimes called "galvanized pipe" or mistakenly called "wrought-iron pipe." Wrought-iron pipe has not been manufactured in the

United States in more than 30 years. All pipe referred to as "wrought-iron pipe" is steel pipe.

POLYETHYLENE (PE) PLASTIC PIPE AND TUBING: PE pipe and tubing is approved for water service. Like polybutylene (PB), PE is an inert polyolefin material. It is resistant to chemical action, making solvent-cement joining impossible. PE pipe is blue or black when supplied for water service. Orange or yellow-colored PE is typically rated only for gas pipe installations. The pipe is continuously marked with the manufacturer's name, the ASTM or the Canadian Standards Association (CSA) standard and the grade of PE.

Various grades of polyethylene resin are used to produce pipe or tubing. In accordance with the ASTM standard, the material is identified by the term "PE," followed by four digits. The first digit indicates the type of material based on the density. The second digit is the category identifying the extrusion flow rate. The last two digits show the hydrostatic design stress in hundreds of pounds per square inch. The hydrostatic design stress is not an indication of the pressure rating of the pipe.

ASTM D2737 is for PE tubing with an SDR of 7.3, 9 and 11. The SDR for tubing is based on the average outside diameter. The grades of polyethylene permitted are PE 2305, PE 2306, PE 3306, PE 3406 and PE 3408.

CROSS-LINKED POLYETHYLENE (PEX) PLASTIC TUBING: PEX tubing is approved for both water service and water distribution, except that PEX tubing that conforms to ASTM F876 may not be used for water distribution. The material is available in sizes [National Pipe Size (NPS)] $1/_4$ inch through 2 inches (6 mm through 51 mm) and a standard dimension ratio (SDR 9) [see Commentary Figure P2906.4(2)].

Unlike most other pipe materials, PEX can be produced by several different manufacturing processes, such as the Engel, Silane, Monosil, Point a'Mousson, Daoplas and electron beam methods. Although each manufacturing process can yield tubing that complies with the referenced standards, each process produces a tube that possesses unique characteristics. The cross-linked molecular structuring gives the pipe additional resistance to rupture over a wider range of temperatures and pressures than that of other polyolefin plastics [PE, PB, and polypropylene (PP)]. Because of PEX pipe's unique molecular structure and resistance to heat, heat fusion is not permitted as a joining method. Because PEX is a member of the polyolefin plastic family, it is resistant to solvents. Therefore, the pipe cannot be joined by solvent cementing.

PEX pipe is flexible and can be bent using two methods: hot and cold bending. For hot bending, a hot-air gun with a diffuser nozzle is used. The pipe cannot be exposed to an open flame. The hot air meeting the pipe surface must not exceed 338°F (170°C), and the heat-up time must not exceed 5 minutes. The pipe is bent using conventional methods and is allowed to cool to room temperature before removal from the bending tool. The minimum hot-bending radius is two and one-half times the outside diameter. PEX can be bent at room temperature (cold bending) to a minimum radius of six times the outside diameter.

Mechanical connectors and fittings for PEX pipe are proprietary and must only be used with the pipe for which they have been designed and tested. Fitting systems currently available include compression fittings with inserts and ferrules or O-rings, insert fittings using a metallic lock (crimp) ring, and insert fittings with compression collars using a cold flaring tool. It is important to consult the manufacturer's installation instructions to identify the fittings authorized for use with PEX piping.

CROSS-LINKED POLYETHYLENE/ALUMINUM/ CROSS-LINKED POLYETHYLENE (PEX-AL-PEX) PIPE: PEX-AL-PEX is approved for both water distribution and water service. PEX-AL-PEX is a composite pipe made of an aluminum tube laminated to interior and exterior layers of PE. The layers are bonded together with an adhesive.

The cross-linked molecular structure gives the pipe additional resistance to rupture that is superior to that of PE (see the commentary to "cross-linked polyethylene plastic tubing" in this section). Therefore, the pipe is suitable for hot and cold water distribution and is pressure rated for 125 psi at 180°F (862 kPa at 82°C).

Although it is partially plastic, the PEX-AL-PEX pipe resembles metal tubing in that it can be bent by hand or with a suitable bending device and maintains its shape without fittings or supports. The minimum bending radius specified by manufacturers is five times the outside diameter.

Mechanical joints are the only methods currently available to join PEX-AL-PEX pipe. A number of pro-prietary mechanical-compression-type connectors have been developed for use with the composite pipe to permit transition to other pipes and fittings. Such fittings must be installed in accordance with the manufacturer's instructions.

POLYETHYLENE/ALUMINUM/POLYETHYLENE (PE-AL-PE) PIPE: PE-AL-PE is identical to PEX-AL-PEX composite pipe except for the physical properties of PE. PE-AL-PE displays the same resistance to temperature and pressure as PEX-AL-PEX does; therefore, it is suitable for hot and cold water distribution and is pressure rated for 125 psi at 180°F (862 kPa at 82°C) (see the commentary to "PEX-AL-PEX" in this section).

POLYVINYL CHLORIDE (PVC) PLASTIC PIPE: PVC water service pipe is a different material from PVC drainage pipe, although both materials are white. The pipe must be continuously marked with the manufacturer's name, ASTM standard and the grade of PVC material.

A number of grades of PVC material are used to produce pipe. In accordance with the ASTM standards, the compounds are identified as PVC 12454-B, 12454-C and 14333-D. The first digit indicates the base resin, with the following four digits identifying the impact strength, modulus of elasticity and deflection temperature. The letter suffix indicates the material's chemical resistance. The compounds were previously identified by type and grade: 12454-B was Type 1; 12454-C was Type 1, Grade 2; and 14333-D was Type 2, Grade 1. The marking on the pipe lists the material grade by the term "PVC," followed by four digits. The first two digits use the previous type and grade numbers to identify the compound. The last two digits indicate the hydrostatic design stress in hundreds of pounds per square inch. The hydrostatic design stress is not an indication of the pressure rating of the pipe.

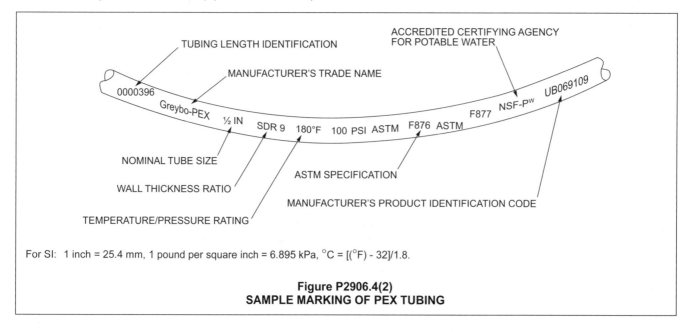

For SI: 1 inch = 25.4 mm, 1 pound per square inch = 6.895 kPa, °C = [(°F) - 32]/1.8.

Figure P2906.4(2)
SAMPLE MARKING OF PEX TUBING

ASTM D1785 is the referenced standard for Schedule 40, 80 and 160 PVC plastic pipe. These dimensions are the same as Schedule 40, 80 and 160 steel pipe. ASTM D2241 is the referenced standard for SDR-PR PVC plastic pipe. SDRs of the pipe are 13.5, 17, 21, 26, 32.5, 41 and 64.

STAINLESS STEEL: Stainless steel is a family of iron-based alloys that must contain at least 10 percent or more chromium. The presence of chromium creates an invisible surface film that resists oxidation and makes the material corrosion resistant. The fact that stainless steel has a great resistance to corrosion means that it will have a very long life compared to mild steel. The benefits of stainless steel includes corrosion resistance, fire and heat resistance, hygiene (easy cleaning ability), aesthetic appearance, strength-to-weight advantage, ease of fabrication, impact resistance and long-term value.

Wrought-austenitic stainless steels combine a useful combination of outstanding corrosion and heat resistance with good mechanical properties over a wide temperature range.

P2906.4.1 Separation of water service and building sewer. Trenching, pipe installation and backfilling shall be in accordance with Section P2604. Where water service piping is located in the same trench with the building sewer, such sewer shall be constructed of materials listed in Table P3002.1(2). Where the building sewer piping is not constructed of materials indicated in Table P3002.1(2), the water service pipe and the building sewer shall be horizontally separated by not less than 5 feet (1524 mm) of undisturbed or compacted earth. The required separation distance shall not apply where a water service pipe crosses a sewer pipe, provided the water service is sleeved to a point not less than 5 feet (1524 mm) horizontally from the sewer pipe centerline on both sides of such crossing. The sleeve shall be of pipe materials indicated in Table P2906.4, P3002.1(2) or P3002.2.

The required separation distance shall not apply where the bottom of the water service pipe that is located within 5 feet (1524 mm) of the sewer is not less than 12 inches (305 mm) above the highest point of the top of the building sewer.

❖ This section was completely reworded for the 2015 edition; however, the requirements were not changed. The language in the previous editions was complex and seemed difficult to understand in an exception format.

The section begins with the most common method to install water service and sewer piping to a building: one trench is excavated and both water service and the building sewer are installed in the same trench. In the majority of those situations, the building sewer is constructed of materials that are indicated in Table P3002.1(2). Materials in that table are suitable for installation inside of a building (those materials are also indicated in Table P3002.2 for building sewer piping). As Table P3002.1(2) materials are more robust than the piping materials indicated only for building sewers, there is a low probability of sewage leaking from the building sewer [see Commentary Figure P2906.4.1(1)].

The second sentence concerns building sewer piping that is of a material not indicated in Table P3002.1(2). In other words, materials indicated only in Table P3002.2. These piping materials are not as robust as Table P3002.1(2) piping materials and could leak sewage into the ground. Water service piping cannot be located in the same trench as these types of building sewer materials. Independent trenches must be excavated for the water service line and the building sewer so that not less than 5 feet of undisturbed earth are between the trenches. Compacted earth is considered to be the same as undisturbed earth [see Commentary Figure P2906.4.1(1)].

The third sentence is for the situation of a water service pipe crossing (either above or below) a building

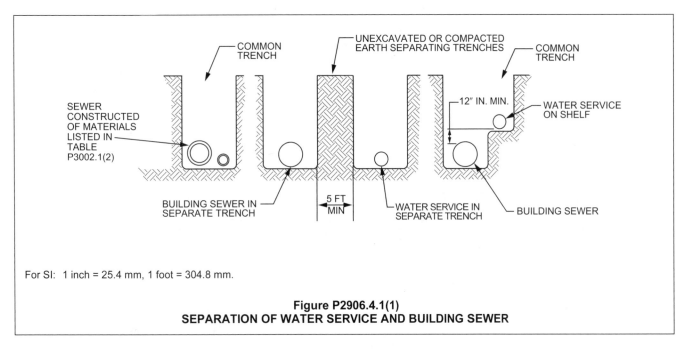

For SI: 1 inch = 25.4 mm, 1 foot = 304.8 mm.

Figure P2906.4.1(1)
SEPARATION OF WATER SERVICE AND BUILDING SEWER

sewer that is of piping materials indicated only in Table P3002.2. Although the code does not directly state that the building sewer piping material referred to in this sentence is only one of those indicated in Table P3002.2, it should be obvious because if the building sewer was of materials indicated in Table P3002.1(2), the water service piping could be in the same trench or anywhere around a building sewer material indicated in Table P3002.1(2). The water service piping has to be sleeved for a length of not less than 5 feet on either side of the building sewer so that if the building sewer did leak sewage into the earth, the water service piping is not directly exposed. The sleeve material can be any of the materials indicated in Table P2906.4, P3002.1(2) or P3002.1(2).

The last sentence concerns water service piping that is at an elevation that is not less than 12 inches above the highest part of the building sewer [where that sewer is of materials indicated only in Table P3002.1(2)]. If sewage leaks from a building sewer, gravity will cause the sewage to seep downwards into the earth and not upwards. Therefore, if the water service piping is above the highest part of the building sewer, there is little probability of sewage contamination of the earth around the water service piping [see Commentary Figure P2906.4.1(2)].

P2906.5 Water-distribution pipe. Water-distribution piping within *dwelling units* shall conform to NSF 61 and shall conform to one of the standards indicated in Table P2906.5. Hot-water-distribution pipe and tubing shall have a pressure rating of not less than 100 psi at 180°F (689 kPa at 82°C).

❖ This section refers to Table P2906.5 for approved piping material inside the building for both hot and cold water. The referenced table specifies materials that are suitable for water distribution piping. Hot water distribution piping must be rated at 100 psi (689 kPa) at 180°F (82°C) as opposed to 160 psi (1103 kPa) at 73°F (23°C), for water service piping. Note that all potable water piping and fittings must conform to NSF 61 so that all toxicity limitations imposed by that standard are met.

TABLE P2906.5. See below.

❖ See the commentary to Section P2906.5.

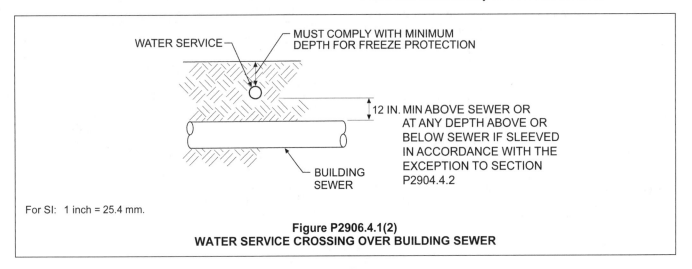

For SI: 1 inch = 25.4 mm.

Figure P2906.4.1(2)
WATER SERVICE CROSSING OVER BUILDING SEWER

TABLE P2906.5
WATER DISTRIBUTION PIPE

MATERIAL	STANDARD
Chlorinated polyvinyl chloride (CPVC) plastic pipe and tubing	ASTM D2846; ASTM F441; ASTM F442; CSA B 137.6
Chlorinated polyvinyl chloride/aluminum/chlorinated polyvinyl chloride (CPVC/AL/CPVC) plastic pipe	ASTM F2855
Copper or copper-alloy pipe	ASTM B42; ASTM B43; ASTM B302
Copper or copper-alloy tubing (Type K, WK, L, WL, M or WM)	ASTM B75; ASTM B88; ASTM B251; ASTM B447
Cross-linked polyethylene (PEX) plastic tubing	ASTM F876; ASTMF 877; CSA B137.5
Cross-linked polyethylene/aluminum/cross-linked polyethylene (PEX-AL-PEX) pipe	ASTM F1281; ASTM F2262; CSA B137.10
Cross-linked polyethylene/aluminum/high-density polyethylene (PEX-AL-HDPE) pipe	ASTM F1986
Galvanized steel pipe	ASTM A53
Polyethylene/aluminum/polyethylene (PE-AL-PE) composite pipe	ASTM F1282
Polyethylene of raised temperature (PE-RT) plastic tubing	ASTM F2769
Polypropylene (PP) plastic pipe or tubing	ASTM F2389; CSA B137.11
Stainless steel (Type 304/304L) pipe	ASTM A312; ASTM A778

P2906.6 Fittings. Pipe fittings shall be *approved* for installation with the piping material installed and shall comply with the applicable standards indicated in Table P2906.6. Pipe fittings used in water supply systems shall comply with NSF 61.

❖ Table 2906.6 can be used to determine the applicable standards for pipe fittings. In some cases, especially with certain plastic pipe or tubing systems, the manufacturer will specify the type of fittings to be used. Where such proprietary conditions exist, they must be adhered to so as not to violate the conditions of the material listing.

TABLE P2906.6. See below.

❖ This table specifies acceptable materials for pipe fittings along with the appropriate standards to which the fittings must comply. Each fitting is designed to be installed in a particular system with a given material or combination of materials. For example, drainage pattern fittings must be installed in drainage systems and vent fittings are limited to the venting system. Many fittings are intended for use only for water distribution systems. Fittings that may be installed in any type of plumbing system are also available.

Fittings must be of the same material as the pipe and must be compatible with the pipe. This avoids any chemical or corrosive action between dissimilar materials. Although a number of standards strictly regulate fittings, many pipe standards include them, as well. All pipe fittings must comply with the limitations set forth in NSF 61.

P2906.7 Flexible water connectors. Flexible water connectors, exposed to continuous pressure, shall conform to ASME A112.18.6/CSA B125.6. Access shall be provided to flexible water connectors.

❖ ASME A112.18.6/CSA B125.3 is the referenced standard for flexible water connectors used under continuous pressure. Such connectors typically consist of a flexible elastomeric tube covered by a metal or plastic braided jacket with termination fittings on each end. They are commonly used to connect the fixture fitting to the fixture supply pipe. Plastic materials coming into contact with potable water must comply with the applicable sections of NSF 14. Flexible connectors for delivery of drinking water must comply with NSF 61. Solder and fluxes containing lead in excess of 0.2 percent must not be used. Metal alloys must not exceed an 8-percent lead content. Copper-alloy components used with the assembled connector must contain not less than 58-percent copper.

Flexible water connectors are designed to function at water pressures up to 125 psi (862 kPa) and supply temperatures from 40°F to 180°F (4°C to 82°C).

TABLE P2906.6
PIPE FITTINGS

MATERIAL	STANDARD
Acrylonitrile butadiene styrene (ABS) plastic	ASTM D2468
Cast-iron	ASME B16.4
Chlorinated polyvinyl chloride (CPVC) plastic	ASSE 1061; ASTM D2846; ASTM F437; ASTM F438; ASTM F439; CSA B137.6
Copper or copper alloy	ASSE 1061; ASME B16.15; ASME B16.18; ASME B16.22; ASME B16.26; ASME B16.51
Cross-linked polyethylene/aluminum/high-density polyethylene (PEX-AL-HDPE)	ASTM F1986
Fittings for cross-linked polyethylene (PEX) plastic tubing	ASSE 1061; ASTM F877; ASTM F1807; ASTM F1960; ASTM F2080; ASTM F2098; ASTM F2159; ASTM F2434; ASTM F2735; CSA B137.5
Gray iron and ductile iron	AWWA C110/A21.10; AWWA C153/A21.53
Malleable iron	ASME B16.3
Insert fittings for Polyethylene/aluminum/polyethylene (PE-AL-PE) and cross-linked polyethylene/aluminum/cross-linked polyethylene (PEX-AL-PEX)	ASTM F1974; ASTM F1281; ASTM F1282; CSA B137.9; CSA B137.10
Polyethylene (PE) plastic	ASTM D2609; CSA B137.1
Fittings for polyethylene of raised temperature (PE-RT) plastic tubing	ASTM F1807; ASTM F2098; ASTM F2159; ASTM F2735; ASTM F2769
Polypropylene (PP) plastic pipe or tubing	ASTM F2389; CSA B137.11
Polyvinyl chloride (PVC) plastic	ASTM D2464; ASTM D2466; ASTM D2467; CSA B137.2; CSA B137.3
Stainless steel (Type 304/304L) pipe	ASTM A312; ASTM A778
Stainless steel (Type 316/316L) pipe	ASTM A312; ASTM A778
Steel	ASME B16.9; ASME B16.11; ASME B16.28

P2906.8 Joint and connection tightness. Joints and connections in the plumbing system shall be gas tight and water tight for the intended use or required test pressure.

❖ This section simply states that joints in a plumbing system must be air and water tight (see commentary, Section P2503.6).

P2906.9 Plastic pipe joints. Joints in plastic piping shall be made with *approved* fittings by solvent cementing, heat fusion, corrosion-resistant metal clamps with insert fittings or compression connections. Flared joints for polyethylene pipe shall be permitted in accordance with Section P2906.3.

❖ Plastic pipe may be joined in numerous ways depending on the material and the methods allowed by the manufacturer. This section provides the different acceptable methods of joining plastic pipe and specifies the various standards applicable in each case. Solvent cementing, heat fusion, clamps with insert fittings, compression connections and flared joints are all permitted methods of joining plastic pipe.

P2906.9.1 Solvent cementing. Solvent-cemented joints shall comply with Sections P2906.9.1.1 through P2906.9.1.4.

❖ Depending on the type of pipe, material, standards and methods to be applied are outlined in Sections P2906.9.1.1 through P2906.9.1.4.

P2906.9.1.1 ABS plastic pipe. Solvent cement for ABS plastic pipe conforming to ASTM D2235 shall be applied to all joint surfaces.

❖ Solvent cementing is the most common method of joining ABS plastic pipe and fittings. It is also one of the most misunderstood methods of joining pipe. The ASTM referenced standard indicates how to handle solvent cement and recommends a practice for making joints.

The pipe and fittings must be at approximately the same temperature prior to solvent cementing. The pipe end and socket fitting must be clean, dry and free of grease, oil and other foreign materials. The pipe end must be prepared as required by the manufacturer. Chamfering of the pipe end will help prevent binding as the pipe is inserted into the fitting and will also help prevent the cement from being scraped off the fitting socket during insertion. The solvent cement must be uniformly applied to the socket of the fitting and to the end of the pipe to the depth of the socket. Excessive solvent cement must be avoided to prevent a bead from forming on the inside of the pipe joint. The joint is made by twisting the pipe into the fitting socket immediately after applying the cement. The pipe must be held in place for a few seconds until the solvent cement begins to set. The pipe will tend to back out unless it is restrained. A small bead of solvent cement forms between the pipe exterior and fitting. The bead must be removed with a clean cloth to avoid the possibility of weakening the pipe wall. A solvent-cement joint must not be exposed to the working pressure for a period of 24 hours. The pipe must not be handled roughly for at least 1 hour after the joint is made.

Because each type of solvent cement is specifically designed for a given piping material, all-purpose solvent cement or universal solvent cement cannot be used to join ABS pipe or fittings unless it conforms to ASTM D2235. Solvent cement for each plastic pipe material requires a unique mixture or combination of solvents and dissolved plastic resins to conform to the standard. Although all-purpose cement may dissolve the surfaces of ABS, PVC and CPVC, it may be too aggressive or not aggressive enough to keep a joint strong enough to meet the minimum joint strength requirements of the referenced standard. Solvent cement must be identified on the can as complying with ASTM D2235.

P2906.9.1.2 CPVC plastic pipe. Joint surfaces shall be clean and free from moisture. Joints shall be made in accordance with the pipe, fitting or solvent cement manufacturer's installation instructions. Where such instructions require a primer to be used, an *approved* primer shall be applied, and a solvent cement, orange in color and conforming to ASTM F493, shall be applied to joint surfaces. Where such instructions allow for a one-step solvent cement, yellow or red in color and conforming to ASTM F493, to be used, the joint surfaces shall not require application of a primer before the solvent cement is applied. The joint shall be made while the cement is wet, and in accordance with ASTM D2846 or ASTM F493. Solvent cement joints shall be permitted above or below ground.

❖ Solvent cementing of CPVC piping must be performed in accordance with the pipe manufacturer's instructions using a solvent cement that complies with ASTM F493. Some pipe manufacturers may require specific brands or types of solvent cement be used for certain types and sizes of piping. The pipe and fitting surfaces to be joined must be free of dirt, grease, oil and any other foreign substances. Especially critical is the need for the joint surfaces to be dry, as solvent cement quickly congeals in the presence of water.

Some pipe manufacturer's instructions may require that a primer liquid be used to prepare the joint surfaces. Primer liquids break down the glossy surface finish of the plastic, help remove any remaining foreign substances from the joint area and begin to soften the plastic surfaces. Because of its softening capability, primer liquids must not be allowed to puddle inside of fittings. Some pipe manufacturers may require specific brands or types of primer liquids be used for certain types and sizes of piping. This section does not require primer liquids to be tinted with a color (such as purple or blue); however, a colored primer could be useful for determining if all joint surfaces have been primed.

Where priming of joint surfaces is required by the pipe manufacturer, the color of the solvent cement to be used must be orange. In general, solvent cement is applied to all joint surfaces immediately after the joint surfaces have been primed. Immediately after application of the solvent cement, the joint is assembled. The assembled joint is held in the assembled position for a period of time as required by the pipe manufacturer. Different sizes of piping require very specific assembly procedures and holding times. The piping manufacturer's instructions must be followed in every detail to ensure a proper joint.

Where the pipe manufacturer allows for a one-step solvent cementing process (in other words, priming of the joint surfaces is not required), the color of the solvent cement to be used must be either red or yellow. One-step solvent cementing can be in accordance with either ASTM D2846 or ASTM F493.

The two colors of one-step cement, red or yellow, deserve further explanation. It is not the intent that red or yellow solvent cements be used interchangeably on any type of CPVC piping. Red solvent cement is required by the piping manufacturer for CPVC fire sprinkler piping that is orange in color. Yellow solvent cement is required by the piping manufacturer for CPVC piping (other than CPVC fire sprinkler piping) where specified by the piping manufacturer that a one-step solvent cement can be used. Where CPVC multipurpose sprinkler systems are installed in a building, and the potable water distribution system of the building is also of CPVC material, there is an inherent need to connect the regular CPVC piping material to the orange CPVC fire sprinkler piping material at numerous locations throughout the building. CPVC adapters for solvent cementing on both ends are available to connect the Schedule 40 pipe-sized sprinkler piping to the copper tube size tubing of the distribution system. The adapter is all of the same material, typically regular (not orange) CPVC material. A dilemma sometimes arises as to which color of solvent cement to use at this connection. The intent of allowing either red- or yellow-colored solvent cement is to accommodate these solvent cemented connections in multipurpose sprinkler systems.

Assembled and cured joints can be located above or below the ground.

P2906.9.1.3 CPVC/AL/CPVC pipe. Joint surfaces shall be clean and free from moisture, and an *approved* primer shall be applied. Solvent cement, orange in color and conforming to ASTM F493, shall be applied to all joint surfaces. The joint shall be made while the cement is wet, and in accordance with ASTM D2846 or ASTM F493. Solvent-cement joints shall be installed above or below ground.

> **Exception:** A primer shall not be required where all of the following conditions apply:
>
> 1. The solvent cement used is third-party certified as conforming to ASTM F493.
>
> 2. The solvent cement used is yellow in color.
>
> 3. The solvent cement is used only for joining $^1/_2$-inch (12.7 mm) through 1-inch (25 mm) diameter CPVC/AL/CPVC pipe and CPVC fittings.
>
> 4. The CPVC fittings are manufactured in accordance with ASTM D2846.

❖ Solvent cementing of CPVC/AL/CPVC piping must be performed in accordance with the pipe manufacturer's instructions using a solvent cement that complies with

ASTM F493. Some pipe manufacturers may require specific brands or types of solvent cement be used for certain types and sizes of piping. The pipe and fitting surfaces to be joined must be free of dirt, grease, oil and any other foreign substances. Especially critical is the need for the joint surfaces to be dry, as solvent cement quickly congeals in the presence of water.

Priming of joint surfaces is required. Primer liquids break down the glossy surface finish of the plastic, help remove any remaining foreign substances from the joint area and begin to soften the plastic surfaces. Because of its softening capability, primer liquids must not be allowed to puddle inside of fittings. Some pipe manufacturers may require specific brands or types of primer liquids be used for certain types and sizes of piping. The color of the solvent cement to be used must be orange.

In general, solvent cement is applied to all joint surfaces immediately after the joint surfaces have been primed. Immediately after application of the solvent cement, the joint is assembled. The assembled joint is held in the assembled position for a period of time as required by the pipe manufacturer. Different sizes of piping require very specific assembly procedures and holding times. The piping manufacturer's instructions must be followed in every detail to ensure a proper joint.

A one-step solvent cementing process (in other words, priming of the joint surfaces is not required) can be used provided that all of the conditions of the exceptions are met.

Assembled and cured joints can be located above or below the ground.

P2906.9.1.4 PVC plastic pipe. A purple primer that conforms to ASTM F656 shall be applied to PVC solvent-cemented joints. Solvent cement for PVC plastic pipe conforming to ASTM D2564 shall be applied to all joint surfaces.

❖ Like CPVC, PVC plastic pipe requires a primer prior to solvent cementing. ASTM F656 recommends that PVC primer not be orange, insofar as this is the recommended color for CPVC solvent cement. The required purple coloring allows the installer and the inspector to verify that the required primer has been applied.

To distinguish the primer from the solvent cement, the solvent cement must be a different color than the primer. Solvent cement for PVC is typically clear.

The PVC solvent cement is applied to the pipe and socket fitting. The joint is made with a slight twisting motion until the pipe reaches the full depth of the socket fitting. It is held in place until the solvent cement begins to set. The installer should wear impervious gloves to prevent solvent cement and primer from coming into contact with the skin. The solvent-cementing procedure must be done in accordance with the referenced standard. To minimize the health and fire hazards associated with solvent cements and primers, the manufacturer's instructions for the use and handling of these materials must be followed.

P2906.9.1.5 Cross-linked polyethylene plastic (PEX). Joints between cross-linked polyethylene plastic tubing or fittings shall comply with Section P2906.9.1.5.1 or Section P2906.9.1.5.2.

❖ This section speaks specifically to PEX systems. Two methods of joining such material are outlined in the following subsections. Often, this type of material, and manufacturer's the fittings and joining methods are proprietary. The installation instructions must be followed and the specified fittings and tools must be used.

P2906.9.1.5.1 Flared joints. Flared pipe ends shall be made by a tool designed for that operation.

❖ Flared joints must be made with tools specific to the design of the connector system. The assembly method must be in accordance with the manufacturer's instructions specific to the type of connectors being used. Flare configurations will vary depending on the tool used. Using a cold flaring tool, wherein the tubing outside diameter is expanded mechanically, the flared end is then secured between the fitting components. The flared surface serves as the sealing surface between the tubing and fitting.

P2906.9.1.5.2 Mechanical joints. Mechanical joints shall be installed in accordance with the manufacturer's instructions. Fittings for cross-linked polyethylene (PEX) plastic tubing shall comply with the applicable standards indicated in Table P2906.6 and shall be installed in accordance with the manufacturer's instructions. PEX tubing shall be factory marked with the applicable standards for the fittings that the PEX manufacturer specifies for use with the tubing.

❖ ASTM F1807, ASTM F1960 and ASTM F2080 are the standards for metal insert fittings and copper crimp rings or cold-expansion fittings with PEX reinforcing rings or metal compression sleeves for use with PEX plastic tubing or pipe. These fittings are intended for use in 100 psi (689.5 kPa) cold and hot water distribution systems operating at temperatures up to and including 180°F (82°C). These fittings must be installed in accordance with the manufacturer's instructions. Insert fittings must be joined to PEX tubing by the compression of a copper crimp ring around the outer circumference of the tubing, forcing the tubing material into annular spaces formed by ribs on the fittings.

The crimping procedure involves sliding the crimp ring onto the tubing, then inserting the ribbed end of the fitting into the end of the tubing until the tubing contacts the shoulders of the fitting or tube stop. The crimp ring must then be positioned on the tubing so that the edge of the crimp ring is the required distance from the end of the tube. The jaws of the crimping tool must be centered over the crimp ring, and the tool must be held so that the crimping jaws are approximately perpendicular to the axis of the barb. The jaws of the crimping tool must be closed around the crimp ring, compressing the crimp ring onto the tubing. The crimp ring must not be crimped more than once.

P2906.10 Polypropylene (PP) plastic. Joints between polypropylene plastic pipe and fittings shall comply with Section P2906.10.1 or P2906.10.2.

❖ Like PEX, polypropylene (PP) cannot be solvent welded, therefore, mechanical and heat-fusion joints are the only options.

P2906.10.1 Heat-fusion joints. Heat fusion joints for polypropylene pipe and tubing joints shall be installed with socket-type heat-fused polypropylene fittings, butt-fusion polypropylene fittings or electrofusion polypropylene fittings. Joint surfaces shall be clean and free from moisture. The joint shall be undisturbed until cool. Joints shall be made in accordance with ASTM F2389.

❖ Heat fusion is the same as thermal welding and involves the use of special tools to heat and soften a surface of a pipe and fitting. Electrofusion fittings have a single-use heating element built into the socket for use with proprietary power supply units.

P2906.10.2 Mechanical and compression sleeve joints. Mechanical and compression sleeve joints shall be installed in accordance with the manufacturer's instructions.

❖ This section covers any type of joint that involves the mechanical application of force around the circumference of the pipe. Such joints generally use an elastomeric sealing element.

P2906.11 Cross-linked polyethylene/aluminum/cross-linked polyethylene. Joints between polyethylene/aluminum/polyethylene (PE-AL-PE) and cross-linked polyethylene/aluminum/cross-linked polyethylene (PEX-AL-PEX) pipe and fittings shall comply with Section P2906.11.1.

❖ See the commentary to Section P2905.11.1.

P2906.11.1 Mechanical joints. Mechanical joints shall be installed in accordance with the manufacturer's instructions. Fittings for PE-AL-PE and PEX-AL-PEX as described in ASTM F1974, ASTM F1281, ASTM F1282, CSA B137.9 and CSA B137.10 shall be installed in accordance with the manufacturer's instructions.

❖ The ASTM standards listed are for insert fittings with either crimp rings or stainless steel clamps. Manufacturer's instructions must be followed for the installation of the fittings to the pipe.

P2906.12 Stainless steel. Joints between stainless steel pipe and fittings shall comply with Section P2906.12.1 or P2906.12.2.

❖ Acceptable joints for stainless steel pipe or fittings are identified in the referenced sections.

P2906.12.1 Mechanical joints. Mechanical joints shall be installed in accordance with the manufacturer's instructions.

❖ Mechanical joints, such as compression joints, grooved couplings, hydraulic pressed fittings or flange designs, can be used to join stainless steel pipe and fittings, and are specified by the manufacturer. Such joints must be assembled and installed in compliance with the manufacturer's instructions.

P2906.12.2 Welded joints. Joint surfaces shall be cleaned. The joint shall be welded autogenously or with an *approved* filler metal in accordance with ASTM A312.

❖ The two basic methods for welding stainless steels are fusion welding and resistance welding. In fusion welding, heat is provided by an electric arc struck between an electrode and the metal to be welded. In resistance welding, bonding is the result of heat and pressure. Heat is produced by the resistance to the flow of electric current through the parts to be welded and pressure is applied by the electrodes.

Welds and the surrounding area should be thoroughly cleaned to avoid impairment of corrosion resistance. Weld spatter, flux or scale may become focal points for corrosive attack if not properly removed, especially in aggressive environments.

P2906.13 Threaded pipe joints. Threaded joints shall conform to American National Taper Pipe Thread specifications. Pipe ends shall be deburred and chips removed. Pipe joint compound shall be used only on male threads.

❖ Threaded pipe joints must conform to the American National Taper Pipe Thread requirements (ASME B1.20.1, referred to under the standards chapter). After cutting the threads, chips must be removed and the pipe must be reamed so that any remaining material that might obstruct the flow within the pipe is removed. It is imperative that the threads not exceed the maximum required for the size of the pipe. If there are too many threads, the improper amount of taper will not allow the threads to engage as intended. If there are not enough threads, the integrity of the joint is diminished.

Pipe joint compound must be applied to the male threads only. This prevents the compound from entering the piping system when the joints are made.

P2906.14 Soldered and brazed joints. Soldered joints in copper and copper alloy tubing shall be made with fittings approved for water piping and shall conform to ASTM B828. Surfaces to be soldered shall be cleaned bright. Fluxes for soldering shall be in accordance with ASTM B813. Brazing fluxes shall be in accordance with AWS A5.31M/A5.31. Solders and fluxes used in potable water-supply systems shall have a lead content of not greater than 0.2 percent.

❖ A soldered joint is the most common method of joining copper pipe and tubing. Pipe must be cut square for proper alignment, adequate surface area for joining and an interior free from obstructions.

When the pipe is cut, an edge (burr) is left protruding into the pipe. The pipe must be reamed to properly remove the burr. Chamfering is required to bevel the outer edge of the cut end of the pipe. Undercutting (excessive reaming) can reduce pipe wall thickness (see Commentary Figure P2906.14).

ASTM B828 governs the procedures for making joints by soldering of copper and copper tubing and fittings. Flux must be applied to all joint surfaces. Flux is a chemically active material that prevents metal oxides from forming in the joint area during heating, allowing

the melted solder to spread out onto the surfaces to be joined. The flux material must conform to ASTM B813, which establishes minimum performance criteria. It also limits the maximum lead content to 0.2 percent. Any solder used in connection with potable water systems must also be limited to a maximum lead content of 0.2 percent.

The solder will flow by capillary action toward the heat. The temperatures of the joint surfaces play a major role in the making of a properly soldered joint. The fitting and pipe must be approximately the same temperature when the solder is applied. Heating the copper liquefies the flux and solder, and expands the surface voids of the material, allowing the solder to penetrate. A strong bond is created when the solder solidifies. If the temperatures are incorrect or uneven, the solder will run down the inside of the pipe and fitting or onto the outside of the pipe. The result is a solder-lined pipe.

Lead-bearing solder can sometimes be visually identified because the solder tends to darken with age. Test kits are available that can be used to identify lead-bearing solders with absolute certainty.

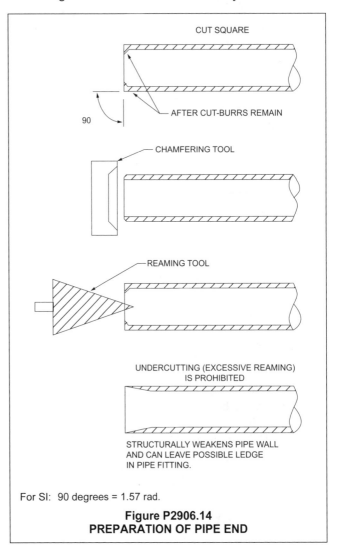

CUT SQUARE

AFTER CUT-BURRS REMAIN

90

CHAMFERING TOOL

REAMING TOOL

UNDERCUTTING (EXCESSIVE REAMING) IS PROHIBITED

STRUCTURALLY WEAKENS PIPE WALL AND CAN LEAVE POSSIBLE LEDGE IN PIPE FITTING.

For SI: 90 degrees = 1.57 rad.

Figure P2906.14
PREPARATION OF PIPE END

P2906.15 Flared joints. Flared joints in water tubing shall be made with *approved* fittings. The tubing shall be reamed and then expanded with a flaring tool.

❖ Outside of PEX materials, copper tube is the most common material that can be joined by the use of a flared joint. See the commentaries to Sections P2904.3 and P2904.9.1.4.1 for flared joint requirements pertaining to PE and PEX materials.

Because the pipe end is expanded in a flared joint, only annealed and bending-tempered drawn copper tubing may be flared. There are two types of flaring tools: hammered and a screw yoke and block. The hammer-type flaring tool is inserted into the pipe and hammered. With the screw yoke and block, the end of the pipe is firmly held by the block assembly. The screw yoke is attached, and a flaring surface is pressed into the pipe end, creating the flare. The flared surface of the pipe is compressed against the mating surface of the fitting by the flare nut to form a water-tight seal. Commonly, pipe-joint compound is placed on flare-fitting threads to lubricate the assembly (see Commentary Figure P2906.15).

Because hard-drawn pipe is subject to splitting when flared, a flared joint is restricted to soft (annealed) copper tubing. A flared joint is readily dismantled and is, in effect, a type of union connection.

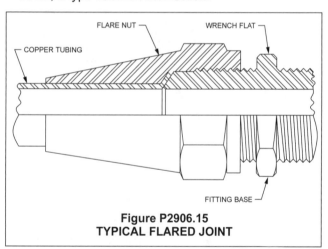

Figure P2906.15
TYPICAL FLARED JOINT

P2906.16 Above-ground joints. Joints within the building between copper pipe or CPVC tubing, in any combination with compatible outside diameters, shall be permitted to be made with the use of *approved* push-in mechanical fittings of a pressure-lock design.

❖ Push-in mechanical fittings of a pressure-lock design are allowed to be used to join copper pipe, or CPVC tubing, where piping is within the building.

In addition, joints between different materials must be installed in accordance with Sections P2905.17.1 through P2405.17.5 (see commentary, Sections P2905.17.1 through P2905.17.3).

P2906.17 Joints between different materials. Joints between different piping materials shall be made in accordance with Section P2906.17.1, P2906.17.2 or P2906.17.3, or with a mechanical joint of the compression or mechanical sealing type having an elastomeric seal conforming to ASTM D1869 or ASTM F477. Joints shall be installed in accordance with the manufacturer's instructions.

❖ Mechanical joints can be used to join dissimilar piping materials. Because many piping materials have the same outside diameter, a number of fittings are designed to join different piping materials. This section requires mechanical joints to be of the compression or mechanical sealing type having an elastomeric seal conforming to ASTM D1869 or ASTM F477. These types of joints must be capable of handling the required test pressure and must also be compatible with the material and its intended use (see commentary, Section P2503.6).

P2906.17.1 Copper or copper-alloy tubing to galvanized steel pipe. Joints between copper or copper-alloy tubing and galvanized steel pipe shall be made with a copper alloy fitting or dielectric fitting. The copper tubing shall be joined to the fitting in an *approved* manner, and the fitting shall be screwed to the threaded pipe.

❖ This section requires a method to protect against galvanic action when joining copper or copper-alloy tubing with galvanized steel pipe. Galvanic action occurs when two dissimilar metals come into contact with each other in the presence of an electrolyte, resulting in an electrical current that causes corrosion of one of the two metals. To stop galvanic action, this section requires either a dielectric coupling that includes an insulator (barrier) between the two metals or a copper alloy fitting that acts as a buffer and retards the galvanic action (see Commentary Figure P2906.17.1 for an illustration of the former).

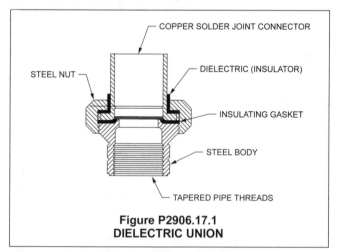

Figure P2906.17.1
DIELECTRIC UNION

P2906.17.2 Plastic pipe or tubing to other piping material. Joints between different types of plastic pipe or between plastic pipe and other piping material shall be made with an *approved* adapter fitting.

❖ The joining of different types of plastic pipe, and of plastic pipe to other piping materials, requires the use of adapter fittings specifically designed for such connections. The adaptor fitting must be compatible with the two dissimilar materials being joined.

Adapter fittings must be evaluated individually and approved by the code official. Some key items to review when evaluating adaptor fittings are joint tightness, compatibility of the adaptor with the dissimilar materials being joined, and the presence of any ledges, shoulders or reductions that could retard or obstruct the flow of water in the piping.

P2906.17.3 Stainless steel. Joints between stainless steel and different piping materials shall be made with a mechanical joint of the compression or mechanical-sealing type or a dielectric fitting.

❖ In theory, all metals immersed in an electrolyte, such as water, have a voltage potential. The relative activity of the metal is determined by voltage potential with those metals having negative voltages being the most active and more likely to corrode. Dielectric sleeves may be used with flange bolting or as gaskets for connections between different types of materials to prevent such corrosion. The plastic or rubber gasket engages the thread of the fitting attached, thereby insulating the metal from the water. This prevents corrosion on either part.

P2906.18 Press-connect joints. Press-connect joints shall conform to one of the standards indicated in Table P2906.6. Press-type mechanical joints in copper tubing shall be made in accordance with the manufacturer's instructions. Cut tube ends shall be reamed to the full inside diameter of the tube end. Joint surfaces shall be cleaned. The tube shall be fully inserted into the press connect fitting. Press-connect joints shall be pressed with a tool certified by the manufacturer.

❖ Press-connect joints are a specific type of mechanical joints in which the copper pipe is inserted into the fitting and a proprietary motorized tool deforms (presses) the fitting and the pipe creating a nonreversible joint. The fittings use an O-ring and are rated for a working pressure of 200 psi (1379 kPa) with a temperature limitation of 250°F (121°C).

P2906.19 Polyethylene of raised temperature plastic. Joints between polyethylene of raised temperature plastic tubing and fittings shall be in accordance with Section P2906.19.1.

❖ Polyethylene of raised temperature tubing can only be connected by using the mechanical joining method in Section P2906.19.1.

P2906.19.1 Mechanical joints. Mechanical joints shall be installed in accordance with the manufacturer's instructions. Fittings for polyethylene of raised temperature plastic tubing shall comply with the applicable standards listed in Table P2906.6 and shall be installed in accordance with the manufacturer's instructions. Polyethylene of raised temperature plastic tubing shall be factory marked with the applicable standards for the fittings that the manufacturer of the tubing specifies for use with the tubing.

❖ Mechanical joints for these materials include insert-type fittings, metallic lock-ring fittings, compression fittings and crimp-type fittings. Each fitting is intended for a specific pipe material and cannot be interchanged

unless specified by the pipe manufacturer. In all cases, the fittings connecting any type of polyethylene pipe must be assembled and installed in accordance with the manufacturer's instructions.

SECTION P2907
CHANGES IN DIRECTION

P2907.1 Bends. Changes in direction in copper tubing shall be permitted to be made with bends having a radius of not less than four diameters of the tube, provided that such bends are made by use of forming equipment that does not deform or create loss in cross-sectional area of the tube.

❖ Annealed copper and tempered drawn copper may be formed by bending. Whereas in most cases copper fittings will be used to change direction, "soft" copper can be bent to make the desired change in direction. Care must be exercised so that the minimum bending radius of four times the diameter of the copper tube is not exceeded. Although such a sharp bend is possible, the use of proper bending tools will be required to avoid a kink in the copper that would reduce both the cross-sectional area and the integrity of the material.

SECTION P2908
SUPPORT

P2908.1 General. Pipe and tubing support shall conform to Section P2605.

❖ This section points the reader to Section P2605, which is titled "Support" and contains provisions pertaining to the horizontal and vertical support of piping materials used within water distribution systems (see commentary to Section P2605.1 and Table P2605.1).

SECTION P2909
DRINKING WATER TREATMENT UNITS

P2909.1 Design. Drinking water treatment units shall meet the requirements of NSF42, NSF 44, NSF 53, NSF 60 or CSA B483.1.

❖ Water delivered to a building is expected to be within established guidelines for health, odor and chemical content. These guidelines are established by the U.S. Department of Health and Human Services and are applicable to all potable water sources, whether these sources are public municipal water treatment facilities or individual private wells. Although the guidelines make certain that the water is safe to consume, occasionally there are times when the water supply has an objectionable odor or dissolved materials within it that are noticeable to the occupants of the building, particularly in the drinking water.

NSF 42 contains the requirements for units that are designed to improve the perceived quality of the drinking water supplied to the occupants of the building.

These units are designed to reduce the amount of specific contaminants present in the water supply.

These contaminants affect the aesthetic quality of the water. Most water treatment units are designed to remove certain odors or dissolved materials in the drinking water supply.

The user or installer of the treatment unit is responsible for the selection of the proper unit based on the offensive characteristics of the water and the unit's reported performance capabilities. Treatment units require periodic maintenance so they will continue to function as designed. Because of this requirement, the treatment unit must be installed in a location where it is accessible for maintenance.

NSF 53, similar to the NSF 42 contains the standards for units designed to improve the quality of drinking water by removing dissolved materials or substances that could have a negative effect on the health of the occupants of the building. These materials or substances may be microbiological, chemical or particulate and are often considered to be either an established or potential health hazard.

NSF 44 contains requirements pertaining to cation (ion) exchange water softeners, which use charged ions in removing dissolved minerals.

P2909.2 Reverse osmosis drinking water treatment units. Point-of-use reverse osmosis drinking water treatment units, designed for residential use, shall meet the requirements of NSF 58 or CSA B483.1. Waste or discharge from reverse osmosis drinking water treatment units shall enter the drainage system through an *air gap* or an *air gap* device that meets the requirements of NSF 58.

❖ Reverse osmosis systems manufactured in accordance with NSF 58 are intended to remove specific contaminants from public or private drinking water supplies considered to be microbiologically safe. The treatment units designed in conformance to NSF 58 are intended to reduce the total dissolved solids (TDS) and other contaminants present in the drinking water. Many of these dissolved materials are considered to be either established or potential health hazards.

The objectionable dissolved materials in a reverse osmosis system are removed using pressure to force the water through a semipermeable membrane. In normal osmosis, water molecules would move across a membrane by the phenomenon of "osmotic pressure" from a solution of lower contaminant concentration into a solution of higher concentration until the concentrations equalize.

In reverse osmosis, a water pressure differential is used to reverse natural osmosis and this forces water molecules to pass through a semipermeable membrane while leaving behind the undesirable molecules that cannot pass through the membrane. The membrane acts as an extremely fine filter (molecular filter).

Reverse osmosis treatment systems are intended to be installed as a point of use system. This means the treatment unit serves the drinking water fixture located immediately adjacent to the treatment system.

The discharge from these units must be through an air gap or air gap device meeting the requirements of

NSF 58 and Section P2902.2.1. See the commentary to Section P2902.2.1 and Table P2902.2.1.

P2909.3 Connection tubing. The tubing to and from drinking water treatment units shall be of a size and material as recommended by the manufacturer. The tubing shall comply with NSF 14, NSF 42, NSF 44, NSF 53, NSF 58 or NSF 61.

❖ The NSF (National Sanitation Foundation) standards listed address the connections for these units. The tubing used for drinking water treatment units must be of a size and material recommended by the manufacturer. The tubing materials are certified during the evaluation of the equipment for compliance with materials or products that come into contact with drinking water.

SECTION P2910
NONPOTABLE WATER SYSTEMS

P2910.1 Scope. The provisions of this section shall govern the materials, design, construction and installation of systems for the collection, storage, treatment and distribution of nonpotable water. The use and application of nonpotable water shall comply with laws, rules and ordinances applicable in the jurisdiction.

❖ This chapter covers the required details about collecting, storing, treating and distributing nonpotable water. The end use for these waters is controlled by the rules established in each jurisdiction. For example, collecting and using rainwater in Atlanta, Georgia might be allowed whereas that practice might not be allowed in Denver, Colorado.

P2910.2 Water quality. Nonpotable water for each end use application shall meet the minimum water quality requirements as established for the intended application by the laws, rules and ordinances applicable in the jurisdiction. Where nonpotable water from different sources is combined in a system, the system shall comply with the most stringent requirements of this code applicable to such sources.

❖ The quality required for the nonpotable water to be used in a particular application must be established by each jurisdiction. It is hoped that, eventually, quality requirements for nonpotable water used in specific applications will become somewhat standardized, similar to the standardization of the quality for potable drinking water. However, unlike the ultimate use for potable water (human ingestion), the use of nonpotable water can be quite varied such that the setting of an ultimate quality standard for any one use will most likely be impractical. Another compounding issue is the extremely wide (quality) variations of raw sources of nonpotable water. For example, raw water sources for potable water are typically lakes, rivers and underground aquifers which provide "nature cleansed" water for starting the potable water treatment process. Treatment is fairly straightforward because the quality and consistency of the raw water source is known. For nonpotable water, the raw water source can greatly vary depending on what nonpotable waters are collected, the combinations of different nonpotable waters and the volumes of each, all of which can vary over any period of time. Of course, the goal is to not "over treat" the nonpotable water for an end use application, as this can be costly in terms of initial equipment installation and long-term operation. Conversely, there is the inherent obligation for ensuring that the use of the nonpotable water does not create hazards for living organisms that come into contact with the water. For example, some raw nonpotable waters might have a pH characteristic that can be accommodated by flush valves for water closets and urinals, but unsuitable for drip irrigation for landscaping. A treatment process to correct the pH so that the nonpotable water can be used for both the landscaping and the flushing applications might be appropriate. However, why use pH adjusting treatment for that volume of nonpotable water only used for flushing, when the flush valves can accommodate the non pH-adjusted water? Treating all of the (combined) water for the highest quality required for any one use might make sense, depending on what treatments are necessary. However, the code does not prohibit combining the raw sources, treating that water to a certain quality level and then sending part of that water to another location for additional treatment for a specific end use application that requires a "higher" quality level than the initial treatment provided. This provides a better understanding of the last sentence of the section: "Only where the raw water sources are combined, and treated, stored and used as a combined treated source does the code requirement apply for treatment to the most stringent of the requirements of the code."

P2910.2.1 Residual disinfectants. Where chlorine is used for disinfection, the nonpotable water shall contain not more than 4 ppm (4 mg/L) of chloramines or free chlorine. Where ozone is used for disinfection, the nonpotable water shall not contain gas bubbles having elevated levels of ozone at the point of use.

Exception: Reclaimed water sources shall not be required to comply with the requirements of this section.

❖ Chlorinating and ozonizing are two of the most common methods for disinfection of water. Chlorine is inexpensive to obtain and ozone is easy to make on site. They are effective, very controllable and the results (bacteria kill) are well established for the quantities and detention times needed for volumes of water. Some "over disinfection" is necessary to ensure that a minimum level of disinfection product is maintained. But, too much disinfection product in the water is costly in the long run and does not yield any better results (the bacteria is dead—more disinfection is unnecessary). The excess disinfection product off-gasses out of the water and produces undesirable effects in the short term. For chlorine, that effect is an unpleasant chlorine odor, similar to the air around most indoor public swimming pools. For ozone, that effect is an effervescence bubbling up and out of the water along with a sweet, pungent smell that is similar to how the air smells on a calm summer day just prior to the beginning of a thunderstorm. These effects in an outdoor environment are typically of no concern. However, in an indoor environ-

ment, the effects can be quite a nuisance to the building occupants and depending on the concentrations and length of exposure, could impact the health of the occupants. Therefore, this section requires control in the use of these products.

Where ultraviolet light, instead of chlorine or ozone, is to be used for disinfection, this method will require approval by the code official in accordance with Section R104.11.

The exception for reclaimed water recognizes that such water is disinfected at the source, the water utility that provides the water. See the definition for "Reclaimed water" in Chapter 2.

P2910.2.2 Filtration required. Nonpotable water utilized for water closet and urinal flushing applications shall be filtered by a 100 micron or finer filter.

Exception: Reclaimed water sources shall not be required to comply with the requirements of this section.

❖ Water closet fill valves (and flush valves) and urinal flush valves have small diameter water ways that could become clogged with particulate matter sometimes found in nonpotable water. The 100-micron filter requirement ensures that valves will function as the manufacturer intended.

The exception for reclaimed water recognizes that such water is filtered at the source, the water utility that provides the water. See the definition for "Reclaimed water" in Chapter 2.

P2910.3 Signage required. Nonpotable water outlets such as hose connections, open-ended pipes and faucets shall be identified at the point of use for each outlet with signage that reads as follows: "Nonpotable water is utilized for [application name]. CAUTION: NONPOTABLE WATER. DO NOT DRINK." The words shall be legibly and indelibly printed on a tag or sign constructed of corrosion-resistant, waterproof material or shall be indelibly printed on the fixture. The letters of the words shall be not less than 0.5 inches (12.7 mm) in height and in colors contrasting the background on which they are applied. In addition to the required wordage, the pictograph shown in Figure P2910.3 shall appear on the signage required by this section.

❖ This section is identical to Section P2901.2.1. See the commentary to that section.

FIGURE P2910.3
PICTOGRAPH—DO NOT DRINK

P2910.4 Permits. Permits shall be required for the construction, installation, alteration and repair of nonpotable water systems. Construction documents, engineering calculations, diagrams and other such data pertaining to the nonpotable water system shall be submitted with each permit application.

❖ Although Section R105 of the code requires permits for the work covered by the code, this section reinforces that requirement as not everyone considers all nonpotable water system work as needing a permit. For example, rainwater collection and distribution systems have been installed for many decades without a single thought towards compliance to a code. There have been guidelines developed by numerous jurisdictions where water shortages are a significant problem. Some guidelines have been adopted and made part of the laws so that code officials could enforce the guidelines. In other areas, the guidelines are not law. This section makes it clear that this chapter is enforceable when the code is adopted.

P2910.5 Potable water connections. Where a potable system is connected to a nonpotable water system, the potable water supply shall be protected against backflow in accordance with Section P2902.

❖ This section simply points to the backflow prevention section for where nonpotable water might possibly contaminate a potable water supply system when a backflow event (in the potable water system) occurs.

P2910.6 Approved components and materials. Piping, plumbing components and materials used in collection and conveyance systems shall be manufactured of material approved for the intended application and compatible with any disinfection and treatment systems used.

❖ The use of the term "approved" may or may not mean approved by the building official, as the code does not provide a basis for which the materials can be evaluated for some applications. For example, a gray water collection system inside of (and below) the building will be constructed of piping materials that comply with Tables P3002.1(1), P3002.1(2) and P3002.3. However, should the continuation of that piping outside of the building be allowed to comply with Table P3002.2 (building sewer piping)? The building official might agree that it does comply with the code, but manufacturers of systems that "process" the nonpotable waters might not agree.

P2910.7 Insect and vermin control. The system shall be protected to prevent the entrance of insects and vermin into storage tanks and piping systems. Screen materials shall be compatible with contacting system components and shall not accelerate the corrosion of system components.

❖ Large floating "mats" of dead insects and vermin in storage tanks will decompose and create odors through the tank vent. Although it is usually not possible to prevent all insects from entry, those that do enter will hopefully become waterlogged and sink to the bottom of the tank. Other sections provide additional requirements for on-site nonpotable water and rainwater. Filtering to remove insects and vermin from

reclaimed water is performed at the water utility that distributes the reclaimed water.

Screens for storage tank vents must not be of materials that are incompatible with the material that they are in contact with so that the screen does not corrode and fall off the opening.

It is uncertain whether this section requires screens at each vent pipe termination (of the vent piping required for equalizing pressures in a drain piping system) to prevent the entry of insects and vermin into the piping system. Any insect and vermin would be captured by filters installed before the storage tank. The code official will need to make this decision.

P2910.8 Freeze protection. Where sustained freezing temperatures occur, provisions shall be made to keep storage tanks and the related piping from freezing.

❖ Because storage tanks have a large volume of water, it would take an "amount of time" of exposure to freezing temperatures for the water in the tank to begin freezing. The length of time necessary to cause the water to freeze will depend on many factors such as sunlight, wind, movement of water in the tank and the lowest temperature excursion. For these reasons, many nonpotable water tanks are buried underground. However, some climates could have only "mildly freezing" temperatures that last for only a few hours a day. In regions that do have freezing temperatures, storage tanks could be located outdoors under open shelters, or in enclosed shelter buildings that have very little or no heat. Tank contents that are kept moving by a circulation system could endure "mildly freezing" overnight conditions.

The exposure of piping, pumps and other equipment to freezing conditions is a more serious issue because in most cases the water is not moving. See the commentary to Section P2603.5 concerning protection of piping against freezing. Note that while the piping, pumps and other equipment might need protection (insulation, heat or both) against freezing conditions, that is not an indication that the storage tank needs similar protection.

P2909.9 Nonpotable water storage tanks. Nonpotable water storage tanks shall comply with Sections P2910.9.1 through P2910.9.11.

❖ A storage tank can be of any shape and be of a multitude of materials. The following sections provide the requirements for storage tanks.

P2910.9.1 Sizing. The holding capacity of the storage tank shall be sized in accordance with the anticipated demand.

❖ Ideally, all nonpotable water collected and stored will (eventually) be reused for another purpose. However, that is not the case with raw gray water, as Section P2911.6 prohibits untreated gray water from being retained for more than 24 hours. Another factor is that there is not enough demand (use) for the collected nonpotable water. For example, a rainwater collection system collects large volumes of rainwater in a continuously "rainy" region. The rainwater is only used to flush water closets (the installed nonwater urinals not needing water). Does it make sense to install a large storage tank to retain all of the rainwater? Perhaps not. After careful analysis of the rainwater flows (historical weather data) and the demands (anticipated use of the fixtures that will use the water), the designer of the system will make the decision on the size of the storage tank.

P2910.9.2 Location. Storage tanks shall be installed above or below grade. Above-grade storage tanks shall be protected from direct sunlight and shall be constructed using opaque, UV-resistant materials such as, but not limited to, heavily tinted plastic, lined metal, concrete and wood; or painted to prevent algae growth; or shall have specially constructed sun barriers including, but not limited to, installation in garages, crawlspaces or sheds. Storage tanks and their manholes shall not be located directly under any soil piping, waste piping or any source of contamination.

❖ Storage tanks can be located anywhere. Possible locations could be inside of buildings, on roofs or on grade (with proper shelter) or underground (inside of or outside of the building footprint). The designer chooses the location that best suits the needs of the system that he is designing.

Tanks above grade need to be protected so that the material of the tank's outer wall is not degraded by exposure to sunlight because sunlight includes wavelengths in the ultraviolet (UV) range. Most untinted thermoplastic materials, such as polypropylene and polyethylene, can start cracking because the sun's UV radiation causes degradation of the connection points within the molecules of plastic materials. The cracking continues until the material loses enough strength to fail. Although "above grade" usually means outdoors and above grade, the situation would be the same for a plastic storage tank located inside of a standard glass greenhouse (plain glass is able to pass some UV radiation). Does this mean that storage tanks located in a room inside of a building where a wall window allows natural light into the room must be heavily tinted? Perhaps not, as the exposure to sunlight will probably be incidental or only be directly shining on the tank for a very limited amount of time (as the sun moves though the sky). However, if the room has many clear skylights (not tinted) such as might be found in a warehouse, that thought might have to reevaluated. Note that this section is only concerned about exposure of the tank to sunlight. However, the sun is not the only generator of UV radiation. A shop that performs open welding could cause significant UV radiation toward a plastic tank if the tank was not shielded in some manner.

Plastic tanks can be protected against UV radiation by the addition of pigments that absorb the UV rays. As UV absorbers (UVAs) will degrade over long periods of time, significant UVA pigments must be added to compensate for the UVA degradation. Thus, the code requires "heavily tinted" plastic tank materials.

Another reason for keeping direct sunlight off of a translucent tank is to prevent algae growth in the water. Algae need sunlight to propagate. Unchecked algae growth causes significant equipment problems and can create difficult to solve water quality issues.

Lined metal tanks are one type of metal tank. This specifically refers to carbon steel metal which will rust and produce scale (corrode). Typical linings are epoxy coatings that are applied after the carbon steel metal is sandblasted clean. Presumably, stainless steel tanks would not require a lining because stainless steel (if the proper series of stainless steel is chosen) will not corrode.

The requirement for storage tanks (and their manholes) to not be located under soil or waste piping, or any "source of contamination" is to make sure that the water in the tank does not become something other than expected.

P2910.9.3 Materials. Where collected on site, water shall be collected in an *approved* tank constructed of durable, nonabsorbent and corrosion-resistant materials. The storage tank shall be constructed of materials compatible with any disinfection systems used to treat water upstream of the tank and with any systems used to maintain water quality within the tank. Wooden storage tanks that are not equipped with a makeup water source shall be provided with a flexible liner.

❖ Storage tanks must be approved by the code official. There are many types of "tanks" that are not of the durability, nonabsorbency and corrosion resistance needed for long-term use as a permanent system for the intended use. For example, a field-built brick (or block) structure with a rubber lining that was adhered to the walls might hold water for the initial test, but is that an appropriate long-term solution? The code official must make the decision.

The tank materials must be compatible with the disinfection and treatment methods for the water that will be stored in the tank. For example, if alternative approval was obtained for the use of UV lamps in the tank for disinfection of the water, then a tank made of untinted plastic might not be compatible with the disinfection method (see commentary for Section P2910.9.2).

The water tightness of wooden storage tanks depends on the wood planking "swelling" and closing off the gaps between the planks. If the water level in a wood tank is allowed to decrease for long periods of time (such as a day), the planks dry out, shrink and the gaps open up to cause leaks. Makeup water added to the tank keeps the planks wet. The makeup water would not be needed at any time that the water level in the tank becomes low, but only when the tank level remains low for a period of time where the planks would start drying out. If adding makeup water does not make sense in the scheme of collecting as much nonpotable water as possible (and reducing the use of potable water), then a wood tank would have to be lined with a flexible liner so that the water level could remain low for long periods of time. For example, rainwater might be used from rainwater storage at a very slow rate while rainwater is collected in "surges" (rainfall events). Filling the wood tank to the overflow point to keep the wood from drying out would be senseless.

P2910.9.4 Foundation and supports. Storage tanks shall be supported on a firm base capable of withstanding the weight of the storage tank when filled to capacity. Storage tanks shall be supported in accordance with this code.

❖ Water weighs 62.4 pounds per cubic foot or 8.34 pounds per gallon. A 100-gallon tank will weigh 834 pounds plus the weight of the tank itself. The weight of a filled water tank is always significant, but especially so when supported by a building structure. Other parts of this code concerned with the building structure must be reviewed.

P2910.9.4.1 Ballast. Where the soil can become saturated, an underground storage tank shall be ballasted or otherwise secured to prevent the tank from floating out of the ground when empty. The combined weight of the tank and hold-down ballast shall meet or exceed the buoyancy force of the tank. Where the installation requires a foundation, the foundation shall be flat and shall be designed to support the storage tank weight when full, consistent with the bearing capability of adjacent soil.

❖ Uplift forces of an empty underground tank are very significant when the ground around the tank becomes saturated with water. The effect is no different than placing a ping pong ball under water and releasing it. It will return to the surface. The effect was scientifically discovered by Archimedes of Syracuse (Greece) around 250 B.C. His principle states: "Any object, wholly or partially immersed in a fluid, is buoyed up by a force equal to the weight of the fluid displaced by the object." In other words, objects have "buoyancy," however, when the weight of the object is greater than the weight of the fluid displaced, the object is not buoyant (and is not uplifted toward the surface of the water).

Consider a tank of 5,000-gallon capacity. The uplift force of that tank (empty and neglecting the empty weight of the tank) in fully saturated soil is 5,000 gallons × 8.34 pounds per gallon or 41,700 pounds. That amount of force will cause a $1^1/_8$-inch nominal diameter reinforcing bar (40 ksi yield strength) to begin to pull apart. Design of the straps and their connections to the tank and to the concrete foundation must be carefully considered to prevent underground tanks from "popping up and out of the ground."

The (concrete) foundation that the tank rests on must also be properly designed to accept the loads from the tank (filled to capacity) without failure and to not overload the soil that the foundation rests on.

P2910.9.4.2 Structural support. Where installed below grade, storage tank installations shall be designed to withstand earth and surface structural loads without damage and with minimal deformation when empty or filled with water.

❖ An underground tank must be able to support the earth loads against the sides and top of the tank. If the tank is under a road, driveway or parking lot, vehicle loads add to the loads that the tank must withstand. Where the tank is adjacent to a road, driveway or parking lot, some of those loads could transfer at an angle away from directly under those areas to the tank. Water-saturated soils also impart significant loads to the sides of an empty tank.

P2910.9.5 Makeup water. Where an uninterrupted nonpotable water supply is required for the intended application, potable or reclaimed water shall be provided as a source of makeup water for the storage tank. The makeup water supply shall be protected against backflow by means of an *air gap* not less than 4 inches (102 mm) above the overflow or an *approved* backflow device in accordance with Section P2902. A full-open valve located on the makeup water supply line to the storage tank shall be provided. Inlets to the storage tank shall be controlled by fill valves or other automatic supply valves installed to prevent the tank from overflowing and to prevent the water level from dropping below a predetermined point. Where makeup water is provided, the water level shall be prohibited from dropping below the source water inlet or the intake of any attached pump.

❖ Where a constant supply of nonpotable water is depended on for any use, a makeup water supply for the storage tank is necessary. For example, where rainwater or treated gray water is used to flush water closets and urinals, a constant supply of water is necessary. If the volume of raw nonpotable water supply is not sufficient to keep up with the demand, then additional water must be added to the tank to keep the nonpotable water supply system going. Makeup water does not necessarily need to be potable water, but often it is. The connection of makeup water to a storage tank must comply with Section 608 to protect the higher quality makeup water from contamination in the event that there is a backflow event in the makeup water system.

The intent of this section is that the makeup water be supplied to the tank by automatic means. The term "fill valve" implies that the type of mechanical "float type" fill valve is found in a flush tank water closet or at the top of a vat. Or the valve could be the type that is electrically controlled (such as a solenoid valve) through float switches or other water level detection devices. A full open valve is required to be on the makeup water line so that the fill valve or other automatic control valve can be removed, repaired or replaced.

Obviously, the makeup water fill valve must shut before water in the tank begins to empty out through the overflow of the tank. However, the shutoff point does not need to be at the overflow level in order to leave room for the addition of nonpotable water from the original sources. The valve must open before the water in the tank reaches a certain low level, as determined by the designer of the system. In most nonpotable water storage tanks, especially those storing water with some particulate matter, the incoming water (both the original water and makeup) is introduced into the tank at a point that is below the water surface and somewhat above the bottom of the tank. This is to reduce "stirring" of the floating debris on the surface of water in the tank. Entry of water into the tank should be as calm as possible. Allowing the level of water to drop below the inlet level would only introduce turbulence on the surface of the water resulting in the particulate becoming entrained throughout the tank. This could later negatively impact the operation of pumps, valves and filters in the distribution system.

P2910.9.5.1 Inlet control valve alarm. Makeup water systems shall be fitted with a warning mechanism that alerts the user to a failure of the inlet control valve to close correctly. The alarm shall activate before the water within the storage tank begins to discharge into the overflow system.

❖ A makeup water supply that fails to shut off could waste many thousands of gallons of potable water. In a residential building environment, this could be very expensive if water is being paid for or, could affect the longevity of a private well pump. A simple high level alarm will notify the building owner of a condition that needs repair.

P2910.9.6 Overflow. The storage tank shall be equipped with an overflow pipe having a diameter not less than that shown in Table P2910.9.6. The overflow outlet shall discharge at a point not less than 6 inches (152 mm) above the roof or roof drain; floor or floor drain; or over an open water-supplied fixture. The overflow outlet shall be covered with a corrosion-resistant screen of not less than 16 by 20 mesh per inch (630 by 787 mesh per m) and by $^1/_4$-inch (6.4 mm) hardware cloth or shall terminate in a horizontal angle seat check valve. Drainage from overflow pipes shall be directed to prevent freezing on roof walks. The overflow drain shall not be equipped with a shutoff valve. Not less than one cleanout shall be provided on each overflow pipe in accordance with Section P3005.2.

❖ Storage tanks can become filled to capacity and additional water added to the tank has to go somewhere or the collection system will back up and cause problems. The overflow is another location where insects and vermin can enter the storage tank so a screen is necessary. In some situations, such as rainwater storage tanks, the overflow could discharge to a storm sewer or a natural waterway. Because it is rainwater, logically, it would seem that allowing it to flow into the storm sewer or natural waterway would be acceptable. However,

that depends on the storm water runoff requirements of the jurisdiction. If the rainwater has been treated (other than screening for particulate) or potable (or other nonpotable) water is used for the makeup water supply, there could be restrictions against discharging to a storm sewer or natural waterway. If the local jurisdiction does not know, state environmental agencies should be sought out to determine the answer. In the United States, the Environmental Protection Agency (EPA) considers storm water discharges as point sources of pollution because rainwater could contain harmful pollutants that could wash into streams, river, lakes and coastal water. Most states administer the National Pollutant Discharge Elimination System (NPDES) Stormwater program; however, in some states and territories and all tribal lands, the EPA directly administers the program.

Although this section implies that any nonpotable water (other than rainwater) could be allowed to overflow to a storm sewer or waterway, it is doubtful that this would be allowed in any jurisdiction, pursuant to the overriding requirements of the EPA. And, even though reclaimed water is of an equal quality (or perhaps of higher quality) than what exits the tertiary stage of the municipal wastewater treatment facility that distributes the reclaimed water, reclaimed water may not necessarily be allowed to flow into storm sewers or waterways.

If the flow of water from the overflow is allowed to drain onto the ground around the outside of a storage tank, it must be routed to drain away from the tank area and not toward a building or an adjacent lot. Foundations for the tank and a building can be quickly undermined when excess water surcharges the soil at these locations. Adjacent lot owners would not appreciate such overflows.

Where a water tank is on a roof (as many water tanks have been located in New York City), water coming out of an overflow during freezing weather creates hazardous conditions on the roof walkways. Prior planning to put the overflow and walkways in different locations is necessary.

Overflows could become clogged, especially where the overflow has an outlet screen. A cleanout is necessary to clear blockages.

TABLE P2910.9.6
SIZE OF DRAIN PIPES FOR WATER TANKS

TANK CAPACITY (gallons)	DRAIN PIPE (inches)
Up to 750	1
751 to 1500	$1^1/_2$
1501 to 3000	2
3001 to 5000	$2^1/_2$
5001 to 7500	3
Over 7500	4

For SI: 1 gallon = 3.875 liters, 1 inch = 25.4 mm.

P2910.9.7 Access. Not less than one access opening shall be provided to allow inspection and cleaning of the tank interior. Access openings shall have an *approved* locking device or other *approved* method of securing access. Below-grade storage tanks, located outside of the building, shall be provided with a manhole either not less than 24 inches (610 mm) square or with an inside diameter not less than 24 inches (610 mm). Manholes shall extend not less than 4 inches (102 mm) above ground or shall be designed to prevent water infiltration. Finished grade shall be sloped away from the manhole to divert surface water. Manhole covers shall be secured to prevent unauthorized access. Service ports in manhole covers shall be not less than 8 inches (203 mm) in diameter and shall be not less than 4 inches (102 mm) above the finished grade level. The service port shall be secured to prevent unauthorized access.

> **Exception:** Storage tanks under 800 gallons (3028 L) in volume installed below grade shall not be required to be equipped with a manhole, but shall have a service port not less than 8 inches (203 mm) in diameter.

❖ Storage tanks, large and small, need to be accessed for inspection. Above-ground tanks are not required to have a manhole access but need an inspection access. Below-grade tanks must have manhole access except where the tank is less than 800 gallons capacity. All openings must be secured to prevent unauthorized access.

P2910.9.8 Venting. Storage tanks shall be provided with a vent sized in accordance with Chapter 31 and based on the aggregate diameter of all tank influent pipes. The reservoir vent shall not be connected to sanitary drainage system vents. Vents shall be protected from contamination by means of an *approved* cap or a U-bend installed with the opening directed downward. Vent outlets shall extend not less than 4 inches (102 mm) above grade, or as necessary to prevent surface water from entering the storage tank. Vent openings shall be protected against the entrance of vermin and insects in accordance with the requirements of Section P2902.7.

❖ As the water level in the tank rises and drops, air needs to escape and enter the tank to prevent pressure (and especially negative pressure) from occurring in the tank.

P2910.9.9 Drain. A drain shall be located at the lowest point of the storage tank. The tank drain pipe shall discharge as required for overflow pipes and shall not be smaller in size than specified in Table P2910.9.6. Not less than one cleanout shall be provided on each drain pipe in accordance with Section P3005.2.

❖ A tank can be drained by a pump or by a drain near the bottom of the tank. Most the time, draining of the tank is required to clean the tank of accumulated debris. In-tank pumps and level controls are usually accessible by lifting them out of the tank, so draining is not required for those repairs. Where there is a tank drain, this section requires a "tank drain pipe." In other words, there needs to be a pipe connected near the bottom of the tank, presumably with a valve, so that drain flow can be controlled. In other words, having a "drain plug"

would not meet the intent of this section as once the plug was removed there would not be any way to control the discharge.

P2910.10 Marking and signage. Each nonpotable water storage tank shall be labeled with its rated capacity. The contents of storage tanks shall be identified with the words "CAUTION: NONPOTABLE WATER—DO NOT DRINK." Where an opening is provided that could allow the entry of personnel, the opening shall be marked with the words, "DANGER—CONFINED SPACE." Markings shall be indelibly printed on the tank, or on a tag or sign constructed of corrosion-resistant waterproof material that is mounted on the tank. The letters of the words shall be not less than 0.5 inches (12.7 mm) in height and shall be of a color in contrast with the background on which they are applied.

❖ The requirements of this section are straightforward. A capacity label, especially for underground tanks, is useful when the contents have to be pumped out and hauled to a disposal point.

P2910.11 Storage tank tests. Storage tanks shall be tested in accordance with the following:

1. Storage tanks shall be filled with water to the overflow line prior to and during inspection. Seams and joints shall be left exposed and the tank shall remain water tight without leakage for a period of 24 hours.

2. After 24 hours, supplemental water shall be introduced for a period of 15 minutes to verify proper drainage of the overflow system and leaks do not exist.

3. Following a successful test of the overflow, the water level in the tank shall be reduced to a level that is 2 inches (51 mm) below the makeup water trigger point by using the tank drain. The tank drain shall be observed for proper operation. The makeup water system shall be observed for proper operation, and successful automatic shutoff of the system at the refill threshold shall be verified. Water shall not be drained from the overflow at any time during the refill test.

❖ The requirements of this section are straightforward. Although 24 hours seems like a long test period, it might take that amount of time for a small leak to develop enough of a puddle to be observed. For a lined carbon steel tank, even a small leak could become a much larger leak over time (because of corrosion), so it is useful to wait a day to find leaks. For wood tanks, the tank must be filled well ahead of the test period so that the wood can swell and close up any gaps.

P2910.12 System abandonment. If the owner of an on-site nonpotable water reuse system or rainwater collection and conveyance system elects to cease use of or fails to properly maintain such system, the system shall be abandoned and shall comply with the following:

1. System piping connecting to a utility-provided water system shall be removed or disabled.

2. The distribution piping system shall be replaced with an *approved* potable water supply piping system. Where an existing potable water pipe system is already in place, the fixtures shall be connected to the existing system.

3. The storage tank shall be secured from accidental access by sealing or locking tank inlets and access points, or filled with sand or equivalent.

❖ Nonpotable water reuse systems do require maintenance and must be kept in good repair to provide water for the intended fixtures or other uses. However, there will be times that abandonment of some systems will be necessary.

P2910.13 Separation requirements for nonpotable water piping. Nonpotable water collection and distribution piping and reclaimed water piping shall be separated from the building sewer and potable water piping underground by 5 feet (1524 mm) of undisturbed or compacted earth. Nonpotable water collection and distribution piping shall not be located in, under or above cesspools, septic tanks, septic tank drainage fields or seepage pits. Buried nonpotable water piping shall comply with the requirements of Section P2604.

Exceptions:

1. The required separation distance shall not apply where the bottom of the nonpotable water pipe within 5 feet (1524 mm) of the sewer is not less than 12 inches (305 mm) above the top of the highest point of the sewer and the pipe materials conforms to Table P3002.2.

2. The required separation distance shall not apply where the bottom of the potable water service pipe within 5 feet (1524 mm) of the nonpotable water pipe is not less than 12 inches (305 mm) above the top of the highest point of the nonpotable water pipe and the pipe materials comply with the requirements of Table P2906.5.

3. The required separation distance shall not apply where a nonpotable water pipe is located in the same trench with a building sewer that is constructed of materials that comply with the requirements of Table P3002.2.

4. The required separation distance shall not apply where a nonpotable water pipe crosses a sewer pipe provided that the nonpotable water pipe is sleeved to not less than 5 feet (1524 mm) horizontally from the sewer pipe centerline on both sides of such crossing, with pipe materials that comply with Table P3002.2.

5. The required separation distance shall not apply where a potable water service pipe crosses a nonpotable water pipe, provided that the potable water service pipe is sleeved for a distance of not less than 5 feet (1524 mm) horizontally from the centerline of the nonpotable pipe on both sides of such crossing, with pipe materials that comply with Table P3002.2.

6. The required separation distance shall not apply to irrigation piping located outside of a building and downstream of the backflow preventer where nonpotable water is used for outdoor applications.

❖ These are the same requirements (excluding Exception 6) as for the installation of a potable water supply system. Nonpotable water irrigation piping does not need separation from the building sewer because this would be too difficult to accomplish and provide limited benefit.

P2910.14 Outdoor outlet access. Sillcocks, hose bibs, wall hydrants, yard hydrants and other outdoor outlets supplied by nonpotable water shall be located in a locked vault or shall be operable only by means of a removable key.

❖ Rainwater (or possibly reclaimed water) could be used outside of a building for various uses. Rainwater could be used to water gardens and vegetation, or to wash vehicles. Reclaimed water might be able to be used for outdoor heavy equipment washdowns. The outlets to connect to these waters sources need to be controlled so that unauthorized use does not occur.

SECTION P2911
ON-SITE NONPOTABLE WATER REUSE SYSTEMS

P2911.1 General. The provisions of this section shall govern the construction, installation, alteration and repair of on-site nonpotable water reuse systems for the collection, storage, treatment and distribution of on-site sources of nonpotable water as permitted by the jurisdiction.

❖ This section is about the specific details for on-site nonpotable water reuse systems. Refer to the definition "On-site nonpotable water reuse system" in Chapter 2.

P2911.2 Sources. On-site nonpotable water reuse systems shall collect waste discharge only from the following sources: bathtubs, showers, lavatories, clothes washers and laundry trays. Water from other *approved* nonpotable sources including swimming pool backwash operations, air conditioner condensate, rainwater, foundation drain water, fluid cooler discharge water and fire pump test water shall be permitted to be collected for reuse by on-site nonpotable water reuse systems, as approved by the building official and as appropriate for the intended application.

❖ Essentially, an on-site nonpotable water reuse system only collects gray water. However, there are many other kinds of nonpotable waters that could be available at the building site. Some of these might be generated in quantities that overshadow the flow of gray water into the system. The "qualities" of gray water are well understood. Other nonpotable waters are quite different from gray water and any type of nonpotable water could have a significant impact on the suitability of the water for an end-use application. For example, cooling tower blowdown water could have such a high level of total dissolved solids that, for an end use as water closet flushing, the bowls of the water closets become scaled and hard to clean. Problems with water closet fill and flush valves, and clogging of the internal waterways of the water closet could occur. Thus, the need for the code official to approve these other nonpotable water sources for inclusion.

P2911.2.1 Prohibited sources. Reverse osmosis system reject water, water softener backwash water, kitchen sink wastewater, dishwasher wastewater and wastewater containing urine or fecal matter shall not be collected for reuse within an on-site nonpotable water reuse system.

❖ There are certain nonpotable waters that are known to be detrimental to on-site nonpotable water reuse systems. They are prohibited from inclusion.

P2911.3 Traps. Traps serving fixtures and devices discharging waste water to on-site nonpotable water reuse systems shall comply with the Section P3201.2.

❖ Plumbing fixtures (and clothes washer waste receptors) are already required to have traps that comply with Section P3201.2. Where other nonpotable water is collected by the on-site nonpotable water reuse systems, a trap is required between the connection of the piping from the nonpotable water source and the collection piping leading to the on-site nonpotable water reuse systems. The trap is required to comply with Section P3201.2. In most cases, these additional non-potable waters will be connected indirectly to the gray water piping system.

P2911.4 Collection pipe. On-site nonpotable water reuse systems shall utilize drainage piping *approved* for use within plumbing drainage systems to collect and convey untreated water for reuse. Vent piping *approved* for use within plumbing venting systems shall be utilized for vents within the gray-water system. Collection and vent piping materials shall comply with Section P3002.

❖ The piping used for collecting all on-site nonpotable water must comply with Section P3002. Vent piping must also be in accordance with Section P3002.

P2911.4.1 Installation. Collection piping conveying untreated water for reuse shall be installed in accordance with Section P3005.

❖ Installation of the collection piping must comply with Section P3005.

P2911.4.2 Joints. Collection piping conveying untreated water for reuse shall utilize joints *approved* for use with the distribution piping and appropriate for the intended applications as specified in Section P3002.

❖ Joints in collection piping must comply with Section P3002.

P2911.4.3 Size. Collection piping conveying untreated water for reuse shall be sized in accordance with drainage sizing requirements specified in Section P3005.4.

❖ Although this section indicates that the drain piping is to be sized in accordance with Section P3005.4, the flows from nonpotable water sources (other than the allowed plumbing fixtures) are not measured in drainage fixture units (dfu). In most cases, a maximum flow rate in gallons per minute can be computed for these flows.

P2911.4.4 Marking. Additional marking of collection piping conveying untreated water for reuse shall not be required

beyond that required for sanitary drainage, waste and vent piping by the Chapter 30.

❖ As the code does not require labeling/marking of sanitary drainage piping, including piping that conveys only gray water, the code does not require labeling/marking for nonpotable water collection piping.

P2911.5 Filtration. Untreated water collected for reuse shall be filtered as required for the intended end use. Filters shall be accessible for inspection and maintenance. Filters shall utilize a pressure gauge or other *approved* method to provide indication when a filter requires servicing or replacement. Filters shall be installed with shutoff valves immediately upstream and downstream to allow for isolation during maintenance.

❖ Filters are typically installed to filter the flow before entering the storage tank so that debris is kept out of the tank. Some type of alert system is required to indicate when the filter needs to be cleaned or changed.

P2911.6 Disinfection. Nonpotable water collected on site for reuse shall be disinfected, treated or both to provide the quality of water needed for the intended end-use application. Where the intended end-use application does not have requirements for the quality of water, disinfection and treatment of water collected on site for reuse shall not be required. Nonpotable water collected on site containing untreated gray water shall be retained in collection reservoirs for not more than 24 hours.

❖ Some nonpotable water, such as gray water, will need disinfection before being used for certain applications. Other nonpotable waters, such as rainwater, will not require disinfection for certain applications. The designer decides what waters need disinfection in order to meet the jurisdiction's water quality requirements for the end use application. Note that untreated gray water cannot be stored for longer than 24 hours because of the significant bacteria present in the water.

P2911.6.1 Gray water used for fixture flushing. Gray water used for flushing water closets and urinals shall be disinfected and treated by an on-site water reuse treatment system complying with NSF 350.

❖ Untreated gray water, even though disinfected, is too much of a health risk to be used inside of a building for the purposes of flushing water closets and urinals. Gray water must be processed through treatment equipment that complies with NSF 350.

P2911.7 Storage tanks. Storage tanks utilized in on-site nonpotable water reuse systems shall comply with Section P2910.9 and Sections P2911.7.1 through P2911.7.3.

❖ In addition to Section P2909.9, there are additional requirements for nonpotable water tanks.

P2911.7.1 Location. Storage tanks shall be located with a minimum horizontal distance between various elements as indicated in Table P2911.7.1.

❖ Table P22911.7.1 indicates the minimum required clearance between nonpotable water storage tanks and sources of possible contamination, water services that could be contaminated, adjacent lot lines and the critical root zone of trees that must be protected.

TABLE P2911.7.1
LOCATION OF NONPOTABLE WATER REUSE STORAGE TANKS

ELEMENT	MINIMUM HORIZONTAL DISTANCE FROM STORAGE TANK (feet)
Critical root zone (CRZ) of protected trees	2
Lot line adjoining private lots	5
Seepage pits	5
Septic tanks	5
Water wells	50
Streams and lakes	50
Water service	5
Public water main	10

❖ For SI: 1 foot = 304.8 mm

P2911.7.2 Inlets. Storage tank inlets shall be designed to introduce water into the tank with minimum turbulence, and shall be located and designed to avoid agitating the contents of the storage tank.

❖ The introduction of gray water into a storage tank (already having some water in the tank) must be in a manner that limits agitation of the water. Typically, the water is introduced below a certain water "low point" level but above the bottom of the tank.

P2911.7.3 Outlets. Outlets shall be located not less than 4 inches (102 mm) above the bottom of the storage tank, and shall not skim water from the surface.

❖ The "outlet" referred to is not the drain outlet of the tank. The purpose of the section is to allow some volume for sediment to settle in the bottom of the tank and to not have pump intakes or intake piping (or floating intake mechanisms) pulling water (and sediment) from this area. This is the same purpose for prohibiting water from being pulled from the surface of the water: to avoid pulling in floating debris from the water surface.

P2911.8 Valves. Valves shall be supplied on on-site nonpotable water reuse systems in accordance with Sections P2911.8.1 and P2911.8.2.

❖ There are two valves that must be provided on nonpotable water reuse systems: a bypass valve and a backwater valve for the overflow and drain of the storage tank.

P2911.8.1 Bypass valve. One three-way diverter valve certified to NSF 50 or other *approved* device shall be installed on collection piping upstream of each storage tank, or drainfield, as applicable, to divert untreated on-site reuse sources to the sanitary sewer to allow servicing and inspection of the system. Bypass valves shall be installed downstream of fixture traps and vent connections. Bypass valves shall be labeled to indicate the direction of flow, connection and storage tank or drainfield connection. Bypass valves shall be installed in

accessible locations. Two shutoff valves shall not be installed to serve as a bypass valve.

❖ There will be situations where flow to the nonpotable water storage tank must be diverted. One of those times will be during cleaning of the tank. The flow must be positively diverted so that it is known that the flow is not backing up in the building but is being diverted to an appropriate location. The only way to positively know is by installing a three-way diverter valve where the flow can only be diverted from one outlet to the other outlet (without the incoming flow being blocked). Note the last sentence of the section that prohibits two independent valves from being used to divert the flow. Opening and closing two independent valves could result in both valves being inadvertently closed. The requirement for listing and labeling of the three-way valve to NSF 50 ensures that the valve is of quality construction and meets durability requirements, as this valve might be installed outdoors.

P2911.8.2 Backwater valve. Backwater valves shall be installed on each overflow and tank drain pipe. Backwater valves shall be in accordance with Section P3008.

❖ Contamination of a nonpotable water storage tank by any type of external water must be avoided. A back water valve is a check valve for nonpotable water in gravity flow piping. Although the reference is to Section P3008, Section P3008.1 does not apply because this section always requires backwater valves. Sections P3008.2 through P3008.5 do apply.

P2911.9 Pumping and control system. Mechanical equipment including pumps, valves and filters shall be accessible and removable in order to perform repair, maintenance and cleaning. The minimum flow rate and flow pressure delivered by the pumping system shall be appropriate for the application and in accordance with Section P2903.

❖ Equipment must be maintained, repaired or replaced, so access will be needed to perform that work. How easy that access will be is up to the designer of the system and the area that he or she has to fit the system into. Every effort should be made to think about how components can be taken apart while still connected to the system. Where components need to be replaced, the designer needs to think about the clearances necessary for removal of the component. For example, many underground storage tanks have submersible pumps attached to discharge piping such that the entire vertical section of discharge piping has to be lifted out to bring the pump up to, and out of the tank. Will there be sufficient overhead clearance at the completion of the entire project? Sometimes, at the "rough-in" installation of a large tank, it does appear that there is sufficient clearance. However, afterwards a large duct or a fire water line may be inappropriately routed across the area counted on for removal of the pump.

The pumps must have the capacity to provide the necessary flow and pressure for proper operation of the plumbing fixtures (water closets and urinals).

P2911.10 Water-pressure-reducing valve or regulator.
Where the water pressure supplied by the pumping system exceeds 80 psi (552 kPa) static, a pressure-reducing valve shall be installed to reduce the pressure in the nonpotable water distribution system piping to 80 psi (552 kPa) static or less. Pressure-reducing valves shall be specified and installed in accordance with Section P2903.3.1.

❖ Most pumps have a variation in pressure output that changes with the flow rate. Typically, the highest output pressure is when the flow rate is zero (or nearly zero). As outlets in the distribution system open, the flow rate from the pump increases to match the demand, and the pressure that the pump generates is somewhat less. As more outlets open, the flow becomes greater and the outlet pressure of the pump becomes even less. Because of this effect, a pump model is chosen to supply the minimum required pressure of the system when the demand (flow rate) of the system is at the greatest. Thus, at low demands (flow rates), the pressure in the system can easily exceed 80 psi and a pressure regulator is required to keep the system pressure from exceeding 80 psi. Note that in distribution systems of multistory buildings, the pressure in the riser(s) might need to be significantly greater than 80 psi to counter the effect of pressure loss caused by elevation change. Pressure reducers are installed for the branches for the fixtures on lower floor levels to maintain the pressure in those branches at or below 80 psi for those floors.

Variable speed pumps for water distribution can resolve the problem of needing to choose a pump that has a much higher pressure at low/no-flow conditions so that there is adequate pressure at peak flow conditions. A variable speed pump changes the rotational speed pump impeller to better match the pump's outlet pressure in response to the flow rate. However, for multistory buildings, there will still be the need for reducing pressures for the branches serving the lower floors.

P2911.11 Distribution pipe. Distribution piping utilized in on-site nonpotable water reuse systems shall comply with Sections P2911.11.1 through P2911.11.3.

Exception: Irrigation piping located outside of the building and downstream of a backflow preventer.

❖ Distribution piping begins at the treated water storage tank and ends at the termination of the fixture supply pipe, such as at a water closet or urinal. Where the end use of the nonpotable water is not a plumbing fixture, then the piping ends at a control valve for the system that utilizes the nonpotable water.

The exception for irrigation systems allows for piping not complying with this section to begin after the backflow preventer for the irrigation system.

P2910.11.1 Materials, joints and connections. Distribution piping shall conform to the standards and requirements specified in Section P2906 for nonpotable water.

❖ To maintain consistency in the pipe sizes, materials and connections of water distribution piping, both pota-

ble and nonpotable, must comply with the requirements of Section P2906. Although there is very little difference between Tables P2906.4 and P2906.5, one notable difference is that PVC piping is not indicated in Table P2906.5. For a potable water supply system, PVC is not allowed to be used for water distribution piping (piping that is downstream of the main water shutoff valve for the building) primarily because the material is not rated for hot water service. Where piping potable water systems inside of a building, there is the concern that PVC piping could be inadvertently used for a hot water line and that the piping could fail under that service condition. Because nonpotable water would not be heated prior to use in a building, it is questionable as to whether this (potable water system) limitation for use of PVC piping for nonpotable water distribution (in a building) is necessary or appropriate. The building official will have to make this decision.

P2911.11.2 Design. On-site nonpotable water reuse distribution piping systems shall be designed and sized in accordance with Section P2903 for the intended application.

❖ Section P2903 requires that water distribution systems be designed in accordance with an accepted engineering practice and provides basic design criteria. A nonpotable water distribution system serving fixtures (water closets and urinals) needs to provide the same pressures and flow rates to the fixtures as if those fixtures were supplied by a potable water system.

P2911.11.3 Marking. On-site nonpotable water distribution piping labeling and marking shall comply with Section P2901.2.

❖ See the commentary to Section P2901.2. The term "labeling" in this section is not to be confused with requirements for third-party inspection and certification of products (e.g., the "listing and labeling" to a standard).

P2911.12 Tests and inspections. Tests and inspections shall be performed in accordance with Sections P2911.12.1 through P2911.12.6.

❖ Piping must be tested after it is installed to ensure the integrity of the installation. Note that the tests of piping are not necessarily attempting to replicate an actual condition of service. This section is also concerned with water quality at the point of use.

P2911.12.1 Collection pipe and vent test. Drain, waste and vent piping used for on-site water reuse systems shall be tested in accordance with Section P2503.

❖ Section P2503 covers testing of drainage and vent piping inside a building (including under the building). Note that Section P2503.5.1 covering air testing of piping, prohibits air testing of plastic piping. Where drainage piping is outside of a building, either above or below ground, Section P2503.4 appears to apply even though the piping is not sewer piping (carrying sewage with fecal matter).

P2911.12.2 Storage tank test. Storage tanks shall be tested in accordance with Section P2910.11.

❖ This section references the storage tank tests in Section P2910.11. In some situations, vent piping for storage tanks will be completely outdoors. The storage tank test does not include testing of the vent piping because the vent piping is completely outdoors, including the vent terminal. However, in other situations, the storage tank vent piping will be indoors (primarily because the storage tank is inside of the building or under the building). This vent piping should be tested separately, not as a part of testing the storage tank. Although it would be possible to temporarily block the tank overflow and fill the storage tank completely full until the water spilled out of the top of the vent pipe, storage tanks might not be designed for more than the pressure of just the tank being filled to capacity (tank overflow level). Adding the head pressure of a vertical run of vent piping, perhaps several stories or more, could overstress the storage tank walls.

This does not necessarily mean that storage tank (indoor) vent piping should not be tested. Testing of the tank vent piping might require that a valve be installed in the vent piping just above the tank, with the valve closed and the vent piping filled with water for the leak check. After the test, the valve could be opened, the water in the piping drained to the tank and the valve removed so that final vent-to-tank connections can be completed.

P2911.12.3 Water supply system test. The testing of makeup water supply piping and distribution piping shall be conducted in accordance with Section P2503.7.

❖ Makeup water supply piping could be another nonpotable water supply but is more often a potable water supply. In either situation, the piping must be leak free. Note that air testing of plastic piping is prohibited. And where piping other than plastic is tested, test air pressure is only required to be 50 psi. Where water testing is performed, the test pressure must be the working pressure of the system.

However, in service, the piping is only required to be leak free at the normal water pressures encountered. This does not imply that all water piping to the storage tank will experience the same pressure. For example, the makeup water control valve might be a 1-inch motorized ball valve located at some distance away from the storage tank. The valve might even be below the top of the storage tank. When the valve opens, the piping downstream of the valve will only be subjected to the pressure of the flow resistance in the downstream piping and any difference in the elevation between the valve and the outlet of the piping. That pressure will most likely be much less than the static pressure upstream of the makeup water valve. For the piping from the makeup water valve to the storage tank, the "test" of the piping should simply be a visual test. There is no need to temporarily install a test valve right at the tank for testing this piping at full line pressure.

Testing of distribution piping is not any different than for potable water.

P2911.12.4 Inspection and testing of backflow prevention assemblies. The testing of backflow preventers and backwater valves shall be conducted in accordance with Section P2503.8.

❖ Section P2503.8 covers the inspection and testing of backflow prevention assemblies and hose connection backflow preventers. The section also covers the inspection of air gaps. A backflow prevention assembly is a type of backflow preventer that can be field tested using sensitive, calibrated gauges to determine if the assembly is operational. Backflow prevention assemblies have "test cocks" for attachment of the gauge's test hoses.

A hose connection backflow preventer is a very simple device to test. All that is necessary is to hold the open outlet of a connected hose above the elevation of the hose connection backflow preventer (the valve before the backflow preventer should be in the "off" position). The contents of the hose will flow back to the backflow preventer and out of the backflow preventer's ports.

An air gap can be visually inspected to verify that the required open air distance is provided.

P2911.12.5 Inspection of vermin and insect protection. Inlets and vents to the system shall be inspected to verify that each is protected to prevent the entrance of insects and vermin into the storage tank and piping systems in accordance with Section P2910.7.

❖ This section points back to Section P2910.7 so that verification can be made that protection has been provided.

All other types of backflow preventers cannot be field tested nor should any attempts be made to perform a field test, as this could damage the device such that it might not perform the intended function under service conditions.

P2911.12.6 Water quality test. The quality of the water for the intended application shall be verified at the point of use in accordance with the requirements of the jurisdiction.

❖ The jurisdiction allowed the installation of the nonpotable water reuse system based on the premise that a certain quality level would be achieved by the treatment and disinfection of the system. The installer has to perform tests to show that the system provides that quality.

P2911.13 Operation and maintenance manuals. Operation and maintenance materials shall be supplied with nonpotable on-site water reuse systems in accordance with Sections P2910.13.1 through P2910.13.4.

❖ The following sections indicate the minimum requirements for operation and maintenance manuals supplied with on-site nonpotable water reuse systems. As these systems can be very diverse and some are very customized, more than these minimum requirements are encouraged. For example, photographs of components, photograph sequences of specialized repair procedures, dimensional drawings including orientation references, common component manufacturer's part numbers and electronic files of the entire O&M package would be well received by the system owner. These systems can be complex and because numerous repair and maintenance firms will be required to work on these systems, perhaps for many generations, information must be complete and clear. And with the current diversity and number of firms providing these systems, there are bound to be future mergers and acquisitions that will make future identification of particular systems difficult as dwindling numbers of people will remain who will remember those "older" systems, let alone have access to the original records of those systems.

P2911.13.1 Manual. A detailed operations and maintenance manual shall be supplied in hard-copy form for each system.

❖ A hard copy (a booklet or binder) of the manual is most basic for anyone to comprehend and use.

P2911.13.2 Schematics. The manual shall include a detailed system schematic, the location of system components and a list of system components that includes the manufacturers and model numbers of the components.

❖ A schematic can take many forms, but in terms of these types of systems, the schematic would most likely be in the form of primarily showing tanks, pumps and treatment units in their approximate "physical forms," with the piping shown as lines connecting those components. Valves, meters and controls are typically shown in a common "legend" format for quick comprehension of their functionality. A numbered list, similar to a "bill of material," identifies every item on the schematic.

P2911.13.3 Maintenance procedures. The manual shall provide a schedule and procedures for system components requiring periodic maintenance. Consumable parts including filters shall be noted along with part numbers.

❖ An on-site nonpotable water reuse system is much more complex than a potable water system for a building. Scheduled maintenance is paramount to making sure that the nonpotable water delivered for an end use does not become a health hazard for the occupants. The other reason for performing required maintenance of these systems is that where the end use for the water is for flushing water closets and urinals, the building owner can ill afford to not have these fixtures operational. If those fixtures do not work, the building will be condemned and evacuated until such time that the system is repaired and the fixtures are operational. This is a totally different concept for most building owners who are unaware of the implications of poor or no maintenance of these systems. If the public water supply to a building is shut off because of a utility piping problem (e.g., water line break in the street, fire hydrant broken), the building owner is not to blame and the occupants "make do" with the inconvenience. However, with an on-site nonpotable water reuse system,

the building owner is fully responsible for providing the nonpotable water to the water closets and urinals. The health department and code enforcement department might not look favorably at a building owner's lack of nonpotable water for his building.

P2911.13.4 Operations procedures. The manual shall include system startup and shutdown procedures. The manual shall include detailed operating procedures for the system.

❖ This requirement addresses indicating what the system components do over a period of time so that the operator of the system knows when the system is not properly operating. That period of time includes startup and shut down of the system, where an operator is usually necessary to override certain automatic functions.

SECTION P2912
NONPOTABLE RAINWATER COLLECTION AND DISTRIBUTION SYSTEMS

P2912.1 General. The provisions of this section shall govern the construction, installation, alteration, and repair of rainwater collection and conveyance systems for the collection, storage, treatment and distribution of rainwater for nonpotable applications, as permitted by the jurisdiction.

❖ Rainwater is very plentiful in many areas of the United States and so reliance on this water source is strong. However, other areas do not have much rainfall so there is no advantage to capturing what little does fall. There are also areas of the country that have legal restrictions about capturing and reusing rainwater. These areas are concerned about yearly replenishment of lakes and underground aquifers that provide the raw water source for making potable water. The more rainwater that is captured and used in a building, the less of that water finds its way back into the natural waterways that feed the lakes and rivers. It is a very delicate problem for some areas. Thus, the jurisdiction decides whether rainwater collection and distribution systems can be installed depending on the end use application. For example, a rainwater collection system for outdoor irrigation might be approved because the water will drain to the natural waterways. However, that same rainwater collection system might not be approved where the system is used to supply water closets and urinals.

P2912.2 Collection surface. Rainwater shall be collected only from above-ground impervious roofing surfaces constructed from *approved* materials. Collection of water from vehicular parking or pedestrian walkway surfaces shall be prohibited except where the water is used exclusively for landscape irrigation. Overflow and bleed-off pipes from roof-mounted *appliances* including, but not limited to, evaporative coolers, water heaters and solar water heaters shall not discharge onto rainwater collection surfaces.

❖ Before it comes in contact with a surface, rainwater is certainly of much higher "quality" than gray water. One can drink collected rainwater out of a clean drinking cup that was put outside during a rainfall event and be fairly certain that he or she would not experience ill effects from doing so. People have been doing this for thousands of years and some people still capture rainwater for drinking and cooking. However, rainwater could become nonpotable before reaching a collection surface because of airborne pollutants that the rainwater picks up as it falls from the sky. Imagine the condition of rainwater in an area around a large coal-fired boiler such as found in a power plant. Furthermore, the collection of rainwater (contacting a roof surface) adds more pollutants to the water as collection surfaces (roofs) have bird droppings, airborne particulates, dirt with pesticides and a host of many other things that would make the rainwater unsafe for drinking.

The requirement for roofing surfaces (used for the collection of rainwater) to be of materials approved by the code official is to make sure that the roof material itself does not add any more contamination to the rainwater that the end use application cannot have or that the treatment processes for the rainwater cannot remove before end use. For example, a roof made of copper with lead-soldered joints might not be appropriate for the end use of the rainwater to be drinking water because of the potential for copper and lead contamination that might not be removed by the proposed treatment process.

Any other potential sources of contamination, such as bleed offs and overflows from mechanical systems and relief valves from water heaters, must not be allowed to discharge to the roof surface for the same reasons as making sure that the roof material does not add contaminants to the rainwater.

P2912.3 Debris excluders. Downspouts and leaders shall be connected to a roof washer and shall be equipped with a debris excluder or equivalent device to prevent the contamination of collected rainwater with leaves, sticks, pine needles and similar material. Debris excluders and equivalent devices shall be self-cleaning.

❖ The debris excluder can be an integral part of the roof washer or it can be separate. The debris excluder screens out sticks, leaves and other large particles from the rainwater flow. As debris captured in the debris excluder can build up quickly, these devices are required to be self-cleaning.

P2912.4 Roof washer. An amount of rainwater shall be diverted at the beginning of each rain event, and not allowed to enter the storage tank, to wash accumulated debris from the collection surface. The amount of rainfall to be diverted shall be field adjustable as necessary to minimize storage tank water contamination. The roof washer shall not rely on manually operated valves or devices, and shall operate automatically. Diverted rainwater shall not be drained to the roof surface, and shall be discharged in a manner consistent with the storm water runoff requirements of the jurisdiction. Roof washers shall be accessible for maintenance and service.

❖ The term "roof washer" can be misleading. The roof washer is actually a device that wastes the initial flow

of rainwater coming from the roof so that the highly contaminated rainwater is diverted from the storage tank. This device must be adjustable so that the amount of water wasted can be varied to meet the actual conditions of service. The basis of adjustment is usually water testing to determine the level of contaminants in the water going to the storage tank. Once set, the roof washer may not need adjustment unless conditions of the captured rainwater change.

P2912.5 Roof gutters and downspouts. Gutters and downspouts shall be constructed of materials that are compatible with the collection surface and the rainwater quality for the desired end use. Joints shall be watertight.

❖ The construction materials of gutters and downspouts must be "compatible" with the raw rainwater quality that is needed for the end uses. The intent is the same as for roof materials that collect rainwater. As gutters are often in direct contact with roofing materials (the collection surface), material compatibility between the two surfaces must be considered. For example, an aluminum collection surface (roofing) might not be compatible with galvanized steel gutters.

P2912.5.1 Slope. Roof gutters, leaders and rainwater collection piping shall slope continuously toward collection inlets and shall be free of leaks. Gutters and downspouts shall have a slope of not less than $^1/_8$ inch per foot (10.4 mm/m) along their entire length. Gutters and downspouts shall be installed so that water does not pool at any point.

❖ Once rainwater is collected by gutters, the rainwater must be kept moving toward the storage tank so that particulate matter and debris do not drop out of the flow. Pockets and pools of nonmoving water will begin to create bacteria "nests" that will contaminate the stored rainwater after the roof washer has wasted the highly contaminated initial flow from the roof.

P2912.5.2 Cleanouts. Cleanouts shall be provided in the water conveyance system to allow access to filters, flushes, pipes and downspouts.

❖ There needs to be access to the collection and conveyance systems and filters for conveying rainwater so that debris can be periodically cleared from those systems.

P2912.6 Drainage. Water drained from the roof washer or debris excluder shall not be drained to the sanitary sewer. Such water shall be diverted from the storage tank and shall discharge to a location that will not cause erosion or damage to property. Roof washers and debris excluders shall be provided with an automatic means of self-draining between rain events and shall not drain onto roof surfaces.

❖ The first sentence prohibits wasted storm water (rainwater) from entering a drain system for sewage only. Rainwater does not need sewage treatment processes, whether they be public or private systems, and adding this water unnecessarily taxes those systems. Control of rainwater runoff from a roof washer or debris excluder is required to avoid erosion and other possible negative effects.

Roof washers and debris excluders must be self-draining so that there is not captive water that could harbor mosquitos. Also, captive water during freezing temperatures could freeze and damage these devices.

The water and debris that is excluded from going to the storage tank should not be wasted to the roof surface, as this will only reintroduce the contaminants to the collection system after the roof washer has cycled for the initial rainwater flow.

P2912.7 Collection pipe. Rainwater collection and conveyance systems shall utilize drainage piping *approved* for use within plumbing drainage systems to collect and convey captured rainwater. Vent piping *approved* for use within plumbing venting systems shall be utilized for vents within the rainwater system. Collection and vent piping materials shall comply with Section P3002.

❖ Where piping is used to convey rainwater or to vent rainwater systems, the piping must comply with materials that are in accordance with Section P3002.

P2912.7.1 Installation. Collection piping conveying captured rainwater shall be installed in accordance with Section P3005.3.

❖ See the commentary to Section P2911.4.1.

P2912.7.2 Joints. Collection piping conveying captured rainwater shall utilize joints *approved* for use with the distribution piping and appropriate for the intended applications as specified in Section P3003.

❖ See the commentary to Section P2911.4.2.

P2912.7.3 Size. Collection piping conveying captured rainwater shall be sized in accordance with drainage-sizing requirements specified in Section P3005.4.

❖ See the commentary to Section P2911.4.3.

P2912.7.4 Marking. Additional marking of collection piping conveying captured rainwater for reuse shall not be required beyond that required for sanitary drainage, waste, and vent piping by Chapter 30.

❖ See the commentary to Section P2911.4.4.

P2912.8 Filtration. Collected rainwater shall be filtered as required for the intended end use. Filters shall be accessible for inspection and maintenance. Filters shall utilize a pressure gauge or other *approved* method to provide indication when a filter requires servicing or replacement. Filters shall be installed with shutoff valves installed immediately upstream and downstream to allow for isolation during maintenance.

❖ See the commentary to Section P2911.5.

P2912.9 Disinfection. Where the intended application for rainwater requires disinfection or other treatment or both, it shall be disinfected as needed to ensure that the required water quality is delivered at the point of use.

❖ Collected rainwater will not require disinfection for many end use applications. The designer decides whether there is need for disinfection in order to meet the jurisdiction's water quality requirements for the end use application.

P2912.10 Storage tanks. Storage tanks utilized in nonpotable rainwater collection and conveyance systems shall comply with Section P2910.9 and Sections P2912.10.1 through P2912.10.3.

❖ In addition to Section P2910.9, there are additional requirements for nonpotable water tanks.

P2912.10.1 Location. Storage tanks shall be located with a minimum horizontal distance between various elements as indicated in Table P2912.10.1.

❖ This section refers to Table P2912.10.1, which is similar to Table P2911.7.1, except there is not a concern for the rainwater contaminating other water sources. Rainwater does not offer the tree roots as much "incentive" to travel towards the storage tank where the tank is underground.

P2912.10.2 Inlets. Storage tank inlets shall be designed to introduce collected rainwater into the tank with minimum turbulence, and shall be located and designed to avoid agitating the contents of the storage tank.

❖ The introduction of rainwater into a storage tank (already having some water in the tank) must be in a manner that limits agitation of the water. Typically, the water is introduced below a certain water "low point" level but above the bottom of the tank. Although there is usually not too much floating debris on the surface of the water, a biofilm does develop on the surface of the water and the bottom of the tank. The biofilms are helpful in maintaining the water in good condition. Therefore, avoiding unnecessary turbulence contributes to the health of these biofilms.

P2912.10.3 Outlets. Outlets shall be located not less than 4 inches (102 mm) above the bottom of the storage tank and shall not skim water from the surface.

❖ The purpose of the section is to allow some volume for sediment to settle in the bottom of the tank and to not have pump intakes or intake piping (or floating intake mechanisms) pulling water (and sediment) from this area.

TABLE P2912.10.1
LOCATION OF RAINWATER STORAGE TANKS

ELEMENT	MINIMUM HORIZONTAL DISTANCE FROM STORAGE TANK (feet)
Critical root zone (CRZ) of protected trees	2
Lot line adjoining private lots	5
Seepage pits	5
Septic tanks	5

For SI: 1 foot = 304.8 mm

P2912.11 Valves. Valves shall be supplied on rainwater collection and conveyance systems in accordance with Sections P2912.11.1 and P2912.11.2.

❖ Sections P2912.11.1 and P2912.11.2 describe the required inlet and outlet valves for the system.

P2912.11.1 Influent diversion. A means shall be provided to divert storage tank influent to allow for maintenance and repair of the storage tank system.

❖ The code does not require a valve for the inlet of the tank; however, valves might be used for this purpose so that maintenance of the storage tank can be performed. Such valves would not necessarily need to be a three-way valve. One valve could be used for the tank inlet and the inlet piping arranged so that the inlet piping (before the valve) "over-flows" to a storm sewer or alternative discharge point while the tank is out of service.

P2912.11.2 Backwater valve. Backwater valves shall be installed on each overflow and tank drain pipe. Backwater valves shall be in accordance with Section P3008.

❖ See the commentary to Section P2911.8.2.

P2912.12 Pumping and control system. Mechanical equipment including pumps, valves and filters shall be easily accessible and removable in order to perform repair, maintenance and cleaning. The minimum flow rate and flow pressure delivered by the pumping system shall appropriate for the application and in accordance with Section P2903.

❖ See the commentary to Section P2903.

P2912.13 Water-pressure-reducing valve or regulator. Where the water pressure supplied by the pumping system exceeds 80 psi (552 kPa) static, a pressure-reducing valve shall be installed to reduce the pressure in the rainwater distribution system piping to 80 psi (552 kPa) static or less. Pressure-reducing valves shall be specified and installed in accordance with Section P2903.3.1.

❖ See the commentary to Section P2911.10.

P2912.14 Distribution pipe. Distribution piping utilized in rainwater collection and conveyance systems shall comply with Sections P2912.14.1 through P2912.14.3.

Exception: Irrigation piping located outside of the building and downstream of a backflow preventer.

❖ See the commentary to Section P2911.11.

P2912.14.1 Materials, joints and connections. Distribution piping shall conform to the standards and requirements specified in Section P2906 for nonpotable water.

❖ See the commentary to Section P2912.11.1.

P2912.14.2 Design. Distribution piping systems shall be designed and sized in accordance with the Section P2903 for the intended application.

❖ See the commentary to Section P2912.11.1.

P2912.14.3 Labeling and marking. Nonpotable rainwater distribution piping labeling and marking shall comply with Section P2901.2.

❖ See the commentary to Section P2912.11.3.

P2912.15 Tests and inspections. Tests and inspections shall be performed in accordance with Sections P2912.15.1 through P2912.15.8.

❖ Piping must be tested after it is installed to ensure the integrity of the installation. Note that the tests of piping are not necessarily attempting to replicate an actual condition of service. Gutters need to be inspected to make sure they have proper slope. This section is also concerned with water quality at the point of use.

P2912.15.1 Roof gutter inspection and test. Roof gutters shall be inspected to verify that the installation and slope is in accordance with Section P2912.5.1. Gutters shall be tested by pouring not less than one gallon of water (3.8 L) into the end of the gutter opposite the collection point. The gutter being tested shall not leak and shall not retain standing water.

❖ Water collected from roof surfaces cannot be collected unless the water can make it to the storage tank. Improperly sloped gutters (usually backwardly sloped or without slope) will cause the gutters to overflow during peak rainfall events resulting in loss of rainwater. Standing water in gutters will evaporate (another loss) or create pools where mosquitos can proliferate. Leaky gutters also waste rainwater. Pouring water into the most remote end of a gutter system is the simplest way to determine proper slope and to find leaks. Note that the minimum quantity of 1 gallon is to prevent someone from pouring a smaller amount of water into the gutter and because the water did not leak out of the gutter (because the water did not arrive at all points in the gutter system), the gutter is considered "leak free." The intent of this section is to introduce sufficient water to the gutter to test the gutter system for leaks, flow direction and standing water.

P2912.15.2 Roofwasher test. Roofwashers shall be tested by introducing water into the gutters. Proper diversion of the first quantity of water in accordance with the requirements of Section P2912.4 shall be verified.

❖ At the same time that the gutters are being tested, the roof washer operation can be tested. The initial amount of rainwater diversion is chosen by the designer. Knowing that the roof washer is adjustable, verification of the amount for this test is not necessary. Only a verification that the device wastes the initial flow and returns to a position to allow the flow to be sent to the storage tank is necessary.

P2912.15.3 Collection pipe and vent test. Drain, waste and vent piping used for rainwater collection and conveyance systems shall be tested in accordance with Section P2503.

❖ See the commentary to Section P2911.12.1.

P2912.15.4 Storage tank test. Storage tanks shall be tested in accordance with the Section P2910.11.

❖ See the commentary to Section P2911.12.2.

P2912.15.5 Water supply system test. The testing of makeup water supply piping and distribution piping shall be conducted in accordance with Section P2503.7.

❖ Makeup water supply piping will be a potable water supply. The piping must be leak free. Note that air testing of plastic piping is prohibited. And where piping other than plastic is tested, test air pressure is only required to be 50 psi. Where water testing is performed, the test pressure must be the working pressure of the system.

However, in service, the piping is only required to be leak free at the normal water pressures encountered. This does not mean that all water piping to the storage tank will experience the same pressure. For example, in a large nonpotable water reuse system, the makeup water control valve might be a 1-inch motorized ball valve located some distance away from the storage tank. The valve might even be below the top of the storage tank. When the valve opens, the piping downstream of the valve will only be subjected to the pressure of the flow resistance in the downstream piping and any difference in the elevation between the valve and the outlet of the piping. That pressure will most likely be much less than the static pressure upstream of the makeup water valve. For the piping from the makeup up water valve to the storage tank, the "test" of the piping should simply be a visual test. There is no need to "temporarily install a test valve" right at the tank for testing this piping at full line pressure. In this example, this would be a 3-inch valve, which would be a needless expenditure.

Testing of distribution piping is not any different than for potable water.

P2912.15.6 Inspection and testing of backflow prevention assemblies. The testing of backflow preventers and backwater valves shall be conducted in accordance with Section P2503.8.

❖ See the commentary to Section P2911.12.4.

P2912.15.7 Inspection of vermin and insect protection. Inlets and vents to the system shall be inspected to verify that each is protected to prevent the entrance of insects and vermin into the storage tank and piping systems in accordance with Section P2910.7.

❖ This section refers back to Section P2510.7 so that verification can be made that protection has been provided.

P2912.15.8 Water quality test. The quality of the water for the intended application shall be verified at the point of use in accordance with the requirements of the jurisdiction.

❖ The jurisdiction allows the installation of the nonpotable rainwater reuse system based on the premise that a certain quality level, at the end use point, would be achieved by the treatment and disinfection systems of the rainwater reuse system. The installer has to perform tests to show that the system provides that quality at the end use point. This does not mean the collected rainwater in the storage tank must meet these requirements, because the treatment of rainwater occurs after the storage tank.

P2912.16 Operation and maintenance manuals. Operation and maintenance manuals shall be supplied with rainwater collection and conveyance systems in accordance with Sections P2912.16.1 through P2912.16.4.

❖ Sections P2912.16.1 through P2912.16.4 describe the details for operation and maintenance manuals.

P2912.16.1 Manual. A detailed operations and maintenance manual shall be supplied in hard-copy form for each system.

❖ See the commentary to Section P2911.13.1.

P2912.16.2 Schematics. The manual shall include a detailed system schematic, the location of system components and a list of system components that includes the manufacturers and model numbers of the components.

❖ See the commentary to Section P2911.13.2.

P2912.16.3 Maintenance procedures. The manual shall provide a maintenance schedule and procedures for system components requiring periodic maintenance. Consumable parts, including filters, shall be noted along with part numbers.

❖ See the commentary to Section P2911.13.3.

P2912.16.4 Operations procedures. The manual shall include system startup and shutdown procedures, and detailed operating procedures.

❖ See the commentary to Section P2911.13.4.

SECTION P2913
RECLAIMED WATER SYSTEMS

P2913.1 General. The provisions of this section shall govern the construction, installation, alteration and repair of systems supplying nonpotable reclaimed water.

❖ In most situations, reclaimed water is supplied to a building site by a public water supply utility. See the definition of "Reclaimed water" in Chapter 2. The systems that provide reclaimed water to the building site are similar to how a public utility supplies potable water to a building site. The public water supply utility goes to great lengths to make sure that these nonpotable water supplies are well marked/identified in order to avoid inadvertent connections to those systems. In most situations, piping, valves, backflow preventers and piping are colored purple.

However, the code does not limit the use of reclaimed water to only that which is supplied by a public water supply utility. A building site could have a wastewater facility (licensed by authorities having jurisdiction over such facilities) where the effluent is of a quality that meets the jurisdiction's quality requirements for reclaimed water.

P2913.2 Water-pressure-reducing valve or regulator. Where the reclaimed water pressure supplied to the building exceeds 80 psi (552 kPa) static, a pressure-reducing valve shall be installed to reduce the pressure in the reclaimed water distribution system piping to 80 psi (552 kPa) static or less. Pressure-reducing valves shall be specified and installed in accordance with Section P2903.3.1.

❖ See the commentary to Section P2911.10.

P2913.3 Reclaimed water systems. The design of the reclaimed water systems shall conform to accepted engineering practice.

❖ As reclaimed systems will be custom engineered for the specific building site, accepted engineering practices must be used for these systems.

P2913.3.1 Distribution pipe. Distribution piping shall comply with Sections P2913.3.1.1 through P2913.3.1.3.

> **Exception:** Irrigation piping located outside of the building and downstream of a backflow preventer.

❖ See the commentary to Section P2911.11.

P2913.3.1.1 Materials, joints and connections. Distribution piping conveying reclaimed water shall conform to standards and requirements specified in Section P2905 for nonpotable water.

❖ See the commentary to Section P2911.11.1.

P2913.3.1.2 Design. Distribution piping systems shall be designed and sized in accordance with Section P2903 for the intended application.

❖ See the commentary to Section P2911.11.2.

P2913.3.1.3 Labeling and marking. Nonpotable rainwater distribution piping labeling and marking shall comply with Section P2901.2.

❖ See the commentary to Section P2911.11.3.

P2913.4 Tests and inspections. Tests and inspections shall be performed in accordance with Sections P2913.4.1 and P2913.4.2.

❖ Piping must be tested after it is installed to ensure the integrity of the installation and that the piping does not leak.

P2913.4.1 Water supply system test. The testing of makeup water supply piping and reclaimed water distribution piping shall be conducted in accordance with Section P2503.7.

❖ See the commentary to Section P2503.7.

P2913.4.2 Inspection and testing of backflow prevention assemblies. The testing of backflow preventers shall be conducted in accordance with Section P2503.8.

❖ See the commentary to Section P2503.8.

Bibliography

The following resource materials were used in the preparation of the commentary for this chapter of the code:

ASTM D2282–99e01, *Specification for Acrylonitrile-Butadiene-Styrene (ABS) Plastic Pipe (SDR-PR)*. West Conshohocken, PA: ASTM International, 2001.

ASTM D2737–03, *Specification for Polyethylene (PE) Plastic Tubing*. West Conshohocken, PA: ASTM International, 2003.

ASTM D2846/D2846M–99, *Specification for Chlorinated Poly (Vinyl Chloride) (CPVC) Plastic Hot- and Cold-Water Distribution Systems*. West Conshohocken, PA: ASTM International, 1999.

ASTM F441/F441M–02, *Specification for Chlorinated Poly (Vinyl Chloride) (CPVC) Plastic Pipe, Schedules 40 and 80*. West Conshohocken, PA: ASTM International, 2002.

ASTM F876–04, *Specification for Cross-linked Polyethylene (PEX) Tubing*. West Conshohocken, PA: ASTM International, 2004.

Chapter 30:
Sanitary Drainage

User note: Code change proposals to this chapter will be considered by the IRC – Plumbing and Mechanical Code Development Committee during the 2015 (Group A) Code Development Cycle.

General Comments

This chapter contains the requirements pertaining to the design, sizing and installation of sanitary drainage systems. Because gray water is a subset of sanitary drainage, by default, a gray water collection system must be designed in the same manner as a sanitary drainage system except the discharge is collected in a gray water tank.

Material requirements for sanitary drainage systems and applicable standards for the various materials are given in a table format for ease of reference. This chapter discusses joining methods of similar or different types of material and includes a table the designer can use to determine the drainage fixture unit (d.f.u) values as they apply to various fixtures and appliances.

It also addresses the appropriate use of various fittings, provisions for future fixtures or connections, adequate placement of cleanouts, sizing requirements pertaining to horizontal and vertical drain lines, and limitations pertaining to vertical and horizontal offsets in vertical drainage stacks. It includes requirements for drainage fixtures and drains that cannot be drained by gravity to the sewer.

This chapter regulates the use of pumps. It provides the requirement for, and the manufacture of, backwater valves.

Purpose

This chapter states the minimum requirements for achieving a properly sized sanitary drainage system. It addresses the proper use of fittings for directing the flow and its contents effectively. It also outlines materials and provisions necessary for servicing the drainage system.

A properly working drainage system will depend on adequate air to balance the pressure fluctuations that are common in such systems; hence, there is a need for properly sized and located vents to the atmosphere. Venting requirements are located in Chapter 31.

In the 19th century, typhoid fever, cholera and dysentery were common. The elimination of these diseases in industrialized nations has been due in large part to modern plumbing systems with proper drainage piping. Medical professionals give much of the credit for improvements in health and longevity to the plumbing profession. Medicine alone, without improvements in sanitation practices, would have had only a marginal effect.

SECTION P3001
GENERAL

P3001.1 Scope. The provisions of this chapter shall govern the materials, design, construction and installation of sanitary drainage systems. Plumbing materials shall conform to the requirements of this chapter. The drainage, waste and vent (DWV) system shall consist of piping for conveying wastes from plumbing fixtures, appliances and appurtenances, including fixture traps; above-grade drainage piping; below-grade drains within the building (*building drain*); below- and above-grade venting systems; and piping to the public sewer or private septic system.

❖ This chapter covers both sanitary drainage and sewers. Because gray water is a subset of sanitary drainage, by default, a gray water collection system must be designed in the same manner as a sanitary drainage system except the discharge is collected in a gray water tank.

All portions of the drain, waste and vent (DWV) system are listed here. Although this chapter primarily addresses drainage and waste lines, venting is also an important part of the entire drainage system in that it provides the air movement needed for the drainage system to function properly.

P3001.2 Protection from freezing. No portion of the above-grade DWV system, other than vent terminals, shall be located outside of a building, in *attics* or crawl spaces, concealed in outside walls, or in any other place subjected to freezing temperatures unless adequate provision is made to protect them from freezing by insulation or heat or both, except in localities having a winter design temperature greater than 32°F (0°C) (ASHRAE 97.5 percent column, winter, see Chapter 3).

❖ In areas where freezing can occur, care must be exercised not to install drainage systems where their contents can freeze. To determine whether the locality is subject to freezing, the ASHRAE 97.5 percent column of the winter design condition table is referenced. The code user is referred to Chapter 3. A review of Table R301.2(1) for the winter design temperature will show that this determination can be made by the use of Appendix D of the *International Plumbing Code®* (IPC®). If the temperature is 32°F (0°C) or lower, this section will apply. In this case, all drainage and vent piping must not be allowed outside the building (unless buried and below the frost level), in attics or crawl spaces, concealed in outside walls or any other place subject to freezing. If the designer has no alternative but to place the piping in these places, provisions must be made to either insulate or heat those sections so that the contents will not freeze. The concern is not only that a stoppage can occur in the pipe, but also that the pipe can burst because freezing water expands in volume. No amount of insulation by itself (without a heat source) can prevent freezing. Insulation can only delay freezing by slowing the rate of heat loss. Pipe heating devices, such as a self-limiting heat tape, can be used. Care is especially necessary where portions of the system that contain standing water, such as a trap, are installed; for example, where the trap is installed within a cantilever and may extend below the floor insulation (see Commentary Figure P2603.6).

P3001.3 Flood-resistant installation. In flood hazard areas as established by Table R301.2(1), drainage, waste and vent systems shall be located and installed to prevent infiltration of floodwaters into the systems and discharges from the systems into floodwaters.

❖ See the commentary to Section P2602.2.

SECTION P3002
MATERIALS

P3002.1 Piping within buildings. Drain, waste and vent (DWV) piping in buildings shall be as indicated in Tables P3002.1(1) and P3002.1(2) except that galvanized wrought-iron or galvanized steel pipe shall not be used underground and shall be maintained not less than 6 inches (152 mm) above ground. Allowance shall be made for the thermal expansion and contraction of plastic piping.

❖ Above-ground DWV piping within buildings must be of the materials specified in Table P3002.1(1). Underground DWV piping must be of the materials specified in Table P3002.1(2). Because galvanized wrought iron and steel pipe are very susceptible to exterior corrosion where buried in the earth, these materials are prohibited from being installed underground or within 6 inches (152 mm) above the ground. Where plastic pipe is used, there must be consideration given to the thermal expansion and contraction of the pipe because of

TABLE P3002.1(1)
ABOVE-GROUND DRAINAGE AND VENT PIPE

MATERIAL	STANDARD
Acrylonitrile butadiene styrene (ABS) plastic pipe in IPS diameters, including schedule 40, DR 22 (PS 200) and DR 24 (PS 140); with a solid, cellular core or composite wall	ASTM D2661; ASTM F628; ASTM F1488; CSA B181.1
Cast-iron pipe	ASTM A74; CISPI 301; ASTM A888
Copper or copper-alloy pipe	ASTM B42; ASTM B43; ASTM B302
Copper or copper-alloy tubing (Type K, L, M or DWV)	ASTM B75; ASTM B88; ASTM B251; ASTM B306
Galvanized steel pipe	ASTM A53
Polyolefin pipe	CSA B181.3
Polyvinyl chloride (PVC) plastic pipe in IPS diameters, including schedule 40, DR 22 (PS 200) and DR 24 (PS 140); with a solid, cellular core or composite wall	ASTM D2665; ASTM F891; CSA B181.2; ASTM F1488
Polyvinyl chloride (PVC) plastic pipe with a 3.25 inch O.D. and a solid, cellular core or composite wall	ASTM D2949; ASTM F1488
Stainless steel drainage systems, Types 304 and 316L	ASME A112.3.1

For SI: 1 inch = 25.4 mm.

TABLE P3002.1(2)
UNDERGROUND BUILDING DRAINAGE AND VENT PIPE

PIPE	STANDARD
Acrylonitrile butadiene styrene (ABS) plastic pipe in IPS diameters, including schedule 40, DR 22 (PS 200) and DR 24 (PS 140); with a solid, cellular core or composite wall	ASTM D2661; ASTM F628; ASTM F1488; CSA B181.1
Cast-iron pipe	ASTM A74; CISPI 301; ASTM A888
Copper or copper alloy tubing (Type K, L, M or DWV)	ASTM B75; ASTM B88; ASTM B251; ASTM B306
Polyolefin pipe	ASTM F1412; CSA B181.3
Polyvinyl chloride (PVC) plastic pipe in IPS diameters, including schedule 40, DR 22 (PS 200) and DR 24 (PS 140); with a solid, cellular core or composite wall	ASTM D2665; ASTM F891; ASTM F1488; CSA B181.2
Polyvinyl chloride (PVC) plastic pipe with a 3.25 inch O.D. and a solid, cellular core or composite wall	ASTM D2949; ASTM F1488
Stainless steel drainage systems, Type 316L	ASME A112.3.1

For SI: 1 inch = 25.4 mm.

the much higher rate of expansion in comparison to metallic pipe. Where expansion and contraction of plastic pipe will cause damage to the pipe, a means for controlling the change of length must be incorporated into the installation.

P3002.2 Building sewer. *Building sewer* piping shall be as shown in Table P3002.2. Forced main sewer piping shall conform to one of the standards for ABS plastic pipe, copper or copper-alloy tubing, PVC plastic pipe or pressure-rated pipe indicated in Table P3002.2.

❖ Building sewer piping must be of the pipe materials listed in Table P3002.2. These materials are suitable for installation underground and have appropriately designed (sanitary pattern) fittings and smooth interior surfaces so as not to impede gravity flow.

The term "forced main" actually means a completely full, pressurized sewer pipe (as opposed to a partially full gravity flow sewer pipe). A pump or ejector is used to "push" the waste through a pipe up to a higher elevation. The amount of pressure that the pressurized sewer pipe must handle is the maximum discharge pressure of the pump or ejector. The following is an example of this scenario:

Consider the outlet of a building sewer from a building at an elevation that is 12 feet (3658 mm) below the elevation of the gravity public sewer main. A pump basin is installed to collect the sewage from the gravity building sewer. A sewage pump is installed in the basin to pump the sewage up to the public sewer main. If the pump discharge outlet is 2 feet (610 mm) below the

building drain elevation, the head that the pump must generate to begin to cause flow in the pipe is 14 feet (4267 mm) (12 + 2 = 14). As flow develops in the pressurized sewer pipe, pipe friction losses require the pump to develop additional head to be able to sustain the flow at the desired flow rate. For simplicity of this example, assume that the pipe friction loss requires an additional 1 foot (305 mm) of head (pressure). Therefore, the total dynamic head that the pump must generate is 15 feet (4572 mm) (14 + 1 = 15). A pump model is selected to produce 15 feet (4572 mm) of head (pressure) and the flow rate is verified to be adequate.

A general characteristic of a majority of sewage pumps is that they develop the highest discharge head when the flow rate is zero or approaches zero. A flow rate of zero would occur if the discharge pipe became blocked such as might occur if the public sewer main became surcharged. To illustrate this, assume that the pump curve (or chart) for the model of pump chosen indicates a maximum discharge head of 20 feet (6096 mm). This head is equivalent to 20 feet (6096 mm) × 0.433 pounds per square inch (psi)/foot of head = 8.6 psi (59 kPa). Therefore, the discharge piping at the outlet of the pump must be suitable for 8.6 psi (59 kPa). However, at higher elevations along the run of the discharge pipe, the pressure in the pipe becomes less (see Commentary Figure P3002.2).

The point of the preceding example is to illustrate how the pressures are generally determined for pressurized sewer piping ("forced mains") and the potential magnitudes of those pressures. Although this section

TABLE P3002.2
BUILDING SEWER PIPE

MATERIAL	STANDARD
Acrylonitrile butadiene styrene (ABS) plastic pipe in IPS diameters, including schedule 40, DR 22 (PS 200) and DR 24 (PS 140); with a solid, cellular core or composite wall	ASTM D2661; ASTM F628; ASTM F1488
Cast-iron pipe	ASTM A74; ASTM A888; CISPI 301
Acrylonitrile butadiene styrene (ABS) plastic pipe in sewer and drain diameters, including SDR 42 (PS 20), PS35, SDR 35 (PS 45), PS50, PS100, PS140, SDR 23.5 (PS 150) and PS200; with a solid, cellular core or composite wall	ASTM F1488; ASTM D2751
Polyvinyl chloride (PVC) plastic pipe in sewer and drain diameters, including PS 25, SDR 41 (PS 28), PS 35, SDR 35 (PS 46), PS 50, PS 100, SDR 26 (PS 115), PS140 and PS 200; with a solid, cellular core or composite wall	ASTM F891; ASTM F1488; ASTM D3034; CSA B182.2; CSA B182.4
Concrete pipe	ASTM C14; ASTM C76; CSA A257.1M; CSA A257.2M
Copper or copper-alloy tubing (Type K or L)	ASTM B75; ASTM B88; ASTMB 251
Polyethylene (PE) plastic pipe (SDR-PR)	ASTM F714
Polyolefin pipe	ASTM F1412; CSA B181.3
Polyvinyl chloride (PVC) plastic pipe in IPS diameters, including schedule 40, DR 22 (PS 200) and DR 24 (PS 140); with solid, cellular core or composite wall	ASTM D2665; ASTM D2949; ASTM D3034; ASTM F1412; CSA B182.2; CSA B182.4
Polyvinyl chloride (PVC) plastic pipe with a 3.25 inch O.D. and a solid, cellular core or composite wall	ASTM D2949, ASTM F1488
Stainless steel drainage systems, Types 304 and 316L	ASME A112.3.1
Vitrified clay pipe	ASTM C425; ASTM C700

For SI: 1 inch = 25.4 mm.

specifically requires pressurized sewer piping (or tubing) be made of ABS, PVC, copper, copper alloy or pressure-rated pipes that are listed in Table P3002.2, it is the designer's or installer's responsibility to choose piping and components from that set of materials that are suitable for the intended pressures. Where the maximum pump head (pressure) is very low, it might not be necessary to use pressure-rated ABS and PVC pipes and fittings since pipes for gravity drainage and sanitary drainage fittings might be adequate for the pressures involved.

P3002.2.1 Building sewer pipe near the water service. The proximity of a *building sewer* to a water service shall comply with Section P2905.4.2.

❖ See the commentary to Section P2905.4.2.

P3002.3 Fittings. Pipe fittings shall be *approved* for installation with the piping material installed and shall comply with the applicable standards indicated in Table P3002.3.

❖ Each fitting is designed for installation in a particular system with a given material. Drainage pattern fittings are typically "directional" to guide the flow derived from gravitational force. The required radius of curvature of the fittings will vary with the location where the fittings are used. Flows changing direction from horizontal to vertical, vertical to horizontal and horizontal to horizontal require progressively longer radiuses to allow the flow to change direction without slowing down to a

point where solids can separate from the liquid. When the velocity of the flow slows below a certain threshold [typically 2 feet (610 mm) per second], solids may separate and adhere to the pipe, thus creating blockages. Therefore, the type of fittings used in drainage systems not only have to be directional but must also be correctly selected for the specific location where they are to be installed. Vent fittings must never be used as drainage fittings. Vent fittings have less radius than drainage fittings because they only convey air.

P3002.3.1 Drainage. Drainage fittings shall have a smooth interior waterway of the same diameter as the piping served. Fittings shall conform to the type of pipe used. Drainage fittings shall not have ledges, shoulders or reductions that can retard or obstruct drainage flow in the piping. Threaded drainage pipe fittings shall be of the recessed drainage type, black or galvanized. Drainage fittings shall be designed to maintain one-fourth unit vertical in 12 units horizontal (2-percent slope) grade. This section shall not be applicable to tubular waste fittings used to convey vertical flow upstream of the trap seal liquid level of a fixture trap.

❖ Successful gravity flow of waste water, especially waste water with solids, requires smooth internal piping surfaces that do not impede flow. The combinations of pipe and pipe fittings for drainage systems must be carefully chosen so that ledges, shoulders or reductions in the direction of flow are not created. A simple example of an inappropriate combination

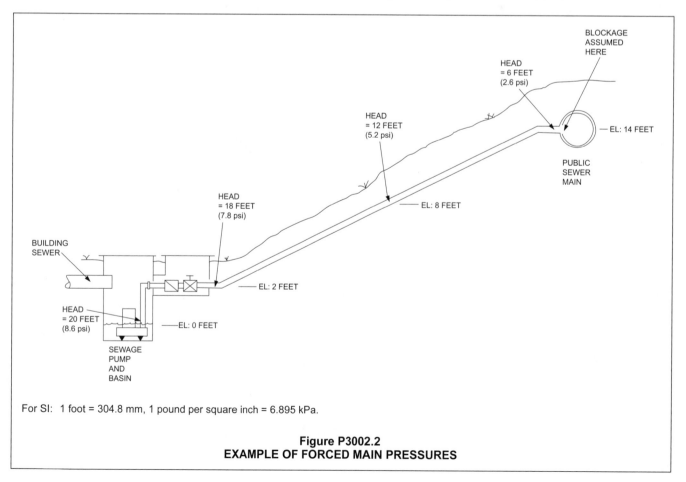

For SI: 1 foot = 304.8 mm, 1 pound per square inch = 6.895 kPa.

Figure P3002.2
EXAMPLE OF FORCED MAIN PRESSURES

would be Schedule 80 wall pipe with commonly available drainage fittings. Drainage fittings (plastic and cast iron) are intended for use only with Schedule 40 wall pipe. Schedule 80 wall pipe has a smaller inside diameter, which would create a ledge in the direction of flow. Another example of an inappropriate combination would be the use of pressure-pipe fittings for a drainage application. The inside diameter of the body of a pressure-pipe fitting is not reduced such that it is the same diameter as the Schedule 40 drainage pipe that connects to the fitting. Only drainage fittings must be used in gravity-flow piping systems.

The last line of this section clarifies that the requirement for not having ledges or obstructions in the flow path does not apply to tubular waste fittings installed in a vertical flow orientation. For example, tubular waste fittings such as a disposal connect tee with an internal baffle or a pop-up actuator rod tailpiece both create a restriction to gravity flow. It is not the intent of this section to prohibit these tubular waste fittings when they are used in a vertical flow orientation; however, it is inappropriate to use such fittings in a horizontal flow orientation. For example, a dishwasher connection tubular fitting with an internal baffle must not be installed in a horizontal flow orientation.

P3002.4 Other materials. Sheet lead, lead bends, lead traps and sheet copper shall comply with Sections P3002.4.1 through P3002.4.3.

❖ This section regulates the material requirements for sheet lead, lead bends, lead traps and sheet copper. Each of these are addressed in the subsections herein.

P3002.4.1 Sheet lead. Sheet lead shall weigh not less than indicated for the following applications:

1. Flashing of vent terminals, 3 psf (15 kg/m²).

2. Prefabricated flashing for vent pipes, $2^1/_2$ psf (12 kg/m²).

❖ Flashing materials are used to close off the opening between a vent pipe and the hole in the roof made for the vent pipe to terminate outdoors. Where sheet lead is used to field fabricate a vent flashing, the sheet lead must have a weight of not less than 3 pounds per square foot (psf) (15 kg/m²). Where such flashings are factory fabricated, the sheet lead must have a weight of not less than $2^1/_2$ psf (12 kg/m²) (see commentary, Section P2606).

P3002.4.2 Lead bends and traps. Lead bends and lead traps shall be not less than $1/_8$-inch (3 mm) wall thickness.

❖ Although lead is not one of the materials referenced in Table P3002.1(1), P3002.1(2) or P3002.3, on those rare occasions where repairs must be made to existing installations having lead bends and lead traps, the wall thickness of the material must be not less than 0.125 inch (3.2 mm).

P3002.4.3 Sheet copper. Sheet copper shall weigh not less than indicated for the following applications:

1. General use, 12 ounces per square feet (4 kg/m²).

2. Flashing for vent pipes, 8 ounces per square feet (2.5 kg/m²).

❖ Sheet copper is sometimes used as a lining or flashing material. Where used as a flashing material, the copper must not weigh less than 8 ounces per square foot (3.7 kg/m²). Where used for other purposes (such as lining material), the copper must weigh not less than 12 ounces per square foot (3.7 kg/m²). See Section P2709.3.1 for the requirements pertaining to the use of sheet copper for shower lining material.

TABLE P3002.3
PIPE FITTINGS

PIPE MATERIAL	FITTING STANDARD
Acrylonitrile butadiene styrene (ABS) plastic pipe in IPS diameters	ASTM D2661; ASTM D3311; ASTM F628; CSA B181.1
Cast-iron	ASME B16.4; ASME B16.12; ASTM A74; ASTM A888; CISPI 301
Acrylonotrile butadiene styrene (ABS) plastic pipe in sewer and drain diameters	ASTM D2751
Polyvinyl chloride (PVC) plastic pipe in sewer and drain diameters	ASTM D 3034
Copper or copper alloy	ASME B16.15; ASME B16.18; ASME B16.22; ASME B16.23; ASME B16.26; ASME B16.29
Gray iron and ductile iron	AWWA C 110/A21.10
Polyolefin	ASTM F1412; CSA B181.3
Polyvinyl chloride (PVC) plastic in IPS diameters	ASTM D2665; ASTM D3311; ASTM F1866
Polyvinyl chloride (PVC) plastic pipe with a 3.25 inch O.D.	ASTM D2949
PVC fabricated fittings	ASTM F1866
Stainless steel drainage systems, Types 304 and 316L	ASME A112.3.1
Vitrified clay	ASTM C700

For SI: 1 inch = 25.4 mm.

SECTION P3003
JOINTS AND CONNECTIONS

P3003.1 Tightness. Joints and connections in the DWV system shall be gas tight and water tight for the intended use or pressure required by test.

❖ DWV systems carry liquid waste and odors (gases) that must be prevented from leaking into the building. Therefore, joints and connections in those systems must be gas tight and liquid tight under test pressure conditions so that they are leak free during normal system operation (see commentary, Section 2503.5).

P3003.1.1 Threaded joints, general. Pipe and fitting threads shall be tapered.

❖ Tapered threads are required at all threaded connections so that they can be tightened to produce a leak free connection under service and test conditions. This is one of those requirements that was always intended and understood, but not actually stated in the code.

P3003.2 Prohibited joints. Running threads and bands shall not be used in the drainage system. Drainage and vent piping shall not be drilled, tapped, burned or welded.

The following types of joints and connections shall be prohibited:

1. Cement or concrete.

2. Mastic or hot-pour bituminous joints.

3. Joints made with fittings not *approved* for the specific installation.

4. Joints between different diameter pipes made with elastomeric rolling O-rings.

5. Solvent-cement joints between different types of plastic pipe.

6. Saddle-type fittings.

❖ Running threads are threads that have no taper and, therefore, will not tighten to produce leak-free joints. Drilling, tapping, burning and welding to create fittings for connections are prohibited because these means do not create the smooth transitions like those found in factory-cast or molded directional fittings.

This section specifically prohibits certain types of joints. Cement and concrete are not effective in sealing a pipe joint because such materials are inflexible and susceptible to cracking and displacement. Mastic or hot pour bituminous joints could dry out and crack. A rolling O-ring has no resistance to being pushed or rolled out of a joint when exposed to pressure or the movement of pipe caused by expansion and contraction. Because a solvent-cement joint actually fuses the plastic pipe and plastic fitting materials, if the materials are different, proper fusion will not occur and the strength and integrity of the joint will be adversely affected. For example, solvent cementing is prohibited between ABS and PVC, PVC and CPVC or ABS and CPVC. Saddle-type fittings can be too easily moved out of alignment. The hole cut into the side of the pipe can weaken the pipe and, typically, does not form a drainage pattern connection. However, saddle-type fit-

tings are commonly used for connecting a building sewer to an existing public sewer.

P3003.3 ABS plastic. Joints between ABS plastic pipe or fittings shall comply with Sections P3003.3.1 through P3003.3.3.

❖ Acceptable joints for ABS plastic pipe are identified in the referenced sections.

P3003.3.1 Mechanical joints. Mechanical joints on drainage pipes shall be made with an elastomeric seal conforming to ASTM C1173, ASTM D3212 or CSA B602. Mechanical joints shall be installed only in underground systems unless otherwise *approved*. Joints shall be installed in accordance with the manufacturer's instructions.

❖ Mechanical joints are predominantly used in underground applications (e.g., bell or hub compression gasket joints and elastomeric transition couplings). Mechanical joints are typically of the unrestrained type. Above-ground mechanical joints are limited primarily to rehabilitation installations and connections to existing systems when it is impractical to make another type of joint. Mechanical joints are sometimes used where other joining means cannot be used. For example, underground PVC pipe may be joined with an elastomeric coupling where, because of water in the trench, solvent cementing is not possible. The referenced standards are addressed only with respect to the elastomeric material used to make the seal in a mechanical joint, not the entire coupling or fitting assembly.

P3003.3.2 Solvent cementing. Joint surfaces shall be clean and free from moisture. Solvent cement that conforms to ASTM D2235 or CSA B181.1 shall be applied to joint surfaces. The joint shall be made while the cement is wet. Joints shall be made in accordance with ASTM D2235, ASTM D2661, ASTM F628 or CSA B181.1. Solvent-cement joints shall be permitted above or below ground.

❖ Solvent cementing is the most common method of joining ABS plastic pipe and fittings. This section clarifies that solvent cementing is an acceptable joining method for both above-ground and below-ground installations. Although it is the most common method, solvent cementing is one of the most misunderstood methods of joining pipe. ASTM International (ASTM) and Canadian Standards Association (CSA) referenced standards indicate how to handle solvent cement safely and recommend a practice for making joints. ABS solvent cement is a flammable liquid and a source of flammable vapors. It must be kept a safe distance from any source of ignition. Solvent cement should also be used in a well-ventilated area. Prolonged breathing of solvent-cement vapors can have adverse health effects. Solvent cement is also an eye and skin irritant. If contact with the eyes is possible, protective goggles should be worn. When frequent contact with the skin is likely, impervious gloves should be worn. The pipe and fittings should be approximately the same temperature prior to solvent cementing. The pipe end and socket fitting must be clean, dry and free from grease, oil and other foreign materials. The solvent cement must be uniformly applied to the socket of the fitting and to the end of the pipe to the depth of the socket. Excessive solvent cement should be avoided to prevent a puddle of cement from forming on the inside of the pipe joint, which may deform or structurally weaken the fitting or the pipe. The joint is made by twisting the pipe into the fitting socket immediately after applying the cement. The pipe will tend to back out unless it is restrained; therefore, it must be held in place for a few seconds until the solvent cement begins to set. A small bead of solvent cement forms between the pipe exterior and fitting. The bead must be removed with a clean cloth to avoid the possibility of weakening the pipe wall. A solvent-cement joint should remain for a period of 24 hours before being exposed to working pressure. There should be no rough handling of the pipe for at least 1 hour after the joint is made. Because each type of solvent cement is specifically designed for a given piping material, all-purpose solvent cement or universal solvent cement cannot be used to join ABS pipe or fittings unless it conforms to ASTM D2235 or CSA B181.1. Solvent cement for each plastic pipe material requires a unique mixture or combination of solvents and dissolved plastic resins to conform to the applicable standard. Although all-purpose cement may dissolve the surfaces of ABS, PVC and CPVC, it may be too aggressive or not aggressive enough to make a joint strong enough to meet the minimum strength requirements of the referenced standards. Solvent cement must be identified on its container as complying with the appropriate ASTM or CSA standard.

P3003.3.3 Threaded joints. Threads shall conform to ASME B1.20.1. Schedule 80 or heavier pipe shall be permitted to be threaded with dies specifically designed for plastic pipe. *Approved* thread lubricant or tape shall be applied on the male threads only.

❖ ASME B1.20.1 identifies pipe dimensions as "NPT," which is often thought of as an abbreviation. Some of the terms from which NPT has been shortened are: National Pipe Thread, Nominal Pipe Thread and National Pipe Tapered Thread. NPT is a coded designation: N stands for USA Standard, P indicates pipe and T means that the threads are tapered. Thread lubricant (pipe-joint compound) must be specifically designed for plastic pipe. It is not the same lubricant that is commonly used for metallic pipe, such as steel. Thread lubricant and tape are designed for two purposes: to lubricate the joint for proper thread mating and to fill in small imperfections on the surfaces of the threads. Field threading is limited to solid-core Schedule 80 pipe because the wall thickness is sufficient to accommodate threads without affecting pipe strength.

P3003.4 Cast iron. Joints between cast-iron pipe or fittings shall comply with Sections P3003.4.1 through P3003.4.3.

❖ Acceptable joints for cast-iron pipe or fittings are identified in the referenced sections.

P3003.4.1 Caulked joints. Joints for hub and spigot pipe shall be firmly packed with oakum or hemp. Molten lead shall be poured in one operation to a depth of not less than 1 inch (25 mm). The lead shall not recede more than $^1/_8$ inch (3 mm) below the rim of the hub and shall be caulked tight. Paint, varnish or other coatings shall not be permitted on the jointing material until after the joint has been tested and *approved.* Lead shall be run in one pouring and shall be caulked tight.

❖ A caulked joint is the oldest form of joining cast-iron pipe. The joint is sealed using tightly packed oakum or hemp. The oakum or hemp expands the first time it becomes wet, thus sealing the joint. Lead is poured in place behind the oakum to prevent the oakum from expanding out of the joint. Quite often, a caulked joint will leak when it is first tested because the oakum has not yet expanded in place. Once the oakum has expanded, the joint will not leak.

The joint is made by packing the oakum or hemp into the hub fitting around the pipe to be joined. The oakum and joint surfaces must be dry to prevent the oakum from expanding prematurely. Once in place, the oakum is packed tight with a packing iron and hammer. Molten lead is poured in one continuous operation, securing the oakum. If the joint surfaces or oakum are wet when the lead is poured, rapidly expanding steam can cause the lead to pop or explode. After the lead solidifies, it must be packed tight. A properly made caulked joint will bear marks on the lead caused by a finishing iron used to pack the lead. Because caulked joints are labor intensive and involve the risk of worker injury, and because lead is a health hazard, such joints are becoming increasingly rare.

The code does not recognize a substitute for lead in caulked joints. In the case of remodeling and repair work where one or more caulked joints must be made to allow transitions to different types of piping, building officials have on occasion approved the use of lead substitutes, such as cement-based compounds. Elastomeric seals are also available for the purpose of making transition joints with hubbed cast-iron pipe.

P3003.4.2 Compression gasket joints. Compression gaskets for hub and spigot pipe and fittings shall conform to ASTM C564. Gaskets shall be compressed when the pipe is fully inserted.

❖ The elastomeric gasket is often called a "rubber gasket" or "rubber joint." The gasket is inserted into the hub of the pipe. A lubricant is used to coat the end of the joining pipe and the gasket. The pipe is inserted into the hub using a special tool that pulls the pipe and hub together, compressing the gasket in the annular space between the pipe and hub.

The pipes must be completely pulled together to properly make the joint. A typical compression joint has two main compression points in the seal.

P3003.4.3 Mechanical joint coupling. Mechanical joint couplings for hubless pipe and fittings shall consist of an elastomeric sealing sleeve and a metallic shield that comply with CISPI 310, ASTM C1277 or ASTM C1540. The elastomeric sealing sleeve shall conform to ASTM C564 or CSA B602 and shall have a center stop. Mechanical joint couplings shall be installed in accordance with the manufacturer's instructions.

❖ Mechanical joint couplings refer to couplings that are used to join hubless cast-iron DWV pipe and fittings. Mechanical joint couplings are manufactured in a variety of types, all of which include an elastomeric seal and a clamp assembly that compresses the seal around the outside circumference of the pipe or fitting. Most designs include a metal shield (sleeve) that fits between the inside diameter of the clamps and the outside diameter of the sealing element. The shield protects the elastomeric seal from external punctures and abuse. Thicker shields contribute shear-force-resistance and rigidity to the assembled joint.

Mechanical joint couplings are usually designed to resist the shear forces that can cause misalignment of pipe and fittings. ASTM C1540 and CSA B602 apply only to the elastomeric seal component of mechanical joint couplings. CISPI 310, ASTM C1277 and ASTM C1540 address entire coupling assemblies. The purpose of the center stop in the elastomeric part of the coupling is to prevent the pipe or fitting from penetrating too far into the coupling.

Mechanical joints and mechanical joint couplings must be installed in accordance with the manufacturer's instructions. Manufacturer's instructions require that a small torque wrench be used to tighten the clamps so that clamping force is even but not excessive which could result in damage to the elastomeric seal.

P3003.5 Concrete joints. Joints between concrete pipe and fittings shall be made with an elastomeric seal conforming to ASTM C443, ASTM C1173, CSA A257.3M or CSA B602.

❖ The only acceptable method of joining concrete pipe is the pipe to form a seal. The ASTM and CSA standards regulate the design, material and performance of the elastomeric gasket.

P3003.6 Copper and copper-alloy pipe and tubing. Joints between copper or copper-alloy pipe tubing or fittings shall comply with Sections P3003.6.1 through P3003.6.4.

❖ Acceptable joints for copper or copper-alloy pipe or fittings are identified in the referenced sections.

P3003.6.1 Brazed joints. All joint surfaces shall be cleaned. An *approved* flux shall be applied where required. Brazing materials shall have a melting point in excess of 1,000°F (538°C). Brazing alloys filler metal shall be in accordance with AWS A5.8.

❖ Although brazed joints are similar to soldered joints, brazed joints are joined at a higher temperature [in excess of 1,000°F (538°C)]. Because of the brazed joint's inherent resistance to elevated temperatures, it is much stronger than a soldered joint. Brazing is often referred to as "silver soldering." Silver soldering is more accurately described as silver brazing and uses high-silver alloys primarily composed of silver, copper and zinc. Silver soldering (brazing) typically requires temperatures in excess of 1,000°F (538°C), and such solders are classified as "hard" solders. Confusion has always been present with respect to the distinction between "silver solder" and "silver-bearing solder." Silver solders are unique and may be further subdivided into soft and hard categories, which are determined by the percentages of silver and the other component elements of the particular alloy. The distinction must be made that silver-bearing solders [with a melting point less than 600°F (316°C)] are used in soft-soldered joints, whereas silver solders [with a melting point greater than 1,000°F (538°C)] are used in silver-brazed joints. The two common series of filler metals used for brazing are phosphorous copper brazing rods (BCuP) and silver brazing rods (BAg). BCuP contains phosphorus and copper alloys. BAg contains 30 to 60 percent silver with zinc and copper alloys. Some of the BAg filler metals also contain cadmium. During brazing, such filler metal produces toxic fumes. Brazing should always be done in well-ventilated areas, and protective eye gear must also be worn. The joint surfaces to be brazed must be cleaned, which is typically accomplished by polishing the pipe with an emery cloth. The socket fitting is cleaned with a specially designed brush. Brazing flux is applied to the joint surfaces. Certain types of BCuP filler metals do not require flux for certain types of joints. The manufacturer's recommendations should be consulted to determine when flux is necessary.

The joints are typically heated with an oxyacetylene torch, though an air acetylene torch may be used on a smaller-diameter pipe. The pipe and fitting must be at approximately the same temperature when making the joint. The filler metal flows to the heat with the aid of the flux. The pipe must first be heated above the fitting. When the flux becomes clear, the heat must be directed to the base of the fitting socket. When the flux becomes clear and quiet, the filler metal is applied to the joint. The heat draws the filler metal into the socket. Any remaining flux residue must be removed from the pipe and joint surfaces.

P3003.6.2 Mechanical joints. Mechanical joints shall be installed in accordance with the manufacturer's instructions.

❖ Mechanical joints for copper or copper-alloy pipe are similar to those used to join brass pipe. Because many different types of mechanical joints are available and most couplings are unique to each manufacturer, the joining method must be in accordance with the manufacturer's installation instructions. Such joints are highly resistant to being pulled apart.

P3003.6.3 Soldered joints. Copper and copper-alloy joints shall be soldered in accordance with ASTM B828. Cut tube ends shall be reamed to the full inside diameter of the tube end. All joint surfaces shall be cleaned. Fluxes for soldering shall be in accordance with ASTM B813 and shall become noncorrosive and nontoxic after soldering. The joint shall be soldered with a solder conforming to ASTM B32.

❖ Soldering is the most common method of joining copper pipe and tubing. ASTM B828 governs the procedures for making capillary joints by soldering copper and copper-alloy tube and fittings. To consistently make satisfactory joints, the following sequence of joint preparation and operations must be followed: measuring and cutting, reaming, cleaning, fluxing, assembly and support, heating, applying the solder, and cooling and cleaning. The joint surfaces must be accurately measured because if the tube is too short, it will not reach all the way into the fitting and a proper joint cannot be made. If the tube is too long, the system strain may affect service life.

The tubing may be cut in a number of different ways to produce a square end. The tubing may be cut with a tubing cutter, a hacksaw, an abrasive wheel or a band saw. Regardless of the method, the cut must be square with the run of the tube so it will seat properly in the fitting. The tubing ends, outside of the tube ends to the inside diameter of the tube, must be reamed to remove the burrs created by the cutting operation. The joint surfaces for a soldered joint must be cleaned to remove all oxides and surface soil from the tube ends and fittings. Cleaning can be accomplished using a sand cloth, abrasive pad and fitting brush.

The space between the tube and fitting is the gap the solder metal fills by capillary action. The spacing is critical for the solder metal to flow into the gap and form a strong joint. Uniformity of capillary space will ensure metal capillary flow. Once the joint surfaces are cleaned, the cleaned area must not be touched with bare hands or oily gloves. Skin oils, lubricating oils or grease will impair the solder metal.

Apply a thin, even coating of flux to both the tubing and fitting that will dissolve and remove traces of oxide from the cleaned surfaces. Care must be taken in applying flux because excessive amounts will cause corrosion that may perforate the wall of the tubing, the fitting or both. The flux material must conform to ASTM B813, which establishes minimum performance criteria. This also limits the maximum lead content to 0.2 percent.

Begin heating with the flame alternating from the fitting cup back onto the tube. Touch the solder to the joint and if the solder does not melt, remove it and continue the heating process. Once the melting temperatures are reached, apply heat to the base of the fitting to aid in capillary action in drawing the molten solder into the cup toward the heat source. Molten solder metal is drawn into the joint by capillary action regard-

less of the direction that the metal is being fed (horizontal or vertical). If excessive heat is used, the flux will burn out and char the pipe, and where the temperature is too uneven, the solder will run down either the inside or the outside of the pipe. Either condition will prevent capillary action. Allow the completed joint to cool naturally; cooling it with water will cause unnecessary shock and stress on the joint. When the joint is cool, clean off any remaining flux.

P3003.6.4 Threaded joints. Threads shall conform to ASME B1.20.1. Pipe-joint compound or tape shall be applied on the male threads only.

❖ Before assembly, the threads must be cleaned with a wire brush to remove burrs or chips resulting from the cutting process. Pipe-joint compound or Teflon tape is applied on the male thread to ensure a tight seal. If pipe-joint compound is applied to the female thread, it will enter the system. The male thread is placed inside the female thread and the joint is tightened using pipe wrenches. It is generally limited to smaller pipe diameters because of the amount of effort required to turn a pipe of larger size in making the joint.

P3003.7 Steel. Joints between galvanized steel pipe or fittings shall comply with Sections P3003.7.1 and P3003.7.2.

❖ Acceptable joints for galvanized steel pipe or fittings are identified in the referenced sections.

P3003.7.1 Threaded joints. Threads shall conform to ASME B1.20.1. Pipe-joint compound or tape shall be applied on the male threads only.

❖ The pipe-joint compound used for steel pipe will also be used for both copper and brass pipe. The application of pipe-joint compound or Teflon tape to only male threads prevents the compound from entering the piping system.

P3003.7.2 Mechanical joints. Joints shall be made with an *approved* elastomeric seal. Mechanical joints shall be installed in accordance with the manufacturer's instructions.

❖ There is a wide variety of mechanical joints available to join steel pipe, including plain or beveled ends, cut or rolled grooves. Each requires different pipe-end preparation in making a joint. Rolled grooves are used where the wall of the pipe is not thick enough to have a groove formed around the pipe near the ends. The coupling assembly for both types is the same. The joint consists of an inner elastomeric gasket and an outer split metallic sleeve with an integral bolt used for tightening. The assembly is placed over both ends and joined with the grooves in the pipe. The bolt is tightened to the torque requirements listed in the manufacturer's installation instructions.

P3003.8 Lead. Joints between lead pipe or fittings shall comply with Sections P3003.8.1 and P3003.8.2.

❖ Acceptable joints for lead pipe or fittings are identified in the referenced sections.

P3003.8.1 Burned. Burned joints shall be uniformly fused together into one continuous piece. The thickness of the joint shall be not less than the thickness of the lead being joined. The filler metal shall be of the same material as the pipe.

❖ A burned joint is formed by fitting the end of one lead pipe into the flared end of another lead pipe. Heat is then applied evenly around the joint, melting the overlapping edges and fusing them together. Lead burning is a method used for joining sheet lead or making lead pans.

P3003.8.2 Wiped. Joints shall be fully wiped, with an exposed surface on each side of the joint not less than $^3/_4$ inch (19 mm). The joint shall be not less than $^3/_8$ inch (9.5 mm) thick at the thickest point.

❖ A wiped joint is formed by fitting the end of one pipe into the flared end of the other. Molten lead is poured onto the joint and wiped with a hand-held pad, providing a minimum dimension of $^3/_4$ inch (19 mm) and a maximum thickness of $^3/_8$ inch (9.5 mm). With the aid of modern technology and the introduction of new materials, lead pipe is rarely used in drainage systems.

P3003.9 PVC plastic. Joints between PVC plastic pipe or fittings shall comply with Sections P3003.9.1 through P3003.9.3.

❖ Acceptable joints for PVC plastic pipe or fittings are identified in the referenced sections.

P3003.9.1 Mechanical joints. Mechanical joints on drainage pipe shall be made with an elastomeric seal conforming to ASTM C1173, ASTM D3212 or CSA B602. Mechanical joints shall not be installed in above-ground systems, unless otherwise *approved*. Joints shall be installed in accordance with the manufacturer's instructions.

❖ The typical mechanical joint must be restrained to prevent separation; therefore, such joints are limited to underground installations where restraint is accomplished by burial of the piping.

P3003.9.2 Solvent cementing. Joint surfaces shall be clean and free from moisture. A purple primer that conforms to ASTM F656 shall be applied. Solvent cement not purple in color and conforming to ASTM D2564, CSA B137.3 or CSA B181.2 shall be applied to all joint surfaces. The joint shall be made while the cement is wet, and shall be in accordance with ASTM D2855. Solvent-cement joints shall be installed above or below ground.

> **Exception:** A primer shall not be required where all of the following conditions apply:
>
> 1. The solvent cement used is third-party certified as conforming to ASTM D2564.
>
> 2. The solvent cement is used only for joining PVC drain, waste and vent pipe and fittings in non-pressure applications in sizes up to and including 4 inches (102 mm) in diameter.

❖ Generally speaking, PVC plastic pipe requires the application of a primer prior to solvent cementing. Although ASTM F656 only recommends that PVC primer be purple, purple primer is a code requirement.

The purple coloring is intended to allow the installer and the inspector to verify that the required primer has been applied. To distinguish the difference between primer and solvent cement, the solvent cement must be a color other than purple. Solvent cement for PVC is typically clear. First, primer is applied to both the fitting socket and the pipe. The PVC solvent cement is then applied to the pipe and socket fitting. The joint is made with a slight twisting until the pipe reaches the full depth of the socket fitting. It is then held in place until the solvent cement begins to set. The solvent-cementing procedure must be done in accordance with the referenced standard.

To minimize the health and fire hazards associated with solvent cements and primers, the manufacturer's instructions for the use and handling of these materials must be followed, in addition to the requirements found in ASTM D2855.

Recent testing by the National Sanitation Foundation determined that for PVC pipe sizes 4 inches and smaller, a primer is not required for solvent cement joints provided that the solvent cement used complies with ASTM D2564 and the DWV piping is only for non-pressure applications. Even though the code requires all plumbing products required to comply with a standard to be third-party certified (see Section P2609.4), one condition of the exception is for the solvent cement that is used (without primer being first applied to the joint surfaces) to be third-party certified to ASTM D2564 to ensure that the solvent cement is a product that meets the requirements of that standard.

P3003.9.3 Threaded joints. Threads shall conform to ASME B1.20.1. Schedule 80 or heavier pipe shall be permitted to be threaded with dies specifically designed for plastic pipe. *Approved* thread lubricant or tape shall be applied on the male threads only.

❖ The requirements regarding threaded joints for PVC plastic pipe are identical to those of ABS and CPVC plastic pipe.

P3003.10 Vitrified clay. Joints between vitrified clay pipe or fittings shall be made with an elastomeric seal conforming to ASTM C425, ASTM C1173 or CSA B602.

❖ An acceptable joint for vitrified clay pipe is a mechanical joint sealed with an elastomeric material. This is a type of compression joint similar to that used with other hubbed pipe and fittings.

P3003.11 Polyolefin plastic. Joints between polyolefin plastic pipe and fittings shall comply with Sections P3003.11.1 and P3003.11.2.

❖ The term polyolefin refers to a family of thermoplastic materials made from the chemical compounds named alkenes (olefins). Alkenes are hydrocarbon monomers that can be molecularly bonded to form long chains of molecules known as polymers. The polyolefin family includes polyethylene (PE), polypropylene (PP), polybutylene (PB) and cross-linked polyethylene (PEX). PE materials cannot be solvent welded.

P3003.11.1 Heat-fusion joints. Heat-fusion joints for polyolefin pipe and tubing joints shall be installed with socket-type heat-fused polyolefin fittings or electrofusion polyolefin fittings. Joint surfaces shall be clean and free from moisture. The joint shall be undisturbed until cool. Joints shall be made in accordance with ASTM F1412 or CSA B181.3.

❖ Heat fusion of plastic pipe and fittings is the equivalent of welding of steel pipe and fittings. Heating irons or heating elements embedded in the fitting are used to thermally soften (melt) the materials to allow them to fuse (weld) together. ASTM F1412 addresses Schedule 40 and 80 PE and polypropylene pipe and fittings for corrosive waste drainage systems.

P3003.11.2 Mechanical and compression sleeve joints. Mechanical and compression sleeve joints shall be installed in accordance with the manufacturer's instructions.

❖ Mechanical joints are commonly proprietary and installed in accordance with specific installation instructions unique to that fitting and pipe material. Mechanical joints include bolted or clamped sleeve couplings with elastomeric seals.

P3003.12 Polyethylene plastic pipe. Joints between polyethylene plastic pipe and fittings shall be underground and shall comply with Section P3003.12.1 or P3003.12.2.

❖ PE pipe is a type of polyolefin pipe and is used for underground building sewers and for insertion renewal of existing sewers (see commentary, Section P3003.16).

P3003.12.1 Heat fusion joints. Joint surfaces shall be clean and free from moisture. Joint surfaces shall be cut, heated to melting temperature and joined using tools specifically designed for the operation. Joints shall be undisturbed until cool. Joints shall be made in accordance with ASTM D2657 and the manufacturer's instructions.

❖ See the commentary to Section P3003.16.1.

P3003.12.2 Mechanical joints. Mechanical joints in drainage piping shall be made with an elastomeric seal conforming to ASTM C1173, ASTM D3212 or CSA B602. Mechanical joints shall be installed in accordance with the manufacturer's instructions.

❖ See the commentary to Section P3003.16.2.

P3003.13 Joints between different materials. Joints between different piping materials shall be made with a mechanical joint of the compression or mechanical-sealing type conforming to ASTM C1173, ASTM C1460 or ASTM C1461. Connectors and adapters shall be *approved* for the application and such joints shall have an elastomeric seal conforming to ASTM C425, ASTM C443, ASTM C564, ASTM C1440, ASTM D1869, ASTM F477, CSA A257.3M or CSA B602, or as required in Sections P3003.13.1 through P3003.13.6. Joints between glass pipe and other types of materials shall be made with adapters having a TFE seal. Joints shall be installed in accordance with the manufacturer's instructions.

❖ Mechanical joints are typically used to join dissimilar piping materials. Because many piping materials have the same outside diameters, many fittings are designed by manufacturers for use with a number of different piping materials. Some mechanical joints are manufactured specifically for joining pipes having different outside diameters. Transition couplings use bushings (inserts) to match the outside diameters of the pipes or are simply designed to fit over pipe ends of different outside diameter. ASTM C1173, ASTM C1460 or ASTM C1461 are the standards for the complete coupling assembly, which includes an elastomeric gasket, clamp assembly or both.

P3003.13.1 Copper or copper-alloy tubing to cast-iron hub pipe. Joints between copper or copper-alloy tubing and cast-iron hub pipe shall be made with a copper-alloy ferrule or compression joint. The copper or copper-alloy tubing shall be soldered to the ferrule in an *approved* manner, and the ferrule shall be joined to the cast-iron hub by a caulked joint or a mechanical compression joint.

❖ An adapter must be used when joining copper tubing to a cast-iron hub. The adapter is called by a wide variety of names, the most common being "soil adapter" or "caulk adapter." The adapter creates a spigot end similar to that of cast-iron pipe and fittings.

P3003.13.2 Copper or copper-alloy tubing to galvanized steel pipe. Joints between copper or copper-alloy tubing and galvanized steel pipe shall be made with a copper-alloy fitting or dielectric fitting. The copper tubing shall be soldered to the fitting in an *approved* manner, and the fitting shall be screwed to the threaded pipe.

❖ When joining copper or copper-alloy tubing to galvanized steel pipe, a method of protecting against galvanic corrosion is required. Galvanic corrosion occurs when two different metals come in contact in the presence of an electrolyte, such as water. Galvanic corrosion accelerates the natural corrosion process that occurs in metals, and because certain metals corrode faster than others, they have been placed in a hierarchy in order of rate of corrosion. The more reactive metal at a connection is called the "anode" (galvanized steel pipe), and the less reactive metal is called the "cathode" (copper tubing). When the copper tubing and galvanized steel pipe are coupled together, the galvanized steel pipe will tend to dissolve in the electrolyte, thereby generating an electric current flow between the metals. This section prescribes two methods of protection against galvanic corrosion: dielectric fittings and copper alloy fittings. The fittings are designed to provide a barrier or buffer between the two materials.

P3003.13.3 Cast-iron pipe to galvanized steel or brass pipe. Joints between cast-iron and galvanized steel or copper-alloy pipe shall be made by either caulked or threaded joints or with an *approved* adapter fitting.

❖ Galvanized steel pipe and copper alloy pipe may be caulked directly into the hub of cast-iron pipe without an adapter fitting. Certain cast-iron fittings have a female tapped inlet to permit a direct threaded connection.

P3003.13.4 Plastic pipe or tubing to other piping material. Joints between different types of plastic pipe or between plastic pipe and other piping material shall be made with an *approved* adapter fitting. Joints between plastic pipe and cast-iron hub pipe shall be made by a caulked joint or a mechanical compression joint.

❖ ABS and PVC pipe may connect directly into the hub of cast-iron pipe with a caulked joint or the use of a compression gasket without the use of an adapter fitting. Special caulk-adapter plastic fittings are available that provide a spigot end for use with hubbed pipe and fittings. Where different grades of plastic pipe are joined, or where plastic pipe and cast-iron spigot-end pipe, Schedule 40 steel pipe or copper pipe are joined, an adapter fitting must be used.

P3003.13.5 Lead pipe to other piping material. Joints between lead pipe and other piping material shall be made by a wiped joint to a caulking ferrule, soldering nipple, or bushing or shall be made with an *approved* adapter fitting.

❖ The common method of adapting to lead is by a piped joint to a ferrule, nipple or bushing; however, a mechanical adapter fitting using an elastomeric material may also be used.

P3003.13.6 Stainless steel drainage systems to other materials. Joints between stainless steel drainage systems and other piping materials shall be made with *approved* mechanical couplings.

❖ Joints between stainless steel drainage systems must be made with a mechanical coupling and have an elastomeric seal conforming to ASTM C425, ASTM C443, ASTM C564, ASTM C1173, ASTM D1869, ASTM F477, CSA A257.3M or CSA B602. Because many different types of mechanical joints are available and most couplings are unique to each manufacturer, the joining method must be in accordance with the manufacturer's installation instructions.

P3003.14 Joints between drainage piping and water closets. Joints between drainage piping and water closets or similar fixtures shall be made by means of a closet flange or a waste connector and sealing gasket compatible with the drainage system material, securely fastened to a structurally firm base. The joint shall be bolted, with an *approved* gasket flange to fixture connection complying with ASME A112.4.3 or setting compound between the fixture and the closet flange or waste connector and sealing gasket. The waste connector and sealing gasket joint shall comply with the joint-tightness test of ASME A112.4.3 and shall be installed in accordance with the manufacturer's instructions.

❖ The flange connection method is the most common water closet-to-piping connection method used in North America. The bottom outlet of the water closet forms one-half of the flange and the floor flange forms the other half of the connection. The setting compound most commonly used is a wax ring made of beeswax or a synthetic wax. In the past, putty was commonly used, but this practice is rarely used as putty dries out and cracks. Elastomeric gaskets are also used between fixture outlets (horns) and flanges [see Commentary Figure P3003.14(2)]. Corrosion-resistant bolts or screws are required in this application so that there is not a reduction of strength of the fastener over time due to corrosion. The flange-to-fixture connection complying with ASME A112.4.3 is a special plastic fitting that is only suitable for floor-mounted, bottom discharge water closets. The device seals to the inside of the drain pipe with an elastomeric O-ring and seals to the outlet of the water closet bowl with an elastomeric gasket. The gasket does not adhere to the bowl but seals to the bowl when backpressure in the drain pipe occurs. Where this device is used, the floor flange only serves as an attachment point for the closet bowl bolts.

Another type of water closet-to-piping system connection method is the gasketed waste tube connection method [see Commentary Figure P3003.14(1)]. This is a relatively new method for the North American plumbing industry; however, the gasketed waste tube connection method has been used in European countries (and countries using European standards) for decades. Note that this method does not require a floor flange. In some cases, the presence of a floor flange could interfere with the water closet installation. Without a floor flange to support the pipe at the floor, it is necessary for the piping installer to adequately support and provide

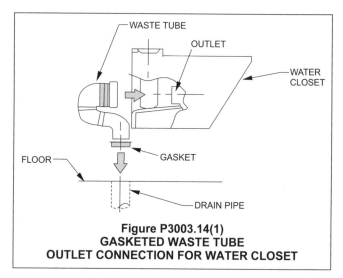

Figure P3003.14(1)
GASKETED WASTE TUBE
OUTLET CONNECTION FOR WATER CLOSET

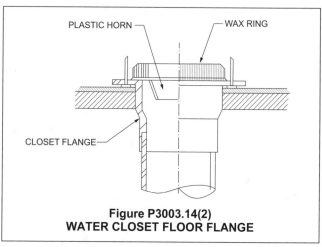

Figure P3003.14(2)
WATER CLOSET FLOOR FLANGE

blocking for the pipe to remain stationary when the sealing gasket/waste tube assembly is pressed into the pipe opening (and for when the water closet is removed). Note that the rough-in dimension for gasketed waste tube connected water closets may not necessarily be standardized in the familiar 14-, 12- and 10-inch (356, 305 and 254 mm) center-of-outlet to face-of-finished-wall dimensions. Piping installers will need to know the exact model of water closet to be installed prior to installing the receiving drainage pipe to ensure that the gasketed waste tube connection arrangement will work as intended.

SECTION P3004
DETERMINING DRAINAGE FIXTURE UNITS

P3004.1 DWV system load. The load on DWV-system piping shall be computed in terms of drainage fixture unit (d.f.u.) values in accordance with Table P3004.1.

❖ This section regulates the maximum loads that can be imposed on a drainage system. Table P3004.1 is used for the determination of d.f.u. The conventional method of sizing a sanitary drainage system is by d.f.u. load values. The fixture unit approach takes into consideration the probability of load on a drainage system. The probability method of sizing drainage systems was developed largely by Dr. Roy B. Hunter. Through his research, Dr. Hunter helped standardize and simplify design principles, which reduced the cost of plumbing systems.

Fixture unit values were determined based on the average rate of discharge by a fixture, the time of a single operation and the frequency of use or interval between operations. The theoretical approach considers a large group of fixtures being connected to the plumbing system with only a small fraction of the total number of fixtures in use simultaneously. The probability method has also been effective in the design of smaller plumbing systems because of the overdesign factor added by Dr. Hunter. Dr. Hunter sought an adequate design methodology to provide satisfactory service without interruption or inconvenience to the user.

TABLE P3004.1
DRAINAGE FIXTURE UNIT (d.f.u.) VALUES FOR VARIOUS PLUMBING FIXTURES

TYPE OF FIXTURE OR GROUP OF FIXTURES	DRAINAGE FIXTURE UNIT VALUE (d.f.u.)[a]
Bar sink	1
Bathtub (with or without a shower head or whirlpool attachments)	2
Bidet	1
Clothes washer standpipe	2
Dishwasher	2
Floor drain[b]	0
Kitchen sink	2
Lavatory	1
Laundry tub	2
Shower stall	2
Water closet (1.6 gallons per flush)	3
Water closet (greater than 1.6 gallons per flush)	4
Full-bath group with bathtub (with 1.6 gallon per flush water closet, and with or without shower head and/or whirlpool attachment on the bathtub or shower stall)	5
Full-bath group with bathtub (water closet greater than 1.6 gallon per flush, and with or without shower head and/or whirlpool attachment on the bathtub or shower stall)	6
Half-bath group (1.6 gallon per flush water closet plus lavatory)	4
Half-bath group (water closet greater than 1.6 gallon per flush plus lavatory)	5
Kitchen group (dishwasher and sink with or without food-waste disposer)	2
Laundry group (clothes washer standpipe and laundry tub)	3
Multiple-bath groups[c]: 1.5 baths 2 baths 2.5 baths 3 baths 3.5 baths	7 8 9 10 11

For SI: 1 gallon = 3.785 L.

a. For a continuous or semicontinuous flow into a drainage system, such as from a pump or similar device, 1.5 fixture units shall be allowed per gpm of flow. For a fixture not listed, use the highest d.f.u. value for a similar listed fixture.

b. A floor drain itself does not add hydraulic load. Where used as a receptor, the fixture unit value of the fixture discharging into the receptor shall be applicable.

c. Add 2 d.f.u. for each additional full bath.

Because the fixture unit values have a built-in probability factor, they cannot be directly translated into flow rates of discharge. For example, 1 d.f.u. is equivalent to a discharge rate of 1 cubic foot per minute (cfm) (0.47 L/s) or approximately 7.5 gallons per minute (gpm) (29 L/m). Two independent fixtures, however, each with a value of 1 drainage fixture unit, cannot be considered to have a combined discharge rate of 15 gpm (57 L/m), because the drainage fixture unit value incorporates the element of probability.

TABLE P3004.1. See page 30-13.

❖ This table identifies common plumbing fixtures with their corresponding d.f.u. values. The d.f.u. values are used to determine the load on a drain, which in turn is used to determine the minimum drainage pipe size requirement for the load served. In determining the fixture unit values, each fixture was evaluated for its impact on the sanitary drainage system. The original development of fixture unit values was based on an arbitrary scale established by Dr. Hunter.

The table takes into consideration the unlikely probability of multiple fixtures being used simultaneously. Therefore, sizing can be based on d.f.u. values for individual fixtures or for fixtures combined with adjacent fixtures within the same bathroom or grouping. For this purpose, the table assigns combined d.f.u. values for groups of bathroom fixtures, kitchen fixtures, appliances and laundry fixtures. These groups are adequately defined within the table itself.

Additional information is provided in the footnotes to the table for the purpose of sizing drains that receive the discharge of pumps or similar devices, floor drains, multiple-bathroom groups and other types of fixtures or drains not specifically listed within the table.

SECTION P3005
DRAINAGE SYSTEM

P3005.1 Drainage fittings and connections. Changes in direction in drainage piping shall be made by the appropriate use of sanitary tees, wyes, sweeps, bends or by a combination of these drainage fittings in accordance with Table P3005.1. Change in direction by combination fittings, heel or side inlets or increasers shall be installed in accordance with Table P3005.1 and Sections P3005.1.1 through P3005.1.4. based on the pattern of flow created by the fitting.

❖ This section refers the user to Table P3005.1 and the appropriate subsections for the determination of the type of drainage fitting that can be used depending on the installation. Consideration is given to the direction of flow, type of fixtures installed and point of connection as it relates to stacks, vents, drains or other connections. Drainage fittings and connections must provide a smooth transition of flow without creating obstructions or causing interference. The use of proper fittings helps to maintain the required flow velocities and reduces the possibility of stoppage in the drainage system.

Combination fittings are commonly used in drainage systems and must be evaluated for their pattern of flow. Combination fittings include fittings made up of two or more fittings and one-piece fittings, such as combination wye and eighth bends or tee-wyes.

TABLE P3005.1. See below.

❖ Where flow is from the horizontal to the horizontal, long-pattern fittings are necessary to prevent excessive reduction in flow velocity. Where flow is from the horizontal to the vertical or from the vertical to the horizontal, a short-pattern fitting can be used because the acceleration of flow caused by gravity will maintain higher flow velocities. An X in the column indicates that a particular fitting is acceptable for the described change in direction. For example, the only X shown for a sanitary tee is in the column for a horizontal-to-vertical change in direction.

The table has been developed to identify where use of a particular drainage pattern fitting is allowed. The terminology used to identify drainage pattern fittings was originally developed by the cast-iron soil pipe industry.

TABLE P3005.1
FITTINGS FOR CHANGE IN DIRECTION

TYPE OF FITTING PATTERN	CHANGE IN DIRECTION		
	Horizontal to vertical[c]	Vertical to horizontal	Horizontal to horizontal
Sixteenth bend	X	X	X
Eighth bend	X	X	X
Sixth bend	X	X	X
Quarter bend	X	X[a]	X[a]
Short sweep	X	X[a,b]	X[a]
Long sweep	X	X	X
Sanitary tee	X[c]	—	—
Wye	X	X	X
Combination wye and eighth bend	X	X	X

For SI: 1 inch = 25.4 mm.

a. The fittings shall only be permitted for a 2-inch or smaller fixture drain.

b. Three inches and larger.

c. For a limitation on multiple connection fittings, see Section P3005.1.1.

The pattern of the fittings is regulated by the fitting standards referenced in Table P3002.3. A fitting of one material will not necessarily look the same as a fitting of another material with the same name.

Note b permits the use of short-sweep fittings of 3 inches (76 mm) in diameter or larger where the change in direction is from the vertical to the horizontal. Note a permits the use of 2-inch (51 mm) or smaller short-sweep and quarter-bend fittings for drains serving a single fixture where the change in direction is from the vertical to the horizontal or from the horizontal to the horizontal. The intent of Note a is to allow short-pattern fittings for fixture drains because fixture drains are generally difficult to install using long-pattern fittings in frame construction. Only cast-iron pipe fittings have a

separate quarter bend fitting and short sweep fitting. For other drainage piping materials, a quarter bend and short sweep are the same fitting. A quarter bend with a very short radius, less than that of a short sweep, is identified as a vent elbow or vent quarter bend. These particular fittings are designed for vent system applications only.

Because there is no waste flow in vent piping and because the airflow in vent piping can be in either direction, the types of fittings used in vent piping are not regulated by this section. For example, a sanitary tee placed on its back with the branch inlet connected to a vertical dry vent riser is a common installation practice and is not prohibited. In this application, the only waste flow is through the straight run of the fitting, and the

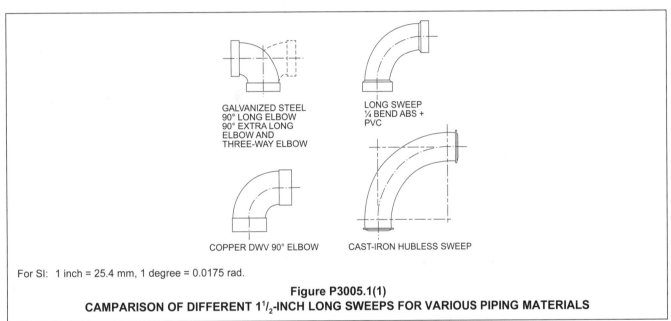

GALVANIZED STEEL
90° LONG ELBOW
90° EXTRA LONG
ELBOW AND
THREE-WAY ELBOW

LONG SWEEP
¼ BEND ABS +
PVC

COPPER DWV 90° ELBOW

CAST-IRON HUBLESS SWEEP

For SI: 1 inch = 25.4 mm, 1 degree = 0.0175 rad.

Figure P3005.1(1)
CAMPARISON OF DIFFERENT 1¹/₂-INCH LONG SWEEPS FOR VARIOUS PIPING MATERIALS

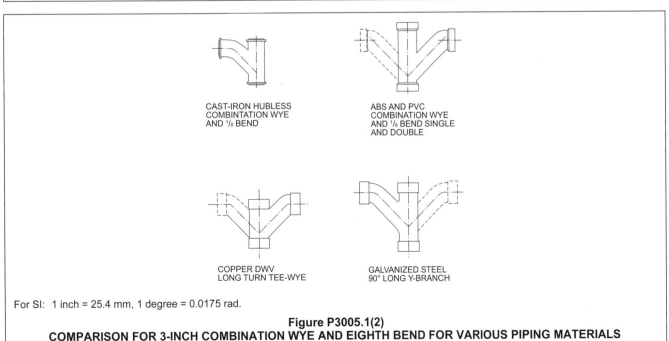

CAST-IRON HUBLESS
COMBINTATION WYE
AND ¹/₈ BEND

ABS AND PVC
COMBINATION WYE
AND ¹/₈ BEND SINGLE
AND DOUBLE

COPPER DWV
LONG TURN TEE-WYE

GALVANIZED STEEL
90° LONG Y-BRANCH

For SI: 1 inch = 25.4 mm, 1 degree = 0.0175 rad.

Figure P3005.1(2)
COMPARISON FOR 3-INCH COMBINATION WYE AND EIGHTH BEND FOR VARIOUS PIPING MATERIALS

branch inlet conducts only air into or out of the horizontal drain pipe. Note that the tee must be oriented to allow compliance with Section P3104.3. The restrictions placed on the use of a sanitary tee are based on its inability to guide waste flow entering through the branch inlet. Such restrictions do not apply where the branch inlet is used for a dry vent connection in compliance with Section P3104.

In all cases, the code requires that fittings be installed in a manner that will encourage high-velocity drainage flow and minimize the chances of blockage [see Commentary Figures P3005.1(1) through P3005.1(7)].

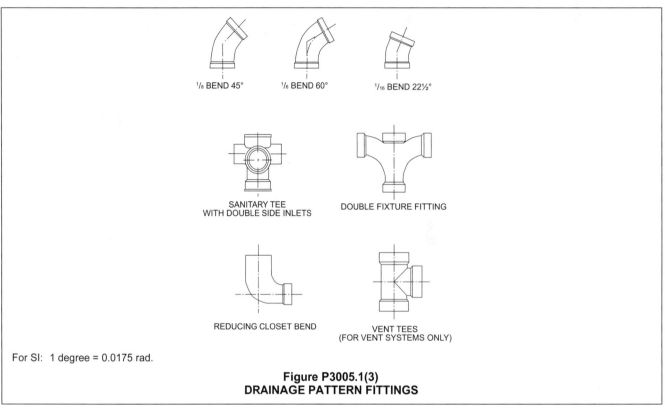

For SI: 1 degree = 0.0175 rad.

Figure P3005.1(3)
DRAINAGE PATTERN FITTINGS

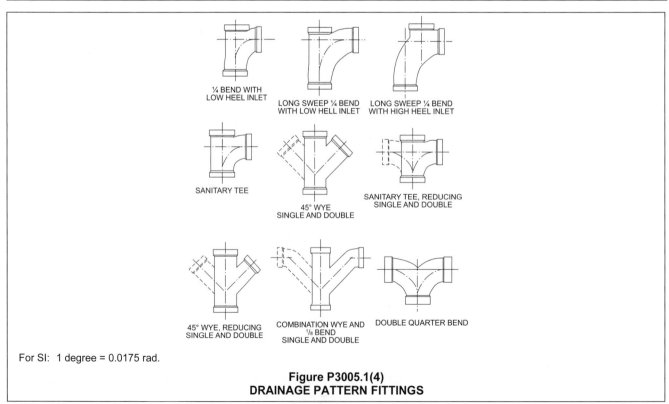

For SI: 1 degree = 0.0175 rad.

Figure P3005.1(4)
DRAINAGE PATTERN FITTINGS

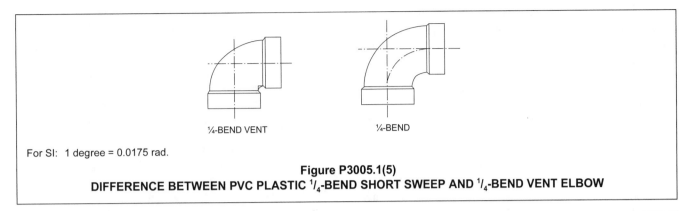

For SI: 1 degree = 0.0175 rad.

Figure P3005.1(5)
DIFFERENCE BETWEEN PVC PLASTIC $^1/_4$-BEND SHORT SWEEP AND $^1/_4$-BEND VENT ELBOW

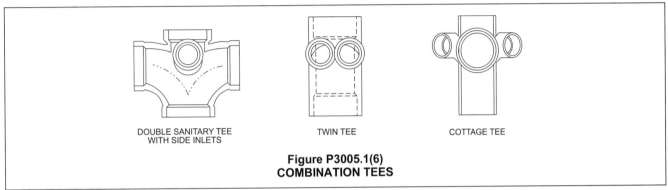

DOUBLE SANITARY TEE
WITH SIDE INLETS

TWIN TEE

COTTAGE TEE

Figure P3005.1(6)
COMBINATION TEES

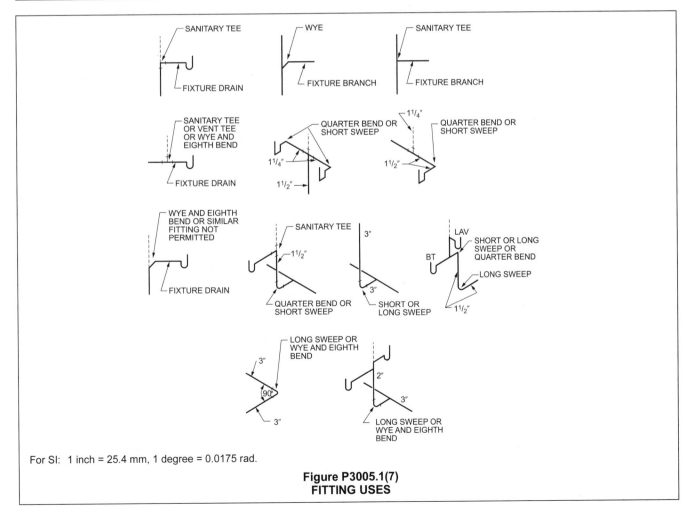

For SI: 1 inch = 25.4 mm, 1 degree = 0.0175 rad.

Figure P3005.1(7)
FITTING USES

P3005.1.1 Horizontal to vertical (multiple connection fittings). Double fittings such as double sanitary tees and tee-wyes or *approved* multiple connection fittings and back-to-back fixture arrangements that connect two or more branches at the same level shall be permitted as long as directly opposing connections are the same size and the discharge into directly opposing connections is from similar fixture types or fixture groups. Double sanitary tee patterns shall not receive the discharge of back-to-back water closets and fixtures or appliances with pumping action discharge.

Exception: Back-to-back water closet connections to double sanitary tee patterns shall be permitted where the horizontal *developed length* between the outlet of the water closet and the connection to the double sanitary tee is 18 inches (457 mm) or greater.

❖ This section addresses the use of double (or multiple) inlet fittings that connect more than one horizontal drain to a vertical drain or stack. This is most common with, but is not limited to, back-to-back fixtures (see Commentary Figure P3005.1.1).

It is mistakenly thought that the use of a double combination fitting, such as tee-wyes and combination wye and $^1/_8$ bends as shown in Commentary Figure P3005.1(4), is not permitted for the water closet arrangement, insofar as it will not permit the vent to connect above the trap weir. However, this arrangement is permitted because the vent connection limitation is designed to prevent self-siphonage and water closets must self-siphon to operate properly (see Section P3105.2).

Double sanitary tees must not be installed where water closets or appliances with pumping action are back-to-back. Washing machines and dishwashers are the typical appliances with pumping action in one-

and two-family dwellings. The exception recognizes that the waste flow in the fixture drain will not have the energy to jump across the fitting branches after it has been conveyed horizontally for $1^1/_2$ feet (457 mm) or more. Only in this case may a double sanitary tee be used for back-to-back water closet installations.

For all other fixture connections using double sanitary tees and tee-wyes or other multiple connection fittings, the back-to-back fixtures must be similar to one another (two lavatories, two bathtubs, etc.), and each opposing branch connecting to the fitting must be the same size.

P3005.1.2 Heel- or side-inlet quarter bends, drainage. Heel-inlet quarter bends shall be an acceptable means of connection, except where the quarter bends serves a water closet. A low-heel inlet shall not be used as a wet-vented connection. Side-inlet quarter bends shall be an acceptable means of connection for both drainage, wet venting and stack venting arrangements.

❖ This section addresses the use of heel-inlet quarter bends and side-inlet quarter bends for the purpose of drainage. There are two types of heel-inlet quarter bends: a heel-inlet and a side-inlet [see Commentary Figures P3005.1.2(1) and P3005.1.2(2)]. A heel-inlet (low or high) quarter bend cannot be used to connect a water closet, and a low-heel-inlet quarter bend, in particular, cannot be used as a wet-vented connection. Side-inlet quarter bends have no usage limitations. This section addresses combination quarter bend fittings where the quarter bend portion is used for vertical to horizontal changes in direction. Note the serious building code violation in Figure P3005.1.2(2) (floor joist cut).

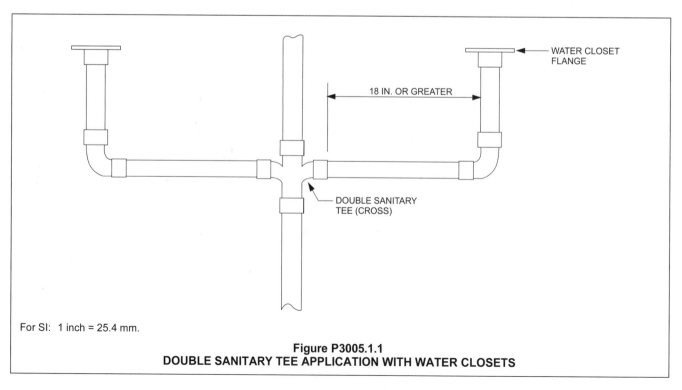

For SI: 1 inch = 25.4 mm.

Figure P3005.1.1
DOUBLE SANITARY TEE APPLICATION WITH WATER CLOSETS

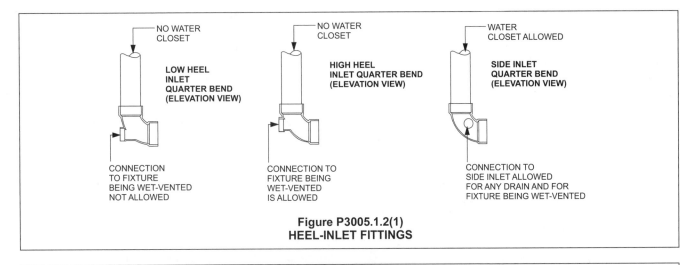

Figure P3005.1.2(1)
HEEL-INLET FITTINGS

Figure P3005.1.2(2)
VIOLATION: I-JOIST SEVERED FOR PIPING INSTALLATION

P3005.1.3 Heel- or side-inlet quarter bends, venting. Heel-inlet or side-inlet quarter bends, or any arrangement of pipe and fittings producing a similar effect, shall be acceptable as a dry vent where the inlet is placed in a vertical position. The inlet is permitted to be placed in a horizontal position only where the entire fitting is part of a dry vent arrangement.

❖ The only application where heel- or side-inlet quarter bends can be used for the purpose of venting a drain is where the inlet is in the vertical position. This means that if waste flows through the quarter bend, the low, high or side inlet would have to point upward. If there is no waste flow through the quarter bend, it is being used only as a vent fitting and the fitting can be installed in any orientation.

P3005.1.4 Water closet connection between flange and pipe. One-quarter bends 3 inches (76 mm) in diameter shall be acceptable for water closet or similar connections, provided that a 4-inch by 3-inch (102 mm by 76 mm) flange is installed to receive the closet fixture horn. Alternately, a 4-inch by 3-inch (102 mm by 76 mm) elbow shall be acceptable with a 4-inch (102 mm) flange.

❖ The reduction from 4 inches (102 mm) to 3 inches (76 mm) by means of a 4-inch by 3-inch (102 mm by 76 mm) closet flange or a 4-inch by 3-inch (102 mm by 76 mm) closet elbow is acceptable and is not to be considered an obstruction, as would be the case with reducing drain sizes downstream of the connection. A 4-inch by 3-inch (102 mm by 76 mm) closet flange is referred to as a reducing flange or simply a 3-inch (76 mm) closet flange.

P3005.1.5 Provisions for future fixtures. Where drainage has been roughed-in for future fixtures, the drainage unit values of the future fixtures shall be considered in determining the required drain sizes. Such future installations shall be terminated with an accessible permanent plug or cap fitting.

❖ Logically, the d.f.u. values for future fixtures must be included in determining the size of the drainage system; otherwise, the system might be undersized such that the future fixtures could not be accommodated. The piping must be permanently plugged or capped to prevent the escape of sewer gases or the intrusion of foreign objects. The point of future connection at the plug or cap must be accessible (see the definition of "Accessible" in Chapter 2). Note that Section P3104.6 also requires a vent to be roughed-in for drainage piping that will serve fixtures intended to be installed.

P3005.1.6 Change in size. The size of the drainage piping shall not be reduced in size in the direction of the flow. A 4-inch by 3-inch (102 mm by 76 mm) water closet connection shall not be considered as a reduction in size.

❖ One of the fundamental requirements of a drainage system is that the size of the piping cannot be reduced in the direction of drainage flow. A size reduction would create an obstruction to flow, possibly resulting in a backup of flow, an interruption of service in the drainage system or stoppage in the pipe. A 4-inch by 3-inch (102 mm by 76 mm) water closet connection is not considered to be a reduction in pipe size.

P3005.2 Cleanouts required. Cleanouts shall be provided for drainage piping in accordance with Sections P3005.2.1 through P3005.2.11.

❖ There must be points of access to drainage systems for clearing blockages. These access points are called cleanouts. See the definition for "Cleanout" in Chapter 2. Blockages in vertical piping rarely, if ever, occur because gravity will cause the blockage to move downward. Therefore, only horizontal piping is required to have cleanouts. In locating cleanouts, the designer or installer should consider the type of equipment that might be necessary for clearing blockages. Typical drain clearing/cleaning equipment includes hand-held cable augers, flat tapes, power-driven cables from $1/_4$-inch diameter to $1^1/_8$-inch diameter (or larger), and water jetting hoses from $1/_4$-inch diameter to 1-inch diameter (or larger). This equipment, especially power driven units, can be sizable and heavy. A large machine could be difficult to maneuver into smaller areas, let alone be safe to operate in a confined area. This section does provide a few limitations for where cleanouts can be located. Otherwise, the experience of the designer or installer is leaned on heavily to determine the best, most logical locations for cleanouts.

A frequently asked question is whether mechanical joints such as elastomeric connections (hubless joint connectors) can be considered as cleanouts. A significant amount of drain cleaning is performed by personnel that, although trained on drain-cleaning methods and equipment, may not have the necessary training on the proper removal and reinstallation of piping. Hangers could be left out, couplings reused or couplings overtightened such that the plumbing system might not work as intended or become a hazard (either environmentally or structurally). In the author's opinion, appropriate cleanout access such as removal of a cleanout plug is a far superior and a safer approach to provide access to the drainage system. However, some code officials might believe that removal of piping connectors would provide the intended access. The code is silent on this topic.

P3005.2.1 Horizontal drains and building drains. Horizontal drainage pipes in buildings shall have cleanouts located at intervals of not more than 100 feet (30 480 mm). *Building drains* shall have cleanouts located at intervals of not more than 100 feet (30 480 mm) except where manholes are used instead of cleanouts, the manholes shall be located at intervals of not more than 400 feet (122 m). The interval length shall be measured from the cleanout or manhole opening, along the *developed length* of the piping to the next drainage fitting providing access for cleaning, the end of the horizontal drain or the end of the *building drain*.

> **Exception:** Horizontal fixture drain piping serving a non-removable trap shall not be required to have a cleanout for the section of piping between the trap and the vent connection for such trap.

❖ Modern drain-cleaning equipment is available with cable (or jetter hose) length to clear blockages in horizontal drainage piping that is 100 feet (or less) in developed piping length from the cleanout opening. Manholes are sometimes used for cleanouts for larger-diameter piping. The larger piping diameter allows for larger-diameter cables (or larger diameter/higher pressure jetter hoses) to be used which are capable of clearing blockages in horizontal drainage piping that is 400 feet in length (or less). A key point to consider in locating the cleanouts is that the lengths indicated in this section are measured from the cleanout opening to the farthest point where the next cleanout allows access to the horizontal drain piping. For example, if the cleanout is on a stack and cleanout opening is 10 feet above the horizontal drain pipe (that the stack connects to), then the length of horizontal drainage piping that can be cleared is only 90 feet. The lengths in this section are for drainage piping that does not have changes in direction any greater than 45 degrees. See Section P3005.2.4 for horizontal changes in direction that are greater than 45 degrees.

The exception covers the special circumstance for horizontal drain piping that begins with a nonremovable trap such as for a floor drain or a shower drain. It is understood that a cable or jetter hose can be negotiated through a trap. However, the convoluted path of the trap creates additional drag on a cable such that running the cable for too far of a distance from the trap can cause high stresses in the cable resulting in a greater potential for cable breakage. Therefore, the cleanout access for these horizontal drains can be located downstream of these traps for a distance of up to the maximum allowable pipe length between the trap and the vent connection for the trap. See Table P3105.1 for trap-to-vent distances.

P3005.2.2 Building sewers. *Building sewers* smaller than 8 inches (203 mm) shall have cleanouts located at intervals of not more than 100 feet (30 480 mm). *Building sewers* 8 inches (203 mm) and larger shall have a manhole located not more than 200 feet (60 960 mm) from the junction of the *building drain* and *building sewer* and at intervals of not more than 400 feet (122 m). The interval length shall be measured from the cleanout or manhole opening, along the *developed length* of the piping to the next drainage fitting providing access for cleaning, a manhole or the end of the *building sewer*.

❖ The commentary for Section P3005.2.1 explains most of the information in this section. Sewers that are 8 inches or larger are required to have manholes for cleanout access because the large cables (or jetter hoses) needed for these pipelines cannot be easily negotiated through the standard pattern wye and one-eighth bend fitting that would be needed to bring the cleanout access opening to grade level. Building sewers could be broken at wye and one-eighth bend entry into the horizontal drainage piping, resulting in earth or stone backfill later migrating into the sewer. Manholes for these larger cables and hoses provide the needed space for access to the piping. Note that the first manhole for an 8-inch or larger diameter sewer is not required to be at the building. A manhole located up to 200 feet away from the building can be used to clear blockages in the sewer line back towards the building.

P3005.2.3 Building drain and building sewer junction. The junction of the *building drain* and the *building sewer* shall be served by a cleanout that is located at the junction or within 10 feet (3048 mm) *developed length* of piping upstream of the junction. For the requirements of this section, removal of a water closet shall not be required to provide cleanout access.

❖ The building sewer begins at the termination of the building drain which ends, by definition, at a point that is 30 inches from the outside wall of the building. Access to the building sewer is very important because blockages can frequently occur as the result of low-flow velocities being unable to transport solids and tree roots that can penetrate the sewer. Blockages might also be the result of other underground utilities that were installed by underground horizontal directional boring methods or "slit" trenching, both of which can sever/penetrate a building sewer without the knowledge of the equipment operator.

The allowance to have the cleanout located upstream of the building drain/building sewer juncture, but only up to 10 feet, provides for building sewer cleanouts to be in basements, crawl spaces, interior rooms or corridors. The prohibition of not allowing the removal of a water closet to provide cleanout access has a twofold purpose: 1. Water closets are either in a water closet compartment or a small single-user toilet room.

Although removal of the water closet could be a cleanout (see "Cleanout" definition in Chapter 2), such spaces are usually not large enough to facilitate the safe operation of cable machines that are powerful enough to have root-cutting capability. 2. The location of a wall (stack) or floor cleanout that requires removal of the water closet to gain safe access to the cleanout defeats the purpose of having installed the cleanout that is to facilitate drain clearing. For example, a cleanout might be located directly behind a water closet, on the (close) wall side of a water closet or on the floor next to the water closet such that the water closet would need to be removed to enable positioning of a large drain cleaning machine near the cleanout opening. A heavy machine near or on the closet flange could cause damage to the flange that might not be properly repaired. The location of a building sewer cleanout on the interior of a building needs to be well thought out.

A note needs to be said about safety and situational awareness when attempting to clear building sewer blockages. As stated previously, underground horizontal directional boring or slit trenching are frequently being used to install many types of underground utilities. Because building sewers are usually not marked on drawings, are frequently made of nonmetallic materials and are rarely installed with detectable tapes or wires such that detection equipment can locate them, a building sewer could be inadvertently penetrated by a utility service. This may cause sewer blockage. Thus, drain cleaning personnel need to be aware of any evidence (including information from the building occupant/owner) of recent construction activities in the area that could have impacted the sewer line, especially if those activities involve natural gas lines or electric power cabling. There have been injuries and deaths associated with sewer cleaning work where gas lines or electric cables penetrated a sewer and were cut by the sewer cleaning operations.

P3005.2.4 Changes of direction. Where a horizontal drainage pipe, a *building drain* or a *building sewer* has a change of horizontal direction greater than 45 degrees (0.79 rad), a cleanout shall be installed at the change of direction. Where more than one change of horizontal direction greater than 45 degrees (0.79 rad) occurs within 40 feet (12 192 mm) of *developed length* of piping, the cleanout installed for the first change of direction shall serve as the cleanout for all changes in direction within that 40 feet (12 192 mm) of developed length of piping.

❖ Table P3005.1 covers what types of fittings can be used for changes in direction of drainage piping. However, where there are many changes in direction greater than 45 degrees, it becomes more difficult for the drain-cleaning equipment (cables or jetter hoses) to negotiate the turns. The length of 40 feet is based on the drain-cleaning equipment's ability to negotiate all of the turns without significant equipment problems.

Where the length of drainage piping to be cleared is longer than 40 feet, some accommodations are necessary to ease the way for the cable or jetter hose for these longer pipelines. The accommodation is to not use single fitting patterns for changes in direction greater than 45 degrees.

A single pattern fitting or a fitting made up from "combining" fittings to make a pattern that is virtually identical to a single fitting must be avoided. Examples are: 1. A quarter bend long sweep fitting can be duplicated by assembling a one-eighth bend spigot end elbow into the hub end of a one-eighth bend elbow; and 2. A combination wye and one-eighth bend fitting can be duplicated by assembling a one-eighth bend spigot end elbow into the branch of a wye fitting. A frequent question is whether two one-eighth bend pattern fittings can be connected together by a short length of pipe and not be considered to be duplicating a single pattern fitting (and not requiring a cleanout for that change in direction). An example would be making a 90 degree change in direction using two one-eighth bend fittings (hubs on both ends) connected by a "make up" length of pipe (such that the hub end of each fitting touches the other). The code does allow 45 degree changes in direction and does not state how far apart those changes are required to be. The logical conclusion is that the length of pipe between the two fittings could be minimal and the made up fitting still be in compliance such that a cleanout would not be required. This might not necessarily be the same conclusion that the local code official might make. And the type of pipe material (and how the fittings are made for that type of pipe such as hubless cast iron fittings versus plastic fittings) might alter the code official's viewpoint in the matter. If a length of pipe needs to be used to make up a fitting to accomplish a change of direction greater than 45 degrees, then it would be better to plan for and use a length that clearly separates the two fittings into two separate changes in direction rather than causing the question to be raised in the first place.

P3005.2.5 Cleanout size. Cleanouts shall be the same size as the piping served by the cleanout, except cleanouts for piping larger than 4 inches (102 mm) need not be larger than 4 inches (102 mm).

Exceptions:

1. A removable P-trap with slip or ground joint connections can serve as a cleanout for drain piping that is one size larger than the P-trap size.

2. Cleanouts located on stacks can be one size smaller than the stack size.

3. The size of cleanouts for cast-iron piping can be in accordance with the referenced standards for cast iron fittings as indicated in Table P3002.3.

❖ Cleanout openings for $1^1/_4$-, $1^1/_2$-, 2-, $2^1/_2$-, 3- and 4-inch pipes must not be smaller than the pipe served, thus allowing use of full-bore cleaning equipment. For larger piping, a 4-inch (102 mm) opening is considered adequate for today's cleaning equipment, and full-bore cleaning equipment is typically not used for such piping.

Manufactured cleanout fittings, like pipe, are sized by nominal dimensions.

Section P3005.2.5, Exception 2 allows a smaller nominal size fitting to serve as a cleanout for a larger nominal pipe size. This exception is very specific and limited to removable "P" traps. Exception 2 is intended to permit the removal of fixture traps to provide access to the connecting drains. Exception 3 accommodates some stack fitting designs that have always had the "one pipe size smaller" cleanout opening.

P3005.2.6 Cleanout plugs. Cleanout plugs shall be copper alloy, plastic or other *approved* materials. Cleanout plugs for borosilicate glass piping systems shall be of borosilicate glass. Brass cleanout plugs shall conform to ASTM A74 and shall be limited for use only on metallic piping systems. Plastic cleanout plugs shall conform to the referenced standards for plastic pipe fittings as indicated in Table P3002.3. Cleanout plugs shall have a raised square head, a countersunk square head or a countersunk slot head. Where a cleanout plug will have a trim cover screw installed into the plug, the plug shall be manufactured with a blind end threaded hole for such purpose.

❖ Metallic cleanout plugs must be brass to provide for easy removal. If the cleanout plug is corroded in place, the brass plug is a soft enough material to be chiseled out. Brass cleanout plugs are limited to metallic fittings because the metal plug threads may damage the softer plastic threads of a fitting. Plastic plugs are intended for use with plastic fittings; however, this section does not prohibit the use of plastic plugs with metallic fittings. Like brass plugs, plastic plugs are less likely to seize in metallic fittings. The cleanout plug must have a square turning surface to allow for ease of removal while minimizing the possibility of stripping the surface during removal.

P3005.2.7 Manholes. Manholes and manhole covers shall be of an approved type. Manholes located inside of a building shall have gas-tight covers that require tools for removal.

❖ There are many designs for manholes and manhole covers. Frequently, jurisdictions have specifications on manhole design and require specially embossed covers to be used. The code official must approve the design.

Where manholes are installed inside of buildings, the manhole covers must be of gas-tight design to prevent sewer gases from entering the building. Covers need to be bolted down or otherwise require tools to remove to cause limited access.

The designer specifying the installation of manholes in areas subject to flooding must consider the protection of manholes from the effects of floodwaters. Floodwaters may damage or completely displace manholes and covers if they are not properly sealed and secured. Additionally, submerged manhole openings (covers) may allow floodwaters to enter the drainage system. The manhole cover may be located above the base flood level elevation or designed to resist hydrostatic forces when submerged in water and dynamic forces resulting from wave action or high-velocity water.

P3005.2.8 Installation arrangement. The installation arrangement of a cleanout shall enable cleaning of drainage piping only in the direction of drainage flow.

Exceptions:

1. Test tees serving as cleanouts.

2. A two-way cleanout installation that is *approved* for meeting the requirements of Section P3005.2.3.

❖ Many fixtures and buildings have been damaged by drain cleaning work that allowed the cable or jetter hose to be routed in an upstream direction. Drainage piping is frequently smaller in diameter at upstream locations and augers can break or permanently jam in the piping. Almost every plumbing contractor has heard of stories or has personally witnessed the aftermath of a cleaning cable that found its way out through a water closet, breaking the water closet and causing significant wall, floor and ceiling damage in a toilet room. Cleanouts must be well thought out ensuring that the preferred cleanout arrangement is for cleaning in a downstream direction.

Exception 1 recognizes that a test tee might be required to be installed for a cleanout. The intent is not for enabling drain cleaning in the upstream direction. The drainage piping upstream of the test tee (serving as a cleanout) must have the required cleanout. Though not required by the code, some permanent indication of flow should be indicated on test tees serving as cleanouts, except those where the direction of flow is obvious, such as on a vertical pipe.

Exception 2 recognizes a specialized two-way cleanout fitting that is constructed with a branch throat that curves toward both the upstream and downstream directions of the run of the fitting. This fitting is only allowed to be used for the building drain/building sewer cleanout requirement of Section P3005.2.3. This fitting is sometimes used as a retrofit for existing buildings where the building drain was not provided with an upstream cleanout. A question that is often asked is how far below grade can a two-way cleanout fitting be located? The concern is about being able to reliably guide the cleaning cable (or jetter hose) in the intended direction (and not in an unintended direction). Or it should be obvious that the cleanout is a specialized two-way cleanout fitting so that a drain cleaner doesn't just assume that the fitting will always direct the cable downstream. The code is silent on this issue.

P3005.2.9 Required clearance. Cleanouts for 6-inch (153 mm) and smaller piping shall be provided with a clearance of not less than 18 inches (457 mm) from, and perpendicular to, the face of the opening to any obstruction. Cleanouts for 8-inch (203 mm) and larger piping shall be provided with a clearance of not less than 36 inches (914 mm) from, and perpendicular to, the face of the opening to any obstruction.

❖ The usability of cleanouts is not to be compromised by their placement in the building. The required clearances in this section are the minimum conditions that will allow the cleanout to be used for the intended purpose. The designer is encouraged to provide addi-tional clearance whenever possible and practical. The required clearances are to be measured perpendicular to the face of the cleanout opening to any obstruction. Obstructions could mean fixed equipment, plumbing fixtures and equipment, appliances that cannot be moved by hand only (such as an ice machine), or permanent building construction such as water closet compartment panels that are fixed to the building.

P3005.2.10 Cleanout access. Required cleanouts shall not be installed in concealed locations. For the purposes of this section, concealed locations include, but are not limited to, the inside of plenums, within walls, within floor/ceiling assemblies, below grade and in crawl spaces where the height from the crawl space floor to the nearest obstruction along the path from the crawl space opening to the cleanout location is less than 24 inches (610 mm). Cleanouts with openings at a finished wall shall have the face of the opening located within $1^1/_2$ inches (38 mm) of the finished wall surface. Cleanouts located below grade shall be extended to grade level so that the top of the cleanout plug is at or above grade. A cleanout installed in a floor or walkway that will not have a trim cover installed shall have a counter-sunk plug installed so the top surface of the plug is flush with the finished surface of the floor or walkway.

❖ A cleanout cannot be located in permanently concealed areas, be concealed by permanent finish materials or be located in an area where cleaning equipment could not be obviously moved to perform the work. A cleanout that cannot be found is a useless cleanout. A cleanout that is poorly located (such as in floor/ceiling area) and presents difficulties resulting in building damage is not an intelligent choice.

P3005.2.10.1 Cleanout plug trim covers. Trim covers and access doors for cleanout plugs shall be designed for such purposes. Trim cover fasteners that thread into cleanout plugs shall be corrosion resistant. Cleanout plugs shall not be covered with mortar, plaster or any other permanent material.

❖ Trim covers and access panels are required to be designed for such purposes so that they are recognizable as a potential location for a cleanout. Such covers and panels must be approved by the code official so that obscure and inappropriate methods for covering up cleanouts are not chosen.

P3005.2.10.2 Floor cleanout assemblies. Where it is necessary to protect a cleanout plug from the loads of vehicular traffic, cleanout assemblies in accordance with ASME A112.36.2M shall be installed.

❖ Load-rated cleanout assemblies are required for vehicular traffic areas.

P3005.2.11 Prohibited use. The use of a threaded cleanout opening to add a fixture or extend piping shall be prohibited except where another cleanout of equal size is installed with the required access and clearance.

❖ If a cleanout opening is eliminated by a permanent piping connection, a new approved cleanout of equal access and capacity must be installed. In existing structures, cleanouts provide a convenient opening for the connection of new piping in remodeling, addition and alteration work. The cleanout fitting is commonly removed to allow a new connection; however, a substitute cleanout must be provided to serve in the same capacity as the cleanout that was eliminated. Many threaded cleanout openings have only a few threads and, therefore, are not intended to receive threaded pipe or male adapters.

P3005.3 Horizontal drainage piping slope. Horizontal drainage piping shall be installed in uniform alignment at uniform slopes not less than $^{1}/_{4}$ unit vertical in 12 units horizontal (2-percent slope) for $2^{1}/_{2}$ inch (64 mm) diameter and less, and not less than $^{1}/_{8}$ unit vertical in 12 units horizontal (1-percent slope) for diameters of 3 inches (76 mm) or more.

❖ The minimum desired velocity in a horizontal drain pipe is approximately 2 feet per second (0.609 m/s). This velocity is often referred to as the "scouring velocity." This minimum velocity is intended to keep solids in suspension. If velocity is too low because the pipe slope is too shallow, the solids will tend to drop out of suspension, settling at the bottom of the pipe. This will ultimately lead to a drain blockage. The slope requirements of this section produce the minimum required drainage flow velocity. Two-percent slope is required for $2^{1}/_{2}$-inch (64 mm) diameter and smaller pipe, and 1-percent slope is required for 3-inch (76 mm) diameter and larger pipe. Drainage pipe can always be installed with greater slopes.

P3005.4 Drain pipe sizing. Drain pipes shall be sized according to drainage fixture unit (d.f.u.) loads. The size of the drainage piping shall not be reduced in size in the direction of flow. The following general procedure is permitted to be used:

1. Draw an isometric layout or riser diagram denoting fixtures on the layout.

2. Assign d.f.u. values to each fixture group plus individual fixtures using Table P3004.1.

3. Starting with the top floor or most remote fixtures, work downstream toward the *building drain* accumulating d.f.u. values for fixture groups plus individual fixtures for each branch. Where multiple bath groups are being added, use the reduced d.f.u. values in Table P3004.1, which take into account probability factors of simultaneous use.

4. Size branches and stacks by equating the assigned d.f.u. values to pipe sizes shown in Table P3005.4.1.

5. Determine the pipe diameter and slope of the *building drain* and *building sewer* based on the accumulated d.f.u. values, using Table P3005.4.2.

❖ This section outlines in detail the method the designer can use to effectively size the drainage portion of the DWV system.

The first step is to draw an isometric layout of the drain lines. This drawing must show each fixture (or group of fixtures) and must indicate all points of connection (see Commentary Figure P3005.4).

Step two is to assign d.f.u. values to each fixture or group of fixtures. This information is found in Table P3004.1 (see commentaries, Section P3004.1 and Table P3004.1).

The third step is to add up the d.f.u.s for each section of piping. Start with the most remote fixture (or group) and follow downstream, adding the total for each section of pipe (see Commentary Figure P3005.4).

Step four is to size the branches and stacks. This requires the use of Table P3005.4.1, and the size of drainage piping must not be reduced in the direction of flow (see commentaries, Section P3005.4.1, Table P3005.4.1, and the definitions of "Branch" and "Stack").

The fifth and final step is to size the building drain and building sewer based on the d.f.u.s as determined in the previous steps. This will require the use of Table P3005.4.2 (see commentaries, Section P3005.4.2, Table P3005.4.2, and the definitions of "Building drain" and "Building sewer").

P3005.4.1 Branch and stack sizing. Branches and stacks shall be sized in accordance with Table P3005.4.1. Below grade drain pipes shall be not less than $1^{1}/_{2}$ inches (38 mm) in diameter. Drain stacks shall be not smaller than the largest horizontal branch connected.

Exceptions:

1. A 4-inch by 3-inch (102 mm by 76 mm) closet bend or flange.

2. A 4-inch (102 mm) closet bend connected to a 3-inch (76 mm) stack tee shall not be prohibited.

❖ This section outlines the method for sizing branches and stacks (see Chapter 2 for the definitions of "Branch" and "Stack"). Refer to Table P3005.4.1 for determining the size of pipe based on the d.f.u. value. It is important to understand the definitions of "Branch" and "Stack," especially as they relate to the definition of a "Building Drain" (see commentaries for the Chapter 2 definitions of "Branch," "Building Drain" and "Stack"). A closer look at these definitions, as they relate to Commentary Figure P3005.4, shows that none of the drain lines in the diagram can be defined as a stack or branch. Rather, they are part of building drain and the building drain branches, which are sized in accordance with Table P3005.4.2. In addition, Table P3201.7 dictates the trap size required for each fixture.

A $1^{1}/_{4}$-inch (32 mm) pipe is limited to use as a fixture drain and is not allowed to be installed underground.

TABLE P3005.4.1
MAXIMUM FIXTURE UNITS ALLOWED TO BE
CONNECTED TO BRANCHES AND STACKS

NOMINAL PIPE SIZE (inches)	ANY HORIZONTAL FIXTURE BRANCH	ANY ONE VERTICAL STACK OR DRAIN
$1^1/_4$ [a]	—	—
$1^1/_2$ [b]	3	4
2 [b]	6	10
$2^1/_2$ [b]	12	20
3	20	48
4	160	240

For SI: 1 inch = 25.4 mm.

a. $1^1/_4$-inch pipe size limited to a single-fixture drain or trap arm. See Table P3201.7.

b. No water closets.

❖ This table lists the maximum number of fixture units that are allowed to discharge into a pipe of a given size. As the number of drainage fixture units increases in the direction of flow, the pipe might also have to be increased in size.

Note that a d.f.u. load is not given for the use of $1^1/_4$-inch (32 mm) pipe. Note a shows that a $1^1/_4$-inch (32 mm) drain can serve only one fixture drain or trap (see the definition of "Fixture drain"). Of course, the fixture must be such that it can be served by a $1^1/_4$-inch (32 mm) drain or trap in accordance with Table P3201.7.

P3005.4.2 Building drain and sewer size and slope. Pipe sizes and slope shall be determined from Table P3005.4.2 on the basis of drainage load in fixture units (d.f.u.) computed from Table P3004.1.

❖ This table provides the size of pipe to be used, based on the accumulated d.f.u. load. As indicated in the table, the d.f.u. capacity increases for a greater slope, providing the designer with greater flexibility in sizing and layout. Where a d.f.u. load is not given, this is due to a minimum slope required for certain sizes of drains or the limitation on fixtures as outlined in Note a. As with Table P3005.4.1, Note b clarifies that piping with diameters of less than 3 inches (76 mm) cannot serve a water closet.

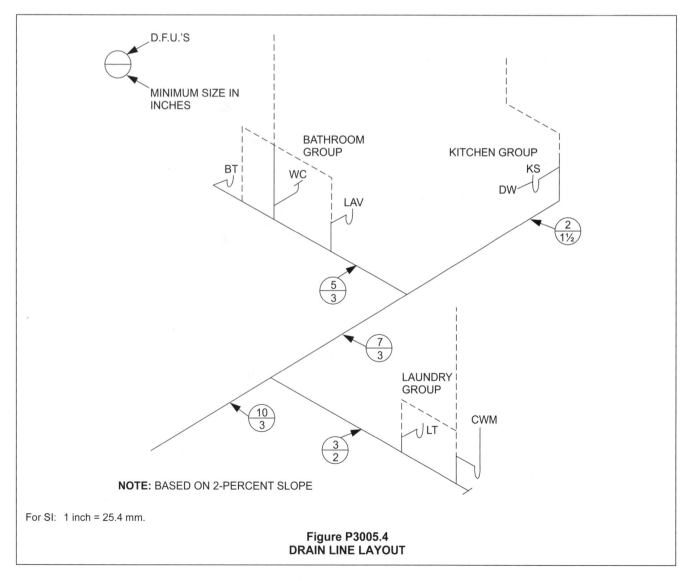

Figure P3005.4
DRAIN LINE LAYOUT

For SI: 1 inch = 25.4 mm.

TABLE P3005.4.2
MAXIMUM NUMBER OF FIXTURE UNITS ALLOWED
TO BE CONNECTED TO THE BUILDING DRAIN,
BUILDING DRAIN BRANCHES OR THE BUILDING SEWER

DIAMETER OF PIPE (inches)	SLOPE PER FOOT		
	$^1/_8$ inch	$^1/_4$ inch	$^1/_2$ inch
$1^1/_2$ a, b	—	Note a	Note a
2 b	—	21	27
$2^1/_2$ b	—	24	31
3	36	42	50
4	180	216	250

For SI: 1 inch = 25.4 mm, 1 foot = 304.8 mm.

a. $1^1/_2$-inch pipe size limited to a building drain branch serving not more than two waste fixtures, or not more than one waste fixture if serving a pumped discharge fixture or food waste disposer discharge.

b. No water closets.

❖ See the commentary to Section P3005.4.2.

P3005.5 Connections to offsets and bases of stacks. Horizontal branches shall connect to the bases of stacks at a point located not less than 10 times the diameter of the drainage stack downstream from the stack. Horizontal branches shall connect to horizontal stack offsets at a point located not less than 10 times the diameter of the drainage stack downstream from the upper stack.

❖ Within 10 pipe diameters of a stack, hydraulic jump can occur causing pressure fluctuations that affect traps and also causing waste and solids to enter the branch fitting.

SECTION P3006
SIZING OF DRAIN PIPE OFFSETS

P3006.1 Vertical offsets. An offset in a vertical drain, with a change of direction of 45 degrees (0.79 rad) or less from the vertical, shall be sized as a straight vertical drain.

❖ An offset is created by two fittings that each cause a change of direction. It brings one section of the pipe out of line and into a line parallel with the other section. Though the drain line changes direction, it is sized as though it were still vertical. The offset (or the section of piping out of line with the run of pipe) portion must not exceed 45 degrees (0.79 rad) with the connecting lines on a vertical installation. Where the offset exceeds an angle of 45 degrees (0.79 rad) it is considered to be horizontal and must be sized in accordance with Section P3006.2 or P3006.3, whichever applies.

P3006.2 Horizontal offsets above the lowest branch. A stack with an offset of more than 45 degrees (0.79 rad) from the vertical shall be sized as follows:

1. The portion of the stack above the offset shall be sized as for a regular stack based on the total number of fixture units above the offset.

2. The offset shall be sized as for a *building drain* in accordance with Table P3005.4.2.

3. The portion of the stack below the offset shall be sized as for the offset or based on the total number of fixture units on the entire stack, whichever is larger.

❖ This section applies to offsets that are installed in a stack, are located above the lowest branch and are not defined as "Vertical offsets" (see commentary, Section P3006.1 and the definition of "Offset").

The portion of the stack above the horizontal offset must be sized for a regular stack in accordance with Table P3005.4.1. The horizontal offset portion must be sized in accordance with Table P3005.4.2. This will then determine the minimum size of the remainder of the stack below the offset, except that additional d.f.u.s for fixtures and branches connecting to it must be included in the sizing of the stack; this could necessitate a further increase in size beyond that required for the horizontal offset. Offsets in stacks create turbulent flow by disrupting the annular flow that occurs in straight stacks. This turbulent flow can, in turn, create pressure fluctuations that can negatively affect trap seals.

P3006.3 Horizontal offsets below the lowest branch. In soil or waste stacks below the lowest horizontal branch, a change in diameter shall not be required if the offset is made at an angle not greater than 45 degrees (0.79 rad) from the vertical. If an offset greater than 45 degrees (0.79 rad) from the vertical is made, the offset and stack below it shall be sized as a *building drain* (see Table P3005.4.2).

❖ Vertical offsets that are below the lowest horizontal branch connection to the stack are not required to increase in size because of the offset. Offsets in stacks create turbulent flow by disrupting the annular flow that occurs in straight stacks. This turbulent flow can, in turn, create pressure fluctuations that can negatively affect trans or seals.

The pressure differences created by horizontal offsets mainly affect the horizontal branches downstream of the offset. Horizontal offsets that are below the lowest horizontal branch connection are required to be sized as a building drain in accordance with Table P3005.4.2.

SECTION P3007
SUMPS AND EJECTORS

P3007.1 Building subdrains. Building subdrains that cannot be discharged to the sewer by gravity flow shall be discharged into a tightly covered and vented sump from which the liquid shall be lifted and discharged into the building gravity drainage system by automatic pumping equipment or other *approved* method. In other than existing structures, the sump shall not receive drainage from any piping within the building capable of being discharged by gravity to the *building sewer*.

❖ Where the drainage system or portions of it cannot discharge by gravity to the sewer, the drainage is collected in a tightly sealed and vented sump and pumped to a gravity sewer or drain. The sump must be sized to provide adequate holding capacity and to limit the retention period of the waste. Though not specifically required by the code, the intent of this section is that the capacity of the sump must not exceed one-half of a day's (12 hours) discharge load from the piping system connected to the sump under normal use. The intent is to keep the waste retention period short to prevent the sump from acting as a waste decomposition (septic) tank (see Commentary Figures P3007.1 and P3007.2).

The minimum capacity of the sump must be such that the pumping equipment operates for at least 15 seconds per pumping cycle to prevent short cycling, thereby extending the life of the equipment.

The cover for the sump must be gas tight to prevent the escape of sewer gas into the building (see commentary, Section P3113.4). Sumps, other than pneumatic ejectors, must be vented in accordance with Section P3113.4.1.

Sumps receiving water closet and urinal discharge use a sewage ejector to pump the soil and waste to a gravity drain. Quite often, the pumping equipment is referred to as a sewage pump by technical standards. A pneumatic ejector is a special type of sewage "pump" that operates by air pressure, which forces the sewage from the pressurized receiver instead of mechanically

pumping it. Pneumatic ejectors require special relief vents to instantaneously relieve the pressure in the receiver (see commentary, Section P3113.4.2).

P3007.2 Valves required. A check valve and a full open valve located on the discharge side of the check valve shall be installed in the pump or ejector discharge piping between the pump or ejector and the gravity drainage system. Access shall be provided to such valves. Such valves shall be located above the sump cover required by Section P3007.3.2 or, where the discharge pipe from the ejector is below grade, the valves shall be accessibly located outside the sump below grade in an access pit with a removable access cover.

❖ The discharge piping must have a check valve to prevent the previously pumped waste from returning to the sump pit when the pump or ejector shuts off. This prolongs the life of the pumping equipment and conserves energy. The check valve will also prevent the gravity drainage system from backing up into the sump or receiver.

The requirement for the installation of a full-open valve on the discharge side of the check valve indicates a similar concern. There will be some point over the service life of the sump or check valve when replacement or repairs are necessary. This full-open valve is there to prevent waste from running back into the discharge piping when maintenance is performed

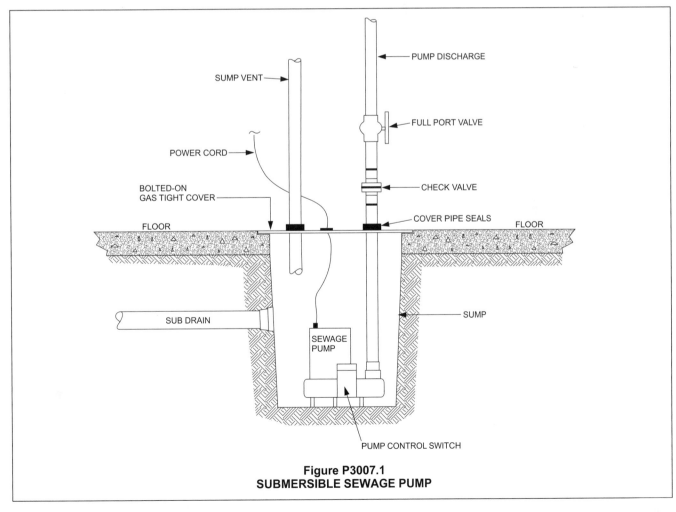

Figure P3007.1
SUBMERSIBLE SEWAGE PUMP

on the check valve or pump/ejector. It also allows isolation of the drainage system to prevent backups and sewer gas leakage during pump replacement and maintenance. Both valves must be accessible for maintenance and must be installed on the discharge pipe between the pump or ejector and the gravity drainage system.

P3007.3 Sump design. The sump pump, pit and discharge piping shall conform to the requirements of Sections P3007.3.1 through P3007.3.5.

❖ This is the main section, under which Sections P3007.3.1 through P3007.3.5 will give the specific requirements for the construction of the sump pump, sump pit and discharge piping related to sumps and ejectors.

P3007.3.1 Sump pump. The sump pump capacity and head shall be appropriate to anticipated use requirements.

❖ The sewage pump or pneumatic ejector must be sized properly to accommodate the peak flow into the receiver or pit, to provide the head pressure required to lift or eject waste, to prevent excessive cycling of the pumping equipment and to handle the type of waste that discharges to the sump.

Sewage pumps and ejectors must be able to handle, without creating blockages, the types of solid waste discharged into the sump pit or receiver. Pumps and ejectors serving water closets must be able to handle the solids associated with soil drainage.

Grinder pumps and grinder ejectors that serve water closets are exempt from the 2-inch (51 mm) spherical solids-handling requirements of this section. These pumps and ejectors reduce solids to a near-liquid state and pump the slurry to the drainage system. The liquidized solids pass readily through the pump and into

the drainage system, thus reducing the possibility of creating a stoppage in either the pump or the receiving drainage line.

In certain instances, the design professional may choose to install duplex pumping equipment. Duplex pumping installations consist of two pumps with special controls installed to alternate pump duty. If the inflow into the sump is at a high rate or one pump fails to operate, the second pump provides assistance.

The intent behind duplex pumping equipment is to reduce the possibility of a disruption of service of the plumbing system. If a pump fails, it may be repaired while the drainage system functions normally with the remaining pump. Duplex pumping equipment is not required by the code.

The sewage pump or ejector must have a minimum capacity that is based on the size of the discharge pipe. The required capacity provides a full-flow velocity of at least 2 feet per second (0.61 m/s) in the discharge pipe (see Table P3007.6).

P3007.3.2 Sump pit. The sump pit shall be not less than 18 inches (457 mm) in diameter and 24 inches (610 mm) deep, unless otherwise *approved*. The pit shall be accessible and located so that drainage flows into the pit by gravity. The sump pit shall be constructed of tile, concrete, steel, plastic or other *approved* materials. The pit bottom shall be solid and provide permanent support for the pump. The sump pit shall be fitted with a gas-tight removable cover that is installed above grade level or floor level, or not more than 2 inches (51 mm) below grade or floor level. The cover shall be adequate to support anticipated loads in the area of use. The sump pit shall be vented in accordance with Chapter 31.

❖ As discussed in the commentary to Section 3007.1, the sump pit must be properly sized to receive waste

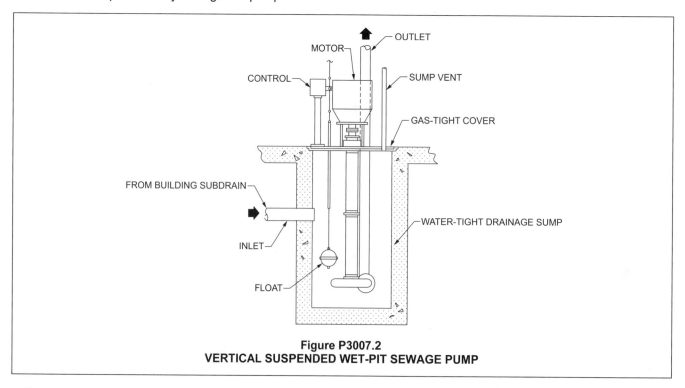

Figure P3007.2
VERTICAL SUSPENDED WET-PIT SEWAGE PUMP

and sewage from the plumbing fixtures that it will serve and allow proper operation of the pump.

In this section, the minimum dimensions of 18 inches (457 mm) in diameter and 24 inches (610 mm) in depth are given, unless the designer or manufacturer presents calculations and other supporting data that will allow the code official to determine that other dimensions work for the situation being addressed. This approval is strictly on a case-by-case basis by the code official. This section states that the sump pit is to be accessible for maintenance and requires that all drainage flow into the pit must be by the force of gravity.

The sump pit must be constructed of a durable material, such as tile, concrete, steel, plastic or other approved materials. The bottom of the sump pit must be solid and structurally capable of supporting the sump pump. The sump pit must have a gas-tight removable cover to prevent the escape of sewer gas. This cover is to be structurally capable of supporting the weight of loads it will receive based on the location of the sump pit. The sump pit must be vented in accordance with Chapter 31. The sump pit cover must be readily removable; therefore, the method used to connect the vent to the sump pit must be designed for disassembly.

The cover to the sump must be above grade or floor level, or no more than 2 inches below grade or floor level. The limitation for below-grade or floor-level distance is so that a repair person will not have difficulty in reaching a cover that is significantly below grade or floor level.

P3007.3.3 Discharge pipe and fittings. Discharge pipe and fittings serving sump pumps and ejectors shall be constructed of materials in accordance with Sections P3007.3.3.1 and P3007.3.3.2 and shall be *approved*.

❖ Section P3007.3.3.1 specifies the allowable types of materials for pipes and fittings. Section P3007.3.3.2 requires suitability of pipes and fittings for the pressure, temperature and burial conditions. Section P3007.6 regulates the size of the discharge piping. Section P3007.2 requires valves in the discharge piping; however, because there are no other sections of the code that regulate pump discharge piping, the code official is required to specifically approve the piping.

P3007.3.3.1 Materials. Pipe and fitting materials shall be constructed of copper alloy, copper, CPVC, ductile iron, PE, or PVC.

❖ The materials listed by this section offer maximum durability for pump discharge service.

P3007.3.3.2 Ratings. Pipe and fittings shall be rated for the maximum system operating pressure and temperature. Pipe fitting materials shall be compatible with the pipe material. Where pipe and fittings are buried in the earth, they shall be suitable for burial.

❖ The pipe and fittings must be able to withstand the pressure developed by the pump. If the pump discharges hot water, the pipe must be rated for the temperature at the maximum pressure. If the pipe and fittings are buried, the materials must be suitable for burial.

P3007.3.4 Maximum effluent level. The effluent level control shall be adjusted and maintained to at all times prevent the effluent in the sump from rising to within 2 inches (51 mm) of the invert of the gravity drain inlet into the sump.

❖ This section states that the control mechanism that starts and stops the pump must be maintained and adjusted to limit the level of effluent in the sump to not higher than 2 inches (51 mm) below the invert level of the gravity drainage pipe(s) entering the sump. This requirement is to reduce the chance that the gravity drainage piping will become flooded or clogged because of standing effluent in the sump that may back up into the drain piping.

P3007.3.5 Ejector connection to the drainage system. Pumps connected to the drainage system shall connect to a *building sewer, building drain*, soil stack, waste stack or horizontal branch drain. Where the discharge line connects into horizontal drainage piping, the connection shall be made through a wye fitting into the top of the drainage piping and such wye fitting shall be located not less than 10 pipe diameters from the base of any soil stack, waste stack or fixture drain.

❖ Section P3007 is titled "Sumps and ejectors." As such, it may be interpreted to be applicable to both sewage pumps (including the "grinder type") and pneumatic sewage ejectors. The terms "sewage pump" and "ejector" are sometimes used interchangeably to describe the same device. This section imposes certain restrictions on how and where sewage pumps and pneumatic sewage ejectors may connect to the gravity drainage system.

This section clarifies that pumps and ejectors are allowed to connect to any drainage pipe of the sanitary drainage system, not just to building sewers or building drain. If the discharge pipe is connecting to a horizontal drain pipe, a wye fitting must be used and it must be located a minimum of 10 pipe diameters away from other pipe connections or fixtures to reduce the likelihood that discharge from the pump/ejector will interfere with the gravity flow in the drainage system. The 10 pipe diameters distance allows the pumped waste flow to settle (flatten out) in the invert of the pipe without creating backups and pressure surges. In a related requirement, the wye fitting must be oriented with the wye branch toward the top of the drainage piping. This requirement is to help prevent disruption of the flow of waste inside the drainage piping because the normal flow is on the bottom of the pipe. Therefore, introducing discharge flow into the top half of the pipe will not cause blockages of the pipe or disruption of other flow inside the pipe.

P3007.4 Sewage pumps and sewage ejectors. A sewage pump or sewage ejector shall automatically discharge the contents of the sump to the building drainage system.

❖ This section is redundant with Section P3007.1 and requires such equipment to operate automatically.

P3007.5 Macerating toilet systems and pumped waste systems. Macerating toilet systems and pumped waste systems shall comply with ASME A112.3.4/CSA B45.9 and shall be installed in accordance with the manufacturer's instructions.

❖ Macerating toilet systems and macerating pumping systems are unique in that there are many acceptable types available. ASME A112.3.4/CSA B45.9 is the standard for macerating toilet systems and pumping systems. Such systems must be installed in accordance with the manufacturer's instructions for their proper use and function. A macerating/pumping system encompasses three components: the container that houses the operating mechanisms; the pressure chamber that automatically activates and deactivates the macerating pump; and an induction motor that drives the shredder blades and pump assembly. Such systems collect waste from a single water closet, lavatory, bathtub or any combination thereof located in the same room, and grind and pump the collected wastes to some point in the sanitary drainage system [see Commentary Figure P3007.5(1)]. These systems are designed for above-the-floor installation and incorporate either a directly affixed macerating tank and integral grinder/pump mechanism to a specially designed rear discharge water closet; or a separate system with a built-in grinder/pump mechanism [see Commentary Figure P3007.5(2)].

P3007.6 Capacity. Sewage pumps and sewage ejectors shall have the capacity and head for the application requirements. Pumps and ejectors that receive the discharge of water closets shall be capable of handling spherical solids with a diameter of up to and including 2 inches (51 mm). Other pumps or ejectors shall be capable of handling spherical solids with a diameter of up to and including 1 inch (25.4 mm). The minimum capacity of a pump or ejector based on the diameter of the discharge pipe shall be in accordance with Table 3007.6.

Exceptions:

1. Grinder pumps or grinder ejectors that receive the discharge of water closets shall have a discharge opening of not less than $1^1/_4$ inches (32 mm).

2. Macerating toilet assemblies that serve single water closets shall have a discharge opening of not less than $^3/_4$ inch (19 mm).

❖ The sewage pump or ejector must be sized properly to accommodate the peak flow into the receiver, to provide the head pressure required for the elevation of lift, to prevent excessive cycling of the pumping equipment and to handle the type of waste that discharges to the sump. Sewage pumps and ejectors must be able to handle, without creating blockages, the types of solid waste discharged into the sump or receiver. Pumps and ejectors serving water closets must be able to handle the solids associated with soil drainage. Grinder pumps/ejectors and macerating toilet assemblies that serve water closets are exempt from the 2-inch (51 mm) spherical solids-handling requirements of this section. These pumps and ejectors reduce solids to a near-liquid state and pump the slurry to the drainage system. The $1^1/_4$-inch (32 mm) minimum-size discharge for grinder pumps and ejectors and $^3/_4$-inch (19 mm) minimum-size discharge for macerating grinder/pump systems will permit solids that are completely reduced to pass without creating a stoppage.

TABLE 3007.6
MINIMUM CAPACITY OF SEWAGE PUMP OR SEWAGE EJECTOR

DIAMETER OF THE DISCHARGE PIPE (inches)	CAPACITY OF PUMP OR EJECTOR (gpm)
2	21
$2^1/_2$	30
3	46

For SI: 1 inch = 25.4 mm, 1 gallon per minute = 3.785 L/m.

❖ Table P3007.6 specifies sewage pump capacity in gallons per minute (L/m) with respect to the pump's discharge opening size. This relationship between pipe size and pump capacity will maintain flow velocity in the piping to help prevent blockage and restriction.

SECTION P3008
BACKWATER VALVES

P3008.1 Sewage backflow. Where the flood level rims of plumbing fixtures are below the elevation of the manhole cover of the next upstream manhole in the public sewer, the fixtures shall be protected by a backwater valve installed in the *building drain*, branch of the *building drain* or horizontal branch serving such fixtures. Plumbing fixtures having flood level rims above the elevation of the manhole cover of the next upstream manhole in the public sewer shall not discharge through a backwater valve.

Exception: In existing buildings, fixtures above the elevation of the manhole cover of the next upstream manhole in the *public sewer* shall not be prohibited from discharging through a backwater valve.

❖ A backwater valve is required in areas where the public sewer might back up into the building through the sanitary drainage system [see Commentary Figure P3008.1(1)]. Where plumbing fixtures are located above the next upstream manhole cover from the building sewer connection to the public sewer, the sewer will back up through the street manhole before entering the building.

Public sewers might become blocked or overloaded, which will result in sewage backing up into the manholes and any laterals (taps) connected to the sewer system. The point of overflow for the public sewer will be the top of the manholes in the backed-up portion of the system.

Fixtures or drains located at an elevation below that of the tops of the manholes for the relative portion of the sewer system are subject to backflow and must be protected by backwater valves [see Commentary Figure P3008.1(2)]. Plumbing fixtures that are not subject to backflow are not permitted to discharge through a

backwater valve. In theory, limiting the fixtures that discharge through a backwater valve will prevent waste from up-stream fixtures from backing up through downstream fixtures because it cannot pass through the backwater valve when the public sewer is blocked or overloaded. Additionally, the valve will be protected

Figure P3007.5(1)
MACERATING TOILET SYSTEMS
(Photo courtesy of SFA SANIFLO)

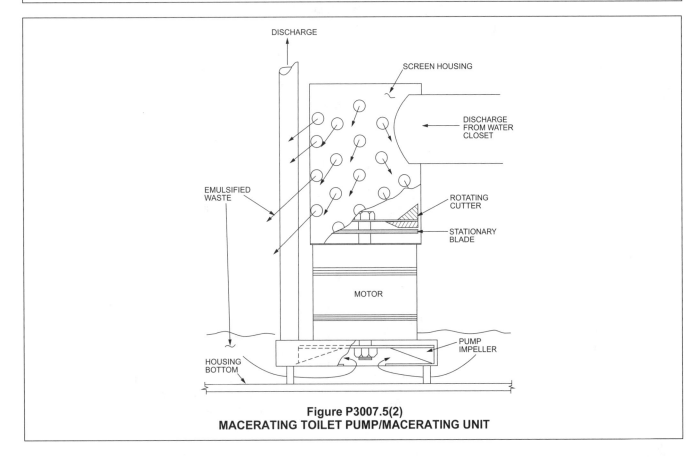

Figure P3007.5(2)
MACERATING TOILET PUMP/MACERATING UNIT

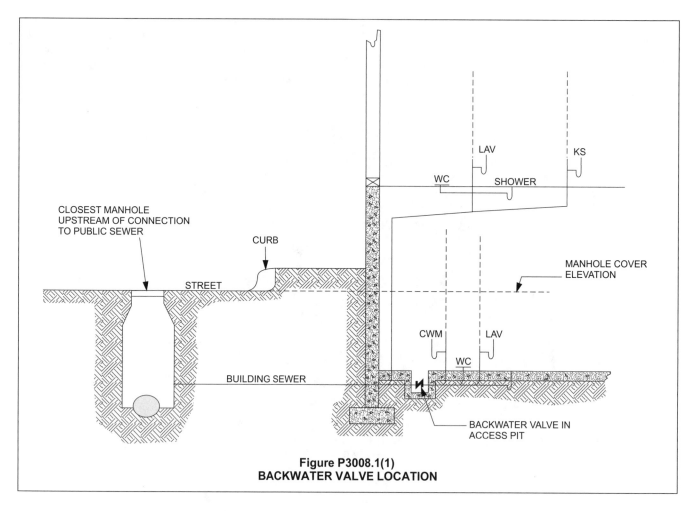

Figure P3008.1(1)
BACKWATER VALVE LOCATION

from excess wear and potential failure resulting from debris and accumulations.

The last sentence of this section, if taken literally, would apply to fixtures having a flood level rim a fraction of an inch above the upstream manhole cover. Such fixtures might still be subject to the backup of sewage, which may be insanitary and present the risk of sewage entering the structure through fixture connections, traps and other "weak links" in the piping system.

The exception to this section allows for retrofit of a backwater valve to the drainage system of an existing building without the need to separate the drainage piping of the building into "fixtures above the next upstream manhole cover elevation" and "fixtures below that elevation." Where an existing building owner is concerned about the public sewer backing up into his (or her) building, the easiest solution is to install a backwater valve in the building sewer. Although such an installation does present the risk of the building owner flooding his building by use fixtures on a floor level above a lower floor with fixtures, he does have control over the building's fixture use. Without the backwater valve, the building owner has no control over flooding of his building from a public sewer backup.

P3008.2 Material. Bearing parts of backwater valves shall be of corrosion-resistant material. Backwater valves shall comply with ASME A112.14.1, CSA B181.1 or CSA B181.2.

❖ This section requires that bearing parts of the backwater valve be made of corrosion-resistant materials so that the valve remains operable for its expected life, thus protecting the building it serves. This section also gives the industry standards to which the backwater valve must be manufactured. As required in Section P2608.4, these backwater valves are to be inspected and verified by an independent third-party quality-assurance facility.

P3008.3 Seal. Backwater valves shall be constructed to provide a mechanical seal against backflow.

❖ A backwater valve is designed to prevent the backflow of drainage in a piping system. The valve incorporates a swing check on the inlet side of the body. This section requires that the swing check have a positive seal against the backflow of drainage. Some back-water valves are also equipped with a gate valve that is manually operated to close the outlet side of the device. Backwater valves equipped with gate valves are typically used in areas subject to severe flooding conditions [see Commentary Figure P3008.1(2)].

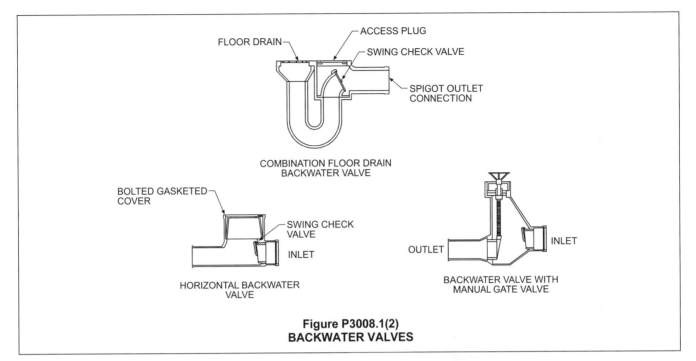

Figure P3008.1(2)
BACKWATER VALVES

P3008.4 Diameter. Backwater valves, when fully opened, shall have a capacity not less than that of the pipes in which they are installed.

❖ This section simply requires that the capacity of the pipe in which the backwater valve is located not be reduced by the valve. This would include components of the backwater valve, including any optional manual gate valves. Backwater valves are designed so as to not create a restriction in the piping.

P3008.5 Location. Backwater valves shall be installed so that the working parts are accessible for service and repair.

❖ Because a backwater valve has movable parts, there is a possibility of stoppage or malfunction. This section requires that backwater valves be located to be accessible to permit the necessary maintenance or repairs.

SECTION P3009
SUBSURFACE LANDSCAPE IRRIGATION SYSTEMS

P3009.1 Scope. The provisions of this section shall govern the materials, design, construction and installation of subsurface landscape irrigation systems connected to nonpotable water from on-site water reuse systems.

❖ The chapter provides specific details for designing and constructing subsurface irrigation systems.

P3009.2 Materials. Above-ground drain, waste and vent piping for subsurface landscape irrigation systems shall conform to one of the standards indicated in Table P3002.2(1). Subsurface landscape irrigation, underground building drainage and vent pipe shall conform to one of the standards indicated in Table P3002.1(2).

❖ The piping for above-ground landscape systems must comply with the standards in Table P3002.2(1). The piping for above-ground landscape systems must comply with the standards in Table P3002.2(2). The appropriate pipe fittings from Table P3002.2.3 must be used to make connections in the pipe.

P3009.3 Tests. Drain, waste and vent piping for subsurface landscape irrigation systems shall be tested in accordance with Section P2503.

❖ Section P2503 covers testing of drainage and vent piping inside a building (including under the building). Note that air testing of piping is prohibited.

P3009.4 Inspections. Subsurface landscape irrigation systems shall be inspected in accordance with Section R109.

❖ The installation of systems needs to be performed prior to concealment to make sure that the system is installed according to the requirements of the code.

P3009.5 Disinfection. Disinfection shall not be required for on-site nonpotable reuse water for subsurface landscape irrigation systems.

❖ Water sent to a subsurface irrigation system is not required to be disinfected as there is not human contact with the water.

P3009.6 Coloring. On-site nonpotable reuse water used for subsurface landscape irrigation systems shall not be required to be dyed.

❖ Coloring of the water is not required because the water is not visible to humans.

P3009.7 Sizing. The system shall be sized in accordance with the sum of the output of all water sources connected to the subsurface irrigation system. Where gray-water collection piping is connected to subsurface landscape irrigation systems, gray-water output shall be calculated according to the gallons-per-day-per-occupant (liters per day per occupant) number based on the type of fixtures connected. The gray-water discharge shall be calculated by the following equation:

$$C = A \times B \qquad \text{(Equation 30-1)}$$

where:

A = Number of occupants:

Number of occupants shall be determined by the actual number of occupants, but not less than two occupants for one bedroom and one occupant for each additional bedroom.

B = Estimated flow demands for each occupant:

25 gallons (94.6 L) per day per occupant for showers, bathtubs and lavatories and 15 gallons (56.7 L) per day per occupant for clothes washers or laundry trays.

C = Estimated gray-water discharge based on the total number of occupants.

❖ The volume of gray-water discharge from residential buildings is the estimated number of occupants living in the building times a fixed amount of 25 gallons per day (94.6 L/day). An additional 15 gallons per day (56.7 L/day) per occupant is added if clothes washers or laundry trays exist. The number of occupants for the building must be not less than two for the first bedroom and one additional for each additional bedroom.

P3009.8 Percolation tests. The permeability of the soil in the proposed absorption system shall be determined by percolation tests or permeability evaluation.

❖ In order to determine how large the subsurface irrigation system (the absorption system) must be in order to accommodate the maximum daily flow of gray water, tests of the subsurface soil must be performed to determine the soil's capacity for accepting the flow. Soil testing can be by the percolation test method or permeability evaluation method.

P3009.8.1 Percolation tests and procedures. Not less than three percolation tests in each system area shall be conducted. The holes shall be spaced uniformly in relation to the bottom depth of the proposed absorption system. More percolation tests shall be made where necessary, de-pending on system design.

❖ A percolation test involves digging at least three holes to a depth where the proposed absorption system is planned to be installed. The undisturbed soil at the bottom of the hole will be representative of the soil that must absorb the gray-water flow.

P3009.8.1.1 Percolation test hole. The test hole shall be dug or bored. The test hole shall have vertical sides and a horizontal dimension of 4 inches to 8 inches (102 mm to 203 mm). The bottom and sides of the hole shall be scratched with a sharp-pointed instrument to expose the natural soil. Loose material shall be removed from the hole and the bottom shall be covered with 2 inches (51 mm) of gravel or coarse sand.

❖ The percolation test holes must be of the same size and shape so that the test results from multiple holes can be compared. The surfaces of holes must be scratched to break up any soil smearing that could

block the flow of test water. Test holes are made purposely small in horizontal dimension so a limited amount of water is necessary to perform the test. The placement of gravel or coarse sand in the bottom of the hole prevents scouring of the bottom of the test hole as test water is poured into the hole.

P3009.8.1.2 Test procedure, sandy soils. The hole shall be filled with clear water to not less than 12 inches (305 mm) above the bottom of the hole for tests in sandy soils. The time for this amount of water to seep away shall be determined, and this procedure shall be repeated if the water from the second filling of the hole seeps away in 10 minutes or less. The test shall proceed as follows: Water shall be added to a point not more than 6 inches (152 mm) above the gravel or coarse sand. Thereupon, from a fixed reference point, water levels shall be measured at 10-minute intervals for a period of 1 hour. Where 6 inches (152 mm) of water seeps away in less than 10 minutes, a shorter interval between measurements shall be used. The water depth shall not exceed 6 inches (152 mm). Where 6 inches (152 mm) of water seeps away in less than 2 minutes, the test shall be stopped and a rate of less than 3 minutes per inch (7.2 s/mm) shall be reported. The final water level drop shall be used to calculate the percolation rate. Soils not meeting these requirements shall be tested in accordance with Section P3009.8.1.3.

❖ To prepare sandy soil for the percolation test, 12 inches (305 mm) of water is poured into the test hole and allowed to drain out. Another 12 inches (305 mm) of water is added to the hole and, if the time that it takes to drain out is 10 minutes or less, then the hole is repeatedly filled with 12 inches (305 mm) of water until the time that it takes for the water to drain away is greater than 10 minutes.

For the sandy soil percolation test, the hole is filled with 6 inches (152 mm) of water and the level measured from a fixed reference point is taken every 10 minutes. If the 6 inches (152 mm) of water completely drains from the hole in less than 10 minutes, then the measurement interval must be shortened so that some water remains in the hole at the end of the measurement interval. If the 6 inches (152 mm) of water completely drains out in 2 minutes or less, then a percolation rate of 3 minutes per inch is assigned. If the 6 inches (152 mm) of water drains out slower than 2 minutes, then the percolation rate is determined from the final water level drop in the test hole. For example, assume that the measurement interval is 5 minutes and the last water level drop (before the hole becomes completely empty) was $^3/_4$ inch (19.1 mm). The percolation rate would be 5 minutes/0.75 inches (19.1 mm) = 6.7 minutes per inch.

P3009.8.1.3 Test procedure, other soils. The hole shall be filled with clear water, and a minimum water depth of 12 inches (305 mm) shall be maintained above the bottom of the hole for a 4-hour period by refilling whenever necessary or by use of an automatic siphon. Water remaining in the hole after 4 hours shall not be removed. Thereafter, the soil shall be allowed to swell not less than 16 hours or more than 30 hours. Immediately after the soil swelling period, the measurements for determining the percolation rate shall be made as follows: any soil sloughed into the hole shall be removed and the water level shall be adjusted to 6 inches (152 mm) above the gravel or coarse sand. Thereupon, from a fixed reference point, the water level shall be measured at 30-minute intervals for a period of 4 hours, unless two successive water level drops do not vary by more than $^1/_{16}$ inch (1.59 mm). Not less than three water level drops shall be observed and recorded. The hole shall be filled with clear water to a point not more than 6 inches (152 mm) above the gravel or coarse sand whenever it becomes nearly empty. Adjustments of the water level shall not be made during the three measurement periods except to the limits of the last measured water level drop. When the first 6 inches (152 mm) of water seeps away in less than 30 minutes, the time interval between measurements shall be 10 minutes and the test run for 1 hour. The water depth shall not exceed 5 inches (127 mm) at any time during the measurement period. The drop that occurs during the final measurement period shall be used in calculating the percolation rate.

❖ For soils other than sandy soils, the soil in the test hole has to be prepared by soaking with not less than 12 inches (305 mm) of water for 4 hours. After the soaking period, the water is left in the hole and the soil allowed to swell for not less than 16 hours nor more than 30 hours. Immediately after the swelling period, the water in the hole must be adjusted to a 6-inch (152 mm) depth above the bottom of the hole and measurements of the water level drop taken at 30-minute intervals for 4 hours. If two consecutive water level drops do not vary more than $^1/_{16}$ inch (1.6 mm), then the test can be stopped and the percolation rate can be determined from the last water level drop.

If the water in the hole drains out in less than 30 minutes, then water level drop measurements must be taken at 10-minute intervals for a total of 1 hour. After each measurement, the hole must be refilled to at least 5 inches (127 mm) above the bottom of the hole.

P3009.8.1.4 Mechanical test equipment. Mechanical percolation test equipment shall be of an *approved* type.

❖ Another way to determine soil percolation rates is by using mechanical percolation test equipment such as a permeameter. Such equipment must be approved by the code official.

P3009.8.2 Permeability evaluation. Soil shall be evaluated for estimated percolation based on structure and texture in accordance with accepted soil evaluation practices. Borings shall be made in accordance with Section P3009.8.1.1 for evaluating the soil.

❖ The soil from the bored percolation test hole must be evaluated for the type of soil structure and texture to determine if the soil is suitable for absorbing water. Evidence of soil mottling could indicate that a seasonal high water table exists, even though the test hole soil is dry.

P3009.9 Subsurface landscape irrigation site location. The surface grade of soil absorption systems shall be located at a point lower than the surface grade of any water well or reservoir on the same or adjoining lot. Where this is not possible, the site shall be located so surface water drainage from the site is not directed toward a well or reservoir. The soil absorption system shall be located with a minimum horizontal distance between various elements as indicated in Table P3009.9. Private sewage disposal systems in compacted areas, such as parking lots and driveways, are prohibited. Surface water shall be diverted away from any soil absorption site on the same or neighboring lots.

❖ Because gray water contains bacteria, it is important to make sure that the soil-absorption system be located away and preferably downhill from water reservoirs and water wells. Table P3009.9 indicates the minimum distances that tanks and absorption systems must be located from certain features on the lot.

TABLE P3009.9
LOCATION OF SUBSURFACE IRRIGATION SYSTEM

ELEMENT	MINIMUM HORIZONTAL DISTANCE	
	STORAGE TANK (feet)	IRRIGATION DISPOSAL FIELD (feet)
Buildings	5	2
Lot line adjoining private property	5	5
Water wells	50	100
Streams and lakes	50	50
Seepage pits	5	5
Septic tanks	0	5
Water service	5	5
Public water main	10	10

For SI: 1 foot = 304.8 mm.

P3009.10 Installation. Absorption systems shall be installed in accordance with Sections P3009.10.1 through P30091.10.5 to provide landscape irrigation without surfacing of water.

❖ This section directs the reader to the subsections that regulate the installation of absorption systems.

P3009.10.1 Absorption area. The total absorption area required shall be computed from the estimated daily gray-water discharge and the design-loading rate based on the percolation rate for the site. The required absorption area equals the estimated gray-water discharge divided by the design loading rate from Table P3009.10.1.

❖ The area of the absorption system is determined by dividing the estimated number of gray water gallons per day by the design loading rate from Table P3009.10.1. The design loading rate is a function of the percolation test results.

**TABLE P3009.10.1
DESIGN LOADING RATE**

PERCOLATION RATE (minutes per inch)	DESIGN LOADING FACTOR (gallons per square foot per day)
0 to less than 10	1.2
10 to less than 30	0.8
30 to less than 45	0.72
45 to 60	0.4

For SI: 1 minute per inch = min/25.4 mm,
1 gallon per square foot = 40.7 L/m².

P3009.10.2 Seepage trench excavations. Seepage trench excavations shall be not less than 1 foot (304 mm) in width and not greater than 5 feet (1524 mm) in width. Trench excavations shall be spaced not less than 2 feet (610 mm) apart. The soil absorption area of a seepage trench shall be computed by using the bottom of the trench area (width) multiplied by the length of pipe. Individual seepage trenches shall be not greater than 100 feet (30 480 mm) in developed length.

❖ One method for making a subsurface irrigation system is to dig a trench 1 to 5 feet (304 to 1524 mm) in width and place a distribution pipe along the center of the width of the trench. Trenches are dug to a depth of where the percolation tests were performed. Where multiple trenches are required, they must be not less than 2 feet (610 mm) apart. Trenches are limited to 100 feet (30 480 mm) in length as this is a practical limit for keeping the entire trench bottom level. The absorption area of the trench is the width of the trench multiplied by the length of the pipe to be placed in the trench.

P3009.10.3 Seepage bed excavations. Seepage bed excavations shall be not less than 5 feet (1524 mm) in width and have more than one distribution pipe. The absorption area of a seepage bed shall be computed by using the bottom of the trench area. Distribution piping in a seepage bed shall be uniformly spaced not greater than 5 feet (1524 mm) and not less than 3 feet (914 mm) apart, and greater than 3 feet (914 mm) and not less than 1 foot (305 mm) from the sidewall or headwall.

❖ Another method for making a subsurface irrigation system is to excavate an area that is not less than 5 feet (1524 mm) wide and place multiple distribution pipes, equally spaced from 3 to 5 feet (914 to 1524 mm) apart in the excavated area. The distribution pipes must be not closer than 1 foot (305 mm) and not more than 3 feet (914 mm) from the headwall and sidewalls. The absorption area of the trench is the width of the excavated bed area multiplied by the length of the bed area.

P3009.10.4 Excavation and construction. The bottom of a trench or bed excavation shall be level. Seepage trenches or beds shall not be excavated where the soil is so wet that such material rolled between the hands forms a soil wire. Smeared or compacted soil surfaces in the sidewalls or bottom of seepage trench or bed excavations shall be scarified to the depth of smearing or compaction and the loose material removed. Where rain falls on an open excavation, the soil shall be left until sufficiently dry so a soil wire will not form when soil from the excavation bottom is rolled between the hands. The bottom area shall then be scarified and loose material removed.

❖ Care must be taken to not excavate or work in trenches or beds when the soil is wet so that a soil wire can be formed between the hands. Smeared and compacted soil must be scarified (scratched) to break up the smearing and compaction. Loose soil must be removed from the trench or bed area. The bottom of bed and trench areas must be level.

P3009.10.5 Aggregate and backfill. Not less than 6 inches (150 mm) in depth of aggregate ranging in size from ¹/₂ to 2¹/₂ inches (12.7 mm to 64 mm) shall be laid into the trench below the distribution piping elevation. The aggregate shall be evenly distributed not less than 2 inches (51 mm) in depth over the top of the distribution pipe. The aggregate shall be covered with *approved* synthetic materials or 9 inches (229 mm) of uncompacted marsh hay or straw. Building paper shall not be used to cover the aggregate. Not less than 9 inches (229 mm) of soil backfill shall be provided above the covering.

❖ The bottom of the bed or trench area must have aggregate placed to a depth of 6 inches (152 mm). Then the distribution pipe is laid on top of the aggregate and covered with more aggregate so that the distribution pipe is covered with 2 inches (51 mm) of aggregate. The installer has the choice of covering the aggregate with synthetic materials or 9 inches (229 mm) of marsh hay or straw. The trench or bed is then covered by not less than 9 inches (229 mm) of soil backfill.

P3009.11 Distribution piping. Distribution piping shall be not less than 3 inches (76 mm) in diameter. Materials shall comply with Table P3009.11. The top of the distribution pipe shall be not less than 8 inches (203 mm) below the original surface. The slope of the distribution pipes shall be not less than 2 inches (51 mm) and not greater than 4 inches (102 mm) per 100 feet (30 480 mm).

❖ Distribution piping must be at least 3 inches (76 mm) in diameter and be of one of the materials listed in Table P3009.11. The elevation of the top of the pipe in the trench or bed must be not less than 8 inches (203 mm) below the original grade elevation. The distribution pipe must be sloped at least 2 inches (51 mm) per 100 feet (30 480 mm) but not more than 4 inches (102 mm) per 100 feet (30 480 mm).

**TABLE P3009.11
DISTRIBUTION PIPE**

MATERIAL	STANDARD
Polyethylene (PE) plastic pipe	ASTM F405
Polyvinyl chloride (PVC) plastic pipe	ASTM D2729
Polyvinyl chloride (PVC) plastic pipe with a 3.5-inch O.D. and solid cellular core or composite wall	ASTM F1488

For SI: 1 inch = 25.4 mm.

P3009.11.1 Joints. Joints in distribution pipe shall be made in accordance with Section P3003 of this code.

❖ Distribution piping must be joined by one of the methods in Section P3003, as applicable to the type of pipe being used.

SECTION P3010
REPLACEMENT OF UNDERGROUND SEWERS BY PIPE BURSTING METHODS

P3010.1 General. This section shall govern the replacement of existing *building sewer* piping by pipe-bursting methods.

❖ This is a method for replacing existing building piping that does not require complete excavation of the existing piping. This method uses an expanding mandrel that is attached to the end of the replacement pipe. The mandrel (with the replacement pipe in tow) is pulled by a cable through the existing sewer line. The mandrel has a hydraulically activated expanding shell that expands and breaks (bursts) the existing pipe to make room for the new pipe as it is pulled into position.

The method requires that launching and receiving pits be excavated at the beginning and ending points of the pipe to be replaced. If there are any lateral connections that need to be made between the beginning and the end, excavated pits are required at those locations as well.

P3010.2 Applicability. The replacement of building sewer piping by pipe bursting methods shall be limited to gravity drainage piping of sizes 6 inches (150 mm) and smaller. The replacement piping shall be of the same nominal size as the existing piping.

❖ Although the pipe-bursting method of underground pipe replacement is used for large pipe sizes, the code limits its use to the replacement of pipes that are 6 inches and smaller. The code also limits this method to use on gravity sewers and a replacement using the same size as existing piping.

P3010.3 Preinstallation inspection. The existing piping sections to be replaced shall be inspected internally by a recorded video camera survey. The survey shall include notations of the position of cleanouts and the depth of connections to the existing piping.

❖ An internal video camera survey of the existing installation is required to identify problem areas and location of connections and cleanouts.

P3010.4 Pipe. The replacement pipe shall be made of a high-density polyethylene (HDPE) that conforms to cell classification number PE3608, PE4608 or PE4710 as indicated in ASTM F714. The pipe fittings shall be manufactured with an SDR of 17 and in compliance with ASTM F714.

❖ The only material that the code allows for sewer replacement is polyethylene plastic pipe (ASTM F714) having an SDR 17 wall thickness. This pipe is an outer-diameter- and wall-thickness-controlled product. SDR 17 provides for an inside diameter that is very close to the inside diameter of PVC and cast iron sewer piping of the same nominal pipe size.

P3010.5 Pipe fittings. Pipe fittings to be connected to the replacement piping shall be made of high-density polyethylene (HDPE) that conforms to cell classification number PE3608, PE4608 or PE4710 as indicated in ASTM F714. The pipe fittings shall be manufactured with an SDR of 17 and in compliance with ASTM D2683.

❖ Fittings for this material must be compatible with the pipe. The requirement for compliance to ASTM D2683 avoids the use of butt-welded fittings. A butt-welded fitting could have internal edges that could catch and restrict flow. Use of a socket fitting allows for pipe ends to be deburred/chamfered before assembly.

P3010.6 Cleanouts. Where the existing *building sewer* did not have cleanouts meeting the requirements of this code, cleanout fittings shall be installed as required by this code.

❖ Cleanouts are required in accordance with the code, whether the existing sewer had the cleanouts or not.

P3010.7 Post-installation inspection. The completed replacement piping section shall be inspected internally by a recorded video camera survey. The video survey shall be reviewed and *approved* by the building official prior to pressure testing of the replacement piping system.

❖ An internal video camera survey of the completed installation is required to be reviewed by the code official prior to pressure testing of replacement pipe, including all connections to the replacement pipe.

P3010.8 Pressure testing. The replacement piping system and the connections to the replacement piping shall be tested in accordance with Section P2503.4.

❖ As with any installation of a new sewer, pressure testing is required.

Chapter 31: Vents

User note: Code change proposals to this chapter will be considered by the IRC – Plumbing and Mechanical Code Development Committee during the 2015 (Group A) Code Development Cycle.

General Comments

Venting protects the trap seals within the plumbing system. When installed in accordance with this chapter, the system will function properly. Failure to adhere to these provisions can result in the loss of trap seals and the introduction of sewer gases into the building.

This chapter includes basic venting requirements and the purpose for them. It also provides installation guidelines that address sizing, penetrations, termination, protection and provisions for future vents. It outlines maximum distances and sizing limitations for trap and fixture drain sizes.

This chapter details various venting concepts, including individual venting for each fixture, common venting, waste stack venting, circuit venting, combination waste and vent systems, and the use of air admittance valves (AAV). Some of these vents are sized to allow a drain to serve as a vent, providing design options that may result in economical material use. This chapter also includes a technique specific to the installation of island sinks.

Purpose

Venting protects the trap seal of each trap. The vents are designed to maintain maximum differential pressures at each trap to less than 1 inch of water column (wc) (249 Pa). Waste flow in the drainage system creates pressure fluctuations that can negatively affect traps. Venting is not intended to provide for the circulation of air within the drainage system. If there were no traps in a drainage system, venting would not be required. The system would function adequately because it would be open to the atmosphere at the fixture connections, thereby allowing airflow.

Chapter 31 covers the requirements for vents and venting. Knowing why venting is required makes it easier to understand this chapter. Venting protects every trap against the loss of a trap seal. All of the provisions set forth in this chapter are intended to limit the pressure differentials in the drainage system to a maximum of 1 inch w.c. (249 Pa) above or below atmospheric pressure (i.e., positive or negative pressures).

SECTION P3101
VENT SYSTEMS

P3101.1 General. This chapter shall govern the selection and installation of piping, tubing and fittings for vent systems. This chapter shall control the minimum diameter of vent pipes, circuit vents, branch vents and individual vents, and the size and length of vents and various aspects of vent stacks and stack vents. Additionally, this chapter regulates vent grades and connections, height above fixtures and relief vents for stacks and fixture traps, and the venting of sumps and sewers.

❖ This section established the scope of this chapter, which governs the selection and installation of vent systems. The chapter also includes requirements for sizing vents, circuit vents, branch vents and individual vents. It regulates the size and length of vents, as well as stating vent stack and stack vent requirements. In addition, it specifies and regulates vent grades, height above fixtures, relief vents for stacks and fixture traps, and venting of sumps and ejectors.

P3101.2 Trap seal protection. The plumbing system shall be provided with a system of vent piping that will allow the admission or emission of air so that the liquid seal of any fixture trap shall not be subjected to a pressure differential of more than 1 inch of water column (249 Pa).

❖ Protection of the trap seal means that during the normal operation of the plumbing system, the water seal

remains in the trap. Each trap has a minimum trap seal depth of 2 inches (51 mm) translating to a hydro-static pressure equal to a 2-inch w.c. (498 Pa). If exposed to a 1-inch w.c. (249 Pa) pressure differential, a 1-inch (25 mm) water seal remains in the trap. The vent methods identified in this chapter limit the pressure differential at trap seals to 1-inch w.c. (249 Pa) or less.

A lower pressure on the outlet side of the trap is detrimental because this could cause all or part of the liquid seal to be lost down the drain. A higher pressure on the outlet side of the trap could result in sewer gases being pushed through the trap seal. For example, bubbles rising through the trap seal of a water closet are an indication of excessive pressure on the outlet side of the trap.

P3101.2.1 Venting required. Every *trap* and trapped fixture shall be vented in accordance with one of the venting methods specified in this chapter.

❖ This section establishes that all traps and trapped fixtures must be vented. It also indicates that the method of venting can be any of the methods described in the chapter. Proper application of the venting options will allow the designer latitude that can be applied to numerous situations encountered in building construction and may also prove cost effective.

P3101.3 Use limitations. The plumbing vent system shall not be used for purposes other than the venting of the plumbing system.

❖ The venting system is designed for a specific function and cannot be used for any other purposes. Any other use could interfere with the proper operation of the plumbing system or could create a hazard. For example, vent piping must not serve as a drain for condensate from an air-conditioning unit. Such use would introduce waste into piping not intended to convey waste and might allow sewer gases to enter the building through the air-conditioning unit.

P3101.4 Extension outside a structure. In climates where the 97.5-percent value for outside design temperature is 0°F (-18°C) or less (ASHRAE 97.5-percent column, winter, see Chapter 3), vent pipes installed on the exterior of the structure shall be protected against freezing by insulation, heat or both. Vent terminals shall be protected from frost closure in accordance with Section P3103.2.

❖ In areas where the outdoor design temperature [in accordance with the American Society of Heating, Refrigeration and Air Conditioning Engineers (ASHRAE) data] is 0°F (-18°C) or less, vent pipes installed on the exterior of the building must be protected from freezing by either insulation or heat, or both. Insulation alone will not prevent freezing temperatures in pipes unless there is a heat source for the pipe. In addition, where subject to frost, the vent terminals must be protected against closure as outlined in Section P3103.2 [see commentary, Table R301.2(1)].

P3101.5 Flood resistance. In flood hazard areas as established by Table R301.2(1), vents shall be located at or above the elevation required in Section R322.1 (flood hazard areas including A Zones) or R322.2 (coastal high-hazard areas including V Zones).

❖ See the commentary to Section P2602.2.

SECTION P3102
VENT STACKS AND STACK VENTS

P3102.1 Required vent extension. The vent system serving each *building drain* shall have not less than one vent pipe that extends to the outdoors.

❖ Each drainage system must have not less than one vent pipe open to the outdoors. This is consistent with Section P3114.7. Section P3102.2 dictates the nature of the required vent. This modernized section has evolved from the traditional "main vent" text and intends to provide for the relief of pressures that might develop in the building drain. This section does not require anything that has not already been done throughout the history of modern plumbing.

P3102.2 Installation. The required vent shall be a dry vent that connects to the *building drain* or an extension of a drain

that connects to the *building drain*. Such vent shall not be an island fixture vent as permitted by Section P3112.

❖ The required vent can be a dry vent taken off of the building drain or it can be the dry vent extension of any drain that connects to the building drain. For example, a vent stack that rises from the building drain meets the intent of this section, as would any dry-vented fixture drain or fixture branch that connects to the building drain. As can be seen, this section will typically be satisfied by any plumbing installation that otherwise complies with the code because almost all systems will have one or more dry-vented drains that connect to the building drain and comply with the sizing of Section P3102.3.

An island fixture vent (see Section P3112) is the only type of vent that cannot satisfy this section because of the inherent risk of blockage in a vent that is allowed to have a horizontal component below the flood level rim of the fixture served (see Commentary Figure P3102.2).

P3102.3 Size. The required vent shall be sized in accordance with Section P3113.1 based on the required size of the *building drain*.

❖ Section P3113.1 requires the vent to have a diameter of not less than one-half of the diameter of the building drain. For example, a 4-inch (102 mm) building drain must have connected to it a 2-inch (51 mm) or larger dry vent extending to the outdoors or a 2-inch (51 mm) or larger drain with a 2-inch (51 mm) or larger dry vent extension to the outdoors.

SECTION P3103
VENT TERMINALS

P3103.1 Roof extension. Open vent pipes that extend through a roof shall be terminated not less than 6 inches (152 mm) above the roof or 6 inches (152 mm) above the anticipated snow accumulation, whichever is greater. Where a roof is to be used for assembly, as a promenade, observation deck or sunbathing deck or for similar purposes, open vent pipes shall terminate not less than 7 feet (2134 mm) above the roof.

❖ Where the roof is occupied for any purpose, such as occurs with a recreational roof deck, plumbing vents must extend at least 7 feet (2134 mm) above the roof to prevent harmful sewer gases from contaminating the area. The sewer gases will tend to disperse into the air rather than accumulate near the roof surface. The minimum termination height must be determined by the local jurisdiction based on the local snowfall averages, but 6 inches (152 mm) above the roof is the minimum termination height necessary to protect the terminal from being blocked by snow and to allow a sufficient length of pipe for a proper roof flashing (see Commentary Figure P3103.1).

The vent terminal through the roof must tolerate movement resulting from the expansion and contraction of the piping material. This is extremely important when the vent piping material has high rates of expansion and contraction, such as acrylonitrilebutadiene-styrene (ABS) and polyvinyl chloride (PVC).

P3103.2 Frost closure. Where the 97.5-percent value for outside design temperature is 0°F (-18°C) or less, vent extensions through a roof or wall shall be not less than 3 inches (76 mm) in diameter. Any increase in the size of the vent shall be

made not less than 1 foot (304.8 mm) inside the thermal envelope of the building.

❖ The possibility of frost closure occurs only in areas of the country having cold climates with outdoor design temperatures of 0°F (-18°C) and less. A vent 3 inches (76 mm) in diameter or greater is not likely to completely close up from frost formation. This section requires vents smaller than 3 inches (76 mm) in diam-

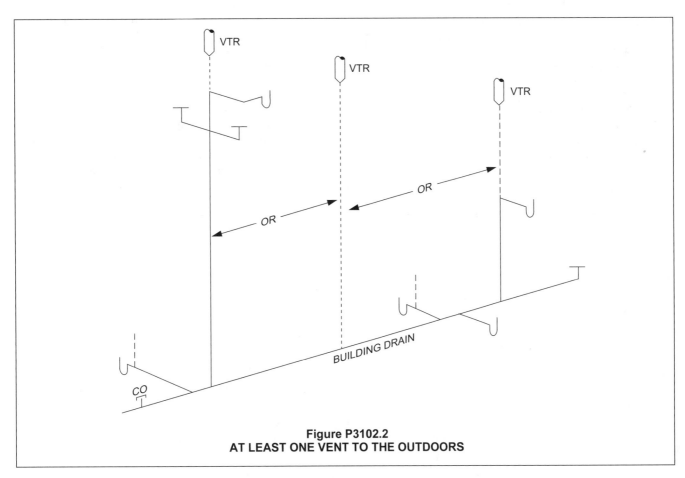

Figure P3102.2
AT LEAST ONE VENT TO THE OUTDOORS

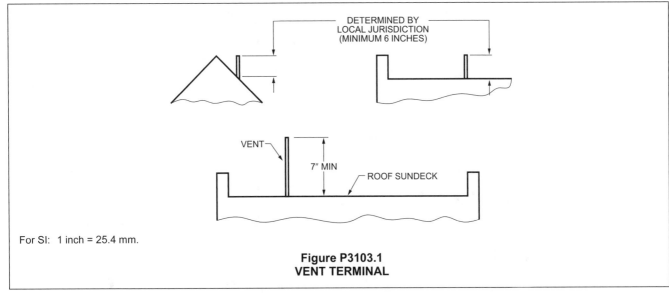

For SI: 1 inch = 25.4 mm.

Figure P3103.1
VENT TERMINAL

eter to be increased in size where frost enclosure is possible. The increase must occur within the thermal envelope of the building to protect the smaller pipe from direct exposure to the outside atmosphere [see commentary, Table R301.2(1)].

P3103.3 Flashings and sealing. The juncture of each vent pipe with the roof line shall be made water tight by an *approved* flashing. Vent extensions in walls and soffits shall be made weather tight by caulking.

❖ Water penetration of the roof or wall, at the point where the vent exits, can cause decay of wood and degradation of other building materials. Flashing must be installed at roof penetrations, and caulking must be applied at wall penetrations to protect the interior of the building from moisture and to prevent the intrusion of insects or vermin (see Sections P2606 and P3002.4.1 for additional provisions).

P3103.4 Prohibited use. A vent terminal shall not be used for any purpose other than a vent terminal.

❖ The use of the vent terminal to support any imposed loads is prohibited.

P3103.5 Location of vent terminal. An open vent terminal from a drainage system shall not be located less than 4 feet (1219 mm) directly beneath any door, openable window, or other air intake opening of the building or of an adjacent building, nor shall any such vent terminal be within 10 feet (3048 mm) horizontally of such an opening unless it is not less than 3 feet (914 mm) above the top of such opening.

❖ To prevent sewer gases from entering the building through gravity openings or forced air intakes, this section requires that vents be terminated not less than 4 feet (1219 mm) below, 10 feet (3048 mm) horizontally from or 3 feet (914 mm) above any opening into the building or adjacent buildings (see Commentary Figure P3103.5).

P3103.6 Extension through the wall. Vent terminals extending through the wall shall terminate not less than 10 feet (3048 mm) from the *lot line* and 10 feet (3048 mm) above the highest adjacent *grade* within 10 feet (3048 mm) horizontally of the vent terminal. Vent terminals shall not terminate under the overhang of a structure with soffit vents.

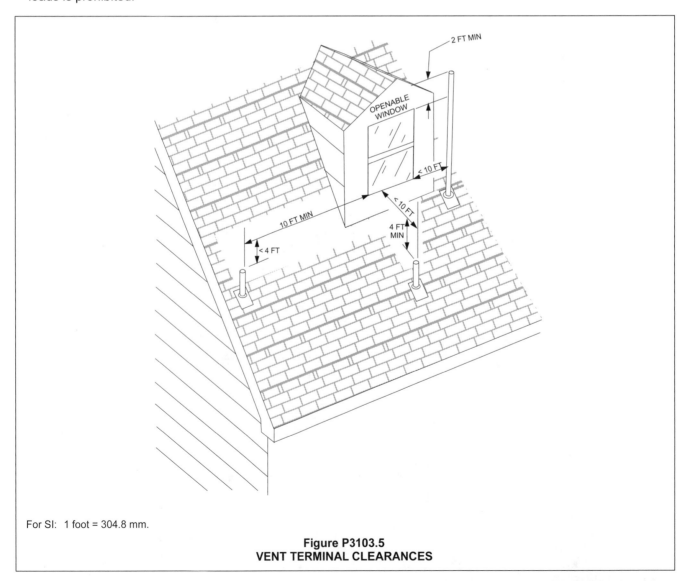

For SI: 1 foot = 304.8 mm.

Figure P3103.5
VENT TERMINAL CLEARANCES

Side wall vent terminals shall be protected to prevent birds or rodents from entering or blocking the vent opening.

❖ Side-wall vent terminations are an alternative to roof penetrations. This alternative might provide significant cost savings and a more aesthetically pleasing installation. For example, a side-wall vent may be preferred to penetrating membrane, built-up, slate or tile roofs, especially for existing buildings. Such roof penetrations are difficult to make leakproof, can be expensive and are often unsightly.

In multiple-story buildings where remodeling work is done, a side wall vent termination can serve as an alternative to running vent piping through finished or occupied stories above. When a side-wall vent is installed, the vent opening must be protected with a screen or louver to prevent birds from building a nest in the pipe. The screen or louver also prevents rodents

from entering the vent system (see Commentary Figure P3103.6). Such vents must not terminate where the emissions can cause structural damage or enter the building envelope through soffit vents. Also, such vents must comply with the location requirements of Section P3103.5.

SECTION P3104
VENT CONNECTIONS AND GRADES

P3104.1 Connection. Individual branch and circuit vents shall connect to a vent stack, stack vent or extend to the open air.

Exception: Individual, branch and circuit vents shall be permitted to terminate at an *air admittance valve* in accordance with Section P3114.

❖ This section further emphasizes the requirement that all vents must extend to the outdoors. This can be

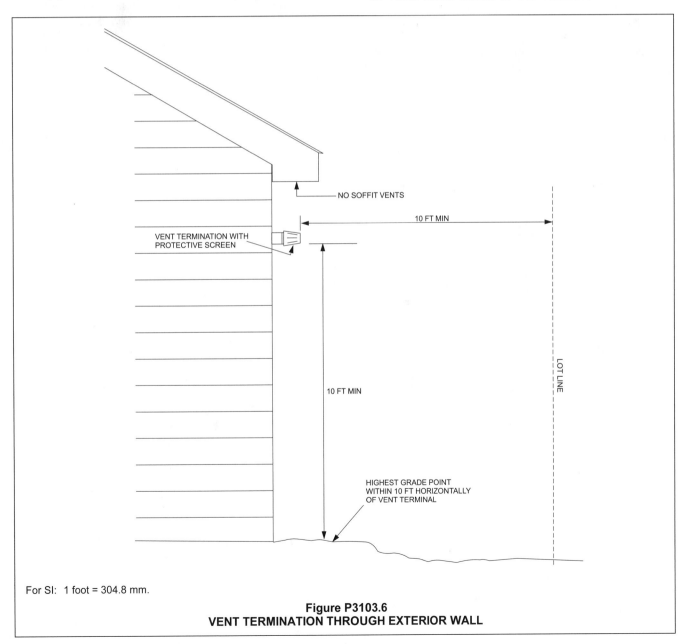

For SI: 1 foot = 304.8 mm.

Figure P3103.6
VENT TERMINATION THROUGH EXTERIOR WALL

accomplished by connecting to a vent stack or stack vent that extends outdoors and terminates to the open air or by extending these vents independently to the outdoors. The intent is to prevent sewer gases from escaping into the building. The exception recognizes the use of air admittance valves as an alternative to terminating a vent outdoors. The exception does not apply to stack vents and vent stacks (see Sections P3102.1 and P3114).

P3104.2 Grade. Vent and branch vent pipes shall be graded, connected and supported to allow moisture and condensate to drain back to the soil or waste pipe by gravity.

❖ Vents commonly convey moisture vapor from the drainage system, and such vapor can condense in the vent piping. Also, rainwater can enter the vent system at the vent terminal. Therefore, the vent must be graded to drain any moisture back into the drainage piping, preventing accumulation of condensate or rainwater. A vent pipe that is not sloped or that has a reverse slope can be partially or completely filled with water, which lessens the efficiency of the vent or makes it useless (see Commentary Figure P3104.2).

P3104.3 Vent connection to drainage system. A dry vent connecting to a horizontal drain shall connect above the centerline of the horizontal drain pipe.

❖ When drainage enters a pipe, the liquid is assumed to proceed down the pipe in the direction of flow. Being a liquid, however, the drainage seeks its own level and can move against the direction of flow before reversing and draining down the pipe. When this occurs, the solids may drop out of suspension to the bottom of the pipe. In a drain, the solids will again move down the pipe with the next discharge of liquid to the drain.

If the horizontal pipe serves only as a dry vent, the solids have no possibility of being brought back into suspension and, over a period of time, the horizontal vent pipe could be completely obstructed by waste. To avoid these blockages, the vent must connect to horizontal drains above the centerline (see Commentary Figure P3104.3). This prevents waste from entering a dry vent by connecting the vent above the flow line of the drain. Such connections will result in vent piping that forms a 45-degree (0.79 rad) or greater angle with the horizontal.

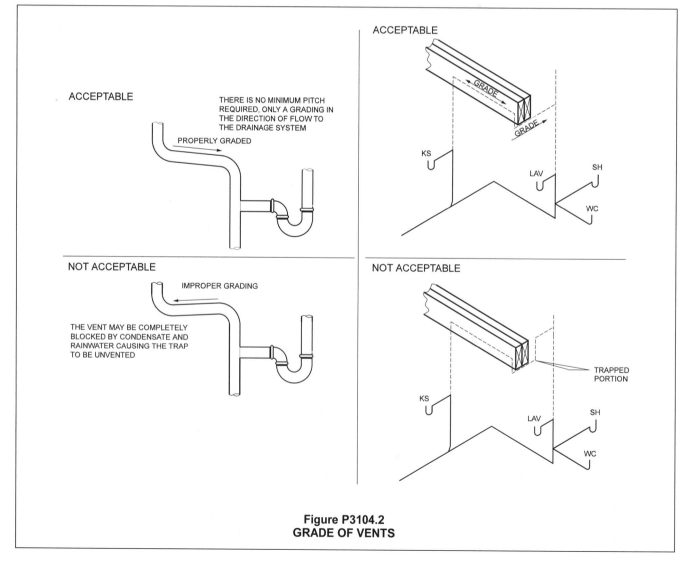

Figure P3104.2
GRADE OF VENTS

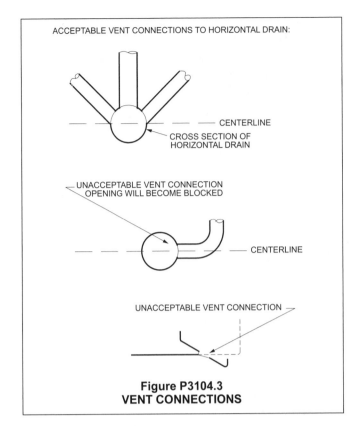

ACCEPTABLE VENT CONNECTIONS TO HORIZONTAL DRAIN:

CENTERLINE

CROSS SECTION OF HORIZONTAL DRAIN

UNACCEPTABLE VENT CONNECTION OPENING WILL BECOME BLOCKED

CENTERLINE

UNACCEPTABLE VENT CONNECTION

**Figure P3104.3
VENT CONNECTIONS**

P3104.4 Vertical rise of vent. A dry vent shall rise vertically to not less than 6 inches (152 mm) above the flood level rim of the highest trap or trapped fixture being vented.

❖ This section prohibits horizontal vent piping from being located where it will be subject to waste flow. A vertical rise of the vent piping will reduce the possibility of having a blockage in the vent pipe. Drainage "backup" resulting from a system stoppage will flow over the fixture flood rim before reaching the horizontal vent line. Drain stoppage is immediately apparent, but vent stoppage is not; however, the latter can cause trap seal loss and allow sewer gases to enter the building. A vertical rise would meet the definition of a vertical pipe, which is an angle of 45 degrees (0.79 rad) or more with the horizontal (see Commentary Figure P3104.4).

P3104.5 Height above fixtures. A connection between a vent pipe and a vent stack or stack vent shall be made not less than 6 inches (152 mm) above the flood level rim of the highest fixture served by the vent. Horizontal vent pipes forming branch vents shall be not less than 6 inches (152 mm) above the flood level rim of the highest fixture served.

❖ This section expresses the same concern as the previous section. In the event of drain stoppage, waste could rise into the vent piping. Over time, the vent piping could become blocked with solids that have settled out of the waste liquids. Additionally, individual, common or branch vents that improperly connect to vent stacks or stack vents could become drains where the fixture drain or drains served are blocked or restricted. It is entirely possible for a vent to serve unintentionally as a drain, and this condition could go unnoticed for a

long period [see Commentary Figures P3104.5(1) and P3104.5(2)].

P3104.6 Vent for future fixtures. Where the drainage piping has been roughed-in for future fixtures, a rough-in connection for a vent, not less than one-half the diameter of the drain, shall be installed. The vent rough-in shall connect to the vent system or shall be vented by other means as provided in this chapter. The connection shall be identified to indicate that the connection is a vent.

❖ When future fixture rough-in drainage piping is installed, vents must also be installed. The vent rough-in must be tied into the vent system or must extend to a vent terminal. The rough-in connection must be identified as a vent so that the purpose of the original installation is evident when the future plumbing is completed.

SECTION P3105
FIXTURE VENTS

P3105.1 Distance of trap from vent. Each fixture trap shall have a protecting vent located so that the slope and the *developed length* in the *fixture drain* from the trap weir to the vent fitting are within the requirements set forth in Table P3105.1.

Exception: The *developed length* of the *fixture drain* from the trap weir to the vent fitting for self-siphoning fixtures, such as water closets, shall not be limited.

❖ The distance from a trap to its vent is limited to reduce the possibility of the trap self-siphoning, which is the siphoning caused by the discharge from the fixture the trap serves. Excessive distance results in excessive fall resulting from the pipe slope and this could cause the drain to resemble an S trap in function. S traps and piping arrangements similar to S traps are prohibited because of the tendency to self-siphon the trap [see Commentary Figures P3105.1(1) and P3105.1(2)].

An exception is made for the trap-to-vent distance for fixtures, such as water closets, because these fixtures rely on self-siphonage to operate properly, and the trap is resealed after each use. The water closet must be vented in all cases, but limiting the distance between the vent connection and the water closet serves no purpose.

**TABLE P3105.1
MAXIMUM DISTANCE OF FIXTURE TRAP FROM VENT**

SIZE OF TRAP (inches)	SLOPE (inch per foot)	DISTANCE FROM TRAP (feet)
$1^1/_4$	$^1/_4$	5
$1^1/_2$	$^1/_4$	6
2	$^1/_4$	8
3	$^1/_8$	12
4	$^1/_8$	16

For SI: 1 inch = 25.4 mm, 1 foot = 304.8 mm,
　　　1 inch per foot = 83.3 mm/m.

❖ The distances listed in Table P3105.1 are based on laboratory testing and prevent the weir of the trap from being located above the highest inlet to the vent. As the total change in elevation of a fixture drain (resulting from the fixture drain slope) approaches the inside diameter of the drain, the weir of the trap approaches the highest level of the vent connection and self-siphoning is more likely (see commentary, Section P3105.2).

P3105.2 Fixture drains. The total fall in a *fixture drain* resulting from pipe slope shall not exceed one pipe diameter,

nor shall the vent pipe connection to a *fixture drain*, except for water closets, be below the weir of the trap.

❖ This section's requirements reinforce those of the previous section. They substantiate the basic rule applied in Table P3105.1 and require that the total fall in a fixture drain caused by slope not to exceed one pipe diameter (based on inside diameter). For example, a 2-

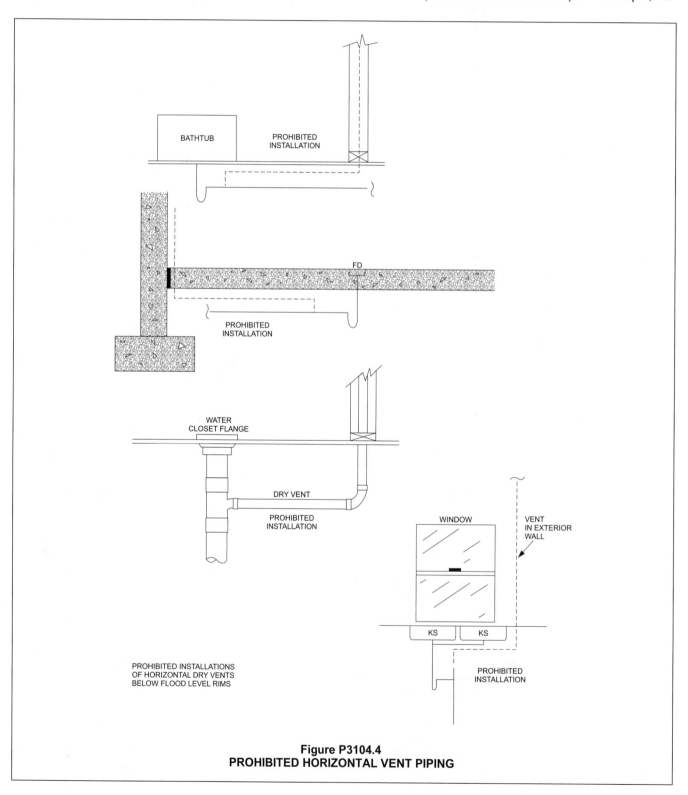

Figure P3104.4
PROHIBITED HORIZONTAL VENT PIPING

inch (51 mm) trap discharges into a 2-inch fixture drain installed at a slope of $^1/_4$ inch per foot (21 mm/m). Eight feet (2438 mm) of pipe run will produce a total fall of 2 inches ($8 \times {}^1/_4$) and this is equal to the drain pipe's internal diameter. Because a water closet depends on trap siphonage in its operation to remove the contents, water closets are exempted from these requirements.

Chapter 32 outlines trap size requirements (see Commentary Figure P3105.2).

P3105.3 Crown vent prohibited. A vent shall not be installed within two pipe diameters of the trap weir.

❖ Crown venting is an arrangement in which a vent connects at the top of the weir (crown) of a trap. The original crown vent was a crown-vented, factory-built S trap, which is no longer manufactured. The problem with this type of connection is that the vent opening can become blocked, thus closing the vent and allowing

the trap to siphon. The blockage is a result of the action of the drainage flowing through the trap. Flow direction and velocity will force waste up into the vent connection. It has been found that when the vent is connected at least two pipe diameters downstream from the trap weir, this problem can be avoided [see Commentary Figures P3105.3(1) and P3105.3(2)].

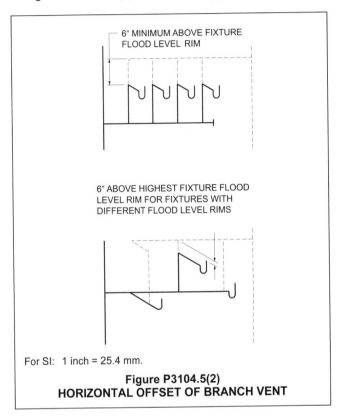

For SI: 1 inch = 25.4 mm.

Figure P3104.5(2)
HORIZONTAL OFFSET OF BRANCH VENT

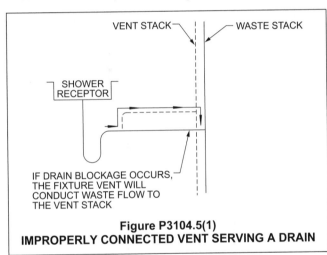

Figure P3104.5(1)
IMPROPERLY CONNECTED VENT SERVING A DRAIN

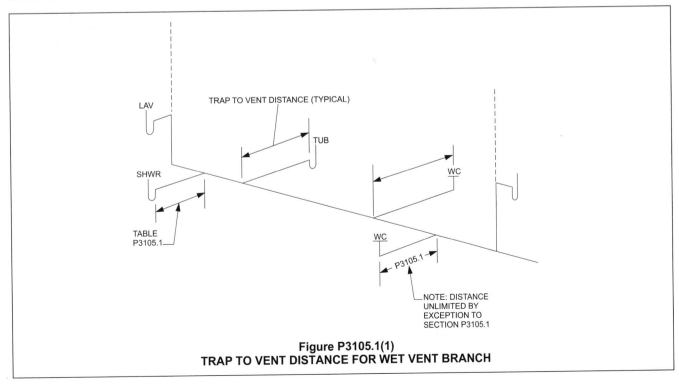

Figure P3105.1(1)
TRAP TO VENT DISTANCE FOR WET VENT BRANCH

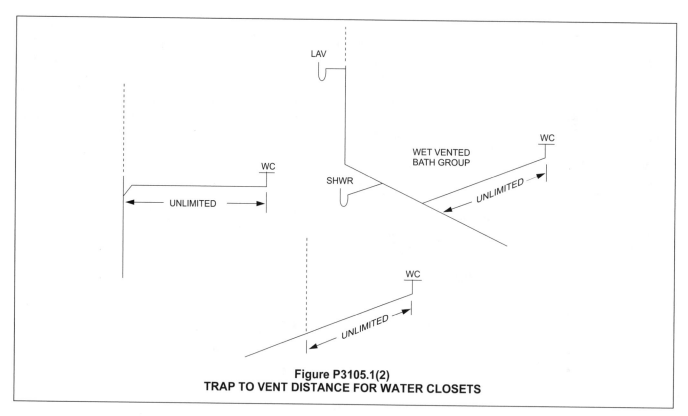

Figure P3105.1(2)
TRAP TO VENT DISTANCE FOR WATER CLOSETS

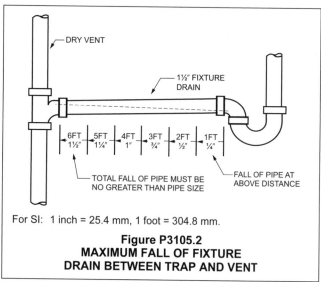

For SI: 1 inch = 25.4 mm, 1 foot = 304.8 mm.

Figure P3105.2
MAXIMUM FALL OF FIXTURE
DRAIN BETWEEN TRAP AND VENT

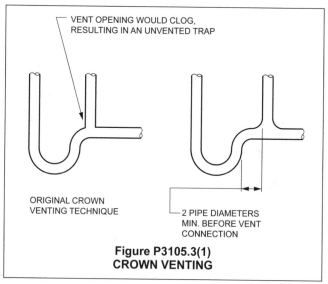

Figure P3105.3(1)
CROWN VENTING

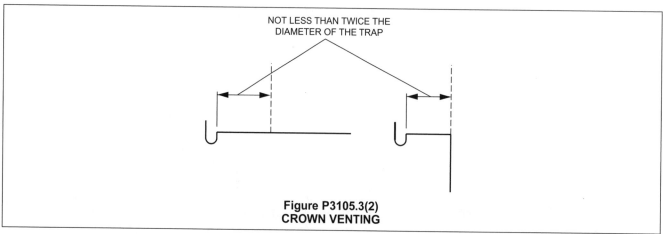

Figure P3105.3(2)
CROWN VENTING

SECTION P3106
INDIVIDUAL VENT

P3106.1 Individual vent permitted. Each trap and trapped fixture shall be permitted to be provided with an individual vent. The individual vent shall connect to the *fixture drain* of the trap or trapped fixture being vented.

❖ The simplest form of venting a trap or trapped fixture is an individual vent for each trap. A single vent pipe is connected between the trap of the fixture and the connection to the drainage system. An individual vent is always placed between the trap it is protecting and the drainage system. With a properly installed individual vent, only the drainage of the fixture served is flowing past the vent. The sizes of individual vents are to be in accordance with Section P3113. See the commentary to Sections P3113.1 and P3113.2. Also, see Commentary Figures P3106.1(1) and P3106.1(2).

SECTION P3107
COMMON VENT

P3107.1 Individual vent as common vent. An individual vent shall be permitted to vent two traps or trapped fixtures as a common vent. The traps or trapped fixtures being common vented shall be located on the same floor level.

❖ Common venting allows an individual vent installed in accordance with Section P3107.1 to serve two fixtures. Sections P3107.2 and P3107.3 contain the provisions that pertain to the connection of fixture drains at the same level or at different levels, respectively. In either case, the fixtures vented by the common vent must be located on the same floor level.

P3107.2 Connection at the same level. Where the *fixture drains* being common vented connect at the same level, the vent connection shall be at the interconnection of the *fixture drains* or downstream of the interconnection.

❖ One form of common venting connects two fixtures at the same level. A typical installation consists of two horizontal drains connecting to a vertical drain through a double-pattern fitting. The extension of the vertical drain serves as the vent. Two drains connecting horizontally to a common horizontal drain through a double-pattern fitting is also a form of common venting and the vent is allowed to connect at the interconnection of the fixture drains or downstream along the horizontal drain. The distance from the trap to the vent for each fixture drain must be in accordance with Table P3105.1 (see Commentary Figure P3107.2).

The fittings used in such installations must be as outlined in Section P3005.

P3107.3 Connection at different levels. Where the *fixture drains* connect at different levels, the vent shall connect as a vertical extension of the vertical drain. The vertical drain pipe connecting the two *fixture drains* shall be considered the vent for the lower *fixture drain*, and shall be sized in accordance with Table P3107.3. The upper fixture shall not be a water closet.

❖ Two fixtures within a single story can connect at different levels to a vertical drain and be considered common vented, but the vent remains an individual vent serving as a common vent. The drain pipe between the upper and lower fixtures must meet the sizing requirements of Table 3107.3 (see commentary, Table P3107.3). The upper fixture cannot be a water closet because the pressures created by the rapid discharge of a water closet could adversely affect the venting of the lower fixture.

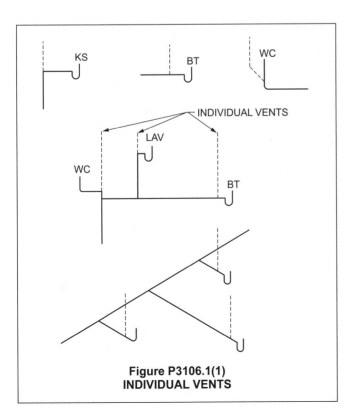

Figure P3106.1(1)
INDIVIDUAL VENTS

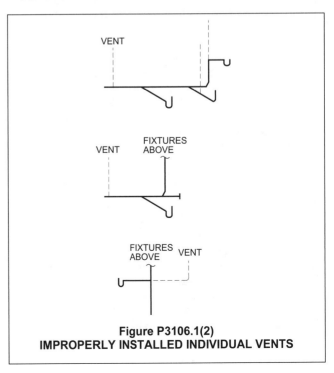

Figure P3106.1(2)
IMPROPERLY INSTALLED INDIVIDUAL VENTS

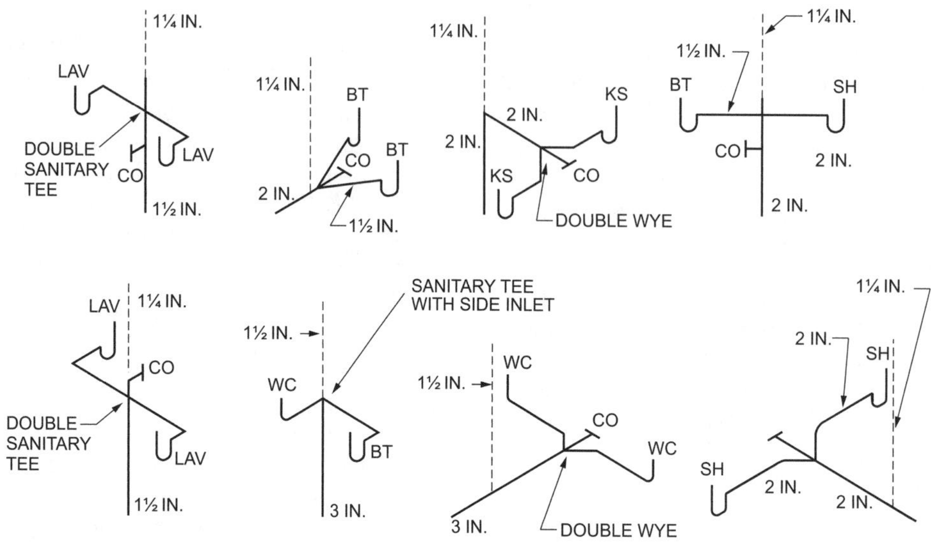

For SI: 1 inch = 25.4 mm.

Figure P3107.2
COMMON VENTS WITH FIXTURE DRAINS CONNECTING AT SAME LEVEL

TABLE P3107.3
COMMON VENT SIZES

PIPE SIZE (inches)	MAXIMUM DISCHARGE FROM UPPER FIXTURE DRAIN (d.f.u.)
$1^1/_2$	1
2	4
$2^1/_2$ to 3	6

For SI: 1 inch = 25.4 mm.

❖ This table lists the maximum allowable drainage fixture units (d.f.u.) that can be served by the common vent, which is that portion of piping between the upper fixture drain and lower fixture drain connection. Note the resemblance to Table P3108.3. In principle, this section is describing a form of vertical wet venting. The common vent is sized based on the d.f.u. of the upper fixture drain. The drain pipe downstream of the lower fixture drain must be sized in accordance with Section P3005.4, based on the combined d.f.u. of both the upper and lower fixture drains connecting to the common vent (see commentary, Section P3005.4 and Commentary Figure P3107.3).

SECTION P3108
WET VENTING

P3108.1 Horizontal wet vent permitted. Any combination of fixtures within two *bathroom groups* located on the same floor level shall be permitted to be vented by a horizontal wet vent. The wet vent shall be considered the vent for the fixtures and shall extend from the connection of the dry vent along the direction of the flow in the drain pipe to the most downstream *fixture drain* connection. Each *fixture drain* shall connect horizontally to the horizontal branch being wet vented or shall have a dry vent. Each wet-vented *fixture drain* shall connect independently to the horizontal wet vent. Only the fixtures within the *bathroom groups* shall connect to the wet-vented horizontal branch drain. Any additional fixtures shall discharge downstream of the horizontal wet vent.

❖ A wet vent is a vent pipe that conveys drainage. It can also be said that a wet vent is a drain that acts as a vent. Wet venting is a method of venting any combination of fixtures within a single or double bathroom group (see the commentary to the definition of "Bathroom group"). Wet venting uses oversized piping to allow for the flow of air above the waste flow. Reliance is placed on two factors for establishing that there will always be adequate volume within the wet vent pipe to permit required airflow (the low probability of simultaneous fixture discharge and the low-flow velocity that results from requiring the fixtures to be on the same floor level). A single properly sized vent will provide airflow when fixtures are being discharged to relieve pressures that develop in the drain piping.

The wet vent is that portion of the drainage system receiving discharge from a fixture and serving as a vent for other fixtures. Only the fixture types and quantities addressed in this section are allowed to connect to or be served by a wet vent. For example, three water closets cannot be vented by the same wet vent system because the set of fixtures from two bathroom groups includes only two water closets. Because of distances or rough-in heights, more than one fixture might be vented by an individual dry vent. However, this does not prohibit such fixtures from discharging into the wet vent. Fixtures that are part of the two bathroom groups are allowed to discharge to the wet vent, even if such

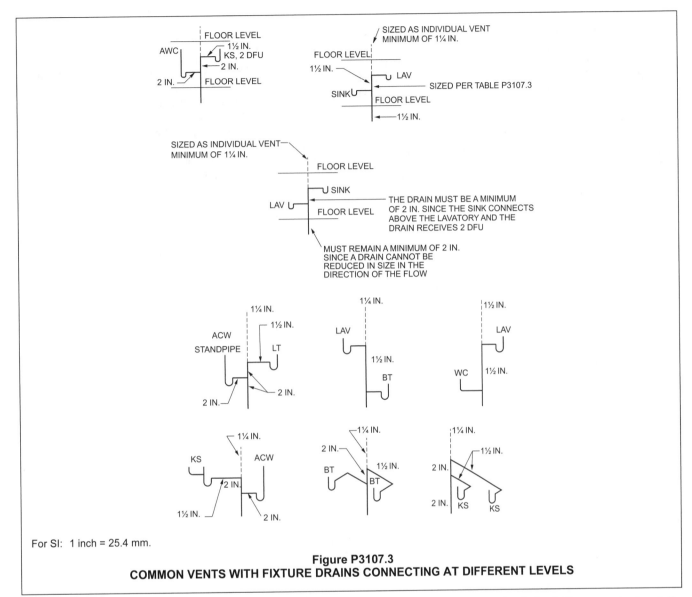

For SI: 1 inch = 25.4 mm.

Figure P3107.3
COMMON VENTS WITH FIXTURE DRAINS CONNECTING AT DIFFERENT LEVELS

fixtures are not being wet vented. Additional fixtures must connect downstream of the wet-vented section. Each fixture drain that is being wet vented must connect to the wet vent in a horizontal plane consistent with the intent of Section P3105.2. Also, each fixture drain must individually connect to the wet vent whether it is wet vented or not [see Commentary Figures P3108.1(1) and P3108.1(2)].

P3108.2 Dry vent connection. The required dry-vent connection for wet-vented systems shall comply with Sections P3108.2.1 and P3108.2.2.

❖ Sections P3108.2.1 and P3108.2.2 cover the dry-vent connections to horizontal wet-vent systems and vertical wet-vent systems, respectively.

P3108.2.1 Horizontal wet vent. The dry-vent connection for a horizontal wet-vent system shall be an individual vent or a common vent for any *bathroom group* fixture, except an emergency floor drain. Where the dry vent connects to a

water closet *fixture drain*, the drain shall connect horizontally to the horizontal wet vent system. Not more than one wet-vented *fixture drain* shall discharge upstream of the dry-vented *fixture drain* connection.

❖ Section P3108.2.1 regulates the dry-vent connection to the horizontal wet vent. The dry vent for a horizontal wet vent must connect to a vent stack, stack vent, air admittance valve or terminate outdoors.

The required vent could be a dry vent directly connected to the main section of the horizontal wet-vent pipe downstream of the most upstream fixture [see Commentary Figure P3108.2.1(1) and (2)]. In this arrangement, the dry vent is classified as an individual vent for the most upstream fixture.

The last sentence of Section P3108.2.1 attempts to reinforce the definition of a horizontal wet vent so that improper fixture connections are avoided. A horizontal wet vent begins at the dry-vent connection and extends downstream to the last fixture being wet vented. The horizontal wet vent does not extend upstream of the connection that leads to the dry-vent connection. Thus, multiple interconnected fixture drains that connect upstream of the dry-vented fixture connection or the dry-vent connection are not vented by the horizontal wet vent. Commentary Figures P3108.2.1(2), (3) and (4) are examples of prohibited arrangements.

An emergency floor drain is not allowed to be the upstream dry-vented fixture because there is a concern that debris could build up in the horizontal drain downstream of the vent connection. Emergency floor drains rarely, if ever, receive any flow that would keep debris clear of the vent connection.

P3108.2.2 Vertical wet vent. The dry-vent connection for a vertical wet-vent system shall be an individual vent or common vent for the most upstream *fixture drain*.

❖ Section P3108.2.2 regulates how the dry-vent connection is made to the vertical wet vent and Table 3108.3 regulates the size of the dry vent. The dry vent is an extension of the vertical wet vent and is connected to a vent stack, stack vent, air admittance valve or point of termination outside the building. The dry-vent size must be at least one-half the largest required pipe size served by the wet vent. For example, if the largest pipe served by the vertical wet vent is 3 inches (76 mm), the dry vent is required to be a minimum of $1^1/_2$ inches (38

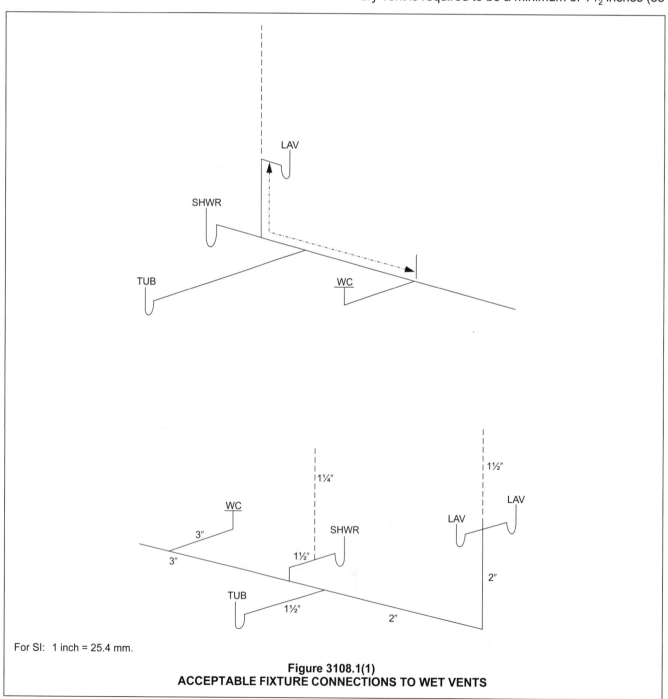

For SI: 1 inch = 25.4 mm.

Figure 3108.1(1)
ACCEPTABLE FIXTURE CONNECTIONS TO WET VENTS

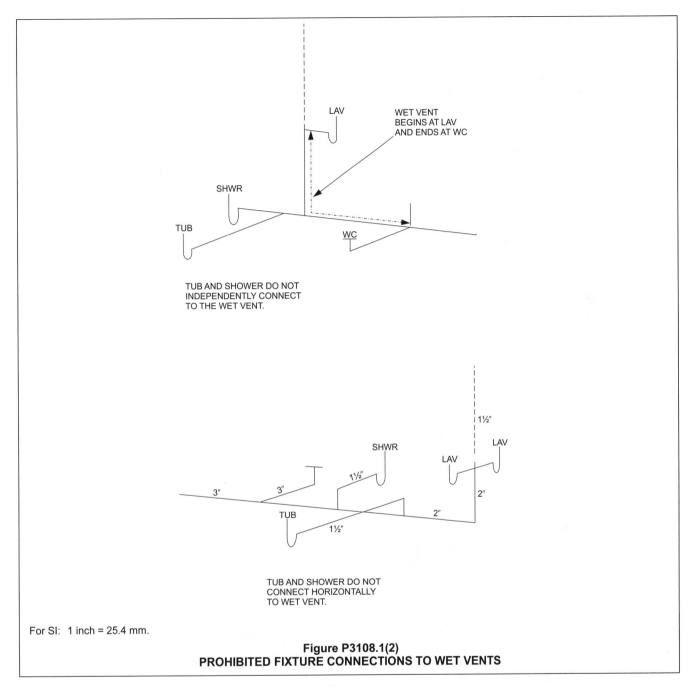

For SI: 1 inch = 25.4 mm.

Figure P3108.1(2)
PROHIBITED FIXTURE CONNECTIONS TO WET VENTS

mm). If the run of the dry vent from the start of the dry-vent fixture connection to the termination of the vent (to a vent stack, stack vent, AAV or termination outdoors) exceeds 40 feet (12 192 mm) developed length, Section P3113.3 requires the dry-vent size to be increased by one nominal pipe size for its entire length.

P3108.3 Size. Horizontal and vertical wet vents shall be not less than the size as specified in Table P3108.3, based on the fixture unit discharge to the wet vent. The dry vent serving the wet vent shall be sized based on the largest required diameter of pipe within the wet-vent system served by the dry vent.

❖ This section refers the reader to Table P3108.3 for sizing the wet-vented section of piping. The table applies to both vertical and horizontal wet venting.

TABLE P3108.3
WET VENT SIZE

WET VENT PIPE SIZE (inches)	FIXTURE UNIT LOAD (d.f.u.)
$1^1/_2$	1
2	4
$2^1/_2$	6
3	12
4	32

For SI: 1 inch = 25.4 mm.

❖ This table provides minimum sizes of wet-vent piping, ranging from $1^1/_2$ inches (38 mm) through 4 inches (102 mm) in diameter.

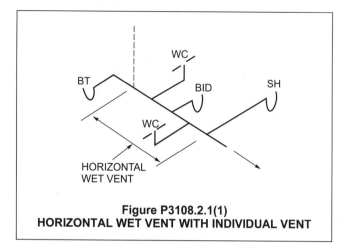

Figure P3108.2.1(1)
HORIZONTAL WET VENT WITH INDIVIDUAL VENT

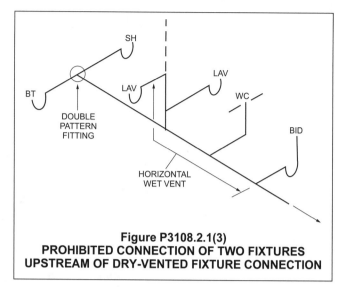

Figure P3108.2.1(3)
PROHIBITED CONNECTION OF TWO FIXTURES
UPSTREAM OF DRY-VENTED FIXTURE CONNECTION

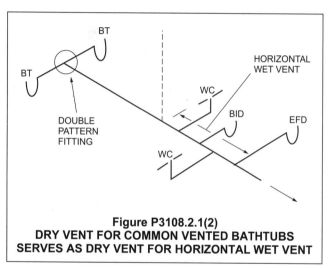

Figure P3108.2.1(2)
DRY VENT FOR COMMON VENTED BATHTUBS
SERVES AS DRY VENT FOR HORIZONTAL WET VENT

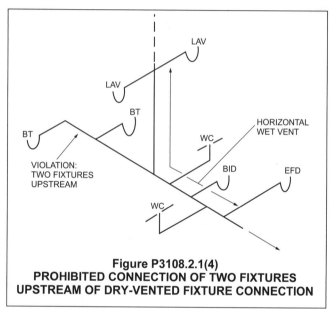

Figure P3108.2.1(4)
PROHIBITED CONNECTION OF TWO FIXTURES
UPSTREAM OF DRY-VENTED FIXTURE CONNECTION

P3108.4 Vertical wet vent permitted. A combination of fixtures located on the same floor level shall be permitted to be vented by a vertical wet vent. The vertical wet vent shall be considered the vent for the fixtures and shall extend from the connection of the dry vent down to the lowest *fixture drain* connection. Each wet-vented fixture shall connect independently to the vertical wet vent. All water closet drains shall connect at the same elevation. Other *fixture drains* shall connect above or at the same elevation as the water closet *fixture drains*. The dry vent connection to the vertical wet vent shall be an individual or common vent serving one or two fixtures.

❖ This section contains the requirements for vertical wet venting, which is the same basic idea as that of horizontal wet venting, as provided for in Section P3108.1. In a vertical wet-vent configuration, the connection of a water closet must be the lowest connection to the vertical wet-vented section. In addition, where two water closets are part of the wet-vented system and connect to the vertical drain independently, they must connect at the same level. All other connections to the vertical drain must be above or at the same height as the water closet connection. Table P3108.3 shows vertical wet-vent sizing.

P3108.5 Trap weir to wet vent distances. The maximum *developed length* of wet-vented *fixture drains* shall comply with Table P3105.1.

❖ Fixture drains that are wet vented end at the point of connection to the wet vent (vertical and horizontal). Thus, the distance between the fixture trap and wet vent is limited in the same manner that all other fixture drains are limited (see Sections P3105.1 and P3105.2).

SECTION P3109
WASTE STACK VENT

P3109.1 Waste stack vent permitted. A waste stack shall be considered a vent for all of the fixtures discharging to the stack where installed in accordance with the requirements of this section.

❖ The term "waste stack" is used because water closets are prohibited in this system and thus, the stack is truly a waste-only stack (see the definition of "Waste"). The principles of waste stack venting are based on some of the original research that was performed in plumbing. The system has been identified by a variety of names, including vertical wet venting, single-stack venting and multifloor stack venting. Common venting and vertical wet-venting installations are allowed only when the various fixtures connecting into the stack are all located on the same floor. Waste stack venting, not to be confused with common venting and vertical wet venting, is allowed to serve fixtures on multiple floors.

P3109.2 Stack installation. The waste stack shall be vertical, and both horizontal and vertical offsets shall be prohibited between the lowest *fixture drain* connection and the highest *fixture drain* connection to the stack. Every *fixture drain* shall connect separately to the waste stack. The stack shall not receive the discharge of water closets or urinals.

❖ Because the stack is also functioning as a vent, the annular flow of waste down the interior pipe wall must be maintained to avoid the pressure fluctuations that are created when the cross-sectional area of the stack is filled with waste at any point. Offsets disrupt the

annular flow and water closets create a surge flow, both of which can create full cross sectional flow in the stack which, in turn, creates positive and negative pressures in the stack that can have adverse effects on traps (see Commentary Figure P3109.2). All fixture drains connecting to the stack must be connected separately; therefore, only fixture drains (as opposed to fixture branches) connect to the stack.

P3109.3 Stack vent. A stack vent shall be installed for the waste stack. The size of the stack vent shall be not less than the size of the waste stack. Offsets shall be permitted in the stack vent and shall be located not less than 6 inches (152 mm) above the flood level of the highest fixture, and shall be in accordance with Section P3104.5. The stack vent shall be permitted to connect with other stack vents and vent stacks in accordance with Section P3113.3.

❖ The waste stack vent serves as both a drain and a vent; therefore, the stack must be vented to the outdoors to neutralize pressures in the stack. To limit flow resistance in the stack vent, it must be no smaller than the waste stack that it serves. The stack vent must be the same diameter as or larger than the stack.

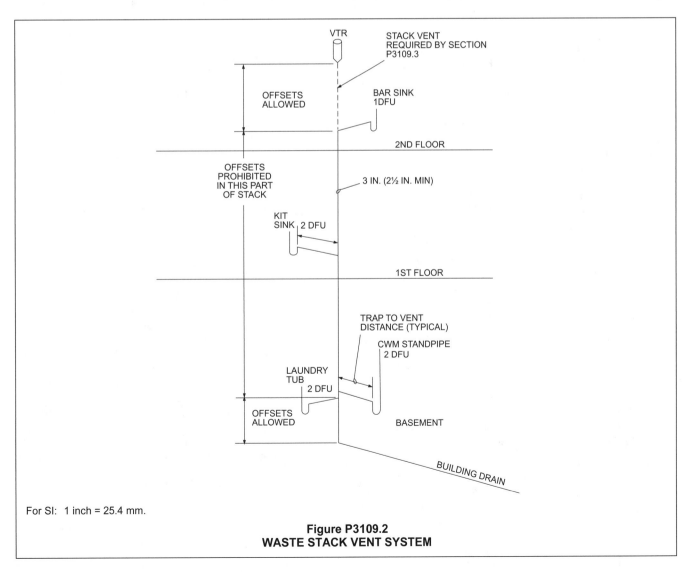

For SI: 1 inch = 25.4 mm.

Figure P3109.2
WASTE STACK VENT SYSTEM

P3109.4 Waste stack size. The waste stack shall be sized based on the total discharge to the stack and the discharge within a *branch interval* in accordance with Table P3109.4. The waste stack shall be the same size throughout the length of the waste stack.

❖ In Commentary Figure P3109.2, the total d.f.u. of all the fixtures could be handled by a 2-inch (51 mm) diameter vertical drain (see Tables P3004.1 and P3005.4.1); however, the total d.f.u. load is 7, which exceeds the loading allowed by Table P3109.4. Thus, Table P3109.4 will require a 2¹/₂-inch (64 mm) minimum diameter stack.

TABLE P3109.4
WASTE STACK VENT SIZE

STACK SIZE (inches)	MAXIMUM NUMBER OF FIXTURE UNITS (d.f.u.)	
	Total discharge into one branch interval	Total discharge for stack
1¹/₂	1	2
2	2	4
2¹/₂	No limit	8
3	No limit	24
4	No limit	50

For SI: 1 inch = 25.4 mm.

❖ This table states the minimum size requirements for waste stack vents based on the d.f.u. totals for all fixture drains served (see commentary, Section P3109.4). Note that the table refers to branch intervals, which are measured between connections of horizontal branches. Recall that Section P3109.2 requires that only fixture drains connect to the stack (see the definition of "Branch interval").

SECTION P3110
CIRCUIT VENTING

P3110.1 Circuit vent permitted. Not greater than eight fixtures connected to a horizontal branch drain shall be permitted to be circuit vented. Each *fixture drain* shall connect horizontally to the horizontal branch being circuit vented. The horizontal branch drain shall be classified as a vent from the most downstream *fixture drain* connection to the most upstream *fixture drain* connection to the horizontal branch.

❖ Circuit venting is based on many of the same principles as wet venting. However, it will allow for the connection of other types or totals of similar fixtures that may be outside of the definition of a bathroom group. The maximum number of fixtures allowed to be circuit vented on one branch is eight. In addition, unlike wet venting, each fixture drain being circuit vented must connect horizontally to the horizontal circuit-vented branch drain.

P3110.2 Vent connection. The circuit vent connection shall be located between the two most upstream *fixture drains*. The vent shall connect to the horizontal branch and shall be installed in accordance with Section P3104. The circuit vent pipe shall not receive the discharge of any soil or waste.

❖ The circuit vent itself must connect to the horizontal branch between the two uppermost fixture drains so that the point of connection is washed by the upstream fixture and is not a dead end. It must conform to Section P3104 and must be a dry vent.

P3110.3 Slope and size of horizontal branch. The slope of the vent section of the horizontal branch drain shall be not greater than one unit vertical in 12 units horizontal (8-percent slope). The entire length of the vent section of the horizontal branch drain shall be sized for the total drainage discharge to the branch in accordance with Table P3005.4.1.

❖ To maintain a relatively flat flow channel depth, the maximum slope for a circuit-vented horizontal branch drain is 1 inch per foot (83 mm/m) (8-percent slope). Unlike wet venting, circuit venting does not depend on a unique sizing table; rather, the circuit-vented horizontal drain branch is to be sized in accordance with Table P3005.4.1. The horizontal branch must be the same size throughout its length.

P3110.4 Additional fixtures. Fixtures, other than the circuit vented fixtures shall be permitted to discharge, to the horizontal branch drain. Such fixtures shall be located on the same floor as the circuit vented fixtures and shall be either individually or common vented.

❖ Although a maximum of eight fixture drains is allowed to connect horizontally to the circuit vent, other fixtures may connect also if they are individually or common vented and are located on the same floor level as the circuit-vented fixtures. Circuit-vented branches are allowed at any elevation, above or below a floor.

SECTION P3111
COMBINATION WASTE AND VENT SYSTEM

P3111.1 Type of fixtures. A combination waste and vent system shall not serve fixtures other than floor drains, sinks and lavatories. A combination waste and vent system shall not receive the discharge of a food waste disposer.

❖ A combination waste and vent system is another type of venting system that allows a drain to serve as a vent. Floor drains, sinks and lavatories discharge clear- or gray-water waste. Because of the low-flow velocities and volumes that occur in these systems, any solids introduced would accumulate, causing blockages and reduction of the internal cross-sectional area of the piping. Therefore, sinks equipped with food waste disposers and fixtures that discharge solid wastes are not allowed to be served by a combination drain and vent. Ground food waste can collect in the drain, causing blockages. Combination drain and vent systems are oversized drains; therefore, the flow velocity will be low, which reduces drain line carry and scouring action.

P3111.2 Installation. The only vertical pipe of a combination waste and vent system shall be the connection between the fixture drain and the horizontal combination waste and vent pipe. The vertical distance shall be not greater than 8 feet (2438 mm).

❖ A combination waste and vent system is a horizontal piping system and the only vertical pipe allowed in the system is the pipe (drop) between the fixture trap and the horizontal waste and vent pipe. This vertical component (drop) is necessary for fixtures that rough-in above the floor and traps that must be installed above the horizontal system. Commentary Figures P3111.2(1) and P3111.2(2) show combination waste and vent systems.

P3111.2.1 Slope. The horizontal combination waste and vent pipe shall have a slope of not greater than $^1/_2$ unit vertical in 12 units horizontal (4-percent slope). The minimum slope shall be in accordance with Section P3005.3.

❖ To control flow velocity and maintain a nearly flat flow channel depth, this section specifies the maximum slope to be $^1/_2$ inch per foot (42 mm/m) or a 4-percent slope.

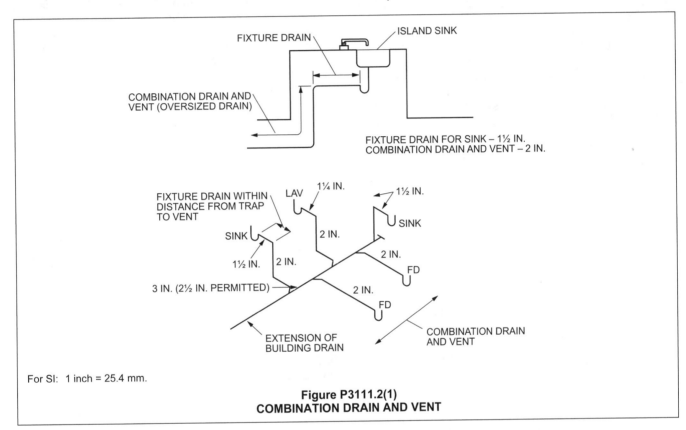

For SI: 1 inch = 25.4 mm.

Figure P3111.2(1)
COMBINATION DRAIN AND VENT

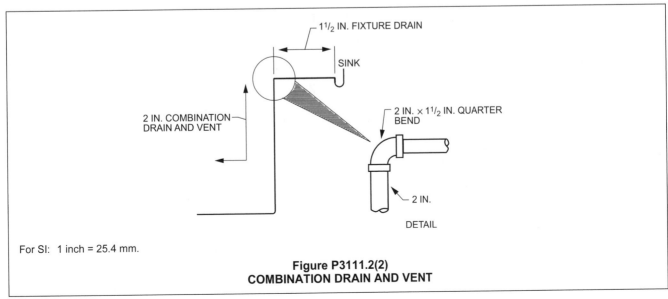

For SI: 1 inch = 25.4 mm.

Figure P3111.2(2)
COMBINATION DRAIN AND VENT

P3111.2.2 Connection. The combination waste and vent system shall be provided with a dry vent connected at any point within the system or the system shall connect to a horizontal drain that serves vented fixtures located on the same floor. Combination waste and vent systems connecting to *building drains* receiving only the discharge from one or more stacks shall be provided with a dry vent. The vent connection to the combination waste and vent pipe shall extend vertically to a point not less than 6 inches (152 mm) above the flood level rim of the highest fixture being vented before offsetting horizontally.

❖ Where a combination waste and vented horizontal branch extends laterally from a vented horizontal drain, an additional dry vent is not required; otherwise, the combination waste and vent system will require a dry vent, as outlined in this section and in Section P3111.2.3. A vented horizontal drain could be a horizontal wet-vent branch, a circuit-vented branch or a combination waste and vent system (see Commentary Figure P3111.2.2).

P3111.2.3 Vent size. The vent shall be sized for the total fixture unit load in accordance with Section P3113.1.

❖ See the commentary to Section P3113.1.

P3111.2.4 Fixture branch or drain. The fixture branch or *fixture drain* shall connect to the combination waste and vent within a distance specified in Table P3105.1. The combination waste and vent pipe shall be considered the vent for the fixture.

❖ The combination waste and vent system is oversized and vented so that the entire system allows sufficient air movement to maintain the trap seals for the fixtures being connected. Because the combination waste and vent system is the vent for the fixture, the fixture drains or branches connecting to the system need not be oversized; therefore, the size and length limitations of Table P3105.1 apply. If the fixture drains are sized in accordance with Table P3111.3, they are, themselves, combination waste and vent pipes; thus, the distance limits between the trap and its vent are not relevant.

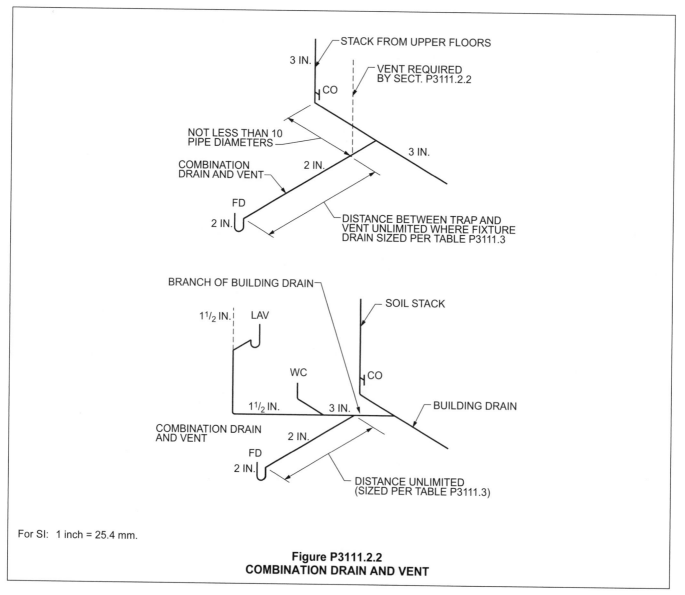

For SI: 1 inch = 25.4 mm.

Figure P3111.2.2
COMBINATION DRAIN AND VENT

P3111.3 Size. The size of a combination drain and vent pipe shall be not less than that specified in Table 3111.3. The horizontal length of a combination drain and vent system shall be unlimited.

❖ There is not a limit as to the horizontal length of a combination waste and vent system because, as long as the pipe is sized according to the Table P3111.3, there will always be sufficient airspace above the flow line to vent the fixtures connected (see commentary, Table P3111.3).

TABLE P3111.3
SIZE OF COMBINATION WASTE AND VENT PIPE

DIAMETER PIPE (inches)	MAXIMUM NUMBER OF FIXTURE UNITS (d.f.u.)	
	Connecting to a horizontal branch or stack	Connecting to a building drain or building subdrain
2	3	4
2½	6	26
3	12	31
4	20	50

For SI: 1 inch = 25.4 mm.

❖ This table outlines the maximum d.f.u. loading for various sizes of pipe. The loading depends on whether the combination waste and vent branch extends from a building drain or sub-drain, or from a horizontal branch or stack (see the definitions of "Building drain" and "Horizontal branch, drainage"). The differences between the second and third columns of this table parallel the differences between Tables P3005.4.1 and P3005.4.2.

SECTION P3112
ISLAND FIXTURE VENTING

P3112.1 Limitation. Island fixture venting shall not be permitted for fixtures other than sinks and lavatories. Kitchen sinks with a dishwasher waste connection, a food waste disposer, or both, in combination with the kitchen sink waste, shall be permitted to be vented in accordance with this section.

❖ Because of the many diverse designs in construction, a fixture might be located without the nearby walls or partitions necessary for conventional venting methods to be used. Other options for island fixtures are as permitted by Sections P3111 and P3114.

P3112.2 Vent connection. The island fixture vent shall connect to the *fixture drain* as required for an individual or common vent. The vent shall rise vertically to above the drainage outlet of the fixture being vented before offsetting horizontally or vertically downward. The vent or branch vent for multiple island fixture vents shall extend not less than 6 inches (152 mm) above the highest island fixture being vented before connecting to the outside vent terminal.

❖ The vent rises to above the outlet of the fixture before turning downward so that in the event of a drain stoppage, waste will rise and stand in the bottom of the future before discharging into the vent piping (see Commentary Figure P3112.2).

P3112.3 Vent installation below the fixture flood level rim. The vent located below the flood level rim of the fixture being vented shall be installed as required for drainage piping in accordance with Chapter 30, except for sizing. The vent shall be sized in accordance with Section P3113.1. The lowest point of the island fixture vent shall connect full size to the drainage system. The connection shall be to a vertical drain pipe or to the top half of a horizontal drain pipe. Cleanouts shall be provided in the island fixture vent to permit rodding of all vent piping located below the flood level rim of the fixtures. Rodding in both directions shall be permitted through a cleanout.

❖ This section contains further provisions for the vent discussed in Section P3112.2. Any portion of the vent that is below the flood level rim of the island sink must be installed with drainage fittings and slope in accordance with Chapter 30 because waste-flow invasion into the vent is possible when the drain piping is blocked.

This will prevent the low (trapped) portion of the vent from becoming filled with waste. Because the vent is installed horizontally below the fixture flood-level rim, it could be subject to waste flow; therefore, a cleanout must be provided (see Commentary Figure P3112.2).

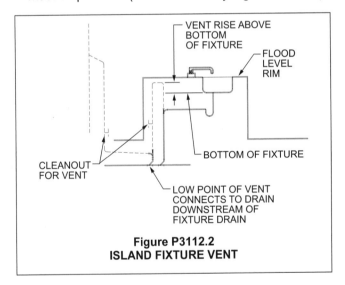

Figure P3112.2
ISLAND FIXTURE VENT

SECTION P3113
VENT PIPE SIZING

P3113.1 Size of vents. The required diameter of individual vents, branch vents, circuit vents, vent stacks and stack vents shall be not less than one-half the required diameter of the drain served. The required size of the drain shall be determined in accordance with Chapter 30. Vent pipes shall be not less than 1¼ inches (32 mm) in diameter. Vents exceeding 40 feet (12 192 mm) in *developed length* shall be increased by one nominal pipe size for the entire *developed length* of the vent pipe.

❖ This section provides for the sizing of different types of vents. Sizing is based on the minimum required diameter of the drain served and the total developed length. All vents must be a minimum of one-half the diameter of the drain that they serve. No vent may be less than 1¼ inches (32 mm) in diameter. This does not exempt vents from the specific provisions for vent sizing outlined in the sections addressing some of the venting concepts, such as those found in Sections P3108.3, P3109.4, P3110.3 and P3111.3.

Research has shown that a vent length over 40 feet (12 192 mm) will have too much friction loss to allow the air movement necessary to maintain control over pressure excursions in the drainage system. Recall that vents are designed to move air under the influence of plus or minus 2 inches of w.c. pressure (498 Pa). If the size of a vent is increased by one pipe size, the allowable developed length will be longer than any

vent run likely to be installed in any dwelling (see Section P3113.2).

P3113.2 Developed length. The *developed length* of individual, branch, and circuit vents shall be measured from the farthest point of vent connection to the drainage system, to the point of connection to the vent stack, stack vent or termination outside of the building.

❖ The developed length of a vent must be determined to apply the 40-foot (12 192 mm) developed length limitation of Section P3113.1 (see Commentary Figure P3113.2).

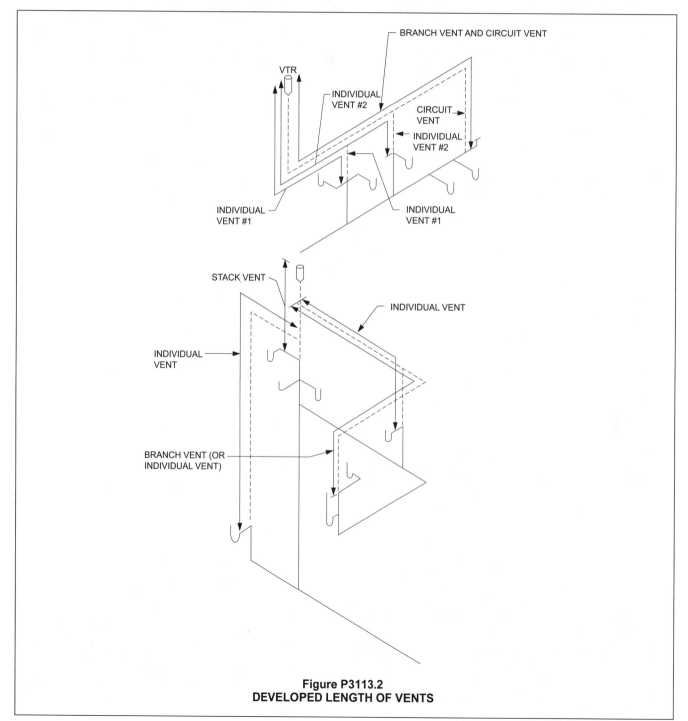

Figure P3113.2
DEVELOPED LENGTH OF VENTS

P3113.3 Branch vents. Where branch vents are connected to a common branch vent, the common branch vent shall be sized in accordance with this section, based on the size of the common horizontal drainage branch that is or would be required to serve the total drainage fixture unit (d.f.u.) load being vented.

❖ Multiple branch vents could connect to a common branch vent, thus serving a large number of fixture units. In other words, one can have a number of horizontal branches, each with a vent, all of which connect together to a common branch vent prior to termination. To determine the proper size of the vents, the d.f.u. total for each of the horizontal branches must be added and then the drain size would be determined for this total in accordance with Section P3005.4. The vent must then be sized in accordance with Section P3113 based on the required drain size for a drain that could serve the d.f.u. load.

P3113.4 Sump vents. Sump vent sizes shall be determined in accordance with Sections P3113.4.1 and P3113.4.2.

❖ The following two subsections state venting requirements for sewage pumps and ejectors.

P3113.4.1 Sewage pumps and sewage ejectors other than pneumatic. Drainage piping below sewer level shall be vented in the same manner as that of a gravity system. Building sump vent sizes for sumps with sewage pumps or sewage ejectors, other than pneumatic, shall be determined in accordance with Table P3113.4.1.

❖ Sumps receiving sanitary drainage must be vented to allow air to enter and exit as the liquid level falls and rises, respectively. Where a sewage pump is used, the vent can connect to the building venting system. Table P3113.4.1 specifies the minimum vent pipe size based on the discharge capacity of the pump and the developed length of vent pipe. Building sub-drain systems, including their venting systems, are designed like gravity drainage systems, including their venting systems, except that they cannot drain to the sewer by gravity and, thus, require a sewage pump or ejector. The vents for gravity systems and subdrain ejector systems can be combined if pneumatic ejectors are not used.

TABLE P3113.4.1. See below.

❖ This table specifies the minimum vent pipe size based on the discharge capacity of the pump and the developed length of the vent.

P3113.4.2 Pneumatic sewage ejectors. The air pressure relief pipe from a pneumatic sewage ejector shall be connected to an independent vent stack terminating as required for vent extensions through the roof. The relief pipe shall be sized to relieve air pressure inside the ejector to atmospheric pressure, but shall be not less than $1^1/_4$ inches (32 mm) in size.

❖ Pneumatic sewage ejectors operate at high air pressure, forcing the sewage from the ejector receiver to the gravity drainage system. Such ejectors are probably not used in dwelling units. An independent vent is necessary to relieve the pressure in the receiver at the completion of the ejector cycle. If such a vent were connected to the building venting system, it would create tremendous pressure differentials, disrupting the normal operation of the drainage system.

SECTION P3114
AIR ADMITTANCE VALVES

P3114.1 General. Vent systems using *air admittance valves* shall comply with this section. Individual and branch-type air admittance valves shall conform to ASSE 1051. Stack-type air admittance valves shall conform to ASSE 1050.

❖ AAVs can be used in place of other vent termination methods provided for in this chapter (see the definition of "Air admittance valve" in Chapter 2).

There are two standards to which AAVs are manufactured. These are ASSE 1051 for individual and branch-type valves and ASSE 1050 for stack-type valves. The standards require that the AAV be marked with the name of the manufacturer, model number or device description, and device classification.

P3114.2 Installation. The valves shall be installed in accordance with the requirements of this section and the manufacturer's instructions. *Air admittance valves* shall be installed

TABLE P3113.4.1
SIZE AND LENGTH OF SUMP VENTS

DISCHARGE CAPACITY OF PUMP (gpm)	MAXIMUM DEVELOPED LENGTH OF VENT (feet)[a]				
	Diameter of vent (inches)				
	$1^1/_4$	$1^1/_2$	2	$2^1/_2$	3
10	No limit[b]	No limit	No limit	No limit	No limit
20	270	No limit	No limit	No limit	No limit
40	72	160	No limit	No limit	No limit
60	31	75	270	No limit	No limit

For SI: 1 inch = 25.4 mm, 1 foot = 304.8 mm, 1 gallon per minute (gpm) = 3.785 L/m.

a. Developed length plus an appropriate allowance for entrance losses and friction caused by fittings, changes in direction and diameter. Suggested allowances shall be obtained from NBS Monograph 31 or other approved sources. An allowance of 50 percent of the developed length shall be assumed if a more precise value is not available.

b. Actual values greater than 500 feet.

after the DWV testing required by Section P2503.5.1 or P2503.5.2 has been performed.

❖ Both the code (see Section P3114) and the manufacturer's installation instructions apply. Where there is a conflict between this section and the manufacturer's installations, the provisions of this section apply.

Section P2503.5.1 provides for testing of the DWV system. That test will produce pressure differentials greater than those for which AAVs are designed. Therefore, the AAVs must not be installed until after the testing has been completed.

P3114.3 Where permitted. Individual vents, branch vents, circuit vents and stack vents shall be permitted to terminate with a connection to an *air admittance valve*. Individual and branch

type air admittance valves shall vent only fixtures that are on the same floor level and connect to a horizontal branch drain.

❖ This section establishes the specific locations within the vent system where AAVs can be installed. Individual vents, branch vents and circuit vents can terminate to an AAV conforming to ASSE 1051, and a stack vent or vent stack can terminate to an AAV conforming to ASSE 1050. An ASSE 1051 AAV cannot be used to vent fixtures that are on multiple floors. Sump vents cannot terminate to an AAV because an AAV can only admit air and the sump vent must both admit and emit air. Commentary Figures P3114.3(1) through P3114.3(4) show various AAV applications. Commentary Figure P3114.3(5) shows improper use of an AAV.

P3114.4 Location. Individual and branch *air admittance valves* shall be located not less than 4 inches (102 mm) above the horizontal branch drain or *fixture drain* being vented.

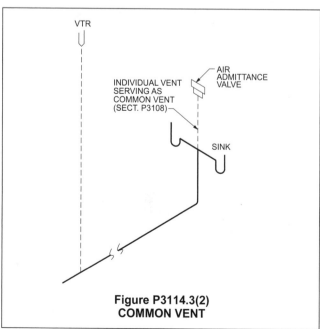

Figure P3114.3(1)
AIR ADMITTANCE VALVES-INDIVIDUAL VENT

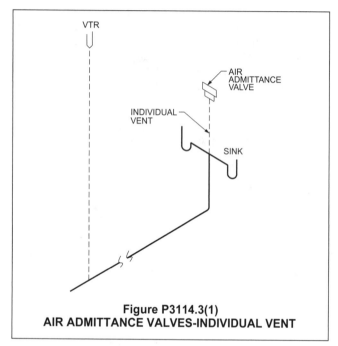

Figure P3114.3(2)
COMMON VENT

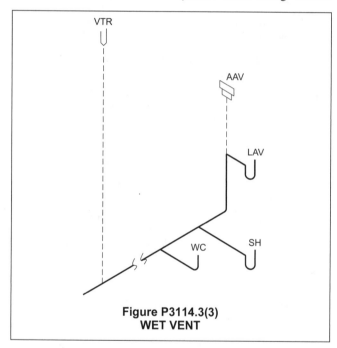

Figure P3114.3(3)
WET VENT

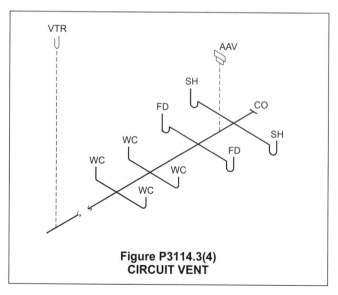

Figure P3114.3(4)
CIRCUIT VENT

Stack-type air admittance valves shall be located not less than 6 inches (152 mm) above the flood level rim of the highest fixture being vented. The *air admittance valve* shall be located within the maximum *developed length* permitted for the vent. The *air admittance valve* shall be installed not less than 6 inches (152 mm) above insulation materials where installed in *attics*.

❖ The height requirements of this section are intended to prevent waste from contaminating an AAV. There will always be an air pocket trapped in the riser beneath an AAV that will prevent waste from reaching the AAV. A vent terminating at an AAV is subject to the same friction losses as any other vent; therefore, the same length limitations apply to such vents (see Commentary Figure P3114.4).

P3114.5 Access and ventilation. Access shall be provided to *air admittance valves*. Such valves shall be installed in a location that allows air to enter the valve.

❖ Because an AAV is a device with a single moving part, it must be accessible for inspection, service, repair or replacement. The code defines accessible as being able to be reached, but first the removal of a panel, door or similar obstruction might be required. Access to an AAV is not achieved if any portion of the structure's permanent finish materials, such as drywall, plaster, paneling, built-in furniture or cabinets or any other similar permanently affixed building component,

must be removed.

Where located in confined spaces, ventilation openings are required in those spaces. AAVs admit air into the drainage system; therefore, an unimpeded air supply must be available to the devices at all times. Additionally, the manufacturer's installation instructions address access and ventilation requirements. The space within a cabinet under a sink counter has an unimpeded air supply and is accessible.

P3114.6 Size. The *air admittance valve* shall be rated for the size of the vent to which the valve is connected.

❖ The size of the vent pipe is regulated by Section P3113.1 or as provided for in the more specific venting provisions found elsewhere in Chapter 31. The AAV is sized based on the required size of the vent pipe and the d.f.u. load limitations found in the manufacturer's instructions. The d.f.u. loading is based on testing performed in accordance with ASSE 1050 and 1051.

P3114.7 Vent required. Within each plumbing system, not less than one stack vent or a vent stack shall extend outdoors to the open air.

❖ The plumbing system must have at least one vent stack or a stack vent extending to the outdoors to relieve positive pressure excursions that might occur, such as from pressure-cleaning operations in the public sewer. Most likely, this vent is also the vent required by Section P3102.1.

P3114.8 Prohibited installations. *Air admittance valves* shall not be used to vent sumps or tanks except where the vent system for the sump or tank has been designed by an engineer.

❖ AAVs must not be used to vent sumps or tanks as there is positive pressure generated when the tank or sump is filling. An AAV cannot relieve positive pressure (see Commentary Figure P3114.8). AAVs can be used as part of an engineered design for tank or sump venting.

Figure P3114.3(5)
VENT CONNECTION NOT PERMITTED

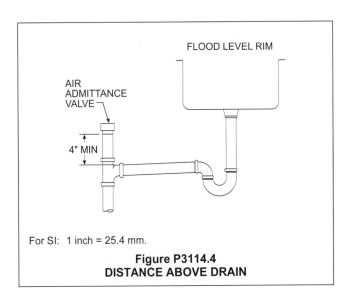

For SI: 1 inch = 25.4 mm.

Figure P3114.4
DISTANCE ABOVE DRAIN

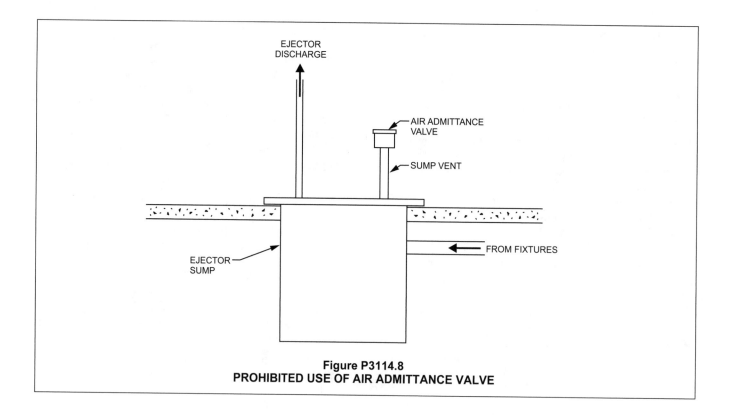

Figure P3114.8
PROHIBITED USE OF AIR ADMITTANCE VALVE

Chapter 32:
Traps

User note: Code change proposals to this chapter will be considered by the IRC – Plumbing and Mechanical Code Development Committee during the 2015 (Group A) Code Development Cycle.

General Comments

Chapter 32 covers fixture traps including criteria for material, design, installation and minimum size of traps as determined by the fixture served, and outlines the approved methods available for maintaining the trap seal. It also specifies types of traps that are prohibited.

Purpose

Traps prevent sewer gas from escaping from the drainage piping into the building. Liquid seal traps are the simplest and most reliable means of preventing sewer gas from entering the interior environment.

SECTION P3201
FIXTURE TRAPS

P3201.1 Design of traps. Traps shall be of standard design, shall have smooth uniform internal waterways, shall be self-cleaning and shall not have interior partitions except where integral with the fixture. Traps shall be constructed of lead, cast iron, copper or copper alloy or *approved* plastic. Copper or copper alloy traps shall be not less than No. 20 gage (0.8 mm) thickness. Solid connections, slip joints and couplings shall be permitted to be used on the trap inlet, trap outlet, or within the trap seal. Slip joints shall be accessible.

❖ This section outlines the types of material that can be used for the construction of fixture traps. Most traps are factory made, though a trap can be fabricated with approved pipe and fittings. When slip joints are used, they must be provided with access as required by Section P2704.1 [see Commentary Figure P3201.1(1)].

A trap must have a pattern allowing unobstructed flow to the drain. Some fixtures, such as water closets, have an integral trap and, thus, are not designed to be connected to an additional trap [see Commentary Figure P3201.1(2)].

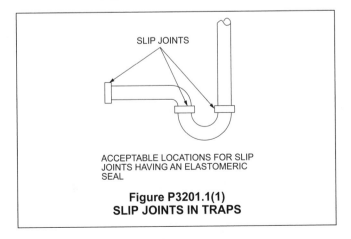

SLIP JOINTS

ACCEPTABLE LOCATIONS FOR SLIP JOINTS HAVING AN ELASTOMERIC SEAL

Figure P3201.1(1)
SLIP JOINTS IN TRAPS

P3201.2 Trap seals. Each fixture trap shall have a liquid seal of not less than 2 inches (51 mm) and not more than 4 inches (102 mm).

❖ A liquid seal of 2 inches (51 mm) is standard for most traps. Some larger pipes, 3 through 6 inches (76 through 152 mm), have a greater seal of up to 4 inches (102 mm) to construct a smooth pattern of flow for the given pipe.

A trap seal must be deep enough to resist the pressures that can develop in a properly vented drainage system, but not deep enough to promote the retention of solids or the growth of bacteria.

Traps that do not periodically receive waste discharge will eventually lose their seal as a result of evaporation. The rate of trap seal evaporation is somewhat dependent on the location of the trap. For example, water in fixture traps in environments with high ambient temperatures or high-volume air movement will evaporate rapidly. Fixtures, such as floor drains, typically have trap seals that are subject to loss by evaporation and, therefore, must be protected by trap seal primers. A deep seal will not prevent the loss of a trap seal, but will simply lengthen the time it will take for it to evaporate.

P3201.2.1 Trap seal protection. Traps seals of emergency floor drain traps and traps subject to evaporation shall be protected by one of the methods in Sections P3201.2.1.1 through P3201.2.1.4.

❖ There are two methods to protect the trap seal against evaporation: 1) a trap seal primer, 2) a trap seal protection device. Sections P3201.2.1.1 through P3201.2.4.1 cover these methods.

P3201.2.1.1 Potable water-supplied trap seal primer valve. A potable water-supplied trap seal primer valve shall supply water to the trap. Water-supplied trap seal primer valves shall conform to ASSE 1018. The discharge pipe from the trap seal primer valve shall connect to the trap above the trap seal on the inlet side of the trap.

❖ The trap primer concept is simple: add a small amount of water to the trap every so often and that trap seal will remain intact. A potable-water-supplied trap primer connects to the potable water distribution system and supplies a small amount of potable water to the trap when a triggering event occurs. For one type of potable water trap seal primer, the triggering event is the use of cold water at a fixture. The cold water distribution piping near the fixture experiences a slight drop of pressure when the cold water valve at the fixture is opened. The trap seal primer device senses the drop in pressure and briefly opens a valve to "waste" a small amount of potable water into a tube that is connected to the trap. Frequent use of the fixture ensures that the trap seal doesn't evaporate. The discharge tube must connect at a level higher than the trap seal so that the tube doesn't become blocked by waste debris floating in or on top of the trap seal.

P3201.2.1.2 Reclaimed or gray-water-supplied trap seal primer valve. A reclaimed or gray-water-supplied trap seal primer valve shall supply water to the trap. Water-supplied

trap seal primer valves shall conform to ASSE 1018. The quality of reclaimed or gray water supplied to trap seal primer valves shall be in accordance with the requirements of the manufacturer of the trap seal primer valve. The discharge pipe from the trap seal primer valve shall connect to the trap above the trap seal on the inlet side of the trap.

❖ The commentary for Section P3201.2.1 is applicable, except the valve is connected to pressurized reclaimed water source or a pressurized (and treated) gray-water source. The triggering fixture could be a water closet or a urinal that is supplied with reclaimed or gray water for flushing. The discharge tube must connect at a level higher than the trap seal so that the tube does not become blocked by waste debris floating in or on top of the trap seal. Note that trap seal primer valve manufacturers provide the water quality requirements for the nonpotable water to be used in the trap seal primer valve.

P3201.2.1.3 Waste-water-supplied trap primer device. A waste-water-supplied trap primer device shall supply water to the trap. Waste-water-supplied trap primer devices shall conform to ASSE 1044. The discharge pipe from the trap seal

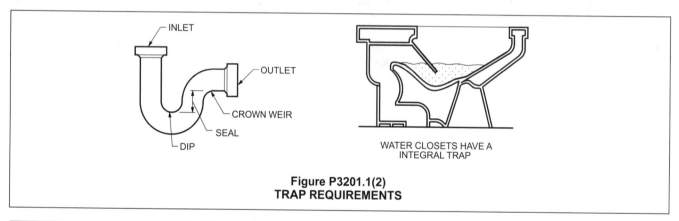

Figure P3201.1(2)
TRAP REQUIREMENTS

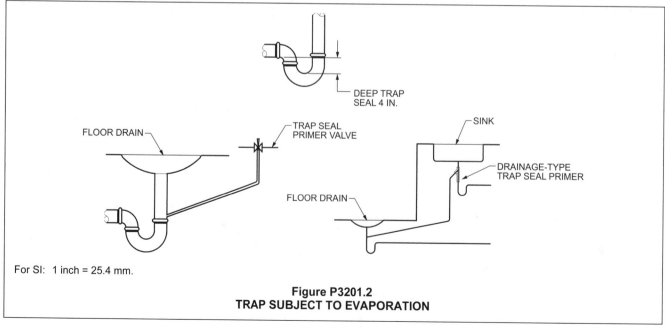

For SI: 1 inch = 25.4 mm.

Figure P3201.2
TRAP SUBJECT TO EVAPORATION

primer device shall connect to the trap above the trap seal on the inlet side of the trap.

❖ This type of device simply diverts some raw gray water from a fixture that discharges gray water. The fixture is typically a lavatory. Usually, such a lavatory has a grid strainer outlet to keep debris from being discharged from the fixture; however, a grid strainer is not a code requirement.

P3201.2.1.4 Barrier-type trap seal protection device. A barrier-type trap seal protection device shall protect the floor drain trap seal from evaporation. Barrier-type floor drain seal protection devices shall conform to ASSE 1072. The devices shall be installed in accordance with the manufacturer's instructions.

❖ This device is an elastomeric "check valve" that is inserted into the pipe opening of a floor drain. When the floor drain is accepting water, the flapper allows the water to flow into the pipe. Otherwise, the flapper returns to its closed position to prevent trap seal exposure to the air and evaporation.

P3201.3 Trap setting and protection. Traps shall be set level with respect to their water seals and shall be protected from freezing. Trap seals shall be protected from siphonage, aspiration or back pressure by an *approved* system of venting (see Section P3101).

❖ Traps must be set level with their seal to perform as designed (see Commentary Figure P3201.3). Because they hold water at all times, traps must be protected where they are subject to freezing. In colder climates, freezing can occur in unheated areas, such as garages or workshops, and also in cantilevered construction where the trap for a bathtub or shower may be installed below the floor insulation.

Where properly installed to code, the venting portion of the drainage system compensates for both negative and positive pressures that can occur in the plumbing system. This helps maintain the water seals in the traps. The venting provisions of Chapter 31 must be adhered to for a properly vented system. This is addressed in Section P3101.2.

P3201.4 Building traps. Building traps shall be prohibited.

❖ Originally, building traps were installed on the building drain, before it was connected to the building sewer, as

a form of rat control. In major cities, rats would breed in the sewer system and enter a building through the sewer connection. A water trap seal was believed to prevent rats from entering the building. Such traps are considered to be high maintenance because of the potential for stoppages.

Building traps may still be necessary in rare cases where the public sewer exerts a positive back pressure on the connected building sewers; however, the installation of a building trap would require special approval. The back-pressure can cause the building vent terminations to emit strong sewer gases, which can create a serious odor problem around the building.

P3201.5 Prohibited trap designs. The following types of traps are prohibited:

1. Bell traps.
2. Separate fixture traps with interior partitions, except those lavatory traps made of plastic, stainless steel or other corrosion-resistant material.
3. "S" traps.
4. Drum traps.
5. Trap designs with moving parts.

❖ A trap is a simple U-shaped piping arrangement that offers minimal resistance to flow. Certain traps or arrangements are prohibited.

Item 1 refers to bell traps, which, because of their design, tend to clog with debris. The trap seal has a larger exposed surface area, which accelerates evaporation.

Item 2 refers to separate fixture traps with interior partitions, except those lavatory traps made of plastic, stainless steel or other corrosion-resistant material. Such lavatory traps are sometimes referred to as bottle traps. This item refers to independent traps that are not integral with the fixture. For example, water closet traps are a form of partition trap, but because they are part of the fixture they are not prohibited. Partition traps are single-wall traps that rely on a single partition to separate the house side of the trap from the sewer side of the trap. Although they may be made of corrosion-resistant material, they are still subject to corrosion and can fail structurally. Because such failure does not result in leakage, there will be no indication of the lost trap seal.

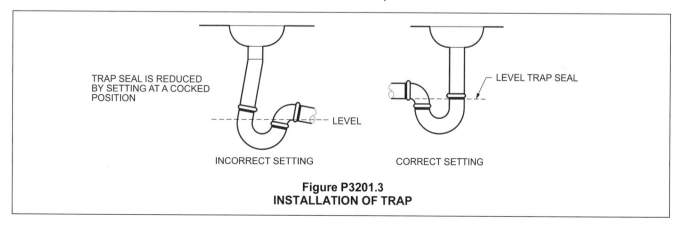

Figure P3201.3
INSTALLATION OF TRAP

Item 3 refers to "S" traps. "S" configuration traps tend to create a siphon effect, which causes loss of the trap seal.

Item 4 refers to drum traps, which, because they are not self-cleaning, are subject to clogging. Although in certain cases they may be desired in nonresidential occupancies, there is no special need for them in dwellings; thus, they are prohibited by the code.

Item 5 refers to trap designs with moving parts. These are mechanical traps that use moving parts, such as floats or flappers. Because of corrosion, clogging and waste deposits that interfere with the operation of moving components and seals, such designs are not dependable.

P3201.6 Number of fixtures per trap. Each plumbing fixture shall be separately trapped by a water seal trap. The vertical distance from the fixture outlet to the trap weir shall not exceed 24 inches (610 mm) and the horizontal distance shall not exceed 30 inches (762 mm) measured from the center line of the fixture outlet to the centerline of the inlet of the trap. The height of a clothes washer standpipe above a trap shall conform to Section P2706.1.2. Fixtures shall not be double trapped.

Exceptions:

1. Fixtures that have integral traps.

2. A single trap shall be permitted to serve two or three like fixtures limited to kitchen sinks, laundry tubs and lavatories. Such fixtures shall be adjacent to each other and located in the same room with a continuous waste arrangement. The trap shall be installed at the center fixture where three fixtures are installed. Common trapped fixture outlets shall be not more than 30 inches (762 mm) apart.

3. Connection of a laundry tray waste line into a standpipe for the automatic clothes-washer drain shall be permitted in accordance with Section P2706.1.2.1.

❖ Double trapping a fixture will cause obstruction of flow and potential stoppages because air is trapped between the two trap seals.

Except for a clothes washer standpipe, which has a maximum vertical distance of 42 inches (1067 mm), the maximum vertical distance of 24 inches (610 mm) from a fixture outlet to the trap weir is required for all other fixtures for two reasons. First, buildup on the wall of the pipe can breed bacteria and cause odors. It is, therefore, best to locate the trap as close as possible to the fixture to minimize the amount of drain pipe that occurs on the inlet (house) side of the trap. Second, the velocity of the drainage flow must be controlled. If the vertical distance is too great, a self-siphoning effect could occur. The horizontal distance limit between the fixture outlet and the trap intends to prevent excessive lengths of pipe on the house side of the trap in which bacteria can grow.

This section allows a single trap to serve a set of not more than three like fixtures (lavatories, laundry tubs or kitchen sinks). In residential applications, this could apply to a double lavatory vanity. Where fixtures are connected to a single trap, the trap must be installed at

the center fixture. This minimizes the amount of piping on the inlet (house) side of the trap, thus reducing the likelihood of bacteria growth and subsequent odors.

Reference is made to Section P2706.2.1, where the provision for connecting a laundry tray to a clothes washer standpipe is allowed.

P3201.7 Size of fixture traps. Trap sizes for plumbing fixtures shall be as indicated in Table P3201.7. Where the tailpiece of a plumbing fixture is larger than that indicated in Table P3201.7, the trap size shall be the same nominal size as the fixture tailpiece. A trap shall not be larger than the drainage pipe into which the trap discharges.

❖ A properly sized trap allows the fixture to drain rapidly and function as intended. The code does not prescribe a maximum trap size; however, if a trap is oversized, it will not scour (cleanse) itself and will be prone to clogging. A trap that is larger than the drainage pipe into which it discharges is further prone to clogging because the reduced size outlet pipe does not allow a water velocity high enough to scour the trap.

TABLE P3201.7
SIZE OF TRAPS FOR PLUMBING FIXTURES

PLUMBING FIXTURE	TRAP SIZE MINIMUM (inches)
Bathtub (with or without shower head and/or whirlpool attachments)	$1^1/_2$
Bidet	$1^1/_4$
Clothes washer standpipe	2
Dishwasher (on separate trap)	$1^1/_2$
Floor drain	2
Kitchen sink (one or two traps, with or without dishwasher and food waste disposer)	$1^1/_2$
Laundry tub (one or more compartments)	$1^1/_2$
Lavatory	$1^1/_4$
Shower (based on the total flow rate through showerheads and bodysprays) Flow rate: 5.7 gpm and less / More than 5.7 gpm up to 12.3 gpm / More than 12.3 gpm up to 25.8 gpm / More than 25.8 gpm up to 55.6 gpm	$1^1/_2$ / 2 / 3 / 4

For SI: 1 inch = 25.4 mm.

❖ This table shows minimum trap sizes for most fixtures. The provisions of Section P3201.6, which allow a set of not more than three similar fixtures (lavatories, laundry trays or kitchen sinks) to be drained by one trap, do not require an increase in the size of the trap.

Because of the low flow rate for showers [2.5 gpm (9.5 L/m) in accordance with Section P2903.2], a $1^1/_2$-inch (38 mm) shower drain is more than adequate and will require less maintenance than a 2-inch (51 mm) drain because the higher flow velocity in the smaller drain will scour the trap and pipe walls. Where multiple shower heads or body spray heads are installed, the drain and trap size may have to be increased to accommodate the increased load.

Chapter 33:
Storm Drainage

User note: Code change proposals to this chapter will be considered by the IRC – Plumbing and Mechanical Code Development Committee during the 2015 (Group A) Code Development Cycle.

General Comments

Chapter 33 covers criteria for material, design and installation of subsoil drainage systems typically used around the perimeter of foundations. Poor or subsoil drainage can affect the structural stability of the foundation walls and, where basements are below grade, cause water damage to finished area and property.

Purpose

This chapter defines the requirements for subsoil drainage systems, including pumping systems for removal of the storm water collected by those systems.

SECTION P3301
GENERAL

P3301.1 Scope. The provisions of this chapter shall govern the materials, design, construction and installation of storm drainage.

❖ This chapter covers the requirements for storm drainage systems.

SECTION P3302
SUBSOIL DRAINS

P3302.1 Subsoil drains. Subsoil drains shall be open-jointed, horizontally split or perforated pipe conforming to one of the standards listed in Table P3302.1. Such drains shall be not less than 4 inches (102 mm) in diameter. Where the building is subject to backwater, the subsoil drain shall be protected by an accessibly located backwater valve. Subsoil drains shall discharge to a trapped area drain, sump, dry well or *approved* location above ground. The subsoil sump shall not be required to have either a gas-tight cover or a vent. The sump and pumping system shall comply with Section P3303.

❖ The drains in a subsoil drainage system consist of collection piping having holes, slots or end-to-end gaps for the ground water to infiltrate into the pipe to then be carried away from the collection area. The requirement for piping to be at least 4 inches (102 mm) in diameter

ensures that adequate flow area will exist after sediments are deposited along the bottom of the pipes. Although not stated by this section, subsoil drain piping is typically installed with zero slope as the water entering the subsoil drainage piping will naturally flow towards an outlet in the piping system.

If the project site is sloped such that finished grade elevation is below the subsoil drainage system, the collected water can flow by gravity to a suitable location on the project site, a dry well, or an area drain. If the area drain is connected to a combined storm sanitary drainage system, the area drain must be trapped so that sewer gases cannot flow into the subsoil drainage system. Where the outlet of a gravity-flow subsoil drainage system could be submerged by high levels of surface water, the subsoil drainage system must have a backwater valve installed (see Commentary Figure P3302.1). Access must be provided for maintaining the backwater valve.

Where the water from a subsoil drainage system cannot flow away from the collection area by gravity, a sump with a sump pump must be provided for the collected water to flow into. The requirements for the sump and sump pump are located in Section 3303. Sumps are not required to have a gas-tight cover; however, Section 3303.1.2 requires the sump to have a cover that can support the anticipated loads. If a gas-

TABLE P3302.1
SUBSOIL DRAIN PIPE

MATERIAL	STANDARD
Cast-iron pipe	ASTM A74; ASTM A888; CISPI 301
Polyethylene (PE) plastic pipe	ASTM F405; CSA B182.1; CSA B182.6; CSA B182.8
Polyvinyl chloride (PVC) Plastic pipe (type sewer pipe, SDR 35, PS25, PS50 or PS100)	ASTM D2729; ASTM D3034; ASTM F891; CSA B182.2; CSA B182.4
Stainless steel drainage systems, Type 316L	ASME A112.3.1
Vitrified clay pipe	ASTM C4; ASTM C700

tight cover is installed, a vent-to-atmosphere is recommended. The vent must not connect to the venting system of a sanitary drainage system.

TABLE P3302.1. See page 33-1.

❖ Most of the standards for the piping materials in Table 3302.1 include requirements for the number, size and location of perforations. Where a pipe of a material listed in the table is not available in a perforated form, a common practice is to install short sections [approximately 2 feet long (610 mm)] of pipe end-to-end, with a gap between the ends. An archaic practice was to cover the gap between the ends of the pipe with asphaltic roofing paper to keep backfill material from flowing into the gaps; however, modern "geotextiles" are available for this purpose. The "open-jointed" installation method is only appropriate for pipe materials supplied in straight lengths and not in coiled form.

SECTION P3303
SUMPS AND PUMPING SYSTEMS

P3303.1 Pumping system. The sump pump, pit and discharge piping shall conform to Sections P3303.1.1 through P3303.1.4.

P3303.1.1 Pump capacity and head. The sump pump shall be of a capacity and head appropriate to anticipated use requirements.

❖ The amount of subsoil drainage flow is dependent on many project site factors, such as the length of subsoil piping system, rainfall duration, porosity of soil and the elevation of the water table. There is no easy method to determine what flow capacity of sump pump is required for any given application. It is easier to determine the required pump head pressure and check the pump flow rate to see if it is reasonable.

The amount of head pressure that the pump must overcome can be approximated from the difference in elevation between the pump and the highest portion of the discharge piping. For example, consider a pump located in a 2-foot-deep (610 mm) sump pit in the floor of a basement that is 9 feet (2743 mm) below finished grade. The desired discharge point for the pumped flow is 1 foot (305 mm) above finished grade. The sump pump must develop at least 12 feet (3658 mm) (2 + 9 + 1 = 12) of head pressure in order to "lift" water from the sump to the discharge point.

Although the flow of water in discharge piping creates a resistance to flow that the pump must also overcome, it is usually not taken into account because most

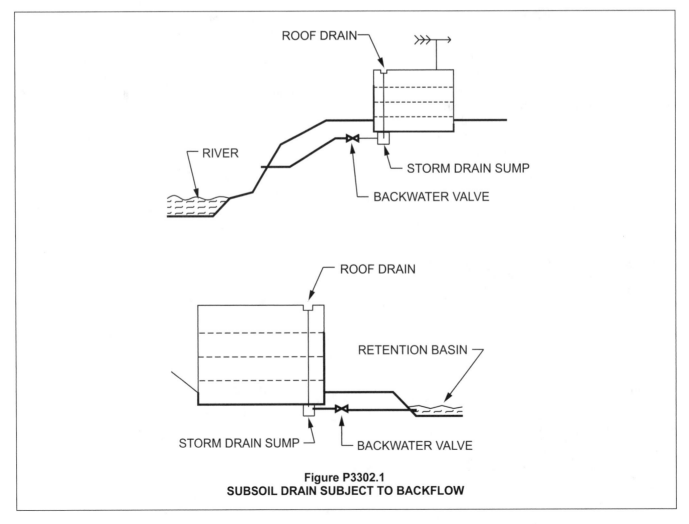

Figure P3302.1
SUBSOIL DRAIN SUBJECT TO BACKFLOW

sump pump discharge piping lengths are relatively short [e.g., 20 feet (6096 mm)].

Sump pumps are typically rated for so many gallons per minute (or hour) at a so many feet of "head" pressure. A pump can be chosen by finding a pump model that can produce the head pressure (determined previously), while delivering at least 20 gpm (1200 L/m) gallons per hour. The flow rate of 20 gpm (1200 L/m) provides for a pump run time of at least 30 seconds when the pump is installed in the minimum size sump pit, as required by Section P3303.1.2.

Although not stated in the code, the assumption is that the sump pump is equipped with an integral float switch or have some other means for automatic control.

P3303.1.2 Sump pit. The sump pit shall be not less than 18 inches (457 mm) in diameter and 24 inches (610 mm) deep, unless otherwise *approved*. The pit shall be accessible and located so that all drainage flows into the pit by gravity. The sump pit shall be constructed of tile, steel, plastic, cast-iron, concrete or other *approved* material, with a removable cover adequate to support anticipated loads in the area of use. The pit floor shall be solid and provide permanent support for the pump.

❖ An 18-inch-diameter (457mm) pit allows adequate room for any typical sump pump with a typical float switch arrangement. The depth of the pit allows for a reasonable pump run time given typical sump pump designs.

P3303.1.3 Electrical. Electrical outlets shall meet the requirements of Chapters 34 through 43.

❖ Although not specifically stated, the intent of the code is for the sump pump power cord to be plugged into a receptacle. The receptacle must meet the requirements of the electrical portion of the code (see Chapters 34 though 43). Given that sump pumps are usually located in unfinished portions of basements, receptacles will be protected by ground fault circuit interrupter devices.

P3303.1.4 Piping. Discharge piping shall meet the requirements of Sections P3002.1, P3002.2, P3002.3 and P3003. Discharge piping shall include an accessible full flow check valve. Pipe and fittings shall be the same size as, or larger than, the pump discharge tapping.

❖ Because a sump pump "lifts" water to another location, a check valve in the discharge line is necessary to prevent the pumped water from flowing backward into the sump. The check valve must have access for cleaning or replacement. The discharge piping must not be smaller than the outlet size of the pump in order to keep piping friction losses at a minimum.

Chapter 34:
General Requirements

This Electrical Part (Chapters 34 through 43) is produced and copyrighted by the National Fire Protection Association (NFPA) and is based on the 2014 *National Electrical Code®* (NEC®) (NFPA 70®-2014), copyright 2013, National Fire Protection Association, all rights reserved. Use of the Electrical Part is pursuant to license with the NFPA.

The title *National Electrical Code®*, the acronym NEC® and the document number NFPA 70® are registered trademarks of the National Fire Protection Association, Quincy, Massachusetts. The section numbers appearing in parentheses or brackets after IRC text are the section numbers of the corresponding text in the *National Electrical Code* (NFPA 70).

IMPORTANT NOTICE AND DISCLAIMER CONCERNING THE NEC AND THIS ELECTRICAL PART.

This Electrical Part is a compilation of provisions extracted from the 2014 edition of the NEC. The NEC, like all NFPA codes and standards, is developed through a consensus standards development process approved by the American National Standards Institute. This process brings together volunteers representing varied viewpoints and interests to achieve consensus on fire and other safety issues. While the NFPA administers the process and establishes rules to promote fairness in the development of consensus, it does not independently test, evaluate or verify the accuracy of any information or the soundness of any judgments contained in its codes and standards.

The NFPA disclaims liability for any personal injury, property or other damages of any nature whatsoever, whether special, indirect, consequential or compensatory, directly or indirectly resulting from the publication, use of, or reliance on the NEC or this Electrical Part. The NFPA also makes no guaranty or warranty as to the accuracy or completeness of any information published in these documents.

In issuing and making the NEC and this Electrical Part available, the NFPA is not undertaking to render professional or other services for or on behalf of any person or entity. Nor is the NFPA undertaking to perform any duty owed by any person or entity to someone else. Anyone using these documents should rely on his or her own independent judgment or, as appropriate, seek the advice of a competent professional in determining the exercise of reasonable care in any given circumstances.

The NFPA has no power, nor does it undertake, to police or enforce compliance with the contents of the NEC and this Electrical Part. Nor does the NFPA list, certify, test, or inspect products, designs, or installations for compliance with these documents. Any certification or other statement of compliance with the requirements of these documents shall not be attributable to the NFPA and is solely the responsibility of the certifier or maker of the statement.

For additional notices and disclaimers concerning NFPA codes and standards see www.nfpa.org/disclaimers.

General Comments

This chapter contains broadly applicable requirements including provisions for the protection of the structural elements of a building, inspection of work, general installation provisions and conductor identification. Section E3403.3 requires that all electrical system components be listed and labeled by an approved agency.

The portions of the National Electrical Code® (NEC®) text that are not applicable to dwelling construction have been excluded; therefore, the provisions herein are appropriate for the scope of coverage of the code. The NEC text has been reformatted, reorganized and, in some cases, restated for the purpose of making the provisions easier to locate, understand and apply. The code is formatted in a manner consistent with the other International Codes® (I-Codes®), designed to be a "cookbook" for dwelling construction and meant to be user friendly for both builders and code officials.

The electrical code provisions of the code are identical to the intent of the NEC provisions except that the code requires all electrical system components be listed and labeled; the NEC does not. The code does not contain unique electrical requirements. A dwelling built to the code will have electrical systems identical to those required by the respective edition of the NEC. The code addresses only those electrical systems that are common to dwelling construction, and the NEC is referenced for any subject not addressed in the code.

The code electrical chapters are produced and maintained by the National Fire Protection Association (NFPA) in accordance with a contract between the International Code Council® (ICC®) and NFPA. The contract requires NFPA to update the code to the new edition of the NEC, and the code and the NEC to remain technically consistent. ICC's technical staff works with NFPA's technical staff to present the updated text in ICC style and format.

Purpose

This chapter establishes the scope of coverage of the code and provides reference to NFPA 70 where the code lacks coverage of any wiring method or material.

The provisions of this chapter intend to protect life, limb and property from the hazards of electricity, attempting to make sure that electrical equipment and conductors are properly located, guarded and identified.

SECTION E3401
GENERAL

E3401.1 Applicability. The provisions of Chapters 34 through 43 shall establish the general scope of the electrical system and equipment requirements of this code. Chapters 34 through 43 cover those wiring methods and materials most commonly encountered in the construction of one- and two-family dwellings and structures regulated by this code. Other wiring methods, materials and subject matter covered in NFPA 70 are also allowed by this code.

❖ These chapters cover the wiring methods, materials, equipment, fixtures and appliances that are most commonly used in residential buildings. Where special equipment, wiring methods or materials are used in dwellings or on residential property, the provisions of the NEC apply (see Section R101.2).

E3401.2 Scope. Chapters 34 through 43 shall cover the installation of electrical systems, equipment and components indoors and outdoors that are within the scope of this code, including services, power distribution systems, fixtures, appliances, devices and appurtenances. Services within the scope of this code shall be limited to 120/240-volt, 0- to 400-ampere, single-phase systems. These chapters specifically cover the equipment, fixtures, appliances, wiring methods and materials that are most commonly used in the construction or alteration of one- and two-family dwellings and accessory structures regulated by this code. The omission from these chapters of any material or method of construction provided for in the referenced standard NFPA 70 shall not be construed as prohibiting the use of such material or method of construction. Electrical systems, equipment or components not specifically covered in these chapters shall comply with the applicable provisions of NFPA 70.

❖ This section states that any material or method of construction not covered by these chapters must comply with the applicable provisions of the NEC. Section R102.4 states that codes and standards referenced in the code must be considered part of the requirements of this code. This section, in the "Administration" chapter, goes on to state that "where differences occur between provisions of this code and referenced codes and standards, the provisions of this code shall apply."

The provisions of Chapters 34 through 43 cover the electrical installations of wiring and equipment most commonly used for one- and two-family dwellings and associated structures. In the development of these chapters, it was considered that the vast majority of one- and two-family dwellings have a 120/240-volt single-phase service of between 100 and 400 amperes. Therefore, it was deemed adequate for the code to include provisions considered necessary only for one-

and two-family dwellings with a service of up to a maximum of 400 amperes (see Section R101.2). For larger services and methods that are less common for one- and two-family dwellings, the provisions of the NEC would apply.

In residential electrical work, many times the electrician or electrical contractor on the job designs the wiring system. The plans may have only the locations of lighting and receptacle outlets, so the electrician or electrical contractor must calculate the load for the dwelling, as well as the load for the branch circuits and then design the layout of the wiring job. The electrical chapters of the code are written in a way that makes them simple to use while still being technically accurate and complete for the most common methods of installation. They contain minimum provisions that, if followed, will result in a safe and sufficient installation. However, larger conductor sizes, more branch circuits, and more lighting and receptacle outlets than the minimum required by the code are sometimes desirable for added convenience.

The electrical provisions in these chapters are based on the 2014 NEC, so the requirements are the same as in that referenced standard. For consistency with the building, mechanical and plumbing provisions, the electrical chapters of the code are written in a somewhat different style and format from those in the NEC. The requirements of the NEC pertain to many different occupancies and varied installations. In the development of the electrical provisions of the code, it was desired to develop an organization scheme that pertained specifically to dwellings. One result of this is that fewer exceptions are needed because the code covers only residential buildings. Also, several tables are included to summarize the various code rules. For example, many of the requirements for the different wiring methods are consolidated into tables.

An effort has been made to arrange the chapters in a logical order, beginning with general requirements. The start of the wiring system and electrical supply is the service and feeders, so provisions concerning them became the next chapter in the sequence. Grounding as a general subject was separated into two areas: service and feeder grounding and equipment grounding. Because system grounding is part of the electrical service, the service grounding and bonding provisions, including the grounding electrode system are included in Chapter 36. The next chapter contains the branch circuit and feeder requirements. The chapters that follow are on wiring methods and requirements for power and lighting distribution. These include provisions for outlets and equipment grounding which in turn include rules for sizing equipment grounding conductors. The final three chapters contain rules on appliances, swimming pools and low-voltage circuits.

E3401.3 Not covered. Chapters 34 through 43 do not cover the following:

1. Installations, including associated lighting, under the exclusive control of communications utilities and electric utilities.

2. Services over 400 amperes.

❖ This code does not directly cover requirements for the service-drop or service-lateral conductors. In most cases, the service-drop and service-lateral conductors are owned and maintained by the utility company, so it is responsible for these conductors. The clearances for these conductors and the depth of cover for laterals that are beyond the service point are not in the code. A dwelling unit with a service rated over 400 amperes may require equipment not covered by this code and is not within the scope of this code.

E3401.4 Additions and alterations. Any addition or alteration to an existing electrical system shall be made in conformity to the provisions of Chapters 34 through 43. Where additions subject portions of existing systems to loads exceeding those permitted herein, such portions shall be made to comply with Chapters 34 through 43.

❖ This section correlates with the administrative provisions of Chapter 1, Section R101.2, which states that the code applies to the construction, alteration, movement, enlargement, replacement, repair, removal and demolition of one- and two-family dwellings. It is important to consider the provisions of Chapter 1 with the electrical chapters. For example, Section R105 has the requirement to obtain a permit from the code official before work is started. Electrical work that is exempt from a permit is also listed in Section R105.

Additions, alterations or repairs to any structure must conform to this code, but they do not require that the entire structure be updated or brought into current code compliance. Where the additions, alterations or repairs do not cause the existing electrical system to be unsafe, upgrading the old system is not required. For example, a simple bedroom addition may not result in a big difference in the electrical load on the service or feeders. The addition of a circuit for a bedroom light and four or five receptacle outlets would probably not be a reason to replace the electrical service equipment with one of larger ampacity. However, in a very old house with a 60-ampere service, the addition of one circuit could well cause an overload on the service, and the service would need to be upgraded. This is not to imply that the electrical load required for a small addition of lighting, outlets or equipment can be neglected. The load for additions to existing dwellings should always be calculated. If the existing service and/or feeder conductors are found to be inadequate to handle the new load, an upgrade would be necessary. Chapters 36 and 37 contain the load calculation requirements.

It is important to look not only at the electrical chapters when considering electrical requirements. A bedroom addition would require that a smoke alarm be installed in the bedroom. Where smoke alarms are wired and not just battery powered, they are installed as part of the wiring job and are in the domain of the electrician and the electrical inspector. Although the requirement to install smoke alarms is in Part III-Building Planning and Construction and not in Part VIII-Electrical, the provisions for the materials and methods for wiring the smoke alarms are included in the electrical chapters. Section R314.2.2 requires that when one or more sleeping rooms are added or created in an existing dwelling, the dwelling unit must have smoke alarms as required for new dwellings. If this dwelling did not already have smoke alarms installed, the code requires that they be installed in all bedrooms and outside each separate sleeping area in the immediate vicinity of the bedrooms and in each additional story of the dwelling including basements and cellars. All of these smoke alarms must be interconnected and hard wired, including the one in the new bedroom. The exception to this rule is that where the wall or ceiling finish is not removed during the construction or remodeling and there is no access to install the wiring, smoke alarms would not have to be interconnected. However, the smoke alarm in the new bedroom would have to be hard wired. Whether and how to interconnect additional smoke alarms in the existing part of the dwelling may be decisions reached between the code official and the installer of the electrical wiring. See the exception to Section R314.4.

Whenever interior alterations, repairs or additions are done that require a permit, smoke alarms must be in-stalled in the dwelling in accordance with the smoke alarm requirements for new construction. This provision indicates that in a house with no existing smoke alarms, where the kitchen is remodeled for example, the kitchen wiring must be installed according to current code rules, and smoke alarms must be installed in all the required locations in the existing house, hard wired and interconnected where possible.

SECTION E3402
BUILDING STRUCTURE PROTECTION

E3402.1 Drilling and notching. Wood-framed structural members shall not be drilled, notched or altered in any manner except as provided for in this code.

❖ The installation of wire and cable requires drilling and notching of floor joists, studs and other wood-framing members. This provision is another example of the necessity to look in other parts of the code when installing or inspecting the electrical wiring. When drilling and notching, two factors must be remembered: the integrity of the framing member and the protection of the wire or cable. To protect the strength and structural load bearing integrity of wood-framed structural members, the provisions of Sections R502.8 and R602.6 must be followed. Requirements in Table E3802.1 are somewhat different and are for the protection of the wire, cable or other wiring method.

E3402.2 Penetrations of fire-resistance-rated assemblies. Electrical installations in hollow spaces, vertical shafts and ventilation or air-handling ducts shall be made so that the possible spread of fire or products of combustion will not be substantially increased. Electrical penetrations into or through fire-resistance-rated walls, partitions, floors or ceilings shall be protected by approved methods to maintain the fire-resistance rating of the element penetrated. Penetrations of fire-resistance-rated walls shall be limited as specified in Section R302.4. (300.21)

❖ Penetrations through fire-resistance-rated walls or floors must meet the requirements of Section R317.3. Wall and/or floor assemblies of at least a 1-hour fire-resistance rating separate two-family dwellings from each other. For the installation of electrical wire, cable, conduit or electrical outlet boxes, it is often necessary to make a through penetration or a membrane penetration. A through penetration is an opening that passes through an entire assembly, accommodating an item such as a cable or conduit. Where membrane construction is provided, such as gypsum board applied to both sides of a stud wall, a through penetration would pass entirely through both membranes and the cavity of the wall. An electrical box installed in the wall or ceiling/floor assembly separating a two-family dwelling and a recessed light fixture in the ceiling of a dwelling that has a separate dwelling unit on the floor above are examples of membrane penetrations, which are openings through only one membrane of a wall, ceiling or floor. Where nonmetallic electrical outlet boxes are used in a membrane penetration, they must be installed in accordance with their listings. For steel electrical boxes, the specific requirements of Section R302.4.2 must be followed.

E3402.3 Penetrations of firestops and draftstops. Penetrations through fire blocking and draftstopping shall be protected in an approved manner to maintain the integrity of the element penetrated.

❖ See Section R302.11 for requirements regarding the penetration of fireblocking systems. Penetrations through fire blocking and draftstopping by electrical conduit, cables and equipment must be installed in a way that will maintain the integrity of the building element being penetrated. There are many fire-blocking and draftstopping materials available, and both the electrician and the builder must be familiar with these products and methods.

SECTION E3403
INSPECTION AND APPROVAL

E3403.1 Approval. Electrical materials, components and equipment shall be approved. (110.2)

❖ Electrical materials and equipment installed in one- and two-family dwellings must be approved by the code official for the specific use or application.

E3403.2 Inspection required. New electrical work and parts of existing systems affected by new work or alterations shall be inspected by the building official to ensure compliance with the requirements of Chapters 34 through 43.

❖ In small towns and municipalities, the building inspector may perform all of the inspections. However, in many larger jurisdictions, inspectors are hired to work under the direction of the code official who manages the building department from the office and does not spend much time in the field. Where the code requires inspections by the code official, it means that the code official can designate the inspector to perform the actual inspection in the field.

E3403.3 Listing and labeling. Electrical materials, components, devices, fixtures and equipment shall be listed for the application, shall bear the label of an approved agency and shall be installed, and used, or both, in accordance with the manufacturer's installation instructions. [110.3(B)]

❖ This code requires that all electrical system components be listed. Testing laboratories publish lists of materials that have been tested. Laboratories and the testing methods they use are evaluated by an evaluation service. An evaluation report is available on the listed products. Code officials rely on these lists and evaluation reports to approve materials for use in the installation of electrical systems. Electrical materials must be approved and installed in accordance with the manufacturer's installation instructions. It is necessary to know these listing requirements as well as the code rules.

SECTION E3404
GENERAL EQUIPMENT REQUIREMENTS

E3404.1 Voltages. Throughout Chapters 34 through 43, the voltage considered shall be that at which the circuit operates. (110.4)

❖ For the code provisions in Chapters 34 through 43, the voltage to be considered must be the voltage at which the circuit operates. Nominal voltage for residential wiring as stated in Section E3602.4 is 120/240-volt, single phase, with a grounded neutral. However, the actual voltage at which the circuit operates can vary from the nominal voltage designation.

E3404.2 Interrupting rating. Equipment intended to interrupt current at fault levels shall have a minimum interrupting rating of 10,000 amperes. Equipment intended to interrupt current at levels other than fault levels shall have an interrupting rating at nominal circuit voltage of not less than the current that must be interrupted. (110.9)

❖ See the commentary to Section E3404.3.

E3404.3 Circuit characteristics. The overcurrent protective devices, total impedance, equipment short-circuit current ratings and other characteristics of the circuit to be protected shall be so selected and coordinated as to permit the circuit protective devices that are used to clear a fault to do so without extensive damage to the electrical equipment of the circuit. This fault shall be assumed to be either between two or more of the circuit conductors or between any circuit conductor and the equipment grounding conductors permitted in Section E3908.8. Listed equipment applied in accordance with its listing shall be considered to meet the requirements of this section. (110.10)

❖ It is important that the ratings, intended uses and proper types of materials are selected for the appropriate use. In one- and two-family dwellings, circuit breakers are the most commonly used overcurrent protective device (OPD) for circuits originating at the panelboard. Fuses are often used for protection of wiring and components of air conditioning equipment and other motor circuits. If a short circuit or ground fault occurs, the circuit must be de-energized as soon as possible to prevent fire and/or damage to the equipment. Section E3705.5 mandates that OPDs be placed in the circuit where the conductors receive their supply. These devices are designed to open the circuit and interrupt the current under other than normal current flow conditions without damage to either the device or to items protected by the device.

Insulation Failure

One of the major causes of trouble in wiring is insulation failure. The results of insulation failure are short circuits and ground faults.

A "short circuit" is a conducting connection between any two conductors of an electrical system; it could be between any two ungrounded conductors or between an ungrounded conductor and the grounded conductor.

A "ground fault" is a conducting connection between a conductor of an electrical system and a grounded object such as a conduit, box or enclosure. Where opposite phase conductors are accidentally connected together and then energized, the connection is called a bolted fault, and there is usually no arcing. But where the ungrounded conductor lightly touches another ungrounded conductor or a metal enclosure around the conductors, a fault occurs that would likely be an arcing ground fault.

Purpose of the OPD

OPDs serve two separate functions. They protect against: (1) current overload or current higher than normal operating levels for the circuit conductors and (2) short circuits and ground faults. Accordingly, OPDs have two different interrupting ratings.

Current Overload Interrupting Rating

The standard ampere rating of an OPD is for the maximum current flow under normal operating conditions. In common practice in the field, this rating is not usually referred to as an interrupting rating, although that is what it is. Electricians simply call it the ampere rating of the breaker. For example, they would install a 20-ampere breaker on the circuit supplying the clothes washer. The 20-ampere rating is an interrupting rating in that the breaker will open

and interrupt the current flow if the load increases to over 20 amperes. The laundry circuit often has a convenience outlet in the same room in addition to the clothes washer receptacle. If a high current use appliance such as a space heater is plugged into this receptacle, the total load could possibly be over 20 amperes. A current flow greater than 20 amperes, if not interrupted, could cause such a high temperature that the conductor insulation or the wire itself would be damaged. Section E3705.6 lists the standard ampere rating of overcurrent protective devices, which determines the rating of the branch circuit and is coordinated with the ampacity of the circuit conductors (see Section E3702.2).

Short-circuit or Ground-fault Interrupting Rating

This section requires that all OPD have a short-circuit and ground-fault interrupting rating of at least 10,000 amperes. If the available short-circuit or ground-fault current is higher than 10,000 amperes, the installed OPD must have sufficient interrupting ratings above the 10,000 ampere level. The electrician or electrical contractor must have this information and be able to provide it to the inspector or the code official (the authority having jurisdiction).

The Amount of Current Flow in a Short-circuit or Ground-fault Circuit

A high-level short circuit or ground fault could destroy the OPD and/or the panelboard, as well as lead to the ignition of a fire. To determine the amount of current (let-through current) in a short circuit or ground fault, it is necessary to know the amount of available short-circuit current in amperes from the supply system and take into account the impedance of the short-circuit or ground-fault circuit. The impedance is very low in a short circuit, and therefore the current flow is very high.

In a bolted short circuit, there is high current flow. The current can be almost all of what the system can deliver. In most residential systems, the available short-circuit current is usually under 10,000 amperes but could be more, depending on several factors including the transformer size, transformer impedance, how close the transformer is to the dwelling and the wire size.

E3404.4 Enclosure types. Enclosures, other than surrounding fences or walls, of panelboards, meter sockets, enclosed switches, transfer switches, circuit breakers, pullout switches and motor controllers, rated not over 600 volts nominal and intended for such locations, shall be marked with an enclosure-type number as shown in Table E3404.4.

Table E3404.4 shall be used for selecting these enclosures for use in specific locations other than hazardous (classified) locations. The enclosures are not intended to protect against conditions such as condensation, icing, corrosion, or contamination that might occur within the enclosure or enter through the conduit or unsealed openings. (110.28)

❖ Enclosures are the housing or cabinet that enclose wiring terminals, overcurrent devices, switches and other electrical components. Enclosures serve to protect what they enclose from damage and also protect persons by shielding them from energized parts. Table E3404.4 lists the enclosure type designations established by the equipment manufacturers, often referred to as National Electrical Manufacturers Association (NEMA) designations. An enclosure must be suitable for the environment to which it will be exposed. In residential applications, Type 3R enclosures are typically used for outdoor installations such as service and equipment disconnects and panelboards.

E3404.5 Protection of equipment. Equipment not identified for outdoor use and equipment identified only for indoor use, such as "dry locations," "indoor use only" "damp locations," or enclosure Type 1, 2, 5, 12, 12K and/or 13, shall be pro-

tected against damage from the weather during construction. (110.11)

❖ During construction, equipment delivered to the job site that is identified for indoor use or for dry locations only must be protected from the weather. The electrical equipment manufacturing industry has designated equipment with different type numbers. A Type 1 enclosure is designated for indoor use only. Equipment should be stored in a dry and clean location until it is installed.

E3404.6 Unused openings. Unused openings, other than those intended for the operation of equipment, those intended for mounting purposes, and those permitted as part of the design for listed equipment, shall be closed to afford protection substantially equivalent to the wall of the equipment.

TABLE E3404.4 (Table 110.28)
ENCLOSURE SELECTION

PROVIDES A DEGREE OF PROTECTION AGAINST THE FOLLOWING ENVIRONMENTAL CONDITIONS	FOR OUTDOOR USE									
	Enclosure-type Number									
	3	3R	3S	3X	3RX	3SX	4	4X	6	6P
Incidental contact with the enclosed equipment	X	X	X	X	X	X	X	X	X	X
Rain, snow and sleet	X	X	X	X	X	X	X	X	X	X
Sleet[a]	—	—	X	—	—	X	—	—	—	—
Windblown dust	X	—	X	X	—	X	X	X	X	X
Hosedown	—	—	—	—	—	—	X	X	X	X
Corrosive agents	—	—	—	X	X	X	—	X	—	X
Temporary submersion	—	—	—	—	—	—	—	—	X	X
Prolonged submersion	—	—	—	—	—	—	—	—	—	X

PROVIDES A DEGREE OF PROTECTION AGAINST THE FOLLOWING ENVIRONMENTAL CONDITIONS	FOR INDOOR USE									
	Enclosure-type Number									
	1	2	4	4X	5	6	6P	12	12K	13
Incidental contact with the enclosed equipment	X	X	X	X	X	X	X	X	X	X
Falling dirt	X	X	X	X	X	X	X	X	X	X
Falling liquids and light splashing	—	X	X	X	X	X	X	X	X	X
Circulating dust, lint, fibers and flyings	—	—	X	X	—	X	X	X	X	X
Settling airborne dust, lint, fibers and flings	—	—	X	X	X	X	X	X	X	X
Hosedown and splashing water	—	—	X	X	—	X	X	—	—	—
Oil and coolant seepage	—	—	—	—	—	—	—	X	X	X
Oil or coolant spraying and splashing	—	—	—	—	—	—	—	—	—	X
Corrosive agents	—	—	—	X	—	—	X	—	—	—
Temporary submersion	—	—	—	—	—	X	X	—	—	—
Prolonged submersion	—	—	—	—	—	—	X	—	—	—

a. Mechanism shall be operable when ice covered.

Note 1: The term raintight is typically used in conjunction with Enclosure Types 3, 3S, 3SX, 3X, 4, 4X, 6 and 6P. The term rainproof is typically used in conjunction with Enclosure Types 3R and 3RX. The term watertight is typically used in conjunction with Enclosure Types 4, 4X, 6 and 6P. The term driptight is typically used in conjunction with Enclosure Types 2, 5, 12, 12K and 13. The term dusttight is typically used in conjunction with Enclosure Types 3, 3S, 3SX, 3X, 5, 12, 12K and 13.

Note 2: Ingress protection (IP) ratings are found in ANSI/NEMA 60529, *Degrees of Protection Provided by Enclosures.* IP ratings are not a substitute for enclosure-type ratings.

Where metallic plugs or plates are used with nonmetallic enclosures they shall be recessed at least $^1/_4$ inch (6.4 mm) from the outer surface of the enclosure. [110.12(A)]

❖ Unused openings include knockouts in device boxes, threaded raceway sockets in conduit bodies and boxes, cabinet and enclosure concentric and eccentric knockouts and circuit breaker knockouts in panelboard covers. Boxes and enclosures must have the integrity to keep out foreign objects, protect the conductors and devices therein, contain any arcing or hot materials resulting from faults and splice failures in the box or enclosure and protect persons from accidental contact with energized components. Metal plugs used with nonmetallic boxes will not be grounded, and recessing them in the outer surface will limit contact.

Knockouts being left open and unused in boxes and other cabinets is a very common problem. A knockout opening in a box is sometimes covered with tape, which is not equivalent to the wall of the box. This procedure does not comply with the code. Unused openings must be closed with appropriate materials for the protection of the circuit conductors and devices as well as for protection against shock by someone inserting a finger or tool into the enclosure. Listed knockout plugs are available for this purpose.

This section is not intended to be applicable to holes provided for fasteners for mounting purposes, cooling slots and similar openings that are part of the design of the listed equipment. In the context of this section, unused opening are openings that are drilled, punched or "knocked out" of the enclosure.

E3404.7 Integrity of electrical equipment. Internal parts of electrical equipment, including busbars, wiring terminals, insulators and other surfaces, shall not be damaged or contaminated by foreign materials such as paint, plaster, cleaners or abrasives, and corrosive residues. There shall not be any damaged parts that might adversely affect safe operation or mechanical strength of the equipment such as parts that are broken; bent; cut; deteriorated by corrosion, chemical action, or overheating. Foreign debris shall be removed from equipment. [110.12(B)]

❖ After installation, electrical materials must be protected. A panelboard installed during the rough wiring phase of the job could result in having the bus bars coated with paint while the walls were being spray painted. Thus, the panel must be covered during construction. Also, the inspector should watch for a router tool being used to cut openings for the electrical boxes after the drywall is hung. In many cases, the openings for the electrical boxes are not premeasured and cut before the wallboard is installed. First the drywall or wallboard is placed on the stud wall, then a router is used to cut the openings where a switch or receptacle box has been installed behind the wallboard. This results in damage to the conductors in the box and can also damage the box; it is a clear violation of the code. Often the conductors are cut and the insulation is scraped from the conductors. Later, after the walls are

finished and painted, it is difficult to repair the wiring within the outlet boxes. It is often thought that code violations in the electrical wiring system are caused only by the electrician as a result of improper installation or improper wiring methods. But the electrical system can be installed in total compliance with code requirements and later be found to violate the code through improper construction practices. Correlation between all construction trades is needed to prevent these types of problems and is required by the code.

E3404.8 Mounting. Electrical equipment shall be firmly secured to the surface on which it is mounted. Wooden plugs driven into masonry, concrete, plaster, or similar materials shall not be used. [110.13(A)]

❖ Equipment must be mounted securely. This provision is an example of a code requirement that is not exactly prescriptive, meaning that it does not spell out such things as the weight of the equipment or the number and size of fasteners. It is more of a performance requirement in that it simply requires that the equipment be mounted securely so that it will not come loose, displace or move when utilized. For example, on a brick house the service equipment is often installed on the surface of the brick wall. Holes lined up with the mounting holes in the back of the meter base or panelboard are drilled into the brick or mortar. If wooden plugs are used, the panelboard may look secure at the time of installation, but the wooden plugs will deteriorate, and the equipment will eventually sag or fall off the wall. Wooden plugs can decay and shrink, causing fasteners to pull out. Proper masonry anchors and fasteners must be used.

For relatively lightweight boxes and enclosures, plastic or metal plugs with lag screws can work well. The weight of the equipment should be taken into account. In some cases, it may be necessary to install additional supports such as lag screws that attach to the framing members. Where service equipment is supplied by a service lateral, conduit is often used from below the ground up to the bottom of the service equipment. This conduit must be securely attached to the house. In new housing developments where the area next to the foundation has been excavated to place the footing and foundation and then backfilled, the ground may settle over the first several months and can pull the conduit containing the service-lateral conductors out of the service equipment. This can easily happen where, for example, a heavy 2-inch (51 mm) rigid metal conduit with a 90-degree (1.57 rad) elbow underground is installed into a connector through concentric knockouts at the bottom of the outdoor panelboard. A one-hole strap attached with a plastic screw anchor may not be enough to prevent the conduit from pulling away from the equipment. This code requirement is not only for service equipment but pertains to all electrical equipment. Sections E4004.5 and E4101.6 cover the provisions for mounting boxes for lighting fixtures weighing over 50 pounds (22.7 kg) and boxes for fans.

E3404.9 Energized parts guarded against accidental contact. Approved enclosures shall guard energized parts that are operating at 50 volts or more against accidental contact. [110.27(A)]

❖ Any electrical circuit of 50 volts or more must have its components enclosed. A doorbell circuit transformer has the line terminals or leads that connect to a 120-volt circuit enclosed in a box, but the load terminals, that supply less than 50 volts (typically 24 volts), are permitted to be in the open.

E3404.10 Prevent physical damage. In locations where electrical equipment is likely to be exposed to physical damage, enclosures or guards shall be so arranged and of such strength as to prevent such damage. [110.27(B)]

❖ Locations where electrical equipment is likely to be exposed to physical damage are determined by the code official in the enforcement of this code. These locations are not spelled out, but many are obvious, such as on the outside of the dwelling where a vehicle or a lawn mower could possibly hit or damage a box or other wiring or equipment. Guards or enclosures may be required in such locations.

E3404.11 Equipment identification. The manufacturer's name, trademark or other descriptive marking by which the organization responsible for the product can be identified shall be placed on all electric equipment. Other markings shall be provided that indicate voltage, current, wattage or other ratings as specified elsewhere in Chapters 34 through 43. The marking shall have the durability to withstand the environment involved. [110.21(A)]

❖ In addition to the name of the manufacturer, markings must be provided on electrical equipment regarding volt-age, current, wattage and other information necessary to define the correct use of such equipment. The equipment must be identified so that it can be determined if it is being used in compliance with code provisions and according to its listing requirements.

E3404.12 Field-applied hazard markings. Where caution, warning, or danger signs or labels are required by this code, the labels shall meet the following requirements:

1. The marking shall adequately warn of the hazard using effective words, colors, or symbols or combinations of such.

2. Labels shall be permanently affixed to the equipment or wiring method.

3. Labels shall not be hand written except for portions of labels or markings that are variable, or that could be subject to changes. Labels shall be legible.

4. Labels shall be of sufficient durability to withstand the environment involved. [110.21(B)]

❖ Obviously, the code does not intend to allow poorly constructed signs and labels, such as those of duct tape and marker pens or marker pens alone. The required markings must use universally understood pictograms, symbols or colors; or must use words that clearly convey the message. Any effective combina-tion of words, symbols, colors, pictograms, etc., are allowed. Labels must be suitable for the environment in which they are located so that they will continue to serve their purpose for the life of the equipment. Hand written labels are not allowed because they are typically illegible and because they lack an authoritative and professional appearance. Caution, warning and danger signs and labels are life safety protection features that must remain in place and convey their messages for as long as the equipment, system, wiring, etc., remains in service.

E3404.13 Identification of disconnecting means. Each disconnecting means shall be legibly marked to indicate its purpose, except where located and arranged so that the purpose is evident. The marking shall have the durability to withstand the environment involved. [110.22(A)]

❖ The main disconnecting means for the dwelling must be marked "Main Disconnect." In the case of a fire, fire department personnel must be able to determine whether the disconnect they are switching off is the main disconnect or not. In outdoor panels especially the marking may fade with time, so the code official may consider a requirement for a durable type of marking. Included in this section is the requirement to label the branch circuit panelboard (see also commentary, Section E3706.2).

Branch circuits must be legibly marked or identified. In many cases, panelboards are marked with illegible handwriting or may have circuits marked with only a simple description. The code does not spell out how detailed the marking must be; that is, the code does not specify whether the marking must indicate the specific rooms a lighting circuit serves or only that it is a lighting circuit. These would be determined by the code official in the enforcement of the code.

SECTION E3405
EQUIPMENT LOCATION AND CLEARANCES

E3405.1 Working space and clearances. Access and working space shall be provided and maintained around all electrical equipment to permit ready and safe operation and maintenance of such equipment in accordance with this section and Figure E3405.1. (110.26)

❖ Enough space should be provided to work on electrical equipment without jeopardizing the safety of the worker. It is not left to the arbitrary decision of the electrician or architect to state how much clearance is needed; the code is very specific with regard to the minimum clearances required for electrical equipment. More clearance is always better.

The code does not simply state that sufficient space must be provided. Figure E3405.1 provides a very complete picture of the basic requirements for location and clearances for panelboards. This also includes the requirements for electrical equipment located outdoors such as the meter base and main disconnect and air conditioner disconnect.

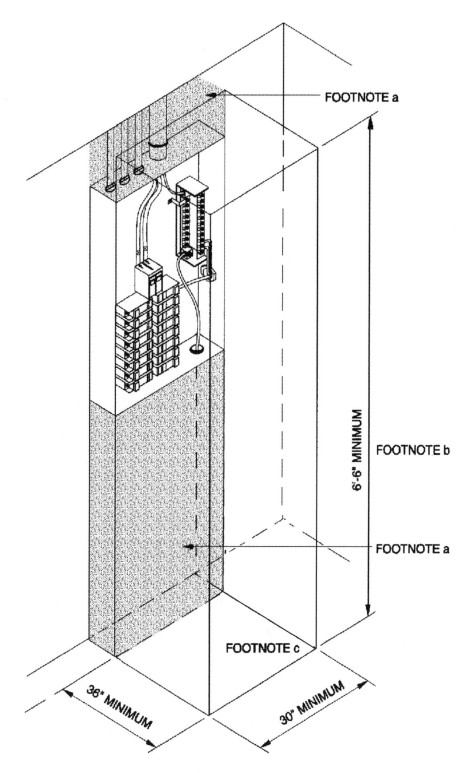

For SI: 1 inch = 25.4 mm, 1 foot = 304.8 mm.

a. Equipment, piping and ducts foreign to the electrical installation shall not be placed in the shaded areas extending from the floor to a height of 6 feet above the panelboard enclosure, or to the structural ceiling, whichever is lower.

b. The working space shall be clear and unobstructed from the floor to a height of 6.5 feet or the height of the equipment, whichever is greater.

c. The working space shall not be designated for storage.

d. Panelboards, service equipment and similar enclosures shall not be located in bathrooms, toilet rooms, clothes closets or over the steps of a stairway.

e. Such work spaces shall be provided with artificial lighting where located indoors and shall not be controlled by automatic means only.

FIGURE E3405.1[a, b, c, d, e]
WORKING SPACE AND CLEARANCES

E3405.2 Working clearances for energized equipment and panelboards. Except as otherwise specified in Chapters 34 through 43, the dimension of the working space in the direction of access to panelboards and live parts of other equipment likely to require examination, adjustment, servicing or maintenance while energized shall be not less than 36 inches (914 mm) in depth. Distances shall be measured from the energized parts where such parts are exposed or from the enclosure front or opening where such parts are enclosed. In addition to the 36-inch dimension (914 mm), the work space shall not be less than 30 inches (762 mm) wide in front of the electrical equipment and not less than the width of such equipment. The work space shall be clear and shall extend from the floor or platform to a height of 6.5 feet (1981 mm) or the height of the equipment, whichever is greater. In all cases, the work space shall allow at least a 90-degree (1.57 rad) opening of equipment doors or hinged panels. Equipment associated with the electrical installation located above or below the electrical equipment shall be permitted to extend not more than 6 inches (152 mm) beyond the front of the electrical equipment. [110.26(A) (1), (2), (3)]

Exceptions:

1. In existing dwelling units, service equipment and panelboards that are not rated in excess of 200 amperes shall be permitted in spaces where the height of the working space is less than 6.5 feet (1981 mm). [110.26(A)(3) Exception No. 1]

2. Meters that are installed in meter sockets shall be permitted to extend beyond the other equipment. Meter sockets shall not be exempt from the requirements of this section. [110.26(A)(3) Exception No. 2]

❖ When a worker is standing in front of a panelboard or air conditioner disconnect switch, for example, he or she should be able to walk up to it and stand in a clear space with no obstructions without having to lean or climb over another piece of electrical equipment or other obstruction. The last sentence of this section states that "equipment associated with the electrical installation located above or below the electrical equipment shall be permitted to extend not more that 6 inches (152 mm) beyond the front of the electrical equipment." At first reading, this may be difficult to understand if taken out of context. The working space in front of any electrical equipment must be clear from the floor or ground to at least 6.5 feet (1981 mm) (see Exception 1) above the floor or ground. Equipment that extends more than 6 inches (152 mm) past the front of other equipment is considered to protrude into the required clear working space (see Exception 2).

On residential properties, such things as the locations of a lawn sprinkler valve box that sits above the surface of the ground, a telephone connection box, a junction box, a subpanel or a current transformer enclosure should be carefully considered when locating the panelboard and service equipment. If these items extend more than 6 inches (152 mm) farther out than the front of the adjacent equipment, they are not permitted in the 36-inch (914 mm) depth or 30-inch (762 mm) width space in front of the panelboard or service equipment.

E3405.3 Indoor dedicated panelboard space. The indoor space equal to the width and depth of the panelboard and extending from the floor to a height of 6 feet (1829 mm) above the panelboard, or to the structural ceiling, whichever is lower, shall be dedicated to the electrical installation. Piping, ducts, leak protection apparatus and other equipment foreign to the electrical installation shall not be installed in such dedicated space. The area above the dedicated space shall be permitted to contain foreign systems, provided that protection is installed to avoid damage to the electrical equipment from condensation, leaks and breaks in such foreign systems (see Figure E3405.1).

Exception: Suspended ceilings with removable panels shall be permitted within the 6-foot (1829 mm) dedicated space.

❖ This section is specific to indoor spaces and Section E3405.4 addresses outdoor spaces. A dedicated space above and below the panelboard equal to the depth and width of the panelboard must be maintained clear of all ducts, piping and other foreign items. In basements, the heating and air conditioning ducts are often run against a wall and the ceiling. To locate the panelboard on a wall with a duct installed above and tight to the wall is a code violation, unless the duct is 6 feet (1829 mm) or more above the top of the panelboard. That is not likely in most dwellings. Many times, the electrician arrives at the job site first and installs the panelboard and the wiring, and later the plumbing and heating, ventilation and air-conditioning (HVAC) contractor installs the piping and ducts. This may seem to be acceptable if the wiring is installed first; however, no such provision is in the code. It is necessary for the electrician to know the location of the ducts and pipes before they are installed or for the installer of the ducts and pipes to avoid the required clear space above the panelboard (see Figure E3405.1).

E3405.4 Outdoor dedicated panelboard space. The outdoor space equal to the width and depth of the panelboard, and extending from grade to a height of 6 feet (1.8 m) above the panelboard, shall be dedicated to the electrical installation. Piping and other equipment foreign to the electrical installation shall not be located in this zone.

❖ See the commentary to Section E3405.3.

E3405.5 Location of working spaces and equipment. Required working space shall not be designated for storage. Panelboards and overcurrent protection devices shall not be located in clothes closets, in bathrooms, or over the steps of a stairway. [110.26(B), 240.24(D), (E), (F)]

❖ This section specifies that the clear working space required in front of electrical equipment not be in a storage area and that OPDs and panelboards must not be located in clothes closets or bathrooms.

Closets are obviously designed for storage, and combustible material in a closet could be stored close to or in contact with the panelboard. To avoid the possibility of easily ignitable material being placed next to a panelboard and to avoid the storage of materials in front of a panelboard, thus blocking access to it, overcurrent protection devices must not be installed in clothes closets. Although the code specifically prohibits the panelboard in a clothes closet, it seems wise to not locate a panelboard in any closet, because no one knows what the occupants of the dwelling will store there.

OPDs in a bathroom are a problem because of the location of plumbing fixtures and moisture present in the room.

The required clear space cannot be part of an area designated for storage. One common problem occurs where a panelboard or other equipment is installed on the wall of a garage. When the house is being built and there are no shelves along the wall of the garage, it is easy to put the panel on that wall. Later, a workbench is built under the panelboard, or shelves are installed all around it. Although this is not a code violation at the time of the panelboard installation, it can be a major problem later and becomes a code violation, even though the house passed inspection initially. It is difficult to foresee that a home owner may install a cabinet over a panelboard in the garage, cutting out the back of the cabinet to apparently allow access to the panel. Care should be taken by the electrician to select a location for the panelboard that will not likely be used as a storage area later.

Locating equipment above stair steps is an obvious fall hazard to occupants and service personnel who would have to access such equipment. The loss of one's footing could also result in a shock hazard when working on the equipment.

E3405.6 Access and entrance to working space. Access shall be provided to the required working space. [110.26(C)(1)]

❖ The minimum working space is defined, but access dimensions such as the specific dimensions of a door or entrance to the working space are not given. The access to the working space must be sufficient.

E3405.7 Illumination. Artificial illumination shall be provided for all working spaces for service equipment and panelboards installed indoors and shall not be controlled by automatic means only. Additional lighting outlets shall not be required where the work space is illuminated by an adjacent light source or as permitted by Exception 1 of Section E3903.2 for switched receptacles. [110.26(D)]

❖ Although the code does not specify how close to the equipment the light source must be installed, it must provide enough light to work safely on the equipment. If the lighting was controlled by automatic means only, such as occupant sensors, the lighting might turn off unexpectedly or might not turn on at all if the control device is inoperative.

SECTION E3406
ELECTRICAL CONDUCTORS AND CONNECTIONS

E3406.1 General. This section provides general requirements for conductors, connections and splices. These requirements do not apply to conductors that form an integral part of equipment, such as motors, appliances and similar equip-

ment, or to conductors specifically provided for elsewhere in Chapters 34 through 43. (310.1)

❖ This section deals specifically with residential wiring.

E3406.2 Conductor material. Conductors used to conduct current shall be of copper except as otherwise provided in Chapters 34 through 43. Where the conductor material is not specified, the material and the sizes given in these chapters shall apply to copper conductors. Where other materials are used, the conductor sizes shall be changed accordingly. (110.5)

❖ Copper conductors are used almost exclusively for branch circuits of 15 and 20 amperes. Nonmetallic sheathed cable with aluminum conductors was used in the 1970s in some regions for these smaller circuits in residential wiring, but now it is not readily available. Aluminum conductors, even in these smaller sizes, are not prohibited by the code. Table E3705.1, which lists the allowable ampacities for conductors, begins with size 12 AWG for aluminum conductors. For repair and remodeling work, the ampacities are included for these smaller aluminum conductors. For larger conductors such as for services and feeders, aluminum is commonly used.

E3406.3 Minimum size of conductors. The minimum size of conductors for feeders and branch circuits shall be 14 AWG copper and 12 AWG aluminum. The minimum size of service conductors shall be as specified in Chapter 36. The minimum size of Class 2 remote control, signaling and power-limited circuits conductors shall be as specified in Chapter 43. [310.106(A)]

❖ The absolute minimum size of conductors is listed here, but for the minimum ampacities or allowable ampacities, refer to Table E3705.1. For example, in that table, size 12 AWG aluminum is listed as having an allowable ampacity of 20 amperes but is limited to a 15 ampere circuit by Section E3705.5.3.

E3406.4 Stranded conductors. Where installed in raceways, conductors 8 AWG and larger shall be stranded. A solid 8 AWG conductor shall be permitted to be installed in a raceway only to meet the requirements of Sections E3610.2 and E4204. [310.106(C)]

❖ This requirement limits the size of solid conductors installed in raceways to 10 AWG and smaller. Solid conductors of 8 AWG and larger are usually very difficult to pull into conduit or tubing and may be damaged in the process. Section E4204 requires a minimum 8 AWG solid conductor to be used to bond all the metallic parts of a swimming pool. Where installed in a raceway, the raceway should be sized to allow sufficient room to hold this conductor alone. An exception to this section is the case where a single solid grounding electrode conductor is pulled into a conduit sleeve for physical protection. For example, a solid number 8 AWG grounding electrode conductor could be installed in conduit. Table E3603.4, Note d requires No. 8 grounding electrode conductors to be protected by means such as metal or nonmetallic conduit. The reference to Section E3610.2 could imply that larger solid grounding electrode conductors could not be installed in a raceway, however, installing a solid No. 6 or 4 copper conductor in a short length of $^1/_2$-inch (12 mm) or $^3/_4$-inch (19 mm) conduit with no bends is common practice and would not likely damage the conductor, especially considering that it is not required to be insulated.

E3406.5 Individual conductor insulation. Except where otherwise permitted in Sections E3605.1 and E3908.9, and E4303, current-carrying conductors shall be insulated. Insulated conductors shall have insulation types identified as RHH, RHW, RHW-2, THHN, THHW, THW, THW-2, THWN, THWN-2, TW, UF, USE, USE-2, XHHW or XHHW-2. Insulation types shall be approved for the application. [310.106(C), 310.104]

❖ The common specific conductor insulation types listed are the conductors included in Table E3705.1. The provisions of this code cover the installations in which these conductor types would be used in residential wiring. Conductor insulation types have different temperature ratings, indicating the temperature the conductor insulation can withstand before being damaged. Wherever conductors are used in wet locations, such as service entrance conductors that are exposed at the service head, the conductors must have a "W" in the insulation type. The following are letters in the conductor insulation types and the corresponding use or meaning:

R — Rubber

T — Thermoplastic

H — Rated for a 15°C (27° F) temperature rise over the 60°C (140° F) rating

HH — Rated for a 30°C (54° F) temperature rise over the 60°C (140° F) rating

N — Outer nylon jacket or equivalent

W — Rated for wet location

X — Cross-linked synthetic polymer

E3406.6 Conductors in parallel. Circuit conductors that are connected in parallel shall be limited to sizes 1/0 AWG and larger. Conductors in parallel shall: be of the same length; consist of the same conductor material; be the same circular mil area and have the same insulation type. Conductors in parallel shall be terminated in the same manner. Where run in separate raceways or cables, the raceway or cables shall have the same physical characteristics. Where conductors are in separate raceways or cables, the same number of conductors shall be used in each raceway or cable. [310.10(H)]

❖ Parallel conductors are typically installed in conduit or raceways, and for one- and two-family dwellings within the scope of this code (dwellings with a service size of up to 400 amperes), only very large circuits or the service or feeder conductors would likely be run in parallel. Although permitted for conductors of 1/0 AWG and above, paralleling conductors in residential wiring would normally be used only for service conductors. The size of the raceway or cable, labor involved and

the cost of the conductors are major factors in the decision to use paralleled conductors.

E3406.7 Conductors of the same circuit. All conductors of the same circuit and, where used, the grounded conductor and all equipment grounding conductors and bonding conductors shall be contained within the same raceway, cable or cord. [300.3(B)]

❖ This provision requires that all conductors of a circuit be in the same raceway, cable or cord to prevent magnetic fields from creating heat in metal enclosures and raceways.

E3406.8 Aluminum and copper connections. Terminals and splicing connectors shall be identified for the material of the conductors joined. Conductors of dissimilar metals shall not be joined in a terminal or splicing connector where physical contact occurs between dissimilar conductors such as copper and aluminum, copper and copper-clad aluminum, or aluminum and copper-clad aluminum, except where the device is listed for the purpose and conditions of application. Materials such as inhibitors and compounds shall be suitable for the application and shall be of a type that will not adversely affect the conductors, installation or equipment. (110.14)

❖ Copper and aluminum conductors must not be joined unless the device or connector is listed for the purpose. Some wire nut connectors are listed for this purpose. If the connector is marked "AL/CU," aluminum and copper conductors could be connected together in the same pressure connector. It is very important to use an approved oxide-inhibiting compound on aluminum conductors. Where aluminum conductors are exposed to the atmosphere, oxidation can occur. Oxidation of aluminum can cause a thin film or layer, which looks like a powder, on the conductor. It will result in heat buildup by impeding current flow. Wire connectors that are listed and approved for connection of aluminum to aluminum or for copper to aluminum conductors are available. Most wire nuts are not specifically listed for use with aluminum conductors, and one cannot assume that they can be used. If they are identified for this application, the container will so indicate with a marking such as "AL/CU" and may indicate that an approved oxide inhibitor should be used where aluminum conductors are connected. Many of these approved wire nuts are made with a specially formulated corrosion-resistant compound to help prevent aluminum corrosion. There is a need for these connectors, for example where an old house wired with aluminum is being remodeled or a room is being added. If any of the old aluminum conductors are spliced, the installer should be careful to use approved connectors. Where connections of larger conductors are made, for example feeder or service conductors, aluminum is commonly used. Again, there may be a need in a retrofit job to make connections between the existing aluminum and copper conductors. Split bolts connectors may be used for this connection. These connectors will have a separator between the two openings where the wire is inserted which functions to prevent contact between the conductors. The bolts must have markings on them such as "AL/CU," indicating they may be used for aluminum-to-copper connections.

E3406.9 Fine stranded conductors. Connectors and terminals for conductors that are more finely stranded than Class B and Class C stranding as shown in Table E3406.9, shall be identified for the specific conductor class or classes. (110.14)

❖ Conductors are more finely stranded as the number of strands increases and the diameter of each strand decreases. Stranded conductors are typically used only where flexibility is desired and are not commonly found in residential construction. The finer the strands, the more difficult it is to make good terminations and splices. The device, wire connector or terminal being used with stranded wire must be suitable for the application. Some devices and terminals state that they are for use with solid wire only and some devices or terminals might not be suitable for use with very finely stranded wire. Other than what is contained in the product listing information, the manufacturer's instructions and on product markings, the code is silent on the proper termination of stranded wire with devices such as snap switches and receptacles. Most workmen agree that crimp-on spade terminals should be used or the device should be of the "backwired" type, which employs a screw to compress the stranded conductor between two pressure plates. Stranded wire tends to deform and "spread out" when placed under the head of a wire-binding screw, creating a questionable termination. There are tricks of the trade, such as reverse twisting and creative wire stripping techniques, that are used to make "side-wiring" with wire-binding screws

more viable, but choosing a more suitable type of device or using spade terminals is believed to be the better way. Of course, there is always the option of splicing the stranded conductor to a solid conductor pigtail.

E3406.10 Terminals. Connection of conductors to terminal parts shall be made without damaging the conductors and shall be made by means of pressure connectors, including set-screw type, by means of splices to flexible leads, or for conductor sizes of 10 AWG and smaller, by means of wire binding screws or studs and nuts having upturned lugs or the equivalent. Terminals for more than one conductor and terminals for connecting aluminum conductors shall be identified for the application. [110.14(A)]

❖ Connections must be made without damaging the conductors. When stripping the insulation from the ends of conductors, it is easy to nick or cut into the wire. Using improper techniques or tools when stripping the conductors may damage the softer aluminum conductors.

In many installations, two or more conductors are connected under one terminal because it is easy to install them and they will fit. However, terminals for connecting multiple conductors must be listed for this purpose. Most modern equipment such as panelboards has terminal strips with a sufficient number of terminals to connect all necessary circuit conductors. In some cases, it may be necessary to install additional terminal strips to terminate each conductor under a separate terminal. If the terminals are approved for the connection of more than one conductor, the approval will be stated in the listing information or supplied with the packaging of the equipment.

TABLE E3406.9 (Chapter 9, Table 10)
CONDUCTOR STRANDING[c]

CONDUCTOR SIZE		NUMBER OF STRANDS		
		Copper		Aluminum
AWG or kcmil	mm²	Class B	Class C	Class B
24-30	0.20-0.05	a	—	—
22	0.32	7	—	—
20	0.52	10	—	—
18	0.82	16	—	—
16	1.3	26	—	—
14-2	2.1-33.6	7	19	7[b]
1-4/0	42.4-107	19	37	19
250-500	127-253	37	61	37
600-1000	304-508	61	91	61
1250-1500	635-759	91	127	91
1750-2000	886-1016	127	271	127

a. Number of strands vary.

b. Aluminum 14 AWG (2.1 mm²) is not available.

c. With the permission of Underwriters Laboratories, Inc., this material is reproduced from UL Standard 486A-B, Wire Connectors, which is copyrighted by Underwriters Laboratories, Inc., Northbrook, Illinois. While use of this material has been authorized, UL shall not be responsible for the manner in which the information is presented, nor for any interpretations thereof.

E3406.11 Splices. Conductors shall be spliced or joined with splicing devices listed for the purpose. Splices and joints and the free ends of conductors shall be covered with an insulation equivalent to that of the conductors or with an insulating device listed for the purpose. Wire connectors or splicing means installed on conductors for direct burial shall be listed for such use. [110.14(B)]

❖ Conductors must be spliced with devices listed for the purpose. The points in the electrical wiring system where conductors are spliced or connected are the weakest links. Here problems could occur such as deterioration of the conductor, overheating, arcing, breaking, or opening of the connection. Therefore, it is very important that conductors be properly spliced. Splicing and terminating conductors in high-voltage wiring, for example, is a skill that must be acquired and developed. But even in low-voltage wiring, the splicing of small conductors is a skill and critical to the proper operation of the system. Bad connections are the source of many of the problems in an electrical wiring system, including fires.

One unapproved method of making a connection is to rely only on twisting the stripped ends of the conductors together using pliers and then using tape to cover the connection without any approved device. Wire nuts or pressure connectors of the appropriate size and type must be used. Some installers claim that the stripped ends of conductors should be securely twisted together with pliers before applying the wire nut, even though the wire nut manufacturers typically say the stripped wire ends should be held together and then twisted together using only the wire nut. The installer should consult the instructions for how to install the particular type of wire connector to be used. The instructions on the container will state, for example, "Pretwisting unnecessary. Hold the stripped wires together with the ends even." Most wire nuts or pressure connectors that are sold for use in residential wiring are for connecting copper wires only, and the container will have such a message on it. See the commentary to Section E3406.8 for details about joining aluminum and copper conductors.

Where a split bolt or clamp-on pressure connector is used, the code requires that the splice be covered with an insulating material equivalent to the conductor insulation. Some of these bolts are supplied with an approved plastic or insulating cover that snaps into place around the splice, but many are bare, and the electrician must cover the connection with splicing compounds and tape. If the insulating tape is listed for the purpose, this could be an approved method if the thickness of the covering is equivalent to that of the conductor insulation.

E3406.11.1 Continuity. Conductors in raceways shall be continuous between outlets, boxes, and devices and shall be without splices or taps in the raceway.

Exception: Splices shall be permitted within surface-mounted raceways that have a removable cover. [300.13(A)]

❖ Most raceways have neither the room nor the means of access to allow splices to occur within them. Splices in raceways can cause overfill of the cross-sectional area and would prevent the conductors from being withdrawn or the addition of new conductors.

E3406.11.2 Device connections. The continuity of a grounded conductor in multiwire branch circuits shall not be dependent on connection to devices such as receptacles and lampholders. The arrangement of grounding connections shall be such that the disconnection or the removal of a receptacle, luminaire or other device fed from the box does not interfere with or interrupt the grounding continuity. [300.13(B)]

❖ The grounded conductor, which is commonly referred to in the field as the neutral conductor, is often connected or spliced only by means of the receptacle in an outlet box. This is commonly done where two 2-wire or 3-wire cables are brought into an outlet box for the installation of a 120-volt duplex receptacle, and the two grounded conductors are connected only to the receptacle and are not tied together before being connected to the receptacle. A multiwire circuit has two ungrounded conductors (usually one red and one black) with a voltage of 240 between them and a grounded (neutral) conductor (white) with a voltage of 120 volts between it and each of the ungrounded conductors. When the receptacle is replaced or one of the wires is disconnected, an open neutral could result. If this is done when the circuit is energized, it could cause abnormal voltages (both exceedingly high and low) to be imposed on any appliances or equipment connected to the circuit (see Commentary Figure E3406.11.2). Therefore, the code requires that the white grounded conductors be connected together and then connected with a short jumper to the receptacle (pigtailed). This way, the continuity of the grounded conductor is maintained at all times even when the device is removed.

Where multiple equipment grounding conductors are brought into a junction box for the connection of devices, they must be connected together. If a device or luminaire is installed in or on the box, a pigtail must be used so that only one conductor connects to the grounding terminal or lead of the device or luminaire. This is to maintain continuity of the grounding conductor and the protection it provides.

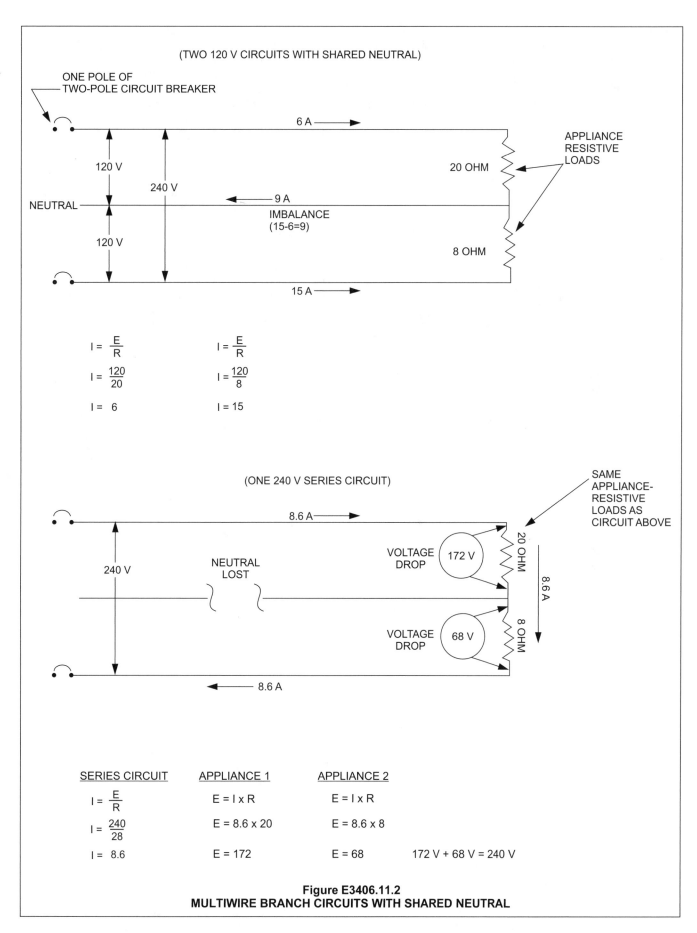

Figure E3406.11.2
MULTIWIRE BRANCH CIRCUITS WITH SHARED NEUTRAL

E3406.11.3 Length of conductor for splice or termination.
Where conductors are to be spliced, terminated or connected to fixtures or devices, a minimum length of 6 inches (152 mm) of free conductor shall be provided at each outlet, junction or switch point. The required length shall be measured from the point in the box where the conductor emerges from its raceway or cable sheath. Where the opening to an outlet, junction or switch point is less than 8 inches (200 mm) in any dimension, each conductor shall be long enough to extend at least 3 inches (75 mm) outside of such opening. (300.14)

❖ This code provision is prescriptive and leaves little room for misinterpretation. It is very difficult to make splices and device terminations while working inside a device box on very short wires. In many cases, especially when remodeling on older houses, it is found that existing wires have been cut or installed very short. When an attempt is made to remove the switch or receptacle, it cannot be pulled out and turned so that the electrician might see or work on the connections. These device boxes are often smaller than would normally be used in modern wiring, which adds to the difficulty. The alternative can also be a problem. If too much wire is left at the outlet, it is difficult to put the extra length back into the box when installing the device. Although the requirement is very precise, care must be taken to tuck the conductors into the outlet box when the switch or receptacle is installed. If the box is too small, the wires could be damaged when they are pushed back into the box to make room for the device. Minimum box sizes are specified in Section E3905, but consideration should be given to the size of the wiring device to be installed. Some devices, such as fan control switches, dimmers and ground-fault circuit interrupter receptacles, take up more room in the box than ordinary switches or receptacles. Therefore, it may be necessary to size the box larger than the minimum required to have room to tuck the required conductor length into the box without damaging the conductors. For junction boxes with an opening of at least 8 inches (203 mm) in any of its dimensions, the requirement is for only 6 inches (152 mm) of free conductor from the point in the box where the conductor emerges from the raceway or cable sheath. It may be easier for electricians to reach into the box and work on conductors in a junction box with openings of at least 8 inches (203 mm). For all other boxes, the conductor must be long enough to extend at least 3 inches (76 mm) beyond the front edge of the box. This provision does not override the requirement for a 6-inch (152 mm) length of free conductor (see Commentary Figure E3406.11.3).

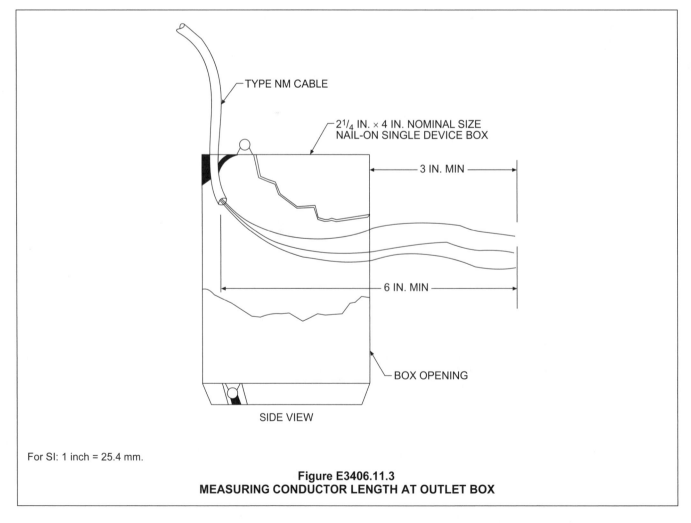

For SI: 1 inch = 25.4 mm.

Figure E3406.11.3
MEASURING CONDUCTOR LENGTH AT OUTLET BOX

E3406.12 Grounded conductor continuity. The continuity of a grounded conductor shall not depend on connection to a metallic enclosure, raceway or cable armor. [200.2(B)]

❖ The grounded conductor, not to be confused with a grounding conductor, is a current carrying conductor and can never utilize sheet metal, conduits or metal armors as any portion of such conductor. Recall that the only place that the neutral (grounded) conductor is bonded to a grounded metal enclosure or raceway is within the service equipment, never downstream of the service equipment. For example, a neutral or grounded conductor can not be terminated on a lug bolted to a service panelboard enclosure because the connection of the conductor to the service neutral terminal bus would depend upon the sheet steel of the enclosure to complete the circuit. Even though the service enclosure is bonded to the neutral conductor, the enclosure must not be utilized as a normal current carrying component for any circuit.

E3406.13 Connection of grounding and bonding equipment. The connection of equipment grounding conductors, grounding electrode conductors and bonding jumpers shall be in accordance with Sections E3406.13.1 and E3406.13.2.

❖ See the commentary to Sections E3406.13.1 and E3406.13.2.

E3406.13.1 Permitted methods. Equipment grounding conductors, grounding electrode conductors, and bonding jumpers shall be connected by one or more of the following means:

1. Listed pressure connectors.

2. Terminal bars.

3. Pressure connectors listed as grounding and bonding equipment.

4. Exothermic welding process.

5. Machine screw-type fasteners that engage not less than two threads or are secured with a nut.

6. Thread-forming machine screws that engage not less than two threads in the enclosure.

7. Connections that are part of a listed assembly.

8. Other listed means. [250.8 (A)]

❖ Item 1 refers to devices such as set screw connectors, split-bolt connectors, clamp-type connectors and compression connectors.

Item 2 refers to terminals bars that are most likely part of the listed electrical equipment being grounded or bonded.

Item 3 refers to devices that are listed for the specific purpose of making grounding and bonding connections such as grounding rod connectors and connectors for bonding metal pipes and reinforcement steel bars.

Item 4 refers to welding kits that include a graphite crucible mold and a thermite powder composed of powdered aluminum and copper oxide along with an ignition powder such as magnesium. When ignited and combusted, the powder yields aluminum oxide and

pure molten copper. The molten copper drops into the mold and welds the conductors together. Exothermic means that the process yields more heat energy than it consumes, thus an external heat source is not needed for such welding.

Item 5 refers to machine screws threaded into factory-provided threaded openings or machine screws with threaded nuts. Sheet metal screws are NOT allowed.

Item 6 refers to machine screws that act as taps and cut threads in the factory-supplied opening in which they are installed. Such screws are part of listed boxes or panelboard enclosures. Sheet metal screws are NOT allowed.

Item 7 refers to any type of connection that is part of an electrical system component, where such component is listed and labeled for the purpose. See Section E3404.3.

Item 8 is a catch-all category for any means of connection that is specifically listed for the application.

E3406.13.2 Methods not permitted. Connection devices or fittings that depend solely on solder shall not be used. [250.8 (B)]

❖ Connections that completely depend on solder to maintain contact with the conductor(s) are prohibited. Solder has a low melting temperature and the joint could come apart under fault conditions. Note that this section does not prohibit solder joints where the conductors are mechanically joined and then soldered. When solder joints were common, the wires were twisted tightly around each other and then soldered. Solder joints are very rare today because of the potential of heat to damage conductor insulation and because they are labor intensive.

SECTION E3407
CONDUCTOR AND TERMINAL IDENTIFICATION

E3407.1 Grounded conductors. Insulated grounded conductors of sizes 6 AWG or smaller shall be identified by a continuous white or gray outer finish or by three continuous white or gray stripes on other than green insulation along the entire length of the conductors. Conductors of sizes 4 AWG or larger shall be identified either by a continuous white or gray outer finish or by three continuous white or gray stripes on other than green insulation along its entire length or at the time of installation by a distinctive white or gray marking at its terminations. This marking shall encircle the conductor or insulation. [200.6(A) & (B)]

❖ Grounded conductors are commonly referred to as neutral conductors. The correct terminology should be used to identify conductors, namely: grounded, grounding and ungrounded. Conductors with green insulation are prohibited from being used as grounded or neutral circuit conductors; white or gray insulation is always used for the grounded conductors. The provision for using a color other than green with three white stripes as a grounded conductor is included for cases where equipment, lighting fixtures or flexible cords may contain conductors with this coloring. One of the conductors of flexible cord and fixture wire must be identified as the grounded conductor.

For grounded conductors up to and including size 6 AWG, the white insulation, gray insulation or three white stripes must be continuous for the length of the grounded conductor. For 4 AWG and larger, any color conductor insulation other than green may be used for the grounded conductor as long as it is identified at the terminations by white markings at the time of installation. The most common method of marking the grounded conductors of colors other than white or gray is the use of white tape, although white paint that would remain permanently on the conductor could be used. Some paints could react with the insulation, causing damage and such paints should be avoided for marking. At junction and outlet boxes, cabinets and panelboards where grounded conductors are identified with white tape, there is no requirement that the grounded conductor be taped or marked white for the entire length of the exposed conductor. Marking the entire length might be assumed and might help to make the grounded conductor more readily identifiable at these locations. However, the length of the white marking at the termination locations is not defined; the only requirement is that the marking encircle the conductor.

In the wiring of one- and two-family dwellings, the installation of grounded conductors with the wrong color of insulation is not common because the smaller conductors used in residential wiring are either part of a multiconductor cable assembly that is manufactured with the proper color identification or the conductors are required to be the proper color because of their size. This is typically an issue for service and feeder conductors only. Using the white conductor of a multiconductor cable as an ungrounded conductor is permitted only under the exceptions to Section E3407.3.

E3407.2 Equipment grounding conductors. Equipment grounding conductors of sizes 6 AWG and smaller shall be identified by a continuous green color or a continuous green color with one or more yellow stripes on the insulation or covering, except where bare. Conductors with insulation or individual covering that is green, green with one or more yellow stripes, or otherwise identified as permitted by this section shall not be used for ungrounded or grounded circuit conductors. (250.119)

Equipment grounding conductors 4 AWG and larger AWG that are not identified as required for conductors of sizes 6 AWG and smaller shall, at the time of installation, be permanently identified as an equipment grounding conductor at each end and at every point where the conductor is accessible, except where such conductors are bare.

The required identification for conductors 4 AWG and larger shall encircle the conductor and shall be accomplished by one of the following:

1. Stripping the insulation or covering from the entire exposed length.

2. Coloring the exposed insulation or covering green at the termination.

3. Marking the exposed insulation or covering with green tape or green adhesive labels at the termination. [250.119(A)]

Exceptions:

1. Conductors 4 AWG and larger shall not be required to be identified in conduit bodies that do not contain splices or unused hubs. [250.119(A)(1) Exception]

2. Power-limited, Class 2 or Class 3 circuit cables containing only circuits operating at less than 50 volts shall be permitted to use a conductor with green insulation for other than equipment grounding purposes. [250.119 Exception No. 1]

❖ Equipment grounding conductors may be bare in most cases; however, the code will indicate when insulation is required, such as for swimming pool grounding. If the conductors are insulated or covered they must be identified by a continuous green color or continuous green color with one or more yellow stripes. An insulated equipment grounding conductor larger than 6 AWG is permitted to be identified at termination points such as junction boxes, cabinets and panelboards by having the insulation stripped from the entire exposed length or by having the insulation colored green or taped with green tape. For example, a 4 AWG aluminum equipment grounding conductor used on a circuit with an overcurrent device of 200 amperes could have insulation of any color if it is marked with green tape where it is exposed to view at all junction boxes and termination points. The bottom line is: if the conductor is small (6 AWG and smaller), the code says that there is no excuse for not installing conductors with the required color of insulation. The code recognizes that larger conductors with green insulation may not be readily available or may require special order. Note that Items 2 and 3 appear to require reidentification of the conductor only at the termination points; however, the second paragraph of this section also requires the conductor to be reidentified at all points where the conductor is accessible, such as in junction and pull boxes. The code official must determine if the specifics of Items 2 and 3 override the general requirement in this regard.

E3407.3 Ungrounded conductors. Insulation on the ungrounded conductors shall be a continuous color other than white, gray and green. [310.110(C)]

Exception: An insulated conductor that is part of a cable or flexible cord assembly and that has a white or gray finish or a finish marking with three continuous white or gray stripes shall be permitted to be used as an ungrounded conductor where it is permanently reidentified to indicate its use as an ungrounded conductor by marking tape, painting, or other effective means at all terminations and at each location where the conductor is visible and accessible. Identification shall encircle the insulation and shall be a color other than white, gray, and green. [200.7(C)(1)]

Where used for single-pole, 3-way or 4-way switch loops, the reidentified conductor with white or gray insulation or three continuous white or gray stripes shall be used only for the supply to the switch, not as a return conductor from the switch to the outlet. [200.7(C)(2)]

❖ Any color of insulation other than white, gray or green is permitted for ungrounded conductors. The only exception occurs where the conductor with white or gray insulation or with white stripes is part of a multiconductor cable or cord assembly and is permanently reidentified by marking or taping with a color other than white, gray, or green at all termination points on the circuit and at all locations where the conductor is visible or accessible. One example of this is a switch loop in which conductors supplying power to a luminaire are brought to the lighting outlet. Only one cable with a black, white, and bare conductor is run to the switch point. The white conductor is connected to the ungrounded conductor at the lighting outlet and supplies power to the switch. When the switch is closed, power is carried back to the lighting outlet on the black conductor of the switch-loop cable. At the device box for the switch and in the lighting outlet box, the white conductor in the switch loop must be colored, marked or taped to indicate that it is an ungrounded conductor. After the installation is completed and the wall is covered, the wiring layout will not be visible. If the switch box has only one cable run to it, and the switch terminals have only the white and black conductors connected, a person opening the box could assume that the white conductor is used as an ungrounded conductor. However, the code requirement is for it to be marked as ungrounded so that such assumptions are not necessary.

A conductor with white or gray insulation installed in conduit or tubing cannot be used as an ungrounded conductor. For example, a white conductor to be used as an ungrounded conductor that is taped with black tape and installed in conduit or tubing would be in violation of the code.

Exception 2 is a very specific application of the same allowance expressed in Exception 1.

E3407.4 Identification of terminals. Terminals for attachment to conductors shall be identified in accordance with Sections E3407.4.1 and E3407.4.2.

❖ See the commentary to Sections E3407.4.1 and E3407.4.2.

E3407.4.1 Device terminals. All devices excluding panelboards, provided with terminals for the attachment of conductors and intended for connection to more than one side of the circuit shall have terminals properly marked for identification, except where the terminal intended to be connected to the grounded conductor is clearly evident. [200.10(A)]

Exception: Terminal identification shall not be required for devices that have a normal current rating of over 30 amperes, other than polarized attachment caps and polarized receptacles for attachment caps as required in Section E3407.4.2. [200.10(A) Exception]

❖ Devices that have provisions for terminating both grounded and ungrounded conductors must have the terminals identified to indicate the connections. This is

required for most of the devices that would be used in residential wiring except for certain specified devices with a normal current rating of over 30 amperes.

E3407.4.2 Receptacles, plugs and connectors. Receptacles, polarized attachment plugs and cord connectors for plugs and polarized plugs shall have the terminal intended for connection to the grounded (white) conductor identified. Identification shall be by a metal or metal coating substantially white in color or by the word "white" or the letter "W" located adjacent to the identified terminal. Where the terminal is not visible, the conductor entrance hole for the connection shall be colored white or marked with the word "white" or the letter "W." [200.10(B)]

❖ Most devices used for residential wiring have the grounded conductor terminals identified by a white (silver) metal composition or metal plating that is lighter in color than the terminals for the ungrounded conductors, which are usually brass colored. Some receptacles and connectors may have corrosion-resistant terminals, which are all white or a light color, or they may have all brass-colored terminals. For these devices especially, the terminals for the connection of the grounded or neutral conductor must be marked with the word "white," marked with the letter "W" or have white paint adjacent to the terminal. Many receptacles have both the distinctive colored terminals and the word "white" to indicate the grounded conductor connection.

Bibliography

The following resource materials were used in the preparation of the commentary for this chapter of the code:

NFPA 70–14, *National Electrical Code.* Quincy, MA: National Fire Protection Association, 2013.

Chapter 35:
Electrical Definitions

General Comments

The electrical trade, like other construction trades, has its vernacular. If people do not understand the language of the code text, they will have difficulty with interpretation of the text. Many words have a unique meaning in the context of a particular code and may be undefined or defined differently in a dictionary.

Purpose

Because much code text meaning depends on the definitions of the terms used in the text, Chapter 35 lists the definitions of terms that do not have everyday, common, universally accepted or dictionary-defined meanings. Chapter 35 is provided as the key to understanding the electrical text.

SECTION E3501
GENERAL

E3501.1 Scope. This chapter contains definitions that shall apply only to the electrical requirements of Chapters 34 through 43. Unless otherwise expressly stated, the following terms shall, for the purpose of this code, have the meanings indicated in this chapter. Words used in the present tense include the future; the singular number includes the plural and the plural the singular. Where terms are not defined in this section and are defined in Section R202 of this code, such terms shall have the meanings ascribed to them in that section. Where terms are not defined in these sections, they shall have their ordinarily accepted meanings or such as the context implies.

❖ In addition to the definitions in this chapter, the code provides definitions at the beginning of Chapter 42, "Swimming Pools," and the beginning of Chapter 43, "Class 2 Remote-Control, Signaling and Power-Limited Circuits." It is necessary that everyone understand the meaning of the terms used when they are using the code provisions. There are many technical terms in the code that would not be understood without a definition. Also, there are many common terms that have a different meaning or very specific meaning when applied to code rules. An individual may try to use terms such as "ground" or "accessible" according to a commonly understood meaning, but the specific definitions in this chapter are necessary to apply the code rules properly. Any definitions not included in this chapter or in Chapter 42 or 43 are to be understood as the commonly used definition in the English language. In addition, there are some terms used in the code such as "service head," "service mast," "threadless coupling," "raceway seal," "multioutlet branch circuit" "opening" and "yoke" that are neither defined herein nor used in their common English sense but are understood with exposure to the electrical industry.

ACCESSIBLE. (As applied to equipment.) Admitting close approach; not guarded by locked doors, elevation or other effective means.

❖ Where equipment must be accessible, it must not be guarded by such things as locked doors or elevation.

This does not mean that the equipment cannot be behind a locked door or installed at a high elevation. The key word of this definition is "guarded"; refer to the definition of "Guarded" in this chapter. Equipment is accessible as long as it is not guarded. Also, equipment may be on a roof and still be accessible by someone climbing a permanent or portable ladder. If it is mounted at a high elevation to prevent people from getting to it, it is guarded.

ACCESSIBLE. (As applied to wiring methods.) Capable of being removed or exposed without damaging the building structure or finish, or not permanently closed in by the structure or finish of the building.

❖ Where wiring methods must be accessible, they must not be concealed behind finished walls or behind or in a part of the dwelling construction that is permanent. The wiring is accessible even if reaching it requires someone to climb a ladder or remove an access door. For example, access to the wiring connections for a hydromassage bathtub may be through an access door that is secured with several screws. The connections are accessible as long as reaching them does not require the removal of a permanent cabinet or finished wall. It may take some time to remove the screws securing the access door, but the door does provide access. However, if the access door were inside a linen closet on the opposite side of a wall where the bathtub is located, there may not be room to physically get to the wiring and adequately and safely work on it. If the hydromassage bathtub pump motor needed to be replaced, it may be practically impossible to reach the enclosure for the wiring connections. Thus, it may technically be accessible by removal of the access door, but the accessible space must be considered also. If it is too small a space, the wiring would not be accessible. Wiring in a crawl space, unfinished basement, or attic is accessible as long as a person can get to the wiring for inspection, maintenance, repair or replacement. See the commentary to Section E3905.1 for making connections and splices accessible.

ACCESSIBLE, READILY. Capable of being reached quickly for operation, renewal or inspections, without requiring those to whom ready access is requisite to take actions such as to use tools, to climb over or remove obstacles or to resort to portable ladders, etc.

❖ A student of or person experienced in using the *International Building Code®* (IBC®) may automatically think of "accessible" and "readily accessible" in terms of something being accessible to physically disabled persons. The definition in this code as it pertains to equipment has not been a part of building codes in the past but has been peculiar to electrical code requirements and also more recently included in mechanical and plumbing codes. The difference between accessible and readily accessible is very important in applying code rules.

The definition here is also in Section R202 of this code. Where the code states that equipment must be readily accessible, it means that a person can walk up to the equipment or device and work on it or operate it without climbing or removing any obstacles. Disconnect switches and panelboards containing overcurrent devices must be readily accessible. They should be located where it is not likely that someone will store materials nearby that could block access (see commentary, Section E3405.5).

AMPACITY. The maximum current in amperes that a conductor can carry continuously under the conditions of use without exceeding its temperature rating.

❖ This term is the combination of two terms, "ampere" and "capacity." It relates to the temperature rating of the conductor insulation material. The maximum current that a conductor can carry continuously without exceeding its temperature rating is not set at a certain value but will vary with the conditions of use. The two conditions of use that must be considered to determine the allowable ampacity are the ambient temperature and the number of conductors in close proximity to each other. See the commentary to Section E3704.5 for more information on ampacity.

APPLIANCE. Utilization equipment, normally built in standardized sizes or types, that is installed or connected as a unit to perform one or more functions such as clothes washing, air conditioning, food mixing, deep frying, etc.

❖ Unlike devices, appliances consume power. Chapter 41 has the general code provisions covering appliances. Chapter 37 contains the rules covering the load calculations and branch circuit conductors for appliances such as clothes dryers and ranges. The provisions for the installation of ceiling fans are in Chapter 41. Many times a ceiling fan is thought of as a lighting fixture, however, it is considered to be an appliance.

APPROVED. Acceptable to the authority having jurisdiction.

❖ The term "approved" is used to describe a specific material or method of installation. It means that the material or method is acceptable to the code official. This term is also defined in Section R202 of this code. That definition states that "approved" means acceptable to the code official. The term "authority having jurisdiction" is used in the *National Electrical Code®*

(NEC®). In the enforcement of the code, the authority having jurisdiction is the same as the code official. As stated in Section R202, it is imperative that the code official (or authority having jurisdiction) bases her or his decision of approval on the result of investigations, tests or accepted principles or practices.

ARC-FAULT CIRCUIT INTERRUPTER. A device intended to provide protection from the effects of arc-faults by recognizing characteristics unique to arcing and by functioning to de-energize the circuit when an arc-fault is detected.

❖ An arc is produced by electricity flowing through an air gap between two conducting surfaces in a circuit. Because of the high resistance offered by the air gap, intense heat is developed in the arc along with radiation of energy including visible light. For comparison, consider the process of arc-welding, which readily melts the base metal and the filler metals. Arcing can result from loose terminals, screws, lugs, splicing devices, plug and receptacle parts and also from damaged or severed conductors.

The type of arc-fault circuit interrupter (AFCI) device currently required by this code is the combination type, which is further explained in Section E3902.11. In the typical installation, the AFCI device consists of a standard circuit breaker with additional electronic circuitry that is capable of recognizing the unique current fluctuations that result from an arcing condition. Like ground fault circuit interrupter (GFCI) circuit breakers, AFCI circuit breakers are designed and installed such that the grounded conductor and ungrounded conductor of the branch circuit both connect to the AFCI device. This is because the devices monitor the balance of current flow between the conductors. Under normal conditions, the current flow on the ungrounded and grounded conductors of a two-wire branch circuit are equal. If the flow on the conductors is not equal, the AFCI will open in a similar manner as a GFCI device except at a higher leakage current threshold.

An AFCI is designed to detect arcing faults that could cause a fire, and then de-energize the circuit thereby preventing a fire from developing. Conventional circuit breakers reduce the risk of fire by protecting the conductors from excess current which causes conductor heating and AFCI devices enhance this protection by also protecting against heating caused by arcing. For example, at a bad (loose) connection at the terminal of a receptacle, series arcing can occur under load. Over time, as the arcing persists, a layer of carbon builds even more resistance in the connection and the result is heat, which could eventually build up enough to cause a fire. AFCIs are required in most rooms of a dwelling. An AFCI is designed to de-energize the circuit before ignition can occur.

Several different AFCI devices have been developed and are available for use at different points in a circuit. A circuit-breaker type of device that mounts in the panelboard at the origin of the branch circuit or feeder provides protection for the entire branch circuit or feeder. An outlet branch circuit AFCI is installed as a receptacle in the branch circuit and provides protection for downstream wiring, including extension cords and power-supply cords plugged into the circuit. Other AFCI devices can be used such as a portable plug-in device with multiple outlets that provides protection for extension cords and power-supply cords plugged into it, and a cord AFCI that has the device built into or attached to the extension cord and is intended to protect only that cord. These devices are designed to clear the arc fault or open the circuit in a specified time of between one second and 200 milliseconds, depending on the current in the arc fault, thus helping to prevent the ignition of a fire.

Note that the code specifies only one type of AFCI device, that being the combination type. Combination type AFCI breakers will be labeled as such and are also distinguished by different color test buttons. For example, one manufacturer used a blue test button for the earlier branch circuit type of AFCI and uses a white button for the combination type.

ATTACHMENT PLUG (PLUG CAP) (PLUG). A device that, by insertion into a receptacle, establishes connection between the conductors of the attached flexible cord and the conductors connected permanently to the receptacle.

❖ Receptacles are often incorrectly referred to as plugs. Plugs are referred to as the male end of a cord assembly whereas a receptacle is the female component. Attachment plugs have voltage and amperage ratings. Specific configurations are designed so that the attachment plug for an appliance such as a clothes dryer, for example, will not fit the receptacle for a range because they have different amperage ratings.

AUTOMATIC. Performing a function without the necessity of human intervention.

❖ Examples of devices that operate automatically are timing switches, occupant sensor controls, ground- and arc-fault interrupters, thermostats and overcurrent devices.

BATHROOM. An area, including a basin, with one or more of the following: a toilet, a urinal, a tub, a shower, a bidet, or similar plumbing fixture.

❖ The definition of "Bathroom" is important in determining where outlets are required and which ones must have GFCI receptacles (see Sections E3601.6.2, E3705.7, E3901.6 and E3902.1).

BONDED (BONDING). Connected to establish electrical continuity and conductivity.

❖ Bonding is necessary to help ensure continuity between all the metallic parts of the electrical wiring system and will provide an effective path for fault current. If the fault current path is interrupted, open or of high enough impedance, circuit breakers and fuses will not operate properly or at the appropriate time. Bonding can be provided with a conductor or a device, as long as one of these provides a solid metallic connection between the parts being bonded. See Section E3509 for bonding provisions and commentary.

BONDING CONDUCTOR OR JUMPER. A reliable conductor to ensure the required electrical conductivity between metal parts required to be electrically connected.

❖ A bonding jumper is required where the continuity between metallic parts is broken, opened or where the impedance to current flow is too high. An example of this is where a metal conduit is installed containing under-ground service conductors and is attached to the bottom of an outdoor panelboard with a fitting through an opening made by breaking out the concentric rings of the prepunched knockout, which is there to adjust the opening to the size of conduit being installed. The electrical conductivity through the remaining two or three tabs holding the remaining concentric rings in place may be negligible, and in the case of ground fault, the fault current would be impeded. Installing the correct size "bonding jumper" between the conduit and the panelboard provides for a solid fault current path. Note that bonding jumper is also defined in Chapter 24 where it has a similar yet unique definition.

BONDING JUMPER (EQUIPMENT). The connection between two or more portions of the equipment grounding conductor.

❖ Where there is a possibility of an incomplete fault current path or a fault current path that would have high impedance, a bonding jumper must be installed to maintain the continuity and current carrying capacity of the fault current path.

BONDING JUMPER, MAIN. The connection between the grounded circuit conductor and the equipment grounding conductor at the service.

❖ The main bonding jumper is a connection between the service grounded conductor and the equipment grounding conductor. Because the code uses the term "jumper" as part of this term, an individual might think of the main bonding jumper only as a conductor or short piece of wire. Indeed, a wire is used in some cases to make the connection, but the connection is often accomplished with a strap, busbar or self-tapping screw.

For example, in the service equipment (and only in the service equipment), at the location of the main disconnect, the grounded service conductor is connected to the same busbar as the equipment grounding conductors and possibly one or more grounding electrode conductors. Therefore, that busbar functions as a main bonding jumper because it is the connection between the grounded service conductor and the equipment grounding conductors. In the typical dwelling service equipment, the grounded (neutral) busbar is bonded to the service panel-board enclosure by means of a special bonding screw or strap provided by the equipment manufacturer. The equipment grounding conductors for the branch circuits and feeders are terminated on the same busbar that is bonded to the service panel-board enclosure or they are terminated on a separate equipment grounding bus bar that is also bonded to the equipment enclosure. If a separate equipment grounding conductor busbar is installed in the service panelboard, it is the steel enclosure that bonds that equipment grounding busbar to the neutral busbar.

The factory supplied bonding jumper screw or strap is shipped disconnected or as a loose component because the manufacturer of the panelboard or enclosed switch cannot anticipate whether or not the equipment will be used as service equipment. If used as service equipment, the neutral conductor bus must be bonded to the enclosure, but, if not used as service equipment, the neutral or grounded conductor bus must be electrically isolated from the enclosure. This is why the bonding jumper screw or strap is not factory installed, it is used in some cases and discarded in others. Where such bonding jumper screws and straps are not required to be installed, they should be discarded to prevent someone from installing them, thinking that they were overlooked.

The purpose of the main bonding jumper is to provide a low-impedance fault-current path back to the power supply to facilitate the tripping of the overcurrent device in case of a ground fault.

BONDING JUMPER, SUPPLY-SIDE. A conductor installed on the supply side of a service or within a service equipment enclosure(s) that ensures the required electrical conductivity between metal parts required to be electrically connected.

❖ Such jumpers are commonly used to bond metallic raceways to equipment enclosures by means of bonding bushings, lugs and short pieces of bare copper conductors.

BRANCH CIRCUIT. The circuit conductors between the final overcurrent device protecting the circuit and the outlet(s).

❖ "Branch circuit" is a general term that includes the conductors run between the final overcurrent device in the panelboard and an outlet such as a receptacle outlet or a lighting fixture. This term helps distinguish a branch circuit from a feeder, which supplies power to the final overcurrent devices.

BRANCH CIRCUIT, APPLIANCE. A branch circuit that supplies energy to one or more outlets to which appliances are to be connected, and that has no permanently connected luminaires that are not a part of an appliance.

❖ This term refers to dedicated branch circuits such as those for laundry appliances and the small-appliance branch circuits required for kitchen and dining areas by Section E3703.2. These are receptacle circuits intended for appliance use only.

BRANCH CIRCUIT, GENERAL PURPOSE. A branch circuit that supplies two or more receptacle outlets or outlets for lighting and appliances.

❖ General-purpose branch circuits supply lighting and receptacle outlets. The receptacle and lighting outlets are not required to be supplied by separate circuits because both are permitted on the same circuit. For dwellings, the number of lighting or receptacle outlets permitted on each general-purpose branch circuit is not specified. The minimum number of general purpose branch circuits is determined from the total computed load (see Section E3703.5).

BRANCH CIRCUIT, INDIVIDUAL. A branch circuit that supplies only one utilization equipment.

❖ Individual branch circuits are used for most major appliances and motors and are sized to provide power to specific loads according to the rating of the equipment, appliance or load served. The circuits supplying power for dishwashers, dryers, ranges and many motor-driven appliances are examples of individual branch circuits. See Section E3702 for rules on sizing individual branch circuits for motor loads. Where an individual branch circuit supplies a receptacle that the appliance is plugged into, it must be a single receptacle. An appliance such as an instant electric water heater installed under a kitchen sink may require an individual branch circuit because of the high current load. If a duplex receptacle were installed under the sink, a disposal could also be connected to that circuit, and if both appliances were operated at once, the overcurrent device could trip. Therefore, this electric water heater should have a single receptacle installed and should be supplied by a separate circuit—an individual branch circuit. Not all individual branch circuits supply receptacles. For example, individual branch circuit conductors are run directly to the furnace and not to a receptacle. Section E3703.1 requires that the furnace be supplied by an individual branch circuit regardless of the horsepower rating of the motor.

BRANCH CIRCUIT, MULTIWIRE. A branch circuit consisting of two or more ungrounded conductors having voltage difference between them, and a grounded conductor having equal voltage difference between it and each ungrounded conductor of the circuit, and that is connected to the neutral or grounded conductor of the system.

❖ See the commentaries for "Branch circuit" and "Branch circuit, individual" (see Commentary Figure E3406.11.2).

CABINET. An enclosure designed either for surface or flush mounting and provided with a frame, mat or trim in which a swinging door or doors are or may be hung.

❖ A cabinet is an enclosure such as a meter base. A panelboard enclosure is a cabinet, which is essentially a metal or plastic box that houses apparatus such as busbars, terminals and overcurrent devices. In everyday usage in the field, many electricians and electrical contractors involved in residential wiring refer to the cabinet and panel-board together as simply the panel, or they use trade jargon terms such as "breaker box" and do not necessarily distinguish between the cabinet and the panelboard as separate items. Perhaps this is because most panel-boards used in one- and two-family dwellings come from the manufacturer already installed in a cabinet. For simplicity, the term "panelboard" is also used in this commentary in many places to refer to the cabinet and panel-board as a unit.

However, it is important to know the definition for cabinet to understand code rules that apply specifically to the panelboard enclosure or other enclosures such as the meter base. See Section E3907 for the rules covering cabinets, which include positioning, mounting, and running cables and raceways into the cabinet. Preassembled cabinet and panelboard units comply with requirements for sizing and space such as providing sufficient wire bending space.

CIRCUIT BREAKER. A device designed to open and close a circuit by nonautomatic means and to open the circuit automatically on a predetermined overcurrent without damage to itself when properly applied within its rating.

❖ Circuit breakers are the most commonly used form of overcurrent protection in residential wiring. Except for service-entrance conductors and some tap conductors, the code requires that an overcurrent protection device be installed in series with each ungrounded conductor at the point where the conductor receives its supply. Conductors are rated to operate at maximum temperatures, and circuit breakers serve to open the circuit and protect the conductor if the temperature in the conductor rises to a level that would damage the conductor. Circuit breakers also protect conductors from short circuits and ground faults. A ground fault is current flowing to something that is grounded, such as a flow between a hot conductor and a metal enclosure. A short circuit is the flow of current between ungrounded conductors or between an ungrounded and a grounded conductor. The short circuit current is usually quite high and flows for a very short time. Circuit breakers must have an interrupting rating sufficient to stop the flow of current before damage occurs from excessive heating or arcing.

CLOTHES CLOSET. A nonhabitable room or space intended primarily for storage of garments and apparel.

❖ A clothes closet is a space or a room that is not considered to be habitable. Its main function is to store clothing, shoes and apparel. The definition of "Habitable space" in Chapter 2 excludes closets, storage rooms, bathrooms and similar spaces, in harmony with this definition and also describes habitable space as that used for living, sleeping, eating or cooking. The question often arises about how to view a walk-in closet that is quite large, provided with cabinetry and furniture, designated as a "dressing room" and that resembles a habitable room more than a "closet." Despite the fact that such a space appears to be more than a closet, its primary purpose is for storage of clothing and apparel and it is not used for living, sleeping, eating or cooking. A dressing room is more akin to a bathroom than a habitable room. This issue is a concern when applying the code provisions relative to receptacle outlet placement and the allowable types of luminaires. If the "main function" of a room is to store clothes, shoes and apparel, it is a clothes closet and the type of luminaires should comply with Section E4003.12.

CONCEALED. Rendered inaccessible by the structure or finish of the building.

❖ Wiring and equipment that is installed within the finished construction is considered to be concealed. Access to it would require the removal of finished construction. The wiring in a crawl space or attic is not concealed if one could get to the wiring to work on it. In some cases, the space in the attic and crawl space or some parts of the attic or crawl space of a dwelling is very limited. For example, where a junction box is visible from one part of a crawl space but the area is too small to actually crawl to the box or reach the box to work on it, is the junction box considered to be concealed? Junction boxes are considered to be concealed if a person would need to remove some finished construction to work on them.

CONDUCTOR.

Bare. A conductor having no covering or electrical insulation whatsoever.

Covered. A conductor encased within material of composition or thickness that is not recognized by this code as electrical insulation.

Insulated. A conductor encased within material of composition and thickness that is recognized by this code as electrical insulation.

❖ An example of a covered conductor is the grounding conductor within the sheathing of Type NM cable. When the sheathing is stripped and the separate conductors are examined, it can be seen that this grounding conductor is covered with paper. This is not an insulated conductor. An improper use of this paper-covered bare conductor would be to use it as an ungrounded or grounded conductor. Insulated conductors are tested to meet standards for temperature and conditions of use and are marked as to the type of insulation. The conductor types most widely used in wiring of one- and two-family dwellings are listed in Section E3406.5 [see Commentary Figure E3501(1)].

CONDUIT BODY. A separate portion of a conduit or tubing system that provides access through a removable cover(s) to the interior of the system at a junction of two or more sections of the system or at a terminal point of the system. Boxes such as FS and FD or larger cast or sheet metal boxes are not classified as conduit bodies.

❖ A conduit body is a fitting designed to provide access to the wiring through a separate cover. It is used mostly where a 90-degree bend with a full radius would be difficult to install. Many different styles and shapes of conduit bodies are available for both metal and nonmetallic conduit systems. Some of the different types are LB, where the opening is on the back; LL and LR, which are intended to lay flat against the sur-face and have the opening on the left or right side respectively; and T, which is a combination of an LL and an LR. A Type C conduit body is used in a straight run of conduit to provide an opening in the conduit run or to make a splice. A splice can be made in a conduit body if it is marked with the cubic-inch capacity.

CONNECTOR, PRESSURE (SOLDERLESS). A device that establishes a connection between two or more conductors or between one or more conductors and a terminal by means of mechanical pressure and without the use of solder.

❖ This term applies to the common bolted-type pressure connector used for terminals in panelboards and other equipment where a screw or bolt binds against the end of a conductor. This term also applies to devices commonly referred to as "wire nuts," which are used to connect two or more conductors. Wire nuts are a screw-on type of connector with an insulated cap made of plastic or other nonconductive material. Most wire nuts contain a spiral metallic spring or coil inside to bind the conductors together. Wire nuts are typically color-coded to indicate the maximum size and number of conductors for which the device is rated. These colors are not determined by the code and may be different for different manufacturers. The proper size connector must be used for the wires to be spliced. See Section E3406.10 for detailed information on making connections.

CONTINUOUS LOAD. A load where the maximum current is expected to continue for 3 hours or more.

❖ Continuous loads could result in higher conductor temperatures. Section E3701.2 requires that conductors for continuous loads be sized at 125 percent of the load.

COOKING UNIT, COUNTER-MOUNTED. A cooking appliance designed for mounting in or on a counter and consisting of one or more heating elements, internal wiring and built-in or separately mountable controls.

❖ This definition must be considered when calculating the feeder neutral load for cooking appliances as prescribed in Section E3704.3 (see also Table E3602.2 and Section E3702.9).

COPPER-CLAD ALUMINUM CONDUCTORS. Conductors drawn from a copper-clad aluminum rod with the copper metallurgically bonded to an aluminum core. The copper forms a minimum of 10 percent of the cross-sectional area of a solid conductor or each strand of a stranded conductor.

❖ Copper-clad aluminum conductors were used in nonmetallic sheathed cable in the late 1950s and early 1960s. It is important to know the terminating and splicing provisions for repair and retrofitting work that involves older dwellings that are wired with copper-clad aluminum conductors.

CUTOUT BOX. An enclosure designed for surface mounting and having swinging doors or covers secured directly to and telescoping with the walls of the box proper (see "Cabinet").

❖ This is a variation of a cabinet defined above, which is an enclosure for electrical devices and equipment such as fuses and circuit breakers, switches, current transformers, etc. A cutout box may be used for a motor disconnect switch with fuses for overcurrent protection. A junction box has a cover attached with screws or bolts, but a cutout box has a swinging door with a hinge. A cutout box is made to contain devices that may need to be worked on, inspected or maintained; the hinged cover makes it easier to get into than a junction box.

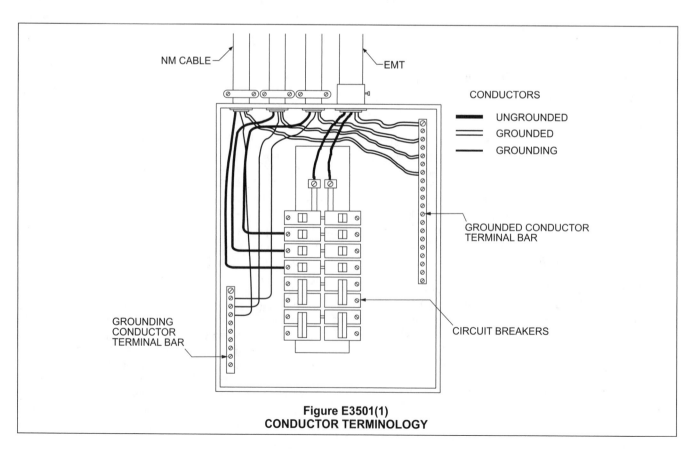

Figure E3501(1)
CONDUCTOR TERMINOLOGY

DEAD FRONT. Without live parts exposed to a person on the operating side of the equipment.

❖ The door or cover of a cabinet or cutout box is a "dead front;" the live parts within are not exposed to a person standing in front of a panelboard or equipment. This term is used in measuring the working clearances for electrical equipment. The clearance from the dead front or front of the enclosure of the cabinet for a panelboard is a minimum of 36 inches (914 mm).

DEMAND FACTOR. The ratio of the maximum demand of a system, or part of a system, to the total connected load of a system or the part of the system under consideration.

❖ Demand factors are used in the calculations for various loads such as service and feeder loads. Demand factors are expressed as a percentage of the calculated load and are used when there are multiple loads that are not likely to be used at the same time. Feeders are not sized to have the current-carrying capacity to supply the total current for all the loads at once. For example, for certain fastened-in-place appliances in a dwelling, the total nameplate rating load must be calculated at 100 percent if the number of appliances is three or fewer. But if there are four or more, the load is calculated at 75 percent of the total load of the appliances. It is unlikely that all four of the appliances would be used at full load at the same time [see Tables E3602.2 and E3704.2(1)].

DEVICE. A unit of an electrical system that carries or controls electrical energy as it principal function.

❖ A switch, receptacle, circuit breaker, fuse, meter and thermostat are all examples of a "device" since they do not use or consume electrical energy; they just control or carry the current. This definition recognizes that some devices such as illuminated switches, timer switches, occupant sensing switches, remote control switches and GFCI receptacles do, in fact, consume some small amount of electrical power, but their primary function is still consistent with this definition.

DISCONNECTING MEANS. A device, or group of devices, or other means by which the conductors of a circuit can be disconnected from their source of supply.

❖ This term is usually used in conjunction with the service, appliances and motors. Disconnecting means can be fused or unfused enclosed switches, enclosed circuit breakers and pull-out devices. Code provisions for disconnecting means are very specific for such things as identification, location, type and ratings. The disconnection of power to a circuit is an obvious safety issue.

DWELLING.

Dwelling unit. A single unit, providing complete and independent living facilities for one or more persons, including permanent provisions for living, sleeping, cooking and sanitation.

One-family dwelling. A building consisting solely of one dwelling unit.

Two-family dwelling. A building consisting solely of two dwelling units.

❖ A guest room in a hotel or motel is not a dwelling unit unless it contains permanent provisions for cooking. A portable microwave is not permanent, so even a motel room with a kitchenette is not a dwelling unit unless the cooking equipment is a built-in range or cook top. Because this code covers only one- and two-family dwellings and townhouses, the definition for a multiple-family dwelling is not included (see Section R101.2).

EFFECTIVE GROUND-FAULT CURRENT PATH. An intentionally constructed, low-impedance electrically conductive path designed and intended to carry current under ground-fault conditions from the point of a ground fault on a wiring system to the electrical supply source and that facilitates the operation of the overcurrent protective device or ground-fault detectors.

❖ See Sections E3908.4 and E3908.5.

ENCLOSED. Surrounded by a case, housing, fence or walls that will prevent persons from accidentally contacting energized parts.

❖ This is a general term that applies to enclosing things such as devices inside a cabinet. It can also apply to equipment inside a room such as an electrical equipment room.

ENCLOSURE. The case or housing of apparatus, or the fence or walls surrounding an installation, to prevent personnel from accidentally contacting energized parts or to protect the equipment from physical damage.

❖ An enclosure contains electrical equipment. All enclosures, from recessed light fixtures to outdoor panelboards, are evaluated by a testing agency to designate proper use and location. The designation is marked on or comes with the equipment. Examples of types of enclosures are Type 1 for general indoor use and Type 3R for outdoor use. A Type 3R enclosure provides a degree of protection against rain, snow and sleet.

ENERGIZED. Electrically connected to, or is, a source of voltage.

❖ Energized conductors and devices have a voltage impressed upon them. Some jurisdictions require that a house be energized before an inspection can be completed. Often, temporary power can be connected to the house for the inspection to check such things as proper polarity of receptacles and the proper functioning of the GFCI receptacles.

EQUIPMENT. A general term including material, fittings, devices, appliances, luminaires, apparatus, machinery and the like used as a part of, or in connection with, an electrical installation.

❖ This is a general catchall term used to refer to anything from large items such as appliances, air-conditioning units, cabinets, panelboards and meter bases to items such as a 4-inch-square junction box or receptacle. Note that "Equipment" is defined differently in Chapter 2.

EXPOSED. (As applied to live parts.) Capable of being inadvertently touched or approached nearer than a safe distance by a person.

❖ The term "exposed" live parts refers to the uninsulated portion of conductors, terminals or devices that are energized and that are not guarded or enclosed. The conductors and overcurrent devices of a panelboard in a cabinet are exposed when the cover is removed.

EXPOSED. (As applied to wiring methods.) On or attached to the surface or behind panels designed to allow access.

❖ Wiring methods that are not concealed in the construction of the dwelling are considered exposed. For example, nonmetallic sheathed cable can be run exposed where not subject to physical damage. The provisions of Section E3802.3 cover the installation of exposed cables.

EXTERNALLY OPERABLE. Capable of being operated without exposing the operator to contact with live parts.

❖ Circuit breakers, for example, are externally operable because a dead front cover is installed over the panelboard containing the breakers so that the operator will not be exposed to live parts.

FEEDER. All circuit conductors between the service equipment, or the source of a separately derived system, or other power supply source and the final branch-circuit overcurrent device.

❖ In residential wiring systems, a feeder is usually a large set of conductors run in cable or in a raceway from the service equipment to the main panel or to a subpanel, which has the final branch circuit overcurrent devices. A feeder carries power to a distribution center or panel that is the origination point for one or more branch circuits. The arrangement and layout of the service equipment can vary, and some houses may not have feeder conductors. Where the meter base, main disconnect and distribution panel are combined into one unit or piece of equipment, this unit may be referred to in common usage or slang as a meter-main combination panel. The main disconnect and all the branch circuit overcurrent devices are located in this equipment, usually on the outside of the house. If all the circuit breakers except for the main disconnect are for branch circuits, and there are no sub-panels anywhere else in or on the house, there is no feeder.

Another arrangement may be where a meter base only is on the outside of the house and the service-entrance conductors are run to the meter base and continue on to a panelboard inside the house. This panel inside the house may be called, in common usage (but not in the code), a main breaker panel because the main breaker is installed in or is part of the panelboard along with the final branch circuit overcurrent devices. Again, if there are no subpanels where additional final overcurrent devices for branch circuits would be installed, there would be no feeder.

Another arrangement of service equipment is where the meter base and main disconnect are located on the outside of the house and the panelboard is separate. The conductors that are run from the main disconnect to the distribution panel are the main power feeder. The main disconnect may consist of up to six switches (which are usually circuit breakers), and the conductors from any of these switches or circuit breakers that feed a distribution panel are feeders (see Section E3706.3). A large house may have two or more distribution panels where final branch circuit overcurrent devices are installed. For example, the conductors from the main service to a panel for a detached garage and to a panel serving swimming pool equipment are feeder conductors. All of the main disconnect switches, which could number up to six, do not necessarily supply feeder conductors. For example, one of the six main disconnects at the service equipment may serve an air conditioning condenser unit near the service equipment. These conductors would be branch circuit conductors, not feeder conductors.

FITTING. An accessory such as a locknut, bushing or other part of a wiring system that is intended primarily to perform a mechanical rather than an electrical function.

❖ Fittings connect cables, raceways, boxes and other equipment together to form a complete wiring system for the wiring method used. Wiring methods for dwellings include cable and raceway systems along with the associated boxes and hardware. Fittings include accessories such as cable and conduit connectors, couplings, adapters, reducers, clamps, bushings, locknuts, nipples, knock-out seals, conduit bodies, capped elbows, pipe and rod ground clamps, brackets, hubs and service heads.

GROUND. The earth.

❖ A "ground" is a conducting connection. Two examples of how this term is used are "ground clamp" and "ground bus," which refer to the conducting connection. This term is not used by the code to describe a wire or conductor. To say the "ground conductor" or the "ground wire" may lead to miscommunication. In talking about a conductor, it is helpful to use either "grounded conductor" or "grounding conductor." In some publications the term "ground wire" is used to indicate the "grounding wire." The terms used to refer to ground and grounding have very specific meanings and should be understood and used properly. "Ground wire" usually means the grounding conductor, but someone could possibly use it to mean the grounded conductor. Many informal words and phrases are used in the electrical trade to describe materials, devices, equipment and conductors. Although not defined in the code or other publications, these informal terms are understood by all individuals with at least some experience. Some terms, however, should be used according to the code definition to make communication accurate and avoid misunderstandings.

GROUNDED (GROUNDING). Connected (connecting) to ground or to a conductive body that extends the ground connection.

❖ Grounded conductors are intentionally connected to the earth by means of an electrode such as a ground rod, concrete reenforcing steel, copper water service pipe or other grounding electrode.

GROUNDED, EFFECTIVELY. Intentionally connected to earth through a ground connection or connections of sufficiently low impedance and having sufficient current-carrying capacity to prevent the buildup of voltages that may result in undue hazards to connected equipment or to persons.

❖ Grounding an object prevents it from being at a voltage potential different from the earth. An electrical system must be effectively grounded by connecting it to the earth in a manner that will provide a current-carrying path of sufficient capacity to limit the voltage to ground (earth), limit the voltage imposed by a lightning strike or a line surge and provide a reference level for the normal system voltage. See Section E3607 for more information on grounding.

GROUNDED CONDUCTOR. A system or circuit conductor that is intentionally grounded.

❖ The grounded conductor of the service, feeder or branch circuits is commonly called the "neutral" and is required to be white or gray. Although the word "neutral" is not included in this chapter and is seldom used in the code, in trade jargon it is used interchangeably with "grounded conductor." The term "neutral" is used in Section E3704.3 regarding the load on a feeder. The code requires that one of the three service conductors supplying the dwelling be grounded for the reasons listed under "Effectively grounded," above. Electrical power is supplied to one- and two-family dwellings from a single-phase, 3-wire transformer, the midpoint of which is grounded at the transformer. The conductor that is connected to the midpoint of the transformer winding is the grounded conductor. The other two service conductors are ungrounded and have a potential difference of 240 volts between them. The grounded conductor has a potential difference of 120 volts between it and each ungrounded conductor. The grounded conductor is the neutral conductor of the system and carries only the unbalanced current between the two phase conductors. According to Section E3607.1, the grounded conductor is grounded at the service disconnect location through its connection to the grounding electrode conductor, which is run to an acceptable grounding electrode. See Section E3608 for the "Grounding Electrode System."

GROUNDING CONDUCTOR, EQUIPMENT (EGC). The conductive path(s) that provides a ground-fault current path and connects normally noncurrent-carrying metal parts of equipment together and, to the system grounded conductor, the grounding electrode conductor or both.

❖ The equipment grounding conductor is not intended to carry current except in case of a ground fault. It must

be sized according to the rating of the overcurrent device serving the feeder or branch circuit. Section E3908 covers equipment grounding. All of the noncurrent carrying metallic parts of the electrical wiring system are tied together to form a continuous low-impedance current path for any fault current that could occur. Equipment is grounded so that in case of a failure in the insulation or for any other reason the metal equipment becomes energized, current will flow along the equipment grounding conductor path back to the source and trip the circuit breaker or fuse. In one- and two-family dwellings, equipment is grounded with the equipment grounding conductor, which is typically the green or bare conductor in the nonmetallic sheathed cable run for the branch circuits. Or, where metal conduit or tubing (such as electrical metallic tubing or EMT) is used, the code permits the conduit or tubing itself to be the equipment grounding conductor. There are several conditions in Section E3908.8.1 for using flexible metal conduit itself as the equipment grounding conductor. In actual practice, it is easier to install the green or bare equipment grounding conductor. The equipment grounding conductor is connected to the grounded conductor only within the service equipment. It must not be tied to the grounded conductor or neutral conductor at any point downstream from the service equipment (see Section E3607.2).

GROUNDING ELECTRODE. A conducting object through which a direct connection to earth is established.

❖ Ground rods, plates, pipes, wire loops, footing reenforcing steel and metal underground water pipes are all examples of grounding electrodes. An electrode in the context of this code is a metal electrically conductive terminal through which current enters the earth.

GROUNDING ELECTRODE CONDUCTOR. A conductor used to connect the system grounded conductor or the equipment to a grounding electrode or to a point on the grounding electrode system.

❖ The grounding electrode conductor is a copper, aluminum or copper-clad aluminum conductor, and it is not in-tended to carry current except in case of a ground fault. It can be bare, and the code does not specify a color for the grounding electrode conductor. The grounding electrode conductor is sized based on the size of the largest service-entrance conductor as shown in Table E3603.4.

The grounding electrode conductor is connected to the grounded service conductor at any accessible point between the load end of the service drop or service lateral and the main disconnect. This connection is usually made in the service equipment where the main disconnect is located or in the meter enclosure. The grounding electrode conductor and the grounded service conductor are bonded together and to the metal service enclosure by the main bonding jumper. The equipment grounding conductors are also brought to this point and tied to the grounded service conductor and to the grounding electrode conductor. The other end of the grounding electrode conductor is connected to one or more grounding electrodes such as an underground water pipe and ground rod.

GROUND-FAULT CIRCUIT-INTERRUPTER. A device intended for the protection of personnel that functions to de-energize a circuit or portion thereof within an established period of time when a current to ground exceeds the value for a Class A device.

❖ A ground-fault circuit-interrupter (GFCI) is a device that will open a circuit when current flows outside of the intended circuit. A current flowing outside the circuit conductors could occur, for example, from a damaged conductor that could create an outside path for current to flow to a grounded object. The small current flow may not be enough to trip the normal circuit breaker. Or it could be a low-resistance current path through an individual who touches something like a cord, box, wire or tool surface that has become "hot" or energized by a ground fault. It has been determined that even a small current flow of 6 milliamperes could be harmful to people under certain circumstances.

The GFCI device contains a coil through which both circuit conductors are routed. If between 4 and 6 milliamperes of current is flowing outside the circuit conductors, this current transformer coil senses the unbalance of current in the two branch circuit conductors and generates a small current that activates the trip mechanism and opens the circuit. A GFCI receptacle provides only ground-fault protection, and the GFCI circuit breaker provides overcurrent protection as well.

GROUND-FAULT CURRENT PATH. An electrically conductive path from the point of a ground fault on a wiring system through normally non-current-carrying conductors, equipment, or the earth to the electrical supply source.

❖ Examples of ground-fault current paths are any combination of equipment grounding conductors, metallic raceways, metallic cable sheaths, electrical equipment, and any other electrically conductive material such as metal, water, and gas piping; steel framing members; stucco mesh; metal ducting; reinforcing steel; shields of communications cables; and the earth itself.

GUARDED. Covered, shielded, fenced, enclosed or otherwise protected by means of suitable covers, casings, barriers, rails, screens, mats or platforms to remove the likelihood of approach or contact by persons or objects to a point of danger.

❖ The code requires live parts to be guarded against accidental contact. Where live parts are installed inside approved boxes, cabinets and other enclosures, they are considered guarded.

IDENTIFIED. (As applied to equipment.) Recognizable as suitable for the specific purpose, function, use, environment, application, etc., where described in a particular code requirement.

❖ Equipment that is identified for a specific purpose is listed and labeled or marked. In this way, it is recognizable as suitable for the use and for the conditions in which it is installed. For example, a cable that is suit-

able for direct burial underground is identified for this use by its listing and labeling as Type UF (Underground Feeder) cable. Another example is the disconnect switch installed on the outside of a house for an air conditioner. It must be identified for outdoor use where it could be subject to rain or snow. A Type 3R enclosure for this disconnect switch would survive the usual outdoor weather conditions. An installer and inspector can rely on the identification of equipment in its listing and labeling because a nationally recognized testing laboratory has tested it.

INTERRUPTING RATING. The highest current at rated voltage that a device is identified to interrupt under standard test conditions.

❖ This rating is the highest current that can flow through a circuit breaker or fuse without damage to the device and its enclosure. This is not the overcurrent rating, such as the identification of a fuse as a 15-ampere or 20-ampere fuse. Where a short circuit occurs as two ungrounded conductors touch, or where a ground fault occurs as an ungrounded conductor comes in contact with a grounded conductor, a very high current can flow. This can damage the overcurrent device and the equipment in which it is mounted. Circuit breakers and fuses are designed to trip without being damaged where the rating is high enough for the current imposed on the device. The amount of short-circuit or ground-fault current available must be calculated and the devices rated accordingly. Most circuit breakers are made with an interrupting rating of 10,000 amperes. See Sections E3404.2 and E3404.3 for more information.

INTERSYSTEM BONDING TERMINATION. A device that provides a means for connecting intersystem bonding conductors for communications systems to the grounding electrode system.

❖ Such devices provide a convenient place for bonding of communications and entertainment systems. All systems serving a dwelling are required to be bonded together. Without the convenience of such bonding devices, bonding is more likely to be overlooked by the installers of the systems. One type of device clamps onto a service enclosure such as a meter box and provides terminals for bonding conductors from telephone, cable and other utilities (see Section E3609.3).

ISOLATED. (As applied to location.) Not readily accessible to persons unless special means for access are used.

❖ This term may be used to refer to some wiring method or equipment that is not readily accessible.

Switches or devices could be considered isolated if they are located inside a locked cabinet. This term is related to location and does not relate to usage such as isolated grounding receptacles.

KITCHEN. An area with a sink and permanent provisions for food preparation and cooking.

❖ This term is used in many locations in the code and requires that three elements be present: a sink, provisions for food preparation and provisions for cooking, all of which must be permanent.

LABELED. Equipment or materials to which has been attached a label, symbol or other identifying mark of an organization acceptable to the authority having jurisdiction and concerned with product evaluation that maintains periodic inspection of production of labeled equipment or materials and by whose labeling the manufacturer indicates compliance with appropriate standards or performance in a specified manner.

❖ The code official must approve the materials and equipment installed in an electrical wiring system. A label of a nationally recognized testing laboratory (NRTL) identifies a product or material and provides other information that the code official uses to determine compliance with the code. Equipment and products that have been labeled have been tested for conformance to a standard and subject to third-party inspection, which verifies that the minimum level of quality required by the applicable standard is maintained. The labeling agency performing the third-party inspection must be approved by the code official, and the basis for this approval may include, but is not necessarily limited to, the capacity and capability of the agency to perform the specific testing and inspection.

The label will have the name and/or symbol of the NRTL and will contain the product category and type of the equipment or product, which has its proper use and application described in the listing publication from the NRTL. The applicable reference standard often states the minimum identifying information that must be on a label. Labels or marks are applied to the tested product by stamping, molding, affixing a sticker, affixing a printed decal or other similar means. If the product is too small or has a shape, texture or is of a material that makes the attachment of a label difficult, the product container will have the label or mark. For this purpose, the installer should make available the product container for the inspector if requested. The product container might have, in addition to the label, installation instructions for the product.

LIGHTING OUTLET. An outlet intended for the direct connection of a lampholder or luminaire.

❖ An outlet box designed for the attachment of a lamp holder, a lighting fixture or a pendant cord terminating in a lamp holder is installed at a lighting outlet, which is an opening in the wiring system. The word "opening" is not defined in the code but is commonly used in the trade. It is the point where junction boxes, device boxes or boxes for the attachment of lighting fixtures are installed. Lighting outlets are required in specific locations in a dwelling as covered in Section E3903.1.

LIGHTING TRACK (Track Lighting). A manufactured assembly designed to support and energize luminaires that are capable of being readily repositioned on the track. Its length can be altered by the addition or subtraction of sections of track.

❖ See Section E4005.

LISTED. Equipment, materials or services included in a list published by an organization that is acceptable to the authority having jurisdiction and concerned with evaluation of products or services, that maintains periodic inspection of production of listed equipment or materials or periodic evaluation of services, and whose listing states either that the equipment, material or services meets identified standards or has been tested and found suitable for a specified purpose.

❖ Nationally recognized testing laboratories (NRTLs) publish lists of electrical materials, components, devices, fixtures and equipment that they have subjected to standardized tests and found suitable for installation and use in a specific manner. The listing information is necessary if there is a question as to the suitability of the product or material for a specific installation. Section E3403.3 requires the listing and labeling of electrical materials, components, devices, fixtures and equipment.

LIVE PARTS. Energized conductive components.

❖ Live parts must be enclosed or guarded to prevent people from receiving electric shock. For example, live parts should not be left exposed during the construction of the dwelling. When the cover (dead front) of a panelboard is removed, live parts will be exposed if the panelboard is energized. Cabinet doors and junction box covers are usually not installed until the finish phase of the project, but means should be taken where necessary to enclose or guard live parts.

LOCATION, DAMP. Location protected from weather and not subject to saturation with water or other liquids but subject to moderate degrees of moisture.

LOCATION, DRY. A location not normally subject to dampness or wetness. A location classified as dry may be temporarily subject to dampness or wetness, as in the case of a building under construction.

LOCATION, WET. Installations underground or in concrete slabs or masonry in direct contact with the earth and locations subject to saturation with water or other liquids, such as vehicle-washing areas, and locations exposed to weather.

❖ The designation of certain areas as a type of location is very important in determining what equipment and materials are permitted to be installed at a specific place in the dwelling. The location must be considered when selecting almost any material such as boxes, cabinets, conductors, lighting fixtures, cables and raceways.

It may seem apparent that all locations inside a dwelling would be "dry locations" for the purpose of installing wire and materials. But, as the code states, some basements may be damp locations because moisture could seep through concrete or masonry walls and because water may be present in some basements. In most cases it is evident that an area might be subject to dampness and the appropriate wiring methods must be used. If the classification of location is questionable, the code official would make the determination.

All conductors used in a wet location must be rated for that use and must have a "W" in the marking of the conductor type such as THW and THWN. The "W" indicates that it is approved for use in a wet location. The in-side of a conduit where the conduit is installed underground is a wet location because of the chance that water could accumulate there. Conductors installed in underground conduits must have the "W" rating.

LUMINAIRE. A complete lighting unit consisting of a light source such as a lamp or lamps together with the parts designed to position the light source and connect it to the power supply. A luminaire can include parts to protect the light source or the ballast or to distribute the light. A lampholder itself is not a luminaire.

❖ The term "luminaire," pronounced "lumenair," has replaced the term "lighting fixture."

MULTIOUTLET ASSEMBLY. A type of surface, or flush, or freestanding raceway; designed to hold conductors and receptacles, assembled in the field or at the factory.

❖ A multioutlet assembly is a raceway type of enclosure that is a long narrow strip with single receptacles spaced at regular intervals such as every 6 inches (152 mm). An 8-foot-long (2438 mm) multioutlet assembly, for example, could be installed on the wall above a workbench where it would be convenient to have several tools plugged in at once. It is designed to be mounted directly on the surface of the wall and is either factory-made or built on the job. It may come as a channel with fittings for couplings, connectors, corner installations and a snap-on cover that holds the receptacles. It could be direct wired by mounting over a receptacle outlet box. If it were used on a garage wall over a workbench, the receptacles would require GFCI protection. A multioutlet assembly may be used in a kitchen where several appliances need to be plugged in or at a desk or built-in counter where there is a need to plug in computers and related equipment.

NEUTRAL CONDUCTOR. The conductor connected to the neutral point of a system that is intended to carry current under normal conditions.

❖ A neutral conductor originates as a center tap of the secondary winding of the utility transformer that supplies power to a 3-wire, single-phase 120/240 volt system. The two ends of the secondary winding are the ungrounded conductors and the midpoint of the winding is the neutral point from which the neutral conductor originates. This neutral conductor serves as the common return conductor for the two ungrounded conductors. The neutral conductor is always a grounded conductor, but a grounded conductor is not always a neutral conductor. This sounds confusing and we need not discuss it further because in the wiring systems covered by this code, the only grounded conductor that exists is the neutral conductor. In systems beyond the scope of this code, there can be grounded phase conductors that are not connected to a neutral point in a transformer.

Most people refer to all grounded (white) conductors in branch circuits in a dwelling as neutral conductors because they all connect to the grounded neutral conductor in the service, although, the code refers to them as grounded conductors (see definition of "Neutral point").

NEUTRAL POINT. The common point on a wye-connection in a polyphase system or midpoint on a single-phase, 3-wire system, or midpoint of a single-phase portion of a 3-phase delta system, or a midpoint of a 3-wire, direct-current system.

❖ The only part of this definition that is applicable in this code is the reference to the midpoint on a single-phase, 3-wire system (see definition of "neutral conductor").

OUTLET. A point on the wiring system at which current is taken to supply utilization equipment.

❖ An "outlet" is not a junction box or any device box such as a switch box. A switch is installed in a device box, and it may be referred to by the electrician as an opening in the wiring system, but it is not an outlet because power is not taken from the distribution system at this point. A box where a receptacle or lighting fixture will be installed is an outlet.

OVERCURRENT. Any current in excess of the rated current of equipment or the ampacity of a conductor. Such current might result from overload, short circuit or ground fault.

❖ Conductors have a rated current which is the allowable ampacity. An overcurrent is a current in excess of the rated current. Overcurrent is caused by overloads and short circuits or ground faults. An overcurrent will cause the temperature of the conductor and/or equipment to rise to a level that could cause damage. In some cases, an overcurrent can exist for a certain length of time without causing damage. For installation in these cases, some overcurrent protection devices are built with a time delay so they will not trip instantaneously.

OVERLOAD. Operation of equipment in excess of normal, full-load rating, or of a conductor in excess of rated ampacity that, when it persists for a sufficient length of time, would cause damage or dangerous overheating. A fault, such as a short circuit or ground fault, is not an overload.

❖ Heat is produced as a result of an overload, which is due to a level of current above the rating of the conductors. The severity of the overload is determined by: 1. The amount of current; 2. The resistance of the conductor; and 3. The amount of time the overload persists. Overloading a conductor results in overcurrent that can cause conductor and device damage, excessive voltage drop and, if severe enough, can cause a fire.

PANELBOARD. A single panel or group of panel units designed for assembly in the form of a single panel, including buses and automatic overcurrent devices, and equipped with or without switches for the control of light, heat or power circuits, designed to be placed in a cabinet or cutout box placed in or against a wall, partition or other support and accessible only from the front.

❖ A panelboard contains the final branch-circuit overcurrent devices, which are the points where the branch circuits originate. The panelboard is an assembly of busbars and mounting hardware designed to be enclosed in a cabinet or cutout box. In residential wiring, the panelboard is usually not distinguished as a separate piece of equipment from its enclosure because it is normally installed inside the cabinet at the factory. Panelboards intended for flush mounting in the wall have adjusting screws so that the dead front or cover can be set to the right depth.

PLENUM. A compartment or chamber to which one or more air ducts are connected and that forms part of the air distribution system.

❖ A plenum is a ductwork component that serves to connect multiple ducts together and is associated with furnaces and air handlers. See Section E3904.7, which addresses cavities within the building construction.

POWER OUTLET. An enclosed assembly that may include receptacles, circuit breakers, fuseholders, fused switches, buses and watt-hour meter mounting means, intended to supply and control power to mobile homes, recreational vehicles or boats, or to serve as a means for distributing power required to operate mobile or temporarily installed equipment.

❖ A pedestal-mounted power outlet is an assembly supplied by an underground feeder or service lateral. A power outlet can also be mounted on a pole or other surface and supplied by an overhead service drop. It has receptacles and overcurrent devices installed in a weatherproof cabinet or box, and can include a meter. It can also be used to provide temporary power during construction.

PREMISES WIRING (SYSTEM). Interior and exterior wiring, including power, lighting, control and signal circuit wiring together with all of their associated hardware, fittings and wiring devices, both permanently and temporarily installed. This includes wiring from the service point or power source to the outlets and wiring from and including the power source to the outlets where there is no service point. Such wiring does not include wiring internal to appliances, luminaires, motors, controllers, and similar equipment.

❖ The premises wiring system includes all of the installed conductors and equipment inside and outside the dwelling. This definition makes it clear that the premises wiring system begins at the utility service point, however, where there is no service point, the premises wiring includes the power source such as solar and wind generation systems. For an overhead service, the service point is at the load end of the utility service drop or at the service head. For an underground service, it is at the load end of the utility service lateral. All of the wiring and equipment from that point on is covered by this code except internal wiring of appliances and factory-made equipment.

QUALIFIED PERSON. One who has the skills and knowledge related to the construction and operation of the electrical equipment and installations and has received safety training to recognize and avoid the hazards involved.

❖ The code may limit access to, operation of or installation of certain equipment to persons who have the ability to work safely with such equipment. For example, NFPA 70 will relax some safety provisions where it can be demonstrated that only qualified persons will be working on or around certain equipment.

RACEWAY. An enclosed channel of metallic or nonmetallic materials designed expressly for holding wires, cables, or busbars, with additional functions as permitted in this code.

❖ Table E3801.2 lists the types of raceways used in residential wiring. Raceways include any channel that encloses conductors including conduit and tubing. The term "conduit" is used in the descriptive name of various types of conduit and is different from "tubing."
EMT, for example, is not conduit. This distinction is important in relation to the rules of the support of boxes with devices and lighting fixtures. The use of EMT for this support is not permitted, but the use of conduit is (see Sections E3906.8.4 and E3906.8.5).

RAINPROOF. Constructed, protected or treated so as to prevent rain from interfering with the successful operation of the apparatus under specified test conditions.

❖ The definition of "Rainproof" is similar to the definition of "Weatherproof" in that the box, enclosure or fitting, is made and installed so that rain will "not interfere with the successful operation" of the contained apparatus. When the enclosure is rainproof, it has been tested to ensure that exposure to a beating rain will not interfere with the operation or result in the live parts and wiring within the enclosure becoming wet. It does not mean that the enclosure is sealed against the entry of water.

RAIN TIGHT. Constructed or protected so that exposure to a beating rain will not result in the entrance of water under specified test conditions.

❖ The definition of "Rain tight" is similar to the definition of "Water tight," which concerns the entrance of water. The testing of equipment designed to be rain tight simulates exposure to a beating rain, and the result must be that water will not enter the enclosure. For example, the weatherhead or service head must be "rain tight" (see Section E3605.9.1).

RECEPTACLE. A receptacle is a contact device installed at the outlet for the connection of an attachment plug. A single receptacle is a single contact device with no other contact device on the same yoke. A multiple receptacle is two or more contact devices on the same yoke.

❖ The distinction between "outlet" and "receptacle" is important. The outlet is the point on the wiring system where a device box and receptacle are installed. A receptacle is the device installed in the box at the outlet. A receptacle can be single, duplex or multiple. The term "yoke" is not defined in this chapter but is used to describe the strap on which the receptacle is mounted and that is made to attach to the device box. A strap or yoke can hold a single device or multiple devices.

Chapter 39 of the code requires receptacle outlets to be installed at specific locations. Duplex receptacles are universally used; however, single and triplex receptacles are also found in some locations.

RECEPTACLE OUTLET. An outlet where one or more receptacles are installed.

❖ See the commentary to "Receptacle."

SERVICE. The conductors and equipment for delivering energy from the serving utility to the wiring system of the premises served.

❖ "Service" is a general term used to refer to all of the service conductors and equipment. A key part of this definition is that the service begins at the load end of the service drop or lateral and ends at the enclosure that houses the main disconnecting means.

SERVICE CABLE. Service conductors made up in the form of a cable.

❖ Service conductors for dwellings are installed in a raceway or in a cable assembly called the service-entrance cable, although almost all modern services are separate conductors in a raceway. Cable Types SE and USE are service entrance cables.

SERVICE CONDUCTORS. The conductors from the service point to the service disconnecting means.

❖ It is clear that service conductors include service-entrance conductors and any other conductors up to the main disconnect. This definition typically excludes the service drop or service lateral conductors because such conductors are on the supply side of the service point. What is included and what is excluded depends on where the service point is. See the definition of "Service point."

SERVICE CONDUCTORS, OVERHEAD. The overhead conductors between the service point and the first point of connection to the service-entrance conductors at the building or other structure.

❖ This definition recognizes that service conductors not owned by the serving utility could exist between the service point and the service-entrance conductors at or on the building. In such cases, the service point is not at the building and is located, for example, on a pole or on an accessory structure on the premises.

SERVICE CONDUCTORS, UNDERGROUND. The underground conductors between the service point and the first point of connection to the service-entrance conductors in a terminal box, meter, or other enclosure, inside or outside of the building wall.

❖ See the commentary to "Service conductors, overhead."

SERVICE DROP. The overhead service conductors between the utility electric supply system and the service point.

❖ Service drop conductors run from the overhead utility line or pole and are connected at the weatherhead to

the service-entrance conductors. They are usually owned and maintained by the utility company. Where utility owned, the size of the service drop conductors is decided by the utility company and is not dictated by this code.

SERVICE-ENTRANCE CONDUCTORS, OVERHEAD SYSTEM. The service conductors between the terminals of the service equipment and a point usually outside of the building, clear of building walls, where joined by tap or splice to the service drop or overhead service conductors.

❖ Overhead service-entrance conductors run from the weatherhead to the main disconnect. The layout and configuration of the service equipment may be such that only a meter is on the outside of the house with the conductors run to the main disconnect on the inside of the house. The code does not dictate that the main disconnect be on the outside of the house except as might be required by Section E3601.6.2. An electric meter installed between the weatherhead and the main disconnect is not relevant. The service-entrance conductors continue inside the house to the main disconnect. Where the meter is in the same box or enclosure as the main disconnect, the service-entrance conductors would run only from the weatherhead down (in raceway or cable) to the meter/main disconnect enclosure.

SERVICE-ENTRANCE CONDUCTORS, UNDERGROUND SYSTEM. The service conductors between the terminals of the service equipment and the point of connection to the service lateral or underground service conductors.

❖ For an underground service, the service-entrance conductors connect to the service lateral and run to the terminals of the service equipment. Where the meter-main combination cabinet is on the outside of the house, there are no service-entrance conductors because the service lateral is defined as the conductors from the utility supply to the first terminal box, meter or enclosure at the structure. If the first enclosure is the location of the main disconnect, which could be an enclosure with a meter and the main disconnect together, the service-lateral conductors connect at the terminals of this equipment, and there are essentially no service-entrance conductors. In this installation, if the panelboard were also in the same enclosure, there would be no feeder to supply the panelboard, and the installer would not need to calculate the size of the service-entrance conductors. If the panel-board were on the inside of the house, a feeder would be installed to serve the panelboard, and the size of feeder would need to be calculated.

SERVICE EQUIPMENT. The necessary equipment, usually consisting of a circuit breaker(s) or switch(es) and fuse(s), and their accessories, connected to the load end of the service conductors to a building or other structure, or an otherwise designated area, and intended to constitute the main control and cutoff of the supply.

❖ See the commentary to "Service."

SERVICE LATERAL. The underground service conductors between the electric utility supply system and the service point.

❖ The utility company may have an underground supply system with transformers on pads at or near the property line. The service lateral runs underground to the building from a transformer or connection box installed at the ground level. Where there is an overhead transmission line, the homeowner may not want a service drop (over-head line) run across the property. In such cases, the conductors can be run down a utility pole in a raceway and then run underground to the house. Both of these installations would be called a service lateral, which is defined as the conductors from the utility main line to the point at which they connect to the service equipment or service-entrance conductors at the house. The service lateral typically terminates at a meter or main disconnect enclosure.

Usually the utility company is responsible for maintaining the conductors of the service lateral.

The electrician or contractor is usually responsible for installing a raceway conductor enclosure from the service equipment down to the required burial depth of the lateral.

SERVICE POINT. The point of connection between the facilities of the serving utility and the premises wiring.

❖ The service point is determined by the electric utility company. This is the point of connection between the electric-utility-owned conductors and the premises wiring-system conductors. If the overhead service drop is installed and owned by the utility company, the service point would be at the connection of the service-entrance conductors to the service drop at a point close to the weatherhead. If the utility company owns the underground service lateral conductors, the service point would be where the service lateral conductors connect to the terminals of the service equipment.

STRUCTURE. That which is built or constructed.

❖ Structures include buildings, houses, towers, fences, sheds and swimming pools.

SWITCHES.

General-use switch. A switch intended for use in general distribution and branch circuits. It is rated in amperes and is capable of interrupting its rated current at its rated voltage.

❖ A general-use switch may be used to control equipment, appliances, or a feeder. These switches are rated in amperes and voltage and are designed and permitted to interrupt up to their rated current at their rated voltage. An example of a general-use switch is the disconnect switch for a condensing unit of an air-conditioner system mounted on the outside of the house. These switches are either fused or nonfused and are often referred to by electricians as safety switches. They are installed in a cabinet or cutout box, which is usually mounted on the surface of the wall and has a hinged door that opens on the side or the top. In many of these installations, the switch handle is on the side and is operated vertically. In some safety switch enclosures, the switch is a pull-out block in the front with a handle. A safety feature of some of these switches is that the door will not open when the switch is in the closed position, and some are built so that when the door is opened, a mechanism moves an insulated cover over the live parts.

General-use snap switch. A form of general-use switch constructed so that it can be installed in device boxes or on box covers or otherwise used in conjunction with wiring systems recognized by this code.

❖ General-use snap switches come from the factory mounted on a yoke or strap, the same as a receptacle yoke, and are installed in a device box with two screws. They also may be used in other ways such as attached to the raised cover of a four-square metal box. General-use snap switches are the common toggle switches used throughout the dwelling for the majority of the switching needs. They are installed to control basically all of the lighting fixtures and also motor loads such as the garbage disposal and pump motors. The switches are rated in amperes and voltage, and these ratings are stamped or marked on the switch. For example, the yoke of the switch might have stamped on it "15A, 120VAC." It is important to use these switches within their ratings.

AC general-use snap switches are designed for and can be used up to a maximum of the full current rating of the switch to control: 1. Resistive loads such as incandescent lighting and heating loads; 2. Inductive loads such as fluorescent lighting; and 3. Tungsten-filament lamp loads of 120 volts. For example, an AC general-use snap switch rated at 15 amperes can control a resistive load of 15 amperes. For stationary motors of 2 horsepower (1.5 kW) or less, an ac general-use snap switch is permitted as the disconnect where the full-load current rating of the motor is a maximum of 80 percent of the rating of the switch. For example, if an ac general-use snap switch were used as the disconnect for a 1-horsepower (0.75 kw) motor with a full-load current rating of 16 amperes, a 20-ampere rated switch would be required. The full-load motor current of 16 amperes times 125 percent is 20 amperes. A 15-ampere rated switch would be permitted to control a motor with a maximum full-load current of 12 amperes or 80 percent of the 15 amperes.

AC/DC general-use snap switches are designed for and can be used up to a maximum of the full current rating of the switch to control resistive loads and tungsten-filament lamp loads. These switches are tested for a tungsten-filament lamp load with a rating of 125 volts and are marked with a "T" where intended for use with this type of lamp, which draws a high current when first turned on. Where an AC/DC general-use snap switch is used for inductive loads, the load cannot exceed 50 percent of the rating of the switch. Some residential lighting circuits could serve relatively large inductive lighting loads such as fluorescent lights, mercury vapor lights, or high-intensity discharge fixtures. If a 20-ampere rated AC/DC general-use snap switch were used for these types of lights, the load could be a maximum of only 10 amperes.

Isolating switch. A switch intended for isolating an electric circuit from the source of power. It has no interrupting rating and is intended to be operated only after the circuit has been opened by some other means.

❖ This type of switch is used mainly for large equipment or for entire parts of an electric wiring system. It is not designed to be operated under load conditions. Such switches are uncommon in residential installations. For a dwelling, a meter socket can serve as an isolation switch by pulling the meter from its socket and placing a cover over the meter opening in the enclosure.

Motor-circuit switch. A switch, rated in horsepower that is capable of interrupting the maximum operating overload current of a motor of the same horsepower rating as the switch at the rated voltage.

❖ A motor-circuit switch must have a horsepower rating of at least the horsepower of the motor and must be able to interrupt the maximum overload current of the motor. A horsepower-rated motor-circuit switch may be a toggle switch and will have the rating stamped or marked on it. But not all toggle switches are motor-circuit switches, and not all motors in a dwelling require a motor-circuit switch. The disconnect switch for a motor could be the branch-circuit overcurrent device if the motor is $^1/_8$ horsepower (93 W) or less. And an ac general-use snap switch could be used for a motor of 2 horsepower (15 kW) or less if the switch has a rating of 125 percent of the motor full-load current.

UNGROUNDED. Not connected to ground or to a conductive body that extends the ground connection.

❖ Ungrounded conductors are typically referred to as "hot" conductors, although this is erroneous because the grounded conductors carry the same voltage and current as the ungrounded conductors in all two-wire circuits. Ungrounded conductors are color coded by any color except white, green and gray. In cable wiring systems, the lone ungrounded conductor in a cable will be black and a second ungrounded conductor will be red and beyond that they will be any other color except whit, green and gray. In dwelling wiring systems, all conductors are ungrounded except the neutral, ungrounded (white), equipment grounding and grounding electrode conductors.

UTILIZATION EQUIPMENT. Equipment that utilizes electric energy for electronic, electromechanical, chemical, heating, lighting or similar purposes.

❖ This equipment, such as a range, exhaust fan, luminaire, or electric furnace, is differentiated by its use from equipment that does not use energy. Junction boxes, switches and raceways are not utilization equipment.

VENTILATED. Provided with a means to permit circulation of air sufficient to remove an excess of heat, fumes or vapors.

❖ Some types of equipment are ventilated such as motors, transformers, recessed lighting fixtures and cabinets. This definition for the electrical chapters does not necessarily refer to the ventilation system for the occupants of the building.

VOLTAGE (OF A CIRCUIT). The greatest root-mean-square (rms) (effective) difference of potential between any two conductors of the circuit concerned.

❖ See the definition of "Voltage (Nominal)."

VOLTAGE, NOMINAL. A nominal value assigned to a circuit or system for the purpose of conveniently designating its voltage class (e.g., 120/240). The actual voltage at which a circuit operates can vary from the nominal within a range that permits satisfactory operation of equipment.

❖ The voltage of the power supplied by the electric utility varies somewhat by regions and by power demand but typically not enough to cause any harm to the components of the premises wiring system or the utilization equipment. The circuit voltage and the voltage to ground are the actual voltages as measured. The nominal voltage is used as a common reference and for calculations [see Commentary Figure E3501(2)].

VOLTAGE TO GROUND. For grounded circuits, the voltage between the given conductor and that point or conductor of the circuit that is grounded. For ungrounded circuits, the greatest voltage between the given conductor and any other conductor of the circuit.

❖ See the definition to "Voltage (nominal)."

WATERTIGHT. Constructed so that moisture will not enter the enclosure under specified test conditions.

❖ See the commentary to "Rain tight."

WEATHERPROOF. Constructed or protected so that exposure to the weather will not interfere with successful operation.

❖ See the commentary to "Rainproof."

Bibliography

The following resource materials were used in the preparation of the commentary for this chapter of the code:

NFPA 70–14, *National Electrical Code*. Quincy, MA: National Fire Protection Association, 2013.

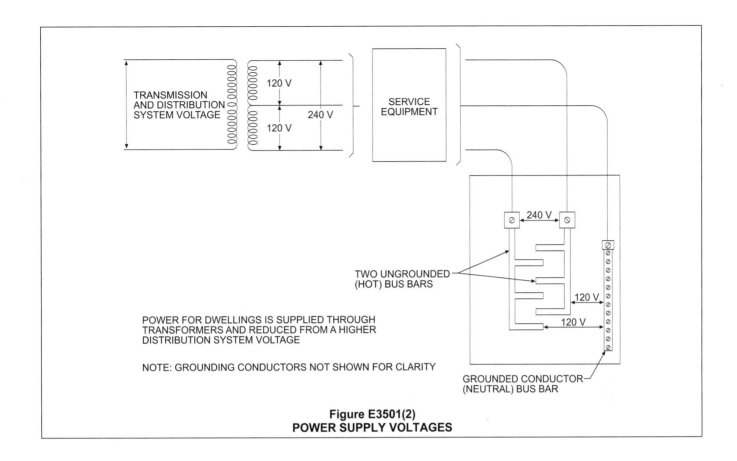

Figure E3501(2)
POWER SUPPLY VOLTAGES

Chapter 36: Services

General Comments

This chapter is the first of the logical order of chapters that mimics the normal sequence of dwelling construction. The first step in dwelling wiring is typically the sizing, design and installation of the service that is the source of power for the building. This chapter addresses the sizing of services, service conductor sizing and installation, system grounding and bonding, overcurrent protection, disconnecting means and the grounding electrode system.

Purpose

This chapter requires services to be properly sized to serve the load. This is intended to prevent overloading and to provide the utility expected by the occupants. This chapter also intends to protect occupants and the building from fire and protect the occupants from electrical shock hazards associated with service conductors and equipment.

SECTION E3601
GENERAL SERVICES

E3601.1 Scope. This chapter covers service conductors and equipment for the control and protection of services and their installation requirements. (230.1)

❖ A logical place to begin when planning the wiring for a house is the service. The size and location of the service equipment and main distribution panelboard are usually determined before the branch circuits are laid out. The service delivers energy from the serving utility to the wiring system of the premises and includes all of the equipment and conductors for this purpose. The electric utility provides power to the dwelling in an overhead service drop or in an underground service lateral. This chapter covers these conductors and all wiring and equipment up to the main disconnect, including the system grounding and bonding requirements. Equipment grounding is covered in later chapters, but the rules for the grounding electrode system are related directly to the service.

E3601.2 Number of services. One- and two-family dwellings shall be supplied by only one service. (230.2)

❖ The code requires that a building be supplied by only one service; that is, one overhead service drop or one underground service lateral run to the one- or two-family dwelling. In some cases, there may be a need for each unit of a two-family dwelling to be considered as a separate building. Generally, where a building is separated into two areas or occupancies, the code requires a 1- or 2-hour fire-resistant wall and/or floor assembly between the two areas. For example, the code defines a townhouse as a single-family dwelling and considers the dwelling as a separate building. See the *International Residential Code®* (IRC®) *Commentary, Volume I*, Sections R202 and R302.2. Because Section R302.3 of the code states that units in two-family dwellings must be separated from each other by wall and/or floor assemblies of not less than 1-hour

fire-resistance rating (note exceptions), the two dwellings in a two-family dwelling would not be considered as separate buildings, and two services would not be permitted. If the need existed for two services, it would be with special permission from the code official, and the fire-resistance rating of the separation between the two dwelling units would be considered in the decision.

Anyone observing existing older dwellings has probably seen a house that was remodeled to create a separate apartment with a separate service drop and separate service equipment. Special written permission of the code official is required for this kind of installation.

E3601.3 One building or other structure not to be supplied through another. Service conductors supplying a building or other structure shall not pass through the interior of another building or other structure. (230.3)

❖ The service entrance conductors are not permitted to be run through one building to serve another. For example, the service equipment for a townhouse could not pass through the crawl space or attic of another townhouse (see Commentary Figure E3601.3). This section does not apply to feeders. Service conductors are not protected from fault currents and pose a risk; whereas, feeder conductors are fully protected from shorts, ground faults and overcurrent, and therefore pose no risk.

E3601.4 Other conductors in raceway or cable. Conductors other than service conductors shall not be installed in the same service raceway or service cable. (230.7)

Exceptions:

1. Grounding electrode conductors and equipment bonding jumpers or conductors.

2. Load management control conductors having overcurrent protection.

❖ The set of fuses or a circuit breaker that is the overcurrent protection for the service conductors is an integral part of the service disconnecting means and is located at the load end of the service lateral or service entrance conductors. By definition, service conductors end at the main disconnect, and any conductors after that point on the system are feeders and branch-circuit conductors. Because service conductors are run directly into the service equipment from the utility company's supply lines, they are protected only at the load end. If feeder or branch-circuit conductors were in the same raceway with the service conductors, they could be subject to fault current.

E3601.5 Raceway seal. Where a service raceway enters from an underground distribution system, it shall be sealed in accordance with Section E3803.6. (230.8)

❖ See the commentary to Section E3803.6.

E3601.6 Service disconnect required. Means shall be provided to disconnect all conductors in a building or other structure from the service entrance conductors. (230.70)

❖ The code requires that there be a means for disconnecting all conductors from the service entrance conductors. The disconnecting means for the ungrounded conductors is almost always an overcurrent device in the form of a circuit breaker or set of fuses. If the service overcurrent device is not an integral part of the service disconnecting means, the main disconnect must be adjacent to the overcurrent device. It is rarely done, but a fuseless safety switch could be installed as the main disconnect; however, it must be next to the service overcurrent device.

The main service disconnect switch, circuit breaker

or set of fuses provides a means to dis-connect the ungrounded conductors from the supply. No overcurrent device should be placed in series with the grounded (neutral) conductor unless that device will disconnect both ungrounded and grounded conductors. Such devices are not used in residential services.

The code requires that the service equipment have a means to disconnect all conductors from the supply, which includes the grounded service conductor or neutral. The neutral is cut and terminated in a bus bar or lugs provided in the service equipment for this purpose. For example, where the main disconnect is located on the outside of the house and the panelboard is on the inside, the neutral could not be run through the service equipment and on to the panelboard without cutting it. Means must be provided to disconnect the neutral or remove it from the terminal at the location of the main disconnect; it need not be disconnected with a switch or as part of the overcurrent device. The code rule is satisfied if the neutral can be removed from the terminal or bus bar.

E3601.6.1 Marking of service equipment and disconnects. Service disconnects shall be permanently marked as a service disconnect. [230.70(B)]

❖ In a house with an overhead service, it may be obvious that the power is supplied through the service drop and continues down through the service-entrance conductors to the enclosure where the main disconnect is located. In an emergency, for example in the case of a fire, the fire department personnel could easily see

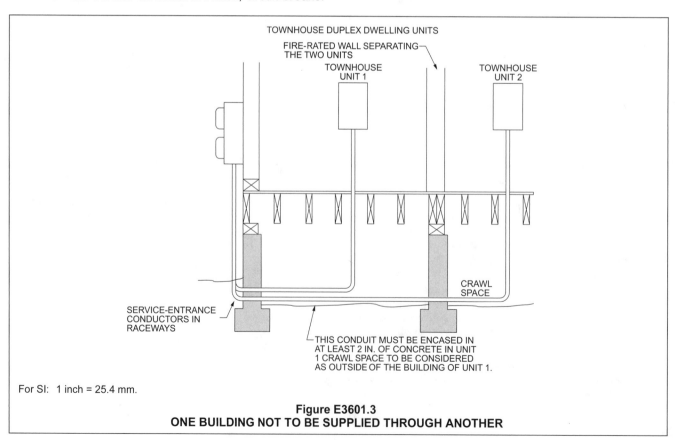

For SI: 1 inch = 25.4 mm.

Figure E3601.3
ONE BUILDING NOT TO BE SUPPLIED THROUGH ANOTHER

where the main disconnect is located and turn off the power. But in many houses there may be enclosures, gutters, raceways and cables on the house. These might be part of the service equipment or other boxes on the load side of the service equipment. Therefore, it is important to have the main disconnect clearly marked. On some older dwellings, the only way to disconnect the power was by pulling the meter. An individual meter socket is not considered service equipment because service equipment constitutes the main control and cutoff of the supply (see the definition of "Service equipment" in Chapter 35). The code does not require a meter, but permits it.

E3601.6.2 Service disconnect location. The service disconnecting means shall be installed at a readily accessible location either outside of a building or inside nearest the point of entrance of the service conductors. Service disconnecting means shall not be installed in bathrooms. Each occupant shall have access to the disconnect serving the dwelling unit in which they reside. [230.70(A)(1), 230.72(C)]

❖ The code permits the service disconnect to be located on either the outside or the inside of a building (see Section E3603.3.3). It need not be attached to or mounted on the house, although it almost always is. Various reasons from practical to personal have been given for locating it on the outside or the inside. Practical reasons may include making it easier for fire service personnel to find it outside the house and turn off the power in case of a fire. In other cases, it is installed inside the house for personal reasons like reducing the likelihood of a burglar turning off the power before attempting to enter.

Wherever the main disconnect is installed, it must be at a readily accessible location (see the definition of "Readily accessible" in Chapter 35). The maximum height permitted for the grip of the operating handle of the disconnect is 6 feet, 7 inches (2007 mm) above the floor or ground (see Section E4001.6). On a house under construction, the final grade level around the house might not be established when the service equipment is installed. After the electrician installs the main disconnect, the contractor might excavate and lower the surrounding grade, making the disconnect out of reach. For an underground service, by grading the yard, the contractor might decrease the burial depth of the service lateral, resulting in a code violation. It is important to consider what the final grade will be before mounting the main disconnect. The same caution also applies to meter placement. Although the code does not specify any dimension, the power company usually has a required range of height above the ground.

Where the main disconnect is located on the inside of the house, a question arises of how far inside the dwelling the unprotected service conductors are allowed to be run before being terminated at the main disconnect. The concern is that the service conductors are unprotected from fault-current and short-current up to the point of the main disconnect, which is usually the main overcurrent device. The available

short-circuit current or fault current is usually very high. In case of a fault current or short circuit, the current flow could be of a magnitude that would heat up and damage the conductors, which could start a fire. If the decision is made to locate the main disconnect inside the house, the code indicates that it must be at a location "nearest the point of entrance of the service conductors."

Many code provisions are very prescriptive and state an exact dimension, but because of the many variables and configurations of laying out the service equipment, the length of the service conductors inside the dwelling is left to the discretion of the code official. Some jurisdictions establish a length limit for this. For instance, the length of the conductors might be permitted to run a maximum of 10 or 15 feet (3048 or 4572 mm) inside the building before being terminated at the load end at the main disconnect and overcurrent device. If the requirement stated that the main disconnect must be placed on the inside wall at the point at which the service conductors enter, the disconnect might be at a location that is either not readily accessible or convenient. Because there are so many different house plans and designs, it is impossible to have detailed prescriptive code provisions that would address every situation [see Commentary Figures E3601.6.2(1) and (2)].

E3601.7 Maximum number of disconnects. The service disconnecting means shall consist of not more than six switches or six circuit breakers mounted in a single enclosure or in a group of separate enclosures. [230.71(A)]

❖ A person must be able to turn off all the power to a house with no more than six motions of the hand while at one location. The main disconnect does not have to be one switch, one circuit breaker or one set of fuses and does not have to be in only one enclosure. For new construction, the service disconnect or disconnects are usually in one enclosure, but on an existing dwelling, for example where a load is being added, an additional main disconnect could be installed in a separate enclosure. However, all the disconnects must be grouped or located together and marked. There may be up to six enclosures that constitute the main disconnecting means for the dwelling, and each one must be clearly marked.

This code provision is usually not a problem in new construction. A house that has only one panelboard with all loads fed from that panelboard would require only one main disconnect, such as a two-pole circuit breaker. Although it is not a code requirement, installing service equipment that has room for additional circuit breakers may be helpful for future expansion. Service equipment enclosures are available for residential wiring with spaces for up to 6 two-pole circuit breakers. With this design, feeders could be run from the main service equipment instead of from the panelboard for such things as a subpanel for swimming pool equipment or for an outbuilding, detached garage or air conditioner.

In an older dwelling, there might be six circuit breakers or fuses that serve the branch circuits with no other main disconnect. If that dwelling is being remodeled and circuits are run for new kitchen appliances or for a room addition, it may be necessary to change the service equipment because the addition of another branch circuit overcurrent device would be over the six maximum permitted for turning off all the power to the house. Obviously, when adding more circuits, the size of the existing service conductors would be a major concern in the decision to install new service equipment, but it would not be the only concern. The number of disconnects could be the deciding factor.

See Section E3603.3.1, which provides for the overload protection of ungrounded service conductors, and Commentary Figure E3601.7.

This section is applied on a per-building basis in cases where one service supplies more than one set of service conductors. For example, in rural locations it is not uncommon to find a service installed on an outbuilding, such as pole barn, and that service supplies two sets of service conductors-one for the pole barn and one for the dwelling. In this case, up to six disconnects would be allowed for each building and they would not be required to be grouped at the service location on the pole barn.

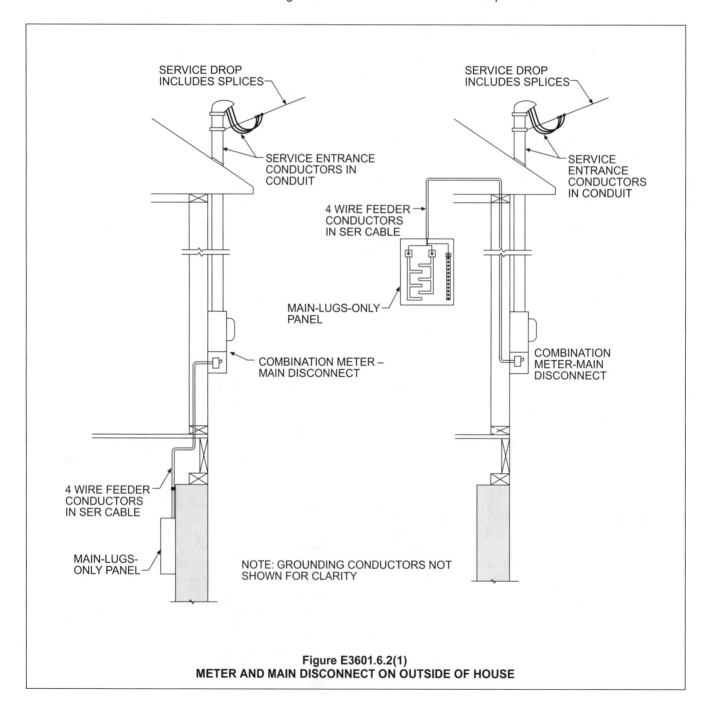

Figure E3601.6.2(1)
METER AND MAIN DISCONNECT ON OUTSIDE OF HOUSE

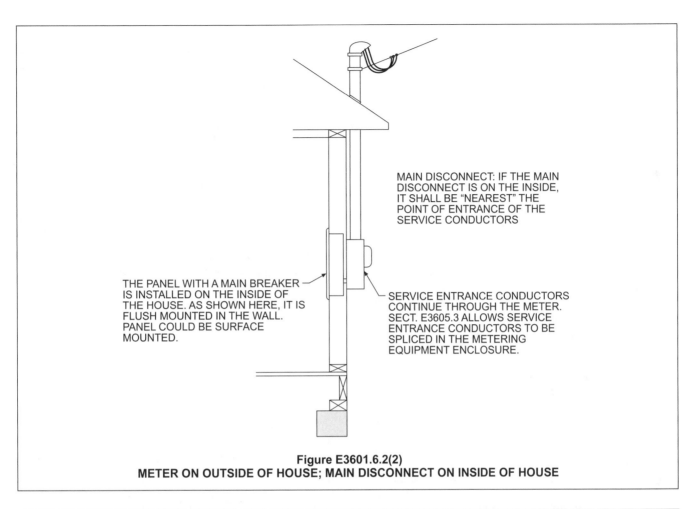

MAIN DISCONNECT: IF THE MAIN DISCONNECT IS ON THE INSIDE, IT SHALL BE "NEAREST" THE POINT OF ENTRANCE OF THE SERVICE CONDUCTORS

THE PANEL WITH A MAIN BREAKER IS INSTALLED ON THE INSIDE OF THE HOUSE. AS SHOWN HERE, IT IS FLUSH MOUNTED IN THE WALL. PANEL COULD BE SURFACE MOUNTED.

SERVICE ENTRANCE CONDUCTORS CONTINUE THROUGH THE METER. SECT. E3605.3 ALLOWS SERVICE ENTRANCE CONDUCTORS TO BE SPLICED IN THE METERING EQUIPMENT ENCLOSURE.

Figure E3601.6.2(2)
METER ON OUTSIDE OF HOUSE; MAIN DISCONNECT ON INSIDE OF HOUSE

Figure E3601.7
SERVICE DISCONNECTING MEANS CONSISTING OF SIX OR FEWER SWITCHES OR CIRCUIT BREAKERS

SECTION E3602
SERVICE SIZE AND RATING

E3602.1 Ampacity of ungrounded conductors. Ungrounded service conductors shall have an ampacity of not less than the load served. For one-family dwellings, the ampacity of the ungrounded conductors shall be not less than 100 amperes, 3 wire. For all other installations, the ampacity of the ungrounded conductors shall be not less than 60 amperes. [230.42(B), 230.79(C) & (D)]

❖ The minimum service rating for any dwelling unit is 100 amperes, but the load must be calculated to determine whether 100 amperes is sufficient. The load depends on many factors, such as use of gas appliances or electric space heating and water heating. The square footage of the house is another factor, but it is only one of the elements that should be taken into account. Some electricians may have a guideline that would require a certain size service corresponding to the square footage of the house, such as a minimum 100-ampere service for a house of 1,000 square feet (93 m²), a 125-ampere rating for a house up to 1,250 square feet (116 m²), a 150-ampere rating for a house of 1,500 square feet (139 m²) and so on. This kind of rule is not a code provision, but it may work in some cases. Because of the various loads that may be present in a dwelling, the code official, in enforcing the code, would require calculations to show that the rating of the service is sufficient.

The provision in this paragraph for a 60-ampere-rated service does not apply to the dwelling but would be for something like a separate service on the residential property such as on a farm where a service might be run to a remote building. Exceptions to the minimum 60-ampere rating are in Section E3603.2.

E3602.2 Service load. The minimum load for ungrounded service conductors and service devices that serve 100 percent of the dwelling unit load shall be computed in accordance with Table E3602.2. Ungrounded service conductors and service devices that serve less than 100 percent of the dwelling unit load shall be computed as required for feeders in accordance with Chapter 37. [220.82(A)]

❖ Where the service conductors serve 100 percent of the load, Table E3602.2 is used to calculate the size of the service conductors. For example, where an overhead service drop feeds the service entrance conductors, which are run to a service equipment cabinet that table must be used. This equipment is referred to by electricians in some areas as a meter-main combination, meaning that the enclosure has a meter base and a panelboard with the main disconnect or main breaker in the panel. In other words, it is a main-breaker panel combined with a meter base. In this case, the service entrance conductors terminated at the main breaker serve 100 percent of the load.

The service conductors must be of sufficient size to carry the load. The procedure set forth in this table is used to calculate the minimum load, in amperes, required for the service conductors. Once the load is

TABLE E3602.2
MINIMUM SERVICE LOAD CALCULATION [220.82(B) & (C)]

LOADS AND PROCEDURE
3 volt-amperes per square foot of floor area for general lighting and general use receptacle outlets.
Plus
1,500 volt-amperes multiplied by total number of 20-ampere-rated small appliance and laundry circuits.
Plus
The nameplate volt-ampere rating of all fastened-in-place, permanently connected or dedicated circuit-supplied appliances such as ranges, ovens, cooking units, clothes dryers not connected to the laundry branch circuit and water heaters.
Apply the following demand factors to the above subtotal:
The minimum subtotal for the loads above shall be 100 percent of the first 10,000 volt-amperes of the sum of the above loads plus 40 percent of any portion of the sum that is in excess of 10,000 volt-amperes.
Plus the largest of the following:
One-hundred percent of the nameplate rating(s) of the air-conditioning and cooling equipment.
One hundred percent of the nameplate rating(s) of the heat pump where a heat pump is used without any supplemental electric heating.
One-hundred percent of the nameplate rating of the electric thermal storage and other heating systems where the usual load is expected to be continuous at the full nameplate value. Systems qualifying under this selection shall not be figured under any other category in this table.
One-hundred percent of nameplate rating of the heat pump compressor and sixty-five percent of the supplemental electric heating load for central electric space-heating systems. If the heat pump compressor is prevented from operating at the same time as the supplementary heat, the compressor load does not need to be added to the supplementary heat load for the total central electric space-heating load.
Sixty-five percent of nameplate rating(s) of electric space-heating units if less than four separately controlled units.
Forty percent of nameplate rating(s) of electric space-heating units of four or more separately controlled units.
The minimum total load in amperes shall be the volt-ampere sum calculated above divided by 240 volts.

determined, the conductor size can be selected in accordance with Section E3603.1. For many smaller houses, especially where gas is used for space heating and water heating, the calculation using Table E3602.2 will usually result in a load of 100 amperes or less. Of course, the minimum rating required for ungrounded service conductors is 100 amperes as stated in Section E3602.1, even if the calculated load is less. The code requirements to calculate the minimum load in order to select the size conductors that are necessary for safety. However, the minimum conductor size and service rating may not be adequate for additional loads that could be added later. A good practice is to calculate the minimum load as required and then consider the feasibility and costs of installing a service larger than required to provide for future expansion.

The following is a description of the loads listed in the table.

General Lighting and Convenience Receptacle Loads:

A unit load of not less than 3 volt-amperes per square foot constitutes the minimum lighting load for each square foot of floor area. The floor area must be computed from the outside dimensions of the dwelling and does not include open porches, garages, or unused spaces not adaptable for future use. However, in all cases these unit values are based on minimum load conditions with a 100-percent power factor, and they may not provide sufficient capacity for the installation contemplated. Included in the lighting load are general use (convenience) receptacle outlets of 15- and 20-ampere rating. The bathroom receptacle outlets supplied by the separate branch circuit required by Section E3703.4 are also included in the general lighting load.

Small Appliance Branch Circuit Loads:

A load of 1,500 volt-amperes is included in the calculation for each two-wire small appliance branch circuit for portable appliances supplied by 15- or 20-ampere receptacles on 20-ampere branch circuits in the kitchen, pantry, dining room and breakfast room. Since the code requires a minimum of two small appliance circuits regardless of the size of the house, the minimum load to be included in the calculation is 3,000 volt-amperes. Where a dwelling has more than two small appliance branch circuits, a load of 1,500 volt-amperes must be included for each circuit. An electrician may choose to supply the refrigerator receptacle on a separate circuit. Because this is not required by the code, an additional 1,500 volt-amperes would not need to be added to the load calculation for this circuit.

Laundry Circuit Load:

A load of 1,500 volt-amperes is included for the circuit supplying the clothes washer and gas-fired dryer. Where the clothes washer is in a laundry room and additional receptacle outlets are provided, such as an outlet for the iron, these receptacles are permitted on the laundry circuit, and only one unit load of 1,500 volt-amperes is included.

Appliance Load:

This is the nameplate rating of all appliances that are fastened in place, permanently connected, or connected to a dedicated circuit including ranges, dryers, and water heaters. The reference to clothes dryers is addressing clothes dryers that are not served by the required 20-amp laundry branch circuit. In other words, the load for a gas-fired clothes dryer served by the 20-amp laundry branch circuit is already included in the 1,500 volt-

amperes counted for this circuit. An electric clothes dryer load must be added in this step because it is not accounted for in the previous step. For the calculation, an appliance that has the nameplate rating given only in amperes must have its rating converted to volt-amperes by multiplying the ampere rating by the rated voltage.

Air-conditioning and Heating Loads:

It is unlikely that air-conditioning (cooling equipment) and heating loads will be used simultaneously. Therefore, only the largest load is used in the calculation. The bottom half of Table E3602.2 lists cooling equipment loads and various combinations of space-heating loads. The largest of any of these loads present in the dwelling is used for the calculation.

Sample Calculation

Description of loads in a dwelling:

Volt-amperes are indicated as VA. Depending on the power factor, VA is not always equivalent to wattage. However, wattage is considered the same as volt-amperes for purposes of calculating service loads.

Sample Calculation: A dwelling with no gas supply for space or water heating has the following:

- Outside dimensions of 30 feet × 65 feet, which does not include the garage, porches, attic or cellar.
- Two small appliance circuits in the kitchen and related rooms.
- One laundry circuit to serve the clothes washer and outlets in the laundry room.
- Water heater rated at 4,500 watts.
- Clothes dryer rated at 5,000 VA.
- Range rated at 12,000 VA.
- Dishwasher rated at 1,200 VA.
- Disposal rated at 1,000 VA.
- Microwave oven rated at 1,800 VA.
- Air-conditioner condensing unit with a nameplate rating of 28 amperes, 240 volts.
- Gas furnace rated at 7 amperes, 120 volts for heating and cooling, but on a separate circuit.

Performing the calculation:

1. General lighting and general purpose receptacle loads:

   30 × 65 = 1,950 sq. ft. × 3 VA/sq. ft. = 5,850 VA

2. Small appliance and laundry circuit loads:

   Two small appliance branch circuits
   at 1,500 VA each = 3,000 VA
   One laundry branch circuit
   at 1,500 VA = 1,500 VA
   4,500 VA

3. Nameplate rating of appliances:

Water heater	4,500 VA
Clothes dryer	5,000 VA
Range	12,000 VA
Dishwasher	1,200 VA
Disposal	1,000 VA
Microwave oven	1,800 VA
Gas furnace: 7 amperes × 120 volts	840 VA
	26,340 VA

4. Combine loads from steps 1, 2, and 3:   36,690 VA

5. Apply demand factor:

Add VA in
this column

Up to first 10,000 at 100%: 10,000 VA

Remainder at 40% =

36,690 - 10,000 =

26,690 × 0.4   10,676 VA
Total   20,676 VA

6. Cooling/Heating Loads:

Gas furnace rated at 840 VA

Cooling equipment load is 28 amperes ×
240 volts =   6,720 VA

The largest of these is the
cooling load: 6,720 VA
Total 27,396 VA

Sizing the ungrounded service conductors:

27,396 VA/240 volts = 114.15 amperes. The standard ampere ratings of fuses and circuit breakers include 100, 110 and 125. Therefore, the main disconnect would need to be at least 125 amperes because the calculated load is over 110 amperes and below 125 amperes (see Sections E3705.5.2 and E3705.6). From Section E3603.1, the minimum size permitted for an ungrounded service conductor would be a size 2 AWG copper conductor or a size 1/0 AWG aluminum conductor from the 75°C column of Table E3705.1. (Per Section E3603.1.1, 0.83 × 125 amps = 104 amps.) The main breaker or set of fuses would be 125 amperes.

Sizing the grounded (neutral) service conductor:

The neutral conductor must be of a size to adequately carry the maximum unbalanced load in accordance with Sections E3603.1 and E3704.3. The size of the service ungrounded conductors that are part of the "main power feeder" is dictated by Section E3603.1. Because Section E3603.1.4 does not directly address the

sizing of grounded service conductors, Section E3704.3 provides for the sizing of such conductors by applying the feeder load calculation method. Section E3704.3 provides the definition of "maximum unbalanced load," which is applicable for any feeder neutral and for the grounded service conductor. It states: "The maximum unbalanced load shall be the maximum net computed load between the neutral and any one ungrounded conductor." Accordingly, the maximum unbalanced load for the 120/240-volt, three-wire range and dryer branch circuits is 70 percent of the load calculated for the ungrounded conductors serving these two appliances. Two-wire 240-volt loads, such as electric water heaters, do not add to the neutral load. The cooling equipment is not counted because it is a 240-volt load that does not have a neutral. Where a blower motor is used for air handling in air conditioning, it is usually a 120-volt motor and must be included in the calculation of the neutral load. Of course, when the circuit breakers are installed in the panelboard, the loads should be balanced as much as possible between the two ungrounded "hot" legs of the service. The following example assumes that the 120-volt loads are equally distributed between the two ungrounded service conductors. If they were not equally distributed across the two ungrounded service conductors (legs), the load calculation would have to be performed twice, once for each leg of the service based on the actual distribution as installed.

Calculating the Service Neutral Load Using Table E3704.2(1) and Section E3704.3:

Add VA in
this column

1. General lighting and convenience receptacle loads:

3 VA per sq. ft. 30 65 =
1950 × 3   =   5,850 VA

2. Small appliance and laundry circuit loads

Two small appliance branch circuit loads
1,500 VA each   =   3,000 VA

One laundry ranch circuit
at 1,500 VA   =   1,500 VA
     =   10,350 VA

Applying demand factor,
first 3,000 at 100%   =   3,000 VA

10,350 - 3,000 = 7,350
at 35%   =   2,573 VA
     =   5,573 VA

Total demand for lighting, receptacle, small appliances and laundry is: 5,573 VA

3. Appliance loads:

Water heater (two-wire, 240 volts,

no neutral load)	=	0 VA
Dishwasher	=	1,200 VA
Disposer	=	1,000 VA
Largest motor load: Disposer: 1000 × 25%	=	250 VA
Microwave Oven	=	1,800 VA
Gas furnace, 7 amperes × 120 volts	=	840 VA
Clothes dryer 5,000 VA × 70%	=	3,500 VA
Range at 8,000 VA × 70%	=	5,600 VA
Subtotal 14,190 VA		14,190 VA
	Total	19,763 VA

4. Cooling/heating: No neutral loads other than furnace.

The branch circuit load for the range is permitted to be calculated in accordance with Table E3704.2(2), which lists the demand load for one range with a rating of 12 kVA or less, as 8,000 VA. This demand load is not used for the range load when calculating the load for the service in accordance with Section E3602.2.

19,763 VA/240 volts = 82.34 amperes.

From Section E3603.1.4 and Table E3705.1, at 75°C (167°F), the minimum size permitted for the grounded service conductor would be a size 4 AWG copper conductor or a size 2 AWG aluminum conductor.

The total load was divided by 240 volts rather than 120 volts because this example uses the common calculation procedure, which assumes that the 120 volt loads are equally distributed between the two ungrounded conductors. The only alternative would be to add all of the 120-volt loads that are actually connected to one ungrounded conductor and divide the total by 120 volts, and this procedure must be repeated for the other ungrounded conductor to determine the greatest load (maximum unbalance). Of course, the lengthy exercise of calculating the neutral conductor size could be avoided by simply making the neutral conductor the same size as the ungrounded conductors. A "full-size" neutral has the same ampacity as the ungrounded conductors and will certainly be able to carry the maximum unbalanced load. Unless there is some economy of scale, the installation of a neutral conductor that is smaller than the ungrounded conductors is generally avoided because of the necessity to provide calculations to justify such an installation.

E3602.2.1 Services under 100 amperes. Services that are not required to be 100 amperes shall be sized in accordance with Chapter 37. [230.42(A), (B), and (C)].

❖ The load calculation method in Table E3602.2 is only for services of 100 amperes and larger. Therefore, smaller services must be sized by the method in Chapter 37. Where a service supplies a separate load, such

as a separate workshop or a panelboard at a small building that houses swimming pool equipment and a bath house, the conductors would be calculated as required for feeders according to Chapter 37 (see Section E3602.1).

E3602.3 Rating of service disconnect. The combined rating of all individual service disconnects serving a single dwelling unit shall be not less than the load determined from Table E3602.2 and shall be not less than as specified in Section E3602.1. (230.79 & 230.80)

❖ The service load is calculated to determine the ampacity rating of the service conductors and service equipment. When designing the service and performing the load calculation, it is commonly assumed that there will be only one switch for the main disconnecting means in the service equipment. However, Section E3601.7 permits the service disconnecting means to consist of up to six switches or circuit breakers, which could all be in the same enclosure or in a group of separate enclosures. The combined rating of the switches or circuit breakers must be at least equal to the calculated load and not less than the minimums dictated by Section E3602.1. If the load calculation for a dwelling is under 100 amperes, then the minimum required ampere rating for the service disconnect would be 100 amperes according to this section (see Section E3603.3.1).

E3602.4 Voltage rating. Systems shall be three-wire, 120/240-volt, single-phase with a grounded neutral. [220.82(A)]

❖ For purposes of performing load calculations, the voltage used must be 120/240 volts. The services considered under the provisions of this code are three-wire, single phase with a grounded neutral.

SECTION E3603
SERVICE, FEEDER AND GROUNDING ELECTRODE CONDUCTOR SIZING

E3603.1 Grounded and ungrounded service conductor size. Service and feeder conductors supplied by a single-phase, 120/240-volt system shall be sized in accordance with Sections E3603.1.1 through E3603.1.4 and Table 3705.1.

❖ Dwellings are supplied by 120/240-volt single-phase systems (see Section E3602.4) and the service and main feeder conductors are sized in accordance with Sections E3603.1.1 through E3603.1.4. The main power feeder is the feeder between the main disconnect and the panelboard that supplies, either by branch circuits or by feeders, or both, all loads that are part of or are associated with the dwelling unit.

E3603.1.1 Ungrounded service conductors. For a service rated at 100 through 400 amperes, the service conductors supplying the entire load associated with a one-family dwelling, or the service conductors supplying the entire load associated with an individual dwelling unit in a two-family dwelling, shall have an ampacity of not less than 83 percent of the service rating.

❖ Note that the code requires services to be at least 100 ampere rated for dwellings and the scope of the electrical part of the code cuts off at 400 amp rate services (see Sections E3401.2 and E3602.1). In previous editions of the code, ungrounded service conductors were sized from a table that allowed smaller conductors than would otherwise be allowed by Table E3705.1, based on assumed load diversity and the duration of loads for the typical dwelling. The ampacity of conductors in the 75°C column of Table E3705.1 is very close to 83 percent of the rating of corresponding conductors in the former Table E3603.1 for a specific service rating. The 75°C column is chosen because the terminals on service equipment will almost certainly have a maximum temperature rating of 75°C. It is apparent that Table E3603.1 from the 2012 IRC has been replaced with a simple calculation for determining 83 percent. For example, if the service rating is 100 ampere, then the ungrounded service conductors must have an ampacity of at least 83 amps (100 amp × 83% = 100 × 0.83 = 83 amp). Table E3705.1 indicates that a number 4 copper conductor under the 75°C column has an ampacity of 85. Likewise, if a service has a 200 amp rating, the ungrounded conductors must have an ampacity of at least 166 amps (200 × 0.83 = 166 amps). A number 2/0 copper conductor has an ampacity of 175 amps under the 75°C column in Table E3705.1.

The "rating" of a service is the calculated load for the service. The calculated service load equals the service rating. Service equipment is built in standard sizes, so the "service rating" translates to be the standard size of equipment that is the closest match to the calculated load for the service. The intent of the code is to apply the 83-percent calculation to the rating of the service disconnect (see Section E3602.3). The service disconnect equipment is built in standard sizes, as are circuit breakers. Because the service disconnect in dwellings is almost always a circuit breaker (or pair of breakers with handle ties), the service disconnect and service overcurrent device are typically one in the same. This means that the 83-percent calculation is applied to the rating of the main overcurrent device/disconnect device (see Sections E3705.5.2 and E3705.6).

E3603.1.2 Ungrounded feeder conductors. For a feeder rated at 100 through 400 amperes, the feeder conductors supplying the entire load associated with a one-family dwelling, or the feeder conductors supplying the entire load associated with an individual dwelling unit in a two-family dwelling, shall have an ampacity of not less than 83 percent of the feeder rating.

❖ This section is essentially the same as Section E3603.1.1 because feeder conductors that carry the entire load for the dwelling are acting no differently than the service conductors, as both service and "main feeder" conductors are carrying the entire load for the dwelling. Recall that the code used to refer to the conductors that carry the entire load for the dwelling as "main feeder" conductors. "Main feeder" conductors are those that, for example, extend from an outdoor main disconnect/overcurrent device and feed an indoor panelboard from which all branch circuits originate. The feeder rating is that which is necessary to serve the calculated load.

E3603.1.3 Feeder size relative to service size. A feeder for an individual dwelling unit shall not be required to have an ampacity greater than that specified in Sections E3603.1.1 and E3603.1.2.

❖ Service loads and feeder loads are calculated differently (see Sections E3602.2 and E3704.2); thus, the size of the conductors could vary by calculation. Sections E3603.1.1 and E3603.1.2 both require the same minimum 83-percent ampacity, but the loads for the feeder and the service could be different based on the two different load calculation methods for services and feeders. This section is simply stating that regardless of the calculations, feeder conductors do not have to be larger in size than the service conductors, for the logical reason that both sets of conductors carry the same loads. See Section E3602.2. The code is saying that main power feeders are treated as service conductors for purposes of sizing, despite the fact that they are actually feeders by definition.

E3603.1.4 Grounded conductors. The grounded conductor ampacity shall be not less than the maximum unbalance of the load and the size of the grounded conductor shall be not smaller than the required minimum grounding electrode conductor size specified in Table E3603.4. [310.15(B)(7)]

❖ The grounded service conductor (neutral) is sometimes reduced in size compared to the ungrounded conductors, but it should not be assumed that the neutral can always be smaller than the ungrounded conductors. It must be large enough to carry the maximum unbalance of current between it and either of the ungrounded conductors. Also, it must not be smaller than the minimum grounding conductor electrode size because it must carry any fault current back to the supply transformer. The neutral becomes the low-impedance path back to the power source. To determine the maximum unbalance of the load, add all of the 120-volt loads that are actually connected to one ungrounded conductor and divide the total by 120 volts. This procedure must be repeated for the other ungrounded conductor to determine the greatest load (maximum unbalance). Of course, the lengthy exercise of calculating the neutral conductor size could be avoided by simply making the neutral conductor the same size as the ungrounded conductors. A "full-size" neutral has the same ampacity as the ungrounded conductors and will certainly be able to carry the maximum unbalanced load. Unless there is some economy of scale, the installation of a neutral conductor smaller than the ungrounded conductors is generally avoided because of the necessity to provide calculations to justify such an installation. Also, there is the concern that the unbalance (or balance) could be changed as a result of rearranging circuits after the initial installation. A full-sized, ungrounded conductor greatly simplifies the system design.

E3603.2 Ungrounded service conductors for accessory buildings and structures. Ungrounded conductors for other than dwelling units shall have an ampacity of not less than 60 amperes and shall be sized as required for feeders in Chapter 37. [230.79(D)]

Exceptions:

1. For limited loads of a single branch circuit, the service conductors shall have an ampacity of not less than 15 amperes. [230.79(A)]

2. For loads consisting of not more than two two-wire branch circuits, the service conductors shall have an ampacity of not less than 30 amperes. [230.79(C)]

❖ Power supplied to accessory buildings and detached structures is usually by a feeder, although in some cases it could be a separate service. This section specifies the size of service conductors for such structures. Section E3607.3 contains important rules for system grounding for an accessory building.

E3603.3 Overload protection. Each ungrounded service conductor shall have overload protection. (230.90)

❖ See the commentary to Section E3603.3.1.

E3603.3.1 Ungrounded conductor. Overload protection shall be provided by an overcurrent device installed in series with each ungrounded service conductor. The overcurrent device shall have a rating or setting not higher than the allowable service or feeder rating specified in Section E3603.1. A set of fuses shall be considered to be all of the fuses required to protect all of the ungrounded conductors of a circuit. Single pole circuit breakers, grouped in accordance with Section E3601.7, shall be considered as one protective device. [230.90(A)]

Exception: Two to six circuit breakers or sets of fuses shall be permitted as the overcurrent device to provide the overload protection. The sum of the ratings of the circuit breakers or fuses shall be permitted to exceed the ampacity of the service conductors, provided that the calculated load does not exceed the ampacity of the service conductors. [230.90(A) Exception No. 3]

❖ The overload protection must be located at the service equipment and is typically integral with the main disconnect in the form of a circuit breaker or set of fuses. The rating of the overcurrent protection device must be not more than the allowable ampacity of the service conductors as determined by Section E3603.1, except as allowed by the exception. Section E3705.5 contains the general rule for overload protection of conductors and states that overcurrent protection be installed at the point where the conductors receive their supply, which is the power company's transformer. On the line side of the main disconnect, such conductors are not protected against short-circuit or fault current. This requirement for overload protection of the ungrounded service conductors correlates with Section E3601.7.

The main disconnecting means may consist of any combination of single- or two-pole circuit breakers, or single fuses or a set of fuses. A set of fuses used as the main disconnect or as one of the six disconnecting means would be used on a 240-volt circuit and would have a handle tie or means to disconnect both hot legs of the circuit. The set of fuses would count as one of the maximum six disconnects. Two-pole circuit breakers

are manufactured with a master handle or a handle tie that connects the handles of each pole together. Thus, a two-pole circuit breaker would be one of the six disconnects. Where two single-pole switches or circuit breakers are tied together with a handle tie, they would count as one of the six disconnecting means. The intent of this code requirement is that all ungrounded service conductors can be disconnected from the source of supply with no more than six operations of the hand.

E3603.3.2 Not in grounded conductor. Overcurrent devices shall not be connected in series with a grounded service conductor except where a circuit breaker is used that simultaneously opens all conductors of the circuit. [230.90(B)]

❖ The grounded service conductor must not be switched because it also must serve as the low-impedance ground path in case of a fault to facilitate the operation of the overcurrent device.

E3603.3.3 Location. The service overcurrent device shall be an integral part of the service disconnecting means or shall be located immediately adjacent thereto. (230.91)

❖ Most often, the service overcurrent device and the service disconnecting means are the same (a main circuit breaker). These devices could be separate but grouped together, such as a set of fuses or a circuit breaker and an unfused switch, although this would be rare today. The key to the intent of this provision lies in Section E3601.6.2; the code intends to limit the length of service-entrance conductors that are both inside of a building and on the line side of the service overcurrent device. Service-entrance conductors on the line side of the service overcurrent device are not protected against short circuits and ground faults. Conductors on the load side of the service overcurrent device are feeders and are not limited in length (see Commentary Figure E3603.3.3).

Figure E3603.3.3
SERVICE OVERCURRENT DEVICE INTEGRAL WITH
SERVICE DISCONNECTING MEANS

SERVICES

2015 INTERNATIONAL RESIDENTIAL CODE® COMMENTARY   **36-11**

E3603.4 Grounding electrode conductor size. The grounding electrode conductors shall be sized based on the size of the service entrance conductors as required in Table E3603.4. (250.66)

❖ This is a prescriptive rule based on the size of the largest ungrounded service-entrance conductor for the dwelling. The size of the grounding electrode conductor is not based on the rating of the overcurrent device, nor is it based on the calculated load; rather, it is based on the size of the installed ungrounded service conductors. Service-entrance conductor size is determined from the calculation, but if a larger service-entrance conductor is used, for example, to provide capacity for future expansion, the grounding electrode conductor would be based on the size installed.

TABLE E3603.4
GROUNDING ELECTRODE CONDUCTOR SIZE[a, b, c, d, e, f]

SIZE OF LARGEST UNGROUNDED SERVICE-ENTRANCE CONDUCTOR OR EQUIVALENT AREA FOR PARALLEL CONDUCTORS (AWG/kcmil)		SIZE OF GROUNDING ELECTRODE CONDUCTOR (AWG/kcmil)	
Copper	Aluminum or copper-clad aluminum	Copper	Aluminum or copper-clad aluminum
2 or smaller	1/0 or smaller	8	6
1 or 1/0	2/0 or 3/0	6	4
2/0 or 3/0	4/0 or 250	4	2
Over 3/0 through 350	Over 250 through 500	2	1/0
Over 350 through 600	Over 500 through 900	1/0	3/0

a. If multiple sets of service-entrance conductors connect directly to a service drop, set of overhead service conductors, set of underground service conductors, or service lateral, the equivalent size of the largest service-entrance conductor shall be determined by the largest sum of the areas of the corresponding conductors of each set.

b. Where there are no service-entrance conductors, the grounding electrode conductor size shall be determined by the equivalent size of the largest service-entrance conductor required for the load to be served.

c. Where protected by a ferrous metal raceway, grounding electrode conductors shall be electrically bonded to the ferrous metal raceway at both ends. [250.64(E)(1)]

d. An 8 AWG grounding electrode conductor shall be protected with rigid metal conduit, intermediate metal conduit, rigid polyvinyl chloride (Type PVC) nonmetallic conduit, rigid thermosetting resin (Type RTRC) nonmetallic conduit, electrical metallic tubing or cable armor. [250.64(B)]

e. Where not protected, 6 AWG grounding electrode conductor shall closely follow a structural surface for physical protection. The supports shall be spaced not more than 24 inches on center and shall be within 12 inches of any enclosure or termination. [250.64(B)]

f. Where the sole grounding electrode system is a ground rod or pipe as covered in Section E3608.3, the grounding electrode conductor shall not be required to be larger than 6 AWG copper or 4 AWG aluminum. Where the sole grounding electrode system is the footing steel as covered in Section E3608.1.2, the grounding electrode conductor shall not be required to be larger than 4 AWG copper conductor. [250.66(A) and (B)]

E3603.5 Temperature limitations. Except where the equipment is marked otherwise, conductor ampacities used in determining equipment termination provisions shall be based on Table E3705.1. [110.14(C)(1)]

❖ The ampacities specified by Table E3705.1 are to be used in determining the maximum operating temperature of equipment connections to the service-entrance conductors. For example, a 4-AWG copper service conductor is limited to a load of 85 amps by Table E3705.1 where the operating temperature must be not greater than 75°C. See Section E3705.4.

SECTION E3604
OVERHEAD SERVICE AND SERVICE-ENTRANCE CONDUCTOR INSTALLATION

E3604.1 Clearances on buildings. Open conductors and multiconductor cables without an overall outer jacket shall have a clearance of not less than 3 feet (914 mm) from the sides of doors, porches, decks, stairs, ladders, fire escapes and balconies, and from the sides and bottom of windows that open. See Figure E3604.1. [230.9(A)]

❖ This code requirement pertains to overhead service-entrance conductors. Where service-entrance conductors are installed as SE cable or in a raceway, there is no minimum distance specified between openings and doors and the cable or raceway. The main concern here is that an individual standing on a balcony or similar place or reaching out of an openable window should not be able to reach open conductors (see Figure E3604.1 and Commentary Figure E3604.1).

E3604.2 Vertical clearances. Overhead service conductors shall not have ready access and shall comply with Sections E3604.2.1 and E3604.2.2. (230.24)

❖ The installer is responsible for the required clearance for the overhead service conductors. The sag in the overhead span of conductors will change with the temperature and with loading from ice and wind. The conductors will contract in colder temperatures and expand in warmer temperatures. The conductors must be installed such that when they are extended because of the increased sag, required clearances will still be maintained.

E3604.2.1 Above roofs. Conductors shall have a vertical clearance of not less than 8 feet (2438 mm) above the roof surface. The vertical clearance above the roof level shall be maintained for a distance of not less than 3 feet (914 mm) in all directions from the edge of the roof. See Figure E3604.2.1. [230.24(A)]

Exceptions:

1. Conductors above a roof surface subject to pedestrian traffic shall have a vertical clearance from the roof surface in accordance with Section E3604.2.2. [230.24(A) Exception No. 1]

2. Where the roof has a slope of 4 inches (102 mm) in 12 inches (305 mm), or greater, the minimum clearance shall be 3 feet (914 mm). [230.24(A) Exception No. 2]

3. The minimum clearance above only the overhanging portion of the roof shall not be less than 18 inches (457 mm) where not more than 6 feet (1829 mm) of overhead service conductor length passes over 4 feet (1219 mm) or less of roof surface measured horizontally and such conductors are terminated at a through-the-roof raceway or approved support. [230.24(A) Exception No. 3]

4. The requirement for maintaining the vertical clearance for a distance of 3 feet (914 mm) from the edge of the roof shall not apply to the final conductor span where the service drop is attached to the side of a building. [230.24(A) Exception No. 4]

5. Where the voltage between conductors does not exceed 300 and the roof area is guarded or isolated, a reduction in clearance to 3 feet (914 mm) shall be permitted. [230.24(A) Exception No. 5]

❖ Most roofs of one- and two-family dwellings are not subject to pedestrian traffic, but may be walked on. This code provision considers that where the slope of the roof is less than 4 inches (102 mm) vertical in 12 inches (305 mm) horizontal, is it easier to walk on; therefore, it is more likely that an individual will be walking on the roof at some point in time. A roof that has a slope of 4 or more inches (102 mm) vertical in 12 inches (305 mm) horizontal is more difficult to walk on, and the conductor clearance may be reduced to 3 feet (914 mm), recognizing that walking on the roof is less likely.

Exception 3 pertains only to the overhanging portion of the roof; that is, where the roof extends past the wall of the dwelling. Where the service equipment is mounted on the wall with the riser extending through the roof the conductors will usually pass over only the overhanging portion of the roof. The clearance can be reduced to 18 inches (457 mm) above the roof surface if a length of 6 feet (1829 mm) or less of service conductor passes over a maximum of 4 feet (1219 mm) of overhanging roof. The 4 feet (1219 mm) is measured in any direction, not just perpendicular to the wall of the house.

Exception 4 is for an installation where the service drop is not attached to the service mast but to the building, and the service riser does not extend through the roof. The service drop conductors may run near the edge of the roof, and the vertical clearance for a distance of 3 feet (914 mm) from the edge of the roof is not required for this installation.

Exception 5 addresses those conditions where human contact with the conductors is extremely unlikely or impossible.

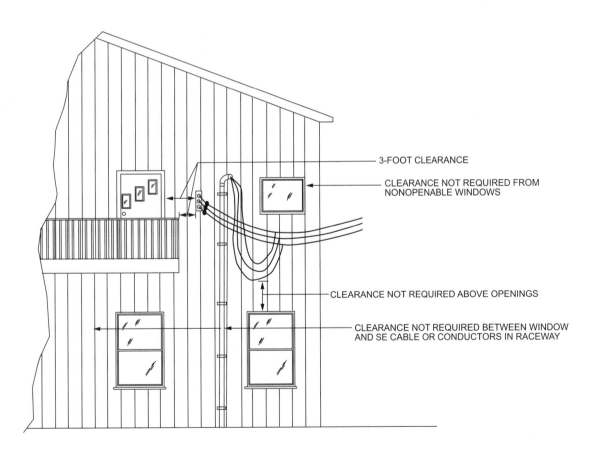

For SI: 1 foot = 304.8 mm.

FIGURE E3604.1
CLEARANCES FROM BUILDING OPENINGS

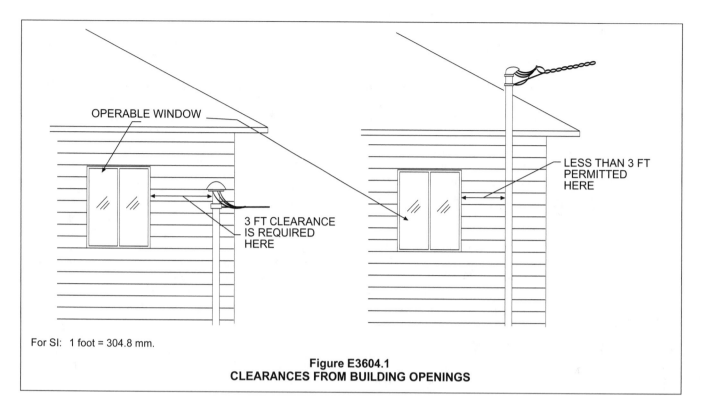

For SI: 1 foot = 304.8 mm.

Figure E3604.1
CLEARANCES FROM BUILDING OPENINGS

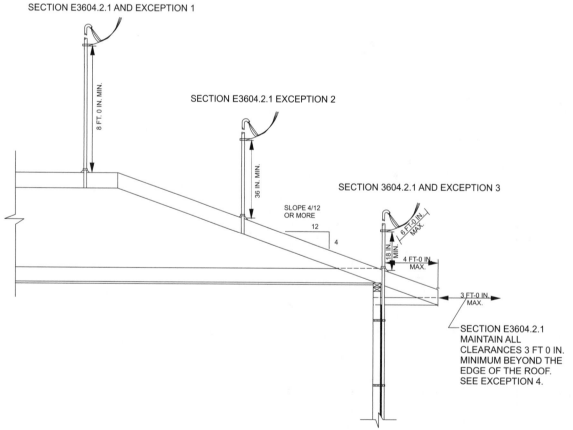

For SI: 1 inch = 25.4 mm, 1 foot = 304.8 mm.

FIGURE E3604.2.1
CLEARANCES FROM ROOFS

E3604.2.2 Vertical clearance from grade. Overhead service conductors shall have the following minimum clearances from final grade:

1. For conductors supported on and cabled together with a grounded bare messenger wire, the minimum vertical clearance shall be 10 feet (3048 mm) at the electric service entrance to buildings, at the lowest point of the drip loop of the building electric entrance, and above areas or sidewalks accessed by pedestrians only. Such clearance shall be measured from final grade or other accessible surfaces.

2. Twelve feet (3658 mm)—over residential property and driveways.

3. Eighteen feet (5486 mm)—over public streets, alleys, roads or parking areas subject to truck traffic. [(230.24(B)(1), (2), and (4)]

❖ The 10-foot (3048 mm) vertical clearance requirement from final grade applies to the drip loop as well as the service conductors.

E3604.3 Point of attachment. The point of attachment of the overhead service conductors to a building or other structure shall provide the minimum clearances as specified in Sections E3604.1 through E3604.2.2. The point of attachment shall be not less than 10 feet (3048 mm) above finished grade. (230.26)

❖ This code section indicates that the height above grade of the point of attachment governs the clearance of the service drop conductors. Several factors must be considered in determining the height of the point of attachment, such as the length of conductor left to form the drip loop; the length of span and sag of the service drop conductors and the 10-, 12- or 18-foot clearance rules. The point of attachment could be 10 feet (3048 mm) above grade at the lowest.

E3604.4 Means of attachment. Multiconductor cables used for overhead service conductors shall be attached to buildings or other structures by fittings approved for the purpose. (230.27)

❖ Only fittings that are approved and designed for the purpose can be used for the attachment of the service conductors. Installation of the supports or fittings to the framing members before the outside or inside walls are finished may be necessary.

E3604.5 Service masts as supports. A service mast used for the support of service-drop or overhead service conductors shall comply with Sections E3604.5.1 and E3604.5.2. Only power service drop or overhead service conductors shall be attached to a service mast.

❖ The code does not contain a prescriptive requirement for the size and type of conduit to be used or for the height above the roof it can extend before being supported by braces or guy wires where overhead service drop conductors are attached to the service mast. In many juris-dictions, the power utility company specifies that if the service riser is installed through the roof, a minimum 2-inch (51 mm) rigid conduit must be installed to provide adequate strength where the con-duit connects the overhead service drop. The size of the mast depends on the length and size of the service drop conductors, the height that the mast extends above the last point of support and loading resulting from wind and ice build-up. The overhead service mast must not support such things as cable TV and telephone wires.

E3504.5.1 Strength. The service mast shall be of adequate strength or shall be supported by braces or guys to safely withstand the strain imposed by the service-drop or overhead service conductors. Hubs intended for use with a conduit that serves as a service mast shall be identified for use with service-entrance equipment.

❖ See Section E3604.5. Threaded hubs on the top of meter enclosures and service equipment enclosures must be identified for the duty. Typically, such conduit hubs bear the full force created by the mast acting as a lever across the fulcrum that occurs at the point of support at the roof.

E3604.5.2 Attachment. Service-drop or overhead service conductors shall not be attached to a service mast at a point between a coupling and a weatherhead or the end of the conduit, where the coupling is located above the last point of securement of the building or other structure or is located above the building or other structure. [230.28(A) & (B)]

❖ This section prohibits the attachment of service conductors to masts at a point on the mast that is above a conduit coupling where the coupling is above the last point of support for the mast. Where a bending force is applied to conduit, a coupling is a weak point and stress failure is likely to occur at the coupling before the mast deforms. The threaded portion of conduit is thinner than the conduit wall and the coupling is a hinge-point not designed to resist bending forces.

E3604.6 Supports over buildings. Service conductors passing over a roof shall be securely supported. Where practicable, such supports shall be independent of the building. (230.29)

❖ Where it is necessary for a service drop to pass over the roof of a building, the service drop should be supported independently of the building over which it crosses.

SECTION E3605
SERVICE-ENTRANCE CONDUCTORS

E3605.1 Insulation of service-entrance conductors. Service-entrance conductors entering or on the exterior of buildings or other structures shall be insulated in accordance with Section E3406.5. (230.41)

Exceptions:

1. A copper grounded conductor shall not be required to be insulated where it is:

   1.1. In a raceway or part of a service cable assembly,

   1.2. Directly buried in soil of suitable condition, or

1.3. Part of a cable assembly listed for direct burial without regard to soil conditions.

2. An aluminum or copper-clad aluminum grounded conductor shall not be required to be insulated where part of a cable or where identified for direct burial or utilization in underground raceways. (230.41 Exception)

❖ Service-entrance conductors are almost always installed in raceways or they are part of a multiwire cable assembly. The grounded conductor need not be insulated if it is in a raceway or cable. Some SE cables are made with a bare conductor that is used as the neutral.

E3605.2 Wiring methods for services. Service-entrance wiring methods shall be installed in accordance with the applicable requirements in Chapter 38. (230.43)

❖ Table E3801.4 includes eight different types of wiring methods as allowable for service-entrance wiring.

E3605.3 Spliced conductors. Service-entrance conductors shall be permitted to be spliced or tapped. Splices shall be made in enclosures or, if directly buried, with listed underground splice kits. Conductor splices shall be made in accordance with Chapters 34, 37, 38 and 39. (230.33, 230.46)

❖ Service-entrance conductors are spliced for the service of a two-family dwelling. On an over-head service, the riser is run down to a gutter where the service-entrance conductors are spliced and run to the two separate main disconnects.

E3605.4 Protection of underground service entrance conductors. Underground service-entrance conductors shall be protected against physical damage in accordance with Chapter 38. (230.32)

❖ See the commentary to Section E3803 for rules on the protection of underground service-entrance conductors.

E3605.5 Protection of all other service cables. Above-ground service-entrance cables, where subject to physical damage, shall be protected by one or more of the following: rigid metal conduit, intermediate metal conduit, Schedule 80 PVC conduit, electrical metallic tubing or other approved means. [230.50(1)]

❖ Where above-ground SE cables are deemed by the code official to be subject to physical damage, they must be protected by installing them in one of the types of conduit listed or they must be installed in electrical metallic tubing. For new work, separate service-entrance conductors are normally installed in one of the listed wiring methods. If SE cable is used, it could be run exposed. Installation in one of the raceways on the list is not required except at any point where it could be damaged.

E3605.6 Locations exposed to direct sunlight. Insulated conductors and cables used where exposed to direct rays of the sun shall comply with one of the following:

1. The conductors and cables shall be listed, or listed and marked, as being sunlight resistant.

2. The conductors and cables are covered with insulating material, such as tape or sleeving, that is listed, or listed and marked, as being sunlight resistant. [310.10(D)]

❖ Ultraviolet radiation from sunlight will attack and degrade many plastic materials such as those used for conductor insulation and cable jackets (sheaths). Therefore, any conductor exposed to sunlight, such as the exposed service-entrance conductors at the drip loop at the connection to a service drop, must be "sunlight resistant." Cables run on the exterior surfaces of buildings, such as SE cables, must be "sunlight resistant" as well. A sunlight-resistant covering, listed and labeled as such, can be used to protect conductors and cable assemblies.

E3605.7 Mounting supports. Service-entrance cables shall be supported by straps or other approved means within 12 inches (305 mm) of every service head, gooseneck or connection to a raceway or enclosure and at intervals not exceeding 30 inches (762 mm). [230.51(A)]

❖ Where installed on the surface of a wall, service entrance cable must be supported at least every 30 inches (762 mm) to keep the cable flat against the wall surface and protected from damage.

E3605.8 Raceways to drain. Where exposed to the weather, raceways enclosing service-entrance conductors shall be suitable for use in wet locations and arranged to drain. Where embedded in masonry, raceways shall be arranged to drain. (230.53)

❖ Although a weatherhead installed at the top of an overhead riser will keep out precipitation, differences in temperature can cause condensation to form inside a raceway. The raceway and the associated equipment have drain holes for this reason, which must not be plugged or covered during installation.

E3605.9 Overhead service locations. Connections at service heads shall be in accordance with Sections E3605.9.1 through E3605.9.7. (230.54)

❖ See the commentary to Sections E3605.9.1 through E3605.9.7.

E3605.9.1 Rain-tight service head. Service raceways shall be equipped with a service head at the point of connection to service-drop or overhead conductors. The service head shall be listed for use in wet locations. [230.54(A)]

❖ See the commentary to Section E3605.9.7.

E3605.9.2 Service cable, service head or gooseneck. Service-entrance cable shall be equipped with a service head or shall be formed into a gooseneck in an approved manner. The service head shall be listed for use in wet locations. [230.54(B)]

❖ See the commentary to Section E3605.9.7.

E3605.9.3 Service-head location. Service heads, and goosenecks in service-entrance cables, shall be located above the point of attachment of the service-drop or overhead service conductors to the building or other structure. [230.54(C)]

Exception: Where it is impracticable to locate the service head or gooseneck above the point of attachment, the service head or gooseneck location shall be not more than 24 inches (610 mm) from the point of attachment. [230.54(C) Exception]

❖ See the commentary to Section E3605.9.7.

E3605.9.4 Separately bushed openings. Service heads shall have conductors of different potential brought out through separately bushed openings. [230.54(E)]

❖ See the commentary to Section E3605.9.7.

E3605.9.5 Drip loops. Drip loops shall be formed on individual conductors. To prevent the entrance of moisture, service-entrance conductors shall be connected to the service-drop or overhead conductors either below the level of the service head or below the level of the termination of the service-entrance cable sheath. [230.54(F)]

❖ See the commentary to Section E3605.9.7.

E3605.9.6 Conductor arrangement. Service-entrance and overhead service conductors shall be arranged so that water will not enter service raceways or equipment. [230.54(G)]

❖ See the commentary to Section E3605.9.7.

E3605.9.7 Secured. Service-entrance cables shall be held securely in place. [230.54(D)]

❖ A weatherhead or service head is installed at the top of the conduit or electrical metallic tubing used as the riser for the service entrance conductors. There is a separate hole in the weather-head for each conductor. The weatherhead is rain tight, which means that when it is installed properly, rain water will not enter the raceway through the weatherhead. Some condensation will occur within the raceway for which drain holes are provided. The point of attachment of the service-drop conductors is below the weather head so that rain water running down along the service conductors will not enter the service raceway and get into the service equipment. At the point the service-entrance conductors emerge from the weatherhead, they are bent downward to form a drip loop so that water will not run along these conductors and enter the service equipment.

SECTION E3606
SERVICE EQUIPMENT—GENERAL

E3606.1 Service equipment enclosures. Energized parts of service equipment shall be enclosed. (230.62)

❖ All energized parts are enclosed within the panelboard or service equipment (see the commentary to Section E3404.9).

E3606.2 Working space. The working space in the vicinity of service equipment shall be not less than that specified in Chapter 34. (110.26)

❖ See the commentary to Section E3405.

E3606.3 Available short-circuit current. Service equipment shall be suitable for the maximum fault current available at its supply terminals, but not less than 10,000 amperes. (110.9)

❖ See the commentary to Section E3404.3.

E3606.4 Marking. Service equipment shall be marked to identify it as being suitable for use as service equipment. Service equipment shall be listed. Individual meter socket enclosures shall not be considered as service equipment. (230.66)

❖ See the commentary to Section E3601.6.1.

SECTION E3607
SYSTEM GROUNDING

E3607.1 System service ground. The premises wiring system shall be grounded at the service with a grounding electrode conductor connected to a grounding electrode system as required by this code. Grounding electrode conductors shall be sized in accordance with Table E3603.4. [250.20(B)(1) and 250.24(A)]

❖ The electrical system must be connected to earth to limit voltage imposed by lightning, line surges or unintentional contact with higher voltage lines and to stabilize voltage to earth during normal operation. The electrical system is connected to earth or grounded by connecting the neutral, referred to as the grounded service conductor, to a grounding electrode conductor, which is run to the grounding electrode system. This is part of the fault-current path and must have sufficiently low impedance to facilitate the operation of the overcurrent device and clear the fault. An electrically continuous and permanent path must be established that will allow enough current to flow during a fault condition to trip the main circuit breaker or set of fuses. When a fault occurs at some point in the electrical wiring system, the current will return to the source of the power supply, which is the utility company transformer. The grounding electrode conductor is sized according to the size of the ungrounded service conductors. The size of the grounding electrode conductor is not based on the rating of the main service disconnect, and it is not based strictly on the calculated load. Although the calculated load determines the minimum acceptable ungrounded conductor size, a larger conductor could be selected and installed. The amount of current that will flow in a ground-fault (line-to-ground) condition is determined by the impedance in the circuit from the source (ungrounded conductor) to the fault and the impedance in the fault-current return path. Obviously, the larger the size of the source or ungrounded conductor, the lower the impedance will be, and the larger the grounding electrode conductor must be.

The grounding electrode conductor sizes in Table E3603.4 are appropriate to carry the fault current likely to be imposed on the grounding electrode conductor.

E3607.2 Location of grounding electrode conductor connection. The grounding electrode conductor shall be connected to the grounded service conductor at any accessible point from the load end of the overhead service conductors, service drop, underground service conductors, or service lateral to and including the terminal or bus to which the grounded service conductor is connected at the service disconnecting means. A grounding connection shall not be made to any grounded circuit conductor on the load side of the service disconnecting means, except as provided in Section E3607.3.2. [250.24(A)(1) and (A)(5)]

❖ Commonly, the grounded conductor is connected to the grounding electrode conductor at a grounding terminal in the service equipment at the location of the main disconnect (see Commentary Figure E3607.2). The code permits this connection to be made at any accessible location between the load end of the service drop or service lateral and the main disconnect and it is often made in a meter cabinet. The meter cabinet is an accessible location for this connection. For an overhead service, the connection could be made at the weatherhead, the meter cabinet, a current transformer cabinet or the main disconnect. If the connection is made in the meter cabinet, it may be considered inaccessible and therefore not permitted by the utility company. The main disconnect is usually the most convenient place to make the connection.

It is extremely important that no connection be made between the grounded conductors and the grounding conductors downstream of the main disconnect. Such a connection on the load side of the main disconnect could create a hazardous parallel path for fault-current

and normal current. An exception to this rule is for the circuits to existing ranges and dryers. Another exception, for existing installations only, is where two or more buildings are supplied from a common service and an equipment grounding conductor is not run with the supply conductors to the other building (see the commentary to Section E3607.3.2). See also Sections E3908.5 and E3908.6.

E3607.3 Buildings or structures supplied by feeder(s) or branch circuit(s). Buildings or structures supplied by feeder(s) or branch circuit(s) shall have a grounding electrode or grounding electrode system installed in accordance with Section E3608. The grounding electrode conductor(s) shall be connected in a manner specified in Section E3607.3.1 or, for existing premises wiring systems only, Section E3607.3.2. Where there is no existing grounding electrode, the grounding electrode(s) required in Section E3608 shall be installed. [250.32(A)]

Exception: A grounding electrode shall not be required where only one branch circuit, including a multiwire branch circuit, supplies the building or structure and the branch circuit includes an equipment grounding conductor for grounding the noncurrent-carrying parts of all equipment. For the purposes of this section, a multiwire branch

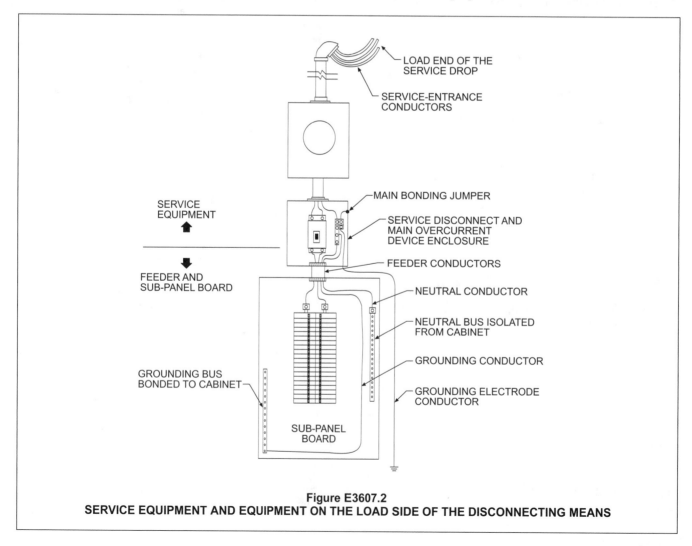

Figure E3607.2
SERVICE EQUIPMENT AND EQUIPMENT ON THE LOAD SIDE OF THE DISCONNECTING MEANS

circuit shall be considered as a single branch circuit. [250.32(A) Exception]

❖ Where two or more buildings are supplied from one service, each building must have a grounding electrode. For example, a dwelling has a service supplied from the power company to service equipment on the outside of the dwelling and has two 2-pole circuit breakers as main service disconnects. One 2-pole circuit breaker serves a feeder run to the panelboard inside the dwelling, and the other serves a feeder run to a panelboard at a separate workshop building 100 feet (30 480 mm) away from the dwelling. Both the dwelling and the workshop must have a grounding electrode as described in Section E3608. If there is no grounding electrode at the workshop, one of the grounding electrodes listed in Section E3608 must be installed. A ground rod, concrete encased electrode, or ground ring could be installed. If water is supplied to the separate workshop building through metal pipe, an underground metal water pipe could serve as the grounding electrode. This section states that the grounding electrode at the separate building must be connected in the building disconnecting means in one of the ways described in the two subsections that follow. Note that Section E3607.3.2 applies only to existing premises wiring systems. A separate building must have a main disconnect and not simply a panelboard. A main disconnect could be a main circuit breaker in the panelboard that would disconnect all the power to the panelboard.

A grounding electrode is not required in a separate building if it is served by only one branch circuit. If a separate building supplied by only one branch circuit does not have a grounding electrode, it does not have to have one installed. If a metal underground water pipe is already installed at the building, it does not have to be bonded to the equipment grounding conductor of the one branch circuit. A building like a small tool shed, for example, may have only a light and receptacle. The equipment grounding conductor run with the branch circuit conductors is sufficient to ground any equipment for this separate building. Of course, a receptacle in a tool storage shed or accessory building at grade level would require ground-fault circuit-interrupter (GFCI) protection, but, the grounding conductor need not be connected to a grounding electrode such as a ground rod.

E3607.3.1 Equipment grounding conductor. An equipment grounding conductor as described in Section E3908 shall be run with the supply conductors and connected to the building or structure disconnecting means and to the grounding electrode(s). The equipment grounding conductor shall be used for grounding or bonding of equipment, structures or frames required to be grounded or bonded. The equipment grounding conductor shall be sized in accordance with Section E3908.12. Any installed grounded conductor shall not be connected to the equipment grounding conductor or to the grounding electrode(s). [250.32(B) and Table 250.122]

❖ Where the feeder that supplies the separate building is run with an equipment grounding conductor, the rules

of this subsection apply. Consider the example of the separate workshop building in the commentary to Section E3607.3. The panelboard at the workshop is served by a feeder in a raceway or cable consisting of four conductors: two ungrounded, one grounded (neutral) and one equipment grounding conductor of the appropriate sizes. The equipment grounding conductor run along with the supply conductors will be connected to the separate building disconnecting means and to the separate building grounding electrode(s). The feeder neutral or grounded conductor of the supply must be insulated and is not bonded to the panelboard cabinet. It is kept separate from the grounding conductors and is not bonded to the separate building grounding electrode(s). The panelboard cabinet of the separate building has one terminal bus for the grounded conductors and another for the grounding conductors. The bonding screw supplied for the purpose of bonding the neutral terminal strip or bus to the metal cabinet must not be installed, and this neutral terminal strip or bus is insulated from the metal cabinet. The grounding electrode is connected to the equipment grounding terminal, which is bonded to the metal cabinet. If the neutral terminal bus and the equipment grounding conductor terminal bus were bonded together, the current intended to return via the neutral will also flow on the equipment grounding conductor. This is a fire- and life-safety hazard and violates Section E3607.2, among others.

The equipment-grounding conductor run with the feeder could be any of the types listed in Section E3908.8, including the conduit or tubing enclosing the circuit conductors where the installation complies with the provisions of Section E3908 (see Commentary Figure E3607.3.1).

E3607.3.2 Grounded conductor, existing premises. For installations made in compliance with previous editions of this code that permitted such connection and where an equipment grounding conductor is not run with the supply conductors to the building or structure, there are no continuous metallic paths bonded to the grounding system in both buildings or structures involved, and ground-fault protection of equipment has not been installed on the supply side of the feeder(s), the grounded conductor run with the supply to the buildings or structure shall be connected to the building or structure disconnecting means and to the grounding electrode(s) and shall be used for grounding or bonding of equipment, structures, or frames required to be grounded or bonded. Where used for grounding in accordance with this provision, the grounded conductor shall be not smaller than the larger of:

1. That required by Section E3704.3.

2. That required by Section E3908.12. [250.32(B)(1) Exception]

❖ Where the feeder that supplies the separate building is run without an equipment grounding conductor, the rules of this subsection apply. Note that this section applies only to existing wiring systems that were installed in compliance with previous editions of the code that allowed such connections. If, for example,

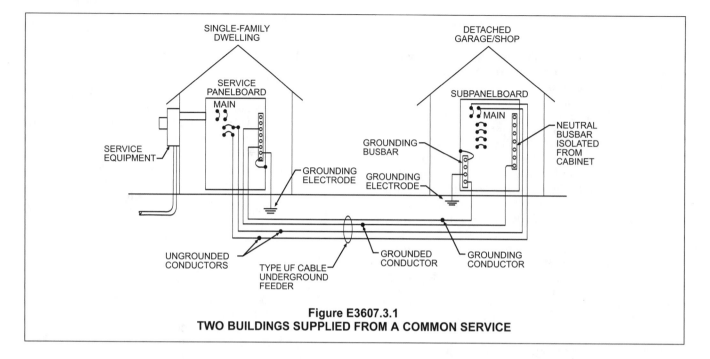

Figure E3607.3.1
TWO BUILDINGS SUPPLIED FROM A COMMON SERVICE

new wiring was to be installed to supply a building that did not have power run to it, the code intends to require the installation to comply with Section E3607.3.1. In previous editions of the code, this section applied to new and existing installations and this was considered to be an unnecessarily risky practice. Recall that the code recently underwent similar revisions regarding the grounding of electric clothes dryers and ranges, where the long-standing practice of grounding the appliance frames to the grounded (neutral) conductor was banned, except for existing installations.

The preferred method is covered in Section E3607.3.1 and for new installations, there is no excuse for not complying with the preferred safer method, thus this section (E3607.3.2) was revised to apply only to existing installations. Let's assume that the panelboard at the workshop discussed in the previous section is served by an existing underground feeder cable consisting of three conductors: two ungrounded and one neutral conductor of the appropriate sizes. The neutral or grounded conductor of the feeder is grounded again at the separate building by a grounding electrode conductor connected to the separate building grounding electrode(s). The panelboard cabinet of the separate building may have one terminal bus for the grounded conductors and another for the grounding conductors; if so, they must be bonded together or bonded to the cabinet by a main bonding jumper. Or there may be only one terminal bus to connect all grounded and grounding conductors. In this case, the bonding screw or strap supplied for the purpose of bonding the neutral terminal strip to the metal cabinet is installed, and this bonding screw or strap is considered as a main bonding jumper. The grounding electrode is also bonded to the equipment-grounding terminal, which is bonded to the metal cabinet.

An important rule that applies where only a three-conductor feeder is run to the separate building and the feeder grounded conductor is bonded to the grounding electrode system at the separate building is that there must be no common continuous metallic paths bonded to the grounding electrode system in both buildings. Bonding at both ends of such objects as a metal water pipe or gas pipe that was run from the house to the separate building could result in a parallel path for neutral current as well as for fault current. If the neutral were broken or became disconnected, the return current could be through the metal pipe, which would cause a dangerous potential on noncurrent carrying metal surfaces (see Commentary Figure E3607.3.2).

E3607.4 Grounding electrode conductor. A grounding electrode conductor shall be used to connect the equipment grounding conductors, the service equipment enclosures, and the grounded service conductor to the grounding electrode(s). This conductor shall be sized in accordance with Table E3603.4. [250.24(D)]

❖ The grounded service conductor, the grounding conductors and the metal service enclosures must be stabilized at a potential voltage to earth of zero volts. The grounding electrode conductor is used to connect these components to the grounding electrode buried in the earth.

E3607.5 Main bonding jumper. An unspliced main bonding jumper shall be used to connect the equipment grounding conductor(s) and the service-disconnect enclosure to the grounded conductor of the system within the enclosure for each service disconnect. [250.24(B)]

❖ The grounded service conductor, the grounding conductors and the metal service enclosures must be connected together by the main bonding jumper. This connection is made at the main disconnect location.

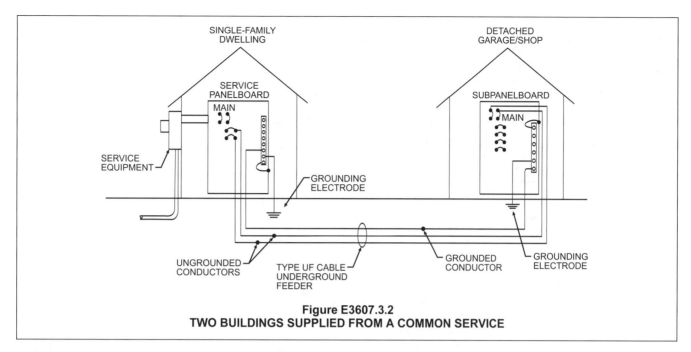

Figure E3607.3.2
TWO BUILDINGS SUPPLIED FROM A COMMON SERVICE

E3607.6 Common grounding electrode. Where an ac system is connected to a grounding electrode in or at a building or structure, the same electrode shall be used to ground conductor enclosures and equipment in or on that building or structure. Where separate services, feeders or branch circuits supply a building and are required to be connected to a grounding electrode(s), the same grounding electrode(s) shall be used. Two or more grounding electrodes that are effectively bonded together shall be considered as a single grounding electrode system. (250.58)

❖ This section requires that the grounding electrode used to ground the service neutral conductor also be used to ground all the metallic enclosures in the electrical system. With one grounding electrode used to ground all of these components, there will not be a different potential; all enclosures will be at the same zero voltage potential as the grounded service conductor. Where two or more grounding electrodes are used at the premises, they must be effectively bonded together and considered as one grounding electrode. This keeps all noncurrent carrying metal surfaces in the electrical system at earth potential.

SECTION E3608
GROUNDING ELECTRODE SYSTEM

E3608.1 Grounding electrode system. All electrodes specified in Sections E3608.1.1, E3608.1.2, E3608.1.3, E3608.1.4 E3608.1.5 and E3608.1.6 that are present at each building or structure served shall be bonded together to form the grounding electrode system. Where none of these electrodes are present, one or more of the electrodes specified in Sections E3608.1.3, E3608.1.4, E3608.1.5 and E3608.1.6 shall be installed and used. (250.50)

Exception: Concrete-encased electrodes of existing buildings or structures shall not be required to be part of the grounding electrode system where the steel reinforcing

bars or rods are not accessible for use without disturbing the concrete. (250.50 Exception)

❖ The code requires that the premises wiring system, which includes the wiring for the house, any outbuildings and appurtenances, be connected to the earth or grounded at the service. The term "grounded" as defined in Chapter 35 means connected to the earth. Grounding will cause the conductors and equipment so connected to have the same voltage potential as the earth. In other words, there would be no voltage difference between a grounded conductor or equipment and the earth. The electrical system must be connected to the earth to limit voltage imposed by lightning, line surges or unintentional contact with higher voltage lines and to stabilize voltage to earth during normal operation.

The electrical system is connected to the earth through the grounding electrode system. An electrode is a conductor or object serving as a conductor terminal placed in an electrolyte or substance through which electricity could enter or leave. In this case, the electrode is placed in the earth, which contains electrolytes such as mineral and salt solutions that are at different potencies depending on the moisture content of the soil. All of the available grounding electrodes at the premises that are identified by the code must be bonded together to form a "grounding electrode system." To put it simply, the code intends to take advantage of any and all electrodes that are on the site in order to create to the best electrode system possible. The code does not specify that these electrodes be bonded together all at one point or location, or at a specific point or location; it just specifies that they be bonded together. The grounded service conductor or neutral is connected to the grounding electrode system with a grounding electrode conductor. This conductor is run from the service equipment, usually at the point of the main disconnect or on the line side of the main disconnect, to any convenient point on the grounding electrode system, convenient meaning at any of the electrodes or at any point along the grounding electrode conductor. The size of the grounding electrode conductor depends on the size required for the largest of the available grounding electrodes.

Because a metal underground water pipe is one of the electrodes that must be connected to the grounding electrode system, it would be convenient to run a short grounding electrode conductor to the nearest cold water pipe, but this is prohibited. In many houses, there is a cold water pipe just inside the house near the service equipment. The general concept would be to let the water piping system serve as the conductor all the way through the house to where the water pipe enters the house and is buried underground. This type of grounding was a common practice in some areas before the early 1990s when code provisions were changed to prohibit using the water-piping system as a grounding electrode conductor. For various reasons, the water piping system may not always remain continuous or remain the ideal conductor, and the wiring system could be left in an ungrounded condition. Piping could be disconnected during repairs of the plumbing system and a section of nonmetallic water piping could be installed.

The code permits only the first 5 feet (1524 mm) of the water piping system from the point where the pipe enters the building to be used to connect the electrodes together. There are usually not any or not very many interconnections of the piping system in the first 5 feet (1524 mm). In most new houses, if the underground water pipe is metal for 10 feet (3048 mm) to the point where it enters the building, it is an available grounding electrode, and the grounding electrode conductor is run continuously from the service to the point where the water pipe enters the house.

Where any or all of the electrodes mentioned under Section E3608.1 are present (existing) on the premises, they must be bonded together as part of the system. Concrete-encased grounding electrodes consisting of concrete-reinforcing steel must be used as part of the electrode system. The code does not specify that a reinforcing bar be made available or left exposed above the concrete for connection of a grounding electrode conductor, or that 20 feet (6096 mm) of size 4-AWG bare copper be encased in the concrete and left exposed to run to the service equipment. However, such steel must be connected to the grounding electrode system if it is present on the job site. The concrete-encased electrode has been proven to be one of the most effective electrodes in any type of soil (see Commentary Figure E3608.1). The exception further clarifies the intent to require steel reinforcing elements to be bonded to the service, because only where the steel is inaccessible in an existing building would bonding not be required.

E3608.1.1 Metal underground water pipe. A metal underground water pipe that is in direct contact with the earth for a distance of

10 feet (3048 mm) or more, including any well casing effectively bonded to the pipe and that is electrically continuous, or made electrically continuous by bonding around insulating joints or insulating pipe to the points of connection of the grounding electrode conductor and the bonding conductors, shall be considered as a grounding electrode (see Section E3608.1). [250.52(A)(1)]

❖ Any metal water pipe in contact with the earth for a distance of 10 feet (3048 mm) or more is considered to be a grounding electrode. It is often thought that only the main water service piping fits this description, but any metal water pipe, such as an irrigation pipe or pipe from a water well, if buried in the earth 10 feet (3048 mm) or more, must be bonded as part of the grounding electrode system. The pipe could be of any material such as copper or steel because the code does not mention the type of metal. In many houses, a water meter, water-pressure reducing valve, or similar equipment is installed in the water supply line. A bonding jumper of the same size as the grounding electrode conductor is installed around such devices because many of these devices are made of or utilize nonconductive materials. In many cases, when the device is removed and/or replaced, the grounding electrode is disconnected. A useful practice is to leave enough slack in the bonding jumper around the device that the jumper will not have to be taken off the piping in the event the device is replaced.

The code assumes that the first 5 feet (1524 mm) of water piping, measured from the point that the piping penetrates an outside wall or floor slab on grade, will not be disturbed or altered by plumbing work. Any piping beyond 5 feet (1524 mm) into the building is more likely to be altered such that electrical continuity is lost. This alteration could take the form of the installation of plastic piping, nonconductive components (e.g., water filters), dielectric fittings or the removal of grounding clamps.

A supplemental grounding electrode is always required when a water pipe electrode is used. The most commonly used supplemental grounding electrode is the ground rod. That is why in so many dwellings a grounding electrode conductor is run from the service equipment to the entry point of the water pipe and another grounding electrode conductor is run to a ground rod. The ground rod is quite often driven into the ground close to the service equipment. A grounding electrode conductor that connects the service equipment to a ground rod, pipe or plate electrode and connects to no other electrodes is not required to be larger than size 6-AWG copper or 4-AWG aluminum (see Section E3610.2).

E3608.1.1.1 Interior metal water piping. Interior metal water piping located more than 5 feet (1524 mm) from the entrance to the building shall not be used as a conductor to interconnect electrodes that are part of the grounding electrode system. [250.68(C)(1)]

❖ See the commentary to Section E3608.1.1.

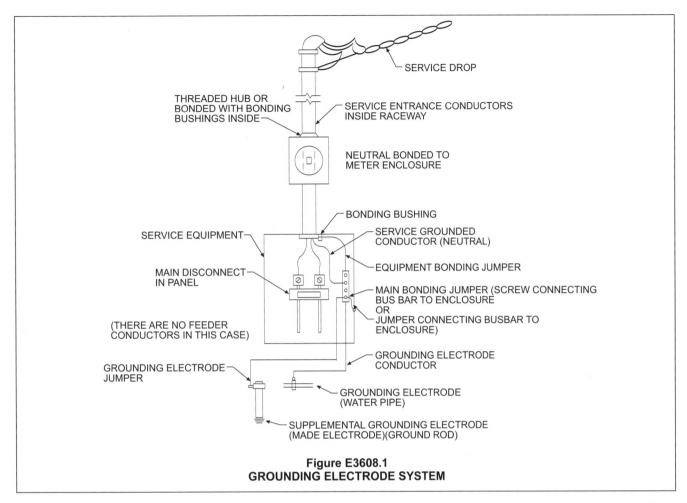

Figure E3608.1
GROUNDING ELECTRODE SYSTEM

E3608.1.1.2 Installation. Continuity of the grounding path or the bonding connection to interior piping shall not rely on water meters, filtering devices and similar equipment. A metal underground water pipe shall be supplemented by an additional electrode of a type specified in Sections E3608.1.2 through E3608.1.6. The supplemental electrode shall be bonded to the grounding electrode conductor, the grounded service-entrance conductor, a nonflexible grounded service raceway, any grounded service enclosure or to the equipment grounding conductor provided in accordance with Section E3607.3.1. Where the supplemental electrode is a rod, pipe or plate electrode in accordance with Section E3608.1.4 or E3608.1.5, it shall comply with Section E3608.4.

Where the supplemental electrode is a rod, pipe or plate electrode in accordance with Section E3608.1.4 or E3608.1.5, that portion of the bonding jumper that is the sole connection to the supplemental grounding electrode shall not be required to be larger than 6 AWG copper or 4 AWG aluminum wire. [250.53(D) and (E)]

❖ See the commentary to Section E3608.1.1.

E3608.1.2 Concrete-encased electrode. A concrete-encased electrode consisting of at least 20 feet (6096 mm) of either of the following shall be considered as a grounding electrode:

1. One or more bare or zinc galvanized or other electrically conductive coated steel reinforcing bars or rods not less than $1/2$ inch (13 mm) in diameter, installed in one continuous 20-foot (6096 mm) length, or if in mul-

tiple pieces connected together by the usual steel tie wires, exothermic welding, welding, or other effective means to create a 20-foot (6096 mm) or greater length.

2. A bare copper conductor not smaller than 4 AWG.

Metallic components shall be encased by at least 2 inches (51 mm) of concrete and shall be located horizontally within that portion of a concrete foundation or footing that is in direct contact with the earth or within vertical foundations or structural components or members that are in direct contact with the earth.

Where multiple concrete-encased electrodes are present at a building or structure, only one shall be required to be bonded into the grounding electrode system. [250.52(A)(3)]

❖ Encasing an electrode in concrete must be done, of course, before the concrete is placed for the footings. The 20 feet (6096 mm) of size 4-AWG bare copper conductor is run in the forms close to the bottom of the footing before the concrete is placed. Enough conductor is left extending out of the footing to be run to the service equipment without a splice. At the point where the copper conductor exits the concrete, it could be broken off by being bent back and forth several times before and during construction. To help protect this conductor, a nonmetallic conduit around the conductor where it leaves the footing would be helpful. In some installations, the end of a horizontal reinforcing bar in the concrete is bent upward so as to create an above-ground exposed connection point for a grounding electrode conductor. In other cases, concrete contractors have been known to provide a short piece of steel-reinforcing rod bent at a 90-degree (1.57 rad) angle and tie-wire it to a footing reinforcing bar for the purpose of providing a stub out for the electrician to connect to. The code does not appear to address these practices and it is the author's opinion that the code does not anticipate that connections will be made that are solely dependent upon a scrap of rebar that is scabbed (tie-wired) onto the functional rebar. It is noted, however, that the code does allow multiple pieces of rebar to be tied together to create a 20-foot (6096 mm) or longer electrode, assuming that such pieces are placed horizontally in the concrete as required by this section. The best conducting path will likely be achieved where the grounding electrode conductor connects directly to the functional reinforcing steel. The code does not assume that the reinforcing steel bars are present because they are required as part of the structural design, although that is typically why they would be present. Reinforcing bars could be installed for the sole purpose of creating an electrode.

This section allows horizontally oriented electrodes to be placed in vertical concrete foundation components that are in direct contact with the earth, such as below-grade foundation walls. Within the footing is an ideal location because it is the deepest part and the moisture levels will likely be greater at greater depths.

Note that Section E3608.1 requires connection to footing steel where footing steel exists; therefore, the electrical contractor and concrete contractor must coordinate their work to allow for the required bonding. The excuse that the electrician was not able to make the required connection to the footing steel before the concrete was poured is not valid.

This section raises the question of what to do with rebar smaller than $1/2$-inch (12.7 mm) in diameter. For example, $3/8$-inch (9.53 mm) rod does not qualify as an electrode, but if it is used in the footing, does the code intend to require it to be bonded regardless? It appears that the code would not require bonding to rebar smaller than $1/2$ inch (12.7 mm), even though doing so would enhance the overall electrode system.

E3608.1.3 Ground rings. A ground ring encircling the building or structure, in direct contact with the earth at a depth below the earth's surface of not less than 30 inches (762 mm), consisting of at least 20 feet (6096 mm) of bare copper conductor not smaller than 2 AWG shall be considered as a grounding electrode. [250.52(A)(4)]

❖ A ground ring is installed around the building as an intentionally constructed grounding electrode. It may be installed: because someone prefers it to other electrodes; as an enhancement to an electrode system or because electrodes, such as footing rebar and copper water service pipes, are not present. A ground ring may be necessary because of the resistivity of the soil, the type of equipment present at the house or other conditions.

E3608.1.4 Rod and pipe electrodes. Rod and pipe electrodes not less than 8 feet (2438 mm) in length and consisting of the following materials shall be considered as a grounding electrode:

1. Grounding electrodes of pipe or conduit shall not be smaller than trade size $3/4$ (metric designator 21) and, where of iron or steel, shall have the outer surface galvanized or otherwise metal-coated for corrosion protection.

2. Rod-type grounding electrodes of stainless steel and copper or zinc-coated steel shall be at least $5/8$ inch (15.9 mm) in diameter unless listed. [250.52(A)(5)]

❖ Rod and pipe electrodes must be at least 8 feet (2438 mm) long, and must be made of steel, stainless steel or nonferrous material. If not of stainless steel or nonferrous material, the rod must have a coating for corrosion protection. Pipe electrodes are rarely used today. The most common electrode is a listed $5/8$-inch (15.9 mm) copper coated steel rod.

E3608.1.4.1 Installation. The rod and pipe electrodes shall be installed such that at least 8 feet (2438 mm) of length is in contact with the soil. They shall be driven to a depth of not less than 8 feet (2438 mm) except that, where rock bottom is encountered, electrodes shall be driven at an oblique angle not to exceed 45 degrees (0.79 rad) from the vertical or shall be buried in a trench that is at least 30 inches (762 mm) deep. The upper end of the electrodes shall be flush with or below ground level except where the aboveground end and the grounding electrode conductor attachment are protected against physical damage. (250.53(G)]

❖ The previous section requires that the ground rod must be at least 8 feet (2438 mm) long. This section requires that at least 8 feet (2438 mm) of the ground rod be in contact with the earth. Therefore, an 8-foot (2438 mm) long ground rod cannot be installed with the top exposed for the connection of the grounding electrode conductor. The ground rod clamp must be suitable for direct burial, or an exothermic welding process must be used. In some cases, the installers leave the upper end of the rod exposed so that the connection of the grounding electrode conductor can be inspected. The only way this would be acceptable is if the rod were longer than 8 feet (2438 mm), and then the attachment of the grounding electrode conductor must be protected against physical damage. In some cases the rod may be driven away from the house to miss the footing as it is being driven. As a result, it is several inches away from the wall of the house in a flower bed or in a future driveway or patio. Some electricians leave it exposed rather than cover it with concrete for a sidewalk or patio. Where the rod is installed in such a location, the code official would need to inspect the connection and the length of the driven or buried ground rod before the concrete slab was placed.

Where the upper end of the rod and connection clamp are exposed above ground, they must be protected against damage that could result in connector or conductor failure.

E3608.1.5 Plate electrodes. A plate electrode that exposes not less than 2 square feet (0.186 m²) of surface to exterior soil shall be considered as a grounding electrode. Electrodes of bare or conductively coated iron or steel plates shall be at least ¹/₄ inch (6.4 mm) in thickness. Solid, uncoated electrodes of nonferrous metal shall be at least 0.06 inch (1.5 mm) in thickness. Plate electrodes shall be installed not less than 30 inches (762 mm) below the surface of the earth. [250.52(A)(7)]

❖ Plate electrodes are now manufactured and are becoming more commonplace as alternatives to rods and pipes. They are constructed of copper or steel and are sized to provide adequate surface area in contact with the soil.

E3608.1.6 Other electrodes. In addition to the grounding electrodes specified in Sections E3608.1.1 through E3608.1.5, other listed grounding electrodes shall be permitted. [250.52(A)(6)]

❖ This section simply recognizes the fact that other listed electrodes could be manufactured now or in the future. Currently rod and plate electrodes are available as listed products.

E3608.2 Bonding jumper. The bonding jumper(s) used to connect the grounding electrodes together to form the grounding electrode system shall be installed in accordance with Sections E3610.2, and E3610.3, shall be sized in accordance with Section E3603.4, and shall be connected in the manner specified in Section E3611.1. [250.53(C)]

❖ See the commentary to Sections E3610.2, E3610.3, E3603.4 and E3611.1.

E3608.3 Rod, pipe and plate electrode requirements. Where practicable, rod, pipe and plate electrodes shall be embedded below permanent moisture level. Such electrodes shall be free from nonconductive coatings such as paint or enamel. Where more than one such electrode is used, each electrode of one grounding system shall be not less than 6 feet (1829 mm) from any other electrode of another grounding system. Two or more grounding electrodes that are effectively bonded together shall be considered as a single grounding electrode system. That portion of a bonding jumper that is the sole connection to a rod, pipe or plate electrode shall not be required to be larger than 6 AWG copper or 4 AWG aluminum wire. [250.53(A)(1), 250.53(B), 250.53(C)]

❖ Moist soil (earth) is more conductive than dry soil and therefore provides for better grounding. Multiple electrodes placed less than 6 feet (1829 mm) apart are not much more effective than a single electrode. This relates to the concept of earth resistance as addressed in Section E3608.4.

E3608.4 Supplemental electrode required. A single rod, pipe, or plate electrode shall be supplemented by an additional electrode of a type specified in Sections E3608.1.2 through E3608.1.6. The supplemental electrode shall be bonded to one of the following:

1. A rod, pipe, or plate electrode.
2. A grounding electrode conductor.
3. A grounded service-entrance conductor.
4. A nonflexible grounded service raceway.
5. A grounded service enclosure.

Where multiple rod, pipe, or plate electrodes are installed to meet the requirements of this section, they shall not be less than 6 feet (1829 mm) apart. [250.53(A)(2) and (A)(3)]

Exception: Where a single rod, pipe, or plate grounding electrode has a resistance to earth of 25 ohms or less, the supplemental electrode shall not be required. [250.53(A)(2) Exception]

❖ The requirement of this section for a supplemental electrode or a maximum of 25 ohms of resistance between the grounding electrode and the earth is only for "made" electrodes (rods, pipes and plates). Note that the fundamental requirements of this section have not changed from previous editions of the code, but the approach has changed. The code now requires the supplemental electrode as the primary rule and allows an exception for low-resistance electrode installations, whereas previously, the code first required a low-resistance electrode installation and mandated a supplemental electrode only where the low resistance could not be demonstrated. The code comes at it from a different angle, but the results are the same.

If the resistance is over 25 ohms for an electrode, the code requires that an additional electrode be installed at least 6 feet (1829 mm) away from the first one. There is no requirement to keep adding electrodes until the resistance is reduced to 25 ohms or less. There are many different types of soil, and the resistivity or conductivity of the soil changes with the amount of moisture in the soil and even with the temperature of the soil. This code rule provides a low-resistance path to ground to limit the voltage imposed on the system by lightning or ground faults. The earth must never be used as the sole grounding conductor or ground-fault path because it has such high resistance. The current flow through the earth would not facilitate the operation of the overcurrent device under fault conditions.

Earth resistance can be measured with "fall-of-potential"-type instruments and modern clamp-on hand-held instruments. At some distance from the electrode under a fall-of-potential test, the earth resistance will peak and level off for subsequent test points before rising again at farther test points. This "leveled-off" resistance is taken as the actual resistance between the electrode and the earth [see Commentary Figures E3608.4(1) and (2)].

The code does not state specifically that the resistance measurement must be taken for each house and submitted. In some areas, it may be well known that the soil resistivity is high and there will almost always be over 25 ohms of resistance to ground from the ground rod, meaning that the exception to this section would never apply. Historically, many code officials have simply taken the approach that two grounding electrodes must always be installed unless the resistance is measured and shown to be 25 ohms or less, and now the code states just that.

on

For SI: 1 foot = 304.8 mm.

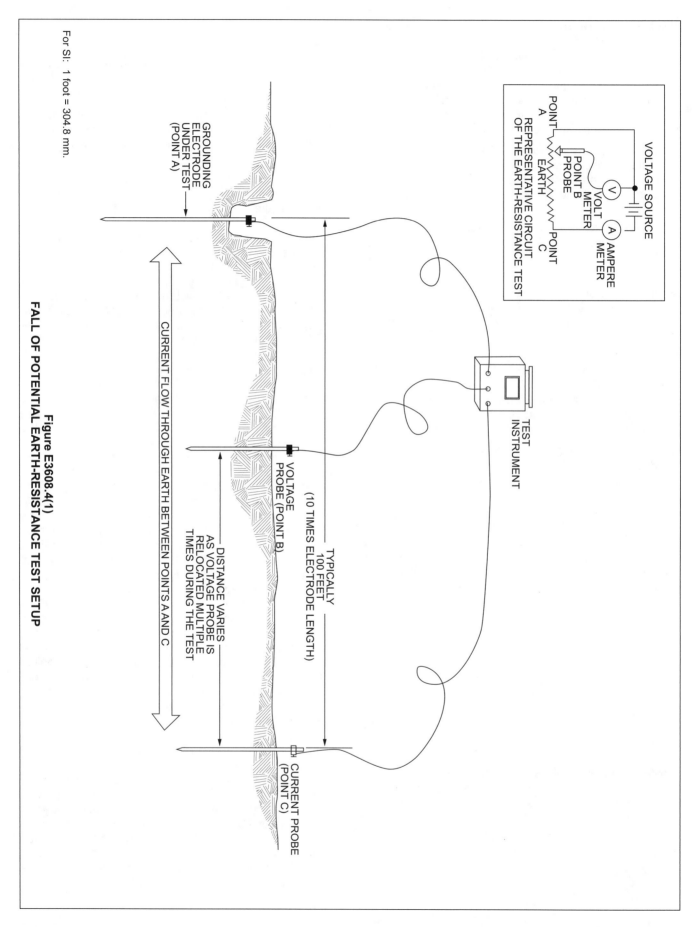

Figure E3608.4(1)
FALL OF POTENTIAL EARTH-RESISTANCE TEST SETUP

GROUNDING
ELECTRODE
UNDER TEST
(POINT A)

REPRESENTATIVE CIRCUIT
OF THE EARTH-RESISTANCE TEST

POINT
A
EARTH
POINT B
PROBE
POINT
C

VOLTAGE SOURCE

VOLT
METER

AMPERE
METER

TEST
INSTRUMENT

TYPICALLY
100 FEET
(10 TIMES ELECTRODE LENGTH)

VOLTAGE
PROBE (POINT B)

DISTANCE VARIES
AS VOLTAGE PROBE IS
RELOCATED MULTIPLE
TIMES DURING THE TEST

CURRENT FLOW THROUGH EARTH BETWEEN POINTS A AND C

CURRENT PROBE
(POINT C)

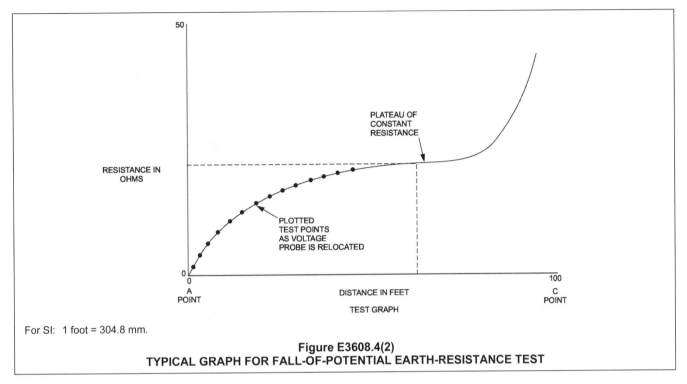

For SI: 1 foot = 304.8 mm.

Figure E3608.4(2)
TYPICAL GRAPH FOR FALL-OF-POTENTIAL EARTH-RESISTANCE TEST

E3608.5 Aluminum electrodes. Aluminum electrodes shall not be permitted. [250.52(B)(2)]

❖ See the commentary to Section E3610.1.

E3608.6 Metal underground gas piping system. A metal underground gas piping system shall not be used as a grounding electrode. [250.52(B)(1)]

❖ Where a dwelling is served with a metal underground gas piping system, the gas utility usually installs an insulating fitting in the line at or near the point the gas line enters the house. This interrupts the continuity of the metal pipe to the earth. Gas piping is not recognized as an electrode and such use could be hazardous in some circumstances.

SECTION E3609
BONDING

E3609.1 General. Bonding shall be provided where necessary to ensure electrical continuity and the capacity to conduct safely any fault current likely to be imposed. (250.90)

❖ Bonding is the connection of metallic parts to make one continuous current path to ensure the fault current path has low impedance. Bonding and grounding are often thought of as the same general subject because bonding is related to grounding as far as the fault-current path is concerned.

E3609.2 Bonding of equipment for services. The noncurrent-carrying metal parts of the following equipment shall be effectively bonded together:

1. Raceways or service cable armor or sheath that enclose, contain, or support service conductors.

2. Service enclosures containing service conductors, including meter fittings, and boxes, interposed in the service raceway or armor. [250.92(A)]

❖ All of the service equipment, raceways, and metallic surfaces must be bonded together to ensure a low impedance fault-current path of electrical continuity. The enclosures and cabinets such as the service mast, weatherhead, conduit to and from the meter base, meter base, cable armor, boxes and the can for the main disconnecting means all contain conductors that have overcurrent protection only on the load end. The service equipment and conductors may be subject to a high fault current, which would require the utility company transformer switch or overcurrent device to open to clear the fault. Where metal conduit or electrical metallic tubing is used to enclose the grounding electrode conductor, the conduit or tubing must be bonded to the grounding electrode conductor to make one continuous fault-current path.

E3609.3 Bonding for other systems. An intersystem bonding termination for connecting intersystem bonding conductors required for other systems shall be provided external to enclosures at the service equipment or metering equipment enclosure and at the disconnecting means for any additional buildings or structures. The intersystem bonding termination shall comply with all of the following:

1. It shall be accessible for connection and inspection.

2. It shall consist of a set of terminals with the capacity for connection of not less than three intersystem bonding conductors.

3. It shall not interfere with opening of the enclosure for a service, building or structure disconnecting means, or metering equipment.

4. Where located at the service equipment, it shall be securely mounted and electrically connected to an enclosure for the service equipment, to the meter enclosure, or to an exposed nonflexible metallic service raceway, or shall be mounted at one of these enclosures and connected to the enclosure or to the grounding electrode conductor with a 6 AWG or larger copper conductor.

5. Where located at the disconnecting means for a building or structure, it shall be securely mounted and electrically connected to the metallic enclosure for the building or structure disconnecting means, or shall be mounted at the disconnecting means and connected to the metallic enclosure or to the grounding electrode conductor with a 6 AWG or larger copper conductor.

6. It shall be listed as grounding and bonding equipment. (250.94)

❖ Communications systems, cable TV and similar systems must be grounded. If the enclosures, raceways and noncurrent carrying metal surfaces of these "other systems" are not bonded to the premises wiring system, they could operate at a difference in potential. Intersystem bonding is required to reduce the shock hazard and minimize the possible fire danger from arcing. Bonding systems together will eliminate any voltage difference between them. Where the service equipment cabinet and meter base are flush with a brick or stucco wall, the raceways are concealed within the wall, the grounding electrode conductor is not accessible and the ground rod has been buried, there is nothing available to which to bond the metallic jacket of the communications cable. If the installer drives a separate ground rod for bonding the cable TV cable shield and boxes, it would be a code violation if not bonded to the service electrode system. An accessible means external to the service equipment enclosures must be provided.

This section requires a method for the convenient bonding of other systems, and such a method must involve a terminal bar or set of terminals that will accommodate at least three intersystem bonding conductors. Providing convenient bonding terminals will increase the likelihood that the communication and entertainment system installers will actually install the required bonding conductors.

E3609.4 Method of bonding at the service. Bonding jumpers meeting the requirements of this chapter shall be used around impaired connections, such as reducing washers or oversized, concentric, or eccentric knockouts. Standard locknuts or bushings shall not be the only means for the bonding required by this section but shall be permitted to be installed to make mechanical connections of raceways. Electrical continuity at service equipment, service raceways and service conductor enclosures shall be ensured by one or more of the methods specified in Sections E3609.4.1 through E3609.4.4.

❖ The metal parts of the service equipment that do not normally carry current must be bonded together. The methods and means to do this are covered in the fol-

lowing four subsections. Concentric and eccentric knockouts greatly reduce the current carrying capacity or conductivity of the connection because very little metal is left intact between the removable sections of the knockout. This is not the case with single-size knockouts. Fault current will flow along the raceway and be impeded at the point where the raceway fitting is attached to the cabinet through a concentric or eccentric knockout. In some cases where part of (but not the entire) knockout is removed, the mechanical connection is so weak that reducing washers must be installed just to strengthen the connection. There certainly is not a solid electrical connection at this point. In addition, standard locknuts are not permitted to provide the only bonding connection of the service equipment. Commonly, a special bonding bushing is used on the inside of the enclosure for the bonding of the raceway or fitting to the enclosure. Such bonding bushings have lugs for jumper connections and may also have set screws that bite into the metal of the enclosure and/or the threads on the conduit or fitting.

E3609.4.1 Grounded service conductor. Equipment shall be bonded to the grounded service conductor in a manner provided in this code.

❖ Bonding to the grounded service conductor is done by using listed pressure connectors such as a terminal bar or bus in a panelboard cabinet, listed lugs, clamps or exothermic welding.

E3609.4.2 Threaded connections. Equipment shall be bonded by connections using threaded couplings or threaded hubs on enclosures. Such connections shall be made wrench tight.

❖ Raceways joined by threaded couplings and connectors must be wrench tight. High impedance in the ground-fault current path can occur if the installer only hand tightens threaded couplings. "Wrench tight" is another term that is not defined in Chapter 35, but the meaning is obvious. Tightening the threaded couplings with a wrench is necessary not only for mechanical strength but to ensure that the fault-current path has good electrical continuity. For overhead service risers, a typical installation uses a 2-inch (51 mm) rigid conduit. A hub is provided with the service equipment that has predrilled holes for the attachment of the conduit. It is important that the conduit be threaded into the hub wrench tight.

E3609.4.3 Threadless couplings and connectors. Equipment shall be bonded by threadless couplings and connectors for metal raceways and metal-clad cables. Such couplings and connectors shall be made wrench tight. Standard locknuts or bushings shall not be used for the bonding required by this section.

❖ Threadless couplings and connectors are permitted for bonding the parts of service equipment together, but these must be rain tight where installed outdoors. For example, a set-screw connector is permitted to connect the conduit of an underground service lateral to the service equipment cabinet where it is run up to the bottom of the cabinet. Also, where the cabinet is installed in the wall and is flush with the surface, set-screw connectors and couplings are permitted, but they must be made wrench tight.

E3609.4.4 Other devices. Equipment shall be bonded by other listed devices, such as bonding-type locknuts, bushings and bushings with bonding jumpers. [250.92(B)]

❖ To ensure electrical continuity throughout the interconnection of the various components of the service equipment, special devices such as bonding bushings are required where raceways and fittings are connected to concentric and eccentric knockouts punched in boxes and cabinets. Bonding bushings have a set screw that positively bonds to the threads of the raceway or fitting, a lug for the connection of a bonding jumper and sometimes a set screw that bites into the enclosure wall.

E3609.5 Sizing supply-side bonding jumper and main bonding jumper. The bonding jumper shall not be smaller than the sizes shown in Table E3603.4 for grounding electrode conductors. Where the service-entrance conductors are paralleled in two or more raceways or cables, and an individual supply-side bonding jumper is used for bonding these raceways or cables, the supply-side bonding jumper for each raceway or cable shall be selected from Table E3603.4 based on the size of the ungrounded supply conductors in each raceway or cable. A single supply-side bonding jumper installed for bonding two or more raceways or cables shall be sized in accordance with Table E3603.4 based on the largest set of parallel ungrounded supply conductors. [250.102(C)]

❖ Paralleled service conductors are rarely used in dwelling unit services. For example, for a 200-ampere service using size 2/0-AWG copper conductors, the table indicates that bonding jumpers would be a minimum of size 4-AWG copper.

E3609.6 Metal water piping bonding. The metal water piping system shall be bonded to the service equipment enclosure, the grounded conductor at the service, the grounding electrode conductor where of sufficient size, or to the one or more grounding electrodes used. The bonding jumper shall be sized in accordance with Table E3603.4. The points of attachment of the bonding jumper(s) shall be accessible. [250.104(A) and 250.104(A)(1)]

❖ Because the interior metal water piping might become energized, it must be bonded to the grounding electrode system. Bonding metal piping systems together reduces the possibility of a voltage potential between them and the associated shock hazard. This is not the same requirement that the metal underground water pipe be used as a grounding electrode. If the water service is a metal pipe for 10 feet (3048 mm) underground and metal where it enters the house, it is used as the grounding electrode. In some dwellings, the water piping system changes to plastic inside the house, and much of the water supply piping is plastic; however, any piping that is metal must be bonded to one of the parts of the grounding electrode system listed in this section. If the entire piping system is metal, it may seem obvious that the grounding electrode conductor being run to the first 5 feet (1524 mm) of the point at which the water pipe enters the house would bond the entire water piping system. In many cases there is a dielectric union at the entry of the water piping system, and a jumper is required from the line side of the union to the downstream side of the water piping. It may be necessary to bond between the hot and cold water pipes because these could have nonconductive materials at the water heater. The bonding jumper must be at least the size of the grounding electrode conductor determined for the service from Table E3603.4. The pipe clamp or fitting used to attach the bonding jumper to the water piping system must be accessible. The water heater is a convenient place for this since the attachment point will always be accessible. Also, this may be the best place to bond between the hot and cold water piping, if that is necessary.

E3609.7 Bonding other metal piping. Where installed in or attached to a building or structure, metal piping systems, including gas piping, capable of becoming energized shall be bonded to the service equipment enclosure, the grounded conductor at the service, the grounding electrode conductor where of sufficient size, or to the one or more grounding electrodes used. The bonding conductor(s) or jumper(s) shall be sized in accordance with Table E3908.12 using the rating of the circuit capable of energizing the piping. The equipment grounding conductor for the circuit that is capable of energizing the piping shall be permitted to serve as the bonding means. The points of attachment of the bonding jumper(s) shall be accessible. [250.104(B)]

❖ The above-ground metal gas piping upstream from equipment shutoff valves must be bonded because it could become energized. The rating of the circuit serving the gas appliance or equipment determines the size of the bonding jumper in accordance with Table E3908.12. Under this practice, the size of a bonding conductor for a gas pipe serving a furnace supplied with a 20 ampere circuit would be 12 AWG according to the table. A three-conductor nonmetallic sheathed cable with size 12-AWG conductors (12-2 with ground) that is to run to the furnace is adequate for bonding the metal gas pipe since the grounding conductor of the cable is bonded to the grounding electrode system. The gas piping is connected to the appliance by metal-to-metal joints; therefore, the piping is bonded to the appliance and the appliance grounding conductor. The gas piping is considered to be bonded where connected to the equipment-grounding conductor for a circuit capable of energizing the piping. This is specifically stated in the 2008 edition of NFPA 70, Section 250.104(B). See Section G2411 of the code, which addresses the bonding of gas piping and corrugated stainless steel tubing (CSST) used for gas. The bonding requirements for CSST are unique and extend beyond the requirements in this section.

SECTION E3610
GROUNDING ELECTRODE CONDUCTORS

E3610.1 Continuous. The grounding electrode conductor shall be installed in one continuous length without splices or joints and shall run to any convenient grounding electrode available in the grounding electrode system where the other electrode(s), if any, are connected by bonding jumpers in accordance with Section E3608.2, or to one or more grounding electrode(s) individually. The grounding electrode conductor shall be sized for the largest grounding electrode conductor required among all of the electrodes connected to it. [250.64(C)]

> **Exception:** Splicing of the grounding electrode conductor by irreversible compression-type connectors listed as grounding and bonding equipment or by the exothermic welding process shall not be prohibited. [250.64(C)(1)]

❖ Except as allowed by the exception, grounding electrode conductors are to be run in continuous lengths without splices. Note that the exception allows two extremely reliable methods of splicing. Unspliced conductors will have the greatest reliability and continuity. Where electrodes are connected together to a common electrode conductor, the largest required size of the conductors will dictate the size of the common conductor. For example, if a 6-AWG conductor ran from the service to a ground rod and a 4-AWG conductor was required to connect the ground rod to the footing steel, the 6-AWG conductor would have to be upsized to a 4 AWG.

E3610.2 Securing and protection against physical damage. Where exposed, a grounding electrode conductor or its enclosure shall be securely fastened to the surface on which it is carried. Grounding electrode conductors shall be permitted to be installed on or through framing members. A 4 AWG or larger conductor shall be protected where exposed to physical damage. A 6 AWG grounding conductor that is free from exposure to physical damage shall be permitted to be run along the surface of the building construction without metal covering or protection where it is and securely fastened to the construction; otherwise, it shall be in rigid metal conduit, intermediate metal conduit, rigid polyvinyl chloride (PVC), nonmetallic conduit, reinforced thermosetting resin (RTRC) nonmetallic conduit, electrical metallic tubing or cable armor. Grounding electrode conductors smaller than 6 AWG shall be in rigid metal conduit, intermediate metal conduit, rigid polyvinyl chloride (PVC) nonmetallic conduit, reinforced thermoseting resin (RTRC) nonmetallic conduit, electrical metallic tubing or cable armor. Grounding electrode conductors and grounding electrode bonding jumpers shall not be required to comply with Section E3803. [250.64(B)]

Bare aluminum or copper-clad aluminum grounding electrode conductors shall not be used where in direct contact with masonry or the earth or where subject to corrosive conditions. Where used outside, aluminum or copper-clad aluminum grounding electrode conductors shall not be installed within 18 inches (457 mm) of the earth. [250.64(A)]

❖ This section provides rules for the physical protection of the grounding electrode conductors. It is up to the code official to decide whether the grounding electrode conductors as installed are exposed to physical damage.

Where run on the interior of the dwelling, it is unlikely that these conductors would be exposed to physical damage. Where installed along the outside wall or surface of the house, the possibility of physical damage is often a concern. In most new houses, the grounding electrode conductor is run on the interior, and another grounding electrode conductor is run to a made electrode on the outside of the building. The conductor that is the sole connection to the supplemental grounding electrode or ground rod is considered one of the grounding electrode conductors, and the requirements of this section apply to it as well as to a grounding electrode conductor run to the entry point of the metal underground water pipe.

Quite often a solid bare copper conductor is installed on the outside of the house as the grounding electrode conductor and is run from the service equipment enclosure to grade along the wall of the house. The smallest size grounding conductor permitted to run exposed without any covering or protection is 6 AWG, but only if it is free from exposure to physical damage. Although it can be done, it is difficult to install these conductors in a neat and workmanlike manner where they are run exposed. Many electricians install these surface conductors in one of the raceways listed in this section.

The code does not permit the installation of aluminum conductors within 18 inches (457 mm) of the earth where used outdoors because of moisture and the potential for corrosion.

E3610.3 Raceways and enclosures for grounding electrode conductors. Ferrous metal raceways and enclosures for grounding electrode conductors shall be electrically continuous from the point of attachment to cabinets or equipment to the grounding electrode, and shall be securely fastened to the ground clamp or fitting. Nonferrous metal raceways and enclosures shall not be required to be electrically continuous. Ferrous metal raceways and enclosures shall be bonded at each end of the raceway or enclosure to the grounding electrode or to the grounding electrode conductor. Bonding methods in compliance with Section E3609.4 for installations at service equipment locations and with E3609.4.2 through E3609.4.4 for other than service equipment locations shall apply at each end and to all intervening ferrous raceways, boxes, and enclosures between the cabinets or equipment and the grounding electrode. The bonding jumper for a grounding electrode conductor raceway shall be the same size or larger than the required enclosed grounding electrode conductor.

Where a raceway is used as protection for a grounding conductor, the installation shall comply with the requirements of Chapter 38. [250.64(E)(4)]

❖ Where grounding electrode conductors are installed inside a metal raceway, the metal conduit or tubing must be bonded at each end to the conductor so that it is electrically continuous. If this is not done and a high-fault current flows through the grounding electrode conductor, the current could create a magnetic flux field around the raceway enclosing the conductor and, depending on the amount of fault current, this would create a high impedance in the fault current return path. One solution is to install nonmetallic conduit where permitted for protection of the grounding electrode conductor.

E3610.4 Prohibited use. An equipment grounding conductor shall not be used as a grounding electrode conductor. (250.121)

> **Exception:** A wire-type equipment grounding conductor shall be permitted to serve as both an equipment grounding conductor and a grounding electrode conductor where installed in accordance with the applicable requirements for both the equipment grounding conductor and the grounding electrode conductor in Chapters 36 and 39. Where used as a grounding electrode conductor, the wire-type equipment grounding conductor shall be installed and arranged in a manner that will prevent objectionable current. [250.121 Exception, 250.6(A)]

❖ Grounding electrode conductors are dedicated purpose conductors that comply with the requirements of Section E3610. Equipment grounding conductors comply with Section E3908. Conductors used to ground equipment such as appliances and panelboard enclosures cannot serve as grounding electrode conductors, except where the conductor is a wire and complies with all of the code provisions for both types of conductors. In dwelling unit construction, the opportunity to utilize a combination conductor would be rare.

SECTION E3611
GROUNDING ELECTRODE CONDUCTOR CONNECTION TO THE GROUNDING ELECTRODES

E3611.1 Methods of grounding conductor connection to electrodes. The grounding or bonding conductor shall be connected to the grounding electrode by exothermic welding, listed lugs, listed pressure connectors, listed clamps or other listed means. Connections depending on solder shall not be used. Ground clamps shall be listed for the materials of the grounding electrode and the grounding electrode conductor and, where used on pipe, rod or other buried electrodes, shall also be listed for direct soil burial or concrete encasement. Not more than one conductor shall be connected to the grounding electrode by a single clamp or fitting unless the clamp or fitting is listed for multiple conductors. One of the methods indicated in the following items shall be used:

1. A pipe fitting, pipe plug or other approved device screwed into a pipe or pipe fitting.

2. A listed bolted clamp of cast bronze or brass, or plain or malleable iron.

3. For indoor communications purposes only, a listed sheet metal strap-type ground clamp having a rigid metal base that seats on the electrode and having a strap of such material and dimensions that it is not likely to stretch during or after installation.

4. Other equally substantial approved means. (250.70)

❖ This section lists items that can be used to connect the grounding electrode conductor to the grounding electrode.

E3611.2 Accessibility. All mechanical elements used to terminate a grounding electrode conductor or bonding jumper to the grounding electrodes that are not buried or concrete encased shall be accessible. [250.68(A) and 250.68(A) Exception]

❖ Devices that connect a conductor to an electrode must be accessible except where such devices are buried or concrete encased. Such connections are often located underground so as to provide a degree of protection from physical damage. Any device used to connect to an electrode must be listed for the environment to which it is exposed, such as listed for direct burial or concrete encasement.

E3611.3 Effective grounding path. The connection of the grounding electrode conductor or bonding jumper shall be made in a manner that will ensure a permanent and effective grounding path. Where necessary to ensure effective grounding for a metal piping system used as a grounding electrode, effective bonding shall be provided around insulated joints and sections and around any equipment that is likely to be disconnected for repairs or replacement. Bonding jumpers shall be of sufficient length to permit removal of such equipment while retaining the integrity of the grounding path. [250.68(B)]

❖ Where a metal piping system is used as the grounding electrode, a permanent and effective ground-fault current path must be maintained. Water meters, pressure reducing devices and the like may be installed in the piping system, and a bonding jumper must be installed around such devices. The bonding jumper must be of sufficient length that the item can be repaired or replaced without disconnecting the bonding jumper.

E3611.4 Interior metal water piping. Where grounding electrode conductors and bonding jumpers are connected to interior metal water piping as a means to extend the grounding electrode conductor connection to an electrode(s), such piping shall be located not more than 5 feet (1524 mm) from the point of entry into the building.

Where interior metal water piping is used as a conductor to interconnect electrodes that are part of the grounding electrode system, such piping shall be located not more than 5 feet (1524 mm) from the point of entry into the building. [250.68(C)(1)]

❖ See the commentary to Section E3608.1.1.

E3611.5 Protection of ground clamps and fittings. Ground clamps or other fittings shall be approved for applications without protection or shall be protected from physical damage by installing them where they are not likely to be damaged or by enclosing them in metal, wood or equivalent protective coverings. (250.10)

❖ If not protected from physical damage by their location, ground clamps and fittings must be protected with approved enclosures. The code official must decide whether physical damage is likely to occur.

E3611.6 Clean surfaces. Nonconductive coatings (such as paint, enamel and lacquer) on equipment to be grounded shall be removed from threads and other contact surfaces to ensure good electrical continuity or shall be connected by fittings that make such removal unnecessary. (250.12)

❖ Nonconductive coatings such as paint must be removed to help ensure the continuity of a grounding circuit that may need to carry high levels of fault current. Fittings that make the removal of paint unnecessary may not be available or may be very hard to find.

Bibliography

The following resource materials were used in the preparation of the commentary for this chapter of the code:

NFPA 70–14, *National Electrical Code*. Quincy, MA: National Fire Protection Association, 2013.

Chapter 37:
Branch Circuit and Feeder Requirements

General Comments

Chapter 37 addresses the sizing of conductors for feeders and branch circuits, specifies the required branch circuits, provides for overcurrent protection of such conductors, specifies limitations for branch circuit loading, and addresses panel board ratings and protection. After the service is designed or installed, the next stage in the design or installation is the feeder and branch circuit design or installation.

Purpose

The primary purpose of this chapter is to provide the required number and type/rating of branch circuits to serve the anticipated loads and to protect conductors from overload (overcurrent). This translates to protection from fire and property damage.

SECTION E3701
GENERAL

E3701.1 Scope. This chapter covers branch circuits and feeders and specifies the minimum required branch circuits, the allowable loads and the required overcurrent protection for branch circuits and feeders that serve less than 100 percent of the total dwelling unit load. Feeder circuits that serve 100 percent of the dwelling unit load shall be sized in accordance with the procedures in Chapter 36. [310.15(B)(7)(2)]

❖ Feeders carry power from the service to the branch circuit overcurrent protective devices or the panelboard, which is where the branch circuits begin. Figure E3607.2 shows that the feeder runs from the service equipment to supply power to the panelboard, in accordance with the definition given in Chapter 35. The terms "subfeeder" and "subpanel" are not defined in the code but are terms commonly used in the field. A "subfeeder" runs from a panelboard to serve another panel or "subpanel." In other words, feeders do not serve loads directly but serve a panelboard from which a branch circuit is run to supply the appliances or equipment. Overcurrent protection is provided at the supply end of the feeder or "subfeeder," and the panel or "subpanel" it serves is not required to have a main circuit breaker or set of main fuses if located within the same building. "Main-lug only" panels located within the same building as the service equipment are allowed to be supplied by feeders. See Sections 225.31 through 225.39 of the *National Electrical Code®* (NEC®).

The conductor size for a feeder that carries 100 percent of the load is calculated under the provisions of Chapter 36. An example of a feeder that carries 100 percent of the load is one where the service consists of only one main disconnect/overcurrent device and where a single set of conductors from such equipment to the panelboard (the feeder conductors) supplies the power for all of the loads of the dwelling.

Where the feeder conductors do not serve 100 percent of the total dwelling unit load, the provisions of this chapter are used to calculate the conductor size.

E3701.2 Branch-circuit and feeder ampacity. Branch-circuit and feeder conductors shall have ampacities not less than the maximum load to be served. Where a branch circuit or a feeder supplies continuous loads or any combination of continuous and noncontinuous loads, the minimum branch-circuit or feeder conductor size, before the application of any adjustment or correction factors, shall have an allowable ampacity equal to or greater than the noncontinuous load plus 125 percent of the continuous load. [210.19(A)(1)(a) and 215.2(A)(1)(a)]

Exception: The grounded conductors of feeders that are not connected to an overcurrent device shall be permitted to be sized at 100 percent of the continuous and noncontinuous load. [215.1(A)(1) Exception No. 2]

❖ This requirement pertains to the size or the allowable ampacity of the conductors and not to the rating of the branch circuit. The branch circuit rating does not always correspond to the conductor size. See Section E3702 for branch circuit ratings. The branch circuit and feeder loads must be calculated to determine the appropriate size conductors and overcurrent protection devices to be used. Demand factors may be used in sizing conductors. For demand factors, see the commentary to Section E3704.3.

In calculating conductor ampacity, the type of load also must be taken into account (see commentary, Section E3705.1). A load that is in operation for three hours or more is a continuous load. When a load is operated continuously, the heating effect on the conductors is a concern because, without time to dissipate, the heat can build up. Although it may seem that the lights are sometimes on for many hours, there is usually much diversity in the operation of residential lighting. Generally, dwelling unit lighting loads are not considered as continuous loads for calculation purposes. Some loads in residential wiring may need to be considered as continuous and be calculated at 125 percent of the full load. However, the code does not specify certain loads as being continuous. For residential wiring, it may be a judgement call as to how long the load will operate at any one time.

For any load considered to be continuous, the allowable conductor ampacity must be at least 125 percent of the calculated load. This is the same as saying that the continuous load on a conductor should not be more than 80 percent of its allowable ampacity.

Even for noncontinuous loads, this is a good practice, although not required. For example, a 20-ampere-rated branch circuit serving a noncontinuous load is permitted to be loaded at up to 20 amperes in accordance with Table E3702.14. A good rule of thumb is to try to estimate a maximum load of 16 amperes for a 20-ampere circuit. However, for general purpose branch circuits in a dwelling supplying receptacles and lighting outlets, it is impossible to know at the time the wiring is being installed what load will be connected. Even for a circuit consisting of only lighting outlets, it is difficult to know what the load will be, as light fixtures and lamps can be changed.

The exception allows the sizing of grounded conductors based on the actual load where there is no heating effect from device terminals.

E3701.3 Selection of ampacity. Where more than one calculated or tabulated ampacity could apply for a given circuit length, the lowest value shall be used. [310.15(A)(2)]

Exception: Where two different ampacities apply to adjacent portions of a circuit, the higher ampacity shall be permitted to be used beyond the point of transition, a distance equal to 10 feet (3048 mm) or 10 percent of the circuit length figured at the higher ampacity, whichever is less. [310.15(A)(2) Exception]

❖ If the conductors of a circuit are exposed to different ambient temperatures (a cable run from a basement through an attic, for example), they could have different ampacities based on the application of Section E3705.2. The portion of the circuit in the higher ambient temperature location is the weakest link; therefore, the lowest calculated ampacity must apply. The exception relaxes this requirement slightly by allowing the higher ampacity to apply for conductors that extend only a short distance into an area of higher ambient temperature.

E3701.4 Branch circuits with more than one receptacle. Conductors of branch circuits supplying more than one receptacle for cord-and-plug-connected portable loads shall have ampacities of not less than the rating of the branch circuit. [210.19(A)(2)]

❖ For multioutlet branch circuits, the rating of the branch circuit overcurrent protection device determines the minimum size of conductors that can be used. Conductors must have an ampacity of not less than the rating of the branch circuit. For example, the use of a 20-ampere circuit breaker requires that the conductors have an ampacity of at least 20 amperes. This is differ-

ent for a dedicated circuit to a specific load such as a large motor or appliance. The conductors must have an ampacity sufficient for the load, but this ampacity for the motor or appliance may not correspond to a standard size of overcurrent device. In this case, the next higher size circuit breaker could be used, and therefore the conductor ampacity could be less than the rating of the branch circuit. The rating of the branch circuit is set by the size of the overcurrent protection device. However, on multioutlet branch circuits that supply receptacles for cord- and plug-connected loads, the conductor ampacity must be matched exactly to the branch circuit rating because it is unknown what load might be plugged into the circuit.

E3701.5 Multiwire branch circuits. All conductors for multiwire branch circuits shall originate from the same panelboard or similar distribution equipment. Except where all ungrounded conductors are opened simultaneously by the branch-circuit overcurrent device, multiwire branch circuits shall supply only line-to-neutral loads or only one appliance. [210.4(A) and 210.4(C)]

❖ In residential wiring, a multiwire branch circuit is a three-wire circuit that consists of two ungrounded conductors that have a potential of 240 volts between them and a grounded conductor that has a potential of 120 volts between it and each ungrounded conductor. The circuit that supplies an electric range or electric clothes dryer is a multiwire branch circuit (see Commentary Figure E3406.10.2).

A multiwire branch circuit could also be two separate circuits that share a neutral conductor. In commercial locations where conduit is used, this type of circuit is used very commonly: three conductors are pulled into the raceway, and the neutral carries only the unbalanced current of the two circuits. However, in residential work where cable is primarily used, it is not as advantageous; the perceived advantage being a savings in materials. Depending on how the receptacle circuits are routed, there may be no advantage to using a multiwire circuit and the potential risks could be avoided. Some electricians like to install a three-conductor cable. It actually contains four conductors: a red and black for the ungrounded or hot conductors, a white for the shared neutral (grounded conductor) and a bare or green for the grounding conductor. This installation can result in serious problems, and care must be taken to ensure that all code provisions are complied with. One concern is that in dwellings a homeowner or individual who does not understand the special code provisions sometimes rearranges the circuit breakers in the panelboard by placing both hot conductors of the multiwire circuit on the same leg (ungrounded conductor) of the service, resulting in the neutral conductor carrying the sum of the return current of both circuits instead of only the unbalanced current of the two circuits. On a 20-ampere multiwire circuit, it is conceivable that the neutral conductor would be subjected to up to 40 amperes on a size 12-AWG conductor. This could heat up the neutral con-

BRANCH CIRCUIT AND FEEDER REQUIREMENTS

ductor and melt or burn the insulation and possibly result in a fire.

One use where this is encountered is for the two kitchen small appliance circuits. Duplex receptacles with breakoff tabs are used so that the two separate receptacles on the same yoke or strap can be supplied by the two hot conductors of the multiwire circuit. A person working on the circuit might check one of the "split-wired" receptacles to determine that the power is turned off at the circuit breaker but not turn off the power to both circuits. The worker is then subject to the risk of shock. This is why multiwire circuits must be supplied by a two-pole circuit breaker or two single-pole circuit breakers with an approved handle tie. An approved handle tie is a clip made specifically for and listed for clamping over the handles of the two single-pole circuit breakers. If one of the circuit breakers trips open, it will operate the handle of the other breaker. Some circuit breakers are made with a hole in the handle through which a nail can be inserted. A nail holding the handles of two breakers together should not be approved. Where the handle of the circuit breaker does not already have a hole, attempts have been made to drill a small hole through the handle to insert a nail or wire. This is an obvious code violation. Not tying the handles together is certainly a code violation. The most foolproof installation would use a two-pole circuit breaker with a single handle, thus avoiding the need to use field-installed ties. Note that although the last sentence of this section alludes to the possibility of having independent overcurrent devices serving the two ungrounded circuit conductors, Section E3701.5.1 strictly prohibits this (see Section E3701.5.1).

The shared neutral conductor of the multiwire circuit must be pigtailed in each outlet box where the duplex receptacle is installed. If the duplex receptacle was used to splice or connect the neutral conductor of the circuit, the circuit would be broken when the receptacle is removed or worked on. If this is done when the circuit is energized, it would place appliances plugged into the circuit in a 240-volt series circuit. The voltage drop across each appliance would depend on its resistance and could cause abnormal voltages and thus damage to the appliances (see commentary, Section E3406.10.2).

E3701.5.1 Disconnecting means. Each multiwire branch circuit shall be provided with a means that will simultaneously disconnect all ungrounded conductors at the point where the branch circuit originates. [210.4(B)]

❖ In its practical effect, this section requires that multiwire circuits be supplied through two-pole circuit breakers to eliminate the possibility of opening one ungrounded conductor while the other remains energized. The reason for this is discussed in the previous section's commentary.

E3701.5.2 Grouping. The ungrounded and grounded circuit conductors of each multiwire branch circuit shall be grouped by cable ties or similar means in at least one location within the panelboard or other point of origination. [210.4(D)]

Exception: Grouping shall not be required where the circuit conductors enter from a cable or raceway unique to the circuit, thereby making the grouping obvious, or where the conductors are identified at their terminations with numbered wire markers corresponding to their appropriate circuit number. [210.4(D) Exception].

❖ As discussed in the commentary to Section E3701.5, multiwire circuits can be hazardous if incorrectly wired. The intent of this section is to make it less likely that multiwire circuits will be incorrectly connected at the time of installation and at any later time when work is performed in the panelboard enclosure. A visual "red flag" will alert someone that the grouped conductors are part of a multiwire circuit. If the only conductors in a conduit are those that serve the load to which the conduit is run, and such load requires a multiwire branch circuit, the circuit will be recognizable when looking inside the panelboard. The same is true for cables such as Type NM, AC and MC because the presence of a white, black and red conductor emerging from the cable is recognizable as a multiwire circuit.

SECTION E3702
BRANCH CIRCUIT RATINGS

E3702.1 Branch-circuit voltage limitations. The voltage ratings of branch circuits that supply luminaires or receptacles for cord-and-plug-connected loads of up to 1,400 volt-amperes or of less than $^1/_4$ horsepower (0.186 kW) shall be limited to a maximum rating of 120 volts, nominal, between conductors.

Branch circuits that supply cord-and-plug-connected or permanently connected utilization equipment and appliances rated at over 1,440 volt-amperes or $^1/_4$ horsepower (0.186 kW) and greater shall be rated at 120 volts or 240 volts, nominal. [210.6(A), (B), and (C)]

❖ The code requires that lighting fixtures in dwelling units be supplied by circuits with a maximum voltage of 120 volts. Lighting fixtures designed for higher voltages are for use in commercial and industrial occupancies. Receptacles supplying cord- and plug-connected loads of up to 1,440 volt-amperes or 12 amperes are also limited to 120 volts. A multiwire circuit with a potential of 240 volts between the two ungrounded conductors that supplies a range or dryer is permitted under the second paragraph because the load is well over 12 amperes (see Section E3701.5). A multiwire circuit connected to one duplex receptacle with the break-off tab removed so that it is a split-wired receptacle is a 240-volt branch circuit, and the cord- and plug-connected load will usually be less than 1,440 volt-amperes; however, this multiwire circuit would be considered as two separate 120-volt circuits. See the commentary to Section E3701.5, which states that a multiwire branch circuit could also be two separate circuits that share a neutral conductor.

E3702.2 Branch-circuit ampere rating. Branch circuits shall be rated in accordance with the maximum allowable ampere rating or setting of the overcurrent protection device. The rating for other than individual branch circuits shall be 15, 20, 30, 40 and 50 amperes. Where conductors of higher ampacity are used, the ampere rating or setting of the specified over-current device shall determine the circuit rating. (210.3)

❖ The rating of the branch circuit is set by the size of the overcurrent protection device.

The rating of the overcurrent device, not the size or ampacity of the branch circuit conductors, determines the rating of the branch circuit. The conductor size is sometimes increased to compensate for voltage drop such as where a size 10-AWG copper conductor is used on a 20-ampere branch circuit where a size 12 AWG would normally be used. It is a 20-ampere branch circuit because a 20-ampere circuit breaker is used.

Most of the loads for residential applications can be supplied by branch circuits of the standard 15-, 20-, 30-, 40- or 50-ampere ratings given in this section. An individual branch circuit (a circuit that supplies only one piece of utilization equipment) is permitted to be of any rating necessary to supply the load it serves.

E3702.3 Fifteen- and 20-ampere branch circuits. A 15- or 20-ampere branch circuit shall be permitted to supply lighting units, or other utilization equipment, or a combination of both. The rating of any one cord-and-plug-connected utilization equipment not fastened in place shall not exceed 80 percent of the branch-circuit ampere rating. The total rating of utilization equipment fastened in place, other than luminaires, shall not exceed 50 percent of the branch-circuit ampere rating where lighting units, cord-and-plug-connected utilization equipment not fastened in place, or both, are also supplied. [210.23(A)(1) and (2)]

❖ This section covers combination loads for 15- and 20-ampere branch circuits that supply multiple outlets, appliances, and equipment.

In this code section, appliances and equipment are referred to as two types: (1) as cord- and plug-connected and not fastened in place and (2) as fastened in place. An appliance or equipment that is fastened in place or connected by permanent wiring methods is sometimes called a fixed appliance or fixed equipment. For this discussion, the term "fixed" will be used for appliances and equipment fastened in place and connected with permanent wiring methods. The term "appliance" will be used to discuss appliances and utilization equipment.

A 15- or 20-ampere branch circuit is permitted to supply a combination of loads except for the two 20-ampere small appliance branch circuits required for the kitchen area receptacles in accordance with Sections E3703.2 and E3901.3 through E3901.3.2. It is common practice to run a separate branch circuit to appliances such as a dishwasher, refrigerator, disposal, etc. It is also permissible to supply two or more loads of different types on the same circuit. A furnace,

considered an appliance, is required to be supplied by a separate individual branch circuit in accordance with Section E3703.1.

Where a separate circuit is installed for appliances, the circuit is called an appliance branch circuit or individual branch circuit. Appliances can also be supplied by a general purpose branch circuit, but the rating of the appliance is restricted. The total ampere rating of fixed appliances must not exceed 50 percent of the rating of the branch circuit if other loads are also served on the same branch circuit. A 120-volt, 20-ampere rated branch circuit that supplies a room air conditioner, for example, could also supply light fixtures or a receptacle for cord- and plug-connected loads, as long as the air conditioner load does not exceed 10 amperes. More than one fixed appliance could be served by the mixed-load circuit. For example, consider a 20-ampere general purpose branch circuit that supplies several receptacles, light fixtures and two fixed attic fans. The rating of this circuit would comply with code as long as the combined rating of the two attic fans was not more than 10 amperes. The number of receptacle outlets on the mixed load circuit is not restricted.

In summary, for a 15- or-20 ampere-rated general purpose branch circuit that serves a combination of loads, the maximum rating of a fixed appliance served by that circuit is 50 percent of the circuit rating. Any cord- and plug-connected appliance not fastened in place and served by this general purpose branch circuit must have an ampere rating not more than 80 percent of the circuit rating.

E3702.4 Thirty-ampere branch circuits. A 30-ampere branch circuit shall be permitted to supply fixed utilization equipment. A rating of any one cord-and-plug-connected utilization equipment shall not exceed 80 percent of the branch-circuit ampere rating. [210.23(B)]

❖ A 30-ampere branch circuit is used for fixed utilization equipment, which is considered equipment fastened in place or connected by permanent wiring methods. Water heaters and condensing units are examples of 30-ampere circuits that would be installed in a dwelling. The load of the appliance should be checked to make sure it does not exceed 80 percent of the branch circuit rating or 5,760 volt-amperes or 24 amperes. In most cases, when a new house is being wired, the exact rating of the appliances is not known. For a clothes dryer, it is common practice to run size 10-AWG copper conductors on a 30-ampere circuit. For calculating a feeder load, a clothes dryer for which the nameplate rating is not known is considered as 5,000 volt-amperes. This would require a 240 volt, 30-ampere rated circuit (5,000 volt-amperes/240 volts = 20.8 amperes, which would require a 30-ampere-rated circuit).

E3702.5 Branch circuits serving multiple loads or outlets. General-purpose branch circuits shall supply lighting outlets, appliances, equipment or receptacle outlets, and combinations of such. Multioutlet branch circuits serving lighting or receptacles shall be limited to a maximum branch-circuit rating of 20 amperes. [210.23(A), (B), and (C)]

❖ The code permits circuits serving general purpose convenience receptacles to also serve lighting outlets. In commercial and industrial buildings, it is common practice to install receptacles on one circuit and lighting loads on another circuit and not to mix the two. But in residential wiring, it is very common to mix the two loads on the same circuit (see commentary, Section E3702.3). The number of general purpose (convenience) receptacle outlets on a branch circuit is not limited.

E3702.6 Branch circuits serving a single motor. Branch-circuit conductors supplying a single motor shall have an ampacity not less than 125 percent of the motor full-load current rating. [430.22(A)]

❖ Some examples of motor loads in residential wiring are an attic fan, a sump pump, a circulating pump for hot water space heating, a garage door opener and a garbage disposal. Such loads are typically served by individual dedicated branch circuits. Circuits that run to motors such as these must have conductors rated for at least 125 percent of the full load rating of the motor to allow for starting current.

E3702.7 Branch circuits serving motor-operated and combination loads. For circuits supplying loads consisting of motor-operated utilization equipment that is fastened in place and that has a motor larger than $^1/_8$ horsepower (0.093 kW) in combination with other loads, the total calculated load shall be based on 125 percent of the largest motor load plus the sum of the other loads. [220.18(A)]

❖ To correctly size branch circuit conductors, the load must be calculated. Branch circuits that supply fastened-in-place utilization equipment having a motor larger than $^1/_8$ horsepower must have the branch circuit load calculated at 125 percent of the motor load plus the other loads served. If none of the motors on a circuit is larger than $^1/_8$ horsepower, it is not necessary to add the extra 25 percent.

E3702.8 Branch-circuit inductive and LED lighting loads. For circuits supplying luminaires having ballasts or LED drivers, the calculated load shall be based on the total ampere ratings of such units and not on the total watts of the lamps. [220.18(B)]

❖ On circuits serving light fixtures that contain a ballast such as fluorescent, mercury-vapor, high-intensity discharge and high-pressure sodium, there is an inductive load. In addition to the wattage or volt-amperes of the lamps, the inductive load must be considered. Therefore, the total ampere rating of the light fixtures must be used for determining the load on the circuit. Such light fixtures are marked with a rating that includes the ballast and lamp load.

E3702.9 Branch-circuit load for ranges and cooking appliances. It shall be permissible to calculate the branch-circuit load for one range in accordance with Table E3704.2(2). The branch-circuit load for one wall-mounted oven or one counter-mounted cooking unit shall be the nameplate rating of the appliance. The branch-circuit load for a counter-mounted cooking unit and not more than two wall-mounted ovens all supplied from a single branch circuit and located in the same room shall be calculated by adding the nameplate ratings of the individual appliances and treating the total as equivalent to one range. (220.55 Note 4)

❖ The oven and all of the heating elements of a range are usually not used all at the same time. Because of this diversity, the demand factors of Table E3704.2(2) can be used. A counter-mounted cooking unit (a stove top) and one or two ovens are treated as one range. If only a separate counter-mounted cooking unit or oven is installed, the nameplate rating must be used as the load for the branch circuit. See the sample feeder load calculations for Section E3704.3.

E3702.9.1 Minimum branch circuit for ranges. Ranges with a rating of 8.75 kVA or more shall be supplied by a branch circuit having a minimum rating of 40 amperes. [210.19(A)(3)]

❖ A minimum rating of 40 amperes is required for any range rated 8.75 kVA or more. Typically, 60-ampere circuits are installed for ranges.

E3702.10 Branch circuits serving heating loads. Electric space-heating and water-heating appliances shall be considered to be continuous loads. Branch circuits supplying two or more outlets for fixed electric space-heating equipment shall be rated 15, 20, 25 or 30 amperes. [424.3(A)]

❖ Such appliance loads are considered to be continuous loads and, thus, cannot be greater than 80 percent of the rating of the circuit (see definition of "Continuous load"). A 30-ampere circuit is the maximum permitted for fixed electric space heating such as baseboard resistance heaters.

E3702.11 Branch circuits for air-conditioning and heat pump equipment. The ampacity of the conductors supplying multimotor and combination load equipment shall be not less than the minimum circuit ampacity marked on the equipment. The branch-circuit overcurrent device rating shall be the size and type marked on the appliance. [440.4(B), 440.35, 440.62(A)]

❖ For these circuits, it is necessary to know the nameplate information in order to determine the proper size conductors. The equipment label will dictate the required branch circuit ampacity and rating of the overcurrent device.

E3702.12 Branch circuits serving room air conditioners. A room air conditioner shall be considered as a single motor unit in determining its branch-circuit requirements where all the following conditions are met:

1. It is cord- and attachment plug-connected.

2. The rating is not more than 40 amperes and 250 volts; single phase.

3. Total rated-load current is shown on the room air-conditioner nameplate rather than individual motor currents.

4. The rating of the branch-circuit short-circuit and ground-fault protective device does not exceed the ampacity of the branch-circuit conductors, or the rating of the branch-circuit conductors, or the rating of the receptacle, whichever is less. [440.62(A)]

❖ Dwelling unit air conditioning is usually provided by whole-house systems and room air conditioners, such as the window type or the through-the-wall type used in motels and hotels, usually cool or heat only a small room or area. Room air conditioners are considered to be fastened-in-place appliances but are subject to specific rules that are different from those for whole-house systems, mainly because these units are connected with cord and attachment plugs. These appliances can be supplied either by an individual branch circuit or by a branch circuit that also supplies other loads including lighting and/or other appliances. In the planning stages of wiring a dwelling unit, an individual branch circuit is commonly used; however, the room air conditioner could be installed later and connected to a branch circuit also serving other loads.

A room air conditioner usually has a compressor motor and a fan motor, but the code rule indicates that for calculating the branch circuit conductor size and overcurrent device rating, it is considered to be a unit. The total rated-load current must be shown on the nameplate. Where these units are served by a 15- or 20-ampere branch circuit, they are subject to the provisions of Section E3702.3. Those same provisions are reiterated specifically for room air conditioners in the two sections below.

E3702.12.1 Where no other loads are supplied. The total marked rating of a cord- and attachment plug-connected room air conditioner shall not exceed 80 percent of the rating of a branch circuit where no other appliances are also supplied. [440.62(B)]

❖ Where supplied by an individual branch circuit, the unit as a whole, which includes both the compressor and the fan motor, must have a rating not more than 80 percent of the branch circuit rating. For a 15-ampere circuit, the rating can be a maximum of 12 amperes, and for a 20-ampere circuit, a maximum of 16 amperes.

E3702.12.2 Where lighting units or other appliances are also supplied. The total marked rating of a cord- and attachment plug-connected room air conditioner shall not exceed 50 percent of the rating of a branch circuit where lighting or other appliances are also supplied. Where the circuitry is interlocked to prevent simultaneous operation of the room air conditioner and energization of other outlets on the same branch circuit, a cord- and attachment-plug-connected room air conditioner shall not exceed 80 percent of the branch-circuit rating. [440.62(C)]

❖ Where the branch circuit serves other loads, the total rating of the room air conditioner, whether it has been planned from the initial stages of the wiring job or installed in a window with the cord being plugged into an existing circuit, must not exceed 50 percent of the branch circuit rating, that is, 50 percent of the size of the overcurrent protection device, not 50 percent of the ampacity of the branch circuit conductors. For example, one room air conditioner with a total unit rating of up to 10 amperes could be plugged into a receptacle on a 20-ampere circuit. An interlock arrangement that prevents other outlets on the same circuit from being used when the room air conditioner is running would allow the 50-percent limit to increase to 80 percent.

E3702.13 Electric vehicle branch circuit. Outlets installed for the purpose of charging electric vehicles shall be supplied by a separate branch circuit. Such circuit shall not supply other outlets. (210.17)

❖ Charging equipment for electric automobiles operates for long periods, up to 8 hours, depending on the input voltage to the charger. Also, the load on the branch circuit is substantial, therefore, the branch circuit has little capacity for other loads. Nuisance tripping of circuit breakers could lead to the inconvenience of having the vehicle not charged when it is needed, plus the loss of power to any other load served by the circuit. It is a wise precaution to have continuous loads served by dedicated branch circuits.

E3702.14 Branch-circuit requirement—summary. The requirements for circuits having two or more outlets, or receptacles, other than the receptacle circuits of Sections E3703.2, E3703.3 and E3703.4, are summarized in Table E3702.14. Branch circuits in dwelling units shall supply only loads within that dwelling unit or loads associated only with that dwelling unit. Branch circuits installed for the purpose of lighting, central alarm, signal, communications or other purposes for public or common areas of a two-family dwelling shall not be supplied from equipment that supplies an individual dwelling unit. (210.24 and 210.25)

❖ Notice from Table E3702.14 that a 20-ampere rated receptacle is not permitted to be installed on a 15-ampere rated circuit, but a 15-ampere rated receptacle is permitted on a 20-ampere circuit. Equipment may have a 20-ampere rated attachment plug intended for use on a 20-ampere circuit. If a circuit is not really a 20-ampere circuit, it may not provide the required amount of current. Section E4002.1.1 requires that a single receptacle on an individual branch be rated equal to the rating of the branch circuit (see commentary for Section E4002.1.1).

A two-family dwelling unit may have areas common to both residences. For example, yard lighting, carport lighting or entry lighting could be used by occupants of both dwellings. If this is the case, the power for the common area loads must be supplied from a separate service and panelboard. Even where a two-family dwelling is owned or managed by one occupant, the common power needs cannot be supplied from the panel in the manager's unit, for example. Common area lighting and other power needs are more prevalent in multiple-family dwellings and are not often seen in two-family dwellings.

TABLE E3702.14 (Table 210.24)
BRANCH-CIRCUIT REQUIREMENTS-SUMMARY[a, b]

	CIRCUIT RATING		
	15 amp	20 amp	30 amp
Conductors: Minimum size (AWG) circuit conductors	14	12	10
Maximum overcurrent-protection device rating Ampere rating	15	20	30
Outlet devices: Lampholders permitted Receptacle rating (amperes)	Any type 15 maximum	Any type 15 or 20	N/A 30
Maximum load (amperes)	15	20	30

a. These gages are for copper conductors.

b. N/A means not allowed.

❖ See Section E3702.13.

SECTION E3703
REQUIRED BRANCH CIRCUITS

E3703.1 Branch circuits for heating. Central heating equipment other than fixed electric space heating shall be supplied by an individual branch circuit. Permanently connected air-conditioning equipment, and auxiliary equipment directly associated with the central heating equipment such as pumps, motorized valves, humidifiers and electrostatic air cleaners, shall not be prohibited from connecting to the same branch circuit as the central heating equipment. (422.12 and 422.12 Exceptions No. 1 and No. 2)

❖ For most furnaces, the motor load, even with the extra 25 percent, would not exceed the maximum 80 percent of the branch circuit rating; however, if the furnace was connected to the same circuit serving other loads and an overload occurred, opening the circuit breaker, the heat supply for the house would be out of order. In some climates, this could be a very serious problem. Thus, the heating equipment must be supplied from a separate circuit. Associated auxiliary equipment and controls may be connected to the same branch circuit.

E3703.2 Kitchen and dining area receptacles. A minimum of two 20-ampere-rated branch circuits shall be provided to serve all wall and floor receptacle outlets located in the kitchen, pantry, breakfast area, dining area or similar area of a dwelling. The kitchen countertop receptacles shall be served by a minimum of two 20-ampere-rated branch circuits, either or both of which shall also be permitted to supply other receptacle outlets in the same kitchen, pantry, breakfast and dining area including receptacle outlets for refrigeration appliances. [210.11(C)(1) and 210.52(B)(1) and (B)(2)]

Exception: The receptacle outlet for refrigeration appliances shall be permitted to be supplied from an individual branch circuit rated 15 amperes or greater. [210.52(B)(1) Exception No. 2]

❖ This section requires at least two 20-ampere-rated circuits for the kitchen and dining area receptacles to serve the many high wattage countertop appliances likely to be used in a dwelling. Section E3901 provides

rules on location and spacing of the receptacles. The code specifies that there must be at least two circuits but does not specify or limit the number of receptacle outlets. The countertop receptacle branch circuits are permitted to supply receptacles installed for refrigerators. So as not to rob capacity from the small appliance circuits, the refrigerators are commonly served by a dedicated branch circuit. These required circuits are intended to serve related/associated spaces and cannot serve receptacle outlets in kitchen, pantry, breakfast and dining areas located in different parts or stories of a dwelling, such as in the case of a home with more than one kitchen and its related areas.

E3703.3 Laundry circuit. A minimum of one 20-ampere-rated branch circuit shall be provided for receptacles located in the laundry area and shall serve only receptacle outlets located in the laundry area. [210.11(C)(2)]

❖ A dedicated branch circuit is required for the laundry area. This circuit can serve receptacles that, for example, are used for ironing, clothes washers and gas dryers.

E3703.4 Bathroom branch circuits. A minimum of one 20-ampere branch circuit shall be provided to supply bathroom receptacle outlet(s). Such circuits shall have no other outlets. [210.11(C)(3)]

Exception: Where the 20-ampere circuit supplies a single bathroom, outlets for other equipment within the same bathroom shall be permitted to be supplied in accordance with Section E3702. [210.11(C)(3) Exception]

❖ This section requires a dedicated branch circuit for the bathroom receptacle outlet(s). Section E3901.6 provides rules for the location of the receptacle(s). Two choices are available for the branch circuit serving the bathroom(s) in the house.

1. One 20-ampere-rated circuit may serve the required receptacle(s) in all the bathrooms in the house. In this case, no other outlets may be served by this circuit. The bathroom lighting and any fan or heat lamps would be served by a distinct general purpose branch circuit.

2. A 20-ampere circuit may serve only one bathroom. For additional bathrooms in the house, additional separate circuits would be run. The one circuit for the single bathroom supplies the required receptacle(s) and is also permitted to serve lighting and equipment in the bathroom such as an exhaust fan and/or heat lamp. Because this circuit is subject to the provisions of Section E3702, the load of the other equipment can be a maximum of 50 percent of the rating of the branch circuit.

The bathroom receptacle must be ground-fault circuit interrupter (GFCI) protected, but lighting and other equipment is not required to be GFCI protected unless otherwise stated as a condition of the listing of a fixture or equipment. Some electricians supply the luminaires and fans in a bathroom from a GFCI-protected circuit to provide additional safety for the occupants who come in contact with the luminaires, fans and switches.

E3703.5 Number of branch circuits. The minimum number of branch circuits shall be determined from the total calculated load and the size or rating of the circuits used. The number of circuits shall be sufficient to supply the load served. In no case shall the load on any circuit exceed the maximum specified by Section E3702. [210.11(A)]

❖ The minimum number of branch circuits required for various specific loads such as the laundry, furnace, bathroom, and small appliance circuits, is clearly prescribed in Section E3703. However, a simple calculation is necessary to find the minimum number of branch circuits required for the general lighting load.

The general lighting load includes all of the general purpose convenience receptacles and lighting outlets and is calculated at 3 volt-amperes per square foot (33 VA/m^2) in accordance with Section E3704.4. The square footage of the house used for this calculation does not include open porches, garages or unused or unfinished areas that are not adaptable for future use. Certainly, these areas have receptacle and lighting outlets, and even areas such as an attic or crawl space could have lighting and receptacles. See Section 3903.4 for lighting required in storage spaces. Although these outlets are part of the general lighting load, these rooms and spaces are not part of the floor area used in the general lighting load calculation. In many jurisdictions, new houses are completed with an unfinished basement, which is intended for future habitable space. The square footage of such basements must be included in the general lighting load calculation.

To find the number of general purpose lighting and receptacle circuits needed, multiply the dwelling square footage by 3 volt-amperes. This is the total lighting and receptacle load. Divide this load by the circuit voltage to find the total amperage needed. The total amperage is then divided by the rating of the circuit breaker to find the minimum number of circuits needed.

For a 2,200-square-foot house, the total lighting load is 3 VA × 2,200 = 6,600 volt-amperes. To get the total amperage load for the 120-volt branch circuits, divide the volt-amperes by 120 volts, which is 55 amperes. For a 15-ampere-rated circuit, divide the 55 amperes by 15. The result is 3.67 circuits, which must be rounded up to four circuits. For a 20-ampere-rated circuit, divide the 55 amperes by 20. The result is 2.75 circuits, which must be rounded up to three 20-ampere circuits to serve all the general purpose convenience receptacles and lighting load of the house. This is the minimum number of circuits required; however, many more are typically installed as a matter of convenience and good planning.

The 2,200-square-foot house in this example could have three or four bedrooms, a family room and a living room. With the addition of hallway, entry hall and out-

door receptacle outlets, it may easily contain 40 or more general purpose receptacle outlets. (This does not include the laundry, bathroom, kitchen and dining area receptacles on the small appliance circuits.) Add to this all the lighting outlets, (which do include the laundry, kitchen and dining room, and could include the bathroom lights, depending on how the bathroom branch circuits are run). See Section E3703.4 for bathroom branch circuits. The outside porch lights, garage lights, attic storage area lights, etc., are included in the general lighting branch circuits, although these areas are not included in the 3 volt-amperes per square foot of floor area. The total number of lighting outlets could easily be 20 or more. For instance, if there were 40 general purpose receptacle outlets and 15 lighting outlets for this house, the total of 55 outlets would be supplied by the minimum required number of four 15-ampere or three 20-ampere circuits. This lighting load must be evenly proportioned among the multioutlet branch circuits in accordance with Section E3703.6. Therefore, there could be approximately 17 to 18 outlets per 20-ampere circuit or 13 to 14 outlets per 15-ampere circuit for this house; however, the number of outlets per circuit is not limited by the code. Evenly proportioned loading of the required number of branch circuits is required.

This calculation covers the first sentence of the section: "The minimum number of branch circuits shall be determined from the total computed load and the size or rating of the circuits used." However, it is very important not to overlook the rest of the paragraph, which requires that the number of circuits be sufficient to supply the load served and that the load not exceed the limitations of Section E3702.

Where the loads are known, such as the number and wattage of the lighting fixtures to be installed or the cord- and plug-connected appliances or equipment to be used, additional circuits will need to be installed according to the load. For good practice, the wiring system should provide for increased loads and the use of added future electrical appliances. These factors should be taken into account in addition to the mere minimum number of circuits.

Many electricians use a rule of thumb method based on experience and practice for deciding on the number of branch circuits for a dwelling. In most cases, these traditional practices result in the actual number of branch circuits installed exceeding the minimum number required by the code calculation. Because the number and wattage of lamps to be used in light fixtures is not known at the time of deciding the number of circuits, and the load of the plug- and cord-connected appliances cannot be known, one rule of thumb method used in some areas is a maximum of 8 to 10 outlets included on any one general purpose branch circuit. This is not a code rule, but this kind of installation would provide a safety and convenience margin so that circuit breakers are not as likely to open on overload; for example, during a cold season when a homeowner may use portable space heaters in the house.

Limiting of outlets per circuit is not a code rule unless the loads are known.

One reason that some electricians limit the number of outlets per circuit may be because of the influence of the method used to determine the number of circuits for commercial buildings. The number of outlets per circuit is limited on commercial buildings. For other than residential wiring, each receptacle outlet is assigned a value of 1.5 amperes, which is 180 volt-amperes divided by 120 volts. Under this method, a 20-ampere circuit could have 13 outlets, and a 15-ampere circuit could have 10 outlets. This is sometimes confused with the requirements for dwellings, and if the commercial method is applied to calculating the number of outlets and branch circuits for dwellings, then the number of circuits would most likely satisfy the residential code requirements. In residential wiring, as long as the minimum number of circuits based on the square footage is provided, there is no limit on the number of outlets per branch circuit. The spacing and location requirements of receptacle outlets for dwellings result in the installation of many outlets that most definitely will not all be used at the same time. Because of the intermittent use and great diversity in the loads and duration of operation of the loads connected to the receptacle outlets in a house, any number of outlets can be installed on each general purpose branch circuit.

Sample Calculation

Consider a three-bedroom, two-bathroom, 1,500-square-foot (139 m²) house served with fuel gas for space and water heating and with the following lighting and receptacle outlets. The house has a 500-square-foot (46 m²) garage.

Number of Lighting Outlets

Three Bedrooms	3
Two Bathrooms	2
Hall	1
Living Room	1
Kitchen/Dining Rooms	2
Laundry Room	1
Outside Doors and Garage	3
Total Lighting Outlets*	13

Number of Receptacle Outlets

Three Bedrooms, four receptacles each	12
Hall	1
Outdoors (front and back)	2
Garage	2
Living Room	6
Total Receptacles*	23

*These numbers are used for this example only. See Chapter 39 for required outlets.

What is the minimum number of branch circuits needed for the total of 36 lighting and receptacle outlets for this house?

To determine the minimum required number of circuits, from Table E3704.2(1) find 3 volt-amperes per square foot for lighting and receptacles.

1,500 square feet × 3 VA = 4,500 VA/120 Volts = 37.5 amps minimum required for the house.

For 15-ampere circuits, divide 15 into 37.5 amperes 37.5/15 = 2.5, round up to 3

For 20-ampere circuits, divide 20 into 37.5 amperes 37.5/20 =1.8, round up to 2

Another method is to take the voltage times the rating of the circuit in amperes to find how many volt-amperes are allowed on the circuit.

A 15-ampere circuit has a capacity for 1,800 VA (120 × 15 = 1,800 VA)

A 20-ampere circuit has a capacity for 2,400 VA (120 × 20 = 2,400 VA)

The minimum lighting load required for this house is 3 × 1,500 or 4,500 VA

Divide the VA that each circuit can handle into 4,500 VA to find the minimum number of general lighting circuits.

Using 15-ampere circuits:

$$4,500/1,800 = 2.5, \text{ round up to } 3$$

Using 20-ampere circuits:

$$4,500/2,400=1.8, \text{ round up to } 2$$

The 36 lighting and receptacle outlets should be evenly proportioned between the two or three general purpose branch circuits. More circuits could be needed, such as a separate circuit to a receptacle in the garage to serve a freezer.

In addition to the general lighting branch circuits, the minimum required circuits for this house are as follows:

- Kitchen and dining area receptacles served by two small appliance branch circuits (see Sections E3703.2 and E3901.3).
- Refrigerator. The refrigerator is permitted to be on one of the kitchen small appliance circuits, or it can be on a separate 15-ampere circuit (see Section E3901.3 and Exception 2).
- Laundry branch circuit (see Sections E3703.3 and E3901.8).
- Bathroom circuit. One 20-ampere circuit could serve the receptacles in both bathrooms (see Sections E3703.4 and E3901.6).
- Central heating or furnace circuit (see Section E3703.1).
- Individual or appliance branch circuits. Section E3702 requires that circuits be provided to supply such loads as a disposal, dishwasher, range

and dryer sized according to the rating of the appliance or equipment.

The code does not prescribe or state in words that the dishwasher and disposal should each be supplied by a separate circuit as it does for the furnace, laundry and bathroom circuits. This must be determined by a simple calculation. Because the dishwasher and disposal are closely associated in proximity and function, some electricians wire both appliances on the same 20-ampere circuit. This might be permitted if all code requirements are satisfied. Because the appliances to be installed are not usually available on the job when the circuits are being installed at the rough-wiring stage, and because the appliances could be upgraded at a future time using the same branch circuits, it is best to run a separate 20-ampere circuit to each appliance. Section E3702.3 requires that if two appliances are supplied by the same branch circuit, the rating of any one appliance (utilization equipment), in this case the dishwasher or disposal, not exceed 80 percent of the branch circuit rating.

Section E3702.7, for combination loads, requires that the branch circuit total computed load be based on the 125 percent of the largest motor plus the sum of the other loads. A dishwasher has a pump motor and heating element, but they usually do not operate at the same time. A typical dishwasher nameplate lists the total load as 11.0 amperes and the motor load as 6.5 amperes. This dishwasher could be supplied by a separate 15-ampere circuit. The disposal is rated at 7.3 amperes and could be supplied by a separate 15-ampere circuit because the computed load is 7.3 × 125 percent, or 9.125 amperes. Could these two appliances be supplied on the same 20-ampere branch circuit? For combination loads, take 125 percent of the largest motor load plus the sum of the other loads. The disposal is the largest motor load at 7.3 × 125 percent = 9.125 amperes + the dishwasher at 11.0 amperes = 20.125 amperes. Neither one exceeds 80 percent of the 20-ampere branch circuit rating, but the total computed load is over 20 amperes. Therefore, two separate circuits are needed for these two appliances. Two separate 15-ampere circuits would satisfy code requirements, but good practice would indicate that two 20-ampere circuits be installed to provide sufficient capacity for future upgrades to appliances of higher load ratings.

E3703.6 Branch-circuit load proportioning. Where the branch-circuit load is calculated on a volt-amperes-per-square-foot (m²) basis, the wiring system, up to and including the branch-circuit panelboard(s), shall have the capacity to serve not less than the calculated load. This load shall be evenly proportioned among multioutlet branch circuits within the panelboard(s). Branch-circuit overcurrent devices and circuits shall only be required to be installed to serve the connected load. [210.11(B)]

❖ What if the actual connected load on the branch circuits was known and was less than the calculated load? The first sentence of this section requires that the wiring system have the capacity to serve at least the calculated load even if the actual load is less.

The branch circuit load calculated for the general lighting and general purpose receptacles must be evenly proportioned among the number of multioutlet branch circuits. Although distributing the loads over a greater number of circuits than the required minimum would provide for future load increases, the code requires only the number of branch circuits necessary to carry the connected load.

SECTION E3704
FEEDER REQUIREMENTS

E3704.1 Conductor size. Feeder conductors that do not serve 100 percent of the dwelling unit load and branch-circuit conductors shall be of a size sufficient to carry the load as determined by this chapter. Feeder conductors shall not be required to be larger than the service-entrance conductors that supply the dwelling unit. The load for feeder conductors that serve as the main power feeder to a dwelling unit shall be determined as specified in Chapter 36 for services. [310.15(B)(7)(2) and (3)]

❖ Feeder conductors that are run to a subpanel serving a detached garage, swimming pool equipment building, detached shop or upper story are examples of feeder conductors that do not serve 100 percent of the dwelling unit load.

But how are the main power feeder conductors sized? The term "main power feeder" refers to the feeder conductors between the main disconnect at the service and the panelboard serving branch circuits for lighting, receptacles and appliances. The main power feeder can be, but is not always, the only feeder in the wiring system and as such might not serve 100 percent of the dwelling unit load. In the example feeder load calculation for Section E3704.3, the main power feeder is not the only feeder in the system.

Use Chapter 36 for sizing the main power feeder conductors. Because of the diversity in the use of the dwelling unit lighting, receptacle and appliance loads, the demand on the main power feeder conductors is lower than the demand on the other feeders. In addition to the demand factors allowed in the actual calculation, Section E3603.1 has a built-in demand factor or reduction in the conductor size. Feeders other than the main power feeder can be expected to be more fully loaded and thus are sized by a different, more conservative method.

E3704.2 Feeder loads. The minimum load in volt-amperes shall be calculated in accordance with the load calculation procedure prescribed in Table E3704.2(1). The associated table demand factors shall be applied to the actual load to determine the minimum load for feeders. (220.40)

❖ Table E3704.2(1) is used to calculate the load for a feeder that supplies power to a sub-panelboard serving branch circuits for lighting, receptacles, motors and appliances. Once the feeder load is determined, the conductor size is selected. For sizing a main power feeder (serves 100 percent of the dwelling-unit load), see Section E3603.1.

The code provisions for calculating the feeder loads allow for the fact that in most cases all of the lighting, receptacles, motors and appliances served by the feeder will not be in use at the same time. Because of this diversity in the operation of the various loads, demand factors can be used. For example, a demand factor of 75 percent can be applied to the total load of four or more appliances served by a feeder. If it is known that the load will be operated 100 percent of the time, the demand factor would be 100 percent.

Sample Calculation

Using Table E3704.2(1), the following is an example of the calculation for a feeder to a subpanelboard serving branch circuits for lighting, receptacles, motors and appliances. In this example, the feeder is protected by Circuit Breaker #4.

A house that measures 30 feet by 60 feet (9144 mm by 18 288 mm) exclusive of the garage and porches has service equipment containing four main disconnects, which are two-pole, 240-volt circuit breakers. The four disconnects serve the following:

- Disconnect 1 serves a feeder to a subpanelboard in a separate bathhouse building used to house swimming pool equipment,

- Disconnect 2 serves a branch circuit to a 240-volt, 18,000-volt-ampere heat pump,

- Disconnect 3 serves a branch circuit to a 240-volt, 8,500-volt-ampere house air conditioner and

- Disconnect 4 serves a feeder to a subpanelboard inside the house, which supplies the following loads:

Appliance or Equipment	Rating
240-volt water heater:	6,000 volt-amperes
120-volt dishwasher:	16,000 volt-amperes

(motor rated 6.5 amperes included in 1,600 VA total rating)

120-volt disposal:	8.3 amperes
120-volt microwave oven:	1,400 volt-amperes
240-volt dryer:	5,500 volt-amperes
240-volt range:	11,000 volt-amperes
120-volt pump motor for a hydromassage bathtub:	10 amperes
120-volt room air conditioner:	1,400 volt-amperes
120-volt garage door opener:	7.2 amperes
120-volt clothes washer:	1,500 volt-amperes
Three 20-ampere kitchen and dining area receptacle branch circuits:	4,500 volt-amperes

Step 1: Lighting load, kitchen and dining area receptacle branch circuits and laundry branch circuit

Total floor area 30 feet × 60 feet = 1,800 × 3 VA/sq. ft. =	5,400 volt-amperes
Three 20-ampere kitchen and dining area receptacle branch circuits:	4,500 volt-amperes*

(Only two are required, but this example indicates three installed.)

One 20-ampere laundry branch circuit:	1,500 volt amperes
	11,400 volt-amperes
Applying the demand factor: 3,000 at 100 percent =	3,000
35 percent of remainder: 8,400 × 0.35 =	2,940
Total for Step 1:	5,940 volt-amperes

Step 2: Fastened-in-place appliances

Water heater:	6,000 volt-amperes
Dishwasher:	1,600 volt-amperes
Disposal: 120 volts × 8.3 amperes =	996 volt-amperes
Microwave oven:	1,400 volt-amperes
Hydromassage bathtub pump motor: 129 volts × 10 amperes =	1,200 volt-amperes
Room air conditioner:	1,400 volt-amperes
Garage door opener: 120 volts × 7.2 amperes =	864 volt-amperes
Total rating of appliances:	13,460 volt-amperes
Applying the demand factor: 13,460 × 0.75 =	10,095
Total for Step 2:	10,095 volt-amperes

Step 3: Motors

25 percent of largest motor:

Hydromassage bathtub pump motor: 120 volts × 10 amperes =	1,200 × 0.25 =
Total for Step 3:	300 volt-amperes

Step 4: Clothes Dryer

Greater of 5,000 VA or nameplate rating.

Total for Step 4:	5,500 volt-amperes

Step 5: Cooking Appliances

8,000 VA for the range per Column A of Table E3604.2(2)

Total for Step 5:	8,000 volt-amperes
Total computed feeder load:	29,835 volt-amperes

The minimum ampacity of the ungrounded feeder conductors is 29,835/240 volts = 124.3 amperes. A two-pole, 125-ampere circuit breaker can be used. Referring to Table E3705.1, copper conductors of size 1 AWG with a 75°C-rated insulation type can be used for ungrounded conductors for this feeder.

The heating and cooling loads for the house in this example are not served by the feeder from Disconnect #4, but would be used in calculating the load for the service conductors. If the heating and cooling loads were served by this feeder, it would not be necessary to consider both loads in the calculation. As shown in Table E3704.2(1), only the largest of these two loads would be used, because it is unlikely that the heating and cooling loads will be used simultaneously.

TABLES E3704.2(1) and E3704.2(2). See below.

❖ See the commentary to Section E3704.2.

E3704.3 Feeder neutral load. The feeder neutral load shall be the maximum unbalance of the load determined in accordance with this chapter. The maximum unbalanced load shall be the maximum net calculated load between the neutral and any one ungrounded conductor. For a feeder or service supplying electric ranges, wall-mounted ovens, counter-mounted cooking units and electric dryers, the maximum unbalanced load shall be considered as 70 percent of the load on the ungrounded conductors. [220.61(A) and (B)]

❖ For a three-wire, 120/240-volt single-phase system, the grounded (neutral) conductor run from the transformer is also grounded at the service equipment. The voltage between the grounded neutral and each ungrounded conductor is 120 volts. A 240-volt branch circuit to a motor or space heater, for example, has no neutral conductor because the load is supplied by the two ungrounded conductors. There is no current in the feeder neutral from the 240-volt load [see Commentary Figure E3401(2)].

TABLE E3704.2(1)
(Table 220.12, 220.14, Table 220.42, 220.50, 220.51, 220.52, 220.53, 220.54, 220.55, and 220.60)
FEEDER LOAD CALCULATION

LOAD CALCULATION PROCEDURE	APPLIED DEMAND FACTOR
Lighting and receptacles: A unit load of not less than 3 VA per square foot of total floor area shall constitute the lighting and 120-volt, 15- and 20-ampere general use receptacle load. 1,500 VA shall be added for each 20-ampere branch circuit serving receptacles in the kitchen, dining room, pantry, breakfast area and laundry area.	100 percent of first 3,000 VA or less and 35 percent of that in excess of 3,000 VA.
Plus	
Appliances and motors: The nameplate rating load of all fastened-in-place appliances other than dryers, ranges, air-conditioning and space-heating equipment.	100 percent of load for three or less appliances. 75 percent of load for four or more appliances.
Plus	
Fixed motors: Full-load current of motors plus 25 percent of the full load current of the largest motor.	
Plus	
Electric clothes dryer: The dryer load shall be 5,000 VA for each dryer circuit or the nameplate rating load of each dryer, whichever is greater.	
Plus	
Cooking appliances: The nameplate rating of ranges, wall-mounted ovens, counter-mounted cooking units and other cooking appliances rated in excess of 1.75 kVA shall be summed.	Demand factors shall be as allowed by Table E3704.2(2).
Plus the largest of either the heating or cooling load	
Largest of the following two selections: 1. 100 percent of the nameplate rating(s) of the air conditioning and cooling, including heat pump compressors. 2. 100 percent of the fixed electric space heating.	

For SI: 1 square foot = 0.0929 m^2.

TABLE E3704.2(2) (220.55 and Table 220.55)
DEMAND LOADS FOR ELECTRIC RANGES, WALL-MOUNTED OVENS, COUNTER-MOUNTED COOKING UNITS AND OTHER COOKING APPLIANCES OVER 1¾ kVA RATING[a, b]

NUMBER OF APPLIANCES	MAXIMUM DEMAND[b, c]	DEMAND FACTORS (percent)[d]	
	Column A maximum 12 kVA rating	Column B less than 3½ kVA rating	Column C 3½ to 8¾ kVA rating
1	8 kVA	80	80
2	11 kVA	75	65

a. Column A shall be used in all cases except as provided for in Footnote d.

b. For ranges all having the same rating and individually rated more than 12 kVA but not more than 27 kVA, the maximum demand in Column A shall be increased 5 percent for each additional kVA of rating or major fraction thereof by which the rating of individual ranges exceeds 12 kVA.

c. For ranges of unequal ratings and individually rated more than 8.75 kVA, but none exceeding 27 kVA, an average value of rating shall be computed by adding together the ratings of all ranges to obtain the total connected load (using 12 kVA for any ranges rated less than 12 kVA) and dividing by the total number of ranges; and then the maximum demand in Column A shall be increased 5 percent for each kVA or major fraction thereof by which this average value exceeds 12 kVA.

d. Over 1.75 kVA through 8.75 kVA. As an alternative to the method provided in Column A, the nameplate ratings of all ranges rated more than 1.75 kVA but not more than 8.75 kVA shall be added and the sum shall be multiplied by the demand factor specified in Column B or C for the given number of appliances.

To calculate the feeder neutral load from the example in Section E3704.2, do not include the water heater load, because it is 240 volts and does not use a neutral conductor. The maximum unbalanced load for the dryer and range is considered as 70 percent of the load used for the ungrounded conductor load calculation.

Appliance or Equipment	Rating
120-volt dishwasher:	1,600 volt-amperes

(motor rated 6.5 amperes included in 1,600 VA total rating)

120-volt disposal:	8.3 amperes
120-volt microwave oven:	1,400 volt-amperes
240-volt dryer:	5,500 volt-amperes
240-volt range:	11,000 volt-amperes
120-volt pump motor for a hydromassage bathtub:	10 amperes
120-volt room air conditioner:	1,400 volt-amperes
120-volt garage door opener:	7.2 amperes
120-volt clothes washer (laundry circuit):	1,500 volt-amperes
Three 20-ampere kitchen and dining area receptacle branch circuits:	4,500 volt-amperes

Step 1: Lighting load, kitchen and dining area receptacle branch circuits and laundry branch circuit

Total floor area 30 feet 60 feet = 1,800 × 3 VA/sq. ft. =	5,400 volt-amperes
Three 20-ampere kitchen and dining area receptacle branch circuits:	45,00 volt-amperes*

*(Only two are required, but this example indicates three installed.)

One 20-ampere laundry branch circuit:	1,500 volt-amperes
	11,400 volt-amperes
Applying the demand factor: 3,000 at 100 percent =	3,000
35 percent of remainder: 8,400 × 0.35 =	2,940
Total for Step 1:	5,940 volt-amperes

Step 2: Fastened-in-place appliances that use a neutral conductor

Dishwasher:	1,600 volt-amperes
Disposal: 120 volts 8.3 amperes =	996 volt-amperes
Microwave oven:	1,400 volt-amperes
Hydromassage bathtub pump motor: 120 volts × 10 amperes =	1,200 volt-amperes
Room air conditioner:	1,400 volt-amperes
Garage door opener: 120 volts × 7.2 amperes =	864 volt-amperes
Total rating of appliances:	7,460 volt-amperes

Applying the demand factor: 7,460 × 0.75 =	10,095
Total for Step 2:	5,595 volt-amperes

Step 3: Motors

25 percent of largest motor:

Hydromassage bathtub pump motor: 120 volts × 10 amperes = 1,200 × 25 percent =	300 volt-amperes
Total for Step 3:	300 volt-amperes

Step 4: Clothes Dryer

70 percent of rating: 5,500 VA × 70 percent =	3,850
Total for Step 4:	3,850 volt-amperes

Step 5: Cooking Appliances

8,000 VA × 70 percent =	5,600
Total for Step 5:	5,600 volt-amperes
Total computed Feeder load:	21,285 volt-amperes

The minimum ampacity of the neutral feeder conductor is 21,285/240 volts = 88.68 amperes. Using the 75°C column of Table E3705.1, a size 3-AWG Type THWN copper conductor is the minimum size permitted.

E3704.4 Lighting and general use receptacle load. A unit load of not less than 3 volt-amperes shall constitute the minimum lighting and general use receptacle load for each square foot of floor area (33 VA for each square meter of floor area). The floor area for each floor shall be calculated from the outside dimensions of the building. The calculated floor area shall not include open porches, garages, or unused or unfinished spaces not adaptable for future use. [220.12, Table 220.12, and 220.14(J)]

❖ See the commentary to Section E3704.2.

E3704.5 Ampacity and calculated loads. The calculated load of a feeder shall be not less than the sum of the loads on the branch circuits supplied, as determined by Section E3704, after any applicable demand factors permitted by Section E3704 have been applied. (220.40)

❖ A feeder must have an ampacity at least equal to the calculated feeder load. Feeder load is calculated in accordance with Section E3704.

E3704.6 Equipment grounding conductor. Where a feeder supplies branch circuits in which equipment grounding conductors are required, the feeder shall include or provide an equipment grounding conductor that is one or more or a combination of the types specified in Section E3908.8, to which the equipment grounding conductors of the branch circuits shall be connected. Where the feeder supplies a separate building or structure, the requirements of Section E3607.3.1 shall apply. (215.6)

❖ Feeders typically supply a subpanelboard which in turn supplies two or more branch circuits. Because the branch circuits will require grounding conductors, there must be a grounding conductor included as part of the feeder, to which the branch circuit grounding conductors will connect (see Section E3607.3.1 relative to feeders to outbuildings). If the subpanel serves 240-volt loads, a four-wire feeder will be necessary. Recall that multiple sections of the code repeat the very important rule that the grounded conductors in such subpanels be electrically isolated from the grounding conductors, thus requiring separate terminal bus bars for the grounded and grounding conductor terminations.

SECTION E3705
CONDUCTOR SIZING
AND OVERCURRENT PROTECTION

E3705.1 General. Ampacities for conductors shall be determined based in accordance with Table E3705.1 and Sections E3705.2 and E3705.3. [310.15(A)]

❖ Table E3705.1 covers the allowable ampacities for other than service conductors and main power feeder conductors. Section E3603.1 covers the conductor sizes for services and feeders that serve 100 percent of the dwelling-unit load.

The selection of conductors for dwelling unit wiring begins with finding the allowable conductor ampacity. Several variables are involved in determining the ampacity of a conductor. These include the ambient temperature, the number of conductors in close proximity, the temperature rating of the terminations and the maximum overcurrent protection ratings for specific conductor sizes such as given in Table E3705.5.3.

Table E3705.1 cannot be used to determine conductor ampacity without knowing the "conditions of use" of the conductor. It is important here to review the definition of ampacity from Chapter 35: "The current in amperes a conductor can carry continuously under the conditions of use without exceeding its temperature rating." This definition applies to continuous use. There is not one ampacity for continuous use and another for noncontinuous use. Where a load is operated continuously (for 3 hours or more), the ampacity of the branch circuit conductors must be 125 percent of the continuous load. If the load is not operated continuously, the branch circuit conductors must have an ampacity at least equal to the load being served (see Section E3701.2).

Because of resistance, current passing through a conductor causes heat to build up in it, which must be dissipated into the ambient air. Ampacity is the amount of current that a conductor can carry without exceeding the thermal limit of its insulation material, which is the temperature rating. Conductors are classified by the different types of insulation that are capable of withstanding a certain level of heat before the insulation degrades, melts or burns. The conductor temperature ratings of 60°C, 75°C and 90°C at the top of the table indicate the maximum temperature a given size conductor with the type of insulation shown can withstand (60°C = 140°F; 75°C = 167°F; 90°C = 194°F). The

maximum current the conductor can carry without raising the temperature higher than 60°C, 75°C or 90°C is shown in the table as the ampacity for that size of conductor with the type of insulation shown at the top of the column. Ampacities shown in Table E3705.1 are based on an ambient temperature of 86°F (30°C). Look at the first column in Table E3705.2 for ambient temperature. The 5th row is 26°C - 30°C, and there is no adjustment factor for the ampacity of conductors with this ambient temperature (a mulitplier of 1 has no effect).

For example, a size 8-AWG copper conductor with Type XHHW insulation can carry 50 amperes continuously at 30°C without raising the temperature of the conductor over 167°F (75°C). If this conductor carries a higher current, a temperature higher than 75°C could develop in the conductor, thus causing damage.

TABLE E3705.1. See page 37-15.

❖ See the commentary to Section E3705.1.

E3705.2 Correction factor for ambient temperatures. For ambient temperatures other than 30°C (86°F), multiply the allowable ampacities specified in Table E3705.1 by the appropriate correction factor shown in Table E3705.2. [310.15(B)(2)]

❖ The ampacity values shown in Table E3705.1 must be reduced according to the factors in Table E3705.2. This table is usually referred to for a reduction in allowable ampacity; however, the ambient temperature correction factors could result in a slight increase in ampacity if the ambient temperature where the conductors are installed never exceeds 77°F (25°C) as indicated in the first row of the table.

It is reasonable to assume that attic temperatures will reach 105°F (41°C) and greater. Where conductors run in attics are exposed to the possibility of such high temperatures, a correction factor for conductor ampacity must be applied. Consider an air-conditioner supplied by a 40-ampere branch circuit with copper size 8-AWG, Type THHN conductors run through an attic where temperatures in summer months reach 125°F (52°C). To find the ampacity of these conductors under this condition of use, find Type THHN conductors in the 90°C column (on the left half of the table, for copper) of Table E3705.1. Reading down to the row for size 8 AWG, these conductors would have an ampacity of 55 amperes if installed in a maximum ambient temperature of 30°C (86°F). However, read down the far right hand column of Table E3705.2 to the row that includes 125°F (52°C). In this row, move left to the column for Type THHN conductors. THHN is at the top of Tables E3705.1 and E3705.2. The correction factor is 0.76, which indicates a correction factor of 76 percent of the 55 amperes, or 41.8 amperes. The maximum allowable ampacity for the size 8-AWG conductors is 41.8 amperes.

If these size 8-AWG Type THHN copper conductors supplying the air conditioner are in nonmetallicsheathed cable, the code simply states that the ampacity shall be that of the 60°C column or 40 amperes (see Section E3705.4.4). So why start with the 90°C column and have to apply the correction factor? If we started with the 60°C column, the correction factor would be 0.41, which would reduce the ampacity of the size 8-AWG conductors to 16.4 amperes (40 A x 0.41). Section E3705.4.4 states that the 90°C (194°F) rating is permitted to be used for ampacity correction and adjustment purposes.

Nonmetallic sheathed cable manufactured today has Type THHN insulation on the conductors for this very purpose. The most commonly used Type NM will be marked "NM-B" indicating that it contains conductors rated for 90°C (see Section E3705.4.4). The derating is done from the ampacity values in the 90°C column. The lowest value, of either the 60°C column or the calculated ampacity using the correction factor, is then used to determine the ampacity. If the ambient temperature was 132°F (55°C), size 8-AWG Type THHN conductors would have an ampacity of 39.05 (55 × 0.71) and could not be used on a 40-ampere circuit. For this installation, size 6-AWG Type THHN conductors could be used. The ampacity would be (75 0.71) 53.25 amperes. This would be acceptable for the 40-ampere branch circuit under the conditions of use

where the conductors pass through the 132°F (55°C) ambient temperature.

TABLE E3705.2. See page 37-16.

❖ See the commentary to Section E3705.2.

E3705.3 Adjustment factor for conductor proximity. Where the number of current-carrying conductors in a raceway or cable exceeds three, or where single conductors or multiconductor cables are stacked or bundled for distances greater than 24 inches (610 mm) without maintaining spacing and are not installed in raceways, the allowable ampacity of each conductor shall be reduced as shown in Table E3705.3. [310.15(B)(3)]

Exceptions:

1. Adjustment factors shall not apply to conductors in nipples having a length not exceeding 24 inches (610 mm). [310.15(B)(3)(2)]

2. Adjustment factors shall not apply to underground conductors entering or leaving an outdoor trench if those conductors have physical protection in the form of rigid metal conduit, intermediate metal conduit, or rigid nonmetallic conduit having a length not

TABLE E3705.1
ALLOWABLE AMPACITIES

CONDUCTOR SIZE	CONDUCTOR TEMPERATURE RATING						CONDUCTOR SIZE
	60°C	75°C	90°C	60°C	75°C	90°C	
AWG kcmil	Types TW, UF	Types RHW, THHW, THW, THWN, USE, XHHW	Types RHW-2, THHN, THHW, THW-2, THWN-2, XHHW, XHHW-2, USE-2	Types TW, UF	Types RHW, THHW, THW, THWN, USE, XHHW	Types RHW-2, THHN, THHW, THW-2, THWN-2, XHHW, XHHW-2, USE-2	AWG kcmil
	Copper			Aluminum or copper-clad aluminum			
14[a]	15	20	25	—	—	—	—
12[a]	20	25	30	15	20	25	12[a]
10[a]	30	35	40	25	30	35	10[a]
8	40	50	55	35	40	45	8
6	55	65	75	40	50	55	6
4	70	85	95	55	65	75	4
3	85	100	115	65	75	85	3
2	95	115	130	75	90	100	2
1	110	130	145	85	100	115	1
1/0	125	150	170	100	120	135	1/0
2/0	145	175	195	115	135	150	2/0
3/0	165	200	225	130	155	175	3/0
4/0	195	230	260	150	180	205	4/0
250	215	255	290	170	205	230	250
300	240	285	320	195	230	260	300
350	260	310	350	210	250	280	350
400	280	335	380	225	270	305	400
500	320	380	430	260	310	350	500
600	350	420	475	285	340	385	600
700	385	460	520	315	375	425	700
750	400	475	535	320	385	435	750
800	410	490	555	330	395	445	800
900	435	520	585	355	425	480	900

For SI: °C = [(°F) - 32]/1.8.

a. See Table E3705.5.3 for conductor overcurrent protection limitations.

exceeding 10 feet (3048 mm) and the number of conductors does not exceed four.[310.15(B)(3)(3)]

3. Adjustment factors shall not apply to type AC cable or to type MC cable without an overall outer jacket meeting all of the following conditions:

   3.1. Each cable has not more than three current-carrying conductors.

   3.2. The conductors are 12 AWG copper.

   3.3. Not more than 20 current-carrying conductors are bundled, stacked or supported on bridle rings. [310.15(B)(3)(4)]

4. An adjustment factor of 60 percent shall be applied to Type AC cable and Type MC cable where all of the following conditions apply:

   4.1. The cables do not have an overall outer jacket.

   4.2. The number of current-carrying conductors exceeds 20.

   4.3. The cables are stacked or bundled longer than 24 inches (607 mm) without spacing being maintained. [310.15(B)(3)(5)]

❖ The allowable ampacity of branch circuit conductors varies not only according to the ambient temperature surrounding the conductors but also according to the number of current-carrying conductors in a raceway or cable or in close proximity. Where multiconductor cables (such as nonmetallic-sheathed cables used in residential wiring) are stacked or bundled together for 24 inches (610 mm) or more without maintaining spacing, the allowable ampacity is reduced according to the factors in Table E3705.3. For example, the allowable ampacity of size 12-AWG copper THHN conductors (found in nonmetallic-sheathed cable) used with 90°C

terminals is 30 amperes. If several of these cables were bundled together for 2 feet (610 mm) or more such that seven current-carrying conductors were in the bundle, then the allowable ampacity would be reduced 70 percent to 21 amperes (see Commentary Figure E3705.3).

In some cases where the allowable ampacity is reduced, either a smaller circuit breaker would need to be installed or the conductors would need to be installed with sufficient spacing between them and not bundled. Some judgement may enter into the interpretation and enforcement of this code rule because in residential loads there is so much diversity in the amount of current use in the branch circuit conductors and in the duration of operation of the loads. To drill one hole in the top plate of the framing above the panelboard and run all the branch circuit cables in a bundle for 2 or 3 feet (610 or 914 mm) into the panel would be a code violation. This is not a good practice, because the heat developed in the current-carrying conductors cannot readily dissipate. Spacing should be maintained or the allowable ampacity of the conductors would be reduced.

The correction factors for ambient temperature and for the number of current-carrying conductors must both be applied. For example, if a 30-ampere electric water heater branch circuit using size 10 AWG copper conductors of Type THHN in nonmetallic sheathed cable was run through an attic where the temperature reached 123°F (51°C), the ampacity would be 30.4 amperes. If this cable was bundled together for over 2 feet (610 mm) with other cables containing six current-carrying conductors, a further reduction would be required. The 30.4 amperes would be reduced by 20 percent (indicated in Table E3705.3) to 24.32 amperes, and the size 10-AWG conductors would be a violation if protected by a 30-ampere circuit breaker.

TABLE E3705.2 [Table 310.15(B)(2)(a)]
AMBIENT TEMPERATURE CORRECTION FACTORS

Ambient Temperature (°C)	Temperature Rating of Conductor			Ambient Temperature (°F)
	For ambient temperatures other than 30°C (86°F), multiply the allowable ampacities specified in the ampacity tables by the appropriate correction factor shown below.			
	60°C	75°C	90°C	
10 or less	1.29	1.20	1.15	50 or less
11-15	1.22	1.15	1.12	51-59
16-20	1.15	1.11	1.08	60-68
21-25	1.08	1.05	1.04	69-77
26-30	1.00	1.00	1.00	78-86
31-35	0.91	0.94	0.96	87-95
36-40	0.82	0.88	0.91	96-104
41-45	0.71	0.82	0.87	105-113
46-50	0.58	0.75	0.82	114-122
51-55	0.41	0.67	0.76	123-131
56-60	—	0.58	0.71	132-140
61-65	—	0.47	0.65	141-149
66-70	—	0.33	0.58	150-158
71-75	—	—	0.50	159-167
76-80	—	—	0.41	168-176
81-85	—	—	0.29	177-185

For SI: 1 °C = [(°F) - 32]/1.8.

**TABLE E3705.3 [Table 310.15(B)(3)(a)]
CONDUCTOR PROXIMITY ADJUSTMENT FACTORS**

NUMBER OF CURRENT-CARRYING CONDUCTORS IN CABLE OR RACEWAY	PERCENT OF VALUES IN TABLE E3705.1
4-6	80
7-9	70
10-20	50
21-30	45
31-40	40
41 and above	35

❖ See the commentary to Section E3705.3.

E3705.4 Temperature limitations. The temperature rating associated with the ampacity of a conductor shall be so selected and coordinated to not exceed the lowest temperature rating of any connected termination, conductor or device. Conductors with temperature ratings higher than specified for terminations shall be permitted to be used for ampacity adjustment, correction, or both. Except where the equipment is marked otherwise, conductor ampacities used in determining equipment termination provisions shall be based on Table E3705.1. [110.14(C)]

❖ The allowable conductor ampacity is not the only consideration in determining what size conductor to use. In addition to the temperature rating of the conductor, the temperature rating of the terminations or terminals the conductor connects to at each end of the conductor is critical. The terminals on devices such as receptacles, switches, circuit breakers and panelboards all have maximum temperature ratings. Current passing through these terminations can heat up the device just as it does in a conductor. The lowest temperature rating of any of the devices in the circuit determines the allowable ampacity. Many switches and receptacles used in residential work have a temperature rating of 60°C (140°F). Circuit breakers might have temperature ratings of 75°C (167°F).

E3705.4.1 Conductors rated 60°C. Except where the equipment is marked otherwise, termination provisions of equipment for circuits rated 100 amperes or less, or marked for 14

**Figure E3705.3
MULTICONDUCTOR CABLES BUNDLED TOGETHER**

AWG through 1 AWG conductors, shall be used only for one of the following:

1. Conductors rated 60°C (140°F);

2. Conductors with higher temperature ratings, provided that the ampacity of such conductors is determined based on the 60°C (140°F) ampacity of the conductor size used;

3. Conductors with higher temperature ratings where the equipment is listed and identified for use with such conductors; or

4. For motors marked with design letters B, C, or D conductors having an insulation rating of 75°C (167°F) or higher shall be permitted to be used provided that the ampacity of such conductors does not exceed the 75°C (167°F) ampacity. [110.14(C)(1)(a)]

❖ See the commentary to Section E3705.4.2.

E3705.4.2 Conductors rated 75°C. Termination provisions of equipment for circuits rated over 100 amperes, or marked for conductors larger than 1 AWG, shall be used only for:

1. Conductors rated 75°C (167°F).

2. Conductors with higher temperature ratings provided that the ampacity of such conductors does not exceed the 75°C (167°F) ampacity of the conductor size used, or provided that the equipment is listed and identified for use with such conductors. [110.14(C)(1)(b)]

❖ Where the temperature rating of the termination on the device is 60°C (140°F), the allowable conductor ampacity is found in the 60°C (140°F) column of Table E3705.1. Conductors with higher temperature rated insulation are allowed to be used, and this is often the case when de-rating is necessary for ambient temperature or for the number of conductors in close proximity.

An example of this would be THHN copper conductors connected to a 70-ampere 2-pole circuit breaker that is listed and marked at 60/75°C (140/167°F). The conductors feed equipment that has terminals listed and marked at 60°C (140°F). Assuming an ambient temperature of 80°F, and not over three current carrying conductors in the raceway, the smallest size THHN copper conductors that may be used in this installation are size 4 AWG, determined from the 60°C column.

E3705.4.3 Separately installed pressure connectors. Separately installed pressure connectors shall be used with conductors at the ampacities not exceeding the ampacity at the listed and identified temperature rating of the connector. [110.14(C)(2)]

❖ The temperature rating of wire nuts and other pressure connectors is one factor in determining the allowable ampacity of the conductors. Make sure that the temperature rating of the connectors is suitable for the temperature rating associated with the conductor ampacity.

E3705.4.4 Conductors of Type NM cable. Conductors in NM cable assemblies shall be rated at 90°C (194°F). Types NM, NMC, and NMS cable identified by the markings NM-B, NMC-B, and NMS-B meet this requirement. The allowable ampacity of Types NM, NMC, and NMS cable shall not exceed that of 60°C (140°F) rated conductors and shall comply with Section E3705.1 and Table E3705.5.3. The 90°C (194°F) rating shall be permitted to be used for ampacity adjustment and calculations provided that the final corrected or adjusted ampacity does not exceed that for a 60°C (140°F) rated conductor. Where more than two NM cables containing two or more current-carrying conductors are installed, without maintaining spacing between the cables, through the same opening in wood framing that is to be sealed with thermal insulation, caulk or sealing foam, the allowable ampacity of each conductor shall be adjusted in accordance with Table E3705.3. Where more than two NM cables containing two or more current-carrying conductors are installed in contact with thermal insulation without maintaining spacing between cables, the allowable ampacity of each conductor shall be adjusted in accordance with Table E3705.3. (334.80 and 334.112)

❖ The code requires that the individual conductors in nonmetallic sheathed cable have an insulation rating of 90°C (194°F), but the ampacity is that of the 60°C column of Table E3705.1. This conservative approach of limiting the ampacity to that of 60°C (140°F) conductors accounts for the fact that NM cable is often embedded in thermal insulation (see commentary, Sections E3705.2 and E3705.3). The adjustment factor for conductor proximity heating effect must be applied where three or more NM cables pass through the same bored hole in wood members and such holes are to be sealed by fire and/or draftstopping materials. Experiments have shown that excessive temperatures can develop in conductors installed in this manner. If cables must be pulled through bored holes through which one or more other cables already pass, extreme caution must be used to prevent friction damage caused by the cables rubbing and abrading each other. Although "one cable per bored hole" is a safe practice, it is not a code requirement.

Where three or more NM cables are installed in contact with thermal insulation, such as cables would be where installed in outside wall cavities, and such cables are bundled or otherwise lack spacing between them, their ampacity must be adjusted in accordance with Table E3705.3. In such cases, the conductors will be subject to additional heating caused by the lack of "breathing room" and the heat retention property of the thermal insulation material.

E3705.4.5 Conductors of Type SE cable. Where used as a branch circuit or feeder wiring method within the interior of a building and installed in thermal insulation, the ampacity of the conductors in Type SE cable assemblies shall be in accordance with the 60°C (140°F) conductor temperature rating.

The maximum conductor temperature rating shall be permitted to be used for ampacity adjustment and correction purposes, provided that the final derated ampacity does not exceed that for a 60°C (140°F) rated conductor. [338.10(B)(4)(a)]

❖ This requirement is similar to that for Type NM cable (see Section E3705.4.4). Type SE (service entrance) cable conductors typically have XHHW insulation or can have THHN or THWN insulation. This section applies to SE cable that meets both conditions: 1. It is used for a feeder or branch circuit; and 2. It is installed within thermal insulation. This section does not apply to SE cable used for service conductors. Note that SE-U (flat U-shaped) cable is limited in application because it has a bare neutral conductor. Type SE-U can be used only for service entrance conductors or for branch or feeder conductors where the bare conductor is used as an equipment grounding conductor. Type SE-R (round shape) cable has an insulated neutral and a bare or insulated grounding conductor and is commonly used for feeders and branch circuits (see Section E3801.4).

E3705.5 Overcurrent protection required. All ungrounded branch-circuit and feeder conductors shall be protected against overcurrent by an overcurrent device installed at the point where the conductors receive their supply. Overcurrent devices shall not be connected in series with a grounded conductor. Overcurrent protection and allowable loads for branch circuits and for feeders that do not serve as the main power feeder to the dwelling unit load shall be in accordance with this chapter.

Branch-circuit conductors and equipment shall be protected by overcurrent protective devices having a rating or setting not exceeding the allowable ampacity specified in Table E3705.1 and Sections E3705.2, E3705.3 and E3705.4 except where otherwise permitted or required in Sections E3705.5.1 through E3705.5.3. [240.4, 240.21, and 310.15(B)(7)(2)]

❖ The overcurrent devices consist of circuit breakers or fuses and are installed at the beginning of the circuit where the conductors receive their supply. The overcurrent protection devices prevent the conductors from carrying currents above the allowable ampacity determined from Table E3705.1 and the associated de-rating provisions. The overcurrent protection device will automatically open and protect the conductors if the current is high enough to cause a dangerous temperature.

E3705.5.1 Cords. Cords shall be protected in accordance with Section E3909.2. [240.5(B)]

❖ See the commentary to Section E3909.2.

E3705.5.2 Overcurrent devices of the next higher rating. The next higher standard overcurrent device rating, above the ampacity of the conductors being protected, shall be permitted to be used, provided that all of the following conditions are met:

1. The conductors being protected are not part of a branch circuit supplying more than one receptacle for cord- and plug-connected portable loads.

2. The ampacity of conductors does not correspond with the standard ampere rating of a fuse or a circuit breaker without overload trip adjustments above its rating (but that shall be permitted to have other trip or rating adjustments).

3. The next higher standard device rating does not exceed 400 amperes. [240.4(B)]

❖ Where a circuit supplies only a specific load, the maximum load current is known. For multioutlet branch circuits, it is not known what the load will be for cord- and plug-connected equipment and appliances. But where the load is known, and it does not match one of the standard ratings of overcurrent devices, an overcurrent device of the next higher rating can be used. In these cases, the conductor ampacity could be less than the rating of the overcurrent device, which is the branch circuit rating. For example, if the ampacity of the conductors serving only a single appliance was calculated to be 27 amperes, a 30-ampere circuit breaker could be used to protect such conductors.

E3705.5.3 Small conductors. Except as specifically permitted by Section E3705.5.4, the rating of overcurrent protection devices shall not exceed the ratings shown in Table E3705.5.3 for the conductors specified therein. [240.4(D)]

❖ Referring to Table E3705.1 on conductor ampacity, the allowable ampacity of size 14 AWG is 20 to 25 amperes, depending on the type of insulation and the temperature ratings. Various ampacities are also given for sizes 12 and 10 AWG. However, Table E3705.5.3 provides the maximum overcurrent device rating for these small size conductors regardless of the allowable ampacity given in Table E3705.1. This section provides an extra level of safety by placing a conservative cap on the overcurrent device ratings for the smaller conductors used throughout the dwelling.

A 15-ampere circuit breaker is used on circuits with size 14-AWG conductors, a 20-ampere circuit breaker with size 12-AWG conductors, and a 30-ampere circuit breaker with size 10 AWG conductors. The only exception is as allowed by Section E3705.5.4.

Do not overlook the note to Table E3705.5.3, which addresses the fact that an even lower overcurrent device rating could be required where the conductors are derated as required by Sections E3705.2 and E3705.3.

TABLE E3705.5.3 [240.4(D)]
OVERCURRENT-PROTECTION RATING

COPPER		ALUMINUM OR COPPER-CLAD ALUMINUM	
Size (AWG)	Maximum overcurrent-protection-device rating[a] (amps)	Size (AWG)	Maximum overcurrent-protection-device rating[a] (amps)
14	15	12	15
12	20	10	25
10	30	8	30

a. The maximum overcurrent-protection-device rating shall not exceed the conductor allowable ampacity determined by the application of the correction and adjustment factors in accordance with Sections E3705.2 and E3705.3.

❖ See the commentary to Section E3705.5.3.

E3705.5.4 Air-conditioning and heat pump equipment. Air-conditioning and heat pump equipment circuit conductors shall be permitted to be protected against overcurrent in accordance with Section E3702.11. [240.4(G)]

❖ The overcurrent protection device rating for air-conditioning and heat-pump circuits is stated on the equipment label.

E3705.6 Fuses and fixed trip circuit breakers. The standard ampere ratings for fuses and inverse time circuit breakers shall be considered to be 15, 20, 25, 30, 35, 40, 45, 50, 60, 70, 80, 90, 100, 110, 125, 150, 175, 200, 225, 250, 300, 350 and 400 amperes. (240.6)

❖ Although 25 amperes is a standard size circuit breaker, and the allowable ampacity from Table E3705.1 shows 25 amperes for size 12-AWG Type THW, it would be a code violation to install a 25-ampere circuit breaker on a branch circuit of this size conductor. An electrician might try to solve a problem of a circuit breaker tripping from overload by looking at Table E3705.1 and, seeing that a size 12-THHN conductor has an ampacity of 30 amperes, installing a 25- or 30-ampere circuit breaker. This has been done and is a code violation and a serious fire hazard (see commentary, Section E3705.5.3).

E3705.7 Location of overcurrent devices in or on premises. Overcurrent devices shall:

1. Be readily accessible. [240.24(A)]
2. Not be located where they will be exposed to physical damage. [240.24(C)]
3. Not be located where they will be in the vicinity of easily ignitible material such as in clothes closets. [240.24(D)]
4. Not be located in bathrooms. [240.24(E)]
5. Not be located over steps of a stairway.
6. Be installed so that the center of the grip of the operating handle of the switch or circuit breaker, when in its

highest position, is not more than 6 feet 7 inches (2007 mm) above the floor or working platform. [240.24(A)]

Exceptions:

1. This section shall not apply to supplementary overcurrent protection that is integral to utilization equipment. [240.24(A)(2)]
2. Overcurrent devices installed adjacent to the utilization equipment that they supply shall be permitted to be accessible by portable means. [240.24(A)(4)]

❖ The panelboard (which contains overcurrent protection devices) is not permitted to be installed in a clothes closet or in any closet. It might be determined that a pantry closet in a kitchen would not have "easily ignitable material." Whether the code official would permit a panelboard to be installed in such a closet is a matter of interpretation. Because of the dampness and humidity, and potential shock hazard, overcurrent devices are not permitted in bathrooms. A panelboard would be permitted to be installed in a laundry room or utility room.

Panelboards cannot be located over stairs because this section prohibits the installation of overcurrent devices over stairs. Such a location would obviously place the user or worker at risk of injury from falling.

E3705.8 Ready access for occupants. Each occupant shall have ready access to all overcurrent devices protecting the conductors supplying that occupancy. [240.24(B)]

❖ In a two-family dwelling, each occupant must have access to all overcurrent protective devices for their own dwelling. This includes the service disconnecting means. The overcurrent protection devices must not be locked such as to prevent occupant access.

E3705.9 Enclosures for overcurrent devices. Overcurrent devices shall be enclosed in cabinets, cutout boxes, or equipment assemblies. The operating handle of a circuit breaker shall be permitted to be accessible without opening a door or cover. [240.30(A) and (B)]

❖ Panelboards are installed in a cabinet or cutout box. See the commentary to the definitions for these items in Chapter 35.

SECTION E3706
PANELBOARDS

E3706.1 Panelboard rating. All panelboards shall have a rating not less than that of the minimum service or feeder capacity required for the calculated load. (408.30)

❖ Panelboards are located in a cabinet or cutout box and hold the circuit breakers for the branch circuits. The busbar to which the circuit breakers are connected must have an ampere rating at least equal to the rating of the feeder serving the panelboard. Some common panelboard ratings are 60-, 100-, 125-, 150- and 200-ampere ratings.

E3706.2 Panelboard circuit identification. All circuits and circuit modifications shall be legibly identified as to their clear, evident, and specific purpose or use. The identification shall include an approved degree of detail that allows each circuit to be distinguished from all others. Spare positions that contain unused overcurrent devices or switches shall be described accordingly. The identification shall be included in a circuit directory located on the face of the panelboard enclosure or inside the panel door. Circuits shall not be described in a manner that depends on transient conditions of occupancy. [408.4(A)]

❖ The panelboard circuit directory must be permanent and legible. In many homes, the door of the panelboard cabinet reveals illegible markings. The outlets that a circuit serves are a mystery to an observer because the markings on the panelboard have become faded or are incomplete. Some electricians and installers write only such things as "lights" or "plugs" or "kitchen plugs" to satisfy this section. Each circuit identification must be detailed enough to distinguish the circuit from all other circuits. Obviously, 10 circuits labeled "lights and receptacles" cannot be distinguished from each other and such labels are a violation of this section. Factory-made panel labeling kits

are available that will facilitate compliance with the intent. Of course, acceptance of such markings is according to the judgment of the code official, but if the markings are not legible, a code violation exists. Correct terminology must be used; for example, branch circuits do supply receptacles, but not "plugs." Obviously, the circuits for appliances such as dishwashers, disposals, furnaces, etc., should be marked. Commentary Figures E3601.7 and E3706.2 illustrate panelboard directories on the inside of the doors. A potential hazard exists where circuit breakers are mislabeled or ambiguously labeled because the occupants or electrical workers might not actually be de-energizing what they intend to de-energize. Installed circuit breakers that serve no load are spares and must be labeled as such. Circuits must not be labeled in a manner that would be understood only by certain occupants of the dwelling. For example, if a circuit was labeled "Uncle Waldo's room" or "Guest bedroom," that designation would be meaningless to any occupants other than the occupants who originally labeled the circuits.

E3706.3 Panelboard overcurrent protection. In addition to the requirement of Section E3706.1, a panelboard shall be protected by an overcurrent protective device having a rating

Figure E3706.2
PANELBOARD CIRCUIT IDENTIFICATION

not greater than that of the panelboard. Such overcurrent protective device shall be located within or at any point on the supply side of the panelboard. (408.36)

❖ The overcurrent device supplying the panelboard must have a rating not more than the ampere rating of the panelboard busbars. The last sentence indicates that the panelboard can be protected by the main breaker installed in the panelboard or at the supply end of the feeder supplying the panelboard.

E3706.4 Grounded conductor terminations. Each grounded conductor shall terminate within the panelboard on an individual terminal that is not also used for another conductor, except that grounded conductors of circuits with parallel conductors shall be permitted to terminate on a single terminal where the terminal is identified for connection of more than one conductor. (408.41 and 408.41 Exception)

❖ This section applies to grounded conductors (see definition), which are typically referred to as neutrals. Panelboard neutral bus bars are usually not listed for use with more than one wire per terminal. This rule can be thought of as the "one wire per hole" rule. If multiple grounded conductors are terminated at one terminal, loosening the terminal screw to remove one conductor would loosen all of the conductors temporarily. Unless all of the affected circuits are de-energized, this will result in arcing and/or loss of the grounded or neutral conductor for all of the circuits associated with that terminal. In the case of multiwire circuits, loss of the neutral connection, even momentarily, could cause a hazardous condition and appliance damage (see commentary, Sections E3701.5 through E3701.5.2). With only one wire per terminal, it is assumed that the wire being removed is part of a circuit that has been de-energized.

E3706.5 Back-fed devices. Plug-in-type overcurrent protection devices or plug-in-type main lug assemblies that are back-fed and used to terminate field-installed ungrounded supply conductors shall be secured in place by an additional fastener that requires other than a pull to release the device from the mounting means on the panel. [408.36(D)]

❖ Back-fed circuit breakers are commonly used in small main lug only subpanels and in generator transfer switch subpanels. Instead of receiving power through their plug-in terminals from bus parts in the panelboard, back-fed breakers receive power through their load terminals, hence, the name "back-fed." The circuit breakers appear to be wired in reverse, but the devices do not know the difference and function normally. The only problem with this arrangement is that if the device was unplugged from the panelboard, the exposed plug-in terminals on the back of the device are still hot (energized) because the device is back-fed through its load terminals. This poses a shock and arcing hazard. Such breakers require the additional securing means (typically a clamp and screw) to warn personnel and prevent accidental unplugging.

Bibliography

The following resource materials were used in the preparation of the commentary for this chapter of the code:

NFPA 70–14, *National Electrical Code*. Quincy, MA: National Fire Protection Association, 2013.

Chapter 38:
Wiring Methods

General Comments

This chapter provides installation details for the wiring methods commonly found in dwelling unit construction, and it dictates where and under what conditions specific wiring methods can be used.

Purpose

The purpose of this chapter is to protect life and property by making sure that the wiring in a building is suitable for the environment to which it is exposed and that the wiring is properly supported and protected from damage.

SECTION E3801
GENERAL REQUIREMENTS

E3801.1 Scope. This chapter covers the wiring methods for services, feeders and branch circuits for electrical power and distribution. (300.1)

❖ Wiring methods consist of the cable, tubing or conduit in which the individual wiring is installed. Cable comes with the wire already installed. Tubing and conduit are installed first, with the wire inserted after. This chapter includes the provisions for installing cable, tubing or conduit. Chapter 39 covers the number of conductors permitted in the tubing or conduit, and boxes. This chapter covers how and where the cable, tubing or conduit may be installed.

E3801.2 Allowable wiring methods. The allowable wiring methods for electrical installations shall be those listed in Table E3801.2. Single conductors shall be used only where part of one of the recognized wiring methods listed in Table E3801.2. As used in this code, abbreviations of the wiring-method types shall be as indicated in Table E3801.2. [110.8, 300.3(A)]

❖ Single insulated conductors are not permitted as a wiring method without being part of a cable assembly or installed in tubing or conduit. In many older houses, a wiring method known as "knob-and-tube" was used, consisting of single insulated conductors supported in free air on ceramic insulators (knobs) and, where run through a wood-framing member, installed through ceramic or fabric insulating tubes. Where remodeling, retrofitting or repair is being done in an old house that has this type of wiring method, it is permitted to connect into or tap into one of these circuits to make an extension of the circuit. But instead of running new individual conductors, the transition to the new wiring should be to cable. In early years, this type of wiring method was installed in the hollow spaces of walls, ceilings and attics. It was not installed directly on the wood framing as cable is today, but supported in free air, thus allowing for the dissipation of any heat buildup in the conductors. Care should be taken if insulation is blown in the wall or attic or if foam insulation is installed

around these wires, thereby affecting heat dissipation. In many older houses, the wiring is still sound and in good shape, but where additions or extensions are made, the possibility for arcing, bad connections, heating and overloading comes into play.

TABLE E3801.2
ALLOWABLE WIRING METHODS

ALLOWABLE WIRING METHOD	DESIGNATED ABBREVIATION
Armored cable	AC
Electrical metallic tubing	EMT
Electrical nonmetallic tubing	ENT
Flexible metal conduit	FMC
Intermediate metal conduit	IMC
Liquidtight flexible conduit	LFC
Metal-clad cable	MC
Nonmetallic sheathed cable	NM
Rigid polyvinyl chloride conduit (Type PVC)	RNC
Rigid metallic conduit	RMC
Service entrance cable	SE
Surface raceways	SR
Underground feeder cable	UF
Underground service cable	USE

E3801.3 Circuit conductors. All conductors of a circuit, including equipment grounding conductors and bonding conductors, shall be contained in the same raceway, trench, cable or cord. [300.3(B)]

❖ Current flow in a conductor produces a magnetic flux around the conductor. Through this induction process, heating of the surrounding metal enclosure occurs, and impedance increases. In alternating current circuits, the polarity of the magnetic field is constantly reversing, which causes induction heating. Where both conductors of the circuit are run together in close proximity, the magnetic field of one conductor is canceled or balanced by that of the other.

E3801.4 Wiring method applications. Wiring methods shall be applied in accordance with Table E3801.4. (Chapter 3 and 300.2)

❖ Table E3801.4 is a summary of the applications or locations permitted for the various wiring methods. The footnotes of the table are provisions of the code and should not be overlooked. The most common wiring method for interior wiring is nonmetallic-sheathed cable (NM). Type NM and NMC cables are permitted for one- and two-family dwellings regardless of the number of stories.

SECTION E3802
ABOVE-GROUND INSTALLATION REQUIREMENTS

E3802.1 Installation and support requirements. Wiring methods shall be installed and supported in accordance with Table E3802.1. (Chapter 3 and 300.11)

❖ Table E3802.1 covers the requirements for the support of above-ground wiring methods used in dwellings. Where run parallel to a framing member, cables and nonmetallic wiring methods must be held back at least 1¹/₄ inches (32 mm) from the edge. This is so that when the drywall is applied, a nail or screw that misses the stud of the framing member cannot penetrate and damage the cable or conduit.

TABLE E3801.4 (Chapter 3 and 300.2)
ALLOWABLE APPLICATIONS FOR WIRING METHODS[a, b, c, d, e, f, g, h, i, j, k]

ALLOWABLE APPLICATIONS (application allowed where marked with an "A")	AC	EMT	ENT	FMC	IMC RMC RNC	LFC[a, g]	MC	NM	SR	SE	UF	USE
Services	—	A	A[h]	A[i]	A	A[i]	A	—	—	A	—	A
Feeders	A	A	A	A	A	A	A	A	—	A[b]	A	A[b]
Branch circuits	A	A	A	A	A	A	A	A	A	A[c]	A	—
Inside a building	A	A	A	A	A	A	A	A	A	A	A	—
Wet locations exposed to sunlight	—	A	A[h]	—	A	A	A	—	—	A	A[e]	A[e]
Damp locations	—	A	A	A[d]	A	A	A	—	—	A	A	A
Embedded in noncinder concrete in dry location	—	A	A	—	A	A[j]	—	—	—	—	—	—
In noncinder concrete in contact with grade	—	A[f]	A	—	A[f]	A[j]	—	—	—	—	—	—
Embedded in plaster not exposed to dampness	A	A	A	A	A	A	A	—	—	A	A	—
Embedded in masonry	—	A	A	—	A[f]	A	A	—	—	—	—	—
In masonry voids and cells exposed to dampness or below grade line	—	A[f]	A	A[d]	A[f]	A	A	—	—	A	A	—
Fished in masonry voids	A	—	—	A	—	A	A	A	—	A	A	—
In masonry voids and cells not exposed to dampness	A	A	A	A	A	A	A	A	—	A	A	—
Run exposed	A	A	A	A	A	A	A	A	A	A	A	—
Run exposed and subject to physical damage	—	—	—	—	A[g]	—	—	—	—	—	—	—
For direct burial	—	A[f]	—	—	A[f]	A	A[f]	—	—	—	A	A

For SI: 1 foot = 304.8 mm.

a. Liquid-tight flexible nonmetallic conduit without integral reinforcement within the conduit wall shall not exceed 6 feet in length.

b. Type USE cable shall not be used inside buildings.

c. The grounded conductor shall be insulated.

d. Conductors shall be a type approved for wet locations and the installation shall prevent water from entering other raceways.

e. Shall be listed as "Sunlight Resistant."

f. Metal raceways shall be protected from corrosion and approved for the application. Aluminum RMC requires approved supplementary corrosion protection.

g. RNC shall be Schedule 80.

h. Shall be listed as "Sunlight Resistant" where exposed to the direct rays of the sun.

i. Conduit shall not exceed 6 feet in length.

j. Liquid-tight flexible nonmetallic conduit is permitted to be encased in concrete where listed for direct burial and only straight connectors listed for use with LFNC are used.

k. In wet locations under any of the following conditions:

    1. The metallic covering is impervious to moisture.

    2. A lead sheath or moisture-impervious jacket is provided under the metal covering.

    3. The insulated conductors under the metallic covering are listed for use in wet locations and a corrosion-resistant jacket is provided over the metallic sheath.

Where holes are bored in studs and vertical framing members for cables and nonmetallic wiring methods, the nearest edge of the hole must be at least $1^1/_4$ inches (32 mm) from the edge of the framing member. This is to help prevent a nail or screw from penetrating the wiring installed through the holes. Where this is impossible, a steel plate at least $^1/_{16}$ inch (1.5 mm) thick must be installed over the framing member where the wiring passes through. However, protection plates that are listed for the purpose may be more or less than $^1/_{16}$ inch in thickness.

Electricians know the $1^1/_4$-inch rule because this is an electrical code provision for protection of cables and nonmetallic wiring methods. But for drilling and notching of framing members, structural integrity must be considered also. Electrical code provisions cover protection of wiring but not structural framing rules.

Building code provisions require that the outer edges of holes drilled in horizontal framing members must be at least 2 inches (51 mm) back from the member edge. This requirement is to prevent weakening of the structural members such as floor joists. See Section R502.8 and Figure R502.8 of the code for details of drilling and notching provisions for structural floor members. This requirement is included in Table E3802.1.

Detailed provisions for drilling and notching of studs in an exterior wall or bearing partition are in Section R602.6. In addition to following the electrical code provisions for protection of wiring, these provisions must be followed for structural strength (see Commentary Figure E3802.1).

TABLE E3802.1 (Chapter 3)
GENERAL INSTALLATION AND SUPPORT REQUIREMENTS FOR WIRING METHODS[a, b, c, d, e, f, g, h, i, j, k]

INSTALLATION REQUIREMENTS (Requirement applicable only to wiring methods marked "A")	AC MC	EMT IMC RMC	ENT	FMC LFC	NM UF	RNC	SE	SR[a]	USE
Where run parallel with the framing member or furring strip, the wiring shall be not less than $1^1/_4$ inches from the edge of a furring strip or a framing member such as a joist, rafter or stud or shall be physically protected.	A	—	A	A	A	—	A	—	—
Bored holes in framing members for wiring shall be located not less than $1^1/_4$ inches from the edge of the framing member or shall be protected with a minimum 0.0625-inch steel plate or sleeve, a listed steel plate or other physical protection.	A[k]	—	A[k]	A[k]	A[k]	—	A[k]	—	—
Where installed in grooves, to be covered by wallboard, siding, paneling, carpeting, or similar finish, wiring methods shall be protected by 0.0625-inch-thick steel plate, sleeve, or equivalent, a listed steel plate or by not less than $1^1/_4$-inch free space for the full length of the groove in which the cable or raceway is installed.	A	—	A	A	A	—	A	A	A
Securely fastened bushings or grommets shall be provided to protect wiring run through openings in metal framing members.	—	—	A[j]	—	A[j]	—	A[j]	—	—
The maximum number of 90-degree bends shall not exceed four between junction boxes.	—	A	A	A	—	A	—	—	—
Bushings shall be provided where entering a box, fitting or enclosure unless the box or fitting is designed to afford equivalent protection.	A	A	A	A	—	A	—	A	—
Ends of raceways shall be reamed to remove rough edges.	—	A	A	A	—	A	—	A	—
Maximum allowable on center support spacing for the wiring method in feet.	4.5[b, c]	10[l]	3[b]	4.5[b]	4.5[i]	3[d, l]	2.5[e]	—	2.5
Maximum support distance in inches from box or other terminations.	12[b, f]	36	36	12[b, g]	12[h, i]	36	12	—	—

For SI: 1 inch = 25.4 mm, 1 foot = 304.8 mm, 1 degree = 0.0175 rad.
a. Installed in accordance with listing requirements.
b. Supports not required in accessible ceiling spaces between light fixtures where lengths do not exceed 6 feet.
c. Six feet for MC cable.
d. Five feet for trade sizes greater than 1 inch.
e. Two and one-half feet where used for service or outdoor feeder and 4.5 feet where used for branch circuit or indoor feeder.
f. Twenty-four inches for Type AC cable and thirty-six inches for interlocking Type MC cable where flexibility is necessary.
g. Where flexibility after installation is necessary, lengths of flexible metal conduit and liquidtight flexible metal conduit measured from the last point where the raceway is securely fastened shall not exceed: 36 inches for trade sizes $^1/_2$ through $1^1/_4$, 48 inches for trade sizes $1^1/_2$ through 2 and 5 feet for trade sizes $2^1/_2$ and larger.
h. Within 8 inches of boxes without cable clamps.
i. Flat cables shall not be stapled on edge.
j. Bushings and grommets shall remain in place and shall be listed for the purpose of cable protection.
k. See Sections R502.8 and R802.7 for additional limitations on the location of bored holes in horizontal framing members.

For SI: 1 inch = 25.4 mm.

Figure E3802.1
CABLES HELD 1¹/₄ INCHES BACK FROM EDGE OF FRAMING MEMBER

E3802.2 Cables in accessible attics. Cables in attics or roof spaces provided with access shall be installed as specified in Sections E3802.2.1 and E3802.2.2. (320.3 and 334.23)

❖ An attic can be totally enclosed or accessible through a small 22-inch by 30-inch (559 mm by 762 mm) opening or provided with a ladder or stairs. Section R807 of the code requires an attic access opening for attic areas over 30 square feet (3 m²) and a vertical height of at least 30 inches (762 mm). Where cables are run in an attic that has an access opening, specific rules apply as stated in the two following sections.

E3802.2.1 Across structural members. Where run across the top of floor joists, or run within 7 feet (2134 mm) of floor or floor joists across the face of rafters or studding, in attics and roof spaces that are provided with access, the cable shall be protected by substantial guard strips that are at least as high as the cable. Where such spaces are not provided with access by permanent stairs or ladders, protection shall only be required within 6 feet (1829 mm) of the nearest edge of the attic entrance. [330.23(A) and 334.23]

❖ If the attic is accessed using a ladder or stairs, any exposed cable run must be protected as stated. These code provisions are intended as protection for the cables where it is likely that people will be in the attic. Any cable run exposed along the surface of rafters or studs could easily be pulled on or used to hang things. Cables that are exposed and run across the surface of joists could be walked on or have storage items placed on them, causing damage to the cables. Thus, they are required to be protected or kept at least 7 feet (2134 mm) above the floor of the attic. Some new dwellings do not have an unfinished attic accessible by stairs or a ladder where cables are exposed, but most do have an attic accessible by a scuttle hole (a term used in the industry for the small attic entrance). In attics with only the scuttle hole entrance, cables run across the face of rafters or studs must be protected with guard strips for a distance of 6 feet (1829 mm) from the entrance. This provision should be inspected during the rough inspection. On the finish inspection a ladder may be necessary to look into the attic, and the wiring is usually covered by insulation.

E3802.2.2 Cable installed through or parallel to framing members. Where cables are installed through or parallel to the sides of rafters, studs or floor joists, guard strips and running boards shall not be required, and the installation shall comply with Table E3802.1. [330.23(B) and 334.23]

❖ In an attic where cable is run along the sides of rafters, studs or joists, no extra protection, such as guard strips, is required because the cable is not subject to damage by being walked on or having storage items placed on it as would be the case if it was run over the tops of joists. Table E3802.1 requires that cable be kept back at least 1.25 inches (32 mm) from the face of the rafter, stud or joist.

E3802.3 Exposed cable. In exposed work, except as provided for in Sections E3802.2 and E3802.4, cable assemblies shall be installed as specified in Sections E3802.3.1 and E3802.3.2. (330.15 and 334.15)

❖ See the commentary to Section E3802.3.2 and Commentary Figures E3802.3.1(1) and E3802.3.1(2).

E3802.3.1 Surface installation. Cables shall closely follow the surface of the building finish or running boards. [334.15(A)]

❖ This section prevents exposed cables from being in the open away from a solid surface except as allowed

by Sections E3802.4, E3802.2.1, E3802.2.2 and Table E3802.1. See the commentary to Section E3802.3.2 and Commentary Figures E3802.3.1(1), E3802.3.1(2) and E3907.8.

E3802.3.2 Protection from physical damage. Where subject to physical damage, cables shall be protected by rigid metal conduit, intermediate metal conduit, electrical metallic tubing, Schedule 80 PVC rigid nonmetallic conduit, or other approved means. Where passing through a floor, the cable shall be enclosed in rigid metal conduit, intermediate metal conduit, electrical metallic tubing, Schedule 80 PVC rigid nonmetallic conduit or other approved means extending not less than 6 inches (152 mm) above the floor. [334.15(B)]

❖ The provisions of this section apply to exposed cable in other than attics and joist spaces of unfinished basements. They apply to cables installed in such places as on the wall of an unfinished basement. Although it may not look good to have cable installed on the surface of a wall, it is permitted where the cable is not subject to physical damage. Running it through conduit is one common protection method.

For example, in an unfinished basement, a switch is

Figure E3802.3.1(1)
PROHIBITED APPLICATION OF TYPE NM CABLE

installed on the concrete or masonry wall at the entrance to the outside stairway. The nonmetallic sheathed cable run to the switch must be protected. One way to do this is to install a short, 3- or 4-foot (914 or 1219 mm) length of conduit or EMT from the ceiling or joist space down to an appropriate switch box attached to the concrete wall. This makes for a neat looking installation along the wall and provides protection for the cable. However, the cable could be subject to damage where it enters the conduit or EMT. The part of the cable in the basement ceiling or joist area could be pulled to gain slack for various reasons, such as when installing heating ducts. At the top of the short run of conduit or EMT, a fitting must be installed to protect the cable where it enters the EMT.

Where cable is installed through a floor, it must be protected such that it cannot be damaged by being walked on or kicked or having objects placed against it [see Commentary Figures E3802.3.1(1) and E3802.3.1(2)].

E3802.3.3 Locations exposed to direct sunlight. Insulated conductors and cables used where exposed to direct rays of the sun shall be listed or listed and marked, as being "sunlight resistant," or shall be covered with insulating material, such as tape or sleeving, that is listed or listed and marked as being "sunlight resistant." [310.10(D)]

❖ See the commentary to Section E3605.6.

E3802.4 In unfinished basements and crawl spaces. Where type NM or SE cable is run at angles with joists in unfinished basements and crawl spaces, cable assemblies containing two or more conductors of sizes 6 AWG and larger and assemblies containing three or more conductors of sizes 8 AWG and larger shall not require additional protection where

attached directly to the bottom of the joists. Smaller cables shall be run either through bored holes in joists or on running boards. Type NM or SE cable installed on the wall of an unfinished basement shall be permitted to be installed in a listed conduit or tubing or shall be protected in accordance with Table E3802.1. Conduit or tubing shall be provided with a suitable insulating bushing or adapter at the point where the cable enters the raceway. The sheath of the Type NM or SE cable shall extend through the conduit or tubing and into the outlet or device box not less than $^1/_4$ inch (6.4 mm). The cable shall be secured within 12 inches (305 mm) of the point where the cable enters the conduit or tubing. Metal conduit, tubing, and metal outlet boxes shall be connected to an equipment grounding conductor complying with Section E3908.13. [334.15(C)]

❖ Although there are options, in most cases in unfinished basement ceilings, cables are installed through holes in the joists. Note that this section also applies to crawl spaces requiring the same protection for cables as required for basements. Cables that are run across the bottom of floor joists are subject to damage resulting from occupants hanging things from or storing things on the cables. If the basement is left unfinished in a new house, the electrician is usually expected to run the cables within the joist area instead of on the lower surface of the joists because it is expected that at a future time a ceiling membrane will be attached directly to the bottom edge of the floor joists to finish the basement ceiling. Even larger cables such as feeders to a panelboard are run through drilled holes. But to accommodate piping and ducts, the ceiling of an unfinished basement may need to be furred down below the surface of the floor joists to prepare for applying the ceiling membrane. For whatever reason, if the electri-

Figure E3802.3.1(2)
PROHIBITED APPLICATION OF TYPE NM CABLE

cian chooses to run the cables on the surface of the lower edge of the joists, these specific requirements apply.

Only the specified cables can be installed across the bottom of joists without further protection. Cables with conductor sizes 14, 12 and 10 AWG must be either run through drilled holes or installed on running boards. Note that the building portion of the code limits the size and location of bored holes in joists (see Commentary Figure E3802.4). Commonly, the installer makes the mistake of boring holes in joists too close to the bottom of the joist. Holes located in the center of the joist have the least effect on the strength of the joist.

E3802.5 Bends. Bends shall be made so as not to damage the wiring method or reduce the internal diameter of raceways.

For types NM and SE cable, bends shall be so made, and other handling shall be such that the cable will not be damaged and the radius of the curve of the inner edge of any bend shall be not less than five times the diameter of the cable. (334.24 and 338.24)

❖ Where installing through drilled holes in studs and joists, it is important to avoid kinks and knots in the cable as it is pulled off the spool or coil. Although a slang term used in some areas for cable is "rope," it should not be installed as if it were rope. Care must be used when pulling cables through holes at various angles. Cables must not be bent too sharply as this can damage the insulation on the conductors by compressing it on the inside of the bend and stretching it on the outside of the bend.

E3802.6 Raceways exposed to different temperatures. Where portions of a raceway or sleeve are known to be subjected to different temperatures and where condensation is known to be a problem, as in cold storage areas of buildings or where passing from the interior to the exterior of a building, the raceway or sleeve shall be filled with an approved

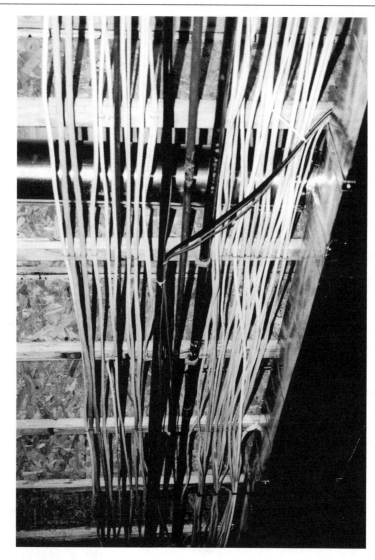

Figure E3802.4
PROHIBITED INSTALLATION OF TYPE MM CABLE ACROSS THE BOTTOM OF FLOOR JOISTS

material to prevent the circulation of warm air to a colder section of the raceway or sleeve. [300.7(A)]

❖ Pressure differentials caused by wind, stack effect and exhaust fans can cause air to pass through various openings in the building envelope. If warm humid air from a dwelling interior passes through a raceway that extends into an unconditioned space or outdoors, condensation can occur within the raceway, causing damage to electrical components and/or the structure. This can be observed where a conduit extends from an interior switch box to an exterior luminaire box. Liquid water can form in the raceway and in the exterior box. This can also be seen where raceways pass through attic spaces and connect to ceiling outlet boxes. Electrical putty such as "duct seal putty" could be used to seal the raceway or sleeve.

E3802.7 Raceways in wet locations above grade. Where raceways are installed in wet locations above grade, the interior of such raceways shall be considered to be a wet location. Insulated conductors and cables installed in raceways in wet locations above grade shall be listed for use in wet locations. (300.9)

❖ Raceways installed above grade in wet locations and all raceways installed below grade are subject to internal moisture; therefore, the conductors in such raceways must be listed as suitable for wet locations.

SECTION E3803
UNDERGROUND INSTALLATION REQUIREMENTS

E3803.1 Minimum cover requirements. Direct buried cable or raceways shall be installed in accordance with the minimum cover requirements of Table E3803.1. [300.5(A)]

❖ Underground wiring must be protected by burial at a sufficient depth. For example, a 120-volt, 20-ampere branch circuit with ground-fault circuit interrupter (GFCI) protection serving a yard light must be buried at least 12 inches (305 mm) deep where it is under the residential drive-way. A service lateral installed in a PVC conduit under the backyard lawn must be buried at least 18 inches (457 mm) deep. Many utility companies require the service lateral to be installed in conduit. Where the service lateral is a direct buried cable under the backyard lawn with no protection such as a concrete slab, it must be buried at least 24 inches (610 mm) deep, according to Table E3803.1.

TABLE E3803.1. See page 38-9.

❖ See the commentary to Sections E3802.1 and E3803.1.

E3803.2 Warning ribbon. Underground service conductors that are not encased in concrete and that are buried 18 inches (457 mm) or more below grade shall have their location identified by a warning ribbon that is placed in the trench not less than 12 inches (305 mm) above the underground installation. [300.5(D)(3)]

❖ A warning ribbon, typically red plastic tape with printed warnings, is required for service conductor installa-

tions to help prevent excavators from damaging the conductors and risking injury or death. Table E3803.1 dictates the conditions under which service conductors must be buried at least 18 inches (457 mm). The ribbon must be at least 12 inches (305 mm) above the conductors to provide warning well in advance of reaching the conductor depth.

E3803.3 Protection from damage. Direct buried conductors and cables emerging from the ground shall be protected by enclosures or raceways extending from the minimum cover distance below grade required by Section E3803.1 to a point at least 8 feet (2438 mm) above finished grade. In no case shall the protection be required to exceed 18 inches (457 mm) below finished grade. Conductors entering a building shall be protected to the point of entrance. Where the enclosure or raceway is subject to physical damage, the conductors shall be installed in rigid metal conduit, intermediate metal conduit, Schedule 80 rigid nonmetallic conduit or the equivalent. [300.5(D)(1)]

❖ Where a light installed on a pole is fed by direct buried cable such as a UF cable, the cable must be protected from at least 18 inches (457 mm) below the ground surface to a height of at least 8 feet (2438 mm) up the pole.

Most service laterals to a dwelling would not be encased in concrete (such as a service lateral enclosed in nonmetallic conduit that is encased in concrete) unless under special circumstances. In all other cases, whether direct-buried or enclosed in conduit, the warning ribbon must be placed in the trench 12 inches (305 mm) or more above the service lateral to help avoid damage or electrocution when digging occurs near the buried conductors. The service lateral is not protected against fault current caused by a short circuit.

E3803.4 Splices and taps. Direct buried conductors or cables shall be permitted to be spliced or tapped without the use of splice boxes. The splices or taps shall be made by approved methods with materials listed for the application. [300.5(E)]

❖ Splicing kits are available for use on direct-buried conductors. These are used in most cases where the conductors have been damaged and it becomes necessary to repair the conductors rather than replace the entire underground run.

E3803.5 Backfill. Backfill containing large rock, paving materials, cinders, large or sharply angular substances, or corrosive material shall not be placed in an excavation where such materials cause damage to raceways, cables or other substructures or prevent adequate compaction of fill or contribute to corrosion of raceways, cables or other substructures. Where necessary to prevent physical damage to the raceway or cable, protection shall be provided in the form of granular or selected material, suitable boards, suitable sleeves or other approved means. [300.5(F)]

❖ Unless the soil is suitable to prevent damage to the buried conductors or raceways, sand must be brought in and placed in the trench below and on top of the underground wiring method, or other approved means must be used. In cases where this is not done, direct buried cable can be damaged upon backfilling by rocks falling into the trench on top of the cable.

Where aluminum cable is used, a small nick or cut in the conductor insulation can cause moisture to penetrate the sheath or insulation, which in turn can cause the aluminum to oxidize and deteriorate. For underground service laterals, many utility companies have equipment to locate the exact point of the damaged or open conductor. It can then be repaired by digging down to the cable and making a splice allowed by Section E3803.4. This is a very costly and time-consuming process and requires digging. Therefore, in landscaped areas, the utility companies often require that service laterals be run in conduit from the point of supply to the service equipment.

E3803.6 Raceway seals. Conduits or raceways shall be sealed or plugged at either or both ends where moisture will enter and contact live parts. [300.5(G)]

❖ Where an underground conduit is run through the basement wall, for example, the temperature difference will cause condensation, and moisture could enter an electrical enclosure such as a box or cabinet. An example is a service lateral that is run directly into service equipment located on the inside wall of a basement. The conduit must be plugged or a sealing compound used to prevent moisture from entering and contacting energized parts.

TABLE E3803.1 (Table 300.5)
MINIMUM COVER REQUIREMENTS, BURIAL IN INCHES[a, b, c, d, e]

LOCATION OF WIRING METHOD OR CIRCUIT	TYPE OF WIRING METHOD OR CIRCUIT				
	1 Direct burial cables or conductors	2 Rigid metal conduit or intermediate metal conduit	3 Nonmetallic raceways listed for direct burial without concrete encasement or other approved raceways	4 Residential branch circuits rated 120 volts or less with GFCI protection and maximum overcurrent protection of 20 amperes	5 Circuits for control of irrigation and landscape lighting limited to not more than 30 volts and installed with type UF or in other identified cable or raceway
All locations not specified below	24	6	18	12	6
In trench below 2-inch-thick concrete or equivalent	18	6	12	6	6
Under a building	0 (In raceway only or Type MC identified for direct burial)	0	0	0 (In raceway only or Type MC identified for direct burial)	0 (In raceway only or Type MC identified for direct burial)
Under minimum of 4-inch-thick concrete exterior slab with no vehicular traffic and the slab extending not less than 6 inches beyond the underground installation	18	4	4	6 (Direct burial) 4 (In raceway)	6 (Direct burial) 4 (In raceway)
Under streets, highways, roads, alleys, driveways and parking lots	24	24	24	24	24
One- and two-family dwelling driveways and outdoor parking areas, and used only for dwelling-related purposes	18	18	18	12	18
In solid rock where covered by minimum of 2 inches concrete extending down to rock	2 (In raceway only)	2	2	2 (In raceway only)	2 (In raceway only)

For SI: 1 inch = 25.4 mm.

a. Raceways approved for burial only where encased concrete shall require concrete envelope not less than 2 inches thick.

b. Lesser depths shall be permitted where cables and conductors rise for terminations or splices or where access is otherwise required.

c. Where one of the wiring method types listed in columns 1 to 3 is combined with one of the circuit types in columns 4 and 5, the shallower depth of burial shall be permitted.

d. Where solid rock prevents compliance with the cover depths specified in this table, the wiring shall be installed in metal or nonmetallic raceway permitted for direct burial. The raceways shall be covered by a minimum of 2 inches of concrete extending down to the rock.

e. Cover is defined as the shortest distance in inches (millimeters) measured between a point on the top surface of any direct-buried conductor, cable, conduit or other raceway and the top surface of finished grade, concrete, or similar cover.

E3803.7 Bushing. A bushing, or terminal fitting, with an integral bushed opening shall be installed on the end of a conduit or other raceway that terminates underground where the conductors or cables emerge as a direct burial wiring method. A seal incorporating the physical protection characteristics of a bushing shall be considered equivalent to a bushing. [300.5(H)]

❖ An example of this requirement is where direct buried service lateral conductors are installed in rigid or intermediate metal conduit from a point underground up to the service equipment. At the point where the conductors emerge from the conduit, a bushing must be installed to protect the conductors. The ground around the conductors and the conduit will settle and pull the conductors against the conduit, which, if installed in compliance with code provisions, is fastened solidly to the wall of the house. Where the conduit is not fastened solidly to the wall of the house, the settling of the ground has been observed to pull the conduit connector away from the service equipment enclosure, exposing the conductor under the service equipment cabinet. It is important to leave slack in the direct buried conductors of the service lateral where they enter the riser conduit in the trench. Where the earth has been backfilled around a new house, settlement of the ground will occur that can also cause the conductors to move.

E3803.8 Single conductors. All conductors of the same circuit and, where present, the grounded conductor and all equipment grounding conductors shall be installed in the same raceway or shall be installed in close proximity in the same trench. [300.5(I)]

Exception: Conductors shall be permitted to be installed in parallel in raceways, multiconductor cables, and direct-buried single conductor cables. Each raceway or multiconductor cable shall contain all conductors of the same circuit, including equipment grounding conductors. Each direct-buried single conductor cable shall be located in close proximity in the trench to the other single conductor cables in the same parallel set of conductors in the circuit, including equipment grounding conductors. [300.5(I) Exception No.1]

❖ Conductors must be grouped together in the same raceway or trench to reduce induction heating in raceways and cables. Where conductors are run in parallel (not often encountered in dwelling electrical systems), each raceway must contain a complete set of the circuit conductors. For example, a parallel set of 4-wire feeders is supplying a 120/240-volt panelboard, and in this case, each raceway would contain two ungrounded conductors (with a voltage of 240 volts between them), one grounded conductor and one grounding conductor, the same as if there was only a single raceway.

E3803.9 Earth movement. Where direct buried conductors, raceways or cables are subject to movement by settlement or frost, direct buried conductors, raceways or cables shall be arranged to prevent damage to the enclosed conductors or to equipment connected to the raceways. [300.5(J)]

❖ See the commentary to Section E3803.8. Conductors in a raceway rising from underground and entering the bottom of an enclosure can be looped in the enclosure in such manner that will allow them to flex when the soil rises and falls and pushes or pulls the conductors.

E3803.10 Wet locations. The interior of enclosures or raceways installed underground shall be considered to be a wet location. Insulated conductors and cables installed in such enclosures or raceways in underground installations shall be listed for use in wet locations. Connections or splices in an underground installation shall be approved for wet locations. [300.5(B)]

❖ This section is a reminder of what is already understood in the code; that is, the interior of raceways and enclosures located underground must be treated as wet locations. It cannot be assumed that the interior of any enclosure or raceway is free of moisture. Moisture will enter in liquid form and/or in vapor form and condense in the raceway or enclosure (see Section E3802.7).

E3803.11 Under buildings. Underground cable and conductors installed under a building shall be in a raceway. [300.5(C)]

Exception: Type MC Cable shall be permitted under a building without installation in a raceway where the cable is listed and identified for direct burial or concrete encasement and one or more of the following applies:

1. The metallic covering is impervious to moisture.

2. A moisture-impervious jacket is provided under the metal covering.

3. The insulated conductors under the metallic covering are listed for use in wet locations, and a corrosion-resistant jacket is provided over the metallic sheath. [300.5(C) Exception No.2]

❖ A common code violation is the installation of Type UF cable under a floor slab on grade to supply floor receptacle boxes. Raceways provide physical protection for cables and conductors under slabs and also allow conductors to be removed and reinstalled in a location that is permanently concealed under concrete. If it complies with all of the provisions of the exception to this section, MC cable is also allowed under a slab because of the physical protection afforded by the metal covering.

Bibliography

The following resource materials were used in the preparation of the commentary for this chapter of the code:

NFPA 70–14, *National Electrical Code.* Quincy, MA: National Fire Protection Association, 2013.

Chapter 39:
Power and Lighting Distribution

General Comments

Chapter 39 addresses the "rough-in" stage of construction in which the wiring system is installed to distribute receptacle and lighting outlets throughout the dwelling. This chapter covers receptacle outlet spacing, GFCI (ground-fault circuit interrupter) and AFCI (arc-fault circuit interrupter) protection, lighting outlet locations, raceway and box fill limitations, box and panel board installation, equipment grounding and flexible cords.

Purpose

The intent of Chapter 39 is to protect occupants from fire, electrical shock and accidents. By eliminating the need for extension cords, by providing adequate lighting and by requiring equipment grounding and GFCI and AFCI protection, the code substantially increases the level of safety in homes.

SECTION E3901
RECEPTACLE OUTLETS

E3901.1 General. Outlets for receptacles rated at 125 volts, 15- and 20-amperes shall be provided in accordance with Sections E3901.2 through E3901.11. Receptacle outlets required by this section shall be in addition to any receptacle that is:

1. Part of a luminaire or appliance;

2. Located within cabinets or cupboards;

3. Controlled by a wall switch in accordance with Section E3903.2, Exception 1; or

4. Located over 5.5 feet (1676 mm) above the floor.

Permanently installed electric baseboard heaters equipped with factory-installed receptacle outlets, or outlets provided as a separate assembly by the baseboard manufacturer shall be permitted as the required outlet or outlets for the wall space utilized by such permanently installed heaters. Such receptacle outlets shall not be connected to the heater circuits. (210.52)

❖ Section E3901 covers the requirements for the locations of receptacle outlets. A receptacle that is built-in or is an integral part of a light fixture, appliance or cabinet is not counted as one of the outlets required by this section. However, a factory-installed receptacle in a baseboard heater is permitted to serve as one of the required outlets, but such a receptacle must not be connected to the circuit supplying the heater.

A baseboard heater must not be installed on the wall below a receptacle outlet unless specifically listed for such installations. Where a cord is plugged into a receptacle, it could drape over the baseboard heater or come into contact with it; thus, the cord would become hot, which could melt or weaken the insulation and create a possible fire or shock hazard. If the plan calls for baseboard heaters, the required receptacles must be either integral with the heater or be positioned so that cords will not contact the heater.

A receptacle outlet is controlled by a light switch does not satisfy the requirements of this section for convenience and appliance receptacle distribution. Switched receptacles make convenient lighting outlets, but are not suitable for most other plug loads because of unintended switch operation by the occupants and because the occupants may not be aware that the receptacle is switched. Note that a duplex receptacle is actually two receptacles, therefore, half of the duplex device could be switched and the other half could be unswitched, thus complying with the intent of this section. If both halves of the duplex receptacle are switched, that outlet is not counted as fulfilling the intent of this section.

E3901.2 General purpose receptacle distribution. In every kitchen, family room, dining room, living room, parlor, library, den, sun room, bedroom, recreation room, or similar room or area of dwelling units, receptacle outlets shall be installed in accordance with the general provisions specified in Sections E3901.2.1 through E3901.2.3 (see Figure E3901.2).

❖ Simply stated, no matter where a lamp or appliance is placed along a wall, a receptacle must be within reach of the approximately 6-foot (1829 mm) long cord of the lamp or appliance. The receptacles in all rooms of a dwelling unit except bathrooms and laundry rooms are referred to as general use (convenience) receptacles. This section and the following four subsections cover the location and spacing requirements for general use receptacles. There is no minimum or maximum number of receptacles for a room. There is no height requirement for general use (convenience) receptacles except that any receptacle located over $5^1/_2$ feet (1676 mm) above the floor is not counted. There are, however, certain height requirements for accessibility standards that must be observed.

Although there are specific requirements for countertop receptacles in a kitchen, the location and spacing of all other receptacles in kitchens and dining areas are included in the provisions of this section.

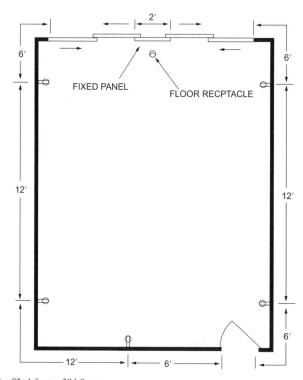

For SI: 1 foot = 304.8 mm.

FIGURE E3901.2
GENERAL USE RECEPTACLE DISTRIBUTION

E3901.2.1 Spacing. Receptacles shall be installed so that no point measured horizontally along the floor line of any wall space is more than 6 feet (1829 mm), from a receptacle outlet. [210.52(A)(1)]

❖ This requirement is intended to eliminate the need for extension cords. In any of the rooms listed (or a similar area of a dwelling), a lamp, appliance, radio, TV or other electrical appliance can be placed at any point along the floor line, and a receptacle must be available within 6 feet (1829 mm) of that location. It is important to not "cut corners" when measuring. The floor line is measured along the wall all the way into and around corners. Where a door opens against a wall, it may seem reasonable to begin the measurement at the end of the door swing, about 30 to 36 inches (762 to 914 mm) from the hinged side of the door casing, but the code requires that the wall space be counted all the way to the doorjamb. The logic of this section is simply to provide enough conveniently located receptacles to prevent the occupants from creating hazards by running or draping cords in poor locations. Poorly located cords could become tripping hazards, could cause appliances to fall, could be covered by rugs, walked on, pinched in doors, etc., all of which could cause fires or injuries. The code never assumes that any wall space meeting the criteria of Section E3901.2.2 will be an "unlikely" place for a lamp or appliance. If it is reasonably possible, an occupant will utilize every spot along a wall space at some point in the life of the building.

E3901.2.2 Wall space. As used in this section, a wall space shall include the following: [210.52(A)(2)]

1. Any space that is 2 feet (610 mm) or more in width, including space measured around corners, and that is unbroken along the floor line by doorways and similar openings, fireplaces, and fixed cabinets.

2. The space occupied by fixed panels in exterior walls, excluding sliding panels.

3. The space created by fixed room dividers such as railings and freestanding bar-type counters.

❖ An appliance or lamp, for example, could be placed in front of the fixed portion of a sliding door, which must be counted as wall space for determining the spacing requirements of receptacle outlets. An example of a fixed room divider is a partial wall that subdivides a space or that separates a bar area from the rest of a room. A guardrail that separates spaces at different elevations could also qualify as a room divider and appliances, lamps, entertainment equipment, etc., could also be placed along such dividers. Appliances or lamps could be placed by the homeowner along any partial wall or railing, and a receptacle must be available at a point not farther than 6 feet (1829 mm) from any point along such wall or railing. Floor outlets may be necessary in situations where the room divider is an open guardrail.

E3901.2.3 Floor receptacles. Receptacle outlets in floors shall not be counted as part of the required number of receptacle outlets except where located within 18 inches (457 mm) of the wall. [210.52(A)(3)]

❖ Where the furniture layout is known, floor receptacles are sometimes located in the middle of a room. For example, power may be required for a floor lamp to be placed near a couch that is located in the center of a living room. Such receptacles are not counted as part of the required number of receptacles as measured along the wall line. In the example under Section E3901.2.2, a floor outlet must be installed not more than 18 inches (457 mm) from the railing to count as the required receptacle.

E3901.2.4 Countertop receptacles. Receptacles installed for countertop surfaces as specified in Section E3901.4 shall not be considered as the receptacles required by Section E3901.2. [210.52(A)(4)]

❖ Receptacles installed to serve a countertop surface are not recognized as also serving any wall space. For example, a bar counter could be installed against a partial wall that subdivides a room. The side of such wall that is opposite of the bar countertop would require receptacles in accordance with Section E3901.2.2 and the receptacles provided to serve the bar countertop cannot also serve this wall space. Keep in mind the intent to prevent occupants from running or draping cords over obstacles such that a hazard is created.

E3901.3 Small appliance receptacles. In the kitchen, pantry, breakfast room, dining room, or similar area of a dwelling unit, the two or more 20-ampere small-appliance branch circuits required by Section E3703.2, shall serve all wall and floor receptacle outlets covered by Sections E3901.2 and E3901.4 and those receptacle outlets provided for refrigeration appliances. [210.52(B)(1)]

Exceptions:

1. In addition to the required receptacles specified by Sections E3901.1 and E3901.2, switched receptacles supplied from a general-purpose branch circuit as defined in Section E3903.2, Exception 1 shall be permitted. [210.52(B)(1) Exception No. 1]

2. The receptacle outlet for refrigeration appliances shall be permitted to be supplied from an individual branch circuit rated at 15 amperes or greater. [210.52(B)(1) Exception No. 2]

❖ Small appliance receptacles are supplied by at least two 20-ampere branch circuits and include all of the receptacles in the kitchen, pantry, dining and similar areas of a house. The countertop receptacles as well as all of the other "low" receptacles [usually installed around 12 to 18 inches (305 to 457 mm) above the floor] in these rooms are served by the small appliance receptacle branch circuits. In a dining room, a receptacle controlled by a wall switch can serve as the required lighting outlet. Such a receptacle would not be served by one of the small appliance branch circuits but would be served by a general-purpose lighting circuit.

The receptacle serving a refrigerator can be included on one of the small appliance circuits or can be supplied by a separate individual branch circuit. Where it is an individual branch circuit, it can be a 15- or 20-ampere-rated branch circuit.

E3901.3.1 Other outlets prohibited. The two or more small-appliance branch circuits specified in Section E3901.3 shall serve no other outlets. [210.52(B)(2)]

Exceptions:

1. A receptacle installed solely for the electrical supply to and support of an electric clock in any of the rooms specified in Section E3901.3. [210.52(B)(2) Exception No.1]

2. Receptacles installed to provide power for supplemental equipment and lighting on gas-fired ranges, ovens, and counter-mounted cooking units. [210.52(B)(2) Exception No.2]

❖ Receptacles for specific fixed-in-place appliances such as a disposal, dishwasher or trash compactor cannot be served by the small-appliance branch circuits. The small-appliance circuits cannot supply anything else, such as the hood fan over the range.

A half century ago, a receptacle for an electric clock installed high on the wall was a standard feature in the kitchen in some areas of the country. Special recessed receptacles were used for clocks. The power for a clock is minimal, and although this type of receptacle

outlet is not as common now, including it on one of the small appliance branch circuits has been permitted for many years. Where a gas range is installed in the kitchen, a 120-volt receptacle outlet is needed to operate such items on the range as the timer, lights and ignition system. This receptacle can be connected to one of the small appliance branch circuits.

E3901.3.2 Limitations. Receptacles installed in a kitchen to serve countertop surfaces shall be supplied by not less than two small-appliance branch circuits, either or both of which shall also be permitted to supply receptacle outlets in the same kitchen and in other rooms specified in Section E3901.3. Additional small-appliance branch circuits shall be permitted to supply receptacle outlets in the kitchen and other rooms specified in Section E3901.3. A small-appliance branch circuit shall not serve more than one kitchen. [210.52(B)(3)]

❖ Small appliance branch circuits can serve receptacles in a dining room. In a formal dining room, where no cooking is done and it is not likely that any appliances will be used, it may seem that the receptacles could be served by a general-purpose lighting branch circuit, but these receptacles must be included on one of the two or more small-appliance branch circuits.

There is not a limit on the number of outlets served by a small appliance branch circuit. Some electrical contractors have traditional ways of laying out these circuits. For example, one practice is that all countertop receptacles on one side of the kitchen sink are wired on one circuit, and another circuit is used for all receptacles on the other side of the sink. Another method is to alternate countertop receptacles on two different circuits so that no two adjacent receptacles are on the same circuit. Yet another method is to use three-conductor cable and wire the top half of the duplex receptacle on one circuit and the bottom half on another circuit. These methods are not code requirements but only preferences for some installers. The code simply requires that all receptacles in the kitchen, pantry and dining areas be served by two or more small appliance branch circuits.

E3901.4 Countertop receptacles. In kitchens pantries, breakfast rooms, dining rooms and similar areas of dwelling units, receptacle outlets for countertop spaces shall be installed in accordance with Sections E3901.4.1 through E3901.4.5 (see Figure E3901.4). [210.52(C)]

❖ This section introduces five subsections. The first four subsections cover the countertop spaces where receptacles are required, and the fifth covers the location of receptacles in those required spaces.

E3901.4.1 Wall countertop space. A receptacle outlet shall be installed at each wall countertop space 12 inches (305 mm) or wider. Receptacle outlets shall be installed so that no point along the wall line is more than 24 inches (610 mm), measured horizontally from a receptacle outlet in that space. [210.52(C)(1)]

Exception: Receptacle outlets shall not be required on a wall directly behind a range, counter-mounted cooking unit or sink in the installation described in Figure E3901.4.1. [210.52(C)(1) Exception]

❖ In many kitchens there is a short section of countertop (between the refrigerator and the range) on which it is helpful to place items when getting them in and out of the refrigerator. The literal code requirement is that if this countertop is 12 inches (305 mm) or more wide, a receptacle is required along the wall at this space. Some installers have omitted the receptacle where the countertop space is only 9 or 10 inches (229 or 254 mm). This is a case where good judgment should prevail. For example, a toaster could be placed on such a countertop space, and if there were not a receptacle available, the toaster cord could be placed across the range to be plugged in. This would create an unsafe situation. Providing a receptacle at a 10-inch (254 mm) countertop space, for example, is a good idea, although not a code requirement.

As indicated in Figure E3901.4, receptacle outlets must not be spaced more than 48 inches (1219 mm) apart along the wall line of the countertop. For this measurement, cutting corners is not permitted. Notice the measurement at the left of the sink in the figure. The edge of the sink is considered to be the beginning of the countertop, and an outlet must be available within 24 inches (610 mm) of that point. Continuing to the left from that point, the measurement follows the wall line.

Using Figure E3901.4, it is incorrect, for example, to draw an imaginary line along the front edge of the countertop in front of the sink all the way to the wall left of the sink, consider that to be one "countertop," then begin at that point to measure to the refrigerator. The wall line in the corner must be included in the measurement.

The exception allows receptacles to be omitted in the wall space or spaces directly behind the sink, range or counter-mounted cooking unit because such space is not likely to be used for countertop appliances since they would be out of the reach of the user, and the cords could be dragged over hot cooking surfaces.

E3901.4.2 Island countertop spaces. At least one receptacle outlet shall be installed at each island countertop space with a long dimension of 24 inches (610 mm) or greater and a short dimension of 12 inches (305 mm) or greater. [210.52(C)(2)]

❖ At an island counter space, one receptacle is required for each counter space. Where the island contains no sink, range or cooktop, only one receptacle is required. More may be installed, for example, at each end. The 24-inch (610 mm) spacing requirement does not apply to the island countertop space because there is no wall line. Depending on the size of the space behind the sink, range or cooktop, an island might be divided into two counter spaces by the presence of the sink or cooktop, thus requiring a receptacle for each counter space so created. In Commentary Figure E3901.4.2, the higher countertop surface is used as a dining table

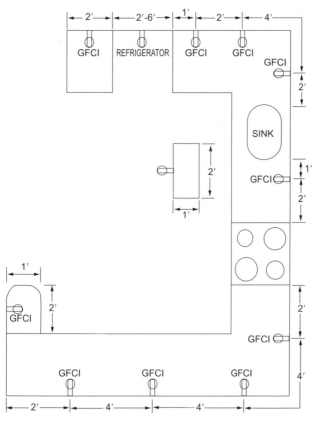

For SI: 1 foot = 304.8 mm.

FIGURE E3901.4
COUNTERTOP RECEPTACLES

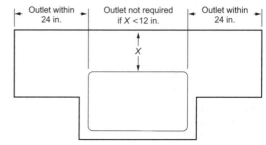

Sink, range or counter-mounted cooking unit extending from face of counter

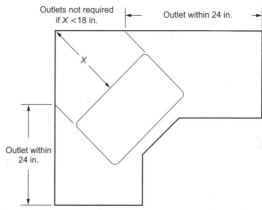

Sink, range or counter-mounted cooking unit mounted in corner

For SI: 1 inch = 25.4 mm.

FIGURE E3901.4.1
DETERMINATION OF AREA BEHIND SINK OR RANGE

Figure E3901.4.2
ISLAND WITH TWO COUNTER SPACES

where chairs are placed along the back side of the island. Also the higher surface is out of reach and unhandy for use as typical countertop workspace. For these reasons, one could interpret that two countertop spaces are present requiring at least two receptacles, regardless of the depth of space of the upper surface.

E3901.4.3 Peninsular countertop space. At least one receptacle outlet shall be installed at each peninsular countertop space with a long dimension of 24 inches (610 mm) or greater and a short dimension of 12 inches (305 mm) or greater. A peninsular countertop is measured from the connecting edge. [210.52(C)(3)]

❖ The connecting edge of a peninsular countertop is an imaginary line along the front edge of the adjacent countertop. At least one receptacle is required at the peninsular countertop space. It need not be at the end of the peninsula (see the commentary to and the text of Section E3901.4).

E3901.4.4 Separate spaces. Countertop spaces separated by range tops, refrigerators, or sinks shall be considered as separate countertop spaces in applying the requirements of Sections E3901.4.1, E3901.4.2 and E3901.4.3. Where a range, counter-mounted cooking unit, or sink is installed in an island or peninsular countertop and the depth of the countertop behind the range, counter-mounted cooking unit, or sink is less than 12 inches (305 mm), the range, counter-mounted cooking unit, or sink has divided the countertop space into two separate countertop spaces as defined in Section E3901.4.4. Each separate countertop space shall comply with the applicable requirements of this section. [210.52(C)(4)]

❖ The intent of this code requirement is to prevent cords from being stretched or laid across the range, cooktop

or sink that divides the countertop space. The left and right edges of the kitchen sink or cooktop, as one faces it, are the points where the measurement begins for the countertop space (see Section E3901.4).

Where there is not much countertop surface extending away from the back of a sink or counter-recessed cooking unit, the code considers such sinks and cooking units as effective dividers that split the island or peninsular counter space into separate counter spaces. If the counter spaces created by this split meet the dimensional requirements of Section E3901.4.2 or E3901.4.3, each space must have its own receptacle(s).

E3901.4.5 Receptacle outlet location. Receptacle outlets shall be located not more than 20 inches (508 mm) above the countertop. Receptacle outlet assemblies installed in countertops shall be listed for the application. Receptacle outlets shall not be installed in a face-up position in the work surfaces or countertops. Receptacle outlets rendered not readily accessible by appliances fastened in place, appliance garages, sinks or rangetops as addressed in the exception to Section E3901.4.1, or appliances occupying dedicated space shall not be considered as these required outlets. [210.52(C)(5)]

Exception: Receptacle outlets shall be permitted to be mounted not more than 12 inches (305 mm) below the countertop in construction designed for the physically impaired and for island and peninsular countertops where the countertop is flat across its entire surface and there are no means to mount a receptacle within 20 inches (508 mm) above the countertop, such as in an overhead cabinet. Receptacles mounted below the countertop in accordance with this exception shall not be located where the countertop extends more than 6 inches (152 mm) beyond its support base. [210.52(C)(5) Exception]

❖ The intent of the code rule is that receptacles be available for small appliances used on the countertop. The code lists several locations for receptacles that do not comply with the intent to provide accessible receptacles. It can be a hazard to have appliances plugged in below the countertop surface because the cord would be draped over the edge of the countertop, and the cord could easily be snagged, pulling the appliance off of the countertop. Where an island or peninsular countertop does not overhang the cabinet more than 6 inches (152 mm) and there is no other location option, the required receptacle is permitted to be installed below the countertop if not more than 12 inches (305 mm) below it. Appliance garages are storage spaces for small countertop appliances. Such spaces have side walls and doors and are accessed at countertop level. Receptacles must not be mounted face-up in a countertop surface because liquids and debris would enter the device. The code recognizes receptacle assemblies that are designed for installation in countertop surfaces and listed for this application. Such assemblies consist of a device box, cover assembly and a pop-up receptacle. These assemblies are hardwired with permanent raceway wiring methods from under the countertop.

E3901.5 Appliance receptacle outlets. Appliance receptacle outlets installed for specific appliances, such as laundry equipment, shall be installed within 6 feet (1829 mm) of the intended location of the appliance. (210.50(C)]

❖ Where the outlet is not more than 6 feet (1829 mm) from the location of an appliance, it is assumed that the appliance cord will reach the outlet. The intent is that extension cords will not have to be used. Common sense should be used here, and the 6-foot (1829 mm) measurement may not be adequate if the measured distance is simply from the edge of the appliance space. For example, if a 120-volt receptacle for a gas clothes dryer is 6 feet (1829 mm) from the edge of the dryer space, the receptacle placement may satisfy the code rule, but the dryer cord at the rear of a gas clothes dryer may be at the opposite side of the space closest to the receptacle. Consideration should be given to the location of the appliances such as garage door openers, sump pumps, clothes washers, etc., so that extension cords will not have to be used.

E3901.6 Bathroom. At least one wall receptacle outlet shall be installed in bathrooms and such outlet shall be located within 36 inches (914 mm) of the outside edge of each lavatory basin. The receptacle outlet shall be located on a wall or partition that is adjacent to the lavatory basin location, located on the countertop, or installed on the side or face of the basin cabinet. The receptacle shall be located not more than 12 inches (305 mm) below the top of the basin.

Receptacle outlets shall not be installed in a face-up position in the work surfaces or countertops in a bathroom basin

location. Receptacle outlet assemblies installed in countertops shall be listed for the application. [210.52(D)]

❖ Because of the popularity of full-height mirrors, it is becoming more difficult to find a suitable location for the receptacle outlets; therefore, the code allows installation of such receptacles on the face or sides of vanity cabinets. The receptacle must not be more than 12 inches (305 mm) below the top of the lavatory because a hazard is created by having appliance cords protruding from the cabinet sides or face. Also, the lower the receptacle, the less appliance cord that will be available for using the appliance. A raceway wiring method will be necessary within the cabinet to protect and support the conductors for such receptacles. Wiring methods in the cabinet must be located so as to be clear of drawers and plumbing components. Type NM cable should not be allowed exposed within any cabinet. Ideally, such receptacles would be supplied from a junction box behind the cabinet back, thereby allowing the cabinet to be replaced easily (see Section E3802.3). Receptacles must not be installed within the cabinet such that they cannot be accessed from the cabinet exterior. Some home owners have requested that the receptacles be installed within the cabinet interior and this would be a serious fire hazard because the temptation would be to leave grooming appliances plugged in while stored in the cabinet. Where a bathroom lighting fixture contains a receptacle, it does not satisfy the requirement for the bathroom wall receptacle as stated in Section E3901.1 because required receptacles are in addition to any that are part of a lighting fixture. It is possible for one receptacle outlet to serve two lavatories if it is placed between the lavatories, for example.

Receptacles must not be mounted face-up in a countertop surface because liquids and debris would enter the device.

The code recognizes receptacle assemblies that are designed for installation in countertop surfaces and listed for this application. Such assemblies consist of a device box, cover assembly and a pop-up receptacle. These assemblies are hard-wired with permanent raceway wiring methods from under the countertop.

E3901.7 Outdoor outlets. Not less than one receptacle outlet that is readily accessible from grade level and located not more than 6 feet, 6 inches (1981 mm) above grade, shall be installed outdoors at the front and back of each dwelling unit having direct access to grade level. Balconies, decks, and porches that are accessible from inside of the dwelling unit shall have at least one receptacle outlet installed within the perimeter of the balcony, deck, or porch. The receptacle shall be located not more than 6 feet, 6 inches (1981 mm) above the balcony, deck, or porch surface. [210.52(E)]

❖ Any dwelling unit that has direct access to grade requires an outdoor receptacle both at the front and back of the dwelling. A receptacle that is part of a yard light mounted on a post, for example, would not satisfy this requirement. Where a dwelling unit has a balcony or deck accessible only through a door from the inside of the house, an outdoor receptacle at the balcony or deck location may not satisfy this requirement if the receptacle is not accessible at grade level. If it is not over 6¹/₂ feet (1981 mm) from the ground and can be reached from standing on the ground, it counts as the required receptacle.

Balconies, decks and porches having or exceeding the minimum usable area and that can be accessed from within the dwelling must have a receptacle to allow the use of appliances, lighting fixtures, radios, computers and other plug loads likely to be found in such spaces. In the case of the typical deck area, the required outdoor receptacle and the required deck receptacle could be the same receptacle if all of the access conditions are met.

E3901.8 Laundry areas. Not less than one receptacle outlet shall be installed in areas designated for the installation of laundry equipment.

❖ Section E3703.3 requires at least one 20-amp-rated branch circuit for the laundry area. Typically, this circuit supplies a duplex receptacle to serve a clothes washer and a gas-fired clothes dryer, however a single (simplex) receptacle would be allowed where the clothes dryer is an electric appliance (see Section E3901.5 for receptacle location requirements). "Areas designated for the installation of laundry equipment" is not defined but is interpreted as including an ironing area within the laundry area, meaning that multiple receptacle locations could be supplied by the required laundry branch circuit. A laundry area could be a separate room or an alcove or any portion of another room. Bear in mind that a 20-amp-rated branch circuit is nearly at its full capacity when serving both a washer and dryer (see Section E3902.9 for GFCI protection).

E3901.9 Basements, garages and accessory buildings. Not less than one receptacle outlet, in addition to any provided for specific equipment, shall be installed in each separate unfinished portion of a basement, in each attached garage, and in each detached garage or accessory building that is provided with electrical power. The branch circuit supplying the receptacle(s) in a garage shall not supply outlets outside of the garage and not less than one receptacle outlet shall be installed for each motor vehicle space. [210.52(G)(1), (2), and (3)]

❖ The code does not require that a detached garage or an accessory building be supplied with electrical power. But if it has power, it must have at least one receptacle that is not dedicated for specific equipment. Because there are many uses and activities in attached garages and unfinished basements that require power, an attached garage, an unfinished basement and each separate unfinished portion of a basement must have at least one receptacle to avoid the overuse or unsafe use of extension cords. This section is not satisfied by a receptacle provided specifically for appliances or equipment, such as sump pumps, laundry appliances, water treatment equipment, security systems, overhead door operators, etc. Section E3909.1 prohibits the use of flexible cord as a substitute for fixed wiring in the dwelling and indicates that cords must not be run through holes in walls, ceilings or floors. Where receptacles are installed at appropriate locations, the receptacles will reduce the need for extension cords. In garages, at least one receptacle outlet must be installed for each vehicle space, thus a two car garage would require at least two receptacle outlets. Although not specifically required by code, the obvious intent is to locate the receptacles so that they can be accessed near each vehicle without the use of extension cords, considering the use of battery chargers, shop vacuums, trouble lights, engine block warmers, etc. Extension cords are likely to be run over by vehicles or draped across vehicles. The philosophy of the code is consistent, that is: Do not make it tempting for people to do unsafe things with electricity and extension cords.

E3901.10 Hallways. Hallways of 10 feet (3048 mm) or more in length shall have at least one receptacle outlet. The hall length shall be considered the length measured along the centerline of the hall without passing through a doorway. [210.52(H)]

❖ The length of an L-shaped hallway is also measured along the centerline and includes the total L-shaped length. The purpose for this requirement is to mandate supply power for a vacuum cleaner. This results in convenience and safety by helping avoid the use of extension cords.

E3901.11 Foyers. Foyers that are not part of a hallway in accordance with Section E3901.10 and that have an area that is greater than 60 ft² (5.57 m²) shall have a receptacle(s) located in each wall space that is 3 feet (914 mm) or more in width. Doorways, door-side windows that extend to the floor, and similar openings shall not be considered as wall space. [210.52(H)]

❖ The code now addresses foyers, which are not covered in Section E3901.2. Foyers can be quite large and receptacles are necessary to serve lamps, cleaning and maintenance appliances, holiday lighting, etc. The spacing requirements are less stringent than they are for the spaces covered by Section E3901.2. Doors, door side-lights that extend to the floor and windows that extend to the floor do not count as wall space under this section (see Section E3901.2.2 for comparison).

E3901.12 HVAC outlet. A 125-volt, single-phase, 15- or 20-ampere-rated receptacle outlet shall be installed at an accessible location for the servicing of heating, air-conditioning and refrigeration equipment. The receptacle shall be located on the same level and within 25 feet (7620 mm) of the heating, air-conditioning and refrigeration equipment. The receptacle outlet shall not be connected to the load side of the HVAC equipment disconnecting means. (210.63)

Exception: A receptacle outlet shall not be required for the servicing of evaporative coolers. (210.63 Exception)

❖ When servicing or repairing HVAC equipment, a technician usually needs electrically powered tools, instruments or equipment. The intent of this section is to avoid the necessity of a technician dragging an extension cord through the attic or crawl space or across a roof.

This applies to both indoor and outdoor equipment. Evaporative coolers are exempt because power tools, diagnostic instruments, trouble lights and vacuum equipment are typically not needed to service them. Evaporative coolers are typical only in dry climates and are sometimes referred to as "swamp coolers."

SECTION E3902
GROUND-FAULT AND ARC-FAULT CIRCUIT-INTERRUPTER PROTECTION

E3902.1 Bathroom receptacles. 125-volt, single-phase, 15- and 20-ampere receptacles installed in bathrooms shall have ground-fault circuit-interrupter protection for personnel. [210.8(A)(1)]

❖ There are many unique layouts for a dwelling unit bath-room. This section and the requirement for bath-room receptacles and GFCI protection are the reasons that there is a code definition of bathroom. A bathroom is not necessarily a room. It is an area that includes a wash basin (lavatory) and also has a water closet, a tub or a shower. The water closet, tub or shower could be in a separate room from the wash basin, and the area would still be defined as a bathroom. There is no exception to the requirement for GFCI protection of receptacles in bathrooms. All receptacles would include, for example, a receptacle in a light fixture, a receptacle for a clothes washer, or a receptacle for any other appliance installed in a bathroom (see Commentary Figure E3902.1).

E3902.2 Garage and accessory building receptacles. 125-volt, single-phase, 15- or 20-ampere receptacles installed in garages and grade-level portions of unfinished accessory buildings used for storage or work areas shall have ground-fault circuit-interrupter protection for personnel. [210.8(A)(2)]

❖ A receptacle in an accessory building, an implement shed or even a storage shed also requires GFCI pro-

tection. Sometimes an appliance such as a food freezer or refrigerator is located in a garage, and where such an appliance is plugged into a GFCI receptacle, nuisance tripping of the GFCI device could cause food to spoil. Therefore, in the past, it was common to have some appliances supplied by non-GFCI protected circuits where the appliances and the receptacles that supplied them were in a space dedicated for the cord- and plug-connected appliances and such appliances were "not easily moved," however, this practice is no longer allowed. The concern for nuisance tripping is not valid for modern appliances that have far less leakage currents than their predecessors. If an appliance such as older model refrigerator trips a GFCI device, it proves that the appliance has high leakage current and should be retired and recycled. The code believes that there is no valid excuse for not having GFCI protection in any location where the occupants could be at risk of electrical shock. There are no exceptions to this section's requirements.

Note that electric utility companies often claim that it is energy wasteful to relocate old inefficient freezers and refrigerators to the garage where they will continue to waste energy in their new location and also labor under extreme temperature conditions.

E3902.3 Outdoor receptacles. 125-volt, single-phase, 15- and 20-ampere receptacles installed outdoors shall have ground-fault circuit-interrupter protection for personnel. [210.8(A)(3)]

Exception: Receptacles as covered in Section E4101.7. [210.8(A)(3) Exception]

❖ The only outdoor receptacles that are exempt from GFCI protection are those located in the eave of the dwelling for cord- and plug-connected deicing equipment. These receptacle outlets are not readily accessible, are supplied from a dedicated circuit especially for the deicing equipment or cable and are usually sep-

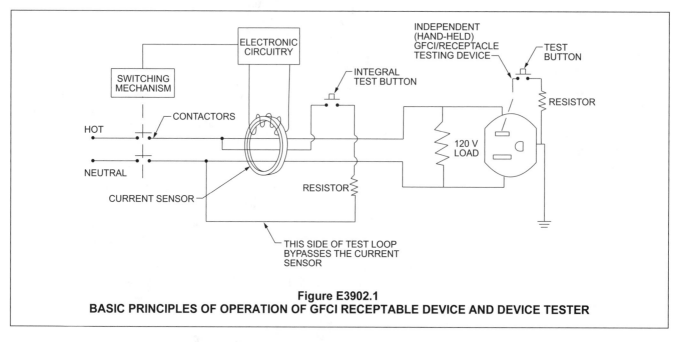

Figure E3902.1
BASIC PRINCIPLES OF OPERATION OF GFCI RECEPTABLE DEVICE AND DEVICE TESTER

arate from eave receptacles that would be used for decorative holiday lighting.

E3902.4 Crawl space receptacles. Where a crawl space is at or below grade level, 125-volt, single-phase, 15- and 20-ampere receptacles installed in such spaces shall have ground-fault circuit-interrupter protection for personnel. [210.8(A)(4)]

❖ This requirement is for GFCI protection of a crawl space receptacle outlet. According to Section E3901.12, a crawl space is not required to have a receptacle outlet unless there is equipment such as HVAC appliances that require servicing located there. Plumbing pipes and electrical cable run in the crawl space area are not considered to be equipment that would need servicing, so a receptacle would not be required because of the piping or wiring. However, if the choice is made to install a crawl space receptacle, it must be GFCI protected.

E3902.5 Unfinished basement receptacles. 125-volt, single-phase, 15- and 20-ampere receptacles installed in unfinished basements shall have ground-fault circuit-interrupter protection for personnel. For purposes of this section, unfinished basements are defined as portions or areas of the basement not intended as habitable rooms and limited to storage areas, work areas, and similar areas. [210.8(A)(5)]

> **Exception:** A receptacle supplying only a permanently installed fire alarm or burglar alarm system. Receptacles installed in accordance with this exception shall not be considered as meeting the requirement of Section E3901.9. [210.8(A)(5) Exception]

❖ Masonry and concrete can provide a conductive path to earth. Bare concrete or masonry walls might also be subject to moisture. Judgment must enter into the decision of whether to consider a basement space as finished or unfinished. Clearly the intent is to protect the occupants from the potential shock hazard associated with using electricity in a location where they are in contact with conductive surfaces that provide an electrical path to the earth. Concrete floors, walls and bare earth are of particular concern. It is the author's opinion that painted concrete walls and floors do not make a basement "finished" and because the electrical shock hazard is not mitigated, GFCI protection would be required. Basements with carpet and padding and furred and paneled walls lessen the concern for exposed conductive surfaces. Considering that GFCI protection is very inexpensive, complete protection for all areas of the basement will eliminate the need to determine what is or is not "finished."

A single receptacle (not a duplex) supplying a plug-in transformer for an alarm system is exempt because the receptacle is occupied, and nuisance tripping could disable the alarm system. This exception states that a receptacle for an alarm system does not satisfy the requirement of Section E3901.9, which is more appropriately the subject matter of Section E3901.9 rather than this section. It is understood, however, that the intent of this section and Section E3901.9 is to not allow an alarm system receptacle to count as the receptacle(s) required for basements, garages and accessory buildings. The alarm receptacle is dedicated for that purpose, therefore, another convenience receptacle must be installed for general purpose use. Besides, the alarm receptacle will be a single, occupied receptacle in order to avoid being GFCI protected, which is the entire premise of the exception.

E3902.6 Kitchen receptacles. 125-volt, single-phase, 15- and 20-ampere receptacles that serve countertop surfaces shall have ground-fault circuit-interrupter protection for personnel. [210.8(A)(6)]

❖ In a kitchen, all receptacles serving any countertop must be GFCI protected. In the past, the only kitchen countertop receptacles that required GFCI protection were the ones close to the kitchen sink. Now, any countertop receptacle, even in an area far away from the sink, is required to have GFCI protection.

E3902.7 Sink receptacles. 125-volt, single-phase, 15- and 20-ampere receptacles that are located within 6 feet (1829 mm) of the outside edge of a sink shall have ground-fault circuit-interrupter protection for personnel. Receptacle outlets shall not be installed in a face-up position in the work surfaces or countertops. [210.8(A)(7)]

❖ This requirement is different from the kitchen countertop receptacles in that only receptacles within 6 feet (1829 mm) measured in any direction from the sink must be GFCI protected. With regards to how the 6-foot (1829 mm) criterion is measured, the term "within" is defined as inside the limits of a distance; therefore, it could be argued that a receptacle that is exactly 6 feet (1829 mm) from a sink is exempt because it is not "within" the 6 foot (1829 mm) distance. The opposing view is that the receptacle would be exempt only if it was over 6 feet (1829 mm) from the sink. Note that this section is applicable to laundry sinks (tubs), utility sinks and bar sinks and is applicable to all receptacles, including those provided for laundry appliances.

E3902.8 Bathtub or shower stall receptacles. 125-volt, single phase, 15- and 20-ampere receptacles that are located within 6 feet (1829 mm) of the outside edge of a bathtub or shower stall shall have ground-fault circuit interrupter protection for personnel. [210.8(A)(8)]

❖ Section E4002.11 prohibits receptacles within bathtub and shower spaces, but receptacles are allowed to be outside of those spaces, with no minimum distance requirement. This section requires GFCI protection for receptacles that are less than 6 feet from the edge of the bathtub or shower space. A user of a shower or bathtub could reach a receptacle that is less than 6 feet from the edge of the shower or bathtub, thus, a shock hazard exists.

E3902.9 Laundry areas. 125-volt, single-phase, 15- and 20-ampere receptacles installed in laundry areas shall have ground-fault circuit interrupter protection for personnel. [210.8(A)(9)]

❖ Laundry areas provide opportunities for occupants to contact a grounded conducting surface, grounded water piping, grounded appliance housings, etc. A potential hazard exists where occupants put themselves between a voltage source and a grounded object or surface. See Section E3901.8.

E3902.10 Kitchen dishwasher branch circuit. Ground-fault circuit-interrupter protection shall be provided for outlets that supply dishwashers in dwelling unit locations. [210.8(D)]

❖ The kitchen provides the opportunity for occupants to place themselves between a voltage source and a grounded surface or object creating a shock hazard. The branch circuits that supply dishwashers must have GFCI protection whether the appliance is hardwired or cord- and plug-connected. This is accomplished by means of a GFCI-type circuit breaker or a GFCI receptacle device. Dishwashers are commonly supplied by a dedicated branch circuit and the appliance is commonly fitted with a cord and plug and plugged into a receptacle mounted in the back of an adjacent kitchen sink base cabinet. Such an arrangement allows the dishwasher to be easily slid in and out of the space in which it is installed and satisfies the disconnect requirement of Table E4101.5 (see Section E4101.3). It would be good practice to provide a single (not duplex) receptacle for the dishwasher so as to discourage occupants from using an open receptacle mounted inside of a cabinet.

E3902.11 Boathouse receptacles. 125-volt, single-phase, 15- or 20-ampere receptacles installed in boathouses shall have ground-fault circuit-interrupter protection for personnel. [210.8(A)(8)]

❖ The potential shock hazards in buildings having unfinished concrete floors on grade and wet surfaces are comparable to shock hazards outdoors and in kitchens, bathrooms, garages, basements and workshops.

E3902.12 Boat hoists. Ground-fault circuit-interrupter protection for personnel shall be provided for 240-volt and less outlets that supply boat hoists. [210.8(C)]

❖ Generally, boat hoists are permanently installed on a boat dock and used to lift a boat out of the water.

E3902.13 Electrically heated floors. Ground-fault circuit-interrupter protection for personnel shall be provided for electrically heated floors in bathrooms, kitchens and in hydromassage bathtub, spa and hot tub locations. [424.44(G)]

❖ Heat-producing cables and grids can be embedded in concrete floors to warm the floor for occupant comfort. The possibility of current leakage combined with wet surfaces and grounded plumbing fixtures justifies this requirement; the branch circuit overcurrent device serving the heating cables must provide GFCI protection.

E3902.14 Location of ground-fault circuit interrupters. Ground-fault circuit interrupters shall be installed in a readily accessible location. [210.8(A)]

❖ See the definition of "Accessible, readily."

E3902.15 Location of arc-fault circuit interrupters. Arc-fault circuit interrupters shall be installed in readily accessible locations.

❖ See the definition of "Accessible, readily."

E3902.16 Arc-fault circuit-interrupter protection. Branch circuits that supply 120-volt, single-phase, 15- and 20-ampere outlets installed in kitchens, family rooms, dining rooms, living rooms, parlors, libraries, dens, bedrooms, sunrooms, recreations rooms, closets, hallways, laundry areas and similar rooms or areas shall be protected by any of the following: [210.12(A)]

1. A listed combination-type arc-fault circuit interrupter, installed to provide protection of the entire branch circuit. [210.12(A)(1)]

2. A listed branch/feeder-type AFCI installed at the origin of the branch-circuit in combination with a listed outlet branch-circuit type arc-fault circuit interrupter installed at the first outlet box on the branch circuit. The first outlet box in the branch circuit shall be marked to indicate that it is the first outlet of the circuit. [210.12(A)(2)]

3. A listed supplemental arc protection circuit breaker installed at the origin of the branch circuit in combination with a listed outlet branch-circuit type arc-fault circuit interrupter installed at the first outlet box on the branch circuit where all of the following conditions are met:

   3.1. The branch-circuit wiring shall be continuous from the branch-circuit overcurrent device to the outlet branch-circuit arc-fault circuit interrupter.

   3.2. The maximum length of the branch-circuit wiring from the branch-circuit overcurrent device to the first outlet shall not exceed 50 feet (15.2 m) for 14 AWG conductors and 70 feet (21.3 m) for 12 AWG conductors.

   3.3. The first outlet box on the branch circuit shall be marked to indicate that it is the first outlet on the circuit. [210.12(A)(3)]

4. A listed outlet branch-circuit type arc-fault circuit interrupter installed at the first outlet on the branch circuit in combination with a listed branch-circuit overcurrent protective device where all of the following conditions are met:

   4.1. The branch-circuit wiring shall be continuous from the branch-circuit overcurrent device to the outlet branch-circuit arc-fault circuit interrupter.

   4.2. The maximum length of the branch-circuit wiring from the branch-circuit overcurrent device to the first outlet shall not exceed 50 feet (15.2 m) for 14 AWG conductors and 70 feet (21.3 m) for 12 AWG conductors.

   4.3. The first outlet box on the branch circuit shall be marked to indicate that it is the first outlet on the circuit.

   4.4. The combination of the branch-circuit overcurrent device and outlet branch-circuit AFCI shall be identified as meeting the requirements for a system combination-type AFCI and shall be listed as such. [210.12(A)(4)]

5. Where metal outlet boxes and junction boxes and RMC, IMC, EMT, Type MC or steel-armored Type AC cables meeting the requirements of Section E3908.8, metal

wireways or metal auxiliary gutters are installed for the portion of the branch circuit between the branch-circuit overcurrent device and the first outlet, a listed outlet branch-circuit type AFCI installed at the first outlet shall be considered as providing protection for the remaining portion of the branch circuit. [210.12(A)(5)]

6. Where a listed metal or nonmetallic conduit or tubing or Type MC cable is encased in not less than 2 inches (50.8 mm) of concrete for the portion of the branch circuit between the branch-circuit overcurrent device and the first outlet, a listed outlet branch-circuit type AFCI installed at the first outlet shall be considered as providing protection for the remaining portion of the branch circuit. [210.12(A)(6)]

Exception: AFCI protection is not required for an individual branch circuit supplying only a fire alarm system where the branch circuit is wired with metal outlet and junction boxes and RMC, IMC, EMT or steel-sheathed armored cable Type AC or Type MC meeting the requirements of Section E3908.8.

❖ This section requires that AFCIs be installed to protect 15- and 20-ampere 120-volt branch circuits that supply outlets installed in the listed dwelling unit rooms. There are two distinct provisions of this requirement to keep in mind (see Commentary Figure E3902.16). First, the section refers to "outlets," which by definition, includes receptacle outlets, lighting outlets and other outlets such as smoke alarms (see Section E3501 for the definition of "Outlet"). The second provision is that the protection is for the "branch circuit" and not just the outlet itself. A branch circuit begins at the last overcurrent device (circuit breaker) and runs to the outlet (see the definition of "Branch circuit" in Section E3401).

AFCIs combine a standard circuit breaker with electronic circuitry that can detect the characteristics of hazardous arcing in an ac circuit [see the commentary to the definition of "Arc-fault circuit interrupter (AFCI)"]. An arc will produce some unique characteristics including high frequency noise, voltage spikes on the waveform and fluctuating arcing current lower than normal current. These and other factors are analyzed by the AFCI to determine whether an arc is present and whether it is an arc that should be interrupted. The AFCI device is designed to ignore normal arcing, such as when switches open and close and when plugs are inserted or removed from receptacles.

Bedrooms were chosen as the initial location in which to introduce AFCI protection requirements. Because arcing can occur in any circuit in a dwelling, the code has greatly expanded the requirement for AFCI protection of 120-volt branch circuits. In fact, the expanded list of rooms and spaces appears to include everything but bathrooms, unfinished basement spaces, garages and outdoors.

Section R314 requires smoke alarm outlets in bedrooms, and such outlets must also be on an AFCI circuit. These smoke-alarm outlets must be interconnected and would conveniently be served by the same branch circuit. Although not specifically required by this code, smoke-alarm circuits powered by a branch circuit that also supplies lighting will be monitored because power loss to the smoke alarms will also mean loss of power to the lighting. In such cases, the occupant will restore power to the alarms in their effort to restore lighting. It is also recognized that there are requirements for smoke alarms to have a secondary power source should the primary power source be interrupted for any reason.

Branch/feeder-type AFCIs are designed to detect and clear (deenergize the circuit) higher energy parallel arc faults of about 75 amperes and higher for line-to-neutral faults and low energy parallel faults of 5 amperes and higher for line-to-ground faults. Many AFCIs on the market today can detect line-to-ground faults at currents as low as 50 milliamperes. Series arcing involves much lower current flows and, depending on the level, might not be detected by branch/feeder-type AFCIs.

A new requirement of this section is that the AFCI device must be the combination type. A combination AFCI integrates the arc detection of a branch/feeder AFCI with low level series arc detection (as low as 5 amperes). One point of confusion to avoid is that a "combination AFCI" is NOT a combination AFCI/GFCI device. UL 1699 is the applicable standard for AFCIs and can provide more detail on the types of AFCIs. Even though AFCI circuit breaker devices and GFCI circuit breaker devices look similar and are installed in a similar manner, they are very different devices with different functions and they cannot be substituted for one another to satisfy code requirements. Devices that

Photo courtesy of Square D Company.

Figure E3902.6
AFCI TYPE CIRCUIT BREAKER

combine a circuit breaker, a GFCI and an AFCI all in one package may be available in the marketplace now or in the future.

Branch/feeder-type AFCI devices (typically in the form of a circuit breaker) protect the branch circuit wiring primarily from parallel arcing between conductors and provide some limited protection for branch circuit extension wiring such as cord sets and appliance cords. Outlet branch circuit-type AFCI devices are generally for installation in the first outlet box downstream of the branch circuit overcurrent device and provide protection from series arcing such as would occur in a broken or severed conductor in an appliance or fixture cord, but also provide protection of upstream and downstream branch circuit conductors. A combination AFCI device incorporates both kinds of protection.

The UL product categories are as follows for the three types of AFCI devices addressed in this code section.

AVZQ	Arc Fault Circuit Interrupters, Branch/Feeder Type
AWAH	Arc Fault Circuit Interrupters, Combination Type
AWBZ	Arc Fault Circuit Interrupters, Outlet Branch Circuit Type

This section allows an AFCI to be located outside of the distribution panelboard. This provision handles unique installations where an existing system has circuits being added that require AFCI protection. Where the older service panelboard or sub-panelboard is fusible or of an older type for which an AFCI device is not available, this provision will allow locating an AFCI outside of the panelboard in a separate enclosure. Outlets supplied by this branch circuit cannot occur upstream of the AFCI device. The conductors between the existing panelboard and the AFCI must have the additional protection from physical damage because they are upstream of the AFCI and, therefore, are not protected from arcing faults.

Items 1 through 6 provide the options for the required protection. Option 1 simply requires a combination-type AFCI that would be installed in the panelboard at the source of the branch circuit. This option is the most simple and likely the most economical approach and would be chosen for new panelboards in new construction.

Option 2 requires a branch/feeder-type AFCI device to protect the circuit conductors between the panelboard and the first outlet box on the circuit. It also requires an outlet branch circuit AFCI device in the first outlet box. Note that the code says "first outlet box," not the first receptacle outlet. The first outlet box on a circuit may or may not be a receptacle outlet box. The branch circuit conductors are not required to have any means of physical protection above and beyond what is otherwise required by the code. An outlet branch circuit AFCI device is typically a receptacle device or a blind device designed for installation in a receptacle box.

Option 3 requires a standard AFCI circuit breaker (older product line; not a combination type) supplemented by an outlet branch circuit-type AFCI device. Because the AFCI device in the panelboard is the less capable type, the code requires that there be no splices in wiring between the two AFCI devices and limits the conductor lengths.

Option 4 requires an outlet branch circuit AFCI device as the only AFCI device provided that such device in combination with the circuit breaker is listed as meeting all of the requirements for a system combination-type AFCI. This appears to be addressing devices made by the same manufacturer and the manufacturer has submitted the device combination to a testing agency to determine if the device combination meets the applicable criteria.

Option 5 requires only an outlet branch circuit-type AFCI device and requires exceptional physical protection of the branch circuit conductors between the overcurrent device and the first outlet. This item could apply in cases where AFCI circuit breakers are not available for the type of existing panelboard, and in such cases, a new sub-panelboard could be placed near the service panelboard. The new sub-panelboard would be supplied by a feeder protected by the wiring methods specified in this item and the required AFCI devices would be installed in the new sub-panelboard.

Option 6 is similar in approach to Option 5.

The exception to this code section exempts dedicated branch circuits for fire alarm systems where the circuit conductors are provided with exceptional physical protection.

E3902.17 Arc-fault circuit interrupter protection for branch circuit extensions or modifications. Where branch-circuit wiring is modified, replaced, or extended in any of the areas specified in Section E3902.16, the branch circuit shall be protected by one of the following:

1. A combination-type AFCI located at the origin of the branch circuit

2. An outlet branch-circuit type AFCI located at the first receptacle outlet of the existing branch circuit. [210.12(B)]

Exception: AFCI protection shall not be required where the extension of the existing conductors is not more than 6 feet (1.8 m) in length and does not include any additional outlets or devices. [210.12(B) Exception]

❖ This section addresses existing buildings and will require AFCI protection where wiring is replaced, altered and extended. For example, if an existing branch circuit serving receptacle and/or lighting outlets is extended to serve a new receptacle, that existing circuit must be protected at its origin or such protection must be provided by an outlet-type AFCI device installed at the first outlet so as to protect all downstream outlets. Any work on branch circuit wiring that would require a permit would trigger the requirement for AFCI protection of the affected branch circuit wiring. The exception applies where the branch circuit extension is very short in length and does not involve any new outlets or devices, thus, the amount of wiring subject to faults is negligible.

SECTION E3903
LIGHTING OUTLETS

E3903.1 General. Lighting outlets shall be provided in accordance with Sections E3903.2 through E3903.4. [210.70(A)]

❖ See the commentary to Sections E3903.2 and E3903.4.

E3903.2 Habitable rooms. At least one wall switch-controlled lighting outlet shall be installed in every habitable room and bathroom. [210.70(A)(1)]

Exceptions:

1. In other than kitchens and bathrooms, one or more receptacles controlled by a wall switch shall be considered equivalent to the required lighting outlet. [210.70(A)(1) Exception No. 1]

2. Lighting outlets shall be permitted to be controlled by occupancy sensors that are in addition to wall switches, or that are located at a customary wall switch location and equipped with a manual override that will allow the sensor to function as a wall switch. [210.70(A)(1) Exception No. 2]

❖ According to Chapter 2, a space for living, sleeping, eating or cooking is a habitable space and does not include a bathroom. Therefore, a bathroom is specifically referred to as an area where a wall-switch-controlled lighting outlet is required. In all rooms other than kitchens and bathrooms, the lighting outlet may be a switched receptacle. Switched receptacles serve floor and table lamps and such lamps are not found in and are not appropriate for kitchens and bathrooms.

It seems unlikely that the dining room lighting outlet would be a switched receptacle. However, if this is the case, such a receptacle must not be supplied by one of the small appliance branch circuits. This is clear from the exception to Section E3901.3, which allows for a switched receptacle to be supplied from a general purpose branch circuit. Thus, all of the receptacles in a dining room must be supplied by a small appliance circuit except the one that is switched and considered to be the lighting outlet (see commentary, Section E3901.1).

E3903.3 Additional locations. At least one wall-switch-controlled lighting outlet shall be installed in hallways, stairways, attached garages, and detached garages with electric power. At least one wall-switch-controlled lighting outlet shall be installed to provide illumination on the exterior side of each outdoor egress door having grade level access, including outdoor egress doors for attached garages and detached garages with electric power. A vehicle door in a garage shall not be considered as an outdoor egress door. Where one or more lighting outlets are installed for interior stairways, there shall be a wall switch at each floor level and landing level that includes an entryway to control the lighting outlets where the stairway between floor levels has six or more risers. [210.70(A)(2)]

Exception: In hallways, stairways, and at outdoor egress doors, remote, central, or automatic control of lighting shall be permitted. [210.70(A)(2) Exception]

❖ Additional areas that are not habitable and would not be included in Section E3903.2 are hallways, stairways and garages. There is no requirement for a three-way switch at each end of a hallway or, for that matter, in any room. The only requirement for three-way switches or four-way switches (where a lighting outlet is controlled from two or three separate switches at different locations) is at a stairway that has at least six risers. Installing three-way and four-way switches at separate entrances of a large room does provide for convenience and also an added measure of safety, but it is not required by the code (see Section R303.6.1). Where intermediate landings occur in stairways and those landings are served by egress doors, a wall switch must also be installed on such landings.

E3903.4 Storage or equipment spaces. In attics, under-floor spaces, utility rooms and basements, at least one lighting outlet shall be installed where these spaces are used for storage or contain equipment requiring servicing. Such lighting outlet shall be controlled by a wall switch or shall have an integral switch. At least one point of control shall be at the usual point of entry to these spaces. The lighting outlet shall be provided at or near the equipment requiring servicing. [210.70(A)(3)]

❖ In attics or crawl spaces that are used for storage or have equipment that requires servicing, a lighting outlet is required with a switch at the point of entrance. If an attic or crawl space is not used for storage and contains no equipment, no lighting outlet is required. A utility room and a basement would have equipment and/or be used for storage and would require a lighting outlet. Similar requirements are found in Section M1305. The required lighting outlet is for general illumination to allow the space to be safely navigated and is not intended to provide all of the illumination necessary for servicing of equipment. Such equipment includes furnaces, boilers, water heaters, well system components, exhaust fans, etc.

SECTION E3904
GENERAL INSTALLATION REQUIREMENTS

E3904.1 Electrical continuity of metal raceways and enclosures. Metal raceways, cable armor and other metal enclosures for conductors shall be mechanically joined together into a continuous electric conductor and shall be connected to all boxes, fittings and cabinets so as to provide effective electrical continuity. Raceways and cable assemblies shall be mechanically secured to boxes, fittings cabinets and other enclosures. (300.10)

Exception: Short sections of raceway used to provide cable assemblies with support or protection against physical damage. (300.10 Exception No. 1)

❖ Where armored cable is used, the armor must be continuous to be able to serve as a conductor to carry fault current. In a raceway system, the metal raceway must serve as a conductor to carry fault current. Where the mechanical connection at a box or fitting is not tight, high resistance will occur and prevent the flow of fault current necessary to trip the overcurrent protective device. Poor connections can also cause arcing under fault conditions creating a fire hazard. Good workmanship is essential to providing raceway electrical continuity. It is not uncommon to find raceways such as conduit entering a box on an angle such that the connector and its locknut contact the box at only a single point, thus making a poor connection that has poor continuity and that is likely to work loose. Proper workmanship must be enforced to achieve the intent of this section.

E3904.2 Mechanical continuity—raceways and cables. Metal or nonmetallic raceways, cable armors and cable sheaths shall be continuous between cabinets, boxes, fittings or other enclosures or outlets.

Exception: Short sections of raceway used to provide cable assemblies with support or protection against physical damage. (300.12 Exception No. 1)

❖ Cable is permitted to be protected from physical damage by running it through a section of raceway (see Section E3904.1).

E3904.3 Securing and supporting. Raceways, cable assemblies, boxes, cabinets and fittings shall be securely fastened in place. (300.11)

❖ Raceways and cables are wiring methods covered in Chapter 38. The supporting requirements are covered in Table E3802.1. Boxes and cabinets must be securely supported according to provisions in Section E3906.8. Detailed requirements are included for support of boxes.

E3904.3.1 Prohibited means of support. Cable wiring methods shall not be used as a means of support for other cables, raceways and nonelectrical equipment. [300.11(C)]

❖ Cables, raceways and electrical and nonelectrical equipment must not be supported by other cables because this would be poor support at best and could cause damage to the cables that are used for supports. This could also be a violation of Section E3802.3.1.

E3904.4 Raceways as means of support. Raceways shall be used as a means of support for other raceways, cables or non-electric equipment only under the following conditions:

1. Where the raceway or means of support is identified as a means of support; or

2. Where the raceway contains power supply conductors for electrically controlled equipment and is used to support Class 2 circuit conductors or cables that are solely for the purpose of connection to the control circuits of the equipment served by such raceway; or

3. Where the raceway is used to support boxes or conduit bodies in accordance with Sections E3906.8.4 and E3906.8.5. [300.11(B)]

❖ The use of raceways to support other raceways, cables or wires is permitted only under the limited circumstances listed. Where wires, cables and other raceways are "piggybacked" on other raceways, the additional weight and tension forces could strain the host raceway and its supports. Other installers tend to attach their wiring to whatever is conveniently located where they are working and this practice can get out of hand.

A thermostat wire for the control of heating and air conditioning may be supported by the raceway supplying power to the equipment. Control wiring is typically supported by the conduit or raceway that supplies furnaces, boilers and condensing units. Where cables such as telephone wiring, IT cables and cable TV wiring are supported by a conduit or electrical metallic tubing, the cables could prevent the dissipation of heat produced inside the raceway and also add additional stresses from the weight of the wires and cables.

E3904.5 Raceway installations. Raceways shall be installed complete between outlet, junction or splicing points prior to the installation of conductors. (300.18)

Exception: Short sections of raceways used to contain conductors or cable assemblies for protection from physical damage shall not be required to be installed complete between outlet, junction, or splicing points. (300.18 Exception)

❖ The proper way to install wiring in raceways is to assemble the complete raceway system first and properly support it. After the entire system is in place and secured, the wire can be pulled into the raceway with less risk of damaging it. Trying to install the wire piecemeal as the conduit is installed will likely result in damage to the conductor insulation.

E3904.6 Conduit and tubing fill. The maximum number of conductors installed in conduit or tubing shall be in accordance with Tables E3904.6(1) through E3904.6(10). (300.17, Chapter 9, Table 1 and Annex C)

❖ The code includes 10 tables to indicate how many conductors of the various types and sizes are allowed in the various sizes and types of raceway.

TABLES E3904.6(1) through E3904.6(1). See page 39-15.

❖ See the commentary to Section E3904.6.

TABLE E3904.6(1) (Annex C, Table C.1)
MAXIMUM NUMBER OF CONDUCTORS IN ELECTRICAL METALLIC TUBING (EMT)[a]

TYPE LETTERS	CONDUCTOR SIZE AWG/kcmil	TRADE SIZES (inches)					
		1/2	3/4	1	1 1/4	1 1/2	2
RHH, RHW, RHW-2	14	4	7	11	20	27	46
	12	3	6	9	17	23	38
	10	2	5	8	13	18	30
	8	1	2	4	7	9	16
	6	1	1	3	5	8	13
	4	1	1	2	4	6	10
	3	1	1	1	4	5	9
	2	1	1	1	3	4	7
	1	0	1	1	1	3	5
	1/0	0	1	1	1	2	4
	2/0	0	1	1	1	2	4
	3/0	0	0	1	1	1	3
	4/0	0	0	1	1	1	3
TW, THHW, THW, THW-2	14	8	15	25	43	58	96
	12	6	11	19	33	45	74
	10	5	8	14	24	33	55
	8	2	5	8	13	18	30
RHH[a], RHW[a], RHW-2[a]	14	6	10	16	28	39	64
	12	4	8	13	23	31	51
	10	3	6	10	18	24	40
	8	1	4	6	10	14	24
RHH[a], RHW[a], RHW-2[a], TW, THW, THHW, THW-2	6	1	3	4	8	11	18
	4	1	1	3	6	8	13
	3	1	1	3	5	7	12
	2	1	1	2	4	6	10
	1	1	1	1	3	4	7
	1/0	0	1	1	2	3	6
	2/0	0	1	1	1	3	5
	3/0	0	1	1	1	2	4
	4/0	0	0	1	1	1	3
THHN, THWN, THWN-2	14	12	22	35	61	84	138
	12	9	16	26	45	61	101
	10	5	10	16	28	38	63
	8	3	6	9	16	22	36
	6	2	4	7	12	16	26
	4	1	2	4	7	10	16
	3	1	1	3	6	8	13
	2	1	1	3	5	7	11
	1	1	1	1	4	5	8
	1/0	1	1	1	3	4	7
	2/0	0	1	1	2	3	6
	3/0	0	1	1	1	3	5
	4/0	0	1	1	1	2	4
XHH, XHHW, XHHW-2	14	8	15	25	43	58	96
	12	6	11	19	33	45	74
	10	5	8	14	24	33	55
	8	2	5	8	13	18	30
	6	1	3	6	10	14	22
	4	1	2	4	7	10	16
	3	1	1	3	6	8	14
	2	1	1	3	5	7	11
	1	1	1	1	4	5	8
	1/0	1	1	1	3	4	7
	2/0	0	1	1	2	3	6
	3/0	0	1	1	1	3	5
	4/0	0	1	1	1	2	4

For SI: 1 inch = 25.4 mm.

a. Types RHW, and RHW-2 without outer covering.

TABLE E3904.6(2) (Annex C, Table C.2)
MAXIMUM NUMBER OF CONDUCTORS IN ELECTRICAL NONMETALLIC TUBING (ENT)[a]

TYPE LETTERS	CONDUCTOR SIZE AWG/kcmil	TRADE SIZES (inches)					
		$^1/_2$	$^3/_4$	1	$1^1/_4$	$1^1/_2$	2
RHH, RHW, RHW-2	14	3	6	10	19	26	43
	12	2	5	9	16	22	36
	10	1	4	7	13	17	29
	8	1	1	3	6	9	15
	6	1	1	3	5	7	12
	4	1	1	2	4	6	9
	3	1	1	1	3	5	8
	2	0	1	1	3	4	7
	1	0	1	1	1	3	5
	1/0	0	0	1	1	2	4
	2/0	0	0	1	1	1	3
RHH, RHW, RHW-2	3/0	0	0	1	1	1	3
	4/0	0	0	1	1	1	2
TW, THHW, THW, THW-2	14	7	13	22	40	55	92
	12	5	10	17	31	42	71
	10	4	7	13	23	32	52
	8	1	4	7	13	17	29
RHH[a], RHW[a], RHW-2[a]	14	4	8	15	27	37	61
	12	3	7	12	21	29	49
	10	3	5	9	17	23	38
	8	1	3	5	10	14	23
RHH[a], RHW[a], RHW-2[a], TW, THW, THHW, THW-2	6	1	2	4	7	10	17
	4	1	1	3	5	8	13
	3	1	1	2	5	7	11
	2	1	1	2	4	6	9
	1	0	1	1	3	4	6
	1/0	0	1	1	2	3	5
	2/0	0	1	1	1	3	5
	3/0	0	0	1	1	2	4
	4/0	0	0	1	1	1	3
THHN, THWN, THWN-2	14	10	18	32	58	80	132
	12	7	13	23	42	58	96
	10	4	8	15	26	36	60
	8	2	5	8	15	21	35
	6	1	3	6	11	15	25
	4	1	1	4	7	9	15
	3	1	1	3	5	8	13
	2	1	1	2	5	6	11
THHN, THWN, THWN-2	1	1	1	1	3	5	8
	1/0	0	1	1	3	4	7
	2/0	0	1	1	2	3	5
	3/0	0	1	1	1	3	4
	4/0	0	0	1	1	2	4
XHH, XHHW, XHHW-2	14	7	13	22	40	55	92
	12	5	10	17	31	42	71
	10	4	7	13	23	32	52
	8	1	4	7	13	17	29
	6	1	3	5	9	13	21
	4	1	1	4	7	9	15
	3	1	1	3	6	8	13
	2	1	1	2	5	6	11
	1	1	1	1	3	5	8
	1/0	0	1	1	3	4	7
	2/0	0	1	1	2	3	6
	3/0	0	1	1	1	3	5
	4/0	0	0	1	1	2	4

For SI: 1 inch = 25.4 mm.

a. Types RHW, and RHW-2 without outer covering.

TABLE E3904.6(3) (Annex C, Table C.3)
MAXIMUM NUMBER OF CONDUCTORS IN FLEXIBLE METALLIC CONDUIT (FMC)[a]

TYPE LETTERS	CONDUCTOR SIZE AWG/kcmil	TRADE SIZES (inches)					
		½	¾	1	1¼	1½	2
RHH, RHW, RHW-2	14	4	7	11	17	25	44
	12	3	6	9	14	21	37
	10	3	5	7	11	17	30
	8	1	2	4	6	9	15
	6	1	1	3	5	7	12
	4	1	1	2	4	5	10
	3	1	1	1	3	5	7
RHH, RHW, RHW-2	2	1	1	1	3	4	7
	1	0	1	1	1	2	5
	1/0	0	1	1	1	2	4
	2/0	0	1	1	1	1	3
	3/0	0	0	1	1	1	3
TW, THHW, THW, THW-2	14	9	15	23	36	53	94
	12	7	11	18	28	41	72
	10	5	8	13	21	30	54
	8	3	5	7	11	17	30
RHH[a], RHW[a], RHW-2[a]	14	6	10	15	24	35	62
	12	5	8	12	19	28	50
	10	4	6	10	15	22	39
	8	1	4	6	9	13	23
RHH[a], RHW[a], RHW-2[a], TW, THW, THHW, THW-2	6	1	3	4	7	10	18
	4	1	1	3	5	7	13
	3	1	1	3	4	6	11
	2	1	1	2	4	5	10
	1	1	1	1	2	4	7
	1/0	0	1	1	1	3	6
	2/0	0	1	1	1	3	5
	3/0	0	1	1	1	2	4
	4/0	0	0	1	1	1	3
	4/0	0	0	1	1	1	2
THHN, THWN, THWN-2	14	13	22	33	52	76	134
	12	9	16	24	38	56	98
	10	6	10	15	24	35	62
	8	3	6	9	14	20	35
THHN, THWN, THWN-2	6	2	4	6	10	14	25
	4	1	2	4	6	9	16
	3	1	1	3	5	7	13
	2	1	1	3	4	6	11
	1	1	1	1	3	4	8
	1/0	1	1	1	2	4	7
	2/0	0	1	1	1	3	6
	3/0	0	1	1	1	2	5
	4/0	0	1	1	1	1	4
XHH, XHHW, XHHW-2	14	9	15	23	36	53	94
	12	7	11	18	28	41	72
	10	5	8	13	21	30	54
	8	3	5	7	11	17	30
	6	1	3	5	8	12	22
	4	1	2	4	6	9	16
	3	1	1	3	5	7	13
	2	1	1	3	4	6	11
	1	1	1	1	3	5	8
	1/0	1	1	1	2	4	7
	2/0	0	1	1	2	3	6
	3/0	0	1	1	1	3	5
	4/0	0	1	1	1	2	4

For SI: 1 inch = 25.4 mm.

a. Types RHW, and RHW-2 without outer covering.

TABLE E3904.6(4) (Annex C, Table C.4)
MAXIMUM NUMBER OF CONDUCTORS IN INTERMEDIATE METALLIC CONDUIT (IMC)[a]

TYPE LETTERS	CONDUCTOR SIZE AWG/kcmil	TRADE SIZES (inches)					
		1/2	3/4	1	1 1/4	1 1/2	2
RHH, RHW, RHW-2	14	4	8	13	22	30	49
	12	4	6	11	18	25	41
	10	3	5	8	15	20	33
	8	1	3	4	8	10	17
	6	1	1	3	6	8	14
	4	1	1	3	5	6	11
	3	1	1	2	4	6	9
	2	1	1	1	3	5	8
	1	0	1	1	2	3	5
	1/0	0	1	1	1	3	4
	2/0	0	1	1	1	2	4
	3/0	0	0	1	1	1	3
	4/0	0	0	1	1	1	3
TW, THHW, THW, THW-2	14	10	17	27	47	64	104
	12	7	13	21	36	49	80
	10	5	9	15	27	36	59
	8	3	5	8	15	20	33
RHH[a], RHW[a], RHW-2[a]	14	6	11	18	31	42	69
	12	5	9	14	25	34	56
	10	4	7	11	19	26	43
	8	2	4	7	12	16	26
RHH[a], RHW[a], RHW-2[a], TW, THW, THHW, THW-2	6	1	3	5	9	12	20
	4	1	2	4	6	9	15
	3	1	1	3	6	8	13
	2	1	1	3	5	6	11
	1	1	1	1	3	4	7
	1/0	1	1	1	3	4	6
	2/0	0	1	1	2	3	5
	3/0	0	1	1	1	3	4
	4/0	0	1	1	1	2	4
THHN, THWN, THWN-2	14	14	24	39	68	91	149
	12	10	17	29	49	67	109
	10	6	11	18	31	42	68
	8	3	6	10	18	24	39
	6	2	4	7	13	17	28
	4	1	3	4	8	10	17
	3	1	2	4	6	9	15
	2	1	1	3	5	7	12
	1	1	1	2	4	5	9
	1/0	1	1	1	3	4	8
	2/0	1	1	1	3	4	6
THHN, THWN, THWN-2	3/0	0	1	1	2	3	5
	2/0	0	1	1	1	2	4
XHH, XHHW, XHHW-2	14	10	17	27	47	64	104
	12	7	13	21	36	49	80
	10	5	9	15	27	36	59
	8	3	5	8	15	20	33
	6	1	4	6	11	15	24
	4	1	3	4	8	11	18
	3	1	2	4	7	9	15
	2	1	1	3	5	7	12
	1	1	1	2	4	5	9
	1/0	1	1	1	3	5	8
	2/0	1	1	1	3	4	6
	3/0	0	1	1	2	3	5
	4/0	0	1	1	1	2	4

For SI: 1 inch = 25.4 mm.

a. Types RHW, and RHW-2 without outer covering.

TABLE E3904.6(5) (Annex C, Table C.5)
MAXIMUM NUMBER OF CONDUCTORS IN LIQUID-TIGHT FLEXIBLE NONMETALLIC CONDUIT (FNMC-B)[a]

TYPE LETTERS	CONDUCTOR SIZE AWG/kcmil	TRADE SIZES (inches)						
		$^3/_8$	$^1/_2$	$^3/_4$	1	$1^1/_4$	$1^1/_2$	2
RHH, RHW, RHW-2	14	2	4	7	12	21	27	44
	12	1	3	6	10	17	22	36
	10	1	3	5	8	14	18	29
	8	1	1	2	4	7	9	1
	6	1	1	1	3	6	7	12
	4	0	1	1	2	4	6	9
RHH, RHW, RHW-2	3	0	1	1	1	4	5	8
	2	0	1	1	1	3	4	7
	1	0	0	1	1	1	3	5
	1/0	0	0	1	1	1	2	4
	2/0	0	0	1	1	1	1	3
	3/0	0	0	0	1	1	1	3
	4/0	0	0	0	1	1	1	2
TW, THHW, THW, THW-2	14	5	9	15	25	44	57	93
	12	4	7	12	19	33	43	71
	10	3	5	9	14	25	32	53
	8	1	3	5	8	14	18	29
RHH[a], RHW[a], RGW-2[a]	14	3	6	10	16	29	38	62
	12	3	5	8	13	23	30	50
	10	1	3	6	10	18	23	39
	8	1	1	4	6	11	14	23
RHH[a], RHW[a], RHW-2[a], TW, THW, THHW, THW-2	6	1	1	3	5	8	11	18
	4	1	1	1	3	6	8	13
	3	1	1	1	3	5	7	11
	2	0	1	1	2	4	6	9
	1	0	1	1	1	3	4	7
	1/0	0	0	1	1	2	3	6
	2/0	0	0	1	1	2	3	5
	3/0	0	0	1	1	1	2	4
	4/0	0	0	0	1	1	1	3
THHN, THWN, THWN-2	14	8	13	22	36	63	81	133
	12	5	9	16	26	46	59	97
	10	3	6	10	16	29	37	61
	8	1	3	6	9	16	21	35
THHN, THWN, THWN-2	6	1	2	4	7	12	15	25
	4	1	1	2	4	7	9	15
	3	1	1	1	3	6	8	13
	2	1	1	1	3	5	7	11
	1	0	1	1	1	4	5	8
	1/0	0	1	1	1	3	4	7
	2/0	0	0	1	1	2	3	6
	3/0	0	0	1	1	1	3	5
	4/0	0	0	1	1	1	2	4
XHH, XHHW, XHHW-2	14	5	9	15	25	44	57	93
	12	4	7	12	19	33	43	71
	10	3	5	9	14	25	32	53
	8	1	3	5	8	14	18	29
	6	1	1	3	6	10	13	22
	4	1	1	2	4	7	9	16
	3	1	1	1	3	6	8	13
	2	1	1	1	3	5	7	11
	1	0	1	1	1	4	5	8
	1/0	0	1	1	1	3	4	7
	2/0	0	0	1	1	2	3	6
	3/0	0	0	1	1	1	3	5
	4/0	0	0	1	1	1	2	4

For SI: 1 inch = 25.4 mm.

a. Types RHW, and RHW-2 without outer covering.

TABLE E3904.6(6) (Annex C, Table C.6)
MAXIMUM NUMBER OF CONDUCTORS IN LIQUID-TIGHT FLEXIBLE NONMETALLIC CONDUIT (FNMC-A)[a]

TYPE LETTERS	CONDUCTOR SIZE AWG/kcmil	TRADE SIZES (inches)						
		3/8	1/2	3/4	1	1 1/4	1 1/2	2
RHH, RHW, RHW-2	14	2	4	7	11	20	27	45
	12	1	3	6	9	17	23	38
	10	1	3	5	8	13	18	30
	8	1	1	2	4	7	9	16
	6	1	1	1	3	5	7	13
	4	0	1	1	2	4	6	10
	3	0	1	1	1	4	5	8
	2	0	1	1	1	3	4	7
	1	0	0	1	1	1	3	5
	1/0	0	0	1	1	1	2	4
	2/0	0	0	1	1	1	1	4
	3/0	0	0	0	1	1	1	3
	4/0	0	0	0	1	1	1	3
TW, THHW, THW, THW-2	14	5	9	15	24	43	58	96
	12	4	7	12	19	33	44	74
	10	3	5	9	14	24	33	55
	8	1	3	5	8	13	18	30
RHH[a], RHW[a], RHW-2[a]	14	3	6	10	16	28	38	64
	12	3	4	8	13	23	31	51
	10	1	3	6	10	18	24	40
	8	1	1	4	6	10	14	24
RHH[a], RHW[a], RHW-2[a], TW, THW, THHW, THW-2	6	1	1	3	4	8	11	18
	4	1	1	1	3	6	8	13
	3	1	1	1	3	5	7	11
	2	0	1	1	2	4	6	10
	1	0	1	1	1	3	4	7
RHH[a], RHW[a], RHW-2[a], TW, THW, THHW, THW-2	1/0	0	0	1	1	2	3	6
	2/0	0	0	1	1	1	3	5
	3/0	0	0	1	1	1	2	4
	4/0	0	0	0	1	1	1	3
THHN, THWN, THWN-2	14	8	13	22	35	62	83	137
	12	5	9	16	25	45	60	100
	10	3	6	10	16	28	38	63
	8	1	3	6	9	16	22	36
	6	1	2	4	6	12	16	26
	4	1	1	2	4	7	9	16
	3	1	1	1	3	6	8	13
	2	1	1	1	3	5	7	11
	1	0	1	1	1	4	5	8
	1/0	0	1	1	1	3	4	7
	2/0	0	0	1	1	2	3	6
	3/0	0	0	1	1	1	3	5
	4/0	0	0	1	1	1	2	4
XHH, XHHW, XHHW-2	14	5	9	15	24	43	58	96
	12	4	7	12	19	33	44	74
	10	3	5	9	14	24	33	55
	8	1	3	5	8	13	18	30
	6	1	1	3	5	10	13	22
	4	1	1	2	4	7	10	16
	3	1	1	1	3	6	8	14
	2	1	1	1	3	5	7	11
	1	0	1	1	1	4	5	8
XHH, XHHW, XHHW-2	1/0	0	1	1	1	3	4	7
	2/0	0	0	1	1	2	3	6
	3/0	0	0	1	1	1	3	5
	4/0	0	0	1	1	1	2	4

For SI: 1 inch = 25.4 mm.

a. Types RHW, and RHW-2 without outer covering.

TABLE E3904.6(7) (Annex C, Table C.7)
MAXIMUM NUMBER OF CONDUCTORS IN LIQUID-TIGHT FLEXIBLE METAL CONDUIT (LFMC)[a]

TYPE LETTERS	CONDUCTOR SIZE AWG/kcmil	TRADE SIZES (inches)					
		1/2	3/4	1	1 1/4	1 1/2	2
RHH, RHW, RHW-2	14	4	7	12	21	27	44
	12	3	6	10	17	22	36
	10	3	5	8	14	18	29
	8	1	2	4	7	9	15
	6	1	1	3	6	7	12
	4	1	1	2	4	6	9
	3	1	1	1	4	5	8
	2	1	1	1	3	4	7
	1	0	1	1	1	3	5
	1/0	0	1	1	1	2	4
	2/0	0	1	1	1	1	3
	3/0	0	0	1	1	1	3
	4/0	0	0	1	1	1	2
TW, THHW, THW, THW-2	14	9	15	25	44	57	93
	12	7	12	19	33	43	71
	10	5	9	14	25	32	53
	8	3	5	8	14	18	29
RHH[a], RHW[a], RHW-2[a], THHW, THW, THW-2	14	6	10	16	29	38	62
	12	5	8	13	23	30	50
	10	3	6	10	18	23	39
	8	1	4	6	11	14	23
RHH[a], RHW[a], RHW-2[a], TW, THW, THHW, THW-2	6	1	3	5	8	11	18
	4	1	1	3	6	8	13
	3	1	1	3	5	7	11
	2	1	1	2	4	6	9
	1	1	1	1	3	4	7
	1/0	0	1	1	2	3	6
	2/0	0	1	1	2	3	5
	3/0	0	1	1	1	2	4
	4/0	0	0	1	1	1	3
THHN, THWN, THWN-2	14	13	22	36	63	81	133
	12	9	16	26	46	59	97
	10	6	10	16	29	37	61
	8	3	6	9	16	21	35
	6	2	4	7	12	15	25
	4	1	2	4	7	9	15
	3	1	1	3	6	8	13
	2	1	1	3	5	7	11
	1	1	1	1	4	5	8
	1/0	1	1	1	3	4	7
	2/0	0	1	1	2	3	6
	3/0	0	1	1	1	3	5
	4/0	0	1	1	1	2	4
XHH, XHHW, XHHW-2	14	9	15	25	44	57	93
	12	7	12	19	33	43	71
	10	5	9	14	25	32	53
	8	3	5	8	14	18	29
XHH, XHHW, XHHW-2	6	1	3	6	10	13	22
	4	1	2	4	7	9	16
	3	1	1	3	6	8	13
	2	1	1	3	5	7	11
	1	1	1	1	4	5	8
	1/0	1	1	1	3	4	7
	2/0	0	1	1	2	3	6
	3/0	0	1	1	1	3	5
	4/0	0	1	1	1	2	4

For SI: 1 inch = 25.4 mm.

a. Types RHW, and RHW-2 without outer covering.

TABLE E3904.6(8) (Annex C, Table C.8)
MAXIMUM NUMBER OF CONDUCTORS IN RIGID METAL CONDUIT (RMC)[a]

TYPE LETTERS	CONDUCTOR SIZE AWG/kcmil	TRADE SIZES (inches)					
		$\frac{1}{2}$	$\frac{3}{4}$	1	$1\frac{1}{4}$	$1\frac{1}{2}$	2
RHH, RHW, RHW-2	14	4	7	12	21	28	46
	12	3	6	10	17	23	38
	10	3	5	8	14	19	31
	8	1	2	4	7	10	16
	6	1	1	3	6	8	13
	4	1	1	2	4	6	10
	3	1	1	2	4	5	9
	2	1	1	1	3	4	7
	1	0	1	1	1	3	5
	1/0	0	1	1	1	2	4
	2/0	0	1	1	1	2	4
	3/0	0	0	1	1	1	3
	4/0	0	0	1	1	1	3
TW, THHW, THW, THW-2	14	9	15	25	44	59	98
	12	7	12	19	33	45	75
	10	5	9	14	25	34	56
	8	3	5	8	14	19	31
RHH[a], RHW[a], RHW-2[a]	14	6	10	17	29	39	65
	12	5	8	13	23	32	52
	10	3	6	10	18	25	41
	8	1	4	6	11	15	24
RHH[a], RHW[a], RHW-2[a], TW, THW, THHW, THW-2	6	1	3	5	8	11	18
	4	1	1	3	6	8	14
	3	1	1	3	5	7	12
	2	1	1	2	4	6	10
	1	1	1	1	3	4	7
	1/0	0	1	1	2	3	6
	2/0	0	1	1	2	3	5
	3/0	0	1	1	1	2	4
	4/0	0	0	1	1	1	3
THHN, THWN, THWN-2	14	13	22	36	63	85	140
	12	9	16	26	46	62	102
	10	6	10	17	29	39	64
	8	3	6	9	16	22	37
	6	2	4	7	12	16	27
	4	1	2	4	7	10	16
	3	1	1	3	6	8	14
	2	1	1	3	5	7	11
	1	1	1	1	4	5	8
THHN, THWN, THWN-2	1/0	1	1	1	3	4	7
	2/0	0	1	1	2	3	6
	3/0	0	1	1	1	3	5
	4/0	0	1	1	1	2	4
XHH, XHHW, XHHW-2	14	9	15	25	44	59	98
	12	7	12	19	33	45	75
	10	5	9	14	25	34	56
	8	3	5	8	14	19	31
	6	1	3	6	10	14	23
	4	1	2	4	7	10	16
	3	1	1	3	6	8	14
	2	1	1	3	5	7	12
	1	1	1	1	4	5	9
	1/0	1	1	1	3	4	7
	2/0	0	1	1	2	3	6
	3/0	0	1	1	1	3	5
	4/0	0	1	1	1	2	4

For SI: 1 inch = 25.4 mm.

a. Types RHW, and RHW-2 without outer covering.

TABLE E3904.6(9) (Annex C, Table C.9)
MAXIMUM NUMBER OF CONDUCTORS IN RIGID PVC CONDUIT, SCHEDULE 80 (PVC-80)[a]

TYPE LETTERS	CONDUCTOR SIZE AWG/kcmil	TRADE SIZES (inches)					
		¹/₂	³/₄	1	1¹/₄	1¹/₂	2
RHH, RHW, RHW-2	14	3	5	9	17	23	39
	12	2	4	7	14	19	32
	10	1	3	6	11	15	26
	8	1	1	3	6	8	13
	6	1	1	2	4	6	11
	4	1	1	1	3	5	8
RHH, RHW, RHW-2	3	0	1	1	3	4	7
	2	0	1	1	3	4	6
	1	0	1	1	1	2	4
	1/0	0	0	1	1	1	3
	2/0	0	0	1	1	1	3
	3/0	0	0	1	1	1	3
	4/0	0	0	0	1	1	2
TW, THHW, THW, THW-2	14	6	11	20	35	49	82
	12	5	9	15	27	38	63
	10	3	6	11	20	28	47
	8	1	3	6	11	15	26
RHH[a], RHW[a], RHW-2[a]	14	4	8	13	23	32	55
	12	3	6	10	19	26	44
	10	2	5	8	15	20	34
	8	1	3	5	9	12	20
RHH[a], RHW[a], RHW-2[a], TW, THW, THHW, THW-2	6	1	1	3	7	9	16
	4	1	1	3	5	7	12
	3	1	1	2	4	6	10
	2	1	1	1	3	5	8
	1	0	1	1	2	3	6
	1/0	0	1	1	1	3	5
RHH[a], RHW[a], RHW-2[a], TW, THW, THHW, THW-2	2/0	0	1	1	1	2	4
	3/0	0	0	1	1	1	3
	4/0	0	0	1	1	1	3
THHN, THWN, THWN-2	14	9	17	28	51	70	118
	12	6	12	20	37	51	86
	10	4	7	13	23	32	54
	8	2	4	7	13	18	31
	6	1	3	5	9	13	22
	4	1	1	3	6	8	14
	3	1	1	3	5	7	12
	2	1	1	2	4	6	10
	1	0	1	1	3	4	7
	1/0	0	1	1	2	3	6
	2/0	0	1	1	1	3	5
	3/0	0	1	1	1	2	4
	4/0	0	0	1	1	1	3
XHH, XHHW, XHHW-2	14	6	11	20	35	49	82
	12	5	9	15	27	38	63
	10	3	6	11	20	28	47
	8	1	3	6	118	15	26
	6	1	2	4		11	19

(continued)

TABLE E3904.6(9) (Annex C, Table C.9)—continued
MAXIMUM NUMBER OF CONDUCTORS IN RIGID PVC CONDUIT, SCHEDULE 80 (PVC-80)[a]

TYPE LETTERS	CONDUCTOR SIZE AWG/kcmil	TRADE SIZES (inches)					
		$^1/_2$	$^3/_4$	1	$1^1/_4$	$1^1/_2$	2
XHH, XHHW, XHHW-2	14	6	11	20	35	49	82
	12	5	9	15	27	38	63
	10	3	6	11	20	28	47
	8	1	3	6	11	15	26
	6	1	2	4	8	11	19
	4	1	1	3	6	8	14
	3	1	1	3	5	7	12
	2	1	1	2	4	6	10
	1	0	1	1	3	4	7
	1/0	0	1	1	2	3	6
	2/0	0	1	1	1	3	5
	3/0	0	1	1	1	2	4
	4/0	0	0	1	1	1	3

For SI: 1 inch = 25.4 mm.

TABLE E3904.6(10) (Annex C, Table C.10)
MAXIMUM NUMBER OF CONDUCTORS IN RIGID PVC CONDUIT SCHEDULE 40 (PVC-40)[a]

TYPE LETTERS	CONDUCTOR SIZE AWG/kcmil	TRADE SIZES (inches)					
		¹/₂	³/₄	1	1¹/₄	1¹/₂	2
RHH, RHW, RHW-2	14	4	7	11	20	27	45
	12	3	5	9	16	22	37
	10	2	4	7	13	18	30
	8	1	2	4	7	9	15
	6	1	1	3	5	7	12
	4	1	1	2	4	6	10
	3	1	1	1	4	5	8
	2	1	1	1	3	4	7
RHH, RHW, RHW-2	1	0	1	1	1	3	5
	1/0	0	1	1	1	2	4
	2/0	0	0	1	1	1	3
	3/0	0	0	1	1	1	3
	4/0	0	0	1	1	1	2
TW, THHW, THW, THW-2	14	8	14	24	42	57	94
	12	6	11	18	32	44	72
	10	4	8	13	24	32	54
	8	2	4	7	13	18	30
RHH[a], RHW[a], RHW-2[a]	14	5	9	16	28	38	63
	12	4	8	13	22	30	50
	10	3	6	10	17	24	39
	8	1	3	6	10	14	23
RHH[a], RHW[a], RHW-2[a], TW, THW, THHW, THW-2	6	1	2	4	8	11	18
	4	1	1	3	6	8	13
	3	1	1	3	5	7	11
	2	1	1	2	4	6	10
	1	0	1	1	3	4	7
	1/0	0	1	1	2	3	6
	2/0	0	1	1	1	3	5
	3/0	0	1	1	1	2	4
	4/0	0	0	1	1	1	3
THHN, THWN, THWN-2	14	11	21	34	60	82	135
	12	8	15	25	43	59	99
	10	5	9	15	27	37	62
	8	3	5	9	16	21	36
THHN, THWN, THWN-2	6	1	4	6	11	15	26
	4	1	2	4	7	9	16
	3	1	1	3	6	8	13
	2	1	1	3	5	7	11
	1	1	1	1	3	5	8
	1/0	1	1	1	3	4	7
	2/0	0	1	1	2	3	6
	3/0	0	1	1	1	3	5
	4/0	0	1	1	1	2	4
XHH, XHHW, XHHW-2	14	8	14	24	42	57	94
	12	6	11	18	32	44	72
	10	4	8	13	24	32	54
	8	2	4	7	13	18	30
	6	1	3	5	10	13	22
	4	1	2	4	7	9	16
	3	1	1	3	6	8	13
	2	1	1	3	5	7	11
	1	1	1	1	3	5	8
	1/0	1	1	1	3	4	7
	2/0	0	1	1	2	3	6
	3/0	0	1	1	1	3	5
	4/0	0	1	1	1	2	4

For SI: 1 inch = 25.4 mm.

a. Types RHW, and RHW-2 without outer covering.

E3904.7 Air handling-stud cavity and joist spaces. Where wiring methods having a nonmetallic covering pass through stud cavities and joist spaces used for air handling, such wiring shall pass through such spaces perpendicular to the long dimension of the spaces. [300.22(C) Exception]

❖ Limiting the wiring in air-handling spaces and cavities is intended to minimize the production and rapid spreading of toxic fumes/smoke created during a fire. In many situations it might be convenient or necessary to run nonmetallic sheathed cable through a stud or joist space that is intended to convey return air. The code permits this for dwelling wiring on a limited basis only where the cables are run perpendicular to the stud or joist (i.e., run horizontally through stud cavities). In a new house under construction that has been framed and is ready for installation of the rough wiring, plumbing and duct work, it is important to know which stud or joist spaces will be used for return air. If the heating and air-conditioning contractor arrives first and installs the duct work system, and the electrician comes along after that, the wiring can be run correctly. That is, installing cable through a return air space or cavity perpendicular to the long dimension of such space is allowed. But, if the electrician arrives first and has no information on which, if any, stud cavities or joist spaces may be used for air handling, wiring might be installed parallel to the stud or joist in the space or cavity. The mechanical contractor might then enclose the space with a sheet metal pan or otherwise create the air handling space, which would cause the wiring to be in violation of the code. The installers of the wiring and of the heating and air-conditioning systems must coordinate. Note that only wiring that is not perpendicular to the long dimension is prohibited. There is no limit to the amount of wiring that can pass through perpendicular to the long dimension. In other words, any number of NM cables can pass horizontally through a stud cavity, but not one can pass through vertically. This might seem illogical, but, in reality only one or two cables are likely to pass through a stud cavity horizontally, with a much greater likelihood for many cables to pass through the cavity vertically. Note that other sections of this code restrict or eliminate the use of framing cavities for conveying airflow, therefore this practice may be disappearing.

SECTION E3905
BOXES, CONDUIT BODIES AND FITTINGS

E3905.1 Box, conduit body or fitting—where required. A box or conduit body shall be installed at each conductor splice point, outlet, switch point, junction point and pull point except as otherwise permitted in Sections E3905.1.1 through E3905.1.6.

Fittings and connectors shall be used only with the specific wiring methods for which they are designed and listed. (300.15)

❖ The point in the wiring system that is the most vulnerable for possible problems from overheating caused by

arcing or sparking is a splice or connection. Except for some cases of underground wiring, a box or enclosure must always be used to make a splice or connection. A splice or connection made outside of a box is a violation and a serious safety concern. Nonmetallic sheathed cable must be used in continuous lengths from box to box. There must be no splices outside of an enclosure listed for the purpose, and the enclosures must be accessible (see Section E3905.11).

E3905.1.1 Equipment. An integral junction box or wiring compartment that is part of listed equipment shall be permitted to serve as a box or conduit body. [300.15(B)]

❖ An example of this is on a hood-fan unit installed above a range. If a box were installed and splices made, the box would probably be covered with the hood. Because the exact height of the hood is not always known, a sufficient length of the cable is left stapled to a stud on the rough wiring without any box and brought through a hole in the drywall. The hood-fan unit has a built-in box where the splices can be made. This is similar for a bathroom exhaust fan, a dishwasher, and many other appliances. It is important to use a proper fitting such as a listed cable connector where the cable enters the equipment junction box. If not protected with a proper connector, the cable could be pulled out and/or damaged by the metal edge of the knockout on the equipment. The proper installation is to use a connector and not simply to wrap tape around the cable where it enters the box.

E3905.1.2 Protection. A box or conduit body shall not be required where cables enter or exit from conduit or tubing that is used to provide cable support or protection against physical damage. A fitting shall be provided on the end(s) of the conduit or tubing to protect the cable from abrasion. [300.15(C)]

❖ A box is not required because there are no splices. An example of the protection required is where a switch box is attached to a concrete or masonry wall in a basement and a short run of EMT is used to protect the cable along the wall. At the top of the short run of EMT, a connector with a bushing or a cable transition fitting must be installed (see commentary, Section E3802.3.2).

E3905.1.3 Integral enclosure. A wiring device with integral enclosure identified for the use, having brackets that securely fasten the device to walls or ceilings of conventional on-site frame construction, for use with nonmetallic-sheathed cable, shall be permitted in lieu of a box or conduit body. [300.15(E)]

❖ An example of an integral enclosure is a receptacle for a range or dryer outlet. It has a base section that is attached to the wall and a cover that is attached by screws after the connections are made to terminals. This type of device is usually for surface mounting.

E3905.1.4 Fitting. A fitting identified for the use shall be permitted in lieu of a box or conduit body where such fitting is accessible after installation and does not contain spliced or terminated conductors. [300.15(F)]

❖ A pulling elbow with an access cover is an example of such fitting. Where a change is made in the type of wiring method, a fitting is permitted as long as it does not contain splices. Where there is a transition from non-metallic-sheathed cable to EMT, a box is not needed if a proper fitting is provided at the transition.

E3905.1.5 Buried conductors. Splices and taps in buried conductors and cables shall not be required to be enclosed in a box or conduit body where installed in accordance with Section E3803.4.

❖ See the commentary to Section E3803.4.

E3905.1.6 Luminaires. Where a luminaire is listed to be used as a raceway, a box or conduit body shall not be required for wiring installed therein. [300.15(J)]

❖ To make the connection of the circuit wires to the fixtures wire, generally a box is required where lighting fixtures are installed. In some older homes cable was brought through the plaster ceiling and the connection made without the use of a box. The light fixture was attached directly to the ceiling with long wood screws, and the canopy of the light fixture contained the splices. This is a code violation because a box is required.

According to this section, however, if a fixture is listed to be used as a box to contain the splices, it would not be a code violation. If the cable or cables are run through the drywall ceiling and brought into the wiring section of such fixture through a knockout opening that is usually provided, the installation would be acceptable and would serve as the enclosure for splicing the circuit conductors to the fixture wires. Where the nonmetallic sheathed cable extends through the finished ceiling wallboard, it is permissible to omit the junction box in the ceiling and use the wiring compartment of the listed fixture as the junction box to make the connections.

E3905.2 Metal boxes. Metal boxes shall be grounded. (314.4)

❖ See Section E3908 on grounding.

E3905.3 Nonmetallic boxes. Nonmetallic boxes shall be used only with cabled wiring methods with entirely nonmetallic sheaths, flexible cords and nonmetallic raceways. (314.3)

Exceptions:

1. Where internal bonding means are provided between all entries, nonmetallic boxes shall be permitted to be used with metal raceways and metal-armored cables. (314.3 Exception No. 1)

2. Where integral bonding means with a provision for attaching an equipment grounding jumper inside the box are provided between all threaded entries in nonmetallic boxes listed for the purpose, nonmetallic boxes shall be permitted to be used with metal raceways and metal-armored cables. (314.3 Exception No. 2)

❖ A nonmetallic device box (used for directly attaching switches or receptacles) is designed for use with non-

metallic-sheathed cable. A nonmetallic 4-inch square box can be used, but a plaster ring would be needed for the attachment of switches and receptacles. Nonmetallic 4-inch square boxes come with knockouts for the entry of nonmetallic-sheathed cable and also have standard size knockouts for conduit and tubing fittings. To use a nonmetallic box in a run of metal conduit or tubing requires special attention to the grounding and bonding requirements.

A metal raceway is permitted to serve as a grounding conductor. Where a nonmetallic box is inserted between two metal raceway entries, the continuity of the grounding conductor is interrupted. Therefore, bonding jumpers are required.

E3905.3.1 Nonmetallic-sheathed cable and nonmetallic boxes. Where nonmetallic-sheathed cable is used, the cable assembly, including the sheath, shall extend into the box not less than $^1/_4$ inch (6.4 mm) through a nonmetallic-sheathed cable knockout opening. (314.7(C)]

❖ The individual conductors should not be exposed outside of the box. Where the cable enters the box, the clamp should not bind against the individual conductors but against the sheathing. Some electricians strip the sheathing off the conductors before installing the cable in the nonmetallic device box, and this sometimes results in the cable sheath being too short to extend into the box. On the rough inspection, it is an indication that the job has been done in a careless manner where the sheathing has been stripped short of entering the box.

The amount of sheathing left on the conductors inside the box affects the box fill, because it takes more room to tuck the conductors back into the box when there is too much sheathing left. This leaves less room to install the device. The length of sheathing left in the box, the length of the conductors, the size of the box and the type of device all should be considered, not just the minimum requirements of the code for box size.

E3905.3.2 Securing to box. Wiring methods shall be secured to the boxes. [314.17(C)]

Exception: Where nonmetallic-sheathed cable is used with boxes not larger than a nominal size of $2^1/_4$ inches by 4 inches (57 mm by 102 mm) mounted in walls or ceilings, and where the cable is fastened within 8 inches (203 mm) of the box measured along the sheath, and where the sheath extends through a cable knockout not less than $^1/_4$ inch (6.4 mm), securing the cable to the box shall not be required. [314.17(C) Exception]

❖ The basic rule here is that the cable, such as armored cable and nonmetallic-sheathed cable, should be secured to the box. Where metal boxes are used, this is always required. All metal device boxes come with clamps, usually inside the box. If the metal box used is not a device box, such as a four-square box with a plaster ring installed to make it possible to attach devices, the cable is secured to the box by approved cable connectors through the knockouts or the box may have integral clamps.

This exception refers to a single-gang nonmetallic box, which is usually not larger than $2^{1}/_{4}$ inches by 4 inches (57 mm by 102 mm). For single-gang nonmetallic boxes, the cable need not be secured to the box itself if the cable is stapled to the stud within 8 inches (203 mm) of cable length outside the box.

The reason for this requirement is that the electrician may run the cable against the stud and then, at the top of the box, make a bend and run the cable over to a knockout opening farthest away from the stud. When the wires are tucked back into the box or when the device is installed, the cable could be pushed up and the conductors exposed outside the box where the sheathing has been stripped off. Many nonmetallic single-gang boxes have internal clamps that work like a spring, putting pressure against the cable so that it is difficult if not impossible to remove the cable simply by pulling it back out of the box. But the cable sheathing should be long enough so that a nonmetallic internal clamp puts pressure against the sheathing and not against the individual conductors. In other words, in addition to extending into the box at least $^{1}/_{4}$ inch (6.3 mm), the sheathing should extend into the box beyond any cable clamp. Running multiple cables through a single knockout opening in such plastic boxes is permitted (see Section E3906.3).

E3905.3.3 Conductor rating. Nonmetallic boxes shall be suitable for the lowest temperature-rated conductor entering the box. [314.17(C)]

❖ All conductors have a temperature rating according to the insulation type, as indicated in Table E3805.1.

Nonmetallic boxes have a temperature rating also, and it should be suitable for the lowest temperature rating of any conductor installed in the box because the lowest temperature rated conductor will set the temperature limit for all conductors within the box. Recall that Type NM cable contains 90°C (194°F) rated conductors but such conductors are sized based on a 60°C (149°F) conductor rating (see Section E3705.4.4).

E3905.4 Minimum depth of boxes for outlets, devices, and utilization equipment. Outlet and device boxes shall have an approved depth to allow equipment installed within them to be mounted properly and without the likelihood of damage to conductors within the box. (314.24)

❖ Boxes that contain devices, including switches, timers, dimmers, receptacles, GFCI devices and electronic controllers must have the necessary volume to prevent crowding of the conductors which can cause damage to conductor insulation. For example, extra deep boxes and 4-inch square boxes are typically used where large devices such as GFCI receptacles are to be installed. Some devices such as receptacles are

made with 6-inch-long stranded conductor leads permanently welded to the device and some are made with special mating connectors with such 6-inch leads. Such devices are meant to save installation time, but can also reduce the chances of damaging conductors in boxes while forcing devices into boxes.

E3905.4.1 Outlet boxes without enclosed devices or utilization equipment. Outlet boxes that do not enclose devices or utilization equipment shall have an internal depth of not less than $^{1}/_{2}$ inch (12.7 mm). [314.24(A)]

❖ An outlet box could be for a lighting fixture or for a device, such as a receptacle or switch. A box for a lighting fixture that is an outlet box but does not contain devices could be as shallow as $^{1}/_{2}$ inch (12.7 mm). A term used for this type of round shallow box is "pancake box," and it can be used, for example, on the surface of a joist or truss where the lighting fixture must be centered in a room or over some feature. Where a device is installed, the box must be at least $^{15}/_{16}$ inch (24 mm) deep. This provision for such a shallow box does not supercede the box fill requirements.

E3905.4.2 Utilization equipment. Outlet and device boxes that enclose devices or utilization equipment shall have a minimum internal depth that accommodates the rearward projection of the equipment and the size of the conductors that supply the equipment. The internal depth shall include that of any extension boxes, plaster rings, or raised covers. The internal depth shall comply with all of the applicable provisions that follow. [314.24(B)]

Exception: Utilization equipment that is listed to be installed with specified boxes.

1. Large equipment. Boxes that enclose devices or utilization equipment that projects more than $1^{7}/_{8}$ inches (48 mm) rearward from the mounting plane of the box shall have a depth that is not less than the depth of the equipment plus $^{1}/_{4}$ inch (6.4 mm). [314.24(B)(1)]

2. Conductors larger than 4 AWG. Boxes that enclose devices or utilization equipment supplied by conductors larger than 4 AWG shall be identified for their specific function. [314.24(B)(2)]

3. Conductors 8, 6, or 4 AWG. Boxes that enclose devices or utilization equipment supplied by 8, 6, or 4 AWG conductors shall have an internal depth that is not less than $2^{1}/_{16}$ inches (52.4 mm). [314.24(B)(3)]

4. Conductors 12 or 10 AWG. Boxes that enclose devices or utilization equipment supplied by 12 or 10 AWG conductors shall have an internal depth that is not less than $1^{3}/_{16}$ inches (30.2 mm). Where the equipment projects rearward from the mounting plane of the box by more than 1 inch (25.4 mm), the box shall have a depth that is not less than that of the equipment plus $^{1}/_{4}$ inch (6.4 mm). [314.24(B)(4)]

5. Conductors 14 AWG and smaller. Boxes that enclose devices or utilization equipment supplied by 14 AWG or smaller conductors shall have a depth that is not less than $^{15}/_{16}$ inch (23.8 mm). [314.24(B)(5)]

❖ More and more devices and pieces of equipment are being designed for installation in boxes, so the code has evolved to provide additional coverage for these increasingly common installations. Examples include: timers, occupant sensors, dimmers and lighting controllers. The concern again is for conductor overcrowding. Overcrowding can cause overheating and can cause damage to conductor insulation from sharp edges, compression and sharp bending.

E3905.5 Boxes enclosing flush-mounted devices. Boxes enclosing flush-mounted devices shall be of such design that the devices are completely enclosed at the back and all sides and shall provide support for the devices. Screws for supporting the box shall not be used for attachment of the device contained therein. (314.19)

❖ Many devices require a box of greater depth than the minimum $^{15}/_{16}$ inch (24 mm) because of the size of the device. For example, where a GFCI receptacle or a dimmer is installed, the box must have sufficient depth and capacity to completely enclose the device on the sides and back. If a cover is installed but the sides of the device are exposed, it would be a code violation. Fasteners that secure the box to the structure must not be used for any other purpose. Devices are to be supported by the box by screws that are independent of fasteners used to support the box itself.

E3905.6 Boxes at luminaire outlets. Outlet boxes used at luminaire or lampholder outlets shall be designed for the support of luminaires and lampholders and shall be installed as required by Section E3904.3. [314.27(A)]

❖ Luminaires (lighting fixtures) are typically mounted to round or octagonal boxes, which are usually close to 4 inches (102 mm) in diameter. Two sets of mounting holes are provided in some round fixture boxes. The holes, machined to accept 8-32 mounting screws (size No. 8 in diameter with 32 threads per inch), are $2^3/_4$ inches (70 mm) and $3^1/_2$ inches (89 mm) apart. This is a standard measurement. Many nonmetallic round boxes have the same two sets of mounting holes. Metal round or octagonal boxes usually have only one set of mounting holes, which are $3^1/_4$ inches (89 mm) apart. Some luminaires are secured by two 8-32 screws directly through the base of the fixture into the mounting holes of the box, and the holes in the fixture base are exactly $3^1/_2$ inches (89 mm) apart to match the holes on the round box. Porcelain keyless lamp holders and pull chain type fixtures are examples of fixtures having the mounting holes the same as those on the round "fixture boxes," as they are sometimes referred to in the field. Most luminaires come with a bracket designed to be attached to the outlet box. The bracket serves as a means to adjust the alignment of the holes for mounting the fixture. The bracket must be used as indicated in the instructions provided with the luminaire.

E3905.6.1 Vertical surface outlets. Boxes used at luminaire or lampholder outlets in or on a vertical surface shall be identified and marked on the interior of the box to indicate the maximum weight of the luminaire or lamp holder that is permitted to be supported by the box if other than 50 pounds (22.7 kg). [314.27(A)(1)]

Exception: A vertically-mounted luminaire or lampholder weighing not more than 6 pounds (2.7 kg) shall be permitted to be supported on other boxes or plaster rings that are secured to other boxes, provided that the luminaire or its supporting yoke is secured to the box with not fewer than two No. 6 or larger screws. [314.27(A)(1) Exception]

❖ Wall-mounted lighting outlet boxes (i.e., installed in or on vertical surfaces) do not have to be rated for supporting at least 50 pounds (22.7 kg), but if they are not so rated, they must be marked to indicate their maximum support capacity.

The code does not specify that fixtures must be mounted only on a round or octagonal box. A device box could be used for a small fixture that weighs not over 6 pounds (2.7 kg) if two 6-32 or larger screws are used. Device boxes are used for switches and receptacles and have mounting holes machined to accept 6-32 screws, which obviously would not provide as much support as the 8-32 screws that are supplied with fixtures. A problem is that the mounting holes on a device box are $3^1/_4$ inches (83 mm) apart, and they do not line up with those on the luminaires. They also might not line up with the mounting holes on many of the brackets supplied with luminaires.

A device box could be installed in the rough-wiring stage with the intention that a small fixture weighing 6 pounds (2.7 kg) or less will be installed at that location in the finish stage. But in the future, the small fixture could be changed for a larger and heavier fixture, thus making the device box not suitable and not in compliance with the code. In some areas, single-gang device boxes have been used in a brick wall for mounting outdoor fixtures. On a brick house, it is easy to install a device box in one of the courses of brick without cutting any brick. If a round box is used, the brick mason has to cut or notch one or more bricks to install the round box flush with the wall. If the luminaire weighs 6 pounds (2.7 kg) or less, use of a device box would be acceptable, but it would be good practice to install a round or octagonal box in anticipation of heavier fixtures. The intent of this section is to require that the boxes be marked to indicate the maximum weight that they can support, except as allowed by the exception (see Section E4003 for more provisions for luminaire installation). The weight capacity of a box can be meaningless if the box is improperly attached to the structure. Be wary of nails and drywall screws that might not have the ability to withstand the pullout or shear forces involved (see Section E3906.8). Because it is an unknown what fixtures the owner will choose to install, it would be wise to provide stout outlet boxes and supports at all locations intended for lighting fixtures.

E3905.6.2 Ceiling outlets. For outlets used exclusively for lighting, the box shall be designed or installed so that a luminaire or lampholder can be attached. Such boxes shall be capable of supporting a luminaire weighing up to 50 pounds (22.7 kg). A luminaire that weighs more than 50 pounds (22.7 kg) shall be supported independently of the outlet box, unless the outlet box is listed and marked on the interior of the box to indicate the maximum weight that the box is permitted to support. [314.27(A)(2)]

❖ This section requires lighting outlet boxes installed in a ceiling to be capable of supporting at least 50 pounds (22.7 kg) to account for the fact that the weight of the luminaire may not be known when the box is installed. For those of us that have ever carried a 50-pound bag of grain, cement or animal feed, it is evident how heavy a 50-pound lighting fixture is. The box and the box supports must be substantial to safely hold a 50-pound weight indefinitely. It is not uncommon to find ceiling outlet boxes attached with roofing nails driven up through the bottom of the box as the only means of support and this is obviously less than desirable because the weight is tending to pull out the nails. Ceiling-mounted luminaires are more likely to be heavier than those that are wall-mounted. Luminaires that weigh more than 50 pounds, such as some chandeliers, must be directly supported by the building structure, except where the outlet box is marked to indicate that it can support the weight of the subject luminaire. All ceiling outlet boxes should be carefully selected based on the anticipated load they might have to support. Ceiling outlet boxes in many locations, such as in parlors, entryways, atriums, great rooms, dining rooms, etc., should be chosen with the expectation that heavy fixtures will be supported from them at some point in time. Roughing in a stout box and supports is much easier than trying to retro fit a box after the building is finished.

E3905.7 Floor boxes. Where outlet boxes for receptacles are installed in the floor, such boxes shall be listed specifically for that application. [314.27(B)]

❖ Floor boxes are specially designed boxes that must be capable of the heavy duty demanded of them. For example, an ordinary device box is typically not listed for installation in a floor and used for a receptacle flush with the floor. Obviously, the support of a floor box should be considered as needing more than just the two nails or plaster ears with which most device boxes are mounted, because the floor receptacle could be walked on, and furniture could be placed on it. The finish or front edge of floor boxes is usually adjustable because the thickness of the flooring material is not known at the rough stage of the wiring. Even if a regular device box were approved for a floor outlet, it would be difficult to estimate the thickness of the flooring so that the box could be mounted at the required height.

Both metal and plastic floor boxes listed for the purpose are available.

E3905.8 Boxes at fan outlets. Outlet boxes and outlet box systems used as the sole support of ceiling-suspended fans (paddle) shall be marked by their manufacturer as suitable for this purpose and shall not support ceiling-suspended fans (paddle) that weigh more than 70 pounds (31.8 kg). For outlet boxes and outlet box systems designed to support ceiling-suspended fans (paddle) that weigh more than 35 pounds (15.9 kg), the required marking shall include the maximum weight to be supported.

Where spare, separately switched, ungrounded conductors are provided to a ceiling-mounted outlet box and such box is in a location acceptable for a ceiling-suspended (paddle) fan, the outlet box or outlet box system shall be listed for sole support of a ceiling-suspended (paddle) fan. [314.27(C)]

❖ In no case is an outlet box allowed to support a fan weighing over 70 pounds (31.8 kg) (see Section E4101.6). Logically, an outlet box that is mounted in a location that would be a likely location for a ceiling fan should be a box that is listed and rated for such application, but this section does not mandate such for that reason alone. If, however, that same box had an extra ungrounded (hot) conductor run to it from a switch location, it would be obvious that the outlet is set up for the future installation of a ceiling fan; and, in this case, the box is required to be a type that is listed and rated for the application. For example, it is common to run a 12/3 or 14/3 cable to a ceiling outlet box for the purpose of supplying a separately switched luminare and separately switched fan. This allows the owner to install a luminare only or a fan and luminare combination at any point in time. There is no excuse for not installing a fan-rated box where the wiring clearly anticipates the installation of a fan. Bear in mind that where EMT is the wiring method, a separate ungrounded conductor could be pulled in later.

The weight capacity of a box can be meaningless if the box is improperly attached to the structure. Be wary of nails and drywall screws that might not have the ability to withstand the pullout or shear forces involved (see Section E3906.8). Many fan-rated boxes have appropriate fasteners included with the box or in its packaging. Box mounting instructions must be followed precisely.

E3905.9 Utilization equipment. Boxes used for the support of utilization equipment other than ceiling-suspended (paddle) fans shall meet the requirements of Sections E3905.6.1 and E3905.6.2 for the support of a luminaire that is the same size and weight. [314.27(D)]

Exception: Utilization equipment weighing not more than 6 pounds (2.7 kg) shall be permitted to be supported on other boxes or plaster rings that are secured to other boxes, provided that the equipment or its supporting yoke is secured to the box with not fewer than two No. 6 or larger screws. [314.27(D) Exception]

❖ See the commentary to Sections E3905.6 and E3905.7.

E3905.10 Conduit bodies and junction, pull and outlet boxes to be accessible. Conduit bodies and junction, pull and outlet boxes shall be installed so that the wiring therein can be accessed without removing any part of the building or structure or, in underground circuits, without excavating sidewalks, paving, earth or other substance used to establish the finished grade. (314.29)

Exception: Boxes covered by gravel, light aggregate or noncohesive granulated soil shall be listed for the application, and the box locations shall be effectively identified and access shall be provided for excavation. (314.29 Exception)

❖ Boxes must be accessible and have an appropriate cover. A junction box could have a blank cover in a finished wall or ceiling. The homeowner may not want a blank cover showing on a finished wall or ceiling, but it would be a code violation to "bury" or hide the box behind the drywall. In many locations of the country, dwellings are constructed with a basement that is left unfinished when the house is purchased and moved into by the owner. At a later date, when the basement is finished, conductors are spliced in the existing boxes and then covered with drywall. This would be a code violation insofar as the boxes would then be concealed behind the ceiling or wall, and the only way to get to the connection would be to cut open or remove the drywall. If the splices are not in a box, that would be a separate code violation.

This code provision requiring a box where a splice is made is not only for the convenience of being able to work on the circuit in the future, but is a safety concern because overheating could occur from arcing in the splice. If it were not possible or feasible to avoid a splice in a concealed space, the homeowner may simply have to have a junction box in the ceiling of the basement with a blank cover on the finished ceiling. That would make the splice accessible. No matter what choice is made, the cable must be continuous from box to box, and the box must be accessible. Another example is in the remodeling of a kitchen. The old outlets may not be in the desired or required location for the new kitchen cabinet layout. It would be a code violation to permanently cover or conceal junction boxes or splices.

E3905.11 Damp or wet locations. In damp or wet locations, boxes, conduit bodies and fittings shall be placed or equipped so as to prevent moisture from entering or accumulating within the box, conduit body or fitting. Boxes, conduit bodies and fittings installed in wet locations shall be listed for use in wet locations. Where drainage openings are installed in the field in boxes or conduit bodies listed for use in damp or wet locations, such openings shall be approved and not larger than $^1/_4$ inch (6.4 mm). For listed drain fittings, larger openings are permitted where installed in the field in accordance with the manufacturer's instructions. (314.15)

❖ Where an outdoor receptacle is installed and a box is surface mounted, it must be listed for the purpose. Commonly used boxes for this purpose are Type FS or FD. The cover must be installed with a gasket to seal the interior of the box from moisture.

E3905.12 Number of conductors in outlet, device, and junction boxes, and conduit bodies. Boxes and conduit bodies shall be of an approved size to provide free space for all enclosed conductors. In no case shall the volume of the box, as calculated in Section E3905.12.1, be less than the box fill calculation as calculated in Section E3905.12.2. The minimum volume for conduit bodies shall be as calculated in Section E3905.12.3. The provisions of this section shall not apply to terminal housings supplied with motors or generators. (314.16)

❖ Selecting the proper size box requires two tasks. One is to determine the cubic inch capacity of the box, and the second is to determine the box fill allowed or needed. This section deals with determining the cubic-inch capacity or volume of the box.

E3905.12.1 Box volume calculations. The volume of a wiring enclosure (box) shall be the total volume of the assembled sections, and, where used, the space provided by plaster rings, domed covers, extension rings, etc., that are marked with their volume in cubic inches or are made from boxes the dimensions of which are listed in Table E3905.12.1. [314.16(A)]

❖ See the commentary to Section E3905.12.1.2.

E3905.12.1.1 Standard boxes. The volumes of standard boxes that are not marked with a cubic-inch capacity shall be as given in Table E3905.12.1. [314.16(A)(1)]

❖ See the commentary to Section E3905.12.1.2.

E3905.12.1.2 Other boxes. Boxes 100 cubic inches (1640 cm³) or less, other than those described in Table E3905.12.1, and nonmetallic boxes shall be durably and legibly marked by the manufacturer with their cubic-inch capacity. Boxes described in Table E3905.12.1 that have a larger cubic inch capacity than is designated in the table shall be permitted to have their cubic-inch capacity marked as required by this section. [314.16(A)(2)]

❖ Section E3905.12.1 and the two subsequent subsections are included in the code to establish the criteria to determine the volume or cubic-inch capacity of the box. The code requires the cubic-inch capacity to be marked on all nonmetallic boxes and any metal box of over 100 cubic inches. For standard size metal boxes listed in Table E3905.12.1 that do not have the cubic inch capacity marked, the volume or cubic inch capacity used for conductor fill is that listed in Table E3905.12.1. Because most boxes used in residential wiring do have the cubic inch capacity marked, Table E3905.12.1 is not often needed to determine the box volume.

E3905.12.2 Box fill calculations. The volumes in Section E3905.12.2.1 through Section E3905.12.2.5, as applicable, shall be added together. No allowance shall be required for small fittings such as locknuts and bushings. [314.16(B)]

❖ The various components of box fill are described in the five subsections below (Sections E3905.12.2.1 through E9805.12.2.5). These components are: 1. Conductors; 2. Internal clamps; 3. Support fittings; 4. Devices; and 5. Grounding conductors. The volume allowance for each of these components must be added and multiplied by the cubic inch volume in Table E3905.12.2.1

for the size conductor used. In Table E3905.12.2.1, the volume allowances are listed in the second column as free space for each conductor. The volume allowance for all nonconductor components (clamps, fittings, devices and grounding conductors) is based on the largest conductor in the box. For devices, two volume allowances are counted.

The exception to Section E3905.12.2.1 allows for fixtures that have several fixture wires from different lampholders on the fixture terminating in the fixture canopy to be omitted from the calculation. Some fixtures may attach to the box with a support fitting like a threaded stud or hickey, which is a stud with openings for fasteners on the sides. These were used in the distant past but are not common in modern wiring.

Example of Box Fill Calculation

An inspector finds that a device box with no internal cable clamps is marked with a volume of 20 cubic inches (328 cm³) and contains two size 12-AWG nonmetallic-sheathed cables of 12-3 with ground and a duplex receptacle. (Each cable has two ungrounded conductors, one grounded conductor, and one grounding conductor). Does this box comply with code provisions?

Determine the minimum cubic-inch capacity required for each component of the box from Table E3905.12.2.1. The table indicates that the volume

allowance for size 12-AWG conductors is 2.25 cubic inches (37 cm³).

Six conductors (ungrounded and grounded conductors)	$6 \times 2.25 = 13.5$
All grounding conductors counted as one	$1 \times 2.25 = 2.25$
One receptacle counts as two volume allowances	$2 \times 2.25 = 4.5$
Total cubic inches required	20.25

The use of this box is a code violation because it does not have the volume required.

E3905.12.2.1 Conductor fill. Each conductor that originates outside the box and terminates or is spliced within the box shall be counted once, and each conductor that passes through the box without splice or termination shall be counted once. Each loop or coil of unbroken conductor having a length equal to or greater than twice that required for free conductors by Section E3406.11.3, shall be counted twice. The conductor fill, in cubic inches, shall be computed using Table E3905.12.2.1. A conductor, no part of which leaves the box, shall not be counted. [314.16(B)(1)]

Exception: An equipment grounding conductor or not more than four fixture wires smaller than No. 14, or both, shall be permitted to be omitted from the calculations where such conductors enter a box from a domed fixture

TABLE E3905.12.1 [Table 314.16(A)]
MAXIMUM NUMBER OF CONDUCTORS IN METAL BOXES[a]

BOX DIMENSIONS (inches trade size and type)	MAXIMUM CAPACITY (cubic inches)	MAXIMUM NUMBER OF CONDUCTORS[a]						
		18 Awg	16 Awg	14 Awg	12 Awg	10 Awg	8 Awg	6 Awg
$4 \times 1^1/_4$ round or octagonal	12.5	8	7	6	5	5	4	2
$4 \times 1^1/_2$ round or octagonal	15.5	10	8	7	6	6	5	3
$4 \times 2^1/_8$ round or octagonal	21.5	14	12	10	9	8	7	4
$4 \times 1^1/_4$ square	18.0	12	10	9	8	7	6	3
$4 \times 1^1/_2$ square	21.0	14	12	10	9	8	7	4
$4 \times 2^1/_8$ square	30.3	20	17	15	13	12	10	6
$4^{11}/_{16} \times {}^{11}/_4$ square	25.5	17	14	12	11	10	8	5
$4^{11}/_{16} \times {}^{11}/_2$ square	29.5	19	16	14	13	11	9	5
$4^{11}/_{16} \times 2^1/_8$ square	42.0	28	24	21	18	16	14	8
$3 \times 2 \times 1^1/_2$ device	7.5	5	4	3	3	3	2	1
$3 \times 2 \times 2$ device	10.0	6	5	5	4	4	3	2
$3 \times 2 \times 2^1/_4$ device	10.5	7	6	5	4	4	3	2
$3 \times 2 \times 2^1/_2$ device	12.5	8	7	6	5	5	4	2
$3 \times 2 \times 2^3/_4$ device	14.0	9	8	7	6	5	4	2
$3 \times 2 \times 3^1/_2$ device	18.0	12	10	9	8	7	6	3
$4 \times 2^1/_8 \times 1^1/_2$ device	10.3	6	5	5	4	4	3	2
$4 \times 2^1/_8 \times 1^7/_8$ device	13.0	8	7	6	5	5	4	2
$4 \times 2^1/_8 \times 2^1/_8$ device	14.5	9	8	7	6	5	4	2
$3^3/_4 \times 2 \times 2^1/_2$ masonry box/gang	14.0	9	8	7	6	5	4	2
$3^3/_4 \times 2 \times 3^1/_2$ masonry box/gang	21.0	14	12	10	9	8	7	4

For SI: 1 inch = 25.4 mm, 1 cubic inch = 16.4 cm³.

a. Where volume allowances are not required by Sections E3905.12.2.2 through E3905.12.2.5.

or similar canopy and terminate within that box. [314.16(B)(1) Exception]

❖ See the commentary to Section E3905.12.2.

TABLE E3905.12.2.1 [Table 314.16(B)]
VOLUME ALLOWANCE REQUIRED PER CONDUCTOR

SIZE OF CONDUCTOR	FREE SPACE WITHIN BOX FOR EACH CONDUCTOR (cubic inches)
18 AWG	1.50
16 AWG	1.75
14 AWG	2.00
12 AWG	2.25
10 AWG	2.50
8 AWG	3.00
6 AWG	5.00

For SI: 1 cubic inch = 16.4 cm³.

E3905.12.2.2 Clamp fill. Where one or more internal cable clamps, whether factory or field supplied, are present in the box, a single volume allowance in accordance with Table E3905.12.2.1 shall be made based on the largest conductor present in the box. An allowance shall not be required for a cable connector having its clamping mechanism outside of the box. A clamp assembly that incorporates a cable termination for the cable conductors shall be listed and marked for use with specific nonmetallic boxes. Conductors that originate within the clamp assembly shall be included in conductor fill calculations provided in Section E3905.12.2.1 as though they entered from outside of the box. The clamp assembly shall not require a fill allowance, but, the volume of the portion of the assembly that remains within the box after installation shall be excluded from the box volume as marked.

❖ See the commentary to Section E3905.12.2.

E3905.12.2.3 Support fittings fill. Where one or more fixture studs or hickeys are present in the box, a single volume allowance in accordance with Table E3905.12.2.1 shall be made for each type of fitting based on the largest conductor present in the box. [314.16(B)(3)]

❖ See the commentary to Section E3905.12.2.

E3905.12.2.4 Device or equipment fill. For each yoke or strap containing one or more devices or equipment, a double volume allowance in accordance with Table E3905.12.2.1 shall be made for each yoke or strap based on the largest conductor connected to a device(s) or equipment supported by that yoke or strap. For a device or utilization equipment that is wider than a single 2-inch (51 mm) device box as described in Table E3905.12.1, a double volume allowance shall be made for each ganged portion required for mounting of the device or equipment. [314.16(B)(4)]

❖ See the commentary to Section E3905.12.2.

E3905.12.2.5 Equipment grounding conductor fill. Where one or more equipment grounding conductors or equipment bonding jumpers enters a box, a single volume allowance in accordance with Table E3905.12.2.1 shall be made based on

the largest equipment grounding conductor or equipment bonding jumper present in the box. [314.16(B)(5)]

❖ See the commentary to Section E3905.12.2.

E3905.12.3 Conduit bodies. Conduit bodies enclosing 6 AWG conductors or smaller, other than short-radius conduit bodies, shall have a cross-sectional area not less than twice the cross-sectional area of the largest conduit or tubing to which they can be attached. The maximum number of conductors permitted shall be the maximum number permitted by Section E3904.6 for the conduit to which it is attached. [314.16(C)(1)]

❖ A listed conduit body will have the minimum required cross-sectional area. This is a manufacturing requirement. Following the minimum fill requirements of Table E3904.6 for the type and size of conduit or tubing used will fulfill the requirements for maximum conductor fill of conduit bodies.

E3905.12.3.1 Splices, taps or devices. Only those conduit bodies that are durably and legibly marked by the manufacturer with their cubic inch capacity shall be permitted to contain splices, taps or devices. The maximum number of conductors shall be calculated using the same procedure for similar conductors in other than standard boxes. [314.16(C)(2)]

❖ It is difficult to calculate the cubic inch capacity of a conduit body. If the body is not marked with its capacity, it cannot contain any splices. Conduit bodies used in residential wiring are usually installed to provide access to pull conductors and make a splice.

E3905.12.3.2 Short-radius conduit bodies. Conduit bodies such as capped elbows and service-entrance elbows that enclose conductors 6 AWG or smaller and that are only intended to enable the installation of the raceway and the contained conductors, shall not contain splices, taps, or devices and shall be of sufficient size to provide free space for all conductors enclosed in the conduit body. [314.16(C)(3)]

❖ Conduit bodies such as LBs, LRs and LLs are commonly used with service conductors installed in raceways. These types of elbows have coverplates held in place with two or more screws. The interior volume is only large enough to accommodate the installation of the conductors. The same is true of capped elbows, except that capped elbows are typically used with branch-circuit and feeder conductors and have a removable coverplate or screwed cap that is in a plane 45 degrees from the plane of the raceway connections (i.e., on the outside corner of the elbow).

SECTION E3906
INSTALLATION OF BOXES, CONDUIT BODIES AND FITTINGS

E3906.1 Conductors entering boxes, conduit bodies or fittings. Conductors entering boxes, conduit bodies or fittings shall be protected from abrasion. (314.17)

❖ See the commentary to Section E3906.1.1.

E3906.1.1 Insulated fittings. Where raceways contain 4 AWG or larger insulated circuit conductors and these conductors enter a cabinet, box enclosure, or raceway, the conductors shall be protected by an identified fitting providing a smoothly rounded insulating surface, unless the conductors are separated from the fitting or raceway by identified insulating material securely fastened in place. [300.4(G)]

> **Exception:** Where threaded hubs or bosses that are an integral part of a cabinet, box enclosure, or raceway provide a smoothly rounded or flared entry for conductors. [300.4(G) Exception]

Conduit bushings constructed wholly of insulating material shall not be used to secure a fitting or raceway. The insulating fitting or insulating material shall have a temperature rating not less than the insulation temperature rating of the installed conductors. [330.4(G)]

❖ Conduit entering a panelboard cabinet or box is usually attached or connected with a fitting that enters the box. A conduit could be threaded and have one locknut on the outside of the box and one on the inside to secure it in place. Whether this method is used or a fitting is used to attach the raceway to the box or cabinet, it is a good practice to install an insulating bushing on all raceway entries. The code requires this for conductors of size 4 AWG or larger because there is more chance of damaging the insulation because these conductors are more difficult to bend than the smaller conductors.

E3906.2 Openings. Openings through which conductors enter shall be closed in an approved manner. [314.17(A)]

❖ The fitting used to secure the cable into the box should close the knockout opening sufficiently. If a small conductor is installed in a large knockout with a connector or fitting too large for the conductor, it will leave an opening through which foreign items could enter the cabinet or box.

E3906.3 Metal boxes and conduit bodies. Where raceway or cable is installed with metal boxes, or conduit bodies, the raceway or cable shall be secured to such boxes and conduit bodies. [314.17(B)]

❖ The exception in Section E3905.3.2 pertains to nonmetallic boxes only. For metal boxes, there are no exceptions. The cable must be secured to the box. It is obvious that conduit and tubing installed with metal boxes must be secured to the box.

E3906.4 Unused openings. Unused openings other than those intended for the operation of equipment, those intended for mounting purposes, or those permitted as part of the design for listed equipment, shall be closed to afford protection substantially equivalent to that of the wall of the equipment. Metal plugs or plates used with nonmetallic boxes or conduit bodies shall be recessed at least $^{1}/_{4}$ inch (6.4 mm) from the outer surface of the box or conduit body. [110.12(A)]

❖ Where a knockout on a box or cabinet is removed by mistake or is otherwise not used, a plug designed for the purpose of closing the opening must be installed.

Taping over the opening and other makeshift methods are not allowable methods of closing unused openings. Knockout plugs are available that snap into the opening or that consist of two plates that are joined by a bolt and nut.

E3906.5 In wall or ceiling. In walls or ceilings of concrete, tile or other noncombustible material, boxes employing a flush-type cover or faceplate shall be installed so that the front edge of the box, plaster ring, extension ring, or listed extender will not be set back from the finished surface more than $^{1}/_{4}$ inch (6.4 mm). In walls and ceilings constructed of wood or other combustible material, boxes, plaster rings, extension rings and listed extenders shall be flush with the finished surface or project therefrom. (314.20)

❖ Boxes installed in a noncombustible surface are permitted to be recessed a maximum of $^{1}/_{4}$ inch (6.4 mm) from the surface. In a wall or ceiling of combustible material, the front edge of the box must be flush with the surface. A problem could occur where the electrician installs the box for $^{1}/_{2}$-inch (12.7 mm) drywall and then later wood paneling or trim is installed on top of the drywall. The box then is recessed and must have an extension installed. It may seem sufficient to use longer 6-32 screws than are provided with the devices, but whenever longer screws are needed, the box is probably recessed too far from the surface.

E3906.6 Noncombustible surfaces. Openings in noncombustible surfaces that accommodate boxes employing a flush-type cover or faceplate shall be made so that there are no gaps or open spaces greater than $^{1}/_{8}$ inch (3.2 mm) around the edge of the box. (314.21)

❖ The maximum gap in the wall opening from the edge of the box is $^{1}/_{8}$ inch (3.2 mm). Usually the drywall installer will repair all gaps with drywall joint compound. Gaps could allow sparks or burning materials to escape from the box into the wall or ceiling cavity.

E3906.7 Surface extensions. Surface extensions shall be made by mounting and mechanically securing an extension ring over the box. (314.22)

> **Exception:** A surface extension shall be permitted to be made from the cover of a flush-mounted box where the cover is designed so it is unlikely to fall off, or be removed if its securing means becomes loose. The wiring method shall be flexible for an approved length that permits removal of the cover and provides access to the box interior and shall be arranged so that any bonding or grounding continuity is independent of the connection between the box and cover. (314.22 Exception)

❖ A box extension must be secured to the box and not just to the wall or ceiling surface. It is acceptable to attach a flexible conduit to a box cover rather than using a box extension. In this case, a grounding conductor must be installed in the conduit to maintain grounding continuity when the cover is removed from the box.

E3906.8 Supports. Boxes and enclosures shall be supported in accordance with one or more of the provisions in Sections E3906.8.1 through E3906.8.6. (314.23)

❖ The six subsections of Section E3906.8 cover the code provisions for the support of boxes and enclosures. There is no specific provision for the number of nails or screws to be used to attach boxes and enclosures, and there are no exact provisions pertaining to the weight of the box. How securely the box is supported is a judgment call by the installer and the inspector. Many variables enter into consideration here including the type of surface or structure the box is mounted on. In some cases, special support must be constructed in addition to the building structure or components. This code section is similar to Section E3404.8 (see the commentary to that section).

E3906.8.1 Surface mounting. An enclosure mounted on a building or other surface shall be rigidly and securely fastened in place. If the surface does not provide rigid and secure support, additional support in accordance with other provisions of Section E3906.8 shall be provided. [314.23(A)]

❖ See Sections E3906.8 and E3404.8.

E3906.8.2 Structural mounting. An enclosure supported from a structural member or from grade shall be rigidly supported either directly, or by using a metal, polymeric or wood brace. [314.23(B)]

❖ See Sections E3906.8 and E3404.8.

E3906.8.2.1 Nails and screws. Nails and screws, where used as a fastening means, shall be attached by using brackets on the outside of the enclosure, or they shall pass through the interior within $1/4$ inch (6.4 mm) of the back or ends of the enclosure. Screws shall not be permitted to pass through the box except where exposed threads in the box are protected by an approved means to avoid abrasion of conductor insulation. [314.23(B)(1)]

❖ Screw threads exposed inside of a box present sharp edges that can easily damage wiring (see Sections E3906.8 and E3404.8).

E3906.8.2.2 Braces. Metal braces shall be protected against corrosion and formed from metal that is not less than 0.020 inch (0.508 mm) thick uncoated. Wood braces shall have a cross section not less than nominal 1 inch by 2 inches (25.4 mm by 51 mm). Wood braces in wet locations shall be treated for the conditions. Polymeric braces shall be identified as being suitable for the use. [314.23(B)(2)]

❖ See Sections E3906.8 and E3404.8.

E3906.8.3 Mounting in finished surfaces. An enclosure mounted in a finished surface shall be rigidly secured there to by clamps, anchors, or fittings identified for the application. [314.23(C)]

❖ Where the house is finished and the homeowner wants to add an outlet, the provisions of this section apply. Support of a box solely by the wallboard, for a switch or receptacle for example, is permitted if clamps, anchors and/or fittings identified for the purpose are used. Boxes specifically designed and approved for this kind

of installation are available. In some areas, these boxes are called an "old work box" from the idea that outlets are being added in an existing house or in an old house. Another term is "cut-in box" from the idea that the wall is being cut into for the installation of the box. A hole is cut in the wall exactly to the shape of the box. The box has brackets that seat against the face or surface of the drywall and brackets that tighten against the inside of the drywall with screws.

Another approved anchor is called a "hold it" or "F" strap, which is a thin piece of metal that slips into the wall along each side of the box and is bent into the box (see also Sections E3906.8 and E3404.8).

E3906.8.4 Raceway supported enclosures without devices or fixtures. An enclosure that does not contain a device(s), other than splicing devices, or support a luminaire, lampholder or other equipment, and that is supported by entering raceways shall not exceed 100 cubic inches (1640 cm³) in size. The enclosure shall have threaded entries or identified hubs. The enclosure shall be supported by two or more conduits threaded wrenchtight into the enclosure or hubs. Each conduit shall be secured within 3 feet (914 mm) of the enclosure, or within 18 inches (457 mm) of the enclosure if all entries are on the same side of the enclosure. [314.23(E)]

Exception: Rigid metal, intermediate metal, or rigid polyvinyl chloride nonmetallic conduit or electrical metallic tubing shall be permitted to support a conduit body of any size, provided that the conduit body is not larger in trade size than the largest trade size of the supporting conduit or electrical metallic tubing. [314.23(E) Exception]

❖ Where boxes are used outdoors in landscaping, the provisions of this section apply to the support of the boxes, where outdoor receptacles or decorative lighting are installed around patios or in flower beds, for example. This section pertains to junction boxes only or boxes that do not support fixtures or contain any devices. The box must be supported by at least two conduits that are threaded into the box or enclosure. If the two conduits are run into the same side of the box, such as in the installation of boxes for landscape wiring where the two conduits emerge from the ground and are run into the bottom of the box for support, the box must be no more than 18 inches (457 mm) above the ground or from the secure support. These types of threaded conduits are rigid metal conduit, intermediate metal conduit and Schedule 80 nonmetallic conduit. EMT, sometimes referred to as "thin wall," is not considered to be conduit and cannot be used to support a box or enclosure in this manner.

It would be a code violation to install the box on only one conduit that emerges from the ground, even if it was a rigid metal conduit and secured by concrete. The code requires two conduits. In some cases, at the end of a circuit, a second conduit is installed just for support of the box and is stubbed into the ground for a short distance but not connected to any other box and has no wire in it.

The exception allows a conduit body to be supported by rigid metal conduit, intermediate metal conduit, EMT or rigid nonmetallic conduit. Conduit bodies are usually not supported independently from the conduit but are intended to be part of the conduit run and are similar to a fitting in this respect.

E3906.8.5 Raceway supported enclosures, with devices or luminaire. An enclosure that contains a device(s), other than splicing devices, or supports a luminaire, lampholder or other equipment and is supported by entering raceways shall not exceed 100 cubic inches (1640 cm³) in size. The enclosure shall have threaded entries or identified hubs. The enclosure shall be supported by two or more conduits threaded wrench-tight into the enclosure or hubs. Each conduit shall be secured within 18 inches (457 mm) of the enclosure. [314.23(F)]

Exceptions:

1. Rigid metal or intermediate metal conduit shall be permitted to support a conduit body of any size, provided that the conduit bodies are not larger in trade size than the largest trade size of the supporting conduit. [314.23(F) Exception No. 1]

2. An unbroken length(s) of rigid or intermediate metal conduit shall be permitted to support a box used for luminaire or lampholder support, or to support a wiring enclosure that is an integral part of a luminaire and used in lieu of a box in accordance with Section E3905.1.1, where all of the following conditions are met:

   2.1. The conduit is securely fastened at a point so that the length of conduit beyond the last point of conduit support does not exceed 3 feet (914 mm).

   2.2. The unbroken conduit length before the last point of conduit support is 12 inches (305 mm) or greater, and that portion of the conduit is securely fastened at some point not less than 12 inches (305 mm) from its last point of support.

   2.3. Where accessible to unqualified persons, the luminaire or lampholder, measured to its lowest point, is not less than 8 feet (2438 mm) above grade or standing area and at least 3 feet (914 mm) measured horizontally to the 8-foot (2438 mm) elevation from windows, doors, porches, fire escapes, or similar locations.

   2.4. A luminaire supported by a single conduit does not exceed 12 inches (305 mm) in any direction from the point of conduit entry.

   2.5. The weight supported by any single conduit does not exceed 20 pounds (9.1 kg).

   2.6. At the luminaire or lampholder end, the conduit(s) is threaded wrenchtight into the box, conduit body, or integral wiring enclosure, or into hubs identified for the purpose. Where a box or conduit body is used for support, the luminaire shall be secured directly to the box or conduit body, or through a threaded conduit nipple

not over 3 inches (76 mm) long. [314.23(F) Exception No. 2]

❖ The main difference in this section and Section E3906.8.4 is that if the box contains a switch or receptacle or supports a luminaire, the conduit must be secured within 18 inches (457 mm) of the enclosure, and the provision for being secured within 36 inches (914 mm) from the box is not allowed. Note that the exceptions omit EMT and rigid nonmetallic conduit.

E3906.8.6 Enclosures in concrete or masonry. An enclosure supported by embedment shall be identified as being suitably protected from corrosion and shall be securely embedded in concrete or masonry. [314.23(G)]

❖ Where boxes and enclosures are supported by being encased in concrete or masonry, they must be identified for the application.

E3906.9 Covers and canopies. Outlet boxes shall be effectively closed with a cover, faceplate or fixture canopy. Screws used for the purpose of attaching covers, or other equipment to the box shall be either machine screws matching the thread gauge or size that is integral to the box or shall be in accordance with the manufacturer's instructions. (314.25)

❖ All outlets must have a cover, whether it is a blank cover, the faceplate of a switch or receptacle or the canopy of a luminaire. A residence should not pass inspection if a junction box is left open with the conductors exposed, even if they are tucked or folded back into the box. Where a luminaire is to be installed in the future, the box must be covered with a blank cover.

E3906.10 Covers and plates. Covers and plates shall be nonmetallic or metal. Metal covers and plates shall be grounded. [314.25(A)]

❖ Metal covers are grounded by the metal attachment screws that screw into the metal box or metal yoke of the device. Receptacles and switches have a grounding terminal that is connected to the yoke or strap on which the device is mounted. Therefore, the metal plate is grounded through the mounting screws into the yoke and through the grounding terminal. If a metal box is used and the yoke seats against the metal, as in the case of a surface-mounted box, there will be direct metal-to-metal contact between the yoke and the metal box. This is an approved method of grounding the device and any metal screws or metal cover plates. If there is no metal-to-metal contact, as where a nonmetallic box is used, the mounting strap of a switch must have a grounding terminal to ensure that a metal face plate will be grounded.

See the code requirements on grounding terminals of wiring devices in Section E3908.14 and switch faceplate grounding in Section E4001.11.1.

E3906.11 Exposed combustible finish. Combustible wall or ceiling finish exposed between the edge of a fixture canopy or pan and the outlet box shall be covered with noncombustible material where required by Section E4004.2. [314.25(B)]

❖ Note that this section is related to Section E4004.2 which contains a threshold of 180 square inches for the combustible finish area. This code provision applies where the fixture has a dome, canopy or other shape that creates a void space behind it and would not apply where the fixture has a flat metal back that fits entirely flat against the wall or ceiling surface. An example would be where a round 4-inch (102 mm) box for a wall-hung luminaire is used on a wall with a wood paneling finish. If installed properly, the front edge of the box will be flush with the finished wall surface. A luminaire is installed having a canopy 8 inches (203 mm) in diameter that seats against the wall, leaving a wall surface area 2 inches (51 mm) wide around the box exposed inside the canopy. This area becomes part of the space where the connections are made between the circuit wires and the fixture wires. If arcing occurs or the fixture overheats, heat buildup could cause the wood wall surface around the box to ignite. Thus, that area of combustible wall must be covered with a noncombustible material. Where walls and ceilings are finished with combustible materials such as wood paneling and boards, and fixtures are mounted on such surfaces, the electrical inspector should verify that the surfaces are protected under the fixture pans, domes and canopies. The method of protection is not specified but would likely consist of a metal ring or plate or some noncombustible fiber pad attached to the outlet box such that it is not visible beyond the body of the fixture. Such metal protection means should be grounded to the outlet box or equipment grounding conductor.

SECTION E3907
CABINETS AND PANELBOARDS

E3907.1 Switch and overcurrent device enclosures with splices, taps, and feed-through conductors. Where the wiring space of enclosures for switches or overcurrent devices contains conductors that are feeding through, spliced, or tapping off to other enclosures, switches, or overcurrent devices, all of the following conditions shall apply:

1. The total area of all conductors installed at any cross section of the wiring space shall not exceed 40 percent of the cross-sectional area of that space.

2. The total area of all conductors, splices, and taps installed at any cross section of the wiring space shall not exceed 75 percent of the cross-sectional area of that space.

3. A warning label shall be applied to the enclosure that identifies the closest disconnecting means for any feed-through conductors. (312.8)

❖ This section allows splices within panelboards and switch and circuit breaker enclosures if there is ample room to make and contain such splices without crowding the conductors. It is commonly necessary to splice branch circuit conductors within panelboards as a result of relocating circuit breakers in the panel or replacing standard circuit breakers with GFCI- or AFCI-type breakers. Splices are also necessary when the panelboard is replaced as part of a service upgrade.

E3907.2 Damp and wet locations. In damp or wet locations, cabinets and panelboards of the surface type shall be placed or equipped so as to prevent moisture or water from entering and accumulating within the cabinet, and shall be mounted to provide an air-space not less than $^1/_4$ inch (6.4 mm) between the enclosure and the wall or other supporting surface. Cabinets installed in wet locations shall be weatherproof. For enclosures in wet locations, raceways and cables entering above the level of uninsulated live parts shall be installed with fittings listed for wet locations. (312.2)

> **Exception:** Nonmetallic enclosures installed on concrete, masonry, tile, or similar surfaces shall not be required to be installed with an air space between the enclosure and the wall or supporting surface. (312.2 Exception)

❖ Cabinets and other boxes and enclosures that are mounted in wet or damp locations must have at least a $^1/_4$-inch (6.4 mm) airspace between the wall, cabinet or box so that water or moisture is not trapped against the metal. In some areas and locations, corrosion will result from moisture behind the cabinet. Concentric knockouts factory punched in cabinets for wet locations cannot be made water tight which is why such knockouts are found only below the level of live parts within the cabinet.

E3907.3 Position in wall. In walls of concrete, tile or other noncombustible material, cabinets and panelboards shall be installed so that the front edge of the cabinet will not set back of the finished surface more than $^1/_4$ inch (6.4 mm). In walls constructed of wood or other combustible material, cabinets shall be flush with the finished surface or shall project therefrom. (312.3)

❖ See Section E3906.5.

E3907.4 Repairing noncombustible surfaces. Noncombustible surfaces that are broken or incomplete shall be repaired so that there will not be gaps or open spaces greater than $^1/_8$ inch (3.2 mm) at the edge of the cabinet or cutout box employing a flush-type cover. (312.4)

❖ See Section E3906.6.

E3907.5 Unused openings. Unused openings, other than those intended for the operation of equipment, those intended for mounting purposes, and those permitted as part of the design for listed equipment, shall be closed to afford protection substantially equivalent to that of the wall of the equipment. Metal plugs and plates used with nonmetallic cabinets shall be recessed at least $^1/_4$ inch (6.4 mm) from the outer surface. Unused openings for circuit breakers and switches shall be closed using identified closures, or other approved means that provide protection substantially equivalent to the wall of the enclosure. (110.12(A)]

❖ Closures identified for the purpose must be used to cover unused circuit breaker openings. Such openings would allow fingers or metal objects to contact uninsulated bus bars in the panelboard which is obviously dangerous. Duct tape and sheet metal scraps are not equivalent (see Section E3906.4). Panelboard enclosure manufacturers provide closure plugs for the purpose of closing openings that are not closed by a circuit breaker.

E3907.6 Conductors entering cabinets. Conductors entering cabinets and panelboards shall be protected from abrasion and shall comply with Section E3906.1.1. (312.5)

❖ See Section E3906.1.1.

E3907.7 Openings to be closed. Openings through which conductors enter cabinets, panelboards and meter sockets shall be closed in an approved manner. [312.5(A)]

❖ See Section E3906.2.

E3907.8 Cables. Where cables are used, each cable shall be secured to the cabinet, panelboard, cutout box, or meter socket enclosure. [312.5(C)]

> **Exception:** Cables with entirely nonmetallic sheaths shall be permitted to enter the top of a surface-mounted enclosure through one or more sections of rigid raceway not less than 18 inches (457 mm) nor more than 10 feet (3048 mm) in length, provided that all the following conditions are met:
>
> 1. Each cable is fastened within 12 inches (305 mm), measured along the sheath, of the outer end of the raceway.
>
> 2. The raceway extends directly above the enclosure and does not penetrate a structural ceiling.
>
> 3. A fitting is provided on each end of the raceway to protect the cable(s) from abrasion and the fittings remain accessible after installation.
>
> 4. The raceway is sealed or plugged at the outer end using approved means so as to prevent access to the enclosure through the raceway.
>
> 5. The cable sheath is continuous through the raceway and extends into the enclosure beyond the fitting not less than $^1/_4$ inch (6.4 mm).
>
> 6. The raceway is fastened at its outer end and at other points in accordance with Section E3802.1.
>
> 7. The allowable cable fill shall not exceed that permitted by Table E3907.8. A multiconductor cable having two or more conductors shall be treated as a single conductor for calculating the percentage of conduit fill area. For cables that have elliptical cross sections, the cross-sectional area calculation shall be based on the major diameter of the ellipse as a circle diameter. [312.5(C) Exception]

❖ Nonmetallic cable must be secured with cable connectors to the cabinet (see Commentary Figure E3907.8). Where a cable enters a cabinet through a knockout, running the cable directly through the hole without any type of connector would be a code violation. Where the connector is omitted, the sharp edge of the knockout could damage the sheathing on the cable and the cable could be pulled out. Installers have been known to wrap the cable with several layers of tape at the point where it enters the cabinet through the knockout. This method is not an acceptable alternative to cable connectors.

The exception recognizes what has been common practice in some localities, and that is running Type NM cables through a conduit riser and into the top of a panelboard cabinet. This practice can overcome a shortage of $^1/_2$-inch (12.7 mm) knockouts on the top of a cabinet and provides physical protection for the cables. It also saves labor in some cases and makes a neat appearance. Section E3705.3 applies where the raceway is longer than 24 inches (610 mm). Note that this installation is limited to surface-mounted cabinets and panelboards only.

TABLE E3907.8 (Chapter 9, Table 1)
PERCENT OF CROSS SECTION OF
CONDUIT AND TUBING FOR CONDUCTORS

NUMBER OF CONDUCTORS	MAXIMUM PERCENT OF CONDUIT AND TUBING AREA FILLED BY CONDUCTORS
1	53
2	31
Over 2	40

❖ Each Type NM cable is treated as a conductor when using this table. The fill limits prevent overcrowding and cable damage.

E3907.9 Wire-bending space within an enclosure containing a panelboard. Wire-bending space within an enclosure containing a panelboard shall comply with the requirements of Sections E3907.9.1 through E3907.9.3.

❖ Recall that a panelboard is defined as the buses, terminals, insulators, switches and overcurrent devices (i.e., the guts) that are housed in a cabinet. Loosely speaking, the entire assembly of a panelboard and cabinet is referred to in the field as the "panel."

E3907.9.1 Top and bottom wire-bending space. The top and bottom wire-bending space for a panelboard enclosure shall be sized in accordance with Table E3907.9.1(1) based on the largest conductor entering or leaving the enclosure. [408.55 (A)]

> **Exceptions:**
>
> 1. For a panelboard rated at 225 amperes or less and designed to contain not more than 42 overcurrent devices, either the top or bottom wire-bending space shall be permitted to be sized in accordance with Table E3907.9.1(2). For the purposes of this exception, a 2-pole or a 3-pole circuit breaker shall be considered as two or three overcurrent devices, respectively. [408.55(A) Exception No. 1]
>
> 2. For any panelboard, either the top or bottom wire-bending space shall be permitted to be sized in accordance with Table E3907.9.1(2) where the wire-bending space on at least one side is sized in accordance with Table E3907.9.1(1) based on the largest conductor to be terminated in any side wire-bending space. [408.55(A) Exception No. 2]
>
> 3. Where the panelboard is designed and constructed for wiring using only a single 90-degree bend for each conductor, including the grounded circuit conductor, and the wiring diagram indicates and specifies the method of wiring that must be used, the top

and bottom wire-bending space shall be permitted to be sized in accordance with Table E3907.9.1(2). [408.55(A) Exception No. 3]

4. Where there are no conductors terminated in that space, either the top or the bottom wire-bending space, shall be permitted to be sized in accordance with Table E3907.9.1(2). [408.55(A) Exception No. 4]

❖ The code requires adequate space to allow conductors to be bent and terminated without excessive bending. Excessive bending can damage conductor insulation by thinning of insulation on the outer radius of the bend and by compressing insulation against raceway fittings and cabinet walls. Adequate space at terminals allows long radius bends to be made and prevents conductors from being compressed against cabinet walls and fittings. Of course, adequate space makes installation easier and safer as well. Note that the wire bending spaces within the typical cabinet (panel) installed in residential buildings is not something that the installer has control over. It is what it is, so to speak, as provided by the manufacturer; however, the installer must plan where to place raceways and overcurrent devices in the cabinet so as to be in code compliance. If there is not the required space in one area of a cabinet, then the raceway needs to enter at a location where there is adequate space. With thoughtful planning, raceways and overcurrent devices can be located where there will be no need to bend the conductors other than slightly. Less space is required for "lay-in" terminals because the conductor does not have to be pulled back and then inserted into a set screw-type terminal. "Lay-in" terminals open up to allow the conductors to be laid into a saddle without requiring an insertion movement. "Lay-in" terminals are common in meter enclosures and other equipment where large conductors are used.

E3907.9.2 Side wire-bending space. Side wire-bending space shall be in accordance with Table E3907.9.1(2) based on the largest conductor to be terminated in that space. [408.55(B)]

❖ The largest conductor in the space governs the space requirements because it represents the worst case scenario. See Note 1 in Table E 3907.9.1(2), which explains how the space dimension is measured. For example, if 2 AWG conductors are the largest conductors in the side space of a panel and they are to be terminated on a 100-amp two-pole circuit breaker, the minimum required space must be $2^1/_2$ inches (64 mm) in width measured from the outside end of the circuit breaker terminals to the wall of the cabinet that is opposite the terminals. The table recognizes that for very large conductors, they might be run in parallel, thus there is a column for two conductors per terminal.

E3907.9.3 Back wire-bending space. The distance between the center of the rear entry and the nearest termination for the entering conductors shall be not less than the distance given in Table E3907.9.1(1). Where a raceway or cable entry is in the wall of the enclosure, opposite a removable cover, the distance from that wall to the cover shall be permitted to comply with the distance required in Table E3907.9.1(2). [408.55 (C)]

❖ Rear wall knockouts in cabinets (panels) might be located such that the center of the KO opening is very close to the terminals intended for the conductors, thus providing inadequate space to maneuver the conductors. Again, thoughtful planning on where to enter the raceway is important. The typical cabinet and panelboard assembly will have a removable cover opposite of the rear entry KOs, therefore, Table E3907.9.1(2) can be used, which requires less space.

Figure E3907.8
CABLES SECURED TO PANELBOARD ENCLOSURES

TABLE E3907.9.1(1) [Table 312.6(B)]
MINIMUM WIRE-BENDING SPACE AT TERMINALS (see note 1)

WIRE SIZE (AWG or kcmil)		WIRES PER TERMINAL			
		One (see note 2)		Two	
All other conductors	Compact stranded AA-8000 aluminum alloy conductors (see Note 3)	inches	mm	inches	mm
14-10	12-8	Not specified	Not specified	—	—
8	6	$1^1/_2$	38.1	—	—
6	4	2	50.8	—	—
4	2	3	76.2	—	—
3	1	3	76.2	—	—
2	1/0	$3^1/_2$	88.9	—	—
1	2/0	$4^1/_2$	114	—	—
1/0	3/0	$5^1/_2$	140	$5^1/_2$	140
2/0	4/0	6	152	6	152
3/0	250	$6^1/_2$[a]	165[a]	$6^1/_2$[a]	165[a]
4/0	300	7[b]	178[b]	$7^1/_2$[c]	190[c]
250	350	$8^1/_2$[d]	216[d]	$8^1/_2$[d]	229[d]
300	400	10[c]	254[c]	10[d]	254[d]
350	500	12[c]	305[c]	12[c]	305[c]
400	600	13[c]	330[c]	13[c]	330[c]
500	700-750	14[c]	356[c]	14[c]	356[c]
600	800-900	15[c]	381[c]	16[c]	406[c]
700	1000	16[c]	406[c]	18[c]	457[c]

1. Bending space at terminals shall be measured in a straight line from the end of the lug or wire connector in a direction perpendicular to the enclosure wall.

2. For removable and lay-in wire terminals intended for only one wire, bending space shall be permitted to be reduced by the following number of inches (millimeters):

   a. $^1/_2$ inches (12.7 mm)

   b. 1 inches (25.4 mm)

   c. $1^1/_2$ inches (38.1 mm)

   d. 2 inches (50.8 mm)

3. This column shall be permitted to determine the required wire-bending space for compact stranded aluminum conductors in sizes up to 1000 kcmil and manufactured using AA-8000 series electrical grade aluminum alloy conductor material.

TABLE E3907.9.1(2) [Table 312.6(A)]
MINIMUM WIRE-BENDING SPACE AT TERMINALS
AND MINIMUM WIDTH OF WIRING GUTTERS (see note 1)

WIRE SIZE (AWG or kcmil)	WIRES PER TERMINAL			
	One		Two	
	inches	mm	inches	mm
14-10	Not specified	Not specified	—	—
8-6	$1^1/_2$	38.1	—	—
4-3	2	50.8	—	—
2	$2^1/_2$	63.5	—	—
1	2	76.2	—	—
1/0-2/0	$3^1/_2$	88.9	5	127
3/0-4/0	4	102	6	152
250	$4^1/_2$	114	6	152
300-350	5	127	8	203
400-500	6	152	8	203
600-700	8	203	10	254

1. Bending space at terminals shall be measured in a straight line from the end of the lug or wire connector in the direction that the wire leaves the terminal to the wall, barrier, or obstruction.

SECTION E3908
GROUNDING

E3908.1 Metal enclosures. Metal enclosures of conductors, devices and equipment shall be connected to the equipment grounding conductor. (250.86)

Exceptions:

1. Short sections of metal enclosures or raceways used to provide cable assemblies with support or protection against physical damage. (250.86 Exception No. 2)

2. A metal elbow that is installed in an underground installation of rigid nonmetallic conduit and is isolated from possible contact by a minimum cover of 18 inches (457 mm) to any part of the elbow or that is encased in not less than 2 inches (51 mm) of concrete. (250.86 Exception No. 3)

❖ Two general subjects of grounding are system grounding and equipment grounding. This section covers provisions for equipment grounding. In general, all metal enclosures must be grounded. Grounding serves the purpose of preventing a voltage potential from existing be-tween a conductive surface and the earth or surfaces in contact with the earth. A voltage potential between a surface and the earth could result in a shock hazard where living creatures, humans included, put themselves in contact with the earth and an energized surface. The voltage potential would cause current to flow through living tissue. Surfaces that are tied to the earth electrically (grounded) will have no voltage potential between them and the earth. The surfaces and the earth will be at the same potential with no differences in voltage between them.

E3908.2 Equipment fastened in place or connected by permanent wiring methods (fixed). Exposed, normally non-current-carrying metal parts of fixed equipment supplied by or enclosing conductors or components that are likely to become energized shall be connected to the equipment grounding conductor where any of the following conditions apply:

1. Where within 8 feet (2438 mm) vertically or 5 feet (1524 mm) horizontally of earth or grounded metal objects and subject to contact by persons;

2. Where located in a wet or damp location and not isolated; or

3. Where in electrical contact with metal. (250.110)

❖ Examples of equipment fastened in place include furnaces, water heaters, dishwashers, condensing units and evaporative coolers. There is almost no equipment served by a residential wiring system that is exempt from grounding requirements. Section E3908.9 lists the methods of grounding this type of equipment.

E3908.3 Specific equipment fastened in place (fixed) or connected by permanent wiring methods. Exposed, normally noncurrent-carrying metal parts of the following equipment and enclosures shall be connected to an equipment grounding conductor:

1. Luminaires as provided in Chapter 40. [250.112(J)]

2. Motor-operated water pumps, including submersible types. Where a submersible pump is used in a metal well casing, the well casing shall be connected to the pump circuit equipment grounding conductor. [250.112(L)]

❖ Although a submersible water pump might not be attached to the metal well casing, the code requires that the casing be bonded to the pump circuit equipment grounding conductor (see Section E 4003.3).

E3908.4 Effective ground-fault current path. Electrical equipment and wiring and other electrically conductive material likely to become energized shall be installed in a manner that creates a low-impedance circuit facilitating the operation of the overcurrent device or ground detector for high-impedance grounded systems. Such circuit shall be capable of safely carrying the maximum ground-fault current likely to be imposed on it from any point on the wiring system where a ground fault might occur to the electrical supply source. [250.(A)(5)]

❖ When a ground-fault occurs at some point in the electrical wiring system, the current will return to the source of the power supply if there is an electrically continuous path. Such path must be able to conduct the current necessary to open the overcurrent devices and clear the fault. For example, if a ground fault occurs because failed insulation allowed a conductor to contact the metal raceway in which it is installed, sufficient fault current must be conducted through the raceway all the way back to the source of the circuit to allow the overcurrent protective device to trip. Where an equipment grounding conductor is used, it must be sized according to Table E3908.12.

E3908.5 Earth as a ground-fault current path. The earth shall not be considered as an effective ground-fault current path. [250.4(A)(5)]

❖ Equipment must be grounded by connection to a grounding conductor and not only to the earth. For example, driving a ground rod at the equipment location and grounding only to the ground rod would not create the electrically continuous and permanent path that must be established to allow enough current to flow during a fault condition to trip the circuit breaker or set of fuses. Some current flows through the earth, but not enough to trip the circuit breaker.

E3908.6 Load-side grounded conductor neutral. A grounded conductor shall not be connected to normally non-current-carrying metal parts of equipment, to equipment grounding conductor(s), or be reconnected to ground on the load side of the service disconnecting means. [250.24(A)(5)]

❖ The grounding and grounded conductors are connected together only within the service equipment. At any and all points downstream of the service, meaning on the load side of the main disconnect, the grounding conductor and the grounded conductor must be electrically isolated. The only exception is the specific allowance in Section E3607.3.2. A shock hazard and fire hazard is created where grounding conductors, raceways and enclosures are connected to a grounded conductor at any point other than in or ahead of the service equipment (see Commentary Figures E3607.2 and E3908.7).

E3908.7 Load-side equipment. A grounded circuit conductor shall not be used for grounding noncurrent-carrying metal parts of equipment on the load side of the service disconnecting means. [250.142(B)]

❖ See Commentary Figure E3908.7 and the commentary to Sections E3607.2 and E3908.6. Commentary Figure E3908.7 illustrates what can happen when a grounded circuit conductor is used to ground an appliance or when the grounded circuit conductor and the grounding conductor are not electrically isolated from each other. Because the grounded circuit conductor is carrying current, the entire appliance housing will be energized, creating a potential shock hazard. Also, any metallic raceway will be energized, causing it to become a current carrier in parallel with the grounded conductor and a potential shock hazard as well. A fire hazard is also introduced because of possible arcing at any weak links in the continuity of the raceway. If the grounded conductor were lost for some reason, the raceway would carry the entire load, and the appliance would continue to operate even though it would be hazardous. Voltage drop along the circuit conductors and raceway can cause voltage potentials to exist between the appliance housing or raceway and any other grounded surface.

E3908.8 Types of equipment grounding conductors. The equipment grounding conductor run with or enclosing the circuit conductors shall be one or more or a combination of the following:

1. A copper, aluminum or copper-clad conductor. This conductor shall be solid or stranded; insulated, covered or bare; and in the form of a wire or a busbar of any shape. [250.118(1)]

2. Rigid metal conduit. [250.118(2)]

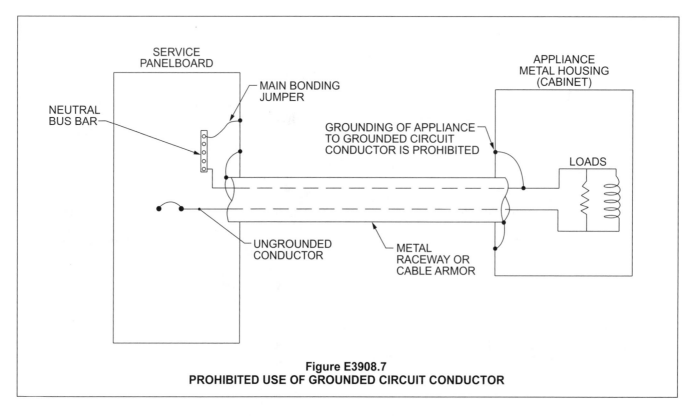

Figure E3908.7
PROHIBITED USE OF GROUNDED CIRCUIT CONDUCTOR

3. Intermediate metal conduit. [250.118(3)]

4. Electrical metallic tubing. [250.118(4)]

5. Armor of Type AC cable in accordance with Section E3908.4. [250.118(8)]

6. Type MC cable that provides an effective ground-fault current path in accordance with one or more of the following:

    6.1. It contains an insulated or uninsulated equipment grounding conductor in compliance with Item 1 of this section.

    6.2. The combined metallic sheath and uninsulated equipment grounding/bonding conductor of interlocked metal tape-type MC cable that is listed and identified as an equipment grounding conductor.

    6.3. The metallic sheath or the combined metallic sheath and equipment grounding conductors of the smooth or corrugated tube-type MC cable that is listed and identified as an equipment grounding conductor. [250.118(10)]

7. Other electrically continuous metal raceways and auxiliary gutters. [250.118(13)]

8. Surface metal raceways listed for grounding. [250.118(14)]

❖ Where conduit is used as the grounding conductor, it is very important that all connections be made properly so that there is an electrically continuous path for any fault current. The conduit system can be considered as a grounding conductor, but it is really a conductor that is "manufactured" or assembled on the job. A grounding conductor can be pulled into the conduit system where there is a lack of confidence in the integrity of all the connectors and couplings of the assembled system. If the electrician forgets to tighten one set screw or locknut on a fitting, for example, a high resistance could be created in the grounding return path.

E3908.8.1 Flexible metal conduit. Flexible metal conduit shall be permitted as an equipment grounding conductor where all of the following conditions are met:

1. The conduit is terminated in listed fittings.

2. The circuit conductors contained in the conduit are protected by overcurrent devices rated at 20 amperes or less.

3. The combined length of flexible metal conduit and flexible metallic tubing and liquid-tight flexible metal conduit in the same ground return path does not exceed 6 feet (1829 mm).

If used to connect equipment where flexibility is necessary to minimize the transmission of vibration from equipment or to provide flexibility for equipment that requires movement after installation, an equipment grounding conductor shall be installed. [250.118(5)]

❖ Flexible metal conduit is suitable as a grounding conductor to a certain extent, but the conditions are very limiting. It can be used only for circuits of 20 amperes

or less, only in total return path lengths of 6 feet (1829 mm) or less, and only with listed fittings. All flexible metal conduit runs over 6 feet in length require that a grounding conductor be installed within the conduit. It is usually easier to install a grounding conductor in a run of 6 feet or less than find the listed grounding fittings and comply with all the conditions. Where flexible metal conduit is used to provide flexibility for the equipment being served, such as a motor that may need to be replaced or equipment that may vibrate, the flexible metal conduit cannot be relied on for grounding, and a grounding conductor must be installed within the conduit.

Because flexible conduit is formed as a spiral, fault current would have to travel along the spiral path and not straight along the length of the conduit. The fault current return path is longer following the long spiral path, and the spiral path acts as an inductor coil; thus, the impedance could be too high to allow the current flow necessary to trip the overcurrent device.

E3908.8.2 Liquid-tight flexible metal conduit. Liquid-tight flexible metal conduit shall be permitted as an equipment grounding conductor where all of the following conditions are met:

1. The conduit is terminated in listed fittings.

2. For trade sizes $^3/_8$ through $^1/_2$ (metric designator 12 through 16), the circuit conductors contained in the conduit are protected by overcurrent devices rated at 20 amperes or less.

3. For trade sizes $^3/_4$ through $1^1/_4$ (metric designator 21 through 35), the circuit conductors contained in the conduit are protected by overcurrent devices rated at not more than 60 amperes and there is no flexible metal conduit, flexible metallic tubing, or liquid-tight flexible metal conduit in trade sizes $^3/_8$ inch or $^1/_2$ inch (9.5 mm through 12.7 mm) in the ground fault current path.

4. The combined length of flexible metal conduit and flexible metallic tubing and liquid-tight flexible metal conduit in the same ground return path does not exceed 6 feet (1829 mm).

If used to connect equipment where flexibility is necessary to minimize the transmission of vibration from equipment or to provide flexibility for equipment that requires movement after installation, an equipment grounding conductor shall be installed. [250.118(6)]

❖ The same conditions apply to liquid-tight flexible conduit as to flexible metal conduit mentioned above, with the exception of size. For a circuit up to 60 amperes, the code specifies the larger conduit sizes indicated for the liquid-tight flexible metal conduit that is to be used as a grounding conductor. Again, the conditions are so limiting that for residential wiring most electricians choose to run an appropriate size grounding conductor within the conduit. If the conduit is used as a grounding conductor, the conditions must be strictly adhered to.

E3908.8.3 Nonmetallic sheathed cable (Type NM). In addition to the insulated conductors, the cable shall have an insulated, covered, or bare equipment grounding conductor.

Equipment grounding conductors shall be sized in accordance with Table E3908.12. (334.108)

❖ The equipment grounding conductor in a nonmetallic sheathed cable is uninsulated (bare) and must be used only as an equipment grounding conductor.

E3908.9 Equipment fastened in place or connected by permanent wiring methods. Noncurrent-carrying metal parts of equipment, raceways and other enclosures, where required to be grounded, shall be grounded by one of the following methods: (250.134)

1. By any of the equipment grounding conductors permitted by Sections E3908.8 through E3908.8.3. [250.134(A)]

2. By an equipment grounding conductor contained within the same raceway, cable or cord, or otherwise run with the circuit conductors. Equipment grounding conductors shall be identified in accordance with Section E3407.2. [250.134(B)]

❖ This section simply requires that equipment be grounded by one of the metallic wiring methods in Section E3908.8 or by a separate conductor run in the conduit, tubing or cable. If the grounding conductor is insulated, it must be green. See Section E3407 for conductor color requirements.

E3908.10 Methods of equipment grounding. Fixtures and equipment shall be considered grounded where mechanically connected to an equipment grounding conductor as specified in Sections E3908.8 through E3908.8.3. Wire type equipment grounding conductors shall be sized in accordance with Section E3908.12. (250 Part VII)

❖ See Section E3908.8.

E3908.11 Equipment grounding conductor installation. Where an equipment grounding conductor consists of a raceway, cable armor or cable sheath or where such conductor is a wire within a raceway or cable, it shall be installed in accordance with the provisions of this chapter and Chapters 34 and 38 using fittings for joints and terminations approved for installation with the type of raceway or cable used. All connections, joints and fittings shall be made tight using suitable tools. (250.120)

❖ See Section E3908.8.

E3908.12 Equipment grounding conductor size. Copper, aluminum and copper-clad aluminum equipment grounding conductors of the wire type shall be not smaller than shown in Table E3908.12, but they shall not be required to be larger than the circuit conductors supplying the equipment. Where a raceway or a cable armor or sheath is used as the equipment grounding conductor, as provided in Section E3908.8, it shall comply with Section E3908.4. Where ungrounded conductors are increased in size from the minimum size that has sufficient ampacity for the intended installation, wire type equipment grounding conductors shall be increased proportionally according to the circular mil area of the ungrounded conductors. [250.122(A) and (B)]

❖ The rule that the equipment grounding conductor does not have to be larger than the circuit conductors may apply where the rating of the circuit breaker has been increased to allow for motor starting. In such a case, the minimum size equipment grounding conductor listed in the table may be larger than the circuit conductors. This is not usually a concern in residential wiring but could be in some rare situations.

Where the equipment grounding conductor is an actual conductor and not a raceway system, using the table to size this conductor is simple because it is prescriptive. But where a raceway like a conduit or the sheath of an armored cable is used as the grounding conductor as allowed in Section E3908.8, the code states here that it must comply with Section E3908.4, which is actually a performance requirement. How is it determined that the conduit or cable sheath is of a sufficient size or has a low enough impedance as an equipment grounding conductor? It is usually assumed that the conduit has enough metal to serve as a conductor and will do so as long as the conduit system is assembled properly or performance is verified through testing. This is why the use of flexible conduit as a grounding conductor is limited by strict conditions.

As an example of sizing the equipment conductor, a 240-volt, 70-ampere circuit to an air-conditioner run with two size 4-AWG copper ungrounded conductors and no grounded conductor would be required to have a minimum size 8-AWG copper equipment grounding conductor.

TABLE E3908.12 (Table 250.122)
EQUIPMENT GROUNDING CONDUCTOR SIZING

RATING OR SETTING OF AUTOMATIC OVERCURRENT DEVICE IN CIRCUIT AHEAD OF EQUIPMENT, CONDUIT, ETC., NOT EXCEEDING THE FOLLOWING RATINGS (amperes)	MINIMUM SIZE	
	Copper wire No. (AWG)	Aluminum or copper-clad aluminum wire No. (AWG)
15	14	12
20	12	10
60	10	8
100	8	6
200	6	4
300	4	2
400	3	1

❖ See the commentary to Section E3908.12.

E3908.12.1 Multiple circuits. Where a single equipment grounding conductor is run with multiple circuits in the same raceway or cable, it shall be sized for the largest overcurrent device protecting conductors in the raceway or cable. [250.122(C)]

❖ It is not necessary to run multiple equipment grounding conductors in the same conduit or cable that supplies multiple equipment. But the equipment grounding conductor must be sized for the largest rated circuit in the conduit or cable. For example, where a 20-ampere circuit and a 30-ampere circuit are run in the same conduit, the equipment grounding conductor must be at least size 10-AWG copper. Another example would be where a three-conductor cable is used. A cable marked "12-3 w/ground" may have a red, a black, a white and a bare conductor. The red and black conductors could supply two separate circuits to two different appliances or equipment, as part of a multiwire branch circuit. The one size 12-AWG bare conductor serves as the equipment grounding conductor for both circuits.

E3908.13 Continuity and attachment of equipment grounding conductors to boxes. Where circuit conductors are spliced within a box or terminated on equipment within or supported by a box, any equipment grounding conductors associated with the circuit conductors shall be connected within the box or to the box with devices suitable for the use. Connections depending solely on solder shall not be used. Splices shall be made in accordance with Section E3406.10 except that insulation shall not be required. The arrangement of grounding connections shall be such that the disconnection or removal of a receptacle, luminaire or other device fed from the box will not interfere with or interrupt the grounding continuity. [250.146(A) and (C)]

❖ Where several nonmetallic sheathed cables are installed in a box, the equipment grounding conductors must be joined together with a listed device. These conductors must not be twisted together with pliers as the only means of splicing. A listed connector such as a wire nut must be used. A jumper or "pigtail" must be used to connect to the grounding terminal of the device so that when the device is removed it will not interrupt the continuity of the grounding return path. Special pressure connectors (e.g., wire nuts with a hole in the end and wire nuts with attached lead wires) are available that provide a lead wire (jumper) for connection to devices. See Sections E3908.15 and E3908.16 for equipment grounding conductors in metal boxes and in nonmetallic boxes.

E3908.14 Connecting receptacle grounding terminal to box. An equipment bonding jumper, sized in accordance with Table E3908.12 based on the rating of the overcurrent device protecting the circuit conductors, shall be used to connect the grounding terminal of a grounding-type receptacle to a grounded box except where grounded in accordance with one of the following: (250.146)

1. Surface mounted box. Where the box is mounted on the surface, direct metal-to-metal contact between the device

yoke and the box shall be permitted to ground the receptacle to the box. At least one of the insulating washers shall be removed from receptacles that do not have a contact yoke or device designed and listed to be used in conjunction with the supporting screws to establish the grounding circuit between the device yoke and flush-type boxes. This provision shall not apply to cover-mounted receptacles except where the box and cover combination are listed as providing satisfactory ground continuity between the box and the receptacle. A listed exposed work cover shall be considered to be the grounding and bonding means where the device is attached to the cover with at least two fasteners that are permanent, such as a rivet or have a thread locking or screw locking means and where the cover mounting holes are located on a flat non-raised portion of the cover. [250.146(A)]

2. Contact devices or yokes. Contact devices or yokes designed and listed for the purpose shall be permitted in conjunction with the supporting screws to establish equipment bonding between the device yoke and flush-type boxes. [250.146(B)]

3. Floor boxes. The receptacle is installed in a floor box designed for and listed as providing satisfactory ground continuity between the box and the device. [250.146(C)]

❖ Where a metal box is used, a receptacle must be grounded by a jumper from the device grounding terminal to the box unless there is metal-to-metal contact between the yoke and the box. A receptacle having a contact device listed for the purpose of equipment grounding could be used. The so-called "self-grounding" receptacles have a spring type of clip that binds against the 6-32 supporting screw of the device to provide a positive connection between the receptacle yoke and the metal box.

E3908.15 Metal boxes. A connection shall be made between the one or more equipment grounding conductors and a metal box by means of a grounding screw that shall be used for no other purpose, equipment listed for grounding or by means of a listed grounding device. Where screws are used to connect grounding conductors or connection devices to boxes, such screws shall be one or more of the following: [250.148(C)]

1. Machine screw-type fasteners that engage not less than two threads.

2. Machine screw-type fasteners that are secured with a nut.

3. Thread-forming machine screws that engage not less than two threads in the enclosure. [250.8(5) and (6)]

❖ Where a metal box is used, the equipment grounding conductor must be bonded or connected to the metal box by means of a listed grounding screw or grounding device. It may be handy to wrap the equipment grounding conductor around the screw used for attaching the box to the structure, but this would be a poor, unreliable connection and a code violation. Screws that support the box or that are part of a cable clamp cannot be used to ground metal boxes. Most, if not all, metal boxes are provided with factory tapped openings for 10-32 machine screws for grounding purposes.

E3908.16 Nonmetallic boxes. One or more equipment grounding conductors brought into a nonmetallic outlet box shall be arranged to allow connection to fittings or devices installed in that box. [250.148(D)]

❖ See Section E3908.13.

E3908.17 Clean surfaces. Nonconductive coatings such as paint, lacquer and enamel on equipment to be grounded shall be removed from threads and other contact surfaces to ensure electrical continuity or the equipment shall be connected by means of fittings designed so as to make such removal unnecessary. (250.12)

❖ See Section E3908.18.

E3908.18 Bonding other enclosures. Metal raceways, cable armor, cable sheath, enclosures, frames, fittings and other metal noncurrent-carrying parts that serve as equipment grounding conductors, with or without the use of supplementary equipment grounding conductors, shall be effectively bonded where necessary to ensure electrical continuity and the capacity to conduct safely any fault current likely to be imposed on them. Any nonconductive paint, enamel and similar coating shall be removed at threads, contact points and contact surfaces, or connections shall be made by means of fittings designed so as to make such removal unnecessary. [250.96(A)]

❖ Paint is an insulator, and a positive connection cannot be made without removing the paint at the point where the equipment grounding conductor is attached (see also Sections E3908.9 and E3906.10).

E3908.19 Size of equipment bonding jumper on load side of an overcurrent device. The equipment bonding jumper on the load side of an overcurrent devices shall be sized, as a minimum, in accordance with Table E3908.12, but shall not be required to be larger than the circuit conductors supplying the equipment. An equipment bonding conductor shall be not smaller than No. 14 AWG.

A single common continuous equipment bonding jumper shall be permitted to connect two or more raceways or cables where the bonding jumper is sized in accordance with Table E3908.12 for the largest overcurrent device supplying circuits therein. [250.102(D) and 250.122]

❖ Bonding jumpers can be used where two or more raceways enter a junction box to ensure the continuity of the conduit as an equipment grounding conductor. A common bonding jumper can be used as mentioned in the commentary to Section E3908.12.1.

E3908.20 Installation equipment bonding jumper. Bonding jumpers or conductors and equipment bonding jumpers shall be installed either inside or outside of a raceway or an enclosure in accordance with Sections E3908.20.1 and E3908.20.2. [250.102(E)]

❖ An equipment bonding jumper is limited to 6 feet (1829 mm) if installed on the outside of the raceway, but the length is not limited if installed on the inside and run

with the circuit conductors. Note the exception for outdoor poles.

E3908.20.1 Inside raceway or enclosure. Where installed inside a raceway or enclosure, equipment bonding jumpers and bonding jumpers or conductors shall comply with the requirements of Sections E3407.2 and E3908.13. [250.102(E)(1)]

❖ See the commentary to Sections E3407.2 and E3908.13.

E3908.20.2 Outside raceway or enclosure. Where installed outside of a raceway or enclosure, the length of the bonding jumper or conductor or equipment bonding jumper shall not exceed 6 feet (1829 mm) and shall be routed with the raceway or enclosure. [250.102(E)(2)]

Equipment bonding jumpers and supply-side bonding jumpers installed for bonding grounding electrodes and installed at outdoor pole locations for the purpose of bonding or grounding isolated sections of metal raceways or elbows installed in exposed risers of metal conduit or other metal raceway, shall not be limited in length and shall not be required to be routed with a raceway or enclosure. [250.102(E)(2) Exception]

❖ An equipment bonding jumper is limited to 6 feet (1829 mm) if installed on the outside of the raceway, but the length is not limited if installed on the inside and run with the circuit conductors. Note the special allowances for bonding grounding electrodes and outdoor pole risers.

E3908.20.3 Protection. Bonding jumpers or conductors and equipment bonding jumpers shall be installed in accordance with Section E3610.2. [250.102(E)(3)]'

❖ See the commentary to Section E3610.2.

SECTION E3909
FLEXIBLE CORDS

E3909.1 Where permitted. Flexible cords shall be used only for the connection of appliances where the fastening means and mechanical connections of such appliances are designed to permit ready removal for maintenance, repair or frequent interchange and the appliance is listed for flexible cord connection. Flexible cords shall not be installed as a substitute for the fixed wiring of a structure; shall not be run through holes in walls, structural ceilings, suspended ceilings, dropped ceilings or floors; shall not be concealed behind walls, floors, ceilings or located above suspended or dropped ceilings. (400.7 and 400.8)

❖ A flexible cord could not be used for the power supply to a furnace or water heater, but could be used for a range. Where a flexible cord is used for an appliance, a receptacle must be placed within reach of the cord in the same area as the appliance so the cord does not have to be run through a wall, ceiling or floor (see Section E4101.3).

E3909.2 Loading and protection. The ampere load of flexible cords serving fixed appliances shall be in accordance with

Table E3909.2. This table shall be used in conjunction with applicable end use product standards to ensure selection of the proper size and type. Where flexible cord is approved for and used with a specific listed appliance, it shall be considered to be protected where applied within the appliance listing requirements. [240.4, 240.5(A), 240.5(B)(1), 400.5, and 400.13]

❖ Where a flexible cord is field installed, it is important to follow the ampere load ratings in Table E3909.2. Many cord- and plug-connected appliances and equipment come with a cord already installed, which is sized to carry the load.

TABLE E3909.2 [Table 400.5(A)(1)]
MAXIMUM AMPERE LOAD FOR FLEXIBLE CORDS

CORD SIZE (AWG)	CORD TYPES S, SE, SEO, SJ, SJE, SJEO, SJO, SJOO, SJT, SJTO, SJTOO, SO, SOO, SRD, SRDE, SRDT, ST, STD, SV, SVO, SVOO, SVTO, SVTOO	
	Maximum ampere load	
	Three current-carrying conductors	Two current-carrying conductors
18	7	10
16	10	13
14	15	18
12	20	25

E3909.3 Splices. Flexible cord shall be used only in continuous lengths without splices or taps. (400.9)

❖ No provision is made for splicing or making tap connections to a flexible cord. Flexible cord is not to be used in any way as a substitute for the fixed wiring of the dwelling.

E3909.4 Attachment plugs. Where used in accordance with Section E3909.1, each flexible cord shall be equipped with an attachment plug and shall be energized from a receptacle outlet. [400.7(B)]

❖ Attachment plugs are used to receive power, not to deliver power to a circuit. Plugs are received by receptacles.

Bibliography

The following resource materials were used in the preparation of the commentary for this chapter of the code:

NFPA 70–14, *National Electrical Code*. Quincy, MA: National Fire Protection Association, 2013.

Chapter 40:
Devices and Luminaires

General Comments

This chapter addresses the "trim-out" (final) stage of construction in which devices and fixtures are installed and connected to the installed wiring system.

Chapter 40 covers receptacle ratings and installation, lighting fixture installation, construction and location, and grounding of devices and fixtures.

Purpose

Chapter 40 is intended to prevent fire and electrical shock hazards associated with the exposed devices and fixtures to which occupants and dwelling contents are exposed.

SECTION E4001
SWITCHES

E4001.1 Rating and application of snap switches. General-use snap switches shall be used within their ratings and shall control only the following loads:

1. Resistive and inductive loads not exceeding the ampere rating of the switch at the voltage involved.

2. Tungsten-filament lamp loads not exceeding the ampere rating of the switch at 120 volts.

3. Motor loads not exceeding 80 percent of the ampere rating of the switch at its rated voltage. [404.14(A)]

❖ "Snap switch" is the term for the common types of switches used in dwelling units for lighting control, small appliances and motor-driven equipment. Most snap switches installed in dwellings are rated 15 amperes and 20 amperes. For resistive loads and non-motor inductive loads, such as fluorescent lighting, the switch must have an ampere rating of at least 100 percent of the load. A routine practice is to install 15-ampere switches because the switch often controls only one light fixture in a room. However, where a switch controls many lights of heavy wattage, it is important to consider the total load and make sure that the ampere rating is adequate. A switch controlling a motor must have a rating of 125 percent of the motor load.

E4001.2 CO/ALR snap switches. Snap switches rated 20 amperes or less directly connected to aluminum conductors shall be marked CO/ALR. [404.14(C)]

❖ Aluminum conductors are not used in modern wiring for 15- and 20-ampere branch circuits. However, this code provision is important because it also covers remodeling and rewiring work in older homes where aluminum conductors may be found. If a snap switch is replaced on aluminum wire, the switch must be marked CO/ALR or otherwise indicate "suitable for copper or aluminum conductors." If a switch is not available marked in this way, one alternative is to connect a copper pigtail to the end of each aluminum conductor to be terminated on the switch. A splicing device listed for the application must be used to connect the pigtail. In some cases, a decision is made to go through the entire house and add copper pigtails and replace all receptacles and switches where there is evidence of terminal overheating (see also Section E3406.8).

E4001.3 Indicating. General-use and motor-circuit switches and circuit breakers shall clearly indicate whether they are in the open OFF or closed ON position. Where single-throw switches or circuit breaker handles are operated vertically rather than rotationally or horizontally, the up position of the handle shall be the closed (on) position.

❖ It is assumed that a vertically operated switch will be "on" in the up position and "off" in the down position, because this is the convention. It would not be wise to possibly confuse occupants by changing the convention to which they are accustomed. Similar logic in the code requires the hot water control to correspond with the left side of a faucet.

E4001.4 Time switches and similar devices. Time switches and similar devices shall be of the enclosed type or shall be mounted in cabinets or boxes or equipment enclosures. A barrier shall be used around energized parts to prevent operator exposure when making manual adjustments or switching. (404.5)

❖ Some modern time switches are electronic, and the setting buttons and controls are not near the energized parts, but many time switches have the dial or controls near the live terminals. It is important that the insulating barrier not be removed to help prevent shock or electrocution when adjusting the settings.

E4001.5 Grounding of enclosures. Metal enclosures for switches or circuit breakers shall be connected to an equipment grounding conductor. Metal enclosures for switches or circuit breakers used as service equipment shall comply with the provisions of Section E3609.4. Where nonmetallic enclosures are used with metal raceways or metal-armored cables, provisions shall be made for connecting the equipment grounding conductor.

Nonmetallic boxes for switches shall be installed with a wiring method that provides or includes an equipment grounding conductor. (404.12)

❖ The exposed metal parts of a cabinet or switch box must be grounded by connection to the equipment grounding conductor by approved devices. Where nonmetallic boxes are used, the equipment grounding conductors, whether they are wires or raceways, must be bonded together to assure continuity of the ground fault return path (see Section E3609.4).

E4001.6 Access. Switches and circuit breakers used as switches shall be located to allow operation from a readily accessible location. Such devices shall be installed so that the center of the grip of the operating handle of the switch or circuit breaker, when in its highest position, will not be more than 6 feet 7 inches (2007 mm) above the floor or working platform. [404.8(A)]

> **Exception:** This section shall not apply to switches and circuit breakers that are accessible by portable means and are installed adjacent to the motors, appliances and other equipment that they supply. [404.8(A) Exception]

❖ This provision for access to switches includes access to circuit breakers, so when the contractor is considering the location for the panelboard and the location of the main disconnect, this rule must be followed. For example, if the main disconnect is installed on the outside of the house and the level of the grade is lowered, the height of the main breaker will be raised. Most ordinary switches are installed lower than 6 feet, 7 inches (2007 mm). However, in some cases for special equipment, this rule will come into play. One should not need a ladder or stool to operate any switch (see definition of "Readily accessible").

The exception recognizes that accessing some switches by portable ladders, for example, is adequate where the switch or circuit breaker is located adjacent to the motor, appliance or equipment that it serves. An example is an HVAC unit mounted on a roof where the switch is installed next to the unit. Such switches require only occasional or seasonal use and need not be readily accessible.

E4001.7 Damp or wet locations. A surface mounted switch or circuit breaker located in a damp or wet location or outside of a building shall be enclosed in a weatherproof enclosure or cabinet. A flush-mounted switch or circuit breaker in a damp or wet location shall be equipped with a weatherproof cover. Switches shall not be installed within wet locations in tub or shower spaces unless installed as part of a listed tub or shower assembly. [404.8(A), (B), and (C)]

❖ There is no exception to the rule that a common light switch wired to control a fan, lighting outlet, or other load cannot be installed in the bathtub or shower space; however, the code does not provide an exact definition or dimension of tub or shower space. It would be wise to locate the switch out of the reach of someone in the tub or shower, even though this is not stated in the code (see Section E3902.8).

E4001.8 Grounded conductors. Switches or circuit breakers shall not disconnect the grounded conductor of a circuit except where the switch or circuit breaker simultaneously disconnects all conductors of the circuit. [404.2(B)]

❖ Switching the grounded conductor of a two-wire circuit will turn the load on and off, but a safety hazard is created because the ungrounded conductor is not opened. When someone is working on the wiring in a lighting outlet or on a luminaire, the ungrounded or hot conductor must be deenergized when the switch is off.

E4001.9 Switch connections. Three- and four-way switches shall be wired so that all switching occurs only in the ungrounded circuit conductor. Color coding of switch connection conductors shall comply with Section E3407.3. Where in metal raceways or metal-jacketed cables, wiring between switches and outlets shall be in accordance with Section E3406.7. [404.2(A)]

> **Exception:** Switch loops do not require a grounded conductor. [404.2(A) Exception]

❖ One standard way to wire three-way switches is to have an ungrounded and a grounded conductor feed into one switch box (sometimes called the power end), then run "travelers" along with the grounded conductor to the other switch box (sometimes called the switch leg end), and then from there, run the switched ungrounded conductor along with the grounded conductor up to the light fixture. In this circuit layout, it is very simple to keep the grounded conductor from being switched. But there are many different variations on designing a three-way switch circuit and some of these over the years have involved switching the grounded conductor. In older homes, lighting fixtures were sometimes wired in the circuit between two three-way switches such that the polarity at the lamp holder changed depending on the position of the switches. This circuit is difficult to diagnose and creates a shock hazard. This provision is included in the code to specifically prohibit the switching of grounded conductors.

This section also requires, by reference to Section E3406.7, that all conductors of a circuit be in the same raceway or cable. This prohibits running the travelers without the grounded conductor and picking up the grounded conductor on the switch leg from a close-by outlet box that has a grounded conductor such as from a receptacle.

In a switch loop, the conductors supplying power to a light are brought first to the lighting outlet. Only one cable with a black, a white and a bare conductor is run to the switch point. The white conductor is connected to the ungrounded conductor at the lighting outlet and supplies power down to the switch. When the switch is closed, power is carried back to the lighting fixture on the black conductor of the switch-loop cable. This hook-up of a switch involves a "loop" of ungrounded conductor and does not require the grounded conductor to be run with the "hot" switch conductors (see Section E3407.3).

E4001.10 Box mounted. Flush-type snap switches mounted in boxes that are recessed from the finished wall surfaces as covered in Section E3906.5 shall be installed so that the extension plaster ears are seated against the surface of the wall. Flush-type snap switches mounted in boxes that are flush with the finished wall surface or project therefrom shall be installed so that the mounting yoke or strap of the switch is seated against the box. Screws used for the purpose of attaching a snap switch to a box shall be of the type provided with a listed snap switch, or shall be machine screws having 32 threads per inch or part of listed assemblies or systems, in accordance with the manufacturer's instructions. [404.10(B)]

❖ Following the rules of this code provision will result in the switch being secure so that it will not move, become loose or cause accidental shorting to a metal box. The ears of the switch are designed to seat solidly against the wall surface or the box itself. Where the wall opening at the outlet is too big because the drywall or plaster has been cut too large, the switch might "float" with only the cover-plate screws holding it in place. This section intends to prevent the use of improper mounting screws for switches. Sheet metal, drywall and wood screws are prohibited because they can loosen and will damage the threads in the box. Also, the wrong screws will have sharp points and likely project into the box and contact wires.

E4001.11 Snap switch faceplates. Faceplates provided for snap switches mounted in boxes and other enclosures shall be installed so as to completely cover the opening and, where the switch is flush mounted, seat against the finished surface. [404.9(A)]

❖ The faceplate is designed to seat against the finished wall surface when the switch is installed properly with its extension plaster ears also seated against the wall surface. If the wall surface around the outlet is not finished properly, the faceplate will not fit properly. An oversized face-plate might cover the opening that has been left too large for the box, but Section E3906.6 requires that the openings in plaster, gypsum board, or plasterboard surfaces around boxes be made so that there is no gap more than $^1/_8$ inch (3.2 mm) from the edge of the box.

E4001.11.1 Faceplate grounding. Snap switches, including dimmer and similar control switches, shall be connected to an equipment grounding conductor and shall provide a means to connect metal faceplates to the equipment grounding conductor, whether or not a metal faceplate is installed. Snap switches shall be considered to be part of an effective ground-fault current path if either of the following conditions is met:

1. The switch is mounted with metal screws to a metal box or metal cover that is connected to an equipment grounding conductor or to a nonmetallic box with integral means for connecting to an equipment grounding conductor.

2. An equipment grounding conductor or equipment bonding jumper is connected to an equipment grounding termination of the snap switch. [404.9(B)]

Exceptions:

1. Where a means to connect to an equipment grounding conductor does not exist within the snap-switch enclosure or where the wiring method does not include or provide an equipment grounding conductor, a snap switch without a grounding connection to an equipment grounding conductor shall be permitted for replacement purposes only. A snap switch wired under the provisions of this exception and located within 8 feet (2438 mm) vertically or 5 feet (1524 mm) horizontally of ground or exposed grounded metal objects, shall be provided with a faceplate of nonconducting noncombustible material with nonmetallic attachment screws, except where the switch-mounting strap or yoke is nonmetallic or the circuit is protected by a ground-fault circuit interrupter. [404.9(B) Exception No.1]

2. Listed kits or listed assemblies shall not be required to be connected to an equipment grounding conductor if all of the following conditions apply:

    2.1. The device is provided with a nonmetallic faceplate that cannot be installed on any other type of device.

    2.2. The device does not have mounting means to accept other configurations of faceplates.

    2.3. The device is equipped with a nonmetallic yoke.

    2.4. All parts of the device that are accessible after installation of the faceplate are manufactured of nonmetallic materials. [404.9(B) Exception No. 2]

3. Connection to an equipment grounding conductor shall not be required for snap switches that have an integral nonmetallic enclosure complying with Section E3905.1.3. [404.9(B) Exception No. 3]

❖ Note that where Exception 1 requires a nonconducting faceplate, the faceplate attachment screws must be nonmetallic. Nylon 6-32 screws are available for this purpose. The theme of the exceptions to this section is to reduce the potential shock hazard by eliminating conducting materials that will come in contact with the user of the switches and controls (see commentary, Sections E3906.10 and E3908.14).

E4001.12 Dimmer switches. General-use dimmer switches shall be used only to control permanently installed incandescent luminaires (lighting fixtures) except where listed for the control of other loads and installed accordingly. [404.14(E)]

❖ General-use dimmer switches are designed for use with only simple resistive loads (incandescent lamps).

Dimmer controls must be listed for the purpose if used for controlling inductive loads, such as fluorescent ballasts. Dimmer switches must not be used to control switched receptacles because it will not be known what type of load will be plugged into such receptacles.

E4001.13 Multipole snap switches. A multipole, general-use snap switch shall not be fed from more than a single circuit unless it is listed and marked as a two-circuit or three-circuit switch. [404.8(C)]

❖ Multipole snap switches contain more than one single-throw or double-throw switch in a single device (e.g., a double-pole, single-throw switch or a double-pole, double-throw switch).

E4001.14 Cord-and-plug-connected loads. Where snap switches are used to control cord-and-plug-connected equipment on a general-purpose branch circuit, each snap switch controlling receptacle outlets or cord connectors that are supplied by permanently connected cord pendants shall be rated at not less than the rating of the maximum permitted ampere rating or setting of the overcurrent device protecting the receptacles or cord connectors, as provided in Sections E4002.1.1 and E4002.1.2. [404.14(F)]

❖ In cases where snap switches control receptacles for cord- and plug-connected equipment, the connected load will be unpredictable; therefore, the switch needs to be rated to handle the maximum current allowed by the branch-circuit overcurrent device. Otherwise, the switch might be overloaded.

E4001.15 Switches controlling lighting loads. The grounded circuit conductor for the controlled lighting circuit shall be provided at the location where switches control lighting loads that are supplied by a grounded general-purpose branch circuit for other than the following:

1. Where conductors enter the box enclosing the switch through a raceway, provided that the raceway is large enough for all contained conductors, including a grounded conductor.

2. Where the box enclosing the switch is accessible for the installation of an additional or replacement cable without removing finish materials.

3. Where snap switches with integral enclosures comply with E3905.1.3.

4. Where the switch does not serve a habitable room or bathroom.

5. Where multiple switch locations control the same lighting load such that the entire floor area of the room or space is visible from the single or combined switch locations.

6. Where lighting in the area is controlled by automatic means.

7. Where the switch controls a receptacle load. [404.2(C)]

❖ Many switching devices such as illuminated or indicating snap switches, electronic timers, occupant sensors, lighting control centers and other box-mounted devices/equipment consume power to operate; therefore, the grounded conductor will need to be present to supply the intended or future devices.

The first two exceptions recognize the conditions where extending the grounded conductor to the box at a later date is easily achieved. Exceptions 4 through 7 address situations where devices needing a grounded conductor are not likely to be installed.

SECTION E4002
RECEPTACLES

E4002.1 Rating and type. Receptacles and cord connectors shall be rated at not less than 15 amperes, 125 volts, or 15 amperes, 250 volts, and shall not be a lampholder type. Receptacles shall be rated in accordance with this section. [406.3(B)]

❖ The minimum rating allowed for receptacles in dwelling units is 15 amperes. Receptacles are not permitted to be of a type suitable for use as lampholders (i.e., a lampholder socket with a receptacle adapter screwed into it).

E4002.1.1 Single receptacle. A single receptacle installed on an individual branch circuit shall have an ampere rating not less than that of the branch circuit. [210.21(B)]

❖ A receptacle device can be single, duplex or multiple. The term "yoke" describes the strap that the receptacle is mounted on; the yoke is used to attach to the device box. A single receptacle is a device on a strap or yoke that can receive only one plug. The rating of the circuit breaker or overcurrent device is the rating of branch circuit, and on an individual branch circuit serving only a single receptacle, all of the circuit current flow might be taken through it. Therefore, the single receptacle must be rated the same as the branch circuit overcurrent device (see definition of "Receptacle").

E4002.1.2 Two or more receptacles. Where connected to a branch circuit supplying two or more receptacles or outlets, receptacles shall conform to the values listed in Table E4002.1.2. [210.21(B)(3)]

❖ Where two or more receptacles are fed by a 20-ampere branch circuit, Table E4002.1.2 indicates that the receptacles could be rated 15 amperes. This could mean that where a 20-ampere branch circuit feeds only one duplex receptacle, that duplex receptacle could be rated 15 amperes because there is more than a single receptacle on the circuit.

TABLE E4002.1.2 [Table 210.21(B)(3)]
RECEPTACLE RATINGS FOR
VARIOUS SIZE MULTI-OUTLET CIRCUITS

CIRCUIT RATING (amperes)	RECEPTACLE RATING (amperes)
15	15
20	15 or 20
30	30
40	40 or 50
50	50

❖ See the commentary to Section E4002.1.2.

E4002.2 Grounding type. Receptacles installed on 15- and 20-ampere-rated branch circuits shall be of the grounding type. [406.4(A)]

❖ All receptacles must be of the grounding type. That is, they must accept a three-prong grounding plug (cord cap).

E4002.3 CO/ALR receptacles. Receptacles rated at 20 amperes or less and directly connected to aluminum conductors shall be marked CO/ALR. [406.3(C)]

❖ See Section E4001.2.

E4002.4 Faceplates. Metal face plates shall be grounded. [406.6(B)]

❖ See the commentary to Sections E3906.10 and E3908.14.

E4002.5 Position of receptacle faces. After installation, receptacle faces shall be flush with or project from face plates of insulating material and shall project a minimum of 0.015 inch (0.381 mm) from metal face plates. Faceplates shall be installed so as to completely cover the opening and seat against the mounting surface. Receptacle faceplates mounted inside of a box having a recess-mounted receptacle shall effectively close the opening and seat against the mounting surface. [406.5(D), 406.6]

Exception: Listed kits or assemblies encompassing receptacles and nonmetallic faceplates that cover the receptacle face, where the plate cannot be installed on any other receptacle, shall be permitted. [406.5(D) Exception]

❖ This section makes a distinction between the face of the actual receptacle and the faceplate or cover plate. The face of the receptacle must project out from a metal faceplate at least $^1/_{64}$ inch (0.4 mm). If the face of the receptacle does not project out from a metal faceplate where a metal faceplate is used, the blades of the cord attachment plug could touch the metal faceplate and create a ground fault or short. See also Section E4001.11. If a receptacle is recessed from the surface of the faceplate, the plug may not engage the receptacle at the proper depth.

E4002.6 Receptacle mounted in boxes. Receptacles mounted in boxes that are set back from the finished wall surface as permitted by Section E3906.5 shall be installed so that the mounting yoke or strap of the receptacle is held rigidly at the finished surface of the wall. Screws used for the purpose of attaching receptacles to a box shall be of the type provided with a listed receptacle, or shall be machine screws having 32 threads per inch or part of listed assemblies or systems, in accordance with the manufacturer's instructions. Receptacles mounted in boxes that are flush with the wall surface or project therefrom shall be so installed that the mounting yoke or strap is seated against the box or raised cover. [406.5(A) and (B)]

❖ The way the drywall (gypsum board) is cut out around the box has an effect on compliance with this code provision. If the drywall is cut out too far at the top and bottom of the box so that the extension plaster ears of the receptacle do not seat firmly against the finished wall surface, the receptacle will "float" between the front

edge of the box and the finished wall surface, and the face of the receptacle will be recessed from the faceplate. The one screw in the center of the faceplate will be the only thing holding the receptacle in place and will likely break. If washers or spacers are used to space the receptacle mounting yoke out from the face of the box, the faceplate will then be anchored to the receptacle and seated firmly against the wall surface. Because the edges of gypsum board (drywall) around a box opening are fragile, it is likely that the receptacle will become loose if it relies on the plaster ears seated against the drywall. This is why it is important to install boxes so that they will be flush with the finished wall surface, thus allowing the devices to seat against the box. Of course, flush box mounting is not always accomplished because of bowed studs and other factors, so electrical contractors will often use spacers where necessary to make the devices seat against the box instead of depending on fragile drywall edges to support the device. This section intends to prevent the use of improper mounting screws for receptacles. Sheet metal, drywall and wood screws are prohibited because they can loosen and will damage the threads in the box. Also, the wrong screws will have sharp points and likely project into the box and contact wires.

E4002.7 Receptacles mounted on covers. Receptacles mounted to and supported by a cover shall be held rigidly against the cover by more than one screw or shall be a device assembly or box cover listed and identified for securing by a single screw. [406.5(C)]

❖ An example of this is where a 4-inch (102 mm) square metal box is used, and the receptacle is installed in the raised box cover. A raised cover, for example, may have a $^1/_2$-inch (12.7 mm) rise. It is the same size as the four-square box and fits directly on the face of the box. A switch or receptacle is secured to the raised cover and not to the box, and then the cover is secured to the box. A switch installed in the cover would be secured by two 6-32 screws, the same as regular coverplate screws, but the cover of a receptacle is normally secured by only one screw. The problem is that when attachment plugs are inserted, the receptacle will break away from the cover. This provision was put into the code to require that in these types of installations, two or more screws must be used to support the receptacle in the cover. The 4-inch (102 mm) box covers are now manufactured with two device mounting holes for receptacles and are supplied with screws and locking nuts (see Section E3908.14).

E4002.8 Damp locations. A receptacle installed outdoors in a location protected from the weather or in other damp locations shall have an enclosure for the receptacle that is weatherproof when the receptacle cover(s) is closed and an attachment plug cap is not inserted. An installation suitable for wet locations shall also be considered suitable for damp locations. A receptacle shall be considered to be in a location protected from the weather where located under roofed open porches, canopies and similar structures and not subject to rain or water runoff. Fifteen- and 20-ampere, 125- and 250-

volt nonlocking receptacles installed in damp locations shall be listed a weather-resistant type. [406.9(A)]

❖ In this section and the following, the distinction is made between receptacles in damp locations and in wet locations. The type of receptacle cover used is the main issue here. The conventional type of receptacle faceplate used on the interior of the dwelling is not permitted outdoors or in any damp or wet location. Several different types of receptacle covers are available for use in damp and wet locations. This section requires that the receptacle have a cover that will make it weatherproof when the cover is closed and the receptacle is not in use. On a receptacle installed on the outside of the house, the receptacle terminals could become corroded from condensation, and the corrosion causes resistance. The result might be that because of corrosion and the development over time of a thin insulating layer on the contacts and terminals, continuity of the circuit might be lost. Damage could result also from the build-up of heat as current is drawn through the poor connection caused by corrosion. The weatherproof cover will help prevent the entrance of moisture to the interior of the receptacle.

The definition of the term "weatherproof" is important to remember: "So constructed or protected that exposure to the weather will not interfere with successful operation." If the receptacle is in a damp location as described here and in Chapter 35, the cover must make the receptacle weatherproof when the cover is closed and no attachment plug is inserted. Note that the receptacle itself must be listed as a special type that is resistant to the environment.

E4002.9 Fifteen- and 20-ampere receptacles in wet locations. Where installed in a wet location, 15- and 20-ampere, 125- and 250-volt receptacles shall have an enclosure that is weatherproof whether or not the attachment plug cap is inserted. An outlet box hood installed for this purpose shall be listed and identified as "extra-duty." Fifteen- and 20-ampere, 125- and 250-volt nonlocking receptacles installed in wet locations shall be a listed weather-resistant type. [406.9(B)(1)]

❖ This section applies to receptacles that are in a wet location either outdoors or indoors. The receptacle box and cover must be weatherproof in all cases, thus requiring what is commonly referred to as a "bubble" cover. Such covers provide the space necessary to allow the cover to close fully over an inserted plug and cord. This provision should end the debate over whether or not the item plugged into the receptacle will be attended while in use (see Section E4002.10). Note that the receptacle itself must be listed as a special type that is resistant to the environment.

E4002.10 Other receptacles in wet locations. Where a receptacle other than a 15- or 20-amp, 125- or 250-volt receptacle is installed in a wet location and where the product intended to be plugged into it is not attended while in use, the receptacle shall have an enclosure that is weatherproof both when the attachment plug cap is inserted and when it is removed. Where such receptacle is installed in a wet location and where the product intended to be plugged into it will be attended while in use, the

receptacle shall have an enclosure that is weatherproof when the attachment plug cap is removed. [406.9(B)(2)]

❖ Where a receptacle other than a 15- or 20-amp receptacle is installed in a wet location, two conditions or situations are described: 1. Where an attachment plug is left connected to the receptacle continually; and 2. Where the attachment plug is connected only when attended. In the first situation, the cover used must make the receptacle weatherproof all the time. There are covers that completely blanket the receptacle and the attachment plug when it is inserted. If the receptacle will be used only when attended, the cover could be one that makes the receptacle weatherproof only when not in use. Note that this section will rarely apply because it does not address 15- or 20-amp receptacles. For other than ranges and clothes dryers, 30-amp and larger receptacles are not commonly used in dwellings.

E4002.11 Bathtub and shower space. A receptacle shall not be installed within or directly over a bathtub or shower stall. [406.9(C)]

❖ Although not stated by the code, it would be wise to locate all receptacles beyond the reach of a person using a bathtub or shower (see Sections E3902.8 and E4001.7).

E4002.12 Flush mounting with faceplate. In damp or wet locations, the enclosure for a receptacle installed in an outlet box flush-mounted in a finished surface shall be made weatherproof by means of a weatherproof faceplate assembly that provides a water-tight connection between the plate and the finished surface. [406.9(E)]

❖ Weatherproof covers on outdoor receptacles must seat completely on the wall surface. A receptacle installed in a brick wall can be difficult to seal and must have a gasket or sealant between the wall and the receptacle cover. On a house that has horizontal or vertical siding that overlaps, the receptacle cover must be completely in one of the sections of siding so that there is not a gap where moisture could enter. Siding manufacturers make components for the siding system that can be installed at the receptacle location to provide a flat surface for the installation of a receptacle cover.

E4002.13 Exposed terminals. Receptacles shall be enclosed so that live wiring terminals are not exposed to contact. [406.5(G)]

❖ Receptacles are required to be enclosed in device boxes or must have integral enclosures (see Section E3905.1).

E4002.14 Tamper-resistant receptacles. In areas specified in Section E3901.1, 125-volt, 15- and 20-ampere receptacles shall be listed tamper-resistant receptacles. [406.12(A)]

Exception: Receptacles in the following locations shall not be required to be tamper resistant:

1. Receptacles located more than 5.5 feet (1676 mm) above the floor.

2. Receptacles that are part of a luminaire or appliance.

3. A single receptacle for a single appliance or a duplex receptacle for two appliances where such receptacles are located in spaces dedicated for the appliances served and, under conditions of normal use, the appliances are not easily moved from one place to another. The appliances shall be cord-and-plug-connected to such receptacles in accordance with Section E3909.4. [406.12(A) Exception]

❖ Section 3901.1 addresses all 15- and 20-amp 125-volt receptacles in a dwelling; therefore, all such devices must be listed tamper-resistant receptacle devices. These receptacle devices have an internal mechanism that blocks access to the plug prong openings except when a plug is inserted into the receptacle. The intent is to protect children who often insert objects into receptacles out of curiosity. The author recalls more than one experience where, as a youngster, a hair pin or nail was inserted into a receptacle and, fortunately, he lived to tell about it.

The contact openings (slots) in receptacles pose a shock or electrocution hazard for anyone who purposely or accidentally inserts a metallic object into such openings in a receptacle. The Consumer Product Safety Commission (CPSC) estimates that there are approximately 3,900 injuries treated by hospital emergency rooms every year that are the result of shock or burns from receptacles. Children are the most at risk because of their curiosity and lack of knowledge of the danger and nearly one-third of the reported injuries occur to children. Tamper-resistant receptacles were developed to reduce injuries and are now required for nearly all receptacle locations. Obviously, the intent is to require tamper-resistant receptacles in all locations where there is a possibility of tampering. Note the three exempt locations in the exception. This section applies to 125-volt, 15- and 20-amp receptacles only.

A typical tamper-resistant receptacle is shown in Commentary Figure E4002.14. Note the embossed "TR" on the face of the receptacle to indicate that it is different from a standard receptacle. Inserting a two-prong (or three-prong) plug into a tamper-resistant receptacle is no different than it is for a standard receptacle. Mechanisms within the receptacle sense that the two parallel prongs of a plug are being inserted simultaneously, and this action unlocks an internal shutter to the electrical contacts. If the internal mechanism senses only one slot being penetrated, such as with a screwdriver, nail, key, pin or other object, the shutter does not release to open the slot to the electrical contacts. Each receptacle manufacturer has developed its own proprietary shutter mechanisms to comply with the operation requirements.

It would be easier for installers to use "TR" receptacles in all locations rather than trying to determine what locations, if any, are exempt (see Exception). Based on the extensive coverage of Section E3901.1, it is hard to determine that there are any exempt locations, other than those addressed in the exception. For example, the exception could be applied to allow a non-TR receptacle in a cabinet above a range hood but the exception would not apply to a receptacle mounted in a sink base cabinet to serve a dishwasher. A single (simplex) receptacle behind a refrigerator would be exempt and a receptacle installed in the garage ceiling for the garage door opener would be exempt.

E4002.15 Dimmer-controlled receptacles. A receptacle supplying lighting loads shall not be connected to a dimmer except where the plug and receptacle combination is a nonstandard configuration type that is specifically listed and identified for each such unique combination.

❖ Dimmer switch controls must not serve switched receptacle outlets because the fixture or appliance plugged into the receptacle might not be compatible with the dimmer output. An incandescent lamp is one thing but a motor operated appliance or a computer is

Figure E4002.14
TYPICAL TAMPER-RESISTANT RECEPTACLE

an entirely different thing. The exception to this prohibition is where the switched receptacle is an unusual configuration, such as a twist lock configuration, and only the fixture or appliance that has a mating plug can utilize the receptacle. The receptacle and plug combination must be listed for the application. Such an arrangement would be very unlikely in a dwelling unit. Switched receptacle outlets in dwellings are used almost exclusively for lighting loads, and because the type of lighting load is unknown, dimmers are prohibited.

SECTION E4003
LUMINAIRES

E4003.1 Energized parts. Luminaires, lampholders, and lamps shall not have energized parts normally exposed to contact. (410.5)

❖ Energized parts of luminaires must be enclosed so that they will not be exposed to accidental contact. Suitable covers must be installed to prevent accidental contact with energized parts.

E4003.2 Luminaires near combustible material. Luminaires shall be installed or equipped with shades or guards so that combustible material will not be subjected to temperatures in excess of 90°C (194°F). (410.11)

❖ A significant amount of heat is developed in some incandescent luminaires and luminaires containing ballasts. If there is no means to dissipate the heat, it could cause damage to the surface on which it is mounted and also could cause a fire. It may be necessary to hold the luminaire away from a combustible ceiling or wall surface if the light fixture will create heat in excess of 90°C (194°F). Compliance with this code provision will help prevent fires in combustible structural elements as well as combustible storage.

E4003.3 Exposed conductive parts. The exposed metal parts of luminaires shall be connected to an equipment grounding conductor or shall be insulated from the equipment grounding conductor and other conducting surfaces. Lamp tie wires, mounting screws, clips and decorative bands on glass spaced at least $1^1/_2$ inches (38 mm) from lamp terminals shall not be required to be grounded. (410.42)

❖ All luminaires must be grounded. In older houses, only luminaires installed in a bathroom or outdoors were grounded. A remodeler can observe that some interior luminaires for a bedroom or living room do not contain a grounding wire, although the wiring method has a grounding conductor. When these fixtures are replaced, the grounding wire must be connected. All new luminaires are supplied with a grounding wire or grounding means.

E4003.4 Screw-shell type. Lampholders of the screw-shell type shall be installed for use as lampholders only. (410.90)

❖ The installation or use of lampholders with a screw-in plug adapter is not permitted. This was common practice in the 1920s and 1930s. Lampholders must be used only for lamps.

E4003.5 Recessed incandescent luminaires. Recessed incandescent luminaires shall have thermal protection and shall be listed as thermally protected. [410.115(C)]

Exceptions:

1. Thermal protection shall not be required in recessed luminaires listed for the purpose and installed in poured concrete. [410.115(C) Exception No.1]

2. Thermal protection shall not be required in recessed luminaires having design, construction, and thermal performance characteristics equivalent to that of thermally protected luminaires, and such luminaires are identified as inherently protected. [410.115(C) Exception No. 2]

❖ The buildup of heat in an incandescent recessed fixture can be a serious problem. A recessed fixture should have a clearance from insulation installed around or over it unless it is listed for no clearance. When the heat builds up to a level that could be dangerous, the required thermal protection device automatically turns off the lamp. After the heat in the fixture has dissipated, the device closes the circuit and the light automatically turns back on. Where this happens repeatedly, the fixture might have a higher wattage lamp than the maximum allowed. The problem could be solved by installing lower-wattage lamps or switching to nonincandescent types of lamps. The recessed fixture that cycles off to allow the heat to dissipate is doing what it is designed to do; however, the cause of the over-heating must be eliminated.

E4003.6 Thermal protection. The ballast of a fluorescent luminaire installed indoors shall have integral thermal protection. Replacement ballasts shall also have thermal protection integral with the ballast. A simple reactance ballast in a fluorescent luminaire with straight tubular lamps shall not be required to be thermally protected. [410.130(E)(1)]

❖ In addition to incandescent recessed fixtures, fluorescent fixtures must have thermal protection. The protection required must be part of the ballast. The thermal trip protection integral with the ballast responds to an abnormally high temperature in the ballast.

E4003.7 High-intensity discharge luminaires. Recessed high-intensity luminaires designed to be installed in wall or ceiling cavities shall have thermal protection and be identified as thermally protected. Thermal protection shall not be required in recessed high-intensity luminaires having design, construction and thermal performance characteristics equivalent to that of thermally protected luminaires, and such luminaires are identified as inherently protected. Thermal protection shall not be required in recessed high-intensity discharge luminaires installed in and identified for use in poured concrete. A recessed remote ballast for a high-intensity discharge luminaire shall have thermal protection that is integral with the ballast and shall be identified as thermally protected. [110.130(F)(1), (2), (3), and (4)]

❖ High-intensity discharge fixtures such as metal-halide, sodium and mercury-vapor operate at high temperatures and must have thermal protection unless identified as inherently protected.

E4003.8 Metal halide lamp containment. Luminaires that use a metal halide lamp other than a thick-glass parabolic reflector lamp (PAR) shall be provided with a containment barrier that encloses the lamp, or shall be provided with a physical means that allows the use of only a lamp that is Type O. [(110.130(F)(5)]

❖ Type O lamps are designed for use in open fixtures where the lamp is not fully enclosed. Such lamps have a shrouded arc tube which protects the outer glass bulb from damage in the event that the inner arc tube ruptures, the intent of the provision is to fully contain all lamps components in the event of lamp failure. This will prevent white hot metal fragments from falling from the lamp and causing a fire. Luminaires are available with lamp sockets that will accept only Type O lamps to prevent use of the wrong lamp. This is similar to Type S tamper proof plug fuses.

E4003.9 Wet or damp locations. Luminaires installed in wet or damp locations shall be installed so that water cannot enter or accumulate in wiring compartments, lampholders or other electrical parts. All luminaires installed in wet locations shall be marked SUITABLE FOR WET LOCATIONS. All luminaires installed in damp locations shall be marked SUITABLE FOR WET LOCATIONS or SUITABLE FOR DAMP LOCATIONS. (410.10)

❖ Installation of an indoor fixture is not permitted outdoors, even if it is under a porch or on a porch ceiling where the roof overhangs several feet. A partially protected location under canopies and roofed open porches is considered a damp location, as stated in Chapter 35. If a homeowner chooses lighting fixtures for a house and the ones for the ceiling of the open porch are the same as the ones for the entryway just inside the door, it would be necessary to check the fixtures to determine whether they are marked as suitable for a damp location. If not, it would be a code violation to install the fixture outdoors.

A wall porch light that is exposed to weather and unprotected by the roof overhang is in a wet location, and the fixture must be marked as suitable for wet locations.

Installation of a fixture that is suitable only for damp locations is not permitted in wet locations. But a fixture that is marked as suitable for wet locations could also be installed in damp locations. Installation of an indoor light fixture is not permitted outdoors.

E4003.10 Lampholders in wet or damp locations. Lampholders installed in wet locations shall be listed for use in wet locations. Lampholders installed in damp locations shall be listed for damp locations or shall be listed for wet locations. (410.96)

❖ An example of a weatherproof type of lampholder is floodlights installed under the eaves of a house. These weatherproof fixtures have a canopy with a gasket that seats against the surface of the underside of the eave (if that is where it is installed). They have one or more

sockets that swivel for adjusting to the desired position. The lamps used are rated for direct exposure to the weather, and the lampholder has a gasket that helps prevent moisture from entering the socket. Fixtures used for this type of application must be weatherproof, and the lampholder must be weatherproof as well. Lampholders listed for wet locations are also suitable for damp locations, but not the other way around.

E4003.11 Bathtub and shower areas. Cord-connected luminaires, chain-, cable-, or cord-suspended-luminaires, lighting track, pendants, and ceiling-suspended (paddle) fans shall not have any parts located within a zone measured 3 feet (914 mm) horizontally and 8 feet (2438 mm) vertically from the top of a bathtub rim or shower stall threshold. This zone is all encompassing and includes the space directly over the tub or shower. Luminaires within the actual outside dimension of the bathtub or shower to a height of 8 feet (2438 mm) vertically from the top of the bathtub rim or shower threshold shall be marked for damp locations and where subject to shower spray, shall be marked for wet locations. [410.4(D)]

❖ In most homes, a bathroom ceiling is not high enough to permit any fixture other than a surface-mounted or recessed-type fixture above a bathtub or shower. Fixtures such as chain-hung or pendant types where a cord is exposed above a bathtub or shower could be a safety hazard, and there is more of a danger of electrocution where water is present. The code does not specifically list a bathroom as a damp or wet location; however, this section requires a light in the ceiling of a shower to be suitable for a wet location or, at a minimum, suitable for a damp location if it is not subject to splashing or direct shower spray. A recessed light with a gasketed glass cover is quite often the fixture used in the ceiling of a shower. There is not a code definition for "out of reach," but a hanging fixture or paddle fan installed at least 8 feet (2438 mm) above the rim of the bathtub would, in most cases, be out of reach of someone standing in or on the bathtub. Note that the requirement for luminaires to be marked for damp locations applies only to the horizontal area bounded by the actual bathtub or shower dimensions, not the entire "zone" horizontal area described in the first part of the code section.

E4003.12 Luminaires in clothes closets. For the purposes of this section, storage space shall be defined as a volume bounded by the sides and back closet walls and planes extending from the closet floor vertically to a height of 6 feet (1829 mm) or the highest clothes-hanging rod and parallel to the walls at a horizontal distance of 24 inches (610 mm) from the sides and back of the closet walls respectively, and continuing vertically to the closet ceiling parallel to the walls at a horizontal distance of 12 inches (305 mm) or the width of the shelf, whichever is greater. For a closet that permits access to both sides of a hanging rod, the storage space shall include the volume below the highest rod extending 12 inches (305 mm) on either side of the rod on a plane horizontal to the floor extending the entire length of the rod (see Figure E4003.12). (410.2)

The types of luminaires installed in clothes closets shall be limited to surface-mounted or recessed incandescent or LED luminaires with completely enclosed light sources, surface-mounted or recessed fluorescent luminaires, and surface-mounted fluorescent or LED luminaires identified as suitable for installation within the closet storage area. Incandescent luminaires with open or partially enclosed lamps and pendant luminaires or lamp-holders shall be prohibited. The minimum clearance between luminaires installed in clothes closets and the nearest point of a closet storage area shall be as follows: [410.16(A) and (B)]

1. Surface-mounted incandescent or LED luminaires with a completely enclosed light source shall be installed on the wall above the door or on the ceiling, provided that there is a minimum clearance of 12 inches (305 mm) between the fixture and the nearest point of a storage space.

2. Surface-mounted fluorescent luminaires shall be installed on the wall above the door or on the ceiling, provided that there is a minimum clearance of 6 inches (152 mm).

3. Recessed incandescent luminaires or LED luminaires with a completely enclosed light source shall be installed in the wall or the ceiling provided that there is a minimum clearance of 6 inches (152 mm).

4. Recessed fluorescent luminaires shall be installed in the wall or on the ceiling provided that there is a minimum

clearance of 6 inches (152 mm) between the fixture and the nearest point of a storage space.

5. Surface-mounted fluorescent or LED luminaires shall be permitted to be installed within the closet storage space where identified for this use. [410.16(C)]

❖ In a clothes closet, a fixture with an exposed (incandescent) lightbulb (lamp) is a definite fire hazard and has been strictly prohibited for many years. Some older houses still have a porcelain fixture in closets, with an exposed lamp such as a keyless lampholder or lampholder with a pullchain for the switch. A luminaire designed for use with an exposed incandescent lamp is prohibited even if a compact fluorescent type or LED lamp is screwed into it. Storage items piled or stacked on closet shelves could easily come into contact with the bare lamp and cause a fire. Also, a naked incandescent lamp could be broken and the hot filament fragments could start a fire. This section clearly defines the storage space and provides minimum clearances from the permitted light fixtures to the storage space. For a small closet where this clearance would not be possible, a lighting fixture would be prohibited. Note that there may be special luminaires on the market such as LED fixtures that are listed and identified for installation within the storage space.

E4003.13 Luminaire wiring—general. Wiring on or within luminaires shall be neatly arranged and shall not be exposed to physical damage. Excess wiring shall be avoided. Conductors shall be arranged so that they are not subjected to temperatures above those for which the conductors are rated. (410.48)

❖ Excessive heat will damage fixture wires and the wall or ceiling surface and can cause a fire. Most ceiling-mounted incandescent luminaires for indoor use for general lighting are marked with a statement indicating the maximum wattage of the lamp. Many typical luminaires indicate a maximum lamp wattage of 60. Where higher wattage lamps are installed, the heat produced will likely damage the fixture wires and the wall finish within the canopy of the fixture. Many fixtures come with thermal insulation inside the canopy, which must not be removed because it helps protect both the circuit wires and the fixture wires from excessive heat. Fortunately, from a safety and energy standpoint, incandescent lamps are becoming more rare.

E4003.13.1 Polarization of luminaires. Luminaires shall be wired so that the screw shells of lampholders will be connected to the same luminaire or circuit conductor or terminal. The grounded conductor shall be connected to the screw shell.

❖ The grounded conductor of the circuit must be connected to the screw shell of the socket. The "hot" conductor is connected to the contact in the bottom of the socket. When a person is replacing the lamp in a luminaire that is energized and touches the screw shell part of the lamp, the person could be electrocuted if the screw shell of the lampholder was connected to the ungrounded (hot) conductor.

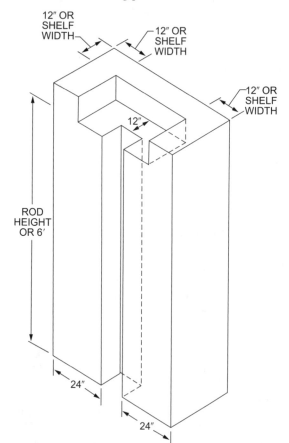

For SI: 1 inch = 25.4 mm, 1 foot = 304.8 mm.

FIGURE E4003.12
CLOSET STORAGE SPACE

E4003.13.2 Luminaires as raceways. Luminaires shall not be used as raceways for circuit conductors except where such luminaires are listed and marked for use as a raceway or are identified for through-wiring. Luminaires designed for end-to-end connection to form a continuous assembly, and luminaires connected together by recognized wiring methods, shall not be required to be listed as a raceway where they contain the conductors of one 2-wire branch circuit or one multiwire branch circuit and such conductors supply the connected luminaires. One additional 2-wire branch circuit that separately supplies one or more of the connected luminaires shall also be permitted. [410.64(A), (B), and (C)]

❖ This is usually a concern only where several fluorescent fixtures are installed end-to-end and the supply conductors run through one fixture to another. Where two or more fluorescent fixtures are installed end-to-end, the supply conductors may enter at one fixture, and the switched circuit conductors are continued to all other fixtures. This is permitted if the fixture wireway is listed as a raceway and if the conductors are rated for the temperature in the fixture. Heat is created by the ballast, and the circuit wires will likely be run next to the ballast.

SECTION E4004
LUMINAIRE INSTALLATION

E4004.1 Outlet box covers. In a completed installation, each outlet box shall be provided with a cover except where covered by means of a luminaire canopy, lampholder or device with a faceplate. (410.22)

❖ See the commentary to Section E3906.9. Note that the requirement of this section is triggered by a minimum surface area of 180 square inches. Such area will likely be doughnut shaped and the calculation of the area in square inches will require that the area of the inner (smaller) circle for the outlet box be calculated and then deducted from the area of the larger circle representing the size of the fixture canopy.

E4004.2 Combustible material at outlet boxes. Combustible wall or ceiling finish exposed between the inside edge of a luminaire canopy or pan and the outlet box and having a surface area of 180 in.^{2} (116 129 mm^2) or more shall be covered with a noncombustible material. (410.23)

❖ See the commentary to Section E3906.11.

E4004.3 Access. Luminaires shall be installed so that the connections between the luminaire conductors and the circuit conductors can be accessed without requiring the disconnection of any part of the wiring. Luminaires that are connected by attachment plugs and receptacles meet the requirement of this section. (410.8)

❖ To access connections to circuit wiring, a person might need to move a lighting fixture; therefore, sufficient slack in the luminaire and/or circuit wires is necessary. The luminaire must be hung or supported independently of the wiring.

E4004.4 Supports. Luminaires and lampholders shall be securely supported. A luminaire that weighs more than 6 pounds (2.72 kg) or exceeds 16 inches (406 mm) in any dimension shall not be supported by the screw shell of a lampholder. [410.30(A)]

❖ A luminaire with a lampholder may be designed to have a decorative part attached and supported directly by the lampholder. The part supported by the lampholder is limited in size and weight. The entire fixture that is supported directly by the outlet box is covered in the next section.

E4004.5 Means of support. Outlet boxes or fittings installed as required by Sections E3905 and E3906 shall be permitted to support luminaires. [410.36(A)]

❖ Most luminaires are designed to be supported directly from the outlet box. Any light fixture weighing over 50 pounds (23 kg) must be supported from the building structure or by a box listed for the application. However, for a large fixture such as a chandelier that may be a little less than 50 pounds (23 kg), it may be well to check how securely the outlet box is supported before assuming that it will be acceptable simply because the code allows it. The ceiling outlet box might be attached to the structure with only two nails or by a weak bracket between two framing members, which may not provide the support necessary for a heavy fixture. It would be wise to provide heavy duty boxes and attachment means where it is anticipated that the owner will want a chandelier-type luminaire.

E4004.6 Exposed components. Luminaires having exposed ballasts, transformers, LED drivers or power supplies shall be installed so that such ballasts, transformers, LED drivers or power supplies are not in contact with combustible material unless listed for such condition. [410.136(A)]

❖ Ballasts, transformers and the circuitry in LED lamp assemblies often operate at high temperatures under normal conditions and even higher temperatures under abnormal conditions. Clearance to combustible materials is a precaution against these devices starting a fire. Because winding failure in such devices can cause extreme temperatures and/or open flames, ballasts are equipped with internal thermal overload protection devices (see Section E4003.6).

E4004.7 Combustible low-density cellulose fiberboard. Where a surface-mounted luminaire containing a ballast, transformer, LED driver or power supply is installed on combustible low-density cellulose fiberboard, the luminaire shall be marked for this purpose or it shall be spaced not less than 1^1/$_2$ inches (38 mm) from the surface of the fiberboard. Where such luminaires are partially or wholly recessed, the provisions of Sections E4004.8 and E4004.9 shall apply. [410.136(B)]

❖ Where a fluorescent lighting fixture is installed directly on combustible low-density cellulose fiberboard, the fixture must be marked very clearly for this installation. The heat from the ballast could exceed the temperature rating of this type of ceiling finish. Thermal insulation installed above a ceiling of this type will add to the buildup of heat. Combustible low-density cellulose fiberboard includes sheets, panels, and tiles that have a density of 20 pounds per cubic foot (320 kg/m³) or less and are formed from bonded plant fiber material.

E4004.8 Recessed luminaire clearance. A recessed luminaire that is not identified for contact with insulation shall have all recessed parts spaced at least $^1/_2$ inch (12.7 mm) from combustible materials. The points of support and the finish trim parts at the opening in the ceiling, wall or other finished surface shall be permitted to be in contact with combustible materials. A recessed luminaire that is identified for contact with insulation, Type IC, shall be permitted to be in contact with combustible materials at recessed parts, points of support, and portions passing through the building structure and at finish trim parts at the opening in the ceiling or wall. [410.116(A)(1) and (A)(2)]

❖ A recessed fixture that is designed to allow the insulation above the ceiling to be in contact with the fixture is referred to as Type IC. Type IC is commonly thought of as standing for "insulation contact", however, it is believed to refer to "in ceiling" or "insulation ceiling."

E4004.9 Recessed luminaire installation. Thermal insulation shall not be installed above a recessed luminaire or within 3 inches (76 mm) of the recessed luminaire's enclosure, wiring compartment, ballast, transformer, LED driver or power supply except where such luminaire is identified for contact with insulation, Type IC. [410.116(B)]

❖ Only Type IC recessed fixtures can have insulation installed above and within 3 inches (76 mm) of the fixture. There should be coordination between the electrician and the insulation installer to be sure that insulation is not placed above and is kept at least 3 inches (76 mm) away from the recessed lights, unless Type IC fixtures are used. Heat will be trapped and can cause damage or a fire.

SECTION E4005
TRACK LIGHTING

E4005.1 Installation. Lighting track shall be permanently installed and permanently connected to a branch circuit having a rating not more than that of the track. [410.151(A) and (B)]

❖ Lighting track cannot be supplied by a cord and plug; rather, it must be hard wired as a permanent fixture. To prevent overloading of the track, the branch circuit serving the track must not have a higher current rating than the track.

E4005.2 Fittings. Fittings identified for use on lighting track shall be designed specifically for the track on which they are to be installed. Fittings shall be securely fastened to the track, shall maintain polarization and connection to the equipment grounding conductor, and shall be designed to be suspended directly from the track. Only lighting track fittings shall be installed on lighting track. Lighting track fittings shall not be equipped with general-purpose receptacles. [410.151(A) and (B)]

❖ Track fittings and the track to which they connect must be designed to mate. Ill-fitting, mismatched or improp-

erly installed components could cause poor electrical connections and could cause lighting units to fall. Tracks are intended to function as a component of a lighting system and are not intended to serve as a substitute for convenience receptacles.

E4005.3 Connected load. The connected load on lighting track shall not exceed the rating of the track. [410.151(B)]

❖ If too many lighting units are connected to a track, the total load could exceed the rating of the track, resulting in a hazard. Section E4005.1 also requires that the circuit rating not exceed the track rating; thus, the overcurrent device will be sized to protect the track.

E4005.4 Prohibited locations. Lighting track shall not be installed in the following locations:

1. Where likely to be subjected to physical damage.

2. In wet or damp locations.

3. Where subject to corrosive vapors.

4. In storage battery rooms.

5. In hazardous (classified) locations.

6. Where concealed.

7. Where extended through walls or partitions.

8. Less than 5 feet (1524 mm) above the finished floor except where protected from physical damage or the track operates at less than 30 volts rms open-circuit voltage.

9. Where prohibited by Section E4003.11. [410.151(C)]

❖ Track lighting must not be located where it is subject to damage from impact or moisture. It must not be concealed by structural or finish elements of a building and must not pass through walls or partitions. Item 9 is consistent with Section E4003.11.

E4005.5 Fastening. Lighting track shall be securely mounted so that each fastening will be suitable for supporting the maximum weight of luminaires that can be installed. Except where identified for supports at greater intervals, a single section 4 feet (1219 mm) or shorter in length shall have two supports and, where installed in a continuous row, each individual section of not more than 4 feet (1219 mm) in length shall have one additional support. (410.154)

❖ Support for track lighting is very important. The code does not state that the track must be supported directly into a truss or framing member, only that it must be securely mounted. It is usually not sufficient to support the track lighting solely by attaching it to the drywall of the ceiling. When track lighting is first installed, it may well support the type of fixture initially installed. However, the quantity and type of fixtures that will be attached to the track at a future time is unknown.

E4005.6 Grounding. Lighting track shall be grounded in accordance with Chapter 39, and the track sections shall be securely coupled to maintain continuity of the circuitry, polarization and grounding throughout. [410.155(B)]

❖ Where several sections of lighting track are joined together, it is important that the grounding continuity be secured. The track has the hot or ungrounded conductor on one side and the grounded conductor on the other side. If a section is reversed, the polarity could be reversed.

Bibliography

The following resource materials were used in the preparation of the commentary for this chapter of the code:

NFPA 70–14, *National Electrical Code*. Quincy, MA: Na-tional Fire Protection Association, 2013.

Chapter 41:
Appliance Installation

General Comments

Chapter 41 covers appliance installation, which is typically the final stage of construction after all wiring, devices and fixtures are installed. This chapter covers flexible cords, overcurrent protection, disconnecting means and installation provisions.

Purpose

This chapter intends to make appliance installations safe by regulating power cords, overcurrent/overload protection support and means of disconnecting from the power supply.

SECTION E4101
GENERAL

E4101.1 Scope. This section covers installation requirements for appliances and fixed heating equipment. (422.1 and 424.1)

❖ Appliances are also referred to as utilization equipment. The classification includes small kitchen appliances such as a food mixer and deep fryer that are cord- and plug-connected as well as larger equipment that is fixed in place such as a dishwasher, water heater, air conditioner or central heating furnace.

E4101.2 Installation. Appliances and equipment shall be installed in accordance with the manufacturer's installation instructions. Electrically heated appliances and equipment shall be installed with the required clearances to combustible materials. [110.3(B) and 422.17]

❖ The code does not have prescriptive provisions for the minimum clearance between appliances and combustible materials. Required clearances for various appliances are given in the manufacturer's installation instructions.

E4101.3 Flexible cords. Cord-and-plug-connected appliances shall use cords suitable for the environment and physical conditions likely to be encountered. Flexible cords shall be used only where the appliance is listed to be connected with a flexible cord. The cord shall be identified as suitable in the installation instructions of the appliance manufacturer. Receptacles for cord-and-plug-connected appliances shall be accessible and shall be located to avoid physical damage to the flexible cord. Except for a listed appliance marked to indicate that it is protected by a system of double-insulation, the flexible cord supplying an appliance shall terminate in a grounding-type attachment plug. A receptacle for a cord-and-plug-connected range hood shall be supplied by an individual branch circuit. Specific appliances have additional requirements as specified in Table E4101.3 (see Section E3909). [422.16(B)(1), (B)(2)]

❖ Flexible cords are permitted for appliances to allow easy removal or replacement of the unit, but only if the appliance is listed to be cord- and plug-connected. Kitchen appliances and range hoods listed in Table

E4101.3 are specifically permitted to be connected with flexible cord. A disposal is connected to the sink and to the drain piping, but these connections are designed to permit disconnection and removal of the disposal. In many areas, power to a dishwasher is provided by direct connection of nonmetallic sheathed cable in the junction box or termination box at the base of the dishwasher; however, flexible cord is allowed. Type NM cable cannot be directly connected to a waste disposal unit because the cable would not be protected from damage. Flexible metal conduit and flexible cord are the two most commonly installed wiring methods for disposal units. In fact, flexible cord is the ideal method for wiring both dishwashers and disposals. Flexible cord allows for ease of installation and removal of the appliances. The cord and plug connection provides a convenient disconnecting means, also. Furnaces, air handlers and most storage-type electric water heaters are not known to be listed for connection by flexible cords and, therefore, must be hard wired. It is a common code violation in some localities where furnaces, boilers and air handling units are wired with cords and plugs.

TABLE E4101.3
FLEXIBLE CORD LENGTH

APPLIANCE	MINIMUM CORD LENGTH (inches)	MAXIMUM CORD LENGTH (inches)
Electrically operated in-sink waste disposal	18	36
Built-in dishwasher	36	48
Trash compactor	36	48
Range hoods	18	36

For SI: 1 inch = 25.4 mm.

❖ See the commentary to Section E4101.3.

E4101.4 Overcurrent protection. Each appliance shall be protected against overcurrent in accordance with the rating of the appliance and its listing. [110.3(B), 422.11(A)]

❖ Some appliances come with installation instructions or are marked with the minimum and maximum ratings of the overcurrent device for the branch circuit feeding the appliance.

E4101.4.1 Single nonmotor-operated appliance. The overcurrent protection for a branch circuit that supplies a single nonmotor-operated appliance shall not exceed that marked on the appliance. Where the overcurrent protection rating is not marked and the appliance is rated at over 13.3 amperes, the overcurrent protection shall not exceed 150 percent of the appliance rated current. Where 150 percent of the appliance rating does not correspond to a standard overcurrent device ampere rating, the next higher standard rating shall be permitted. Where the overcurrent protection rating is not marked and the appliance is rated at 13.3 amperes or less, the overcurrent protection shall not exceed 20 amperes. [422.11(E)]

❖ The overcurrent device protects the conductors, but it also serves to protect the appliance or equipment itself. An appliance marked with the rated ampere load might not provide information on the rating of branch circuit overcurrent protection. If this was the case and the appliance is rated over 13.3 amperes, the provisions of this section limit the size of the overcurrent device on the branch circuit to 150 percent of the appliance rated current. For example, a typical electric water heater with a maximum load of 4,500 watts draws 18.75 amperes and 150 percent of that is 28.1 amperes, therefore, the typically used 30-ampere overcurrent device is too large. A 25 amp fuse or circuit breaker can serve a continuous load of 18.75 amps and is less than 150 percent of the appliance rating. See Sections E3705.5.2 and E3705.6.

E4101.5 Disconnecting means. Each appliance shall be provided with a means to disconnect all ungrounded supply conductors. For fixed electric space-heating equipment, means shall be provided to disconnect the heater and any motor controller(s) and supplementary overcurrent-protective devices. Switches and circuit breakers used as a disconnecting means shall be of the indicating type. Disconnecting means shall be as set forth in Table E4101.5. (422.30, 422.35, and 424.19)

❖ In Table E4101.5, permanently connected appliances are those with fixed wiring (as opposed to cord- and plug-connected appliances). A circuit breaker in a panelboard could serve as a disconnect if it serves a rated load of $^1/_8$ horsepower (93 W) or less or 300 VA or less. If it is rated over $^1/_8$ horsepower (93 W), or over 300 VA, the circuit breaker is permitted to serve as the disconnecting means if it is within sight of the appliance or if it can be locked in the open position. The second from bottom row in Table E4101.5 indicates that if the appliance has a unit switch with a marked OFF position, it can serve as the disconnect if there is an additional redundant disconnect. In residential wiring, the main service disconnect is permitted to serve as the additional disconnect for this purpose. Air-conditioning condensing units and heat pump units must have a disconnect within sight from the unit as the only allowable means (see Commentary Figure E4101.5). Note that on the typical heat pump or condensing unit there is not likely to be any surface on which a disconnect can be mounted because any flat surface will probably be an access panel or the location of the equipment label (see Note a to Table E4101.5).

TABLE E4101.5. See page 41-3.

❖ See the commentary to Section E4101.5.

Figure E4101.5
TYPICAL APPLIANCE DISCONNECTING MEANS

TABLE E4101.5
DISCONNECTING MEANS
[422.31(A), (B), and (C); 422.34; 422.35; 424.19; 424.20; and 440.14]

DESCRIPTION	ALLOWED DISCONNECTING MEANS
Permanently connected appliance rated at not over 300 volt-amperes or $1/8$ horsepower.	Branch-circuit overcurrent device.
Permanently connected appliances rated in excess of 300 volt-amperes.	Branch circuit breaker or switch located within sight of appliance or such devices in any location that are capable of being locked in the open position. The provision for locking or adding a lock to the disconnecting means shall be installed on or at the switch or circuit breaker used as the disconnecting means and shall remain in place with or without the lock installed.
Motor-operated appliances rated over $1/8$ horsepower.	For permanently connected motor-operated appliances with motors rated over $1/8$ horsepower, the branch circuit switch or circuit breaker shall be permitted to serve as the disconnecting means where the switch or circuit breaker is within sight from the appliance. Where the branch circuit switch is not located within sight from the appliance, the disconnecting means shall be one of the following types: a listed motor-circuit switch rated in horsepower, a listed molded case circuit breaker, a listed molded case switch, a listed manual motor controller additionally marked "Suitable as Motor Disconnect" where installed between the final motor branch-circuit short-circuit protective device and the motor. For stationary motors rated at 2 hp or less and 300 volts or less, the disconnecting means shall be permitted to be one of the following devices: 1. A general-use switch having an ampere rating not less than twice the full-load current rating of the motor. 2. On AC circuits, a general-use snap switch suitable only for use on AC, not general-use AC–DC snap switches, where the motor full-load current rating is not more than 80 percent of the ampere rating of the switch. 3. A listed manual motor controller having a horsepower rating not less than the rating of the motor and marked "Suitable as Motor Disconnect". The disconnecting means for motor circuits rated 600 volts, nominal, or less shall have an ampere rating not less than 115 percent of the full-load current rating of the motor except that a listed unfused motor-circuit switch having a horsepower rating not less than the motor horsepower shall be permitted to have an ampere rating less than 115 percent of the full-load current rating of the motor. The disconnecting means shall be installed within sight of the appliance. **Exception:** A unit switch with a marked-off position that is a part of an appliance and disconnects all ungrounded conductors shall be permitted as the disconnecting means and the switch or circuit breaker serving as the other disconnecting means shall be permitted to be out of sight from the appliance.
Appliances listed for cord-and-plug connection.	A separable connector or attachment plug and receptacle provided with access.

(continued)

TABLE E4101.5—continued
DISCONNECTING MEANS

DESCRIPTION	ALLOWED DISCONNECTING MEANS
Permanently installed heating equipment with motors rated at not over $^1/_8$ horsepower with supplementary overcurrent protection.	Disconnect, on the supply side of fuses, in sight from the supplementary overcurrent device, and in sight of the heating equipment or, in any location, if capable of being locked in the open position.
Heating equipment containing motors rated over $^1/_8$ horsepower with supplementary overcurrent protection.	Disconnect permitted to serve as required disconnect for both the heating equipment and the controller where, on the supply side of fuses, and in sight from the supplementary overcurrent devices, if the disconnecting means is also in sight from the controller, or is capable of being locked off and simultaneously disconnects the heater, motor controller(s) and supplementary overcurrent protective devices from all ungrounded conductors. The provision for locking or adding a lock to the disconnecting means shall be installed on or at the switch or circuit breaker used as the disconnecting means and shall remain in place with or without the lock installed. The disconnecting means shall have an ampere rating not less than 125 percent of the total load of the motors and the heaters.
Heating equipment containing no motor rated over $^1/_8$ horsepower without supplementary overcurrent protection.	Branch-circuit switch or circuit breaker where within sight from the heating equipment or capable of being locked off and simultaneously disconnects the heater, motor controller(s) and supplementary overcurrent protective devices from all ungrounded conductors. The provision for locking or adding a lock to the disconnecting means shall be installed on or at the switch or circuit breaker used as the disconnecting means and shall remain in place with or without the lock installed. The disconnecting means shall have an ampere rating not less than 125 percent of the total load of the motors and the heaters.
Heating equipment containing motors rated over $^1/_8$ horsepower without supplementary overcurrent protection.	Disconnecting means in sight from motor controller or as provided for heating equipment with motor rated over $^1/_8$ horsepower with supplementary overcurrent protection and simultaneously disconnects the heater, motor controller(s) and supplementary overcurrent protective devices from all ungrounded conductors. The provision for locking or adding a lock to the disconnecting means shall be installed on or at the switch or circuit breaker used as the disconnecting means and shall remain in place with or without the lock installed. The disconnecting means shall have an ampere rating not less than 125 percent of the total load of the motors and the heaters.
Air-conditioning condensing units and heat pump units.	A readily accessible disconnect within sight from unit as the only allowable means.[a]
Appliances and fixed heating equipment with unit switches having a marked OFF position.	Unit switch where an additional individual switch or circuit breaker serves as a redundant disconnecting means.
Thermostatically controlled fixed heating equipment.	Thermostats with a marked OFF position that directly open all ungrounded conductors, which when manually placed in the OFF position are designed so that the circuit cannot be energized automatically and that are located within sight of the equipment controlled.

For SI: 1 horsepower = 0.746 kW.

a. The disconnecting means shall be permitted to be installed on or within the unit. It shall not be located on panels designed to allow access to the unit or located so as to obscure the air-conditioning equipment nameplate(s).

E4101.6 Support of ceiling-suspended paddle fans. Ceiling-suspended fans (paddle) shall be supported independently of an outlet box or by a listed outlet box or outlet box system identified for the use and installed in accordance with Section E3905.8. (422.18)

❖ The provisions for paddle fans are included in this chapter because they are considered to be appliances and not luminaires by the code. The outlet boxes permitted by this section for the support of paddle fans are not the standard luminaire boxes that are used for most lighting outlets (see Section E3905.8). Although not required by the code, it would be wise to install fan support boxes for all outlets where it is anticipated that the owner will want to add a fan in the future. Section E3905.8 will require fan-rated boxes where wiring is provided in anticipation of a fan installation.

E4101.7 Snow-melting and deicing equipment protection. Outdoor receptacles that are not readily accessible and are supplied from a dedicated branch circuit for electric snow-melting or deicing equipment shall be permitted to be installed without ground-fault circuit-interrupter protection for personnel. However, ground-fault protection of equipment shall be provided for fixed outdoor electric deicing and snow-melting equipment. [210.8(A)(3) Exception, 426.28]

❖ Receptacles can be installed under the eaves for the connection of deicing cables installed along the roof edge. These receptacles need not be ground-fault circuit interrupter (GFCI) protected. However, permanently installed deicing and snow melting equipment must have ground-fault protection. Circuit breakers are available with ground-fault equipment protection which will open the circuit if leakage current is detected and will help prevent fires caused by arcing from low-level fault currents.

Bibliography

The following resource materials were used in the preparation of the commentary for this chapter of the code:

NFPA 70–14, *National Electrical Code.* Quincy, MA: Na-tional Fire Protection Association, 2013.

Chapter 42: Swimming Pools

General Comments

This chapter addresses all aspects of wiring, fixtures, motors and electrical accessories for swimming pools, wading pools, hot tubs, spas and hydromassage bathtubs.

Purpose

This chapter focuses on protection of occupants from electrical shock. The dangers of using electricity around water, wet surfaces, grounded surfaces and plumbing are well known, and this chapter is intended to minimize or eliminate those hazards.

SECTION E4201
GENERAL

E4201.1 Scope. The provisions of this chapter shall apply to the construction and installation of electric wiring and equipment associated with all swimming pools, wading pools, decorative pools, fountains, hot tubs and spas, and hydromassage bathtubs, whether permanently installed or storable, and shall apply to metallic auxiliary equipment, such as pumps, filters and similar equipment. Sections E4202 through E4206 provide general rules for permanent pools, spas and hot tubs. Section E4207 provides specific rules for storable pools and storable/portable spas and hot tubs. Section E4208 provides specific rules for spas and hot tubs. Section E4209 provides specific rules for hydromassage bathtubs. (680.1)

❖ Because of a unique potential of electrical shock hazard around pools, a separate chapter is included in the code to provide specific detailed requirements regarding the wiring in and around swimming pools, wading pools, spas and hot tubs and hydromassage tubs. A conductive path for fault current can easily be established through the water to the earth. When a person in a pool touches a metallic surface that is energized, the fault-current path through the individual can be fatal. Also, a person in the pool not touching anything but the water could become part of a fault-current path. For example, if an energized device such as an electric appliance drops into the pool, an electrical potential could be established in the pool resulting in a voltage gradient that could cause the person to be surrounded by different levels of voltage in the water acting like a conductor. Because of these types of hazards, more stringent and unique code provisions apply to wiring around swimming pools.

E4201.2 Definitions. (680.2)

CORD-AND-PLUG-CONNECTED LIGHTING ASSEMBLY. A lighting assembly consisting of a cord-and-plug-connected transformer and a luminaire intended for installation in the wall of a spa, hot tub, or storable pool.

❖ This is a definition for a lighting fixture used in the wall of a storable swimming pool, spa, or hot tub. It is made with a nonmetallic housing and is supplied with power from a cord- and plug-connected transformer.

DRY-NICHE LUMINAIRE. A luminaire intended for installation in the floor or wall of a pool, spa or fountain in a niche that is sealed against the entry of water.

❖ A dry-niche luminaire is installed behind a clear window below the water level and does not allow pool water to enter the forming shell or the area around the lamp housing. For swimming pool wiring in dwellings, access to the lamp would typically be through a deck box, and relamping is done from behind the waterproof lens or glass window. The dry-niche area where the lamp is installed must have drainage for any water that may enter from the deck area of the pool when the cover is off for servicing. The glass window is sealed against the entry of pool water.

FORMING SHELL. A structure designed to support a wet-niche luminaire assembly and intended for mounting in a pool or fountain structure.

❖ A forming shell is built into the wall of the pool and is the supporting part of the wet-niche luminaire.

FOUNTAIN. Fountains, ornamental pools, display pools, and reflection pools. The definition does not include drinking fountains.

❖ Decorative pools and fountains fall under the scope of this chapter.

HYDROMASSAGE BATHTUB. A permanently installed bathtub equipped with a recirculating piping system, pump, and associated equipment. It is designed so it can accept, circulate and discharge water upon each use.

❖ A hydromassage bathtub is similar to a normal bathtub in that the water is drained after each use. The hydromassage bathtub does not store water and is not left filled with water like a spa or hot tub. It has a pump and piping system for jetting water when filled. The pump must be plugged into a ground-fault circuit interrupter (GFCI) protected receptacle. All receptacles in the same bathroom with the hydromassage bathtub must be GFCI protected.

LOW VOLTAGE CONTACT LIMIT. A voltage not exceeding the following values:

1. 15 volts (RMS) for sinusoidal AC
2. 21.2 volts peak for nonsinusoidal AC
3. 30 volts for continuous DC
4. 12.4 volts peak for DC that is interrupted at a rate of 10 to 200 Hz

MAXIMUM WATER LEVEL. The highest level that water can reach before it spills out.

❖ In the plumbing part of this code the term "flood-level rim" is used; the meaning is the same as "maximum water level."

NO-NICHE LUMINAIRE. A luminaire intended for installation above or below the water without a niche.

❖ This is an underwater luminaire installed on the surface of the wall of the pool on a mounting bracket. For changing the lamp, the cord should be long enough to detach the fixture from the bracket underwater and bring it out of the water. A no-niche luminaire is typically used on an aboveground pool that does not have a forming shell.

PACKAGED SPA OR HOT TUB EQUIPMENT ASSEMBLY. A factory-fabricated unit consisting of water-circulating, heating and control equipment mounted on a common base, intended to operate a spa or hot tub. Equipment may include pumps, air blowers, heaters, luminaires, controls and sanitizer generators.

❖ The packaged unit comes already assembled from the manufacturer, minimizing the need to design and install each component in the field. The spa or hot tub vessel comes as a separate component and is installed on the job along with the packaged assembly of equipment.

PERMANENTLY INSTALLED SWIMMING, WADING, IMMERSION AND THERAPEUTIC POOLS. Those that are constructed in the ground or partially in the ground, and all others capable of holding water with a depth greater than 42 inches (1067 mm), and all pools installed inside of a building, regardless of water depth, whether or not served by electrical circuits of any nature.

❖ Special provisions are included in the code for swimming pools because of the high risk of electric shock resulting from the conductive path from a human body to the water surrounding a human body through the concrete and reinforcing steel in the wall of the pool to the earth.

POOL. Manufactured or field-constructed equipment designed to contain water on a permanent or semipermanent basis and used for swimming, wading, immersion, or therapeutic purposes.

POOL COVER, ELECTRICALLY OPERATED. Motor-driven equipment designed to cover and uncover the water surface of a pool by means of a flexible sheet or rigid frame.

❖ Because of the close proximity of the motor of an electrically operated pool cover to the edge of the pool, the code has special provisions regarding these pool covers.

SELF-CONTAINED SPA OR HOT TUB. A factory-fabricated unit consisting of a spa or hot tub vessel with all water-circulating, heating and control equipment integral to the unit. Equipment may include pumps, air blowers, heaters, luminaires, controls and sanitizer generators.

❖ The equipment needed to operate the spa or hot tub such as pumps, heaters, controls, etc., is manufactured together with the spa or hot tub vessel and comes as a complete unit.

SPA OR HOT TUB. A hydromassage pool, or tub for recreational or therapeutic use, not located in health care facilities, designed for immersion of users, and usually having a filter, heater, and motor-driven blower. They are installed indoors or outdoors, on the ground or supporting structure, or in the ground or supporting structure. Generally, a spa or hot tub is not designed or intended to have its contents drained or discharged after each use.

❖ A hot tub is typically built with wood, such as redwood, but a spa is usually made of fiberglass, concrete or tile. They may have much the same equipment that a swimming pool could have, such as stainless-steel handrails and electric equipment for heating and circulating the water.

STORABLE SWIMMING, WADING OR IMMERSION POOLS; OR STORABLE/PORTABLE SPAS AND HOT TUBS. Those that are constructed on or above the ground and are capable of holding water with a maximum depth of 42 inches (1067 mm), or a pool with nonmetallic, molded polymeric walls or inflatable fabric walls regardless of dimension.

❖ This definition was included in the code to help differentiate between a storable pool and a permanent pool built in the ground. Also, the code provisions and definition for storable pools are included in the code because these structures have most of the same equipment that an in-ground pool could have. They are not built with concrete and reinforcing steel, but similar electrical hazards exist. Although storable pool are intended to be temporary, and many storable pools are disassembled during the cold months of the year, an on-ground pool that is left installed all year could still be defined as a storable pool.

THROUGH-WALL LIGHTING ASSEMBLY. A lighting assembly intended for installation above grade, on or through the wall of a pool, consisting of two interconnected groups of components separated by the pool wall.

❖ Such assemblies allow the installations of pool lighting in above-ground pools. The components house the luminaire and also serve to seal the opening made in the wall of the pool.

WET-NICHE LUMINAIRE. A luminaire intended for installation in a forming shell mounted in a pool or fountain structure where the luminaire will be completely surrounded by water.

❖ A wet-niche luminaire is designed to have the pool water completely surround the fixture within the forming shell. A wet-niche luminaire is connected with a cord that extends to a junction box, typically at the pool deck. It has enough cord so that when the fixture is relamped, it is brought up out of the water without the cord being disconnected and then the cord is recoiled and placed behind the fixture in the forming shell.

SECTION E4202
WIRING METHODS FOR POOLS, SPAS, HOT TUBS AND HYDROMASSAGE BATHTUBS

E4202.1 General. Wiring methods used in conjunction with permanently installed swimming pools, spas, hot tubs or hydromassage bathtubs shall be installed in accordance with Table E4202.1 and Chapter 38 except as otherwise stated in this section. Storable swimming pools shall comply with Section E4207. [680.7; 680.21(A); 680.23(B) and (F); 680.25(A); 680.42; 680.43; and 680.70]

❖ Table E4202.1 covers the rules for the raceways and cables that can be used in wiring associated with swimming pools. The rules in Chapter 38 for wiring methods apply except as modified by the provisions of this chapter. It is important to consider the notes of Table E4202.1. For example, Note c indicates that EMT can be used only on or within buildings.

E4202.2 Flexible cords. Flexible cords used in conjunction with a pool, spa, hot tub or hydromassage bathtub shall be installed in accordance with the following:

1. For other than underwater luminaires, fixed or stationary equipment shall be permitted to be connected with a flexible cord to facilitate removal or disconnection for maintenance or repair. For other than storable pools, the flexible cord shall not exceed 3 feet (914 mm) in length. Cords that supply swimming pool equipment shall have a copper equipment grounding conductor not smaller than 12 AWG and shall terminate in a grounding-type attachment plug. [680.7(A), (B), and (C); 680.21(A)(5)]

TABLE E4202.1
ALLOWABLE APPLICATIONS FOR WIRING METHODS[a, b, c, d, e, f, g, h, k]

WIRING LOCATION OR PURPOSE (Application allowed where marked with an "A")	AC, FMC, NM, SR, SE	EMT	ENT	IMC[i], RMC[i], RNC[h]	LFMC	LFNMC	UF	MC[i]	FLEX CORD
Panelboard(s) that supply pool equipment: from service equipment to panelboard	A[b, e] SR not permitted	A[c]	A[b]	A	—	A	A[e]	A[e]	—
Wet-niche and no-niche luminaires: from branch circuit OCPD to deck or junction box	AC[b] only	A[c]	A[b]	A	—	A	—	A[b]	—
Wet-niche and no-niche luminaires: from deck or junction box to forming shell	—	—	—	A[d]	—	A	—	—	A[g]
Dry niche: from branch circuit OCPD to luminaires	AC[b] only	A[c]	A[b]	A	—	A	—	A[b]	—
Pool-associated motors: from branch circuit OCPD to motor	A[b]	A[c]	A[b]	A	A[e]	A[e]	A[b]	A	A[g]
Packaged or self-contained outdoor spas and hot tubs with underwater luminaire: from branch circuit OCPD to spa or hot tub	AC[b] only	A[c]	A[b]	A	A[f]	A[f]	—	A[b]	A[g]
Packaged or self-contained outdoor spas and hot tubs without underwater luminaire: from branch circuit OCPD to spa or hot tub	A[b]	A[c]	A[b]	A	A[f]	A[f]	A[b]	A	A[g]
Indoor spas and hot tubs, hydromassage bathtubs, and other pool, spa or hot tub associated equipment: from branch circuit OCPD to equipment	A[b]	A[c]	A[b]	A	A	A	A	A	A[g]
Connection at pool lighting transformers or power supplies	AC[b] only	A[c]	A[b]	A	A[l, f]	A[f]	—	A[b]	—

For SI: 1 foot = 304.8 mm.

a. For all wiring methods, see Section E4205 for equipment grounding conductor requirements.
b. Limited to use within buildings.
c. Limited to use on or within buildings.
d. Metal conduit shall be constructed of brass or other approved corrosion-resistant metal.
e. Limited to where necessary to employ flexible connections at or adjacent to a pool motor.
f. Sections installed external to spa or hot tub enclosure limited to individual lengths not to exceed 6 feet. Length not limited inside spa or hot tub enclosure.
g. Flexible cord shall be installed in accordance with Section E4202.2.
h. Nonmetallic conduit shall be rigid polyvinyl chloride conduit Type PVC or reinforced thermosetting resin conduit Type RTRC.
i. Aluminum conduits shall not be permitted in the pool area where subject to corrosion.
j. Where installed as direct burial cable or in wet locations, Type MC cable shall be listed and identified for the location.
k. See Section E4202.3 for listed, double-insulated pool pump motors.
l. Limited to use in individual lengths not to exceed 6 feet. The total length of all individual runs of LFMC shall not exceed 10 feet.

2. Other than listed low-voltage lighting systems not requiring grounding, wet-niche luminaires that are supplied by a flexible cord or cable shall have all exposed noncurrent-carrying metal parts grounded by an insulated copper equipment grounding conductor that is an integral part of the cord or cable. Such grounding conductor shall be connected to a grounding terminal in the supply junction box, transformer enclosure, or other enclosure and shall be not smaller than the supply conductors and not smaller than 16 AWG. [680.23(B)(3)]

3. A listed packaged spa or hot tub installed outdoors that is GFCI protected shall be permitted to be cord-and-plug-connected provided that such cord does not exceed 15 feet (4572 mm) in length. [680.42(A)(2)]

4. A listed packaged spa or hot tub rated at 20 amperes or less and installed indoors shall be permitted to be cord-and-plug-connected to facilitate maintenance and repair. (680.43 Exception No. 1)

5. For other than underwater and storable pool lighting luminaire, the requirements of Item 1 shall apply to any cord-equipped luminaire that is located within 16 feet (4877 mm) radially from any point on the water surface. [680.22(B)(5)]

❖ Flexible cords are used to facilitate easy removal or repair of fixed or stationary equipment. For example, a pool cover motor could be connected by a flexible cord not over 3 feet (914 mm) long. Limiting the length of the cord will help prevent excessive cord length from being in the pool water or otherwise exposed. Underwater fixtures such as no-niche and wet-niche fixtures are permitted to have longer cords so that the fixture can be removed and brought up on the deck for servicing. Some equipment on a storable pool could have cords longer than 3 feet (914 mm) because it is not fixed or stationary, being installed so that it can be removed and stored.

An example of a permitted flexible cord in the area of a swimming pool is one for a luminaire within 16 feet (4877 mm) of the edge of the water. It is permitted to be cord- and plug-connected if the cord is a maximum of 3 feet (914 mm) long, has a grounding conductor of at least size 12 AWG and has a grounding-type attachment plug (cap).

E4202.3 Double insulated pool pumps. A listed cord and plug-connected pool pump incorporating an approved system of double insulation that provides a means for grounding only the internal and nonaccessible, noncurrent-carrying metal parts of the pump shall be connected to any wiring method recognized in Chapter 38 that is suitable for the location. Where the bonding grid is connected to the equipment grounding conductor of the motor circuit in accordance with Section E4204.2, Item 6.1, the branch circuit wiring shall comply with Sections E4202.1 and E4205.5. [680.21(B)]

❖ See Sections E4202.1, E4204.2 and E4205.5.

SECTION E4203
EQUIPMENT LOCATION AND CLEARANCES

E4203.1 Receptacle outlets. Receptacles outlets shall be installed and located in accordance with Sections E4203.1.1 through E4203.1.5. Distances shall be measured as the shortest path that an appliance supply cord connected to the receptacle would follow without penetrating a floor, wall, ceiling, doorway with hinged or sliding door, window opening, or other effective permanent barrier. [680.22(A)(5)]

❖ The measurement between the edge of the pool and a receptacle is the shortest unobstructed route of an appliance supply cord plugged into the receptacle. The code considers such things as doors, windows, walls, floors and ceilings to be "effective permanent barriers" in determining the distance of a receptacle from the edge of the pool. When measuring the distance from the pool water to the receptacle, if the measurement is, for example, a total of 9 feet (2743 mm) from the pool to the receptacle, but the measurement is through a doorway that has a sliding or hinged door installed (not an opening without a door), the door is considered a barrier and the receptacle would not be considered 9 feet (2743 mm) from the edge of the pool and would not require GFCI protection because of the proximity of the swimming pool. The receptacle in this example could be required to have GFCI protection for a different reason, such as if it were in a bathroom (see commentary, Section E3902.1).

E4203.1.1 Location. Receptacles that provide power for water-pump motors or other loads directly related to the circulation and sanitation system shall be permitted to be located between 6 feet and 10 feet (1829 mm and 3048 mm) from the inside walls of pools and outdoor spas and hot tubs, where the receptacle is single and of the grounding type and protected by ground-fault circuit interrupters.

Other receptacles on the property shall be located not less than 6 feet (1829 mm) from the inside walls of pools and outdoor spas and hot tubs. [680.22(A)(2) and (A)(3)]

❖ Only receptacles for specific equipment are permitted between 6 and 10 feet (1524 mm and 3048 mm) from the inside wall of the pool, and they must be a single receptacle of the locking and grounding type so that a typical radio, for example, could not be plugged into it. The code prohibits receptacles that supply power to appliances from being installed within 6 feet (3048 mm) of the inside wall of the pool.

E4203.1.2 Where required. At least one 125-volt, 15- or 20-ampere receptacle supplied by a general-purpose branch circuit shall be located a minimum of 6 feet (1829 mm) from and not more than 20 feet (6096 mm) from the inside wall of pools and outdoor spas and hot tubs. This receptacle shall be located not more than 6 feet, 6 inches (1981 mm) above the floor, platform or grade level serving the pool, spa or hot tub. [680.22(A)(1)]

❖ At least one receptacle is required in the area between 6 and 20 feet (3048 mm and 6096 mm) from the inside wall of a swimming pool and outdoor spas and hot tubs. This requirement is for indoor and outdoor swimming pools but only for outdoor spas and hot tubs. This provides a power source for appliances to be used near the pool without resorting to the use of extension cords. Because a typical appliance supply cord is about 6 feet (1829 mm), it is less likely that an appliance would be bumped and knocked into the pool where it is plugged into a receptacle located at least 6 feet from the pool water. This required receptacle could be on a wall above a raised deck, for example, as long as the receptacle is not over 78 inches (1981 mm) above the swimming pool deck (see Section E4203.1.3).

E4203.1.3 GFCI protection. All 15- and 20-ampere, single phase, 125-volt receptacles located within 20 feet (6096 mm) of the inside walls of pools and outdoor spas and hot tubs shall be protected by a ground-fault circuit-interrupter. Outlets supplying pool pump motors supplied from branch circuits rated at 120 volts through 240 volts, single phase, whether by receptacle or direct connection, shall be provided with ground-fault circuit-interrupter protection for personnel. [680.21(C) and 680.22(A)(4)]

❖ Pool pump motor receptacles must be GFCI protected regardless of their location. All receptacles within 20 feet (6096 mm) of the pool water must have GFCI protection including any single-, locking- and grounding-type receptacles installed for specific equipment. This applies to storable pools as well as permanently installed indoor and outdoor pools. It may not be common to have an indoor pool with a deck over 20 feet (6096 mm) wide, [20 feet (6096 mm) from the inside wall of the pool to the wall of the building], but where this occurs, any receptacle over 20 feet (6096 mm) from the pool would not have to be GFCI protected. Where a pool is installed outdoors, the 20-foot (6096 mm) rule is irrelevant since all outdoor receptacles at a dwelling require GFCI protection per Section E3902.3.

E4203.1.4 Indoor locations. Receptacles shall be located not less than 6 feet (1829 mm) from the inside walls of indoor spas and hot tubs. A minimum of one 125-volt receptacle shall be located between 6 feet (1829 mm) and 10 feet (3048 mm) from the inside walls of indoor spas or hot tubs. [680.43(A) and 680.43(A)(1)]

❖ This code provision is specifically for indoor spas and hot tubs. The rule to have at least one receptacle between 6 feet and 10 feet (1524 mm and 3048 mm) from the inside wall of the spa or hot tub is to help prevent the hazards inherent with the use of extension cords.

E4203.1.5 Indoor GFCI protection. All 125-volt receptacles rated 30 amperes or less and located within 10 feet (3048 mm) of the inside walls of spas and hot tubs installed indoors, shall be protected by ground-fault circuit-interrupters. [680.43(A)(2)]

❖ All receptacles (30 amp and smaller) within 10 feet (3048 mm) of the inside wall of the spa or hot tub require GFCI protection. The code provisions for locating receptacles around hydromassage bathtubs are different from those for spas and hot tubs. A receptacle is not prohibited in the area within 6 feet (1524 mm) and is not required between 6 feet and 10 feet (1524 and 3048 mm) from a hydromassage bathtub, but any receptacles located within 6 feet (1524 mm) of the inside wall of a hydromassage bathtub must be GFCI protected.

E4203.2 Switching devices. Switching devices shall be located not less than 5 feet (1524 mm) horizontally from the inside walls of pools, spas and hot tubs except where separated from the pool, spa or hot tub by a solid fence, wall, or other permanent barrier or the switches are listed for use within 5 feet (1524 mm). Switching devices located in a room or area containing a hydromassage bathtub shall be located in accordance with the general requirements of this code. [680.22(C); 680.43(C); and 680.72]

❖ This section refers not only to ordinary light switches but also to such devices as a timer, panelboard, etc. By requiring these devices to be at least 5 feet (1524 mm) from the inside walls of the pool, spa or hot tub, the code intends that they cannot be reached by someone in the pool. People will need to get out of the pool, spa or hot tub to operate the switching device, thus reducing the possible shock hazard. This specific rule does not pertain to switching devices around hydromassage bathtubs.

E4203.3 Disconnecting means. One or more means to simultaneously disconnect all ungrounded conductors for all utilization equipment, other than lighting, shall be provided. Each of such means shall be readily accessible and within sight from the equipment it serves and shall be located at least 5 feet (1524 mm) horizontally from the inside walls of a pool, spa, or hot tub unless separated from the open water by a permanently installed barrier that provides a 5-foot (1524 mm) or greater reach path. This horizontal distance shall be measured from the water's edge along the shortest path required to reach the disconnect. (680.12)

❖ The disconnect must be at least 5 feet (1524 mm) away from the inside walls of pools spas and hot tubs and must be within sight from the pool, spa or hot tub. (Note the exception for an effective barrier) This means it must be visible and not more than 50 feet (15 240 mm) away. This allows the power to equipment such as motors and heaters to be turned off while some-one is working on these units.

E4203.4 Luminaires and ceiling fans. Lighting outlets, luminaires, and ceiling-suspended paddle fans shall be installed and located in accordance with Sections E4203.4.1 through E4203.4.6. [680.22(B)]

❖ Compliance with these prescriptive code provisions will help reduce the increased hazards of electrocution near water.

E4203.4.1 Outdoor location. In outdoor pool, outdoor spas and outdoor hot tubs areas, luminaires, lighting outlets, and ceiling-suspended paddle fans shall not be installed over the pool or over the area extending 5 feet (1524 mm) horizontally from the inside walls of a pool except where no part of the luminaire or ceiling-suspended paddle fan is less than 12 feet (3658 mm) above the maximum water level. [680.22(B)(1)]

❖ For an outdoor pool, spa or hot tub; luminaires and paddle fans are not permitted in the zone that is over the water and including the area extending 5 feet (1524 mm) horizontally from the inside edges of the pool and extending upward to a distance of 12 feet (3658 mm) above the water level. There are no exceptions to this rule. See Section E4203.4.3.

E4203.4.2 Indoor locations. In indoor pool areas, the limitations of Section E4203.4.1 shall apply except where the luminaires, lighting outlets and ceiling-suspended paddle fans comply with all of the following conditions:

1. The luminaires are of a totally enclosed type;

2. Ceiling-suspended paddle fans are identified for use beneath ceiling structures such as porches and patios.

3. A ground-fault circuit interrupter is installed in the branch circuit supplying the luminaires or ceiling-suspended paddle fans; and

4. The distance from the bottom of the luminaire or ceiling-suspended paddle fan to the maximum water level is not less than 7 feet, 6 inches (2286 mm). [680.22(B)(2)]

❖ Where provided with GFCI protection, fixtures of the totally enclosed type and paddle fans identified for use on porches and patios, are permitted as close as 7$\frac{1}{2}$ feet (2286 mm) above the water level. A totally enclosed luminaire is one that completely conceals the lamp and lampholder from view and contact.

E4203.4.3 Low-voltage luminaires. Listed low-voltage luminaires not requiring grounding, not exceeding the low-voltage contact limit, and supplied by listed transformers or power supplies that comply with Section E4206.1 shall be permitted to be located less than 1.5 m (5 ft) from the inside walls of the pool. [680.22(B)(6)]

❖ This section is written as an exception to the previous Sections E4203.4.1 and E4203.4.2, and essentially states that low-voltage luminaires meeting all of the stated conditions can be located at any distance from the inside walls of the pool. The product listing and manufacturer's installation instructions might dictate a minimum distance. Because this section only addresses the area within 5 feet horizontally of the inside wall of the pool, there appears to be no allowance to locate such luminaires over the area of the pool.

E4203.4.4 Existing lighting outlets and luminaires. Existing lighting outlets and luminaires that are located within 5 feet (1524 mm) horizontally from the inside walls of pools and outdoor spas and hot tubs shall be permitted to be located not less than 5 feet (1524 mm) vertically above the maximum water level, provided that such luminaires and outlets are rigidly attached to the existing structure and are protected by a ground-fault circuit-interrupter. [680.22(B)(3)]

❖ This section covers luminaires that are already installed on a house or structure when the construction

of the pool begins. This rule on existing lighting outlets and fixtures is for indoor and outdoor pools but only for outdoor spas and hot tubs. If the existing luminaires are rigidly attached to the house or structure and end up being within 5 feet (1524 mm) horizontally from the inside walls, they must be at least 5 feet (1524 mm) above the maximum water level, and the supply circuit must have GFCI protection.

E4203.4.5 Indoor spas and hot tubs.

1. Luminaires, lighting outlets, and ceiling-suspended paddle fans located over the spa or hot tub or within 5 feet (1524 mm) from the inside walls of the spa or hot tub shall be not less than 7 feet, 6 inches (2286 mm) above the maximum water level and shall be protected by a ground-fault circuit interrupter. [680.43(B)(1)(b)]

   Luminaires, lighting outlets, and ceiling-suspended paddle fans that are located 12 feet (3658 mm) or more above the maximum water level shall not require ground-fault circuit interrupter protection. [680.43(B)(1)(a)]

2. Luminaires protected by a ground-fault circuit interrupter and complying with Item 2.1 or 2.2 shall be permitted to be installed less than 7 feet, 6 inches (2286 mm) over a spa or hot tub.

   2.1. Recessed luminaires shall have a glass or plastic lens and nonmetallic or electrically isolated metal trim, and shall be suitable for use in damp locations.

   2.2. Surface-mounted luminaires shall have a glass or plastic globe and a nonmetallic body or a metallic body isolated from contact. Such luminaires shall be suitable for use in damp locations. [680.43(B)(1)(c)]

❖ The general rules for luminaires and paddle fans over indoor spas and hot tubs are the same as for indoor swimming pools found in Section E4203.4.2. However, where lighting is desired above an indoor spa or hot tub, the ceiling is often not high enough to meet the 7$\frac{1}{2}$ foot (2286 mm) rule. The room containing the spa or hot tub would need to have a high enough ceiling so that the bottom of a luminaire or fan would be at least 7$\frac{1}{2}$ feet above the water level. Recessed or surface mounted light fixtures are permitted to be less than 7$\frac{1}{2}$ feet above the water level of the spa or hot tub but must have GFCI protection, and the fixtures must be suitable for use in damp locations, must be all nonmetallic or any metallic parts must be electrically isolated from the fixture and must have a suitable lens or cover so that the lamp is not exposed. Cord-hung or pendant fixtures and paddle fans are not permitted above an indoor spa or hot tub at a height less than 7$\frac{1}{2}$ feet above the water level.

E4203.4.6 GFCI protection in adjacent areas. Luminaires and outlets that are installed in the area extending between 5 feet (1524 mm) and 10 feet (3048 mm) from the inside walls of pools and outdoor spas and hot tubs shall be protected by ground-fault circuit-interrupters except where such fixtures and outlets are installed not less than 5 feet (1524 mm) above the maximum water level and are rigidly attached to the structure. [680.22(B)(4)]

❖ In the area extending horizontally between 5 feet and 10 feet from the inside wall of a pool or outdoor spaoutlets or hot tub, luminaires and lighting outlets (not receptacle) are permitted, but they must have GFCI protection. In the same area, but at a height of over 5 feet (1524 mm) above the water level, the circuits feeding rigidly attached light fixtures need not have GFCI protection.

E4203.5 Other outlets. Other outlets such as for remote control, signaling, fire alarm and communications shall be not less than 10 feet (3048 mm) from the inside walls of the pool. Measurements shall be determined in accordance with Section E4203.1. [680.22(D)]

❖ This section addresses the outlets for telephone, cable TV, internet, security systems, thermostats, fire alarm, etc., all of which can pose some electrical shock hazard.

E4203.6 Overhead conductor clearances. Except where installed with the clearances specified in Table E4203.6, the following parts of pools and outdoor spas and hot tubs shall not be placed under existing service-drop conductors, overhead service conductor, or any other open overhead wiring; nor shall such wiring be installed above the following:

1. Pools and the areas extending not less than 10 feet, (3048 mm) horizontally from the inside of the walls of the pool.

2. Diving structures and the areas extending not less than 10 feet (3048 mm) horizontally from the outer edge of such structures.

3. Observation stands, towers, and platforms and the areas extending not less than 10 feet (3048 mm) horizontally from the outer edge of such structures.

Overhead conductors of network-powered broadband communications systems shall comply with the provisions in Table E4203.6 for conductors operating at 0 to 750 volts to ground.

Utility-owned, -operated and -maintained communications conductors, community antenna system coaxial cables and the supporting messengers shall be permitted at a height of not less than 10 feet (3048 mm) above swimming and wading pools, diving structures, and observation stands, towers, and platforms. [680.8(A), (B), and (C)]

❖ Service drop conductors are not permitted above a pool or within 10 feet (3048 mm) horizontally from the edge of the pool unless they are at least 22.5 feet (6858 mm) above the water in any direction and 14.5 feet (4420 mm) above a diving platform. Telephone and cable TV lines must be at least 10 feet (3048 mm) above a pool or diving platform. Typically, swimming pools are not installed under the service drop lines because this minimum height is difficult to maintain. Overhead conductors of network-powered broadband communications systems must comply with the column in the table that is applicable to conductors operating at 0 to 750 volts to ground.

E4203.7 Underground wiring. Underground wiring shall not be installed under or within the area extending 5 feet (1524 mm) horizontally from the inside walls of pools and outdoor hot tubs and spas except where the wiring is installed to supply pool, spa or hot tub equipment or where space limitations prevent wiring from being routed 5 feet (1524 mm) or more horizontally from the inside walls. Where installed within 5 feet (1524 mm) of the inside walls, the wiring method shall be a complete raceway system of rigid metal conduit, intermediate metal conduit or a nonmetallic raceway system. Metal conduit shall be corrosion resistant and suitable for the location. The minimum cover depth shall be in accordance with Table E4203.7. (680.10)

❖ Where underground wiring is installed, corrosion of the raceway is possible that could result in a hazardous situation under ground-fault conditions. Because of dampness and the presence of chemicals such as chlorine in the area, corrosion of metallic parts is more likely in a pool area and in the ground around a pool. For this reason, where it is necessary to install wiring underground within 5 feet (1524 mm) of the pool, it must be in rigid metal conduit, intermediate metal conduit or nonmetallic raceway. A typical corrosion-resistant rigid metal conduit has an outer nonmetallic covering. The other option that is commonly used is to run the wiring in a nonmetallic raceway such as PVC conduit. Such raceway/conduits must be complete systems as opposed to only sections of conduit that are within 5 feet of the pool, spa or tub.

TABLE E4203.6 [Table 680.8(A)]
OVERHEAD CONDUCTOR CLEARANCES

| | INSULATED SUPPLY OR SERVICE DROP CABLES, 0-750 VOLTS TO GROUND, SUPPORTED ON AND CABLED TOGETHER WITH AN EFFECTIVELY GROUNDED BARE MESSENGER OR EFFECTIVELY GROUNDED NEUTRAL CONDUCTOR (feet) | ALL OTHER SUPPLY OR SERVICE DROP CONDUCTORS (feet) | |
| | | Voltage to ground | |
		0-15 kV	Greater than 15 to 50 kV
A.Clearance in any direction to the water level, edge of water surface, base of diving platform, or permanently anchored raft	22.5	25	27
B.Clearance in any direction to the diving platform	14.5	17	18

For SI: 1 foot = 304.8 mm.

TABLE E4203.7 (680.10)
MINIMUM BURIAL DEPTHS

WIRING METHOD	UNDERGROUND WIRING (inches)
Rigid metal conduit	6
Intermediate metal conduit	6
Nonmetallic raceways listed for direct burial and under concrete exterior slab not less than 4 inches in thickness and extending not less than 6 inches (162 mm) beyond the underground installation	6
Nonmetallic raceways listed for direct burial without concrete encasement	18
Other approved raceways[a]	18

For SI: 1 inch = 25.4 mm.

a. Raceways approved for burial only where concrete-encased shall require a concrete envelope not less than 2 inches in thickness.

SECTION E4204
BONDING

E4204.1 Performance. The equipotential bonding required by this section shall be installed to reduce voltage gradients in the prescribed areas of permanently installed swimming pools and spas and hot tubs other than the storable/portable type.

❖ "Equipotential" means equal potential (the same voltage). A difference in voltage between two bodies, objects or surfaces is referred to as a potential because of the potential energy that exists from the voltage (i.e., pressure). Potential energy is converted to current flow when the two bodies, objects or surfaces are electrically connected. The intent of such bonding is to eliminate or reduce voltage gradients (differences) between any and all conductive parts between which a living creature could place itself and thereby endanger itself. Voltage differences of only a few volts can be dangerous to occupants in the water or near the water.

E4204.2 Bonded parts. The parts of pools, spas, and hot tubs specified in Items 1 through 7 shall be bonded together using insulated, covered or bare solid copper conductors not smaller than 8 AWG or using rigid metal conduit of brass or other identified corrosion-resistant metal. An 8 AWG or larger solid copper bonding conductor provided to reduce voltage gradients in the pool, spa, or hot tub area shall not be required to be extended or attached to remote panelboards, service equipment, or electrodes. Connections shall be made by exothermic welding, by listed pressure connectors or clamps that are labeled as being suitable for the purpose and that are made of stainless steel, brass, copper or copper alloy, machine screw-type fasteners that engage not less than two threads or are secured with a nut, thread-forming machine screws that engage not less than two-threads, or terminal bars. Connection devices or fittings that depend solely on solder shall not be used. Sheet metal screws shall not be used to connect bonding conductors or connection devices: [680.26(B)]

1. Conductive pool shells. Bonding to conductive pool shells shall be provided as specified in Item 1.1 or 1.2. Poured concrete, pneumatically applied or sprayed concrete, and concrete block with painted or plastered coatings shall be considered to be conductive materials because of their water permeability and porosity. Vinyl liners and fiberglass composite shells shall be considered to be nonconductive materials.

1.1. Structural reinforcing steel. Unencapsulated structural reinforcing steel shall be bonded together by steel tie wires or the equivalent. Where structural reinforcing steel is encapsulated in a nonconductive compound, a copper conductor grid shall be installed in accordance with Item 1.2.

1.2. Copper conductor grid. A copper conductor grid shall be provided and shall comply with Items 1.2.1 through 1.2.4:

1.2.1. It shall be constructed of minimum 8 AWG bare solid copper conductors bonded to each other at all points of crossing.

1.2.2. It shall conform to the contour of the pool.

1.2.3. It shall be arranged in a 12-inch (305 mm) by 12-inch (305 mm) network of conductors in a uniformly spaced perpendicular grid pattern with a tolerance of 4 inches (102 mm).

1.2.4. It shall be secured within or under the pool not more than 6 inches (152 mm) from the outer contour of the pool shell. [680.26(B)(1)]

2. Perimeter surfaces. The perimeter surface shall extend for 3 feet (914 mm) horizontally beyond the inside walls of the pool and shall include unpaved surfaces, poured concrete surfaces and other types of paving. Perimeter surfaces that extend less than 3 feet (914 mm) beyond the inside wall of the pool and that are separated from the pool by a permanent wall or building 5 feet (1524 mm) or more in height shall require equipotential bonding on the pool side of the permanent wall or building. Bonding to perimeter surfaces shall be provided as specified in Item 2.1 or 2.2 and shall be attached to the pool, spa, or hot tub reinforcing steel or copper conductor grid at a minimum of four points uniformly spaced around the perimeter of the pool, spa, or hot tub. For nonconductive pool shells, bonding at four points shall not be required.

Exceptions:

1. Equipotential bonding of perimeter surfaces shall not be required for spas and hot tubs where all of the following conditions apply:

1.1. The spa or hot tub is listed as a self-contained spa for aboveground use.

1.2. The spa or hot tub is not identified as suitable only for indoor use.

1.3. The installation is in accordance with the manufacturer's instructions and is located on or above grade.

1.4. The top rim of the spa or hot tub is not less than 28 in. (711 mm) above all perimeter surfaces that are within 30 in. (762 mm), measured horizontally from the spa or hot tub. The height of nonconductive external steps for entry to or exit from the self-contained spa is not used to reduce or increase this rim height measurement.

2. The equipotential bonding requirements for perimeter surfaces shall not apply to a listed self-contained spa or hot tub located indoors and installed above a finished floor.

2.1. Structural reinforcing steel. Structural reinforcing steel shall be bonded in accordance with Item 1.1.

2.2. Alternate means. Where structural reinforcing steel is not available or is encapsulated in a nonconductive compound, a copper conductor(s) shall be used in accordance with Items 2.2.1 through 2.2.5:

2.2.1. At least one minimum 8 AWG bare solid copper conductor shall be provided.

2.2.2. The conductors shall follow the contour of the perimeter surface.

2.2.3. Splices shall be listed.

2.2.4. The required conductor shall be 18 to 24 inches (457 to 610 mm) from the inside walls of the pool.

2.2.5. The required conductor shall be secured within or under the perimeter surface 4 to 6 inches (102 mm to 152 mm) below the subgrade. [680.26(B)(2)]

3. Metallic components. All metallic parts of the pool structure, including reinforcing metal not addressed in Item 1.1, shall be bonded. Where reinforcing steel is encapsulated with a nonconductive compound, the reinforcing steel shall not be required to be bonded. [680.26(B)(3)]

4. Underwater lighting. All metal forming shells and mounting brackets of no-niche luminaires shall be bonded. [680.26(B)(4)]

   Exception: Listed low-voltage lighting systems with nonmetallic forming shells shall not require bonding. [680.26(B)(4) Exception]

5. Metal fittings. All metal fittings within or attached to the pool structure shall be bonded. Isolated parts that are not over 4 inches (102 mm) in any dimension and do not penetrate into the pool structure more than 1 inch (25.4 mm) shall not require bonding. [680.26(B)(5)]

6. Electrical equipment. Metal parts of electrical equipment associated with the pool water circulating system, including pump motors and metal parts of equipment associated with pool covers, including electric motors, shall be bonded. [680.26(B)(6)]

   Exception: Metal parts of listed equipment incorporating an approved system of double insulation shall not be bonded. [680.26(B)(6) Exception]

   6.1. Double-insulated water pump motors. Where a double-insulated water pump motor is installed under the provisions of this item, a solid 8 AWG copper conductor of sufficient length to make a bonding connection to a replacement motor shall be extended from the bonding grid to an accessible point in the vicinity of the pool pump motor. Where there is no connection between the swimming pool bonding grid and the equipment grounding system for the premises, this bonding conductor shall be connected to the equipment grounding conductor of the motor circuit. [680.26(B)(6)(a)]

   6.2. Pool water heaters. For pool water heaters rated at more than 50 amperes and having specific instructions regarding bonding and grounding, only those parts designated to be bonded shall be bonded and only those parts designated to be grounded shall be grounded. [680.26(B)(6)(b)]

7. All fixed metal parts including, but not limited to, metal-sheathed cables and raceways, metal piping, metal awnings, metal fences and metal door and window frames. [680.26(B)(7)]

   Exceptions:

   1. Those separated from the pool by a permanent barrier that prevents contact by a person shall not be required to be bonded. [680.26(B)(7) Exception No. 1]

   2. Those greater than 5 feet (1524 mm) horizontally from the inside walls of the pool shall not be required to be bonded. [680.26(B)(7) Exception No. 2]

   3. Those greater than 12 feet (3658 mm) measured vertically above the maximum water level of the pool, or as measured vertically above any observation stands, towers, or platforms, or any diving structures, shall not be required to be bonded. [680.26(B)(7) Exception No. 3]

❖ Bonding is the joining of metallic parts to form an electrically conductive path that will result in electrical continuity between components of a common grid to ensure that the electrical potential will be the same throughout. Keeping the electrical potential at the same level will reduce the shock hazard created by any stray currents in the pool or in the ground around the pool. The risk of electric shock is great in damp locations, and it is increased where a person is in a pool of water. An electrical potential may occur in a pool because of such things as lightning, ground faults or the operation of electrical equipment in close proximity. Bonding together of all of the metallic items in and around a pool will help eliminate the voltage gradients or differences in electrical potential from one part of the pool to another or from metallic equipment to the pool water. The intent is to prevent the occupants/users of the pool, spa and hot tub from being subjected to voltage potentials that could cause current to flow through their bodies. The various items enumerated in the seven parts of this section provide a clear description of what parts must be bonded together and what parts are exempt from bonding.

Many of the bonding connections, such as the tie wires connecting the reinforcing steel together, must be inspected prior to concealment. Item 2 requires the bonding of walking surfaces around a pool, spa and hot tub, whether paved or unpaved bare earth. Where a permanent wall or building having a height of 5 feet or more is located such that it separates a portion of the 3-foot perimeter from the inside wall of the pool, only the portion of the wall or building on the poolside is required to be bonded.

Spas and hot tubs that meet all of the conditions of Exception 1 of Item 2, "perimeter surfaces," are exempt from the perimeter surface bonding requirements because the listing of the unit has evaluated the potential risks. Steps for the unit that are constructed of nonconductive material are not to be used as a reference elevation to determine compliance with the required 28-inch minimum height above perimeter surfaces. Steps of conductive material would impact the rim height measurement because such steps are another type of surface that could have a voltage potential difference. Note that wood, stone and concrete steps would be considered as conductive.

E4204.3 Pool water. Where none of the bonded parts is in direct connection with the pool water, the pool water shall be in direct contact with an approved corrosion-resistant conductive surface that exposes not less than 9 in.^{2} (5800 mm^2) of surface area to the pool water at all times. The conductive surface shall be located where it is not exposed to physical damage or dislodgement during usual pool activities, and it shall be bonded in accordance with Section E4204.2.

❖ This section requires that the water itself be bonded utilizing a conductive surface as an electrode. The electrode can be metal parts that are submerged in the water such as ladders, lighting components and drain bodies. The electrode can also be a metal body that is constructed specifically for that purpose.

E4204.4 Bonding of outdoor hot tubs and spas. Outdoor hot tubs and spas shall comply with the bonding requirements of Sections E4204.1 through E4204.3. Bonding by metal-to-metal mounting on a common frame or base shall be permitted. The metal bands or hoops used to secure wooden staves

shall not be required to be bonded as required in Section E4204.2. [680.42 and 680.42(B)]

❖ Wooden staves are used to build wooden tubs in the same manner as wooden barrels are made (see Sections E4204.1 through E4204.3).

This section applies to indoor hot tubs and spas. Item 3 requires copper and galvanized steel water supply piping to be bonded if within 5 feet of the inside walls. This would also apply to copper, steel and cast iron drain and vent piping. Item 5 applies to electrical devices that are not related to the spa or hot tub, but happen to be located within 5 feet of the inside walls.

E4204.5 Bonding of indoor hot tubs and spas. The following parts of indoor hot tubs and spas shall be bonded together:

1. All metal fittings within or attached to the hot tub or spa structure. [680.43(D)(1)]

2. Metal parts of electrical equipment associated with the hot tub or spa water circulating system, including pump motors unless part of a listed self-contained spa or hot tub. [680.43(D)(2)]

3. Metal raceway and metal piping that are within 5 feet (1524 mm) of the inside walls of the hot tub or spa and that are not separated from the spa or hot tub by a permanent barrier. [680.43(D)(3)]

4. All metal surfaces that are within 5 feet (1524 mm) of the inside walls of the hot tub or spa and that are not separated from the hot tub or spa area by a permanent barrier. [680.43(D)(4)]

   Exception: Small conductive surfaces not likely to become energized, such as air and water jets and drain fittings, where not connected to metallic piping, towel bars, mirror frames, and similar nonelectrical equipment, shall not be required to be bonded. [680.43(D)(4) Exception]

5. Electrical devices and controls that are not associated with the hot tubs or spas and that are located less than 5 feet (1524 mm) from such units. [680.43(D)(5)]

❖ See the commentary to Sections E4204.1 and E4204.2.

E4204.5.1 Methods. All metal parts associated with the hot tub or spa shall be bonded by any of the following methods:

1. The interconnection of threaded metal piping and fittings. [680.43(E)(1)]

2. Metal-to-metal mounting on a common frame or base. [680.43(E)(2)]

3. The provision of an insulated, covered or bare solid copper bonding jumper not smaller than 8 AWG. It shall not be the intent to require that the 8 AWG or larger solid copper bonding conductor be extended or attached to any remote panelboard, service equipment, or any electrode, but only that it shall be employed to eliminate voltage gradients in the hot tub or spa area as prescribed. [680.43(E)(3)]

❖ The methods of bonding are the same as allowed in Sections E4204.2 and E4204.4.

E4204.5.2 Connections. Connections to bonded parts shall be made in accordance with Section E3406.13.1.

❖ Connection devices used for bonding must be listed and labeled for the application, such as listed for direct burial in concrete or soil or listed for submersion in water.

SECTION E4205
GROUNDING

E4205.1 Equipment to be grounded. The following equipment shall be grounded:

1. Through-wall lighting assemblies and underwater luminaires other than those low-voltage lighting products listed for the application without a grounding conductor.

2. All electrical equipment located within 5 feet (1524 mm) of the inside wall of the pool, spa or hot tub.

3. All electrical equipment associated with the recirculating system of the pool, spa or hot tub.

4. Junction boxes.

5. Transformer and power supply enclosures.

6. Ground-fault circuit-interrupters.

7. Panelboards that are not part of the service equipment and that supply any electrical equipment associated with the pool, spa or hot tub. (680.7)

❖ Grounding and bonding are required for different reasons. Bonding is required for all metal parts of the electrical equipment and for nonelectrical metal parts of the structure to establish an equipotential grid. Bonding conductors in many cases are permitted to be externally clamped or connected to the noncurrent-carrying metal parts of equipment. The attachment need not be accessible. Bonding of metal parts of electrical equipment is required to ensure a low-impedance path for fault current back to the source of the circuit to facilitate the operation of the overcurrent device. For equipment grounding, a separate grounding conductor is connected to the equipment's grounding terminal.

E4205.2 Luminaires and related equipment. Other than listed low-voltage luminaires not requiring grounding, all through-wall lighting assemblies, wet-niche, dry-niche, or no-niche luminaires shall be connected to an insulated copper equipment grounding conductor sized in accordance with Table E3908.12 but not smaller than 12 AWG. The equipment grounding conductor between the wiring chamber of the secondary winding of a transformer and a junction box shall be sized in accordance with the overcurrent device in such circuit. The junction box, transformer enclosure, or other enclosure in the supply circuit to a wet-niche or no-niche luminaire and the field-wiring chamber of a dry-niche luminaire shall be grounded to the equipment grounding terminal of the panelboard. The equipment grounding terminal shall be directly connected to the panelboard enclosure. The equipment grounding conductor shall be installed without joint or splice. [680.23(F)(2) and 680.23(F)(2) Exception]

Exceptions:

1. Where more than one underwater luminaire is supplied by the same branch circuit, the equipment grounding conductor, installed between the junction boxes, transformer enclosures, or other enclosures in the supply circuit to wet-niche luminaires, or between the field-wiring compartments of dry-niche luminaires, shall be permitted to be terminated on grounding terminals. [680.23(F)(2)(a)]

2. Where an underwater luminaire is supplied from a transformer, ground-fault circuit-interrupter, clock-operated switch, or a manual snap switch that is located between the panelboard and a junction box connected to the conduit that extends directly to the underwater luminaire, the equipment grounding conductor shall be permitted to terminate on grounding terminals on the transformer, ground-fault circuit-interrupter, clock-operated switch enclosure, or an outlet box used to enclose a snap switch. [680.23(F)(2)(b)]

❖ Of particular note here is the requirement that the equipment grounding conductor must be insulated and sized in accordance with Table E3908.12. An insulated conductor run within the metal raceway is used because the conduit could corrode and open the conductive grounding path. The junction box or enclosure in the supply circuit to an underwater luminaire or related equipment must be grounded to the equipment grounding terminal of the panelboard. This is done by running a minimum size 12-AWG insulated equipment grounding conductor directly between the terminals of the junction box or enclosure and the panelboard. This conductor must not be confused with the size 8-AWG bonding conductor used to connect the forming shell to the bonding grid.

E4205.3 Nonmetallic conduit. Where a nonmetallic conduit is installed between a forming shell and a junction box, transformer enclosure, or other enclosure, a 8 AWG insulated copper bonding jumper shall be installed in this conduit except where a listed low-voltage lighting system not requiring grounding is used. The bonding jumper shall be terminated in the forming shell, junction box or transformer enclosure, or ground-fault circuit-interrupter enclosure. The termination of the 8 AWG bonding jumper in the forming shell shall be covered with, or encapsulated in, a listed potting compound to protect such connection from the possible deteriorating effect of pool water. [680.23(B)(2)(b)]

❖ In this section, the size 8-AWG conductor that must be run within the nonmetallic-conduit is not the same bonding conductor used to connect the forming shell to the common bonding grid. The size 8-AWG or larger bonding conductor connects to the forming shell and must be insulated. Note that Section E3406.4 requires 8-AWG and larger conductors to be stranded where installed in raceways and there is no exemption indicated for Section E4205. Stranded conductors are far less likely to have the insulation damaged during installation. The size 8-AWG conductor run within the nonmetallic-conduit provides electrical continuity between the forming shell and the junction box or transformer enclosure. The 8-AWG conductor must have green insulation in accordance with Section

E3407.2. Corrosion can occur when the connection is exposed to pool water, and thus the connection must be encapsulated in a listed potting compound.

E4205.4 Flexible cords. Other than listed low-voltage lighting systems not requiring grounding, wet-niche luminaires that are supplied by a flexible cord or cable shall have all exposed noncurrent-carrying metal parts grounded by an insulated copper equipment grounding conductor that is an integral part of the cord or cable. This grounding conductor shall be connected to a grounding terminal in the supply junction box, transformer enclosure, or other enclosure. The grounding conductor shall not be smaller than the supply conductors and not smaller than 16 AWG. [680.23(B)(3)]

❖ See Section E4205.2.

E4205.5 Motors. Pool-associated motors shall be connected to an insulated copper equipment grounding conductor sized in accordance with Table E3908.12, but not smaller than 12 AWG. Where the branch circuit supplying the motor is installed in the interior of a one-family dwelling or in the interior of accessory buildings associated with a one-family dwelling, using a cable wiring method permitted by Table E4202.1, an uninsulated equipment grounding conductor shall be permitted provided that it is enclosed within the outer sheath of the cable assembly. [680.21(A)(1) and (A)(4)]

❖ Motors used with pool equipment must be grounded with a minimum size 12-AWG equipment grounding conductor. This conductor must be insulated or within an overall sheath such as it would be in nonmetallic-sheathed cable.

E4205.6 Feeders. An equipment grounding conductor shall be installed with the feeder conductors between the grounding terminal of the pool equipment panelboard and the grounding terminal of the applicable service equipment. The equipment grounding conductor shall be insulated, shall be sized in accordance with Table E3908.12, and shall be not smaller than 12 AWG.

❖ Where a separate panelboard that supplies swimming pool equipment is fed from service equipment, it must have an insulated equipment grounding conductor not smaller than size 12-AWG run with the feeders from the service equipment.

E4205.6.1 Separate buildings. A feeder to a separate building or structure shall be permitted to supply swimming pool equipment branch circuits, or feeders supplying swimming pool equipment branch circuits, provided that the grounding arrangements in the separate building meet the requirements of Section E3607.3. The feeder equipment grounding conductor shall be an insulated conductor. (680.25(B)(2))

❖ See Section E3607.3.

E4205.7 Cord-connected equipment. Where fixed or stationary equipment is connected with a flexible cord to facilitate removal or disconnection for maintenance, repair, or storage, as provided in Section E4202.2, the equipment grounding conductors shall be connected to a fixed metal part

of the assembly. The removable part shall be mounted on or bonded to the fixed metal part. [680.7(C)]

❖ The equipment grounding conductor is required in order to provide a path of low impedance that helps limit the voltage to ground and serves to carry enough current to trip the overcurrent protective device. Flexible cord is used in some cases to supply power to equipment that may need to be removed for replacement or repair, and the equipment grounding conductor must be connected to a part of the equipment that is not removable.

E4205.8 Other equipment. Other electrical equipment shall be grounded in accordance with Section E3908. (Article 250, Parts V, VI, and VII; and 680.6)

❖ Equipment not addressed specifically in this section must comply with the grounding provisions of Section E3908.

SECTION E4206
EQUIPMENT INSTALLATION

E4206.1 Transformers and power supplies. Transformers and power supplies used for the supply of underwater luminaires, together with the transformer or power supply enclosure, shall be listed for swimming pool and spa use. The transformer or power supply shall incorporate either a transformer of the isolated-winding type with an ungrounded secondary that has a grounded metal barrier between the primary and secondary windings, or a transformer that incorporates an approved system of double insulation between the primary and secondary windings. [680.23(A)(2)]

❖ Transformers used to supply power for underwater swimming pool and spa luminaires are of the independent two-winding type and must be listed for such application. They are of the isolated-winding type, meaning they have a shield or metal barrier between the primary and secondary windings. The metal barrier is grounded by an equipment grounding conductor connected to the equipment grounding terminal in the transformer housing. Without a metal barrier or shield, deterioration of the insulation of the primary and secondary windings could cause the two windings to come into contact with each other, creating a higher voltage on the secondary windings. As an alternative to isolated-winding-type transformers, the code official can approve transformers that employ a double insulation system that separates the primary and secondary windings, thereby providing a level of safety equivalent to the isolated-winding-type transformers.

E4206.2 Ground-fault circuit-interrupters. Ground-fault circuit-interrupters shall be self-contained units, circuit-breaker types, receptacle types or other approved types. (680.5)

❖ A GFCI is a device designed to interrupt or open the circuit when a fault current to ground exceeds a certain value that is less than that required to trip the circuit breaker supplying the branch circuit. A Class A GFCI trips when the ground-fault current is from 4 to 6 milliamperes. Class A GFCIs are suitable for use in branch circuits feeding swimming pool equipment. How-ever, existing branch circuits in older swimming pool installations might have sufficient leakage current to cause a Class A GFCI to trip.

E4206.3 Wiring on load side of ground-fault circuit-interrupters and transformers. For other than grounding conductors, conductors installed on the load side of a ground-fault circuit-interrupter or transformer used to comply with the provisions of Section E4206.4, shall not occupy raceways, boxes, or enclosures containing other conductors except where the other conductors are protected by ground-fault circuit interrupters or are grounding conductors. Supply conductors to a feed-through type ground-fault circuit interrupter shall be permitted in the same enclosure. Ground-fault circuit interrupters shall be permitted in a panelboard that contains circuits protected by other than ground-fault circuit interrupters. [680.23(F)(3)]

❖ Except within panelboard cabinets, the load side conductors of a GFCI circuit are not permitted to share the same wiring space with other conductors that are not GFCI protected other than for equipment grounding conductors. Note the exception for conductors that supply feed-through type GFCI devices where the supply conductors and the GFCI device are allowed to be in the same box. The magnetic inductance caused by current flow in adjacent conductors could cause current to flow in circuit conductors on the load side of GFCI device. Also, insulation failure could energize GFCI protected conductors.

E4206.4 Underwater luminaires. The design of an underwater luminaire supplied from a branch circuit either directly or by way of a transformer or power supply meeting the requirements of Section E4206.1, shall be such that, where the fixture is properly installed without a ground-fault circuit-interrupter, there is no shock hazard with any likely combination of fault conditions during normal use (not relamping). In addition, a ground-fault circuit-interrupter shall be installed in the branch circuit supplying luminaires operating at more than the low-voltage contact limit, such that there is no shock hazard during relamping. The installation of the ground-fault circuit-interrupter shall be such that there is no shock hazard with any likely fault-condition combination that involves a person in a conductive path from any ungrounded part of the branch circuit or the luminaire to ground. Compliance with this requirement shall be obtained by the use of a listed underwater luminaire and by installation of a listed ground-fault circuit-interrupter in the branch circuit or a listed transformer or power supply for luminaires operating at more than the low-voltage contact limit. Luminaires that depend on submersion for safe operation shall be inherently protected against the hazards of overheating when not submerged. [680.23(A)(1), (A)(3), (A)(7) and (A)(8)]

❖ An underwater light fixture must be listed and of such a design that will not pose an electric shock hazard during operation of the fixture without a GFCI device in the circuit. Any light fixture operating at over the low-

voltage contact limit must have GFCI protection. This will provide safety from electric shock during relamping. Wet-niche and no-niche fixtures are detached and brought above the water level for relamping (see definition of "Low voltage contact limit").

E4206.4.1 Maximum voltage. Luminaires shall not be installed for operation on supply circuits over 150 volts between conductors. [680.23(A)(4)]

❖ Where luminaires are designed to operate at 120 volts, the circuit must have GFCI protection.

E4206.4.2 Luminaire location. Luminaires mounted in walls shall be installed with the top of the fixture lens not less than 18 inches (457 mm) below the normal water level of the pool, except where the luminaire is listed and identified for use at a depth of not less than 4 inches (102 mm) below the normal water level of the pool. A luminaire facing upward shall have the lens adequately guarded to prevent contact by any person or shall be listed for use without a guard. [680.23(A)(5) and (A)(6)]

❖ If a person is hanging onto the edge in the deep end of a pool, directly in front of a wet-niche luminaire with a broken bulb or damaged lens, fault current could leak into the pool directly to the person. Where the luminaire is at least 18 inches (457 mm) below the normal water level, the person's chest, the most sensitive area for shock hazard, would most likely be above water or above the luminaire and not be directly in line with the fixture.

E4206.5 Wet-niche luminaires. Forming shells shall be installed for the mounting of all wet-niche underwater luminaires and shall be equipped with provisions for conduit entries. Conduit shall extend from the forming shell to a suitable junction box or other enclosure located as provided in Section E4206.9. Metal parts of the luminaire and forming shell in contact with the pool water shall be of brass or other approved corrosion-resistant metal. [680.23(B)(1)]

The end of flexible-cord jackets and flexible-cord conductor terminations within a luminaire shall be covered with, or encapsulated in, a suitable potting compound to prevent the entry of water into the luminaire through the cord or its conductors. If present, the grounding connection within a luminaire shall be similarly treated to protect such connection from the deteriorating effect of pool water in the event of water entry into the luminaire. [680.23(B)(4)]

Luminaires shall be bonded to and secured to the forming shell by a positive locking device that ensures a low-resistance contact and requires a tool to remove the luminaire from the forming shell. [680.23(B)(5)]

❖ These fixtures are typically installed in the wall of a pool or field-fabricated spa. A forming shell is installed before concrete is placed. The forming shell is not sealed from pool water, and the fixture is designed to be surrounded by water. Wet-niche fixtures that are UL listed have a permanently attached, factory installed flexible cord at least 12 feet (3658 mm) long. With this length of flexible cord, the fixture can be removed from the housing or forming shell and brought up to the pool deck for relamping without lowering the water level and without disconnecting the cord from the junction box above the pool deck.

E4206.5.1 Servicing. All wet-niche luminaires shall be removable from the water for inspection, relamping, or other maintenance. The forming shell location and length of cord in the forming shell shall permit personnel to place the removed luminaire on the deck or other dry location for such maintenance. The luminaire maintenance location shall be accessible without entering or going into the pool water. [680.23(B)(6)]

❖ This requirement protects the person performing the maintenance and also helps prevent water from entering the luminaire as it is serviced.

E4206.6 Dry-niche luminaires. Dry-niche luminaires shall have provisions for drainage of water. Other than listed low-voltage luminaires not requiring grounding, a dry-niche luminaire shall have means for accommodating one equipment grounding conductor for each conduit entry. Junction boxes shall not be required but, if used, shall not be required to be elevated or located as specified in Section E4206.9 if the luminaire is specifically identified for the purpose. [680.23(C)(1) and (C)(2)]

❖ These underwater luminaires are designed to be serviced or relamped from the rear in a passageway behind the pool wall. This is not very common in residential pools. A no-niche lighting fixture could be installed in the wall of a pool or spa that is partially above ground level where an elevated deck is built above ground to match the top edge of the pool. Some dry-niche fixtures are installed with a box flush with the deck (where listed for use with the dry-niche fixture, it need not be raised above the deck) from which the lamp can be replaced.

E4206.7 No-niche luminaires. No-niche luminaires shall be listed for the purpose and shall be installed in accordance with the requirements of Section E4206.5. Where connection to a forming shell is specified, the connection shall be to the mounting bracket. [680.23(D)]

❖ See Section E4206.5.

E4206.8 Through-wall lighting assembly. A through-wall lighting assembly shall be equipped with a threaded entry or hub, or a nonmetallic hub, for the purpose of accommodating the termination of the supply conduit. A through-wall lighting assembly shall meet the construction requirements of Section E4205.4 and be installed in accordance with the requirements of Section E4206.5 Where connection to a forming shell is specified, the connection shall be to the conduit termination point. [680.23(E)]

❖ See Sections E4205.4 and E4206.5.

E4206.9 Junction boxes and enclosures for transformers or ground-fault circuit interrupters. Junction boxes for underwater luminaires and enclosures for transformers and ground-fault circuit-interrupters that supply underwater luminaires shall comply with the following: [680.24(A)]

❖ The branch-circuit wiring for underwater swimming pool fixtures is brought to a junction box, transformer or

enclosure installed a relatively short distance from the pool or spa. A flexible cord from the fixture is installed in a raceway and connected in the junction box or transformer enclosure, which must be listed specifically for use with underwater luminaires. It must be labeled as a swimming pool junction box.

E4206.9.1 Junction boxes. A junction box connected to a conduit that extends directly to a forming shell or mounting bracket of a no-niche luminaire shall be:

1. Listed as a swimming pool junction box; [680.24(A)(1)]

2. Equipped with threaded entries or hubs or a nonmetallic hub; [680.24(A)(1)(1)]

3. Constructed of copper, brass, suitable plastic, or other approved corrosion-resistant material; [680.24(A)(1)(2)]

4. Provided with electrical continuity between every connected metal conduit and the grounding terminals by means of copper, brass, or other approved corrosion-resistant metal that is integral with the box; and [680.24(A)(1)(3)]

5. Located not less than 4 inches (102 mm), measured from the inside of the bottom of the box, above the ground level, or pool deck, or not less than 8 inches (203 mm) above the maximum pool water level, whichever provides the greatest elevation, and shall be located not less than 4 feet (1219 mm) from the inside wall of the pool, unless separated from the pool by a solid fence, wall or other permanent barrier. Where used on a lighting system operating at the low-voltage contact limit or less, a flush deck box shall be permitted provided that an approved potting compound is used to fill the box to prevent the entrance of moisture; and the flush deck box is located not less than 4 feet (1219 mm) from the inside wall of the pool. [680.24(A)(2)]

❖ The junction box must be sized according the provisions of Section E3905. The junction box must be large enough to accommodate the raceways connected to it. A nonmetallic conduit between a wet-niche luminaire and the deck junction box must be sized large enough for both a size 8 AWG insulated copper bonding conductor and the flexible cord that supplies the fixture. Terminals must be provided within the box for connection of the bonding conductors. A box that is listed under swimming pool junction boxes will have the correct number of integral grounding and bonding terminals for the number of conduit entries it has. The box must have at least one grounding terminal for each conduit entry plus one. Field installation of a grounding terminal or grounding bar is not permitted (see Section E4206.9.4).

E4206.9.2 Other enclosures. An enclosure for a transformer, ground-fault circuit-interrupter or a similar device connected to a conduit that extends directly to a forming shell or mounting bracket of a no-niche luminaire shall be:

1. Listed and labeled for the purpose, comprised of copper, brass, suitable plastic, or other approved corrosion-resistant material; [680.24(B)(1)]

2. Equipped with threaded entries or hubs or a nonmetallic hub; [680.24(B)(2)]

3. Provided with an approved seal, such as duct seal at the conduit connection, that prevents circulation of air between the conduit and the enclosures; [680.24(B)(3)]

4. Provided with electrical continuity between every connected metal conduit and the grounding terminals by means of copper, brass or other approved corrosion-resistant metal that is integral with the enclosures; and [680.24(B)(4)]

5. Located not less than 4 inches (102 mm), measured from the inside bottom of the enclosure, above the ground level or pool deck, or not less than 8 inches (203 mm) above the maximum pool water level, whichever provides the greater elevation, and shall be located not less than 4 feet (1219 mm) from the inside wall of the pool, except where separated from the pool by a solid fence, wall or other permanent barrier. [680.24(B)(2)]

❖ A transformer enclosure that has conduits that extend directly to the underwater luminaire must have the conduit entries sealed against air circulation. Corrosion of internal components of the enclosure could occur at a faster rate if not sealed.

E4206.9.3 Protection of junction boxes and enclosures. Junction boxes and enclosures mounted above the grade of the finished walkway around the pool shall not be located in the walkway unless afforded additional protection, such as by location under diving boards or adjacent to fixed structures. [680.24(C)]

❖ Junction boxes must not be installed in the open at walkways and similar locations where subject to damage from foot traffic, carts, patio furniture, etc. Above-grade boxes would also present a serious tripping hazard.

E4206.9.4 Grounding terminals. Junction boxes, transformer and power supply enclosures, and ground-fault circuit-interrupter enclosures connected to a conduit that extends directly to a forming shell or mounting bracket of a no-niche luminaire shall be provided with grounding terminals in a quantity not less than the number of conduit entries plus one. [680.24(D)]

❖ See the commentary to Section E4206.9.1.

E4206.9.5 Strain relief. The termination of a flexible cord of an underwater luminaire within a junction box, transformer or power supply enclosure, ground-fault circuit-interrupter, or other enclosure shall be provided with a strain relief. [680.24(E)]

❖ A flexible cord installed from an underwater luminaire within a raceway must be connected within the box with a strain relief device. When the fixture is removed from the housing or forming shell for relamping, the cord is pulled out of the forming shell and it could be pulled against the connections in the junction box, resulting in bad connections or damage to the cord.

E4206.10 Underwater audio equipment. Underwater audio equipment shall be identified for the purpose. [680.27(A)]

❖ Underwater audio equipment can pose an electrical shock hazard and is installed under provisions similar to those for underwater luminaires (see commentary, Section E4206.4).

E4206.10.1 Speakers. Each speaker shall be mounted in an approved metal forming shell, the front of which is enclosed by a captive metal screen, or equivalent, that is bonded to and secured to the forming shell by a positive locking device that ensures a low-resistance contact and requires a tool to open for installation or servicing of the speaker. The forming shell shall be installed in a recess in the wall or floor of the pool. [680.27(A)(1)]

❖ Underwater speakers are installed in a listed forming shell that is recessed into the wall of the pool. A metal screen must be installed over the forming shell to prevent contact with the speaker. The screen must require a tool to open. Typically, the screen is attached with screws so that it is not easily removed.

E4206.10.2 Wiring methods. Rigid metal conduit of brass or other identified corrosion-resistant metal, rigid polyvinyl chloride conduit, rigid thermosetting resin conduit or liquid-tight flexible nonmetallic conduit (LFNC-B) shall extend from the forming shell to a suitable junction box or other enclosure as provided in Section E4206.9. Where rigid nonmetallic conduit or liquid-tight flexible nonmetallic conduit is used, an 8 AWG solid or stranded insulated copper bonding jumper shall be installed in this conduit with provisions for terminating in the forming shell and the junction box. The termination of the 8 AWG bonding jumper in the forming shell shall be covered with, or encapsulated in, a suitable potting compound to protect such connection from the possible deteriorating effect of pool water. [680.27(A)(2)]

❖ The wiring method of extending a conduit from the speaker forming shell to a deck junction box is the same as for a wet-niche lighting fixture.

E4206.10.3 Forming shell and metal screen. The forming shell and metal screen shall be of brass or other approved corrosion-resistant metal. Forming shells shall include provisions for terminating an 8 AWG copper conductor. [680.27(A)(3)]

❖ Listed equipment is available for installation of underwater audio equipment.

E4206.11 Electrically operated pool covers. The electric motors, controllers, and wiring for pool covers shall be located not less than 5 feet (1524 mm) from the inside wall of the pool except where separated from the pool by a wall, cover, or other permanent barrier. Electric motors installed below grade level shall be of the totally enclosed type. The electric motor and controller shall be connected to a branch circuit protected by a ground-fault circuit-interrupter. The device that controls the operation of the motor for an electrically operated pool cover shall be located so that the operator has full view of the pool. [680.27(B)(1) and (B)(2)]

❖ Pool cover motors and their controllers are typically located at the side of the pool in a box or recessed with a cover flush with the pool deck. Obviously, the box or recessed enclosure must be provided with means to drain water that will accumulate, and the motor must be totally enclosed and served by a GFCI-protected circuit.

The operating control for an electrically operated pool cover must be located to allow the person operating the device to see that people are clear of moving parts. This is a common-sense safety approach for the operation of all machinery.

E4206.12 Electric pool water heaters. Electric pool water heaters shall have the heating elements subdivided into loads not exceeding 48 amperes and protected at not more than 60 amperes. The ampacity of the branch-circuit conductors and the rating or setting of overcurrent protective devices shall be not less than 125 percent of the total nameplate load rating. (680.9)

❖ See Sections E4101.4.1 and E3702.10.

E4206.13 Pool area heating. The provisions of Sections E4206.13.1 through E4206.13.3 shall apply to all pool deck areas, including a covered pool, where electrically operated comfort heating units are installed within 20 feet (6096 mm) of the inside wall of the pool. [680.27(C)]

❖ Where electric heating is installed in an area within 20 feet (6096 mm) of the inside wall of the pool, specific code rules apply to the wiring for heating units.

E4206.13.1 Unit heaters. Unit heaters shall be rigidly mounted to the structure and shall be of the totally enclosed or guarded types. Unit heaters shall not be mounted over the pool or within the area extending 5 feet (1524 mm) horizontally from the inside walls of a pool. [680.27(C)(1)]

❖ Where unit heaters are used in a pool, spa or hot tub area, they must be kept at least 5 feet (1524 mm) horizontally from the inside walls of the pool. They must be guarded or enclosed. These two terms have clear definitions, given in Chapter 35.

E4206.13.2 Permanently wired radiant heaters. Electric radiant heaters shall be suitably guarded and securely fastened to their mounting devices. Heaters shall not be installed over a pool or within the area extending 5 feet (1524 mm) horizontally from the inside walls of the pool and shall be mounted not less than 12 feet (3658 mm) vertically above the pool deck. [680.27(C)(2)]

❖ Radiant heaters in a pool area must be at least 12 feet (3058 mm) above the pool deck. Although these heating units must be securely fastened, there is no exception that would permit installation over the pool. They must be at least 5 feet horizontally from the inside walls of the pool.

E4206.13.3 Radiant heating cables prohibited. Radiant heating cables embedded in or below the deck shall be prohibited. [680.27(C)(3)]

❖ Installation of heating cables is prohibited in the deck of the pool area because of the possible breakdown of the insulation on the cable causing a ground fault or current leakage.

SECTION E4207
STORABLE SWIMMING POOLS, STORABLE SPAS, AND STORABLE HOT TUBS

E4207.1 Pumps. A cord and plug-connected pool filter pump for use with storable pools shall incorporate an approved system of double insulation or its equivalent and shall be provided with means for grounding only the internal and nonaccessible noncurrent-carrying metal parts of the appliance.

The means for grounding shall be an equipment grounding conductor run with the power-supply conductors in a flexible cord that is properly terminated in a grounding-type attachment plug having a fixed grounding contact. Cord and plug-connected pool filter pumps shall be provided with a ground-fault circuit interrupter that is an integral part of the attachment plug or located in the power supply cord within 12 inches (305 mm) of the attachment plug. (680.31)

❖ Filter pumps must be protected by a system of double insulation or the equivalent. The means for grounding the internal noncurrent-carrying metal parts must be an equipment grounding conductor run with the power-supply conductors in the flexible cord terminated with a grounding-type attachment plug.

E4207.2 Ground-fault circuit-interrupters required. Electrical equipment, including power-supply cords, used with storable pools shall be protected by ground-fault circuit-interrupters. 125-volt, 15- and 20-ampere receptacles located within 20 feet (6096 mm) of the inside walls of a storable pool, storable spa, or storable hot tub shall be protected by a ground-fault circuit interrupter. In determining these dimensions, the distance to be measured shall be the shortest path that the supply cord of an appliance connected to the receptacle would follow without passing through a floor, wall, ceiling, doorway with hinged or sliding door, window opening, or other effective permanent barrier. (680.32)

❖ Similar requirements are found in Section E4203.1.3. The 20-foot distance is measured for the worst case condition along the shortest path that the cord could be laid in order to reach the receptacle.

E4207.3 Luminaires. Luminaires for storable pools, storable spas, and storable hot tubs shall not have exposed metal parts and shall be listed for the purpose as an assembly. In addition, luminaires for storable pools shall comply with the requirements of Section E4207.3.1 or E4207.3.2. (680.33)

❖ Luminaires for storable pools, spas and hot tubs are intended for temporary installation on or in the wall of the pool, generally between 8 and 10 inches (203 and 254 mm) below the top of the pool wall. These luminaires are manufactured as an assembly with flexible cords of 25 feet (7620 mm) or more so that they can be routed away from the pool to an enclosure for a transformer or GFCI enclosure. The transformer or GFCI assembly typically has a 3- to 6-foot-long (914 to 1829 mm) power supply cord, and the unit is designed for temporary installation on a structure near a receptacle.

E4207.3.1 Within the low-voltage contact limit. A luminaire installed in or on the wall of a storable pool shall be part of a cord and plug-connected lighting assembly. The assembly shall:

1. Have a luminaire lamp that is suitable for the use at the supplied voltage;

2. Have an impact-resistant polymeric lens, luminaire body, and transformer enclosure;

3. Have a transformer meeting the requirements of section E4206.1 with a primary rating not over 150 volts; and

4. Have no exposed metal parts. [680.33(A)]

❖ For use on a storable pool, a cord-and-plug-connected low-voltage light fixture is a listed assembly that receives its power through a flexible cord plugged into a transformer. The primary of the transformer is supplied from a 120-volt branch circuit.

E4207.3.2 Over the low-voltage contact limit but not over 150 volts. A lighting assembly without a transformer or power supply, and with the luminaire lamp(s) operating at over the low-voltage contact limit, but not over 150 volts, shall be permitted to be cord and plug-connected where the assembly is listed as an assembly for the purpose and complies with all of the following:

1. It has an impact-resistant polymeric lens and luminaire body.

2. A ground-fault circuit interrupter with open neutral conductor protection is provided as an integral part of the assembly.

3. The luminaire lamp is permanently connected to the ground-fault circuit interrupter with open-neutral protection.

4. It complies with the requirements of Section E4206.4.

5. It has no exposed metal parts. [680.33(B)]

❖ These fixtures are connected through flexible cords plugged into GFCI devices that have open-neutral protection. A typical GFCI device monitors the current in the ungrounded and grounded conductors, and when it is not balanced, it will trip. A GFCI with open-neutral protection will trip if the neutral is opened or lost.

E4207.4 Receptacle locations. Receptacles shall be located not less than 6 feet (1829 mm) from the inside walls of a storable pool, storable spa or storable hot tub. In determining these dimensions, the distance to be measured shall be the shortest path that the supply cord of an appliance connected to the receptacle would follow without passing through a floor, wall, ceiling, doorway with hinged or sliding door, window opening, or other effective permanent barrier. (680.34)

❖ This section is more stringent in that all receptacles must be at least 10 feet from the inside walls of the storable pool, spa and hot tub (see Sections E4203.1 through E4203.1.2).

E4207.5 Clearances. Overhead conductor installations shall comply with Section E4203.6 and underground conductor installations shall comply with Section E4203.7.

❖ See Sections E4203.6 and E4203.7.

E4207.6 Disconnecting means. Disconnecting means for storable pools and storable/portable spas and hot tubs shall comply with Section E4203.3.

❖ See Section E4203.3.

E4207.7 Ground-fault circuit interrupters. Ground-fault circuit interrupters shall comply with Section E4206.2.

❖ See Section E4206.2.

E4207.8 Grounding of equipment. Equipment shall be grounded as required by Section E4205.1.

❖ See Section E4205.1.

E4207.9 Pool water heaters. Electric pool water heaters shall comply with Section E4206.12.

❖ See Section E4206.12.

SECTION E4208
SPAS AND HOT TUBS

E4208.1 Ground-fault circuit-interrupters. The outlet(s) that supplies a self-contained spa or hot tub, or a packaged spa or hot tub equipment assembly, or a field-assembled spa or hot tub with a heater load of 50 amperes or less, shall be protected by a ground-fault circuit-interrupter. (680.44)

A listed self-contained unit or listed packaged equipment assembly marked to indicate that integral ground-fault circuit-interrupter protection is provided for all electrical parts within the unit or assembly, including pumps, air blowers, heaters, lights, controls, sanitizer generators and wiring, shall not require that the outlet supply be protected by a ground-fault circuit interrupter. [680.44(A)]

❖ The definitions provided at the beginning of this chapter for "Packaged spa or hot tub equipment assembly" and "Self-contained spa or hot tub" are useful in understanding and applying the rules on GFCI protection. Where these units are listed and have integral GFCI protection, the flexible cord need not be plugged into a GFCI receptacle.

E4208.2 Electric water heaters. Electric spa and hot tub water heaters shall be listed and shall have the heating elements subdivided into loads not exceeding 48 amperes and protected at not more than 60 amperes. The ampacity of the branch-circuit conductors, and the rating or setting of overcurrent protective devices, shall be not less than 125 percent of the total nameplate load rating. (680.9)

❖ The ampere load on the circuits supplying water heaters for spas and hot tubs is limited to 80 percent of the branch circuit rating. The maximum load that can be protected by a single over-current device is 48 amperes, therefore, 60 amperes ($0.8 \times 60 = 48$ or $1.25 \times 48 = 60$) is the maximum rating for an overcurrent device protecting this water heater circuit. If the water heater total load consists of more than 48 amperes, the load must be divided into subgroups such that no subgroup load is more than 48 amperes. Heaters with loads over 48 amperes will have multiple circuit breakers or sets of fuses integral with the heater.

E4208.3 Underwater audio equipment. Underwater audio equipment used with spas and hot tubs shall comply with the provisions of Section E4206.10. [680.43(G)]

❖ See Section E4206.10.

E4208.4 Emergency switch for spas and hot tubs. A clearly labeled emergency shutoff or control switch for the purpose of stopping the motor(s) that provides power to the recirculation system and jet system shall be installed at a point that is readily accessible to the users, adjacent to and within sight of the spa or hot tub and not less than 5 feet (1524 mm) away from the spa or hot tub. This requirement shall not apply to single-family dwellings. (680.41)

❖ The emergency switch must be located to allow quick access in an emergency, but must also be a safe distance from the spa or hot tub. This section applies only to spas and hot tubs intended for use by the occupants of more than one dwelling.

SECTION E4209
HYDROMASSAGE BATHTUBS

E4209.1 Ground-fault circuit-interrupters. Hydromassage bathtubs and their associated electrical components shall be supplied by an individual branch circuit(s) and protected by a readily accessible ground-fault circuit-interrupter. All 125-volt, single-phase receptacles not exceeding 30 amperes and located within 6 feet (1829 mm) measured horizontally of the inside walls of a hydromassage tub shall be protected by a ground-fault circuit interrupter(s). (680.71)

❖ Hydromassage bathtubs typically have a cord- and plug-connected motor as part of the unit package. A flexible cord, usually from 18 inches to 3 feet (457 mm to 914 mm) long, is provided for the power supply to the motor. A GFCI-protected receptacle is typically located in the accessible area near the pump, behind an access panel; however, based on the definition of "Readily accessible," one could interpret this section as requiring the GFCI device to be exterior to the bathtub enclosure so that access panels would not have to be removed to gain access to the device. This may necessitate the use of a GFCI circuit breaker in the panelboard. Note that this section requires a dedicated circuit for these bathtubs.

Where located farther than 6 feet from the inside walls of a bathtub, receptacles not supplying the bathtub do not need GFCI protection because they are not easily accessed by and are not within the immediate reach of bathers. Receptacles located where the occupants would be able to use electrical appliances and electronic products in close proximity to the bathtub

should be avoided; however, if they are installed within 6 feet, they must be GFCI protected.

Section E3902.1 requires that all receptacles in a bathroom be GFCI protected; therefore, if a hydromassage bathtub is located in a room defined as a bathroom, the distance between the receptacles and the bathtub is irrelevant.

E4209.2 Other electric equipment. Luminaires, switches, receptacles, and other electrical equipment located in the same room, and not directly associated with a hydromassage bathtub, shall be installed in accordance with the requirements of this code relative to the installation of electrical equipment in bathrooms. (680.72)

❖ Receptacles and lighting fixtures in the area are subject to the code provisions for bathrooms. There are no specific rules for locating receptacles at a certain distance from the hydromassage bathtub as there are for pool, spas or hot tubs.

E4209.3 Accessibility. Hydromassage bathtub electrical equipment shall be accessible without damaging the building structure or building finish. Where the hydromassage bathtub is cord- and plug-connected with the supply receptacle accessible only through a service access opening, the receptacle shall be installed so that its face is within direct view and not more than 12 inches (305 mm) from the plane of the opening. (680.73)

❖ The equipment, usually at the base of the hydromassage bathtub, must be accessible. This access is usually provided through an access panel. The cover or door must not be permanently sealed by tile, etc. See Section P2720.1 for specific pump access requirements. The receptacle serving cord- and plug-connected units is typically located under the unit and behind an access panel. This receptacle must be within easy reach and in plain view to allow the cord to be unplugged, to allow replacement of the receptacle and to allow resetting or replacement of any GFCI device.

E4209.4 Bonding. Both metal piping systems and grounded metal parts in contact with the circulating water shall be bonded together using an insulated, covered or bare solid copper bonding jumper not smaller than 8 AWG. The bonding jumper shall be connected to the terminal on the circulating pump motor that is intended for this purpose. The bonding jumper shall not be required to be connected to a double insulated circulating pump motor. The 8 AWG or larger solid copper bonding jumper shall be required for equipotential bonding in the area of the hydromassage bathtub and shall not be required to be extended or attached to any remote panelboard, service equipment, or any electrode. Where a double-insulated circulating pump motor is used, the 8 AWG or larger solid copper bonding jumper shall be long enough to terminate on a replacement nondouble-insulated pump motor and shall be terminated to the equipment grounding conductor of the branch circuit for the motor. (680.74)

❖ Metal parts of the hydromassage bathtub must be bonded together with at least a size 8-AWG solid copper conductor. This is usually done at the base of the unit where the equipment is located. Because the typical hydromassage bathtub is constructed entirely of nonconductive materials, such as plastics and fiberglass, and the water supply and drain piping is commonly plastic, there may not be any components that require bonding to the pump motor. However, metal water supply and drain piping serving the bathtub would be required to be bonded.

Bibliography

The following resource materials were used in the preparation of the commentary for this chapter of the code:

NFPA 70–14, *National Electrical Code*. Quincy, MA: National Fire Protection Association, 2013.

Chapter 43:
Class 2 Remote-control, Signaling and Power-limited Circuits

General Comments

This chapter covers low-voltage power-limited circuits such as alarm, door bell, remote control and signaling circuits.

Purpose

By regulating installation methods, conductor types and locations and power supplies, this chapter intends to mitigate the potential fire and shock hazards associated with low-voltage power-limited circuits.

SECTION E4301
GENERAL

E4301.1 Scope. This chapter contains requirements for power supplies and wiring methods associated with Class 2 remote-control, signaling, and power-limited circuits that are not an integral part of a device or appliance. Other classes of remote-control, signaling and power-limited conductors shall comply with Article 725 of NFPA 70. (725.1)

❖ Wiring for heating and cooling controls, security systems, intercom systems and door bell chimes are examples of Class 2 wiring in one- and two-family dwellings. Low-voltage circuits that are part of an appliance or equipment are not covered in this chapter. For example, the low-voltage wiring that is integral with a furnace is not covered, but the thermostat wiring is, because it is run in the dwelling exterior to the furnace. For a garage door opener, the wiring to the safety sensors at the base of the door and the push button control wiring is subject to the provisions of this chapter.

E4301.2 Definitions.

CLASS 2 CIRCUIT. That portion of the wiring system between the load side of a Class 2 power source and the connected equipment. Due to its power limitations, a Class 2 circuit considers safety from a fire initiation standpoint and provides acceptable protection from electric shock. (725.2)

❖ There is little or no shock or fire hazard from the operation of devices on power-limited low-voltage circuits. Class 2 circuits typically operate at 30 volts or less and use power supplies that limit current flow to reasonably safe levels.

REMOTE-CONTROL CIRCUIT. Any electrical circuit that controls any other circuit through a relay or an equivalent device. (Article 100)

❖ A low-voltage remote-control circuit is used to open or close a switch to control current flow in another circuit. Current is supplied by a low-voltage power source to a relay, which is used to control the energy supply to equipment, appliances or luminaires. A thermostat is part of a remote-control circuit for the HVAC system that it controls.

SIGNALING CIRCUIT. Any electrical circuit that energizes signaling equipment. (Article 100)

❖ Examples of signaling circuits in one- and two-family dwellings include wiring for security systems, fire alarm systems, chimes, bells and buzzers. For the hearing impaired, a doorbell button is used to energize a relay to turn on a dedicated lamp or to initiate a sequence of strobe lights. Also for the hearing impaired, signaling devices are available that cause the normal lighting fixtures to flash on and off rapidly for a short period. The signaling devices are usually controlled by low-voltage circuits.

SECTION E4302
POWER SOURCES

E4302.1 Power sources for Class 2 circuits. The power source for a Class 2 circuit shall be one of the following:

1. A listed Class 2 transformer.

2. A listed Class 2 power supply.

3. Other listed equipment marked to identify the Class 2 power source.

4. Listed information technology (computer) equipment limited power circuits.

5. A dry-cell battery provided that the voltage is 30 volts or less and the capacity is equal to or less than that available from series connected No. 6 carbon zinc cells. [725.121(A)]

❖ Class 2 low-voltage circuits in residential wiring usually have the voltage and current inherently limited by a transformer. These transformers have inherent protection against overcurrent. Low-voltage transformers are designed to limit current flow to safe values even when a short circuit occurs. Low-voltage transformers are rated in volt-amperes, and a typical step-down transformer for a 24-volt system is rated at 40 volt-amperes or less. The power rating of Class 2 low-voltage circuits in typical residential applications is less than 100 volt-amperes.

E4302.2 Interconnection of power sources. A Class 2 power source shall not have its output connections paralleled or otherwise interconnected with another Class 2 power source except where listed for such interconnection. [725.121(B)]

❖ Power sources must not be interconnected in a way that allows excessive current flow in Class II circuits. The power supply for a remote-control circuit is usually a step-down transformer, from 120 volts to 24 volts for example.

SECTION E4303
WIRING METHODS

E4303.1 Wiring methods on supply side of Class 2 power source. Conductors and equipment on the supply side of the power source shall be installed in accordance with the appropriate requirements of Chapters 34 through 41. Transformers or other devices supplied from electric light or power circuits shall be protected by an over-current device rated at not over 20 amperes. The input leads of a transformer or other power source supplying Class 2 circuits shall be permitted to be smaller than 14 AWG, if not over 12 inches (305 mm) long and if the conductor insulation is rated at not less than 600 volts. In no case shall such leads be smaller than 18 AWG. (725.127 and 725.127 Exception)

❖ The maximum rating of a branch circuit used to supply power to Class 2 low-voltage circuits is 20 amperes. The input leads to a transformer must be at least size 18 AWG. All of the code provisions covering branch circuits must be followed for the wiring up to and including the primary side of the transformer or power supply. The primary leads of a transformer are installed in the same box or enclosure with the branch-circuit conductors and must have at least the same voltage rating as that of the branch-circuit wires.

E4303.2 Wiring methods and materials on load side of the Class 2 power source. Class 2 cables installed as wiring within buildings shall be listed as being resistant to the spread of fire and listed as meeting the criteria specified in Sections E4303.2.1 through E4303.2.3. Cables shall be marked in accordance with Section E4303.2.4. Cable substitutions as described in Table E4303.2 and wiring methods covered in Chapter 38 shall also be permitted. (725.130 (B); 725.135 (A), (C), (G) and (M); 725.154; Table 725.154; Figure 725.154 (A); and 725.179)

❖ Class 2 low-voltage wiring on the load side of the power source is limited to the cable types listed in Table E4303.2 and wiring methods covered in Chapter 38. The conductors for Class 2 low-voltage circuits are listed based on their flame spread characteristics.

They are typically rated as conductors having low smoke-producing characteristics during standardized tests for fire and smoke. The cable types in Table E4303.2 are listed in descending order of resistance to spread of fire.

TABLE E4303.2
CABLE USES AND PERMITTED SUBSTITUTIONS
[Figure 725.154(A)]

CABLE TYPE	USE	PERMITTED SUBSTITUTIONS[a]
CL2P	Class 2 Plenum Cable	CMP, CL3P
CL2R	Class 2 Plenum Cable	CMP, CL3P, CL2P, CMR, CL3R
CL2	Class 2 Cable	CMP, CL3P, CL2P, CMR, CL3R, CL2R CMG, CM, CL3
CL2X	Class 2 Cable, Limited Use	CMP, CL3P CL2P, CMR, CL3R, CL2R, CMG, CM, CL3, CL2, CMX, CL3X

a. For identification of cables other than Class 2 cables, see NFPA 70.

E4303.2.1 Type CL2P cables. Cables installed in ducts, plenums and other spaces used to convey environmental air shall be Type CL2P cables listed as being suitable for the use and listed as having adequate fire-resistant and low smoke-producing characteristics. [725.179(A)]

❖ Under standardized test methods, Class 2 plenum cable (CL2P) is determined to have adequate fire-resistant and low smoke-producing characteristics. The Underwriters Laboratories (UL) *Marking Guide for Wire and Cable* lists the testing criteria for power-limited circuit cable. Cable Type CL2P exhibits a maximum peak optical density of 0.5, a maximum optical density of 0.15, and maximum flame-spread distance of 5 feet (1524 mm).

E4303.2.2 Type CL2 cables. Cables for general-purpose use, shall be listed as being resistant to the spread of fire and listed for the use. [725.179 (C)]

❖ The technical criterion for listing cable Type CL2 as being resistant to the spread of fire is based on a vertical tray flame test where the cable "does not spread flame to the top of the tray" as stated in the (UL) *Marking Guide for Wire and Cable.*

E4303.2.3 Type CL2X cables. Type CL2X limited-use cable shall be listed as being suitable for use in dwellings and for the use and in raceways and shall also be listed as being flame retardant. Cables with a diameter of less than $^1/_4$ inch (6.4 mm) shall be permitted to be installed without a raceway. [725.179 (D)]

❖ According to the (UL) *Marking Guide for Wire and Cable,* cable Type CL2X is intended for use in Class 2 circuits in one- and two-family dwellings or multifamily dwellings where the cable diameter is less than 0.25 inch (6.4 mm). This cable complies with "vertical wire" flame test requirements in UL 1581.

E4303.2.4 Type CL2R cables. Cables installed in a vertical run in a shaft or installed from floor to floor shall be listed as suitable for use in a vertical run in a shaft or from floor to floor and shall also be listed as having fire-resistant characteristics capable of preventing fire from being conveyed from floor to floor. [725.179(B)]

> **Exception:** CL2X and CL3X cables with a diameter of less than $\frac{1}{4}$ inch (6.4 mm) and CL2 and CL3 cables shall be permitted in risers in one- and two-family dwelling units. [725.154 (G)]

❖ Cables can be tested to a standard that determines whether the cables will convey fire upward through a shaft or chase or upward on exposed cable risers. The code is concerned about combustible cables that will burn and carry fire from one story to the next, thereby accelerating the spread of fire.

E4303.2.5 Marking. Cables shall be marked in accordance with Table E4303.2.5. Voltage ratings shall not be marked on cables.

❖ The voltage rating is not marked on cables listed for use in Class 2 low-voltage circuits because a voltage rating could be misinterpreted to mean that the cable is suitable for use in Class 1 circuits.

Table E4303.2.5 [Table 725.179(K)]
CABLE MARKING

CABLE MARKING	TYPE
CL2P	Class 2 plenum cable
CL2R	Class 2 riser cable
CL2	Class 2 cable
CL2X	Class 2 cable, limited use

SECTION E4304
INSTALLATION REQUIREMENTS

E4304.1 Separation from other conductors. In cables, compartments, enclosures, outlet boxes, device boxes, and raceways, conductors of Class 2 circuits shall not be placed in any cable, compartment, enclosure, outlet box, device box, raceway, or similar fitting with conductors of electric light, power, Class 1 and nonpower-limited fire alarm circuits. (725.136)

Exceptions:

1. Where the conductors of the electric light, power, Class 1 and nonpower-limited fire alarm circuits are separated by a barrier from the Class 2 circuits. In enclosures, Class 2 circuits shall be permitted to be installed in a raceway within the enclosure to separate them from Class 1, electric light, power and nonpower-limited fire alarm circuits. [725.136(B)]

2. Class 2 conductors in compartments, enclosures, device boxes, outlet boxes and similar fittings where electric light, power, Class 1 or nonpower-limited fire alarm circuit conductors are introduced solely to connect to the equipment connected to the Class 2 circuits. The electric light, power, Class 1 and nonpower-limited fire alarm circuit conductors shall be routed to maintain a minimum of $\frac{1}{4}$ inch (6.4 mm) separation from the conductors and cables of the Class 2 circuits; or the electric light power, Class 1 and nonpower-limited fire alarm circuit conductors operate at 150 volts or less to ground and the Class 2 circuits are installed using Types CL3, CL3R, or CL3P or permitted substitute cables, and provided that these Class 3 cable conductors extending beyond their jacket are separated by a minimum of $\frac{1}{4}$ inch (6.4 mm) or by a nonconductive sleeve or nonconductive barrier from all other conductors. [725.136(D)]

❖ Class 2 low-voltage wiring must not be run in the same raceway or installed in the same box or enclosure with 120-volt branch-circuit wiring. For example, the 120-volt branch circuit conductors for a furnace are run in conduit to the furnace junction box. The Class 2 thermostat wiring controlling the furnace cannot be run in the same raceway.

Class 2 low-voltage conductors could be installed in an enclosure with higher voltage power conductor if a barrier or partition is installed to separate the wiring of each system. This does not apply to conductors such as the leads to the primary side of the transformer as indicated in Section E4303.1. A common code violation is the installation of a Class 2 transformer inside of a panelboard enclosure.

E4304.2 Other applications. Conductors of Class 2 circuits shall be separated by not less than 2 inches (51 mm) from conductors of any electric light, power, Class 1 or nonpower-limited fire alarm circuits except where one of the following conditions is met:

1. All of the electric light, power, Class 1 and nonpower-limited fire alarm circuit conductors are in raceways or in metal-sheathed, metal-clad, nonmetallic-sheathed or Type UF cables.

2. All of the Class 2 circuit conductors are in raceways or in metal-sheathed, metal-clad, nonmetallic-sheathed or Type UF cables. [725.136(I)]

❖ Class 2 circuit conductors must be separated by at least 2 inches (51 mm) from any power and lighting branch circuit conductors except where either the Class 2 or power and lighting conductors are enclosed in raceways, sheathed cable or metal-clad cable. Low-voltage door bell wiring, for example, can be run through the same bored holes as used for the Type NM cable.

E4304.3 Class 2 circuits with communications circuits. Where Class 2 circuit conductors are in the same cable as communications circuits, the Class 2 circuits shall be classified as communications circuits and shall meet the requirements of Article 800 of NFPA 70. The cables shall be listed as communications cables or multipurpose cables.

Cables constructed of individually listed Class 2 and communications cables under a common jacket shall be permitted to be classified as communications cables. The fire-resistance rating of the composite cable shall be determined by the performance of the composite cable. [725.139(D)]

❖ In some cases, a composite cable is used that has telephone wire and Class 2 signaling wire. For these installations, the code provisions for communication circuits apply.

E4304.4 Class 2 cables with other circuit cables. Jacketed cables of Class 2 circuits shall be permitted in the same enclosure or raceway with jacketed cables of any of the following:

1. Power-limited fire alarm systems in compliance with Article 760 of NFPA 70.

2. Nonconductive and conductive optical fiber cables in compliance with Article 770 of NFPA 70.

3. Communications circuits in compliance with Article 800 of NFPA 70.

4. Community antenna television and radio distribution systems in compliance with Article 820 of NFPA 70.

5. Low-power, network-powered broadband communications in compliance with Article 830 of NFPA 70. [725.139(E)]

❖ Jacketed or sheathed Class 2 cables are permitted in the same raceway or enclosure with the cables listed because they are protected from faults by the jackets on the cables, and the cables listed in this code section operate at low voltage.

E4304.5 Installation of conductors and cables. Cables and conductors installed exposed on the surface of ceilings and sidewalls shall be supported by the building structure in such a manner that they will not be damaged by normal building use. Such cables shall be supported by straps, staples, hangers, cable ties or similar fittings designed so as to not damage the cable. Nonmetallic cable ties and other nonmetallic accessories used to secure and support cables located in stud cavity and joist space plenums shall be listed as having low smoke and heat release properties. The installation shall comply with Table E3802.1 regarding cables run parallel with framing members and furring strips. The installation of wires and cables shall not prevent access to equipment nor prevent removal of panels, including suspended ceiling panels. Raceways shall not be used as a means of support for Class 2 circuit conductors, except where the supporting raceway contains conductors supplying power to the functionally associated equipment controlled by the Class 2 conductors. [300.22 (C) (1) and 725.24]

❖ Although Class 2 low-voltage cables can be installed in the same bored holes as Type NM cable, they cannot be taped, strapped or attached by any means to a race-way. An exception to this is where a low-voltage control cable is run along with and strapped to the raceway to a furnace or condensing unit. In this case, the Class 2 cable must be part of the control circuitry for the equipment supplied by the raceway wiring method. Low-voltage wiring is sometimes run in a careless manner because there is not as much concern for shock or fire hazard. However, low-voltage wiring deserves the same respect as higher voltage wiring and should be supported from the building struc-

ture and run in a manner that will protect it from damage. Extreme care is necessary to avoid damaging a cable anytime that more than one cable is pulled through the same bored/drilled hole. It is common to find Class 2 cables fastened to ductwork, furnaces and air handlers using scraps of sheet metal with sharp edges that can cut the cable insulation. Class 2 cables are often damaged outdoors where they are run through the air to heat pumps and condensing units, and subsequently damaged by animals, weed trimmers and yard tools. Damage to such cables can lead to the loss of heat in buildings in freezing climates, for example.

Bibliography

The following resource materials were used in the preparation of the commentary for this chapter of the code:

NFPA 70–14, *National Electrical Code*. Quincy, MA: National Fire Protection Association, 2013.

Chapter 44:
Referenced Standards

This chapter lists the standards that are referenced in various sections of this document. The standards are listed herein by the promulgating agency of the standard, the standard identification, the effective date and title, and the section or sections of this document that reference the standard. The application of the referenced standards shall be as specified in Section R102.4.

General Comments

Chapter 44 contains a comprehensive list of standards that are referenced in the code. It is organized to make locating specific document references easy.

It is important to understand that not every document related to residential design, installation and construction is qualified to be a "referenced standard." The International Code Council® (ICC®) has adopted a criterion that standards referenced in the *International Codes*® and standards intended for adoption into the *International Codes* must meet to qualify as a referenced standard. The policy is summarized as follows:

- Code references: The scope and application of the standard must be clearly identified in the code text.
- Standard content: The standard must be written in mandatory language and be appropriate for the subject covered. The standard cannot have the effect of requiring proprietary materials or prescribing a proprietary testing agency.
- Standard promulgation: The standard must be readily available and developed and maintained in a consensus process such as those used by ASTM or ANSI.

The ICC Code Development Procedures, of which the standards policy is a part, are updated periodically. A copy of the latest version can be obtained from the ICC offices.

Once a standard is incorporated into the code through the code development process, it becomes an enforceable part of the code. When the code is adopted by a jurisdiction, the standard also is part of that jurisdiction's adopted code. It is for this reason that the criteria were developed. Compliance with this policy means that documents incorporated into the code are developed through the use of a consensus process, are written in mandatory language, and do not mandate the use of proprietary materials or agencies. The requirement for a standard to be developed through a consensus process means that the standard is representative of the most current body of available knowledge on the subject as determined by a broad range of interested or affected parties without dominance by any single interest group. A true consensus process has many attributes, including but not limited to:

- An open process that has formal (published) procedures that allow for the consideration of all viewpoints;
- A definitive review period that allows for the standard to be updated and/or revised;
- A process of notification to all interested parties; and
- An appeals process.

Many available documents related to system design, installation and construction, though useful, are not "standards" and are not appropriate for reference in the code. Often, these documents are developed or written with the intention of being used for regulatory purposes and are unsuitable for use as a standard because of extensive use of recommendations, advisory comments and nonmandatory terms. Typical examples include installation instructions, guidelines and practices.

The objective of ICC's standards policy is to provide regulations that are clear, concise and enforceable; thus, the requirement for standards to be written in mandatory language. This requirement is not intended to mean that a standard cannot contain informational or explanatory material that will aid the user of the standard in its application. When the standard's promulgating agency wants such material to be included, however, the information must appear in a nonmandatory location, such as an annex or appendix, and be clearly identified as not being part of the standard.

Overall, standards referenced by the code must be authoritative, relevant, up to date and, most important, reasonable and enforceable. Standards that comply with ICC's standards policy fulfill these expectations.

Purpose

This code contains numerous references to documents that are used to regulate materials and methods of construction. The references to these documents within the code text consist of the promulgating agency's acronym and its publication designation (for example, AWS A5.8) and a further indication that the document being referenced is the one that is listed in Chapter 44. Chapter 44 contains all of the information that is necessary to identify the specific referenced document. Included is the following information on a document's promulgating agency:

- The promulgating agency (the agency's title);
- The promulgating agency's acronym; and
- The promulgating agency's address.

For example, a reference to an ASME standard within the code indicates that the document is promulgated by the American Society of Mechanical Engineers (ASME), which is located in New York, New York. Chapter 44 lists the standards agencies alphabetically for ease of identification.

Chapter 44 also includes the following information on the referenced document itself:

- The document's publication designation;
- The document's edition year;
- The document's title;

- Any addenda or revisions to the document known at the time of the code's publication; and
- Every section of the code in which the document is referenced.

For example, a reference to ASME A112.1.3 indicates that this document can be found in Chapter 44 under the heading ASME. The specific standard's designation is A112.1.3. For convenience, these designations are listed in alphanumeric order. Chapter 44 shows that ASME A112.1.3 is titled Air Gap Fittings for Use with Plumbing Fixtures, Appliances, and Appurtenances; the applicable edition (that is, its year of publication) is 2000; and it is referenced in two specifically identified sections of the code (see Commentary Figure 44).

Chapter 44 also indicates when a document has been discontinued or replaced by its promulgating agency. When a document is replaced by a different one, a note appears to tell the user the designation and title of the new document.

The key aspect of the manner in which standards are referenced by the code is that a specific edition of a specific standard is clearly identified. The requirements necessary for compliance can be readily determined. The basis for code compliance is, therefore, established and available to the code official, the mechanical contractor, the designer and the owner.

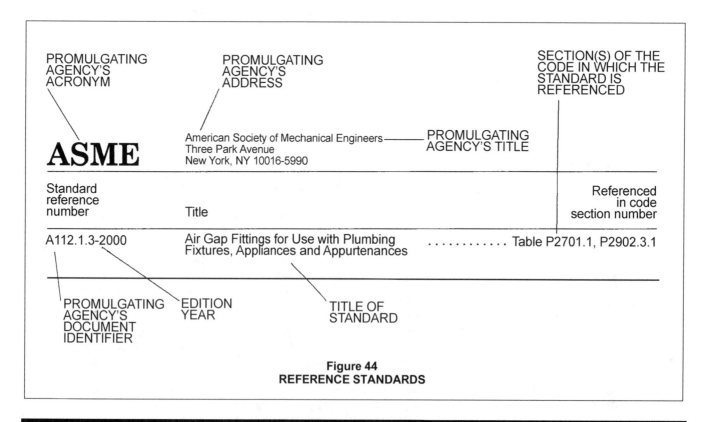

Figure 44
REFERENCE STANDARDS

AAMA

American Architectural Manufacturers Association
1827 Walden Office Square, Suite 550
Schaumburg, IL 60173

Standard reference number	Title	Referenced in code section number
AAMA/WDMA/CSA 101/I.S.2/A440—11	North American Fenestration Standards/Specifications for Windows, Doors and Skylights	R308.6.9, R609.3, N1102.4.3
450—10	Voluntary Performance Rating Method for Mulled Fenestration Assemblies	R609.8
506—11	Voluntary Specifications for Hurricane Impact and Cycle Testing of Fenestration Products	R609.6.1
711—13	Voluntary Specification for Self-adhering Flashing Used for Installation of Exterior Wall Fenestration Products	R703.4
712—11	Voluntary Specification for Mechanically Attached Flexible Flashing	R703.4
714—12	Voluntary Specification for Liquid Applied Flashing Used to Create a Water-resistive Seal around Exterior Wall Openings in Buildings	R703.4
AAMA/NPEA/ NSA 2100—12	Specifications for Sunrooms	R302.2.1.1

ACI

American Concrete Institute
38800 Country Club Drive
Farmington Hills, MI 48331

Standard reference number	Title	Referenced in code section number
318—14	Building Code Requirements for Structural Concrete	R301.2.2.2.4, R301.2.2.3.4, R402.2, Table R404.1.2(2), Table R404.1.2(5), Table R404.1.2(6), Table R404.1.2(7), Table R404.1.2(8), Table R404.1.2(9), R404.1.3, R404.1.3.1, R404.1.3.3, R404.1.3.4, R404.1.4.2, R404.5.1, R608.1, R608.1.1, R608.1.2, R608.2, R608.5.1, R608.6.1, R608.8.2, R608.9.2, R608.9.3
332—14	Code Requirements for Residential Concrete Construction	R402.2, R403.1, R404.1.3, R404.1.3.4, R404.1.4.2, R506.1
530—13	Building Code Requirements for Masonry Structures	R404.1.2, R606.1, R606.1.1, R606.12.1, R606.12.2.3.2, R606.12.2.3.1, R703.12
R606.12.3.1530.1—13	Specification for Masonry Structures	R404.1.2, R606.1, R606.1.1, R606.2.9, R606.2.12, R606.12.1, R606.12.2.3.2, R606.12.3.1, 703.12

ACCA

Air Conditioning Contractors of America
2800 Shirlington Road, Suite 300
Arlington, VA 22206

Standard reference number	Title	Referenced in code section number
Manual D—2011	Residential Duct Systems	M1601.1, M1602.2
Manual J—2011	Residential Load Calculation—Eighth Edition	N1103.6, M1401.3
Manual S—13	Residential Equipment Selection	N1103.6, M1401.3

AISI

American Iron and Steel Institute
25 Massachusetts Avenue, NW Suite 800
Washington, DC 20001

Standard reference number	Title	Referenced in code section number
AISI S100—12	North American Specification for the Design of Cold-formed Steel Structural Members, 2012	R505.1.3, R603.6, R608.9.2, R608.9.3, R804.3.6
AISI S200—12	North American Standard for Cold-formed Steel Framing—General Provisions 2012	R702.3.3
AISI S220—11	North American Standard for Cold-formed Steel Framing—Nonstructural Members	R702.3.3

AISI—continued

AISI S230—07/ S3-12 (2012)	Standard for Cold-formed Steel Framing—Prescriptive Method for One- and Two-family Dwellings, 2007 with Supplement 3, dated 2012 (Reaffirmed 2012) . . . . R301.1.1, R301.2.1.1, R301.2.2.3.1, R301.2.2.3.5, R603.6, R603.9.4.1, R603.9.4.2, R608.9.2, R608.9.3, Figure 608.9(11), R608.10

AMCA

Air Movement and Control Association
300 West University
Arlington Heights, IL 60004

Standard reference number	Title	Referenced in code section number
ANSI/AMCA 210- ANSI/ASHRAE 51—07	Laboratory Methods of Testing Fans for Aerodynamic Performance Rating	Table M1506.2

AMD

Association of Millwork Distributors Standards
10047 Robert Trent Parkway
New Port Richey, FL 34655-4649

Standard reference number	Title	Referenced in code section number
AMD 100—2013	Structural Performance Ratings of Side Hinged Exterior Door Systems and Procedures for Component Substitution	R609.3

ANCE

Association of the Electric Sector
Av. Lázaro Cardenas No. 869
Col. Nueva Industrial Vallejo
C.P. 07700 México D.F.

Standard reference number	Title	Referenced in code section number
UL/CSA/ANCE 60335-2—2012	Standard for Household and Similar Electric Appliances, Part 2: Particular Requirements for Motor-compressors M1403.1, M1412.1, M1413.1

ANSI

American National Standards Institute
25 West 43rd Street, Fourth Floor
New York, NY 10036

Standard reference number	Title	Referenced in code section number
A108.1A—99	Installation of Ceramic Tile in the Wet-set Method, with Portland Cement Mortar	R702.4.1
A108.1B—99	Installation of Ceramic Tile, Quarry Tile on a Cured Portland Cement Mortar Setting Bed with Dry-set or Latex Portland Mortar	R702.4.1
A108.4—99	Installation of Ceramic Tile with Organic Adhesives or Water-Cleanable Tile-setting Epoxy Adhesive.	R702.4.1
A108.5—99	Installation of Ceramic Tile with Dry-set Portland Cement Mortar or Latex Portland Cement Mortar	R702.4.1
A108.6—99	Installation of Ceramic Tile with Chemical-resistant, Water-cleanable Tile-setting and -grouting Epoxy	R702.4.1
A108.11—99	Interior Installation of Cementitious Backer Units.	R702.4.1
A118.1—99	American National Standard Specifications for Dry-set Portland Cement Mortar	R702.4.1
A118.3—99	American National Standard Specifications for Chemical-resistant, Water-cleanable Tile-setting and -grouting Epoxy, and Water-cleanable Tile-setting Epoxy Adhesive	R702.4.1

ANSI—continued

A118.4—99	American National Standard Specifications for Latex-Portland Cement Mortar	R606.2.10
A118.10—99	Specification for Load-bearing, Bonded, Waterproof Membranes for Thin-set Ceramic Tile and Dimension Stone Installation	P2709.2, P2709.2.4
A136.1—99	American National Standard Specifications for Organic Adhesives for Installation of Ceramic Tile	R702.4.1
A137.1—2012	American National Standard Specifications for Ceramic Tile	R702.4.1
LC1/CSA 6.26—13	Fuel Gas Piping Systems Using Corrugated Stainless Steel Tubing (CSST)	G2414.5.3
LC4/CSA 6.32—12	Press-connect Metallic Fittings for Use in Fuel Gas Distribution Systems	G2414.10.2
Z21.1—2010	Household Cooking Gas Appliances	G2447.1
Z21.5.1/CSA 7.1—14	Gas Clothes Dryers—Volume I—Type I Clothes Dryers	G2438.1
Z21.8—94 (R2002)	Installation of Domestic Gas Conversion Burners	G2443.1
Z21.10.1/CSA 4.1—12	Gas Water Heaters—Volume I—Storage Water Heaters with Input Ratings of 75,000 Btu per hour or Less	G2448.1
Z21.10.3/CSA 4.3—11	Gas Water Heaters—Volume III—Storage Water Heaters with Input Ratings above 75,000 Btu per hour, Circulating and Instantaneous	G2448.1
Z21.11.2—11	Gas-fired Room Heaters—Volume II—Unvented Room Heaters	G2445.1
Z21.13/CSA 4.9—11	Gas-fired Low-pressure Steam and Hot Water Boilers	G2452.1
Z21.15/CSA 9.1—09	Manually Operated Gas Valves for Appliances, Appliance Connector Valves and Hose End Valves	Table G2420.1.1
Z21.22—99 (R2003)	Relief Valves for Hot Water Supply Systems—with Addenda Z21.22a—2000 (R2003) and 21.22b—2001 (R2003)	P2804.2, P2804.7
Z21.24/CGA 6.10—06	Connectors for Gas Appliances	G2422.1
Z21.40.1/ CSA 2.91—96 (R2011)	Gas-fired, Heat-activated Air-conditioning and Heat Pump Appliances	G2449.1
Z21.40.2/ CSA 2.92—96 (R2011)	Air-conditioning and Heat Pump Appliances (Thermal Combustion)	G2449.1
Z21.42—2014	Gas-fired Illuminating Appliances	G2450.1
Z21.47/CSA 2.3—12	Gas-fired Central Furnaces	G2442.1
Z21.50/CSA 2.22—12	Vented Gas Fireplaces	G2434.1
Z21.56/CSA 4.7—13	Gas-fired Pool Heaters	G2441.1
Z21.58—95/CSA 1.6—13	Outdoor Cooking Gas Appliances	G2447.1
Z21.60/CSA 2.26—12	Decorative Gas Appliances for Installation in Solid Fuel-burning Fireplaces	G2432.1
Z21.75/CSA 6.27—07	Connectors for Outdoor Gas Appliances and Manufactured Homes	G2422.1
Z21.80—11	Line Pressure Regulators	G2421.1
ANSI/CSA America FCI—12	Stationary Fuel Cell Power Systems	M1903.1
Z21.84—12	Manually Listed, Natural Gas Decorative Gas Appliances for Installation in Solid Fuel-burning Fireplaces	G2432.1, G2432.2
Z21.86—08	Gas-fired Vented Space Heating Appliances	G2436.1, G2437.1, G2446.1
Z21.88/CSA 2.33—14	Vented Gas Fireplace Heaters	G2435.1
Z21.91—07	Ventless Firebox Enclosures for Gas-fired Unvented Decorative Room Heaters	G2445.7.1
Z21.97—12	Outdoor Decorative Appliances	G2454.1
Z83.6—90 (R1998)	Gas-fired Infrared Heaters	G2451.1
Z83.8/CSA 2.6—09	Gas-fired Unit Heaters, Gas Packaged Heaters, Gas Utility Heaters and Gas-fired Duct Furnaces	G2444.1
Z83.19—01 (R2009)	Gas-fuel High-intensity Infrared Heaters	G2451.1
Z83.20—08	Gas-fired Low-intensity Infared Heaters Outdoor Decorative Appliances	G2451.1
Z97.1—2014	Safety Glazing Materials Used in Buildings—Safety Performance Specifications and Methods of Test	R308.1.1, R308.3.1

APA

APA—The Engineered Wood Association
7011 South 19th
Tacoma, WA 98466

Standard reference number	Title	Referenced in code section number
ANSI/A190.1—12	Structural Glued-laminated Timber	R502.1.3, R602.1.3, R802.1.2
ANSI/APA PRP 210—08	Standard for Performance-rated Engineered Wood Siding	R604.1, Table R703.3(1), R703.3.3
ANSI/APA PRG 320—2012	Standard for Performance-rated Cross Laminated Timber	R502.1.6, R602.1.6, R802.1.6
ANSI/APA PRR 410—2011	Standard for Performance-rated Engineered Wood Rim Boards	R502.1.7, R602.1.7, R802.1.7
APA E30—11	Engineered Wood Construction Guide	Table R503.2.1.1(1), R503.2.2, R803.2.2, R803.2.3

ASCE/SEI

American Society of Civil Engineers
Structural Engineering Institute
1801 Alexander Bell Drive
Reston, VA 20191

Standard reference number	Title	Referenced in code section number
5—13	Building Code Requirements for Masonry Structures	R404.1.2, R606.1, R606.1.1, R606.12.1, R606.12.2.3.1, R606.12.2.3.2, R606.12.3.1, R703.12
6—13	Specification for Masonry Structures	R404.1.2, R606.1, R606.1.1, R606.2.9, R606.2.12, R606.12.1, R606.12.2.3.1, R606.12.2.3.2, R606.12.3.1, R703.12
7—10	Minimum Design Loads for Buildings and Other Structures with Supplement No. 1	R301.2.1.1, R301.2.1.2, R301.2.1.2.1, R301.2.1.5, R301.2.1.5.1, Table R608.6(1), Table R608.6(2), Table R608.6(3), Table R608.6(4), Table R608.7(1A), Table R608.7(1B), Table R608.7(1C), R608.9.2, R608.9.3, R609.2, AH107.4.3
24—14	Flood-resistant Design and Construction	R301.2.4, R301.2.4.1, R322.1, R322.1.1, R322.1.6, R322.1.9, R322.2.2, R322.3.3
32—01	Design and Construction of Frost-protected Shallow Foundations	R403.1.4.1

ASHRAE

American Society of Heating, Refrigerating
and Air-Conditioning Engineers, Inc.
1791 Tullie Circle, NE
Atlanta, GA 30329

Standard reference number	Title	Referenced in code section number
ASHRAE—2013	ASHRAE Handbook of Fundamentals	N1102.1.5, Table N1105.5.2(1), P3001.2, P3101.4
ASHRAE 193—2010	Method of Test for Determining Air Tightness of HVAC Equipment	N1103.3.2.1
34—2013	Designation and Safety Classification of Refrigerants	M1411.1

ASME

American Society of Mechanical Engineers
Three Park Avenue
New York, NY 10016-5990

Standard reference number	Title	Referenced in code section number
ASME/A17.1/ CSA B44—2013	Safety Code for Elevators and Escalators	R321.1
A18.1—2008	Safety Standard for Platforms and Stairway Chair Lifts	R321.2
A112.1.2—2004	Air Gaps in Plumbing Systems	P2717.1, Table P2902.3, P2902.3.1
A112.1.3—2000 (Reaffirmed 2011)	Air Gap Fittings for Use with Plumbing Fixtures, Appliances and Appurtenances	Table P2701.1, P2717.1, Table P2902.3, P2902.3.1
A112.3.1—2007	Stainless Steel Drainage Systems for Sanitary, DWV, Storm and Vacuum Applications Above and Below Ground	Table P3002.1(1), Table P3002.1(2), Table P3002.2, Table P3002.3, Table P3302.1
A112.3.4—2013/ CSA B45.9—13	Macerating Toilet Systems and Related Components	Table P2701.1, P3007.5
A112.4.1—2009	Water Heater Relief Valve Drain Tubes	P2804.6.1
A112.4.2—2009	Water-closet Personal Hygiene Devices	P2722.5
A112.4.3—1999 (R2010)	Plastic Fittings for Connecting Water Closets to the Sanitary Drainage System	P3003.14
A112.4.14—2004 (R2010)	Manually Operated, Quarter-turn Shutoff Valves for Use in Plumbing Systems	Table P2903.9.4
A112.6.1M—1997 (R2008)	Floor-affixed Supports for Off-the-floor Plumbing Fixtures for Public Use	Table P2701.1, P2702.4
A112.6.2—2000 (R2010)	Framing-affixed Supports for Off-the-floor Water Closets with Concealed Tanks	Table P2701.1, P2702.4
A112.6.3—2001 (R2007)	Floor and Trench Drains	Table P2701.1
A112.14.1—03 (Reaffirmed 2008)	Backwater Valves	P3008.2

ASME—continued

A112.18.1—2012/ CSA B125.1—2012	Plumbing Supply Fittings	Table P2701.1, P2708.5, P2722.1, P2722.3, P2727.2, P2902.2, Table P2903.9.4
A112.18.2—2011/ CSA B125.2—2011	Plumbing Waste Fittings	Table P2701.1, P2702.2
A112.18.3—2002 (Reaffirmed 2008)	Performance Requirements for Backflow Protection Devices and Systems in Plumbing Fixture Fittings	P2708.5, P2722.3
A112.18.6/ CSA B125.6—2009	Flexible Water Connectors	P2906.7
A112.19.1—2013/ CSA B45.2—2013	Enameled Cast-iron and Enameled Steel Plumbing Fixtures	Table P2701.1, P2711.1
A112.19.2—2013/ CSA B45.1—2013	Ceramic Plumbing Fixtures	Table P2701.1, P2705.1, P2711.1, P2712.1, P2712.2, P2712.9
A112.19.3—2008/ CSA B45.4—08 (R2013)	Stainless Steel Plumbing Fixtures	Table P2701.1, P2705.1, P2711.1, P2712.1
A112.19.5—2011/ CSA B45.15—2011	Flush Valves and Spuds for Water-closets, Urinals and Tanks	Table P2701.1
A112.19.7/2012 CSA B45.10—2012	Hydromassage Bathtub Systems	Table P2701.1
A112.19.12—2006	Wall-mounted and Pedestal-mounted, Adjustable and Pivoting Lavatory and Sink Carrier Systems	Table P2701.1, P2711.4, P2714.2
A112.19.14—2006 (R2011)	Six-Liter Water Closets Equipped with Dual Flushing Device	P2712.1
A112.19.15—2005	Bathtub/Whirlpool Bathtubs with Pressure-sealed Doors	Table P2701.1, P2713.2
A112.36.2m—1991 (R2008)	Cleanouts	P3005.2.10.2
B1.20.1—1983 (R2006)	Pipe Threads, General-purpose (Inch)	G2414.9, P3003.6.4, P3003.17.1, P3003.9.3
B16.3—2011	Malleable-iron-threaded Fittings, Classes 150 and 300	Table P2906.6
B16.4—2011	Gray-iron-threaded Fittings	Table P2906.6, Table P3002.3
B16.9—2007	Factory-made, Wrought-steel Buttwelding Fittings	Table P2906.6
B16.11—2011	Forged Fittings, Socket-welding and Threaded	Table P2906.6
B16.12—2009	Cast-iron-threaded Drainage Fittings	Table P3002.3
B16.15—2011	Cast-bronze-threaded Fittings	Table P2906.6, Table P3002.3
B16.18—2012	Cast-copper-alloy Solder Joint Pressure Fittings	Table P2906.6, Table P3002.3
B16.22—2001 (R2010)	Wrought-copper and Copper-alloy Solder Joint Pressure Fittings	Table P2906.6, Table P3002.3
B16.23—2002 (R2011)	Cast-copper-alloy Solder Joint Drainage Fittings (DWV)	Table P3002.3
B16.26—2011	Cast-copper-alloy Fittings for Flared Copper Tubes	Table P2906.6, Table P3002.3
B16.28—1994	Wrought-steel Buttwelding Short Radius Elbows and Returns	Table P2906.6
B16.29—2012	Wrought-copper and Wrought-copper-alloy Solder Joint Drainage Fittings (DWV)	Table P3002.3
B16.33—2012	Manually Operated Metallic Gas Valves for Use in Gas Piping Systems up to 125 psig (Sizes $^1/_2$ through 2)	Table G2420.1.1
B16.34—2009	Valves—Flanged, Threaded and Welding End	Table P2903.9.4
B16.44—2002 (Reaffirmed 2007)	Manually Operated Metallic Gas Valves for Use in Above-ground Piping Systems up to 5 psi	Table G2420.1.1
B16.51—2011	Copper and Copper Alloy Press-Connect Pressure Fittings	Table 2906.6
B36.10M—2004	Welded and Seamless Wrought-steel Pipe	G2414.4.2
BPVC—2010/2011 addenda	ASME Boiler and Pressure Vessel Code (2007 Edition)	M2001.1.1, G2452.1
CSD-1—2011	Controls and Safety Devices for Automatically Fired Boilers	M2001.1.1, G2452.1
ASSE 1016/ASME 112.1016/ CSA B125.16—2011	Performance Requirements for Automatic Compensating Valves for Individual Showers and Tub/Shower Combinations	Table P2701.1, P2708.4, P2722.2

ASSE

American Society of Sanitary Engineering
901 Canterbury, Suite A
Westlake, OH 44145

Standard reference number	Title	Referenced in code section number
1001—2008	Performance Requirements for Atmospheric-type Vacuum Breakers	Table P2902.3, P2902.3.2
1002—2008	Performance Requirements for Anti-siphon Fill Valves for Water Closet Flush Tank	Table P2701.1, Table P2902.3, P2902.4.1
1003—2009	Performance Requirements for Water-pressure-reducing Valves for Domestic Water Distribution Systems	P2903.3.1

ASSE—continued

1008—2006	Performance Requirements for Plumbing Aspects of Residential Food Waste Disposer Units	Table P2701.1
1010—2004	Performance Requirements for Water Hammer Arresters	P2903.5
1011—2004	Performance Requirements for Hose Connection Vacuum Breakers	Table P2902.3, P2902.3.2
1012—2009	Performance Requirements for Backflow Preventers with Intermediate Atmospheric Vent	Table P2902.3, P2902.3.3, P2902.5.1, P2902.5.5.3
1013—2009	Performance Requirements for Reduced Pressure Principle Backflow Preventers and Reduced Pressure Principle Fire Protection Backflow Preventers	Table P2902.3, P2902.3.5, P2902.5.1, P2902.5.5.3
1015—2009	Performance Requirements for Double Check Backflow Prevention Assemblies and Double Check Fire Protection Backflow Prevention Assemblies	Table P2902.3, P2902.3.6
ASSE 1016/ASME 112.1016/ CSA B125.16—2011	Performance Requirements for Automatic Compensating Valves for Individual Showers and Tub/Shower Combinations	Table P2701.1, P2708.4, P2722.2
1017—2010	Performance Requirements for Temperature-actuated Mixing Valves for Hot Water Distribution Systems	P2724.1, P2802.1, P2803.2
1018—2010	Performance Requirements for Trap Seal Primer Valves; Potable Water Supplied	P3201.2.1, P3201.2.2
1019—2010	Performance Requirements for Freeze-resistant, Wall Hydrants, Vacuum Breaker, Draining Types	Table P2701.1, P2902.3.2
1020—2004	Performance Requirements for Pressure Vacuum Breaker Assembly	Table P2902.3, P2902.3.4
1023—2010	Performance Requirements for Hot Water Dispensers, Household-storage-type—Electrical	Table P2701.1
1024—2004	Performance Requirements for Dual Check Backflow Preventers, Anti-siphon-type, Residential Applications	Table P2902.3, P2902.3.7
1035—2008	Performance Requirements for Laboratory Faucet Backflow Preventers	Table P2902.3, P2902.3.2
1037—2010	Performance Requirements for Pressurized Flushing Devices (Flushometer) for Plumbing Fixtures	Table P2701.1
1044—2010	Performance Requirements for Trap Seal Primer Devices Drainage Types and Electronic Design Types	P3201.2.3
1047—2009	Performance Requirements for Reduced Pressure Detector Fire Protection Backflow Prevention Assemblies	Table P2902.3, P2902.3.5
1048—2009	Performance Requirements for Double Check Detector Fire Protection Backflow Prevention Assemblies	Table P2902.3, P2902.3.6
1050—2009	Performance Requirements for Stack Air Admittance Valves for Sanitary Drainage Systems	P3114.1
1051—2009	Performance Requirements for Individual and Branch-type Air Admittance Valves for Plumbing Drainage Systems	P3114.1
1052—2004	Performance Requirements for Hose Connection Backflow Preventers	Table P2701.1, Table P2902.3, P2902.3.2
1056—2010	Performance Requirements for Spill-resistant Vacuum Breakers	Table P2902.3, P2902.3.4
1060—2006	Performance Requirements for Outdoor Enclosures for Fluid-conveying Components	P2902.6.1
1061—2010	Performance Requirements for Removable and Nonremovable Push Fit Fittings	Table P2906.6
1062—2006	Performance Requirements for Temperature-actuated, Flow Reduction (TAFR) Valves for Individual Supply Fittings	Table P2701.1, P2724.2
1066—2009	Performance Requirements for Individual Pressure Balancing In-line Valves for Individual Fixture Fittings	P2722.4
1070—2004	Performance Requirements for Water-temperature-limiting Devices	P2713.3, P2721.2, P2724.1
1072—07	Performance Requirements for Barrier-type Floor Drain Trap Seal Protection Devices	P3201.2.4

ASTM

ASTM International
100 Barr Harbor Drive
West Conshohocken, PA 19428

Standard reference number	Title	Referenced in code section number
A36/A36M—08	Specification for Carbon Structural Steel	R606.15, R608.5.2.2
A53/A53M—12	Specification for Pipe, Steel, Black and Hot-dipped, Zinc-coated Welded and Seamless	R407.3, Table M2101.1, G2414.4.2, Table P2906.4, Table P2906.5, Table P3002.1(1)
A74—13A	Specification for Cast-iron Soil Pipe and Fittings	Table P3002.1(1), Table P3002.1(2), Table P3002.2, Table P3002.3, P3005.2.6, Table P3302.1
A106/A106M—11	Specification for Seamless Carbon Steel Pipe for High-temperature Service	Table M2101.1, G2414.4.2
A126—(2009)	Gray Iron Castings for Valves, Flanges and Pipe Fittings	Table P2903.9.4

ASTM—continued

A153/A153M—09	Specification for Zinc Coating (Hot Dip) on Iron and Steel Hardware	R317.3, Table R606.3.4.1, 703.6.3, R905.7.5, R905.8.6
A167—99 (2009)	Specification for Stainless and Heat-resisting Chromium-nickel Steel Plate, Sheet and Strip	Table R606.3.4.1
A240/A240M—13A	Standard Specification for Chromium and Chromium-nickel Stainless Steel Plate, Sheet and Strip for Pressure Vessels and for General Applications	Table R905.10.3(1)
A254—97 (2007)	Specification for Copper Brazed Steel Tubing	Table M2101.1, G2414.5.1
A307—12	Specification for Carbon Steel Bolts and Studs, 60,000 psi Tensile Strength	R608.5.2.2
A312/A312M—13A	Specification for Seamless, Welded and Heavily Cold Worked Austenitic Stainless Steel Pipes	Table P2906.4, Table P2906.5, Table P2906.6, P2905.12.2
A463/A463M—10	Standard Specification for Steel Sheet, Aluminum-coated by the Hot-dip Process	Table R905.10.3(2)
A539—99	Specification for Electric-resistance-welded Coiled Steel Tubing for Gas and Fuel Oil Lines	M2202.1
A615/A615M—12	Specification for Deformed and Plain Billet-steel Bars for Concrete Reinforcement	R402.3.1, R403.1.3.5.1, R404.1.3.3.7.1, R608.5.2.1
A641/A641M—09a	Specification for Zinc-coated (Galvanized) Carbon Steel Wire	Table R606.3.4.1
A653/A653M—11	Specification for Steel Sheet, Zinc-coated (Galvanized) or Zinc-iron Alloy-coated Galvanized) by the Hot-dip Process	R317.3.1, R505.2.2, R603.2.2, Table R606.3.4.1, R608.5.2.3, R804.2.2, R804.2.3, Table R905.10.3(1), Table R905.10.3(2), M1601.1.1
A706/A706M—09B	Specification for Low-alloy Steel Deformed and Plain Bars for Concrete Reinforcement	R402.3.1, R403.1.3.5.1, R404.1.3.3.7.1, R608.5.2.1
A755/A755M—2011	Specification for Steel Sheet, Metallic Coated by the Hot-dip Process and Prepainted by the Coil-coating Process for Exterior Exposed Building Products	Table R905.10.3(2)
A778—01 (2009e1)	Specification for Welded Unannealed Austenitic Stainless Steel Tubular Products	Table P2905.4, Table P2905.5, Table P2905.6
A792/A792M—10	Specification for Steel Sheet, 55% Aluminum-zinc Alloy-coated by the Hot-dip Process	R505.2.2, R603.2.2, R608.5.2.3, R804.2.2, Table 905.10.3(2)
A875/A875M—13	Specification for Steel Sheet, Zinc-5%, Aluminum Alloy-coated by the Hot-dip Process	R608.5.2.3, Table R905.10.3(2)
A888—13A	Specification for Hubless Cast Iron Soil Pipe and Fittings for Sanitary and Storm Drain, Waste and Vent Piping Application	Table P3002.1(1), Table P3002.1(2), Table P3002.2, Table P3002.3, Table P3302.1
A924/A924M—13	Standard Specification for General Requirements for Steel Sheet, Metallic-coated by the Hot-Dip Process	Table R905.10.3(1)
A996/A996M—2009b	Specifications for Rail-steel and Axle-steel Deformed Bars for Concrete Reinforcement	Table R404.1.2(9), R402.3, R403.1.3.5.1, R404.1.3.3.7.1, R608.5.2.1, Table R608.5.4(2)
A1003/A1003M—13A	Standard Specification for Steel Sheet, Carbon, Metallic and Nonmetallic-coated for Cold-formed Framing Members	R505.2.1, R505.2.2, R603.2.1, R603.2.2, R804.2.1
B32—08	Specification for Solder Metal	P3003.6.3
B42—10	Specification for Seamless Copper Pipe, Standard Sizes	Table M2101.1, Table P2906.4, Table P2906.5, Table P3002.1(1)
B43—09	Specification for Seamless Red Brass Pipe, Standard Sizes	Table M2101.1, Table P2906.4, Table P2906.5, Table P3002.1(1)
B75—11	Specification for Seamless Copper Tube	Table M2101.1, Table P2905.4, Table P2906.5, Table P3002.1(1), Table P3002.1(2), Table P3002.2
B88—09	Specification for Seamless Copper Water Tube	Table M2101.1, G2414.5.2, Table, P2906.4, Table P2905.6, Table P3002.1(1), Table P3002.1(2), Table P3002.2
B101—12	Specification for Lead-coated Copper Sheet and Strip for Building Construction	Table R905.2.8.2, Table R905.10.3(1)
B135—10	Specification for Seamless Brass Tube	Table M2101.1
B209—10	Specification for Aluminum and Aluminum-alloy Sheet and Plate	Table 905.10.3(1)
B227—10	Specification for Hard-drawn Copper-clad Steel Wire	R606.15
B251—10	Specification for General Requirements for Wrought Seamless Copper and Copper-alloy Tube	Table M2101.1, Table P2906.4, Table P2906.5, Table P3002.1(1), Table P3002.1(2), Table P3002.2
B302—12	Specification for Threadless Copper Pipe, Standard Sizes	Table M2101.1, Table P2906.4, Table P2906.5, Table P3002.1(1)
B306—09	Specification for Copper Drainage Tube (DWV)	Table M2101.1, Table P3002.1(1), Table P3002.1(2)

ASTM—continued

B370—12	Specification for Copper Sheet and Strip for Building Construction	Table R905.2.8.2, Table R905.10.3(1), Table P2701.1
B447—12a	Specification for Welded Copper Tube	Table P2906.4, Table P2906.5
B695—04 (2009)	Standard Specification for Coatings of Zinc Mechanically Deposited on Iron and Steel	R317.3.1, R317.3.3
B813—10	Specification for Liquid and Paste Fluxes for Soldering Applications of Copper and Copper Alloy Tube	Table M2101.1, M2103.3, P2906.14, P3003.6.3
B828—02 (2010)	Practice for Making Capillary Joints by Soldering of Copper and Copper Alloy Tube and Fittings	M2103.3, P2906.14, P3003.6.3
C4—2009	Specification for Clay Drain Tile and Perforated Clay Drain Tile	Table P3302.1
C5—10	Specification for Quicklime for Structural Purposes	R702.2.1
C14—11	Specification for Non-reinforced Concrete Sewer, Storm Drain and Culvert Pipe	Table P3002.2
C22/C22M—00 (2010)	Specification for Gypsum	R702.2.1, R702.3.1
C27—98 (2008)	Specification for Standard Classification of Fireclay and High-alumina Refractory Brick	R1001.5
C28/C28M—10	Specification for Gypsum Plasters	R702.2.1
C33/C33M—13	Specification for Concrete Aggregates	R403.4.1
C34—12	Specification for Structural Clay Load-bearing Wall Tile	Table R301.2(1), R606.2.2
C35/C35M—1995 (2009)	Specification for Inorganic Aggregates for Use in Gypsum Plaster	R702.2.1
C55—2011	Specification for Concrete Building Brick	R202, Table R301.2(1), R606.2.1
C56—12	Standard Specification for Structural Clay Nonloadbearing Tile	R606.2.2
C59/C59M—00 (2011)	Specification for Gypsum Casting Plaster and Molding Plaster	R702.2.1
C61/C61M—00 (2011)	Specification for Gypsum Keene's Cement	R702.2.1
C62—13	Standard Specification for Building Brick (Solid Masonry Units Made from Clay or Shale)	R202, Table R301.2(1), R606.2.2
C73—10	Specification for Calcium Silicate Face Brick (Sand Lime Brick)	R202, Table R301.2(1), R606.2.1
C76—13A	Specification for Reinforced Concrete Culvert, Storm Drain and Sewer Pipe	Table P3002.2
C90—13	Specification for Load-bearing Concrete Masonry Units	Table R301.2(1), 606.2.1
C91—12	Specification for Masonry Cement	R702.2.2
C94/C94M—13	Standard Specification for Ready-mixed Concrete	R404.1.3.3.2, R608.5.1.2
C126—13	Standard Specification for Ceramic Glazed Structural Clay Facing Tile, Facing Brick, and Solid Masonry Units	R606.2.2
C129—11	Specification for Nonload-bearing Concrete Masonry Units	Table R301.2(1)
C143/C143M—12	Test Method for Slump of Hydraulic Cement Concrete	R404.1.3.3.4, R608.5.1.9
C145—85	Specification for Solid Load-bearing Concrete Masonry Units	R202, Table R301.2(1)
C150—12	Specification for Portland Cement	R608.5.1.1, R702.2.2
C199—84 (2011)	Test Method for Pier Test for Refractory Mortar	R1001.5, R1001.8, R1003.12
C203—05a (2012)	Standard Test Methods for Breaking Load and Flexural Properties of Block-type Thermal Insulation	Table R610.3.1
C207—06 (2011)	Specification for Hydrated Lime for Masonry Purposes	Table R606.2.7
C208—12	Specification for Cellulosic Fiber Insulating Board	R602.1.10, Table R602.3(1), Table R906.2
C212-10	Standard Specification for Structural Clay Facing Tile	R602.2.2
C216—13	Specification for Facing Brick (Solid Masonry Units Made from Clay or Shale)	R202, Table R301.2(1), R606.2.2
C270—12a	Specification for Mortar for Unit Masonry	R606.2.7, R606.2.10
C272/C272M—12	Standard Test Method for Water Absorption of Core Materials for Sandwich Constructions	Table R610.3.1
C273/C273M—11	Standard Test Method for Shear Properties of Sandwich Core Materials	Table R610.3.1
C315—07 (2011)	Specification for Clay Flue Liners and Chimney Pots	R1001.8, R1003.11.1, Table R1003.14(1), G2425.12
C406/C406M—2010	Specifications for Roofing Slate	R905.6.4
C411—11	Test Method for Hot-surface Performance of High-temperature Thermal Insulation	M1601.3
C425—04 (2009)	Specification for Compression Joints for Vitrified Clay Pipe and Fittings	Table P3002.2, P3003.10, P3003.13
C443—12	Specification for Joints for Concrete Pipe and Manholes, Using Rubber Gaskets	P3003.5, P3003.13
C475/C475—12	Specification for Joint Compound and Joint Tape for Finishing Gypsum Wallboard	R702.3.1
C476—10	Specification for Grout for Masonry	R606.2.11
C503—10	Standard Specification for Marble Dimension Stone	R606.2.4
C514—04 (2009)el	Specification for Nails for the Application of Gypsum Wallboard	R702.3.1
C552—12b	Standard Specification for Cellular Glass Thermal Insulation	Table R906.2
C557—03 (2009)e01	Specification for Adhesives for Fastening Gypsum Wallboard to Wood Framing	R702.3.1
C564—12	Specification for Rubber Gaskets for Cast Iron Soil Pipe and Fittings	P3003.4.2, P3003.4.3, P3003.13

ASTM—continued

C568—10	Standard Specification for Limestone Dimension Stone	R606.2.4
C578—12b	Specification for Rigid, Cellular Polystyrene Thermal Insulation	R316.8, R403.3, R610.3.1, R703.11.2.1, Table R703.15.1, Table R703.15.2, Table R703.16.1, Table R703.16.2, Table R906.2
C587—04(2009)	Specification for Gypsum Veneer Plaster	R702.2.1
C595/C595M—13	Specification for Blended Hydraulic Cements	R608.5.1.1, R702.2.2
C615/C615M—11	Standard Specification for Granite Dimension Stone	R606.2.4
C616/C615M—10	Standard Specification for Quartz Dimension Stone	R606.2.4
C629—10	Standard Specification for Slate Dimension Stone	R606.2.4
C631—09	Specification for Bonding Compounds for Interior Gypsum Plastering	R702.2.1
C645—13	Specification for Nonstructural Steel Framing Members	R702.3.3
C652—13	Specification for Hollow Brick (Hollow Masonry Units Made from Clay or Shale)	R202, Table R301.2(1), R606.2.2
C685/C685M—11	Specification for Concrete Made by Volumetric Batching and Continuous Mixing	R404.1.3.3.2, R608.5.1.2
C700—13	Specification for Vitrified Clay Pipe, Extra Strength, Standard Strength and Perforated	Table P3002.2, Table P3002.3, Table P3302.1
C728—05 (2013)	Standard Specification for Perlite Thermal Insulation Board	Table R906.2
C744—11	Standard Specification for Prefaced Concrete and Calcium Silicate Masonry Units	R606.2.1
C836/C836M—12	Specification for High Solids Content, Cold Liquid-applied Elastomeric Waterproofing Membrane for Use with Separate Wearing Course	R905.15.2
C843—99 (2012)	Specification for Application of Gypsum Veneer Plaster	R702.2.1
C844—04 (2010)	Specification for Application of Gypsum Base to Receive Gypsum Veneer Plaster	R702.2.1
C847—12	Specification for Metal Lath	R702.2.1, R702.2.2
C887—05 (2010)	Specification for Packaged, Dry, Combined Materials for Surface Bonding Mortar	R406.1, R606.2.8
C897—05 (2009)	Specification for Aggregate for Job-mixed Portland Cement-based Plasters	R702.2.2
C920—11	Standard Specification for Elastomeric Joint Sealants	R406.4.1
C926—13	Specification for Application of Portland Cement-based Plaster	R702.2.2, R702.2.2.1, R703.7, R703.7.2.1, R703.7.4
C933—13	Specification for Welded Wire Lath	R702.2.1, R702.2.2
C946—10	Standard Practice for Construction of Dry-Stacked, Surface-Bonded Walls	R606.2.8
C954—11	Specification for Steel Drill Screws for the Application of Gypsum Panel Products or Metal Plaster Bases to Steel Studs from 0.033 in (0.84 mm) or to 0.112 in. (2.84 mm) in Thickness	R505.2.5, R603.2.5, R702.3.5.1, R804.2.5
C955—11C	Specification for Load-bearing (Transverse and Axial) Steel Studs, Runners (Tracks), and Bracing or Bridging for Screw Application of Gypsum Panel Products and Metal Plaster Bases	R702.3.3
C957—10	Specification for High-solids Content, Cold Liquid-applied Elastomeric Waterproofing Membrane for Use with Integral Wearing Surface	R905.15.2
C1002—07	Specification for Steel Drill Screws for the Application of Gypsum Panel Products or Metal Plaster Bases	R702.3.1, R702.3.5.1
C1029—13	Specification for Spray-applied Rigid Cellular Polyurethane Thermal Insulation	R905.14.2
C1032—06 (2011)	Specification for Woven Wire Plaster Base	R702.2.1, R702.2.2
C1047—10a	Specification for Accessories for Gypsum Wallboard and Gypsum Veneer Base	R702.2.1, R702.2.2, R702.3.1
C1063—12D	Specification for Installation of Lathing and Furring to Receive Interior and Exterior Portland Cement-based Plaster	R702.2.2, R703.7
C1088—13	Standard Specification for Thin Veneer Brick Units Made from Clay or Shale	R606.2.2
C1107/C1107M—13	Standard Specification for Packaged Dry, Hydraulic-cement Grout (Nonshrink)	R402.3.1
C1116/C116M—10A	Standard Specification for Fiber-reinforced Concrete and Shotcrete	R402.3.1
C1157/C1157M—11	Standard Performance Specification for Hydraulic Cement	R608.5.1.1
C1167—11	Specification for Clay Roof Tiles	R905.3.4
C1173—10e1	Specification for Flexible Transition Couplings for Underground Piping Systems	P3003.3.1, P3003.5, P3003.6.1, P3003.10.1, P3003.12.1, P3003.12.2, P3003.13
C1177/C1177M—08	Specification for Glass Mat Gypsum Substrate for Use as Sheathing	R702.3.1, Table 906.2
C1178/C1178M—11	Specification for Glass Mat Water-resistant Gypsum Backing Panel	R702.3.1, R702.3.7, Table R702.4.2
C1186—08 (2012)	Specification for Flat Nonasbestos Fiber Cement Sheets	R703.10.1, R703.10.2
C1261—10	Specification for Firebox Brick for Residential Fireplaces	R1001.5, R1001.8
C1277—12	Specification for Shielded Couplings Joining Hubless Cast Iron Soil Pipe and Fittings	P3003.4.3
C1278/ C1278M—07a (2011)	Specification for Fiber-reinforced Gypsum Panels	R702.3.1, R702.3.7, Table R702.4.2, Table R906.2
C1283—11	Practice for Installing Clay Flue Lining	R1003.9.1, R1003.12

ASTM—continued

C1288—99 (2010)	Standard Specification for Discrete Nonasbestos Fiber-cement Interior Substrate Sheets	Table R503.2.1.1(1), Table R503.2.1.1(2), Table 602.3(2), Table R702.4.2
C1289—13e1	Standard Specification for Faced Rigid Cellular Polyisocyanurate Thermal Insulation Board	R316.8, R703.11.2.1, Table R703.15.1, Table R703.15.2, Table R703.16.1, Table R703.16.2, Table R906.2
C1325—08b	Standard Specification for Nonasbestos Fiber-mat Reinforced Cement Interior Substrate Sheets Backer Units	Table R702.4.2
C1328/C1328M—12	Specification for Plastic (Stucco) Cement	R702.2.2
C1363—11	The Standard Test Method for Thermal Performance of Building Materials and Envelope Assemblies by Means of a Hot Box Apparatus	N1101.10.4.1
C1364—10B	Standard Specification for Architectural Cast Stone	R606.2.5
C1386—07	Standard Specification for Precast Autoclaved Aerated Concrete (AAC) Wall Construction Units	R606.2.3
C1396/C1396M—2013	Specification for Gypsum Board	Table R602.3(1), R702.2.1, R702.2.2, R702.3.1, R702.3.7
C1405—12	Standard Specification for Glazed Brick (Single Fired, Brick Units)	R606.2.2
C1440—08	Specification for Thermoplastic Elastomeric (TPE) Gasket Materials for Drain, Waste and Vent (DWV), Sewer, Sanitary and Storm Plumbing Systems	P3003.13
C1460—08	Specification for Shielded Transition Couplings for Use with Dissimilar DWV Pipe and Fittings Above Ground	P3003.18
C1461—08	Specification for Mechanical Couplings Using Thermoplastic Elastomeric (TPE) Gaskets for Joining Drain, Waste and Vent (DWV) Sewer, Sanitary and Storm Plumbing Systems for Above and Below Ground Use	P3003.13
C1492—03 (2009)	Specification for Concrete Roof Tile	R905.3.5
C1513—2013	Standard Specification for Steel Tapping Screws for Cold-formed Steel Framing Connections	R505.2.5, R603.2.5, R702.3.8.1, Table R703.3(2), R804.2.5 Table R703.16.1, Table R703.16.2
C1540—11	Specification for Heavy Duty Shielded Couplings Joining Hubless Cast-iron Soil Pipe and Fittings	P3003.4.3
C1634—11	Standard Specification for Concrete Facing Brick	R606.2.1
C1658/C1658M—12	Standard Specification for Glass Mat Gypsum Panels	R702.3.1
C1668—12	Standard Specification for Externally Applied Reflective Insulation Systems on Rigid Duct in Heating, Ventilation, and Air Conditioning (HVAC) Systems	M1601.3
D41—05	Specification for Asphalt Primer Used in Roofing, Dampproofing and Waterproofing	Table R905.9.2, Table R905.11.2
D43—00 (2006)	Specification for Coal Tar Primer Used in Roofing, Dampproofing and Waterproofing	Table R905.9.2
D226/D226M—09	Specification for Asphalt-saturated (Organic Felt) Used in Roofing and Waterproofing	R703.2, R905.1.1, Table R905.1.1(1), R905.8.4, Table R905.9.2
D227/D227M—03 (2011)e1	Specification for Coal Tar Saturated (Organic Felt) Used in Roofing and Waterproofing	Table R905.9.2
D312—00 (2006)	Specification for Asphalt Used in Roofing	Table R905.9.2
D422—63 (2007)	Test Method for Particle-size Analysis of Soils	R403.1.8.1
D449—03 (2008)	Specification for Asphalt Used in Dampproofing and Waterproofing	R406.2
D450—07	Specification for Coal-tar Pitch Used in Roofing, Dampproofing and Waterproofing	Table R905.9.2
D1227—95 (2007)	Specification for Emulsified Asphalt Used as a Protective Coating for Roofing	Table R905.9.2, Table R905.11.2, R905.15.2
D1248—12	Specification for Polyethylene Plastics Extrusion Materials for Wire and Cable	M1601.1.2
D1527—99 (2005)	Specification for Acrylonitrile-butadiene-styrene (ABS) Plastic Pipe, Schedules 40 and 80	Table P2906.4
D1621—10	Standard Test Method for Compressive Properties of Rigid Cellular Plastics	Table R610.3.1
D1622—08	Standard Test Method for Apparent Density of Rigid Cellular Plastics	Table R610.3.1
D1623—09	Standard Test Method for Tensile and Tensile Adhesion Properties of Rigid Cellular Plastics	Table R610.3.1
D1693—2013	Test Method for Environmental Stress-cracking of Ethylene Plastics	Table M2101.1
D1784—11	Standard Specification for Rigid Poly (Vinyl Chloride) (PVC) Compounds and Chlorinated Poly (Vinyl Chloride) (CPVC) Compounds	M1601.1.2
D1785—12	Specification for Poly (Vinyl Chloride) (PVC) Plastic Pipe, Schedules 40, 80 and 120	Table P2906.4, Table AG101.1
D1863/D1863M—05 (2011)e1	Specification for Mineral Aggregate Used in Built-up Roofs	Table R905.9.2
D1869—95 (2010)	Specification for Rubber Rings for Asbestos-cement Pipe	P2906.17, P3003.13
D1970/D1970M—2013	Specification for Self-adhering Polymer Modified Bitumen Sheet Materials Used as Steep Roofing Underlayment for Ice Dam Protection	R905.1.1, R905.2.8.2, R905.16.4
D2104—03	Specification for Polyethylene (PE) Plastic Pipe, Schedule 40	Table P2905.4

ASTM—continued

D2126—09	Standard Test Method for Response of Rigid Cellular Plastics to Thermal and Humid Aging	Table R610.3.1
D2178—04	Specification for Asphalt Glass Felt Used in Roofing and Waterproofing	Table R905.9.2
D2235—04 (2011)	Specification for Solvent Cement for Acrylonitrile-butadiene-styrene (ABS) Plastic Pipe and Fittings	P2906.9.1.1, P3003.3.2
D2239—12A	Specification for Polyethylene (PE) Plastic Pipe (SIDR-PR) Based on Controlled Inside Diameter	Table P2906.4, Table AG101.1
D2241—09	Specification for Poly (Vinyl Chloride) (PVC) Pressure-rated Pipe (SDR-Series)	Table P2905.4, Table AG101.1
D2282—99 (2005)	Specification for Acrylonitrile-butadiene-styrene (ABS) Plastic Pipe (SDR-PR)	Table P2905.4
D2412—11	Test Method for Determination of External Loading Characteristics of Plastic Pipe by Parallel-plate Loading	M1601.1.2
D2447—03	Specification for Polyethylene (PE) Plastic Pipe Schedules 40 and 80, Based on Outside Diameter	Table M2101.1
D2464—06	Specification for Threaded Poly (Vinyl Chloride) (PVC) Plastic Pipe Fittings, Schedule 80	Table P2906.6
D2466—06	Specification for Poly (Vinyl Chloride) (PVC) Plastic Pipe Fittings, Schedule 40	Table P2906.6
D2467—06	Specification for Poly (Vinyl Chloride) (PVC) Plastic Pipe Fittings, Schedule 80	Table P2906.6
D2468—96a	Specification for Acrylonitrile-butadiene-styrene (ABS) Plastic Pipe Fittings, Schedule 40	Table P2906.6
D2513—2013e1	Specification for Thermoplastic Gas Pressure Pipe, Tubing and Fittings	Table M2101.1, G2414.6, G2414.6.1, G2414.11, G2415.17.2
D2559—12A	Standard Specification for Adhesives for Bonded Structural Wood Products for Use Under Exterior (West Use) Exposure Conditions	R610.3.3
D2564—12	Specification for Solvent Cements for Poly (Vinyl Chloride) (PVC) Plastic Piping Systems	P2906.9.1.4, P3003.8.2, P3003.9.2
D2609—02 (2008)	Specification for Plastic Insert Fittings for Polyethylene (PE) Plastic Pipe	Table P2906.6
D2626/ D2626M—04 (2012)e1	Specification for Asphalt-saturated and Coated Organic Felt Base Sheet Used in Roofing	R905.3.3, Table R905.9.2
D2657—07	Standard Practice for Heat Fusion-joining of Polyolefin Pipe Fittings	M2105.11.1, P2906.3.1, P3003.12.1
D2661—11	Specification for Acrylonitrile-butadiene-styrene (ABS) Schedule 40 Plastic Drain, Waste, and Vent Pipe and Fittings	Table P3002.1(1), Table P3002.1(2), Table P3002.2, Table P3002.3, P3003.3.2
D2665—12	Specification for Poly (Vinyl Chloride) (PVC) Plastic Drain, Waste and Vent Pipe and Fittings	Table P3002.1(1), Table P3002.1(2), Table P3002.2, Table P3002.3, Table AG101.1
D2672—96a (2009)	Specification for Joints for IPS PVC Pipe Using Solvent Cement	Table P2906.4
D2683—2010e1	Specification for Socket-type Polyethylene Fittings for Outside Diameter-controlled Polyethylene Pipe and Tubing	Table M2105.5, M2105.11.1, P3010.5
D2729—11	Specification for Poly (Vinyl Chloride) (PVC) Sewer Pipe and Fittings	P3009.11, Table P3302.1, Table AG101.1
D2737—2012A	Specification for Polyethylene (PE) Plastic Tubing	Table P2906.4, Table AG101.1
D2751—05	Specification for Acrylonitrile-butadiene-styrene (ABS) Sewer Pipe and Fittings	Table P3002.2, Table P3002.3
D2822/ D2822M—05 (2011)e1	Specification for Asphalt Roof Cement, Asbestos Containing	Table R905.9.2
D2823/ D2823—05 (2011)e1	Specification for Asphalt Roof Coatings, Asbestos Containing	Table R905.9.2
D2824—06 (2012)e1	Specification for Aluminum-pigmented Asphalt Roof Coatings, Nonfibered, Asbestos Fibered and Fibered without Asbestos	Table R905.9.2, Table R905.11.2
D2846/D2846M—09BE1	Specification for Chlorinated Poly (Vinyl Chloride) (CPVC) Plastic Hot- and Cold-water Distribution Systems	Table M2101.1, Table P2906.4, Table P2906.5, Table P2906.6, P2906.9.1.2, P2906.9.1.3, Table AG101.1
D2855—96 (2010)	Standard Practice for Making Solvent-cemented Joints with Poly (Vinyl Chloride) (PVC) Pipe and Fittings	P3003.9.2
D2898—10	Test Methods for Accelerated Weathering of Fire-retardant-treated Wood for Fire Testing	R802.1.5.4, R802.1.5.8
D2949—10	Specification for 3.25-in. Outside Diameter Poly (Vinyl Chloride) (PVC) Plastic Drain, Waste and Vent Pipe and Fittings	Table P3002.1(1), Table P3002.1(2), Table P3002.2, Table P3002.3, Table AG101.1
D3019—08	Specification for Lap Cement Used with Asphalt Roll Roofing, Nonfibered, Asbestos Fibered and Nonasbestos Fibered	Table R905.9.2, Table R905.11.2

ASTM—continued

D3034—08	Specification for Type PSM Poly (Vinyl Chloride) (PVC) Sewer Pipe and Fittings.	Table P3002.2, Table P3002.3, Table P3202.1, Table AG101.1
D3035—2012e1	Specification for Polyethylene (PE) Plastic Pipe (DR-PR) Based On Controlled Outside Diameter	Table M2105.4, Table AG101.1
D3161/D3161M—2013	Test Method for Wind Resistance of Asphalt Shingles (Fan Induced Method)	R905.2.4.1, Table R905.2.4.1, R905.16.7
D3201—2013	Test Method for Hygroscopic Properties of Fire-retardant Wood and Wood-base Products	R802.1.5.9
D3212—07	Specification for Joints for Drain and Sewer Plastic Pipes Using Flexible Elastomeric Seals	P3003.3.1, P3003.9.1, P3003.12.2
D3261—12	Specification for Butt Heat Fusion Polyethylene (PE) Plastic Fittings for Polyethylene (PE) Plastic Pipe and Tubing	Table M2105.5, M2105.11.1
D3309—96a (2002)	Specification for Polybutylene (PB) Plastic Hot- and Cold-water Distribution System	Table M2101.1
D3311—11	Specification for Drain, Waste and Vent (DWV) Plastic Fittings Patterns	P3002.3
D3350—2012e1	Specification for Polyethylene Plastic Pipe and Fitting Materials	Table M2101.1
D3462/D3462M—10A	Specification for Asphalt Shingles Made From Glass Felt and Surfaced with Mineral Granules	R905.2.4
D3468—99 (2006)e01	Specification for Liquid-applied Neoprene and Chlorosulfanated Polyethylene Used in Roofing and Waterproofing	R905.15.2
D3679—11	Specification for Rigid Poly (Vinyl Chloride) (PVC) Siding	R703.11
D3737—2012	Practice for Establishing Allowable Properties for Structural Glued Laminated Timber (Glulam)	R502.1.3, R602.1.3, R802.1.2
D3747—79 (2007)	Specification for Emulsified Asphalt Adhesive for Adhering Roof Insulation	Table R905.9.2, Table R905.11.2
D3909/D3909M—97b (2012)e1	Specification for Asphalt Roll Roofing (Glass Felt) Surfaced with Mineral Granules	R905.2.8.2, R905.5.4, Table R905.9.2
D4022/D4022M—2007 (2012)e1	Specification for Coal Tar Roof Cement, Asbestos Containing	Table R905.9.2
D4068—09	Specification for Chlorinated Polyethylene (CPE) Sheeting for Concealed Water Containment Membrane	P2709.2, P2709.2.2
D4318—10	Test Methods for Liquid Limit, Plastic Limit and Plasticity Index of Soils	R403.1.8.1
D4434/D4434M—12	Specification for Poly (Vinyl Chloride) Sheet Roofing	R905.13.2
D4479/D4479M—07 (2012)e1	Specification for Asphalt Roof Coatings-asbestos-free	Table R905.9.2
D4551—12	Specification for Poly (Vinyl) Chloride (PVC) Plastic Flexible Concealed Water-containment Membrane	P2709.2, P2709.2.1
D4586/D4586M—07 (2012)e1	Specification for Asphalt Roof Cement-asbestos-free	Table R905.9.2
D4601/D4601M—04 (2012)e1	Specification for Asphalt-coated Glass Fiber Base Sheet Used in Roofing	Table R905.9.2
D4637/D4637M—2013	Specification for EPDM Sheet Used in Single-ply Roof Membrane	R905.12.2
D4829—11	Test Method for Expansion Index of Soils	R403.1.8.1
D4869/D4869M—05 (2011)e01	Specification for Asphalt-saturated (Organic Felt) Underlayment Used in Steep Slope Roofing	R905.1.1, Table R905.1.1(1), R905.16.3, R905.16.4.2
D4897/ D4897M—01 (2009)	Specification for Asphalt Coated Glass-fiber Venting Base Sheet Used in Roofing	Table R905.9.2
D4990—97a (2005)e01	Specification for Coal Tar Glass Felt Used in Roofing and Waterproofing	Table R905.9.2
D5019—07a	Specification for Reinforced Nonvulcanized Polymeric Sheet Used in Roofing Membrane	R905.12.2
D5055—2013	Specification for Establishing and Monitoring Structural Capacities of Prefabricated Wood I-joists	R502.1.2
D5456—2013	Standard Specification for Evaluation of Structural Composite Lumber Products	R502.1.5, R602.1.5, R802.1.7
D5516—09	Test Method for Evaluating the Flexural Properties of Fire-retardant-treated Softwood Plywood Exposed to the Elevated Temperatures	R802.1.5.7
D5643/D5643M—06 (2012)e1	Specification for Coal Tar Roof Cement Asbestos-free	Table R905.9.2
D5664—10	Test Methods For Evaluating the Effects of Fire-retardant Treatments and Elevated Temperatures on Strength Properties of Fire-retardant-treated Lumber	R802.1.5.7
D5665—99a (2006)	Specification for Thermoplastic Fabrics Used in Cold-applied Roofing and Waterproofing	Table R905.9.2
D5726—98 (2005)	Specification for Thermoplastic Fabrics Used in Hot-applied Roofing and Waterproofing	Table R905.9.2

ASTM—continued

D6083—05e01	Specification for Liquid-applied Acrylic Coating Used in Roofing	Table R905.9.2, Table R905.11.2, Table R905.14.3, R905.15.2
D6162—2000a (2008)	Specification for Styrene Butadiene Styrene (SBS) Modified Bituminous Sheet Materials Using a Combination of Polyester and Glass Fiber Reinforcements	Table R905.11.2
D6163—00 (2008)	Specification for Styrene Butadiene Styrene (SBS) Modified Bituminous Sheet Materials Using Glass Fiber Reinforcements	Table R905.11.2
D6164/D6164M—11	Specification for Styrene Butadiene Styrene (SBS) Modified Bituminous Sheet Materials Using Polyester Reinforcements	Table R905.11.2
D6222/D6222M—11	Specification for Atactic Polypropylene (APP) Modified Bituminous Sheet Materials Using Polyester Reinforcements	Table R905.11.2
D6223/D6223M—02 (2011)e1	Specification for Atactic Polypropylene (APP) Modified Bituminous Sheet Materials Using a Combination of Polyester and Glass Fiber Reinforcement	Table R905.11.2
D6298—05e1	Specification for Fiberglass-reinforced Styrene Butadiene Styrene (SBS) Modified Bituminous Sheets with a Factory Applied Metal Surface	Table R905.11.2
D6305—08	Practice for Calculating Bending Strength Design Adjustment Factors for Fire-retardant-treated Plywood Roof Sheathing	R802.1.5.6
D6380—03 (2009)	Standard Specification for Asphalt Roll Roofing (Organic Felt)	Table R905.1.1(1), R905.2.8.2, R905.5.4
D6694—08	Standard Specification for Liquid-applied Silicone Coating Used in Spray Polyurethane Foam Roofing Systems	Table R905.14.3, R905.15.2
D6754/D6745M—10	Standard Specification for Ketone-ethylene-ester-based Sheet Roofing	R905.13.2
D6757—2013	Standard Specification for Inorganic Underlayment for Use with Steep Slope Roofing Products	Table R905.1.1(1), R905.1.1, R905.16.3, R905.16.4.2
D6841—08	Standard Practice for Calculating Design Value Treatment Adjustment Factors for Fire-retardant-treated Lumber	R802.1.5.7
D6878/D6878—11a	Standard Specification for Thermoplastic-polyolefin-based Sheet Roofing	R905.13.2
D6947—07	Standard Specification for Liquid Applied Moisture Cured Polyurethane Coating Used in Spray Polyurethane Foam Roofing System	Table R905.14.3, R905.15.2
D7032—10a	Standard Specification for Establishing Performance Ratings for Wood-plastic Composite Deck Boards and Guardrail Systems (Guards or Handrails)	R507.3, R507.3.1, 507.3.4, 507.3.4
D7158—D7158M—2011	Standard Test Method for Wind Resistance of Sealed Asphalt Shingles (Uplift Force/Uplift Resistance Method)	R905.2.4.1, Table R905.2.4.1
D7254—07	Standard Specification for Polypropylene (PP) siding	Table R703.3(1), R703.14
D7425/D7425M—11	Standard Specification for Spray Polyurethane Foam Used for Roofing Application	R905.14.2
D7672—2012	Standard Specification for Evaluating Structural Capacities of Rim Board Products and Assemblies	R502.1.7, R602.1.7, R802.1.7
D7793—13	Standard Specification for Insulated Vinyl Siding	R703.13, Table R703.3(1)
E84—2013a	Test Method for Surface Burning Characteristics of Building Materials	R202, R302.9.3, R302.9.4, R302.10.1, R302.10.2, R316.3, R316.5.9, R316.5.11, R507.3.2, R802.1.5, M1601.3, M1601.5.2
E96/E96M—2013	Test Method for Water Vapor Transmission of Materials	R202, Table R610.3.1, M1411.6 M1601.4.6
E108—2011	Test Methods for Fire Tests of Roof Coverings	R302.2.2, R902.1
E119—2012a	Test Methods for Fire Tests of Building Construction and Materials	Table R302.1(1), Table R302.1(2), R302.2, R302.2.2, R302.3, R302.4.1, R302.11.1
E136—2012	Test Method for Behavior of Materials in a Vertical Tube Furnace at 750°C	R202, R302.11
E283—04	Test Method for Determining the Rate of Air Leakage Through Exterior Windows, Curtain Walls and Doors Under Specified Pressure Differences Across the Specimen	N1102.4.5
E330—02	Test Method for Structural Performance of Exterior Windows, Curtain Walls and Doors by Uniform Static Air Pressure Difference	R609.4, R609.5, R703.1.2
E331—00 (2009)	Test Method for Water Penetration of Exterior Windows, Skylights, Doors and Curtain Walls by Uniform Static Air Pressure Difference	R703.1.1
E779—10	Standard Test Method for Determining Air Leakage Rate by Fan Pressurization	N1102.4.1.2
E814—2013	Test Method for Fire Tests of Through-penetration Firestops	R302.4.1.2
E970—2010	Test Method for Critical Radiant Flux of Exposed Attic Floor Insulation Using a Radiant Heat Energy Source	R302.10.5
E1509—12	Standard Specification for Room Heaters, Pellet Fuel-burning Type	M1410.1
E1602—03 (2010)e1	Guide for Construction of Solid Fuel Burning Masonry Heaters	R1002.2
E1827—11	Standard Test Methods for Determining Airtightness of Building Using an Orifice Blower Door	N1102.4.1.2
E1886—05	Test Method for Performance of Exterior Windows, Curtain Walls, Doors and Storm Shutters Impacted by Missile(s) and Exposed to Cyclic Pressure Differentials	R301.2.1.2, R609.6.1

ASTM—continued

E1996—2012a   Standard Specification for Performance of Exterior Windows, Curtain
Walls, Doors and Impact Protective Systems Impacted by
Windborne Debris in Hurricanes R301.2.1.2, R301.2.1.2.1, R609.6.1

E2178—2013   Standard Test Method for Air Permeance of Building Materials. R202

E2231—09   Standard Practice for Specimen Preparation and Mounting of Pipe and
Duct Insulation Materials to Assess Surface Burning Characteristics M1601.3

E2273—03 (2011)   Standard Test Method for Determining the Drainage Efficiency of Exterior
Insulation and Finish Systems (EIFS) Clad Wall Assemblies R703.9.2

E2568—09e1   Standard Specification for PB Exterior Insulation and Finish Systems R703.9.1, R703.9.2

E2570—07   Standard Test Methods for Evaluating Water-resistive Barrier (WRB) Coatings
Used Under Exterior Insulation and Finish Systems (EIFS) or EIFS with Drainage. R703.9.2

E2634—11   Standard Specification for Flat Wall Insulating Concrete Form (ICF) Systems R404.1.3.3.6.1, R608.4.4

F405—05   Specification for Corrugated Polyethylene (PE) Pipe and Fittings. . . . . Table P3009.11, Table P3302.1,
Table AG101.1

F409—12   Specification for Thermoplastic Accessible and Replaceable Plastic Tube and
Tubular Fittings. . . . . .Table P2701.1, P2702.2, P2702.3

F437—09   Specification for Threaded Chlorinated Poly (Vinyl Chloride)
(CPVC) Plastic Pipe Fittings, Schedule 80. . . . . Table P2906.6

F438—09   Specification for Socket-type Chlorinated Poly (Vinyl Chloride) (CPVC)
Plastic Pipe Fittings, Schedule 40. . . . . Table P2906.6

F439—12   Specification for Chlorinated Poly (Vinyl Chloride) (CPVC)
Plastic Pipe Fittings, Schedule 80. . . . . Table P2906.6

F441/F441M—13   Specification for Chlorinated Poly (Vinyl Chloride) (CPVC)
Plastic Pipe, Schedules 40 and 80. Table P2906.4, Table P2906.5, Table AG101.1

F442/F442M—13   Specification for Chlorinated Poly (Vinyl Chloride)
(CPVC) Plastic Pipe (SDR-PR) . . . . Table P2906.4, Table P2906.5, Table AG101.1

F477—10   Specification for Elastomeric Seals (Gaskets) for Joining Plastic Pipe P2906.17, P3003.13

F493—10   Specification for Solvent Cements for Chlorinated Poly (Vinyl Chloride)
(CPVC) Plastic Pipe and Fittings P2906.9.1.2, P2906.9.1.3

F628—08   Specification for Acrylonitrile-butadiene-styrene (ABS) Schedule 40 Plastic
Drain, Waste and Vent Pipe with a Cellular Core . . . . Table P3002.1(1), Table P3002.1(2),
Table P3002.2, Table P3002.3, P3003.3.2, Table AG101.1

F656—10   Specification for Primers for Use in Solvent Cement Joints of Poly (Vinyl Chloride)
(PVC) Plastic Pipe and Fittings P2906.9.1.4, P3003.9.2

F714—13   Specification for Polyethylene (PE) Plastic Pipe (SDR-PR)
Based on Outside Diameter. . . . . Table P3002.2, P3010.4

F876—13   Specification for Cross-linked Polyethylene (PEX) Tubing. . . . . Table M2101.1, Table P2906.4,
Table P2906.5, Table AG101.1

F877—11A   Specification for Cross-linked Polyethylene (PEX) Plastic Hot- and
Cold-water Distribution Systems . . . . Table M2101.1,
Table P2906.4, Table P2906.5, Table P2906.6

F891—10   Specification for Coextruded Poly (Vinyl Chloride)
(PVC) Plastic Pipe with a Cellular Core. . . . . Table P3002.1(1), Table P3002.1(2),
Table P3002.2, Table P3302.1, Table AG101.1

F1055—13   Specification for Electrofusion Type Polyethylene Fittings for Outside
Diameter Controlled Polyethylene and Crosslinked Polyethylene Pipe and Tubing . . . . Table M2105.5,
M2105.11.2

F1281—11   Specification for Cross-linked Polyethylene/Aluminum/Cross-linked
Polyethylene (PEX-AL-PEX) Pressure Pipe . . . . .Table M2101.1, Table P2906.4, Table P2906.5,
Table P2906.6, P2506.11.1, Table AG101.1

F1282—10   Specification for Polyethylene/Aluminum/Polyethylene (PE-AL-PE)
Composite Pressure Pipe. . . . . .Table M2101.1, Table P2906.4, Table P2906.5,
Table P2906.6, P2906.11.1, Table AG101.1

F1412—09   Specification for Polyolefin Pipe and Fittings for Corrosive
Waste Drainage. . . . . .Table P3002.1(2), Table P3002.2,
Table P3002.3, P3003.11.1

F1488—09e1   Specification for Coextruded Composite Pipe . . . . Table P3002.1(1), Table P3002.1(2),
Table P3002.2, Table P3009.11

F1554—07a   Specification for Anchor Bolts, Steel, 36, 55 and 105-ksi Yield Strength. R608.5.2.2

F1667—11A e1   Specification for Driven Fasteners, Nails, Spikes and Staples. R317.3, R703.3.2, R703.6.3,
Table R703.15.1, Table R703.15.2, R905.2.5

F1807—13   Specification for Metal Insert Fittings Utilizing a Copper Crimp Ring
for SDR9 Cross-linked Polyethylene (PEX) Tubing and SDR9
Polyethylene of Raised Temperature (PE-RT) Tubing. . . . . .Table M2101.1, Table P2906.6

F1866—07   Specification for Poly (Vinyl Chloride) (PVC) Plastic Schedule 40 Drainage and
DWV Fabricated Fittings . . . . Table P3002.3

ASTM—continued

F1924—12	Standard Specification for Plastic Mechanical Fittings for Use on Outside Diameter Controlled Polyethylene Gas Distribution Pipe and Tubing	M2105.11.1
F1960—12	Specification for Cold Expansion Fittings with PEX Reinforcing Rings for Use with Cross-linked Polyethylene (PEX) Tubing	Table M2101.1, Table P2906.6
F1970—12	Standard Specification for Special Engineered Fittings, Appurtenances or Valves for Use in Poly (Vinyl Chloride) (PVC) or Chlorinated Poly (Vinyl Chloride) (CPVC) Systems	M2105.5, Table 2903.9.4
F1973—08	Standard Specification for Factory Assembled Anodeless Risers and Transition Fittings in Polyethylene (PE) and Polyamide 11 (PA 11) Fuel Gas Distribution Systems	G2415.15.2
F1974—09	Specification for Metal Insert Fittings for Polyethylene/Aluminum/Polyethylene and Cross-linked Polyethylene/Aluminum/Cross-linked Polyethylene Composite Pressure Pipe	P2506.11.1, Table P2906.6
F1986—01 (2011)	Multilayer Pipe Type 2, Compression Joints for Hot and Cold Drinking Water Systems	Table P2906.4, Table P2906.5, Table P2906.6
F2080—12	Specification for Cold-expansion Fittings with Metal Compression-sleeves for Cross-linked Polyethylene (PEX) Pipe	P2906.6
F2090—10	Specification for Window Fall Prevention Devices—with Emergency Escape (Egress) Release Mechanisms	R310.1.1, R312.2.1, R312.2.2
F2098—08	Standard Specification for Stainless Steel Clamps for Securing SDR9 Cross-linked Polyethylene (PEX) Tubing to Metal Insert and Plastic Insert Fittings	Table M2101.1, Table P2906.6
F2159—11	Standard Specification for Plastic Insert Fittings Utilizing a Copper Crimp Ring for SDR9 Cross-linked Polyethylene (PEX) Tubing and SDR9 Polyethylene of Raised Temperature (PE-RT) Tubing	Table P2906.6
F2262—09	Standard Specification for Cross-linked Polyethylene/Aluminum/Cross-linked Polyethylene Tubing OD Controlled SDR9	Table P2906.4, Table P2906.5
F2389—10	Standard for Pressure-rated Polypropylene (PP) Piping Systems	Table M2105.12.1, Table P2906.4, Table P2906.5, Table P2906.6, P2906.10.1, Table AG101.1
F2434—09	Standard Specification for Metal Insert Fittings Utilizing a Copper Crimp Ring for Polyethylene/Aluminum/Cross-linked Polyethylene (PEX-AL-PEX) Tubing	Table P2906.6
F2623—08	Standard Specification for Polyethylene of Raised Temperature (PE-RT) SDRG Tubing	Table M2101.1, Table AG101.1
F2735—09	Standard Specification for Plastic Insert Fittings for SDR9 Cross-linked Polyethylene (PEX) and Polyethylene of Raised Temperature (PE-RT) Tubing	Table M2101.1, Table P2906.6
F2769—10	Polyethylene or Raised Temperature (PE-RT) Plastic Hot and Cold-Water Tubing and Distribution Systems	Table M2101.1, Table P2906.4, Table P2906.5, Table P2906.6, Table AG101.1
F2806—10	Standard Specification for Acrylonitrile-butadiene-styrene (ABS) Plastic Pipe (Metric SDR-PR)	Table M2101.1
F2855—12	Standard Specification for Chlorinated Poly (Vinyl Chloride)/Aluminum/Chlorinated Poly (Vinyl Chloride) (CPVC AL CPVC) Composite Pressure Tubing	Table P2906.4, Table P2906.5, Table AG101.1
F2969—12	Standard Specification for Acrylonitrile-butadiene-styrene (ABS) IPS Dimensioned Pressure Pipe	Table M2101.1

AWC

American Wood Council
222 Catocin Circle, Suite 201
Leesburg, VA 20175

Standard reference number	Title	Referenced in code section number
AWC STJR—2015	Span Tables for Joists and Rafters	R502.3, R802.4, R802.5
AWC WFCM—2015	Wood Frame Construction Manual for One- and Two-family Dwellings	R301.1.1, R301.2.1.1, R602.10.8.2, R608.9.2, Figure R608.9(9), R608.10
ANSI AWC NDS—2015	National Design Specification (NDS) for Wood Construction—with 2015 Supplement	R404.2.2, R502.2, Table R503.1, R602.3, R608.9.2, Table R703.15.1, Table R703.15.2, R802.2
ANSI/AWC PWF—2015	Permanent Wood Foundation Design Specification	R317.3.2, R401.1, R404.2.3

AWPA

American Wood Protection Association
P.O. Box 361784
Birmingham, AL 35236-1784

Standard reference number	Title	Referenced in code section number
C1—03	All Timber Products—Preservative Treatment by Pressure Processes R902.2	
M4—11	Standard for the Care of Preservative-treated Wood Products R317.1.1, R318.1.2	
U1—14	USE CATEGORY SYSTEM: User Specification for Treated Wood Except Section 6 Commodity Specification H .... R317.1, R402.1.2, R504.3, R703.6.3, R905.7.5, Table R905.8.5, R905.8.6	

AWS

American Welding Society
8669 NW 36 Street, #130
Doral, FL 33166

Standard reference number	Title	Referenced in code section number
A5.8M/A5.8—2011 ANSI/AWS	Specifications for Filler Metals for Brazing and Braze Welding P3003.6.1	
A5.31M/A5.31—2012	Specification for Fluxes for Brazing and Braze Welding Edition: 2nd M2103.3, M2202.2, P2906.14, M2103.3	

AWWA

American Water Works Association
6666 West Quincy Avenue
Denver, CO 80235

Standard reference number	Title	Referenced in code section number
C104/A21.4—08	Cement-mortar Lining for Ductile-iron Pipe and Fittings for Water P2906.4	
C110/A21.10—12	Ductile-iron and Gray-iron Fittings .... Table P2906.6,	
C115/A21.15—11	Flanged Ductile-iron Pipe with Ductile-iron or Gray-iron Threaded Flanges .... Table P2906.4	
C151/A21.51—09	Ductile-iron Pipe, Centrifugally Cast, for Water .... Table P2906.4	
C153/A21.53—11	Ductile-iron Compact Fittings for Water Service........................... Table P2906.6	
C500—09	Standard for Metal-seated Gate Valves for Water Supply Service.....................Table P2903.9.4	
C504—10	Standard for Rubber-seated Butterfly Valves.....................................Table P2903.9.4	
C507—11	Standard for Ball Valves, 6 In. Through 60 In...................................Table P2903.9.4	
C510—07	Double Check Valve Backflow Prevention Assembly........................ Table P2902.3, P2902.3.6	
C511—07	Reduced-pressure Principle Backflow Prevention Assembly.........................Table P2902.3, P2902.3.5, P2902.5.1	
C901—08	Polyethylene (PE) Pressure Pipe and Tubing $^1/_2$ in. (13 mm) through 3 in. (76 mm) for Water Service ....P2906.4, Table AG101.1	
C903—05	Polyethylene-aluminum-polyethylene & Crosslinked Polyethylene Composite Pressure Pipe, $^1/_2$ in. (12 mm) through 2 in. (50 mm), for Water Service .... Table M2101	
C904—06	Cross-linked Polyethylene (PEX) Pressure Pipe, $^1/_2$ in. (12 mm) through 3 in. (76 mm) for Water Service ....P2906.4, Table AG101.1	

CEN

European Committee for Standardization (EN)
Central Secretariat
Rue de Stassart 36
B-10 50 Brussels

Standard reference number	Title	Referenced in code section number
EN 15250-2007	Slow Heat Release Appliances Fired by Solid Fuel Requirements and Test Methods............... R1002.5	

CGSB

Canadian General Standards Board
Place du Portage 111, 6B1
11 Laurier Street
Gatineau, Quebec, Canada KIA 1G6

Standard reference number	Title	Referenced in code section number
CAN/CGSB-37.54—95	Polyvinyl Chloride Roofing and Waterproofing Membrane	R905.13.2
37-GP-52M—(1984)	Roofing and Waterproofing Membrane, Sheet Applied, Elastomeric	R905.12.2
37-GP-56M—(1980)	Membrane, Modified Bituminous, Prefabricated and Reinforced for Roofing—with December 1985 Amendment	Table R905.11.2

CISPI

Cast Iron Soil Pipe Institute
5959 Shallowford Road, Suite 419
Chattanooga, TN 37421

Standard reference number	Title	Referenced in code section number
301—04a	Standard Specification for Hubless Cast Iron Soil Pipe and Fittings for Sanitary and Storm Drain, Waste and Vent Piping Applications	Table P3002.1(1), Table P3002.1(2), Table P3002.2, Table P3002.3, Table P3302.1
310—04	Standard Specification for Coupling for Use in Connection with Hubless Cast Iron Soil Pipe and Fittings for Sanitary and Storm Drain, Waste and Vent Piping Applications	P3003.4.3

CPA

Composite Panel Association
19465 Deerfield Avenue, Suite 306
Leesburg, VA 20176

Standard reference number	Title	Referenced in code section number
ANSI A135.4—2012	Basic Hardboard	Table R602.3(2)
ANSI A135.5—2012	Prefinished Hardboard Paneling	R702.5
ANSI A135.6—2012	Engineered Wood Siding	R703.5
ANSI A135.7—2012	Engineered Wood Trim	R703.5
A208.1—2009	Particleboard	R503.3.1, R602.1.9, R605.1

CPSC

Consumer Product Safety Commission
4330 East West Highway
Bethesda, MD 20814-4408

Standard reference number	Title	Referenced in code section number
16 CFR, Part 1201—(2002)	Safety Standard for Architectural Glazing	R308.1.1, R308.3.1, Table R308.3.1(1)
16 CFR, Part 1209—(2002)	Interim Safety Standard for Cellulose Insulation	R302.10.3
16 CFR, Part 1404—(2002)	Cellulose Insulation	R302.10.3

CSA

CSA Group
8501 East Pleasant Valley Road
Cleveland, OH 44131-5516

Standard reference number	Title	Referenced in code section number
AAMA/WDMA/CSA 101/I.S.2/A440—11	North American Fenestration Standard/Specification for Windows, Doors and Unit Skylights	R308.6.9, R609.3, N1102.4.3
ANSI/CSA America FCI—2012	Stationary Fuel Cell Power Systems	M1903.1
ASME A112.3.4—2013/ CSA B45.9—13	Macerating Toilet Systems and Related Components	Table P2701.1, P3007.5
ASME A112.18.1—2012/ CSA B125.1—2012	Plumbing Supply Fittings	Table P2701.1, P2708.4, P2708.5, P2722.1, P2722.2, P2722.3, P2902.2, Table P2903.9.4
ASME A112.18.2—2011/ CSA B125.2—2011	Plumbing Waste Fittings	Table P2701.1, P2702.2
A112.18.6/ CSA B125.6—2009	Flexible Water Connectors	P2906.7
ASME A112.19.1—2013/ CSA B45.2—13	Enameled Cast-iron and Enameled Steel Plumbing Fixtures	Table 2701.1, P2711.1
ASME A112.19.2—2013/ CSA B45.1—13	Ceramic Plumbing Fixtures	Table P2701.1, P2705.1, P2711.1, P2712.1, P2712.2, P2712.9
ASME A112.19.3—2008/ CSA B45.4—08 (R2013)	Stainless Steel Plumbing Fixtures	Table P2701.1, P2705.1, P2711.1, P2712.1
ASSE 1016/ASME 112.1016/ CSA B125.16—2011	Performance Requirements for Automatic Compensating Valves for Individual Showers and Tub/Shower Combinations	Table P2701.1, P2708.4, P2722.2
A112.19.5—2011/ CSA B45.15—2011	Flush Valves and Spuds for Water-closets, Urinals and Tanks	Table P2701.1
A112.19.7—2012/ CSA B45.10—2012	Hydromassage Bathtub Systems	Table P2701.1
ASME A17.1/ CSA B44—2013	Safety Code for Elevators and Escalators	R321.1
CSA 8—93	Requirements for Gas Fired Log Lighters for Wood Burning Fireplaces—with revisions through January 1999	G2433.1
A257.1M—2009	Circular Concrete Culvert, Storm Drain, Sewer Pipe and Fittings	Table P3002.2
A257.2M—2009	Reinforced Circular Concrete Culvert, Storm Drain, Sewer Pipe and Fittings	Table P3002.2, P3003.13
A257.3M—2009	Joints for Circular Concrete Sewer and Culvert Pipe, Manhole Sections and Fittings Using Rubber Gaskets	P3003.5, P3003.18
B64.1.1—11	Vacuum Breakers, Atmospheric Type (AVB)	Table P2902.3, P2902.3.2
B64.1.2—11	Pressure Vacuum Breakers (PVB)	Table P2902.3, P2902.3.4
B64.1.3—11	Spill Resistant Pressure Vacuum Breakers (SRPVB)	P2902.3.2
B64.2—11	Vacuum Breakers, Hose Connection Type (HCVP)	Table P2902.3, P2902.3.2
B64.2.1—11	Hose Connection Vacuum Breakers (HCVB) with Manual Draining Feature	Table P2902.3, P2902.3.2
B64.2.1.1—11	Hose Connection Dual Check Vacuum Breakers (HCDVB)	Table P2902.3, P2902.3.2
B64.2.2—11	Vacuum Breakers, Hose Connection Type (HCVP) with Automatic Draining Feature	Table P2902.3, P2902.3.2
B64.3—11	Dual Check Backflow Preventers with Atmospheric Port (DCAP)	Table P2902.3, P2902.5.1
B64.4—11	Backflow Preventers, Reduced Pressure Principle Type (RP)	Table P2902.3, P2902.3.5, P2903.5.1
B64.4.1—11	Reduced Pressure Principle for Fire Sprinklers (RPF)	Table P2902.3, P2902.3.5
B64.5—11	Double Check Backflow Preventers (DCVA)	Table P2902.3, P2902.3.6
B64.5.1—11	Double Check Valve Backflow Preventers, Type for Fire Systems (DCVAF)	Table P2902.3, P2902.3.6
B64.6—11	Dual Check Valve Backflow Preventers (DuC)	Table P2902.3, P2902.3.7
B64.7—11	Laboratory Faucet Vacuum Breakers (LFVB)	Table P2902.3, P2902.3.2
B125.3—12	Plumbing Fittings	Table 2701.1, P2713.3, P2721.2, Table P2902.3, P2902.4.1, Table P2903.9.4
B137.1—13	Polyethylene (PE) Pipe, Tubing and Fittings for Cold Water Pressure Services	Table P2906.4, Table P2906.6
B137.2—13	Polyvinylchloride PVC Injection-moulded Gasketed Fittings for Pressure Applications	Table P2906.6
B137.3—13	Rigid Poly (Vinyl Chloride) (PVC) Pipe for Pressure Applications	Table P2906.4, Table 2906.6, P3003.9.2, Table AG101.1

CSA—continued

B137.5—13	Cross-linked Polyethylene (PEX) Tubing Systems for Pressure Applications . . .	Table P2906.4, Table P2906.5, Table P2906.6, Table AG101.1
B137.6—13	Chlorinated polyvinylchloride CPVC Pipe, Tubing and Fittings For Hot- and Cold-water Distribution Systems	Table P2906.4, Table P2906.5, Table 2906.6, Table AG101.1
B137.9—13	Polyethylene/Aluminum/Polyethylene (PE-AL-PE) Composite Pressure Pipe Systems	Table M2101.1, P2506.11.1, Table P2906.4
B137.10—13	Cross-linked Polyethylene/Aluminum/Cross-linked Polyethylene (PE-AL-PE) Composite Pressure Pipe Systems	Table M2101.1, Table P2906.4, Table P2906.5, Table P2906.6, P2906.11.1
B137.11—13	Polypropylene (PP-R) Pipe and Fittings for Pressure Applications	Table P2906.4, Table 2906.5, Table P2906.6, Table AG101.1
B181.1—11	Acrylonitrile-butadiene-styrene (ABS) Drain, Waste and Vent Pipe and Pipe Fittings	Table P3002.1(1), Table P3002.1(2), Table P3002.3, P3003.3.2, P3003.8.2
B181.2—11	Polyvinylchloride (PVC) and chlorinated polyvinylchloride (CPVC) Drain, Waste and Vent Pipe and Pipe Fittings	Table P3002.1(1), Table P3002.1(2), P3003.9.2, P3003.14.2, P3008.2, Table P3302.1
B181.3—11	Polyolefin and polyvinylidene (PVDF) Laboratory Drainage Systems	Table P3002.1(1), Table P3002.1(2), Table P3002.2, Table P3002.3, P3003.11.1
B182.2—11	PSM Type polyvinylchloride (PVC) Sewer Pipe and Fittings	Table P3002.2, Table P3002.3, Table P3302.1
B182.4—11	Profile polyvinylchloride (PVC) Sewer Pipe & Fittings	Table P3002.2, Table P3002.3, Table P3302.1
B182.6—11	Profile Polyethylene (PE) Sewer Pipe and Fittings for leak-proof Sewer Applications	Table P3302.1
B182.8—11	Profile Polyethylene (PE) Storm Sewer and Drainage Pipe and Fittings.	Table P3302.1
B356—10	Water Pressure Reducing Valves for Domestic Water Supply Systems	P2903.3.1
B483.1—14	Drinking Water Treatment Systems	P2909.1, P2909.2
B602—10	Mechanical Couplings for Drain, Waste and Vent Pipe and Sewer Pipe.	P3003.3.1, P3003.4.3, P3003.5, P3003.9.1, P3003.10.1, P3003.12.2, P3003.13
CSA B45.5—11/ IAPMO Z124—11	Plastic Plumbing Fixtures.	Table P2701.1, P2711.1, P2711.2, P2712.1
CSA C448 Series-02-CAN/CSA—2002	Design and Installation of Earth Energy Systems— First Edition; Update 2: October 2009; Consolidated Reprint 10/2009	Table M2105.4, Table M2105.5
O325—07	Construction Sheathing	R503.2.1, R602.1.8, R604.1, R803.2.1
O437-Series—93	Standards on OSB and Waferboard (Reaffirmed 2006).	R503.2.1, R602.1.8, R604.1, R803.2.1
UL/CSA/ANCE 60335-2-40—2012	Standard for Household and Similar Electrical Appliances, Part 2: Particular Requirements for Motor-compressors	M1403.1, M1412.1, M1413.1

CSSB

Cedar Shake & Shingle Bureau
P. O. Box 1178
Sumas, WA 98295-1178

Standard reference number	Title	Referenced in code section number
CSSB—97	Grading and Packing Rules for Western Red Cedar Shakes and Western Red Shingles of the Cedar Shake and Shingle Bureau	R702.6, R703.6, Table R905.7.4, Table R905.8.5

DASMA

Door and Access Systems Manufacturers Association International
1300 Summer Avenue
Cleveland, OH 44115-2851

Standard reference number	Title	Referenced in code section number
108—12	Standard Method for Testing Garage Doors: Determination of Structural Performance Under Uniform Static Air Pressure Difference.	R609.14
115—12	Standard Method for Testing Garage Doors: Determination of Structural Performance Under Missile Impact and Cyclic Wind Pressure	R301.2.1.2

DOC

United States Department of Commerce
1401 Constitution Avenue, NW
Washington, DC 20230

Standard reference number	Title	Referenced in code section number
PS 1—09	Structural Plywood	R404.2.1, Table R404.2.3, R503.2.1, R602.1.8, R604.1, R610.3.2, R803.2.1
PS 2—10	Performance Standard for Wood-based Structural-use Panels	R404.2.1, Table R404.2.3, R503.2.1, R602.1.8, R604.1, R610.3.2, Table 610.3.2, R803.2.1
PS 20—05	American Softwood Lumber Standard	R404.2.1, R502.1.1, R602.1.1, R802.1.1

DOTn

Department of Transportation
1200 New Jersey Avenue SE
East Building, 2nd floor
Washington, DC 20590

Standard reference number	Title	Referenced in code section number
49 CFR, Parts 192.281(e) & 192.283 (b) (2009)	Transportation of Natural and Other Gas by Pipeline: Minimum Federal Safety Standards	G2414.6.1

FEMA

Federal Emergency Management Agency
500 C Street, SW
Washington, DC 20472

Standard reference number	Title	Referenced in code section number
FEMA TB-2—08	Flood Damage-resistant Materials Requirements	R322.1.8
FEMA TB-11—01	Crawlspace Construction for Buildings Located in Special Flood Hazard Area	R408.7

FM

Factory Mutual Global Research
Standards Laboratories Department
1301 Atwood Avenue, P. O. Box 7500
Johnson, RI 02919

Standard reference number	Title	Referenced in code section number
4450—(1989)	Approval Standard for Class 1 Insulated Steel Deck Roofs—with Supplements through July 1992	R906.1
4880—(2010)	American National Standard for Evaluating Insulated Wall or Wall and Roof/Ceiling Assemblies, Plastic Interior Finish Materials, Plastic Exteriorv Building Panels, Wall/Ceiling Coating Systems, Interior and Exterior Finish Systems	R316.6

GA

Gypsum Association
6525 Belcrest Road, Suite 480
Hyattsville, MD 20782

Standard reference number	Title	Referenced in code section number
GA-253—12	Application of Gypsum Sheathing	Table R602.3(1)

HPVA

Hardwood Plywood & Veneer Association
1825 Michael Faraday Drive
Reston, Virginia 20190-5350

Standard reference number	Title	Referenced in code section number
ANSI/HP-1—2013	Standard for Hardwood and Decorative Plywood	R702.5

IAPMO

IAPMO
4755 E. Philadelphia Street
Ontario, CA 91761-USA

Standard reference number	Title	Referenced in code section number
CSA B45.5—11/ IAPMO Z124—11	Plastic Plumbing Fixtures	Table P2701.1, P2711.1, P2711.2, P2712.1

ICC

International Code Council, Inc.
500 New Jersey Avenue, NW
6th Floor
Washington, DC 20001

Standard reference number	Title	Referenced in code section number
IBC—15	International Building Code®	R101.2, R110.2, R202, R301.1.1, R301.1.3, R301.2.2.1.1, R301.2.2.1.2, R301.2.2.4, R301.3, R308.5, R320.1, R320.1.1, R403.1.8, Table R602.10.3(3), Table R606.12.2.1, R609.2, R802.1.5.4, R905.10.3, N1107.4, G2402.3
ICC/ANSI A117.1—09	Accessible and Usable Buildings and Facilities	R321.3
ICC 400—12	Standard on the Design and Construction of Log Structures	R301.1.1, 502.1.4, R602.1.4, R703.1, R802.1.3
ICC 500—14	ICC/NSSA Standard on the Design and Construction of Storm Shelters	R323.1
ICC 600—14	Standard for Residential Construction in High-wind Regions	R301.2.1.1
IEBC—15	International Existing Building Code®	R110.2
IECC—15	International Energy Conservation Code®	N1101.2, N1101.5, N1101.13.1
IFC—15	International Fire Code®	R102.7, R324.2, M2201.7, G2402.3, G2412.2
IFGC—15	International Fuel Gas Code®	G2401.1, G2402.3, G2423.1
IMC—15	International Mechanical Code®	N1103.2.1, N1103.6, G2402.3
IPC—15	International Plumbing Code®	Table R301.2(1), R903.4.1, G2402.3, R2601.1, Table P2902.3, P2902.5.5,
IPMC—15	International Property Maintenance Code®	R102.7
IPSDC—15	International Private Sewage Disposal Code®	R322.1.7
ISPSC—15	International Swimming Pool and Spa Code™	R326.1

ISO

International Organization for Standardization
1, ch. de la Voie - Creuse
Case postale 56
CH-1211 Geneva 20, Switzerland

Standard reference number	Title	Referenced in code section number
8336—2009	Fibre-cement Flat Sheets-product Specification and Test Methods	Table R503.2.1.1(1), Table R503.2.1.1(2), Table R602.3(2), Table R702.4.2, R703.10.1, R703.10.2
15874—2002	Polypropylene Plastic Piping Systems for Hot and Cold Water Installations	Table M2101.1

MSS

Manufacturers Standardization Society of the Valve and Fittings Industry
127 Park Street, Northeast
Vienna, VA 22180

Standard reference number	Title	Referenced in code section number
SP-42—09	Corrosion Resistant Gate, Globe, Angle and Check Valves with Flanged and Butt Weld Ends (Glasses 150, 300 & 600)	Table P2903.9.4
SP-58—09	Pipe Hangers and Supports—Materials, Design, Manufacture, Selection, Application and Installation	G2418.2
SP-67—11	Butterfly Valves	Table P2903.9.4
SP-70—11	Gray Iron Gate Valves, Flanged and Threaded Ends	Table P2903.9.4
SP-71—11	Gray Iron Swing Check Valves, Flanged and Threaded Ends	Table P2903.9.4
SP-72—10	Ball Valves with Flanged or Butt-Welding Ends for General Service	P2903.9.4
SP-78—11	Cast Iron Plug Valves, Flanged and Threaded Ends	Table P2903.9.4
SP-80—08	Bronze Gate, Globe, Angle and Check Valves	Table P2903.9.4
SP-110—10	Ball Valves, Threaded, Socket Welded, Solder Joint, Grooved and Flared Ends	Table P2903.9.4

NAIMA

North American Insulation Manufacturers Association
44 Canal Center Plaza, Suite 310
Alexandria, VA 22314

Standard reference number	Title	Referenced in code section number
AH 116—09	Fibrous Glass Duct Construction Standards, Fifth Edition	M1601.1.1

NFPA

National Fire Protection Association
1 Batterymarch Park
Quincy, MA 02269

Standard reference number	Title	Referenced in code section number
13—13	Installation of Sprinkler Systems	R302.3
13D—13	Standard for the Installation of Sprinkler Systems in One- and Two-family Dwellings and Manufactured Homes	R302.13, R313.1.1, R313.2.1, R325.5, P2904.1, P2904.6.1
13R—13	Standard for the Installation of Sprinkler Systems in Low Rise Residential Occupancies	R325.5
31—11	Standard for the Installation of Oil-burning Equipment	M1701.1, M1801.3.1, M1805.3
58—14	Liquefied Petroleum Gas Code	G2412.2, G2414.6.2
70—14	National Electrical Code	E3401.1, E3401.2, E4301.1, Table E4303.2, E4304.3, E4304.4, R324.3
72—13	National Fire Alarm and Signaling Code	R314.1, R314.7.1
85—15	Boiler and Combustion Systems Hazards Code	G2452.1
211—13	Standard for Chimneys, Fireplaces, Vents and Solid Fuel Burning Appliances	R1002.5, G2427.5.5.1
259—13	Standard for Test Method for Potential Heat of Building Materials	R316.5.7, R316.5.8
275—13	Standard Method of Fire Tests for the Evaluation of Thermal Barriers	R316.4
286—15	Standard Methods of Fire Tests for Evaluating Contribution of Wall and Ceiling Interior Finish to Room Fire Growth	R302.9.4, R316.6
501—13	Standard on Manufactured Housing	R202
720—15	Standard for the Installation of Carbon Monoxide (CO) Detectors and Warning Equipment	R315.6.1, R315.6.2
853—15	Standard on the Installation of Stationary Fuel Cell Power Systems	M1903.1

NSF

NSF International
789 N. Dixboro
Ann Arbor, MI 48105

Standard reference number	Title	Referenced in code section number
14—2011	Plastics Piping System Components and Related Materials	M1301.4, P2609.3, P2909.3
41—2011	Nonliquid Saturated Treatment Systems (Composting Toilets)	P2725.1
42—2011	Drinking Water Treatment Units—Anesthetic Effects	P2909.1, P2909.3
44—2012	Residential Cation Exchange Water Softeners	P2909.1, P2909.3
50—2012	Equipment for Swimming Pools, Hot Tubs and Other Recreational Water Facilities.	P2911.8.1
53—2011A	Drinking Water Treatment Units—Health Effects	P2909.1, P2909.3
58—2012	Reverse Osmosis Drinking Water Treatment Systems	P2909.2, P2909.3
61—2012	Drinking Water System Components—Health Effects	P2609.5, P2722.1, P2903.9.4, P2906.4, P2906.5, P2906.6, P2908.3
350—2011	Onsite Residential and Commercial Water Reuse Treatment Systems	P2910.6.1
358-1—2011	Polyethylene Pipe and Fittings for Water-based Ground Source "Geothermal" Heat Pump Systems	M2105.4, M2105.5, Table AG101.1
358-2—2012	Polypropylene Pipe and Fittings for Water-based Ground Source "Geothermal" Heat Pump Systems	M2105.5
359—2012	Valves for Crosslinked Polyethylene (PEX) Water Distribution Tubing Systems	Table P2903.9.4
372—2010	Drinking Water Systems Components—Lead Content	P2906.2.1

PCA

Portland Cement Association
5420 Old Orchard Road
Skokie, IL 60077

Standard reference number	Title	Referenced in code section number
100—12	Prescriptive Design of Exterior Concrete Walls for One- and Two-family Dwellings (Pub. No. EB241)	R301.2.2.2.4, R301.2.2.3.4, R404.1.3, R404.1.3.2.1, R404.1.3.2.2, R404.1.3.4, R404.1.4.2, R608.1, R608.2, R608.5.1, R608.9.2, R608.9.3

SBCA

Structural Building Components Association
6300 Enterprise Lane
Madison, WI 53719

Standard reference number	Title	Referenced in code section number
BCSI—2013	Building Component Safety Information Guide to Good Practice for Handling, Installing, Restraining & Bracing of Metal Plate Connected Wood Trusses	502.11.2, 802.10.3
CFS-BCSI—2008	Cold-formed Steel Building Component Safety Information (CFSBCSI) Guide to Good Practice for Handling, Installing & Bracing of Cold-formed Steel Trusses	505.1.3, 804.3.6
FS100—12	Standard Requirements for Wind Pressure Resistance of Foam Plastic Insulating Sheathing Used in Exterior Wall Covering Assemblies	R316.8

SMACNA

Sheet Metal & Air Conditioning Contractors National Assoc. Inc.
4021 Lafayette Center Road
Chantilly, VA 22021

Standard reference number	Title	Referenced in code section number
SMACNA—10	Fibrous Glass Duct Construction Standards (2003)	M1601.1.1, M1601.4.1
SMACNA—15	HVAC Duct Construction Standards, Metal and Flexible (2005)	M1601.4.1

SRCC

Solar Rating & Certification Corporation
400 High Point Drive, Suite 400
Cocoa, FL 32926

Standard reference number	Title	Referenced in code section number
SRCC 100—13	Standard 100 for Solar Collectors M2301.3.1	
SRCC 300—13	Standard 300 for Solar Water Heating Systems M2301.2.3, M2301.4, M2301.2.6, M2301.2.8	
SRCC 600—13	Standard 600 for Solar Concentrating Collectors M2301.3.1	

TMS

The Masonry Society
105 South Sunset Street, Suite Q
Longmont, CO 80501

Standard reference number	Title	Referenced in code section number
402—2013	Building Code Requirements for Masonry Structures . . . . .R404.1.2, R606.1, R606.1.1, R606.2.3.2, R606.12.1, R606.12.2.3.1, R606.12.3.1, Table R703.4, 703.12	
403—2013	Direct Design Handbook for Masonry Structures. R606.1, R606.1.1, R606.12.1, R606.12.3.1	
602—2013	Specification for Masonry Structures R404.1.2, R606.2.9, R606.2.12, R606.12.3.1, R703.12	

TPI

Truss Plate Institute
218 N. Lee Street, Suite 312
Alexandria, VA 22314

Standard reference number	Title	Referenced in code section number
TPI 1—2014	National Design Standard for Metal-plate-connected Wood Truss Construction R502.11.1, R802.10.2	

UL

UL LLC
333 Pfingsten Road
Northbrook, IL 60062

Standard reference number	Title	Referenced in code section number
17—2008	Vent or Chimney Connector Dampers for Oil-fired Appliances— with revisions through January 2010 . . . . M1802.2.2	
55A—04	Materials for Built-up Roof Coverings. R905.9.2	
58—14	Liquefied Petroleum Gas Code. M2201.1	
80—2007	Steel Tanks for Oil-burner Fuel—with revisions August 2009 . . . . M2201.1	
103—2010	Factory-built Chimneys for Residential Type and Building Heating Appliances—with revisions through July 2012 . . . . .R202, R1005.3, G2430.1	
127—2011	Factory-built Fireplaces R1001.11, R1004.1, R1004.4, R1004.5, R1005.4, G2445.7	
174—04	Household Electric Storage Tank Water Heaters— with revisions through September 2012 . . . . M2005.1	
180—2012	Liquid-level Indicating Gauges for Oil Burner Fuels and Other Combustible Liquids M2201.5	
181—05	Factory-made Air Ducts and Air Connectors—with revisions through May 2003 . . . . . . M1601.1.1, M1601.4.1	
181A—2013	Closure Systems for Use with Rigid Air Ducts and Air Connectors— with revisions through December 1998 . . . . .M1601.2, M1601.4.1	
181B—2013	Closure Systems for Use with Flexible Air Ducts and Air Connectors— with revisions through August 2003. . . . . M1601.4.1	
217—06	Single- and Multiple-station Smoke Alarms—with revisions through April 2012 . . . . R314.1.1, R315.1.1	
263—2011	Standards for Fire Test of Building Construction and Materials. . . . . Table 302.1(1), Table R302.1(2), R302.2, R302.3, R302.4.1, R302.11.1, Table R312.1(1), R606.2.2	

UL—continued

268—2009	Smoke Detectors for Fire Alarm Systems	R314.7.1, R314.7.4, R315.6.4
325—02	Door, Drapery, Gate, Louver and Window Operations and Systems— with revisions through June 2013	R309.4
343—2008	Pumps for Oil-burning Appliances—with revisions through June 2013	M2204.1
378—06	Draft Equipment—with revisions through January 2010	M1804.2.6
441—10	Gas Vents	G2426.1
508—99	Industrial Control Equipment—with revisions through March 2013	M1411.3.1
536—97	Flexible Metallic Hose—with revisions through June 2003	M2202.3
641—2010	Type L, Low-temperature Venting Systems— with revisions through May 2013	R202, R1003.11.5, M1804.2.4, G2426.1
651—2011	Schedule 40 and Schedule 80 Rigid PVC Conduit and Fittings— with revisions through March 2012	G2414.6.3
705—04	Standard for Power Ventilators—with revisions through March 2012	M1502.4.4
723—08	Standard for Test for Surface Burning Characteristics of Building Materials—with revisions through September 2010	R202, R302.9.3, R302.9.4, R302.10.1, R302.10.2, R316.3, R316.5.9, R316.5.11, R507.3.2, R802.1.5, M1601.3, M1601.5.2
726—95	Oil-fired Boiler Assemblies—with revisions through April 2011	M2001.1.1, M2006.1
727—06	Oil-fired Central Furnaces—with revisions through April 2010	M1402.1
729—03	Oil-fired Floor Furnaces—with revisions through August 2012	M1408.1
730—03	Oil-fired Wall Furnaces—with revisions through August 2012	M1409.1
732—95	Oil-fired Storage Tank Water Heaters—with revisions through April 2010	M2005.1
737—2011	Fireplaces Stoves	M1414.1, M1901.2
790—04	Standard Test Methods for Fire Tests of Roof Coverings— with revisions through October 2008	R302.2.2, R902.1
795—2011	Commercial-industrial Gas Heating Equipment— with Revisions through September 2012	G2442.1, G2452.1
834—04	Heating, Water Supply and Power Boilers—Electric—with revisions through January 2013	M2001.1.1
842—07	Valves for Flammable Fluids—with revisions through October 2012	M2204.2
858—05	Household Electric Ranges—with revisions through April 2012	M1901.2
875—09	Electric Dry-bath Heaters with revisions through November 2011	M1902.2
896—93	Oil-burning Stoves—with revisions through August 2012	M1410.1
923—2013	Microwave Cooking Appliances	M1504.1
959—2010	Medium Heat Appliance Factory-built Chimneys	R1005.6
1026—2012	Electric Household Cooking and Food Serving Appliances	M1901.2
1040—96	Fire Test of Insulated Wall Construction—with revisions through October 2012	R316.6
1042—2009	Electric Baseboard Heating Equipment—with Revisions through June 2013	M1405.1
1256—02	Fire Test of Roof Deck Construction—with revisions through January 2007	R906.1
1261—01	Electric Water Heaters for Pools and Tubs—with revisions through July 2012	M2006.1
1479—03	Fire Tests of Through-Penetration Firestops—with revisions through October 2012	R302.4.1.2
1482—2011	Solid-Fuel-type Room Heaters	R1002.2, R1002.5, M1410.1
1618—09	Wall Protectors, Floor Protectors, and Hearth Extensions— with revisions through May 2013	R1004.2, M1410.2
1693—2010	Electric Radiant Heating Panels and Heating Panel Sets— with revisions through October 2011	M1406.1
1703—02	Flat-plate Photovoltaic Modules and Panels— with revisions through November 2014	R324.3.1, R902.4, R905.16.5, R907.5
1715—97	Fire Test of Interior Finish Material—with revisions through January 2013	R316.6
1738—2010	Venting Systems for Gas-burning Appliances, Categories II, III and IV— with revisions Through May 2011	G2426.1
1741—2010	Inverters, Converters, Controllers and Interconnection System Equipment for use with Distributed Energy Resources	R324.3
1777—07	Chimney Liners—with revisions through July 2009	R1003.11.1, R1003.18, G2425.12, G2425.15.4, M1801.3.4
1995—2011	Heating and Cooling Equipment	M1402.1, M1403.1, M1407.1, M1412.1, M1413.1
1996—2009	Electric Duct Heaters—with revisions through November 2011	M1402.1, M1407.1
2034—08	Standard for Single- and Multiple-station Carbon Monoxide Alarms— with revisions through February 2009	R314.1.1, R315.1.1
2075—2013	Standard for Gas and Vapor Detectors and Sensors	R314.7.4, R315.6.1, R315.6.4
2158A—2010	Outline of Investigation for Clothes Dryer Transition Duct	M1502.4.3
2523—09	Standard for Solid Fuel-fired Hydronic Heating Appliances, Water Heaters and Boilers— with revisions through February 2013	M2005.1, M2001.1.1

UL—continued

UL/CSA/ANCE
60335-2-40—2012   Standard for Household and Similar Electrical Appliances,
Part 2: Particular Requirements for Motor-compressors M1403.1, M1412.1, M1413.1

ULC

ULC
7 Underwriters Road
Toronto, Ontario, Canada M1R 3B4

Standard reference number	Title	Referenced in code section number
CAN/ULC S 102.2—2010	Standard Methods for Test for Surface Burning Characteristics of Building Materials and Assemblies	R302.10.1, R302.10.2

WDMA

Window & Door Manufacturers Association
2025 M Street, NW Suite 800
Washington, DC 20036-3309

Standard reference number	Title	Referenced in code section number
AAMA/WDMA/CSA 101/I.S2/A440—11	North American Fenestration Standard/ Specifications for Windows, Doors and Skylights	R308.6.9, R609.3, N1102.4.3
I.S. 11—13	Industry Standard Analytical Method for Design Pressure (DP) Ratings of Fenestration Products	R308.6.9.1, R609.3.1

Appendix A:
Sizing and Capacities of Gas Piping

(This appendix is informative and is not part of the code. This appendix is an excerpt from the 2015 *International Fuel Gas Code*, coordinated with the section numbering of the *International Residential Code*.)

A.1 General piping considerations. The first goal of determining the pipe sizing for a fuel gas *piping* system is to make sure that there is sufficient gas pressure at the inlet to each *appliance*. The majority of systems are residential and the appliances will all have the same, or nearly the same, requirement for minimum gas pressure at the *appliance* inlet. This pressure will be about 5-inch water column (w.c.) (1.25 kPa), which is enough for proper operation of the *appliance* regulator to deliver about 3.5-inches water column (w.c.) (875 kPa) to the burner itself. The pressure drop in the *piping* is subtracted from the source delivery pressure to verify that the minimum is available at the *appliance*.

There are other systems, however, where the required inlet pressure to the different appliances may be quite varied. In such cases, the greatest inlet pressure required must be satisfied, as well as the farthest *appliance*, which is almost always the critical *appliance* in small systems.

There is an additional requirement to be observed besides the capacity of the system at 100-percent flow. That requirement is that at minimum flow, the pressure at the inlet to any *appliance* does not exceed the pressure rating of the *appliance* regulator. This would seldom be of concern in small systems if the source pressure is $^1/_2$ psi (14-inch w.c.) (3.5 kPa) or less but it should be verified for systems with greater gas pressure at the point of supply.

To determine the size of *piping* used in a gas *piping* system, the following factors must be considered:

(1) Allowable loss in pressure from *point of delivery* to *appliance*.

(2) Maximum gas demand.

(3) Length of *piping* and number of fittings.

(4) Specific gravity of the gas.

(5) Diversity factor.

For any gas *piping* system or special *appliance*, or for conditions other than those covered by the tables provided in this code such as longer runs, greater gas demands or greater pressure drops, the size of each gas *piping* system should be determined by standard engineering practices acceptable to the code official.

A.2 Description of tables.

A.2.1 General. The quantity of gas to be provided at each *outlet* should be determined, whenever possible, directly from the manufacturer's gas input Btu/h rating of the *appliance* that will be installed. In case the ratings of the appliances to be installed are not known, Table 402.2 shows the approximate consumption (in Btu per hour) of certain types of typical household appliances.

To obtain the cubic feet per hour of gas required, divide the total Btu/h input of all appliances by the average Btu heating value per cubic feet of the gas. The average Btu per cubic feet of the gas in the area of the installation can be obtained from the serving gas supplier.

A.2.2 Low pressure natural gas tables. Capacities for gas at low pressure [less than 2.0 psig (13.8 kPa gauge)] in cubic feet per hour of 0.60 specific gravity gas for different sizes and lengths are shown in Tables 402.4(1) and 402.4(2) for iron pipe or equivalent rigid pipe; in Tables 402.4(8) through 402.4(11) for smooth wall semirigid tubing; and in Tables 402.4(15) through 402.4(17) for corrugated stainless steel tubing. Tables 402.4(1) and 402.4(6) are based upon a pressure drop of 0.3-inch w.c. (75 Pa), whereas Tables 402.4(2), 402.4(9) and 402.4(15) are based upon a pressure drop of 0.5-inch w.c. (125 Pa). Tables 402.4(3), 402.4(4), 402.4(10), 402.4(11), 402.4(16) and 402.4(17) are special low-pressure applications based upon pressure drops greater than 0.5-inch w.c. (125 Pa). In using these tables, an allowance (in equivalent length of pipe) should be considered for any *piping* run with four or more fittings (see Table A.2.2).

A.2.3 Undiluted liquefied petroleum tables. Capacities in thousands of Btu per hour of undiluted liquefied petroleum gases based on a pressure drop of 0.5-inch w.c. (125 Pa) for different sizes and lengths are shown in Table 402.4(28) for iron pipe or equivalent rigid pipe, in Table 402.4(30) for smooth wall semi-rigid tubing, in Table 402.4(32) for corrugated stainless steel tubing, and in Tables 402.4(35) and 402.4(37) for polyethylene plastic pipe and tubing. Tables 402.4(33) and 402.4(34) for corrugated stainless steel tubing and Table 402.4(36) for polyethylene plastic pipe are based on operating pressures greater than $1^1/_2$ pounds per square inch (psi) (3.5 kPa) and pressure drops greater than 0.5-inch w.c. (125 Pa). In using these tables, an allowance (in equivalent length of pipe) should be considered for any *piping* run with four or more fittings [see Table A.2.2].

A.2.4 Natural gas specific gravity. Gas *piping* systems that are to be supplied with gas of a specific gravity of 0.70 or less can be sized directly from the tables provided in this code, unless the code official specifies that a gravity factor be applied. Where the specific gravity of the gas is greater than 0.70, the gravity factor should be applied.

Application of the gravity factor converts the figures given in the tables provided in this code to capacities for another gas of different specific gravity. Such application is accomplished by multiplying the capacities given in the tables by the multipliers shown in Table A.2.4. In case the exact specific gravity does not appear in the table, choose the next higher value specific gravity shown.

TABLE A.2.2
EQUIVALENT LENGTHS OF PIPE FITTINGS AND VALVES

		SCREWED FITTINGS[1]				90° WELDING ELBOWS AND SMOOTH BENDS[2]					
		45°/Ell	90°/Ell	180°close return bends	Tee	R/d = 1	R/d = 1$\frac{1}{3}$	R/d = 2	R/d = 4	R/d = 6	R/d = 8
k factor =		0.42	0.90	2.00	1.80	0.48	0.36	0.27	0.21	0.27	0.36
L/d' ratio[4] n =		1 4	30	67	60	16	12	9	7	9	12
Nominal pipe size, inches	Inside diameter d, inches, Schedule 40[6]	L = Equivalent Length In Feet of Schedule 40 (Standard-weight) Straight Pipe[6]									
$\frac{1}{2}$	0.622	0.73	1.55	3.47	3.10	0.83	0.62	0.47	0.36	0.47	0.62
$\frac{3}{4}$	0.824	0.96	2.06	4.60	4.12	1.10	0.82	0.62	0.48	0.62	0.82
1	1.049	1.22	2.62	5.82	5.24	1.40	1.05	0.79	0.61	0.79	1.05
1$\frac{1}{4}$	1.380	1.61	3.45	7.66	6.90	1.84	1.38	1.03	0.81	1.03	1.38
1$\frac{1}{2}$	1.610	1.88	4.02	8.95	8.04	2.14	1.61	1.21	0.94	1.21	1.61
2	2.067	2.41	5.17	11.5	10.3	2.76	2.07	1.55	1.21	1.55	2.07
2$\frac{1}{2}$	2.469	2.88	6.16	13.7	12.3	3.29	2.47	1.85	1.44	1.85	2.47
3	3.068	3.58	7.67	17.1	15.3	4.09	3.07	2.30	1.79	2.30	3.07
4	4.026	4.70	10.1	22.4	20.2	5.37	4.03	3.02	2.35	3.02	4.03
5	5.047	5.88	12.6	28.0	25.2	6.72	5.05	3.78	2.94	3.78	5.05
6	6.065	7.07	15.2	33.8	30.4	8.09	6.07	4.55	3.54	4.55	6.07
8	7.981	9.31	20.0	44.6	40.0	10.6	7.98	5.98	4.65	5.98	7.98
10	10.02	11.7	25.0	55.7	50.0	13.3	10.0	7.51	5.85	7.51	10.0
12	11.94	13.9	29.8	66.3	59.6	15.9	11.9	8.95	6.96	8.95	11.9
14	13.13	15.3	32.8	73.0	65.6	17.5	13.1	9.85	7.65	9.85	13.1
16	15.00	17.5	37.5	83.5	75.0	20.0	15.0	11.2	8.75	11.2	15.0
18	16.88	19.7	42.1	93.8	84.2	22.5	16.9	12.7	9.85	12.7	16.9
20	18.81	22.0	47.0	105.0	94.0	25.1	18.8	14.1	11.0	14.1	18.8
24	22.63	26.4	56.6	126.0	113.0	30.2	22.6	17.0	13.2	17.0	22.6

(continued)

TABLE A.2.2—continued
EQUIVALENT LENGTHS OF PIPE FITTINGS AND VALVES

		MITER ELBOWS[3] (No. of miters)					WELDING TEES		VALVES (screwed, flanged, or welded)			
		1-45°	1-60°	1-90°	2-90[o5]	3-90[o5]	Forged	Miter[3]	Gate	Globe	Angle	Swing Check
k factor =		0.45	0.90	1.80	0.60	0.45	1.35	1.80	0.21	10	5.0	2.5
L/d' ratio[4] n =		15	30	60	20	15	45	60	7	333	167	83
Nominal pipe size, inches	Inside diameter d, inches, Schedule 40[6]	L = Equivalent Length In Feet of Schedule 40 (Standard-weight) Straight Pipe[6]										
1/2	0.622	0.78	1.55	3.10	1.04	0.78	2.33	3.10	0.36	17.3	8.65	4.32
3/4	0.824	1.03	2.06	4.12	1.37	1.03	3.09	4.12	0.48	22.9	11.4	5.72
1	1.049	1.31	2.62	5.24	1.75	1.31	3.93	5.24	0.61	29.1	14.6	7.27
1 1/4	1.380	1.72	3.45	6.90	2.30	1.72	5.17	6.90	0.81	38.3	19.1	9.58
1 1/2	1.610	2.01	4.02	8.04	2.68	2.01	6.04	8.04	0.94	44.7	22.4	11.2
2	2.067	2.58	5.17	10.3	3.45	2.58	7.75	10.3	1.21	57.4	28.7	14.4
2 1/2	2.469	3.08	6.16	12.3	4.11	3.08	9.25	12.3	1.44	68.5	34.3	17.1
3	3.068	3.84	7.67	15.3	5.11	3.84	11.5	15.3	1.79	85.2	42.6	21.3
4	4.026	5.04	10.1	20.2	6.71	5.04	15.1	20.2	2.35	112.0	56.0	28.0
5	5.047	6.30	12.6	25.2	8.40	6.30	18.9	25.2	2.94	140.0	70.0	35.0
6	6.065	7.58	15.2	30.4	10.1	7.58	22.8	30.4	3.54	168.0	84.1	42.1
8	7.981	9.97	20.0	40.0	13.3	9.97	29.9	40.0	4.65	22.0	111.0	55.5
10	10.02	12.5	25.0	50.0	16.7	12.5	37.6	50.0	5.85	278.0	139.0	69.5
12	11.94	14.9	29.8	59.6	19.9	14.9	44.8	59.6	6.96	332.0	166.0	83.0
14	13.13	16.4	32.8	65.6	21.9	16.4	49.2	65.6	7.65	364.0	182.0	91.0
16	15.00	18.8	37.5	75.0	25.0	18.8	56.2	75.0	8.75	417.0	208.0	104.0
18	16.88	21.1	42.1	84.2	28.1	21.1	63.2	84.2	9.85	469.0	234.0	117.0
20	18.81	23.5	47.0	94.0	31.4	23.5	70.6	94.0	11.0	522.0	261.0	131.0
24	22.63	28.3	56.6	113.0	37.8	28.3	85.0	113.0	13.2	629.0	314.0	157.0

For SI: 1 foot = 305 mm, 1 degree = 0.01745 rad.

Note: Values for welded fittings are for conditions where bore is not obstructed by weld spatter or backing rings. If appreciably obstructed, use values for "Screwed Fittings."

1. Flanged fittings have three-fourths the resistance of screwed elbows and tees.

2. Tabular figures give the extra resistance due to curvature alone to which should be added the full length of travel.

3. Small size socket-welding fittings are equivalent to miter elbows and miter tees.

4. Equivalent resistance in number of diameters of straight pipe computed for a value of $(f - 0.0075)$ from the *relation* $(n - k/4f)$.

5. For condition of minimum resistance where the centerline length of each miter is between d and $2\frac{1}{2}d$.

6. For pipe having other inside diameters, the equivalent resistance can be computed from the above n values.

Source: Crocker, S. *Piping Handbook*, 4th ed., Table XIV, pp. 100-101. Copyright 1945 by McGraw-Hill, Inc. Used by permission of McGraw-Hill Book Company.

TABLE A.2.4
MULTIPLIERS TO BE USED WITH TABLES 402.4(1)
THROUGH 402.4(22) WHERE THE SPECIFIC GRAVITY
OF THE GAS IS OTHER THAN 0.60

SPECIFIC GRAVITY	MULTIPLIER	SPECIFIC GRAVITY	MULTIPLIER
0.35	1.31	1.00	0.78
0.40	1.23	1.10	0.74
0.45	1.16	1.20	0.71
0.50	1.10	1.30	0.68
0.55	1.04	1.40	0.66
0.60	1.00	1.50	0.63
0.65	0.96	1.60	0.61
0.70	0.93	1.70	0.59
0.75	0.90	1.80	0.58
0.80	0.87	1.90	0.56
0.85	0.84	2.00	0.55
0.90	0.82	2.10	0.54

A.2.5 Higher pressure natural gas tables. Capacities for gas at pressures 2.0 psig (13.8 kPa) or greater in cubic feet per hour of 0.60 specific gravity gas for different sizes and lengths are shown in Tables 402.4(5) through 402.4(7) for iron pipe or equivalent rigid pipe; Tables 402.4(12) to 402.4(14) for semirigid tubing; Tables 402.4(18) and 402.4(19) for corrugated stainless steel tubing; and Table 402.4(22) for polyethylene plastic pipe.

A.3 Use of capacity tables.

A.3.1 Longest length method. This sizing method is conservative in its approach by applying the maximum operating conditions in the system as the norm for the system and by setting the length of pipe used to size any given part of the *piping* system to the maximum value.

To determine the size of each section of gas *piping* in a system within the range of the capacity tables, proceed as follows (also see sample calculations included in this Appendix):

(1) Divide the *piping* system into appropriate segments consistent with the presence of tees, branch lines and main runs. For each segment, determine the gas load (assuming all appliances operate simultaneously) and its overall length. An allowance (in equivalent length of pipe) as determined from Table A.2.2 shall be considered for *piping* segments that include four or more fittings.

(2) Determine the gas demand of each *appliance* to be attached to the *piping* system. Where Tables 402.4(1) through 402.4(24) are to be used to select the *piping* size, calculate the gas demand in terms of cubic feet per hour for each *piping* system *outlet*. Where Tables 402.4(25) through 402.4(37) are to be used to select the *piping* size, calculate the gas demand in terms of

thousands of Btu per hour for each *piping* system *outlet*.

(3) Where the *piping* system is for use with other than undiluted liquefied petroleum gases, determine the design system pressure, the allowable loss in pressure (pressure drop), and specific gravity of the gas to be used in the *piping* system.

(4) Determine the length of *piping* from the *point of delivery* to the most remote *outlet* in the building/*piping* system.

(5) In the appropriate capacity table, select the row showing the measured length or the next longer length if the table does not give the exact length. This is the only length used in determining the size of any section of gas *piping*. If the gravity factor is to be applied, the values in the selected row of the table are multiplied by the appropriate multiplier from Table A.2.4.

(6) Use this horizontal row to locate ALL gas demand figures for this particular system of *piping*.

(7) Starting at the most remote *outlet*, find the gas demand for that *outlet* in the horizontal row just selected. If the exact figure of demand is not shown, choose the next larger figure left in the row.

(8) Opposite this demand figure, in the first row at the top, the correct size of gas *piping* will be found.

(9) Proceed in a similar manner for each *outlet* and each section of gas *piping*. For each section of *piping*, determine the total gas demand supplied by that section.

Where a large number of *piping* components (such as elbows, tees and valves) are installed in a pipe run, additional pressure loss can be accounted for by the use of equivalent lengths. Pressure loss across any *piping* component can be equated to the pressure drop through a length of pipe. The equivalent length of a combination of only four elbows/tees can result in a jump to the next larger length row, resulting in a significant reduction in capacity. The equivalent lengths in feet shown in Table A.2.2 have been computed on a basis that the inside diameter corresponds to that of Schedule 40 (standard-weight) steel pipe, which is close enough for most purposes involving other schedules of pipe. Where a more specific solution for equivalent length is desired, this can be made by multiplying the actual inside diameter of the pipe in inches by $n/12$, or the actual inside diameter in feet by n (n can be read from the table heading). The equivalent length values can be used with reasonable accuracy for copper or brass fittings and bends although the resistance per foot of copper or brass pipe is less than that of steel. For copper or brass valves, however, the equivalent length of pipe should be taken as 45 percent longer than the values in the table, which are for steel pipe.

A.3.2 Branch length method. This sizing method reduces the amount of conservatism built into the traditional Longest Length Method. The longest length as measured from the meter to the furthest remote *appliance* is only used to size the initial parts of the overall *piping* system. The Branch Length Method is applied in the following manner:

(1) Determine the gas load for each of the connected appliances.

(2) Starting from the meter, divide the *piping* system into a number of connected segments, and determine the length and amount of gas that each segment would carry assuming that all appliances were operated simultaneously. An allowance (in equivalent length of pipe) as determined from Table A.2.2 should be considered for piping segments that include four or more fittings.

(3) Determine the distance from the *outlet* of the gas meter to the *appliance* furthest removed from the meter.

(4) Using the longest distance (found in Step 3), size each *piping* segment from the meter to the most remote *appliance outlet*.

(5) For each of these *piping* segments, use the longest length and the calculated gas load for all of the connected appliances for the segment and begin the sizing process in Steps 6 through 8.

(6) Referring to the appropriate sizing table (based on operating conditions and *piping* material), find the longest length distance in the first column or the next larger distance if the exact distance is not listed. The use of alternative operating pressures and/or pressure drops will require the use of a different sizing table, but will not alter the sizing methodology. In many cases, the use of alternative operating pressures and/or pressure drops will require the approval of both the code official and the local gas serving utility.

(7) Trace across this row until the gas load is found or the closest larger capacity if the exact capacity is not listed.

(8) Read up the table column and select the appropriate pipe size in the top row. Repeat Steps 6, 7 and 8 for each pipe segment in the longest run.

(9) Size each remaining section of branch *piping* not previously sized by measuring the distance from the gas meter location to the most remote *outlet* in that branch, using the gas load of attached appliances and following the procedures of Steps 2 through 8.

A.3.3 Hybrid pressure method. The sizing of a 2 psi (13.8 kPa) gas *piping* system is performed using the traditional Longest Length Method but with modifications. The 2 psi (13.8 kPa) system consists of two independent pressure zones, and each zone is sized separately. The Hybrid Pressure Method is applied as follows:

The sizing of the 2 psi (13.8 kPa) section (from the meter to the line regulator) is as follows:

(1) Calculate the gas load (by adding up the name plate ratings) from all connected appliances. (In certain circumstances the installed gas load can be increased up to 50 percent to accommodate future addition of appliances.) Ensure that the line regulator capacity is adequate for the calculated gas load and that the required pressure drop (across the regulator) for that capacity does not exceed $^3/_4$ psi (5.2 kPa) for a 2 psi (13.8 kPa) system. If the pressure drop across the regulator is too high (for the connected gas load), select a larger regulator.

(2) Measure the distance from the meter to the line regulator located inside the building.

(3) If there are multiple line regulators, measure the distance from the meter to the regulator furthest removed from the meter.

(4) The maximum allowable pressure drop for the 2 psi (13.8 kPa) section is 1 psi (6.9 kPa).

(5) Referring to the appropriate sizing table (based on *piping* material) for 2 psi (13.8 kPa) systems with a 1 psi (6.9 kPa) pressure drop, find this distance in the first column, or the closest larger distance if the exact distance is not listed.

(6) Trace across this row until the gas load is found or the closest larger capacity if the exact capacity is not listed.

(7) Read up the table column to the top row and select the appropriate pipe size.

(8) If there are multiple regulators in this portion of the *piping* system, each line segment must be sized for its actual gas load, but using the longest length previously determined above.

The low pressure section (all *piping* downstream of the line regulator) is sized as follows:

(1) Determine the gas load for each of the connected appliances.

(2) Starting from the line regulator, divide the piping system into a number of connected segments or independent parallel piping segments, and determine the amount of gas that each segment would carry assuming that all appliances were operated simultaneously. An allowance (in equivalent length of pipe) as determined from Table A.2.2 should be considered for piping segments that include four or more fittings.

(3) For each piping segment, use the actual length or longest length (if there are sub-branchlines) and the calculated gas load for that segment and begin the sizing process as follows:

    (a) Referring to the appropriate sizing table (based on operating pressure and piping material), find the longest length distance in the first column or the closest larger distance if the exact distance is not listed. The use of alternative operating pressures and/or pressure drops will require the use of a different sizing table, but will not alter the sizing methodology. In many cases, the use of alternative operating pressures and/or pressure drops can require the approval of the code official.

    (b) Trace across this row until the appliance gas load is found or the closest larger capacity if the exact capacity is not listed.

    (c) Read up the table column to the top row and select the appropriate pipe size.

    (d) Repeat this process for each segment of the piping system.

A.3.4 Pressure drop per 100 feet method. This sizing method is less conservative than the others, but it allows the designer to immediately see where the largest pressure drop occurs in the system. With this information, modifications can be made to bring the total drop to the critical *appliance* within the limitations that are presented to the designer.

Follow the procedures described in the Longest Length Method for Steps (1) through (4) and (9).

For each *piping* segment, calculate the pressure drop based on pipe size, length as a percentage of 100 feet (30 480 mm) and gas flow. Table A.3.4 shows pressure drop per 100 feet (30 480 mm) for pipe sizes from $^1/_2$ inch (12.7 mm) through 2 inches (51 mm). The sum of pressure drops to the critical *appliance* is subtracted from the supply pressure to verify that sufficient pressure will be available. If not, the layout can be examined to find the high drop section(s) and sizing selections modified.

Note: Other values can be obtained by using the following equation:

$$\text{Desired Value} = MBH \times \sqrt{\frac{\text{Desired Drop}}{\text{Table Drop}}}$$

For example, if it is desired to get flow through $^3/_4$-inch (19.1 mm) pipe at 2 inches/100 feet, multiply the capacity of $^3/_4$-inch (19.1 mm) pipe at 1 inch/100 feet by the square root of the pressure ratio:

$$147\ MBH \times \sqrt{\frac{2''\ \text{w.c.}}{1''\ \text{w.c.}}} = 147 \times 1.414 = 208\ MBH$$

$$(MBH = 1000\ \text{Btu/h})$$

A.4 Use of sizing equations. Capacities of smooth wall pipe or tubing can also be determined by using the following formulae:

(1) High Pressure [1.5 psi (10.3 kPa) and above]:

$$Q = 181.6 \sqrt{\frac{D^5 \cdot (P_1^2 - P_2^2) \cdot Y}{C_r \cdot fba \cdot L}}$$

$$= 2237\ D^{2.623} \left[\frac{(P_1^2 - P_2^2) \cdot Y}{C_r \cdot L}\right]^{0.541}$$

(2) Low Pressure [Less than 1.5 psi (10.3 kPa)]:

$$Q = 187.3 \sqrt{\frac{D^5 \cdot \Delta H}{C_r \cdot fba \cdot L}}$$

$$= 2313 D^{2.623} \left(\frac{\Delta H}{C_r \cdot L}\right)^{0.541}$$

where:

Q = Rate, cubic feet per hour at 60°F and 30-inch mercury column

D = Inside diameter of pipe, in.

P_1 = Upstream pressure, psia

P_2 = Downstream pressure, psia

Y = Superexpansibility factor = 1/supercompressibility factor

C_r = Factor for viscosity, density and temperature*

    = $0.00354\ ST\left(\dfrac{Z}{S}\right)^{0.152}$

Note: See Table 402.4 for Y and C_r for natural gas and propane.

TABLE A.3.4
THOUSANDS OF BTU/H (MBH) OF NATURAL GAS PER 100 FEET OF PIPE AT VARIOUS PRESSURE DROPS AND PIPE DIAMETERS

PRESSURE DROP PER 100 FEET IN INCHES W.C.	PIPE SIZES (inch)					
	$^1/_2$	$^3/_4$	1	$1^1/_4$	$1^1/_2$	2
0.2	31	64	121	248	372	716
0.3	38	79	148	304	455	877
0.5	50	104	195	400	600	1160
1.0	71	147	276	566	848	1640

For SI: 1 inch = 25.4 mm, 1 foot = 304.8 mm.

S = Specific gravity of gas at 60°F and 30-inch mercury column (0.60 for natural gas, 1.50 for propane), or = 1488μ

T = Absolute temperature, °F or = $t + 460$

t = Temperature, °F

Z = Viscosity of gas, centipoise (0.012 for natural gas, 0.008 for propane), or = 1488μ

fba = Base friction factor for air at 60°F (CF = 1)

L = Length of pipe, ft

ΔH = Pressure drop, in. w.c. (27.7 in. H$_2$O = 1 psi)

(For SI, see Section 402.4)

A.5 Pipe and tube diameters. Where the internal diameter is determined by the formulas in Section 402.4, Tables A.5.1 and A.5.2 can be used to select the nominal or standard pipe size based on the calculated internal diameter.

TABLE A.5.1
SCHEDULE 40 STEEL PIPE STANDARD SIZES

NOMINAL SIZE (inch)	INTERNAL DIAMETER (inch)	NOMINAL SIZE (inch)	INTERNAL DIAMETER (inch)
$^1/_4$	0.364	$1^1/_2$	1.610
$^3/_8$	0.493	2	2.067
$^1/_2$	0.622	$2^1/_2$	2.469
$^3/_4$	0.824	3	3.068
1	1.049	$3^1/_2$	3.548
$1^1/_4$	1.380	4	4.026

For SI: 1 inch = 25.4 mm.

A.6 Examples of piping system design and sizing.

A.6.1 Example 1: Longest length method. Determine the required pipe size of each section and *outlet* of the *piping* system shown in Figure A.6.1, with a designated pressure drop of 0.5-inch w.c. (125 Pa) using the Longest Length Method. The gas to be used has 0.60 specific gravity and a heating value of 1,000 Btu/ft^3 (37.5 MJ/m^3).

Solution:

(1) Maximum gas demand for *Outlet* A:

$$\frac{\text{Consumption (rating plate input, or Table 402.2 if necessary}}{\text{Btu of gas}} =$$

$$\frac{35,000 \text{ Btu per hour rating}}{1,000 \text{ Btu per cubic foot}} = 35 \text{ cubic feet per hour} = 35 \text{ cfh}$$

Maximum gas demand for *Outlet* B:

$$\frac{\text{Consumption}}{\text{Btu of gas}} = \frac{75,000}{1,000} = 75 \text{ cfh}$$

Maximum gas demand for *Outlet* C:

$$\frac{\text{Consumption}}{\text{Btu of gas}} = \frac{35,000}{1,000} = 35 \text{ cfh}$$

Maximum gas demand for *Outlet* D:

$$\frac{\text{Consumption}}{\text{Btu of gas}} = \frac{100,000}{1,000} = 100 \text{ cfh}$$

(2) The length of pipe from the *point of delivery* to the most remote *outlet* (A) is 60 feet (18 288 mm). This is the only distance used.

(3) Using the row marked 60 feet (18 288 mm) in Table 402.4(2):

(a) *Outlet* A, supplying 35 cfh (0.99 m^3/hr), requires $^1/_2$-inch pipe.

TABLE A.5.2
COPPER TUBE STANDARD SIZES

TUBE TYPE	NOMINAL OR STANDARD SIZE (inches)	INTERNAL DIAMETER (inches)
K	$^1/_4$	0.305
L	$^1/_4$	0.315
ACR (D)	$^3/_8$	0.315
ACR (A)	$^3/_8$	0.311
K	$^3/_8$	0.402
L	$^3/_8$	0.430
ACR (D)	$^1/_2$	0.430
ACR (A)	$^1/_2$	0.436
K	$^1/_2$	0.527
L	$^1/_2$	0.545
ACR (D)	$^5/_8$	0.545
ACR (A)	$^5/_8$	0.555
K	$^5/_8$	0.652
L	$^5/_8$	0.666
ACR (D)	$^3/_4$	0.666
ACR (A)	$^3/_4$	0.680
K	$^3/_4$	0.745
L	$^3/_4$	0.785
ACR	$^7/_8$	0.785
K	1	0.995
L	1	1.025
ACR	$1^1/_8$	1.025
K	$1^1/_4$	1.245
L	$1^1/_4$	1.265
ACR	$1^3/_8$	1.265
K	$1^1/_2$	1.481
L	$1^1/_2$	1.505
ACR	$1^5/_8$	1.505
K	2	1.959
L	2	1.985
ACR	$2^1/_8$	1.985
K	$2^1/_2$	2.435
L	$2^1/_2$	2.465
ACR	$2^5/_8$	2.465
K	3	2.907
L	3	2.945
ACR	$3^1/_8$	2.945

For SI: 1 inch = 25.4 mm.

(b) *Outlet* B, supplying 75 cfh (2.12 m³/hr), requires ³/₄-inch pipe.

(c) Section 1, supplying *Outlets* A and B, or 110 cfh (3.11 m³/hr), requires ³/₄-inch pipe.

(d) Section 2, supplying *Outlets* C and D, or 135 cfh (3.82 m³/hr), requires ³/₄-inch pipe.

(e) Section 3, supplying *Outlets* A, B, C and D, or 245 cfh (6.94 m³/hr), requires 1-inch pipe.

(4) If a different gravity factor is applied to this example, the values in the row marked 60 feet (18 288 mm) of Table 402.4(2) would be multiplied by the appropriate multiplier from Table A.2.4 and the resulting cubic feet per hour values would be used to size the *piping*.

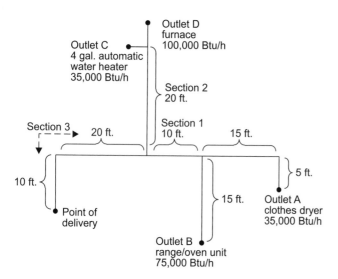

FIGURE A.6.1
PIPING PLAN SHOWING A STEEL PIPING SYSTEM

A.6.2 Example 2: Hybrid or dual pressure systems. Determine the required CSST size of each section of the *piping* system shown in Figure A.6.2, with a designated pressure drop of 1 psi (6.9 kPa) for the 2 psi (13.8 kPa) section and 3-inch w.c. (0.75 kPa) pressure drop for the 13-inch w.c. (2.49 kPa) section. The gas to be used has 0.60 specific gravity and a heating value of 1,000 Btu/ft³ (37.5 MJ/ m³).

Solution:

(1) Size 2 psi (13.8 kPa) line using Table 402.4(18).

(2) Size 10-inch w.c. (2.5 kPa) lines using Table 402.4(16).

(3) Using the following, determine if sizing tables can be used.

(a) Total gas load shown in Figure A.6.2 equals 110 cfh (3.11 m³/hr).

(b) Determine pressure drop across regulator [see notes in Table 402.4(18)].

(c) If pressure drop across regulator exceeds ³/₄ psig (5.2 kPa), Table 402.4(18) cannot be used. Note: If pressure drop exceeds ³/₄ psi (5.2 kPa), then a larger regulator must be selected or an alternative sizing method must be used.

(d) Pressure drop across the line regulator [for 110 cfh (3.11 m³/hr)] is 4-inch w.c. (0.99 kPa) based on manufacturer's performance data.

(e) Assume the CSST manufacturer has tubing sizes or EHDs of 13, 18, 23 and 30.

(4) Section A [2 psi (13.8 kPa) zone]

(a) Distance from meter to regulator = 100 feet (30 480 mm).

(b) Total load supplied by A = 110 cfh (3.11 m³/hr) (furnace + water heater + dryer).

(c) Table 402.4(18) shows that EHD size 18 should be used.

Note: It is not unusual to oversize the supply line by 25 to 50 percent of the as-installed load. EHD size 18 has a capacity of 189 cfh (5.35 m³/ hr).

(5) Section B (low pressure zone)

(a) Distance from regulator to furnace is 15 feet (4572 mm).

(b) Load is 60 cfh (1.70 m³/hr).

(c) Table 402.4(16) shows that EHD size 13 should be used.

(6) Section C (low pressure zone)

(a) Distance from regulator to water heater is 10 feet (3048 mm).

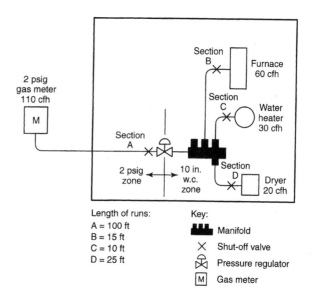

FIGURE A.6.2
PIPING PLAN SHOWING A CSST SYSTEM

(b) Load is 30 cfh (0.85 m³/hr).

(c) Table 402.4(16) shows that EHD size 13 should be used.

(7) Section D (low pressure zone)

(a) Distance from regulator to dryer is 25 feet (7620 mm).

(b) Load is 20 cfh (0.57 m³/hr).

(c) Table 402.4(16) shows that EHD size 13 should be used.

A.6.3 Example 3: Branch length method. Determine the required semirigid copper tubing size of each section of the *piping* system shown in Figure A.6.3, with a designated pressure drop of 1-inch w.c. (250 Pa) (using the Branch Length Method). The gas to be used has 0.60 specific gravity and a heating value of 1,000 Btu/ft³ (37.5 MJ/m³).

Solution:

(1) Section A

(a) The length of tubing from the *point of delivery* to the most remote *appliance* is 50 feet (15 240 mm), A + C.

(b) Use this longest length to size Sections A and C.

(c) Using the row marked 50 feet (15 240 mm) in Table 402.4(10), Section A, supplying 220 cfh (6.2 m³/hr) for four appliances requires 1-inch tubing.

(2) Section B

(a) The length of tubing from the *point of delivery* to the range/oven at the end of Section B is 30 feet (9144 mm), A + B.

(b) Use this branch length to size Section B only.

(c) Using the row marked 30 feet (9144 mm) in Table 402.4(10), Section B, supplying 75 cfh (2.12 m³/hr) for the range/oven requires ¹/₂-inch tubing.

(3) Section C

(a) The length of tubing from the *point of delivery* to the dryer at the end of Section C is 50 feet (15 240 mm), A + C.

(b) Use this branch length to size Section C.

(c) Using the row marked 50 feet (15 240 mm) in Table 402.4(10), Section C, supplying 30 cfh (0.85 m³/hr) for the dryer requires ³/₈-inch tubing.

(4) Section D

(a) The length of tubing from the *point of delivery* to the water heater at the end of Section D is 30 feet (9144 mm), A + D.

(b) Use this branch length to size Section D only.

(c) Using the row marked 30 feet (9144 mm) in Table 402.4(10), Section D, supplying 35 cfh (0.99 m³/hr) for the water heater requires ³/₈-inch tubing.

(5) Section E

(a) The length of tubing from the *point of delivery* to the furnace at the end of Section E is 30 feet (9144 mm), A + E.

(b) Use this branch length to size Section E only.

(c) Using the row marked 30 feet (9144 mm) in Table 402.4(10), Section E, supplying 80 cfh (2.26 m³/hr) for the furnace requires ¹/₂-inch tubing.

A.6.4 Example 4: Modification to existing piping system. Determine the required CSST size for Section G (retrofit application) of the *piping* system shown in Figure A.6.4, with a designated pressure drop of 0.5-inch w.c. (125 Pa) using the branch length method. The gas to be used has 0.60 specific gravity and a heating value of 1,000 Btu/ft³ (37.5 MJ/m³).

Solution:

(1) The length of pipe and CSST from the *point of delivery* to the retrofit *appliance* (barbecue) at the end of Section G is 40 feet (12 192 mm), A + B + G.

(2) Use this branch length to size Section G.

(3) Assume the CSST manufacturer has tubing sizes or EHDs of 13, 18, 23 and 30.

(4) Using the row marked 40 feet (12 192 mm) in Table 402.4(15), Section G, supplying 40 cfh (1.13 m³/hr) for the barbecue requires EHD 18 CSST.

(5) The sizing of Sections A, B, F and E must be checked to ensure adequate gas carrying capacity since an *appliance* has been added to the *piping* system (see A.6.1 for details).

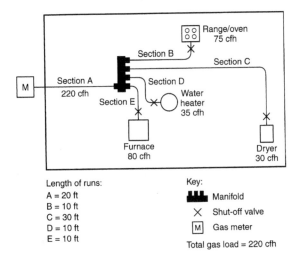

FIGURE A.6.3
PIPING PLAN SHOWING A COPPER TUBING SYSTEM

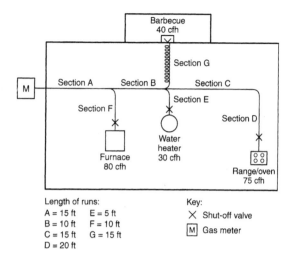

Length of runs:

A = 15 ft   E = 5 ft
B = 10 ft   F = 10 ft
C = 15 ft   G = 15 ft
D = 20 ft

Key:

✕ Shut-off valve
Ⓜ Gas meter

**FIGURE A.6.4
PIPING PLAN SHOWING A MODIFICATION
TO EXISTING PIPING SYSTEM**

A.6.5 Example 5: Calculating pressure drops due to temperature changes. A test *piping* system is installed on a warm autumn afternoon when the temperature is 70°F (21°C). In accordance with local custom, the new *piping* system is subjected to an air pressure test at 20 psig (138 kPa). Overnight, the temperature drops and when the inspector shows up first thing in the morning the temperature is 40°F (4°C).

If the volume of the *piping* system is unchanged, then the formula based on Boyle's and Charles' law for determining the new pressure at a reduced temperature is as follows:

$$\frac{T_1}{T_2} = \frac{P_1}{P_2}$$

where:

T_1 = Initial temperature, absolute (T_1 + 459)

T_2 = Final temperature, absolute (T_2 + 459)

P_1 = Initial pressure, psia (P_1 + 14.7)

P_2 = Final pressure, psia (P_2 + 14.7)

$$\frac{(70 + 459)}{(40 + 459)} = \frac{(20 + 14.7)}{(P_2 + 14.7)}$$

$$\frac{529}{499} = \frac{34.7}{(P_2 + 14.7)}$$

$$(P_2 + 14.7) \times \frac{529}{499} = 34.7$$

$$(P_2 + 14.7) \times \frac{34.7}{1.060}$$

P_2 = 32.7 - 14.7

P_2 = 18 *psig*

Therefore, the gauge could be expected to register 18 psig (124 kPa) when the ambient temperature is 40°F (4°C).

A.6.6 Example 6: Pressure drop per 100 feet of pipe method. Using the layout shown in Figure A.6.1 and ΔH = pressure drop, in w.c. (27.7 in. H_2O = 1 psi), proceed as follows:

(1) Length to A = 20 feet, with 35,000 Btu/hr.

For $^1/_2$-inch pipe, $\Delta H = {}^{20\ feet}/_{100\ feet} \times 0.3$ inch w.c. = 0.06 in w.c.

(2) Length to B = 15 feet, with 75,000 Btu/hr.

For $^3/_4$-inch pipe, $\Delta H = {}^{15\ feet}/_{100\ feet} \times 0.3$ inch w.c. = 0.045 in w.c.

(3) Section 1 = 10 feet, with 110,000 Btu/hr. Here there is a choice:

For 1-inch pipe: $\Delta H = {}^{10\ feet}/_{100\ feet} \times 0.2$ inch w.c. = 0.02 in w.c.

For $^3/_4$-inch pipe: $\Delta H = {}^{10\ feet}/_{100\ feet} \times [0.5$ inch w.c. + $^{(110,000\ Btu/hr-104,000\ Btu/hr)}/_{(147,000\ Btu/hr-104,000\ Btu/hr)} \times (1.0$ inches w.c. - 0.5 inch w.c.)] = 0.1 × 0.57 inch w.c.≈ 0.06 inch w.c.

Note that the pressure drop between 104,000 Btu/hr and 147,000 Btu/hr has been interpolated as 110,000 Btu/hr.

(4) Section 2 = 20 feet, with 135,000 Btu/hr. Here there is a choice:

For 1-inch pipe: $\Delta H = {}^{20\ feet}/_{100\ feet} \times [0.2$ inch w.c. + $^{(14,000\ Btu/hr)}/_{(27,000\ Btu/hr)} \times 0.1$ inch w.c.] = 0.05 inch w.c.

For $^3/_4$-inch pipe: $\Delta H = {}^{20\ feet}/_{100\ feet} \times 1.0$ inch w.c. = 0.2 inch w.c.

Note that the pressure drop between 121,000 Btu/hr and 148,000 Btu/hr has been interpolated as 135,000 Btu/hr, but interpolation for the ¾-inch pipe (trivial for 104,000 Btu/hr to 147,000 Btu/hr) was not used.

(5) Section 3 = 30 feet, with 245,000 Btu/hr. Here there is a choice:

For 1-inch pipe: $\Delta H = {}^{30\ feet}/_{100\ feet} \times 1.0$ inches w.c. = 0.3 inch w.c.

For $1^1/_4$-inch pipe: $\Delta H = {}^{30\ feet}/_{100\ feet} \times 0.2$ inch w.c. = 0.06 inch w.c.

Note that interpolation for these options is ignored since the table values are close to the 245,000 Btu/hr carried by that section.

(6) The total pressure drop is the sum of the section approaching A, Sections 1 and 3, or either of the following, depending on whether an absolute minimum is needed or the larger drop can be accommodated.

Minimum pressure drop to farthest *appliance*:

ΔH = 0.06 inch w.c. + 0.02 inch w.c. + 0.06 inch w.c. = 0.14 inch w.c.

Larger pressure drop to the farthest *appliance*:

ΔH = 0.06 inch w.c. + 0.06 inch w.c. + 0.3 inch w.c. = 0.42 inch w.c.

Notice that Section 2 and the run to B do not enter into this calculation, provided that the appliances have similar input pressure requirements.

For SI units: 1 Btu/hr = 0.293 W, 1 cubic foot = 0.028 m³, 1 foot = 0.305 m, 1 inch w.c. = 249 Pa.

Appendix B:
Sizing of Venting Systems Serving Appliances Equipped with Draft Hoods, Category I Appliances, and Appliances Listed for Use with Type B Vents

(This appendix is informative and is not part of the *code*. This appendix is an excerpt from the 2015 *International Fuel Gas Code*, coordinated with the section numbering of the *International Residential Code*.)

EXAMPLES USING SINGLE APPLIANCE VENTING TABLES

Example 1: Single draft-hood-equipped appliance.

An installer has a 120,000 British thermal unit (Btu) per hour input *appliance* with a 5-inch-diameter draft hood outlet that needs to be vented into a 10-foot-high Type B vent system. What size vent should be used assuming (a) a 5-foot lateral single-wall metal vent connector is used with two 90-degree elbows, or (b) a 5-foot lateral single-wall metal vent connector is used with three 90-degree elbows in the vent system?

Solution:

Table 504.2(2) should be used to solve this problem, because single-wall metal vent connectors are being used with a Type B vent.

(a) Read down the first column in Table 504.2(2) until the row associated with a 10-foot height and 5-foot lateral is found. Read across this row until a vent capacity greater than 120,000 Btu per hour is located in the shaded columns *labeled* "NAT Max" for draft-hood-equipped appliances. In this case, a 5-inch-diameter vent has a capacity of 122,000 Btu per hour and can be used for this application.

(b) If three 90-degree elbows are used in the vent system, then the maximum vent capacity listed in the tables must be reduced by 10 percent (see Section 504.2.3 for single *appliance* vents). This implies that the 5-inch-diameter vent has an adjusted capacity of only 110,000 Btu per hour. In this case, the vent system must be increased to 6 inches in diameter (see calculations below).

> 122,000 (0.90) = 110,000 for 5-inch vent
> From Table 504.2(2), Select 6-inch vent
> 186,000 (0.90) = 167,000; This is greater than the required 120,000. Therefore, use a 6-inch vent and connector where three elbows are used.

Example 2: Single fan-assisted appliance.

An installer has an 80,000 Btu per hour input fan-assisted *appliance* that must be installed using 10 feet of lateral connector attached to a 30-foot-high Type B vent. Two 90-degree elbows are needed for the installation. Can a single-wall metal vent connector be used for this application?

Solution:

Table 504.2(2) refers to the use of single-wall metal vent connectors with Type B vent. In the first column find the row associated with a 30-foot height and a 10-foot lateral. Read across this row, looking at the FAN Min and FAN Max columns, to find that a 3-inch-diameter single-wall metal vent connector is not recommended. Moving to the next larger size single wall connector (4 inches), note that a 4-inch-diameter single-wall metal connector has a recommended minimum vent capacity of 91,000 Btu per hour and a recommended maximum vent capacity of 144,000 Btu per hour. The 80,000 Btu per hour fan-assisted *appliance* is outside this range, so the conclusion is that a single-wall metal vent connector cannot be used to vent this *appliance* using 10 feet of lateral for the connector.

However, if the 80,000 Btu per hour input *appliance* could be moved to within 5 feet of the vertical vent, then a 4-inch single-wall metal connector could be used to vent the *appliance*. Table 504.2(2) shows the acceptable range of vent capacities for a 4-inch vent with 5 feet of lateral to be between 72,000 Btu per hour and 157,000 Btu per hour.

If the *appliance* cannot be moved closer to the vertical vent, then Type B vent could be used as the connector material. In this case, Table 504.2(1) shows that for a 30-foot-high vent with 10 feet of lateral, the acceptable range of vent capacities for a 4-inch-diameter vent attached to a fan-assisted *appliance* is between 37,000 Btu per hour and 150,000 Btu per hour.

Example 3: Interpolating between table values.

An installer has an 80,000 Btu per hour input *appliance* with a 4-inch-diameter draft hood outlet that needs to be vented into a 12-foot-high Type B vent. The vent connector has a 5-foot lateral length and is also Type B. Can this *appliance* be vented using a 4-inch-diameter vent?

Solution:

Table 504.2(1) is used in the case of an all Type B vent system. However, since there is no entry in Table 504.2(1) for a height of 12 feet, interpolation must be used. Read down the 4-inch diameter NAT Max column to the row associated with 10-foot height and 5-foot lateral to find the capacity value of 77,000 Btu per hour. Read further down to the 15-foot height, 5-foot lateral row to find the capacity value of 87,000 Btu per hour. The difference between the 15-foot height capacity value and the 10-foot height capacity value is 10,000 Btu per hour. The capacity for a vent system with a 12-foot height is equal to the capacity for a 10-foot height plus $^2/_5$ of the difference between the 10-foot and 15-foot height values, or $77,000 + {}^2/_5 (10,000) = 81,000$ Btu per hour. Therefore, a 4-inch-diameter vent can be used in the installation.

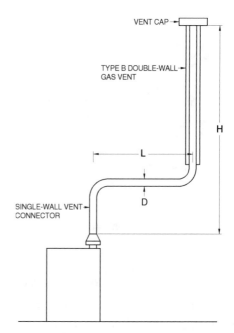

For SI: 1 foot = 304.8 mm, 1 British thermal unit per hour = 0.2931 W.

Table 504.2(2) is used when sizing a single-wall metal vent connector attached to a Type B double-wall gas vent.

Note: The appliance may be either Category I draft hood equipped or fan-assisted type.

FIGURE B-2
TYPE B DOUBLE-WALL VENT SYSTEM SERVING A SINGLE APPLIANCE WITH A SINGLE-WALL METAL VENT CONNECTOR

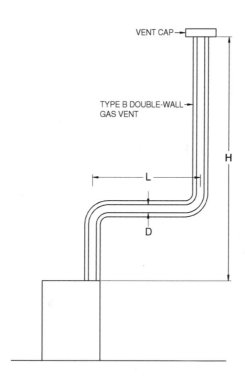

For SI: 1 foot = 304.8 mm, 1 British thermal unit per hour = 0.2931 W.

Table 504.2(1) is used when sizing Type B double-wall gas vent connected directly to the appliance.

Note: The appliance may be either Category I draft hood equipped or fan-assisted type.

FIGURE B-1
TYPE B DOUBLE-WALL VENT SYSTEM SERVING A SINGLE APPLIANCE WITH A TYPE B DOUBLE-WALL VENT

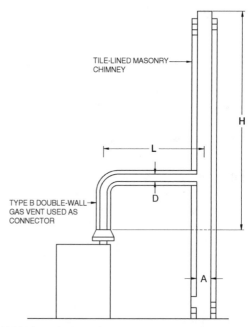

Table 504.2(3) is used when sizing a Type B double-wall gas vent connector attached to a tile-lined masonry chimney.

Note: "A" is the equivalent cross-sectional area of the tile liner.

Note: The appliance can be either Category I draft hood equipped or fan-assisted type.

FIGURE B-3
VENT SYSTEM SERVING A SINGLE APPLIANCE WITH A MASONRY CHIMNEY OF TYPE B DOUBLE-WALL VENT CONNECTOR

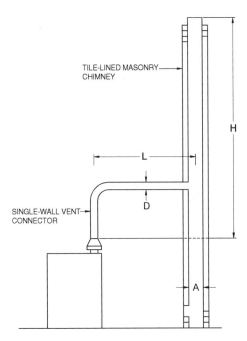

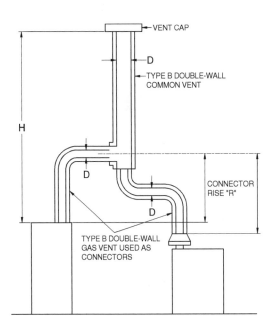

Table 504.2(4) is used when sizing a single-wall vent connector attached to a tile-lined masonry chimney.

Note: "A" is the equivalent cross-sectional area of the tile liner.

Note: The appliance can be either Category I draft hood equipped or fan-assisted type.

FIGURE B-4
VENT SYSTEM SERVING A SINGLE APPLIANCE
USING A MASONRY CHIMNEY AND A
SINGLE-WALL METAL VENT CONNECTOR

Table 504.3(1) is used when sizing Type B double-wall vent connectors attached to a Type B double-wall common vent.

Note: Each appliance can be either Category I draft hood equipped or fan-assisted type.

FIGURE B-6
VENT SYSTEM SERVING TWO OR MORE APPLIANCES
WITH TYPE B DOUBLE-WALL VENT AND TYPE B
DOUBLE-WALL VENT CONNECTOR

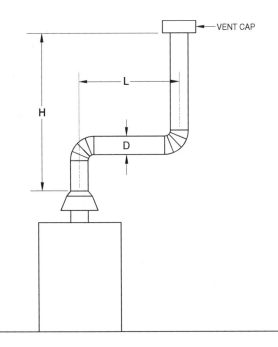

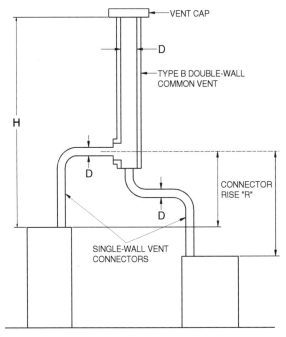

Asbestos cement Type B or single-wall metal vent serving a single draft-hood-equipped appliance [see Table 504.2(5)].

FIGURE B-5
ASBESTOS CEMENT TYPE B OR SINGLE-WALL
METAL VENT SYSTEM SERVING A SINGLE
DRAFT-HOOD-EQUIPPED APPLIANCE

Table 504.3(2) is used when sizing single-wall vent connectors attached to a Type B double-wall common vent.

Note: Each appliance can be either Category I draft hood equipped or fan-assisted type.

FIGURE B-7
VENT SYSTEM SERVING TWO OR MORE APPLIANCES
WITH TYPE B DOUBLE-WALL VENT AND
SINGLE-WALL METAL VENT CONNECTORS

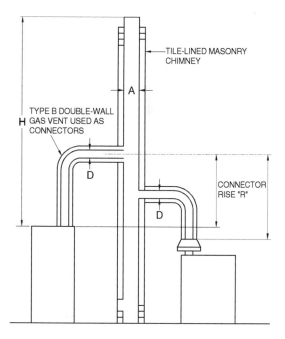

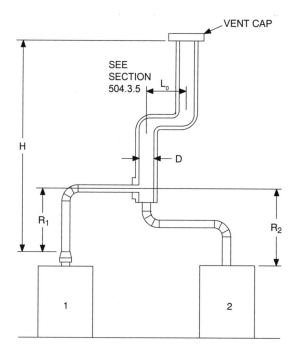

Table 504.3(3) is used when sizing Type B double-wall vent connectors attached to a tile-lined masonry chimney.

Note: "A" is the equivalent cross-sectional area of the tile liner.

Note: Each appliance can be either Category I draft hood equipped or fan-assisted type.

FIGURE B-8
MASONRY CHIMNEY SERVING TWO OR MORE APPLIANCES WITH TYPE B DOUBLE-WALL VENT CONNECTOR

Asbestos cement Type B or single-wall metal pipe vent serving two or more draft-hood-equipped appliances [see Table 504.3(5)].

FIGURE B-10
ASBESTOS CEMENT TYPE B OR SINGLE-WALL METAL VENT SYSTEM SERVING TWO OR MORE DRAFT-HOOD-EQUIPPED APPLIANCES

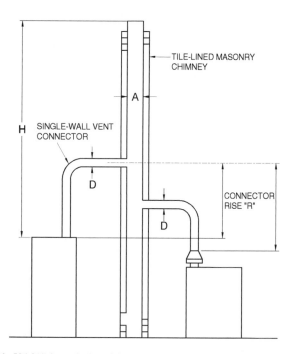

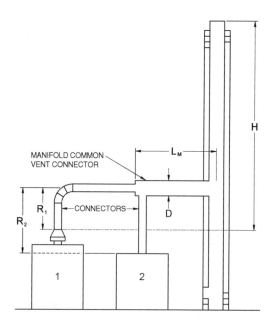

Table 504.3(4) is used when sizing single-wall metal vent connectors attached to a tile-lined masonry chimney.

Note: "A" is the equivalent cross-sectional area of the tile liner.

Note: Each appliance can be either Category I draft hood equipped or fan-assisted type.

FIGURE B-9
MASONRY CHIMNEY SERVING TWO OR MORE APPLIANCES WITH SINGLE-WALL METAL VENT CONNECTORS

Example: Manifolded Common Vent Connector L_u shall be no greater than 18 times the common vent connector manifold inside diameter; i.e., a 4-inch (102 mm) inside diameter common vent connector manifold shall not exceed 72 inches (1829 mm) in length (see Section 504.3.4).

Note: This is an illustration of a typical manifolded vent connector. Different appliance, vent connector, or common vent types are possible. Consult Section 502.3.

FIGURE B-11
USE OF MANIFOLD COMMON VENT CONNECTOR

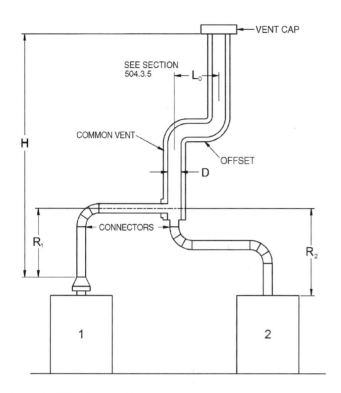

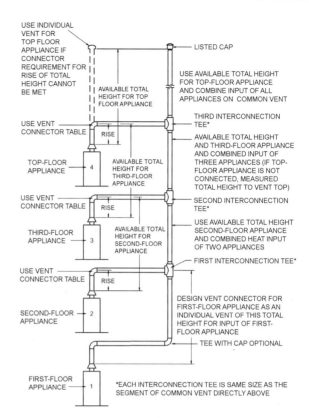

Example: Offset Common Vent

Note: This is an illustration of a typical offset vent. Different appliance, vent connector, or vent types are possible. Consult Sections 504.2 and 504.3.

FIGURE B-12
USE OF OFFSET COMMON VENT

Principles of design of multistory vents using vent connector and common vent design tables (see Sections 504.3.11 through 504.3.17).

FIGURE B-14
MULTISTORY VENT SYSTEMS

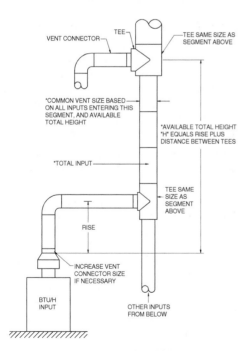

Vent connector size depends on:
- Input
- Rise
- Available total height "H"
- Table 504.3(1) connectors

Common vent size depends on:
- Combined inputs
- Available total height "H"
- Table 504.3(1) common vent

FIGURE B-13
MULTISTORY GAS VENT DESIGN PROCEDURE FOR EACH SEGMENT OF SYSTEM

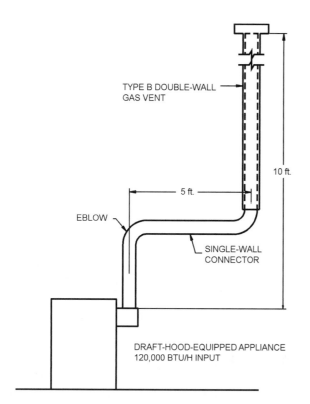

For SI: 1 foot = 304.8 mm, 1 British thermal unit per hour = 0.2931 W.

FIGURE B-15 (EXAMPLE 1)
SINGLE DRAFT-HOOD-EQUIPPED APPLIANCE

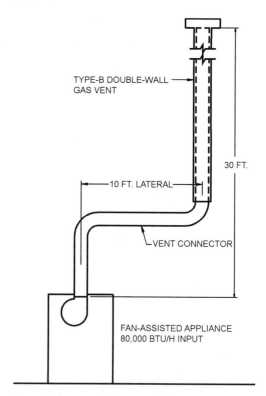

For SI: 1 foot = 304.8 mm, 1 British thermal unit per hour = 0.2931 W.

FIGURE B-16 (EXAMPLE 2)
SINGLE FAN-ASSISTED APPLIANCE

EXAMPLES USING COMMON VENTING TABLES

Example 4: Common venting two draft-hood-equipped appliances.

A 35,000 Btu per hour water heater is to be common vented with a 150,000 Btu per hour furnace using a common vent with a total height of 30 feet. The connector rise is 2 feet for the water heater with a horizontal length of 4 feet. The connector rise for the furnace is 3 feet with a horizontal length of 8 feet. Assume single-wall metal connectors will be used with Type B vent. What size connectors and combined vent should be used in this installation?

Solution:

Table 504.3(2) should be used to size single-wall metal vent connectors attached to Type B vertical vents. In the vent connector capacity portion of Table 504.3(2), find the row associated with a 30-foot vent height. For a 2-foot rise on the vent connector for the water heater, read the shaded columns for draft-hood-equipped appliances to find that a 3-inch-diameter vent connector has a capacity of 37,000 Btu per hour. Therefore, a 3-inch single-wall metal vent connector can be used with the water heater. For a draft-hood-equipped furnace with a 3-foot rise, read across the appropriate row to find that a 5-inch-diameter vent connector has a maximum capacity of 120,000 Btu per hour (which is too small for the furnace) and a 6-inch-diameter vent connector has a maximum vent capacity of 172,000 Btu per hour. Therefore, a 6-inch-diameter vent connector should be used with the 150,000 Btu per hour furnace. Since both vent connector horizontal lengths are less

than the maximum lengths *listed* in Section 504.3.2, the table values can be used without adjustments.

In the common vent capacity portion of Table 504.3(2), find the row associated with a 30-foot vent height and read over to the NAT + NAT portion of the 6-inch-diameter column to find a maximum combined capacity of 257,000 Btu per hour. Since the two appliances total only 185,000 Btu per hour, a 6-inch common vent can be used.

Example 5a: Common venting a draft-hood-equipped water heater with a fan-assisted furnace into a Type B vent.

In this case, a 35,000 Btu per hour input draft-hood-equipped water heater with a 4-inch-diameter draft hood *outlet*, 2 feet of connector rise, and 4 feet of horizontal length is to be common vented with a 100,000 Btu per hour fan-assisted furnace with a 4-inch-diameter flue collar, 3 feet of connector rise, and 6 feet of horizontal length. The common vent consists of a 30-foot height of Type B vent. What are the recommended vent diameters for each connector and the common vent? The installer would like to use a single-wall metal vent connector.

Solution: [Table 504.3(2)].

Water Heater Vent Connector Diameter. Since the water heater vent connector horizontal length of 4 feet is less than the maximum value listed in Section 504.3.2, the venting table values can be used without adjustments. Using the Vent Connector Capacity portion of Table 504.3(2), read down the

Total Vent Height (*H*) column to 30 feet and read across the 2-foot Connector Rise (*R*) row to the first Btu per hour rating in the NAT Max column that is equal to or greater than the water heater input rating. The table shows that a 3-inch vent connector has a maximum input rating of 37,000 Btu per hour. Although this is greater than the water heater input rating, a 3-inch vent connector is prohibited by Section 504.3.21. A 4-inch vent connector has a maximum input rating of 67,000 Btu per hour and is equal to the draft hood *outlet* diameter. A 4-inch vent connector is selected. Since the water heater is equipped with a draft hood, there are no minimum input rating restrictions.

Furnace Vent Connector Diameter. Using the Vent Connector Capacity portion of Table 504.3(2), read down the Total Vent Height (*H*) column to 30 feet and across the 3-foot Connector Rise (*R*) row. Since the furnace has a fan-assisted combustion system, find the first FAN Max column with a Btu per hour rating greater than the furnace input rating. The 4-inch vent connector has a maximum input rating of 119,000 Btu per hour and a minimum input rating of 85,000 Btu per hour. The 100,000 Btu per hour furnace in this example falls within this range, so a 4-inch connector is adequate. Since the furnace vent connector horizontal length of 6 feet does not exceed the maximum value listed in Section 504.3.2, the venting table values can be used without adjustment. If the furnace had an input rating of 80,000 Btu per hour, then a Type B vent connector [see Table 504.3(1)] would be needed in order to meet the minimum capacity limit.

Common Vent Diameter. The total input to the common vent is 135,000 Btu per hour. Using the Common Vent Capacity portion of Table 504.3(2), read down the Total Vent

Height (*H*) column to 30 feet and across this row to find the smallest vent diameter in the FAN + NAT column that has a Btu per hour rating equal to or greater than 135,000 Btu per hour. The 4-inch common vent has a capacity of 132,000 Btu per hour and the 5-inch common vent has a capacity of 202,000 Btu per hour. Therefore, the 5-inch common vent should be used in this example.

Summary. In this example, the installer can use a 4-inch-diameter, single-wall metal vent connector for the water heater and a 4-inch-diameter, single-wall metal vent connector for the furnace. The common vent should be a 5-inch-diameter Type B vent.

Example 5b: Common venting into a masonry chimney.

In this case, the water heater and fan-assisted furnace of Example 5a are to be common vented into a clay tile-lined masonry chimney with a 30-foot height. The chimney is not exposed to the outdoors below the roof line. The internal dimensions of the clay tile liner are nominally 8 inches by 12 inches. Assuming the same vent connector heights, laterals, and materials found in Example 5a, what are the recommended vent connector diameters, and is this an acceptable installation?

Solution:

Table 504.3(4) is used to size common venting installations involving single-wall connectors into masonry chimneys.

Water Heater Vent Connector Diameter. Using Table 504.3(4), Vent Connector Capacity, read down the Total Vent Height (*H*) column to 30 feet, and read across the 2-foot Connector Rise (*R*) row to the first Btu per hour rating in the

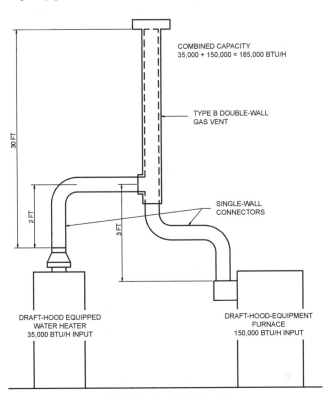

FIGURE B-17 (EXAMPLE 4)
COMMON VENTING TWO DRAFT-HOOD-EQUIPPED APPLIANCES

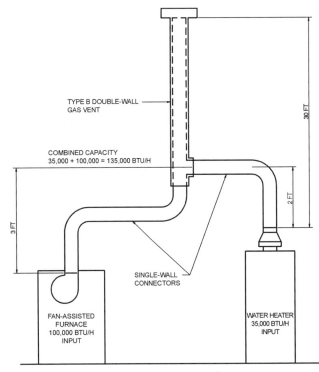

FIGURE B-18 (EXAMPLE 5A)
COMMON VENTING A DRAFT HOOD WITH A FAN-ASSISTED FURNACE INTO A TYPE B DOUBLE-WALL COMMON VENT

NAT Max column that is equal to or greater than the water heater input rating. The table shows that a 3-inch vent connector has a maximum input of only 31,000 Btu per hour while a 4-inch vent connector has a maximum input of 57,000 Btu per hour. A 4-inch vent connector must therefore be used.

Furnace Vent Connector Diameter. Using the Vent Connector Capacity portion of Table 504.3(4), read down the Total Vent Height (*H*) column to 30 feet and across the 3-foot Connector Rise (*R*) row. Since the furnace has a fan-assisted combustion system, find the first FAN Max column with a Btu per hour rating greater than the furnace input rating. The 4-inch vent connector has a maximum input rating of 127,000 Btu per hour and a minimum input rating of 95,000 Btu per hour. The 100,000 Btu per hour furnace in this example falls within this range, so a 4-inch connector is adequate.

Masonry Chimney. From Table B-1, the equivalent area for a nominal liner size of 8 inches by 12 inches is 63.6 square inches. Using Table 504.3(4), Common Vent Capacity, read down the FAN + NAT column under the Minimum Internal Area of Chimney value of 63 to the row for 30-foot height to find a capacity value of 739,000 Btu per hour. The combined input rating of the furnace and water heater, 135,000 Btu per hour, is less than the table value, so this is an acceptable installation.

Section 504.3.17 requires the common vent area to be no greater than seven times the smallest *listed appliance* categorized vent area, flue collar area, or draft hood outlet area. Both appliances in this installation have 4-inch-diameter outlets. From Table B-1, the equivalent area for an inside diameter of 4 inches is 12.2 square inches. Seven times 12.2 equals 85.4, which is greater than 63.6, so this configuration is acceptable.

Example 5c: Common venting into an exterior masonry chimney.

In this case, the water heater and fan-assisted furnace of Examples 5a and 5b are to be common vented into an exterior masonry chimney. The chimney height, clay tile liner dimensions, and vent connector heights and laterals are the same as in Example 5b. This system is being installed in Charlotte, North Carolina. Does this exterior masonry chimney need to be relined? If so, what corrugated metallic liner size is recommended? What vent connector diameters are recommended?

Solution:

In accordance with Section 504.3.20, Type B vent connectors are required to be used with exterior masonry chimneys. Use Tables 504.3(7a), (7b) to size FAN+NAT common venting installations involving Type-B double wall connectors into exterior masonry chimneys.

The local 99-percent winter design temperature needed to use Table 504.3(7b) can be found in the ASHRAE *Handbook of Fundamentals*. For Charlotte, North Carolina, this design temperature is 19°F.

Chimney Liner Requirement. As in Example 5b, use the 63 square inch Internal Area columns for this size clay tile liner. Read down the 63 square inch column of Table 504.3(7a) to the 30-foot height row to find that the combined *appliance* maximum input is 747,000 Btu per hour. The combined input rating of the appliances in this installation, 135,000 Btu per hour, is less than the maximum value, so this criterion is satisfied. Table 504.3(7b), at a 19°F design temperature, and at the same vent height and internal area used above, shows that the minimum allowable input rating of a space-heating appliance is 470,000 Btu per hour. The furnace input rating of 100,000 Btu per hour is less than this minimum value. So this criterion is not satisfied, and an alternative venting design needs to be used, such as a Type B vent shown in Example 5a or a *listed* chimney liner system shown in the remainder of the example.

In accordance with Section 504.3.19, Table 504.3(1) or 504.3(2) is used for sizing corrugated metallic liners in masonry chimneys, with the maximum common vent capacities reduced by 20 percent. This example will be continued assuming Type B vent connectors.

Water Heater Vent Connector Diameter. Using Table 504.3(1), Vent Connector Capacity, read down the Total Vent Height (*H*) column to 30 feet, and read across the 2-foot Connector Rise (*R*) row to the first Btu/h rating in the NAT Max column that is equal to or greater than the water heater input rating. The table shows that a 3-inch vent connector has a maximum capacity of 39,000 Btu/h. Although this rating is greater than the water heater input rating, a 3-inch vent connector is prohibited by Section 504.3.21. A 4-inch vent connector has a maximum input rating of 70,000 Btu/h and is equal to the draft hood outlet diameter. A 4-inch vent connector is selected.

Furnace Vent Connector Diameter. Using Table 504.3(1), Vent Connector Capacity, read down the Vent Height (*H*) column to 30 feet, and read across the 3-foot Connector Rise (*R*) row to the first Btu per hour rating in the FAN Max column that is equal to or greater than the furnace input rating. The 100,000 Btu per hour furnace in this example falls within this range, so a 4-inch connector is adequate.

Chimney Liner Diameter. The total input to the common vent is 135,000 Btu per hour. Using the Common Vent Capacity Portion of Table 504.3(1), read down the Vent Height (*H*) column to 30 feet and across this row to find the smallest vent diameter in the FAN+NAT column that has a Btu per hour rating greater than 135,000 Btu per hour. The 4-inch common vent has a capacity of 138,000 Btu per hour. Reducing the maximum capacity by 20 percent (Section 504.3.19) results in a maximum capacity for a 4-inch corrugated liner of 110,000 Btu per hour, less than the total input of 135,000 Btu per hour. So a larger liner is needed. The 5-inch common vent capacity *listed* in Table 504.3(1) is 210,000 Btu per hour, and after reducing by 20 percent is 168,000 Btu per hour. Therefore, a 5-inch corrugated metal liner should be used in this example.

Single-Wall Connectors. Once it has been established that relining the chimney is necessary, Type B double-wall vent connectors are not specifically required. This example could be redone using Table 504.3(2) for single-wall vent connectors. For this case, the vent connector and liner diameters would be the same as found above with Type B double-wall connectors.

TABLE B-1
MASONRY CHIMNEY LINER DIMENSIONS
WITH CIRCULAR EQUIVALENTS[a]

NOMINAL LINER SIZE (inches)	INSIDE DIMENSIONS OF LINER (inches)	INSIDE DIAMETER OR EQUIVALENT DIAMETER (inches)	EQUIVALENT AREA (square inches)
4 × 8	$2^1/_2 \times 6^1/_2$	4	12.2
		5	19.6
		6	28.3
		7	38.3
8 × 8	$6^3/_4 \times 6^3/_4$	7.4	42.7
		8	50.3
8 × 12	$6^1/_2 \times 10^1/_2$	9	63.6
		10	78.5
12 × 12	$9^3/_4 \times 9^3/_4$	10.4	83.3
		11	95
12 × 16	$9^1/_2 \times 13^1/_2$	11.8	107.5
		12	113.0
		14	153.9
16 × 16	$13^1/_4 \times 13^1/_4$	14.5	162.9
		15	176.7
16 × 20	13 × 17	16.2	206.1
		18	254.4
20 × 20	$16^3/_4 \times 16^3/_4$	18.2	260.2
		20	314.1
20 × 24	$16^1/_2 \times 20^1/_2$	20.1	314.2
		22 ×	380.1
24 × 24	$20^1/_4 \times 20^1/_4$	22.1	380.1
		24	452.3
24 × 28	$20^1/_4 \times 20^1/_4$	24.1	456.2
28 × 28	$24^1/_4 \times 24^1/_4$	26.4	543.3
		27	572.5
30 × 30	$25^1/_2 \times 25^1/_2$	27.9	607
		30	706.8
30 × 36	$25^1/_2 \times 31^1/_2$	30.9	749.9
		33	855.3
36 × 36	$31^1/_2 \times 31^1/_2$	34.4	929.4
		36	1017.9

For SI: 1 inch = 25.4 mm, 1 square inch = 645.16 m^2.

a. Where liner sizes differ dimensionally from those shown in Table B-1, equivalent diameters can be determined from published tables for square and rectangular ducts of equivalent carrying capacity or by other engineering methods.

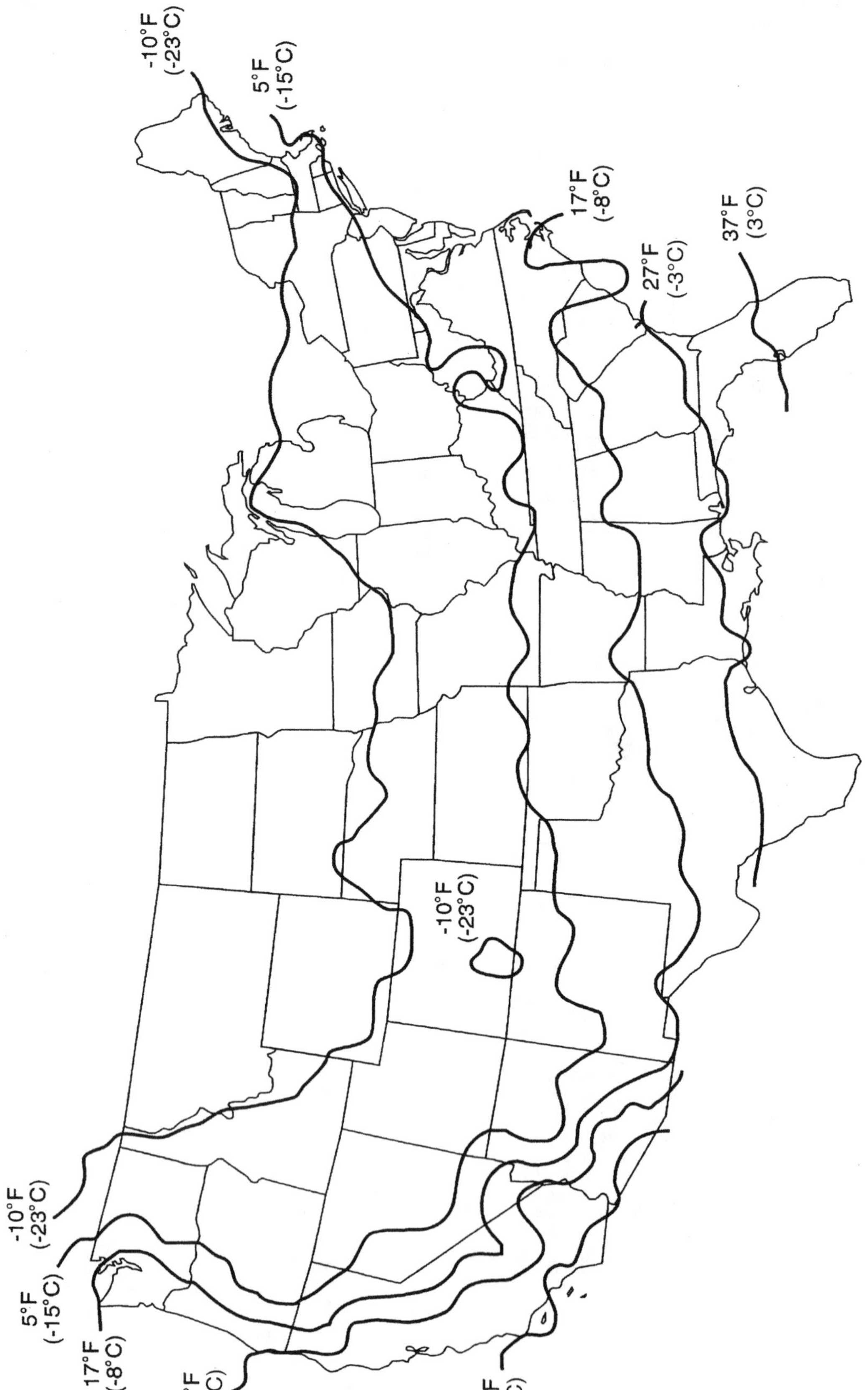

FIGURE B-19

Appendix C:
Exit Terminals of Mechanical Draft and Direct-vent Venting Systems

(This appendix is informative and is not part of the code. This appendix is an excerpt from the 2015 International Fuel Gas Code, *coordinated with the section numbering of the* International Residential Code.)*

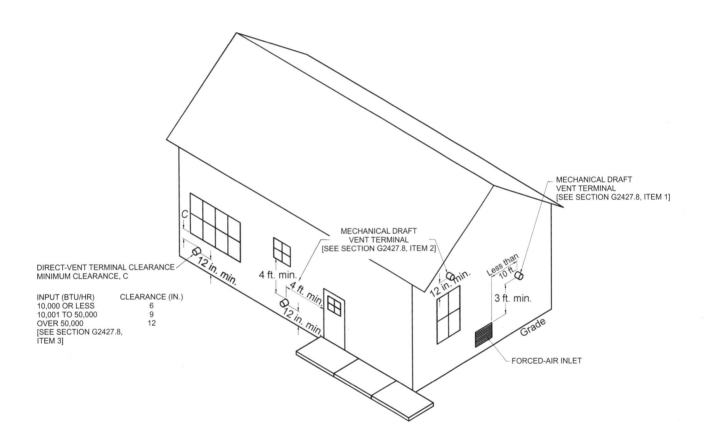

DIRECT-VENT TERMINAL CLEARANCE
MINIMUM CLEARANCE, C

INPUT (BTU/HR)	CLEARANCE (IN.)
10,000 OR LESS	6
10,001 TO 50,000	9
OVER 50,000	12
[SEE SECTION G2427.8, ITEM 3]	

Labels on figure:
- MECHANICAL DRAFT VENT TERMINAL [SEE SECTION G2427.8, ITEM 1]
- MECHANICAL DRAFT VENT TERMINAL [SEE SECTION G2427.8, ITEM 2]
- 12 in. min.
- 4 ft. min.
- 4 ft. min.
- 12 in. min.
- 12 in. min.
- Less than 10 ft.
- 3 ft. min.
- Grade
- FORCED-AIR INLET
- C

For SI: 1 inch = 25.4 mm, 1 foot = 304.8 mm, 1 British thermal unit per hour = 0.2931 W.

APPENDIX C
EXIT TERMINALS OF MECHANICAL DRAFT AND DIRECT-VENT VENTING SYSTEMS

Appendix D:
Recommended Procedure for Safety Inspection of an Existing Appliance Installation

(This appendix is not a part of the requirements of this code and is included for informational purposes only. This appendix is an excerpt from the 2015 *International Fuel Gas Code*, coordinated with the section numbering of the *International Residential Code*.)

D.1 General. The following procedure is intended as a guide to aid in determining that an appliance is properly installed and is in a safe condition for continued use. Where a gas supplier performs an inspection, their written procedures should be followed.

D.1.1 Application. This procedure is intended for existing residential installations of a furnace, boiler, room heater, water heater, cooking appliance, fireplace appliance and clothes dryer. This procedure should be performed prior to any attempt to modify the appliance installation or building envelope.

D.1.2 Weatherization Programs. Before a building envelope is to be modified as part of a weatherization program, the existing appliance installation should be inspected in accordance with these procedures. After all unsafe conditions are repaired, and immediately after the weatherization is complete, the appliance inspections in D.5.2 are to be repeated.

D.1.3 Inspection Procedure. The safety of the building occupant and inspector are to be determined as the first step as described in D.2. Only after the ambient environment is found to be safe should inspections of gas piping and appliances be undertaken. It is recommended that all inspections described in D.3, D.4, and D.6, where the appliance is in the off mode, be completed and any unsafe conditions repaired or corrected before continuing with inspections of an operating appliance described in D.5 and D.6.

D.1.4 Manufacturer Instructions. Where available, the manufacturer's installation and operating instructions for the installed appliances should be used as part of these inspection procedures to determine if it is installed correctly and is operating properly.

D.1.5 Instruments. The inspection procedures include measuring for fuel gas and carbon monoxide (CO) and will require the use of a combustible gas detector (CGD) and a CO detector. It is recommended that both types of detectors be listed. Prior to any inspection, the detectors should be calibrated or tested in accordance with the manufacturer's instructions. In addition, it is recommended that the detectors have the following minimum specifications.

(1) Gas Detector: The CGD should be capable of indicating the presence of the type of fuel gas for which it is to be used (e.g. natural gas or propane). The combustible gas detector should be capable of the following:

    a. *PPM:* Numeric display with a parts per million (ppm) scale from 1ppm to 900 ppm in 1 ppm increments.

    b. *LEL:* Numeric display with a percent lower explosive limit (% LEL) scale from 0 percent to 100 percent in 1 percent increments.

    c. *Audio:* An audio sound feature to locate leaks.

(2) CO Detector: The CO detector should be capable of the following functions and have a numeric display scale as follows:

    a. *PPM:* For measuring ambient room and appliance emissions a display scale in parts per million (ppm) from 0 to 1,000 ppm in 1 ppm increments.

    b. *Alarm:* A sound alarm function where hazardous levels of ambient CO is found (see D.2 for alarm levels)

    c. *Air Free:* Capable of converting CO measurements to an air free level in ppm. Where a CO detector is used without an air free conversion function, the CO air free can be calculated in accordance with footnote 3 in Table D.6.

D.2 Occupant and Inspector Safety. Prior to entering a building, the inspector should have both a combustible gas detector (CGD) and CO detector turned on, calibrated, and operating. Immediately upon entering the building, a sample of the ambient atmosphere should be taken. Based on CGD and CO detector readings, the inspector should take the following actions:

(1) The CO detector indicates a carbon monoxide level of 70 ppm or greater [1]. The inspector should immediately notify the occupant of the need for themselves and any building occupant to evacuate; the inspector shall immediately evacuate and call 911.

(2) Where the CO detector indicates a reading between 30 ppm and 70 ppm[1]. The inspector should advise the occupant that high CO levels have been found and

---

[1] U.S. Consumer Product Safety Commission, *Responding to Residential Carbon Monoxide Incidents, Guidelines For Fire and Other Emergency Response Personnel,* Approved 7/23/02

recommend that all possible sources of CO should be turned off immediately and windows and doors opened. Where it appears that the source of CO is a permanently installed appliance, advise the occupant to keep the appliance off and have the appliance serviced by a qualified servicing agent.

(3) Where CO detector indicates CO below 30 ppm[1] the inspection can continue.

(4) The CGD indicates a combustible gas level of 20% LEL or greater. The inspector should immediately notify the occupant of the need for themselves and any building occupant to evacuate; the inspector shall immediately evacuate and call 911.

(5) The CGD indicates a combustible gas level below 20% LEL, the inspection can continue.

If during the inspection process it is determined a condition exists that could result in unsafe appliance operation, shut off the appliance and advise the owner of the unsafe condition. Where a gas leak is found that could result in an unsafe condition, advise the owner of the unsafe condition and call the gas supplier to turn off the gas supply. The inspector should not continue a safety inspection on an operating appliance, venting system, and piping system until repairs have been made.

D.3 Gas Piping and Connection Inspections.

(1) Leak Checks. Conduct a test for gas leakage using either a non-corrosive leak detection solution or a CGD confirmed with a leak detection solution.

The preferred method for leak checking is by use of gas leak detection solution applied to all joints. This method provides a reliable visual indication of significant leaks.

The use of a CGD in its audio sensing mode can quickly locate suspect leaks but can be overly sensitive indicating insignificant and false leaks. All suspect leaks found through the use of a CGD should be confirmed using a leak detection solution.

Where gas leakage is confirmed, the owner should be notified that repairs must be made. The inspection should include the following components:

a. All gas piping fittings located within the appliance space.

b. Appliance connector fittings.

c. Appliance gas valve/regulator housing and connections.

(2) *Appliance Connector.* Verify that the appliance connection type is compliant with Section G2422 of the *International Fuel Gas Code.* Inspect flexible appliance connections to determine if they are free of cracks, corrosion and signs of damage. Verify that there are no uncoated brass connectors. Where connectors are determined to be unsafe or where an uncoated brass connector is found, the appliance shut-

off valve should be placed in the off position and the owner notified that the connector must be replaced.

(3) *Piping Support.* Inspect piping to determine that it is adequately supported, that there is no undue stress on the piping, and if there are any improperly capped pipe openings.

(4) *Bonding.* Verify that the electrical bonding of gas piping is compliant with Section G2411 of the *International Fuel Gas Code.*

D.4 Inspections to be performed with the Appliance Not Operating. The following safety inspection procedures are performed on appliances that are not operating. These inspections are applicable to all appliance installations.

(1) *Preparing for Inspection.* Shut off all gas and electrical power to the appliances located in the same room being inspected. For gas supply, use the shutoff valve in the supply line or at the manifold serving each appliance. For electrical power, place the circuit breaker in the off position or remove the fuse that serves each appliance. A lock type device or tag should be installed on each gas shutoff valve and at the electrical panel to indicate that the service has been shut off for inspection purposes.

(2) *Vent System Size and Installation.* Verify that the existing venting system size and installation are compliant with Chapter 5 of the *International Fuel Gas Code.* The size and installation of venting systems for other than natural draft and Category I appliances should be in compliance with the manufacturer's installation instructions. Inspect the venting system to determine that it is free of blockage, restriction, leakage, corrosion, and other deficiencies that could cause an unsafe condition. Inspect masonry chimneys to determine if they are lined. Inspect plastic venting system to determine that it is free of sagging and it is sloped in an upward direction to the outdoor vent termination.

(3) *Combustion Air Supply.* Inspect provisions for combustion air as follows:

a. *Non-Direct Vent Appliances.* Determine that non-direct vent appliance installations are compliant with the combustion air requirements in Section G2407 of the *International Fuel Gas Code.* Inspect any interior and exterior combustion air openings and any connected combustion air ducts to determine that there is no blockage, restriction, corrosion or damage. Inspect to determine that the upper horizontal combustion air duct is not sloped in a downward direction toward the air supply source.

b. *Direct Vent Appliances.* Verify that the combustion air supply ducts and pipes are securely fastened to direct vent appliance and determine that there are no separations, blockage, restriction,

---

[1] U.S. Consumer Product Safety Commission, *Responding to Residential Carbon Monoxide Incidents, Guidelines For Fire and Other Emergency Response Personnel,* Approved 7/23/02

corrosion or other damage. Determine that the combustion air source is located in the outdoors or to areas that freely communicate to the outdoors.

c. *Unvented Appliances.* Verify that the total input of all unvented room heaters and gas-fired refrigerators installed in the same room or rooms that freely communicate with each other does not exceed 20 Btu/hr/ft³.

(4) *Flooded Appliances.* Inspect the appliance for signs that the appliance may have been damaged by flooding. Signs of flooding include a visible water submerge line on the appliance housing, excessive surface or component rust, deposited debris on internal components, and mildew-like odor. Inform the owner that any part of the appliance control system and any appliance gas control that has been under water must be replaced. All flood-damaged plumbing, heating, cooling and electrical appliances should be replaced.

(5) *Flammable Vapors.* Inspect the room/space where the appliance is installed to determine if the area is free of the storage of gasoline or any flammable products such as oil-based solvents, varnishes or adhesives. Where the appliance is installed where flammable products will be stored or used, such as a garage, verify that the appliance burner(s) is a minimum of 18" above the floor unless the appliance is listed as flammable vapor ignition resistant.

(6) *Clearances to Combustibles.* Inspect the immediate location where the appliance is installed to determine if the area is free of rags, paper or other combustibles. Verify that the appliance and venting system are compliant with clearances to combustible building components in accordance with Sections G2408.5, G2425.15.4, G2426.5, G2427.6.1, G2427.10.5 and other applicable sections of Section G2427.

(7) *Appliance Components.* Inspect internal components by removing access panels or other components for the following:

a. Inspect burners and crossovers for blockage and corrosion. The presence of soot, debris, and signs of excessive heating may indicate incomplete combustion due to blockage or improper burner adjustments.

c. Metallic and non-metallic hoses for signs of cracks, splitting, corrosion, and lose connections.

d. Signs of improper or incomplete repairs

e. Modifications that override controls and safety systems

f. Electrical wiring for loose connections; cracks, missing or worn electrical insulation; and indications of excessive heat or electrical shorting. Appliances requiring an external electrical supply should be inspected for proper electrical connection in accordance with the National Electric Code.

(8) *Placing Appliances Back in Operation.* Return all inspected appliances and systems to their preexisting state by reinstalling any removed access panels and components. Turn on the gas supply and electricity to each appliance found in safe condition. Proceed to the operating inspections in D.5 through D.6.

D.5 Inspections to be performed with the Appliance Operating. The following safety inspection procedures are to be performed on appliances that are operating where there are no unsafe conditions or where corrective repairs have been completed.

D.5.1 General Appliance Operation.

(1) *Initial Startup.* Adjust the thermostat or other control device to start the appliance. Verify that the appliance starts up normally and is operating properly.

Determine that the pilot(s), where provided, is burning properly and that the main burner ignition is satisfactory, by interrupting and re-establishing the electrical supply to the appliance in any convenient manner. If the appliance is equipped with a continuous pilot(s), test all pilot safety devices to determine whether they are operating properly by extinguishing the pilot(s) when the main burner(s) is off and determining, after 3 minutes, that the main burner gas does not flow upon a call for heat. If the appliance is not provided with a pilot(s), test for proper operation of the ignition system in accordance with the appliance manufacturer's lighting and operating instructions.

(2) *Flame Appearance.* Visually inspect the flame appearance for proper color and appearance. Visually determine that the main burner gas is burning properly (i.e., no floating, lifting, or flashback). Adjust the primary air shutter as required. If the appliance is equipped with high and low flame controlling or flame modulation, check for proper main burner operation at low flame.

(3) *Appliance Shutdown.* Adjust the thermostat or other control device to shut down the appliance. Verify that the appliance shuts off properly.

D.5.2 Test for Combustion Air and Vent Drafting for Natural Draft and Category I Appliances. Combustion air and vent draft procedures are for natural draft and category I appliances equipped with a draft hood and connected to a natural draft venting system.

(1) *Preparing for Inspection.* Close all exterior building doors and windows and all interior doors between the space in which the appliance is located and other spaces of the building that can be closed. Turn on any clothes dryer. Turn on any exhaust fans, such as range hoods and bathroom exhausts, so they will operate at maximum speed. Do not operate a summer exhaust fan. Close fireplace dampers and any fireplace doors.

(2) *Placing the Appliance in Operation.* Place the appliance being inspected in operation. Adjust the thermostat or control so the appliance will operate continuously.

(3) *Spillage Test.* Verify that all appliances located within the same room are in their standby mode and ready for operation. Follow lighting instructions for each appliance as necessary. Test for spillage at the draft hood relief opening as follows:

    a. After 5 minutes of main burner operation, check for spillage using smoke.

    b. Immediately after the first check, turn on all other fuel gas burning appliances within the same room so they will operate at their full inputs and repeat the spillage test.

    c. Shut down all appliances to their standby mode and wait for 15 minutes.

    d. Repeat the spillage test steps a through c on each appliance being inspected.

(4) Additional Spillage Tests: Determine if the appliance venting is impacted by other door and air handler settings by performing the following tests.

    a. Set initial test condition in accordance with D.5.2 (1).

    b. Place the appliance(s) being inspected in operation. Adjust the thermostat or control so the appliance(s) will operate continuously.

    c. Open the door between the space in which the appliance(s) is located and the rest of the building. After 5 minutes of main burner operation, check for spillage at each appliance using smoke.

    d. Turn on any other central heating or cooling air handler fan that is located outside of the area where the appliances are being inspected. After 5 minutes of main burner operation, check for spillage at each appliance using smoke. The test should be conducted with the door between the space in which the appliance(s) is located and the rest of the building in the open and in the closed position.

(5) Return doors, windows, exhaust fans, fireplace dampers, and any other fuel gas burning appliance to their previous conditions of use.

(6) If, after completing the spillage test it is believed sufficient combustion air is not available, the owner should be notified that an alternative combustion air source is needed in accordance with Section G2407 of the *International Fuel Gas Code.* Where it is believed that the venting system does not provide adequate natural draft, the owner should be notified that alternative vent sizing, design or configuration is needed in accordance with Chapter 24 of the *International Fuel Gas Code.* If spillage occurs, the owner should be notified as to its cause, be instructed as to which position of the door (open or closed) would lessen its impact, and that corrective action by a HVAC professional should be taken.

D.6 Appliance-Specific Inspections. The following appliance-specific inspections are to be performed as part of a complete inspection. These inspections are performed either with the appliance in the off or standby mode (indicated by "*OFF*") or on an appliance that is operating (indicated by "*ON*"). The CO measurements are to be undertaken only after the appliance is determined to be properly venting. The CO detector should be capable of calculating CO emissions in ppm air free.

(1) Forced Air Furnaces:

    a. OFF. Verify that an air filter is installed and that it is not excessively blocked with dust.

    b. OFF. Inspect visible portions of the furnace combustion chamber for cracks, ruptures, holes, and corrosion. A heat exchanger leakage test should be conducted.

    c. ON. Verify both the limit control and the fan control are operating properly. Limit control operation can be checked by blocking the circulating air inlet or temporarily disconnecting the electrical supply to the blower motor and determining that the limit control acts to shut off the main burner gas.

    d. ON. Verify that the blower compartment door is properly installed and can be properly re-secured if opened. Verify that the blower compartment door safety switch operates properly.

    e. *ON.* Check for flame disturbance before and after blower comes on which can indicate heat exchanger leaks.

    f. *ON.* Measure the CO in the vent after 5 minutes of main burner operation. The CO should not exceed threshold in Table D.6.

(2) Boilers:

    a. OFF and ON. Inspect for evidence of water leaks around boiler and connected piping.

    b. ON. Verify that the water pumps are in operating condition. Test low water cutoffs, automatic feed controls, pressure and temperature limit controls, and relief valves in accordance with the manufacturer's recommendations to determine that they are in operating condition.

    c. ON. Measure the CO in the vent after 5 minutes of main burner operation. The CO should not exceed threshold in Table D.6.

(3) Water Heaters:

    a. *OFF.* Verify that the pressure-temperature relief valve is in operating condition. Water in the heater should be at operating temperature.

    b. OFF. Verify that inspection covers, glass, and gaskets are intact and in place on a flammable vapor ignition resistant (FVIR) type water heater.

    c. ON. Verify that the thermostat is set in accordance with the manufacturer's operating instructions and measure the water temperature at the closest tub or sink to verify that it is no greater than 120°F.

    d. OFF. Where required by the local building code in earthquake prone locations, inspect that the water heater is secured to the wall studs in two locations (high and low) using appropriate metal strapping and bolts.

    e. ON. Measure the CO in the vent after 5 minutes of main burner operation. The CO should not exceed threshold in Table D.6.

(4) Cooking Appliances

    a. *OFF.* Inspect oven cavity and range-top exhaust vent for blockage with aluminum foil or other materials.

    b. *OFF.* Inspect cook top to verify that it is free from a build-up of grease.

    c. *ON.* Measure the CO above each burner and at the oven exhaust vents after 5 minutes of burner operation. The CO should not exceed threshold in Table D.6.

(5) Vented Room Heaters

    a. OFF. For built-in room heaters and wall furnaces, inspect that the burner compartment is free of lint and debris.

    b. OFF. Inspect that furnishings and combustible building components are not blocking the heater.

    a. ON. Measure the CO in the vent after 5 minutes of main burner operation. The CO should not exceed threshold in Table D.6.

(6) Vent-Free (unvented) Heaters

    a. OFF. Verify that the heater input is a maximum of 40,000 Btu input, but not more than 10,000 Btu where installed in a bedroom, and 6,000 Btu where installed in a bathroom.

    b. OFF. Inspect the ceramic logs provided with gas log type vent free heaters that they are properly located and aligned.

    c. OFF. Inspect the heater that it is free of excess lint build-up and debris.

    c. OFF. Verify that the oxygen depletion safety shutoff system has not been altered or bypassed.

    d. ON. Verify that the main burner shuts down within 3 minutes by extinguishing the pilot light. The test is meant to simulate the operation of the oxygen depletion system (ODS).

    e. ON. Measure the CO after 5 minutes of main burner operation. The CO should not exceed threshold in Table D.6.

(7) Gas Log Sets and Gas Fireplaces

    a. OFF. For gas logs installed in wood burning fireplaces equipped with a damper, verify that the fireplace damper is in a fixed open position.

    b. ON. Measure the CO in the firebox (log sets installed in wood burning fireplaces or in the vent (gas fireplace) after 5 minutes of main burner operation. The CO should not exceed threshold in Table D.6.

(8) Gas Clothes Dryer

    a. *OFF.* Where installed in a closet, verify that a source of make-up air is provided and inspect that any make-up air openings, louvers, and ducts are free of blockage.

    b. OFF. Inspect for excess amounts of lint around the dryer and on dryer components. Inspect that there is a lint trap properly installed and it does not have holes or tears. Verify that it is in a clean condition.

    c. OFF. Inspect visible portions of the exhaust duct and connections for loose fittings and connections, blockage, and signs of corrosion. Verify that the duct termination is not blocked and that it terminates in an outdoor location. Verify that only approved metal vent ducting material is installed (plastic and vinyl materials are not approved for gas dryers).

    d. ON. Verify mechanical components including drum and blower are operating properly.

    e. ON. Operate the clothes dryer and verify that exhaust system is intact and exhaust is exiting the termination.

    f. ON. Measure the CO at the exhaust duct or termination after 5 minutes of main burner operation. The CO should not exceed threshold in Table D.6.

TABLE D.6
CO THRESHOLDS

Central Furnace (all categories)	400 ppm[1] air free[2,3]
Floor Furnace	400 ppm air free
Gravity Furnace	400 ppm air free
Wall Furnace (BIV)	200 ppm air free
Wall Furnace (Direct Vent)	400 ppm air free
Vented Room Heater	200 ppm air free
Vent-Free Room Heater	200 ppm air free
Water Heater	200 ppm air free
Oven/Boiler	225 ppm as measured
Top Burner	25 ppm as measured (per burner)
Clothes Dryer	400 ppm air free
Refrigerator	25 ppm as measured
Gas Log (gas fireplace)	25 ppm as measured in vent
Gas Log (installed in wood burning fireplace)	400 ppm air free in firebox

1. Parts per million
2. Air free emission levels are based on a mathematical equation (involving carbon monoxide and oxygen or carbon dioxide readings) to convert an actual diluted flue gas carbon monoxide testing sample to an undiluted air free flue gas carbon monoxide level utilized in the appliance certification standards. For natural gas or propane, using as-measured CO ppm and O2 percentage:

$$CO_{AFppm} = \left(\frac{20.9}{20.9 - O_2}\right) \times CO_{ppm}$$

(continued)

TABLE D.6—continued
CO THRESHOLDS

Where:

CO_{AFppm} = Carbon monoxide, air-free ppm

CO_{ppm} = As-measured combustion gas carbon monoxide ppm

$O2$ = Percentage of oxygen in combustion gas, as a percentage

3. An alternate method of calculating the CO air free when access to an oxygen meter is not available:

$$CO_{AFppm} = \left(\frac{UCO_2}{CO_2}\right) \times CO$$

Where:

UCO_2 = Ultimate concentration of carbon dioxide for the fuel being burned in percent for natural gas (12.2 percent) and propane (14.0 percent)

CO_2 = Measured concentration of carbon dioxide in combustion products in percent

CO = Measured concentration of carbon monoxide in combustion products in percent

Appendix E:
Manufactured Housing Used as Dwellings

(The provisions contained in this appendix are not mandatory unless specifically referenced in the adopting ordinance.)

General Comments

Manufactured housing in the United States represents a major portion of housing construction. The criteria for the construction of the structure are governed by the National Manufactured Housing Construction and Safety Act. This act, in effect for quite some time, provides a minimum baseline for such structures no matter where they are installed; manufactured housing is constructed in a factory and not subject to inspection by the local jurisdiction when installed.

While this act may seem to cover the bulk of the construction of manufactured housing, it does not cover those areas related to the placement of the housing on the property. This appendix works in conjunction with the act in providing a complete set of regulations that can assure the owner and the jurisdiction alike that the structure is safe.

Section AE101 is the scope of the appendix. Section AE102 addresses the application of these criteria to existing manufactured housing. Section AE201 contains the definitions necessary for the proper interpretation of the regulations. Section AE301 discusses the type of work that requires permits for such installations. Section AE302 identifies the information needed on such permits. Section AE303 addresses permit issuance. Section AE304 contains the criteria for the establishment of fees for such permits. Section AE305 is the material on required permits. Section AE306 addresses when special inspections are necessary for the work regulated by this appendix, and Section AE307 discusses how to handle the connection to the utilities.

Section AE401 addresses the issues related to the occupancy of this type of housing. Section AE402 contains the provisions for locating the structure on the property. Section AE501 discusses the design of the foundation systems for manufactured housing. Section AE502 contains the specific criteria for the foundation system chosen. Section AE503 addresses the enclosure of the perimeter of the underfloor area. Section AE504 discusses the structural concerns of adding to such housing. Section AE505 identifies which set of regulations applies to building service equipment. Section AE506 is concerned with the path of egress that occupants will need to use in the event of an emergency. Section AE507 provides a cross reference to the HUD Safety Standards. Section AE601 provides more specific criteria for the use of various foundation systems. Section AE602 addresses the use of piers as foundation material. Section AE603 discusses the allowable height of such piers. Section AE604 refers the reader to the manufacturer's installation instructions for the use of anchors for the foundation system, and Section AE605 addresses the use of ties.

Purpose

This appendix provides a complete set of regulations in conjunction with federal law for the installation of manufactured housing. Application of the law and the provisions together provides safe housing.

SECTION AE101
SCOPE

AE101.1 General. These provisions shall be applicable only to a *manufactured home* used as a single *dwelling unit* installed on privately owned (nonrental) lots and shall apply to the following:

1. Construction, *alteration* and repair of any foundation system that is necessary to provide for the installation of a *manufactured home* unit.

2. Construction, installation, *addition*, *alteration*, repair or maintenance of the building service *equipment* that is necessary for connecting *manufactured homes* to water, fuel, or power supplies and sewage systems.

3. *Alterations*, *additions* or repairs to existing *manufactured homes*. The construction, *alteration*, moving, demolition, repair and use of accessory buildings and

structures, and their building service *equipment*, shall comply with the requirements of the codes adopted by this *jurisdiction*.

These provisions shall not be applicable to the design and construction of *manufactured homes* and shall not be deemed to authorize either modifications or *additions* to *manufactured homes* where otherwise prohibited.

Exception: In addition to these provisions, new and replacement *manufactured homes* to be located in flood hazard areas as established in Table R301.2(1) of the *International Residential Code* shall meet the applicable requirements of Section R322 of the *International Residential Code*.

❖ This appendix discusses the use of the *International Residential Code*® (IRC®), which is to be used in conjunction with federal laws that govern the construction of manufactured housing. Since some federal laws only cover the design and construction of the housing itself, other provisions must be used to address the design and construction of the foundation system, the connection to utilities, and the construction of accessory buildings. These issues have typically fallen under the adopted codes of the locality.

For installations in flood hazard areas, criteria in Section R322 apply, depending on whether the manufactured home is to be placed in a coastal high hazard area (V Zone) or Coastal A Zone where foundations will be exposed to wave impacts, or in flood hazard areas referred to as A Zones. Some riverine floodplains have identified floodways. Section R322.1.9 specifies that foundations and anchorage of manufactured homes located in floodways must be designed in accordance with ASCE 24. See FEMA P-85 for additional guidance and some pre-engineered foundation designs.

HUD's October 2007 final rule for the Model Manufactured Home Installation Standard (24 CFR, Part 3285) includes some provisions related to flood hazard areas. Manufacturers must clearly specify that their installation instructions and foundation specifications have either (a) been designed for flood-resistant considerations, in which case the conditions are to be listed (velocities, depths or wave action) and the design must be certified by a registered professional engineer or architect, or (b) not been designed to address flood loads, in which case the instructions must direct the installer to "obtain an alternate design prepared and certified by a registered professional engineer or registered architect for the support and anchorage." HUD places the burden on the installer to determine whether a home site is wholly or partly in a flood hazard area and to obtain additional designs, if needed.

SECTION AE102
APPLICATION TO EXISTING MANUFACTURED HOMES AND BUILDING SERVICE EQUIPMENT

AE102.1 General. *Manufactured homes* and their building service *equipment* to which *additions*, *alterations* or repairs are made shall comply with all the requirements of these provisions for new facilities, except as specifically provided in this section.

❖ This section discusses the use of this appendix for existing manufactured housing. As a general rule, this section requires that the activity must comply with the criteria for new construction; however, the following subsections contain many variances to that rule.

AE102.2 Additions, alterations or repairs. *Additions* made to a *manufactured home* shall conform to one of the following:

1. Be certified under the National Manufactured Housing Construction and Safety Standards Act of 1974 (42 U.S.C. Section 5401, et seq.).

2. Be designed and constructed to comply with the applicable provisions of the National Manufactured Housing Construction and Safety Standards Act of 1974 (42 U.S.C. Section 5401, et seq.).

3. Be designed and constructed in compliance with the code adopted by this *jurisdiction*.

Additions shall be structurally separated from the *manufactured home*.

Exception: A structural separation need not be provided when structural calculations are provided to justify the omission of such separation.

Alterations or repairs may be made to any *manufactured home* or to its building service *equipment* without requiring the existing *manufactured home* or its building service *equipment* to comply with all the requirements of these provisions, provided the *alteration* or repair conforms to that required for new construction, and provided further that no hazard to life, health or safety will be created by such *additions*, *alterations* or repairs.

Alterations or repairs to an existing *manufactured home*, which are nonstructural and do not adversely affect any structural member or any part of the building or structure having required fire protection, may be made with materials equivalent to those of which the *manufactured home* structure is constructed, subject to approval by the *building official*.

Exception: The installation or replacement of glass shall be required for new installations.

Minor *additions*, *alterations* and repairs to existing building service *equipment* installations may be made in accordance with the codes in effect at the time the original installation was made, subject to the approval of the *building official*, and provided such *additions*, *alterations* and repairs will not cause the existing building service *equipment* to become unsafe, insanitary or overloaded.

❖ This section discusses alterations, additions, and repairs to manufactured homes and is important because of confusion over the application of federal regulations to these topics. In many areas of the country, manufactured housing is ordered and installed, and a patio cover or a carport is added later. The jurisdiction then has to make a determination of which set of criteria applies. This section allows the use of any one of the three specified alternative methods of compliance for additions, making it clear that the National Manufactured Housing Construction and Safety Standards Act can apply, or the additions can be designed using the code that is adopted by the jurisdiction. See Section AE507.1 for alterations made to a manufactured home.

AE102.3 Existing installations. Building service *equipment* lawfully in existence at the time of the adoption of the applicable codes may have their use, maintenance or repair continued if the use, maintenance or repair is in accordance with the original design and no hazard to life, health or property has been created by such building service *equipment*.

❖ Existing building service equipment is generally "grandfathered" as an allowable condition when a newer edition of a code is adopted, provided that a minimum level of safety is maintained. In most cases, that level is the set of criteria or the code under which the original building service equipment was installed. Unfortunately, there are a number of areas where codes have only recently been adopted. In those cases, the code official must apply those criteria that can be reasonably applied to an existing installation.

AE102.4 Existing occupancy. *Manufactured home*s that are in existence at the time of the adoption of these provisions may have their existing use or occupancy continued if such use or occupancy was legal at the time of the adoption of these provisions, provided such continued use is not dangerous to life, health and safety.

The use or occupancy of any existing *manufactured home* shall not be changed unless evidence satisfactory to the *building official* is provided to show compliance with all applicable provisions of the codes adopted by this *jurisdiction*. Upon any change in use or occupancy, the *manufactured home* shall cease to be classified as such within the intent of these provisions.

❖ Very much like the previous section, this section's provisions describe a minimum level of safety for the occupants of an existing structure. Also, if the occupancy is changed, evidence must be provided indicating that all applicable codes and standards are complied with by the new use.

AE102.5 Maintenance. All *manufactured homes* and their building service *equipment*, existing and new, and all parts thereof, shall be maintained in a safe and sanitary condition. All devices or safeguards which are required by applicable codes or by the *Manufactured Home* Standards shall be maintained in conformance to the code or standard under which it was installed. The owner or the owner's designated agent shall be responsible for the maintenance of *manufactured homes*, accessory buildings, structures and their building service *equipment*. To determine compliance with this section, the *building official* may cause any *manufactured home*, accessory building or structure to be reinspected.

❖ As with all structures, building service equipment, and parts, proper maintenance is vital to maintaining a safe and sanitary environment for the occupants. This section specifically designates the owner or her or his representative as being responsible for such maintenance. In order to determine if the structure and associated equipment are properly maintained, the code official can have an inspection done.

AE102.6 Relocation. *Manufactured home*s which are to be relocated within this *jurisdiction* shall comply with these provisions.

❖ This section is a clarification that even relocated manufactured housing must comply with these provisions.

SECTION AE201
DEFINITIONS

AE201.1 General. For the purpose of these provisions, certain abbreviations, terms, phrases, words and their derivatives shall be construed as defined or specified herein.

❖ These terms are necessary for the proper design and enforcement of this appendix. Any terms not listed herein will be found in Chapter 2 of this code, in the other *International Codes*® (I-Codes®), or will use their normally accepted meanings.

ACCESSORY BUILDING. Any building or structure or portion thereto, located on the same property as a *manufactured home*, which does not qualify as a *manufactured home* as defined herein.

❖ This definition addresses those structures that are not manufactured housing. An example would be a patio cover or shed.

BUILDING SERVICE EQUIPMENT. Refers to the plumbing, mechanical and electrical *equipment*, including piping, wiring, fixtures and other accessories which provide sanitation, lighting, heating, ventilation, cooling, fire protection and facilities essential for the habitable occupancy of a *manufactured home* or accessory building or structure for its designated use and occupancy.

❖ This term covers the plumbing, mechanical, and electrical equipment serving the manufactured home.

MANUFACTURED HOME. A structure transportable in one or more sections which, in the traveling mode, is 8 body feet (2438 body mm) or more in width or 40 body feet (12 192 body mm) or more in length or, when erected on site, is 320 or more square feet (30 m²), and which is built on a permanent chassis and designed to be used as a *dwelling* with or without a permanent foundation when connected to the required utilities, and includes the plumbing, heating, air-conditioning and electrical systems contained therein; except that such term shall include any structure which meets all the requirements of this paragraph, except the size requirements and with respect to which the manufacturer voluntarily files a certification required by the Secretary of the U.S. Department of Housing and Urban Development (HUD) and complies with the standards established under this title.

For mobile homes built prior to June 15, 1976, a *label* certifying compliance with the *Standard for Mobile Homes*, NFPA 501, ANSI 119.1, in effect at the time of manufacture, is required. For the purpose of these provisions, a mobile home shall be considered a *manufactured home*.

❖ The term "Manufactured home" in the IRC is consistent with the HUD definition. An existing manufactured home requires a label certifying compliance with the Standard for Mobile Homes, NFPA 501, ANSI 119.1 if built prior to June 15, 1976. Included in this definition is everything that is built into the home.

MANUFACTURED HOME INSTALLATION. Construction which is required for the installation of a *manufactured home*, including the construction of the foundation system, required structural connections thereto and the installation of on-site water, gas, electrical and sewer systems and connections thereto which are necessary for the normal operation of the *manufactured home*.

❖ The components covered are those that have to be installed on the site prior to the arrival of the manufactured home. Included are the foundation systems, connections to the utilities, and anything else necessary for the operation of the home.

MANUFACTURED HOME STANDARDS. The *Manufactured Home Construction and Safety Standards* as promulgated by the HUD.

❖ This definition gives the title of the federal standard that is applicable for the construction of the manufactured home.

PRIVATELY OWNED (NONRENTAL) LOT. A parcel of real estate outside of a *manufactured home* rental community (park) where the land and the *manufactured home* to be installed thereon are held in common ownership.

❖ This is one of the few instances that an I-Code addresses ownership. This definition makes it clear that both the land and manufactured home are to be held by the same owner or owners.

SECTION AE301
PERMITS

AE301.1 Initial installation. A *manufactured home* shall not be installed on a foundation system, reinstalled or altered without first obtaining a *permit* from the *building official*. A separate *permit* shall be required for each *manufactured home* installation. When *approved* by the *building official*, such *permit* may include accessory buildings and structures, and their building service *equipment*, when the accessory buildings or structures will be constructed in conjunction with the *manufactured home* installation.

❖ This section specifies that a permit is needed for the installation, reinstallation or alteration of each manufactured home. See Section AE302.1 for the information that is required to apply for a permit. The permit process gives the jurisdiction the opportunity to verify that the manufactured home installation complies with the specified requirements in this appendix chapter.

AE301.2 Additions, alterations and repairs to a manufactured home. A *permit* shall be obtained to alter, remodel, repair or add accessory buildings or structures to a *manufactured home* subsequent to its initial installation. *Permit* issuance and fees therefor shall be in conformance to the codes applicable to the type of work involved.

An *addition* made to a *manufactured home*, as defined in these provisions, shall comply with these provisions.

❖ Any alteration, remodeling, repair, or addition to a manufactured home must have a permit. This section

also coordinates with Section AE102.2, which discusses the application of the appendix provisions.

AE301.3 Accessory buildings. Except as provided in Section AE301.1, *permits* shall be required for all accessory buildings and structures, and their building service *equipment*. *Permit* issuance and fees therefor shall be in conformance to the codes applicable to the types of work involved.

❖ This section requires a separate permit for the construction/installation of all accessory structures. Any fee that is charged is intended to cover those services provided by the jurisdiction for the enforcement of the code.

AE301.4 Exempted work. A *permit* shall not be required for the types of work specifically exempted by the applicable codes. Exemption from the *permit* requirements of any of said codes shall not be deemed to grant authorization for any work to be done in violation of the provisions of said codes or any other laws or ordinances of this *jurisdiction*.

❖ Certain types of work are exempted from permit in most codes. See Section R105.2 for items that do not require a permit. This does not mean, however, that the exempted work does not have to meet the code. Such work must still comply with all applicable codes and standards.

SECTION AE302
APPLICATION FOR PERMIT

AE302.1 Application. To obtain a *manufactured home* installation *permit*, the applicant shall first file an application, in writing, on a form furnished by the *building official* for that purpose. At the option of the *building official*, every such application shall:

1. Identify and describe the work to be covered by the *permit* for which application is made.

2. Describe the land on which the proposed work is to be done by legal description, street address or similar description that will readily identify and definitely locate the proposed building or work.

3. Indicate the use or occupancy for which the proposed work is intended.

4. Be accompanied by plans, diagrams, computations and specifications, and other data as required in Section AE302.2.

5. Be accompanied by a soil investigation when required by Section AE502.2.

6. State the valuation of any new building or structure; or any *addition*, remodeling or *alteration* to an existing building.

7. Be signed by the permittee, or permittee's authorized agent, who may be required to submit evidence to indicate such authority.

8. Give such other data and information as may be required by the *building official*.

❖ This section lists the items that must be found on an application for a permit. The purpose of requiring this information is so that the installation can be verified to comply with this appendix chapter.

AE302.2 Plans and specifications. Plans, engineering calculations, diagrams and other data as required by the *building official* shall be submitted in not less than two sets with each application for a *permit*. The *building official* may require plans, computations and specifications to be prepared and designed by an engineer or architect licensed by the state to practice as such.

Where no unusual site conditions exist, the *building official* may accept *approved* standard foundation plans and details in conjunction with the manufacturer's *approved* installation instructions without requiring the submittal of engineering calculations.

❖ The submittal of plans and other supporting information is necessary to verify that the manufactured home installation complies with all necessary codes and standards. Under certain circumstances the code official has the discretion to waive the submittal of engineering calculations.

AE302.3 Information on plans and specifications. Plans and specifications shall be drawn to scale on substantial paper or cloth, and shall be of sufficient clarity to indicate the location, nature and extent of the work proposed and shown in detail that it will conform to these provisions and all relevant laws, ordinances, rules and regulations. The *building official* shall determine what information is required on plans and specifications to ensure compliance.

❖ This section requires that the information necessary to verify the compliance of the installation of the manufactured home be provided on the submittal documents.

SECTION AE303
PERMITS ISSUANCE

AE303.1 Issuance. The application, plans and specifications, and other data filed by an applicant for *permit* shall be reviewed by the *building official*. Such plans may be reviewed by other departments of this *jurisdiction* to verify compliance with any applicable laws under their *jurisdiction*. If the *building official* finds that the work described in an application for a *permit*, and the plans, specifications and other data filed therewith, conform to the requirements of these provisions, and other data filed therewith conform to the requirements of these provisions and other pertinent codes, laws and ordinances, and that the fees specified in Section AE304 have been paid, the *building official* shall issue a *permit* therefor to the applicant.

When the *building official* issues the *permit* where plans are required, the *building official* shall endorse in writing or stamp the plans and specifications *APPROVED*. Such *approved* plans and specifications shall not be changed, modified or altered without authorization from the *building official*, and all work shall be done in accordance with the *approved* plans.

❖ This section requires that the code official issue a permit when the information submitted has been shown to

comply with the applicable codes and standards. These plans are the approved set and there must be no alteration to them; otherwise, authorization from the code official is necessary.

AE303.2 Retention of plans. One set of *approved* plans and specifications shall be returned to the applicant and shall be kept on the site of the building or work at all times during which the work authorized thereby is in progress. One set of *approved* plans, specifications and computations shall be retained by the *building official* until final approval of the work.

❖ A set of the approved plans and specifications must be kept by the jurisdiction for a period of time that is typically set out in state law. These retained documents are very useful in resolving disputes that can arise during the construction process and beyond.

AE303.3 Validity of permit. The issuance of a *permit* or approval of plans and specifications shall not be construed to be a *permit* for, or an approval of, any violation of any of these provisions or other pertinent codes of any other ordinance of the *jurisdiction*. No *permit* presuming to give authority to violate or cancel these provisions shall be valid.

The issuance of a *permit* based on plans, specifications and other data shall not prevent the *building official* from thereafter requiring the correction of errors in said plans, specifications and other data, or from preventing building operations being carried on thereunder when in violation of these provisions or of any other ordinances of this *jurisdiction*.

❖ This section states the fundamental premise that the permit is only a license to proceed with the intended work. It is not a license to violate, cancel, or set aside any provisions of this code or any other law or regulation. This places the burden of complying with all codes and regulations on the permit applicant and not on the building official.

AE303.4 Expiration. Every *permit* issued by the *building official* under these provisions shall expire by limitation and become null and void if the work authorized by such *permit* is not commenced within 180 days from the date of such *permit*, or if the work authorized by such *permit* is suspended or abandoned at any time after the work is commenced for a period of 180 days. Before such work can be recommenced, a new *permit* shall be first obtained, and the fee therefor shall be one-half the amount required for a new *permit* for such work, provided no changes have been made or will be made in the original plans and specifications for such work, and provided further that such suspension or abandonment has not exceeded 1 year. In order to renew action on a *permit* after expiration, the permittee shall pay a new full *permit* fee.

Any permittee holding an unexpired *permit* may apply for an extension of the time within which work may commence under that *permit* when the permittee is unable to commence work within the time required by this section for good and satisfactory reasons. The *building official* may extend the time for action by the permittee for a period not exceeding 180 days upon written request by the permittee showing that circumstances beyond the control of the permittee have prevented action from being taken. No *permit* shall be extended more than once.

❖ A permit becomes invalid under two separate conditions. The first is when work is not commenced within 180 days; the second is when such work has been suspended for 180 days or more. To reinstate the permit, a new permit must first be obtained, and one-half of the fees paid. If the builders know that they can run into a problem with having work delayed for 180 days or longer, then they can submit a request for an extension to the building official if the reason is a good one. The building official can extend the permit for up to an additional 180 day period. This can only happen once; after that a new permit must be obtained.

AE303.5 Suspension or revocation. The *building official* may, in writing, suspend or revoke a *permit* issued under these provisions whenever the *permit* is issued in error or on the basis of incorrect information supplied, or in violation of any ordinance or regulation or any of these provisions.

❖ As with any license to do work, a permit can be revoked or suspended if it has been shown that the permit has been issued in error or that erroneous information has been provided. A permit to do work cannot violate any codes, laws, or regulations imposed on such a project.

**SECTION AE304
FEES**

AE304.1 Permit fees. The fee for each *manufactured home* installation *permit* shall be established by the *building official*.

When *permit* fees are to be based on the value or valuation of the work to be performed, the determination of value or valuation under these provisions shall be made by the *building official*. The value to be used shall be the total value of all work required for the *manufactured home* installation plus the total value of all work required for the construction of accessory buildings and structures for which the *permit* is issued, as well as all finish work, painting, roofing, electrical, plumbing, heating, air conditioning, elevators, fire-extinguishing systems and any other permanent *equipment* which is a part of the accessory building or structure. The value of the *manufactured home* itself shall not be included.

❖ Each jurisdiction must establish a set of fees for permit issuance. The fees collected cover only the services provided by the city in regard to the enforcement of the code and other laws in relation to the work anticipated. There is additional guidance for the building officials in establishing the valuation of a project found in this section.

AE304.2 Plan review fees. When a plan or other data are required to be submitted by Section AE302.2, a plan review fee shall be paid at the time of submitting plans and specifications for review. Said plan review fee shall be as established

by the *building official*. Where plans are incomplete or changed so as to require additional plan review, an additional plan review fee shall be charged at a rate as established by the *building official*.

❖ If the review of a set of construction documents is required to verify that the project is in compliance with the code, then an additional fee is imposed to cover the service of that review. If the plans are incomplete, then an additional fee is imposed to cover the additional plan review of the project. Each jurisdiction must establish a set of fees to be applied where plans are incomplete or changed.

AE304.3 Other provisions.

AE304.3.1 Expiration of plan review. Applications for which no *permit* is issued within 180 days following the date of application shall expire by limitation, and plans and other data submitted for review may thereafter be returned to the applicant or destroyed by the *building official*. The *building official* may extend the time for action by the applicant for a period not exceeding 180 days upon request by the applicant showing that circumstances beyond the control of the applicant have prevented action from being taken. No application shall be extended more than once. In order to renew action on an application after expiration, the applicant shall resubmit plans and pay a new plan review fee.

❖ This section is similar to the description of the expiration of permit in Section AE303.4. The applicant is given a period of time to complete the plan review process and obtain a permit for the work. There is an opportunity to extend this time period since it is understood that problems can be encountered while obtaining a permit.

AE304.3.2 Investigation fees—work without a permit.

AE304.3.2.1 Investigation. Whenever any work for which a *permit* is required by these provisions has been commenced without first obtaining said *permit*, a special investigation shall be made before a *permit* may be issued for such work.

❖ This fee is imposed to cover the services of reviewing a project after it has been commenced without a permit. Oftentimes work is covered or concealed by the time the inspection takes place, and the fees involved may have to reflect the difficulty of investigations to verify compliance with the codes and other regulations.

AE304.3.2.2 Fee. An investigation fee, in addition to the *permit* fee, shall be collected whether or not a *permit* is then or subsequently issued. The investigation fee shall be equal to the amount of the *permit* fee required. The minimum investigation fee shall be the same as the minimum fee established by the *building official*. The payment of such investigation fee shall not exempt any person from compliance with all other provisions of either these provisions or other pertinent codes or from any penalty prescribed by law.

❖ This section indicates that the building official must determine the fees involved for such investigations.

AE304.3.3 Fee refunds.

AE304.3.3.1 Permit fee erroneously paid or collected. The *building official* may authorize the refunding of any fee paid hereunder which was erroneously paid or collected.

❖ Any fees that were collected in error can be refunded to the applicants.

AE304.3.3.2 Permit fee paid when no work done. The *building official* may authorize the refunding of not more than 80 percent of the *permit* fee paid when no work has been done under a *permit* issued in accordance with these provisions.

❖ This section addresses the situation where a permit was issued and fees paid and the work was not begun. Eighty percent of such fees can be refunded to applicants by the building official. Twenty percent of the permit fee is retained to cover the administrative work that has already been done.

AE304.3.3.3 Plan review fee. The *building official* may authorize the refunding of not more than 80 percent of the plan review fee paid when an application for a *permit* for which a plan review fee has been paid is withdrawn or canceled before any plan reviewing is done.

The *building official* shall not authorize the refunding of any fee paid, except upon written application by the original permittee not later than 180 days after the date of the fee payment.

❖ If a plan review is not done and the permit withdrawn or canceled, then the building official can authorize the reimbursement of most of the fees paid. A refund request must be in writing.

SECTION AE305
INSPECTIONS

AE305.1 General. All construction or work for which a *manufactured home* installation *permit* is required shall be subject to inspection by the *building official*, and certain types of construction shall have continuous inspection by special inspectors as specified in Section AE306. A survey of the *lot* may be required by the *building official* to verify that the structure is located in accordance with the *approved* plans.

It shall be the duty of the *permit* applicant to cause the work to be accessible and exposed for inspection purposes. Neither the *building official* nor this *jurisdiction* shall be liable for expense entailed in the removal or replacement of any material required to allow inspection.

❖ This is one of the most important functions of a building department. This section authorizes the building official to have a project inspected for compliance with the approved documents and codes. The work must be exposed and accessible to the inspectors. If the work to be inspected is not exposed, the jurisdiction is not responsible for any expenses incurred for making the work exposed and accessible.

AE305.2 Inspection requests. It shall be the duty of the person doing the work authorized by a *manufactured home*

installation *permit* to notify the *building official* that such work is ready for inspection. The *building official* may require that every request for inspection be filed at least one working day before such inspection is desired. Such request may be in writing or by telephone at the option of the *building official*.

It shall be the duty of the person requesting any inspections required, either by these provisions or other applicable codes, to provide access to and means for proper inspection of such work.

❖ It is the responsibility of the people doing the work to ask for inspections at the proper intervals. Also, the people doing the work need to provide the means for making such inspections. For example, this can be in the form of a ladder or other such device.

AE305.3 Inspection record card. Work requiring a *manufactured home* installation *permit* shall not be commenced until the *permit* holder or the *permit* holder's agent shall have posted an inspection record card in a conspicuous place on the premises and in such position as to allow the *building official* conveniently to make the required entries thereon regarding inspection of the work. This card shall be maintained in such position by the *permit* holder until final approval has been issued by the *building official*.

❖ The purpose of keeping the permit in a conspicuous location on the job site is to make it available for the inspector to make notes and keep track of the previous inspections.

AE305.4 Approval required. Work shall not be done on any part of the *manufactured home* installation beyond the point indicated in each successive inspection without first obtaining the approval of the *building official*. Such approval shall be given only after an inspection has been made of each successive step in the construction as indicated by each of the inspections required in Section AE305.5. There shall be a final inspection and approval of the *manufactured home* installation, including connections to its building service *equipment*, when completed and ready for occupancy or use.

❖ No work on the project can progress beyond a required inspection without having approval from the building official.

AE305.5 Required inspections.

AE305.5.1 Structural inspections for the manufactured home installation. Reinforcing steel or structural framework of any part of any *manufactured home* foundation system shall not be covered or concealed without first obtaining the approval of the *building official*. The *building official*, upon notification from the *permit* holder or the *permit* holder's agent, shall make the following inspections and shall either approve that portion of the construction as completed or shall notify the *permit* holder or the *permit* holder's agent wherein the same fails to comply with these provisions or other applicable codes:

1. Foundation inspection: To be made after excavations for footings are completed and any required reinforcing steel is in place. For concrete foundations, any required forms shall be in place prior to inspection. All materials

for the foundation shall be on the job, except where concrete from a central mixing plant (commonly termed "transit mixed") is to be used, the concrete materials need not be on the job. Where the foundation is to be constructed of *approved* treated wood, additional framing inspections as required by the *building official* may be required.

2. Concrete slab or under-floor inspection: To be made after all in-slab or under-floor building service *equipment*, conduit, piping accessories and other ancillary *equipment* items are in place but before any concrete is poured or the *manufactured home* is installed.

3. Anchorage inspection: To be made after the *manufactured home* has been installed and permanently anchored.

❖ The inspections specified in Section AE305.5 are the minimum thought necessary to verify the compliance of the installation. This section lists the structural inspections for the foundation, underfloor equipment, concrete slab, and anchorage of the manufactured home.

AE305.5.2 Structural inspections for accessory building and structures. Inspections for accessory buildings and structures shall be made as set forth in this code.

❖ This section suggests that the jurisdiction have the same inspections for accessory structures that it would have for any building.

AE305.5.3 Building service equipment inspections. All building service *equipment* which is required as a part of a *manufactured home* installation, including accessory buildings and structures authorized by the same *permit*, shall be inspected by the *building official*. Building service *equipment* shall be inspected and tested as required by the applicable codes. Such inspections and testing shall be limited to site construction and shall not include building service *equipment* which is a part of the *manufactured home* itself. No portion of any building service *equipment* intended to be concealed by any permanent portion of the construction shall be concealed until inspected and *approved*. Building service *equipment* shall not be connected to a water, fuel or power supply, or sewer system, until authorized by the *building official*.

❖ All equipment serving the manufactured home must be inspected. This section also requires that no connection be made to the utilities without the building official's approval.

AE305.5.4 Final inspection. When finish grading and the *manufactured home* installation, including the installation of all required building service *equipment*, is completed and the *manufactured home* is ready for occupancy, a final inspection shall be made.

❖ When all work is complete, this section requires one final inspection to verify that the approved plans have been followed. Once this inspection is approved, then the home is ready for occupancy.

AE305.6 Other inspections. In addition to the called inspections specified in Section AE305.5.4, the *building official* may make or require other inspections of any construction work to ascertain compliance with these provisions or other codes and laws which are enforced by the code enforcement agency.

❖ This section allows a building official to impose other inspections as may be necessary to verify compliance with this set of criteria or other laws and regulations.

SECTION AE306
SPECIAL INSPECTIONS

AE306.1 General. In addition to the inspections required by Section AE305, the *building official* may require the owner to employ a special inspector during construction of specific types of work as described in this code.

❖ A special inspector may be needed to verify compliance with the code. This could be due to the use of high strength concrete, special soil conditions or anchorage, or even flood plain installations.

SECTION AE307
UTILITY SERVICE

AE307.1 General. Utility service shall not be provided to any building service *equipment* which is regulated by these provisions or other applicable codes, and for which a *manufactured home* installation *permit* is required by these provisions, until *approved* by the *building official*.

❖ The building official must approve the final inspection before connection can be made to any utilities that serve the manufactured home.

SECTION AE401
OCCUPANCY CLASSIFICATION

AE401.1 Manufactured homes. A *manufactured home* shall be limited in use to a single *dwelling unit*.

❖ There can be instances of applications to change the use of the structure to something other than a dwelling. This chapter was not developed with that in mind, and in those cases the *International Building Code*® (IBC®) should be used.

AE401.2 Accessory buildings. Accessory buildings shall be classified as to occupancy by the *building official* as set forth in this code.

❖ The use of accessory structures is indeed accessory to the home, so the structures allowed would be those normally attached to such dwellings. Examples would be tool sheds, patio covers, carports, and gazebos.

SECTION AE402
LOCATION ON PROPERTY

AE402.1 General. *Manufactured homes* and accessory buildings shall be located on the property in accordance with applicable codes and ordinances of this *jurisdiction*.

❖ This section refers to other codes and laws that may be in effect for locating the manufactured home on the site. Section R302 of the IRC also applies, as well as any zoning ordinance of the jurisdiction.

SECTION AE501
DESIGN

AE501.1 General. A *manufactured home* shall be installed on a foundation system which is designed and constructed to sustain within the stress limitations specified in this code and all loads specified in this code.

> **Exception:** When specifically authorized by the *building official*, foundation and anchorage systems which are constructed in accordance with the methods specified in Section AE600 of these provisions, or in the HUD, *Permanent Foundations for Manufactured Housing,* 1984 Edition, Draft, shall be deemed to meet the requirements of this appendix.

❖ This section specifies that an appropriate foundation system be used for manufactured housing. The provided exception allows the use of special foundations when approved by the building official. These systems are called out in Section AE600 of this appendix chapter and in the United States Department of Housing and Urban Development Handbook: *Permanent Foundations for Manufactured Housing.*

AE501.2 Manufacturer's installation instructions. The installation instructions as provided by the manufacturer of the *manufactured home* shall be used to determine permissible points of support for vertical loads and points of attachment for anchorage systems used to resist horizontal and uplift forces.

❖ The installation instructions from the manufacturer of the housing must be followed. The code requirements of Section AE501.1 always apply, and where those foundations requirements are more stringent than the manufacturer's requirements, the code overrides the manufacturer's installation instructions. For example, the HUD Permanent Foundations document would override the manufacturer's installation instructions.

AE501.3 Rationality. Any system or method of construction to be used shall submit to a rational analysis in accordance with well-established principles of mechanics.

❖ Any structural system must comply with standard engineering practice.

SECTION AE502
FOUNDATION SYSTEMS

AE502.1 General. Foundation systems designed and constructed in accordance with this section may be considered a permanent installation.

❖ Foundation systems designed in accordance with this section are permanent installations. That is, they are not subject to any time limitations that may be imposed on temporary uses.

AE502.2 Soil classification. The classification of the soil at each *manufactured home* site shall be determined when

required by the *building official*. The *building official* may require that the determination be made by an engineer or architect licensed by the state to conduct soil investigations.

The classification shall be based on observation and any necessary tests of the materials disclosed by borings or excavations made in appropriate locations. Additional studies may be necessary to evaluate soil strength, the effect of moisture variation on soil-bearing capacity, compressibility and expansiveness.

When required by the *building official*, the soil classification design-bearing capacity and lateral pressure shall be shown on the plans.

❖ The determination of the classification of soil at the manufactured-home site is the same as for buildings and structures that are constructed under any building code. It is left to the discretion of the local building official to determine if an investigation is necessary.

AE502.3 Footings and foundations. Footings and foundations, unless otherwise specifically provided, shall be constructed of materials specified by this code for the intended use and in all cases shall extend below the frost line. Footings of concrete and masonry shall be of solid material. Foundations supporting untreated wood shall extend at least 8 inches (203 mm) above the adjacent finish *grade*. Footings shall have a minimum depth below finished *grade* of 12 inches (305 mm) unless a greater depth is recommended by a foundation investigation.

Piers and bearing walls shall be supported on masonry or concrete foundations or piles, or other *approved* foundation systems which shall be of sufficient capacity to support all loads.

❖ This section requires that footings and foundations supporting manufactured housing extend both above and below ground a specified amount. These numbers are consistent with those found in IRC Sections R323.1 and R403.1.4. A footing that is not below the frost line would result in uneven frost heave movement and possible damage to the manufactured home.

AE502.4 Foundation design. When a design is provided, the foundation system shall be designed in accordance with the applicable structural provisions of this code and shall be designed to minimize differential settlement. Where a design is not provided, the minimum foundation requirements shall be as set forth in this code.

❖ This section states that a foundation design must follow the structural provisions of the IRC. If a design is not provided, then the prescriptive criteria in the appendix must be followed.

AE502.5 Drainage. Provisions shall be made for the control and drainage of surface water away from the *manufactured home*.

❖ Manufactured housing must meet the same criteria for drainage of surface water away from other buildings and structures as other structures in the IRC. Such water directed toward a structure would cause the decay of that structure.

AE502.6 Under-floor clearances—ventilation and access. A minimum clearance of 12 inches (305 mm) shall be maintained beneath the lowest member of the floor support framing system. Clearances from the bottom of wood floor joists or perimeter joists shall be as specified in this code.

Under-floor spaces shall be ventilated with openings as specified in this code. If combustion air for one or more heat-producing *appliance* is taken from within the under-floor spaces, ventilation shall be adequate for proper *appliance* operation.

Under-floor access openings shall be provided. Such openings shall be not less than 18 inches (457 mm) in any dimension and not less than 3 square feet (0.279 m²) in area, and shall be located so that any water supply and sewer drain connections located under the *manufactured home* are accessible.

❖ This section requires a nominal amount of space for access to the underfloor area and for its proper ventilation. This section also specifies that all underfloor connections of the water supply system and the drainage system be accessible.

SECTION AE503
SKIRTING AND PERIMETER ENCLOSURES

AE503.1 Skirting and permanent perimeter enclosures. Skirting and permanent perimeter enclosures shall be installed only where specifically required by other laws or ordinances. Skirting, when installed, shall be of material suitable for exterior exposure and contact with the ground. Permanent perimeter enclosures shall be constructed of materials as required by this code for regular foundation construction.

Skirting shall be installed in accordance with the skirting manufacturer's installation instructions. Skirting shall be adequately secured to ensure stability, minimize vibration and susceptibility to wind damage, and compensate for possible frost heave.

❖ Skirting is not required by this appendix chapter. However, the skirting of the perimeter of the underfloor area of manufactured housing is required by other laws or ordinances in many areas of the United States. Such skirting is intended to hide unsightly areas and to prevent the intrusion of rodents. The manufacturer's installation instructions must be followed, and the materials themselves must be able to resist the effects of being installed outdoors.

AE503.2 Retaining walls. Where retaining walls are used as a permanent perimeter enclosure, they shall resist the lateral displacements of soil or other materials and shall conform to this code as specified for foundation walls. Retaining walls and foundation walls shall be constructed of *approved* treated wood, concrete, masonry or other *approved* materials or combination of materials as for foundations as specified in this code. Siding materials shall extend below the top of the exterior of the retaining or foundation wall, or the joint between

the siding and enclosure wall shall be flashed in accordance with this code.

❖ This section applies where a continuous foundation wall is used, and the elevation of the exterior grade is different from the grade within the crawl space.
   Also, this section requires that when skirting is used as a retaining wall, the material must be designed to withstand the loads imposed and the exposure to the earth itself.

SECTION AE504
STRUCTURAL ADDITIONS

AE504.1 General. Accessory buildings shall not be structurally supported by or attached to a *manufactured home* unless engineering calculations are submitted to substantiate any proposed structural connection.

Exception: The *building official* may waive the submission of engineering calculations if it is found that the nature of the work applied for is such that engineering calculations are not necessary to show conformance to these provisions.

❖ Additions to manufactured homes must be structurally independent of the housing itself. If not, an engineered design consistent with that of the manufacturer of the dwelling is acceptable. An exception is provided to allow for simple structures such as carports and some sheds to be added to the housing without engineering calculations.

SECTION AE505
BUILDING SERVICE EQUIPMENT

AE505.1 General. The installation, *alteration*, repair, replacement, *addition* to or maintenance of the building service *equipment* within the *manufactured home* shall conform to regulations set forth in the *Manufactured Home* Standards. Such work which is located outside the *manufactured home* shall comply with the applicable codes adopted by this *jurisdiction*.

❖ This section specifies which set of regulations are to be used for the installation, alteration, repair, replacement, addition to, or maintenance of any building service equipment that serves the home. The HUD Manufactured Home Construction and Safety Standards govern when the equipment is within the home.

SECTION AE506
EXITS

AE506.1 Site development. Exterior stairways and ramps which provide egress to the public way shall comply with the applicable provisions of this code.

❖ The path of egress from the home must comply with the means of egress criteria in the body of the IRC. See IRC Sections R311 through R316 for stairway and ramp requirements. Any means of egress element within the structure of the manufactured housing itself is covered by the HUD safety standards.

AE506.2 Accessory buildings. Every accessory building or portion thereof shall be provided with exits as required by this code.

❖ Accessory buildings must have adequate means of egress for the occupants.

SECTION AE507
OCCUPANCY, FIRE SAFETY AND ENERGY CONSERVATION STANDARDS

AE507.1 General. *Alterations* made to a *manufactured home* subsequent to its initial installation shall conform to the occupancy, fire safety and energy conservation requirements set forth in the *Manufactured Home* Standards.

❖ This is a reference to the HUD safety standards for any alteration to manufactured housing.

SECTION AE600
SPECIAL REQUIREMENTS FOR FOUNDATION SYSTEMS

AE600.1 General. This section is applicable only where specifically authorized by the *building official.*

❖ This gives the building official the discretion to use the prescriptive criteria found in the following sections for the foundation system of manufactured housing. See Section AE501.1 for alternative HUD criteria that are commonly used.

SECTION AE601
FOOTINGS AND FOUNDATIONS

AE601.1 General. The capacity of individual load-bearing piers and their footings shall be sufficient to sustain all loads specified in this code within the stress limitations specified in this code. Footings, unless otherwise *approved* by the *building official*, shall be placed level on firm, undisturbed soil or an engineered fill which is free of organic material, such as weeds and grasses. Where used, an engineered fill shall provide a minimum load-bearing capacity of not less than 1,000 pounds per square foot (48 kN/m²). Continuous footings shall conform to the requirements of this code. Section AE502 of these provisions shall apply to footings and foundations constructed under the provisions of this section.

❖ These are the general criteria for the design of foundation systems, similar to those found in Chapter 4. If fill is to be used for foundation support, the load-bearing capacity should be verified by field testing prior to foundation construction. Section AE502.3 requires the foundation to extend below the frost line.

SECTION AE602
PIER CONSTRUCTION

AE602.1 General. Piers shall be designed and constructed to distribute loads evenly. Multiple-section homes may have concentrated roof loads which will require special consideration. Load-bearing piers may be constructed utilizing one of the following methods listed. Such piers shall be considered to resist only vertical forces acting in a downward direction. They shall not be considered as providing any resistance to horizontal loads induced by wind or earthquake forces.

1. A prefabricated load-bearing device that is listed and *labeled* for the intended use.

2. Mortar shall comply with ASTM C270, Type M, S or N; this may consist of one part Portland cement, one-half part hydrated lime and four parts sand by volume. Lime shall not be used with plastic or waterproof cement.

3. A cast-in-place concrete pier with concrete having specified compressive strength at 28 days of 2,500 pounds per square inch (17 225 kPa).

Alternative materials and methods of construction may be used for piers which have been designed by an engineer or architect licensed by the state to practice as such.

Caps and leveling spacers may be used for leveling of the *manufactured home.* Spacing of piers shall be as specified in the manufacturer's installation instructions, if available, or by an *approved* designer.

❖ This section allows the use of devices that have been developed to provide the foundation for manufactured housing. These devices and other such products can only resist gravity loads. As always, an engineered foundation or a foundation that meets the requirements of the HUD Permanent Foundations document referenced in Section AE501.1 is acceptable.

SECTION AE603
HEIGHT OF PIERS

AE603.1 General. Piers constructed as indicated in Section AE602 may have heights as follows:

1. Except for corner piers, piers 36 inches (914 mm) or less in height may be constructed of masonry units, placed with cores or cells vertically. Piers shall be installed with their long dimension at right angles to the main frame member they support and shall have a minimum cross-sectional area of 128 square inches (82 560 mm²). Piers shall be capped with minimum 4-inch (102 mm) *solid masonry* units or equivalent.

2. Piers between 36 and 80 inches (914 and 2032 mm) in height and all corner piers greater than 24 inches (610 mm) in height shall be at least 16 inches by 16 inches (406 mm by 406 mm) consisting of interlocking masonry units and shall be fully capped with minimum 4-inch (102 mm) *solid masonry* units or equivalent.

3. Piers greater than 80 inches (2032 mm) in height may be constructed in accordance with the provisions of Item 2, provided the piers shall be filled solid with grout and reinforced with four continuous No. 5 bars. One bar shall be placed in each corner cell of hollow masonry unit piers or in each corner of the grouted space of piers constructed of *solid masonry* units.

4. Cast-in-place concrete piers meeting the same size and height limitations of Items 1, 2 and 3 may be substituted for piers constructed of masonry units.

❖ Masonry or concrete piers have specific height limitations for stability. Minimum cross-sectional areas are also specified for adequate load-bearing capacity.

SECTION AE604
ANCHORAGE INSTALLATIONS

AE604.1 Ground anchors. Ground anchors shall be designed and installed to transfer the anchoring loads to the ground. The load-carrying portion of the ground anchors shall be installed to the full depth called for by the manufacturer's installation instructions and shall extend below the established frost line into undisturbed soil.

Manufactured ground anchors shall be listed and installed in accordance with the terms of their listing and the anchor manufacturer's instructions, and shall include the means of attachment of ties meeting the requirements of Section AE605. Ground anchor manufacturer's installation instructions shall include the amount of preload required and load capacity in various types of soil. These instructions shall include tensioning adjustments which may be needed to prevent damage to the *manufactured home*, particularly damage that can be caused by frost heave. Each ground anchor shall be marked with the manufacturer's identification and listed model identification number which shall be visible after installation. Instructions shall accompany each listed ground anchor specifying the types of soil for which the anchor is suitable under the requirements of this section.

Each *approved* ground anchor, when installed, shall be capable of resisting an allowable working load at least equal to 3,150 pounds (14 kN) in the direction of the tie plus a 50-percent overload [4,725 pounds (21 kN) total] without failure. Failure shall be considered to have occurred when the anchor moves more than 2 inches (51 mm) at a load of 4,725 pounds (21 kN) in the direction of the tie installation. Those ground anchors which are designed to be installed so that loads on the anchor are other than direct withdrawal shall be designed and installed to resist an applied design load of 3,150 pounds (14 kN) at 40 to 50 degrees from vertical or within the angle limitations specified by the home manufacturer without displacing the tie end of the anchor more than 4 inches (102 mm) horizontally. Anchors designed for the connection of multiple ties shall be capable of resisting the combined working load and overload consistent with the intent expressed herein.

When it is proposed to use ground anchors and the *building official* has reason to believe that the soil characteristics at a given site are such as to render the use of ground anchors advisable, or when there is doubt regarding the ability of the ground anchors to obtain their listed capacity, the *building official* may require that a representative field installation be made at the site in question and tested to demonstrate ground-anchor capacity. The *building official* shall approve the test procedures.

❖ Approved devices for anchoring manufactured housing is allowed by this section. The manufacturer's installation instructions must be followed. This section describes the conditions by which the anchors will be tested and designed. Building officials can either verify the information themselves or ask for a third party evaluation of the product to verify its compliance with these criteria.

   The site soil condition must be suitable for the use of ground anchors. The capacity of a ground anchor is significantly affected by the soil moisture content. Pull-out and lateral load capacity should be verified by tests.

AE604.2 Anchoring equipment. Anchoring *equipment*, when installed as a permanent installation, shall be capable of resisting all loads as specified within these provisions. When the stabilizing system is designed by an engineer or architect licensed by the state to practice as such, alternative designs may be used, providing the anchoring *equipment* to be used is capable of withstanding a load equal to 1.5 times the calculated load. All anchoring *equipment* shall be listed and *labeled* as being capable of meeting the requirements of these provisions. Anchors as specified in this code may be attached to the main frame of the *manufactured home* by an *approved* $^3/_{16}$-inch-thick (4.76 mm) slotted steel plate anchoring device. Other anchoring devices or methods meeting the requirements of these provisions may be permitted when *approved* by the *building official*.

Anchoring systems shall be so installed as to be permanent. Anchoring *equipment* shall be so designed to prevent self-disconnection with no hook ends used.

❖ This section applies to the devices that hold the manufactured home to its ground anchors. This set of equipment should also be addressed in the approved manufacturer's installation instructions.

AE604.3 Resistance to weather deterioration. All anchoring *equipment*, tension devices and ties shall have a resistance to deterioration as required by this code.

❖ All anchorage devices and equipment must be able to withstand exposure to weather without deterioration.

AE604.4 Tensioning devices. Tensioning devices, such as turnbuckles or yoke-type fasteners, shall be ended with clevis or welded eyes.

❖ This section requires that the ends of tensioning devices for the anchorage system must have either clevises or welded eyes. Such eyes constitute a substantial connection and are required to resist a substantial tension load.

SECTION AE605
TIES, MATERIALS AND INSTALLATION

AE605.1 General. Steel strapping, cable, chain or other *approved* materials shall be used for ties. All ties shall be fastened to ground anchors and drawn tight with turnbuckles or other adjustable tensioning devices or devices supplied with the ground anchor. Tie materials shall be capable of resisting an allowable working load of 3,150 pounds (14 kN) with no more than 2-percent elongation and shall withstand a 50-percent overload [4,750 pounds (21 kN)]. Ties shall comply with the weathering requirements of Section AE604.3. Ties shall connect the ground anchor and the main structural frame. Ties shall not connect to steel outrigger beams which fasten to and intersect the main structural frame unless specifically stated in the manufacturer's installation instructions. Connection of cable ties to main frame members shall be $^5/_8$-inch (15.9 mm) closed-eye bolts affixed to the frame member in an *approved* manner. Cable ends shall be secured with at least two U-bolt cable clamps with the "U" portion of the clamp installed on the short (dead) end of the cable to ensure strength equal to that required by this section.

Wood floor support systems shall be fixed to perimeter foundation walls in accordance with provisions of this code. The minimum number of ties required per side shall be sufficient to resist the wind load stated in this code. Ties shall be as evenly spaced as practicable along the length of the *manufactured home* with the distance from each end of the home and the tie nearest that end not exceeding 8 feet (2438 mm). When continuous straps are provided as vertical ties, such ties shall be positioned at rafters and studs. Where a vertical tie and diagonal tie are located at the same place, both ties may be connected to a single anchor, provided the anchor used is capable of carrying both loads. Multiple-section *manufactured homes* require diagonal ties only. Diagonal ties shall be installed on the exterior main frame and slope to the exterior at an angle of 40 to 50 degrees from the vertical or within the angle limitations specified by the home manufacturer. Vertical ties which are not continuous over the top of the *manufactured home* shall be attached to the main frame.

❖ All tie material used must meet the minimum criteria found in this section. The fact that it does meet the criteria should be a part of the information submitted by the proponent of an approved evaluation by a third party. Tie systems are commonly pretested systems that are expressly manufactured for this purpose.

SECTION AE606
REFERENCED STANDARDS

ASTM C270—04   Specification for Mortar for Unit Masonry AE602

NFPA 501—03   Standard on Manufactured Housing AE201

Bibliography

The following resource materials were used in the preparation of the commentary for this appendix of the code:

ASTM C270–04, *Standard Specification for Mortar for Unit Masonry.* West Conshohocken, PA: ASTM International, 2004.

National Manufactured Housing Construction and Safety Standards Act of 1974. (42 United States Code, Section 5401, et seq.).

NFPA 501–03, *Standard on Manufactured Housing.* Quincy, MA: National Fire Protection Association, 2003.

United States Department of Housing and Urban Development. *Permanent Foundations for Manufactured Housing.* Draft. Washington, D.C.: United States Department of Housing and Urban Development, 1984.

Appendix F
Radon Control Methods

(The provisions contained in this appendix are not mandatory unless specifically referenced in the adopting ordinance.)

General Comments

Radon is a radioactive gas that has been identified as a cancer-causing agent. According to the Environmental Protection Agency (EPA), it is estimated to cause many thousands of deaths each year and increases the potential for lung cancer. Radon comes from the natural (radioactive) breakdown of uranium in soil, rock, and water and finds its way into the air. The primary concern of this appendix is the transfer of radon gases from the soil into the dwelling through openings in the floor system.

The provisions of this appendix regulate the design and construction of radon-resistant measures intended to reduce the entry of radon gases into the living space of residential buildings.

Section AF101 establishes the scope of Appendix F, Section AF102 defines the specific terms related to the appendix, and Section AF103 discusses the construction techniques for radon-resistant construction.

Purpose

In the case of residential construction, radon is created in the soil beneath the house. Varying from one area of the United States to another, even from one house to another, the amount of radon gas in the soil is based on the soil chemistry. Since the movement of radon from the soil into the living area of a residence is enhanced as the house warms, the areas of high radon potential are typically found in portions of the United States with colder climates. The construction of an effective and efficient radon mitigation system is necessary where the radon potential reaches a point considered unacceptable. This appendix establishes prescriptive provisions to reduce the amount of radon entering a dwelling unit from the soil beneath the residence.

SECTION AF101
SCOPE

AF101.1 General. This appendix contains requirements for new construction in *jurisdictions* where radon-resistant construction is required.

Inclusion of this appendix by jurisdictions shall be determined through the use of locally available data or determination of Zone 1 designation in Figure AF101 and Table AF101(1).

❖ Where adopted by the jurisdiction, the provisions of this appendix provide regulations for radon-resistant construction. The jurisdiction may choose to adopt this chapter based on available data, or, alternatively, through designation as a Zone 1 structure based on Figure AF101. Zone 1 areas have a relatively high potential for radon contamination, deemed to measure at more than 4 pCi/L. See Figure AF102 for illustrations of the four basic construction methods utilized in the code for radon mitigation.

SECTION AF102
DEFINITIONS

AF102.1 General. For the purpose of these requirements, the terms used shall be defined as follows:

❖ This section clarifies the terminology used in this appendix. The terms take on unique and specific meanings, with many of the terms used solely in the context of radon-resistant construction.

DRAIN TILE LOOP. A continuous length of drain tile or perforated pipe extending around all or part of the internal or external perimeter of a *basement* or crawl space footing.

❖ Much like a drainage system for moving water away from a foundation or basement wall, a drain tile loop can consist of either drain tile or perforated pipe. It can be located on the interior or exterior side of a crawl space or basement footing.

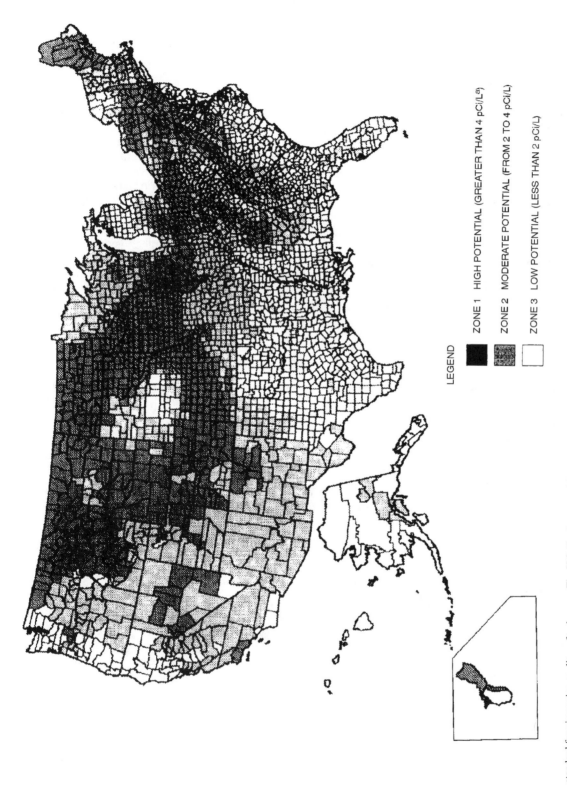

LEGEND

ZONE 1 HIGH POTENTIAL (GREATER THAN 4 pCi/L[a])

ZONE 2 MODERATE POTENTIAL (FROM 2 TO 4 pCi/L)

ZONE 3 LOW POTENTIAL (LESS THAN 2 pCi/L)

FIGURE AF101
EPA MAP OF RADON ZONES

a. pCi/L standard for picocuries per liter of radon gas. The U.S. Environmental Protection Agency (EPA) recommends that homes that measure 4 pCi/L and greater be mitigated. The EPA and the U.S. Geological Survey have evaluated the radon potential in the United States and have developed a map of radon zones designed to assist *building officials* in deciding whether radon-resistant features are applicable in new construction.

The map assigns each of the 3,141 counties in the United States to one of three zones based on radon potential. Each zone designation reflects the average short-term radon measurement that can be expected to be measured in a building without the implementation of radon-control methods. The radon zone designation of highest priority is Zone 1. Table AF101 lists the Zone 1 counties illustrated on the map. More detailed information can be obtained from state-specific booklets (EPA-402-R-93–021 through 070) available through State Radon Offices or from EPA Regional Offices.

TABLE AF101(1)
HIGH RADON-POTENTIAL (ZONE 1) COUNTIES[a]

ALABAMA
Calhoun
Clay
Cleburne
Colbert
Coosa
Franklin
Jackson
Lauderdale
Lawrence
Limestone
Madison
Morgan
Talladega

CALIFORNIA
Santa Barbara
Ventura

COLORADO
Adams
Arapahoe
Baca
Bent
Boulder
Chaffee
Cheyenne
Clear Creek
Crowley
Custer
Delta
Denver
Dolores
Douglas
El Paso
Elbert
Fremont
Garfield
Gilpin
Grand
Gunnison
Huerfano
Jackson
Jefferson
Kiowa
Kit Carson
Lake
Larimer
Las Animas
Lincoln
Logan
Mesa
Moffat
Montezuma
Montrose
Morgan
Otero
Ouray
Park
Phillips
Pitkin
Prowers
Pueblo
Rio Blanco
San Miguel
Summit
Teller
Washington
Weld
Yuma

CONNECTICUT
Fairfield
Middlesex
New Haven
New London

GEORGIA
Cobb
De Kalb
Fulton
Gwinnett

IDAHO
Benewah
Blaine
Boise
Bonner
Boundary
Butte
Camas
Clark
Clearwater
Custer
Elmore
Fremont
Gooding
Idaho
Kootenai
Latah
Lemhi
Shoshone
Valley

ILLINOIS
Adams
Boone
Brown
Bureau
Calhoun
Carroll
Cass
Champaign
Coles
De Kalb
De Witt
Douglas
Edgar
Ford
Fulton
Greene
Grundy
Hancock
Henderson
Henry
Iroquois
Jersey
Jo Daviess
Kane
Kendall
Knox
La Salle
Lee
Livingston
Logan
Macon
Marshall
Mason
McDonough
McLean
Menard
Mercer

Morgan
Moultrie
Ogle
Peoria
Piatt
Pike
Putnam
Rock Island
Sangamon
Schuyler
Scott
Stark
Stephenson
Tazewell
Vermilion
Warren
Whiteside
Winnebago
Woodford

INDIANA
Adams
Allen
Bartholomew
Benton
Blackford
Boone
Carroll
Cass
Clark
Clinton
De Kalb
Decatur
Delaware
Elkhart
Fayette
Fountain
Fulton
Grant
Hamilton
Hancock
Harrison
Hendricks
Henry
Howard
Huntington
Jay
Jennings
Johnson
Kosciusko
LaGrange
Lawrence
Madison
Marion
Marshall
Miami
Monroe
Montgomery
Noble
Orange
Putnam
Randolph
Rush
Scott
Shelby
St. Joseph
Steuben
Tippecanoe
Tipton
Union
Vermillion

Wabash
Warren
Washington
Wayne
Wells
White
Whitley

IOWA
All Counties

KANSAS
Atchison
Barton
Brown
Cheyenne
Clay
Cloud
Decatur
Dickinson
Douglas
Ellis
Ellsworth
Finney
Ford
Geary
Gove
Graham
Grant
Gray
Greeley
Hamilton
Haskell
Hodgeman
Jackson
Jewell
Johnson
Kearny
Kingman
Kiowa
Lane
Leavenworth
Lincoln
Logan
Marion
Marshall
McPherson
Meade
Mitchell
Nemaha
Ness
Norton
Osborne
Ottawa
Pawnee
Phillips
Pottawatomie
Pratt
Rawlins
Republic
Rice
Riley
Rooks
Rush
Saline
Scott
Sheridan
Sherman
Smith
Stanton
Thomas

Trego
Wallace
Washington
Wichita
Wyandotte

KENTUCKY
Adair
Allen
Barren
Bourbon
Boyle
Bullitt
Casey
Clark
Cumberland
Fayette
Franklin
Green
Harrison
Hart
Jefferson
Jessamine
Lincoln
Marion
Mercer
Metcalfe
Monroe
Nelson
Pendleton
Pulaski
Robertson
Russell
Scott
Taylor
Warren
Woodford

MAINE
Androscoggin
Aroostook
Cumberland
Franklin
Hancock
Kennebec
Lincoln
Oxford
Penobscot
Piscataquis
Somerset
York

MARYLAND
Baltimore
Calvert
Carroll
Frederick
Harford
Howard
Montgomery
Washington

MASS.
Essex
Middlesex
Worcester

MICHIGAN
Branch
Calhoun
Cass

Hillsdale
Jackson
Kalamazoo
Lenawee
St. Joseph
Washtenaw

MINNESOTA
Becker
Big Stone
Blue Earth
Brown
Carver
Chippewa
Clay
Cottonwood
Dakota
Dodge
Douglas
Faribault
Fillmore
Freeborn
Goodhue
Grant
Hennepin
Houston
Hubbard
Jackson
Kanabec
Kandiyohi
Kittson
Lac Qui Parle
Le Sueur
Lincoln
Lyon
Mahnomen
Marshall
Martin
McLeod
Meeker
Mower
Murray
Nicollet
Nobles
Norman
Olmsted
Otter Tail
Pennington
Pipestone
Polk
Pope
Ramsey
Red Lake
Redwood
Renville
Rice
Rock
Roseau
Scott
Sherburne
Sibley
Stearns
Steele
Stevens
Swift
Todd
Traverse
Wabasha
Wadena
Waseca
Washington

Watonwan
Wilkin
Winona
Wright
Yellow Medicine

MISSOURI
Andrew
Atchison
Buchanan
Cass
Clay
Clinton
Holt
Iron
Jackson
Nodaway
Platte

MONTANA
Beaverhead
Big Horn
Blaine
Broadwater
Carbon
Carter
Cascade
Chouteau
Custer
Daniels
Dawson
Deer Lodge
Fallon
Fergus
Flathead
Gallatin
Garfield
Glacier
Granite
Hill
Jefferson
Judith Basin
Lake
Lewis and Clark
Madison
McCone
Meagher
Missoula
Park
Phillips
Pondera
Powder River
Powell
Prairie
Ravalli
Richland
Roosevelt
Rosebud
Sanders
Sheridan
Silver Bow
Stillwater
Teton
Toole
Valley
Wibaux
Yellowstone

(continued)

TABLE AF101(1)—continued
HIGH RADON-POTENTIAL (ZONE 1) COUNTIES[a]

NEBRASKA
Adams
Boone
Boyd
Burt
Butler
Cass
Cedar
Clay
Colfax
Cuming
Dakota
Dixon
Dodge
Douglas
Fillmore
Franklin
Frontier
Furnas
Gage
Gosper
Greeley
Hamilton
Harlan
Hayes
Hitchcock
Hurston
Jefferson
Johnson
Kearney
Knox
Lancaster
Madison
Nance
Nemaha
Nuckolls
Otoe
Pawnee
Phelps
Pierce
Platte
Polk
Red Willow
Richardson
Saline
Sarpy
Saunders
Seward
Stanton
Thayer
Washington
Wayne
Webster
York

NEVADA
Carson City
Douglas
Eureka
Lander
Lincoln
Lyon
Mineral
Pershing
White Pine

NEW HAMPSHIRE
Carroll

NEW JERSEY
Hunterdon
Mercer
Monmouth
Morris
Somerset
Sussex
Warren

NEW MEXICO
Bernalillo
Colfax
Mora
Rio Arriba
San Miguel
Santa Fe
Taos

NEW YORK
Albany
Allegany
Broome
Cattaraugus
Cayuga
Chautauqua
Chemung
Chenango
Columbia
Cortland
Delaware
Dutchess
Erie
Genesee
Greene
Livingston
Madison
Onondaga
Ontario
Orange
Otsego
Putnam
Rensselaer
Schoharie
Schuyler
Seneca
Steuben
Sullivan
Tioga
Tompkins
Ulster
Washington
Wyoming
Yates

N. CAROLINA
Alleghany
Buncombe
Cherokee
Henderson
Mitchell
Rockingham
Transylvania
Watauga

N. DAKOTA
All Counties

OHIO
Adams
Allen
Ashland
Auglaize
Belmont
Butler
Carroll
Champaign
Clark
Clinton
Columbiana
Coshocton
Crawford
Darke
Delaware
Fairfield
Fayette
Franklin
Greene
Guernsey
Hamilton
Hancock
Hardin
Harrison
Holmes
Huron
Jefferson
Knox
Licking
Logan
Madison
Marion
Mercer
Miami
Montgomery
Morrow
Muskingum
Perry
Pickaway
Pike
Preble
Richland
Ross
Seneca
Shelby
Stark
Summit
Tuscarawas
Union
Van Wert
Warren
Wayne
Wyandot

PENNSYLVANIA
Adams
Allegheny
Armstrong
Beaver
Bedford
Berks
Blair
Bradford
Bucks
Butler
Cameron
Carbon
Centre
Chester
Clarion
Clearfield
Clinton
Columbia
Cumberland
Dauphin
Delaware
Franklin
Fulton
Huntingdon
Indiana
Juniata
Lackawanna
Lancaster
Lebanon
Lehigh
Luzerne
Lycoming
Mifflin
Monroe
Montgomery
Montour
Northampton
Northumberland
Perry
Schuylkill
Snyder
Sullivan
Susquehanna
Tioga
Union
Venango
Westmoreland
Wyoming
York

RHODE ISLAND
Kent
Washington

S. CAROLINA
Greenville

S. DAKOTA
Aurora
Beadle
Bon Homme
Brookings
Brown
Brule
Buffalo
Campbell
Charles Mix
Clark
Clay
Codington
Corson
Davison
Day
Deuel
Douglas
Edmunds
Faulk
Grant
Hamlin
Hand
Hanson
Hughes
Hutchinson
Hyde
Jerauld
Kingsbury
Lake
Lincoln
Lyman
Marshall
McCook
McPherson
Miner
Minnehaha
Moody
Perkins
Potter
Roberts
Sanborn
Spink
Stanley
Sully
Turner
Union
Walworth
Yankton

TENNESSEE
Anderson
Bedford
Blount
Bradley
Claiborne
Davidson
Giles
Grainger
Greene
Hamblen
Hancock
Hawkins
Hickman
Humphreys
Jackson
Jefferson
Knox
Lawrence
Lewis
Lincoln
Loudon
Marshall
Maury
McMinn
Meigs
Monroe
Moore
Perry
Roane
Rutherford
Smith
Sullivan
Trousdale
Union
Washington
Wayne
Williamson
Wilson

UTAH
Carbon
Duchesne
Grand
Piute
Sanpete
Sevier
Uintah

VIRGINIA
Alleghany
Amelia
Appomattox
Augusta
Bath
Bland
Botetourt
Bristol
Brunswick
Buckingham
Buena Vista
Campbell
Chesterfield
Clarke
Clifton Forge
Covington
Craig
Cumberland
Danville
Dinwiddie
Fairfax
Falls Church
Fluvanna
Frederick
Fredericksburg
Giles
Goochland
Harrisonburg
Henry
Highland
Lee
Lexington
Louisa
Martinsville
Montgomery
Nottoway
Orange
Page
Patrick
Pittsylvania
Powhatan
Pulaski
Radford
Roanoke
Rockbridge
Rockingham
Russell
Salem
Scott
Shenandoah
Smyth
Spotsylvania
Stafford
Staunton
Tazewell
Warren
Washington
Waynesboro
Winchester
Wythe

WASHINGTON
Clark
Ferry
Okanogan
Pend Oreille
Skamania
Spokane
Stevens

W. VIRGINIA
Berkeley
Brooke
Grant
Greenbrier
Hampshire
Hancock
Hardy
Jefferson
Marshall
Mercer
Mineral
Monongalia
Monroe
Morgan
Ohio
Pendleton
Pocahontas
Preston
Summers
Wetzel

WISCONSIN
Buffalo
Crawford
Dane
Dodge
Door
Fond du Lac
Grant
Green
Green Lake
Iowa
Jefferson
Lafayette
Langlade
Marathon
Menominee
Pepin
Pierce
Portage
Richland
Rock
Shawano
St. Croix
Vernon
Walworth
Washington
Waukesha
Waupaca
Wood

WYOMING
Albany
Big Horn
Campbell
Carbon
Converse
Crook
Fremont
Goshen
Hot Springs
Johnson
Laramie
Lincoln
Natrona
Niobrara
Park
Sheridan
Sublette
Sweetwater
Teton
Uinta
Washakie

a. The EPA recommends that this county listing be supplemented with other available State and local data to further understand the radon potential of a Zone 1 area.

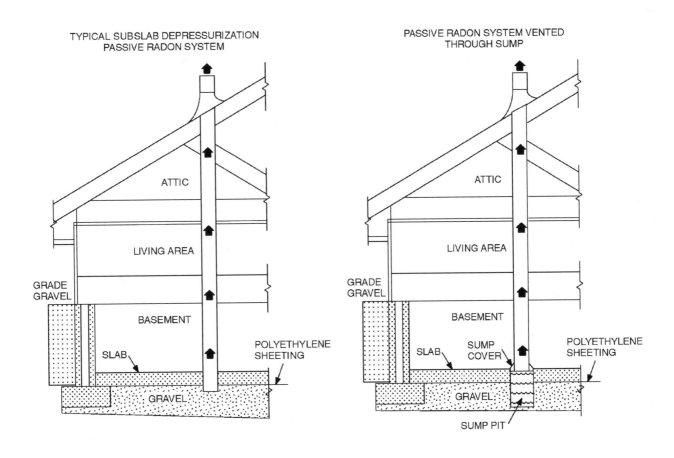

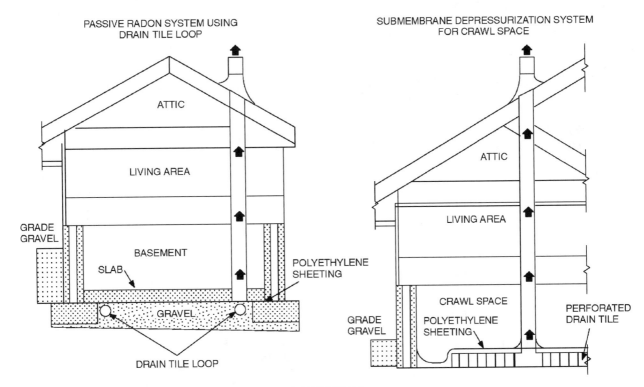

FIGURE AF102
RADON-RESISTANT CONSTRUCTION DETAILS FOR FOUR FOUNDATION TYPES

RADON GAS. A naturally occurring, chemically inert, radioactive gas that is not detectable by human senses. As a gas, it can move readily through particles of soil and rock, and can accumulate under the slabs and foundations of homes where it can easily enter into the living space through construction cracks and openings.

❖ A radioactive gas, radon occurs naturally. It is not detectable by sight, smell, or other human senses. As a gas, it can move readily through particles of soil, aggregate and small cracks and openings in foundation and slab-on-grade construction. It can accumulate under the slabs and foundations of homes where it can easily enter the living space through construction cracks and openings.

SOIL-GAS-RETARDER. A continuous membrane of 6-mil (0.15 mm) polyethylene or other equivalent material used to retard the flow of soil gases into a building.

❖ A membrane of 6-mil (0.15 mm) polyethylene is specifically identified as an acceptable material for retarding the flow of gas from the soil into a structure, if the polyethylene is applied in a continuous manner. Other membrane materials can aslo be used as soil-gas retarders if they provide equal or better protection.

SUBMEMBRANE DEPRESSURIZATION SYSTEM. A system designed to achieve lower submembrane air pressure relative to crawl space air pressure by use of a vent drawing air from beneath the soil-gas-retarder membrane.

❖ Where a basement or crawl space is present, this system can be used in much the same manner as a subslab depressurization system. Shown in Figure AF102, this method draws air from beneath the soil-gas-retarder membrane and vents it to the exterior of the building.

SUBSLAB DEPRESSURIZATION SYSTEM (Active). A system designed to achieve lower subslab air pressure relative to indoor air pressure by use of a fan-powered vent drawing air from beneath the slab.

SUBSLAB DEPRESSURIZATION SYSTEM (Passive). A system designed to achieve lower subslab air pressure relative to indoor air pressure by use of a vent pipe routed through the *conditioned space* of a building and connecting the subslab area with outdoor air, thereby relying on the convective flow of air upward in the vent to draw air from beneath the slab.

❖ One of several methods available for mitigating radon entry into a dwelling unit, this passive system uses the convective movement of air to remove radon from the area below the slab. Illustrated in Figure AF102, this method uses a vertical vent pipe between the subslab area and the exterior of the building to draw air from the subslab area to the outside. By use of a vent pipe routed through the *conditioned* space of a building that connects the subslab area with outdoor air, the system relies on the convective flow of air upward in the vent to draw air from beneath the slab.

SECTION AF103
REQUIREMENTS

AF103.1 General. The following construction techniques are intended to resist radon entry and prepare the building for post-construction radon mitigation, if necessary (see Figure AF102). These techniques are required in areas where designated by the *jurisdiction*.

❖ This section sets forth construction details designed to reduce radon movement from the soil to the interior of the building. Also see the four drawings in Figure AF102 that illustrate several basic construction methods for radon mitigation.

AF103.2 Subfloor preparation. A layer of gas-permeable material shall be placed under all concrete slabs and other floor systems that directly contact the ground and are within the walls of the living spaces of the building, to facilitate future installation of a subslab depressurization system, if needed. The gas-permeable layer shall consist of one of the following:

1. A uniform layer of clean aggregate, a minimum of 4 inches (102 mm) thick. The aggregate shall consist of material that will pass through a 2-inch (51 mm) sieve and be retained by a $^1/_4$-inch (6.4 mm) sieve.

2. A uniform layer of sand (native or fill), a minimum of 4 inches (102 mm) thick, overlain by a layer or strips of geotextile drainage matting designed to allow the lateral flow of soil gases.

3. Other materials, systems or floor designs with demonstrated capability to permit depressurization across the entire subfloor area.

AF103.3 Soil-gas-retarder. A minimum 6-mil (0.15 mm) [or 3-mil (0.075 mm) cross-laminated] polyethylene or equivalent flexible sheeting material shall be placed on top of the gas-permeable layer prior to casting the slab or placing the floor assembly to serve as a soil-gas-retarder by bridging any cracks that develop in the slab or floor assembly, and to prevent concrete from entering the void spaces in the aggregate base material. The sheeting shall cover the entire floor area with separate sections of sheeting lapped at least 12 inches (305 mm). The sheeting shall fit closely around any pipe, wire or other penetrations of the material. All punctures or tears in the material shall be sealed or covered with additional sheeting.

❖ An acceptable sheeting material must be installed on top of the gas-permeable base layer to serve as s soil-gas-retarder. See the definitions for gas-permeable layer and soil-gas-retarder in Section AF102.1. In accordance with the definition of soil-gas-retarder, the sheeting material is to be a minimum 6-mil (0.15 mm) polyethylene membrane or any other flexible sheeting that provides equivalent protection. The soil-gas-retarder resists the vertical flow of radon gas into the slab or other type of floor assembly. Therefore, the membrane must cover the entire floor area of the building, with joints adequately lapped and penetrations tightly sealed. Any tears, rips, or punctures are to be adequately repaired with additional sheeting material.

AF103.4 Entry routes. Potential radon entry routes shall be closed in accordance with Sections AF103.4.1 through AF103.4.10.

❖ This section identifies the various points at which radon may enter a building and specifies the appropriate methods for sealing or otherwise protecting the potential entry routes.

AF103.4.1 Floor openings. Openings around bathtubs, showers, water closets, pipes, wires or other objects that penetrate concrete slabs, or other floor assemblies, shall be filled with a polyurethane caulk or equivalent sealant applied in accordance with the manufacturer's recommendations.

❖ It is typical for a floor slab or other type of floor assembly to be penetrated by underslab or underfloor plumbing, mechanical and electrical components. Polyurethane caulk or an equivalent sealant material must be installed at all penetrations created by the passage of piping, vents, conduit, cable, or other items penetrating the floor. The sealant is to be installed in accordance with the recommendations of the manufacturer.

AF103.4.2 Concrete joints. All control joints, isolation joints, construction joints, and any other joints in concrete slabs or between slabs and foundation walls shall be sealed with a caulk or sealant. Gaps and joints shall be cleared of loose material and filled with polyurethane caulk or other elastomeric sealant applied in accordance with the manufacturer's recommendations.

AF103.4.3 Condensate drains. Condensate drains shall be trapped or routed through nonperforated pipe to daylight.

AF103.4.4 Sumps. Sump pits open to soil or serving as the termination point for subslab or exterior drain tile loops shall be covered with a gasketed or otherwise sealed lid. Sumps used as the suction point in a subslab depressurization system shall have a lid designed to accommodate the vent pipe. Sumps used as a floor drain shall have a lid equipped with a trapped inlet.

❖ A gasketed or sealed lid must be provided on any sump pit that serves as the end point for a subslab or exterior drain tile loop system. Such a lid is also required if the sump pit is open to the soil. The sump lid must be designed to accommodate the vent pipe where the sump is used as the suction point in a subslab decompression system. Where used as a floor drain, the sump pit lid is to be equipped with a trapped inlet.

AF103.4.5 Foundation walls. Hollow block masonry foundation walls shall be constructed with either a continuous course of *solid masonry*, one course of masonry grouted solid, or a solid concrete beam at or above finished ground surface to prevent the passage of air from the interior of the wall into the living space. Where a brick veneer or other masonry ledge is installed, the course immediately below that ledge shall be sealed. Joints, cracks or other openings around all penetrations of both exterior and interior surfaces of masonry block or wood foundation walls below the ground

surface shall be filled with polyurethane caulk or equivalent sealant. Penetrations of concrete walls shall be filled.

❖ Where the foundation is made up of hollow masonry units, it is necessary to provide a means to prohibit the flow of air and potential soil gas within the cavities of the block masonry. Several methods are identified, including the use of solid masonry or solid-grouted masonry for a minimum of one course. The solid barrier must be located at or above the finished ground surface to prevent gases that enter the wall cavity from traveling up and into the living spaces. In those situations where a ledge for brick, stone, or other masonry material is provided, the barrier must be located directly below the ledge.

All penetrations, joints (including the joints where foundation walls meet concrete slab-on-grade construction), cracks and other openings that occur below ground level in masonry, concrete, wood and other types of foundation walls are to be filled with polyurethane caulk or a similar type of flexible sealant. The required penetration and opening protection must be provided on both the interior and exterior sides of the foundation walls.

AF103.4.6 Dampproofing. The exterior surfaces of portions of concrete and masonry block walls below the ground surface shall be dampproofed in accordance with Section R406.

❖ Dampproofing of the exterior surfaces of concrete and masonry block walls located below ground level must be done in accordance with the provisions of Section R406. A variety of methods are established for dampproofing concrete and masonry foundations.

AF103.4.7 Air-handling units. Air-handling units in crawl spaces shall be sealed to prevent air from being drawn into the unit.

Exception: Units with gasketed seams or units that are otherwise sealed by the manufacturer to prevent leakage.

❖ Unless sealed by the manufacturer or provided with gasketed seams to prevent leakage, air-conditioning systems located in crawl spaces must be field-sealed to eliminate the potential for air and gas to be drawn into the unit and distributed throughout the building.

AF103.4.8 Ducts. Ductwork passing through or beneath a slab shall be of seamless material unless the air-handling system is designed to maintain continuous positive pressure within such ducting. Joints in such ductwork shall be sealed to prevent air leakage.

Ductwork located in crawl spaces shall have seams and joints sealed by closure systems in accordance with Section M1601.4.1.

❖ Where ductwork passes through or is installed beneath a concrete floor slab, the ducts must be free of seams that may allow air and gas to enter the duct system. Seams are only permitted where it can be demonstrated that the air-handling equipment will maintain continuous positive pressure within the ducting. In such situations, the seams must be sealed to eliminate any air leakage.

The provision allows ductwork passing through a crawl space to have seams and joints, provided they are sealed by one of the methods prescribed in Section M106.4.1. This will allow the use of fibrous glass and seamed metal ducts and field-fabricated ductwork.

AF103.4.9 Crawl space floors. Openings around all penetrations through floors above crawl spaces shall be caulked or otherwise filled to prevent air leakage.

AF103.4.10 Crawl space access. Access doors and other openings or penetrations between *basements* and adjoining crawl spaces shall be closed, gasketed or otherwise filled to prevent air leakage.

❖ The provisions of Section R408.4 mandate a minimum of one 18-inch by 24-inch (457 mm by 610 mm) opening to access a crawl space. Section M1305.1.4 addresses access to under-floor mechanical equipment. Where such openings or any other access points to the crawl space are provided, the doors or panels must be closed and gasketed to create an airtight separation.

AF103.5 Passive submembrane depressurization system. In buildings with crawl space foundations, the following components of a passive submembrane depressurization system shall be installed during construction.

Exception: Buildings in which an *approved* mechanical crawl space ventilation system or other equivalent system is installed.

AF103.5.1 Ventilation. Crawl spaces shall be provided with vents to the exterior of the building. The minimum net area of ventilation openings shall comply with Section R408.1.

AF103.5.2 Soil-gas-retarder. The soil in crawl spaces shall be covered with a continuous layer of minimum 6-mil (0.15 mm) polyethylene soil-gas-retarder. The ground cover shall be lapped not less than 12 inches (305 mm) at joints and shall extend to all foundation walls enclosing the crawl space area.

❖ An acceptable sheeting material must be installed on top of the gas-permeable base layer to serve as a soil-gas-retarder. In accordance with the requirements of the definition of "Soil-gas-retarder" in Section AF102.1, the material is to be a minimum 6-mil (0.15 mm) polyethylene membrane or any other flexible sheeting that provides equivalent protection. The soil-gas-retarder resists the vertical flow of radon gas into the slab or other type of floor assembly. Therefore, the membrane must cover the entire floor area of the building, with joints adequately lapped and penetrations tightly sealed. Any tears, rips, or punctures are to be adequately repaired with additional sheeting material.

AF103.5.3 Vent pipe. A plumbing tee or other *approved* connection shall be inserted horizontally beneath the sheeting and connected to a 3- or 4-inch-diameter (76 or 102 mm) fitting with a vertical vent pipe installed through the sheeting. The vent pipe shall be extended up through the building floors, and terminate not less than 12 inches (305 mm) above the roof in a location not less than 10 feet (3048 mm) away from any window or other opening into the *conditioned spaces* of the building that is less than 2 feet (610 mm) below the exhaust point, and 10 feet (3048 mm) from any window or other opening in adjoining or adjacent buildings.

AF103.6 Passive subslab depressurization system. In *basement* or slab-on-grade buildings, the following components of a passive subslab depressurization system shall be installed during construction.

AF103.6.1 Vent pipe. A minimum 3-inch-diameter (76 mm) ABS, PVC or equivalent gas-tight pipe shall be embedded vertically into the subslab aggregate or other permeable material before the slab is cast. A "T" fitting or equivalent method shall be used to ensure that the pipe opening remains within the subslab permeable material. Alternatively, the 3-inch (76 mm) pipe shall be inserted directly into an interior perimeter drain tile loop or through a sealed sump cover where the sump is exposed to the subslab aggregate or connected to it through a drainage system.

The pipe shall be extended up through the building floors, and terminate at least 12 inches (305 mm) above the surface of the roof in a location at least 10 feet (3048 mm) away from any window or other opening into the *conditioned spaces* of the building that is less than 2 feet (610 mm) below the exhaust point, and 10 feet (3048 mm) from any window or other opening in adjoining or adjacent buildings.

AF103.6.2 Multiple vent pipes. In buildings where interior footings or other barriers separate the subslab aggregate or other gas-permeable material, each area shall be fitted with an individual vent pipe. Vent pipes shall connect to a single vent that terminates above the roof or each individual vent pipe shall terminate separately above the roof.

❖ An individual vent pipe is required for each unique under-slab area that defines a separate gas-permeable layer, such as those spaces separated by interior footings. The vent pipes may terminate individually above the roof or may be connected to a single vent. Also see the definition for "Gas-permeable layer" in Section AF102.1 and its commentary.

AF103.7 Vent pipe drainage. Components of the radon vent pipe system shall be installed to provide positive drainage to the ground beneath the slab or soil-gas-retarder.

❖ The manner of installation of a radon vent pipe system must be such that positive drainage is created to the ground beneath the floor slab or soil-gas-retarder.

AF103.8 Vent pipe accessibility. Radon vent pipes shall be accessible for future fan installation through an *attic* or other area outside the *habitable space*.

Exception: The radon vent pipe need not be accessible in an *attic* space where an *approved* roof-top electrical supply is provided for future use.

AF103.9 Vent pipe identification. Exposed and visible interior radon vent pipes shall be identified with not less than one *label* on each floor and in accessible *attics*. The *label* shall read: "Radon Reduction System."

❖ Interior vent pipes installed as a portion of the radon venting system must be adequately identified to reduce the potential for improper use or modification of the venting system. The identification is required for every floor level and in all accessible attics where the radon vents are exposed and visible. At a minimum, the identification label must state: "Radon Reduction System."

AF103.10 Combination foundations. Combination *basement*/crawl space or slab-on-grade/crawl space foundations shall have separate radon vent pipes installed in each type of foundation area. Each radon vent pipe shall terminate above the roof or shall be connected to a single vent that terminates above the roof.

❖ Where the design of the structure combines a basement with a crawl space foundation, or a slab-on-grade floor with a crawl space foundation, separate radon vent pipes are to be provided for each individual type of foundation system. The vent piping must extend above the roof, either as individual vent terminations or as a single vent termination connected to the multiple vents.

AF103.11 Building depressurization. Joints in air ducts and plenums in un*conditioned spaces* shall meet the requirements of Section M1601. Thermal envelope air infiltration requirements shall comply with the energy conservation provisions in Chapter 11. Fireblocking shall meet the requirements contained in Section R302.11.

AF103.12 Power source. To provide for future installation of an active submembrane or subslab depressurization system, an electrical circuit terminated in an *approved* box shall be installed during construction in the *attic* or other anticipated location of vent pipe fans. An electrical supply shall also be accessible in anticipated locations of system failure alarms.

❖ It is possible that a passive depressurization system will be converted to an active system at some future time. In anticipation of such an occurrence, an electrical circuit must be provided to an approved box. The box should be located in the attic or other location that provides access to the vent pipe fans and an access opening to the location must be provided. This section provides specific minimum dimensions for the access opening. An additional electrical supply must be provided at the anticipated future locations of system failure alarms.

Appendix G:
Piping Standards for Various Applications

General Comments

This appendix is a valuable reference for designers to use when developing various piping systems for an application. This appendix is meant to be used as a starting point so that extensive research on piping is not necessary to begin a system design. Some applications require determination of friction losses of large piping networks before equipment sizing can be performed and as such, the internal sizes of piping (and the types of fittings that can be used for the selected piping) can greatly impact the size of equipment. This appendix provides a list of standards for various types of piping so that the details for the chosen piping can be easily found.

Note that most, but not all, of the standards in the table are recognized as part of the "base code" (see Chapter 44). Section AG102 indicates the standards that are not referenced in the code text of Chapters 1 through 43. Some of the standards in the table in this appendix might not be referenced in the code text of Chapters 1 through 43 for the application that the user has chosen the piping for. Any standards that are not specifically referenced in Chapters 1 through 43 code language for the specific application will have to be approved in accordance with Section R104.11. The mere existence of the "listing" of a standard in this table is not an indication that the material (as indicated by the standard) complies with the requirements of the code. This appendix is provided as only a convenience and does not imply code approval of any of these standards for any the indicated applications.

SECTION AG101
PLASTIC PIPING STANDARDS

AG101.1 Plastic piping. Table AG101.1 provides a list of plastic piping product standards for various applications.

❖ See the commentary to "General comments" for this appendix.

TABLE AG101.1. See page G-2.

❖ See the commentary to "General comments" for this appendix.

TABLE AG101.1
PLASTIC PIPING STANDARDS FOR VARIOUS APPLICATIONS[a,b]

APPLICATION	LOCATION	TYPE OF PLASTIC PIPING								
		ABS	CPVC	PE	PE-AL-PE	PE-RT	PEX	PEX-AL-PEX	PP	PVC
Central vacuum	System piping	—	—	—	—	—	—	—	—	ASTM F2158
Foundation drainage	System piping	ASTM F628	—	ASTM F405	—	—	—	—	—	ASTM D2665 ASTM D2729 ASTM D3034
Geothermal ground loop	System piping	—	ASTM D2846 ASTM F441 ASTM F442 ASTM F2855 CSA B137.6	ASTM D2239 ASTM D2737 ASTM D3035	ASTM F1282	ASTM F2623 ASTM F2769	ASTM F876 CSA B137.5	ASTM F1281	ASTM F2389 CSA B137.11	ASTM D1785 ASTM D2241 CSA B137.3
	Loop piping	—	—	ASTM D2239 ASTM D2737 ASTM D3035 NSF 358-1	ASTM F1282	ASTM F2623 ASTM F2769	ASTM F876 CSA B137.5	—	ASTM F2389 CSA B137.11	—
Gray water	Nonpressure distribution/ collection	ASTM F628	—	ASTM D2239 ASTM D2737 ASTM D3035 ASTM F2306	—	—	—	—	ASTM F2389 CSA B137.11	ASTM D1785 ASTM D2729 ASTM D2949 ASTM D3034 ASTM F891 ASTM F1760 CSA B137.3

(continued)

TABLE AG101.1—continued
PLASTIC PIPING STANDARDS FOR VARIOUS APPLICATIONS[a,b]

APPLICATION	LOCATION	TYPE OF PLASTIC PIPING								
		ABS	CPVC	PE	PE-AL-PE	PE-RT	PEX	PEX-AL-PEX	PP	PVC
Gray water	Pressure/distribution	—	ASTM D2846 ASTM F441 ASTM F442 ASTM F2855 CSA B137.6	ASTM D2239 ASTM D2737 ASTM D3035	ASTM F1282	ASTM F2623 ASTM F2769	ASTM F876 CSA B137.5	ASTM F1281	ASTM F2389 CSA B137.11	ASTM D1785 ASTM D2241 CSA B137.3
Radiant cooling	Loop piping	—	ASTM D2846 ASTM F441 ASTM F442 ASTM F2855	ASTM D2239 ASTM D2737 ASTM D3035	ASTM F1282	ASTM F2623 ASTM F2769	ASTM F876 CSA B137.5	ASTM F1281	ASTM F2389 CSA B137.11	—
Radiant heating	Loop piping	—	ASTM D2846 ASTM F441 ASTM F442 ASTM F2855	—	ASTM F1282	ASTM F2623 ASTM F2769	ASTM F876 CSA B137.5	ASTM F1281	ASTM F2389 CSA B137.11	—
Rainwater harvesting	Nonpressure/collection	ASTM F628	—	ASTM F1901	—	—	—	—	ASTM F2389 CSA B137.11	ASTM D1785 ASTM D2729 ASTM D2949 ASTM F891 ASTM F1760 CSA B137.3
	Pressure/distribution	—	ASTM D2846 ASTM F441 ASTM F442 ASTM F2855 CSA B137.6	ASTM D2239 ASTM D2737 ASTM D3035	ASTM F1282	ASTM F2623 ASTM F2769	ASTM F876 CSA B137.5	ASTM F1281	ASTM F2389 CSA B137.11	ASTM D1785 ASTM D2241 CSA B137.3

(continued)

TABLE AG101.1—continued
PLASTIC PIPING STANDARDS FOR VARIOUS APPLICATIONS[a,b]

APPLICATION	LOCATION	TYPE OF PLASTIC PIPING								
		ABS	CPVC	PE	PE-AL-PE	PE-RT	PEX	PEX-AL-PEX	PP	PVC
Radon venting	System piping	ASTM F628	—	—	—	—	—	—	—	ASTM D1785 ASTM F891 ASTM F1760
Reclaimed water	Main to building service	—	ASTM D2846 ASTM F441 ASTM F442 ASTM F2855 CSA B137.6	ASTM D3035 AWWA C901 CSA B137.1	ASTM F1282	ASTM F2623 ASTM F2769	ASTM F876 AWWA C904 CSA B137.5	—	ASTM F2389 CSA B137.11	ASTM D1785 ASTM D2241 AWWA C905 CSA B137.3
	Pressure/ distribution/ irrigation	—	ASTM D2846 ASTM F441 ASTM F442 ASTM F2855 CSA B137.6	ASTM D2239 ASTM D2737 ASTM D3035	ASTM F1282	ASTM F2623 ASTM F2769	ASTM F876 CSA B137.5	ASTM F1281	ASTM F2389 AWWA C900 CSA B137.11	ASTM D1785 ASTM D2241 AWWA C900
Residential fire sprinklers[c]	Sprinkler piping	—	ASTM F441 ASTM F442 CSA B137.6 UL 1821	—	—	ASTM F2769	ASTM F876 CSA B137.5 UL 1821	—	ASTM F2389 CSA B137.11	—
Solar heating	Pressure/ distribution	—	ASTM D2846 ASTM F441 ASTM F442 ASTM F2855	—	—	ASTM F2623 ASTM F2769	ASTM F876 CSA B137.5	ASTM F1281	ASTM F2389 CSA B137.11	—

a. This table indicates manufacturing standards for plastic piping materials that are suitable for use in the applications indicated. Such applications support green and sustainable building practices. The system designer or the installer of piping shall verify that the piping chosen for an application complies with local codes and the recommendations of the manufacturer of the piping.

b. Fittings applicable for the piping shall be as recommended by the manufacturer of the piping.

c. Piping systems for fire sprinkler applications shall be listed for the application.

SECTION AG102
REFERENCED STANDARDS

AG102.1 General.

ASTM

F1760—01(2011)	Standard Specification for Coextruded Poly (Vinyl Chloride) (PVC) Non-Pressure Plastic Pipe Having Reprocessed-Recycled Content
F1901—10	Standard Specification for Polyethylene (PE) Pipe and Fittings for Roof Drain Systems
F2158—08	Standard Specification for Residential Central-Vacuum Tube and Fittings
F2306—08	12" to 60" Annular Corrugated Profile-wall Polyethylene (PE) Pipe and Fittings for Gravity Flow Storm Sewer and Sub-surface Drainage Applications

AWWA

900—07	Polyvinyl chloride (PVC) Pressure Pipe and Fabricated Fittings, 4 in. through 12 in. (350 mm through 1200 mm), for Water Transmission and Distribution
905—10	Polyvinyl chloride (PVC) Pressure Pipe and Fabricated Fittings, 14 in. through 48 in. (100 mm through 300 mm)

UL

1821—2011	Standard for Thermoplastic Sprinkler Pipe and Fittings for Fire Protection Service

Appendix H: Patio Covers

(The provisions contained in this appendix are not mandatory unless specifically referenced in the adopting ordinance.)

General Comments

Patio covers and similar structures are regulated under the general provisions of the code unless this appendix chapter is adopted to address their design and construction. The provisions of this appendix address several specific areas where the general criteria of the code would establish more stringent requirements. The floor area below the patio cover is viewed independently from the remainder of the dwelling unit, with only a limited number of provisions. This appendix sets forth the regulations and limitations for patio covers. Section AH101 establishes the scope of Appendix H and the limitations on the uses of patio covers. Sections AH103 and AH104 describe the limitations for structures designated as patio covers as well as the effect a patio cover has on light and ventilation, emergency escape and rescue, and egress requirements. Section AH105 establishes the structural requirements for patio structures, includ-

ing design load and footings. Section AH106 addresses the construction of patio covers without footings. Section AH107 contains the requirements for aluminum screen enclosures in hurricane-prone regions.

Purpose

In many climates, the interior living space of a dwelling unit is extended to create an outdoor living area. At times, a separate structure is constructed to create an outdoor area that is protected from the environment. Such structures are patio covers, with a limited number of requirements as set forth by this appendix. The provisions address those uses permitted in patio cover structures, the minimum design loads to be assigned for structural purposes, and the effect of the patio cover on egress and emergency escape or rescue from sleeping rooms.

SECTION AH101
GENERAL

AH101.1 Scope. Patio covers shall conform to the requirements of Sections AH101 through AH106.

❖ This appendix chapter regulates the design and construction of patio covers that are accessory to residential dwelling units.

AH101.2 Permitted uses. Patio covers shall be permitted to be detached from or attached to *dwelling units*. Patio covers shall be used only for recreational, outdoor living purposes, and not as carports, garages, storage rooms or habitable rooms.

❖ A patio cover may be constructed as a freestanding structure or may be attached directly to a dwelling unit. In either case, the acceptable uses of the floor area beneath the patio cover are limited. This section allows a patio cover to be used for recreational use only, as an outdoor living area. A further provision specifically prohibits use of the area under a patio cover for vehicle parking, general storage or habitable space.

SECTION AH102
DEFINITION

AH102.1 General. The following word and term shall, for the purposes of this appendix, have the meaning shown herein.

❖ Definitions of terms are sometimes needed to help in clarifying code requirements. The purpose of including

the definition of patio covers here is that the term is used in this appendix. Therefore, for convenience, the definition is provided here as well as in Chapter 2.

PATIO COVER. A structure with open or glazed walls that is used for recreational, outdoor living purposes associated with a dwelling unit.

❖ Patio covers are open structures. As such, the special rules provided in this appendix deal with the concerns related to patio covers while at the same time providing less stringent requirements.

SECTION AH103
EXTERIOR WALLS AND OPENINGS

AH 103.1 Enclosure walls. Enclosure walls shall be permitted to be of any configuration, provided the open or glazed area of the longer wall and one additional wall is equal to at least 65 percent of the area below a minimum of 6 feet, 8 inches (2032 mm) of each wall, measured from the floor. Openings shall be permitted to be enclosed with any of the following:

1. Insect screening.

2. Approved translucent or transparent plastic not more than 0.125 inch (3.2 mm) in thickness.

3. Glass conforming to the provisions of Section R308.

4. Any combination of the foregoing.

❖ It is common for a portion of the covered area to be enclosed by solid wall construction, with the remainder of the enclosure provided by glazing or screening. This section indicates that the minimum opening length of 65 percent only need be provided at a height of 80 inches (2032 mm) and below. Insect screening, approved minimum $^1/_8$-inch (3.2 mm) plastic panels, and glazing materials in compliance with Section R308 are permitted as enclosure materials.

AH103.2 Light, ventilation and emergency egress. Exterior openings required for light and ventilation shall be permitted to open into a patio structure conforming to Section AH101, provided that the patio structure shall be unenclosed if such openings are serving as emergency egress or rescue openings from sleeping rooms. Where such exterior openings serve as an exit from the *dwelling unit*, the patio structure, unless unenclosed, shall be provided with exits conforming to the provisions of Section R311 of this code.

❖ A common type of patio cover is one that has one wall of the dwelling unit as the enclosure. This is acceptable, as long as adequate light and ventilation are brought into the building and emergency egress is not blocked. In addition, if this is the exit to the dwelling unit, then the patio enclosure must provide the exit as well. The sections of R311 apply, including proper width of the exit, stairs, and level surface of the means of egress.

Due to the general openness required of patio structures, it is permissible to have the exterior openings required for light and ventilation purposes by Section R303 open into a patio cover structure. If the exterior openings are also required for emergency escape or rescue by the provisions of Section R310, the patio cover structure cannot be enclosed. Therefore, the allowances for full enclosure by insect screening, light-transmitting plastics, or glazing are not applicable. In addition, where an exit passes through the patio cover structure, the structure must be either unenclosed or provided with at least one complying exit.

SECTION AH104
HEIGHT

AH104.1 Height. Patio covers are limited to one-story structures not exceeding 12 feet (3657 mm) in height.

❖ In order to provide requirements that are limited or less stringent applications of the requirements otherwise imposed by the code, patio covers are limited to a height to one story and 12 feet.

SECTION AH105
STRUCTURAL PROVISIONS

AH105.1 Design loads. Patio covers shall be designed and constructed to sustain, within the stress limits of this code, all dead loads plus a vertical live load of not less than 10 pounds per square foot (0.48 kN/m^2), except that snow loads shall be used where such snow loads exceed this minimum. Such cov-

ers shall be designed to resist the minimum wind loads set forth in Section R301.2.1.

❖ In addition to the dead loads imposed on the structure, a minimum design live load of 10 pounds per square foot (0.48 kN/m^2) is to be used for vertical loading on patio covers. If the snow load provides for a higher design load, then it must be used as the design criteria. Any wind loads must also be considered in the design and construction of a patio cover.

AH105.2 Footings. In areas with a frostline depth of zero as specified in Table R301.2(1), a patio cover shall be permitted to be supported on a slab-on-*grade* without footings, provided the slab conforms to the provisions of Section R506, is not less than 3.5 inches (89 mm) thick and the columns do not support live and dead loads in excess of 750 pounds (3.34 kN) per column.

❖ Footings supporting a patio cover must extend below the frostline depth established by the jurisdiction. Where the frostline does not extend below ground level and is established as zero, the patio cover structure need only be supported by a slab on grade. Use of a concrete slab without footings is only permitted if the slab is at least 3.5 inches (89 mm) thick and in compliance with Section R506. In addition, the maximum loading supported by each column is limited to 750 pounds (3.34 kN), including both live loads and dead loads.

SECTION AH106
SPECIAL PROVISIONS FOR ALUMINUM SCREEN ENCLOSURES IN HURRICANE-PRONE REGIONS

AH106.1 General. Screen enclosures in *hurricane-prone regions* shall be in accordance with the provisions of this section.

❖ Because screened enclosures are open structures and many of them have solid roofs, they must be designed to withstand the wind loads in hurricane-prone regions of the country. Sections AH106.1.1 through AH106.5 provide guidance and minimum requirements for the design of these enclosures.

AH106.1.1 Habitable spaces. Screen enclosures shall not be considered *habitable spaces*.

❖ Screened enclosures are classified as nonhabitable spaces to avoid having to comply with some of the requirements for habitable spaces such as space heating. Obviously, if the owner encloses the space at a later date, compliance with all applicable code provisions for habitable spaces would be required.

AH106.1.2 Minimum ceiling height. Screen enclosures shall have a ceiling height of not less than 7 feet (2134 mm).

❖ The minimum ceiling height of 7 feet (2134 mm) is necessary because these enclosures are typically an addition to a house, and tying an enclosure into an existing roof while maintaining an adequate slope for run-off would make a 7$^1/_2$-foot (2286 mm) ceiling difficult to achieve. The code already permits other spaces such as bathrooms and kitchens to have 7-foot (2134 mm) ceilings, so applying this exception to screened enclosures, which are classified as nonhabitable, is reasonable.

AH106.2 Definition. The following word and term shall, for the purposes of this appendix, have the meaning shown herein.

SCREEN ENCLOSURE. A building or part thereof, in whole or in part self-supporting, and having walls of insect screening, and a roof of insect screening, plastic, aluminum or similar lightweight material.

❖ The definition of screen enclosures is provided to clarify that they do not have to be completely self-supporting and that the roof may be of screen or a lightweight, solid material such as aluminum.

AH106.3 Screen enclosures. Screen enclosures shall comply with Sections AH106.3.1 and AH106.3.2.

AH106.3.1 Thickness. Actual wall thickness of extruded aluminum members shall be not less than 0.040 inch (1.02 mm).

❖ This section specifies a minimum acceptable wall thickness for extruded aluminum members to provide the structural integrity of the screened enclosure during high windload periods.

AH106.3.2 Density. Screen density shall be not more than 20 threads per inch by 20 threads per inch mesh.

❖ This section specifies a maximum screen mesh density of 20 threads per inch by 20 threads per inch. A more dense screen mesh could develop more resistance to wind and result in failure of the structure due to increased loads. The minimum screen thread count is 16 per inch for window openings in the *International Property Maintenance Code*® (IPMC®).

AH106.4 Design. The structural design of screen enclosures shall comply with Sections AH106.4.1 through AH106.4.3.

AH106.4.1 Wind load. Structural members supporting screen enclosures shall be designed to support the minimum wind loads given in Tables AH106.4(1) and AH106.4(2) for the ultimate design wind speed, V_{ult}, determined from Figure AH106.4.1. Where any value is less than 10 pounds per square foot (psf) (0.479 kN/m²) use 10 pounds per square foot (0.479 kN/m²).

❖ The ultimate design wind speed used for design is determined from Table R301.2(1) or Figure R301.2(4). Wind exposure category is established on a site-specific basis in accordance with Section R301.2.1.4. Once the wind speed and exposure category is determined, the wind load from Table AH106.4(1), adjusted for height and exposure in accordance with Table AH106.4(2), is used to design the structural members of the screen enclosure.

TABLE AH106.4(1)
DESIGN WIND PRESSURES FOR SCREEN ENCLOSURE FRAMING[a, b, e, f, g, h]

LOAD CASE	WALL	ULTIMATE DESIGN WIND SPEED, V_{ult} (mph)									
		100	105	110	120	130	140	150	160	170	180
		Exposure Category B Design Pressure (psf)									
A[c]	Windward and leeward walls (flow thru) and windward wall (nonflow thru) L/W = 0-1	6	7	8	9	11	13	14	16	18	21
A[c]	Windward and leeward walls (flow thru) and windward wall (nonflow thru) L/W = 2	7	8	9	11	12	14	16	19	21	24
B[d]	Windward: Nongable roof	9	10	11	13	15	18	21	23	26	30
B[d]	Windward: Gable roof	11	13	14	16	19	22	26	29	33	37
	ROOF										
All[e]	Roof-screen	2	3	3	3	4	4	5	6	7	7
All[e]	Roof-solid	7	8	8	10	12	13	15	18	20	22

For SI: 1 mile per hour = 0.44 m/s, 1 pound per square foot = 0.0479 kPa, 1 foot = 304.8 mm.

a. Design pressure shall be not less than 10 psf in accordance with Section AH106.4.1.

b. Loads are applicable to screen enclosures with a mean roof height of 30 feet or less in Exposure B. For screen enclosures of different heights or exposure, the pressures given shall be adjusted by multiplying the table pressure by the adjustment factor given in Table AH106.4(2).

c. For Load Case A flow thru condition, the pressure given shall be applied simultaneously to both the upwind and downwind screen walls acting in the same direction as the wind. The structure shall also be analyzed for wind coming from the opposite direction. For the nonflow thru condition, the screen enclosure wall shall be analyzed for the load applied acting toward the interior of the enclosure.

d. For Load Case B, the table pressure multiplied by the projected frontal area of the screen enclosure is the total drag force, including drag on screen surfaces parallel to the wind, that must be transmitted to the ground. Use Load Case A for members directly supporting the screen surface perpendicular to the wind. Load Case B loads shall be applied only to structural members that carry wind loads from more than one surface.

e. The roof structure shall be analyzed for the pressure given occurring both upward and downward.

f. Table pressures are MWFRS loads. The design of solid roof panels and their attachments shall be based on component and cladding loads for enclosed or partially enclosed structures as appropriate.

g. Table pressures apply to 20-inch by 20-inch by 0.013-inch mesh screen. For 18-inch by 14-inch by 0.013-inch mesh screen, pressures on screen surfaces shall be permitted to be multiplied by 0.88. For screen densities greater than 20 inches by 20 inches by 0.013 inch, pressures for enclosed buildings shall be used.

h. Linear interpolation shall be permitted.

TABLE AH106.4(2)
ADJUSTMENT FACTOR FOR
BUILDING HEIGHT AND EXPOSURE

MEAN ROOF HEIGHT (feet)	EXPOSURE		
	B	C	D
15	1.00	1.21	1.47
20	1.00	1.29	1.55
25	1.00	1.35	1.61
30	1.00	1.40	1.66
35	1.05	1.45	1.70
40	1.09	1.49	1.74
45	1.12	1.53	1.78
50	1.16	1.56	1.81
55	1.19	1.59	1.84
60	1.22	1.62	1.87

For SI: 1 foot = 304.8 mm.

AH106.4.2 Deflection limit. For members supporting screen surfaces only, the total load deflection shall not exceed $l/60$. Screen surfaces shall be permitted to include not more than 25-percent solid flexible finishes.

❖ The deflection limit of l/60 is from the Aluminum Association, Standard ADM - 1, *Aluminum Design Manual*. The limit on solid flexible finishes that is permitted to be included in screen surfaces addresses elements such as kick plates.

AH106.4.3 Roof live load. The roof live load shall be not less than 10 psf (0.479 kN/m^2).

❖ The 10 psf minimum roof live load is the same as the roof live load for patio covers as specified in the IBC.

AH106.5 Footings. In areas with a frost line depth of zero, a screen enclosure shall be permitted to be supported on a concrete slab-on-*grade* without footings, provided the slab conforms to the provisions of Section R506, is not less than $3^1/_2$ inches (89 mm) thick and the columns do not support loads in excess of 750 pounds (3.36 kN) per column.

❖ This allowance, which permits the enclosures to be constructed without foundations, is due to the minor value of these structures and the lack of any substantive safety issues involved. This provision addresses the exemption of footings in regard to frost penetration; it does not address allowable load-bearing capacity of the soil.

Bibliography

The following resource materials were used in the preparation of the commentary for this appendix of the code.

IPMC–15 *International Property Maintenance Code.* Washington, DC: International Code Council, 2015.

ASCE 7–05, *Minimum Design Loads for Buildings and Other Structures.* Reston, VA: American Society of Civil Engineers, 2005.

Reinhold, T. A., J. D. Belcher, C. O. Everly and David Miller. *Wind Loads on Screen Enclosures.* Clemson, SC: Department of Civil Engineering, Clemson University, 2000.

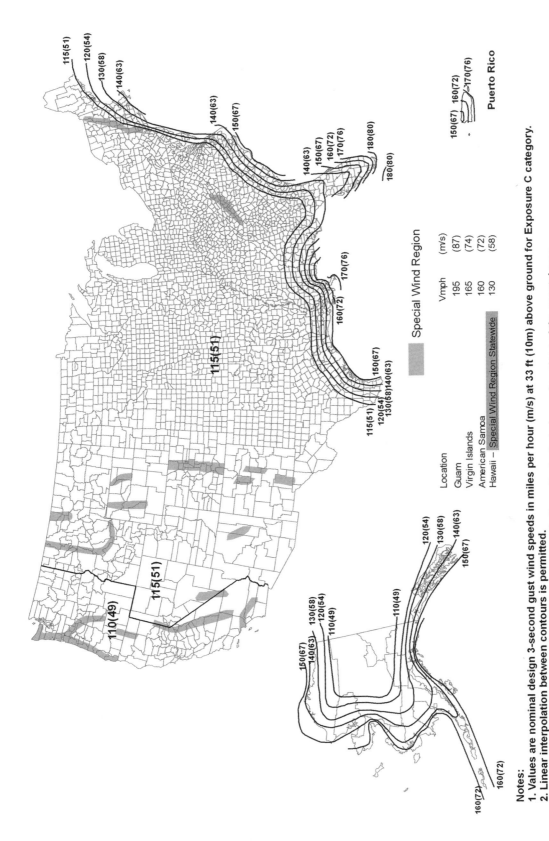

Special Wind Region

Special Wind Region Statewide

Location	Vmph	(m/s)
Guam	195	(87)
Virgin Islands	165	(74)
American Samoa	160	(72)
Hawaii – Special Wind Region Statewide	130	(58)

Puerto Rico

Notes:
1. Values are nominal design 3-second gust wind speeds in miles per hour (m/s) at 33 ft (10m) above ground for Exposure C category.
2. Linear interpolation between contours is permitted.
3. Islands and coastal areas outside the last contour shall use the last wind speed contour of the coastal area.
4. Mountainous terrain, gorges, ocean promontories, and special wind regions shall be examined for unusual wind conditions.
5. Wind speeds correspond to approximately a 7% probability of exceedance in 50 years (Annual Exceedance Probability = 0.00143, MRI = 700 Years).

FIGURE AH106.4.1
ULTIMATE DESIGN WIND SPEEDS FOR PATIO COVERS AND SCREEN ENCLOSURES

Appendix I:
Private Sewage Disposal

(The provisions contained in this appendix are not mandatory unless specifically referenced in the adopting ordinance.)

SECTION AI101
GENERAL

AI101.1 Scope. Private sewage disposal systems shall conform to the *International Private Sewage Disposal Code.*

Appendix J:
Existing Buildings and Structures

(The provisions contained in this appendix are not mandatory unless specifically referenced in the adopting ordinance.)

General Comments

Appendix J contains the provisions for the repair, renovation, alteration and reconstruction of existing buildings and structures that are within the scope of the International Residential Code® (IRC®). The appendix, when adopted, includes or refers to all code requirements for existing buildings. The format of the appendix readily identifies the alternative methods of code compliance, developed to provide consistency with the purpose of the code.

Section AJ101 sets forth the intent and purpose of the appendix. Section AJ102 contains methods to obtain compliance with the appendix. Section AJ103 establishes the use of a preliminary meeting to discuss the proposed work. Section AJ104 sets forth the potential means for investigating and evaluating an existing building. Section AJ105 requires the identification of a work area on the permit. Section AJ201 establishes definitions for specific terms used in the appendix. Section AJ301 regulates repairs of existing buildings and structures. Section AJ401 addresses renovations of existing buildings and structures. Section AJ501 regulates alterations to existing buildings and structures, and Section AJ601 establishes requirements for reconstruction work in existing buildings and structures.

Purpose

Many existing residential buildings do not meet the code requirements that currently apply to new construction. Although these buildings are potentially salvageable, rehabilitation is often cost prohibitive because the building may not comply with all aspects of new construction requirements. At the same time, it is necessary to regulate construction activity in existing buildings that undergo repair, renovation, alteration, or reconstruction. Such activity represents an opportunity for the jurisdiction to determine both that new work conforms to the code (or the intent of the code) and that existing conditions either remain at their current level of compliance or are improved. To accomplish this objective and to make the rehabilitation process more available, this chapter allows for a controlled departure from full code compliance without compromising minimum life safety, fire safety, structural and environmental features of the rehabilitated building.

SECTION AJ101
PURPOSE AND INTENT

AJ101.1 General. The purpose of these provisions is to encourage the continued use or reuse of legally existing buildings and structures. These provisions are intended to permit work in existing buildings that is consistent with the purpose of this code. Compliance with these provisions shall be deemed to meet the requirements of this code.

❖ This section allows for residential structures to continue their existing use while permitting repair, renovation, alteration or reconstruction work that may not be in full compliance with the general provisions of the code. Although strict compliance with the general requirements in the code may be modified by the provisions of this chapter, the resultant work will be consistent with the intent and purpose criteria set forth in Chapter 1. As a rule, the general requirements of the code are applicable unless modified by the provisions of this appendix. Upon compliance with the provisions of this appendix, the work meets the requirements of the code.

AJ101.2 Classification of work. For purposes of this appendix, work in existing buildings shall be classified into the categories of repair, renovation, *alteration* and reconstruction. Specific requirements are established for each category of work in these provisions.

❖ This appendix addresses four unique and specific types of work that may occur on an existing residential structure. Each type is regulated independently based on the provisions of the chapter. Repairs, where work is performed to restore a building element or system back to working order, are regulated by Section AJ301. Renovation, the act of restoring a building element to its original condition, is addressed in Section AJ401. Section AJ501 deals with alterations, where elements, components or systems of the building are modified using new materials and methods, and Section AJ601 addresses reconstruction of existing building elements. Although these areas seem to overlap in their scope, each has unique characteristics that are identified and regulated by this appendix.

AJ101.3 Multiple categories of work. Work of more than one category shall be part of a single work project. Related work permitted within a 12-month period shall be considered to be a single work project. Where a project includes one category of work in one building area and another category of work in a separate and unrelated area of the building, each project area shall comply with the requirements of the respective category of work. Where a project with more than one category of work is performed in the same area or in related areas of the building, the project shall comply with the requirements of the more stringent category of work.

❖ It is not uncommon for two or more types of construction activity to be taking place at the same time, as part of a single work project. Under such conditions, there are two situations that may occur: 1. Where the categories of work are isolated in different areas of the building and unrelated to each other, each project area is regulated by the specific requirements of the work being performed; or 2. Where the multiple categories of work occur in the same project area, the more restrictive provisions of the categories apply.

SECTION AJ102
COMPLIANCE

AJ102.1 General. Regardless of the category of work being performed, the work shall not cause the structure to become unsafe or adversely affect the performance of the building; shall not cause an existing mechanical or plumbing system to become unsafe, hazardous, insanitary or overloaded; and unless expressly permitted by these provisions, shall not make the building any less compliant with this code or to any previously *approved* alternative arrangements than it was before the work was undertaken.

❖ The general limitation on any work done under the provisions of this chapter is that the level of safety, health and public welfare of the existing building must not be reduced by any work being performed. In the case of structural stability, the existing degree of structural strength must be maintained or increased. In general terms, the structure is not to be made unsafe. This requirement can be broadly interpreted because its application can vary case by case. To the extent that existing mechanical and plumbing systems are involved, the level of protection or sanitation must not be reduced.

Where this appendix chapter does not specifically modify the general provisions of the code, such provisions are applicable and conformance is required. Only in those areas where this chapter addresses specific materials or methods for specific elements, components or systems do the general code requirements not apply.

AJ102.2 Requirements by category of work. Repairs shall conform to the requirements of Section AJ301. Renovations shall conform to the requirements of Section AJ401. *Alterations* shall conform to the requirements of Section AJ501

and the requirements for renovations. Reconstructions shall conform to the requirements of Section AJ601 and the requirements for *alterations* and renovations.

❖ Depending on the category of work being performed, conformance to one or more groups of requirements is necessary. If the work is simply a repair job or a renovation, only the provisions of Section AJ301 or AJ401, respectively, are applicable. If the building is to undergo an alteration, the requirements of both Sections AJ401 and AJ501 must be applied. Section AJ601 must be used for reconstruction work.

AJ102.3 Smoke detectors. Regardless of the category of work, smoke detectors shall be provided where required by Section R314.3.1.

❖ The benefit of smoke detectors in residential buildings is unquestioned. Occupants of such buildings spend about one-third of the day asleep. The potential for a fire getting out of control before the occupants are awakened is quite probable, as reflected in statistics that indicate residential fire deaths far exceed those of any other building classification. As a result, the general provisions of Section R313.2.1 for smoke alarms are applicable in all cases and for all categories of work. This includes specific requirements where a dwelling unit undergoes interior alterations or repairs.

AJ102.4 Replacement windows. Regardless of the category of work, where an existing window, including the sash and glazed portion, or safety glazing is replaced, the replacement window or safety glazing shall comply with the requirements of Sections AJ102.4.1 through AJ102.4.3, as applicable.

❖ Replacement windows must comply with the applicable provisions for energy efficiency, safety glazing and emergency escape and rescue openings contained in Sections AJ102.4.1, AJ102.4.2 and AJ102.4.3.

AJ102.4.1 Energy efficiency. Replacement windows shall comply with the requirements of Chapter 11.

❖ Under all circumstances, a replacement window must comply with the energy efficiency provisions of Chapter 11. The provision is applicable only where the entire window is changed out, including the sash and glazing. Where such conditions exist, the new window is to have a maximum fenestration *U*-factor in compliance with Chapter 11. If the replacement window is in a hazardous location, safety glazing complying with Section R308 must be installed.

AJ102.4.2 Safety glazing. Replacement glazing in hazardous locations shall comply with the safety glazing requirements of Section R308.

❖ Just as for windows used in new construction, the glazing in replacement windows must comply with the safety glazing requirements of Section R308.

AJ102.4.3 Emergency escape and rescue openings. Where windows are required to provide emergency escape and rescue openings, replacement windows shall be exempt from the maximum sill height requirements of Section R310.1 and the requirements of Sections R310.1.1, R310.1.2, R310.1.3 and R310.2 provided that the replacement window meets the following conditions:

1. The replacement window is the manufacturer's largest standard size window that will fit within the existing frame or existing rough opening. The replacement window shall be permitted to be of the same operating style as the existing window or a style that provides for an equal or greater window opening area than the existing window.

2. The replacement window is not part of a change of occupancy.

3. Window opening control devices complying with ASTM F2090 shall be permitted for use on windows required to provide emergency escape and rescue openings.

❖ The emergency escape and rescue opening requirements for replacement windows are not as stringent as those for windows that are used in new construction. Most of these differences are due to the fact that it is often difficult and cost prohibitive to change the opening size of windows in order to comply with current opening requirements that may differ significantly from the code under which a dwelling was originally built. Replacement windows are not required to comply with the sill height requirements of Section R310.1 and the requirements of Sections R310.1.2, R310.1.3 and R310.2 if all of the provisions of Items 1 through 3 to Section AJ102.4.3 are satisfied.

AJ102.4.4 Window control devices. Where window fall prevention devices complying with ASTM F2090 are not provided, window opening control devices complying with ASTM F2090 shall be installed where an existing window is replaced and where all of the following apply to the replacement window:

1. The window is operable.

2. The window replacement includes replacement of the sash and the frame.

3. The top of the sill of the window opening is at a height less than 24 inches (610 mm) above the finished floor.

4. The window will permit openings that will allow passage of a 4-inch-diameter (102 mm) sphere where the window is in its largest opened position.

5. The vertical distance from the top of the sill of the window opening to the finished grade or other surface below, on the exterior of the building, is greater than 72 inches (1829 mm).

The window opening control device, after operation to release the control device allowing the window to fully open, shall not reduce the minimum net clear opening area of the window unit.

❖ Similar to the provisions for replacement windows related to emergency escape and rescue opening requirements, the provisions for replacement windows as related to window control devices are not as stringent as those for windows that are used in new con-

struction. This is due to the fact that it is often difficult and cost prohibitive to change the opening size of windows in order to comply with current opening requirements that may differ significantly from the code under which a dwelling was originally built. Replacement window installations are required to comply with this section only if all of the items listed in Section AJ102.4.4 are applicable. Put another way, if the one or more of the items listed are not applicable for a particular replacement window installation, a window control device is not required to be installed.

AJ102.5 Flood hazard areas. Work performed in existing buildings located in a flood hazard area as established by **Table** R301.2(1) shall be subject to the provisions of Section R105.3.1.1.

❖ Long-term reduction in exposure to flood hazards is one of the reasons floodplain development is regulated. If additions, alterations or repairs of an existing building, including repair of damage from any cause, constitute substantial improvement or repair of substantial damage, the existing building is to be brought into compliance as required in Section R105.3.1.1 (see commentary, Section R105.3.1.1). Improvements are deemed to be "substantial" if the cost, including the value of labor as well as donated labor and materials, equals or exceeds 50 percent of the market value of the building before the improvements are made. Damage is considered to be "substantial" where the cost of restoring a damaged building to its predamage condition equals or exceeds 50 percent of the market value of the building before the damage occurred (for additional guidance, refer to FEMA P-259 and FEMA P-758).

AJ102.6 Equivalent alternatives. Work performed in accordance with the *International Existing Building Code* shall be deemed to comply with the provisions of this appendix. These provisions are not intended to prevent the use of any alternative material, alternative design or alternative method of construction not specifically prescribed herein, provided that any alternative has been deemed to be equivalent and its use authorized by the *building official*.

❖ Work that complies with the *International Existing Building Code*© (IEBC®) is deemed to comply with this appendix.

A comprehensive regulatory document such as a construction code cannot envision and then address all future innovations in the industry. As a result, the code must be applicable to, and provide a basis for, the approval of a number of newly developed, innovative materials, systems and methods for which no code text or referenced standards yet exist. The building official is expected to exercise sound technical judgment in accepting materials, systems or methods that, while not anticipated by the code text, can be demonstrated to offer equivalent performance. By virtue of this section, the code regulates new and innovative construction practices while contributing to the safety of building occupants.

AJ102.7 Other alternatives. Where compliance with these provisions or with this code as required by these provisions is technically infeasible or would impose disproportionate costs because of construction or dimensional difficulties, the building official shall have the authority to accept alternatives. These alternatives include materials, design features and operational features.

❖ The building official may accept alternative solutions involving materials, design features or operational features where compliance with the code provisions creates practical difficulties. A practical difficulty, in this context, means it is technically infeasible to meet the code requirements, or the costs involved in providing code compliance are highly disproportionate to the overall cost of the work. It is up to the building official to evaluate the legitimacy of these two exemptions from the general requirements. This section identifies construction or dimensional difficulties as a basis for using these modifications.

AJ102.8 More restrictive requirements. Buildings or systems in compliance with the requirements of this code for new construction shall not be required to comply with any more restrictive requirement of these provisions.

❖ Where the existing building is in compliance with the requirements of the code for new construction, it is not necessary to comply with any higher level requirements as set forth in this appendix. If the work to be performed is in compliance with the general provisions of the code, as would be required of a newly constructed building, there is no reason to require a higher level of performance for existing structures.

AJ102.9 Features exceeding code requirements. Elements, components and systems of existing buildings with features that exceed the requirements of this code for new construction, and are not otherwise required as part of *approved* alternative arrangements or deemed by the *building official* to be required to balance other building elements not complying with this code for new construction, shall not be prevented by these provisions from being modified as long as they remain in compliance with the applicable requirements for new construction.

❖ Unless used as a portion of an alternative solution as an equivalency, modifications to existing elements, components, or systems need only meet the requirements of the code for new construction. Even though the modification may actually reduce the previous level of compliance, the modification is acceptable if it does not reduce compliance to a level below that accepted for new work.

SECTION AJ103
PRELIMINARY MEETING

AJ103.1 General. If a building *permit* is required at the request of the prospective *permit* applicant, the *building official* or his or her designee shall meet with the prospective applicant to discuss plans for any proposed work under these provisions prior to the application for the *permit*. The purpose of this preliminary meeting is for the *building official* to gain an understanding of the prospective applicant's intentions for the proposed work, and to determine, together with the prospective applicant, the specific applicability of these provisions.

❖ Where the provisions of the code are performance provisions, such as those set forth in this appendix for existing buildings, the party or parties involved with the anticipated work must meet with the building official at the onset of the project to discuss any concerns or questions. Those issues that may be interpreted or applied in various ways must be addressed so that the position of the jurisdiction is understood.

This section indicates that the issues must be discussed prior to application for a building permit, with an emphasis on the scope of the work and the applicability of the code provisions for existing buildings. The effort should be cooperative, with input and discussion by both the building official and the potential permit applicant.

SECTION AJ104
EVALUATION OF AN EXISTING BUILDING

AJ104.1 General. The *building official* shall have the authority to require an existing building to be investigated and evaluated by a registered *design professional* in the case of proposed reconstruction of any portion of a building. The evaluation shall determine the existence of any potential nonconformities to these provisions, and shall provide a basis for determining the impact of the proposed changes on the performance of the building. The evaluation shall use the following sources of information, as applicable:

1. Available documentation of the existing building.

    1.1. Field surveys.

    1.2. Tests (nondestructive and destructive).

    1.3. Laboratory analysis.

Exception: Detached one- or two-family dwellings that are not irregular buildings under Section R301.2.2.2.5 and are not undergoing an extensive reconstruction shall not be required to be evaluated.

❖ It is possible that the nature of the reconstruction work requires an evaluation of the existing building by a registered design professional. Based on the investigation and evaluation of the existing conditions in the building or a portion thereof, the building official is able to analyze the work that must take place to comply with the provisions of the code. The need for an evaluation is determined by the building official based on each situation. In addition, the code recognizes that such an evaluation is unnecessary for single- or two-family dwelling units of conventional construction unless the building is undergoing a major reconstruction.

The evaluation should focus on those existing conditions that are not in conformance with the code and its intended purpose, as well as determining how the proposed modifications will address building performance. All resources that may assist in the investigation and evaluation of the existing conditions must be considered along with the three general sources identified in the code.

SECTION AJ105
PERMIT

AJ105.1 Identification of work area. The work area shall be clearly identified on the *permits* issued under these provisions.

❖ Once the limits of one or more work areas have been established based on the scope of the work involved, the work areas must be properly identified and clearly described. The provisions of this appendix are typically limited to those portions of the building characterized as the work areas. The work areas must be described in satisfactory detail on any permits so that all parties have a consistent understanding of the area limits.

SECTION AJ201
DEFINITIONS

AJ201.1 General. For purposes of this appendix, the terms used are defined as follows.

❖ This section clarifies the terminology used in this appendix. The terms take on very specific meanings that are often different from the way they are typically used.

ALTERATION. The reconfiguration of any space; the *addition* or elimination of any door or window; the reconfiguration or extension of any system; or the installation of any additional *equipment*.

❖ Where the work is extensive enough to cause a change in the layout of a building or a portion thereof, the work is an alteration. This activity could include taking an existing space and changing the configuration of the floor plan by adding or removing walls or partitions. Additional rooms may be created or multiple rooms could be transformed into a single room or area. As a result of such an alteration, the exiting system may be modified or the occupant load revised.

Where an additional door or window is installed, or where an existing door or window is removed, the work is an alteration. An alteration is also an increase to any system of the building, including the expansion of any electrical wiring, plumbing piping or mechanical ducts. If additional equipment, such as another electrical panel or HVAC unit, is installed in an existing building, the installation must be regulated as an alteration.

CATEGORIES OF WORK. The nature and extent of construction work undertaken in an existing building. The categories of work covered in this appendix, listed in increasing order of stringency of requirements, are repair, renovation, *alteration* and reconstruction.

❖ This term describes the four types of construction activity that are regulated by this appendix chapter. The code sets forth specific criteria for each of the defined types of work, with each category placed in a hierarchical position based on the mandated level of requirements. The least restrictive category of work is defined as repair, with an increased degree of regulation for renovations and an even higher level for alterations. The most stringent requirements are applicable for work categorized as reconstruction.

DANGEROUS. Where the stresses in any member; the condition of the building, or any of its components or elements or attachments; or other condition that results in an overload exceeding 150 percent of the stress allowed for the member or material in this code.

❖ Although this term is recognized in general terms as a situation that is unsafe, perilous or likely to cause injury or death, for the purposes of this appendix it is much more narrow in scope. It is limited to a structural condition where the structural member or material is subjected to a load significantly higher than that which it is designed to support.

EQUIPMENT OR FIXTURE. Any plumbing, heating, electrical, ventilating, air-conditioning, refrigerating and fire protection *equipment*; and elevators, dumb waiters, boilers, pressure vessels, and other mechanical facilities or installations that are related to building services.

❖ A multitude of building service components are included in the definition of equipment or fixtures. These include elements of the plumbing, mechanical and electrical systems, as well as fire protective features that are installed in the building.

LOAD-BEARING ELEMENT. Any column, girder, beam, joist, truss, rafter, wall, floor or roof sheathing that supports any vertical load in addition to its own weight, or any lateral load.

❖ The structural components of a building that carry loads other than their own are load-bearing elements. These include members that carry gravity loads, lateral loads or both. The definition lists the various elements that are load-bearing where they support more than their own weight.

MATERIALS AND METHODS REQUIREMENTS. Those requirements in this code that specify material standards; details of installation and connection; joints; penetrations; and continuity of any element, component or system in the building. The required quantity, fire resistance, flame spread, acoustic or thermal performance, or other performance attribute is specifically excluded from materials and methods requirements.

❖ Materials standards, installation details and similar specific requirements make up the materials and methods requirements. On the other hand, performance-related requirements tend to be excluded by definition.

RECONSTRUCTION. The reconfiguration of a space that affects an exit, a renovation or *alteration* where the work area is not permitted to be occupied because existing means-of-egress and fire protection systems, or their equivalent, are not in place or continuously maintained; or there are extensive *alterations* as defined in Section AJ501.3.

❖ The most stringent requirements for the four categories of work are assigned to reconstruction activities. Reconstruction causes a high level of concern because of its impact on fire and life safety. Where an existing exit is affected, the existing level of egress or fire protection is not acceptable or extensive alterations are to be undertaken, the work is considered as reconstruction.

REHABILITATION. Any repair, renovation, *alteration* or reconstruction work undertaken in an existing building.

❖ This is the general term for describing any or all of the four categories of work involved on an existing building.

RENOVATION. The change, strengthening or *addition* of load-bearing elements; or the refinishing, replacement, bracing, strengthening, upgrading or extensive repair of existing materials, elements, components, *equipment* or fixtures. Renovation does not involve reconfiguration of spaces. Interior and exterior painting are not considered refinishing for purposes of this definition, and are not renovation.

❖ A renovation can occur for either structural or non-structural work. The definition includes structural members that are replaced, modified or added to a building, as well as nonstructural materials that are changed in some form. The act of renovation focuses on specific elements rather than the reconfiguration of floor area.

REPAIR. The patching, restoration or minor replacement of materials, elements, components, *equipment* or fixtures for the purposes of maintaining those materials, elements, components, *equipment* or fixtures in good or sound condition.

❖ The least-stringent requirements for the four categories of work on existing building regulate repairs. Repair work maintains elements and systems of buildings in sound condition. Repair activities do not change the configuration of the space, nor do they address new construction or equipment.

WORK AREA. That portion of a building affected by any renovation, *alteration* or reconstruction work as initially intended by the owner and indicated as such in the *permit*. Work area excludes other portions of the building where incidental work entailed by the intended work must be performed, and portions of the building where work not initially intended by the owner is specifically required by these provisions for a renovation, *alteration* or reconstruction.

❖ The work area must be correctly identified to define the extent of the rehabilitation activities. Most of the requirements in this appendix are applicable only to the work area under consideration and do not apply to

other portions of the building. The work area, determined at the time of permitting, is limited to that part of the building affected by the rehabilitation work.

SECTION AJ301
REPAIRS

AJ301.1 Materials. Except as otherwise required herein, work shall be done using like materials or materials permitted by this code for new construction.

❖ There are two possible options for materials used in repair work on an existing building. Unless prohibited by other provisions of the section, it is acceptable to use materials consistent with those that are already present. This allowance follows the general concept that the repair work is making the building no more unsafe or hazardous than it was prior to the work being done. Instead of using the same type of materials, the code permits the use of any materials currently allowed by the code.

AJ301.1.1 Hazardous materials. Hazardous materials no longer permitted, such as asbestos and lead-based paint, shall not be used.

❖ It is generally possible to repair a structure, its components and its systems with materials consistent with those materials that were used previously. However, where materials that are now considered hazardous are involved in the repair work, they may no longer be used. For example, the code identifies asbestos and lead-based paint as two hazardous materials that cannot be used in the repair process. Certain materials previously considered acceptable for building construction are a threat to the health of the occupants.

AJ301.1.2 Plumbing materials and supplies. The following plumbing materials and supplies shall not be used:

1. All-purpose solvent cement, unless *listed* for the specific application.

2. Flexible traps and tailpieces, unless *listed* for the specific application.

3. Solder having more than 0.2 percent lead in the repair of potable water systems.

❖ Specific methods and materials of plumbing installations are identified as no longer acceptable because of their negative impact on public health and safety. Where such existing materials and supplies are a part of the repair of a plumbing element or system, alternate materials must be used. The use of all-purpose solvent cement is permitted for plumbing repair work only if listed for the specific application. The same is true for flexible traps and flexible tailpieces.

AJ301.2 Water closets. Where any water closet is replaced with a newly manufactured water closet, the replacement water closet shall comply with the requirements of Section P2903.2.

❖ Where a new water closet replaces an existing water closet, the new fixture must be designed for a maximum water consumption of 1.6 gallons (6 L) for each flushing cycle. Addressing environmental concerns, this limitation assists in reducing the amount of water consumed during the ongoing use of the building.

AJ301.3 Electrical. Repair or replacement of existing electrical wiring and *equipment* undergoing repair with like material shall be permitted.

Exceptions:

1. Replacement of electrical receptacles shall comply with the requirements of Chapters 34 through 43.

2. Plug fuses of the Edison-base type shall be used for replacements only where there is not evidence of overfusing or tampering in accordance with the applicable requirements of Chapters 34 through 43.

3. For replacement of nongrounding-type receptacles with grounding-type receptacles and for branch circuits that do not have an *equipment* grounding conductor in the branch circuitry, the grounding conductor of a grounding-type receptacle outlet shall be permitted to be grounded to any accessible point on the grounding electrode system, or to any accessible point on the grounding electrode conductor, as allowed and described in Chapters 34 through 43.

❖ Under most conditions, it is acceptable to repair existing electrical installations with the same types of wiring materials and electrical equipment as were used previously. However, this section identifies three conditions where additional criteria must be considered. Electrical receptacles must comply as for new construction as described in Chapters 34 through 43. Edison-base-type plug fuses can be replaced with like fuses only if it can be shown that no tampering or overfusing has occurred. Alternate methods of grounding are also set forth when nongrounding-type receptacles are replaced.

SECTION AJ401
RENOVATIONS

AJ401.1 Materials and methods. The work shall comply with the materials and methods requirements of this code.

❖ The general provisions of the code are to be used for renovation work. There are several modifications to these provisions for door and window dimensions, interior finish materials, and the parapets of unreinforced masonry buildings assigned to a high seismic design category.

AJ401.2 Door and window dimensions. Minor reductions in the clear opening dimensions of replacement doors and windows that result from the use of different materials shall be allowed, whether or not they are permitted by this code.

❖ During many renovation projects, it is common for existing doors and windows to be removed and replaced. Quite often the new doors and windows are of different materials and do not provide the same clear opening dimensions as the originals. Reductions in opening dimensions beyond those allowed by the code are permitted if they are minor. Even though not spe-

cifically defined, a minor reduction would remain consistent with the intent and purpose of the provisions.

AJ401.3 Interior finish. Wood paneling and textile wall coverings used as an interior finish shall comply with the flame spread requirements of Section R302.9.

❖ Where wood paneling or textile wall covering materials are being replaced in a renovation project, the new materials must be in compliance with the provisions of Section R302.9 R315. The new interior finishes are to be regulated for flame spread and smoke development as they are for new construction.

AJ401.4 Structural. Unreinforced masonry buildings located in Seismic Design Category D_2 or E shall have parapet bracing and wall anchors installed at the roofline whenever a reroofing *permit* is issued. Such parapet bracing and wall anchors shall be of an *approved* design.

❖ When reroofing work requiring a building permit takes place on an unreinforced masonry building located in Seismic Design Category D_2 or E, it may be necessary to strengthen the parapet by providing bracing in an approved manner. In addition, wall anchors must be provided at the roof line. The existing parapet and wall anchors should be evaluated to determine the extent of the structural strengthening needed. The scope of the structural work and the methods of compliance are to be approved by the building official.

SECTION AJ501
ALTERATIONS

AJ501.1 Newly constructed elements. Newly constructed elements, components and systems shall comply with the requirements of this code.

Exceptions:

1. Openable windows may be added without requiring compliance with the light and *ventilation* requirements of Section R303.

2. Newly installed electrical *equipment* shall comply with the requirements of Section AJ501.5.

❖ Where the alteration of any portion of a building includes new construction, the work must be accomplished in accordance with the requirements of the code. An exception permits the installation of openable windows as additional features without requiring adherence to the light and ventilation provisions of Section R303. In addition, new electrical components and equipment need comply only with the requirements of Section AJ501.5.

AJ501.2 Nonconformities. The work shall not increase the extent of noncompliance with the requirements of Section AJ601, or create nonconformity to those requirements that did not previously exist.

❖ The extent of noncompliance is limited in regard to stairways, handrails, guards, interior finish materials and dwelling separation walls as addressed in Section AJ601. In addition, nonconformity must not be created regarding any requirements that did not previously exist.

AJ501.3 Extensive alterations. Where the total area of all of the work areas included in an *alteration* exceeds 50 percent of the area of the *dwelling unit*, the work shall be considered to be a reconstruction and shall comply with the requirements of these provisions for reconstruction work.

> **Exception:** Work areas in which the *alteration* work is exclusively plumbing, mechanical or electrical shall not be included in the computation of the total area of all work areas.

❖ If the amount of construction activity in a dwelling unit involves more than 50 percent of the unit's floor area, the stringency of the requirements is increased. The category of work is reconstruction, with the requirements for both alterations and reconstruction to be followed. This increase to a higher-level category is not required where the alteration consists only of plumbing, mechanical or electrical work.

AJ501.4 Structural. The minimum design loads for the structure shall be the loads applicable at the time the building was constructed, provided that a dangerous condition is not created. Structural elements that are uncovered during the course of the *alteration* and that are found to be unsound or dangerous shall be made to comply with the applicable requirements of this code.

❖ As building codes have progressed over the years, structural design values have been reviewed and modified. Unless a dangerous condition will be created, it is permissible to use the minimum structural design loads in place at the time the building was constructed.

As the alteration of the building is progressing, there may be occasions where structural elements are exposed and found to be damaged, unsound or otherwise dangerous. In such situations, it is mandatory that structural integrity of the building components be achieved. All necessary steps must be taken to ensure that the applicable structural requirements of the code are met.

AJ501.5 Electrical equipment and wiring.

AJ501.5.1 Materials and methods. Newly installed electrical *equipment* and wiring relating to work done in any work area shall comply with the materials and methods requirements of Chapters 34 through 43.

> **Exception:** Electrical *equipment* and wiring in newly installed partitions and ceilings shall comply with the applicable requirements of Chapters 34 through 43.

❖ In any work area, electrical equipment and wiring installed must comply with the requirements of Chapters 33 through 42. Such requirements are also applicable in the construction of new walls, partitions and ceiling systems.

AJ501.5.2 Electrical service. Service to the *dwelling unit* shall be not less than 100 ampere, three-wire capacity and service *equipment* shall be dead front having no live parts

exposed that could allow accidental contact. Type "S" fuses shall be installed where fused *equipment* is used.

> **Exception:** Existing service of 60 ampere, three-wire capacity, and feeders of 30 ampere or larger two- or three-wire capacity shall be accepted if adequate for the electrical load being served.

❖ It is typical that minimum 100-ampere service be provided to each dwelling unit that is undergoing alteration of electrical service. However, where it can be determined that the service loading does not exceed 60 amperes, an existing 60-ampere service is acceptable.

AJ501.5.3 Additional electrical requirements. Where the work area includes any of the following areas within a *dwelling unit*, the requirements of Sections AJ501.5.3.1 through AJ501.5.3.5 shall apply.

❖ This section sets forth additional requirements for specified enclosed spaces, kitchens, laundry rooms, ground-fault circuit interruption and lighting outlets in conjunction with work areas of dwelling units.

AJ501.5.3.1 Enclosed areas. Enclosed areas other than closets, kitchens, *basements*, garages, hallways, laundry areas and bathrooms shall have not less than two duplex receptacle outlets, or one duplex receptacle outlet and one ceiling- or wall-type lighting outlet.

❖ Those areas and rooms typically viewed as habitable spaces must be provided with at least two duplex receptacle outlets or one duplex receptacle outlet and a ceiling- or wall-type lighting outlet. A minimum number of receptacles must be available to reduce the potential for dangerous electrical conditions.

AJ501.5.3.2 Kitchen and laundry areas. Kitchen areas shall have not less than two duplex receptacle outlets. Laundry areas shall have not less than one duplex receptacle outlet located near the laundry *equipment* and installed on an independent circuit.

❖ To reduce the possibility that dangerous conditions are created in kitchen and laundry areas, a minimum number of receptacle outlets is mandated. In the kitchen, at least two duplex receptacle outlets are required. Near laundry equipment, a minimum of one duplex receptacle outlet is mandated. The receptacle outlet in the laundry area must be located on an independent circuit.

AJ501.5.3.3 Ground-fault circuit-interruption. Ground-fault circuit-interruption shall be provided on newly installed receptacle outlets if required by Chapters 34 through 43.

❖ The installation of ground-fault circuit interruption for all new receptacle outlets is covered by the provisions of Chapters 34 through 43. Those locations in new construction identified by the code to be protected by ground-fault circuit interruption are the same in alterations to existing buildings, but interruption is required only in those locations where new receptacle outlets are installed.

AJ501.5.3.4 Lighting outlets. Not less than one lighting outlet shall be provided in every bathroom, hallway, stairway, attached garage and detached garage with electric power to illuminate outdoor entrances and exits, and in utility rooms and *basements* where these spaces are used for storage or contain *equipment* requiring service.

❖ Lighting outlets are to be installed in a variety of specific locations identified by this section. Areas where at least one lighting outlet is required include bathrooms, hallways, stairways and attached garages. Detached garages provided with electrical power must have facilities for illuminating exterior entrances and exits. Basements and utility rooms used as storage areas or to house equipment that must be serviced must also be provided with at least one lighting outlet.

AJ501.5.3.5 Clearance. Clearance for electrical service *equipment* shall be provided in accordance with Chapters 34 through 43.

❖ All electrical equipment is to be provided with the necessary working space and other clearances as set forth in Chapters 34 through 43.

AJ501.6 Ventilation. Reconfigured spaces intended for occupancy and spaces converted to habitable or occupiable space in any work area shall be provided with *ventilation* in accordance with Section R303.

❖ Adequate ventilation, by either natural or mechanical means, must be provided in occupiable spaces located within a work area. The requirement is applicable to spaces that are altered in shape or size, as well as those areas that are converted to habitable space. The ventilation requirements of Section R303 provide the minimum ventilation criteria that are to be used.

AJ501.7 Ceiling height. *Habitable spaces* created in existing *basements* shall have ceiling heights of not less than 6 feet, 8 inches (2032 mm), except that the ceiling height at obstructions shall be not less than 6 feet 4 inches (1930 mm) from the *basement* floor. Existing finished ceiling heights in nonhabitable spaces in *basements* shall not be reduced.

❖ This section permits owners of older homes to create habitable space in basements. In existing basements, many times it is technically and structurally infeasible to modify an existing ceiling height to comply with the code for new construction. This section recognizes that the minimum ceiling heights easily achievable in new construction are often impossible to provide in basements of older homes and will allow homeowners to finish off their basements.

AJ501.8 Stairs.

❖ Sections AJ501.8.1 through AJ501.8.3 permit maintaining current width, headroom and landing size on existing basement stairways when alterations are made to the stairway or to other portions of the basement. In existing basements, many times it is technically and structurally infeasible to modify an existing stairway to comply with the code for new construction. The provisions of this section do not require a homeowner to replace or modify their existing basement stairs simply because the space at the bottom of the

stairs is being renovated. Because Section R102.7.1 specifically states that alterations shall not result in an unsafe building, this section does not preclude the building official from requiring the replacement of stairs considered substandard or hazardous.

AJ501.8.1 Stair width. Existing *basement* stairs and handrails not otherwise being altered or modified shall be permitted to maintain their current clear width at, above and below existing handrails.

❖ See the commentary to Section AJ501.8.

AJ501.8.2 Stair headroom. Headroom height on existing *basement* stairs being altered or modified shall not be reduced below the existing stairway finished headroom. Existing *basement* stairs not otherwise being altered shall be permitted to maintain the current finished headroom.

❖ See the commentary to Section AJ501.8.

AJ501.8.3 Stair landing. Landings serving existing *basement* stairs being altered or modified shall not be reduced below the existing stairway landing depth and width. Existing *basement* stairs not otherwise being altered shall be permitted to maintain the current landing depth and width.

❖ See the commentary to Section AJ501.8.

SECTION AJ601
RECONSTRUCTION

AJ601.1 Stairways, handrails and guards.

AJ601.1.1 Stairways. Stairways within the work area shall be provided with illumination in accordance with Section R303.6.

❖ Stairways located in the work area of a building undergoing reconstruction activities must be illuminated in accordance with Section R303.7. The provisions set forth two options for locating the light sources. In addition, the required locations of lighting control switches are specified.

AJ601.1.2 Handrails. Every required exit stairway that has four or more risers, is part of the means of egress for any work area, and is not provided with at least one handrail, or in which the existing handrails are judged to be in danger of collapsing, shall be provided with handrails designed and installed in accordance with Section R311 for the full length of the run of steps on not less than one side.

❖ The handrail requirements of Section R311 apply to reconstruction work only in certain circumstances. The stairway under consideration must be a required exit element for a work area of the building and consist of at least four risers. Either there are no handrails provided for the stairway, or the existing rail or rails are unsafe because of the possibility of collapse. If all of these conditions exist, a complying handrail is to be installed on at least one side of the stairway.

Handrail height, continuity, termination, clearance, and gripping surface must all be reviewed for conformance with the provisions of Section R311.

AJ601.1.3 Guards. Every open portion of a stair, landing or balcony that is more than 30 inches (762 mm) above the floor or *grade* below, is part of the egress path for any work area, and does not have *guards*, or in which the existing *guards* are judged to be in danger of collapsing, shall be provided with *guards* designed and installed in accordance with Section R312.

❖ The guard requirements of Section R312 apply to reconstruction work only in certain circumstances. The guard under consideration must be serving a walking surface at least 30 inches (762 mm) above the floor below and located along an exit path for a work area of the building. Either there is no guard provided or the existing guard creates an unsafe condition because of the potential for collapse. If all of these conditions exist, a complying guard is to be installed as protection at the elevation change.

Guard height and opening limitations must be reviewed for compliance with the provisions of Section R312.

AJ601.2 Wall and ceiling finish. The interior finish of walls and ceilings in any work area shall comply with the requirements of Section R302.9. Existing interior finish materials that do not comply with those requirements shall be removed or shall be treated with an *approved* fire-retardant coating in accordance with the manufacturer's instructions to secure compliance with the requirements of this section.

❖ In a reconstruction work area, all wall and ceiling finish materials must be in compliance with the flame spread and smoke-development limitations of Section R302.9 R315. The flame spread classification is limited to 200, with a maximum smoke-developed index of 450. Where the existing interior finish materials do not comply with these requirements, the materials are to be removed or treated with an approved fire retardant coating. Where the treatment method is used to obtain compliance, the manufacturer's instructions must be followed.

AJ601.3 Separation walls. Where the work area is in an attached *dwelling unit*, walls separating *dwelling units* that are not continuous from the foundation to the underside of the roof sheathing shall be constructed to provide a continuous fire separation using construction materials consistent with the existing wall or complying with the requirements for new structures. Performance of work shall be required only on the side of the wall of the *dwelling unit* that is part of the work area.

❖ Where reconstruction work takes place in a dwelling unit that is attached to one or more additional dwelling units, the wall or walls separating the units are to be continuous from the foundation to the underside of the roof deck. If such conditions do not exist, the wall must be extended to the roof sheathing to maintain the necessary separation of the units. The materials used in the wall construction must be at least equivalent to those of the existing wall, or alternatively, as required for new construction. There is no requirement for work to be done outside the work area; therefore, the wall

construction need be provided only on the side of the dwelling where the actual work area occurs.

AJ601.4 Ceiling height. *Habitable spaces* created in existing *basements* shall have ceiling heights of not less than 6 feet, 8 inches (2032 mm), except that the ceiling height at obstructions shall be not less than 6 feet 4 inches (1930 mm) from the *basement* floor. Existing finished ceiling heights in nonhabitable spaces in *basements* shall not be reduced.

❖ This section permits owners of older homes to create habitable space in basements. In existing basements, many times it is technically and structurally infeasible to modify an existing basement ceiling height to comply with the code for new construction. This section recognizes that the minimum ceiling heights easily achievable in new construction are often impossible to provide in basements of older homes and will allow homeowners to finish off their basements.

Bibliography

The following resource materials were used in the preparation of the commentary for this appendix of the code:

FEMA P-758, *Substantial Improvement/Substantial Damage Desk Reference*. Frederick, MD: Federal Emergency Management Agency, 2010.

FEMA P-259, *Engineering Principles and Practices for Retrofitting Flood Prone Residential Structures*. Frederick, MD: Federal Emergency Management Agency, 2012.

Appendix K:
Sound Transmission

(The provisions contained in this appendix are not mandatory unless specifically referenced in the adopting ordinance.)

General Comments

Regulation of sound transmission bears directly on the psychological and long-term physical well-being of a building's occupants. By implementing the minimum sound transmission criteria set forth in this appendix, a jurisdiction can assist in providing a more livable environment for the occupants.

Section AK101 is a general overview of the requirements for limiting sound transmission. Section AK102 establishes the specific provisions for air-borne sound insulation in wall and floor/ceiling assemblies. Section AK103 establishes the specific provisions for structure-borne sound transfer in floor/ceiling assemblies.

Purpose

In an effort to appropriately regulate the environmental and physiological aspects of habitable spaces within dwelling units, the code addresses a variety of conditions. In addition to the fundamental requirements for adequate lighting, ventilation and sanitation, regulations are often desirable to control the transmission of sound and noise between individual dwelling units within a multiple-family residential building. The provisions of this appendix set forth a minimum Sound Transmission Class (STC) rating for common walls and floor-ceiling assemblies between dwelling units. In addition, a minimum Impact Insulation Class (IIC) rating is also established to limit structure-borne sound through common floor-ceiling assemblies separating dwelling units.

SECTION AK101
GENERAL

AK101.1 General. Wall and floor-ceiling assemblies separating *dwelling units*, including those separating adjacent *townhouse* units, shall provide air-borne sound insulation for walls, and both air-borne and impact sound insulation for floor-ceiling assemblies.

❖ Because noise transmission can be quantified, and noise affects the quality of life, the code incorporates regulations that address noise transmission in multiple-family residential construction. To provide for an acceptable degree of livability in structures housing two or more dwelling units, the code mandates a minimum level of sound insulation between units. In such residential structures, the occupants of a dwelling unit typically have little control over noise generated in other units. Where the dwelling units are separated by a common floor-ceiling assembly, both air-borne sound and structure-borne sound are present. Where a common wall system separates the units, only air-borne sound is regulated. The components of construction regulated are those through which noise is primarily transmitted.

SECTION AK102
AIR-BORNE SOUND

AK102.1 General. Air-borne sound insulation for wall and floor-ceiling assemblies shall meet a sound transmission class (STC) rating of 45 when tested in accordance with ASTM E90. Penetrations or openings in construction assemblies for piping; electrical devices; recessed cabinets; bathtubs; soffits;

or heating, ventilating or exhaust ducts shall be sealed, lined, insulated or otherwise treated to maintain the required ratings. *Dwelling unit* entrance doors, which share a common space, shall be tight fitting to the frame and sill.

❖ The code requires common walls and floor-ceiling assemblies between dwelling units to have a minimum sound transmission. The higher the number (rating), the higher the resistance (less sound transmission). Standard architectural wall and floor-ceiling construction assemblies have been tested for sound transmission ratings, and reference to the construction specifications will yield such information. Air-borne sound, such as a voice or music, is transferred through the air.

As a rule, vertical assemblies meeting the requirements of this section consist of double walls or walls containing insulation similar to exterior walls. Horizontal assemblies typically contain some type of insulating materials within the assembly or resilient furring channels to isolate the ceiling membrane from the structural members of the floor construction.

AK102.1.1 Masonry. The sound transmission class of concrete masonry and clay masonry assemblies shall be calculated in accordance with TMS 0302 or determined through testing in accordance with ASTM E90.

❖ This section provides an alternate to testing masonry walls to ASTM E90. This section permits calculation of the air-borne STC for various concrete masonry and clay masonry wall assemblies. The calculation must be in accordance with TMS 0302. The methods in TMS 0302 are derived from testing of masonry wall assemblies in accordance with ASTM E90.

SECTION AK103
STRUCTURAL-BORNE SOUND

AK103.1 General. Floor/ceiling assemblies between *dwelling units*, or between a *dwelling unit* and a public or service area within a structure, shall have an impact insulation class (IIC) rating of not less than 45 when tested in accordance with ASTM E492.

❖ The IIC rating is a measure of an assembly's ability to resist sound transmission. The higher the number (rating), the higher the resistance (less sound transmission). Floors between dwelling units are required to have a minimum IIC rating of 45.

  Usually, carpeted floor assemblies meet the minimum requirement for an IIC rating of 45. Other areas with hard-surfaced finishes may require additional treatment or insulation to comply with these requirements.

  Various resource documents containing STC ratings and IIC ratings are listed in the bibliography. These sources, or any other similar sources, can be submitted as a basis for approval once it is demonstrated to the building official that the data are based on ASTM E90 and ASTM E492.

SECTION AK104
REFERENCED STANDARDS

ASTM

ASTM E90—04   Test Method for Laboratory Measurement of Air-borne Sound Transmission Loss of Building Partitions and Elements........AK102

ASTM E492—09   Specification for Laboratory Measurement of Impact Sound Transmission through Floor-ceiling Assemblies Using the Tapping Machine..................AK103

The Masonry Society

TMS 0302—12   Standard for Determining the Sound Transmission Class Rating for Masonry Walls.......AK102.1.1

Bibliography

The following resource materials were used in the preparation of the commentary for this appendix of the code:

ASTM E90–04, *Standard Test Method for Laboratory Measurement of Air-borne Sound Transmission Loss of Building Partitions and Elements*. West Conshohocken, PA: ASTM International, 2004.

ASTM E492–04, *Standard Specification for Laboratory Measurement of Impact Sound Transmission through Floor-ceiling Assemblies using the Tapping Machine*. West Conshohocken, PA: ASTM International, 2004.

BIA TN 5A–83, *Sound Insulation-clay Masonry Walls*. Reston, VA: Brick Institute of America, 1983.

Gypsum Association. GA 600-97: *Fire-resistance Design Manual*. 15th Edition. Washington, D.C.: Gypsum Association, 1997.

National *Concrete Masonry Association. NCMA TEK 69A–78: New Data on Sound Reduction with Concrete Masonry Walls*. Herndon, VA: National Concrete Masonry Association, 1978.

TMS 302–07, *Standard for Determining the Sound Transmission Class Rating for Masonry Walls*. Boulder, CO: The Masonry Society, 2007.

Appendix L:
Permit Fees

(The provisions contained in this appendix are not mandatory unless specifically referenced in the adopting ordinance.)

TOTAL VALUATION	FEE
$1 to $ 500	$24
$501 to $2,000	$24 for the first $500; plus $3 for each additional $100 or fraction thereof, up to and including $2,000
$2,001 to $40,000	$69 for the first $2,000; plus $11 for each additional $1,000 or fraction thereof, up to and including $40,000
$40,001 to $100,000	$487 for the first $40,000; plus $9 for each additional $1,000 or fraction thereof, up to and including $100,000
$100,001 to $500,000	$1,027 for the first $100,000; plus $7 for each additional $1,000 or fraction thereof, up to and including $500,000
$500,001 to $1,000,000	$3,827 for the first $500,000; plus $5 for each additional $1,000 or fraction thereof, up to and including $1,000,000
$1,000,001 to $5,000,000	$6,327 for the first $1,000,000; plus $3 for each additional $1,000 or fraction thereof, up to and including $5,000,000
$5,000,001 and over	$18,327 for the first $5,000,000; plus $1 for each additional $1,000 or fraction thereof

Appendix M:
Home Day Care—R-3 Occupancy

(The provisions contained in this appendix are not mandatory unless specifically referenced in the adopting ordinance.)

General Comments

This appendix provides means of egress and smoke detection requirements for a Group R-3 Occupancy that is to be used as a home day care for more than five children who receive custodial care for less than 24 hours.

This appendix sets forth the means of egress requirements that are unique to a home day care, such as two exits required, fenced yards, type of fencing, guards for decks and type of locks and latches for exits. This appendix also requires installation of smoke detectors.

Section AM101 describes a home day care operated within a dwelling and defines the limitation of the occupancy. Section AM102 addresses definitions. Section AM103 establishes the specific provisions for the means of egress unique to a home day care. Section AM104 contains the provisions for smoke detectors.

Purpose

This appendix is strictly for guidance and/or adoption by those jurisdictions that have Licensed Home Care Pro-

vider laws and statutes that allow more than five children to be cared for in a person's home.

Several states have laws for Licensed Home Care Providers that will allow more than five children to be cared for in a person's home. Many state jurisdictions require documentation from the local jurisdictions stating that the applicant has complied with the local planning and building department zoning and building codes. Typically, a state jurisdiction will allow six full-time children and two or more part-time children. The part-time child care is usually before and after school hours. The *International Building Code®* (IBC®) regulates Group I-4 day care facilities in Section 308.5 and states, "A facility such as the above with five or fewer persons shall be classified as a Group R-3," Section 308.5.2 child care facility works with more than five children. When directed to the code, one finds that there are no requirements for day care operations in a Group R-3. When a jurisdiction adopts this appendix, the provisions in IBC Sections 308.5 and 308.5.2 should be considered also.

SECTION AM101
GENERAL

AM101.1 General. This appendix shall apply to a home day care operated within a *dwelling*. It is to include buildings and structures occupied by persons of any age who receive custodial care for less than 24 hours by individuals other than parents or guardians or relatives by blood, marriage, or adoption, and in a place other than the home of the person cared for.

SECTION AM102
DEFINITION

EXIT ACCESS. That portion of a means-of-egress system that leads from any occupied point in a building or structure to an exit.

SECTION AM103
MEANS OF EGRESS

AM103.1 Exits required. If the occupant load of the residence is more than nine, including those who are residents, during the time of operation of the day care, two exits are required from the ground-level *story*. Two exits are required from a home day care operated in a *manufactured home* regardless of the occupant load. Exits shall comply with Section R311.

AM103.1.1 Exit access prohibited. An exit access from the area of day care operation shall not pass through bathrooms, bedrooms, closets, garages, fenced rear *yards* or similar areas.

> **Exception:** An exit may discharge into a fenced *yard* if the gate or gates remain unlocked during day care hours. The gates may be locked if there is an area of refuge located within the fenced *yard* and more than 50 feet (15 240 mm) from the *dwelling*. The area of refuge shall be large enough to allow 5 square feet (0.5 m²) per occupant.

AM103.1.2 Basements. If the *basement* of a *dwelling* is to be used in the day care operation, two exits are required from the *basement* regardless of the occupant load. One of the exits may pass through the *dwelling* and the other must lead directly to the exterior of the *dwelling*.

> **Exception:** An emergency and escape window complying with Section R310 and which does not conflict with Section AM103.1.1 may be used as the second means of egress from a *basement*.

AM103.1.3 Yards. If the *yard* is to be used as part of the day care operation it shall be fenced.

AM103.1.3.1 Type of fence and hardware. The fence shall be of durable materials and be at least 6 feet (1529 mm) tall, completely enclosing the area used for the day care operations. Each opening shall be a gate or door equipped with a self-closing and self-latching device to be installed at a minimum of 5 feet (1528 mm) above the ground.

> **Exception:** The door of any *dwelling* which forms part of the enclosure need not be equipped with self-closing and self-latching devices.

AM103.1.3.2 Construction of fence. Openings in the fence, wall or enclosure required by this section shall have intermediate rails or an ornamental pattern that do not allow a sphere 4 inches (102 mm) in diameter to pass through. In addition, the following criteria must be met:

1. The maximum vertical clearance between *grade* and the bottom of the fence, wall or enclosure shall be 2 inches (51 mm).

2. Solid walls or enclosures that do not have openings, such as masonry or stone walls, shall not contain indentations or protrusions, except for tooled masonry joints.

3. Maximum mesh size for chain link fences shall be $1^1/_4$ inches (32 mm) square, unless the fence has slats at the top or bottom which reduce the opening to no more than $1^3/_4$ inches (44 mm). The wire shall be not less than 9 gage [0.148 inch (3.8 mm)].

AM103.1.3.3 Decks. Decks that are more than 12 inches (305 mm) above *grade* shall have a guard in compliance with Section R312.

AM103.2 Width and height of an exit. The minimum width of a required exit is 36 inches (914 mm) with a net clear width of 32 inches (813 mm). The minimum height of a required exit is 6 feet, 8 inches (2032 mm).

AM103.3 Type of lock and latches for exits. Regardless of the occupant load served, exit doors shall be openable from the inside without the use of a key or any special knowledge or effort. When the occupant load is 10 or less, a night latch, dead bolt or security chain may be used, provided such devices are openable from the inside without the use of a key or tool, and mounted at a height not to exceed 48 inches (1219 mm) above the finished floor.

AM103.4 Landings. Landings for stairways and doors shall comply with Section R311, except that landings shall be required for the exterior side of a sliding door when a home day care is being operated in a Group R-3 occupancy.

SECTION AM104
SMOKE DETECTION

AM104.1 General. Smoke detectors shall be installed in *dwelling* units used for home day care operations. Detectors shall be installed in accordance with the approved manufacturer's instructions. If the current smoke detection system in the *dwelling* is not in compliance with the currently adopted code for smoke detection, it shall be upgraded to meet the currently adopted code requirements and Section AM103 before day care operations commence.

AM104.2 Power source. Required smoke detectors shall receive their primary power from the building wiring when that wiring is served from a commercial source and shall be equipped with a battery backup. The detector shall emit a signal when the batteries are low. Wiring shall be permanent and without a disconnecting switch other than those required for overcurrent protection. Required smoke detectors shall be interconnected so if one detector is activated, all detectors are activated.

AM104.3 Location. A detector shall be located in each bedroom and any room that is to be used as a sleeping room, and centrally located in the corridor, hallway or area giving access to each separate sleeping area. When the *dwelling* unit has more than one *story*, and in *dwellings* with *basements*, a detector shall be installed on each *story* and in the *basement*. In *dwelling* units where a *story* or *basement* is split into two or more levels, the smoke detector shall be installed on the upper level, except that when the lower level contains a sleeping area, a detector shall be installed on each level. When sleeping rooms are on the upper level, the detector shall be placed at the ceiling of the upper level in close proximity to the stairway. In *dwelling* units where the ceiling height of a room open to the hallway serving the bedrooms or sleeping areas exceeds that of the hallway by 24 inches (610 mm) or more, smoke detectors shall be installed in the hallway and the adjacent room. Detectors shall sound an alarm audible in all sleeping areas of the *dwelling* unit in which they are located.

Appendix N:
Venting Methods

(This appendix is informative and is not part of the code. This appendix provides examples of various venting methods.)

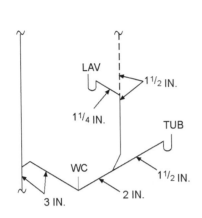

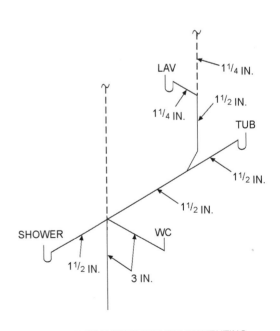

A. TYPICAL SINGLE-BATH ARRANGEMENT

B. TYPICAL POWDER ROOM

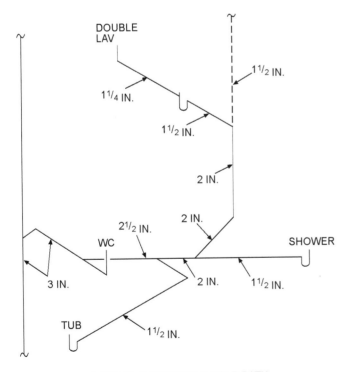

C. MORE ELABORATE SINGLE-BATH
ARRANGEMENT

D. COMBINATION WET AND STACK VENTING
WITH STACK FITTING

For SI: 1 inch = 25.4 mm.

FIGURE N1
TYPICAL SINGLE-BATH WET-VENT ARRANGEMENTS

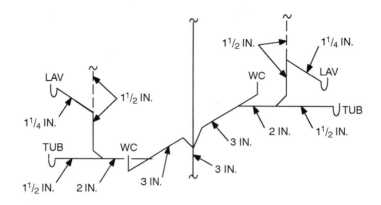

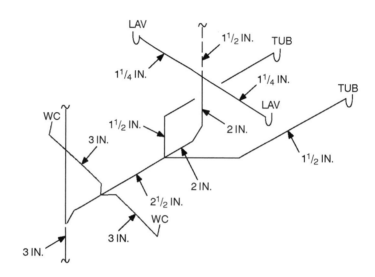

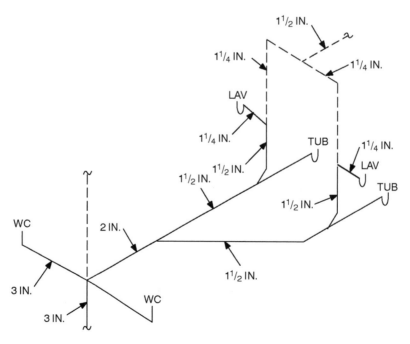

For SI: 1 inch = 25.4 mm.

FIGURE N2
TYPICAL DOUBLE-BATH WET-VENT ARRANGEMENTS

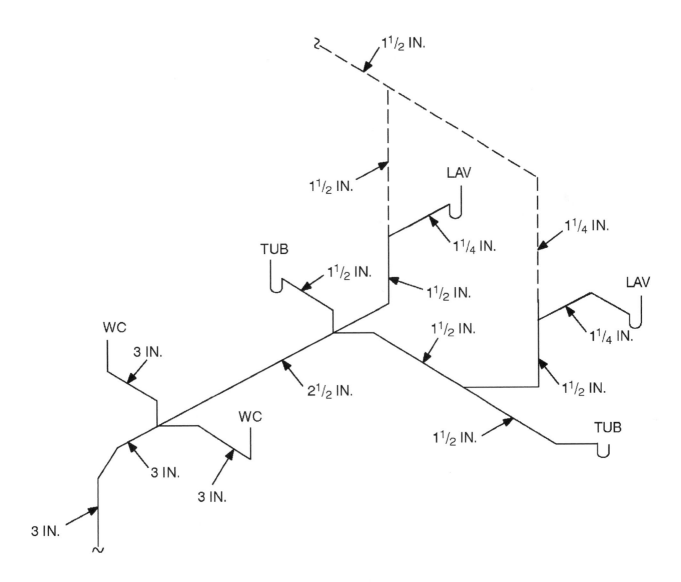

For SI: 1 inch = 25.4 mm.

FIGURE N3
TYPICAL HORIZONTAL WET VENTING

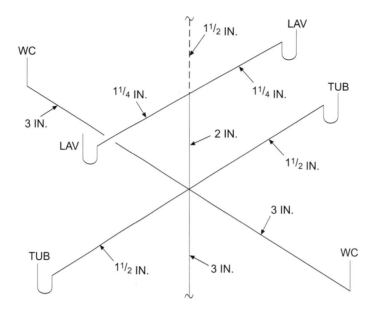

A. VERTICAL WET VENTING

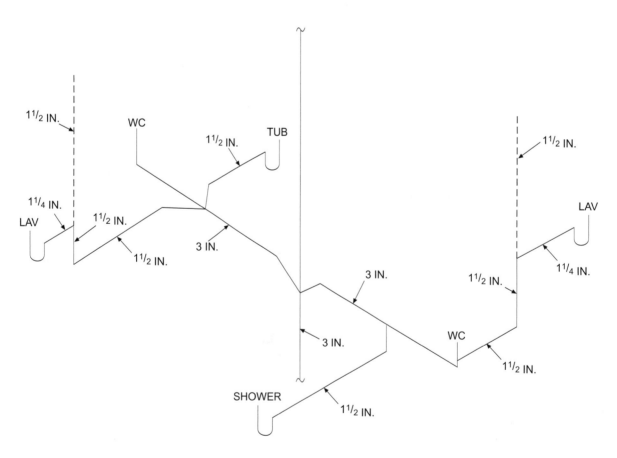

B. HORIZONTAL WET VENTING

For SI: 1 inch = 25.4 mm.

FIGURE N4
TYPICAL METHODS OF WET VENTING

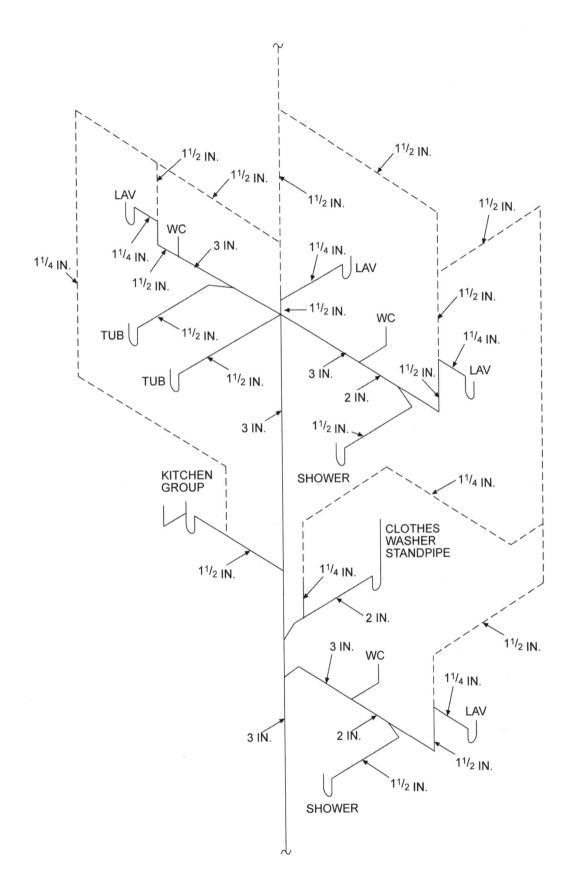

For SI: 1 inch = 25.4 mm.

FIGURE N5
SINGLE STACK SYSTEM FOR A TWO-STORY DWELLING

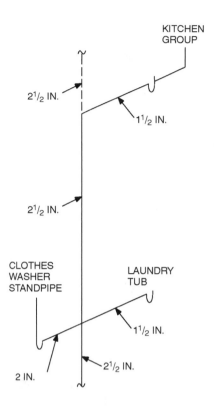

For SI: 1 inch = 25.4 mm.

FIGURE N6
WASTE STACK VENTING

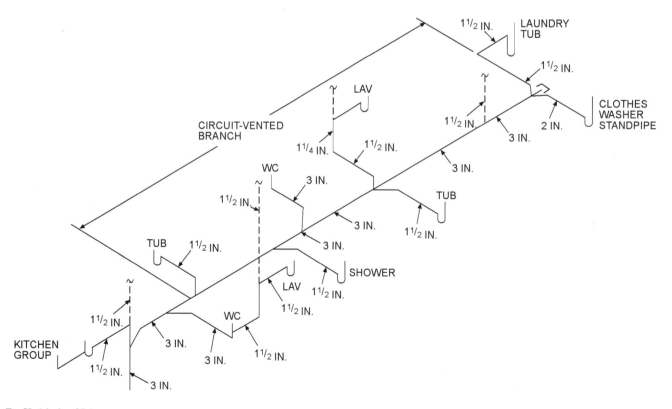

For SI: 1 inch = 25.4 mm.

FIGURE N7
CIRCUIT VENT WITH ADDITIONAL NONCIRCUIT-VENTED BRANCH

Appendix O:
Automatic Vehicular Gates

(The provisions contained in this appendix are not mandatory unless specifically referenced in the adopting ordinance.)

SECTION AO101
GENERAL

AO101.1 General. The provisions of this appendix shall control the design and construction of automatic vehicular gates installed on the lot of a one- or two-family dwelling.

SECTION AO102
DEFINITION

AO102.1 General. For the purposes of these requirements, the term used shall be defined as follows and as set forth in Chapter 2.

VEHICULAR GATE. A gate that is intended for use at a vehicular entrance or exit to the lot of a one- or two-family dwelling, and that is not intended for use by pedestrian traffic.

SECTION AO103
AUTOMATIC VEHICULAR GATES

AO103.1 Vehicular gates intended for automation. Vehicular gates intended for automation shall be designed, constructed and installed to comply with the requirements of ASTM F 2200.

AO103.2 Vehicular gate openers. Vehicular gate openers, where provided, shall be listed in accordance with UL 325.

Appendix P:
Sizing of Water Piping System

(The provisions contained in this appendix are not mandatory unless specifically referenced in the adopting ordinance.)

SECTION AP101
GENERAL

AP101.1 Scope.

AP101.1.1 This appendix outlines two procedures for sizing a water piping system (see Sections AP103.3 and AP201.1). The design procedures are based on the minimum static pressure available from the supply source, the head changes in the system caused by friction and elevation, and the rates of flow necessary for operation of various fixtures.

AP101.1.2 Because of the variable conditions encountered in hydraulic design, it is impractical to specify definite and detailed rules for sizing of the water piping system. Accordingly, other sizing or design methods conforming to good engineering practice standards are acceptable alternatives to those presented herein.

SECTION AP102
INFORMATION REQUIRED

AP102.1 Preliminary. Obtain the necessary information regarding the minimum daily static service pressure in the area where the building is to be located. If the building supply is to be metered, obtain information regarding friction loss relative to the rate of flow for meters in the range of sizes likely to be used. Friction loss data can be obtained from most manufacturers of water meters.

AP102.2 Demand load.

AP102.2.1 Estimate the supply demand of the building main and the principal branches and risers of the system by totaling the corresponding demand from the applicable part of Table AP103.3(3).

AP102.2.2 Estimate continuous supply demands, in gallons per minute (gpm) (L/m), for lawn sprinklers, air conditioners, etc., and add the sum to the total demand for fixtures. The result is the estimated supply demand for the building supply.

SECTION AP103
SELECTION OF PIPE SIZE

AP103.1 General. Decide from Table P2903.1 what is the desirable minimum residual pressure that should be maintained at the highest fixture in the supply system. If the highest group of fixtures contains flushometer valves, the pressure for the group should be not less than 15 pounds per square inch (psi) (103.4 kPa) flowing. For flush tank supplies, the available pressure should be not less than 8 psi (55.2 kPa) flowing, except blowout action fixtures must not be less than 25 psi (172.4 kPa) flowing.

AP103.2 Pipe sizing.

AP103.2.1 Pipe sizes can be selected using the following procedure or by use of other design methods conforming to acceptable engineering practice that are *approved* by the *building official*. The sizes selected must not be less than the minimum required by this code.

AP103.2.2 Water pipe sizing procedures are based on a system of pressure requirements and losses, the sum of which must not exceed the minimum pressure available at the supply source. These pressures are as follows:

1. Pressure required at fixture to produce required flow. See Sections P2903.1 of this code and Section 604.3 of the *International Plumbing Code*.

2. Static pressure loss or gain (due to head) is computed at 0.433 psi per foot (9.8 kPa/m) of elevation change.

   Example: Assume that the highest fixture supply outlet is 20 feet (6096 mm) above or below the supply source. This produces a static pressure differential of 8.66 psi (59.8 kPa) loss [20 feet by 0.433 psi per foot (2096 mm by 9.8 kPa/m)].

3. Loss through water meter. The friction or pressure loss can be obtained from meter manufacturers.

4. Loss through taps in water main.

5. Loss through special devices, such as filters, softeners, backflow prevention devices and pressure regulators. These values must be obtained from the manufacturer.

6. Loss through valves and fittings. Losses for these items are calculated by converting to the *equivalent length* of piping and adding to the total pipe length.

7. Loss caused by pipe friction can be calculated where the pipe size, pipe length and flow through the pipe are known. With these three items, the friction loss can be determined. For piping flow charts not included, use manufacturers' tables and velocity recommendations.

Note: For all examples, the following metric conversions are applicable.

   1 cubic foot per minute = 0.4719 L/s.

   1 square foot = 0.0929 m².

   1 degree = 0.0175 rad.

   1 pound per square inch = 6.895 kPa.

   1 inch = 25.4 mm.

   1 foot = 304.8 mm.

   1 gallon per minute = 3.785 L/m.

AP103.3 Segmented loss method. The size of water service mains, branch mains and risers by the segmented loss method, must be determined by knowing the water supply demand [gpm (L/m)], available water pressure [psi (kPa)] and friction loss caused by the water meter and *developed length* of pipe [feet (m)], including the *equivalent length* of fittings. This design procedure is based on the following parameters:

1. The calculated friction loss through each length of pipe.

2. A system of pressure losses, the sum of which must not exceed the minimum pressure available at the street main or other source of supply.

3. Pipe sizing based on estimated peak demand, total pressure losses caused by difference in elevation, equipment, *developed length* and pressure required at the most remote fixture; loss through taps in water main; losses through fittings, filters, backflow prevention devices, valves and pipe friction.

Because of the variable conditions encountered in hydraulic design, it is impractical to specify definite and detailed rules for the sizing of the water piping system. Current sizing methods do not address the differences in the probability of use and flow characteristics of fixtures between types of occupancies. Creating an exact model of predicting the demand for a building is impossible and final studies assessing the impact of water conservation on demand are not yet complete. The following steps are necessary for the segmented loss method.

1. **Preliminary.** Obtain the necessary information regarding the minimum daily static service pressure in the area where the building is to be located. If the building supply is to be metered, obtain information regarding friction loss relative to the rate of flow for meters in the range of sizes to be used. Friction loss data can be obtained from manufacturers of water meters. Enough pressure must be available to overcome all system losses caused by friction and elevation so that plumbing fixtures operate properly. Section 604.6 of the *International Plumbing Code* requires that the water distribution system be designed for the minimum pressure available taking into consideration pressure fluctuations. The lowest pressure must be selected to guarantee a continuous, adequate supply of water. The lowest pressure in the public main usually occurs in the summer because of lawn sprinkling and supplying water for air-conditioning cooling towers. Future demands placed on the public main as a result of large growth or expansion should be considered. The available pressure will decrease as additional loads are placed on the public system.

2. **Demand load.** Estimate the supply demand of the building main and the principal branches and risers of the system by totaling the corresponding demand from the applicable part of Table AP103.3(3). When estimating peak demand, sizing methods typically use water supply fixture units (w.s.f.u.) [see Table AP103.3(2)]. This numerical factor measures the load-producing effect of a single plumbing fixture of a given kind. The use of fixture units can be applied to a single basic

probability curve (or table), found in the various sizing methods [see Table AP103.3(3)]. The fixture units are then converted into a gpm (L/m) flow rate for estimating demand.

2.1. Estimate continuous supply demand in gpm (L/m) for lawn sprinklers, air conditioners, etc., and add the sum to the total demand for fixtures. The result is the estimated supply demand for the building supply. Fixture units cannot be applied to constant-use fixtures, such as hose bibbs, lawn sprinklers and air conditioners. These types of fixtures must be assigned the gpm (L/m) value.

3. **Selection of pipe size.** This water pipe sizing procedure is based on a system of pressure requirements and losses, the sum of which must not exceed the minimum pressure available at the supply source. These pressures are as follows:

3.1. Pressure required at the fixture to produce required flow. See Section P2903.1 of this code and Section 604.3 of the *International Plumbing Code*.

3.2. Static pressure loss or gain (because of head) is computed at 0.433 psi per foot (9.8 kPa/m) of elevation change.

3.3. Loss through a water meter. The friction or pressure loss can be obtained from the manufacturer.

3.4. Loss through taps in water main [see Table AP103.3(4)].

3.5. Loss through special devices, such as filters, softeners, backflow prevention devices and pressure regulators. These values must be obtained from the manufacturers.

3.6. Loss through valves and fittings [see Tables AP103.3(5) and AP103.3(6)]. Losses for these items are calculated by converting to the *equivalent length* of piping and adding to the total pipe length.

3.7. Loss caused by pipe friction can be calculated where the pipe size, pipe length and flow through the pipe are known. With these three items, the friction loss can be determined using Figures AP103.3(2) through AP103.3(7). Where using charts, use pipe inside diameters. For piping flow charts not included, use manufacturers' tables and velocity recommendations. Before attempting to size any water supply system, it is necessary to gather preliminary information including available pressure, piping material, select design velocity, elevation differences and *developed length* to the most remote fixture. The water supply system is divided into sections at major changes in elevation or where branches lead to fixture groups. The peak demand must be determined in each part of the hot and cold water supply system. The expected flow through each section is determined in w.s.f.u. and converted to gpm (L/m) flow rate. Sizing methods require determination of the "most hydraulically remote" fixture to compute the pressure loss caused by pipe and fittings. The hydraulically remote fixture represents the most downstream fixture along the circuit of piping requiring the most available pressure to operate properly. Consideration must be given to all pressure demands and losses, such as friction caused by pipe, fittings and equipment; elevation; and the residual pressure required by Table P2903.1. The two most common and frequent complaints about water supply system operation are lack of adequate pressure and noise.

Problem: What size Type L copper water pipe, service and distribution will be required to serve a two-story factory building having on each floor, back-to-back, two toilet rooms each equipped with hot and cold water? The highest fixture is 21 feet above the street main, which is tapped with a 2-inch corporation cock at which point the minimum pressure is 55 psi. In the building *basement*, a 2-inch meter with a maximum pressure drop of 11 psi and 3-inch reduced pressure principle backflow preventer with a maximum pressure drop of 9 psi are to be installed. The system is shown in Figure AP103.3(1). To be determined are the pipe sizes for the service main, and the cold and hot water distribution pipes.

Solution: A tabular arrangement such as shown in Table AP103.3(1) should first be constructed. The steps to be followed are indicated by the tabular arrangement itself as they are in sequence, Columns 1 through 10 and Lines A through L.

Step 1

Columns 1 and 2: Divide the system into sections breaking at major changes in elevation or where branches lead to fixture groups. After Point B [see Figure AP103.3(1)], separate consideration will be given to the hot and cold water piping. Enter the sections to be considered in the service and cold water piping in Column 1 of the tabular arrangement. Column 1 of Table AP103.3(1) provides a line-by-line, recommended tabular arrangement for use in solving pipe sizing.

The objective in designing the water supply system is to ensure an adequate water supply and pressure to all fixtures and equipment. Column 2 provides the psi (kPa) to be considered separately from the minimum pressure available at the main. Losses to take into consideration are the following: the differences in elevations between the water supply source and the highest water supply outlet; meter pressure losses; the tap in main loss; special fixture devices, such as water softeners and backflow prevention devices; and the pressure required at the most remote fixture outlet.

The difference in elevation can result in an increase or decrease in available pressure at the main. Where the water supply outlet is located above the source, this results in a loss in the available pressure and is subtracted from the pressure at the water source. Where the highest water supply outlet is located below the water supply source,

there will be an increase in pressure that is added to the available pressure of the water source.

Column 3: Using Table AP103.3(3), determine the gpm (L/m) of flow to be expected in each section of the system. These flows range from 28.6 to 108 gpm. Load values for fixtures must be determined as w.s.f.u. and then converted to a gpm rating to determine peak demand. Where calculating peak demands, the w.s.f.u. are added and then converted to the gpm rating. For continuous flow fixtures, such as hose bibbs and lawn sprinkler systems, add the gpm demand to the intermittent demand of fixtures. For example, a total of 120 w.s.f.u. is converted to a demand of 48 gpm. Two hose bibbs × 5 gpm demand = 10 gpm. Total gpm rating = 48.0 gpm + 10 gpm = 58.0 gpm demand.

Step 2

Line A: Enter the minimum pressure available at the main source of supply in Column 2. This is 55 psi (379.2 kPa). The local water authorities generally keep records of pressures at different times of the day and year. The available pressure can also be checked from nearby buildings or from fire department hydrant checks.

Line B: Determine from Table P2903.1 the highest pressure required for the fixtures on the system, which is 15 psi (103.4 kPa), to operate a flushometer valve. The most remote fixture outlet is necessary to compute the pressure loss caused by pipe and fittings, and represents the most downstream fixture along the circuit of piping requiring the available pressure to operate properly as indicated by Table P2903.1.

Line C: Determine the pressure loss for the meter size given or assumed. The total water flow from the main through the service as determined in Step 1 will serve to aid in the meter selected. There are three common types of water meters; the pressure losses are determined by the American Water Works Association Standards for displacement type, compound type and turbine type. The maximum pressure loss of such devices takes into consideration the meter size, safe operating capacity [gpm (L/m)] and maximum rates for continuous operations [gpm (L/m)]. Typically, equipment imparts greater pressure losses than piping.

Line D: Select from Table AP103.3(4) and enter the pressure loss for the tap size given or assumed. The loss of pressure through taps and tees in psi (kPa) is based on the total gpm (L/m) flow rate and size of the tap.

Line E: Determine the difference in elevation between the main and source of supply and the highest fixture on the system. Multiply this figure, expressed in feet (mm), by 0.43 psi. Enter the resulting psi (kPa) loss on Line E. The difference in elevation between the water supply source and the highest water supply outlet has a significant impact on the sizing of the water supply system. The difference in elevation usually results in a loss in the available pressure because the water supply outlet is generally located above the water supply source. The loss is caused by the pressure required to lift the water to the outlet. The pressure loss is subtracted from the pressure at the water source. Where the highest water supply outlet is located below the water source, there will be an increase in pressure that is added to the available pressure of the water source.

Lines F, G and H: The pressure losses through filters, backflow prevention devices or other special fixtures must be obtained from the manufacturer or estimated and entered on these lines. Equipment, such as backflow prevention devices, check valves, water softeners, instantaneous, or tankless water heaters, filters and strainers, can impart a much greater pressure loss than the piping. The pressure losses can range from 8 to 30 psi.

Step 3

Line I: The sum of the pressure requirements and losses that affect the overall system (Lines B through H) is entered on this line. Summarizing the steps, all of the system losses are subtracted from the minimum water pressure. The remainder is the pressure available for friction, defined as the energy available to push the water through the pipes to each fixture. This force can be used as an average pressure loss, as long as the pressure available for friction is not exceeded. Saving a certain amount for available water supply pressures as an area incurs growth, or because of the aging of the pipe or equipment added to the system is recommended.

Step 4

Line J: Subtract Line I from Line A. This gives the pressure that remains available from overcoming friction losses in the system. This figure is a guide to the pipe size that is chosen for each section, incorporating the total friction losses to the most remote outlet (measured length is called *developed length*).

> **Exception:** Where the main is above the highest fixture, the resulting psi (kPa) must be considered a pressure gain (static head gain) and omitted from the sums of Lines B through H and added to Line J.

The maximum friction head loss that can be tolerated in the system during peak demand is the difference between the static pressure at the highest and most remote outlet at no-flow conditions and the minimum flow pressure required at that outlet. If the losses are within the required limits, every run of pipe will be within the required friction head loss. Static pressure loss is at the most remote outlet in feet × 0.433 = loss in psi caused by elevation differences.

Step 5

Column 4: Enter the length of each section from the main to the most remote outlet (at Point E). Divide the water supply system into sections breaking at major changes in elevation or where branches lead to fixture groups.

Step 6

Column 5: Where selecting a trial pipe size, the length from the water service or meter to the most remote fixture

outlet must be measured to determine the *developed length*. However, in systems having a flushometer valve or temperature-controlled shower at the topmost floors, the *developed length* would be from the water meter to the most remote flushometer valve on the system. A rule of thumb is that size will become progressively smaller as the system extends farther from the main source of supply. A trial pipe size can be arrived at by the following formula:

Line J: (Pressure available to overcome pipe friction) × 100/*equivalent length* of run total *developed length* to most remote fixture × percentage factor of 1.5 (Note: a percentage factor is used only as an estimate for friction losses imposed for fittings for initial trial pipe size) = psi (average pressure drop per 100 feet of pipe).

For trial pipe size, see Figure AP103.3(3) (Type L copper) based on 2.77 psi and 108 gpm = $2^{1}/_{2}$ inches. To determine the *equivalent length* of run to the most remote outlet, the *developed length* is determined and added to the friction losses for fittings and valves. The *developed lengths* of the designated pipe sections are as follows:

A-B	54 feet
B-C	8 feet
C-D	13 feet
D-E	150 feet

Total *developed length* = 225 feet

The *equivalent length* of the friction loss in fittings and valves must be added to the *developed length* (most remote outlet). Where the size of fittings and valves is not known, the added friction loss should be approximated. A general rule that has been used is to add 50 percent of the *developed length* to allow for fittings and valves. For example, the *equivalent length* of run equals the *developed length* of run (225 feet × 1.5 = 338 feet). The total *equivalent length* of run for determining a trial pipe size is 338 feet.

> **Example:** 9.36 (pressure available to overcome pipe friction) × 100/338 (*equivalent length* of run = 225 × 1.5) = 2.77 psi (average pressure drop per 100 feet of pipe).

Step 7

Column 6: Select from Table AP103.3(6) the *equivalent lengths* for the trial pipe size of fittings and valves on each pipe section. Enter the sum for each section in Column 6. (The number of fittings to be used in this example must be an estimate). The *equivalent length* of piping is the *developed length* plus the *equivalent lengths* of pipe corresponding to the friction head losses for fittings and valves. Where the size of fittings and valves is not known, the added friction head losses must be approximated. An estimate for this example is found in Table AP.1.

Step 8

Column 7: Add the figures from Columns 4 and 6, and enter in Column 7. Express the sum in hundreds of feet.

Step 9

Column 8: Select from Figure AP103.3(3) the friction loss per 100 feet of pipe for the gpm flow in a section (Column 3) and trial pipe size (Column 5). Maximum friction head loss per 100 feet is determined on the basis of the total pressure available for friction head loss and the longest *equivalent length* of run. The selection is based on the gpm demand, uniform friction head loss and maximum design velocity. Where the size indicated by the hydraulic table indicates a velocity in excess of the selected velocity, a size must be selected that produces the required velocity.

Step 10

Column 9: Multiply the figures in Columns 7 and 8 for each section and enter in Column 9.

Total friction loss is determined by multiplying the friction loss per 100 feet for each pipe section in the total *developed length* by the pressure loss in fittings expressed as *equivalent length* in feet (mm). Note: Section C-F should be considered in the total pipe friction losses only if greater loss occurs in Section C-F than in pipe Section D-E. Section C-F is not considered in the total *developed length*. Total friction loss in *equivalent length* is determined in Table AP.2.

Step 11

Line K: Enter the sum of the values in Column 9. The value is the total friction loss in *equivalent length* for each designated pipe section.

Step 12

Line L: Subtract Line J from Line K and enter in Column 10.

The result should always be a positive or plus figure. If it is not, repeat the operation using Columns 5, 6, 8 and 9 until a balance or near balance is obtained. If the difference between Lines J and K is a high positive number, it is an indication that the pipe sizes are too large and should be reduced, thus saving materials. In such a case, the operations using Columns 5, 6, 8 and 9 should be repeated.

The total friction losses are determined and subtracted from the pressure available to overcome pipe friction for the trial pipe size. This number is critical because it provides a guide to whether the pipe size selected is too large and the process should be repeated to obtain an economically designed system.

Answer: The final figures entered in Column 5 become the design pipe size for the respective sections. Repeating this operation a second time using the same sketch but considering the demand for hot water, it is possible to size the hot water distribution piping. This has been worked up as a part of the overall problem in the tabular arrangement used for sizing the service and water distribution piping. Note that consideration must be given to the pressure losses from the street main to the water heater (Section A-B) in determining the hot water pipe sizes.

TABLE AP.1

COLD WATER PIPE SECTION	FITTINGS/VALVES	PRESSURE LOSS EXPRESSED AS EQUIVALENT LENGTH OF TUBE (feet)	HOT WATER PIPE SECTION	FITTINGS/VALVES	PRESSURE LOSS EXPRESSED AS EQUIVALENT OF TUBE (feet)
A-B	3 – 2¹/₂″ Gate valves	3	A-B	3 – 2¹/₂″ Gate valves	3
	1 – 2¹/₂″ Side branch tee	12	—	1 – 2¹/₂″ Side branch tee	12
B-C	1 – 2¹/₂″ Straight run tee	0.5	B-C	1 – 2″ Straight run tee	7
	—	—	—	1 – 2″ 90-degree ell	0.5
C-F	1 – 2¹/₂″ Side branch tee	12	C-F	1 – 1¹/₂″ Side branch tee	7
C-D	1 – 2¹/₂″ 90-degree ell	7	C-D	1 – ¹/₂″ 90-degree ell	4
D-E	1 – 2¹/₂″ Side branch tee	12	D-E	1 – 1¹/₂″ Side branch tee	7

For SI: 1 inch = 25.4 mm, 1 foot = 304.8 mm, 1 degree = 0.01745 rad.

TABLE AP.2

PIPE SECTIONS	FRICTION LOSS EQUIVALENT LENGTH (feet)	
	Cold Water	Hot Water
A-B	0.69 × 3.2 = 2.21	0.69 × 3.2 = 2.21
B-C	0.085 × 3.1 = 0.26	0.16 × 1.4 = 0.22
C-D	0.20 × 1.9 = 0.38	0.17 × 3.2 = 0.54
D-E	1.62 × 1.9 = 3.08	1.57 × 3.2 = 5.02
Total pipe friction losses (Line K)	5.93	7.99

For SI: 1 foot = 304.8 mm.

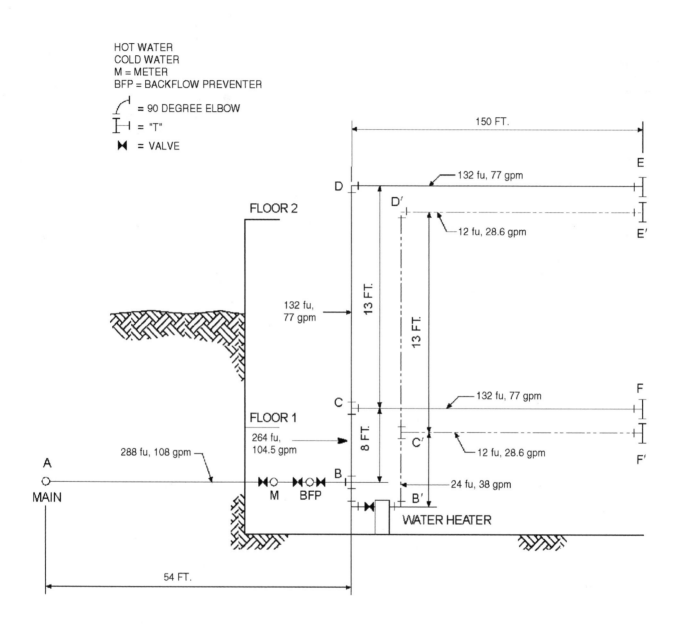

HOT WATER
COLD WATER
M = METER
BFP = BACKFLOW PREVENTER

= 90 DEGREE ELBOW

= "T"

= VALVE

For SI: 1 foot = 304.8 mm, 1 gallon per minute = 3.785 L/m.

FIGURE AP103.3(1)
EXAMPLE—SIZING

TABLE AP103.3(1)
RECOMMENDED TABULAR ARRANGEMENT FOR USE IN SOLVING PIPE SIZING PROBLEMS

COLUMN	1		2	3	4	5	6	7	8	9	10
Line		Description	Pounds per square inch	Gallons per min through section	Length of section (feet)	Trial pipe size (inches)	Equivalent length of fittings and valves (feet)	Total equivalent length [(Col. 4 + Col. 6)/100 feet)]	Friction loss per 100 feet of trial size pipe (psi)	Friction loss in equivalent length Column 8 x Column 7 (psi)	Excess pressure over friction losses (psi)
A	Service and cold water distribution piping[a]	Minimum pressure available at main	55.00								
B		Highest pressure required at a fixture (see Table P2903.1)	15.00								
C		Meter loss 2" meter	11.00								
D		Tap in main loss 2" tap [see Table AP103.3(4)]	1.61								
E		Static head loss 21 ft × 0.43 psi/ft	9.03								
F		Special fixture loss backflow preventer	9.00								
G		Special fixture loss—Filter	0.00								
H		Special fixture loss—Other	0.00								
I		Total overall losses and requirements (Sum of Lines B through H)	45.64								
J		Pressure available to overcome pipe friction (Line A minus Line I)	9.36								
	Pipe section (from diagram) cold water distribution piping	A-B	288	108.0	54	2½	15.00	0.69	3.2	2.21	—
		B-C	264	104.5	8	2½	0.5	0.085	3.1	0.26	—
		C-D	132	77.0	13	2½	7.00	0.20	1.9	0.38	—
		C-F[b]	132	77.0	150	2½	12.00	1.62	1.9	3.08	—
		D-E[b]	132	77.0	150	2½	12.00	1.62	1.9	3.08	—
K	Total pipe friction losses (cold)			—	—	—	—	—	—	5.93	
L	Difference (Line J minus Line K)			—	—	—	—	—	—	—	3.43
	Pipe section (from diagram) Hot water Distribution Piping	A'B'	288	108.0	54	2½	12.00	0.69	3.3	2.21	—
		B'C'	24	38.0	8	2	7.5	0.16	1.4	0.22	—
		C'D'	12	28.6	13	1½	4.0	0.17	3.2	0.54	—
		C'F'[b]	12	28.6	150	1½	7.00	1.57	3.2	5.02	—
		D'E'[b]	12	28.6	150	1½	7.00	1.57	3.2	5.02	—
K	Total pipe friction losses (hot)			—	—	—	—	—	—	7.99	
L	Difference (Line J minus Line K)			—	—	—	—	—	—	—	1.37

For SI: 1 inch = 25.4 mm, 1 foot = 304.8 mm, 1 pound per square inch = 6.895 kPa, 1 gallon per minute = 3.785 L/m.

a. To be considered as pressure gain for fixtures below main (to consider separately, omit from "I" and add to "J").

b. To consider separately, in Line K use Section C-F only if greater loss than the loss in Section D-E.

TABLE AP103.3(2)
LOAD VALUES ASSIGNED TO FIXTURES[a]

FIXTURE	OCCUPANCY	TYPE OF SUPPLY CONTROL	LOAD VALUES, IN WATER SUPPLY FIXTURE UNITS (w.s.f.u.)		
			Cold	Hot	Total
Bathroom group	Private	Flush tank	2.7	1.5	3.6
Bathroom group	Private	Flushometer valve	6.0	3.0	8.0
Bathtub	Private	Faucet	1.0	1.0	1.4
Bathtub	Public	Faucet	3.0	3.0	4.0
Bidet	Private	Faucet	1.5	1.5	2.0
Combination fixture	Private	Faucet	2.25	2.25	3.0
Dishwashing machine	Private	Automatic	—	1.4	1.4
Drinking fountain	Offices, etc.	$^3/_8$" valve	0.25	—	0.25
Kitchen sink	Private	Faucet	1.0	1.0	1.4
Kitchen sink	Hotel, restaurant	Faucet	3.0	3.0	4.0
Laundry trays (1 to 3)	Private	Faucet	1.0	1.0	1.4
Lavatory	Private	Faucet	0.5	0.5	0.7
Lavatory	Public	Faucet	1.5	1.5	2.0
Service sink	Offices, etc.	Faucet	2.25	2.25	3.0
Shower head	Public	Mixing valve	3.0	3.0	4.0
Shower head	Private	Mixing valve	1.0	1.0	1.4
Urinal	Public	1" flushometer valve	10.0	—	10.0
Urinal	Public	$^3/_4$" flushometer valve	5.0	—	5.0
Urinal	Public	Flush tank	3.0	—	3.0
Washing machine (8 lb)	Private	Automatic	1.0	1.0	1.4
Washing machine (8 lb)	Public	Automatic	2.25	2.25	3.0
Washing machine (15 lb)	Public	Automatic	3.0	3.0	4.0
Water closet	Private	Flushometer valve	6.0	—	6.0
Water closet	Private	Flush tank	2.2	—	2.2
Water closet	Public	Flushometer valve	10.0	—	10.0
Water closet	Public	Flush tank	5.0	—	5.0
Water closet	Public or private	Flushometer tank	2.0	—	2.0

For SI: 1 inch = 25.4 mm, 1 pound = 0.454 kg.

a. For fixtures not listed, loads should be assumed by comparing the fixture to one listed using water in similar quantities and at similar rates. The assigned loads for fixtures with both hot and cold water supplies are given for separate hot and cold water loads, and for total load. The separate hot and cold water loads are three-fourths of the total load for the fixture in each case.

TABLE AP103.3(3)
TABLE FOR ESTIMATING DEMAND

SUPPLY SYSTEMS PREDOMINANTLY FOR FLUSH TANKS			SUPPLY SYSTEMS PREDOMINANTLY FOR FLUSHOMETERS		
Load	Demand		Load	Demand	
(w.s.f.u.)	(gpm)	(cfm)	(w.s.f.u.)	(gpm)	(cfm)
1	3.0	0.04104	—	—	—
2	5.0	0.0684	—	—	—
3	6.5	0.86892	—	—	—
4	8.0	1.06944	—	—	—
5	9.4	1.256592	5	15.0	2.0052
6	10.7	1.430376	6	17.4	2.326032
7	11.8	1.577424	7	19.8	2.646364
8	12.8	1.711104	8	22.2	2.967696
9	13.7	1.831416	9	24.6	3.288528
10	14.6	1.951728	10	27.0	3.60936
11	15.4	2.058672	11	27.8	3.716304
12	16.0	2.13888	12	28.6	3.823248
13	16.5	2.20572	13	29.4	3.930192
14	17.0	2.27256	14	30.2	4.037136
15	17.5	2.3394	15	31.0	4.14408
16	18.0	2.90624	16	31.8	4.241024
17	18.4	2.459712	17	32.6	4.357968
18	18.8	2.513184	18	33.4	4.464912
19	19.2	2.566656	19	34.2	4.571856
20	19.6	2.620128	20	35.0	4.6788
25	21.5	2.87412	25	38.0	5.07984
30	23.3	3.114744	30	42.0	5.61356
35	24.9	3.328632	35	44.0	5.88192
40	26.3	3.515784	40	46.0	6.14928
45	27.7	3.702936	45	48.0	6.41664
50	29.1	3.890088	50	50.0	6.684
60	32.0	4.27776	60	54.0	7.21872
70	35.0	4.6788	70	58.0	7.75344
80	38.0	5.07984	80	61.2	8.181216
90	41.0	5.48088	90	64.3	8.595624
100	43.5	5.81508	100	67.5	9.0234
120	48.0	6.41664	120	73.0	9.75864
140	52.5	7.0182	140	77.0	10.29336
160	57.0	7.61976	160	81.0	10.82808
180	61.0	8.15448	180	85.5	11.42964
200	65.0	8.6892	200	90.0	12.0312
225	70.0	9.3576	225	95.5	12.76644

(continued)

TABLE AP103.3(3)—continued
TABLE FOR ESTIMATING DEMAND

SUPPLY SYSTEMS PREDOMINANTLY FOR FLUSH TANKS			SUPPLY SYSTEMS PREDOMINANTLY FOR FLUSHOMETERS		
Load	Demand		Load	Demand	
(w.s.f.u.)	(gpm)	(cfm)	(w.s.f.u.)	(gpm)	(cfm)
250	75.0	10.026	250	101.0	13.50168
275	80.0	10.6944	275	104.5	13.96956
300	85.0	11.3628	300	108.0	14.43744
400	105.0	14.0364	400	127.0	16.97736
500	124.0	16.57632	500	143.0	19.11624
750	170.0	22.7256	750	177.0	23.66136
1,000	208.0	27.80544	1,000	208.0	27.80544
1,250	239.0	31.94952	1,250	239.0	31.94952
1,500	269.0	35.95992	1,500	269.0	35.95992
1,750	297.0	39.70296	1,750	297.0	39.70296
2,000	325.0	43.446	2,000	325.0	43.446
2,500	380.0	50.7984	2,500	380.0	50.7984
3,000	433.0	57.88344	3,000	433.0	57.88344
4,000	535.0	70.182	4,000	525.0	70.182
5,000	593.0	79.27224	5,000	593.0	79.27224

For SI: 1 gallon per minute = 3.785 L/m, 1 cubic foot per minute = 0.000471 m^3/s.

TABLE AP103.3(4)
LOSS OF PRESSURE THROUGH TAPS AND TEES IN POUNDS PER SQUARE INCH (psi)

GALLONS PER MINUTE	SIZE OF TAP OR TEE (inches)						
	$^5/_8$	$^3/_4$	1	$1^1/_4$	$1^1/_2$	2	3
10	1.35	0.64	0.18	0.08	—	—	—
20	5.38	2.54	0.77	0.31	0.14	—	—
30	12.10	5.72	1.62	0.69	0.33	0.10	—
40	—	10.20	3.07	1.23	0.58	0.18	—
50	—	15.90	4.49	1.92	0.91	0.28	—
60	—	—	6.46	2.76	1.31	0.40	—
70	—	—	8.79	3.76	1.78	0.55	0.10
80	—	—	11.50	4.90	2.32	0.72	0.13
90	—	—	14.50	6.21	2.94	0.91	0.16
100	—	—	17.94	7.67	3.63	1.12	0.21
120	—	—	25.80	11.00	5.23	1.61	0.30
140	—	—	35.20	15.00	7.12	2.20	0.41
150	—	—	—	17.20	8.16	2.52	0.47
160	—	—	—	19.60	9.30	2.92	0.54
180	—	—	—	24.80	11.80	3.62	0.68
200	—	—	—	30.70	14.50	4.48	0.84
225	—	—	—	38.80	18.40	5.60	1.06
250	—	—	—	47.90	22.70	7.00	1.31
275	—	—	—	—	27.40	7.70	1.59
300	—	—	—	—	32.60	10.10	1.88

For SI: 1 inch = 25.4 mm, 1 pound per square inch = 6.895 kPa, 1 gallon per minute = 3.785 L/m.

TABLE AP103.3(5)
ALLOWANCE IN EQUIVALENT LENGTHS OF PIPE FOR FRICTION LOSS IN VALVES AND THREADED FITTINGS (feet)

FITTING OR VALVE	PIPE SIZE (inches)							
	$^1/_2$	$^3/_4$	1	$1^1/_4$	$1^1/_2$	2	$2^1/_2$	3
45-degree elbow	1.2	1.5	1.8	2.4	3.0	4.0	5.0	6.0
90-degree elbow	2.0	2.5	3.0	4.0	5.0	7.0	8.0	10.0
Tee, run	0.6	0.8	0.9	1.2	1.5	2.0	2.5	3.0
Tee, branch	3.0	4.0	5.0	6.0	7.0	10.0	12.0	15.0
Gate valve	0.4	0.5	0.6	0.8	1.0	1.3	1.6	2.0
Balancing valve	0.8	1.1	1.5	1.9	2.2	3.0	3.7	4.5
Plug-type cock	0.8	1.1	1.5	1.9	2.2	3.0	3.7	4.5
Check valve, swing	5.6	8.4	11.2	14.0	16.8	22.4	28.0	33.6
Globe valve	15.0	20.0	25.0	35.0	45.0	55.0	65.0	80.0
Angle valve	8.0	12.0	15.0	18.0	22.0	28.0	34.0	40.0

For SI: 1 inch = 25.4 mm, 1 foot = 304.8 mm, 1 degree = 0.0175 rad.

TABLE AP103.3(6)
PRESSURE LOSS IN FITTINGS AND VALVES EXPRESSED AS EQUIVALENT LENGTH OF TUBE[a] (feet)

NOMINAL OR STANDARD SIZE (inches)	FITTINGS				Coupling	VALVES			
	Standard Ell		90-degree Tee			Ball	Gate	Butterfly	Check
	90 Degree	45 Degree	Side Branch	Straight Run					
$^3/_8$	0.5	—	1.5	—	—	—	—	—	1.5
$^1/_2$	1	0.5	2	—	—	—	—	—	2
$^5/_8$	1.5	0.5	2	—	—	—	—	—	2.5
$^3/_4$	2	0.5	3	—	—	—	—	—	3
1	2.5	1	4.5	—	—	0.5	—	—	4.5
$1^1/_4$	3	1	5.5	0.5	0.5	0.5	—	—	5.5
$1^1/_2$	4	1.5	7	0.5	0.5	0.5	—	—	6.5
2	5.5	2	9	0.5	0.5	0.5	0.5	7.5	9
$2^1/_2$	7	2.5	12	0.5	0.5	—	1	10	11.5
3	9	3.5	15	1	1	—	1.5	15.5	14.5
$3^1/_2$	9	3.5	14	1	1	—	2	—	12.5
4	12.5	5	21	1	1	—	2	16	18.5
5	16	6	27	1.5	1.5	—	3	11.5	23.5
6	19	7	34	2	2	—	3.5	13.5	26.5
8	29	11	50	3	3	—	5	12.5	39

For SI: 1 inch = 25.4 mm, 1 foot = 304.8 mm, 1 degree = 0.01745 rad.

a. Allowances are for streamlined soldered fittings and recessed threaded fittings. For threaded fittings, double the allowances shown in the table. The equivalent lengths presented in the table are based on a C factor of 150 in the Hazen-Williams friction loss formula. The lengths shown are rounded to the nearest half-foot.

Note: Fluid velocities in excess of 5 to 8 feet per second are not usually recommended.

For SI: 1 inch = 25.4 mm, 1 foot = 304.8 mm, 1 gallon per minute = 3.785 L/m, 1 pound per square inch = 6.895 kPa, 1 foot per second = 0.305 m/s.

a. This figure applies to smooth new copper tubing with recessed (streamline) soldered joints and to the actual sizes of types indicated on the diagram.

FIGURE AP103.3(2)
FRICTION LOSS IN SMOOTH PIPE[a]
(TYPE K, ASTM B88 COPPER TUBING)

WATER FLOW RATE, GALLONS PER MINUTE

PRESSURE DROP PER 100 FEET OF TUBE, POUNDS PER SQUARE INCH

Note: Fluid velocities in excess of 5 to 8 feet per second are not usually recommended.

For SI: 1 inch = 25.4 mm, 1 foot = 304.8 mm, 1 gallon per minute = 3.785 L/m, 1 pound per square inch = 6.895 kPa, 1 foot per second = 0.305 m/s.

a. This figure applies to smooth new copper tubing with recessed (streamline) soldered joints and to the actual sizes of types indicated on the diagram.

FIGURE AP103.3(3)
FRICTION LOSS IN SMOOTH PIPE[a]
(TYPE L, ASTM B88 COPPER TUBING)

WATER FLOW RATE, GALLONS PER MINUTE

PRESSURE DROP PER 100 FEET OF TUBE, POUNDS PER SQUARE INCH

Note: Fluid velocities in excess of 5 to 8 feet per second are not usually recommended.

For SI: 1 inch = 25.4 mm, 1 foot = 304.8 mm, 1 gallon per minute = 3.785 L/m, 1 pound per square inch = 6.895 kPa, 1 foot per second = 0.305 m/s.

a. This figure applies to smooth new copper tubing with recessed (streamline) soldered joints and to the actual sizes of types indicated on the diagram.

FIGURE AP103.3(4)
FRICTION LOSS IN SMOOTH PIPE[a]
(TYPE M, ASTM B88 COPPER TUBING)

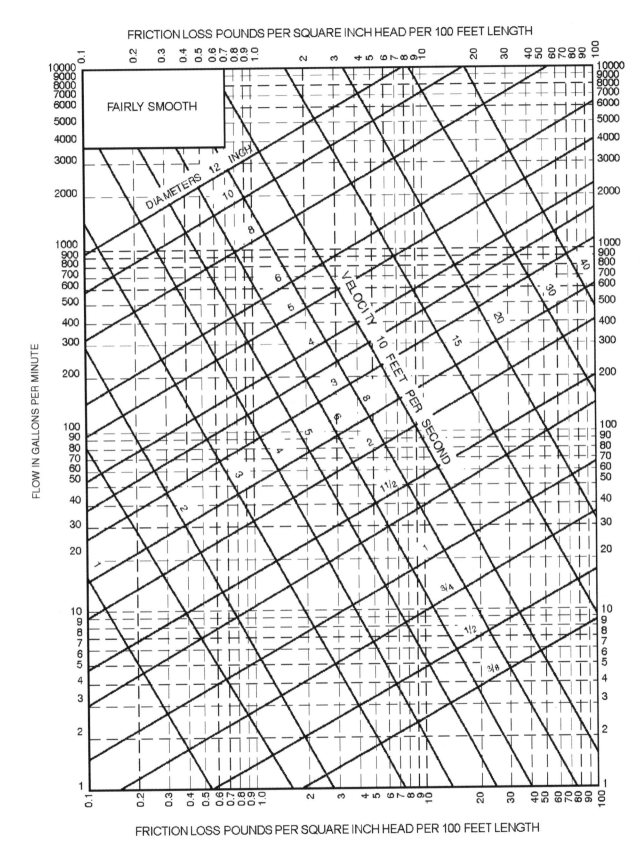

FRICTION LOSS POUNDS PER SQUARE INCH HEAD PER 100 FEET LENGTH

FAIRLY SMOOTH

FLOW IN GALLONS PER MINUTE

FRICTION LOSS POUNDS PER SQUARE INCH HEAD PER 100 FEET LENGTH

For SI: 1 inch = 25.4 mm, 1 foot = 304.8 mm, 1 gallon per minute = 3.785 L/m, 1 pound per square inch = 6.895 kPa, 1 foot per second = 0.305 m/s.

a. This figure applies to smooth new steel (fairly smooth) pipe and to actual diameters of standard-weight pipe.

FIGURE AP103.3(5)
FRICTION LOSS IN FAIRLY SMOOTH PIPE[a]

FRICTION LOSS POUNDS PER SQUARE INCH HEAD PER 100 FEET LENGTH

FRICTION LOSS POUNDS PER SQUARE INCH HEAD PER 100 FEET LENGTH

For SI: 1 inch = 25.4 mm, 1 foot = 304.8 mm, 1 gallon per minute = 3.785 L/m, 1 pound per square inch = 6.895 kPa, 1 foot per second = 0.305 m/s.

a. This figure applies to fairly rough pipe and to actual diameters which, in general, will be less than the actual diameters of the new pipe of the same kind.

FIGURE AP103.3(6)
FRICTION LOSS IN FAIRLY ROUGH PIPE[a]

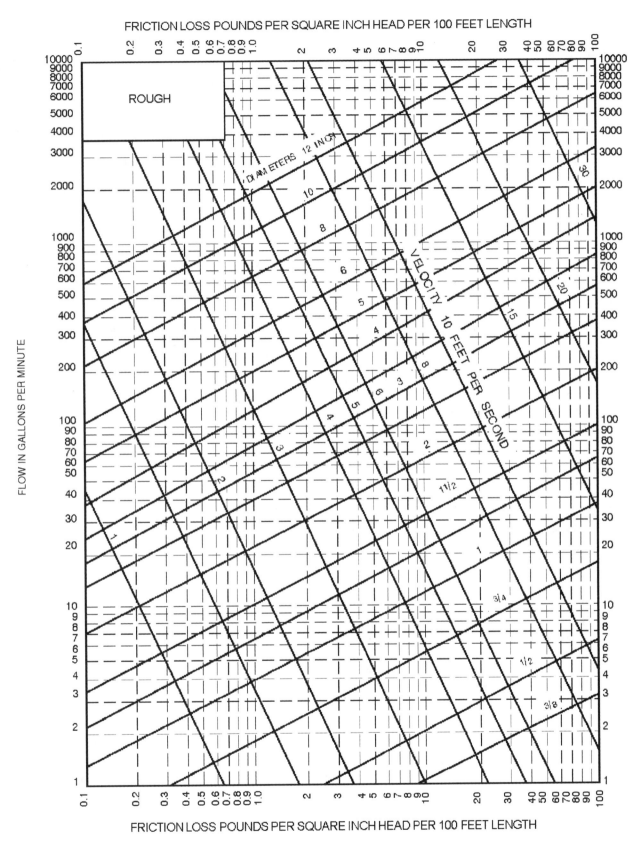

FRICTION LOSS POUNDS PER SQUARE INCH HEAD PER 100 FEET LENGTH

FLOW IN GALLONS PER MINUTE

FRICTION LOSS POUNDS PER SQUARE INCH HEAD PER 100 FEET LENGTH

For SI: 1 inch = 25.4 mm, 1 foot = 304.8 mm, 1 gallon per minute = 3.785 L/m, 1 pound per square inch = 6.895 kPa, 1 foot per second = 0.305 m/s.

a. This figure applies to very rough pipe and existing pipe, and to their actual diameters.

FIGURE AP103.3(7)
FRICTION LOSS IN ROUGH PIPE[a]

SECTION AP201
SELECTION OF PIPE SIZE

AP201.1 Size of water-service mains, branch mains and risers. The minimum size water service pipe shall be $^3/_4$ inch (19.1 mm). The size of water service mains, branch mains and risers shall be determined according to water supply demand [gpm (L/m)], available water pressure [psi (kPa)] and friction loss caused by the water meter and *developed length* of pipe [feet (m)], including the *equivalent length* of fittings. The size of each water distribution system shall be determined according to the procedure outlined in this section or by other design methods conforming to acceptable engineering practice and *approved* by the *building official*:

1. Supply load in the building water distribution system shall be determined by the total load on the pipe being sized, in terms of w.s.f.u., as shown in Table AP103.3(2). For fixtures not listed, choose a w.s.f.u. value of a fixture with similar flow characteristics.

2. Obtain the minimum daily static service pressure [psi (kPa)] available (as determined by the local water authority) at the water meter or other source of supply at the installation location. Adjust this minimum daily static pressure [psi (kPa)] for the following conditions:

   2.1. Determine the difference in elevation between the source of supply and the highest water supply outlet. Where the highest water supply outlet is located above the source of supply, deduct 0.5 psi (3.4 kPa) for each foot (0.3 m) of difference in elevation. Where the highest water supply outlet is located below the source of supply, add 0.5 psi (3.4 kPa) for each foot (0.3 m) of difference in elevation.

   2.2. Where a water pressure-reducing valve is installed in the water distribution system, the minimum daily static water pressure available is 80 percent of the minimum daily static water pressure at the source of supply or the set pressure downstream of the water pressure-reducing valve, whichever is smaller.

   2.3. Deduct all pressure losses caused by special equipment, such as a backflow preventer, water filter and water softener. Pressure loss data for each piece of equipment shall be obtained through the manufacturer of the device.

   2.4. Deduct the pressure in excess of 8 psi (55 kPa) resulting from the installation of the special plumbing fixture, such as temperature-controlled shower and flushometer tank water closet. Using the resulting minimum available pressure, find the corresponding pressure range in Table AP201.1.

3. The maximum *developed length* for water piping is the actual length of pipe between the source of supply and the most remote fixture, including either hot (through the water heater) or cold water branches multiplied by a factor of 1.2 to compensate for pressure loss through

fittings. Select the appropriate column in Table AP201.1 equal to or greater than the calculated maximum *developed length*.

4. To determine the size of the water service pipe, meter and main distribution pipe to the building using the appropriate table, follow down the selected "maximum *developed length*" column to a fixture unit equal to or greater than the total installation demand calculated by using the "combined" w.s.f.u. column of Table AP201.1. Read the water service pipe and meter sizes in the first left-hand column and the main distribution pipe to the building in the second left-hand column on the same row.

5. To determine the size of each water distribution pipe, start at the most remote outlet on each branch (either hot or cold branch) and, working back toward the main distribution pipe to the building, add up the w.s.f.u. demand passing through each segment of the distribution system using the related hot or cold column of Table AP201.1. Knowing demand, the size of each segment shall be read from the second left-hand column of the same table and the maximum *developed length* column selected in Steps 1 and 2, under the same or next smaller size meter row. In no case does the size of any branch or main need to be larger than the size of the main distribution pipe to the building established in Step 4.

TABLE AP201.1
MINIMUM SIZE OF WATER METERS, MAINS AND DISTRIBUTION PIPING BASED ON WATER SUPPLY FIXTURE UNIT VALUES (w.s.f.u.)

METER AND SERVICE PIPE (inches)	DISTRIBUTION PIPE (inches)	MAXIMUM DEVELOPMENT LENGTH (feet)									
Pressure Range 30 to 39 psi		40	60	80	100	150	200	250	300	400	500
$^3/_4$	$^1/_2$ [a]	2.5	2	1.5	1.5	1	1	0.5	0.5	0	0
$^3/_4$	$^3/_4$	9.5	7.5	6	5.5	4	3.5	3	2.5	2	1.5
$^3/_4$	1	32	25	20	16.5	11	9	7.8	6.5	5.5	4.5
1	1	32	32	27	21	13.5	10	8	7	5.5	5
$^3/_4$	$1^1/_4$	32	32	32	32	30	24	20	17	13	10.5
1	$1^1/_4$	80	80	70	61	45	34	27	22	16	12
$1^1/_2$	$1^1/_4$	80	80	80	75	54	40	31	25	17.5	13
1	$1^1/_2$	87	87	87	87	84	73	64	56	45	36
$1^1/_2$	$1^1/_2$	151	151	151	151	117	92	79	69	54	43
2	$1^1/_2$	151	151	151	151	128	99	83	72	56	45
1	2	87	87	87	87	87	87	87	87	87	86
$1^1/_2$	2	275	275	275	275	258	223	196	174	144	122
2	2	365	365	365	365	318	266	229	201	160	134
2	$2^1/_2$	533	533	533	533	533	495	448	409	353	311

METER AND SERVICE PIPE (inches)	DISTRIBUTION PIPE (inches)	MAXIMUM DEVELOPMENT LENGTH (feet)									
Pressure Range 40 to 49 psi		40	60	80	100	150	200	250	300	400	500
$^3/_4$	$^1/_2$ [a]	3	2.5	2	1.5	1.5	1	1	0.5	0.5	0.5
$^3/_4$	$^3/_4$	9.5	9.5	8.5	7	5.5	4.5	3.5	3	2.5	2
$^3/_4$	1	32	32	32	26	18	13.5	10.5	9	7.5	6
1	1	32	32	32	32	21	15	11.5	9.5	7.5	6.5
$^3/_4$	$1^1/_4$	32	32	32	32	32	32	32	27	21	16.5
1	$1^1/_4$	80	80	80	80	65	52	42	35	26	20
$1^1/_2$	$1^1/_4$	80	80	80	80	75	59	48	39	28	21
1	$1^1/_2$	87	87	87	87	87	87	87	78	65	55
$1^1/_2$	$1^1/_2$	151	151	151	151	151	130	109	93	75	63
2	$1^1/_2$	151	151	151	151	151	139	115	98	77	64
1	2	87	87	87	87	87	87	87	87	87	87
$1^1/_2$	2	275	275	275	275	275	275	264	238	198	169
2	2	365	365	365	365	365	349	304	270	220	185
2	$2^1/_2$	533	533	533	533	533	533	533	528	456	403

(continued)

MINIMUM SIZE OF WATER METERS, MAINS AND DISTRIBUTION PIPING BASED ON WATER SUPPLY FIXTURE UNIT VALUES (w.s.f.u.)

METER AND SERVICE PIPE (inches)	DISTRIBUTION PIPE (inches)	MAXIMUM DEVELOPMENT LENGTH (feet)									
Pressure Range 50 to 60 psi		40	60	80	100	150	200	250	300	400	500
3/4	1/2 a	3	3	2.5	2	1.5	1	1	1	0.5	0.5
3/4	3/4	9.5	9.5	9.5	8.5	6.5	5	4.5	4	3	2.5
3/4	1	32	32	32	32	25	18.5	14.5	12	9.5	8
1	1	32	32	32	32	30	22	16.5	13	10	8
3/4	1 1/4	32	32	32	32	32	32	32	32	29	24
1	1 1/4	80	80	80	80	80	68	57	48	35	28
1 1/2	1 1/4	80	80	80	80	80	75	63	53	39	29
1	1 1/2	87	87	87	87	87	87	87	87	82	70
1 1/2	1 1/2	151	151	151	151	151	151	139	120	94	79
2	1 1/2	151	151	151	151	151	151	146	126	97	81
1	2	87	87	87	87	87	87	87	87	87	87
1 1/2	2	275	275	275	275	275	275	275	275	247	213
2	2	365	365	365	365	365	365	365	329	272	232
2	2 1/2	533	533	533	533	533	533	533	533	533	486

METER AND SERVICE PIPE (inches)	DISTRIBUTION PIPE (inches)	MAXIMUM DEVELOPMENT LENGTH (feet)									
Pressure Range Over 60		40	60	80	100	150	200	250	300	400	500
3/4	1/2 a	3	3	3	2.5	2	1.5	1.5	1	1	0.5
3/4	3/4	9.5	9.5	9.5	9.5	7.5	6	5	4.5	3.5	3
3/4	1	32	32	32	32	32	24	19.5	15.5	11.5	9.5
1	1	32	32	32	32	32	28	28	17	12	9.5
3/4	1 1/4	32	32	32	32	32	32	32	32	32	30
1	1 1/4	80	80	80	80	80	80	69	60	46	36
1 1/2	1 1/4	80	80	80	80	80	80	76	65	50	38
1	1 1/2	87	87	87	87	87	87	87	87	87	84
1 1/2	1 1/2	151	151	151	151	151	151	151	144	114	94
2	1 1/2	151	151	151	151	151	151	151	151	118	97
1	2	87	87	87	87	87	87	87	87	87	87
1 1/2	2	275	275	275	275	275	275	275	275	275	252
2	2	365	368	368	368	368	368	368	368	318	273
2	2 1/2	533	533	533	533	533	533	533	533	533	533

For SI: 1 inch = 25.4, 1 foot = 304.8 mm, 1 pound per square inch = 6.895 kPa.
a. Minimum size for building supply is a 3/4-inch pipe.

Appendix Q:
Reserved

Appendix R:
Light Straw-clay Construction

(The provisions contained in this appendix are not mandatory unless specifically referenced in the adopting ordinance.)

General Comments

Mixtures of clay and straw known as clay-straw, straw-clay, and light straw-clay have been successfully used as a nonload-bearing wall infill material since 1950 in Europe and 1990 in the United States. The use of light straw-clay wall systems has seen a surge in popularity recently because of its attributes. These include light straw-clay's thermal performance (with its balance of mass and insulation) and its low environmental impact, using local, minimally processed materials that ultimately return to the earth without harm. Other beneficial qualities include the low cost of the raw materials and the easily-learned skills needed for installing the material.

Two heavier forms of wall construction using clay and straw have been used in various parts of the world for thousands of years. A monolithic clay and straw wall system known as cob has been used in what are now the United Kingdom, Africa, the Middle East and eastern North America. A system of woven wood or woven bamboo with applied clay and straw known as wattle and daub was in common use for centuries throughout Europe, Africa, Asia and North and South America. Both cob and wattle and daub continue to be utilized today. Many thousands of existing structures using cob or wattle and daub, dating back 300–400 years, have been continuously occupied, attesting to the durability of the materials of clay and straw.

Purpose

In the United States, residential and nonresidential structures using light straw-clay have been completed in 17 states, and most of those have been constructed with full permits and inspections. However, many of these structures have been constructed on a case-by-case basis without the benefit of a building code for light-straw clay construction.

As of 2014, only the states of New Mexico and Oregon had adopted guidelines or codes for light straw-clay construction. This appendix provides prescriptive and performance requirements for all aspects of nonbearing light straw-clay construction. This is especially important because most design professionals, code officials and builders are currently unfamiliar with this wall system.

The purpose of this appendix and its commentary is to provide requirements and guidance regarding construction utilizing straw and unfired clay, which are materials previously not addressed in the code. The nature of light straw-clay as a material and system demands compliance with certain requirements that differ from established practices with conventional materials and systems in the code. The commentary to this appendix is particularly useful in these instances because it explains the reasons for requirements that are new to users of the code and this appendix.

SECTION AR101
GENERAL

AR101.1 Scope. This appendix shall govern the use of light straw-clay as a nonbearing building material and wall infill system in Seismic Design Categories A and B.

❖ In Seismic Design Categories A and B, lateral loads due to wind govern the requirements for wall bracing. In Seismic Design Categories C, D_0, D_1 and D_2 seismic loading often governs. A building's performance when subjected to lateral wind forces is independent of the building's weight, whereas a building's performance when subjected to seismic forces is a function of the building's weight. A wood frame wall with light straw-clay infill is significantly heavier than a conventional wood frame wall for which the code's bracing requirements are formulated. Thus, this appendix is limited to use in Seismic Design Categories A and B, where the relative increased weight of light straw-clay walls is not relevant, and the bracing requirements in the code text and tables remain applicable.

Buildings using light straw-clay construction in Seismic Design Categories C, D_0, D_1 and D_2 are allowed if the building official finds that the proposed design satisfies the requirements of Section R104.11, Alternative materials, design and methods of construction and equipment. This would likely require an approved engineered design for wall bracing that accounts for the weight of light straw-clay walls. However, because this appendix is limited to Seismic Design Categories A and B for wall bracing requirements only, all other requirements of this appendix, where adopted, would apply to a building using light straw-clay construction in Seismic Design Categories C, D_0, D_1 and D_2.

In general, many variations of light straw-clay construction are possible and many have been practiced historically. Designs that do not comply with certain requirements in this appendix may be approved if the building official finds the proposed design satisfies Section R104.11 by means of equivalent alternatives to these requirements.

SECTION AR102
DEFINITIONS

AR102.1 General. The following words and terms shall, for the purposes of this appendix, have the meanings shown herein. Refer to Chapter 2 of the *International Residential Code* for general definitions.

CLAY. Inorganic soil with particle sizes of less than 0.00008 inch (0.002 mm) having the characteristics of high to very high dry strength and medium to high plasticity.

CLAY SLIP. A suspension of clay soil in water.

CLAY SOIL. Inorganic soil containing 50 percent or more clay by volume.

❖ The word "soil" is commonly associated with topsoil, which contains organic matter. Clay soil used for construction purposes is a nonfertile (inorganic) mineral subsoil that contains clay, silt and sand. Sections AR103.3.2 and AR103.4.1 and Commentary Figure AR103.3.2 describe methods for determining the suitability of this material for light straw-clay construction.

INFILL. Light straw-clay that is placed between the structural members of a building.

LIGHT STRAW-CLAY. A mixture of straw and clay compacted to form insulation and plaster substrate between or around structural and nonstructural members in a wall.

❖ The density range of light straw-clay is 10 to 50 pounds per cubic foot (160 to 801 kg/m³).

NONBEARING. Not bearing the weight of the building other than the weight of the light straw-clay itself and its finish.

STRAW. The dry stems of cereal grains after the seed heads have been removed.

VOID. Any space in a light straw-clay wall in which a 2-inch (51 mm) sphere can be inserted.

SECTION AR103
NONBEARING LIGHT STRAW-CLAY
CONSTRUCTION

AR103.1 General. Light straw-clay shall be limited to infill between or around structural and nonstructural wall framing members.

AR103.2 Structure. The structure of buildings using light straw-clay shall be in accordance with the *International Residential Code* or shall be in accordance with an *approved* design by a registered *design professional.*

AR103.2.1 Number of stories. Use of light straw-clay infill shall be limited to buildings that are not more than one *story above grade plane.*

> **Exception:** Buildings using light straw-clay infill that are greater than one *story above grade plane* shall be in accordance with an approved design by a registered *design professional.*

AR103.2.2 Bracing. Wind shall be in accordance with Section R602.10 and shall use Method LIB. Walls with light straw-clay infill shall not be sheathed with solid sheathing.

❖ The "solid sheathing" prohibition in this section refers only to the application of sheathing that will inhibit air-drying of newly installed straw-clay material. Because light straw-clay is installed wet and is not fully dried to equilibrium (see Section AR103.5.1) before the application of plaster, any solid impermeable permanent sheathing that would inhibit its drying is prohibited.

AR103.2.3 Weight of light straw-clay. Light straw-clay shall be deemed to have a design dead load of 40 pounds per cubic foot (640 kg per cubic meter) unless otherwise demonstrated to the *building official.*

❖ Densities of light straw-clay can vary depending on wall design, project location and desired thermal properties. Historically, densities ranging from 10 to 50 pounds per cubic foot (160 to 801 kg/m³) have been utilized. Structural engineering is not required for one-story light straw-clay structures that comply with the prescriptive structural requirements of the code and this appendix. However, structural calculations may be needed when departing from these requirements or may be otherwise advantageous.

  Densities of light straw-clay stated in this appendix are for the straw-clay material only and do not include other elements such as the wall's plaster or other finish and its framing.

AR103.2.4 Reinforcement of light straw-clay. Light straw-clay shall be reinforced as follows:

1. Vertical reinforcing shall be not less than nominal 2-inch by 6-inch (51 mm by 152 mm) wood members at not more than 32 inches (813 mm) on center where the vertical reinforcing is nonload bearing and at 24 inches (610 mm) on center where it is load bearing. The vertical reinforcing shall not exceed an unrestrained height of 10 feet (3048 mm) and shall be attached at top and bottom in accordance with Chapter 6 of the this code. In lieu of these requirements, vertical reinforcing shall be in accordance with an *approved* design by a registered *design professional.*

2. Horizontal reinforcing shall be installed in the center of the wall at not more than 24 inches (610 mm) on center and shall be secured to vertical members. Horizontal reinforcing shall be of any of the following: ³/₄-inch (19.1 mm) bamboo, ¹/₂-inch (12.7 mm) fiberglass rod, 1-inch (25 mm) wood dowel or nominal 1-inch by 2-inch (25 mm by 51 mm) wood.

❖ The term "reinforcement" in this section means a component of the wall system that contains or keeps the light straw-clay infill in place during construction and in service. This includes containment of the light straw-clay where the wall is subjected to lateral out-of-plane wind or seismic forces as required by the code.

Item 1: Vertical reinforcing can be composed of nonload-bearing elements, with containment of the light straw-clay as their sole function, or elements that also serve a structural function, such as load-bearing studs. Vertical reinforcing and wall frame construction are based on the conventional wood framing requirements of Section R602, with similar methods and requirements for studs, plates and headers, etc. The main departure from the requirements in Section R602 in light straw-clay construction is the constraint on the use of solid sheathing, as noted in the commentary to Section AR103.2.2. This section is not intended to prohibit commonly accepted vertical reinforcing methods that are structurally equivalent to 2-inch by 6-inch solid wood members, such as "Larsen trusses" or similar composite members. Commentary Figures AR103.2.4(1), AR103.2.4(2) and AR103.2.4(3) show examples of wall framing for light straw-clay construction.

Item 2: Unlike the placement of steel reinforcing bar in typical concrete construction, which is used for tensile and shear reinforcing, horizontal reinforcing members in light straw-clay are used primarily for the doweling effects of the reinforcing within the wall system. The horizontal reinforcing helps: (1) Maintain overall integrity of the wall assembly during con-

struction and in service; (2) Transfer of out-of-plane wind or seismic forces from the light straw-clay mass to the vertical reinforcing framing members and; (3) Control the vertical shrinkage and settling of the light straw-clay material as it dries. In practice, the horizontal reinforcing is placed while slightly compressing the moist straw-clay mix in order to limit "rebound" and help control shrinkage.

AR103.3 Materials. The materials used in light straw-clay construction shall be in accordance with Sections AR103.3.1 through AR103.3.4.

AR103.3.1 Straw. Straw shall be wheat, rye, oats, rice or barley, and shall be free of visible decay and insects.

❖ Straw, like wood, is composed primarily of cellulose and lignin. Construction-grade straw can be readily identified by visual examination. The best straw is composed of the dry (i.e., not having been wetted since cut) intact stems of wheat, rye, oats, rice or barley plants and must be free of visible decay and contaminants such as insects, fertile topsoil and green plant material, since mold and mildew can grow in the presence of these readily-digested food sources.

Straw is best harvested dry and kept dry until it is mixed with clay slip and installed. The straw will have a golden color and a glossy surface. Excessive exposure to moisture will result in straw with reduced color and gloss. This straw is not optimal for light straw-clay construction.

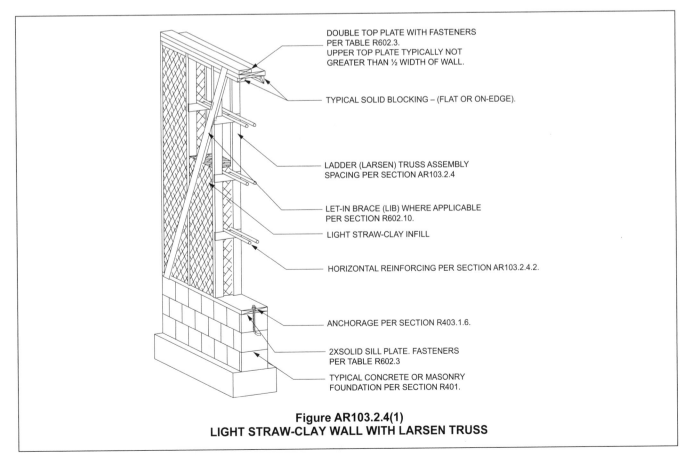

Figure AR103.2.4(1)
LIGHT STRAW-CLAY WALL WITH LARSEN TRUSS

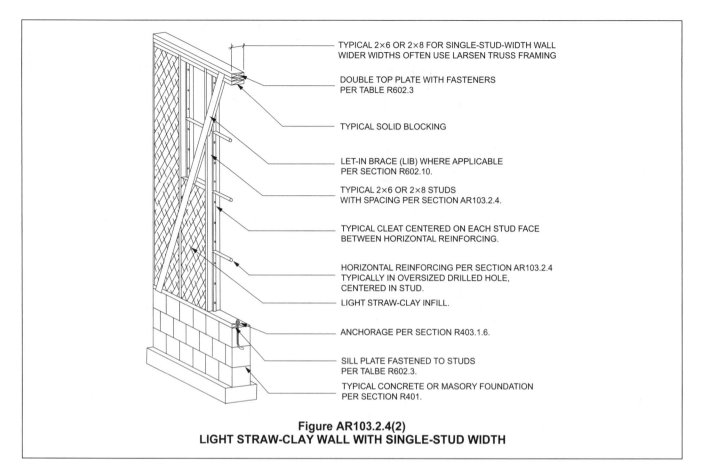

TYPICAL 2×6 OR 2×8 FOR SINGLE-STUD-WIDTH WALL WIDER WIDTHS OFTEN USE LARSEN TRUSS FRAMING

DOUBLE TOP PLATE WITH FASTENERS PER TABLE R602.3

TYPICAL SOLID BLOCKING

LET-IN BRACE (LIB) WHERE APPLICABLE PER SECTION R602.10.

TYPICAL 2×6 OR 2×8 STUDS WITH SPACING PER SECTION AR103.2.4.

TYPICAL CLEAT CENTERED ON EACH STUD FACE BETWEEN HORIZONTAL REINFORCING.

HORIZONTAL REINFORCING PER SECTION AR103.2.4 TYPICALLY IN OVERSIZED DRILLED HOLE, CENTERED IN STUD.

LIGHT STRAW-CLAY INFILL.

ANCHORAGE PER SECTION R403.1.6.

SILL PLATE FASTENED TO STUDS PER TALBE R602.3.

TYPICAL CONCRETE OR MASORY FOUNDATION PER SECTION R401.

Figure AR103.2.4(2)
LIGHT STRAW-CLAY WALL WITH SINGLE-STUD WIDTH

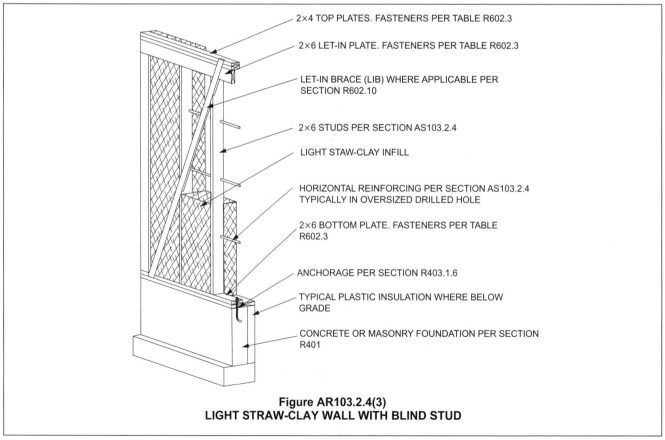

2×4 TOP PLATES. FASTENERS PER TABLE R602.3

2×6 LET-IN PLATE. FASTENERS PER TABLE R602.3

LET-IN BRACE (LIB) WHERE APPLICABLE PER SECTION R602.10

2×6 STUDS PER SECTION AS103.2.4

LIGHT STAW-CLAY INFILL

HORIZONTAL REINFORCING PER SECTION AS103.2.4 TYPICALLY IN OVERSIZED DRILLED HOLE

2×6 BOTTOM PLATE. FASTENERS PER TABLE R602.3

ANCHORAGE PER SECTION R403.1.6

TYPICAL PLASTIC INSULATION WHERE BELOW GRADE

CONCRETE OR MASONRY FOUNDATION PER SECTION R401

Figure AR103.2.4(3)
LIGHT STRAW-CLAY WALL WITH BLIND STUD

AR103.3.2 Clay soil. Suitability of clay soil shall be determined in accordance with the Figure 2 Ribbon Test or the Figure 3 Ball Test of the Appendix to ASTM E2392/E2392M.

❖ This section is intended to ensure that the quality of the clay slip used to make light straw-clay is suitable for the intended application. Clay soil appropriate for light straw-clay construction contains particles of sand and silt in addition to microscopic particles of clay; however, only the microscopic clay particles that form the protective bonds around and between the individual strands of straw ensure stability and longevity. The percentage of clay in the clay slip will determine the effectiveness of the coating and binding of the straw. See Commentary Figure AR103.3.2 for recommended mixture ratios and testing methods.

A ribbon or ball test, as stated in this section, is sufficient to determine suitability for wall densities between 30 and 50 pounds per cubic foot (480 to 801 kg/m³). A lab test known as a "soil texture test" or "physical analysis test" yields an even more accurate analysis, giving precise percentages of clay, sand and silt.

A lab test that measures percentages of sand, silt and clay is strongly recommended for lower-density, higher-insulating light straw-clay mixes. This is because there is much less clay soil added to the mix, so the little clay soil present must have a relatively high percentage of clay to provide the protective bonds that ensure stability and longevity of the wall. Inexpensive soil texture or physical analysis tests are readily available through university cooperative extension and agricultural soil laboratories in most parts of the U.S. and Canada.

AR103.3.3 Clay slip. Clay slip shall be of sufficient viscosity such that a finger dipped in the slip and withdrawn remains coated with an opaque coating.

❖ This requirement is intended to provide a simple field method to verify that enough clay is present in the slip to achieve adequate coverage of the straw and ensure cohesive integrity of the installed light straw-clay mix. This test works for all densities of light straw-clay.

AR103.3.4 Light straw-clay mixture. Light straw-clay shall contain not less than 65 percent and not more than 85 percent straw, by volume of bale-compacted straw to clay soil. Loose straw shall be mixed and coated with clay slip such that there is not more than 5 percent uncoated straw.

❖ The intent of this section is to describe straw and clay percentages for densities in the 30 to 50 pounds per cubic foot range (480 to 801 kg/m³). The percentage of straw described in this section means straw-plus-air in the wall. These percentages are not meant to limit higher percentages of straw-plus-air used in lighter mixes where the subsoil content in the wall is lower and meets the recommended testing in Commentary Figure AR103.3.2. Commentary Figure AR103.3.4 is useful for clarifying the percentages of components used to achieve the range of light straw-clay densities.

AR103.4 Wall construction. Light straw-clay wall construction shall be in accordance with the requirements of Sections AR103.4.1 through AR103.4.7.

AR103.4.1 Light straw-clay maximum thickness. Light straw-clay shall be not more than 12 inches (305 mm) thick, to allow adequate drying of the installed material.

❖ Historically, 12 inches (305 mm) has been considered the maximum wall thickness for straw-clay mixtures 30 pcf and greater. Current research suggests that a higher percentage of clay in the slip allows for lighter

DENSITY (pcf)	MINIMUM % CLAY IN SUBSOIL (1)	MINIMUM CLAY: SILT RATIO	CLAY TESTING METHOD (2)	MAXIMUM LIGHT STRAW-CLAY (thickness, inches)
10	70	3.5:1	A	15
12	46	1.7:1	A	15
13	40	1.33:.1	A	15
15	35	0.95:1	A	15
20	30	0.60:1	A	12
30	25	0.42:1	A or B	12
40	Test Method B	Test Method B	B	12
50	Test Method B	Test Method B	B	12

Courtesy of Douglas Piltingsrud@StrawClay.org. Used by permission.

1. The minimum percentage criteria may be met by using unprocessed raw clay subsoil or subsoil refined by removal of sand and silt.

2. Subsoil Testing Methods:
   a. Lab test for percent of clay, silt and sand via hydrometer method.
   b. Ribbon Test or the Figure 3 Ball Test in the appendix of ASTM E2392/E2392M.

**Figure AR103.3.2
RATIOS AND TESTING OF LIGHT STRAW-CLAY MIXTURES**

density straw-clay mixtures (with higher *R*-values) while still achieving adequate drying in walls up to 15 inches (381 mm) thick. Section AR103.4.1 is intended to ensure safe drying parameters for denser mixes (30 pcf and greater), and is not intended to limit the wall thickness for lighter mixtures where a history of successful drying of walls with lower densities that exceed 12 inches (305 mm) is demonstrated to the building official.

See Commentary Figure AR103.3.2 for the recommended maximum thicknesses for light straw-clay of various densities. This figure also shows the relationships between light straw-clay components and the recommended clay soil testing method for those densities.

AR103.4.2 Distance above grade. Light straw-clay and its exterior finish shall be not less than 8 inches (203 mm) above exterior finished *grade*.

AR103.4.3 Moisture barrier. An *approved* moisture barrier shall separate the bottom of light straw-clay walls from any masonry or concrete foundation or slab that directly supports the walls. Penetrations and joints in the barrier shall be sealed with an *approved* sealant.

❖ Moisture barriers to protect the straw-clay from the rising dampness inherent to masonry or concrete foundations include materials historically used for this purpose such as asphalt-impregnated roofing felt, plastic sheeting, elastomeric gasketing, clay, tar, and liquid-applied products manufactured for use as a moisture barrier over concrete or masonry.

AR103.4.4 Contact with wood members. Light straw-clay shall be permitted to be in contact with untreated wood members.

❖ Light straw-clay installation is similar to the installation of wet-blown cellulose insulation in that both are installed damp, dry in the following weeks, and then

remain dry throughout their service life. In addition, clay has a long history of preserving wood. Clay is hydrophilic (i.e., water-attracting) and is highly water-blocking due to its strong surface bonding with any available water. Also, because of clay's very large total surface area (due to its layered microscopic structure), it has considerable capacity to capture and store water. Even when straw-clay is applied wet and in full contact with wood, the clay functions to pull water away from adjacent materials as the composite dries, including wood and its own straw fibers. In this way, and with its ability to act as a water barrier (e.g., clay is used to line ponds to prevent leakage), the clay keeps the moisture in adjacent materials below the levels necessary to support microbe growth (i.e., rot) and, therefore, acts as a preservative (see commentary, Section AR103.5.3).

AR103.4.5 Contact with nonwood structural members. Nonwood structural members in contact with light straw-clay shall be resistant to corrosion or shall be coated to prevent corrosion with an *approved* coating.

❖ A light straw-clay installation is similar to a wet-blown cellulose installation in that the material is damp only temporarily. This limited wetting is a common construction condition and standard coatings such as primer paint and electro-galvanizing are acceptable to prevent corrosion of nonwood structural members. In this section, the term "nonwood" is intended to mean metal that is susceptible to corrosion with prolonged exposure to moisture.

AR103.4.6 Installation. Light straw-clay shall be installed in accordance with the following:

1. Formwork shall be sufficiently strong to resist bowing where the light straw-clay is compacted into the forms.

2. Light straw-clay shall be uniformly placed into forms and evenly tamped to achieve stable walls free of voids. Light straw-clay shall be placed in lifts of not more

LIGHT STRAW-CLAY DENSITY (pcf)	STRAW (pcf)	SUBSOIL (pcf)	SUBSOIL (1) VOLUME %	STRAW (2) VOLUME %	AIR VOLUME %	STRAW + AIR VOLUME %	*R*-VALUE (hr/F°/cu.ft./ BTU/inch)
10	6.7	3.3	2.0	7.4	90.6	98.0	1.80
12	6.7	5.3	3.1	7.4	89.4	96.9	1.72
13	6.7	6.3	3.7	7.4	88.9	96.3	1.69
15	6.7	8.3	4.9	7.4	87.7	95.1	1.63
20	6.7	13.3	7.9	7.4	84.7	92.1	per AR104.1
30	6.7	23.3	13.8	7.4	78.7	86.2	per AR104.1
40	6.7	33.3	19.9	7.4	72.8	80.2	per AR104.1
50	6.7	43.3	25.7	7.4	66.9	74.3	per AR104.1

Courtesy of Douglas Piltingsrud@StrawClay.org. Used by permission.

1. Uses 168 pcf for subsoil specific (2.7 g/cc).

2. Uses 90 pcf for straw gravity (1.45 g/cc). Straw Volume % and associated columns may increase for 20 pcf walls and above due to the weight from additional subsoil.

Figure AR103.3.4
CHARACTERISTICS OF LIGHT STRAW-CLAY AT VARYING DENSITIES

than 6 inches (152 mm) and shall be thoroughly tamped before additional material is added.

3. Formwork shall be removed from walls within 24 hours after tamping, and walls shall remain exposed until moisture content is in accordance with Section AR103.5.1. Visible voids shall be patched with light straw-clay prior to plastering.

❖ **Item 1:** Commonly used formwork for light straw-clay installations includes plywood, OSB panels and solid wood boards; often in practice these materials are then repurposed as subflooring and roof sheathing. In contrast to poured concrete, which is liquid and extremely heavy, light straw-clay is semi-solid, lightweight and is placed using only a person's body weight to tamp it into place. Accordingly, the formwork for light straw-clay is typically constructed much more lightweight than concrete formwork. For example, it often uses only 2 × 4s for whalers and is attached temporarily with screws to the permanent wall framing as part of the formwork's structure and bracing.

Typically, formwork is first installed completely on one side of the wall, then incrementally on the other side from the bottom and then upward as the straw-clay is also incrementally placed. This allows a person to tamp the material standing partially inside the formwork, and allows for possible reuse of the lower forms ("slip forming") once the material in the form above is tamped in place (see commentary, Section AR103.4.6, Items 2 and 3).

Item 2: This item is intended to ensure the mix is compacted using light tamping only. It is also intended to achieve the desired density, consolidate the mix, ensure adhesion between stalks of straw and achieve wall integrity without crushing the straw stalks. Intact straw stalks and the air spaces inside and between them ensure that the maximum wall R-value is achieved for that density of mix. For this reason, compaction is not normally accomplished by mechanical means. Rather, it is done by a person standing inside the formwork and "walking in the forms" as the straw-clay mix is inserted. The installer ensures complete distribution of material to all corners, making sure that no voids are left, and uses his or her body weight to lightly compress the straw-clay to the desired density.

Item 3: This item is intended to allow the light straw-clay to begin drying as soon as possible after placement and to avoid the possibility of mold growth on the surface of the straw-clay beneath the forms. Forms can generally be removed as soon as the straw has "set" (i.e., moist stalks have relaxed into place). Optimum time is generally 1 to 4 hours after placement.

AR103.4.7 Openings in walls. Openings in walls shall be in accordance with the following:

1. Rough framing for doors and windows shall be fastened to structural members in accordance with the *International Residential Code*. Windows and doors shall be flashed in accordance with the *International Residential Code*.

2. An *approved* moisture barrier shall be installed at window sills in light straw-clay walls prior to installation of windows.

AR103.5 Wall finishes. The interior and exterior surfaces of light straw-clay walls shall be protected with a finish in accordance with Sections AR103.5.1 through AR103.5.5.

AR103.5.1 Moisture content of light straw-clay prior to application of finish. Light straw-clay walls shall be dry to a moisture content of not more than 20 percent at a depth of 4 inches (102 mm), as measured from each side of the wall, prior to the application of finish on either side of the wall. Moisture content shall be measured with a moisture meter equipped with a probe that is designed for use with baled straw or hay.

❖ This provision is intended to require sufficient drying of the straw-clay in order to achieve dimensional stability and a stable surface for a noncracking plaster application; however, a "full-dry" condition (in full equilibrium with ambient conditions) is not needed for successful plastering because plasters as required in Section AR103.5.2 are highly permeable and allow the straw-clay to continue drying after plastering.

It is not always possible to obtain consistent moisture meter readings if only a few locations are sampled. Sampling multiple locations yields more reliable information about the overall dryness, particularly as the less-dense mixes offer less surface contact area for the probes of electrical moisture meters. As an additional measure, vertical wall shrinkage should be observed as the material dries and plastering should begin only when shrinkage has stopped and any voids are filled. Depending on wind, temperature and humidity conditions, the time required for straw-clay to reach a plaster-ready state may vary from a few weeks to several months.

AR103.5.2 Plaster finish. Exterior plaster finishes shall be clay plaster or lime plaster. Interior plaster finishes shall be clay plaster, lime plaster or gypsum plaster. Plasters shall be permitted to be applied directly to the surface of the light straw-clay walls without reinforcement, except that the juncture of dissimilar substrates shall be in accordance with Section AR103.5.4. Plasters shall have a thickness of not less than $^1/_2$ inch (12.7 mm) and not more than 1 inch (25 mm) and shall be installed in not less than two coats. Exterior clay plaster shall be finished with a lime-based or silicate-mineral coating.

❖ The purpose of a lime-based or silicate-mineral coating over exterior clay plaster is to improve the plaster's resistance to erosion from weather. This section does not require exterior plaster, but rather describes its characteristics when used.

Some admixtures in clay plasters have a history of imparting desirable plaster characteristics, such as enhanced moisture resistance, durability and crack resistance. These admixtures include wheat paste, manure, casein and prickly pear juice. It is not the intent of this section to exclude the use of these or other admixtures that can demonstrate erosion resistance from weather equivalent to lime-based or silicate-mineral coatings.

The centuries-old European predecessors and the light straw-clay buildings built to date in North America have all been constructed without the use of a water-resistive barrier. The light straw-clay materials in this appendix are highly vapor permeable, which allows passage of water vapor without harm to the wall. Code precedents exist for construction systems such as adobe construction and log construction, which do not require the use of a water-resistive barrier. In these systems, as in light straw-clay construction, there is sufficient moisture buffer capacity to hold and re-release moisture without damage to structural members or degradation of the wall due to weather-related moisture. For these reasons, a water-resistive barrier is not required in light straw-clay construction.

AR103.5.3 Separation of wood and plaster. Where wood framing occurs in light straw-clay walls, such wood surfaces shall be separated from exterior plaster with No.15 asphalt felt, Grade D paper or other approved material except where the wood is preservative treated or naturally durable.

> **Exception:** Exterior clay plasters shall not be required to be separated from wood.

❖ Clay is hydrophilic (i.e., water-attracting, drawing moisture to it and away from adjacent materials) and is highly water-blocking due to its strong surface bonding with any available water. In addition, because of clay's very large total surface area (due to its layered microscopic structure), it has considerable capacity to capture and store water. Historical examples include the centuries-old half-timbered buildings still in use in Germany, with straw-clay infill and clay plasters in direct contact with the original wood frame structure. This combination of hydrophilic properties, water-blocking properties and water storage capacity protects adjacent materials that might otherwise be harmed by extended exposure to moisture. Because of this, clay plasters are not required to be separated from wood (see also commentary, Section AR103.4.4).

AR103.5.4 Bridging across dissimilar substrates. Bridging shall be installed across dissimilar substrates prior to the application of plaster. Acceptable bridging materials include: expanded metal lath, woven wire mesh, welded wire mesh, fiberglass mesh, reed matting or burlap. Bridging shall extend not less than 4 inches (102 mm), on both sides of the juncture.

❖ This section is intended to prevent cracking of the plaster where it spans across wall framing members or substrates that interrupt the face of the installed light straw-clay. The bridging materials provide greater tensile and shear strength for plasters in areas that are prone to differential movement and provide a material to which the plasters can mechanically bond across substrates that otherwise offer little or no means for bonding.

AR103.5.5 Exterior siding. Exterior wood, metal or composite material siding shall be spaced not less than $3/_4$ inch (19.1 mm) from the light straw-clay such that a ventilation space is created to allow for moisture diffusion. The siding shall be fastened to wood furring strips in accordance with the manufacturer's instructions. Furring strips shall be spaced not more than 32 inches (813 mm) on center, and shall be securely fastened to the vertical wall reinforcing or structural framing. Insect screening shall be provided at the top and bottom of the ventilation space. An air barrier consisting of not more than $3/_8$-inch-thick (9.5 mm) clay plaster or lime plaster shall be applied to the light straw-clay prior to the application of siding.

❖ Plaster installed in a continuous layer over the interior or exterior light straw-clay surface meets the intent of this section, which is to provide the required air barrier.

SECTION AR104
THERMAL INSULATION

AR104.1 R-value. Light straw-clay, where installed in accordance with this appendix, shall be deemed to have an *R*-value of 1.6 per inch.

❖ Current research indicates that *R*-value in light straw-clay varies predictably with density. Higher percentages of clay in the slip and improved quality control of the light straw-clay mix allows for lighter density straw-clay mixtures with higher *R*-values. These lighter mixtures are particularly well suited to colder climate regions. This appendix does not intend to limit the use of these lighter and more insulating mixtures. Commentary Figure AR103.3.4 summarizes the *R*-value findings for densities from 10 to 15 pounds per cubic foot from the 2004 USDA Forest Products Laboratory *Engineering Report of Light Clay Specimens* (see the bibliography) and may be used for calculating thermal performance for walls using these lighter density mixes. The *R*-value stated in Section AR104.1 and those in Commentary Figure AR103.3.4 are for the light straw-clay material only, and exclude any additional thermal resistance provided by plaster or other finish over the light straw-clay.

SECTION AR105
REFERENCED STANDARD

ASTM E2392/
E2392M—10   Standard Guide for Design of Earthen Wall
              Building Systems AR103.3.2

Bibliography

The following resource materials were used in the preparation of the commentary for this appendix of the code:

Baker-Laporte, Paula, and Robert Laporte. *Econest, Creating Sustainable Sanctuaries of Clay, Straw and Timber*. Layton, UT: Gibbs Smith, 2005.

Duncan, Richard, PE. *Resistance to Out-Of-Plane-Lateral Forces of Light Straw Clay Wall Infill.*http://www.econesthomes.com/wp-content/uploads/2013/01/Light-Straw-Clay-Out-of-Plane-Study-FINAL_merged.pdf, 2013.

Host-Jablonski, L. *Typical Outline Specifications for a Northern Light Straw-Clay House.* In *The Affordable Natural House Contractor Training Reference Manual,* 2nd Edition. S. Thering ed. Madison, WI: University of Wisconsin–Extension, 2008.

Piltingsrud, D., and L. Host-Jablonski. *An Introduction to the Science of Northern Light Straw-Clay Construction.* In *The Affordable Natural House Contractor Training Reference Manual,* 2nd Edition. S. Thering, ed. Madison, WI: University of Wisconsin–Extension, 2008.

State of New Mexico Construction Industries Division Clay Straw Guidelines. Santa Fe, NM: State of NM CID publication, 2001.

StrawClay.org. Madison, Wisconsin: Design Coalition, Inc., 2014.

Thornton, J. *Initial Material Characterization of Straw Light Clay.* Ottawa, Ontario: Canada Mortgage and Housing Corporation, 2004.

USDA Forest Products Laboratory *Engineering Report of Light Clay Specimens: Thermal Conductivities for Design Coalition's Straw/Clay Formulations Extend Volhard's K-Value vs. Density Curve in Low Conductivity End.* Madison, WI: USDA Forest Products Laboratory, 2004.

2011 Oregon Reach Code—Section 1307. Country Club Hills, IL: International Code Council, Inc., 2012.

Appendix S:
Strawbale Construction

(The provisions contained in this appendix are not mandatory unless specifically referenced in the adopting ordinance.)

General Comments

Strawbale construction, a wall system using stacked bales of straw that are covered with plaster or other finish, originated in Nebraska in the late 1800s, as made possible by the invention of the baling machine. It was practiced regionally into the 1940s, and buildings from that first era, some over 100 years old, are still in service. After decades of nonuse, strawbale construction was rediscovered in the 1980s and utilized again in the American southwest. Since then it has been further developed and explored, subjected to considerable testing and researched regarding structural performance (under vertical loads, and lateral wind and seismic loads), moisture issues, fire resistance and thermal and acoustic properties.

Since the 1980s, the use of strawbale construction has steadily increased and there are now strawbale buildings in all 50 U.S. states, as well as in more than 50 countries throughout the world. It is estimated that there are over 1,000 strawbale buildings in California alone. Strawbale construction has been used primarily in the construction of residences, but it has also been used for schools, office buildings, wineries, retail buildings, a municipal police station and a federal post office. Most strawbale buildings have been one story, though many two- or three-story buildings have also been constructed and as many as eight-stories for a nonload-bearing strawbale apartment building in France. Over both its early and modern history, strawbale construction has proven, through testing and in practice, to be a safe, durable, resource- and energy-efficient and fully viable wall system (see Commentary Figure AS101 for typical load-bearing and typical post-and-beam strawbale wall systems with their basic components).

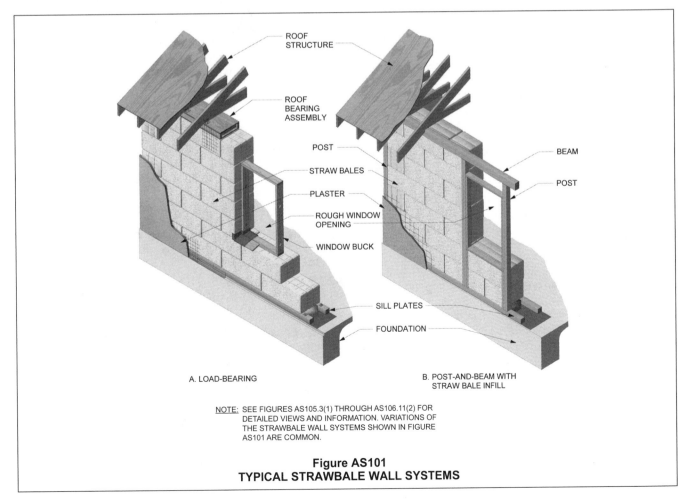

NOTE: SEE FIGURES AS105.3(1) THROUGH AS106.11(2) FOR DETAILED VIEWS AND INFORMATION. VARIATIONS OF THE STRAWBALE WALL SYSTEMS SHOWN IN FIGURE AS101 ARE COMMON.

Figure AS101
TYPICAL STRAWBALE WALL SYSTEMS

Purpose

As of 2014, only New Mexico, Oregon and North Carolina had adopted statewide strawbale building codes. California has legislated strawbale construction guidelines for voluntary adoption by local jurisdictions. In addition, nine U.S. cities or counties have adopted strawbale building codes. Three countries outside of the United States—Germany, France, and Belarus—have limited strawbale building codes. New Zealand has official guidelines for strawbale construction. In jurisdictions in the United States without a strawbale building code, strawbale buildings have been permitted on a case-by-case basis, often with little reliable guidance for building officials, builders and owners. This has been an impediment to strawbale construction's proper and broader use.

Most strawbale building codes that exist in the U.S. are derived from the first load-bearing strawbale code created for and adopted in 1996 by Tucson and Pima County, Arizona. One jurisdiction based its code on the State of New Mexico's nonload-bearing code, also adopted in 1996. Field experience, testing and research since 1996 have shown these codes to be deficient. They are often either too restrictive or not restrictive enough, and in some cases do not address important issues at all. The purpose of this appendix is to bring the practice of strawbale construction into alignment with current understanding and to unify disparate strawbale building codes, while providing flexibility to allow time-tested local and regional variations, as well as viable new variations as strawbale construction evolves.

SECTION AS101
GENERAL

AS101.1 Scope. This appendix provides prescriptive and performance-based requirements for the use of baled straw as a building material. Other methods of strawbale construction shall be subject to approval in accordance with Section R104.11 of this code. Buildings using strawbale walls shall comply with the this code except as otherwise stated in this appendix.

❖ Historically, many variations of strawbale construction have been practiced, influenced by climate, level of high wind or seismic risk, available materials, local building practices and regional architecture. Still more variations are possible. The intent of this appendix, through both prescriptive and performance-based requirements, is to be inclusive of as many safe and durable methods of strawbale construction as possible. (See Commentary Figure AS101 for typical load-bearing and post-and-beam strawbale walls with their basic components.)

Variations of strawbale construction that are not explicitly allowed by this appendix may or may not be viable and acceptable. Such methods should be evaluated by the local building official in accordance with Section R104.11 of the code.

All components and aspects of strawbale buildings other than their strawbale walls, including foundations, nonstrawbale walls, roof structure, energy efficiency and mechanical, plumbing and electrical systems, must comply with the code, unless this appendix states otherwise. See the commentary to Section AS105.6 for recommendations regarding plumbing in strawbale walls as related to moisture control.

Although there are no provisions in this appendix related to electrical wiring systems in strawbale walls, these installations must address the same code criteria as for other wall systems in the code, such as protection of wiring from damage after construction (e.g., separation of wiring from the surface of finishes to protect from inadvertent nailing), secure attachment of wiring or conduit systems and boxes and air sealing of electrical boxes and penetrations.

Decades of electrical installations in strawbale walls have yielded common practices, some of which have been codified in local or state strawbale building codes. Nonmetallic sheathed cable has been allowed with and without conduit, with frequency of attachment complying with the electrical code, to either wood framing members or to the bales using long wire "staples". Electrical boxes have been secured to either wood framing members or to 12-inch (305 mm) wooden stakes driven into the bales. Plaster mesh has also been used as a means of securing electrical boxes in strawbale walls. All wiring unprotected by conduit and armored cable has generally been required to be set back from the face of the strawbales at least $1^1/_2$ inches (38 mm), except where entering an electrical box.

SECTION AS102
DEFINITIONS

AS102.1 Definitions. The following words and terms shall, for the purposes of this appendix, have the meanings shown herein. Refer to Chapter 2 of the *International Residential Code* for general definitions.

BALE. Equivalent to straw bale.

❖ Many agricultural and nonagricultural materials are baled. However, in this appendix, the term bale is used to mean straw bale.

CLAY. Inorganic soil with particle sizes less than 0.00008 inch (0.002 mm) having the characteristics of high to very high dry strength and medium to high plasticity.

❖ Clay has been used for thousands of years as a building material. This includes unfired clay in adobe bricks, rammed earth walls, cob walls, earthen plasters and earthen floors, and fired clay in bricks, roofing tiles and floor and wall tiles. In all of these materials, clay is the binder, sometimes along with another binder such as lime or cement, that holds together other materials such as sand or straw. In this appendix, clay appears as a component of clay slip, straw-clay, clay plaster, and soil-cement plaster.

Clay for these purposes is obtained from inorganic subsoil with sufficient clay content, or can be obtained as a commercially bagged and quarried material. Clay subsoil typically contains particles of sand and silt in addition to clay. Two ways of determining the suitability of clay subsoil, for use in the materials in this appendix that contain clay, are the Figure 2 Ribbon Test and the Figure 3 Ball Test of the appendix to ASTM E2392/E2392M, *Standard Guide for Design of Earthen Wall Building Systems.*

CLAY SLIP. A suspension of clay particles in water.

❖ Clay slip is used to make straw-clay. It is also sometimes used to coat the bottom of the first course of bales to protect the straw against potential moisture intrusion. For walls that will receive a clay plaster finish, clay slip is sometimes applied to the face of a bale wall during construction to provide temporary protection against light rain, or immediately before application of the first coat of clay plaster for improved bonding to the straw. A simple way to determine if clay slip contains sufficient clay and is of sufficient viscosity for these purposes is to dip a finger in the slip. If upon withdrawal the finger remains covered with an opaque coating, the slip contains sufficient clay.

FINISH. Completed compilation of materials on the interior or exterior faces of stacked *bales.*

❖ See Section AS104 for acceptable finishes on straw-bale walls.

FLAKE. An intact section of compressed *straw* removed from an untied *bale.*

❖ Industrialized baling machines push clumps of straw taken from the straw collection chamber tight toward the end of the bale chamber until the set bale length is achieved. The bale is then tied by the machine and the process is repeated.

In this process, the clumps of straw are flattened into 3- to 4-inch-thick (76 to 102 mm) mats known as flakes. When a bale is untied and pulled apart, the separation tends to occur between flakes, and the flakes tend to remain compressed and intact. Flakes of straw are often used in strawbale construction to fill voids between bales, or between bales and framing members in strawbale walls.

LAID FLAT. The orientation of a bale with its largest faces horizontal, its longest dimension parallel with the wall plane, its *ties* concealed in the unfinished wall and its *straw* lengths oriented across the thickness of the wall.

❖ See Commentary Figure AS102.1.

LOAD-BEARING WALL. A strawbale wall that supports more than 100 pounds per linear foot (1459 N/m) of vertical load in addition its own weight.

❖ The definition of this term is consistent with the definition of "Wall, load-bearing" for stud walls in the *International Building Code® (IBC®).*

MESH. An openwork fabric of linked strands of metal, plastic, or natural or synthetic fiber, embedded in plaster.

NONSTRUCTURAL WALL. Walls other than load-bearing walls or shear walls.

❖ The definition of this term is consistent with the definition in ASCE 7. ASCE 7 is one of the standards used as the basis for structural loads on buildings in the code and in the IBC.

In nonstructural strawbale walls, the straw bales are infill only and serve as enclosure, insulation and as a substrate for plaster. The straw bales and their finish carry no superimposed vertical or in-plane lateral loads, and the bales can be in any orientation including laid flat, on-edge, or on-end (see Commentary Figure AS102.1). However, other elements in the same wall, such as wood or steel framing, or structural panels, may also carry such structural loads.

ON-EDGE. The orientation of a *bale* with its largest faces vertical, its longest dimension parallel with the wall plane, its *ties* on the face of the wall and its *straw* lengths oriented vertically.

❖ See Commentary Figure AS102.1.

PIN. A vertical metal rod, wood dowel or bamboo, driven into the center of stacked bales, or placed on opposite surfaces of stacked bales and through-tied.

❖ See the commentary to Section AS105.4.2.

PLASTER. Gypsum or cement plaster, as defined in Sections R702 and AS104, or clay plaster, soil-cement plaster, lime plaster or cement-lime plaster as defined in Section AS104.

PRECOMPRESSION. Vertical compression of stacked bales before the application of finish.

❖ See the commentary to Section AS106.12.1.

REINFORCED PLASTER. A plaster containing mesh reinforcement.

RUNNING BOND. The placement of *straw bales* such that the head joints in successive courses are offset not less than one-quarter the bale length.

❖ The definition of this term is consistent with the definition in the body of the code, except that straw bales are used in place of masonry units.

SHEAR WALL. A strawbale wall designed and constructed to resist lateral seismic and wind forces parallel to the plane of the wall in accordance with Section AS106.13.

❖ The term "shear wall" is used interchangeably with the term "braced wall panel" in this appendix.

SKIN. The compilation of plaster and reinforcing, if any, applied to the surface of stacked bales.

STRUCTURAL WALL. A wall that meets the definition for a load-bearing wall or shear wall.

❖ The definition of this term is consistent with the definition in ASCE 7. ASCE 7 is one of the standards used as the basis for structural loads on buildings in the code and in the IBC.

STACK BOND. The placement of straw bales such that head joints in successive courses are vertically aligned.

❖ The definition of this term is consistent with its definition in the code, except that straw bales are used in place of masonry units.

STRAW. The dry stems of cereal grains after the seed heads have been removed.

❖ See the commentary to Section AS103.3.7.

STRAW BALE. A rectangular compressed block of straw, bound by ties.

STRAWBALE. The adjective form of straw bale.

❖ Strawbale construction has historically been referred to as "straw bale", "straw-bale", or "strawbale" construction. Applying accepted English grammar, when used as an adjective, "straw-bale" is correct. However, the English language contains many compound adjectives that originated as two words, but became one word because of frequency and simplicity use. For these reasons the compound adjective "strawbale" is used in this appendix.

STRAW-CLAY. Loose straw mixed and coated with clay slip.

❖ Straw-clay is a material in the context of this appendix and should not be confused with the term light straw-clay construction, which is a nonload-bearing wall system described in Appendix R.

TIE. A synthetic fiber, natural fiber or metal wire used to confine a straw bale.

❖ See the commentary to Section AS103.3.

TRUTH WINDOW. An area of a strawbale wall left without its finish, to allow view of the straw otherwise concealed by its finish.

SECTION AS103
BALES

AS103.1 Shape. Bales shall be rectangular in shape.

❖ Straw bales as referred to in this appendix are rectangular in shape. Rectangular bales created by common industrialized baling machines fall into one of three size categories: "two-string," "three-string," and "jumbo." See the commentary to Sections AS103.2 and AS103.3 for information regarding bale dimensions in these categories. Round (cylindrical) bales have also been used for strawbale building, but are not covered in this appendix.

AS103.2 Size. Bales shall have a height and thickness of not less than 12 inches (305 mm), except as otherwise permitted or required in this appendix. Bales used within a continuous wall shall be of consistent height and thickness to ensure even distribution of loads within the wall system.

❖ This section states the minimum height and thickness (width) for straw bales used in strawbale construction, except that a larger minimum thickness is stated in Section AS106.13.1 for strawbale braced panels. There is no length requirement.

The most commonly used straw bales for building are known as "two-string" and "three-string" bales, named for the number of ties that hold each bale together (see Commentary Figure AS103.2). Their width and height dimensions are determined by the compression chamber of the baler. These two types of bales typically have dimensions as shown in Commentary Figure AS103.2. Of the three dimensions of a bale, the length can vary the most. Two-string and three-string bales are usually small and light enough to be handled by most able-bodied persons.

"Jumbo" bales have also been used for building. They are typically 3 or 4 feet × 4 feet × 8 feet (910 or

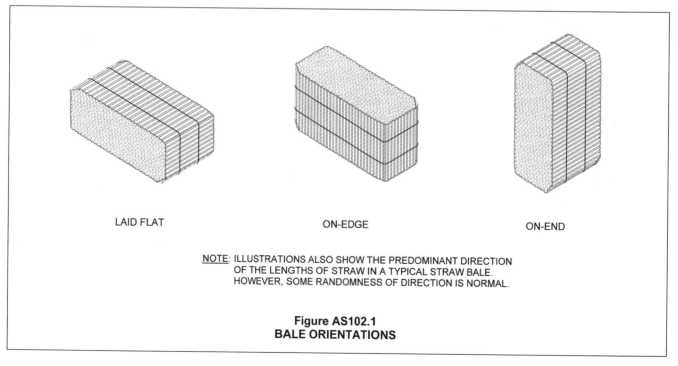

LAID FLAT   ON-EDGE   ON-END

NOTE: ILLUSTRATIONS ALSO SHOW THE PREDOMINANT DIRECTION OF THE LENGTHS OF STRAW IN A TYPICAL STRAW BALE. HOWEVER, SOME RANDOMNESS OF DIRECTION IS NORMAL.

Figure AS102.1
BALE ORIENTATIONS

1220 mm × 1220 mm × 2440 mm), and can only be moved by mechanized equipment. Bales as small as 1 foot × 1 foot × 2 feet (305 mm × 305 mm × 610 mm) have been successfully used in strawbale construction, including for structural walls in one-story buildings.

This section also requires bales in a continuous wall to be of consistent height and thickness (again, no requirement for length). Like unit masonry, it is advantageous for the length of rectangular bales to be approximately twice their width. Thus, when stacking bales in a running bond, corners can be efficiently turned and half-bales efficiently used.

Although straw bales in industrialized countries are almost universally made with industrialized baling machines, manually compressed bales have been used in strawbale construction and are acceptable if the bales meet the requirements of Section AS103.

AS103.3 Ties. Bales shall be confined by synthetic fiber, natural fiber or metal ties sufficient to maintain required bale density. Ties shall be not less than 3 inches (76 mm) and not more than 6 inches (152 mm) from the two faces without ties and shall be spaced not more than 12 inches (305 mm) apart. Bales with broken ties shall be retied with sufficient tension to maintain required bale density.

❖ In strawbale construction, the ties that confine a bale serve two major functions: (1) To enable the bale to be easily moved and handled during construction, and (2) To keep the bale at its required density. If ties are removed (such as when making half-bales) or broken before installation, the bales must be retied to the density required by Section AS103.5.

Straw bales in industrialized agriculture are now tied almost exclusively with polypropylene ties, though steel wire ties are sometimes found. Any synthetic fiber, natural fiber (e.g., sisal or hemp) or metal wire ties that can maintain the required bale density is acceptable. The most commonly used bales for building are "two-string" and "three-string" bales, named for the number of ties used (see Commentary Figure AS103.2).

Some practitioners of strawbale construction intentionally cut ties on bales laid flat in nonstructural walls before plastering after courses of bales are sufficiently confined at their ends. The purpose is to allow the bales to expand along their length to fill gaps between bales or between bales and vertical framing, creating a more consistent insulating core and lathing surface. When ties are cut, a bale tends to expand only lengthwise and change in the other two dimensions is negligible, due to the nature of how baling machines create bales (see the commentary to the definition of "Flake").

AS103.4 Moisture content. The moisture content of bales at the time of application of the first coat of plaster or the installation of another finish shall not exceed 20 percent of the weight of the bale. The moisture content of bales shall be determined by use of a moisture meter designed for use with baled straw or hay, equipped with a probe of sufficient length to reach the center of the bale. Not less than 5 percent and not less than 10 bales used shall be randomly selected and tested.

❖ Field experience and laboratory testing have shown that as long as the moisture content of straw does not exceed 20 percent for prolonged periods of time, the straw will not degrade. Some studies have shown that degradation does not occur until 25 percent moisture content is reached. 20 percent moisture content is widely considered a safe maximum for straw in strawbale construction. However, practitioners of strawbale construction prefer a working range of 7-15 percent moisture content, which allows a margin for modest future wetting without degradation.

It is required that the bales have a moisture content of less than 20 percent at the time the walls are plas-

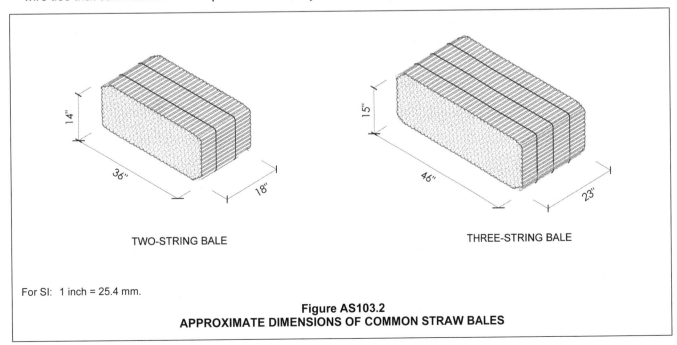

For SI: 1 inch = 25.4 mm.

Figure AS103.2
APPROXIMATE DIMENSIONS OF COMMON STRAW BALES

tered or receive another finish, but should also be below 20 percent at all stages. These stages include: at time of baling, in storage, in transport, at the job site and in service. Knowing or controlling the moisture content of bales before procurement or arrival on site can be difficult or impossible. Therefore, at procurement, bales that measure above 20 percent moisture content or that show visible signs of previous wetting, such as dark staining from microbial growth, should not be used for construction (see commentary, Section AS105.6).

The required method of determining whether bales are acceptable for construction is to test the quantity stated in this section with a moisture meter designed for use with baled straw or hay (see Commentary Figure AS103.4 for pictures of example moisture meters). The building official may specify how this is demonstrated. The bales tested must not exceed 20 percent moisture content, and their centers should be checked. If any of the bales tested show a moisture content higher than 20 percent, such bales should not be used for construction, and the building official may then specify how many more bales must be tested before accepting the entire lot of bales for construction.

Once the lot of bales is demonstrated to be acceptable, if bales with moisture content above 20 percent are subsequently discovered, such bales should not be used for construction.

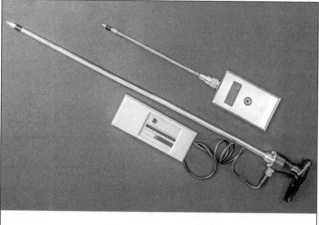

Figure AS103.4
EXAMPLE MOISTURE METERS FOR STRAW BALES

AS103.5 Density. Bales shall have a dry density of not less than 6.5 pounds per cubic foot (104 kg/cubic meter). The dry density shall be calculated by subtracting the weight of the moisture in pounds (kg) from the actual bale weight and dividing by the volume of the bale in cubic feet (cubic meters). Not less than 2 percent and not less than five bales to be used shall be randomly selected and tested on site.

❖ The required method of determining whether bales are of sufficient density for construction is to check the quantity stated in this section. The building official may specify how this is demonstrated.

The dry density is determined as follows: 1. Weigh the bale; 2. Find its percent moisture with a moisture meter designed for use with baled straw or hay; and 3. Calculate the volume of the bale in cubic feet by multiplying its length x width x height in feet.

Enter those numbers in the following equation, where A = total weight (lbs), B = percent moisture content, C = Volume (cu.ft.) and D = Dry Density (lbs/cu.ft):

$$[A - (B \times A)] \div C = D$$

For example: A two-string bale measuring 14 inches × 18 inches × 36 inches is determined to have a 12 percent moisture content and weighs 47 lbs. Its volume is 1.17 feet × 1.5 feet x 3 feet = 5.26 cu.ft. Therefore A = 47 lbs, B = 12, and C = 5.25 cu.ft. Substituting the numbers in the equation: [47 - (.12 × 47)]/5.26] = 7.86 lb per cu.ft. dry density.

Field experience and testing have shown the density required in this section to be a minimum threshold for proper performance of strawbale walls in terms of insulation and structure and as a substrate for plaster. Most bales made by industrialized balers significantly exceed this density. See the commentary to Section AS103.3 for information regarding the practice of intentionally cutting the ties of installed bales.

AS103.6 Partial bales. Partial bales made after original fabrication shall be retied with ties complying with Section AS103.3.

❖ Partial or custom-size bales, most commonly half bales, are typically made on site when sizes other than a full-size bale are required. Also see the commentary to Section AS103.3.

AS103.7 Types of straw. Bales shall be composed of straw from wheat, rice, rye, barley or oat.

❖ Bales made of straw from the five named cereal grains have been successfully used in modern strawbale buildings on six continents since 1990. Straw is an agricultural byproduct baled after nutrient grains have been harvested and is commonly used for livestock bedding and erosion control.

Baled grasses such as hay or alfalfa must not be used. They are cultivated and baled as livestock feed, are baled green, contain nutrients that would support active decomposition by micro-organisms and are unacceptable for building.

AS103.8 Other baled material. The dry stems of other cereal grains shall be acceptable where approved by the building official.

❖ In addition to straw from the five cereal grains named in Section AS103.7, dry stalks from other plants such as flax have been successfully used for bales in strawbale buildings in North America. The building official can accept other baled dry plant material for construction where such material is demonstrated to be equivalent in performance to the types of straw allowed in Section AS103.7. Also see the commentary to Section AS103.7 for information regarding baled plants such as hay and alfalfa that are not acceptable for use in buildings.

SECTION AS104
FINISHES

AS104.1 General. Finishes applied to strawbale walls shall be any type permitted by this code, and shall comply with this section and with Chapters 3 and 7 of this code unless stated otherwise in this section.

❖ Historically, plaster has been the most common finish on strawbale walls. However, other cladding materials permitted by the code have also been used and are not prohibited by this appendix.

AS104.2 Purpose, and where required. Strawbale walls shall be finished so as to provide mechanical protection, fire resistance and protection from weather and to restrict the passage of air through the bales, in accordance with this appendix and this code. Vertical strawbale wall surfaces shall receive a coat of plaster not less than $^3/_8$ inch (10 mm) thick, or greater where required elsewhere in this appendix, or shall fit tightly against a solid wall panel. The tops of strawbale walls shall receive a coat of plaster not less than $^3/_8$ inch (10 mm) thick where straw would otherwise be exposed.

Exception: Truth windows shall be permitted where a fire-resistance rating is not required. Weather-exposed truth windows shall be fitted with a weather-tight cover. Interior truth windows in Climate Zones 5, 6, 7, 8 and Marine 4 shall be fitted with an air-tight cover.

❖ The requirement that vertical surfaces of strawbale walls receive a coat of plaster or fit tightly against a wall panel or other abutting material is intended to restrict air movement. Restriction of air movement across the surface of the bales, especially vertical movement, is important for inhibiting the potential spread of fire and for minimizing convective heat loss (see commentary, Section AS107). Restriction of air movement through the bales reduces the potential for convective heat transfer and migration of moist air through the wall assembly.

The requirement that exposed tops of strawbale walls receive a coat of plaster (where not covered tightly by materials such as wood framing, plywood, or gypsum board) is intended to protect the straw against fire and to inhibit air and moisture movement into the wall. Clay plaster has proven to be especially effective in managing moisture at the top of strawbale walls.

The exception to Section AS104.2 allows truth windows where a fire-resistance rating is not required. See the definition of "Truth window." The purpose of the air-tight cover requirement for interior truth windows in the indicated climate zones is to prevent condensation in the wall from warm, moist interior air bypassing the air barrier of the interior finish. Weather-tight and air-tight covers can be operable and can include a transparent material, such as glass or acrylic, to maintain the truth window's purpose, which is to display the straw in the wall. In the cover's closed position it must be weather-tight, or air-tight, as required. Also see Section AS105.6.2 and its commentary.

AS104.3 Vapor retarders. Class I and II vapor retarders shall not be used on a strawbale wall, nor shall any other material be used that has a vapor permeance rating of less than 3 perms, except as permitted or required elsewhere in this appendix.

❖ High vapor permeability of both interior and exterior finishes is desirable for strawbale walls to allow moisture migration through the wall and dispersion of any moisture that enters the wall. For this reason, Class I and Class II vapor retarders (≤1 perm, and ≤1 perm, and thus considered vapor barriers) are not allowed on strawbale walls, except in extreme situations as required in this appendix, where the importance of keeping water vapor from entering the wall exceeds the importance of enabling moisture to exit the wall.

There are two sections in this appendix where such a situation is defined. In Section AS105.6.2, a Class I or Class II vapor retarder is required on strawbale walls enclosing showers or steam rooms. In Section AS105.6.5, a Class II vapor retarder is required to separate straw bales from adjacent concrete or masonry.

The minimum specified permeance rating is intended to be the "wet bulb" perm rating of the material because it is under wet conditions that it is most important for the materials to be vapor permeable. Some materials have identical wet bulb and dry bulb perm ratings, whereas others, such as plywood, have substantially different wet bulb and dry bulb perm ratings. Unless identical, the dry bulb perm rating of a material is always lower than the wet bulb rating, and can be used to comply with this section if the wet bulb rating is unknown because the dry bulb rating will be conservative.

AS104.4 Plaster. Plaster applied to bales shall be any type described in this section, and as required or limited in this appendix. Plaster thickness shall not exceed 2 inches (51 mm).

❖ Plaster has been the most common finish on strawbale walls and as such is given particular attention in this appendix. This appendix contains provisions for two plaster types that also appear in the body of the code —gypsum plaster and cement plaster—and contains provisions for four plaster types that do not appear in the body of the code: clay plaster, soil-cement plaster, lime plaster and cement-lime plaster. Plasters used on structural walls must comply with Section AS106.6.

All of the plaster types contained in this appendix have been successfully used in strawbale construction. Each has comparative advantages and disadvantages in terms of compressive strength, durability, vapor permeability, moisture management, availability, cost, ease of application, acoustic properties, fire resistance and aesthetic qualities. The choice of plaster should be carefully considered based on the context and requirements of the building project. Such considerations are typically influenced by climate and the required or desired structural performance of the strawbale walls. Similar consideration should be given when choosing between plaster and other finish materials or assemblies for strawbale walls.

The reason for limiting plaster thickness to 2 inches (51 mm) is that the additional weight of thicker plaster has potential structural consequences, especially in areas of high seismic risk. Thicker plaster may also unacceptably reduce vapor permeability depending on the plaster used; however, there are also potential benefits of thicker plaster, including improved thermal performance in some climates due to the increased mass. The building official may allow plaster thicker than 2 inches (51 mm) if deemed to present no significant additional risk. Also see the commentary to Section AS106.2.

AS104.4.1 Plaster and membranes. Plaster shall be applied directly to strawbale walls to facilitate transpiration of moisture from the bales, and to secure a mechanical bond between the skin and the bales, except where a membrane is allowed or required elsewhere in this appendix.

❖ Strawbale construction has historically been practiced with plaster applied directly to the stacked straw bales, without any membrane air barrier or water-resistive barrier between the plaster and the straw. This allows the plaster to mechanically bond with the straw, which is especially important for structural strawbale walls (see commentary, Section AS106.8). Also, under certain conditions, the presence of a membrane may impede the dispersion of moisture through the plaster to the interior or exterior.

It is important to note that while the code requires a water-resistive barrier for exterior cement plaster over wood-frame construction in accordance with Section R703.7.3, a water-resistive barrier is not required for cement plaster, or any exterior plaster allowed by this appendix, over strawbale walls. The moisture management characteristics of strawbale construction compared with wood-frame construction account for the differing requirements. In strawbale construction, the many tubular lengths of straw in the bale core give a strawbale wall considerably more capacity than a wood-frame wall to safely absorb, store and disperse moisture.

There are situations where a membrane is allowed or required by this appendix. For example, Section AS105.6.2 requires a Class I or Class II vapor retarder on strawbale walls enclosing showers or steam rooms. One means of achieving this is the installation of a membrane vapor retarder. See the commentary to Section AS105.6.1 for situations where a water-resistive barrier membrane is sometimes used between straw and plaster.

Installation of a membrane between the straw bales and plaster always requires mesh or lath that is adequately attached to the bales or through-tied to mesh or lath on the other side, including on nonstructural walls. This mesh or lath substitutes for the typical lathing bond between plaster and straw (see Section AS104.4.2) that is interrupted where a membrane is

installed. Where mesh is required structurally, such attachment must be of an approved engineered design (see Section AS106.8).

AS104.4.2 Lath and mesh for plaster. The surface of the straw bales functions as lath, and other lath or mesh shall not be required, except as required for out-of-plane resistance by Table AS105.4 or for structural walls by Tables AS106.12 and AS106.13(1).

AS104.4.3 Clay plaster. Clay plaster shall comply with Sections AS104.4.3.1 through AS104.4.3.6.

AS104.4.3.1 General. Clay plaster shall be any plaster having a clay or clay-soil binder. Such plaster shall contain sufficient clay to fully bind the plaster, sand or other inert granular material, and shall be permitted to contain reinforcing fibers. Acceptable reinforcing fibers include chopped straw, sisal and animal hair.

❖ The relative amounts of soil and added sand in the plaster mix depend on the amount of clay, sand and silt in the clay soil. Experimentation or experience is necessary to determine mixes that are workable and that yield a cured plaster with minimal or no cracking and sufficient compressive strength for the application.

See the commentary to the definition of "Clay" regarding methods used to determine the suitability of clay soil for clay plasters.

In addition to the acceptable reinforcing fibers listed, hemp fiber and agave fibers also have a history of successful use.

AS104.4.3.2 Lath and mesh. Clay plaster shall not be required to contain reinforcing lath or mesh except as required in Tables AS105.4 and AS106.13(1). Where provided, mesh shall be natural fiber, corrosion-resistant metal, nylon, high-density polypropylene or other approved material.

AS104.4.3.3 Thickness and coats. Clay plaster shall be not less than 1 inch (25 mm) thick, except where required to be thicker for structural walls as described elsewhere in this appendix, and shall be applied in not less than two coats.

❖ The purposes of requiring a minimum of two coats for clay plaster include limiting through-cracks (where the second coat fills in or bridges any cracks in the first coat) and aiding in the drying of the plaster.

AS104.4.3.4 Rain-exposed. Clay plaster, where exposed to rain, shall be finished with lime wash, lime plaster, linseed oil or other *approved* erosion-resistant finish.

❖ Exception 2 in the definition of "Weather-exposed surfaces" in the IBC might be considered when determining whether a clay plaster should be regulated as "exposed to rain," subject to the evaluation of the building official, along with consideration of the local climate. That exception reads: "Walls or portions of walls beneath an unenclosed roof area, where located a horizontal distance from an open exterior opening equal to at least twice the height of the opening."

Lime plaster and lime wash over clay plaster should only be used with careful consideration of local climate, materials and experience, as spalling can occur under certain conditions. Lime in the bonding coat of the clay plaster has been shown to decrease the chance of spalling of these lime-based protective coats.

AS104.4.3.5 Prohibited finish coat. Plaster containing Portland cement shall not be permitted as a finish coat over clay plasters.

❖ Finish coats of Portland cement plaster over clay plasters have a history of spalling and, therefore, are prohibited. In addition, cement plaster has a significantly lower vapor permeability than clay plaster and, without sufficient lime, risks reducing the vapor permeability of the plaster below the minimum 3 perm requirement of Section AS104.3.

As a matter of practice, an often-desired attribute of clay plaster is its high vapor permeability (16-19 perms per inch). Applying cement plaster, or any finish that significantly reduces its vapor permeability, runs counter to the intention. Other finishes that protect clay plaster from erosion without significantly reducing its vapor permeability are available and commonly used (see Section AS104.4.3.4 and its commentary).

AS104.4.3.6 Plaster additives. Additives shall be permitted to increase plaster workability, durability, strength or water resistance.

AS104.4.4 Soil-cement plaster. Soil-cement plaster shall comply with Sections AS104.4.4.1 through AS104.4.4.3.

AS104.4.4.1 General. Soil-cement plaster shall be composed of soil (free of organic matter), sand and not less than 10 percent and not more than 20 percent Portland cement by volume, and shall be permitted to contain reinforcing fibers.

❖ The relative amounts of soil and added sand in the plaster mix depend on the amount of clay, sand and silt in the soil. Experimentation or experience is necessary to determine mixes that are workable and that yield a cured plaster with minimal or no cracking and sufficient compressive strength for the application.

AS104.4.4.2 Lath and mesh. Soil-cement plaster shall use any corrosion-resistant lath or mesh permitted by this code, or as required in Section AS106 where used on structural walls.

AS104.4.4.3 Thickness. Soil-cement plaster shall be not less than 1 inch (25 mm) thick.

AS104.4.5 Gypsum plaster. Gypsum plaster shall comply with Section R702. Gypsum plaster shall be limited to use on interior surfaces of nonstructural walls, and as an interior finish coat over a structural plaster that complies with this appendix.

❖ The subsection of Section R702 that pertains to gypsum plaster is Section R702.2.1. Gypsum plaster, like all other plasters described in this appendix, can be applied directly to strawbale walls in accordance with Section AS104.4.2.

AS104.4.6 Lime plaster. Lime plaster shall comply with Sections AS104.4.6.1 and AS104.4.6.3.

AS104.4.6.1 General. Lime plaster is any plaster with a binder that is composed of calcium hydroxide (CaOH) including Type N or S hydrated lime, hydraulic lime, natural hydraulic lime or quicklime. Hydrated lime shall comply with ASTM C206. Hydraulic lime shall comply with ASTM C1707. Natural hydraulic lime shall comply with ASTM C141 and EN 459. Quicklime shall comply with ASTM C5.

AS104.4.6.2 Thickness and coats. Lime plaster shall be not less than $^7/_8$ inch (22 mm) thick, and shall be applied in not less than three coats.

AS104.4.6.3 On structural walls. Lime plaster on strawbale structural walls in accordance with Table AS106.12 or Table AS106.13(1) shall use a binder of hydraulic or natural hydraulic lime.

❖ For structural strawbale walls using lime plaster, the binder is required to be hydraulic or natural hydraulic lime because they develop the minimum compressive strength required by Table AS106.6.1 more reliably than nonhydraulic limes. However, the building official may decide to allow Type N or S hydrated lime, or quicklime for lime plasters on structural walls where such plasters are demonstrated to achieve the minimum required compressive strength through testing in accordance with Section AS106.6.1.

AS104.4.7 Cement-lime plaster. Cement-lime plaster shall be plaster mixes CL, F or FL, as described in ASTM C926.

AS104.4.8 Cement plaster. Cement plaster shall conform to ASTM C926 and shall comply with Sections R703.7.2, R703.7.4 and R703.7.5, except that the amount of lime in plaster coats shall be not less than 1 part lime to 6 parts cement to allow a minimum acceptable vapor permeability. The combined thickness of plaster coats shall be not more than $1^1/_2$ inches (38 mm) thick.

❖ Cement plaster is required to contain lime in the proportion stated in order to achieve a vapor permeability that is consistent with the minimum 3 perm requirement of Section AS104.3.

SECTION AS105
STRAWBALE WALLS—GENERAL

AS105.1 General. Strawbale walls shall be designed and constructed in accordance with this section. Strawbale structural walls shall be in accordance with the additional requirements of Section AS106.

❖ The provisions of this section apply to all strawbale walls, including nonstructural and structural walls (see the definitions of "Nonstructural wall" and "Structural wall"), except where a subsection states that the provision(s) apply only to nonstructural walls. In addition, structural strawbale walls must be designed and constructed in accordance with Section AS106.

AS105.2 Building requirements for use of strawbale nonstructural walls. Buildings using strawbale nonstructural walls shall be subject to the following limitations and requirements:

1. Number of stories: not more than one, except that two stories shall be allowed with an *approved* engineered design.

2. Building height: not more than 25 feet (7620 mm).

3. Wall height: in accordance with Table AS105.4.

4. Braced wall panel length, and increase in Seismic Design Categories C, D_0, D_1 and D_2: the required length of bracing for buildings using strawbale nonstructural walls shall comply with Section R602.10.3 of this code, with the additional requirements that Table R602.10.3(3) shall be applicable to buildings in Seismic Design Category C, and that the minimum total length of braced wall panels in Table R602.10.3(3) shall be increased by 60 percent.

❖ All buildings with nonstructural strawbale walls must employ a lateral force-resisting system in accordance with Section R602.10.3. For such buildings in Seismic Design Categories C, D_0, D_1 and D_2, the minimum total length of braced wall panels in Table R602.10.3(3) must be increased by 60 percent. This increase is due to the additional weight of strawbale walls (especially when finished with plaster on both sides) relative to wood-framed walls. This weight imposes additional seismic load on the lateral force-resisting system.

It is also possible for buildings with load-bearing strawbale walls to employ a conventional lateral force-resisting system. In Seismic Design Categories C, D_0, D_1 and D_2, the same increase in minimum total length of conventional braced wall panels described in Item 4 should apply.

Buildings employing strawbale braced wall panels as their lateral force-resisting system must comply with Section AS106.13. When used in Seismic Design Categories C, D_0, D_1 and D_2, the required total lengths of these strawbale braced wall panels in Table AS106.13(3) already account for the seismic load due to the weight of the plastered strawbale walls, therefore, no increase is necessary.

The 60-percent increase in Item 4 assumes a wall dead load of up to 60 psf. This is consistent with the wall dead load limit stipulated in Section AS106.2. Smaller wall dead loads, especially for strawbale walls with a nonplaster finish, could justify a smaller percentage increase if demonstrated in accordance with accepted engineering practice.

AS105.3 Sill plates. Sill plates shall support and be flush with each face of the straw bales above and shall be of naturally durable or preservative-treated wood where required by this code. Sill plates shall be not less than nominal 2 inches by 4 inches (51 mm by 102 mm) with anchoring complying with Section R403.1.6 and the additional requirements of Tables AS105.4 and AS106.6(1), where applicable.

❖ Although not required by this appendix, it is common and good practice to install continuous flashing that covers the joint between the bottom of the sill plate and its supporting material (e.g., concrete slab or foundation) and as a means of providing a capillary break where the plaster is supported by concrete or masonry. Such flashing should drain to the exterior. Examples of this flashing are shown in Commentary Figures AS105.3(1) and AS105.3(2).

The purpose of this flashing is to protect against intrusion of moisture that might accumulate near the bottom of the exterior plaster. If a water-resistive barrier is installed over the bales (e.g., under a finish other than plaster, between wood framing and plaster, or between the straw and plaster in some applications), it should lap over the flashing.

AS105.4 Out-of-plane resistance and unrestrained wall dimensions. Strawbale walls shall employ a method of out-of-plane resistance in accordance with Table AS105.4, and comply with its associated limits and requirements.

❖ Every strawbale wall must employ an acceptable method of out-of-plane resistance in accordance with this section. The maximum unrestrained height of any strawbale wall, both nonstructural and structural, is a function of its method of out-of-plane resistance, as well as other parameters included in Table AS105.4. The overall height of a strawbale wall can exceed the unrestrained height limits if it employs an approved horizontal restraint (see Table AS105.4, Note b) at an intermediate height between the sill plates and the top plate or roof bearing assembly.

One method of providing out-of-plane resistance for strawbale walls that is not listed in Table AS105.4 is the use of wood posts or studs, to which a sufficient number and distribution of bales are attached via wood stakes, gussets or other means, to provide out-of-plane stability under the design wind and seismic loads for the building and its location. This method is most commonly employed in post-and-beam structures with strawbale infill, using plasters without reinforcing mesh, or using nonplaster finishes. This method should be used only with an approved engineered design, unless a local history of acceptable experience exists.

TABLE AS105.4. See page S-13.

❖ See the commentary to Section AS105.4.2.

AS105.4.1 Determination of out-of-plane loading. Out-of-plane loading for the use of Table AS105.4 shall be in terms of the design wind speed and seismic design category as determined in accordance with Sections R301.2.1 and R301.2.2 of this code.

AS105.4.2 Pins. Pins used for out-of-plane resistance shall comply with the following or shall be in accordance with an *approved* engineered design. Pins shall be external, internal or a combination of the two.

1. Pins shall be $^1/_2$-inch-diameter (12.7 mm) steel, $^3/_4$-inch-diameter (19.1 mm) wood or $^1/_2$-inch-diameter (12.7 mm) bamboo.

2. External pins shall be installed vertically on both sides of the wall at a spacing of not more than 24 inches (610

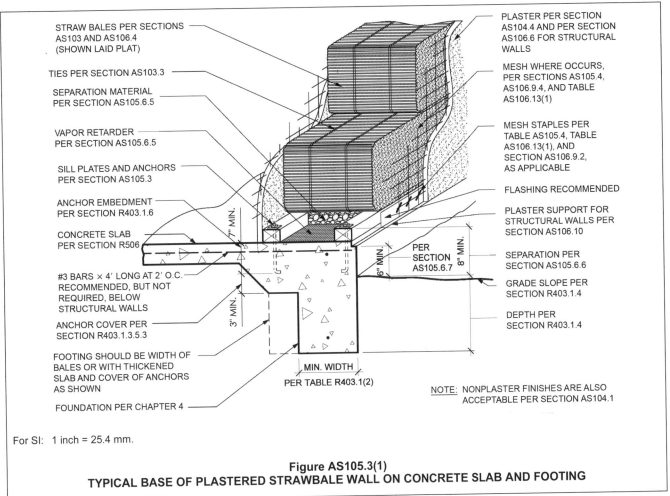

STRAW BALES PER SECTIONS AS103 AND AS106.4 (SHOWN LAID PLAT)

TIES PER SECTION AS103.3

SEPARATION MATERIAL PER SECTION AS105.6.5

VAPOR RETARDER PER SECTION AS105.6.5

SILL PLATES AND ANCHORS PER SECTION AS105.3

ANCHOR EMBEDMENT PER SECTION R403.1.6

CONCRETE SLAB PER SECTION R506

#3 BARS × 4' LONG AT 2' O.C. RECOMMENDED, BUT NOT REQUIRED, BELOW STRUCTURAL WALLS

ANCHOR COVER PER SECTION R403.1.3.5.3

FOOTING SHOULD BE WIDTH OF BALES OR WITH THICKENED SLAB AND COVER OF ANCHORS AS SHOWN

FOUNDATION PER CHAPTER 4

PLASTER PER SECTION AS104.4 AND PER SECTION AS106.6 FOR STRUCTURAL WALLS

MESH WHERE OCCURS, PER SECTIONS AS105.4, AS106.9.4, AND TABLE AS106.13(1)

MESH STAPLES PER TABLE AS105.4, TABLE AS106.13(1), AND SECTION AS106.9.2, AS APPLICABLE

FLASHING RECOMMENDED

PLASTER SUPPORT FOR STRUCTURAL WALLS PER SECTION AS106.10

SEPARATION PER SECTION AS105.6.6

GRADE SLOPE PER SECTION R403.1.4

DEPTH PER SECTION R403.1.4

7" MIN.

3' MIN.

6" MIN.

PER SECTION AS105.6.7

8" MIN.

MIN. WIDTH PER TABLE R403.1(2)

NOTE: NONPLASTER FINISHES ARE ALSO ACCEPTABLE PER SECTION AS104.1

For SI: 1 inch = 25.4 mm.

Figure AS105.3(1)
TYPICAL BASE OF PLASTERED STRAWBALE WALL ON CONCRETE SLAB AND FOOTING

mm) on center. External pins shall have full lateral bearing on the sill plate and the top plate or roof-bearing element, and shall be tightly tied through the wall to an opposing pin with ties spaced not more than 32 inches (813 mm) apart and not more than 8 inches (203 mm) from each end of the pins.

3. Internal pins shall be installed vertically within the center third of the bales, at spacing of not more than 24 inches (610 mm) and shall extend from top course to bottom course. The bottom course shall be similarly connected to its support and the top course shall be similarly connected to the roof- or floor-bearing member above with pins or other *approved* means. Internal pins shall be continuous or shall overlap through not less than one bale course.

❖ The term "pin" (see definition of "Pin") comes from the once-common use of internal pins, typically made of steel reinforcing bar, in modern stawbale construction. In this method, the bales are skewered with steel bar as a means of stabilizing the walls during construction and to provide resistance to out-of-plane wind and seismic forces. The practice of internal pinning (with steel or other materials) has fallen into disuse by most practitioners of strawbale construction due to its greater difficulty and cost compared with other means

of achieving the same or better wall stability. There is also debate among strawbale building practitioners regarding whether steel pins may be a location of potential condensation in some conditions.

The term "external pin" might be considered a misnomer relative to general use or understanding of the word "pin" (typically being inside another material), but it is commonly used terminology in strawbale construction.

AS105.5 Connection of light-framed walls to strawbale walls. *Light-framed* walls perpendicular to, or at an angle to a straw bale wall assembly, shall be fastened to the bottom and top wood members of the strawbale wall in accordance with requirements for wood or cold-formed steel *light-framed* walls in this code, or the abutting stud shall be connected to alternating straw bale courses with a $^1/_2$-inch diameter (12.7 mm) steel, $^3/_4$-inch-diameter (19.1 mm) wood or $^5/_8$-inch-diameter (15.9 mm) bamboo dowel, with not less than 8-inch (203 mm) penetration.

❖ The connection methods stated in this section are not meant to preclude the use of other common methods, such as an all-thread rod inserted through the abutting stud and bale wall, with a plywood washer on the opposite side of the bale and a steel nut and washer on both ends of the threaded rod.

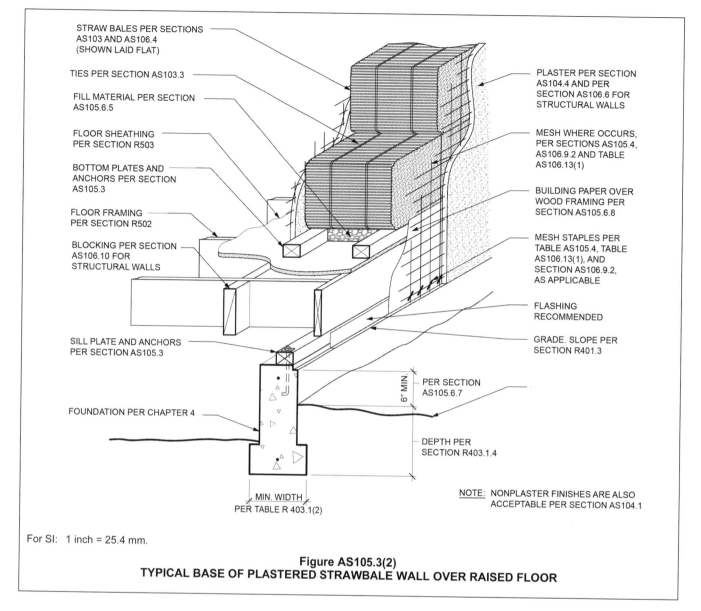

STRAW BALES PER SECTIONS AS103 AND AS106.4 (SHOWN LAID FLAT)

TIES PER SECTION AS103.3

FILL MATERIAL PER SECTION AS105.6.5

FLOOR SHEATHING PER SECTION R503

BOTTOM PLATES AND ANCHORS PER SECTION AS105.3

FLOOR FRAMING PER SECTION R502

BLOCKING PER SECTION AS106.10 FOR STRUCTURAL WALLS

SILL PLATE AND ANCHORS PER SECTION AS105.3

FOUNDATION PER CHAPTER 4

PLASTER PER SECTION AS104.4 AND PER SECTION AS106.6 FOR STRUCTURAL WALLS

MESH WHERE OCCURS, PER SECTIONS AS105.4, AS106.9.2 AND TABLE AS106.13(1)

BUILDING PAPER OVER WOOD FRAMING PER SECTION AS105.6.8

MESH STAPLES PER TABLE AS105.4, TABLE AS106.13(1), AND SECTION AS106.9.2, AS APPLICABLE

FLASHING RECOMMENDED

GRADE. SLOPE PER SECTION R401.3

PER SECTION AS105.6.7

6" MIN.

DEPTH PER SECTION R403.1.4

MIN. WIDTH PER TABLE R 403.1(2)

NOTE: NONPLASTER FINISHES ARE ALSO ACCEPTABLE PER SECTION AS104.1

For SI: 1 inch = 25.4 mm.

Figure AS105.3(2)
TYPICAL BASE OF PLASTERED STRAWBALE WALL OVER RAISED FLOOR

AS105.6 Moisture control. Strawbale walls shall be protected from moisture intrusion and damage in accordance with Sections AS105.6.1 through AS105.6.8.

❖ Preventing intrusion of moisture into strawbale walls is important for maximizing service life and for maintaining the insulating value of the straw.

There is a colloquial metaphor commonly used by practitioners of strawbale construction that summarizes the basic principles for keeping a wall dry: "Good boots, a good hat, and a coat that breathes." This translates into keeping the bottom of the wall protected from ground and weather-related moisture, providing ample roof overhangs (especially in wet climates) to shield the wall and its openings from rain and providing a protective wall finish that is vapor permeable. This dictum is also considered to be wise advice for buildings constructed of wood frame and other materials.

Straw, like wood, is composed primarily of cellulose, hemicellulose and lignin. Straw and wood can last indefinitely as long as there is insufficient free moisture (moisture not bound in the material) to cause deterioration. If free moisture becomes available (generally when the moisture content of the material exceeds 20 percent) and other necessary environmental conditions exist—including temperatures above 50°F (10°C) and the presence of oxygen—then microbial degradation of the straw can occur. In buildings, the only practically controllable condition is the availability of free moisture.

Though straw and wood have similar material makeup, straw is more susceptible to damage from moisture over a shorter period of time. This is due to the greater surface area of the many lengths of tubular straw, compared with only the perimeter surface area of wood framing members. However, the greater surface area of straw also has the comparative advantage of potentially absorbing and storing more moisture before it becomes free moisture and causes degradation. This, combined with the much larger storage capacity of a continuous core of straw compared to the spaced framing members in a wood-framed wall, gives a strawbale wall significantly greater moisture management capacity.

Subsections AS105.6.1 through AS105.6.8 contain requirements to minimize the possibility of moisture intrusion into strawbale walls from common sources such as rain or snow and condensation (from uncontrolled flow of relatively warm, moist air into the wall). See the additional commentary after each subsection.

Potential sources that are not addressed in these subsections are leaks and condensation from plumbing pipes in the walls. While there is no requirement regarding plumbing pipes in this appendix, it is recommended and common practice to minimize the installation of supply and waste pipes in strawbale walls, or to install them at the bottom of the wall (between sill plates) or in a continuous sleeve that drains to the exterior.

In addition to the importance of preventing moisture from entering a strawbale wall, moisture must also be allowed to readily exit the wall. Thus, the finish on each side should be as vapor permeable as possible within the requirements of this appendix and the code. Sections AS105.6.1, AS105.6.2 and AS104.3 include minimum vapor permeability requirements for finishes and any associated vapor retarder or water-resistive barrier.

AS105.6.1 Water-resistant barriers and vapor permeance ratings. Plastered bale walls shall be constructed without any membrane barrier between straw and plaster to facilitate tran-

spiration of moisture from the bales, and to secure a structural bond between straw and plaster, except as permitted or required elsewhere in this appendix. Where a water-resistant barrier is placed behind an exterior finish, it shall have a vapor permeance rating of not less than 5 perms, except as permitted or required elsewhere in this appendix.

❖ This section is intended to both facilitate transpiration of moisture out of the strawbale core through its finish and to allow a mechanical bond between plaster finishes and the straw. Some strawbale designs incorporate water-resistive barriers. This typically includes water-resistive barriers on the exterior face of bales below window sills, on the first course(s) of bales that are particularly exposed to weather-related moisture, behind nonplaster finishes, over sheathing, or as part of a rain screen. Where a water-resistive barrier is used, its vapor permeance rating must be at least 5 perms (which is the definition of vapor permeable in the code).

Where a water-resistive barrier is used behind a plaster finish on a structural wall, its mesh must be through-tied to the plaster skin on the other side by means of an approved engineered design (see Section AS106.8). Though not addressed in this appendix, where a water-resistive barrier is used behind plaster on a nonstructural wall, it should contain mesh or other lath that is adequately attached to the bale core or to the finish on the other side of the wall to ensure that the plaster will remain in place in normal service.

AS105.6.2 Vapor retarders. Wall finishes shall have an equivalent vapor permeance rating of a Class III vapor

TABLE AS105.4
OUT-OF-PLANE RESISTANCE AND UNRESTRAINED WALL DIMENSIONS

METHOD OF OUT-OF-PLANE RESISTANCE[a]	FOR WIND DESIGN SPEEDS (mph)	FOR SEISMIC DESIGN CATEGORIES	UNRESTRAINED WALL DIMENSIONS, H[b]		MESH STAPLE SPACING AT BOUNDARY RESTRAINTS
			Absolute limit in feet	Limit based on bale thickness T[c] in feet (mm)	
Nonplaster finish or unreinforced plaster	≤ 100	A, B, C, D_0	$H \leq 8$	$H \leq 5T$	None required
Pins per Section AS105.4.2	≤ 100	A, B, C, D_0	$H \leq 12$	$H \leq 8T$	None required
Pins per Section AS105.4.2	≤ 110	A, B, C, D_0, D_1, D_2	$H \leq 10$	$H \leq 7T$	None required
Reinforced[c] clay plaster	≤ 110	A, B, C, D_0, D_1, D_2	$H \leq 10$	$H \leq 8T^{0.5}$ ($H \leq 140T^{0.5}$)	≤ 6 inches
Reinforced[c] clay plaster	≤ 110	A, B, C, D_0, D_1, D_2	$10 < H \leq 12$	$H \leq 8T^{0.5}$ ($H \leq 140T^{0.5}$)	≤ 4 inches[e]
Reinforced[c] cement, cement-lime, lime or soil-cement plaster	≤ 110	A, B, C, D_0, D_1, D_2	$H \leq 10$	$H \leq 9T^{0.5}$ ($H \leq 157T^{0.5}$)	≤ 6 inches
Reinforced[c] cement, cement-lime, lime or soil-cement plaster	≤ 120	A, B, C, D_0, D_1, D_2	$H \leq 12$	$H \leq 9T^{0.5}$ ($H \leq 157T^{0.5}$)	≤ 4 inches[e]

For SI: 1 inch = 25.4 mm, 1 foot = 304.8 mm.

a. Finishes applied to both sides of stacked bales. Where different finishes are used on opposite sides of a wall, the more restrictive requirements shall apply.

b. *H* = Stacked bale height in feet (mm) between sill plate and top plate or other *approved* horizontal restraint, or the horizontal distance in feet (mm) between *approved* vertical restraints. For load-bearing walls, *H* refers to vertical height only.

c. *T* = Bale thickness in feet (mm).

d. Plaster reinforcement shall be any mesh allowed in Table AS106.16 for the matching plaster type, and with staple spacing in accordance with this table. Mesh shall be installed in accordance with Section AS106.9.

e. Sill plate attachment shall be with $^5/_8$-inch anchor bolts or approved equivalent at not more than 48 inches on center where staple spacing is required to be ≤ 4 inches.

retarder on the interior side of exterior strawbale walls in Climate Zones 5, 6, 7, 8 and Marine 4, as defined in Chapter 11. Bales in walls enclosing showers or steam rooms shall be protected on the interior side by a Class I or Class II vapor retarder.

❖ The interior wall finish of exterior strawbale walls requires an equivalent vapor permeance rating of a Class III vapor retarder (> 1 perm and ≤ 10 perms) in the stated climate zones. This requirement is intended to prevent condensation in the wall in cold winter climates by retarding the passage of water vapor into the wall from interior air. However, sufficient vapor permeability of the interior finish is also important to allow migration of moisture from the wall to the interior under some conditions. This includes conditions that occur in many locations in the stated climate zones where interiors are mechanically cooled during warm, humid seasons.

Because of potentially conflicting seasonal demands in the stated climate zones, a balance between the vapor-retarding ability and the vapor permeability of the interior finish of exterior strawbale walls is desirable. Field experience of strawbale practitioners, coupled with principles of building science regarding moisture migration, dictate a perm rating in the upper half of a Class III vapor retarder (between 5 and 10 perms) is optimal for the interior finish of exterior strawbale walls in these climate zones.

In cold climates, the comparative vapor permeance of the interior and exterior finishes are at least as important as the vapor permeability of the interior finish alone. Although not a requirement in this appendix, it is recommended that the vapor permeance of the exterior finish of strawbale walls be greater than that of the interior finish. This will allow water vapor that is driven from the relatively warm and moist interior air toward the colder and drier exterior to continue out of the wall before accumulating and condensing.

The requirement for a Class I or Class II vapor retarder on walls enclosing showers or steam rooms applies to all climate zones and interior and exterior strawbale walls. The term "shower" means a shower or a combination tub/shower. Walls enclosing tubs without showers are not intended to be subject to this requirement.

Also see the commentary to Section AS104.3.

AS105.6.3 Penetrations in exterior strawbale walls. Penetrations in exterior strawbale walls shall be sealed with an *approved* sealant or gasket on the exterior side of the wall in all climate zones, and on the interior side of the wall in Climate Zones 5, 6, 7, 8 and Marine 4, as defined in Chapter 11.

❖ In all climate zones, sealing penetrations (e.g., by piping, conduit, electrical boxes or ventilation caps) in the exterior finish of strawbale walls helps minimize or prevent the intrusion of weather-related moisture into the straw core of the wall. In climate zones with cold winters, and in marine climate zones where the air is especially laden with moisture, sealing penetrations on the interior side of exterior walls is important to prevent

warm, moist interior air from bypassing the air barrier of the interior finish of the wall, which could potentially condense in the bale core.

Window and door openings are considered large penetrations in exterior walls and are subject to the requirements of this section. The interfaces between window units and door units and exterior strawbale walls must be sealed on the exterior to protect against weather-related moisture intrusion. They must also be sealed on the interior in the listed climate zones in order to protect against intrusion of moisture from interior air. Window sills are subject to the requirements of Section AS105.6.4.

Windows and doors require particular attention to minimize moisture intrusion into the wall. Appropriate flashing measures should be taken and detailed according to the exterior finish (plaster or other cladding), the profile of any window or door trim and the location of the unit within the thickness of the strawbale wall. Common practice is to seal the plaster to the rough opening, which is in turn sealed to the window or door unit. Conventional "peel-and-stick" window flashings are often used to bridge from the window or door unit onto the surrounding bales at the jambs, head and sill.

Penetrations in strawbale walls that are not weather-exposed are not intended to require sealing on the exterior. However, penetrations in such walls must be sealed on the interior in Climate Zones 5, 6, 7, 8 and Marine 4. Exception 2 in the definition of "Weather-exposed surfaces" in the IBC might be used to determine whether a penetration in a strawbale wall is considered weather exposed, subject to the evaluation of the local building official, with additional consideration for the local climate. That exception reads, "Walls or portions of walls beneath an unenclosed roof area, where located a horizontal distance from an open exterior opening equal to at least twice the height of the opening."

AS105.6.4 Horizontal surfaces. Bale walls and other bale elements shall be provided with a water-resistant barrier at weather-exposed horizontal surfaces. The water-resistant barrier shall be of a material and installation that will prevent water from entering the wall system. Horizontal surfaces shall include exterior window sills, sills at exterior niches and buttresses. The finish material at such surfaces shall be sloped not less than 1 unit vertical in 12 units horizontal (8-percent slope) and shall drain away from bale walls and elements. Where the water-resistant barrier is below the finish material, it shall be sloped not less than 1 unit vertical in 12 units horizontal (8-percent slope) and shall drain to the outside surface of the bales wall's vertical finish.

❖ As with wood-frame walls, horizontal (or nearly horizontal) surfaces in or on strawbale walls can be vulnerable to weather-related moisture entering the wall. Window sills for windows that are recessed into the wall are especially vulnerable and should be carefully detailed and constructed to meet the prescriptive and performance criteria of this section. It is not uncommon for the water-resistive barrier under window sills to lap over a vapor permeable water-resistive barrier that continues down the exterior face of the bales below in order to provide additional protection for those potentially vulnerable bales.

Horizontal surfaces in or on strawbale walls that are not weather-exposed do not require a water-resistive barrier. Exception 2 in the definition of "Weather-exposed surfaces" in the IBC might be used to determine whether a horizontal surface in a strawbale wall is considered weather-exposed, subject to the evaluation of the local building official, with additional consideration of the local climate. That exception reads, "Walls or portions of walls beneath an unenclosed roof area, where located a horizontal distance from an open exterior opening equal to at least twice the height of the opening."

Ample roof overhangs and wrap-around porch roofs are often employed by strawbale building practitioners, especially in wet climates and heavy snow fall areas, to protect walls and their openings from weather-related moisture intrusion.

AS105.6.5 Separation of bales and concrete. A sheet or liquid-applied Class II *vapor retarder* shall be installed between bales and supporting concrete or masonry. The bales shall be separated from the vapor retarder by not less than $^3/_4$ inch (19.1 mm), and that space shall be filled with an insulating material such as wood or rigid insulation, or a material that allows vapor dispersion such as gravel, or other approved insulating or vapor dispersion material. Sill plates shall be installed at this interface in accordance with Section AS105.3. Where bales abut a concrete or masonry wall that retains earth, a Class II vapor retarder shall be provided between such wall and the bales.

❖ A Class II vapor retarder is required to separate bales from supporting concrete or masonry to prevent "rising damp" from reaching the bales and potentially causing mold or mildew on the straw. Further, a minimum $^3/_4$-inch space is required between the vapor retarder and the straw and insulating or vapor dispersion material. Insulating materials help keep the underside of the first course of bales above dew point to avoid condensation on the straw. Vapor dispersion materials also allow safe dispersion of any moisture that might condense or otherwise reach the area at the base of the wall.

See Commentary Figure AS106(1) for a typical base-of-wall detail for slab-on-grade condition.

AS105.6.6 Separation of bales and earth. Bales shall be separated from earth by not less than 8 inches (203 mm).

❖ Straw, like wood, is subject to decay with prolonged exposure to excessive free moisture; therefore, straw bales are required to be separated from potentially moisture-laden earth by at least 8 inches (203 mm). This separation also reduces exposure of straw at the bottom of the wall to weather-related moisture, such as rain splash-back and snow melt. Greater separation is recommended in situations where excessive moisture exposure is expected due to local climate conditions.

AS105.6.7 Separation of exterior plaster and earth. Exterior plaster applied to straw bales shall be located not less

than 6 inches (102 mm) above earth or 3 inches (51 mm) above paved areas.

❖ Section R703.7.2.1 requires the bottom of weep screeds for exterior cement plaster to be not less than 4 inches above the earth and 2 inches (51 mm) above paved areas. For exterior plaster applied to straw bales, Section AS105.6.7 requires the plaster to be not less than 6 inches (152 mm) above earth and 3 inches (51 mm) above paved areas. The additional comparative separation is to give greater assurance that exterior moisture at grade will not reach the straw. This separation requirement is for all exterior plasters allowed in this appendix, not only for cement plaster as referenced in Section R703.7.2.1.

Greater separation is recommended where excessive, weather-related moisture exposure, including rain splash-back, is expected due to local climatic conditions, and for weather-exposed clay plasters, even where protected by an erosion-resistant finish as required by Section AS104.4.3.4.

AS105.6.8 Separation of wood and plaster. Where wood framing or wood sheathing occurs on the exterior face of strawbale walls, such wood surfaces shall be separated from exterior plaster with two layers of Grade D paper, No. 15 asphalt felt or other *approved* material in accordance with Section R703.7.3.

Exceptions:

1. Where the wood is preservative treated or *naturally durable* and is not greater than $1^1/_2$ inches (38 mm) in width.

2. Clay plaster shall not be required to be separated from untreated wood that is not greater than $1^1/_2$ inches (38 mm) in width.

❖ A water-resistive barrier such as two layers of Grade D paper or No. 15 asphalt felt is required to separate exterior plaster from wood framing or sheathing due to the possibility of moisture penetrating the plaster and reaching the framing or sheathing. The water-resistive barrier protects the framing or sheathing from potential decay.

Also see Sections AS104.4.1 and AS106.8 and their commentary for information regarding where a water-resistive barrier between plaster and straw bales is not required.

A water-resistive barrier is not required under the conditions noted in Exception 1 because such materials are not susceptible to decay in the presence of moisture.

In Exception 2, a water-resistive barrier is not required between clay plaster and untreated wood because clay is hydrophilic (i.e., water-attracting, drawing moisture away from adjacent materials) and blocks passage of water due to its strong surface bonding with available moisture. Also, clay has considerable capacity to capture and store water due to the large surface area of its microscopic layers. This combination of hydrophilic properties, water-blocking properties and water storage capacity protects adjacent materials that might otherwise be harmed by prolonged exposure to moisture.

The $1\frac{1}{2}$-inch (38 mm) maximum width in Exceptions 1 and 2 relates to the prevention of plaster cracking. Plasters (all types identified in this appendix) on strawbale walls have demonstrated the ability to span across $1\frac{1}{2}$-inch (38 mm) wide wood members without significant cracking, even when unreinforced and without the "slip sheet" effect of a water-resistive barrier between the plaster and the wood member. For wood members greater than $1\frac{1}{2}$ inches (38 mm) wide, a two-layered water-resistive barrier, along with an appropriate reinforcing mesh that extends at least 1 inch (25 mm) beyond the edges of the wood, has been shown to greatly reduce or eliminate plaster cracking in these locations.

AS105.7 Inspections. The *building official* shall inspect the following aspects of strawbale construction in accordance with Section R109.1:

1. Sill plate anchors, as part of and in accordance with Section R109.1.1.

2. Mesh placement and attachment, where mesh is required by this appendix.

3. *Pins*, where required by and in accordance with Section AS105.4.

❖ Items 2 and 3, regarding inspections for mesh and pins, are necessary only where their use is required by this appendix. Mesh or pins are considered required where used as a means of compliance with Section AS105.4 for out-of-plane resistance and for mesh in structural applications in accordance with Section AS106. Where mesh or pins are installed voluntarily, no inspection of their installation is required.

SECTION AS106
STRAWBALE WALLS—STRUCTURAL

AS106.1 General. Plastered strawbale walls shall be permitted to be used as structural walls in one-story buildings in accordance with the prescriptive provisions of this section.

❖ The provisions of Section AS106 apply to structural strawbale walls (load-bearing and/or shear walls) in one-story buildings. Some provisions in this section are also considered good practice for nonstructural walls and are identified as such in this commentary.

There are many examples of two-story buildings constructed with structural strawbale walls in the U.S and other countries. This section is not meant to prohibit such structures. However, they are outside of the prescriptive and performance structural requirements of this appendix and, as such, should be accompanied by an approved engineered design. Note that Section AS105.2, Item 4 allows nonstructural strawbale walls in two-story buildings with an approved engineered design.

Structural strawbale walls require a plaster finish (as prescribed in this section), as the plaster skins together with the strawbale core provide the structural capacity of the composite wall system. The plaster must be reinforced if the wall is used to resist shear. Shear walls (braced wall panels) may be utilized in load-bearing or nonload-bearing applications. In both load-bearing walls and shear walls, the strawbale core provides lateral bracing for the relatively thin plaster skin, while also serving as a backup load-carrying system should the plaster skins degrade.

AS106.2 Loads and other limitations. Live and dead loads and other limitations shall be in accordance with Section R301 of the *International Residential Code*. Strawbale wall dead loads shall not exceed 60 psf (2872 N/m^2) per face area of wall.

❖ The amount of bracing required for seismic loads is directly related to the weight of the structure. The dead load limit of 60 psf for strawbale walls is used to avoid potential overburden of the lateral load-resisting system in Seismic Design Categories C, D$_0$, D$_1$ and D$_2$. A larger wall dead load could be acceptable with an approved engineered design, but may require total braced wall panel lengths that are greater than the 60 percent stipulated in Section AS105.2, Item 4 for conventional braced wall panels.

The weight "per face area of wall" means the weight of all materials within a horizontally projected area from one face of the wall to the other. This includes finishes on both sides and the straw bale core.

For example, the weight per square foot of a strawbale wall with two-string bales laid flat and 1-inch-thick lime plaster on both sides would be determined as follows: First, determine the weight per cubic foot of a typical bale for the project (total bale weight / total bale volume). For this example, say: 47 lb/(1.16 feet × 1.5 feet × 3 feet) = 8.5 pcf. For this example assume the lime plaster weighs 130 pcf. Therefore the weight per square foot of the strawbale wall in this example is: (1 foot × 1 foot × 1.5 foot × 8.5 pcf) + 2(1 foot × 1 foot × $\frac{1}{12}$ foot × 130 pcf) = 12.75 lb + 21.67 lb = 34.42 psf.

AS106.3 Foundations. Foundations for plastered strawbale walls shall be in accordance with Chapter 4.

❖ Foundations that satisfy the requirements of Chapter 4 are acceptable for buildings with strawbale walls.

Three tables in the code are used to determine footing width and thickness, depending on the type of wall construction. Strawbale walls, with any finish allowed in this appendix, are closest to the weight of light-frame walls with brick veneer in Table R403.1(2); therefore, that table should be used.

See Commentary Figure AS105.3(1) for typical concrete slab-on-grade construction with monolithic foundation supporting a strawbale wall. The figure shows a strawbale structural wall with its required anchor bolts for exterior and interior sill plates and a thickened slab below the interior sill plate. This is necessary to provide the 7-inch (178 mm) embedment for the anchor bolts as required by Section R403.1.6 and the minimum 3-inch (76 mm) cover for reinforcement as required by Section R403.1.3.5.3. Nonstructural strawbale walls may also require anchor bolts for their sill plates, and therefore also may require a similar thickened slab.

For structural strawbale walls on a raised floor, Section AS106.10 requires an approved engineered design. See the commentary to that section for recommendations for raised floors supporting structural strawbale walls. Also, Commentary Figure AS105.3(2) illustrates the typical elements and configuration for a strawbale wall on a raised floor.

AS106.4 Configuration of bales. Bales in strawbale structural walls shall be laid flat or on-edge and in a running bond or stack bond, except that bales in structural walls with unreinforced plasters shall be laid in a running bond only.

AS106.5 Voids and stuffing. Voids between bales in strawbale *structural walls* shall not exceed 4 inches (102 mm) in width, and such voids shall be stuffed with flakes of straw or straw-clay, before application of finish.

❖ Stuffing of voids between bales and between bales and framing members is considered necessary for the structural performance of strawbale structural walls, as well as for the thermal performance of all exterior strawbale walls. The density of straw or straw-clay that fills voids should achieve the minimum required density for bales (see Section AS103.5) to the degree practicable. Also see the commentary to Section AS108.

AS106.6 Plaster on structural walls. Plaster on *load-bearing* walls shall be in accordance with Table AS106.12. Plaster on shear walls shall be in accordance with Table AS106.13(1).

❖ Requirements for plaster on strawbale structural walls are more stringent than for nonstructural applications. The term "shear wall" is used interchangeably with the term "braced wall panel" in this appendix.

AS106.6.1 Compressive strength. For plaster on strawbale structural walls, the building official is authorized to require a 2-inch (51mm) cube test conforming to ASTM C109 to demonstrate a minimum compressive strength in accordance with Table AS106.6.1.

❖ Minimum compressive strengths in Table AS106.6.1 are provided to clarify minimum expectations for builders and building officials. These minimum values relate (with appropriate factors of safety) to the allowable bearing capacities for walls with plaster types indicated in Table AS106.12, and to the minimum braced wall panel lengths for wall types indicated in Tables AS106.13(2) and AS106.13(3).

The minimum compressive strengths in Table AS106.6.1 are on the low end of the range of compressive strengths typical for each plaster type. As such, a building official may expect these values to be achieved without testing for a plaster made with good-quality materials and workmanship. However, the building official is authorized to require an ASTM C109 2-inch cube test to confirm a plaster's minimum compressive strength for structural strawbale walls.

Since ASTM C109 applies to the compressive strength testing of hydraulic cement mortars, the fol-lowing modifications are needed for the testing of soil-cement and clay plasters:

(a) Soil-cement plaster samples must be cured in a moist environment in accordance with Section 10.5 of ASTM C109, and must not be immersed in lime-saturated water in accordance with the same section.

(b) Clay plaster samples must be dried to the approximate ambient moisture conditions of the project site, and must not be not cured in the moist or wet curing environments described in Section 10.5 of ASTM C109.

Rationale: Hydraulic cement mortar and the cement, cement-lime, lime, and soil-cement plasters described in this appendix all contain sand and hydraulic cement and/or hydraulic lime. These mixes develop their strength by hydration (a water-based chemical reaction, sometimes called curing) over time. Clay plaster develops its strength by drying, not by moist curing. Soil-cement plaster contains both hydraulic cement that must be moist cured and clay that must dry to an acceptable moisture content. As such it should be moist cured, but not immersed in water.

TABLE AS106.6.1
MINIMUM COMPRESSIVE STRENGTH FOR PLASTERS ON STRUCTURAL WALLS

PLASTER TYPE	MINIMUM COMPRESSIVE STRENGTH (psi)
Clay	100
Soil-cement	1000
Lime	600
Cement-lime	1000
Cement	1400

For SI: 1 pound per square inch = 6894.76 N/m².

AS106.7 Straightness of plaster. Plaster on strawbale structural walls shall be straight, as a function of the bale wall surfaces they are applied to, in accordance with all of the following:

1. As measured across the face of a bale, straw bulges shall not protrude more than $^3/_4$ inch (19.1 mm) across 2 feet (610 mm) of its height or length.

2. As measured across the face of a bale wall, straw bulges shall not protrude from the vertical plane of a bale wall more than 2 inches (51 mm) over 8 feet (2438 mm).

3. The vertical faces of adjacent bales shall not be offset more than $^3/_8$ inch (9.5 mm).

❖ The plaster on structural strawbale walls carries in-plane forces. Significant deviation from straightness may compromise a wall's strength and structural performance. The requirements of Section AS106.7 are intended to ensure a minimum degree of straightness for the expected structural performance of the wall.

Note that some practitioners have constructed large-radius strawbale walls by curving bales laid flat. Such a wall can be considered to comply with this section and be used as a load-bearing wall if the wall's plaster is straight vertically, and if it meets all other requirements in this appendix for load-bearing strawbale walls.

APPENDIX S

AS106.8 Plaster and membranes. Strawbale structural walls shall not have a membrane between straw and plaster, or shall have attachment through the bale wall from one plaster skin to the other in accordance with an *approved* engineered design.

❖ The application of plaster with no membrane between the straw and plaster allows the plaster to bond directly to the straw, thus allowing composite structural performance of the wall assembly. Specifically it allows: (a) The plaster to be well braced by the straw-bale core; (b) the plaster to resist delamination from the straw core under load; and (c) the strawbale core to transfer shear forces from one plaster skin to the other. In circumstances where a membrane is required or desired, the mesh in one plaster skin must be through-tied to the mesh in the opposite plaster skin in accordance with an approved engineered design.

See the commentary to Section AS104.4.1 for information regarding membranes and issues of moisture management and the support of plaster on nonstructural walls.

AS106.9 Mesh. Mesh in plasters on strawbale structural walls, and where required by Table AS105.4, shall be installed in accordance with Sections AS106.9.1 through AS106.9.4.

❖ Mesh required on structural strawbale walls and mesh used as a method of out-of-plane resistance (in accordance with Table AS105.4) must be installed in accordance with all subsections of Section AS106.9. Where mesh is installed but not required, these provisions are recommended as good practice.

AS106.9.1 Mesh laps. Mesh required by Table AS105.4 or AS106.12 shall be installed with not less than 4-inch (102 mm) laps. Mesh required by Table AS106.13(1) or in walls designed to resist wind uplift of more than 100 plf (1459 N/m), shall run continuous vertically from sill plate to the top plate or roof-bearing element, or shall lap not less than 8 inches (203 mm). Horizontal laps in such mesh shall be not less than 4 inches (102 mm).

❖ Mesh "required by Table AS106.13(1)" refers to the wall types described in that table where used as braced wall panels to resist wind or seismic forces in accordance with Table AS106.13(2) or AS106.13(3). Mesh "in walls designed to resist uplift of more than 100 plf" refers to the use of mesh in accordance with Section AS106.14.

AS106.9.2 Mesh attachment. Mesh shall be attached with staples to top plates or roof-bearing elements and to sill plates in accordance with all of the following:

1. **Staples.** Staples shall be pneumatically driven, stainless steel or electro-galvanized, 16 gage with 1 1/2-inch (38 mm) legs, 7/16-inch (11.1 mm) crown; or manually driven, galvanized, 15 gage with 1-inch (25 mm) legs. Other staples shall be permitted to be used as designed by a registered design professional. Staples into preservative-treated wood shall be stainless steel.

2. **Staple orientation.** Staples shall be firmly driven diagonally across mesh intersections at the required spacing.

3. **Staple spacing.** Staples shall be spaced not more than 4 inches (102 mm) on center, except where a lesser spacing is required by Table AS106.13(1) or Section AS106.14, as applicable.

❖ Mesh is required to be attached to a wall's horizontal boundary elements, such as roof-bearing elements and sill plates. Attachment to posts or studs (where present) is not required, but is recommended, particularly at vertical boundaries, such as wall corners or where a wall is interrupted by a window or door opening.

Staples are required to be stainless steel or galvanized steel to minimize or prevent corrosion of staples in the event that significant moisture is present. Where galvanized, Section R317.3.1 requires that the fasteners be hot-dipped, zinc coated galvanized steel. Staples are required to be stainless steel where driven into preservative-treated wood.

Installing staples diagonally across mesh intersections is important for strawbale braced wall panels (shear walls) in order to minimize slip of the mesh during load transfer between the mesh and the top and bottom plates, regardless of the direction of movement. This can also be important for transfer of loads for mesh used on load-bearing walls, or to resist out-of-plane loads or wind uplift loads. Diagonal stapling also helps limit splitting of the wood member, thus helping to maintain the strength of the connection. Diagonally placed staples are required for structural strawbale walls and are recommended good practice where mesh is used on strawbale walls.

AS106.9.3 Steel mesh. Steel mesh shall be galvanized, and shall be separated from preservative-treated wood by Grade D paper, No. 15 roofing felt or other *approved* barrier.

❖ Where steel mesh is used in structural strawbale plasters, it is required to be galvanized to reduce the potential for corrosion. Where galvanized, Section R317.3.1 requires that the fasteners be hot-dipped, zinc coated galvanized steel. The required separation of steel mesh from preservative-treated wood in this section is intended to prevent corrosion in the mesh.

AS106.9.4 Mesh in plaster. Required mesh shall be embedded in the plaster except where staples fasten the mesh to horizontal boundary elements.

❖ Required mesh must be embedded in the plaster so that they work together as a structural composite to resist in-plane forces. The plaster provides resistance to compressive forces and the mesh provides resistance to tensile and shear forces. The plaster helps resist deformation of the embedded mesh under load, while resolving diagonal tensile forces into alignment with the principal directions of the mesh.

Ideally, mesh is embedded in the middle third of the overall thickness of the plaster. In practice, deviation from ideal embedment can occur due to the irregular surface of the bales and normal construction practices. Various spacers have been used to hold mesh away from the bales, and U-shaped wire "staples" or wire "through-ties" have been used to bring mesh closer to the bales and to help locate the mesh in the middle third of the plaster application.

Where the mesh is stapled to the top-of-wall member or assembly and to the sill plates, the mesh cannot be fully embedded in the plaster. Tested strawbale wall assemblies used as the basis for the structural provisions of this appendix were constructed using direct fastening of mesh to their horizontal boundary elements and not more than reasonable effort to locate the mesh in the middle third of their plasters.

AS106.10 Support of plaster skins. Plaster *skins* on strawbale structural walls shall be continuously supported along their bottom edge. Acceptable supports include: a concrete or masonry stem wall, a concrete slab-on-grade, a wood-framed floor blocked with an *approved* engineered design or a steel angle anchored with an *approved* engineered design. A weep screed as described in Section R702.7.2.1 is not an acceptable support.

❖ Plaster skins are much stiffer than the strawbale core of a strawbale wall and thus carry most of the applied load. The strawbale core provides lateral bracing to the relatively thin plaster skins.

Both vertical and lateral loads are delivered through the plaster skins to the skins' bottom edges. It is vital that the bottom edge of the plaster skins are supported in a way that is capable of delivering these loads to the foundation. Section AS106.10 describes acceptable ways of supporting plaster skins for structural strawbale walls. See Commentary Figure AS105.3(1) for typical plaster support on a concrete slab-on-grade with monolithic foundation, and Commentary Figure AS105.3(2) for typical plaster support for a raised floor condition.

Where structural strawbale walls are supported by a wood-framed floor, an approved engineered design is required. The requirement that these floors be blocked refers to blocking between joists that supports the interior plaster where floor joists run perpendicular to the wall. Where the joists run parallel to the wall, a joist directly below the interior plaster provides that support. Such blocking or joists should also support the interior sill plate of the wall, and all sill plates should have adequate attachment to the floor framing, so that the system maintains a continuous load path to the foundation. Where strawbale braced wall panels are used, the engineered design should also address any associated uplift in the floor system.

Note that the commonly used weep screed for cement plaster over wood sheathing, as described in Section R703.7.2.1, is not an acceptable structural support for strawbale construction. Conventional weep screeds are not designed to support structural loads; however, such a weep screed is acceptable for nonstructural strawbale walls.

AS106.11 Transfer of loads to and from plaster skins. Where plastered strawbale walls are used to support superimposed vertical loads, such loads shall be transferred to the plaster *skins* by continuous direct bearing or by an *approved* engineered design. Where plastered strawbale walls are used to resist in-plane lateral loads, such loads shall be transferred to the reinforcing mesh from the structural member or assembly above and to the sill plate in accordance with Table AS106.13(3).

❖ Section AS106.11 contains criteria for the transfer of loads to the top of load-bearing strawbale walls, to or from the top-of-wall structural members and to the sill plates of strawbale shear walls (braced wall panels). Prescriptive requirements for the sill plates are contained in Section AS105.3. There are no prescriptive requirements for the top-of-wall members or assemblies for structural or nonstructural strawbale walls.

Roof loads imposed on load-bearing strawbale walls require continuous direct bearing on the top edges of the plaster, or an engineered means of transferring the loads to the plaster skins. Commentary Figure AS106.11(1) shows an example of direct bearing onto the top edge of the plaster. Other configurations are also possible. Roof loads should be transferred approximately equally to both plaster skins, or only half the allowable bearing capacity should be used. See Section AS106.12.2 for information regarding concentrated loads.

One example of an engineered means of load transfer without direct bearing is from the roof bearing assembly through attached mesh into a plaster-mesh matrix, with approved calculations or test results demonstrating that the allowable capacity of the load-transfer system is greater than the design loads. Any system other than direct bearing should demonstrate a load path with sufficient capacity.

For strawbale walls used to resist in-plane lateral loads (wind or seismic), the load transferred from the structural member or assembly at the top of the wall through the required mesh and staples and into the required sill plate shall be as stipulated in Table AS106.13(1).

Historically, load-bearing strawbale walls have used some variation of a wooden box beam as a roof- or floor-bearing assembly at the top of the wall, though other members or assemblies have been used. Commentary Figure AS106.11(1) shows a generic box beam at the top of a load-bearing strawbale wall. Box beams and other roof-bearing assemblies can also serve as headers over wall openings, as indicated in Commentary Figure AS106.11(1). The figure also shows an example of locations for mesh attachment to the box beam when the strawbale wall is used as a braced wall panel (shear wall).

For nonstructural strawbale walls, the members at the top of the walls are typically the beams of a post-and-beam frame, where the top course of straw bales is infilled tightly against the underside of the beams, or notched to accommodate them. Commentary Figure AS106.11(2) shows a generic configuration of the top of a post-and-beam wall with strawbale infill. The figure also shows example locations for mesh attachment to the beams and roof assembly where the strawbale wall is used as a braced wall panel.

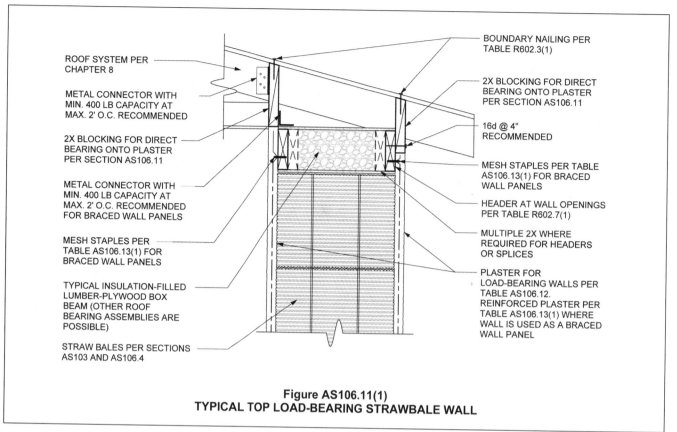

ROOF SYSTEM PER CHAPTER 8

METAL CONNECTOR WITH MIN. 400 LB CAPACITY AT MAX. 2' O.C. RECOMMENDED

2X BLOCKING FOR DIRECT BEARING ONTO PLASTER PER SECTION AS106.11

METAL CONNECTOR WITH MIN. 400 LB CAPACITY AT MAX. 2' O.C. RECOMMENDED FOR BRACED WALL PANELS

MESH STAPLES PER TABLE AS106.13(1) FOR BRACED WALL PANELS

TYPICAL INSULATION-FILLED LUMBER-PLYWOOD BOX BEAM (OTHER ROOF BEARING ASSEMBLIES ARE POSSIBLE)

STRAW BALES PER SECTIONS AS103 AND AS106.4

BOUNDARY NAILING PER TABLE R602.3(1)

2X BLOCKING FOR DIRECT BEARING ONTO PLASTER PER SECTION AS106.11

16d @ 4" RECOMMENDED

MESH STAPLES PER TABLE AS106.13(1) FOR BRACED WALL PANELS

HEADER AT WALL OPENINGS PER TABLE R602.7(1)

MULTIPLE 2X WHERE REQUIRED FOR HEADERS OR SPLICES

PLASTER FOR LOAD-BEARING WALLS PER TABLE AS106.12. REINFORCED PLASTER PER TABLE AS106.13(1) WHERE WALL IS USED AS A BRACED WALL PANEL

Figure AS106.11(1)
TYPICAL TOP LOAD-BEARING STRAWBALE WALL

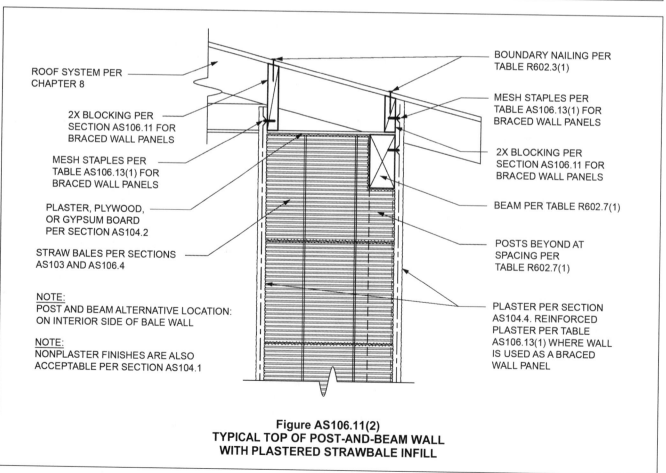

ROOF SYSTEM PER CHAPTER 8

2X BLOCKING PER SECTION AS106.11 FOR BRACED WALL PANELS

MESH STAPLES PER TABLE AS106.13(1) FOR BRACED WALL PANELS

PLASTER, PLYWOOD, OR GYPSUM BOARD PER SECTION AS104.2

STRAW BALES PER SECTIONS AS103 AND AS106.4

NOTE:
POST AND BEAM ALTERNATIVE LOCATION: ON INTERIOR SIDE OF BALE WALL

NOTE:
NONPLASTER FINISHES ARE ALSO ACCEPTABLE PER SECTION AS104.1

BOUNDARY NAILING PER TABLE R602.3(1)

MESH STAPLES PER TABLE AS106.13(1) FOR BRACED WALL PANELS

2X BLOCKING PER SECTION AS106.11 FOR BRACED WALL PANELS

BEAM PER TABLE R602.7(1)

POSTS BEYOND AT SPACING PER TABLE R602.7(1)

PLASTER PER SECTION AS104.4. REINFORCED PLASTER PER TABLE AS106.13(1) WHERE WALL IS USED AS A BRACED WALL PANEL

Figure AS106.11(2)
TYPICAL TOP OF POST-AND-BEAM WALL
WITH PLASTERED STRAWBALE INFILL

AS106.12 Load-bearing walls. Plastered strawbale walls shall be permitted to be used as load-bearing walls in one-*story* buildings to support vertical loads imposed in accordance with Section R301, in accordance with and not more than the allowable bearing capacities indicated in Table AS106.12.

❖ See Commentary Figure AS106.11(1) for the top of a typical load-bearing wall. See Commentary Figures AS105.3(1) and AS105.3(2) for the base of a typical load-bearing wall over a slab-on-grade and for a raised wood-framed floor, respectively.

TABLE AS106.12. See page S-22.

❖ The allowable bearing capacities in Table AS106.12 are for strawbale walls with plaster on both sides. Plaster can be installed on one side only; however, the allowable bearing capacity is then half of that shown in the table.

AS106.12.1 Precompression of load-bearing strawbale walls. Prior to application of plaster, walls designed to be load bearing shall be precompressed by a uniform load of not less than 100 plf (1459 N/m).

❖ The purpose of requiring precompression of load-bearing strawbale walls is to accelerate settling of the stacked straw bales before plaster is applied so that potential cracking of the plaster is minimized. The required 100 plf (1459 N/m) precompression is intended to be permanent and can be achieved by a variety of means, including vertically tensioned steel mesh fastened to sill plates and top plates, or tensioned packaging straps, or other tensile material looped under sill plates and over top plates. In many cases, the dead load of the roof on the wall (prior to plaster) may meet the 100 plf (1459 N/m) precompression requirement.

AS106.12.2 Concentrated loads. Concentrated loads shall be distributed by structural elements capable of distributing the loads to the bearing wall within the allowable bearing capacity listed in Table AS106.12 for the plaster type used.

❖ Roof-bearing members or assemblies must be used to distribute concentrated loads to avoid exceeding the allowable bearing capacity listed in Table AS106.12 for the plaster type used. The load transfer to the plaster skins must be by continuous direct bearing or with an approved engineered design, as required in Section AS106.11.

Concentrated loads commonly occur at both sides of a window or door opening. For openings not exceeding 4 feet, the header (typically as part of a continuous roof-bearing assembly) may be used to transfer the load over the first 2 feet on each side of the opening. For larger openings, studs or posts on each side of the opening are recommended. Headers must comply with Tables R602.7(1) and R602.7(2) and may be part of a continuous roof-bearing assembly.

Concentrated loads from a beam or post resting on the roof-bearing assembly can be considered to be carried by a 4-foot length of wall (2 feet on each side of the load) where the wall is unbroken by an opening.

Any uniform loads carried by a wall where a concentrated load is supported must be taken into account when determining whether the allowable bearing capacity of the wall is exceeded. Concentrated loads should be transferred approximately equally to both plaster skins, or only half the allowable bearing capacity should be used.

AS106.13 Braced panels. Plastered strawbale walls shall be permitted to be used as braced wall panels for one-story buildings in accordance with Section R602.10 of the *International Residential Code*, and with Tables AS106.13(1), AS106.13(2) and AS106.13(3). Wind design criteria shall be in accordance with Section R301.2.1. Seismic design criteria shall be in accordance with Section R301.2.2.

❖ Plastered strawbale walls can be used as braced wall panels (shear walls) to resist in-plane wind and/or seismic loads where they are constructed in accordance with Table AS106.13(1), and meet the minimum total lengths stipulated in Table AS106.13(2) for wind loads and Table AS106.13(3) for seismic loads.

Testing of plastered strawbale walls has demonstrated their ability to absorb and dissipate energy under extreme in-plane loading. The reinforced plaster provides an in-plane lateral load-resisting element with stiffness comparable to wood panels on light wood framing. The strawbale core braces the relatively thin plaster skins and provides significant backup capacity, including the ability to support superimposed gravity loads.

Conventional hold-downs are not required on strawbale braced wall panels to resist overturning from seismic or wind forces. The required reinforcing mesh and the weight of the roof and wall assemblies provide adequate resistance to overturning forces from seismic and wind loads.

TABLE AS106.13(1). See page S-22.

❖ The strawbale braced wall panel types in Table AS106.13(1) are with plaster on both sides. Plaster is permitted to be installed on one side only if the minimum total braced wall panel lengths shown in Tables AS106.13(2) and AS106.13(3) are doubled.

For strawbale braced wall panels used to resist wind forces in accordance with Table AS106.13(2) or seismic forces in accordance with Table AS106.13(3), it is recommended that the mesh required in Table AS106.13(1) be supplemented by a 1-foot wide band of mesh along the sill plate with the same fastening as for the required mesh. This improves the transfer of forces to the sill plate.

Although strawbale braced wall panel type A1 appears in Table AS106.13(1), it is not listed as a braced wall panel type in Table AS106.13(2) for resisting wind loads or in Table AS106.13(3) for resisting seismic loads. Only mesh-reinforced plasters are used to resist in-plane lateral loads in this appendix.

TABLE AS106.13(2). See page S-23.

❖ The minimum total lengths of strawbale braced wall panels in Table AS106.13(2) are based on the basic wind speed at the building location (see the definition of "Basic wind speed" in the code). The basic wind speed can be determined by converting the ultimate design wind speed found in Figure R301.2(4)A to the nominal design wind speed in accordance with Section R301.2.1.3 and Table R301.2.1.3. Nominal design speed is equivalent to basic wind speed, therefore the converted value can be used as the basic wind speed in Table AS106.13(2). The required total length of strawbale braced wall panels may be determined by linear interpolation between the tabulated basic wind speeds, or by using the length for a higher basic wind speed. Alternatively, the building official may allow the basic wind speed to be determined from another source, such as Figure R301.2(4)A of the 2012 code.

Note c clarifies minimum total lengths of braced wall panel along each braced wall line in Table AS106.13(2). Where the required length is satisfied with a single braced wall panel, the aspect ratio (H:L) of the braced wall panel must not exceed 2:1. The required length may also be satisfied using multiple braced wall panels. If each of those braced wall panels has an aspect ratio not exceeding 1:1, the sum of their lengths must simply satisfy the required total length in the table. If any of those braced wall panels has an aspect ratio exceeding 1:1, the required total length must be increased by multiplying it by the largest aspect ratio (H:L) of the braced wall panels in that line.

Where a strawbale braced wall panel ends at a corner, the lengths of its plaster skins on opposite sides of the wall may differ. In such cases the braced wall panel length should be taken as the average of the two lengths.

TABLE AS106.12
ALLOWABLE SUPERIMPOSED VERTICAL LOADS (LBS/FOOT) FOR PLASTERED LOAD-BEARING STRAWBALE WALLS

WALL DESIGNATION	PLASTER[a] (both sides) Minimum thickness in inches each side	MESH[b]	STAPLES[c]	ALLOWABLE BEARING CAPACITY[d] (plf)
A	Clay $1^1/_2$	None required	None required	400
B	Soil-cement 1	Required	Required	800
C	Lime $^7/_8$	Required	Required	500
D	Cement-lime $^7/_8$	Required	Required	800
E	Cement $^7/_8$	Required	Required	800

For SI: 1 inch = 25.4mm, 1 pound per foot = 14.5939 N/m.

a. Plasters shall conform to Sections AS104.4.3 through AS104.4.8, AS106.7 and AS106.10.

b. Any metal mesh allowed by this appendix and installed in accordance with Section AS106.9.

c. In accordance with Section AS106.9.2, except as required to transfer roof loads to the plaster skins in accordance with Section AS106.11.

d. For walls with a different plaster on each side, the lower value shall be used.

TABLE AS106.13(1)
PLASTERED STRAWBALE BRACED WALL PANEL TYPES

WALL DESIGNA-TION	PLASTER[a] (both sides) Type	Thickness (minimum in inches each side)	SILL PLATES[b] (nominal size in inches)	ANCHOR BOLT[c] SPACING (inches on center)	MESH[d] (inches)	STAPLE SPACING[e] (inches on center)
A1	Clay	1.5	2 × 4	32	None	None
A2	Clay	1.5	2 × 4	32	2 × 2 high-density polypropylene	2
A3	Clay	1.5	2 × 4	32	2 × 2 × 14 gage	4
B	Soil-cement	1	4 × 4	24	2 × 2 × 14 gage	2
C1	Lime	$^7/_8$	2 × 4	32	17-gage woven wire	3
C2	Lime	$^7/_8$	4 × 4	24	2 × 2 × 14 gage	2
D1	Cement-lime	$^7/_8$	4 × 4	32	17 gage woven wire	2
D2	Cement-lime	$^7/_8$	4 × 4	24	2 × 2 × 14 gage	2
E1	Cement	$^7/_8$	4 × 4	32	2 × 2 × 14 gage	2
E2	Cement	1.5	4 × 4	24	2 × 2 × 14 gage	2

SI: 1 inch = 25.4 mm

a. Plasters shall conform with Sections AS104.4.3 through AS104.4.8, AS106.7, AS106.8 and AS106.12.

b. Sill plates shall be Douglas fir-larch or southern pine and shall be *preservative treated* where required by the *International Residential Code*.

c. Anchor bolts shall be in accordance with Section AS106.13.3 at the spacing shown in this table.

d. Installed in accordance with Section AS106.9.

e. Staples shall be in accordance with Section AS106.9.2 at the spacing shown in this table.

TABLE AS103.13(3). See page S-24.

❖ For buildings located in Seismic Design Categories A and B, the minimum total length of braced wall panels is governed by the basic wind speed in accordance with Table AS106.13(2).

Buildings located in Seismic Design Categories C, D_0, D_1, and D_2 are also subject to wind loads. The required length of their bracing along each braced wall line is the greater value determined from Table AS106.13(2) or AS106.13(3), with applicable adjustment factors.

Soil Class D is assumed in these tables and in Table R602.10.3(3). Table R602.10.3(3) applies to wood-frame construction. For site-specific soil classes other than D, adjustments can be made to the minimum total length of strawbale braced wall panels, as is allowed for braced wall panels in wood-frame construction in accordance with Note b of Table R602.10.3(3).

Note c clarifies the minimum required total lengths of braced wall panels along each braced wall line in Table AS106.13(3). Where the required length is satisfied with a single braced wall panel, the aspect ratio (H:L) of the braced wall panel must not exceed 2:1. The required length may also be satisfied using multiple braced wall panels. If each of those braced wall panels has an aspect ratio not exceeding 1:1, the sum of their lengths must simply satisfy the required total length in the table. If any of those braced wall panels has an aspect ratio exceeding 1:1, the required total length must be increased by multiplying it by the largest aspect ratio (H:L) of the braced wall panels in that line.

Where a strawbale braced wall panel ends at a corner, the lengths of its plaster skins on opposite sides of the wall may differ. In such cases, the braced wall panel length should be taken as the average of the two lengths.

TABLE AS106.13(2)
BRACING REQUIREMENTS FOR STRAWBALE BRACED WALL PANELS BASED ON WIND SPEED

• EXPOSURE CATEGORY B[d] • 25-FOOT MEAN ROOF HEIGHT • 10-FOOT EAVE-TO-RIDGE HEIGHT[d] • 10-FOOT WALL HEIGHT[d] • 2 BRACED WALL LINES[d]			MINIMUM TOTAL LENGTH (FEET) OF STRAWBALE BRACED WALL PANELS REQUIRED ALONG EACH BRACED WALL LINE[a, b, c, d]		
Basic wind speed (mph)	Story location	Braced wall line spacing (feet)	Strawbale braced wall panel[e] A2, A3	Strawbale braced wall panel[e] C1, C2, D1	Strawbale braced wall panel[e] D2, E1, E2
≤ 85	One-story building	10	6.4	3.8	3.0
		20	8.5	5.1	4.0
		30	10.2	6.1	4.8
		40	13.3	6.9	5.5
		50	16.3	7.7	6.1
		60	19.4	8.3	6.6
≤ 90	One-story building	10	6.4	3.8	3.0
		20	9.0	5.4	4.3
		30	11.2	6.4	5.1
		40	15.3	7.4	5.9
		50	18.4	8.1	6.5
		60	21.4	8.8	7.0
≤ 100	One-story building	10	7.1	4.3	3.4
		20	10.2	6.1	4.8
		30	14.3	7.2	5.7
		40	18.4	8.1	6.5
		50	22.4	9.0	7.1
		60	26.5	9.8	7.8
≤ 110	One-story building	10	7.8	4.7	3.7
		20	12.2	6.6	5.3
		30	17.3	7.9	6.3
		40	22.4	9.0	7.1
		50	26.5	9.8	7.8
		60	31.6	11.4	8.5

For SI: 1 inch = 25.4 mm, 1 foot = 305 mm, 1 mile per hour = 0.447 m/s.

a. Linear interpolation shall be permitted.

b. All *braced wall panels* shall be without openings and shall have an aspect ratio (H:L) ≤ 2:1.

c. Tabulated minimum total lengths are for *braced wall lines* using single braced wall panels with an aspect ratio (H:L) ≤ 2:1, or using multiple *braced wall panels* with *aspect ratios* (H:L) ≤ 1:1. For *braced wall lines* using two or more *braced wall panels* with an aspect ratio (H:L) > 1:1, the minimum total length shall be multiplied by the largest *aspect ratio* (H:L) of *braced wall panels* in that line.

d. Subject to applicable wind adjustment factors associated with "All methods" in Table R602.10.3(2)

e. Strawbale braced panel types indicated shall comply with Sections AS106.13.1 through AS106.13.3 and with Table AS106.13(1).

Braced wall panel designation A3 is not intended to be excluded from use in resisting seismic forces in accordance with Table AS106.13(3). If this panel type is used, the minimum total lengths should be 25 percent greater than the lengths required for braced wall panel designation A2.

AS106.13.1 Bale wall thickness. The thickness of the stacked bale wall without its plaster shall be not less than 15 inches (381 mm).

❖ The thickness of a strawbale wall contributes to its stability under strong ground shaking because: (a) The width of the bale ensures overlying bales are well supported as gaps open during in-plane rocking; (b) The width of the bale enhances out-of-plane strength and stiffness; and (c) lateral restoring forces are developed as walls rock out-of-plane. This section is not meant to restrict the use of two-string bales on edge [which typically result in a bale wall thickness of 14 inches (356 mm)] where a history of good experience exists and where wall-height-to-thickness requirements in Table AS105.4 are satisfied, particularly in Seismic Design Categories A and B. Walls with 12-inch-wide (305 mm) bales for use in small residential buildings have performed well in the field and in full-scale structural tests (Donovan, 2014, see the bibliography).

AS106.13.2 Sill plates. Sill plates shall be in accordance with Table AS106.13(1).

AS106.13.3 Sill plate fasteners. Sill plates shall be fastened with not less than $5/_8$-inch-diameter (15.9 mm) steel anchor bolts with 3-inch by 3-inch by $3/_{16}$-inch (76.2 mm by 76.2 mm by 4.8 mm) steel washers, with not less than 7-inch (177.8 mm) embedment in a concrete or masonry foundation, or shall be an approved equivalent, with the spacing shown in Table AS106.13(1). Anchor bolts or other fasteners into framed floors shall be of an approved engineered design.

AS106.14 Resistance to wind uplift forces. Plaster mesh in *skins* of strawbale walls that resist uplift forces from the roof assembly, as determined in accordance with Section R802.11, shall be in accordance with all of the following:

1. Plaster shall be any type and thickness allowed in Section AS104.

2. Mesh shall be any type allowed in Table AS106.13(1), and shall be attached to top plates or roof-bearing elements and to sill plates in accordance with Section AS106.9.2.

TABLE AS106.13(3)
BRACING REQUIREMENTS FOR STRAWBALE BRACED WALL PANELS BASED ON SEISMIC DESIGN CATEGORY

• SOIL CLASS D[d] • WALL HEIGHT = 10 FEET[d] • 15 PSF ROOF-CEILING DEAD LOAD[d] • BRACED WALL LINE SPACING ≤ 25 FEET[d]			MINIMUM TOTAL LENGTH (FEET) OF STRAWBALE BRACED WALL PANELS REQUIRED ALONG EACH BRACED WALL LINE[a, b, c, d]	
Seismic Design Category	Story location	Braced wall line length (feet)	Strawbale Braced Wall Panel[e] A2, C1, C2, D1	Strawbale Braced Wall Panel[e] B, D2, E1, E2
C	One-story building	10 20 30 40 50	5.7 8.0 9.8 12.9 16.1	4.6 6.5 7.9 9.1 10.4
D₀	One-story building	10 20 30 40 50	6.0 8.5 10.9 14.5 18.1	4.8 6.8 8.4 9.7 11.7
D₁	One-story building	10 20 30 40 50	6.3 9.0 12.1 16.1 20.1	5.1 7.2 8.8 10.4 13.0
D₂	One-story building	10 20 30 40 50	7.1 10.1 15.1 20.1 25.1	5.7 8.1 9.9 13.0 16.3

For SI: 1 inch = 25.4 mm, 1 foot = 305 mm, 1 pound per square foot = 0.0479 kPa.

a. Linear interpolation shall be permitted.

b. *Braced wall panels* shall be without openings and shall have an *aspect ratio* (H:L) ≤ 2:1.

c. Tabulated minimum total lengths are for *braced wall lines* using single *braced wall panels* with an *aspect ratio* (H:L) ≤ 2:1, or using multiple *braced wall panels* with *aspect ratios* (H:L) ≤ 1:1. For *braced wall lines* using two or more *braced wall panels* with an aspect ratio (H:L) > 1:1, the minimum total length shall be multiplied by the largest *aspect ratio* (H:L) of *braced wall panels* in that line.

d. Subject to applicable seismic adjustment factors associated with "All methods" in Table R602.10.3(4), except "Wall dead load."

e. Strawbale *braced wall panel* types indicated shall comply with Sections AS106.13.1 through AS106.13.3 and Table AS106.13(1).

3. Sill plates shall be not less than nominal 2-inch by 4-inch (51 mm by 102 mm) with anchoring complying with Section R403.1.6.

4. Mesh attached with staples at 4 inches (51 mm) on center shall be considered to be capable of resisting uplift forces of 100 plf (1459 N/m) for each plaster skin.

5. Mesh attached with staples at 2 inches (51 mm) on center shall be considered to be capable of resisting uplift forces of 200 plf (2918 N/m) for each plaster skin.

❖ Section R802.11 gives wind-uplift-resistance requirements for roof assemblies. Mesh used to satisfy plaster reinforcement requirements for strawbale load-bearing walls or braced wall panels may also be used to resist wind uplift forces. Supplemental mesh may also be provided in the plaster of structural or nonstructural walls for this purpose alone.

Mesh must be adequately anchored as part of a complete load path in accordance with this section. Also see Section AS106.9.1, which indicates that mesh used in walls designed to resist wind uplift of more than 100 plf must run continuous vertically from sill plate to the top plate or roof-bearing element, or shall lap not less than 8 inches.

SECTION AS107
FIRE RESISTANCE

AS107.1 Fire-resistance rating. Strawbale walls shall be considered to be nonrated, except for walls constructed in accordance with Section AS107.1.1 or AS107.1.2. Alternately, fire-resistance ratings of strawbale walls shall be determined in accordance with Section R302 of the *International Residential Code*.

❖ Many variations of strawbale wall assemblies have been used in practice, but only two assemblies, tested in accordance with one of the two standards required for determining fire-resistance ratings of walls, are included in the appendix. These two assemblies, as described in Sections AS107.1.1 and AS107.1.2, can be used as fire-resistance-rated walls with their stated ratings, where a fire-resistance-rated wall is required by the code. All other strawbale wall assemblies, whether finished with plaster or another acceptable finish, are considered nonrated without further justification.

This section also refers to Sections R302 and 302.1 and Tables R302.1(1) and R302.1(2), which give minimum fire-resistance ratings for exterior walls in certain circumstances and state that such fire-resistance-rated walls must be tested in accordance with ASTM E119 or UL 263. The rated wall assemblies described in Sections AS107.1.1 and AS107.1.2 were tested in accordance with ASTM E119 at a certified testing laboratory.

Although only two strawbale wall assemblies are given fire-resistance ratings in this appendix (based on specific ASTM E119 tests), two decades of field experience with and testing of other assemblies, including European tests to German Institute for Standardization

(DIN) standards, have more broadly shown plastered strawbale walls to be highly resistant to fire.

As with wood-frame construction or other assemblies that are flammable prior to being finished, straw bale walls are vulnerable to fire while the straw is exposed. However, because of the density of the bales and the limited oxygen within the compressed straw, even unplastered strawbale wall assemblies have shown good resistance to fire. Their burning characteristics have been compared to large wood timbers where charring occurs at the surface but the lack of oxygen and the insulating effect of the charred material greatly slows combustion beyond the charred surface.

An important fire-related issue that is not addressed in this section, but is addressed in other sections of this appendix, is that strawbale walls must not contain vertical spaces that could act as a chimney to spread fire across unprotected straw where an ignition source is present. This is mostly a matter of preventing fire from spreading to wood roof or floor framing above the wall because, although fire can spread quickly across the face of bales, the compressed bales generally do not contain sufficient oxygen to sustain a flame. Vertical spaces in a strawbale wall that must be eliminated include: (a) Space between the face of the stacked bales and their finish; (b) Space between bales (especially in a stack bond wall); and (c) Spaces between vertical framing members and the straw where straw bales are used as infill in a wood frame wall.

Section AS104.2 requires either a plaster finish on strawbale walls or that the bales "fit tightly against a solid wall panel" (such as plywood, fiberboard, or gypsum board). This is intended to address the condition described in "a" in the preceding paragraph. Section AS106.5 requires stuffing of voids for structural walls between bales, but this is important for all walls as a matter of fire blocking to address the conditions described in "b" and "c" in the preceding paragraph, both between bales and between bales and any vertical framing. Bales should be notched around framing members wherever possible, or the spaces should be stuffed with straw or straw-clay.

Section R302.10.1 provides requirements for the flame spread index and smoke-developed index for insulation in wall assemblies, roof-ceiling assemblies, crawl spaces and attics. Such materials must have a flame spread index not exceeding 25 and a smoke-developed index not exceeding 450, as tested in accordance with ASTM E84 or UL 723. Straw bales in exterior strawbale walls typically function as insulation (among other functions) and thus are subject to this section. Although this issue for straw bales is not addressed in this appendix, an ASTM E84 test of unplastered straw bales conducted in an accredited testing laboratory in the year 2000 indicated a flame spread index of 10 and a smoke-developed index of 350, which satisfies the requirements of Section R302.10.1.

AS107.1.1 One-hour rated clay plastered wall. One-hour fire-resistance-rated nonload-bearing clay plastered strawbale walls shall comply with all of the following:

1. Bales shall be laid flat or on-edge in a running bond.

2. Bales shall maintain thickness of not less than 18 inches (457 mm).

3. Gaps shall be stuffed with straw-clay.

4. Clay plaster on each side of the wall shall be not less than 1 inch (25 mm) thick and shall be composed of a mixture of 3 parts clay, 2 parts chopped straw and 6 parts sand, or an alternative approved clay plaster.

5. Plaster application shall be in accordance with Section AS104.4.3.3 for the number and thickness of coats.

❖ See the commentary to Section AS107.1.

AS107.1.2 Two-hour rated cement plastered wall. Two-hour fire-resistance-rated nonload-bearing cement plastered strawbale walls shall comply with all of the following:

1. Bales shall be laid flat or on-edge in a running bond.

2. Bales shall maintain a thickness of not less than 14 inches (356 mm).

3. Gaps shall be stuffed with straw-clay.

4. $1^1/_2$-inch (38 mm) by 17-gage galvanized woven wire mesh shall be attached to wood members with $1^1/_2$-inch (38 mm) staples at 6 inches (152 mm) on center. 9 gage U-pins with not less than 8-inch (203 mm) legs shall be installed at 18 inches (457 mm) on center to fasten the mesh to the bales.

5. Cement plaster on each side of the wall shall be not less than 1 inch (25 mm) thick.

6. Plaster application shall be in accordance with Section AS104.4.8 for the number and thickness of coats.

❖ See the commentary to Section AS107.1.

AS107.2 Openings in rated walls. Openings and penetrations in bale walls required to have a fire-resistance rating shall satisfy the same requirements for openings and penetrations as prescribed in the *International Residential Code.*

AS107.3 Clearance to fireplaces and chimneys. Strawbale surfaces adjacent to fireplaces or chimneys shall be finished with not less than $^3/_8$-inch (10 mm) thick plaster of any type permitted by this appendix. Clearance from the face of such plaster to fireplaces and chimneys shall be maintained as required from fireplaces and chimneys to combustibles in Chapter 10, or as required by manufacturer's instructions, whichever is more restrictive.

❖ Section R1001.11 addresses the required clearances from masonry fireplaces to combustible materials, Section R1002.5 addresses the required clearances from masonry heaters to combustible materials and Section R1003.18 addresses the required clearances from masonry chimneys to combustible materials.

Strawbale walls must maintain those clearances to the face of the strawbale wall's required plaster. For factory-built fireplaces and chimneys, strawbale walls must maintain the clearances to combustible materials required by the manufacturer's installation instructions, from the fireplace or chimney to the face of the strawbale wall's required plaster.

These clearance requirements are more conservative for strawbale construction than for wood-frame construction, adding both a finish requirement and the resulting increased distance to the combustible material. Whereas wood framing is allowed to be directly exposed in these locations, this appendix requires the straw to be finished with plaster, and the distances are required to be to the face of the plaster, thereby increasing the distance to combustible material by the thickness of the plaster.

The requirement for a plaster finish is not intended to preclude the use of materials that afford similar protection, such as gypsum board, wood framing, or plywood, where those materials are tight against the straw.

SECTION AS108
THERMAL INSULATION

AS108.1 R-value. The unit *R*-value of a strawbale wall with bales laid flat is R-1.3 per inch of bale thickness. The unit *R*-value of a strawbale wall with bales on-edge is R-2 per inch of bale thickness.

❖ The unit *R*-values in this section are used for exterior strawbale walls to determine compliance with Section N1102.1.1 and Table N1102.1.1. Following the principles of Section N1102.1.2, only the *R*-value of the straw (as the cavity insulation) and any insulating sheathing, should be used when computing the *R*-value, not finishes such as plaster or siding. The computed *R*-value for the strawbale wall must be at least the *R*-value in Table N1102.1.1 in the column "Wood Frame Wall *R*-value," for the climate zone where the structure is located.

The unit *R*-values in this section reflect results from what is considered the most reliable thermal testing for strawbale walls to date—the 1998 guarded hot-box test at Oak Ridge National Laboratory. (See the bibliography at the end of this appendix for *Thermal Performance of Straw Bale Wall Systems.*)

Strawbale walls have been shown to have different unit *R*-values depending upon the orientation of the bales (i.e., laid flat or on-edge). This is because industrialized baling machines typically create bales where the strands of straw are predominantly oriented across the "width" of the bale (across its strings). Straw conducts heat more readily along than across its length (where more air pockets interrupt the flow of heat). Thus, the thermal resistance per inch of a bale on-edge is typically greater than that of a bale laid flat.

The unit *R*-values stated in this section should be considered accurate for typical bales. However, particular bales that show a predominant straw orientation differing from that described here might be reasonably considered to have unit *R*-values between the R-1.3 and R-2, respectively, for bales laid flat or on-edge.

Other factors that affect the thermal performance of a straw bale wall include bale density, adhesion of the facing materials (e.g., plaster) to the straw bales and homogeneity of the wall system. Although a minimum dry density of 6.5 lbs/cu.ft. is required, higher-density bales generally provide higher thermal performance. Additionally, good adhesion of a plaster finish to the straw bales reduces air infiltration, which helps maintain or increase the *R*-value of the assembly.

Finally, any gaps between bales, or between a bale and other wall elements such as framing members, should be tightly filled with straw or straw-clay. Otherwise a significant negative impact on the overall thermal performance of the wall assembly can occur. See Section AS106.5 for voids and stuffing. Stuffing of voids is mandated for structural walls between bales. Further, it is important for thermal performance of all exterior walls, between bales and between bales and framing members.

Though not explicitly addressed in this appendix, straw bales have been and may be used as insulation in nonstructural strawbale walls in any orientation, including laid flat, on-edge, and on-end (see Commentary Figure AS103.2). In nonstructural strawbale walls, elements in the same wall other than the bales and their plaster, such as wood or steel framing or structural panels, may carry structural loads.

In addition to their function as thermal insulation in walls, straw bales have been used as insulation in raised floors and ceilings. However, issues of structure and of protection from fire must be addressed where straw bales are used in these locations. The floor or ceiling structure must be adequate for carrying the additional weight of the bales, for both static and seismic loading. A complete load path to the foundation must be provided for these loads. For fire protection the bales must be tightly confined on all sides (similar to their required confinement in a wall system) by materials such as plaster, wood framing, plywood or gypsum board.

See the commentary to Section AS107.1 for information regarding the use of straw bales as insulation and how they satisfy the requirements of Section R302.10.1 for the flame spread index and smoke-developed index of insulation in wall assemblies, roof-ceiling assemblies, crawl spaces and attics.

SECTION AS109
REFERENCED STANDARDS

ASTM

C5—10	Standard Specification for Quicklime for Structural Purposes	AS104.4.6.1
C109/C 109M—12	Standard Test Method for Compressive Strength of Hydraulic Cement Mortars	AS106.6.1
C141/C 141M—09	Standard Specification for Hydrated Hydraulic Lime for Structural Purposes	AS104.4.6.1
C206—03	Standard Specification for Finishing Hydrated Lime	AS104.4.6.1
C926—12a	Standard Specification for Application of Portland Cement Based Plaster	AS104.4.7, AS104.4.8
C1707—11	Standard Specification for Pozzolanic Hydraulic Lime for Structural Purposes	AS104.4.6.1

EN

459—2010	Part 1: Building Lime. Definitions, Specifications and Conformity Criteria; Part 2: Test Methods	AS104.4.6.1

Bibliography

The following resource materials were used in the preparation of the commentary for this appendix of the code:

ASCE 7–10, *Minimum Design Loads for Buildings and Other Structures*. Reston, VA: American Society of Civil Engineers, 2010.

Aschheim, Mark, S. Jalali, C. Ash, K. Donahue, and M. Hammer. "Allowable Shears for Plastered Straw-Bale Walls." *Journal of Structural Engineering*. New York, NY: American Society of Civil Engineers, July 1, 2014.

Aschheim, Mark, Kevin Donahue, and Martin Hammer. *Strawbale Construction—Parts 1 & 2*, Structure Magazine. Chicago, IL: National Council of Structural Engineers Associations, September & October 2012.

ASTM C109–12, *Standard Test Method for Compressive Strength of Hydraulic Cement Mortars*. West Conshohocken, PA: ASTM International, 2012.

CMHC Technical Series 08-107, *Effect of Mesh and Bale Orientation on Strength of Straw Bale Walls*. Ottowa, Ontario: Canada Mortgage and Housing Corporation, 2008.

DOE-G010094–01, *House of Straw*. Washington DC: U.S. Department of Energy, Office of Energy Efficiency and Renewable Energy, 1995.

Donovan, Darcey. *Seismic Performance of Innovative Straw Bale Wall Systems*, NEES Research Report 2009-0666. Network for Earthquake Engineering Simulation, 2014. https://nees.org/warehouse/project/666

Eisenberg, David, and Martin Hammer. "Strawbale Construction and Its Evolution in Building Codes." *Building Safety Journal*. Washington, DC: International Code Council, Inc., February 2014.

Hammer, Martin. "The Status of Straw-bale Codes and Permitting Worldwide," *The Last Straw Journal*. Lincoln, NB: Green Prairie Foundation, Issue #54, April 2006.

King, Bruce, et al. *Design of Straw Bale Buildings*. San Rafael, CA: Green Building Press, 2006.

King, Bruce. "Straw-Bale Construction: A Review of Testing and Lessons Learned to Date." *Building Safety Journal*. Washington, DC: International Code Council, Inc., May-June 2004.

King, Bruce. "Straw-Bale Construction: What Have We Learned?" *The Last Straw Journal*. Lincoln, NB: Green Prairie Foundation, Issue #53, Spring 2006.

Listiburek, Joseph. *Builder's Guide to Mixed Climates*. Newtown, CT: Taunton Press, Inc., 2000.

Lstiburek, Joseph. *Insulations, Sheathings and Vapor Retarders*, Research Report 0412. Westford, MA: Building Science Press, 2004.

Magwood, Chris. "Nylon Strapping for Pre-tensioning Bale Walls." *The Last Straw Journal*. Lincoln, NB: Green Prairie Foundation, Issue #53, Spring 2006.

Parker, Andrew, M. Aschheim, D. Mar, M. Hammer, and B. King. "Recommended Mesh Anchorage Details For Straw Bale Walls." *Journal of Green Building*. Glen Allen, VA: College Publishing, Vol.1, No.4, 2004.

Racusin, Jacob and Ace McArleton, *The Natural Building Companion*. White River Junction, VT: Chelsea Green Publishing, 2012.

Stone, Nehemiah. *Thermal Performance of Straw Bale Wall Systems*. San Rafael, CA: Ecological Building Network, 2003.

Straube, John. "Moisture Basics and Straw-Bale Moisture Basics." *The Last Straw Journal*. Lincoln, NB: Green Prairie Foundation, Issue #53, Spring 2006.

The Straw Bale Alternative Solutions Resource. The Province of British Columbia, Canada: Alternative Solutions Resource Initiative, 2013.

Appendix T:
Recommended Procedure for Worst-case Testing of Atmospheric Venting Systems under N1102.4 or N1105 Conditions $\le 5ACH_{50}$

(This appendix is informative and is not part of the code.)

General Comments

Energy efficiency improvements often have a direct impact on the building pressure boundary affecting the safe operation of combustion equipment. Under certain conditions, reduced natural air-leakage coupled with the installation of atmospheric combustion appliances contribute to poor indoor air quality and possible health problems due to spillage, inadequate draft, or carbon monoxide concerns.

Purpose

This appendix is intended to provide guidance to builders, code officials and home performance contractors for worst-case testing of atmospheric venting systems to identify problems that weaken draft and restrict combustion air in existing buildings. Worst-case vent testing uses the home's exhaust fans, air-handling appliances and chimneys to create worst-case depressurization in the combustion appliance zone (CAZ). This appendix is basically a distilled version of predominant combustion safety test procedures for atmospherically vented appliances found in home performance programs across the country, such as the Enivironmental Protection Agency's Healthy Indoor Environments Protocols, and Home Performance with Energy Star; the Department of Energy's Workforce Guidelines for Home Energy Upgrades; Housing and Urban Development's Community Development Block Grants and Weatherization Assistance Programs; the Building Performance Institute's Technical Standards for the Building Analyst Professional; and the Residential Energy Services Network's Interim Guidelines for Combustion Appliance Testing and Writing Work Scopes. This is intended to take the combustion safety test procedures that are used most commonly by these home performance, weatherization, and beyond code programs, and reduce them to their simplest and most straightforward form for the purpose of combustion safety in field assessment through the use of building diagnostic tools.

SECTION T101
SCOPE

T101.1 General. This appendix is intended to provide guidelines for worst-case testing of atmospheric venting systems. Worst-case testing is recommended to identify problems that weaken draft and restrict combustion air.

SECTION T202
GENERAL DEFINITIONS

COMBUSTION APPLIANCE ZONE (CAZ). A contiguous air volume within a building that contains a Category I or II atmospherically vented appliance or a Category III or IV direct-vent or integral vent appliance drawing combustion air from inside the building or dwelling unit. The CAZ includes, but is not limited to, a mechanical closet, a mechanical room or the main body of a house or dwelling unit.

DRAFT. The pressure difference existing between the *appliance* or any component part and the atmosphere that causes a continuous flow of air and products of *combustion* through the gas passages of the *appliance* to the atmosphere.

Mechanical or induced draft. The pressure difference created by the action of a fan, blower or ejector that is located between the *appliance* and the *chimney* or vent termination.

Natural draft. The pressure difference created by a vent or *chimney* because of its height and the temperature difference between the *flue* gases and the atmosphere.

SPILLAGE. Combustion gases emerging from an appliance or venting system into the combustion appliance zone during burner operation.

SECTION T301
TESTING PROCEDURE

T301.1 Worst-case testing of atmospheric venting systems. Buildings or dwelling units containing a Category I or II atmospherically vented appliance; or a Category III or IV direct-vent or integral vent appliance drawing combustion air from inside of the building or dwelling unit, shall have the Combustion Appliance Zone (CAZ) tested for spillage, acceptable draft and carbon monoxide (CO) in accordance with this Section. Where required by the *code official*, testing shall be conducted by an *approved* third party. A written report of the results of the test shall be signed by the party conducting the test and provided to the *code official*. Testing shall be performed at any time after creation of all penetrations of the *building thermal envelope* and prior to final inspection.

Exception: Buildings or dwelling units containing only Category III or IV direct-vent or integral vent appliances that do not draw combustion air from inside of the building or dwelling unit.

The enumerated test procedure as follows shall be complied with during testing:

1. Set combustion appliances to the pilot setting or turn off the service disconnects for combustion appliances. Close exterior doors and windows and the fireplace damper. With the building or dwelling unit in this configuration, measure and record the baseline ambient pressure inside the building or dwelling unit CAZ. Compare the baseline ambient pressure of the CAZ to that of the outside ambient pressure and record the difference (Pa).

2. Establish worst case by turning on the *clothes dryer* and all exhaust fans. Close interior doors that make the CAZ pressure more negative. Turn on the air handler, where present, and leave on if, as a result, the pressure in the CAZ becomes more negative. Check interior door positions again, closing only the interior doors that make the CAZ pressure more negative. Measure net change in pressure from the CAZ to outdoor ambient pressure, correcting for the base ambient pressure inside the home. Record "worst case depressurization" pressure and compare to Table T301.1(1).

   Where CAZ depressurization limits are exceeded under worst-case conditions in accordance with Table T301.1(1), additional combustion air shall be provided or other modifications to building air-leakage performance or exhaust appliances such that depressurization is brought within the limits prescribed in Table T301.1(1).

3. Measure worst-case spillage, acceptable draft and carbon monoxide (CO) by firing the fuel-fired appliance with the smallest Btu capacity first.

   a. Test for spillage at the draft diverter with a mirror or smoke puffer. An appliance that continues to spill flue gases for more than 60 seconds fails the spillage test.

   b. Test for CO measuring undiluted flue gases in the throat or flue of the appliance using a digital gauge in parts per million (ppm) at the 10-minute mark. Record CO ppm readings to be compared with Table T301.1(3) upon completion of Step 4. Where the spillage test fails under worst case, go to Step 4.

   c. Where spillage ends within 60 seconds, test for acceptable draft in the connector not less than 1 foot (305 mm), but not more than 2 feet (610 mm) downstream of the draft diverter. Record draft pressure and compare to Table T301.1(2).

   d. Fire all other CONNECTED appliances simultaneously and test again at the draft diverter of each appliance for spillage, CO and acceptable draft using procedures 3a through 3c.

4. Measure spillage, acceptable draft, and carbon monoxide (CO) under natural conditions—without *clothes dryer* and exhaust fans on—in accordance with the procedure outlined in Step 3, measuring the net change in pressure from worst case condition in Step 3 to natural in the CAZ to confirm the worst case depressurization taken in Step 2. Repeat the process for each appliance, allowing each vent system to cool between tests.

5. Monitor indoor ambient CO in the breathing zone continuously during testing, and abort the test where indoor ambient CO exceeds 35 ppm by turning off the appliance, ventilating the space, and evacuating the building. The CO problem shall be corrected prior to completing combustion safety diagnostics.

6. Make recommendations based on test results and the retrofit action prescribed in Table T301.1(3).

TABLE T301.1(1)
CAZ DEPRESSURIZATION LIMITS

VENTING CONDITION	LIMIT (Pa)
Category I, atmospherically vented water heater	−2.0
Category I or II atmospherically vented boiler or furnace common vented with a Category I atmospherically vented water heater	−3.0
Category I or II atmospherically vented boiler or furnace, equipped with a flue damper, and common vented with a Category I atmospherically vented water heater	−5.0
Category I or II atmospherically vented boiler or furnace alone	
Category I or II atmospherically vented, fan-assisted boiler or furnace common vented with a Category I atmospherically vented water heater	
Decorative vented, gas appliance	
Power vented or induced-draft boiler or furnace alone, or fan-assisted water heater alone	−15.0
Category IV direct-vented appliances and sealed combustion appliances	−50.0

For SI: 6894.76 Pa = 1.0 psi.

TABLE T301.1(2)
ACCEPTABLE DRAFT TEST CORRECTION

OUTSIDE TEMPERATURE (°F)	MINIMUM DRAFT PRESSURE REQUIRED (Pa)
< 10	−2.5
10 – 90	(Outside Temperature ÷ 40) – 2.75
> 90	−0.5

For SI: 6894.76 Pa = 1.0 psi.

TABLE T301.1(3)
ACCEPTABLE DRAFT TEST CORRECTION

CARBON MONOXIDE LEVEL (ppm)	AND OR	SPILLAGE AND ACCEPTABLE DRAFT TEST RESULTS	RETROFIT ACTION
0 – 25	and	Passes	Proceed with work
25 < × ≤ 100	and	Passes	Recommend that CO problem be resolved
25 < × ≤ 100	and	Fails in worst case only	Recommend an appliance service call and repairs to resolve the problem
100 < × ≤ 400	or	Fails under natural conditions	**Stop!** Work shall not proceed until appliance is serviced and problem resolved
> 400	and	Passes	**Stop!** Work shall not proceed until appliance is serviced and problem resolved
> 400	and	Fails under any condition	**Emergency!** Shut off fuel to appliance and call for service immediately

Appendix U:
Solar-ready Provisions—Detached One- and Two-family Dwellings, Multiple Single-family Dwellings (Townhouses)

(The provisions contained in this appendix are not mandatory unless specifically referenced in the adopting ordinance.)

General Comments

This appendix is intended to support future potential improvements for detached one- and two-family dwellings, and multiple single-family dwellings for solar electric and solar thermal systems. This appendix does not require the installation of conduit, prewiring, or preplumbing. It does not require any specific physical orientation of the residential building. It does not require any increased load capacities for residential roofing systems. It does not require the redesign of plans.

Many building departments have been mandated by local regulations to accelerate permits and inspections for solar installation. Having important information and documentation available to the building department, solar contractor and homeowner will assist in supporting the accelerated working environment many municipalities have mandated.

The U.S. Department of Energy's (DOE) SunShot Initiative has set a goal to make solar energy cost competitive with other forms of energy by the year 2021, which would reduce installed costs of solar energy systems by about 75 percent. This initiative, combined with increased pressures on our energy supply and demand, should encourage and drive greater adoption of renewable energy systems on residential buildings.

Purpose

This appendix is intended to identify the areas of a residential building roof, called the solar ready zone, for potential future installation of renewable energy systems. The ability to plan ahead for possible future solar equipment starts with documenting necessary solar ready zone information on the plans, some of which may already be mandated in permit construction requirements. This appendix also requires the builder to post specific information about the home, for use by the homeowner(s).

The documentation of solar ready zones and roof load calculations (already performed during the design phase) will assist building departments as well as any future solar contractors seeking to install renewable energy systems on the roof. The builder/designer is knowledgeable on the intricacies of each model and plan and can easily identify unobstructed roof areas as well as spaces where conduit, wiring and plumbing can be routed from the roof to the respective utility areas. This will save building departments and solar designers time and effort when installing future solar systems. If a homeowner wishes to install a solar energy system later, this documentation can save thousands of dollars in labor, installation, design and integration of the solar system into the house.

SECTION U101
SCOPE

U101.1 General. These provisions shall be applicable for new construction where solar-ready provisions are required.

SECTION U102
GENERAL DEFINITIONS

SOLAR-READY ZONE. A section or sections of the roof or building overhang designated and reserved for the future installation of a solar photovoltaic or solar thermal system.

SECTION U103
SOLAR-READY ZONE

U103.1 General. New detached one- and two-family dwellings, and multiple single-family dwellings (townhouses) with

not less than 600 square feet (55.74 m^2) of roof area oriented between 110 degrees and 270 degrees of true north shall comply with sections U103.2 through U103.8.

Exceptions:

1. New residential buildings with a permanently installed on-site renewable energy system.

2. A building with a solar-ready zone that is shaded for more than 70 percent of daylight hours annually.

U103.2 Construction document requirements for solar ready zone. Construction documents shall indicate the solar-ready *zone*.

U103.3 Solar-ready zone area. The total solar-ready *zone* area shall be not less than 300 square feet (27.87 m^2) exclusive of mandatory access or set back areas as required by the *International Fire Code*. New multiple single-family dwell-

---

ings (townhouses) three stories or less in height above grade plane and with a total floor area less than or equal to 2,000 square feet (185.8 m^2) per dwelling shall have a solar-ready *zone* area of not less than 150 square feet (13.94 m^2). The solar-ready *zone* shall be composed of areas not less than 5 feet (1.52 m) in width and not less than 80 square feet (7.44 m^2) exclusive of access or set back areas as required by the *International Fire Code*.

U103.4 Obstructions. Solar-ready *zones* shall be free from obstructions, including but not limited to vents, chimneys, and roof-mounted equipment.

U103.5 Roof load documentation. The structural design loads for roof dead load and roof live load shall be clearly indicated on the construction documents.

U103.6 Interconnection pathway. Construction documents shall indicate pathways for routing of conduit or plumbing from the solar-ready *zone* to the electrical service panel or service hot water system.

U103.7 Electrical service reserved space. The main electrical service panel shall have a reserved space to allow installation of a dual pole circuit breaker for future solar electric installation and shall be labeled "For Future Solar Electric." The reserved space shall be positioned at the opposite (load) end from the input feeder location or main circuit location.

U103.8 Construction documentation certificate. A permanent certificate, indicating the solar-ready *zone* and other requirements of this section, shall be posted near the electrical distribution panel, water heater or other conspicuous location by the builder or registered design professional.

Index

A

ABSORPTION COOLING EQUIPMENT M1412

ACCESS

To appliances M1305
To attic R807
To crawl space R408.4
To equipment M1401.2
To floor furnace M1408.4
To plumbing connections P2704
To plumbing fixtures .P2705
To whirlpool pumps P2720.1

ACCESSIBLE

Definition. R202
Readily accessible, definition R202

ACCESSORY STRUCTURE

Definition. R202

ADDRESS (SITE) R319

ADMINISTRATIVE . . . . Chapter 1

Authority R104
Entry R104.6
Inspections R109
Permits R105
Purpose R101.3
Violations R113

AIR

Combustion . . . . .Chapter 17, G2407
Combustion air, definition R202
Return. M1602

AIR CONDITIONERS

Branch circuits E3702.11
Room air conditioners. E3702.12

ALLOWABLE SPANS

Of floor joists. R502.3, R505.3.2
Of headers R602.7, R603.6, R610.10
Of rafters and ceiling joists R802.4, R802.5,
R804.3.1, R804.3.2

**ALTERNATIVE MATERIALS
(see MATERIALS)** R104.11

AMPACITY E3501

ANCHOR BOLTS .R403.1.6

APPEAL

Board of R112.1
Right of R112

APPLIANCE

Access to M1305
Attic. M1305.1.3
Clearance for M1306

Connectors, fuel-gas. . . . . Chapter 24
Definition. R202
Definition applied to electrical
equipment E3501
Electrical appliance
disconnection means. E4101.5
Electrical appliance installation. E4101
Equipment (general) . . . . Chapter 14
Floor furnace. M1408
Flue area. R1003.14
Fuel-burning . . . . Chapter 24
Heating and cooling . . . . Chapter 14
Installation. M1307
Labeling M1303
Open-top broiler units M1505.1
Outdoor. G2454
Ranges M1901
Room heaters M1410
Special fuel-burning equipment . . . . Chapter 19
Vented (decorative). . . . . Chapter 24
Wall furnace M1409
Warm-air furnace M1402
Water heaters . . . . Chapter 20, Chapter 24

APPLICATION

Plywood R703.3

APPROVAL

Mechanical M1302

APPROVED

Definition. R202
Definition applied to electrical equipment E3501

ARC-FAULT CIRCUIT-INTERRUPTER E3902

AREA

Disposal, private sewage disposal P2602
Flue (appliances) R1003.14
Flue masonry R1003.15

ARMORED CABLE . . . . Table E3801.2

ASPHALT SHINGLES. R905.2

ATTACHMENT PLUG (PLUG CAP) (CAP)

Definition. E3501

ATTIC

Access R807

AUTOMATIC FIRE SPRINKLER SYSTEMS . . . R313

B

BACKFILL

For piping P2604

BACKFLOW, DRAINAGE
Definition R202
BACKWATER VALVE R202
BASEMENT WALL
Definition R202
BATH AND SHOWER SPACES R307
BATHROOM
Exhaust M1507.4
Group. . . . . R202, Table P3004.1
BATHTUB
Enclosure. P2713.2
Hydromassage. E4209
Outlets for P2713.1
Whirlpool P2720
BEAM SUPPORTS. R606.14
BEARING
Of joists R502.6
BIDETS. P2721
BOILER
Definition R202
Requirements . . . . Chapter 20
BONDING. E3609, E4204
Definition applied to
electrical installations E3501
BONDING JUMPER. E3501
Bonding of service equipment E3609.2
Bonding to other systems E3609.3
Main bonding jumper. E3607.5
Metal water piping bonding E3608.1.1
Sizing bonding jumpers E3609.5
BORED HOLES (see NOTCHING)
BOXES E3906
Nonmetallic boxes E3905.3
Support of boxes E3904.3
Where required .E3905.1
BRANCH CIRCUIT. . . . . E3501, Chapter 37
Branch circuit ampacity E3701.2
Branch circuit ratings E3702
Branch circuits required E3703
BUILDING
Definition R202
Drain, definition R202
Existing, definition R202
Sewer, definition R202
BUILDING OFFICIAL
Definition R202
Inspection and tests of
fuel-gas piping . . . . Chapter 24
BUILDING PLANNING
Automatic fire sprinkler systems R313
Carbon monoxide alarms. R315
Ceiling height R305
Decay protection R317
Design criteria R301
Emergency escape R310
Exterior wall R302.1
Fire-resistant construction R302
Foam plastic. R316
Garages and carports R309
Glazing. R308
Guardrails. R312
Handrails R311.7.8, R311.8.3
Insulation R302.10
Landing R311.3, R311.3.1, R311.3.2, R311.5.1
Light, ventilation and heating R303
Means of egress. R311
Minimum room area R304
Planning. . . . . Chapter 3
Plumbing fixture clearances R307
Radon protection . . . . Appendix F
Ramps R311.8
Sanitation. R306
Site address R319
Smoke alarms R314
Stairways R311.7
Storm shelters R323
Termite protection R318
BUILDING THERMAL ENVELOPE
Definition R202
BUILT-UP GIRDERS (see GIRDERS)
BUILT-UP ROOFING (see ROOFING)

C

CABINETS AND PANELBOARDS E3907
CAPACITY
Expansion tank. M2003.2
CARBON MONOXIDE ALARMS R315
CARPORT. R309.4
CEILING
Finishes R805
Height. R305
CEILING FANS. E4203.4
CENTRAL FURNACES (see FURNACES)
CHASES R606.8
CHIMNEYS
And fireplaces . . . . Chapter 10
Caps. R1003.9.1
Clearance. R1003.18
Corbeling R1003.5
Crickets R1003.20
Design (masonry). R1003.1

Factory-built R1005
Fireblocking R1003.19
Flue area R1003.14, R1003.15
Flue lining R100311
Load R1003.8
Masonry and factory built, size M1805
Multiple flue R1003.14
Rain caps 1003.9.3
Spark arrestors R1003.9.2
Termination R1003.9
Wall thickness R1003.10

CIRCUIT BREAKER
Definition E3501

CLASS 2 CIRCUITS
Class 2 remote-control, signaling
and power-limited circuits . . . . Chapter 43

CLAY
Tiles R905.3

CLEANOUT
Definition R202
Drainage P3005.2
Masonry chimney R1003.17

CLEARANCE
Above cooking top M1901.1
For appliances M1306.1
For chimneys R1003.18
Reduction methods M1306.2
Vent connector M1803.3.4

CLEARANCES
Around electrical equipment E3405.1, E3604.1,
E3604.2, E3604.3

CLOTHES CLOSETS
Lighting fixtures E4003.12

CLOTHES DRYERS . . . . .M1502, Chapter 24

CLOTHES WASHING MACHINES P2718

COLUMNS R407

COMBUSTIBLE
Materials R202

COMBUSTION AIR
Air . . . . Chapter 17
Definition R202

COMMON VENT
Definition R202
Requirements M1801.11

CONCRETE
Compressive Strength R402.2
Floors (on ground) .R506
Tile (roof) R905.3
Weathering . . . . Figure R301.2(3), R402.2

CONCRETE-ENCASED ELECTRODE E3608.1.2

CONDUCTOR E3406
Ampacity E3705.1
Definition E3501
Identification E3407
Insulation E3406.5
Material E3406.2
Parallel E3406.6
Size E3406.3, E3704.1
Ungrounded E3603.1

CONDUIT BODY
Definition E3501

CONNECTION
Access to connections P2704
For fuel-burning appliances . . . . Chapter 24
For fuel-oil piping M2202
Joints P2904
Plumbing fixture P2601.2
To water supply P2902.1

CONNECTIONS
Aluminum E3406.8
Device E3406.11.2

CONNECTOR
Chimney and vent M1803
Vent, definition R202

CONNECTOR, PRESSURE (SOLDERLESS)
Definition E3501

CONSTRUCTION
Cavity wall masonry R608
Flood-resistant R322
Floors . . . . Chapter 5
Footings R403
Foundation material R402
Foundation walls R404
Foundations . . . . Chapter 4
Masonry R606, R607,
R608, R610
Roofs . . . . Chapter 8
Steel framing R505, R603, R804
Wood framing R502, R602, R802
Walls . . . . Chapter 6

CONTINUOUS LOAD
Definition E3501

CONTINUOUS WASTE
Definition R202

CONTROL
Devices . . . . Chapter 24

CONTROLS
For forced-air furnaces . . . . Chapter 24

COOKING UNIT, COUNTER-MOUNTED
Definition E3501

COOLING

Absorption cooling equipment M1412

Access to equipment M1401.2

Evaporative cooling equipment M1413

Installation M1401.1

Refrigeration cooling equipment M1404

Return air-supply source M1602

COPPER-CLAD ALUMINUM CONDUCTORS

Definition E3501

CORDS

Flexible E4101.3, E4202.2

CORROSION

Protection of metallic piping P2603.3

COURT

Definition R202

COVER REQUIREMENTS E3803.1

COVERING

Exterior R703

Interior R702

Roof . . . . Chapter 9

Wall . . . . Chapter 7

CRAWL SPACE R408

CRITERIA

Design R301

CROSS CONNECTION

Definition R202

D

DAMPER, VOLUME

Definition R202

DECAY

Protection against R317

DECK

Supported by exterior wall R507

Wood/plastic composite
boards R507.3

DECORATIVE APPLIANCES

Outdoor G2454

Vented . . . . Chapter 24

DEFINITIONS

Building R202

Electrical E3501

Mechanical system R202

Plumbing R202

DESIGN

Criteria R301

DIRECTIONAL

Fittings, plumbing P2707

DISCONNECTING MEANS

Definition E3501

DISHWASHING MACHINES P2717

DOORS

Egress R311.2

Exterior R311.3.2, R609

DRAFT HOOD

Definition R202

DRAFTSTOPPING R302.12, R502.12

DRAIN

Floor P2719

Shower receptors P2709

DRAINAGE

Cleanouts P3005.2

Foundation R405

Inspection and tests P2503

Storm drainage . . . . Chapter 33

DRILLING AND NOTCHING (see NOTCHING)

DRINKING WATER TREATMENT P2909

DRIP LOOPS E3605.9.5

DRYERS

Domestic clothes M1502

DUCTS . . . . Chapter 16

Installation M1601.4

Insulation N1103.2, M1601.3

Material M1601

System, definition R202

DWELLING

Definition R202, E3501

DWELLING UNIT

Definition R202, E3501

Separation R302.2, R302.3

E

EJECTORS (see SUMPS AND EJECTORS)

ELECTRICAL

Appliance (labeling) M1303

Inspection E3403

Vehicle charging E3702.13

ELECTRICAL METALLIC TUBING . . . . Table E3904.6

**ELECTRICAL NONMETALLIC
TUBING** . . . . Table E3904.6

ELECTRICAL RESISTANCE HEATERS

Baseboard convectors M1405

Duct heaters M1407

Radiant heating M1406

ELECTRODES

Grounding E3608

EMERGENCY ESCAPE R202, R310

ENCLOSURE

Definition E3501

ENERGY CONSERVATION . . . . Chapter 11

ENTRY R104.6

EQUIPMENT
Definition applied to electrical equipment E3501
General, mechanical. . . . . Chapter 14
Heating and cooling . . . . Chapter 14

EXCAVATIONS
For appliance installation M1305.1.4.2

EXISTING PLUMBING SYSTEMS P2502

EXTERIOR
Covering. R703
Insulation finish systems. R703.9
Lath. R703.7.1
Plaster R703.7

EXTERIOR WALL
Construction R302.1
Definition. R202
Fire-resistance rating R302.1

F

FACEPLATES. E4001.11, E4002.4

FACTORY BUILT
Chimneys R1005, M1805
Fireplace stoves R1005.3
Fireplaces. R1004

FASTENING . . . . Table R602.3(1)

FAUCETS P2701

FAUCETS
Handle orientation P2722.2
Requirements P2701

FEEDER
Ampacity. E3704.5
Conductor size E3704.1
Feeder neutral load E3704.3
Loads E3704.2
Requirements E3704

FENESTRATION
Definition. R202

FINISHES
Flame spread and smoke density. R302.9
For ceilings R805
Interior .R302.9, R315, R702

FIRE-RESISTANT CONSTRUCTION R302

FIRE SPRINKLER SYSTEM
Inspections of P2904.8
Sizing of P2904.6
Sprinkler location P2904.1.1

FIREBLOCKING
Barrier between stories. R302.11, R602.8
Chimney R1003.19
Fireplace. R1001.12

FIREPLACES . . . . Chapter 10
Clearance R1001.11
Corbeling. R1001.8
Factory-built R1004
Fireblocking. R1001.12
Walls. R1001.5

FITTING
Definition applied to electrical installations E3501

FITTINGS
DWV piping. P3002.3
Prohibited joints P3003.2
Water supply. P2906.6

FIXTURE UNIT
Drainage, definition. R202
Unit valves P2903.9.3
Water supply, definition. R202

FIXTURES
Plumbing fixture, definition R202
Plumbing fixture, general . . . . Chapter 27
Trap seals. P3201.2

FLAME SPREAD INDEX. R302.9, R302.10

FLASHING. R703.8.5, R703.4, R903.2, R905

FLEXIBLE CORDS E3909, E4101.3, E4202.2

FLEXIBLE METAL CONDUIT. . . . E3801.1, E3908.8.1

FLOOD HAZARD AREA
Construction R322
Plumbing systems. P2601.3

FLOOR DRAIN P2719

FLOOR FURNACE M1408

FLOORS
Concrete (on ground) R506
Steel framing. R505
Treated-wood (on ground) R504
Wood framing R502

FLUE
Area R1003.14, R1003.15
Lining R1003.11, R1003.12
Multiple R1003.13

FOAM PLASTICS R316

FOOD-WASTE DISPOSERS
General. P2716
Discharge connection to sink tailpiece P2707.1

FOOTINGS
Excavation for piping near P2604.4
Requirements R403

FORCED MAIN SEWER
Materials P3002.2
Testing P2503.4

FOOTINGS R403

FOUNDATIONS . . . . Chapter 4
  Cripple walls R602.9, R602.10.9, R602.11.2
  Frost protection R403.1.4.1
  Inspection R109.1.1
  Walls R404
FRAME
  Inspection R109.1.4
FREEZE PROTECTION OF PLUMBING P2603.6
FUEL-BURNING APPLIANCES
  (see APPLIANCE, definition)
  Identification . . . . Chapter 24
FUEL GAS . . . . Chapter 24
FUEL OIL
  Oil tanks M2201
  Piping, fittings and connections M2202
  Pumps and valves M2204
FURNACES
  Clearance of warm-air furnaces M1402.2
  Exterior M1401.4
  Floor (see Floor Furnace)
  Wall (see Wall Furnace)
  Warm-air . . . . M1402, Chapter 24
FUSES E3705.6

G

GARAGES R309
GAS
  Appliance labeling . . . . Chapter 24
GAS PIPING SYSTEM NOT
TO BE USED AS GROUNDING
ELECTRODE . . . . Chapter 24
GIRDERS R502.5
GLAZING R308
  Aggregate R303.1
  Protection of openings R301.2.1.2
GRADE
  Definition R202
  Of lumber R502.1.1, R602.1.1, R802.1.1
  Plane, definition R202
  Slope of piping, definition R202
GRAYWATER
  For subsurface irrigation P3009
  Identification for P2901.2
  Supply to fixtures P2901.1
GROUND
  Definition of electrical E3501
  Floors (on ground) R504, R506
  Joint connection .P2906.8
GROUND-FAULT CIRCUIT-
INTERRUPTER PROTECTION E3902, E4203,
                            E4206, E4207.2, E4208

GROUND SOURCE HEAT PUMP M2105
GROUNDED
  Definition E3501
GROUNDED CONDUCTOR
  Definition E3501
  Identification E3407
GROUNDED, EFFECTIVELY
  Definition E3501
GROUNDING
  Effective grounding path E3611.3
  Of equipment E3908, E4205.1
GROUNDING CONDUCTOR
  Definition E3501
  Equipment E3501, E3607.3.1, E3908
GROUNDING ELECTRODE
CONDUCTOR E3607.4
  Connection E3607.2, E3611
  Definition E3501
  Size E3603.4
GROUNDING ELECTRODE SYSTEM E3608
GROUNDING ELECTRODES E3608, E3608.1
  Resistance of E3608
GUARDED
  Definition applied
    to electrical equipment E3501
GUARDING OF ENERGIZED PARTS E3404.9
GUARDS R312
  Definition R202
GYPSUM
  Wallboard R702.3

H

HABITABLE SPACE
  Definition R202
HALLWAYS R311.6
HANDRAILS R311.7.8, R311.8.3
  Definition R202
HAND SHOWER (see SHOWER)
HEADERS
  SIP R610.10
  Steel R603.6
  Wood R602.7
HEARTH R1001.9
  Extension R1001.9, R1001.10
HEATERS
  Baseboard M1405
  Central furnaces M1402
  Duct M1407
  Heat pumps M1403
  Masonry R1002
  Pool .M2006, Chapter 24

Radiant M1406
Sauna . . . . Chapter 24
Unvented G2445
Vented room . . . . .M1410, Chapter 24
Water . . . . Chapter 20, Chapter 28

HEATING. R303.9

HEATING EQUIPMENT . . . . Chapter 14

HEIGHT
Ceiling. R305

HOLLOW-UNIT MASONRY
(see MASONRY) R606.3.2.2, 606.13.1.2

HORIZONTAL
Pipe, definition R202
Vent length M1803.3.2

HOT TUBS. . . . . E4201, Appendix G

HOT WATER
Definition. R202
Distribution pipe rating P2906.5
Distribution systems P2905
Heaters. .M2005, Chapter 24
Supplied to bathtubs. P2713
Temperature actuated valves for P2724

HUB DRAIN (See WASTE RECEPTOR)

HYDROGEN GENERATING
AND REFUELING M1307.4

HYDRONIC HEATING SYSTEMS
Baseboard convectors M2102
Boilers. M2001
Expansion tanks M2003
Floor heating systems M2103
Operating and safety controls. M2002
Piping systems installation M2101

I

IDENTIFIED
Definition applied to electrical equipment E3501

INDIVIDUAL
Branch circuit, definition E3501
Sewage disposal systems R202, P2602
Water supply and sewage
disposal systems. P2602

INLET
To masonry chimneys R1003.16

INLINE SANITARY WASTE
VALVE P3101.2.1, P3201.6.1

INSPECTION
Card AE305.3
Excavation for piping P2604.1
Fuel-supply system. . . . . Chapter 24
Of plumbing system P2503
On site R109.1

INSTALLATION
Existing . . . . Appendix J
Of appliances M1307
Of cooling and heating equipment M1401.1
Of ducts. M1601.4
Of flue liners R1003.12
Of plumbing fixtures P2705
Of wall furnaces M1409.3

INSULATION. R302.10

INTERIOR
Lath. R702.2.3
Other finishes R702.5
Plaster. R702.2
Wall covering . . . . Chapter 7

INTERMEDIATE METAL CONDUIT . . . E3801, E3904

INTERRUPTING RATING E3404.2
Definition E3501

IRRIGATION (see SUBSURFACE IRRIGATION)

J

JOINTS, PIPE
And connections P2906, P3003
Slip P2704.1, P3005.2.9

JOIST
Bearing R502.6, R606.6.3.1

JUMPERS
Bonding. E3609

JUNCTION BOXES E3905, E4206.9.1

K

KITCHEN
Definition. R202

KITCHEN RECEPTACLES E3703.2, E3901

L

L VENT TERMINATION M1804.2.4

LABELED
Definition. R202
Definition applied to electrical equipment E3501

LABELING
Appliances M1303
Definition. R202

LAMPHOLDERS E4003, E4004

LANDINGS R311.3, R311.3.1,
R311.7.6, R311.8.2

LATERAL SUPPORT . . . R502.7, R606.6.4, R607.5.2

LATH
Exterior R703.7.1
Interior. R702.2.3

LAUNDRY CIRCUIT E3703.3

LAUNDRY TUBS

Drain connection to standpipe P2706.1.2.1

Requirements P2715

LAVATORIES P2711

Clearances R307

Waste outlets P2711.3

LEAD

Allowable amount, potable water
    components P2906.2

Bends and traps P3002.4.2

Caulked joints P3003.6.1

Flashing P3002.4.1

LIABILITY R104.8

LIGHT, VENTILATION AND HEATING R303

LIGHTING

Luminaire in clothes closets E4003.12

Luminaire installation E4004

Luminaries E4003, E4004, E4206

LIGHTING OUTLETS

Definition E3501

Required lighting outlets E3903

LINING

Flue R1003.11, R1003.12

LINTEL R606.10, R608.8,
    R703.8.3, R1001.7

**LIQUID-TIGHT FLEXIBLE
CONDUIT** E3801.4, E3908.8.2

LISTED

Definition applied to electrical equipment E3501

LISTED AND LISTING

Definition applied to building and
    mechanical provisions R202

Mechanical appliances M1302.1

Potable water components P2609

LOAD

Additional R1003.8

Roof R301.6

Seismic risk map . . . . Figure R301.2(2)

Snow load map . . . . Figure R301.2(5)

Wind speed map . . . . Figure R301.2(4)A

LOADS

Branch circuit loads E3702, E3703.6

Dead load R301.4

Feeder load E3704.2

Feeder neutral load E3704.3

Horizontal load AE602.1

Live load R301.5

Service load E3602.2

LOADS, LIVE AND DEAD

Definition R202

LOCATION

Of furnaces M1408.3, M1409.2

LOCATION (DAMP) (DRY) (WET)

Definitions E3501

LUMBER

Grade R502.1.1, R602.1.1, R802.1.1

LUMINAIRE E4004

Clearances E4004.8

M

MACERATING TOILET P2723

MAKEUP AIR M1503.4

MANUFACTURED HOME

Definition R202

Provisions . . . . Appendix E

MASONRY

Anchorage R606.11

Cavity wall R606.13

Chases R606.8

General R606

Hollow unit R202, R606.3.2.2, R606.13.1.2

Inspection R109.1.4

Seismic requirements R606.12

Solid, definition R202

Veneer R703.8

Veneer attachment R703.3

MATERIALS

Alternative R104.11

Combustible R202, R1001.11, R1001.12,
    R1003.18, R1003.19

For ducts M1601

For fixture accessories P2702

For flue liners R1003.11

For fuel-supply systems . . . . Chapter 24

For hearth extension R1001.9

For siding R703.5

Hydronic pipe M2101, M2103, M2104

Plumbing pipe P2906, P3002

MECHANICAL

Inspection R109.1.2

System requirements . . . . Chapter 13

**MEDIUM PRESSURE FUEL-GAS
PIPING SYSTEM** . . . . Chapter 24

MEMBRANE

Penetration R302.4.2

Polyethylene R504.2.2, R506.2.3

Water-resistive R703.2

Waterproofing R406.2

METAL

Ducts . . . . Chapter 16

Roof panels R905.10

Roof shingles R905.4

METAL-CLAD CABLE . . . . Table E3701.2
METHODS
Water distribution pipe sizing . . . . Appendix P
MODIFICATIONS R104.10
MOISTURE CONTROL R702.7
MORTAR
Joints R606.3.1
MOTORS
Motor branch circuits E3702.6, E3702.7
MULTIPLE
Appliance venting systems M1801.11
Flues R1003.13
MULTIWIRE BRANCH CIRCUITS E3701.5

N

NONCOMBUSTIBLE MATERIAL
Definition R202
NONMETALLIC BOXES E3905.3, E3908.16
NONMETALLIC RIGID CONDUIT E3801.4
NONMETALLIC-SHEATHED
CABLE E3705.4.4, E3801.4,
E3905.3.1, E3908.8.3
NONPOTABLE WATER SYSTEMS . . . P2910, P2911,
P2912, P2913
NOTCHING
Electrical E3402.1
For mechanical systems M1308
For plumbing piping P2603.2
Steel joists R505.2.5, R505.3.5,
R804.2.5, R804.3.3
Steel studs R603.2.6, R603.3.4
Wood joists R502.8, R802.7.1
Wood studs R602.6
Wood top plates R602.6.1

O

OCCUPIED SPACE
Definition R202
OIL
Piping and connections M2202
Supply pumps and valves M2204
Tanks M2201
OPEN-TOP GAS BROILER UNITS M1505.1
OPENING
Requirements,
combustion air . . . . Chapter 17, Chapter 24
Waterproofing of piping penetrations P2606.1
OUTLET
Definition . . . . Chapter 24, E3501

OVEN, WALL-MOUNTED
Definition applied to electrical provisions E3501
OVERCURRENT
Definition E3501
OVERCURRENT PROTECTION E3705
OVERCURRENT-PROTECTION RATING E3705
OVERLOAD
Definition E3501

P

PANELBOARD
Definition E3501
PANELBOARDS E3706
Clearance and dedicated space E3405.2,
E3405.3
Headroom E3405.3
PARALLEL PATH
CONNECTION N1102.2.6 (R402.2.6)
PARAPETS R302.2.2, R606.4.4
PARTICLEBOARD
Floor R503.3
Walls R605
PERMITS R105
PIERS R606.7
Masonry R404.1.9
PIPE
Fittings P2702.2
Materials listing . . . . Table M2101.1, P2609
Protection P2603
Standards, drain,
waste and vent . . . . Table P3002.1(1),
Table P3002.1(2)
Standards, sewer . . . . Table P3002.2
Standards, water supply P2906
PIPING
Connections and materials . . . Table M2101.1, P2905
Drain, waste and vent P3002.1
Excavations next to footings P2604.4
Fuel-gas size determination . . . . Chapter 24
Penetrations P2606
Protection P2603
Sizing methods, water distribution P2903.7
Support . . . . M2101.9, Chapter 24, P2605
PLANNING
Building . . . . Chapter 3
PLANS R106
PLASTER
Exterior R703.7
Interior R702.2

PLENUM

Definition R202

Definition, electrical installations E3501

PLUMBING

Fixture clearances R307

Fixtures . . . . Chapter 27

Inspection R109.1.2

Materials P2702.2, P2905, P3002

Requirements and definitions R202

System, definition R202

Third-party certification P2609.4

Traps . . . . Chapter 32

Vents . . . . Chapter 31

PLYWOOD

Application R703.5

Materials, walls R604

POTABLE WATER

Definition R202

Piping listing. P2609.4

PRECAST CONCRETE

Footings. R403.4

Foundation material. R402.3.1

Foundation walls R404.5

PRESSURE, WATER SUPPLY

Maximum P2903.3.1

Minimum P2903.3

PRIVATE

Sewage disposal system . . . . Appendix I

PROHIBITED

Receptors P2706.3

Return air sources M1602.2

Traps P3201.5

Valves below grade P2903.9.5

Water closets. P2712.1

PROTECTION

Against decay and termites R319, R320

Against radon. . . . . Appendix F

Of backflow preventers P2902.6

Of ferrous gas piping . . . . Chapter 24

Of potable water supply P2902

PURLINS R802.5.1

PURPOSE. R101.3

R

***R*-VALUE**

Definition R202

RACEWAY

Definition E3501

Raceway installations E3904

Raceway seals. E3601.5, E3803.6

Raceways as means of support. E3904.4

RADON

Map . . . . Appendix F

RAFTERS

Grade of lumber. R802.1

Spans. R802.1.1,

Tables R802.5.1(1) – R802.5.1(8)

RAIN TIGHT

Definition applied to electrical

provisions E3501

RAINPROOF

Definition applied to electrical

provisions E3501

RAINWATER

Identification for P2901.2

Supply to fixtures .P2901.1

RAMPS. R311.8

RANGES

Branch circuits for ranges E3702.9

RANGES AND OVENS

Vertical clearance above cooking top M1901.1

READILY ACCESSIBLE

Definition R202

Definition, electrical installations E3501

RECEPTACLE

Definition E3501

RECEPTACLE OUTLET

Definition E3501

Required outlets. E3901

RECEPTACLES

Rating, type and installation. E4002

RECEPTORS

Plumbing fixtures and traps . . . . Chapter 27

Site-built materials P2709.2

Waste. P2706

RECESSED LUMINAIRES E4003.5, E4003.12,

E4004.8, E4004.9

RECLAIMED WATER

Identification for P2901.2

Supply to fixtures .P2901.1

REQUIREMENTS

Connections for fuel-burning

appliances. . . . . Chapter 24

Return air M1602

RESISTANCE TO GROUND

Electrodes E3608.4

RESISTANT SIDING MATERIAL (see MATERIALS)

RETURN-AIR M1602

RIDGE BOARD. R802.3

RIGID METALLIC CONDUIT E3908
RIGID NONMETALLIC CONDUIT E3801.4
ROOF
Coverings R905
Drainage. R903.4
Flashing R703.4, R903.2, R905
Steel framing R804
Wood framing R802
ROOF-CEILING CONSTRUCTION
(see CONSTRUCTION) . . . . Chapter 8
Wind uplift. R802.11, R804.3.8
ROOFING
Built-up R905.9
Liquid-applied coating. R905.15
Modified bitumen R905.11
Sprayed polyurethane foam R905.14
Thermoplastic single-ply. R905.13
Thermoset single-ply R905.12
ROOM
Heaters, vented . . . . .M1410, Chapter 24
Minimum Sizes R304

S

SANITATION. R306
SEISMIC RISK MAP . . . . Figure R301.2(2)
SEPTIC TANK
Definition. R202
SERVICE
Definition E3501
SERVICE CABLE
Definition. E3501
SERVICE CONDUCTORS
Definition. E3501
Drip loops E3605.9.5
Insulation E3605.1
Overload protection E3603.3
Rating of ungrounded
service conductors E3602.1
Size. E3603.1
SERVICE DISCONNECT
Location E3601.6.2
Marking of. E3601.6.1
Maximum number of. E3601.7
Rating of E3602.3
Required E3601.6
SERVICE DROP
Clearances E3604.1, E3604.2
Definition. E3501
Point of attachment. E3604.3

SERVICE-ENTRANCE CONDUCTORS,
OVERHEAD SYSTEM
Definition. E3501
SERVICE-ENTRANCE CONDUCTORS,
UNDERGOUND SYSTEM
Definition. E3501
SERVICE EQUIPMENT
Definition. E3501
SERVICE LATERAL
Definition. E3501
SERVICE LOAD E3602.2
SERVICE POINT
Definition. E3501
SERVICE VOLTAGE RATING E3602.4
SERVICES. . . . . .Chapter 36
SEWAGE
Disposal, private P2602
SEWER, BUILDING
Definition. R202
Separation from water service . . P2604.2, P2906.4.1
Size. P3005.4.2
Testing P2503.4
SHAKES
Wood. R702.6, R703.6, R905.8
SHINGLE
Asphalt shingles R905.2
Metal. R905.4
Slate R905.6
Wood. R905.7
SHOWER
Control valve. P2708.4
Control valve riser to shower head P2708.3
Compartment R307.2
Drain size P2708.2
Hand P2708.5
Receptor P2709
Stall dimensions P2708
Walls P2710
SHUTOFF VALVE (see VALVES)
SIDING
Exterior coverings. R703
SINKS P2714
SITE
Address. R319
Preparation R408.5, R504.2, R506.2
SIZE
Of combustion air openings . . . . Chapter 17
Of rooms. R304
Of trap. P3201.7

SIZING METHODS
Water piping P2903.7
SKYLIGHTS R308.6
SLATE SHINGLES R905.6
SMOKE ALARMS R314
SMOKE-DEVELOPED INDEX R302.9, R302.10
SNOW LOAD MAP . . . . Figure R301.2(5)
SOLAR ENERGY SYSTEMS M2301
SPANS
Steel (allowable) R505.3.2, R804.3.2.1
Wood (allowable) R502.3, R802.5
SPAS E4208
SPLICES E3406.11
SPLICES AND TAPS E3803.4
SPRINKLER (see FIRE SPRINKLER SYSTEMS)
STAIRWAYS R311.7
STANDARDS . . . . Chapter 44, Appendix G
STANDPIPE (see WASTE RECEPTOR)
STEEL
Fireplace units R1001.5.1
Floor construction R505
Roof-ceiling construction R804
Walls R603
STORM SHELTERS R323
STORY
Definition R202
STOVES
Factory-built fireplace M1414
STRUCTURAL AND PIPING PROTECTION . . . P2603
STRUCTURAL INSULATED PANEL (SIP) R610
STUDS
Wood R602.2, R602.3
Spacing R602.3.1
Steel R603.2, R603.3
SUBSURFACE IRRIGATION SYSTEMS P3009
SUMPS AND EJECTORS P3007
SUMP PUMPS P3007.3.2, P3303
SUPPLY
Combustion air . . . . .Chapter 17, Chapter 24
Fuel systems . . . . Chapter 24
Oil supply pumps and valves M2204
Required gas . . . . Chapter 24
Return air M1602
Water . . . . Chapter 29
SUPPORT
Of decks R507.2.2.1
Of ducts M1601.4.4
Of floor joists R502.6, R505.3.2
Of masonry chimneys R1003.2,
R1003.3, R1003.4
Of pipe . . . . M2101.9, Chapter 24, P2605

SUPPORT REQUIREMENTS
FOR WIRING METHODS E3802.1
SWIMMING POOLS . . . . R326, Chapter 42
SWITCHES
Definition E3501
Rating and application E4001
SYSTEMS
Mechanical venting . . . . Chapter 18, Chapter 24
Plumbing, drainage, waste
and venting . . . . Chapter 25, Chapter 30,
Chapter 31, Chapter 33

T

TAIL PIECES, PLUMBING FIXTURE P2703
TANK
For fuel oil-burning appliances M2201
Septic, definition R202
TEMPERATURE ISOLINES . . . . . . .Figure R301.2(1)
TEMPERATURE LIMITATIONS E3705.4
TERMINALS E3406.10
TERMINATION
Of chimneys R1003.9
Of vents (general) M1804.2
TERMITES
Infestation probability map . . . . . . .Figure R301.2(6)
Protection R318
TEST
For leaks in supply piping . . . . Chapter 24
Of backflow preventers P2503.8
Of building sewers P2503.4
Of plumbing systems P2503
Of shower liner .P2503.6
THICKNESS
Of chimney walls R1003.10
TIES
Veneer R703.8.4
TILE
Shingles (clay and concrete) R905.3
TOWNHOUSE
Definition R202
Scope R101.2
Separation R302.2
TRACK LIGHTING E4005
TRAP . . . . Chapter 32
Arm, definition R202
Building P3201.1
Seal protection .P3201.2.1
TRENCHING AND BACKFILLING P2604
TRUSSES
Steel R505.1.3, R804.3.6
Wood R502.11, R802.10

TYPE OF VENTING
SYSTEMS REQUIRED .... Chapter 18, Chapter 24

U

U-FACTOR
  Definition.................................R202
UNDER FLOOR
  Access to furnaces M1305.1.4
  Plenums M1601.5
  Space R408
UNDERGROUND INSTALLATION
REQUIREMENTS
  Duct systems M1601.1.2
  Electrical cable E3803
  Wiring E4203.7
UNVENTED ATTIC ASSEMBLIES R806.5
UTILIZATION EQUIPMENT
  Definition.................................E3401

V

VALVES
  Backwater.............................P3008
  Meter .... Chapter 24
  Oil-supply M2204
  Relief, water heater P2804
  Shutoff, fuel-gas .... Chapter 24
  Shutoff, fuel-oil M2204.2
  Water heaters.......................P2903.9.2
  Water supply........................ P2903.9
VAPOR RETARDERS................... R702.7
  Definition.................................R202
VENEER
  Masonry R703.7
VENT
  B or BW vent .... Chapter 24
  Definition......................... Chapter 24
  L vent ....M1804.2.4, Chapter 24
  Plumbing system, definition R202
  Termination.......................M1804.2,
                          M2203.5, P3103.5
VENTED
  Decorative appliances .... Chapter 24
  Floor furnaces......................... M1408
  Room heaters M1410
  Wall furnaces M1409
VENTILATION
  Bathroom and kitchen.................. M1507
  For hydrogen systems M1307.4
  Roof R806

  Under floor R408.1
  Whole house.......................M1507.3
VENTING (MECHANICAL)
  Chimney and vent connectors .. M1803, Chapter 24
  Components .... M1802, Chapter 24
  General..................... M1801, Chapter 24
  Required .... M1801, M2203.4, Chapter 24
  Systems .... Chapter 18, Chapter 24
VENTING (PLUMBING)
  Air admittance valves P3114
  Circuit P3110
  Combination waste and vent system P3111
  Common P3107
  Connections and grades..................P3104
  Fixture.............................P3105
  Individual...........................P3106
  Island fixture P3112
  Pipe sizing..........................P3113
  Stacks and stack vents P3102
  Systems P3101
  Terminals P3103
  Waste stack P3109
  Wet P3108
VERTICAL
  Clearances above cooking top M1901.1
  Pipe, definition R202
VIOLATIONS
  And penalties R113
VOLTAGE
  Definition.................................E3501
VOLTAGE RATING (SERVICES) E3602.4
VOLTAGE TO GROUND
  Definition.................................E3501
VOLTAGES................................E3404.1

W

WALL FURNACE (see APPLIANCE)
  General...............................M1409.1
  Installation...........................M1409.3
  Location M1409.2
WALLBOARD
  Gypsum.............................R702.3
WALLS
  Bathtub compartments P2710
  Bracing, steel R603.9
  Bracing, wood R602.10
  Construction .... Chapter 6
  Covering .... Chapter 7
  Cripple............................R602.9
  Deflection R301.7

Exterior covering R703
Finishes R302.9, R702
Fireplace R1001.5
Foundation. R404
Insulating concrete form R608.3, R608.4,
R608.5.3
Shower compartments. P2710
Steel framing R603
Structural insulated panels (SIP) R610
Thickness, masonry chimneys. R1003.10
Wood framing R602

WARM-AIR FURNACES M1402
Access to M1401.2
Clearance from M1402.2

WASTE
Clear water P2792.1
Continuous, definition R202

WASTE RECEPTOR
Hub drain P2706.1
Standpipe. P2706.1
Prohibited types. .P2706.3

WATER
Distribution system, definition R202
Heater, definition R202
Heaters . . . . . . .Chapter 20, Chapter 24, Chapter 28
Individual supply and sewage disposal P2602
Piping sizing methods P2903
Pressure P2903.3
Service, separation from
sewer P2604.2, P905.4.1
Supply and distribution . . . . Chapter 29
Supply system, definition. R202

WATER CLOSET
Carriers for wall hung. P2702.4
Clearances for R306.1
Dual flushing P2712.1
Personal hygiene devices P2722.5
Seats P2712.7
Standards P2712.1

WATER HEATER
Drain water heater heat recovery units P2903.11
In garages P2801.5.2
Location P2801.3, P2801.4, P2801.5
Pan P2801.5
Plumbing requirements . . . . Chapter 28
Relief valves P2804
Seismic bracing P2801.7
Solar thermal P2802, P2902.5.5
Thermal expansion from P2903.4
Used for space heating P2803

WATER TIGHT
Definition applied to electrical provisions E3501
WATERPROOFING
And dampproofing R406
Of openings through roofs and walls P2607
WEATHERPROOF
Definition applied to electrical provisions E3501
WELLS (see INDIVIDUAL WATER SUPPLY)
WHIRLPOOL BATHTUBS P2720
WHOLE-HOUSE VENTILATION M1507.3
WIND SPEED MAP. . . . . Figure R301.2(4)A
WINDOW. R609
Fall prevention R312.2
Opening limiting devices R312.2.2
WIRING METHODS . . . . Chapter 37
WOOD
Floor construction. R502
Foundation walls R404.2
Roof-ceiling construction R802
Shakes. R905.8
Shingles R905.7
Trusses R502.11, R802.10
Walls R602
WORKING SPACE .E3606.2
Around electrical equipment. E3405.1, E3405.2
WORKMANSHIP, PLUMBING P2608

Y

YARD
Definition R202

INTERNATIONAL
CODE
COUNCIL®

People Helping People Build a Safer World®

GET IMMEDIATE DOWNLOADS OF THE STANDARDS YOU NEED

Browse hundreds of industry standards adopted by reference. Available to you 24/7!

Count on ICC for standards from a variety of publishers, including:

ACI	CPSC	GYPSUM
AISC	CSA	HUD
ANSI	DOC	ICC
APA	DOJ	ISO
APSP	DOL	NSF
ASHRAE	DOTn	SMACNA
ASTM	FEMA	USC
AWC	GBI	

DOWNLOAD YOUR STANDARDS TODAY!
SHOP.ICCSAFE.ORG

15-11221